$2400

Textbook of Organic Medicinal
and Pharmaceutical Chemistry

Authors

T. C. Daniels, Ph.D.

Emeritus Professor of Pharmaceutical Chemistry, University of California

Robert F. Doerge, Ph.D.

Professor of Pharmaceutical Chemistry and Chairman of the Department of Pharmaceutical Chemistry, School of Pharmacy, Oregon State University

Ole Gisvold, Ph.D.

Professor of Medicinal Chemistry, College of Pharmacy, University of Minnesota

E. C. Jorgensen, Ph.D.

Professor of Chemistry and Pharmaceutical Chemistry, School of Pharmacy, University of California

Edward E. Smissman, Ph.D.

Professor of Medicinal Chemistry, The School of Pharmacy, University of Kansas

Taito O. Soine, Ph.D.

Professor of Medicinal Chemistry and Chairman of the Department of Medicinal Chemistry, College of Pharmacy, University of Minnesota

Abraham Taub, A.M.

Emeritus Distinguished Service Professor of Pharmaceutical Chemistry, College of Pharmaceutical Sciences, Columbia University

Allen I. White, Ph.D.

Dean and Professor of Pharmaceutical Chemistry, College of Pharmacy, Washington State University

Robert E. Willette, Ph.D.

Associate Professor of Medicinal Chemistry, School of Pharmacy, University of Connecticut

Charles O. Wilson, Ph.D.

Dean and Professor of Pharmaceutical Chemistry, School of Pharmacy, Oregon State University

Textbook of Organic Medicinal and Pharmaceutical Chemistry

Edited by

Charles O. Wilson, Ph.D.

*Dean and Professor of Pharmaceutical Chemistry, School of Pharmacy,
Oregon State University*

Ole Gisvold, Ph.D.

*Professor of Medicinal Chemistry, College of Pharmacy,
University of Minnesota*

and

Robert F. Doerge, Ph.D.

*Professor of Pharmaceutical Chemistry,
Chairman of the Department of Pharmaceutical Chemistry,
School of Pharmacy, Oregon State University*

SIXTH EDITION

J. B. LIPPINCOTT COMPANY
Philadelphia · Toronto

5   7   8   6   4

Preface

In this textbook, it is our purpose to describe, within certain chemical and pharmacologic classifications, those substances used as therapeutic agents and as pharmaceutical aids. This volume is written for the undergraduate pharmacy student who has previously completed a regular year's course in the fundamentals of organic chemistry. The information assembled is that which is of practical value to present-day pharmacists, since it now is understood quite generally that a pharmacist should have a thorough knowledge of what a drug is, its limitations, applications, forms and uses, as well as of those characteristics which pertain strictly to its compounding.

Products used in medicine and in pharmacy have been discussed under chapter headings which are understood readily by those in the pharmaceutical profession. Since many pharmacists and teachers of pharmacists classify medicinals by pharmacologic action, some of the chapters have been prepared with this in mind. Examples of these are: Antimalarials, Central Nervous System Depressants, Adrenergic Agents and Analgesic Agents. As each class of oganic compounds is introduced, there is a discussion of the basic principles of organic chemistry, and the discussion serves to orient the student and to provide a review. Certain terms are used in pharmacy to identify groups of pharmaceuticals; therefore, there are chapters on subjects such as Antibiotics, Steroids, Proteins and Vitamins.

To bring about a better understanding of why certain organic compounds have been selected as pharmaceuticals and to impress on the student the importance of physical properties, a chapter is included on Physicochemical Properties in Relation to Biologic Action. Usually, the chemical properties of a compound are responsible for the method of detoxication in the body. This area is covered in a chapter on Metabolic Changes of Drugs and Related Organic Compounds.

The authors of this book have attempted to include a discussion of all products described in the *U.S.P. XVIII, N.F. XIII, New Drugs,* and *Accepted Dental Remedies,* as well as the most important pharmaceuticals reported in the periodical literature. A compound is identified as "U.S.P." or "N.F." when it is accepted in the current edition of the official book (*U.S.P. XVIII, N.F. XIII*).

CHARLES O. WILSON
OLE GISVOLD
ROBERT F. DOERGE

Contents

16. CENTRAL NERVOUS SYSTEM STIMULANTS 449
T. C. Daniels and E. C. Jorgensen

17. ADRENERGIC AGENTS 473
Edward E. Smissman

18. CHOLINERGIC AGENTS AND RELATED DRUGS 495
Ole Gisvold

19. AUTONOMIC BLOCKING AGENTS AND RELATED DRUGS 513
Taito O. Soine

1

Introduction

Ole Gisvold, Ph.D.,
Professor of Medicinal Chemistry, College of Pharmacy,
University of Minnesota

and

Charles O. Wilson, Ph.D.,
Dean and Professor of Pharmaceutical Chemistry, School of Pharmacy,
Oregon State University

The turn of the century has seen the development of most of the organic medicinal agents, the great majority of which have been produced in the past 35 years.

Pure organic compounds, natural or synthetic, together with the so-called organometallics are the chief source of agents for the cure, the mitigation or the prevention of disease today. These remedial agents have had their origin in essentially three ways: (1) from naturally occurring materials of both plant and animal origin, (2) from the synthesis of organic compounds whose structures are closely related to those of naturally occurring compounds (e.g., morphine, atropine, cortisone and cocaine) that have been shown to possess useful medicinal properties, and (3) from pure synthesis in which no attempt has been made to pattern after a known, naturally occurring compound exhibiting some activity. Many of these are antihistamines, barbiturates, diuretics, tranquilizers and antiseptics.

Many of the compounds from the first group are prepared synthetically today as a financially expedient measure. Examples of these are most of the vitamins, some sex hormones, corticometric principles, methyl salicylate, camphor and menthol. On the other hand, cardiac glycosides, quinine, atropine, antibiotics and insulin either cannot be synthesized or can be isolated from natural sources at a cost that can compete with synthetic methods. Examples of compounds found in the second group are the numerous sympathomimetic drugs and local anesthetics, antispasmodics, mydriatic and myotic drugs. Examples of the third group include the synthetic antimalarials, dyes, some analgesics, diuretics, phenols, barbiturates and surface active agents.

Even though isolation and synthesis of many active constituents from animal and plant sources have been accomplished, there are new problems yet to be solved—for example, the synthesis of vitamin B_{12}, the very active antipernicious anemia factor found in liver. These accomplishments, together with others, will add to the present large number of useful organic medicinal compounds and bring about a more nearly complete complement of drugs.

In many cases, at present, there is no simple and direct correlation between the activity of organic compounds and their chemical structure beyond the broad generalization that compounds similarly constituted may be expected to have similar ac-

1

tivities. This is not always true, for often a fine shade of difference in chemical structure may lie between a very active compound on the one hand and a completely inactive one on the other or even one whose activity may be antagonistic to the original model. The last-mentioned compounds have received intensive research attention in recent years and usually are referred to as metabolic antagonists. The very useful sulfa drugs were shown to be metabolic antagonists. These studies are useful to the biochemist and the physiological chemist, and it is hoped that new and useful medicinal agents also will be developed as a result of the extension of these studies. In other cases, each member of a whole series of compounds more or less related in structure to one another may have some activity. This can well be illustrated by sympathomimetic drugs, which include a series of compounds from the simple 2-aminoheptane to epinephrine. The fact that a series of compounds, the members of which are structurally related to one another, exhibits a similarity in activity does not preclude the possibility that some other compounds, unrelated structurally, can have similar activity. For example, anesthetic properties are present not only in the cocaine or procaine type of molecule but also in benzyl alcohol, quinine, nupercaine, phenacaine, plasmochin and other compounds. Nevertheless, a convenient method for the study of organic medicinal agents according to a hybrid chemical classification is, in part, a desirable approach, because it allows the student to familiarize himself with the chemical, the physical and the biochemical properties of such groups. It is well to remember that the chemical, the physical and conformational and, now, the biochemical properties of organic compounds are functions of their structures. Therefore, much can be gained by studying medicinal agents from these points, noting the changes in activities that are affected by the changes in these factors.

Sometimes the activity of a drug is dependent chiefly on its physical and chemical properties, whereas in other instances the arrangement, the position and the size of the groups in a given molecule also are important and lead to a high degree of specificity. In the latter case, this high degree of specificity is usually associated with the mode of action of the drug, involving enzymes or enzyme systems.

As our knowledge of enzyme systems increases, the development of more nearly perfect organic drugs will be made possible. Furthermore, this increased knowledge will help us to explain the mode of action of a number of valuable organic drugs now in use.

Many antimetabolites of amino acids that exhibit a reversible competitive activity have been prepared. None is recognized as an official drug. Some are incorporated into proteins and true enzymes in vivo to produce the phenomenon called lethal synthesis. Some of the naturally occurring antibiotics such as cycloserine, etc., are antimetabolites of certain amino acids. Thus, nature has been successful in the production of antimetabolites of amino acids just as it has in the case of atropine which blocks the action of acetylcholine.

In the case of the steroid hormones a tremendous number of modifications has been reported in the literature. These modifications were prepared in order to develop more potent steroids, particularly those that emphasize one activity at the expense of another—for example, the accentuation of the anti-inflammatory activity of the corticoids, the anabolic activity of the androgens, etc. An attempt has been made in this revision to include the important changes that have been made in the steroid structures and to correlate these changes with their effects upon the biologic activities.

Many modifications of the purines, the pyrimidines, the nucleosides and the nucleotides have been reported in the literature in the quest for antimetabolites that might prove to be useful in the treatment of cancer. Very few of these antimetabolites have survived clinical trial, and even these are not curative agents. No attempt has been made to discuss the changes in the structures of this group of metabolites that have proved to be effective in the production of antimetabolites. Suffice it to say that not only has success been encountered with the preparation of reversible competitive antimetabolites but some of these modified compounds are actually incorporated in vivo into nucleic acids. The latter phenomenon is called lethal synthesis and probably would preclude the use of such antimetabolites as medicinal agents.

The chemical properties frequently determine the locale of absorption. For example, weakly acidic, feebly basic and neutral drugs are absorbed from both the stomach and the intestines. Basic drugs of the order $K_b = 5.4 \times 10^{-4}$ are not absorbed significantly from the stomach but are well absorbed from the intestines provided that the other factors are favorable. On the other hand, strong bases and strong acids are poorly absorbed, if at all, from the G.I. tract. Although strong bases such as streptomycin, curare, etc., are not absorbed orally, many such drugs are absorbed when they are administered by injection.

Among many other factors that no doubt influence the absorption and the distribution of drugs, particle size can play a significant role in the rate of absorption of drugs from the site of administration. In the case of medicinal agents that have a high solubility in both water and lipoids, particle size is much less significant than it is in those cases in which the partition coefficient is very high. Between these extremes an intermediate situation exists.

Advancement in the synthesis of polypeptides has proceeded at such a rapid pace that some very low molecular weight polypeptide hormones such as oxytocin, vaso-pressin, etc., have been synthesized. In addition, some analogs of these polypeptides also have been synthesized in order to delineate the contribution to activity of some of the amino acid residues. The amino acid sequence also has been determined in the complex polypeptide chains A and B of insulin, corticotropin, glucagon, etc. In some cases it has been shown that the biologic activity of these protein hormones is not a function of the molecule as a whole; rather, the activity is due to a portion of the molecule (active center). In these instances fragments of the parent polypeptide chain are active. This poses the possibility that these fragments may be synthesized and analogous synthetic studies carried out as with oxytocin, vasopressin, etc.

In order to appreciate and understand organic medicinal products more fully, the student should be well grounded in such fields as organic, physical, biologic and physiologic chemistry, bacteriology, physiology and zoology.

It is not possible in a textbook of reasonable size to include a complete discussion of all subjects mentioned. Furthermore, the inclusion of commercially available forms of the medicinal compounds per se or in combination is not considered because of their nonfundamental nature. The material presented here should provide a basis for a more detailed study of the scientific literature. It is hoped that the student will use some of the references for this purpose.

2

Physicochemical Properties in Relation to Biologic Action

T. C. Daniels, Ph.D.,
*Emeritus Professor of Pharmaceutical Chemistry, School of Pharmacy,
University of California*

and

E. C. Jorgensen, Ph.D.,
*Professor of Chemistry and Pharmaceutical Chemistry, School of Pharmacy,
University of California*

INTRODUCTION

During the past century, a period which completely encompasses the era of development and growth of the field of synthetic drugs, much consideration has been given to the possible relationships existing between chemical constitution, physical properties and biologic action. During the 19th century, medicinally useful agents were developed most often by isolation from the natural sources known and tested by folk medicine. As the science of chemistry developed, chemical structures were elucidated for these natural products, and it was observed that similar structural units were sometimes present in compounds possessing related biologic activity. For example, in 1869, Crum-Brown and Fraser[1] observed that tertiary amines with varying pharmacologic properties tended to show similar pharmacologic properties when quaternized.

It has been hoped that general principles could be established which would guide the rational development of useful new agents, with select and discrete physiologic properties. Attempts to relate a particular functional group to a particular biologic response have been largely unsuccessful. There are many examples of a wide variety of biologic activities shown by compounds with the same functional groups.[2]

There are also many compounds unrelated chemically which show the same general pharmacologic properties. As examples may be mentioned the relatively large number of different chemical classes of compounds that have local anesthetic properties or may serve as hypnotics or as analgesics. The frequent failure to find a simple relationship between chemical structure, physical properties and biologic action may be accounted for in terms of the complex nature of biologic systems.

Many competing events take place between the introduction of a drug and its final interaction with a specific receptor or organized tissues in which the desired response is to be initiated. Structural features which contribute the proper physical properties must be present to shepherd the drug through these devious pathways, and an adequate number of molecules must survive the passage to bring about a significant reaction with the receptor, or to disrupt the order within organized tissues. In order to exert their biologic effects, drugs must be soluble in and transported by the body fluids, pass various membrane barriers, escape excessive distribution into inert body depots, endure metabolic attack, penetrate to the sites of action and there orient and interact in a specific fashion, causing the alteration of function termed the action of the drug.

The physical and chemical properties of a molecule are determined by the number, the kind and the arrangement of the atoms. Both properties are closely interrelated, and for this reason the term "physicochemical" is a better expression of the properties that relate to biologic action than physical properties per se. In some cases the properties involved relate to a specific chemical function (e.g., dissociation constants of acids and bases), or they may relate to an over-all property such as the solubility of a molecule in a polar or nonpolar solvent.

The physicochemical properties of a compound are measurable characteristics by which the compound may interact with other systems. Biologic response to a drug is a consequence of the interaction of that drug with the living system, causing some change in the biologic processes present before the drug was administered. Since physicochemical properties relate to the processes by which drugs reach and are concentrated at their sites of action, it is important to examine the extent to which any one property correlates with the observed

biologic activity. The possible importance of such properties as solubility, partition coefficients, adsorption, surface activity, degree of dissociation at the pH of the body fluids, interatomic distances between functional groups, redox-potentials (reduction-oxidation), resonance, hydrogen bonding, dimensional factors, chelation and the spatial configuration of the molecule are worthy of consideration. Moreover, the physicochemical properties of the cells on which the drug acts are of considerable theoretic importance and interest. The physical properties of drugs can be varied through chemical modification more or less at will, but for the most part the properties of the normal cell remain constant. Living matter in a very general sense consists of an organized, highly complex colloidal system in which the cell or the body fluids function as the continuous phase, and organized proteins, emulsified fats, etc., are dispersed phases. The living cell surfaces* are immersed in extracellular fluids, and those compounds reaching the cell surface must do so by means of transport through the extracellular fluids. Compounds insoluble in the extracellular fluids and not solubilized by enzyme or other chemical action cannot be transported effectively by these fluids to the cell surface and normally can be expected to show no significant activity. Therefore, all biologically active substances must have or acquire through chem-

* Plant and bacterial cells are surrounded by an outer *nonliving* wall.

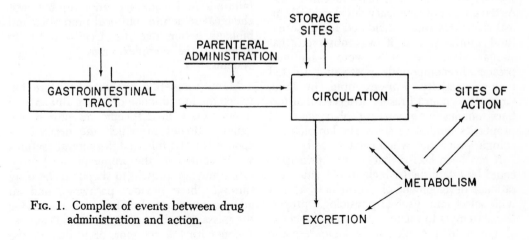

FIG. 1. Complex of events between drug administration and action.

ical modification some minimum solubility in the polar extracellular fluids.

COMPLEX OF EVENTS BETWEEN DRUG ADMINISTRATION AND DRUG ACTION

Following introduction into the body, a drug must pass many barriers and survive competing alternate pathways before it finally reaches the site of action (cell receptors) where the useful biologic response is developed. Before consideration of factors and forces important to the interaction of the drug and the receptor, physicochemical properties affecting the absorption, the distribution and the metabolism of a drug should be considered. A simplified diagram of this system is presented in Figure 1.

INFLUENCE OF ROUTE OF ADMINISTRATION

Parenteral administration of a drug involves no absorption complications, if carried out by the intravenous, the intraarterial, the intraspinal or the intracerebral routes, for these place the drug directly into the body fluids. However, the subcutaneous, the intramuscular, the intradermal, the intraperitoneal, etc., routes produce a depot from which the drug must reach the blood or the lymph in order to produce systemic effects. Factors of importance in determining the rate at which this takes place are those which determine the dissolution rate of the drug and its transfer from one phase to another.[3] These factors are similar to those which have been studied in far greater detail for absorption of a drug from the gastrointestinal tract.

Following oral administration of a drug, dissolution rate is of primary importance in determining eventual levels attained in the blood and the tissues.[4, 5] If the drug is too insoluble in the environment of the gastrointestinal tract to dissolve at an appreciable rate, it cannot diffuse to the gastrointestinal wall and be absorbed. Variation in particle size and surface area, coating, chemical modification, etc., producing differences in rates of dissolution, provide an important

approach to "prolonged action" medication.[6]

Following dissolution of a drug, its action in tissues outside of the gastrointestinal tract must be preceded by passage through the membranes separating the lumen of the stomach and the intestines from the mucosal blood supply. The same factors influencing the penetration of the gastrointestinal-plasma barrier are important in the penetration of other membranes such as the blood-brain barrier and select cells and tissues. The major differences are in variations in acidity for various body compartments: the high acidity of the stomach (pH 1 to 3.5), the far less acidic environment of the lumen of the intestine (duodenal contents, pH 5 to 7; duodenum to ileo-cecal valve, pH 6 to 7; lower ileum, pH about 8), and the essentially neutral environment of the circulating fluids of the body and of the tissues and organs supplied by these fluids (plasma, cerebrospinal fluid, pH 7.4).

Absorption from the gastrointestinal tract, as well as penetration of other membrane barriers, may be passive or active. Substances which are normal cellular metabolites, or close chemical relatives, may pass the membrane by a process of active absorption, energy being used by the body to effect the transfer of such normal food stuff as glucose and amino acids from the gastrointestinal tract to the plasma. The preferential cellular uptake of potassium ion over sodium ion is assumed to involve such a carrier system. Enzymatic systems distinct from those involved in metabolic activity have been found in bacteria to be associated with membrane penetration and given the general name "permeases."[7] Some lipid-insoluble substances penetrate cell membranes by a passive diffusion process, so that the cell wall is thought of as sievelike with a connection via the pores between aqueous phases on both sides of the membrane.[8] The rate of such passive diffusion through aqueous channels depends on the size of the pores, the molecular volume of the solute, and the differences in transmembrane concentrations. Penetration through pores has been shown to be important for absorption from the intestine only for those drugs with molecular weights less than 100.[9] Pores of other sizes must be present in other tissues;

for example, the glomerulus of Bowman's capsule in the kidney is permeable to molecules smaller in size than albumin (molecular weight 70,000).

However, although there are specialized transport mechanisms for natural cell substances and diffusion through pores for small polar molecules, most organic compounds foreign to the body penetrate tissue cells as though the boundaries were lipid in nature, with passage across these barriers predictable from the lipid solubilities of the molecules. This concept was advanced in 1901 by Overton[10] following a study of the lipid solubility of organic compounds and the relationship between this property and ease of penetration of cells. The fatlike nature of cellular membranes has since stood the test of considerable experimentation.

The lipid solubility of drugs has been shown to be an important physical property in governing the rate of passage through a variety of membrane barriers.[11] Examples include passage across the mucosal membranes of the oral cavity (*buccal* and *sublingual* absorption), the gastrointestinal membranes (stomach, small intestine, colon), through the skin, across the renal tubule epithelium, into the bile, the central nervous system and tissue cells. Most drugs are weak acids or bases, and the degree of their ionization, as determined by the dissociation constant (pKa) of the drug and pH of the environment, influences their lipid solubilities. The un-ionized molecule possesses the higher lipid solubility and passes most membrane barriers more readily than does the ionized molecule. The highly charged nature

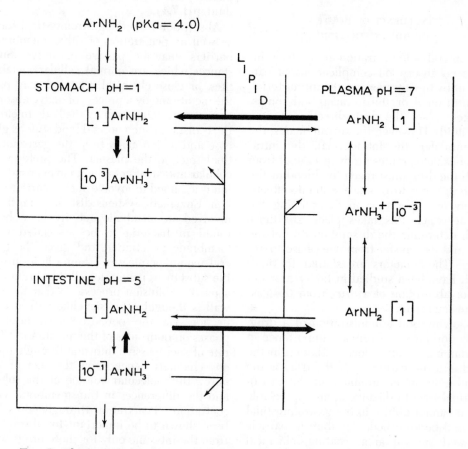

FIG. 2. Theoretical distribution between gastrointestinal tract and plasma for a lipid soluble aromatic amine with pKa = 4.0. Data from Brodie and Hogben.[14]

of the lipoprotein making up the cell wall accounts for this difference in the ability of undissociated molecules and their ions to penetrate the cell. The electrostatic forces interacting between the ion and the cell wall serve to repel or bind the ion, thus decreasing cell penetration. Also, hydration of the ion results in a species larger in size than the undissociated molecule, and this may interfere with diffusion through pores.

The dissociation constant (pKa) is the negative log of the acidic dissociation constant and is the preferred expression for both acids and bases.

for acids $R-COOH = R-COO^- + H^+$

$$pKa = pH + \log \frac{\text{un-ionized acid}}{\text{ionized acid}}$$

$$= pH + \log \frac{[RCOOH]}{[RCOO^-]}$$

for bases $RNH_3^+ = R-NH_2 + H^+$

$$pKa = pH + \log \frac{\text{ionized base}}{\text{un-ionized base}}$$

$$= pH + \log \frac{[RNH_3^+]}{[RNH_2]}$$

An acid with a small pKa (pKa = 1) placed in an environment with pH 7 would be almost completely ionized

$$(R-COO^-/RCOOH = 10^6/1);$$

it would be classed as a strong acid. Weak acids have a large pKa; weak bases, a small pKa.

Absorption From the Stomach

It was noted in 1940[12] that large doses of alkaloids were not toxic in the presence of a highly acidic gastric content. When the gastric contents were made alkaline, the animals rapidly died. This indicated that the gastric epithelium is selectively permeable to the undissociated alkaloidal bases, which produce their toxic effects upon absorption. Later, the observation by Shore[13] that certain parenterally administered weak bases concentrated in the gastric juice led to the development of a general pH–partition hypothesis explaining the rate and the extent of absorption of ionizable drugs from the gastrointestinal tract. A simplified example of this process is presented in Figure

2. When a lipid-soluble, moderately weak base such as an aromatic amine ($ArNH_2$; pKa = 4.0) is administered orally and passes into the strongly acidic environment of the stomach (pH = 1), the base will exist largely (1,000/1) in the poorly lipid soluble ionic form ($ArNH_3^+$) and will be only slowly absorbed through the gastric epithelium. As a specific example, absorption of aniline (pKa = 4.6) from the rat stomach during a 1-hour exposure was measured as 6 per cent of the administered dose.[14] Weaker bases (pKa < 2.5), such as acetanilide (pKa = 0.3), caffeine (pKa = 0.8) and antipyrine (pKa = 1.4), are absorbed better (36 to 14%), since they are significantly un-ionized even in this strongly acidic environment. By decreasing the acidity of the rat stomach to pH 8 with sodium bicarbonate, even aniline is absorbed to the extent of 56 per cent. This approaches a maximum value under the experimental conditions used, limited by the rate of blood flow in the gastric mucosa. These results are consistent with the concept of the gastric mucosa being selectively permeable to the lipid-soluble undissociated form of drugs.

Weak acids (pKa 2.5 to 10) exist largely in the un-ionized form in the stomach and are well absorbed. For example, salicylic acid (pKa = 3.0) is 61 per cent absorbed, benzoic acid (pKa = 4.2) is 55 per cent absorbed from the rat stomach following 1-hour exposure. That lipid solubility is the physical property governing the passage of uncharged molecules across membrane barriers is supported by the observation that 3 barbiturates with similar pKa values were absorbed at rates proportional to the lipid/water partition coefficients (K = chloroform/water) of the un-ionized forms.[15] Thiopental (pKa = 7.6; K = > 100) was absorbed very rapidly, the less lipid-soluble secobarbital (pKa = 7.9; K = 23.3) less rapidly, and barbital with its poor lipid solubility (pKa = 7.8; K = 0.7) was absorbed very slowly. The same pattern of absorption has been observed in man.[16]

Substances completely ionized, and therefore poorly lipid soluble, at the pH of the stomach (or the intestines), such as the strongly acidic sulfonic acids ($R-SO_3H$)

and strongly basic quaternary ammonium compounds (R_4N^+), are not well absorbed.

The stomach also serves as a "site of loss" for weak bases administered intravenously. Since these exist largely in the un-ionized form in the blood (Fig. 2), they penetrate cellular membranes readily, including the lipid barrier between the mucosal blood supply and the stomach, where they may be trapped as the ions. This site of loss has been confirmed for many basic drugs,[13] since they have been found concentrated in gastric juice following intravenous administration.

ABSORPTION FROM THE INTESTINES

When weak bases pass from the strongly acidic environment of the stomach into the less-acidic intestinal lumen, the extent of ionization decreases as shown in Figure 2. The concentration of un-ionized species for a base with $pKa = 4.0$, is about 10 times that of the ionized species, and, since the neutral molecule freely diffuses through the intestinal mucosa, the drug is well absorbed.

When aniline ($pKa = 4.6$) was perfused through the small intestine of the rat over a wide concentration range, and at the fairly rapid rate of 1.5 ml. per minute, 53 to 59 per cent of the administered dose was absorbed, even though the time of drug contact was only about 7 minutes. Although the measured pH of the content of the lumen of the intestine is 6.5, absorption studies are more consistent with a "virtual" pH of 5.3 for the absorbing surface of the intestinal mucosa.[17] This is derived principally from the observation that a drastic reduction in the extent of absorption from the intestine occurred for acids with a $pKa < 2.5$ (strong acids), and for bases with a $pKa > 8.5$ (strong bases). The rat colon has been shown[18] to follow the same pattern as the intestine, with lipid solubility being the primary physical property governing the rate of drug absorption.

Studies of physicochemical factors related to absorption following oral administration of quaternary ammonium ions illustrate additional complexities in the problem of absorption from the gastrointestinal tract. Some appear to be unabsorbed, e.g., pyrvinium pamoate (Povan) and dithiazanine io-

dide, anthelmintic drugs whose lack of absorption prevents undesirable systemic effects and preserves high gastrointestinal concentrations for toxic effects against intestinal parasites. Others, in spite of their permanent ionic character, cross the intestinal epithelium but at a very slow rate as compared with most uncharged molecules. This rate falls with time following administration, suggesting the formation of nonabsorbable complexes with the charged carboxyl and sulfonic acid residues of intestinal mucosa. Mucin added to the intestinal loop has been shown to decrease the rate of absorption of quaternary compounds.[19]

When the relatively inactive trimethylene-bis(trimethylammonium) dichloride was administered orally together with an active hypotensive bis-quaternary compound, IN 292, a marked enhancement of effect has been shown, although none was shown following concomitant intravenous administration.[20] It is postulated that the inactive quaternary competes with the active quaternary for mucosal binding sites, allowing enhanced absorption of the active molecule.

$$(CH_3)_3\overset{+}{N} - (CH_2)_3 - \overset{+}{N}(CH_3)_3 \cdot 2\,Cl^-$$

Trimethylene-bis(trimethylammonium) dichloride

IN 292

ABSORPTION OF DRUGS INTO THE EYE

When a drug is applied topically to the conjunctival sac, a portion will pass directly through the conjunctiva membrane into the blood and the remainder passes through the cornea at rates dependent on the degree of ionization and the partition coefficient of the drug.[21] Since the un-ionized molecule possesses the higher lipid solubility, weak acids penetrate more rapidly from solutions having a low pH and weak bases from solutions buffered at high pH values.

Drugs may pass from the bloodstream into the ocular fluid by two general routes: (1) through the epithelium of the ciliary body, and (2) through the capillary walls and connective tissue of the iris. The rate of absorption into the aqueous humor appears to parallel closely the partition coefficient of the drug.

SITES OF LOSS

Relatively few drug molecules will survive to reach the site of action in a complex biologic system. The sites to which a drug may be lost may be reversible storage depots, or enzyme systems which produce metabolic alteration to a more or less active form, or the drug may be excreted before or after metabolism (Fig. 1). All of these ways in which the drug may be lost rather than react at the normal site have been collectively described as "sites of loss." The distribution between the sites of loss and of action is largely dependent on the physicochemical properties of the drug, including solubility, degree of ionization, and the nature and the strength of the forces binding the drug at these sites.

Storage Sites

Body compartments exist in a variety of types, each characterized by the nature of the physicochemical factors which retain the drug in competition with other sites. The gastrointestinal tract has already been mentioned as such a site, retaining molecules which lack adequate lipid solubility, small size or special transport systems.

Protein Binding. Binding of drugs by plasma protein is usually readily reversible, with most drugs bound to proteins in the albumin fraction.[14] Binding resembles salt formation, for generally the ionic form of a drug interacts with plasma protein, aided by secondary binding by nonpolar portions of the molecule. The latter forces may be adequate alone, for drugs which are not electrolytes are also protein bound, e.g., hydrocortisone. The resulting protein binding may act as a transport system for the drug, which, while bound, is hindered in its access to the sites of metabolism, action and excretion. The drug-protein complex is too large to pass

through the renal glomerular membranes and therefore remains in the circulating blood, thereby prolonging the duration of action. Protein binding not only may prevent rapid excretion of the drug; also, it limits the amount of free drug available for metabolism and for interaction with specific receptor sites. For example, the trypanocide, suramin, remains in the body in the protein-bound form for several months following a single intravenous injection. Its slow dissociation releases enough free drug for protection against sleeping sickness.

There exists a high degree of structural specificity for the interaction between plasma proteins and many small molecules. Drug binding by protein is generally more dependent on detailed chemical structure than are the competing events of drug absorption and localization in lipoidal tissues, which are most dependent on solubility in nonpolar solvents. For example, specific structural requirements for binding of thyroxine analogs to a thyroxine-binding fraction of serum albumin have been established.[22] For maximal binding, the molecular features of thyroxine are most favorable: a diphenyl ether nucleus, 4 iodine atoms, a free phenolic hydroxyl group and an alanine side chain, or an anionic group separated by 3 carbon atoms from the aromatic nucleus. The resulting association constant of 500,000 is significantly reduced if any of these structural features are altered (see Table 1). Compounds possessing single aromatic rings have very low association constants ranging from 8 to 11,000, the strongest of these being salicylic acid and 2,4-dinitrophenol.

In general, structural requirements for plasma protein binding are related, but not as specific as those for the biologic receptor. Thus, salicylate may give some thyroxine-like effects in high doses by displacing thyroxine from serum protein, but there is no indication that salicylate will elicit a thyroxinelike effect at the cell receptor.

Other drugs may exert an indirect biologic effect by displacing active substances from protein binding. Thus, the sulfonylurea antidiabetic agents are thought to exert some activity by displacing insulin from its com-

TABLE 1. RELATIONSHIP BETWEEN STRUCTURE OF THYROXINE ANALOGS
AND BINDING BY THYROXINE-BINDING ALBUMIN

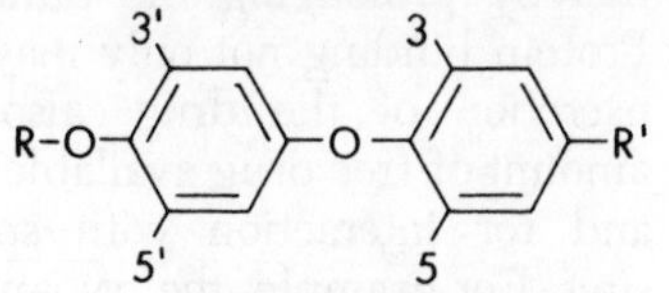

| | SUBSTITUENTS | | | ASSOCIATION |
R	3′, 5′	3, 5	R′	CONSTANT
H	I_2	I_2	$CH_2CH(NH_2)COOH$	500,000
CH_3	I_2	I_2	$CH_2CH(NH_2)COOH$	20,000
H	I, H	I_2	$CH_2CH(NH_2)COOH$	24,600
H	I_2	I_2	CH_2CH_2COOH	160,000
H	I_2	I_2	CH_2COOH	100,000
H	I_2	I_2	COOH	72,000
H	I_2	I_2	$CH_2CH_2NH_2$	32,000
H	Cl_2	Cl_2	$CH_2CH(NH_2)COOH$	23,400
H	$(NO_2)_2$	$(NO_2)_2$	$CH_2CH(NH_2)COOH$	6,600
H	H_2	I_2	$CH_2CH(NH_2)COOH$	6,400
H	H_2	I, H	$CH_2CH(NH_2)COOH$	5,060
H	H_2	H_2	$CH_2CH(NH_2)COOH$	660

Data from Sterling.[22]

plex with plasma protein, as well as by releasing insulin from pancreatic β-cells.

The protein-bound anticoagulants bishydroxycoumarin (Dicoumarol) and warfarin (Coumadin) are displaced from protein by many other drugs, including phenylbutazone (Butazolidin), clofibrate (Atromid-S), norethandrolone (Nilevar), sulfonamides, etc. This displacement of the anticoagulants potentiates their action by increasing the amount of free drug available for competitive inhibition of vitamin K in the clotting process. The resulting increase in clotting time may lead to hemorrhaging. Displacement of a protein-bound drug by administration of another drug is more generally a therapeutic hazard than has been commonly recognized.

Tissue proteins or related tissue constituents may also bind drugs, thus providing depots outside of the plasma. For example, the antimalarial drug quinacrine (Atabrine) shows a 2,000-fold concentration in liver over plasma 4 hours after administration; in 14 days of daily administration, the concentration in liver is 20,000 times that in plasma. Similar concentration of drug occurs in other body tissues such as lung, spleen and muscle.

Neutral Fat. Neutral fat constitutes some 20 to 50 per cent of body weight and as such makes up a depot of considerable importance. Drugs with high partition coefficients (lipid/water) are concentrated in these inert depots. Physical solution in lipid has been suggested[23, 24] as the principal reason for the rapid disappearance of ultrashort-acting barbiturates from the plasma (see Chap. 15). Thiopental, a thiobarbiturate with pKa = 7.6, is approximately 50 per cent ionized in the plasma (pH = 7.4). However, the high lipid solubility for the undissociated

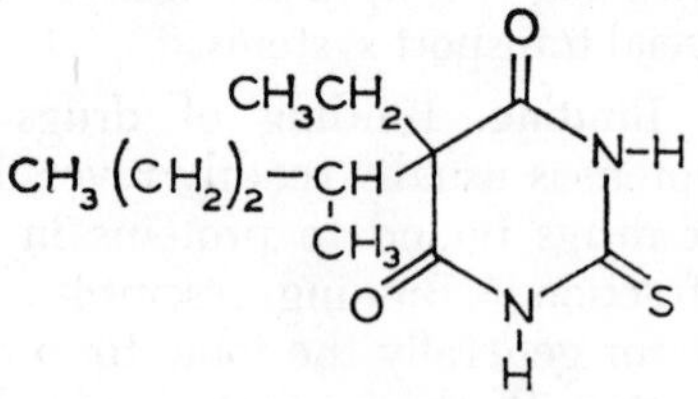

Thiopental pKa = 7.6

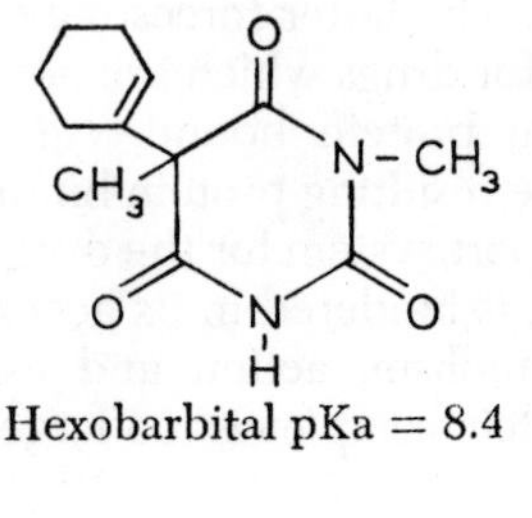

Hexobarbital pKa = 8.4

molecule as compared with its oxygen analog causes this to partition rapidly into neutral fat, thus decreasing blood levels below those adequate for the maintenance of anesthesia. Some N-methyl bartituric acid analogs (e.g., hexobarbital) also have a short duration of anesthetic activity. This has been attributed to the influence of the N-methyl group on acid strength; these have a pKa of about 8.4 compared with 7.6 for those without the N-methyl substituent. At physiologic pH hexobarbital exists largely in the undissociated form which would distribute rapidly into the lipid depots.

An alternate explanation has been proposed,[25] since the blood supply to neutral fat is too poor to account for the initial depletion from the plasma which takes place within a few minutes. The lean body tissues such as viscera and muscle are well perfused with blood, and their cells possess the lipid-permeable membrane which allows these to serve as the initial depot. Redistribution to body fat, metabolism and excretion appear to occur more slowly, and these are related to prolongation of depression rather than rapid recovery from anesthesia.

Lipid accumulation has also been implicated in the long duration of action of the adrenergic blocking agents (see Chap. 19), dibenamine and dibenzyline.[26, 27] Some 20 per cent of the initial dose is rapidly deposited in fat, followed by a slow return to the bloodstream. The strength of the covalent bond formed at the active sites with these alkylating agents also appears to play a role in their long duration of action.[28]

As an example of a different type relating to either neutral fat depots or "lipophilic receptor sites," the neuromuscular blocking agent hexafluorenium (Mylaxen) is relatively ineffective in the unanesthetized animal.[29]

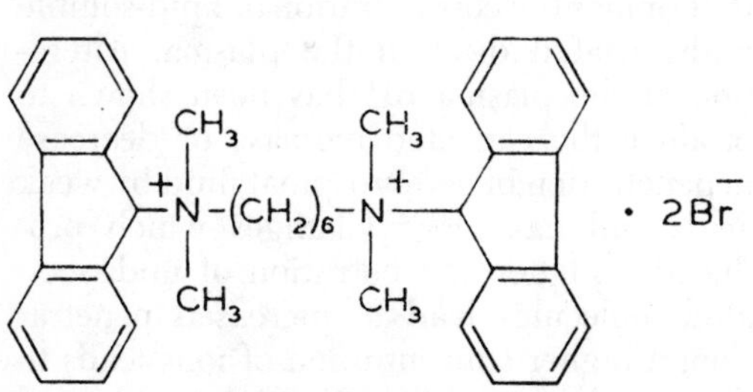

However, in the presence of anesthesia with cyclopropane, ether and related lipophilic anesthetics, potent blockade of muscular function occurs. It was assumed that the lipophilic fluorene groups were absorbed by biologically inert sites of loss which are lipid in nature, and that concomitant administration of a nonpolar anesthetic would result in greater saturation of these lipophilic receptors. This would result in a decrease in available sites of loss with a resulting increase in the hexafluorenium available at the neuromuscular junction. As a test for this hypothesis, relatively inert compounds more closely related in structure to the lipophilic portion of the drug were administered: dibenzylamine, 9-dimethylaminofluorene and its quaternary derivative. All caused significant potentiation of the hexafluorenium effect in the order:

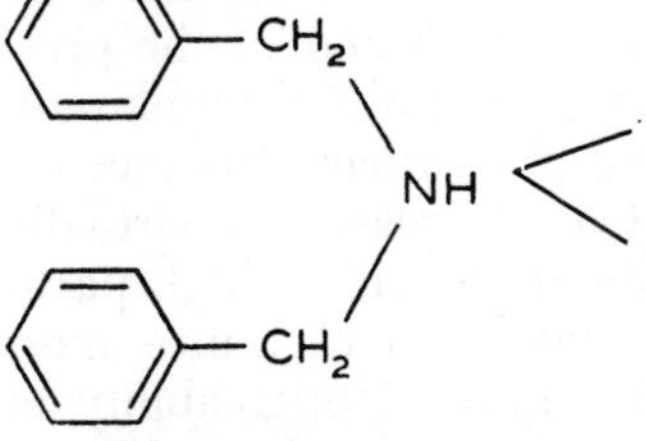

Dibenzylamine

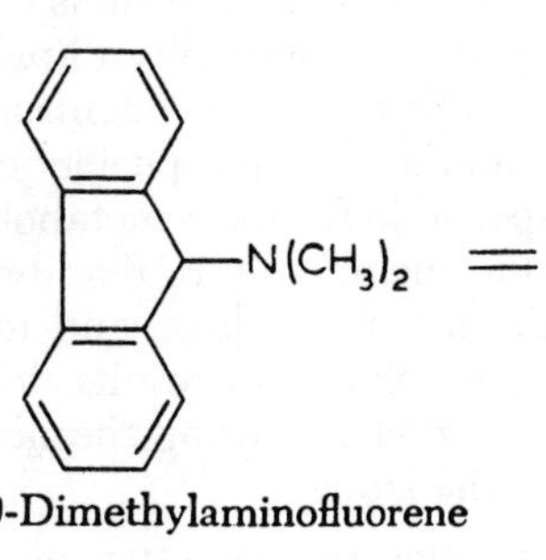

9-Dimethylaminofluorene

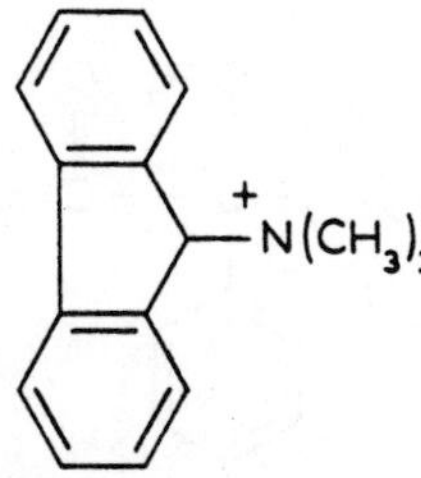

Hexafluorenium

Quaternary Derivative

The coplanarity of the fluorene rings could also permit enhanced binding by van der Waals' forces and could account for an increased effect over dibenzylamine. The equal effect for the tertiary and the quaternary bases demonstrates that the lipid portion, rather than the ionic, is involved.

This property of synergistic activity for biologically inert substances, involving the blockage of sites of loss and conservation of drugs, has further examples at other sites.

METABOLISM AND EXCRETION

The nature of the processes involved in drug metabolism and excretion are discussed in detail in Chapter 3. Excretion, either of the unaltered drug or of its metabolites, is an irreversible site of loss. However, metabolic alteration of the drug may lead to a metabolite with enhanced, reduced or essentially unchanged biologic activity.

One of the major routes of excretion is by way of the kidney, which implies the presence or the formation of a water-soluble substance. Following glomerular filtration, tubular reabsorption into plasma is virtually complete for substances with a high partition coefficient (lipid/water). Since most active drugs (by virtue of their ability to penetrate lipid cellular membranes) are lipid soluble, metabolic conversion, usually in the liver, to a more polar form is essential for their excretion. Presumably, a lipid membrane surrounds the liver microsomes in which are found the nonspecific enzyme systems responsible for most metabolic conversions. This membrane is readily penetrated by the lipophilic drug, and metabolism to a more polar form results, followed by increased excretion during the next passage through the kidney.

The potentiation of the action of a wide variety of drugs, such as analgesics, central nervous system stimulants and depressants, etc., by the compound SKF 525 (β-diethylaminoethyl 2,2-diphenylvalerate) has been accounted for on the basis of its inhibition of many metabolic reactions.[30] This is another example of a synergistic effect by blocking a site of loss.

SITES OF ACTION

After a drug reaches the bloodstream, and a portion of it survives distribution to sites of loss, other cell boundaries must be crossed before it reaches its site of action.

Penetration of Drugs into Tissue Cells

The capillary wall is sufficiently porous to permit the passage of water-soluble molecules of relatively large size. The boundaries of organ tissue cells present a barrier of lipid character to the passage of foreign substances. A completely ionized molecule such as hexamethonium does not enter tissue cells,[31] but lipid-soluble, un-ionized molecules possessing high partition coefficients readily penetrate a variety of cells and tissues. Thus, in dogs, phenobarbital has been shown[32] to be increased in concentration in body tissues (brain, fat, liver and muscle) when the plasma pH was lowered. Presumably, the increased concentration of undissociated molecules facilitated cell membrane penetration, providing a shift of drug from extracellular to intracellular fluids.

A wide variation occurs in the rate at which various drugs penetrate cerebrospinal fluid and the brain. Here, too, lipid solubility of the un-ionized molecule is the physical property largely governing the rate of entry.[33] The dissociation constant of a weak acid or base is of importance insofar as this determines the concentration of lipid-soluble undissociated drug in the plasma. Alteration of the plasma pH has been shown to produce the expected increase or decrease in penetration of cerebrospinal fluid by weak acids and bases.[33, 34] Changes which produced a higher concentration of undissociated molecules lead to increased penetration; a higher concentration of ions leads to decreased penetration. Sulfonic acids and

SKF 525

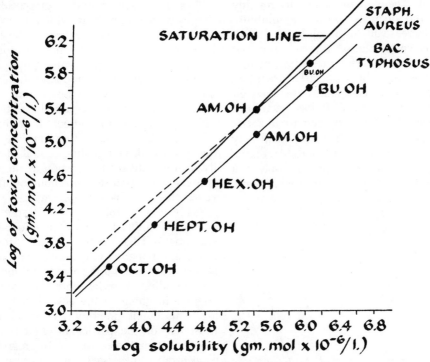

FIG. 3. Bactericidal concentration vs. solubility for normal primary alcohols. (Ferguson, J.: Proc. Roy. Soc. (Ser. B) 127:387, 1939)

quaternary ammonium compounds do not penetrate the cerebrospinal fluid in any significant amount.

The same characteristic lipid-barrier is found in a variety of other cells, including those of some bacteria. The toxic effects of certain chelating agents, such as 8-hydroxy-quinoline (oxine), are best explained in terms of bacteria cell penetration by the lipid-soluble saturated Fe (oxine)$_3$ species (see this chapter, chelating agents).

Thus solubility and partition coefficients play a role of primary importance in determining the presence, the absence or the relative intensity of biologic action. This is well illustrated by a consideration of this physical property and its influence within certain homologous series.

SOLUBILITY AND PARTITION COEFFICIENTS

In homologous series of undissociated or slightly dissociated compounds[35] in which the change in structure involves only an increase in the length of the carbon chain, gradations in the intensity of action have been observed for a number of unrelated pharmacologic groups of compounds, e.g., normal alcohols (antibacterial), alkyl resorcinols, alkyl hydrocupreines, alkyl phenols and cresols[36] (antibacterial), esters of *p*-aminobenzoic acid (anesthetics), alkyl 4,4'-stilbenediols[37] (estrogenic). It is well recognized that increasing the nonpolar portion of a molecule by increasing the length of a hydrocarbon side-chain produces certain rather definite changes in the physical properties: the boiling point increases, solubility in water decreases, the partition coefficient increases, surface activity increases, viscosity increases. Normally, the lower members of a homologous series show a low order of biologic activity; with the increasing length of the carbon chain (nonpolar portion of the molecule), the activity increases, passing through a maximum. Further increase in the length of the carbon chain results in a rapid decrease in the activity. The increase in activity roughly parallels the decrease in water solubility and the increase in lipid solubility (partition coefficient). It also may be associated with an increase in surface activity. Changes of this type may be associated with the availability of the compound for the cell where the action occurs. It has been suggested[35] that the observed decrease in activity with increasing length of the chain may be due to the diminishing solubility of the compounds in water or, more precisely, the lowered solubility in the extracellular fluid, in which the cell is immersed, that serves as a medium of transport to the cell surface.

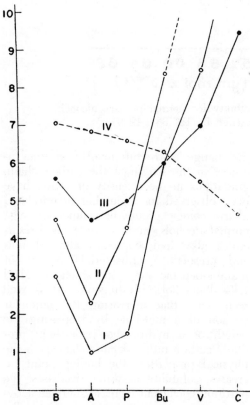

FIG. 4. B = Benzoic, A = Phenylacetic, P = β-Phenylpropionic, Bu = γ-Phenylbutyric, V = δ-Phenylvaleric, C = ϵ-Phenylcaproic. I. Solubility (concentration $1/m \times 10^{-1}$). II. Distribution coefficient (cottonseed oil/H_2O). III. Bactericidal activities (concentration $1/m \times 10^{-3}$). IV. Surface tension (dynes/cm. $\times 10^{-1}$).

This has been illustrated graphically by Ferguson in Figure 3, in which are plotted the log of toxic concentration $v.$ the log of solubility of the normal alcohols for two organisms. Also given on the graph is the "saturation line." A compound falling on this line would have to be present at the concentration of its saturated solution to show the bactericidal effect. If a line for a series crosses this saturation line, then the series will have a sharp cutoff of activity at the point of intersection as the series is ascended, because those compounds beyond the crossover point which would appear on the dotted line will not have enough solubility to give a bactericidal concentration. This neatly accounts for the observation made with a number of substances that the biologic activity increases on ascending a series and then abruptly falls off in going to the next higher homolog. This cutoff point will depend on the resistance of the particular organism. The more resistant the organism, the higher the concentration necessary for killing and the earlier the cutoff will appear in the series.

The ω-phenyl substituted carboxylic acids illustrate the relationship of solubility, partition coefficients and antibacterial activity for a homologous series of compounds.[38] Benzoic acid, although not truly a member of the homologous series, is included in Figure 4 which compares the water solubilities, the distribution coefficients, the surface tension lowering and the bactericidal properties for these homologs. The series falls short of showing the length of the carbon chain necessary for decreased activity, but it can be assumed that with continued increase in the length of the aliphatic side chain the activity will pass through a maximum then diminish rapidly. In homologous series of this type the polar portion (hydrophilic carboxyl group) remains constant, and the nonpolar portion of the molecule (hydrocarbon) is increased. As a result of this type of change the water solubility decreases, lipid solubility increases, and the partition coefficients increase as the series is ascended.

The 4-n-alkylresorcinols and the n-alcohols also illustrate the relationship of bio-

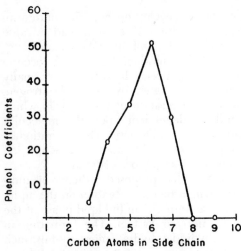

FIG. 5. Phenol coefficients of 4-*n*-alkyl-resorcinols against *B. typhosus*.

TABLE 2. PHENOL AND PARTITION
COEFFICIENTS OF ESTERS OF
p-HYDROXYBENZOIC ACID

ESTER	INHIBITION OF FERMEN-TATION	Staph. aureus BACTERI-CIDAL	PARTITION Co-EFFICIENT*
Methyl	3.7	2.6	1.2
Ethyl	5.3	7.1	3.4
n-Propyl	25.0	15.0	13.0
i-Propyl	15.0	13.0	7.3
n-Butyl	40.0	37.0	17.0
Amyl	53.0	......	150.0
Allyl	15.0	12.0	7.6
Benzyl	69.0	83.0	119.0
Phenol	1.0	1.0	......

* Lipid/water.

logic activity in homologous series differing only in the length of the carbon chain. The phenol coefficients of 4-*n*-alkylresorcinols against *B. typhosus* are shown in Figure 5. Against this organism, a maximum is reached when the alkyl side-chain has 6 carbons. Schaffer and Tilley[39] studied the same series of resorcinols and observed that the phenol coefficients for *Staph. aureus* continued to increase through the 4-*n*-nonylresorcinol. This variation in sensitivity of different organisms to members of a homologous series of compounds (selective toxicity) may be due to a number of factors. For a more extensive discussion of these, *see* Albert, A.: Selective Toxicity, 1968.

The normal aliphatic alcohols show a regular increase in antibacterial activity as the homologous series is ascended from methyl through octyl alcohols.[39] The branched chain alcohols are more water-soluble and have lower partition coefficients than the corresponding primary normal alcohols; the branched chain alcohols are also less active as antibacterial agents. *n*-Hexyl alcohol is more than twice as active as the secondary hexyl alcohol and 5 times as active as the tertiary hexyl alcohol. The higher molecular weight alcohols (e.g., cetyl) are inactive as antibacterial agents; therefore, the alcohols have the same type of curve as the alkylresorcinols against *Staph. aureus*, i.e., activity increasing with increase in the length of the carbon side-chain through a maximum.

Esters of a number of aromatic carboxylic acids have been studied, and it has been observed that as the carbon length of the alcohol is increased the antibacterial activity rapidly increases.[40] The phenol coefficients of the esters of vanillic acid are reported as follows: methyl, 1.7; ethyl, 7.3; *n*-propyl, 33.4. The isopropyl ester, which has a lower partition coefficient than the *n*-propyl ester, has a phenol coefficient of only 11.2.

The partition coefficients and observed antibacterial activity in a series of esters of *p*-hydroxybenzoic acid are quite parallel, as shown in Table 2.[41]

The discussion of partition coefficients thus far has dealt largely with antibacterial agents. Historically, partition coefficients were first correlated with the biologic activity of hypnotic and narcotic drugs by Overton[10] and Meyer.[42] The importance of the physical properties of the central nervous system depressants is now generally recognized. Many of the compounds of this class, such as the volatile anesthetics, are relatively inert chemically.

The theory of narcosis, first expressed by Meyer in 1899, involved three postulates. As expressed by Falk,[43] they are as follows:

1. All chemically indifferent substances which are soluble in fats and fatlike bodies must exert a narcotic action on living proto-

plasm insofar as they can become distributed in it.

2. The effect must manifest itself first and most markedly in those cells in which fatty or lipid substances predominate in the chemical structure and, presumably, form essential participants of the cell function, viz., the nerve cells.

3. The relative efficiency of such narcotic agents must be dependent on their mechanical affinity for lipid substances on the one hand and the remaining body constituents, i.e., principally water, on the other hand. Their efficiency is therefore dependent on the partition coefficient which determines their distribution in a mixture of water and lipid substances.

This theory has been supported in that excellent correlation between hypnotic activity and the partition coefficient has been observed[44, 45] for many compounds. However, this correlation is not proof that the above proposed mechanism is correct. It should be recognized that the theory relates only to the availability of the compound for the site of action and does not suggest a mechanism of action for the hypnotic drugs. Furthermore, it does not explain why all compounds with high partition coefficients do not show hypnotic properties.

A possible explanation for the latter problem may be related to molecular size. The recognition that the chemically unreactive rare gas xenon was capable of producing general anesthesia[46] led Wulf and Featherstone[47] to call attention to the relationship between some fundamental properties of molecules and their depressant effects. They pointed out that a correlation exists between the constants "a" and "b" in the van der Waals' equation (these terms measure the sphere of influence of a molecule) and the presence or the absence of anesthetic potency. In general, a critical "size" (van der Waals' "b," relating to molecular volume) was found necessary for the anesthetic molecule. This was larger than that for substances (such as H_2O, "b" = 3.05; O_2, "b" = 3.18; N_2, "b" = 3.91) which might normally occupy the lateral space separating lipid and protein molecules of the cell. Molecular volumes ("b" values) for some of the anesthetic agents are: N_2O, 4.4; Xe, 5.1;

ethylene, 5.7; cyclopropane, 7.5; chloroform, 10.2; ethyl ether, 13.4. None had a value lower than 4.4. Wulf and Featherstone suggest that the anesthetic agents may occupy the space between lipid layers normally occupied by water, oxygen and nitrogen, causing a separation of these layers which would be dependent upon the molecular volume of the anesthetic. This alteration in cell structure could produce a depression of function leading to anesthesia.

Pauling[48] has proposed a theory of anesthesia which focuses attention on the aqueous phase, rather than the lipid phase of the central nervous system. The formation in the brain fluid of hydrate microcrystals, such as those known in vitro for chloroform, xenon and other anesthetic agents, is suggested. The anesthetic agents, together with side chains of proteins and other solutes in the encephalonic fluid, could occupy and stabilize by van der Waals forces chambers made up of water molecules. The resulting microcrystalline hydrates could alter the conductivity of electrical impulses necessary for maintenance of mental alertness, leading to narcosis or anesthesia.

Most of the current theories of general anesthesia are based on positive correlations obtained between the partial pressures of agents required to produce anesthesia and such physical properties as solubility in oil, the distribution between oil and water, the vapor pressure ("thermodynamic activity") of the pure liquid or the partial pressure of hydrate crystals. All of these physical properties which correlate with anesthetic activity are related to the van der Waals attraction of the molecules of anesthetic agent for other molecules, and all are interrelated, since the energy of intermolecular attraction is approximately proportional to the polarizability (mole refraction) of the molecules of anesthetic agent. As yet, no direct experimental evidence uniquely supports one theory of mode of anesthetic action at the molecular level.

A quantitative measure of the importance of partition behavior on drug action has been introduced. Hansch[49] has determined the effect of substituent groups on distribution between water and the nonpolar solvent, 1-octanol. The distribution coefficients for the parent compound, e.g., phenoxyacetic acid ($C_6H_5OCH_2COOH$) and a derivative, e.g., 3-trifluoromethylphenoxyacetic acid ($3\text{-}CF_3\text{-}C_6H_5OCH_2COOH$) are measured, and a value, π, for the substituent trifluoromethyl group is determined by the difference between the logarithm of the distribution coefficients:

$$\pi_{CF_3} = \log P_{CF_3} - \log P_H,$$

where P_{CF_3} is the partition coefficient of the 3-trifluoromethyl derivative, and P_H that of the unsubstituted parent compound. The π values, taken in conjunction with the Hammett sigma (σ) values[50] (measures of the electronic contributions of substituents relative to hydrogen) have been used effectively in correlating chemical structure, physical properties and biologic activities.

Hansch has assumed that a rate-limiting condition for many biologic responses involves the movement of the drug through a large number of cellular compartments made up of essentially aqueous or organic phases. The molecule possessing solubility and structural characteristics such that the sum of the free energy changes is minimal for the many partitionings made between phases, including adsorption-desorption steps at solid surfaces, will have ideal lipohydrophilic character and will most easily reach its site of action. The π value is a measure of the substituent's contribution to solubility behavior in such a series of partitions.

Table 3 lists some typical substituent constants[51] arranged in order of decreasing contribution to lipophilic character when substituted in the 3-position of phenoxyacetic acid. Values of π and σ are approximately constant and additive in a variety of different aromatic systems, as long as no strong group interactions occur. Therefore, the substituent constants for a polysubstituted aromatic compound are approximately equal to the sum of the π and σ values for individual substituents. The additive character of these constants has been demonstrated by good correlations obtained from the action of polysubstituted phenols on gram-negative and gram-positive organisms, the action of thyroxine analogs on rodents and the carcinogenic activity of derivatives of dimethyl-

TABLE 3. CONSTANTS FOR SOLUBILITY (π) AND ELECTRONIC (σ) EFFECTS OF 3-SUBSTITUENTS IN PHENOXYACETIC ACID[*]

R	π[†]	σ[‡]
$n\text{-}C_4H_9$	$+1.90$	-0.15
SCF_3	$+1.58$	$+0.51$
SF_5	$+1.50$	$+0.68$
$n\text{-}C_3H_7$	$+1.43$	-0.15
OCF_3	$+1.21$	$+0.35$
I	$+1.15$	$+0.28$
CF_3	$+1.07$	$+0.55$
C_2H_5	$+0.97$	-0.15
Br	$+0.94$	$+0.23$
SO_2CF_3	$+0.93$	$+0.93$
Cl	$+0.76$	$+0.23$
SCH_3	$+0.62$	-0.05
CH_3	$+0.51$	-0.17
OCH_3	$+0.12$	-0.27
NO_2	$+0.11$	$+0.78$
H	0	0
$COOH$	-0.15	$+0.27$
$COCH_3$	-0.28	$+0.52$
CN	-0.30	$+0.63$
OH	-0.49	-0.36
$NHCOCH_3$	-0.79	-0.02
SO_2CH_3	-1.26	$+0.73$

[*] Data from Hansch.[51, 53]
[†] $\pi = \log P_x - \log P_H$, where P_x and P_H are the partition coefficients between 1-octanol and water.
[‡] $\sigma =$ Hammett sigma constant for 4-substituents.

aminoazobenzene and of aromatic hydrocarbons and benzacridines.[49]

A different set of π values has been obtained for substituents not attached to an aromatic nucleus.[52] In a homologous series, if functional groups are removed by two or more methylene (CH_2) groups, interaction is small and values may usually be determined additively. Both the methyl and methylene ($-CH_2-$) groups have an additive π value of about $+0.50$; thus π values for a homologous series substituted in the 3-position of phenoxyacetic acid are: $H = 0$; $CH_3 = 0.51$; $C_2H_5 = 0.97$; $n\text{-}C_3H_7 = 1.43$; $n\text{-}C_4H_9 = 1.90$.

Relative to hydrogen $= 0$, a positive value for π means that the group enhances solubility in nonpolar solvents, a negative value

TABLE 4. ANALOGS OF CHLOROMYCETIN TESTED AGAINST STAPHYLOCOCCUS AUREUS[*]

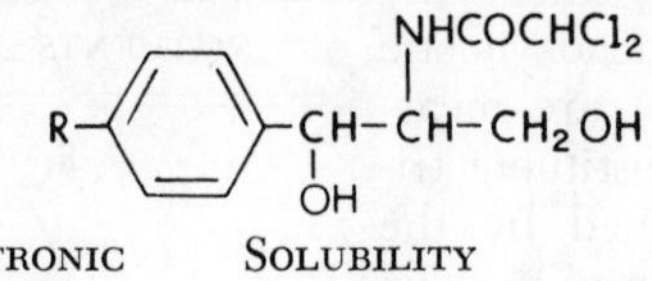

$$R-\underset{\underset{OH}{|}}{\underset{|}{\overset{NHCOCHCl_2}{|}}}-CH-CH-CH_2OH$$

SUBSTITUENT R	ELECTRONIC σ[†]	SOLUBILITY π	LOG A[‡] CALCULATED	LOG A[‡] OBSERVED
NO$_2$	0.71	0.06	1.77	2.00
CN	0.68	−0.31	1.47	1.40
SO$_2$CH$_3$	0.65	−0.47	1.27	1.04
COOCH$_3$	0.32	−0.04	0.89	1.00
Cl	0.37	0.70	1.08	1.00
N≡N−C$_6$H$_5$	0.58	1.72	0.69	0.78
OCH$_3$	0.12	−0.04	0.46	0.74
NHCOC$_6$H$_5$	0.22	0.72	0.76	0.40
NHCOCH$_3$	0.10	−0.79	−0.28	−0.30
OH	0	−0.62	−0.29	< −0.40
COOH	0.36	−0.16	0.90	< −0.40

[*] Data from Hansch.[53]
[†] σ = Hammett sigma constant for 3-substituents.
[‡] A = activity relative to chloromycetin = 100.

that solubility in polar solvents is enhanced. A positive value for σ denotes an electron-attracting effect; a negative value denotes electron-donation by the group. Thus, the methyl group is typical of alkyl groups, in enhancing nonpolar solubility ($\pi = +0.51$) and is electron donating ($\sigma = -0.17$). By contrast, the acetamido group (CH$_3$-CONH−) as a substituent strongly enhances water solubility ($\pi = -0.79$) and is a weak electron acceptor ($\sigma = +0.10$).

Particularly noteworthy is the exceptionally strong lipophilic character of fluoro-substituted groups as compared with the hydrogen-substituted analog; e.g., CF$_3$ > CH$_3$; SCF$_3$ > SCH$_3$; OCF$_3$ > OCH$_3$; SO$_2$CF$_3$ > SO$_2$CH$_3$. The frequent enhancement of biologic activity when a hydrogen, a methyl or a halogen is replaced by the trifluoromethyl group may be related to the significant contribution to lipophilic character.

In relating the application of the pi (π) and sigma (σ) substituent constants to biologic activity, Hansch has derived the equation:[53]

$$\log (1/C) = -k\pi^2 + k'\pi + \rho\sigma + k''$$

where C is the concentration of drug necessary to produce the biologic response

(log A, the logarithm of relative biologic activities is equally applicable), k, k′ and k″ are constants for the system being studied, ρ (rho) is a reaction constant, π is the substituent constant for solubility contribution, and σ is the substituent constant for electronic contributions. In this form, contributions by steric factors are assumed to be constant as substituents are varied.

As an example of an application of these substituent constants, the relative antibacterial activities of chloromycetin (R = NO$_2$) and a series of its analogs have been compared,[53] in which the 4-nitro group has been varied in its nature.

When substituent constants, π and σ, and relative observed antibacterial activities are substituted in the equation $\log A = -k\pi^2 + k'\pi + \rho\sigma + k''$, the system and the reaction constants which best fit the experimental data are: $\log A = -0.54\pi^2 + 0.48\pi + 2.14\sigma + 0.22$. A comparison of observed antibacterial activities and those calculated from the derived equation (Table 4) shows excellent correlation. From these data, it is concluded that a strong electron-attracting group enhances activity ($\sigma_{NO_2} = +0.71$), as does a moderately lipophilic group ($\pi_{NO_2} = +0.06$). The great potential such correlations hold for directing the course of structure-activity studies is apparent.

Benzeneboronic acids ($X–C_6H_5–B-(OH)_2$ are carriers of boron, which, if localized in tumor tissue, could be useful in the treatment of cancer. Radiation with neutrons would lead to neutron capture by boron and release local high concentrations of high energy alpha radiation capable of destroying the tumor. The problem of structural factors leading to selective localization of compounds in tumor tissue has been evaluated by the Hansch method,[54] and it has been found that penetration of the brain is highly dependent on π, while localization of boronic acids in tumor tissue is dependent on electron-releasing substituents ($-\sigma$). Since the compounds are not significantly ionized at physiologic pH, it is suggested that an electron-releasing group, which would facilitate cleavage to boric acid, might release this polar molecule inside the tumor, where it would be trapped by lipophilic barriers. Alternatively, electron release might enhance binding of the boronic acid with an electron deficient component of the tumor tissue.

The hypnotic activities of a variety of drugs, including barbiturates, tertiary alcohols, carbamates, amides, and N,N-diacylureas (see Chap. 15) have been correlated with their distribution behavior using the model nonpolar-polar system, octanol-water.[55] The most active depressant drugs, of all classes, have partition coefficients of about $100/1(\log P = 2)$ in the octanol/water system. All effective hypnotics contain a very polar nonionic portion of the molecule, as illustrated by their large negative π values: 5,5-unsubstituted barbituric acid, -1.35; hydroxyl (–OH), -1.16; carbamate (–OCONH$_2$), -1.16; carboxamide (–CONH$_2$), -1.71; N,N-diacylurea (–CONHCONHCO–), -1.68. In addition, they possess hydrocarbon or halogenated hydrocarbon residues which are sufficiently lipophilic to provide the intact molecule with nonionic surface-active character, and a distribution coefficient ($\log P$) in the usual range of 1 to 3.

Examples of the additive nature of the Hansch substituent constants (π) in estimating the partition coefficient ($\log P$), and the closeness of this value to the ideal coefficient

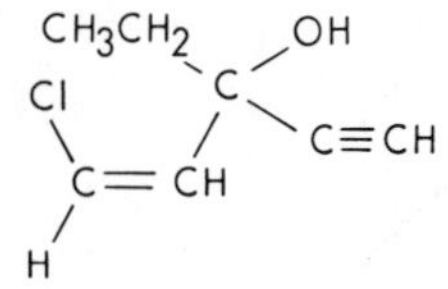

Ethchlorvynol
(Placidyl)

Substituent	π
C – OH	-1.16
C≡CH	0.84
CH$_3$CH$_2$	1.00
ClHC = CH	1.32

$\Sigma \pi = 2.00 = \log P$

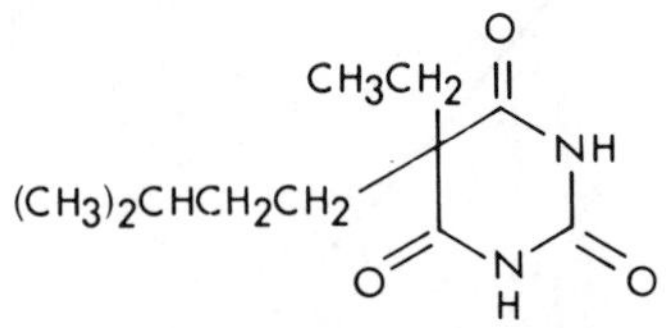

Amobarbital
(Amytal)

Substituent	π
–CCONHCONHCO	-1.35
CH$_3$CH$_2$	1.00
(CH$_3$)$_2$CHCH$_2$CH$_2$	2.30

$\Sigma \pi = 1.95 = \log P$

for hypnotics ($\log P = 2$), are illustrated with calculations for the hypnotic-sedative drugs amobarbital, a barbiturate, and ethchlorvynol, an acetylenic tertiary alcohol.

In addition to correlations based upon electronic and solubility constants for substituents, parameters for steric contributions of substituents have been applied.[56] Steric constants (Es) derived from substituent effects on the rates of hydrolysis of aliphatic esters or *ortho*-substituted benzoic acid esters, or calculated values based upon van der Waals radii, have been used to correlate structure-activity relationships in substituted phenoxyethylcyclopropylamine [R–C$_6$H$_4$OCH$_2$CH$_2$-N–CH(CH$_2$)$_2$] monoamine oxidase inhibitors. The reduced activity produced by *meta* substitution was best correlated with steric inhibition of fit to the enzyme surface.

Although partition coefficients may be correlated with the observed biologic activ-

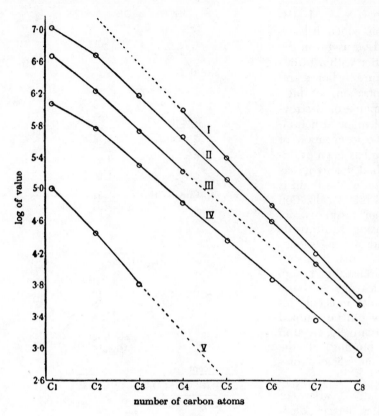

FIG. 6. Properties of normal primary alcohols. I. Solubility (mole $\times$ $10^{-6}/$ liter). II. Toxic concentration for *B. typhosus* (mole $\times$ $10^{-6}/$liter). III. Concentrations reducing surface tension of water to 50 dynes/cm. (mole $\times$ $10^{-6}/$ liter). IV. Vapor pressure at 25° (mm. $\times$ 10^4). V. Partition coefficient between water and cotton-seed oil ($\times$ 10^3). (Ferguson, J.: Proc. Roy. Soc. (Ser. B) 127:387, 1939)

ity, it should be remembered that other properties also are involved and that the biologic action must be due to the over-all properties of the molecule and not to any one physical or chemical property.

FERGUSON PRINCIPLE

The observation that many compounds containing diverse chemical groups show narcotic action is itself indicative that mainly physical rather than chemical properties are involved. The fact that narcotic action is attained rapidly and remains at the same level as long as a reservoir or critical concentration of the drug is maintained but quickly disappears when the supply of drug is removed suggests that an equilibrium exists between the external phase and the phase at the site of action in the organism designated the *biophase*. The numerous relationships between biologic action and physical properties in homologous series are said to be indicative of physical equilibria.

In many such series the toxicity increases; or, stated another way, the equitoxic concentrations decrease as the series is ascended, a generalization known as Richardson's rule. Fühner,[57] in 1904, found this decrease to proceed according to a geometric progression, 1, 1/3, $1/3^2$, $1/3^3$, . . . , as the number of carbon atoms increases arithmetically. This finding holds for a number of series, including alcohols, ketones, amines, esters, urethanes and hydrocarbons. Certain, but not all, physical properties change according to a geometric progression in ascending a homologous series. These include vapor pressure, water solubility, capillary activity and distribution between immiscible phases. Since logarithms represent a geometric progression, a plot of the logarithms of the value of these various properties against the number of carbon atoms gives straight lines (Fig. 6). The attribute these physical properties have in common is that they involve a distribution between heterogeneous phases, e.g., solubility involves distribution between solid or liquid and saturated solution; surface activity the distribution between solution and surface; vapor pressure the distribution between liquid and vapor, etc.; and this log relationship is a consequence of that common attribute. Toxic action of this type also must involve such an equilibrium between the agent in the biophase and the agent in the extracellular fluids.

This logarithmic change in distribution coefficiency, which is the common denominator involved in each of these properties, results from the relation

$$\log k = (\overline{F}°_2 - \overline{F}°_1) / RT$$

in which the distribution coefficient k is a log function of the difference in the partial molal free energies $\overline{F}°_1$ and $\overline{F}°_2$ of the substance in its standard states in phases 1 and 2. Each additional CH_2 group gives rise to a constant increment in the difference between the partial molal free energies.

The fact that the biologic effect parallels some physical property, such as the oil/water distribution ratio, is in itself not evidence that a particular mechanism is involved; for example, that narcosis takes place in a lipid medium. On the contrary, it may relate only to the fact that both the biologic effect and the oil/water distribution ratio have in common a heterogeneous phase distribution.

Ferguson[58] advanced the concept that it is unnecessary to define the nature of the biophase, or receptor, nor is it necessary to measure the concentration at this site. If equilibrium conditions exist between the drug in the biophase and that in the extracellular fluids, although the concentration in each phase is different, the tendency for the drug to escape from each phase is the same. In such a system the partial molal free energy for the substance must be equal in each phase ($\overline{F}_1 = \overline{F}_2$), since this serves as a quantitative measure of the escaping tendency from that phase. The degree of saturation of each phase is a reasonable approximation of the tendency to escape from that phase, and this may be called the *thermodynamic activity*. Since this thermodynamic activity is the same in both the biophase and the extracellular phase, measurements made in the latter, which is

TABLE 5. ISONARCOTIC CONCENTRATIONS OF GASES AND VAPORS FOR MICE AT 37° C.[*]

SUBSTANCE	VAPOR PRESSURE (mm.) p_s	NARCOTIC CONCENTRATION % BY VOLUME c	PARTIAL PRESSURE (mm.) AT NARCOTIC CONCENTRATION $(760xc/100) = p_t$	APPROXIMATE THERMODYNAMIC ACTIVITY p_t/p_s
Nitrous oxide	59,300	100	760	0.01
Acetylene	51,700	65	494	0.01
Methyl ether	6,100	12	91	0.02
Methyl chloride	5,900	14	106	0.01
Ethylene oxide	1,900	5.8	44	0.02
Ethyl chloride	1,780	5.0	39	0.02
Diethyl ether	830	3.4	26	0.03
Methylal	630	2.8	21	0.03
Ethyl bromide	725	1.9	14	0.02
Dimethylacetal	288	1.9	14	0.05
Diethylformal	110	1.0	8	0.07
Dichlorethylene	450	0.95	7	0.02
Carbon disulfide	560	1.1	8	0.02
Chloroform	324	0.5	4	0.01

[*] Adapted from a table by Ferguson.[58]

TABLE 6. ISOANESTHETIC CONCENTRATION OF GASES AND VAPORS IN MAN AT 37°[*]

SUBSTANCE	VAPOR PRESSURE[*] mm. p_s	ANESTHETIC[†] CONC. IN VOL. % c	PARTIAL PRESSURE AT ANESTHETIC CONC. $(760xc/100) = p_t$	APPROXIMATE THERMODYNAMIC ACTIVITY p_t/p_s
Nitrous oxide	59,300	100	760	0.01
Ethylene	49,500	80	610	0.01
Acetylene	51,700	65	495	0.01
Ethyl chloride	1,780	5	38	0.02
Ethyl ether	830	5	38	0.05
Vinyl ether	760	4	30	0.04
Ethyl bromide	725	1.9	14	0.02
1,2-Dichloroethylene	450	0.95	7	0.02
Chloroform	324	0.5	4	0.01

[*] From data in Table 5 and the Handbook of Chemistry and Physics, Chemical Rubber Company.
[†] From data in Goodman and Gilman.[59]

accessible, may be directly equated with the former, which is inaccessible.

Evidence that toxic action of various types closely parallels the thermodynamic activity of compounds has been given by Ferguson. The relationship has been established for the homologous alcohols in such diverse actions as the inhibition of development of sea urchin eggs, bactericidal action against *B. typhosus*, tadpole narcosis and hemolysis of ox blood. It also holds for the action of various types of compounds with respect to narcosis, bactericidal and insecticidal action. Thus while the bactericidal concentration of an alcohol varies from 10.8 to 0.0034 moles per liter in going from methyl to octyl alcohol, the thermodynamic activity varies from only 0.33 to 0.88.[58] For a variety of depressant gases and vapors (Table 5), the isonarcotic concentrations varied from 100 to 0.5 per cent by volume, but the ratio of the partial vapor pressure to saturation pressure, which gives the approximate thermodynamic activity, varied from only 0.01 to 0.07.

Similar data may be compiled, using the data for partial pressures of gases and vapors required to produce anesthesia in man. As shown in Table 6, the approximate thermodynamic activities (degree of saturation of the vapors) range between 0.01 and 0.05, while the concentration range was 200-fold.

Where the biologic activity parallels thermodynamic activity, the compounds are said to be *structurally nonspecific*. Such compounds exhibit "physical toxicity" and show the following characteristics:

1. Biologic action is related directly to the thermodynamic activity.

2. Compounds widely different chemically give the same effect if their thermodynamic activity is the same.

3. Depressants of cellular function, e.g., anesthetics, belong to this class.

4. An equilibrium appears to exist between the internal biophase and the external phase.

5. If such an equilibrium exists, the thermodynamic activity must be the same in each phase.

6. Measuring the thermodynamic activity in the external phase will then give the thermodynamic activity in the biophase.

7. The thermodynamic activity is, as a first approximation, equal to the ratio p_t/p_s, where p_t is the partial pressure of a gas necessary to give a biologic effect and p_s the saturated vapor pressure.

8. In case of a solution, the thermodynamic activity is approximately equal to S_t/S_o, where S_t is the concentration neces-

sary to give a biologic effect and S_o is the solubility of the substance.

9. In general, substances which are present in the same proportional saturation have the same thermodynamic activity and the same degree of biologic action.

10. Therefore, saturated solutions of different substances should have the same biologic effect.

In contrast with those compounds which are structurally nonspecific and are characterized by a wide variety of chemical types giving a like biologic response, dependent only on their thermodynamic activity, there are those compounds that are said to be *structurally specific.*

Compounds that are structurally specific usually are effective in lower concentrations than those that are nonspecific. However, equilibria are involved with the former as well as with the latter. This may involve equilibria between an external phase and the biophase, or equilibria between the drug and the receptors or the enzymes on or within the cell. The bonds involved may be any of the known types, including covalent, ionic, ion-dipole, dipole-dipole, hydrogen, van der Waals' and hydrophobic bonds. In cases of the structurally specific agents the bonds are likely to be stronger and the equilibrium shifted over to the side favoring maximum biologic activity. The law of mass action and its equations are applicable to such situations. However, it should be realized that physical properties may be important in determining the action of both structurally specific and structurally nonspecific compounds.

Whether a drug is structurally specific or nonspecific is of fundamental importance in developing new therapeutic agents. This can be decided by determining the thermodynamic activity necessary to produce the useful biologic effect ("Ferguson Value," p_t/p_s, S_t/S_o) being studied, using several different chemical types and comparing the values for the new compounds. If the new agents have comparable values, they most probably are structurally nonspecific, and minute variations in structure usually would not be expected to produce marked changes in biologic action.

DRUG–RECEPTOR INTERACTIONS

CHARACTERISTICS OF THE DRUG

Most drugs that belong to the same pharmacologic class have certain structural features in common. These frequently include, for example, a basic nitrogen atom, an aromatic ring, an ester or amide group, a phenolic or alcoholic hydroxyl group, or an aliphatic or alicyclic portion of the molecule. Structural features usually are present in the molecule which permit these "functional" groups to be oriented in a similar pattern in space.

Paul Ehrlich's introduction of the receptor concept provided the basis for relating structural similarities in molecules with similarities in biologic activity. The drug receptor is conceived as a small chemically reactive area which combines with complementary functional groups of a drug to produce a biologic response. The concept of specifically oriented functional areas forming a receptor leads directly to specific structural requirements for functional groups of a drug which must be complementary to the receptor.

Regardless of the ultimate mechanism by which the drug and the receptor interact, the drug must approach the receptor and fit closely to its surface. Steric factors determined by the stereochemistry of the receptor site surface and that of the drug molecules are therefore of primary importance in determining the nature and the efficiency of the drug-receptor interaction. Unless the drug is of the structurally nonspecific cellular depressant type discussed under the Ferguson Principle, it must possess a high degree of structural specificity to initiate a response at a particular receptor.

Some structural features contribute a high degree of structural rigidity to the molecule. For example, aromatic rings are planar, and the atoms attached directly to these rings are held in the plane of the aromatic ring. Thus, the quaternary nitrogen and carbamate oxygen attached directly to the benzene ring in the cholinesterase inhibitor neostigmine are restricted to the plane of the ring, and, consequently, the spatial arrangement of at least these atoms is established.

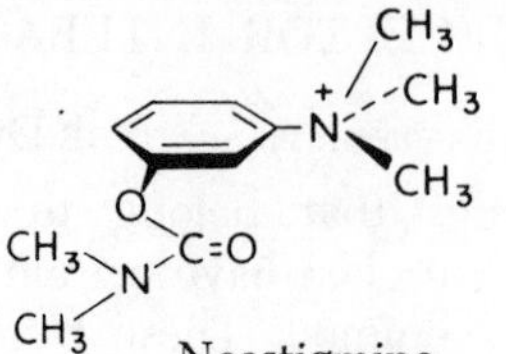

Neostigmine

The relative positions of atoms attached directly to multiple bonds is also fixed. In the case of the double bond, *cis* and *trans* isomers result. For example, diethylstilbestrol exists in two fixed stereoisomeric forms. *Trans*-diethylstilbestrol is estrogenic, while the *cis*-isomer is only 7 per cent as active. In *trans*-diethylstilbestrol, resonance interactions and minimal steric interference tend to hold the two aromatic rings and connecting ethylene carbon atoms in the same plane.

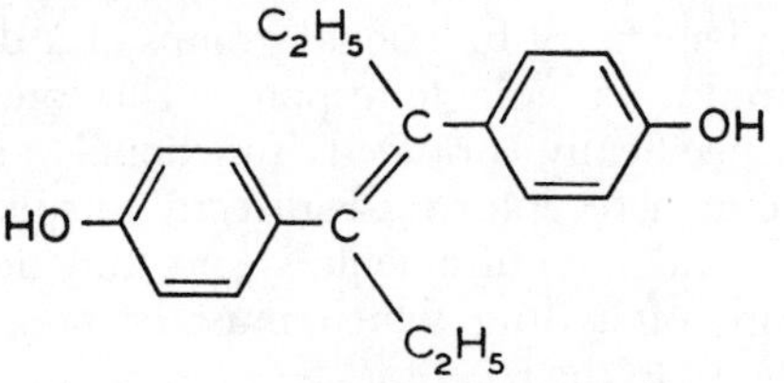

Trans-diethylstilbestrol

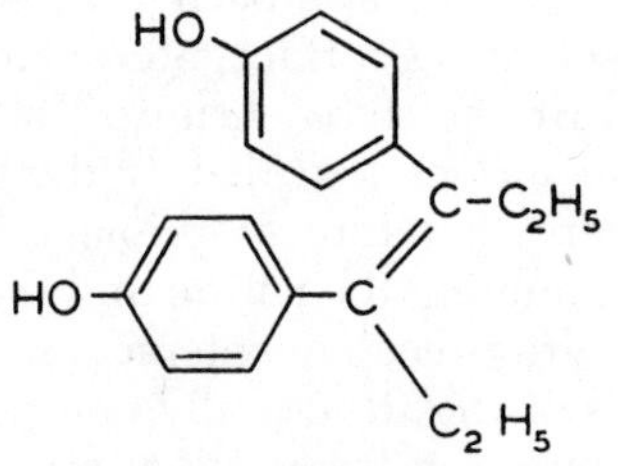

Cis-diethylstilbestrol

Geometric isomers, such as the *cis* and the *trans* isomers, hold structural features at different relative positions in space. These isomers also have significantly different physical and chemical properties. Therefore, their distributions in the biologic medium are different, as well as their capabilities for interacting with a biologic receptor in a structurally specific manner.

More subtle differences exist for *conformational* isomers. Like geometric isomers, these exist as different arrangements in space for the atoms or groups in a single

classic structure. Rotation about bonds allows interconversion of conformational isomers; however, an energy barrier between isomers is often sufficiently high for their independent existence and reaction. Differences in reactivity of functional groups, or interaction with biologic receptors, may be due to differences in steric requirements. In certain semirigid ring systems, such as the steroids, conformational isomers show significant differences in biologic activities (see Chap. 25).

The principles of conformational analysis have established some generalizations in regard to the more stable structures for reduced (nonaromatic) ring systems. In the case of cyclohexane derivatives, bulky groups tend to be held approximately in the plane of the ring, the *equatorial* position. Substituents attached to bonds perpendicular to the general plane of the ring (*axial* position) are particularly susceptible to steric crowding. Thus, 1, 3-diaxial substituents larger than hydrogen may repel each other, twisting the flexible ring and placing the substituents in the less crowded equatorial conformation.

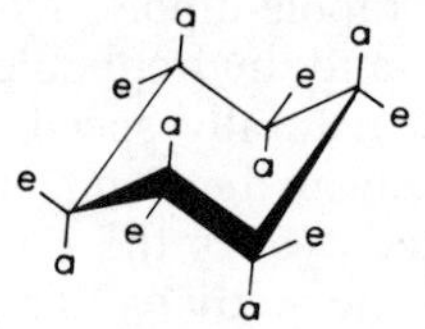

Equatorial (*e*) and *axial* (*a*) substitution in the chair form of Cyclohexane.

Similar calculations may be made for reduced heterocyclic ring systems, such as substituted piperidines. Generally, an equilibrium mixture of conformers may exist. For example, the potent analgesic trimeperidine (see Chap. 24) has been calculated to exist largely in the form in which the bulky phenyl group is in the *equatorial* position, this form being favored by 7 kcal/mole over the *axial* species. The ability of a molecule to produce potent analgesia has been related to the relative spatial positioning of a flat aromatic nucleus, a connecting aliphatic or alicyclic chain, and a nitrogen atom which exists largely in the ionized form at physiologic pH.[60] It might be expected that one of the conformers would be responsible for the analgesic activity; however, in this case it

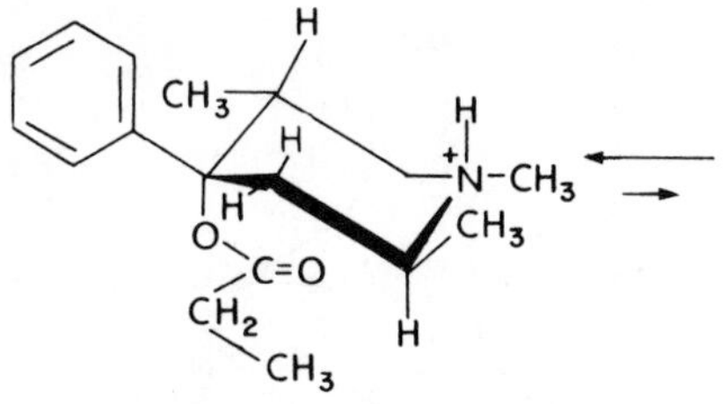

Trimeperidine (*equatorial*-phenyl)

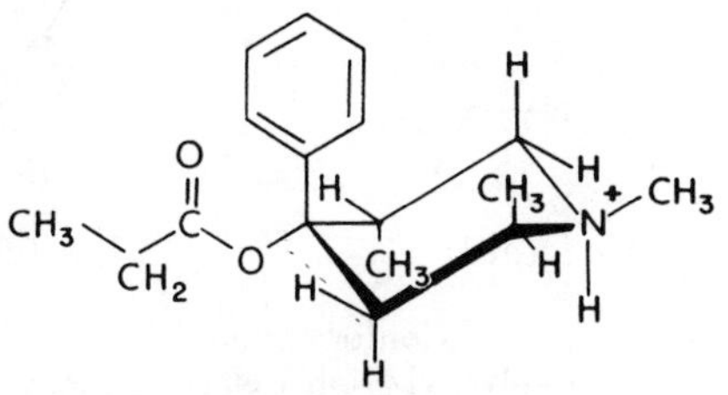

Trimeperidine (*axial*-phenyl)

appears that both the *axially* and the *equatorially* oriented phenyl group may contribute. In structurally related isomers whose conformations are fixed by the fusion of an additional ring, both compounds in which the phenyl group is in the *axial* and those in which it is in the *equatorial* position have equal analgetic potency.[61]

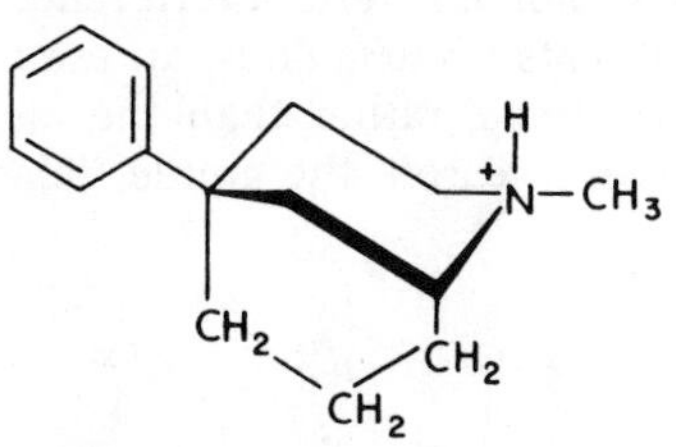

Equatorial-phenyl (Analgesic E.D.$_{50}$ 18.4 mg./kg.)

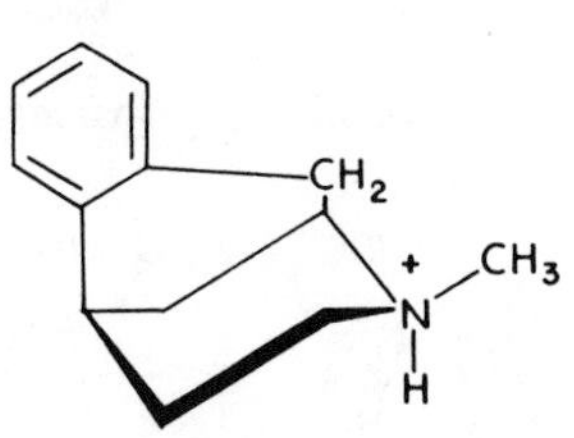

Axial-phenyl (Analgesic E.D.$_{50}$ 18.7 mg./kg.)

Ring-fused analgesics

In a related study of conformationally rigid diastereoisomeric analogs of meperidine, the *endo*-phenyl epimer was found to be more potent than was the *exo*-isomer.[62] However, the *endo*-isomer was shown to penetrate brain tissue more effectively due to slight differences in pK$_a$ values and partition coefficients between the isomers. This emphasizes the importance of considering differences in physical properties of closely related compounds before interpreting differences in biologic activities solely on steric grounds and relative spatial positioning of functional groups.

Open chains of atoms, which form an important part of many drug molecules, are not equally free to assume all possible conformations, there being some which are sterically preferred.[63] Energy barriers to free rotation of the chains are present, due to interactions of nonbonded atoms. For example, the atoms tend to position themselves in space so as to occupy staggered positions, with no two atoms directly facing (eclipsed). Thus, for butane at 37°, the calculated relative probabilities for four possible conformations show that the maximally extended *trans* form is favored 2-to-1 over the two equivalent bent (skew) forms. The *cis* form, in which all of the atoms are facing or *eclipsed*, is much hindered, and only about 1 molecule in 1,000 may be expected to be in this conformation at normal temperatures.

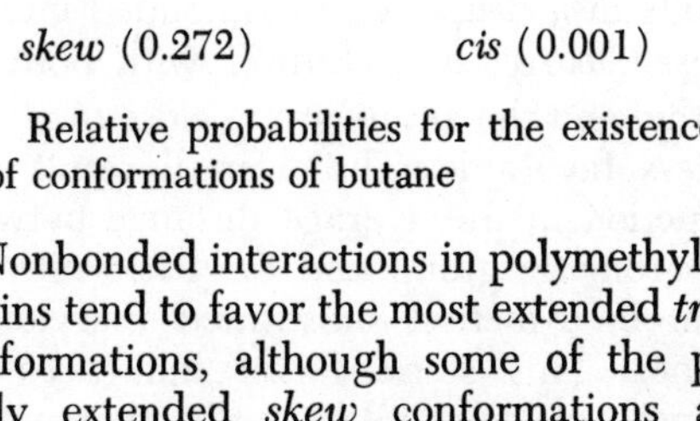

trans (1.0)

skew (0.272)

skew (0.272)

cis (0.001)

Relative probabilities for the existence of conformations of butane

Nonbonded interactions in polymethylene chains tend to favor the most extended *trans* conformations, although some of the partially extended *skew* conformations also exist. A branched methyl group reduces somewhat the preference for the *trans* form in that portion of the chain, and therefore the probability distribution for the length of the chain is shifted toward the shorter distances. This situation is present in substituted chains which contain the elements of many drugs, such as the β-phenylethyl amines. It should be noted that such amines are largely protonated at physiologic pH, and exist in a charged tetra-covalent form. Thus, their stereochemistry closely resembles that of carbon, although in the following diagrams, the hydrogen atoms attached to nitrogen are not shown. As may be expected, the fully extended *trans* form, with

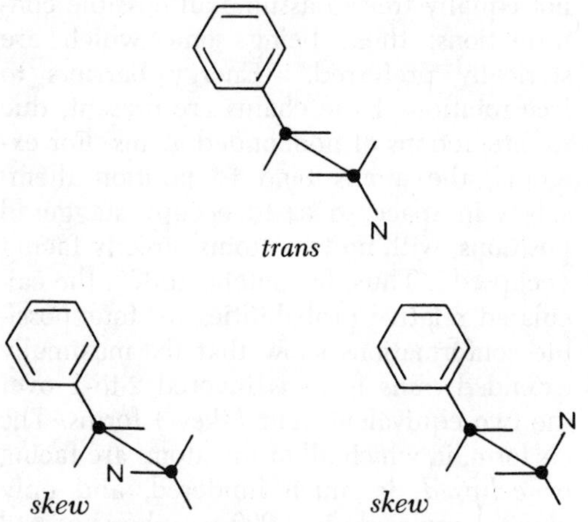

trans

skew form 1   *skew* form 2

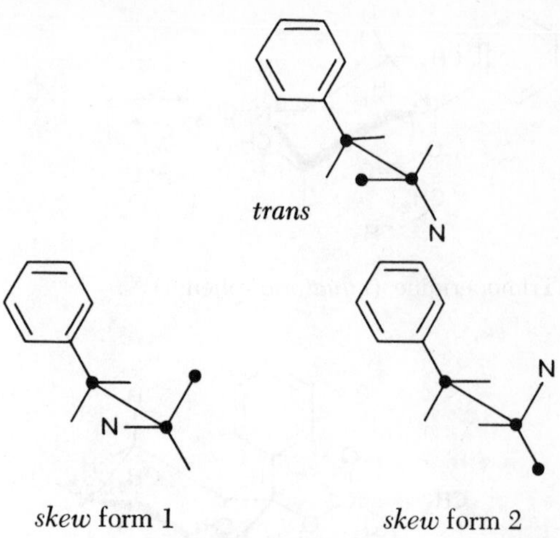

Conformations of
α-Methyl-β-phenylethylamines

trans

skew   *skew*

Conformations of β-phenylethylamines

maximal separation of the phenyl ring and the nitrogen atom, is favored and a smaller population of the two equivalent *skew* forms, in which the ring and the nitrogen are closer together, exists in solution. Introduction of an α-methyl group alters the favored position of the *trans* form, since positioning of the bulky methyl group away from the phenyl group (*skew* form 2) also results in a decrease in nonbonded interactions. Clearly, *skew* form 1 with both the methyl and the amine group close to phenyl is less favorable. The over-all result is a reduction in the average distance between the aromatic group and the basic nitrogen atom in α-methyl substituted β-arylethylamines. This steric factor influences the strength of the binding interaction with a biologic receptor required to produce a

given pharmacologic effect. It is possible that the altered stereochemistry of α-methyl β-arylethylamines may partially account for their slow rate of metabolic deamination (see Chap. 17).

The introduction of atoms other than carbon into a chain strongly influences the conformation of the chain. Due to resonance contributions of forms in which a double bond occupies the central bonds of esters and amides, a planar configuration is favored, in which minimal steric interference of bulky substituents occurs. Thus, an ester is mainly in the *trans*, rather than the *cis* form. For the same reason, the amide link-

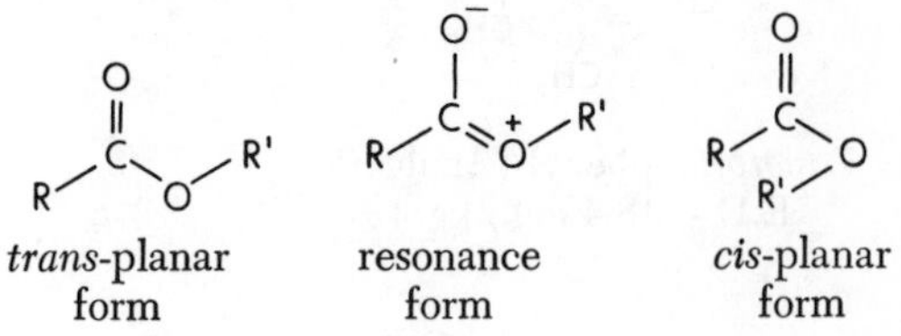

trans-planar   resonance   *cis*-planar
form   form   form

Stabilizing planar structure of esters

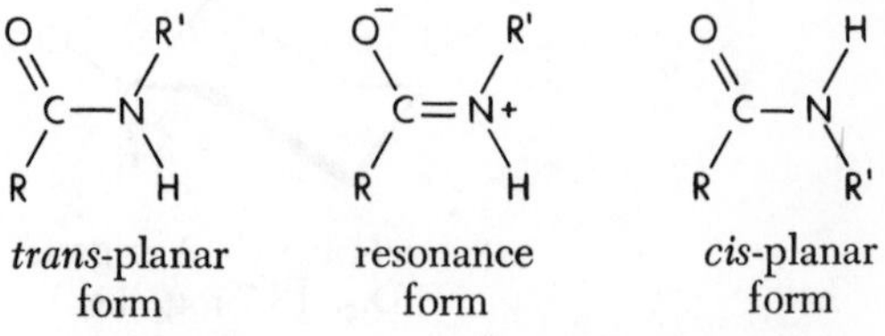

trans-planar   resonance   *cis*-planar
form   form   form

Stabilizing planar structure of amides

age is essentially planar, with the more bulky substituents occupying the *trans* position. Therefore, ester and amide linkages in a chain tend to hold bulky groups in a plane and to separate them as far as possible. As components of the side chains of drugs, ester and amide groups favor fully extended chains and, also, add polar character to that segment of the chain.

The above considerations make it clear that the ester linkages in succinyl choline provide both a polar segment which is readily hydrolyzed by plasma cholinesterase (see Chap. 19), and additional stabilization to the fully extended form. This form is also favored by repulsion of the positive charges at the ends of the chain.

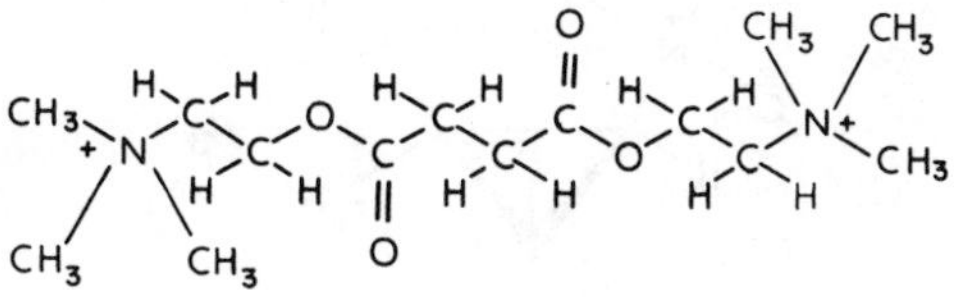

Extended form of succinyl choline

The conformations favored by stereochemical considerations may be further influenced by *intramolecular interactions* between specific groups in the molecule. *Electrostatic forces,* involving attractions by groups of opposite charge, or repulsion by groups of like charge, may alter molecular size and shape. Thus, the terminal positive charges on the polymethylene bis-quaternary ganglionic blocking agent hexamethonium and the neuromuscular blocking agent decamethonium make it most likely that the ends of these molecules are maximally separated in solution.

$$(CH_3)_3\overset{+}{N}{-}(CH_2)_n{-}\overset{+}{N}(CH_3)_3$$
Hexamethonium n=6
Decamethonium n=10

In some cases *dipole-dipole interactions* appear to influence structure in solution. Methadone may exist partially in a cyclic form in solution, due to dipolar attractive forces between the basic nitrogen and carbonyl group.[64] In such a conformation, it

closely resembles the conformationally more rigid potent analgesics, morphine, meperi-

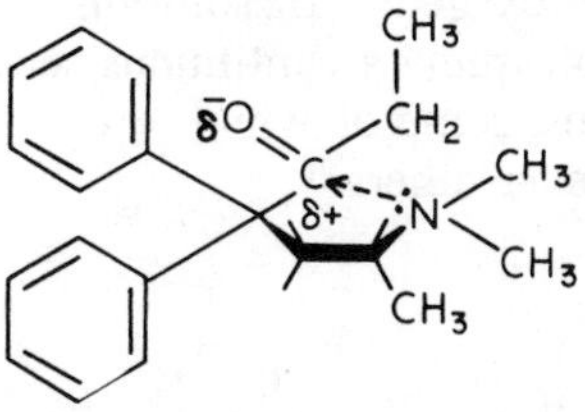

Ring conformation of methadone by dipolar interactions

dine and their analogs (see Chap. 24), and it may be this form which interacts with the analgesic receptor.

An intramolecular *hydrogen bond,* usually formed between donor —OH and =NH groups, and acceptor oxygen (:Ö=) and nitrogen (:N≡) atoms, might be expected to add stability to a particular conformation of a drug in solution. However, in aqueous solution donor and acceptor groups tend to be bonded to water, and little gain in free energy would be achieved by the formation of an intramolecular hydrogen bond, particularly if unfavorable steric factors involving nonbonded interactions were introduced in the process. Therefore, it is likely that internal hydrogen bonds play only a secondary role to steric factors in determining the conformational distribution of flexible drug molecules.

Conformational Flexibility and Multiple Modes of Action. It has been proposed that the conformational flexibility of most open-chain neurohormones, such as acetylcholine, epinephrine, serotonin, and related physiologically active biomolecules, such as histamine, permits multiple biologic effects to be produced by each molecule, by virtue of the ability to interact in a different and unique conformation with different biologic receptors. Thus, it has been suggested that acetylcholine may interact with the muscarinic receptor of postganglionic parasympathetic nerves and with acetylcholinesterase in the fully extended conformation, and in a different, more folded structure, with the nicotinic receptors at ganglia and at neuromuscular junctions.[65, 66] Acetylcholine bromide exists in a quasi-ring form in the crystal, with an N-methyl hydrogen atom close to, and perhaps forming a hydrogen bond with, the backbone oxygen.[67] In solution, however, it is able to assume a continuous series of conformations, some of which are energetically favored over others.[66]

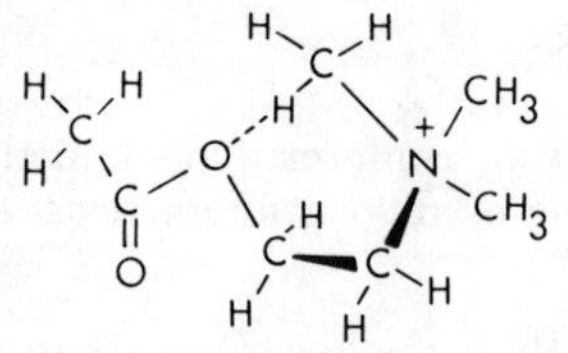

Quasi-ring form of
acetylcholine

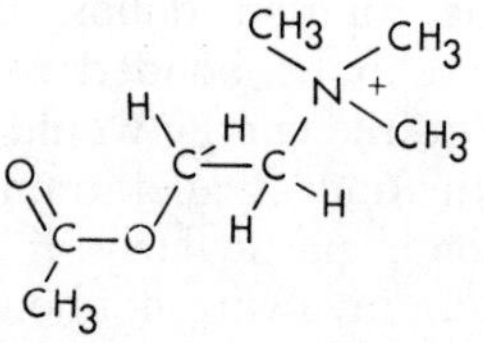

Extended conformation of
acetylcholine

Conformationally rigid acetylcholine-like molecules have been used to study the relationships between these various possible conformations of acetylcholine and their biologic effects. (+)-*Trans*-2-acetoxy cyclopropyl trimethylammonium iodide, in which the quaternary nitrogen atom and acetoxyl groups are held apart in a conformation approximating that of the extended conformation of acetylcholine, was about 5 times as active as acetylcholine in its muscarinic effect on dog blood pressure, and equiactive to acetylcholine in its muscarinic effect on the guinea pig ileum.[68] The (+)-*trans*-isomer was hydrolyzed by acetylcholinesterase at a rate equal to the rate of hydrolysis of acetylcholine. In contrast, the (−)-*trans*-isomer

and the mixed (+),(−)-*cis*-isomers were 1/500 and 1/10,000 as active as acetylcholine in muscarinic tests on guinea-pig ileum. Similarly, the *trans*-diaxial relationship be-

Trans–2-Acetoxy Cyclopropyl
Trimethylammonium Iodide

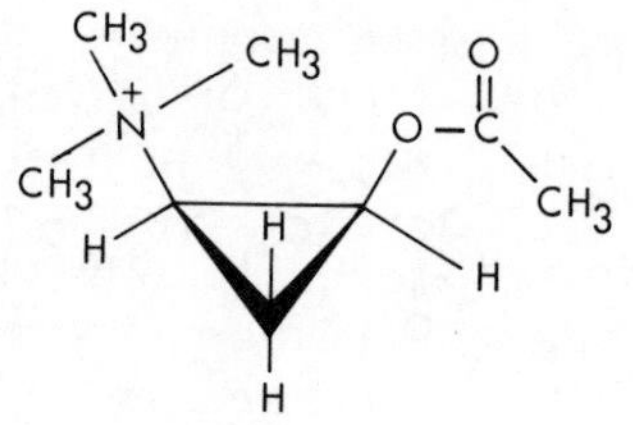

Cis–2-Acetoxy Cyclopropyl
Trimethylammonium Iodide

tween the quaternary nitrogen and acetoxyl group led to maximal muscarinic response and rate of hydrolysis by true acetylcholinesterase in a series of isomeric 3-trimethylammonium-2-acetoxyl decalins.[69] The nico-

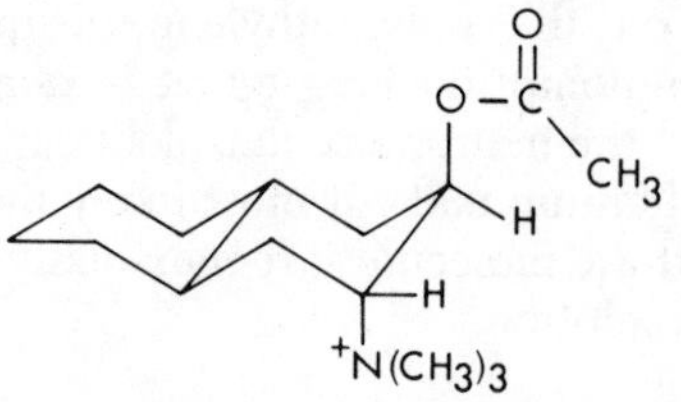

Trans-diaxial 3-Trimethylammonium-2-acetoxy Decalin

tinic response appears to be highly sensitive to steric effects of substituents on the acetylcholine molecule. Even the minimally substituted *cis*- and *trans*-isomers of 2-acetoxy cyclopropyl trimethylammonium iodide showed negligible nicotinic activity, pre-

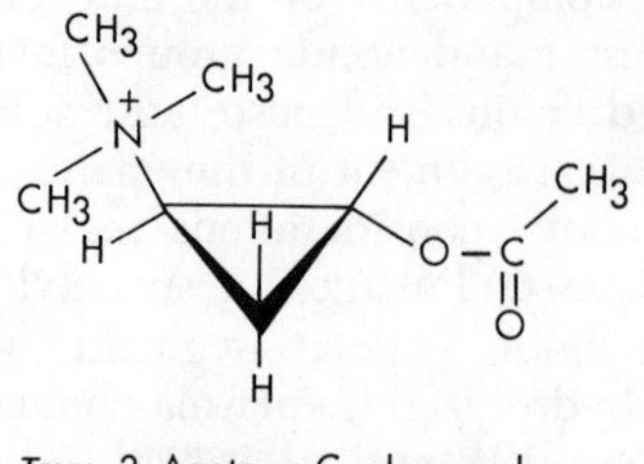

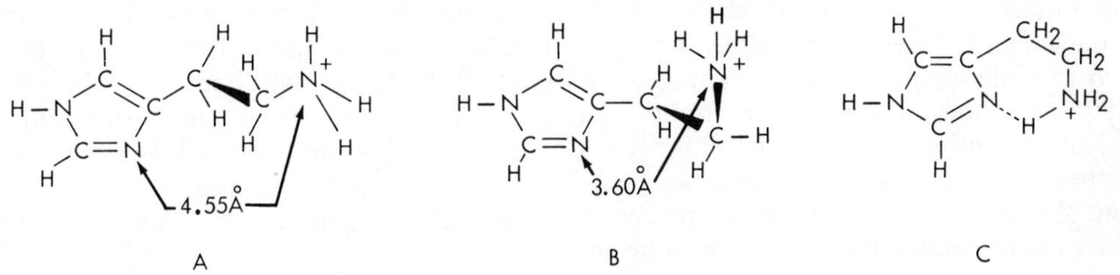

Conformations of Histamine

sumably due to the 1,3-interaction of the methylene groups of the cyclopropane ring with the carbonyl oxygen, which is believed to be required for nicotinic activity.

Using an approach which focuses on the parent molecule, rather than on conformationally fixed analogs, molecular orbital calculations have indicated that histamine may exist in two extended conformations (A,B) of nearly equal and minimal energy[70] rather than the earlier predicted coiled form (C) involving intramolecular hydrogen bonds.[71] In one extended conformation (A), one imidazole ring nitrogen atom is about 4.55 Å from the side chain nitrogen, while in conformation B this distance is about 3.60 Å. Histamine receptors have been differentiated into at least two classes, there being different structural requirements for stimulation of smooth muscle, such as the guinea-pig ileum (blocked by antihistamines), and for the

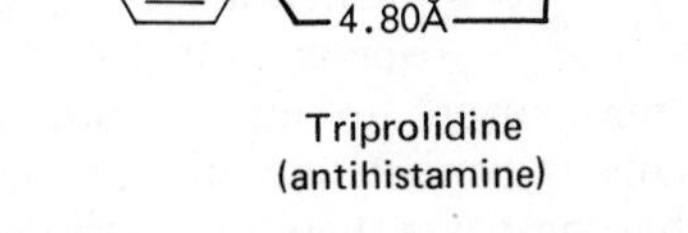

Triprolidine
(antihistamine)

stimulation of secretion of gastic acid (not blocked by antihistamines). It is proposed, on the basis of the internitrogen distance of closest approach of 4.8 ± 0.2 Å for the relatively rigid antihistamine triprolidine, that histamine acts on smooth muscle in confor-

mation A, in which the internitrogen distance of 4.55 Å closely approximates the spacing found in the specific antagonist. It is further presumed that the histamine-induced release of gastric acid may be brought about by a histamine–receptor interaction in an alternate conformation of closer internitrogen spacing, such as conformation B. No specific antagonists to the gastric acid stimulation of histamine are available to test this hypothesis.

Optical Isomerism and Biologic Activity. The widespread occurrence of differences in biologic activities for *optical isomers* has been of particular importance in the development of theories in regard to the nature of drug-receptor interactions. *Diastereoisomers*, compounds with two or more asymmetric centers, have the same functional groups and, therefore, can undergo the same types of chemical reactions. However, the diastereoisomers (e.g., ephedrine, *pseudo*-ephedrine, see Chap. 17) have different physical properties, undergo different rates of reactions, substituent groups occupy different relative positions in space, and the different biologic properties shown by such isomers may be accounted for by the influence of any of these factors on drug distribution, metabolism or interaction with the drug receptor.

However, *optical enantiomorphs*, also called *optical antipodes* (mirror images) present a very different case, for they are compounds whose physical and chemical properties are usually considered identical except for their ability to rotate the plane of polarized light. Here one might expect the compounds to have the same biologic activity. However, such is not the case with many of the enantiomorphs that have been investigated.

As examples of compounds whose optical isomers show different activities may be cited the following: (−)-hyoscyamine is 15 to 20 times more active as a mydriatic than (+)-hyoscyamine; (−)-hyoscine is 16 to 18 times as active as (+)-hyoscine; (−)-epinephrine is 12 to 15 times more active as a vasoconstrictor than (+)-epinephrine; (+)-norhomoepinephrine is 160 times more active as a pressor than (−)-norhomoepinephrine; (−)-synephrine has 60 times the pressor activity of (+)-synephrine; (−)-amino acids are either tasteless or bitter, while (+)-amino acids are sweet; (−)-ascorbic acid has good antiscorbutic properties, while (+)-ascorbic acid has none.

Although it is well established that optical antipodes have different physiologic activities, there are different interpretations as to why this is so. Differences in distribution of isomers, without considering differences in action at the receptor site, could account for different activities for optical isomers. Diastereoisomer formation with optically active components of the body fluids (e.g., plasma proteins) could lead to differences in absorption, distribution and metabolism. Distribution could also be affected by preferential metabolism of one of the optical antipodes by a stereospecific enzyme (e.g., D-amino acid oxidase). Preferential adsorption could also occur at a stereospecific site of loss (e.g., protein binding). Cushny[72] accounted for this difference by assuming that the optical antipodes reacted with an optically active receptor site to produce diastereoisomers with different physical and chemical properties. Easson and Stedman,[73] taking a somewhat different view, point out that optical antipodes can in theory have different physiologic effects for the same reason that structural isomers can have different effects, i.e., because of different molecular arrangements, one antipode can react with a hypothetical receptor while the other cannot. Assuming a receptor

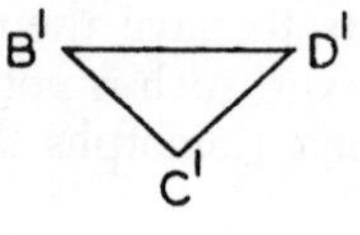

in tissues to which a drug can be attached and have activity only if the complementary parts B, D, C are superimposed, it is apparent that of the two enantiomorphs only I can be so superimposed. Under these conditions, I therefore would be active, and II would show no activity. This interpreta-

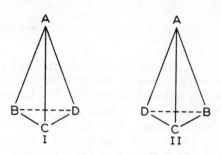

tion in a sense is not greatly different from that given by Cushny, because the receptor has a unique configuration not much different from that of an optically active compound. Both theories demand a structure of unique configuration in the body, but in the one theory only one enantiomorph reacts, while in the other they both react, with one combination having greater biologic activity than the other.

Easson and Stedman[73] have also postulated that the optical antipodes of epinephrine owe their differences in activity to a difference in ease of attachment to the receptor surface. This is illustrated below for the pressor activity of (−)- and (+)-epinephrine.[74]

Thus, only in (−)-epinephrine can the three groups essential for maximal pressor activity in sympathomimetic amines—the positively charged nitrogen, the aromatic ring and the alcoholic hydroxyl group—attach to the complementary receptor surface. In the (+)-isomer, any two binding groups may orient to attach, but not all three. This is consistent with the observation[75] that desoxyepinephrine, which lacks the alcoholic hydroxyl and therefore may only bind in two positions, has about the same pressor effect as (+)-epinephrine.

Belleau[76] has suggested a more detailed model for the adrenergic receptor, in which the anionic site is a phosphate group. For-

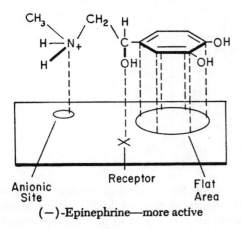

(−)-Epinephrine—more active

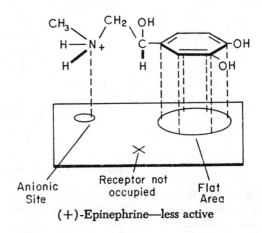

(+)-Epinephrine—less active

mation of an ion-pair at the anionic site would trigger the excitatory "α-response," while chelation of the phenolic hydroxyl group with a metal would lead to the inhibitory "β-response." Bulky groups on nitrogen, such as methyl, isopropyl, and larger, would inhibit ion-pair formation and favor the inhibitory response as seen with isoproterenol.

In the field of potent analgesics (morphine, meperidine, methadone, etc.) Beckett[74] has described a receptor surface made up of three regions. These are: (1) a flat surface providing binding to an aromatic ring, (2) a cavity into which the connecting chain between the aromatic group and nitrogen may fit and be held by van der Waals' forces and (3) an anionic site which binds the cationic nitrogen. Configurational studies have shown that all potent analgesics thus far studied either possess or may adopt the conformations which allow ready association with this receptor. Optical antipodes to the potent analgesics are less active as analgesics, although some, e.g., dextromethorphan (Romilar), retain antitussive properties.

THE DRUG RECEPTOR

The drug receptor is a component of the cell whose interaction with the drug initiates a chain of events leading to an observable biologic response. Primarily by analogy with the well-studied substrate–enzyme interactions, it is usually assumed that those drug receptors which are not enzymes resemble enzymes in their general nature but are, in contrast, an integral part of the organized structure of the cell and, therefore, may not be isolable by presently available technics. There are a number of specific examples of drug-enzyme interactions which are related to pharmacologic effects. The best established of these are inhibitors of acetylcholinesterase acting as cholinergic agents (e.g., physostigmine, isoflurophate; see Chap. 18), inhibitors of carbonic anhydrase acting as diuretics (e.g., acetazolamide; see Chap. 20), inhibitors of monoamine oxidase acting as central nervous system stimulants (e.g., tranylcypromine; see Chap. 16), and the aldehyde oxidase inhibitor disulfiram, used to discourage the chronic consumption of alcohol. However, most drug actions appear to take place on or within the cell in regions which have not been isolated and characterized as enzymes. Preliminary optimism that a protein fraction from the electric organ of the eel, *Electrophorus electricus,* represented the isolated acetylcholine receptor[77] has not been supported by subsequent studies.[78] This illustrates the difficulties involved in separating drug receptors from tissue proteins. Extraction processes must break both weak and strong bonds binding the receptor to cellular material. Even if a macromolecule survives this operation and is isolated with its functional groups intact, functionality may be lost if the macromolecule changes its structural form in solution.

The cell membrane is one region of the cell which contains organized components

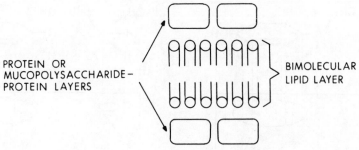

PROTEIN OR
MUCOPOLYSACCHARIDE-
PROTEIN LAYERS

BIMOLECULAR
LIPID LAYER

Schematic representation of the cell membrane

which can interact with small molecules in a specific manner. The structural unit of the cell membrane is thought to consist of a bimolecular layer of lipid molecules about 25 Å thick, held between two layers, each about 25 Å thick, which are at least partially protein. The lipid layers may consist of cholesterol and phospholipids, with the non-polar hydrocarbon chains held together at the center by van der Waals attraction, the polar heads being oriented outward, and associated by polar bonds with the protein sheaths. High-molecular-weight charged mucopolysaccharides, with their constituent carboxylic and sulfate ester groups acting as solvated anions, may be associated with the protein in one or both of the outer layers. Water filled pores, lined by the polar side-chains of protein molecules, are assumed to permit passage of small polar molecules. The proteins constitute a potentially highly organized region of the cell membrane. Molecular specificity is well known in such proteins as enzymes and antibodies, and it is generally believed that proteins are an important component of the drug receptor. The nature of the amide link in proteins provides a unique opportunity for the formation of multiple internal hydrogen bonds, as well as internal formation of hydrophobic, van der Waals and ionic bonds by side-chain groups, leading to such organized structures as the α-helix, which contains about four amino acid residues for each turn of the helix. An organized protein structure would hold the amino-acid side-chains at relatively fixed positions in space and available for specific interactions with a small molecule.

Proteins have the potential to adopt many different conformations in space without breaking their covalent amide linkages. They may shift from highly coiled structures to partially disorganized structures, with parts of the molecule existing in random chain, or to folded sheet structures, depending on the environment. In the mono-layer of a cell membrane, the interaction of a foreign small molecule with an organized protein may lead to a drastic change in the structural and physical properties of the membrane. Such changes could well be the initiating events in the production of a tissue or organ response to a drug.

The large body of information now available on relationships between chemical structure and biologic activity strongly supports the concept of flexible receptors. The fit of drugs onto or into macromolecules is only rarely an all-or-none process as pictured by the earlier "lock and key" concept of a receptor. Rather, the binding or partial insertion of groups of moderate size onto or into a macromolecular pouch appears to be a continuous process, even though over a limited range, as indicated by the frequently occurring regular increase and decrease in biologic activity as one ascends a homologous series of drugs. A range of productive associations between drug and receptor may be pictured, which lead to agonist responses, such as those produced by cholinergic drugs. Similarly, strong associations may lead to unproductive changes in the configuration of the macromolecule, leading to an antagonistic or blocking response, such as that produced by anticholinergic agents. The fundamental structural unit of the drug receptor is generally considered to be protein in nature, although this may be supplemented by its associations with other units, such as mucopolysaccharides and nucleic acids.

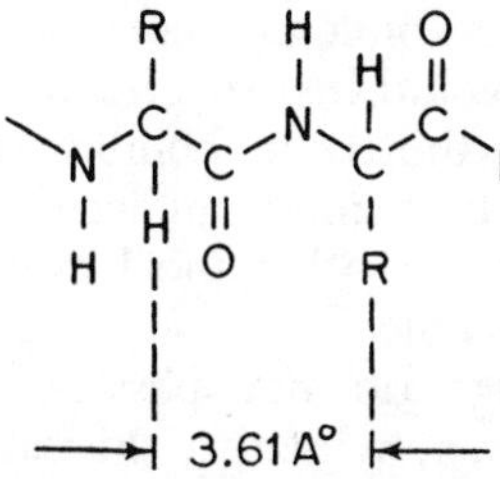

Identity Distance in Extended Protein

In the maximally extended protein, the distance between peptide bonds ("identity distance") is 3.61 Å. For many types of biologic activity, the distance between functional groups leading to maximal activity approximates this identity distance or some whole number multiple of it. Many parasympathomimetic (acetylcholinelike) and parasympatholytic (cholinergic blocking) agents have a separation of 7.2 Å (2×3.6) between the ester carbonyl group and nitrogen.[79] This distance is doubled between quaternary nitrogens of curarelike drugs; 14.5 Å (4×3.61).[80] The preferred separation of hydrogen bonding groups in estrogenic compounds (e.g., hydroxyls of diethylstilbestrol) is 14.5 Å (4×3.61).[81]

A related spacing of 5.5 Å, which corresponds to two turns of the α-helical structure common to proteins, is found between functional groups of many drugs. The most frequently occurring of these is the R—X—CH$_2$—CH$_2$—NR$_2'$ (X = N; X = O, N) structure which is present in local anesthetics, antihistamines, adrenergic blocking agents and others.[82]

Studies involving the relative effectiveness of various molecules of well-defined structural and functional types have contributed to an understanding of the stereochemical and physicochemical properties of their biologic receptors. Pfeiffer[79] concluded that parasympathomimetic stimulant action depends on two adjacent oxygen atoms at distances of approximately 5.0 Å. and 7.0 Å. from a methyl group or groups attached to nitrogen. Since these compounds (acetylcholine, methacholine, urecholine, etc.) do not have rigid structures, the actual distance between the oxygen and the methyl groups varies; however, the more extended conformations would be favored in solution.

Welsh and Taub[83] have concluded that a carbonyl group at a maximum distance of 7 Å from the quaternary nitrogen is an important linking group with the acetylcholine receptor protein of the *Venus* heart. They suggest that some type of bond forms between the carbonyl carbon or ketone oxygen and an appropriate group in the protein molecule.

The nature of the acetylcholinesterase receptor site probably has been investigated more thoroughly than the reactive site of any other enzyme. On the basis of studies

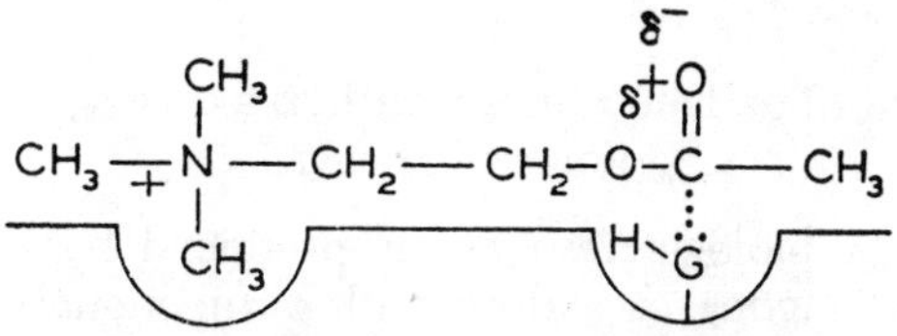

Anionic Site
Acetylcholinesterase

Esteratic Site
(Nachmansohn and Wilson)

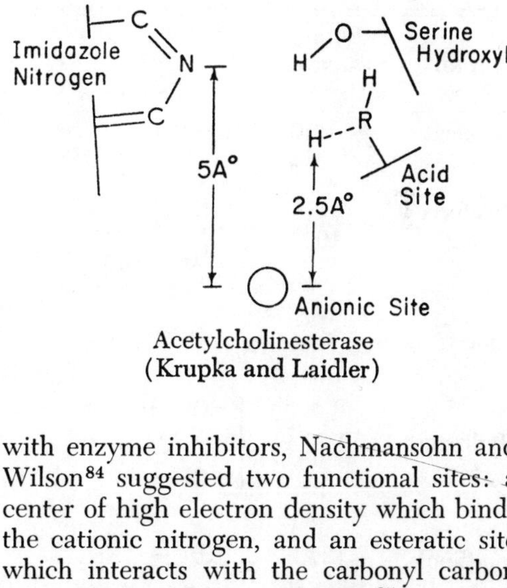

Acetylcholinesterase
(Krupka and Laidler)

with enzyme inhibitors, Nachmansohn and Wilson[84] suggested two functional sites: a center of high electron density which binds the cationic nitrogen, and an esteratic site which interacts with the carbonyl carbon atom. Friess and his co-workers[85] attempted to define the distance between the anionic and the esteratic sites by studying enzyme inhibition with cyclic aminoalcohols (e.g., *cis*-2-dimethylaminocyclohexanol) and their esters. The *cis*-isomers were more active than the *trans*, and a distance of about 2.5 Å was indicated as separating the nitrogen and the oxygen and, by inference, the receptors which bind these on the enzyme.

Krupka and Laidler[86] correlated previous stereochemical studies with kinetic data and described a complex esteratic site made up of three components: a basic site (imidazole nitrogen, 5 Å from the anionic site), an acid site, 2.5 Å from the anionic site, and a serine hydroxyl group. Following stereospecific binding of acetylcholine, the serine hydroxyl is acetylated to effect ester cleavage. Subsequently, a water molecule is held in the proper position through hydrogen bonds with imidazole, serine is deacetylated (hydrolyzed) and the reactive enzyme is regenerated.

THE DRUG RECEPTOR INTERACTION; FORCES INVOLVED

A biologic response is produced by the interaction of a drug with a functional or organized group of molecules which may be called the biologic receptor site. This interaction would be expected to take place by utilizing the same bonding forces involved as when simple molecules interact. These, together with typical examples, are collected in Table 7.

Most drugs do not possess functional groups of a type which would lead to ready formation of the strong and essentially irreversible covalent bonds between drug and biologic receptors. In most cases it is desirable that the drug leave the receptor site when the concentration decreases in the extracellular fluids; therefore, most useful drugs are held to their receptors by ionic or weaker bonds. However, in a few cases where relatively long-lasting or irreversible effects are desired (e.g., antibacterial, anticancer), drugs which form covalent bonds are effective and useful.

The alkylating agents, such as the nitrogen mustards (e.g., mechlorethamine) used in cancer chemotherapy, furnish an example of drugs which act by formation of covalent bonds. These are believed to form the reactive immonium ion intermediates, which alkylate and thus link together proteins or nucleic acids, preventing their normal participation in cell division.

Covalent bond formation between drug and receptor is the basis of Baker's[87] concept of *"active-site-directed irreversible inhibition."* Considerable experimental evidence on the nature of enzyme inhibitors has supported this concept. Compounds studied possess appropriate structural features for reversible and highly selective association with an enzyme. If, in addition, the compounds carry reactive groups capable of forming covalent bonds, the substrate may be irreversibly bound to the drug-receptor complex by covalent bond formation with reactive groups adjacent to the active site. In studies with reversibly-binding antimetabolites that carried additional alkylating and acylating groups of varying reactivities, selective irreversible binding by the related enzymes lactic dehydrogenase and glutamic dehydrogenase has been demonstrated. The selectivity of response has been attributed to the formation of a covalent bond between the carbophenoxyamino substituent of 5-(carbophenoxyamino)sali-

TABLE 7. TYPES OF CHEMICAL BONDS[*]

BOND TYPE	BOND STRENGTH kcal./mole	EXAMPLE
Covalent	40–140	$CH_3 - OH$
Reinforced ionic	10	$R-N-H \cdots O$ ... $\overset{\oplus}{N} \cdots \overset{\ominus}{O} C-R'$
Ionic	5	$R_4N^{\oplus} \cdots {}^{\ominus}I$
Hydrogen	1–7	$-OH \cdots O=$; $-OH \cdots \overset{\parallel}{C}$
Ion-dipole	1–7	$R_4N^{\oplus} \cdots :NR_3$
Dipole-dipole	1–7	$O=C \cdots :NR_3$ (δ^- δ^+)
Van der Waals'	0.5–1	$C \cdots C$
Hydrophobic	1	See Text

[*] Adapted from a table *in* Albert, A.: Selective Toxicity, p. 183, New York, Wiley, 1968.

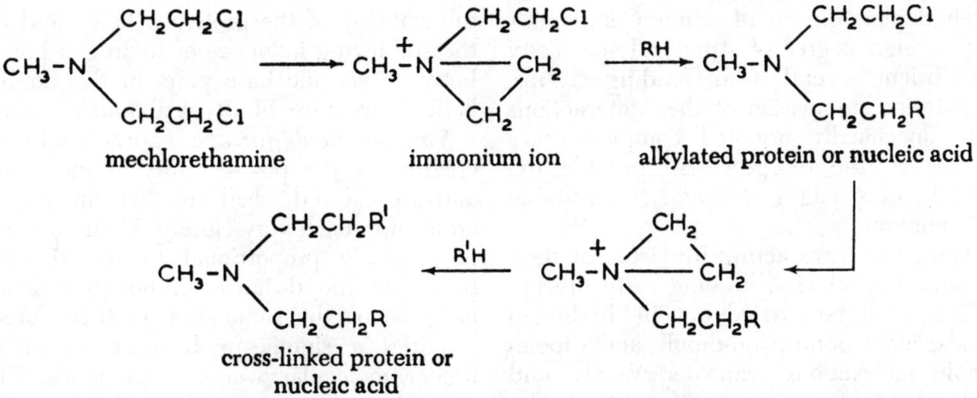

mechlorethamine   immonium ion   alkylated protein or nucleic acid

cross-linked protein or
nucleic acid

R, R' = free amino groups of proteins, adenyl or phosphate groups of nucleic acids.

cylic acid and a primary amino group in glutamic dehydrogenase[88] and between the maleamyl substituent of 4-(maleamyl)salicylic acid and a sulfhydryl group in lactic dehydrogenase.[89] Assignments of covalent bond formation with specific groups in the enzymes are based on the fact that the α,β-unsaturated carbonyl system of maleamyl groups reacts most rapidly with sulfhydryl groups, much more slowly with amino groups and extremely slowly with hydroxyl groups. In contrast, the carbophenoxy group will react only with a primary amino group on a protein.

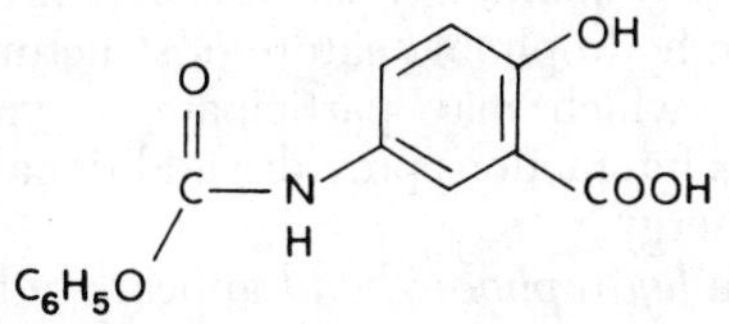

5-(Carbophenoxyamino)salicylic acid

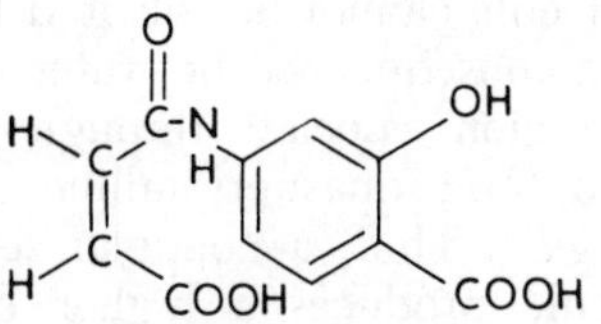

4-(Maleamyl)salicylic acid

In the purine series, similar studies[90] have led to the rational development of an active-site-directed inhibitor of adenosine deaminase. Studies on 9-alkyladenines showed that hydrophobic interactions between the 9-alkyl substituent and a nonpolar region

of the enzyme were important in the formation of the reversible drug-inhibitor complex. A nonpolar aromatic group, containing the active but nonselective bromoacetamido group, was substituted in the 9-position, and the resulting 9-(p-bromoacetamidobenzyl)adenine was shown to form initially a reversible enzyme-inhibitor complex, followed by formation of an irreversible complex, presumably by alkylation.

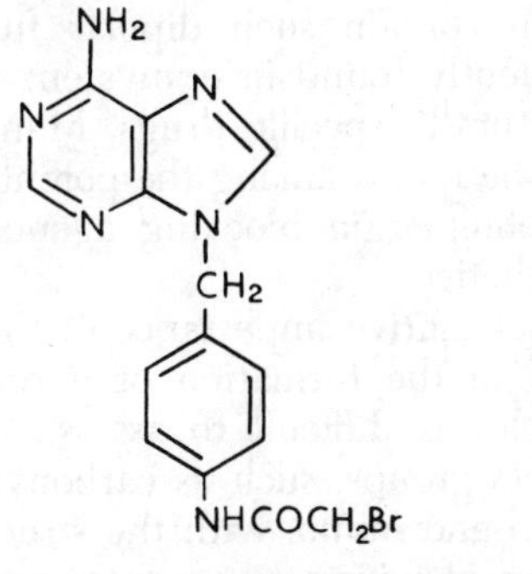

9-(p-Bromoacetamidobenzyl)adenine

Other examples of covalent bond formation between drug and biologic receptor site include the reaction of arsenicals and mercurials with essential sulfhydryl groups, the acylation of bacterial cell-wall constituents by penicillin and the inhibition of cholinesterase by the organic phosphates.

It is desirable that most drug effects be reversible. For this to occur, relatively weak forces must be involved in the drug-receptor complex, and yet strong enough so that other binding sites of loss will not competitively deplete the site of action. Compounds with a high degree of structural specificity may orient several weak binding groups, so that the summation of their interactions with specifically oriented complementary groups on the receptor will provide the total bond strength sufficient for a stable combination.

Thus, for drugs acting by virtue of their structural specificity, binding to the receptor site will be carried out by hydrogen bonds, ionic bonds, ion-dipole and dipole-dipole interactions, van der Waals' and hydrophobic forces. Ionization at physiologic pH would normally occur with the carboxyl, sulfonamido and aliphatic amino groups, as well as the quaternary ammonium group at any pH. These sources of potential ionic bonds are frequently found in active drugs. Differences in electronegativity between carbon and other atoms such as oxygen and nitrogen lead to an unsymmetrical distribution of electrons (dipoles) which are also capable of forming weak bonds with regions of high or low electron density, such as ions or other dipoles. Carbonyl, ester, amide, ether, nitrile and related groups which contain such dipolar functions are frequently found in equivalent locations in structurally specific drugs. Many examples may be found among the potent analgesics, the cholinergic blocking agents and local anesthetics.

The relative importance of the *hydrogen bond* in the formation of a drug-receptor complex is difficult to assess. Many drugs possess groups, such as carbonyl, hydroxyl, amino and imino, with the structural capabilities of acting as acceptors or donors in the formation of hydrogen bonds. However, such groups would usually be solvated by water, as would the corresponding groups on a biologic receptor. Relatively little net change in free energy would be expected in exchanging a hydrogen bond with a water molecule for one between drug and receptor. However, in a drug-receptor combination, a number of forces could be involved, including the hydrogen bond which would contribute to the stability of the interaction. Where multiple hydrogen bonds may be formed, the total effect may be a sizeable one, such as that demonstrated by

the stability of the protein α-helix, and by the stabilizing influence of hydrogen bonds between specific base pairs in the double helical structure of deoxyribonucleic acid.

Van der Waals' forces are attractive forces created by the polarizability of molecules and are exerted when any two uncharged atoms approach very closely. Their strength is inversely proportional to the seventh power of the distance. Although individually weak, the summation of their forces provides a significant bonding factor in higher molecular weight compounds. For example, it is not possible to distil normal alkanes with more than 80 carbon atoms, since the energy of about 80 kcal. per mole required to separate the molecules is approximately equal to the energy required to break a carbon-carbon covalent bond. Flat structures, such as aromatic rings, permit close approach of atoms. With van der Waals' force approximately 0.5 to 1.0 kcal./ mole for each atom, about 6 carbons (a benzene ring) would be necessary to match the strength of a hydrogen bond. The aromatic ring is frequently found in active drugs, and a reasonable explanation for its requirement for many types of biologic activity may be derived from the contributions of this flat surface to van der Waals' binding to a correspondingly flat receptor area.

The hydrophobic nature of structural elements which may participate in van der Waals interactions provides additional binding energy.

The *hydrophobic bond* appears to be one of the more important forces of association between nonpolar regions of drug molecules and biologic receptors. A nonpolar region of a molecule cannot be solvated by water, and, as a consequence, the water molecules in that region associate through hydrogen bonds to form quasi-crystalline structures ("icebergs"). Thus, a nonpolar segment of a molecule produces a higher degree of order in surrounding water molecules than is present in the bulk phase. If two nonpolar regions, such as hydrocarbon chains of a drug and a receptor, should come close together, these regions would be shielded to a greater extent from interaction with water molecules. As a result some of the quasi-crystalline water structures would collapse,

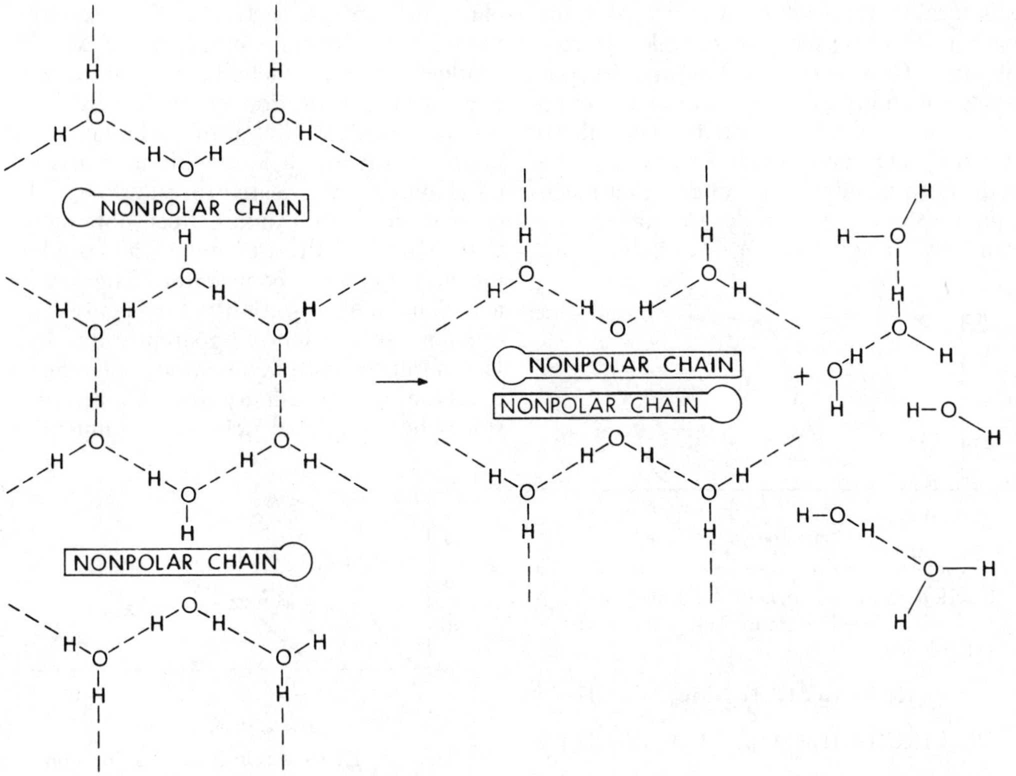

Isolated nonpolar chains in an ordered aqueous environment

Association of nonpolar chains displacing ordered water structures

Schematic representation of hydrophobic bond formation

producing a gain in entropy relative to the isolated nonpolar structures. The gain in free energy achieved through a decrease in the ordered state of many water molecules stabilizes the close contact of nonpolar regions, this association being called "hydrophobic bonding."

THE DRUG-RECEPTOR INTERACTION AND SUBSEQUENT EVENTS

Once bound at the receptor site, drugs may act, either to initiate a response (*stimulant* or *agonist* action), or to decrease the activity potential of that receptor (*antagonist* action) by blocking access to it by active molecules. The chain of events leading to an observable biologic response must be initiated in some fashion by either the process of formation or the nature of the drug-receptor complex. Current theories in regard to the mechanism of action of drugs at the receptor level are based primarily on the studies of Clark[91] and Gaddum,[92] whose work supports the assumption that the tissue response is proportional to the number of receptors occupied. The "occupancy theory" of drug action has been modified by Ariens[93] and Stephenson,[94] who have divided the drug-receptor interaction into two steps: (a) combination of drug and receptor, and (b) production of effect. Thus, any drug may have structural features which contribute independently to the *affinity* for the receptor, and to the efficiency with which the drug-receptor combination initiates the response (*intrinsic activity* or *efficacy*). The Ariens-Stephenson concept retains the assumption that the response is related to the number of drug-receptor complexes.

In the Ariens-Stephenson theory, both agonist and antagonist molecules possess structural features which would enable formation of a drug-receptor complex (strong affinity). However, only the agonist possesses the ability to cause a stimulant action, i.e., possesses intrinsic activity. The affinity of a drug may be estimated by comparison of the dose required to produce a pharmacologic response with the dose required by a standard drug. Thus, acetylcholine pro-

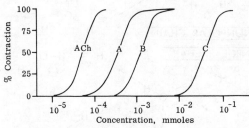

FIG. 7. Dose-response curves for contraction produced by acetylcholine (ACh) and alkyltrimethylammonium salts on the rat jejunum.

A. $CH_3CH_2OCH_2CH_2\overset{+}{N}Me_3$. B. CH_3-$CH_2CH_2CH_2CH_2\overset{+}{N}Me_3$. C. $CH_3CH_2CH_2$-$\overset{+}{N}Me_3$. (Modified from Ariens, E. J., and Simonis, A. M.[93])

duces a normal "S"-shaped curve if the logarithm of the dose is plotted against the per cent contraction of the rat jejunum (a segment of the small intestine). A series of related alkyl trimethylammonium salts (ethoxyethyl trimethylammonium, pentyl trimethylammonium, propyl trimethylammonium; Fig. 7) are able to produce the same degree of contraction of the tissue as can acetylcholine, but higher doses are required. The shape of the dose-response curve is the same, but the series of parallel curves are shifted to higher dose levels. Therefore, the alkyl trimethylammonium compounds are said to possess the same intrinsic activity as acetylcholine, being able to produce the same maximal response, but to show a lower affinity for the receptor, since larger amounts of drug are required.

By contrast, structural change of a molecule can lead to a gradual decline in the maximal height and slope of the log dose-response curves (Fig. 8), in which case the loss in activity may be attributed to a de-

cline in intrinsic activity. For example, pentyl trimethylammonium ion is able to produce a full acetylcholinelike contraction. Successive substitution of methyl by ethyl groups (pentyl ethyl dimethylammonium, pentyl diethyl methylammonium, pentyl triethylammonium) leads to successive decreases in the maximal effect obtainable, with pentyl triethylammonium ion producing no observable contraction. The loss in acetylcholinelike activity for pentyl triethylammonium ion is apparently due to a loss in intrinsic activity, without a significant decrease in the affinity for the receptor, since the compound acts as a competitive

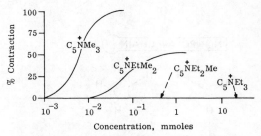

FIG. 8. Dose-response curves for contraction produced by pentyl trialkylammonium salts on the rat jejunum. (Modified from Ariens[93])

inhibitor (antagonist) for active derivatives of the same series.

In the case of an antagonist, it is desirable to have high affinity and low or zero intrinsic activity—that is, to bind firmly to the receptor, but to be devoid of activity. Many examples are available where structural modifications of an agonist molecule lead successively to compounds with decreasing agonist and increasing antagonist activity. Such modifications on acetylcholinelike structures, usually by addition of bulky nonpolar groups to either end (or both ends) of the molecule, may lead to the complete antagonistic activity found in the parasympatholytic compounds (e.g., atropine) discussed in Chapter 19.

In contrast to the occupancy theory, Croxatto[95] and Paton[96] have proposed that excitation by a stimulant drug is proportional to the *rate* of drug-receptor combination rather than to the number of receptors occupied. The *rate theory* of drug action proposes that the rate of association and dissociation of an agonist is rapid, and this leads to the production of numerous impulses per unit time. An antagonist, with strong receptor binding properties, would have a high rate of association but a low rate of dissociation. The occupancy of receptors by antagonists, assumed to be a nonproductive situation, prevents the productive events of association by other molecules. This concept is supported by the fact that even blocking molecules are known to cause a brief stimulatory effect before blocking action develops. During the initial period of drug-receptor contact when few receptors are occupied, the rate of association would be at a maximum. When a significant number of sites are occupied, the rate of association would fall below the level necessary to evoke a biologic response.

The *occupation* and the *rate* theories of drug action do not provide specific models at the molecular level to account for a drug acting as agonist or antagonist. The *induced-fit* theory of enzyme-substrate interaction,[97] in which combination with the substrate induces a change in conformation of the enzyme, leading to an enzymatically active orientation of groups, provides the basis for similar explanations of mechanisms of drug action at receptors. Assuming that protein constituents of membranes play a role in regulating ion flow, it has been proposed[98] that acetylcholine may interact with the protein and alter the normal forces which stabilize the structure of the protein, thereby producing a transient rearrangement in the membrane structure and a consequent change in its ion regulating properties. If the structural change of the protein led to a configuration in which the stimulant drug was bound less firmly and dissociated, the conditions of the *rate* theory would be satisfied. A drug-protein combination which did not lead to a structural change would result in a stable binding of the drug and a blocking action.

A related hypothesis (the *macromolecular perturbation theory*) of the mode of acetylcholine action at the muscarinic (postganglionic parasympathetic) receptor has been advanced by Belleau.[99] It is proposed that interaction of small molecules (substrate or drug) with a macromolecule (such as the protein of a drug receptor) may lead either to *specific conformational perturbations* (SCP) or to *nonspecific conformational perturbations* (NSCP). A SCP (specific change in structure or conformation of a protein molecule) would result in the specific response of an agonist (i.e., the drug-receptor would possess intrinsic activity). If a NSCP occurs, no stimulant response would be obtained, and an antagonistic or blocking action may be produced. If a drug possesses features which contribute to formation of both a SCP and a NSCP, an equilibrium mixture of the two complexes may result, which would account for a partial stimulant action.

The alkl trimethylammonium ions ($R-\overset{+}{N}Me_3$), in which the alkyl group, R, is varied from 1 to 12 carbon atoms, provide a homologous series of muscarinic drugs which serve as models for the macromolecular perturbation theory of events which may occur at the drug receptor. With these simple analogs, hydrophobic forces, in addition to ion-pair formation, are considered to be the most important in contributing to receptor binding. Lower alkyl trimethylammonium ions (C_1 to C_6) stimulate the muscarinic receptor and are considered to possess a chain length which is able to form a hydrophobic bond with nonpolar regions of the receptor, altering receptor structure in a specific perturbation (Fig. 9; e.g., $C_5\overset{+}{N}Me_3$). With a chain of 8 to 12 carbon atoms, the antagonistic action observed is considered to result from a nonspecific conformational perturbation (NSCP) of a network of nonpolar residues at the periphery of the catalytic surface (Fig. 9; e.g., $C_9\overset{+}{N}Me_3$). The intermediate heptyl and octyl derivatives act as partial agonists, and it is considered that they may form an equilibrium mixture of drug-receptor combinations, with both active SCP forms and inactive NSCP forms present (Fig. 9; e.g., $C_7\overset{+}{N}Me_3$).

The events initiated by specific conformational changes of the receptor are un-

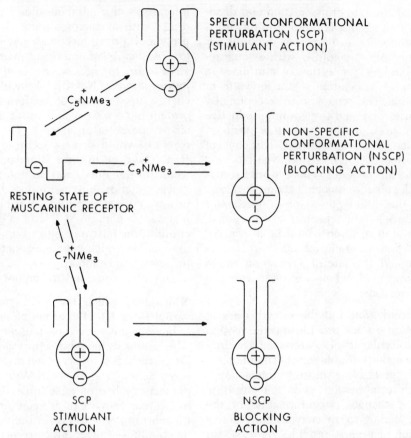

SPECIFIC CONFORMATIONAL
PERTURBATION (SCP)
(STIMULANT ACTION)

$C_5NMe_3^+$

NON-SPECIFIC
CONFORMATIONAL
PERTURBATION (NSCP)
(BLOCKING ACTION)

$C_9NMe_3^+$

RESTING STATE OF
MUSCARINIC RECEPTOR

$C_7NMe_3^+$

SCP
STIMULANT
ACTION

NSCP
BLOCKING
ACTION

FIG. 9. Schematic representation of alkyl trimethylammonium ions reacting with the muscarinic receptor. (Modified from Belleau[99])

known. However, it is suggested[99] that water freed of its ordered structure during the formation of a hydrophobic bond may be available for hydration of Na^+ and K^+ during transport. An alternate concept is that a certain chain length (C_1-C_6) or hydrophobic character for a portion of the drug is effective in bringing order to protein strands, so that groups whose interaction contributes to the energy necessary for ion transport are brought in proximity. A longer hydrophobic chain (C_9-C_{12}) could bring about disorder by associating with an additional segment of protein, either altering the ordered state required or screening the site from a necessary interaction with other molecules. A specific conformational perturbation of the enzymatic drug receptor monoamine oxidase during association with

its substrates also has been proposed by Belleau.[100]

In general, drugs which reduce the activity of hormones (such as acetylcholine, epinephrine, serotonin, various steroids and histamine) are called *antagonists* or *blocking agents*. If the substrate is a compound normally required in the metabolism of the organism (e.g., vitamins, coenzymes), the drug that blocks its use is called an *antimetabolite*. The best example of an effective and useful antimetabolite is the antibacterial sulfonamides (see Chap. 11). These drugs, close structural relatives to *p*-aminobenzoic acid, interfere with the incorporation of the latter into pteroylglutamic acid (folic acid) required by some bacteria. Since mammals obtain their folic acid preformed from food sources, a toxic effect selective for *p*-aminobenzoic-acid-requiring bacteria is achieved by the sulfonamide antimetabolites.

In the event that antagonism is shown between two drugs, the term *therapeutic interference* is used. For example, N-allyl-*nor*-morphine (nalorphine) has a weak analgetic effect alone but antagonizes that produced by morphine. In terms of a receptor hypothesis, nalorphine is bound as firmly as morphine to the receptor but is less efficient in initiating the events which lead to analgesia.[101]

ISOSTERISM

In the search for novel, more potent, less toxic and more selectively acting drugs, those associated with pharmaceutical research have developed considerable intuition, based on a large body of experimental knowledge, in selecting appropriate structural modifications. As understanding of the stereochemical and physicochemical nature of molecular features has increased, intuition has been strengthened by the application of modern structural theory. The term, *isosterism* has been widely used to describe the selection of structural components whose steric, electronic and solubility characteristics make them interchangeable in drugs.

The concept of isosterism has evolved and changed significantly in the years since its introduction by Langmuir[102] in 1919. Langmuir, while seeking a correlation which would explain similarities in physical properties for nonisomeric molecules, defined *isosteres* as compounds or groups of atoms having the same number and arrangement of electrons. Those isosteres which were isoelectric, i.e., with the same total charge as well as same number of electrons, would possess similar physical properties. For example, the molecules N_2 and CO, both possessing 14 total electrons and no charge, show similar physical properties. Related examples described by Langmuir were CO_2 and N_2O, and N_3^- and NCO^-.

In 1925, Grimm[103] provided a significant advancement to the concept when he introduced his "Hydride Displacement Law." The combination resulting from the addition of a hydrogen atom and its lone electron to an atom, the presence of the small H atom in the orbital then being ignored, was described as a "pseudoatom." Some physical properties for this "pseudoatom" resemble those of the atom which has one more electron than that from which the "pseudoatom" was derived. CH and N, both with 7 total electrons, are examples of "pseudoatoms" and atoms which are isosterically related. The atoms and combinations compared in this way by Grimm are shown below.

The recognition that the electrons in the outermost orbital are involved in most

AN EXPANDED TABLE OF ISOSTERES

Grimm's Hydride Displacement

Total electrons	6	7	8	9	10	11
	C	N	O	F	Ne	Na^+
		CH	NH	OH	FH	–
			CH_2	NH_2	OH_2	FH_2^+
				CH_3	NH_3	OH_3^+
					CH_4	NH_4^+

Atoms and Groups of Atoms with the Same Number of Peripheral Electrons

Peripheral electrons	4	5	6	7	8
	N^+	P	S	Cl	ClH
	P^+	As	Se	Br	BrH
	S^+	Sb	Te	I	IH
	As^+	–	PH	SH	SH_2
	Sb^+	–	–	PH_2	PH_3

chemical reactions led Erlenmeyer to redefine isosteres as "Atoms, ions or molecules in which the peripheral layers of electrons can be considered to be identical."[104] Thus, elements in the same group of the periodic table (e.g., O, S, Se, Te) and the "pseudoatoms" related to these with the same number of peripheral electrons (e.g., NH, CH_2) may be considered as isosteres.

Other groups of atoms which impart similar physical or chemical properties to a molecule, due to similarities in size, electronegativity or stereochemistry, are now frequently referred to under the general term of *isostere*. The early recognition that benzene and thiophene were alike in many of their properties led to the term "ring equivalents" for the vinylene group ($-CH=CH-$) and divalent sulfur ($-S-$). The vinylene group in an aromatic ring system may be replaced by other atoms isosteric to sulfur, such as oxygen (furan) or NH (pyrrole); however, in such cases, aromatic character is significantly decreased.

With increased understanding of the structures of molecules, less emphasis has been placed on the number of electrons involved, for variations in hybridization during bond formation may lead to considerable differences in the angles, the lengths and the polarities of bonds formed by atoms with the same number of peripheral electrons. Even the same atom may vary widely in its structural and electronic characteristics when it forms a part of a different functional group. Thus, nitrogen is part of a planar structure in the nitro group but forms the apex of a pyramidal structure in ammonia and the amines.

For this reason, a less rigorous definition of isosteres has evolved which includes those groups which possess similar steric and electronic configurations, regardless of the number of electrons involved.[105] Examples of such isosteric pairs are: the carboxylate ($-COO-$) and sulfonamido ($-SO_2-NR-$); ketone ($-CO-$) and sulfone ($-SO_2-$); chlorine ($-Cl$) and trifluromethyl (CF_3) groups.

Because of the frequent application of the concept of isosterism in studies relating to the actions of structurally related compounds on biologic systems, Friedman has applied the term "bio-isosteric" to compounds which "fit the broadest definition for isosteres, and have the same type of biological activity."[106] Antagonistic activity found for structural analogs is also included in this definition.

The general principles of isosterism are frequently applied in the structural alteration of compounds known to possess biologic activity. Such compounds may be altered by isosteric replacements of atoms or groups, in order to develop analogs with select biologic effects, or to act as antagonists to normal metabolites. Each series of compounds showing a specific biologic effect must be considered separately, for there are no general rules which will predict whether biologic activity will be increased or decreased. It appears that when isosteric replacement involves the bridge connecting groups necessary for a given response, a gradation of like effects results, with steric factors (bond angles) being important. Some examples of this type are:

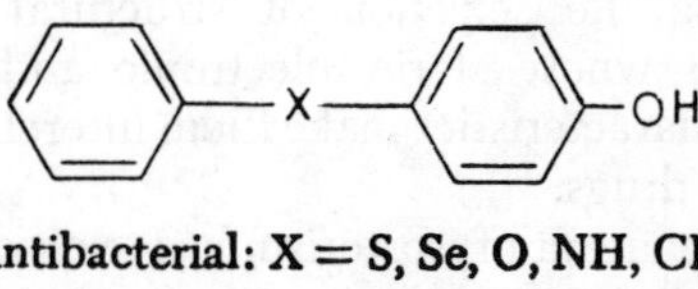

Antibacterial: X = S, Se, O, NH, CH_2

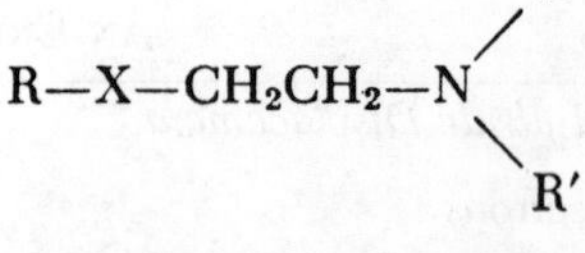

Antihistamines: X=O, NH, CH_2.
Cholinergic blocking agents: X=$-COO-$, $-CONH-$, $-COS-$.

When a group is present in a part of a molecule where it may be involved in an essential interaction or may influence the reactions of neighboring groups, isosteric replacement sometimes produces analogs which act as antagonists. Some examples from the field of cancer chemotherapy are:

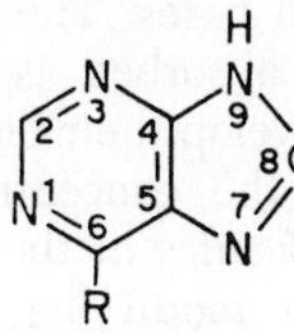

	R	
Adenine	NH_2	⎫
Hypoxanthine	OH	⎬ Metabolites
6-Mercaptopurine	SH	Antimetabolite

The 6-NH_2 and 6-OH groups appear to play essential roles in the hydrogen-bonding interactions of base pairs during nucleic acid replication in cells. The substitution of the significantly weaker hydrogen bonding isosteric sulfhydryl groups results in a partial blockage of this interaction, and a decrease in the rate of cellular synthesis.

In a similar fashion, replacement of the hydroxyl group of pteroylglutamic acid (folic acid) by the amino group leads to aminopterin, an antagonist useful in the treatment of certain types of cancer.

As a better understanding develops of the nature of the interactions between drug and biologic receptor, selection of isosteric groups with particular electronic, solubility and steric properties should permit the rational preparation of more selectively acting drugs. But in the meanwhile, results obtained by the systematic application of the principles of isosteric replacement are aiding in the understanding of the nature of these receptors.

The Hansch[53] approach, discussed previously, which correlates the contributions to biologic activity made by selected physicochemical properties of substituents, has greatly facilitated the systematic alteration of groups in the design of more useful drugs. In addition, the purely empirical mathematical approach of Free and Wilson[107] provided structure-activity data which have been used to correlate and to predict activity. This method assumes a constant and additive contribution to biologic activity by substituents on a fundamental structural unit. Therefore, the series studied must be on a linear portion of the usual parabolic curve of a homologous series. The summation of

group contributions must not provide greater than maximal solubility characteristics to the molecule, and place the compound on the negative slope of the biologic response curve. This method has been used in studies of substituent effects on the ability of the parent phenylethylamine structure to inhibit uptake of norepinephrine by the isolated rat heart.[108] It has also been used to correlate and predict butyrylcholinesterase inhibitory activity by 1-substituted-3-(N-ethyl-N-methylcarbamoyl)piperidine hydrobromide.[109]

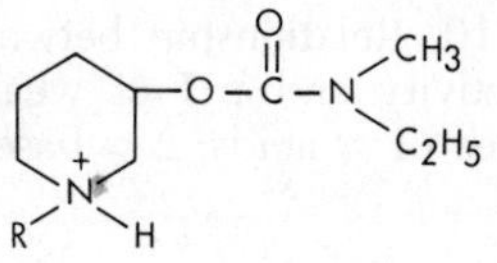

1- Substituted-3-(N-ethyl-N-methylcarbamoyl)piperidine
Hydrobromide

A large number of examples of isosteric replacement are to be found among currently distributed drugs. Within many pharmacologic classes, such as the local anesthetics, antihistaminic, parasympathomimetic and parasympatholytic classes in particular, examples are readily apparent.

SELECTED PHYSICOCHEMICAL PROPERTIES

Factors which influence the passage of a drug from its site of administration to its site of action have been described in the preceding sections. Some specific physicochemical properties which are important to drug action will now be discussed in greater detail.

IONIZATION

Acids and bases may be responsible for biologic action as undissociated molecules or in the form of their respective ions. A great many compounds, particularly the weak acids and bases, appear to act as undissociated molecules. It is probable that when action takes place inside the cell or within cell membranes, molecules gain entrance to the cell in the undissociated form and after reaching the site of action may

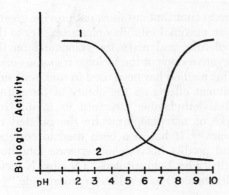

FIG. 10. Relationship between biologic activity and pH of weak acids and bases (1 = acids; 2 = bases.)

then function as ions. Numerous examples of this type are known, and several have been described in previous sections. A general relationship between biologic activity and pH for weak acids and bases which require a high concentration of undissociated molecules for maximal effect is shown in Figure 10.

For example, the antibacterial activity of benzoic acid, salicylic acid, mandelic acid and other acids of this type is greatest in acid media. The efficiency of these acids as antibacterial agents may be increased as much as 100 times in going from neutral to acid solutions (pH 3).[110]

The antibacterial action of the phenols is greatest at a pH below 4.5, and the activity again increases at a pH of 10 and above.[111] This increase at high pH has been attributed to partial oxidation of the phenol to a more active quinone.

The solubility and partition coefficients of acids and bases may be altered greatly by changes in pH. Thus, cocaine hydrochloride is freely soluble in water (1:0.4), while the free base has only low water-solubility (1:600). On the other hand, the solubility of the free base in chloroform, ether and vegetable oils is rather high, and the partition between these solvents and water will favor the nonpolar solvents (high partition coefficients). The partition coefficients of the salt between water and all of the nonpolar solvents will, of course, be extremely low. The same is true in general for the salts

of other acids and bases. The salts of acids or bases that are absorbed as undissociated molecules will develop their biologic activity in proportion to the concentration of free undissociated molecules in the solution.

In addition to modifying the physical properties of solutions, changes in pH may also affect the reactivity of acidic and basic groups on the cell surface or within the cell. At the isoelectric point, potential anions and cations in a protein or cell exist as "zwitterions." The effect of modifying the pH above or below the isoelectric point may be shown as follows:

Increasing the pH of the medium will increase the concentration of the anions on the cell, thereby increasing the activity for

TABLE 8. IONIZATION AND BACTERIOSTATIC EFFECTS OF AMINOACRIDINES[*]

ACRIDINE	MIN. BACTERIO-STATIC CONC., STREPT. PYOG.	PER CENT IONIZED (pH 7.3; 37°)
3-NH$_2$	1/80,000	73
9-NH$_2$	1/160,000	99
3,6-diNH$_2$	1/160,000	99
3,7-diNH$_2$	1/160,000	76
3,9-diNH$_2$	1/160,000	100
4,9-diNH$_2$	1/80,000	98
4-NH$_2$	1/5,000	< 1
2-NH$_2$	1/10,000	2
1-NH$_2$	1/10,000	2
4,5-diNH$_2$	< 1/5,000	< 1
2,7-diNH$_2$	1/20,000	4

[*] Adapted from Albert, A.: Selective Toxicity, p. 262, New York, Wiley, 1968.

biologically active cations. Decreasing the pH of the medium will increase the concentration of cations on the cell and thereby increase the activity for biologically active anions.

Active Ions. Some compounds show increased biologic activity when their degree of ionization is increased. Because of the difficulty with which ions penetrate membranes, it is most likely that compounds of this type exert their effect on the outside of the cell.

The per cent of ionization at the body pH (7.3) for a large number of acridine compounds has been recorded by Albert and co-workers.[112] It was observed that a basicity sufficient to induce at least 75 per cent ionization at pH 7.3 at 20° (or 60% at 37°) is necessary for effective antibacterial action in the series. It also was shown that the acridine cations are largely responsible for the activity. The undissociated molecules, anions or zwitterions have an insignificant effect on activity. Representative members of the group are shown in Table 8.

Amino group substitution has a marked influence on the base strength of the heterocyclic nitrogen. Resonance stabilization of the ion by an amino group in the 3, 6 and the 9 positions will increase base strength, leading to a higher concentration of ion at pH 7.3 and increased bacteriostatic activity.

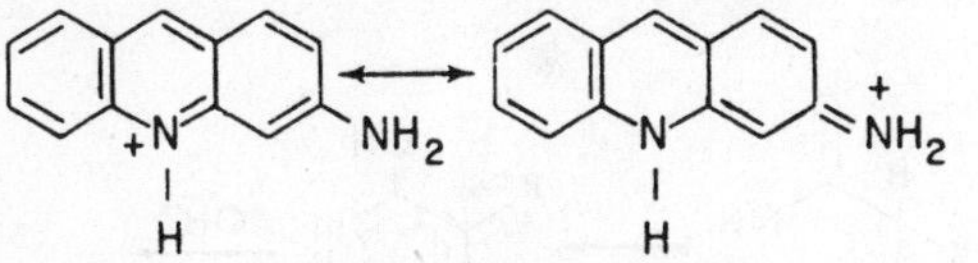

3-aminoacridinium ions

Substitution of an amino group in the 4-position leads to base-weakening intramolecular hydrogen bonding; substitution in the 1 and the 2 positions permits no resonance stabilization of the ion, and base strength remains low.

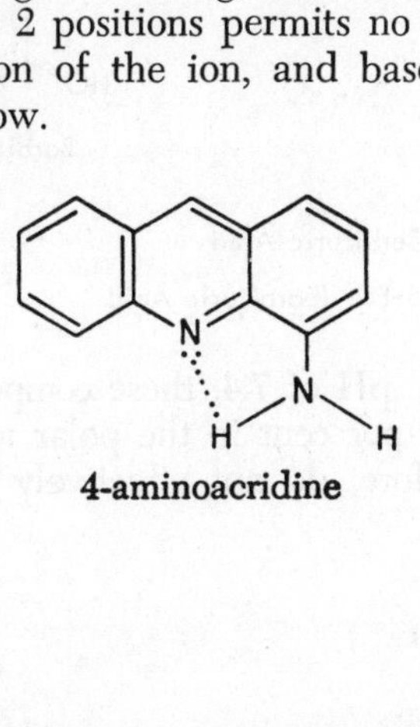

4-aminoacridine

Thus, substitution to produce a biologically active cation at physiologic pH is the most important structural feature of the aminoacridines. Other features of the molecule are also of importance, for if the total flat surface of the molecule is reduced below about 38 square angstroms, antibacterial activity is largely lost. Examples are 9-aminoacridine with one ring reduced (9-aminotetrahydroacridine; antibact. conc. 1/5,000); and 4-aminoquinoline (antibact. conc. <1/5,000).

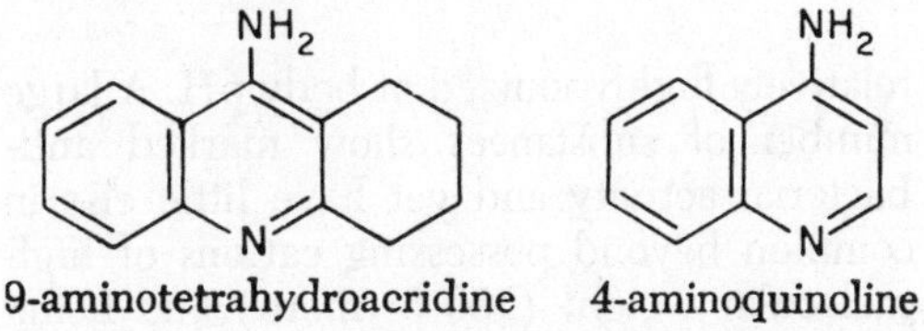

9-aminotetrahydroacridine   4-aminoquinoline

It is postulated that a sufficiently flat surface is necessary to supplement, by van der Waals' forces, the ionic bond between the drug cation and the receptor anion.

The basic dyes (e.g., triphenylmethanes, acridines, etc.) appear to function as antibacterial cations by reacting with essential anions (acid groups) of the bacterial cell to give slightly dissociated compounds. This may be illustrated by the following equation:

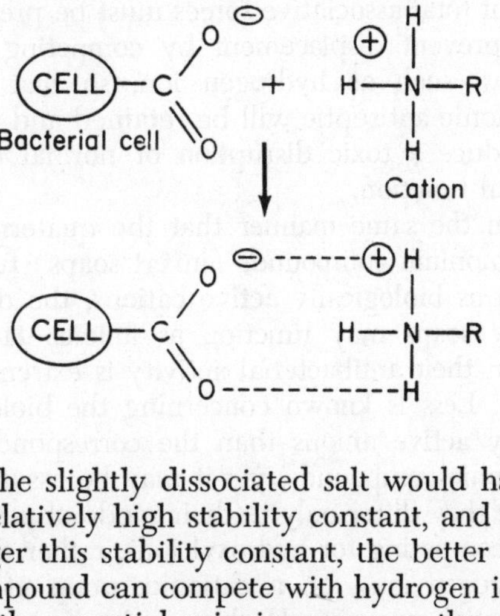

The slightly dissociated salt would have a relatively high stability constant, and the larger this stability constant, the better the compound can compete with hydrogen ions for the essential anionic groups on the cell. In this manner, functional groups of the organism can be blocked and cellular metabolism inhibited (bacteriostasis). This means that active compounds of this type must be

$$K_s = \frac{\left[\left(\text{CELL}\right)-C\begin{smallmatrix}O\\||\\O^{\ominus}\end{smallmatrix}\cdots\cdots\begin{smallmatrix}\oplus H\\|\\H-N-R\\|\\H\end{smallmatrix}\right]}{\left[\left(\text{CELL}\right)-C\begin{smallmatrix}O\\||\\O^{\ominus}\end{smallmatrix}\right]\left[\begin{smallmatrix}H\\|\\{}^{\oplus}N-R\\|\\H\end{smallmatrix}\right]}$$

relatively highly ionized at body pH. A large number of substances show marked antibacterial activity and yet have little else in common beyond possessing cations of high molecular weight (150 or more) and being highly ionized at pH of 7. Among these may be listed the aliphatic amines, quaternary ammonium compounds, diamines, amidines, diamidines, guanidines, biguanidines, pyridinium compounds, etc.

The relatively high molecular weight lipophilic residues joined to the cationic head must contribute hydrophobic or van der Waals binding to complimentary nonpolar regions of the bacteria, in addition to the interaction between the drug cation and anionic or polar regions of the bacteria. Sufficient total associative forces must be present to prevent displacement by competing cations, such as hydrogen ion, so that the cationic antiseptic will be retained and will produce a toxic disruption of normal bacterial function.

In the same manner that the quaternary ammonium compounds (invert soaps) function as biologically active cations, the ordinary soaps may function as anions. However, their antibacterial activity is extremely low. Less is known concerning the biologically active anions than the corresponding cationic compounds, yet it may be assumed that they differ only in their mode of action by competing for hydroxyl ions, rather than hydrogen ions, on a cationic group of an essential enzyme. Highly ionized anionic compounds normally show no significant biologic activity, and this may be attributed to the predominantly anionic nature of living cells. For example, most bacteria have an iso-

electric point of approximately 4 and, at a pH of 7 or more, are anionic in character.

It has been pointed out that in certain cases biologic activity increases with increased ionization. In other cases, where undissociated molecules are responsible for the biologic effect, the activity decreases with increased ionization. However, it should be kept in mind that in the case of biologically active acids and bases the concentration of ions and undissociated molecules is determined by the pKa of the acid or the base and the pH value of the environment in which action occurs.

Minor changes in structure can produce significant changes in the degree of ionization of a weak acid or base. This change may be the primary reason for the presence or absence of biologic activity in closely related compounds. Thus, all barbituric acid derivatives that are useful central-nervous-system depressants are 5,5-disubstituted, whereas barbituric acid and its 5-monosubstituted derivatives are inactive. It is most likely that the relatively high acidity of barbituric acid (pKa 4.0) and of its 5-monosubstituted derivatives (e.g., 5-ethylbarbituric acid; pKa 4.4) is responsible for the lack of hypnotic-sedative properties. These compounds are stronger acids, since they are able to assume a completely aromatic structure which can stabilize the barbiturate ion by delocalization of the extra pair of electrons formed. At

Barbiturate Ion

R = H: Barbituric Acid

R = C$_2$H$_5$:   5-Ethylbarbituric Acid

physiologic pH of 7.4, these compounds are about 99.9 per cent in the polar ionic form and, therefore, do not effectively penetrate the lipoidal barriers to the central nervous system. In contrast, 5,5-disubstituted barbituric acid derivatives cannot assume fully aromatic character and are much weaker acids, usually ranging in pKa from 7.0 to 8.5 (e.g., 5,5-diethylbarbituric acid, barbital; pKa 7.4). Therefore, at physiologic pH, these compounds exist about 50 per cent or higher in the nonpolar, unionized form, capable of ready passage into the lipoidal tissues of the central nervous system.

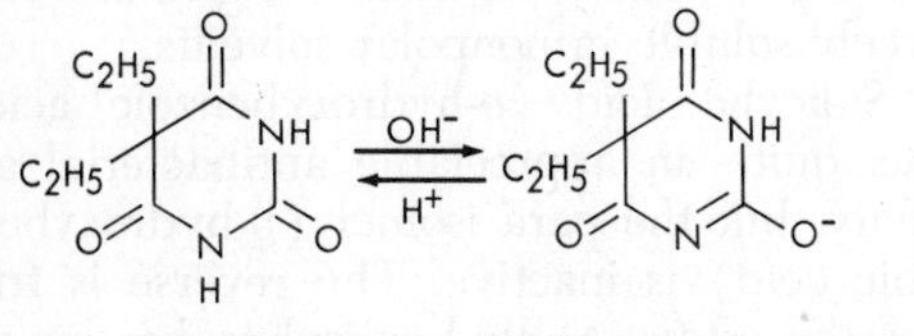

5,5-Diethylbarbituric Acid   5,5-Diethylbarbiturate Ion
(Barbital)

HYDROGEN BONDING AND BIOLOGIC ACTION

The hydrogen bond, which is sometimes referred to as a hydrogen bridge, is a bond in which a hydrogen atom serves to hold two other atoms together. The two bonds attached to hydrogen cannot both be covalent, for hydrogen has only one stable orbital and can form only one covalent bond. The other bond must be ionic. Consequently, the hydrogen bond is formed with electronegative atoms, which are those that form ionic bonds. Atoms capable of forming hydrogen bonds have at least one unshared electron pair together with a complete octet, and these include F, O, N, and to a lesser degree Cl and S.

The strength of the hydrogen bond varies from 1 to 10 kilocal. per mole and usually is about 5 kilocal. per mole. Thus it is only about one tenth as strong as most covalent bonds, which range from about 35 to 110 kilocal. per mole. In spite of the relative weakness of the bond, it may have a profound effect on the properties of substances, such as changes in boiling points, melting points, heats of vaporization, solubilities, dielectric constants, infrared absorption spectra and other properties.

The distance between two atoms held together by a hydrogen bond is consider-ably greater than the distance between the same two atoms held together by an ordinary chemical bond. Thus, the distance between oxygen atoms that are hydrogen-bonded varies from 2.5 to 2.8 Å, while the distance between the oxygen atoms in the oxygen molecule is 1.20 Å. On the other hand, the distance between hydrogen-bonded atoms is less than the distance between nonbonded atoms. Like ordinary chemical bonds, the strength of the hydrogen bond is an inverse function of the distance between the atoms.

Considerable evidence has been accumulated to show that proteins are held in specific configurations by hydrogen bonds and that the denaturation of proteins involves the breaking of some of these bonds. It is significant that virtually all reagents that denature proteins are reagents capable of breaking hydrogen bonds.

The most common hydrogen bonds are the following: O—H $\cdots$ O, N—H $\cdots$ O, N—H $\cdots$ N, F—H $\cdots$ F, O—H $\cdots$ N and N—H $\cdots$ F. If such bonds occur within a molecule they are termed *intramolecular*; if they occur between two molecules they are called *intermolecular* hydrogen bonds. Molecules are known that form both intramolecular and intermolecular hydrogen bonds simultaneously, an example being salicylic acid. *o*-Nitrophenol is an example of a molecule that forms an intramolecular hydrogen bond, while *p*-nitrophenol can form only intermolecular hydrogen bonds.

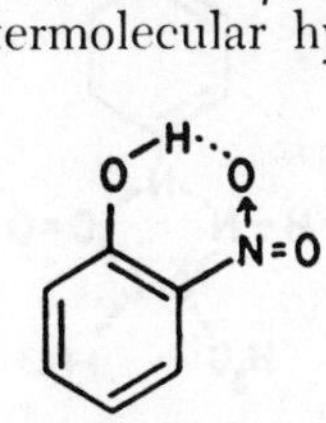

o-Nitrophenol

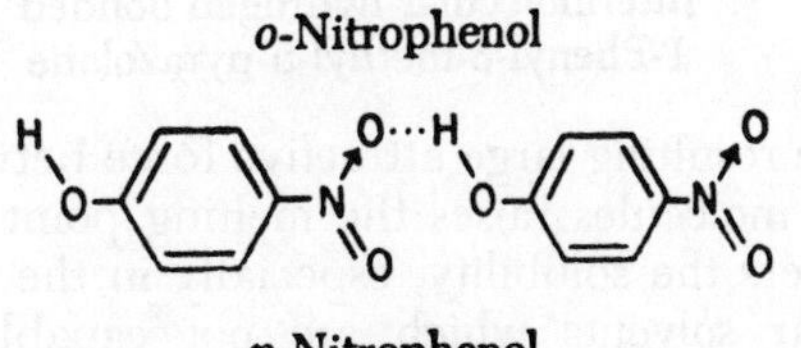

p-Nitrophenol

Intermolecular bonds are frequently much weaker than the intramolecular bonds. Strong intramolecular hydrogen bonds usually are found in 5-membered rings (6-membered, counting the hydrogen atom).

Since the physical and chemical properties of a compound may be greatly altered by hydrogen bonding, it is reasonable to expect that this may also have a significant effect and show some correlation with biologic properties. To be sure, the finding of such correlations is not proof that the hydrogen bond as such is directly responsible for observed differences in biologic activity. Concomitant or associated properties may be partly responsible, and the extent to which each property contributes to the over-all biologic action cannot be determined precisely. In a number of cases hydrogen bonding may be correlated with the observed biologic activity.

1-Phenyl-3-methyl-5-pyrazolone shows no analgesic properties; on the other hand, 1-phenyl-2,3-dimethyl-5-pyrazolone (antipyrine) is a well-known analgesic agent. The former has a melting point of 127° and is comparatively insoluble at ordinary temperatures in water and only slightly soluble in ether. The latter has a lower melting point (112°), is soluble in water (1:1) and moderately soluble in ether (1:43). It is unusual for a methyl group to bring about such large changes. The effect appears to be best explained by the fact that the first compound through intermolecular hydrogen bonding forms a linear polymer.

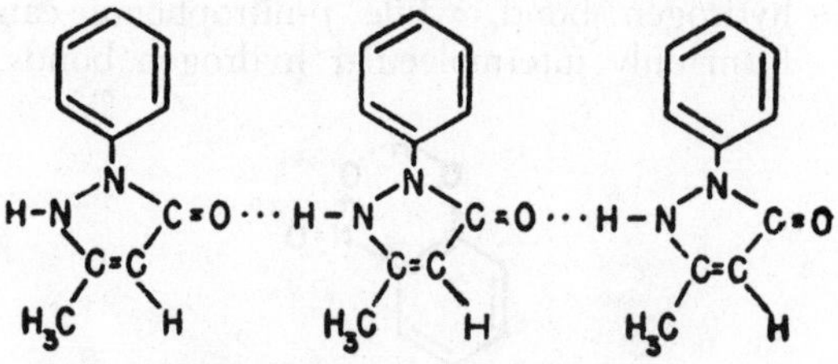

Intermolecular hydrogen bonded
1-Phenyl-3-methyl-5-pyrazolone

The resulting large attractive force between the molecules raises the melting point and lowers the solubility, especially in the nonpolar solvents which are not capable of breaking the hydrogen bonds. On the other hand, the methyl compound (antipyrine) cannot form hydrogen bonds and has only comparatively weak attractive forces between its molecules.

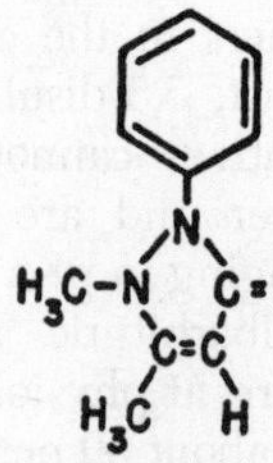

1-Phenyl-2,3-dimethyl-5-pyrazolone
(Antipyrine)

Its melting point is lower in spite of its having a higher molecular weight, and it is freely soluble in nonpolar solvents.

Salicylic acid (o-hydroxybenzoic acid) has quite an appreciable antibacterial activity, but the para isomer (p-hydroxybenzoic acid) is inactive. The reverse is true for the esters. Methyl salicylate has an extremely weak antibacterial action, but methyl p-hydroxybenzoate shows good action. A number of the esters of p-hydroxybenzoic acid (especially methyl and propyl) are used as preservatives in various pharmaceutical and cosmetic preparations. The difference in antibacterial action of the free acids and their esters may be accounted for through hydrogen bond formation. Only the ortho isomer (salicylic acid) shows analgesic and antipyretic properties. Likewise, salicylic acid is the only one of the three isomers that can form intramolecular hydrogen bonds. The m- and the p-isomers can form only intermolecular hydrogen bonds.

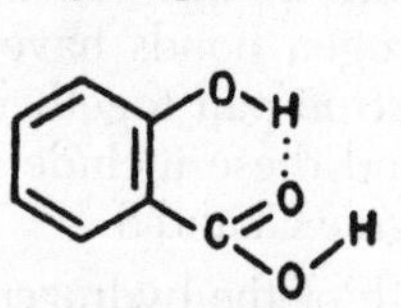

Salicylic Acid

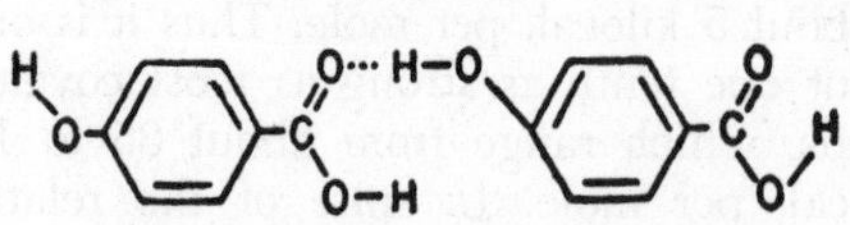

p-Hydroxybenzoic Acid (dimer)

Salicylic acid is a much stronger acid (pKa = 3.0) than p-hydroxybenzoic acid (pKa = 4.5). Salicylic acid is less soluble in water than the p-isomer, but its partition coefficient (benzene/water) is approximately 300 times greater. The higher melting point of *p*-hydroxybenzoic acid may be associated with intermolecular hydrogen bonding, which can also account for the low partition coefficient and low bactericidal action. It should be noted that salicylic acid with an intramolecular hydrogen bond has the phenolic hydroxyl masked, but the carboxylic acid group is free and can function as an antibacterial agent similar to benzoic acid. *p*-Hydroxybenzoic acid, on the other hand, must form intermolecular hydrogen bonds leading to a high degree of association and thereby lowering its antibacterial activity. In the case of the esters of salicylic and *p*-hydroxybenzoic acid, the opposite effect in bactericidal power is observed. Methyl salicylate is without significant antibacterial activity, but the esters of *p*-hydroxybenzoic acid show useful antibacterial properties. Methyl salicylate, through intramolecular hydrogen bond formation, has the phenolic hydroxyl group masked.

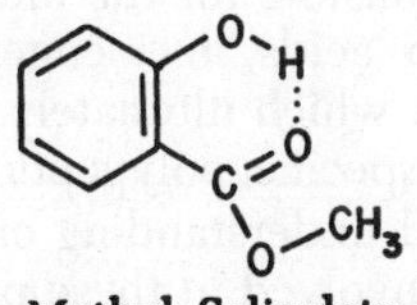

Methyl Salicylate

Methyl *p*-hydroxybenzoate and other esters of *p*-hydroxybenzoic acid can form only intermolecular hydrogen bonds, which may be illustrated with the following structures:

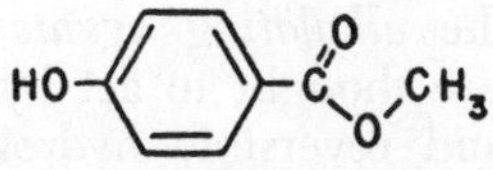

Methyl *p*-Hydroxybenzoate

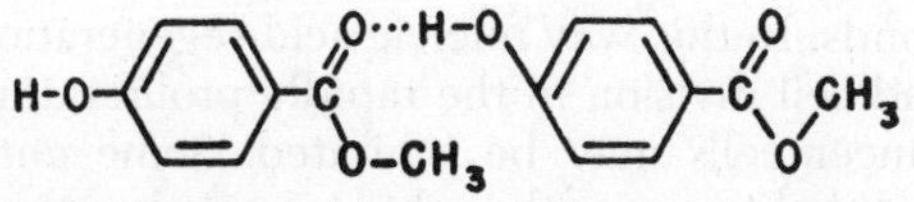

Methyl *p*-Hydroxybenzoate (dimer)

Association through hydrogen bonding to form the dimer or higher polymers may occur, but the partition coefficient and the antibacterial activity data suggest that the esters of *p*-hydroxybenzoic acid are not highly associated and that they may function as substituted phenols.

The structures of tetracycline and chlortetracycline are of interest in that they may form intramolecular hydrogen bonds.

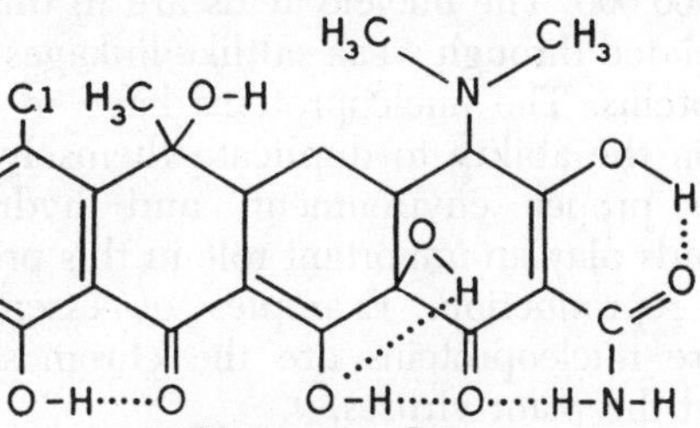

Chlortetracycline

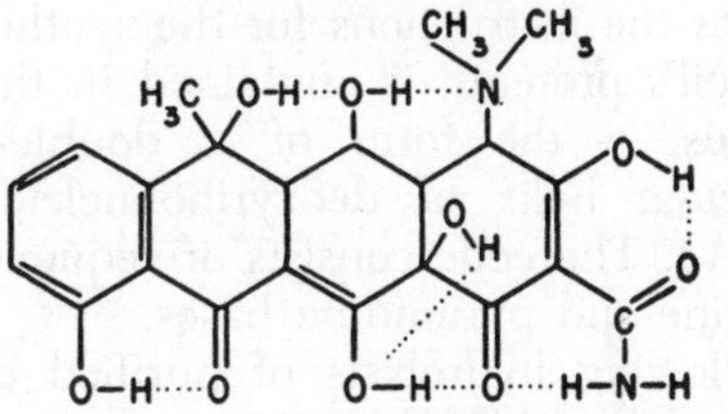

Tetracycline

Such compounds also chelate with metal ions. Whether the antibiotic properties of these compounds are in any way associated with the formation of intramolecular hydrogen bonds and chelates is not established.

It appears quite possible that certain types of biologically active molecules may first attach themselves to the site of action in the organism (such as the functional group of an enzyme system) through the initial formation of a hydrogen bond. Indeed, such bonding may be primarily responsible for the observed biologic effect. Molecules may form more than one hydrogen bond as a cross link where two or more hydrogen bonding atoms are available in the same molecule. The distance between the electron bonding and the reactive hydrogens of the drug and the substrate must coincide to give the multiple bonding. Such linkages are quite specific.

The nucleic acids, fundamental reproductive units of cells, provide an important example of molecules held together by specific hydrogen bonds.[113] Nucleic acids are composed of purines and pyrimidines in glycosidic combination with ribose or 2-deoxyribose forming nucleosides, which are in turn phosphorylated to form nucleotides. The nucleotides are linked together through phosphate bridges forming long chains, and these chains associated with other chains through hydrogen bonding form the nucleic acids of molecular weight 200,000 to 2,000,000. The nucleic acids are in turn associated through weak saltlike linkages with proteins. The nucleoproteins have in common the ability to duplicate themselves in the proper environment, and hydrogen bonds play an important role in this process of reproduction. Examples of essentially pure nucleoproteins are the chromosomes and the plant viruses.

The genetic code of the cell, which constitutes the instructions for the synthesis of the cell's proteins, is contained in the cell nucleus in the form of a double-chain molecular helix of deoxyribonucleic acid (DNA). The code consists of sequences of 4 purine and pyrimidine bases.

Following hydrolysis of purified deoxyribonucleic acid, adenine and thymine are found in nearly equimolar amounts; guanine and cytosine form a similar 1:1 pair. These purine-pyrimidine pairs are thought to occupy adjacent positions on neighboring nucleic acid strands, and to be held together by specific hydrogen bonds. It now appears that a triplet code is involved: a sequence of 3 purine or pyrimidine bases is needed to specify which amino acid will be incorporated in a specific location in the protein.

Adenine   Thymine

Guanine   Cytosine

When a cell divides, half of the nucleoprotein from the mother cell can be found in each daughter cell. It is possible that the paired strands separate, each going into a different daughter cell to act as a matrix for the formation of a new strand, but more likely the paired strands are equally distributed.

The large DNA molecule does not take part directly in protein synthesis, instead the genetic code is transcribed into shorter single chains of ribonucleic acid (RNA), called "messenger RNA," which is presumed to be a copy of the bases in one strand of DNA. Still smaller units of RNA, called "transfer RNA," which are specific for each amino acid, pick up and activate the amino acid by acylation of the ribose portion of the RNA. The activated amino acid is deposited at a position in the polypeptide chain specified by messenger RNA, the site of protein synthesis within the cell being a particle called the ribosome.

Hydrogen bonds play a key role in maintaining the structural integrity of the base pairs of DNA. Similar weak chemical bonds must be responsible for the interactions between amino acids, messenger RNA and transfer RNA which ultimately result in the synthesis of specific polypeptide chains. A more detailed understanding of the specific interactions involved in these processes will aid in understanding and attacking many problems of abnormal cellular function.

Mutagenic agents may cause rupture of the nucleic acid chain, thus disrupting the self-duplicating sequence. Many current attempts at chemotherapeutic control of cancer are directed at this level of cellular function. The *alkylating agents* (nitrogen mustards) are thought to act by replacing the weak and reversible hydrogen bonds between adjacent nucleic acid strands with strong and relatively irreversible covalent bonds. In this way nucleic acid regeneration and cell division in the rapidly proliferating cancer cells may be inhibited. Some *antimetabolites* are thought to act by their structural similarity to the purine and pyrimidine constituents of nucleic acids. An example is mercaptopurine, an analog of adenine in which the 6-amino group, normally responsible for specific hydrogen bonding, is replaced by a 6-thiol group. This is believed to be a close enough analog to fit into pathways leading to normal nucleic acid formation but incapable of replacing the functions of adenine which contribute to cellular reproduction.

CHELATION AND BIOLOGIC ACTION

The term *chelate* is applied to those compounds that result from a combination of an electron donor with a metal ion to form a ring structure. The compounds capable of forming a ring structure with a metal are designated as *ligands*. If the metal is bonded to carbon, the ring structure is not a chelate but an organometallic compound which has different properties. If the metal is not in a ring, the compound is called simply a metal complex. Nearly all the metals can form chelates and complexes. However, the electron donor atoms in the chelating agent are limited almost entirely to N, O and S. If the complex-forming ligand supplies both electrons for chelation, then the bond is classified as a co-ordinate covalent bond and by convention is represented as M←X, where M is the metal and X the ligand. If one electron is supplied by the metal and one by the ligand (normal covalent bond), the bond is shown as M—X.

A ligand molecule containing only 2 electron-donating groups is designated as "bidentate" and is able to form only a single ring; if it contains 3 electron-donating groups, it is "tridentate" and may form 2 rings in an interlocked complex; the porphyrins are able to form a number of interlocked ring systems with metals and are designated as "polydentate" structures.

The size of the rings in chelate compounds is of interest with respect both to stability and occurrence. Three-membered rings have not been identified, but 4-membered rings are known, and 4-membered rings containing sulfur may be quite stable. The 5- and the 6-membered chelate rings are most common and usually show the greatest stability. The chelates are identified by a number of properties, no one of which constitutes positive identification, so evidence from as many different sources as possible is desirable.

The "normal" chemical reactions of a metal ion in solution disappear if a chelate is formed. The chelate may serve to prevent precipitation of an ion which normally would precipitate. For example, the cupric ion in basic Fehling's solution normally would be precipitated as cupric hydroxide but is prevented from doing so by the formation of the copper tartrate chelate. In this case, the formation of the chelate results in an increased water solubility. In other cases, the chelate may be insoluble in water and soluble in organic solvents. Water-soluble chelating agents, called *sequestering agents*, often are used to remove objectionable metal ions by combining with them to form stable water-soluble chelates.

A number of naturally occurring chelates are present in biologic systems. The amino acids, proteins and acids of the tricarboxylic acid cycle are the principal *ligands*; and the metals involved are iron, magnesium, manganese, copper, cobalt and zinc, among others. A group with iron present is the heme-proteins, such as hemoglobin, which is found in the red blood cells of vertebrates and is involved in oxygen transport. In the heme-proteins, an iron atom is covalently bound to a porphine derivative. Other heme-proteins are myoglobin, an intercellular pigment involved in oxygen storage; catalase, a compound present in the tissues of plants and animals which catalyzes the decomposition of H_2O_2; peroxidases, which are enzymes involved in the oxidation of a substrate with peroxides; cytochromes, which play a part in cellular oxidation. Because of the change of magnetic character of iron compounds with different types of bonding, a study of magnetic susceptibilities has given a rather detailed insight into the way iron is bound in these compounds. Although this tool is not available in studying most of the other naturally occurring metal compounds, it is fairly certain that these metals also are present as chelates. The evidence for this is that the bonding of the metal is so strong that it could result only from the formation of chelate rings.

Copper-containing enzymes include the oxidases, ascorbic acid oxidase, tyrosinase, polyphenoloxidase and laccase. Magnesium is present in chlorophyll, but it is involved also in the action of some proteolytic enzymes, phosphatases and carboxylases. Manganese activates most carboxylases and some proteolytic enzymes. Zinc, which is present in insulin, has an activating effect on some carboxylases, proteolytic enzymes and phosphatases. Cobalt activates some enzymes belonging to each of the above classes and is present in vitamin B_{12}.

The fact that a number of biologically important compounds are chelates opens up other approaches to chemotherapy. One such approach depends on the use of an unnatural chelating agent to reduce or eliminate the toxic effects of a metal. To serve in this capacity the chelating agent (ligand) must effectively compete with the chemical systems in the body to which the excess metal is bound.

The agent, because of its greater affinity for the metal, forms a more stable chelate, thereby decreasing the concentration of the toxic metal ion in the tissues by binding it as a soluble chelate for excretion by the kidneys. The stability of chelates is expressed as a constant $(\log K_s)$* which represents the over-all equilibrium between the metal and the chelates which it forms with the ligand. In the case of a trivalent metal (e.g., Fe^{+++}), $\log K_1$ represents the stability constant for the 1:1 chelate; $\log K_2$ the 2:1 chelate; $\log K_3$ the 3:1 chelate, and the over-all constant $(\log K_s)$ is the product of the individual constants,[114] that is,

$$\log K_s = \log K_1 + \log K_2 + \log K_3.$$

A chelate varies with the ligand and the metal to which it is bound. For example, the *stability constants* $(\log K_s)$ of glycine 2:1 chelates with several divalent metals are

as follows: Cu^{++} 15; Ni^{++} 11; Co^{++} 9; Fe^{++} 8; and Mn^{++} 5.5.

Dimercaprol was first introduced in 1945 under the name BAL (British anti-lewisite) as an antidote for the organic arsenical "lewisite." Subsequent studies have revealed its effectiveness for the treatment of poisoning due to antimony, gold and mercury, as well as arsenic[115] (see Chap. 3).

($\pm$)Penicillamine, a hydrolysis product of the various penicillins, is an effective antidote for the treatment of poisoning by copper.[116] Hepatolenticular degeneration (Wilson's disease), a familial disorder in which there is decreased excretion of copper and, sometimes, increased excretion of amino acids, may be treated with penicillamine to promote the removal of protein-bound copper and its excretion in the urine.

Penicillamine also has been used with some success as an antidote for the treatment of mercury and lead (divalent metals) poisoning. The 2:1 copper chelate, possessing two ionizable and solubilizing carboxyl groups, acts as a sequestering agent.

Deferoxamine mesylate (Desferal) a trihydroxamic acid compound made up of the elements of 3 moles of a 1,5-pentamethylene diamine, 2 moles of succinic acid, and one of acetic acid, is isolated from *Streptomyces pilosus*.[117] The compound combines with Fe^{+++} to form a water-soluble chelate which is excreted by the kidneys. The agent removes excess iron from the tissues but does not displace it from essential proteins (e.g., transferrin) involved in the iron trans-

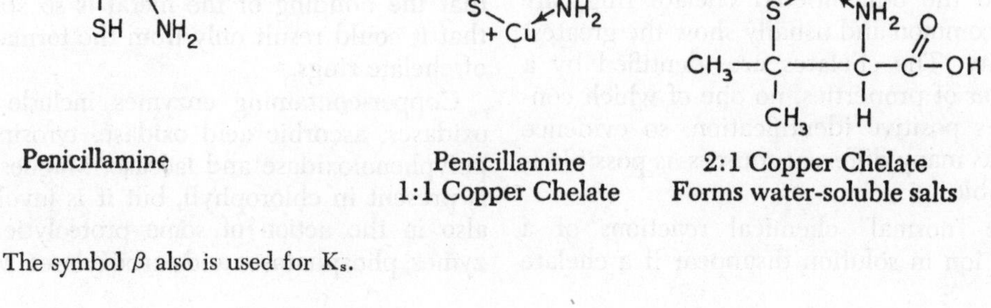

Penicillamine	Penicillamine 1:1 Copper Chelate	2:1 Copper Chelate Forms water-soluble salts

* The symbol β also is used for K_s.

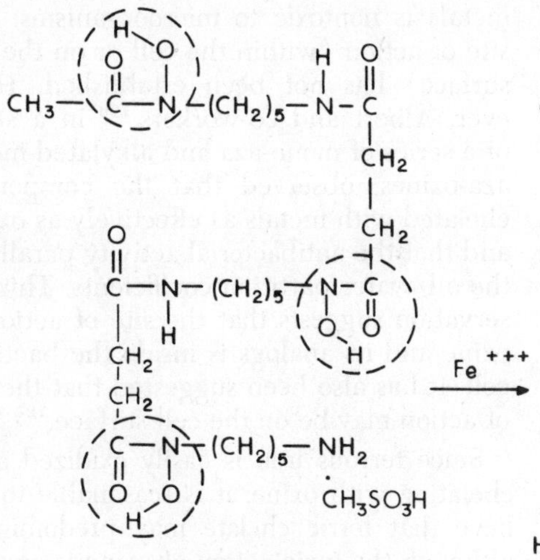

Deferoxamine
(Chelating groups circled)

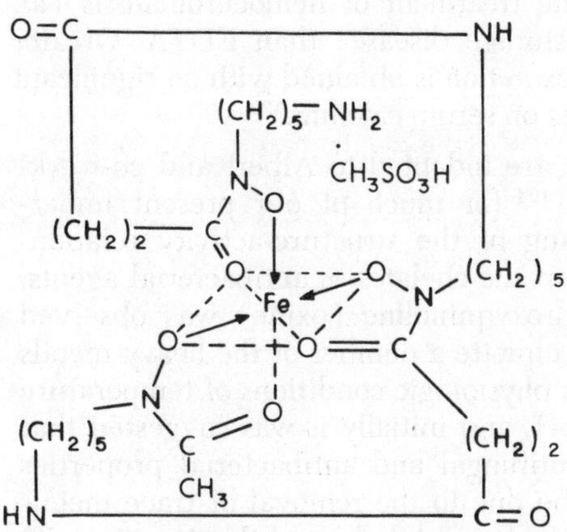

Deferoxamine—iron chelate

port mechanism. The compound is reported to be selective for iron with little or no affinity for calcium, copper and other metals.[118, 119] It is nontoxic and has been used successfully in the treatment of primary (hereditary) and secondary hemochromatosis and as an effective antidote for the treatment of acute iron poisoning in children.

The reddish-colored iron chelate of deferoxamine has a high stability constant (log $K_s = 30.7$) which may be attributed to its unique chemical structure in which the iron is octahedrally bound by the hydroxamic acid oxygen atoms and carbonyl oxygens of the ligand.

The compound ethylenediamine tetraacetic acid (EDTA) forms water-soluble stable chelates with many metals and is an important sequestering agent which has received wide application. Among these may be mentioned: its use as an antioxidant for the stabilization of drugs which rapidly deteriorate in the presence of trace metals (e.g., ascorbic acid, epinephrine and penicillin); prevention of rancidity in detergents; purifying oils; clarifying wines and soap solutions; removal of lead arsenate spray residues from fruits; removal of radioactive contaminants; titration of metals; and as an antidote for heavy metal poisoning.

Like penicillamine, EDTA is able to form water-soluble, stable metal chelates in the body which may be excreted readily. The free acid and sodium salts of EDTA, when administered to mammals, produce an excessive loss of essential body calcium and are quite toxic. The calcium-disodium salt (edathamil) is comparatively nontoxic and serves as an effective antidote for the treatment of lead poisoning. It is also reported to be effective as an antidote for other heavy metals, including copper, chromium, iron and nickel.[120]

The general structure given below is assigned to the water-soluble metal chelates of EDTA. It should be noted that the metal (M) is bound by 2 coordinate-covalent

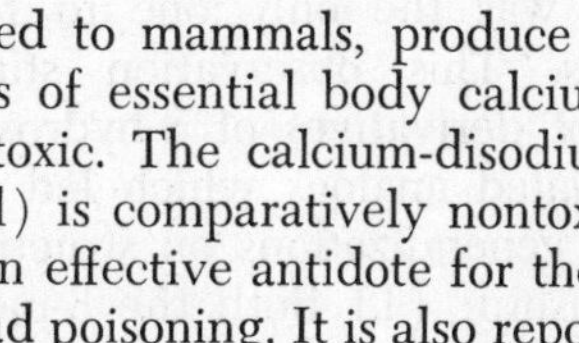

Disodium salt of metal chelates with ethylenediamine tetraacetic acid
M = Bound metal

bonds ($\rightarrow$) and 2 normal covalent bonds (—). The strength of the metal binding (stability constant) varies with each metal. For the divalent ions the order is as follows:

$$Cu^{++} > Ni^{++} > Pb^{++} > Co^{++}, Zn^{++}$$
$$> Fe^{++} > Mn^{++} > Mg^{++}, > Ca^{++}.[121]$$

Trisodium calcium diethylenetriamine pentaacetic acid is reported as more effective for the treatment of hemochromatosis (an iron storage disease) than EDTA. Greater iron excretion is obtained with no significant effects on serum calcium.[122]

We are indebted to Albert and co-workers[123, 124] for much of our present understanding of the structure-activity relationships of the chelates as antibacterial agents. 8-Hydroxyquinoline (oxine) was observed to precipitate a number of the heavy metals under physiologic conditions of temperature and pH, and initially it was suggested that the antifungal and antibacterial properties may be due to the removal of trace metals essential for metabolism of the organisms.[125] A study of the 7 isomeric monohydroxyquinolines demonstrated that only the 8-hydroxy isomer was active in inhibiting the growth of microorganisms and that the same isomer was the only one to form metal chelates. This observation stimulated a study of derivatives of 8-hydroxyquinoline and related analogs which led to the following generalizations on structure-activity relationship: (1) Both the 8-methyl ether and the 1-methyl derivatives of 8-hydroxyquinoline which are unable to form chelates show no antibacterial effect. (2) Substitution of a mercapto for the hydroxy group in oxine gives an active chelating agent which is also active as an antibacterial. (3) The substitution of a methyl group in the 2-position of oxine gives an active chelating agent in vitro, but the compound is relatively inactive as an antibacterial. This decreased activity is attributed to lack of penetration of the cell, or interaction with the cell receptor, due to steric hindrance. (4) The introduction of a highly ionizable group in oxine (e.g., 8-hydroxyquinoline-5-sulfonic acid) does not alter the chelating property in vitro, but the antibacterial activity is lost, presumably due to the inability of the ion to penetrate the cell wall. A high partition coefficient appears to be essential for antibacterial activity.

It has been well established that oxine and its analogs act as antibacterial and antifungal agents by complexing with iron or copper. Oxine, in the absence of these metals is nontoxic to microorganisms. The site of action (within the cell or on the cell surface) has not been established. However, Albert and co-workers,[126] in a study of a series of mono-aza and alkylated mono-aza-oxines, observed that the compounds chelated with metals as effectively as oxines and that the antibacterial activity paralleled the oil/water partition coefficients. This observation suggests that the site of action of oxine and its analogs is inside the bacterial cell. It has also been suggested that the site of action may be on the cell surface.[127]

Since ferrous iron is easily oxidized after chelating with oxine, it is reasonable to believe that ferric chelate may predominate, although the toxic action of oxine is equally developed by the addition of ferrous or ferric salts. The addition of an excess of either iron or oxine inhibits the antibacterial action.[128] Thus, the growth of *Staph. aureus* in untreated meat broth is completely inhibited by oxine (M/100,000), but this concentration has no effect on the organism when suspended in distilled water. The toxic effect of oxine is due to its combination with trace amounts of iron in the meat broth; when the concentration of oxine was increased (M/800) the inhibition of growth (antibacterial effect) disappears, due to a "concentration quenching" effect which is attributed to a shifting of the equilibrium from the unsaturated (1:1- and 2:1-complexes) to the saturated, nontoxic 3:1-oxine-iron complex; and inhibition of growth again occurs when the concentration of iron is increased (M/800), since the equilibrium is shifted from the saturated 3:1-complex to the unsaturated 1:1- and 2:1-complexes which are toxic.

If the site of action is within the cell, it is reasonable to assume that only the saturated (3:1-oxine-ferric complex) will be able to penetrate the cell membrane. The unsaturated 1:1- and 2:1-complexes as cations cannot penetrate. It is postulated that the nontoxic 3:1-complex which is able to penetrate the cell membrane breaks down inside the cell to form the toxic unsaturated 1:1- or 2:1-complexes.[126] Although it appears less likely that the site of action is outside the cell membrane, it must be

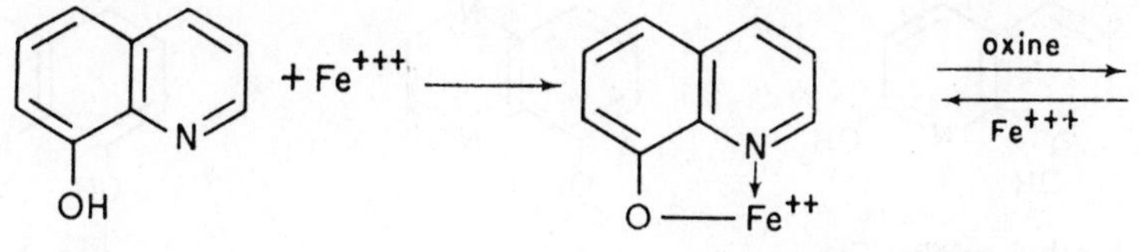

1:1-oxine-ferric chelate
unsaturated: active

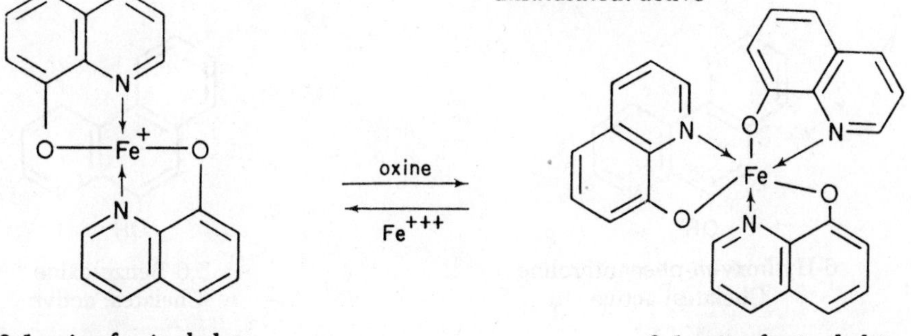

2:1-oxine-ferric chelate
unsaturated: active

3:1-oxine-ferric chelate
saturated: inactive

assumed that, if such is the case, the unsaturated 2:1-complex would be responsible for the toxic effect. Since excess iron or oxine decreases this toxic effect, the equilibrium would be shifted to form primarily the 1:1- or the 3:1-complexes, respectively, which would bring about a concomitant decrease in toxicity.

The antibacterial properties of oxine-iron complexes are antagonized by metals that form more stable complexes. The addition of low concentrations of cobaltous sulfate

(M/25,000) completely inactivates the antibacterial action of (M/100,000) oxine-iron solutions.[128]

The structures below are representative of the biologically active and inactive analogs of oxine. It should be noted that all active compounds (antibacterial) form metal chelates, but not all chelating structures are biologically active.

Cupric salts form 1:1- and 1:2-complexes with oxine. By analogy with the iron chelates it may be assumed that only the

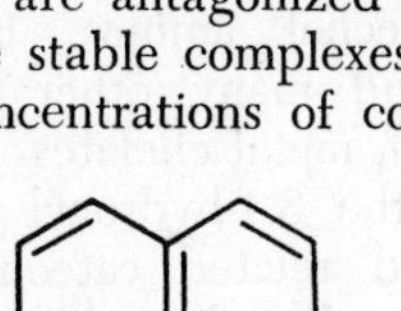

8-Hydroxyquinoline (Oxine)
Chelates: active

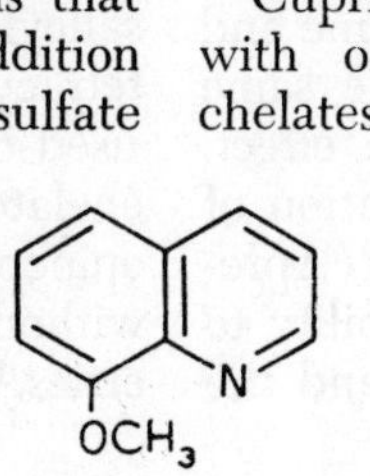

8-Methoxyquinoline
Nonchelating: inactive

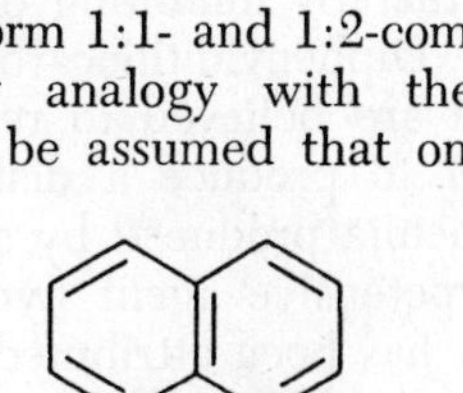

Oxine methochloride
Nonchelating: inactive

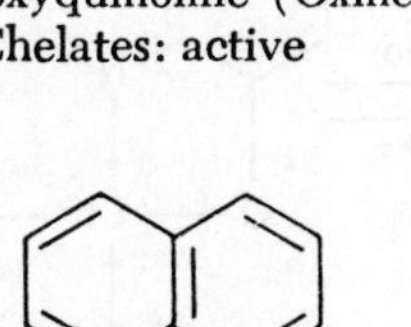

8-Mercaptoquinoline
Chelates: active

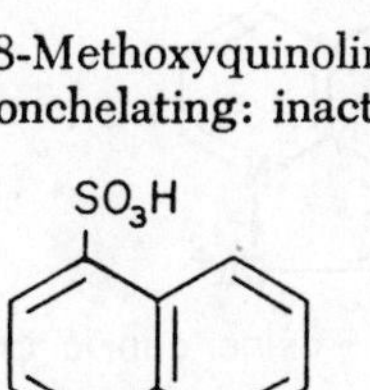

8-Hydroxyquinoline-5-
sulfonic acid
Chelates: inactive

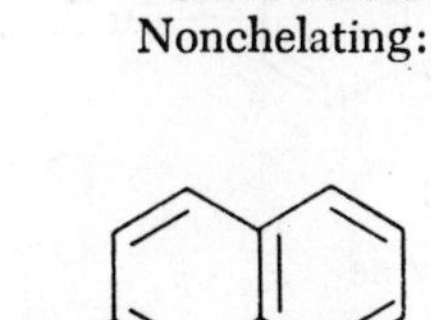

7-Chloro-8-hydroxy-
quinoline
Chelates: active

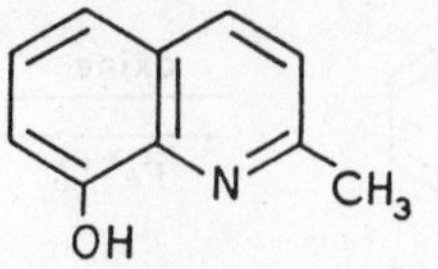

2-Methyl-oxine
Chelates: decreased activity

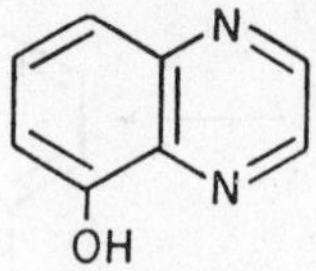

4-aza-oxine
Chelates: active

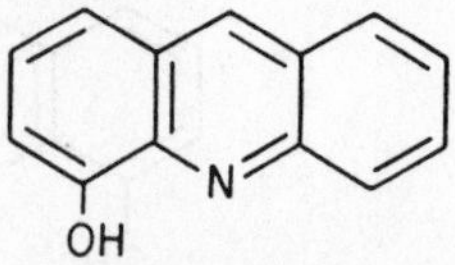

4-Hydroxyacridine
Chelates: active

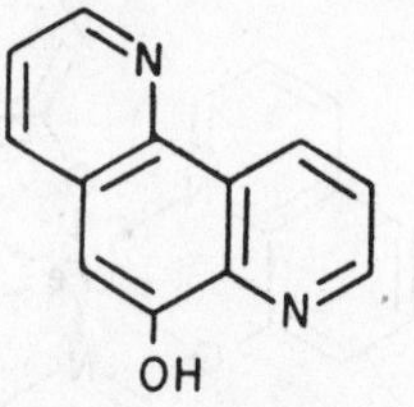

6-Hydroxy-*m*-phenanthroline
Chelates: active

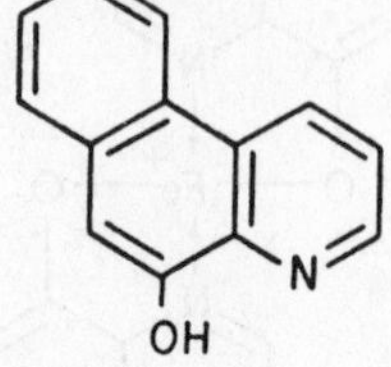

5,6 Benzo-oxine
Chelates: active

unsaturated 1:1-complex would be active as an antibacterial or antifungal agent.

Numerous drugs unrelated to oxine form chelate complexes with metals; and although the complex formation may have no direct relation with the major action of the drug, it may be responsible for significant side-effects. Thus, the antitubercular agent, thiacetazone, may produce an onset of *diabetes mellitus*, and it has been suggested that this may be due to its ability to chelate with zinc in the *beta* cells of the pancreas, thereby inhibiting the production of insulin. Diphenyldithiocarbazone, oxine and alloxan are believed to react in the same manner to produce a diabetogenic effect. The anemia produced by administration of the hypotensive agent hydralazine (Apresoline) has been attributed to its ability to complex with iron.[129] Dimercaprol and the

antitubercular drug isonicotonic acid hydrazide (INH) tend to induce histaminelike actions, and it has been suggested that this may be due to complexing with a copper-catalyzed enzyme responsible for the destruction of histamine.[130] INH may function as a chelating agent in inhibiting the growth of *Mycobacterium tuberculosis*, but the evidence for this mode of action is not conclusive. The drug is an active chelating agent (see p. 59), and derivatives, such as 1-methyl-1-isonicotinoyl hydrazine, which are unable to chelate are inactive.[131] The salicylates, catechol amines, biguanides, tetracyclines and many other commonly used drugs form metal chelates. Boric acid chelates with the 3,4-hydroxyl groups of epinephrine and related catechol amines without altering the pharmacologic properties.[132]

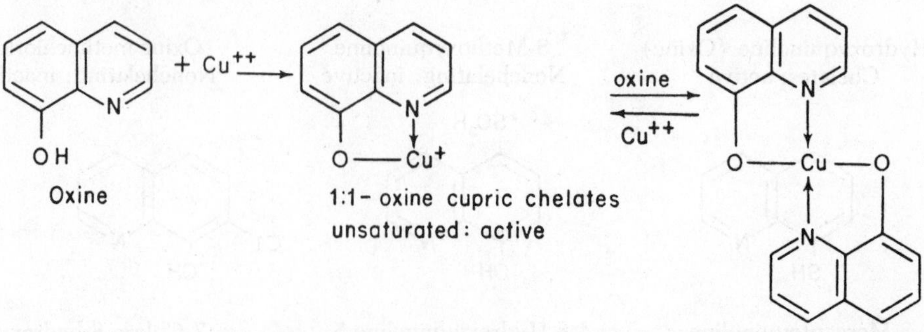

Oxine

1:1- oxine cupric chelates
unsaturated: active

2:1-oxine cupric chelate
saturated: inactive

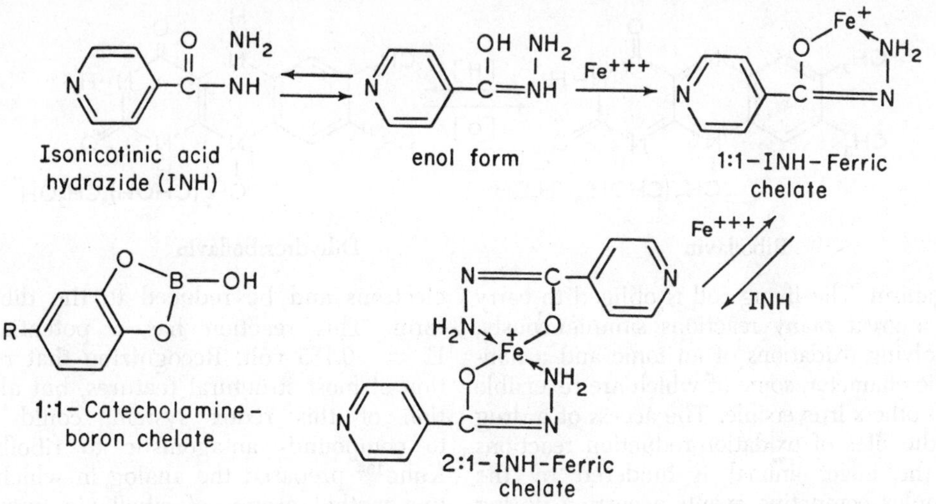

Isonicotinic acid
hydrazide (INH)

enol form

1:1-INH-Ferric
chelate

1:1-Catecholamine-
boron chelate

2:1-INH-Ferric
chelate

In summary, chelation may be used for a variety of purposes, including (1) sequestration of metals to control the concentration of metal ions (e.g., buffer systems); (2) stabilization of drugs (e.g., epinephrine); (3) elimination of toxic metals from intact organisms (e.g., EDTA as an antidote for treatment of lead poisoning); (4) improvement of metal absorption, which has been demonstrated in plants and, by analogy, may also be true in mammals (e.g., EDTA-iron complex increases uptake of iron in plants); and (5) increasing the toxic effects of a metal (e.g., antibacterial activity of the unsaturated oxine-iron chelates).

The full significance and importance of chelation in biology and medicine remains to be established, but from the knowledge available it is evident that the chelates represent an extremely important group of naturally occurring compounds and that foreign ligands may play an increasingly important role in chemotherapy. A review of the role of metal-binding in the biologic activities of drugs has been presented by Foye (see Selected Reading).

OXIDATION-REDUCTION POTENTIALS AND BIOLOGIC ACTION

The oxidation-reduction potential (redox potential) may be defined as a quantitative expression of the tendency that a compound has to give or to receive electrons. It is re-lated to the concentrations of the oxidant and reductant present under the conditions of the measurement and may be expressed mathematically by the following equation:

$$E_h = E'_o - \frac{0.06}{n} \log \left(\frac{reductant}{oxidant} \right)$$

E_h = The oxidation-reduction potential of the system being measured.

E'_o = The standard electrode potential at a specific hydrogen-ion concentration.

n = The number of electrons transferred. 0.06 is a thermodynamic constant for 1 electron transfer at 30°.

It should be noted that when the concentration of reductant equals the concentration of the oxidant, E_h is equal to E'_o.

The oxidation-reduction potential may be compared with an acid-base reaction. The latter case may be regarded as the transfer of a proton from an atom in one molecule to the atom in another, while in the case of an oxidation-reduction reaction, there is an electron transfer. Since living organisms function at an optimum redox potential range which varies with the organism, it might be assumed that the oxidation-reduction potentials of compounds of a certain type would correlate with the observed biologic effect.[133] However, there are a number of reasons why few satisfactory correlations have been observed. The oxidation-reduction potential applies to a single reversible ionic equilibrium which does not exist in a living

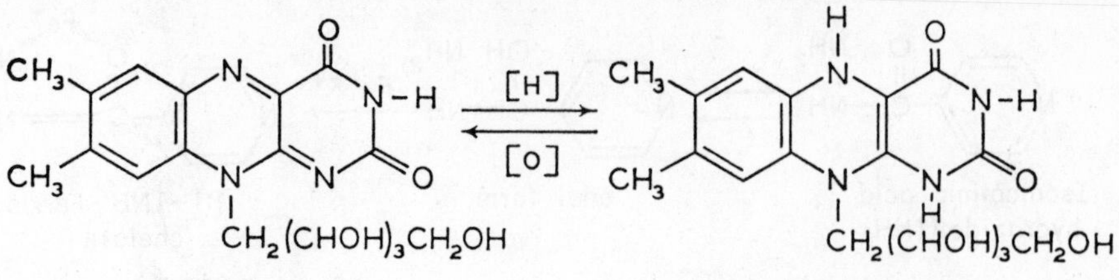

Riboflavin   Dihydroriboflavin

organism. The living cell is obliged to carry on a great many reactions simultaneously, involving oxidations of an ionic and a non-ionic character, some of which are reversible and others irreversible. The access of a drug to the sites of oxidation-reduction reactions in the intact animal is hindered by the complex competing events occurring during absorption, distribution, metabolism and excretion. Included among these may be multiple competing biologic redox systems. Therefore, it is to be expected that correlations between redox potential and biologic activity generally hold only for compounds of very similar structure and physical properties. In such series, variations in routes of distribution and in steric factors which might modify the drug redox system interaction would be minimized. Only a few series studied have met these criteria. Page and Robinson[134] have studied the relation between the bacteriostatic activity and the normal redox potentials of substituted quinones. They find no simple relation between the reduction potentials (E_o') of 20 substituted quinones and their observed bacteriostatic activity. The quinones showing marked activity against *Staph. aureus* gave reduction potentials falling between -0.10 and $+0.15$ volt, with optimum activity associated with a potential of approximately $+0.03$ volt. The same authors failed to find a similar relation between oxidation-reduction potentials of 18 commercially distributed oxidation-reduction indicators of varied structure and their activity against *Staph. aureus*.

The reduction potentials of a number of acridines have been studied.[135] The more active compounds have an $E_{1/2h}$ of less than -0.4, but no detailed correlation with the antibacterial activity was observed.

Riboflavin, in its cofactor form, owes its biologic activity to its ability to accept electrons and be reduced to the dihydro form. This reaction has a potential of $E_o = -0.185$ volt. Recognizing that retention of most structural features, but alteration of this redox system, could lead to compounds antagonistic to riboflavin, Kuhn[136] prepared the analog in which the two methyl groups of riboflavin were replaced by chlorines. This compound had a potential of $E_o = -0.095$ volt, and its antagonistic properties were suggested as being due to the dichloro-dihydro form being a weaker reducing agent than the dihydro form of riboflavin. It may be absorbed at the specific receptor site but not have a negative enough potential to carry out the biologic reductions of riboflavin. More recently "nonredox analogs of riboflavin" have been proposed as potential anticancer agents,[137] and compounds have been prepared which alter the redox potential or fix the molecule in the nonoxidizable dihydro form.

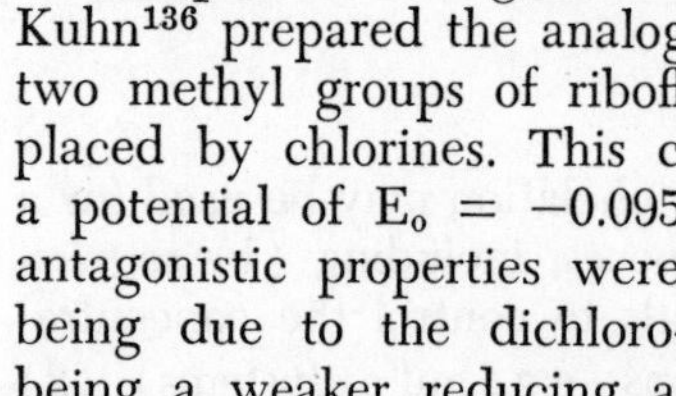

Altered redox potential

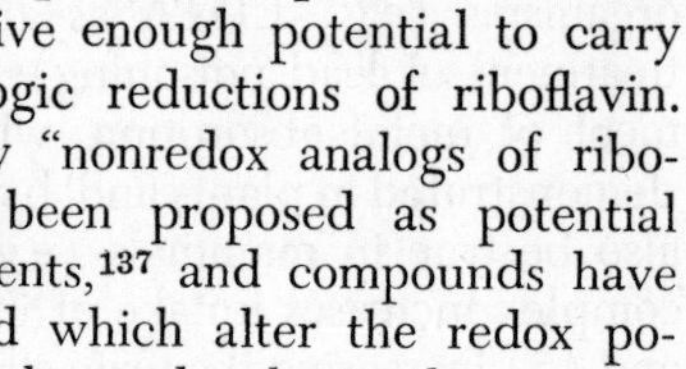

Fixed in dihydro form

Riboflavin analogs

Phenothiazine   Semiquinone Ion   Phenozothionium Ion
(Active species)

The anthelmintic activities of a series of substituted phenothiazines have been correlated with the possession of a redox potential which could lead to maximal formation of semiquinone ion (a radical ion) at physiologic pH.[138] Against mixed infestations of *Syphacia obvelata* and *Aspirculurus tetraptera* in mice, anthelmintic activity was present in unsubstituted phenothiazine and those substituted derivatives (3–EtO–; 3–MeO–; 3–Me; 2–Cl–7–MeO; 4–Cl–7–MeO–; 3–F–; 3–Cl–; 3–Br–) with E_m values* which were within 0.1 volts of the value 0.583 V (acetic acid–water). At a potential similar to that of the biologic oxidation-reduction system involved, semiquinone concentration would be maximal and thus facilitate or compete with an essential biologic electron transfer reaction, producing a toxic or paralyzing effect. When corrected for solvent effects, the active potential is in the range of that of isolated cytochromes.

The additional requirement of a free 3 or 7 position in the phenothiazine nucleus for significant anthelmintic activity, and the inactivity of phenothiazine tranquilizing drugs (2-substituted 10-dimethylaminopropylphenothiazines) again point up the difficulty of correlating redox potential and activity for compounds with differing structural and solubility characteristics.

SURFACE ACTIVITY: ADSORPTION AND ORIENTATION AT SURFACES

The molecules in a pure liquid are attracted equally on all sides by neighboring molecules, but the molecules on the surface are attracted inward without any opposing outside force. This contracting force is known as the surface tension of the liquid and is measured in dynes/cm. For water at 20° the surface tension is 72.75 dynes/cm.

* E_m = bivalent midpoint electrode potential.

When a solute is added to water it will increase or decrease the surface tension by interaction with the hydrogen-bonded water molecules. Certain types of molecules have the property of concentrating and orienting in a definite configuration at the interface or surface of a solution and thereby lowering the surface energy.

In general, compounds less polar than water will be positively adsorbed in water and concentrate on the surfaces of the solution. Compounds more polar than water are negatively adsorbed, and the solute will be uniformly dispersed in the bulk of the solution.

The orientation of surface-active molecules at the surface of water or at the interface of polar and nonpolar liquids takes place with the nonpolar (e.g., hydrocarbon) portion of the molecule oriented toward the vapor phase or nonpolar liquid and the polar groups (e.g., $-COOH$, $-OH$, $-NH_2$, $-NO_2$, etc.) toward the polar liquid. Three forces are involved in orientations of this type: namely, van der Waals' forces, hydrogen bonds and ion dipoles. Van der Waals' forces represent mainly the forces of attraction between the nonpolar groups tending to hold the hydrocarbon groups together. An aliphatic hydrocarbon chain of 18 methylene groups has an attractive force for another hydrocarbon chain of the same length (both in *trans* configuration) of approximately 15 kcal. per mole. This force is of the same order of magnitude as an ion dipole association. Oleic acid on water forms a monomolecular film in which the hydrocarbon chain tends to stand vertically on the water surface, pointing into the vapor of the solution, and the $-COOH$ groups are pointed into the water. If mineral oil or another nonpolar liquid such as benzene is now added to the water solution to form a liquid/liquid interface, the hydrocarbon chains of oleic acid will be pointed into the nonpolar liquid, and the —COOH groups will remain in contact with the water. The oleic acid thereby tends to make the transition from the nonpolar to the polar phase less abrupt. In such cases, the free surface energy is greatly lowered.

Compounds showing pronounced surface activity usually are unsuited for use in the animal body. Such compounds are lost through adsorption on proteins, and they also have the undesirable feature of disorganizing the cell membrane and producing hemolysis of red blood cells. In general, highly surface-active agents are limited in use to topical application as skin disinfectants or for the sterilization of inanimate objects, such as instruments. The surface-active cations of the quaternary ammonium type are used for this purpose. They are characterized by having a hydrophilic cation attached to a long nonpolar group. Compounds of this type are nonspecific antibacterial agents, since they are adsorbed on all tissues as well as on bacteria. Their activity is greatly reduced by body fluids and by high molecular weight anions, such as the ordinary soaps. In common with other surface-active substances, they tend to act destructively on denuded tissue. An important feature of the surface-active compounds is the property of "wetting" shown by their solutions. This property may serve to bring the active agent into more intimate contact with the total surface to which it is applied. This is, of course, a decided advantage for solutions intended for use as topical antiseptics.

A surfactant molecule exhibits two distinct regions of lipophilic and hydrophilic character and such compounds are commonly categorized as *amphiphilic,* or as *amphiphils.* Molecules of this type may vary markedly from predominantly hydrophilic to predominantly lipophilic, depending on the relative ratio of polar to nonpolar groups present. The polar or hydrophilic groups differ widely in their degree of polarity and, as might be expected, the more polar groups (e.g., $-OSO_3^-$, $-SO_3^-$, $-NR_3^+$ etc.) are able to increase the hydrophilic character of a molecule to a greater extent than the weaker polarizing groups such as $-NH_2$, $-OH$, $-COOH$. Thus, water solubility is lost, and lipid solubility is gained in the normal alcohols, amines and carboxylic acids at C_4 to C_6. The higher molecular weight members are miscible with oil and immiscible with water. However, the more active n-alkylated anions and cations (e.g., $-OSO_3^-$ and $-NR_3^+$) increase the hydrophilic character, and water solubility is not lost until the number of carbon atoms reaches C_{14} or higher. This difference in the hydrophilic properties of polar groups reflects their ability to interact with water. Water molecules are held together by hydrogen bonds (see section on hydrogen bonding), and a compound dissolves in water by breaking the hydrogen bonds and forming new hydrogen bonds with water molecules (e.g., alcohol hydrogen bonded to water, $R-O-H \cdots OH_2$). If the group is strongly ionized, the interaction with water is much greater, and the hydrophilic character of the compound is increased. Compounds of intermediate molecular weight in each of the above series will normally be miscible with both organic solvents and water.

There are four general classes of surface-active agents: (1) *anionic* compounds, such as the ordinary soaps, salts of bile acids, salts of the sulfate or phosphate esters of alcohols, and salts of sulfonic acids; (2) *cationic* compounds, such as the high molecular weight aliphatic amines, and quaternary ammonium derivatives; (3) *nonionic* compounds (e.g., polyoxyethylene ethers and glycol esters of fatty acids); and *amphoteric* surfactants.

The surface-active ions of intermediate to high molecular weight (ca. 150 to 300) show the same electrical and osmotic properties in dilute solutions as equivalent concentrations of inorganic electrolytes. Therefore, the ions in dilute solution are distributed in the monomeric state. However, with increasing concentration of the surfactant, a critical point is reached, at which the molecules associate (become polymeric) in an oriented fashion to form *micelles.* The concentration at which the

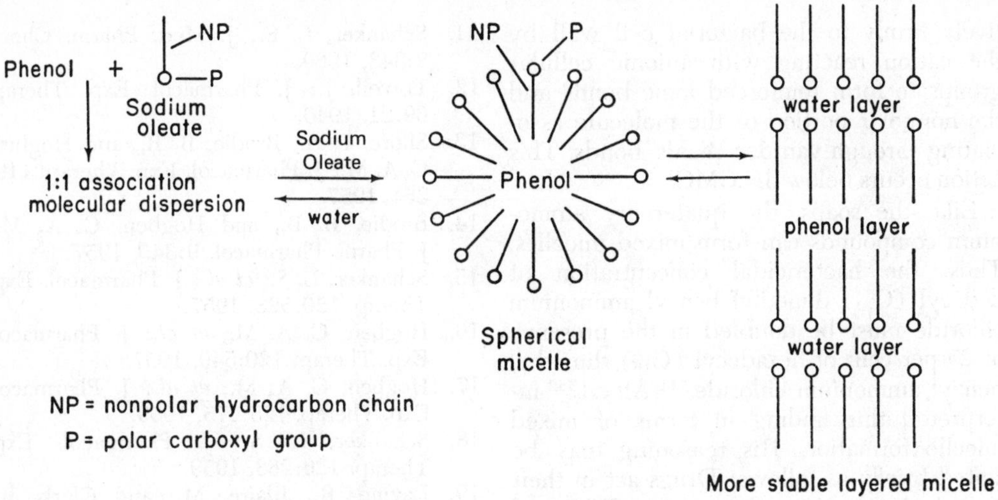

Phenol + Sodium oleate

NP = nonpolar hydrocarbon chain
P = polar carboxyl group

1:1 association molecular dispersion

Spherical micelle

More stable layered micelle

polymeric species develops is commonly designated as the *critical micelle concentration* (CMC) and differs for each surfactant.

At the critical micelle concentration large polymers (macromolecules) begin to form, and the solution becomes colloidal in nature. This is a reversible process; therefore, a micelle on dilution will revert to the monomeric state. The *solubilization* of organic compounds (insoluble in water) begins at the CMC and increases rapidly with increasing concentration of the solubilizing agent. The simplified structures above show the micelles of sodium oleate solubilizing an insoluble phenol. It should be mentioned that the size, the shape and the properties of micelles vary with each surfactant. The critical micelle concentration decreases rapidly as the molecular weight increases. Thus, the CMC (moles/liter) for potassium myristate (C_{14}) is only one fourth that of potassium laurate (C_{12}), and when the two soaps are mixed it requires only 15 per cent of potassium myristate to reduce by 50 per cent the CMC of potassium laurate. The decrease in the CMC is attributed to *mixed micelle* formation.[139] When phenols are solubilized by soap, mixed micelles (soap and phenol) are formed, and the activity of the phenol may be enhanced or reduced depending on the ratio of soap to phenol used.

The anthelmintic activity of hexylresorcinol is reported to be increased by low concentrations and decreased by high concentrations of soap. If the soap concentration is kept below the CMC, a 1:1 association of the phenol and soap occurs which facilitates the penetration of phenol through the surface of the worm. If the CMC is exceeded the micelle competes favorably with the worms for the phenol and there is decreased activity.[140] The antibacterial activity of the high molecular weight quaternary ammonium compounds (cationic soaps) appears to be dependent on two or more factors, such as: (1) the charge density on the nitrogen atom; (2) the size and the length of the nonpolar groups attached to the nitrogen; and (3) the lipophilic-hydrophilic balance. The fact that the lower and excesssively high molecular weight compounds are inactive indicates that more than a charged nitrogen is necessary for antibacterial activity. The active compounds are those in which the charged nitrogen is unsymmetrically positioned in the molecule; therefore, long nonpolar chains appear to be a necessary structural feature. This suggests that the most active compounds will be those having a maximum charge on an unsymmetrically positioned nitrogen and lipophilic-hydrophilic balance to impart optimum surface activity. Molecules of this type will be attracted and held comparatively firmly to the bacterial cell wall by the cation reacting with anionic cellular groups to form reinforced ionic bonds and the nonpolar portion of the molecule associating through van der Waals bonds. This action occurs below the CMC.

Like the soaps, the quaternary ammonium compounds can form mixed micelles. Thus, the bactericidal concentration of dodecyl (C_{12}) dimethyl benzyl ammonium chloride must be doubled in the presence of 25 per cent of hexadecyl (C_{16}) dimethyl benzyl ammonium chloride.[141] Albert[142] interpreted this finding in terms of mixed micelle formation. His reasoning may be stated briefly as follows: Drugs act in their monomolecular form, but at the CMC and above the micelle competes with the microorganism for the monomers, thereby reducing the effective antibacterial concentration. He has also suggested that mixed micelle formation may be responsible for the phenomenon of therapeutic interference.

Surface-active agents can be expected to have a pronounced effect on the permeability of the cell. Mildly surface-active agents may be adsorbed as a monolayer on the cell membrane and thereby interfere with the absorption of other compounds through this membrane. On the other hand, highly surface-active agents may cause a disintegration or lysis of the cell membrane.

REFERENCES CITED

1. Crum-Brown, A., and Fraser, T.: Trans. Roy. Soc. Edinburgh 25:151, 693, 1868-9.
2. Burger, A.: J. Chem. Educ. 35:142, 1958.
3. Wagner, J. B.: J. Pharm. Sci. 50:359, 1961.
4. Wurster, D. E., and Taylor, P. W.: J. Pharm. Sci. 54:169, 1965.
5. Levy, G.: J. Pharm. Sci. 50:388, 1961.
6. Lazarus, J., and Cooper, J.: J. Pharm. Pharmacol. 11:257, 1959.
7. Cohen, G. N., and Monod, J.: Bact. Rev. 21:169, 1957.
8. Collander, R., and Bärlund, H.: Acta bot. fenn. 11:1, 1933.
9. Schanker, L. S., et al.: J. Pharmacol. Exp Therap. 123:81, 1958.
10. Overton, E.: Studien über Narkose, Jena, Fischer, 1901.
11. Schanker, L. S.: J. Med. Pharm. Chem. 2:343, 1960.
12. Travell, J.: J. Pharmacol. Exp. Therap. 69:21, 1940.
13. Shore, P. A., Brodie, B. B., and Hogben, C. A. M.: J. Pharmacol. Exp. Therap. 119:361, 1957.
14. Brodie, B. B., and Hogben, C. A. M.: J. Pharm. Pharmacol. 9:345, 1957.
15. Schanker, L. S., et al.: J. Pharmacol. Exp. Therap. 120:528, 1957.
16. Hogben, C. A. M., et al.: J. Pharmacol. Exp. Therap. 120:540, 1957.
17. Hogben, C. A. M., et al.: J. Pharmacol. Exp. Therap. 125:275, 1959.
18. Schanker, L. S.: J. Pharmacol. Exp. Therap. 126:283, 1959.
19. Levine, R., Blaire, M., and Clark, B.: J. Pharmacol. Exp. Therap. 121:63, 1957.
20. Cavallito, C. J., and O'Dell, T. B.: J. Am. Pharm. A. (Sci. Ed.) 47:169, 1958.
21. Schanker, L. W.: Physiological Transport of Drugs, in Harper, N. J., and Simmonds, A. B. (eds.): Advances in Drug Research, pp. 71-106, London, Academic Press, 1964.
22. Sterling, K. L.: J. Clin. Invest. 43:1721, 1964.
23. Brodie, B. B., et al.: J. Pharmacol. Exp. Therap. 48:85, 1950.
24. Brodie, B. B. Bernstein, E., and Mark, L. C.: J. Pharmacol. Exp. Therap. 105:421, 1952.
25. Price, H. L., et al.: J. Clin. Pharmacol. Therap. 1:16, 1960.
26. Axelrod, J., Aronow, L., and Brodie, B. B.: J. Pharmacol. Exp. Therap. 106:166, 1952.
27. Brodie, B. B., Aronow, L., and Axelrod, J.: J. Pharmacol. Exp. Therap. 111:21, 1954.
28. Agarwal, S. L., and Harvey, S. C.: J. Pharmacol. Exp. Therap. 117:106, 1956.
29. Cavallito, C., et al.: Anesthesiology, 17:547, 1956.
30. Fouts, J. R., and Brodie, B. B.: J. Pharmacol. Exp. Therap. 115:68, 1955.
31. Paton, W. D. M., and Zaimis, E. J.: Pharmacol. Rev. 4:219, 1952.
32. Waddell, W. J., and Butler, T. C.: J. Clin. Invest. 36:1217, 1957.
33. Brodie, B. B., Kurz, H., and Schanker, L. S.: J. Pharmacol. Exp. Therap. 130:20, 1960.
34. Roll, D. P., Stabenau, J. R., and Zubrod, C. G.: J. Pharmacol. Exp. Therap. 125:185, 1959.
35. Daniels, T. C.: Ann. Rev. Biochem. 12:447, 1943.

36. Coulthard, C. E., Marshall, J., and Pyman, F. L.: J. Chem. Soc. 280, 1930.
37. Dodds, E. C., *et al.*: Proc. Roy. Soc. London, s. B 127:140, 1939.
38. Daniels, T. C., and Lyons, R. E.: J. Phys. Chem. 35:2049, 1931.
39. Schaffer, J. M., and Tilley, F. W.: J. Bact. 12:303, 1926; 14:259, 1927.
40. Sabalitschka, Th., and Tietz, H.: Arch. Pharm. 269:545, 1931.
41. Sabalitschka, Th., and Tietz, H.: Pharm. acta helv. 5:286, 1930.
42. Meyer, H.: Arch. exp. Path. Pharmakol. 42:109, 1899; 46:338, 1901.
43. Falk, J. E.: Australian J. Sc. 7:48, 1944.
44. Winterstein, H.: Die Narkose, ed. 2, Berlin, Springer, 1926.
45. Meyer, K. H., and Gottlieb-Billroth, H.: Ztschr. physiol. Chem. 112:6, 1920.
46. Cullen, S. C., and Gross, E. G.: Science 113:580, 1951.
47. Wulf, R. J., and Featherstone, R. M.: Anesthesiology 18:97, 1957.
48. Pauling, L.: Science 134:15, 1961.
49. Hansch, C., and Fujita, T.: J. Am. Chem. Soc. 86:1616, 1964.
50. Jaffé, H. H.: Chem. Rev. 53:191, 1953.
51. Fujita, T., Iwasa, J., and Hansch, C.: J. Am. Chem. Soc. 86:5175, 1964.
52. Iwasa, J., Fujita, T., and Hansch, C.: J. Med. Chem. 8:150, 1965.
53. Hansch, C., *et al.*: J. Am. Chem. Soc. 85:2817, 1963.
54. Hansch, C., Steward, A. R., and Iwasa, J.: Molecular Pharmacol. 1:87, 1965.
55. Hansch, C., Steward, A. R., Anderson, S. M., and Bentley, D.: J. Med. Chem. 11:1, 1968.
56. Kutter, E., and Hansch, C.: J. Med. Chem. 12:647, 1969.
57. Fühner, H.: Arch. exp. Path. Pharmakol. 51:1, 52:69, 1904.
58. Ferguson, J.: Proc. Roy. Soc. London, s. B 127:387, 1939.
59. Goodman, L. S., and Gilman, A.: The Pharmacological Basis of Therapeutics, New York, Macmillan, 1965.
60. Beckett, A. H., and Casy, A. F.: J. Pharm. Pharmacol. 6:986, 1954.
61. Eddy, N. B.: Chem. Ind. 1959, 1462.
62. Portoghese, P. S., Mikhail, A. A., and Kupferberg, H. J.: J. Med. Chem. 11:219, 1968.
63. Gill, E. W.: Progr. Med. Chem. 4:39, 1965.
64. Beckett, A. H.: J. Pharm. Pharmacol. 8:848, 1956.

65. Martin-Smith, M., Smail, G. A., and Stenlake, J. B.: J. Pharm. Pharmacol. 19:561, 1967.
66. Kier, L. B.: Mol. Pharmacol. 3:487, 1967; 4:70, 1968.
67. Chothia, C., and Pauling, P.: Nature 219:1156, 1968.
68. Chiou, C. Y., Long, J. P., Cannon, J. G., and Armstrong, P. D.: J. Pharmacol. Exp. Ther. 166:243, 1969.
69. Smissman, E., Nelson, W., Day, J., and LaPidus, J.: J. Med. Chem. 9:458, 1966.
70. Kier, L. B.: J. Med. Chem. 11:441, 1968.
71. Niemann, C. C., and Hayes, J. T.: J. Am. Chem. Soc. 64:2288, 1942.
72. Cushny, A. R.: Biological Relations of Optically Active Isomeric Substances, Baltimore, Williams and Wilkins, 1926.
73. Easson, L. H., and Steadman, E.: Biochem. J. 27:1257, 1933.
74. Beckett, A.: Prog. Drug. Res. 1:455-530, 1959.
75. Blaschko, H.: Proc. Roy. Soc. London 137(B):307, 1950.
76. Belleau, B.: *in* Uvnas, B. (ed.): Proc. First Intern. Pharmacol. Meeting, Vol. 7, p. 75, New York, Pergamon, 1963.
77. Ehrenpreis, S.: Biochem. biophys. acta 44:561, 1960.
78. Ehrenpreis, S.: Science 136:175, 1962.
79. Pfeiffer, C.: Science 107:94, 1948.
80. Barlow, R. B., and Ing, H. R.: Brit. J. Pharmacol. 3:298, 1948.
81. Fisher, A., Keasling, H., and Schueler, F.: Proc. Soc. Exp. Biol. Med. 81:439, 1952; *see* Schueler, Selected Reading, p. 410.
82. Gero, A., and Reese, V. J.: Science 123:100, 1956.
83. Welsh, J. H., and Taub, R.: J. Pharmacol. Exp. Therap. 103:62, 1951.
84. Nachmansohn, D., and Wilson, I. B.: Advances Enzym. 12:259, 1951.
85. Friess, S. L., *et al.*: J. Am. Chem. Soc. 76:1363, 1954; 78: 199, 1956; 79:3269, 1957; 80:5687, 1958.
86. Krupka, R. M., and Laidler, K. J.: J. Am. Chem. Soc. 83:1458, 1961.
87. Baker, B. R.: J. Pharm. Sci. 53:347, 1964.
88. Baker, B. R., and Patel, R. P.: J. Pharm. Sci. 52:927, 1963.
89. Baker, B. R., and Alumaula, P. I.: J. Pharm. Sci. 52:915, 1963.
90. Schaeffer, H. J.: J. Pharm. Sci. 54:1223, 1965.
91. Clark, A. J.: J. Physiol. 61:530, 547, 1926.

92. Gaddum, J. H.: J. Physiol. 61:141, 1926; 89:7P, 1937.
93. Ariëns, E. J., and Simonis, A. M.: J. Pharm. Pharmacol. 16:137, 289, 1964.
94. Stephenson, R. P.: Brit. J. Pharmacol. 11: 379, 1956.
95. Croxatto, R., and Huidobro, F.: Arch. int. Pharmacodyn. 106:207, 1956.
96. Paton, W. D. M.: Proc. Roy. Soc. (London) B154:21, 1961.
97. Koshland, D. E.: Proc. Nat. Acad. Sci. 44:98, 1958.
98. Nachmansohn, D.: Chemical and Molecular Basis of Nerve Activity, New York, Academic Press, 1959.
99. Belleau, B.: J. Med. Chem. 7:776, 1964.
100. Belleau, B., and Moran, J.: Ann. N.Y. Acad. Sci. 107:822, 1963.
101. Beckett, A. H., Casey, A. F., and Harper, N. J.: J. Pharm. Pharmacol. 8:874, 1956.
102. Langmuir, I.: J. Am. Chem. Soc. 41: 1543, 1919.
103. Grimm, H. G.: Z. Electrochem. 31:474, 1925; Naturwiss. 17:557, 1929.
104. Erlenmeyer, H.: Bull. Soc. chim. biol. 30:792, 1948.
105. Bovet, D.: Science 129:1255, 1959.
106. Friedman, H. L.: Symposium on Chemical-Biological Correlation, National Research Council, Pub. No. 206, p. 295, Washington, D. C., 1950.
107. Free, S. H., Jr., and Wilson, J. W.: J. Med. Chem. 7:395, 1964.
108. Ban, T., and Fujita, T.: J. Med. Chem. 12:353, 1969.
109. Purcell, W. P.: Biochim. biophys. acta 105:201, 1965.
110. Rahn, O., and Conn, J. E.: Ind. Eng. Chem. 36:185, 1944.
111. Kuroda, T.: Biochem. Ztschr. 169:261, 1926.
112. Albert, A., et al.: Brit. J. Exp. Path. 26: 160, 1945.
113. Overend, W., and Peacocke, A.: Endeavor 16:90, 1957.
114. Albert, A.: Biochem. J. 47:531, 1950; 50:690, 1952.
115. Stocken, L. A., and Thompson, R. H. S.: Physiol. Rev. 29:168, 1949.
116. Walshe, J.: Am. J. Med. 21:487, 1956.
117. Bickel, H.: Experientia 16:129, 1960.
118. Moeschlin, S., and Schnider, U.: New Eng. J. Med. 269:57, 1963.
119. Brannerman, R. M., et al.: Brit. Med. J. 1573, 1962.
120. Chenoweth, M. B.: Pharmacol. Rev. 8:57, 1956.
121. Mellor, D. P., and Maley, L.: Nature 161:436, 1948.
122. Fahey, J. L., et al.: J. Lab. Clin. Med. 57:436, 1961.
123. Albert, A., and Magrath, D.: Biochem. J. 41:534, 1947.
124. Albert, A., Gibson, M. I., and Rubbo, S. D.: Brit. J. Exp. Path. 34:119, 1953.
125. Albert, A.: M. J. Australia 1:245, 1944.
126. Albert, A., et al: Brit. J. Exp. Path. 35: 75, 1954.
127. Beckett, A. H., et al.: J. Pharm. Pharmacol. 10:160T, 1958.
128. Rubbo, S., Albert, A., and Gibson, M.: Brit. J. Exp. Path. 31:425, 1950.
129. Perry, H. M., and Schroeder, H. A.: Am. J. Med. 16:606, 1954.
130. Bruns, F., and Stüttgen, G.: Biochem. Ztschr. 322:68, 1951.
131. Cymerman-Craig, J., et al.: Nature 176: 34, 1955.
132. Trautner, E. M., and Messer, M.: Nature 169:31, 1952.
133. Goldacre, R. J.: Australian J. Sc. 6:112, 1944.
134. Page, J. E., and Robinson, F. A.: Brit. J. Exp. Path. 24:89, 1943.
135. Breyer, B., Buchanan, G. S., and Duewell, H.: J. Chem. Soc. 360, 1944.
136. Kuhn, R., Weygand, F., Möller, E.: Chem. Ber. 76(2):1044, 1943.
137. Reist, E. J., et al.: J. Org. Chem. 25: 1368, 1455, 1960.
138. Tozer, T. N., Tuck, L. D., and Craig, J. C.: J. Med. Chem. 12:294, 1969.
139. Klevens, H. B.: J. Physical Colloid Chem. 52:130, 1948.
140. Alexander, A. E., and Trim, A. R.: Proc. Roy. Soc. London 133(B):220, 1946.
141. Valko, E., and Dubois, A.: J. Bact. 47: 15, 1944.
142. Albert, A.: Selective Toxicity, p. 97, New York, Wiley, 1965.

SELECTED READING

Albert, A.: Selective Toxicity, ed. 4, New York, Wiley, 1968.
Ariëns, E. J.: Molecular Pharmacology, Vol. 1, New York, Academic Press, 1964.
Barlow, R. B.: Introduction to Chemical Pharmacology, ed. 2, New York, Wiley, 1964.
Beckett, A.: Stereochemical factors in biological activity, Progr. Drug Res. 1:455-530, 1959.

Bloom, B. M.: Receptor Theories, *in* Burger, A. (ed.): Medicinal Chemistry, ed. 3, p. 108, New York, Wiley-Interscience, 1970.

Bloom, B. M., and Laubach, G. D.: The relationship between chemical structure and pharmacological activity, Ann. Rev. Pharmacol. 2:62, 1962.

Brodie, Bernard B., and Hogben, Adrian M.: Some physico-chemical factors in drug action, J. Pharm. Pharmacol. 9:345, 1957.

Burger, A.: Relation of Chemical Structure and Biological Activity, *in* Burger, A. (ed.): Medicinal Chemistry, ed. 3, p. 64, New York, Wiley-Interscience, 1970.

Cammarata, A., and Martin, A. N.: Physical Properties and Biological Activity, *in* Burger, A. (ed.): Medicinal Chemistry, ed. 3, p. 118, New York, Wiley-Interscience, 1970.

Foye, W. O.: Role of metal-binding in the biological activities of drugs, J. Pharm. Sci. 50:93, 1961.

Gill, E. W.: Drug receptor interactions, Progr. Medicinal Chem. 4:39, 1965.

Gourley, D. R. H.: Basic mechanisms of drug action, Progr. Drug Res. 7:11, 1964.

Schueler, F. W.: Chemobiodynamics and Drug Design, New York, Blakiston, 1960.

Wooley, D. W.: A Study of Antimetabolites, New York, Wiley, 1952.

3

Metabolic Changes of Drugs and Related Organic Compounds (Detoxication)

T. C. Daniels, Ph.D.,

Emeritus Professor of Pharmaceutical Chemistry, School of Pharmacy,
University of California

and

E. C. Jorgensen, Ph.D.,

Professor of Chemistry and Pharmaceutical Chemistry, School of Pharmacy,
University of California

INTRODUCTION

The term *detoxication* originally referred to a reduction in toxicity resulting from changes which foreign substances undergo in the body. The implication was that every foreign substance was converted into a less toxic substance. However, it became apparent that a number of substances were not metabolized in the body and that some were converted into metabolites considerably more toxic than the original compounds. These include, among others, the pentavalent arsenic compounds, which are converted to the more toxic trivalent compounds, and picric acid, which is transformed to the more toxic picramic acid. As a result of such findings the term detoxication has been broadened to include all the metabolic changes which foreign organic compounds undergo in the animal body. It is in such a sense that detoxication is used in this chapter.

Many foreign substances have been observed to pass through the body without undergoing any appreciable change. For the most part, these compounds appear to belong to one of two general classes.

1. Compounds that are insoluble in the body fluids and show no biologic action. Examples are mineral oil and barium sulfate.

2. Compounds freely soluble in body fluids, relatively insoluble in nonpolar solvents and resistant to chemical change. Compounds of this type, with the notable exception of the active cations and anions, are usually relatively nontoxic and rapidly excreted. Examples of this type are the aromatic and aliphatic sulfonic acids and certain of the carboxylic acids, such as mandelic.

A relatively large number of specific and nonspecific enzyme systems have been shown to be responsible for the metabolism of organic compounds, but only a small number of types of chemical change (pathways) are involved. Compounds that are foreign to the body, such as most drugs, may react with enzymes of normal metabolic processes. If the foreign compound closely resembles the natural substrate in

69

structure, it may be metabolized by the enzyme or, in some cases, it may act as an enzyme inhibitor. Examples of foreign compounds acting as substrates for enzymes normally concerned with natural substrates are: procaine and succinylcholine (nonspecific plasma cholinesterase), and benzylamine (monoamine oxidase).

In addition to the enzymes known to act on natural substrates, other enzymes are present, mainly in the liver microsomes, which seem to metabolize foreign compounds primarily. Thus, the foreign compound, acetanilid, is hydroxylated by liver microsomes (see p. 76), whereas this system has no effect on the related natural compound, L-phenylalanine. A specific L-phenylalanine hydroxylase, present in another part of the liver cell, is responsible for the conversion of L-phenylalanine to its *p*-hydroxyl derivative, L-tyrosine.

It is possible that accessibility to the microsomal enzyme system is a determining factor. In general, only lipid-soluble substances, presumably capable of passive diffusion through lipoidal membrane barriers, are metabolized by the nonspecific liver microsomal enzyme systems. An alternate possibility is that microsomal enzyme systems interact best with substrates of nonpolar character, for example, through preliminary formation of hydrophobic bonds. The metabolites formed are almost invariably less lipid-soluble than the parent compound. As a result, they are usually less toxic, since the more polar (water soluble) metabolites are more soluble in body fluids, are more readily excreted and fail to penetrate or penetrate less readily the lipoidal barriers to sites of biologic action.

Highly lipid-soluble molecules are passively reabsorbed from the glomerular filtrate (see Chap. 20) by the kidney tubules and are poorly excreted. Such compounds tend to remain in the body until they are converted by metabolic processes into more polar compounds. This is due to the lipid character of the epithelial cells of the kidney tubules which serve as an effective barrier for the passive reabsorption of polar molecules.[1]

Quick[2] has associated the "detoxication" process with the conversion of "a weak acid which the body cannot excrete to a strong acid which it can eliminate." While acid strength may be an important factor, it does not appear to be the sole determining factor in drug metabolism. As an example, benzoic acid (K_a 6.5×10^{-5}) has a partition coefficient (ether/water) of approximately 30 and is excreted by man largely as hippuric acid (K_a 2.3×10^{-4}), which has a partition coefficient in the same solvents of less than 0.4. Thus, while the acid strength of hippuric acid is more than three times that of benzoic acid, its partition coefficient is less by a factor of 75. Gentisic acid (2,5-dihydroxybenzoic acid) has an acid strength (K_a 1.03×10^{-3}) more than four times greater than that of hippuric acid; yet it is excreted up to 50 per cent as the sulfate and glucuronide in man. Gentisic acid has a partition coefficient (ether/water) of approximately 12, and the partition coefficient of gentisic acid sulfate although it is not recorded, can be assumed to be extremely low. Thus, metabolic reactions tend to convert relatively lipophilic compounds, whose solubility properties favor retention in the body, into more hydrophilic compounds with properties favoring urinary excretion. Conversion into derivatives with increased acid strength, leading to more highly ionized compounds at the physiologic pH of 7.4, is only one mechanism for the formation of more polar metabolites.

In addition to acid strength and partition coefficient, a number of other factors determine the degree and the type of chemical change foreign compounds undergo in the body. Among these may be mentioned the functional groups present, steric effects and other properties that may affect the availability of the molecule for the cell receptors where metabolic reactions occur.

In general, toxic compounds tend to undergo chemical change. The introduction of carboxyl or other solubilizing groups into phenols, piperidine compounds and other biologically active molecules tends to lower the toxicity. For example, ester drugs, such as atropine, cocaine, procaine, and arecoline, are biologically active and toxic until hydrolysis occurs. The acids and aminoalcohols formed from the hydrolysis, are, without exception, relatively nontoxic. Likewise, the hydrolysis products have lower partition coefficients than the esters from which they are obtained.

The nature of the chemical change involved in drug metabolism may be similar or quite different in different species of animals. Even in the same species individual variation leads to relatively large differences in the extent of specific metabolic reactions.

Phenylacetic acid conjugates differently in various species. In man it conjugates with glutamine, in the fowl with ornithine, and in the dog with glycine. The phenylacetyl derivatives of glutamine, glycine and ornithine when fed to man, dog and fowl are excreted unchanged. The fact that this species difference does not apply to the conjugates may indicate that the conjugated compounds, because of their physical properties, are not available at sites of action where degradation and reconjugation may occur.

The great majority of organic compounds undergo some metabolic change in the body, although some compounds appear to be slowly excreted largely unchanged. Metabolism may involve: (1) a single-step conversion of a biologically active compound to an inactive compound which is excreted (e.g., conjugation of a phenol with sulfate or glucuronic acid); (2) a two-step conversion in which there is first inactivation followed by conjugation with glucuronic acid (e.g., hydroxylation and conjugation of phenobarbital); (3) a two-step process in which an inactive compound is converted to a biologically active compound followed by inactivation and excretion (e.g., phenacetin, arseno compounds and some azo dyes); and (4) a two-step process in which there is first a change in activity of an active compound followed by inactivation (e.g., *O*-demethylation of codeine to give morphine which then conjugates with sulfate and glucuronic acid).[3]

The metabolic changes drugs undergo are of considerable interest and frequently of great practical value in the search for new and improved medicinals. The discovery by French workers[4] that the azo dye Prontosil, which is inactive in vitro, is converted by reduction in the body to the active

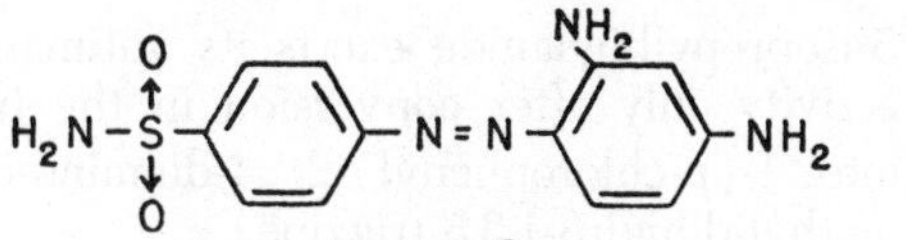

Prontosil

4-Sulfonamido-2′,4′-diaminoazobenzene

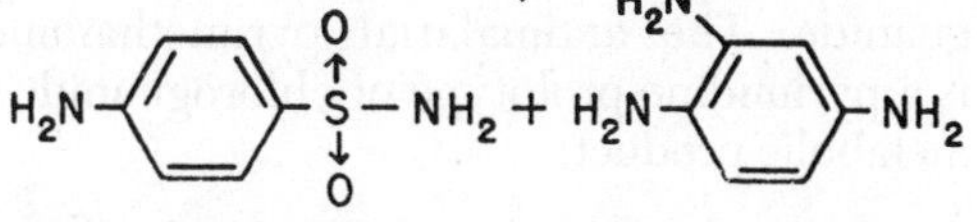

Sulfanilamide   1,2,4-Triaminobenzene

sulfanilamide, led to the rapid development of the sulfonamides as therapeutic agents. Later studies on the metabolic acetylation of the sulfonamides aided in the development of compounds that are acetylated to a lesser extent or whose acetylated derivatives are more soluble and therefore minimize kidney damage due to crystallization in the renal tubules. Likewise, studies on absorption of the sulfonamides from the intestine led to the introduction of the poorly absorbed N^4-acylated derivatives such as Sulfasuxidine for the treatment of intestinal infections. The introduction of mandelic acid as a genito-urinary antiseptic followed the observation that it is excreted unchanged and, in an acid environment in the urine, possesses significant bactericidal activity.

The therapeutically useful "arsine oxides" resulted from the observation that arseno compounds, $-As=As-$, are oxidized to arsenoxides, $-As=O$, which although more toxic are superior therapeutic agents. Chloroguanide (Paludrine) 1-(*p*-chlorophenyl)-

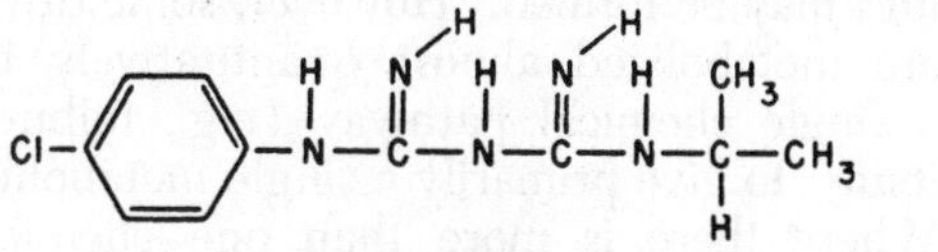

Chloroguanide

↓ −2H

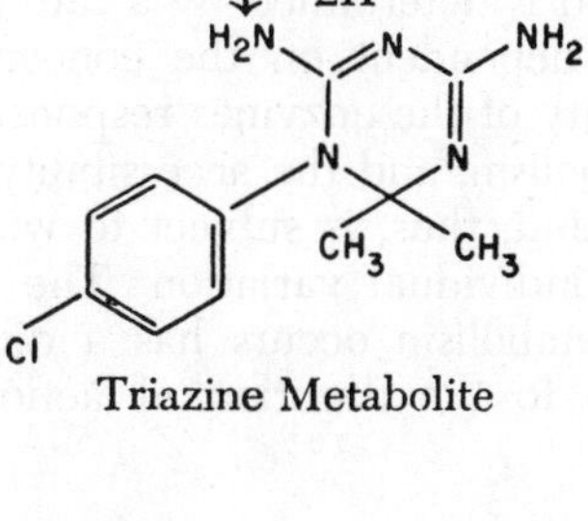

Triazine Metabolite

5-isopropylbiguanide exerts its antimalarial activity only after conversion in the body into 1-(*p*-chlorophenyl)-2, 4-diamino-6-di-methyl-dihydro-1,3,5-triazine.

The latter compound has been synthesized and shown to be considerably more active in experimental animals than chloroguanide. The antimalarial pyrimethamine is a pyrimidine prototype of chloroguanide's metabolic product.

Pyrimethamine
5-(*p*-Chlorophenyl)-2,6-diamino-4-ethylpyrimidine

Other examples of drug developments related to metabolism include the recognition that the analgesic properties of phenacetin (*p*-ethoxyacetanilide) are dependent on its conversion by O-dealkylation to form the active metabolite, acetaminophen (*p*-hydroxyacetanilide). Also, the antidepressant properties of imipramine and amitriptyline, both tertiary amines, appear to be mediated by their secondary amine metabolites, desipramine and nortriptyline.

FACTORS INFLUENCING THE METABOLISM OF DRUGS

Drug metabolism frequently involves a number of chemical pathways and therefore a relatively large number of metabolites may be formed.[5] However, some drugs are metabolized almost quantitatively by a single chemical pathway (e.g., tolbutamide) to give primarily a single metabolite. Where there is more than one chemical pathway involved, the amount of a metabolite formed is determined by a rate process which is dependent on the concentration and activity of the enzymes responsible for the metabolism, and the accessibility to the enzymes and, thus, is subject to wide species and individual variation. The rate at which metabolism occurs has a direct relationship to the duration of action of a drug and varies with species and is also subject to individual variation within the same species. The activity of a drug may be greatly modified by the concurrent administration of another drug or foreign compound which may alter the sensitivity of receptor sites for the drug or inhibit the enzymes or access to the enzymes responsible for its metabolism.

The major factors which have been observed to influence drug metabolism are the following: genetic factors; species and strain; sex; age; enzyme inhibitors, and stimulators.

Genetic Factors: Pharmacogenetics. The biochemical as well as the morphologic characteristics of living organisms are genetically controlled. The absence of an allelic gene responsible for the synthesis of an enzyme essential for normal metabolism leads to an enzyme deficiency and an inherent metabolic abnormality, such as phenylketonuria. Less apparent are the variations in the synthesis of enzymes, under genetic control, which lead to marked individual variations in the intensity and duration of action of drugs. *Pharmacogenetics* is the study of genetically determined variations as revealed by the effects of drugs.[6, 7]

The presence or absence in rabbits of an esterase capable of hydrolyzing atropine is under genetic control. The subcutaneous administration of atropine to a rabbit deficient in the esterase produces a pupillary dilation for as long as three days, while the same dose administered to a rabbit which has the esterase produces pupillary dilation for only a few hours.

Studies on the rates of acetylation of the antitubercular drug isoniazid (isonicotinic acid hydrazide) have shown that some individuals are rapid acetylators and others at a much slower rate. This variation is determined by the genetically controlled concentrations of acetyl coenzyme A. Acetylisoniazid is only one of several metabolic products excreted following administration of the drug (see p. 107), but it is of special interest since it may serve as an intermediate in the metabolic formation of other, more polar compounds. Since the antitubercular activity of isoniazid is dependent on the concentration of free drug in the blood serum, the rate of acetylation and the rates of other metabolic pathways must determine both the intensity and the duration of action of the drug.[8] Many other drugs are metabolized in part as acetylated derivatives and those individuals who are genetically supplied with higher concentrations of acetyl coenzyme A for the acetylation process (fast acetylators) may be unresponsive or less responsive to a drug metabolized by this pathway.[7, 8]

Species and Strain Differences.[9-11] Jaffe[12] in 1877 was the first to observe that benzoic acid when fed to hens was excreted as ornithuric acid but in dogs was excreted as hippuric acid. Since that time many observations have been recorded as to species difference in drug metabolism. The differences observed are both qualitative (type of reaction involved) and quantitative (same type of reaction but with variation in rate of metabolism), with the latter predominating. The following indicates some of the major species differences which have been recorded:[10] (a) Cats, in contrast to other species, are unable to form glucuronides in significant amounts; (b) dogs are unable to acetylate aromatic amines such as the sulfonamides; (c) in rabbits, amphetamine is primarily deaminated, but in dogs it is hydroxylated in the aromatic ring; and (d) in dogs, acetanilid is hydroxylated in the *para* position, and in cats in the *ortho* position. A good example of species difference in the rate of metabolism has been observed with hexobarbital, which at 50 mg./kg. anesthetizes man or dog for more than 5 hours; yet at twice this dosage level (100 mg/kg.) it anesthetizes mice for only 12 minutes[13] (approximately).

Wide differences in the rates of metabolism and duration of action also have been observed in different strains of the same species; this observation led to the interesting suggestion by Brodie[14] that a specific species and strain of experimental animal which metabolizes a given class of drugs in the same manner and at the same rate as man should be searched for.

Sex Differences. Many experimental animals show no sex variations in the rates of drug metabolism (mice, guinea pigs,

rabbits and dogs), but the rat is a noteworthy exception. Female rats after puberty metabolize a variety of drugs (e.g., barbiturates, narcotics, sulfonamides) at a significantly lower rate than the males.[9] Up to the age of puberty, the sex variations are not present and the difference observed after puberty has been attributed to the production of androgens and related anabolic steroids which increase the activity of the liver microsomal metabolizing enzymes. The increase in enzymatic activity parallels the increase in anabolic activity more closely than the increase in androgenic activity.

Age Differences. The newborn in a number of species of experimental animals (mice, guinea pigs and rabbits) have been shown to be deficient in the liver microsomal enzymes necessary for the metabolism of a variety of drugs. In experimental animals the metabolizing enzyme systems appear during the first week and develop to a maximum activity approximately 8 weeks after birth.[15] The implication of this finding is obvious with respect to the hazards of administration of drugs to young children who are deficient in the enzymes essential for metabolism.

Inhibitors of Drug Metabolism (Prolonging Duration of Action). The concurrent administration of a compound which inhibits or blocks the enzymes responsible for metabolism of a drug will increase the duration of action of the drug.[9] A number of enzyme inhibitors serving this function have been discovered, including among them, the following representative compounds:

$$CH_3CH_2CH_2-C-\overset{\overset{\displaystyle O}{\|}}{C}-O-CH_2CH_2-N\overset{\displaystyle C_2H_5}{\underset{\displaystyle C_2H_5}{}}$$

β-Diethylaminoethyl-α,α-diphenylvalerate
(SKF 525)

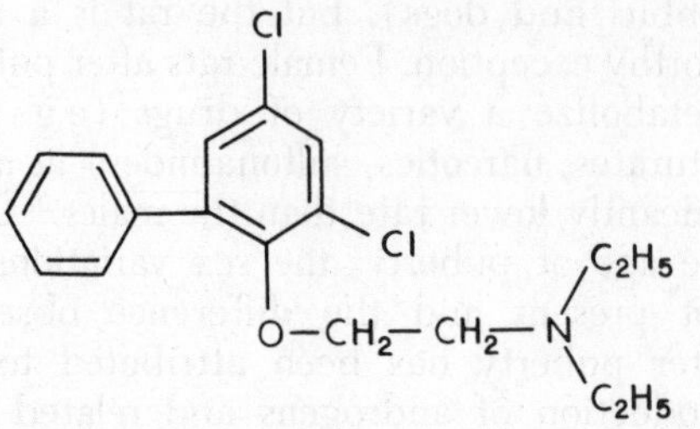

2,4-Dichloro-6-phenylphenoxyethyl
diethylamine (Lilly 18947)

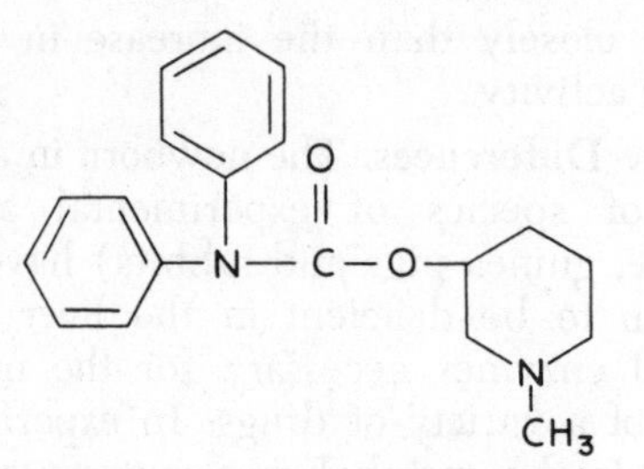

N-methyl-3-piperidyl Diphenylcarbamate

Stimulation of Drug Metabolism (Shortening the Duration of Action). Pretreatment of experimental animals with a variety of compounds has been shown to increase the rate of drug metabolism and to shorten the duration of drug action. Gillette[9] suggests that the stimulators function by increasing "the amount of the drug metabolizing enzymes or a component of these systems, and not by altering the permeability of microsomes or by blocking inhibitory reactions." There is evidence to suggest that the metabolic stimulation may involve more than one mechanism.

Representative of compounds which have been demonstrated to stimulate drug metabolism are the following: (a) Polycyclic hydrocarbons such as 3-methylcholanthrene and 3,4-benzpyrene stimulate the metabolism of aminopyrine, hexobarbital and other drugs; (b) phenobarbital and other long-acting barbiturates shorten the duration of action of hexobarbital in the rat; (c) nikethamide, a respiratory stimulant, and the anesthetic gases diethyl ether and nitrous oxide (CNS depressants) increase the rate of metabolism of the barbiturates; (d) imi-

pramine increases the rate of metabolism of acetanilid and aminopyrine, and (e) chlordane and many other insecticides greatly increase the metabolic rate of hexobarbital, but the stimulatory effect develops slowly over a period of several days.[9]

Many drugs, when administered chronically, have been observed to stimulate their own metabolism, as well as the metabolism of other drugs, e.g., aminopyrine, chlorpromazine, meprobamate, hexobarbital, pentobarbital, phenobarbital, phenylbutazone, probenecid, tolbutamide, etc. This increase in metabolic activity appears to be due to an increased concentration of the hepatic microsomal enzymes and is referred to as "enzyme induction."[16]

It has been shown that alcohol is metabolized more than twice as rapidly in alcoholic patients who have recently been drinking as it is in nonalcoholic subjects. It has also been established that the half-life of warfarin, diphenylhydantoin and tolbutamide in the blood of alcoholic test subjects is significantly shorter than in nonalcoholic subjects. This increased rate of clearance from the circulation indicates a more rapid metabolism of the drugs by the hepatic microsomal enzymes and suggests that continued intake of alcohol induces nonspecifically an increase in the concentration of the enzymes.[16a] Acceleration of the rate of metabolism of a drug leads to a gradual decrease in the plasma concentration of free drug and a corresponding decrease in its pharmacologic activity. The lowering of the plasma level of a drug by "enzyme induction" may also explain why some drugs, when first administered, produce signs of toxicity which disappear on continued use of the drug because of the accelerated rate of metabolism.

Miscellaneous Factors. A number of other factors have been observed to influence markedly drug metabolism, including the nutritional state of the animal, hormone levels, ascorbic acid deficiencies, pathologies of the liver, such as hepatic tumors and obstructive jaundice, and nonspecific protein binding of the drug or its storage in fatty depots.

SITE OF METABOLISM

The chemical changes substances undergo in the body take place principally in the intestines, the liver and the kidneys, with the liver serving as the dominant organ. The role of the liver in detoxication has been supported with a wealth of data obtained from studies on hepatectomized animals, liver slices, homogenates and with perfusion studies using the intact organ. A relationship between the lipid solubility of drugs and their oxidation by liver microsomes has been established.[17] In a series of alkylamines, only compounds with high chloroform/water partition coefficients are oxidized by rabbit liver microsomes in vitro.

The enzymes responsible for oxidation, as well as other metabolic reactions, are presumed to be located within the lipoidal membrane of the microsomes, and accessible only to compounds capable of passive diffusion into the lipid phase.

To illustrate the importance of the liver in the metabolism of drugs and other foreign compounds, metabolic transformations shown[18] to be mediated by liver microsomal enzyme systems include deamination, N-dealkylation, O-dealkylation, oxidation of thioethers, hydroxylation of aromatic ring systems, aromatization of hydroaromatic compounds, oxidation of alcohols and aldehydes, reduction of the nitro and azo groups, conjugation with sulfuric, glucuronic, mercapturic, amino and acetic acids, N- and O-methylations, hydrolysis of esters and amides, dehalogenation, and replacement of sulfur by oxygen.

METABOLIC CHANGES IN THE GASTROINTESTINAL TRACT

Action of the Salivary Secretions. With the exception of troches, gargles and sublingual medication, most drugs administered orally have only superficial contact with the salivary secretions and are not altered significantly. However, the salivary secretions contain a number of enzymes that are capable of destroying certain drugs.

Catalase, which catalyzes the decomposition of hydrogen peroxide, is present. Urease converts urea into ammonia and carbonic acid. Other amides, such as salicylamide, appear to be converted into the corresponding acid and ammonia. Ptyalin, "salivary amylase," hydrolyzes starch and glycogen.

Action of the Gastric Secretions. The acid juices of the stomach contain the enzymes pepsin, rennin and lipase. Hydrochloric acid is present in approximately 0.5 per cent concentration. The salts of biologically active amines are poorly absorbed from the normal stomach. However, the salts of most weak organic acids will be converted to the free acids (undissociated) and may be well absorbed from the stomach. The acid secretions of the stomach play an important role in facilitating the absorption of salts of weak acids, for example, the barbiturates (see Chap. 2). The basic ester drugs, atropine, Syntropan, etc., normally pass through the stomach unchanged.

Penicillin is quite unstable in the strong acid of the stomach, and penicillin products intended for oral administration usually are protected by an alkaline buffer in order to minimize the decomposition. Insulin and other biologically active substances containing protein may be decomposed (proteolysis), completely, or nearly so.

Changes Occurring in the Intestines. The intestinal secretions, including bile and pancreatic juice, are responsible for a number of significant changes, such as the hydrolysis of the aliphatic and aromatic esters, the glyceryl esters (fats), the anilids and related compounds.

All but the most resistant esters (e.g., atropine and other amino alcohol esters) appear to be more or less completely hydrolyzed. Most absorption of foodstuffs and drugs takes place in the small intestine. In the large intestine occur putrefactive changes which are due primarily to bacterial activity. The bacterial flora normally present in the intestine are capable of metabolizing drugs in the same general manner as the drug-metabolizing enzymes present in the liver microsomes and other body tissues. However, drugs that are administered orally are usually well absorbed from the stomach or small intestine and have only limited contact with intestinal bacteria.

TYPES OF METABOLIC REACTIONS

The chemical reactions drugs and other foreign organic compounds undergo in the body may be divided into the following classes:

    I. Oxidations

    II. Reductions

    III. Replacement reactions
        A. Hydrolysis
        B. Acetylation
        C. Methylation
        D. "Conjugation" (condensations)
            1. Sulfuric acid
            2. Glucuronic acid
            3. Glycine
            4. Glutamine
            5. Ornithine
            6. Cysteine and acetylcysteine

    IV. Thiocyanate formation

Representative types of metabolic reactions are outlined below. Details on the reaction mechanisms have been presented by Gillette[9] and Williams.[3]

OXIDATIVE REACTIONS

The enzyme systems catalyzing metabolic oxidation of compounds are found principally in liver microsomes. The additional general requirement for reduced triphosphopyridine nucleotide (TPNH) and oxygen has led to the suggestion that TPNH reduces a component in the microsomes; this reacts with oxygen to form the reactive oxidizing intermediate, the reaction being catalyzed by a group of nonspecific microsomal enzymes.[19]

Hydroxylation of Aromatic Compounds

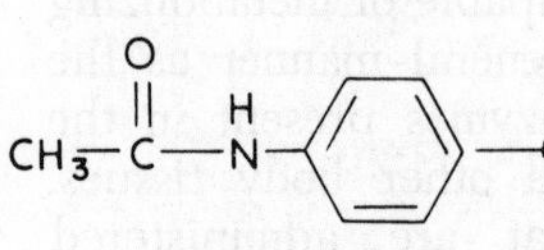

Acetanilide

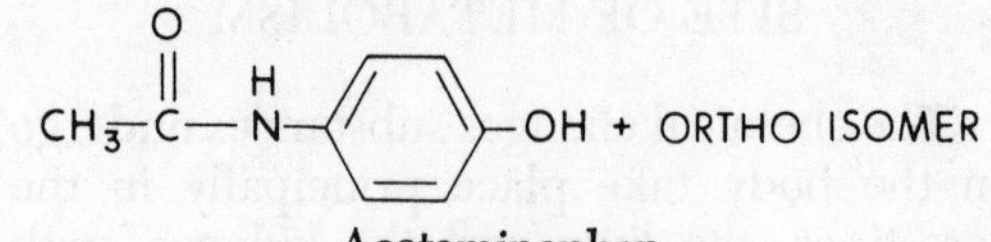

Acetaminophen

Hydroxylation of Aliphatic Side Chains and Aliphatic Compounds

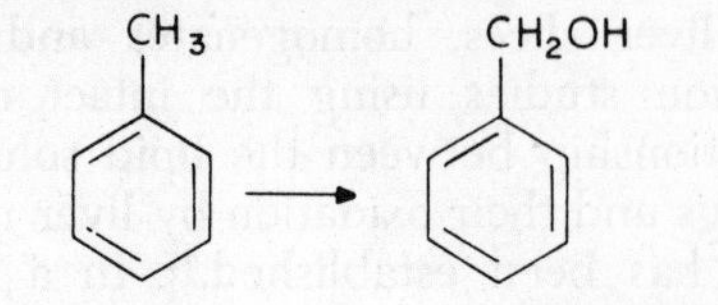

Toluene   Benzyl alcohol

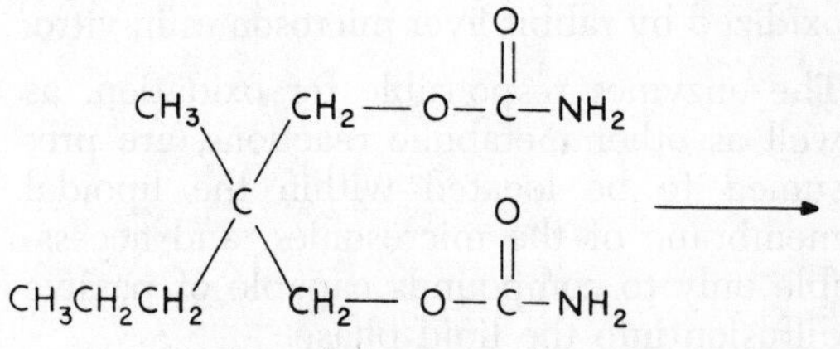

Meprobamate

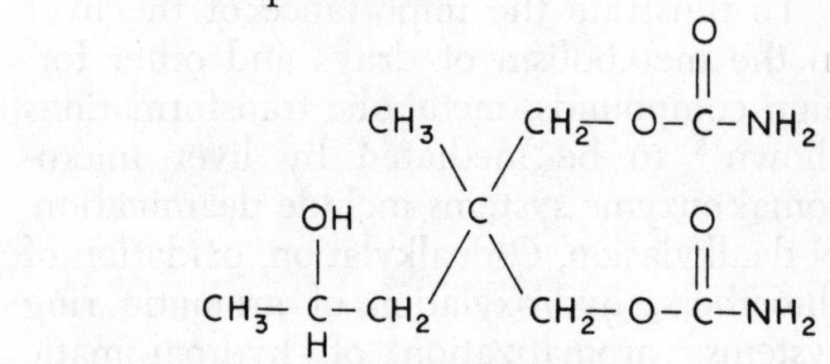

Hydroxymeprobamate

Oxidative O-Dealkylation

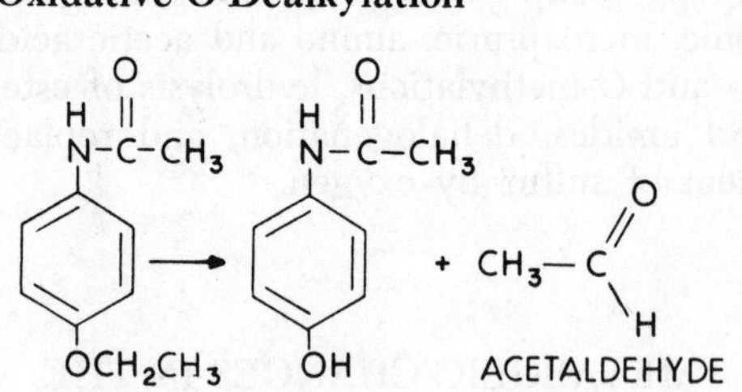

Phenac-   Acetam-   ACETALDEHYDE
etin   inophen

Oxidative N-Dealkylation

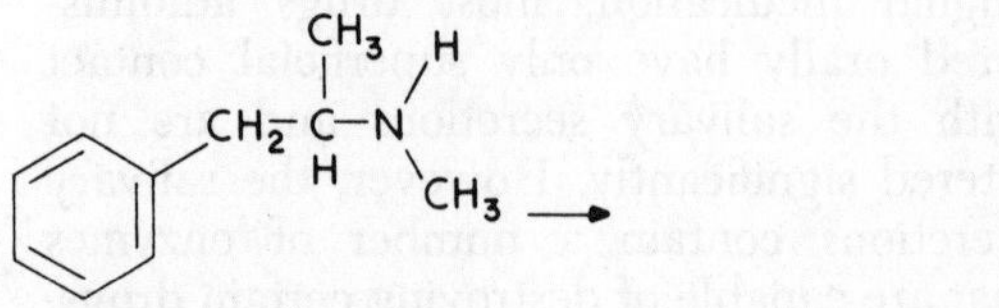

Methamphetamine

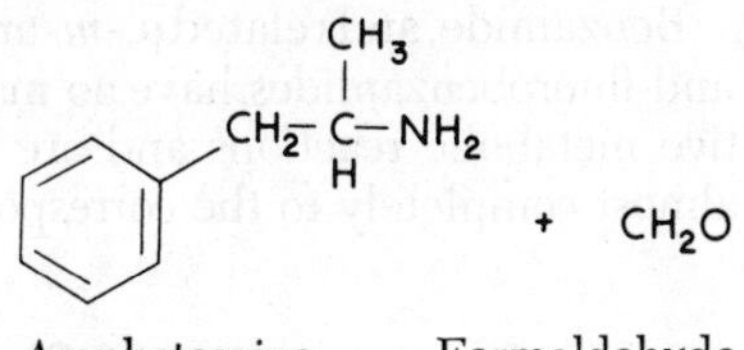

Amphetamine   Formaldehyde

Oxidative Deamination

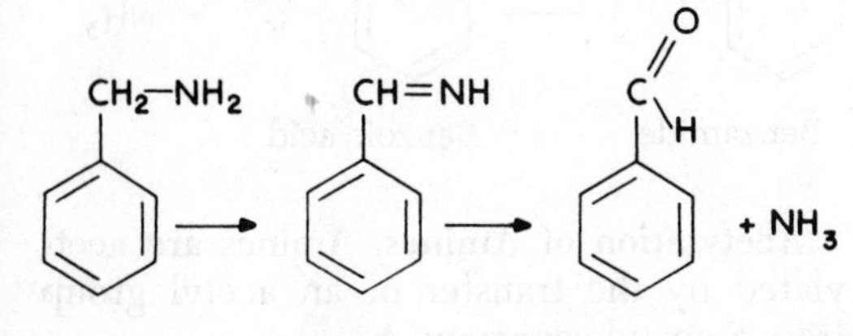

Benzyl-   Benzyl-   Benzalde-   Ammonia
amine   imine   hyde

Nitrogen Oxidation

$$(CH_3)_3N \longrightarrow (CH_3)_3N \longrightarrow O$$

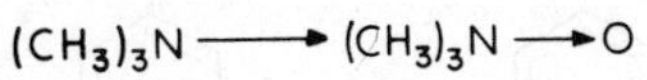

Trimethylamine   Trimethylamine Oxide

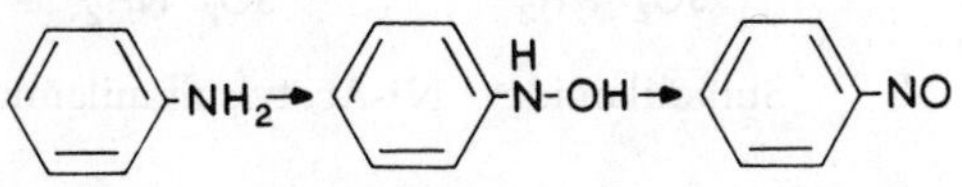

Aniline   Phenylhydroxyl-   Nitrosobenzene
                amine

Oxidation of Thioethers (sulfoxidation)

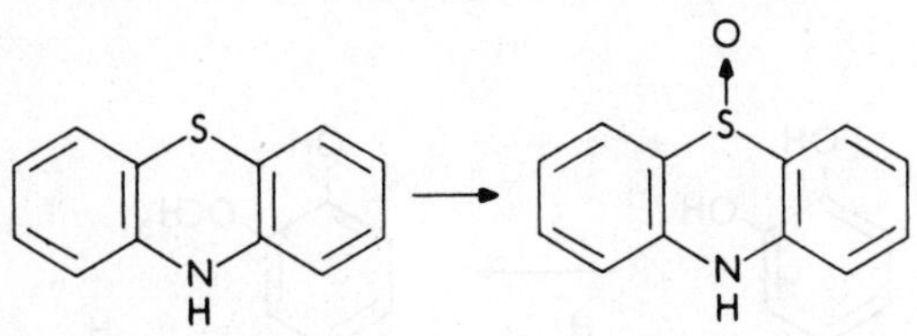

Phenothiazine   Phenothiazine
                           Sulfoxide

Ring Aromatization

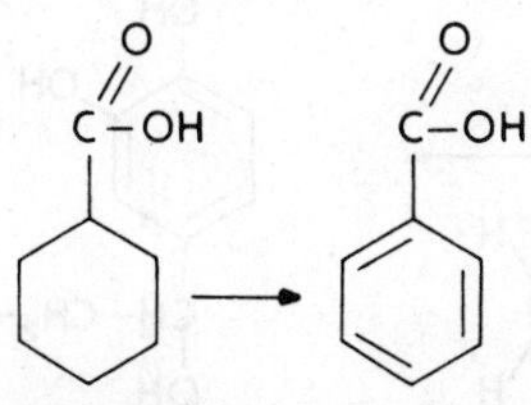

Hexahydrobenzoic   Benzoic Acid
Acid

**Oxidation of Primary Alcohols
and Aldehydes**

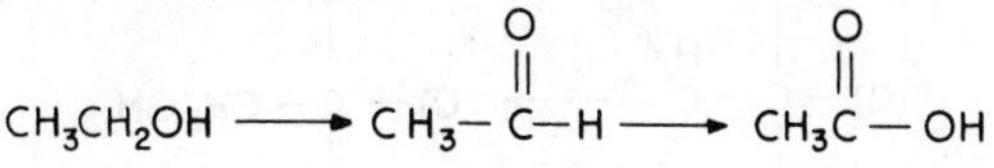

Ethanol   Acetaldehyde   Acetic Acid

**Oxidation of Arseno Compounds
to Arsenoxides**

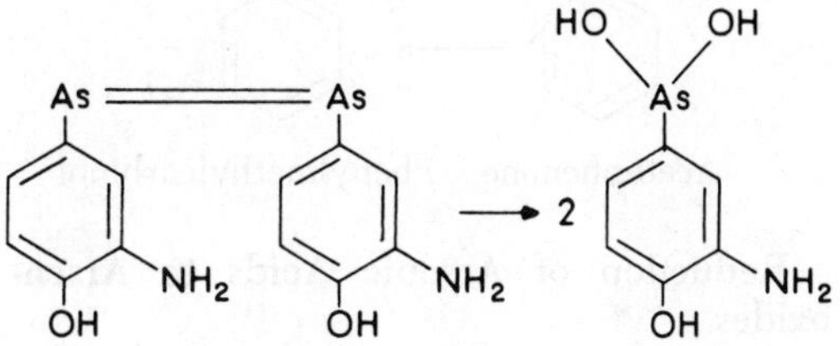

Arsphenamine   Oxophenarsine

REDUCTIVE REACTIONS

Metabolic reductions, such as those which act on azo and nitro compounds, are carried out by enzyme systems which may use triphosphopyridine nucleotide (TPNH) as a hydrogen donor.

Reduction of Azo Compounds

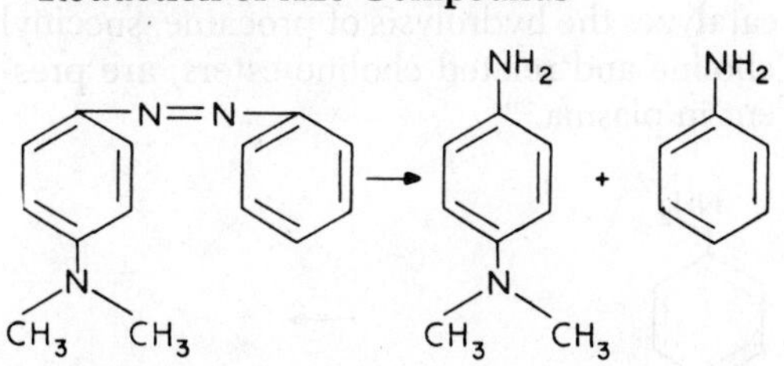

4-Dimethylamino-   4-Dimethylamino-   Aniline
azobenzene   aniline
(Butter yellow)

Reduction of Aromatic Nitro Compounds

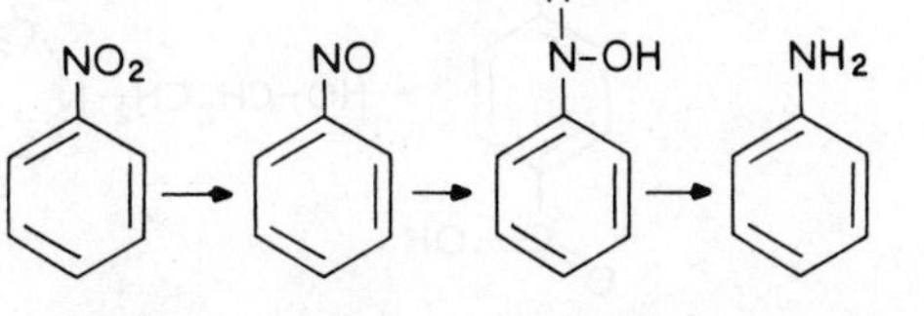

Nitrobenzene   Nitroso-   Phenylhy-   Aniline
                benzene   droxylamine

Reduction of Aldehydes and Ketones

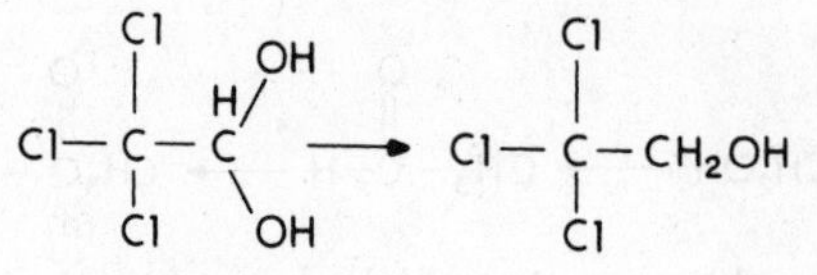

Chloral Hydrate   2,2,2-Trichloroethanol

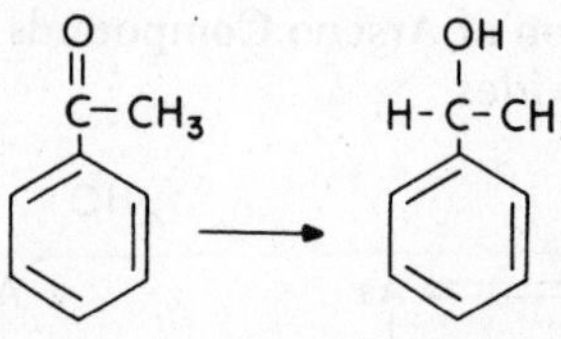

Acetophenone   Phenylmethylcarbinol

Reduction of Arsonic Acids to Arsenoxides

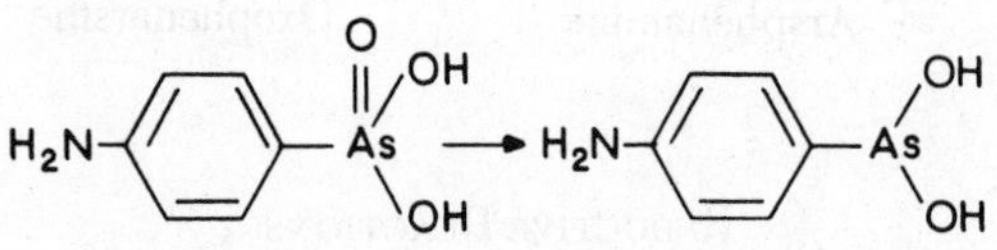

Arsanilic Acid   *p*-Aminophenylarsinic acid

REPLACEMENT REACTIONS

Hydrolysis of Esters and Amides. Esterases, such as the pseudocholinesterase which catalyzes the hydrolysis of procaine, succinyl choline and related choline esters, are present in plasma.[20]

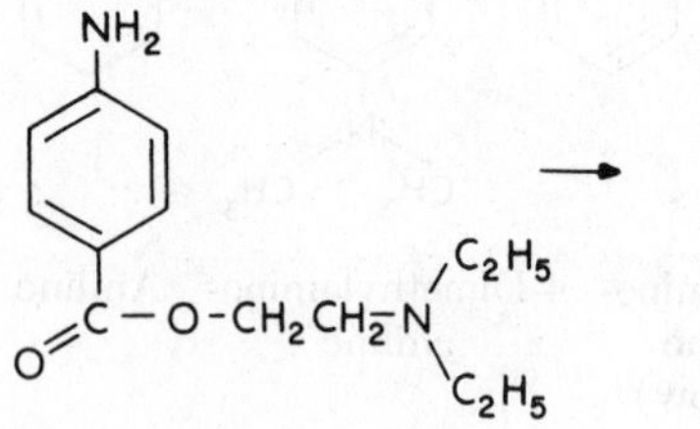

Procaine

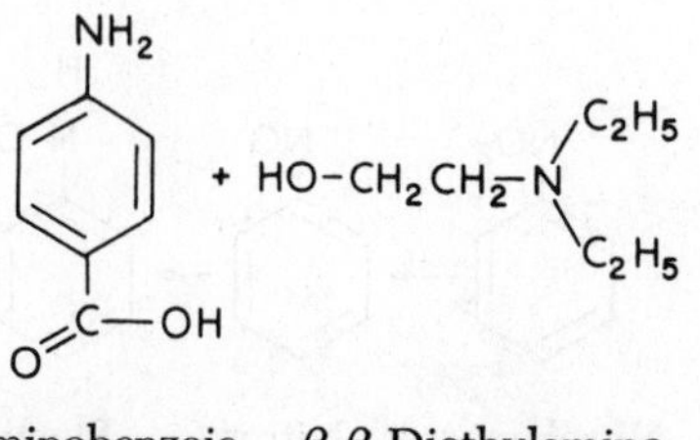

p-Aminobenzoic   β,β-Diethylamino-
acid   ethanol

Benzamide, and related *o*-, *m*- and *p*-chloro- and-fluorobenzamides have no major alternative metabolic reactions and are hydrolyzed almost completely to the corresponding acid.

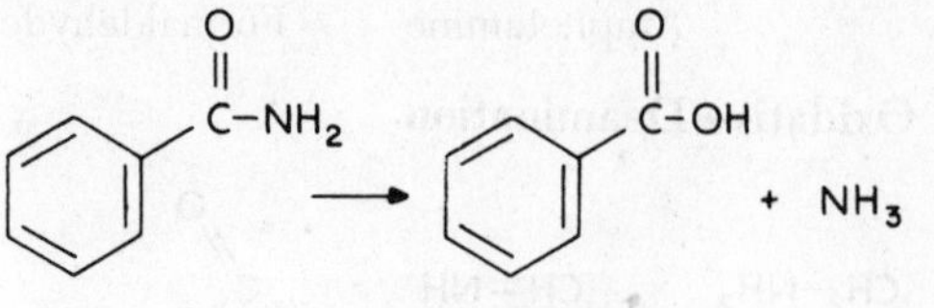

Benzamide   Benzoic acid

Acetylation of Amines. Amines are acetylated by the transfer of an acetyl group from S-acetyl coenzyme A.

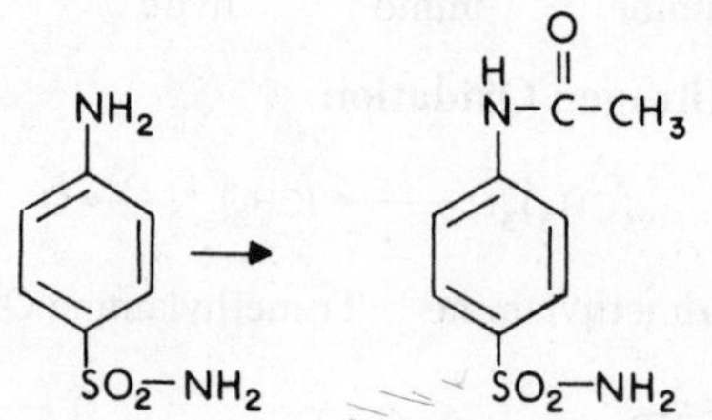

Sulfanilamide   N⁴-Acetylsulfanilamide

O and N Methylations. Enzymes transfer the methyl group of S-adenosylmethionine to the nitrogen of a variety of amines and to the oxygen of catechols.

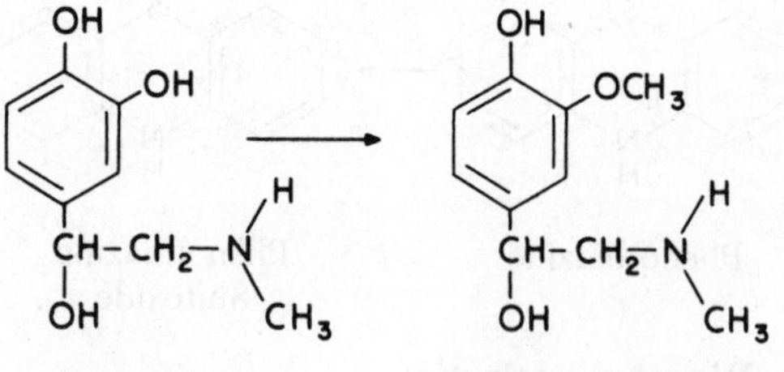

Epinephrine   Metanephrine

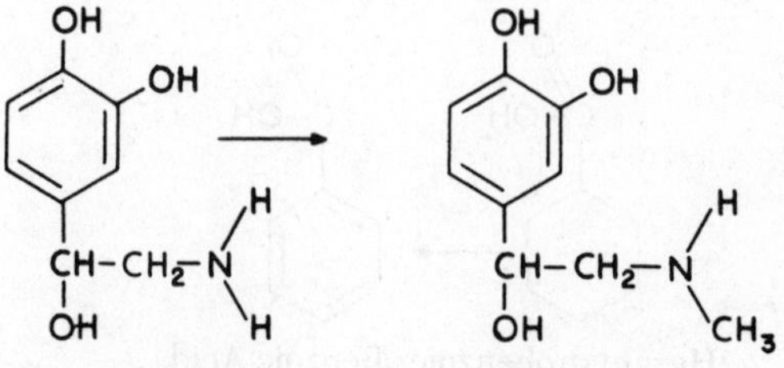

Norepinephrine   Epinephrine

Conjugation (Condensations). Conjugation or condensation reactions follow a common pattern in which the conjugating agent first forms a reactive intermediate, followed by enzymatic transfer to the substrate.

SULFATE CONJUGATION. Inorganic sulfate is first converted to a high energy, nucleotide-bound form, 3'-phosphoadenosine-5'-phosphosulfate. The sulfate group is then transferred to a variety of aromatic and aliphatic hydroxyl or amino groups, the transfer being mediated by enzymes called sulfokinases.

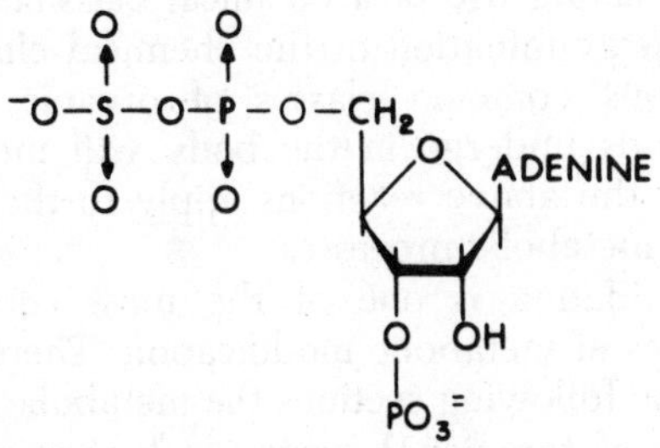

3'-Phosphoadenosine-5'-phosphosulfate

Alkyl and Aryl Sulfate Esters

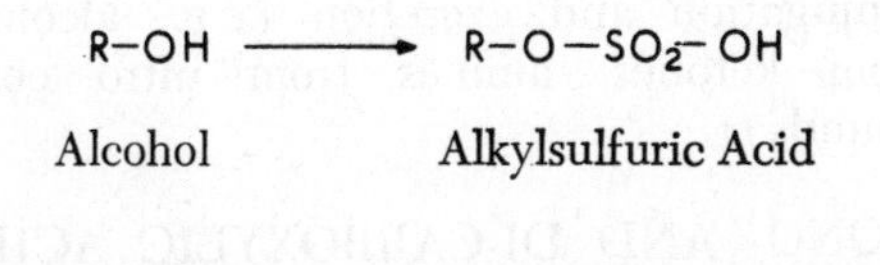

Phenol   Phenylsulfuric Acid

$$R-OH \longrightarrow R-O-SO_2^- OH$$

Alcohol   Alkylsulfuric Acid

Aromatic Amines Form Sulfamic Acids

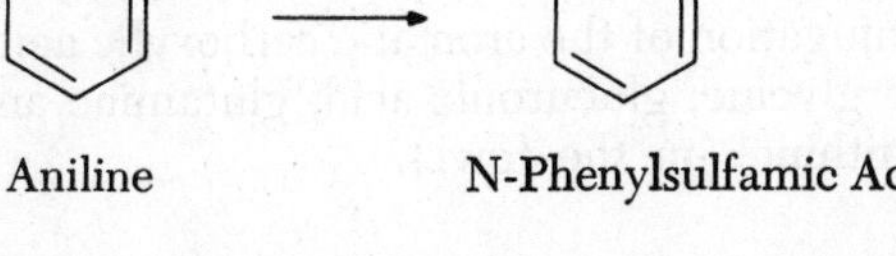

Aniline   N-Phenylsulfamic Acid

GLUCURONIDES. Uridine diphosphate glucuronic acid serves as the active species, transfer of glucuronic acid to substrate being catalyzed by the enzyme, uridine diphosphate-transglucuronylase.

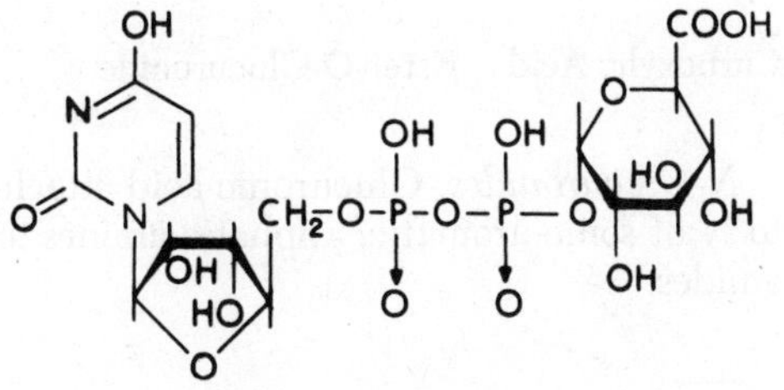

Uridine Diphosphate Glucuronic Acid

Five types of glucuronide conjugation have been established:[21]

O-Ether Glucuronides. Glucuronic acid attached to oxygen of phenols or primary, secondary and tertiary alcohols.

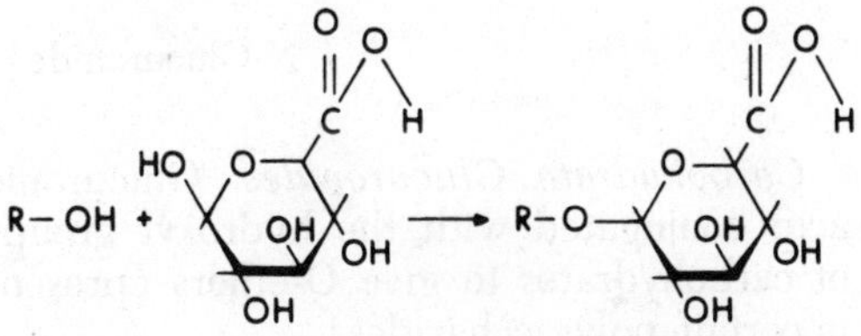

Alcohol   Glucuronic   Ether
or Phenol   Acid   O-Glucuronide

S-Glucuronides. Glucuronic acid attached to S of sulfhydryl groups such as thiophenols and mercaptobenzothiazole.

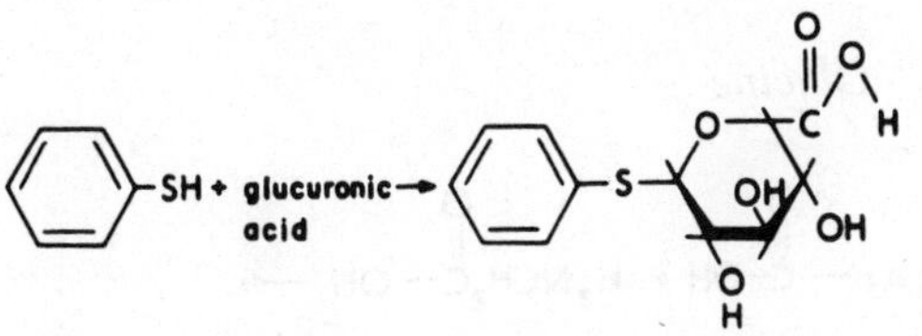

Thiophenol   Thiophenyl Glucuronide

Ester Glucuronides. Glucuronic acid attached to oxygen of aromatic and certain aliphatic carboxylic acids.

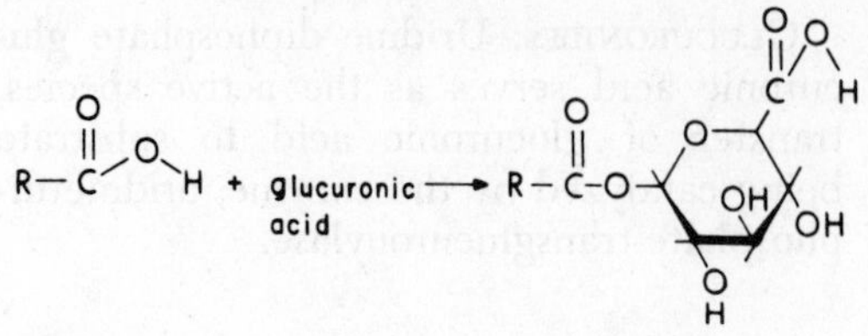

Carboxylic Acid   Ester-O-Glucuronide

N-Glucuronides. Glucuronic acid attached to N of some aromatic, aliphatic amines and amides.

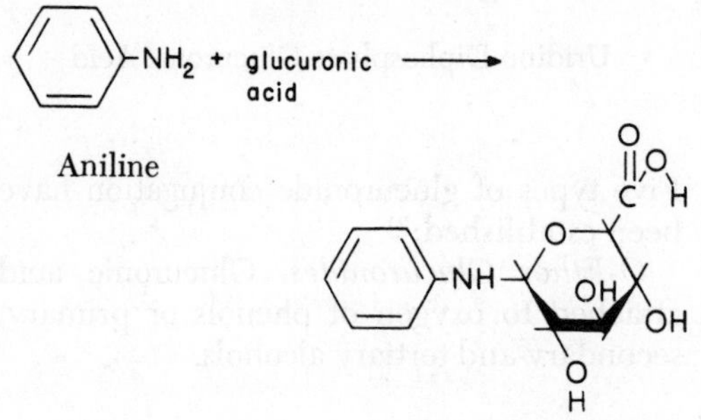

Aniline

N-Glucuronide

Carbohydrate Glucuronides. Glucuronic acid conjugated with the hydroxyl groups of carbohydrates to give O-ethers (present in certain polysaccharides).

CONJUGATION WITH AMINO ACIDS. The condensation of carboxylic acids with the amino group of the amino acids glycine, glutamine and ornithine is preceded by the conversion of the acid into an active ester, S-acyl coenzyme A. The acyl group is transferred to the amino group by the appropriate enzyme.

Glycine

Aromatic   Glycine
Carboxylic Acid

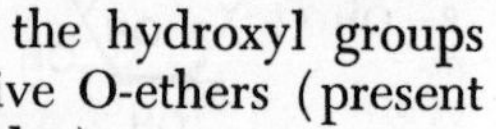

Glycine Conjugate

THIOCYANATE FORMATION

The enzyme rhodanese catalyzes the conversion of the toxic cyanide ion into thiocyanate,[22] which is less than 1 per cent as toxic. Sulfur may be transferred from a variety of sources, including thiosulfate.

$$HCN + Na_2S_2O_3 \longrightarrow$$
Hydrocyanic
Acid

$$HS-CN + Na_2SO_3$$
Thiocyanic
Acid

The metabolism of drugs and other compounds foreign to the body will be reviewed by making use of a chemical classification.

An examination of the chemical changes various common classes of organic compounds undergo in the body will indicate how the above reactions apply to the general metabolic processes.

Oxidation is one of the most common routes of metabolic modification. Therefore, in the following sections the metabolic reactions of functional groups at higher oxidation levels are usually presented before those of groups at lower oxidation levels. Following their metabolic oxidation, compounds frequently undergo conjugation reactions characteristic of the new functional group formed. Similarly, the metabolic reactions of groups formed by hydrolysis (acids, alcohols, phenols) are presented before the more complex group from which they may be formed.

Where compounds are resistant to metabolic oxidation, reduction frequently takes place, particularly if the reduction produces a more polar group or one susceptible to conjugation and excretion (e.g., alcohols from ketones, amines from nitro compounds).

MONO- AND DI-CARBOXYLIC ACIDS

Aromatic carboxylic acids tend to conjugate in the body by one or more of several metabolic pathways. The conjugates formed and excreted vary with the species. The metabolites commonly involved in the conjugation of the aromatic carboxylic acids are glycine, glucuronic acid, glutamine and ornithine (in the fowl).

The excretion of benzoic acid has been studied carefully in a number of animals. There has been sustained interest in this acid because of its permitted use as a food preservative and because of its value as a test agent for liver function. All mammals thus far studied excrete benzoic acid in part as hippuric acid, and a number (man, sheep, pig, dog and rabbit) also excrete it as an ester of glucuronic acid. The following reactions, making use of benzoic and phenylacetic acids, illustrate the general metabolic reactions of the aromatic acids.

Diphenylacetic acid conjugates in part with glucuronic acid, as does benzoic acid.[23] Phenylacetic acid does not conjugate with glycine in man, but is excreted as a conjugate of glutamine as shown below.

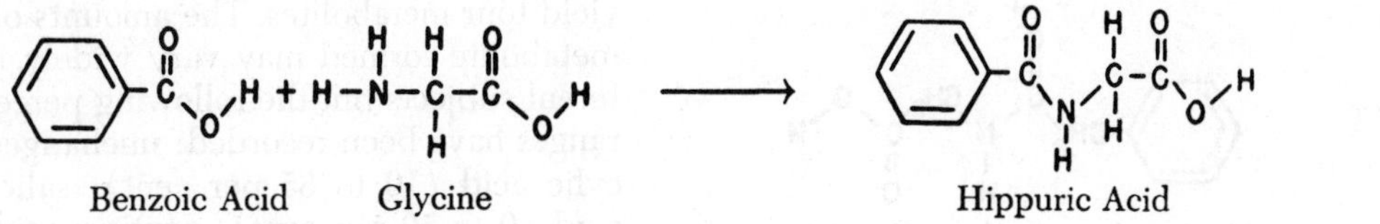

Benzoic Acid   Glycine   Hippuric Acid

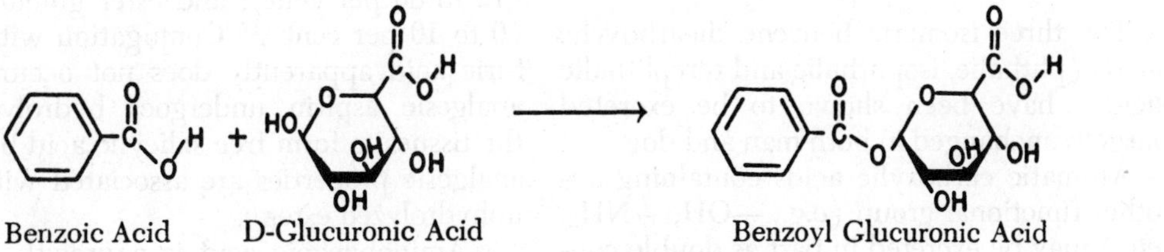

Benzoic Acid   D-Glucuronic Acid   Benzoyl Glucuronic Acid

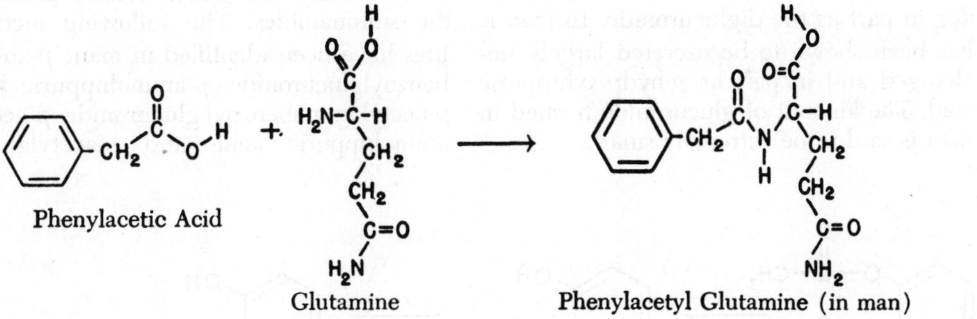

Phenylacetic Acid   Glutamine   Phenylacetyl Glutamine (in man)

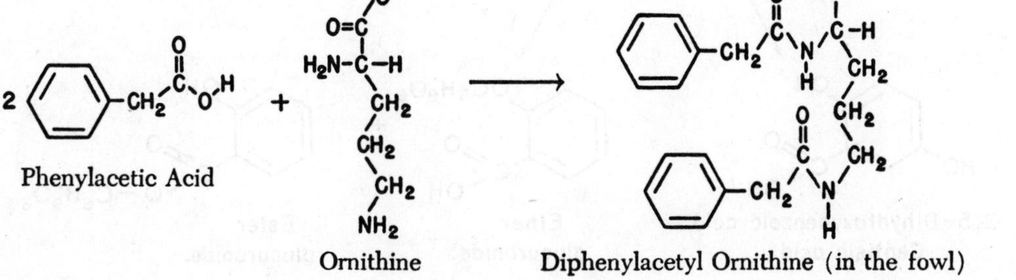

Phenylacetic Acid   Ornithine   Diphenylacetyl Ornithine (in the fowl)

In the fowl, phenylacetic acid conjugates with ornithine.

In the dog, phenylacetic acid is converted to phenaceturic acid.

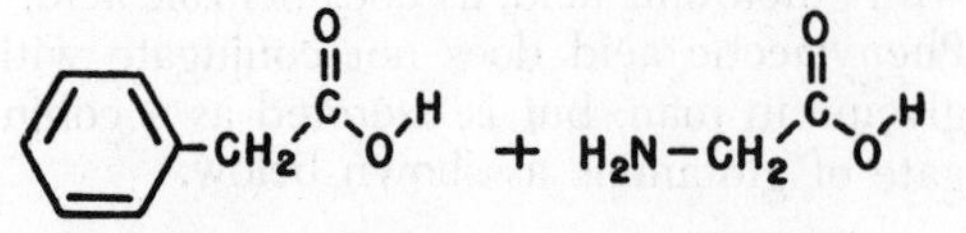

Phenylacetic Acid   Glycine

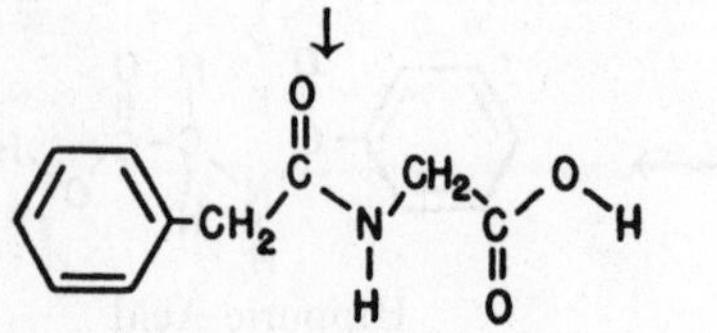

↓

Phenaceturic Acid (in the dog)

The three isomeric benzene dicarboxylic acids (phthalic, isophthalic and terephthalic acids) have been shown to be excreted largely unchanged in both man and dog.

Aromatic carboxylic acids containing another functional group (e.g., —OH, —NH₂, etc.) may be excreted in part as double conjugates. As an example, p-hydroxybenzoic acid has been shown[24] to be excreted by the dog in part as the diglucuronide. In man it has been shown to be excreted largely unchanged and in part as p-hydroxyhippuric acid. The amount of glucuronide formed in man is said to be extremely small.

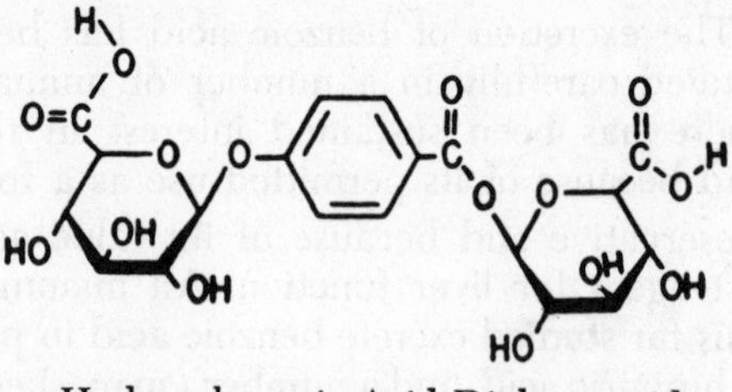

p-Hydroxybenzoic Acid Diglucuronide

Salicylates in man have been observed to yield four metabolites. The amounts of each metabolite formed may vary widely in different subjects but the following percentage ranges have been recorded: unchanged salicylic acid (10 to 85 per cent); salicyluric acid (0 to 50 per cent); gentisic acid (approximately 1 per cent); ether glucuronide (12 to 30 per cent); and ester glucuronide (0 to 10 per cent).[25] Conjugation with sulfuric acid apparently does not occur. The analgesic aspirin undergoes hydrolysis in the tissues to form free salicylic acid but the analgesic properties are associated with the unhydrolyzed ester.

p-Aminobenzoic acid is a growth factor for certain microorganisms and competitively inhibits the bacteriostatic action of the sulfonamides. The following metabolites have been identified in man: p-aminobenzoylglucuronide; p-aminohippuric acid; p-acetylaminobenzoyl glucuronide; p-acetylaminohippuric acid; and p-acetylamino-

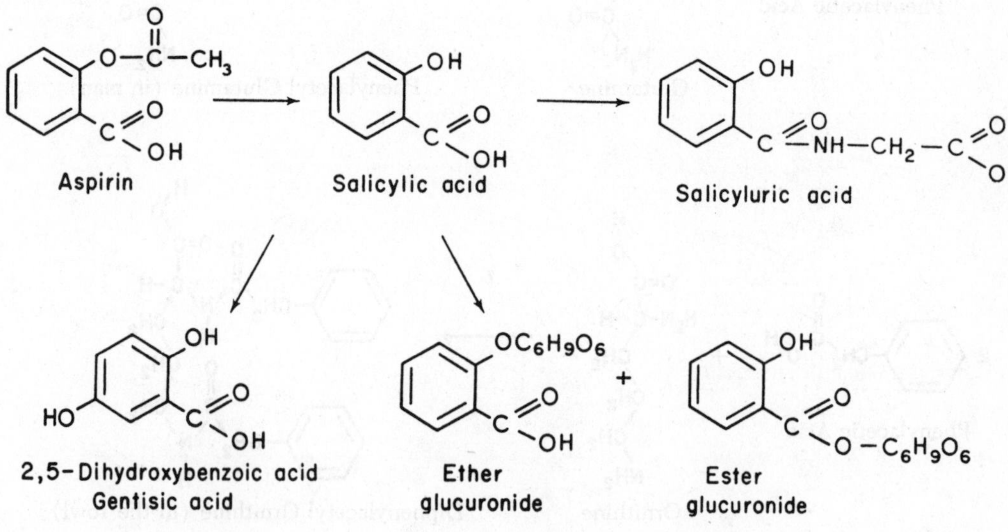

Aspirin   Salicylic acid   Salicyluric acid

2,5–Dihydroxybenzoic acid   Ether   Ester
Gentisic acid   glucuronide   glucuronide

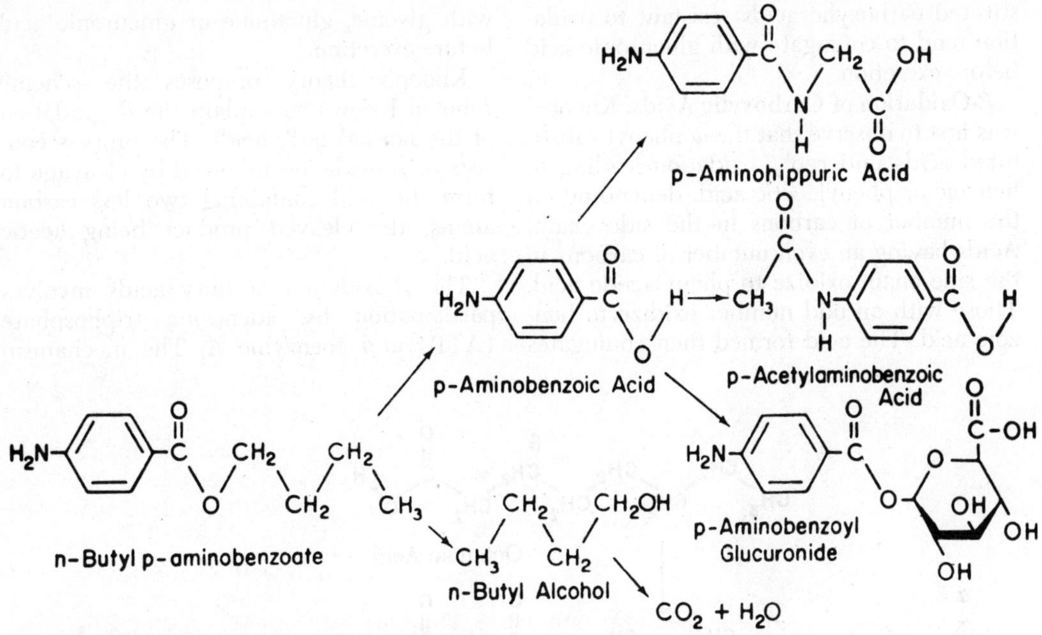

p-Aminohippuric Acid

p-Aminobenzoic Acid

p-Acetylaminobenzoic Acid

p-Aminobenzoyl Glucuronide

n-Butyl p-aminobenzoate

n-Butyl Alcohol

$CO_2 + H_2O$

benzoic acid. The first two metabolites predominate in the compounds excreted.[26] N-glucuronide formation also is possible but has not been identified. A number of the esters of *p*-aminobenzoic acid (ethyl, β-diethylaminoethyl, etc.) are employed as local anesthetics. In the body, the esters are hydrolyzed rapidly; this is followed by the oxidation of the alcohols or amino-alcohols and the acetylation and conjugation of the *p*-aminobenzoic acid. This is shown for butyl aminobenzoate in the above equations.

p-Aminosalicylic acid (PAS) is widely used in combination with other agents for the treatment of tuberculosis. The acid is rapidly absorbed and is partly excreted unchanged but several metabolites have been isolated from man, including *p*-aminosalicyluric acid, *p*-acetylaminosalicylic acid and *p*-acetylaminosalicylic acid glucuronide.

The normal aliphatic acids are largely metabolized to carbon dioxide and water. The iso acids (methyl substituted) usually form acetone as one of the oxidation products and this is eliminated unchanged. Sub-

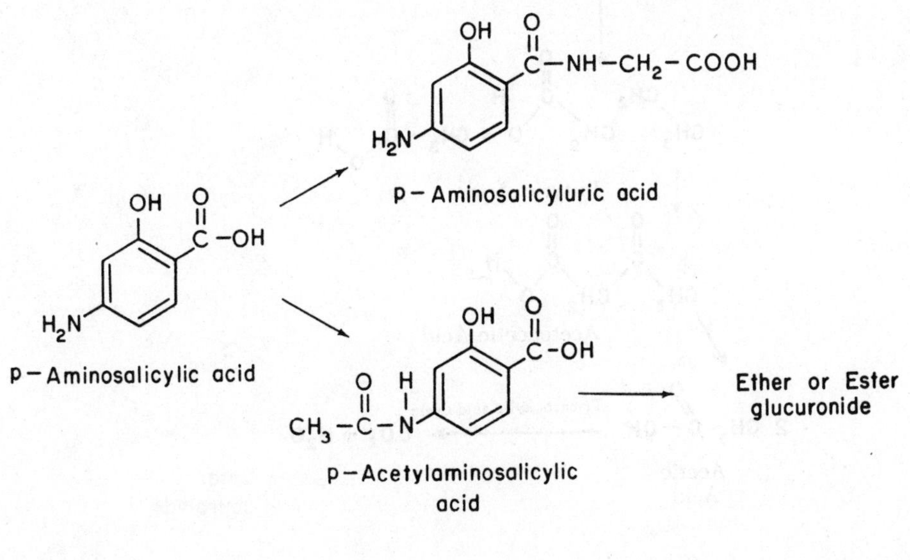

p — Aminosalicyluric acid

p — Aminosalicylic acid

p—Acetylaminosalicylic acid

Ether or Ester glucuronide

stituted carboxylic acids resistant to oxidation tend to conjugate with glucuronic acid before excretion.

β-Oxidation of Carboxylic Acids. Knoop[27] was first to observe that the ω-phenyl substituted acids undergo β-oxidation leading to benzoic or phenylacetic acid, depending on the number of carbons in the side chain. Acids having an even number of carbons in the side chain oxidize to phenylacetic acid. Those with an odd number oxidize to benzoic acid. The acid formed then conjugates with glycine, glutamine or glucuronic acid before excretion.

Knoop's theory proposes the scheme (shown below) to explain the degradation of the normal fatty acids. The process consists of β-oxidation followed by cleavage to form the acid containing two less carbon atoms, the cleaved product being acetic acid.

The β-oxidation of fatty acids involves participation by adenosine triphosphate (ATP) and coenzyme A. The mechanism

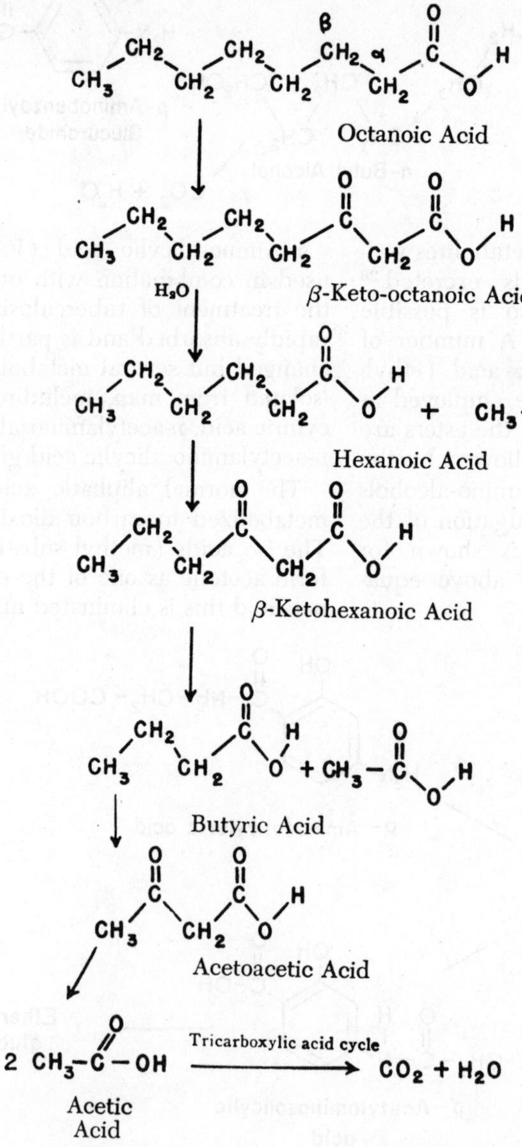

Octanoic Acid

β-Keto-octanoic Acid

Hexanoic Acid

β-Ketohexanoic Acid

Butyric Acid

Acetoacetic Acid

$$2 \ CH_3-C-OH \xrightarrow{\text{Tricarboxylic acid cycle}} CO_2 + H_2O$$

Acetic
Acid

for the removal of a two carbon fragment by β-oxidation of a fatty acid may be described as follows:

$$R-CH_2-CH_2-COOH$$

$$\downarrow ATP + CoA-SH$$

$$R-CH_2-CH_2-CO-SCoA$$

$$\downarrow -2H \text{ (by FAD)}$$

$$R-CH = CH-CO-SCoA$$

$$\downarrow +H_2O$$

$$R-CHOH-CH_2-CO-SCoA$$

$$\downarrow -2H \text{ (by DPN)}$$

$$R-CO-CH_2-CO-SCoA$$

$$\downarrow CoA-SH$$

$$R-CO-SCoA + CH_3CO-SCoA$$

In addition to β-oxidation, multiple alternate oxidation[28] (simultaneous β-oxidation)[29] may occur, and oxidation of the terminal carbon (methyl group)-"ω-oxidation" of fatty acids[30] also is known to occur. ω-Oxidation leads to the formation of dicarboxylic acids. The feeding of the triglyceride of undecanoic (C_{11}) acid, as an example, is found to form three dicarboxylic acids: undecandioic (C_{11}) azelaic (C_9) and pimelic (C_7). Likewise, the feeding of undecandioic acid to dogs shows that part of the compound is excreted unchanged and part is oxidized to azelaic and pimelic acids.

The lower molecular weight aliphatic dicarboxylic acids, oxalic and malonic are largely excreted unchanged. However, oxalic acid is quite toxic and is excreted slowly, due to the formation of an insoluble calcium salt which may be distributed throughout the tissues. Malonic acid is much less toxic than oxalic acid but is an effective inhibitor of succinic acid dehydrogenase and therefore may cause succinic acid to be excreted.

ALDEHYDES

Most of the simple aliphatic aldehydes appear to be oxidized in the body to carbon dioxide and water. The halogenated aldehydes (e.g., chloral) and tertiary substituted aldehydes are resistant to oxidation and may be reduced to the corresponding alcohols, which are then conjugated with glucuronic acid and excreted.

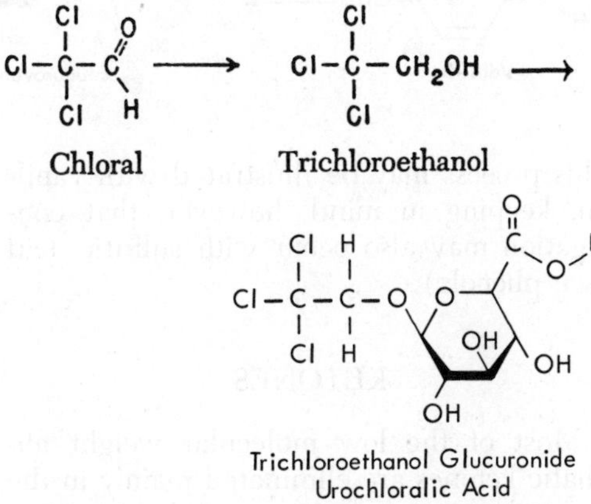

Chloral → Trichloroethanol →

Trichloroethanol Glucuronide
Urochloralic Acid

Most aromatic aldehydes are readily oxidized to the corresponding carboxylic acids.

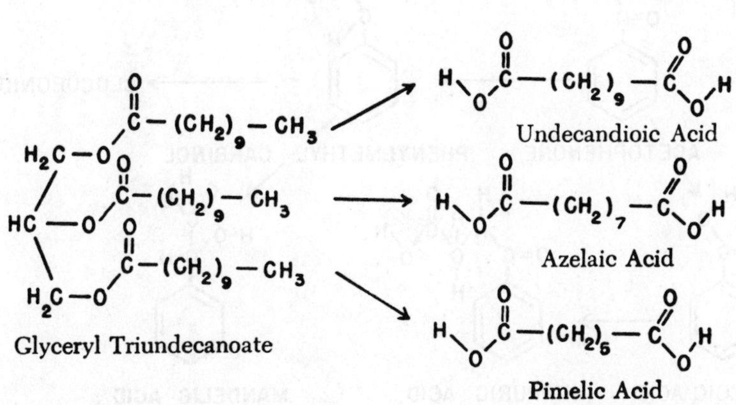

Glyceryl Triundecanoate

Undecandioic Acid

Azelaic Acid

Pimelic Acid

Benzaldehyde, administered orally or parenterally, is converted to benzoic acid.

Cinnamic aldehyde is converted to cinnamic acid. The presence of a phenolic group in the molecule tends to hinder the oxidation of an aldehyde group. Thus with *p*-hydroxybenzaldehyde, some of the compound is excreted in the rabbit with the aldehyde group unchanged and the hydroxyl group conjugated with glucuronic acid. Vanillin also is partly excreted as glucurovanillin. Vanillin undergoes oxidation in the body to vanillic acid, which is then conjugated with glucuronic acid. The conjugation may precede the oxidation.[31]

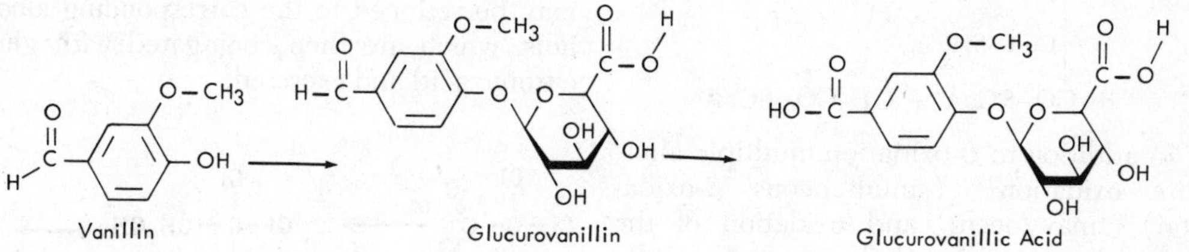

expired air and to a lesser extent in the urine unchanged. Thus 50 to 60 per cent of acetone; 30 to 35 per cent of 2-butanone (ethylmethyl ketone); 38 to 54 per cent of 2-pentanone (methyl-n-propyl ketone); and 3-pentanone (diethyl ketone) are eliminated in the expired air or excreted unchanged.[32] The main metabolic pathway of the aliphatic ketones (with the exception of acetone) is reduction to the corresponding secondary alcohols followed by conjugation with glucuronic acid.

Vanillin   Glucurovanillin   Glucurovanillic Acid

This process may be illustrated with vanillin, keeping in mind, however, that conjugation may also occur with sulfuric acid (see phenols).

KETONES

Most of the low molecular weight aliphatic ketones are eliminated mainly in the

The mixed aromatic-aliphatic ketones are partially reduced to the secondary alcohols and in part oxidized to carboxylic acids.[33] Acetophenone has been shown to undergo the following changes.

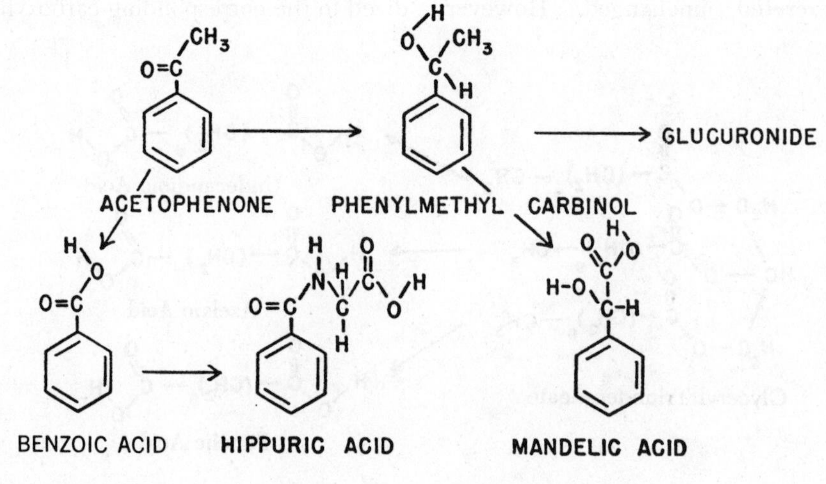

ACETOPHENONE   PHENYLMETHYL CARBINOL   GLUCURONIDE

BENZOIC ACID   HIPPURIC ACID   MANDELIC ACID

Benzyl methyl ketone is oxidized to benzoic acid; phenylethyl methyl ketone and phenylbutyl methyl ketone are oxidized to phenylacetic acid. This is in accord with the views of Dakin[34] that "most aromatic methyl ketones primarily undergo oxidation in the body, so as to yield acids with two less carbons, except in the case of acetophenone in which the carbonyl group is directly attached to the nucleus." Phenylbutyl methyl ketone in man presumably successively forms γ-phenylbutyric acid, phenylacetic acid and phenylacetyl glutamine, which is excreted (see carboxylic acids).

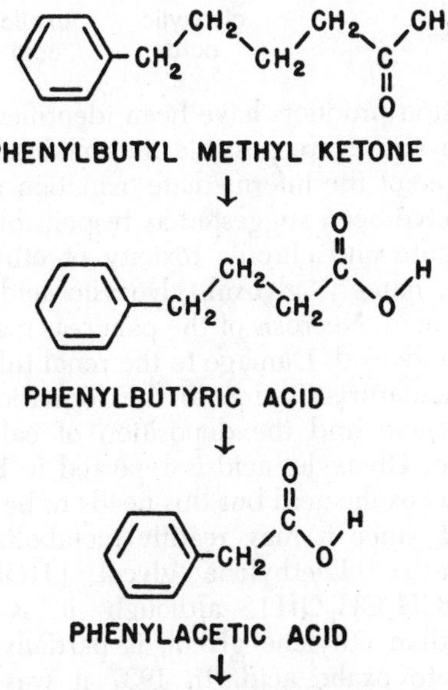

Camphor   5-Hydroxy-camphor   5-Hydroxy-camphor Glucuronide

PHENYLBUTYL METHYL KETONE
↓
PHENYLBUTYRIC ACID
↓
PHENYLACETIC ACID
↓
PHENYLACETYL GLUTAMINE (IN MAN)

Most of the quinones thus far studied appear to be resistant to oxidation in the body, and some are known to undergo reduction. 1,4-Benzoquinone (quinone) is reduced in part to hydroquinone, which is then conjugated with sulfuric and glucuronic acid. Menadione (2-methyl-1,4-naphthoquinone) is oxidized in part to phthalic acid.[35]

Camphor is metabolized to form a hydroxy derivative which is then conjugated with glucuronic acid and excreted. The main oxidation product is 5-hydroxycamphor but there is also formed some of the 3-hydroxy derivative.[36]

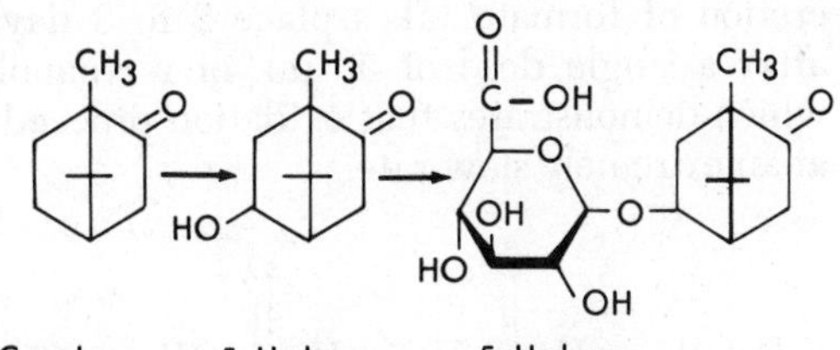

ALCOHOLS AND GLYCOLS

Many aliphatic alcohols are oxidized in the body to carbon dioxide and water. Glucuronide formation has long been known as a conjugation reaction for aliphatic alcohols, but only recently has it been shown that they are also excreted in the urine as sulfates.[37]

Methyl alcohol is not in line with other members of its homologous series, since it is considerably more toxic than the higher homologs, ethyl and propyl alcohols. Methanol is metabolized at only about one fifth the rate of ethanol in rabbits. This slow rate gives rise to the possibility of accumulation in various tissues and delayed toxic effects. The narcotic effect of methanol is thought to be due to the compound itself, and this effect is less than its higher homologs, possibly due to its lower fat solubility. The fate of methanol in the body is probably as follows: some is excreted unchanged by way of the kidneys and lungs, the rest is oxidized to formaldehyde, some of this is combined with protein and the rest oxidized to formic acid, some of this is excreted and the rest oxidized to carbon dioxide and water. One danger associated with methanol poisoning is blindness. The much greater toxicity of methanol as compared with ethanol is apparently dependent on the greater toxicity of the intermediate metabolites, formaldehyde and formic acid, as compared with acetaldehyde and acetic acid. Formaldehyde has been detected in the vitreous humor of methanol-poisoned animals, and this tissue also increases in acidity in such animals. Rabbits are able to metabolize formic acid and are much less susceptible to methanol poisoning. In man formate is found in both the blood and the urine. The maximum excretion of formate takes place 2 to 3 days after a single dose of 50 ml. of methanol, which demonstrates that oxidation proceeds at an extremely slow rate.

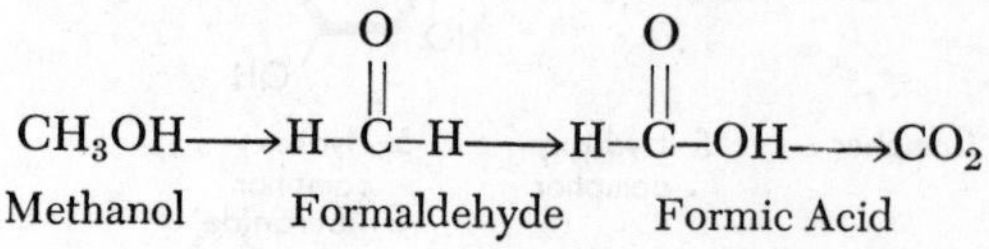

$$CH_3OH \longrightarrow \underset{\text{Formaldehyde}}{H-\overset{\displaystyle O}{\overset{\|}{C}}-H} \longrightarrow \underset{\text{Formic Acid}}{H-\overset{\displaystyle O}{\overset{\|}{C}}-OH} \longrightarrow CO_2$$
Methanol

The glycols tend to be oxidized in the body to the corresponding mono- or dicarboxylic acids. Those failing to undergo oxidation may conjugate in part with glucuronic acid or be excreted unchanged. Ethylene glycol is the most toxic of the glycols and the toxicity appears to be dependent on the oxidation products formed in the body. Not all of the intermediate

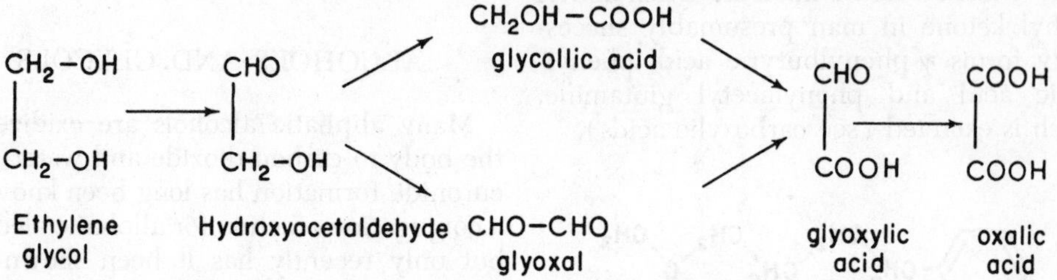

$$\underset{\substack{\text{Ethylene}\\\text{glycol}}}{\overset{\displaystyle CH_2-OH}{\underset{\displaystyle CH_2-OH}{|}}} \longrightarrow \underset{\text{Hydroxyacetaldehyde}}{\overset{\displaystyle CHO}{\underset{\displaystyle CH_2-OH}{|}}} \begin{array}{c} \nearrow \ \underset{\text{glycollic acid}}{CH_2OH-COOH} \searrow \\ \\ \searrow \underset{\substack{\text{glyoxal}}}{CHO-CHO} \nearrow \end{array} \underset{\substack{\text{glyoxylic}\\\text{acid}}}{\overset{\displaystyle CHO}{\underset{\displaystyle COOH}{|}}} \longrightarrow \underset{\substack{\text{oxalic}\\\text{acid}}}{\overset{\displaystyle COOH}{\underset{\displaystyle COOH}{|}}}$$

When ethanol is administered with methanol, the ethanol appears to be preferentially oxidized, leading to an increased rate of excretion of unchanged methanol and a suppression of formate excretion.[38] This observation has led a number of workers to suggest the use of ethanol as an antidote for methanol poisoning.[39-41] Its effectiveness as an antidote remains to be established.

Ethanol is rapidly oxidized in the body to acetaldhyde, acetic acid and carbon dioxide. The administration of disulfiram (Antabuse) interferes with the second step of the metabolism of ethanol, the oxidation of acetaldhyde to acetic acid, producing an accumulation of acetaldehyde and this may lead to alarming symptoms.

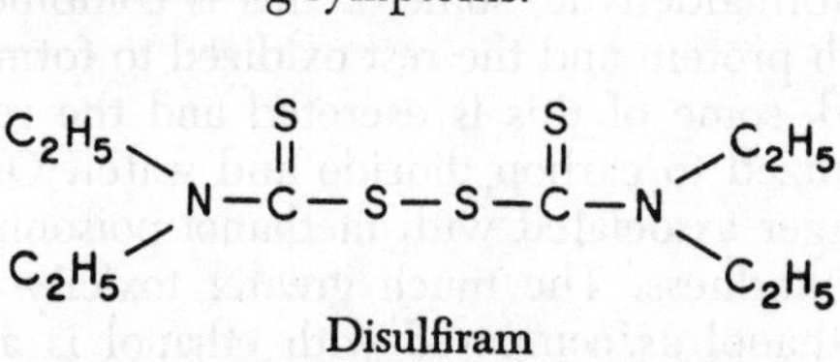

$$\underset{C_2H_5}{\overset{\displaystyle C_2H_5}{\diagdown}} N-\overset{\displaystyle\overset{\displaystyle S}{\|}}{C}-S-S-\overset{\displaystyle\overset{\displaystyle S}{\|}}{C}-N \overset{\displaystyle C_2H_5}{\underset{\displaystyle C_2H_5}{\diagup}}$$

Disulfiram
Bis(diethylthiocarbamyl) disulfide

The unsubstituted primary and secondary aliphatic alcohols of low molecular weight are, generally, more or less completely oxidized to carbon dioxide and water.

Tertiary alcohols and halogenated alcohols (e.g., tertiary butyl alcohol, tribromoethanol, trichloroethanol, etc.) are resistant to oxidation and are excreted largely as the conjugates of glucuronic acid.

oxidation products have been identified but shown above are possible intermediates.

Three of the intermediate oxidation products have been suggested as responsible for the acute and chronic toxicity of ethylene glycol, namely, glyoxal, glyoxylic acid and oxalic acid. Necrosis of the pancreas may be due to glyoxal. Damage to the renal tubules and hematuria is due to the formation of oxalic acid and the deposition of calcium oxalate. Glyoxylic acid is reported to be as toxic as oxalic acid but this needs to be confirmed, since it may readily metabolize to the latter. Diethylene glycol ($HOCH_2$-$CH_2OCH_2CH_2OH$), although it is less toxic than ethylene glycol, is partially oxidized to oxalic acid. In 1937 it was employed as a vehicle in the preparation of an elixir of sulfanilamide and was the cause of many deaths.[42] Triethylene glycol ($HOCH_2CH_2OCH_2CH_2OCH_2CH_2OH$) is less toxic than diethylene glycol and it appears that this compound is not oxidized to oxalic acid. The polyethylene glycols resulting from the polymerization of ethylene oxide (e.g., Carbowaxes) are poorly absorbed from the gastrointestinal tract and are relatively nontoxic.

Propane-1,2-diol (propylene glycol) is nontoxic and is relatively widely used as a solvent in pharmaceutical preparations. It is partially excreted unchanged and in part oxidized to lactic acid.[43] In the rabbit it has also been shown to be excreted as the monoglucuronide. Propane-1,3-diol is somewhat more toxic than propylene glycol and this is believed due to its metabolism to malonic acid which is a recognized enzyme inhibitor. Phenylglycol is reported to be largely metabolized to mandelic acid and there is also excreted a monoglucuronide of the glycol. 3-*o*-Tolyloxypropane-1,2-diol (mephenesin) is metabolized largely to 3-*o*-tolyloxylactic acid. In dogs, from 30 to 40 per cent of

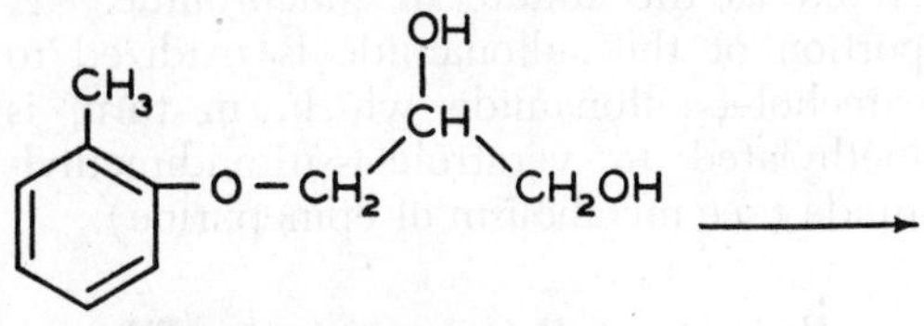

3-*o*-Tolyloxypropane-1,2-diol
(Mephenesin)

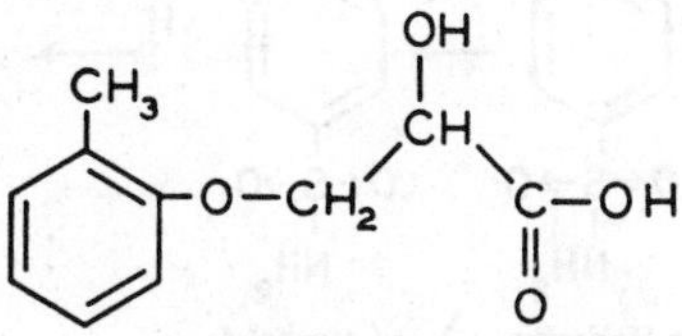

3-*o*-Tolyloxylactic Acid

the drug is excreted as the conjugate, presumably of glucuronic acid.

Chloramphenicol (Chloromycetin) D-(−) *threo*-1-*p*-nitrophenyl-2-dichloroacetamido-1,3-propanediol is largely metabolized in man, with only 5 to 10 per cent of the drug excreted unchanged. The two major metabolic products excreted are the 3-glucuronide and the deacylated chloramphenicol. There is also some reduction of the aromatic nitro group but in man it is excreted largely unchanged.[44, 45]

The primary aromatic alcohols are oxidized in part to benzoic or phenylacetic acid, depending on the number of carbons in the side chain. Benzyl alcohol, as an example, is oxidized to benzoic acid, β-phenylethyl alcohol to phenylacetic acid. γ-Phenylpropanol may undergo β-oxidation to benzoic acid, which will then conjugate with glycine (see acids).

The secondary aromatic alcohols (e.g., phenylmethyl carbinol) are more resistant to oxidation and are excreted in part as the conjugates of glucuronic acid.[33] The tertiary aromatic alcohols, such as triphenylcarbinol, are excreted in part unchanged.[46] The failure of triphenylcarbinol to conjugate may be due to steric hindrance. Alicyclic alcohols, such as menthol and borneol, are excreted largely as the glucuronides.

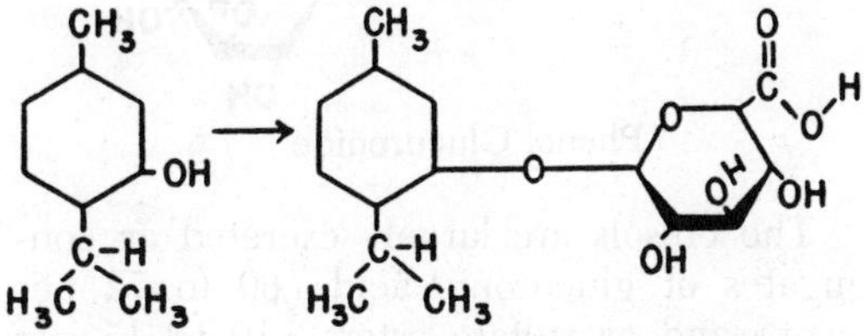

Menthol   Menthol Glucuronide

PHENOLS

Three types of chemical changes are associated with the phenols in the organism, namely: oxidation, conjugation with glucuronic acid and conjugation with sulfates to form the sulfuric acid esters. Some phenolic compounds are excreted partially unchanged.

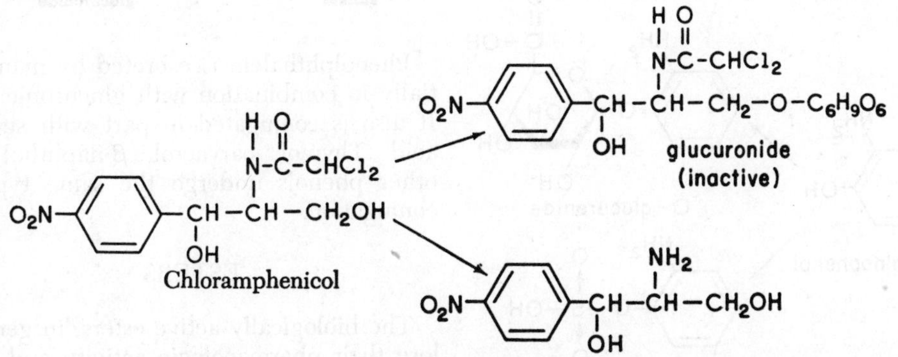

glucuronide
(inactive)

Chloramphenicol

Deacylated Chloramphenicol (inactive)

Phenol will serve to illustrate the types of chemical change characteristic of the phenols.

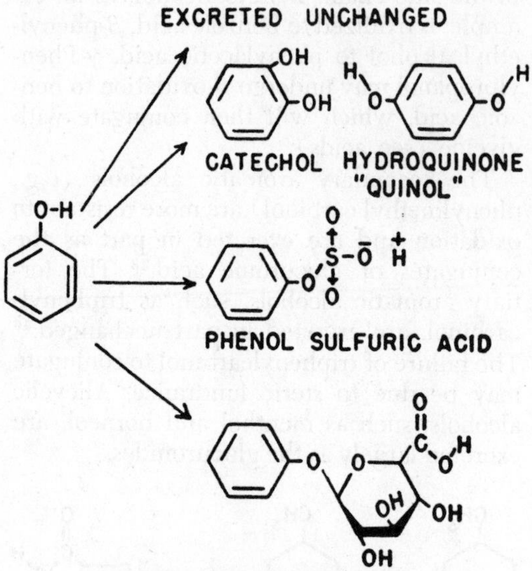

EXCRETED UNCHANGED

CATECHOL HYDROQUINONE "QUINOL"

PHENOL SULFURIC ACID

Phenol Glucuronide

The cresols are largely excreted as conjugates of glucuronic acid (60 to 72 per cent) and as sulfate esters (10 to 15 per cent). *p*-Cresol is partially oxidized in rabbits to *p*-hydroxybenzoic acid which is excreted both free and as the glucuronide. Oxidation of the *o*- and *m*-cresols to the corresponding benzoic acids has not been observed but all three isomers are metabolized into small amounts of the dihydroxytoluenes.[47] The isomeric monoaminophenols are excreted primarily as the glucuronide ethers (60 to 70 per cent) and as sulfate esters (12 to 15 per cent).[48] Acetylation of

o – aminophenol

O – glucuronide

sulfate ester

the amino group may also occur; thus *m*-aminophenol is excreted as the *m*-acetylaminophenyl sulfate and glucuronide but the *o*- and *p*-isomers appear not to be acetylated[49] (see amino compounds this chapter). Like aniline, the aminophenols may form N-glucuronides and sulfamates.

Following the administration of *p*-hydroxybenzenesulfonamide to rabbits, approximately 80 per cent of the compound is excreted as the sulfate or glucuronide.[50] A portion of the sulfonamide is oxidized to catechol-4-sulfonamide which, in turn, is methylated to veratrole-4-sulfondimethylamide (see metabolism of epinephrine).

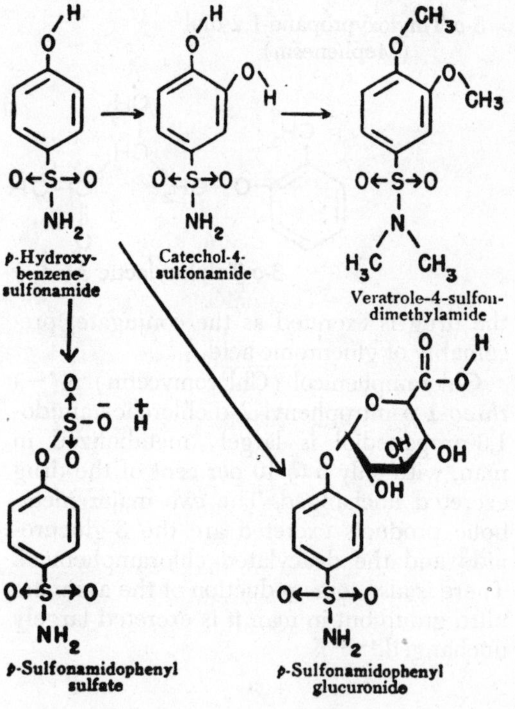

p-Hydroxy-benzene-sulfonamide

Catechol-4-sulfonamide

Veratrole-4-sulfon-dimethylamide

p-Sulfonamidophenyl sulfate

p-Sulfonamidophenyl glucuronide

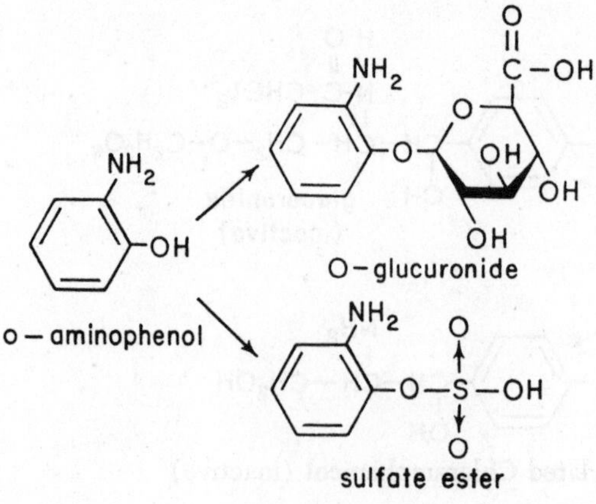

Phenolphthalein is excreted by man partially in combination with glucuronic acid. It also is conjugated in part with sulfuric acid. Thymol, carvacrol, β-naphthol and other phenols undergo the same type of conjugation.

ESTERS

The biologically active esters, in general, lose their pharmacologic activity and most of their toxicity when hydrolyzed. This is the most likely method for the detoxication of the large number of ester drugs in use (e.g., local anesthetics, antispasmodics, parasympathomimetics, etc.). Arecoline, an alkaloid ester occurring in the seeds of *Areca catechu*, will serve to illustrate the influence of hydrolysis on the toxicity of biologically active esters.

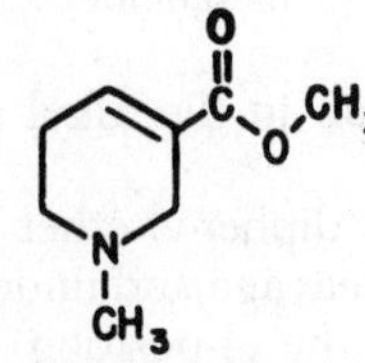

Arecoline (toxic) 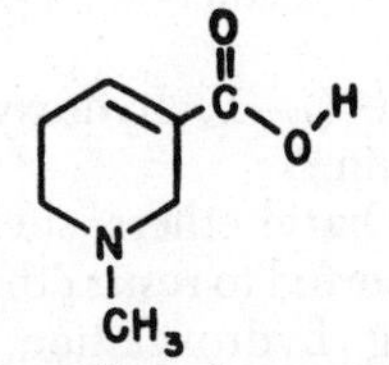 Arecaidine (nontoxic)

Methylphenidate (see Chap. 16) has been observed to undergo predominantly hydrolysis. The major metabolite in rats is phenyl-(2-piperidyl) acetic acid.[51] Hydrolysis is also a major metabolic pathway for the ester drugs meperidine, anileridine and ethoheptazine (see Chap. 24).

Esters of phenols are rapidly hydrolyzed and subsequently undergo the metabolic reactions of the component carboxylic acid and phenol. For example, acetylsalicylic acid (aspirin) is rapidly hydrolyzed to salicylic acid and acetic acid by several body tissues, including the plasma. Using C^{14}-labeled aspirin, intact drug in concentrations less than 2 mg. per 100 ml. was found in plasma for up to 2 hours following an oral dose of 1.2 g.[52] The only metabolite apparent in plasma at that time was salicylic acid, which appeared to arise from the hydrolysis of aspirin after absorption. The urinary metabolites of aspirin are the same as those of salicylic acid (see Acids).

The nitrite and nitrate esters of the aliphatic alcohols (e.g., glyceryl trinitrate, erythrityl tetranitrate, amyl nitrite, etc.) are detoxified by hydrolysis.

Whitemore[53] suggests that the nitrate esters undergo an intramolecular oxidation-reduction in accordance with the following reaction:

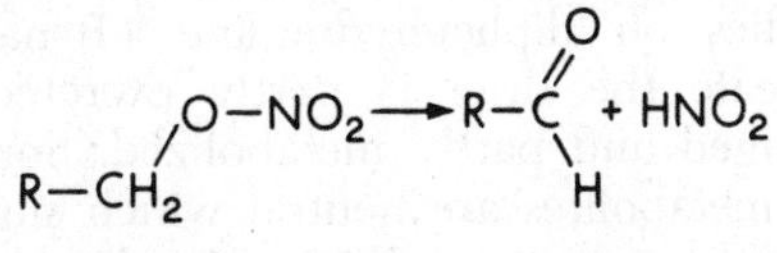

$$R-CH_2-O-NO_2 \longrightarrow R-\overset{O}{\overset{\|}{C}}-H + HNO_2$$

The nitrite esters, by a different mechanism (hydrolysis followed by oxidation), may give the same products in the body as the nitrate esters. This may be shown for ethyl nitrite as follows.

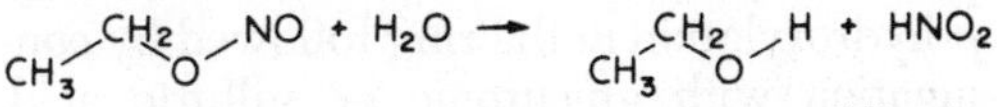

$$CH_3CH_2O-NO + H_2O \longrightarrow CH_3CH_2O-H + HNO_2$$

Ethyl Nitrite Ethyl Alcohol Nitrous Acid

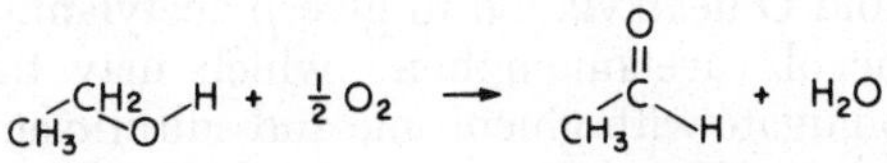

$$CH_3CH_2O-H + \tfrac{1}{2}O_2 \longrightarrow CH_3\overset{O}{\overset{\|}{C}}H + H_2O$$

Ethyl Alcohol Acetaldehyde

The aldehyde will undergo further oxidation to give carbon dioxide and water.

Krantz et al.[54] have published evidence to show that the action of the nitrate esters may be due to the unhydrolyzed molecule and not to products of hydrolysis. They find in a series of nitrates of glycollic acid and its esters that the observed activity parallels the partition coefficients. As an alternate explanation, it is possible that the partition coefficient influences the rate of penetration to the enzyme systems which catalyze the formation of nitrite ion.

TABLE 9. ACTIVITY AND PARTITION COEFFICIENTS OF DERIVATIVES OF GLYCOLLIC ACID*

COMPOUND	EFFECTIVE DE-PRESSOR (MOLAR) CONCENTRATION	PARTITION (OIL/WATER) COEFFICIENT
Sodium glycollate nitrate	0.10	0.9
Ethyl glycollate nitrate	0.013	17.0
n-Propyl glycollate nitrate	0.008	24.0
n-Butyl glycollate nitrate	0.003	108.0
n-Heptyl glycollate nitrate	0.001	142.00

* After Krantz, J. C., Jr., Carr, C. J., Forman, S., and Cone, N.: J. Pharmacol. Exp. Ther. 70:323, 1940.

ETHERS

The aliphatic ethers (ethyl ether, vinyl ether) employed as anesthetics appear not to be metabolized and are eliminated, primarily in the expired air, unchanged. Two of the commonly used antihistaminic drugs (see Chap. 23) (diphenhydramine and doxylamine) contain an aliphatic ether linkage. Studies on diphenhydramine (Benadryl) indicate the drug is partly excreted unchanged and partly metabolized. Some of the metabolites are neutral which suggests the ether group is split to give benzhydrol and β-dimethylaminoethanol as intermediates.[55] Doxylamine may undergo a similar cleavage but this has not been established.

Mixed arylalkyl ethers (e.g., anisole, phenetole) have been observed to undergo p-hydroxylation in the ring followed by conjugation with glucuronic or sulfuric acid. The ether group remains unchanged. Phenacetin (p-acetylaminophenetol) undergoes rapid O-dealkylation to give p-acetylaminophenol (acetaminophen) which may then conjugate with glucuronic and sulfuric acid. In man 80 to 90 per cent of the drug is metabolized in this manner.[56] The analgesic activity of phenacetin in man is dependent on the relative rates of oxidative de-ethylation to the active metabolite acetaminophen (p-acetylaminophenol) and its subsequent conversion to the pharmacologically inactive glucuronide and sulfate conjugates.

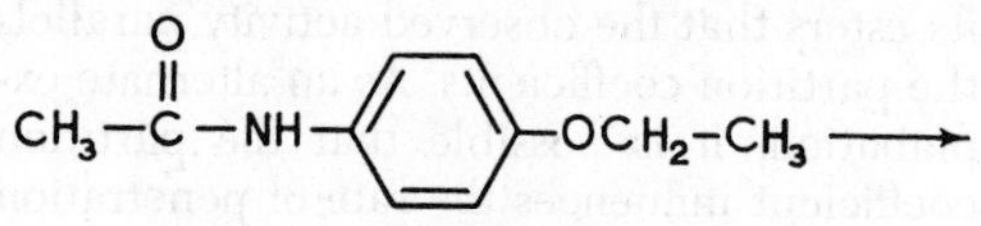

Phenacetin
(*p*-Acetylaminophenetol)

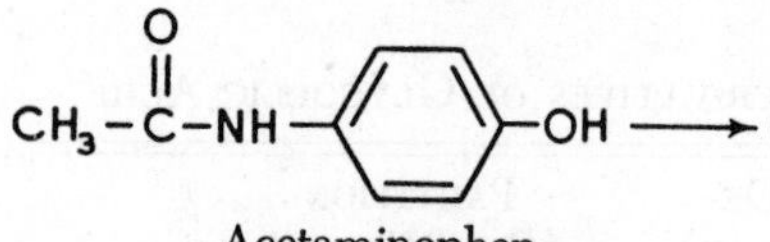

Acetaminophen
(*p*-Acetylaminophenol)

O-Demethylation to form morphine is reported to be an important metabolic pathway for codeine.[57] In addition to O-dealkylation and ring hydroxylation to form phenols, arylalkyl ethers have been observed to

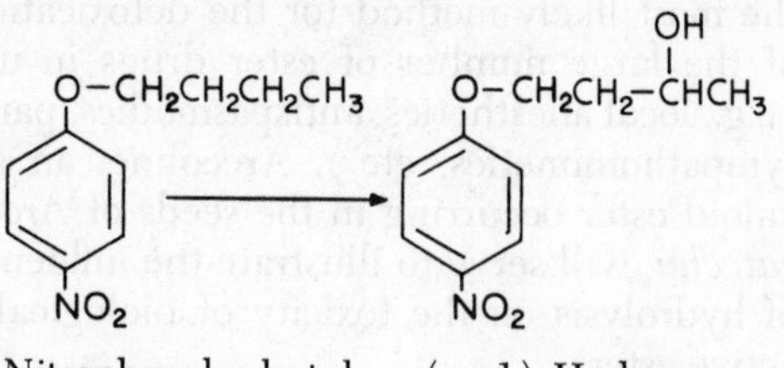

p-Nitrophenyl-*n*-butyl   (ω−1)-Hydroxy
ethers   metabolite

give (ω−1) hydroxylation in the alkyl side chain.[58]

Diaryl ethers, such as diphenyl ether, are reported to resist ether cleavage and undergo ring hydroxylation in the 4-position followed by excretion as the glucuronide and sulfate ester.[59]

HYDROCARBONS

The saturated aliphatic hydrocarbons of high molecular weight are poorly absorbed and pass through the body unchanged. However, n-hexadecane is absorbed by the rat, partially stored in the fat depots and in part oxidized to fatty acids. Some of the unsaturated hydrocarbons have been observed to undergo hydration in the body to form the corresponding alcohol, which may be conjugated with glucuronic acid.

The aromatic hydrocarbons, although resistant to oxidation, are oxidized in part. Benzene is partially oxidized to phenol, which in turn is oxidized to catechol, hydroquinone and muconic acid.

The aromatic hydrocarbons with short, normal aliphatic side chains, for example, toluene, ethylbenzene, xylene and mesitylene, are oxidized to carboxylic acids or hydroxy carboxylic acids. Toluene is oxidized to benzoic acid and excreted largely as hippuric acid. Ethylbenzene in the rabbit is oxidized to phenylmethylcarbinol, phenylacetic acid, benzoic acid and in part to mandelic acid. The phenylmethylcarbinol is excreted largely as the glucuronide, phenylacetic acid as phenaceturic acid and benzoic acid as hippuric acid. The mandelic acid formed is excreted unchanged. n-Propylbenzene in the rabbit is oxidized in part to benzoic acid which is excreted as hippuric acid, but the main metabolic products are 1-phenylpropanol and 2-phenylpropanol, both of which are excreted as the glucuronides.

Alkylbenzenes with branched side chains such as isopropylbenzene and *tert*-butylbenzene in the rabbit are mainly oxidized to alcohols but there are also some acids formed. Both the alcohols and acids are excreted as glucuronides. The metabolism of isopropylbenzene will illustrate the changes involved:

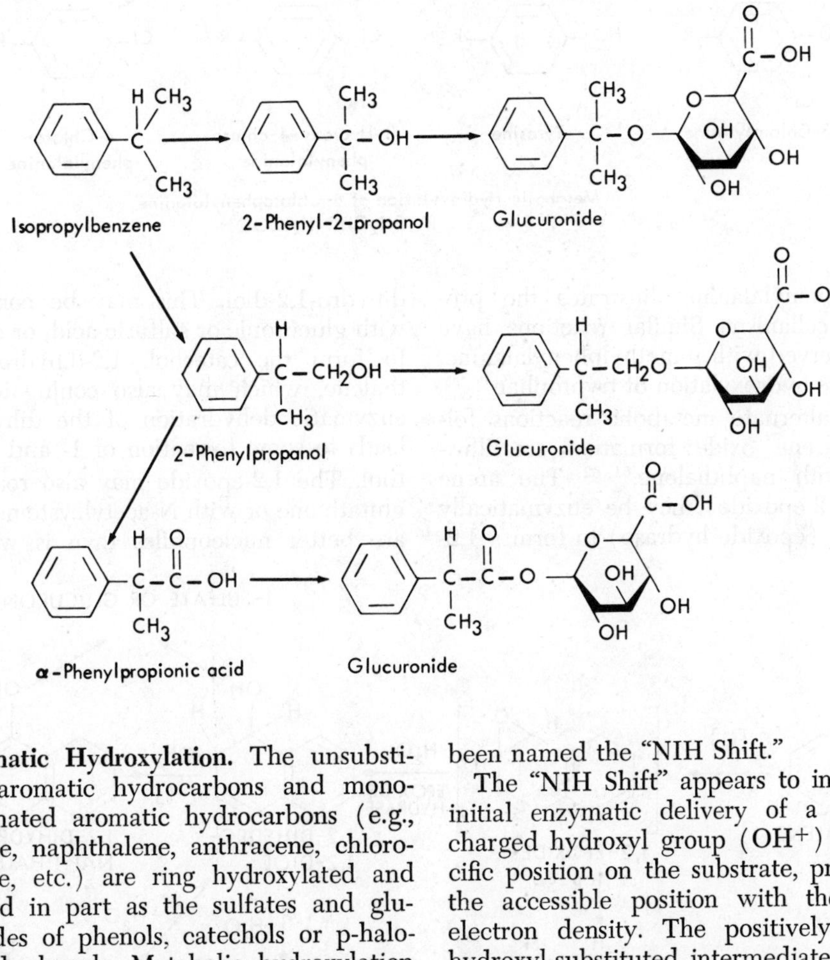

mechanism for aromatic hydroxylation.[60] The unusual reaction was discovered when 4-tritiophenylalanine was tested for potential use in a phenylketonuria assay. Instead of the expected release of tritium on hydroxylation with phenylalanine hydroxylase, most of the tritium was retained in the molecule and was found to have shifted to the 3-position. In recognition of the group at the National Institutes of Health who discovered this process of hydroxylation-induced intramolecular migration, it has

Isopropylbenzene   2-Phenyl-2-propanol   Glucuronide

2-Phenylpropanol   Glucuronide

α-Phenylpropionic acid   Glucuronide

Aromatic Hydroxylation. The unsubstituted aromatic hydrocarbons and monohalogenated aromatic hydrocarbons (e.g., benzene, naphthalene, anthracene, chlorobenzene, etc.) are ring hydroxylated and excreted in part as the sulfates and glucuronides of phenols, catechols or p-halogenated phenols. Metabolic hydroxylation is a common reaction for a large variety of aromatic compounds. A mechanism for the biologic hydroxylation process, in which aromatic substituents may undergo an intramolecular migration, has been determined and is considered to be a general

been named the "NIH Shift."

The "NIH Shift" appears to involve the initial enzymatic delivery of a positively charged hydroxyl group (OH^+) to a specific position on the substrate, presumably the accessible position with the highest electron density. The positively charged hydroxyl-substituted intermediate may follow alternate pathways, involving either a substituent or hydrogen shift to an adjacent position on the aromatic ring, or arene oxide formation and ring opening, which may result in the hydroxyl group being shifted. The action of phenylalanine hydroxylase on

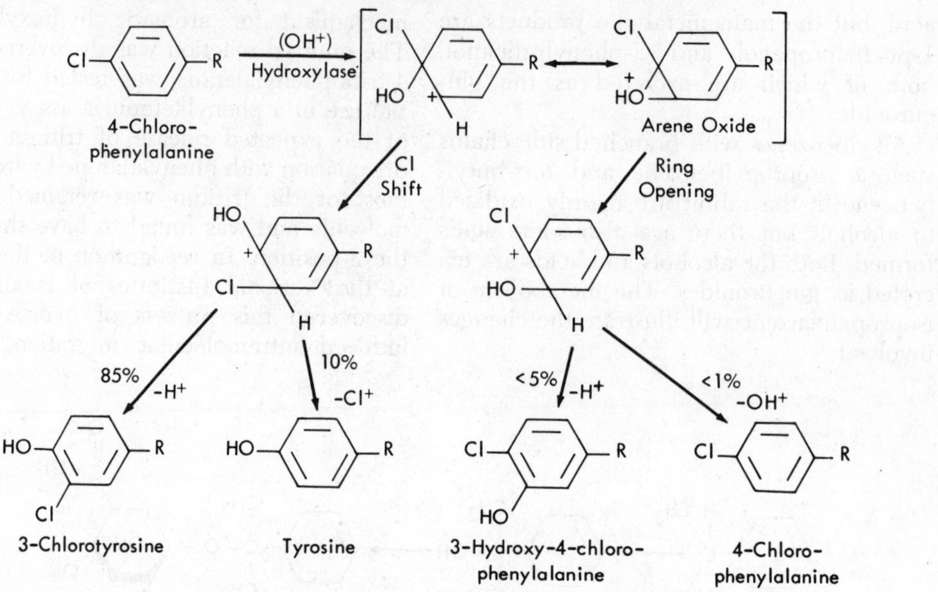

Metabolic Hydroxylation of 4-Chlorophenylalanine
R=CH₂CH(NH₂)COOH

4-chlorophenylalanine illustrates the proposed mechanism. Similar reactions have been observed with *p*-methylphenylalanine, and in the hydroxylation of tryptophan.

Some alternate metabolic reactions following arene oxide formation are illustrated with naphthalene.[61, 62] The arene oxide (1,2-epoxide) may be enzymatically hydrated (epoxide hydrase) to form a 1,2-dihydro-1,2-diol. This may be conjugated with glucuronic or sulfuric acid, or oxidized to form the catechol, 1,2-dihydroxynaphthalene, which may also conjugate. Nonenzymatic dehydration of the dihydrodiol leads to some formation of 1- and 2-naphthol. The 1,2-epoxide may also react with glutathione or with N-acetylcysteine, which are better nucleophiles than is water, to

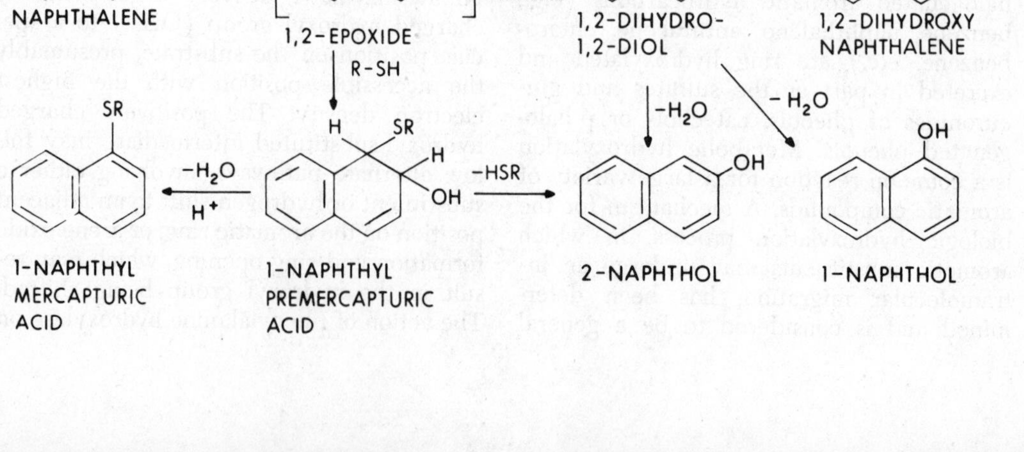

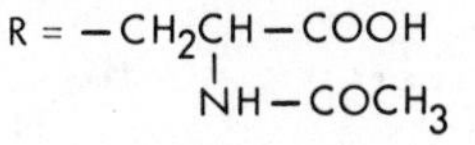

$$R = -CH_2CH-COOH$$
$$\qquad\;\; | $$
$$\qquad\;\; NH-COCH_3$$

N-Acetylcysteine

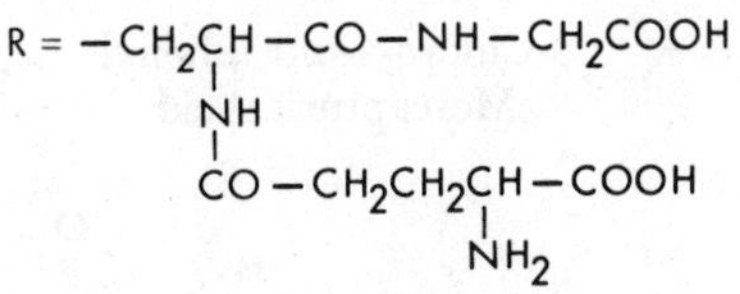

$$R = -CH_2CH-CO-NH-CH_2COOH$$
$$\qquad\;\; | $$
$$\qquad\;\; NH$$
$$\qquad\;\; | $$
$$\qquad\;\; CO-CH_2CH_2CH-COOH$$
$$\qquad\qquad\qquad\qquad | $$
$$\qquad\qquad\qquad\qquad NH_2$$

Glutathione

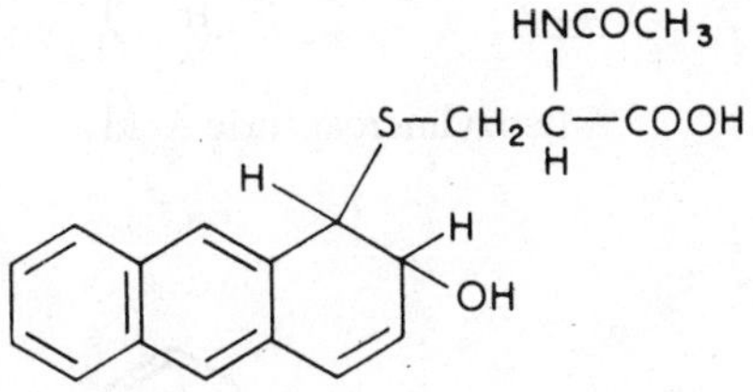

$$HNCOCH_3$$
$$\qquad | $$
$$S-CH_2\,C-COOH$$

N-acetyl-S-(2-hydroxy-1,2-
dihydroanthranil)-L-cysteine
(1-Anthryl Premercapturic acid)

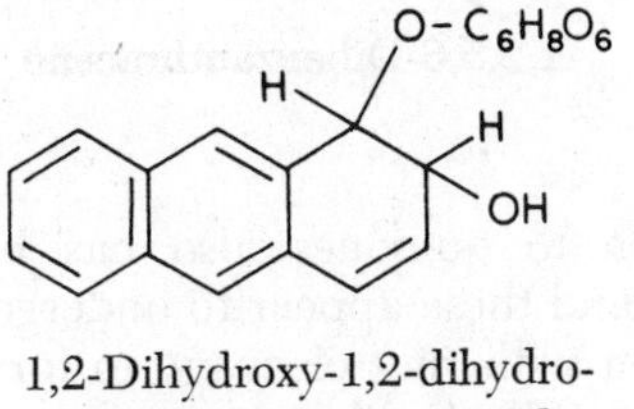

$$O-C_6H_8O_6$$

1,2-Dihydroxy-1,2-dihydro-
anthracene-1-glucuronide

form the premercapturic acids. Mercapturic acids are not normal metabolites, but are formed by dehydration of premercapturic acids upon exposure to an acidic environment during the usual extraction procedure from urine. It is not known whether acetylcysteine or cysteine is involved directly in the conjugation, or whether glutathione is the conjugating species, followed by hydrolysis and acetylation to form the premercapturic acids.

Anthracene is metabolized in the same general manner as naphthalene, leading to the formation of the excretion products 1-anthryl premercapturic acid and 1,2-dihydroxy-1,2-dihydroanthracene-1-glucuronide.

Acid-labile premercapturic acids have also been isolated from the urine of animals given benzene, and the monohalogenated benzenes. By reactions analogous to those of naphthalene, bromobenzene has been shown to form glucuronide and sulfate con-

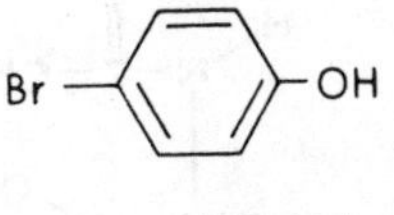

p-Bromophenol

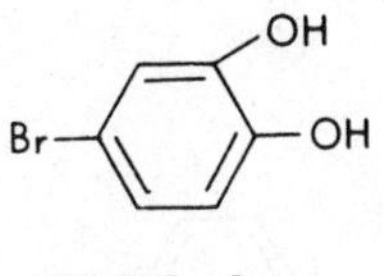

3,4-Dihydroxy-
bromobenzene

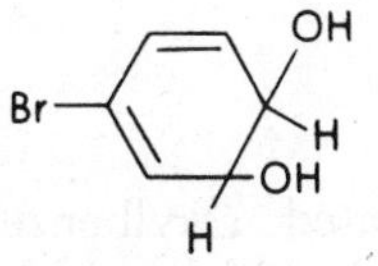

3,4-Dihydroxy-3,4-dihydro-bromobenzene

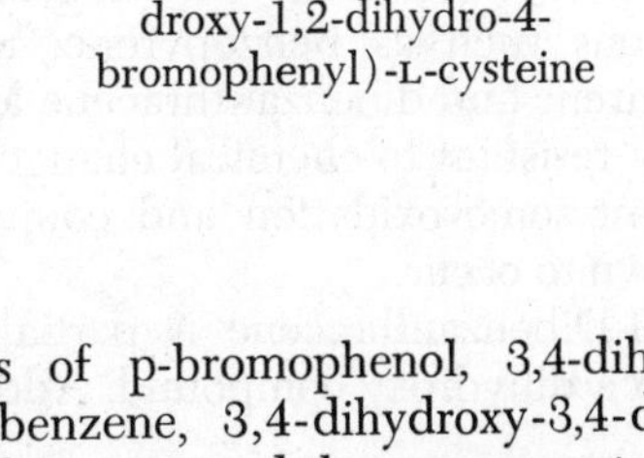

$$NH-COCH_3$$
$$\qquad | $$
$$S-CH_2\,C-COOH$$

N-Acetyl-S-(2-hy-
droxy-1,2-dihydro-4-
bromophenyl)-L-cysteine

jugates of p-bromophenol, 3,4-dihydroxybromobenzene, 3,4-dihydroxy-3,4-dihydrobromobenzene, and the premercapturic acid derivative, N-acetyl-S-(2-hydroxy-1,2-dihydro-4-bromophenyl)-L-cysteine.

Aromatic hydrocarbons containing a labile halogen appear to conjugate directly with acetylcysteine to form a mercapturic acid derivative. For example, 2,4-dichloronitrobenzene forms 3-chloro-4-nitrophenyl mercapturic acid.[63]

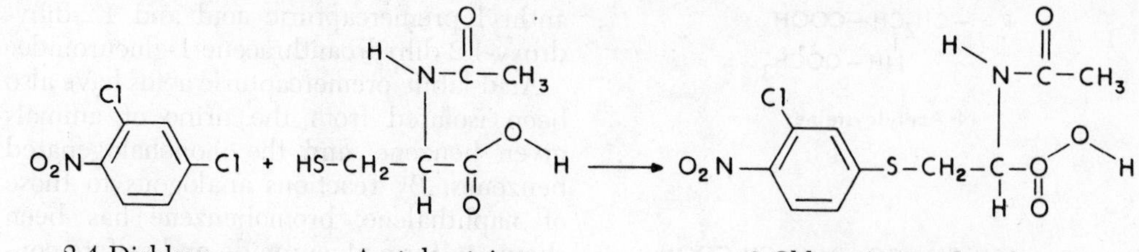

2,4-Dichloro-
nitrobenzene   Acetylcysteine   3-Chloro-4-nitrophenyl
                                          Mercapturic Acid

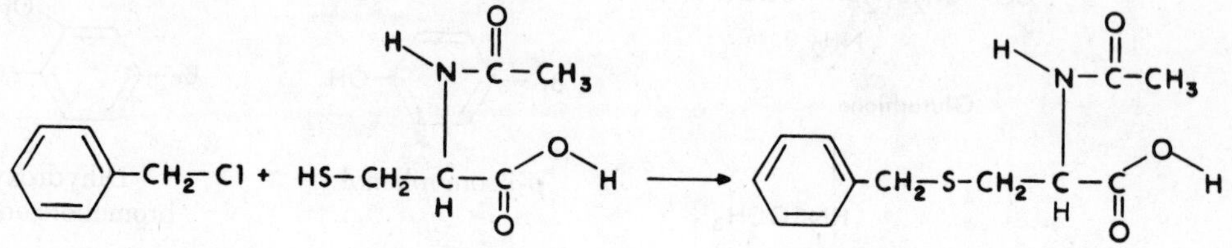

Benzyl Chloride   Acetylcysteine   S-Benzylmercapturic Acid

α-Halogenated alkylbenzenes, such as benzyl chloride and benzyl bromide conjugate directly with acetylcysteine and are excreted mainly as S-benzylmercapturic acid, without prior formation of the premercapturic acid.[64] ω-Halogenated alkylbenzenes such as phenylethyl bromide, 3-bromopropylbenzene, 4-bromobutylbenzene, although forming some mercapturic acid, undergo ring hydroxylation to an increasing extent as the halogen is further removed from the ring. The phenols thus formed may then undergo conjugation with glucuronic and sulfuric acids.

The cancer-producing (carcinogenic) hydrocarbons such as benzopyrene, methylcholanthrene and dibenzanthracene are presumably resistant to chemical change in the body, but some oxidation and conjugation are known to occur.

1,2,5,6-Dibenzanthracene is partially oxidized to a dihydroxy compound. Additional

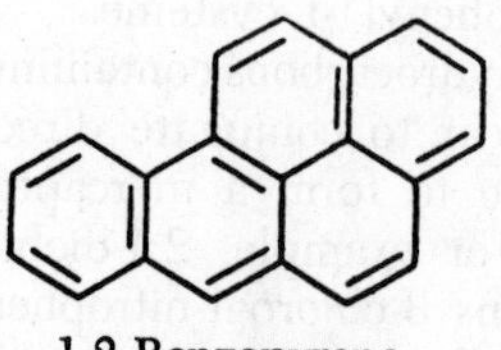

1,2,5,6-Dibenzanthracene

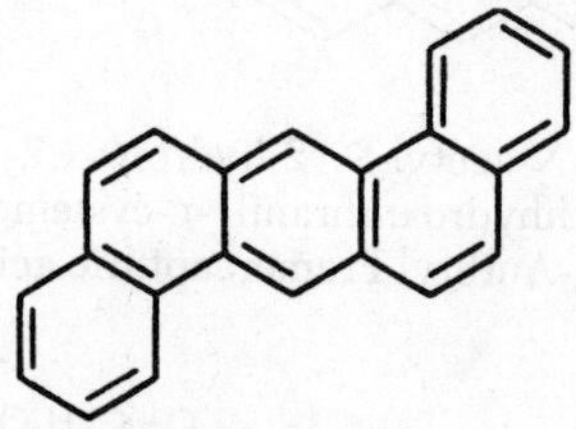

oxidation to quinones also has been observed, and these appear to undergo further oxidation with ring cleavage to form dicarboxylic acids. Unlike the noncarcinogenic hydrocarbons such as naphthalene, mercapturic acid metabolites have not been found with dibenzanthracene. The polar metabolites of carcinogens, so far as is known, are noncarcinogenic.

Metabolism of the anticonvulsant drug phenaceturea (Phenurone) is of special interest because of its extreme toxicity. In rabbits, only 7 per cent of the administered drug is excreted unchanged. Most of the drug is metabolized by aromatic hydroxylation and O-methylation to form 3-methoxy-4-hydroxyphenaceturea, accompanied by minor hydrolysis and conjugation reactions. It has been suggested that the aromatic hydroxyl metabolites may be responsible for some toxic effects of the drug.[65]

1,2-Benzopyrene

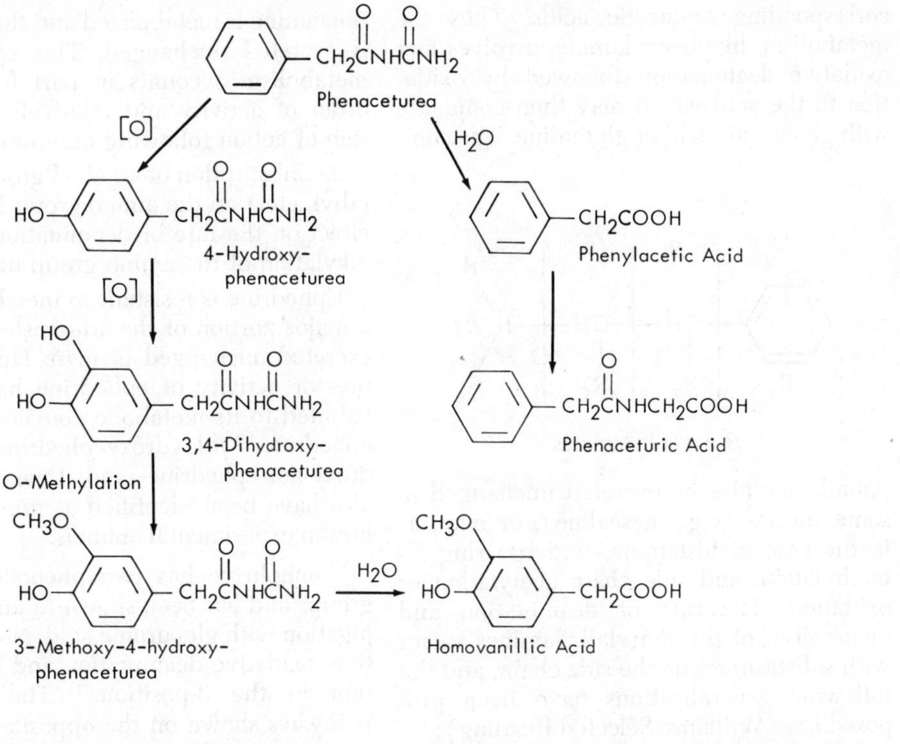
CH₂CNHCNH₂
Phenaceturea
[O]
H₂O
HO
CH₂CNHCNH₂
4-Hydroxy-
phenaceturea
CH₂COOH
Phenylacetic Acid
[O]
HO
HO
CH₂CNHCNH₂
3,4-Dihydroxy-
phenaceturea
CH₂CNHCH₂COOH
Phenaceturic Acid
O-Methylation
CH₃O
HO
CH₂CNHCNH₂
H₂O
3-Methoxy-4-hydroxy-
phenaceturea
CH₃O
HO
CH₂COOH
Homovanillic Acid

NITROGENOUS COMPOUNDS

Aliphatic Amines. The primary aliphatic amines are for the most part metabolized, although some (e.g., ethylamine) are mainly excreted unchanged. The metabolism first involves oxidative deamination by mono- and diamine oxidase which are normally present in the liver, kidney and intestinal mucosa. Following deamination they are oxidized to carboxylic acids and urea.

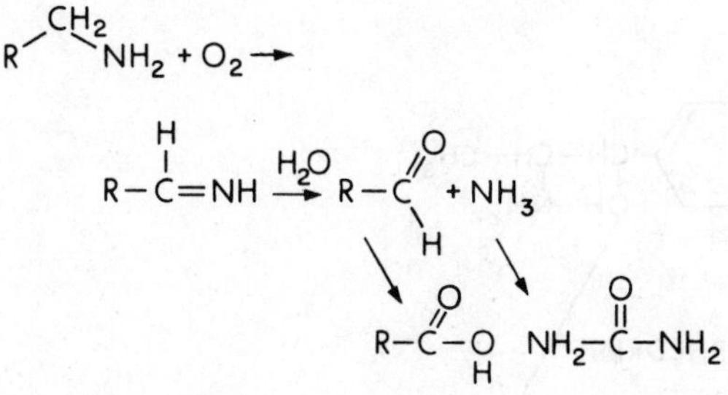
R
CH₂
NH₂ + O₂
R–C=NH
H₂O
R–C
+ NH₃
R–C–O
NH₂–C–NH₂

Benzylamine is oxidized in the dog to benzoic acid and excreted as hippuric acid. *p*-Hydroxybenzylamine undergoes a similar oxidation to *p*-hydroxybenzoic acid (see acids). However, N-acetyl-*p*-hydroxybenzylamine is not appreciably deacetylated and oxidized but is excreted mainly as the glucuronide ether.

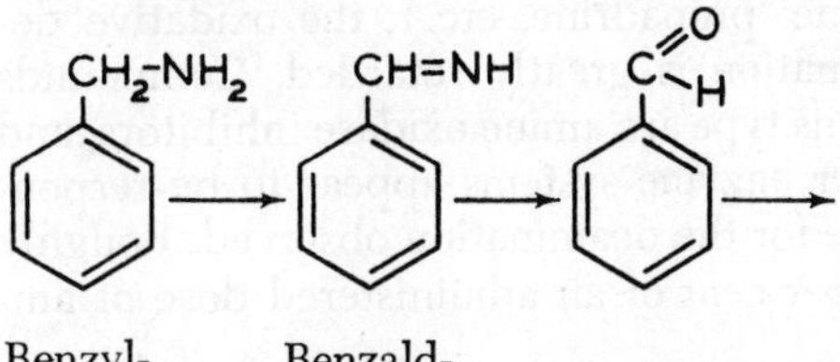
CH₂-NH₂
CH=NH
C
H

Benzyl-amine Benzald-imine Benzaldehyde

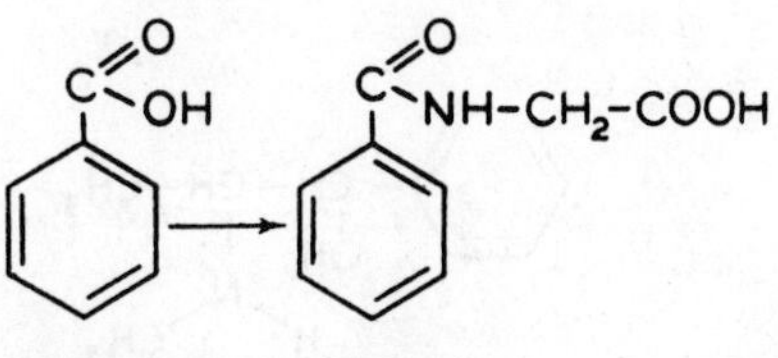
C
OH
C
NH-CH₂-COOH

Benzoic Acid Hippuric Acid

The β-arylalkylamines show significant differences in metabolism. The primary amines of this class such as β-phenylethylamine, tyramine, mescaline, β-indolethylamine and histamine, although showing species differences, are oxidized in part to the corresponding arylacetic acids. Thus the metabolism, like benzylamine, involves first oxidative deamination followed by oxidation to the acid which may then conjugate with glucuronic acid or glutamine. The com-

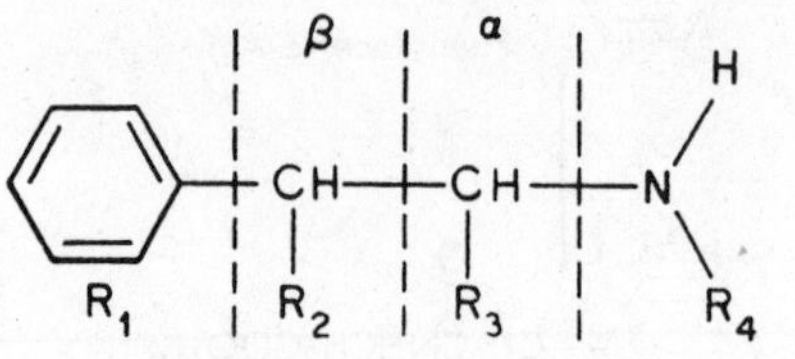

β-Arylalkylamines

pound may also be excreted unchanged in some species (e.g., mescaline), or may, as in the case of histamine, undergo ring N-methylation, and side chain acetylation or oxidation. The rate of deamination and metabolism of the β-arylalkylamines varies with substitutions on the side chain, and the following generalizations have been proposed (See Williams, Selected Reading):

1. Compounds in which a methyl group is substituted on the β-carbon R_2 show a decrease in the rate of oxidative deamination;

2. When a methyl group is substituted on the α-carbon R_3 (e.g., ephedrine, amphetamine, propadrine, etc.), the oxidative deamination is greatly retarded. Compounds of this type are amine oxidase inhibitors, and other enzyme systems appear to be responsible for the deamination observed. Roughly 50 per cent of an administered dose of am-

phetamine is metabolized and the remainder is excreted unchanged. This resistance to metabolism accounts in part for the high order of activity and relatively long duration of action following oral administration;

3. Substitution of an alkyl group (methyl, ethyl, etc.) on the amino group R_4 has little effect on the rate of deamination, since dealkylation of the amino group occurs.

Ephedrine is resistant to metabolism, and a major portion of the administered drug is excreted unchanged in man. However, the pressor activity of ephedrine has been attributed to its metabolic conversion to norephedrine.[66] *p*-Hydroxyephedrine and *p*-hydroxy-nor-ephedrine and their conjugates also have been identified as minor metabolites in experimental animals.

Epinephrine has two phenolic hydroxyl groups and has been shown to undergo conjugation with glucuronic acid, N-demethylation, oxidative deamination and O-methylation in the 3-position.[67] The metabolic pathways shown on the opposite page have been suggested.

Methyldopa (Aldomet), (−)-3-(3,4-dihydroxyphenyl)-2-methylalanine, a drug widely used for the treatment of hypertension, is metabolized in a manner similar to that of the structurally related epinephrine except that side chain oxidation is stopped at the ketone stage due to the presence of a branched methyl group. Sulfate or glucuronide conjugates of the phenolic groups also form (see facing page).

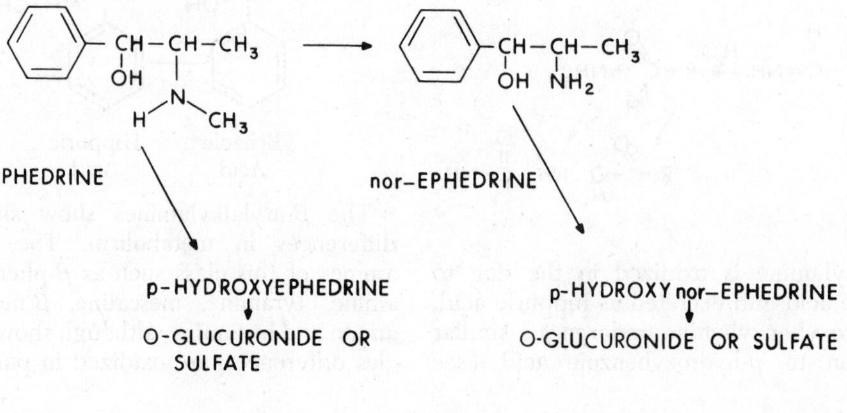

EPHEDRINE

nor–EPHEDRINE

p–HYDROXYEPHEDRINE
↓
O–GLUCURONIDE OR SULFATE

p–HYDROXY nor–EPHEDRINE
↓
O–GLUCURONIDE OR SULFATE

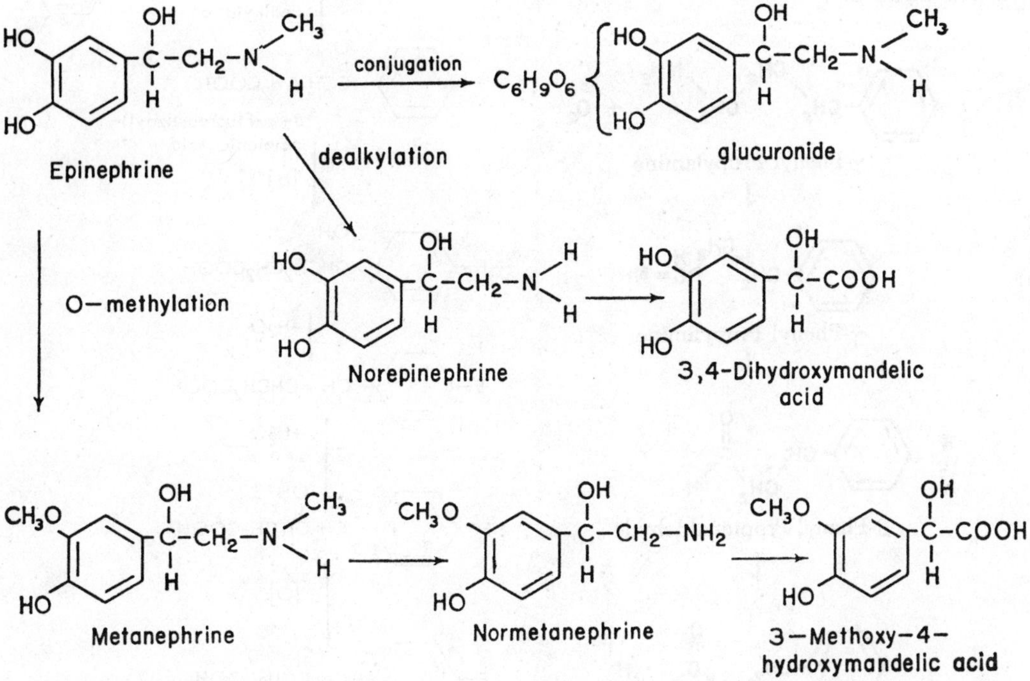

Epinephrine

conjugation

$C_6H_9O_6$ { glucuronide

dealkylation

O–methylation

Norepinephrine

3,4-Dihydroxymandelic acid

Metanephrine

Normetanephrine

3–Methoxy–4–hydroxymandelic acid

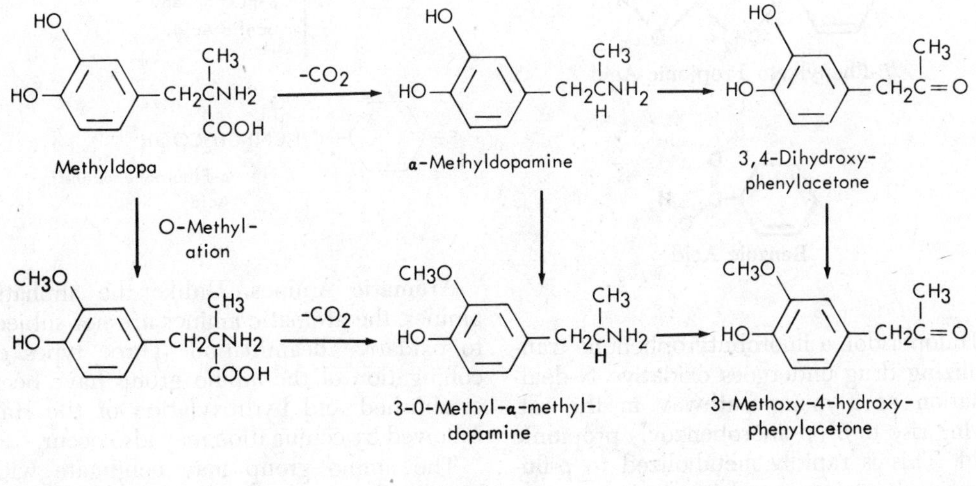

Methyldopa

O–Methylation

$-CO_2$

α–Methyldopamine

3,4-Dihydroxyphenylacetone

$-CO_2$

3-0-Methyl-α-methyldopamine

3-Methoxy-4-hydroxyphenylacetone

γ-Phenyl propylamine also has been shown to be oxidized to benzoic acid, and this may be explained in the same manner, the final step involving β-oxidation (see carboxylic acids).

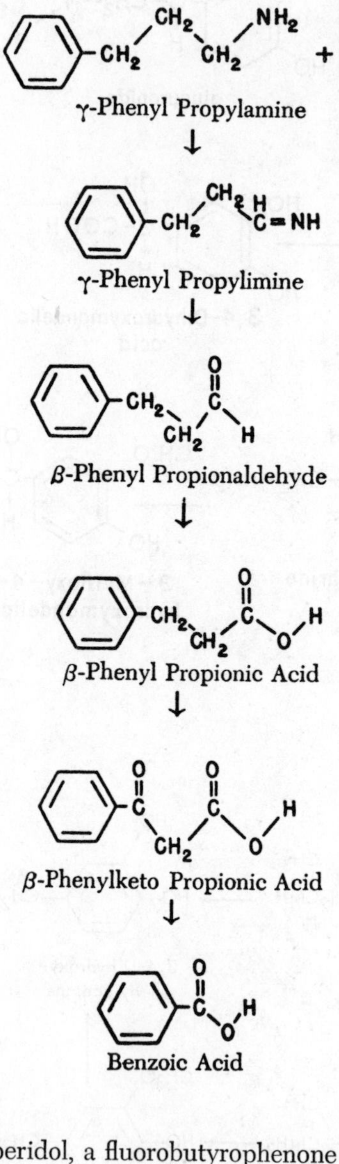

γ-Phenyl Propylamine

↓

γ-Phenyl Propylimine

↓

β-Phenyl Propionaldehyde

↓

β-Phenyl Propionic Acid

↓

β-Phenylketo Propionic Acid

↓

Benzoic Acid

Haloperidol

Oxidative dealkylation

β-(p-Fluorobenzoyl)-propionic Acid

[H]

-H$_2$O

+H$_2$O

[O]

[O]

p-Fluorophenyl-acetic Acid

p-Fluorophenaceturic Acid

Haloperidol, a fluorobutyrophenone tranquilizing drug undergoes oxidative N-dealkylation as a major pathway in the rat, giving rise to β-(p-fluorobenzoyl) propionic acid. This is rapidly metabolized to p-fluorophenylacetic acid and its glycine conjugate, p-fluorophenaceturic acid, presumably by the unique route shown below.[68, 69]

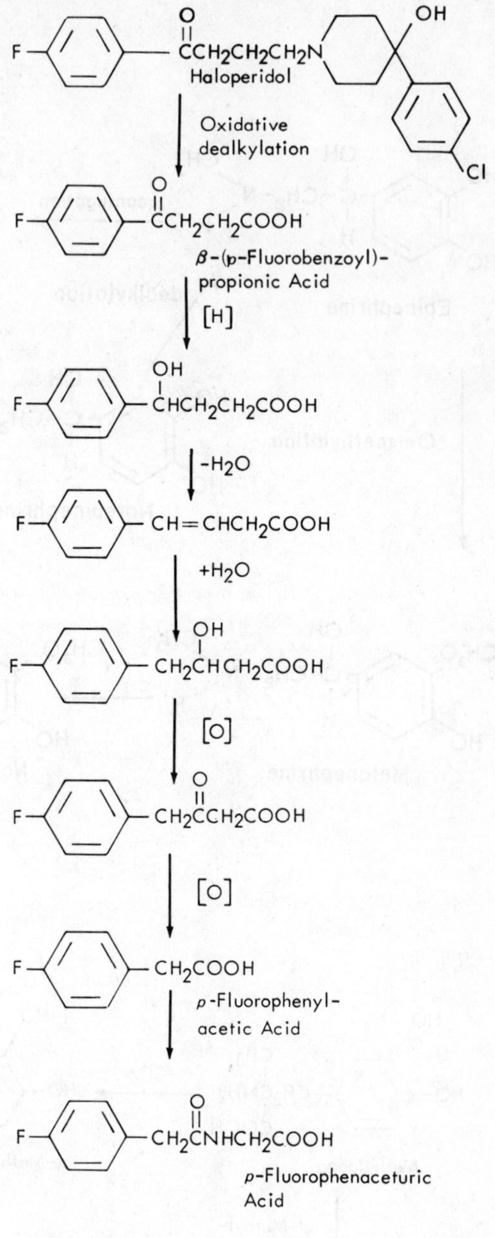

Aromatic Amines. Unlike the aliphatic amines, the aromatic amines are not subject to oxidative deamination. Three types of conjugation of the amino group have been established and hydroxylation of the ring followed by conjugation may also occur.

The amino group may conjugate with acetic, glucuronic or sulfuric acids to form N-acetylated, N-glucurono and N-sulfonic (sulfamic acid) derivatives. The first N-glucuronide was discovered as a labile metabolite of aniline.[70] The N-glucuronides are not hydrolyzed by β-glucuronidase but they are rapidly hydrolyzed in acidic solutions. It is reasonable to believe they may be a common metabolite of the primary aromatic amines. Acetanilid, acetophenetidin, acetotoluidines and other derivatives of aromatic amines have been observed to be excreted in part as N-glucuronides. A large number of aromatic amines have been observed to undergo acetylation in vivo (see sulfonamides). Sulfamate formation ($-NHSO_2-OH$), has been demonstrated and may be regarded as a common metabolic pathway for aromatic amines. 2-Naphthylamine sulfamate is one of the metabolites formed from 2-naphthylamine.

Hydroxylation of the ring also is a common metabolic pathway of the aromatic amines. In aniline the hydroxylation takes place *ortho* or *para* to the amino group. The hydroxyl group introduced usually conjugates to form glucuronides or sulfate esters.

Acetanilid is mainly oxidized in the body to N-acetyl-*p*-aminophenol[71] which is excreted 70 to 85 per cent conjugated as the sulfuric acid ester or the glucuronide. N-Acetyl-*p*-aminophenol is itself an effective analgesic and the analgesic action of acetanilid is thought to be largely due to this oxidation product.

Aromatic amino compounds have been observed to form methemoglobin, a modified form of oxyhemoglobin, in which the oxygen is held so firmly that it does not function in respiration. Acetanilid, phenacetin and other similar compounds form methemoglobin, and the following mechanism involving metabolic products has been proposed. The *p*-aminophenol undergoes oxidation forming a quinoneimine which, by means of a redox system, is able to form methemoglobin and produce methemoglobinemia.

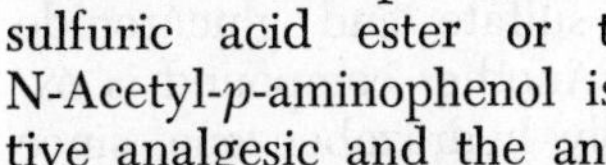

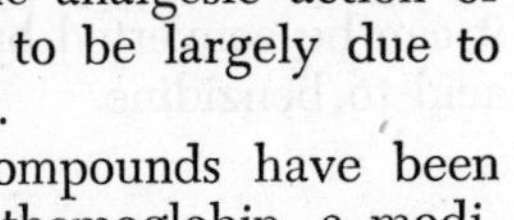

Quinoneimine

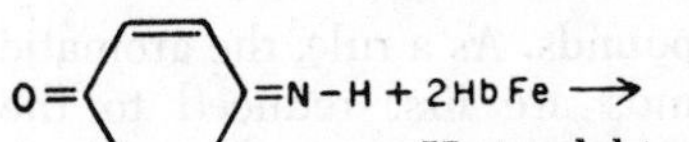

Hemoglobin

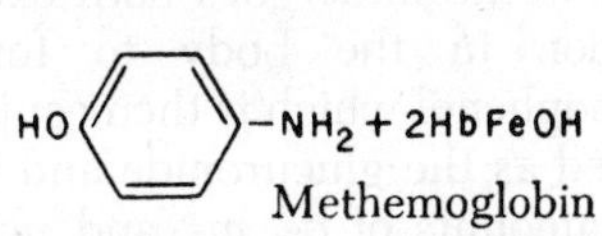

Methemoglobin

Since the aminophenol functions as a catalyst, small amounts can transform a large amount of hemoglobin to methemoglobin. Substances which produce methemoglobin also are said to cause porphyrinuria. Compounds which can be converted in the body to *o*- or *p*-aminophenols may be responsible for methemoglobin formation.

Many alkyl substituted aliphatic, aromatic and heterocyclic amines have been observed to undergo biologic dealkylation. Some amino compounds dealkylate prior to oxidative deamination but many secondary and tertiary amines also dealkylate and the compound containing the free amino group is excreted unchanged. The rate of dealkylation is approximately three times greater for the secondary N-methylamines than for the corresponding mono-substituted ethyl, propyl and butyl amines. The tertiary amines with two alkyl groups may also be dealkylated (step-wise) to a primary amine but at a slower rate than the corresponding secondary amines (see Williams, Selected Reading). The dealkylation of dimethylbenzylamine will serve to illustrate the N-dealkylation process.

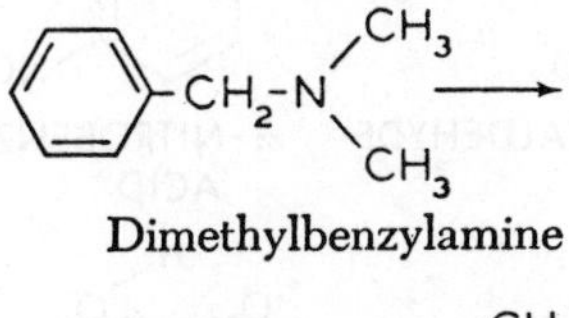

Dimethylbenzylamine

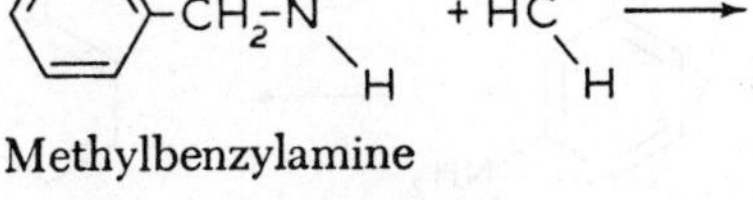

Methylbenzylamine

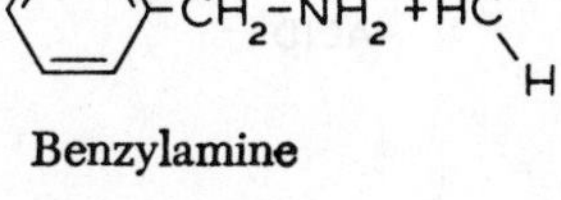

Benzylamine

Nitro Compounds. As a rule, the aromatic nitro compounds are first reduced to the amino derivative and then are acetylated. Nitrobenzene undergoes both oxidation and reduction in the body to form mainly *p*-aminophenol which is then conjugated and excreted as the glucuronide and the sulfate. Small amounts of *o*-, *m*-, and *p*-nitrophenol and *o*-, and *m*-aminophenol also have been found as metabolites in the rabbit. In general, the nitrophenols tend to be reduced to the corresponding aminophenols, which may then conjugate with glucuronic acid.

One of the few naturally occurring nitro compounds, chloramphenicol (see alcohols, this chapter), is principally metabolized by reactions involving an alcoholic hydroxyl. Some reduction of the aromatic nitro to an aromatic amino group occurs, but this appears to play a minor metabolic role.

Aromatic compounds containing more than one nitro group (e.g., dinitrophenol, picric acid, etc.) have only one of the nitro groups reduced. For example, picramic acid

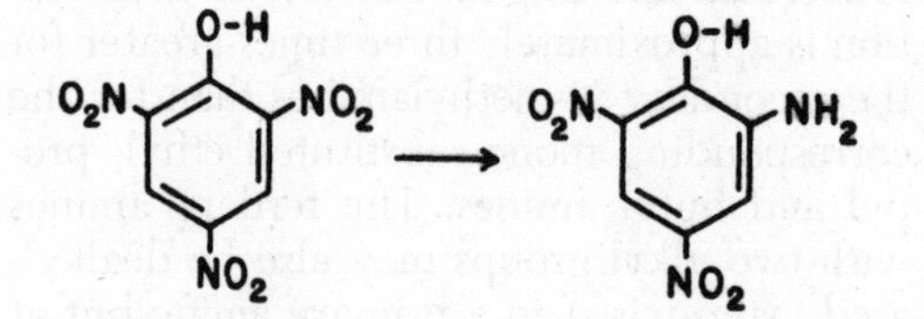

is the main metabolic product of picric acid.

Aromatic nitro compounds, having substituents easily oxidized, may be expected to undergo both oxidation and reduction. Thus, the mononitrobenzaldhydes are oxidized to the corresponding nitrobenzoic acids. The meta and para isomers are then conjugated in part with glycine and excreted as hippurates. It also has been shown that the *m*-nitrobenzaldehyde is oxidized and reduced to *m*-aminobenzoic acid, which then is excreted in part as the acetyl derivative and in part as *m*-acetylaminohippuric acid. The mononitrophenylacetic acids are excreted by man partially unchanged.

Azo Compounds. In general, these are reduced, giving rise to two aromatic amino groups. The amines thus formed then may undergo further change, such as oxidation and acetylation. Reference has been made to the reduction of the prontosils to sulfanilamide and the reductive cleavage of the azo groups in 4-dimethylaminoazobenzene (Butter Yellow). Azobenzene is characteristic of the metabolism of this type compound and it is converted to aniline which is hydroxylated in the para postion and excreted as the sulfate and glucuronide (see acetanilid). Another compound is excreted, presumably hydrazobenzene, since it can be converted by strong hydrochloric acid to benzidine.

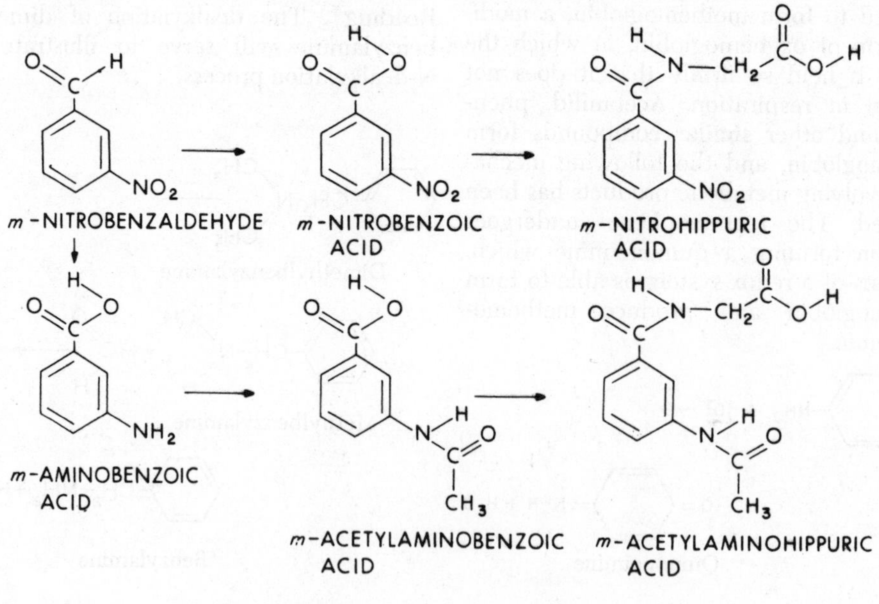

Picric Acid   Picramic Acid

m-NITROBENZALDEHYDE   *m*-NITROBENZOIC ACID   *m*-NITROHIPPURIC ACID

m-AMINOBENZOIC ACID   *m*-ACETYLAMINOBENZOIC ACID   *m*-ACETYLAMINOHIPPURIC ACID

CARBOXYLIC ACID AMIDES
AND CARBAMATES

The carboxylic acid amides are hydrolyzed largely to the free acids. The final metabolic products will depend therefore on the specific acids and amines formed. For example, benzamide is excreted largely as hippuric acid and phenylacetamide as phenaceturic acid. The enzyme benzamidase which occurs in animal tissues catalyzes the reaction

$$C_6H_5COOH + NH_3 \rightleftarrows C_6H_5CONH_2 + H_2O$$

Therefore the enzyme may be involved either in the production or the hydrolysis of amides in the body.

Salicylamide has been shown to be excreted by dogs and rabbits partly as a sulfate ester and as salicyluric acid. The latter is formed following the hydrolysis of the amide to the acid and then conjugation with glycine. The main urinary metabolite of salicylamide, after oral administration to cancer patients, is the glucuronide of salicylamide. A small amount of the original drug was recovered but no other metabolites have been detected.

Like the amides, the related carbamates may undergo hydrolysis. However, they are sufficiently resistant to hydrolysis so that most of the metabolites identified involve attack on other portions of the molecule.

The metabolism of meprobamate, a dicarbamate, remains uncertain, but in man only 10 per cent of the drug is excreted unchanged. In the dog it is metabolized to give hydroxymeprobamate derivatives, one derived from the oxidation of the methyl group and the second from the oxidation of the propyl group. The former (2-hydroxymethyl-2-n-propyl propane-1,3-diol dicarbamate) lacks CNS depressant activity. The hydroxypropyl derivative has not been fully characterized, but the metabolite appears to be primarily the 2-hydroxypropyl derivative.[72] Both the hydroxymethyl and the hydroxypropyl derivatives are excreted in part as the O-glucuronides. There is also evidence of the formation of an N-glucuronide which, following hydrolysis, gives free meprobamate.[73, 74] Recently, it has been shown that the N-mono-glucuronide is relatively

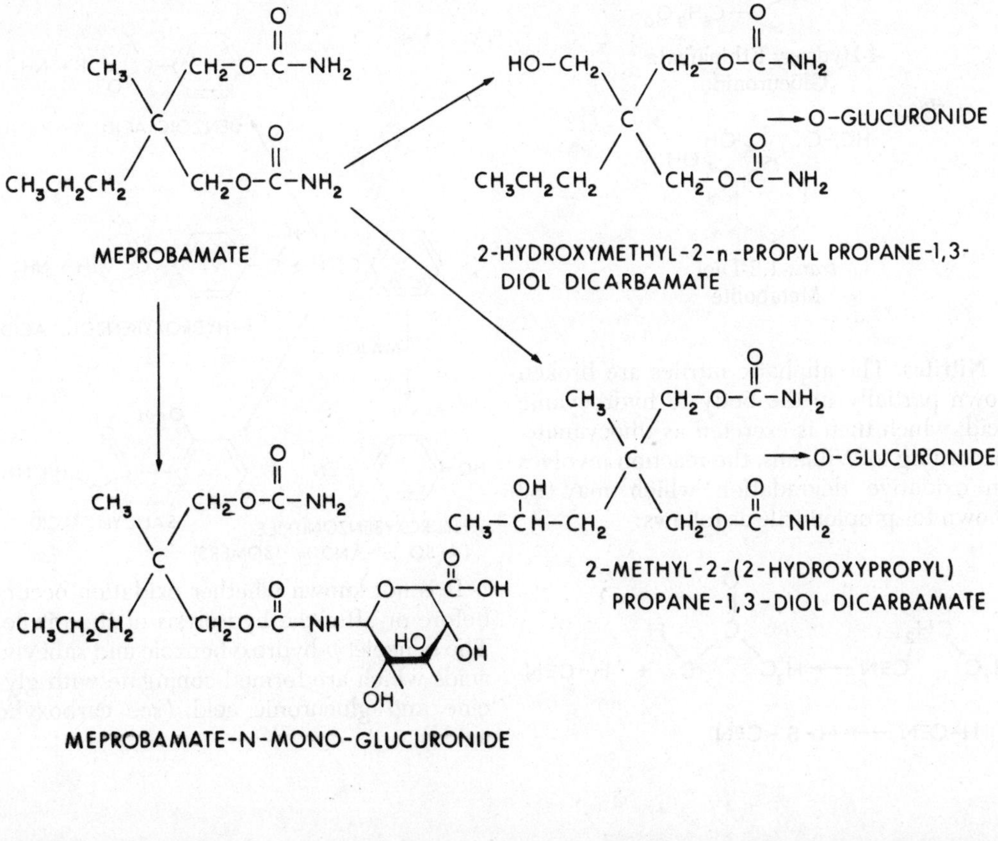

MEPROBAMATE

2-HYDROXYMETHYL-2-n-PROPYL PROPANE-1,3-DIOL DICARBAMATE

2-METHYL-2-(2-HYDROXYPROPYL) PROPANE-1,3-DIOL DICARBAMATE

MEPROBAMATE-N-MONO-GLUCURONIDE

stable to acid and base hydrolysis and it is claimed to be the principal metabolic product of meprobamate in man.[75]

Ethinamate (Valmid) is a CNS depressant carbamate which has been shown to be hydroxylated in the cyclohexyl ring in both the 2- and 4-positions and to undergo hydrolysis and conjugation. The glucuronide of 1-ethynyl-4-hydroxycyclohexyl carbamate (4-hydroxy ethinamate glucuronide) and 1-ethynyl-*trans*-1,2-cyclohexanediol have been isolated as metabolites in man.[76]

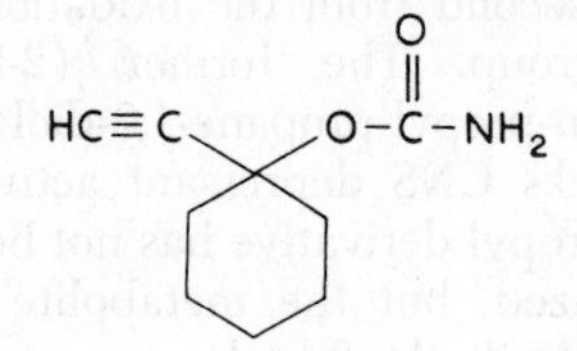

Ethinamate

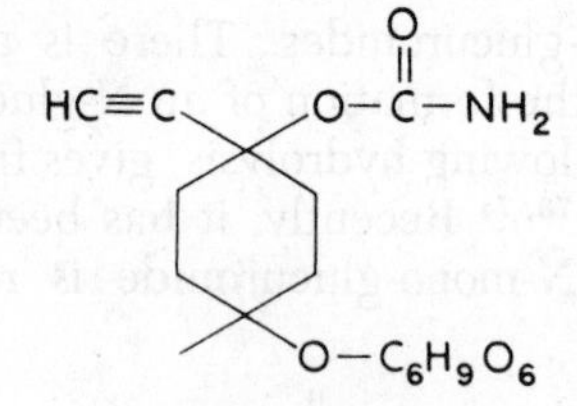

4-Hydroxy Ethinamate
Glucuronide

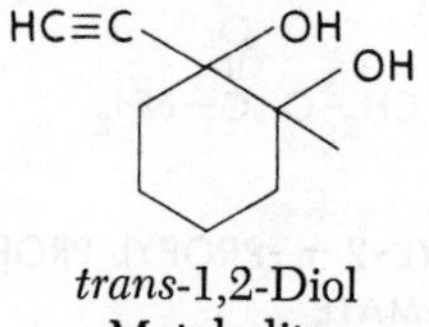

trans-1,2-Diol
Metabolite

Nitriles. The aliphatic nitriles are broken down partially in the body to hydrocyanic acid, which then is excreted as thiocyanate. According to Williams, the reaction involves an oxidative degradation which may be shown for propionitrile as follows:

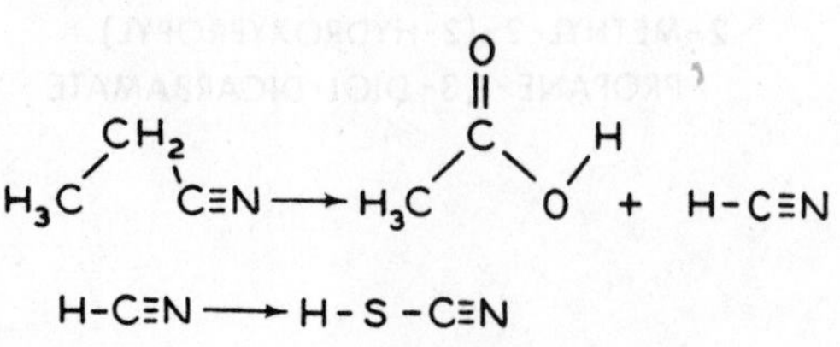

In the body, the formation of thiocyanic acid from hydrocyanic acid is brought about by the enzyme rhodanese, which is widely distributed in animal tissue. Lang[77] has shown that thiocyanic acid is formed from hydrocyanic acid and sodium thiosulfate in vitro and that the reaction does not require oxygen.

$$H-C{\equiv}N + Na_2S_2O_3 \longrightarrow HS-C{\equiv}N + Na_2SO_3$$

In addition to splitting off HCN, the aliphatic nitriles undergo some hydrolysis to the corresponding acids.

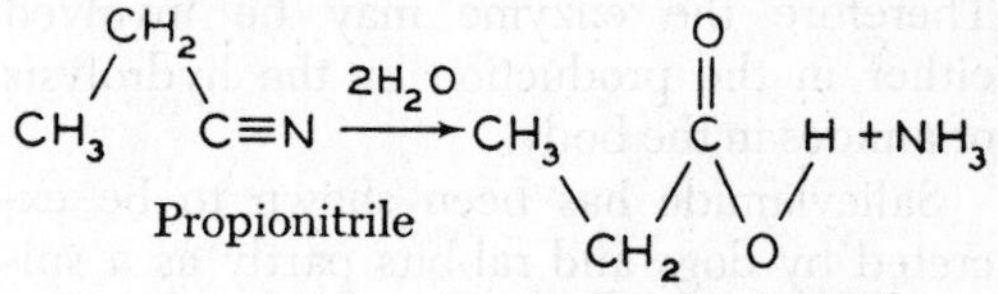

Propionitrile   Propionic Acid

The aromatic nitriles appear to undergo primarily hydroxylation and, to a lesser extent, hydrolysis with or without oxidation. Williams[78] has reported that benzonitrile undergoes in part the following changes in the body.

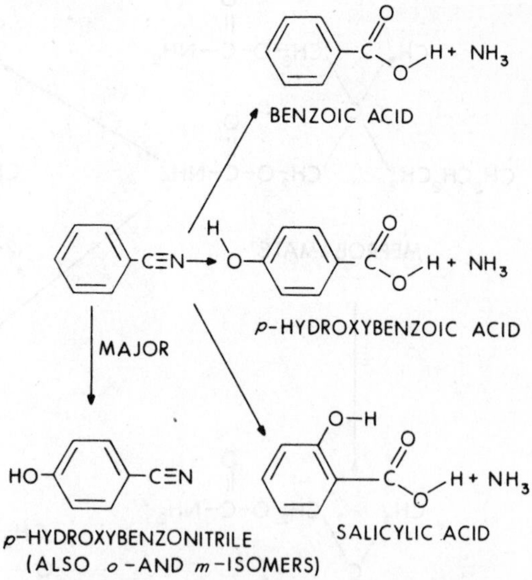

BENZOIC ACID

p-HYDROXYBENZOIC ACID

MAJOR

SALICYLIC ACID

p-HYDROXYBENZONITRILE
(ALSO *o*-AND *m*-ISOMERS)

It is not known whether oxidation occurs before or after the hydrolysis of the nitrile. The benzoic, *p*-hydroxybenzoic and salicylic acids which are formed conjugate with glycine and glucuronic acid (see carboxylic acids).

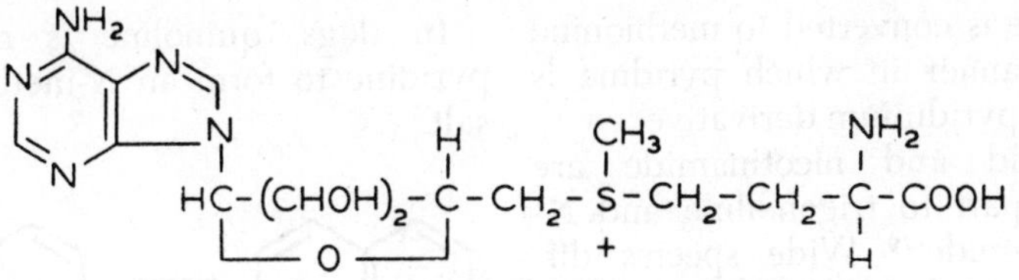

S-Adenosylmethionine

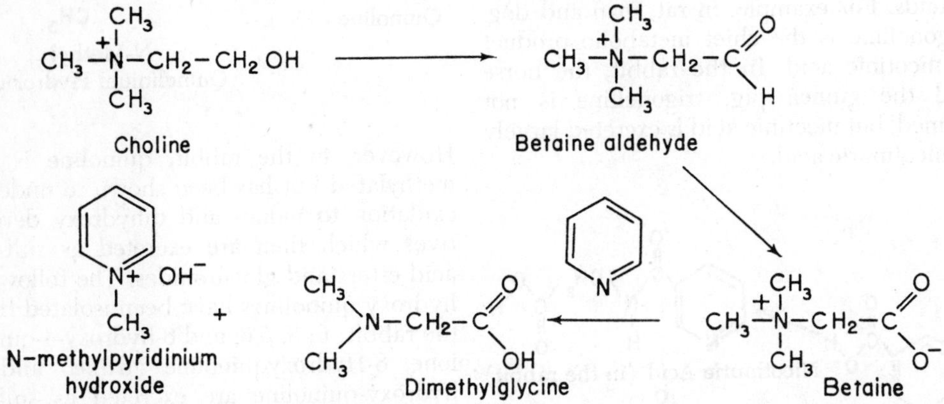

Choline → Betaine aldehyde

N-methylpyridinium hydroxide + Dimethylglycine ← Betaine

HETEROCYCLIC NITROGEN COMPOUNDS

A number of heterocyclic amines are known to undergo methylation in the body, and the methylation process is an important metabolic pathway. To illustrate, pyridine is excreted as a salt of N-methyl pyridinium hydroxide, quinoline as a salt of N-methyl quinolinium hydroxide and nicotinic acid is excreted in part as trigonelline.

Betaine and S-adenosylmethionine have been established as the main methylating agents in the biological methylation process. It should be noted that the methyl groups in both compounds are attached to an "onium" atom (quaternary nitrogen or tertiary sulfur) and it is believed all methyl donors must have similar structures.

Dietary choline is the biological source of betaine, and the above reactions show the main chemical changes involved in the conversion of the relatively toxic pyridine into the nontoxic pyridinium compound.

Dimethylglycine is not a methyl donor but is reconverted to betaine in a cyclic series of reactions.

Besides pyridine, other commonly recognized methyl acceptors are: ethanolamine, methylethanolamine, dimethylethanolamine, nicotinamide, homocysteine and norepinephrine.

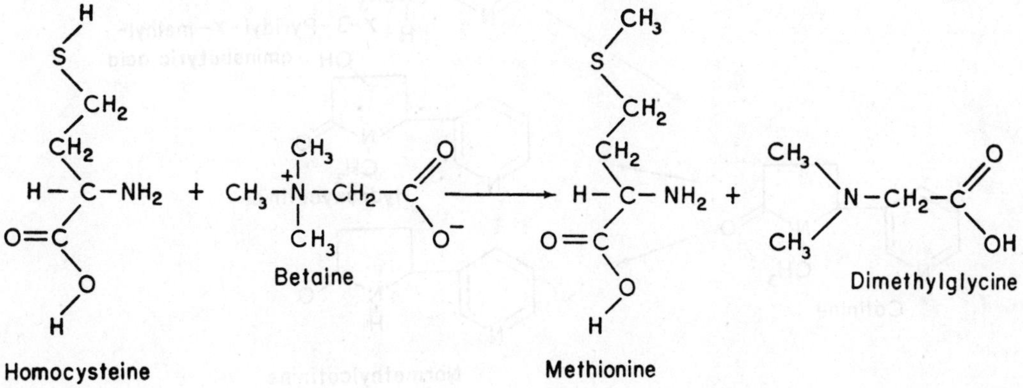

Homocysteine + Betaine → Methionine + Dimethylglycine

Homocysteine is converted to methionine in the same manner in which pyridine is converted to its pyridinium derivative.

Nicotinic acid and nicotinamide are methylated in part to trigonelline and N-methyl nicotinamide.[79] Wide species differences have been observed for the methylation process of heterocyclic nitrogen compounds. For example, in rat, man and dog, trigonelline is the chief metabolic product of nicotinic acid. In the rabbit, the horse and the guinea pig, trigonelline is not formed, but nicotinic acid is excreted largely as nicotinuric acid.

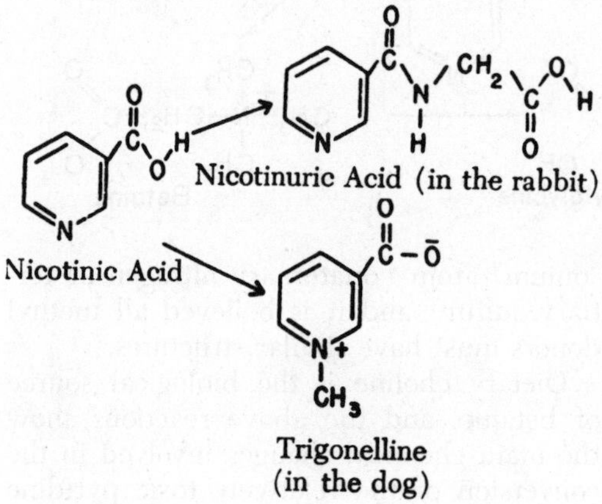

In dogs, quinoline is methylated like pyridine to form an N-methyl quinolinium salt.

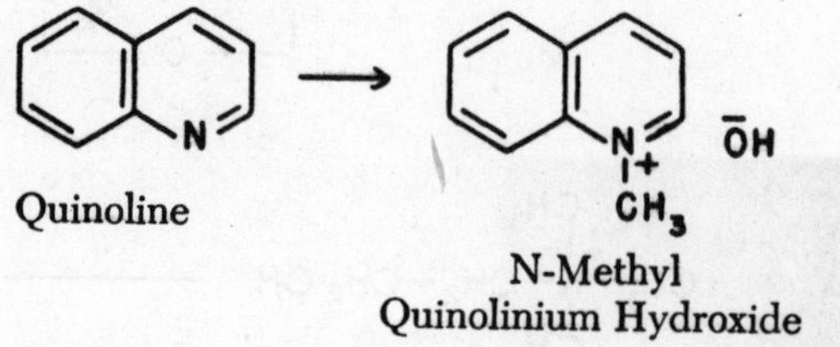

Quinoline → N-Methyl Quinolinium Hydroxide

However, in the rabbit, quinoline is not methylated but has been shown to undergo oxidation to mono- and dihydroxy derivatives which then are excreted as sulfuric acid esters and glucuronides. The following hydroxy quinolines have been isolated from the rabbit: 6; 8; 5,6; and 6-hydroxy-4-quinolone. 8-Hydroxyquinoline (oxine) and 4-hydroxy-quinoline are excreted as sulfate esters or glucuronides.

Nicotine and Related Pyridine Derivatives. Five metabolic products of nicotine have been isolated from dog urine[80, 81] and the same products, with the exception of N-methylnicotine, have been detected in the urine of man. The products isolated from dog urine are shown below.

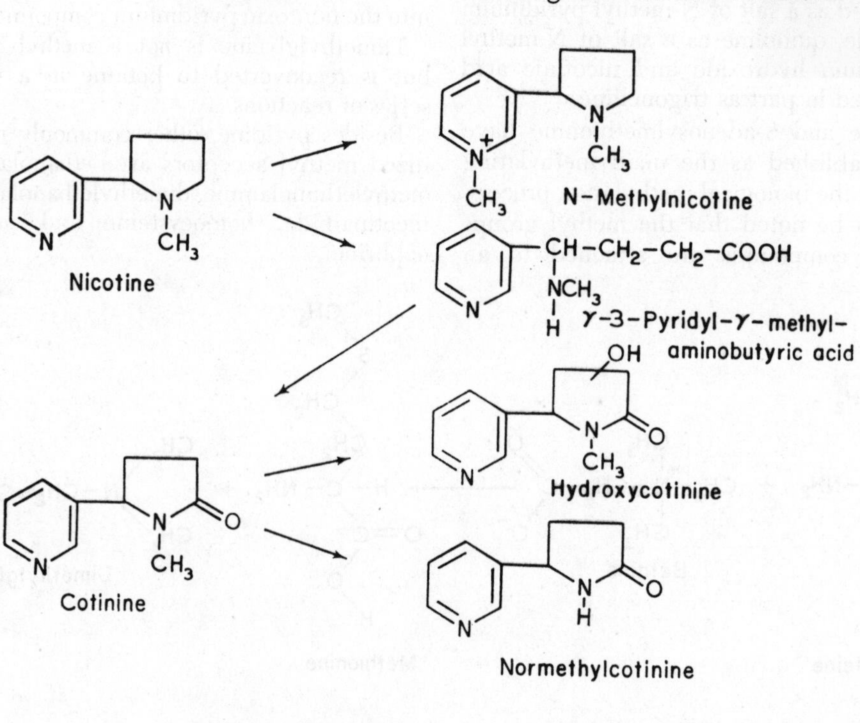

Nicotinuric Acid (in the rabbit)

Nicotinic Acid

Trigonelline (in the dog)

Nicotine

N-Methylnicotine

γ-3-Pyridyl-γ-methyl-aminobutyric acid

Cotinine

Hydroxycotinine

Normethylcotinine

The position of the hydroxyl group in hydroxycotinine has not been determined. N-Methylcotinine has been isolated from dog urine following the administration of cotinine and may therefore be a metabolite of nicotine.

Isoniazid (isonicotinic acid hydrazide, INH), a widely used antitubercular drug, is inactivated by metabolic acetylation, hydrolysis, N-methylation, and the formation of substituted hydrazones. Acetylation is the major pathway, and, since the rate of acetylation may vary markedly in individuals (see under genetic factors, p. 72), response to the drug may also vary due to differences in rates of metabolic inactivation.

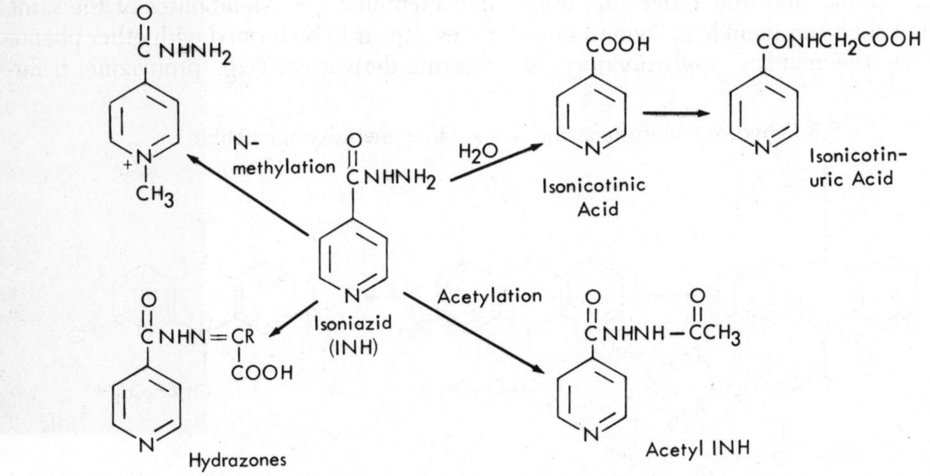

changed, 55 per cent as the carbostyril II and 22 per cent as the dihydroxy derivative III. With cinchonidine, 20 per cent was unchanged, 60 per cent converted to the carbostyril II and 20 per cent converted to the dihydroxy compound III.

Quinine and quinidine gave only small

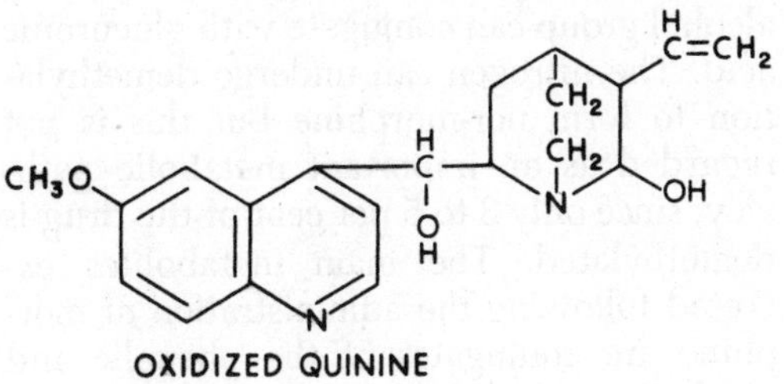

OXIDIZED QUININE

Isonicotinic Acid

Isonicotinuric Acid

Hydrazones

Isoniazid (INH)

Acetyl INH

Cinchona Alkaloids. The cinchona alkaloids contain a quinoline and quinuclidine ring. The principal of metabolic products obtained from cinchonine, cinchonidine, quinine and quinidine are hydroxy derivatives of the alkaloids with the hydroxyl alpha to the nitrogen.[82]

Following the administration of cinchonine to man 4 per cent was recovered un-

amounts of the carbostyril and the main metabolic products had one or two hydroxy groups on the quinuclidine ring.

Morphine and Codeine. Because of addiction liability, the biologic fate of this group of compounds has received much attention and is of wide interest. Only the major metabolites of morphine and codeine will be given. A detailed discussion of the metabo-

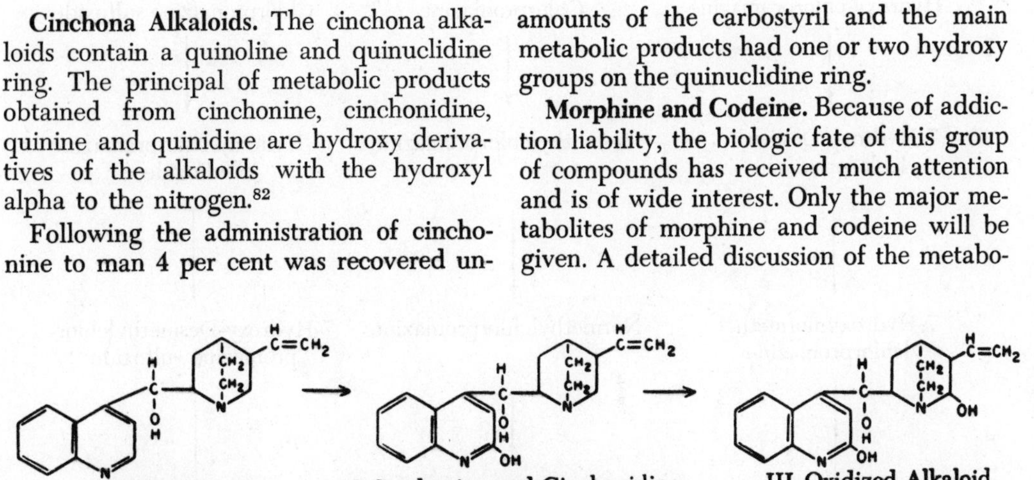

I Cinchonine and Cinchonidine

II Cinchonine and Cinchonidine Carbostyril

III Oxidized Alkaloid Carbostyril

lism of the narcotic analgesics is given by Way and Adler.[83]

The three functional groups primarily involved in the metabolism of morphine include a phenolic hydroxyl, a secondary hydroxyl and an N-methyl. The phenolic hydroxyl group can conjugate with glucuronic and sulfuric acid, the secondary alcohol group can conjugate with glucuronic acid. The nitrogen can undergo demethylation to form nor-morphine but this is not regarded as an important metabolic pathway, since only 3 to 5 per cent of the drug is demethylated. The main metabolites excreted following the administration of morphine are conjugates of the phenolic and alcoholic hydroxyl groups. The conjugates are quite stable and the older literature designated such compounds as "bound morphines." In the monkey approximately 70 per cent of the administered dose is conjugated and up to 12 per cent of the drug is excreted unchanged.

Codeine is the 3-methyl ether of morphine and could be expected to undergo both O-demethylation (see ethers, this chapter) and N-demethylation as well as conjugation with the unsubstituted alcoholic hydroxyl. All of these reactions take place, but the main metabolic pathway is conjugation with glucuronic acid. Approximately 50 per cent of an administered dose of codeine is excreted by man as the glucuronide.

Phenothiazines and Related Compounds. The metabolism of chlorpromazine has received much attention, and many of its hypothesized metabolites have been isolated and identified.[84, 85] Metabolites of the same types appear to be formed with other phenothiazine derivatives (e.g., promazine, triflu-

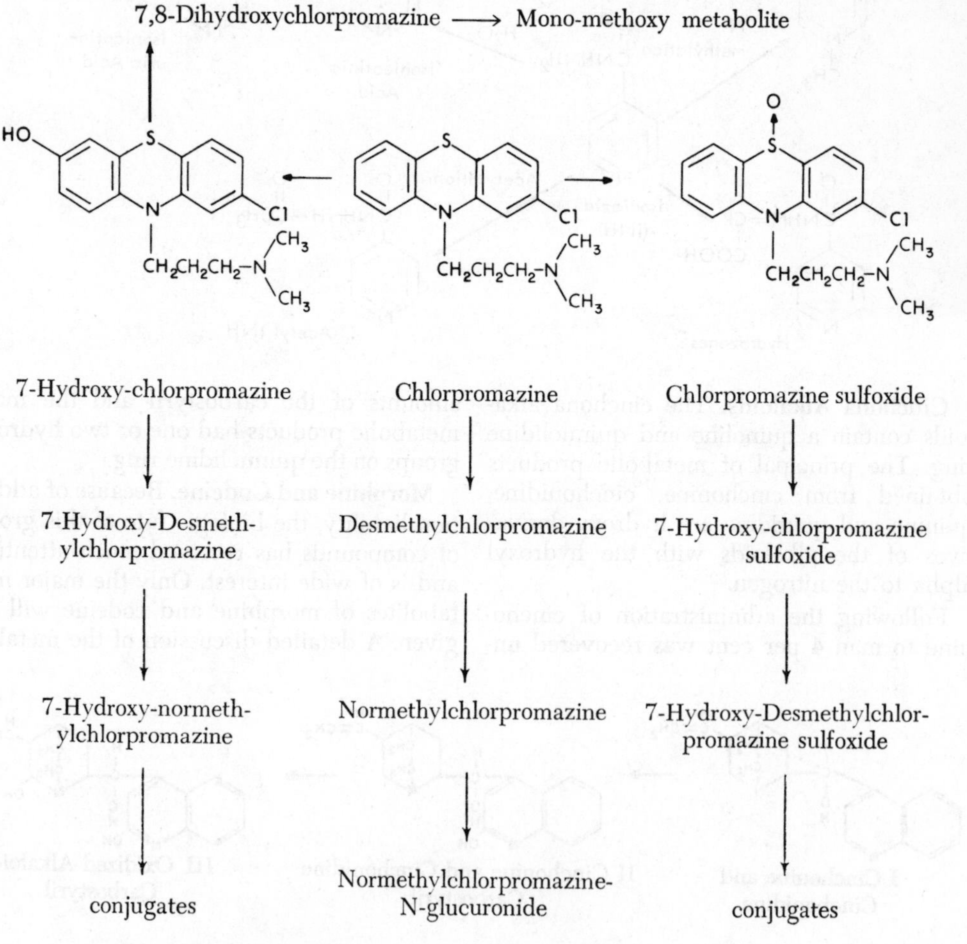

7,8-Dihydroxychlorpromazine ⟶ Mono-methoxy metabolite

| 7-Hydroxy-chlorpromazine | Chlorpromazine | Chlorpromazine sulfoxide |

| 7-Hydroxy-Desmethylchlorpromazine | Desmethylchlorpromazine | 7-Hydroxy-chlorpromazine sulfoxide |

| 7-Hydroxy-normethylchlorpromazine | Normethylchlorpromazine | 7-Hydroxy-Desmethylchlorpromazine sulfoxide |

| conjugates | Normethylchlorpromazine-N-glucuronide | conjugates |

promazine), and chlorpromazine serves as a model drug to illustrate the metabolism. Wide species variation in sulfoxidation of the phenothiazine nucleus has been observed, but in man this appears to be a minor pathway which may account for 5 per cent or less of the metabolites. The major metabolic pathways involve ring hydroxylation and side chain dealkylation, and these may occur in the same molecule, along with sulfoxide formation. 7-Hydroxychlorpromazine, 7-hydroxychlorpromazine sulfoxide, 7-hydroxy-desmethylchlorpromazine and 7-hydroxy-normethylchlorpromazine and their glucuronic acid conjugates have

been characterized[86] among the phenolic metabolites of chlorpromazine in man and in the dog. 7-Hydroxychlorpromazine has been implicated[87] in photosensitive reactions, such as purple skin pigmentation and corneal opacities which have developed in patients receiving large doses of chlorpromazine over long periods of time.

An in-vitro study of the metabolism of 7-hydroxychlorpromazine by rat liver microsomes showed conversion to 7,8-dihydroxychlorpromazine, followed by formation of a mono-methoxy derivative (O-methylation), a reaction similar to that of epinephrine and related catecholamines. The dihydroxy

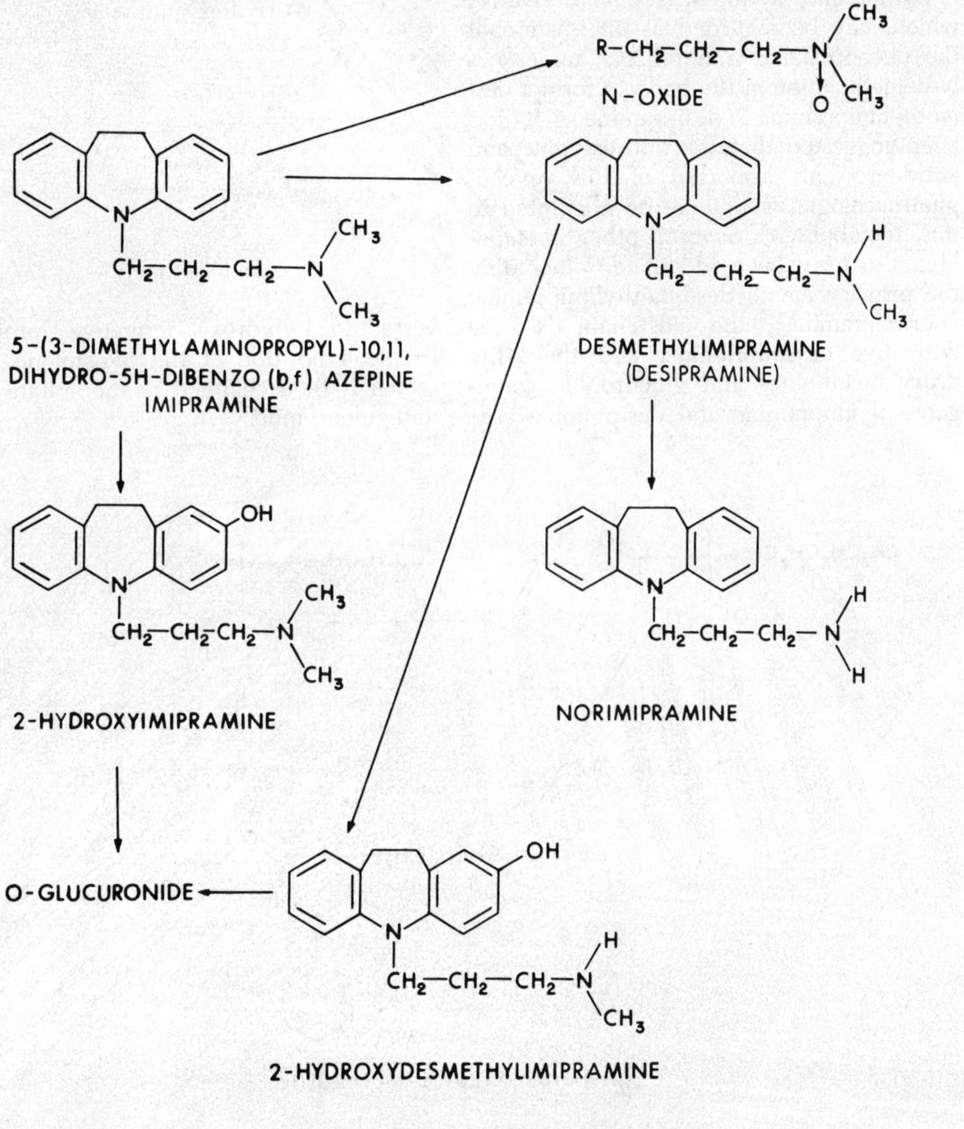

5-(3-DIMETHYLAMINOPROPYL)-10,11, DIHYDRO-5H-DIBENZO (b,f) AZEPINE IMIPRAMINE

N-OXIDE

DESMETHYLIMIPRAMINE (DESIPRAMINE)

2-HYDROXYIMIPRAMINE

NORIMIPRAMINE

O-GLUCURONIDE

2-HYDROXYDESMETHYLIMIPRAMINE

metabolite was not formed from 7-hydroxy-chlorpromazine sulfoxide, and formed less readily from desmethylchlorpromazine. It was suggested that the 7,8-dihydroxymetabolite, a catechol susceptible to ready oxidation and polymerization, may be responsible for some toxic reactions and for skin pigmentation.[88]

The major expected pathways of chlorpromazine metabolism are shown on page 108. All of the hydroxy derivatives may conjugate to form sulfate esters or glucuronides, and the normethyl compounds may also form N-glucuronides. N-oxides also are reported to be minor metabolites.

Imipramine, a dibenzazepine derivative which may be regarded as an isostere of the phenothiazine tranquilizers, undergoes N-demethylation in the body to form a desmethylimipramine (desipramine). It has been suggested that the antireserpine and antidepressant properties of this psychopharmacologic agent may be attributed to this metabolite.[89] Several other metabolites also have been identified,[90] including the primary amine desdimethylimipramine (norimipramine), the side-chain N-oxide derivative of imipramine, and the 2-hydroxy metabolites and glucuronide conjugates of imipramine and desipramine (see

p. 109). Possibly, the ethylene bridge may undergo hydroxylation by analogy to the major reaction which occurs to the closely related antidepressant drugs, amitriptyline and nortriptyline (see Chap. 16).

Pyrazolone Derivatives. Antipyrine is in part excreted unchanged and partly con-

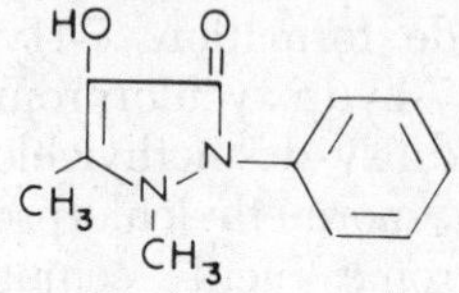

4-Hydroxyantipyrine

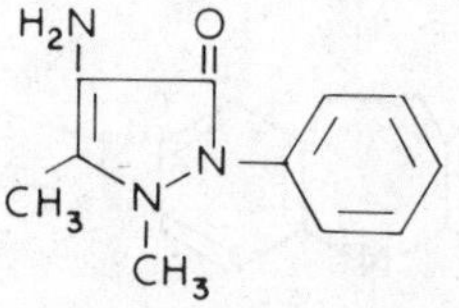

4-Aminoantipyrine

verted to a hydroxy derivative (probably in the 4 position of the pyrazolone ring) which is then changed to the sulfate ester and glucuronide.

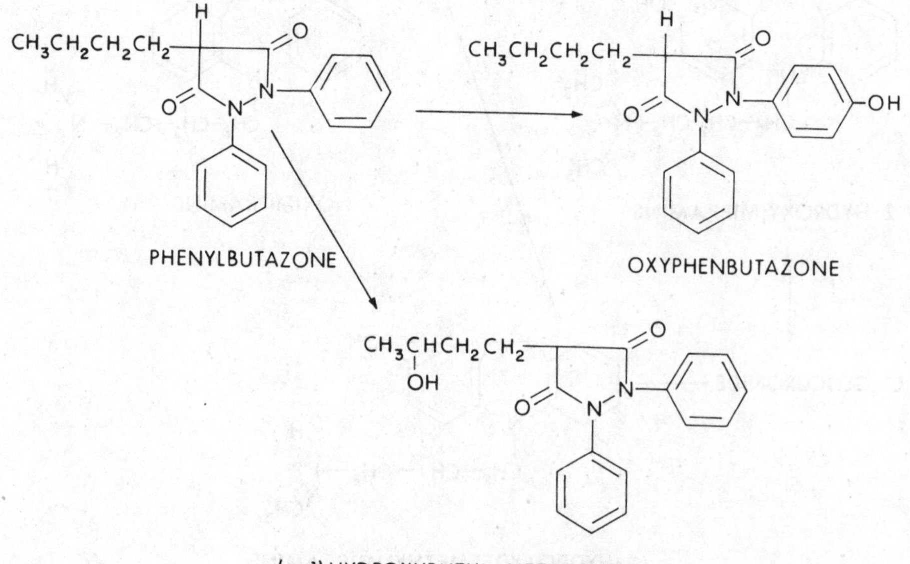

PHENYLBUTAZONE

OXYPHENBUTAZONE

(ω-1) HYDROXYBUTYL METABOLITE

Aminopyrine (Pyramidon) is metabolized in man by demethylation to give the analgetically active 4-aminoantipyrine which is partially converted to an inactive 4-acetylaminoantipyrine. These two products account for about 50 per cent of the administered drug. Small amounts of 4-hydroxy-antipyrine also have been isolated, together with a glucuronide of unknown structure.

Phenylbutazone has been shown to form two major metabolites,[91] one involving ring phenyl hydroxylation in the para position, and the other (ω–1) oxidation of the butyl side chain. The ring hydroxy metabolite (oxyphenbutazone) retains the antipyretic and analgetic properties of the parent drug. Conjugates of these metabolites have not been identified.

Barbiturates and Related Compounds. The metabolism of the barbiturates may be classified under four general headings: (1) Oxidation of groups substituted in the 5-position; (2) removal of N-alkyl radicals (N-demethylation); (3) conversion of thio-barbiturates to their oxygen analogs; and (4) hydrolytic cleavage of the barbiturate ring.[92] The last is believed to be only a minor pathway but has been observed in the dog with both pentobarbital and amobarbital. Oxidation of groups in the 5-position appears to be the most important metabolic pathway. The main metabolic product of phenobarbital is 5-ethyl 5-(p-hydroxy-phenyl)barbituric acid.[93] The metabolism takes place quite slowly and the metabolite is excreted in man unchanged or as the sulfate ester, both of which are inactive as CNS depressants. In dogs the p-hydroxy metabolite is excreted largely conjugated with glucuronic acid.[94] The long-acting barbiturates (e.g., barbital, phenobarbital) are excreted slowly. Barbital is excreted over a period of several days largely unchanged. Oxidation of the terminal (ω) and penultimate (ω–1) carbons have been reported for pentobarbital and thiopental. Pentobarbital forms about equal amounts of an alcohol

OXIDATION OF GROUPS SUBSTITUTED IN THE 5-POSITION

Phenobarbital

p–Hydroxyphenobarbital

ω Oxidation

(ω-1) Oxidation

Pentobarbital

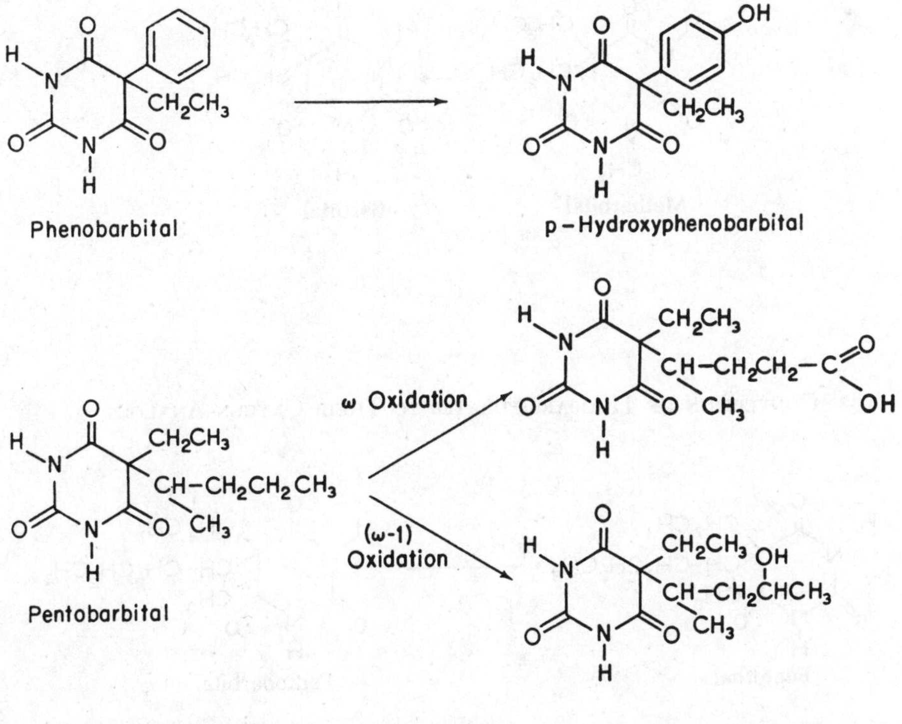

OXIDATION OF GROUPS SUBSTITUTED IN THE 5-POSITION

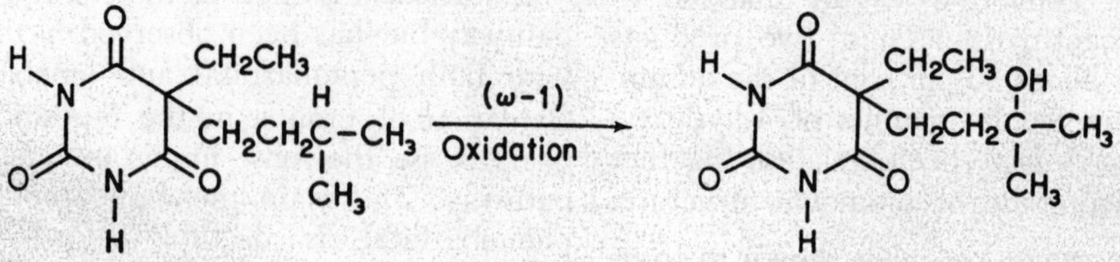

Amobarbital

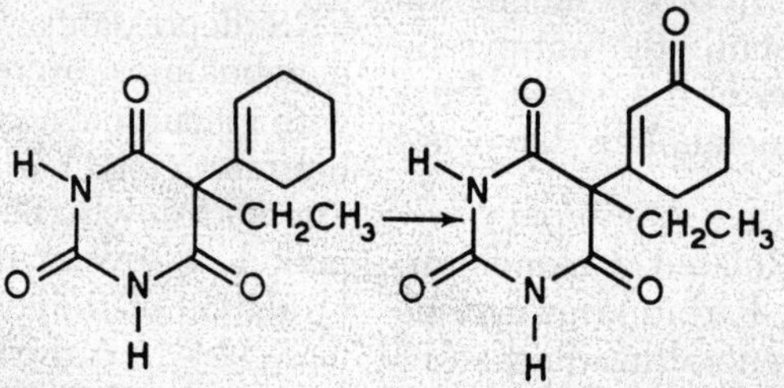

$(\omega - 1)$ Oxidation

Cyclobarbital   3 - Oxocyclobarbital

N-DEMETHYLATION

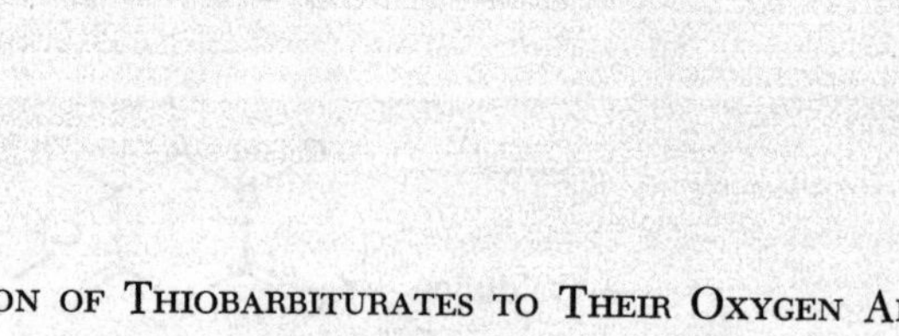

Metharbital   Barbital

CONVERSION OF THIOBARBITURATES TO THEIR OXYGEN ANALOGS

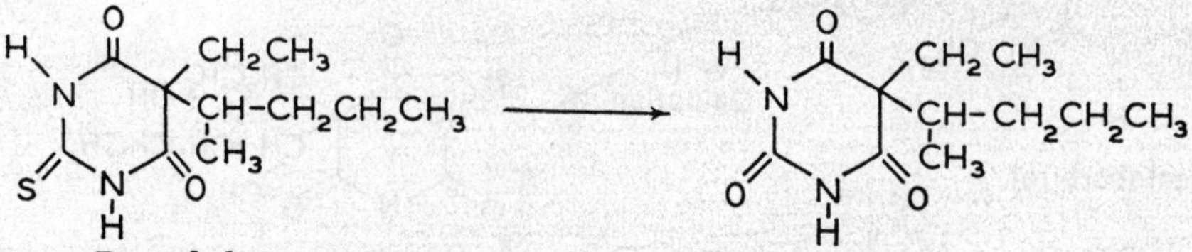

Pentothal   Pentobarbital

HYDROLYTIC CLEAVAGE OF THE BARBITURATE RING

Hexobarbital

Ring Oxidation

3-Oxohexobarbital

Ring cleavage

N-Methyl-N'-[α-(1-cyclohexenyl)-α-methylacetyl]urea $+ CO_2$

N-demethylation

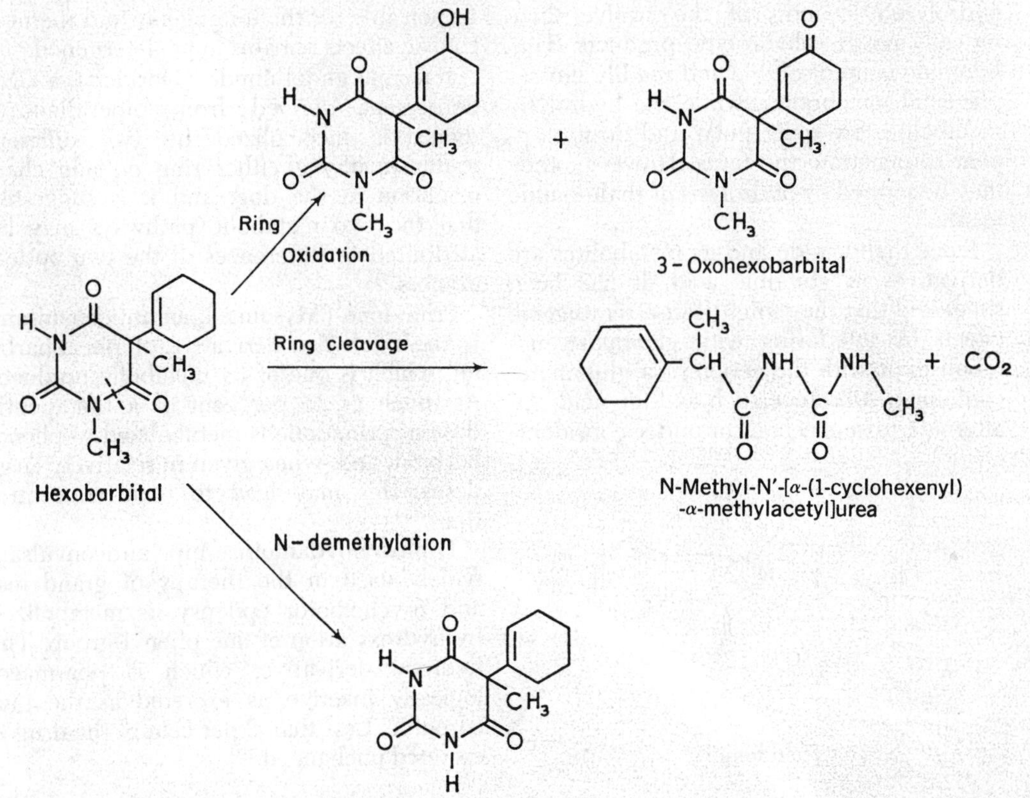

N-Des-methylhexobarbital

(penultimate oxidation) and an acid (ω-oxidation), whereas thiopental appears to be converted primarily to a carboxylic acid (ω-oxidation).[95] Amobarbital primarily undergoes penultimate oxidation in the 5-isoamyl side chain. Unsaturated cyclic substituents in the 5-position (e.g., cyclobarbital, hexobarbital) are metabolically oxidized in the 3- and 6-positions. The examples on pages 111 and 112 illustrate the several types of metabolic changes observed for the barbiturates (see Chap. 15):

Hexobarbital has been reported to undergo cleavage of the barbiturate ring in rabbits and dogs. The barbiturate ring is relatively stable in vitro and cleavage of the ring is regarded as a minor metabolic pathway, since the cleavage metabolites have never represented more than 5 per cent of the dose administered.[96] Cleavage of the ring appears to occur primarily in the N-alkyl derivatives. Hexobarbital has been observed to undergo N-demethylation, oxidation of the cyclohexenyl ring and cleavage of the barbiturate ring. However, the metabolites isolated account for only 10 per cent of the drug administered.

Cyclic Compounds Related to the Barbiturates. The metabolism of thalidomide, the phthalimide derivative of 3-aminopiperidine-2,6-dione, due to neurotoxic and teratogenic effects, has been studied intensively to determine if these effects are produced by the intact drug or by one or more of its metabolic products.[97, 98] Thalidomide is stable in vitro at pH values between 2 and

6, but above pH 6 it is quite unstable and at the physiologic pH of 7.4 it is rapidly hydrolyzed to most of the twelve theoretically possible hydrolytic products. Thalidomide is lipid soluble and readily crosses placental membranes, while the hydrolytic metabolites are quite polar and do not appear to penetrate the fetus. However, they may be formed by hydrolysis of thalidomide in situ.

Since thalidomide and its metabolites are derivatives of glutamic acid, it has been suggested that they might cause teratogenic effects by interfering with glutamate metabolism or with the action of a glutamate-containing substance such as folic acid. An alternate toxic mechanism under considera-

tion is the acylation and inactivation of essential substances in the fetus. The agent(s) responsible for the teratogenic and the neurotoxic effects remains to be determined.

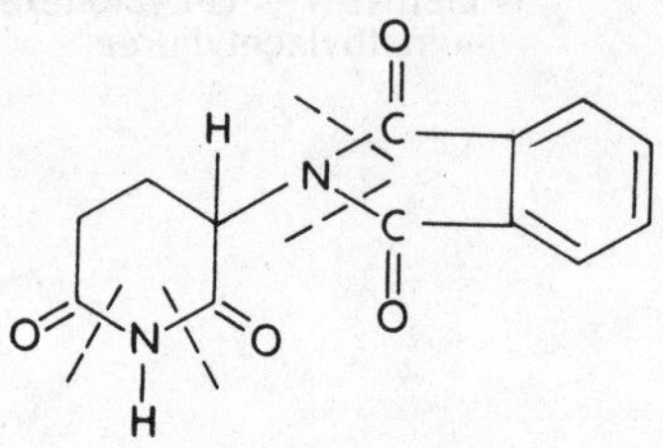

Racemic glutethimide (Doriden), a CNS depressant derived from piperidine-2,6-dione, is metabolized by two different routes involving either ring or side chain oxidation in the dog, and it is suggested that the two metabolic pathways may be attributed to differences in the two optical isomers.[99]

Primidone (Mysoline), an anticonvulsant, is the 2-dihydro derivative of phenobarbital, which is one of its metabolic products. As much as 15 per cent of a therapeutic dose of primidone is metabolized to phenobarbital, and, when given in relatively large doses, this may lead to toxic concentrations.[100]

Diphenylhydantoin, an anticonvulsant widely used in the therapy of grand mal and psychomotor epilepsy, is metabolized by hydroxylation of one phenyl group. The hydroxy derivative, which is pharmacologically inactive, is excreted as the glucuronide. Less than 2 per cent of the drug is excreted unchanged.[101]

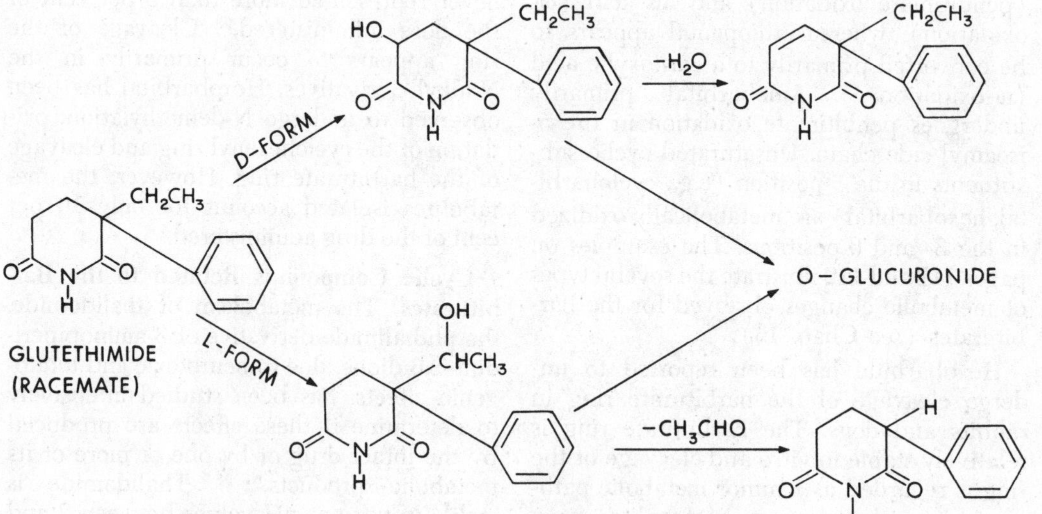

Thalidomide

GLUTETHIMIDE (RACEMATE)

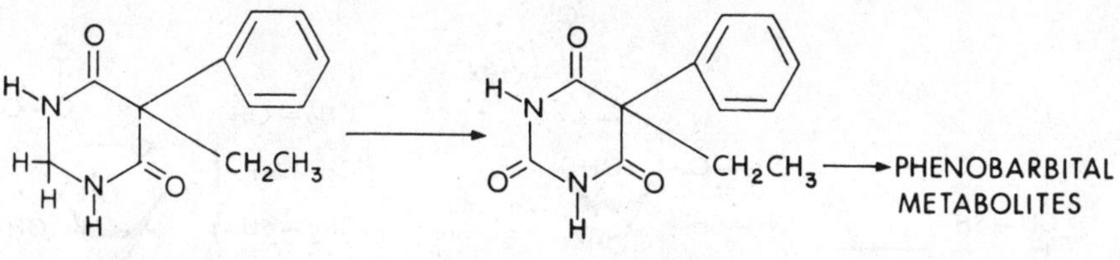

Primidone → Phenobarbital → PHENOBARBITAL METABOLITES

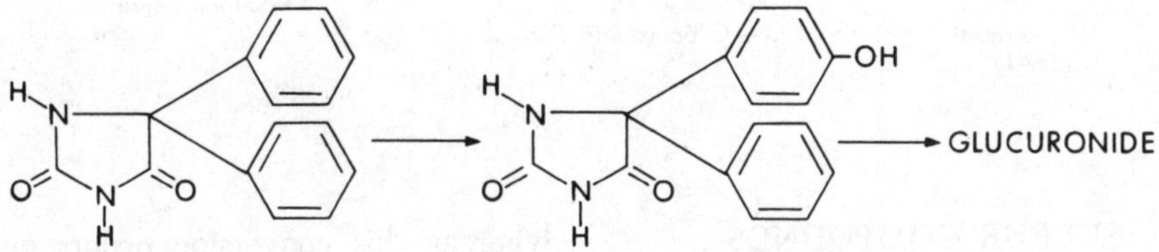

Diphenylhydantoin → 5-*p*-Hydroxyphenyl-5-phenylhydantoin → GLUCURONIDE

Miscellaneous Heterocyclic Nitrogen Compounds. The trichomonacidal agent metronidazole (Flagyl) is metabolized to the extent of about 35 per cent in man by oxidation of the 2-methyl group to the hydroxymethyl derivative. Less than 10 per cent of the 1-acetic acid metabolite is formed also, while about 35 per cent is excreted unchanged.[102]

Diazepam (Valium), a CNS depressant drug, is metabolized in man by N-demethylation and by hydroxylation in the 3-position to form an active derivative, oxazepam (Serax), which is excreted largely as the glucuronide.[103]

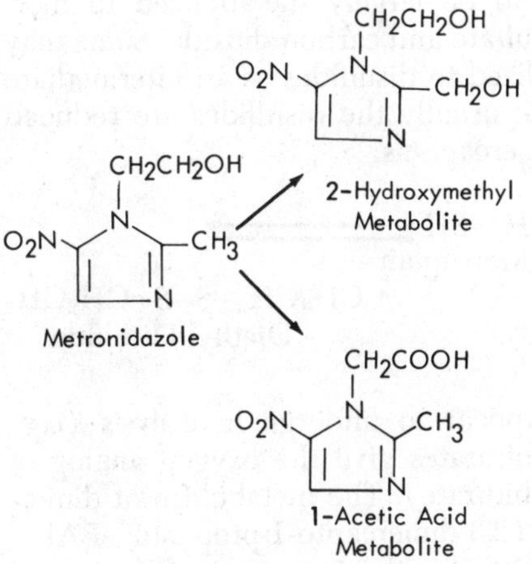

Metronidazole → 2-Hydroxymethyl Metabolite

Metronidazole → 1-Acetic Acid Metabolite

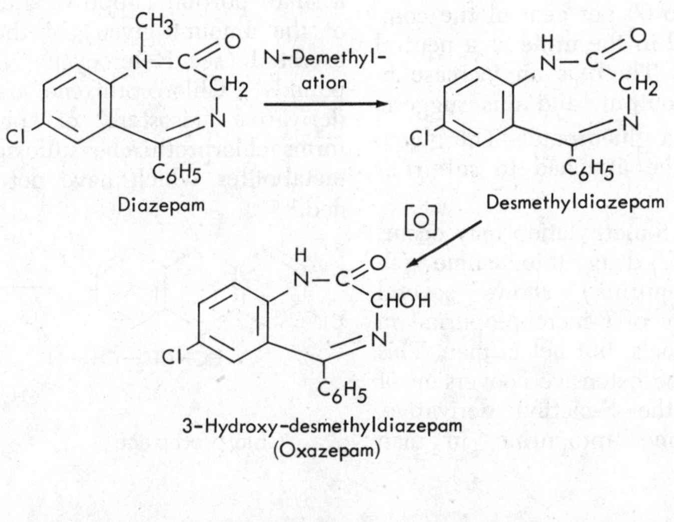

Diazepam → (N-Demethylation) → Desmethyldiazepam

Desmethyldiazepam → [O] → 3-Hydroxy-desmethyldiazepam (Oxazepam)

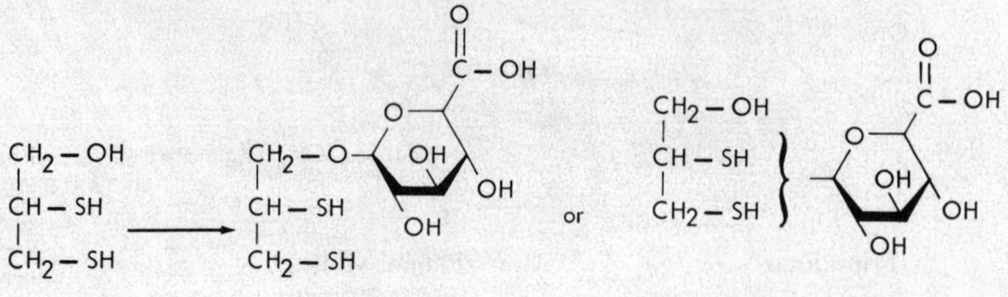

Dimercaprol
(BAL)

O — Glucuronide

or

S — Glucuronide

SULFUR COMPOUNDS

Thiols and Disulfides. The thioalcohols (mercaptans) may be metabolized in a number of ways. Methyl mercaptan is reported to be largely metabolized to inorganic sulfate and carbon dioxide. Some may be oxidized to disulfides as an intermediate step but usually the disulfides are reduced to the mercaptans.

$$2CH_3CH_2-SH \rightleftharpoons$$
Ethyl Mercaptan

$$CH_3CH_2-S-S-CH_2CH_3$$
Diethyl Disulfide

Some appear to undergo hydrolysis (e.g., thiobarbiturates give the oxygen analog of the barbiturate). The metabolism of dimercaprol (2,3-dimercapto-1-propanol; BAL), an antidote for the treatment of acute and chronic poisoning by arsenic, mercury, gold and other heavy metals, is not known, but in rats from 40 to 60 per cent of the compound is excreted in the urine as a neutral sulfur derivative. There is an increase in glucuronic acid output, and this suggests the formation of a glucuronide. The glucuronic acid may be attached to sulfur or oxygen.

In some cases, S-methylation may occur. The antileukemic drug thioguanine (2-amino-6-mercaptopurine) shows several times the potency of 6-mercaptopurine in experimental animals, but not in man. This is attributed to the extensive conversion of thioguanine to the S-methyl derivative, 2-amino-6-methylmercaptopurine, in man

whereas this conversion occurs only to a minor degree in other species studied.[104]

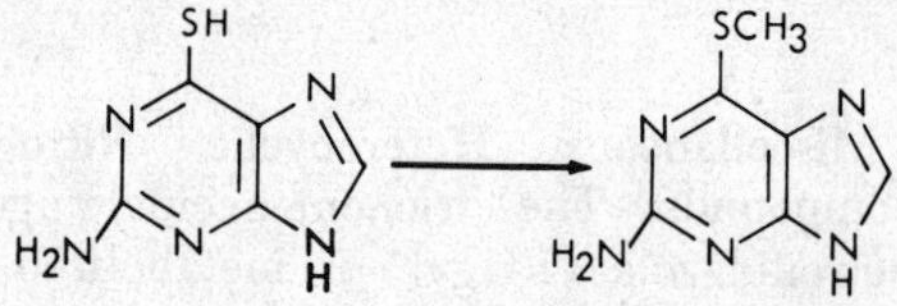

Thioguanine
2-Amino-6-mercaptopurine

2-Amino-6-methyl-
mercaptopurine

Thioethers. The metabolism of thioethers or sulfides may involve cleavage to form hydrogen sulfide or mercaptans but some are known to form sulfoxides and sulfones. It has been suggested that dimethylthioether (methyl sulfide) may be metabolized to dimethyl sulfone.

Heterocyclic Sulfur Compounds. A number of the phenothiazine tranquilizing agents have been observed to form sulfoxides but these appear to account for only a small portion (approximately 5 per cent of the amount given) of the metabolites excreted (see Heterocyclic Nitrogen Compounds). Chlorprothixene, a thiaxanthene derivative (isostere of phenothiazine), forms chlorprothixene sulfoxide and other metabolites which have not been identified.[105]

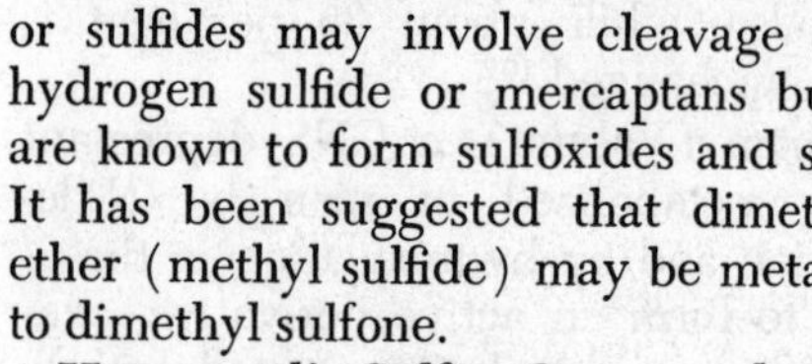

Chlorprothixene

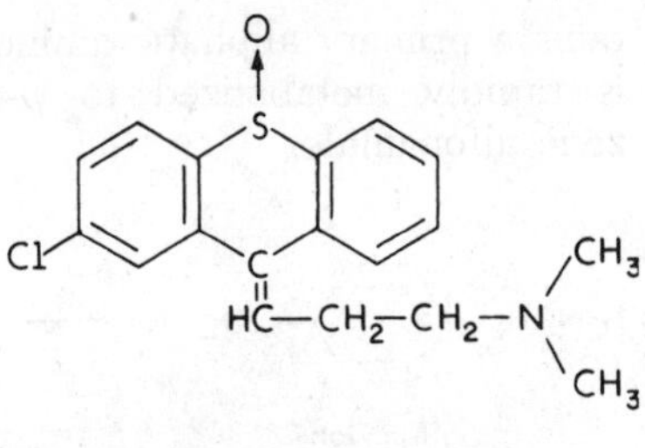

Chlorprothixene sulfoxide

Sulfonic Acids. The aliphatic and aromatic sulfonic acids appear for the most part to be excreted unchanged. Compounds of this type are quite stable and freely soluble in water. Representatives such as benzenesulfonic acid, *p*-hydroxybenzenesulfonic acid, sulfanilic acid and n-octanesulfonic acid are reported to be excreted unchanged.

Sulfamic Acids. The sulfamic acids are normal metabolites (see aromatic amines) and are excreted for the most part unchanged. They are relatively strong acids (pK$_a$ of sulfamic acid 3.2) and have a low order of toxicity. The sodium and calcium salts of cyclohexylsulfamic acid (sodium cyclamate, calcium cyclamate) have been used as noncaloric sweetening agents. The compounds are excreted largely unchanged, although cyclohexylamine has been observed in the urine of some individuals ingesting sodium cyclamate.[106, 107] Based on evidence that the cyclamates are able to produce cancer of the bladder in rats, they have been barred from further use in all food products.

Sulfonamides and Sulfonylureas. Sulfonamide compounds are of special interest because of their activity and wide use as antibacterial, antidiabetic and diuretic agents. The metabolism of the sulfonamides normally does not involve the sulfonamide group and that portion of the molecule usually remains intact in the metabolites excreted. Acetazolamide (see Chap. 20) is for the most part excreted unchanged. In man approximately 70 per cent of the drug is excreted in 24 hours. It has not been established whether the remaining portion of the drug is metabolized or gradually excreted unchanged.

Benzothiazole-2-sulfonamide, in contrast to the other sulfonamides, appears to be metabolized in dogs to a significant degree. The sulfonamido group is reduced to a thiol which then conjugates with glucuronic acid. Approximately 25 per cent of the drug administered to dogs is excreted as the S-glucuronide of 2-mercaptobenzothiazole.[108] The metabolism of benzothiazole-2-sulfonamide is of interest, since it is the only sulfonamide in which the sulfonamide group is known to be reduced to a thiol. It is also the first compound reported to yield an S-glucuronide. Instead of a direct reduction, it has been shown[109] that the sulfonamide group is first replaced by

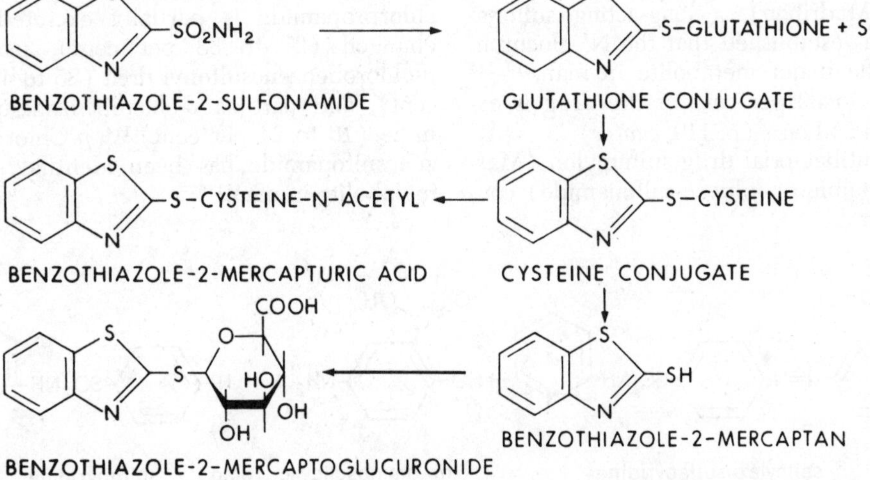

BENZOTHIAZOLE-2-SULFONAMIDE

GLUTATHIONE CONJUGATE

BENZOTHIAZOLE-2-MERCAPTURIC ACID

CYSTEINE CONJUGATE

BENZOTHIAZOLE-2-MERCAPTAN

BENZOTHIAZOLE-2-MERCAPTOGLUCURONIDE

glutathione, the sulfonamide sulfur being excreted as inorganic sulfate. The glutathione conjugate undergoes subsequent hydrolytic removal of two amino acids to form the cysteine conjugate, which is acetylated to yield the mercapturic acid derivative as the major metabolite in the rat, rabbit or dog. Cleavage of the sulfur-carbon bond of the glutathione conjugate (or cysteine derivative) leads to the formation of benzothiazole-2-mercaptan, most of which forms a thioether with glucuronic acid before urinary excretion.

The antibacterial sulfonamides have been observed to metabolize by acetylation of the aromatic amino group and by hydroxylation of the benzene or hetero ring attached to the amide nitrogen (see Chap. 11). As mentioned earlier, the azo sulfonamides (Prontosil, salicylazosulfapyridine) first undergo reduction to give the free sulfa drug and an aromatic amino compound (see below).

Conjugation with acetic acid is the most common metabolic pathway of the sulfonamides but the extent of acetylation varies markedly with the sulfonamide used. Most but not all of the acetyl derivatives are less soluble than the unacetylated compound and this may delay excretion and produce crystalluria. Hydroxylation of one of the rings followed by conjugation with glucuronic or sulfuric acid facilitates the excretion. Sulfapyridine will serve to illustrate the main metabolic pathways (p. 119, *top*).

Studies on the metabolism of sulfadimethoxine (Madribon), a long-acting sulfonamide, has established that the N^1-glucuronide is the major metabolite in man.[110, 111] Only 20 to 30 per cent of the drug is excreted in 24 hours (p. 119, *center*).

The antibacterial drug sulfamylon (Mafenide, α-amino-*p*-toluene-sulfonamide) contains a primary aliphatic amino group and is rapidly metabolized to *p*-carboxybenzenesulfonamide.

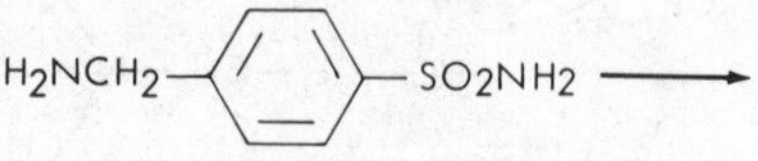

Sulfamylon

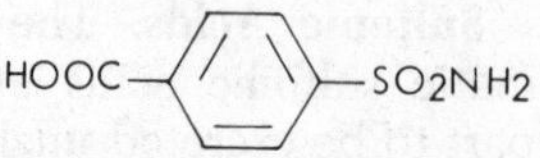

p-Carboxybenzenesulfonamide

The sulfonylureas, which are widely used in the treatment of diabetes mellitus, are mainly metabolized by reactions involving the oxidation of the methyl group (tolbutamide) or the hydrolysis of the ureide to form a sulfonamide (chlorpropamide). The main metabolic pathways of the two drugs are shown on page 119. Tolbutamide is rapidly absorbed and excreted with a mean biological half-life of from five to seven hours. Practically all of the drug is excreted as the carboxytolbutamide in man.[112] In the rat and rabbit, hydroxymethyltolbutamide is the major metabolite, and carboxytolbutamide a minor metabolite. The hydroxymethyl metabolite is approximately one half as active as tolbutamide in its hypoglycemic effect, while the carboxy metabolite is essentially inactive.[113] In the dog chlorpropamide is partially excreted unchanged (27 to 33 per cent), part as *p*-chlorobenzenesulfonyl urea (35 to 40 per cent), and part as *p*-chlorobenzenesulonamide (16 to 24 per cent).[114] *p*-Chlorobenzenesulfonamide has been identified as a metabolite in man.[115]

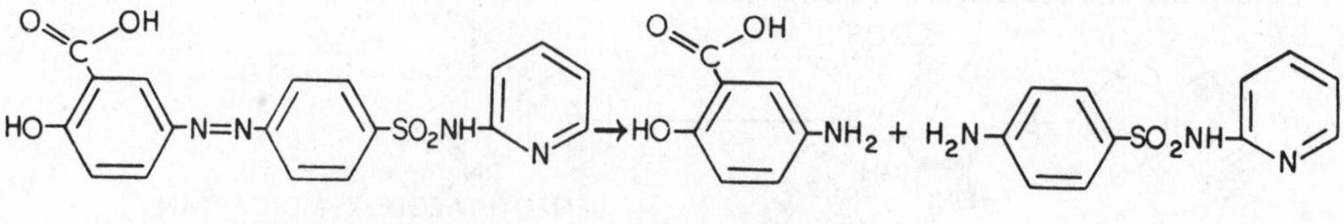

salicylazosulfapyridine   m-aminosalicylic acid   sulfapyridine

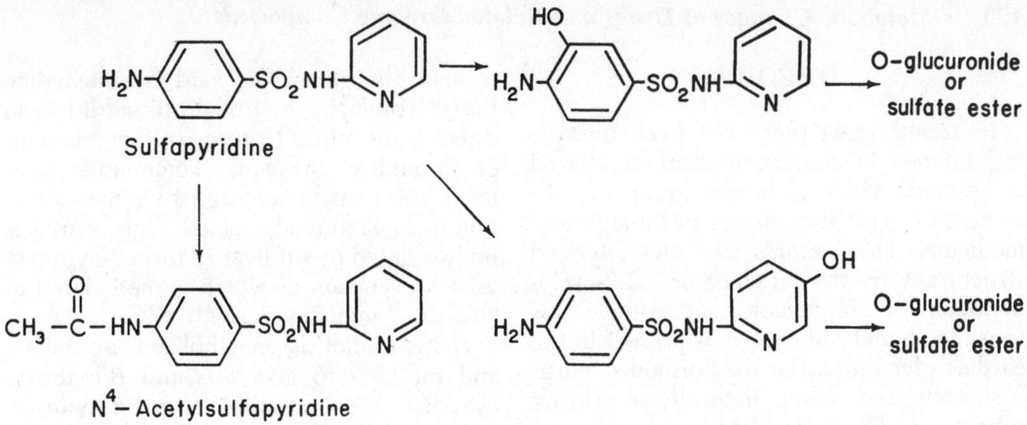

Sulfapyridine

N⁴− Acetylsulfapyridine

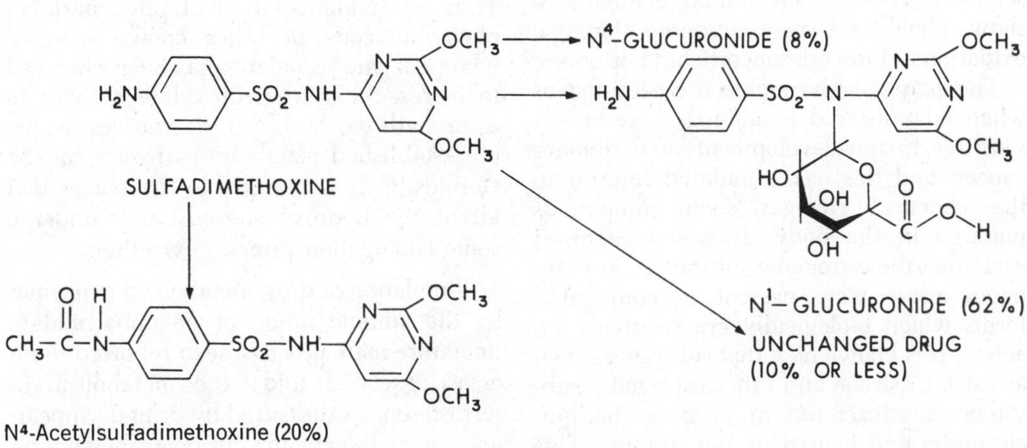

O−glucuronide or sulfate ester

O−glucuronide or sulfate ester

SULFADIMETHOXINE

N⁴-Acetylsulfadimethoxine (20%)

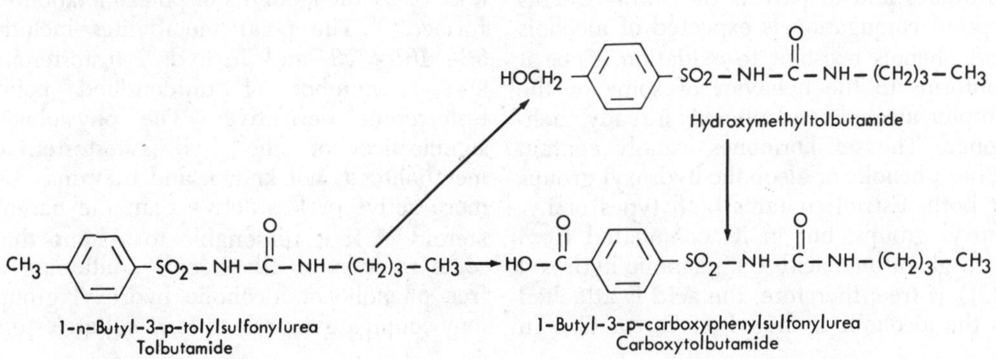

N⁴−GLUCURONIDE (8%)

N¹− GLUCURONIDE (62%)
UNCHANGED DRUG
(10% OR LESS)

Hydroxymethyltolbutamide

1-n-Butyl-3-p-tolylsulfonylurea
Tolbutamide

1-Butyl-3-p-carboxyphenylsulfonylurea
Carboxytolbutamide

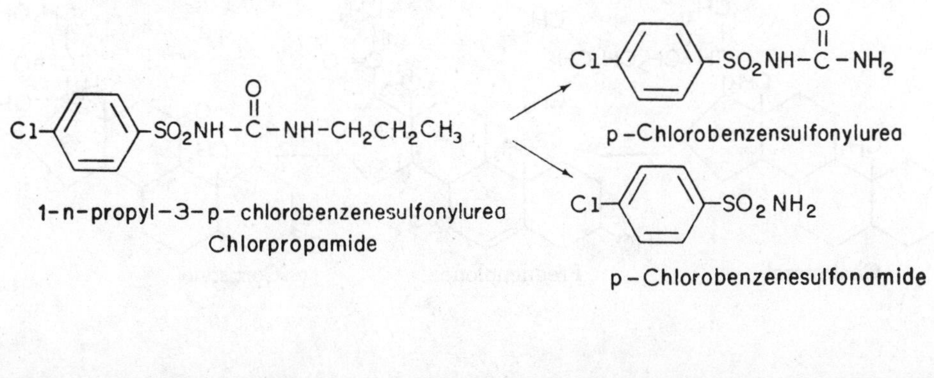

1-n-propyl−3−p−chlorobenzenesulfonylurea
Chlorpropamide

p-Chlorobenzensulfonylurea

p-Chlorobenzenesulfonamide

STEROIDS

In recent years there has been increasing interest in the metabolism of steroid compounds which include a group of substances of great importance in biology and medicine. The steroids are characterized structurally by the presence of a 1,2-cyclopentanoperhydrophenanthrene ring. This four-membered ring system is present in the cardiac glycosides, the sex hormones, corticosteroids and other naturally occurring substances. Cholesterol appears to be an important precursor for other steroids and is implicated in the development of atherosclerosis. There is substantial evidence to show cholesterol may undergo biologic oxidation to form adrenocorticoid hormones.

The estrogenic hormones are carcinogens when administered to animals hereditarily sensitive to the development of mammary cancer, and this has stimulated interest in the chemical changes such compounds undergo in the body. It was recognized early that the estrogenic substances in pregnancy urine were present in conjugated forms which biologically are relatively inactive. It is known now that substances such as estriol, estrone and other estrogenic substances are excreted in part as the glucuronides and in part as the sulfates. This type of conjugation is expected of alcohols and phenols resistant to oxidation, since it conforms to the behavior of some of the simpler alcohols and phenols already mentioned. The sex hormones usually contain either phenolic or alcoholic hydroxyl groups or both. Estriol contains both types of hydroxyl groups, but in its conjugated form with glucuronic acid, the phenolic hydroxyl (C_3) is free; therefore, the acid is attached to the alcoholic hydroxyl on C_{16} or C_{17}. In estrone, the glucuronic acid is attached to the C_3 phenolic hydroxyl. In addition to direct conjugation to form sulfate esters or glucuronides, estrogenic compounds have been observed to undergo ring hydroxylation in some animal species. Thus, estriol is hydroxylated by rat liver to form 2-hydroxyestriol, a portion of which is methylated to give the 2-methoxy derivative.[116]

17β-Estradiol is metabolized by mouse and rat liver to give 6α- and β-hydroxyestradiol, 6-oxoestradiol and 6β-hydroxyestrone.[117, 118] There is evidence to suggest that the steroid hormones are normal body substrates for drug-metabolizing enzymes. Thus, the administration of phenobarbital, phenylbutazone or other known metabolizing enzyme stimulants has been observed to increase steroid hydroxylase activity to form hydroxy and oxo derivatives which are established metabolic pathways for the steroids.[119] It is reasonable to assume that all of the hydroxy steroids may undergo some conjugation prior to excretion.

Stimulation of drug-metabolizing enzymes by the administration of phenobarbital to immature male rats has been reported to increase (several fold) the metabolism of testosterone as measured by drug disappearance or by the amounts of polar metabolites formed.[120] The polar metabolites include 6β-, 16α-, 2β- and 7α-hydroxytestosterone and a number of unidentified polar testosterone derivatives. The physiologic significance of the hydroxytestosterone metabolites is not known and they may be more active or less active than the parent steroid.[119] It is reasonable to assume that some fraction of all steroids containing a free phenolic or alcoholic hydroxyl group may conjugate directly to form sulfate esters

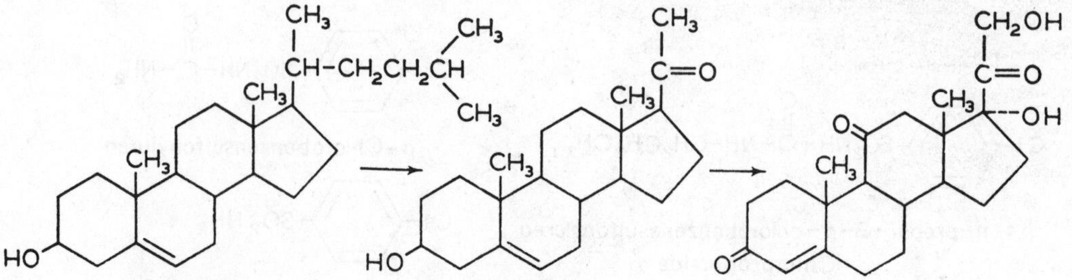

Cholesterol   Pregnenolone   Cortisone

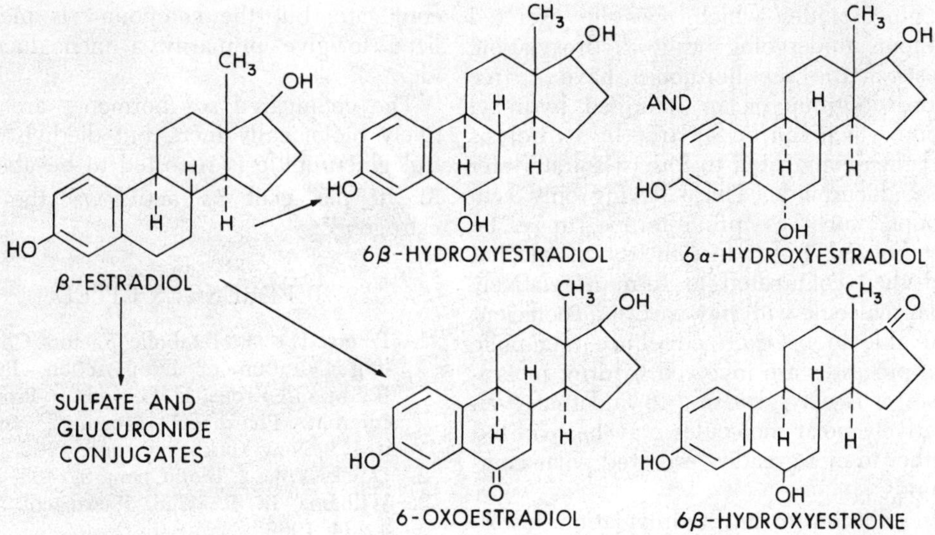

6β-HYDROXYESTRADIOL   AND   6α-HYDROXYESTRADIOL

β-ESTRADIOL

SULFATE AND
GLUCURONIDE
CONJUGATES

6-OXOESTRADIOL   6β-HYDROXYESTRONE

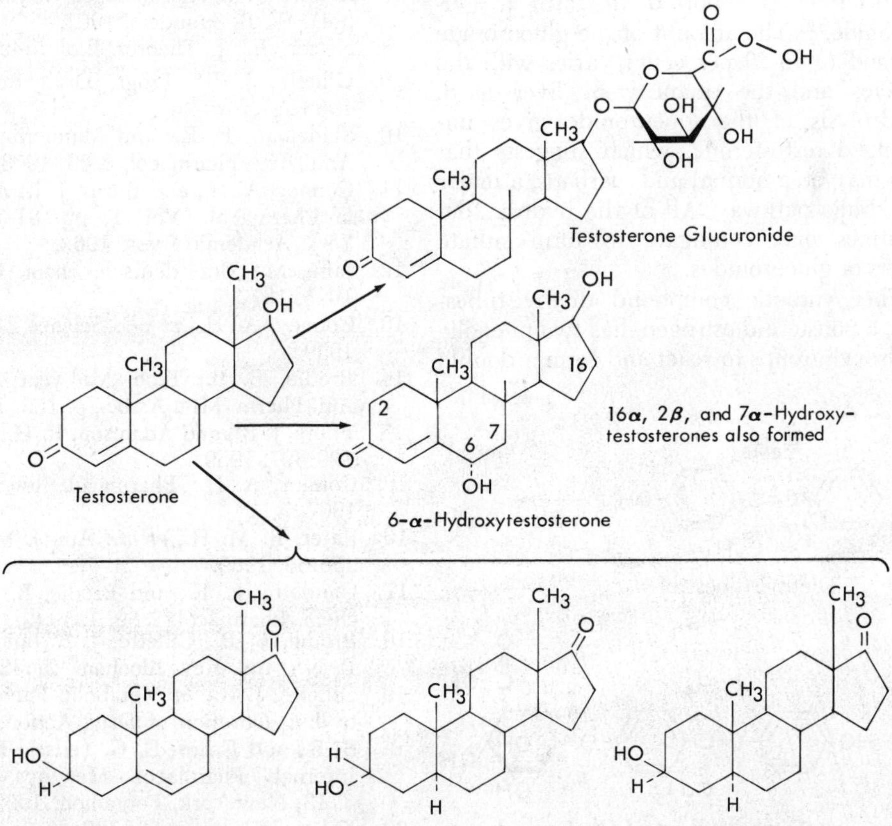

Testosterone Glucuronide

Testosterone

6-α-Hydroxytestosterone

16α, 2β, and 7α-Hydroxy-
testosterones also formed

Dehydroisoandrosterone   Androsterone   Isoandrosterone

or glucuronides which may be excreted without undergoing ring hydroxylation. Most of the sex hormones have a free hydroxyl group or an esterified hydroxyl group which can be set free by hydrolysis and then conjugated to form a sulfate ester or a glucuronide. Those having only keto groups, such as progesterone, may be metabolized by reduction to an alcohol and then conjugated to form a relatively polar molecule with low partition coefficient and able to be excreted. Since metabolic rate processes are involved in drug metabolism, it is not surprising to find that even relatively polar molecules may be oxidized further to more readily excreted polar compounds.

In addition to ring hydroxylation, testosterone conjugates directly to form a glucuronide or a sulfate ester. It is partially converted also to androsterone, dehydroisoandrosterone and isoandrosterone (see Chap. 25). When incubated with liver slices from six different animal species, testosterone is reported to form a glucuronide.[121] The amount of the glucuronide formed (1 to 20 per cent) varies with the species and the quantity of liver used. Hydrolysis of the glucuronide gives unchanged testosterone, which suggests that this may be a normal and, perhaps, a major metabolic pathway. All of the hydroxy derivatives may conjugate to form sulfate esters or glucuronides.

The synthetic compound diethylstilbestrol, a nonsteroid estrogen, has two phenolic hydroxyl groups to react and form a double conjugate, but the compound is metabolized to give primarily a monoglucuronide.

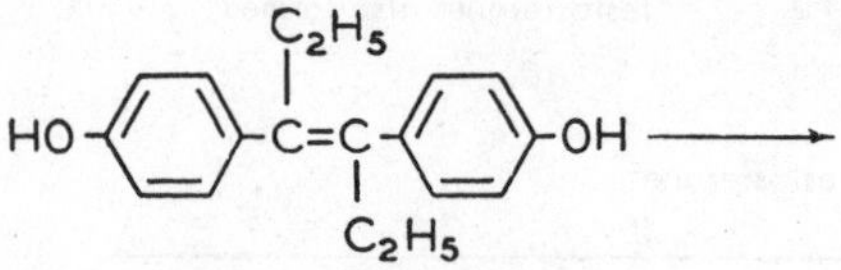

The conjugated sex hormones are relatively biologically inert, but diethylstilbestrol glucuronide is reported to be about 5 to 10 per cent as active as the free estrogen.[122]

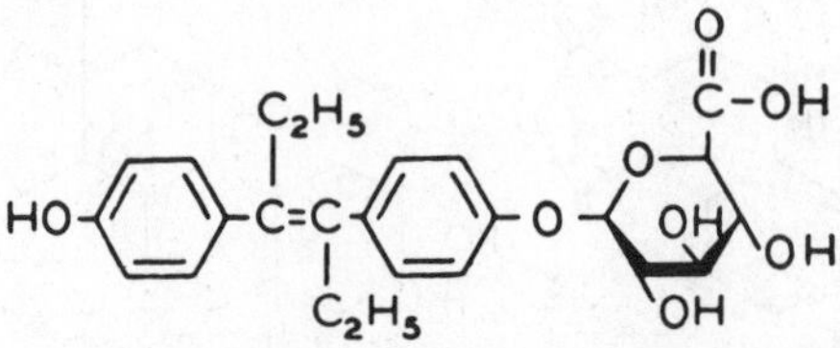

REFERENCES CITED

1. Peters, L.: *in* Metabolic Factors Controlling Duration of Drug Action, Brodie, B. B., and Erdös, E. G. (eds.): Proc. 1st Internat. Pharmacol. Meetings, Vol. 6, p. 179, New York, Pergamon, 1962.
2. Quick, A. J.: J. Biol. Chem. 97:403, 1932.
3. Williams, R. T.: Clin. Pharmacol. Ther. 4:234, 1963.
4. Trefouël, J., *et al.*: Compt. rend. Soc. biol. 120:756, 1935.
5. Brodie, B. B.: J. Pharm. Pharmacol. 8:1, 1956.
6. Meier, H.: Experimental Pharmacogenetics, New York, Academic Press, 1963.
7. Kalow, W.: Pharmacogenetics, Philadelphia, W. B. Saunders, 1962.
8. Nelson, E.: J. Theoret. Biol. 5:493, 1963.
9. Gillette, J. R.: Progr. Drug Res. 6:49, 1963.
10. Shideman, F. E., and Mannering, G. J.: Ann. Rev. Pharmacol. 3:33, 1963.
11. Conney, A. H., and Burns, J. J.: Advances in Pharmacol., Vol. 1, pp. 31-58, New York, Academic Press, 1962.
12. Jaffe, M.: Ber. deutsch. chem. Ges. 10: 1925, 1877.
13. Conney, A. H., *et al.*: Science 130:1478, 1959.
14. Brodie, B. B.: Proc. Mid-year Meeting, Am. Pharm. Mfg. Assoc., p. 122, 1952.
15. Fouts, J. R., and Adamson, R. H.: Science 129:897, 1959.
16. Conney, A. H.: Pharmacol. Rev. 19:317, 1967.
16a. Kater, R. M. H., *et al.*: Am. J. Med. Sci. 258:35, 1969.
17. Gaudette, L. E., and Brodie, B. B.: Biochem. Pharmacol. 2:89, 1959.
18. Brodie, B. B., Gillette, J. R., and LaDu, B. N.: Ann. Rev. Biochem. 27:427, 1958.
19. Gillette, J. R.: *in* Metabolic Factors Controlling Duration of Drug Action, Brodie, B. B., and Erdös, E. G. (eds.): Proc. 1st Internat. Pharmacol. Meetings, vol. 6, p. 13, New York, Pergamon, 1962.
20. Kalow, W.: *op. cit.*, p. 137.

Diethylstilbestrol

Diethylstilbestrol Glucuronide

21. Dutton, G. J.: *op. cit.*, p. 39.
22. Sörbo, B.: *op. cit.*, p. 121.
23. Miriam, S. R., Wolf, J. T., and Sherwin, C. P.: J. Biol. Chem. 71:249, 1927.
24. Quick, A. J.: J. Biol. Chem. 97:403, 1932.
25. Alpen, E. L., *et al.*: J. Pharmacol. Exp. Ther. 102:150, 1951.
26. Tabor, C. W., *et al.*: J. Pharmacol. and Exp. Ther. 102:98, 1951.
27. Knoop, F.: Beitr. chem. physiol. Path. 6:150, 1905.
28. Jowett, M., and Quastel, J. H.: Biochem. J. 29:2159, 1935.
29. Hurtley, W. H.: Quart. J. Med. 9:301, 1915-16.
30. Verkade, P. E., and van der Lee, J.: Hoppe-Seyler Z. physiol. Chem. 227:213, 1934.
31. Sammons, H. G., and Williams, R. T.: Biochem. J. 35:1175, 1941.
32. Haggard, H. W., Miller, D. P., and Greenberg, L. A.: J. Indust. Hyg. 27:1, 1945.
33. Quick, A. J.: J. Biol. Chem. 80:515, 1928.
34. Dakin, H. D.: J. Biol. Chem. 5:173, 1908.
35. Shemiakin, M. M., and Schukina, L. A.: Nature (Lond.) 154:513, 1944.
36. Asahina, Y., and Ishidate, M.: Ber. deutsch. chem. Ges. 66:1673, 1933; 67:71, 1934; 68:947, 1935.
37. Boström, H., and Vestmark, A.: Biochem. Pharmacol. 6:72, 1961.
38. Bartlett, G. R.: Am. J. Physiol. 163:619, 1950.
39. Kendal, L. P., and Ramanathan, A. N.: Biochem. Z. 54:424, 1953.
40. Roe, O.: Acta med. scandinav. 125 (suppl. 182):256, 1946.
41. Leaf, G., and Zatman, L. J.: Brit. J. Industr. Med. 9:19, 1952.
42. Leech, P. N.: JAMA 109:1531, 1937.
43. Newman, H. W., *et al.*: J. Pharmacol. Exp. Ther. 68:194, 1940.
44. Glazko, A. J., Dill, W. A., and Rebstock, M. C.: J. Biol. Chem. 183:679, 1950.
45. Glazko, A. J., *et al.*: J. Pharmacol. Exp. Ther. 96:445, 1949.
46. Miriam, S. R., Wolf, J. T., and Sherwin, C. P.: J. Biol. Chem. 71:695, 1927.
47. Bray, H. G., Thorpe, W. V., and White, K.: Biochem. J. 46:275, 1950.
48. Bray, H. G., Clowes, R. C., and Thorpe, W. V.: Biochem. J. 51:70, 1952.
49. Williams, R. T.: Biochem. J. 37:329, 1943.
50. Williams, R. T.: Biochem. J. 35:557, 1941.
51. Bernhard, K., Bühler, U., and Bickel, M. H.: Helv. chim. acta 42:802, 1959.
52. Mandel, H. G., *et al.*: J. Pharmacol. Exp. Ther. 112:495, 1954.

53. Whitmore, F. C.: Organic Chemistry, pp. 489-490, New York, Van Nostrand, 1937.
54. Krantz, J. C., Jr., *et al.*: J. Pharmacol. Exp. Ther. 70:323, 1940.
55. Glazko, A. J., *et al.*: J. Biol. Chem. 179:417, 1949; 179:409, 1949.
56. Brodie, B. B., and Axelrod, J.: J. Pharmacol. Exp. Ther. 97:58, 1949.
57. Adler, T. K.: J. Pharmacol. Exp. Ther. 106:371, 1952.
58. Tsukamoto, H., *et al.*: Chem. Pharm. Bull. 12:987, 1964.
59. Bray, H. G., *et al.*: Biochem. J. 54:547, 1953.
60. Guroff, G., Daly, J. W., Jerina, D. M., Renson, J., Withop B., and Udenfriend, S.: Science 157:1524, 1967.
61. Boyland, E.: *in* Metabolic Factors Controlling Duration of Drug Action. Brodie, B. B., and Erdös, E. G. (eds.): Proc. 1st Internat. Pharmacol. Meetings, vol. 6, p. 65, New York, Pergamon, 1962.
62. Williams, R. T.: Detoxication Mechanisms, ed. 2, p. 210, New York, Wiley, 1959.
63. Bray, H. G., *et al.*: Biochem. J. 65:483, 1957.
64. Knight, R. H., and Young, L.: Biochem. J. 70:111, 1958.
65. Tatsumi, K., *et al.*: Biochem. Pharmacol. 16:1941, 1967.
66. Axelrod, J.: J. Pharmacol. Exp. Ther. 109:62, 1953.
67. Axelrod, J.: Physiol. Rev. 39:751, 1959.
68. Braun, G. A., *et al.*: European J. Pharmacol. 1:58, 1967.
69. Soudijn, W., *et al.*: European J. Pharmacol. 1:47, 1967.
70. Smith, J. N., and Williams, R. T.: Biochem. J. 44:242, 1949.
71. Brodie, B. B., and Axelrod, J.: J. Pharmacol. Exp. Ther. 94:29, 1948.
72. Yamamoto, A., *et al.*: Chem. Pharm. Bull. 10:522, 1962.
73. Walkenstein, S. S., *et al.*: J. Pharmacol. Exp. Ther. 123:254, 1958.
74. Wiser, R., and Seifter, J.: Fed. Proc. 19:390, 1960.
75. Tsukamoto, H., *et al.*: Chem. Pharm. Bull. 11:421, 1963.
76. Murata, T.: Chem. Pharm. Bull. 9:334, 1961.
77. Lang, K.: Biochem. Ztschr. 259:243, 1933.
78. Smith, J. N., and Williams, R. T.: Biochem. J. 46:243, 1950.
79. Komori, Y., and Sendju, Y.: J. Biochem. 6:163, 1926.

80. McKennis, H., *et al.*: J. Am. Chem. Soc. 79:6342, 1957; 80:6597, 1958; 81:3951, 1951.

81. Turnbull, L. B., *et al.*: Fed. Proc. 19:268, 1960.

82. Brodie, B. B., Baer, J. E., and Craig, L. C.: J. Biol. Chem. 188:567, 1951.

83. Way, E. L., and Adler, T. K.: Pharmacol. Rev. 12:383, 1960.

84. Emmerson, J. L., and Miya, T. S.: J. Pharm. Sci. 52:411, 1963.

85. Beckett, A. H., *et al.*: Biochem. Pharmacol. 12:779, 1963.

86. Fishman, V., and Goldenberg, H.: Proc. Soc. Exp. Biol. Med. 112:501, 1963.

87. Perry, T. L., *et al.*: Science 146:81, 1964.

88. Daly, J. W., and Manion, A. A.: Biochem. Pharmacol. 16:2131, 1967.

89. Gillette, J. R., *et al.*: Experientia 17:417, 1961.

90. Häfliger, F., and Burckhardt, V.: *in* Gordon, M. (ed.): Psychopharmacological Agents, vol. 1, p. 83, New York, Academic Press, 1964.

91. Burns, J. J., *et al.*: J. Pharmacol. 113:481, 1955.

92. Mark, L. C.: Clin. Pharmacol. Ther. 4:504, 1963.

93. Butler, T. C.: Science 120:494, 1954.

94. Butler, T. C.: J. Am. Pharm. Assoc. (Sci. ed.) 116:326, 1956.

95. Cooper, J. R., and Brodie, B. B.: J. Pharmacol. Exp. Ther. 114:409, 1955; 120:75, 1957.

96. Tsukamoto, H., *et al.*: Pharm. Bull. (Tokyo) 3:459, 1955; 3:397, 1955; 4:364, 368, 371, 1956.

97. Faigle, J. W., *et al.*: Experientia 18:389, 1962.

98. Smith, R. L., *et al.*: Life Sciences 1:333, 1962.

99. Keberle, H., *et al.*: Experientia 18:105, 1962.

100. Plaa, G. L., *et al.*: JAMA 168:1769, 1958.

101. Sparberg, M.: Ann. Int. Med. 59:914, 1963.

102. Stambaugh, J. E., *et al.*: Life Sci. 6:1811, 1967.

103. Schwartz, M. A., *et al.*: J. Pharmacol. Exp. Therap. 149:423, 1965.

104. Elion, G. B., *et al.*: Cancer Chemotherapy Rep. 16:197, 1962.

105. Petersen, P. V., and Nielsen, I. M.: *in* Gordon, M. (ed.): Psychopharmacological Agents, vol. 1, p. 319, New York, Academic Press, 1964.

106. Kojima, S., and Ichibagase, H.: Chem. Pharm. Bull. (Tokyo) 14:971, 1966.

107. Leahy, J., Wakefield, M., and Taylor, T.: Food Cosmet. Toxicol. 5:447, 1967.

108. Clapp, J. W.: J. Biol. Chem. 223:207, 1956.

109. Colucci, D. F., and Buyske, D. A.: Biochem. Pharmacol. 14:457, 1965.

110. Bridges, J. W., *et al.*: Biochem. J. 91:12p, 1964.

111. Uno, T., *et al.*: Chem. Pharm. Bull. 13:261, 1965.

112. Nelson, E., and O'Reilly, I.: J. Pharmacol. Exp. Ther. 132:103, 1961.

113. Tagg, J., *et al.*: Biochem. Pharmacol. 16:143, 1967.

114. Welles, J. S., Root, M. A., and Andersen, R. C.: Proc. Soc. Exp. Biol. Med. 101:668, 1959.

115. Johnson, P. C., *et al.*: Ann. N.Y. Acad. Sci. 74:459, 1959.

116. King, R. J. B.: Biochem. J. 79:355, 1961.

117. Brewer, H., *et. al.*: Biochim. biophys. acta 65:1, 1962.

118. Mueller, H. C., and Rumney, G.: J. Am. Chem. Soc. 79:1004, 1957.

119. Conney, A. H., *et al.*: Ann. N.Y. Acad. Sci. 123:98, 1965.

120. Conney, A. H., and Klutch, A.: J. Biol. Chem. 238:1611, 1963.

121. Fishman, W. H., and Sie, H. G.: J. Biol. Chem. 218:335, 1956.

122. Wilder Smith, A. E., and Williams, P. C.: Biochem. J. 42:253, 1948.

SELECTED READING

Boyland, E., and Booth, J.: The Metabolic Fate and Excretion of Drugs, Ann. Rev. Pharmacol. 2:129, 1962.

Brodie, B. B., *et al.*: Ann. Rev. Biochem. 27:427, 1958.

Fishman, W. H.: Chemistry of Drug Metabolism, Springfield, Ill., Thomas, 1961.

McMahon, R. E.: Drug Metabolism, *in* Burger, A. (ed.): Medicinal Chemistry, ed. 3, p. 50, New York, Wiley-Interscience, 1970.

Maynert, E. W.: Metabolic Fate of Drugs, Ann. Rev. Pharmacol. 1:45, 1961.

Parke, D. V.: The Biochemistry of Foreign Compounds, New York, Pergamon, 1968.

Shideman, F. E., and Mannering, G. J.: Metabolic Fate, Ann. Rev. Pharmacol. 3:33, 1963.

Williams, R.: Detoxication Mechanisms, ed. 2, New York, Wiley, 1959.

4

Hydrocarbons

Abraham Taub, A.M.,

Emeritus Distinguished Service Professor of Pharmaceutical Chemistry,
College of Pharmaceutical Sciences,
Columbia University

and

Charles O. Wilson, Ph.D.,

Dean and Professor of Pharmaceutical Chemistry, School of Pharmacy,
Oregon State University

ALKANES (SATURATED HYDROCARBONS)

Alkanes or paraffin hydrocarbons are a very stable group of compounds of carbon and hydrogen comprising methane, CH_4, and its homologs. Their generic formula is C_nH_{2n+2}.

As the series increases in carbon chain length, the physical properties of normal alkanes show a gradation from gases, CH_4 to C_5H_{12}, to liquids up to about $C_{16}H_{34}$, then low-melting solids up to about $C_{23}H_{48}$ and rigid solids up to $C_{94}H_{190}$, the longest alkane synthesized.

For the first 3 members of the series, only normal, straight chain compounds are possible:

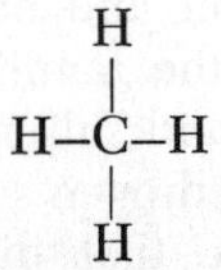

Methane 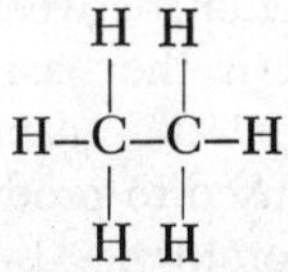   Ethane

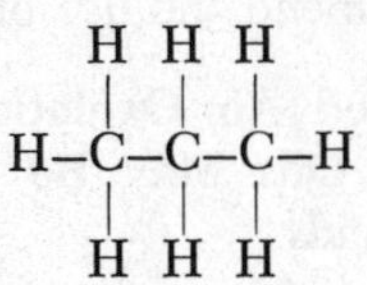

Propane

Beyond this, isomeric forms, having the same empirical formula but different chemical and physical properties, are possible:

Butanes

$$CH_3-CH_2-CH_2-CH_3$$
n-Butane

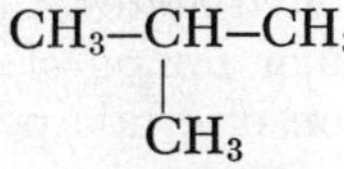

$$CH_3-CH-CH_3$$
$$| $$
$$CH_3$$
Isobutane
or
2-Methylpropane

The name 2-methylpropane is preferred for the isomeric or branched chain form of butane. It is representative of the Geneva system of nomenclature, in which the carbon atoms are arranged in the longest chain possible and numbered consecutively from one end of the chain. The substituents then are located by indicating the number of the carbon atom to which they are attached, the lowest possible number being used as a prefix. Thus, in the 3 isomeric pentanes, the second member is called 2-methylbutane and not 3-methylbutane.

$$CH_3-CH_2-CH_2-CH_2-CH_3$$
n-Pentane

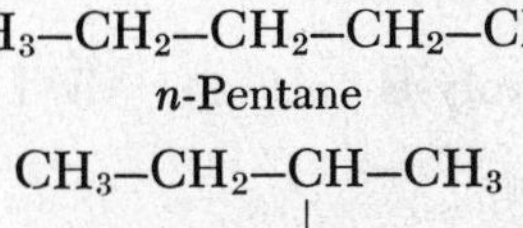

$$CH_3-CH_2-CH-CH_3$$
$$|$$
$$CH_3$$
2-Methylbutane

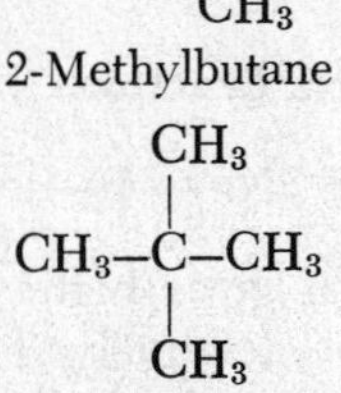

$$CH_3$$
$$|$$
$$CH_3-C-CH_3$$
$$|$$
$$CH_3$$
2,2-Dimethylpropane
(Tetramethylmethane)

125

The prefix methyl, applied to the CH_3 group, is typical of the name applied to a radical derived from an alkane with one hydrogen atom removed, the suffix *yl* replacing *ane*: methane, methyl (CH_3); butane, butyl (C_4H_9); alkane, alkyl, the latter being generic for saturated radicals and designated by the symbol R.

PREPARATION

Alkane hydrocarbons have been found in small amount in vegetable sources[1]: *n*-heptane, C_7H_{16}, in some species of pine; eicosane, $C_{20}H_{42}$, in spinach and in rose buds; nonacosane, $C_{29}H_{60}$, in pears and cabbages and hentriacontane, $C_{31}H_{64}$, in beeswax.

Commercial sources of the alkanes include natural gas and petroleum. The lower members, up to C_5H_{12}, are found in gas wells in the United States. However, petroleum is the major source of hydrocarbons. The composition of crude petroleum is indeed complex. It is made up of hundreds of different compounds. The composition varies with the source. Pennsylvania oils have a high content of paraffinic hydrocarbons with a residue of solid paraffins. Mid-continent oils have a higher naphthenic and aromatic hydrocarbon content, and Russian oils consist mostly of naphthenic hydrocarbons.

The commercial isolation of hydrocarbon fractions consists of a series of distillations of crude petroleum and subsequent washings and purifications.

Individual alkane hydrocarbons may be synthesized by a number of methods. A few of the more widely used syntheses are reviewed briefly.

1. Hydrolysis of an Alkyl Grignard Halide

$$R-Mg-X + H_2O \rightarrow R-H + Mg-X-OH$$

2. The Wurtz Reaction

$$2R-X + 2Na \rightarrow R-R + 2Na\,X$$

This reaction gives better yields for hydrocarbons of relatively high molecular weight.

3. Electrolysis of Carboxylic Acids

$$2R-COONa \rightarrow R-R + 2CO_2 + Na$$

The hydrocarbon and CO_2 are liberated at the anode.

CHEMICAL PROPERTIES

Alkanes undergo pyrolysis, yielding smaller molecules of unsaturated hydrocarbons, hydrogen and carbon.

Alkanes generally do not interact with highly reactive reagents such as concentrated sulfuric acid, nitric acid or potassium permanganate. Nevertheless, it is possible to cause some alkanes to interact under special conditions. The following reactions are illustrative.

1. Photochemical Halogenation

$$CH_4 + Cl_2 \xrightarrow{\text{ultraviolet light}} HCl + \underset{\substack{\text{Methyl} \\ \text{Chloride}}}{CH_3Cl}$$

This reaction may be explosive unless concentrations are kept low. The reaction proceeds to the formation of CH_2Cl_2 (methylene chloride), $CHCl_3$ (chloroform) and CCl_4 (carbon tetrachloride).

2. Sulfur Reduction

$$CH_3(CH_2CH_2)_nCH_3 + S \longrightarrow$$
$$nH_2S + CH_3(CH=CH)nCH_3$$

This equation is a general one expressing the reaction between sulfur and hydrocarbons. It is the basis for the familiar mixture of paraffin, asbestos and sulfur (Aichtu-ess) used to produce hydrogen sulfide in the laboratory. Also, note that directions given for preparing Sulfur Ointment U.S.P. do not recommend the use of heat.

3. Controlled Air Oxidation. High-Boiling Mineral Oils May Be Converted to Carboxylic Acids

$$C_nH_{(2n+1)}-CH_3 \xrightarrow{(O)} C_nH_{2n+1}COOH$$

The acids formed resemble the higher fatty acids, e.g., stearic acid, and have been utilized as fat substitutes in countries deficient in vegetable crops and livestock.

4. Vapor Phase Nitration[2]

$$C_2H_6 \xrightarrow[\text{1 atmos.}]{HNO_3, \, 420°} \underset{\text{Nitroethane}}{CH_3CH_2NO_2}$$

The nitroparaffins,[2] a commercial development of recent years, are valuable industrial solvents and serve as starting materials in the syntheses of unique pharmaceutical emulsifiers and of therapeutic agents such as the sympathomimetic amines.

The lower members of the alkane series are of little use in medicine. Attempts to utilize their slight anesthetic properties have proved unsuccessful. As the molecular weight increases, up to C_7H_{16}, there is a concomitant increase in toxicity. Above this molecular weight, the narcotic properties diminish rapidly.

The alkane products of pharmaceutical interest are mixtures of hydrocarbons derived from petroleum.

The saturated hydrocarbons are nonpolar in character and, therefore, are insoluble in water and other polar solvents. However, the hydrocarbons are completely miscible with all nonpolar solvents. Some of these are ether, chloroform, benzene, petroleum benzin, amyl acetate, fixed oils and carbon disulfide.

PRODUCTS

Solvent Hexane, Petroleum Ether, Purified Benzin, Petroleum Benzin. Solvent hexane is a purified distillate from petroleum, consisting of hydrocarbons, chiefly of the methane series. The first distillate from petroleum consists primarily of butane and pentane, boiling around 20°. It is a clear, volatile liquid containing chiefly pentane and hexane and having an ethereal or faint petroleumlike odor. It has a boiling range of 35° to 80° and a specific gravity range of 0.634 to 0.660. It differs from commercial cleaning benzin in being free from fat or other residue. It is highly flammable, and its vapors are explosive. Because of its excellent solvent prop-

erties for fixed and volatile oils, it is used as. a pharmaceutical extractant and in the defatting of drugs. It is miscible with chloroform, ether, benzene and dehydrated alcohol and is insoluble in water. The latter property permits the clean-cut separation of benzin-soluble active agents from aqueous solutions. It was used at one time as a local anesthetic.

Deodorized Kerosene. The kerosene fraction of petroleum contains aliphatic hydrocarbons from C_9H_{20} to $C_{15}H_{32}$ and also some unsaturated hydrocarbons. By treating commercial kerosene with concentrated sulfric acid and washing, followed by filtration through adsorbent clays such as fuller's earth, a deodorized kerosene is obtained.

$$RCH=CHR + H_2SO_4 \longrightarrow$$

Water soluble

This, because of its freedom from irritant unsaturated compounds, is used in cosmetic preparations such as hair oils and brilliantines and in liniments and antiseptic solutions. It is readily miscible with vegetable and mineral oils and evaporates at a very slow rate. It is used also in the manufacture of insecticidal liquids containing pyrethrum extract and chlorophenothane (D.D.T.).

Light Mineral Oil N.F.; Light Liquid Paraffin; Light White Mineral Oil. Light mineral oil is a mixture of liquid hydrocarbons obtained from petroleum. Mineral oils for pharmaceuticals are purified and freed from sulfur compounds, unsaturated hydrocarbons and solid paraffins. The N.F. permits the addition of an antioxidant to prevent the development of oxidative rancidity. Vitamin E (*dl-α-tocopherol*) is often used in 10 ppm. Often, unsaturated aliphatic compounds are present in minute amounts and will be oxidized to low molecular weight aldehydes and acids which affect both taste and odor. The official oil should be tasteless and odorless when cold and should not develop more than a faint odor of petroleum when heated.

Light mineral oil is colorless, non-fluorescent, odorless and tasteless but is otherwise analogous to lubricating oils in composition and is composed chiefly of C_{15} to C_{20} hydrocarbons. It is miscible with most fixed oils but not with castor oil. It is insoluble in water and in alcohol. Most American oils are chiefly aliphatic hydrocarbons. Russian mineral oils contain naphthenes or cyclic polymethylene hydrocarbons,

$$\overline{CH_2-(CH_2)_n-CH_2,}$$

where n may be 1 to 4, and the hydrogen atoms may be substituted by low molecular weight alkyl groups. Naphthenes tend to give greater viscosity to oils than do paraffins. In either series, the viscosity and the specific gravity, in general, increase with increased molecular weight. This is the basis of the distinction between light and heavy mineral oils. Light white mineral oil has a specific gravity range of 0.828 to 0.880 and a kinematic viscosity, at body temperature (37.8°), of not more than 37 centistokes (172.7 seconds by Saybolt Universal Viscosimeter).

Light white mineral oil is used in cosmetic products such as sun-screen oils, baby oils, hair gloss and cleansing creams. It is used as a pharmaceutical necessity in ointments and nasal preparations. Nose drops and sprays containing mineral oil should be applied cautiously and, preferably, under physician's supervision in infants, because of the possibility of aspiration of the oil and subsequent lipoid pneumonia, since mineral oil is not readily absorbed in the lungs.

Category—vehicle.

Mineral Oil U.S.P.; Liquid Paraffin; White Mineral Oil; Heavy Liquid Petrolatum. Mineral oil is a mixture of liquid hydrocarbons obtained from petroleum. The hydrocarbons usually present range in carbon content from C_{18} to C_{24}. Mineral oil has a specific gravity range of 0.860 to 0.905 and a kinematic viscosity at 37.8° of not less than 38.1 centistokes (177.2 seconds, Saybolt). Heavy Russian mineral oils may have viscosities in excess of 300 seconds.

Although mineral oil is composed of hydrocarbons of marked stability, some oils, particularly those less highly refined, on exposure to light and air develop a kerosene odor and taste. This is believed to be due to peroxide formation. The U.S.P. allows the addition of an antioxidant to prevent peroxide formation. A concentration of 10 ppm. of *dl-α*-tocopherol may be used. There is no official test prescribed for measuring the stability of a mineral oil. Golden[3] has developed a shelf-life test based upon heating the oil for 2 to 15 minutes at 300° F. and testing for peroxide formation with an acetone solution of ferrous thiocyanate. Those oils that remain free of peroxide formation for 15 minutes have an estimated shelf-life of at least a year.

Mineral oil has been used widely as an intestinal lubricant and laxative and for softening the contents of the lower intestine in the treatment of hemorrhoids and other rectal disturbances. Oils of higher viscosity are desirable because they are less likely to leak out from the lower bowel. Petrolatum is sometimes added further to prevent such leakage. Mineral oil also has been used as a noncaloric oil in obesity diets. Some studies[4] have indicated that mineral oil used near mealtime interferes with the absorption of vitamins A, D and K from the digestive tract and, therefore, interferes with the utilization of calcium and phosphorus, leaving the user liable to deficiency diseases; when used during pregnancy it predisposes to hemorrhagic diseases of the newborn. Mineral oil should be prescribed for limited periods and be administered only at bedtime. A recent study revealed that mineral oil in doses up to 30 ml. taken at bedtime over long periods of time did not have any effect on the vitamin A concentration of the blood nor were any other deleterious effects noted. It should be given to infants only upon the advice of a physician. The usual dose is 1 tablespoonful.

OCCURRENCE

Mineral Oil Emulsion N.F.

Petrolatum N.F.; Petrolatum Jelly; Yellow Petrolatum. Petrolatum is a purified, semisolid mixture of hydrocarbons obtained from petroleum. Petrolatum is a colloidal dispersion of aliphatic liquid hydrocarbons (C_{18} to C_{24}) in solid hydrocarbons (C_{25} to C_{30}), the disperser being a plastic material known as protosubstance, consisting of noncrystalline, naturally occurring, branched chain paraffinic type hydrocarbons (C_{25} to C_{30}). Without this, a mixture of mineral oil and paraffin would not be stable; the oil would leak out or "sweat." The fact that petrolatum does not produce an oily stain on paper indicates the presence of the oil in the inner phase of the dispersion.

Petrolatum is yellowish to light amber in color. It has not more than a slight fluorescence even after being melted, and it is transparent in thin layers. It is free or nearly free from odor and taste. It is soluble in benzene, chloroform, ether, petroleum, benzin, carbon disulfide, solvent hexane or in most fixed and volatile oils. It is partly soluble in acetone; the protosubstance, which is precipitated out, may be dissolved by the addition of amyl acetate. Petrolatum is insoluble in alcohol or water.

The melting point ranges from 38° to 60°, and the specific gravity between 0.815 and 0.880 at 60°. These should not be confused with consistency, measured by a penetrometer, which characterizes the firmness of texture of a petrolatum. Consistency is dependent on the microscopic fibers which make up a petrolatum. If these are tough and stiff, a product very firm in consistency results. Fibers also may vary in length. If they are too short, the product will be soupy at summer temperatures. If they are too long, the petrolatum is too tacky. Medium length is preferred for most pharmaceutical products. Petrolatums of soft consistency generally are used when ease of spreading is desired. Those of medium consistency are used most widely for ointments and cosmetic creams. Those of hard consistency are used in lipsticks and when

considerable mineral oil is to be incorporated.

Petrolatum is miscible with a relatively small proportion of water. However, the addition of small amounts of such ingredients as cetyl alcohol, lanolin and cholesteryl esters[4] may increase the water absorption properties to nearly 10 times the weight of the petrolatum base.

Category—ointment base.

White Petrolatum U.S.P.; White Petroleum Jelly. White Petrolatum is a purified mixture of semisolid hydrocarbons obtained from petroleum and is wholly or nearly decolorized. This differs from Petrolatum N.F. only in respect to color. It is white or faintly yellowish and transparent in thin layers. This type is preferred as a household topical dressing.

Although dressings of many types have been advocated for burns, petrolatum has been suggested as a simple and effective application. Healing of uncomplicated burns has been reduced from 7 to 2 days by the use of petrolatum on sterile gauze covered with a compression elastic bandage, compared with other methods of topical treatment.[5] White petrolatum is widely used as an oleaginous ointment base.

OCCURRENCE
Hydrophilic Petrolatum U.S.P.
Hydrophilic Ointment U.S.P.
Petrolatum Gauze U.S.P.

Plastibase; Squibb Base; Jelene Ointment Base. This is a combination of 95 per cent Mineral Oil U.S.P. and 5 per cent high molecular weight polyethylene (approximately 1,300). These are mixed at high temperature with high-speed stirring and then shock-cooled to an ointment state by a patented process (U.S. Nos. 2,628,187 and 2,628,205).

Plastibase is a soft, smooth, homogeneous, neutral and nonirritating ointment base. In most respects it is chemically similar to petrolatum but differs somewhat in physical properties. The consistency remains nearly unchanged within a temperature range of 40° to 180° F. Ointments are soft; they spread uniformly and compound easily.

Paraffin N.F.; Petrolatum Wax. Paraffin is a purified mixture of solid hydrocarbons obtained from petroleum. It is a white, translucent solid of crystalline structure composed of C_{24} to C_{30} hydrocarbons. Its solubility is similar to that of petrolatum. Its melting range is 47° to 65°. It is used in pharmacy mainly to raise the melting point of ointment bases.

The paraffins of commerce are, mainly, interlaced plate-type crystals representing straight chain hydrocarbons. Within recent years, paraffinic fractions which consist chiefly of branched chain hydrocarbons have been separated. Physically, they are composed of minute interlacing needles; they are known as microcrystalline waxes. They are plastic; they have a higher melting point, and they are tougher and more flexible than regular paraffin. They are used in polishes, paper coatings, and laminated boards. A pharmaceutical grade of somewhat lower m.p. (55°), known as Protowax, is useful to prevent leakage of oils, or sweating, in lipsticks, ointments and cosmetics.

Category—stiffening agent.

Microcrystalline wax is obtained from petroleum and has a much finer crystalline structure than paraffin. Its melting range is from 85° to 90°. The main pharmaceutical uses are in cosmetics, ointment bases and as a component of some of the wax-fat coatings used in sustained release products.

Ozokerite, earth wax, is naturally occurring solid saturated and unsaturated aliphatic hydrocarbons of high molecular weight (m.p. 80°). It is found in Austria and Poland and in this country in Texas and Utah. It comes in yellowish-brown to green color and in bleached yellow and white forms.

ALKENES (OLEFINS, UNSATURATED ALIPHATIC HYDROCARBONS)

Alkenes have the generic formula C_nH_{2n}. They resemble the paraffins in physical properties but, in marked contrast, are highly reactive. Structurally, this is indicated by the double bond, R=R. The lowest member is ethylene or ethene, $CH_2=CH_2$. Only one form exists of ethlyene and also of the next member of the homologous series, propene, or propylene, $CH_2=CH-CH_3$. As the molecular weight increases, two types of isomeric compounds are possible—isomers with respect to the position of the branched chain alkyl substituent and with respect to the position of the double bond or unsaturated linkage. Alkenes are named as are paraffins, the suffix *ane* being replaced by *ylene* or preferably *ene*. The position of the double bond is located by number, using the lowest number of the carbon chain which the bond follows.

$$CH_2=CH-CH_2-CH_3$$
Butene-1

$$CH_3-CH=CH-CH_3$$
Butene-2

$$CH_2=CH-CH_3$$
$$|$$
$$CH_3$$
2-Methylpropene (Isobutylene)

PREPARATION

Alkenes are found in natural gas and petroleum. They are obtained in larger quantities from the latter by "cracking." In the laboratory they usually are prepared by removing hydrogen halide from a halogenated hydrocarbon with alcoholic potassium hydroxide solution.

$$CH_3-CHI-CH_3(\text{ or } CH_2I-CH_2-CH_3)$$

$$\xrightarrow[80-100°]{\text{alc. KOH}} CH_2=CH-CH_3$$

The yield of propene from 2-iodopropane is 94 per cent, in contrast to 36 per cent from 1-iodopropane.

Commercial synthesis is accomplished by the dehydration of alcohols at temperatures of 150° or higher in the presence of H_2SO_4 or H_3PO_4 or by alcohols with superheated steam at 360° over a catalyst.

$$CH_3-CH_2OH \xrightarrow[360°]{Al_2O_3} CH_2=CH_2 + H_2O$$

CHEMICAL PROPERTIES

Alkenes are characterized by the following reactions, which are typical:

Addition at the Double Bond

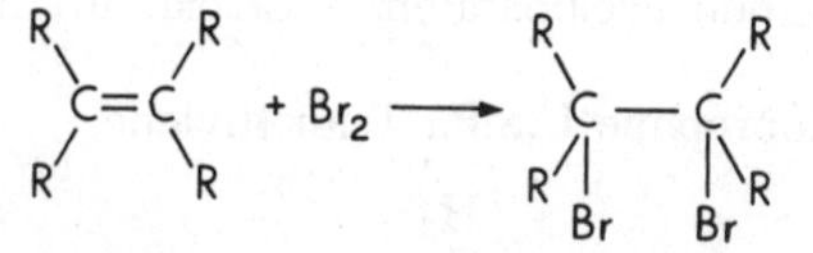

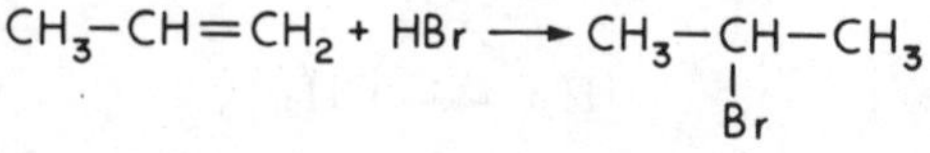

$$\underset{R}{\overset{R}{>}}C=C\underset{R}{\overset{R}{<}} + Br_2 \longrightarrow \underset{R\ Br}{\overset{R}{>}}C-C\underset{Br\ R}{\overset{R}{<}}$$

$$CH_3-CH=CH_2 + HBr \longrightarrow CH_3-\underset{Br}{CH}-CH_3$$

In this reaction the halogen becomes affixed to the carbon carrying the smaller number of hydrogen atoms.

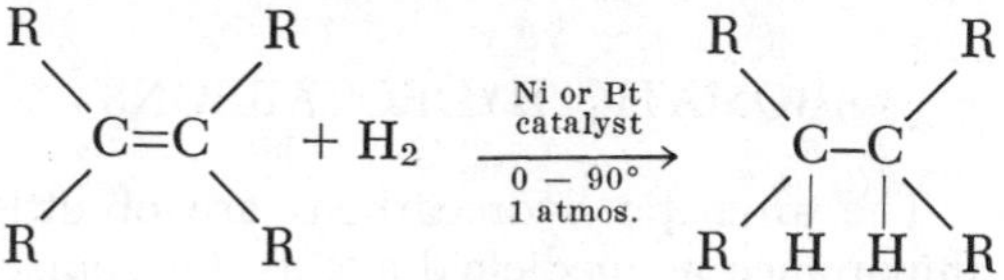

$$\underset{R}{\overset{R}{>}}C=C\underset{R}{\overset{R}{<}} + H_2 \xrightarrow[\substack{0-90° \\ 1\ atmos.}]{\substack{Ni\ or\ Pt \\ catalyst}} \underset{R\ H}{\overset{R}{>}}C-C\underset{H\ R}{\overset{R}{<}}$$

$$CH_2=CH_2 \xrightarrow[0-15°]{conc.\ H_2SO_4} C_2H_5\ OSO_2OH$$

Ethylsulfuric
Acid

$\downarrow$ H₂O

CH₃ CH₂OH
Ethanol

Oxidation to Glycols

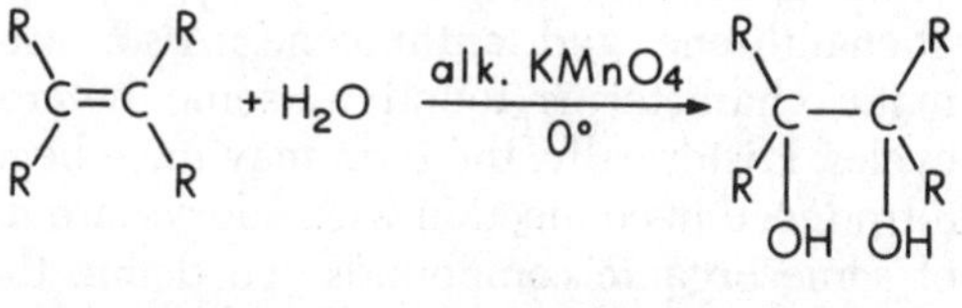

$$\underset{R}{\overset{R}{>}}C=C\underset{R}{\overset{R}{<}} + H_2O \xrightarrow[0°]{alk.\ KMnO_4} \underset{R\ OH}{\overset{R}{>}}C-C\underset{OH\ R}{\overset{R}{<}}$$

Ethylene N.F. (Ethene). The formula for ethylene is $CH_2=CH_2$. Ethylene, first isolated in 1669, is obtained by petroleum "cracking" or catalytic dehydrogenation; the ethylene is purified by fractional distillation of the gas, liquefied at 50 atmospheres and then washed to remove toxic by-products. Later it is compressed in steel cylinders.

Ethylene is a colorless gas of slightly sweet odor and taste. One liter under standard conditions weighs 1.260 g. It is soluble in 9 volumes of water and in 0.5 volume alcohol at 25°. It is *highly flammable, and*

mixtures with air or oxygen are explosive in the presence of a flame or spark. Ethylene for medicinal use as an anesthetic (*see* Chap. 15) must be free from acetylene, aldehydes, hydrogen sulfide, phosphine and carbon monoxide and from all but traces of acids, alkalies or carbon dioxide. It is assayed gasometrically by absorption in bromine water.

In addition to its use as an anesthetic, ethylene has been utilized in the controlled ripening of bananas, citrus fruits, bulbs and potatoes, permitting the shipment of green or unripe crops. It enables starches to hydrolyze to sugars and develops the matured coloring principles.

Category—general anesthetic.
Application—by inhalation as required.

ALKYNES (ACETYLENES, ETHINES)

These unsaturated hydrocarbons have the generic formula C_nH_{2n-2}. Their chemical reactivity is represented structurally by a triple bond, $R-C\equiv C-R$. The nomenclature is analogous to that of the alkenes, the suffix *yne* being used.

Acetylene (Ethnine, Narcylene), $HC\equiv CH$, the lowest member of the group, typifies many of the reactions of the group found in illuminating gas. Acetylene may be synthesized from 1,2-dibromoethane.

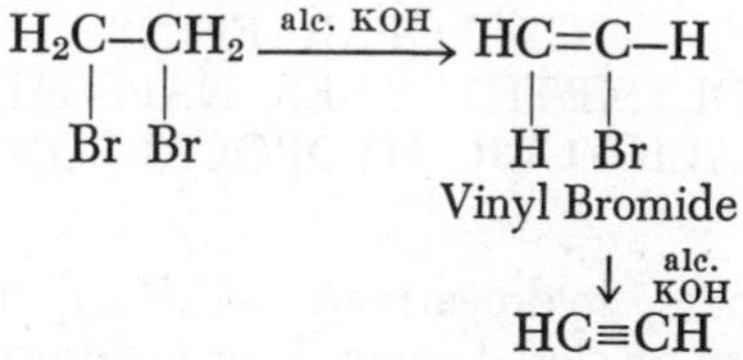

$$\underset{Br\ Br}{\overset{}{H_2C-CH_2}} \xrightarrow{alc.\ KOH} \underset{H\ Br}{\overset{}{HC=C-H}}$$

Vinyl Bromide

$$\downarrow \overset{alc.}{KOH}$$

$$HC\equiv CH$$

Commercially, it is made by the action of water on calcium carbide, CaC_2, or by passing hydrogen through an electric arc.

It reacts stepwise with hydrogen or halogen:

$$HC\equiv CH \xrightarrow[catalyst]{H_2} CH_2=CH_2$$

$$\downarrow \overset{H_2}{catalyst}$$

$$CH_3-CH_3$$

Acetylene forms explosive acetylides with salts of some heavy metals such as copper or silver (H–C≡C–Ag) (see test for acetylene in Ethylene N.F.).

Acetylene is a colorless gas with a garlic-like odor. Because of its tendency to explode when compressed, it is made available, commercially, dissolved under 200 lbs. pressure in acetone; 300 volumes dissolve in 1 volume of acetone.

Acetylene is the basis for a host of industrial chemicals and intermediates, including polymerized products such as Neoprene, a synthetic rubber.

DIENES OR DIOLEFINS

These aliphatic hydrocarbons (C_nH_{2n-2}) that contain two unsaturated linkages are classified on the basis of the position of the double bonds: (1) those in which the double bonds are adjacent, such as allene $(CH_2=C=CH_2)$ and its homologs; (2) those in which the double bonds are conjugated or separated by a single bond, such as butadiene-1,3($CH_2=CH–CH=CH_2$); (3) those in which at least two single bonds separate the double bonds, such as pentadiene-1,4($CH_2=CH–CH_2–CH=CH_2$).

Dienes do not include any compounds of pharmaceutical interest, although diene linkages are encountered as side chains in some drug components.

CYCLOPARAFFINS
(POLYMETHYLENES, NAPHTHENES, ALICYCLIC HYDROCARBONS)

These hydrocarbons (C_nH_{2n}), though ringed or closed chain, bear a closer resemblance to the paraffins than to the benzenoid hydrocarbons. The carbon valences are fully saturated. Based on the Bayer strain theory, the 5- and 6-carbon rings are most stable; the 3- and 4-membered rings are more reactive and deviate in properties from the other members of the series.

Nomenclature resembles that of the paraffins, the prefix *cyclo* being used to designate the closed part of the molecule. Typical isomeric compounds are represented as follows:

$$
\begin{array}{cc}
\begin{array}{c} CH_2\text{—}CH_2 \\ | \qquad\quad | \\ CH_2\text{—}CH_2 \\ \text{Cyclobutane} \end{array}
&
\begin{array}{c} CH_2 \\ | \quad\diagdown \\ \quad\quad CH\text{–}CH_3 \\ | \quad\diagup \\ CH_2 \\ \text{Methylcyclopropane} \end{array}
\end{array}
$$

Only one cycloparaffin is official in the *U.S.P.*

Cyclopropane U.S.P.; Trimethylene.

$$
\begin{array}{c}
H_2 \\
C \\
\diagup \quad \diagdown \\
H_2C\text{——}CH_2
\end{array}
$$

This was first isolated in 1882, but it was not until 1929 that its study as an impurity in propane led to its use as a valuable inhalation type anesthetic (see Chap. 15).

AROMATIC HYDROCARBONS

The aromatic hydrocarbons are of little importance as medicinal agents. Of course, in many cases they are the nuclei for more active and desirable compounds, but as yet no medicinal application has been found for the aromatic hydrocarbons. Carcinogenic properties have been observed and studied for some of the more complex aromatic hydrocarbons.

Aromatic character and aromaticity are terms that are used in reference to cyclic compounds, such as benzene, naphthalene, phenanthrene and anthracene. Also, aromatic character is found in some heterocycles. Historically, the term may have been introduced in connection with odor or aroma of some organic compounds. To define the term "aromatic character" clearly and quickly is most difficult, since such a characteristic is associated with several chemical phenomena. When the chemical properties of benzene are compared with those of the unsatured alkenes and alkynes, the diminished unsaturation reactivity and the tendency to the formation and the preservation of structural type are impressive. That benzene is unsaturated has been established unquestionably, yet it is so much less reactive than hexatriene ($CH_2=CH–CH=CH–CH=CH_2$) that the unsaturation seems to be of a modified special character. In contrast with the simple alkenes and alkynes, benzene reacts only slowly with bromine and is inert toward hydrogen bromide. It decolorizes weak, cold, acidic solutions of potassium permanganate but does not decolorize alkaline permanganate solutions, whereas olefins quickly decolorize acidic or basic permanganate. The stability of the unsaturated ring system of benzene and its ease of formation is exhibited by the fact that aromatic type compounds are produced by the pyrolysis of coal, the dehydration of camphor with phosphorous pentoxide to *p*-cymene and dehydrogenation reactions with sulfur, selenium and palladium.

A specific character of these aromatic compounds is that most of their reactions are substitutions; this is in contrast with the reactions of alkenes and alkynes, which are additions. Picric acid, under suitable conditions, reacts with the condensed ring polynuclear hydrocarbons to form complexes that are called picrates.[6] These complexes, due to specific melting points, are useful as derivatives for identification and, since the hydrocarbon can be regenerated, are important for purification.

Benzene (Benzol). (*See* Reagents U.S.P.) The chief source of benzene (C_6H_6) is coal tar, from which it is obtained in yields of 1 to 2 per cent. Sometimes impurities such as toluene, carbon disulfide and thiophene may be present.

It is a colorless, transparent liquid having a burning taste and an agreeable aromatic odor. Benzene is miscible with alcohol, chloroform, ether, acetone, glacial acetic acid or oils and is immiscible with water. Thus, it is a good solvent for fats, resins or certain alkaloids. The vapor of benzene is highly flammable, producing much soot, and forms an explosive mixture with air in concentrations of 1.5 to 8.0 per cent.

External contact with benzene should be avoided because of the defatting effect and irritant action. Repeated exposure may establish serious, abnormal skin conditions.

Naphthalene (Tar Camphor) ($C_{10}H_8$) occurs as glistening white crystalline plates, having a sharp taste and a characteristic odor. The crystals are soluble in organic solvents (hot alcohol; very slightly soluble in cold alcohol) and insoluble in water. The solid sublimes readily and will vaporize slowly under normal conditions (an exposed moth ball vaporizes in about three months).

Anthracene ($C_{14}H_{10}$). This hydrocarbon is recovered from anthracene oil (with a boiling point between 270° and 360°) obtained by the distillation of coal tar.

Anthracene occurs as colorless crystalline plates that have a blue fluorescence. The crystals dissolve in most organic solvents, are less soluble in alcohol or ether and are insoluble in water.

Anthracene is an ingredient in synthetic coal-tar ointment.

Phenanthrene ($C_{14}H_{10}$) is a structural isomer of anthracene isolated in the same oil fraction from the distillation of coal tar. The hydrocarbon occurs as white glistening plates, soluble in ether, benzene, alcohol, glacial acetic acid or carbon disulfide, but insoluble in water. Solutions of it possess a blue fluorescence.

SILICONES*

The term *silicone* is used to designate the organosilicon-oxide polymers. These same structures also are called *organopolysiloxanes*. Silicones may be considered as being related to glass, in that they are a chemical hybrid of organic and inorganic substances. They are organic derivatives of silica, SiO_2, and possess properties never before obtained in any known compound.

Chemically, the "silicones" are very similar to the "silicic acids." First, let us review some structural aspects of the silicic acids. Silicon dioxide, SiO_2, on hydration, should yield orthosilicic acid (actually, this does not occur). $SiO_2 + 2H_2O \rightarrow H_4SiO_4$. However, if silicon tetrachloride is treated with water, hydrolysis takes place, with the formation of a polymer that is gelatinous and forms a colloidal state upon dilution.

$$SiCl_4 + 4H_2O \rightarrow H_4SiO_4 + 4HCl$$

---

* Silicones are not hydrocarbons but do have similar properties and pharmaceutic uses.

The structure of orthosilicic acid may be considered as:

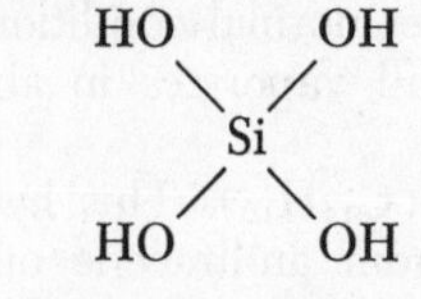

HO
OH
Si
HO
OH

It readily polymerizes by splitting out water from two or more molecules to form a large number of "silicic acids." Examples of these are orthodisilicic acid, a compound of which is kaolinite (aluminum orthodisilicate);

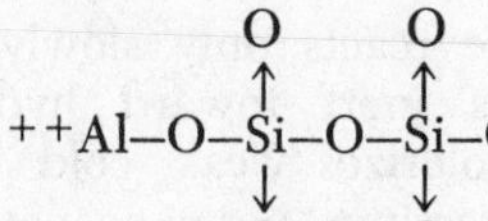

O
O
++Al–O–Si–O–Si–O–Al++
O
O

and Magnesium Trisilicate U.S.P. ($Mg_2Si_3O_8$) which may be thought of as a derivative of trisilicic acid ($H_4Si_3O_8$).

Mg++
⁻O
O
O
O⁻
Si
Si
Si
⁻O
O
O
O⁻

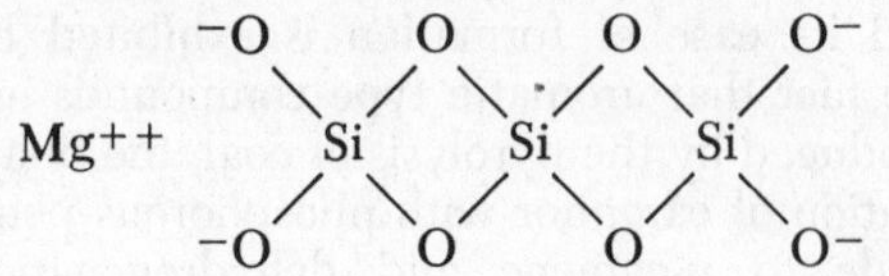

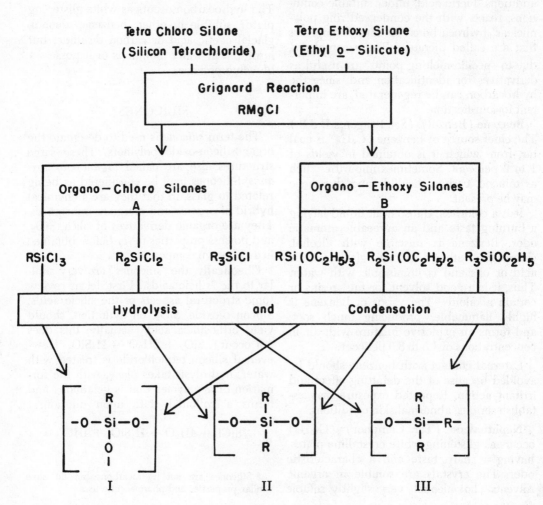

$$SiO_2 + 2C + 2Cl_2 \longrightarrow SiCl_4 + 2CO$$

$$SiCl_4 + 4C_2H_5{-}O{-}C_2H_5 \longrightarrow (C_2H_5O)_4Si$$

Tetra Chloro Silane
(Silicon Tetrachloride)
Tetra Ethoxy Silane
(Ethyl o-Silicate)
Grignard Reaction
RMgCl
Organo-Chloro Silanes
A
Organo-Ethoxy Silanes
B
RSiCl3
R2SiCl2
R3SiCl
RSi(OC2H5)3
R2Si(OC2H5)2
R3SiOC2H5
Hydrolysis
and
Condensation
R
-O-Si-O-
O
R
-O-Si-O-
R
R
-O-Si-R-
R

I II III

Silicones are polymers similar in structure to the silicic acids, differing only in organic radicals (methyl, phenyl, etc.) attached by a carbon-silicon bond to silicon. The synthesis of these compounds is accomplished indirectly. The scheme shown on page 134 outlines a possible process.[7]

The organosilanes represented by the 3 compounds of A and the 3 of B are hydrolyzable to organo-substituted silicic acids called silanols. A condensation rapidly takes place between the silanols, splitting out water, which results in the silicon-oxygen-silicon polymer. A study of the types of silanols, I, II and III, will reveal quickly the unlimited number of combinations. Type I polymers produce resinous silicones; II yields a linear liquid product; and III is used to stop the polymerization of types I and II and, thus, control the size of the organopolysiloxane (silicone) molecule.

Silicone fluids are linear molecules where n = 0 to 2,000 and R = methyl (DC 200); in the more viscous fluids (DC 702 and DC 550) some phenyl groups are present.

$$(CH_3)_3Si-\left[\begin{array}{c} CH_3 \\ | \\ O-Si- \\ | \\ CH_3 \end{array}\right]_n O-Si(CH_3)_3$$

Silicone compounds are methyl or methylphenylpolysiloxanes containing various fillers and sometimes a greater or smaller number of cross-linked molecules. Two common ones are the petrolatumlike DC 4, having a siliceous filler, and antifoam A, also greaselike.

Silicone resins are methyl or methylphenylpolysiloxanes of high molecular weight, with a large amount of cross linking. These resins are inert chemically and are used as insulating varnishes, protective coatings and releasing agents on bakery pans. Examples are DC 2102 and DC Pan Glaze.

Silicones appear to have a future as agents for varied therapeutic, diagnostic and laboratory uses. They are practically inert physiologically, being nontoxic[8] in all methods of administration except intravenous. These compounds are stable under practically every conceivable condition, being resistant to oxidation. All are nonvolatile and noncrystalline. The fluidity or viscosity for all silicones is nearly constant over a temperature range of −70° to 200°. Solubility characteristics vary with the type and the molecular size of the silicone, but, in general, for silicones, hydrocarbon or chlorinated hydrocarbon solvents are best (benzene, toluene, petroleum fractions, carbon tetrachloride and methylene chloride); solvents containing oxygen (ether is an exception) have poor solvent properties. Silicone completely repels water.

Silicones have no effect on rubber and are not corrosive to metals; in fact, some are applied to surgical instruments as a coating, to protect them from rusting. Glass vials for injectables are coated on the inside to induce 100 per cent drainage. The best and most permanent lubrication for syringes is DC 200. An available product is DOW Corning Stopcock Grease, for use on all glass-to-glass contacts. In many cases of surgery, or in experiments involving blood, all glass articles are silicone coated. Silicote[9] (Pro-Derma, Covicone) is a formulation of silicone and petrolatum and is used for skin protection in diaper rash, ulcers and irritation from colostomies. Dental ointments designed for oral use containing silicones are well retained because of inertness and water insolubility.

PRODUCTS

Simethicone, also known as dimethicone and methylpolysiloxane, is a silicone oil consisting of dimethylsiloxane polymers and technically designated as dimethylpolysiloxane of the DC 200 series of fluids. This compound is the silicone used most frequently in pharmaceutical products. Simethicone is a mixture of liquid dimethyl polysiloxanes containing 4.45 per cent of silica aerogel, based on the weight of the mixture, processed to take on a defoaming characistic. The mixture is homogeneous and has a greaselike consistency.

A number of studies utilizing silicones in ointments,[10] liniments and as media for intramuscular injections give promise for future uses.

HALOGENATED HYDROCARBONS

One or more of the hydrogen atoms of paraffin hydrocarbons may be replaced by chlorine, bromine, iodine or fluorine, although not necessarily by direct substitution. Halogenated hydrocarbons do not occur naturally. In addition to the methods of direct substitution of saturated hydrocarbons and of addition of halides or hydrogen halides to unsaturated hydrocarbons (see p. 131), halogenated hydrocarbons may be prepared by the interaction of alcohols, aldehydes or ketones with hydrogen halides, phosphorus halides or thionyl chloride ($SOCl_2$).

$$ROH + HCl \xrightarrow[\text{catalyst}]{ZnCl_2} RCl + H_2O$$

$$ROH + PBr_5 \longrightarrow RBr + POBr_3 + HBr$$

$$3\ ROH + PI_3 \longrightarrow 3\ RI + H_3PO_3$$

$$ROH + SOCl_2 \longrightarrow RCl + SO_2 + HCl$$

$$RCHO + PCl_5 \longrightarrow RCHCl_2 + POCl_3$$

The addition of halogens to hydrocarbons increases the specific gravity and raises the boiling points of the hydrocarbons. The iodine derivatives have the highest boiling points and the bromine derivatives have intermediate boiling points compared with those of the chlorinated hydrocarbons of analogous structure. Corresponding fluorinated hydrocarbons have the lowest boiling points. All are immiscible with water, although the presence of the halogen atom confers some polar properties on these compounds. They are less flammable than the corresponding hydrocarbons and become noninflammable as the degree of halogenation reaches a maximum.

Their reactions in forming saturated and unsaturated hydrocarbons have been discussed previously. Other characteristic reactions include the formation of (1) the Grignard reagent (R-Mg-X), (2) nitriles (organic cyanides), (3) alcohols, (4) ethers from metallic alcoholates and (5) amines.

1. $RBr + Mg \rightarrow R-Mg-Br$

2. $RCl + NaCN \rightarrow R-C\equiv N + NaCl$

3. $RI + aq.\ KOH \rightarrow ROH + KI$

4. $R-O-Na + RBr \rightarrow R-O-R + NaBr$

5. $RBr + NH_3 \rightarrow RNH_3\ Br \xrightarrow{NaOH} RNH_2 + NaBr + H_2O$

ALKANE DERIVATIVES

Methyl Chloride (Monochloromethane). Methyl chloride (CH_3Cl) is prepared industrially by the vapor phase chlorination of methane above $200°$. It is a gas of ethereal odor, having a boiling point of $-23.7°$. It burns with a greenish flame. It is compressed easily to a liquid of specific gravity 1.782. When evaporated rapidly in a current of air, a temperature of $-55°$ may be obtained. Upon this property depends its rather limited use as a local anesthetic. To avoid blister formation, it generally is mixed with ethyl chloride and sprayed directly on the skin or applied with cotton.

The narcotic potency of hydrocarbons increases with the extent of halogenation. Methyl chloride has the lowest and carbon tetrachloride the highest narcotic potency of the chlorinated methanes.

Methylene Chloride (Dichloromethane). Methylene chloride (CH_2Cl_2) is a liquid with a boiling point of $40.8°$ and a specific gravity of 1.336. Its vapors are not inflammable, although weakly combustile mixtures are formed with air.

Although methylene chloride is more effective and less toxic as a general anesthetic than methyl chloride, it does not compare favorably with chloroform.

Methylene chloride is an excellent solvent for oils, rubber, some resins and cellulose plastics. It enters the composition of many paint and varnish removers.

Chloroform N.F. (Trichloromethane). Chloroform ($CHCl_3$) was first synthesized by Leibig in 1831; it was introduced as an obstetric anesthetic in 1847 by Simpson, an Edinburgh surgeon, within a year of the introduction of ether as an anesthetic in this country.

Chloroform is prepared industrially by the haloform reaction, the starting materials being alcohol or acetone and an alkali chlorine compound such as bleaching powder or sodium hypochlorite.

$$CH_3CH_2OH \xrightarrow{\text{NaOCl}} CH_3CHO \xrightarrow{\text{NaOCl}}$$

Ethanol   Acetaldehyde

$$CCl_3CHO \xrightarrow{\text{NaOH}} CHCl_3 + \overset{\displaystyle O}{\overset{\|}{HC}}-ONa$$

Chloral   Chloro-   Sodium
                     form   Formate

Another commercial synthesis involves the reduction of carbon tetrachloride with iron and water.

$$CCl_4 + 2H \; (H_2O, Fe) \rightarrow CHCl_3 + HCl$$

Chloroform is a colorless, mobile liquid of ethereal odor and sweet taste. It has a boiling point of 61° and a specific gravity of 1.475. It is soluble in about 200 volumes of water and miscible with most organic solvents and oils. Its heated vapors burn with a green flame; the liquid itself is not inflammable. In the presence of air, sunlight or open flames, it is oxidized to phosgene,

$$Cl-C=O$$
$$|$$
$$Cl$$

a highly reactive gas which hydrolyzes in the lung tissues to hydrogen chloride, producing pulmonary edema. To minimize the existence of phosgene in chloroform during storage, the *N.F.* prescribes the presence of 0.5 to 1 per cent of alcohol. Alcohol reacts with phosgene to form nontoxic diethylcarbonate. Also expected would be the oxidation of alcohol to acetaldehyde. Other impurities that should be absent from anesthetic grade chloroform are chlorine and chlorinated decomposition products, acids, aldehydes, ketones and readily carbonizable substances. Chloroform should be stored in airtight, light-resistant containers at a temperature not above 30°. When corks are used, they should be covered with tin foil.

Chloroform is used as a hypnotic and a sedative in cough liquids and lozenges and as a counterirritant and an analgesic in liniments. Chloroform water is used as a carminative in a dose of 15 ml. It is used also as a preservative for crude drugs and galenicals. It serves as a solvent for fats, resins and some plastics and is an excellent extractant for alkaloids and other soluble medicinal agents in their manufacture and assay.

Category—solvent.

Carbon Tetrachloride. (See Chapter 9.)

Methyl Bromide. Methyl bromide (CH_3Br) is a colorless gas; when compressed, it forms a colorless noninflammable liquid, boiling at 4.5°. It is used in fire extinguishers; in the presence of flames or hot surfaces, hydrogen bromide and bromine are liberated. Because of the excellent penetration of the gas, it has been used effectively as a fumigant to combat insect pests in grains and other foodstuffs and in crude drugs. It is used also as a rodenticide. Its vapors are toxic to the nervous system.

Iodoform N.F. (See Chapter 9.)

Fluorohydrocarbons. Freon®, Genetron®, Isotron®. Compounds in common use are the chlorofluoromethanes and the chlorofluoroethanes. Difluordichloromethane (CF_2Cl_2) is an example. Because of their low boiling point ($-30°$), noninflammability and unexpected freedom from toxicity, Midgley and Henne introduced the Freon group as refrigerants in 1930. Tests by the U. S. Bureau of Mines have shown them to be nontoxic at concentrations of 20 per cent in air for exposures as long as 8 hours. They have been adopted widely in the closed systems of mechanical refrigerators.

In 1943, the "aerosol bomb" (U. S. Pat. 2,321,023) was developed for insecticidal sprays to be used by the Armed Forces.

The aerosols consist of a solution of active ingredients and of a propellant in a sealed container with a specially designed valve and a standpipe.

Therapeutic application of aerosols seems to be a well-established medical practice. Today we find them designed for inhalation products that include antibiotics, vasoconstrictor amines, endocrines, antihistamines, local anesthetics, radiologically opaque solutions, radioactive isotopes and sulfonamides. Most uses are limited to localized action.

Products that are of a cosmetic nature include aerosols of personal deodorants, colognes, sun-screen lotions, shampoos, and other such products.

Compounds recognized by the N.F. are:
Dichlorodifluoromethane N.F.
   Category—aerosol propellant.
Dichlorotetrafluoroethane N.F.
   Category—aerosol propellant.
Trichloromonofluoromethane N.F.
   Category—aerosol propellant.

Ethyl Chloride N.F. (Monochloroethane, Kelene). Although gaseous at room temperature (b.p. 12°), ethyl chloride (CH_3CH_2Cl) is available in liquefied form in valved glass or metal cylinders. In its liquefied state, it has a specific gravity of 0.921 at 0°. It has an ethereal odor and burning taste. It is slightly soluble in water and soluble in alcohol and in ether. It is flammable and explosive when present in as little as a 4 per cent concentration in air.

Ethyl chloride was introduced by Carlson in 1894 as a general inhalation anesthetic in dentistry but is now used mainly as a local anesthetic (*see* Chap. 15).

Ethyl Bromide. Ethyl bromide (CH_3CH_2Br) is a colorless liquid of specific gravity 1.43 and a boiling point of 38°. It has an ethereal odor and a sharp, burning taste. It is only sparingly soluble in water but is soluble in alcohol and in ether. It is used as a refrigerant topical anesthetic, like ethyl chloride; it also is used for temporary relief from neuralgia. In admixture with methyl and ethyl chlorides, it has been used for general anesthesia.

Chlorinated Paraffin, Chlorcosane. Chlorinated Paraffin is a liquid paraffin which has been treated with chlorine.

It is an odorless, viscous liquid, yellow to amber in color, with a specific gravity of 1.03. It is not miscible with water, is slightly soluble in alcohol and is miscible with ether, chloroform or benzene.

Chlorcosane was introduced by Dakin in 1918 as a special solvent for Dichloramine T.

Alkene Derivatives

Trichloroethylene U.S.P. (1,1,2-Trichloroethene), Trimar®, Trilene®.

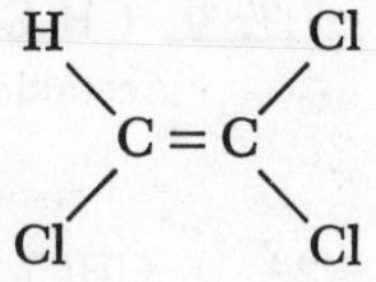

Trichloroethylene is made by elimination of hydrogen chloride from symmetrical tetrachlorethane ($CHCl_2 \cdot CHCl_2$) by passing the latter over hydrated lime, or over pumice at 500°.

Trichloroethylene is a colorless, mobile, highly volatile liquid with a specific gravity of 1.46 and a boiling point of 88°. (See Chap. 15.)

Tetrachloroethylene U.S.P. See Chapter 9.

Aromatic Derivatives

The halogenated aromatic hydrocarbons have not been found useful as medicaments, although many are used. They are quite toxic and vary in physiologic response according to which one of the halogens is present and whether the halogen is attached to the aromatic ring or is in a side chain. Substituting a halogen for a hydrogen on the nucleus of an aromatic hydrocarbon (chlorobenzene) increases the toxicity and does not impart any sedative or anesthetic effect. The order of toxicity is approximately the following: toluene < *o*-chlorotoluene < *m*-chlorotoluene < *p*-chlorotoluene. Increasing the number of halogen atoms in the hydrocarbon lessens the toxicity. This is explained by the decrease in solubility of the halogenated hydrocarbons. Thus, the order of toxicity is: chlorobenzene > 1,4-dichlorobenzene > 1,3,5-trichlorobenzene. Where water solubility is not a limiting factor, an increase in the number of halogens contained in the aromatic compound will give rise to increased toxicity.

The halogenated aromatic hydrocarbons that have the halogen in a side chain, e.g., benzyl bromide and chloromethyltetralin, have decided lacrimatory properties. Exposed skin also is attacked, with resulting redness and discomfort.

REFERENCES CITED

1. Kremers, E.: Phytochem III, Madison, Wisconsin, Bull. Univ. Wisconsin, 1934.
2. Hass, H. B., and Riley, E. F.: Chem. Rev. 32:373, 1943.
3. Golden, M. J.: J. Am. Pharm. A. (Sci. Ed.) 34:76, 1945.
4. Cataline, F. L., Jeffries, S. F., and Reinish, F.: J. Am. Pharm. A. (Sci. Ed.) 34:33, 1945.
5. Stubenbord, J. G.: Aviation Med. 16:192, 1945.
6. Fieser, L. F., and Fieser, M.: Organic Chemistry, ed. 3, p. 585, New York, Reinhold, 1956.
7. Rowe, V. K., *et al.*: J. Indust. Hyg. Toxicol. 30:332, 1948.
8. Bardondes, R. de R., *et al.*: Mil. Surgeon 106:379, 1950.
9. Talbot, S. R., *et al.*: J. Invest. Derm. 17:125, 1951.
10. Bickmore, J., and Plein, E. M.: J. Am. Pharm. A. (Sci. Ed.) 42:79, 1953.

SELECTED READING

Bennett, H.: Commercial Waxes, New York, Chem. Pub. Co., 1944.

Boyd, J. H., Jr.: Petroleum, source of raw materials for chemical industry, Chem. Eng. News 23:345, 1943.

Hyde, J. F.: Chemical backgrounds of silicones, Science 147:829-836, 1965.

Mather, K. F.: Petroleum—today and tomorrow, Science 106:603, 1947.

Meyer, E.: White mineral oil and petroleum in pharmaceuticals and cosmetics, J. Am. Pharm. A. (Sci. Ed.) 24:319, 1935.

Riegel, E. R.: Industrial Chemistry, Chapter 24, New York, Reinhold, 1942.

Schindler, H.: Petrolatum for drugs and cosmetics, Drug and Cosm. Ind. 81 (1):36, July, 1961.

Shreve, R. N.: The Chemical Process Industries, Chapter 37, New York, McGraw-Hill, 1945.

5

Alcohols and Ethers

Charles O. Wilson, Ph.D.,
Dean and Professor of Pharmaceutical Chemistry, School of Pharmacy,
Oregon State University

ALCOHOLS

INTRODUCTION

The aliphatic hydroxy-containing compounds included in this section are the monohydric alcohols, the dihydric alcohols, the trihydric alcohols and the polyhydric alcohols.

Alcohols possess properties similar to those of both hydrocarbons and water and may be considered as derivatives of either. Water (H-O-H) with one of its hydrogen atoms replaced by a hydrocarbon group yields a monohydric alcohol. A hydrocarbon with one of its hydrogen atoms replaced by a hydroxyl group yields a monohydric alcohol.

Hydrocarbons undergo a decided change in physical, chemical and physiologic properties when one or more hydroxyl groups are introduced. The physical properties approach those of water, especially in the alcohols of low molecular weight. These are readily water soluble (methyl, ethyl, propyl and isopropyl alcohols). However, water solubility decreases as the molecular weight of the alcohol increases but increases with an increase in the number of hydroxyl groups and the branching of the chain. Therefore, it is of interest to note that secondary alcohols are more soluble in water than are the corresponding primary alcohols and that tertiary alcohols are more soluble than secondary ones. Solubility of some alcohols in

water given as grams per 100 ml. are: n-butyl, 7.9; isobutyl, 12.5; and n-amyl, 2.7. Alcohols have solvent powers similar to those of water and hydrocarbons; this is particularly true of methanol, ethanol, propanol-2, glycol and glycerin. The similarity to the solvent powers of hydrocarbons increases with an increase in the molecular weight. In a given carbon chain, as the hydroxyl groups are increased, so are the boiling point and sweetness and solubility in water.

The hydroxyl group of the alcohols permits hydrogen bonding with water molecules. But, as pointed out, when the carbon content of alcohols increases, the molecule becomes more nonpolar and, thus, loses water solubility. Not only water but other compounds with $-OH$, $-NH_2$ or $=NH$ groups show good solubility in most common alcohols. This is the basis for the solubility of many organic compounds in alcohol.

There are 3 types of monohydric alcohols, depending on the number of chemical groups attached to the same carbon atom bearing the hydroxyl group. A primary alcohol ($R-CH_2OH$) has one R group and two hydrogens attached to the hydroxyl carbon; a secondary alcohol (R_2CHOH) has two R groups and one hydrogen attached to the hydroxyl carbon; a tertiary alcohol (R_3COH) does not have any hydrogen attached to the carbon atom to which the hydroxyl is attached.

Alicyclic alcohols are formed from cyclic hydrocarbons by replacing one or more hydrogen atoms by a hydroxyl group. These are usually secondary alcohols. A phenol is an aromatic compound that has a hydroxyl group attached to an aromatic ring. In any chemical structure, a carbon atom rarely bears two hydroxyl groups. Glycols are aliphatic alcohols containing two hydroxyl groups, e.g., ethylene glycol. There occur in nature compounds having two or more hydroxyl groups (glycerin and mannitol), as well as compounds containing several hydroxyl groups along with other chemical groups, such as aldehyde (glucose), carboxyl (uronic acids) and amino (glucosamine).

Alcohols in the form of esters are distributed widely in nature. Volatile oils, fixed oils, fats and waxes are the most common sources of these alcohols.

MONOHYDRIC ALCOHOLS

PREPARATION

The availability of alcohols is most varied: some are obtained from natural sources (glycerin from fat); some are prepared from complex natural material by fermentation processes (ethanol from starch); some are prepared by synthesis from simple chemicals (methanol from hydrogen and carbon monoxide), others from standard procedures in organic chemistry (isopropyl alcohol by the reduction of acetone).

Methods applicable to the preparation of dihydric alcohols are discussed in connection with ethylene glycol. Procedures used to obtain trihydric and polyhydric alcohols are limited in number, and examples are given for the specific alcohols later described.

CHEMICAL PROPERTIES

In chemical properties, the alcohols resemble water by reacting with sodium and phosphorus halides. They are neutral to litmus paper and often form, with organic salts, a complex called "alcohol of crystallization."

Their chemical properties approach those of the hydrocarbons as the molecular weight increases and those of water as the number of hydroxyl groups increases. A hydroxyl group introduced into a hydrocarbon produces a class of compounds, alcohols, that exhibit individual chemical properties. Some of these properties are: reaction with metals to form metal alkylates, dehydration to form olefins, oxidation to form aldehydes and ketones and combination with acids to form esters.

Oxidation of alcohols may be carried out by several procedures, using a variety of oxidizing agents.

A. *Primary alcohols* are oxidized to aldehydes and then to organic acids.

$$C_2H_5OH \xrightarrow{(O)} CH_3CHO \xrightarrow{(O)} CH_3COOH$$

B. *Secondary alcohols* are oxidized to ketones containing the same number of carbon atoms.

$$(CH_3)_2CHOH \xrightarrow{(O)} (CH_3)_2CO$$

C. *Tertiary alcohols* are more difficult to oxidize; they produce ketones or acids containing fewer carbon atoms than does the original alcohol.

$$(CH_3)_3COH \xrightarrow[(acid)]{} (CH_3)_2C=CH_2$$
$$\downarrow (O)$$
$$(CH_3)_2CO + CO_2$$

ANTIBACTERIAL ACTION AND CHEMICAL STRUCTURE

The antibacterial values of the straight chain alcohols increase with an increase in molecular weight, but as the molecular weight increases the water-solubility decreases so that beyond C_8 the activity begins to fall off. The isomeric alcohols show a drop in activity from primary to secondary to tertiary. Thus, n-propyl alcohol has a phenol coefficient against *Staph. aureus* of 0.082 as compared with 0.054 for isopropyl alcohol. Of course, because the latter is commercially available at a lower price it is more widely used than n-propyl alcohol. Isopropyl alcohol is slightly more effective than ethyl alcohol against the vegetative phase, but both are rather ineffective against the spore phase.

Methyl Alcohol; Methanol, Carbinol, Wood Alcohol, Columbian Spirit, Wood Spirit, Wood Naphtha, Methyl Hydroxide. Methyl alcohol rarely is found free in nature, but it is found often in ester form (methyl salicylate) and in ether form (methoxyl group). Prior to 1923, methanol was obtained chiefly by the destructive distillation of wood, a method which now supplies only about 10 per cent of the total yearly production. The hydrogenation of carbon monoxide, carried out in the presence of zinc and chromium oxides under about 3,000 lbs. pressure at from 350° to 400°, is the procedure most used now. Other alcohols (*n*-propyl and isobutyl) also are formed by this reaction and may be increased or decreased in amount by modifying the reaction condition.

$$CO + 2H_2 \xrightarrow[\substack{3,000 \text{ lbs.} \\ 350°\text{-}400°}]{ZnO \ CrO} CH_3OH$$

Methyl alcohol is a colorless liquid with a peculiar odor and a hot, sharp taste. It is miscible with ether, water and ethanol; its general solvent powers resemble those of ethyl alcohol. The vapors are irritating to the eyes, and, like ethanol, it burns easily.

The pharmaceutical use of methanol *per se* is prohibited, and under no conditions may it be used as a menstruum or solvent for medicinals. It may be used as a denaturant in ethyl alcohol which is employed for the preparation of solid extracts. However, in industry it is used extensively as a solvent and chemical agent.

Ingestion of even small quantities of methanol produces a drunkenness similar to that of ethanol, a state which leads to eye injury, blindness and death. Some 20 per cent of it is slowly oxidized to formic acid by the body.

Alcohol U.S.P.; Ethanol; Ethyl Alcohol, Spiritus Vini Rectificatus (Cologne Spirit, Wine Spirit). Ethanol has been known since earliest times as a fermentation product of carbohydrates. An important source today is from the fermentation of molasses. The steps of the process are quite complicated, but essentially they are:

$$C_{12}H_{22}O_{11} + H_2O \xrightarrow{\text{invertase}} 2C_6H_{12}O_6$$

$$C_6H_{12}O_6 \xrightarrow{\text{zymase}} 2C_2H_5OH + 2CO_2$$

A synthetic method of preparation using acetylene or ethylene has been employed, although only the ethylene procedure has shown commercial possibilities. Ethylene, $CH_2=CH_2$, plus 1 mole of water, H_2O, provides the essentials for a mole of alcohol. However, direct hydration of ethylene is difficult and gives low yields. By using sulfuric acid on ethylene to form ethylsulfuric acid and diethyl sulfate,

$$CH_2=CH_2 + HOSO_2OH$$
$$\downarrow$$
$$CH_3CH_2OSO_2OH$$
Ethyl Sulfuric Acid

$$2CH_3CH_2OSO_2OH \xrightarrow[H_2SO_4]{} (CH_3CH_2O)_2SO_2$$
Diethyl Sulfate

which are diluted with an equal volume of water, alcohol is formed and removed by distillation.

$$C_2H_5OSO_2OH + H_2O$$
$$\downarrow$$
$$C_2H_5OH + H_2SO_4$$

$$(C_2H_5O)_2SO_2 + H_2O$$
$$\downarrow$$
$$C_2H_5OH + C_2H_5OSO_2OH$$

The commercial product is about 95 per cent alcohol by volume, because this concentration of alcohol (92.3 per cent w/w) and water forms a constant-boiling mixture at 78.2°. Pure alcohol boils at 78.3° and cannot be obtained by direct distillation.

Ethanol is a clear, colorless, volatile liquid having a burning taste and a characteristic odor. It is flammable and miscible with water, ether, chloroform and most alcohols. Its chemical properties are characteristic of primary alcohols. Most incompatibilities associated with it are due to solubility characteristics. It does not dissolve most inorganic and organic salts, gums or proteins. Due to the aldehydes sometimes present in alcohol, the following chemical changes are often observed: the reduction of mercuric chloride to mercurous chloride, the formation of explosive mixtures with silver salts in the presence of nitric acid and the development of a dark color with alkalies.

Ethanol suspected of containing methanol is treated with resorcinol and concentrated sulfuric acid. A pink color denotes presence of methanol. Detection of isopropanol in ethanol is facilitated by a 1 per cent solution of *p*-dimethylaminobenzaldehyde in concentrated sulfuric acid. Positive test is a brilliant red-violet ring which slowly decomposes. Similar red-brown color is given by *n*-propanol.

The Treasury Department of the U. S. Government oversees the use of alcohol and provides definitions and information pertaining thereto.*

"The term 'alcohol' means that substance known as ethyl alcohol, hydrated oxide of ethyl, or spirit of wine, from whatever source or whatever process produced, having a proof of 160 or more, and not including the substances commonly known as whisky, brandy, rum, or gin."

Besides alcohol available as ethyl alcohol, there are two other forms: (1) completely denatured alcohol and (2) specially denatured alcohol. Denatured alcohol is ethyl alcohol to which has been added such denaturing materials as render the alcohol unfit for use as an intoxicating beverage. It is free of tax and is solely for use in the arts and industries.

Completely denatured alcohol, is prepared according to one of two formulas:

A. Contains ethyl alcohol, wood alcohol and benzene. This is not suitable even for external use.

B. Contains ethyl alcohol, methanol, aldehol† and benzene. This mixture is usually used as an antifreeze.

Specially denatured alcohol is ethyl alcohol treated with one or more acceptable denaturants so that its use may be permitted for special purposes in the arts and industries. Examples are: menthol in alcohol intended for use in dentifrices or mouth washes; iodine in alcohol intended for prep-

aration of tincture of iodine; phenol, methyl salicylate or sucrose octa-acetate in alcohol intended for bathing or as an antiseptic, and methanol in alcohol to be used in the preparation of solid drug extracts.

Ethyl alcohol has a low narcotic potency. It seldom is used in medical practice as a therapeutic agent but almost always is employed as a solvent, preservative, mild counterirritant or antiseptic. It may be injected near nerves and ganglia to alleviate pain or ingested as a source of food energy, for hypnotic effect, as a carminative or as a mild vasodilator. The body readily oxidizes ethanol, first to acetaldehyde and then to carbon dioxide and water. (*See* Disulfiram.)

Externally, it is refrigerant, astringent, rubefacient and slight anesthetic (Rubbing Alcohol N.F.).

The specific uses of alcohol in pharmacy are extremely varied and numerous. Spirits are a class of pharmaceuticals using alcohol exclusively as the solvent, whereas elixirs are hydro-alcoholic preparations. Most fluid extracts contain a small percentage of alcohol as a preservative and solvent.

A concentration of 70 per cent has long been held to be optimal for bactericidal action, but there is little evidence to support it. The rate of kill of organisms suspended in alcohol concentration between 60 and 95 per cent is always so rapid that it is difficult to establish a significant difference.[1] Lower concentrations are also effective, but longer contact times are necessary, e.g., a period of 24 hours is required for a 15 per cent solution to kill *Staph. albus.*[2] It has been reported that concentrations over 70 per cent can be used safely for preoperative treatment of the skin.[3]

It also is the initial material used for the production of many important medicinal agents, such as chloroform, ether and iodoform.

Category—local anti-infective (as 70% w/v solution).

OCCURRENCE	PER CENT ALCOHOL
Diluted Alcohol U.S.P.	50.0
Rubbing Alcohol N.F.	70.0

Dehydrated Alcohol; Dehydrated Ethanol; Absolute Alcohol. Absolute or dehy-

---

* Regulation No. 3, Industrial and Denatured Alcohol, published by U. S. Treasury Department 1927, 1938.

† Aldehol is an oxidation product of kerosene (b.p., 340° to 370°), having a boiling point of 200° to 240°, composed of glycols, aldehydes and acids.

drated alcohol is ethyl hydroxide in a form as pure as it is possible to obtain. It contains not less than 99 per cent by weight of C_2H_5OH.

There are many laboratory procedures available for the preparation of anhydrous ethanol. Some of the compounds used in these methods are calcium oxide, anhydrous calcium sulfate, anhydrous sodium sulfate, aluminum ethoxide, diethyl phthalate and diethyl succinate. Commercially, absolute alcohol is prepared by azeotropic distillation of an ethanol and benzene mixture. Because the ethanol contains about 5 per cent water, the resultant combination, ethyl alcohol-water-benzene, first distills at 64.8° (C_6H_6 74 per cent, water 7.5 per cent and C_2H_5OH 18.5 per cent). All of the water is removed at this temperature, and then the remaining ethyl alcohol and benzene distill at 68.2°. The ethyl alcohol is always in great excess; thus, when all the benzene has been removed, pure ethyl alcohol is collected at 78.3°.

Dehydrated alcohol has a great affinity for water and must be stored in tightly closed containers. It is used primarily as a chemical agent but has been injected for the relief of pain in carcinoma and in other conditions where pain is local.

Isopropyl Alcohol N.F., 2-Propanol. Isopropyl alcohol[2] became recognized about 1935 as a suitable substitute for ethyl alcohol in many external uses, but it must not be taken internally.

$$CH_3COCH_3 \xrightarrow[Ni]{H} CH_3CHOHCH_3$$

It is prepared, with Raney nickel as a catalyst, by the high pressure hydrogenation of acetone, or, with sulfuric acid, from propylene by hydration.

$$CH_3CH{=}CH_2 \xrightarrow[H_2SO_4]{H_2O} CH_3CHOHCH_3$$

Isopropyl alcohol is a colorless, clear, volatile liquid having a slightly bitter taste and a characteristic odor. It is miscible with water, ether and chloroform.

It is used to remove creosote from the skin and as a disinfectant for the skin and surgical instruments. A 40-per-cent solution is approximately equal in antiseptic power to a 60-per-cent solution of ethyl alcohol. The effective concentrations are between 50 and 95 per cent by weight. A 91 per cent solution in water forms a constant boiling point mixture and is thus the most economical concentration. It is frequently used by diabetics for cold sterilization of their syringes and needles.

In recent years it has been used in many toiletries and pharmaceuticals as a solvent and preservative and to replace, in some cases, ethyl alcohol.

Internally, isopropyl alcohol is oxidized to acetone and eliminated in much the same manner as ethyl alcohol. However, it is about twice as toxic and may cause depression. Unlike methanol, no effects on the eyes have been observed.

Category—solvent.

OCCURRENCE	PER CENT ISOPROPYL ALCOHOL
Isopropyl Rubbing Alcohol N.F. . .	70

A number of other alcohols are discussed in Chapter 15.

DIHYDRIC ALCOHOLS (GLYCOLS)

Glycols[1] are alcohols with two hydroxyl groups in the molecule. The lower molecular weight glycols are colorless, somewhat hygroscopic liquids. They have solubility properties between those of water and of hydrocarbons, i.e., they are miscible with water yet will dissolve volatile oils, some dyes, resins and a few gums. Toxicity studies on the glycols[5] indicate that they vary according to the number of carbon atoms and the position of the hydroxyls and that their derivatives also vary with the substituent. They resemble the monohydric alcohols in chemical properties.

The glycols can be prepared with potassium acetate and subsequent hydrolysis from an olefin or by oxidation with potassium permanganate.

PRODUCTS

Ethylene Glycol; Glycol, Ethanediol-1,2. Ethylene glycol, the simplest stable member of the dihydric series, is not used in pharmacy. When ingested, it is oxidized to poisonous oxalic acid. A major commercial use of ethylene glycol is as an antifreeze for liquid-cooled motors.

Cellosolves are represented by the formula $ROCH_2CH_2OH$, the most common ones being those in which R ranges from methyl to amyl. The compounds are usually hygroscopic liquids miscible with water and most organic solvents. Pharmaceutic use is found in ointments, creams and lotions. A phenyl cellosolve tincture is available.

Diethylene Glycol, β,β'-dihydroxy-diethylether, $HOCH_2CH_2OCH_2CH_2OH$, is a colorless hygroscopic liquid with properties similar to the other glycols and is used extensively as an antifreeze. This glycol was used, in 1937, as a solvent for an elixir of sulfanilamide[3] and was the cause of many deaths. Carbitols are compounds of the formula $ROCH_2CH_2OCH_2CH_2OH$ and are derivatives of diethylene glycol. They are quite numerous where R is aliphatic or aromatic. The methyl and the ethyl Carbitols are useful in ointments, creams and lotions. Ethyl Carbitol (Dioxitol, Ethyl Diglycol) is used as a substitute for glycerin.

Propylene Glycol U.S.P. (Methylethylene Glycol, 1,2-Propanediol). Propylene glycol ($CH_3CHOHCH_2OH$) is a clear, viscous, nontoxic liquid similar to glycol and is hygroscopic. It may be prepared from propylene by forming the chlorohydrin and treating with sodium carbonate or distilling a mixture of glycerin and sodium hydroxide. In properties, it is typical of the series. It is miscible with water or organic solvents (acetone and chloroform). Although it will dissolve volatile oils, it is immiscible *with fixed oils.* Compare these properties with those of glycerin. As a replacement for glycerin, it is used as a solvent in pharmaceuticals, perfumes and flavoring extracts. It is nontoxic on internal administration and has a strong bactericidal activity. In fact, it is one of the best of all of the glycols. A recent use is the sterilization of air in hospitals and office buildings by vaporization from electric heaters or from saturated cloth rolls. As a fumigant, it is excellent against streptococci, staphylococci and similar organisms.

Propylene glycol has become widely used as a solvent, extractant and preservative. It is acceptable to the Food and Drug Administration for use in foods and to both the medical and dental professions for use in pharmaceuticals. It is a better general solvent than glycerin and dissolves a wide variety of compounds, such as phenols, sulfa drugs, barbiturates, vitamins (A and D), most alkaloids and many local anesthetics. As an antiseptic it is equal to ethyl alcohol and against mold growth is equal to glycerin and only slightly less efficient than ethanol.

Category—pharmaceutic aid.

OCCURRENCE	PER CENT
Hydrophilic Ointment U.S.P.	12
Amobarbital Elixir N.F.	31

POLYETHYLENE GLYCOLS

Polyethylene glycols are polymers represented by the formula $HOCH_2(CH_2OCH_2)_n$-CH_2OH. The n may range from one to a larger number that produces solid, waxy materials having a molecular weight as high as 10,000. The liquid polyethylene glycols cover the molecular weight range from 150 to 700, while the solid polyethylene glycols, the Carbowax (Union Carbide) compounds, vary from 1,000 to 10,000. Those that are readily available are: polyethylene glycol 200, 300, 400 and 600. Also, a triethylene glycol, $HOCH_2(CH_2OCH_2)_2CH_2$-OH, is distributed for use in glycol vaporizers as an aid in air sanitation. The solid polymers, Carbowax, come as 1,000, 1,540, 1,500, 4,000 and 6,000 (see products, below).

Two methods of preparation are used commonly.

1. Polymerization of ethylene oxide:

$$H_2C\overset{\displaystyle O}{\underset{\displaystyle}{\diagup\!\!\diagdown}}CH_2 \xrightarrow[\Delta \text{ pressure}]{\text{trace } H_2O} H(OCH_2CH_2)_nOH$$

2. Careful dehydration of ethylene glycol:

$$nHOCH_2CH_2OH \xrightarrow{-H_2O} H(OCH_2CH_2)_nOH$$

All of the polymers are characterized by ether linkages and terminal primary alcohol groups. They are heat stable and may be considered chemically inert for pharmaceutic application. The alcohol group may be esterified to form desirable emulsifying agents (see Polyethylene Glycol Monostearate in Chapter 12). Diesterified products are more resinlike. When esterfied with only one mole of fatty acid, excellent emulsifying and detergent agents are formed that are not affected by acids or hard water.

Liquid polyethylene glycols are all light-colored, viscous, slightly hygroscopic liquids that are miscible with water and many organic solvents, such as alcohols, acetone, ethyl acetate, benzene and toluene. They are incompatible with the aliphatic hydrocarbons, kerosene, liquid petrolatum and petrolatum.

In pharmacy these liquid glycols find wide use as suspending agents (calamine lotion), solvents (see Furacin Solution), emulsifiers and dispersants (see Chapter 12).

PRODUCTS

Polyethylene Glycol 300 N.F.
Polyethylene Glycol 400 U.S.P.
Polyethylene Glycol 1540 N.F.
Polyethylene Glycol 4000 U.S.P.

Poloxalene, Pluronics®, are compounds of a similar type, composed of polyoxyethylene-polyoxypropylene-polyoxyethylene glycols, $[HO(C_2H_4O)_n(C_3H_6O)_n(C_2H_4O)_nH]$.

TRIHYDRIC ALCOHOLS

Glycerin U.S.P.; Glycerol, Propanetriol, Trihydroxypropane. Glycerin, an important pharmaceutical for many years, was isolated in 1779; its structure was determined in 1835. For over 100 years, the principal source of glycerin was the hydrolysis of fats (q.v.). In 1938, a method was devised to produce glycerin from propylene, a petroleum product:

$$\underset{\substack{CH_3 \\ | \\ CH \\ \| \\ CH_2}}{} \xrightarrow{Cl_2} \underset{\substack{CH_2Cl \\ | \\ CH \\ \| \\ CH_2}}{} \xrightarrow{OH^-} \underset{\substack{CH_2OH \\ | \\ CH \\ \| \\ CH_2}}{} \xrightarrow{HOCl}$$

$$\underset{\substack{CH_2OH \\ | \\ CHOH \\ | \\ CH_2Cl}}{} \xrightarrow[Lime]{Soda} \underset{\substack{CH_2OH \\ | \\ CH \\ \diagdown \\ | \quad O \\ \diagup \\ CH_2}}{} \xrightarrow[H^+]{H_2O} \underset{\substack{CH_2OH \\ | \\ CHOH \\ | \\ CH_2OH}}{}$$

A fermentation process also has been used for the production of glycerol by a method similar to that employed for ethyl alcohol. The reduction of acetaldehyde is inhibited by the use of sodium bisulfite or sodium carbonate. The glyceraldehyde, an intermediate, is then converted into glycerol.

Glycerin is a clear, colorless, viscous liquid with a faint odor and a sweet taste. It is highly hygroscopic and is miscible with alcohol or water but insoluble in organic solvents. It is a fair solvent for inorganic salts and is better than alcohol. Due to alcohol groups, it is a sequestering agent, which accounts for glycerin solutions of cupric hydroxide (see Haines' Reagent, N.F. XIII, p. 943) and calcium salts, including calcium oxide and calcium hydroxide. The chemical properties are closely allied with those of alcohols. By strong heating or treatment with dehydrating agents, it yields acrolein. Glycerin combines with boric acid or borates to produce a stronger acid solution than boric acid and, thus, is incompatible with carbonates or other materials sensitive to acid. Most oxidizing agents react to form oxalic acid and carbon dioxide. Also, in the formation of Boroglycerin Glycerite a reaction occurs that involves the loss of 3 moles of water to give $C_3H_5BO_3$.

The applications of glycerin in pharmacy are extremely varied, and its uses in industry alone would require a book[4] for complete discussion. In the classes of preparations known as glycerites and fluidglycerates, it is employed as a vehicle. The solvent and the preservative properties of glycerin are used widely, while the consistency and sweet taste make it suitable for use in cough remedies. As an emollient and demulcent, it is an ingredient in lotions and hand creams. An irritant property, perhaps due to dehydration, is made use of in Glycerin Suppositories N.F., which stimulate bowel movements. In prescription compounding, it is used frequently as a stabilizer, to retard precipitation by decreasing ionization in many cases and as a softening agent.

Category—humectant, pharmaceutic aid (solvent).

OCCURRENCE
Glycerin Suppositories N.F.

POLYHYDRIC ALCOHOLS

Erythritol (Tetrahydroxybutane, Erythrol, Erythrite). Erythritol is a tetrahydric alcohol having two asymmetric carbon atoms.

Erythritol has a sweet taste, is readily soluble in water and is insoluble in alcohol. In the form of erythritol tetranitrate it is a useful vasodilator for arterial hypertension (see Chap. 21).

Mannitol U.S.P. The chewable tablet as a dosage form is used frequently for a wide range of therapeutic agents. Mannitol, because of its properties, is primarily responsible for the effectiveness of chewable tablets. It is nonhygroscopic and chemically stable, as well as having a pleasant taste and mouth feel. Mannitol also provides a smooth disintegration. (See Chap. 20 for use as kidney function test.)

Sorbitol U.S.P., D-Glucitol, is found in a number of fruits and berries and is prepared readily by the hydrogenation of glucose. Sorbitol is a white powder. It is available commercially as Alex,® a noncrystallizable liquid containing 83 per cent sorbitol. This form is used in cosmetics and galenicals. Sorbitol is isomeric with mannitol and has similar properties. However, it is metabolized by the body as a source of food; mannitol is not. An outstanding property of sorbitol is

its narrow humectant range. It serves as a starting material for the synthesis of vitamin C, and its solutions are used widely in industry and in pharmaceutical manufacturing. It has advantages over glycerin for many purposes.

Category—flavor.

OCCURRENCE	PER CENT (w/w) SORBITOL
Sorbitol Solution U.S.P.	70

TERPENE ALCOHOLS

Terpene alcohols frequently are found, either free or in ester combination, in volatile oils. They contribute in large measure to the odor and the perfume value of oils. Two, menthol and santalol, are used as medicinal agents.

Linalool is present free and as the acetate ester in bergamot oil (not less than 36 per cent), bitter orange oil, Coriander Oil U.S.P. and Lavender Oil N.F. (not less than 30 per cent). It is a tertiary alcohol having two double bonds. Thus, it is difficult to esterify and reacts characteristically with reagents that add to double bonds.

Menthol U.S.P., p-Methane-3-ol, is a terpene secondary alcohol containing three asymmetric carbon atoms. This structure allows the formation of 8 stereoisomeric men-

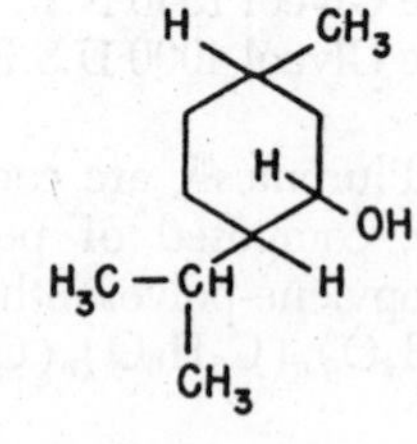

thols of which 4, (+)- and (−)-menthol and (+)- and (−)-neomenthol, are known. It has been a medicament for many years and occurs primarily in peppermint oils, as free (−)-menthol (see formula), and to a lesser extent, as menthyl acetate. The alcohol may be prepared synthetically by the reduction of thymol and the resolution of the racemate to obtain the levo and the dextro forms. The U.S.P. recognizes the levorotatory form and the racemic form.

A synthetic process in present use starts with *m*-cresol and, in the difficult step of reduction, uses copper chromite as the catalyst.[5]

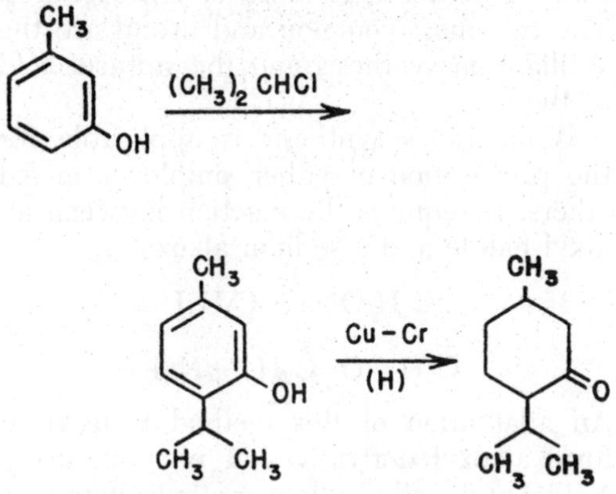

The ketones, a mixture of menthone and isomenthone, are treated with sodium hydroxide to convert them to the enol form. By reduction of the enol form, racemic menthol is obtained.

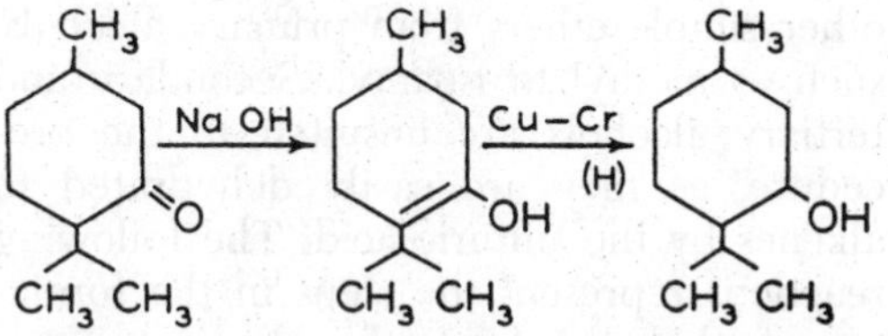

The mixture of isomeric menthols for most purposes presents no difficulty and is equally as usable as natural (−)-menthol or the pure synthetic (±)-menthol, in regard to toxicity and pharmacologic properties. The difference is, however, that all the stereoisomers except (±)-menthol have a disagreeable odor and an unpleasant taste.

Menthol usually occurs as needlelike crystals or in fused masses. It has a sharp, peppermintlike odor, is only slightly soluble in water but is soluble in organic solvents or in fixed and volatile oils. Eutectic mixtures are formed with a number of substances, such as camphor, phenol, chloral hydrate and thymol.

Menthol is used[6] primarily in external preparations or in sprays and solutions designed for use in the nasal passages or mouth. A decided cooling effect is noticed on application to the skin and a slight anesthetic action accompanies this. It also possesses some antiseptic action, as well as the action of a counterirritant. These combined properties make menthol a useful ingredient in lotions, nasal preparations, vapors for breathing, analgesic balms, cough drops and the like. Internally, the value of menthol is questionable, but it is used for nausea and dyspepsia.

Category—topical antipruritic.

For external use—0.1 to 2 per cent in preparations for use on skin.

Terpin Hydrate N.F., Terpinol, *cis-p*-menthane-1,8-diol Hydrate, is a dihydroxy terpene alcohol that may be present in some fresh volatile oils and is developed in some

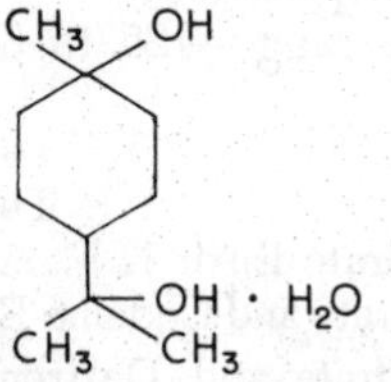

oils on standing. It is prepared commercially by treating turpentine oil with sulfuric acid, nitric acid and alcohol and allowing the mixture to stand for several days. (The reactions involved in this process are shown on page 150.)

The α-pinene is the constituent which is converted to terpin hydrate. By acid treatment, dipentene, geraniol and linalool also produce terpin hydrate. The removal of a mole of water from terpin hydrate forms the anhydride known as cineole or eucalyptol (q.v.). Phosphoric acid treatment may remove a mole of water at four different positions, and this gives rise to a group of terpeneols that are valuable in perfumery because of their lilac odor.

Terpin hydrate exists usually as colorless crystals or as a white powder that is slightly soluble in water (1:200), soluble in alcohol (1:13) or slightly soluble in organic solvents.

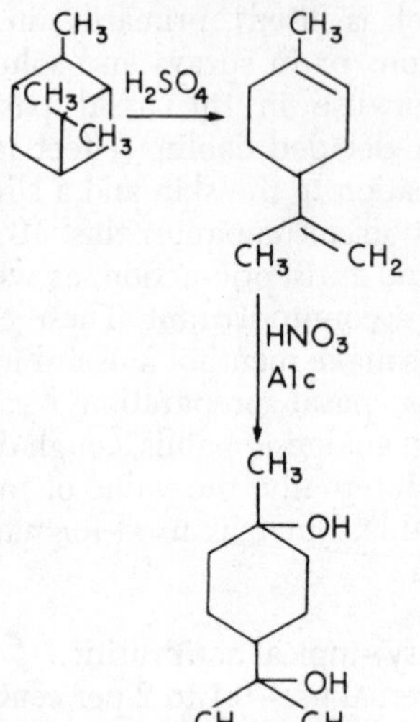

It has been used for many years as an expectorant in cough preparations, although there is little scientific evidence of its value.

Category—expectorant.
Usual dose range—125 to 300 mg. every 6 hours.

OCCURRENCE	PER CENT TERPIN HYDRATE
Terpin Hydrate Elixir N.F.	1.7
Terpin Hydrate and Codeine Elixir N.F.	1.7
Terpin Hydrate and Dextromethorphan Hydrobromide Elixir N.F.	1.7

ETHERS

INTRODUCTION

Ethers, or organic oxides, are characterized by the structure R—O—R and may be considered as derivatives of water (H—O—H) which has had the hydrogen atoms replaced by the same or different chemical groups, such as alkyl, aryl, alicyclic or heterocyclic. Inasmuch as ethers usually are prepared from alcohols, they are more often considered as derivatives of alcohols (R—O—H) in which the hydrogen has been replaced by one of the previously mentioned groups. Less often, they are thought of as a hydrocarbon with a hydrogen replaced by an alkoxyl group (RO—), or as isosteres of hydrocarbons, where a methylene group ($-CH_2-$) is replaced by an oxygen atom.

An ether in which the groups R of R—O—R are the same is known as a simple ether; when the R's differ, it is a mixed or unsymmetrical ether.

PREPARATION

Preparation of ethers can be accomplished by any one of several procedures. The two most common and useful are the Williamson synthesis and the sulfuric acid method.

Williamson's synthesis is applicable for the preparation of either simple or mixed ethers. It requires the reaction between an alkyl halide and a sodium alkoxide.

$$C_2H_5ONa + C_2H_5I$$
$$\downarrow$$
$$C_2H_5-O-C_2H_5 + NaI$$

An adaptation of this method is used to form alkoxyl derivatives of phenols, using NaOH and methylsulfate or ethylsulfate.

$$C_6H_5OH + (CH_3)_2SO_4$$
$$\downarrow \text{ NaOH}$$
$$C_6H_5-O-CH_3 + CH_3 NaSO_4$$

The sulfuric acid method is particularly suitable for the preparation of ethyl ether from ethanol and for the preparation of other simple ethers from primary alcohols, such as methyl to isoamyl. Secondary and tertiary alcohols are unsuited to the procedure, as they are easily dehydrated to alkenes by the sulfuric acid. The following reactions represent the steps in the formation of ethyl ether from ethanol:

$$C_2H_5OH + HOSO_2OH$$
$$\downarrow$$
$$C_2H_5OSO_2OH + H_2O$$
$$C_2H_5OSO_2OH + HOC_2H_5$$
$$\downarrow 140°$$
$$C_2H_5OSO_2OC_2H_5 + H_2O$$
$$C_2H_5OSO_2OC_2H_5 + C_2H_5OH$$
$$\downarrow 140°$$
$$C_2H_5-O-C_2H_5 + C_2H_5OSO_2OH$$

Once the reaction is started, alcohol is added at the same rate that the ether distills off, and the water formed is retained by the sulfuric acid. The process is continuous until the acid becomes too dilute. The name "sulfuric ether" has thus resulted. Ether so produced may be contaminated with thioacids, thioethers, sulfates, sulfur oxides, ethylene, aldehydes, organic acids and alcohol.

CHEMICAL PROPERTIES

Except for the saturated hydrocarbons, ethers are perhaps the most chemically inert class of organic compounds. They react neither with sodium, strong acids or bases at moderate temperatures nor with phosphorus trichloride, cold phosphorus pentachloride, alkali metals or oxidizing agents.

Potassium dichromate in concentrated sulfuric acid, when warmed, will react with ether to produce, successively, acetaldehyde, acetic acid, carbon dioxide and water. The reaction may be used as a basis for quantitative assay.

At low ten.peratures, ethers combine with strong mineral acids to form oxonium salts, which usually are unstable at moderate temperatures.* However, the linkage can be split with strong hydrobromic or hydriodic acid.

$$C_2H_5OC_2H_5 + HBr$$
$$\downarrow$$
$$C_2H_5OH + C_2H_5Br$$

The hydrogen of the ethers is attacked readily by bromine and chlorine to form halogenated ethers. Because of the stability of ethers, they are most useful as solvents; this is particularly true of diethyl ether.

The ethers are more volatile than the isomeric alcohols and they decrease in volatility, inflammability and water solubility

* The structure of oxonium salts may be shown by using ethyl ether and hydrochloric acid. This

reaction is used to distinguish ethers from hydrocarbons. In some cases, the oxonium salts are stable above room temperature. Dioxane, as an example, forms an oxonium compound with sulfuric acid; the oxonium compound has a melting point of 101°.

as the molecular weight increases. All organic solvents are miscible with ethers, as are fixed and volatile oils.

ALIPHATIC ETHERS

Diethyl Ether (Sulfuric Ether, Ethoxyethane). Diethyl ether ($C_2H_5OC_2H_5$), usually referred to as "ether," has been known for more than 400 years, and its structure has been known since 1807. It is prepared readily (q.v.) and occurs as a volatile, inflammable liquid, colorless, lighter than water, slightly soluble in water and having a characteristic odor. By reason of its close chemical relationship to the hydrocarbons, it has become a most useful solvent. The molecular weight of oxygen is 16 and that of a methylene group is 14, thereby allowing the comparison of diethyl ether (b.p. 34.6°) with *n*-pentane (b.p. 36.1°). Ether boils at a much lower temperature than does the corresponding ethyl alcohol (b.p. 78.5°). This difference is due to hydrogen bonding in the alcohol and of the unassociated form in the case of alkanes and ethers. The solvent powers of ether are very great for most organic compounds that do not contain a large number of hydroxyl groups (carbohydrates)—for example, alkaloids, fats, amines, sterols, waxes. Most inorganic complexes or salts are insoluble in ether, but it does dissolve the halogens and mercuric chloride. The low boiling point of ether makes it first choice as an extraction solvent when subsequent evaporation or distillation is necessary. Few substances are injured by heating at this low temperature.

Although ether is immiscible with water, each solvent does dissolve some of the other. At room temperature, ether dissolves about 1 to 1.5 per cent of water and water dissolves about 8 per cent of ether. Ether may be dried in several ways, but, as the first step, the peroxides and alcohol should be removed (q.v.). The wet ether then is treated for several days with a drying agent, such as calcium chloride, after which it is decanted into a clean container, and solid sodium or sodium hydroxide is added. It may be decanted and used, or it may be redistilled. In laboratory operations, ether is a dangerous chemical because it is highly inflammable, forms explosive mixtures with air and has a low flash point. It is 2½ times heavier than air and thus settles in low places. Upon long standing in contact with air (partially filled containers), ether is known to develop peroxides. Their development is favored by ultraviolet light and sunlight. In well-filled containers, either glass or tin-plated iron, the peroxides will form in about 6 months and remain for about 1 year. It appears that in old ether (over 2 years), the content of peroxides decreases and aldehydes are formed from the alcohol present. These peroxides are nonvolatile and remain as a residue when such an ether is evaporated. In such cases, they explode with terrific force when heated to about 100° and constitute a real danger. The peroxides are found not only in ethyl ether, but are possible also in isopropyl and butyl ethers.

$$\begin{array}{c} H \\ | \\ CH_3-C-O-CH_2-CH_3 \\ | \\ O-O-H \end{array}$$

Ether Peroxide

The peroxides may be destroyed by shaking the ether with an aqueous solution of a reducing agent, such as ferrous sulfate, zinc dust and sulfuric acid, sodium sulfite, ferrous chloride, sodium hydrosulfite, catechol, or by filtering it through activated alumina. Ether supplied in glass bottles contains iron wire to prevent peroxide formation. Many studies have been made to find a preservative for ether; the best one found involves keeping ether in copper or copper-plated containers. Mercury, too, has been found to be a satisfactory stabilizer, since, like iron and copper, it is preferentially oxidized. Acetaldehyde stems from the alcohol present and results in the formation of organic acids.

PRODUCTS

Ether U.S.P.; Ethyl Ether; Diethyl Ether. Faraday is credited with first observing (in 1815) the depressant property of ether; it

was used in surgical anesthesia in 1842 by Long and in 1846 by Morton.[7] (See Chapter 15.) Its preparation and properties have been discussed previously. It occurs as a colorless, mobile liquid having a burning, sweetish taste and a characteristic odor. Ether U.S.P. is intended for anesthetic use and thus has rigid specifications as to content and method of handling. It may contain up to 4 per cent of alcohol and water. The alcohol has little value as a preservative but does raise the boiling point and prevent frosting on the anesthetic mask. In the U.S.P., a caution limits the size of container to 3 kilos and permits the ether to be used only up to 24 hours after the container has been opened.

The ether must be free of acids, aldehydes and peroxides. Acids are tested for by using .02 N sodium hydroxide. A test for aldehydes, sensitive to 1 part in 1,000,000, using Nessler's solution (alkaline mercuric-potassium iodide T.S.) shows no yellow color when the test is negative. The peroxide test is carried out on 10 ml. of ether. One ml. of potassium iodide T.S. is added, and the mixture is shaken for 1 hour. If peroxides are present, the potassium iodide, as a reducing agent, has its iodide ion oxidized to free elemental iodine (colored). (*See* Chap. 15 concerning its use as a general anesthetic.)

OCCURRENCE	PER CENT ETHER
Collodion U.S.P.	75
Flexible Collodion U.S.P.	..
Salicylic Acid Collodion U.S.P.	75

Ethyl Oxide, Solvent Ether. This is similar in every respect to Ether U.S.P. except that it is not required to be of such high purity and, therefore, may be produced more cheaply. It is employed as a solvent.

Vinyl Ether N.F.; Divinyl Oxide; Vinethene®. (See Chap 15.)

AROMATIC ETHERS

The aromatic ether structure is present in some pharmaceutical compounds; there are numerous active compounds having alkoxy groups, such as methoxy, ethoxy, propoxy. Since the ether structure is usually of secondary importance in most compounds the classification is based on other groups. (See quinine, guaiacol and eugenol).

Anethole U.S.P. *p*-Propenylanisole (Paramethoxypropenylbenzene). Anethole is para-propenyl anisole. It is obtained from anise oil and other sources, or is prepared synthetically. This aromatic ether is isolated from several volatile oils by fractionating, chilling and crystallizing. It may also be prepared synthetically.[9]

This ether is a colorless or faintly-yellow liquid which congeals at about 20° to 23°. It has the aromatic odor of anise oil, a sweet taste and the characteristic of being affected by light. It is soluble in organic solvents but is insoluble in water.

Category—flavor.

OCCURRENCE	PER CENT ANETHOLE
Diphenhydramine Hydrochloride Elixir U.S.P.	0.0003

Safrole is the methylene ether of 3,4-dihydroxyallylbenzene and is the main constituent of sassafras oil. It is used for flavoring purposes, as a starting material for the synthesis of piperonal and in the preparation of insecticidal material.

TERPENE ETHERS

Eucalyptol (Cineol, Cajuputol, Cineole 1:8) is a terpene inner ether found in oil of eucalyptus and other oils. It may be obtained synthetically by the dehydration of terpin hydrate.

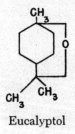

Eucalyptol

Eucalyptol is a clear liquid with a characteristic odor; it is insoluble in water but soluble in oils or organic solvents.

It has been administered internally, but such use is unsound. Primarily, it is used as a flavor and as an antiseptic in mouth washes, nasal sprays and throat preparations.

REFERENCES CITED

1. DuMez, A. G.: J. Am. Pharm. A. 28:416, 1939.
2. Smyth, H. F.: J. Indust. Hyg. & Toxicol. 23:259, 1941.
3. Leech, P. N.: J.A.M.A. 109:1531, 1937.
4. Leffingwell, G., and Lesser, M.: Glycerin, New York Chem. Pub. Co., 1948.
5. Barney, A. L., and Hass, H. B.: Ind. Eng. Chem. 36:85, 1944; Brode, W. R., and Van Dolah, R. W.: Ind. Eng. Chem. 39: 1157, 1947.
6. Bliss, A. R., and Glass, H. B.: Drug and Cos. Ind. 44:289, 1939; 46:160, 1940.
7. Guthrie, D.: A History of Medicine, p. 302, Philadelphia, Lippincott, 1946.
8. Mallinckrodt, E.: J. Am. Chem. Soc. 49: 2655, 1927.
9. Quelet, R.: Bull. Soc. chim. 7:196, 1940; Compt. rend. Soc. biol. 202:956, 1936.

SELECTED READING

Bradshaw, H. H.: Impurities in ether, Am. J. Surg. 45:511, 1939.

Curme, G. O., Jr., and Johnston, F.: Glycols, A.C.S. Monograph 114, New York, Reinhold, 1953.

Gold, H., and Gold, D.: Stability of U.S.P. Ether after metal container is opened, J.A.M.A. 102:817, 1934.

Minor, C. S., and Dalton, N. N.: Glycerol, A.C.S. Monograph 117, New York, Reinhold, 1953.

Nitardy, F. W., and Topley, M. W.: The stability of anesthetic ether, J. Am. Pharm. A. 17:10, 1928.

6

Aldehydes and Ketones

Charles O. Wilson, Ph.D.,
Dean and Professor of Pharmaceutical Chemistry, School of Pharmacy,
Oregon State University

ALDEHYDES

The aldehydes are characterized by the typical group

$$\overset{H}{\underset{}{-C=O,}}$$

which is called the aldehyde group and often is written simply —CHO but never —COH. Because the carbon is primary, this group must be a terminal one and in cyclic compounds can be only a substituent and never an integral part of the ring. The active hydrogen attached to the carbonyl radical gives the aldehydes unique properties as compared with other compounds that contain the carbonyl radical, such as acids, esters and ketones.

The nomenclature of aldehydes follows any one of several systems. Aldehydes may be regarded as derivatives of the corresponding primary alcohols or of the acids to which they will oxidize, or they may be viewed as substitution products of acetaldehyde. In addition, the names may follow the rules that have been evolved from the Geneva system. Thus, CH_3-CH_2-CHO is propyl aldehyde, propionic aldehyde, methylacetaldehyde, propionaldehyde or propanal. In more complicated substances, the aldehyde group may be considered as a substituting one and given the prefix formyl- or the suffix carboxaldehyde; the compound previously cited then would then be formylethane or ethanecarboxaldehyde.

PREPARATION

The aliphatic compounds may be prepared by oxidation of primary alcohols, although it is not easy to prevent further oxidation to the acid.

$$CH_3-CH_2-CH_2OH \xrightarrow{(O)}$$
$$CH_3-CH_2-CHO \; (+ H_2O)$$

A still better process is direct dehydrogenation over metallic copper at about 250°; this gives excellent yields of formaldehyde or acetaldehyde.

$$CH_3 - CH_2OH \xrightarrow[250°]{Cu} CH_3 - CHO + H_2$$

Another method is to hydrolyze α,α-dihalogen compounds; for example, formaldehyde can be produced from methylene iodide.

$$CH_2I_2 + 2KOH \longrightarrow HCHO + 2KI + H_2O$$

A frequently used process involves the Grignard reagent. This substance is added to dry hydrogen cyanide or to formic esters, and the mixture is hydrolyzed as usual.

$$R - Mg - I \xrightarrow{HCN} R - CH = N - Mg - I$$
$$\downarrow H_2O$$
$$R - CHO$$

This method is especially convenient for those compounds having a high molecular weight.

The aldehydes also can be produced by the distillation of an appropriate salt with a formate.

155

R—COOM + H—COOM ⟶

$$M_2CO_3 + R—CHO$$

Then the product must be separated by suitable means from the ketone R—CO—R which is formed simultaneously.

Finally, aldehydes often can be made by reduction of acid derivatives, such as ester or acid chloride, although not from the acids themselves, using hydrogen and a catalyst.

$$R—COOR \xrightarrow[\text{(Pd)}]{H_2} R—CHO$$

Under other conditions of reduction, it is difficult to prevent further change to primary alcohol.

PHYSICAL PROPERTIES

The aliphatic aldehydes with low molecular weight are colorless, neutral gases or liquids with a characteristically pungent odor and are more or less soluble in water. The higher ones and those containing rings are much less soluble, have a more agreeable odor and may be insoluble and odorless liquids or solids.

CHEMICAL PROPERTIES

Most aldehydes can be oxidized easily to the corresponding acids. They will reduce reagents such as Fehling's solution or ammoniacal silver nitrate quite readily when warm. Even oxygen of the air can bring about a gradual change, so that care must be exercised in preserving them. The presence of an aryl group in aldehydes decreases the oxidizability by reagents, although autoxidation brings about a more rapid change by oxygen of the air.

The carbonyl group has great capacity for adding any one of a variety of reagents, especially if these substances contain an active hydrogen, in which case a hydroxyl group is formed.

$$R—CHO + (H—) \rightarrow R—CHOH—$$

1. Hydrogen can be added to reduce the compound to a primary alcohol. This reduction can be accomplished by sodium amalgam, by zinc and acid, by hydrogen and a catalyst, by electrolysis or by other means.

2. Ammonia adds to some of the lower aldehydes to give compounds of the type R—CHOH—NH$_2$, which are crystalline and soluble in water. Such an "aldehyde ammonia" represents a simple intermediate in purification, because the addition process can be reversed by acids. However, the ammonia compounds show a great tendency toward decomposition to complex molecules.

3. Hydrogen cyanide adds to form hydroxynitriles or cyanohydrins, R—CHOH—CN. The reaction may be stopped or reversed by alkalies. The cyanohydrins are useful intermediates in synthesis, because the —CN can be hydrolyzed to a carboxyl group.

4. Sodium bisulfite reacts with aldehydes to form compounds of the formula R—CHOH—SO$_3$Na. These substances are crystalline and more or less soluble in water, but, usually, they can be separated if the salt is used in saturated solution and, particularly, if the aldehyde is dissolved previously in an anhydrous solvent, such as ether. The reaction is easily reversible by acids or alkalies, making it a convenient method of isolation and purification.

The Schiff test for the presence of aldehydes may depend initially on a similar reaction. The reagent is prepared by decolorizing fuchsin with sulfur dioxide. When a small amount of aldehyde is added, a red or purple-red color is produced; this color is removed by acids more or less easily, except in the case of formaldehyde.

5. Water undoubtedly often adds to the carbonyl group to produce hydrates, but these hydrates are seldom stable enough to isolate, except under special circumstances. Analogously, alcohols could be expected to add, forming hemiacetals.

The reaction is highly reversible and such compounds usually are very unstable, but they can be isolated in some cases (chloral alcoholate). The principle of hemiacetal formation is involved in the cyclization of monosaccharides. (*See* Chap. 26.)

In the presence of a trace of acid, aldehydes condense with alcohols to form acetals.

$$R-CHO + 2R'OH \longrightarrow$$
$$R-CH(OR')_2 + H_2O$$

The acetals can be hydrolyzed to aldehydes by boiling with acids and therefore are sometimes very useful.

6. The Grignard reagent will add to aldehydes to give a product that hydrolyzes to an alcohol; from formaldehyde, the alcohol is primary, from other aldehydes, secondary. For example, propanal with methyl magnesium bromide can be used to prepare secondary butyl alcohol.

$$CH_3CH_2-CHO + CH_3-MgBr \longrightarrow$$
$$CH_3CH_2-CH(OMgBr)-CH_3$$
$$CH_3CH_2-CH(OMgBr)-CH_3 \xrightarrow{H_2O}$$
$$CH_3CH_2-CHOH-CH_3$$

7. When acetaldehyde is treated with a dilute alkali, an addition takes place to produce the compound $CH_3-CHOH-CH_2-CHO$. The mechanism of the reaction is rather complex, but the final product is as if there were simple addition of one molecule to the carbonyl group of another. The compound is called aldol because it is both alcohol and aldehyde, and the process is known as aldol condensation. Other aldehydes of the series will undergo a similar polymerization, but not if they do not contain an alpha hydrogen, such as in trimethylacetaldehyde, $(CH_3)_3C-CHO$.

When aldehydes typical of the lower aldehydes of the aliphatic series are boiled with alkalies, they form insoluble orange or brown resins. This reaction, which serves to distinguish these aldehydes from formaldehyde and aldehydes of the aromatic and heterocyclic groups, is probably a further aldol condensation.

In those aldehydes that have no alpha hydrogen, alkalies bring about the so-called Cannizzaro reaction, in which half of the molecules are oxidized and half reduced. Thus, formaldehyde produces methyl alcohol and a formate.

$$2H-CHO + KOH \rightarrow H-COOK + CH_3OH$$

Benzaldehyde and its substitution products behave in a similar manner.

8. On standing or in the presence of catalysts, aliphatic aldehydes polymerize in quite another way. Acetaldehyde with a small amount of acid is converted to a trimer, $(CH_3-CHO)_3$, known as paraldehyde. If this compound is heated with sulfuric acid, the process is reversed. On the other hand, if the original reaction is conducted at temperatures below $0°$, there is produced a solid substance that is known as metaldehyde and is probably a metamer. It also can be reconverted to acetaldehyde by warming with acids. Neither of these polymers shows any of the reactions typical of carbonyl compounds, nor do similar products from other aldehydes, and the structure is essentially that of acetals (see paraldehyde).

The carbonyl oxygen of aldehydes will condense with substances containing active hydrogen, with the splitting off of water, a reaction similar to the one that formed acetals. One common application of this type is with primary amines, such as hydroxylamine, phenylhydrazine or semicarbazide. In all cases, there is probably a primary addition at the carbonyl group, but water eventually is split out to produce oximes, phenylhydrazones and semicarbazones respectively. These products are generally crystalline and have well-defined melting points that are useful in identifying specific aldehydes. For example, the compounds from benzaldehyde would have the formulas $C_6H_5-CH=NOH$, $C_6H_5-CH=N-NH-C_6H_5$, and $C_6H_5-CH=N-NH-CO-NH_2$.

When phosphorus pentachloride and similar agents are triturated or warmed with aldehydes, the carbonyl oxygen is replaced by two chlorine atoms.

$$R-CHO + PCl_5 \rightarrow R-CHCl_2 + POCl_3$$

This is perfectly analogous to the action of such agents on an alcohol, where an $-OH$ group is displaced by a chlorine atom, since the carbonyl oxygen is equivalent to two hydroxy groups.

PHYSIOLOGIC PROPERTIES

The lower aliphatic aldehydes, because of ease of reaction, are more or less irritating to mucous membranes when inhaled or swallowed and are thus inimical to plants and animals. The higher ones or those containing rings are much less toxic. The aldehyde group otherwise confers no useful properties on compounds in which it occurs. Most of the aldehydes that are employed in pharmacy and medicine have advantages because of properties other than those contributed by the aldehyde group. (See Chaps. 9 and 15.)

PRODUCTS

Benzaldehyde N.F., Artificial Oil of Bitter Almond, is found naturally combined in some glycosides, such as amygdalin of bitter almonds and other rosaceous kernels. It may be prepared by the oxidation of benzyl alcohol or by distillation of a benzoate and a

Benzaldehyde

formate, but commercially the usual method is by oxidation of toluene with manganese dioxide or chromyl chloride or by hydrolysis of benzal chloride using ferric benzoate or powdered iron as a catalyst. In any method the benzaldehyde may be purified by distillation with steam. A related aldehyde, cinnamaldehyde, is present in cinnamon oil.

$$C_6H_5CH_3 \xrightarrow{CrO_2Cl_2} C_6H_5CHO$$

$$C_6H_5CH_3 \xrightarrow{Cl_2}$$
$$C_6H_5CHCl_2 \xrightarrow[95°-100°]{Fe} C_6H_5CHO$$

Benzaldehyde is a colorless, strongly refractive liquid, having an odor resembling that of bitter almond oil and a burning, aromatic taste. It dissolves in about 350 volumes of water and is miscible with alcohol, ether or fixed or volatile oils. It is heavier than water, having a specific gravity of about 1.045, and has a high index of refraction, 1.5440 to 1.5465. If made from glycosides,

it may contain some hydrocyanic acid; if made from benzal chloride, it may contain chlorinated products. It is assayed by reaction with hydroxylamine hydrochloride and titration of the released hydrogen chloride.

The aromatic nucleus in benzaldehyde reduces somewhat the reactivity of the aldehyde group, but, qualitatively, it behaves as do the compounds of the aliphatic series. Having no alpha hydrogen, it does not undergo aldol condensation, but, with alkalies, it is converted to benzoic acid and benzyl alcohol by the so-called Cannizzaro process. With alkalies and acetaldehyde, it forms cinnamaldehyde by condensation (Claisen-Schmidt), with aliphatic acid anhydrides and salts, it forms unsaturated acids (Perkin), and, with certain esters in the presence of sodium, it condenses to form unsaturated esters (Claisen). These and other reactions make benzaldehyde very useful in synthesis.

One remarkable property of benzaldehyde is its great tendency to auto-oxidize. While it is affected much less easily by oxidizing agents than aliphatic aldehydes are, oxygen of the air is absorbed rather rapidly to form a peroxide, which decomposes to give benzoic acid.

$$C_6H_5CHO + O_2 \rightarrow C_6H_5CO_3H$$
$$C_6H_5CHO + C_6H_5CO_3 \rightarrow 2C_6H_5COOH$$

Consequently, benzaldehyde must be stored in well-filled, tight, light-resistant containers. The presence of a small amount of an antioxidant, such as hydroquinone, will make the commercial article more stable.

It is used almost entirely as a flavor and in the preparation of synthetics, such as triphenylmethane dyes, cinnamic aldehyde and acid and unsaturated esters. It is somewhat antispasmodic and anesthetic but is not efficient enough for these purposes to compete with similar therapeutic agents.

Category—flavor.

OCCURRENCE	PER CENT BENZALDEHYDE
Compound Benzaldehyde Elixir N.F.	0.05
Ephedrine Sulfate Syrup U.S.P. ...	0.006
Terpine Hydrate Elixir N.F.	0.005
Terpin Hydrate and	
Codeine Elixir N.F.	0.005

Terpine Hydrate and Dextromethorphan Hydrobromide Elixir N.F. . . . 0.005

Vanillin U.S.P., 4-hydroxy-3-methoxy-
benzaldehyde, occurs naturally in a large
number of plants, including vanilla beans,
but it is prepared more conveniently and
economically from lignin waste in the manufacture of paper[1, 2] or from eugenol of
clove oil. In the latter process, the eugenol
is converted by alkalies to isoeugenol, acetylated, oxidized and finally hydrolyzed.

Vanillin occurs as fine, white to slightly
yellow crystals, usually needlelike, having
an odor and taste suggestive of vanilla. It is
soluble in water (1:100), in glycerin (1:20)
and in other organic solvents. Because of
the phenolic group, it dissolves also in fixed
alkali hydroxides and gives a blue color with
ferric chloride. Solutions of vanillin in water
are acid and give a white precipitate with
lead subacetate. Vanillin is rather volatile,
easily oxidized and affected by light, so that
it must be stored in tight, light-resistant
containers.

Category—pharmaceutic acid (flavor).

	PER CENT
OCCURRENCE	VANILLIN
Mineral Oil Emulsion N.F.	0.004
Compound Benzaldehyde Elixir N.F.	0.1

Ethyl Vanillin N.F., 3-ethoxy-4-hydroxy-
benzaldehyde, is synthesized and occurs as
fine white or slightly yellowish crystals. It
has the same general physical and chemical
characteristics as vanillin, but ethyl vanillin
possesses a more delicate and a more intense
vanilla odor and taste.

Vanillin R = CH₃
Ethyl Vanillin R = C₂H₅

Category—flavor.

Citral occurs primarily in oil of lemon
grass (about 80%) and in oil of lemon (8%).
It exists in two geometric forms, α and β.

It is an aliphatic aldehyde containing two
double bonds. Terpeneless lemon oil is
nearly pure citral, α and β; it is more soluble in water than is the natural oil, it has
about 16 times the flavoring ability, and it
does not develop a terebinthinate odor. It
is in a great measure responsible for the
lemon odor.

KETONES

Ketones also are characterized by the
carbonyl group, but they are much less active chemically than aldehydes because they
lack the labile hydrogen. They may be
represented generally by the formula
R—CO—R′.

As with other classes of organic compounds, there are several systems of nomenclature. The more common ketones have
trivial names, but they may be considered
also as derivatives of the simplest one, acetone; the two radicals may be designated
in alphabetical order or the Geneva system may be used with the suffix *-one*.
Thus, CH_3CH_2—CO—CH_3 is methylace-
tone, ethyl methyl ketone or butanone,
while $CH_3CH_2CH_2$—CO—CH_3 is ethylace-
tone, methyl propyl ketone or pentan-2-one.
In more complex compounds, and especially
in rings, the carbonyl group may be designated by the prefix keto- or oxy-.

PREPARATION

The methods of preparation are similar
to those for aldehydes. The oxidation of secondary alcohols proceeds smoothly and with
little danger of carrying the process too far,
but direct dehydrogenation of the alcohol
by heating the vapors over metallic copper is
preferable for acetone and cyclohexanone.

Hydrolysis of dihalides, in which the two
halogen atoms are on the same secondary
carbon atom, is an excellent method, if the
necessary compound is available. Thus, isopropylidene chloride gives acetone.

$$CH_3-CCl_2-CH_3 \xrightarrow{H_2O} CH_3-CO-CH_3$$

A convenient method for those ketones
of higher molecular weight is addition of
the Grignard reagent to amides or nitriles
with subsequent hydrolysis.

$$R-CN \xrightarrow{R'-Mg-I} R-C(R')=N-Mg-I$$

$$\downarrow_{H_2O}$$

$$R-CO-R'$$

For simple, volatile ketones, one of the most practical methods is to distill a salt of barium, calcium or thorium. This can be illustrated by the preparation of dipropyl ketone from barium butyrate.

$$(C_3H_7COO)_2Ba \rightarrow (C_3H_7)_2CO + BaCO_3$$

If a mixed ketone should be desired, such as hexyl methyl ketone, one would need to start with a mixture of salts, and the product would be contaminated with two other compounds.

Finally, ketones frequently can be prepared with advantage from substituted acetoacetic ester by appropriately conducted hydrolysis using dilute acid or alkali. The reactions for producing ethyl methyl ketone will illustrate.

$$CH_3-CO-CH_2-COOEt \xrightarrow{Na}$$

$$CH_3-C(ONa)=CH-COOEt \xrightarrow[\text{warm}]{CH_3I}$$

$$CH_3-CO-CH(CH_3)-COOEt \xrightarrow[\text{dilute}]{KOH}$$

$$CH_3-CO-CH_2-CH_3$$

Physical Properties

The lower aliphatic and alicyclic ketones are volatile liquids that are soluble in water and have a pleasant odor. The higher ones and those with aryl groups are liquids or solids of sparing solubility and sometimes with no odor.

Products

Acetone N.F., 2-propanone, dimethyl ketone, was observed first in the distillate from wood in 1661 by Robert Boyle; it was isolated from the heating of acetates by Macaire and Marcet in 1823. A small amount is present in normal blood and urine; this may be increased greatly in diabetes, apparently by decarboxylation of acetoacetic acid. It was manufactured formerly by heating acetates or by the distillation of wood, but practically all of the annual production (about 120,000 tons) today is made from petroleum or from starches by fermentation.

The fermentation method, developed since 1914, first utilized corn starch and *Clostridium acetobutylicum* Weizmann. The chief products are *n*-butyl alcohol, acetone and ethyl alcohol in the proportion of about 6:3:1. Other organisms also were found for this process, as well as a variety of raw materials, including potatoes, molasses, bran and other waste from plants. By altering the factors, acetone can be obtained as the chief product.

The usual starting material from petroleum is propylene, which can be produced in almost unlimited amount by the cracking process. This compound is catalytically hydrated to isopropyl alcohol, which may in turn be oxidized to acetone.

$$CH_3-CH=CH_2 \xrightarrow{H_2O}$$

$$CH_3-CHOH-CH_3 \xrightarrow{(O)} CH_3-CO-CH_3$$

Other methods are available to utilize ethylene from the cracking process, either through ethyl alcohol or acetaldehyde. A similar method of some commercial importance involves the use of acetylene, which first is converted to acetaldehyde (q.v.) and acetic acid. Vapors of the last are passed over manganous oxide or other catalytic agents at 400° to 500°.

$$2CH_3-COOH \xrightarrow[500°]{MnO} CH_3-CO-CH_3 + CO_2 + H_2O$$

Furthermore, it is possible to convert acetylene directly to acetone by hydration, using zinc oxide as catalyst.[3]

$$2CH\equiv CH + 3H_2O \xrightarrow{ZnO}$$

$$CH_3-CO-CH_3 + CO_2\uparrow + H_2\uparrow$$

Commercial acetone is usually of a high degree of purity, seldom less than 99 per cent, the remainder being practically all water. It is a transparent, colorless, mobile liquid with a characteristic odor, boiling at about 56° but volatile and inflammable at much lower temperatures. It has a specific gravity of about 0.79 and is miscible with water, alcohol or other solvents. Like other methyl ketones, it gives the haloform reaction. With sodium nitroprusside in the presence of alkalies, it gives a red color that is deepened by the addition of excess acetic acid.

Although acetone is not used in medicine, it is of immense value in industry, chiefly as a solvent for fats, waxes, oils, varnishes, lacquers, rubber and like materials. In addition, it is employed in the manufacture of a great number of substances, including chloroform, iodoform, explosives, varnish removers, plastics, rayon and medicinals. It is added to tanks of acetylene to aid in compression and prevent explosion, for acetone will dissolve 24 times its volume of the gas.

Category—solvent.

Occurrence	Per Cent Acetone
Carbol-Fuchsin Solution N.F.	5
Nitromersol Tincture N.F.	10
Thimerosal Tincture N.F.	10

Methyl Isobutyl Ketone N.F., 4-Methyl-2-pentanone, Isopropylacetone. This is a colorless liquid with a faint ketonic and camphor odor. It is miscible with alcohol and other organic solvents.

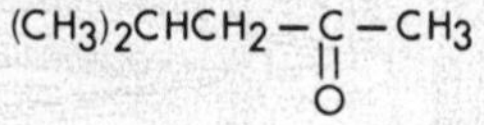

$$(CH_3)_2CHCH_2 - \underset{\underset{O}{\|}}{C} - CH_3$$

Methyl Isobutyl Ketone

These properties make it useful as a component of S.D.A. Formula 23-H (Internal Revenue Service, U. S. Treasury Department), which is used in Rubbing Alcohol N.F.

Category—alcohol denaturant.

Camphor U.S.P., 2-bornanone, 2-camphanone is a terpene ketone and may be obtained from the wood of *Cinnamomum camphora* by distillation or produced synthetically using α-pinene isolated from turpentine oil.

α-Pinene $\xrightarrow{HCl}$ bornyl chloride $\xrightarrow{base}$ camphene

$\xrightarrow{CH_3COOH}$ isobornyl acetate $\xrightarrow{H_2O}$ isoborneol

$\xrightarrow{(O)}$ camphor

The camphor structure has two asymmetric carbon atoms, but only two optically active forms are possible.

Natural camphor is the dextro form and

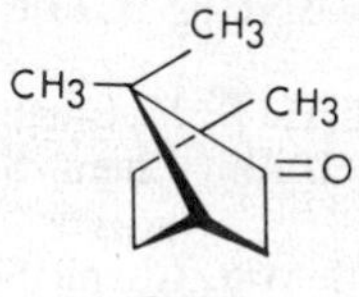

Camphor

synthetic camphor is a racemic form, but for industrial or medicinal use the difference is insignificant. Actually, the greater activity of the levo forms makes synthetic camphor more powerful physiologically than natural camphor.

Usually, camphor is supplied as white, crystalline masses or as white, translucent, small cakes. It has a characteristic odor and a sharp, burning taste. Powdered camphor is prepared readily by rubbing camphor with a small quantity of alcohol, ether or chloroform. It is slightly soluble in water (1:800) and soluble in alcohol or organic solvents. An eutectic mixture is formed with menthol, thymol, chloral hydrate, salol, phenol or resorcinol.

Camphor is employed most commonly locally as a mild antiseptic, analgesic and antipruritic (see official occurrence). The sensation of coolness it creates when applied either to the epidermis or to the nasal passages makes it a pleasant medicinal agent. In paregoric and in cold preparations, it is used because there is thought to be a slight expectorant and diaphoretic action. Its use internally as a circulatory and respiratory stimulant is unsound.

Camphor from the many preparations in use gains entrance into the system, where it is reduced to the alcohol camphorol and then combines with glucuronic acid by a glucosidic linkage and is excreted in the urine.

Category—topical antipruritic.

For external use: topically, 1 to 3 per cent in preparations for use on the skin.

Occurrence	Per Cent Camphor
Flexible Collodion U.S.P.	2
Paregoric U.S.P.	0.4
Camphorated Parachlorophenol N.F.	65

REFERENCES CITED

1. Creighton, R. H. J., McCarthy, J. L., and Hibbert, H.: J. Am. Chem. Soc. 63:3049, 1941.
2. Pearl, I. A.: J. Am. Chem. Soc. 64:1429, 1942.
3. British Patent 192,392; Chem. Abst. 17: 3190, 1923; see also D.R.P. 655,969, Feb. 5, 1938 in Chem. Abst. 32:3773, 1938.

SELECTED READING

McClure, H. B., and Bateman, R. L.: The synthetic aliphatic industry, Chem. Eng. News 25:3208, 3286, 1947.

7

Acids, Their Salts and Esters

Charles O. Wilson, Ph.D.,
Dean and Professor of Pharmaceutical Chemistry, School of Pharmacy,
Oregon State University

INTRODUCTION

Organic acids are characterized by having one or more carboxyl groups (COOH) in the molecule. They often are referred to as monocarboxylic (monobasic) acids, dicarboxylic (dibasic) acids, tricarboxylic (tribasic) acids, etc. The structure of the carboxyl group is such that an organic acid will ionize sufficiently to provide hydrogen (hydronium, H_3O^+) ions.

$$R-C\!\!\!\!\overset{O}{\diagup}\!\!-OH \rightleftharpoons R-\overset{O}{\underset{\|}{C}}-O^- \quad H^+ \xrightarrow{\ H_2O\ } H_3O^+$$

Organic acids are weak acids, since they ionize only slightly in polar solvents, such as water.

Just as the alcohols may be thought of as hydrocarbons having a hydrogen replaced by a hydroxyl group (OH), so may the acids be considered as hydrocarbons having a carboxyl group replacing a hydrogen atom. The simplest acid then would be the methane derivative, acetic acid, CH_3COOH. Formic acid, HCOOH, is a hydrocarbon derivative but is unusual in that the carboxyl group is attached to a hydrogen atom and thus becomes the simplest organic acid.

Organic acids are the final oxidation product of primary alcohols and of aldehydes before the breakdown of the carbon chain to carbon dioxide and water.

The carboxylic acids also may be looked upon as being derived from water (HOH)

$$RCH_2OH \xrightarrow{(O)} RCHO \xrightarrow{(O)}$$
$$RCOOH \xrightarrow{(O)} CO_2 + H_2O$$

by the replacement of a hydrogen atom with an acyl group

$$(R-\overset{O}{\underset{\|}{C}}-).$$

The term *fat acids** applies to the straight chain saturated or unsaturated acids having an even number of carbon atoms from C_4 to C_{30}. They are so called because such acids are found as glyceryl esters in fats and oils. Fat acids may be saturated or unsaturated and may or may not contain hydroxyl groups. Fatty acids* is the name often used to denote the acids of the acetic acid series, which, of course, are all saturated; the series includes acids with even- and odd numbers of carbon atoms.

As a general rule, introduction of the carboxyl group into any organic molecule tends to increase solubility in water and to decrease both toxicity and physiologic action.

SATURATED MONOBASIC ACIDS

The monobasic saturated acids and their esters and salts represent an important group of organic compounds used in pharmacy.

* These names are often used interchangeably, and in pharmacy and medicine the term fatty acid is commonly used to designate an acid obtained from natural glycerides.

Aliphatic straight-chain acids up to C_{10} and those of even-numbered carbon atoms up to C_{18} have been found in nature. Also, the acids of C_{11} and C_{17} have been reported.

Preparation of these acids may be carried out in many ways, of which the following are typical:

1. Oxidation of Alcohols, Aldehydes or Ketones

$$CH_3CH_2OH \xrightarrow{(O)} CH_3COOH$$

$$CH_3CHO \xrightarrow{(O)} CH_3COOH$$

$$CH_3CH_2CHOHCH_3 \xrightarrow{(O)} CH_3CH_2COCH_3$$

$$\downarrow (O)$$

$$CH_3CH_2COOH + HCOOH$$

2. Hydrolysis of Acid Derivatives

ESTER

$$CH_3CH_2COOC_2H_5 + HOH$$

$$\downarrow H^+$$

$$CH_3CH_2COOH + CH_3CH_2OH$$

NITRILE

$$CH_3CH_2CN + 2H_2O \xrightarrow{H^+} CH_3CH_2COO^- + NH_4^+$$

SALT

$$CH_3COONa + H_2O \xrightarrow{H^+} CH_3COO^- + Na^+$$

3. Grignard Reaction

$$CH_3CH_2MgX + CO_2 \longrightarrow$$

$$CH_3CH_2COOMgX + H_2O$$

$$\downarrow HX$$

$$CH_3CH_2COOH + MgX_2$$

The melting points of the aliphatic saturated acids increase with an increase in the number of carbon atoms. They are liquids up to C_8 and C_9, and the higher members are waxy solids. The lower members are soluble in water (formic, acetic and propionic acids), but, as molecular weight increases, the solubility in water decreases. All are soluble in organic solvents, including methyl and ethyl alcohols. They dissociate very slightly in aqueous solution. The sodium and the potassium salts are soluble and are frequently employed as buffers. The salts formed with other metals, such as Ca, Ba, Mg, Sr salts of the fatty acids, with the exception of the low-molecular-weight fat acids, are insoluble in water. Those acids of low molecular weight have characteristic odors, but the higher ones, of 8 carbon atoms or more, are odorless.

A carboxyl group present in an organic compound provides the characteristic chemical properties which are associated with organic acids. These compounds exhibit acid properties, and suitable methods may be used to cause the replacement of the hydrogen atom by metals or acyl groups (RCO): the hydroxyl of the carboxyl groups may be substituted by the groups RO, SH, NH_2, Cl, H or RCOO; the oxygen and hydroxyl may be substituted by hydrogen (reaction known as decarboxylation). In general, the acids are resistant both to oxidation, which results in carbon dioxide and water, and to reduction, which results in aldehydes and alcohols.

The physiologic activity possessed or conferred by the carboxyl group is limited and generally insignificant. As previously noted, the tendency is for a compound to decrease in activity upon the introduction of a carboxyl group, but at the same time it may assume entirely new, useful properties. Note the difference between methane and acetic acid and between acetic acid and malonic acid. The physiologic activity of the straight-chain even-numbered acids is negligible, since they are available from many foods (fats) and are readily oxidized by the body. Lower members are corrosive in a concentrated form but not when diluted. Like the alcohols and hydrocarbons, the branched-chain acids are more active physiologically than are their straight-chain isomers. The hypnotic property of the drug valerian was thought to be due to valeric acid. Several valeric acid esters and urea derivatives, some brominated, have been introduced as sedatives into medicine. The alpha hydrogen atoms of acetic acid also may be substituted by alkyl groups, such as $(CH_3)_3$ and $(CH_3)_2$ (C_2H_5), thus producing an active hypnotic compound.

PRODUCTS

Formic Acid. The simplest organic acid, formic acid, derives its name from the red ant (*Formica rufa*) from which it was isolated first. It also is found in the stinging nettle, pine needles, caterpillars and in some plants and fruits.

It may be prepared by several methods, but it usually is obtained by treating sodium hydroxide with carbon monoxide; then, to liberate formic acid, sulfuric acid is added to the sodium formate thus formed. Formic acid of about 85 per cent purity may be prepared, although a 25 per cent solution usually is used commercially.

The acid is a colorless liquid having a characteristic pungent odor and is soluble in water, alcohol, glycerin or ether. It is the strongest aliphatic acid and, because of the aldehyde group present, possesses the properties of both aldehydes and acids.

Formic acid is corrosive and irritating. Diluted, it is relatively nontoxic when ingested, as it is oxidized readily to carbon dioxide and water. It has some use in industry but is of doubtful value in therapeutics.

Formate salts, such as calcium, magnesium, lithium, potassium and sodium formate, are used in sodium chloride substitutes.

Acetic Acid U.S.P. Ethanoic Acid, contains 36 to 37 per cent of CH_3COOH. Acetic acid and its salts and esters are distributed widely in nature. The acid has been known for over 100 years, first in the form of vinegar. Acetic acid has been obtained from many sources but is now produced from ethanol by oxidation and from acetylene by hydration to acetaldehyde, which then is oxidized.

$$CH_3CH_2OH \xrightarrow{(O)} CH_3CHO \xrightarrow{(O)} CH_3COOH$$

$$CH{\equiv}CH \xrightarrow{HOH} CH_3CHO \xrightarrow{(O)} CH_3COOH$$

Acetic acid is a corrosive, colorless liquid with a pungent odor and a sharp, acid taste. It is nontoxic and of little therapeutic value. In most cases the acid is used in diluted form where a weak and innocuous acid is required (see Triacetin).

Acetates usually employed in pharmacy are Sodium Acetate N.F., Potassium Acetate N.F., and solutions of aluminum acetate and subacetate.

Sodium acetate and potassium acetate are the common salts. In solution their alkalinity is due to hydrolysis where the acetate ion functions as a proton acceptor. The alkalinity is utilized with preparations of theobromine and theophylline.

Most acetates are very soluble in water. The only one not very soluble is silver acetate. Acetates are stable in solution and are oxidized in the body to bicarbonate. Ethyl Acetate N.F. is used as a flavor.

Acetic acid is metabolizable readily and is in a number of biologic transformations.

Category—pharmaceutic aid.

OCCURRENCE	PER CENT ACETIC ACID
Glacial Acetic Acid U.S.P.	99.5
Acetic Acid U.S.P.	36
Diluted Acetic Acid N.F.	6
Aluminum Subacetate Solution U.S.P.	16
Aluminum Acetate Solution U.S.P. ..	1.5

Stearic Acid U.S.P. "Stearic Acid is a mixture of solid acids obtained from fats, and consists chiefly of stearic acid $[CH_3(CH_2)_{16}COOH]$ and palmitic acid $[CH_3(CH_2)_{14}COOH]$." The production of stearic acid is associated with the saponification of fats (beef, tallow), which procedure is carried out by the use of steam, alkali or by the Twitchell method. Fats are composed of the glycerides of fat acids. The acids most frequently found are oleic, linoleic, stearic, palmitic and myristic. Saponification of these fats with sodium hydroxide yields the sodium salts of the acids.

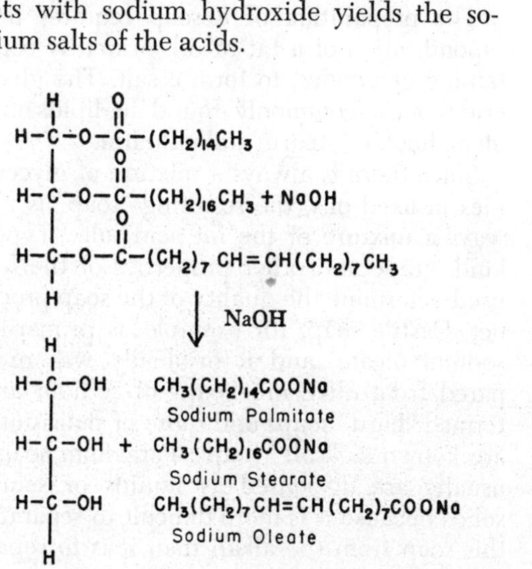

The sodium salts are treated with a mineral acid, thus forming the free fat acids.

$$2 CH_3(CH_2)_n COONa + H_2SO_4 \rightarrow$$

$$2 CH_3(CH_2)_n COOH + Na_2SO_4$$

Oleic acid and any other unsaturated acids are liquid and are separated readily from the solid saturated fat acids. To prepare pure stearic acid is a difficult task and one that is unnecessary for pharmaceutic use.

Stearic Acid U.S.P. is a solid, white, wax-like, crystalline substance having practically no taste or odor; it is insoluble in water but soluble in organic solvents.

It is used to prepare stearates and in ointments, suppositories, creams and cosmetic products. The potassium and sodium salts are of special interest, since these are the common soaps. Esters, such as glycol stearate, are used in ointments, and butyl stearate is satisfactory as an enteric-coating for tablets.

Category—pharmaceutic aid.

SOAPS

A soap is a salt of a fat acid. Most commonly, a soap is considered as a combination of an alkali (NaOH or KOH) with the fat acids derived from animal or vegetable oils. A soap of sodium, potassium or ammonium is soluble in water, but most soaps of other metals, such as calcium, magnesium, iron, barium, strontium, are insoluble in water.

The preparation of a soap requires the saponification of a fat or oil, primarily containing glycerides, to form a salt. The glycerides most commonly found in lipids are olein, linolein, stearin and palmitin.

Since there is always a mixture of glycerides in fixed oils, the resulting "soap" is always a mixture of the fat acid salts. Type, kind, grade and other properties of the fat used determine the quality of the soap product. Castile soap, for example, is primarily sodium oleate, and it originally was prepared from olive oil. Soaps of sodium are termed "hard" soaps and those of potassium are known as "soft" soaps. Potassium soaps usually are dispensed as liquids or semi-solids because it is more difficult to separate this soap from the alkali than it is to sepa-rate the more common sodium soaps (see Green Soap N.F.). The reasons for this are: (1) potassium soaps are more soluble than are the sodium soaps, and (2) it would require a large quantity of potassium chloride, a more expensive salt than sodium chloride, to salt out the potassium soap. However, when a potassium soap is prepared in the conventional manner, it is found to be just as "hard" as a sodium soap. A laundry soap usually contains gritty material; cosmetic or facial soaps contain perfume; medicinal soaps contain a therapeutic agent, such as an antiseptic or fungicide.

All soaps in aqueous solution undergo partial hydrolysis, yield free alkali and free

$$C_{17}H_{35}COONa \underset{\longleftarrow}{\overset{HOH}{\longrightarrow}} C_{17}H_{35}COOH + NaOH$$
Stearic Acid

fat acid. The pH of the usual hand soap is 10 to 11. Other, more crude, soaps are more strongly alkaline because the lack of purification causes some original saponifying alkali to be retained. The detergent effect of soap is due in part to the presence of free alkali, which is responsible for its antiseptic value also. A keratolytic or disinte-grating action on the skin is likewise a characteristic of soap and is one of the reasons why the alkali content is held to a minimum.

PRODUCTS

Hard Soap. This soap is prepared from fats or oils, with sodium hydroxide, and it consists of the sodium salts of the fat acids. It occurs as white flakes or cakes having a faint odor, freedom from rancidity and slow solubility in water or alcohol. Its uses are most varied, but in pharmacy hard soap is employed primarily as a lubricant and less often as a detergent. Application of it is made in suppositories, tooth pastes, tooth powders, liniments, pills and laxatives.

Sodium Stearate N.F. This is a mixture of varying proportions of sodium stearate and sodium palmitate. It is prepared by neutral-izing Stearic Acid U.S.P. with sodium carbonate.

$$CH_3(CH_2)_n COONa^+$$

Sodium Palmitate n=14
Sodium Stearate n=16

The salt is used in preparing some ointments, suppositories, creams and cosmetics.

Category—emulsifying and stiffening agent.

OCCURRENCE	PER CENT SODIUM STEARATE
Glycerin Suppositories N.F.	9

Calcium Stearate N.F. is the calcium salts of stearic and palmitic acids in varying proportions. It is a fine, white to off-white, unctuous powder with a slight characteristic odor. It is insoluble in water, in alcohol, and in ether. Its primary use is in tablet formulation as a solid lubricant for the granules during the tableting process.

Category—tablet lubricant.

Zinc Stearate U.S.P. is prepared by mixing solutions of equal molar amounts of sodium stearate and zinc acetate. Zinc stearate, being insoluble in water, is washed and removed by filtration. The dry salt is a light, fluffy, fine, white powder which is insoluble in organic solvents.

It is used in ointments and powders as a mild astringent and antiseptic. It is also used as a tablet lubricant in the tableting process. The salt is a water-repellent powder and was used widely to replace talcum powder until it was found that inhalation may cause pulmonary inflammation.

Category—pharmaceutic aid (tablet lubricant).

OCCURRENCE	PER CENT ZINC STEARATE
Zinc-Eugenol Cement N.F.	1.0
Compound Iodochlorhydroxyquin Powder N.F.	20.0

Magnesium Stearate U.S.P. is the magnesium salts of stearic acid and palmitic acids in varying proportions. It is a fine white, unctuous powder with a faint characteristic odor. Its properties and uses are similar to those of Calcium Stearate N.F.

Category—pharmaceutic aid (tablet lubricant).

Aluminum Monostearate U.S.P., occurs as a white to off-white bulky powder. It is insoluble in water, alcohol and ether. It is prepared by mixing solutions of a soluble aluminum salt and sodium stearate. Aluminum monostearate is used in Sterile Procaine Penicillin G with Aluminum Stearate Suspension U.S.P. for its action as a dispersing agent and its thixotropic properties. It forms a gel which becomes a free-flowing solution when shaken gently before use.

Category—pharmaceutic aid.

Green Soap N.F., sapo mollis, soft soap "is a potassium soap made by the saponification of vegetable oils, excluding coconut oil and palm-kernel oil, without the removal of glycerin." It is a soft, unctuous, yellow-to-green-colored mass. This is due to the presence of glycerin, formed as a result of saponification. The color is due in part to the vegetable oil used and in part to the age of the soap. These factors also affect the odor.

Green soap is more strongly alkaline than is hard soap, and, therefore, it has greater detergent properties and irritant action, due to excess alkali. It is used to cleanse the skin of keratin elements and to stimulate the tissue in some skin diseases. Very rarely is it used internally.

A potassium soap is formed in the preparation of saponated cresol solution.

OCCURRENCE	PER CENT GREEN SOAP
Green Soap Tincture N.F.	6.5

ESTERS

Next to the salts, the esters are the most important acid derivatives used in pharmacy and medicine. The esters may be considered as organic acids with the hydroxyl of the carboxyl group replaced by an alkoxy group (RO). Numerous examples of ester-type compounds (RCH_2COOR) are to be found among the therapeutic agents; such compounds are atropine, cocaine, fats, waxes and most local anesthetics.

Many procedures are available for the preparation of esters. Some of the most commonly used methods are

1. The action of an alcohol on an acid with the aid of heat and some mineral acid

(HCl or H_2SO_4) as a catalyst (Fischer method):

$$RCOOH + ROH \xrightarrow{H^+} RCOOR$$

2. The action of an alkylating agent on the sodium or silver salt of an acid:

$$RCOOAg + RI \rightarrow RCOOR + AgI$$

3. The action of an acid chloride on an alcohol:

$$RCOCl + ROH \rightarrow RCOOR + HCl$$

4. For the methyl ester, the action of diazomethane on an acid:

$$RCOOH + CH_2N_2 \rightarrow RCOOCH_3 + N_2$$

The carboxylic acid esters are neutral liquids or solids which are hydrolyzed, slowly by water and rapidly by acids or alkalies, into their components. In the case of alkalies, the salt of the acid is obtained. The simpler esters are liquids and possess a fragrant odor, whereas the acids have a pungent or nauseating smell, e.g., ethyl butyrate has the odor of pineapple, while butyric acid has the odor of rancid butter.

A few of the very simple esters are soluble in water, but those composed of more than four carbon atoms are practically insoluble in water.

Certain chemical reactions are characteristic of esters, and the use of these reactions provides methods of producing compounds that are otherwise hard to obtain. Esters are reduced to alcohols by suitable methods, for example, catalytic reductions and the use of sodium and alcohol.

$$RCOOR' \xrightarrow{(H)} RCH_2OH + HOR'$$

Hydrolysis of esters provides a method of obtaining acids or acid salts.

Ester interchange (alcoholysis) is useful in preparing the methyl or ethyl ester of an acid. By using an excess of methyl or ethyl alcohol, an acid catalyst and heat, one can cause most esters, including glycerides, to be a source of the methyl or ethyl ester of the acid.

$$RCOOR' + CH_3OH \xrightarrow{H^+} RCOOCH_3 + R'OH$$

Amides generally can be prepared by the action of ammonia on an ester.

$$RCOOR' + NH_3 \rightarrow RCONH_2 + R'OH$$
$$\text{Amide}$$

Ethyl Acetate N.F., acetic ether, vinegar naphtha, is obtained by the slow distillation of a mixture of ethyl alcohol, acetic acid and sulfuric acid. It is a transparent, colorless liquid, with a fragrant, refreshing, slightly acetous odor and a peculiar, acetous, burning taste. The ester is miscible with ether, alcohol and fixed and volatile oils.

At present, pharmaceutically it is used to impart a pleasant odor and flavor; it has wider application in industry as a solvent.

Category—flavor.

Isopropyl Myristate is a mixture composed principally of the isopropyl ester of myristic acid with lesser amounts of the isopropyl esters of other fatty acids. It is used in cosmetics and pharmaceuticals for its emollient and dispersing properties.

SATURATED POLYBASIC ACIDS

Dibasic acids or their salts are not used too frequently in medicine. A list of the usual ones and a learning aid follow:

Oh	Oxalic
My	Malonic
Such	Succinic
Good	Glutaric
Apple	Adipic
Pie	Pimelic
Sweet	Suberic
As	Azelaic
Sugar	Sebacic

PRODUCTS

Oxalic Acid (Ethanedioic Acid) occurs most commonly as the dihydrate, HOOC-COOH · $2H_2O$, in clear to white crystalline prisms or granules. It is soluble in water (1:7), alcohol (1:25), ether (1:100) or glycerin 1:5.5 and insoluble in other organic solvents. Oxalic acid is a strong reducing agent, which is the basis for much of its industrial application (bleaching agent).

Pharmaceutically, a most frequent reaction is that between calcium ions and oxalic acid. Externally, either the acid or its soluble salts will prevent blood coagulation by removing calcium ions as insoluble calcium oxalate. However, internally, a combination of oxalic and malonic acids (Koagamin) is administered intravenously or intramuscularly to increase blood coagulation. The action is stated to be the removal of some calcium ions that appear to be associated with prothrombin, thus, at the point of bleeding, it quickens the action of prothrombin. Koagamin also renders the platelets more labile than normal. Overdoses will reverse the action.

Succinic Acid (Butandioic Acid) is $HOOCCH_2CH_2COOH$, occurring as colorless or white monoclinic crystals. It is obtained either by hydrolysis of ethylene cyanide or by reductive fermentation of maleic or tartaric acids. The crystals are soluble in water (1:13), alcohol (1:18.5) and glycerin (1:20) and slightly soluble or insoluble in organic solvents.

Succinic acid or succinates enter into the intermediary metabolism of carbohydrates and fats by increasing the oxygen utilization of the tissues. In arthritis, there is considered to be an impairment of tissue oxidation; therefore, succinate salts are used. Succinates are known to decrease the toxicity of salicylates and are used in salicylate-containing pharmaceuticals for arthritis and rheumatism. Sodium succinate (Soduxin) is used intravenously in the treatment of barbiturate poisoning because of its important role in tissue respiration.[1] An analgesic action also is possessed by succinates, presumably by increasing the blood supply to an injured part. Calcium, sodium and magnesium succinates are available (see also benzyl succinate).

Several dicarboxylic acids or their salts possess laxative action,[2] and succinates are effective orally, with the advantage of having no nephrotoxic action.

HYDROXY ACIDS

Hydroxy acids have the properties of alcohols (hydroxyl group) and of acids (carboxyl group). They have both types of groups required for esterification. This gives rise to the possibility of forming polyesters or an internal ester linkage (a lactone). Usually, the hydroxy acids are those that have a hydroxyl group on the α, β, γ, δ or ϵ carbon atom. More than one hydroxyl group may be present, as in the case of gluconic and tartaric acids. Lactone formation occurs most readily with the γ-hydroxy acids producing 5-membered ring systems. These form in acid media and are converted to their salts by warm alkali. Note that the formation of γ-lactones is common among the onic acid of the monosaccharides (see also glucuronolactone).

In the pharmacy the α-hydroxy acids are used mostly, exceptions being citric acid and gluconic acid. An α-hydroxyl increases the acidity, as we note that lactic acid is about pKa 3.87, whereas propionic acid is about pKa 4.87. Hydroxyl groups in beta, gamma, delta and epsilon positions have practically no effect on ionization.

A major function of hydroxy acids and their salts is that of sequestering agent. Sequestration is the state of combining a cation in a large complex soluble ion and thus preventing the original cation from exhibiting its usual chemical properties.[3] Cations commonly sequestered in pharmaceuticals are copper, calcium, magnesium, manganese, antimony, bismuth, ferrous and ferric ions. The acids used are tartaric and citric. Note that glycerin and sucrose also are used as sequestering agents. Thioacids, those containing a mercapto group (HS-), such as thioglycollic and thiomalic acids, function in the same manner (see bismuth sodium thioglycollate and gold sodium thiomalate). The use of BAL depends upon the sequestering action of dimercapto glycerol.

PRODUCTS

Lactic Acid U.S.P. 2-Hydroxypropionic Acid (α-Hydroxypropionic Acid) is a mixture of lactic acid, $CH_3CHOHCOOH$, and lactic anhydride, $CH_3CHOHCO \cdot OCH(CH_3)$ COOH. In 1780, Scheele discovered this product of bacterial fermentation; it is found in products such as sour milk, cheese, buttermilk, wine and sauerkraut.

Although lactic acid may be synthesized, it is produced commercially by the fermentation of molasses, whey or corn sugar, with either *Lactobacillus delbruckii* or *L. bulgaricus.* Because of an asymmetric carbon atom, lactic acid may exist in three forms; it usually is supplied in the DL form. The acid is a clear, slightly yellow, odorless, syrupy liquid, miscible with water, alcohol, or ether but immiscible with chloroform. A 0.1 N solution has a pH of 2.4. The levoform, known as sarcolactic acid, is formed in muscle tissue as a result of work.

The free acid seldom is used because it is caustic in concentrated form. It is added to infant formulas to aid digestion and to decrease the tendency of regurgitation, and it is employed as a spermatocidal agent in contraceptives. It is used to prepare lactates of minerals which are intended for internal administration, e.g., Calcium Lactate N.F., Sodium Lactate Injection U.S.P. and Ferrous Lactate. Ferrous lactate is a desirable form of iron to be given orally. By the use of *Lactobacillus acidophilus*, an acidic intestinal flora which produces lactic acid by fermentation is developed.

Category—pharmaceutic aid.

Occurrence	Per Cent Lactic Acid
Compound Iodochlorhydroxyquin Powder N.F.	2.5

Sodium Lactate occurs as a colorless, thick, odorless liquid that is miscible with water or alcohol to produce neutral solutions. The salt is prepared by mixing equimolar amounts of sodium carbonate and lactic acid in solution and then concentrating, by evaporation or distillation, to the desired strength. This salt is available as a 70 to 80 per cent aqueous solution.

By oral administration, the lactate ion is without specific effect in the body, but, when introduced by injection, the lactate ion is readily oxidized to the bicarbonate ion. Sodium Lactate Injection U.S.P. is a one-sixth molar solution of the racemic salt which is prepared with sterile distilled water. This produces an isotonic solution.

Sodium Lactate Injection U.S.P. may be used in conditions in which sodium bicarbonate is needed for a systemic alkalinizing effect. Lactated Ringer's Injection U.S.P. contains sodium lactate and is a "combined solution" developed by Dr. Alexis F. Hartman for treating mild cases of acidosis or alkalosis. Lactated Potassic Saline Injection N.F. (Darrow's Solution) commonly is used to replenish fluid and electrolytes.

Occurrence
    Sodium Lactate Injection U.S.P.
    Lactated Ringer's Injection U.S.P.

Other lactates, used primarily for the cation portion are Calcium Lactate N.F., and ferrous lactate.

Tartaric Acid N.F. (Dihydroxysuccinic Acid) is a dihydroxy dicarboxylic acid that was first isolated by Scheele. It is obtained from tartar, a crystalline deposit of crude potassium bitartrate occurring in wine.

$$
\begin{array}{c}
H \\
| \\
HO-C-COOH \\
| \\
HO-C-COOH \\
| \\
H
\end{array}
$$

The acid contains 2 asymmetric carbon atoms, and it exists in racemic, levo, dextro and meso forms. In the wine industry, a crude form of potassium bitartrate, called argol, is produced. This is treated with a calcium salt to form insoluble calcium tartrate, which is acidified with sulfuric acid to yield insoluble calcium sulfate and soluble tartaric acid.

The acid occurs as large, colorless crystals or as a fine, white, crystalline powder that is practically 100 per cent pure. It is stable in air, has an acid taste and is odorless. Its pKa values of 3.0 and 4.3 make it useful as a buffer in the pH range of 2.5 to 5.0. It is soluble in water (1:0.75) or in alcohol (1:3), but it is insoluble in most organic solvents. The properties are typical of those of organic acids, and it forms a precipitate with salts of potassium, calcium, barium, strontium, lead, silver or copper. Potassium bitartrate is one of the few relatively water-insoluble (1:165) potassium salts.

Tartrates in general are insoluble or very slightly soluble. Sodium tartrate is soluble,

Potassium Sodium Tartrate N.F. is very soluble, potassium and ammonium dissolve with difficulty, and the others are insoluble. Tartrates and tartaric acid are sequestering agents similar to the citrates. Soluble complex ions are formed in the presence of excess tartrate. Examples of this property are the solubilizing of cupric ions by rochelle salt in Fehling's Solution, the soluble antimony salt, Antimony Potassium Tartrate U.S.P. and bismuth potassium tartrate.

The hydrogen of the alcoholic OH is more active than the OH of ethanol and is easily replaced in basic media by ions of copper, bismuth, antimony or iron. Thus, there is formed a complex that results in a soluble compound. This sequestering property of tartrates is very pronounced. With ferric ions[4] a soluble ferritartaric acid is formed.

Tartrates may be oxidized by ammoniacal silver nitrate to produce a silver mirror. The tartrate perhaps yields pyruvic acid ($CH_3CO-COOH$) and finally acetic acid and carbon dioxide.

Tartaric acid is used in preparing many useful tartrates, and, as the free acid, it is employed in effervescent salts and refrigerant drinks. The acid is not decomposed by the body, as are most organic acids, but it is excreted in the urine and, therefore, increases the acidity of the system. However, most of it is retained within the intestine and functions by osmotic pressure as a saline laxative. Several salts are used as saline cathartics, i.e., Potassium Sodium Tartrate N.F., and potassium bitartrate.

Category—buffer.

OCCURRENCE	PER CENT TARTARIC ACID
Effervescent Sodium Phosphate N.F.	25.2

Citric Acid U.S.P. is a tribasic acid that is distributed widely in nature. The official form is of not less than 99.7 per cent purity calculated on the anhydrous basis.

$$\begin{array}{c} H \\ HC-COOH \\ | \\ HOC-COOH \cdot H_2O \\ | \\ HC-COOH \\ H \end{array}$$

It is the acid present in citrus fruits and berries. Commercially, the acid may be produced from limes or lemons, from the residue from pineapple canning and from the fermentation of beet molasses. Juice from the citrus fruits is treated with chalk, and the precipitated calcium citrate is acidified with sulfuric acid. Calcium sulfate is filtered off and the citric acid is recovered from the filtrate. The fermentation process[5] accounts for the largest amount of citric acid and may be carried out with any one of over nineteen varieties of fungi (*Citromyces, Aspergillus, Penicillium*) to give liquor concentrations of from 10 to 15 per cent citric acid.

The acid occurs as a white, crystalline powder or as large, colorless, translucent crystals. It is efflorescent in air, odorless, sour-tasting and is soluble in water (1:0.5), alcohol (1:2) or ether (1:30) and insoluble in other organic solvents. An aqueous solution of citric acid is unstable because it undergoes slow decomposition. Its chemical reactions are those characteristic of organic acids. Salts are formed readily with all hydroxides, and they produce alkaline, aqueous solutions (sodium, potassium, calcium, magnesium). Citric acid effervesces carbonates, and this property is employed widely in effervescent salts.

A test for the citrate is the formation of calcium citrate in a nearly neutral (pH 7.6) solution.

A test for distinguishing from tartrates consists of adding potassium permanganate T.S. to a hot solution of a citrate, to which has been added mercuric sulfate T.S. (Denige's reagent). A white precipitate is produced from the reaction between mercuric subsulfate and acetone dicarboxylic acid.

In vivo, the citrate ion gives rise to the bicarbonate ion and, thus, contributes alkaline properties.

The citrates of the alkali metals are soluble in water, whereas most other citrate salts are insoluble in water. However, the insoluble citrates may be solubilized by an excess of citric acid or citrate ion (usually furnished by sodium citrate). The citrate ion often is used as a sequestering agent with ions of metals, such as magnesium, manganese, calcium, ferric, bismuth, strontium, barium, copper and silver. The metal ion is held in solution in a complex anion form that is soluble and yet prevents the metal ion from exhibiting its usual properties. This principle is utilized in iron preparations, Benedict's solution and Anticoagulant Sodium Citrate Solution U.S.P. (see Table 10).

TABLE 10

PHARMACEUTICAL PREPARATION	SEQUESTERED ION
SEQUESTERING AGENT—Citrate	
Anticoagulant Sodium Citrate Solution N.F.	Calcium
Anticoagulant Acid Citrate Dextrose Solution U.S.P.	Calcium
Benedict's solution	Cupric
Magnesium Citrate Solution N.F.	Magnesium
Tannic Acid Glycerite N.F.	Ferric
SEQUESTERING AGENT—Tartrate	
Fehling's solution	Cupric
Antimony Potassium Tartrate U.S.P.	Antimony
Bismuth Potassium Tartrate ...	Bismuth

Therapeutic application of free citric acid is rare, but for a variety of reasons it has extensive pharmaceutical use. Many salts of citric acid are available, and they are used for both the metal ion and the citrate portion. Examples are Sodium Citrate U.S.P. and Potassium Citrate N.F.

OCCURRENCE	PER CENT CITRIC ACID
Citric Acid Syrup U.S.P.	1
Effervescent Sodium Phosphate N.F.	20
Ephedrine Sulfate Syrup U.S.P. ...	0.1
Milk of Magnesia U.S.P.	0.1
Citrated Caffeine N.F.	50
Magnesium Citrate Solution N.F. ...	9.4
Orange Syrup U.S.P.	0.5
Anticoagulant Acid Citrate Dextrose Solution U.S.P.	0.8
Sodium Phosphate Solution N.F. ...	13

UNSATURATED ACIDS

A series of unsaturated monobasic acids is possible, of which acrylic acid, $CH_2=CH—COOH$, is the simplest member. Other than crotonic acid, $CH_3CH=CHCOOH$, and tiglic acid, $CH_3CH=C(CH_3)COOH$, one with less than 10 carbon atoms rarely is found in nature. There is one dicarboxylic unsaturated acid, fumaric acid, that for the most part exhibits similar chemical properties. The position of the ethylenic linkage also gives to the unsaturated acids the reactivity of the double bond, whereby they enter into all the addition, oxidation and reduction reactions that are characteristic of olefinic hydrocarbons. The carboxyl group, in general, possesses the same characteristics as it does in other acids.

Solubility in water is similar to that found in the saturated series, but the unsaturated acids and their esters are usually liquids.

PRODUCTS

Fumaric Acid is the simplest unsaturated dicarboxylic acid; it is a trans-butenedioic acid, a geometric isomer of maleic acid, which is the *cis* configuration. The acid occurs naturally in the plant kingdom and is a normal product of carbohydrate metabolism. The acid plays a role in tissue respiratory processes.

$$HOOC—CH$$
$$\|$$
$$HC—COOH$$

It is obtained as a product in a microbiologic chemical synthesis from glucose by the action of fungi such as *Rhizopus nigricans*. Fumaric acid occurs as colorless, needlelike, crystals or a white crystalline powder, having a mild acidic taste. It contains no water of hydration, is quite stable and is nonhygroscopic. Solubility is less than 1 per cent in water, ether or alcohol. It is insoluble in vegetable oils and other organic solvents. During World War II, it was introduced as an antioxidant and a substitute for tartaric acid in food products (baking powders) and pharmaceutical preparations. The salts used mostly are those of sodium, calcium and magnesium, all of which are less toxic than the tartrates. The laxative efficiency is superior to the tartrates, and there are less nephropathic and nephrotoxic hazards. The acid also is used with sodium benzoate as a food preservative.

Fumaric acid is suitable for use in effervescent preparations and as an acidifying agent in liquid or dry pharmaceuticals. Magnesium fumarate ($MgC_4H_2O_4 \cdot 4H_2O$) is most laxative, sodium fumarate ($Na_2C_4-H_2O_4$) is best for anhydrous products and calcium fumarate ($CaC_4H_2O_4 \cdot 3H_2O$) is suitable for tablets. Stannous fumarate is available as a teniacide for poultry.

Oleic Acid U.S.P [$CH_3(CH_2)_7CH=CH(CH_2)_7COOH$] is chiefly *cis*-9-octadecenoic acid present as the glyceride in most animal fats and vegetable oils. A process for preparing the pure acid is based on the fact that the lead salt is soluble in ether. A fat or an oil is saponified with sodium hydroxide, and the sodium salts of the mixed fat acids are treated with sulfuric acid to yield the free fat acids. Lead acetate is added to the free acids, thus forming the water-insoluble lead salts. These are treated then with ether, which dissolves the lead salt of oleic acid and also that of any other unsaturated acid present. Ether is removed, and the residue is acidified to yield the free oleic acid. A purification procedure consists of preparing the barium oleate, which then can be recrystallized from alcohol and treated with sulfuric acid to liberate the free oleic acid.

The acid is a light yellow or brown-colored liquid that becomes darker on exposure to air and on standing due to oxidation. Upon heating, oleic acid decomposes with the production of acrid vapors. It is this unsaturated acid, in the form of olein, which contributes to the low melting point of lard and the fluidity of vegetable oils. The free acid is insoluble in water, but it is miscible with organic solvents. It has a lard-like odor and taste. It congeals below 10°.

$$CH_3(CH_2)_7\underset{\underset{HOOC(CH_2)_7CH}{\|}}{CH} \xrightarrow{HNO_3} CH_3(CH_2)_7\underset{\underset{HC(CH_2)_7COOH}{\|}}{CH}$$

Oleic Acid (*cis*)   Elaidic Acid (*trans*)

Oleic acid exists in the *cis* form and is converted into elaidic acid (the *trans* form) on treatment with nitrous acid, nitrogen

tetroxide or nitric acid. By reduction, oleic acid is converted into stearic acid, and, by oxidation, pelargonic and azelaic acids result.

It is used in pharmacy in the preparation of oleates and in ointments exclusively. Intravenously, it causes hemolysis of the red blood corpuscles, an effect which is characteristic of most unsaturated fat acids.

Oleic acid is used to form soaps in a number of pharmaceuticals whereby emulsifying properties are obtained. Triethanolamine oleate is used in Benzyl Benzoate Lotion N.F.; ammonium oleate is present in a number of preparations.

	PER CENT
OCCURRENCE	OLEIC ACID
Benzyl Benzoate Lotion N.F.	2
Green Soap N.F.	2

Linoleic Acid [$CH_3(CH_2)_4CH=CHCH_2-CH=CH(CH_2)_7COOH$] and **Linolenic Acid** [$CH_3CH_2CH=CHCH_2CH=CHCH_2-CH=CH(CH_2)_7COOH$]. Linoleic and linolenic acids are present in animal tissues. A lack of these acids in the diet leads to eczemas, dry scaly skin and other dermatologic conditions. An external preparation, Linolestrol, containing linoleic and linolenic acids along with lecithin and cholesterol, is available for treatment of the skin conditions. For internal use, there are Linocid and Linolex mixtures of linoleic and linolenic acids.

HALOGENATED ACIDS

Dichloroacetic Acid (Bichloroacetic Acid), $Cl_2CHCOOH$, is a colorless liquid with a characteristic odor. It is a stronger acid than acetic acid, but it is somewhat less acidic than trichloroacetic acid. The acid is available for use as a keratolytic in the removal of corns, warts and calluses.

Trichloroacetic Acid U.S.P. is one of the three chlorinated acetic acids. It has been known since 1838 and is prepared by oxidizing chloral hydrate with either nitric acid or permanganate.

$$CCl_3CHO \xrightarrow{(O)} CCl_3COOH$$

The acid occurs as colorless, deliquescent crystals that have a characteristic odor. Usually, the introduction of a halogen atom into an acid increases the acidic properties, and this is true of acetic acid; the acid properties increase with increase in the number of chlorine atoms. Trichloroacetic acid is a stronger acid (as strong as hydrochloric acid) than acetic acid and is very corrosive to the skin. It is soluble in water (1:0.1) or most organic solvents.

It is astringent, antiseptic and caustic; the caustic properties are the most useful, as in treating forms of keratosis, such as moles and warts. Solutions of trichloroacetic acid are very efficient protein precipitants.

Category—caustic.

For external use—topically, to skin or mucous membranes as required.

FATS AND OILS

Fats and oils are known as lipid substances, a general name applied to animal and plant products that are composed of esters of glycerin and fatty acids. Fats and oils are differentiated on the basis of melting points: fats are solid at 20° and are predominantly mixtures of esters of saturated fat acids of higher molecular weights (C_{10} or more); oils are liquid at 20° and mainly are mixtures of esters of unsaturated fat acids. A contributing factor also may be esters of low molecular weight fatty acids. The fluidity, or liquid state, of most vegetable oils is due generally to the presence of oleic acid ester with glycerol (olein).

In animal fats, the rigidity and the high melting point are due to the presence of palmitin and stearin. The composition of fats is directly responsible for their melting points, i.e., lard 36° to 42°; beef fat 46° and mutton suet 45° to 50°.

Glycerin has 3 hydroxy groups, each of which may be esterified with the same fatty acid, thus forming a symmetrical glyceride (tristearin); they also may be esterified with two or more different fatty acids, thus forming an unsymmetrical glyceride (oleopalmitostearin).

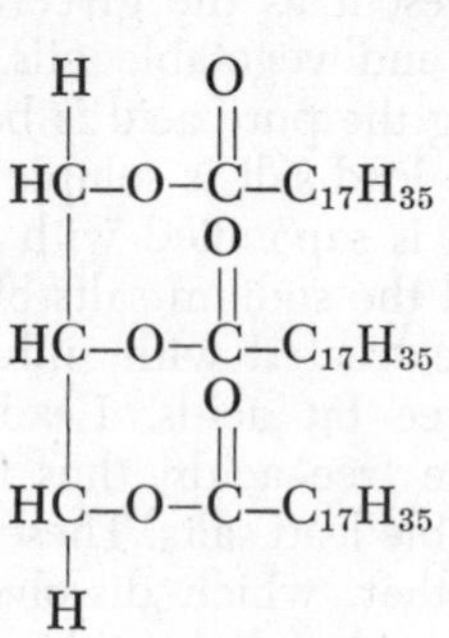

$$
\begin{array}{l}
H \quad\; O \\
| \quad\;\; \| \\
HC-O-C-C_{17}H_{35} \\
| \qquad\; O \\
\qquad\;\; \| \\
HC-O-C-C_{17}H_{35} \\
| \qquad\; O \\
\qquad\;\; \| \\
HC-O-C-C_{17}H_{35} \\
| \\
H
\end{array}
$$

Tristearin

$$
\begin{array}{l}
H \quad\; O \\
| \quad\;\; \| \\
HC-O-C-C_{17}H_{35} \text{ (stearic acid portion)} \\
| \qquad\; O \\
\qquad\;\; \| \\
HC-O-C-C_{15}H_{31} \text{ (palmitic acid portion)} \\
| \qquad\; O \\
\qquad\;\; \| \\
HC-O-C-C_{17}H_{33} \text{ (oleic acid portion)} \\
| \\
H
\end{array}
$$

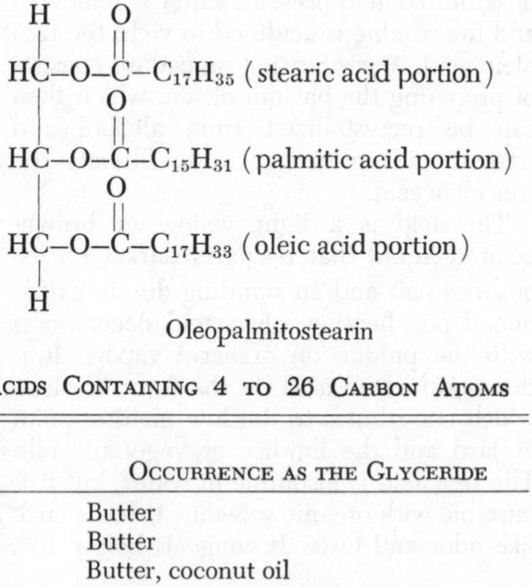

Oleopalmitostearin

TABLE 11. IMPORTANT SATURATED FATTY ACIDS CONTAINING 4 TO 26 CARBON ATOMS

ACID	NUMBER OF CARBON ATOMS	FORMULA	OCCURRENCE AS THE GLYCERIDE
Butyric	4	$CH_3(CH_2)_2COOH$	Butter
Caproic	6	$CH_3(CH_2)_4COOH$	Butter
Caprylic	8	$CH_3(CH_2)_6COOH$	Butter, coconut oil
Capric	10	$CH_3(CH_2)_8COOH$	Butter, coconut oil, palm-nut oil
Lauric	12	$CH_3(CH_2)_{10}COOH$	Coconut oil, spermaceti, palm-kernel oil
Myristic	14	$CH_3(CH_2)_{12}COOH$	Nutmeg fat, coconut oil
Palmitic	16	$CH_3(CH_2)_{14}COOH$	Almost all fats
Stearic	18	$CH_3(CH_2)_{16}COOH$	Almost all animal fats, cocoa butter
Arachidic	20	$CH_3(CH_2)_{18}COOH$	Rape oil, cocoa butter, peanut oil
Behenic	22	$CH_3(CH_2)_{20}COOH$	Rape oil, earth-nut oil, benne oil
Lignoceric	24	$CH_3(CH_2)_{22}COOH$	Peanut oil
Cerotic	26	$CH_3(CH_2)_{24}COOH$	Waxes, wool fat

TABLE 12. IMPORTANT UNSATURATED FATTY ACIDS

ACID	NUMBER OF CARBON ATOMS	FORMULA	OCCURRENCE AS THE GLYCERIDE
One Double Bond			
9, 10 Decylenic	10	$CH_2=CH(CH_2)_7COOH$	......
9, 10 Undecylenic	11	$CH_3CH=CH(CH_2)_7COOH$	......
Dodecylenic	12	$CH_3CH_2CH=CH(CH_2)_7COOH$	......
Palmitoleic	16	$CH_3(CH_2)_5CH=CH(CH_2)_7COOH$	Animal and vegetable oils
Oleic	18	$CH_3(CH_2)_7CH=CH(CH_2)_7COOH$	Animal and vegetable oils
Erucic	22	$CH_3(CH_2)_7CH=CH(CH_2)_{11}COOH$	Rapeseed oil
Two Double Bonds			
Linoleic	18	$CH_3(CH_2)_4CH=CHCH_2CH=$ $CH(CH_2)_7COOH$	Linseed oil Cottonseed oil
Three Double Bonds			
Linolenic	18	$CH_3CH_2CH=CHCH_2CH=$ $CHCH_2CH=CH(CH_2)_7COOH$	Linseed oil
Four Double Bonds			
Arachidonic	20	$CH_3(CH_2)_4(CH=CHCH_2)_4(CH_2)_2COOH$	Lecithin and cephalin
Clupanodonic	18	$CH_3(CH_2)_2(CH=CHCH_2)_4(CH_2)_2COOH$	Fish oils
One Triple Bond			
Tariric	18	$CH_3(CH_2)_7C \equiv C(CH_2)_7COOH$	Picramnia seed fats
Monohydroxy and Double Bond			
Ricinoleic	18	$CH_3(CH_2)_5CH(OH)CH_2CH=$ $CH(CH_2)_7COOH$	Castor oil
Cyclic and One Double Bond			
Hydnocarpic	16	$C_{15}H_{27}COOH$	Chaulmoogra oil
Chaulmoogric	18	$C_{17}H_{31}COOH$	Chaulmoogra oil

In 1823, the constitution of fats was first explained, and, since that time, about fifty fatty acids have been found in nature. All but one are straight-chain acids containing an even number of carbon atoms. The odd-numbered acid is isovaleric acid, which has been isolated from oils of the dolphin and the porpoise.

In Table 11 are listed the important saturated fatty acids containing 4 to 26 carbon atoms.

In Table 12 are given the important unsaturated fatty acids. Those found in fats and oils have an even number of carbon atoms, ranging in carbon content from 10 to 24 atoms.

Fats and oils, as a class, have physical properties that are quite varied, yet quite similar and characteristic. They range from liquids to solids of low melting point; they are insoluble in water, nearly insoluble in alcohol and soluble in most other organic solvents. Castor oil is an exception in that it is soluble in alcohol. All will float on water, as they have a specific gravity of less than one. Because they are nonvolatile, they will leave a greasy stain on cloth or paper. The physical consistency and the solvent properties of fats and oils make them desirable vehicles for therapeutic agents.

Fats and oils are classified chemically as esters of the polyhydric alcohol, glycerin, and fatty acids. These esters are known as glycerides. The chemical properties are those of esters, but the fatty acid portion is responsible for most of their chemical reactions and tests. Glycerides are saponified readily by alkali to form soaps (q.v.) and glycerin. As previously mentioned, the saturated fatty acids form solid glycerides, such as palmitin and stearin, whereas the unsaturated fatty acids form liquid glycerides, such as olein. Due to the presence of the unsaturated fatty acid glyceride, the vegetable oils contain double bonds and may be "hydrogenated," thereby producing a solid fat. This method is used to produce the many hydrogenated cooking fats that are used as substitutes for lard. Iodine may add to the unsaturated linkage of fatty acid glycerides to produce the iodized oils. Double bonds are responsible for the "drying" properties of oils so that they form a hard film when exposed to air. Vegetable oils often are classified according to their drying ability, with the best drying oils containing a large percentage of the glycerides of linoleic and linolenic acids.

1. Nondrying oils containing a large quantity of oleic acid: palm oil, peanut oil, coconut oil, date oil, olive oil, rice oil.

2. Semidrying oils containing oleic and linoleic acids: corn oil, Brazil-nut oil, cottonseed oil, soybean oil, sesame oil, rapeseed oil.

3. Drying oils containing linoleic and linolenic acids: linseed oil, hempseed oil, tung oil, walnut oil, poppyseed oil, sunflower oil.

Rancidity of fats may be due to oxidative or hydrolytic processes. Oxidative rancidity is the most important and results from the oxidation of the unsaturated residue at the double bond and the methylene groups adjacent to the double bond. This produces, in a fat or oil, short-chained acids, aldehydes and ketones which alter the odor and the taste. Many conditions, such as light, heat, moisture and the presence of metals, affect this oxidative process. The methylene

group is, perhaps, first oxidized to a peroxide, which, upon decomposition, oxidizes the fatty acid molecule to yield smaller fragments. The double bond also is oxidized. Ketones are known to result from the action of mold on compounds with double bonds.

These oxidative changes are important in all material containing unsaturated compounds, particularly fats and oils. Antioxidants recognized by the Food and Drug Administration are propyl gallate, nordihydroguaiaretic acid, resin guaiac, tocopherols (vitamin E), lecithin, citric acid, butyl hydroxyanisole (alone or in combination with nordihydroguaiaretic acid or propyl gallate with or without citric acid or phosphoric acid). The exact mode of action is not too well established, but it is not preferential oxidation. Antioxidants protect against or prevent rancidity by interference with oxidative processes. Tocopherol may be added to Light Mineral Oil N.F. and Mineral Oil U.S.P.

Hydrolytic rancidity is the breakdown of a glyceride into the acids and glycerin. This chemical change is a serious problem in those fats which, like butter, contain glycerides of fatty acids of low molecular weight and, also, have a bad odor themselves. Examples of such acids are butyric, caproic, caprylic and capric. Antioxidants are of no value in hydrolytic rancidity. In fats, such as lard, suet or tallow, the acids produced are of high molecular weight, odorless and tasteless.

From a nutrition standpoint, fats are of primary importance as a source of energy and also are transferred directly as fat in the animal body. Vegetable oils often are used as dietary aids for undernourished, convalescent or underweight patients. Usually, aqueous emulsions with sucrose or dextrose are used to provide large doses that are tolerated easily. Two products of this type are Ediol and Lipomul.

Theobroma Oil U.S.P., Cocoa Butter, Cacao Butter, "is the fat obtained from the roasted seed of *Theobroma Cacao* Linné (Fam. *Sterculiaceae*)." It contains, primarily, stearin, palmitin, olein and laurin. This mixture of glycerides produces a substance that melts between 30° and 35°, thus making it very useful as a suppository base.

Category—suppository base.

Lard "is the purified internal fat of the abdomen of the hog." It is a white, soft, unctuous mass containing more olein than do beef fat or mutton suet. It is used for its emollient properties and in the preparation of ointments and cerates.

Benzoinated Lard is lard containing 1 per cent Siam benzoin as an antioxidant and preservative. A catechol, coniferyl alcohol, is the active ingredient.

Olive Oil U.S.P., Sweet Oil, "is the fixed oil obtained from the ripe fruit of *Olea europaea* Linné (Fam. *Oleaceae*)." It contains about 70 per cent olein, and the remainder is mostly palmitin. It has mild laxative properties, but its important uses are as an emollient, a food and in soap manufacture.

Category—emollient.
For external use—topically to skin.

Castor Oil U.S.P., Oleum Ricini, "is fixed oil obtained from the seed of *Ricinus communis* Linné (Fam. *Euphorbiaceae*)." Due to the presence of the glyceride of ricinoleic acid (80%), the oil is used as a laxative. It is the only fixed oil that is soluble in alcohol, so it is added to collodion to increase the flexibility.

Solubility in alcohol is due to the presence of a hydroxyl group in the ricinolein.

Castor oil is quite different in solubility from other fatty oils.[6] It tends to dissolve in oxygenated solvents (alcohols) and be insoluble in hydrocarbon type solvents (benzin), which is opposite to other vegetable oils. It is miscible with dehydrated alcohol, glacial acetic acid, chloroform or ether.

Castorwax is hydrogenated castor oil composed primarily of the glyceride of hydroxystearin. It is useful as a nonionic surface-active agent and, because of a melting point of 84° to 87°, as a stiffening agent, wax extender and so on.

Category—Cathartic.
Usual dose—15 ml.
Usual dose range—15 to 60 ml.

Occurrence	Per Cent Castor Oil
Flexible Collodion U.S.P.	3.0
Aromatic Castor Oil N.F.	96.5
Salicylic Acid Collodion U.S.P.	2.7

Riodine is a liquid preparation of the iodized, unsaturated fat acids occurring in castor oil. The oil is treated with hydrogen iodide until 66 per cent of the unsaturated glycerides are iodized; the result is a product containing about 17 per cent of iodine. It is light amber in color, insoluble in water but soluble in organic solvents.

Riodine is available in pearls and is used for the effect of the iodide ion.

The usual dose is 0.4 to 1.2 g.

Cod Liver Oil N.F. "is the partially destearinated fixed oil obtained from fresh livers of *Gadus morrhua* Linné and other species of the family *Gadidae*." Non-Destearinated Cod Liver Oil N.F. is the complete, untreated oil; it was made official in order to permit its importation from the Scandinavian countries. Due to the presence of stearin, it will congeal on cooling and will form a deposit of stearin. This form rarely is used internally but is used topically. Both grades of cod liver oil are of importance because of the vitamins A and D, but only Cod Liver Oil N.F., is used internally as an aid to nutrition. There is some indication that cod liver oil is of value in ointment form for the treatment of wounds and burns. The vitamins present appear to be responsible for the ability of the oil to promote healing. Suppositories also are available (Desitin and "A and D" products). The sodium salt of the fatty acids, sodium morrhuate (q.v.), has been introduced as a sclerosing agent. The official pharmaceuticals include only those products which are intended for internal administration.

Corn Oil U.S.P., Oleum Maydis, is "the refined fixed oil expressed from the embryo of *Zea mays* Linné (Fam. *Gramineae*)." It is a semidrying oil. Use is made of corn oil as a solvent for vitamins and injections of oil-soluble medicaments, such as estrogenic substances, menadione and diethylstilbestrol.

Category—solvent.

Cottonseed Oil U.S.P., Oleum Gossypii Seminis, "is the refined fixed oil obtained from the seed of cultivated plants of various varieties of *Gossypium hirsutum* Linné or of other species of *Gossypium* (Fam. *Malvaceae*)." It is a liquid containing the glycerides olein and linolein. Exposure to air causes the unsaturated glycerides to be oxidized, thus causing the development of a gummy substance.

Category—solvent.

OCCURRENCE	PER CENT COTTONSEED OIL
Zinc-Eugenol Cement N.F.	15

Sesame Oil U.S.P., Oleum Sesami, Teel Oil, Benne Oil, "is a fixed oil obtained from the seed of one or more cultivated varieties of *Sesamum indicum* Linné (Fam. *Pedaliaceae*)." It contains about 75 per cent olein, together with other glycerides. It is classified as a "semidrying" oil. Sesame oil is used as an emollient, a food and a solvent. Lipoiodine Diagnostic is a 60 per cent solution of lipoiodine in sesame oil.

Category—pharmaceutic aid (solvent for intramuscular injections).

Halibut Liver Oil "is the fixed oil obtained from the fresh or suitably preserved livers of *Hippoglossus hippoglossus* Linné (Fam. *Pleuronectidae*)." Like cod liver oil, it contains vitamins A and D and, therefore, has nutritional uses.

Almond Oil N.F., Sweet Almond Oil, "is the fixed oil obtained from the kernels of varieties of *Prunus amygdalus* Batsch (Fam. *Rosaceae*)." It is a "nondrying" oil composed mainly of olein, with some linolein but no stearin present. The oil is free from odor and taste and has no tendency to become gummy. It is used as an emollient and as an emulsifier of volatile oils.

Category—emollient; perfume.

OCCURRENCE	PER CENT EXPRESSED ALMOND OIL
Rose Water Ointment N.F.	56

Peanut Oil U.S.P., Arachis Oil, Oleum Arachidis, "is the fixed oil obtained by cold pressure from the peeled seeds of one or more of the cultivated varieties of *Arachis hypogaea* Linné (Fam. *Leguminosae*)." It is a "nondrying" oil containing olein, linolein, arachiden and lignocerin. In pharmaceutic uses, it is similar to olive oil. One application is a mixture of wax and peanut oil as a vehicle for penicillin.

Category—pharmaceutic aid (solvent).

Persic Oil N.F. Apricot Kernel Oil, Peach Kernel Oil "is the oil expressed from the kernels of varieties of *Prunus armeniaca* Linné (Apricot Kernel Oil), or from the kernels of varieties of *Prunus persica* Sieb. et Zucc. (Peach Kernel Oil). (Fam. *Rosaceae*)." In physical and chemical properties it is very similar to expressed oil of almond and has the same uses.

Category—vehicle.

WAXES

Yellow Wax N.F.; Beeswax (Yellow Beeswax). Yellow wax "is the purified wax from the honeycomb of the bee, *Apis mellifera* Linné (Fam. *Apidae*)." It is composed primarily of esters, high molecular weight alcohols and acids. The alcohols include those containing 24 to 32 carbon atoms, and the acids include those containing 24 to 34 carbon atoms. Some paraffinic hydrocarbons are also present, for example, hentriacontane, $C_{31}H_{64}$. Beeswax differs from spermaceti in having esters of acids and alcohols of higher molecular weight. These esters contribute to the greater hardness and higher melting point of beeswax (62° to 65°). Beeswax occurs in broken pieces or in flat disks, varying in color from grayish brown to yellow. There is usually a pleasant honeylike odor about the wax. There is practically no unsaturation present.

Beeswax is used for its physical properties, primarily in external pharmaceuticals such as ointments, cerates, suppositories, plasters and dressings.

Category—stiffening agent.

White Wax U.S.P.; Bleached Beeswax; Cera Alba (White Beeswax). White wax is yellow wax that has been treated with a bleaching or oxidizing agent. It is practically the same in composition as yellow wax, except that the bleaching process produces some free fatty acids and causes the wax to be less unctuous in character.

White wax has been used with peanut oil as a vehicle for penicillin or its salts, in preparations designed for intramuscular injection. The purpose is to decrease the rate of absorption of the penicillin. One of the chief constituents of wax, myricin (myricyl palmitate, $C_{15}H_{31}COOC_{31}H_{63}$), is used now to delay absorption of medicinals injected intramuscularly. An example is a tubocurarine chloride injection.

Category—pharmaceutic aid.

Carnauba Wax U.S.P. This is a wax of higher molecular weight which is obtained from the leaves of the Brazilian wax palm tree *Copernicia cerifera*. It occurs as a light brown to pale yellow, moderately coarse powder, or flakes, and has a characteristic, not unpleasant odor when melted. It has a melting point range of 81 to 86°. The hardness and high-polish characteristics make it useful where a high gloss is desired. The primary pharmaceutical use is in polishing mixtures for coated tablets and it is used in the last stage in tablet coating. It is insoluble in water, freely soluble in warm benzene, slightly soluble in boiling alcohol and soluble in chloroform and in toluene.

Category—pharmaceutic aid.

Spermaceti U.S.P. Cetaceum "is a waxy substance obtained from the head of the sperm whale." The primary constituent is cetyl palmitate,

$$CH_3(CH_2)_{14}CO \mid O(CH_2)_{15}CH_3$$
$$\text{Palmitate} \qquad\quad \text{Cetyl}$$

along with varying amounts of other fatty acid esters and alcohols of high molecular weight. Due to its emollient effects and low melting point (42° to 50°), it is used in ointments, creams and cosmetics.

Use of radioactive carbon 14 has shown that spermaceti in ointments or creams does not penetrate the skin.

Category—pharmaceutic aid.

CATIONIC-EXCHANGE RESINS

These resins are of a type which contains an acid group, such as phenolic, carboxylic or sulfonic. The exact composition varies from one type to another, and most are of secret composition. Some are known to be polystyrene polymers containing carboxylic, sulfonic or phenolic groups. Representatives of these classes are: sulfonic-type, Win 3000, Katonium and Dowex; carboxylic-type, Amberlite XE-64, Natrinil. They are insoluble, nontoxic, nonabsorbable but irritant to varying degrees.[7]

In 1946, the cation-exchange resins were suggested by Dock as a means of withdrawing or withholding sodium ions in the intestinal tract and, thus, from the body. Up to this time, sodium ions were controlled by restricting the salt intake to avoid the danger of further accumulations of fluid (low-salt diet) and by increasing the fluid intake to provide greater urine volume, thereby aiding removal of sodium ions. The hydrogen of the acid group readily reacts with sodium ions and/or potassium ions, thus removing them in nonionized compounds and preventing these ions from being metabolized.

$$\text{resin } SO_3H + Na^+ \rightarrow \text{resin } SO_3Na + H^+$$
$$\text{resin—COOH} + Na^+ \rightarrow$$
$$\text{resin—COONa} + H^+$$
$$\text{resin—}C_6H_4OH + Na^+ \rightarrow$$
$$\text{resin—}C_6H_4ONa + H^+$$

The sulfonic-type resin is a strong acid which is not very satisfactory in removing sodium ions from the intestinal tract because it functions best under strong acidic conditions (low pH). Upon oral administration, the acid form removes ions from the mouth, the esophagus, the stomach and so on to such a degree that irritation is noticed and nausea and vomiting often follow.

The carboxylic-type functions best in nearly neutral or mildly alkaline media, such as is usually present in the intestinal tract. It is a weak acid, having no noticeable effect on the oral mucosa or the stomach. Two available products, Carbo-Resin and Natrinil, contain 50 per cent of resin of this type.

The efficiency of the resin in removing cations (sodium ions) will depend on the concentration of the ions, the pH of the medium, the duration of contact and the physical form of the resin in regard to porosity and particle size.

Sodium ions become extremely important to the patient with cardiac edema and hypertension associated with acute nephritis or heart failure.

The treatment of dropsy, or edema, has been a major medical problem for centuries. In most edemata the excretion of sodium ions is impaired and, due to osmosis, the tissues become waterlogged or, more correctly, brine-logged. A cationic-exchange resin permits the enjoyment of a normal diet containing sodium ions, which are removed immediately by the resin and excreted in the feces. Potassium ions appear to be removed by the resin in ratio of their concentrations. To ensure only negligible loss of potassium ions, a portion of the resin is administered as the potassium form. The free acid resin (H-form) has an undesirable taste, and the ammonium resin, which is tasteless, is used currently. It readily gives up ammonium ions to combine with sodium, and the released ammonium ions act as a diuretic to aid in removing excess water and chloride ions.

Of other elements which also could be removed by the resin, only calcium has been studied extensively, and it was found to be slightly altered. However, it seems to be agreed that occasional supplementation of minerals, such as magnesium, copper and iron, may be necessary. No evidence of vitamin deficiency has been observed. Use of resin is contraindicated in renal damage, acidosis, low salt syndrome, potassium deficiency, oliguria and anuria.

Doses of about 50 g. per day usually are required, in conjunction with a moderate salt intake. Since there are theoretical dangers in the treatment, the patients require close supervision.

Sodium Polystyrene Sulfonate U.S.P., Kayexalate,® is a cation-exchange resin prepared in sodium form. It is available as a water-insoluble golden brown, fine powder that is odorless and tasteless. It has an exchange capacity of approximately 3.1 mEq. of potassium per g. The *U.S.P.* states that "each g. exchanges not less than 110 mg. and not more than 135 mg. of potassium, calculated on the anhydrous basis."

Category—ion-exchange resin (potassium).

Usual dose—15 g. up to 4 times a day.

REFERENCES CITED

1. Barrett, R. H.: Ann. Int. Med. 31:739, 1949.
2. Warshaw, L. J., and Gold, H. J.: J. Am. Pharm. A. (Sci. Ed.) 36:56, 1947.
3. Soine, T. O., *et al.*: Inorganic Pharmaceutical Chemistry, ed. 8, pp. 43, 228, Philadelphia, Lea and Febiger, 1961.
4. C. A. 43:5745, 1949; 41:6206, 1947; J. Ind. Chem. Soc. 27:443, 1950.
5. von Loesecke, H. W.: J. Ind. Eng. Chem. (Ind. Ed.) 23:1952, 1945.
6. Gilvert, E. E.: J. Chem. Ed. 18:338, 1941.
7. Dock, W.: Tr. A. Am. Physicians 59:282, 1946.

SELECTED READING

Budowski, P., and Markly, K. S.: Chemistry and physiological properties of sesame oil, Chem. Rev. 48:125, 1951.

Daubert, B. F.: Some recent advances in the chemistry of fats, J. Am. Pharm. A. (Sci. Ed.) 33:321, 1944.

Hilfer, H.: Waxes in cosmetics, Drug and Cos. Ind. 72:178, 1953.

Ralston, A. W.: Fatty Acids and Their Derivatives, New York, Wiley, 1948.

8

Various Nitrogenous Compounds

Charles O. Wilson, Ph.D.,
Dean and Professor of Pharmaceutical Chemistry, School of Pharmacy,
Oregon State University

AMINES

All of the organic nitrogenous compounds may be considered as derived from ammonia by substitution of hydrogen. The amines are often formed by replacing hydrogen with alkyl or aryl radicals and, because there are three atoms of hydrogen, successive substitution yields three kinds of compounds: primary, $R-NH_2$; secondary, R_2NH and tertiary, R_3N, amines. Each of these behaves chemically like ammonia in adding acids to form salts that are similar to ammonium salts; using HA for the acid, there are $R-NH_3A$, R_2NH_2A and R_3NHA. Similarly, there are compounds analogous to ammonium hydroxide: $R-NH_3OH$, R_2NH_2OH and R_3NHOH. Finally, there are two more classes of compounds that are formed by replacing the last hydrogen of ammonium salts and ammonium hydroxide: R_4NA and R_4NOH.

These 11 types are named as derivatives of ammonia (amine) and ammonium compounds. Using methyl as the alkyl and Cl as the acid radical, the compounds given below are possible.

In more complicated amines, the salts often are designated as hydrochlorides, sulfates, nitrates and so on. Thus, $(CH_3)_3NHCl$ would be trimethylamine hydrochloride and might then be written $(CH_3)_3N \cdot HCl$.

PRIMARY

$R-NH_2$	CH_3-NH_2	Methylamine
$R-NH_3Cl$	CH_3-NH_3Cl	Methylammonium chloride
$R-NH_3OH$	CH_3-NH_3OH	Methylammonium hydroxide

SECONDARY

R_2NH	$(CH_3)_2NH$	Dimethylamine
R_2NH_2Cl	$(CH_3)_2NH_2Cl$	Dimethylammonium chloride
R_2NH_2OH	$(CH_3)_2NH_2OH$	Dimethylammonium hydroxide

TERTIARY

R_3N	$(CH_3)_3N$	Trimethylamine
R_3NHCl	$(CH_3)_3NHCl$	Trimethylammonium chloride
R_3NHOH	$(CH_3)_3NHOH$	Trimethylammonium hydroxide

QUATERNARY

R_4NCl	$(CH_3)_4NCl$	Tetramethylammonium chloride
R_4NOH	$(CH_3)_4NOH$	Tetramethylammonium hydroxide

181

In the Geneva system, the primary group (—NH₂) is expressed by the prefix amino-, using the appropriate number. The simple alkyl derivatives are called alkylamino- or dialkylamino- compounds. Thus, *n*-butyl-amine would become 1-aminobutane, while nonyldimethylamine would be 1-dimethyl-aminononane. The number, in each case, denotes the position of the amino group on the parent compound.

The actual structure of these compounds should be noted at this point. In primary, secondary and tertiary amines, as in ammonia, three of the surface electrons are shared; dimethylamine, for example, could be pictured as

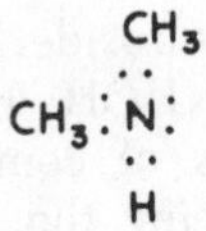

$$CH_3 \overset{\overset{\textstyle CH_3}{\displaystyle ..}}{\underset{\displaystyle H}{:N:}}$$

This leaves 2 of the 5 outer electrons free to be transferred in a polar way or to be shared in the covalent way. When an acid is added to form a salt, the hydrogen ion probably is combined to produce an ammonium ion, such as

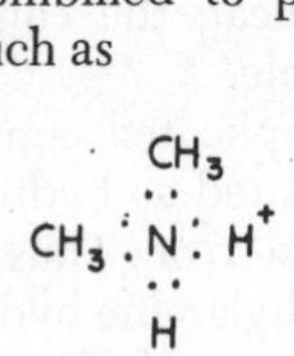

$$CH_3 \overset{\overset{\textstyle CH_3}{\displaystyle ..}}{\underset{\displaystyle H}{:N: H^+}}$$

When water is added to produce the hydroxide, there is a similar process but the hydroxyl ion is probably associated through a hydrogen bond. Thus, a real picture of a quaternary amine such as (CH₃)₄NCl would be (CH₃)₄N⁺ Cl⁻, just as a real picture of sodium chloride would be Na⁺ Cl⁻ and not NaCl, as it usually is written. Some chemists have a tendency to indicate in formulas the electrons and ionic charges, but most authors prefer to write those of the amines in the customary way.

Primary, secondary and tertiary alkyl-amines are all colorless, the lower ones being gases and the higher ones liquids or solids. Most of them will dissolve readily in water, even with alkyl groups having as many as 9 carbon atoms, and they are also soluble in alcohol. The more volatile alkyl-amines have an ammoniacal and fishy odor and, unlike ammonia, are inflammable. The quaternary salts are crystalline solids, and the quaternary hydroxides are usually hygroscopic solids.

PREPARATION

There are many methods for producing the simple amines, and some of these will be noted under particular compounds. At this point, only a few of the methods that are generally applicable will be noted.

Alkylation. All classes may be made by the process of alkylation, which will be described under the chemical properties.

Reduction. The primary and the secondary amines can be made by reducing a great variety of nitrogenous compounds; indeed, the end-product of such a reaction is almost invariably an amine, no matter what substance is used. The most useful compounds for the purpose are nitro and nitroso derivatives, nitriles, oximes, carbylamines and amides. Thus, acetaldoxime gives ethyl-amine.

$$CH_3-CH=NOH \xrightarrow{(H)} CH_3-CH_2-NH_2$$

The choice of reducing agent to be used will depend somewhat on the substance that is to be affected, but usually the more powerful reducing agents are better because the process cannot very well be carried too far. For example, the nitroparaffins can be changed to alkylamines by zinc and dilute acid.

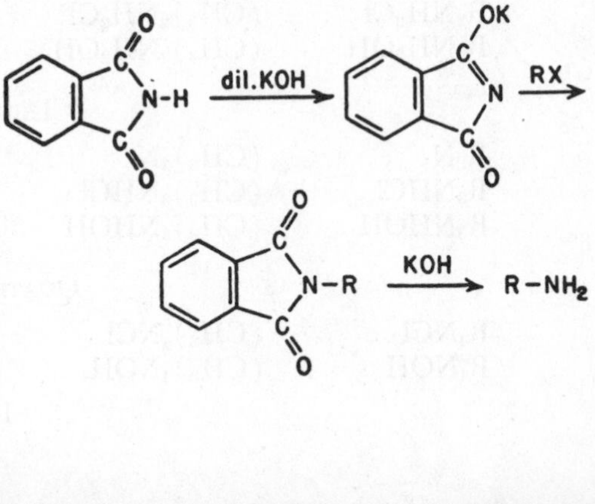

Gabriel's Method. This is an excellent method for the preparation of primary amines. Phthalimide contains a hydrogen atom that is replaceable by potassium or sodium, and the resulting compound will react with alkyl halides to form monoalkyl derivatives of phthalimide. The latter may be hydrolyzed to furnish an amine and a phthalate.

A similar method starts with *p*-nitrosamines of benzene; *p*-nitrosodimethylaniline decomposes to dimethylamine on heating with alkali.

Hofmann Degradation. The primary amines may be prepared by the action of alkaline hypobromites on primary amides in the Hofmann degradation. The end-product from $R-CO-NH_2$ is $R-NH_2$, but the process takes place in 4 steps. This is shown, using the potassium salt.

$$R-CO-NH_2 \xrightarrow{Br_2} R-CO-NHBr \xrightarrow{KOH}$$
Amide $\qquad$ Bromamide

$$R-C(OK) = NBr \rightarrow R-NCO \rightarrow R-NH_2$$
Salt $\qquad$ Isocyanate $\quad$ Amine

Both the bromamide and the isocyanate can be isolated under special conditions. When the radical contains too many carbon atoms, usually above 4, the final product tends to be a nitrile of the next lower acid. Thus, $R-CH_2-CO-NH_2$ may produce largely $R-CN$ instead of $R-CH_2-NH_2$. However, the nitrile subsequently can be reduced to the latter and so give the same end-product. The isocyanates, which are intermediate in this process, and, also, the isothiocyanates can be changed directly to primary amines by heating with alkalies.

Decarboxylation. Amino acids or alkyl derivatives generally can be decarboxylated by heating with barium hydroxide or by the aid of certain microorganisms or enzymes. Thus, $R-CH(NH_2)-COOH$ by such a process gives $R-CH_2-NH_2$.

Many other methods have been described, some of them involving considerable rearrangement, but those that have been noted are the most commonly used. Industrially, several special processes are employed; as examples, the methylamines can be made by heating formaldehyde with alkali and ammonium chloride, and lower alcohols can be converted to primary amines by heating with ammonium chloride and alkali under pressure.

PROPERTIES

In general, the alkylamines are entirely analogous to ammonia in their chemical behavior, although there may be considerable differences in degree under certain conditions. For instance, most of the lower amines are inflammable, while ammonia can be burned only under unusual circumstances. Most of them combine with water to form oily hydrates analogous to ammonium hydroxide, which exists in small amount in solution or in the pure state at low temperature. These hydrates can be decomposed by reaction with strong bases to produce the free amine.

Like ammonia, the amines unite directly with acids to form salts. The alkalinity of aqueous solutions is higher than that of ammonia and sometimes increases with the number of alkyl groups. Quaternary ammonium bases are equivalent in strength to the strong alkalies because they contain no hydrogen to lessen the activity of the hydroxyl ion by hydrogen bonding and will absorb carbon dioxide to form carbonates. The basic properties of any amine in aqueous solution are shown by its ability to precipitate metallic hydroxides and, in excess, to dissolve many of them as does ammonia.

The pKa for primary amines varies from 8.4 to 10.9, and, generally, for secondary as well as tertiary amines the pKa decreases as the molecular weight increases. Examination of the simpler amines indicates that the pKa for a series (methyl or ethyl) increases in going from primary amines to tertiary amines.

The salts that are formed by the addition of acids are colorless, crystalline solids that are generally soluble in water or alcohol and, like other salts, are highy ionized. From them, the bases may be liberated by addition of alkalies. The quaternary salts, however, are not so altered because the corresponding bases are about as strong as the reagents. In these cases, the bases may be produced by shaking the salts with moist silver oxide, forming an insoluble silver salt.

$$2R_4NCl + Ag_2O + H_2O \rightarrow$$
$$2AgCl \downarrow + 2R_4NOH$$

Any one of the alkyl amino halide salts is

$$NH_3 \qquad + \qquad CH_3Cl \qquad \longrightarrow \qquad CH_3NH_3Cl$$

$$CH_3NH_3Cl \qquad + \qquad NH_3^* \qquad \rightleftharpoons \qquad CH_3NH_2 + NH_4Cl$$

$$CH_3NH_2 \qquad + \qquad CH_3Cl \qquad \longrightarrow \qquad (CH_3)_2NH_2Cl$$

$$(CH_3)_2NH_2Cl \qquad + \qquad NH_3^* \qquad \rightleftharpoons \qquad (CH_3)_2NH + NH_4Cl$$

$$(CH_3)_2NH \qquad + \qquad CH_3Cl \qquad \longrightarrow \qquad (CH_3)_3NHCl$$

$$(CH_3)_3NHCl \qquad + \qquad NH_3^* \qquad \rightleftharpoons \qquad (CH_3)_3N + NH_4Cl$$

$$(CH_3)_3N \qquad + \qquad CH_3Cl \qquad \longrightarrow \qquad (CH_3)_4NCl$$

* Any other amine in the system also will react.

decomposed into alkyl halide and an amine when heated. Thus, dimethylammonium chloride produces methylamine and methyl chloride, a reversal of the process of alkylation. In the same way, the quaternary bases will give a tertiary amine and an alcohol. If the radicals are different in the substance being heated, the alcohol or alkyl halide that is split out generally will contain the smallest alkyl group.

All of the salts can combine, as do those of ammonium, with various metallic salts to form double compounds. Examples are:

$$2R-NH_3Cl:PtCl_4 \text{ or } (R-NH_3)_2PtCl_6,$$
$$R_2NH_2Cl:AuCl_3 \text{ or } R_2NH_2AuCl_4 \text{ and }$$
$$2R_3NHCl:HgCl_2 \text{ or } (R_3NH)_2HgCl_4.$$

When ammonia or any simple amine is heated with an inorganic alkyl ester, union takes place to produce a salt of an amine with one or more alkyl groups. As an example, $R-NH_2$ will give $R-N(CH_3)H_2I$ with methyl iodide, and methylamine is converted to $(CH_3)_2NH_2I$. For this purpose, alkyl halides or sulfates are most convenient, but it is noteworthy that compounds of tertiary radicals do not add, they produce olefins instead. Since the process of addition usually is performed in order to introduce alkyl radicals into the amine in place of hydrogen, it is called alkylation (methylation, ethylation).

Unfortunately, however, the reaction is not quite so simple as noted. The ammonium salt that is formed can be converted by bases to the free amine, as has already been mentioned, and all amines are bases. Therefore, while the addition is more or less complete, other reactions go on in equilibrium, so that the final result is to form a conglomerate mixture of compounds. If ammonia is heated with methyl chloride, for example, the foregoing mixture would result. On the other hand, if a strong base is added to the reaction mixture, the equilibrium processes are carried completely to the right and the final product is a quaternary salt.

$$NH_3 + 4CH_3Cl \longrightarrow (CH_3)_4NCl + 3KCl + 3H_2O$$

From this, the tertiary compound may be obtained by distillation.

$$(CH_3)_4NCl \longrightarrow (CH_3)_3N + CH_3Cl$$

Therefore, this alkylation can be employed to prepare tertiary amines, quaternary salts or, from the latter, the corresponding bases. It is not usually efficient for producing the primary or secondary compounds. Under special circumstances and by carefully controlled conditions, it is sometimes possible to get primary or secondary amines in more or less pure form, but, in general, this is not a good method for preparing them.

Acylating agents react with amines to form amides. Such a reaction with the acid is slow and unsatisfactory in some cases, but either the anhydride or the chloride is more rapid and more convenient.

$$2NH_3 + R'-CO-Cl$$
$$\downarrow$$
$$R'-CO-NH_2 + NH_4Cl$$

$$2R-NH_2 + R'-CO-Cl$$
$$\downarrow$$
$$R'-CO-NH-R + R-NH_3Cl$$

$$2R_2NH + R-CO-Cl$$
$$\downarrow$$
$$R-CO-NR_2 + R_2NH_2Cl$$

The agents that are applicable may include the halides of nonmetals or of inorganic acids, such as the chlorides of thionyl, sulfuryl, phosphoryl, sulfur or nitrosyl.

An interesting modification of this process was proposed by Hofmann to distinguish and separate the various classes of amines. When these are treated with ethyl oxalate, the primary forms a dialkyloxamide, R—NH—CO—CO—NH—R, that is soluble in water, and from it the amine can be recovered by distillation with alkali. The secondary compound forms an insoluble dialkyloxamic ester, CH$_3$—CH$_2$—O—CO—CO—NR$_2$, from which the amine can be released by heating with alkali. The unaltered tertiary compound can first be removed by simple distillation.

Another means of distinguishing the amines is by reaction with carbon disulfide in alcoholic solution. Primary amines add to yield salts of alkylthiocarbamic acid, R—NH—CS—S—NH$_3$R, while the secondary amine gives dialkyl-dithiocarbamic acid salts, R$_2$N—CS—S—NH$_2$R$_2$. When, after distilling off the unchanged tertiary amine, the residue is boiled with mercuric or ferric chloride, the primary compound is converted to a mustard oil, RNCS.

One reaction that is very useful, especially for the primary amine, is that brought about by a metallic nitrite and a dilute mineral acid, essentially a reaction of nitrous acid. At room temperature the primary compound is converted to an alcohol, probably through an intermediate diazonium salt.

$$R—NH_2 + NaO—NO + 2HCl$$
$$\downarrow$$
$$R—N_2Cl + 2H_2O + NaCl$$

$$R—N_2Cl + H_2O$$
$$\downarrow$$
$$R—OH + N_2 + HCl$$

This provides a convenient method for transforming the group —NH$_2$ to the alcoholic —OH. The secondary amine is changed to a nitrosamine by condensation.

$$R_2NH + NaO—NO + HCl$$
$$\downarrow$$
$$R_2N—NO + NaCl + H_2O$$

The nitrosamines are oily, yellow liquids that are insoluble in water, stable to heat or reagents generally, but reducible to hydrazines. Tertiary amines are unchanged by the nitrous acid but are converted to salts of the acid that is used.

Primary amines will form isocyanides or carbylamines when boiled with chloroform and alcoholic alkali.

$$R—NH_2 + CHCl_3 + 3KOH$$
$$\downarrow$$
$$R—NC + 3KCl + 3H_2O$$

These are poisonous, ill-smelling liquids; the characteristic odor may be used to identify the presence of a primary amine.

PHYSIOLOGIC PROPERTIES

The amines with low molecular weight are similar to ammonia in physiologic as well as chemical properties. They undoubtedly have a profound effect in the animal body, but, since the irritant action far outweighs any possible other effect, they are useless as drugs. The salts are readily hydrolyzed to the amines, as are the salts of ammonia, but they would have no great advantage over the latter in producing osmosis and systemic acidity. Trimethylamine and its salts have been administered in rheumatic conditions on the doubtful basis that they tend to dissolve urate deposits, and unsuccessful efforts have been made to apply other lower members of the series. Some of them, notably the methyl compounds, are useful in synthesis and otherwise in industry. Those amines with higher molecular weight are less volatile and less irritant and are employed to give more useful effects in medicine; this is particularly true of those resembling ephedrine in physiologic action (see Chap. 17).

AMINE SALTS

The selection of the acid with which to prepare amine salts is dependent upon numerous factors. A compound destined to be a therapeutic agent must be studied not only as a chemical compound but also for chemical and physical properties that permit it to be used as a medicinal agent.

In salt determination the therapeutic use, as well as dosage form and method of administration, are of primary consideration. In many cases the acid selected determines taste, solubility (high or low), rate of absorption, stability, physical form, odor, light sensitivity, hygroscopic properties, and yield. All of the factors are evaluated as they relate to the cost of production and ease of dosage formulation.

The following amine salts are used:

Acid Maleate (Maleate)
Acid Tartrate (Tartrate)
Camphorsulfonate (Camsylate)
Ethanedisulfonate (Edisylate)
Ethanesulfonate (Esylate)
Gluconate
Hydrobromide
Hydrochloride
Hydroiodide
Methanesulfonate (Mesylate)
Methylsulfate
Pamoate
Phosphate
Salicylate
Sulfate

PRODUCTS

Ethylenediamine U.S.P. is prepared from ethylene dichloride and ammonia. It is a clear, colorless, thick, strongly alkaline liquid containing at least 97 per cent of $H_2NCH_2CH_2NH_2$. The *U.S.P.* points out that, due to the alkalinity, it will absorb carbon dioxide from the air to form a non-volatile carbonate. Ethylenediamine must be protected against undue exposure to the atmosphere. It is very irritating to skin and mucous membrane.

The compound is used as a pharmaceutic aid and in aminophylline injection.

	PER CENT ETHYLENE-
OCCURRENCE	DIAMINE
Thimerosal Solution N.F.	0.02
Thimerosal Tincture N.F.	0.02

Tromethamine N.F., Trizma®, Tham®, Talatrol®, 2-amino-2-(hydroxymethyl)-1,3-propanediol occurs as a very pure (99%) white crystalline powder with a slight, characteristic odor. This amine is also a trihydroxy alcohol which is freely soluble in water and in lower aliphatic alcohols. It is insoluble in the usual organic solvents.

Tromethamine is referred to as an organic amine buffer but functions as a weak base (pKa 7.84), and, on entering the body, it becomes a component of a buffer system. It is used to correct systematic acidosis. In vivo it reacts with hydrogen ions and their related acid anions when salts are formed and excreted by the kidney. Combinations with electrolytes are available.

Category—alkalizer.

Usual dose—intravenous, 300 mg. per kg. of body weight as a single dose with the average total dose being about 25 g. given over a period of not less than 1 hour. Doses of up to 500 mg. per kg. may be required, depending on the severity and progression of the acidosis.

OCCURRENCE

Tromethamine for Injection N.F.

ANION EXCHANGE RESINS

In 1935, two English chemists produced the first synthetic "ion-exchange resin."

Synthetic resins are now numerous and the ion-exchange process is now as important to science and industry as are fundamental technics such as distillation, crystallization and fractionation. Two types of ion-exchange resins are used, namely, anion-exchange resins and cation-exchange resins.

Polyamine-Methylene Resin, Resinat®, Resmicon®, Amberlite-IR-4®, is described chemically as a polyethylene polyamine methylene substituted resin of diphenylol dimethylmethane and formaldehyde in basic form. It is a light-amber, granular, free-flowing powder that is insoluble in water, alcohol, ether and aqueous solutions of alkalies or acids.[1]

Many compounds split into charged particles (ions) when they are in solution. Those ions of positive charge are cations (Na^+), while those with a negative charge are anions (Cl^-).

Anion-exchange resins are basic resins usually, containing amines which remove anions only from acid solutions and do so by removing the whole acid molecule (HCl). In 1945, the polyamine-formalde-

hyde resins were suggested to remove hydrochloric acid from the gastric juice in treating peptic ulcer and hyperchlorhydria. The primary amine used in the equation is probably not actually present.

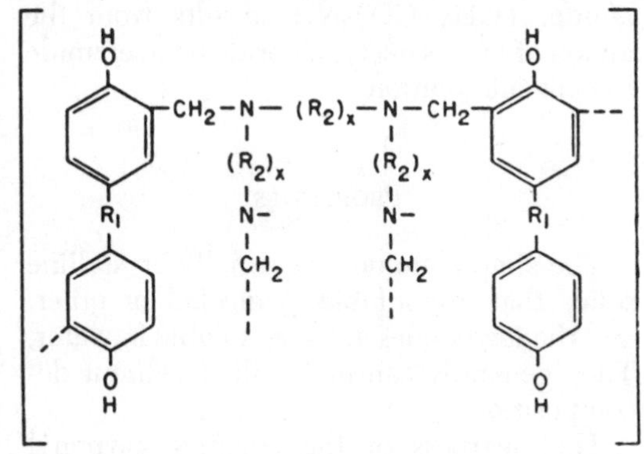

$$Resin-NH_2 + HCl \rightarrow Resin-NH_2 \, HCl$$

The resin should be in a finely divided form (200-mesh) and taken in doses of from 1.5 to 3 g. per day (t.i.d. p.c.). Five g. of resin is equal to 1 g. of dry aluminum hydroxide. The pH of gastric juice is raised immediately to 5, and pepsin and trypsin are inactivated. Advantages of anion-exchange resins are: great speed of action; good neutralizing power; inhibition of trypsin and pepsin; no acid rebound; phosphate radical $(PO_4^{\equiv})$ is not removed; chloride ions are unaffected and constipating property is negligible.

Clinical studies show that the anion-exchange resins reduce pain in the ulcer patient and are equally useful for nonulcer patients. In comparison with the many other substances used as antacids, the amine type resins are the best. Gastric mucin often is admixed with the resin to suspend and disperse it and also to act as a buffering agent. Dosage varies with the severity of the peptic ulcer or hyperacidity and should be adjusted to give adequate relief.

Cholestyramine Resin U.S.P., Cuemid®, Questran®, is the chloride form of a strongly basic anion exchange resin. It is a styrene copolymer with divinylbenzene with quaternary ammonium functional groups. It has an affinity for bile salts so that the ingested resin combines with bile salts in the intestinal tract, leading to their increased fecal excretion. In the process the chloride ion is exchanged for the bile salt anion. This makes the resin useful in pruritis resulting from partial biliary obstruction, a condition that leads to increased serum bile salt levels.

Cholestyramine resin does not bind with drugs that are neutral or with those that are amine salts; however, it is possible that acidic drugs (in the anion form) could be bound. For example, in animal tests it was found the absorption of aspirin given concurrently with the resin was only moderately depressed during the first 30 minutes.

Category—ion-exchange resin (bile salts).

Usual dose—4 g. three times daily.

Usual dose range—10 to 16 g. daily.

AMIDES

The amides or acid amides are derived from acids by substituting an amino or alkylamino group for the hydroxyl; they also may be regarded as derivatives of ammonia or amines with acyl groups instead of hydrogen. The simple compounds have the formula $R-CO-NH_2$, but amides also could be of the types $R-CO-NH-R'$, $R-CO-NR'_2$, $(R-CO)_2NH$ and $(R-CO)_3N$. The chief point of interest in all such substances is that the nitrogen is connected directly to a carbonyl group, a condition that markedly influences the basic character of the nitrogen and is encountered in many medicinal agents.

The simple or primary amides are named from the corresponding acids. The suffix "ic" and the word "acid" are omitted, and the word "amide" is added as a suffix. Thus, formic acid becomes formamide, and acetic acid becomes acetamide; in the Geneva system, "amide" is added to the root, e.g., ethanoic acid gives ethanamide. In more complicated compounds, it seems permissible to append suffixes, such as carbonamide, carboxyamide or carboxylamide.

PREPARATION

Of the various methods that may be used to prepare amides, the most useful are those given below.

Dry Distillation. Dry distillation of ammonium salts or of a mixture of sodium salts and ammonium chloride is efficient for some amides but often leads mostly to cyanide.

$$\underset{\substack{\| \\ O}}{R-C}-O-NH_4 \rightleftarrows \underset{\substack{\| \\ O}}{R-C}-NH_2 \rightleftarrows R-C\equiv N$$

The action is highly reversible and proceeds best if appropriate means are adopted to remove the water as it is formed, an easy task for giving the cyanide but more difficult in case of the amide.

Action of Ammonia or Amines on Esters, Acid Anhydrides or Acid Chlorides.

$$R-NH_2 + \underset{\substack{\| \\ O}}{R'-C}-O-R''$$
$$\downarrow$$
$$\underset{\substack{\| \ \ | \\ O \ \ H}}{R'-C-N}-R + R''-OH$$

The reaction takes place in the cold, but often requires considerable time and a large excess of the amine. Acetamide, for example, may be made in good yield by mixing ethyl acetate and strong ammonia solution and allowing to digest for at least 24 hours.

Hydration of Nitriles. By working under carefully regulated conditions, the primary amides can be prepared by hydration of nitriles. This sometimes can be brought about by mixtures of acids in the cold or by hydrogen peroxide in alkaline solution. If the conditions are not right, ammonium salts may be the products.

From Fatty Acids. Amides can be obtained by the interaction of fatty acids with potassium thiocyanate, with carbylamines or with alkyl isocyanates. For example, the ethyl derivative of acetamide may be made by treating ethyl isocyanate with acetic acid.

$$CH_3-COOH + CH_3-CH_2-NCO$$
$$\downarrow$$
$$\underset{\substack{\| \ \ | \\ O \ \ H}}{CH_3-C-N}-CH_2-CH_3 + CO_2$$

Acylation. More highly acylated amides may be prepared by using an excess of acylating agent or by heating the simple amides with acid anhydrides. Thus, diacetamide, $(CH_3-CO)_2NH$, results from the action of excess acetyl chloride on acetamide in benzene solution.

PROPERTIES

The simple amides are usually crystalline solids that are soluble in alcohol or ether, and the lower ones are very soluble in water. They generally can be distilled without decomposition.

The nearness of the negative carbonyl group weakens the basic nitrogen so much that salts with acids are decidedly unstable. Indeed, hydrogen of the amino group often is influenced enough to be acidic in character and capable of being replaced by metals, a condition that is pronounced especially in the alkyl or aryl derivatives or in the diacyl amides. In general, whenever the combination $-CO-NH-$ is found, the hydrogen may be replaceable to form fairly stable salts of the type $-CONa=N-$; with the grouping $-CO-NH-CO-$, this is certain to be the case, as in the barbitals and hydroxy derivatives of purine. The sulfone group, $-SO_2-$, has a similar influence on the nitrogen in compounds containing $-SO_2NH-$, as is found in the sulfa drugs.

The amides are hydrolyzed easily to

$$\underset{\substack{\| \\ O}}{R-C}-NH_2 \xrightarrow{H_2O}$$

$$\underset{\substack{\| \\ O}}{R-C}-O-NH_4 \xrightarrow{NaOH} \underset{\substack{\| \\ O}}{R-C}-ONa$$

ammonia or amine and the acid. Heating with water may accomplish this, but it is performed more easily by boiling with dilute acid or alkali. On the other hand, dehydration of the primary amides gives the cyanide or nitrile.

With metal nitrites and dilute acid, the primary amides produce the acid and nitrogen.

$$\begin{matrix} & O \\ & \| \\ R-C-NH_2 + NaO-NO + HCl \\ \downarrow \\ R-COOH + N_2 + NaCl + H_2O \end{matrix}$$

Hypobromites act on the primary compounds to give primary amines or cyanides (Hofmann Degradation).

PRODUCTS

Formamide, $HCO-NH_2$, is a thick colorless liquid that is miscible with water, alcohol or ether. When heated rapidly, it decomposes to carbon monoxide and ammonia, but dehydrating agents produce hydrogen cyanide. It combines with chloral to form chloralformamide, $CCl_3-CHOH-NH-CHO$, which has been employed as a hypnotic and sedative.

Acetamide, $CH_3-CO-NH_2$, consists of colorless crystals that melt at 82°, boil at 222° and are very soluble in water or alcohol. It is used as a solvent, a plasticizer and a stabilizer and to manufacture numerous useful derivatives.

The higher compounds of the acetic acid series have more or less sedative and hypnotic action but are of comparatively minor importance.

IMIDES

Imides are represented by compounds of ammonia (NH_3), where 2 hydrogen atoms are replaced by acyl groups. These structures are most common when their formation results in a 5-membered or 6-membered ring, such as succinimide, phthalimide, glutarimide and saccharin. Due to electronegative property of the adjacent carbonyl groups to the nitrogen, imides are acidic in character and form salts with metals (see Sodium Saccharin N.F.). Both succinimide (pKa 10.5) and phthalimide (pKa 8.2) are acidic.

CARBAMATES

Carbamic Acid, NH_2-COOH. Although carbamic acid itself is unknown as a pure compound, it furnishes salts and esters that are important (see Chap. 15). It is the monoamide of carbonic acid and also may be regarded as aminoformic acid. The ammonium salt is formed when ammonium carbonate loses a molecule of water and, with ammonium bicarbonate that can be formed by loss of ammonia, makes up the

$$\begin{matrix} & O & & & O \\ & \| & \overset{-H_2O}{\longleftarrow} & & \| \\ NH_2-C-O-NH_4 & & NH_4-O-C-O-NH_4 \\ & & & \downarrow -NH_3 \\ & & & O \\ & & & \| \\ & & HO-C-O-NH_4 \end{matrix}$$

official ammonium carbonate.

The esters can be prepared by the action of an alcohol on isocyanic acid or its polymer or by treating esters of chloroformic acid with ammonia.

$$\begin{matrix} & & & O \\ & & & \| \\ H-NCO + R-OH & \longrightarrow & R-O-C-NH_2 \end{matrix}$$

$$\begin{matrix} & O \\ & \| \\ R-O-C-Cl + 2NH_3 \\ \downarrow \\ & O \\ & \| \\ R-O-C-NH_2 + NH_4Cl \end{matrix}$$

The ethyl ester is known as "urethane" and this has been adopted as a generic title for any of the esters. Alkyl and aryl derivatives have been made by starting with a corresponding isocyanate, and these compounds also have been called urethanes. The most common reagent for manufacturing them is phenyl isocyanate, and the phenylurethanes that are produced usually have characteristic melting points for identifying the alcohol.

$$C_6H_5-NCO + R-OH \longrightarrow$$

$$\begin{matrix} & H & O \\ & | & \| \\ & C_6H_5-N-C-O-R \end{matrix}$$

UREA AND DERIVATIVES

Urea U.S.P., carbamide, the diamide of carbonic acid, consists of colorless to white, prismatic crystals or a white powder. It is soluble in water (1:15) or in alcohol (1:10), and its aqueous solution is neutral to litmus.

$$H_2N-\overset{\overset{\displaystyle O}{\|}}{C}-NH_2$$
Urea

Even so, urea is a weak base (pKa 0.18) that forms salts with acids, e.g., urea nitrate is important because it is one of the few insoluble nitrates. It will react with acids, esters, acid chlorides and so on to form ureides, just as ammonia reacts to form amides.

Urea is the normal excretion product for nitrogen in the urine of animals and constitutes the chief solid in the urine of man, 5 to 60 g. per day. Its origin in metabolism, involving the amino acids arginine, ornithine and citrulline and the enzyme arginase, will be considered when the proteins are discussed.

Urea is of some importance in medicine, either administered internally or applied locally (see Chap. 20). It is employed locally to treat infected wounds; it has some antiseptic action and is said to promote granulation and healing, as well as to remove dead tissue. Strong solutions have been used to inject into the base of a wart in order to remove it. Urea is said to promote the bacteriostatic activity of the sulfa drugs by inhibiting those substances that oppose them and by increasing the solubility and penetration. For such purpose it may be given by mouth or used topically. Fruit juices or other vehicles may be used to disguise the taste.

Allantoin, Glyoxyldiureide, Cordianine, the ureido derivative of hydantoin, is a white, crystalline powder that is sparingly soluble in cold water but readily soluble in hot water.

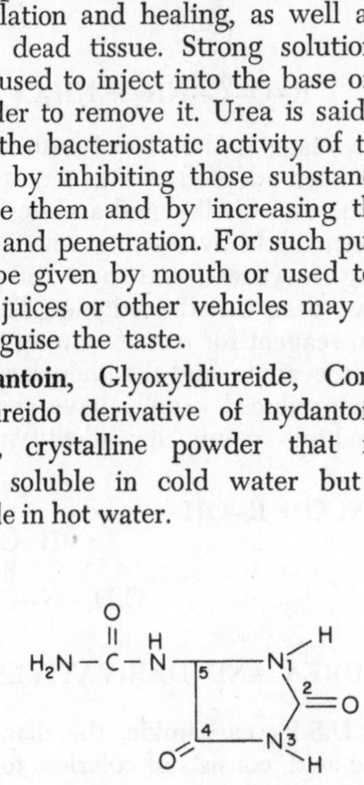

The application of allantoin in medicine followed the use of fly maggots. During World War I it was noted that osteomyelitic abscesses and severe wounds healed faster after being infested with maggots, and these maggots were cultivated commercially for the purpose of treating such conditions. Later, allantoin was found to be present in extracts of the larvae and was recommended as a substitute for the purpose of stimulating cell proliferation in sluggish wounds, especially in osteomyelitis. Still later, it was suggested that the effective agent is urea, which is part of the molecule, and since then the employment of allantoin for the purpose has decreased gradually (see Urea). It is applied as a 0.4 to 2.0 per cent solution, frequently combined with antiseptics, urea and sulfa drugs.

Thiourea, Thiocarbamide, is a white solid.

$$H_2N-\overset{\overset{\displaystyle S}{\|}}{C}-NH_2$$
Thiourea

Thiourea has properties as an antithyroid agent, although it is less effective in this respect than some of its derivatives. It has antioxidant properties which are associated with the sulfur atom, since urea itself has no antioxidant properties. Thiourea possibly may function as an antioxidant, either by being itself oxidized or by inhibiting an enzyme responsible for oxidation. While the evidence is that the equilibrium

$$H_2N-\overset{\overset{\displaystyle S}{\|}}{C}-NH_2 \leftrightarrows H_2N-\overset{\overset{\displaystyle S}{|}}{\underset{}{C}}=NH$$

is far to the left, it has been suggested that there may be enough of the form with a sulfhydryl group to account for the antioxidant properties of the compound. It is used as an antioxidant for epinephrine and to preserve the natural color of dried fruits.

Allythiourea, "Thiosinamin,"

$$H_2C = CH-CH_2-NH-\overset{\overset{\displaystyle S}{\|}}{C}-NH_2,$$

is claimed to have a softening action on certain animal tissues and has been used subcutaneously for the softening of fibrous scars. At the present time it is used comparatively little. Fibrolysin is a 15 per cent solution of a mixture consisting of 2 moles of thiosinamin and 1 mole of sodium salicylate in water.

CHLOROPHYLL

After many years of intensive research on the structure of chlorophyll *a* by a number of chemists a structural formula for it recently was proposed that was verified by its total synthesis. Chlorophyll *a* has the following structural formula.[2]

Chlorophyll *a*

The earliest studies on the therapeutic use of chlorophyll were in about 1851. Sir Joseph Priestly observed that green plants absorb carbon dioxide and yield oxygen in sunlight. The name chlorophyll was given to the pigments, in 1818, by Pelletier and Caventous. Most of the early applications were as a hematopoietic, which were based on the similarity of the porphin structure in chlorophyll and hemoglobin. Etioporphyrin ($C_{32}H_{38}N_4$), which is tetramethyltetraethylporphyrin and does not contain iron or magnesium, has been obtained by the degradation of either hemoglobin or chlorophyll. Likewise, the bile pigment, bilirubin, is obtainable from either one.

Unfortunately, a number of derivatives have been made that also have been designated as chlorophylls. In current pharmaceutic literature, there is considerable confusion regarding the chlorophyll products. First of all, the correct chemical naming of these derivatives is often ignored. See below.

Pheophytins. Both *a* and *b* forms of pheophytin are of the same general structure as the chlorophylls except that 2 hydrogen atoms replace magnesium.

Chlorophyllides. When green leaves are extracted with ethyl or methyl alcohol instead of the usual acetone, the phytyl radical is displaced, through the influence of the enzyme chlorophyllase, by the smaller alkyl group to give chlorophyllides. Ethyl chlorophyllide is water insoluble but is soluble in glyceride-type or hydrocarbon oils. It often is referred to as green chlorophyll dye which is used as a coloring agent in foods, soaps and cosmetics.

Chlorophyllins. Beginning about 1930, Dr. Benjamin Gruskin began working on water-soluble chlorophyll derivatives which culminated in the granting of United States patent 2,120,667, which is controlled by Rystan Company, Inc. When chlorophyll-*a* is hydrolyzed with sodium hydroxide, a water-soluble chlorophyllin is obtained. The ester groups at positions 7 and 10 yield the sodium salts and the respective alcohols. Position 9 represents a beta keto acid formation that under the influence of sodium hydroxide may break from the carbon at 10 to yield a carboxyl group at 9 which, of course, forms a sodium salt. The chlorophyllin thus obtained is either a disodium or trisodium salt. Potassium salts also are used. Usually, the copper complex (Chloresium) or the iron complex is used. All of the present day applications of "chlorophyll" utilize the water-soluble chlorophyllins. Products that contain another metal (copper, iron, cobalt, tin, nickel, silver, gold, mercury, thallium) in place of magnesium are called "metallophyllins." Among those available are: aurophyllin, ferrophyllin, hydrogyrophyllin and manganophyllin. The copper chlorophyllin is by far the most used but primarily for topical application. To avoid toxicity, either magnesium chlorophyllin or ferrophyllin is used internally. The magnesium chlorophyllin is not too stable and is a dull green color. The

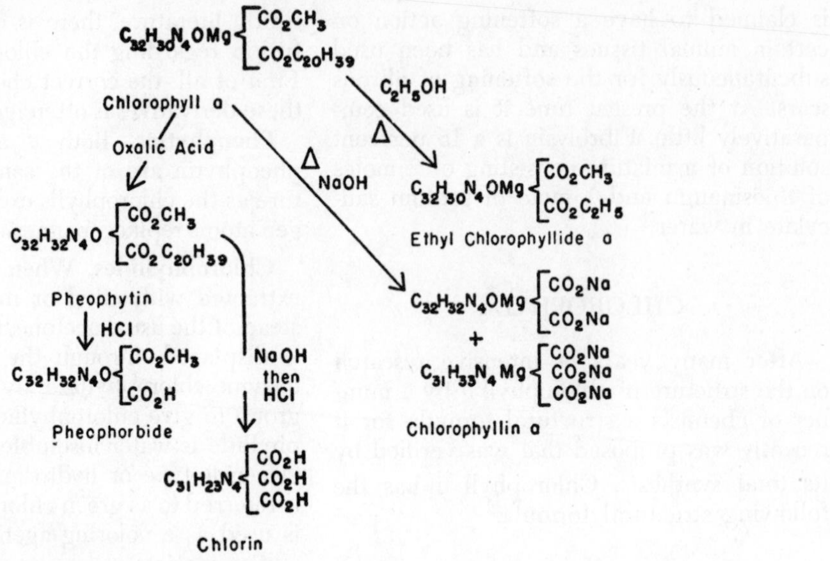

$C_{32}H_{30}N_4OMg\begin{cases}CO_2CH_3\\CO_2C_{20}H_{39}\end{cases}$

Chlorophyll a

C_2H_5OH

Oxalic Acid

Δ

Δ

NaOH

$C_{32}H_{30}N_4OMg\begin{cases}CO_2CH_3\\CO_2C_2H_5\end{cases}$

Ethyl Chlorophyllide a

$C_{32}H_{32}N_4O\begin{cases}CO_2CH_3\\CO_2C_{20}H_{39}\end{cases}$

Pheophytin

$C_{32}H_{32}N_4OMg\begin{cases}CO_2Na\\CO_2Na\end{cases}$

+

$C_{31}H_{33}N_4Mg\begin{cases}CO_2Na\\CO_2Na\\CO_2Na\end{cases}$

Chlorophyllin a

HCl

$C_{32}H_{32}N_4O\begin{cases}CO_2CH_3\\CO_2H\end{cases}$

NaOH then HCl

Pheophorbide

$C_{31}H_{23}N_4\begin{cases}CO_2H\\CO_2H\\CO_2H\end{cases}$

Chlorin

copper compound has an intense green color and is more stable to light and heat.

Current uses of chlorophyll derivatives are primarily in treating wounds and burns, as a hematopoietic agent and as a deodorant. There is some evidence that chlorophyllins have a marked stimulating effect on tissue cells and thus cause the formation of unusually firm and fine textured granulation tissue. The literature gives several favorable reports[3] on the treatment of ulcers of all kinds, wounds, otitis media, sinusitis, skin diseases, burns and so on. The forms employed, usually of the water-soluble variety, are a 0.2 per cent solution in normal saline for irrigations, a 1 per cent ointment containing benzocaine and urea and a 0.5 per cent solution for nasal drops.

A number of authors have reported the effectiveness of chlorophyllins, especially ferrophyllin, in stimulating red blood cell and hemoglobin regeneration. As a deodorant, chlorophyllins appear to act as an oxidation catalyst, causing the oxygen in the air to oxidize the odor-producing material.

The chlorophyllins are effective deodorizers internally or externally and are used in every possible way from tooth pastes to footwear. Effective concentration is between 1:1,000 to 1:10,000 dilution with inert material.

Water-Soluble Chlorophyll Derivatives; Chloresium.® Water-soluble derivatives consist chiefly of the copper chlorophyllin in the form of sodium or potassium salts. It occurs as a blue-black glistening powder having an aminelike odor. Solubility is best in water and decreasing from alcohol and chloroform to ether. A 1 per cent solution has a pH of 9.5 to 10.7 and is a dark green color.

REFERENCES CITED

1. McChesney, E. W., *et al.*: J. Am. Pharm. A. (Sci. Ed.) 40:193, 1951.
2. Woodward, R. B., *et al.*: J. Am. Chem. Soc. 82:3800, 1960; Angw. Chemie, 72: 651, 1960.
3. Bowers, W. F.: Am. J. Surg. 73:37, 1947.

9

Anti-infective Agents

Robert F. Doerge, Ph.D.,
Professor of Pharmaceutical Chemistry,
Chairman of the Department of Pharmaceutical Chemistry,
School of Pharmacy, Oregon State University

INTRODUCTION

Chemotherapy may be defined as the study and the use of agents which are selectively more toxic to the invading organisms than to the host. Paul Ehrlich, the father of chemotherapy, was more absolute in his concept and used the term to describe the cure of an infectious disease *without* injury to the host. This ideal has been rather closely approached by the antibiotic, penicillin. The scientific principles of chemotherapy were established chiefly during the period 1919-1935, but only since this time and especially with the advent of the sulfonamides and the antibiotics have the material benefits in terms of useful medicinal products been realized. The only chemotherapeutic agents known before the time of Ehrlich were cinchona for malaria, ipecac for amebic dysentery and mercury for treating the symptoms of syphilis.

The first 30 years of the 20th century saw the development of useful chemotherapeutic agents, among which were organic compounds containing heavy metals such as arsenic, mercury and antimony, dyes, and a few modifications of the quinine molecule. These agents represented extremely important advances but even so had many drawbacks. The next 30 years of the 20th century comprise the period of greatest advance in the area of chemotherapy. During this time the sulfonamides

and sulfones (see Chap. 11), many phenols and their derivatives (see Chap. 10), the antimalarial agents (see Chap. 13), the surfactants (see Chap. 12) and, of great importance, the antibiotics (see Chap. 14) were studied and introduced into medical practice. The development of these newer drugs has relegated some of the older drugs to positions of minor importance or historical interest only.

The knowledge and the use of chemotherapeutic agents can be classified according to the diseases and the infestations for which they are used; or they can be classified according to separate compounds or groups of related compounds. In this book the chapters covering chemotherapeutic agents are organized by an amalgamation of the two systems. When the knowledge is best expressed and interrelated by the chemical classification this method is used, but where several classes of drugs may be rather specific for a single disease or group of related diseases the medical classification is used.

ANTIBACTERIAL AGENTS

ALCOHOLS AND RELATED COMPOUNDS

Various alcohols and alcohol derivatives have been used as antiseptics. Ethyl alcohol and isopropyl alcohol are widely used for this purpose (see Chap. 5).

PRODUCTS

Chlorobutanol U.S.P., Chloretone®, 1,1,1-Trichloro-2-methyl-2-propanol. Chlorobutanol is tertiary trichlorobutyl alcohol which may be synthesized from acetone and chloroform.

$$CH_3COCH_3 + CHCl_3 \xrightarrow{KOH} \underset{\underset{CCl_3}{|}}{\overset{\overset{CH_3}{|}}{CH_3-C-OH}}$$

It is a white, crystalline solid having a characteristic camphorlike odor and taste. It is available in two forms: the anhydrous form, and the hydrated form containing not over one-half molecule of water of hydration. The anhydrous form is used in preparing oil solutions. Because it volatilizes readily at room temperatures, chloretone is difficult to dry and must be stored carefully. The compound dissolves in water (1:125), alcohol (1:1), glycerin (1:10), all oils or in organic solvents.

Chlorobutanol is widely used as a bacteriostatic agent in pharmaceuticals for injection, ophthalmic use or intranasal administration. When used in aqueous preparations, it has the distinct disadvantage of being only slowly soluble.[1] It is more soluble in boiling water, but when heated the compound hydrolyzes and is lost by volatilization as well. Solutions which are buffered below pH 5 and in closed systems can be autoclaved at 121° for 20 minutes with only slight loss due to hydrolysis.[2]

As part of a thorough kinetic study of the degradation of chlorobutanol in aqueous solution, it was calculated that solutions at pH 5 would lose 13 per cent when heated at 115° for 30 minutes.[3] The solution could then be stored at 25° for well over 5 years before showing a further 10 per cent loss. The hydrolysis of chlorobutanol can be represented as follows:

$$\underset{\underset{CCl_3}{|}}{\overset{\overset{CH_3}{|}}{CH_3COH}} + H_2O \longrightarrow$$

$$\underset{\overset{\|}{O}}{CH_3CCH_3} + CO + 3HCl$$

When chlorobutanol is used in oil solutions these problems of hydrolysis and slow rate of solubility are not met.

Chlorobutanol is also used to some extent as a hypnotic and sedative. The action and effectiveness are similar to chloral hydrate. There is also some local anesthetic action. Advantage is taken of this in making powders for topical use, oil sprays for nasal application and injectable solutions (Bismuth Subsalicylate Injection 3 per cent).

Category—pharmaceutic aid (antimicrobial agent); dental analgesic.

For external use—topically, as a 25 per cent solution in clove oil.

Benzyl Alcohol N.F.; Phenylcarbinol, Phenylmethanol. Benzyl alcohol occurs free in nature (Oil of Jasmine, 6%) and is found as an ester of acetic, cinnamic and benzoic acids in gum benzoin, storax resin, Peru balsam and tolu balsam and in some volatile oils (jasmine and hyacinth). In maize, a glucoside of benzyl alcohol is found. It is synthesized readily from toluene (1) and by the Cannizzaro reaction from benaldehyde (2).

The alcohol is soluble in water (1:25) and in 50 per cent alcohol (1:15). It is miscible with fixed and volatile oils, ether, alcohol or chloroform. Benzyl alcohol is a clear liquid with a faint aromatic odor. It can be boiled without decomposition.

The chemical properties of benzyl alcohol are much the same as those of primary alcohols, since it is phenylmethanol. On oxidation it first yields benzaldehyde and then benzoic acid. It differs from the aliphatic alcohols in being resinified by sulfuric acid, and it does not form the corresponding sulfuric ester. In comparison with hydroxybenzene (phenol), the introduction of the "CH₂" group destroys the phenolic properties, such as the caustic action and the reaction with alkalies or ferric chloride.

Benzyl alcohol commonly is incorporated as a preservative in vials of injectible drugs and also because it exerts a local anesthetic[4] effect when injected or applied on mucous membranes. A saturated piece of cotton is effective for toothache when used in the same manner as clove oil. The concentrations usually employed are 1 to 4 per cent (maximum solubility) in water or saline solution. In such small doses it is nonirritating and nontoxic. Since it is also strongly antiseptic, ointments containing benzyl alcohol up to 10 per cent are useful in preventing secondary infection in the itching of pruritus and other skin conditions. A suitable lotion may be prepared with equal parts of benzyl alcohol, water and alcohol.

Interest in the benzyl group began with the observation that when the isoquinoline ring of the alkaloid papaverine is opened, a β-phenylethyl amine derivative is obtained (see Alverine citrate). Since this secondary amine and its related compounds are useful as antispasmodics, the benzyl group was tried in numerous compounds.

The benzyl group, when present in ester form, exerts an antispasmodic[5] action. In compounds such as the barbiturates when the benzyl group is placed in position five (5), the opposite effect is produced, i.e., it shows convulsive[6] properties. However, the phenylethyl derivatives of the barbiturates lack the convulsive property of the benzyl compounds.

There are a few other aromatic alcohols of minor importance. For a pharmacologic study of some of these see Hirschfelder.[7]

Category—bacteriostatic (injection).

Phenylethyl Alcohol N.F.; Phenethyl Alcohol, 2-Phenylethanol, Orange Oil or Rose Oil, $C_6H_5CH_2CH_2OH$. This compound is useful in perfumery, occurs in oils of rose, orange flowers, pine needles and Neroli. It is prepared by the reduction of ethyl phenylacetate,

$$C_6H_5-CH_2-\overset{O}{\overset{\|}{C}}-O-C_2H_5 \xrightarrow[C_2H_5OH]{Na} C_6H_5 CH_2CH_2OH$$

or with phenylmagnesium chloride and ethylene oxide.

$$C_6H_5MgCl + \underset{\underset{O}{\diagdown\diagup}}{CH_2-CH_2} \longrightarrow C_6H_5CH_2CH_2OH$$

This alcohol is soluble in water (2%). It may be sterilized by boiling, since it boils at 220°.

Hjort[8] found it to be slightly more anesthetic than benzyl alcohol and of the same order of toxicity.

Category—bacteriostatic, preservative.

Ethylene Oxide has been used for many years to sterilize temperature-sensitive medical equipment and more recently has been found to be of value in the sterilization of certain thermolabile pharmaceuticals. As a gas it will diffuse through porous material, it is readily removed by aeration following treatment and effectively destroys all forms of microorganisms at ordinary temperatures.[9] It is a colorless, flammable gas at ordinary room temperature and pressure but can be liquefied at 12°. The gas in air forms explosive mixtures in all proportions from 3 to 80 per cent by volume. The explosion hazard is eliminated when the ethylene oxide is mixed with more than 7.15 times its volume of CO_2. Carboxide is a commercially available product which is 10 per cent ethylene oxide and 90 per cent CO_2; it can be released in the air in any quantity without forming an explosive mixture. Water vapor is a factor in ethylene oxide sterilization. The amount of water vapor which must be added to the gas appears to depend on the amount of moisture absorbed by the material to be sterilized.[10] Plastic intravenous injection equipment can be sterilized in the shipping carton, using ethylene oxide.[11]

Formaldehyde Solution U.S.P., Formalin, Formol. Formaldehyde Solution is a colorless, aqueous solution containing not less than 37 per cent of formaldehyde (CH_2O) with methanol added to prevent polymerization. It is miscible with water or alcohol and has the pungent odor that is typical of the lower members of the aliphatic aldehyde series.

Owing to the ease with which oxidation and polymerization can take place, the chief impurities to be found in the solution are formic acid and paraformaldehyde.

On long standing, especially in the cold, the solution may become cloudy. Therefore, it should be preserved in tightly closed containers at temperatures not below 15°.

Assay is performed by oxidation to sodium formate with hydrogen peroxide and a volumetric solution of sodium hydroxide. After heating the mixture on a water bath for 5 minutes, the excess of alkali is determined by titration with normal sulfuric acid. One of the tests for identity involves the reduction of an ammoniacal solution of silver nitrate to a precipitate or mirror of metallic silver. Another test depends upon the deep red color produced by warming with a solution of salicylic acid in sulfuric acid.

Formaldehyde was prepared first by Hofmann (1868) by passing a hot mixture of methyl alcohol and air over platinum. It still is obtained commercially in the same way, although a variety of catalysts have been utilized, including copper, silver, oxides of iron and molybdenum and vanadium pentoxide. It also can be produced by the oxidation of methane in natural gas.[12] Formaldehyde does not occur naturally in significant amounts. It is frequently found in the aqueous distillate during the preparation of volatile oils from plants.

Formaldehyde differs from typical aliphatic aldehydes in some important reactions. When evaporated with a solution of ammonia, it forms methenamine by condensation.

$$6H-CHO + 4NH_3 \longrightarrow (CH_2)_6N_4 + 6H_2O$$

When warmed with alkali, it yields no resin but partly undergoes the Cannizzaro reaction, which is characteristic of any aldehyde that has no alpha hydrogen atom.

$$2H-CHO + NaOH \longrightarrow$$
$$H-COONa + CH_3OH$$

However, it will give an aldol condensation if allowed to stand for some time with dilute alkali, finally producing a mixture of sugars known as formose; both pentoses and hexoses have been isolated from this. In addition, formaldehyde will undergo an aldol condensation with other aldehydes, and the product usually will give the Cannizzaro reaction. With acetaldehyde and

calcium hydroxide, for example, it furnishes pentaerythritol and a formate.

$$3H-CHO + CH_3-CHO \longrightarrow$$
$$(CH_2OH)_3C-CHO$$

$$(CH_2OH)_3C-CHO + H-CHO + H_2O$$
$$\downarrow$$
$$(CH_2OH)_4C + H-COOH$$

Finally, it shows a remarkable tendency to polymerize, since evaporation of the solution yields a white, friable mass of paraformaldehyde $(CH_2O)_n$. If a strong solution is distilled with 2 per cent sulfuric acid and the vapors are condensed quickly, a crystalline trimer known as trioxane or trioxymethylene is formed.

Either one of these products can be depolymerized by heat, thus giving a convenient source of formaldehyde for synthetic and other processes.

It also will condense with a great variety of other compounds. Any active hydrogen, whether it is attached to carbon, nitrogen or oxygen, will react with the carbonyl group to produce water, either with or without primary addition. In this way, formaldehyde becomes a very useful synthetic agent and frequently is employed for the manufacture of medicinals, dyes, plastics, etc. From the pharmaceutical viewpoint, one of the most important of these reactions is that on amino acids and proteins. Formaldehyde adds to the amino group to mask its basic character, thus making it possible to measure carboxyl value in the so-called formal titration. The nature of this primary effect is not known with certainty, but we may assume it to be addition to yield the grouping $HOCH_2N =$. However, the action does not stop there; undoubtedly, there is an elimination of water and a primary alteration of the protein molecule (denaturing). The final result on animal or vegetable tissue is a coagulation and hardening, an irreversible process.

Formaldehyde, either as a gas or in solution, has a powerful effect on all kinds of tissue; it is irritating to mucous membranes, hardens the skin and kills bacteria or inhibits their growth. It is an excellent germicide, probably equal to phenols or mercury, and its volatility renders it more penetrating. A dilution of 1:5,000 inhibits the growth of any organism, and, in many cases, 1:20,000 will retard any multiplication. Large doses by mouth cause the usual symptoms of gastroenteritis and ultimate collapse. The gas is very irritating when inhaled.

The gas has been employed to disinfect rooms, excreta, instruments and clothing but is little used at present.

Usually, applications to the body are not to be recommended, but, diluted with water or alcohol, the solution has been applied as a hardener of the skin, to prevent excessive perspiration and, also, to disinfect the hands or the site of an operation.

Industrially, formaldehyde has a considerable number of important and varied uses. The mere fact that more than 250,000 tons of the solution are manufactured in the United States annually is a striking illustration. It is indispensable in the manufacture of certain phenolic resins, urea plastics, mirrors, some kinds of rayon, dyes, esters of cellulose, explosives, photographic plates and papers, embalming fluids and many organic chemicals. It also is used in photographic processes, with dyes to improve fastness, to aid the tanning process, to preserve and coagulate rubber latex and to prevent mildew in vegetable products.

Category—disinfectant. Application—full strength or as a 10 per cent solution to inanimate objects.

Paraformaldehyde, obtained by evaporating a formaldehyde solution, also is known as paraform, triformol and, erroneously, trioxymethylene. It is a white powder that is slowly soluble in cold water and more readily soluble in hot water, but with some decomposition, to produce an odor of formaldehyde. Because it can be converted completely to the gas by heating, it is used largely as a convenient form of transportation. It has been used as the active ingredient of contraceptive creams.

Sodium Formaldehydesulfoxylate, $HOCH_2SO_2Na \cdot 2H_2O$, also known as ron-

golite, formopone and hydrolit, is made by reducing the sodium bisulfite addition product of formaldehyde by zinc dust and acetic acid. It consists of white crystals or pieces that are soluble in water but only sparingly so in alcohol. It is almost odorless when freshly prepared but quickly develops a characteristic garliclike odor. It is decomposed rapidly by acids and will reduce even very mild oxidizing agents. It is not used as an antibacterial agent but has been used as a reducing agent with some phenothiazine derivatives.

The compound was formerly used as an antidote for mercury poisoning.

Methenamine N.F., Hexamethylenetetramine, Urotropin,® Uritone,® is actually not an aldehyde at all, but its value depends upon the liberation of formaldehyde, and, therefore, it should be considered at this point. It is manufactured by evaporating a solution of formaldehyde to dryness with strong ammonia water.

The compound consists of colorless crystals or a white crystalline powder without odor. It sublimes at about 260° without melting and burns readily with a smokeless flame. It dissolves in 1.5 ml. of water to make an alkaline solution and in 12.5 ml. of alcohol. Warm acids will liberate formaldehyde, which may be recognized by its odor, and the subsequent addition of alkalies will give an odor of ammonia. The assay of the compound depends upon decomposition with volumetric solution of sulfuric acid and titration of the excess with sodium hydroxide.

Methenamine is used internally as an antiseptic, especially in the urinary tract. In itself it has practically no bacteriostatic power and can be efficacious only when it is acidified to produce formaldehyde. Because concentration in the kidney and the bladder never can become very high, the success in treating infections of the urinary tract usually has not been great. In order to obtain a maximum effect, the administration of the compound generally is accompanied by sodium biphosphate, ammonium chloride or a similar acidifying agent.

Category—antibacterial (urinary tract).

Usual dose—1 g. 4 times daily.

OCCURRENCE   PER CENT

Methenamine Tablets N.F.

Methenamine and

Sodium Biphosphate Tablets N.F.

Amobarbital Elixir N.F. 0.44

Methenamine Mandelate U.S.P., Hexamethylenetetramine Mandelate, Mandelamine®, is a white crystalline powder with a sour taste and practically no odor. It is very soluble in water and has the advantage of furnishing its own acidity, although in its use the custom is to carry out a preliminary acidification of the urine for 24 to 36 hours before administration. It is effective with smaller amounts of mandelic acid and thus avoids the gastric disturbances attributed to the acid when used alone.

Category—antibacterial (urinary).

Usual dose—1 g. 4 times daily.

Usual dose range—3 to 4 g. daily.

OCCURRENCE

Methenamine Mandelate Oral Suspension U.S.P.

Methenamine Mandelate Tablets U.S.P.

Methenamine Hippurate, Hiprex®, is the hippuric acid salt of methenamine. It is readily absorbed after oral administration and is concentrated in the urinary bladder, where it exerts its antibacterial activity. Its activity is increased in acid urine.

Usual dose—for adults and children over 12 years of age, 1 g. twice daily.

ACIDS AND THEIR DERIVATIVES

Benzoic Acid Derivatives

Benzoic Acid U.S.P. Benzoic acid and its esters occur in nature as constituents in gum benzoin, in Peru and Tolu balsams and in cranberries. As hippuric acid, it occurs in combination with glycine in the urine of herbivorous animals.

Benzoic Acid

The acid may be obtained by distillation from a natural product, such as benzoin, or prepared synthetically by several procedures.

$$(1) \quad C_6H_5CH_3 \xrightarrow[\text{H}_2\text{SO}_4]{\text{MnO}_2} C_6H_5COOH$$

$$(2) \, C_6H_5CH_3 \xrightarrow{Cl_2} C_6H_5CCl_3 \xrightarrow[\text{Fe}]{\text{Ca(OH)}_2} C_6H_5COOH$$

Benzoic acid forms white crystals, scales or needles, that are odorless or may have a slight odor of benzoin or benzaldehyde. It sublimes at ordinary temperature and distills with steam. It is slightly soluble in water (0.3%), benzene (1%) and benzin, but it is more soluble in alcohol (30%), chloroform (20%), acetone (30%), ether (30%) and volatile and fixed oils.

All of the reactions characteristic of the carboxyl group are exhibited by benzoic acid. Reactions characteristic of aromatic compounds, as, for instance, nitration, halogenation and sulfonation, take place with benzoic acid. The acid is more strongly acidic (pKa 4.2) than acetic acid (pKa 4.7), and most of the common electron-attracting substituents increase the acidity. Solutions containing the ions of iron, silver, lead or mercury form a precipitate of the respective salt with benzoic acid. The iron salt is a reddish-tan or salmon-colored precipitate.

Benzoic acid is used externally as an antiseptic[13] and is employed in lotions, ointments and mouthwashes. In concentrations over 0.1 per cent, it may produce local irritation. It is employed as a food preservative, especially in the form of its salts (i.e., sodium benzoate). When used as a preservative in foods and in pharmaceutical products, benzoic acid and its salts are more effective as the pH is lowered; thus it is the undissociated benzoic acid molecule which is the effective agent. The pKa of benzoic acid is 4.2, so that at this pH only 50 per cent would be in the undissociated form, while at pH 3.5 over 80 per cent would be in the undissociated form.[14, 15] When benzoic acid or its salts are used in emulsions, the effectiveness as a preservative depends on the distribution between the oil phase and the water phase as well as the pH of the system.[16]

Category—pharmaceutic aid (antifungal preservative).

OCCURRENCE	PER CENT BENZOIC ACID
Kaolin Mixture with Pectin N.F.	0.2
Paregoric U.S.P.	0.38
Starch Glycerite N.F.	0.2

Sodium Benzoate U.S.P. Sodium benzoate is prepared by adding sodium bicarbonate to an aqueous suspension of benzoic acid. The product has a sweet, astringent taste; it is stable in air and is a white, odorless, crystalline substance or a granular powder of 99 per cent purity. It is soluble in water or alcohol.

Sodium benzoate has chemical properties similar to those of all benzoates and the incompatibilities are similar to those of benzoic acid. It is used primarily (0.1%) as a preservative in acid media for the antiseptic effect of benzoic acid. It is not effective in preserving nonacid products. The solubilizing property for caffeine is used in the official physical mixtures (q.v.). This salt is used internally as a test for liver function. The benzoate is conjugated with glycine and excreted as hippuric acid, formation of which takes place in the liver, so that the rate at which this material is excreted determines the functional ability of the liver.

Category—pharmaceutic (antifungal preservative).

OCCURRENCE	PER CENT SODIUM BENZOATE
Acacia Syrup N.F.	0.1
Caffeine and Sodium Benzoate Injection U.S.P.	...

Salicylic Acid U.S.P., *o*-Hydroxybenzoic Acid. This acid has been known for over 130 years, having been discovered in 1839. It is found free in nature and in the form of salts and esters. A very common ester is methyl salicylate (oil of wintergreen). Salicylic acid may be obtained from oil of wintergreen by saponification with sodium hydroxide and then neutralization with hydrochloric acid. This is referred to as "natural salicylic acid" and is used to prepare salts which are preferred by some. The natural acid usually is tinted pink or yellow and has a faint wintergreenlike odor. At one time it was believed that the synthetic salicylic acid was contaminated with some cresotinic acid $[C_6H_3 \cdot CH_3(OH)(COOH)]$ and was thus more toxic, its salts less desirable. It has since been shown, not only that cresotinic acid is absent, but also that cresotinic acid is nontoxic.

In 1859, Kolbe introduced a method for the synthetic preparation of salicylic acid, and, with slight changes, this is still used. Sodium phenolate is prepared and saturated under pressure with carbon dioxide; the resulting product then is acidified and salicylic acid is isolated.

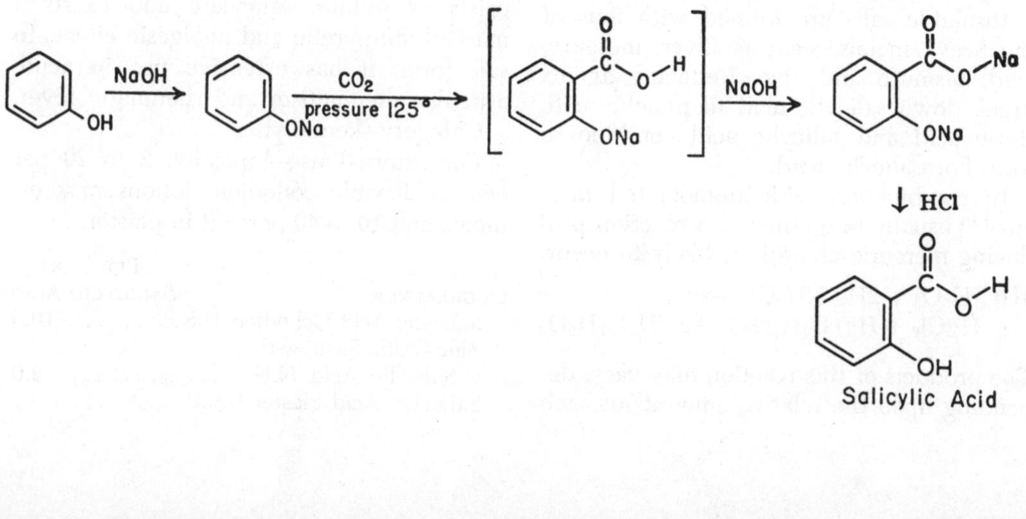

Salicylic Acid

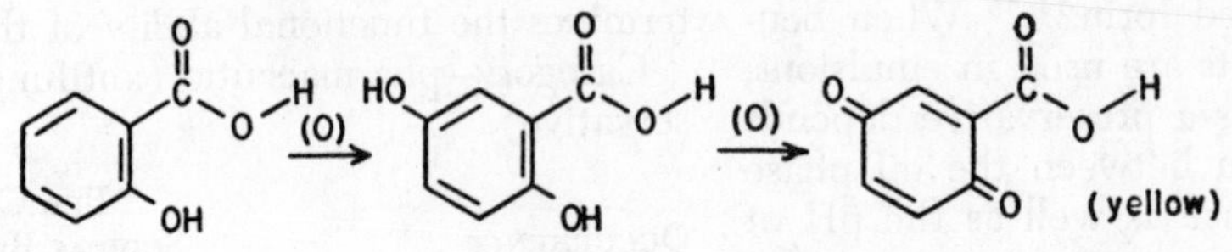

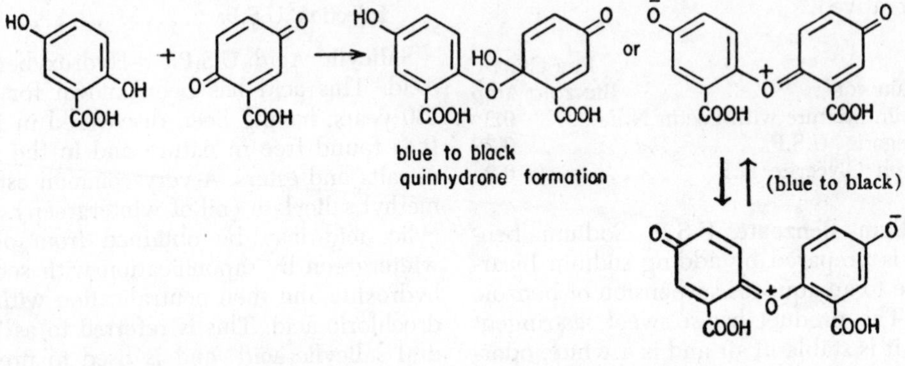

blue to black
quinhydrone formation

(blue to black)

Salicylic acid usually occurs as white, needlelike crystals or as a fluffy, crystalline powder. The synthetic acid is stable in air and is odorless. It is slightly soluble in water (1:460) and is soluble in most organic solvents.

The chemical properties of this acid are due to the phenolic hydroxyl group (see Chap. 10) and to the carboxyl group. Since it is also a phenol, it responds with the reactions of phenols, such as the producing of a violet color with ferric salts, halogenation and oxidation. Oxidizing agents form colored compounds, perhaps of a quinoid[17] type, and destroy the molecule. The colored compounds produced on standing in alkaline solution are due to quinhydrone formation. For examples of quinhydrone formation see the equations above.

Insoluble salts are formed with ions of the heavy metals, such as silver, mercury, lead, bismuth and zinc. Reducing agents break down salicylic acid to pimelic acid. Boric acid and salicylic acid combine to form borosalicylic acid.

In combination with ammoniated mercury,[18] usually in ointments, a reaction producing mercuric chloride is likely to occur.

$$4HC_7H_5O_3 + 2Hg\,NH_2Cl \longrightarrow$$
$$HgCl_2 + Hg(C_7H_5O_3)_2 + 2NH_4C_7H_5O_3$$

The products of this reaction may vary, depending upon the relative amounts of each compound and the presence of moisture. The mercury bichloride formed results in a preparation which is very irritating when topically applied.[18]

Salicylic acid has strong antiseptic and germicidal properties, due to the fact that it is a carboxylated phenol. The presence of the carboxyl group appears to enhance the antiseptic property and to decrease the destructive, escharotic effect. It is used externally as a mild escharotic and antiseptic in ointments and solutions. Many hair tonics and remedies for athlete's foot, corns and warts employ the keratolytic action of salicylic acid. Internally, it seldom is administered as the free acid but is taken in the form of its salts for the effects of the salicylate portion. In the circulatory system, it exists as sodium salicylate and exerts a marked antipyretic and analgesic effect. In salt form, it has extensive use in acute articular rheumatism and rheumatic fever.

Category—keratolytic.

For external use—topically, 2 to 20 per cent in flexible collodion, lotions or ointments and 10 to 40 percent in plaster.

OCCURRENCE	PER CENT SALICYLIC ACID
Salicylic Acid Collodion U.S.P.	10.0
Zinc Oxide Paste with Salicylic Acid N.F.	2.0
Salicylic Acid Plaster U.S.P.	...

Mandelic Acid Derivatives

Mandelic Acid; Racemic Mandelic Acid. Mandelic Acid usually is prepared from benzaldehyde and hydrogen cyanide or sodium cyanide. The solid sodium cyanide has

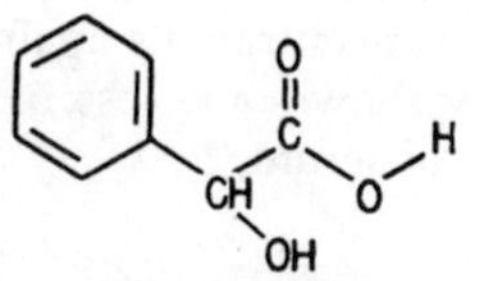

Mandelic Acid

the advantage of being easier to handle. The following equation illustrates the reaction steps:

Mandelonitrile   *rac*-Mandelic Acid

In bitter almonds, there occurs the glucoside, amygdalin, which is the gentiobioside of mandelonitrile. On hydrolysis, this nitrile yields (−)-mandelic acid (see glycosides).

The acid occurs as a white, crystalline powder which may be odorless or have a slight aromatic odor. Upon exposure to light, the crystals slowly darken and decompose.

In 1935,[19] Fuller announced the isolation of β-hydroxybutyric acid from the urine of patients with uncontrolled diabetes and also from urine excreted by individuals on a ketogenic diet. Previous to Fuller's work, the freedom from infections of the urinary tract among diabetics had been observed, as well as the stability of or resistance to putrefaction of diabetic urine. Due to these observations, the ketogenic diet was used in the treatment of upper urinary tract infections. β-Hydroxybutyric acid is an oxidation product of fatty acids in the course of their conversion to carbon dioxide and water and is, therefore, a normal metabolite. Since the acid is metabolized, oral administration was not feasible because it would be further oxidized before excretion in the urine could take place.

Rosenheim,[20] in an effort to find a suitable acid (β-hydroxyacid) found that mandelic acid possessed good bacteriostatic and bactericidal properties (e.g., for *Escherichia coli, Streptococcus faecalis, Salmonella*) and yet, on oral administration, was excreted in the urine, because it was not metabolized. Further studies[21] indicate that in urine of pH 5.5 or less, mandelic acid was most effective as a urinary antiseptic in such conditions as cystitis and pyelitis. Pharmaceutic forms are mandelic acid with ammonium chloride, sodium mandelate with ammonium chloride, calcium mandelate with ammonium chloride, calcium mandelate, methenamine mandelate and mandelates with sodium biphosphate.

Mandelic acid preparations are less extensively used today because of the introduction of the antibacterial sulfonamides. The usual dose is 3 g.

Calcium Mandelate is prepared readily by using a soluble calcium salt and mandelic acid in an aqueous medium. The compound is crystallized from water and occurs as a white, odorless powder which is only slightly soluble in water and is insoluble in alcohol.

The calcium salt, being practically insoluble, is nearly tasteless and, also, produces less gastric irritation than does either the sodium or the ammonium salt. In the intestine, the mandelates all behave the same way, by hydrolyzing, and so liberate free mandelic acid that is excreted in the urine.

Miscellaneous

Nalidixic Acid N.F., NEGram®, is 1-ethyl-1,4-dihydro-7-methyl-4-oxo-1,8-naphthyridine-3-carboxylic acid.

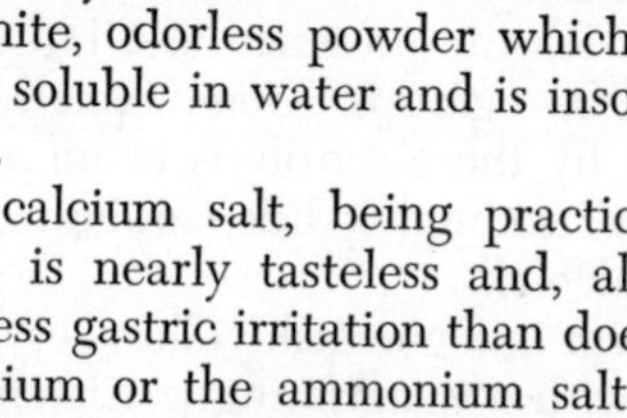

Nalidixic Acid

Nalidixic acid is useful in the treatment of infections of the urinary tract in which gram-negative bacteria are predominant. Gram-positive bacteria are less sensitive to the drug. It is rapidly absorbed and excreted when used orally.

Category—antibacterial.

Usual dose—1 g. 4 times daily for 1 to 2 weeks; for additional treatment, may be reduced to 500 mg. 4 times daily.

OCCURRENCE

Nalidixic Acid Tablets N.F.

IODINE-CONTAINING COMPOUNDS

Not many iodine-containing compounds are widely used as antiseptics in medicine today. However, in spite of the wide choice of antiseptics now available, iodine, as a tincture or in aqueous solution with an iodide, is still widely used. It is also an effective virucide, fungicide and amebicide.

Iodoform N.F., Triiodomethane. The haloform synthesis of iodoform (CH_3) from alcohol or acetone, alkali and iodine is also the basis of the sensitive Lieben iodoform test for the presence of CH_3CHOH- or CH_3CO- groups. Commercially, it is prepared by the electrolysis of an aqueous solution of sodium iodide, sodium carbonate and alcohol.

Iodoform is a greenish-yellow, crystalline powder with a penetrating persistent odor. It is slightly volatile at 25°. It melts at 115° and emits vapors of iodine at higher temperature. It is insoluble in water, sparingly soluble in alcohol or glycerin and soluble in ether, chloroform or fixed oils. Its use as a dusting powder on wounds is dependent upon the slow liberation of iodine. An iodoform ointment and an iodoform paste are used. Attempts to disguise its odor with thymol or paraformaldehyde have not prevented its becoming obsolescent. Odorless antiseptic compounds have replaced it largely.

Category—local anti-infective.

Iodophors

Various surfactants will act as solubilizers or carriers for iodine with the resulting complex possessing antibacterial properties. In practice, the nonionic surfactants along with the addition of an acid to stabilize the product and to enhance the antibacterial properties have been most successful.[22] About 80 per cent of the iodine which dissolves in the carrier remains as bacteriologically active or available iodine. Phosphoric acid is used because of its buffering action in the pH range of 3 to 4.[23] Iodophors have been found to be fungicidal, active against tubercle bacilli and effective in moderate concentrations against *Bacillus subtilis*; they show some loss of activity in the presence of serum.[24]

PRODUCTS

Povidone-Iodine N.F., Betadine®, Isodine®, is a complex of iodine with poly(1-vinyl-2-pyrrolidinone). It is water-soluble and releases iodine slowly, providing a nontoxic, nonstaining antiseptic. It contains about 10 per cent of available iodine.

As an aqueous solution it is useful for skin preparation prior to surgery and injections, for the treatment of wounds and lacerations and for bacterial and mycotic infections of the skin.

Category—local anti-infective.

Application—topically.

OCCURRENCE

Povidone-Iodine Aerosol N.F.
Povidone-Iodine Solution N.F.

CHLORINE-CONTAINING COMPOUNDS

N-Chlorocompounds are represented by amides, imides and amidines in which one or more of the hydrogen atoms attached to nitrogen have been replaced by chlorine. All of these products are designed to liberate hypochlorous acid ($HClO$) and, therefore, simulate the antiseptic action of hypochlorites, such as Sodium Hypochlorite Solution N.F.

In contact with water, the N-chlorocompounds slowly liberate hypochlorous acid.

$$\begin{array}{c} O \quad\; H \\ \| \quad / \\ R-C-N \quad + \; HOH \longrightarrow \\ \backslash Cl \qquad\qquad R-C-NH_2 + HClO \\ \qquad\qquad\qquad \| \\ \qquad\qquad\qquad O \end{array}$$

The antiseptic property is greatest at pH 7 and decreases as the solution becomes more alkaline or acidic. It is known that hypochlorous acid will chlorinate amide nitrogen, and it is assumed to attack bacterial protein by this route. Proteins are chlorinated as follows:

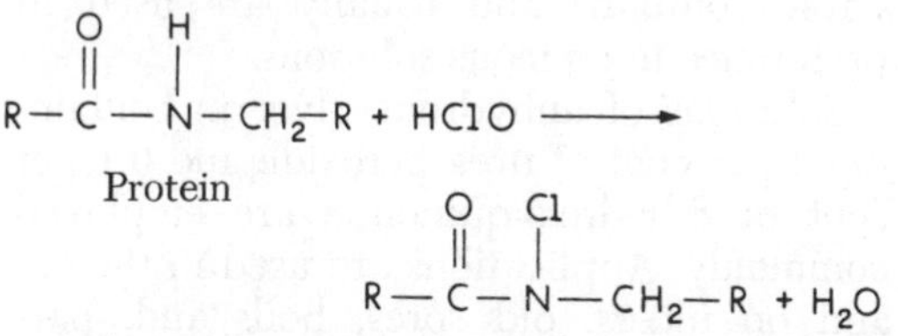

$$R-\overset{\overset{O}{\|}}{C}-\overset{\overset{H}{|}}{N}-CH_2-R + HClO \longrightarrow$$

Protein

$$R-\overset{\overset{O}{\|}}{C}-\overset{\overset{Cl}{|}}{N}-CH_2-R + H_2O$$

The term "active chlorine" is associated with these N-chlorocompounds and hypochlorites, which means the chlorine that is liberated from a substance when treated with an acid.

PRODUCTS

Chloramine-T, Chloramine, Chlorazene, Sodium *p*-toluenesulfonchloramide, occurs as a trihydrate crystalline powder, soluble in water (1:7) and in alcohol but insoluble in ether and chloroform. It has a slight odor of chlorine; on exposure to air, it liberates chlorine and is affected by light. Chloramine-T contains the equivalent of not less than 11.5 per cent and not more than 13 per cent of active chlorine. In solution there is a slow decomposition to yield sodium hypochlorite at a pH between 7 and 8. The solution is alkaline to litmus but does not color phenolphthalein T.S.

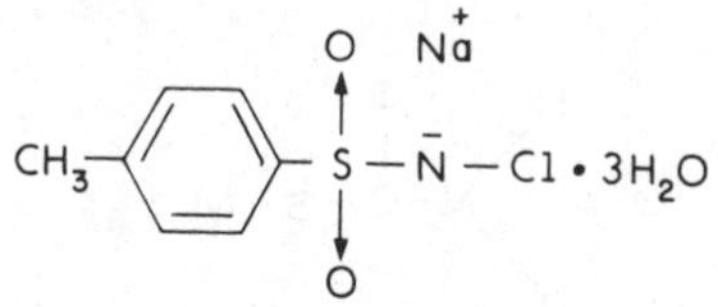

Chloramine-T

Solutions when acidified yield chlorine just as do all hypochlorites. It is used, like the inorganic hypochlorites, as an antiseptic and disinfectant but is less irritant. Note that it is less alkaline than Sodium Hypochlorite Solution N.F. It is applied to mucous membranes as a 0.1 per cent aqueous solution and is used to irrigate or dress wounds as a 1 per cent solution.

Dichloramine-T *p*-Toluenesulfondichloramide occurs as white or greenish-yellow crystals or as a crystalline powder having the odor of chlorine. It is almost insoluble in water but is soluble in alcohol. It is soluble in petroleum benzin (1:1), in chloroform (1:1), in carbon tetrachloride (1:2:5) and in eucalyptol or chlorinated paraffin. Dichloramine-T is used for the same purpose as Chloramine-T. It contains not less than 28 per cent nor more than 30 per cent of active chlorine. A 1 per cent solution in chlorinated paraffin is used for application to mucous surfaces, and a 5 per cent solution in the same solvent is used in dressing wounds.

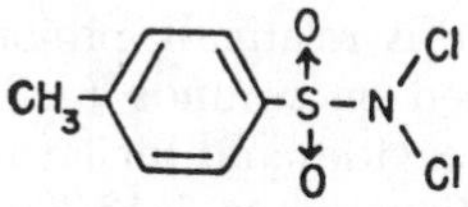

Dichloramine-T

Halazone U.S.P., *p*-Dichlorosulfamoylbenzoic Acid, *p*-Sulfondichloramidobenzoic Acid, *p*-Carboxysulfondichloramide, is a white crystalline powder with a chlorine-like odor. It is affected by light. It is slightly soluble in water and chloroform and is soluble in dilute alkalines. The sodium salt of the compound is used in sterilizing drinking water.

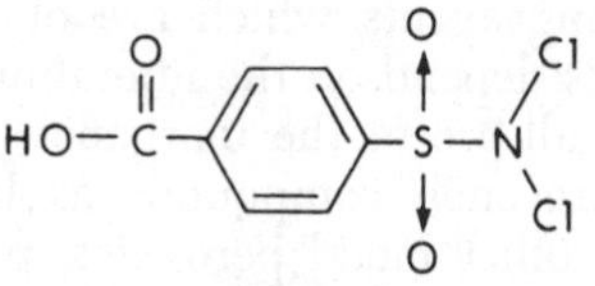

Halazone

Category—disinfectant.
Application—2 to 5 parts per million in drinking water.

OCCURRENCE
Halazone Tablets for Solution U.S.P.

Chloroazodin, Azochloramide®, N,N'-Dichlorodicarbonamidine, contains the equivalent of not less than 37.5 per cent and not more than 39.5 per cent of active chlorine (Cl). It is prepared by treating a solution of guanidine nitrate in dilute acetic acid and sodium acetate with a solution of sodium hypochlorite at 0°.

$$H_2N-\overset{\overset{\displaystyle N^+H_2}{\|}}{\underset{\underset{\displaystyle NH_2}{\diagdown}}{C}} \quad NO_3^- \longrightarrow H_2N-\underset{\underset{\displaystyle NCl}{\|}}{C}-N=N-\underset{\underset{\displaystyle NCl}{\|}}{C}-NH_2$$

It consists of bright yellow needles or flakes with a faint odor of chlorine and a slightly burning taste, and it is explosive at about 155°. It is not very soluble in water or other solvents, including glyceryl triacetate (triacetin), and the solutions decompose on warming or exposure to light.

Chloroazodin is similar to the chloramines and to sodium hypochlorite, but it does not react rapidly with water or reagents, and its action in use is relatively prolonged. Solutions are used on wounds (1:3,300), as a packing for cavities and for lavage and irrigation; dilutions up to 1:13,200 have been proposed for mucous membranes. It often is used in isotonic solutions buffered at pH 7.4. For dressing and packing the stable solution in glyceryl triacetate is employed, and a dilution of this in a vegetable oil (1:2,000) is claimed to be nonirritating to mucous membranes. Tablets of a saline mixture with buffer are available for making solutions.

OXIDIZING AGENTS

Oxidizing agents which are of value as antiseptics depend on the liberation of oxygen, and all are in the inorganic class. Included are such compounds as hydrogen peroxide, other metal peroxides, potassium permanganate and sodium perborate.

Urea Peroxide, Carbamide Peroxide, Perhydrit®, is a stable complex of urea and hydrogen peroxide, $H_2NCONH_2 \cdot H_2O_2$, containing 34 per cent of H_2O_2. This is equivalent to 16 per cent by weight of active oxygen. It occurs as a white crystalline solid that is soluble in water, alcohol, glycerin and propylene glycol and slightly soluble in other organic solvents. The solid complex is reasonably stable stored at room temperature and protected from moisture.

The primary purpose of urea peroxide is to provide a stable form of hydrogen peroxide that upon contact with moisture will liberate oxygen slowly. A 10 per cent aqueous solution decomposes slowly for as long

as 10 days and then deteriorates rapidly. The nonaqueous solvents provide quite stable solutions and usually are used in preference to aqueous solutions.

Solutions of anhydrous glycerin containing 4 per cent of urea peroxide and 0.1 per cent of 8-hydroxyquinoline are employed commonly. Applications are used in the ear and on ulcers, old sores, boils and, particularly, against gram-positive and gram-negative bacteria.[25]

THIURAM DERIVATIVES

Tetramethylthiuram Disulfide. From the study of a series of dithiocarbamate salts, thiuram monosulfides and thiuram disulfides for antibacterial and antifungal activity against human pathogens, it was found that tetramethylthiuram disulfide had a high degree of activity.[26]

Later tests have shown that soap containing 1 per cent of the compound is a highly effective germicidal agent for use on human skin.[27]

Tetramethylthiuram disulfide is also used in a plastic film as a surgical dressing. The aerosol spray is a solution of the antibacterial agent and a methacrylate resin in ethyl acetate with dichlorodifluoro- and trichloromonofluoromethane as the propellant (Rezifilm).

$$\underset{R}{\overset{R}{\diagdown}}N-\overset{\overset{\displaystyle S}{\|}}{C}-S-M$$

Dithiocarbamate salts

$$\underset{R}{\overset{R}{\diagdown}}N-\overset{\overset{\displaystyle S}{\|}}{C}-S-\overset{\overset{\displaystyle S}{\|}}{C}-N\overset{R}{\underset{R}{\diagup}}$$

Thiuram monosulfide

$$\underset{R}{\overset{R}{\diagdown}}N-\overset{\overset{\displaystyle S}{\|}}{C}-S-S-\overset{\overset{\displaystyle S}{\|}}{C}-N\overset{R}{\underset{R}{\diagup}}$$

Thiuram disulfide

Furan Derivatives

Furan or furfuran (from *furfur* = bran) occurs in wood tar and in the products of pyrolysis of many organic materials. It is a colorless liquid with a chloroformlike odor, is slightly soluble in water and boils at about 35°.

It is not used in medicine, but several derivatives are of value. Many natural substances are derivatives of furan, although most of these are γ-lactones whose rings are broken easily by alkalies to furnish salts of γ-hydroxy acids.

Note the relationship of the furfuryl group

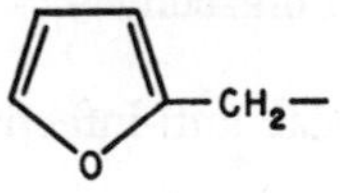

and the furyl group

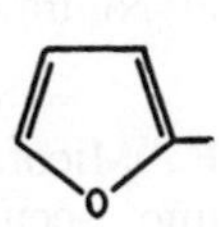

to those of similar structure in the thiophene and benzene series. A furan ring (furyl group) substituted for a phenyl nucleus in a physiologically active compound usually decreases effectiveness. In general, it is less active than compounds containing a thienyl or pyrryl structure.

Furfural (furfurol), the corresponding adehyde, is very important in commerce. It is a colorless, oily liquid that rapidly turns brown to black on standing, has an agreeable odor, is soluble (1:15) in water and boils at about 162°. With reagents, it behaves very much as does benzaldehyde, including the Cannizzaro reaction, the formation of a phenylhydrazone and semicarbazone.

Furfural is manufactured in large quantities by heating any vegetable source of pentosans with dilute acid under pressure. The materials generally used are corn cobs or oat bran, but straw, wood, gums and like materials have enough pentosans to make the process economically profitable.

Furfural is an excellent solvent for resins, for cellulose nitrates and acetates and for gums. It also is used as a fungicide and insecticide, in paint removers, in the manufacture of synthetic resins of the Bakelite type, in shoe dyes and in rubber accelerators.

The partially and completely saturated furan ring is known as dihydrofuran and tetrahydrofuran, respectively. The latter is a good solvent, resembling ethyl ether in its solubilizing properties.

Products

Nitrofurazone N.F., Furacin®, is 5-nitro-2-furaldehyde semicarbazone.

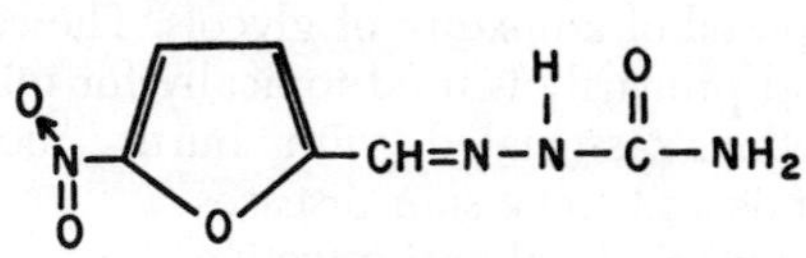

Nitrofurazone

It is an odorless, tasteless, lemon-yellow crystalline solid that is stable at autoclave temperatures for 15 minutes. It decomposes above 227°. In crystalline form or in solution, it darkens on long exposure to light; however, there is no loss in antibacterial activity. Nitrofurazone is very slightly soluble in water and practically insoluble in ether, chloroform and benzene. It is slightly soluble in propylene glycol (1:300,) acetone and alcohol. The best solubility is in the polyethylene glycols. There is no deterioration, either in solution or the dry state. In dispensing preparations of nitrofurazone, light-resistant containers should be used.

Nitrofurazone was first studied in 1944 and was reported to possess good bacteriostatic[28] and bactericidal properties. It is effective against a very wide range of both gram-positive and gram-negative organisms but is not fungistatic. Its action is inhibited by organic matter, such as blood, serum or pus, as well as p-aminobenzoic acid.

Studies on related compounds reveal that no other substitution, either in the 5 or in the 2 position of furan, will reproduce the activity of nitrofurazone. Even analogs of thiophene or pyrrole are inactive. Nitrofurazone is unique, even among other 5-nitrofuran derivatives, in its effect on bacteria. No functional group or specific property has been identified as the key to its activity.

The mode of action of nitrofurazone on the bacterial cell is still obscure. Indications are that it temporarily blocks an energy transfer by the organism necessary for cell division. It is known that the nitro group is reduced, presumably to the 5-hydroxyl-amine (HOHN-), derivative, with total loss of color. The antibacterial action may result from its inhibition of bacterial respiratory enzymes. Since it can be reduced, it may act as a hydrogen acceptor.

Nitrofurazone is available in solutions, ointments and suppositories (usually 0.2%). Water-soluble bases are used which are composed of a mixture of glycols. The compound primarily is used topically for mixed infections associated with burns, ulcers, wounds and some skin diseases.

Category—local anti-infective.

OCCURRENCE

Nitrofurazone Cream N.F.
Nitrofurazone Ointment N.F.
Nitrofurazone Solution N.F.

Nitrofurantoin U.S.P., Furadantin®, 1-[(5-nitrofurfurylidene)amino]hydantoin, is a nitrofuran derivative that is suitable for

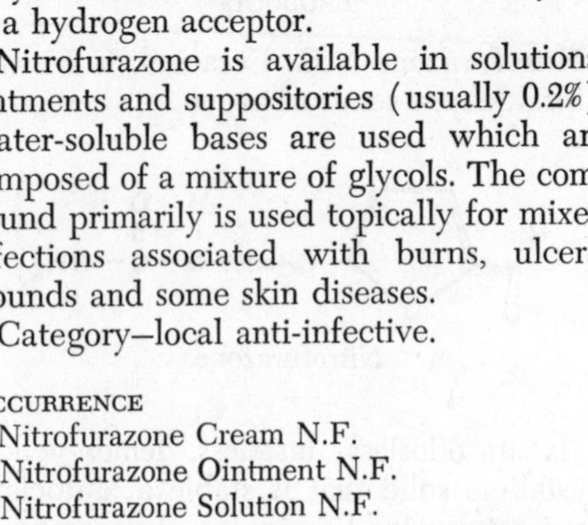

Nitrofurantoin

oral use. The compound has been used successfully in treating infections of the urinary tract. It has been effective for infections that were resistant to antibiotics. Few side-effects, such as diarrhea, pruritus or crystalluria, have been observed.

Category—antibacterial (urinary).
Usual dose—100 mg. 4 times daily.
Usual dose range—200 to 400 mg. daily.

OCCURRENCE

Nitrofurantoin Oral Suspension U.S.P.
Nitrofurantoin Tablets U.S.P.

Furazolidone N.F., Furoxone®, 3-[(5-nitrofurfurylidene)amino]-2-oxazolidinone,

occurs as yellow, odorless, crystalline powder that has a bitter after taste. It is insoluble in water or alcohol. It was found to be effective against a variety of organisms[29] and is used orally in medicine in the treatment of bacterial diarrheal disorders and enteritis. Side-effects, such as nausea and vomiting, may occur, but usually subside when the dosage is reduced.

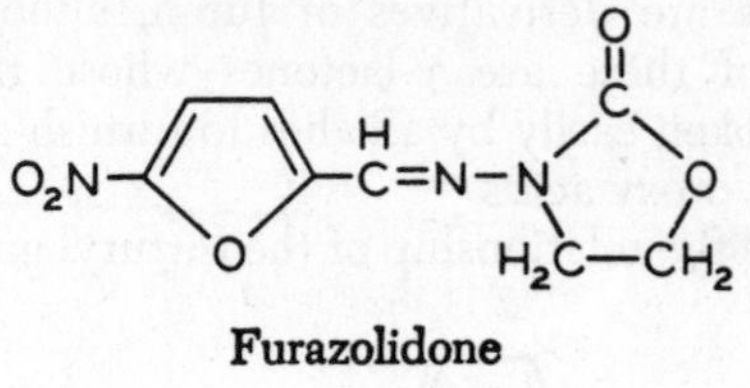

Furazolidone

Category—local anti-infective, local anti-trichomonal.

OCCURRENCE

Furazolidone and Nifuroxime Powder N.F.
Furazolidone and Nifuroxime Suppositories N.F.

Nifuroxime N.F., Micofur®, (Z)5-nitro-2-furaldehyde oxime, occurs as a white to pale yellow crystalline powder when fresh. It darkens upon standing, especially upon contact with bases and metals other than stainless steel or aluminum, and should not be used if darker than a medium tan. It is soluble (1:1000) in water and in alcohol (1:25).

Nifuroxime in combination with furazolidone, as Tricofuron® Vaginal Suppositories and Powder, is used against vaginal infections caused by *Candida albicans* or *Trichomonas vaginalis*.

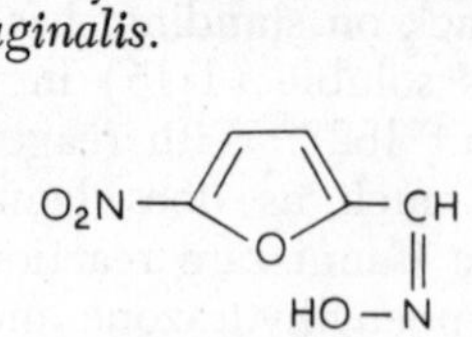

Nitrofuroxime

Category—local anti-infective, local anti-trichomonal.

DYES

The discovery that some dyes would stain certain tissues and not others led Ehrlich to the idea that dyes might be found that would selectively stain, combine with and destroy pathogenic organisms without causing appreciable harm to the host. He and other workers studied a number of dyes with this idea in view and, as a result of these studies, some azo, thiazine, triphenylmethane and acridine dyes came into use as antiseptics and trypanocides and for other medicinal purposes. However, there appears to be no correlation between the dyeing properties of a series of compounds and their antiseptic or bacteriostatic properties.

Prior to the advent of the sulfonamides and the antibiotics the organic dyes were used more extensively as antibacterial agents than they are today. They were used topically for various skin infections. Their chief disadvantage is that they stain the skin and clothing.

The dyes considered in this chapter as well as many of the certified dyes belong to 4 classes: the azo dyes, the acridine dyes, the triphenylmethane dyes and the thiazine dye, methylene blue. They can be further subdivided on the basis of the change on the color nucleus when in aqueous solution. Those that ionize with a negative charge are "acid dyes" and are anionic, while those that ionize with a positive charge are called "basic dyes" in contrast with the acid dyes and are cationic.

The acid dyes are usually sulfonic acids and in the salt form are water-soluble and are generally insoluble in hydrocarbons. They all tend to form slightly water-soluble complexes with the basic or cationic dyes. This may also occur with high molecular weight amine salts. The basic dyes, being cationic, do not combine with metal ions. Metal ions such as Mg^{++}, Ba^{++}, Ca^{++}, Cu^{++} and Fe^{++} will discolor some dyes and may form insoluble precipitates with the acidic dyes. As a general rule, the basic dyes are more resistant to reducing conditions than other dyes. They are considered to be light-sensitive, yet in some cases they may be relatively stable. Light stability of dyes used for coloring sugar-coated tablets is often a problem. The use of insoluble pigments incorporated in a titanium dioxide and syrup suspension will, in many cases, obviate this problem.[30]

Some dyes change color rapidly with the pH and can be used as indicators, while others discolor more slowly and are a stability problem when used as a colorant. Some of the acid dyes may even precipitate at low pH.

Commercial dyes are frequently impure; some of them are mixed with diluents, such as inorganic salts or dextrose. Others may be mixtures of several different colored compounds rather than being composed of one specific compound. Dyes with the same name may vary considerably, depending on the manufacturer.

Some of the confusion in regard to dyes has been removed by standards set up by several official bodies. All dyes used in coloring pharmaceutical products and foods must conform to the Coal Tar Color Regulations established by the United States Food, Drug and Cosmetic Act. Standards for medicinal dyes and food colors also are sanctioned by the *United States Pharmacopeia,* the *National Formulary* and the Dye Certification Division of the United States Department of Agriculture.

Certified colors that are used as colorants in foods and drugs are analyzed and approved by the Food and Drug Administration. To be of certifiable purity each batch of colorant must be virtually free of undesirable by-products and metallic impurities, particularly lead, arsenic and copper. The tolerance for lead is 10 p.p.m. and for arsenic is 1.4 p.p.m. Color certification is controlled to the extent that neither the producer nor the seller may open a container without losing the right to call it certified. If certified dyes are mixed to get a particular shade, the mixture must be recertified. This also applies when repackaged in smaller units without mixing or diluting.

The Food and Drug Administration has classified the certifiable dyes under 3 groups: Group I—Food, Drug and Cosmetic dyes (F. D. & C. dyes), these may be used for coloring foods, drugs and cosmetics; Group II—Drug and Cosmetic dyes (D. & C. dyes), these are designated for use in drugs and cosmetics but not for use in foods; Group III—External Drug and Cosmetic dyes (Ext. D. & C. dyes), these are restricted to use in preparations that will not come in contact with the lips or other mucous membranes and are strictly for use only in externally applied drugs and cosmetics.

Triphenylmethane Dyes

Gentian Violet U.S.P., Pyoktannin®, N,-N, N', N', N'', N''-Hexamethylpararosaniline Chloride, Crystal Violet, Methyl Violet, Methylrosaniline Chloride. The commercial product usually contains small amounts of the closely related compounds, penta- and tetramethylpararosaniline chlorides. Some of the methyl violets of commerce have methyl groups substituted in the ring, and there is considerable lack of uniformity in composition of those being distributed commercially. The pure synthetic crystal violet is presumably free of nuclear methyl groups.

Gentian Violet may be prepared by the reaction shown below.

Gentian Violet occurs as a green powder or as green particles with a metallic luster. The commercial dye frequently contains dextrose and other diluents and should not be used medicinally. It is soluble in water (1:35), in alcohol (1:10) and in glycerin (1:15), but it is insoluble in ether. The dye is much more effective against gram-positive organisms than against gram-negative organisms. It is used topically as a 1 to 3 per cent solution in the treatment of *Monilia albicans* infections, vaginal yeast infections, impetigo and Vincent's angina.

In addition to its use as an antibacterial agent, gentian violet is employed as the dye in indelible pencils. Copying leads contain about 33 per cent of the dye. Eye injuries from indelible pencils are complicated by the toxic effect of the dye which causes local necrosis that may lead to blindness. In making routine examination of such injuries, employing sodium fluorescein, it was observed that the dye surrounding the injured membrane precipitated and could be removed by flushing with the anionic fluorescein solution. By repeated washings, most of the dye can be removed in this manner. At the present time, sodium fluorescein is the agent of choice for treatment of such injuries, although it should be recognized that other anionic agents of high molecular weight may be superior. The mechanism probably involves precipitation of the dye first, fol-

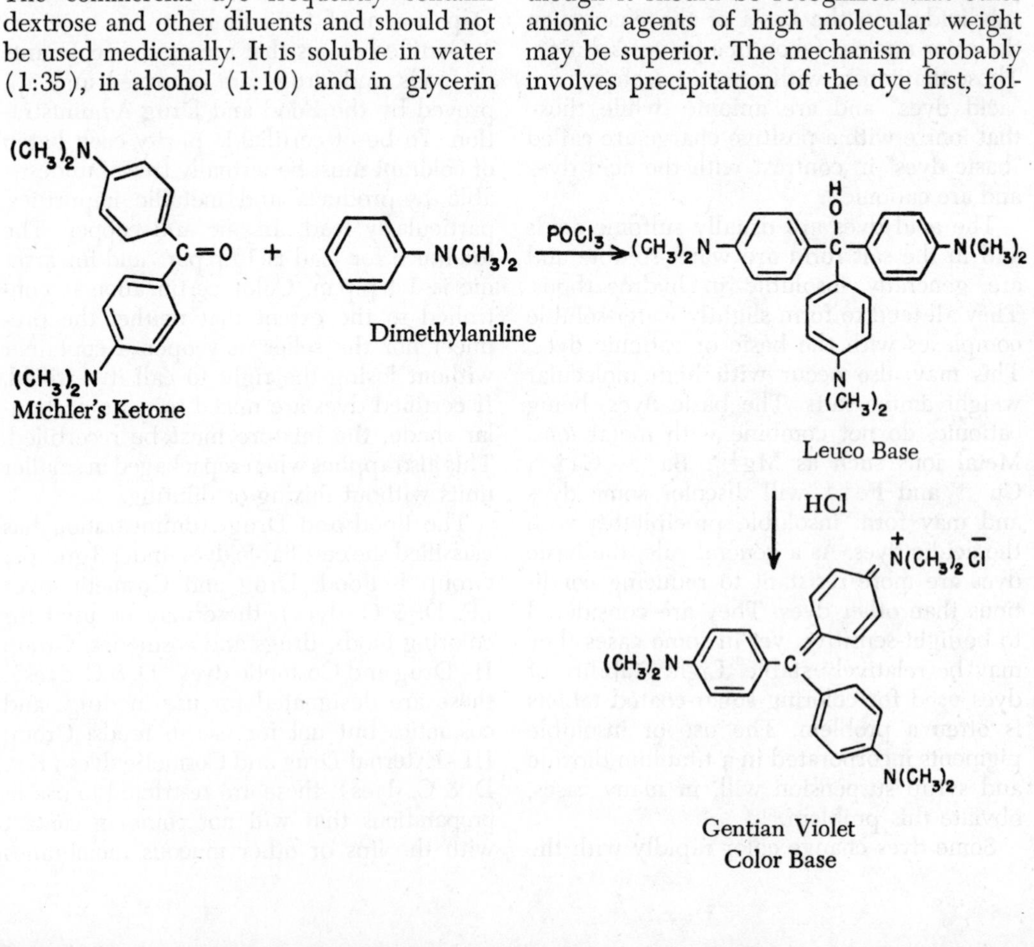

Michler's Ketone   +   Dimethylaniline $\xrightarrow{POCl_3}$ Leuco Base $\xrightarrow{HCl}$ Gentian Violet Color Base

lowed by solubilization with excess of the anionic agent.

Gentian Violet is also used systemically for strongyloidiasis and oxyuriasis (see p. 246).

Category—topical anti-infective.

For external use—topically, in a 1 per cent solution twice daily.

	PER CENT
OCCURRENCE	GENTIAN VIOLET
Gentian Violet Solution U.S.P.	1.0

Basic Fuchsin N.F. is a mixture of the hydrochlorides of rosaniline and pararosaniline. It is a metallic green powder or crystals, soluble in alcohol, with the solution being a carmine red. It is also soluble in water but is insoluble in ether.

Basic Fuchsin is an ingredient of Carbol-Fuchsin Solution N.F. (Castellani's Paint), which is used topically in the treatment of various fungous infections, including ringworm and "athlete's foot."

It is employed also as Schiff's Reagent in testing for aldehydes. This reagent is fuchsin decolorized with sulfur dioxide.

Category—local anti-infective.

Malachite Green, Tetramethyldi-*p*-aminotriphenylcarbinol anhydride chloride, is a green crystalline powder soluble in water and alcohol.

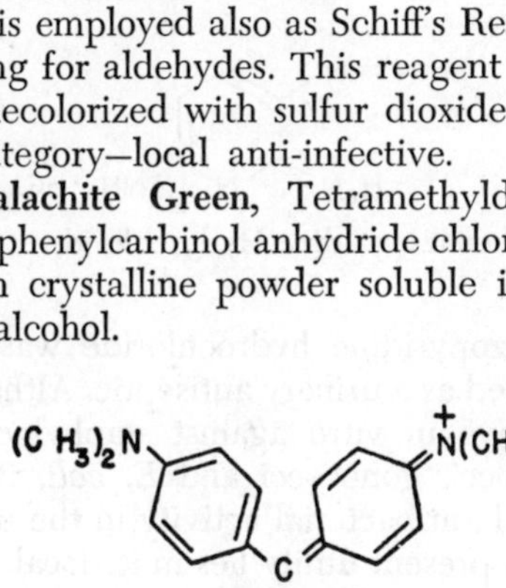

Malachite Green

The commercial dye is used in the form of the $ZnCl_2$ double salt. Only the zinc-free compound should be used medicinally. It has been used in a 1 per cent ointment for the treatment of impetigo and ulcers, in a 1:1,000 water solution as a wound antiseptic and 1:2,000 in normal saline solution by subcutaneous injection for the treatment of trypanosomiasis.

Brilliant Green, Tetraethyldiaminotriphenylcarbinol Anhydride Sulfate. Brilliant green is a green, crystalline powler, soluble in alcohol (1:20) and in water (1:20). It is a homolog of malachite green and can be prepared in the same way by substituting diethylaniline for dimethylaniline. It is used as a surgical antiseptic in a 1:1,000 solution. It is said to promote epithelization.

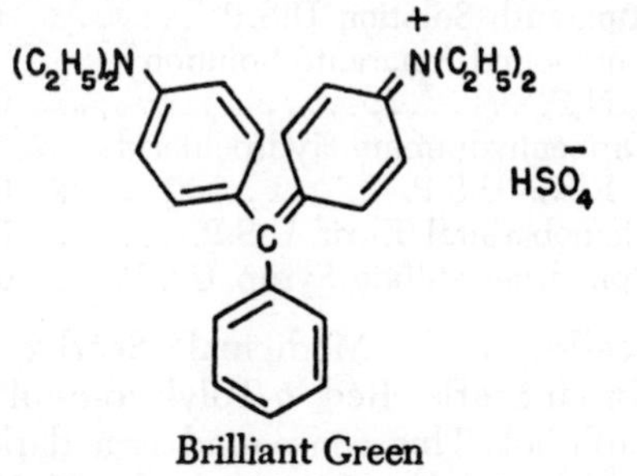

Brilliant Green

It is employed as a 1 to 2 per cent ointment in the treatment of sluggish ulcers.

Acriflavine–Brilliant-Green–Methylrosaniline-Chloride Mixture; Dymixal®. This is a mixture containing 23 per cent of acriflavine, 31 per cent of brilliant green and 46 per cent of methylrosaniline chloride. Such mixtures may be synergistic. It has been used in the treatment of burns.

Azo Dyes

Amaranth U.S.P.; F.D. and C. Red No. 2; Trisodium Salt of 3-Hydroxy-4-[(4-sulfo-1-naphthyl)azo]-2,7-naphthalenedisulfonic

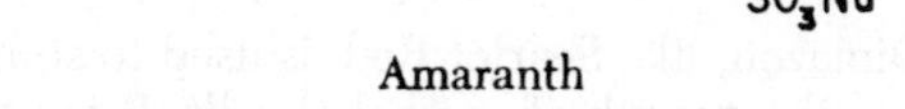

Amaranth

Acid. This compound is a dark, red-brown powder soluble in water (1:15) and very slightly soluble in alcohol. It is used to color pharmaceutical preparations and food products. It is moderately fast to light and in the presence of ferrous ions, but the hue does become darker or duller. It is stable to oxidizing agents but very poorly stable to reducing agents. The dye is reduced by invert sugars such as corn syrup; this reaction is also catalyzed by light.

Category—pharmaceutic aid (color).

OCCURRENCE	PER CENT AMARANTH
Amaranth Solution U.S.P.	1
Compound Amaranth Solution N.F.	0.09
Diphenhydramine Hydrochloride Elixir U.S.P.	0.0016
Phenobarbital Elixir U.S.P.	0.01
Ephedrine Sulfate Syrup U.S.P. . .	0.004

Scarlet Red; Medicinal Scarlet Red; Biebrich Scarlet Red, *o*-Tolylazo-*o*-tolylazo-*β*-naphthol. This compound is a dark red, odorless powder. It is soluble (1:15) in chloroform, readily soluble in oils, fats and phenol, slightly soluble in alcohol, acetone and benzene and almost insoluble in water. The compound is made by coupling diazotized *o*-aminoazotoluene with *β*-naphthol.

It is used to stimulate the growth of epithelial cells in wounds, burns and skin grafting. It usually is applied externally in about a 5 per cent ointment.

Dimazon, Pellidol, 4-Diacetylamino-3-methyl-2′-methyl-azobenzene, is an orange, crystalline powder, readily soluble in alcohol, ether, benzene and mineral oil but insoluble in water. It is prepared by the acetylation of aminoazotoluene.

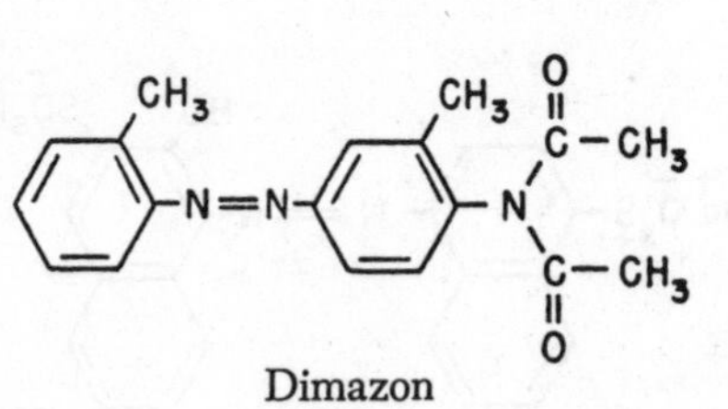

Dimazon

Dimazon, like Scarlet Red, is used to stimulate the growth of epithelial cells. It is used as a 2 per cent solution in petrolatum ointment or in olive oil or as a 5 per cent powder mixed with talc.

Resorcin Brown; D. & C. Brown No. 1; Sodium 4-*p*-Sulfonphenylazo-2-(2,4-xylyl-azo)-1,3-resorcinol. The compound occurs as a deep brown powder, soluble in water, glycerin and alcohol but sparingly soluble in ether and acetone. It is fairly fast to light, and stable toward oxidizing agents, but it is very poorly stable toward reducing agents and is precipitated by ferrous salts.

Resorcin brown is used in coloring drugs and cosmetics.

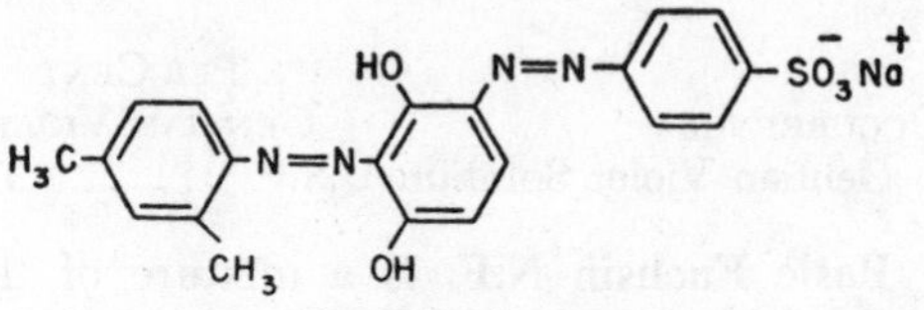

Resorcin Brown

Phenazopyridine Hydrochloride N.F., Pyridium®, 2,3-Diamino-3-(phenylazo)pyridine Monohydrochloride is a brick red, fine crystalline powder. It is slightly soluble in alcohol, in chloroform, and in water.

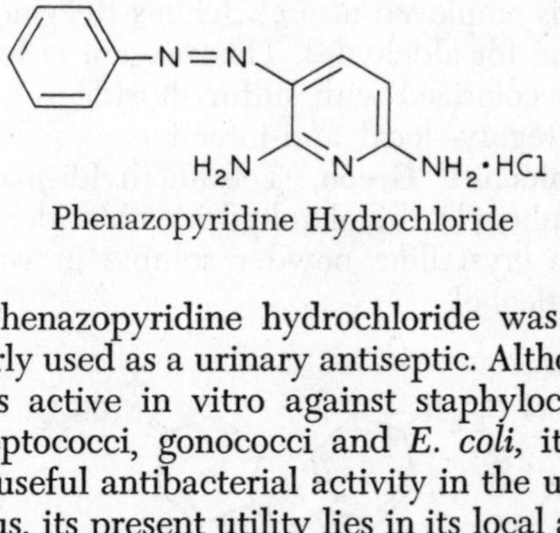

Phenazopyridine Hydrochloride

Phenazopyridine hydrochloride was formerly used as a urinary antiseptic. Although it is active in vitro against staphylococci, streptococci, gonococci and *E. coli*, it has no useful antibacterial activity in the urine. Thus, its present utility lies in its local analgesic effect on the mucosa of the urinary tract. It is now usually given in combination with urinary antiseptics. The drug is rapidly excreted in the urine, to which it gives an orange-red color. Stains in fabrics may be removed by soaking in a 0.25 per cent solution of sodium dithionate.

Category—analgesic (urinary tract).

Usual dose—100 mg. 3 or 4 times daily.

OCCURRENCE
Phenazopyridine Hydrochloride Tablets N.F.
Phenazopyridine Hydrochloride and Sulfisoxazole Tablets N.F.

Ethoxazene Hydrochloride, Serenium®, 4-[(p-Ethoxyphenyl)azo]-*m*-phenylenediamine Hydrochloride, is a dark red, slightly bitter powder, soluble in boiling water (1.100).

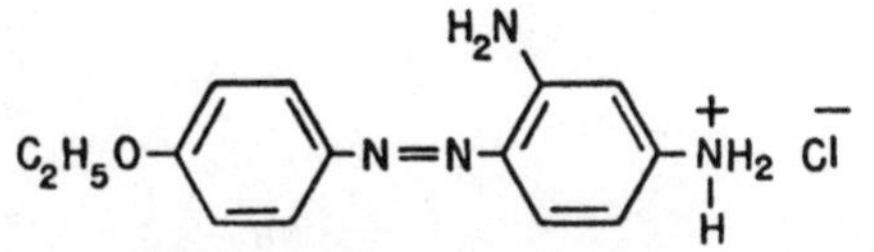

Ethoxazene Hydrochloride

Ethoxazene hydrochloride parallels pyridium hydrochloride in its properties and uses. It has no useful antibacterial action in the urine; it is used for its local analgesic action on urinary tract mucosa and discolors the urine orange-red. It is given orally, with the usual adult dose being 100 mg. 3 times daily before meals.

Acridine Dyes

Acridine has the same relation to anthracene that pyridine has to benzene. It may be considered to be derived from anthracene if one of the middle CH groups is replaced by nitrogen. The numbering on the inside of the formula below is that used by Chemical Abstracts; the outside numbers are those employed by the British. Thus, 9-aminoacridine in the Chemical Abstracts system and 5-aminoacridine in the British system are the same compound. The system used here is that adopted by Chemical Abstracts.

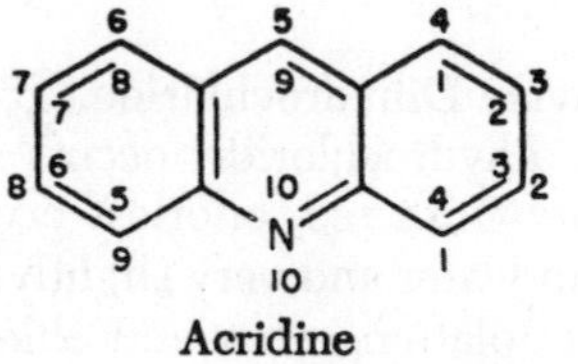

Acridine

Browning was the first to observe (in 1912) that one of the aminoacridines "trypaflavine" (later called acriflavine) was active against trypanosomes. The next year, this compound and proflavine were introduced as general antiseptics and were used extensively in World War I. In 1923, Rivanol, a diaminoethoxy derivative, was introduced. The work of Albert and co-workers[31] on the acridines has led to the discovery that the antibacterial action of these compounds is proportional to the extent to which they are ionized at the biologic pH value 7.3. Evidence also was presented that the cations of the acridine derivatives injure the bacteria by competing with the hydrogen ion, which functions as an essential bacterial cation. As a result of this work, 9-aminoacridine has come into use as an externally applied, nonstaining antibacterial agent. It is not useful for injection, because it is precipitated by saline solutions. 4-Methyl-9-aminoacridine has been introduced as an acridine derivative that is compatible with physiologic saline solution and, therefore, has the possibility of being used by injection. Most of the acridines, however, have a necrotic and inflammatory action on tissues and so are not used intramuscularly or subcutaneously. The aminoacridines maintain their activity in the presence of pus and the body fluids and are claimed to interfere very little with the healing process. However, acriflavine is said to inhibit leukocytic activity in concentrations of 20 per cent of that required to inhibit the growth of bacteria in the bloodstream. Because they, like most cations, are not nearly as active in acid solution as in neutral or basic surroundings, they are effective urinary antiseptics only if the urine is basic. The reason for this is presumably their mode of action, which involves a competition with hydrogen ion. The greater the concentration (law of mass action) of hydrogen ions, the less chance the aminoacridine ions have of displacing the hydrogen ions.

Since some of these compounds have high activity against gram-negative organisms and are more active than the sulfonamides against gram-positive organisms, they may possibly come into more extensive use.

Products

9-Aminoacridine Hydrochloride, Monacrin®, 5-Aminoacridine Hydrochloride, Acramine Yellow, is a pale yellow, crystalline, bitter, odorless powder. It is soluble in

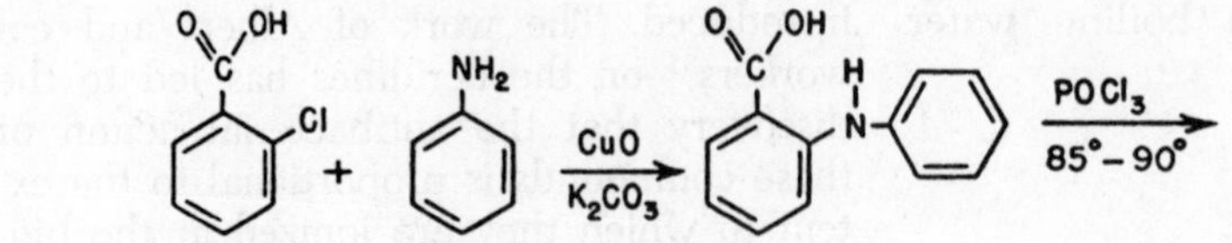

o-Chloro-   Aniline
benzoic
Acid

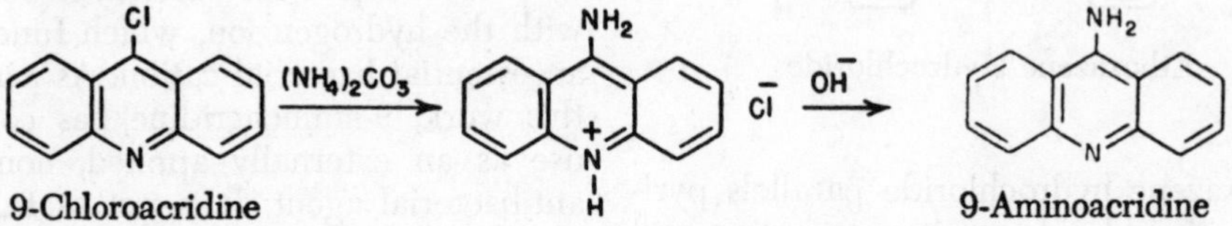

9-Chloroacridine   9-Aminoacridine

water (1:300), in 90 per cent alcohol (1:150) and in glycerol (1:55).

The synthesis shown above of 9-aminoacridine serves to illustrate the synthesis of the acridine antibacterial agents.

The compound is incompatible with acids, alkalis and chlorides. It is not suitable for injection, partially because of the chloride incompatibility, but it is an effective bacteriostatic and bactericidal agent for topical application.

The bacteriostatic activity of the sulfonamides is enhanced when used with 9-aminoacridine. There is approximately a tenfold increase in the activity with the mixed drugs.

Acrisorcin, Akrinol®, 9-Aminoacridinium 4-Hexylresorcinolate. This compound is prepared from 9-aminoacridine and 4-hexylresorcinol and occurs as yellow crystals

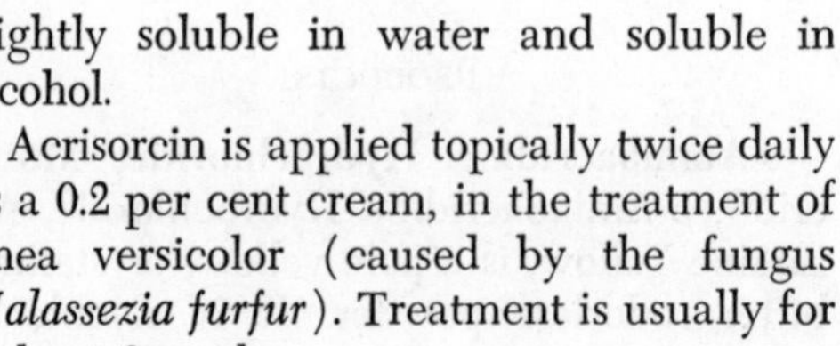

Acrisorcin

slightly soluble in water and soluble in alcohol.

Acrisorcin is applied topically twice daily as a 0.2 per cent cream, in the treatment of tinea versicolor (caused by the fungus *Malassezia furfur*). Treatment is usually for at least 6 weeks.

9-Amino-4-methylacridine Hydrochloride, 5-Amino-1-methylacridine, Neomonoacrin®, Salacrin®, is a yellow, crystalline, bitter powder that is not affected by light. It is soluble in water (1:220) and in absolute alcohol (1:380). Dilute solutions of the compound exhibit a strong, blue fluorescence. This compound is one of the least toxic of some 30 9-aminoacridine derivatives that have been tested. It is nonstaining to skin, is a more active antibacterial agent than 9-aminoacridine and is compatible with physiologic saline solution.

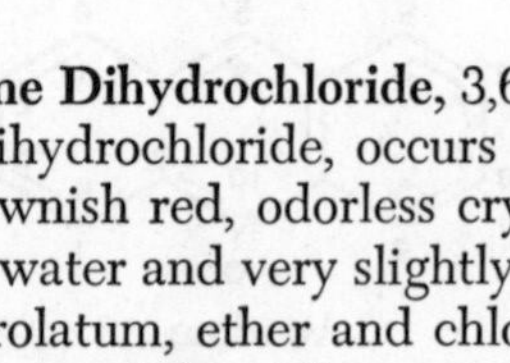

**4-Methyl-9-aminoacridine
Hydrochloride**

Proflavine Dihydrochloride, 3,6-Diaminoacridine dihydrochloride, occurs as orangered to brownish red, odorless crystals. It is soluble in water and very slightly soluble in liquid petrolatum, ether and chloroform. It is sensitive to light. The water solutions sometimes become turbid, in which case they should be discarded.

Proflavine Sulfate, 3,6-Diaminoacridine sulfate, occurs as a reddish brown, odorless powder. It is affected by light and is less soluble in water (1:300) than the corresponding dihydrochloride. It is slightly soluble in alcohol and nearly insoluble in ether, liquid petrolatum and chloroform.

Acriflavine; Acriflavine Base; Neutral Acriflavine; 3,6-Diamino-10-Methylacridinium Chloride. The compound is a deep orange, odorless, granular powder. It is soluble in water (1:3), sparingly soluble in alcohol and nearly insoluble in ether and chloroform. Its solutions exhibit a marked fluorescence and are sensitive to light. The commercial medicinal product is mixed with 3,6-diaminoacridine. Because this compound

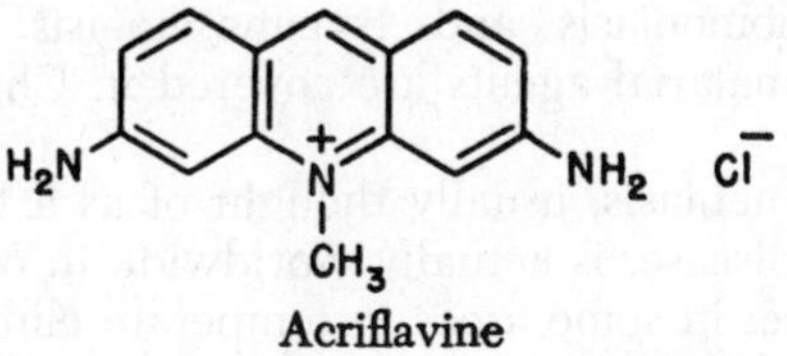

Acriflavine

is more nearly neutral, it is less irritating to tissues than the acriflavine hydrochloride.

Category—antibacterial. Usual dose—locally in 1:1,000 solution; for irrigation in 1:1,000 to 1:10,000 solution.

Acriflavine Hydrochloride, Trypaflavine, 3,6-Diamino-10-methylacridinium chloride

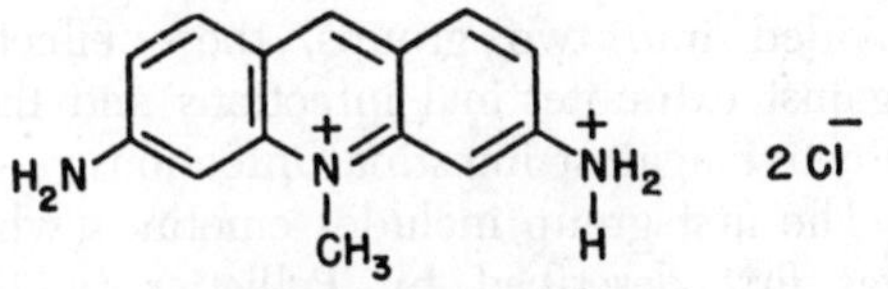

Acriflavine Hydrochloride

hydrochloride, is a reddish brown, odorless, crystalline powder. It is soluble in water but nearly insoluble in ether, chloroform and liquid petrolatum. The solutions of the hydrochloride are acidic, while those of acriflavine base are slightly basic. For some purposes, such as use on mucous surfaces, solutions of the base appear to be less irritating than those of the salt.

4,5 - Dimethylproflavine Hydrochloride; 3,6 - Diamino - 4,5 - dimethylacridine Hydrochloride; 1-9-Dimethylproflavine Hydrochloride; 2,8-Diamino-1,9-Dimethylacridine hydrochloride. This is a brick red powder, soluble in water, sparingly soluble in alcohol and practically insoluble in ether.

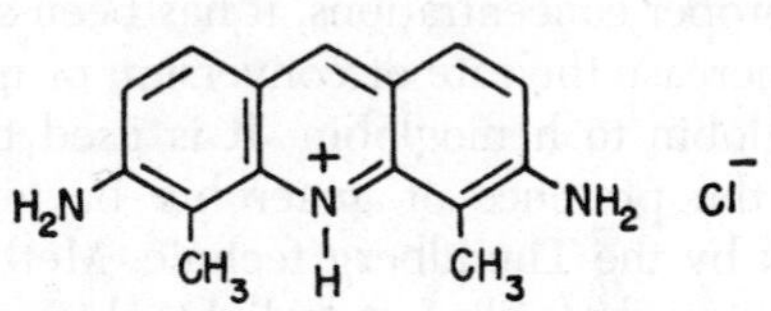

4,5-Dimethylproflavine Hydrochloride

It is made from 2,6-diaminotoluene and anhydrous formic acid. The compound stains cloth and tissues. It has a high bacteriostatic index.

Methylene Blue

Methylene Blue U.S.P., 3,7-bis(dimethylamino)phenazathionium chloride, occurs as dark green crystals or powder with a bronze luster. It is soluble in chloroform, in water (1:25) and in alcohol (1:65). Its solutions may be sterilized by autoclaving.

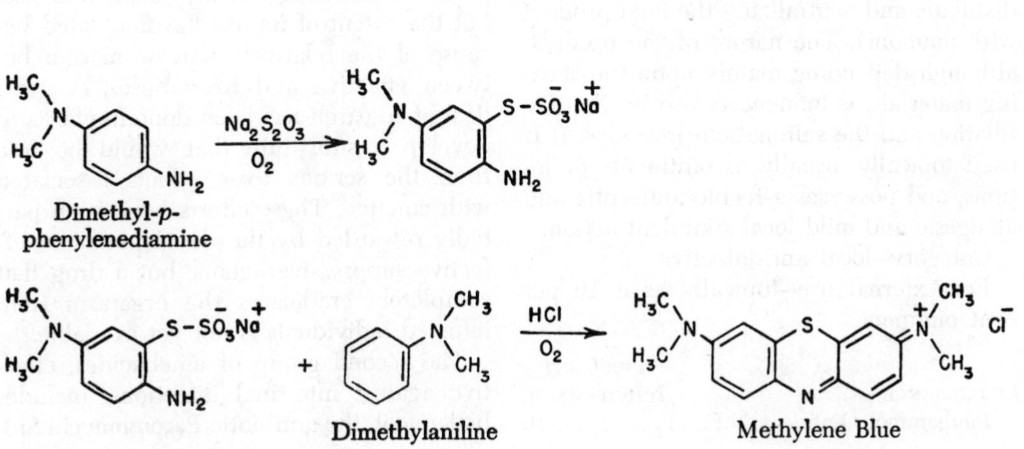

Dimethyl-*p*-phenylenediamine

Dimethylaniline   Methylene Blue

Methylene blue may be synthesized as shown on the preceding page.

It has a comparatively low toxicity and is used to test the renal function of the kidneys and also as a dye in vital nerve staining. It has some action against malaria but is inferior to the cinchona alkaloids, quinacrine and some of the new synthetics in this respect. It is a weak antiseptic that has been used in treating skin diseases and some urinary conditions. Methylene blue also is employed in the treatment of cyanosis resulting from the sulfonamide drugs and as an antidote for cyanide and nitrate poisoning. In proper concentrations, it has been shown to increase the rate of conversion of methemoglobin to hemoglobin. It is used to test for the presence of anaerobic bacteria in milk by the Thundberg technic. Methylene blue is only fairly fast to light, shows moderate stability to oxidizing and reducing agents and good stability to ferrous ions.

Category—antimethemoglobinemic; antidote to cyanide poisoning.

Usual dose—intravenous, 50 ml. of 1 per cent solution.

OCCURRENCE
Methylene Blue Injection U.S.P.

SULFUR COMPOUNDS

Ichthammol N.F., Ammonium Ichthosulfonate, Isarol®, Ichthyol®, Ichthymall®, is obtained by the destructive distillation of certain bituminous schists, sulfonating the distillate, and neutralizing the final product with ammonia. The nature of the product, although depending mainly upon the starting materials, is influenced also by the distillation and the sulfonation processes. It is used topically, usually as ointments or lotions, and possesses a feeble antiseptic and analgesic and mild local stimulant action.

Category—local anti-infective.

For external use—topically as a 10 per cent ointment.

	PER CENT
OCCURRENCE	ICHTHAMMOL
Ichthammol Ointment N.F.	10

ANTIPROTOZOAL AGENTS

Diseases caused by protozoa, especially in the United States and other countries in the temperate zone, are not as widespread as bacterial and viral diseases. Protozoal diseases are more prevalent in the tropical countries of the world, where they occur both in man and in livestock, causing suffering, death and great economic loss. The main protozoal diseases in humans in the United States include malaria, amebiasis, trichomoniasis and trypanosomiasis. The antimalarial agents are covered in Chapter 13.

Amebiasis, usually thought of as a tropical disease, is actually worldwide in occurrence; in some areas in temperate climates, where sanitary conditions are poor, the incidence may be 20 per cent or more. An ideal chemotherapeutic agent against amebiasis would be effective against the causative organism, *Entamoeba histolytica*, irrespective of whether it occurs in the lumen of the colon, the wall of the colon, or extraintestinally, in the liver, lung or other organs, but such an agent is yet to be discovered. The presently available agents are divided into two groups, those effective against extraintestinal infections and those effective against intestinal infections.

The first group includes emetine (which was first described by Pellietier in 1817 and reported to be of value in the chemotherapy of acute amebic dysentery in 1912), and the antimalarial drugs, chloroquine and amodiaquine. After the 1912 report, emetine was quickly taken into use, but the extent of its use has fluctuated because of the relatively narrow margin between effective and toxic doses. A great deal of research has been done in efforts to develop a substitute that would be free from the serious toxic effects associated with emetine. These efforts have been partially rewarded by the development of effective suppressive agents, but a drug that completely eradicates the organism from infected individuals is not yet available.

The second group of amebacides, effective against intestinal infections, includes bialamicol, the antibiotic Paromomycin Sulfate N.F. (see Chap. 14); the arsenicals, Carbarsone N.F. (see p. 230) and Glycobiarsol N.F. (see p. 231); and 3 derivatives of 8-hydroxy-7-iodoquinoline—chiniofon, Iodochlorhydroxyquin U.S.P. and Diiodohydroxyquin U.S.P. (see Chap. 10).

Trichomoniasis, caused by *Trichomonas vaginalis*, is common in the United States. Although it is often considered to be a relatively unimportant affliction, it causes serious physical discomfort and, sometimes, marital problems because of its disruptive effect on sexual relations. Chemotherapeutic research in the past 20 years has led to marked progress in knowledge of the biologic properties of the causative organism, accurate testing methods have been devised and highly effective compounds with systemic activity have been discovered.

Heterocyclic nitro compounds such as Furazolidone N.F., a nitrofuran (see p. 206), and Metronidazole U.S.P., a nitroimidazole, show great promise as drugs in the chemotherapy of human trichomoniasis. Other trichomonacides include Carbarsone N.F. (see p. 230) and the 8-hydroxy-7-iodoquinolines Iodochlorhydroxyquin U.S.P. and Diiodohydroxyquin U.S.P. (see Chap. 10).

Trypanosomiasis, caused by pathogenic members of the family *Trypanosonidae*, occurs both in man and in livestock. The main disease in man, sleeping sickness, is of minor consideration in the United States.

PRODUCTS

Emetine Hydrochloride U.S.P. The nonphenolic alkaloid, emetine, is obtained either by isolation from natural sources or synthetically by methylating naturally occurring cephaëline (phenolic). Emetine is obtained from the crude drug by first extracting the total alkaloids with a suitable solvent and then separating them by a method similar to that outlined for the separation of phenolic and nonphenolic bases. The free base was isolated first by Pelletier and Caventou in 1817.

The free base is levorotatory and occurs as a water-insoluble, light-sensitive, white powder. It is soluble in alcohol or the im-

Emetine Hydrochloride

miscible solvents. It contains 2 basic nitrogens and forms salts quite readily. The *U.S.P.* sets limits for the water of hydration content (8 to 14%). Other than the hydrochloride, the hydrobromide and the camphosulfonate (as a solution) sometimes are used.

The hydrochloride occurs as a white or very slightly yellowish, odorless, crystalline powder. The salt is freely soluble in water (1:4) and in alcohol. Its solutions have an approximate pH of 5.5 but, when prepared for injection, they should be adjusted to a pH of 3.5. Sterilization of solutions may be effected by bacteriologic filtration. Solutions are light-sensitive and should be preserved in light-resistant containers.

As the name implies, emetine possesses emetic action, due to its marked irritation of mucous membranes when ingested orally. However, it is used principally for its amebicidal qualities. Considerable research has shown that, while emetine causes prompt recession of the symptoms of acute intestinal amebiasis, it cures only 10 to 15 per cent of the cases, and is now considered to be the least valuable agent for curing the disease.

The recession of symptoms quite often leads patients to believe that they are cured, although they are still carriers. Therefore, emetine probably is used best for symptomatic control of acute amebic dysentery and should be supplemented by other more effective drugs. However, emetine is said by some investigators to be the only amebicide of value in amebic abscess or amebic hepatitis.

Category—antiamebic.

Usual dose—subcutaneous, 1 mg. per kg. of body weight, but not exceeding 65 mg., daily for 10 days.

OCCURRENCE
Emetine Hydrochloride Injection U.S.P.

Bialamicol Hydrochloride, Camoform® Hydrochloride, is 6,6'-diallyl-α,α'-bis(diethylamino)-4,4'-bi-o-cresol dihydrochloride.

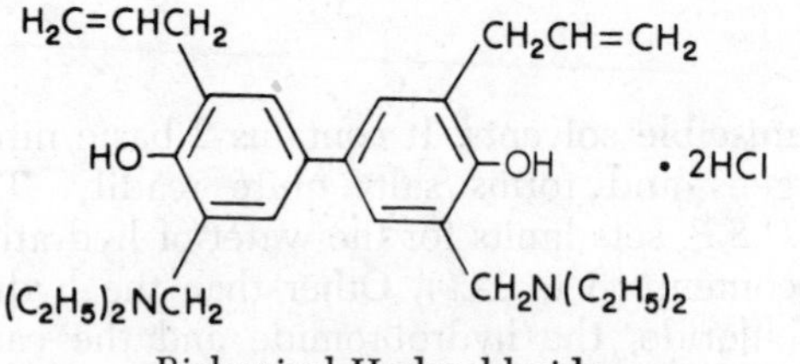

Bialamicol Hydrochloride

This drug is used for the treatment of intestinal amebiasis. It is given orally and should be given with food to minimize gastric irritation The adult dosage is 250 to 500 mg. 3 times daily for 5 days. If a second treatment is required, 3 weeks should elapse before it is begun.

Metronidazole U.S.P., Flagyl®, 2-methyl-5-nitroimidazole-1-ethanol, is a pale yellow, crystalline compound which has limited solubility in water yet is adequately absorbed after oral administration. It is stable in air but darkens on exposure to light. Clinical and experimental studies have shown it to be an effective trichomonacidal agent. After oral administration, the serum and urine levels reach their peaks in about 2 to 3

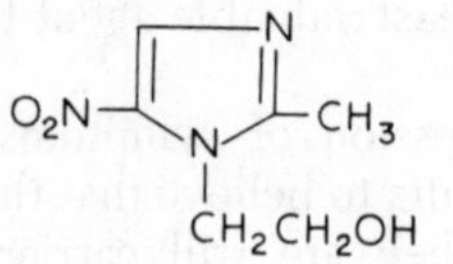

Metronidazole

hours. Darkened urine may occur when the drug is given in doses higher than those generally recommended. The pigment responsible for the darkened urine has not been positively identified, but is probably a metabolite of metronidazole. The darkened urine appears to have no clinical significance.

Category—antitrichomonal.
Usual dose—oral, 250 mg. 3 times daily in the female, or twice daily in the male, for 10 days; vaginal, 500 mg. daily in addition to 250 mg. orally twice daily.

OCCURRENCE
Metronidazole Suppositories U.S.P.
Metronidazole Tablets U.S.P.

Hydroxystilbamidine Isethionate U.S.P.; 2-Hydroxy-4,4'-stilbenedicarboxamidine diisethionate; 2-hydroxy-4,4'-diamidinostilbene. This consists of yellow crystals which are stable in air but are light-sensitive. The pH of a 1 per cent aqueous solution is about 4. Solutions for medicinal use should be freshly prepared and free of any cloudiness. The solution when given by intravenous infusion should be carefully protected from light.

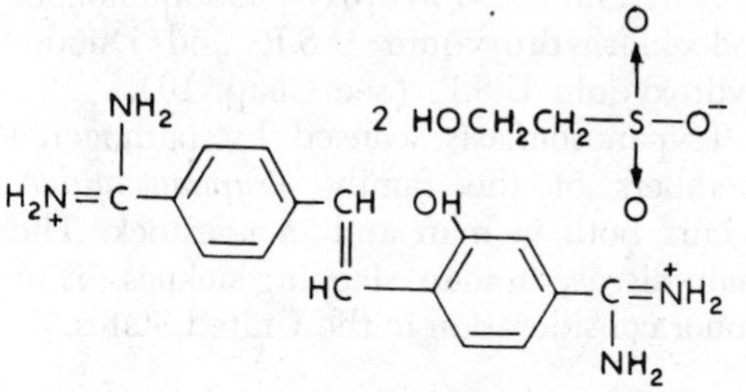

Hydroxystilbamidine Isethionate

Category—antileishmanial.
Usual dose—intravenous infusion, 225 mg. once daily.
Usual dose range—150 to 225 mg. daily every other day, to a total of 6 to 7 g.

OCCURRENCE
Sterile Hydroxystilbamidine Isethionate U.S.P.

Lucathone Hydrochloride U.S.P., 1-{[2-(diethylamino)ethyl]amino}-4-methylthioxanthen-9-one hydrochloride, occurs as a yellowish orange, almost odorless powder, with a bitter taste which is followed by a burning sensation. It is used orally in the treatment of schistosomiasis.

Lucanthone Hydrochloride

The drug is readily absorbed after oral administration. After administration has been discontinued it takes about 3 days for complete elimination from the body. Duration of therapy is usually about 7 days.

Category—antischistosomal.

Usual dose—5 mg. per kg. of body weight 3 times daily for 1 week.

Usual dose range—10 to 20 mg. per kg. of body weight daily for 1 to 3 weeks.

OCCURRENCE
Lucanthone Hydrochloride Tablets U.S.P.

ANTIVIRAL AGENTS

The chemotherapy of viral disease is today at about the same stage of development as was the chemotherapy of bacterial infections prior to the discovery and development of the sulfonamides. Viral diseases such as smallpox and poliomyelitis are at present controlled by public health measures and immunization. With few exceptions, treatment of viral diseases consists of making the condition tolerable for the patient and ensuring that a secondary bacterial infection does not develop.

The two major obstacles to effective antiviral chemotherapy are, first, the close relationship that exists between the multiplying virus and the host cell and, second, the fact that many viral-caused diseases can be diagnosed and recognized only after it is too late for effective treatment. In the first case, an effective antiviral agent must prevent completion of the viral growth cycle in the infected cells without being toxic to the surrounding normal cells. One encouraging development is the discovery that some virus-specific enzymes are elaborated during multiplication of the virus particles and this may be a point of attack by a specific enzyme inhibitor. However, recognition of the disease state too late for effective treatment would render antiviral drugs useless, even if they were available. Thus, until early recognition of the impending disease state is provided, most antiviral chemotherapeutic agents will have their greatest value as prophylactic agents.

From a chemotherapeutic standpoint, viruses that infect animals can be divided into 2 groups. The first and smaller group are the rickettsia and the large viruses (both held by some investigators not to be true viruses) which are more or less effectively controlled by some of the sulfonamides (see Chap. 11) and antibiotics (see Chap. 14). The larger group (true viruses), with the notable exception of the herpes simplex virus, cannot be controlled by chemotherapeutic agents. In 1963 a chemical agent which acts as an antimetabolite was made available commercially. Otherwise, only when secondary bacterial infection is present can the course of disease caused by true viruses be shortened or modified. Most attempts to inhibit virus multiplication without causing damage to the host have been unsuccessful, probably because virus multiplication is so intimately dependent on host cell metabolism.[32]

Idoxuridine U.S.P., Stoxil®, Dendrid®, Herplex®, is 2′-deoxy-5-iodouridine. It is slightly soluble in water and insoluble in chloroform or ether. This product has been found to be an effective antiviral agent in the treatment of dendritic keratitis caused by herpes simplex. Until this discovery was made there was no satisfactory chemotherapy for this infection.

Aqueous solutions are slightly acidic in reaction and are stable for one year if refrigerated.[33] At room temperature there is up to 10 per cent loss in 1 year. Solutions may not be sterilized by autoclaving and must be dispensed in amber bottles to protect from light. The ointment does not re-

Idoxuridine

quire refrigeration and is stable for at least 2 years.

Category—antiviral (ophthalmic).

For external use—topically, as a 0.5 per cent ointment 4 to 6 times daily, or as a 0.1 per cent solution every 1 to 2 hours, to the conjunctiva.

	PER CENT IDOXURIDINE
OCCURRENCE	
Idoxuridine Ophthalmic Ointment U.S.P.	0.5
Idoxuridine Ophthalmic Solution U.S.P.	0.1

Amantadine Hydrochloride N.F., Symmetrel®, 1-adamantanamine hydrochloride, is a white crystalline powder, freely soluble in water and insoluble in alcohol or chloroform; it has a bitter taste. It is useful in the

NH₂ · HCl

Amantadine Hydrochloride

prevention but not the treatment of influenza caused by the A₂ strains of the Asian influenza virus. Aside from vaccination, it is the only prophylactic presently available against any strain of Asian influenza. It appears to exert its effect by preventing penetration of the adsorbed virus into the host cell. It has no therapeutic value, and is in-

effective once the virus has penetrated the host cell. Compared to vaccination, amantadine has two advantages: it is oral medication, and it provides immediate protection. Its chief drawbacks are that protection stops shortly after daily dosage stops and the protection is against only the A₂ strains of the virus.

A recent empirical observation is that the drug may be useful against the crippling disabilities of Parkinson's disease.

Category—antiviral (prophylactic).

Usual dose—200 mg. daily as a single dose or in 2 divided doses.

OCCURRENCE
Amantadine Hydrochloride Capsules N.F.
Amantadine Hydrochloride Syrup N.F.

ANTINEOPLASTIC AGENTS

A tremendous amount of work has been done in the study and the screening of compounds against neoplastic diseases, but it appears that, before a real "breakthrough" can be made, reliable diagnostic tests for the early detection of cancer in man must be developed and, on a more fundamental level, discovery must be made of the origins of spontaneous cancer.

At present there is no known chemical compound which will cure any form of cancer. Therapy is still limited largely to surgery and treatment with ionizing radiation. Nevertheless, there are many anti-cancer agents which are able to produce relief of pain, significant increase in survival time, prevention of metastases following surgery and at least temporary disappearance of tumors. These agents also are capable of producing temporary regressions of certain neoplasms which are not amenable to surgery or radiation. The effective dose is often very close to the toxic dose and all of the drugs have rather low therapeutic indexes. In order to obtain the maximum therapeutic response the drug frequently is administered to the point of systemic toxicity, in the hope that the malignant cells will recover more slowly than the normal cells. Then too, each succeeding course of treat-

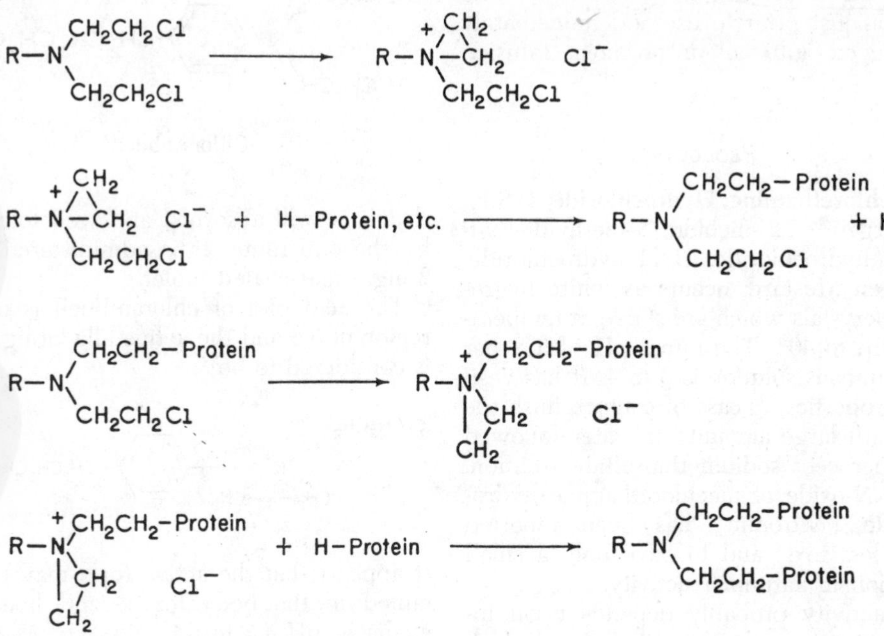

$$R-N\begin{array}{c}CH_2CH_2Cl\\CH_2CH_2Cl\end{array} \longrightarrow R-N^+\begin{array}{c}CH_2\\|\\CH_2\\CH_2CH_2Cl\end{array} Cl^-$$

$$R-N^+\begin{array}{c}CH_2\\|\\CH_2\\CH_2CH_2Cl\end{array} Cl^- + H-Protein, etc. \longrightarrow R-N\begin{array}{c}CH_2CH_2-Protein\\CH_2CH_2Cl\end{array} + H^+$$

$$R-N\begin{array}{c}CH_2CH_2-Protein\\CH_2CH_2Cl\end{array} \longrightarrow R-N^+\begin{array}{c}CH_2CH_2-Protein\\|\\CH_2\\CH_2\end{array} Cl^-$$

$$R-N^+\begin{array}{c}CH_2CH_2-Protein\\|\\CH_2\\CH_2\end{array} Cl^- + H-Protein \longrightarrow R-N\begin{array}{c}CH_2CH_2-Protein\\CH_2CH_2-Protein\end{array}$$

ment is likely to be less effective than the one before.

The chemical compounds used in the chemotherapy of cancer can be divided into: (1) alkylating agents—the nitrogen mustards and related compounds; (2) antimetabolites—folic acid analogs and purine and pyrimidine analogs (see Chap. 27); (3) hormones—ACTH and cortisone and its congeners, and various estrogens and androgens (see Chap. 25); and (4) a miscellaneous group which includes urethane and alkaloids from *Vinca rosea*.

Mustard gas [bis(β-chloroethyl)sulfide], used during World War I, was found to be of some value in the localized lesions of skin cancer. However, the high toxicity, the low solubility in water and its vesicant properties prevented its clinical application. Nitrogen analogs of mustard gas have been found to be easier to handle because the hydrochloride or other salts which were stable solids with a high water-solubility could be formed.

It was discovered that nitrogen mustard (Mechlorethamine) was an effective antineoplastic agent, and, since then, a large number of related compounds and other alkylating agents have been synthesized and evaluated. The antineoplastic activity of the nitrogen mustards has generally been considered to be a function of their property of cyclizing in water to the highly reactive ethyleneimonium ions which react with compounds containing replaceable hydrogens, such as the free amino groups and other groups of proteins, of nucleic acids and others. The nitrogen mustards are relatively stable in the salt form but cyclize rather rapidly in dilute aqueous solutions and at physiologic pH. In the absence of any other reactant they will react with water, with the resulting compound possessing no biologic activity.

If the nitrogen mustard reacts with water in the absence of other reactants, the following inactive compound is formed.

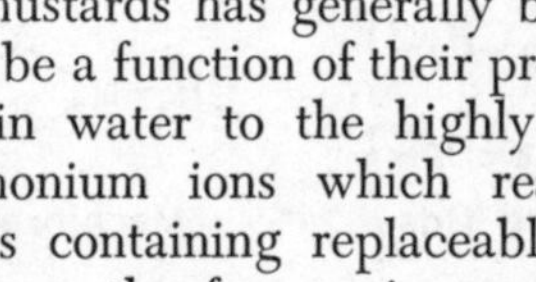

$$R-N\begin{array}{c}CH_2CH_2OH\\CH_2CH_2OH\end{array}$$

For this reason the nitrogen mustards are

marketed in the dry state, to be put into solution just prior to use and immediately introduced into an intravenous infusion fluid.

Products

Mechlorethamine Hydrochloride U.S.P., Mustargen®, 2,2′-dichloro-N-methyldiethylamine hydrochloride, HN2 hydrochloride, Nitrogen Mustard, occurs as white hygroscopic crystals which are stable at temperatures up to 40°. The initial pH of a 2 per cent aqueous solution is 3 to 4. It has vesicant properties; in case of contact, flush the skin with large amounts of water followed by 2 per cent sodium thiosulfate solution.

The N-oxide of mechlorethamine hydrochloride (Nitromin) has been reported to be less toxic and to have only a small reduction in antitumor activity.

Its activity probably depends upon in-vivo reduction prior to its alkylations, which may explain its reported improved therapeutic ratio.

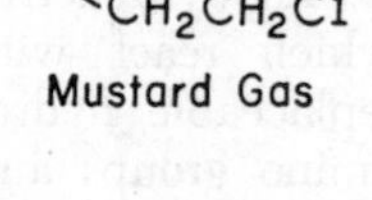

Mustard Gas 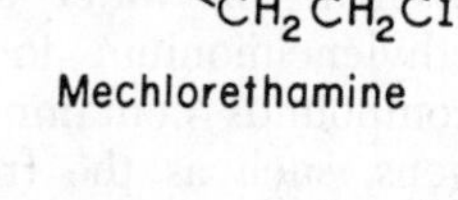   Mechlorethamine

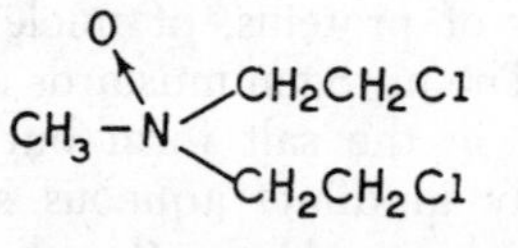

Mechlorethamine N-oxide

Category—antineoplastic.
Usual dose—intravenous, 100 mcg. per kg. of body weight once daily for 4 days.
Usual dose range—200 to 600 mcg. per kg., per course of treatment.

OCCURRENCE
Mechlorethamine Hydrochloride for Injection U.S.P.

Chlorambucil U.S.P., Leukeran®, 4{p-[bis-(2-chloroethyl)amino]phenyl}butyric acid is indicated in the treatment of chronic lymphocytic leukemia, Hodgkin's disease and related conditions. Studies have shown

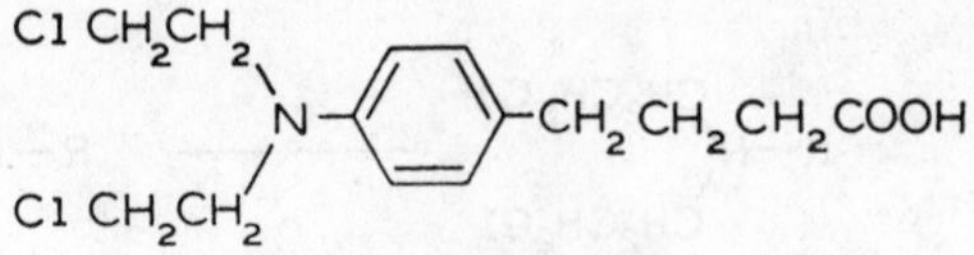

Chlorambucil

that it is well absorbed and well tolerated by the oral route. It is administered as a 2 mg. sugar-coated tablet.

The acid pKa of chlorambucil is in the region of 5.8 and the active alkylating form is considered to be[34]:

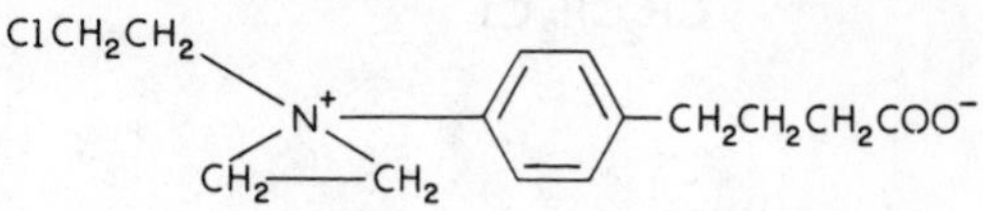

It appears that the active form may be retained in the body for several hours in media of pH 4.5 to 5.5.
Category—antineoplastic.
Usual dose—6 mg. once daily.
Usual dose range—2 to 12 mg. daily.

OCCURRENCE
Chlorambucil Tablets U.S.P.

Uracil Mustard, 5-[bis(2-chloroethyl)-amino]uracil, is used as an alkylating agent belonging to the nitrogen mustard class of compounds. It was synthesized in attempt

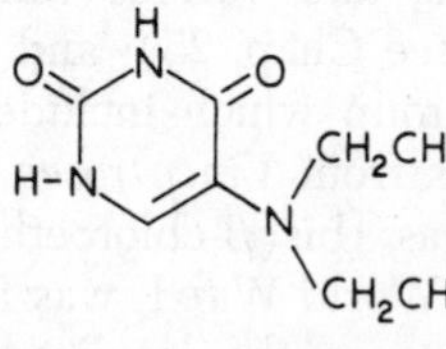

Uracil Mustard

to improve the effectiveness of nitrogen mustard without also increasing toxicity. It is not completely free of side-effects but it compares well with similar drugs.

It is a cream-white, crystalline compound which is unstable in the presence of water. It is supplied as 1-mg. capsules for oral administration.

Melphalan U.S.P., Alkeran®, is L-3-[p-[bis(2-chloroethyl)amino]phenyl]alanine.It is insoluble in water but can be dissolved in a mixture of ethyl alcohol and propylene glycol. It is stable in the dry form but, as is the case with other nitrogen-mustard derivatives, it is not stable in the presence of moisture.

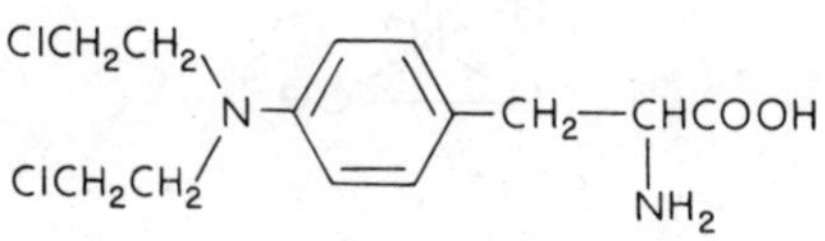

Melphalan

Category—antineoplastic.
Usual dose—initial, 6 mg. once daily; maintenance, 2 mg. once daily.

OCCURRENCE
Mephalan Tablets U.S.P.

Cyclophosphamide U.S.P., Cytoxan®, 2-[bis(2-chloroethyl)amino]tetrahydro-2H,1,3,2-oxazaphosphorine-2-oxide, is a cyclic phosphoramide ester of nitrogen mustard. It exists as the monohydrate and is quite stable in this form.

At temperatures above 35° the material loses the water of hydration and becomes anhydrous; in this form it is not very stable and deteriorates within a few days. The anhydrous form is not very soluble in water.

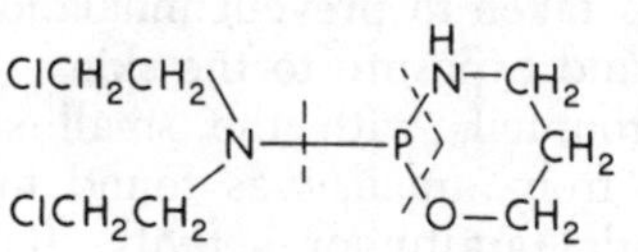

Cyclophosphamide

The monohydrate is soluble to a maximum of 4 per cent in water and in physiologic saline solution at room temperature. The solution for administration should be prepared shortly before use but is satisfactory for use up to 3 hours after preparation. It should be allowed to stand until clear before it is administered.

Cyclophosphamide is not active in vitro, and the in-vivo activity is thought to be due to cleavage of the cyclic P–N bond or P–O bond followed by cleavage of the phosphamide bridge connecting the mustard nitrogen atom and the former phosphamide ring. It is believed that these cleavages are produced by cellular phosphatases and phosphamidases and do not occur in the circulating plasma. Since phosphamidases occur in greater amounts in cancerous cells than in normal cells, the drug may have a more selective action than other cytotoxic agents.

Category—antineoplastic.
Usual dose—oral, 2 to 4 mg. per kg. of body weight daily; intravenous, 5 to 10 mg. per kg. of body weight daily for 4 to 10 days.

OCCURRENCE
Cyclophosphoramide for Injection U.S.P.
Cyclophosphamide Tablets U.S.P.

Pipobroman N.F., Vercyte®, 1,4-bis(3 bromopropionyl)piperazine, is a white, crystalline powder which is slightly soluble in water and sparingly soluble in alcohol. Pipobroman has been used in the treatment of selected patients with polycythemia vera and chronic granulocytic leukemia. The mechanism of action is not known but it has been classified as an alkylating agent by the Cancer Chemotherapy National Service Center (CCNSC). It is readily absorbed from the gastrointestinal tract. The meta-

Pipobroman

bolic fate and route of excretion are unknown.

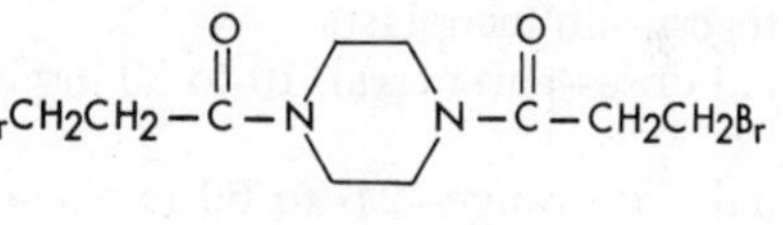

Category—antineoplastic.
Usual dose range—1 to 3 mg. per kg. of body weight daily, depending on condition being treated and patient response.

OCCURRENCE
Pipobroman Tablets N.F.

Triethylenemelamine N.F., TEM®, 2,4,6-tris(1-aziridinyl)-s-triazine, 2,4,6-tris(ethylenimino)-s-triazine, 2,4,6-triethyleneimino-1,3,5-triazine, occurs as water-soluble crystals. This type of compound was investigated as an extension of the study of nitrogen mustards. TEM is transformed in slightly acid media to the active alkylating

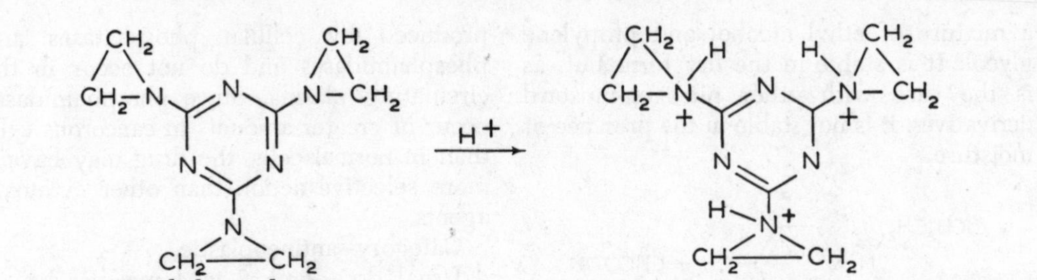

Triethylenemelamine

intermediate. Thus it can be given orally and absorbed in the active form.

It is inactivated by the gastric juice and should be given on an empty stomach along with a mild alkali such as sodium bicarbonate.

Category—antineoplastic.

Usual dose—initial, 2.5 mg. daily for 2 to 3 days; maintenance, 0.5 to 1 mg. weekly to 2.5 to 5 mg. every 2 to 5 days.

OCCURRENCE

Triethylenemelamine Tablets N.F.

Thiotepa U.S.P., tris(1-aziridinyl)phosphine sulfide, occurs as white crystalline flakes which are freely soluble in water. It should be stored in the refrigerator in light-resistant containers. Solutions of the drug may be kept for 5 days in a refrigerator.

Category—antineoplastic.

Usual dose—parenteral, 10 to 30 mg. once a week.

Usual dose range—2.5 to 60 mg. every 5 to 20 days.

OCCURRENCE

Thiotepa for Injection U.S.P.

Busulfan U.S.P., Myleran®, 1,4-butanediol dimethanesulfonate, tetramethylene dimethanesulfonate, occurs as crystals which are practically insoluble in water, but as hydrolysis progresses the material dissolves.

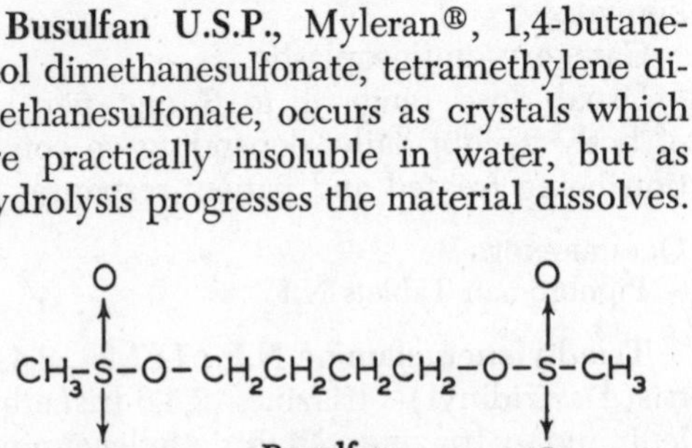

Busulfan

This type of compound functions as an alkylating agent probably with the formation of a cyclic intermediate.[35] The drug is administered orally.

Category—antineoplastic.

Usual dose—2 mg. once daily.

Usual dose range—1 to 4 mg.

OCCURRENCE

Busulfan Tablets U.S.P.

Fluorouracil U.S.P., 5-Fluorouracil. This compound is a white crystalline powder. It is sparingly soluble in water and slightly

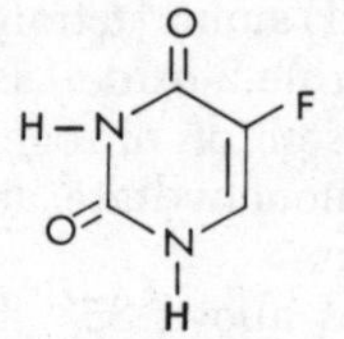

Fluorouracil

soluble in alcohol. It is heat stable but solutions should be protected from light. It should be handled very carefully and precautions taken to prevent inhalation of the power and exposure to the skin.

Fluorouracil, with the small structural change from uracil, was found to possess substantial antitumor activity. Its present use is in the palliative treatment of certain solid tumors for which surgery or irradiation is not possible. There is evidence to indicate that the drug acts by blocking the

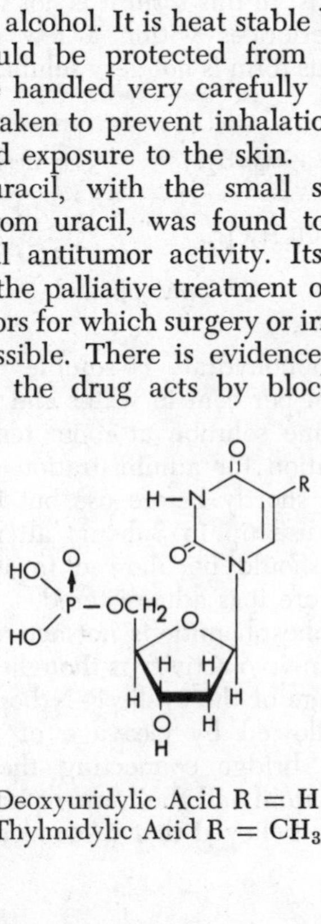

Deoxyuridylic Acid R = H
Thylmidylic Acid R = CH₃

methylation reaction of deoxyuridylic acid to thymidylic acid and, in this way, interferes with the synthesis of deoxyribonucleic acid (DNA).

The fluorouracil injection is an aqueous solution containing 50 mg. per ml. and need not be further diluted. Dosage is based on the patient's actual weight unless he is obese or has fluid retention.

It is also used topically as a solution or cream in the treatment of solar keratoses.

Category—antineoplastic.

Usual dose—intravenous, 12 mg. per kg. of body weight once daily for 4 or 5 days.

Usual dose range—6 to 15 mg. per kg., not exceeding 1 g. daily.

OCCURRENCE
Fluorouracil Injection U.S.P.

Thioguanine N.F., 2-Aminopurine-6-thiol. This compound is a pale yellow, crystalline powder, insoluble in water and in alcohol. It is freely soluble in dilute solutions of fixed bases.

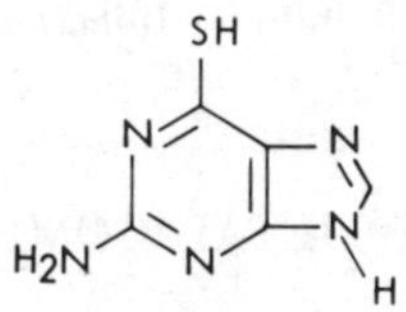

Thioguanine

Thioguanine is a close relative of mercaptopurine and, like mercaptopurine, it is an antimetabolite which interferes with purine metabolism. It is indicated for the treatment of acute leukemia and has been used for the treatment of chronic granulocytic leukemia. It must be used only under close medical supervision.

Category—antineoplastic.

Usual dose—initial, 2 mg. per kg. of body weight per day; maintenance, to be determined according to the needs of the patient.

OCCURRENCE
Thioguanine Tablets N.F.

Mercaptopurine U.S.P., Purinethiol®, purine-6-thiol, 6-mercaptopurine, is the 6-thiol analog of adenine. It occurs as an essentially odorless yellow crystalline powder that is insoluble in water, but soluble in hot alcohol and in dilute alkali solutions.

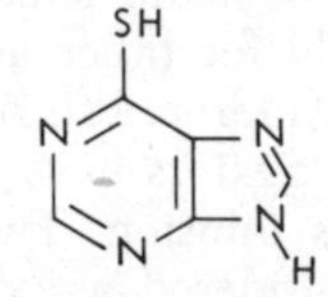

Mercaptopurine

Mercaptopurine is an antimetabolite for adenine in the synthesis of nucleotides by living cells. Therefore, it can repress cell division and at least cause temporary remission of leukemia and chronic myelogenous leukemia. Although it has the same spectrum of activity as thioguanine, mercaptopurine should always take precedence, with thioguanine being used as an alternate drug.

Category—antineoplastic.

Usual dose—2.5 mg. per kg. of body weight once daily.

Usual dose range—2.5 to 5 mg. per kg. daily.

OCCURRENCE
Mercaptopurine Tablets U.S.P.

Cytarabine, Cytosar®, Cytosine Arabinoside, 1-β-D-arabinosyl cytosine. This compound is furnished as a freeze-dried preparation to be reconstituted with Bacteriostatic Water for Injection with 0.9 per cent benzyl alcohol prior to administration. When reconstituted, the pH of the resulting solution is about 5. The solutions should be stored at room temperature and used within 48 hours. Any solution that develops a slight haze should be discarded.

Cytarabine is a synthetic nucleoside in which the sugar is arabinose, rather than

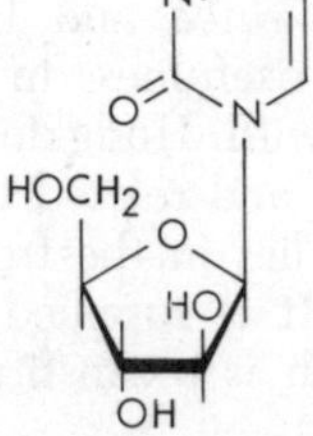

Cytarabine

ribose (as in cytidine) or deoxyribose (in deoxycytidine). It is primarily indicated in the treatment of acute leukemia of adults and secondarily for other acute leukemias of adults and children. It is not active orally and most investigators have given the drug by intravenous infusion. Two mg. per kg. per day is considered a judicious starting dose. The main toxic effect of cytarabine is bone marrow suppression, with leukopenia, thrombocytopenia and anemia.

Urethan N.F., ethyl carbamate, ethyl urethan, occurs as highly water-soluble crystals. It is incompatible with both acids and alkalies. This compound was originally introduced as a hypnotic but is now used to some extent in the treatment of neoplastic syndromes such as leukemia and multiple myeloma. The toxic action on the cells is similar in many ways to that of the nitrogen mustards, but it is a much weaker agent.

Useful antineoplastic activity of the urethan series appears to be limited to this one compound.[36] Changes in the ester group, substitution on the nitrogen atom, or replacement of either of the oxygens by sulfur result in either inactive or less active compounds.

The drug is rapidly metabolized to carbon dioxide, ethyl alcohol and ammonia.

$$H_2N-\overset{\overset{O}{\|}}{C}-OC_2H_5 \xrightarrow{H_2O} CO_2+C_2H_5OH+NH_3$$
Urethan

Category—antineoplastic.
Usual dose—3 g. daily.
Usual dose range—2 to 6 g.

OCCURRENCE
Urethan Tablets N.F.

Vinblastine Sulfate U.S.P., Velban®, vincaleukobastine sulfate, is an alkaloid obtained from *Vinca rosea* and has shown varying degrees of usefulness in the treatment of patients with Hodgkin's disease, monocytic leukemia and related conditions; its main usefulness lies in the treatment of Hodgkin's disease. It is supplied as a lyophilized plug which is reconstituted prior to injection.
Category—antineoplastic.
Usual dose—intravenous, 150 to 200 mcg.

per kg. of body weight per week once a week.
Usual dose range—100 to 500 mcg. per kg. of body weight (the latter amount divided into 2 doses given on successive days) per week.

OCCURRENCE
Sterile Vinblastine Sulfate U.S.P.

Vincristine Sulfate U.S.P., Oncovin®, leurocristine sulfate, is closely related in structure to the vinca alkaloid vinblastine. It is used in the treatment of acute leukemia in children. It is supplied as a lyophilized plug which is reconstituted prior to injection. Both the dry form and the solution are light sensitive and should be protected from light.
Category—antineoplastic.
Usual dose—intravenous, 50 to 75 mcg. per kg. of body weight once a week.
Usual dose range—50 to 150 mcg. per kg. of body weight weekly.

OCCURRENCE
Vincristine Sulfate for Injection U.S.P.

HEAVY METAL COMPOUNDS

SILVER COMPOUNDS

Silver was one of the first metals discovered by man, and although the metal and its compounds have been used since early times for the treatment of various conditions and diseases, their use at present is confined to local applications as antiseptics and germicides.

Silver has an oligodynamic effect resulting in a prolonged activity with very low concentrations. The long-continued use of all silver preparations involves the possibility that the patient may develop argyria —a gray discoloration of the skin, the conjunctiva and the internal organs.

PRODUCTS

Silver Nitrate U.S.P. has been used as a prophylactic against ophthalmia neonatorum, and in some states its use is still compulsory by law. In many infants a conjunctivitis results, so that solutions of antibiotics which are much less irritating are being used.

Category—topical anti-infective.

For external use—topically, 0.1 ml. of a 1 per cent solution, to the conjunctiva; to burned areas of the skin, as a 0.5 per cent solution in a wet dressing.

OCCURRENCE

Silver Nitrate Ophthalmic Solution U.S.P.

Mild Silver Protein, N.F.; Silver Protein; Silvol. This compound occurs as dark brown or black scales or granules. It is odorless, affected by light and is freely soluble in water but almost insoluble in organic solvents. It contains about 20 per cent silver held in colloidal suspension by a protein acting as a protective colloid. The solution should be freshly prepared or contain a suitable stabilizer and should be dispensed in amber-colored bottles. The concentrations used vary from 3 to 50 per cent. Since the discovery of penicillin and the sulfonamide drugs, the use of silver protein compounds has decreased.

Category—local anti-infective.

Application—topically, as a 5 to 25 per cent solution to the skin or mucous membrane several times daily as required.

MERCURY COMPOUNDS

Mercury and its compounds have been used since early times in the treatment of various diseases. Metallic mercury incorporated in ointment bases was applied locally for the treatment of skin infections and syphilis. A few inorganic mercury compounds have been used orally but are no longer commonly used because of the gastrointestinal disturbances and other toxic manifestations resulting therefrom. A number of organic mercury compounds are now in use mainly as antiseptics, disinfectants and diuretics. In some of these, the mercury is attached to carbon and is held rather firmly to the organic portion of the molecule, in others the mercury is attached to oxygen or nitrogen and may be ionized almost completely or partially.

It appears[37] that the antibacterial action of mercury compounds is explained best on the basis of their interfering with SH (sulfhydryl) compounds that are essential cellular metabolites. Large concentrations of SH compounds will inactivate mercury compounds completely as far as their bactericidal or bacteriostatic action is concerned. This reaction is reversible. Apparently, mercury compounds inhibit the growth of bacteria because the mercury combines with SH groups to form a complex of the type R—S—Hg—R′, thus depriving the cell of the SH groups necessary for its metabolism. However, if other SH-containing compounds are introduced which take the mercury away from the bacteria, the latter can grow again.

Experiments have been carried out in which bacteria that have been rendered inactive by mercury compounds have resumed growth when treated with hydrogen sulfide, thioglycollic acid and other sulfhydryl compounds. Thus, apparently, the mercury compounds are not bactericidal but only bacteriostatic in character.

The antibacterial activity of mercurial antiseptics is reduced greatly in the presence of serum and other proteins because the proteins supply SH groups which inactivate the mercury compounds by combining with mercury as they do with arsenic (see BAL). Thus, mercurial antiseptics are more effective on relatively unabraded skin than on highly abraded areas or mucous membranes. Mercurial antiseptics do not kill spores effectively.

Organic compounds of mercury can be made in a number of ways, including the following:

1.

$$CH_3-\overset{\overset{\displaystyle O}{\|}}{C}-O-Hg-O-\overset{\overset{\displaystyle O}{\|}}{C}-CH_3 + C_6H_6 \rightarrow$$
Mercuric Acetate

$$C_6H_5-Hg-O-\overset{\overset{\displaystyle O}{\|}}{C}-CH_3 + CH_3-\overset{\overset{\displaystyle O}{\|}}{C}-OH$$
Phenylmercuric Acetate

2. $C_6H_5SO_2Cl + HgCl_2 \rightarrow$
$$C_6H_5HgCl + SO_2 + HCl$$
Phenylmercuric Chloride

3. $C_6H_5MgBr + HgCl_2 \rightarrow$
Phenylmagnesium Bromide $C_6H_5HgCl + MgClBr$
Phenylmercuric Chloride

4. $2 C_6H_5N_2Cl \cdot HgCl_2 + 6 Cu \xrightarrow{\text{Heat}}$

$C_6H_5Hg-C_6H_5 + Hg + N_2 + 6 CuCl$
Mercury Diphenyl

5. $2 C_6H_5Br + 2 Hg(Na) \xrightarrow{CH_3-\overset{\overset{\displaystyle O}{\|}}{C}-OC_2H_5}$

$C_6H_5-Hg-C_6H_5 + HgBr_2$

The C—Hg linkages are broken by strong acids, halogens and thiocyanogen $(SCN)_2$. The active metals replace mercury; thus, phenyl compounds of sodium, magnesium and zinc, for example, may be made from the mercury compounds. The halides of nonmetals such as phosphorus, arsenic and boron react with mercury diphenyl to give phenyl derivatives of the nonmetal, one or more of the halogen atoms being replaced by phenyl, depending on the conditions. Diphenyl mercury is stable to compounds containing active hydrogen, for example, water, hydrogen sulfide, alcohols and ammonia, which decompose C—Hg linkages.

PRODUCTS

o-Hydroxyphenylmercuric Chloride, Phenol mercuric chloride, is slightly soluble in cold water but freely soluble in alcohol and hot benzene. It is used as an antiseptic in soap and similar products and as an ingredient in Mercresin. It also is employed in the eardrop preparation Myringacaine.

Phenylmercuric Nitrate N.F., merphenyl nitrate, occurs as a white crystalline powder and is a mixture of phenylmercuric nitrate and phenylmercuric hydroxide. It is very slightly soluble in water and slightly soluble in alcohol and glycerin. It

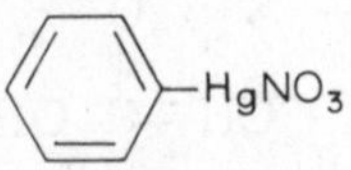

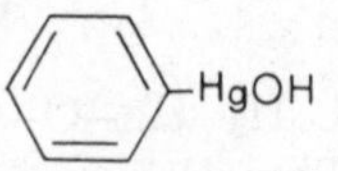

Phenylmercuric   Phenylmercuric
    Nitrate   Hydroxide

is used in 1:1,500 to 1:24,000 aqueous buffered solutions. The former concentration is for disinfection of intact skin or minor abrasions; the more dilute concentration is for application to mucous membranes or for wet dressings or for continuous irrigation of

wounds. It also is used in a concentration of 1:1,500 in an oxycholesterin base ointment for the treatment of superficial infections and minor lesions.

Commercial preparations containing phenylmercuric nitrate are Phemernite® and Merpectogel®.

Category—bacteriostatic.

Phenylmercuric Acetate N.F., acetoxyphenylmercury, occurs in the form of white prisms that are soluble in alcohol and benzene but only slightly soluble in water. It is used for its bacteriostatic properties. It has been used as a herbicide; also as a trichomonicide in the preparation Nylmerate Jelly, used as a vaginal antiseptic.

Category—preservative (bacteriostatic).

Nitromersol N.F., Metaphen®, 6-(hydroxymercuri)-5-nitro-o-cresol inner salt, occurs as a yellow powder that is practically insoluble in water and has a low solubility in alcohol, acetone and ether. It dissolves in alkalies due to the formation of a salt. Two of the structures for the compound given in the literature are shown below. The N.F. gives formula (1). It is very probable that neither of these structures is correct. In (2)

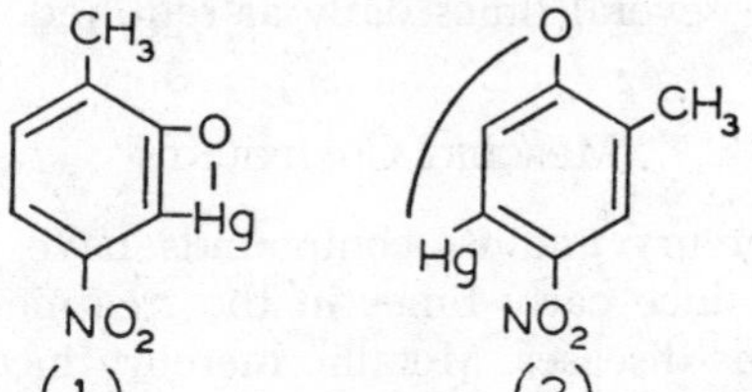

Nitromersol

there is too great a distance between the Hg and the O for the formation of a bond, and in (1) the normal valence angle of 180° for mercury would have to be distorted

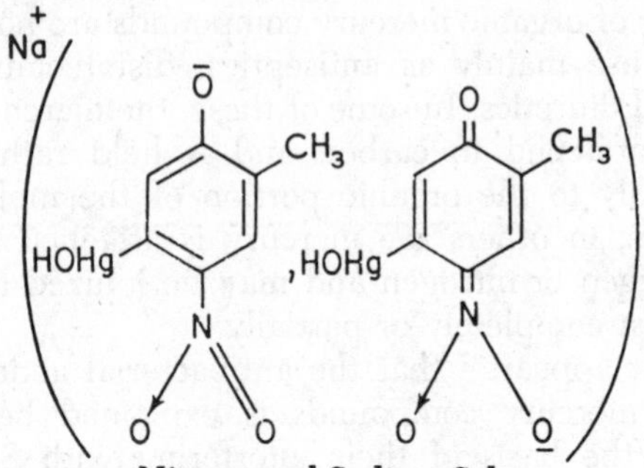

Nitromersol Sodium Salt

greatly to form the 4-membered ring. The forms in which it is supplied most commonly are a 1:500 aqueous solution and a 1:200 alcohol-acetone-aqueous solution, in both of which the compound is present as the sodium salt.

It is used mainly for disinfection of the skin.

Category—local anti-infective.

Application—topically, as a solution or tincture.

Occurrence	Per Cent Nitromersol
Nitromersol Solution N.F.	0.2
Nitromersol Tincture N.F.	0.5

Merbromin, Mercurochrome®, Disodium-2',7'-dibromo-4'-(hydroxymercuri)fluorescein. This compound was one of the first organic mercurials used as a general antiseptic. It is freely soluble in water, nearly insoluble in alcohol and acetone and insoluble in ether and chloroform. It is a non-irritating antiseptic that is used for the disinfection of wounds, the skin and mucous surfaces. It is used as a 2 per cent aqueous solution and as a 2 per cent aqueous-acetone-alcohol solution, called surgical merbromin solution.

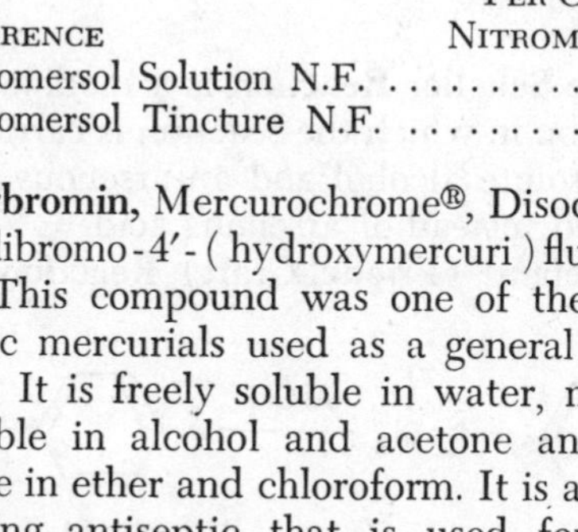

Merbromin

Thimerosal N.F., Merthiolate®, Sodium [(o-Carboxyphenyl)thio]ethylmercury. This occurs as a cream-colored powder. It is soluble in water and is compatible with alcohol, soaps and physiologic salt solution. It does not stain fabric or tissues. It is used as an antiseptic in various ways: 1:1,000 tincture for skin disinfection, 1:1,000 aqueous solution for wounds and denuded surfaces, 1:5,000 in ophthalmic ointment,

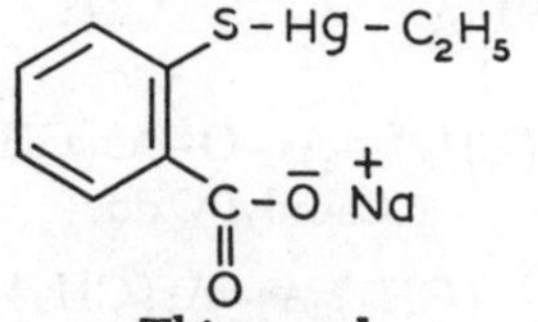

Thimerosal

1:20,000 to 1:5,000 aqueous for urethral irrigation, 1:5,000 to 1:2,000 aqueous for nasal mucous membranes.

Category—local anti-infective.

Occurrence	Per Cent Thimerosal
Thimerosal Aerosol N.F.	0.1
Thimerosal Solution N.F.	0.1
Thimerosal Tincture N.F.	0.1

Sodium Meralein, Merodicein®, mono-hydroxymercuridiiodoresorcin-sulfophthalein sodium, is used as an antiseptic in 0.2 per cent isotonic saline solution for infections of the sinuses. It is also present in Thantis lozenges.

Sodium Hydroxymercuri-o-nitrophenolate, Mercurophen®, Sodium 4-(Hydroxymercuri)-2-nitrophenolate. The compound

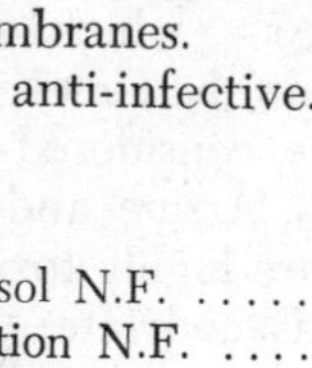

Sodium Hydroxymercuri-o-nitrophenolate

is a red, odorless powder freely soluble in water. It is used as a germicide in sterilizing instruments and skin (1:1,000). For application to mucous membranes and for irrigation, dilutions from 1:2,000 to 1:15,000 are used.

ARSENIC COMPOUNDS

A pharmacist, Louis Cadet, discovered the first organic arsenic compound in 1760. He heated arsenous acid and potassium acetate and obtained a fuming liquid with an offensive odor; this compound later was shown by others to contain cacodyl oxide and cacodyl.

$$(CH_3)_2-As-O-As(CH_3)_2$$
Cacodyl Oxide

$$(CH_3)_2As-As(CH_3)_2$$
Cacodyl, Dicacodyl,
Tetramethyldiarsine

These compounds contain the cacodyl radical $(CH_3)_2As-$, which was so named by Berzelius from the Greek word for stinking. They can be considered as derivatives of arsine, AsH_3. Arsine and almost all of its derivatives are highly toxic.

Sodium Cacodylate; $(CH_3)_2AsOONa$. This is a white, crystalline, deliquescent compound that is very soluble in water and freely soluble in alcohol. It has been used in chronic skin diseases.

Copper Cacodylate, $Cu[(CH_3)_2AsO_2]_2$. This compound is present with iron cacodylate, $Fe[(CH_3)_2AsO_2]_3$, in the preparation Ferro-Cuprin®, which is used as a hematinic.

Aromatic Arsenicals

Several methods are available for preparing aromatic arsenicals.

Bechamps Reaction

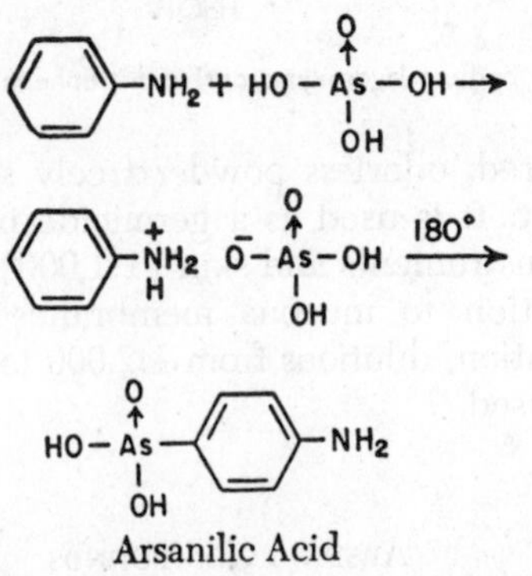

Arsanilic Acid

Other compounds with strong ortho-para directing groups in the ring which can pro-

vide some activation react in a similar fashion, thus phenol can be arsonated.

Bart Reaction

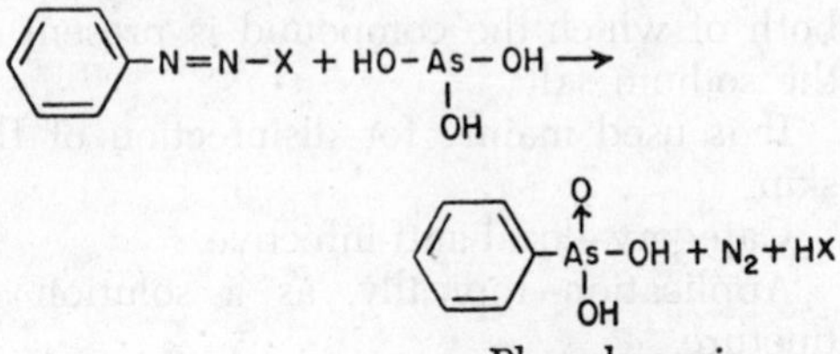

Phenylarsonic
Acid

The Scheller Reaction is a modified Bart reaction in which the reaction is carried out in absolute alcohol and an arsenous halide is used instead of arsenous acid.

Wieland (Friedel-Craft) Reaction

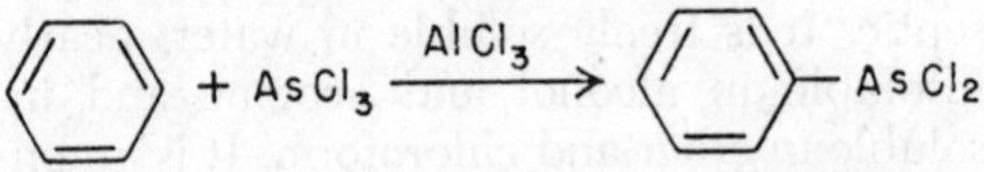

Dichlorophenyl
Arsine

The arsenic compounds most valuable as therapeutic agents belong to the aromatic series. These compounds can be divided into 2 groups: (1) those having arsenic in the pentavalent state, and (2) those having arsenic in the trivalent state. The first compound of this type studied was atoxyl which was prepared by Bechamp.[38] Ehrlich and Bertheim[39] proved that it is *p*-aminobenzenearsonic acid; they named it arsanilic acid. They became interested in arsonic compounds after Thomas, in 1905, used this acid therapeutically in the treatment of sleeping sickness and found it to be trypanocidal in vivo. However, it proved not to be active in vitro, a fact which Ehrlich accounted for by assuming that the inactive pentavalent compound was reduced to an active trivalent "arsenoxide" in the body.

There are 2 types of trivalent arsenic compounds of therapeutic importance, the so-called "arsenoxides," which also are called arsenoso compounds and are thought to be characterized by the structure $-As=O$, and the arseno compounds, which are presumed to have the structure —As=As—. These structures were assigned originally because arsenoso compounds were considered analogous to nitroso compounds and arseno compounds were thought to be similar to the azo compounds. Structures of this type are to be found in standard books of reference and in most textbooks of organic chemistry. However, it is possible that such structures with an ordinary double bond attached to arsenic, antimony and bismuth are incorrect. (See the third edition of this book, pp. 539-540, for a detailed discussion.)

The "arsenoxide" state is the form in which arsenic has the greatest therapeutic activity. This was not appreciated fully until Tatum and Cooper,[40] in 1932, showed that oxophenarsine hydrochloride could be used in small doses for the effective treatment of syphilis. Before this, much larger doses of pentavalent or trivalent arsenicals at the arsenobenzene oxidation level had been used, with the result that, while a small part of the arsenic in these compounds was converted to the effective "arsenoxide" state, the largest part stayed in the ineffective pentavalent or arsenobenzene state, serving to undermine the patient's health and comfort.

Although all arsenic compounds are quite toxic, sodium arsanilate and its derivatives seem to act with particular violence on the optic nerve, causing atrophy which results in blindness. In spite of the fact that the "arsenoxide" compounds are the most toxic, they are, in some respects, the safest to use, because both the arseno compounds and the arsonic acid derivatives are converted to the more toxic compound in the body. Thus, the arseno compounds can undergo oxidation, and the arsonic acids can be reduced to the "arsenoxide" state. The amount of these compounds converted to the most active and most toxic "arsenoxide" form is small but may vary with different individuals and, thus, cannot be predetermined. Also, the arseno compounds, upon slight exposure to air, may be converted in part to the more toxic "arsenoxide" state. To ensure the uniformity of some types of arsenicals, each lot must be released by the

National Institutes of Health before being offered for sale.

However, the pentavalent arsenicals have one advantage; they are highly mobile in the body, penetrating rapidly through all parts and, for the most part, being excreted unchanged. This great mobility made the pentavalent compounds the most suitable for the treatment of neurosyphilis. The trivalent compounds are not effective for this, since they fail to penetrate the central nervous system. Another condition in which the pentavalent arsenical was favored is amebic dysentery, where a large arsenic concentration was desired in the gastrointestinal tract.

Ehrlich made two brilliant guesses in regard to the therapeutic activity of arsenic: the first being that the pentavalent compounds were reduced to the active trivalent compounds in the body; the second being that arsenoxide compounds reacted with thiol groups in the organism. This has been well established by later workers,[41, 42, 43] who demonstrated that the action of arsenic compounds could be nullified by the addition of thiol compounds, such as thioglycollic acid. The large number of thiol groups essentially uses up all the arsenic compound, leaving none or very little of it available to combine with the thiol groups in the organisms. The action of BAL as a detoxifying agent is dependent upon the reaction of its thiol groups with arsenic. About 40 enzymes are known that are inactivated when their thiol groups are blocked. The reaction between an "arsenoxide" and a thiol compound can be demonstrated in the test tube. It involves the splitting-out of water, as indicated by this equation:

$$C_6H_5As\begin{matrix}OH\\ \\OH\end{matrix} + 2RSH \longrightarrow$$

$$C_6H_5As\begin{matrix}S-R\\ \\S-R\end{matrix} + 2H_2O$$

An interesting question in regard to the action of arsenicals is this: since thiol groups are present both in the parasite and the host, how can the arsenical eliminate the parasite without eliminating the host? Several explanations have been suggested[44] to account for this difference. The parasites may be able to concentrate the arsenical on or in their bodies. At any event, it is well known that small organisms have a much higher metabolic rate than man, involving the consumption of roughly 200 times as much food per unit of body weight. This fact could, at least in part, account for the difference in the effect on the parasite as compared with the effect on the host.

PRODUCTS

Carbarsone N.F., N-Carbamoylarsanilic Acid, *p*-Ureidobenzenearsonic Acid. This compound occurs as a white, crystalline powder. It is slightly soluble in water and alcohol, nearly insoluble in ether but soluble in basic aqueous solutions. It is synthesized from urethane and arsanilic acid.

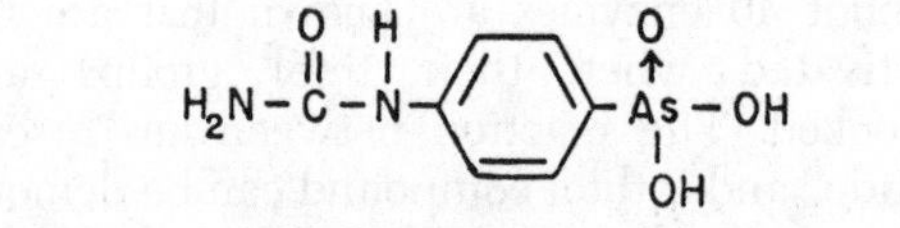

Urethane   Arsanilic Acid

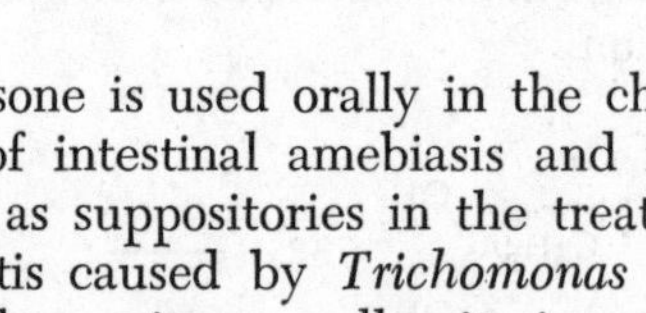

Carbarsone

Carbarsone is used orally in the chemotherapy of intestinal amebiasis and intravaginally as suppositories in the treatment of vaginitis caused by *Trichomonas vaginalis*. When given orally it is readily absorbed from the gastrointestinal tract and is then slowly excreted in the urine. For this reason, rest periods between treatment periods are needed to prevent cumulative poisoning. It is one of the safest arsenicals in use. Dimercaprol (see Chap. 12) is a useful antidote.

Category—antiamebic.

Usual dose—250 mg. two or three times daily for 10 days.
Usual dose range—100 to 250 mg.

OCCURRENCE
Carbarsone Capsules N.F.
Carbarsone Tablets N.F.

Tryparsamide; Monosodium N-(Carbamoylmethyl)arsanilate; Sodium salt of *p*-Arsenophenylglycineamide. This compound is a white, odorless, crystalline powder, soluble in water (1:2), slightly soluble in alcohol and insoluble in ether. It was used in the treatment of syphilis of the central nervous system and is currently used in the treatment of advanced trypanosomiasis due to *Trypanosoma gambiense*. The drug is

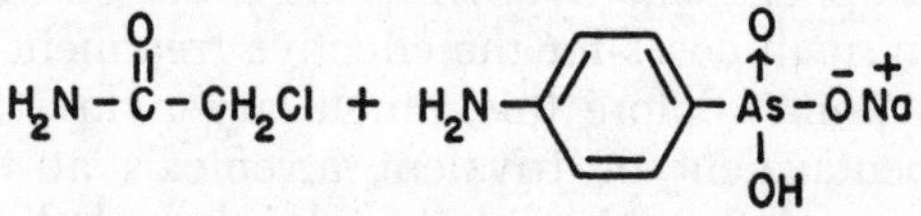

Chloroacetamide

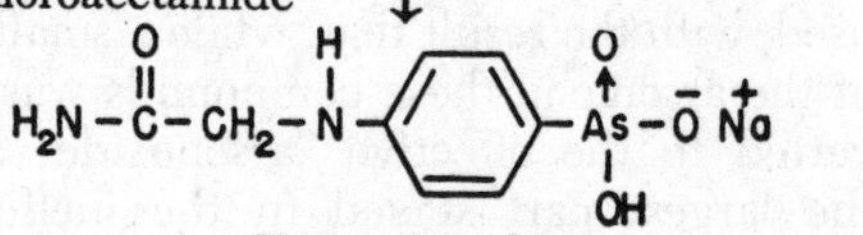

Tryparsamide

administered intravenously, usually in 2 g. doses, at intervals of 5 to 7 days, for 10 or 12 injections.

Acetarsone, Stovarsol®, is 3-acetamido-4-hydroxybenzenearsonic acid. This compound is a white, odorless powder. It is slightly soluble in water, insoluble in alcohol but soluble in basic aqueous solutions. It can be made from *p*-hydroxyphenylarsonic acid by nitrating, reducing and then acetylating. It is used in the treatment of amebiasis, trichomonas vaginitis and Vincent's angina.

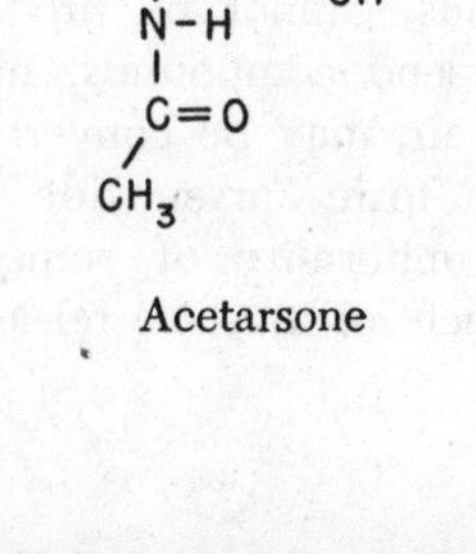

Acetarsone

Glycobiarsol N.F., Milibis®, (Hydrogen N-Glycolylansanilato)oxobismuth, Bismuthyl N-Glycoloylarsanilate. The compound is a yellow to pink powder that decomposes on heating and is very slightly soluble in water or alcohol. The saturated aqueous solution is acidic, the pH being in the range 2.8 to 3.5.

The compound is made from bismuth nitrate and sodium *p*-N-glycoloylarsanilate and is used in the treatment of intestinal amebiasis. It is reported to have low toxicity which may be due to low solubility. It imparts a black color to feces, as a result of the formation of bismuth sulfide. Glycobiarsol is also used as vaginal suppositories in the treatment of trichomonal and monilial vaginitis.

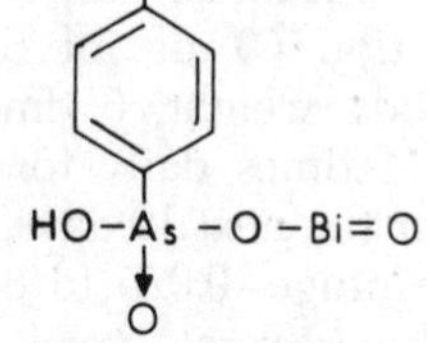

Bismuth Glycolylarsanilate

Category—antiamebic.
Usual dose—500 mg. three times daily for 7 to 10 days.

OCCURRENCE
Glycobiarsol Tablets N.F.

Atoxyl; Sodium Arsanilate; Sodium *p*-Aminophenylarsonate. This was one of the first arsenic compounds used in medicine and is of interest now mainly because of its historical importance, although it is still used to some extent in veterinary medicine for trypanosomiasis.

Oxophenarsine Hydrochloride, Mapharsen®, 2-Amino-4-arsenosophenol Hydrochlo-

Oxophenarsine
Hydrochloride   (Old structure)

ride, occurs as a white, odorless powder, soluble in water. It was used in the treatment of syphilis, and the reactions following its use were said to be less severe than those accompanying the use of the arsphenamines. The compound is distributed as a mixture with buffers and other salts to make its solution isotonic and at a pH not damaging to tissues. It is interesting to note that this compound was made and tested years ago by Ehrlich but discarded because he found it very toxic. However, the important thing with any drug is not the inherent toxicity but the ratio of the therapeutic to the toxic dose, and this ratio with oxophenarsine hydrochloride later was found to be favorable.

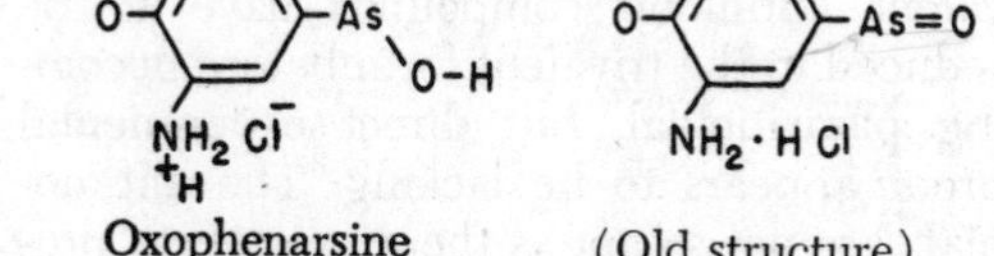

Arsphenamine; 3,3′-Diamino-4,4′-dihydroxyarsenobenzene Dihydrochloride; Diarsenol; Salvarsan; 606. The chemistry of arsphenamine is covered in detail in the third edition of this book, p. 544.

Neoarsphenamine; Sodium 3,3′-diamino-4,4′-dihydroxyarsenobenzene-N-methanol Sulfoxylate; Neodiarsenol; Neosalvarsan; 914. The chemistry of neoarsphenamine is covered in detail in the third edition of this book, p. 544.

Sulfarsphenamine; Disodium 3,3′-diamino-4,4′-dihydroxyarsenobenzene-N-dimethylenesulfate or sulfonate. The chemistry of sulfarsphenamine is covered in detail in the third edition of this book, p. 545.

Thiocarbarsone; 4-Carbamidophenyl bis-(carboxymethylthio)arsenite. The compound is a white, crystalline powder, insoluble in water and acid but readily soluble in alkali.

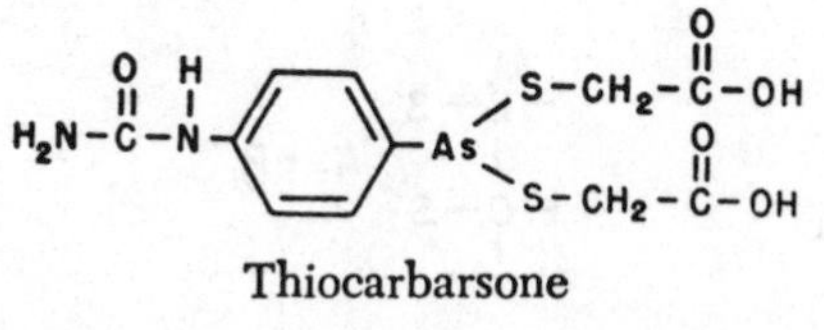

Thiocarbarsone

It has been used as an amebicide. This compound and related compounds are effective against intestinal amebiasis, but the side effects encountered—dermal reactions, nausea and vomiting—have militated against their use.

Arsenic Antidote

Dimercaprol U.S.P., 2,3-dimercapto-1-propanol, BAL (British Anti-Lewisite), dithioglycerol, is a colorless liquid with a mercaptanlike odor. BAL is soluble in water (1:20), in benzyl benzoate and in methanol. It was developed during World War II by the British as an antidote for "Lewisite." The

$$CH_2-CHCH_2OH$$
$$|\quad\quad|$$
$$SH\quad SH$$

name BAL is an abbreviation for British Anti-Lewisite. It is an effective antidote for poisoning with arsenic, gold, antimony, mercury and perhaps other heavy metals. The skin damage resulting from arsenical vesicant agents can be prevented by a previous application of BAL preparations. The damage to the skin by the same agents also can be arrested and perhaps reversed by application of BAL shortly after exposure. In systemic poisoning resulting from various arsenical agents, parenteral administration of BAL in oil has been demonstrated to be quite effective.

The antidote properties of BAL for the metals are associated with the fact that the heavy metal ions tie up the —SH groups in the tissues and thus interfere with the pyruvate oxidase and perhaps other enzyme systems which are dependent on the —SH groups for their activity. The synthetic dithiol compounds, such as BAL, compete effectively with the tissues for the metal, removing the metal by forming a ring compound of the type

$$
\begin{array}{c}
|\\
-C-S\\
|\quad\quad\diagdown\\
\quad\quad\quad As-R\\
|\quad\quad\diagup\\
-C-S\\
|
\end{array}
$$

which is relatively nontoxic and is excreted fairly rapidly. To exhibit the detoxifying effect, it is apparently necessary for the compound to have 2 thiol groups on adjacent carbon atoms or on atoms separated by one other atom so stable 5- or 6-membered ring compounds can be formed.

Monothiol compounds are much less effective.

BAL may be applied locally as an ointment, 5 per cent W/V in a base of lanolin, Lanette wax and diethylphthalate. It is injected intramuscularly as a 5 or 10 per cent solution in peanut oil, to which 2 g. of benzyl benzoate is added for each gram of BAL to make the latter miscible with the peanut oil in all proportions. Solutions of this type can be sterilized in nitrogen-filled ampuls by heating to 170° for 1 hour without having more than 1.5 per cent of the BAL destroyed in the process. Solutions of BAL in water or propylene glycol are reported to be unstable. 1,2,3-Trimercaptopropane has been reported to occur as an impurity in varying amounts in the commercial product.[45]

Category—antidote to arsenic, gold and mercury poisoning.

Usual dose—intramuscular, 0.03 ml. of 10 per cent solution (3 mg. of dimercaprol) per kg. of body weight, 6 times daily for 2 days, then 4 times daily for 1 day, and twice daily for the next 10 days, if required.

Usual dose range—0.025 to 0.05 ml. (2.5 to 5 mg. of dimercaprol) per kg.

OCCURRENCE	PER CENT DIMERCAPROL
Dimercaprol Injection U.S.P.	10

ANTIMONY COMPOUNDS

The use of antimony compounds in place of arsenic for the treatment of trypanosomiasis was suggested at the beginning of the century. The types of compounds used are stibonic acid, $RSbO(OH)_2$, of which the radical prefix is stibono; stibnic acid, $RSb(O)OH$, radical prefix stibnic; stibenoso, $RSb(OH)_2$ or $RSbO$; stibinoso, $R_2Sb(OH)$. The arylstibonic acids can be made by the Scheller reaction which was mentioned under arsenic. It is assumed generally that, like pentavalent arsenic, the pentavalent antimony compounds have to be reduced to the trivalent form before becoming parasiticidal, but direct experimental proof appears to be lacking. The antimonials are not so apt as the arsenicals to produce resistant trypansomal strains.

PRODUCTS

Antimony Potassium Tartrate U.S.P., antimonyl potassium tartrate, tartar emetic, occurs either as colorless, odorless, transparent crystals or as a white powder, depending on whether or not the compound contains water of crystallization. The crystals effloresce when exposed to air. It is soluble in water (1:12), in glycerol (1:15) and is insoluble in alcohol.

The structure frequently given for antimony potassium tartrate is undoubtedly incorrect, first, because antimony does not

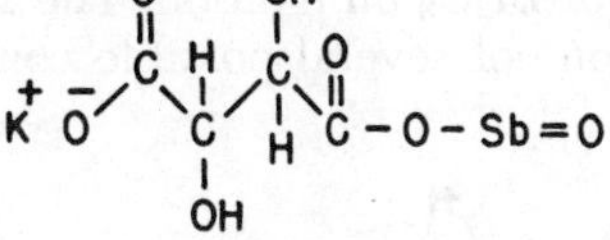

have any appreciable tendency to form double bonds and, secondly, because antimony, like bismuth, arsenic and some other metals, reacts with secondary alcohol groups. The structure proposed by Reihlen and Hezel overcomes this objection,[46] but, if

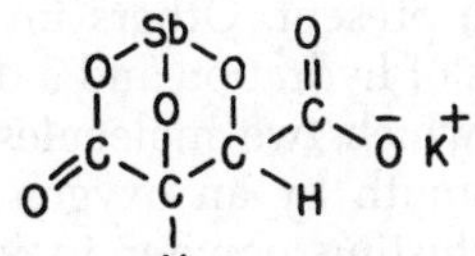

one attempts to construct this molecule with the Fisher-Hirschfelder-Taylor models, it is apparent at once that the second ring cannot be formed without greatly distorting the usual bond distances and angles. A more plausible structure is one in which the third valence of antimony is satisfied by an oxygen on a neighboring molecule or by a hydroxyl group. A dimer held together by an oxygen between 2 antimony atoms also may be a possible structure.

The compound is used orally as an expectorant and an emetic. It is employed intravenously in the treatment of a number of tropical diseases, including leishmaniasis and schistosomiasis. It is considered the drug of choice against *Schistosoma japonicum*. The average oral dose as an expectorant is 3 mg.

Category—antischistosomal.

Usual dose—intravenous, as a 0.5 to 1 per cent solution, initial dose 40 mg.; repeated every 2 days, each dose increased by 20 mg. until 140 mg. is reached, then 140 mg. every other day for a total course-of-treatment dosage of 2 g.

Stibophen U.S.P., Fuadin®; Pentasodium antimony-*bis*[catechol-2,4-disulfonate]. This compound is a white, odorless, crystalline powder. It is freely soluble in water and nearly insoluble in alcohol and ether. It is used in the treatment of schistosomiasis and granuloma inguinale. The compound is sen-

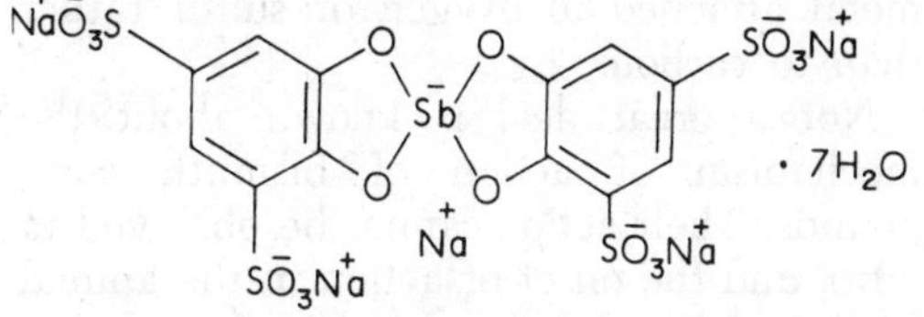

Stibophen

sitive to light and should not be brought into contact with iron. Unused portions of opened ampuls should be discarded because the compound is subject to oxidation. The ampuls usually contain about 0.1 per cent sodium bisulfite to protect the solution during processing and storage.

Category—antischistosomal.

Usual dose—intramuscular or intravenous, 100 mg. on first day, then 300 mg. on alternate days, up to a total dose of 2.5 to 4.6 g.

OCCURRENCE

Stibophen Injection U.S.P.

Antimony Sodium Thioglycollate is a white powder that is odorless or has a faint mercaptanlike odor. It is freely soluble in

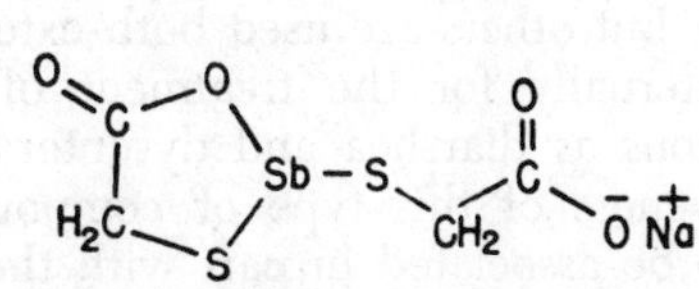

Antimony Sodium Thioglycollate

water, is insoluble in alcohol and is decomposed and discolored by light. It is used in the treatment of schistosomiasis, leishmaniasis and granuloma inguinale. It is somewhat less toxic and less irritating than antimony and postassium tartrate. The intravenous or intramuscular dose is 50 to 100 mg.

BISMUTH COMPOUNDS

Bismuth was used in medicine before 1920 mainly as an opaque substance in radiology, as a paste for tuberculous fistulas and for the treatment of intestinal infections. It has been used extensively for the treatment of syphilis but is being replaced by antibiotics.

The structure of most of the organic bismuth compounds used in medicine has the metal attached to oxygen or sulfur rather than to carbon.

Not a great deal is known about the mechanism of action of bismuth compounds. Their action cannot be observed in vitro, and the onset of action in the animal body is delayed. It has been suggested that they combine with a protein or glutathione to form an active complex. Bismuth is in the same periodic group as arsenic and might be expected to react in a similar fashion with sulfhydryl groups. However, the larger size of the bismuth atom would suggest a much weaker reaction than in the case of arsenic. The administration of bismuth reduces the ascorbic acid level in the animal body, and, to relieve this tendency, the use of bismuth ascorbate has been proposed.

Bismuth compounds may be divided into two classes. The water-insoluble compounds are used for their protective action on inflamed or irritated surfaces. These compounds are mildly astringent and antiseptic, but their mechanical protective action and drying properties are of considerable importance. Some of the compounds are used entirely externally as dusting powder antiseptics, but others are used both externally and internally for the treatment of such conditions as diarrhea and dysentery. The effectiveness of this type of compound is said to be associated in part with the fineness of the state of division of the compound.

Water-soluble bismuth compounds are used for their trypanocidal and spirochetocidal action. These compounds have been used fairly extensively for the treatment of syphilis, alone or, usually, in conjunction with arsenic therapy.

Water-Insoluble Compounds

Bismuth Subgallate, basic bismuth gallate, Dermatol, is a bright yellow, odorless and tasteless powder that is stable in air but sensitive to light. It is insoluble in water, alcohol and ether but dissolves in bases and in warm, moderately strong solutions of mineral acids. A certain amount of decomposition accompanies its solution in these solvents. The composition of the substance is somewhat variable, giving from 52 to 67 per cent of Bi_2O_3 on ignition. The structure given is one of several possible compounds

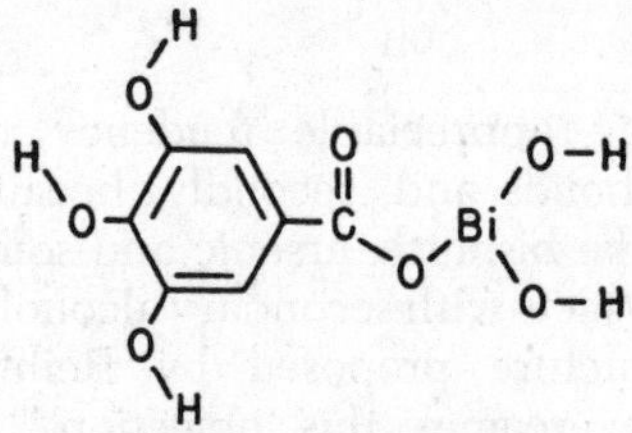

Bismuth Subgallate

that may be present. Others involve different amounts of hydration and a dimer or anhydride in which two molecules are linked through bismuth by an oxygen atom. It is used as a dusting powder in various skin disorders.

Bismuth Subsalicylate; Basic Bismuth Salicylate. This compound is a white odorless powder that is unstable in air and is affected by light. It should be stored in well-closed, light-resistant containers. It is insoluble in water and is of variable composition, giving from 62 to 66 per cent of Bi_2O_3, of the structure given, as well as dimers and hydrates of this structure.

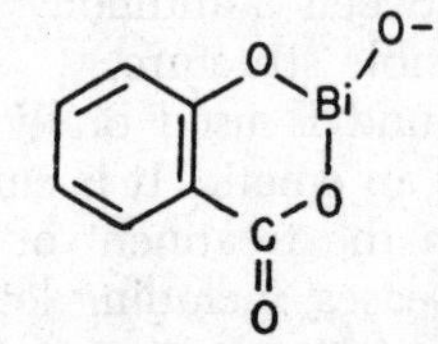

Bismuth Subsalicylate

It is used both externally and internally as an antiseptic and an astringent protective.

It is used also for its systemic effect in the treatment of syphilis, for which purpose it is injected as a suspension in oil. The oral dose for such conditions as diarrhea is 1 g. The intramuscular dose as an antisyphilitic is 100 mg. in oil.

Bismuth Tribromophenate, Xeroform, is a yellow powder, insoluble in water and only slightly soluble in alcohol and chloroform. It is of somewhat variable composition, but the structure given represents one of the possible components.

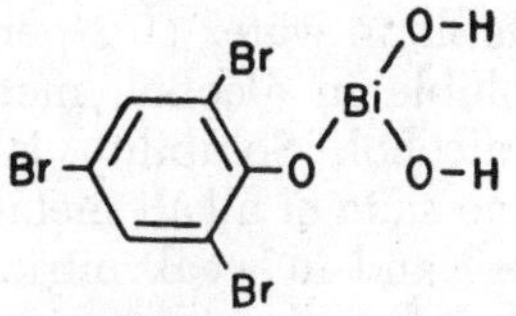

Bismuth Tribromophenate

It is used externally as an antiseptic dusting powder for the treatment of skin infections. Ointments and lotions of from 3 to 10 per cent concentration may be used.

Bismuth Formic Iodide, BFI® Powder, is a mixture of formaldehyde, gelatin, thymol iodide and bismuth oxyiodide. It is used as an antiseptic powder in the same manner as iodoform.

Water-Soluble Compounds

Bismuth Potassium Tartrate; Potassium Bismuth Tartrate; Potassium Bismuthyl Tartrate. This compound is a white, odorless powder that has a sweet taste and is sensitive to light. It is soluble in water (1:2) and is insoluble in alcohol, ether and chloroform. The substance is of variable composition, containing from 60 to 64 per cent of bismuth. A possible structure is the following:

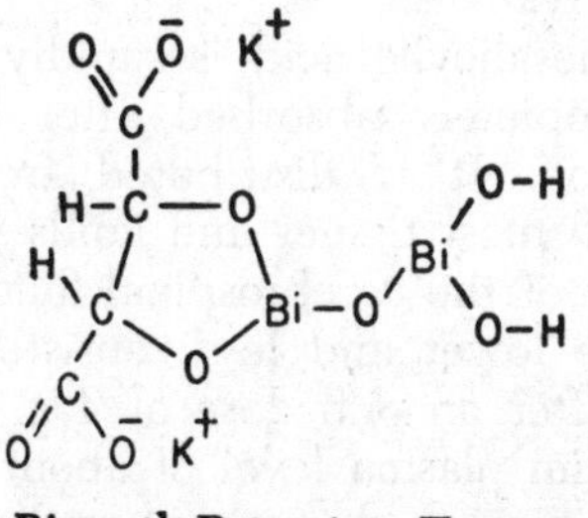

Bismuth Potassium Tartrate

Bismuth potassium tartrate has been used in the treatment of syphilis but is little used at the present time. The dose is 100 mg. intramuscularly.

Bismuth Soluble is a 2.5 per cent aqueous solution of basic bismuth potassium bismuthotartrate which is used as an antisyphilitic.

Bismuth Sodium Tartrate, sodium bismuthyltartrate, is a white, odorless and tasteless powder that is soluble in water (1:3). It contains 72.7 to 73.9 per cent bismuth. It has the same use and general composition as the corresponding potassium salt. The dose is 30 mg. intramuscularly.

Bismuth Sodium Thioglycollate, Thio-Bismol®, is a yellow, hygroscopic, granular compound with a garliclike odor. It is freely soluble in water, but the water solutions are unstable.

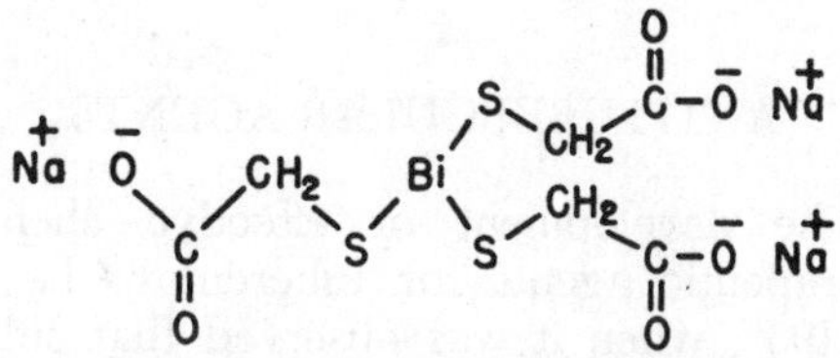

Bismuth Sodium Thioglycollate

It is used in the treatment of syphilis and is said to be absorbed readily, with the production of little local injury. The intramuscular dose is 200 mg. for adults.

Bismuth Sodium Triglycollamate, Bistrimate®. This is a double salt of sodium bismuthyl triglycollamate and disodium triglycollamate. The compound is a derivative of nitrilotriacetic acid. It is readily soluble in water, giving solutions which are near pH 7. The solutions are stable over a pH range of 2.8 to 10 and are compatible with phosphates and chlorides.

The drug is used orally in the treatment

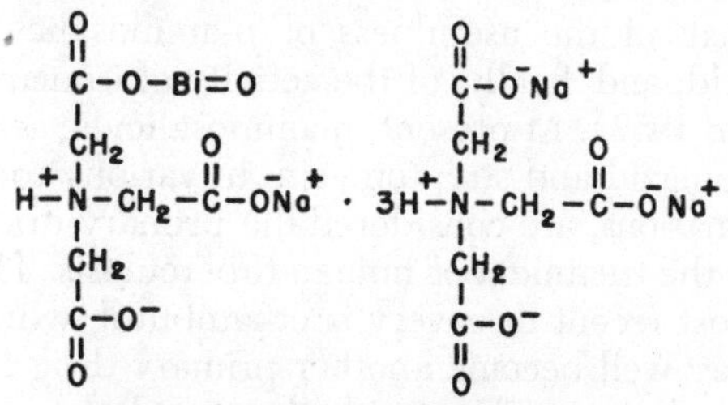

Bismuth Sodium Triglycollamate

of syphilis and certain skin diseases in doses of 410 mg. 2 or 3 times daily after meals. It blackens the feces.

TIN COMPOUNDS

A few tin preparations in which the tin is combined with an organic compound (usually a protein) have been proposed for the treatment of staphylococcic skin infections, such as carbuncles and acne.

Chondro-Stann is a 6 per cent aqueous solution of sclero-sulfhydryl protein combined with tin. The dose is 1 ml. intramuscularly.

Stannine is a tin proteinate preparation in tablet form for oral administration.

Stanno-Yeast is a preparation of tin proteinate and brewers yeast in tablet form for oral administration.

ANTITUBERCULAR AGENTS

The development of effective chemotherapeutic agents for tuberculosis began in 1938, when it was observed that sulfanilamide had a slight inhibitory effect on the course of experimental tuberculosis in guinea pigs. Later, the activity of the sulfones was discovered. Dapsone, 4,4'-diaminodiphenylsulfone, was investigated clinically, but was considered to be too toxic. Later evidence indicates this was probably due to the use of too large doses. Dapsone is now considered one of the most effective drugs for the treatment of leprosy (See Chap. 11). It also appears to be of value in treating certain resistant forms of malaria.

Major advances in the chemotherapy of tuberculosis were, first, the discovery of the antitubercular activity of streptomycin by Waksman and his associates (in 1944); next, of the usefulness of *p*-aminosalicylic acid; and, finally, of the activity of isoniazid (in 1952). At present, *p*-aminosalicylic acid, isoniazid and streptomycin, in various combinations, are considered the primary drugs in the treatment of human tuberculosis. The most recent discovery is ethambutol, which may well become another primary drug for the disease. Thus, with these and the sec-

ondary drugs that are now available, together with public health measures for locating existing cases, successful control of tuberculosis in the United States is now possible.

PRODUCTS

Aminosalicylic Acid U.S.P., 4-Aminosalicylic Acid, PAS, Parasal®, Pamisyl®. The acid is available as practically odorless, white or yellowish white crystals which darken on exposure to light and air. It is slightly soluble in water (0.1 per cent) but is more soluble in alcohol, methanol and isopropyl alcohol. Solubility is increased with alkaline salts of alkali metals (sodium bicarbonate) and in weak nitric acid. The amine salts of hydrochloric and sulfuric acids are insoluble. Aqueous solutions have a pH of about 3.2, and, when heated, the acid decomposes.

p-Aminosalicylic acid aids streptomycin or dihydrostreptomycin in the treatment of tuberculosis. The additional benefit is not as important as its help in preventing the development of bacterial resistance. The acid is taken orally, usually in tablet form. Often, severe gastrointestinal irritation accompanies the use of PAS or its sodium salt. To overcome this disadvantage, coated tablets, capsules and granules are used. Often, an antacid, such as aluminum hydroxide, is prescribed concurrently.

Studies of structural modifications have shown that the maximum activity is obtained when the hydroxyl group is in the 2-position and the free amino group in the 4-position. However, esters and acylation of the amino group, if labile enough to be hydrolyzed in vivo to *p*-aminosalicylic acid, may be used. In fact, advantages of less gastric irritation are claimed for some of these derivatives.

p-Aminosalicylic acid is rapidly and almost completely absorbed after oral administration. It is distributed freely and equally to most tissues and fluids with the exception of the cerebrospinal fluid, where levels are lower and less consistently obtained. After an oral dose of 4 g. in man, a maximum plasma level of about 7.5 mg. per cent is reached in about an hour. It is excreted in the urine, both unchanged and as metabolites and has a biologic half-life of about 2 hours. Up to one third of the dose is excreted unchanged, up to two thirds as acetyl *p*-aminosalicylic acid and up to about one fourth is conjugated with glycine and excreted as *p*-aminosalicyluric acid.

Category—antibacterial (tuberculostatic).
Usual dose—3 g. four times a day.
Usual dose range—10 to 20 g. daily.

OCCURRENCE
Aminosalicylic Acid Tablets U.S.P.

Sodium Aminosalicylate U.S.P., Sodium 4-Aminosalicylate, Parasal® Sodium, Pasara Sodium, Pasem® Sodium, Parapas® Sodium, Paraminose®. This compound is the dihydrate salt, occurring as a yellow-white, odorless powder or in crystals. It is soluble in alcohol and very soluble in water, provided that the solution has a pH of 7.25. Aqueous solutions decompose quite readily, the rate depending on the pH and the temperature. The pH of maximum stability is in the range of 7 to 7.5. Two types of reaction are involved in the decomposition process. The first is decarboxylation to yield *m*-aminophenol. The second involves the oxidation of *p*-aminosalicylic acid or of the *m*-aminophenol, or both, with the formation of brown to black pigments. Freshly prepared solutions of pure sodium *p*-aminosalicylate are nearly colorless, but on standing they develop an amber and eventually a dark brown to black color. The presence of the amber color is not necessarily a sign of extensive decomposition; however, the *U.S.P.* cautions that solutions should be prepared within 24 hours of administration and that in no case should a solution be used if its color is darker than that of a freshly prepared solution. It is generally agreed that solutions for parenteral or topical use should be sterilized by filtration.

Using 4.8 per cent solutions suitable for intravenous infusion, it was found that the following amounts of *m*-aminophenol formed when stored at the conditions indicated[47] in the table below. The addition of 0.1 per cent of sodium sulfite will prevent discoloration (oxidation) but not decarboxylation.[48]

TEMPERATURE, °C.	TIME, DAYS	MG./100 ML
20	1	None
20	2	10
0	7	7
0	30	11
0	45	15
−5	30	9
−5	60	11
−5	90	14
−5	120	15

Category—antibacterial (tuberculostatic).
Usual dose—3 g. five times daily.
Usual dose range—10 to 20 g. daily.

OCCURRENCE
Sodium Aminosalicylate Tablets U.S.P.

Potassium Aminosalicylate U.S.P., Paskalium®, Paskate®, Potassium 4-Aminosalicylate. This salt has properties similar to those of sodium *p*-aminosalicylate. It is reported to cause less gastric irritation than the free acid or the sodium salt. Of course, its use is indicated when the sodium ion intake must be kept at low levels.

Category—antibacterial (tuberculostatic).
Usual dose—3 g. 4 times daily.
Usual dose range—10 to 20 g. daily.

OCCURRENCE
Potassium Aminosalicylate Tablets U.S.P.

Calcium Aminosalicylate N.F., Parasa® Calcium, calcium 4-aminosalicylate, is available in three forms: powder, granules and capsules. It exhibits all the desirable actions of *p*-aminosalicylic acid but materially reduces gastrointestinal irritation.

Category—antibacterial (tuberculostatic).
Usual dose—4 g. 4 times daily.
Usual dose range—10 to 25 g. daily, divided into 4 doses.

OCCURRENCE
Calcium Aminosalicylate Capsules N.F.
Calcium Aminosalicylate Tablets N.F.

PAS Resin, Rezipas®, contains *p*-aminosalicylate ions adsorbed on an anion-exchange resin. It is supplied as a tasteless powder representing 50 per cent of *p*-aminosalicylic acid. In the stomach ion-exchange takes place, freeing the PAS to be absorbed from the intestine. The advantage claimed for this product is that it causes a lower incidence of gastric irritation than either the free acid or the inorganic salts.

Phenyl *p*-Aminosalicylate, Pheny-PAS-Tebamin®, is the phenyl ester of PAS. One gram supplies the equivalent of 0.67 g. of PAS. This product is recommended especially for the tubercular patient who does not tolerate the usual forms of PAS.

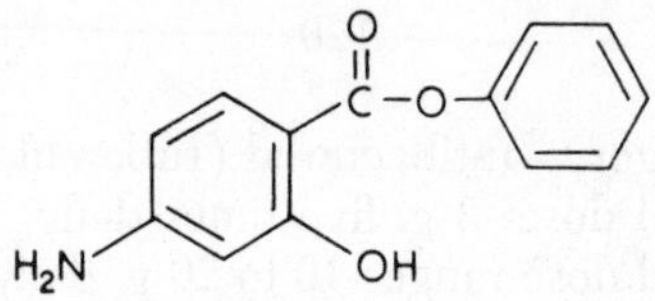

Phenyl *p*-Aminosalicylate

Calcium Benzoylpas N.F., Benzapas®, is the calcium salt of N-benzoylaminosalicylic acid. This derivative of PAS is practically insoluble in water and, when completely hydrolyzed, yields 47.4 per cent of PAS. This modification has been made to decrease the incidence of gastric irritation which may occur when the free acid or the inorganic salts must be administered for long periods.

Category—antibacterial (tuberculostatic).

Usual dose range—10 to 15 g. daily in 2 or 3 divided doses.

OCCURRENCE
Calcium Benzoylpas Tablets N.F.

Isoniazid U.S.P., Rimifon®, INH, isonicotinic acid hydrazide, isonicotinyl hydrazide, occurs as nearly colorless crystals which are very soluble in water. Hydrazides are prepared readily by refluxing a methyl or ethyl ester with a hydrazine.

Antitubercular drugs have been studied ever since Koch identified the tubercle bacillus, *Mycobacterium tuberculosis.* Up to 1952, the primary compounds used to treat tuberculosis were sulfonamides, various sulfones, *p*-aminosalicylic acid, streptomycin, dihydrostreptomycin and tibione.[49] In 1945, and again in 1948, it was pointed out that nicotinamide had tuberculostatic activity equal to that of *p*-aminosalicylic acid. In view of this and the fact that tibione is a thiosemicarbazone of *p*-acetamidobenzalde-hyde, the thiosemicarbazones of alpha, beta and gamma nicotinaldehyde were prepared and studied. Of these pyridine analogs of tibione, the alpha is inactive, and the beta and the gamma are superior to tibione.

In the method used for synthesizing[50] gamma-nicotinaldehyde thiosemicarbazone, isonicotinylhydrazine (isoniazid) was an intermediate. Since the product was available, it was subjected routinely to study on tuberculosis. Experiments on animals and humans revealed no serious or irreversible toxic effects. Reactions included central nervous system stimulation (leg twitching and insomnia) or autonomic activity (dryness of secretions) and dizziness. There is a wide margin of safety between the therapeutic and lethal doses in animals, about twenty times the oral therapeutic dose being necessary to kill 50 per cent of mice.

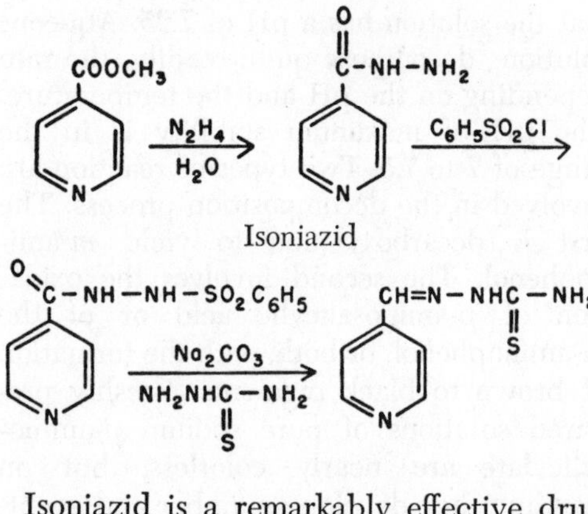

Isoniazid

Isoniazid is a remarkably effective drug and is now considered one of the primary drugs (along with aminosalicylic acid and streptomycin) for chemotherapy of tuberculosis. But, even so, it is not completely effective in all types of the disease. Isoniazid is well absorbed after oral administration and is rather rapidly excreted, with between 50 and 70 per cent of a dose being eliminated in the urine within 24 hours. There was no evidence of elevated plasma levels in patients receiving a dose of 1.5 mg. per kg. twice daily for several weeks. It is excreted unchanged and in several metabolically modified forms, the principal metabolites being the acetylated form and isonicotinic acid. The acetylated drug is much less active than the free drug, which is also the case with the antibacterial sulfonamides.

Isoniazid is freely distributed to all the tissues and fluids of the body, including the cerebrospinal fluid and the placental fluid in the pregnant woman.

The activity of the drug is only on the growing bacilli and not on the resting forms. At present there is no completely satisfactory explanation of the mechanism of action of isoniazid. There is evidence to indicate that it may exert its effect by interference with enzyme systems requiring pyridoxal phosphate as a coenzyme.

The principal toxic reactions are peripheral neuritis and gastrointestinal disturbances such as loss of appetite and constipation. The side-effects are dosage-related, and the incidence may be expected to increase as the dose is increased. The peripheral neuritis resembles that caused by pyridoxine deficiency and it is now current practice for many physicians treating tuberculosis to give the patients fairly large doses of pyridoxine. The mechanism by which isoniazid produces the peripheral neuropathy is not well understood. The pyridoxine does not seem to interfere with the antibacterial action of isoniazid.

The problem often occurs that resistant strains of the tubercle bacillus develop during therapy. For this reason isoniazid is seldom used as the sole chemotherapeutic agent but is usually administered with aminosalicylic acid given orally or with streptomycin administered intramuscularly. In some dosage regimens all three drugs are used.

None of the derivatives of isoniazid is more useful in therapy than the parent compound. Any change in structure leads to a decrease in potency and, in most cases, to a loss of potency. The isopropyl derivative 1-isonicotinyl-2-isopropylhydrazine shows good activity, but clinical trial proved that it was too toxic for use considering that other safer drugs were available. However, the isopropyl derivative, iproniazid, is worthy of special mention because it has led to the development of a group of psychomotor stimulants useful in drug therapy of certain kinds of depression.

Category—antibacterial (tuberculostatic).

Usual dose—oral or intramuscular, 300 mg. once daily.

Usual dose range—300 mg. to 500 mg. daily.

OCCURRENCE
Isoniazid Injection U.S.P.
Isoniazid Syrup U.S.P.
Isoniazid Tablets U.S.P.

Pyrazinamide U.S.P., Aldinamide, Pyrazinecarboxamide, is the pyrazine analog of nicotinamide. It occurs as a white crystalline powder, practically insoluble in water, slightly soluble in acetone, in alcohol and in chloroform. It is a fairly active drug, but it causes a rather significant incidence of liver damage. Because of its hepatotoxic potential, the drug is generally reserved for the treatment of hospitalized patients when the primary drugs and other secondary drugs cannot be used because of bacterial resistance or because the patients cannot tolerate them. Pyrazinamide increases reabsorption of urates and should thus be used with care in patients with a history of gout.

Pyrazinamide

Category—antibacterial (tuberculostatic).

Usual dose—5 to 7.5 mg. per kg. of body weight 4 times daily.

Usual dose range—1 to 3 g. daily.

OCCURRENCE
Pyrazinamide Tablets U.S.P.

Ethionamide, U.S.P., Trecator®, 2-Ethylthioisonicotinamide. This drug occurs as a yellow, crystalline substance, very sparingly soluble in water or ether, and soluble in hot acetone or in dichloroethane. In contrast to ring modifications in the isoniazid series, the 2-alkyl substituted thioisonicotinamides were more active than the parent compound. The 2-ethyl derivative was the most interesting of the group studied.

Ethionamide is a secondary drug in the chemotherapy of tuberculosis and is intended mainly for use in the treatment of pulmonary tuberculosis resistant to isoni-

$$S = C - NH_2$$

Ethionamide

azid or when the patient is intolerant to other drugs. It is administered orally and the highest tolerated dosage is generally recommended.

Category—antibacterial (tuberculostatic).

Usual dose—250 mg. 3 times daily.

Usual dose range—500 mg. to 1 g. daily.

OCCURRENCE

Ethionamide Tablets U.S.P.

Ethambutol, Myambutol®, (+)-2,2′-(ethylene diimino)di-1-butanol dihydrochloride, EMB, is a white crystalline powder freely soluble in water and slightly soluble in alcohol.

$$C_2H_5 - \underset{\underset{H}{|}}{\overset{\overset{CH_2OH}{|}}{C}} - \underset{\underset{H}{|}}{N} - CH_2CH_2 - \underset{|}{N} - \underset{\underset{H}{|}}{\overset{\overset{CH_2OH}{|}}{C}} - C_2H_5 \cdot 2\ HCl$$

Ethambutol Dihydrochloride

This compound is remarkably stereospecific. Tests have shown that although the toxicities of the dextro, levo, and meso isomers are about equal, their activities vary considerably. The dextro isomer is 16 times as active as the meso isomer and the levo isomer is even less active than the meso isomer. In addition, the length of the alkylene chain, the nature of the branching of the alkyl substituents on the nitrogens, and the extent of N-alkylation all have a pronounced effect on the activity.

Ethambutol is rapidly absorbed after oral administration and peak serum levels occur in about 2 hours. It is rapidly excreted, mainly in the urine. Up to 80 per cent is excreted unchanged, with the balance being metabolized and excreted as 2,2′(ethylenediimino)-dibutyric acid and as the corresponding di-aldehyde.

It is recommended not for use alone but in conjunction with other antitubercular drugs in the chemotherapy of pulmonary tuberculosis. The usual dose is of the order of 15 to 25 mg. per kg. of body weight given orally as a single daily dose.

ANTIFUNGAL AGENTS

Many remedies have been used against fungus infections, and research still continues, which would lead one to conclude that the ideal topical antifungal agent has not yet been found. Fatty acids in perspiration have been found to be fungistatic; this has led to the introduction of fatty acids in therapy. They are also used as copper and zinc salts so that the combined antifungal action of the metal ion is obtained. Salicylic acid and some of its derivatives are used for their antifungal activity. Various other structures also have antifungal activity, e.g., hydrocarbon acids, furan derivatives, diamthazole and hexetidine.

Prior to the discovery of griseofulvin (Chap. 14) there was no satisfactory drug that could be used systemically for fungal infections.

PRODUCTS

Propionic Acid has become an important fungicide because it is nontoxic, nonirritant and readily available.

It is a clear, corrosive liquid with a characteristic odor and is soluble in water or alcohol. In 1939 Peck,[51] in his studies on perspiration, observed that it was not the pH of perspiration that was responsible for the fungicidal and fungistatic effect but the presence of fatty acids and their salts. Previous to this, Bruce had found that fatty acids of odd-numbered carbon atoms were bacteriostatic while acids of even-numbered carbon atoms were not. Note, however, that caprylic acid is active. Chemical analysis of sweat showed it to contain, among other ingredients, 0.0091 per cent of propionic acid. The fungicidal action of propionic acid salts, such as those of sodium, ammonium, calcium, zinc and potassium, was found to be the same as that of the free acid. The free acid may be used to treat fungus infections, such as athlete's foot, but usually is employed in the form of its salts because they are more easily handled and are odorless.

A number of fatty acids and their salts are efficient fungicides, but propionic, caprylic and undecylenic acids are used because of availability .

Sodium Propionate N.F.; Mycoban®. In 1943, sodium propionate became important as an effective agent in the treatment of fungus infections.[52] The salt occurs as transparent, colorless crystals that have a faint odor resembling acetic and butyric acids and are deliquescent in moist air. It is soluble in water (1:1) or alcohol (1:24). Sodium propionate is most effective at pH 5.5 and usually is used in a 10 per cent ointment, powder or solution. Other fatty acid salts containing an odd number of carbon atoms are effective as fungicides, but they are more toxic and not so readily available. Undecylenic acid is one that is being used. By 1944 propionate-propionic acid mixtures had been found to be superior, not only as fungicides but also as bactericides. However, they are only slightly better than the undecylenate-undecylenic acid mixture.

Propionic acid salts of sodium, calcium, zinc, potassium and copper are used in preparations for the treatment of fungus infections, such as athlete's foot (tinea pedis).

Category—local antifungal.

For external use—topically in dosage forms containing 5 to 10 per cent of sodium propionate.

Zinc Propionate occurs as plates or as needles in the case of the monohydrate. It is freely soluble in water and sparingly soluble in alcohol. It decomposes in a moist atmosphere, giving off propionic acid. Therefore, it should be kept in well closed containers. It is used as a fungicide, particularly on adhesive tape to reduce irritation caused by fungi and bacterial action.

Propionate Compound, Propion® Gel, is a mixture containing 10 per cent each of calcium and sodium propionates in a jelly for local application in the treatment of vulvovaginal moniliasis.

Sodium Caprylate. Caprylic acid, found in several oils such as coconut and palmkernel, is the acid from which the sodium salt is prepared. Like propionic acid, caprylic acid is an ingredient of perspiration, where it contributes to the antifungal prop-

erties. The sodium salt is soluble in water, sparingly soluble in alcohol and occurs as cream-colored granules.

It is an antifungal agent similar to propionates and undecylenates, being effective against infections due to trichophytons, microsporons and *Candida albicans*. There appears to be no skin sensitivity produced by continuous or repeated use. Sodium caprylate is available as a solution, powder or ointment.

$$CH_3 (CH_2)_5 CH_2COO^- Na^+$$
Sodium Caprylate

Free caprylic acid is a light-amber, oily liquid possessing a disagreeable odor. It is insoluble in water and only slightly soluble in alcohol, as would be expected.

Zinc Caprylate is a fine, white powder that is practically insoluble in water and alcohol. It decomposes on exposure to moist atmosphere, liberating caprylic acid, and, therefore, the container should be kept well closed. It is used as a fungicide as is zinc propionate. Aluminum and copper salts also are used in proprietaries.

Propionate-Caprylate Mixture; Sopronol®; Propionate-Caprylate Compound. These are mixtures made from the free acids, sodium salts and zinc salts. The ingredients and amounts depend on the dosage form. They are employed against superficial fungus infections.

Undecylenic Acid N.F., 10-undecenoic acid, may be represented as CH_2=CH-$(CH_2)_8COOH$. The official form is not less than 95 per cent $C_{11}H_{20}O_2$. It may be obtained by the destructive distillation of castor oil. The ricinoleic acid, present in castor oil as the glyceride, is the source of undecylenic acid.

$$CH_3(CH_2)_5 CHOHCH_2 CH = CH(CH_2)_7 COOH$$

Ricinoleic Acid

↓ Vacuum

$$CH_3(CH_2)_5 CHO \quad + \quad CH_2 = CHCH_2(CH_2)_7COOH$$
n-Heptyl Aldehyde   Undecylenic Acid

It occurs as a yellow liquid having a characteristic odor and a persistent bitter or acrid taste. At lower temperatures (between 21° and 22°) it congeals and at 24° melts. The acid is practically insoluble in water and miscible with alcohol, chloroform, ether, benzene and with both fixed and volatile oils. It possesses the properties of a double bond and is a very weak organic acid.

The higher fatty acids (heptylic, caprylic, pelargonic, capric and undecylenic) have been found to be effective antifungal agents.[53] Undecylenic acid is one of the best fatty acids available as a topical fungistatic agent.[54] It may be used in up to 10 per cent strength in solutions, emulsions, adsorbed on powders or in ointments. Application to eyes, ears, nose or other areas of mucous membrane is not advisable. Even local use as a fungicide may be irritating. Internally, a very pure form is used (dose 7:5 to 10 g. daily) in capsules for the treatment of psoriasis and neurodermatitis.

There are in use a number of undecylenic acid salts, such as zinc undecylenate, copper undecylenate (Undesilin® and Decupryl®), sodium and potassium. Mixtures of the acid and salts are also used.

Category—local antifungal.

For external use—topically as a 1 to 10 per cent ointment as required.

	PER CENT
OCCURRENCE	UNDECYLENIC ACID
Compound Undecylenic Acid	
Ointment N.F.	5

Zinc Undecylenate N.F., zinc 10-undecenoate, is a fine white powder practically insoluble in water and alcohol. It is used as a fungicide in connection with the free acid and other compounds.

Category—local antifungal

For external use—topically as a 20 per cent ointment as required.

Sorbic Acid N.F., 2,4-hexadienoic acid, has been found to be effective for inhibiting the growth of molds and yeasts. It is soluble to the extent of 0.15 per cent in water. The pKa is 4.76. In a test of its fungistatic properties,[55] concentrations as low as 0.05 per cent were found to be effective.

$$CH_3CH=CHCH=CHCOOH$$
Sorbic Acid

Sorbic acid has been found to be useful as a mold inhibitor in various medicinal syrups, elixirs, ointments and lotions containing sugars and other components that support mold growth. It is used in films and other food-packaging materials.

Category—preservative (antimicrobial).

Potassium Sorbate N.F., 2,4-Hexadienoic Acid, Potassium Salt; Potassium 2,4-Hexadienoate. This compound occurs as white crystalline power and has a characteristic odor. It is freely soluble in water and soluble in alcohol. It is used like sorbic acid, especially where greater solubility in water is required. The suggested experimental levels are 0.025 to 0.1 per cent by total weight.

Category—preservative (antimicrobial).

Triacetin N.F., Enzactin®, Fungacetin®, glyceryl triacetate, is an ester of glycerin and acetic acid, prepared by heating a mixture of the two.

It is a colorless, oily liquid having a slight odor and a bitter taste. It is soluble in water (6 to 100), soluble in organic solvents and miscible with alcohol.

Triacetin acts as a topical antifungal agent by virtue of the acetic acid which is formed by slow enzymatic hydrolysis by esterases in the skin. The rate of release is self-limited, because as the pH drops to 4 the esterases are inactivated. It is nonirritating to the skin.

Category—local antifungal.

For external use—topically to affected area twice daily as a 15 per cent aerosol, a 25 per cent cream, or a 33 per cent powder.

OCCURRENCE
Triacetin Aerosol N.F.
Triacetin Cream N.F.
Triacetin Powder N.F.

Salicylanilide N.F., Salinidol,® Shirlan Extra, is the anilide of salicylic acid, or N-phenyl salicylamide. It occurs as white or slightly pink crystals that are slightly soluble in water, alcohol or isopropyl alcohol and in organic solvents. Salicylanilide is an antifungal agent useful in the treatment of tinea capitis.

Due to its irritant action on the skin, the concentration used should be 5 per cent or

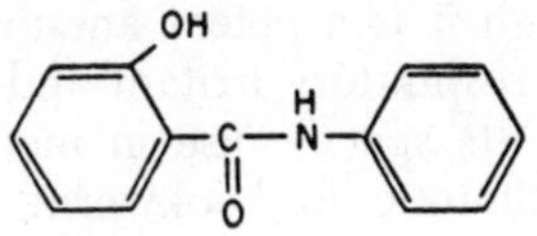

OH

(structure: Salicylanilide)

Salicylanilide

less. It is recommended that its use be limited to ringworm of the scalp. It is used usually in ointment form, although liquid products are available.

Category—local antifungal.

For external use—topically as a 3 to 5 per cent ointment applied to the scalp as required.

Tolnaftate U.S.P., Tinactin®, is a O-2-naphthyl *m*,N-dimethylthiocarbanilate.

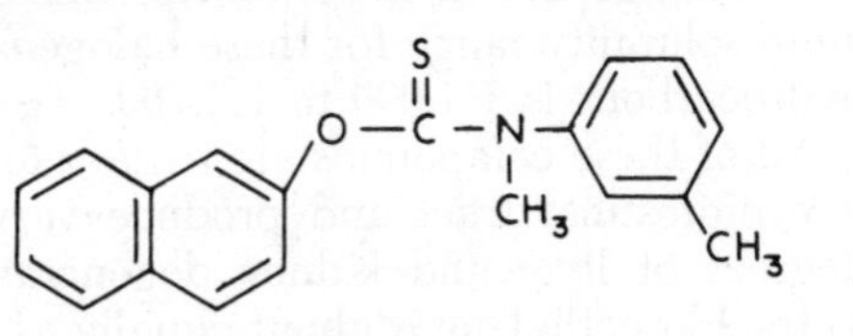

(structure: Tolnaftate)

Tolnaftate

This compound, which is essentially an ester of β-naphthol, is reported to be a potent antifungal agent. Only one or two drops of a 1 per cent solution in a polyethylene glycol is adequate for areas as large as the hand.

Category—antifungal.

For external use—topically, as a 1 per cent ointment, powder, or solution twice daily.

OCCURRENCE

Tolnaftate Cream U.S.P.
Tolnaftate Solution U.S.P.

Coparaffinate, Iso-Par®, is a mixture of water-insoluble isoparaffinic acids partially neutralized with hydroxybenzyl dialkyl amines. The acids are both monocarboxylic

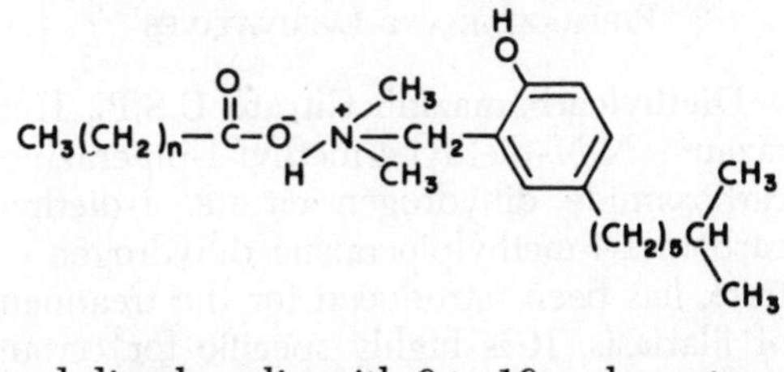

(structure: Coparaffinate)

$$CH_3(CH_2)_n - \overset{O}{\underset{H}{C}} - \bar{O} \overset{+}{\underset{CH_3}{N}} - CH_2$$

and dicarboxylic with 6 to 16 carbon atoms. Coparaffinate is a brown, thick, oily liquid with a characteristic and persistent odor.

It is alcohol-soluble but water-insoluble. It is used as a 17 per cent topical ointment in the treatment of pruritis ani and mycotic infections of the hands and the feet.

Chlordantoin, Sporostacin®, is 5-(1-ethylamyl)-3-trichloromethylthiohydantoin. It is a white, crystalline powder with adequate stability. It is an odorless and nonstaining fungicide. It is reported to be almost entirely free of untoward reactions, such as irritation and sensitization. It is used topically with benzalkonium chloride as a solution, a lotion and a cream.

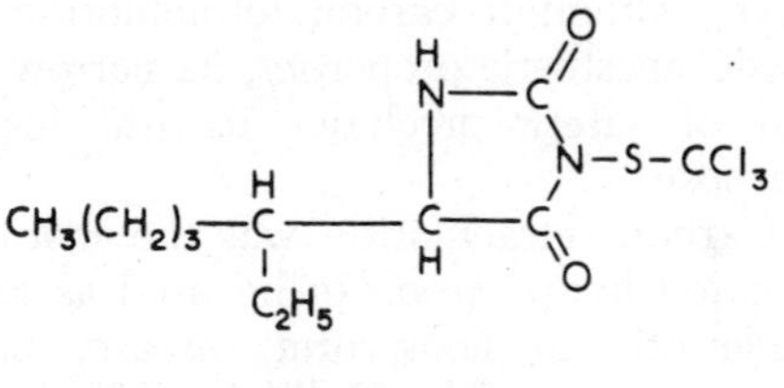

(structure: Chlordantoin)

$$CH_3(CH_2)_3 - \overset{H}{\underset{C_2H_5}{C}} - \overset{H}{C} - C$$

Chlordantoin

ANTHELMINTICS

Anthelmintics are drugs which possess the property of ridding the body of parasitic worms. Several classes of chemicals are used and include: (1) chlorinated hydrocarbons, (2) phenols and derivatives (Chapt. 10), (3) piperazine and derivatives, (4) dyes, (5) antimalarial drugs (Chapt. 13), and (6) alkaloids and other natural products.

CHLORINATED HYDROCARBONS

Carbon Tetrachloride, tetrachloromethane, (CCl_4) is prepared commercially by the chlorination of carbon bisulfide, the latter being prepared by heating sulfur and coke in an electric furnace.

$$C + 2S \rightarrow CS_2 \xrightarrow[\text{SbCl}_5 \text{ (catalyst)}]{+ 3 Cl_2} CCl_4 + S_2Cl_2$$

It is a colorless, mobile liquid, specific gravity 1.590, boiling point 77°. Like chloroform, it is miscible with alcohol, ether and most organic solvents as well as with fixed and volatile oils. The absence of hydrogen from the molecule of this chlorinated hydrocarbon makes it completely noninflammable and accounts for its use as a fire extinguisher. However, it should not be used on hot metal or in confined quarters, since it forms the toxic phosgene. Occasional explosions also have been reported. The vapors of carbon tetrachloride, like those of all chlorinated hydrocarbons, are toxic. When used as a dry cleaner or an industrial solvent, the atmospheric concentration should be kept below 100 p.p.m. for safety. Initial toxic effects are manifested in headache, nausea, abdominal pain and diarrhea. Cumulative effects are marked in the liver, the heart and the kidneys. Although carbon tetrachloride possesses anesthetic properties, its narrow margin of safety precludes its use for this purpose.

Carbon tetrachloride was the first halogenated hydrocarbon to be used as an anthelmintic in hookworm disease, having been introduced by Hall[56] in 1921. Dosage is always followed by a purgative to remove dead worms as well as excess drug. Carbon tetrachloride is capable of causing liver necrosis. This can be combated to an extent by oral or intravenous calcium gluconate. The drug should not be administered during pregnancy because of the liver and kidney damage it produces in the fetus. The main use as an anthelmintic is in veterinary medicine.

Tetrachloroethylene U.S.P., Perchloroethylene, Tetrachloroethene, $Cl_2C=CCl_2$.

Tetrachloroethylene may be synthesized from dry hydrogen chloride and carbon monoxide at 300° and 200 atmospheres pressure in the presence of a nickelous oxide catalyst or by passing symmetrical ethylene dichloride and chlorine over heated pumice at 400°.

Tetrachloroethylene is a colorless, mobile liquid of ethereal odor with a specific gravity of 1.61 and a boiling point of 122°. It is miscible with an equal volume of alcohol and with most organic solvents. Like trichloroethylene, it is unstable to air, moisture and light, decomposing in part to phosgene and hydrochloric acid. The *U.S.P.* permits up to 1 per cent alcohol as preservative. It is noninflammable. It is used industrially as a solvent and as a cleaner of textiles and metals.

Although it is a potent anesthetic, it is a skin and respiratory irritant and difficult to vaporize. Its specific use in medicine is as an anthelmintic in hookworm infestation. Wright and Schaffer,[57] in their attempt to correlate anthelmintic efficiency of chlorinated alkyl hydrocarbons and chemical structure, observed the following trends: in any one homologous series anthelmintic efficiency increases with the lengthening of the carbon chain; correspondingly, there is a decrease in water solubility, about 1:1,000 to 1:5,000, for most effective anthelmintic compounds; the optimum range varies for different homologous series. Substitution of bromine or iodine for chlorine makes less difference in anthelmintic efficiency than does change in water solubility; the optimum solubility range for these halogenated hydrocarbons is 1:1,000 to 1:1,700.

All of these compounds are irritant to the gastrointestinal tract and produce varying degrees of liver and kidney degeneration. Tetrachloroethylene is about equally as efficient as carbon tetrachloride and is preferred in hookworm treatment because it is less toxic and does not raise the guanidine content of the blood, a criterion of importance where calcium deficiency exists.

It may be given on sugar or in gelatin capsules after first emptying the gastrointestinal tract. It is followed by a saline cathartic. Oils, fats and alcohol favor absorption and toxic side-effects and, therefore, should be avoided.

Category—anthelmintic (hookworms and some trematodes).

Usual dose—5 ml. as a single dose.

Usual dose range—3 to 5 ml.

OCCURRENCE

Tetrachloroethylene Capsules U.S.P.

PIPERAZINE AND DERIVATIVES

Diethylcarbamazine Citrate U.S.P., Hetrazan®, N,N-diethyl-4-methyl-1-piperazinecarboxamide dihydrogen citrate, 1-diethylcarbamyl-4-methylpiperazine dihydrogen citrate, has been introduced for the treatment of filariasis. It is highly specific for certain parasites, including filariae and ascaris.

It is a colorless, crystalline solid, highly

soluble in water, alcohol and chloroform but insoluble in most organic solvents. A 1 per cent solution has a pH of 4.1. The drug

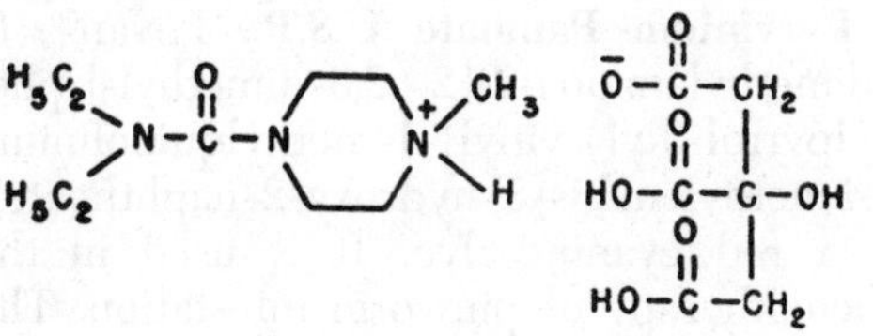

Diethylcarbamazine Citrate

is stable under varied conditions of climate and moisture.

Category—antifilarial.

Usual dose—2 mg. per kg. of body weight 3 times daily for 1 to 3 weeks.

OCCURRENCE

Diethylcarbamazine Citrate Tablets U.S.P.

Piperazine, Arthriticin®, diethylenediamine, dispermine, hexahydropyrazine, occurs as colorless, volatile crystals that are freely soluble in water or glycerol. It crystallizes as a hexahydrate from water. It can be made by warming ethylene chloride with ammonia in alcoholic solution.

$$2CH_2Cl-CH_2Cl + 6NH_3 \rightarrow$$
$$NH(CH_2-CH_2)_2NH + 4NH_4Cl$$

It was introduced into medicine because it will dissolve uric acid in a test tube, and it was hoped that this might be of service in gout and other rheumatic diseases. The clinical results, however, have been almost nil because the distribution of therapeutic doses could not be expected to furnish sufficient concentration. The claim that piperazine is a powerful diuretic has not been confirmed. Piperazine is used as a stabilizing buffer for the estrone sulfate ester (see Sulestrex®, a piperazine estrone sulfate).

After the discovery of the activity of the piperazine derivative, diethylcarbamazine, it was established that piperazine itself was active and is used commonly today as an anthelmintic for the treatment of pinworms (*Enterobius vermicularis; Oxyuris* v.) and roundworms (*Ascaris lumbricoides*) in children and adults. A number of salts of piperazine are available by brand names, usually in the form of a syrup or tablet. It appears

to function by inducing a state of narcosis in the worms. An important aspect of successful treatment is that the worms be voided before the effects of the drug have worn off. Piperazine and its salts have generally replaced gentian violet as the drug of choice in the treatment of human pinworm infections.

Piperazine Citrate U.S.P., Tripiperazine Dicitrate, Antepar® Citrate, Multifuge® Ci-

Piperazine Citrate

trate, Parazine® Citrate, Pipazin® Citrate, occurs as a white crystalline powder with a slight odor. It is insoluble in alcohol and soluble in water, a 10 per cent solution having a pH of 5 to 6.

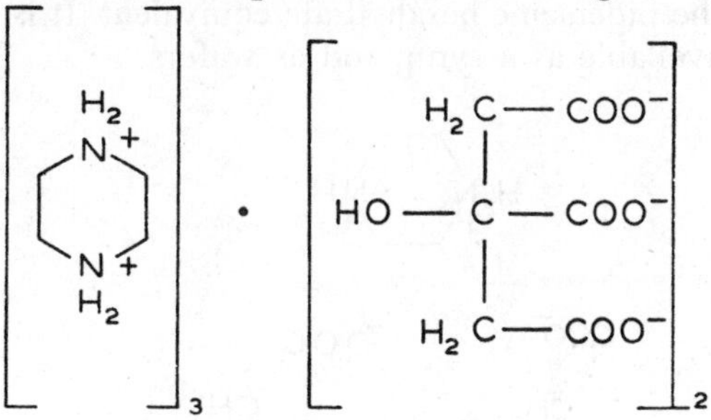

Piperazine citrate is administered orally. In some commercial products, the dose is expressed in terms of the equivalent amount of piperazine hexahydrate, i.e., 550 mg. anhydrous piperazine citrate equivalent to 500 mg. piperazine hexahydrate.

Category—anthelmintic (intestinal pinworms and roundworms).

Usual dose—against *Enterobius,* the equivalent of 2 g. of piperazine hexahydrate once daily for 1 week; against *Ascaris,* 3.5 g. once daily for 2 days.

Usual dose range—for *Enterobius,* up to 2.5 g. daily; for *Ascaris,* up to 3.5 g. daily.

OCCURRENCE

Piperazine Citrate Syrup U.S.P.
Piperazine Citrate Tablets U.S.P.

Piperazine Tartrate, Piperat® Tartrate is formed from piperazine hexahydrate and tartaric acid.

The administration and the dosage are the same as for other salts of the base. It is available as an oral solution and as tablets.

Piperazine Calcium Edathamil, Perin®,

is a chelated compound prepared by the action of ethylenediaminetetraacetic acid on piperazine and calcium carbonate. It occurs as crystals with a slightly saline taste, which are freely soluble in water but very slightly soluble in alcohol or chloroform. The pH of a 20 per cent aqueous solution is about 5.

Piperazine calcium edathamil is administered orally in doses expressed in terms of the piperazine hexahydrate equivalent. It is available as a syrup and as wafers.

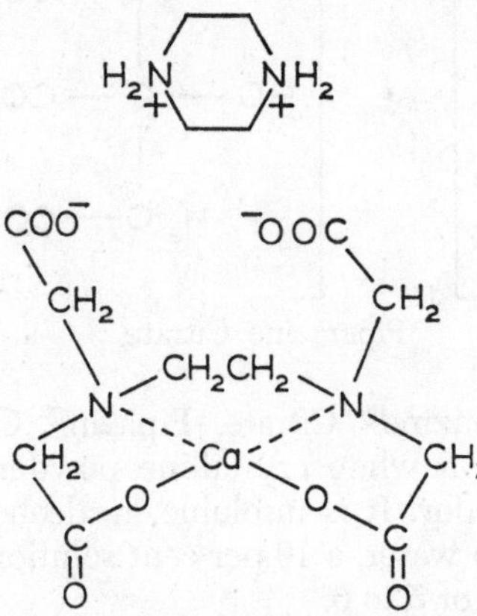

Piperazine Calcium Edathamil

Dyes Used as Anthelmintics

Gentian Violet U.S.P. is used in the treatment of pinworm, but has been largely replaced by piperazine and its salts. It is one of the few drugs effective in strongyloides infestations. For pinworm infestations, it is administered as enteric coated tablets before or with meals 3 times daily for 8 to 10 days, and for 16 to 18 days for strongyloides infestations. The adult dosage is 60 mg. 3 times daily; the dose for children should not exceed 90 mg. total daily dose. The usual size tablets are 10 and 30 mg.

Pyrvinium Pamoate U.S.P., Povan®, 6-(dimethylamino)-2-[2-(2,5-dimethyl-1-phenylpyrrol-3-yl)-vinyl]-1-methylquinolinium 4,4′methylenebis[3-hydroxy-2-naphthoate], is a red cyanine dye. It is used in the chemotherapy of pinworm infestation. The drug is sparingly soluble and poorly absorbed from the intestinal tract and because of its local irritant action, it may cause nausea and vomiting. If vomiting occurs before the drug has left the stomach the vomitus may be red colored; in addition, during treatment with the drug the feces will be frequently stained reddish brown.

Single dose treatment is usually highly effective in eradicating pinworm infestation in children and adults.

Category—anthelmintic (intestinal pinworms).

Usual dose—the equivalent of 5 mg. of pyrvinium per kg. of body weight in a single dose.

Occurrence
Pyrvinium Pamoate Oral Suspension U.S.P.
Pyrvinium Pamoate Tablets U.S.P.

Natural Products Used as Anthelmintics

Many natural products have been used in the past as anthelmintics, but only a few are used today. Even these are being supplanted by the newer drugs now available.

Aspidium, which is used as an anthelmintic, contains a number of compounds that are derived from filicinic acid, 1,1-dimethylcyclohexan-2,4,6-trione. The most im-

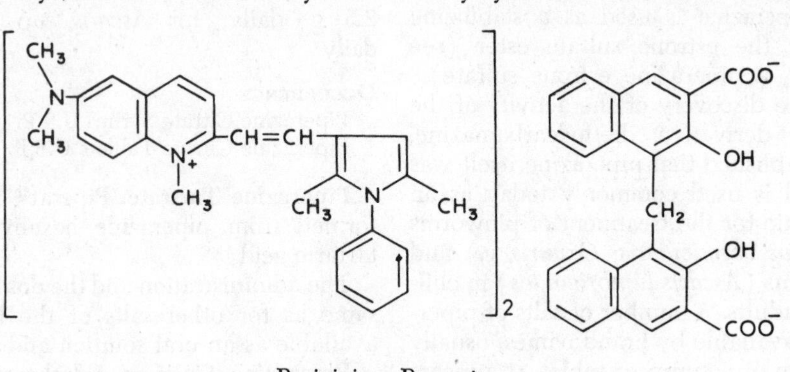

Pyrivinium Pamoate

portant of these are aspidinol, albaspidin and flavaspidic acid, all of which contain butyryl groups. As verified by synthesis, aspidinol[58] has the structural formula

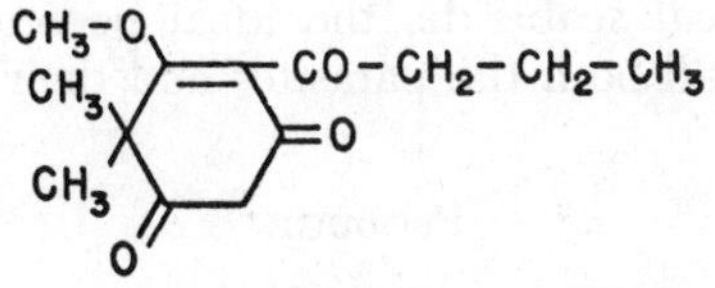

Aspidinol

The anthelmintic action is attributed to the mixture of compounds and none of these substances alone can replace the oleoresin in medicine.

The oleoresin is an ether extract of the dried rhizomes of *Dryopteris filix mas* and is a thick, dark-green liquid with a characteristic odor. It is a useful drug in the treatment of tapeworm infestations. The usual dose is 5 g., with a range of 3 to 5 g., and with a second dose given, if needed, only after a rest period of 7 to 10 days. Only saline laxatives should be used during the treatment.

Santonin; Santolactone. Santonin is the inner anhydride (lactone) of santoninic acid. It occurs as white or colorless crystals which soon become yellow on exposure to

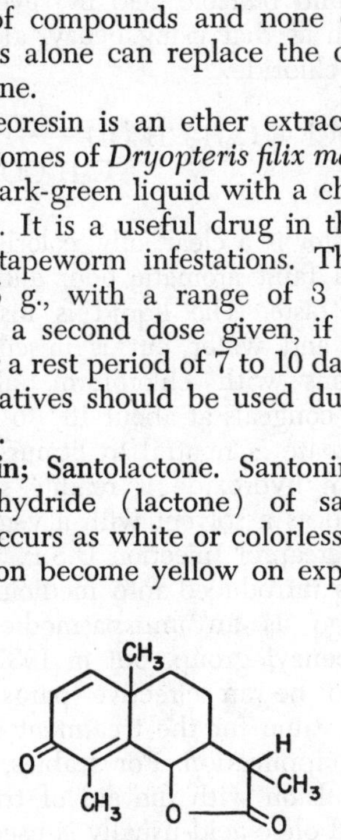

Santonin

light. The lactone is insoluble in water and soluble in organic solvents, forming yellow solutions.

Santonin has a definite anthelmintic action against worms of the ascarides family (roundworms). The action is one of rendering the worm unable to move normally and thus causing it to be carried out of the bowel in the feces. Structures having the lactone ring have been found to possess anthelmintic properties.

The usual dose is 60 mg.

Pelletierine Tannate is not a pure salt but is a mixture of the tannates of all the alkaloids in pomegranate bark. The proportion of alkaloids in the mixture is variable.

Pelletierine tannate exists as a light yellow, odorless, amorphous powder which is affected by light. It has an astringent taste. One gram is soluble in 250 ml. of water, and the compound is soluble in alcohol or warm dilute acids. It is only slightly soluble in ether and is insoluble in chloroform.

Alkali decomposes the preparation to yield the free alkaloids. In a saturated aqueous solution, the tannic acid portion reacts in the normal manner with ferric chloride to give a blue-black color.

The usual dose is 250 mg.

UNCLASSIFIED ANTHELMINTICS

Thiabendazole U.S.P., Mintezol®, is 2-(4-thiazolyl)benzimidazole. Thiabendazole is

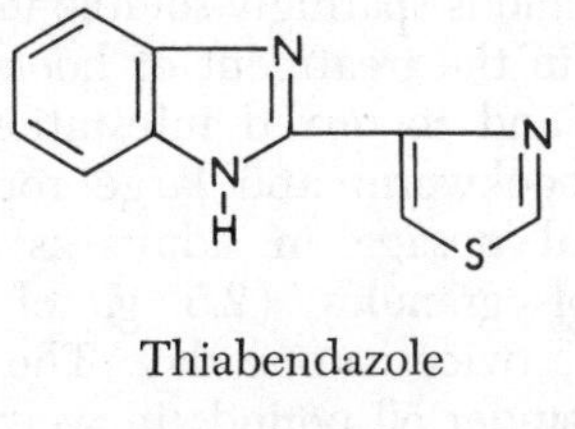

Thiabendazole

a stable compound, both as a solid and in solution. It forms colored complexes with metal ions, such as iron. It has a basic pKa of 4.7 and is only slightly soluble in water but becomes more soluble as the pH is raised or lowered; its maximum solubility is at pH 2.5, at which it will give a 1.5 per cent solution.[59]

Thiabendazole is effective in the treatment of several helminthic diseases. It has shown a high degree of efficacy against threadworm and pinworm, moderate effectiveness against large roundworm and hookworm, and less activity against whipworm. It has been used successfully in the treatment of cutaneous larva migrans (creeping eruption). There have been reports that, in several cases of trichinosis, relief of symptoms and fever has followed its use, but there is no evidence that it will eliminate the adult *Trichinella spiralis*. It is an odorless, tasteless, nonstaining compound and generally is administered as a suspension given after meals.

In addition to its use in human medication, it has been widely accepted for controlling gastrointestinal parasites in livestock. It is also highly active as a fungicide, and a wettable powder has been marketed to control stem-end rot and fruit spoilage in citrus fruit.

Category—anthelmintic (pinworm, threadworm, whipworm, and hookworm and in cutaneous larva migrans).

Usual dose—25 mg. per kg. of body weight twice daily for 2 days.

Usual dose range—up to 3 g. daily for 1 to 4 days.

OCCURRENCE

Thiabendazole Oral Suspension U.S.P.

Bephenium Hydroxynaphthoate, Alcopara®, benzyldimethyl (2-phenoxyethyl)-ammonium 3-hydroxy-2-naphthoate, is a pale yellow crystalline powder, with a bitter taste, and is sparingly soluble in water. It is useful in the treatment of hookworm infestation and in mixed infestations which include hookworm and large roundworm. The usual dosage in adults is one 5-g. packet of granules (2.5 g. of bephenium ion), twice in one day. The dose for children under 50 pounds in weight is one half this amount. Because of the bitter taste, the dose is generally mixed with milk, fruit juice, or carbonated beverage just prior to administration; no food should be taken for at least 2 hours afterwards.

ANTISCABIOUS AND ANTIPEDICULAR AGENTS

Antiscabious agents or scabicides are drugs used against the mite, *Sarcoptes scabiei*, which thrives when personal hygiene is neglected. The ideal scabicide must kill both the parasites and their eggs. Sulfur

preparations have been used for many years but are now being supplanted by more effective and less offensive agents. Antipedicular agents or pediculicides are used to eliminate head, body and crab lice. Like the ideal scabicide, the ideal pediculicide must kill both the parasites and their eggs.

PRODUCTS

Benzyl Benzoate U.S.P. This ester occurs naturally in Peru balsam and in some resins. It is prepared synthetically from benzyl alcohol and benzoic acid by several methods, such as that using benzyl alcohol and benzoyl chloride.

$$C_6H_5COCl + C_6H_5CH_2OH \longrightarrow$$
$$C_6H_5OOCH_2C_6H_5$$

The ester is a clear, oily, colorless liquid, having a faint aromatic odor and a sharp, burning taste. The liquid is insoluble in glycerin and water but is miscible in all proportions with chloroform, alcohol or ether. It congeals at about 18° to 20°. Benzyl benzoate is neutral to litmus and with potassium hydroxide is readily saponified. It is used as a solvent with a vegetable oil for Dimercaprol Injection U.S.P.

It was introduced into medicine several years ago as an antispasmodic because of the benzyl group, but in 1937[60] it was found to be an effective parasiticide of especial value for the treatment of scabies by local application. For scabies, a 25 per cent emulsion with the aid of triethanolamine and oleic acid usually is used. Due to some local anesthetic effect, there is instantaneous relief from itching. A single treatment often produces a complete cure.[61] Other advantages are absence of odor, no staining of clothes and no skin irritation. It is used topically, as a lotion over previously dampened skin of entire body, except the face.

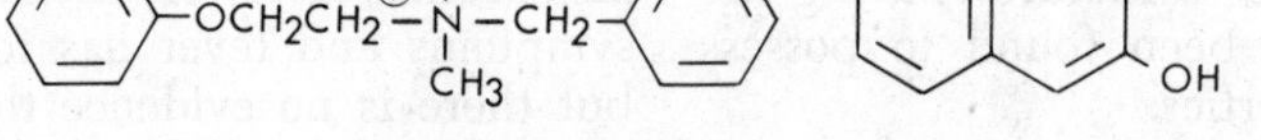

Bephenium Hydroxynaphthoate

PER CENT
OCCURRENCE   BENZYL BENZOATE
Benzyl Benzoate Lotion N.F. 25

Gamma Benzene Hexachloride U.S.P., Lindane®, γ-1,2,3,4,5,6-Hexachlorocyclohexane, Benzene Hexachloride (666, Gamex, B.H.C., Gammexane). This halogenated compound was prepared first in 1825 and has been a subject of research since that time. Bender, [62] in 1935, reported the value of benzene hexachloride as an insecticide in his patent dealing with its preparation by the addition of benzene to liquid chlorine.

Gamma Benzene Hexachloride

Benzene hexachloride is a mixture of a number of isomers, 5 of which have been isolated: alpha, beta, gamma, delta and epsilon. The gamma isomer, which is present to the extent of 10 to 13 per cent in the commercial product, is by far the most active form and is responsible for the insecticidal property. The gamma isomer is extracted with organic solvents and obtained in 99 per cent purity. This is known as Lindane. The formula given above corresponds to one of the optically active inositols which would, of course, have OH groups substituted for the chlorine.

In powder form, it has a light buff to tan color, a persistent musty odor and a bitter taste. It is insoluble in water but readily soluble in many organic solvents[63] such as xylene, carbon tetrachloride, methanol, benzene or kerosene. The compound is unusually stable in neutral or acid environments. It withstands the effect of hot water and may be recrystallized with hot concentrated nitric acid. In the presence of alkalis, such as dry lime or lime water, however, hydrogen chloride is split out readily, leaving a mixture of the isomers of trichlorobenzene.

Benzene hexachloride exhibits three modes of action against insects: (1) contact, (2) fumigant and (3) stomach poison.

Insects which are susceptible to it are affected most by the first two modes of action, since the effect is rapid and the contact or fumigant action is felt before enough material can be eaten to be lethal. Physiologically, the effect upon insects seems to be the same as with DDT, that is, the nervous system is affected first. It is widely used in the destruction of cotton insects, aphids of fruit and vegetables.

Toxicity of this compound to warm-blooded animals is about the same as, or lower than, DDT, and, although it may be irritating to some people, recent studies indicate that a considerable quantity would have to be ingested before any ill effects would be produced.

Pharmaceutically, it is used externally as a parasiticide in the form of lotions and ointments in the treatment of scabies and pediculosis.

Category—pediculicide; scabicide.

For external use—topically, 1 per cent in ointment or lotion, or in an inert base as a dusting powder once or twice a week.

OCCURRENCE
Gamma Benzene Hexachloride Cream U.S.P.
Gamma Benzene Hexachloride Lotion U.S.P.

Chlorophenothane U.S.P., 1,1,1-Trichloro-2,2-(p-chlorophenyl)ethane, DDT. Zeidler,[64] who first prepared the insecticide, DDT, obtained the compound by the reaction between chloral and chlorobenzene in the presence of sulfuric acid. Practically all of the processes used to prepare DDT are modifications[65] of the original method of Zeidler. In 1942,[66] its insecticidal properties were reported.

$$CCl_3\ H$$

DDT

DDT is a white, waxy solid with a very faint, fruity odor. Sometimes it is marketed as a cream or gray powder. It is practically insoluble in water but is soluble in organic solvents, such as xylene, methylnaphthalene and chlorhexanone. Deodorized kerosene solutions (about 4%) are used commonly for household sprays and aerosol sprays. It is stable in aqueous alkali, but in alcoholic alkali it loses hydrogen chloride and, consequently, its effectiveness.

Study of the action of DDT on certain insects indicates that it is both a contact and a stomach poison. The contact action appears to depend upon its effect on the nervous system of insects. It is distinctly toxic to warm-blooded animals when ingested or if absorbed through the skin.[67]

DDT as an insecticide has a wide range of effectiveness, particularly for lice (typhus), mosquitoes (malaria), house flies and moths. It is not effective against mites and thus should not be used alone in the treatment of scabies. There are reports of the emergence of resistant strains of insects; this has become a problem in some cases with the malaria-bearing mosquito.

Category—pediculicide.

For external use—topically, 5 or 10 per cent as a dusting powder once or twice a week.

Crotamitin, Eurax®, N-Ethyl-o-crotontoluide. This compound is a colorless, odorless oily liquid. It is practically insoluble in water, but dissolves in oils, fats, alcohol, acetone and ether. It is stable to light and air.

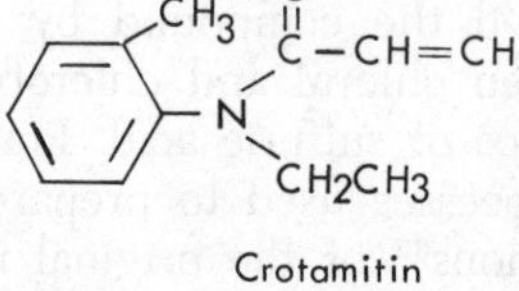

Crotamitin

Crotamitin, in the form of a lotion or in a washable ointment base, is used in the prevention and treatment of scabies. It also has an antipruritic action.

Chlordane, Octa-Klor®, Synklor®, 1,2,4,5,6,7,8,8-Octachloro-2,3,3a,4,7,7a-hexahydro-4,7-methanoindene. This chlorinated hydrocarbon was reported first in 1945.[68] Commercial grades are about 60 per cent $C_{10}H_6Cl_8$ and are dark-colored, viscous liquids that are insoluble in water but miscible with organic solvents. The organic solvents that are used are primarily aliphatic and aromatic hydrocarbons, including deodorized kerosene. Chlordane should not

be formulated with any material which has an alkaline reaction because it loses its chlorine in their presence and is converted to an inactive compound.

It exhibits a high order of toxicity to a wide range of insects and related arthropods. Lethal action on susceptible organisms may result from direct contact, from ingestion or from exposure to its vapor. This product is mild in action to warm-blooded

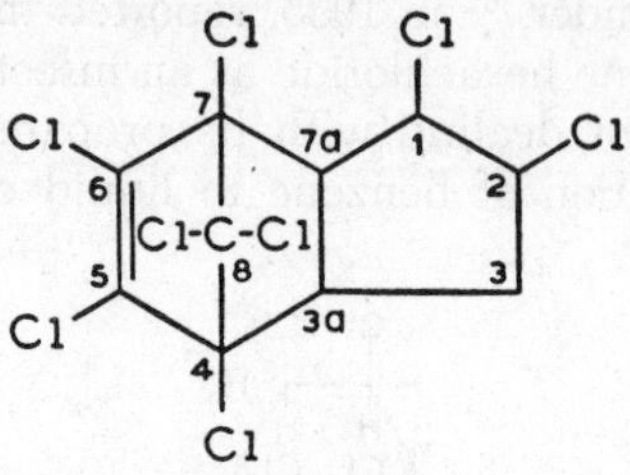

Chlordane

animals and, therefore, may be used safely for insect control under a wide variety of circumstances.

The usual concentration of the insecticide is about 2 per cent. It appears particularly effective against roaches, ants, crickets, ticks, fleas, pill bugs and silverfish. Other insecticides for household use, such as DDT, are quite inefficient against these insects.

A chlordane dust or powder consists of chlordane dispersed in clay or talc. It is used in veterinary medicine against fleas, lice and ticks.

Isobornyl Thiocyanoacetate, Technical; Bornate®. This is an impure form of the compound containing 82 per cent or more of isobornyl thiocyanoacetate with other

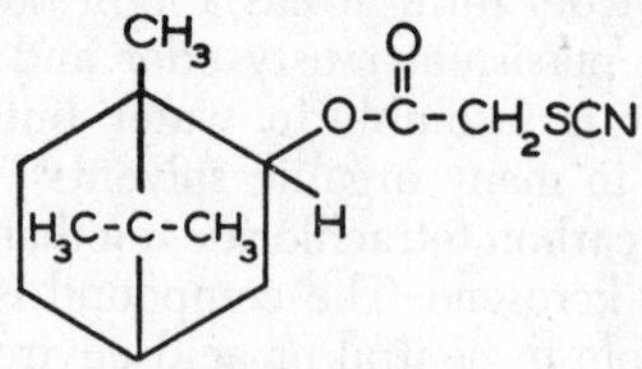

Isobornyl Thiocyanoacetate

terpenes. It has a terpenelike odor and exists as an oily, yellow liquid. Solubility in organic solvents is good, but emulsions are necessary with water.

In the form of about 5 per cent emulsions, it is available as Lotion Bornate® and Cidalon®. Applications externally have been found to be very effective in the control of pediculosis. It eradicates both the adult form and the ova of *Pediculus humanus capitis* (head louse) and *Phthirus pubis* (crab louse.)[69, 70]

Ethohexadiol U.S.P., 2-Ethyl-1,3-hexanediol, Ethylhexanediol (Rutgers 612); $CH_2OHCH(C_2H_5)CHOHCH_2CH_2CH_3$. It is a viscous, clear liquid, sparingly soluble in water and similar in physical and chemical properties to glycerin. This glycol is nonirritating to skin but is irritating to the eyes.

Ethohexadiol is not used as an anti-infective agent but as a repellant for mosquitoes, chiggers, black flies, gnats and most other biting insects.

Category—arthropod repellant.

For external use—topically, to skin and clothing.

Diethyltoluamide U.S.P., N,N-Diethyl-*m*-toluamide, is useful as a repellant for various kinds of inspects, especially mosquitos. It occurs as a colorless liquid with a faint, pleasant odor. Only the *meta*-isomer has activity as repellant.

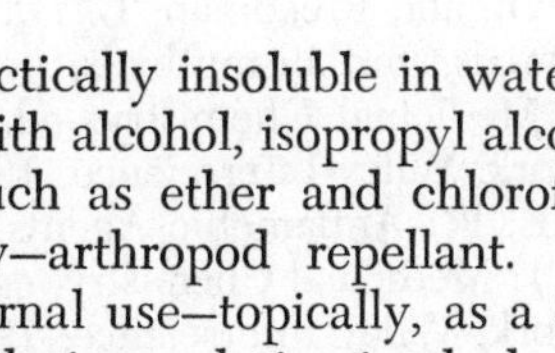

Diethyltoluamide

It is practically insoluble in water but is miscible with alcohol, isopropyl alcohol and solvents such as ether and chloroform.

Category—arthropod repellant.

For external use—topically, as a 50 to 75 per cent solution or lotion in alcohol to skin and clothing.

OCCURRENCE

Diethyltoluamide Solution U.S.P.

REFERENCES CITED

1. Deeb, E. N., and Boenigk, J. W.: J. Am. Pharm. A. (Sci. Ed.) 47:807, 1958.
2. Murphy, J. T., *et al.*: Arch. Ophthal. 53:63, 1955.
3. Nair, A. D., and Lach, J. L.: J. Am. Pharm. A. (Sci. Ed.) 48:390, 1959.
4. Macht, D. I.: J. Pharmacol. Exp. Ther. 11:263, 1918.
5. ——: J. Pharmacol. Exp. Ther. 11:419, 1918.
6. Dox, A. W.: J. Am. Chem. Soc. 46:2843, 1924.
   ——: J. Am. Chem. Soc. 55:1141, 1922.
   Dox, A. W., *et al.*: J. Am. Chem. Soc. 45:245, 1923.
7. Hirschfelder, A. D., *et al.*: J. Pharmacol. Exp. Ther. 15:237, 1920.
8. Hjort, A. M., and Eagan, F. T.: J. Pharmacol, Exp. Ther. 14:211, 1919.
9. Gilbert, G. L., *et al.*: Appl. Microbiol. 12:496, 1964.
10. Opfell, J. B., *et al.*: J. Am. Pharm. A. (Sci. Ed.) 48:617, 1959.
11. Grundy, W. E., *et al.*: J. Am. Pharm. A. (Sci. Ed.) 46:439, 1957.
12. Berl, E.: U. S. Patent 2,270,779, January 20, 1942; through Chem. Abstr. 36:3187[9], 1942.
13. Goshorn, R. H., and Degering, E. F.: Ind. & Eng. Chem. 30:646, 1938.
14. Bandelin, F. J.: J. Am. Pharm. A. (Sci. Ed.) 47:691, 1958.
15. Rahn, O., and Conn, J. E.: Ind. & Eng. Chem. 36:185, 1944.
16. Garrett, E. R., and Woods, O. R.: J. Am. Pharm. A. (Sci. Ed.) 42:736, 1953.
17. Nencki, M.: Arch. exp. Path. Pharmakol. 20:396, 1886.
18. Huyck, C. L.: J. Am. Pharm. A. (Pract. Ed.) 10:568, 1949.
19. Fuller, A. T: Biochem. J. 27:976, 1933; Lancet 1:855, 1935.
20. Rosenheim, M. L.: Lancet 1:1032, 1935.
21. Hemmholtz, H. F., and Osterberg, A. E.: JAMA 107:1794, 1936.
22. Gershenfeld, L.: J. Milk & Food Tech. 18:223, 1955.
23. Brost, G. A., and Krupin, F.: Soap Chem. Specialties 33:93, 1957.
24. Lawrence, C. A., *et al.*: J. Am. Pharm. A. (Sci. Ed.) 46:500, 1957.
25. Brown, E. A., *et al.*: J. Am. Pharm. A. Sci. Ed.) 35:304, 1946.
26. Miller, C. R., and Elson, W. O.: J. Bact. 57:47, 1949.
27. Baer, R. L., and Rosenthal, S. A.: J. Invest. Derm. 23:193, 1954.
28. Dodd, M. C., *et al.*: J. Pharmacol. Exp. Ther. 82:11, 1944.
29. Yurchenco, J. A., *et al.*: Antibiotics & Chemother. 3:1035, 1953.

30. Tucker, S. J., *et al.*: J. Am. Pharm. A. (Sci. Ed.) 47:849, 1958.
31. Albert, A., *et al.*: Brit. J. Exp. Path. 26: 160, 1945.
32. Tamm, I.: Yale J. Biol. Med. 29:33, 1956.
33. Ravin, L. J., and Gulesich, J. J.: J. Am. Pharm. A. NS4:122, 1964.
34. Linford, J. H.: Biochem. Pharmacol. 12: 317, 1963.
35. Parham, W. E., and Wilbur, Jr., J. M.: J. Org. Chem. 26:1569, 1961.
36. Skipper, H. E., and Bryan, C. E.: J. Nat. Cancer Inst. 9:391, 1949.
37. Fildes, P.: Brit. J. Exp. Path. 21:67, 1940.
38. Bechamp, M. A.: Compt. rend. Acad. sc. 56:1173, 1863.
39. Ehrlich, P., and Bertheim, A.: Ber. deutsch. chem. Ges. 40:3292, 1907.
40. Tatum, A., and Cooper, G.: Science 75: 541, 1932.
41. Voegtlin, C., and Smith, H.: Pub. Health Rep. 35:2264, 1920.
42. Rosenthal, S.: Pub. Health Rep. 47:933, 1932.
43. Eagle, H.: J. Pharmacol. Exp. Ther. 66: 436, 1939.
44. Albert, A.: Bull. Post-Grad. Comm. Med. Univ. Sidney 2:1, 1946.
45. Ellin, R. I., and Kondritzer, A. A.: J. Am. Pharm. A. (Sci. Ed.) 47:12, 1958.
46. Rheihlen, H., and Hezel, E.: Ann. der Chemie 487:213, 1931.
46a. Rich, A. R., and Follis, R. H., Jr.: Bull. Johns Hopkins Hosp. 62:77, 1938.
47. Külling, E.: Pharm. Acta Helvet. 34:430, 1959.
48. Schneller, G. H.: Am. Prof. Pharm. 18: 148, 1952.
49. Fox, H. H.: J. Chem. Ed. 29:29, 1952.
50. ——: Science 116:131, 1952.
51. Peck, S. M., *et al.*: Arch. Derm. 39:126, 1939.
52. Kenney, E. L.: Bull. Johns Hopkins Hosp. 73:379, 1943.
53. Keeney, E. L., *et al.*: Bull. Johns Hopkins Hosp. 75:417, 1944.
54. Schwartz, L.: Am. Prof. Pharm. 13:157, 1947.
55. Puls, D. D., *et al.*: J. Am. Pharm. A. (Sci. Ed.) 44:85, 1955.
56. Hall, M. C.: JAMA 77:1641, 1921.
57. Wright, W. H., and Schaffer, J. M.: Am. J. Hyg. 16:325, 1932.
58. Robertson, A. and Sandrock, W. J.: J. Chem. Soc. 1933, 819.
59. Robinson, H. J., *et al.*: Tox. Appl. Pharmacol. 7:53, 1965.
60. Kissmeyer, A.: Lancet 1:21, 1941.

61. MacKenzie, I. F.: Brit. M. J. 2:403, 1941.
62. Bender, H.: U. S. Patent 2,010,841, Aug. 13, 1935; through Chem. Abstr. 29:6607[7], 1935.
63. Chamlin, G. R.: J. Chem. Ed. 23:283, 1945.
64. Zeidler, O.: Ber. deutsch. chem. Ges. 7:1180, 1874.
65. Gunther, F. A.: J. Chem. Ed. 2:238, 1945.
66. Hughes, R. M.: British Patent 547,874, Sept. 15, 1942; through Chem. Abstr. 37: 6400[3], 1943.
67. Woodward, G., *et al.*: J. Pharmacol. Exp. Ther. 82:152, 1944.
68. Kearns, C. W., *et al.*: J. Econ. Ent. 38: 661, 1945.
69. Landis, L., *et al.*: J. Am. Pharm. A. (Sci. Ed.) 40:321, 1951.
70. Shelanski, H. A., *et al.*: Arch. Derm. 51: 179, 1945.

SELECTED READING

Albert, A.: Selective Toxicity, ed. 4, New York, Wiley, 1968.
Baker, J. W., Schumacher, I., and Roman, D. P.: Antiseptics and Disinfectants, *in* Burger, A. (ed.): Medicinal Chemistry, ed. 3, p. 627, New York, Wiley-Interscience, 1970.
Brown, A. W.: Anthelmintics, new and old, Clin. Pharmacol. Therap. 10:5, 1969.
Davis, W., and Larionov, L. F.: Progress in Chemotherapy of Cancer, Bull. W.H.O. 30: 327, 1964.
Doak, G. O., and Freedman, L. D.: Arsenicals, Antimonials, and Bismuthials, *in* Burger, A. (ed.): Medicinal Chemistry, ed. 3, p. 610, New York, Wiley-Interscience, 1970.
Elslager, E. F.: Antiamebic Agents, *in* Burger, A. (ed.): Medicinal Chemistry, ed. 3, p. 522, New York, Wiley-Interscience, 1970.
Haller, H. L.: Wartime development of insecticides, Ind. & Eng. Chem. (Ind. Ed.) 39: 467, 1947.
Jacobs, M. B.: Industrial Poisons, Hazards, Solvents, New York, Interscience, 1941.
Kunin, C. M.: Introduction to Chemotherapy, *in* Burger, A. (ed.): Medicinal Chemistry, ed. 3, p. 246, New York, Wiley-Interscience, 1970.
Lewis, A., and Shepherd, G.: Antimycobacterial Agents, *in* Burger, A. (ed.): Medicinal Chemistry, ed. 3, p. 409, New York, Wiley-Interscience, 1970.

Lubs, H. A.: The Chemistry of Synthetic Dyes and Pigments, pp. 622-686, New York, Reinhold, 1955.

McClure, H. B., and Bateman, R. L.: The synthetic aliphatic industry, Chem. Eng. News 25:3208, 3286, 1947.

Miura, K., and Reckendorf, H. K.: The nitrofurans, Progr. Med. Chem. 5:320, 1967.

Montgomery, J. A., Johnston, T. P., and Shealy, Y. F.: Drugs for Neoplastic Diseases, *in* Burger, A. (ed.): Medicinal Chemistry, ed. 3, p. 680, New York, Wiley-Interscience, 1970.

Montgomery, J. A.: On the chemotherapy of cancer, Progr. Drug Res. 8:431, 1965.

Osdene, T. S.: Antiviral Agents, *in* Burger, A. (ed.): Medicinal Chemistry, ed. 3, p. 662, New York, Wiley-Interscience, 1970.

Osdene, T. S.: Antiviral agents, Topics in Med. Chem. 1:137, 1967.

Paul, H. E., and Paul, M. F.: The nitrofurans —chemotherapeutic properties, *In*: Schnitzer, R. J., and Hawking, F. (eds.): Exper. Chemother., vol. 2, p. 307, New York, Academic Press, 1964.

Peacock, W. H.: The Application Properties of the Certified "Coal Tar" Colorants, Calco Technical Bulletin No. 715, American Cyanamid Company, Calco Chemical Division, Bound Brook, New Jersey.

Ralston, A. W.: Fatty Acids and Their Derivatives, New York, Wiley, 1948.

Robson, J. M., and Sullivan, F. M.: Antituberculosis drugs, Pharmaol. Rev. 15:169, 1963.

Shimkin, M. B.: Cancer chemotherapy, Topics in Med. Chem. 1:79, 1967.

Taylor, E. P.: Antifungal agents, Progr. in Med. Chem. 2:220, 1962.

Tomcufcik, A. S.: Chemotherapeutic Agents for Trypanosomiasis and other Protozoan Diseases, *in* Burger, A. (ed.): Medicinal Chemistry, ed. 3, p. 562, New York, Wiley-Interscience, 1970.

Tomcufcik, A. S., and Hardy, E. M.: Anthelmintics, *in* Burger, A. (ed.): Medicinal Chemistry, ed. 3, p. 583, New York, Wiley-Interscience, 1970.

Weinberg, E. D.: Antifungal Agents, *in* Burger, A. (ed.): Medicinal Chemistry, ed. 3, p. 601, New York, Wiley-Interscience, 1970.

Wheeler, G. P.: Studies related to the mechanisms of action of cytotoxic alkylating agents: a review, Cancer. Res. 22:651, 1962.

Woolfe, G.: The chemotherapy of amoebiasis, Progr. Drug Res. 8:11, 1965.

10

Phenols and Their Derivatives

Ole Gisvold, Ph.D.,
Professor of Medicinal Chemistry, College of Pharmacy,
University of Minnesota

PHENOLS

A phenol is a compound in which a hydrogen atom of an aromatic nucleus has been replaced by a hydroxyl group (OH). As

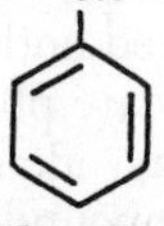

Phenol

such it becomes structurally a special tertiary alcohol. Because aromatic nuclei have a number of similar replaceable hydrogens, more than one hydroxy substitution is possible. For example, benzene has 6 similar replaceable hydrogens, and, thus, the maximum number of hydroxyl groups that can be introduced in benzene would be 6. When monohydroxy substitution in benzene alone is considered, only one product is possible, viz., hydroxybenzene or phenol itself, by virtue of all positions being similar. Obviously, the hydrogens of the aromatic nuclei also can be replaced by one or more of a number of groups such as:

Alkyl . . . . Thymol
Substituted alkyls . . Neosynephrine, Stilbestrol
Carboxyl . . . . Salicylic Acid
Aldehyde . . . . Salicylaldehyde
Halogens . . . . Tribromophenol
Nitro . . . . Picric Acid
Sulfonic acid . . . . . . Phenolsulfonic Acid

It follows that a large number of variously substituted monohydroxybenzenes or phenols are possible. A number of these compounds have been found in nature (e.g., thymol and carvacrol) or have been prepared synthetically (e.g., chlorothymol, chlorohexol and trinitrophenol) and are used as drugs in medicine.

In the case of dihydroxy substitution in benzene, three isomers are possible, for structural reasons, for example:

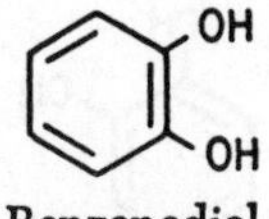

Catechol, 1,2-Benzenediol, *o*-Dihydroxybenzene, Pyrocatechol or 1,2-Dihydroxybenzene

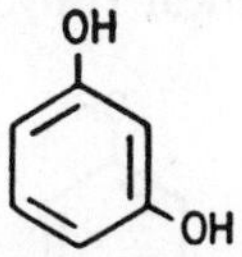

Resorcinol, 1,3-Benzenediol, Resorcin, *m*-Dihydroxybenzene or 1,3-Dihydroxybenzene

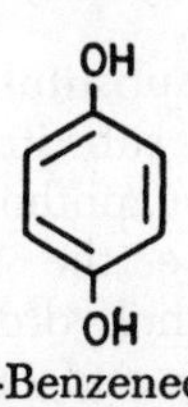

Hydroquinone, 1,4-Benzenediol, 1,4-Dihydroxybenzene, *p*-Dihydroxybenzene or Quinol

The hydrogen attached to the benzene nucleus also may be replaced by the groups as described under phenol, with the production of a number of variously substituted dihydroxybenzenes, such as:

Alkyl ....Eugenol and nordihydro-
                 guaiaretic acid
Substituted alkyl ..Epinephrine and coniferyl
                 alcohol
Carboxyl ....Protocatechuic acid
Aldehyde ....Vanillin
Alkyl ....Hexylresorcinol

A number of these compounds have been found in nature (e.g., eugenol, guaiaretic acid, nordihydroguaiaretic acid and urushiol), have been prepared from natural products (e.g., cresols from coal tar, guaiacol from wood tar and pyrogallol from gallic acid) or have been prepared synthetically (e.g., hexylresorcinol).

Only 3 trihydroxybenzenes also are possible, for structural reasons. These are:

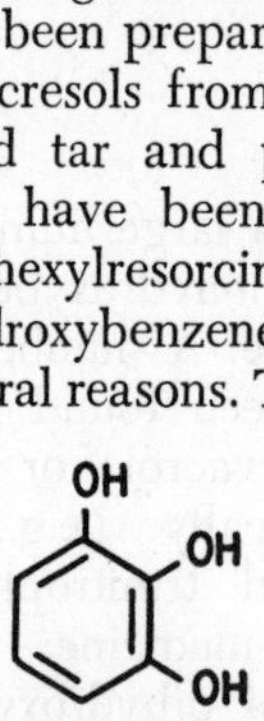

Pyrogallol, 1,2,3-Benzenetriol, 1,2,3-Trihydroxybenzene or Pyrogallic Acid

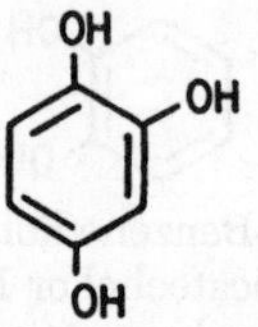

1,2,4-Benzenetriol, 1,2,4-Trihydroxybenzene or Unsymmetrical Trihydroxybenzene

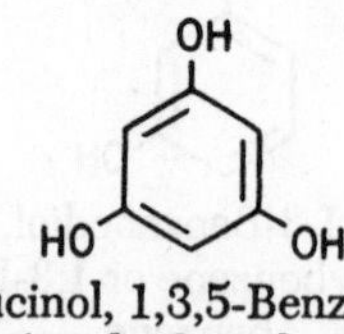

Phloroglucinol, 1,3,5-Benzenetriol, or 1,3,5-Trihydroxybenzene

In the hydroxy substituted naphthalenes, the monohydroxy substituted compounds, namely, 1- and 2-naphthols or alpha- and beta-naphthols, are the most significant. 2-Methyl-1,4-naphthohydroquinone is a precursor to menadione and, on a mole for mole basis, is as physiologically active as mena-

dione. (See section on vitamin K in Chapter 28.) Although many other polyhydroxy naphthols and substituted naphthols are possible, few are encountered in medicine. This is true of other polycyclic compounds also (anthracene and phenanthrene) that contain one or more phenolic hydroxyl groups. 1,8,9-Trihydroxyanthracene is used in medicine.

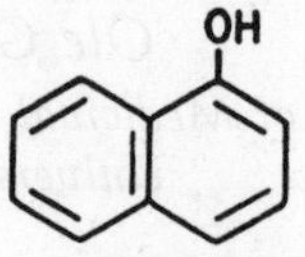

α-Naphthol 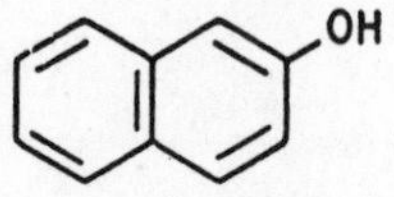   β-Naphthol

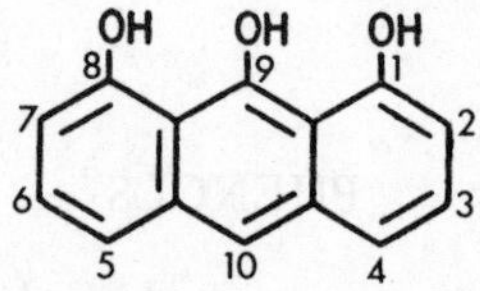

1,8,9-Trihydroxyanthracene

Anthrahydroquinone does not occur in nature. It is oxidized readily to anthraquinone. Substituted anthraquinones do occur in nature; they are phenolic compounds, such as barbaloin, aloin and rhein. No phenanthrene compounds containing phenolic hydroxyl groups are used as medicinal agents. However, partially reduced and substituted phenanthrene structures containing phenolic hydroxyl groups are present in some of the most useful drugs, such as morphine and estradiol.

PROPERTIES

Because phenols are characterized by the OH group, they undergo many of the reactions characteristic of alcohols (see Chapter 5).

Although phenols can be considered, from one standpoint, as tertiary alcohols, they differ from alcohols in that they are weakly acidic. This acidity is a function of the structure of the hydroxyl groups attached directly to the unsaturated nucleus or to the presence of the enolic grouping $-CH = C(OH)-$. The acidic properties of phenols are modified by electron-attracting or -repelling substituents. Thus, for example, phenol ($Ka = 1 \times 10^{-10}$) will form salts with sodium or potassium hydroxide but not

$$R-OH + NaOH \rightleftharpoons R-O^-Na^+$$

with their corresponding carbonates. On the other hand, the introduction of electron-attracting groups, such as nitro groups, increases the acidity of phenols, depending on the number of nitro groups introduced. For example, the mononitrophenols have acid dissociation constants of $1-7 \times 10^{-8}$; 2,4-dinitrophenol, $Ka = 1 \times 10^{-4}$, and trinitrophenol or picric acid is nearly as strongly acidic as a mineral acid and will decompose carbonates.

$$2C_6H_2(NO_2)_3OH + Na_2CO_3 \rightarrow$$
$$2C_6H_2(NO_2)_3ONa + H_2CO_3$$

Although, as a general rule, with the increase in the number of hydroxyl groups an increased solubility in water is obtained, this is not true of the polyhydroxyphenols, as is evidenced by the following solubility data.

COMPOUND	SOLUBILITY (GM. IN 100 ML. H_2O)
Catechol	45
Resorcinol	123
Hydroquinone	8
Pyrogallol	62
Phloroglucinol	7.7

Many phenols give characteristic colors with ferric chloride in very dilute aqueous or alcoholic solutions. Sometimes very strong alcoholic solutions are required to obtain a color test. These colors vary with the phenol and the number and position of the phenolic hydroxy groups. For example: phenol––violet, guaiacol––blue to green, cresols––blue, catechol––green, resorcinol––dark violet, pyrogallol––bluish black.

Phenols are susceptible to attack by oxygen of the air and by oxidizing agents, such as ferric chloride and chromic acid. Electron-attracting groups such as Cl, NO_2, etc., increase the stability of phenols, whereas electron-donating groups decrease their stability toward oxidation by air, etc. The oxidation of the polyhydroxy phenols by air is accelerated very markedly by the presence of alkali. In the case of catechol, it takes place in a matter of minutes, and with pyrogallol the reaction is quantitative. Those polyhydroxyphenols in which the hydroxyl groups are ortho or para to each other are particularly susceptible to oxidation. One form of oxidation as an initial step may consist of the removal of a hydroxylic hydrogen atom, with the formation of a free anion containing univalent oxygen. Such anions (semiquinones in some cases) are usually so unstable and reactive that they are quickly converted to the quinone or to other secondary products, such as quinhydrones, dimers and dehydrodiphenols (see below).

The sensitivity to oxidation of phenols and compounds containing phenolic hydroxyl groups is a problem with many drugs, and special storage conditions and preservatives, such as bisulfites, hydrosulfites and citric and tartaric acids, are used as stabilizing

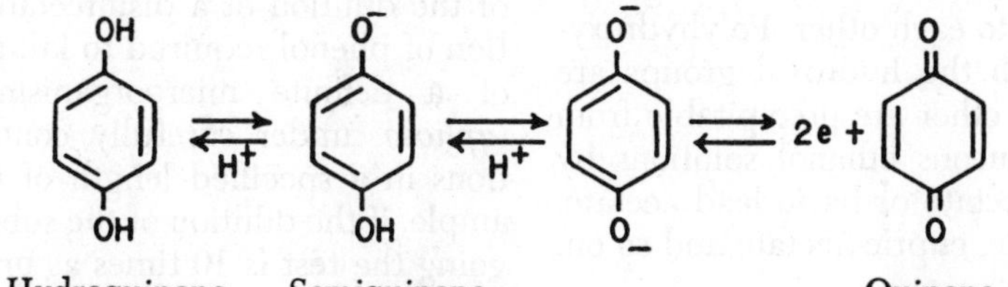

Hydroquinone   Semiquinone (anion)   Quinone (yellow)

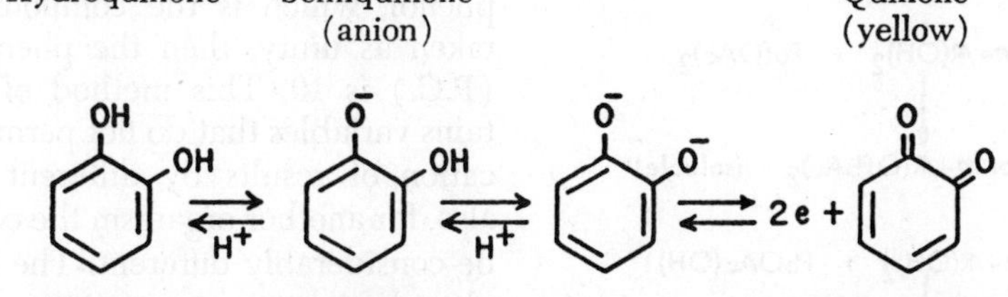

Catechol   Semiquinone (anion)   o-Benzoquinone (red)

agents to retard oxidation or prolong their usefulness. For example, catechol or drugs containing catechol groups (epinephrine) are marketed in solutions containing one of these preservatives to prevent the following change:

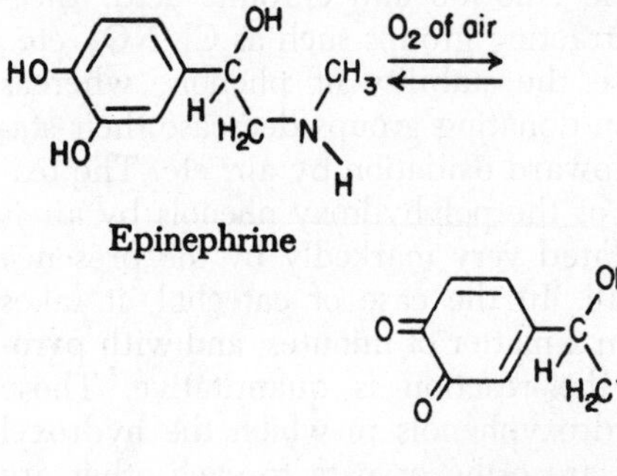

Epinephrine

Epinephrine Quinone

Sympathomimetic and other drugs which are para or ortho-hydroxybenzyl alcohol derivatives react with bisulfite to yield corresponding sulfonic acid derivatives. Degradation of epinephrine solutions stored in an oxygen-free atmosphere occurs at faster rates in the presence of as little as 0.1 per cent bisulfite than in complete absence of bisulfite.[1]

Monohydroxyphenols in general are not precipitable by lead acetate or the acetates of other heavy metals, e.g., barium; however, many are precipitated from solution by basic lead acetate. This is also true of polyhydroxyphenols in which the hydroxyl groups are not adjacent to each other. Polyhydroxyphenols in which the hydroxyl groups are adjacent to each other are precipitable from aqueous and aqueous ethanol solutions by means of lead acetate or basic lead acetate, barium hydroxide, cupric acetate and so on.

1. $AROH$ or $m\text{-}AR(OH)_2$ + $Pb(OAc)_2$
      ↓
      $AROPbAc$ or $m\text{-}AR(OPbAc)_2$ (soluble)

2. $AROH$ or $m\text{-}AR(OH)_2$ + $PbOAc(OH)$
      ↓
      $AROPbOH$ or $m\text{-}AR(OPbOH)_2$ (insoluble)

3. $o\text{-}AR(OH)_2$ + $Pb(OAc)_2$ or $Pb(OH)OAc$

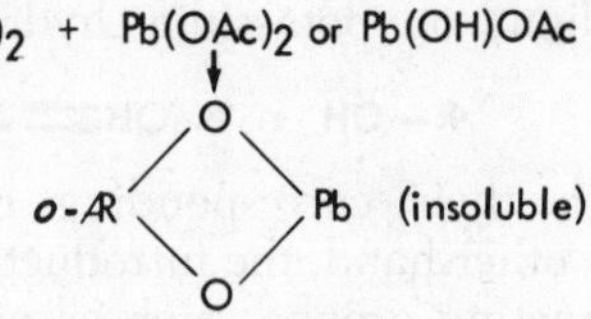

A number of polyhydroxyphenols, particularly those that have three adjacent hydroxyl groups, will combine with and precipitate or coagulate proteins. Catechol type phenols, such as urushiol, the toxic principle of the poison ivies, appear to combine quite firmly with protein. This has been demonstrated with hide powder.*

Some phenols such as 2,4-dihydroxybenzophenone, certain coumarins, salicylic acid, etc., absorb ultraviolet light and have been used as stabilizers, protectants, etc.

Phenols will condense readily with aliphatic and aromatic aldehydes with or without the aid of catalysts and/or heat. This property is particularly true of some of the polyhydric phenols, such as resorcinol, which will combine with formaldehyde in the cold. The extent of polymerization depends upon the ratio of aldehyde to the phenol.

Physiologic Properties

The bactericidal activity of most substances has been compared with that of Phenol U.S.P. as a standard, and this activity is reported as the phenol coefficient.[2, 3] The phenol coefficient is defined as the ratio of the dilution of a disinfectant to the dilution of phenol required to kill a given strain of a definite microorganism, *Eberthella typhosa*, under carefully controlled conditions in a specified length of time. For example, if the dilution of the substance undergoing the test is 10 times as great as that of phenol, which is the compound used and taken as unity, then the phenol coefficient (P.C.) is 10. This method of testing contains variables that do not permit easy duplication of results by different laboratories; also, for another organism the coefficient may be considerably different. The P.C. of many phenols is very temperature dependent.[3a]

* Unpublished experimentation by O. Gisvold.

Almost all phenolic compounds exhibit some antibacterial properties, and this activity is not too specific, although in some cases the phenol coefficients of a given phenol for *Eberthella typhosa* and *Staphylococcus aureus* may differ quite widely. The antimicrobial activity of phenols may be due to structural damage and alteration of permeability mechanisms of microsomes, lysosomes and cell walls. Phenol derivatives have caused leakage of radioactivity from *E. coli.*[4]

TABLE 13. PHENOL COEFFICIENTS OF SOME SUBSTITUTED PHENOLS AND SOME ANTISEPTICS[*]

COMPOUND	ORGANISM		
	E. typhosa	*Staph. aureus*	*Strep. hemolyticus*
Phenol	1.0	...	...
2-Chlorophenol	3.6	3.8	...
2-Bromophenol	3.8	3.7	...
3-Chlorophenol	7.4	5.8	...
4-Chlorophenol	3.9	3.9	...
4-Bromophenol	5.4	4.6	...
2,4-Dichlorophenol	13.0	13.0	...
2,4-Dibromophenol	19.0	22.0	...
2,4,6-Trichlorophenol	23.0	25.0	...
p-Methylphenol[†]	2.5§	...	...
p-Ethylphenol	7.5§	10.0	...
p-n-Propylphenol	20.0§	14.0	...
p-n-Butylphenol	70.0§	21.0	...
p-n-Amylphenol	104.0§	20.0	...
p-n-Hexylphenol	90.0§	...	...
Thymol	...	28.0‖	...
Chlorothymol	...	61.3	...
4-Ethylmetacresol[‡]	12.5§	...	...
4-*n*-Propylcresol	34.0§	...	...
4-*n*-Butylcresol	100.0§	...	...
4-*n*-Amylcresol	280.0§	...	...
4-*n*-Hexylcresol	275.0§	...	...
p-Methyl-*o*-chlorophenol	6.3	7.5	5.6
p-Ethyl-*o*-chlorophenol	17.3	15.7	15.0
p-n-Propyl-*o*-chlorophenol	38.0	32.0	35.0
p-n-Butyl-*o*-chlorophenol	87.0	94.0	89.0
p-n-Hexyl-*o*-chlorophenol	...	714.0	625.0
p-n-Amyl-*o*-chlorophenol	80.0	286.0	222.0
o-n-Amyl-*p*-bromophenol	63.0	571.0	...

[*] After Suter, C. M.: Rev. 28:269, 1941.
[†] Position shown to be unimportant.
[‡] The 4-alkylorthocresols and the 2-alkylparacresols assay much like the 4-alkylmetacresols.
§ 20°.
‖ 25°.

Because phenol itself is antiseptic, early workers (Ehrlich, 1906, and others) sought to improve its activity by modifications of its structure. The introduction of the halogens, chlorine or bromine, into the nuclei of phenols increases their antiseptic activities. This activity increases with the increase in the number of halogens introduced; however, the solubility decreases, thus rendering the polyhalogenated phenols much less useful. Furthermore, phenols, as well as their halogen derivatives, are too toxic for internal use. The introduction of nitro groups increases the antiseptic activity to a moderate degree, whereas carboxyl and sulfonic acid groups are ineffective or moderately effective. The introduction of alkyl groups into phenol, cresols and so on causes a marked increase in antiseptic activity. The structure and the size of the alkyl chain exert marked differences in their effects. Normal alkyl chains are more effective than isoalkyl chains, which in turn are more effective than secondary chains, and the tertiary chains are the least effective; however, the latter do exert considerable activity. Alkoxyl groups also increase the activity of phenols.

The monohydroxybiphenyls, the phenanthrols and the fluorenols have marked bactericidal and fungicidal properties.

Table 13 gives the phenol coefficients of some of the substituted phenols and of some of the better known antiseptics.[5]

The alkyl phenols, although powerful antiseptics, are too toxic for internal use and are used for skin sterilization, skin antiseptics and oral antiseptics.

Phenols exert a definite vermicidal activity which is enhanced by the presence of alkyl groups. The most effective anthelmintics in this group must have a solubility in water of a relatively low order (1:1,000 to 1:2,000) so as to prevent too great absorption from the stomach and intestines.

Phenols and their derivatives have antiseptic, anthelmintic, anesthetic, keratolytic, caustic, vesicant and protein precipitant properties. The extent of activity, in any one or more of the above properties, varies with the type of phenol, i.e., mono, di and trihydroxy substitution, and with the type and extent of substitution, i.e., alkyl, alkoxyl, acetoxy, halogen, nitro, sulfonic, and so on, groups.

As a rule, phenols are inactive in the presence of serum, possibly because they combine with serum albumin and serum globulin and, thus, are not free to act upon the bacteria. A second undesirable feature of phenols for bloodstream infections and other infections involving blood is the inhibitory effect upon leucocytic activity as compared with their activity to inhibit bacterial growth. Table 14 shows these relationships of phenol compared with some of the well-known bactericidal and bacteriostatic agents.[6]

Because phenols have a well-known tendency to bind to proteins, a limited number of substituted phenols have been examined for their serum albumin and mitochondrial protein binding properties. The binding properties depend on the lipophilic character of the substituent, and a linear free-energy relationship exists between the logarithm of the binding constants and substituent π ($\pi = \log P_x/P_H$, where P_H is the partition coefficient of a parent compound between octanol and water, and P_x is that of the derivative).[7]

O-Methylation

Some phenols and substituted phenols are metabolized in part by O-methylation.[8] An enzyme, phenol-O-methyltransferase, highly localized in the microsomes of liver and also present in other tissues, transfers a methyl

TABLE 14. PHENOL COMPARED WITH SOME WELL-KNOWN BACTERICIDAL AND BACTERISTATIC AGENTS*

ANTISEPTIC	LEAST CONC. NECESSARY TO INHIBIT GROWTH OF BACTERIA IN BLOOD	LEAST CONC. NECESSARY TO INHIBIT LEUCOCYTIC ACTIVITY	LEUCOCYTE BACTERIA RATIO
Phenol	1:320	1:1,280	1/4
Eusol	1:2	1:8	1/4
Acriflavine	1:100,000	1:500,000	1/5
Mercuric Chloride	1:2,500	1:20,000	1/8
Cetavlon	1:60,000 *Staph.*	1:2,000	30
Sulfanilamide	1:200,000 *Strep.*	1:200	1,000
Penicillin	1:50,000,000 *Strep.*	1:100	500,000

* After Fleming, A.: Chem. & Ind. 18, 1945.

group from S-adenosylmethionine to some simple alkyl, methoxy- and halophenols as substrates. Greater specificity of O-methylation appears to reside in catechol-O-methyltransferase (COMT), hydroxyindole-O-methyltransferase (found only in the pineal gland) and diiodotyrosine-O-methyltransferase.

Although recent studies have indicated that polyphenols augment the peripheral activities of epinephrine by inhibition of COMT, the central convulsant toxicity of catechol and pyrogallol is not related to their COMT inhibiting properties.[9]

PRODUCTS

Phenol U.S.P., Carbolic Acid. Phenol is monohydroxybenzene obtained from coal tar in 0.7 per cent yield by extraction with alkali. The phenolates are decomposed, and the phenolic fraction subjected to fractional distillation for separation and purification of the individual phenolic fractions. This source of phenol was not sufficient to meet the demand, and phenol is now synthesized on a commercial scale.[10]

Phenol is often called carbolic acid, and this terminology is derived from its weakly acidic properties. Its sodium and potassium salts (sodium and potassium phenolates) are soluble in water.

Phenol occurs as colorless to light pink, interlaced, or separate, needle-shaped crystals or as a white or light pink, crystalline mass that has a characteristic odor. It is soluble 1:15 in water, very soluble in alcohol, glycerin, fixed and volatile oils and is soluble in petrolatum and liquid petrolatums 1:70. Water is soluble 10 per cent in phenol.

Phenol is the most stable member of the group of phenols, although slight oxidation does take place upon exposure to air. It can be sterilized by heat and readily forms eutectic mixtures with a number of compounds, such as thymol, menthol and salol. Phenol is substituted readily by bromine to form the insoluble tribromo derivative.

Phenol is one of the oldest antiseptics, having been introduced in surgery by Sir Joseph Lister in 1867. In addition to its bactericidal activity, which is not very strong, it has a caustic and slight anesthetic action. Phenol is, in general, a protoplasmic poison and is toxic to all types of cells. High concentrations will precipitate proteins, whereas low concentrations denature proteins without coagulating them. This denaturing activity does not firmly bind phenol and, thus, it is free to penetrate the tissues. The action on tissues is a toxic one, and pure phenol is corrosive to the skin, destroying much tissue, and may lead to gangrene. Even the prolonged use of weak solutions of phenol in the form of lotions is apt to cause tissue damage and dermatitis.

Phenol is used commonly in 0.1 to 1 per cent concentrations as an antipruritic in phenolated calamine lotion or as an ointment or simple aqueous solution. Aqueous solutions stronger than 2 per cent should not be applied to the surface of the body. Pure phenol in very small amounts may be used to cauterize small wounds. A 4 per cent solution in glycerin may be used if necessary. Crude phenol is cheap enough to be used for a disinfectant. Phenol is too soluble and too readily absorbed to be of value as an intestinal antiseptic.

Category—pharmaceutic aid (antimicrobial agent).

Liquefied Phenol U.S.P. Liquefied carbolic acid is phenol maintained in a liquid state by the presence of 10 per cent of water. Liquefied phenol is a solution of water in phenol and is a convenient way in which to use it in most pharmaceutic applications. However, its water content precludes its use in fixed oils, petrolatum and liquid petrolatum.

Category—topical antipruritic.

For external use—topically, 0.5 to 2 per cent, in lotion and ointments for use on skin.

OCCURRENCE	PER CENT LIQUEFIED PHENOL
Phenolated Calamine Lotion U.S.P.	1

Parachlorophenol N.F., 4-Chlorophenol. Mono-*p*-chlorophenol is prepared by the chlorination of phenol under conditions which will yield predominantly the para-isomer, which can be separated very easily from the ortho isomer by fractional distillation.

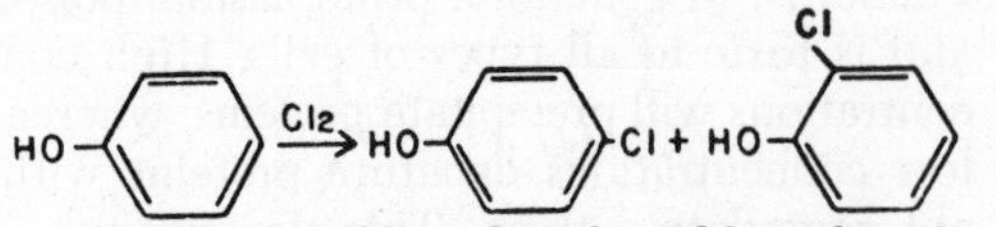

Phenol　　*p*-Chlorophenol　*o*-Chlorophenol

Parachlorophenol occurs as white or pink crystals that have a characteristic phenolic odor. It is sparingly soluble in water or liquid petrolatum, very soluble in alcohol, glycerin, fixed and volatile oils and is soluble in petrolatum.

Parachlorophenol has a P.C. of about 4 and is used in combination with camphor in liquid petrolatum for intratracheal installation in acute chronic laryngitis, tracheitis and tracheobronchitis.

The introduction of chlorine, although it markedly increases the antiseptic value of the parent compound, also decreases the solubility in water. Therefore, the polychlorinated phenols have little application in pharmacy. However, it is worthy of note that pentachlorophenol is an outstanding commercial wood preservative by virtue of its powerful fungicidal properties. It has a phenol coefficient of 50.

Category—local anti-infective (dental).

	PER CENT PARACHLORO-
OCCURRENCE	PHENOL
Camphorated Parachlorophenol N.F.	35

Hexachlorophene U.S.P.; 2,2′-Methylene-bis(3,4,6-trichlorophenol); Gamophen®; Surgi-Cen®; pHisoHex®; Hex-O-San®; Germa-Medica®; 2,2′dihydroxy-3,5,6,3′,5′,6′-hexachlorodiphenylmethane. It is synthesized as follows:

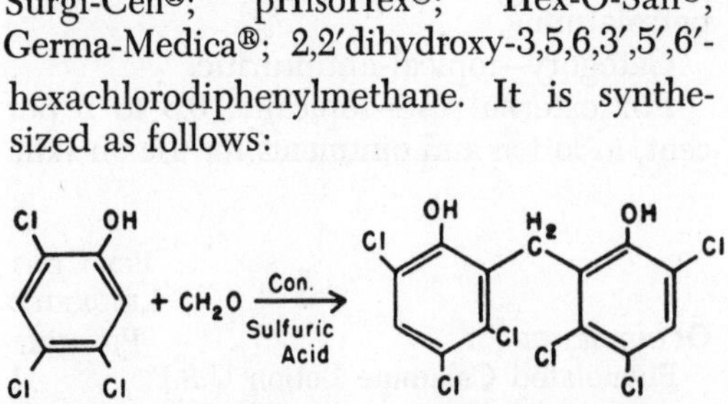

2,4,5-Trichlorophenol　　**Hexachlorophene**

It is a white crystalline powder that is insoluble in water, soluble in alcohol, acetone, the lipoid solvents and is stable in air.

Most bisphenolic compounds are far more effective than the monomers; moreover, their chlorine content further increases the antiseptic activity (P.C., 40 for *S. Aureus* and 15 for *Salm. typhi*).[11] This type of compound is deposited on the skin either through a combination with the epidermis or the sebaceous glands or both. Its usefulness as an antiseptic in low concentrations, is, therefore, prolonged. Hexachlorophene is incorporated in soaps, detergent creams, oils and other suitable vehicles for topical application in 2 to 3 per cent concentrations. It is effective against gram-positive bacteria, whereas gram-negative organisms are much more resistant to its action.

Some surface-active agents such as Tween-80 markedly decrease the activity of hexachlorophene.[12]

OCCURRENCE	PER CENT
Hexachlorophene Liquid Soap U.S.P.	0.225-0.26

Many bisphenols have pronounced fungicidal and bactericidal activities.[13, 14]

Zinc Phenolsulfonate. Zinc sulfocarbolate is essentially a chemically pure zinc salt of *p*-hydroxybenzene sulfonic acid that is prepared by the sulfonation of phenol, followed by neutralization with zinc oxide or zinc carbonate. It occurs as colorless crystals or

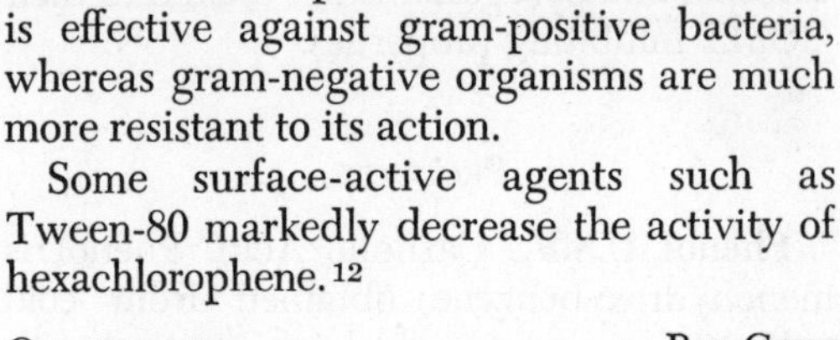

as white granules or a powder that is soluble 1:1.6 in water and 1:1.8 in alcohol at 25°. It has an astringent metallic taste. Zinc phenolsulfonate probably is used primarily for its astringent properties. Phenolsulfonic acid is an important agent in dentistry.

Cresol N.F. Cresol, also called cresylic acid and tricresol, is a mixture of three iso-

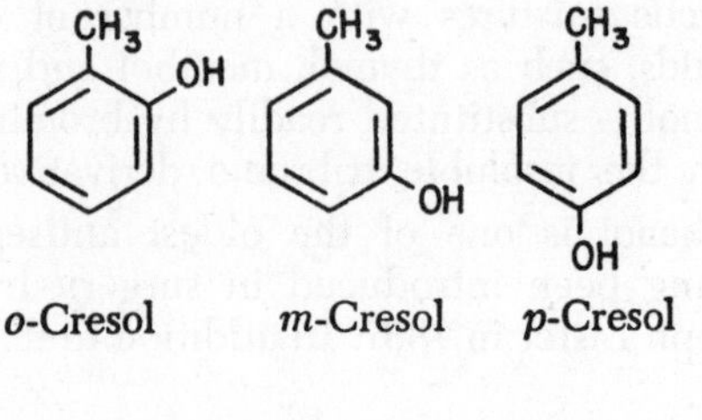

o-Cresol　　　*m*-Cresol　　*p*-Cresol

meric cresols obtained from coal tar or petroleum. The alkali-soluble fraction of coal tar is subjected to fractional distillation, and the three isomeric hydroxytoluenes are obtained as one fraction because they are not readily resolved into pure entities.

Cresol is a colorless, yellowish to brownish-yellow or pinkish, highly refractive liquid that has a phenol-like, sometimes empyreumatic odor. It is soluble in alcohol or glycerin, and 1 ml. is soluble in 50 ml. of water.

By virtue of the methyl groups, the cresols have a P.C. for *E. typhosa* of about 2.5. Cresol supplies the need of a cheap antiseptic and disinfectant, and it is a constituent of Saponated Cresol Solution N.F. Because only 1 g. of cresol is soluble in 50 ml. of water, it is emulsified for this preparation to the extent of 50 per cent by the potassium soaps of vegetable oils. This stock solution can be diluted with water to give a solution, a colloidal suspension or an emulsion of cresol of the desired strength. Cresol may be used as a 2 per cent solution to disinfect the hands. A 1:500 dilution of saponated cresol solution sometimes is used as a mildly antiseptic vaginal douche. Salts of the sulfonic acid derivatives, such as calcium cresol sulfonate, are used in some commercial cough preparations.

Category—disinfectant.

Application—1 to 5 per cent solution on dishes, utensils, and other inanimate objects.

OCCURRENCE	PER CENT CRESOL
Saponated Cresol Solution N.F.	50

Meta-Cresylacetate, Cresatin®, is the acetyl ester of *m*-cresol. It can be prepared by heating *m*-cresol with acetic anhydride.

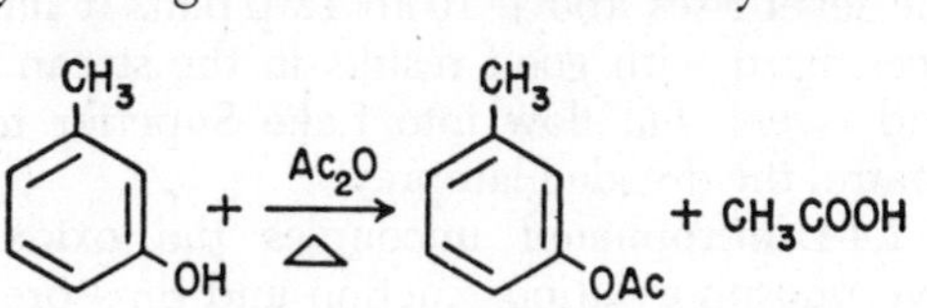

Meta-cresylacetate occurs as a colorless oily liquid, possessing a characteristic odor. It is practically insoluble in water, but it is soluble in the ordinary organic solvents, in liquid petrolatum (not over 5%) and in fixed and volatile oils.

The acetylation of the phenolic group in *m*-cresol produces a compound that slowly liberates the free phenol upon contact with tissues. This property lowers the toxicity and the corrosive action of *m*-cresol so that it can be used for the treatment of infections of the nose, the throat and the ear.

Meta-cresylacetate may be used pure in ointment or in solution in alcohols or oils.

Thymol U.S.P. Thymol or thyme camphor is isopropyl *m*-cresol. It is prepared from the oil of thyme (*Thymus vulgaris*) by extraction with alkali and subsequent acidulation of the alkaline extract to liberate the thymol. It can be distilled to effect further purification if necessary. Thymol also occurs in a number of other plants belonging to the mint family. It can be prepared from (−)-menthone by bromination followed by treatment with quinoline or by the treatment of piperitone (from Australian eucalyptus oils) with ferric chloride or from *m*-cresol and isopropyl alcohol or propylene in the presence of a suitable catalyst.

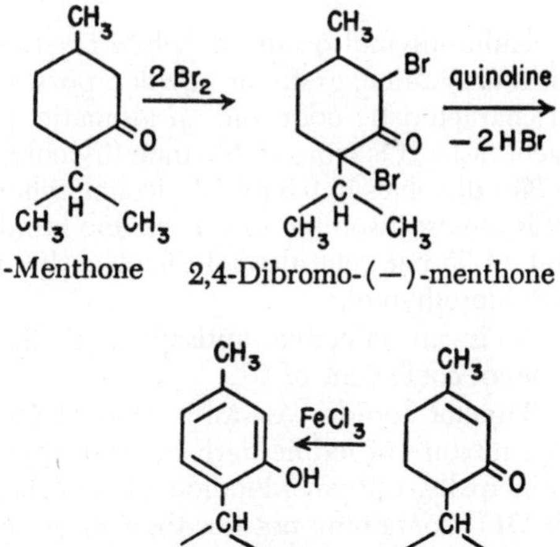

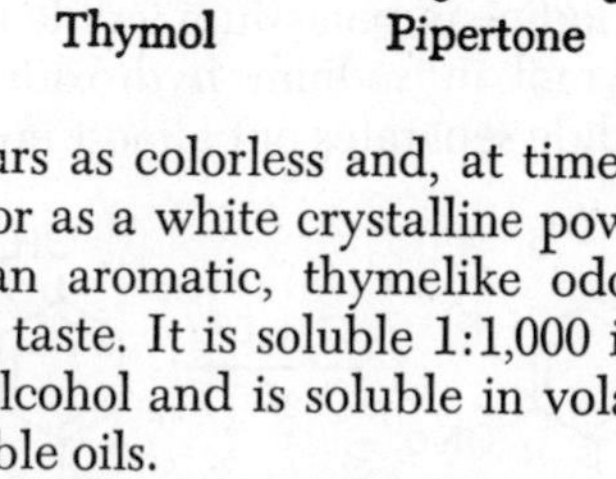

Thymol occurs as colorless and, at times, large crystals, or as a white crystalline powder that has an aromatic, thymelike odor and a pungent taste. It is soluble 1:1,000 in water, 1:1 in alcohol and is soluble in volatile and vegetable oils.

Although thymol is affected by light, it is quite stable and can be sterilized by heat. It forms eutectic mixtures (see Phenol).

Thymol is used in Trichloroethylene U.S.P. at a level of about 0.01 per cent as an antioxidant preservative.

Thymol has fungicidal properties and is effective in controlling dermatitis caused by pathogenic yeasts. It is effective in a 1 per cent alcoholic solution for the treatment of epidermophytosis and in a 2 per cent concentration in dusting powders for ringworm; it can be used both orally and locally for actinomycosis infections. Its use as an intestinal antiseptic is declining in favor of other, superior drugs.

Chlorothymol is 6-chlorothymol and is prepared by the chlorination of thymol.

Thymol   Chlorothymol

Chlorothymol occurs as white crystals or as a crystalline, granular powder, possessing a characteristic odor and an aromatic, pungent taste. It is more stable than thymol. One gram dissolves in 0.5 ml. of alcohol, whereas it is almost insoluble in water. One hundred ml. of 75 per cent alcohol dissolves 100 mg. of chlorothymol.

It is an effective antiseptic. It has a phenol coefficient of 61.

Thymol Iodide; Aristol®. Thymol iodide is a mixture of iodine derivatives of thymol, principally dithymoldiiodide ($C_6H_2CH_3C_3-H_7OI)_2$ containing not less than 43 per cent of iodine. It is prepared by adding a solution of iodine in potassium iodide to a solution of thymol in sodium hydroxide. The thymol iodide separates out almost immediately.

Sodium Thymolate   Thymol Iodide

Thymol iodide occurs as a reddish-brown or reddish-yellow, bulky powder, with a very slight, aromatic odor. It is insoluble in water or glycerin, slightly soluble in alcohol and soluble in fixed and volatile oils and in collodion.

Thymol iodide is affected by light and loses iodine upon heating; this reaction is no doubt due to the unstable hypo-iodide structure.

Thymol iodide is used as an antiseptic dusting powder for local application in dermatosis. It also can be used in ointments, oils and collodions.

o-Phenylphenol, Dowicide® 1, occurs as white or light buff to pink, free flowing flakes that are insoluble in water and are soluble in alcohol 6:1, in propylene glycol 3:1, and in olive oil 0.5:1. Dowicide® A is the sodium salt of Dowicide® 1; it is water-soluble 1.25:1 and is soluble in alcohol 3.35:1. This substituted phenol has a broad spectrum of activity and is effective against gram-positive and gram-negative bacteria as well as fungi, algae and certain viruses. It is one member of a family of substituted phenols that are economically effective antimicrobial agents. It is fungicidal in 0.05 to 0.006 per cent concentrations, depending upon the fungus tested.

Clorophene, Santophen 1(Monsanto), 4-choloro-α-phenyl-O-cresol is an economically useful antimicrobial agent.

Fluorophenols. Fluorine when introduced into alkyl phenols does not exhibit the same uniformity of activity as does chlorine.

3-Trifluoromethyl-4-nitrophenol kills lamprey larvae at concentrations of 0.5 to 4 p.p.m. and does not start to kill fish unless the level rises above 10 to 12 p.p.m. It has been used with good results in the streams and rivers that flow into Lake Superior to control the dreaded lamprey.[15]

2,4-Dinitrophenol uncouples the oxidative phosphorylation reaction and thus prevents the formation of ATP from ADP. It previously was used as a reducing agent but is not considered safe.

A number of substituted phenols are effective uncouplers[16] of oxidative phosphorylation, and a lack of pronounced steric effects suggests that such phenols might play a critical role by accepting a proton at some key enzymatic site. 2,4 Dinitrophenol at toxic levels, when administered locally, has anti-inflammatory activity that has little to do with it uncoupling properties.[17]

Trinitrophenol; Picric Acid. This compound is the 2,4,6-trinitro derivative of phenol. It occurs as pale yellow prisms or scales that are odorless but have an intensely bitter taste. One gram dissolves in 80 ml. of water and in 12 ml. of alcohol. Because trinitrophenol may explode when heated rapidly or when subjected to percussion, it usually is admixed with 10 to 20 per cent of water for safety in transportation. Most, if not all, polynitro compounds are explosive.

Trinitrophenol will decompose carbonates in solution to form the corresponding salt of picric acid. It precipitates proteins, alkaloids, amines and certain aromatic hydrocarbons to form the corresponding picrates. In the case of the nitrogenous substances, these picrates may be looked upon as salts which are not very soluble. It will dye wool and other animal fibers yellow.

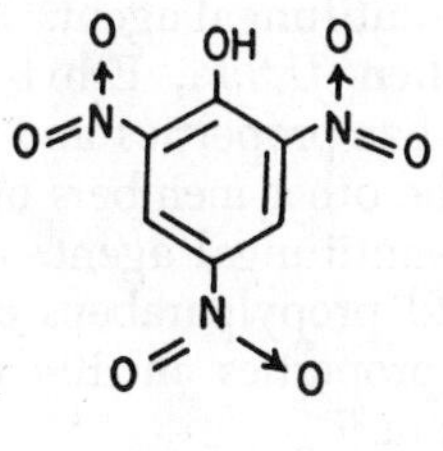

Picric Acid

Picric acid has a P.C. of 5.9 for *E. typhosa*. This antiseptic activity, together with its ability to form insoluble picrates with proteins and nitrogenous bases, makes picric acid useful in the treatment of burns in 1 per cent solutions. However, the uncombined picric acid may be absorbed, and large amounts cause toxic effects. It is also present in Butesin (butyl aminobenzoate) Picrate, a constituent of an ointment that is used for the treatment of burns. The butyl aminobenzoate is present for its local anesthetic activity.

Betanaphthol is betahydroxynaphthalene and is prepared synthetically as follows:

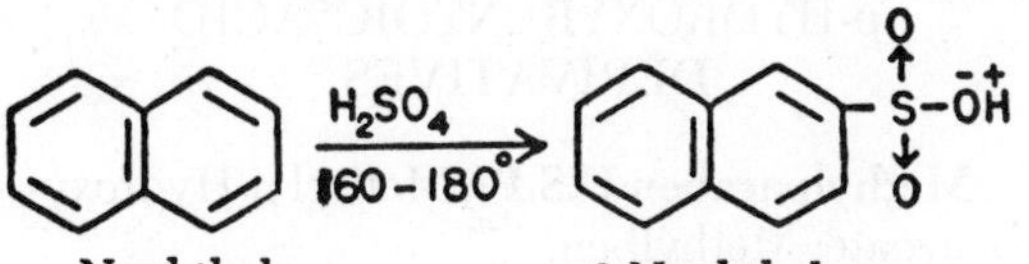

Naphthalene   β-Naphthalene Sulfonic Acid

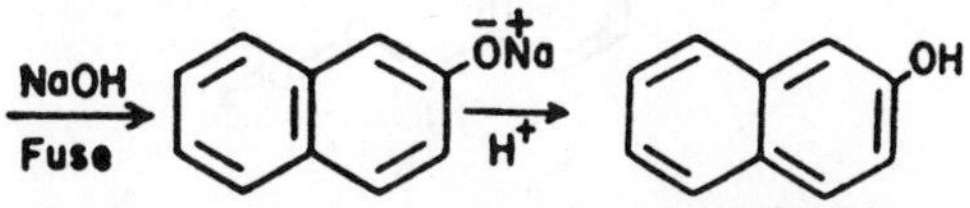

β-Naphthol

Betanaphthol occurs as pale buff-colored, shining crystalline leaflets or as a white or yellowish-white, crystalline powder. One gram is soluble in 1,000 ml. of water or in 1 ml. of alcohol. It is soluble in glycerin and olive oil. It has a faint, phenol-like odor and is stable in air.

Betanaphthol, because of its low solubility in water, is not absorbed readily from the intestinal tract. It is used as an intestinal antiseptic and is safer for internal medication than alphanaphthol, which is more toxic. Nevertheless, betanaphthol is toxic in larger doses and, even though it is effective against hookworm, other drugs are superior. When used for hookworms, it is used in conjunction with purgatives.

When applied locally, betanaphthol has irritant, corrosive, germicidal, fungicidal, parasiticidal and some local anesthetic activity. Five or 10 per cent concentrations in ointment form usually are employed for the treatment of ringworm, psoriasis and pediculosis.

The antibacterial and antifungal properties of β-naphthol derivatives have been reported.[18]

β-Naphthol Benzoate (Benzonaphthol) is a solid ester prepared synthetically from β-naphthol and benzoic acid. It occurs as tasteless, colorless, crystals that are insoluble in water but soluble in organic solvents.

Internally, it has weak antiseptic properties which have led to its use as an intestinal antiseptic in the treatment of diarrhea and dysentery, and it is used as a vermifuge. Externally, in the form of ointments (3 to 10%), it is used as a parasiticide but is less effective against scabies than is benzyl benzoate.

p-HYDROXYBENZOIC ACID DERIVATIVES

Methylparaben U.S.P.; Methyl *p*-Hydroxybenzoate; Methylben.

p-Hydroxybenzoic acid may be prepared by the same procedure used for salicylic acid, except that a temperature of about 200° is necessary when the carbon dioxide reacts with the sodium phenolate. The free acid possesses only slight antiseptic action, but when esterified (for example, with the alcohols methyl, ethyl, propyl or butyl) the resulting compound is very active. Methyl *p*-hydroxybenzoate occurs as small, white crystals or as a crystalline powder which has a slightly burning taste, and a faint, characteristic odor or none at all. It is soluble in water (1:400), alcohol (1:2) and ether (1:10) and slightly soluble in benzene and in carbon tetrachloride.

The *p*-hydroxybenzoic acid esters have a low order of acute toxicity and are less toxic than the esters of salicylic or benzoic acid and increase in toxicity as the molecular weight increases, the butyl ester being about 3 times as toxic as the methyl ester. Hydrolysis in vivo would yield *p*-hydroxybenzoic acid, which has a low order of toxicity.[19]

No irritation was obtained when ointments containing 10 per cent of the methyl or propyl esters were in contact with rabbit skin for 48 hours. The preservative effect of these esters also increases with the molecular weight; the methyl ester is more effective against molds and the propyl ester more effective against yeasts. The ester grouping apparently behaves like an alkyl group (see Hexylresorcinol) in its effect on the antiseptic action. Oil solubility increases along with the size of the ester group, thus making the propyl ester better than the methyl for oils and fats. The ester may be used to preserve almost any pharmaceutical.[20]

Clinical research has indicated that methyl- and propyl-paraben prevented the overgrowth of monilia, the most frequently occurring fungus infection associated with antibiotic therapy.

Category—antifungal agent.

OCCURRENCE	PER CENT METHYLPARABEN
Hydrophilic Ointment U.S.P.	0.025

Propylparaben U.S.P.; Propyl *p*-Hydroxybenzoate; Propylben. This ester is prepared and used in the same manner as methylparaben. It occurs as a white powder or as small, colorless crystals which are only slightly soluble in water (1:2,500) and soluble in most organic solvents. It is used as a preservative, primarily against yeasts (see methylparaben).

Category—antifungal agent.

OCCURRENCE	PER CENT PROPYLPARABEN
Hydrophilic Ointment U.S.P.	0.015

Butylparaben U.S.P., butyl *p*-hydroxybenzoate occurs as small, colorless crystals or white powder. It is very slightly soluble in water or glycerol but is very soluble in alcohols and in propylene glycol.

Category—antifungal agent.

Ethylparaben U.S.P., Ethyl *p*-Hydroxybenzoate. It has properties and uses similar to those of the other members of this group.

Category—antifungal agent.

Methyl and propylparabens exert growth stimulating properties in 10-day-old chick embryo femora.[21]

CATECHOL AND DERIVATIVES

Catechol, pyrocatechin, orthodihydroxybenzene, is a white, crystalline solid that has feeble antiseptic properties, i.e., the P.C. for *E. typhosa* is 0.87 and 0.58 for *Staph. aureus*. It is not used in medicine as such, but some of its derivatives are recognized. These are guaiacol; guaiacol carbonate and eugenol.

Although catechol per se has feeble antiseptic properties, some of its alkyl derivatives are very effective. These compounds are not used because they also have undesirable properties, such as marked instability and vesicant properties. The bis-catechol compounds have very pronounced antiseptic activities but also suffer from some of the above mentioned disadvantages. In addition, they exhibit low solubilities in water. Phenolic compounds that have two or more phenolic hydroxyl groups that are ortho or para to each other, such as catechol, hydroquinone, pyrogallol, esters of gallic acid, norconidendron and nordihydroguaiaretic acid are very effective antioxidants. The last named compound is produced on a commercial scale from the desert shrub *Larrea divaricata* and is used in small amounts for its antioxidant activity in animal fats. This activity is enhanced by synergists, such as citric, tartaric and phosphoric acids. A concentrated solution in propylene glycol or a highly hydrogenated vegetable oil provides a suitable means by which this antioxidant can be incorporated into the desired product.

The antioxidant activity of catechol is significantly reduced by the nuclear substitution with an acyl ester or acid (protocatechuic) group. The propyl ester is slightly less active than protocatechuic acid. The alkyl and allyl catechols are more active than catechol.[22]

Gentisic acid, 5 hydroxysalicylic acid, is a metabolite of salicylic acid and a potentially useful antioxidant. It is excreted in dogs 63 per cent unchanged.[23]

Rhus Preparations (Poison Ivy). These are a number of preparations that are essentially solutions in vegetable oils of the purified vesicant principles found in *Rhus toxicodendron, Rhus diversiloba* and *Rhus venenata.* The active constituents of these plants are related compounds that are exceedingly powerful vesicating agents even in very minute quantities (less than 1 gamma of urushiol over a ¼ inch circular area). These preparations are injected intramuscularly for either prophylactic or curative measures for poisoning by these ivies. The principle of therapy is supposedly based upon the theory of allergies and their methods of treatment. However, their use appears to be quite irrational and this conclusion is supported in the 1945 Council on Pharmacy Report which is quoted in part as follows: "Ivy extracts whether orally or parenterally administered should be vigor-

ously discouraged in severe cases of ivy poisoning rashes. Many patients are made worse by such treatment and practice is not in conformity with theory."*

The vesicant principles of these plants is usually a mixture which, in the case of poison ivy, is called urushiol and recently[24] was shown to be a mixture of four compounds having the carbon skeleton of 3-pentadecyl catechol. These are 3-pentadecylcatechol (hydrourushiol) 1.7 per cent; 3-pentadecenyl-8'-catechol 10.3 per cent; 3-pentadecadienyl-8',11'-catechol 64 per cent and 3-pentadecatrienyl-8',11',14'-catechol 23 per cent.

The local vesicant reactions due to urushiol may range from simple erythema through edema and infiltration to thickly studded vesiculation and even ulceration. The pruritis may be intense, and the fantastic orgies of exquisite, pleasurable, ecstatic itching, as stressed by a psychiatrist, border on dissolute revelry.

Urushiol, cardanol from cashew nut shell liquid and its chemically related vesicating principles are oils that are soluble in alcohol, organic solvents, fixed oils and fixed alkalis. They are precipitated by lead acetate and basic lead acetate, but the precipitates, when brought into contact with skin, are apparently fully as effective as the free compounds.[25]

Ferric chloride and cupric acetate each yield a black precipitate with the vesicant principles, and these precipitates are also exceedingly vesicating. Zinc oxide (calamine), aluminum oxide and aluminum subacetate fail to adsorb these phenols. Magnesium oxides (activated) are very effective adsorbents for urushiol and related compounds. However, the resulting adsorbate, if left in contact with the skin, has vesicant properties.[25] It appears that skin (i.e. protein) will break up these precipitates and adsorbates and form more firmly bound protein complexes. The possibility that the oil of the sweat and sebaceous glands may bring about the same result is not excluded, because it appears that these glands are the

* Stevens, F. A.: A.M.A. Council on Pharmacy & Chemistry Reports, p. 86, 1945.

seat of greatest attack by these vesicants. However, organic solvents are not effective in the removal of the vesicants; this would lead one to believe that protein complexes are formed. A logical approach to chemical treatment is the use of weak, aqueous or dilute alcoholic alkali (about 1% NaOH) solutions which have proven very effective in actual practice. The alkaline wash can be followed with several rinsings of water. Alkali treatment acts in two ways: (1) it will dissolve the ivy phenols which can be removed by washing with water, (2) the ivy phenols which are like catechol in character are very susceptible to oxidation by air in the presence of alkali. Activated magnesium oxides as dusting powders are valuable adjuvants to the alkaline treatment, if they are removed at reasonably frequent intervals.

Sodium hypochlorite solutions are also effective in the treatment of poison ivy.

Guaiacol is a liquid consisting principally of $C_6H_4(OH)(OCH_3)(1:2)$, usually obtained from creosote, or a solid consisting almost entirely of $C_6H_4(OH)(OCH_3)(1:2)$, usually prepared synthetically.

Guaiacol

The term *guaiacol* is used because the compound is a degradation product of one of the constituents of resin of guaiac.

Liquid guaiacol is colorless or yellowish. Solid guaiacol is crystalline and is colorless or yellowish. Guaiacol is soluble in alcohol, glycerin and 1:60 to 70 ml. of water. Methylation of one of the phenolic groups in catechol has increased markedly its stability; however, it darkens gradually upon exposure to light.

Guaiacol is a slightly less active antiseptic than phenol, i.e., the P.C. for *E. typhosa* is 0.91 and for *Staph. aureus* is 0.73. The corresponding *n*-amyl ether has a P.C. of 22 for *E. typhosa* and 23 for *Staph. aureus*. The mono-alkl ethers of hydroquinone and resorcinol exhibit the same order of activity as the mono-alkyl ethers of catechol. However, the mono-*n*-hexyl ethers of resorcinol

and hydroquinone are more active against *Staph. aureus*, i.e., they have a P.C. of 125 and 100 respectively, than the corresponding mono-ether of catechol, which has a P.C. of 28.

Guaiacol is used as an intestinal antiseptic, with doubtful value. It also is used as an expectorant orally and can be vaporized for inhalation purposes.

The usual dose is 0.5 ml.

Potassium Guaiacolsulfonate N.F., Potassium hydroxymethoxybenzenesulfonate, Thiocol.

Guaiacol can be sulfonated at 70° or 80° to yield a monosulfonic acid derivative which is converted to the potassium salt.

Potassium guaiacolsulfonate occurs as white crystals or as a white crystalline powder. It has a slightly aromatic odor and a slightly bitter taste. Its aqueous solutions are neutral or alkaline to litmus paper and it is affected by light. One gram of potassium guaiacolsulfonate dissolves in about 7.5 ml. of water at 25°. It is insoluble in alcohol and in ether.

The introduction of a sulfonic acid group in guaiacol decreases its toxicity and gastrointestinal irritant effects. Claims for the sedative and expectorant properties of this compound are often questioned.

Category—Expectorant.

Usual dose—500 mg.

Creosote, Wood Creosote, Creasote. Creosote is a mixture of phenols obtained from wood tar. The phenolic fraction of wood tar is recovered in the usual fashion for the separation of phenols from tars. Creosote is composed chiefly of guaiacol, creosol and methyl cresol. These constituents are derived by the pyrolysis of the lignin portion of woody tissue. Lignin is a polymer of 3-methoxyl-4-hydroxy benzene (guaiacol) that has a 3 carbon side chain in position 1.

Creosote is an almost colorless or yellowish, highly refractive, oily liquid, having a penetrating, smoky odor and a burning taste. Creosote is slightly soluble in water but is miscible with alcohol, with ether and with fixed or volatile oils.

Creosote is used for its antiseptic, usually intestinal, anesthetic and expectorant properties.

Eugenol U.S.P., 4-allyl-2-methoxyphenol, is a phenol obtained from clove oil and from other sources. Clove oil, which contains not less than 82 per cent of eugenol, is extracted with alkali, and the eugenol is freed subsequently from the separated alkaline extract by acidulation. Further purification can be effected by distillation. A number of other volatile oils also contain eugenol.

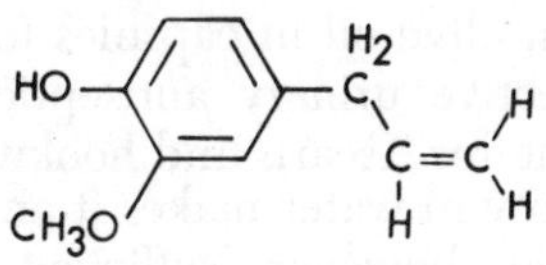

Eugenol

Eugenol is a colorless or pale yellow liquid, having a strongly aromatic odor of clove and a pungent, spicy taste. It is slightly soluble in water, soluble in twice its volume of 70 per cent alcohol and is miscible with alcohol and with fixed and volatile oils.

The para-allyl and ortho-methoxy groups contribute to the antiseptic and anesthetic activity of the phenolic group, so much that eugenol is used for toothaches, lumbago and for its antiseptic activity in mouthwashes. It has a P.C. of 14.4.

Category—dental analgesic.

For external use—topically to dental cavities, in dental protectives.

OCCURRENCE
Zinc-Eugenol Cement N.F.

Vanillic acid esters have good fungicidal properties especially against *Bacillus mycoides*, *Aerobacter aerogenes* and *Aspergillus niger*.[26] Ethyl vanillate is less toxic than sodium benzoate.

RESORCINOL AND DERIVATIVES

Resorcinol U.S.P., *m*-Dihydroxybenzene, resorcin, is prepared synthetically.

Resorcinol occurs as white or nearly white, needle-shaped crystals or powder. It has a faint, characteristic odor and a sweetish, followed by a bitter taste. One gram is

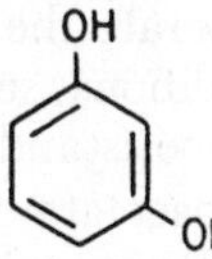

Resorcinol

soluble in 1 ml. of water and in 1 ml. of alcohol, and it is freely soluble in glycerin. Resorcinol should be stored in dark-colored or light-resistant containers. It is much less stable in solution, particularly in the presence of alkaline substances.

Although resorcinol is feebly antiseptic (P.C. = 0.4 for both *E. typhosa* and *Staph. aureus*), it is used in 1 to 3 per cent solutions and in ointments and pastes in 10 to 20 per cent concentrations for its antiseptic and keratolytic action in skin diseases, such as ringworm, parasitic infections, eczema, psoriasis and seborrheic dermatitis. Resorcinol has some fungicidal properties. It is very poorly bound by protein.

Category—keratolytic.

For external use—topically, as a 2 to 20 per cent lotion, ointment, or paste.

OCCURRENCE	PER CENT RESORCINOL
Compound Resorcinol Ointment N.F.	6

Resorcinol Monoacetate N.F., Euresol. This compound is prepared by partial acetylation of resorcinol.

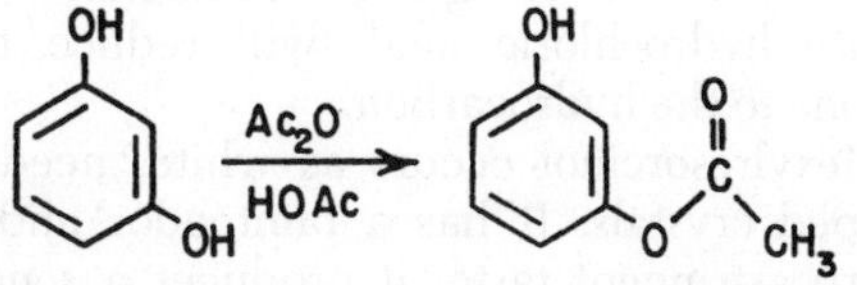

Resorcinol   Resorcinol Monoacetate

Resorcinol monoacetate is a viscous, pale yellow or amber liquid with a faint characteristic odor and a burning taste. It is soluble in alcohol and sparingly soluble in water. Although partial acetylation of resorcinol has increased its stability, it should be stored in tight, light-resistant containers.

Resorcinol is acetylated partially to produce a milder product with a longer lasting action. Prior to hydrolysis, the ester group contributes properties similar to an alkyl group. The resorcinol monoacetate is hydrolyzed slowly to liberate the resorcinol. It is used in skin conditions such as alopecia, seborrhea, acne, sycosis and chilblains.

Resorcinol monoacetate is equal or superior to undecylenic acid for *Candida albicans* and *Microsporum gypseum*. Resorcinol monobenzoate is fungistatic at 0.1 to 0.01 per cent concentrations and is less toxic and more active than resorcinol monoacetate.

It is used in 5 to 20 per cent concentrations in ointments and in 3 to 5 per cent alcoholic solutions for scalp lotions.

As a general rule, the esters of phenols are not as stable as other organic esters. They are hydrolyzed very easily by alkalies or alkaline solutions. They even will hydrolyze very slowly in the solid state in the presence of moisture.

Category—antiseborrheic; keratolytic.

For external use—in topical preparations for application to the scalp.

Hexylresorcinol N.F., Crystoids®, 4-hexylresorcinol, is prepared as shown below.

In the synthesis below the first step takes place in the presence of anhydrous zinc chloride, and the condensation is of the Friedel and Crafts type. Resorcinol is substituted so readily that an acid, rather than an acid chloride, can be used. Also, because of this ease of substitution, the moderately active zinc chloride provides adequate catalysis. In the second step, a typical Clemmensen reduction, using zinc amalgam and dilute hydrochloric acid, will reduce the ketone to the hydrocarbon.

Hexylresorcinol occurs as white, needle-shaped crystals. It has a faint odor and a sharp astringent taste; it produces a sensation of numbness when placed on the tongue. It is freely soluble in alcohol, glycerin and vegetable oils and is soluble 1:2,000 in water. It is sensitive to light.

Although resorcinol is feebly antiseptic, it is less toxic than phenol. This is the basis

for the preparation of numerous alkylated resorcinols, the most effective of which was the 4-*n*-hexyl, which has a P.C. of 46 to 56 against *E. typhosus* and 98 against *Staph. aureus.*

Hexylresorcinol was introduced by Leonard and marketed as a 1:1,000 solution under the name of S.T. 37. It was recommended as a general skin antiseptic, being effective for both gram-positive and gram-negative organisms. It can be administered dissolved in olive oil in capsules to be used for an effective urinary antiseptic and an anthelmintic for Ascaris and hookworms. Its low solubility in water makes it an effective anthelmintic; however, sufficient amounts

TABLE 15. PHENOL COEFFICIENTS OF 4-ALKYLRESORCINOLS*

	PHENOL COEFFICIENT	
4-ALKYLRESORCINOL	*E. typhosa*	*Staph. aureus*
n-Propyl	5	3.7
n-Butyl	22	10.0
Isobutyl	15	..
n-Amyl	33	30
Isoamyl	24	..
n-Hexyl	46-56	98
Isohexyl	27	..
n-Heptyl	30	280
n-Octyl	0†	680
n-Nonyl	..	980

* After Suter, C.M.: Chem. Rev. 28:269, 1941.
† At 45° C., more active than hexyl, heptyl and octyl resorcinols.[3a]

are absorbed to be of value in urinary tract infections.

Hexylresorcinol is irritating to the respiratory tract and to the skin, and an alcoholic solution has vesicant properties. This vesicating effect is a general property of alkylated phenols and reaches a very high degree in urushiol.

The alkyl substituted phenols and resor-

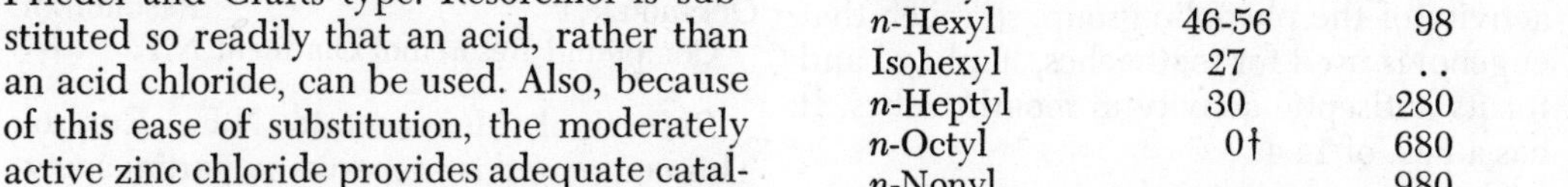

Hexylresorcinol

cinols possess the ability to reduce surface tension (see Surface-Active Agents). It is believed that these compounds may owe at least part of their increased bactericidal activity to this ability to lower surface tension, because many surface-active agents are very effective germicides. Hexylresorcinol, like many of the alkyl phenols, exhibits some local anesthetic activity.

One part of 10 per cent hexylresorcinol in propylene glycol in 100 million parts of air will kill 99.8 per cent of Staph organisms in one hour.

Category—anthelmintic (intestinal roundworm and trematodes).

Usual dose—1 g.; may repeat at weekly intervals if necessary.

OCCURRENCE
Hexylresorcinol Pills N.F.

5-n-Alkylresorcinols have been found in wheat bran and *Anacardium, Ginko* and *Grevillea* species.[27]

The active euphoric component of marihuana, Δ¹-tetrahydrocannabinol, may well be derived from cannabidiolic acid that in turn contains a terpene and a substituted resorcinol moiety.

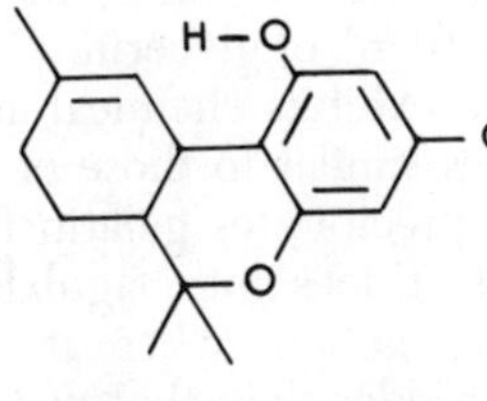

Δ'-Tetrahydrocannabinol

HYDROQUINONE AND DERIVATIVES

Hydroquinone N.F. occurs as fine white needles that darken on exposure to light and air. It is freely soluble in water, in alcohol or in ether. Hydroquinone and its derivative, the monobenzyl ether, possess the property of causing depigmentation of the skin. Hydroquinone is generally used in a 2 per cent concentration in an ointment.

Category—depigmenting agent.

For external use—topically, as a 2 per cent ointment to the affected area once or twice daily at 12-hour intervals.

OCCURRENCE
Hydroquinone Ointment N.F.

Monobenzone N.F., Benoquin®, *p*-(Benzyloxy)phenol. This occurs as a white crystalline, odorless powder that is freely soluble in alcohol but insoluble in water.

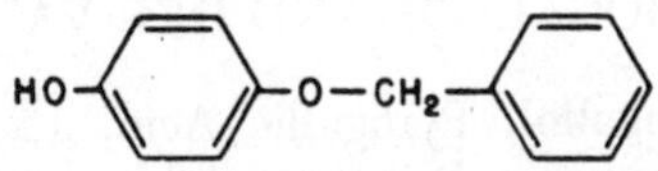

The activity of monobenzone was discovered when people using rubber gloves containing it as an antioxidant developed areas of hypopigmentation.

Long term clinical studies have demonstrated that monobenzone will not produce uniform controllable depigmentation. Depigmented areas persist for years and may be permanent. In 25 per cent of the individuals no reaction occurred.[28]

With proper precautions it is used for the treatment of hyperpigmentation due to increased amount of melanin in the skin.

Category—depigmenting agent.

For external use—topically as a 20 per cent ointment or 5 per cent lotion once or twice daily.

OCCURRENCE
Monobenzone Lotion N.F.
Monobenzone Ointment N.F.

The monalkyl, benzyl and phenethyl ethers of resorcinol, hydroquinone and catechol have a marked increase in antimicrobial properties compared to those of the parent phenols.

The monalkyl ethers of phloroglucinol also are effective antimicrobial agents.[29]

tert-Butyl hydroquinone (TBHQ) is available as a free flowing crystalline material that has significant solubilities in a rather wide range of fats and oils and solvents to which it may be added as a stabilizer or which may be used as carriers for it in various fat, oil, or food applications. TBHQ is at least equivalent to BHA and more effective than BHT or propyl gallate in increasing the oxidative stability of lard, as measured by the Active Oxygen Method.

Like propyl gallate, its antioxidant properties are not carried over into baked products. In the case of vegetable oils, safflower oils especially, TBHQ has equivalent or greater potency as compared with BHA, BHT and propyl gallate.[30] TBHQ is rapidly metabolized and has a low order of toxicity.

PYROGALLOL AND DERIVATIVES

Pyrogallol; Pyrogallic Acid, 1,2,3-Trihydroxybenzene. It is prepared by heating gallic acid, which, in turn, is obtained by the hydrolysis of tannic acid. This property of

Tannic Acid $\xrightarrow[\text{H}^+]{\text{H}_2\text{O}}$

Gallic Acid

$\downarrow \Delta \ 210°$

$+ CO_2$

Pyrogallol

decarboxylation is common to the polyhydroxylated benzoic acids. The ease with which this decarboxylation takes place is dependent upon the number of hydroxyl groups present. The greater the number of hydroxyl groups, the more easily decarboxylation takes place.

Pyrogallol occurs as light, white or nearly white, odorless leaflets or fine needles. One gram is soluble in 2 ml. of water and in 1.5 ml. of alcohol. Pyrogallol when in solution and exposed to air turns brown and acquires an acid reaction due to oxidation by oxygen of the air. If an aqueous solution is rendered alkaline, it can be used in gas analysis to absorb oxygen. It is a strong reducing agent and is used in photography and in engraving.

Pyrogallol is feebly antiseptic (P.C. negligible for *E. typhosa* and *Staph. aureus*).

It is used as an antipruritic, irritant, caustic and keratolytic in ringworm, parasitic diseases, psoriasis, lupus, chronic lichen simplex, trichophyton infections and chronic scalp eczema. As such it is used in ointments in concentrations up to 20 per cent. Pyrogallol stains the skin, and these stains can be removed by lemon juice or sodium bisulfite. Pyrogallol has been used in some hair dyes.

Pyrogallol should not be used on extensive broken areas of the skin because it may be absorbed in quantities large enough to prove toxic. If absorbed, pyrogallol will cause necrosis of the liver and the kidneys and will destroy red blood cells.

Eugallol is pyrogallol monoacetate. It is prepared by the partial acetylation of pyrogallol. This partial acetylation permits a slow liberation of pyrogallol and produces a milder product with a longer lasting action. It is used much the same as pyrogallol and is marketed as a 67 per cent solution in acetone (see Resorcinol Monoacetate).

Gallic Acid is obtained by the hydrolysis of tannic acid. It occurs as white or pale fawn-colored, silky, interlaced needles or in triclinic prisms. It has an astringent, slightly acidulous taste. One gram of gallic acid is soluble in 87 ml. of water, in 4.6 ml. of alcohol or in 10 ml. of glycerin.

Gallic acid exhibits chemical and medicinal properties similar to those of pyrogallol. Although it precipitates gelatin from aqueous solutions, it does not coagulate albumin. Its astringent action, when it is used in diarrhea, is milder than that of tannic acid.

Propyl Gallate is an effective, economically useful antioxidant.

Bismuth Subgallate, basic bismuth gallate, Dermatol, is a basic salt which, when dried at 105° for 3 hours, yields not less than 52 per cent and not more than 57 per cent of Bi_2O_3. Bismuth subgallate occurs as an amorphous, bright yellow powder. It is odorless and tasteless. It is stable in air, but is affected by light. Bismuth subgallate is nearly insoluble in water and in alcohol.

Bismuth subgallate will be decomposed by alkalies, alkali carbonates or mineral acids.

It is used in the treatment of dysentery and diarrhea, with a usual dose of 1 g.

The antiseptic action of bismuth is combined with the astringent action of gallic acid.

Phloroglucinol. Per se phloroglucinol has weak antimicrobial properties and has little application as a medicinal agent.

Aspidin, a derivative of phloroglucinol found in the rhizomes of male fern (*Dryopteris filixmas*), is an active anthelmintic. This led to the preparation of a number of derivatives of phloroglucinol, of which the most active against *Hymenolepis nana* were the diacyl types.[31]

8-HYDROXYQUINOLINE AND DERIVATIVES

8-Hydroxyquinoline, oxine, quinophenol, oxyquinoline, is a white, crystalline powder that is insoluble in water but soluble in alcohol, acids or bases. It is antiseptic if not actually bactericidal. The salts with acids are soluble in water and, therefore, are more usable as antiseptics. Among others that have been employed are the citrate, the tartrate (Termine), the benzoate and the sulfate (Chinosol, Quinosol). The last, which has been the most popular, is a yellow, crystalline powder that is fairly soluble in water to give an acid solution and has a bitter taste. It is presumably an active bactericide, having been given a P.C. of 10 against *Staph. aureus*, but its action against other organisms is variable and in most cases low. In part, the mode of action of oxine and its derivatives is due to its ability to bind, through chelation (q.v.) certain essential trace elements. It is comparatively nontoxic and has been given to guinea pigs orally in doses as high as 3 g. without evident harm. As an antiseptic, it is applied in concentrations of 1:3,000 to 1:1,000, and it has been administered orally in doses of 300 mg. 3 times a day.

Chiniofon, Quinoxyl®, Anayodin®, Yatren®, is a mixture of 7-iodo-8-hydroxy-quinoline-5-sulfonic acid, its sodium salt and sodium carbonate, containing not less than 26.5 per cent and not more than 29.0 per cent of Iodine (I). It is a canary-yellow powder with a slight odor, a bitter taste and a sweet after-taste. When added to water, it effer-

vesces due to reaction with the sodium carbonate and finally dissolves (1:25). With ferric chloride in dilute solution it gives a deep, emerald-green color and with copper sulfate a dense, white precipitate. It is prepared by sulfonating 8-hydroxyquinoline, subjecting the product to potassium iodide and bleaching powder in the presence of potassium carbonate, precipitating the derived phenol with hydrochloric acid and

finally, mixing with sodium carbonate.

Chiniofon is used almost solely as a remedy in amebic dysentery, although it also has been employed, like iodoform, as a surgical dusting powder and in gonorrhea and diphtheria.

With therapeutic doses toxic effects seldom are observed, except that many patients have a diarrhea during the first few days and that iodism may appear in the susceptible. The usual dose is 250 mg. 3 times a day for 7 days; range 250 to 750 mg. It can be given in a retention enema of 3 g. in 100 ml. of water at not over 44°, in which case the oral dose is suspended or reduced during the rectal administration.

Iodochlorhydroxyquin U.S.P., Vioform®, 5-chloro-7-iodo-8-quinolinol, 5-chloro-8-hydroxy-7-iodoquinoline, consists of a spongy, voluminous, yellowish-white powder that has a slight, characteristic odor and is affected by light. It is practically insoluble in water or alcohol but dissolves in hot ethyl acetate and in hot acetic acid. It is decomposed by heating with dilute hydrochloric acid, more easily with concentrated sulfuric acid, giving off vapors of iodine. It is prepared by iodinating 5-chloro-

8-hydroxyquinoline, a direct product of a Skraup synthesis.

The compound originally was introduced as an odorless substitute for iodoform and acts by slow liberation of iodine. It is used as an undiluted powder in surgery, in atopic dermatitis, in eczema of the external auditory canal, in chronic dermatitis, in oil dermatitis and in acute psoriasis and impetigo. It also may be applied as a 2 to 3 per cent ointment, emulsion or paste or used in suppositories for the vagina in the treatment of trichomonas vaginitis.

However, the chief use at present, is as a remedy in amebic dysentery. For this purpose, it is more toxic than chiniofon but also more potent, and it can bring about apparent cures without any unpleasant symptoms, although in a few cases there may be abdominal distress, and the possibility of iodism is always present. Its antiamebic dose is 250 mg. 3 times a day for 10 days. Range, 250 to 500 mg.

Category—topical anti-infective

Usual dose—250 mg. as vaginal suppository once daily.

Usual dose range—250 to 500 mg. daily.

For external use—topically, as a 3 per cent cream or ointment 2 or 3 times daily.

OCCURRENCE
Iodochlorhydroxyquin Cream U.S.P.
Iodochlorhydroxyquin Ointment U.S.P.
Iodochlorhydroxyquin Suppositories U.S.P.
Iodochlorhydroxyquin and Hydrocortisone Cream N.F.
Iodochlorhydroxyquin and Hydrocortisone Lotion N.F.
Iodochlorhydroxyquin and Hydrocortisone Ointment N.F.
Compound Iodochlorhydroxyquin Powder N.F.
Iodochlorhydroxyquin Tablets N.F.

Diiodohydroxyquin U.S.P., Diodoquin®, 5,7-diiodo-8-quinolinol, 8-hydroxy-5,7-diiodoquinoline, is a light yellowish to tan, microcrystalline, odorless powder that is insoluble in water. It is recommended in the treatment of amebic dysentery and the infestation by *Trichomonas hominis* (*intestinalis*) and is claimed to be just as effective as chiniofon and much less toxic. However, sev-

eral cases of acute dermatitis have followed its use.

Category—anti-amebic.

Usual dose—650 mg. 3 times daily for 20 days.

Usual dose range—650 mg. to 2 g. daily.

OCCURRENCE
Diiodohydroxyquin Tablets U.S.P.

Chloroquinaldol, Sterosan®, 5,7-dichloro-8-hydroxyquinaldine, is an effective bac-

teriostatic and fungistatic agent developed after a study of oxyquinoline derivatives. Chloroquinaldol is strongly bacteriostatic for gram-positive staphylococci, streptococci and enterococci. It is more potent than iodochloroxyquinoline or dichloroxyquinoline. The chemical properties are the same as those of the oxyquinoline derivatives. It is available in a 3 per cent ointment or cream.

QUINONES

Quinones are the product of oxidation of ortho or paradihydroxyaromatic compounds. Resorcinol does not form a quinone.

Para-quinone (1,4-Benzoquinone) is a yellow compound which has a strong pungent odor and colors the skin brown. Some

p-Quinone

highly-substituted para-quinones have been found in nature as pigments.

In 1932,[32] it was found that *p*-benzoquinone and toluquinone were toxic to guineapig spermatozoa. Substitution of methoxyl groups[33] (2,6-dimethoxyquinone) on the ring produced a compound that was antagonistic to the growth of *Staph. aureus.* Replacement of OCH_3 by OH always resulted in decreased activity. Naturally occurring pigments, such as fumigatin, spinalosin and phaenecin, found in molds, are related to quinone. Anthelmintic properties have been reported for hydroxybenzoquinones having long chain alkyl groups. An example is embelen (2,5-dihydroxy-3-undecyl-1,4-benzoquinone), which has been isolated from the berries of *Embelia ribes* and Rapanone (2,5-dihydroxy-3-tridenyl-1,4-benzoquinone) found in *Rapanea maximowiczie.* The anthelmintic activity of male fern [(*Asipidium*) *Dryopteris filixmas*] is thought to be due to a complex quinone known as filicinic acid. A series of related compounds were prepared in 1921 by Karrer.[34]

In 1948, several derivatives of 1,4-benzoquinone and naphthoquinone were reported as possessing outstanding antifungal qualities.

o-Quinone, 1,2-Benzoquinone, *o*-Benzoquinone, is the quinone derived by the oxidation of catechol.

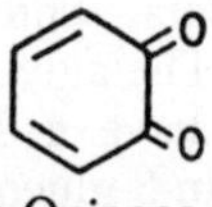

o-Quinone

NAPHTHOQUINONES

Naphthoquinones are of two types, i.e., alpha and beta. Very few derivatives of beta-naphthoquinone have been found in nature and none is of interest in pharmacy.

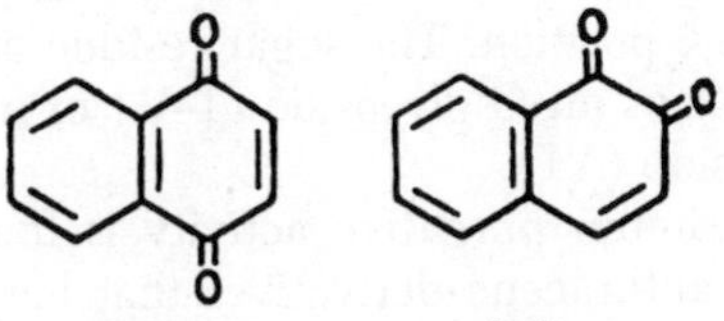

alpha   beta

Alpha-naphthoquinone has derivatives that are found in nature and others, prepared synthetically, that are of value in medicine. (See Vitamin K and Menadione.) Many of these are pigments, such as lawsone, juglone, echinochrome A and lapachol, obtained from either plant or animal sources.

ANTHRAQUINONES

Anthraquinone compounds are distributed widely in nature. They are red, yellow or orange-yellow coloring matters and are used as dyes and laxatives.

Anthraquinone is a tricyclic structure in which the quinoid double bonds have less activity than those in quinone or napthoquinone. This is due perhaps to their inclusion between the aromatic rings. The results in anthraquinone having weak reducing properties and, thus, very useful properties for treating some skin conditions. (See Anthrarobin.)

Anthraquinone is prepared readily from anthracene by oxidation, using nitric acid, dichromate with sulfuric acid or air (oxygen) with vanadium oxide as the catalyst. It occurs as yellow needles which are odorless, only slightly soluble in most solvents

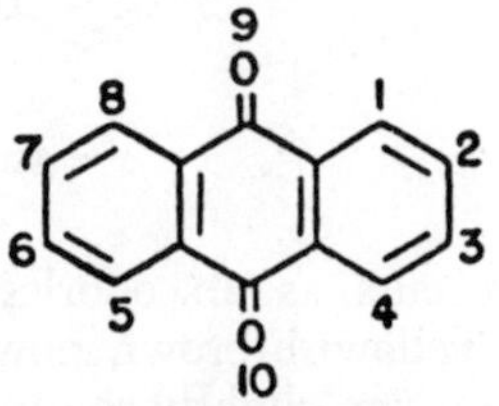

Anthraquinone

and readily sublimed. This compound is very stable and is attacked only with difficulty by nitric acid or other oxidizing agents. Chemically, it resembles the diketones more than the quinones.

TABLE 16. HYDROXYANTHRAQUINONE DYES

NAME	CHEMICAL NAME
Alizarin	1,2-Dihydroxyanthraquinone
Quinizarin	1,4-Dihydroxyanthraquinone
Anthrarufin	1,5-Dihydroxyanthraquinone
Chrysazin	1,8-Dihydroxyanthraquinone
Anthraflavin	2,6-Dihydroxyanthraquinone
Hystazarin	2,3-Dihydroxyanthraquinone
Anthragallol	1,2,3-Trihydroxyanthraquinone
Purpurin	1,2,4-Trihydroxyanthraquinone
Flavopurpurin	1,2,6-Trihydroxyanthraquinone
Anthrapurpurin	1,2,7-Trihydroxyanthraquinone

Several important dyestuffs are derivatives of anthraquinone (see Table 16). Many of these, such as purpurin and alizarin, possess laxative action also but are used only as dyes. Note that the laxative anthraquinones (emodins) have in position 3 another group, such as methyl, carboxyl or hydroxyl methyl (CH_2OH).

Other polyhydroxy anthraquinone derivatives, such as alizarin, Bordeaux, anthracene blue and carminic acid are used as dyes. Of these, carminic acid, because of its presence in cochineal, is used as a coloring material in pharmacy.

Anthralin N.F., 1,8,9-anthracenetriol; 1,8-Dihydroxyanthranol, Cignolin, Dithranol, is prepared as follows:

1,8-Anthraquinone
(Chrysazin)

Anthralin

Anthralin occurs as an odorless, tasteless, crystalline, yellowish-brown powder that is insoluble in water, slightly soluble in alcohol and soluble in most lipoid solvents. Anthralin has antiseptic, irritant and proliferating properties which indicate its use as a substitute for chrysarobin in the treatment of psoriasis, chronic dermatomycosis and chronic dermatoses.

Category—local antieczematic.

For external use—topically, as a 0.1 to 1 per cent ointment.

OCCURRENCE
Anthralin Ointment N.F.

Carminic Acid. This complex hydroxy-anthraquinone is obtained from cochineal. It is extracted readily with water and may be crystallized as bright red needles. The compound is originally red but, due to the presence of one carboxyl group, forms "lakes" with metals, such as tin (cochineal scarlet)

and aluminum, producing colors of crimson, scarlet and violet. It is used primarily as a mordant dye for silk and wool.

Cochineal Solution is essentially an acid solution of carminic-acid-aluminum "lake" and is used to color pharmaceuticals that are acid in reaction. The solution is incompatible with alkali.

Carmine is essentially the carminic-acid-aluminum "lake" prepared from the coloring principle present in cochineal. It is a red, solid material which has both acid- and base-soluble fractions. As previously stated, Cochineal Solution contains the acid-soluble portion, whereas Carmine Solution contains the base-soluble fraction.

Anthraquinone Derivatives. These[35] are distributed widely in plants, particularly in the families Liliaceae, Polygonaceae, Rhamnaceae and Rubiaceae and the genus Cassia. Some also are found in the families Euphorbiaceae, Scrophulariaceae and Hyperiaceae, and they are somewhat common in certain fungi and lichens (30% dry weight of *Helminthosporium gramineum*).

The anthraquinone derivatives occur chiefly as glycosides in purgative drugs such as rhubarb, cascara, senna, frangula, buckthorn and aloes. The aglycon moiety of such glycosides may have a 1,8-quinone structure, aloe-emodin-8-glucoside, rhein-8-glucoside, etc., or a partially reduced form of that structure. One reduced form is of the emodin-oxanthrone type; a second type is found as a dimer, e.g., a dianthrone such as sennoside. The aglycon usually contains hydroxyl groups that vary in number, position and kind. A number of important members contain two phenolic groups in the 1 and the 8 positions and a methyl (III), a hydroxymethyl (I) or a carboxyl (II) group in the 3 position. The sugar residue may be present as an O-glycoside (I–V) or as a C-glycoside (VI).

Maximum purgative activity is found in those anthracene derivatives that have (1) two phenolic° hydroxyl groups or three hydroxyl groups (two being phenolic); (2) an anthrone structure (one C=O of

° One phenolic OH or acetylation of the OH groups of the active compounds leads to inactivity.

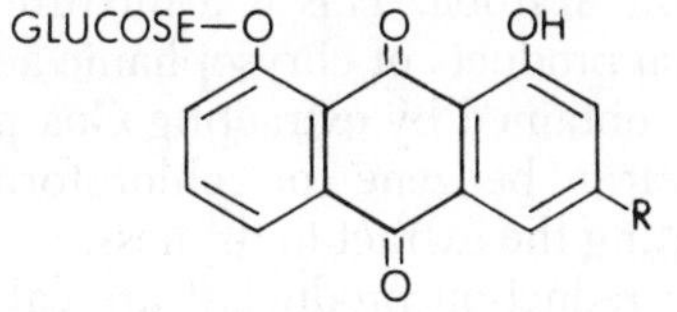

I. Aloe-emodin-8-glucoside, R = CH$_2$OH

II. Rhein-8-glucoside, R = COOH

III. Chrysophanol-8-glucoside, R = CH$_3$

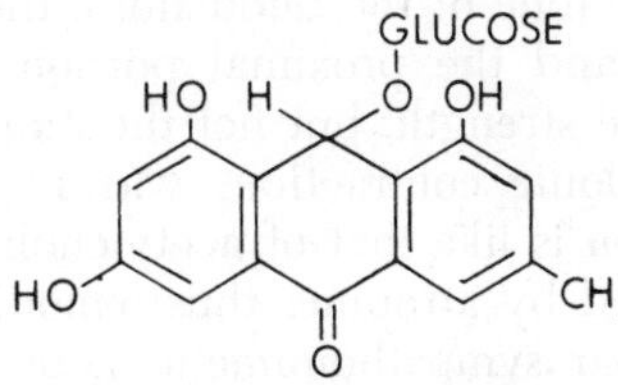

IV. Emodin-oxanthrone glucoside.

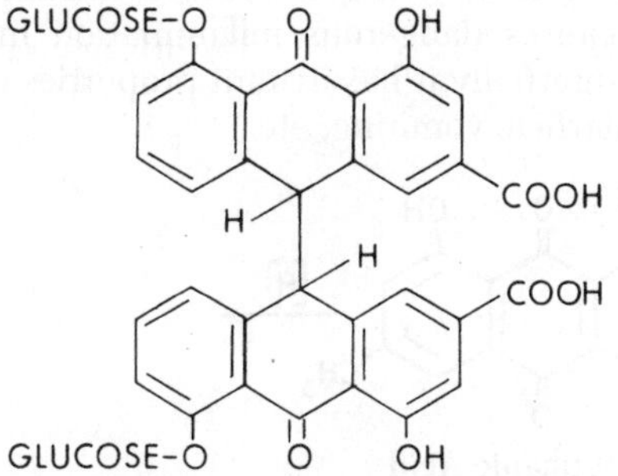

V. Sennoside A.

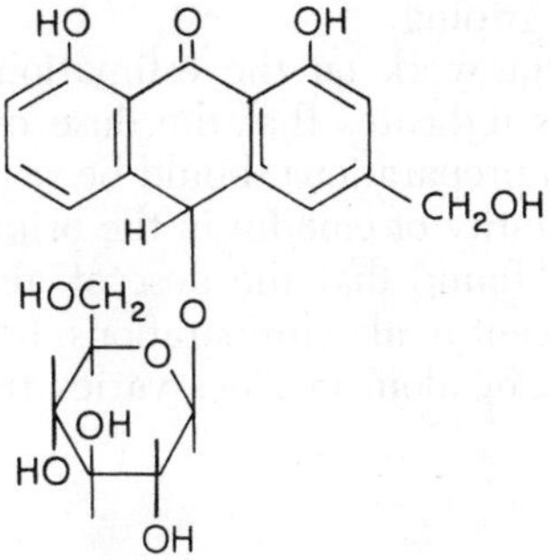

VI. Barbaloin.

the 9,10-anthraquinone reduced to a secondary OH group) and (3) an O-glycoside residue (in some cases a C-glycoside, barbaloin [VI]). A dimer [V] is considerably more active than a monomer [II]. The sugar residue minimizes the absorption of these compounds, so that therapeutic concentrations can reach the large intestine readily. Sennoside C, the 8,8'-diglucoside of rhein-aloe-emodin-dianthrone, is found in senna leaves.[36]

In the case of senna pods, glycosides have been isolated that, on mild hydrolysis, yield sennoside A [V] (a secondary glycoside) and several glucose molecules. In mice, these glycosides are 50 to 120 per cent as active as the sennosides. The sugar-free anthraquinones are relatively inert and their corresponding anthrones only somewhat more active, possibly because they are biologically altered and only small amounts reach the large intestine. Large doses only are effective. With mixed glycosides a synergism of activity has been noted. Mice are resistant to aloin-like [VI] compounds, whereas humans are not.

Of the cascara glycosides, 10 to 20 per cent are normal (O) glycosides (1% of the bark) and 80 to 90 per cent are aloin-like compounds (C-glycosides) (5 to 8% of the bark) which are resistant to hydrolysis in vitro. Barbaloin and deoxybarbaloin (chrysaloin) are present and barbaloin (+) and (−) isomers (at C$_{10}$) occur as components of cascarosides A, B, C and D; the two latter (C and D) occur as both O- and C-glycosides.

Cascarosides are tasteless. Preliminary results suggest that they are more potent than the aloins, which are intensely bitter.

Barbaloin [VI] is an important constituent of aloes and aloin, and certain varieties of aloes contain at least two glycosides of barbaloin called *aloinosides*.

Rhubarb, the dried rhizome of *Rheum palmatum*, contains aloe-emodin, emodin, emodin mono-methyl ether, rhein and chrysophanol. Some studies have suggested that the glycosides of rhein anthrones [II] were the main active components.

Liquid preparations of senna deteriorate rapidly—thus, the growing trend to solid standardized preparations (Senokot®), which are gaining favor.

The mechanism[37, 38] of action of the anthraquinone purgatives was determined on frangula-emodin. It was found to decrease the tone of the duodenum, the small intestine and the proximal portion of the colon. The strength, but not the amplitude, of the colonic contractions was increased. This action is like that of acetylcholine and is inhibited by atropine; thus, emodin may exert a parasympathomimetic action. Perhaps it affixes itself to the muscle protein of the intestinal wall. This appears to be the basis for prescribing belladonna with emodin-containing drugs (i.e., aloes) to reduce griping.

Recent work on the estimation of aloin in aloes indicates that the dose of aloin or emodin preparations should be regulated by the quantity of emodin in the original drug. Stone[39] found that the dose of aloin varied 25 per cent in aloe preparations, because the amount of aloin in aloes varies from lot to lot.

ANTHRAQUINONE REDUCTION PRODUCTS

The partially reduced compounds related to anthraquinone are of value in treating certain skin diseases (eczema and psoriasis). All of these have the carbonyl group in position 10 reduced to methylene (CH_2).

Anthrarobin, Dihydroxyanthranol, is derived from alizarin by reacting it with zinc and ammonia. It is soluble in chloroform or in ether and insoluble in water or in acid media. Anthrarobin is a strong reducing agent which readily absorbs oxygen, thereby being converted to blue alizarin. It is, of course, incompatible with oxidizing agents.

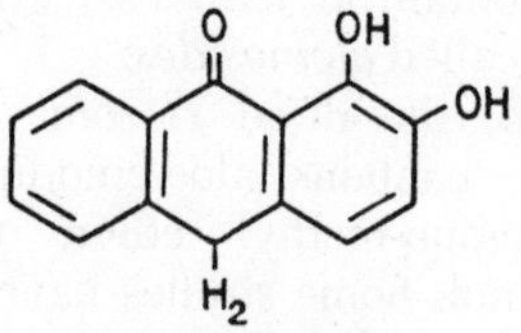

Anthrarobin

It has a mode of action similar to that of chrysarobin and is employed in skin disorders as a 10 per cent solution or ointment.

Chrysarobin. Chrysarobin is a mixture of neutral principles obtained from Goa Powder, a substance deposited in the wood

of Andira Araroba. This is a mixture of the reduction products of chrysophanic acid and emodin, obtained by extracting Goa powder with warm benzene or chloroform and evaporating the extract to dryness.

These reduction products[40] are called anthrones, dianthrones and anthranols and commonly are referred to as "neutral principles." The complex mixture is soluble in ether, chloroform (1:15) or alcohol (1:400) and is slightly soluble in water. It occurs as a brownish to orange-yellow tasteless powder that is irritating to mucous membranes and causes dangerous inflammation in the eye. Internally it has irritant properties causing diarhea, vomiting, etc.

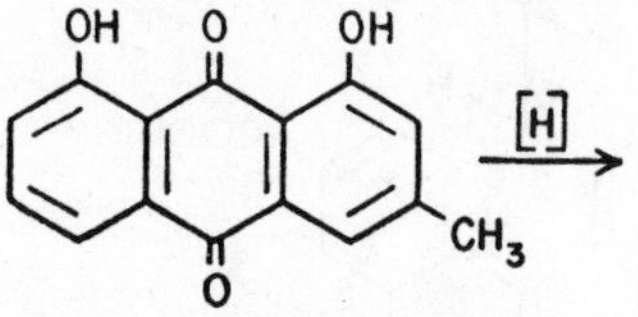

Chrysophanic Acid

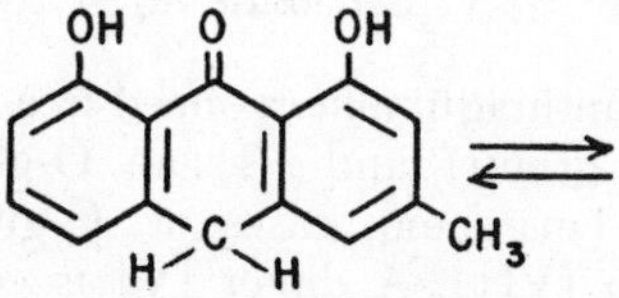

Chrysophanic Acid Anthrone

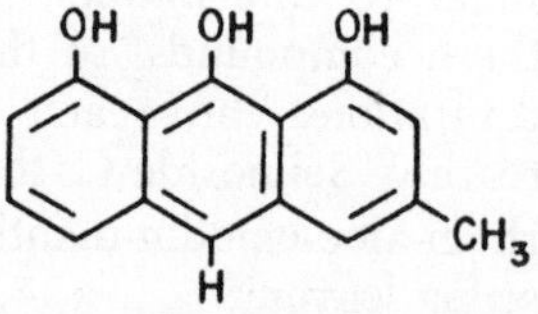

It is used in ointment form and in collodion in about 5 per cent strength. The mode of action in treating skin diseases is thought to be the removal of oxygen from the surrounding tissue, since the "neutral principles" are very active reducing agents. Skin conditions such as eczema, ringworm infections and psoriasis are treated with it.

TANNINS

The term tannin is applied to a group of complex phenolic substances that occur in nature and exhibit similar physical and chemical properties, namely:

1. They are soluble in water or hydro-alcoholic solutions.

2. They are soluble in glycerol or propylene glycol.

3. Usually, they are amorphous in character, and their aqueous or alcoholic solutions are colloidal.

4. Usually, they are insoluble in petroleum ether, ether or benzene.

5. They are astringent to the taste and tissues.

6. They are precipitated from aqueous solutions by gelatin, albumin and other proteins.

7. Their action on hide produces leather.

8. They form precipitates with some alkaloids and some nitrogenous bases.

9. Colored solutions are obtained with iron salts.

Tannins are found in a wide variety of plants and plant tissues. The amount of tannin present in any particular part of a plant may vary greatly from the amount found in another part of the same plant. Tannins are abundant in actively growing parts of the plant, such as buds, immature fruits, root shoots, inner barks, outer barks and galls. The high tannin content of unripe fruits, e.g., choke cherries and persimmons, disappears partially or wholly when the fruit ripens.

Tannins can be classified roughly into two major groups: (1) hydrolyzable tannins and (2) nonhydrolyzable tannins. The members of the first group can be hydrolyzed to simpler molecules by acid, alkali or enzymatic methods, and they usually give a blue to black color with ferric chloride test solution. The nonhydrolyzable tannins usually are precipitated by bromine water, give a green color with ferric chloride and usually contain a phloroglucinol nucleus in part. Many of the nonhydrolyzable tannins polymerize when heated, particularly in the presence of hydrogen ions, to form red substances which are hard to dissolve and are called *phlobaphenes* or *tannin reds*. Polymerization of this type also can take place slowly under ordinary conditions. Light catalyzes this reaction. This phenomenon may account for some of the insoluble materials that separate from some tinctures and fluid extracts upon long standing. Many of the unhydrolyzable tannins (phlobatannins) yield red solutions.

Fear[41] has shown that, contrary to the generally accepted idea, all alkaloids are not precipitated by tannins. Tannins have no effect whatsoever on aconitine, apomorphine, berberine, betaine and many other alkaloids. With atropine, cotarnine, emetine, ephedrine, hydrastine and cocaine hydrochlorides, only a slight opalescence is produced. Fear observed that, if a dilute solution of an alkaloid failed to produce a precipitate with a dilute solution of tannin, no precipitate would be produced by using higher concentrations.

Some of the more carefully characterized hydrolyzable tannins that yield gallic acid on hydrolysis are as follows: gallotannin from nutgalls, acer-tannin from *Acer ginuala*, glucogallin from the Chinese rhubarb and hamamelis-tannin from *Hamamelis*.

Caffetannin yields quinic and caffeic acids upon hydrolysis. Caffeic acid is 3,4-dihydroxycinnamic acid.

Because the condensed, nonhydrolyzable tannins or phlobatannins are of unknown constitution, it is not possible to classify them very readily. Examples of this type of tannin are the tannins found in the barks of the various species of cinchona, wild cherry and most conifers, the best known of which is that found in hemlock. Other examples are the tannins of Indian cutch, cube gambir, rhubarb, hops, tea and the quebrachos.

PRODUCTS

Tannic Acid, gallotannic acid, tannin, is a tannin usually obtained from nutgalls, the excrescences obtained from the young twigs of *Quercus infectoria* Oliver, and other allied species of *Quercus* (Fam. *Fagaceae*). The gallotannin content of the same and different kinds of galls varies considerably. The average gallotannin content of several commercial varieties of galls follows: Aleppo galls, 58 per cent; Chinese galls, 77 per cent; Bassorch galls, 30 per cent and Morea galls, 30 per cent.

Gallotannin obtained from galls can be purified readily by dissolving it in ethyl acetate and extracting this solution with alkali carbonates to remove any free gallic acid.[42] The ethyl acetate then is evaporated to deposit the purified gallotannin.

The constitution of gallotannin is not known definitely. It yields, upon hydrolysis, gallic acid in varying amounts, sometimes as high as 100 per cent. Some authors[43, 44, 45] believe that gallotannin contains glucose and have proposed that gallotannin is a penta-digalloyl glucose. Meta-digalloyl residues have been obtained from gallotannin upon hydrolysis by the enzyme tannase.[46] Meta-digallic acid has been synthesized.[47]

m-Digallic Acid

Tannic acid occurs as an amorphous powder, as glistening scales or as spongy masses, varying in color from yellowish-white to light brown. It is odorless or has a faint, characteristic odor and a strongly astringent taste. Tannic acid is very soluble in water, in acetone and in alcohol. It is freely soluble in diluted alcohol and slightly soluble in dehydrated alcohol. One gram of tannic acid dissolves in about 1 ml. of warm glycerin.

Tannic acid, because it contains gallic acid residues, is very unstable in the presence of alkalies, alkali carbonates and other basic substances.

Tannic acid is used as a hemostatic, an astringent in diarrhea, in the treatment of *Rhus* dermatitis and in the treatment of burns. Its use in the treatment of burns was based upon its ability to precipitate the dead protein tissue to form an eschar. The protein thus combined is not available as a nutrient for bacteria, and the eschar (tanning effect) furnishes a protective covering which reduces the loss of serum (pressure dressing effect) and prevents external contamination. Five to 10 per cent aqueous tannic acid solutions were applied as a wet dressing or sprayed on the burned areas. Alternate applications of silver nitrate often

were used to produce a heavier eschar. Other antiseptics also were used in conjunction with tannic acid.

One disadvantage claimed for the use of tannic acid therapy is the liver damage that possibly may result from the gallic acid liberated upon the hydrolysis of the tannic acid. This problem would not be encountered if a nonhydrolyzable tannin (phlobatannin) were used. It has been demonstrated in rats that a phlobatannin is much less toxic and equally as effective as tannic acid for burns.[48]

Acetyltannic Acid, diacetyltannic acid, tannyl acetate, Acetannin, is a product obtained by the acetylation of gallotannin. The use of this compound in diarrhea is based upon the slow liberation of tannic acid through hydrolysis in the intestines.

A tannic acid whose major constituent is penta-galloylated quinic acid is available in commerce.* It is obtained from the seed pod of a tree or shrub (Tara) widely distributed in northwest South America.

REFERENCES CITED

1. Schroeter, L. C., *et al.*: J. Am. Pharm. A. 48:723, 1958; 48:535, 1959.
2. U.S. Department of Agriculture: Circular 198, December, 1931.
3. J. Am. Pharm. A. 36:129 & 134, 1947.
3a. Reddish, G. F.: Antiseptics, Disinfectants, Fungicides, and Physical Sterilization, ed. 2, p. 537, Philadelphia, Lea and Febiger, 1957.
4. Beckett, A., *et al.*: J. Pharm. Pharmacol. 11:421, 1959; also see J. Pharm. Sci. 52: 126, 1963.
5. Suter, C. M.: Chem. Rev. 28:269, 1941.
6. Fleming, A.: Chem. & Ind. 18, 1945.
7. Hansch, C., *et al.*: J. Am. Chem. Soc. 87:5770, 1965.
8. Axelrod, J., and Daly, J.: Biochim. biophys. acta 159:472, 1968.
9. Angel, H., and Rogers, K. J.: Nature 217:84, 1968.
10. Weiss, J. M.: Chem. and Eng. News 30:4715, 1952.
11. Sykes, G.: Disinfection and Sterilization, ed. 2, p. 320, London, Spon, 1965.

* Mallinckrodt Chemical Works.

12. Erlandson, A. L., and Lawrence, C. A.: Science 118:274, 1953.
13. Marsh, P. B., et al.: Ind. Eng. Chem. 41: 2176, 1949.
14. Florestano, H. J., and Bahler, M. E.: J. Am. Pharm. A. (Sci. Ed.) 42:576, 1953.
15. C. E. N. July 20, 1959.
16. Wedding, R. T., et al.: Arch. Biochem. Biophys. 121:9, 1967.
17. Goldstein, S., et al.: Proc. Soc. Exp. Biol. Med. 128:980, 1968.
18. Baichwal, et al.: J. Am. Pharm. A. 46:603, 1957; 47:537, 1958.
19. Richardson, A., et al.: J. Am. Pharm. A. 45:268, 1956.
20. Neidig, C. P., and Burrell, H.: Drug Cosmet. Ind. 54:408, 1944.
21. White, A. A.: Proc. Soc. Exp. Biol. Med. 126:588, 1967.
22. Seth, S. C., et al.: Ind. Jour. Chem. 1:435, 1963.
23. Astill, D. B., et al.: Biochem. J. 90:194, 1964.
24. Dawson, C. R., and Markiewitz, K. H.: J. Org. Chem. 30:1610, 1965.
25. Gisvold, O.: J. Am. Pharm. A. (Sci. Ed.) 30:17, 1941.
26. Pearl, I. A., and McCoy, J. F.: Food Industries 17:1173, 1458, 1945.
27. Wenkert, E., et al.: J. Org. Chem. 29:435, 1964.
28. Becker, S. W., and Spencer, M. C.: J.A.M.A. 180:279, 1962.
29. Touchstone, J. C., et al.: J. Am. Chem. Soc. 78:5643, 1956.
30. Sherwin, E. R., and Thompson, J. W.: Food Technol. 21:106, 1967.
31. Bowden, K., et al.: Brit. J. Pharmacol. 24: 714, 1965; Bowden, K., and Ross, W. J.: J. Pharm. Pharmacol. 17:239, 1965.
32. Gutland, I. M.: Biochem. J. 26:32, 1932.
33. Oxford, A. E.: Chem. & Ind. 20:189, 1942.
34. Karrer, P., et al.: Helvet. chim. acta 2:407, 466, 1921.
35. Fairbairn, J. W.: Lloydia 27:79, 1964.
36. Schmid, W., and Angliker, E.: Helv. chim. acta 48:1912, 1965.
37. Vallette, G., and Leboefu, H.: Ann Pharm. franc. 5:89, 1947.
38. ——: Compt. rend. Soc. biol. 141:29, 1947.
39. Stone, K. G.: J. Am. Pharm. A. (Sci. Ed.) 36:391, 1947.
40. Gardner, J. H.: J. Am. Pharm. A (Sci. Ed.) 28:143, 1939.
41. Fear, C. M.: Analyst 54:316, 1929.
42. Fischer, E., and Freudenberg, K.: Ber. deutsch. chem. Ges. 45:920, 1912.
43. Karrer, P., et al.: Helvet, chim. acta 6:3, 1923.
44. Freudenberg, K.: Ber. deutsch. chem. Ges. 45:915, 925, 1912.
45. Geake, A., and Nierenstein, M.: Ber. deutsch. chem. Ges. 47:891, 1914.
46. Freudenberg, K.: Z. physiol. Chem. 116: 277, 1921.
47. Fischer, E. et al.: Ber. deutsch. chem. Ges. 51:45, 1918.
48. Bope, F. U., Cranston, E. M., and Gisvold, O.: J. Pharmacol. Exp. Ther. 94:209, 1948.

SELECTED READING

Haslam, E.: Chemistry of Vegetable Tannins, London, Academic Press, 1966.

Kligman, A. M.: Poison ivy (Rhus) dermatitis, Arch. Dermat. 77:149-180, 1958.

Marsh, P. B., et al.: Fungicidal activity of bisphenols, Ind. & Eng. Chem. 41:2176, 1949.

Nierenstein, M.: Tannins, in Allen, A. H.: Commercial Organic Analysis, ed. 5, vol. 5, pp. 1-204, Philadelphia, Blakiston, 1927.

——: The Natural Organic Tannins, Cleveland, Sherwood Press, 1935.

Ostrolenk, M., and Brewer, C. M.: A bactericidal spectrum of some common organisms, J. Am. Pharm. A., 38:95, 1949.

Phenol and Its Derivatives, Nat. Institutes of Health Bull. No. 190, 1949.

Russell, A.: The natural tannins, Chem. Rev. 17:155, 1935.

Suter, C. M.: The relationship between the structure and bactericidal properties of phenols, Chem. Rev. 28:269, 1941.

Research Today, No. 2, 1951.

Reddish, G. F.: Antiseptics, Disinfectants, Fungicides, and Physical Sterilization, ed. 2, Philadelphia, Lea and Febiger, 1957.

Sykes, G.: Disinfection and Sterilization, ed. 2, Philadelphia, Lippincott, 1965.

11

Sulfonamides and Sulfones with Antibacterial Action

Edward E. Smissman, Ph.D.,

Professor of Medicinal Chemistry, The School of Pharmacy,
University of Kansas, Lawrence, Kansas

SULFONAMIDES

In 1935 the discovery that Prontosil (2',4'-diaminoazobenzene-4-sulfonamide) possessed strong bacteriostatic action[1] led to one of the most important developments in the history of chemotherapy. Tréfouel, Tréfouel, Nitti and Bovet[2] reported the sulfonamide portion of Prontosil to be responsible for the observed antibacterial action and that azo dyes of the Prontosil type are reduced in the animal system to sulfanilamide. Sulfanilamide was first synthesized by Gelmo[3] in 1908 and Prontosil by Mietsch and Klarer[4] in 1935. Following the papers on the antibacterial properties of these compounds, thousands of sulfanilamide derivatives and related compounds were synthesized.

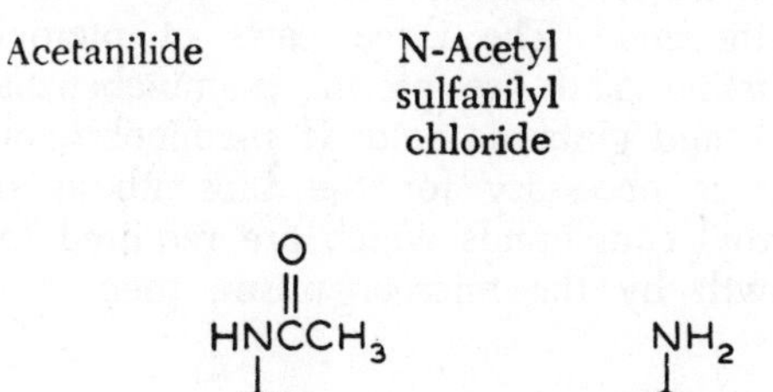

The sulfonamide drugs are relatively easy to prepare. Industrially, as well as in the laboratory, chlorosulfonation of acetanilide is the method of choice. The key intermediate, N-acetylsulfanilyl chloride, can be allowed to react with primary or secondary amines to secure the desired N[1]-substituted sulfonamides.

The nomenclature for sulfanilamide derivatives is based on the following numbering system: substituents on the nitrogen of the sulfonamide nitrogen are called N[1]-

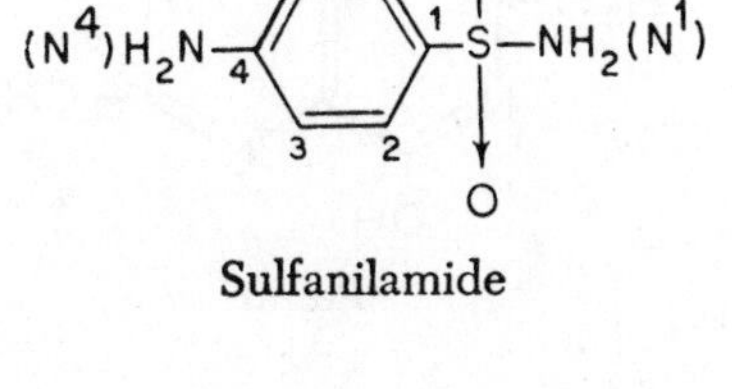

Acetanilide → N-Acetyl sulfanilyl chloride (via $ClSO_3H$) → (via NH_3)

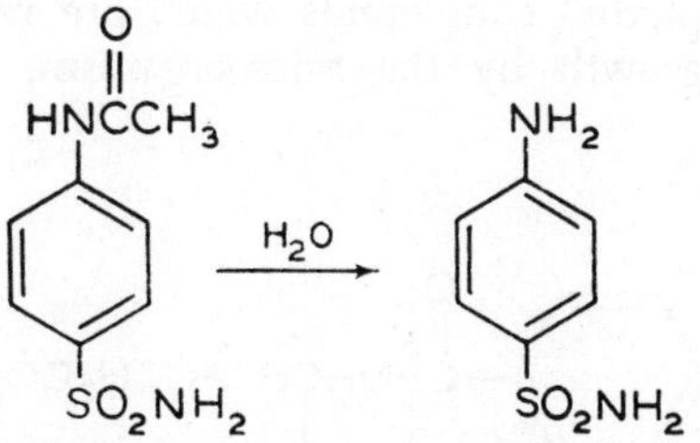

N[4]-Acetyl sulfanilamide → Sulfanilamide (via H_2O)

Sulfanilamide

283

substituents, and those on the amino nitrogen, N^4-substituents. The nomenclature radical is designated as a sulfanilamido-group.

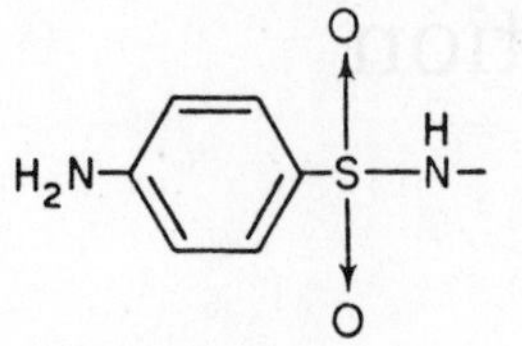

Sulfanilamido–

In naming a heterocyclic substituted sulfonamide, the point of attachment of the hetero ring is given.

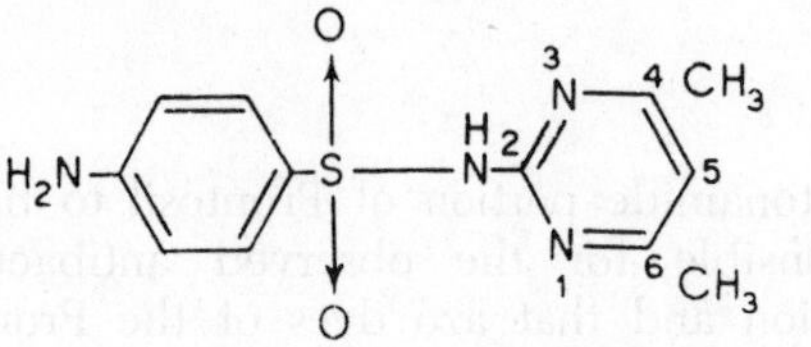

N^1-(4,6-Dimethyl-2-pyrimidyl)sulfanilamide
Sulfamethazine

Many N^4-substituted sulfonamides have been prepared by causing the N^1-substituted sulfanilamide to react with acylchlorides or anhydrides.

The therapeutic effect of sulfonamides is achieved by inhibiting the multiplication of the infectious organism and thus allowing the host to eradicate the infection by normal defense mechanisms. Bacteria will readily become resistant to inadequate concentrations of sulfonamides; therefore, it is necessary to use intensive treatment for short periods. The sulfonamides exhibit bacteriostatic action, not bactericidal effects, on certain types of organisms. They are effective agents against the following microorganisms: *Bacillus coli,* Friedlander's bacillus, *Clostridium septicum, Clostridium welchii,* gonococcus, meningococcus, pneumococcus, *Streptococcus hemolyticus* (beta), *Streptococcus viridans, Shigella dispar,* and *Haemophilus influenzae.* The sulfonamides are not effective agents against virus infections. Most strains of staphylococci are resistant to sulfa drugs. Anaerobic streptococci, the enterococcus group of streptococci, and rheumatoid arthritis infections do not respond to sulfonamides.

In 1940, Woods[5] discovered that *p*-aminobenzoic acid (PABA) prevents the bacteriostatic effect of the sulfonamides. This was of interest to Fildes[6] because he believed that sulfonamide bacteriostasis was due to immobilization of some essential metabolite of the microorganism, either by irreversible combination with it or by blocking some enzyme system required for the utilization of the metabolite. The above workers suggested that sulfanilamide exerts its action by competing with *p*-aminobenzoic acid. Subsequent studies have proved them to be correct; sulfonamides produce their bacteriostasis by a competitive replacement of PABA in an enzyme system essential to the growth of a susceptible organism but not essential to the host animal. Only those bacteriostatic substances antagonized by PABA are considered to possess true sulfanilamide action.

Para-aminobenzoic acid is involved in the synthesis of folic acid coenzymes, which are essential growth factors for some microorganisms. Pteroylglutamic acid, a substance with full folic acid activity, has been isolated and synthesized.[7] The three parts of pteroylglutamic acid are pterin, *p*-aminobenzoic acid, and glutamic acid. If *p*-aminobenzoic acid is necessary for the biosynthesis of pteroyl compounds which are required for growth by the microorganism, then it is

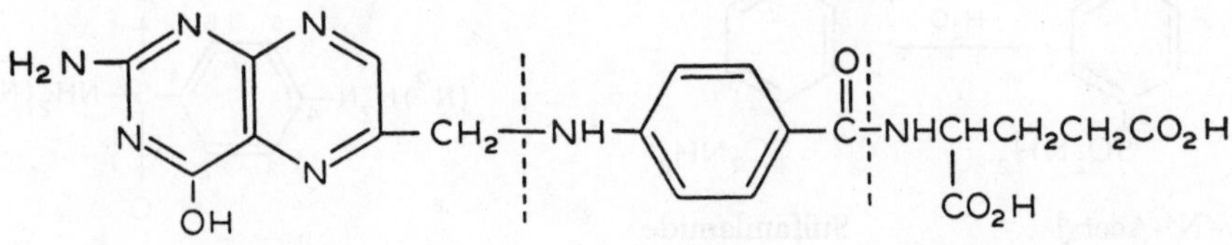

Pterin   PABA   Glutamic Acid

logical to deduce that a sulfonamide, by occupying the enzyme surface required for the biosynthesis, would prevent the growth of the bacteria. Conversely, since man does not synthesize his own folic acid substances, the sulfonamides would not interfere with human cell growth.

Two laboratories[8, 9, 10, 11] have reported experimental data which elucidate the biosynthetic pathway for tetrahydrofolic acid (THFA) in *E. coli*. Seydel[12] discusses this pathway and the sites at which antagonists can interfere with the biosynthesis of the folates (Fig. 11).

All of the requirements for bacteriostatic action are contained in the parent compound sulfanilamide. If the N^4-amino group is replaced by groups which can be converted in the body to a free amino group, activity is maintained.

The fact that no compound with an alkylated aromatic amino group (N^4), in the sulfonamide series, has been found to show sulfonamide activity as such attests to the importance of a free or potentially free aromatic amino group for activity. All of the known N^4-substituted compounds which show activity in vivo break down in the body to produce a free aromatic amino group in the para position to the sulfonamide group. Such groups as $-NO_2$, $-NHOH$, and the azo group can be reduced in the body to a free amino group; an acylamido group can be hydrolyzed to give a free amino group, and thus all of the benzene sulfonamides with the above functions in the para position will exhibit activity in vivo but not in vitro.

If an alkyl, an alkoxy, or other functional group is placed in the para position, no

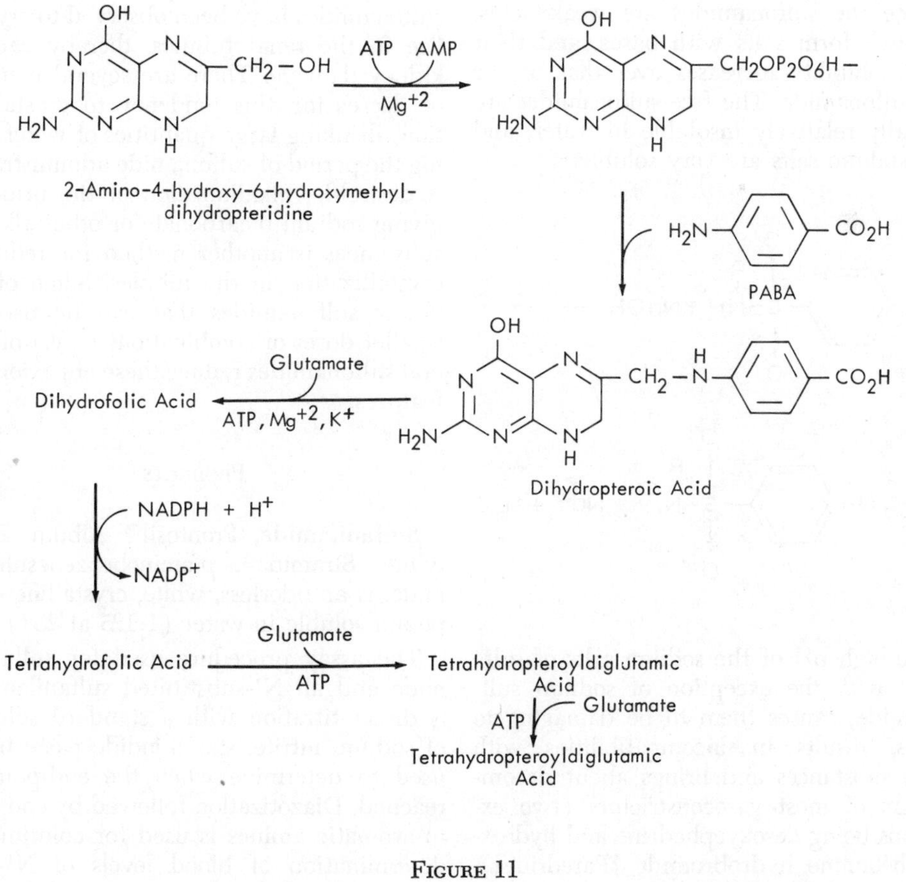

FIGURE 11

activity is observed. If the N^4-amino group is changed to the 2 or 3 position on the benzene ring, giving orthanilamide and metanilamide, inactive compounds result. Substituents on the aromatic ring of the sulfanilamides either diminish or completely destroy bacteriostatic activity.

The substitution of the N^1-amide nitrogen with various groups results in a wide fluctuation in activity. This fluctuation has been correlated with the acid dissociation constant. Sulfonamides which have a nonsubstituted or monosubstituted N^1-amide nitrogen are acidic and will readily form salts. The maximum activity is usually shown by those compounds having a pKa of about 6.7; activity diminishes if the compounds are either more or less acidic.[13] Substitution of a free sulfonic acid ($-SO_3H$) group for the sulfonamide function destroys activity, but replacement by a sulfinic acid group ($-SO_2H$) and acetylation of the N^4-position gave an active compound.[14]

Since the sulfonamides are weak acids, they will form salts with bases, and their water solubility increases over that of the free sulfonamide. The free sulfonamides are generally relatively insoluble in water, and their sodium salts are very soluble.

The high pH of the sodium salts of sulfa drugs, with the exception of sodium sulfacetamide, causes them to be damaging to tissues, results in incompatibilities with acidic substances and brings about decomposition of most vasoconstrictors (two exceptions being desoxyephedrine and hydroxyamphetamine hydrobromide [Paredrine]).

They cannot be used in high concentrations on nasal or other mucous membranes and must be injected intravenously with extreme care. Injections outside the vein may cause necrosis of tissues. The amount of salt formed at a given pH is a function of the acid strength (pKa) of the sulfonamide. The larger the dissociation constant (the smaller the pKa), the more of the sulfonamide that will be in the form of a salt at a given pH. The pKa's of the common sulfonamides vary from 4.77 to 10.43, with the majority of them between 6 and 8. At a pH of 7 a sulfonamide with a pKa of 7 will be 50 per cent in the salt form. At pH 7, sulfanilamide with a pKa of 10.43 would be less than 0.1 per cent in the salt form.

The sulfonamides are converted in the body to N^4-acetyl compounds, and they are in part excreted as such. In general, these acetylated compounds have a lower solubility and a lower pKa than the parent unacetylated compounds. The N^4-acetylated sulfonamides have been observed to crystallize in the renal tubules, thereby causing kidney damage. There are several methods of decreasing this tendency to crystallization: drinking large quantities of water during the period of sulfonamide administration is desirable; alkalinization of the urine by giving sodium bicarbonate or other alkaline substances is another method for reducing crystallization in the tubules. Some of the newer sulfonamides that can be used in smaller doses or combinations (q.v.) of several sulfonamides reduce these objectionable features.

PRODUCTS

Sulfanilamide, Prontosil® Album, Prontylin®, Stramide®, *p*-aminobenzenesulfonamide, is an odorless, white, crystalline compound soluble in water (1:125 at 25°).

The assay procedure used for sulfanilamide and all N^1-substituted sulfanilamides is direct titration with a standard solution of sodium nitrite, starch-iodide paste being used to determine when the end-point is reached. Diazotization followed by coupling to aromatic amines is used for colorimetric determination of blood levels of N^1-substituted sulfonamides, sulfanilamide and N^4-acetylated sulfonamides, after hydrolysis.

Sulfanilamide is structurally the simplest of the sulfonamides. Its potency is low, compared with more complex agents, and, although it has been used in the treatment of a wide variety of infections, in most cases it is no longer the drug of choice.

Absorption from the intestine is rapid and complete, maximum blood concentrations being reached in 1 to 2 hours after a single dose. It is usually given in the form of tablets, the dose depending on the severity and the type of infection. The chief index for the control of therapy is the concentration of sulfanilamide in the blood or other body fluids and not the dose of the drug prescribed. A dose of 4 g. by mouth followed by 1 g. every 4 hours results in a blood level of about 5 mg. per cent, which is suitable for an infection of moderate severity. In severe infection the usual dose is calculated as 100 g. per kg. of body weight for an initial dose and one sixth the initial amount at 4-hour intervals until the temperature remains normal for 72 hours.

Toxic effects are moderate, cyanosis being common, and nausea, vomiting, headache, dizziness, skin rash and drug fever frequent. Acute hemolytic anemia and acidosis are peculiar to sulfanilamide, the former rarely and the latter never being encountered with other sulfonamides. It is common practice to give sodium bicarbonate concurrently to prevent acidosis. It has been found that *p*-aminobenzoic acid in relatively small amounts overcomes the antibacterial action of sulfanilamide and its derivatives. The administration of local anesthetics derived from *p*-aminobenzoic acid, such as procaine, along with local application of sulfanilamide, must be avoided.

Sulfapyridine U.S.P., Dagenan®, M and B 693®, Coccolase®, N^1-2-Pyridylsulfanila-

Sulfapyridine

mide. This compound is a white, crystalline, odorless and tasteless substance. It is stable in air but slowly darkens on exposure to light. It is soluble in water (1:3,500), in alcohol (1:440) and in acetone (1:65) at 25°. It is freely soluble in dilute mineral acids and aqueous solutions of sodium and potassium hydroxide. Its outstanding effect in curing pneumonia was first recognized by Whitby; however, because of its relatively high toxicity it has been supplanted largely by sulfathiazole, sulfadiazine and sulfamerazine. The drug is acetylated readily in the body, and a number of cases of kidney damage have resulted from acetylsulfapyridine crystals deposited in the kidneys. It also causes severe nausea in a majority of patients.

Sulfapyridine was the first drug to have an outstanding curative action on pneumonia. It is considerably more potent than sulfanilamide in the treatment of streptococcal and gonococcal infections. It gave impetus to the study of the whole class of N^1-heterocyclically substituted derivatives of sulfanilamide.

Category—suppressant for dermatitis herpetiformis.

Usual dose—500 mg. to 1 g. 4 times daily.

Usual dose range—2 to 6 g. daily.

OCCURRENCE
Sulfapyridine Tablets U.S.P.

Sodium Sulfapyridine. This compound is a white, odorless, crystalline substance, very soluble in water (1:15) and alcohol (1:10). Its aqueous solutions have a pH of about 11 and absorb carbon dioxide from the air readily with precipitation of sulfapyridine. It is used for intravenous injection as a 5 per cent solution in sterile water for patients requiring an immediate high blood level of the drug. It must not be given subcutaneously or intramuscularly because of tissue damage due to its high alkalinity.

Salicylazosulfapyridine, Azulfidine, 5-[*p*-(2-Pyridylsulfamyl)phenylazo]salicylic Acid. This compound is a brownish-yellow, odorless powder, slightly soluble in alcohol but

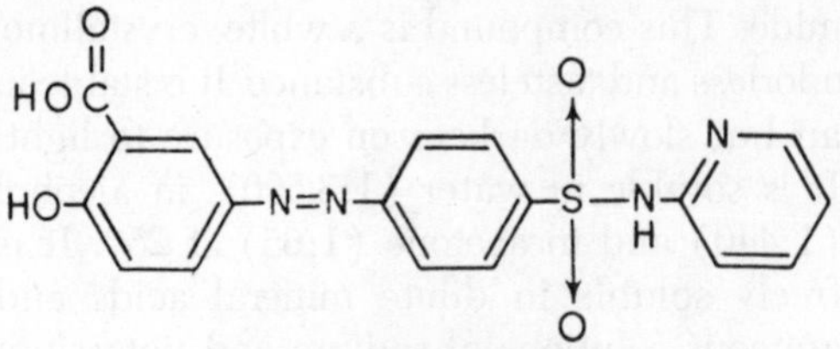

Salicylazosulfapyridine

practically insoluble in water, ether and benzene.

It is broken down in the body to *m*-aminosalicylic acid and sulfapyridine. The drug is excreted through the kidneys and is colorimetrically detectable in the urine, producing an orange-yellow color when the urine is alkaline and no color when the urine is acid. The drug is said to have special affinity for connective tissue and therefore is proposed for use in chronic ulcerative colitis. It has the same potential toxic effects as sulfapyridine, which are unusually high compared with other sulfonamides. The usual dose is 1 g.

Sulfacetamide N.F., Albucid®, Sulamyd®, Sulfacet®, N-Sulfanilylacetamide, N[1]-Acetyl-

Sulfacetamide

sulfanilamide. This compound is a white, crystalline powder, soluble in water (1:62.5 at 37°) and in alcohol. It is very soluble in hot water, and its water solution is acidic.

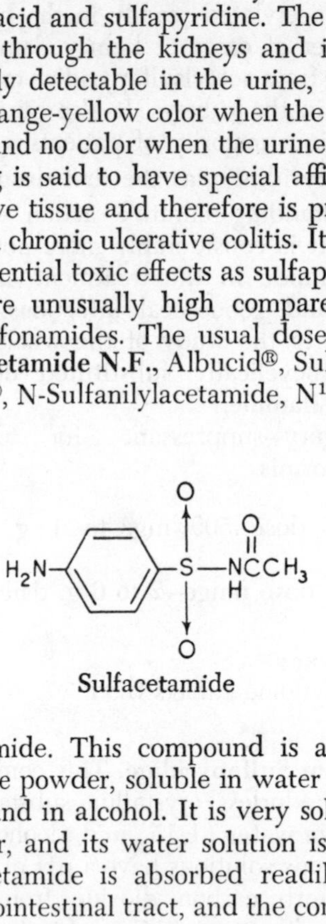

Sulfacetamide is absorbed readily from the gastrointestinal tract, and the concentration in the blood is proportional to the oral dose administered. It is excreted primarily in the urine, in which both free and conjugated forms are found. Because of its high solubility and ready elimination, it has found considerable use in urinary tract infections, since it is possible to maintain high urinary concentrations without danger of kidney damage. It is usually given as an initial dose of 4 g. followed by 1 g. every 4 hours.

Sulfacetamide Tablets N.F.
Sulfacetamide, Sulfadiazine and Sulfamerazine Oral Suspension N.F.
Sulfacetamide, Sulfadiazine and Sulfamerazine Tablets N.F.

Sodium Sulfacetamide U.S.P., Sodium Sulamyd®, N-Sulfanilylacetamide Monosodium Salt.

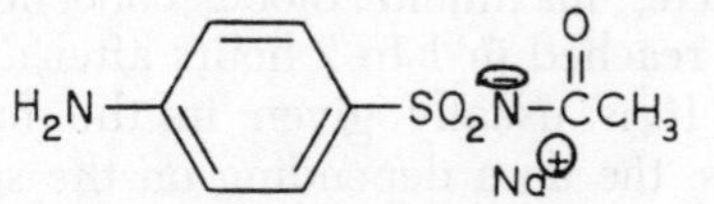

Sodium Sulfacetamide

This compound is obtained as the monohydrate and is a white, odorless, bitter, crystalline powder which is very soluble (1:2.5) in water. Because the sodium salt is highly soluble at the physiologic pH of 7.4, it is especially suited, as a solution, for repeated topical application in the local management of ophthalmic infections susceptible to sulfonamide therapy.

For external use it is employed as a 10 per cent ointment for use in the eyes or as a 30 per cent solution every 2 to 4 hours.

Sodium Sulfacetamide Ophthalmic
    Ointment U.S.P.
Sodium Sulfacetamide Ophthalmic
    Solution U.S.P.

Sulfaguanidine, N[1]-Amidinosulfanilamide, N[1]-Guanylsulfanilamide. This compound is

Sulfaguanidine

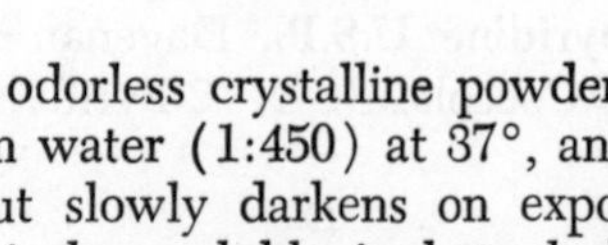

a white, odorless crystalline powder that is soluble in water (1:450) at 37°, and stable in air but slowly darkens on exposure to light. It is less soluble in hot alcohol and acetone and insoluble in benzene, ether and chloroform. Sulfaguanidine is readily soluble in cold dilute mineral acids but insoluble in cold dilute bases. Hot alkaline solutions decompose the compound to sulfanilamide, with the evolution of ammonia. The group

$$\text{NH}$$
$$\text{||}$$
$$\text{--C--NH}_2$$

is essentially an amidine which is a fairly strong base (pKa 12).

Sulfaguanidine was the first sulfonamide which was found to be so poorly absorbed that large amounts could be given orally without the development of high blood levels and toxic side-reactions. As a result, it can be used to counteract infections in the intestine, e.g., bacillary dysentery, by oral administration without developing a high concentration of the drug in the blood.

Sulfaguanidine has a comparatively low toxicity; however, various toxic manifestations, such as vomiting, drug rash and fever, are not uncommon. These reactions appear to occur more frequently than they do from the administration of sulfasuxidine and sulfathalidine, both of which are used for the same purpose as sulfaguanidine.

Sulfaguanidine has been used for the treatment of coccidiosis in chickens and for other veterinary purposes.

The initial dose in man is 50 mg. per kg. of body weight, followed by the same dose every 4 hours until the condition has abated. The total period of administration should not exceed 14 days.

Sulfapyrimidine Group

The sulfapyrimidines, in general, due to their effectiveness and relative lack of toxicity, are among the drugs of choice in the great majority of infections in which sulfonamide therapy is indicated.

Sulfadiazine U.S.P., Pyrimal®, N¹-2-Pyrimidinylsulfanilamide, 2-Sulfanilamidopyrimidine. Sulfadiazine is a white, odorless, crystalline powder soluble in water at 37°

Sulfadiazine

(1:8,100), at 25° (1:13,000) in human serum at 37° (1:620), and sparingly soluble in alcohol and acetone. It is readily soluble in dilute mineral acids and bases.

In vivo, sulfadiazine is slightly less potent than sulfathiazole and is about 8 times as active as sulfanilamide. It exhibits fewer toxic reactions than most of the other sulfonamides. Sulfadiazine is absorbed slowly but completely from the intestine, and it is acetylated to a lesser degree than sulfanilamide, sulfapyridine and sulfathiazole. It is relatively easy to maintain high blood levels which can be reached in about 6 hours. Sulfadiazine and acetylsulfadiazine both are excreted slowly by the kidneys. Only about 65 per cent of the drug appears in the urine and takes 3 to 4 days to be excreted. This low renal clearance is responsible for the facile maintenance of a high blood level.

Sulfadiazine is considered a drug of choice in a number of infections, including pneumococcal, meningococcal, Friedländer's bacillus, *Shigella dispar*, and *H. influenzae* infections. Nausea, dizziness, cyanosis, acidosis, fever, and rash are rarely observed. Hematuria and anuria have been reported but are not common. Since the solubility of sulfadiazine and its acetyl derivative increases rapidly with increased pH, sodium bicarbonate is given to maintain slight alkalinity in the urine in order to prevent precipitation in the kidney tubules.

Usual dose—initial, 4 g. then 1 g. 4 to 6 times daily.
Usual dose range—2 to 8 g. daily.

OCCURRENCE
Sulfadiazine Tablets U.S.P.
Sulfadiazine and Sulfamerazine Tablets N.F.

Sodium Sulfadiazine U.S.P., Soluble Sulfadiazine. This compound is an anhydrous, white, colorless, crystalline powder soluble in water (1:2) and slightly soluble in alcohol. Its water solutions are alkaline (pH 9-10) and absorb carbon dioxide from the air with precipitation of sulfadiazine. It is administered as a 5 per cent solution in sterile water intravenously for patients requiring an immediate high blood level of the sulfonamide.

The usual dose is 4 g. and may be repeated in 8 hours. The usual dose range is 2 to 8 g. daily.

OCCURRENCE
Sodium Sulfadiazine Injection U.S.P.

Sulfamerazine U.S.P., N^1-(4-Methyl-2-pyrimidinyl)sulfanilamide, 2-Sulfanilamido-

Sulfamerazine

4-methylpyrimidine. This compound is a white, crystalline compound with a slightly bitter taste. It slowly darkens on exposure to light. It dissolves in water at 20° (1:6,250) and at 37° (1:3,300). It is readily soluble in dilute acids and bases, sparingly so in acetone, slightly soluble in alcohol, and very slightly soluble in chloroform and ether.

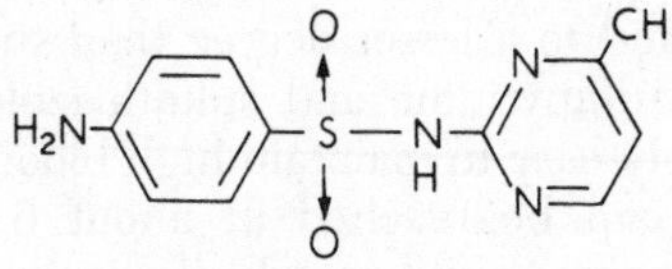

Sulfamerazine is some 6 times more potent than sulfanilamide in vivo and thus is less potent than sulfadiazine. It is absorbed more rapidly but excreted more slowly than sulfadiazine. It is similar in its therapeutic properties to sulfadiazine, but, because of its slow excretion, high blood levels can be maintained with a smaller or less frequent dose. It is less toxic than sulfanilamide, sulfapyridine and sulfathiazole but has a higher incidence of toxic reactions than sulfadiazine, including renal complications, drug fever and rashes. Its usual dose is 4 g. initially followed by 1 g. every 6 hours.

OCCURRENCE
Sulfamerazine and Sulfadiazine Tablets N.F.
Sulfamerazine, Sulfadiazine and Sulfacetamide Oral Suspension N.F.
Sulfamerazine, Sulfadiazine and Sulfacetamide Tablets N.F.

Sodium Sulfamerazine. The water solution of this compound has a pH of about 10 and, on exposure to air, absorbs carbon dioxide with precipitation of sulfamerazine. The compound is used for the same purpose as sodium sulfadiazine and has properties similar to those of that compound.

Sulfamethazine U.S.P., N^1-(4,6-Dimethyl-2-pyrimidinyl)sulfanilamide, 2-Sulfanilamido-4,6-dimethylpyrimidine. This compound is similar in chemical properties to sulfamerazine and sulfadiazine and is found to have greater water solubility than either of the above. Because it is more soluble in acid urine than is sulfamerazine, the possibility of kidney damage from use of the drug is decreased. The human body appears to handle the drug unpredictably; hence, there is some disfavor to its use in this country except in combination sulfa therapy and in veterinary medicine.

OCCURRENCE
Trisulfapyrimidines Oral Suspension U.S.P.
Trisulfapyrimidines Tablets U.S.P.

Sulfisomidine; Elkosin®; N^1-(2,6-Dimethyl-4-pyrimidinyl)sulfanilamide, 4-Sulfanilamido-2,6-dimethylpyrimidine. This compound is the most soluble of the pyrimidine

Sulfisomidine

derivatives of sulfanilamide, having a solubility of 360 to 1,100 mg. per cent in urine at pH 5.5 to 7.5. The acetyl derivative is not as soluble, but, since only 10 per cent of the drug in the urine is in the acetylated form, it is unlikely to cause damage. Hematuria and crystalluria are encountered only rarely. Alkalinization of urine is not necessary to increase solubility of the drug, but fluid intake should be maintained above 1.5 liters daily. Aside from diminished renal toxicity, the side-effects of sulfisomidine are similar to those of other sulfonamides.

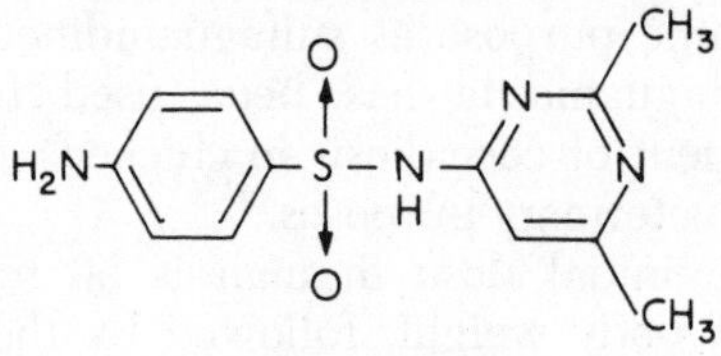

The initial dose in severe infections is calculated on the basis of 100 mg. per kg. of body weight. Subsequent doses of one sixth the initial dose should be given every 4 hours until infection is under control.

Sulfadimethoxine N.F.; Madribon®; N^1-(2,6 - Dimethoxy - 4 - pyrimidinyl)sulfanilamide. This compound is a white, crystalline

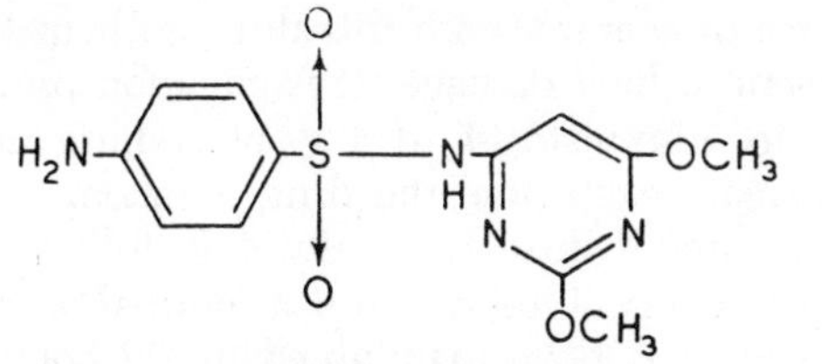

Sulfadimethoxine

powder. It is reported to be absorbed rapidly and the incidence of kidney damage is reported to be low.

Usual dose—initial, 2 g., then 1 g. daily.

Usual dose range—500 mg. to 1 g.

OCCURRENCE

Sulfadimethoxine Oral Suspension N.F.
Sulfadimethoxine Tablets N.F.

Sulfameter, Sulla®, N^1(5-methoxy-2-pyrimidinyl)sulfanilamide, is a long-acting sulfonamide. It occurs as a white, crystalline powder and is insoluble in water and slightly soluble in alcohol. It should be protected from light.

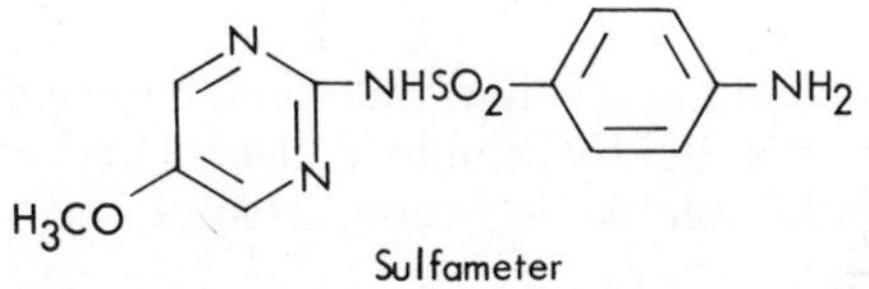

Sulfameter

Sulfameter is readily absorbed from the gastrointestinal tract. Following oral administration, measurable levels of the drug are reached in approximately 2 hours and peak serum levels occur within 4 to 8 hours. Limited data suggest that measurable amounts of the drug are still present in the plasma 96 hours after the drug is administered; approximately 90 per cent of the sulfonamide in the plasma is nonacetylated.

Sulfameter is administered orally as a single daily dose, preferably after breakfast. Because of the long-lasting blood levels of sulfameter, a smaller dosage should be administered than is normally used with shorter-acting sulfonamides.

The usual dosage for adults weighing more than 45 kg. (100 pounds) is 1.5 g. the first day, followed by 500 mg. daily thereafter.

MIXED SULFONAMIDES

The danger of crystal formation in the kidneys on administration of sulfonamides has been reduced greatly through the use of the more soluble sulfonamides such as sulfisoxazole. This danger may be diminished still further by administering mixtures of sulfonamides. When several sulfonamides are administered together, the antibacterial action of the mixture is the summation of the activity of the total sulfonamide concentration present, but the solubilities are independent of the presence of similar compounds. Thus, by giving a mixture of sulfadiazine, sulfamerazine and sulfacetamide, the same therapeutic level can be maintained with much less danger of crystalluria, since only one third the amount of any one compound is present. Some of the mixtures employed are the following:

Sulfacetamide, Sulfadiazine and Sulfamerazine Tablets N.F., Buffonamide®, Cetazine®, Incorposul®. This is a mixture of equal weights of these sulfonamides either with or without an agent to increase the pH of the urine.

Category—antibacterial. The usual dose of total sulfonamides is initial dose 4 g., then 500 mg. to 1 g. every 4 hours.

Sulfacetamide, Sulfadiazine and Sulfamerazine Oral Suspension N.F. The suspension usually available contains 167 mg. of each sulfonamide in each 4 ml.

Category—antibacterial. The usual dose of total sulfonamides is initial dose 4 g., then 500 mg. to 1 g. every 4 hours.

Sulfadiazine and Sulfamerazine Tablets N.F., Duosulf®, Duozine®, Merdisul®, Sulfonamides Duplex. A mixture of equal weights of sulfadiazine and sulfamerazine either with or without an agent to increase the pH of the urine.

Category—antibacterial. The usual dose of combined sulfonamides is initial 4 g., then 2 g. every 4 hours.

Trisulfapyrimidines Oral Suspension U.S.P., Ray-Tri-Mides®, Nectrizine®, Sulfalose®, Syrasulfas®, Terfonyl®, Trifonamide®, Trionamide®, Trisulfazine®. This mixture contains equal weights of Sulfadiazine U.S.P., Sulfamerazine U.S.P., and Sulfamethazine U.S.P., either with or without an agent to increase the pH of the urine.

Category—antibacterial.

Usual dose—initial, 40 ml. (4 g. of trisulfapyrimidines), then 10 ml. 4 to 6 times daily.

Usual dose range—20 to 80 ml. daily.

Trisulfapyrimidines Tablets U.S.P., Neotrizine®, Sulfose®, Truozine®. These tablets contain essentially equal quantities of sulfadiazine, sulfamerazine, and sulfamethazine.

Category—antibacterial.

Usual dose—initial, 4 g.; maintenance, 1 g. 4 to 6 times daily.

Usual dose range—2 to 8 g. daily.

OTHER HETEROCYCLIC SULFONAMIDES

Sulfathiazole, N^1-2-Thiazolylsulfanilamide. This compound is a white, crystalline, odorless substance soluble in water (1:1,700) and alcohol (1:2,000) at 25°. It is soluble in acetone and dilute acids and bases.

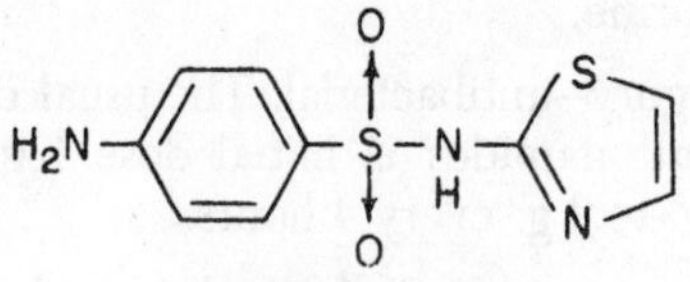

Sulfathiazole

Sulfathiazole is acetylated rapidly in body tissues and excreted rapidly by the kidneys. In some instances it is difficult to maintain adequate blood levels because of this rapid excretion. It is more potent than sulfapyridine in the treatment of streptococcal, staphylococcal, pneumococcal and gonococcal infections and is often the drug of choice in treatment of these infections. It causes less nausea, dizziness and cyanosis than sulfanilamide or sulfapyridine. Drug fever and rash occur in a high percentage of patients treated with this drug, and other less toxic sulfonamides have replaced it in many cases. Hematuria and other kidney damage has been observed, due in some instances to the formation of acetylsulfathiazole crystals in the kidneys. The daily urinary output should be maintained at more than 1,000 ml. in the

course of therapy with this drug, to help to prevent kidney damage. It is common practice to administer 1 to 4 g. of sodium bicarbonate each time the drug is taken.

The initial dose is usually 4 g. followed by 1 g. every 4 hours until a normal temperature has been maintained for 72 hours.

Sodium Sulfathiazole; Soluble Sulfathiazole. This compound is a white, odorless, crystalline substance readily soluble in water or alcohol, giving a highly alkaline solution (pH 9.5 to 10.5) which will precipitate sulfathiazole on absorption of carbon dioxide. The anhydrous salt can be sterilized.

Para-Nitrosulfathiazole; Nisulfazole®; p-Nitro-N-(2-thiazolyl)benzenesulfonamide; 2-(p - Nitrophenylsulfonamido)thiazole. This

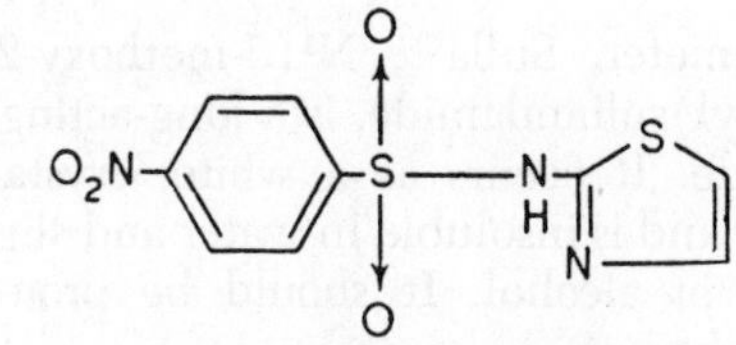

Para-Nitrosulfathiazole

compound is a yellow, odorless, bitter powder. It is slightly soluble in alcohol and very slightly soluble in water, chloroform and ether.

This compound is used only for rectal injection as an adjunct in the local treatment of nonspecific ulcerative colitis. The dose is 10 ml. of a 10 per cent suspension after each stool and at bedtime.

Sulfisoxazole U.S.P.; Gantrisin®; Entusul®; N^1-(3,4-Dimethyl-5-isoxazolyl)sulfanilamide; 5-Sulfanilamido-3,4-dimethylisoxazole. This compound is a white, odorless, slightly bitter, crystalline powder. At pH 6 this sulfonamide has a water solubility of 350 mg. in 100 ml., and its acetyl deriva-

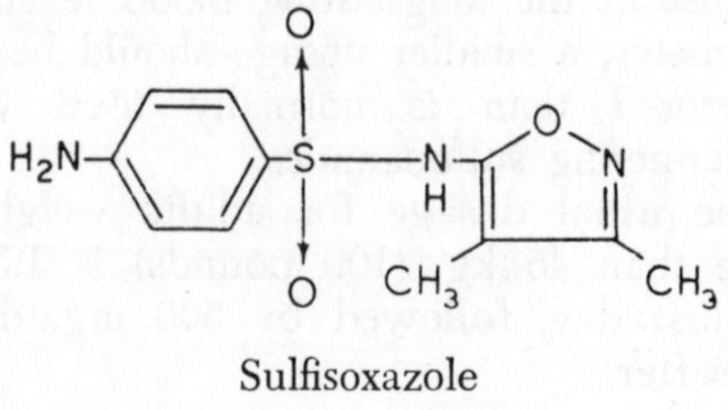

Sulfisoxazole

tive has a solubility of 110 mg. in 100 ml. of water. This solubility is sufficiently high in body fluids so that the drug is not deposited in the kidneys; it has low toxicity and no addition of alkalinizing agents is necessary on administration.

Sulfisoxazole possesses the action and the uses of other sulfonamides and is used for infections involving sulfonamide-sensitive bacteria. It is claimed to be effective in treatment of gram-negative urinary infections. The initial dose is 4 g., followed by 1 g. 4 to 6 times daily until the temperature has been normal for at least 48 hours. The usual dose range is 1 to 8 g. daily.

OCCURRENCE

Sulfisoxazole Tablets U.S.P.
Sulfisoxazole and Phenazopyridine Hydrochloride Tablets N.F.

Acetyl Sulfisoxazole N.F., Gantrisin® Acetyl, N-(3,4-Dimethyl-5-isoxazolyl)-N-sulfanilylacetamide, N^1-Acetyl-N^1-(3,4-dimethyl-5-isoxazolyl)sulfanilamide, shares the actions and the uses of the parent compound, sulfisoxazole. The acetyl derivative is tasteless and, therefore, suitable for oral administration, especially in liquid preparations of the drug. There is evidence that the acetyl compound is split in the intestinal tract and absorbed as sulfisoxazole.

The initial dose is 4.6 g., the equivalent of 4 g. of sulfisoxazole; then 1.15 g., the equivalent of 1 g. of sulfisoxazole, every 4 to 6 hours. The usual dose range is 1.15 g. to 2.3 g. every 4 to 6 hours.

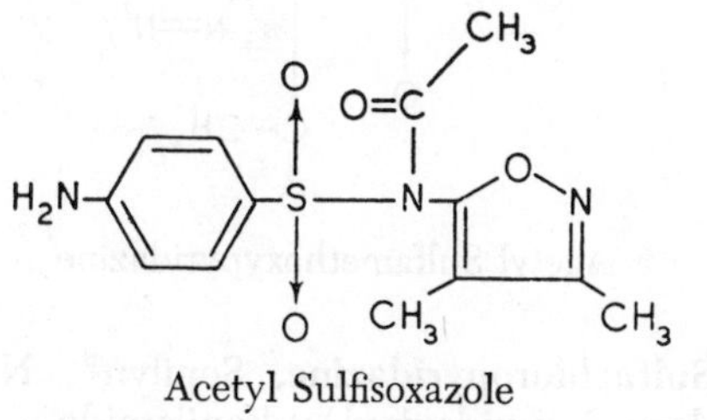

Acetyl Sulfisoxazole

OCCURRENCE
Acetyl Sulfisoxazole Oral Suspension N.F.

Sulfisoxazole Diolamine, Gantrisin® Diethanolamine, 2,2'-Iminodiethanol Salt of N^1-(3,4-Dimethyl-5-isoxazolyl)sulfanilamide.

This salt is prepared by adding enough diethanolamine to a solution of sulfisoxazole to bring pH to about 7.5. It is used as a salt to make the drug more soluble at physiologic pH range of 6.0 to 7.5 and is used in solution for systemic administration of the drug by slow intravenous, intramuscular, or subcutaneous injection when sufficient blood levels cannot be maintained by oral administration alone. It also is used for instillation of drops or ointment in the eye for the local treatment of susceptible infections.

Sulfamethoxazole N.F., Gantanol®, N^1-(5-Methyl-3-isoxazolyl)sulfanilamide.

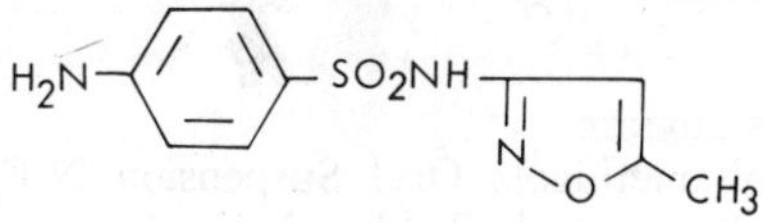

Sulfamethoxazole

Sulfamethoxazole is a sulfonamide drug closely related to sulfisoxazole in chemical structure and antimicrobial activity. It occurs as a tasteless, odorless, almost white, crystalline powder. The solubility of sulfamethoxazole at the pH range of 5.5 to 7.4 is slightly less than that of sulfisoxazole but greater than sulfadiazine, sulfamerazine, or sulfamethazine.

Following oral administration, sulfamethoxazole is not as completely or as rapidly absorbed as sulfisoxazole and its peak blood level is only about 50 per cent as high. Its rate of excretion is somewhat slower than that of sulfisoxazole. Following oral administration of a single 2 g. dose, plasma levels of approximately 10 to 12 mg. per 100 ml. are attained in two hours. For severe infections, the recommended adult dosage is 2 g. initially, followed by 1 g. two to three times daily thereafter.

OCCURRENCE
Sulfamethoxazole Oral Suspension N.F.
Sulfamethoxazole Tablets N.F.

Sulfamethizole N.F.; Thiosulfil®; N^1-(5-Methyl-1,3,4-thiadiazol-2-yl)sulfanilamide, 5-Methyl-2-sulfanilamido-1,3,4-thiadiazole. This compound is a white, crystalline powder soluble 1:2,000 in water. Its high solu-

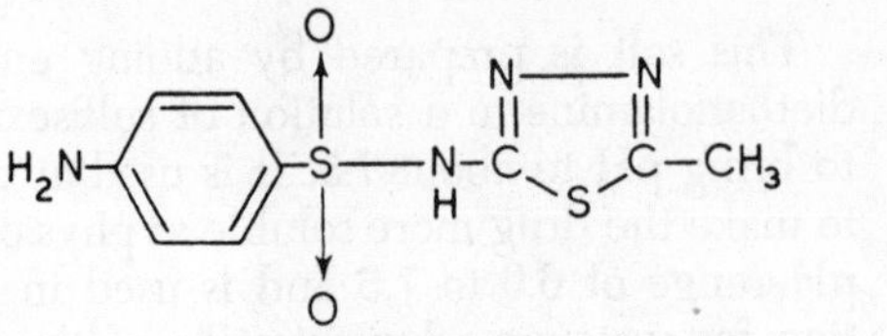

Sulfamethizole

bility makes it useful for the treatment of urinary tract infections. It may be administered to patients with infections who are sensitive to other sulfonamides, since current evidence suggests little cross sensitization to sulfamethizole. The usual dose is 500 mg. 4 times daily. The usual dose range is 500 mg. to 1 g.

OCCURRENCE
Sulfamethizole Oral Suspension N.F.
Sulfamethizole Tablets N.F.

Sulfaethidole, N.F., Sul-Spansion®, Sul-Spantab®,   N¹-(5-Ethyl-1,3,4-thiadiazole-2-yl)sulfanilamide,   5-Ethyl-2-sulfanilamido-1,3,4-thiadiazole. This is a white, crystal-

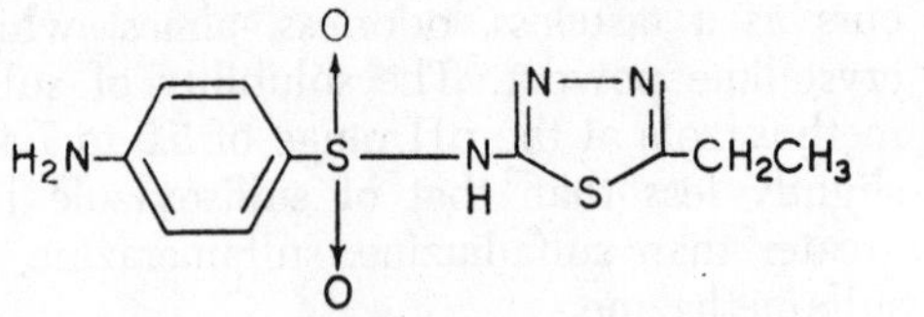

Sulfaethidole

line powder which is soluble 1:4,000 in water. It is identical in chemical structure with sulfamethizole, except for an ethyl rather than a methyl in the 5 position of the thiadiazole ring. It is available as a suspension and in tablets, both formulated to provide a sustained release of the drug. Not only has this drug the lowest degree of acetylation of the common sulfonamides but the acetyl derivative is more soluble than the nonacetylated form. This drug is useful in urologic therapy.

Sulfamethoxypyridazine, Kynex®, Midicel®, N¹-(6-Methoxy-3-pyridazinyl)sulfanilamide. This compound is a white or yellowish-white, bitter, crystalline powder. It is odorless and stable in air, but it darkens slowly on exposure to light. It is sparingly soluble in alcohol and very slightly soluble in water.

Sulfamethoxypyridazine is absorbed rapidly from the gastrointestinal tract but has a very slow rate of excretion. Because of

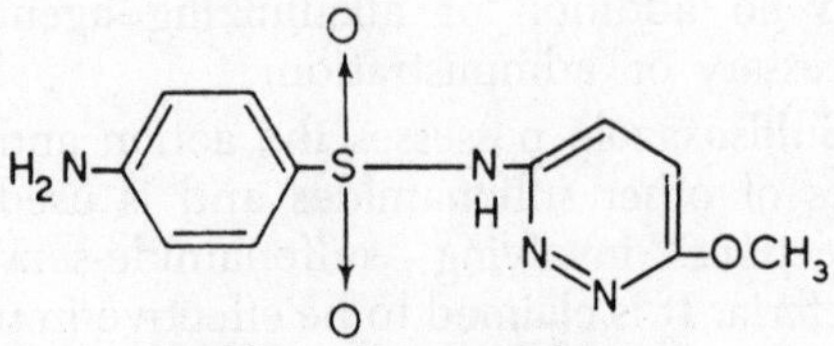

Sulfamethoxypyridazine

this slow rate of excretion, a longer duration of action is obtained and less frequent administration is required than is the case with other sulfonamides. A diminished incidence of renal toxicity has been noted, but, in general, the same side-effects and untoward reactions may be expected as with other sulfonamides.

It is administered orally in small doses: 1 g. the first day, followed by 500 mg. every day thereafter or 1 g. every other day.

Acetyl Sulfamethoxypyridazine; Kynex-Acetyl; N¹-Acetyl-N¹-(6-methoxy-3-pyridazinyl)sulfanilamide, 3-(N¹-Acetylsulfanilamido)-6-methoxypyridazine. Acetyl sulfamethoxypyridazine has the same actions and uses as the parent sulfonamide, sulfamethoxypyridazine, except that it is tasteless and therefore better suited to pediatric use as a liquid medication.

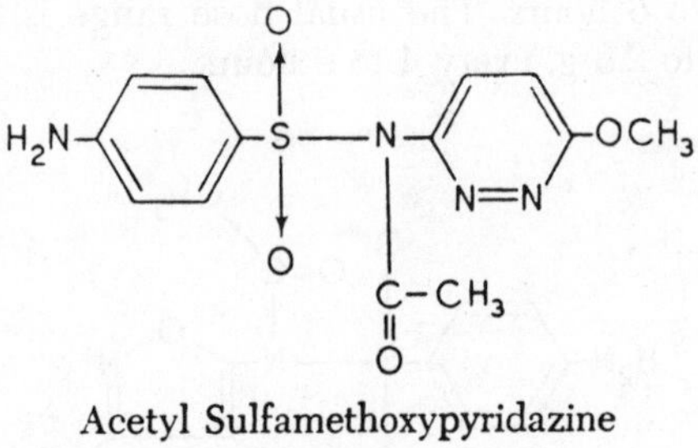

Acetyl Sulfamethoxypyridazine

Sulfachloropyridazine, Sonilyn®, N¹-(6-Chloro-3-pyridazinyl)sulfanilamide. This sulfonamide is well tolerated, rapidly absorbed and excreted rapidly in the urine. In the recommended doses it has been particularly valuable in chronic infections that involve only the urinary tract. It is apparently

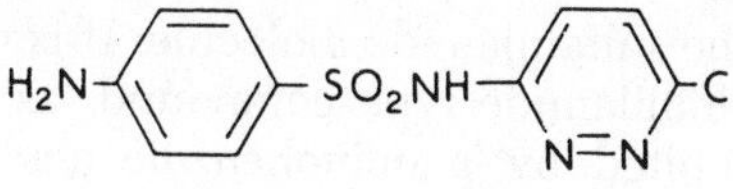

Sulfachloropyridazine

Succinylsulfathiazole

very effective in infections due to *Proteus vulgaris.* Dosage in adults is 1 g. initially and then 1 g. every 8 hours.

Sulfaphenazole, Sulfabid®, 1-Phenyl-5-sulfanilamidopyrazole. This agent is an intermediate-acting sulfonamide which occurs as a tasteless, odorless, white or cream-colored crystalline powder. It is practically insoluble in water and slightly soluble in alcohol.

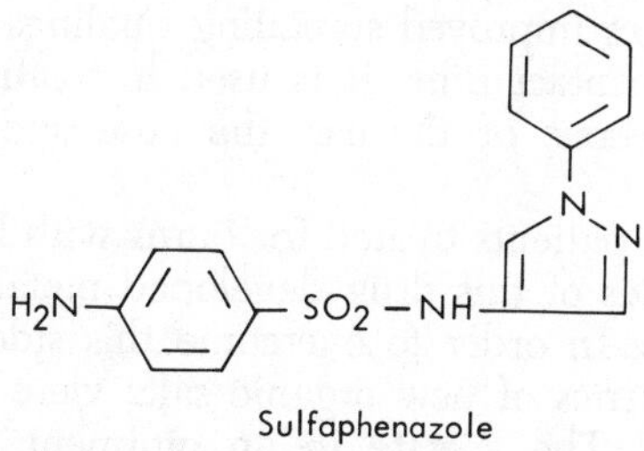

Sulfaphenazole

Sulfaphenazole is readily absorbed from the gastrointestinal tract. Following oral administration of a single 2 g. dose of sulfaphenazole, plasma levels of approximately 10 to 12 mg. per 100 ml. are attained within two hours. A peak level of 10 to 15 mg. per 100 ml. is reached in three to six hours, followed by a gradual decrease to 3 to 8 mg. per 100 ml. within 24 hours. With equal doses of the sulfonamides, sulfaphenazole plasma concentrations are somewhat lower than those attained with sulfamethoxypyridazine or sulfadimethoxine.

Sulfaphenazole is used in the treatment of systemic and urinary tract infections caused by susceptible organisms. For moderate to severe infections, the adult dosage is 3 g. initially, followed by 1 g. every 12 hours thereafter. For mild systemic infections or urinary tract infections in adults, the initial dose is 2 g. followed by 1 g. every 12 hours.

N⁴-Substituted Sulfonamides

Succinylsulfathiazole U.S.P., Sulfasuxidine®, N⁴-Succinylsulfathiazole, 4'-(2-Thi-

azolylsulfamoyl)succinanilic Acid. This compound is a white, odorless crystalline powder which is stable in air but slowly darkens on exposure to light. It is soluble in water (1:4,800), sparingly soluble in alcohol and acetone and insoluble in chloroform and ether. It is soluble in dilute acids and bases and dissolves in sodium bicarbonate solution with evolution of carbon dioxide.

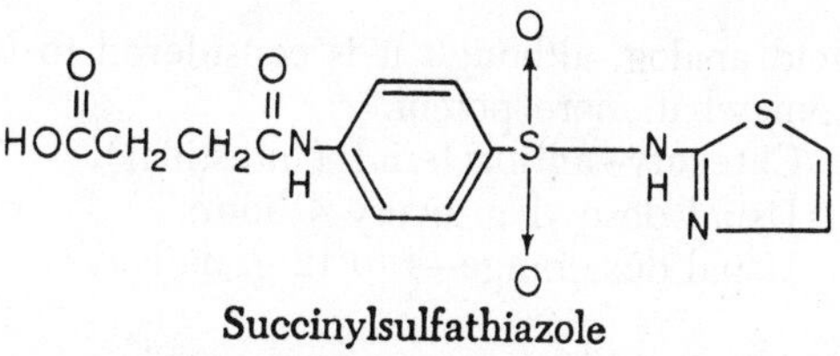

It is absorbed very poorly from the gastrointestinal tract, only 5 per cent being recovered from the urine. Because of its low absorption and low toxicity, compared with sulfaguanidine, it is useful in intestinal infections. This compound is inactive in vitro; therefore, it is probably slowly cleaved to sulfathiazole to give the active form.

Category—antibacterial (intestinal).

Usual dose—3 g. 6 times daily.

Usual dose range—6 to 18 g. daily.

Occurrence

Succinylsulfathiazole Tablets U.S.P.

Phthalylsulfathiazole N.F., Sulfathalidine®; 4'-(2-Thiazolylsulfamoyl)phthalanilic Acid, 2-(N⁴-Phthalylsulfanilamido)thiazole. This compound is an odorless, white,

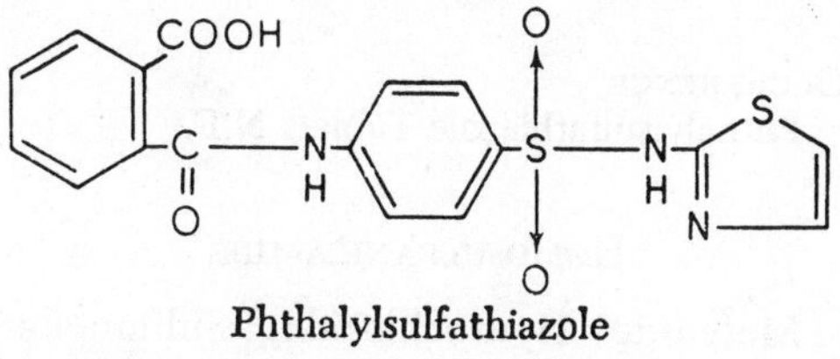

Phthalylsulfathiazole

crystalline powder with a slightly bitter taste. It slowly darkens on exposure to light. It is insoluble in water and chloroform and slightly soluble in alcohol. It is readily soluble in strong acids and bases and liberates carbon dioxide from a solution of sodium bicarbonate.

This compound, like succinylsulfathiazole, is poorly absorbed from the intestinal tract and has properties similar to its succinic acid analog, although it is considered to be somewhat more potent.

Category—antibacterial (intestinal).
Usual dose—1 g. every 4 hours.
Usual dose range—4 to 12 g. daily.

OCCURRENCE
Phthalylsulfathiazole Tablets N.F.

Phthalylsulfacetamide N.F., 4'-(Acetyl-sulfamoyl)phthalanilic Acid, N^1-Acetyl-N^4-phthaloylsulfanilamide. This compound occurs as a white crystalline solid that is very sparingly soluble in water. It possesses the property of diffusing into the intestinal wall, but is absorbed into the bloodstream in amounts too small to give a systemic effect.

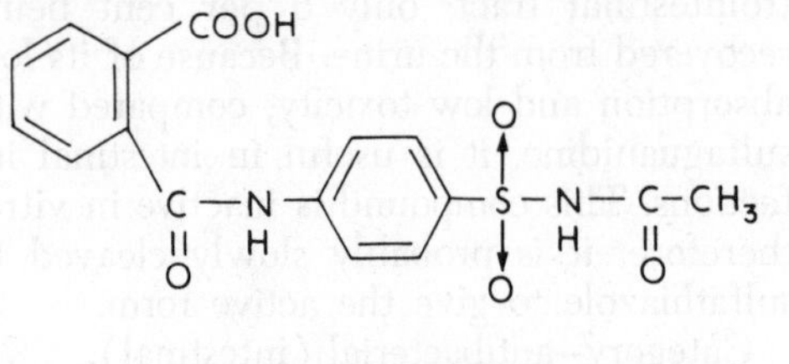

Phthalylsulfacetamide

Its main use is as an intestinal antibacterial agent in gastrointestinal infections and preoperative sterilization of the gastrointestinal tract. It may be used after abdominal surgery.

Category—antibacterial (intestinal).
Usual dose—2 g. three times daily.
Usual dose range—1.5 to 4 g.

OCCURRENCE
Phthalylsulfathiazole Tablets N.F.

HOMOSULFANILAMIDE

Mafenide Hydrochloride, Sulfamylon® Hydrochloride, *p*-Aminomethylbenzenesulfonamide. This compound is a homo-

Sulfbenzamine

log of the sulfanilamide molecule. It is not a true sulfanilamide type compound, as it is not inhibited by *p*-aminobenzoic acid. Its antibacterial action involves a mechanism that is different from that of true sulfanilamide type compounds. This compound is particularly effective against *Clostridium welchii* in topical application and was used during World War II by the German army for prophylaxis of wounds. It is not effective by mouth. It is employed currently alone or with antibiotics, such as streptomycin, in the treatment of slow-healing infected wounds. In combination with (±)-desoxy-ephedrine hydrochloride it is used in nasal drops. A solution of 5 per cent mafenide hydrochloride and 1 per cent methylcellulose gives improved spreading qualities and longer contact time. It is used in treatment of infections of the ear, the nose and the throat.

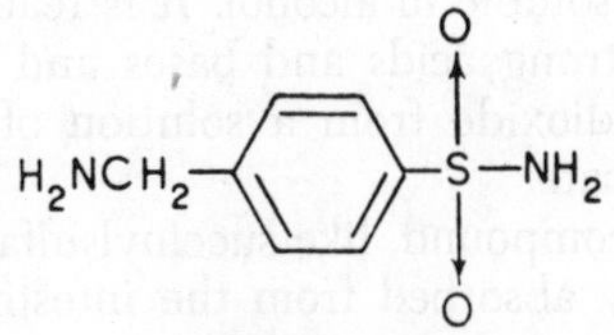

Some patients treated for burns with large quantities of this drug developed metabolic acidosis. In order to overcome this side effect a series of new organic salts were prepared.[15] The acetate in an ointment base proved to be the most efficacious.

SULFONES

The sulfones are primarily of interest as antibacterial agents, although there are some reports of their use in the treatment of malarial and rickettsial infections. They are less effective agents than are the sulfonamides. *p*-Aminobenzoic acid partially antagonizes the action of many of the sulfones, suggesting that the mechanism of action is similar to that of the sulfonamides. It also has been observed that infections which arise in patients being treated with sulfones are cross-resistant to sulfonamides. Some sulfones have found use in the treatment of leprosy.

The search for antileprotic drugs has been hampered by the inability to cultivate *Mycobacterium leprae* on artificial media and by the lack of experimental animals susceptible to human leprosy. Recently a method of isolating and growing *M. leprae* in the foot pads of mice has been reported and may allow for

TABLE 17. SULFONAMIDES

COMPOUND	CHEMICAL NAME	PROPRIETARY NAMES
Sulfanilamide	p-Aminobenzenesulfonamide	Protosil Album, Protylin
Mafenide	p-Aminomethylbenzenesulfonamide	Sulfamylon
N¹-SUBSTITUTED SULFONAMIDES		
Acetyl Sulfamethoxy-pyridazine	N¹-Acetyl-N¹-(6-methoxy-3-pyridazinyl) sulfa-nilamide	Kynex Acetyl
Acetylsulfisoxazole N.F.	N¹-Acetyl-N¹-(3,4-Dimethyl-5-isoxazolyl) sulfa-nilamide	Gantrisin Acetyl
Para-Nitrosulfathiazole	2-(p-Nitrophenylsulfonamido)thiazole	Nisulfazole
Salicylazosulfapyridine	5[p-(2-Pyridylsulfamyl)phenylazo]salicylic acid	Azulfidine
Sulfacetamide N.F.	N¹-Acetylsulfanilamide	Albucid, Sulamyd
Sulfachloropyridazine	N¹-(6-chloro-3-pyridazinyl)sulfanilamide	Sonilyn
Sulfadiazine U.S.P.	N¹-(2-Pyrimidyl)sulfanilamide	Pyrimal
Sulfadimethoxine N.F.	2,4-Dimethoxy-6-sulfanilamido-1,3-diazine	Madribon
Sulfaethidole N.F.	N¹-(5-Ethyl-1,3,4-thiadiazole-2-yl)sulfanilamide	Thiosulfil
Sulfaguanidine	N¹-Guanylsulfanilamide	
Sulfamerazine U.S.P.	N¹-(4-Methyl-2-pyrimidyl)sulfanilamide	
Sulfameter	N¹-(5-Methoxy-2-pyrimidinyl)sulfanilamide	Sulla
Sulfamethazine U.S.P.	N¹-(4,6-Dimethyl-2-pyrimidyl)sulfanilamide	
Sulfamethizole N.F.	N¹-(5-Methyl-1,3,4-thiadiazol-2-yl)sulfanilamide	Thiosulfil
Sulfamethoxazole N.F.	N¹-(5-Methyl-3-isoxazoyl)sulfanilamide	Gantanol
Sulfamethoxypyridazine	N¹-(6-Methoxy-3-pyridazinyl)sulfanilamide	Kynex, Midicel
Sulfaphenazole	1-Phenyl-5-sulfanilamidopyrazole	Sulfabid
Sulfapyridine U.S.P.	N¹-(2-Pyridyl)sulfanilamide	Dagenan
Sulfathiazole	N¹-(2-Thiazolyl)sulfanilamide	Thiazamide Cibazol
Sulfisomidine	N¹-(2,6-Dimethyl-4-pyrimidinyl)sulfanilamide	Elkosin
Sulfisoxazole U.S.P.	N¹-(3,4-Dimethyl-5-isoxazolyl)sulfanilamide	Gantrisin
N⁴-SUBSTITUTED SULFONAMIDES		
Phthalylsulfathiazole N.F.	2-(N⁴-Phthalylsulfanilamido)thiazole	Sulfathalidine
Phthalylsulfacetamide N.F.	N¹-Acetyl-N⁴-phthalylsulfanilamide	
Succinylsulfathiazole U.S.P.	2-(N⁴-Succinylsulfanilamido)thiazole	Sulfasuxidine

the screening of possible antileprotic agents. Sulfones were introduced into the treatment of leprosy after it was found that sodium glucosulfone was effective in experimental tuberculosis in guinea pigs.

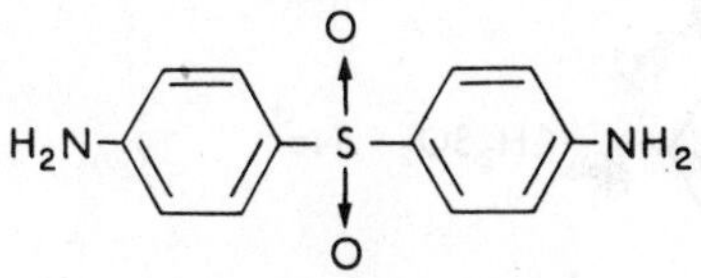

The parent sulfone in the clinically useful area is dapsone (4,4′-sulfonyldianiline).

Four types of variations on this structure have given useful compounds.

Dapsone

1. Substitution on both the 4 and 4′ aminofunctions

2. Monosubstitution on only one of the aminofunctions
3. Nuclear substitution on one of the benzenoid rings
4. Replacement of one of the phenyl rings with a heterocyclic ring

The antibacterial activity and the toxicity of the disubstituted sulfones are thought to be due chiefly to the formation in vivo of dapsone. Hydrolysis of disubstituted derivatives to the parent sulfone apparently occurs readily in the acid medium of the stomach, but only to a very limited extent following parenteral administration. Monosubstituted and nuclear-substituted derivatives are believed to act as entire molecules.

PRODUCTS

Dapsone U.S.P., Avlosulfon®, DDS, 4,4′-Sulfonyldianiline,   *p,p′*-Diaminodiphenyl sulfone. This compound occurs as an odorless, white crystalline powder which is very slightly soluble in water and sparingly soluble in alcohol. The pure compound is light-stable, but the presence of traces of impurities, including water, makes it photosensitive and thus susceptible to discoloration in light. Although no chemical change is detectable following discoloration, the drug should be protected from light.

Dapsone is used in the treatment of both lepromatous and tuberculoid types of leprosy.

Category—antibacterial (leprostatic); suppressant for dermatitis herpetiformis.

Usual dose—leprostatic: initial, 25 mg. twice a week for 1 month, then increased by 25 mg. per dose at monthly intervals to a maximum of 100 mg. 4 times a week; suppressant for dermatitis herpetiformis: 100 to 200 mg. daily.

OCCURRENCE
Dapsone Tablets U.S.P.

Sodium Glucosulfone Injection U.S.P., Promin®, Disodium *p,p′*-Diaminodiphenyl-sulfone-N,N′-di(dextrosesulfonate). This is a sterile solution with no added antimicrobial agents (preservatives). The solution is sufficiently stable so that it can be sterilized by heat. Glucosulfone is effective in the treatment of leprosy. Beneficial effects have been reported in the treatment of tuberculosis but the results have not been exceptional. In tuberculosis treatment glucosulfone is used principally as an adjunct to streptomycin therapy.

It is administered intravenously in a usual dose of 2 g. daily for 6 days of each week. The usual dose range is 2 to 5 g.

Sodium Sulfoxone U.S.P., Diasone® Sodium, Disodium [Sulfonylbis(*p*-phenyleneimino)]dimethanesulfinate. This compound is a white to pale yellow powder with a characteristic odor. It is slightly soluble in alcohol and very soluble in water. It is affected by light.

Sodium sulfoxone is used in the treatment of leprosy. Lesions usually do not progress under therapy, although not all respond favorably.

Usual dose—300 mg. once or twice daily.
Usual dose range—300 mg. to 1 g. daily.

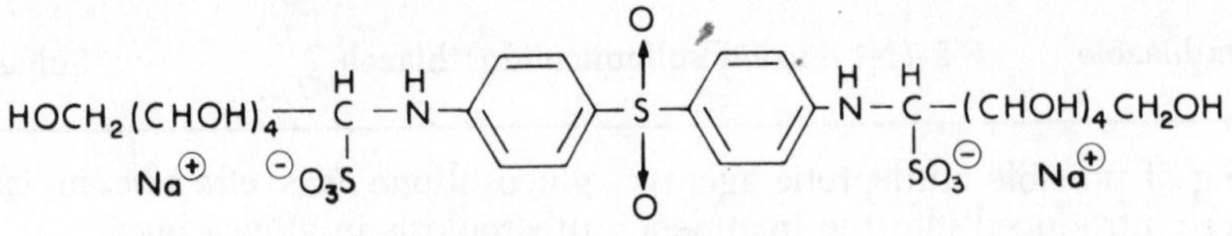

$$HOCH_2(CHOH)_4 - \overset{H}{\underset{\underset{\ominus O_3 S}{Na^{\oplus}}}{C}} - \overset{H}{N} - \underset{}{\overset{O}{\underset{O}{S}}} - N - \overset{H}{\underset{\underset{SO_3^{\ominus}}{Na^{\oplus}}}{C}} - (CHOH)_4 CH_2OH$$

Sodium Glucosulfone

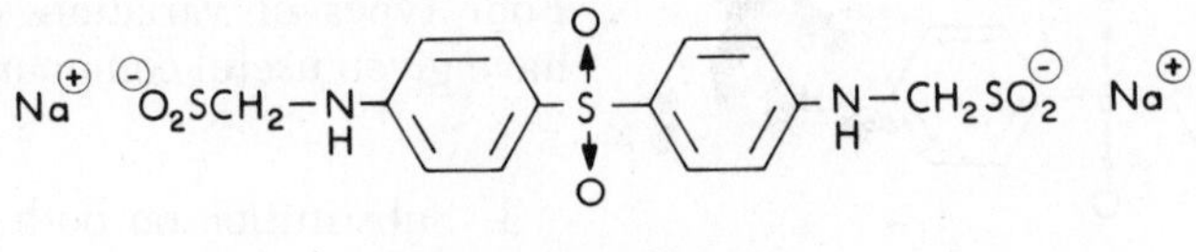

$$Na^{\oplus} \ ^{\ominus}O_2SCH_2 - \underset{H}{N} - \overset{O}{\underset{O}{S}} - \underset{H}{N} - CH_2SO_2^{\ominus} \ Na^{\oplus}$$

Sodium Sulfoxone

OCCURRENCE

Sodium Sulfoxone Tablets U.S.P.

Sodium Acetosulfone, Promacetin®, N-(6-sulfanilylmetanilyl)acetamide Sodium derivative. This agent is used in the treatment

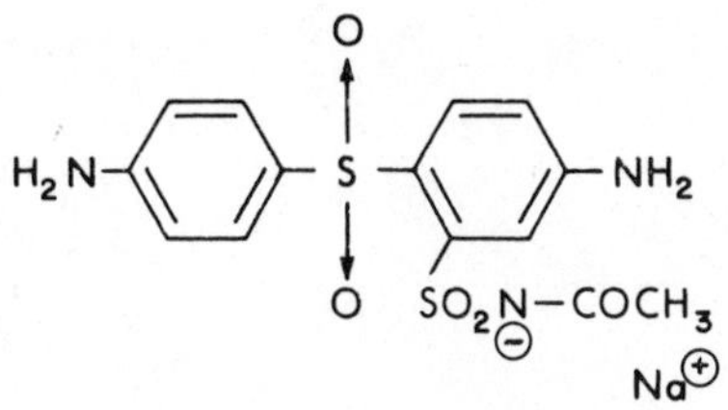

Sodium Acetosulfone

of both lepromatous and tuberculoid types of leprosy. It is also effective in controlling the symptoms of dermatitis herpetiformis.

REFERENCES

1. Domagk, G.: Deutsch. med. Wschr. 61: 250, 1935.
2. Trefouel, J., Trefouel, Mme., Nitti, F., and Bovet, D.: Compt. rend. Soc. biol. 120: 756, 1935.
3. Gelmo, P.: J. prakt. Chem. 77:369, 1908.
4. Mietsch, F., and Klarer, J.: Deutsch. Rep. Pat. 607, 537, 1935.
5. Woods, D. D.: Brit. J. Exp. Path. 21:74, 1940.
6. Fildes, P.: Lancet 1:955, 1940.
7. Angier, R. B., *et al.*: J. Am. Chem. Soc. 70:14, 19, 23, 25, 27, 1948.
8. Jaenicke, L., and Chan, Ph.C.: Angew. Chem. 72:752, 1960.
9. Brown, G. M., Weisman, R. A., and Molnar, D. A.: J. Biol. Chem. 236:2534, 1961.
10. Brown, G. M.: XVII Intern. Kongr. Reine u. Angew. Chem., Verlag Chemie, Weinheim, Germany, 1952.
11. Brown, G. M.: Methods Med. Res. 10: 233, 1964.
12. Seydel, J. K.: J. Pharm. Sci. 57:1455, 1968.
13. Bell, P. H., and Roblin, R. O., Jr.: J. Am. Chem. Soc. 64:2905, 1942.
14. Gray, W. H., Buttle, B. A. H., and Stephenson, D.: Biochem. J. 31:724, 1937.
15. Rakoczy, R., and Nachod, F. C.: J. Med. Chem. 10:273, 1967.

SELECTED READING

Bushby, S. R. M.: The Chemotherapy of Leprosy, Pharmacol. Rev. 10:1, 1958.
Cochrane, R. G.: *in* Lincicome, D. P. (ed.): International Review of Tropical Medicine, p. 1-42, New York, Academic Press, 1961.
Hawking, F., and Lawrence, J. S.: The Sulfonamides, New York, Grune and Stratton, 1961.
Northey, E. H.: Sulfonamides and Allied Compounds, New York, Reinhold, 1948.
Pinder, R. M.: Antimalarials, *in* Burger, A. (ed.): Medicinal Chemistry, ed. 3, p. 492, New York, Wiley-Interscience, 1970.
Schueler, F. W.: Molecular Modification in Drug Design, Am. Chem. Soc. Advances in Chemistry Series, No. 45, Washington, D.C., 1964.
Seydel, J. K.: Molecular basis for the action of chemotherapeutic drugs, structure-activity-studies of sulfonamides, Proc. III Intern. Pharmacol. Congr., São Paulo, New York, Pergamon Press, 1966.
———: Sulfonamides, structure-activity relationship, and mode of action, J. Pharm. Sci. 57: 1455, 1968.
Shepherd, R. G.: Sulfanilamides and other p-Aminobenzoic Acid Antagonists, *in* Burger, A. (ed.): Medicinal Chemistry, ed. 3, p. 255, New York, Wiley-Interscience, 1970.

12

Surfactants and Chelating Agents

Abraham Taub, A.M.,
Emeritus Distinguished Service Professor of Pharmaceutical Chemistry,
College of Pharmaceutical Sciences, Columbia University

SURFACTANTS

Surfactants are a diverse group of chemical compounds which have in common the property of modifying the characteristics of a boundary, surface or interface between two liquids or a liquid and a solid or a liquid and a gas. Such modifying influence becomes manifest in the applicability of these compounds as detergents, wetting agents, foaming agents, dispersing agents, emulsifiers and related industrial and pharmaceutic agents. Overlapping performance of these agents is common. Soap acts as an emulsifier in ammonia liniment; it helps to subdivide the sesame oil into minute droplets enveloped by a continuous aqueous phase. Soap acts as a detergent and a dispersing agent when it lifts dirt from the skin and promotes the separation of dirt particles from each other. Sodium lauryl sulfate acts as a foaming agent in bubble bath powders and as a wetting agent in sulfur hair lotions by reducing the surface tension and permitting the powdered sulfur to sink below the liquid surface. Surfactants generally are used in low concentration and may prove effective in some instances at dilutions of 1:100,000.

Surfactants are characterized by the presence of hydrophilic or water solubilizing groups and lipophilic or fat solubilizing groups. Such groups as $-OSO_2ONa$,

$-COONa$, $-SO_2Na$, $-OSO_2H$ and $-SO_2H$ give strong hydrophilic properties to a compound, the sequence representing the approximate decreasing order of effectiveness. The following groups produce this effect to a lesser degree: $-OH$, $-SH$, $-O-$, $=CO$, $-CHO$, $-NO_2$, $-NH_2$, $-NHR$, $-NR_2$, $-CN$, $-CNS$, $-COOH$, $-COOR$, $-OPO_3H_2$, $-OS_2O_2H$, $-Cl$, $-Br$, $-I$.

Lipophilic groups are typified by aliphatic hydrocarbon chains, aryl-alkyl groups and polycyclic hydrocarbon groups. The presence of unsaturated linkages, such as $-CH=CH-$ and $-C\equiv C-$, helps promote solubility in water and often locates the water solubilizing group.

Surfactants are classified as anionic surfactants when the lipophilic group is present in that part of the molecule which acquires a negative charge upon ionization. They are classed as cationic surfactants when the lipophilic group is in the cation. A third class, nonionic surfactants, is characterized by the presence of weakly hydrophilic groups and lipophilic groups in nonionizable compounds, which render them water dispersible or water soluble. A fourth class, amphoteric surfactants, known also as ampholytic surfactants or ampholytes, contains in addition to the lipophilic group at least one cationic hydrophilic group and at least one anionic hydrophilic group.

The configuration of these classes may be

301

represented by four examples, as follows:
Anionic Surfactant—Sodium laurate:

$$\left[CH_3(CH_2)_{10}COO \right]^{-} \quad Na^+$$

Cationic Surfactant—Lauryl triethyl ammonium chloride:

$$\left[\begin{array}{c} C_2H_5 \\ | \\ C_{12}H_{25}-N-C_2H_5 \\ | \\ C_2H_5 \end{array} \right]^{+} \quad Cl^{-}$$

Nonionic Surfactant—Ethylene glycol monolaurate:

$$CH_3(CH_2)_{10}-COO-CH_2-CH_2-OH$$

Amphoteric Surfactant—Dodecyl glycine:

$$\begin{array}{c} H \\ | \\ C_{12}H_{25}-N^+-CH_2-C-O^- \\ | \quad\quad\quad || \\ H \quad\quad\quad O \end{array}$$

In each case, the underlined part of the molecule is lipophilic; the unmarked parts are hydrophilic.

Surfactants are colloidal electrolytes and thus are able to form ionic micelles. These tend to orient themselves with their long axes perpendicular to the interfacial boundary. At an oil-water interface, the hydrophilic group projects into the water layer and is referred to as a polar group. The lipophilic or nonpolar group projects into the oil layer. It is this phenomenon, whereby the surface-active micelle is directed simultaneously into polar and nonpolar media, which gives rise to such surface activity as dispersion, wetting, detergency and emulsification.

The mere presence of lipophilic and hydrophilic groups is not sufficient to render a compound surface active. There must be a balance between these groups. Thus, CH_3COONa, sodium acetate, is relatively too heavily weighted on the hydrophilic side and is very soluble in water but manifests no surface-active properties. As the hydrocarbon chain is lengthened to about ten carbon atoms, colloidal properties are exhibited,

water and oil solubility of the respective hydrophilic and lipophilic groups are good and surface-active properties are displayed. Sodium laurate, $CH_3(CH_2)_{10}COONa$, is an example of a proper balance between lipophilic and hydrophilic groups. If the chain of carbons becomes too long, namely, about 16, this reduces the solubility in water of the polar group. Thus, sodium stearate, $CH_3(CH_2)_{16}COONa$, is slowly soluble in water and exhibits poor surface-active properties. It is top-heavy on the lipophile side. However, if a double bond is present near the middle of the carbon chain, as in sodium oleate, or if a hydrophilic group, such as an OH, is introduced in the lipophilic group, balance is restored. Sodium ricinoleate,

$$\begin{array}{c} CH_3(CH_2)_5CH-CH_2CH=CH(CH_2)_7COONa, \\ | \\ OH \end{array}$$

is water soluble and is an excellent surface-active agent. The addition of too many double bonds, or OH groups, while making the compound more water soluble, again would upset the balance, this time making it heavy on the hydrophile side, and surface-active properties would be reduced markedly or eliminated.

An attempt has been made to set up a semiquantitative system of classification, based on the relative influence of the hydrophilic and the lipophilic groups, for predicting the best type of emulsifier or other surface activity. It is known as the HLB (hydrophile-lipophile balance) system. On the basis of experimentation, low numbers (4 to 6) are assigned to those compounds which give rise to w/o emulsions and which tend to be of an oil-soluble nature. Higher numbers (8 to 18) are given to compounds which form o/w emulsions. The highest numbers (15 to 18) are for compounds which tend to be water soluble and which have the property of solubilizing oils in water. For any given oil and water, there is one HLB, irrespective of the nature of the individual surface-active agents in a blend, attained by proper proportioning which is optimum for producing a stable emulsion.

The large number of possibilities of introducing groups for the purpose of attaining specific hydrophile-lipophile balances has given rise to thousands of new compounds having surface-active properties.

ANIONIC SURFACTANTS

The history of the development of anionic surfactants is also the history of synthetic detergents. Frémy first sulfonated olive oil in 1831. The first completely synthetic surface-active agent was made on a commercial scale in Germany, in 1916; in this country, the development dates from the 1930's when the alkyl sulfates first were marketed.

There are six major groups of chemical compounds included among the anionic agents: soaps and carboxylic acid derivatives, alkyl sulfates, alkyl sulfonates, alkyl aryl sulfonates, sulfated and sulfonated amides and amines, and sulfated and sulfonated esters and ethers. A selected number of typical examples having pharmaceutic interest will be described.

Products

Sodium Lauryl Sulfate U.S.P., Duponol® C, is a mixture of sodium alkyl sulfates consisting chiefly of sodium lauryl sulfate $[CH_3(CH_2)_{10}CH_2OSO_3Na]$. This is prepared by sulfating long-chain alcohols and neutralizing to form the sodium salts. The alcohols are derived by reduction of coconut oil and other fatty glycerides by high pressure hydrogenation with copper-chromium oxide catalyst, or by sodium reduction.

It occurs as white or light-yellow crystals or flakes having a slight coconut fatty odor. It is soluble in water (1:10). The official form permits up to 8 per cent of combined sodium sulfate and sodium chloride.

Sodium lauryl sulfate is representative of this class of surfactants and accounts for about 20 per cent of the total production of these agents.

It is compatible with alkalies and with soaps, but, unlike the latter, it also is unaffected by dilute acids, calcium or magnesium ions. A 0.1 per cent solution will exert detergent action in a hard water containing 2,500 ppm. of calcium. This makes it widely used in textile and other industrial applications. It is the basis of soapless shampoos and serves as a unique tablet lubricant and as an emulsifier in preparing water-miscible ointment bases.

Category—surfactant.

	PER CENT
OCCURRENCE	SODIUM LAURYL SULFATE
Hydrophilic Ointment U.S.P.	1.0

Sodium Alkyl Sulfoacetate; Sodium Lauryl Sulfoacetate; Lathanol LAL®. A mixture consisting principally of compounds of the general formula $ROCOCH_2SO_3Na$. R designates the alkyl group, which is predominantly lauryl ($C_{12}H_{25}$). Commercial preparations usually contain some inert material.

This is a white powder of slight acrid taste and mild coconutlike odor. It is soluble (1:10) in water to form neutral or slightly alkaline solutions. Because of its detergent and foaming properties, it is utilized in dentifrices and other dental preparations.

Sodium 2-Ethylhexyl Sulfate; Tergemist®,

$$CH_3CH_2CH_2CH_2CHCH_2SO_4Na$$
$$\overset{|}{C_2H_5}$$

is available as an aqueous solution in 0.125 per cent concentration with 0.1 per cent potassium iodide. It is applied topically in a fine mist from a nebulizer and is used for the liquefaction of bronchial secretion in conditions associated with thickened sputum.

Sodium Tetradecyl Sulfate; Sodium Sotradecol®, Sodium 7-Ethyl-2-methyl-4-hendecanol sulfate. Sodium Sotradecol is a

$$CH_3CHCH_2CHCH_2CH_2CHCH_2CH_2CH_2CH_3$$
$$\overset{|}{CH_3}\quad\overset{|}{OSO_3Na}\quad\overset{|}{C_2H_5}$$

white, water-soluble powder made from synthetically prepared fatty acids and, therefore, claimed to be free of allergenic substances of natural origin. It is used in the form of 1 per cent, 3 per cent and 5 per cent aqueous solutions, buffered at pH 7.6, in the obliterative treatment of varicose veins. This highly branched chain type of alkyl sulfate appears to be a more effective vein obliterant than straight chain surfactants, such as sodium morrhuate or sodium lauryl sulfate.

A technical grade, known as Tergitol Penetrant 4, is available as a 25 per cent aqueous solution.

Sodium Heptadecyl Sulfate; Tergitol® Anionic 7, (25 per cent in aqueous solution).

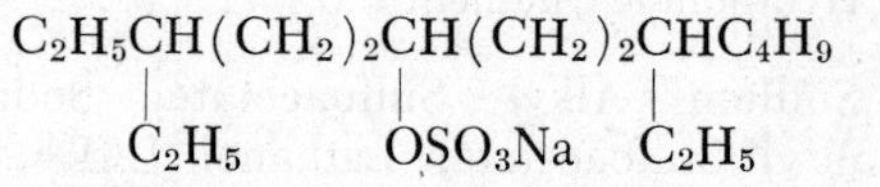

$$C_2H_5CH(CH_2)_2CH(CH_2)_2CHC_4H_9$$
$$\qquad\quad C_2H_5 \qquad OSO_3Na \quad C_2H_5$$

Like the previous compound, this is an alkyl sulfate derived from a secondary alcohol. Although this compound, in concentrations of 1 to 5 per cent, may be used to emulsify water-insoluble, low viscosity solvents, such as petroleum benzin, naphtha, and trichlorethylene, in water, it is used primarily as a wetting agent in low concentrations (0.1 per cent). In the manufacture of penicillin, it has proved to be serviceable in breaking difficult solvent-water emulsions through preferential wetting. The same property may also account for its usefulness in bacteria determinations: when added to bacteriologic media, it inhibits the growth of all bacteria except the coliform group.

Sulfocolaurate. This compound is the lauric acid ester of the potassium salt of sulfoacetic acid amidified with β-amino ethyl alcohol. It is prepared by esterifying

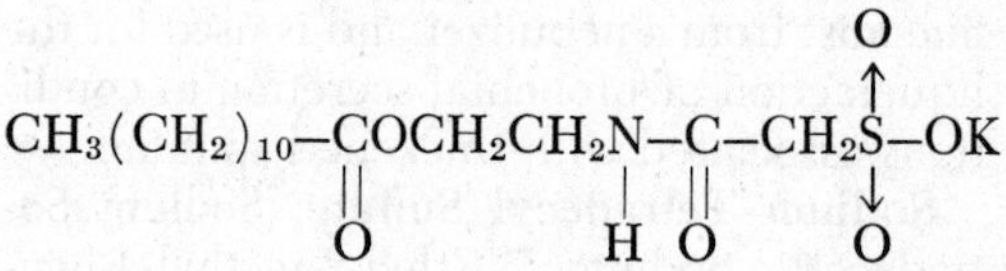

$$CH_3(CH_2)_{10}-COCH_2CH_2N-C-CH_2S-OK$$

Sulfocolaurate

the chloracetamide of β-aminoethyl alcohol with lauric acid and reacting the resulting product with potassium sulfite.

The commercial product contains 8 per cent of potassium chloride. It is a white powder sparingly soluble in cold water and very soluble at 37.5°. It produces copious foam and is used in dentifrices as a foaming agent in concentrations of 1 to 2 per cent. It has been accepted for inclusion in *Accepted Dental Remedies* because of its lack of irritation to the tissues of the mouth in the concentration recommended.

Sodium Lauroyl Sarkosinate, Sarkosyl® NL 30, is one of a group of N-acyl sarkosinates representing modified fatty acids in

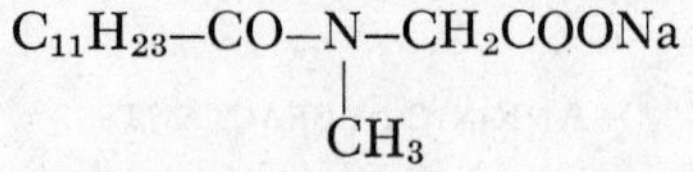

$$C_{11}H_{23}-CO-N-CH_2COONa$$
$$\qquad\qquad\qquad CH_3$$

which the hydrocarbon chain is interrupted by an amidomethyl group. It is made by condensing lauroyl chloride with the sodium salt of sarcosine (N-methyl glycine) and neutralizing with sodium hydroxide. It is colorless, odorless, water-soluble and nearly tasteless; it has mild detergent properties, making it useful in hand creams. Because it adsorbs strongly at interfaces and is markedly antienzymic, it has found special use in dentifrices.

Dioctyl Sodium Sulfosuccinate U.S.P., Aerosol O.T.®, Colace®, Doxinate®, Sodium 1,4-Bis(2-ethylhexyl) Sulfosuccinate. This is prepared by treating the reaction product of maleic anhydride and octyl alcohol with sodium bisulfite.

It is a white, waxlike, plastic solid usually occurring in pellets, with an odor suggestive of octyl alcohol. It is freely soluble in alcohol, glycerin and petroleum benzin. It dissolves slowly in water (1:70). A 1 per cent solution has a pH of 6.5.

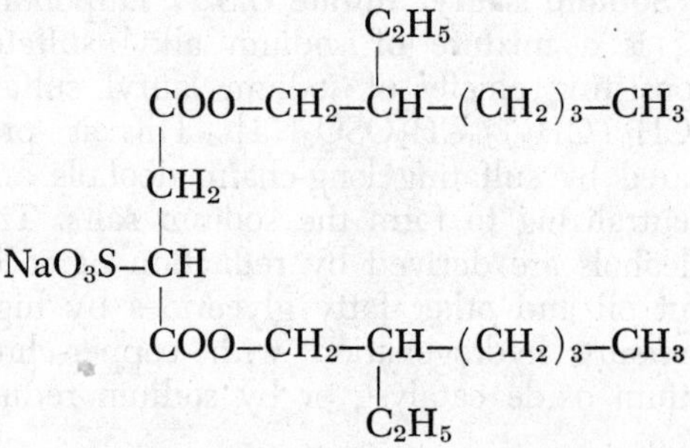

Of the several alkyl substituted sulfosuccinates, the dioctyl is by far the most powerful wetting agent, as measured by the Draves and Clarkson test which measures the sinking time of a skein of yarn under specified conditions.

By contrast with long-chain, sulfated, alcohol-type surfactants, which have terminal polar groups, dioctyl sodium sulfosuc-

$$\qquad\qquad\qquad C_2H_5$$
$$COO-CH_2-CH-(CH_2)_3-CH_3$$
$$CH_2$$
$$NaO_3S-CH$$
$$COO-CH_2-CH-(CH_2)_3-CH_3$$
$$\qquad\qquad\qquad C_2H_5$$

Dioctyl Sodium Sulfosuccinate

cinate carries two polar groups in the middle of a branched chain. In the former, the molecules at an oil-water interface can pack very closely together, with their long axes parallel and perpendicular to the interface. A relatively large number will be required to cover a small surface area. In contrast, the molecules of dioctyl sodium sulfosuccinate, when oriented at the interface, are relatively short but considerably greater in area. A small number of molecules will be required to cover a given area. This illustrates the significance of molecular configuration of surface-active agents in comparing their relative efficiencies.

Dioctyl sodium sulfosuccinate is moderately stable in acid and mild alkali solution; it is decomposed by strong bases. Its diverse uses include the wetting of wax molds to prevent bubbles in dental castings, the cleaning of fruits, the washing of laboratory and ampul glassware, the preparation of hydrophilic ointment bases, the solubilizing of cresol in water and the enhancement of bactericidal properties of antiseptics.

Because it is relatively inert pharmacologically when administered orally, and since it lowers surface tension in the gastrointestinal tract and therefore permits water to permeate the intestinal contents, dioctyl sodium sulfosuccinate is used as a fecal softener as well as a wetting agent.

It is available in oral and suppository dosage forms.

Category—wetting agent; nonlaxative cathartic (fecal matter softener).
Usual dose—100 mg. 2 or 3 times daily.
Usual dose range—50 to 500 mg. daily.

OCCURRENCE
Dioctyl Sodium Sulfosuccinate Capsules U.S.P.
Dioctyl Sodium Sulfosuccinate Solution N.F.
Dioctyl Sodium Sulfosuccinate Syrup N.F.
Dioctyl Sodium Sulfosuccinate Tablets U.S.P.

Dioctyl Calcium Sulfosuccinate N.F., Surfak®, Bis(2-ethylhexyl) S-Calcium Sulfosuccinate. This is a white gelatinous solid similar in action to the sodium compound. It provides fecal softening without peristaltic stimulation. It is especially useful in conditions in which the sodium ion is contraindicated.

Category—wetting agent; nonlaxative fecal matter softener.
Usual dose—240 mg.

OCCURRENCE
Dioctyl Calcium Sulfosuccinate Capsules N.F.

Sodium Stearate and **Hard Soap** are discussed on page 166.
Medicinal Soft Soap (Green Soap) is discussed on page 167.
Sodium Xylene Sulfonate; Naxonate®.

This compound is typical of a class of anionic surfactants formed by the alkylation of an aromatic hydrocarbon, followed by sulfonation. It is a white powder, very soluble in water.

The presence of alkyl groups substituted in the nucleus of this compound tends to increase wetting and penetrating values. In the case of sodium xylene sulfonate, these properties approach a maximum, and the compound becomes a hydrotropic or water-solubilizing agent. Naxonate helps "couple" or make mutually soluble many immiscible liquids, such as cresol and water, and, in general, helps in improving compatibility and speeding up reactions in aqueous solution.

Nacconol® NR and Santomerse® No. 1 are mixed alkyl derivatives of benzene sulfonates. The source of the alkyl chains (C_{12}-C_{18}) is a special fraction of kerosene which places these products in lower cost categories. These two make up a substantial percentage of all detergents produced in this country. In addition to their wide use in industry, they find application in dentrifices, shampoos and cosmetics.

Within recent years, syndets (synthetic detergents) have contributed significantly to water pollution. Two million tons of detergent of all kinds are produced annually in the United States, of which 60 per cent are syndets. Some of the syndets are not biodegradable in sewage disposal systems, and their presence is responsible for the foamy rivers and frothy tap water.

Biodegradability is the capacity of a detergent to be broken down by bacteria into carbon dioxide, water and simple organic molecules which bacteria utilize to make new protoplasm. Conventional fatty acid soaps are degraded readily by river bacteria, but these bacteria do not attack or assimilate syndets that contain branched side-chain hydrocarbons. The petroleum-based alkyl benzene sulfonates, which are compounds of this type, have been largely replaced by biodegradable detergents.

The solution lies in the replacement of the branched chain with straight-chain alkyl benzene sulfonates. Even more completely biodegradable, though more costly, are the sodium salts of alkyl sulfates derived from C_{12} to C_{18} straight-chain alcohols such as tallow alcohols and lauryl alcohol. These are known as "soft" or biodegradable detergents, and they are rapidly displacing the "hard" or nonbiodegradable syndets.

CATIONIC SURFACTANTS

The substances in this class that are most frequently used are aliphatic quaternary ammonium salts, pyridinium salts and picolinium salts, although arsonium, phosphonium and sulfonium salts and primary and secondary amine salts also exhibit properties common to cationic agents. Of the substituent hydrocarbon groups, one is usually a long chain of 8 to 16 carbon atoms, which, however, may be an interrupted chain modified by the inclusion of oxygenated and aryl groups. The other substituents may be short alkyl, aryl, or aryl-short alkyl hydrocarbon groups.

Cationic surfactants, particularly quaternary ammonium and pyridinium salts, are unusual in that they exhibit not only detergent but also germicidal properties. Jacobs and Heidelberger, in 1916, and later Domagk, in 1935, demonstrated the latter qualities. These compounds offer certain advantages over more common germicides, such as phenols, iodine, dyes, mercurial and silver compounds, in that they are more soluble in water, nonstaining, relatively less toxic or caustic, effective in high dilution and noncorrosive to metallic and rubber surgical appliances.

Some anionic agents also have proved germicidal, usually at low pH, but their action for the most part is limited to the inhibition of gram-positive organisms. Cationic agents inhibit both gram-positive and gram-negative bacteria and generally function best in neutral or alkaline media.

Cationic agents are believed to function by exerting a marked effect on the metabolism of the bacterial organism. They are adsorbed upon the surface membranes of the bacteria and denature or disarrange the protein molecules. As a result of the cytologic damage, the cell contents seep out, a process which is analogous to hemolysis. The extent of such bacteriolysis has been correlated with germicidal efficacy.

However, cationic agents are not without disadvantages. They are incompatible with soap and other anionic agents. The large part of the quaternary ammonium salt molecule, being positively charged, combines with the large negatively charged part of the anionic agent. In high concentration, a precipitate or cloudiness is manifest. In low concentration, there is no visible manifestation of incompatibility, but bactericidal properties are reduced or completely lost.

Recently, it has been shown that, in many instances, the incompatibility of cationic germicides with anionic detergents can be overcome by the addition of nonionic solubilizers such as the polyethylene oxide derivative of dibutyl phenol, or other alkyl phenol polyglycol ethers. The nonionic agent must be present in amount equal to at least 10 per cent of the combined weights of the ionic surfactants.

Cationic surfactants do not kill spores and some fungi and viruses. They cannot be relied upon to give absolute sterility of surgical instruments that are heat labile. When used as sanitizing rinses for reduction of bacterial flora on utensils, their effectiveness is lowered by the presence of food residues. In general, increase in pH or in temperature increases their effectiveness.

In appraising the claims of germicidal efficacy appearing in the literature for the several available cationic germicides, or in comparing these with other germicidal agents, values obtained by standard F.D.A. phenol coefficient or agar plate in-vitro tests must to a large extent be discounted. Many factors have been shown to vitiate the results or give erroneous values. Among these are the marked reduction in activity caused by the presence of serum and, also, the bacteriostatic influence of traces of adsorbed cationic agent that have been carried over on bacterial surfaces through several dilutions of culture media, as a result of failure to provide effective inactivating agents. The naphthylamine sulfonic acids offer some promise as cationic inactivating agents. Tamol N, the sodium salt of a condensed aryl sulfonic acid, has proved to be an effective inactivator.

In-vivo tests appear to give a more reliable index. The Nungester and Kempf technic, whereby a freshly segmented mouse tail is impregnated with a virulent bacterial organism, exposed for a predetermined time to the action of the germicide and introduced surgically into the peritoneal cavity of the animal, evaluates the germicide in terms of percentage survival of a number of test animals at different concentration levels of the germicide.

Another factor of importance is the toxicity of the germicide to body tissues. The Welch-Hunter toxicity test measures the ratio of the dilution volume tolerated by tissue (for example, phagocytes) to the weakest dilution volume that is germicidal. A ratio of more than 1 obviously would indicate unfitness for use. Many germicides in use today have indexes above 1.

Until results are published which utilize these more rational technics, any stated specific numeric data representative of germicidal concentrations must be considered in a conservative manner.

Products

Benzalkonium Chloride U.S.P., Zephiran® Chloride, Roccal®, BTC®, Alkyldimethylbenzylammonium Chloride. Benzalkonium Chloride is a mixture of alkyldimethylbenzylammonium chlorides of the general formula $[C_6H_5CH_2N(CH_3)_2R]Cl$, in which R represents a mixture of alkyls from n-C_8H_{17} to $C_{16}H_{33}$.

It is a white, bitter-tasting gel, slightly soluble in benzene and very soluble in water and alcohol. Aqueous solutions are colorless, alkaline to litmus, and foam strongly.

Benzalkonium chloride possesses wetting, detergent, keratolytic and emulsifying actions. It is used as a surface antiseptic for intact skin and mucosa at 1:750 to 1:10,000 concentration. Above 1:1,000, it has proved to be irritant on prolonged contact. It is effective for many pathogenic nonsporulating bacteria and fungi after several minutes exposure. For irrigation, 1:20,000 to 1:40,000 solutions are employed. For sterile storage of surgical instruments, 1:1,000 solutions are used, 0.5 per cent of sodium nitrite being added as an anticorrosive agent.

For presurgical antisepsis, all traces of soap used in preliminary scrubbing must be removed, or inactivation of the cationic detergent will ensue. When tinted solutions are used to help delineate area of operation, cationic-type dyes must in general be selected to avoid incompatibility. The official solution may contain a suitable coloring agent and may be buffered by the addition of ammonium acetate in a quantity not more than 40 per cent of the weight of benzalkonium chloride.

Category—topical anti-infective.

For external use—topically to the conjunctiva, 0.1 ml. of a 0.01 per cent solution or to the skin and mucous membranes as a 0.02 to 0.01 per cent solution.

OCCURRENCE
Benzalkonium Chloride Solution U.S.P.

Cetyl Trimethyl Ammonium Bromide; Cetavlon®, CTAB®. This cationic agent differs from Zephiran in not containing an

$$CH_3(CH_2)_{15}NH_2 \xrightarrow{CH_3Br} CH_3(CH_2)_{15}N(CH_3)_3Br$$

Cetyl Trimethyl Ammonium Bromide

aromatic radical. Cetavlon is made by completely alkylating cetylamine with methyl bromide, and, thus, it contains methyl in place of an aromatic group.

It is a white powder that has a somewhat astringent and bitter taste and is only slightly soluble in cold water (1:200) but soluble (1:2.5) in warm water. It is recommended especially for washing dirty wounds, since it is a good detergent, but the concentration required for different organisms varies considerably. It can be used up to 1 per cent without irritation, but much more dilute solutions may be efficacious.

Cetyl Dimethyl Ethyl Ammonium Bromide is very similar to Cetavlon (above). It is an almost white, crystalline powder that is soluble in water (1:4). Aqueous solutions are neutral. It is suitable for sterilizing instruments and usually is used in hydroalcoholic solution with sodium nitrite as a reducing agent.

Domiphen Bromide; Bradosol® Bromide; Dodecyldimethyl (2-phenoxyethyl) ammonium Bromide is a white crystalline salt soluble in water and alcohol. It is an antiseptic of low toxicity and is germicidal to most organisms found in mouth and throat infections. Dosage is 1.5 mg., combined with benzocaine in a throat lozenge.

Benzethonium Chloride N.F., Phemerol® Chloride, Hyamine® 1622, Benzyldimethyl {2-[2-(*p*-(1,1,3,3,-tetramethylbutyl) phenoxy) ethoxy] ethyl } ammonium Chloride. This compound differs from benzalkonium chloride in that the long alkyl group has been replaced by a group bearing phenyl and ether linkages between the terminal alkyl group and the quaternary nitrogen atom. Many such derivatives have been synthesized for the purpose of enlarging the spectrum of bacterial effectiveness.

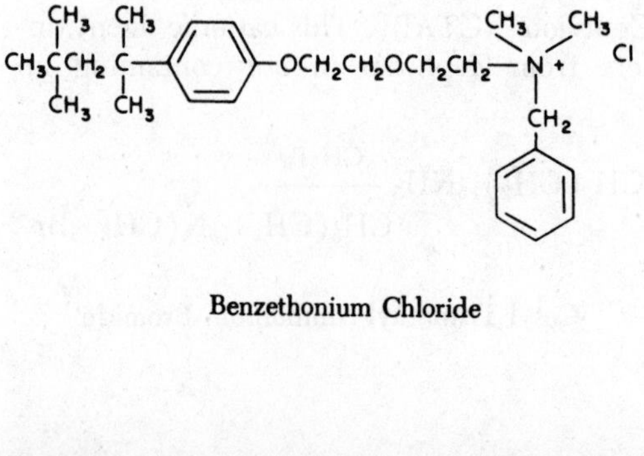

Benzethonium Chloride

Benzethonium chloride is a colorless, odorless, bitter, crystalline powder, soluble in water and slightly soluble in chloroform. A 1 per cent solution has a pH of 5. Solutions above 2 per cent are precipitated by mineral acids and many salts.

Its actions and uses are similar to those of benzalkonium chloride, and it is employed at 1:1,000 concentration for general antisepsis. For irrigation of the eye, nose, or mucous membranes, a 1:5,000 solution is employed. An alcoholic tincture (1:5,000) also is available.

Category—local anti-infective.
Application—topical, as a 1:750 solution to the skin or 1:5,000 solution nasally.

OCCURRENCE
Benzethonium Chloride Solution N.F.

Methylbenzethonium Chloride N.F., Diaparene® Chloride; Bactine®, Hyamine®, 10X; Benzyldimethyl{2-{2-[4-(1,1,3,3-tetramethylbutyl)tolyloxy]ethoxy}ethyl}ammonium Chloride. This is a colorless, crystalline compound, bitter in taste, soluble in water, alcohol and chloroform.

It has a specific use in the bacteriostasis of the intestinal saprophyte, *Bacterium ammoniagenes*, which produces ammonia in decomposed urine and is responsible for diaper dermatitis in infants. It is used in

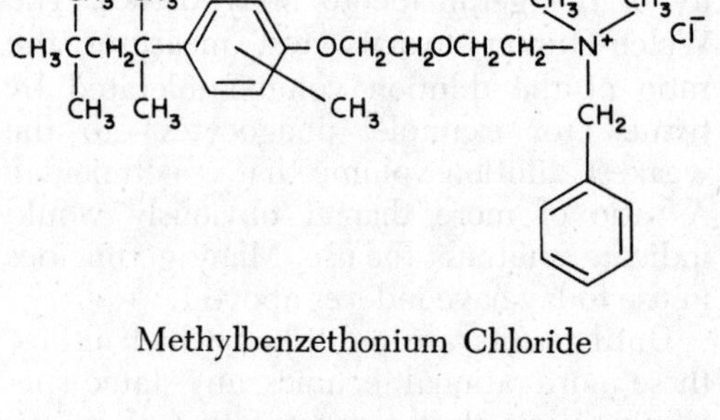

Methylbenzethonium Chloride

aqueous solution, 1:25,000, as a final rinse before drying, for diapers previously washed free of soap cleansers; it offers protection against urine decomposition for 15 hours.

It also is marketed for topical use in the treatment of diaper dermatitis.

Category—local anti-infective.

For external use—topically as a 0.055 per cent powder.

OCCURRENCE

Methylbenzethonium Chloride Powder N.F.

Cetylpyridinium Chloride N.F., Ceepryn® Chloride, 1-Hexadecylpyridinium

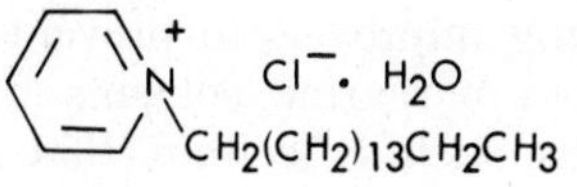

Cetylpyridinium Chloride

Chloride. In this compound the quaternary nitrogen is part of a heterocyclic nucleus. The cetyl derivative has been selected in preference to other alkyl derivatives studied, because of its maximal activity. Also, it is believed that the absence of a benzyl group reduces the toxicity of the compound.

Cetylpyridinium chloride is a white powder freely soluble in water and alcohol.

It is compatible with ephedrine hydrochloride, procaine hydrochloride, urea and allantoin. It is incompatible with tannic acid, picric acid, alum and soap.

It has proved to be germicidally effective over a pH range of 5 to 10. It is available for use as a general antiseptic in 1:100-1:1,000 aqueous solution on intact skin, 1:1,000 for minor lacerations, 1:5,000 to 1:10,000 on mucous membranes, and in tinted tinctures from 1:200 to 1:500 concentration; it is available also in the form of jelly, powder, suppository and as a 1:4,000 phosphate-buffered gargle.

Category—local anti-infective.

Application—topically as a 1:1,000 to 1:100 solution to intact skin; 1:1,000 for minor lacerations; 1:10,000 to 1:2,000 to mucous membranes.

OCCURRENCE

Cetylpyridinium Chloride Solution N.F.

Cetyl bromide is used to form the quaternary ammonium salt of the antihistamine Neohetramine. Neohetramine, N-(2-pyrimi-

dyl)-N-(*p*-methoxybenzyl)-N',N'-dimethylethylenediamine and cetyl bromide form Thonzonium Bromide, an ingredient of Biomydrin used in treating infections and allergic rhinitis.

Emcol-607®. This compound is lauroyl colamin formylmethyl pyridinium chloride.

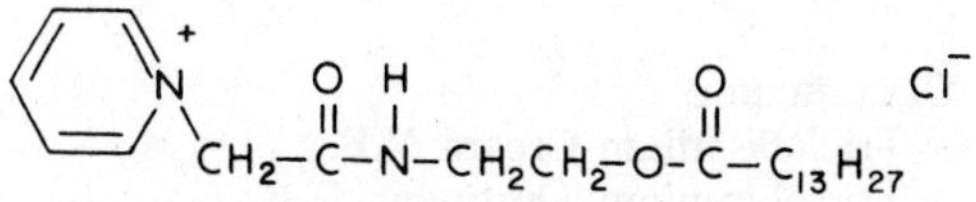

In a long series of compounds studied, this complex long-chain derivative of pyridinium chloride, with a terminal myristate (or alternatively, laurate) grouping, proved the most effective bactericidally. It is illustrative of a careful balance between lipophilic and hydrophilic groups.

It is a white or tan, flaky, odorless powder, very soluble in water, alcohol and acetone, and it has the general properties of the class of cationic agents. Because of a low toxicity index, about 0.6, and relatively high bactericidal and fungicidal activity, it has been advocated for such diversified uses as sanitary laundering of textiles and dish washing, general disinfection, cosmetic preservation, presurgical and other topical antisepsis and in fungous infections. In connection with these uses, possible inactivation by trace anionic agents encountered must be given consideration.

It is used in concentrations of 1:1,000 to 1:30,000.

Triclobisonium Chloride N.F., Triburon Chloride®, Hexamethylenebis[dimethyl (1-methyl-3-(2,2,6-trimethylcyclohexyl)propyl]ammonium Dichloride.

This is an example of terminally substituted alicyclic groups characterizing the quaternary compound.

It is a white or nearly white powder, practically odorless, freely soluble in water, alcohol and chloroform, and almost completely

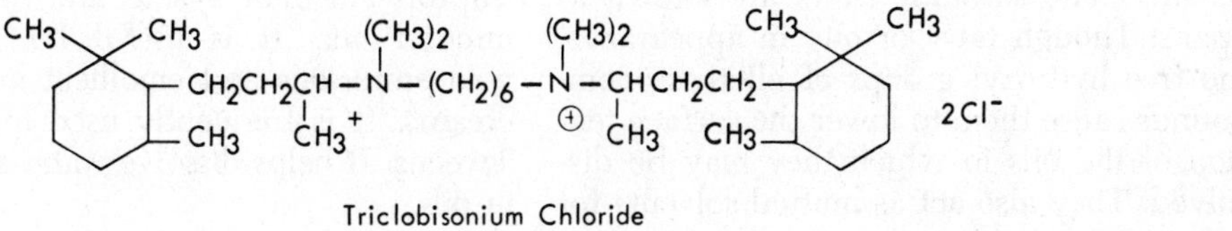

Triclobisonium Chloride

insoluble in ether. It is used as a microbicide having antitrichomonal and antimonilial activity.

Category—local anti-infective.

Application—topically, as a 0.1 per cent ointment 3 or 4 times daily; vaginal, 5 ml. of a 0.1 per cent cream every night for 2 weeks.

OCCURRENCE

Triclobisonium Cream N.F.
Triclobisonium Ointment N.F.

NONIONIC SURFACTANTS

In this class of surfactants, the hydrophilic character usually is obtained from the presence of free hydroxyl or ether oxygen groups and the hydrophobic character from fatty acid hydrocarbon chains of C_{10} to C_{18} or from phenols.

Products

Glyceryl Monostearate N.F., Monostearin.

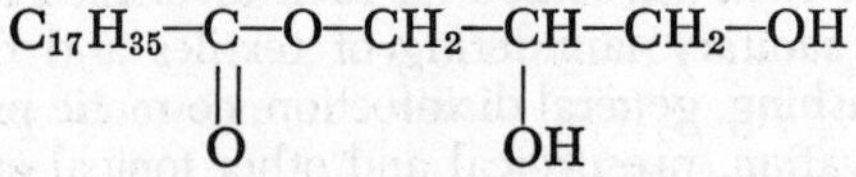

$$C_{17}H_{35}-\underset{\underset{O}{\|}}{C}-O-CH_2-\underset{\underset{OH}{|}}{CH}-CH_2-OH$$

Glyceryl Monostearate occurs as a white, waxlike solid or as white, waxlike beads or flakes. It has a slight, agreeable, fatty odor and taste. It is affected by light.

It melts at 57° and is soluble in hot alcohol, most organic solvents, fixed oils and mineral oil. It is insoluble in water but readily dispersible in hot water with the aid of an anionic or cationic agent.

Glyceryl monostearate is an example of the nondispersible type of nonionic agent. Many compounds of this type are available; the mono-oleates, monoricinoleates, monopalmitates, monolaurates and monostearates of ethylene, diethylene and propylene glycols, glycerol, sorbitol and sorbitan (an inner anhydride of sorbitol) are used most frequently. The sorbitan esters are known as Spans. Though fatty or oily in appearance, the free hydroxyl groups of all these compounds cause them to lower the surface tension of the oils in which they may be dissolved. They also act as mutual solvents for

polar and nonpolar compounds. They may form both w/o and o/w emulsions.

These unique properties give rise to applications such as emulsion stabilizers, dispersing agents of pigments in oils or of food solids in fats, such as cocoa in chocolate fat, shortening improvers to prevent leakage of milk from margarine, solvents for the phospholipids, such as lecithin, that are hard to disperse, and blenders of soap with mineral oil or kerosene. A few, such as propylene glycol monostearate, have been used to prepare hydrophilic suppository bases.

Category—emulsifying aid.

Tegacid®. This is a combination of glyceryl monostearate and a phosphoric acid derivative of an oleic acid amide of diethylethylenediamine. The combination provides an emulsifying agent of value in acidic type creams, such as aluminum sulfate antiperspirant creams.

Acetylated Glyceryl Monostearate is prepared by acetylating one of the hydroxyl groups of glyceryl monostearate. It is not water dispersible, but is easily emulsified by other surfactants and produces emulsions and cosmetic creams which are nongreasy and rub easily into the skin. It is miscible with castor oil and with 20 per cent alcohol. It is used in tablet film coating processes, alone or with polyvinylpyrrolidone.

Sorbitan Monooleate, Span® 80. This is a member of a series of fatty acid esters of sorbitan, an inner ether of sorbitol. It is a

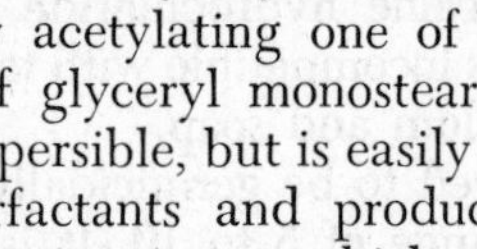

R is $(C_{17}H_{33})COO$

Sorbitan Monooleate

yellow oily liquid, insoluble in water, acetone and propylene glycol, and soluble in isopropyl alcohol, xylene, and vegetable and mineral oils. It is useful for formulating w/o emulsions and emollient and cosmetic creams. It is frequently used together with Tweens. It helps dissolve water-soluble dyes in oils.

Polyethylene Glycol 400 Monostearate; Nonaethylene Glycol Monostearate. This is a soft, white, waxy solid with a melting

$$HOCH_2(CH_2OCH_2)_8CH_2O-\overset{O}{\overset{\|}{C}}-C_{17}H_{35}$$

point of 26°, soluble in many organic solvents and dispersible in hot water. The last-mentioned property characterizes a number of nonionic agents that contain a limited number of ether linkages in the hydrophilic part of the molecule. Other compounds in this water-dispersible group include polyglycerol and sorbitan fatty esters. Sorbitan sesquioleate in petrolatum is marketed as the hydrophilic ointment base, Polysorb.

When dispersed in water, these agents act as thickeners in place of vegetable gums and as self-dispersible lubricants. They form w/o as well as o/w type emulsions. They are used in preparing "soluble oils," clear oily products containing substances such as cresols, kerosene or perfume oils, which form emulsions directly upon addition to water.

Polyoxyl 40 Stearate U.S.P., Polyoxyethylene 40 Monostearate, Stearethate 40, Myrj® 52, is a mixture of the monostearate and distearate esters of mixed polyoxyethylene diols and the corresponding free glycols, the average polymer length being equivalent to about 40 oxyethylene units.

This waxy, nearly odorless solid has a congealing range between 39° and 44°; it is soluble in water, alcohol, ether and acetone and is insoluble in mineral and vegetable oils.

The wide use of polyoxyl 40 stearate as an emulsifier is based on its compatibility with a wide variety of medicinal agents, its inherent stability and its nonirritating qualities. It was previously used as an emulsifier in the official formula for Hydrophilic Ointment (U.S.P. XV).

Category—surfactant.

Polysorbate 80 U.S.P., Sorbitan Monooleate Polyoxyethylene (20) Derivative, Tween® 80. Polysorbate 80 is an oleate ester of sorbitol and its anhydrides copolymerized with approximately 20 moles of ethylene oxide for each mole of sorbitol and sorbitol anhydrides.

This compound differs from the other two examples in that it is completely soluble in water as well as in many fixed oils and organic solvents. The several chains of ether linkages terminating in free hydroxyl groups account for the water solubility. The hydrophilic groups slightly overbalance the lipophilic oleate group. It is not as soluble in some oils as the water-dispersible type of nonionic agent; for example, it is insoluble in mineral oil. Frequently, combinations of water-dispersible, nondispersible and water-soluble nonionic agents are used to produce more stable emulsions. Three related esters, the corresponding monolaurate, monopalmitate and monostearate, are known as Tween 20, 40 and 60, respectively.

Polysorbate 80 has recently been shown to have a marked retardant effect upon metal catalyzed ascorbic acid oxidation. The stabilization effect of this surfactant is attributed to micelle formation; both the size of the micelles and the number of micelles formed increase as the concentration of the polysorbate 80 is increased, up to an optimum concentration of 30 per cent.

Water-soluble nonionic agents are especially useful in the presence of high electrolyte concentration. They also have the unique property of dispersing perfume oils or oil-soluble vitamins in water, producing clear solutions.

Category—surfactant.

OCCURRENCE	PER CENT POLYSORBATE 80
Coal Tar Ointment U.S.P.	0.5
Coal Tar Solution U.S.P.	5

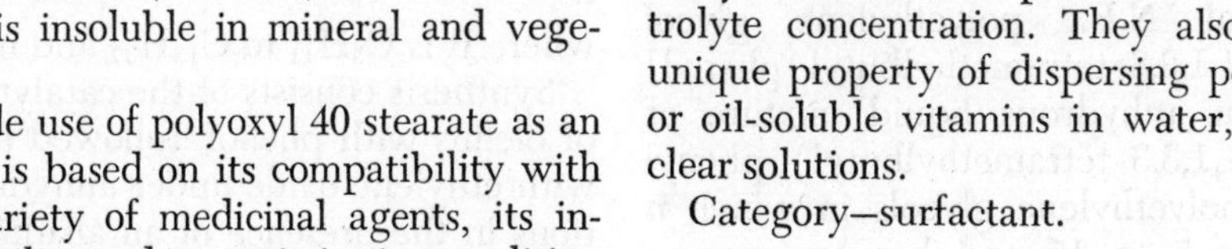

Polysorbate 80

Other compounds in this water-soluble class having similar characteristics are certain partial fatty acid esters of highly polymerized glycol and of long chain polyoxyethylene, polyoxypropylene and methoxy polyoxyethylene glycols.

Triton® X-100; Alkyl Aryl Polyether Alcohol. This is one of a series of related nonionic compounds, differing in structure from the previous compounds in this class, which were all esters. A formula typical of this group is:

$$CH_3(CH_2)_{10}-\underset{\underset{CH_3}{|}}{\overset{\overset{CH_3}{|}}{C}}-C_6H_4-OCH_2(CH_2-O-CH_2)_6CH_2OH$$

Among the chemical variants in this subclass are naphthol in place of phenol, longer or shorter ethylene oxide chains, unsaturation and branched chains in the alkyl group and sulfation of the free alcohol group and sodium salt thereof. These types of compounds also are known under the name Antarox, with special letter and number designations.

They are useful for emulsification, dispersion, wetting and detergency. Their chief virtue lies in their compatibility with cationic as well as anionic agents. Thus, the quaternary cationic germicides, themselves poor emulsifiers, can be made into stable emulsified products by use of specific Triton compounds.

Octoxynol N.F., polyethylene glycol mono[p-(1,1,3,3-tetramethylbutyl)phenyl] ether is an anhydrous liquid mixture of mono-p-(1,1,3,3-tetramethylbutyl) phenyl ethers of polyethylene glycols in which n varies from 5 to 15 and has an average molecular weight of 647. It is a pale yellow liquid with a faint odor and bitter taste. The pH of a 5 per cent solution is between 7 and 9. It is miscible with water, alcohol and acetone, and incompletely miscible with glycerin. It is soluble in benzene and toluene and insoluble in solvent hexane.

$$CH_3-\underset{\underset{CH_3}{|}}{\overset{\overset{CH_3}{|}}{C}}-CH_2-\underset{\underset{CH_3}{|}}{\overset{\overset{CH_3}{|}}{C}}-C_6H_4-O(CH_2CH_2O)_xH$$

Category—surfactant.

Two closely related compounds, whose surfactant properties have been applied to the immobilization of spermatozoa, are di-isobutylphenoxypolyethoxyethanol, used in 1 per cent concentration in Ortho-Gynol Vaginal Jelly, and nonylphenoxypolyethoxyethanol, used in 5 per cent concentration in Delfen Vaginal Cream.

Nonoxynol 9, nonylphenoxypolyethoxyethanol, has been used as a suppository base. It also acts as a unique solubilizing agent or carrier for iodine, the combination being known as an iodophor or surfactant iodine complex. (See Chapter 9.)

Polyoxyalkylene Fatty Ethers, Brij® series, are made by addition of ethylene oxide or other alkylene oxide to fatty alcohols; for example Brij 35 is the lauryl alcohol derivative:

$$C_{12}H_{25}OCH_2CH_2O(CH_2CH_2O)_xCH_2CH_2OH$$

It is a waxy, white solid, soluble in water and alcohol and insoluble in fixed oils and mineral oil. Since it contains no ester linkages, it is stable to acids and alkalies over a broad pH range. This property makes it valuable for preparing hair-wave lotions, depilatory and antiperspirant creams.

Tyloxapol, Triton® WR-1339, Triton® A 20, Superinone®, p-isooctylpolyoxyethylene phenol polymer is a nonionic surfactant of the alkyl phenol class:

$$R-C_6H_4(O-CH_2-CH_2)_nOH$$

where R is C_5H_{11} to $C_{18}H_{37}$ and n = 2 to 30.

Synthesis consists of the catalytic reaction of olefins with phenol, followed by reaction with ethylene oxide under anhydrous conditions in the presence of an alkaline catalyst.

The alkyl phenol nonionics are liquids or waxy solids having excellent color, odor and temperature stability, and a lipophilic-hydrophilic balance suitable for wide applicability as detergents and emulsifiers at both low and high pH values. They are nonirritant and nonsensitizing to the skin. They are less defatting to the skin than many other surfactants.

Poloxalene, Pluronic®, is a group of nonionic compounds classified as block polymer surfactants. They are polymerized propylene glycols of molecular weights ranging from 900 to several thousand, which are insoluble in water, chemically combined with water-soluble polyoxyethylene groups at both ends of the polypropylene glycol. The lipophilic polymer is first synthesized by condensing propylene oxide with propylene glycol until a polymer of desired chain length is produced. Then the condensation is continued with ethylene oxide to produce hydrophilic end groups ranging from 10 to 90 per cent of the molecular weight of the end-product. The type structure may be represented as:

$$HO(CH_2CH_2O)_a \underset{\underset{CH_3}{|}}{(CHCH_2O)_b} (CH_2CH_2O)_cH$$

Since the length of both the lipophilic and the hydrophilic chains is controllable, a series of compounds has been made available having diverse characteristics—liquids, pastes and solids, low and high foam producers, and variable in dispersing, wetting, detergency, rinsability and whiteness retention. The toxicities have been shown to be very low both in feeding tests and in skin irritation tests. Pluronics have been applied in solubilizing antibiotics and vitamins, and in shampoos and cosmetics.

Poloxalkol; Polykol®, is an oxyalkylene polymer of the above structure. It is used orally in 200 mg. dosage as a fecal moistener and softener. It is inert physiologically. Its action depends on its detergency. Its advantage over dioctyl sodium sulfosuccinate of similar action is its relative tastelessness.

Compounds based on ethylenediamine in place of propylene glycol are known as Tetronics.

A problem that is peculiar to many macromolecular compounds and particularly to most nonionic surfactants is the prevention of molding and bacterial decomposition of aqueous dispersions and emulsions made with this class of surfactants in which phenolic preservatives, such as the parabens, are used. Because of hydrogen bonding between the phenolic hydrogen and one of the ether oxygen atoms of the polyalkylene chain present in many nonionic surfactants, there is binding and inactiva-

tion of all or part of such preservatives. The problem may be met by computing the degree of binding and using an experimentally determined excess or "free" concentration of the paraben. Also, the use of sorbic acid ($CH_3CH{=}CHCH{=}CHCOOH$) in approximately 0.2 per cent concentration, either alone or in combination with the paraben, appears to solve the problem of preservation in most instances.

AMPHOTERIC SURFACTANTS

The lipophilic group in this class may be any of those enumerated in the other three classes discussed in the preceding sections. The cationic groups in the hydrophilic part of the molecule are generally quaternary nitrogen types or amine salts. The anionic group of the hydrophilic component may be a carboxylic or sulfonic acid or a sulfated ester group.

At a specific pH the positive and the negative charges of the cationic and the anionic groups of the hydrophilic component may neutralize each other so that the surfactant will then behave as if it were a nonionic type. At a lower pH it exhibits the properties of a cationic, and at a higher pH those of an anionic surfactant. Amphoteric surfactants are of relatively high cost; therefore, their uses are limited to products in which their special properties are significant.

N-Lauryl β-Aminopropionate, Derifat® 170, is one of a series of Derifats, made by reacting lauryl amine and chloropropionic acid.

$$C_{12}H_{25}\overset{+}{N}H_2CH_2CH_2COO^-$$

Its properties are typical of many ampholytic surfactants. It is substantive to skin, hair and fibers, has good hard-water tolerance and has a fair degree of emulsification, wetting and foaming properties. It is most effective in neutral or slightly acidic solutions. It exhibits bacteriostatic and bactericidal properties and low toxicity. Although it exhibits some of the characteristics of cationic surfactants, it is compatible with both the cationic and the anionic compounds. It is used as an adjunct in the production of syndet bars (synthetic detergent bar soap) and as a freeze-thaw stabilizer in emulsions, detergent sanitizers and germicidal shampoos. It has the advantage over cationic germicides in not precipitating proteins as the pH increases.

Because of the polyfunctional properties of the amphoteric surfactant molecule, there are many possibilities for developing compounds of considerable versatility, not available in the other surfactant classes. There can be variation not only in the lipophilic part, but also in the degree of balance of the cationic and the anionic components of the hydrophilic part of the molecule.

A group of compounds developed on this principle of balance are known as Miranols, based on an imidazoline cationic moiety. Miranol 2MCA has the formula:

$$C_{11}H_{23}-C \overset{\overset{\displaystyle CH_2}{\underset{N}{\Big|}}}{\underset{\displaystyle \underset{C_{12}H_{25}OSO_3^-}{N^+}}{\Big|\Big|}} \overset{\displaystyle CH_2}{\underset{\displaystyle CH_2COO^-Na^+}{\Big|}} CH_2CH_2OCH_2COO^-Na^+$$

The presence of the two nitrogen groups gives increased germicidal activity. The balancing of this strongly cationic group with three anionic groupings provides additional properties of marked detergency and foaming, usually lacking in cationic germicidal compounds. It is particularly suitable in shampoos when mixed with anionic detergents; it prevents salt formation and also solubilizes lime soaps, thus eliminating after-rinse residues on hair. It has the further advantage of being non-eye-stinging and the least irritant of shampoo detergents. It is compatible with all types of surfactants.

EMULSIFYING AIDS

Included in this chapter are a number of substances of pharmaceutic application which do not fall categorically into cationic, anionic, nonionic or amphoteric classes but are of assistance in the preparation and stabilization of dispersions and emulsions.

Products

Oleyl Alcohol is a mixture of aliphatic alcohols consisting chiefly of $CH_3(CH_2)_7-CH=CH(CH_2)_7CH_2OH$.

It is prepared by the Bouveault-Blanc method of reduction of butyl oleate

$$CH_3(CH_2)_7CH=CH(CH_2)_7COOC_4H_9$$
$$\downarrow \overset{\displaystyle C_4H_9OH}{Na}$$
$$CH_3(CH_2)_7CH=CH(CH_2)_7CH_2OH$$

or by the hydrogenation of triolein in the presence of zinc chromite. The alcohol exists as a pale yellow liquid having a faint characteristic odor and bland taste. It is soluble in alcohol, in fixed oils or in mineral oil.

It is a useful emulsifying assistant and not an emulsifying agent because the hydroxyl group is sufficiently polar to be oriented toward the aqueous phase in emulsions, while, in the fat phase, the aliphatic chain remains dissolved.

Cetyl Alcohol N.F., 1-Hexadecanol, Palmityl Alcohol. Cetyl alcohol is a mixture of solid alcohols consisting chiefly of cetyl alcohol $CH_3(CH_2)_{14}CH_2OH$. It is found free and esterified in spermaceti and at one time was obtained from this source. The alcohol is prepared now by hydrogenating a mixture of fatty acids having a high percentage of palmitic acid. This saturated alcohol is marketed in various unctuous forms, namely, white flakes, granules, cubes or castings. It has a slight odor and a bland, mild taste. It is soluble in alcohol, ether or vegetable and mineral oils and insoluble in water.

Like oleyl alcohol, it is an emulsifying assistant because of its hydrating properties. Because it is adsorbed and retained by the epidermis of the skin, it is used in lotions and creams. Ointments containing oleyl or cetyl alcohol are washed more easily from the skin (washable ointments).

Category—emulsifying and stiffening agent.

Stearyl Alcohol U.S.P., Stenol. Stearyl alcohol "is a mixture of solid alcohols consisting chiefly of stearyl alcohol, $CH_3(CH_2)_{16}CH_2OH$." It is prepared from stearic acid by catalytic hydrogenation. This alcohol has the same appearance and properties as cetyl alcohol. It is similar in use to cetyl and oleyl alcohol.

Category—pharmaceutic aid.

	PER CENT
OCCURRENCE	STEARYL ALCOHOL
Hydrophilic Ointment U.S.P.	25
Hydrophilic Petrolatum U.S.P.	3

Of the three fatty alcohols, stearyl alcohol causes the greatest potentiation of the "water number" of petrolatum. The "water number" is defined as the largest weight of water which 100 g. of an ointment or fat will hold at 20°.

Polyethylene Glycol 400 U.S.P. "is a polymer of ethylene oxide and water, represented by the formula: $H(OCH_2CH_2)_nOH$, where n varies from 8.2 to 9.1."

The number 400 refers to the average molecular weight. These polyethylene glycols, more correctly called polyoxyethylene glycols, are available as a series of viscous, colorless liquids, ranging from 200 to 600 in average molecular weight. They are characterized by complete solubility in water and in many organic solvents, such as aliphatic alcohols, ketones and esters, and aromatic hydrocarbons but are insoluble in aliphatic hydrocarbons. In addition to their industrial uses as plasticizers and in the synthesis of fatty acid esters, they are used with Carbowaxes to obtain hydrophilic ointments of varying consistency (see Chap. 5).

OCCURRENCE	PER CENT
Polyethylene Glycol Ointment U.S.P. ...	60

Polyethylene Glycol 300 N.F. "is an addition polymer of ethylene oxide and water, represented by the formula $H(OCH_2CH_2)_n$-OH where n varies from 5 to 5.75. It has a molecular weight of not less than 285 and not more than 315." It is a viscous liquid, nearly colorless and having a slight characteristic odor. It is slightly hygroscopic. It freezes between -15 and $-8°$. It is miscible with water, alcohol, acetone, glycols and aromatic hydrocarbons. It is insoluble in ether and aliphatic hydrocarbons. The pH of a 5 per cent aqueous solution is 7. It is useful in compounding water-soluble ointment bases and topical applications.

Category—solvent; dispersing agent.

Polyethylene Glycol 1540 N.F., Polyethylene Glycol 1300-1600, "is an addition polymer of ethylene oxide and water represented by the formula $H(OCH_2CH_2)_nOH$ where n varies from 28 to 36. It has a molecular weight of not less than 1,300 and not more than 1,600." It is like beeswax in consistency. It is white, plastic and waxy. It melts between 43° and 46°. It has a slight characteristic odor. It is very soluble in water and chloroform, slightly soluble in absolute alcohol and insoluble in ether. The pH of a 5 per cent aqueous solution ranges from 4 to 7. It is used in compounding water-soluble ointment vehicles.

Category—vehicle.

Polyethylene Glycol 4000 U.S.P., Carbowax® Compound 4000, differs in composition from the "400" compound in that n varies from 68 to 84. The Carbowaxes are bland nontoxic solids, available from 1,000 to 6,000 in average molecular weight and from semisolid to waxy consistencies. Polyethylene glycol 4,000 is white, waxy, odorless and tasteless and has a melting range of 53° to 56°. It is soluble in water, alcohol and chloroform; it is not hygroscopic. It is an excellent solvent and dispersing agent for many medicinal substances. When made into ointment form with polyethylene glycol 400, it is directly miscible with medicinals that are difficult to incorporate, such as coal tar, ichthammol or balsam of Peru. Carbowax ointments tend to liquefy if more than about 3 per cent of water is added. However, the addition of 5 per cent cetyl alcohol permits the incorporation of water and aqueous solutions up to 10 per cent. In cosmetic emulsions and creams, the Carbowaxes add softening and skin-smoothing properties without undesirable tackiness.

	PER CENT
OCCURRENCE	POLYETHYLENE GLYCOL 4000
Polyethylene Glycol Ointment U.S.P. ...	40

Compound Undecylenic Acid Ointment N.F. contains 75 per cent of Polyethylene Glycol Ointment U.S.P.

For additional information on the polyethylene glycols see Chapter 5, Alcohols.

Cholesterol U.S.P. is discussed in Chapter 25, Steroids. Together with its esters, it gives to lanolin its capacity to absorb water; it forms w/o emulsions. Because there are other substances in lanolin that enhance its emulsifying power, it has been found valuable to utilize lanolin concentrates, obtained by solvent extraction, rather than pure cholesterol, as emulsifying aids. These are known as absorption bases and are composed of cholesterol, with or without its esters and derivatives together with natural sterols and, usually, dissolved in petrolatum, which enhances the water absorption and emulsifying power. Amerchol, Isolan, Falba, Protegin X are a few of the better known absorption bases. They are more stable and less subject to rancidity than lanolin. They are useful in forming stable emulsions in the presence of acids, alkalies or electrolytes. They form w/o ointments and creams that do not dry out readily.

Category—pharmaceutic aid.

	PER CENT
OCCURRENCE	CHOLESTEROL
Hydrophilic Petrolatum U.S.P.	3

Triethanolamine U.S.P. is a mixture consisting largely of triethanolamine, $N(CH_2-CH_2OH)_3$, with smaller amounts of the primary and the secondary compounds, $NH_2-CH_2-CH_2OH$ and $NH(CH_2-CH_2-OH)_2$, respectively. It is manufactured by the action of ammonia on ethylene oxide (q.v.). "It has an alkalinity equivalent to not less than 6.7 ml. and not more than 7.2 ml. of normal acid for each gram of Triethanolamine." The substance is a colorless to pale-yellow, viscous, hygroscopic liquid having a slightly ammoniacal odor. It is miscible with water or alcohol and is soluble in chloroform.

Triethanolamine is an excellent emulsifying agent in the preparation of ointments and creams. It combines with fatty acids to form compounds with good detergent properties, and these are soluble in water and in oils. It is claimed to possess a certain amount of bacteriostatic power.

For the preparation of stable emulsions, the mixture of fatty acid (e.g., oleic) and triethanolamine is added to a portion of the oil, and an equal portion of the water is added while stirring. During continued agitation, the remaining oil and water are added in portions. For vegetable oils, 2.4 per cent of the amine and 11.5 per cent of the acid may be employed; for paraffin oils, the amount of triethanolamine should be increased to 5 per cent. Emulsions so made with 20 to 40 per cent of oil may be diluted with as much as 5 times the volume of water.

Category—pharmaceutic aid.

	PER CENT
OCCURRENCE	TRIETHANOLAMINE
Benzyl Benzoate Lotion N.F.	0.5

Monoethanolamine N.F., 2-aminoethanol, $NH_2-CH_2-CH_2OH$, is a clear, viscous, colorless liquid with an ammoniacal odor; it is affected by light.

Category—surfactant.

	PER CENT
OCCURRENCE	MONOETHANOLAMINE
Thimerosal Solution N.F.	0.1
Thimerosal Tincture N.F.	0.1

β-Isopropanolamine; 2-amino-1-propanol,

$$HO-CH_2-CH-CH_3$$
$$|$$
$$NH_2$$

is prepared by the reduction of acetylcarbinoloxime with sodium amalgam in dilute acetic acid.

This compound is similar in its properties to triethanolamine. It has the advantage of forming fatty acid esters which tend to discolor less rapidly than do the triethanolamine esters; therefore, whiter emulsions may be prepared.

"Amino Glycol"; 2-Amino-2-methyl-1,3-propanediol. This is a white, crystalline, odorless, bland compound, very soluble in water, 250 g. dissolving in 100 ml. of water. The pH of a 0.1 M solution is 10.8.

It is used similarly to the alkanolamines. One part of aminoglycol combines with 2.7 parts of stearic acid. This stearate, made during emulsification of oils, tends to form creams that are softer and remain whiter than when other saponifiers are used.

Polyvinyl Alcohols, PVA, Elvanol,® are long chain polyhydric alcohols prepared from polyvinyl acetates by catalytic alcoholysis with methanol in methyl acetate. The head-to-tail, or 1,3-glycol structure of this compound is represented as

$$\left[\begin{array}{c} \diagdown \\ CH_2-CH-CH_2-CH-CH_2-CH \\ | \qquad | \qquad | \\ OH \qquad OH \qquad OH \end{array} \diagup \right]_x$$

The carbon chain length of the polyvinyl alcohols is dependent upon the degree of polymerization of the polyvinyl acetate used. Commercial grades are dry, white or cream colored powders, insoluble in petroleum solvents, soluble or dispersible either in hot or cold water or in hydroalcoholic solution, depending upon the degree of alcoholysis during synthesis. The pH of 4 per cent aqueous solutions is in the range of 6 to 8. Viscosities of solutions increase with carbon chain length. Aqueous solutions are compatible with lower alcohols. PVA solutions are used for making water-soluble films, and as suspending agents, binders, protective colloids, and emulsifiers either with or without the aid of surfactants. Emulsions can be stabilized with respect to creaming, by use of PVA, either in acid or alkaline range. Borax or boric acid increases the viscosity of PVA solutions, as does an increase in pH. PVA is precipitated from solution by alkali chlorides, carbonates and sulfites, and by aluminum and zinc sulfates. Preservatives are necessary to prevent mold growth in aqueous PVA solutions.

Povidone N.F., Polyvinylpyrrolidone, PVP, Plasdone®, is a synthetic polymer consisting essentially of linear 1-vinyl-2-pyrrolidone groups. It is synthesized from γ-butyrolactone, ammonia and acetylene. The resulting monomer, vinylpyrrolidone, polymerizes by heating in the presence of hydrogen peroxide and ammonia to PVP, its mean molecular weight ranging from 10,000 to 700,000, and its nitrogen content from 12 to 13 per cent.

$$\begin{array}{ccc} H_2C-CH_2 & & \left\{ \begin{array}{c} H_2C-CH_2 \\ | \qquad | \\ H_2C \quad CO \\ \diagdown \diagup \\ N \\ | \\ -C-CH_2- \end{array} \right\}_n \end{array}$$

Vinylpyrrolidone   PVP
(monomer)   (polymer)

It is a white to creamy white odorless hygroscopic powder, soluble in water, alcohol and chloroform, and insoluble in ether. The pH of a 5 per cent aqueous solution ranges from 3 to 7. The viscosity of aqueous solutions up to 10 per cent is the same as that of water; above this concentration the viscosity increases depending upon the concentration and molecular weight of the polymer used.

PVP serves as a protective colloid to stabilize emulsions and to provide an emollient film on the skin. Because of its substantivity to hair, it has been applied in hair sprays to keep hair in place without flaking off on drying, as most gums do. PVP has been used in improving tablet granulations and in simplifying tablet coating procedures, as a clarifying agent, and as a bodying agent in non-nutritive sweetener liquids and vitamin-mineral liquid concentrates. PVP forms a colloidal water-dispersible complex with absorbable gelatin sponge, which has been utilized in aerosol dosage form as a hemostatic wound packing.

Category—dispersing and suspending agent.

Carboxyvinyl Polymers, Carbopols,® are high molecular weight synthetic, water-soluble resins. They are white powders producing clear aqueous or hydroalcoholic solutions when neutralized with inorganic alkalies or with amines. They have certain advantages over natural gum solutions.

They are compatible with nonionic and anionic emulsifiers, and they yield stable emulsions with controlled viscosities over long periods of time when used in concentrations of less than 0.1 per cent. Carbopols are also dispersible in alcohol, glycerin, hydrocarbons, and many other organic solvents when neutralized with ethanolamine or other suitable amine. In aqueous concentrations above 0.2 per cent they act as primary emulsifiers. Emulsions made with Carbopols are not as dependent on a critical oil-soluble/water-soluble balance as when natural or synthetic gums are used. Carbopol dispersions are resistant to bacterial and fungal attack.

Polycarbophil N.F. is a polyacrylic acid cross-linked with divinyl glycol. It occurs as white to off-white granules having a characteristic odor. It is insoluble in water but swells in water, the amount depending on the pH.

Category—absorbent (gastrointestinal).

Usual dose—500 mg. 4 times daily.

CHELATING AGENTS

Chelating agents are substances containing two or more donor groups capable of combining with a metal to form a special type of complex known as a chelate. (See p. 53, Chap. 2.)

Ethylenediaminetetraacetic Acid, EDTA, Versene® Acid, Sequestrene® A. This compound and its alkali salts have the unique property of deactivating or deionizing alkaline earth ions and heavy metal ions in solution. This phenomenon is called sequestering, and the inactivated ion complex is called a chelate (Greek: *khele*, claw). (See Citric Acid.) Sodium and potassium salts of EDTA are among the most powerful chelating agents known. EDTA is made by the action of monochloracetic acid on ethylenediamine.

The removal of Ca^{++} from solution by the tetrasodium salt of EDTA is shown graphically below.

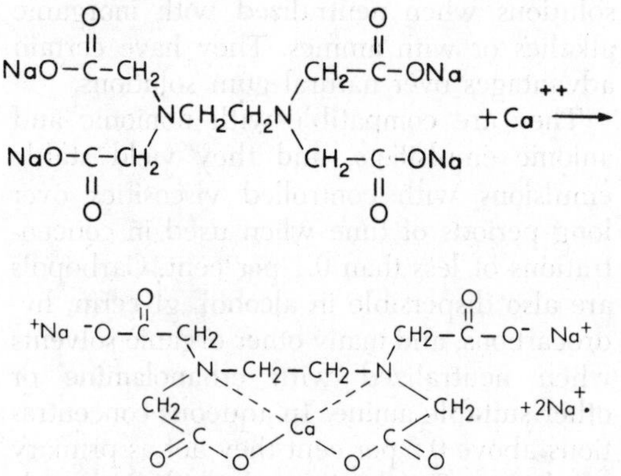

$$NaO-C-CH_2 \quad CH_2-C-ONa$$
$$NCH_2CH_2N \quad + Ca^{++} \longrightarrow$$
$$NaO-C-CH_2 \quad CH_2-C-ONa$$

$$^+Na^-O-C-CH_2 \quad CH_2-C-O^- Na^+$$
$$N-CH_2-CH_2-N$$
$$CH_2 \quad CH_2 \quad + 2Na^+$$
$$C-O \quad Ca \quad O-C$$

The calcium has not been removed physically from solution as is the case when using water softeners. Ring closure has occurred; the calcium has formed a water-soluble stable complex with the EDTA and no longer exhibits its characteristic cationic properties. This property of chelation of EDTA and its salts has proved to be widely applicable in the industrial and pharmaceutic arts.

By inactivating alkaline earth ions, emulsions frequently are prevented from breaking, and their stability is improved through the addition of relatively small amounts (0.1 to 1%) of these complexing agents. The clarity of shampoos is maintained similarly by preventing the formation of insoluble lime and magnesia soaps. By sequestering traces of iron, copper and manganese, so widely prevalent in water supplies and solutions through contact with metallic equipment, discoloration and oxidative decomposition is prevented or retarded in pharmaceutic products containing penicillin, ascorbic acid, epinephrine, salicylate and many other medicinals of normally poor stability.

The di-, tri- and tetrasodium salts, as well as the disodium calcium salts and the ethylenediamine salt of EDTA, are available for specific complexing purposes. A sufficient amount of these agents, based upon stoichiometric proportions, must be used for optimum effectiveness.

PRODUCTS

Disodium Edetate U.S.P., disodium ethylenediaminetetraacetate, is a white crystalline powder, freely soluble in water. It is used in the form of 15 to 20 per cent solution as an antidote for digitalis poisoning, in dosage of 50 mg. per kg. of body weight by slow intravenous infusion over a period of 3 to 4 hours once daily.

Category—metal complexing agent.

OCCURRENCE
Disodium Edetate Injection U.S.P.

Calcium Disodium Edetate U.S.P., Calcium Disodium Versenate®, calcium disodium ethylenediamine tetraacetate, is available in a 20 per cent aqueous solution diluted to 3 per cent for administration by intravenous drip in the treatment of heavy metal poisoning. It is particularly effective in lead poisoning.

Category—metal complexing agent.

Usual dose—intravenous infusion, 1 g. in 250 to 500 ml. of isotonic solution over a period of 1 hour two times a day.

Usual dose range—up to 10 g. per course of treatment.

OCCURRENCE

Calcium Disodium Edetate Injection U.S.P.

Piperazine Calcium Edetate, Perin®, is a chelate produced by reacting EDTA with calcium carbonate and piperazine. Structurally, the piperazine molecule, in the form of a bridge, replaces the two sodium atoms in the formula of calcium disodium edetate. The drug is administered orally, similarly to piperazine, for the treatment of pinworms and roundworms. As a consequence of chelation the drug exhibits a much lower order of toxicity. (See Chap. 9.)

Ferrocholinate, Chel-Iron®, Ferrolip®, is a chelate prepared by reacting equimolecular quantities of freshly precipitated ferric hydroxide with choline dihydrogen citrate.

Because the iron in this water-soluble hematinic compound is chelated, its diffusion rate in the circulation is altered and the toxic properties markedly reduced. It is claimed to be less irritant than ferrous sulfate or ferrous gluconate, and safer in the event of overdosage. Maintenance dosage for adults is 330 to 660 mg. (39 to 78 mg. elemental iron) administered orally as solution, syrup or tablets.

Dextriferron Injection N.F., Astrafer®, is a sterile colloidal solution prepared from ferric hydroxide complexed with partially hydrolyzed dextrin in water for injection. It is nearly neutral in reaction (pH 7.3 to 7.9). It is administered only by slow intravenous injection in severe iron deficiency anemia in conditions not amenable to oral therapy. It is estimated that 70 to 100 per cent of the administered iron is utilized in the synthesis of hemoglobin. The dosage is critically calculated on the basis of the hemoglobin content of the patient. Approximately 25 mg. of iron are needed for women and 35 mg. for men to increase the hemoglobin content by 1 per cent. It is available in ampuls containing 100 mg. iron in 5 ml.

Category—source of iron.

Usual dose—intravenous, initial trial dose of 1.5 ml., then increase dose in increments of 1 to 1.5 ml. per day until a daily dose of 5 ml. is reached.

Iron Dextran Injection U.S.P., Imferon®, is a sterile colloidal solution of ferric hydroxide complexed with partially hydrolyzed dextran of low molecular weight in water for injection. It contains 0.5 per cent phenol as preservative. It is a dark brown, slightly viscous liquid with a pH range of 5.2 to 6. It is administered intramuscularly and is intended for use only in cases in which oral administration of iron is ineffective or impractical.

Category—hematinic.

Usual dose—intramuscular, the equivalent of 100 mg. of iron once daily.

Usual dose range—intramuscular, the equivalent of 50 to 250 mg. of iron daily to every other day.

Iron Sorbitex Injection U.S.P., Jectofer®, is a sterile aqueous solution of a complex of iron, sorbitol and citric acid, stabilized with dextrin and an excess of sorbitol. Each ml. is equivalent to 50 mg. of elemental iron.

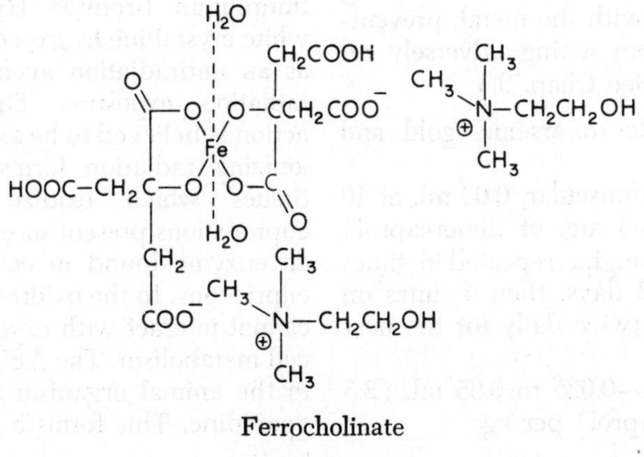

Ferrocholinate

It is a dark brown clear liquid having a pH range of 7.2 to 7.9. The solution is stable for two years after manufacture.

Category—hematinic.

Usual dose—intramuscular, the equivalent of 100 mg. of iron once daily.

Usual dose range—100 to 200 mg. daily.

CDTA. Chel® CD, *trans*-1,2-Diamino-cyclohexanetetraacetic Acid.

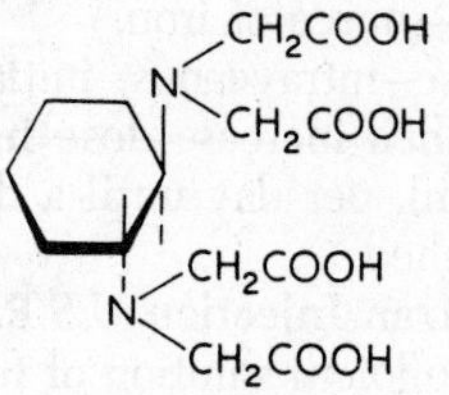

trans-1,2-Diaminocyclohexanetetraacetic Acid

Although EDTA is generally considered one of the most powerful chelating agents, a closely related compound has been introduced recently which produces metal chelates many times more stable. CDTA forms amine salts that are more soluble in many organic solvents. CDTA forms chelates with copper and manganese 100 times more stable than corresponding EDTA chelates. CDTA is also more stable to oxidation.

Dimercaprol U.S.P., 2,3-Dimercaprol, B.A.L., British Anti-Lewisite, acts both as

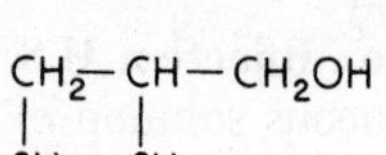

a prophylactic and antidote in poisoning by Lewisite, an organic arsenical, and by other heavy metal compounds. It forms stable ring compounds in situ with the metal, preventing the arsenic from acting adversely on enzyme systems. (See Chap. 9.)

Category—antidote to arsenic, gold and mercury poisoning.

Usual dose—intramuscular, 0.03 ml. of 10 per cent solution (3 mg. of dimercaprol) per kg. of body weight, repeated 6 times daily on the first 2 days, then 4 times on the third day, and twice daily for the next 10 days if needed.

Usual dose range—0.025 to 0.05 ml. (2.5 to 5 mg. of dimercaprol) per kg.

OCCURRENCE	PER CENT DIMERCAPROL
Dimercaprol Injection U.S.P.	10

Penicillamine U.S.P., Cuprimine®, D(−)3-

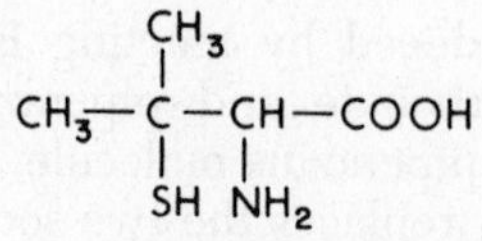

Penicillamine

Mercaptovaline. This degradation product of penicillin is a white crystalline powder soluble in water (pH range 4.5-5.5), slightly soluble in alcohol, and insoluble in ether and chloroform. It is used in Wilson's disease (hepatolenticular degeneration), which is characterized by high tissue copper levels. The drug forms a nondissociable complex with the copper and promotes its urinary excretion. This reduces the copper content of the blood and, consequently, reduces the deposition of copper in the brain, the liver, the kidneys and the eyes.

Category—metal complexing agent.

Usual dose—250 mg. 4 times daily.

Usual dose range—1 to 5 g. daily.

OCCURRENCE

Penicillamine Capsules U.S.P.

AET, Antiradon®, S-(2-Aminoethyl)iso-

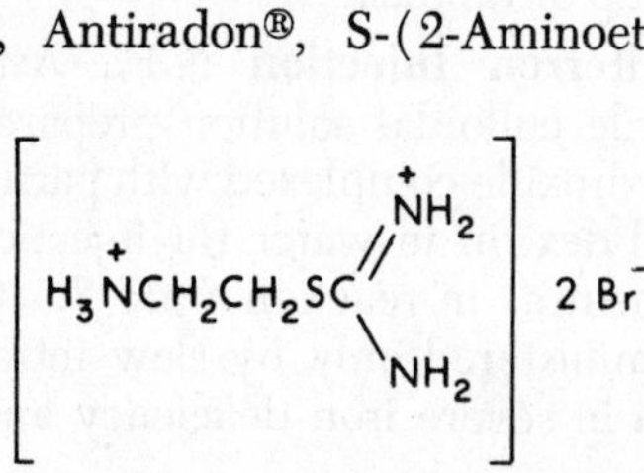

thiuronium Bromide Hydrobromide. This white crystalline hygroscopic powder is used as an antiradiation agent prior to nuclear radiation exposure. The mechanism of action is believed to be as follows: in theory, ionizing radiation forms peroxides in the tissues which oxidize irrevocably the cuprous ions present in cytochrome oxidase, an enzyme found in cell mitochondria, to cupric ions. In the oxidized state the enzyme cannot interact with oxygen, thus impairing cell metabolism. The AET rearranges in situ in the animal organism to 2-mercaptoethyl guanidine. This forms a dissociable chelate

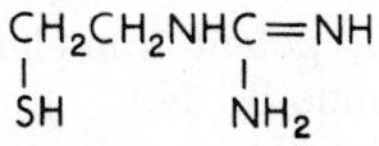

$$CH_2CH_2NHC{=}NH$$
$$SH \qquad NH_2$$

in which the copper is attached to the S and N. In this chelated form the cuprous ions in the enzyme are not oxidized by the effects of ionizing radiation.

Oral dosage is 150 mg. AET per kg. of body weight prior to exposure to irradiation.

Deferoxamine Mesylate, Desferal®, DFOM.

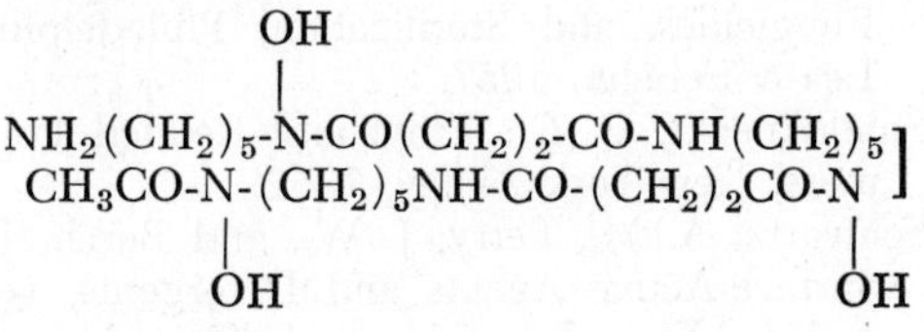

$$OH$$
$$NH_2(CH_2)_5\text{-}N\text{-}CO(CH_2)_2\text{-}CO\text{-}NH(CH_2)_5 \Big]$$
$$CH_3CO\text{-}N\text{-}(CH_2)_5NH\text{-}CO\text{-}(CH_2)_2CO\text{-}N$$
$$OH \qquad\qquad\qquad OH$$

The removal of excessive amounts of iron from the body, whether stored as a result of pathologic conditions or from acute iron poisoning, is a difficult procedure. Chelating agents such as EDTA tend to remove also other essential metals such as copper and magnesium. A biologically derived chemical has recently been developed which has a more specific iron-eliminating action. Deferoxamine is a polymer consisting of three molecules of trihydroxamic acid, having a molecular weight of 597. One mole of DFOM binds 1 mole of Fe^{+++} to form the iron chelate, ferrioxamine.

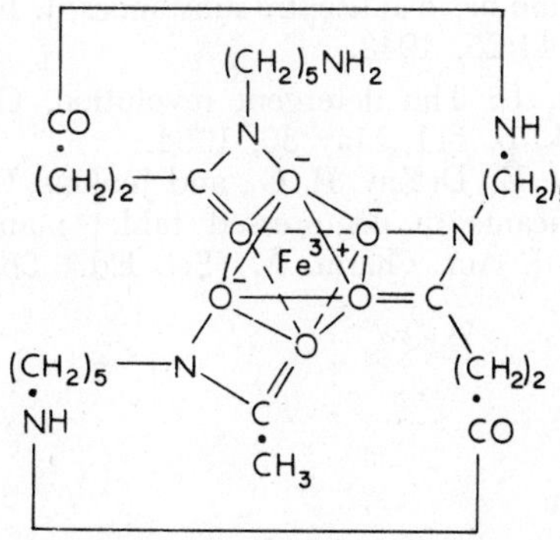

Ferrioxamine

DFOM in the form of the hydrochloride is soluble 5 per cent in water; the methane-sulfonate (mesylate) is soluble 25 per cent in water. DFOM has a very high binding capacity for iron and can withdraw the iron from proteins such as transferrin in blood serum, and from ferritin in the tissues, forming the red colored ferrioxamine which is water soluble and readily excreted by the kidneys. However, DFOM does not bind the iron of hemoglobin and, therefore, does not interfere with the absorption of nutritional hemoglobin; nor does it produce anemia during therapy. In iron poisoning the initial dose is 1 g. dissolved in 2 ml. of sterile water for injection, administered intramuscularly. This is followed by 0.5 g. every 4 to 6 hours, depending upon the clinical response. For intravenous use 0.5 g. is dissolved in 2 ml. of sodium chloride injection or in dextrose injection, and administered at a rate not exceeding 15 mg./kg. per hour. The reconstituted solution may be stored under sterile conditions at room temperature for not longer than two weeks.

Deferoxamine, penicillamine, the iron chelates, CTDA, B.A.L. and AET furnish striking evidence of the diversity of therapeutic application based on formation of chelates within body tissues.

BIBLIOGRAPHY

Ahsan, S. S., and Blaug, S. M.: A study of tablet coating using polyvinylpyrrolidine and acetylated monoglyceride, Drug Standards 26:29, 1958.

Baker, Z., Harrison, R. W., and Miller, B. F.: Action of synthetic detergents on the metabolism of bacteria, J. Exp. Med. 73:249, 1941.

Balabukha, V. S.: Chemical Protection of the Body Against Ionizing Radiation, New York, Macmillan, 1964.

Bean, H. S., Beckett, A. H., and Carless, J. E.: Solubility in systems containing surface-active agents, *In* Advances In Pharmaceutical Sciences, Vol. 1, pp. 86-194, New York, Acad. Press, 1964.

Chaberek, S., and Martel, A. E.: Organic Sequestering Agents, New York, Wiley, 1959.

Charles, R. D., and Carter, P. J.: The effect of sorbic acid and other preservatives on organism growth in typical nonionic emulsified commercial cosmetics, J. Soc. Cosmet. Chem. 10:383, 1959.

Chelation Phenomena, Ann. N. Y. Acad. Sci. 88:281, 1960.

Commercial Surfactants, Am. Perf. Aromat., 1st Documentary ed., New York, Moore, 1960.

Dwyer, F. P., and Mellor, D. P.: Chelating Agents and Metal Chelates, New York, Academic Press, 1964.

Epstein, A. K., Harris, B. R., and Katzman, M.: Relationship of bactericidal potency to length of fatty acid radicals of certain quaternary ammonium derivatives, Proc. Soc. Exp. Biol. Med. 53:238, 1943.

Evans, W. P.: Mobilization and inactivation of preservatives by nonionic detergents, J. Pharm. Pharmacol. 16:323, 1964.

Gershenfeld, L., and Milanick, V. E.: Bactericidal and bacteriostatic properties of surface tension depressants, Am. J. Pharm. 113:306, 1941.

Greenberg, S.: Differentiation test for anionic, non-ionic, and cationic surfactants, Chem. Anal. 51:11, 1962.

Griffin, W. C.: Classification of surface-active agents by "HLB," J. Soc. Cosmet. Chem. 1:311, 1949.

Heald, R. C.: Useful combinations of anionic and cationic surfactants, Am. Perf. Aromat. 75:45, 1960.

Kostenbauder, H. B.: Some physicochemical aspects of phenolic preservatives in the presence of macromolecules, Am. Perf. Aromat. 75:28, 1960.

Lachman, L.: Antioxidants and chelating agents as stabilizers in liquid dosage forms, Drug & Cosmet. Ind. 102:43, 1968. (Feb.)

Lawrence, C. A.: Quaternary Ammonium Germicides, New York, Acad. Press, 1950.

Lehrman, G. P., and Skauen, D. M.: Polyvinylpyrrolidone and other binding agents in tablet formulation, Drug Standards 26:170, 1958.

Lesshafft, C. T., and DeKay, H. G.: A study of nonionic surfactants in a hydrophilic ointment base, Drug Standards 25:45, 51, 1957.

Miller, O. H.: Predicting incompatabilities of new drugs—anionic, cationic, nonionic, J. Am. Pharm. A. (Pract. Ed.) 13:657, 1952.

Moeschlin, S., and Schnider, U.: Treatment of hemochromatosis and acute iron poisoning with a new, potent iron-eliminating agent, desferrioxamine-B, N.E. J. Med. 269:57, 1963.

Moilliet, J. L., and Collie, B.: Surface Activity, New York, Van Nostrand, 1961.

Nungester, W. J., and Kempf, T. H.: An infection-prevention test for the evaluation of skin disinfectants, J. Infect. Dis. 71:174, 1942.

Poust, R. I., and Colaizzi, J. L.: Stability of ascorbic acid in aqueous solutions of polysorbate 80, J. Pharm. Sci. 57:2119, 1968.

Reddish, G. F.: Antiseptics, Disinfectants, Fungicides, and Sterilization, Philadelphia, Lea & Febiger, 1957.

Schildknecht, C. E.: Vinyl and Related Polymers, New York, Wiley, 1952.

Schwartz, A. M., Perry, J. W., and Berch, J.: Surface-Active Agents and Detergents, vol. 2, New York, Interscience, 1958.

Schmitz, I. A., and Harris, W. S.: Germicidal ampholytic surface-active agents, Mfg. Chemist 29:51, 1958.

Shinuda, K., Nakagawa, T., Tamamushi, B., and Isemura, T.: Colloidal Surfactants, New York, Acad. Press, 1963.

Sisley, J. P., and Wood, P. J.: Encyclopedia of Surface-Active Agents, New York, Chem. Pub. Co., 1952.

Slattery, T. W.: A new type of absorbable wound packing in aerosol dosage form, Am. J. Hosp. Pharm. 26:43, 1969. (Jan.)

Smith, R. L.: The Sequestration of Metals, New York, Macmillan, 1959.

Surface-active agents, Ann. New York Acad. Sci. 46:347, 1946.

Welch, H., and Brewer, C. M.: Toxicity indices of some basic antiseptic substances, J. Immunol. 43:25, 1943.

Winton, J.: The detergent revolution, Chem. Week, p. 111, May 30, 1964.

Wolff, J. E., DeKay, H. G., and Jenkins, G. L.: Lubricants in compressed tablet manufacture, J. Am. Pharm. A. (Sci. Ed.) 36:407, 1947.

13

Antimalarials

Allen I. White, Ph.D.,
Dean and Professor of Pharmaceutical Chemistry, College of Pharmacy,
Washington State University

INTRODUCTION

Malaria is the most widespread of all infectious diseases. At least one third of the earth's population is exposed to infection, with its disabling and lethal effects. It has been eradicated almost completely in the United States (although infected military personnel returning from Viet Nam may alter that condition), but malaria continues to be a worldwide problem of great economic, social and political consequence. Research conducted prior to and during World War II strongly indicated that measures for the successful suppression, treatment and cure of malaria had been found. However, in twenty-five years since that time, the synthetic drugs, so successful at first, have proved to be of decreasing effectiveness because of the capability of the malaria organism to develop resistance to them. As a result, there has been an increasing return to the use of the natural product, quinine, as the drug of choice. Also, as a result of resistance which has been developed by vector mosquitoes, insecticides have become less effective in suppression of the spread of the disease than it was assumed they would be after the introduction of DDT. Consequently, there is a renewed interest in research concerned with malaria chemotherapy, insect repellants and the basic biology, particularly the biochemistry, of the plasmodia causing the disease. To date no successful immunologic process that provides protection against malaria has been found, although there is hope that some day a protective vaccination may be developed. The search continues for the ideal antimalarial drug—an inexpensive, palatable, long-acting, nontoxic compound that is prophylactic, suppressive and curative without inducing resistance.

Etiology. Malaria in man may be caused by four species of Plasmodium (protozoan parasites) and in other mammals, birds and reptiles by many other species. The four for which man is the natural host (and the associated type of malaria) are: *Plasmodium vivax* (benign tertian); *Plasmodium falciparum* (estivo-autumnal; malignant tertian); *Plasmodium malariae* (quartan), and *Plasmodium ovale* (ovale tertian). The first three of these are widely distributed and occur most frequently in tropical and subtropical countries. The disease is characterized by successive chills, fever and sweats. The infections are labeled tertian or quartan because the fever tends to recur every third or fourth day, although some variation in time intervals may be observed. Infection with *P. malariae* is characterized by a quartan type of malaria.

All species of Plasmodium have two hosts, a vertebrate and a mosquito that acts as both vector and definitive host. Mosquitoes of the genus Anopheles are the vectors for human malaria. Mosquitoes of the genera Aedes, Culex and Culiseta as well as Anopheles may be vectors for the plasmodia infecting other vertebrates. The malaria organism requires time in both hosts to complete its multistage life cycle, but, in the case of some species, it may remain dormant in one of its several stages in the vertebrate host.

The sexual phase of the life cycle begins when a female mosquito bites an infected vertebrate and ingests blood containing the malarial parasite in the gametocyte stage. In the stomach of the mosquito the sexual phase of development called sporogony occurs. The male and the female gametocytes form gametes. An ookinete is formed by fertilization and penetrates the stomach wall. Outside the stomach, an oocyst is formed which produces sporozoites that are released by the rupturing of the oocyst. The sporozoites travel to the salivary glands of the mosquito, from which they may be transferred to an uninfected vertebrate host by the bite of the mosquito when it starts its blood meal. Injected sporozoites disappear rapidly from the blood of the vertebrate, entering the parenchymal cells of the liver and, perhaps, some other tissues. The parasite now begins the asexual phase of development called schizogony. In this pre-erythrocytic (primary exoerythrocytic) stage, the parasite grows and divides to form a schizont. The schizont segments to form many merozoites, which causes the rupturing of the cell, and the merozoites enter the bloodstream. The merozoites invade the red blood cells, beginning the erythrocytic stage. Within the red blood cells, the merozoites become trophozoites, and multiplication by schizogony occurs. The schizonts that are formed from the trophozoites divide into merozoites and, thus, continuously increase the number of merozoites available to invade more red blood cells, so that, finally, the number of rupturing cells is sufficiently great to initiate the clinical symptoms of the disease. The asexual cycle continues until chemotherapy is initiated, immunity is developed, or death occurs. The continuous invasion and subsequent eruption of erythrocytes lead to the development of another significant symptom of malaria, anemia. It is the chronic anemia of the victim that contributes to the malaise and the general lassitude of the people in malarial countries.

At any time, but particularly when nor-mal reproduction of the erythrocytes becomes unfavorable, some of the trophozoites from the erythrocyte stage develop into male (micro) or female (macro) gametocytes that circulate in the blood to become available for ingestion by another mosquito. Thus, the life cycle is complete. Some species of Plasmodium, notably *P. vivax* (but not *P. falciparum*), are capable of existing in paraerythrocytic (secondary exoerythrocytic) forms that have a variety of patterns but always pass through schizont stages. By this development, the parasite may enter into a dormant state in the tissues of the host during which time it may appear that the infection has been overcome. However, some time later the parasite may return to the bloodstream, thus causing a relapse.

Chemotherapy. Antimalarial drugs may be classified into five different types, depending on which stage of the life cycle of the organism is affected:

1. *Sporozoitocides:* Drugs capable of killing the sporozoites as soon as they are introduced into the bloodstream by the bite of a mosquito. Such drugs would be most desirable, since they would be truly causal prophylactics capable of preventing the development of the disease. Unfortunately, very few compounds with such chemotherapeutic properties have been found.

2. *Exoerythrocytic Schizontocides:* Drugs capable of killing the parasite as it exists in the schizont stage, in either the primary or the secondary exoerythrocytic form. Such drugs, sometimes called *tissue schizontocides*, may be said to be curative, because they are capable of eradicating the organism before it enters the red blood cells or while it is dormant in the host. A favorable effect on relapse rate results. Only a few drugs have been found that possess such activity to a significant degree.

3. *Erythrocytic Schizontocides:* Drugs capable of inhibiting the development of schizonts during the erythrocytic stage. Usually, such drugs keep the number of the blood forms of the organism at a level below that necessary to precipitate the clinical symptoms of the disease. Such drugs, sometimes called *suppressives* or *clinical prophylactics*, are known also as *blood schizonto-*

TABLE 18. ACTIVITY OF SOME ANTIMALARIALS

	Quinine		Chloroquine		Amodiaquine		Primaquine		Quinacrine		Proguanil		Pyrimethamine	
	F	V	F	V	F	V	F	V	F	V	F	V	F	V
Sporozoitocide			(±)*											
Exoerythrocytic schizontocide:														
primary							+				+		(+)†	(+)†
secondary							+							
Erythrocytic schizontocide	±	+	+	+	+	+			+	+	+	+	+	+
Gametocytocide							+	+					+	+
Sporontocide							+	(+)			+	+	+	+

F = *P. falciparum*
V = *P. vivax*
+ = fully active
± = not consistently active
* At excessive dosage
† On release of merozoites

cides. Many of the widely used antimalarials exhibit this kind of activity.

4. *Gametocytocides:* Drugs capable of killing the parasite as it exists in the gametocyte stage. Such drugs help to prevent the spread of the disease, since the mosquito vectors do not become infected. A few of the antimalarial drugs possess such activity.

5. *Sporontocides:* Drugs capable of preventing sporogony in the mosquito by their effect on the gametocytes in the blood of the vertebrate host. Interestingly enough, all sporontocidal drugs show activity as exoerythrocytic schizontocides as well.

The ideal drug would be one that exhibited all five types of activity against all four species of Plasmodium that cause malaria in man. No such broad-spectrum antimalarial has yet been found. Table 18 shows the principal kinds of activity exhibited by the more widely employed antimalarials in the chemotherapy of *P. vivax* and *P. falciparum*. The diagram shown in Figure 12 presents the same information in simplified form.

History. The general antifebrile properties of the bark of the cinchona undoubtedly were known to the Incas before the arrival of the Spaniards early in the 16th century. However, it was probably the observations of early Jesuit missionaries that led to the discovery that infusions of cinchona bark were effective for the treatment of the tertian "ague" that was common in tropical Central and South America even then and to the introduction of the crude drug into Western Europe. The first recorded use in South America was about 1630; and, in Europe, in 1639.[1] Thus began the first era in the chemotherapy of malaria. Cinchona

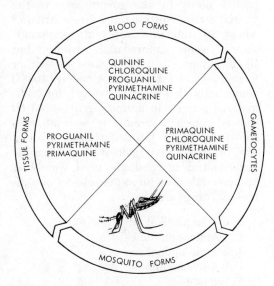

FIGURE 12

and the purified alkaloids obtained from it were to remain the only drugs of significance in the treatment of malaria for 3 centuries. By the time of World War I, the advance of synthetic organic chemistry and the ideas of Paul Ehrlich in regard to chemotherapy provided the Germans, who had been cut off from the world's supply of quinine (then controlled by the Dutch), with the means and the stimulus to seek synthetic antimalarials. They based their research on methylene blue, which was the only compound, excluding the cinchona alkaloids and some toxic arsenicals, known at that time to have any antimalarial activity. That work introduced the modern era of synthetic antimalarials, and the successful development of pamaquine was announced in 1926. In the period that followed, the French joined the Germans in the search for better synthetic antimalarials. During World War II, and now during the Viet Nam war, special impetus was given again to the search for better compounds as fighting took place in tropical and subtropical countries in which malaria was endemic. During the period of 1941 to 1946, approximately 16,000 substances were screened by experimental infection of birds of various species which were used as the principal test animals. About one third of the compounds tested were synthesized in a special program supported jointly by the governments of the United States, Australia and Great Britain.[2] About 70 different classes of compounds showed some antimalarial activity, but significant promise was limited to four or five groups. Of the approximately 16,000 compounds that were screened for antimalarial activity in animals, only about 80 showed sufficient promise to warrant trials for treatment of human malaria. Only a few of these have gained acceptance as significant drugs and are used widely as antimalarials. The important classes of antimalarial drugs are:

1. Cinchona alkaloids
2. 4-Aminoquinolines
3. 8-Aminoquinolines
4. 9-Aminoacridines
5. Biguanides
6. Pyrimidines

7. Sulfones

Other types of compounds that have shown interesting activities (but often disappointing in human trials) include 1,4-naphthoquinones, hydroxynaphthalenes, aryl pteridines, dihydrotriazines, and quinoxalines. Antibiotics have been consistently disappointing as antimalarials, although a few have some degree of activity. Certain sulfonamides, particularly sulfadiazine, sulfisoxazole and the longer-acting sulforthomidine, have been used with varying degrees of success with quinine and pyrimethamine for the treatment of falciparum infections.

The treatment of isolated or relatively few malaria cases presents the clinician with no great problem, as he may select a series of drugs to attack any or all of the phases existing in humans (see Table 18). However, the treatment of endemic malaria throughout a community or country presents the malariologist with a difficult problem. Usually hampered by insufficient economic resources and inadequate public health facilities in malarial countries, he requires the ideal antimalarial to achieve the goal of wiping out the disease. Until that objective is achieved, he seeks ways of achieving the best results with the drugs available. Sometimes, improved treatment is gained by combining drugs. The simultaneous use of drugs that are effective against different stages of the parasite's development has been found to be effective. Also, the use of two drugs that have the same type of activity may produce additive effect.

Recently, attention has been focused on the development of antimalarials with prolonged duration of activity. Malariologists seek as a minimum goal a drug that could be administered as infrequently as twice a year. Such a form of administration would be superior to the mass-scale Pinotti method, in which an antimalarial is mixed into salt that is distributed to the population. Such a procedure has been shown to be effective in reducing endemic malaria, but many problems have been encountered. It seems likely that the most successful form of prolonged-action antimalarials will be depot or repository injections of compounds with very low rates of absorption. The pamoate salt of cycloguanil ("proguanide triazine") is an example of such a drug. It may be possible to develop a drug that will be fixed in body tissues for a prolonged period after oral administration. Also, delaying of the degradation or the excretion of a drug are means of prolonging activity after administration. However, practical applications of prolonged-action forms of the last three types have not been developed yet, and it seems doubtful that they offer promising research objectives. But any dosage form that requires an injection does not meet the optimum goals of malariologists. Therefore, the search for a prolonged-action antimalarial goes forward on all fronts.

The development of plasmodial resistance to antimalarial drugs began to receive serious attention when evidence began to accumulate in 1961 that the efficacy of one of the most widely used and successful erythrocytic schizontocides, chloroquine, was greatly reduced in some areas of the world.[3] Earlier, it had been established that proguanil and pyrimethamine induced the formation of resistant strains in man. Thus, the malaria plasmodium resembles most other pathogenic microorganisms in having the ability to develop resistance to anti-infective agents. The development of resistance has focused attention on the mechanisms of action of antimalarials, since resistance may be related to an ability of the plasmodium to adapt to specific changes in environment caused by the drug. It has been suggested that such a mechanism is involved in the development of microbial resistance to sulfa drugs, and this may be true also in regard to the effect produced by proguanil and pyrimethamine, which appear to inhibit the PABA–folic-acid–folinic-acid sequence. However, the inhibition of malaria organisms by pyrimethamine is not antagonized by administration of either folic acid or folinic acid. Therefore, it has been suggested that pyrimethamine may act by interference with DNA synthesis at a point beyond the utilization of folinic acid.[4]

Other biosynthetic pathways may be inhibited by other antimalarials. It has been suggested that the action of 8-aminoquinolines is a result of the inhibiting effect that their metabolites have on TPNH-linked reduction processes in liver cells or is due to a sensitivity to oxidative damage peculiar to enzymes involved in the pentose phosphate pathway of tissue schizonts as compared with those of blood schizonts.[5] The mechanism for the development of plasmodial resistance to 4-aminoquinolines is uncertain. It has been postulated that chloroquine acts by inhibition of DNA replication and that susceptibility or resistance to it is the result of the capacity of the membranes of the infected cells to permit passage of the drug.[6] Resistance to quinine and other cinchona alkaloids has been observed for some time, but the effect has been minor, and there has been a resurgence in the use of quinine in the 1960's. The mechanism by which quinine exerts its effect is not definitely known. Interference with enzymatic oxidative conversion of pyruvate, formation of lactate from glucose and glycolysis have been implicated. Inhibition of the incorporation of phosphate into RNA and DNA by quinine has been noted.[4] Whether adaptive responses to the inhibition of these and other metabolic systems are responsible for the development of resistance or whether drug-induced mutation is the cause will be decided by further biochemical research. It is worthy of note that the development of cross resistance has not been observed as a result of the simultaneous use of drugs of different types.

CINCHONA ALKALOIDS

The development of cinchona, from its origin as an uncultivated South American plant to its recent extensive cultivation in the Congo, and particularly in Indonesia, is an interesting chapter in the history of the drug.[7] The importance of cinchona as a drug source is attested by the U. S. Department of Commerce reports[8] on the import of the bark and the alkaloids obtained from it. In 1969, 3,589,418 av. oz. of salts and alkaloids obtained from cinchona bark were imported.

The crude drug contains numerous alkaloids, of which quinine, quinidine, cinchonine and cinchonidine are the most abundant and the most important. The average commercial yields of alkaloids from cin-

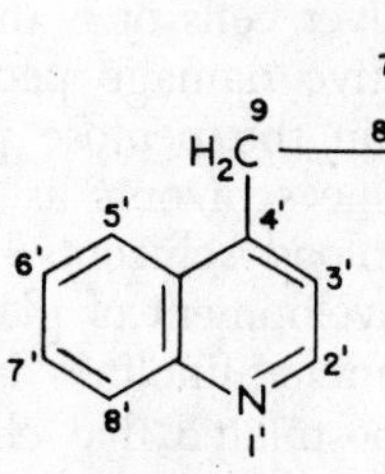

RUBAN

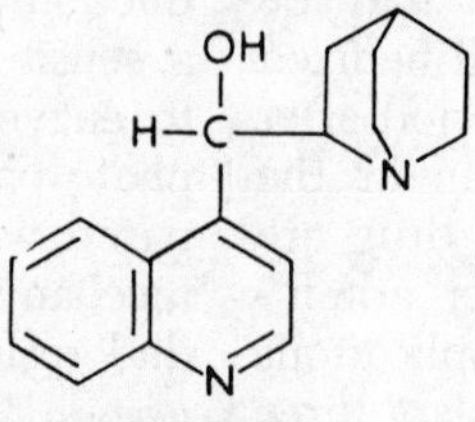

9–RUBANOL

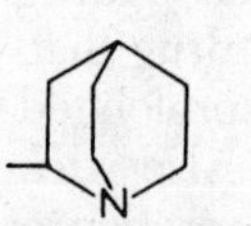

QUINUCLIDINE NUCLEUS

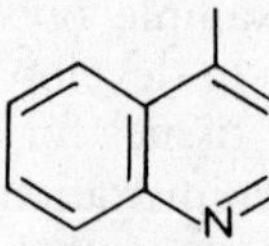

QUINOLINE NUCLEUS

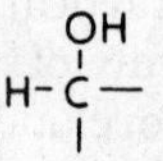

SECONDARY ALCOHOL GROUP

chona bark are: quinine, about 5 per cent; quinidine, 0.1 per cent; cinchonine, 0.3 per cent, and cinchonidine, 0.4 per cent. Another source of cinchona alkaloids is cuprea bark, obtained from *Remijia pedunculata*. One of the numerous alkaloids extracted from it is cupreine, which was the base for the making of ethyl hydrocupreine (Optochin) and isoamylhydrocupreine (Eucupin), derivatives which no longer are of therapeutic importance. These and other cinchona alkaloids and their derivatives are derived from the parent compound, *ruban*. The nomenclature was proposed by Rabe[9] to simplify the naming of these complex compounds and to indicate their origin from the Rubiaceae.

CHEMISTRY

Because of the medicinal and the economic importance of cinchona alkaloids, interest was focused on the chemistry of these compounds at an early date. Quinine was isolated first by Pelletier and Caventou in 1820; they showed also that the alkaloids of cinchona were responsible for its febrifuge properties. In the last quarter of the 19th century, intensive work was carried out by many early organic chemists, such as Skraup and Koenigs, to establish the structure of quinine, but major credit is due to Rabe and his coworkers who postulated its true structure. Rabe's formula was finally corroborated by the synthesis of quinine by Woodward and Doering[10] in 1945. The complete stereochemistry of cinchona alkaloids has been determined (see below).

Examination of the structures of the cinchona alkaloids shows the existence of four asymmetric centers, carbons 3, 4, 8 and 9. All of the cinchona alkaloids are identical in configuration at C-3 and C-4. As may be seen in the structures, four different isomers

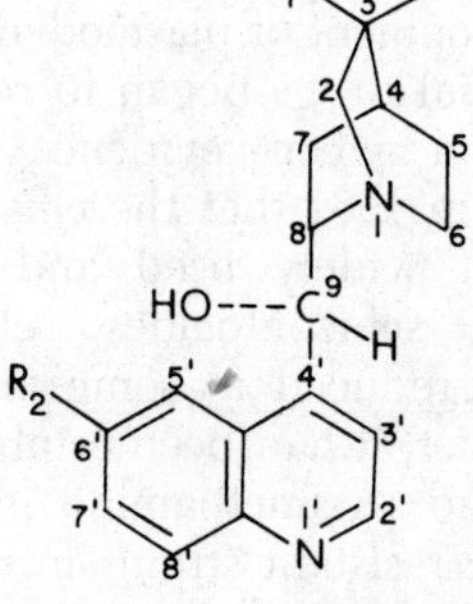

may exist as a result of different configurations at C-8 and C-9. The differences in orientation result in differences in the optical rotation of the alkaloids. Structural relations among some of the important cinchona alkaloids and derivatives is given in Table 19.

STRUCTURE-ACTIVITY RELATIONSHIPS

Quinine has been investigated extensively in an effort either to develop a synthetic substitute or to modify its structure in such a manner as to improve its action. It was noted early that stereochemical changes at C-8 had little effect on antimalarial activity. For the treatment of acute attacks of *P. vivax*, quinine, quinidine, cinchonidine and cinchonine have been found to be about equally successful, but the blood-level concentrations required to produce equivalent effects vary with the drug.[12] On evidence that has been obtained only from non-human malarias, inversion at C-3 appears to have no important effect on activity. Changes in the configuration at C-9 to the *epi* isomers result in inactive compounds.

TABLE 19. STRUCTURAL RELATIONSHIP OF CINCHONA ALKALOIDS

(−) ISOMER	R_1	R_2	(+) ISOMER
Quinine	$-CH=CH_2$	$-OCH_3$	Quinidine
Cinchonidine	$-CH=CH_2$	$-H$	Cinchonine
Cupreine	$-CH=CH_2$	$-OH$	
Hydrocupreine	$-CH_2CH_3$	$-OH$	Hydrocupreidine
Ethylhydrocupreine	$-CH_2CH_3$	$-OCH_2CH_3$	Ethylhydrocupreidine
Isoamylhydrocupreine	$-CH_2CH_3$	$-OC_5H_{11}$ (i)	

Activity has been shown to decrease markedly when the secondary alcohol group was modified in any way. The entire quinuclidine nucleus has been shown to be unnecessary for activity, but the α-aryl-β-tertiary amino alcohol system about the central C-9 atom seems to be essential.[13] Other considerations have been reported,[14] but results have been based on activity determinations with plasmodia other than those infecting man.

ABSORPTION, DISTRIBUTION AND EXCRETION

The absorption, the distribution and the excretion of cinchona alkaloids have been studied extensively.[15] After oral administration, the cinchona alkaloids are absorbed rapidly and nearly completely, with peak blood-level concentrations occurring in 1 to 4 hours. Blood levels fall off very quickly after administration is stopped. A single dose of quinine is disposed of in about 24 hours. Therefore, repeated doses must be administered. Soluble salts of quinine produce higher initial blood levels than are obtained with the free base, but either form is equally satisfactory for maintenance doses. In animals, quinine is concentrated chiefly in pancreas, liver, spleen, lung and kidney. Relatively small amounts are found in the blood, muscle or connective and nervous tissues. Various tissues contain enzymes capable of metabolizing the cinchona alkaloids, but the principal action appears to take place in the liver, where an oxidative process results in the addition of a hydroxyl group to the 2′ position of the quinoline ring. The resulting degradation products, called carbostyrils, are much less toxic, are eliminated more rapidly and possess lower antimalarial activity than the parent compounds. The carbostyrils may be further oxidized to dihydroxy compounds. Attempts to obtain longer acting analogs by blocking the 2′ position with alkyl or aryl groups produced active but not clinically successful compounds.

TOXICITY

The toxic reactions to the cinchona alkaloids have been studied extensively.[16] Acute poisoning with quinine is not common. In one case, a death was reported after administration of 18 g.; in another case, it was reported that the patient recovered after administration of 19.8 g. of quinine. A fatality resulted after the intravenous administration of 1 g. of quinine. The toxic manifestations that are most common are due to hypersensitivity to the alkaloids and are referred to collectively as cinchonism. Frequent reactions are allergic skin reactions, tinnitus, slight deafness, vertigo and slight mental depression. The most serious is amblyopia, which may follow the administration of very large doses of quinine but is not common; usual therapeutic regimens do not produce this effect. When quinine and, possibly, other cinchona alkaloids are used in the treatment of *P. falciparum,*, hemoglobinuria may be produced, with the development of blackwater fever. The synthetic aminoquinolines do not cause this effect. Quinine will pass from the maternal to the fetal circulation, and administration of it during pregnancy may lead to fetal blindness.[17] It has been implicated in congenital deafness.

USES, ROUTES OF ADMINISTRATION AND DOSAGE FORMS

The cinchona alkaloids act only as

erythrocytic (blood) schizontocides on the asexual form of the malaria parasite. They are used in benign tertian and quartan malaria to suppress the development of the clinical symptoms of malaria rather than to provide a radical cure for the disease. Usually, quinine can cure infection caused by *P. falciparum*, since there is no secondary exoerythrocytic form of that parasite. Quinine and related alkaloids suppress attacks due to *P. vivax* and *P. falciparum* equally well but are not as effective against *P. malariae*.

In addition to antimalarial action, cinchona alkaloids are antipyretic. The action of quinine on the central temperature-regulating mechanism causes peripheral vasodilation. This effect accounts for the traditional use of quinine in cold remedies and fever treatments. Quinine has been used as a diagnostic agent for myasthenia gravis (by accentuating the symptoms). Also, it has been used for the treatment of night cramps or "restless legs." Although they possess local anesthetic action, the cinchona alkaloids have no real place in modern therapeutics for such use. The same applies to their use as antiseptics. Quinine salts cause contractions of the uterine muscle, but any reputation as abortifacients based on such activity is undeserved. Aside from the uncertainty of action, the oxytocic use of quinine is attended by toxic dangers to both the mother and the fetus. The antifibrillating effect of quinidine in the treatment of cardiac arrythmias is discussed in Chapter 21.

The antimalarial action of cinchona alkaloids may be obtained by oral, intravenous or intramuscular administration. Administration by injection, particularly intravenous injection, is not without hazard and should be used cautiously. For intramuscular injection, quinine dihydrochloride is usually used, although there is sometimes a preference for the less irritating hydrochloride. Intravenous injections may cause severe cardiovascular depression leading to generalized collapse. Crude extract preparations containing the alkaloids of cinchona have been used widely as economical antimalarials for oral administration. During World War II, a critical re-evaluation[18] of earlier work substantiated the view that the important cinchona alkaloids were roughly equally effective and that it was not necessary for mixtures to contain a high proportion of quinine. A mixture known as quinetum, containing a large amount of quinine, had been used for some time in malaria therapy. As the interest in pure quinine increased, another crude mixture ("cinchona fibrifuge"), composed of the alkaloids remaining after quinine removal, was introduced to replace quinetum. Subsequently, the Malaria Commission of the League of Nations introduced "totaquina." It was accepted in the *British Pharmacopoeia* in 1932 and later in *U.S.P. XII* and *U.S.P. XIII*. The *N.F. X* defined it as containing 7 to 12 per cent of anhydrous crystallizable cinchona alkaloids. Totaquine now is the most widely used of inexpensive antimalarial drugs. The usual dose is 600 mg.

Quinine. Quinine is obtained from quinine sulfate prepared by extraction from the crude drug. To obtain it from solutions of quinine sulfate, a solution of the sulfate is alkalinized with ammonia or sodium hydroxide. Another method is to pour an aqueous solution of quinine bisulfate into excess ammonia water, with stirring. In either procedure, the precipitated base is washed and recrystallized. The pure alkaloid crystallizes with 3 molecules of water of crystallization. It is efflorescent, losing 1 molecule of water at 20° under normal conditions and losing 2 molecules in a dry atmosphere. All water is removed at 100°.

It occurs as a levorotatory, odorless, white crystalline powder possessing an intensely bitter taste. It is only slightly soluble in water (1:1,500), but it is quite soluble in alcohol (1:1), chloroform (1:1) or ether.

It behaves as a diacidic base and forms salts readily. These may be of two types, the *acid* or *bi-salts* and the *neutral* salts. The neutral salts are formed by involvement of only the tertiary nitrogen in the quinuclidine nucleus, and the acid salts are the result of involvement of both basic nitrogens. Inasmuch as the quinoline nitrogen is very much less basic than the quinuclidine nitrogen, involvement of both nitrogens results in a definitely acidic compound. The

TABLE 20. FORMULAS AND SOLUBILITIES OF THE COMMON QUININE SALTS

QUININE SALT AND FORMULA	WATER	ALCOHOL	SOLUBILITY* CHLOROFORM	ETHER	GLYCERIN
Quinine Sulfate U.S.P. $B_2 \cdot H_2SO_4 \cdot 2H_2O$†	500 35(100°)	120 10(80°)	Slightly	Slightly	. . . . . .
Quinine Bisulfate $B \cdot H_2SO_4 \cdot 7H_2O$	10 1(100°)	25 1(80°)	625	. . . . . .	15
Quinine Hydrochloride $B \cdot HCl \cdot 2H_2O$	16 0.5(100°)	1	1	350	7
Quinine Dihydrochloride N.F. $B \cdot 2HCl$	0.6	12	Slightly	Very slightly	. . . . . .
Quinine Hydrobromide $B \cdot HBr \cdot H_2O$	40 3.2(80°)	1	1	25	7

* The solubilities given in this table, unless so indicated, represent the number of cubic centimeters of solvent required to dissolve 1 g. of the salt at a temperature of 25°.

† The letter "B", as used in the formulas given above, represents the quinine base, $C_{20}H_{24}O_2N_2$.

formulas and the solubilities of the commonly used salts of quinine are given in Table 20. With regard to salts of quinine, although they may possess some advantageous features with respect to one another, the speed of absorption and subsequent therapeutic effect seem to be little influenced by the solubility of the compound.[18]

The usual dose is 1 g. daily.

PRODUCTS

Quinine Sulfate U.S.P., Quininium Sulfate. Quinine sulfate is a very common salt of quinine and is ordinarily the "quinine" asked for by the layman.

It is prepared in one of two ways, i.e., from the crude bark or from the free base. When prepared from the crude bark, the powdered cinchona is alkalinized and then extracted with a hot, high-boiling petroleum fraction to remove the alkaloids. By carefully adding diluted sulfuric acid to the extract, the alkaloids are converted to sulfates, the sulfate of quinine crystallizing out first. The crude alkaloidal sulfate is decolorized and recrystallized to obtain the article of commerce. Commercial quinine sulfate is not pure but contains from 2 to 3 per cent of impurities, which consist mainly of hydroquinine and cinchonidine.

To obtain quinine sulfate from the free base, it is neutralized with dilute sulfuric acid. The resulting sulfate, when recrystallized from hot water, forms masses of crystals with the approximate formula $(C_{20}H_{24}O_2N_2)_2 \cdot H_2SO_4 \cdot 8H_2O$. This compound readily effloresces in dry air to the official dihydrate, which occurs as fine, white needles of a somewhat bulky nature.

Quinine sulfate often is prescribed in liquid mixtures. From a taste standpoint, it is better to suspend the salt rather than to dissolve it. However, in the event that a solution is desired, it may be accomplished by the use of alcohol or, more commonly, by addition of a small amount of sulfuric acid to convert it to the more soluble bisulfate. The capsule form of administration is the most satisfactory for masking the taste of quinine when it is to be administered orally.

Category—antimalarial (*Plasmodium falciparum*).

Usual dose—therapeutic, 300 mg. 3 times daily for 9 days.

OCCURRENCE

Quinine Sulfate Capsules U.S.P.
Quinine Sulfate Tablets U.S.P.

The need for the development of salts of quinine other than the sulfate (Table 20) resulted from the special uses to which they were put and from the relatively low solubility of the sulfate.

The dihydrochloride is recognized in the N.F. XIII for use in the official injection. The sulfate salts of cinchonidine and cincho-

TABLE 21. STRUCTURAL RELATIONSHIP OF 4-AMINOQUINOLINES

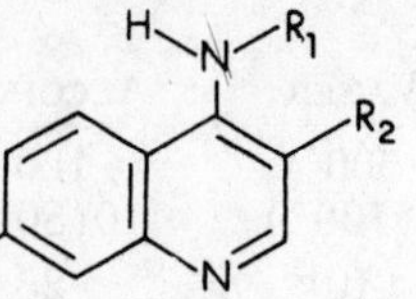

COMPOUND	R₁	R₂
1. Chloroquine	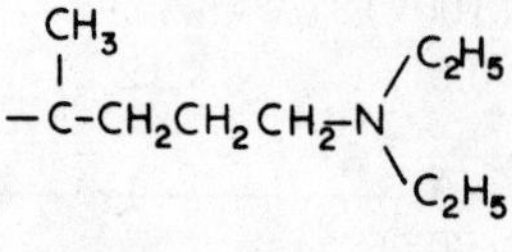 $-C(CH_3)-CH_2CH_2CH_2-N(C_2H_5)(C_2H_5)$	H
2. Hydroxychloroquine	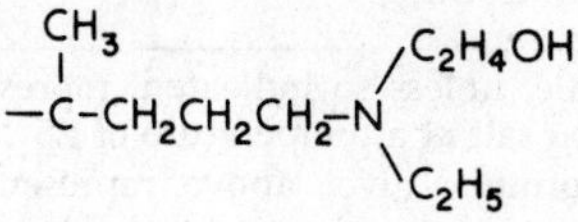 $-C(CH_3)-CH_2CH_2CH_2-N(C_2H_4OH)(C_2H_5)$	H
3. Sontoquine	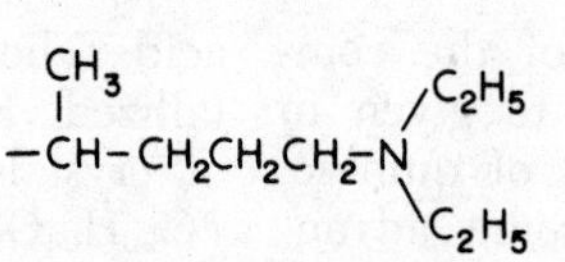 $-CH(CH_3)-CH_2CH_2CH_2-N(C_2H_5)(C_2H_5)$	CH₃
4. Amodiaquine	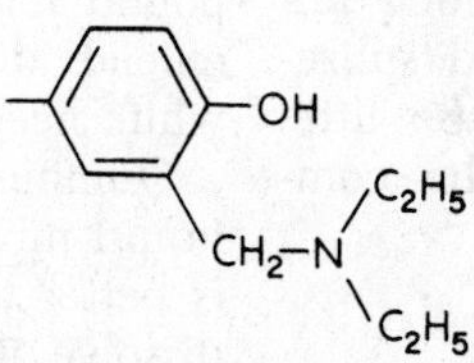 (2-hydroxyphenyl)CH₂-N(C₂H₅)(C₂H₅)	H

nine may be used as antimalarials. The dextrorotatory cinchonine salt is of value in the treatment of patients who display a sensitivity to the levorotatory cinchona alkaloids.

4-AMINOQUINOLINES

The synthesis of 4-aminoquinolines for antimalarial studies was first undertaken by Russian and German workers just prior to World War II. The secrecy that surrounded medical research at that time has obscured the history of the development of this important class of compounds, but no doubt they resulted as an extension of the studies that had led to the development of pamaquine, an 8-aminoquinoline. The general direction that research had been taking may be seen by the German report[19] in 1942 that 4-, 6- and 8-aminoquinolines gave antimalarials when properly substituted. The 4-aminoquinoline first to receive wide study

was sontoquine. The German pharmocologists rated sontoquine as "superior to quinine and 60 to 100 per cent as effective as quinacrine" against avian infections and, apparently, were not impressed with the clinical results of either chloroquine or sontoquine. These conclusions may have been influenced by the difficulties encountered in the synthesis of the intermediate, 3-methyl-4,7-dichloroquinoline. The Vichy French were permitted to subject these compounds to field trials. Sontoquine in tablet form was captured from the Vichy French in North Africa. Prior to this, researches in the United States on 4-aminoquinoline derivatives had been instigated to check the results described in the German patent literature and the favorable reports appearing in the Russian literature on certain members of this group. A reinvestigation by the Office of Scientific Research and Development showed that chloroquine was at least four times as effective as quinacrine against avian malaria and twice as effective against the common human malarias. It also is interesting to note that sontoquine, judged by the Germans as the best of a series, is only about one third as effective against both avian and human infections as chloroquine.

Extensive research has led to the introduction of a number of 4-aminoquinolines into clinical use. The chemical relationships among the members of the group are shown in Table 21.

STRUCTURE-ACTIVITY RELATIONSHIPS

Although the Germans had selected sontoquine as the best of the 4-aminoquinolines available at that time, it is now agreed that substitution of a methyl group on the C-3 reduces activity. It also has been determined that substituting a methyl group on C-8 causes a complete loss of activity. The C-7 position has been found to be best for halogen substitution. The introduction of other groups on the quinoline nucleus reduces antimalarial activity. As may be seen in Table 21, variations in the 4-amino side chain have been studied. Early work indicated that the 4-diethylamino-1-methylbutylamino group present in chloroquine was optimum for activity. However, more recently it has been shown that the introduction of hydroxy groups on the side chain, particularly on an ethyl group of the terminal nitrogen, tends to reduce toxicity and to produce higher blood level concentrations. The substituted anilo compounds were introduced[20] to combine the antimalarial effect found in some α-dialkylamino-*o*-cresols with that of 4-aminoquinoline. Low toxicity for such compounds is claimed, with slightly less activity than chloroquine.

ABSORPTION, DISTRIBUTION AND EXCRETION

The 4-aminoquinolines are absorbed readily from the gastrointestinal tract,[21] but amodiaquin gives lower plasma levels than others in the group. Peak plasma concentrations are reached in 1 to 3 hours, with blood levels falling off rather rapidly after administration is stopped. Normally, 4-aminoquinolines are administered in divided doses over the period of therapy. These drugs tend to concentrate in the liver, the spleen, the heart, the kidney and the brain. The relatively high localization in the liver may be an important factor in their usefulness for the treatment of hepatic amebiasis. Small amounts of 4-aminoquinolines have been found in the skin but probably not in sufficient quantity to account for their suppressant action on polymorphous light dermatoses. These compounds are excreted rapidly, with most of the unmetabolized drug being accounted for in the urine.

TOXICITY

Although the toxicity of 4-aminoquinolines is quite low in the usual antimalarial regimen, both acute and chronic toxic reactions may develop. Acute side-effects include nausea, vomiting, anorexia, abdominal cramps, diarrhea, headache, dizziness, pruritus and urticaria. Interference with accommodation may result in a blurring of vision. Usually, such symptoms are completely reversible on reduction of the dose or complete withdrawal of the drug. Toxic effects that are found less frequently are leukopenia, tinnitus and deafness. Long-term administration or high dosages may have serious effects on the eyes, and ophthalmologic examinations should be carefully carried out. Also, periodic blood examinations should be made. Patients with liver diseases particularly should be watched when 4-aminoquinolines are used.

USES, ROUTES OF ADMINISTRATION AND DOSAGE FORMS

The 4-aminoquinolines are effective erythrocytic schizontocides in infections due to all species of human plasmodia. In infections due to *P. falciparum*, their use can effect complete cure of the disease. However, resistant strains of this species have developed and the 4-aminoquinolines may be without value in such infections. Radical cures are not effected when secondary tissue schizonts are present, as in the case of *P. vivax*.

The 4-aminoquinolines, particularly chloroquine and hydroxychloroquine, are used in the treatment of extraintestinal amebiasis, with very satisfactory results. They are of value in the treatment of chronic discoid lupus erythematosus but are of questionable value in the treatment of the systemic form of the disease. Symptomatic relief has been secured through the use of 4-aminoquinolines in the treatment of rheumatoid arthritis. Although the mechanism for their effect in collagen diseases has not been established, these drugs appear to suppress the formation of antigens that may be responsible for hypersensitivity reactions which cause the symptoms to develop. Long-term therapy of at least 4 to 5 weeks is usually required before beneficial results are obtained in the treatment of collagen diseases.

For the treatment of malaria, these drugs usually are given orally as salts of the amines in tablet form. In case of nausea or vomiting after oral administration, intramuscular injection may be used. For prophylactic treatment, the drugs may be incorporated into table salt. To protect the drugs from the high humidity of tropical climates, coating of the granules with a combination of cetyl and stearyl alcohols has been employed. These drugs are sometimes combined with other drugs such as proguanil or pyrimethamine to obtain a broader spectrum of activity (see Table 18).

Products

Chloroquinine U.S.P., CQ, 7-Chloro-4-[[4-(diethylamino)-1-methylbutyl]amino]-quinoline. Chloroquine occurs as a white or slightly yellow crystalline powder that is odorless and has a bitter taste. It is usually partly hydrated, very slightly soluble in water and soluble in dilute acids, chloroform and ether.

Although several methods are available for the commercial synthesis of chloroquine, the procedure reported by Elderfield[22] and Kenyon[23] has proved to be the most practical.

Category—antimalarial.

Usual dose—200 to 250 mg., repeated in six hours if necessary, as hydrochloride in Chloroquine Hydrochloride Injection.

Usual dose range—200 mg. to 1 g. for one day only as hydrochloride in Chloroquine Hydrochloride Injection.

OCCURRENCE
Chloroquine Hydrochloride Injection U.S.P.

Chloroquine Phosphate U.S.P., Aralen®, Resochin®, 7-Chloro-4-[[4-(diethylamino)-1-methylbutyl]amino]quinoline Phosphate. Chloroquine phosphate occurs as a white, crystalline powder that is odorless, has a bitter taste and slowly discolors on exposure to light. It is freely soluble in water, and aqueous solutions have a pH of about 4.5. It is almost insoluble in alcohol, ether and chloroform. It exists in two polymorphic forms, either of which (or a mixture of both) may be used medicinally.

Category—antimalarial; antiamebic; lupus erythematosus suppressant.

Usual dose—antimalarial: suppressive, 500 mg. weekly; therapeutic, initial, 1 g. then 500 mg. in 6 hours, and 500 mg. on the second and third days; antiamebic (for abscess only): 250 mg. two or three times daily for 2 weeks, up to a total of 11 g.; lupus erythematosus suppressant: 250 mg. twice daily for 1 to 2 weeks, then 250 mg. daily.

OCCURRENCE
Chloroquine Phosphate Tablets U.S.P.

Hydroxychloroquine Sulfate U.S.P., Plaquenil®, 2-[[4-[(7-Chloro-4-quinolyl)amino]-pentyl]ethylamino]ethanol Sulfate (1:1).

Hydroxychloroquine sulfate occurs as a white or nearly white, crystalline powder that is odorless but has a bitter taste. It is freely soluble in water, producing solutions with a pH of about 4.5. It is practically insoluble in alcohol, ether and chloroform.

While successful as an antimalarial, hydroxychloroquine has achieved greater use than chloroquine in the control and the treatment of collagen diseases because it is somewhat less toxic.

Category—suppressant for lupus erythematosus.

Usual dose—400 mg. daily.

Usual dose range—200 to 800 mg. daily.

OCCURRENCE
Hydroxychloroquine Sulfate Tablets, U.S.P.

Amodiaquine Hydrochloride N.F., Camoquin®, 4-[(7-Chloro-4-quinolyl)amino]-α-(diethylamino)-o-cresol Dihydrochloride Dihydrate. Amodiaquine dihydrochloride occurs as a yellow, odorless crystalline powder having a bitter taste. It is soluble in water, sparingly soluble in alcohol and very slightly soluble in ether, chloroform and benzene. The pH of a 1 per cent solution is between 4 and 4.8. The synthesis[20] of amodiaquine is more expensive than that of chloroquine.

This compound is an economically important antimalarial. Amodiaquine is highly suppressive in *P. vivax* and *P. falciparum* infections, being three to four times as active as quinine. However, it has no curative activity except against *P. falciparum*. Amodiaquine is altered rapidly in vivo to yield products which appear to be excreted slowly and have a prolonged suppressive activity.

Category—antimalarial.

Usual dose—suppressive, 520 mg., the equivalent of 400 mg. of amodiaquine base,

every 2 weeks; therapeutic, 780 mg. to 1.3 g., the equivalent of 600 mg. to 1 g. of amodiaquine base.

OCCURRENCE
Amodiaquine Hydrochloride Tablets N.F.

8-AMINOQUINOLINES

As previously stated, an 8-aminoquinoline, pamaquine, was the first truly successful synthetic antimalarial. The developments that eventually led to pamaquine date back to the research conducted by I. G. Farbenindustrie in World War I. Methylene blue, which was the only noteworthy lead to the production of a nonquinine antimalarial, was modified by introducing a dialkylaminoalkyl group in place of one of the aminomethyl groups, to improve its activity. Although no useful compounds were obtained, it appeared worthwhile to introduce the dialkylaminoalkyl side chain into other compounds. Because 6-methoxy-8-amino-

TABLE 22. STRUCTURAL RELATIONSHIPS OF 8-AMINOQUINOLINES

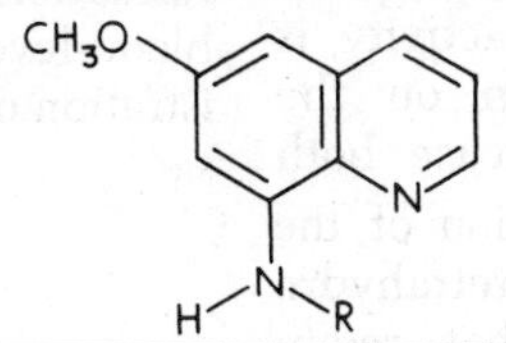

COMPOUND	R
1. Pamaquine	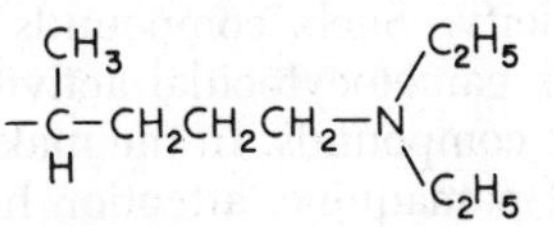 $-\underset{H}{\overset{CH_3}{C}}-CH_2CH_2CH_2-N\underset{C_2H_5}{\overset{C_2H_5}{}}$
2. Primaquine	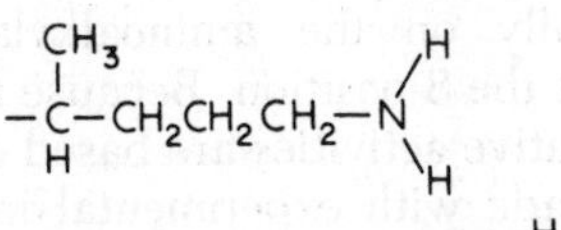 $-\underset{H}{\overset{CH_3}{C}}-CH_2CH_2CH_2-N\underset{H}{\overset{H}{}}$
3. Pentaquine	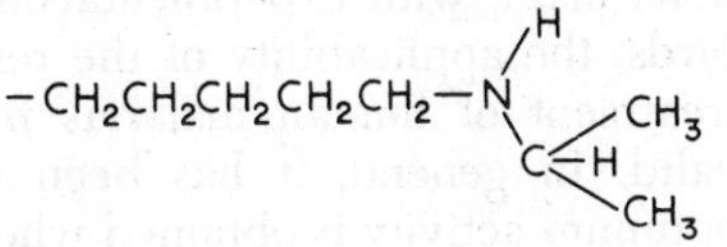 $-CH_2CH_2CH_2CH_2CH_2-N\overset{H}{\underset{C\overset{H}{\underset{CH_3}{\diagdown}}}{\diagup CH_3}}$
4. Isopentaquine	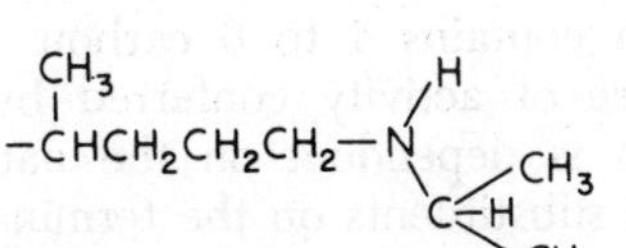 $-\overset{CH_3}{\underset{}{C}}HCH_2CH_2CH_2-N\overset{H}{\underset{C\overset{H}{\underset{CH_3}{\diagdown}}}{\diagup CH_3}}$

quinoline exhibited some antimalarial activity, the dialkylaminoalkyl group was introduced into the amino group at C_8. Initial results were promising, and extension of modifications of the side chain led to the preparation of pamaquine.

The promising lead provided by pamaquine stimulated research that has led to the synthesis and testing of many 8-aminoquinolines. However, their use today is not extensive, and they are employed almost exclusively for their exoerythrocytic schizontocidal activity against *P. vivax* and *P. malariae*. Some of the members of the group exhibit good gametocytocidal activity. The chemical relationships among some of the more important members of the group are shown in Table 22.

Structure-Activity Relationships

Although optimum activity in 8-aminoquinolines is obtained in compounds that have a 6-methoxy substituent, such a group is not essential for antimalarial action.[24] Compounds that have a 6-hydroxy group are quite active, but those with a 6-ethoxy group have little activity. When a 6-methyl group is introduced, complete loss of activity is observed. Additional substitution on the quinoline nucleus tends to decrease both activity and toxicity. The reduction of the quinoline nucleus to 1,2,3,4-tetrahydro analogs produces compounds that retain antimalarial activity, but with lower potency and toxicity. Such compounds appear to have less gametocytocidal activity than the aromatic compounds. In the making of variations of pamaquine, attention has centered principally on the aminoalkylamino side chain at the 8-position. Because most of the comparative activities are based on observations made with experimental infections in birds, the applicability of the results to the treatment of human malarias may not be valid. In general, it has been found that optimum activity is obtained when the alkyl chain contains 4 to 6 carbon atoms. The degree of activity conferred by the alkyl chain is dependent on the nature of the alkyl substituents on the terminal nitrogen. The greatest activity appears to be achieved in alkyl groups containing 5 carbon atoms,

with the normal pentyl group being favored over the 1-methylbutyl group and the 4-methylbutyl group. It may be noted that the terminal nitrogen may be a primary, a secondary or a tertiary amine. In regard to activity, the nature of the amine appears to be of less consequence than the length of the linking alkyl chain.

Absorption, Distribution and Excretion

The 8-aminoquinolines are absorbed rapidly from the gastrointestinal tract, to the extent of 85 to 95 per cent within 2 hours after oral administration.[25] Peak plasma concentration is reached within 2 hours after ingestion, after which the drug rapidly disappears from the blood. The drugs are localized mainly in liver, lung, brain, heart and muscle tissue. Metabolic changes in the drug are produced very rapidly, and, on excretion, metabolic products account for nearly all of the drug. Only about 1 per cent of the drug is eliminated through the urine unchanged. It may be that the antiplasmodial and the toxic properties of these drugs are produced by metabolic transformation products. To maintain therapeutic blood-level concentrations, frequent administration of 8-aminoquinolines may be necessary.

Toxicity

The toxic effects of the 8-aminoquinolines are found principally in the central nervous system and the hemopoietic system. Occasionally, anorexia, abdominal pain, vomiting and cyanosis may be produced. The toxic effects related to the blood system are more common; hemolytic anemia (particularly in dark-skinned people), leukopenia and methemoglobinemia are the usual findings. Toxicity is increased by quinacrine; therefore, the simultaneous use of quinacrine and 8-aminoquinolines must be avoided.

Uses, Routes of Administration and Dosage Forms

Because of the toxic effects, the 8-aminoquinolines are seldom used today except to prevent relapses due to the exoerythrocytic forms of the parasites, particularly *P. vivax*. They may be administered concurrently with 4-aminoquinolines to combine curative effect with suppressive action and to reduce the likelihood of the development of resistant strains of plasmodia. Usually, they are administered orally, in tablet form, as salts such as hydrochlorides or phosphates. Pamaquine is used as the methylene-bis-β-hydroxynaphthoate (naphthoate or pamoate), because this salt is of low solubility and is absorbed slowly, and thus, blood levels are maintained for longer periods and are more uniform.

PRODUCTS

Primaquine Phosphate U.S.P., Primaquinium Phosphate; 8-[(4-Amino-1-methylbutyl)amino]-6-methoxyquinoline Phosphate. Primaquine phosphate is an orange-red, crystalline substance having a bitter taste. It is soluble in water and insoluble in chloroform and ether. Its aqueous solutions are acid to litmus. It may be noted that it is the primary amine homolog of pamaquine.

Primaquine has been found to be the most effective and the best tolerated of the 8-aminoquinolines. Against *P. vivax*, it is 4 to 6 times as active an exoerythrocytic schizontocide as pamaquine and about one half as toxic. When 15 mg. of the base are administered daily for 14 days, radical cure is achieved in most *P. vivax* infections. Success has been achieved against some very resistant strains of *P. vivax* by administering 45 mg. of the base once a week for 8 weeks, with simultaneous administration of 300 mg. of chloroquine base. This regimen also tends to lessen the toxic hemolytic effects produced in primaquine-sensitive individuals.

Category—antimalarial.

Usual dose—26.3 mg., the equivalent of 15 mg. of primaquine base, daily for 14 days.

OCCURRENCE

Primaquine Phosphate Tablets U.S.P.

9-AMINOACRIDINES

The intensive research on synthetic antimalarials on the part of German chemists that had led to the development of pama-

quine continued, and, in 1932, quinacrine was developed.[26] Quinacrine contains the same dialkylaminoalkyl side chain as pamaquine does, and has 2-chloro and 7-methoxy substitutions on the acridine nucleus. This compound did not receive a great deal of attention until the beginning of World War II; at that time it was widely studied, and it was used widely until the introduction of chloroquine. It has now been displaced as an erythrocytic schizontocide by the 4-aminoquinoline derivatives, which are its equal or superior in effectiveness and do not stain the skin yellow, an undesirable but harmless feature of quinacrine therapy.

A number of 9-aminoacridines have been tested for antimalarial activity, but none has been found to be superior to quinacrine. A new acridine compound (CI-423) that is a 10-oxide derivative of quinacrine without the 7-methoxy group has been found to be about 4 times as active as quinacrine in some experimental infections.[27, 28] Other polycyclic structures containing hetero atoms, such as azacridines, have been used to make compounds analogous to quinacrine, but no advantages in such structures have been found.

PRODUCTS

Quinacrine Hydrochloride U.S.P.; Mepacrine Hydrochloride; Atabrine®; Atebrin®; 6-Chloro-9-[[4-(diethylamino)-1-methylbutyl]amino]-2-methoxyacridine Dihydrochloride.

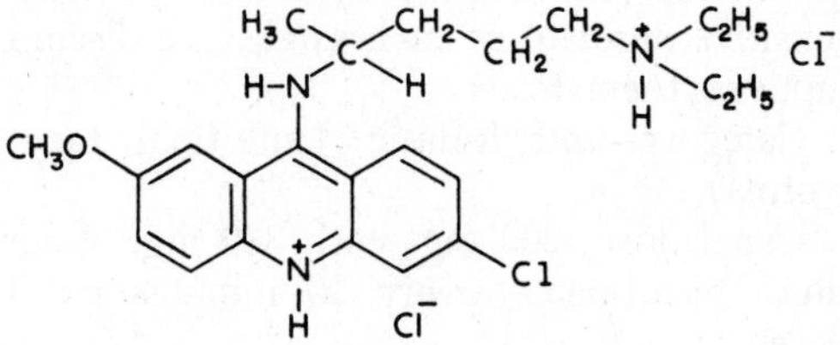

Quinacrine Hydrochloride

The wide use of this compound during the early 1940's resulted in a large number of synonyms for quinacrine in various countries throughout the world.

The dihydrochloride salt is a yellow crystalline powder that has a bitter taste. It is sparingly soluble (1:35) in water and soluble in alcohol. A 1:100 aqueous solution has a pH of about 4.5 and shows a fluorescence. Solutions of the dihydrochloride are not stable and should not be stored. A dimethanesulfonate salt produces somewhat more stable solutions, but they too should not be kept for any length of time.

The yellow color that quinacrine imparts to the urine and the skin is temporary and should not be mistaken for jaundice. Quinacrine may produce toxic effects in the central nervous system, such as headaches, epileptiform convulsions and transient psychoses that may be accompanied by nausea and vomiting. Hemopoietic disturbances such as aplastic anemia may occur. Skin reactions and hepatitis are other symptoms of toxicity. Deaths have occurred from exfoliative dermatitis caused by quinacrine.

As an antimalarial, quinacrine acts as an erythrocytic schizontocide in all kinds of human malaria. It has some effectiveness as a gametocytocide in *P. vivax* and *P. malariae* infections. It may be employed in the treatment of blackwater fever when the use of quinine is contraindicated. It is also an effective curative agent for the treatment of giardiasis due to *Giardia lamblia*, eliminating the parasite from the intestinal tract. It is an important drug for use in the elimination of intestinal cestodes such as *Taenia saginata* (beef tapeworm), *T. solium* (pork tapeworm) and *Hymenolepis nana* (dwarf tapeworm). Like the 4-aminoquinolines, quinacrine may also be used to treat light-sensitive dermatoses such as chronic discoid lupus erythematosus.

Category—anthelmintic (intestinal tapeworms).

Usual dose—200 mg. with 300 mg. of sodium bicarbonate every 10 minutes for 4 doses.

OCCURRENCE
Quinacrine Hydrochloride Tablets U.S.P.

BIGUANIDES

The development of biguanides as antimalarials began in the mid 1940's as a result of a research program of some British scientists[29] who had observed the activity of sulfas, particularly sulfadiazine, against malaria infections. It was thought that the incorporation of certain dialkylaminoalkyl chains onto the pyrimidine ring might lead to significant antimalarial compounds. Although some of their pyrimidine derivatives were active, their studies led them to some open models including certain biguanides.

TABLE 23. STRUCTURAL RELATIONSHIPS OF BIGUANIDES

COMPOUND	X	Y
1. Proguanil	Cl	H
2. Chlorproguanil	Cl	Cl
3. Bromoguanide	Br	H

These compounds showed definite activity, and subsequent chemical modifications led to the production of the compound now called proguanil. The structures of the 3 important biguanides are shown in Table 23.

It has been discovered that the substitution of a halogen on the para position of the phenyl ring significantly increases activity. Chlorine is used in proguanil, but the bromine analog also is very active. Later, it was observed that a second chlorine added to the 3-position of the phenyl ring of proguanil further enhanced activity. However, the dichloro compound, chlorproguanil, is more toxic than proguanil itself.

It has been established[30] that the active forms of biguanides are their metabolic products. For proguanil this is 4,6-diamino-1-*p*-chlorophenyl-1,2-dihydro-2,2-dimethyl-1,3,5-triazine. Because these products are eliminated so rapidly, they are not useful per se in the treatment of human malarias although they are about 10 times as active as their precursors. However, a repository preparation of the metabolite of proguanil (see below) has become available and has achieved spectacular success as an antimalarial with a prolonged duration of activity (cycloguanil pamoate).

Although the biguanides are not sporo-zoitocidal, they are capable of attacking the primary exoerythrocytic stage of *P. falcip-arum* infections and, thus, act as true causal prophylactics. They do not exhibit such activity against *P. vivax*, and some strains of *P. falciparum* show resistance to this action. They possess powerful schizontocidal properties, preventing nuclear division in the early schizont of *P. vivax*, *P. falciparum* and *P. malariae*. A valuable property of these drugs is their ability to inhibit sporogony in the vector. Therefore, they may be used as sporontocidal prophylactics. Unfortunately, the malaria organisms can develop a considerable amount of resist-ance to biguanides, a resistance that persists through the sexual stage in the mosquito. This has led to a considerable decrease in the use of proguanil as an antimalarial. Cross resistance with pyrimethamine is pos-sible.

The biguanides are absorbed from the gastrointestinal tract very quickly, but not as rapidly as quinine or chloroquine. They concentrate in the liver, the lungs, the spleen and the kidney but appear not to cross the blood-brain barrier. They are metabolized in large part in the body and are eliminated very rapidly, principally in the urine. As a result, frequent administration of these drugs is necessary.

The toxic manifestations of biguanides are very mild in man. Some gastrointestinal disturbances may occur if the drugs are taken on an empty stomach but not if they are taken after meals. With excessive doses (1 g. of proguanil), some renal disorders such as hematuria and albuminuria may develop.

PRODUCTS

Proguanil Hydrochloride; Guanatol®; Pal-udrine®; Chlorguanide; 1-(*p*-Chlorophenyl)-5-isopropylbiguanide Hydrochloride. Pro-guanil hydrochloride occurs as a white, crystalline powder or as colorless crystals that are soluble in water (1:75) and alcohol (1:30). It is odorless, has a bitter taste and is stable in air but slowly darkens on expo-sure to light.

Usual dose—suppression of *P. falciparum* and *P. vivax*: 300 mg. weekly; prophylaxis: 100 mg. daily for *P. falciparum;* curative: 300 mg. daily for 10 days for *P. falciparum*.

Cycloguanil Pamoate; Camolar®; CI-501, 4,6-Diamino-1-(*p*-chlorophenyl)-1,2-dihy-dro-2,2-dimethyl-s-triazine (2:1) with 4,4'-Methylenebis[3-hydroxy-2-naphthoic acid]. Among the most exciting developments in antimalarial therapy were the reports[31, 32] in 1963 concerning a repository form of the dihydrotriazine metabolite of proguanil. By combining this very active compound with 4,4'-methylenebis[3-hydroxy-2-naphthoic acid] and by carefully regulating the crystal size, suspensions for intramuscular injec-tion were prepared and found capable of protecting man from *P. vivax* infections for 6 to 19 months after a single dose. Subse-quently, it has been found[33] that cycloguanil pamoate is capable of providing long-term protection against *P. falciparum* infections that are sensitive to proguanil. The dose administered for such long-term protection is the equivalent of 5 mg. of the free base per kg. of body weight. The single intra-muscular injection is administered into the gluteal muscle. In time it will be determined if this preparation affords the malariologist with a preparation meeting all of his ob-jectives.

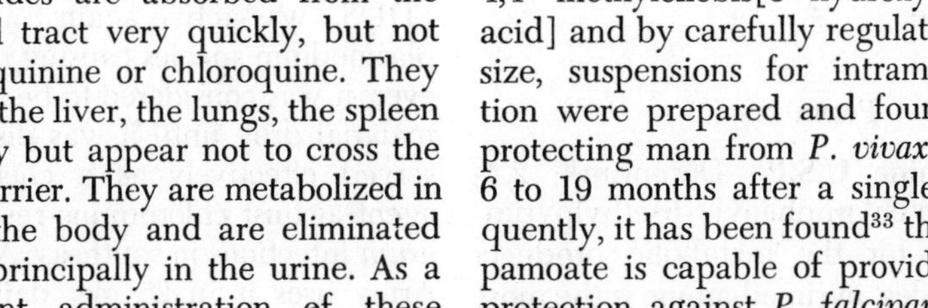

Cycloguanil Pamoate

PYRIMIDINES

Following the observations made in the late 1940's that some 2,4-diaminopyrimidines were capable of interfering with the utilization of folic acid by *Lactobacillus casei*, a property also shown by proguanil, these compounds received intensive study as potential antimalarials. It was noted that certain 2,4-diamino-5-phenoxypyrimidines possessed a structural resemblance to proguanil, and a series of such compounds was synthesized and found to possess good antimalarial action. Subsequently, a large series of 2,4-diamino-5-phenylpyrimidines was prepared and tested for activity.[34] Maximum activity was obtained when an electron-attracting group was present in the 6-position of the pyrimidine ring and when a chlorine atom was present in the *para* position of the phenyl ring. If the two rings were separated by either an oxygen atom or a carbon atom, antimalarial action decreased. The best in the series of compounds was the one that became known as pyrimethamine.

PRODUCTS

Pyrimethamine U.S.P.; Daraprim®; 2,4-Diamino-5-(*p*-chlorophenyl)-6-ethylpyrimidine. Except for the metabolic products of the biguanides, pyrimethamine is the most active antimalarial developed for clinical use. It is an effective erythrocytic schizontocide against all human malarias. It will also

Pyrimethamine

act as an exoerythrocytic schizontocide in most infections due to *P. vivax and P. falciparum*. Sporontocidal action is exhibited by pyrimethamine, thus making it capable of breaking the chain of transmission of malaria. It does not exhibit important gametocytocidal action. In combination with anti-bacterial sulfonamides it has been recommended for the treatment of infection due to resistant strains of *P. falciparum*.[35, 36].

Pyrimethamine is slowly but completely absorbed from the gastrointestinal tract. It is localized in the liver, the lungs, the kidney and the spleen and is excreted through the urine, chiefly in metabolized form. It is relatively nontoxic, but overdoses may lead to depression of cell growth by inhibition of folic acid activity.

It is administered in the form of the free base, a relatively tasteless powder.

Category—antimalarial.

Usual dose—suppressive, 25 mg. weekly; therapeutic, 50 mg. daily for 2 days.

Usual dose range—25 to 75 mg.

OCCURRENCE

Pyrimethamine Tablets U.S.P.

SULFONES

It has been known for some time that 4.4'-diaminodiphenylsulfone, Dapsone U.S.P., (DDS) was active against a number of the plasmodium species causing malaria.[37] However, it was considered to be an inferior antimalarial drug until it was discovered that it served effectively as a chemoprophylactic agent against chloroquine-resistant *P. falciparum* infections in southeast Asia.[38] The U. S. Army uses it in 25 mg. daily doses, along with weekly doses of chloroquine and primaquine, to suppress falciparum malaria. These drugs, along with antimosquito regimens, are credited with keeping a severe malaria problem under control and with preventing fatalities from infections that lead to hospital treatments.

The effectiveness of DDS has prompted the development of programs seeking the synthesis of sulfone compounds of superior activity and with longer duration of action.[39, 40, 41] Among the compounds tested, N,N'-diacetyl-4,4'-diaminodiphenylsulfone (DADDS) has been found to be the most promising. Its more prolonged activity and lower toxicity as compared to DDS is probably related to its slow conversion to either the monoacetyl derivative or DDS itself, both of which act as the antimalarial agents. It is apparent that the antimalarial activity

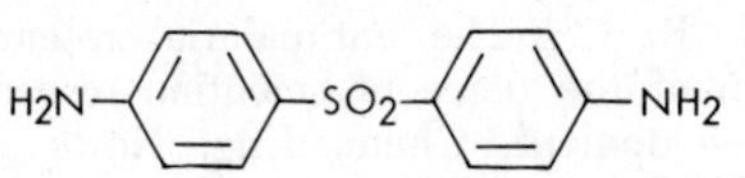

H₂N—⟨ ⟩—SO₂—⟨ ⟩—NH₂

4,4'-Diaminodiphenylsulfone
(DDS)

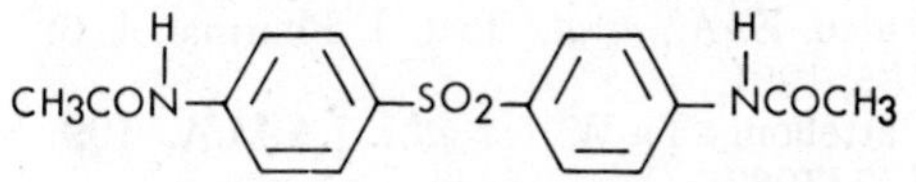

CH₃CON—⟨ ⟩—SO₂—⟨ ⟩—NCOCH₃

4,4'-diacetyl-4,4'-diaminodiphenylsulfone
(DADDS)

of the sulfones is dependent upon an ability to interfere with PABA utilization by the plasmodia. They significantly potentiate drugs known to inhibit the conversion of folic acid to folinic acid, an activity consistent with that mode of action. This sequential blockade of two consecutive steps in the biosynthesis of purine and pyrimidine nucleotides probably accounts for the effectiveness of sulfones in acting against otherwise resistant plasmodia. (For further information about Dapsone, see Chapter 11.)

REFERENCES

1. Suppan, L.: Three Centuries of Cinchona, *in* Proc. Celebration 300th Anniversary of the First Recognized Use of Cinchona, p. 29, St. Louis, 1931.
2. Elderfield, R. C.: The Antimalarial Research Program of the Office of Scientific Research and Development, Chem. Eng. News 24:2598, 1946.
3. Most, H.: Military Medicine 129:587, 1964.
4. Schellenberg, K. A., and Coatney, G. R.: Biochem. Pharmacol. 6:143, 1961.
5. Alving, A. S., *et al.*: Malaria, 8-Aminoquinolines and Haemolysis *in* Goodwin, L. G., and Nimmo-Smith, R. H. (eds.): Drugs, Parasites and Hosts, p. 96, Boston, Little, 1962.
6. Ciak, J., and Hahn, F. E.: Science 151: 347, 1966.
7. Taylor, H.: Cinchona in Java, New York, Greenberg, 1945.
8. United States Imports for Consumption and General Imports, p. 301, Bureau of the Census, Department of Commerce, Washington, D. C., 1969 Annual.
9. Rabe, P.: Ber. deutsch. chem. Ges. 55: 522, 1922.
10. Woodward, R. B., and Doering, W. E.: J. Am. Chem. Soc. 67:860, 1945.
11. Turner, R. B., and Woodward, R. B.: The Chemistry of the Cinchona Alkaloids *in* Manske, R. H. F., and Holmes, H. L.

(eds.): The Alkaloids, vol. 3, p. 24, New York, Acad. Press, 1953.
12. Findlay, G. M.: Recent Advances in Chemotherapy, ed. 3, vol. 2, p. 274, Philadelphia, Blakiston, 1951.
13. Lutz, R. E., *et al.*: J. Org. Chem. 12:617, 1947.
14. Russell, P. B.: Antimalarials, *in* Burger, A. (ed.): Medicinal Chemistry, ed. 2, p. 814, New York, Interscience, 1960.
15. Findlay, G. M.: Recent Advances in Chemotherapy, ed. 3, vol. 2, p. 121, Philadelphia, Blakiston, 1951.
16. ———: Recent Advances in Chemotherapy, ed. 3, vol. 2, p. 187, Philadelphia, Blakiston, 1951.
17. Richardson, S.: South. Med. J. 29:1156, 1936.
18. Shannon, J. A.: J. Am. Pharm. A. (Pract. Ed.) 7:163, 1946.
19. Schonhofer, F., *et al.*: Z. physiol. Chem. 274:1, 1942.
20. Burckhalter, J. F., *et al.*: J. Am. Chem. Soc. 68:1894, 1946.
21. Berliner, R. W., *et al.*: J. Clin. Invest. 27 (Suppl.):98, 1948.
22. Elderfield, R. C.: Chem. and Eng. News 24:2598, 1946.
23. Kenyon, R. L., *et al.*: Ind. Eng. Chem. 41:654, 1949.
24. Fourneau, E., *et al.*: Ann. inst. Pasteur 50:731, 1933.
25. Covell, G., *et al.*: World Health Org. Monograph Ser. No. 27, 1955.
26. Mauss, H., and Mietzsch, F.: Klin. Wschr. 12:1276, 1933.
27. Thompson, P. E., *et al.*: Am. J. Trop. Med. Hyg. 10:335, 1961.
28. Elslager, E. F., *et al.*: J. Med. Pharm. Chem. 5:1159, 1962.
29. Curd, F. H. S., and Rose, F. L.: J. Chem. Soc., p. 343, 1946.
30. Crowther, A. F., and Levi, A. A.: Brit. J. Pharmacol. 8:93, 1953.
31. Thompson, P. E., *et al.*: Am. J. Trop. Med. Hyg. 12:481, 1963.
32. Schmidt, L. H., *et al.*: Am. J. Trop. Med. Hyg. 12:494, 1963.
33. Contacos, P. G., *et al.*: Am. J. Trop. Med. Hyg. 13:386, 1964.

34. Falco, E. A., *et al.*: Brit. J. Pharmacol. 6: 185, 1961.
35. Bartelloni, T. W., *et al.*: J.A.M.A. 199: 173, 1967.
36. Martin, D. C., and Arnold, J. D.: J.A.M.A. 203:476, 1968.
37. Coggeshal, L. T., *et al.*: J.A.M.A. 117: 1077, 1941.
38. Blount, R. E.: Ann. Int. Med. 70:142, 1969.
39. Eslager, E. F., and Worth, D. F.: Nature 206:630, 1965.
40. Popoff, I. C., and Singhal, G. H.: J. Med. Chem. 11:631, 1968.
41. Serafin, B., *et al.*: J. Med. Chem. 12:336, 1969.

SELECTED READING

Bruce-Chwatt, L. J.: Changing tides of chemotherapy of malaria, Brit. Med. J. 1964, 5383, 581.

Elderfield, R. C.: The antimalarial research program of the office of scientific research and development, Chem. Eng. News 24: 2598, 1946.
Findlay, G. M.: Recent Advances in Chemotherapy, ed. 2, vol. 2, Philadelphia, Blakiston, 1951.
Hill, J.: Chemotherapy of Malaria, Part 2. The Antimalarial Drugs, *in* Schnitzer, R. G., and Hawking, F. (eds.): Experimental Chemotherapy, vol. 1, p. 513, New York, Acad. Press, 1963.
Pinder, R. M.: Antimalarials, *in* Burger, A. (ed.): Medicinal Chemistry, ed. 3, p. 492, New York, Wiley-Interscience, 1970.
Powell, R. D.: The chemotherapy of malaria. Clin. Pharmacol. Ther. 7:48, 1966.
Teschan, P. E. (ed.): Panel on malaria, Ann. Int. Med. 70:127, 1969.
Turner, R. B., and Woodward, R. B.: The chemistry of the cinchona alkaloids, *in* Manske, R. H. F., and Holmes, H. L. (eds.): The Alkaloids, vol. 3, p. 1, New York, Acad. Press, 1953.

14

Antibiotics

Allen I. White, Ph.D.,
Dean and Professor of Pharmaceutical Chemistry, College of Pharmacy,
Washington State University

INTRODUCTION

The accidental discovery of penicillin by Sir Alexander Fleming[1] in 1929 was the prime factor in starting the fascinating and fruitful research activities that have produced the amazingly effective anti-infective agents commonly known as antibiotics. However, it was not until Florey and Chain and their associates at Oxford (1940) undertook to apply antibiotics in therapy that Fleming's discovery became meaningful to practical medicine. Long before this, man had learned to use empirically as anti-infective material a number of crude substances which we now assume were effective because of antibiotic substances contained in them. As early as 500 to 600 B.C., the Chinese used a molded curd of soybean to treat boils, carbuncles and similar infections. Vuillemin[2] in 1889 used the term *antibiosis* (literally, against life) to apply to the biologic concept of survival of the fittest in which one organism destroys another to preserve itself. It is from this root that the widely used word *antibiotic* has evolved. So broad has its use become, not only by the lay public but also by the medical professions and science in general, that the term is almost impossible to define satisfactorily. There is no knowledge today that can relate either chemically or biologically all the various substances designated as antibiotics other than by their abilities to antagonize the same or similar microorganisms.

Waksman[3] proposed the widely cited definition that "an antibiotic or an antibiotic substance is a substance produced by microorganisms, which has the capacity of inhibiting the growth and even of destroying other microorganisms." However, the restriction that an antibiotic must be a product of a microorganism is not in keeping with common use. The definition of Benedict and Langlykke[4] more aptly describes the use of the term today. They state that an antibiotic is ". . . a chemical compound derived from or produced by a living organism, which is capable, in small concentrations, of inhibiting the life processes of microorganisms." In this chapter, only those substances of importance to modern medical practice and those that meet the requirements proposed by Baron[5] (points 1, 3 and 4 below) plus one additional provision (point 2) will be included. Today, when the pharmaceutical chemist is so active in synthesizing structural analogs of important naturally occurring medicinal agents, it has become necessary to add the qualification that permits the inclusion of synthetically obtained compounds not known to be products of metabolism. Therefore, a substance is classified as an antibiotic if:

1. It is a product of metabolism (although

it may be duplicated or even have been anticipated by chemical synthesis).

2. It is a synthetic product produced as a structural analog of a naturally occurring antibiotic.

3. It antagonizes the growth and/or the survival of one or more species of microorganisms.

4. It is effective in low concentrations.

The possibility that Nature held the secret to many antibiotic substances in addition to penicillin became a driving force in the search for new compounds with the discovery by Dubos in 1939 that *Bacillus brevis* produced tyrothricin. Under the direction of S. A. Waksman, who later became a Nobel Laureate for his contributions, work leading to the isolation (1944) of streptomycin from *Streptomyces griseus* was undertaken. The discovery that this antibiotic possessed in-vivo activity against *Mycobacterium tuberculosis* as well as gram-negative organisms was electrifying. Evidence was now ample that antibiotics were produced widely in nature. Broad screening programs were set up to find agents that would be effective in the treatment of infections that hitherto had been resistant to chemotherapeutic agents, as well as to provide safer and more rapid therapy for infections for which the previously available treatment had various shortcomings. The development of the broad-spectrum antibiotics such as chloramphenicol and the tetracyclines, the isolation of antifungal antibiotics such as nystatin and griseofulvin and the production of an ever-increasing number of antibiotics that may be used to treat infections that have developed resistance to some of the older antibiotics attest to the success of the many research programs on antibiotics throughout the world.

The natural scientific interest in the field of antibiotics, as well as the commercial success of antibiotics used in therapy, has led to the isolation of antibiotic substances that may now be numbered in the thousands. Of course, only a few of these have been made available for use in medical practice, because, to be useful as a drug, a substance must possess not only the ability to combat the disease process but other attributes as well. For an antibiotic to be successful in

therapy, it should be decisively effective against a pathogen without producing significant toxic side-effects. In addition it should be sufficiently stable so that it can be isolated and processed and then stored for a reasonable length of time without appreciable loss in activity. It is important that it be amenable to processing into desirable dosage forms from which it may be absorbed readily. Finally, the rate of detoxication and elimination from the body should be such as to require relatively infrequent dosage to maintain proper concentration levels, yet be sufficiently rapid and complete that the removal of the drug from the body is accomplished soon after administration has been discontinued.

Relatively few substances that have shown promise as antibiotics have been able to fulfill these requirements to the extent that their commercial production has been warranted. Although the number of antibiotic substances that have shown sufficient promise to be named may be numbered in the hundreds, few of them have been produced in large enough quantities to place them on clinical trial, and only a few more than 3 dozen antibiotics are now released for general medical practice in the United States. To the pharmacist faced with an array of dosage forms and sizes of each antibiotic, not to mention combinations, the number of antibiotics may loom large. When viewed from the standpoint of the number of microorganisms and other living organisms investigated for antibiotic activity, when considered from the standpoint of research activity and cost, and when evaluated from the standpoint of the needs yet remaining for agents that will successfully combat infectious diseases for which there are no satisfactory cures, the number of antibiotics successfuly developed to date is not large. Table 24 lists the antibiotics, together with their sources, that were generally available for clinical use in the United States at the end of 1970.

The spectacular success of antibiotics in the treatment of the diseases of man has prompted the expansion of their use into a number of related fields. Extensive use of their antimicrobial power is made in veterinary medicine. The discovery that low-level administration of antibiotics to meat-pro-

TABLE 24. COMMERCIALLY AVAILABLE ANTIBIOTICS FOR CLINICAL USE

ANTIBIOTIC	SOURCE
Amphotericin B	*Streptomyces nodosus*
Ampicillin	Semisynthetic
Bacitracin	*Bacillus subtilis*
Candicidin	*Streptomyces griseum*
Carbenicillin	Semisynthetic
Cephaloglycin	Semisynthetic
Cephalothin	Semisynthetic
Cephaloridine	Semisynthetic
Chloramphenicol	*Streptomyces venezuelae* (also synthetic)
Chlortetracycline	*Streptomyces aureofaciens*
Clindamycin	Semisynthetic
Cloxacillin	Semisynthetic
Colistin	*Aerobacillus colistinus*
Cycloserine	*Streptomyces venezuelae* (also synthetic)
Dactinomycin	*Streptomyces parvullus*
Demeclocycline	*Streptomyces aureofaciens*
Dicloxacillin	Semisynthetic
Doxycycline	Semisynthetic
Erythromycin	*Streptomyces erythreus*
Gentamicin	*Micromonospora purpurea*
Gramicidin	*Bacillus brevis*
Griseofulvin	*Penicillium griseofulvum Dierckx*
Kanamycin	*Streptomyces kanamyceticus*
Lincomycin	*Streptomyces lincolnensis*
Methacycline	Semisynthetic
Methicillin	Semisynthetic
Nafcillin	Semisynthetic
Neomycin	*Streptomyces fradiae*
Novobiocin	*Streptomyces niveus*
Nystatin	*Streptomyces noursei*
Oleandomycin	*Streptomyces antibioticus*
Oxacillin	Semisynthetic
Oxytetracycline	*Streptomyces rimosus*
Paromomycin	*Streptomyces* species
Penicillin G	*Penicillium* species
Penicillin O	Biosynthetic
Phenethicillin	Semisynthetic
Phenoxymethyl Penicillin	Biosynthetic
Polymyxin B	*Bacillus polymyxa*
Streptomycin	*Streptomyces griseum*
Tetracycline	*Streptomyces* species
Tyrothricin	*Bacillus brevis*
Vancomycin	*Streptomyces orientalis*
Viomycin	*Streptomyces griseus* var. *purpureus*

ducing animals resulted in faster growth, lower mortality rates and better quality has led to extensive use of these products as feed supplements. A number of antibiotics are being used to control bacterial and fungal diseases of plants. Their use in food preservation is being studied carefully. Indeed, such uses of antibiotics have made necessary careful studies of their chronic effects on man and their effect on various commercial processes. For example, foods having low-level amounts of antibiotics may be capable of producing allergic reactions in hypersensitive persons, or the presence of antibiotics in milk may interfere in the manufacture of cheese.

The success of antibiotics in therapy and related fields has made them one of the most important products of the drug industry today. The quantity of antibiotics produced in the United States each year may now be measured in several millions of pounds and valued at several hundreds of million of dollars. With research activity stimulated to find new substances to treat viral infections so far combated with limited success, with the promising discovery that some antibiotics are active against cancers that may be viral in origin, the future development of more antibiotics and increase in the amounts produced seems to be assured.

The commercial production of antibiotics for medicinal use follows a general pattern, differing in detail for each antibiotic. The general scheme may be divided into 6 steps: (1) preparation of a pure culture of the desired organism for use in inoculation of the fermentation medium; (2) fermentation during which the antibiotic is formed; (3) isolation of the antibiotic from the culture media; (4) purification; (5) assay for potency, tests for sterility, absence of pyrogens, other necessary data; (6) formulation into acceptable and stable dosage forms.

The ability of some antibotics such as chloramphenicol and the tetracyclines to antagonize the growth of a large number of pathogens has resulted in their being designated as "broad-spectrum" antibiotics. Others have a high degree of specificity and are classified as "narrow-spectrum" antibiotics.

The manner in which antibiotics exert their actions against susceptible organisms is varied. In many instances, the mechanism of action is not fully known; in a few cases, penicillins, for example, the site of action is known, but precise details of the mechanism are still under investigation. The biochemical processes of microorganisms are lively subjects for research, since an understanding of those mechanisms that are peculiar to the metabolic systems of infectious organisms is the basis for the future development of modern chemotherapeutic agents. Antibiotics that interfere with those metabolic systems found in microorganisms and not in mammalian cells are the most

successful anti-infective agents (e.g., penicillins). The fact that some antibiotics structurally resemble some essential metabolites of microorganisms has suggested that competitive antagonism may be the mechanism by which they exert their effects. Other mechanisms, such as chelation with essential trace elements, have been suggested for some antibiotics. Much work remains to be done in this area, and, as mechanisms of action are revealed, the probability of developing successful structural analogs of effective antibiotics will increase.

The chemistry of antibiotics is so varied that a chemical classification is of little value. However, it is worthy of note that some similarities can be found, indicating, perhaps, that some antibiotics are the products of similar mechanisms in different organisms and that these structurally similar products may exert their activities in a similar manner. For example, a number of important antibiotics have in common a macrolide structure, that is, a large lactone ring. In this group are erythromycin, carbomycin and oleandomycin. The tetracycline family presents a groups of compounds very closely related chemically. A number of compounds contain closely related sugar moieties such as are found in streptomycins, kanamycins, neomycins and paromomycins. The antifungal antibiotics nystatin, the amphotericins and fumagillin are examples of a group of conjugated polyene compounds. The bacitracins, tyrothricin and polymyxin are among a large group of polypeptides that exhibit antibiotic action

The normal biologic processes of microbial pathogens are varied and complex. Thus, it seems reasonable to assume that there are many ways in which they may be inhibited and that different microgranisms that elaborate antibiotics antagonistic to a common "foe" produce compounds that are chemically dissimilar and that act on different processes. In fact, Nature has produced many chemically different antibiotics that are capable of attacking the same microorganism by different pathways. The diversity of structure in antibiotics has proved to be of real value clinically. As the pathogenic cell is called on to combat the effect of one antibiotic and, thus, develops drug resistance, another antibiotic, attacking another metabolic process of the resisting cell, will deal it a crippling blow. The development of new and different antibiotics has been a very important step in providing the means for treating resistant strains of organisms which previously had been susceptible to an older antibiotic. The evolution of so-called "hospital strains" of staphylococci resistant to penicillin and the tetracyclines has become a serious medical problem. Only the development of newer antibiotics that attack the microorganism through different pathways has made the successful treatment of these resistant diseases possible.

THE PENICILLINS

Until 1944, it was assumed that the active principle in penicillin was a single substance and that variation in activity of different products was due to the amount of inert materials in the samples. Now it is known that, during the biologic elaboration of the antibiotic, a number of closely related compounds may be produced. These compounds differ chemically in the acid moiety of the amide side chain. Variations in this moiety produce differences in antibiotic effect and in chemical-physical properties including stability. Thus, it has become proper to speak of penicillins, referring to a group of compounds, and to identify each of the penicillins specifically. As each of the different penicillins was first isolated, letter designations were used in America; the British used Roman numerals.

Over 30 penicillins have been isolated from fermentation mixtures. Some of these occur naturally; others have been biosynthesized by altering the culture media so as to provide certain precursors that may be incorporated as acyl groups Commercial production of penicillins today depends chiefly on various strains of *P. notatum* and *P. chrysogenum*. Recently, many more penicillins have been synthesized, and undoubtedly many more will be added to the list in attempts to find superior products. Table 25 shows the general structure of the penicillins and relates the structures of the more

familiar ones to their various designations. It may be noted that the numbering system shown follows that used by the U.S.P. which assigns to the nitrogen atom the number one position and to the sulfur atom the number four position. The more conventional system is the reverse of that procedure, assigning the number one position to the sulfur atom and the number four position to the nitrogen atom.

The early commercial penicillin was a yellow-to-brown amorphous powder which was so unstable that refrigeration was required to maintain a reasonable level of activity for a short period of time. Improved procedures for purification provide the white crystalline material in use today. The crystalline penicillin must be protected from moisture, but, when kept dry, the salts will remain stable for years without refrigeration.

Because penicillin, when it was first used in chemotherapy, was not a pure compound and exhibited varying activity among samples it was necessary to evaluate it by microbiologic assay. The procedure for assay was developed at Oxford, England, and the value became known as the Oxford Unit. One Oxford Unit is defined as the smallest amount of penicillin that will inhibit, in vitro, the growth of a strain of Staphylococcus in 50 ml. of culture media under specified conditions. Now that pure crystalline penicillin is available, the U.S.P. defines "Unit" as the antibiotic activity of 0.6 microgram of U.S.P. Sodium Penicillin G Reference Standard. The weight-unit relationship of the penicillins will vary with the nature of the acyl substituent and with the salt formed of the free acid. One milligram of sodium penicillin G is equivalent to 1,667 units. One milligram of procaine penicillin G is equivalent to 1,009 units. One milligram of potassium phenoxymethylpenicillin is equivalent to 1,530 units.

The commercial production of pencillin has increased markedly since its introduction. As production increased, the cost of penicillin dropped correspondingly. When penicillin was first available, 100,000 units of it sold for $20. Fluctuations in the production of penicillins reflect the popularity of broad-spectrum antibiotics as compared with penicillins, the development of penicillin-resistant strains of a number of pathogens, the recent introduction of semisynthetic penicillins, the use of penicillins in animal feeds and for veterinary purposes, and the increase in marketing problems in a highly competitive sales area.

Examination of the structure of the penicillin molecule shows it to contain a fused ring system of unusual design, the β-lactam thiazolidine structure. The 5-membered thiazolidine ring appears in other natural compounds, but the 4-membered β-lactam ring is unique. The nature of this ring delayed the elucidation of the structure of penicillin, but its determination was reached as a result of a collaborative research program involving research groups in Great Britain and the United States during the years 1943-1945.[6] Attempts to synthesize these compounds resulted at best only in trace amounts until Sheehan and Henery-Logan[7] adapted technics developed in peptide syntheses to the synthesis of penicillin V. This procedure is not likely to replace the established fermentation processes, because the last step in the

TABLE 25. STRUCTURE OF PENICILLINS

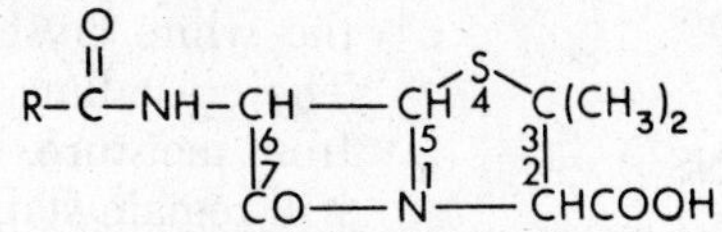

GENERIC NAME	CHEMICAL NAME	R GROUP
Penicillin G	Benzylpenicillin	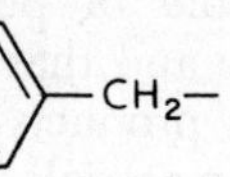
Phenoxymethyl penicillin		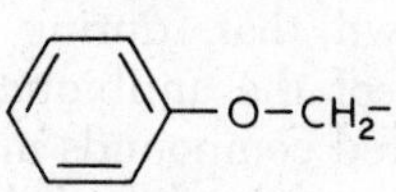
Phenethicillin	Phenoxyethylpenicillin	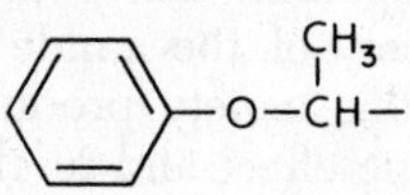
Propicillin	(−)-Phenoxypropylpenicillin	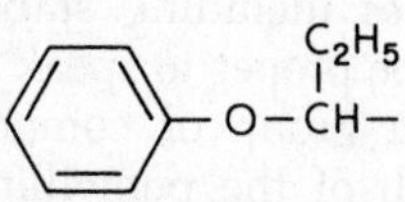
Methicillin	2,6-Dimethoxyphenyl-penicillin	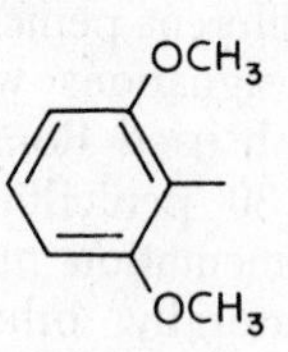
Nafcillin	2-Ethoxy-1-naphthyl-penicillin	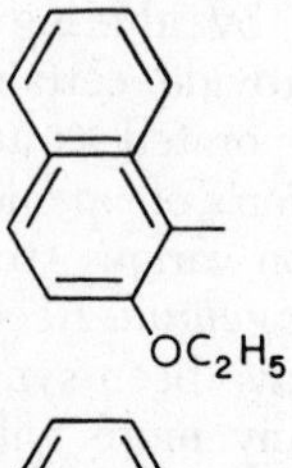
Oxacillin	5-Methyl-3-phenyl-4-isoxazolylpenicillin	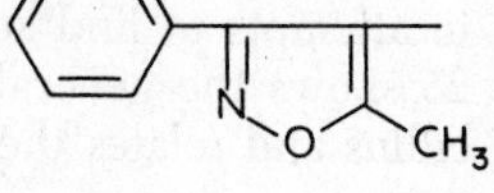

TABLE 25. STRUCTURE OF PENICILLINS—(*Continued*)

GENERIC NAME	CHEMICAL NAME	R GROUP
Cloxacillin	5-Methyl-3-(2-chlorophenyl)-4-isoxazolylpenicillin	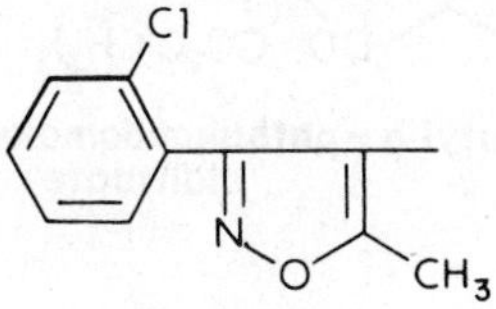
Dicloxacillin	5-Methyl-3-(2,6-dichlorophenyl)-4-isoxazolylpenicillin	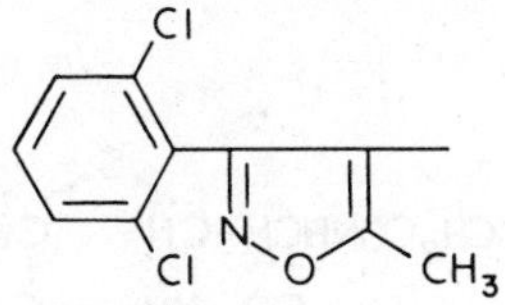
Ampicillin	D-α-Aminobenzylpenicillin	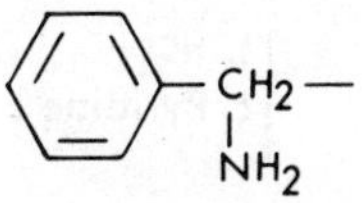
Penicillin O	Allylmercaptomethyl-penicillin	$CH_2=CH-CH_2-S-CH_2-$
Penicillin K	n-Heptylpenicillin	$CH_3(CH_2)_5CH_2-$
Penicillin X	p-Hydroxybenzyl-penicillin	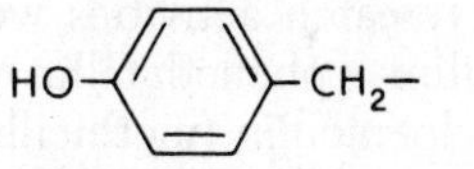
Penicillin F	2-Pentenylpenicillin	$CH_3CH_2CH=CHCH_2-$
Dihydro-penicillin F	n-Amylpenicillin	$CH_3(CH_2)_3CH_2-$

reaction series develops only 10 to 12 per cent of penicillin. It is of advantage in research because it provides a means of obtaining many new amide chains hitherto not possible to achieve by biosynthetic procedures.

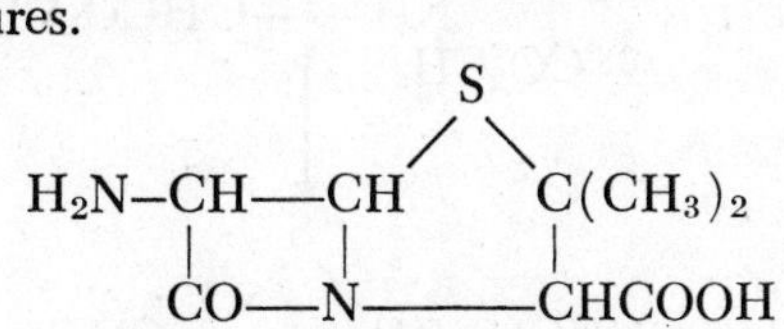

6-Aminopenicillanic Acid

Two other developments have provided additional means for making new penicillins. A group of British scientists, Batch-elor *et al.*,[8] have reported the isolation of 6-aminopenicillanic acid from a culture of *P. chrysogenum*. This compound can be converted to penicillins by acylation of the 6-amino group. Sheehan and Ferris[9] provided another route to synthetic penicillins by converting a natural penicillin such as potassium penicillin G to an intermediate from which the acyl side chain has been removed, which then can be treated to form biologically active penicillins with a variety of new side chains. By these procedures, new penicillins superior in activity and stability to those formerly in wide use have been found, and no doubt, others will be

PHENOXYMETHYLPENICILLIN SYNTHESIS

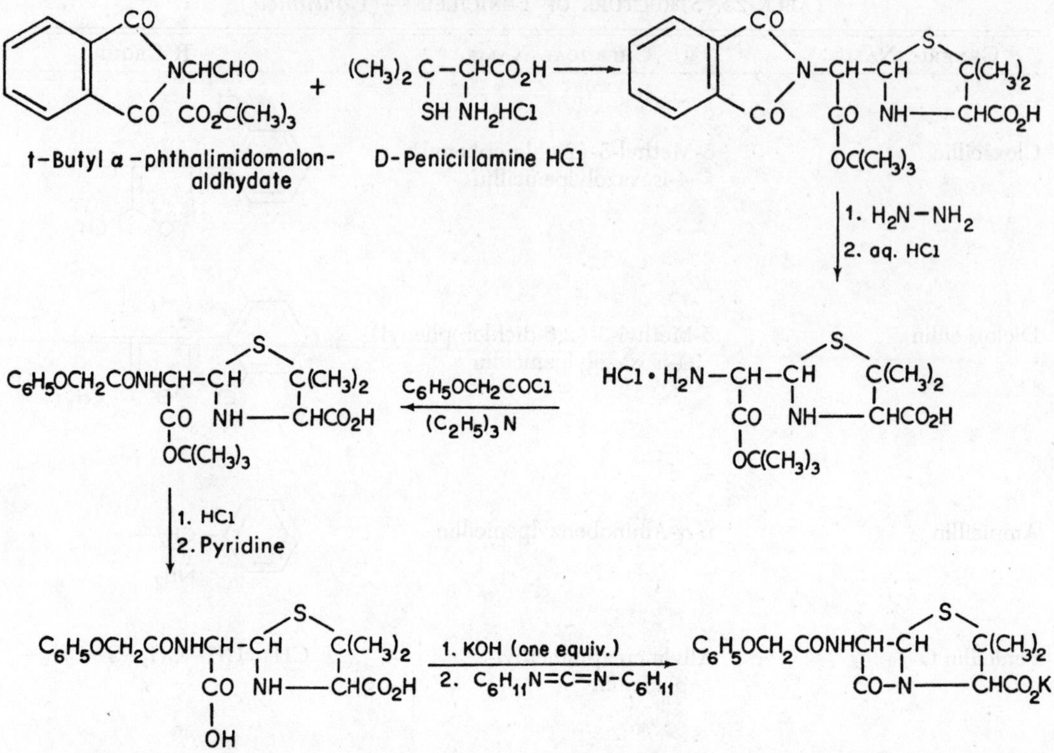

$$\text{t-Butyl } \alpha\text{-phthalimidomalon-aldehydate} \quad + \quad (CH_3)_2C-CHCO_2H \ (\text{with } SH \text{ and } NH_2HCl) \quad \longrightarrow$$

t-Butyl α-phthalimidomalon-aldehydate   D-Penicillamine HCl

1. H_2N-NH_2
2. aq. HCl

$$C_6H_5OCH_2CONHCH-CH \quad C(CH_3)_2 \ (CO, \ OC(CH_3)_3, \ NH-CHCO_2H) \quad \xleftarrow[\ (C_2H_5)_3N\ ]{C_6H_5OCH_2COCl} \quad HCl\cdot H_2N-CH-CH \quad C(CH_3)_2 \ (CO, \ OC(CH_3)_3, \ NH-CHCO_2H)$$

1. HCl
2. Pyridine

$$C_6H_5OCH_2CONHCH-CH \quad C(CH_3)_2 \ (CO, \ OH, \ NH-CHCO_2H) \quad \xrightarrow[\ 2.\ C_6H_{11}N=C=N-C_6H_{11}\ ]{1.\ KOH\ (one\ equiv.)} \quad C_6H_5OCH_2CONHCH-CH \quad C(CH_3)_2 \ (CO-N-CHCO_2K)$$

produced. The first commercial products of these research activities were phenoxyethyl-penicillin (phenethicillin) and dimethoxy-phenylpenicillin (methicillin) (pp. 354, 355).

The purified penicillins are white or slightly yellowish-white crystalline powders without odor. All of the natural penicillins, are strongly dextrorotatory. The solubility and other physiochemical properties of the penicillins are affected by the nature of the acyl side chain and by the cations used to

make salts of the acid. The main cause of deterioration in penicillins is hydrolysis. Some of the crystalline salts of the penicillins are hygroscopic, making it necessary to store them in sealed containers. The course of the hydrolysis (see p. 351) is affected by the pH of the solution. Nucleophilic attack, particularly by hydroxide ion, produces penicilloic acid which loses CO_2 to form penilloic acid. Electrophilic attack, particularly by hydrogen ion, involves the amide side

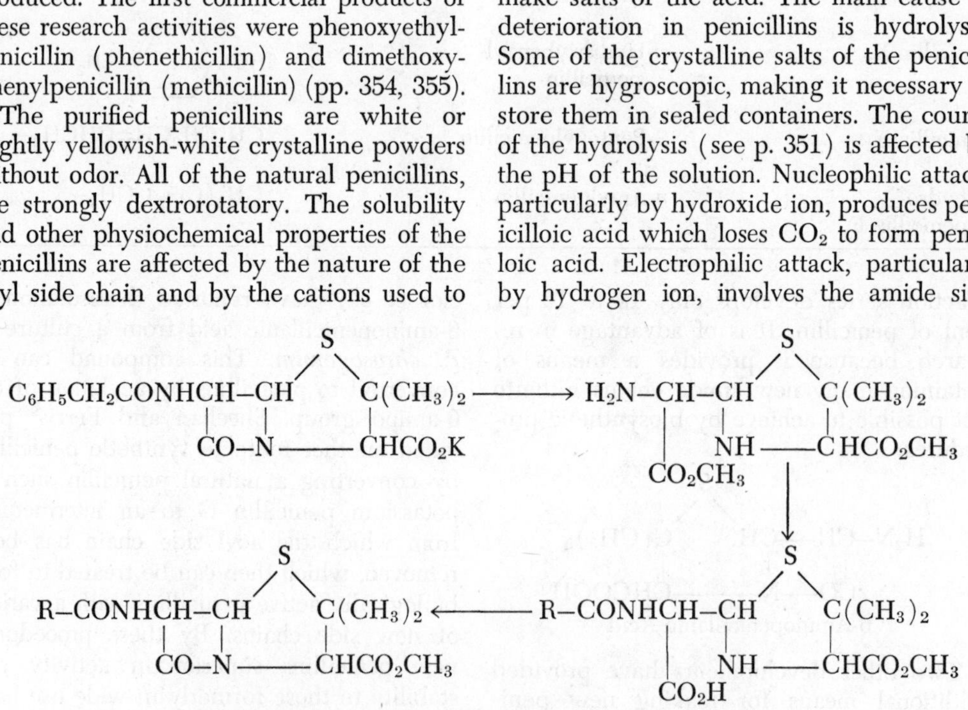

$$C_6H_5CH_2CONHCH-CH \quad C(CH_3)_2 \ (CO-N-CHCO_2K) \quad \longrightarrow \quad H_2N-CH-CH \quad C(CH_3)_2 \ (NH-CHCO_2CH_3, \ CO_2CH_3)$$

$$R-CONHCH-CH \quad C(CH_3)_2 \ (CO-N-CHCO_2CH_3) \quad \longleftarrow \quad R-CONHCH-CH \quad C(CH_3)_2 \ (NH-CHCO_2CH_3, \ CO_2H)$$

Conversion of Natural Penicillin to Synthetic Penicillin

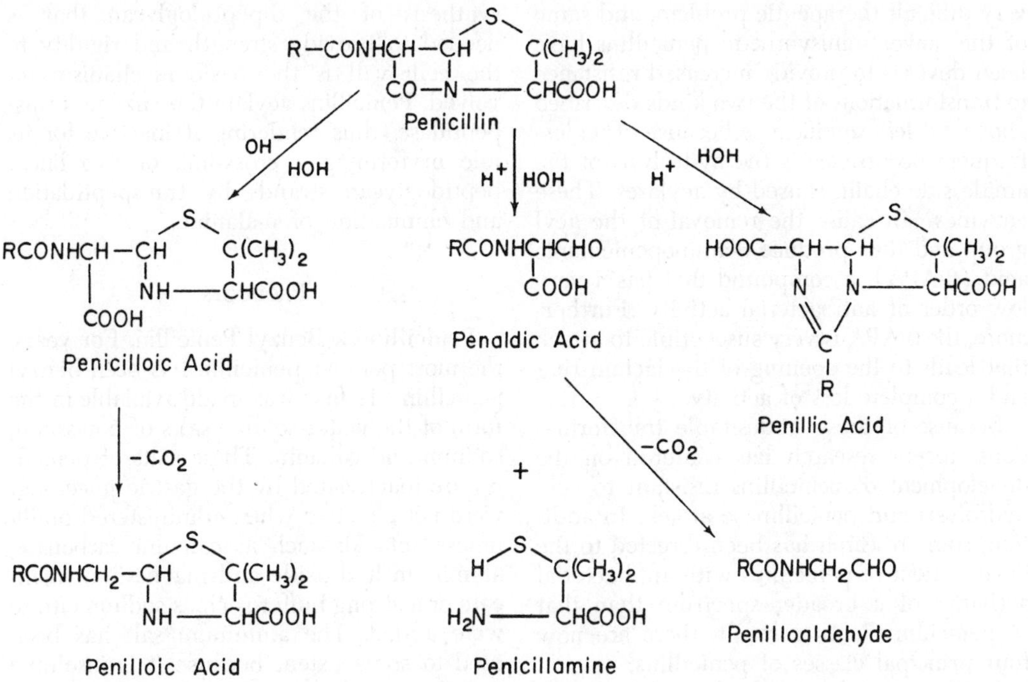

Reaction scheme showing penicillin degradation products: Penicillin, Penicilloic Acid, Penaldic Acid, Penillic Acid, Penilloic Acid, Penicillamine, Penilloaldehyde.

chain but the precise mechanism is in doubt. The introduction of an electron-attracting group, particularly in the alpha position, into the amide side chain, inhibits the electron displacement involving the carbonyl group and the β-lactam ring, thus making such penicillins as phenoxymethl penicillin more acid stable. By controlling the pH of aqueous solutions within a range of 6.0 to 6.8, and by refrigeration of the solutions, aqueous preparations of the soluble penicillins may be stored for periods up to several weeks. The relationship of these properties to the pharmaceutics of penicillins has been reviewed by Schwartz and Buckwalter.[10] It has been noted that some buffer systems, particularly phosphates and citrates, exert a favorable effect on penicillin stability independent of the pH effect. However, Finholt et al.[11] have shown that these buffers may catalyze penicillin degradation if the pH is adjusted to obtain the requisite ions. Hydroalcoholic solutions of potassium penicillin G show about the same degree of instability as do aqueous solutions.[11a] Since penicillins are inactivated by metal ions such as zinc and copper, it has been suggested that the phosphates and the citrates combine with these

metals so as to prevent their existing as ions in solution.

Oxidizing agents also inactivate pencillins, but reducing agents have little effect on them. Temperature affects the rate of deterioration; although the dry salts are stable at room temperature and do not require refrigeration, prolonged heating will inactivate the penicillins.

From a clinical point of view, the most significant transformations of penicillins are caused by gastric acid and by the enzymes generally called penicillinases. The strong acid in the stomach leads to the hydrolysis of the amide side chain and an opening of the lactam ring, with a resulting loss of activity. The term penicillinase is applied to at least two kinds of enzymes, beta-lactamase and acylases, that inactivate penicillins. Beta-lactamase produces an opening of the lactam ring and thus renders the penicillin inactive. The enzyme exists as a natural antagonist to penicillins in many microorganisms and, when it is present in significant amounts, pencillin resistance is produced. During an infection, increase in the numbers of a penicillin-resistant strain (probably a mutant) of a microorganism can lead to a very difficult therapeutic problem, and some of the newer semisynthetic penicillins have been devised to provide increased resistance to transformations of the two kinds described above. Of less significance because of its less frequent occurrence is the hydrolysis of the amide side chain, caused by acylases. These enzymes can cause the removal of the acyl group and thus produce 6-aminopenicillanic acid (6-APA), a compound that has a very low order of antibacterial activity. Furthermore, the 6-APA is very susceptible to attack that leads to the opening of the lactam ring and a complete loss of activity.

Because of these undesirable transformations, recent research has centered on the development of penicillins resistant to acid hydrolysis and penicillinase attack. In addition, some research has been directed to the development of penicillins with antibacterial activities of a broader spectrum than that of penicillin G. As a result, there are now four principal classes of penicillins:

1. The natural penicillins, such as penicillin G, in which the acyl portion of the amide side chain consists of a benzyl group or an alkyl group.

2. The acid-resistant penicillins, such as phenoxymethyl penicillin and phenethicillin, in which a phenoxy group is attached to the alpha carbon of an alkyl group making up the acyl moiety of the amide side chain.

3. The penicillinase-resistant penicillins, such as methicillin, nafcillin, oxacillin, cloxacillin and dicloxacillin, in which a ring structure having aromatic properties is attached directly to the carbonyl carbon of the amide side chain. The aromatic ring is substituted at one or both of the ortho positions with groups that appear to act by sterically blocking the attack on the lactam ring by beta-lactamase.[12]

4. The broad-spectrum penicillins, such as ampicillin, in which various changes of unspecified nature in the acyl portion of the amide side chain produce penicillins capable of inhibiting microorganisms resistant to penicillin G.

As a result of a considerable amount of research, it is now generally concluded that penicillins act on microorganisms by interfering with the development of the cell wall.[13] Specifically, inhibition of the biosynthesis of the dipeptidoglycan that is needed to provide strength and rigidity to the cell wall is the basic mechanism involved. Penicillins acylate the enzyme transpeptidase, thus rendering it inactive for its role in forming a cross-link of two linear peptidoglycan strands by transpeptidation and elimination of D-alanine.

Products

Penicillin G, Benzyl Penicillin. For years, the most popular penicillin has been benzyl penicillin. It first was made available in the form of the water-soluble salts of potassium, sodium and calcium. These salts of penicillin are inactiviated by the gastric juice, and were not effective when administered orally unless antacids such as calcium carbonate, aluminum hydroxide and magnesium trisilicate or a strong buffer such as sodium citrate were added. The aluminum salt has been used to some extent because it is insoluble in solutions at the pH of the gastric juice and, thus, is less susceptible to inactivation. Also, because penicillin is poorly absorbed from the intestinal tract oral doses must be very large—about 5 times the amount necessary with parenteral administration. Only after the production of penicillin had increased sufficiently so that low-priced penicillin was available did the oral dosage forms become popular. The water-soluble potassium and sodium salts are used to achieve rapid high blood-level concentrations of penicillin G.

The rapid elimination of penicillin from the bloodstream through the kidneys and the need for maintaining an effective blood-level concentration have led to the development of "repository" forms of this drug. Suspensions of penicillin in peanut oil or sesame oil with white beeswax added were first employed for prolonging the duration of injected forms of penicillin. This dosage form was replaced by a suspension in vegetable oil to which aluminum monostearate or aluminum distearate was added. Today, most repository forms are suspensions of high-molecular-weight amine salts of penicillin in a similar base.

Category—antibacterial.

Usual dose—Potassium Penicillin G: oral

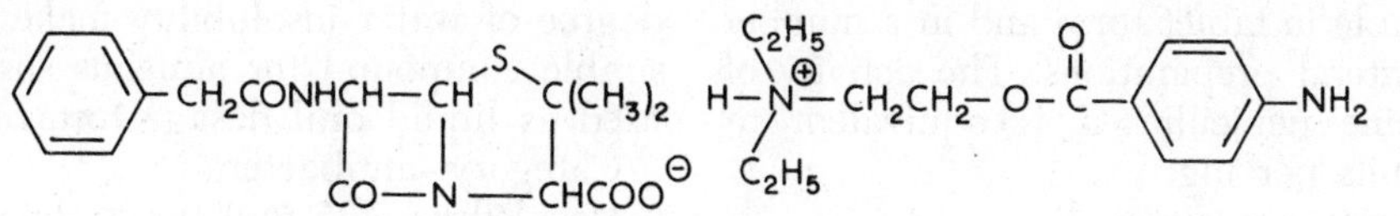

Procaine Penicillin G

or intramuscular, 500,000 U.S.P. units 4 times daily; intravenous, 5,000,000 units one or two times a day.

Usual dose range—oral or intramuscular, 300,000 to 2,000,000 units daily; intravenous, 2,000,000 to 40,000,000 units daily.

OCCURRENCE
Potassium Penicillin G U.S.P.
Potassium Penicillin G for
Injection U.S.P.
Sterile Potassium Penicillin G U.S.P.
Potassium Penicillin G Tablets U.S.P.
Sodium Penicillin G N.F.
Sodium Penicillin G for Injection N.F.

Procaine Penicillin G U.S.P. Abbocillin®, Crysticillin®, Duracillin®, Wycillin®. The first widely used amine salt of penicillin G was made from procaine. It can be made readily from sodium penicillin G by treatment with procaine hydrochloride. This salt is considerably less soluble in water than are the alkaline metal salts, requiring about 250 ml. to dissolve 1 g. The free penicillin is released only as the compound dissolves and then dissociates. It has an activity of 1,009 units per mg. A large number of preparations for injection of procaine penicillin G are commercially available. Most of these are either suspensions in water to which a suitable dispersing or suspending agent, a buffer and a preservative have been added, or suspensions in peanut oil or sesame oil that have been gelled by the addition of 2 percent aluminum monostearate. Some of the commercial products are mixtures of potassium or sodium penicillin G with procaine penicillin G, to provide a rapid development of a high blood-level concentration of penicillin through the use of the water-soluble salt plus the prolonged duration of effect obtained from the insoluble salt. In addition to the injectable forms, procaine penicillin G is available in oral dosage forms. It is claimed that the rate of absorption of this salt is as rapid as that of other forms usually administered orally.

Category—antibacterial.

Usual dose—intramuscular, 600,000 units one or two times a day.

Usual dose range—300,000 to 1,200,000 units daily.

OCCURRENCE
Sterile Procaine Penicillin G Suspension
U.S.P.
Sterile Procaine Penicillin G with
Aluminum Stearate Suspension U.S.P.
Procaine Penicillin G and Sodium
Penicillin G for Injection N.F.

Benzathine Penicillin G U.S.P., Bicillin®, Permapen®, is N,N'-dibenzylethylenediamine dipenicillin G. Since it is the salt of a diamine, two moles of penicillin are available from each molecule of the salt. It is very insoluble in water, requiring about 5,000 ml. to dissolve 1 g. This property gives the compound great stability and prolonged duration of effect. At the pH of gastric juice it is quite stable, and food intake does not interfere with its absorption. It

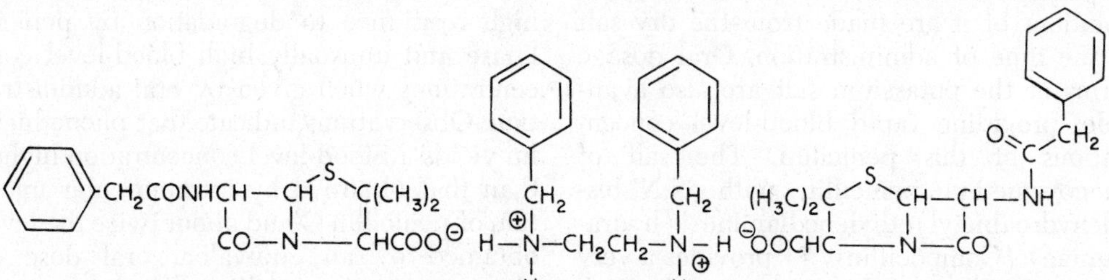

Benzathine Penicillin G

is available in tablet form and in a number of parenteral preparations. The activity of benzathine penicillin G is equivalent to 1,211 units per mg.

Category—antibacterial.

Usual dose—intramuscular, 1,200,000 units once a month.

Usual dose range—600,000 to 3,000,000 units 3 times a week to once a month.

OCCURRENCE

Sterile Benzathine Penicillin G Suspension U.S.P.

A number of other amines have been used to make penicillin salts, and research is continuing to investigate this subject. Other amines that have been used include 2-chloroprocaine, L-N-methyl-1,2-diphenyl-2-hydroxyethylamine (L-ephenamine), dibenzylamine, tripelennamine (Pyribenzamine), and N,N'-bis-(dehydroabietyl)ethylenediamine (hydrabamine).

Phenoxymethyl Penicillin U.S.P., Penicillin-V, Pen-Vee®, V-Cillin®. Phenoxymethyl penicillin was reported by Behrens et al.[14] in 1948 as a biosynthetic product. However, it was not until 1953 that its clinical value was recognized by some Eu-

Phenoxymethyl Penicillin

ropean scientists. Since then it has enjoyed wide use because of its resistance to hydrolysis by gastric juice and its ability to produce uniform concentrations in blood. The free acid requires about 1,200 ml. of water to dissolve 1 g., and it has an activity of 1,695 units per mg. For parenteral solutions, the potassium salt usually is employed. This salt is very soluble in water. Solutions of it are made from the dry salt at the time of administration. Oral dosage forms of the potassium salt are also available, providing rapid blood-level concentrations of this penicillin. The salt of phenoxymethyl penicillin with N,N'-bis-(dehydroabietyl)ethylenediamine (hydrabamine) (Compocillin-V®) provides a very long-acting form of this compound. Its high

degree of water insolubility makes it a desirable compound for aqueous suspensions used as liquid oral dosage forms.

Category—antibacterial.

Usual dose—125 to 250 mg. three or four times a day.

Usual dose range—500 mg. to 2 g. daily.

OCCURRENCE

Phenoxymethyl Penicillin for Oral Suspension U.S.P.

Phenoxymethyl Penicillin Tablets U.S.P.

Phenoxymethyl Penicillin Capsules U.S.P.

Potassium Phenoxymethyl Penicillin U.S.P.

Potassium Phenoxymethyl Penicillin Tablets U.S.P.

Potassium Phenethicillin N.F., Chemipen®, Darcil®, Maxipen®, Ro-Cillin®, Syncillin®, Potassium (1-phenoxyethyl)penicillin. Late in 1959, the first of the penicillins to be produced as a result of synthetic procedures was placed on the market. It is a close structural analog of phenoxymethyl penicillin and has similar properties.

It is interesting to note that the methy-

Potassium Phenethicillin

lene carbon between the carbonyl group and the ether oxygen of the acyl moiety in phenethicillin is asymmetric. The optical isomers have been isolated, and tests have shown (−)-α-phenoxyethyl penicillin is somewhat more active than the (+)-form. However, the racemic mixture is at least as active as the (−)-isomer and is the material available for medicinal use.

The advantages claimed for this product, which differs from phenoxymethyl penicillin only by a methyl group on the acyl moiety, include high stability in acidic solutions, high resistance to degradation by penicillinase and unusually high blood-level concentrations when given by oral administration. Observations indicate that phenethicillin yields a blood-level concentration higher than that obtained by intramuscular injection of penicillin G and about twice the level obtained by an equivalent oral dose of phenoxymethyl penicillin.

Like phenoxymethyl penicillin, phenethicillin is used as the potassium salt. It is recommended for its effect against Streptococci, *Diplococcus pneumoniae*, Neisseria and *Staphylococcus aureus*. Most gram-negative organisms, the Rickettsiae, syphilis and infections resulting in endocarditis or meningitis are resistant to phenethicillin. Of interest is the report that some strains of staphylococci that are resistant to other penicillins have been inhibited by this penicillin in vitro. It appears to have the ability to produce some of the allergic reactions that develop in the use of other penicillins. One mg. is approximately equivalent to 1,600 U.S.P. units.

Category—antibacterial.

Usual dose—125 mg. or 250 mg. 3 times a day.

Usual dose range—125 to 500 mg.

OCCURRENCE

Potassium Phenethicillin for Oral Solution N.F.

Potassium Phenethicillin Tablets N.F.

Sodium Methicillin U.S.P., Staphcillin®, 2,6-Dimethoxyphenyl penicillin Sodium. During 1960, the second penicillin produced as a result of the research that developed

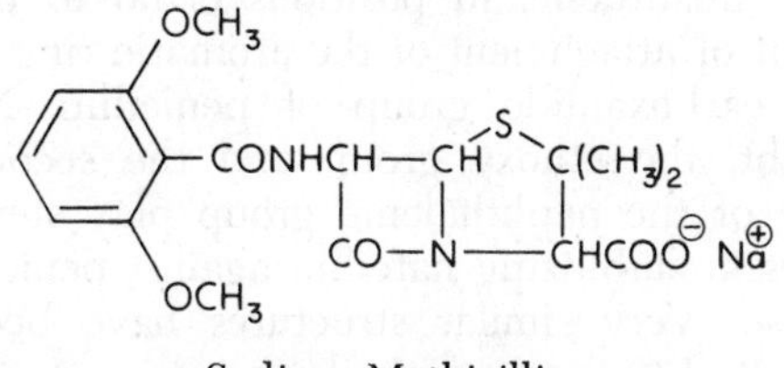

Sodium Methicillin

synthetic analogs was introduced for medicinal use. By reacting 2,6-dimethoxybenzoyl chloride with 6-aminopenicillanic acid, 6-(2,6-dimethoxybenzamido) penicillanic acid forms. The sodium salt is a white crys-

talline solid that is extremely soluble in water, forming clear neutral solutions. Like other penicillins, it is very sensitive to moisture, losing about half of its activity in 5 days at room temperature. Refrigeration at 5° reduces the loss in activity to about 20 percent in the same period. Solutions prepared for parenteral use may be kept as long as 24 hours if refrigerated. It is extremely sensitive to acid, a pH of 2 causing a 50 percent loss in activity in 20 minutes.

Sodium methicillin is particularly resistant to inactivation by penicillinase found in staphylococcal organisms and somewhat more resistant than penicillin G to penicillinase from *B. cereus*. It may be assumed that the absence of the benzyl methylene group of penicillin G and the steric protection afforded by the 2- and 6-methoxy groups makes this compound particularly resistant to enzyme hydrolysis.[12]

Sodium methicillin has been introduced for use in the treatment of staphylococcal infections due to strains found resistant to other penicillins. It is recommended that it not be used in general therapy to avoid the possible widespread development of organisms resistant to it.

Category—antibacterial.

Usual dose—intramuscular or intravenous, 1 to 2 g. 4 to 6 times a day.

Usual dose range—intramuscular, 4 to 8 g. daily; intravenous, 4 to 16 g. daily.

OCCURRENCE

Sodium Methicillin for Injection U.S.P.

Sodium Oxacillin U.S.P., Prostaphlin®, (5-Methyl-3-phenyl-4-isoxazolyl) penicillin Sodium Monohydrate. Sodium oxacillin is the salt of a semisynthetic penicillin that is highly resistant to inactivation by penicillinase. Apparently, the steric effects of the 3-phenyl and 5-methyl groups of the isoxa-

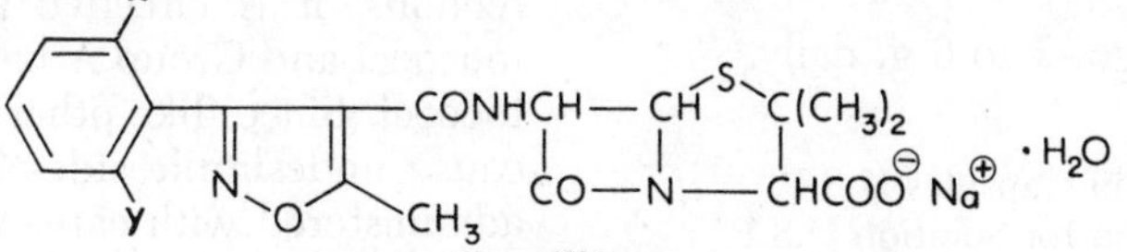

Oxacillins

X, Y = H: Sodium Oxacillin
X = Cl; Y = H: Sodium Cloxacillin
X, Y = Cl: Sodium Dicloxacillin

zolyl ring prevent the binding of this penicillin to the beta-lactamase active site and thus protect the lactam ring from degradation in much the same way as has been suggested for methicillin.[12] It is also relatively resistant to acid hydrolysis and, therefore, may be administered orally with good effect.

Sodium oxacillin, which is available in capsule form, is well absorbed from the gastrointestinal tract, particularly in patients in the fasting state. Effective blood levels are obtained in about 1 hour, but its rapid excretion through the kidneys requires repeated doses about every 4 hours.

The use of oxacillin should be restricted to the treatment of infections caused by staphylococci resistant to penicillin G. Since it may cause undesirable reactions similar to those produced by other pencillins, its use should be carefully observed, particularly in patients known to be penicillin sensitive.

Category—antibacterial.

Usual dose—oral, intramuscular, or intravenous, 500 mg. 4 to 6 times a day.

Usual dose range—1 to 12 g. daily.

OCCURRENCE
Sodium Oxacillin Capsules U.S.P.
Sodium Oxacillin for Injection U.S.P.
Sodium Oxacillin for Solution U.S.P.

Sodium Cloxacillin U.S.P., Tegopen®, [3-(o-Chlorophenyl)-5-methyl-4-isoxazolyl]-penicillin Sodium Monohydrate. The chlorine atom ortho to the position of attachment of the phenyl ring to the isoxazole ring enhances the activity of this compound over that of oxacillin, not by an increase in intrinsic activity or absorption, but by achieving higher blood plasma levels. In almost all other respects it resembles oxacillin.

Category—antibacterial.

Usual dose—500 mg. to 1 g. 4 times a day.

Usual dose range—1 to 6 g. daily.

OCCURRENCE
Sodium Cloxacillin Capsules U.S.P.
Sodium Cloxacillin for Solution U.S.P.

Sodium Dicloxacillin, Dynapen®, Pathocil®, Veracillin®, [3-(2,6-Dichlorophenyl)-5-methyl-4-isoxazolyl] penicillin Sodium Monohydrate. The substitution of chlorine atoms on both carbons ortho to the position of attachment of the phenyl ring to the isoxazole ring is presumed to further enhance the stability of this oxacillin congener and to produce high plasma concentrations of dicloxacillin. Its medicinal properties and use are like those of sodium cloxacillin.

Category—antibacterial.

Usual dose—125 mg. 4 times a day for mild infections; 250 mg. 4 times a day for severe infections.

Sodium Nafcillin U.S.P., Unipen®, 6-(2-Ethoxy-1-naphthyl)penicillin Sodium. Sodium nafcillin is another semisynthetic penicillin produced as a result of the search for penicillinase-resistant compounds. Like oxacillin, it is resistant to acid hydrolysis also. Like methicillin and oxacillin, nafcillin

Sodium Nafcillin

has substituents in positions ortho to the point of attachment of the aromatic ring to the carboxamide group of penicillin. No doubt, the ethoxy group and the second ring of the naphthaleme group play steric roles in stabilizing nafcillin against penicillinase. Very similar structures have been reported to produce similar results in some substituted 2-biphenylpenicillins.[15]

Sodium nafcillin may be used in infections caused solely by penicillin-G—resistant staphylococci or when streptococci are present also. Although it is recommended that it be used exclusively for such resistant infections, it is effective also against pneumococci and Group A beta-hemolytic streptococci. Since, like other penicillins, it may cause undesirable side-effects, it should be administered with care. When given orally, it is absorbed somewhat slowly from the intestine, but satisfactory blood levels are obtained in about 1 hour. Relatively small amounts are excreted through the kidneys, with the major portion excreted in the bile.

Even though some cyclic reabsorption from the gut may thus occur, nafcillin should be readministered every 4 to 6 hours when given orally. This salt is readily soluble in water and may be administered intramuscularly or intravenously to obtain high blood level concentrations quickly for the treatment of serious infections.

Category—antibacterial.

Usual dose—oral, intravenous, and intramuscular, the equivalent of 500 mg. of nafcillin 4 to 6 times a day.

Usual dose range—1 to 6 g. daily.

OCCURRENCE

Sodium Nafcillin Capsules U.S.P.
Sterile Sodium Nafcillin U.S.P.

Ampicillin U.S.P., Penbritin®, Polycillin®, Omnipen®, Alpen®, Amcil®, Principen®, 6-[D - α - Aminophenylacetamido]penicillanic Acid, D-α-Aminobenzylpenicillin.

With ampicillin another goal in the research on semisynthetic penicillins—an antibacterial spectrum broader than that of penicillin G—has been attained. This product is active against the same gram-positive organisms that are susceptible to other penicillins, and it is more active against some gram-negative bacteria and enterococcal in-

Ampicillin

fections. Obviously, the α-amino group plays a significant role in the broader activity, but the mechanism for its action is not known. It has been suggested that the amino group confers an ability to cross cell wall barriers that are impenetrable to other penicillins.

Ampicillin is not resistant to penicillinase, and it produces the allergic reactions and other untoward effects that are found in penicillin-sensitive patients. However, because such reactions are relatively few, it may be used in some infections for which a broad-spectrum antibiotic such as a tetracycline may be indicated but is not preferred because of undesirable reactions. However, ampicillin is not so widely active

that it should be used as a broad-spectrum antibiotic in the same manner as the tetracyclines.

Ampicillin is water-soluble and stable to acid. It is administered orally and is absorbed from the intestinal tract to produce peak blood level concentrations in about 2 hours. Oral doses must be repeated about every 6 hours, because it is rapidly excreted unchanged through the kidneys. It is available as a white crystalline anhydrous powder that is sparingly soluble in water or as the colorless or slightly buff-colored crystalline trihydrate that is soluble in water. Either form may be used for oral administration either in capsules or as a suspension. The white, crystalline sodium salt is very soluble in water and is used to make the solutions used for injections that should be used within one hour after being made.

Category—antibacterial.

Usual dose—500 mg. 4 times a day.

Usual dose range—1 to 4 g. daily.

OCCURRENCE

Ampicillin Capsules U.S.P.
Ampicillin for Oral Suspension U.S.P.
Sodium Ampicillin U.S.P.
Sterile Sodium Ampicillin U.S.P.

Hetacillin, Versapen®, is prepared by the reaction of ampicillin with acetone. In aqueous solution it is rapidly converted back to ampicillin and acetone. The spectrum of antibacterial action is identical with that of ampicillin and probably is due to the hydrolysis product, ampicillin.[15a] Four dosage forms—capsules, chewable tablets, oral suspension and pediatric drops—have been approved by the Food and Drug Administration.

Carbenicillin, Geopen®, Disodium α-carboxybenzyl penicillin.

A new semisynthetic penicillin released in the United States in 1970 is carbenicillin, a product introduced in England and first reported by Acred *et al.*[16] in 1967. Examina-

Carbenicillin

tion of its structure shows that it differs from ampicillin by having an ionizable carboxyl group substituted on the *alpha* carbon atom of the benzyl side chain rather than an amino group. Carbenicillin has a broad range of antimicrobial activity, broader than any other known penicillins, a property attributed to the unique carboxyl group.

Carbenicillin is not stable in acids and is inactivated by penicillinases. It must be administered by injection, the intravenous route being generally used. It is recommended for use in serious infections not susceptible to other penicillins, particularly those due to *Pseudomonas aeruginosa* (pyocyanea), to Proteus species (particularly indole positive strains), and to certain strains of *Escherichia coli*. It may be given in large doses, for example, 200 mg. per kilogram in serious Pseudomonas urinary tract infections or 20 to 30 grams daily for Proteus and *E. coli* infections. Adverse reactions similar to those observed with other penicillins may occur, the most serious possibility being the hypersensitivity reactions.[17b]

Penicillin O, Cer-O-Cillin®, Allylmercaptomethyl Penicillin, Allylthiomethyl Penicillin. Penicillin O is a biosynthetic penicillin that has found a limited use in therapy. Because penicillin G is capable of causing some rather severe allergic responses (e.g., "penicillin tongue," urticaria), penicillin O has been used as a substitute in the treatment of patients hypersensitive to penicillin G. Penicillin O has an activity of 1,630 units per mg. and an antibacterial spectrum similar to that of penicillin G. It is very soluble in water and extremely susceptible to hydrolysis. In the dry state it may be stored for periods up to 3 years or more without significant deterioration. The soluble sodium salt is usually administered parenterally. A more stable and less water-soluble salt made with 2-chloroprocaine is available in an injectable

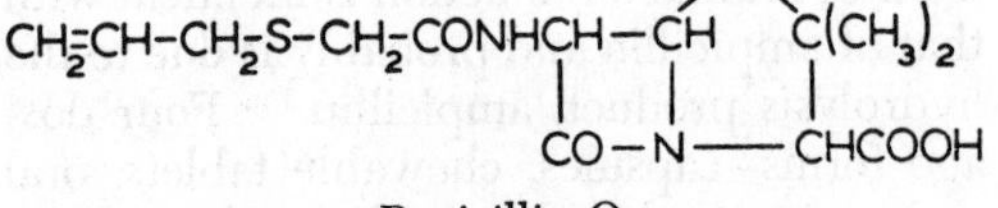

$$CH_2{=}CH{-}CH_2S{-}CH_2CONHCH{-}CH \quad C(CH_3)_2$$
$$CO{-}N{-}CHCOOH$$

Penicillin O

dosage form (Depo-Cer-O-Cillin®). This salt has an activity of about 960 units per mg.

THE CEPHALOSPORINS

The cephalosporins are antibiotics obtained from species of the fungus Cephalosporium and from semisynthetic processes. Although work began on this group of antibiotics in 1945, it has been only since 1964 that they have gained a place in therapy. The earlier developments pertaining to the isolation, the chemistry and the antibacterial properties of the cephalosporins and their relationships to the penicillins have been reviewed by Abraham[17] and by Van Heynigen.[18] Compounds having three different chemical structures have been isolated from Cephalosporium. One of these, cephalosporin P_1, has a steroid structure. It possesses low antibacterial properties and has not been employed in clinical medicine.

Of greater interest is the antibiotic cephalosporin N that was first isolated from *C. salmosynnematum* and was given the

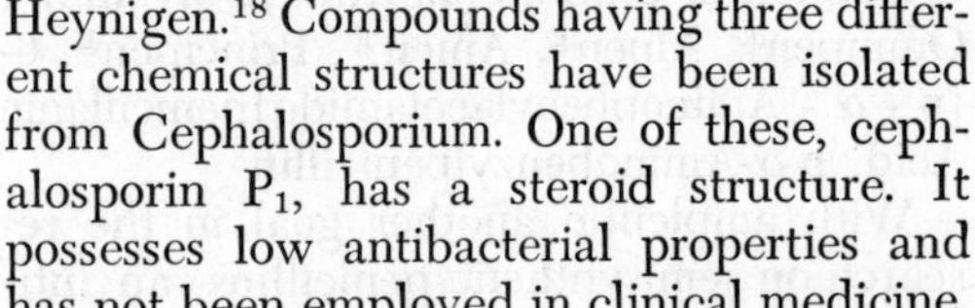

$$H_3\overset{\oplus}{N}$$
$$CHCH_2CH_2CH_2CONHCH{-}CH{-}C(CH_3)_2$$
$$\overset{\ominus}{OOC}$$
$$CO{-}N{-}CHCOOH$$

Penicillin N
(Cephalosporin N, Synnematin B)

name synnematin and then synnematin B. Its structure was determined to be D-(4-amino-4-carboxybutyl)penicillin and it is now frequently referred to as penicillin N.

Its structure shows it to be an acyl derivative of 6-APA and D-α-aminoadipic acid. The unusual zwitterionic side chain produces a compound less effective against gram-positive organisms than are other penicillins. However, it is more active than penicillin G against a number of gram-negative organisms and particularly some of the salmonellae. It has been employed successfully in clinical trials for the treatment of typhoid fever but it has not been released as an approved drug.

The third antibiotic isolated from Cephalosporia is cephalosporin C. Its structure

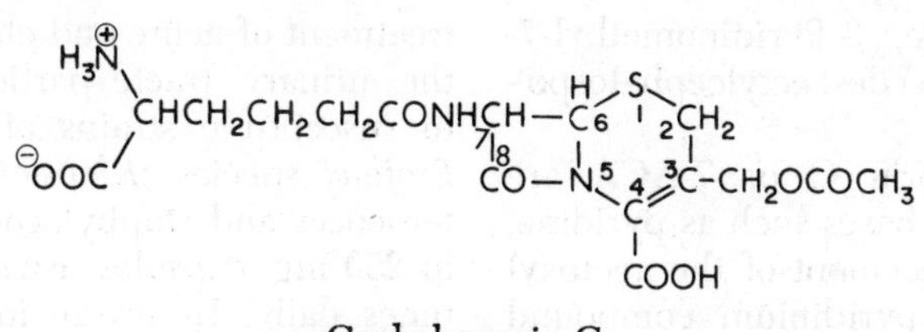

Cephalosporin C

shows it to be a congener of penicillin N, containing a dihydrothiazine ring instead of the thiazolidine ring of the penicillins. Because early studies of the antibacterial properties of cephalosporin C showed it to be similar in spectrum to penicillin N but less active, interest in it was not great in spite of its resistance to degradation by penicillinase. However, the discovery that the α-aminoadipoyl side chain could be hydrolytically removed to produce 7-amino-cephalosporanic acid (7-ACA) prompted investigations that have led to semisynthetic cephalosporins of medicinal value. The relationship of 7-ACA and its acyl derivatives to 6-APA and the semisynthetic penicillins is obvious.

Woodward *et al.*[19] have prepared cephalosporin C by an elegant synthetic procedure, but the commercially available drugs are obtained from Cephalosporia or as semisynthetic products from 7-ACA of natural origin. The chemical nomenclature of cephalosporins is very complex when named as a bicyclic octene. To simplify matters, the bicyclic ring nucleus of cephalosporins with the oxygen atom of the β-lactam ring attached has been given the name cephem. If the double bond occurs between the 2 and the 3 atoms, the compounds are 2-cephems or △²-cephems. If the double bond occurs between the 3 and the 4 atoms, the compounds are 3-cephems or △³-cephems, as shown for sodium cephalothin below. Interestingly, 2-cephems produce inactive compounds.

Sodium Cephalothin U.S.P., Keflin®, Sodium Cephosporn C, Sodium 3-(Hydroxymethyl)-8-oxo-7-[2-(2-thienyl)acetamido]-5-thia-1-azabicyclo[4.2.0]-oct-2-ene-2-carboxylate Acetate (Ester), Sodium 3-Methyl-7-(2-thiophene-2-acetamido)-3-cephem-4-carboxylate, 7-(2-Thienylacetamido)cephalosporanic Acid.

Sodium cephalothin occurs as a white to

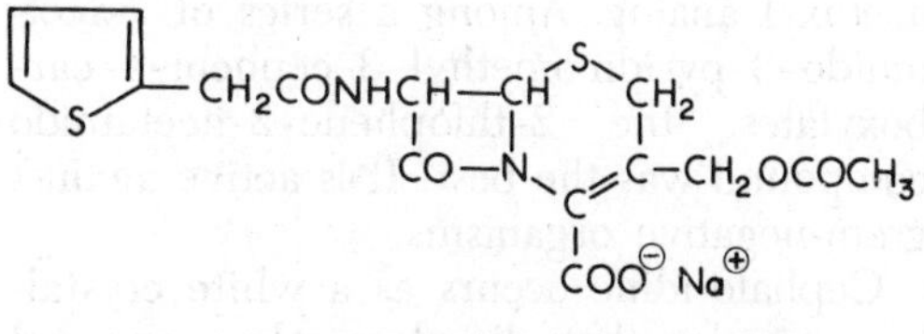

Sodium Cephalothin

off-white crystalline powder that is practically odorless. It is freely soluble in water and is insoluble in most organic solvents. Although it has been described as a broad-spectrum antibacterial compound, it is not in the same class as the tetracyclines. Its spectrum of activity is broader than that of penicillin G, being more like ampicillin in character. It is poorly absorbed from the gastrointestinal tract and must be administered parenterally for systemic infections. It is relatively nontoxic and is acid-stable. It is excreted rapidly through the kidneys, about 60 per cent being lost within 6 hours of administration. Hypersensitivity reactions from cephalothin have been observed and there is some evidence of cross-sensitivity in patients noted previously to be penicillin sensitive.

Category—antibacterial.

Usual dose—parenteral, the equivalent of 500 mg. to 1 g. of cephalothin 4 to 6 times a day.

Usual dose range—2 to 8 g. daily.

OCCURRENCE
Sterile Sodium Cephalothin U.S.P.

Cephaloridine, Loridine®, 3-Pyridinomethyl-7-(2-thiophene-2-acetamido)-3-

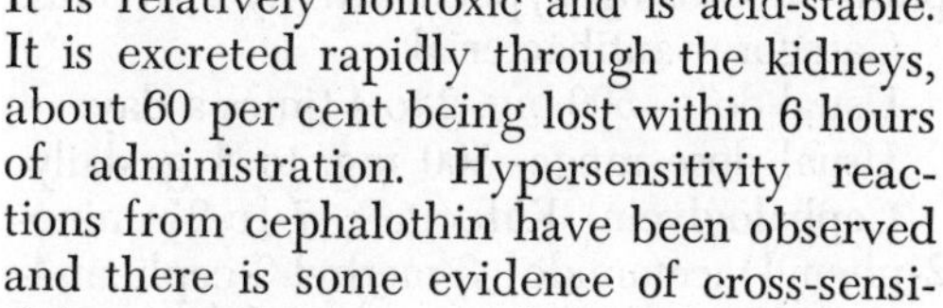

Cephaloridine

cephem-4-carboxylate, 3-Pyridinomethyl-7-(2-thienylacetamido)desacetylcephalosporanic Acid.

When cephalosporin C or 7-ACA are treated with organic bases such as pyridine, a nucleophilic displacement of the acetoxyl group occurs. The pyridinium compound thus produced is more potent than the acetoxyl analog. Among a series of 7-acetamido-3-pyridinomethyl-3-cephem-4-carboxylates, the 2-thiophene-2-acetamido compound was the best. It is active against gram-negative organisms.

Cephaloridine occurs as a white crystalline powder that discolors when exposed to light. It is somewhat unstable and should be stored in a refrigerator. It is very soluble in water and deteriorates rapidly in aqueous solutions which should be used within 24 hours of their preparation and then only if stored at 2° to 15° C.

The intramuscular injection of cephaloridine is less painful than I.M. injection of sodium cephalothin and it is not excreted as rapidly. However, in elevated doses it may produce a nephrotoxicity that makes the control of dosage necessary. It is capable of causing hypersensitivity reactions.

Category—antibacterial.

Usual dose—500 mg. 2 to 4 times a day.

Usual dose range—500 mg. to 4 g. daily.

Cephaloglycin, Kafocin®, 7-[D-2-Amino-2-phenyl)acetamido]-3-methyl-3-cephem-4-carboxylic acid, 7-(D-α-amino-phenylacetamido)cephalosporanic acid.

Cephaloglycin is a congener of ampicillin introduced during 1970. It differs from cephalothin by having a phenylglycine group instead of the 2-thiophenyl-2-acet-

Cephaloglycin

amido function. It occurs as a white to off-white powder that is acid stable and absorbed after oral administration, an advantage over the other cephalosporin compounds. It is recommended for the

treatment of acute and chronic infections of the urinary tract, particularly those due to susceptible strains of *Escherichia coli*, *Proteus* species, *Klebsiella-Aerobacter*, enterococci and staphylococci. It is available in 250 mg. capsules usually administered 4 times daily. In severe infections, the dose may be 500 mg. 4 times daily.

THE STREPTOMYCINS AND RELATED ANTIBIOTICS

The discovery of streptomycin was the result of a planned and deliberate search begun in 1939 and brought to fruition in 1944 by Waksman and his associates.[20] This success stimulated world-wide searches for antibiotics from the actinomycetes and particularly from the genus Streptomyces. Among the many antibiotics isolated from that genus, a number are compounds closely related in structure to streptomycin. Four of them, kanamycin, neomycin, paramomycin, and gentamicin, are of current interest as clinically useful antibiotics. The five structurally related antibiotics are poorly absorbed from the gastrointestinal tract and all but gentamicin are used to treat local infections in that area. Because of the broad-spectrum nature of their antimicrobial activity, they are used also for systemic infections, but their undesirable side-effects, particularly their ototoxicity, have led to restrictions in their employment. When administered for systemic infections, they must be given parenterally, usually by intramuscular injection.

The organism that produces streptomycin, *Streptomyces griseus*, also produces a number of other antibiotic compounds, hydroxystreptomycin, mannisidostreptomycin and cycloheximide (q.v.). None of these has achieved importance as medicinally useful substances. The term *Streptomycin A* has been used to refer to what is commonly called streptomycin, and mannisidostreptomycin has been called Streptomycin B. Hydroxystreptomycin differs from streptomycin in having a hydroxyl group in place of one of the hydrogens of the streptose methyl group. Mannisidostreptomycin has a mannose residue attached by glycosidic

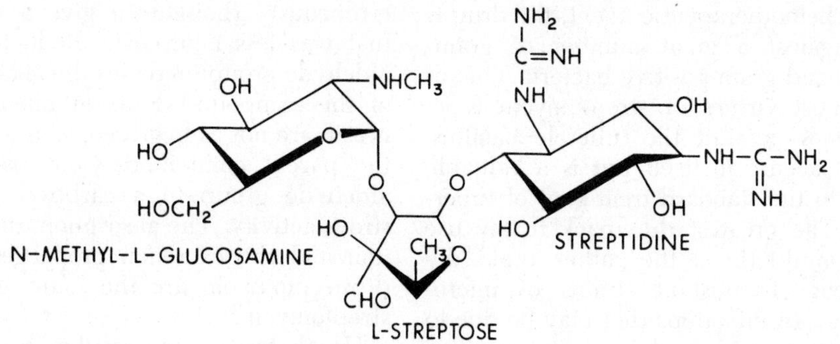

Streptomycin

linkage through the hydroxyl group at carbon four of the N-methyl-L-gucosamine moiety. The work of Dyer *et al.*[21, 22] has established the complete stereochemical structure of streptomycin except for the nature of the two glycosidic linkages.

PRODUCTS

Streptomycin Sulfate U.S.P., Streptomycin sulfate is a white, odorless powder that is hygroscopic but stable toward light and air. It is freely soluble in water, forming solutions that are slightly acidic or nearly neutral. It is very slightly soluble in alcohol and is insoluble in most other organic solvents. Acid hydrolysis of streptomycin yields streptidine and streptobiosamine, the compound that is a combination of L-streptose and N-methyl-L-glucosamine

Streptomycin acts as a triacidic base through the effect of its two strongly basic guanidino groups and the more weakly basic methylamino group. Aqueous solutions may be stored at room temperature for 1 week without any loss of potency, but they are most stable if the pH is adjusted between 4.5 and 7.0. The solutions decompose if sterilized by heating, so sterile solutions are prepared by adding sterile distilled water to the sterile powder. The early salts of streptomycin contained impurities that were difficult to remove and caused a histaminelike reaction. By forming a complex with calcium chloride, it was possible to free the streptomycin from these impurities and to obtain a product that was generally well tolerated. The mechanism by which streptomycin exerts its effect is not

definitely known.[23] It has been noted that streptomycin is bound to ribosomes and it appears to inhibit protein synthesis. It may cause a coding ambiguity for the insertion of amino acids in the primary sequence of protein synthesis, thus leading to the insertion of the wrong amino acid in the required protein molecule. However, the possibility still exists that other factors may be responsible for the antibacterial action of streptomycin. Evidence suggests that the other antibiotics of the aminoglycoside group related to streptomycin act in a similar way.

A clinical problem which develops sometimes with the use of streptomycin is the early development of resistant strains of bacteria, making necessary a change in therapy of the disease. Another factor which limits its therapeutic efficacy is its chronic toxicity. Certain neurotoxic reactions have been observed after the use of streptomycin. They are characterized by vertigo, disturbance of equilibrium and diminished auditory acuity. In an effort to reduce ototoxicity, a mixture of equal parts of streptomycin sulfate and dihydrostreptomycin sulfate has been used. The patient thus gets only one half as much of each drug; this was assumed to reduce the risk of vestibular damage from streptomycin and of hearing loss from dihydrostreptomycin. Evidence has shown that the rationale behind such a combination is not valid and that a greater occurrence of damage to the 8th nerve occurs from the mixture than from streptomycin alone. Minor toxic effects include skin rashes, mild malaise, muscular pains and drug fever.

As a chemotherapeutic agent, the drug is active against a great number of gram-negative and gram-positive bacteria. One of the greatest virtues of streptomycin is its effectiveness against the tubercle bacillus. It is not a cure in itself but is a valuable adjunct to the standard treatment of tuberculosis. The greatest drawback to the use of this antibiotic is the rather rapid development of resistant strains of microorganisms. In infections that may be due to both streptomycin- and penicillin-sensitive bacteria, the combined administraton of the two antibiotics has been advocated. Combinations of penicillin and dihydrostreptomycin also are used. In addition, combinations of a readily soluble salt of penicillin, a slowly soluble amine salt of penicillin, and dihydrostreptomycin sulfate are available. The possible development of damage to the otic nerve by the continued use of streptomycin-containing preparations has led to the discouragement of the use of such products. There is an increasing tendency to reserve the use of streptomycin products for the treatment of tuberculosis. The fact that streptomycin is not absorbed when given orally and is not significantly destroyed in the gastrointestinal tract accounts for the fact that at one time it was rather widely used in the treatment of infections of the intestinal tract. For systemic action, streptomycin usually is given by intramuscular injection.

Category—antibacterial (tuberculostatic).

Usual dose—intramuscular, the equivalent of 1 g. of streptomycin 2 to 7 times a week.

Usual dose range—1 g. weekly to 2 g. daily.

OCCURRENCE

Streptomycin Sulfate Injection U.S.P.

Dihydrostreptomycin Sulfate. Dihydrostreptomycin differs from streptomycin in that the free aldehyde group of the streptose portion of the molecule is converted to a primary alcohol. This change is accomplished by catalytic hydrogenation of streptomycin hydrochloride or of the trihydrochloride—calcium chloride complex. The change in molecular structure has no effect on the activity but, at one time, it was erroneously thought to give a compound that was less neurotoxic. Reduction of the aldehyde group increases the alkali stability of the compound. Salts of dihydrostreptomycin are not as hygroscopic as are those of the parent compound. Conversion of the aldehyde group to a carboxyl group destroys activity. The absorption, the distribution and the excretion properties of dihydrostreptomycin are the same as those of streptomycin.

Usual dose—intramuscular, equivalent of 1 g. of dihydrostreptomycin base.

Usual dose range—300 mg. to 3 g.

Kanamycin Sulfate U.S.P., Kantrex®. Kanamycin was isolated in 1957 in Japan by Umezawa and co-workers[24] from *Streptomyces kanamyceticus*. Its activity against mycobacteria and many intestinal bacteria, as well as a number of pathogens that show resistance to other antibiotics, brought a great deal of attention to this antibiotic. As a result, kanamycin was tested and released for medical use in a very short time.

Research activity has been focused intensively on the determination of the structures of the kanamycins. It has been determined by chromatography that *S. kanamyceticus* elaborates three closely related structures that have been designated kanamycins A, B and C. Commercially available kanamycin is almost pure kanamycin A, the least toxic of the three forms. The kanamycins differ only by the nature of the sugar moieties attached to the glycosidic oxygen on the 4-position of the central deoxystreptamine. The absolute configuration of the deoxystreptamine in kanamycins has been reported as represented above by Tatsuoka *et al.*[25] The chemical relationships among the kanamycins, the neomycins and the paromomycins have been reported by Hichens and Rinehart.[26] It may be noted that the kanamycins do not have the D-ribose molecule that is present in neomycins and paromomycins. Perhaps this structural difference is significant in the lower toxicity observed with kanamycins. The kanosamine fragment linked glycosidically to the 6-position of deoxystreptamine is 3-amino-3-deoxy-D-glucose (3-D-glucosamine) in all three kanamycins. The structures of the kanamycins have been proved by total syn-

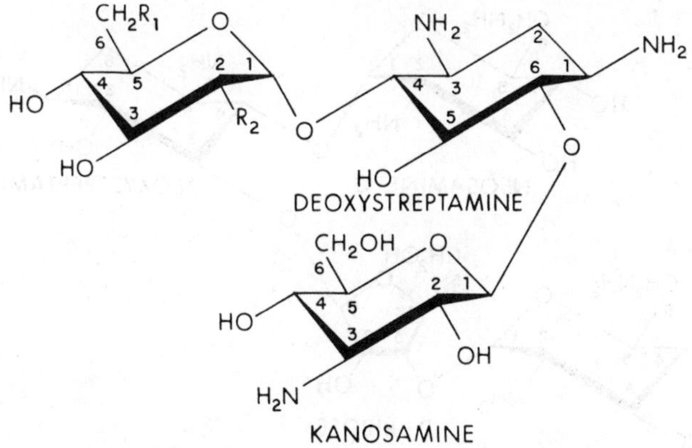

DEOXYSTREPTAMINE

KANOSAMINE

Kanamycin A: $R_1 = NH_2$; $R_2 = OH$
Kanamycin B: $R_1 = NH_2$; $R_2 = NH_2$
Kanamycin C: $R_1 = OH$; $R_2 = NH_2$

thesis.[27, 28] It may be seen that they differ in the nature of the substituted D-glucoses attached glycosidically to the 4-position of the deoxystreptamine ring. Kanamycin A contains 6-amino-6-deoxy-D-glucose; kanamycin B contains 2,6-diamino-2,6-dideoxy-D-glucose; and kanamycin C contains 2-amino-2-deoxy-D-glucose. (See above.)

Kanamycin is basic and forms salts of acids through its amine groups. It is water soluble as the free base but is used in therapy as the sulfate salt, which is very soluble. It is very stable to both heat and chemicals. Solutions resist both acids and alkali within the pH range of 2.0 to 11.0.

The use of kanamycin in the U. S. is usually restricted to infections of the intestinal tract and to systemic infections arising from pathogens that have developed resistance to other antibiotics. It has been recommended also for antisepsis of the bowel preoperatively. It is poorly absorbed from the intestinal tract, so systemic infections must be treated by intramuscular injections. Injections of it are rather painful, and the concomitant use of a local anesthetic is indicated. The use of kanamycin in treatment of tuberculosis has not been widely advocated, since the discovery that mycobacteria develop resistance to it very rapidly. Aoki, Hayashi and Ito[29] have found kanamycin to inhibit oxidative mechanisms in the tubercle bacilli as does streptomycin. Their tests indicated that the interference of kanamycin is not identical with that of streptomycin in the oxidation of benzoic acid, niacin and malonic acid. However, clinical experience as well as experimental work of Morikubo[30] indicates that kanamycin does develop cross resistance in the tubercle bacilli with dihydrostreptomycin, viomycin and other antituberculars. Like streptomycin, kanamycin may cause a decrease in or complete loss of hearing. Upon development of such symptoms, its use should be stopped immediately. Umezawa *et al.*[31] have reported that the N-methane sulfonate salts of kanamycin are considerably less toxic than the monosulfate.

Category—antibacterial.

Usual dose—oral, for therapy in intestinal infections, the equivalent of 1 g. of kanamycin 3 to 4 times daily; for preoperative preparation, 1 g. every hour for 4 doses, then 1 g. every 6 hours for 36 to 72 hours; intramuscular, for therapy in systemic infections, the equivalent of 7.5 mg. of kanamycin per kg. of body weight twice daily.

Usual dose range—oral, 3 to 12 g. daily; intramuscular, on a body-weight basis, up to a total of not more than 1.5 g. daily for up to 5 days.

OCCURRENCE

Kanamycin Sulfate Capsules U.S.P.
Kanamycin Sulfate Injection U.S.P.

Neomycin Sulfate U.S.P., Mycifradin®, Neobiotic®. In a search for less-toxic antibiotics than streptomycin, Waksman and

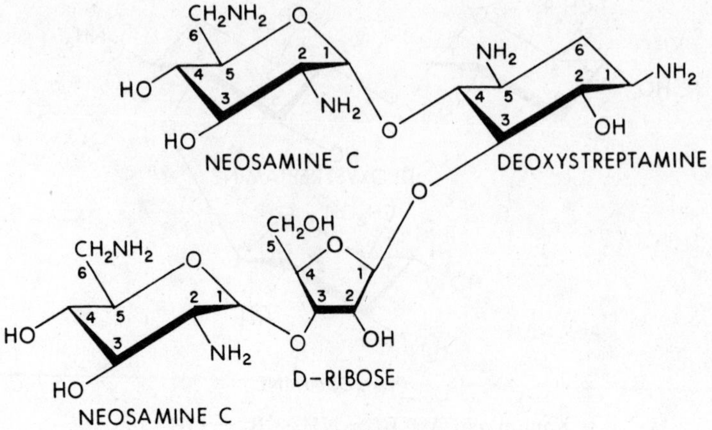

Neomycin C

Lechevalier[32] obtained neomycin in 1949 from *Streptomyces fradiae*. Since that time neomycin has increased steadily in importance, and today it is considered to be one of the most useful antibiotics in the treatment of gastrointestinal infections, dermatologic infections and acute bacterial peritonitis. Also, it is employed in abdominal surgery to reduce or avoid complications due to infections from bacterial flora of the bowel. It has a broad-spectrum activity against a variety of organisms. It shows a low incidence of toxic and hypersensitive reactions. It is very slightly absorbed from the digestive tract, so its oral use does not produce any systemic effect. Neomycin-resistant strains of pathogens have seldom been reported to develop from those organisms against which neomycin is effective. A complete review on neomycin has been edited by Waksman.[33]

Neomycin as the sulfate salt is a white to slightly yellow crystalline powder that is very soluble in water. It is hygroscopic and photosensitive. Neomycin sulfate contains the equivalent of 60 per cent of the free base.

Neomycin, as produced by *S. fradiae*, is a mixture of closely related substances. Included in the "neomycin complex" is neamine (originally designated neomycin A) and neomycins B and C. *S. fradiae* also elaborates another antibiotic called fradicin that has some antifungal properties but no

antibacterial activity. This substance is not present in "pure" neomycin.

The structures of neamine and neomycin B and C are known and the absolute configurational structures of neamine and neomycin have been reported by Hichens and Rinehart.[26] Neamine may be obtained by methanolysis of neomycins B and C during which the glycosodic link between the deoxystreptamine and D-ribose is broken. Therefore, neamine is a combination of deoxystreptamine and neosamine C linked glycosidically (alpha) at the 4-position of deoxystreptamine. According to Hichens and Rinehart, neomycin B differs from neomycin C by the nature of the sugar attached terminally to D-ribose. That sugar, called neosamine B, differs from neosamine C in its stereochemistry. It has been suggested by Rinehart *et al.*[34] that in neosamine B the configuration is that of 2,6-diamino-2,6-dideoxy-L-idose in which the orientation of the 6-aminomethyl group is inverted to that of the 6-amino-6-deoxy-D-glucosamine in neosamine C. In both instances the glycosidic links are assumed to be alpha. However, Huettenrauch[35] more recently has suggested that both of the diamino sugars in neomycin C have the L-idose configuration and that the glycosidic link is beta in the one attached to D-ribose. The proof of these details concerning the absolute configuration of neomycin B is dependent upon further evidence. The combination of neosamine B with D-ribose is called neobiosamine B, and the combination of neosamine C with D-ribose is called neobiosamine C. In both molecules, the glycosidic links at the D-ribose fragment are beta oriented.

Category—antibacterial.

Usual dose—for preoperative preparation, the equivalent of 700 mg. of neomycin base every hour for 4 doses, then 700 mg. every 4 hours for 24 to 72 hours.

Usual dose range—2.8 to 8.4 g. daily.

For external use—topically, as a 0.35 per cent solution, lotion, or ointment, or as a solution in a wet dressing.

OCCURRENCE

Neomycin Sulfate Ointment U.S.P.
Neomycin Sulfate Oral Solution U.S.P.
Neomycin Sulfate Tablets U.S.P.
Neomycin Sulfate and Dexamethasone Sodium Phosphate Cream N.F.
Neomycin Sulfate and Hydrocortisone Acetate Lotion N.F.
Neomycin Sulfate, Polymyxin B Sulfate and Gramicidin Ophthalmic Solution N.F.
Neomycin Sulfate, Polymyxin B Sulfate and Zinc Bacitracin Ointment N.F.

Paromomycin Sulfate, N.F., Humatin®. The isolation of paromomycin was reported in 1956 as an antibiotic obtained from a Streptomyces species (P-D 04998) that is said to resemble closely *S. rimosus*. The parent organism had been obtained from soil samples collected in Colombia. How-

ever, paromomycin more closely resembles neomycin and streptomycin, in antibiotic activity, than oxytetracycline, the antibiotic obtained from *S. rimosus*.

In-vitro tests reported by Coffey *et al.*[36] indicate that paromomycin has a broad spectrum of activity. Oral administration has shown it to be poorly absorbed from the gastrointestinal tract and to be very effective in combating infections of this area due to entamoeba, salmonella, shigella, *Escherichia coli*, proteus and aerobactor. It has been introduced as an antibiotic recommended for the treatment of intestinal amebiasis and bacterial diarrheas and for the suppression of the intestinal flora prior to surgery. High dosage by parenteral administration shows very little toxicity. Diarrhea and an overgrowth of resistant organisms, particularly monilia, has occurred from oral dosage. Cross resistance with neomycin and streptomycin has been shown in vitro.

The general structure of paromomycin was first reported by Haskell *et al.*[37] as one compound. Subsequently, chromatographic determinations have shown paromomycin to consist of two fractions which have been named paromomycin I and paromomycin II. The absolute configurational structures for the paromomycins were suggested by Hichens and Rinehart[26] as shown in the structural formula, and has been confirmed

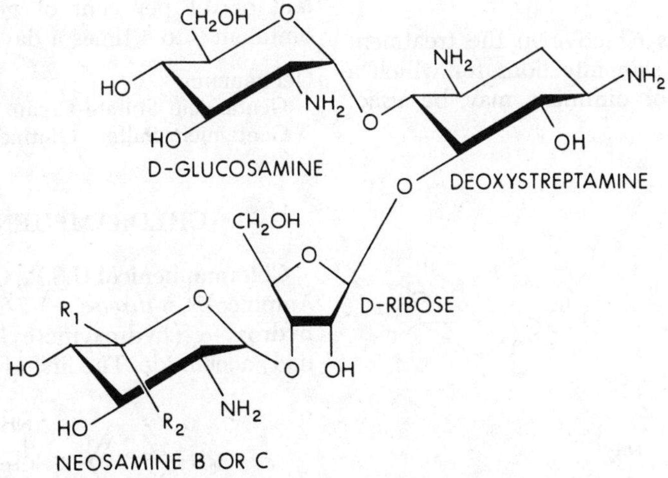

Paromomycin I: $R_1 = H$; $R_2 = CH_2NH_2$
Paromomycin II: $R_1 = CH_2NH_2$; $R_2 = H$

by DeJongh *et al.*[38] by mass spectrometric studies. It may be noted that the structure of paromomycin is the same as that of neomycin B except that paromomycin contains D-glucosamine instead of the 6-amino-6-deoxy-D-glucosamine found in neomycin B. The same relationship in structures is found between paromomycin II and neomycin C. The combination of D-glucosamine with deoxystreptamine is obtained by partial hydrolysis of both paromomycins and is called paromamine [4-(2-amino-2-deoxy-α-4-glucosyl)-deoxystreptamine].

Category—antiamebic.

Usual dose—the equivalent of 500 mg. of paromomycin every 6 hours taken with meals.

Usual dose range—500 mg. to 1 g.

OCCURRENCE

Paromomycin Sulfate Capsules N.F.

Gentamicin Sulfate U.S.P., Garamycin®. Gentamicin was isolated in 1958 and reported in 1963 by Weinstein *et al.*[39] to belong to the streptomycinoid (aminocyclitol) group of antibiotics. It is obtained commercially from *Micromonospora purpurea*. Like the other members of its group, it has a broad spectrum of activity against many common pathogens of both gram-positive and gram-negative types. Of particular interest is its high degree of activity against penicillin-G-resistant strains of *Staphylococcus aureus* and against *Pseudomonas aeruginosa*.

Gentamicin is effective in the treatment of a variety of skin infections for which a topical cream of ointment may be used.

However, since it offers no real advantage over topical neomycin in the treatment of all but pseudomonal infections, it is recommended that topical gentamicin be reserved for use in such infections and in the treatment of burns complicated by pseudomonemia. An injectable solution containing 40 mg. of gentamicin sulfate per ml. may be used for serious systemic and genitourinary tract infections caused by gram-negative bacteria, particularly pseudomonas and proteus species.

Gentamicin sulfate is a mixture of the salts of compounds identified as gentamicins C_1, C_2 and C_{1a}. The structures of these gentamicins have been reported by Cooper *et al.*[40] to have the structures shown although the geometry of the glycosidic links has not been established.

Coproduced but not a part of the commercial product are gentamicins A and B. Their structures have been reported by Maehr and Schaffner[41] and are closely related to the gentamicins C. Although the gentamicin molecules are similar in a number of respects to other aminocyclitols such as streptomycins, they are sufficiently different so that their medical effectiveness is significantly greater.

Gentamicin sulfate is a white to buff colored substance that is soluble in water and insoluble in alcohol, acetone and benzene.

Category—antibacterial.

For external use—topically, the equivalent of 0.1 per cent of gentamicin in an ointment 3 to 4 times a day.

OCCURRENCE

Gentamicin Sulfate Cream U.S.P.
Gentamicin Sulfate Ointment U.S.P.

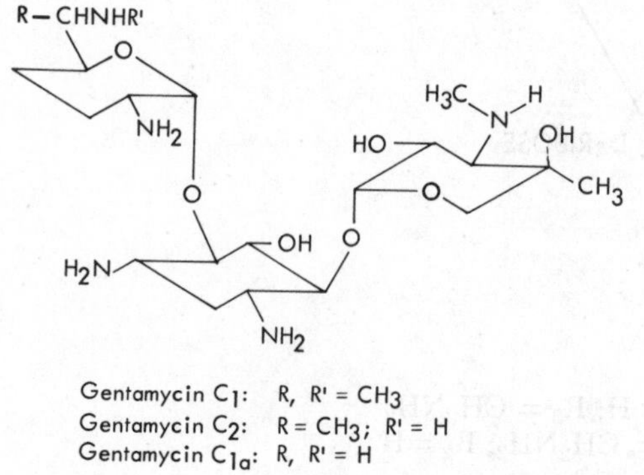

CHLORAMPHENICOL

Chloramphenicol U.S.P., Chloromycetin®, Amphicol®, D-*threo*-(−)-2,2-Dichloro-N-[β-hydroxy-α-(hydroxymethyl)-*p*-nitrophenethyl] acetamide. The first of the widely used

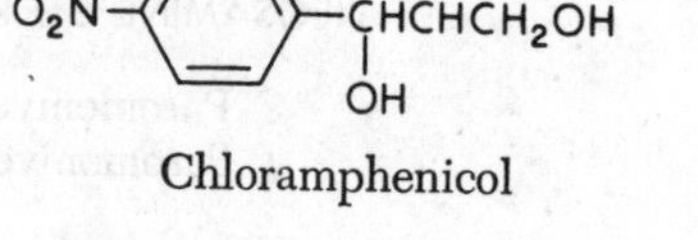

Gentamycin C₁: R, R' = CH₃
Gentamycin C₂: R = CH₃; R' = H
Gentamycin C₁ₐ: R, R' = H

Chloramphenicol

broad spectrum antibiotics, chloramphenicol, was isolated by Ehrlich *et al.*[42] in 1947 They obtained it from *Streptomyces venezuelae,* an organism that has been found in a sample of soil collected in Venezuela. Since that time, chloramphenicol has been isolated as a product of a number of organisms found in soil samples from widely separated places. More important, its chemical structure was soon established, and in 1949, Controulis, Rebstock and Crooks[43] reported its synthesis. This opened the way for the commercial production of chloramphenicol by a totally synthetic route. It was the first and is still the only therapeutically important antibiotic to be so produced in competition with microbiologic processes. A number of synthetic procedures have been developed for chloramphenicol. The commercial process most generally used has started with p-nitroacetophenone.[44]

Chloramphenicol is a white crystalline compound that is very stable. It is very soluble in alcohol and other polar organic solvents but is only slightly soluble in water. It has no odor but has a very bitter taste.

It may be noted that chloramphenicol possesses two asymmetric carbon atoms in the acylamidopropanediol chain. Biologic activity resides almost exclusively in the D-*threo* isomer; the L-*threo* and the D-and L-*erythro* isomers are virtually inactive.

A large number of structural analogs of chloramphenicol have been synthesized to provide a basis for correlation of structure to antibiotic action. It appears that the p-nitrophenyl group may be replaced by other aryl structures without appreciable loss in activity. Substitution on the phenyl ring with several different types of groups for the nitro group, a very unusual structure in biologic products, does not cause a great decrease in activity. However, the maintenance of the aromatic character of this moiety appears to be essential. Modifications of the side chain show it to possess a high degree of specificity in structure for antibiotic action. A conversion of the alcohol group on carbon atom 1 of the side chain to a keto group causes an appreciable loss in activity. The relationship of the structure of chloramphenicol to its antibiotic activity will not be clearly seen until the

mode of action of this compound is known. The review article by Brock[45] reports on the large amount of research that has been devoted to this problem. It has been established that chloramphenicol exerts its bacteriostatic action by a strong inhibition of protein synthesis. The details of such inhibition are as yet undetermined, and the precise point of action is unknown. Some process lying between the attachment of amino acids to soluble RNA and the final formation of protein appears to be involved.

The broad-spectrum activity of chloramphenicol and its singular effectiveness in the treatment of a number of infections not amenable to treatment by other drugs has made it an extremely popular antibiotic. Unfortunately, instances of serious blood dyscrasias and other toxic reactions have resulted from the use of chloramphenicol. Because of these reactions, it is now recommended that it not be used in the treatment of infections for which other antibiotics are as effective and not as hazardous. When properly used with careful observation for untoward reactions, chloramphenicol provides some of the very best therapy for the treatment of serious infections. Because of its bitter taste, this antibiotic is administered orally either in capsules or as the palmitate ester. Chloramphenicol Palmitate U.S.P., is insoluble in water and may be suspended in aqueous bases for liquid dosage forms. The ester forms by reaction with the hydroxyl group on the number 3 carbon atom. In the alimentary tract it is slowly hydrolyzed to the active antibiotic. Parenteral administration of chloramphenicol is made by use of an aqueous suspension of very fine crystals (IM) or by use of a solution of the sodium salt of the succinate ester of chloramphenicol. Sterile chloramphenicol sodium succinate has been used to prepare aqueous solutions for subcutaneous, intramuscular or intravenous injections.

Category—antibacterial and antirickettsial.

Usual dose—oral, 12.5 mg. per kg. of body weight 4 times daily; intravenous, 25 mg. per kg. 2 or 3 times daily.

Usual dose range—50 to 100 mg. per kg. daily.

For external use—topically to the conjunctiva, as a 1 per cent ointment 4 times daily, or 0.1 ml. of a 0.16 to 1 per cent solution 6 to 12 times daily.

OCCURRENCE

Chloramphenicol Capsules U.S.P.

Chloramphenicol Ophthalmic Ointment U.S.P.

Chloramphenicol for Ophthalmic Solution U.S.P.

Sterile Chloramphenicol for Suspension N.F.

Chloramphenicol Palmitate U.S.P.

Chloramphenicol Palmitate Oral Suspension U.S.P.

Chloramphenicol Sodium Succinate U.S.P.

Sterile Chloramphenicol Sodium Succinate U.S.P.

THE TETRACYCLINES

Among the most important broad-spectrum antibiotics are the members of the tetracycline family. Seven such compounds —tetracycline, oxytetracycline, chlortetracycline, demeclocycline, methacycline, doxycycline and rolitetracycline—have been introduced into medical use. A number of others have been shown to possess antibiotic activity. The tetracyclines are obtained by fermentation procedures from streptomyces species or by chemical transformations of the natural products. Their chemical identities have been established by degradation studies and confirmed by the synthesis of two members of the group, oxytetracycline and 6-demethyl-6-deoxy-tetracycline, in their ± forms.[46, 47] The important members of the group are derivatives of an octahydronaphthacene, a hydrocarbon that is made up of a system of 4 fused rings. It is from this system that the group name is obtained. The antibiotic

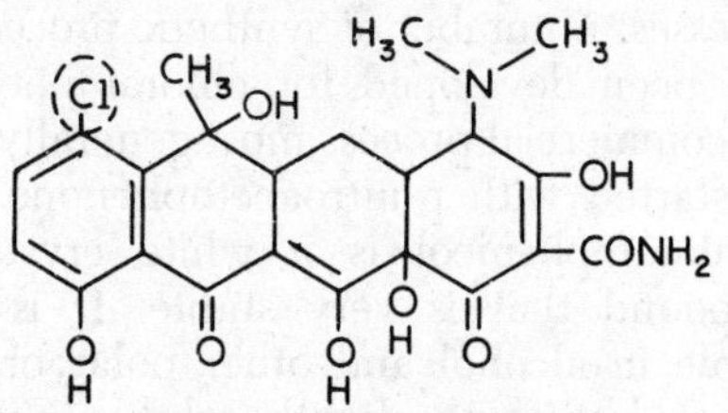

CHLORTETRACYCLINE

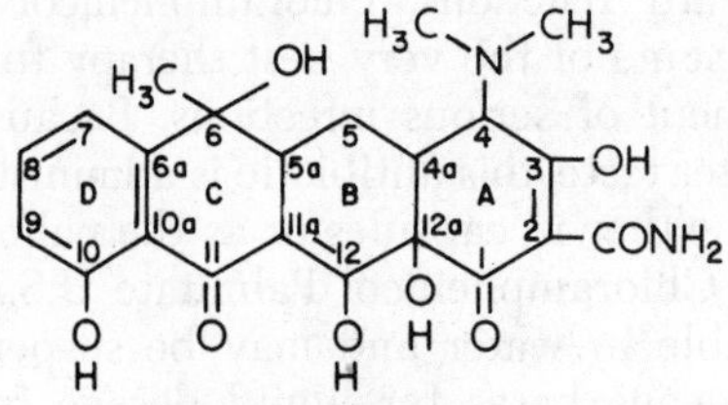

TETRACYCLINE

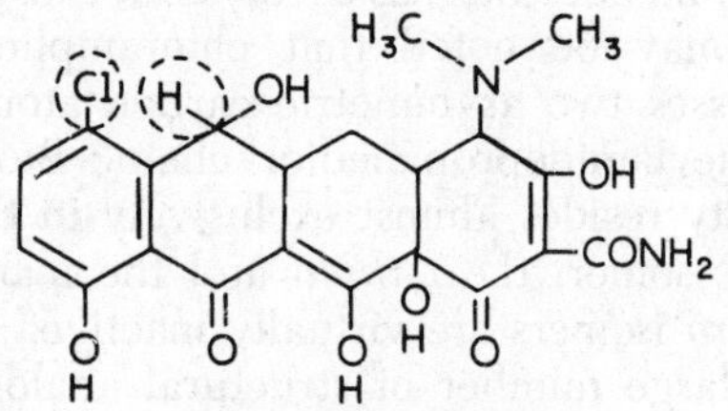

DEMECLOCYCLINE

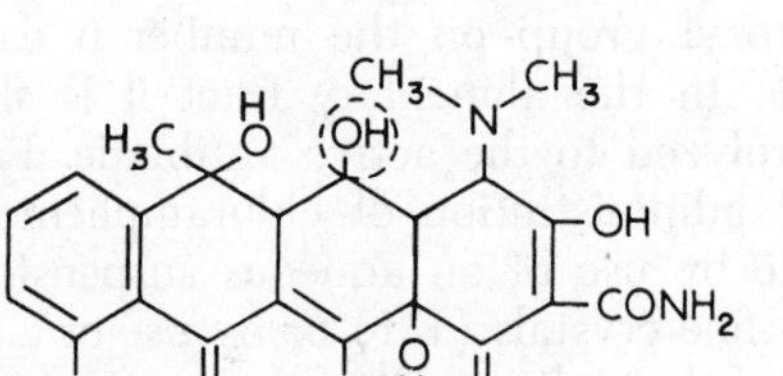

OXYTETRACYCLINE

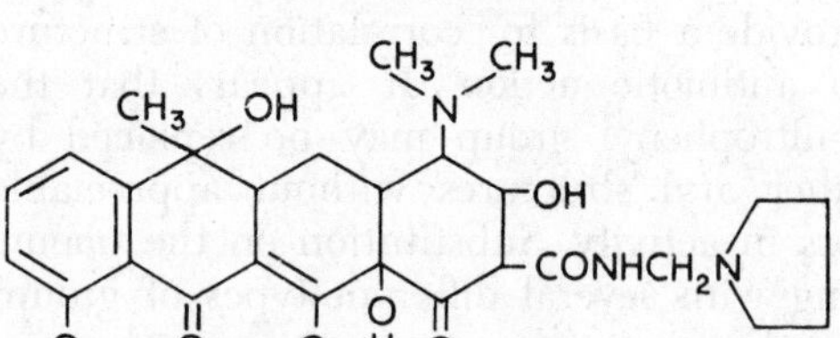

ROLITETRACYCLINE

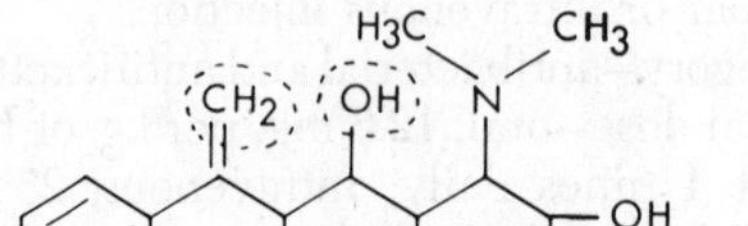

Methacycline

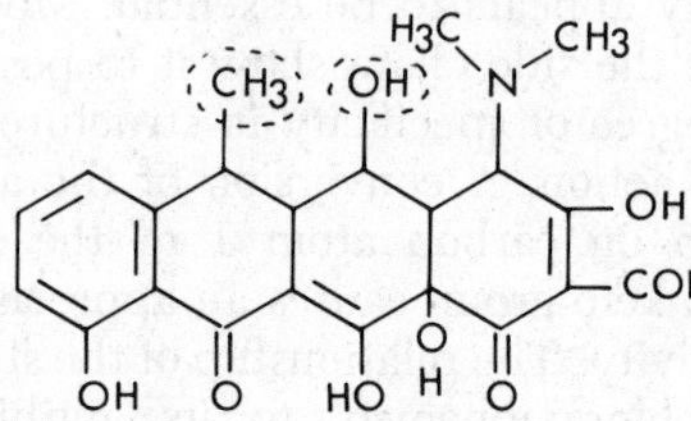

Doxycycline

spectra and the chemical properties of these compounds are very similar but not identical. Their structural relationships are shown on page 368.

The tetracyclines are amphoteric compounds, forming salts with either acids or bases. In neutral solutions these substances exist mainly as zwitter ions. The acid salts, which are formed through protonation of the dimethylamino group on carbon atom 4, exist as crystalline compounds that are very soluble in water. However, these amphoteric antibiotics will crystallize out of aqueous solutions of their salts unless stabilized by an excess of acid. The hydrochloride salts are used most commonly for oral administration and are usually encapsulated because of their bitter taste. Water-soluble salts may be obtained also from bases such as sodium or potassium hydroxides but are not stable in aqueous solutions. Water-insoluble salts are formed with divalent and polyvalent metals. This property accounts for the use of calcium salts to form tasteless suspensions for liquid oral dosage forms.

The unusual structural groupings in the tetracyclines produce three acidity constants in aqueous solutions of the acid salts. The particular functional groups responsible for each of the thermodynamic pKa values has been determined by Leeson et al.[48] to be as shown in the formula below. These groupings had been identified by Stephens et al.[49] previously as the sites for protonation, but their earlier assignments as to which produced the values responsible for pKa2 and pKa3 were opposite to those of Leeson et al. This latter assignment has been substantiated by Rigler et al.[50]

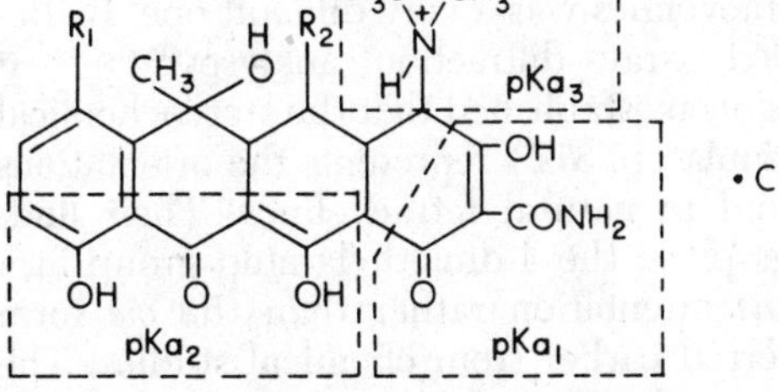

The approximate pKa values for each of these groups in four tetracycline salts in

common use are shown in Table 26. The values are taken from Stephens et al.[49] and from Benet and Goyan.[51]

An interesting property of the tetracyclines is their ability to undergo epimerization at carbon atom 4 in solutions of intermediate pH range. These isomers are called *epi*tetracyclines. Under the influence of the acidic conditions, an equilibrium is established in about a day and consists of approximately equal amounts of the isomers.

Table 26. pKa VALUES (OF HYDROCHLORIDES) IN AQUEOUS SOLUTION AT 25°

	pKA$_1$	pKA$_2$	pKA$_3$
Tetracycline	3.3	7.7	9.5
Chlortetracycline	3.3	7.4	9.3
Demeclocycline	3.3	7.2	9.3
Oxytetracycline	3.3	7.3	9.1

The partial structures below indicate the two forms of the epimeric pair.

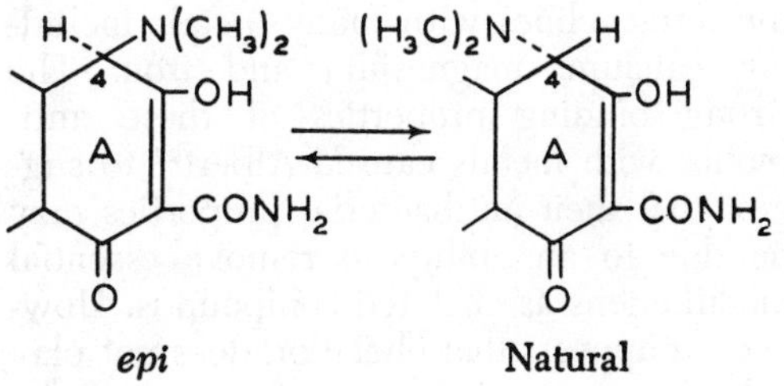

The 4-*epi*tetracyclines have been isolated and characterized. They exhibit much less activity than the "natural" isomers, thus accounting for a decrease in therapeutic value of aged solutions.

Strong acids and strong bases attack the tetracyclines having a hydroxyl group on the number 6 carbon atom, causing a loss in activity through modification of the C ring. Strong acids produce a dehydration through a reaction involving the 6-hydroxyl group and the 5a-hydrogen. The double bond thus formed between positions 5a and 6 induces a shift in the position of the double bond between carbon atoms 11a and 12 to a position between carbon atoms 11 and 11a, forming the more energetically

Tetracycline

acid ← | → base

CH₃   N(CH₃)₂

N(CH₃)₂

— OH

— CONH₂

OH OH O O

OH ||

Anhydrotetracycline

O

— OH

— CONH₂

OH O O

OH ||

Isotetracycline

favored resonant system of the naphthalene group found in the inactive anhydrotetracyclines. Bases promote a reaction between the 6-hydroxyl group and the ketone group at the 11 position, causing the bond between the 11 and 11a atoms to cleave and to form the lactone ring found in the inactive isotetracyclines. These two unfavorable reactions stimulated the research that has led to the development of the more stable and longer acting compounds, 6-deoxytetracycline, methacycline and doxycycline.

Stable chelate complexes are formed by the tetracyclines with many metals including calcium, magnesium and iron. The strong binding properties of these antibiotics with metals caused Albert[52] to suggest that their antibacterial properties may be due to an ability to remove essential metallic ions as chelated compounds. However, it appears that chelation does not play a basic role in the mode of action of the tetracyclines, but it may facilitate transport of the compounds to their sites of action. The conclusion of Jackson,[53] in his review on the mode of action of tetracyclines, is that they act principally by interference with protein synthesis. Maxwell[54] has substantiated the observation that tetracycline prevents the binding of aminoacyl-transfer RNA to messenger RNA-ribosome complex.

As a result of the large amount of research carried out to synthesize structural modifications of the tetracyclines, some interesting structure-activity relationships may be drawn. The synthesis of tetracycline analogs reported up to mid-1962 has been reviewed by Barrett,[55] and the relationships between chemical structure and the biologic activity of tetracyclines was reviewed by

Boothe[56] in 1962. The high antimicrobial power of tetracycline established some time ago that substitutions on the 5 and 7 carbon atoms were not essential. Similarly, the activity of 6-demethyltetracycline (demecycline) and demeclocycline has established that the methyl group on carbon atom 6 may be replaced by hydrogen. The activity of doxycycline and 6-deoxy-6-demethyltetracycline[57] (Minocycline) indicates that hydroxyl substitution on the 6 carbon atom is not essential either. The 6-deoxy-6-methylenetetracyclines [e.g., 6-deoxy-6-demethyl-6-methylene-5-oxytetracycline (meclocycline)] and their mercaptan adducts prepared by Blackwood et al.[58, 59] in 1961 possess typical tetracycline activity and illustrate further the extent of modification possible at the 6-position, with retention of biologic activity. Removal of the 4-dimethylamino group causes a loss of about 75 per cent of the antibiotic effect of the parent tetracyclines. In this connection, it is interesting to note that the 4-*epi*tetracyclines are less active than the dedimethylamino compounds, suggesting that structural conformations in this area may be important in fitting the molecule on a possible enzyme site. The problem of determining the complete stereochemistry of the tetracyclines was a very difficult one. By detailed x-ray diffraction analyses[60, 61, 62] it has been established that the stereochemical formula (p. 371) represents the orientations found in natural tetracyclines. Their findings place the 4-dimethylamino group in a *trans* orientation rather than the *cis* form inferred earlier from chemical studies. The x-ray diffraction studies also prove that a conjugated system exists in the structure from carbons 10 through 12 and that the formula below represents only one of a number of canonical forms existing in that portion of the molecule.

The importance of the shape of the ring system is indicated by a substantial loss in

Tetracycline: X, Z = H; Y = CH₃
Chlortetracycline: X = Cl; Y = CH₃; Z = H
Oxytetracycline: X = H; Y = CH₃; Z = OH
Demeclocycline: X = Cl; Y, Z = H

activity by *epi*merization at carbon atom 5a. Dehydrogenation to form a double bond between carbon atoms 5a and 11a produces a marked decrease in activity. Similarly, dehydration by strong acids of the 6 hydroxyl group and a hydrogen on the 5a carbon produces inactive anhydrotetracycline that has an aromatic C ring. The 2-carboxamide group apparently is relatively free from steric hindrance in the tetracycline molecule, as Gottstein, Minor and Cheyney[63] have shown that substitution of bulky groups for one of the hydrogens on the amide nitrogen does not cause any appreciable loss in activity. In fact, substitution of a pyrrolidinomethyl group at this point increases the water-solubility of tetracycline about 2,500 times without appreciable change in activity.

It appears that biologically there is considerable sensitivity to the chemical reactivity of the various groups on the tetracycline molecule. This may be demonstrated by the fact that changing the 2-carboxamide group to a nitrile or acetyl group produces compounds with little antibiotic power and by the fact that quaternization of the 4-dimethylamino group produces compounds with greatly reduced antibiotic activity.

Until the significance of the substitutions on the 1, 2 and 3 carbon atoms, and the conjugated system existing therein, is understood and until the significance of the oxygen functions on carbon atoms 10, 11, 12

and 12a is clarified, the structural characteristics of these compounds cannot be fully related to their activities.

PRODUCTS

Tetracycline U.S.P., Achromycin®, Cyclopar®, Panmycin®, Steclin®, Tetracyn®. During the chemical studies on chlortetracycline, it was discovered that controlled catalytic hydrogenolysis would selectively remove the 7-chloro atom and thus produce tetracycline. This process was patented by Conover[64] in 1955. Later, tetracycline was obtained from fermentations of streptomyces species but the commercial supply is still chiefly dependent upon the hydrogenolysis of chlortetracycline.

Tetracycline is 4-dimethylamino-1,4,4a,5, 5a,6,11,12a-octahydro-3,6,10,12,12a-pentahydroxy-6-methyl-1,11-dioxo-2-naphthacenecarboxamide. It is a bright-yellow crystalline salt that is stable in air but darkens in color upon exposure to strong sunlight. Tetracycline is stable in acid solutions having a pH higher than 2. It is somewhat more stable in alkaline solutions than chlortetracycline, but, like those of the other tetracyclines, such solutions rapidly lose their potencies. One gram of the base requires 2,500 ml. of water and 50 ml. of alcohol to dissolve it. The hydrochloride salt is most commonly used in medicine, although the free base is absorbed from the gastrointestinal tract about equally well. One gram of the hydrochloride salt dissolves in about 10 ml. of water and in 100 ml. of alcohol. Tetracycline has become the most popular antibiotic of its group, largely because its blood-level concentration appears to be higher and more enduring than that of either oxytetracycline or chlortetracycline. Also, it is found in higher concentration in the spinal fluid than are the other two compounds.

A number of combinations of tetracycline with agents that increase the rate and the height of blood-level concentrations are on the market.

One such adjuvant is citric acid (Achromycin V). Also, an insoluble tetracycline phosphate complex (Panmycin Phosphate, Sumycin, Tetrex) is made by mixing a solution of tetracycline, usually as the hydrochloride, with a solution of sodium metaphosphate. A variety of claims concerning the efficacy of these adjuvants have been made. The mechanisms of their actions is not clear but it has been reported[65, 66] that these agents enhance blood level concentrations over those obtained when tetracycline hydrochloride alone is administered orally. Remmers et al.[67, 68] have reported on the effects that selected aluminum-calcium gluconates complexed with some tetracyclines have on the blood level concentrations when administered orally, intramuscularly or intravenously. Such complexes enhanced blood levels in dogs when injected but not when given orally. They have also observed enhanced blood levels in experimental animals when complexes of tetracyclines with aluminum metaphosphate, with aluminum pyrophosphate and aluminum-calcium phosphinicodilactates were administered orally. As has been noted previously, the tetracyclines are capable of forming stable chelate complexes with metal ions such as calcium and magnesium that would retard absorption from the gastrointestinal tract. The complexity of the systems involved has not permitted unequivocal substantiation of the idea that these adjuvants act by competing with the tetracyclines for substances in the alimentary tract that would otherwise be free to complex with these antibiotics and thus retard their absorption. Certainly, there is no evidence that they act by any virtue they possess as buffers, an idea alluded to sometimes in the literature.

Category—antiamebic; antibacterial; antirickettsial.

Usual dose—oral, the equivalent of 250 mg. of tetracycline hydrochloride 4 times daily; intramuscular, the equivalent of 100 mg. of tetracycline hydrochloride 2 or 3 times daily; intravenous infusion, the equivalent of 250 to 500 mg. of tetracycline hydrochloride over a period of ½ to 1 hour, twice daily.

Usual dose range—oral, the equivalent of 1 to 4 g. of tetracycline hydrochloride daily; intramuscular, the equivalent of 200 to 500 mg. of tetracycline hydrochloride daily; in-travenous, the equivalent of 500 mg. to 2 g. of tetracycline hydrochloride daily.

OCCURRENCE
Tetracycline Oral Suspension U.S.P.
Tetracycline for Oral Suspension U.S.P.
Tetracycline Hydrochloride U.S.P.
Tetracycline Hydrochloride Capsules U.S.P.
Tetracycline Hydrochloride for Injection U.S.P.
Tetracycline Hydrochloride for Ophthalmic Solution N.F.
Tetracycline Hydrochloride Tablets N.F.
Tetracycline Phosphate Complex N.F.
Tetracycline Phosphate Complex Capsules N.F.

Rolitetracycline N.F., Syntetrin®, N-(pyrrolidinomethyl)tetracycline, has been introduced for use by intramuscular and intravenous injection. This derivative is made by condensing tetracycline with pyrrolidine and formaldehyde in the presence of t-butyl

N-(pyrrolidinomethyl)tetracycline

alcohol.[63] It is very soluble in water, 1 g. dissolving in about 1 ml., and provides a means of injecting the antibiotic in a small volume of solution. It is recommended in cases for which the oral dosage forms are not suitable.

Category—antibacterial.

Usual dose range—intramuscular, 150 mg. to 350 mg. every 12 hours; intravenous infusion, 350 mg. to 700 mg. every 12 hours

OCCURRENCE
Rolitetracycline for Injection N.F.

Chlortetracycline Hydrochloride N.F., Aureomycin® Hydrochloride. Chlortetracycline was isolated by Duggar[69] in 1948 from Streptomyces aureofaciens. This compound, which was produced in an extensive search for new antibiotics, was the first of the group of highly successful tetracyclines.

It soon became established as a valuable antibiotic with broad-spectrum activities. It is used in medicine chiefly as the acid salt of the compound whose systematic chemical designation is 7-chloro-4-(dimethylamino)-1,4,4a,5,5a,6,11,12a-octahydro-3,6,10,12,12a-pentahydroxy-6-methyl-1,11-dioxo-2-naphthacenecarboxamide. The hydrochloride salt is a crystalline powder having a bright-yellow color that suggested its brand name, Aureomycin. It is stable in air, but is slightly photosensitive and should be protected from light. It is odorless and has a bitter taste. One gram of the hydrochloride salt will dissolve in about 75 ml. of water, producing a pH of about 3. It is only slightly soluble in alcohol and practically insoluble in other organic solvents.

Chlortetracycline hydrochloride is most generally administered orally in capsules to avoid its bitter taste. It may also be administered parenterally (I.V.). The calcium salt is used in liquid preparations for oral use.

The 7-bromo analog of chlortetracycline has been isolated from streptomyces species grown on special media rich in bromide ion. Bromtetracycline has antibiotic properties very similar to chlortetracycline.

Category—antibacterial; antiprotozoan.

Usual dose—250 mg. 4 times a day.

Usual dose range—250 to 500 mg.

OCCURRENCE

Chlortetracycline Hydrochloride Capsules N.F.

Chlortetracycline Hydrochloride for Injection N.F.

Ophthalmic Chlortetracycline Hydrochloride N.F.

Oxytetracycline Hydrochloride U.S.P., Terramycin®. Early in 1950, Finlay *et al.*[70] reported the isolation of oxytetracycline from *Streptomyces rimosus*. It was soon established that this compound was a chemical analog of chlortetracycline and showed similar antibiotic properties. The structure of oxytetracycline was elucidated by Hochstein *et al.*,[71] and this work provided the basis for the confirmation of the structure of the other tetracyclines.

Oxytetracycline hydrochloride is a crystalline compound having a pale-yellow color and a bitter taste. The amphoteric base is only very slightly soluble in water and slightly soluble in alcohol. It is an odorless substance with a slightly bitter taste. It is stable in air but darkens upon exposure to strong sunlight. The hydrochloride salt is a stable yellow powder having a more bitter taste than the free base. It is much more soluble in water, 1 g. dissolving in 2 ml., and also is more soluble in alcohol. Both compounds are inactivated rapidly by alkali hydroxides and by acid solutions below pH 2. Both forms of oxytetracycline are absorbed from the digestive tract rapidly and equally well, so that the only real advantage the free base offers over the hydrochloride salt is its less bitter taste. Oxytetracycline hydrochloride also is used for parenteral administration (I.V. and I.M.). Insoluble calcium dioxytetracycline is used in water-insoluble suspensions for oral administration.

Category—antibacterial; antirickettsial.

Usual dose—oral, the equivalent of 250 mg. of oxytetracycline 4 times daily; intramuscular, 100 mg. 2 or 3 times daily; intravenous infusion, 250 to 500 mg. over a period of ½ to 1 hour twice daily.

Usual dose range—oral, 1 to 4 g. daily; intramuscular, 200 to 500 mg. daily; intravenous infusion, 500 mg. to 2 g. daily.

OCCURRENCE

Oxytetracycline Hydrochloride Capsules U.S.P.

Oxytetracycline Hydrochloride for Injection U.S.P.

Ophthalmic Oxytetracycline Hydrochloride N.F.

Oxytetracycline Injection N.F.

Calcium Oxytetracycline N.F.

Calcium Oxytetracycline Oral Suspension N.F.

Methacycline Hydrochloride N.F., Rondomycin®, 6-Deoxy-6-demethyl-6-methylene-5-oxytetracycline Hydrochloride, 6-Methylene-5-oxytetracycline Hydrochloride. The synthesis of methacycline, reported by Blackwood *et al.*[58] in 1961, was accomplished by chemical modification of oxytetracycline. It has an antibiotic spectrum similar to the other tetracyclines but has a greater potency; about 600 mg. of methacycline is equivalent to 1 g. of tetracycline. Its particular value lies in its longer serum half-life, doses of 300 mg. producing continuous serum antibacterial activity for 12 hours. Its toxic manifestations and contraindications are similar to the other tetracyclines.

The greater stability of methacycline, both *in vivo* and *in vitro*, is a result of the modification at carbon atom 6. The removal of the 6-hydroxy group markedly increases the stability of ring C to both acids and bases, preventing the formation of anhydrotetracyclines by acids and of isotetracyclines by bases. Methacycline hydrochloride is a yellow to dark yellow crystalline powder that is slightly soluble in water and insoluble in nonpolar solvents. It should be stored in tight, light-resistant containers in a cool place.

Category—antibacterial.

Usual dose—600 mg. (equivalent to 560 mg. of methacycline base) daily in divided doses.

OCCURRENCE

Methacycline Hydrochloride Capsules N.F.
Methacycline Hydrochloride Oral
    Suspension N.F.

Demeclocycline N.F., Declomycin®, 7-Chloro-6-demethyltetracycline, was isolated in 1957 by McCormick *et al.*[72] from a mutant strain of *Streptomyces aureofaciens*. Chemically, it is 7-chloro-4-(dimethyamino)-1,4,4a,5,5a,6,11,12a-octahydro-3,6,10,12,12a-pentahydroxy-1,11-dioxo-2-naphthacenecarboxamide. Thus, it differs from chlortetracycline only by the absence of the methyl group on carbon atom 6. The absence of this methyl group enhances the stability of ring C to both acid and alkali.

Demeclocycline is a yellow, crystalline powder that is odorless and has a bitter taste. It is sparingly soluble in water. A 1 per cent solution has a pH of about 4.8. It has an antibiotic spectrum similar to that of other tetracyclines, but it is slightly more active than the others against most of the micro-organisms for which they are used. This, together with its slower rate of elimination through the kidneys, gives demeclocycline an effectiveness comparable with that of the other tetracyclines, at about three fifths of the dose. Like the other tetracyclines, it may cause infrequent photosensitivity reactions that produce erythema after exposure to sunlight. It appears that demeclocycline may produce the reaction somewhat more frequently than the other tetracyclines. The incidence of discoloration and mottling of the teeth in youths found with demeclocycline appears to be as low as with the other tetracyclines.

Category—antibacterial.

Usual dose—600 mg. daily in 4 divided doses of 150 mg. each or 2 divided doses of 300 mg. each.

Usual dose range—150 to 900 mg. per day.

OCCURRENCE

Demeclocycline for Oral Suspension N.F.
Demeclocycline Oral Suspension N.F.
Demeclocycline Hydrochloride N.F.
Demeclocycline Hydrochloride Capsules
    N.F.
Demeclocycline Hydrochloride Tablets N.F.

Doxycycline, Vibramycin®, α-6-Deoxy-5-oxytetracycline. The most recent addition to the tetracycline group of antibiotics available for antibacterial therapy is doxycycline, first reported by Stephens *et al.*[73] in 1958. It was first obtained in small yields by a chemical transformation of oxytetracycline but it is now produced by catalytic hydrogenation of methacycline or by reduction of a benzyl mercaptan derivative of methacycline with Raney nickel. In the latter process a nearly pure form of the 6-α methyl epimer is produced. It is worthy of note that this isomer has the 6-methyl group oriented differently from that in the tetracyclines bearing also a 6-hydroxy group and that the 6-α methyl epimer is more than three times as active as its β-epimer.[74] Apparently the difference in orientation of the methyl groups, slightly affecting the shapes of the molecules, causes a significant difference in biologic effect. Also, as in methacycline, the absence of the 6-hydroxyl group produces a compound that is very stable to acids and bases and that has a long biologic half-life. In addition, it is very well absorbed from the gastrointestinal tract, thus allowing a smaller dose to be administered.

Doxycycline is available as the hyclate salt, a hydrochloride salt solvated as the hemiethanolate hemihydrate, and as the monohydrate. The hyclate form is sparingly soluble in water and is used in the capsule dosage form; the monohydrate is water-insoluble and is used for aqueous suspensions which are stable periods up to 2 weeks when kept in a cool place.

Category—antibacterial.

Usual dose—100 mg. every 12 hours during the first day of treatment followed by a maintenance dose of 100 mg. daily.

THE MACROLIDES

Among the many antibiotics isolated from the actinomycetes is the group of chemically related compounds called the macrolides. It was in 1950 that picromycin, the first of this group to be identified as a macrolide compound, was first reported. In 1952, erythromycin and carbomycin were reported as new antibiotics and these were followed in subsequent years by other macrolides. At the present time more than three dozen such compounds are known, and new ones are likely to appear in the future. Of all of these, only two, erythromycin and oleandomycin, are available for medical use in the United States. One other, carbomycin, has been available, but, because of its poor and irregular absorption from the gastrointestinal tract and its inferior antibacterial activity when compared to erythromycin, it never enjoyed wide use and was withdrawn from the market. Spiramycin is used in Europe and other parts of the world, but its activity in vitro is inferior to that of erythromycin, and it is difficult to account for its reputed therapeutic success. The spiramycins (also called foromacidins) are elaborated as three closely related compounds by *Streptomyces spectabilis,* but it is spiramycin I (foromocidin A, Trobicin) that has been used in Europe as an antiinfective because it is claimed to have a high affinity for tissue and to be eliminated more slowly than other macrolides. Leucomycin, from *Streptomyces kitasatoensis,* has been used elsewhere in the world but has not achieved important success. Many different leucomycins have been isolated and identified. The macrolides generally are active against gram-positive organisms and inactive against gram-negative organisms, but some exceptions have been noted.

The macrolide antibiotics have three common chemical characteristics: (1) a large lactone ring (which prompted the name *macrolide*), (2) a ketone group, and (3) a glycosidically linked amino sugar. Usually, the lactone ring has 12, 14 or 16 atoms in it and is often partially unsaturated with an olefinic group conjugated with the ketone function. (The polyene macrocyclic lactones, such as pimaricin, and the polypeptide lactones generally are not included among the macrolide antibiotics.) They may have, in addition to the amino sugar, a neutral sugar that is glycosidically linked to the lactone ring (see erythromycin). Because of the presence of the dimethylamino group on the sugar moiety, the macrolides are bases with pKa values between 6.0 and 9.0. This feature has been employed to make clinically useful salts. The free bases are only slightly soluble in water but dissolve in the somewhat polar organic solvents. They are stable in aqueous solutions at or below room temperature but are inactivated by acids, bases and heat. The chemistry of the macrolide antibiotics has been reviewed by Wiley,[75] Miller,[76] and Morin and Gorman.[77]

PRODUCTS

Erythromycin U.S.P., E-Mycin®, Erythrocin®, Ilotycin®. Early in 1952, McGuire et al.[78] reported the isolation of erythromycin from *Streptomyces erythreus.* It achieved a rapid acceptance as a well-tolerated antibiotic of value in the treatment of staphylococcic, beta-hemolytic, streptococcic and pneumococcic infections. It is also useful for the treatment of acute and chronic intestinal amebiasis. It has proved to be effective against a number of organisms that have developed resistance to penicillin, the tetracyclines and streptomycin. Its chemical structure was reported by Wiley et al.[79] in 1957 and its stereochemistry by Celmer[80] in 1965.

The amino sugar attached through a glycosidic link to the number 5 carbon atom is desosamine, a structure found in a number of other macrolide antibiotics. The tertiary amine of desosamine (3,4,6-trideoxy-3-

dimethylamino-D-*xylo*-hexose) confers a basic character to erythromycin and provides the means by which acid salts may be prepared. The other carbohydrate structure linked as a glycoside to carbon atom 3 is called cladinose (2,3,6-trideoxy-3-methoxy-3-C-methyl-L-*ribo*-hexose) and is unique to the erythromycin molecule.

Erythromycin A

Erythromycin is a very bitter, white or yellowish-white crystalline powder. It is soluble in alcohol and in the other common organic solvents but only slightly soluble in water. Saturated aqueous solutions develop an alkaline pH in the range of 8.0 to 10.5. It is extremely unstable at a pH of 4 or lower. The optimum pH for stability of erythromycin is at or near neutrality.

As the free base, erythromycin may be used in oral dosage forms and for topical administration. However, to overcome its bitter taste and to provide more acceptable pharmaceutical forms for its administration, derivatives of erythromycin are commonly used. These derivatives are of two types: acid salts of the dimethylamine such as the glucoheptonate, the lactobionate and the stearate; and esters of the OH group on the desosamine moiety such as the ethyl carbonate, the ethyl succinate and the propionate (estolate). The carbonate ester is hydrolyzed in the gastrointestinal tract to the active base before absorption. However, the ethyl succinate and propionate esters are biologically active themselves and need not be hydrolyzed to show anti-infective action. When administered orally, these compounds may be partially hydrolyzed, but are ab-

sorbed as the esters to an appreciable extent. It is claimed that these esters provide a more rapid onset and higher and more prolonged therapeutic concentration in the blood.

The glucoheptonate (gluceptate) and lactobionate salts are water-soluble, thus providing means for the parenteral administration of erythromycin. The stearate salt is water-insoluble and tasteless and is used in tablets and suspensions. The ethyl carbonate ester is also water-insoluble and is used for pediatric suspensions.

As is common with other macrolide antibiotics, compounds closely related to erythromycin have been obtained from culture filtrates of S. *erythreus*. Two such analogs have been found and are designated as erythromycins B and C. Erythromycin B differs from erythromycin A only at the number 12 carbon atom where a hydrogen has replaced the hydroxyl group. The B analog is more acid stable but has only about 80 per cent of the activity of erythromycin. The C analog differs from erythromycin by the replacement of the methoxyl group on the cladinose moiety by a hydrogen atom. It appears to be as active as erythromycin but is present in very small amounts in fermentation liquors.

The mode of action of erythromycin has not been completely established but it appears to inhibit protein synthesis by binding to the 50 S ribosomal subunit.[81] Strains of staphylococci often develop a resistance to erythromycin. Cross resistance with other macrolides appears to be of a high order. Streptococcal infections are especially amenable to erythromycin therapy.

Category—antibacterial.

Usual dose—oral, as the base, the ethyl carbonate or the stearate, the equivalent of 250 mg. of erythromycin 4 times daily; intravenous, as the gluceptate or the lactobionate, the equivalent of 250 mg. of erythromycin 3 or 4 times daily.

Range—oral or intravenous, 1 to 4 g. daily.

OCCURRENCE
Erythromycin Ethylcarbonate U.S.P.
Erythromycin Ethylcarbonate for Oral
    Suspension U.S.P.
Sterile Erythromycin Gluceptate U.S.P.

Erythromycin Lactobionate for Injection U.S.P.
Erythromycin Stearate U.S.P.
Erythromycin Stearate Tablets U.S.P.
Erythromycin Estolate N.F.
Erythromycin Estolate Capsules N.F.
Erythromycin Estolate for Oral Suspension N.F.
Erythromycin Estolate Tablets N.F.
Erythromycin Ethylsuccinate N.F.
Erythromycin Ethylsuccinate Injection N.F.
Erythromycin Ethylsuccinate for Oral Suspension N.F.
Erythromycin Ethylsuccinate Tablets N.F.

Oleandomycin Phosphate N.F., Matromycin®. A second macrolide antibiotic in medical use is oleandomycin. It was isolated by Sobin, English and Celmer[82] in 1955 from *Streptomyces antibioticus.* By controlling conditions of the fermentation procedure, a closely related compound, oleandomycin B, may be developed. The structure of oleandomycin was first reported by Hochstein *et al.*[83] and its stereochemistry reported by Cellmer.[84] It has been shown to contain 2 sugars and a complex lactone ring designated oleandolide. One of the sugars (left, below) is desosamine, the same amino sugar that occurs in erythromycin. The other sugar is L-oleandrose and, like desosamine, is linked glycosidically to oleandolide.

Oleandolide is a 14-atom ring that contains an exocyclic methylene epoxide on carbon atom 8. Mild alkali causes the loss of the hydroxyl group at carbon atom 11 by β-elimination to produce anhydro-oleandomycin.

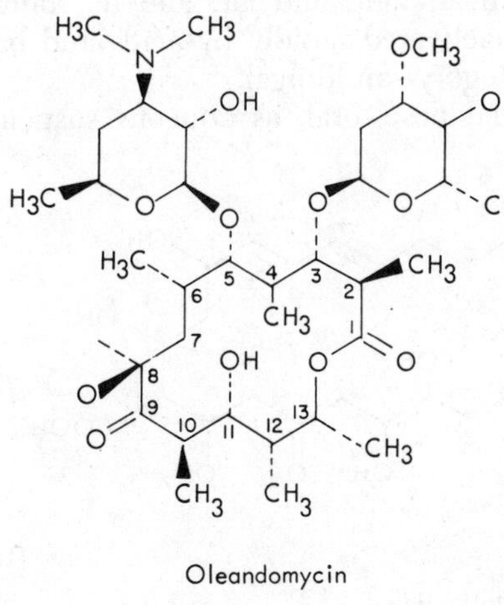

Oleandomycin

Oleandomycin phosphate is a white crystalline powder that is soluble in water. Through the amino group of the desosamine structure a number of acid salts have been prepared, but none of these provides an important increase in the absorption of the antibiotic. The phosphate salt is most commonly employed in medicine and may be administered intravenously and intramuscularly as well as given orally. Its main action is seen against gram-positive bacteria and its main use has been in the treatment of infections refractory to the older and more widely used antibiotics. It has been used as triacetyloleandomycin in combination with tetracycline (Signemycin) on the basis that it provides a synergistic effect and provides protection against resistant microorganisms.

Category—antibacterial.

Usual dose range—intravenous, 1 to 3 g. daily in divided doses; intramuscular, 200 mg. every 6 to 8 hours.

OCCURRENCE
Sterile Oleandomycin Phosphate N.F.

Troleandomycin N.F., Triacetyloleandomycin; Cyclamycin®, TAO®. Oleandomycin contains 3 free hydroxyl groups that are susceptible to acylation. Each of the sugar structures contains one —OH group and there is an additional —OH on the lactone ring. The triacetyl derivative retains the antibiotic activity of the parent compound. Triacetyloleandomycin has been found to achieve more rapid and higher blood-level concentrations than the phosphate salt of the parent antibiotic and has the additional advantage of being practically tasteless. It is stable and practically insoluble in water. Its antibiotic uses are the same as for oleandomycin.

Category—antibacterial.

Usual dose—250 mg. 4 times daily.

Usual dose range—250 to 500 mg.

OCCURRENCE
Troleandomycin Capsules N.F.
Troleandomycin Oral Suspension N.F.

THE POLYENES

A number of antibiotics are known to contain a conjugated polyene system as a characteristic chemical grouping. Rather surprisingly, such antibiotics often show similar antifungal activity, which suggests a structure-activity relationship for which there is not yet a satisfactory explanation. Among the polyenes are a group of macrocyclic lactones that show some degree of chemical relationship. They differ from the macrolide antibiotics of the erythromycin type by having a larger lactone ring in which there is a conjugated polyene system. Many of them contain a glycosidically linked sugar such as the aminodesoxyhexose, mycosamine, that is present in amphotericin B, nystatin, pimaricin and some others. The macrolide polyenes are sometimes classified by the number of double bonds present in the conjugated group, into tetraenes, pentaenes, hexaenes and heptaenes. Characteristic ultraviolet absorption spectra are used as the basis for the classification determination.

The macrolide polyenes include three antibiotics that are used as antifungal agents in the United States. They are amphotericin B, candicidin and nystatin. Others that have received varying amounts of attention by research workers are ascosin, candidin, filipin, fungichromin, perimycin, pimaricin, rimocidin and trichomycin. A complete structure for one of these, pimaricin, has been reported by Golding *et al.*,[85] but, for the most part, the knowledge of their structures is still incomplete. Their general lack of water solubility, their poor stability and their rather toxic properties have contributed to their failure to achieve an important place in therapy. To improve their usefulness, their amphoteric characteristics have been overcome by acylating the amino group of the sugar function and then form-

ing water-soluble salts of the free carboxyl group on the macrolide ring with alkaline bases.[86]

A tetraene that is not a macrolide and does not possess either antibacterial or antifungal activity is fumagillin. However, it possesses significant amebicidal activity.

Nystatin U.S.P.; Mycostatin®. In 1951, Hazen and Brown[87] reported the isolation of nystatin from a strain of *Streptomyces noursei*. It has established itself in human therapy as a valuable agent for the treatment of both internal and local infections of *Candida albicans*. Its success against other monilial infections is less impressive, but it shows in-vitro activity against many yeasts and molds. Its dosage is expressed in terms of units. One mg. of nystatin contains not less than 2,000 U.S.P. units.

The chemical identity of nystatin has not yet been completely elucidated. The U.S.P. gives the molecular formula of nystatin as $C_{46}H_{77}NO_{19}$. However, Ikeda *et al.*[88] assigned a molecular formula of $C_{47}H_{75}NO_{18}$ to it and proposed that nystatinolide, the aglycon portion of the molecule, has the molecular formula $C_{41}H_{64}O_{15}$ and the structure shown. The sugar moiety of the molecule is mycosamine, 3,6-dideoxy-3-amino-D-mannose. The position of the glycosidic link of the sugar to the aglycon is assumed to be through one of the hydroxyl groups near the lactone portion of the aglycon.

Nystatin is a yellow to light tan powder that has a cereal-like odor. It is very slightly soluble in water and only sparingly soluble in nonpolar solvents. It is unstable to moisture, heat, light and air, and its solutions are inactivated rapidly by acids and bases.

Category—antifungal.

Usual dose—oral, as aqueous suspension

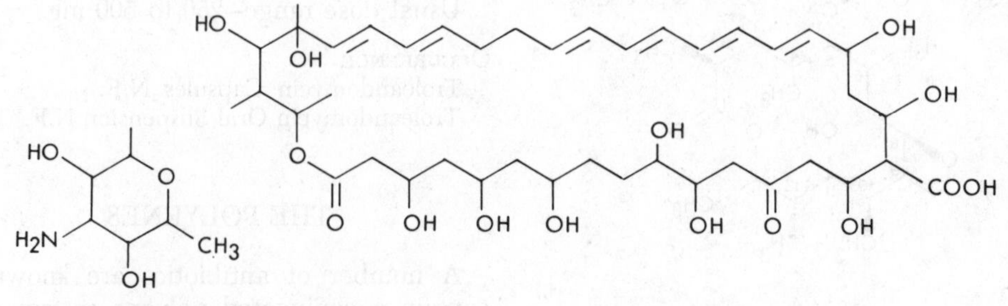

Mycosamine

Nystatinolide

or tablets, 500,000 units 3 times a day; intravaginal, as tablets, 100,000 units 2 times a day.

Usual dose range—oral, 1,500,000 to 6,000,000 units daily.

For external use—topically, as a 100,000-units-per g. ointment or suspension, twice daily.

OCCURRENCE
Nystatin Ointment U.S.P.
Nystatin for Oral Suspension U.S.P.
Nystatin Tablets U..S.P.

Amphotericin B U.S.P., Fungizone®. A polyene antibiotic having potent antifungal action was reported in 1956 by Gold *et al.*[89] to be produced from a Streptomyces species isolated from a sample of soil obtained from the Orinoco River in Venezuela. The species name *Streptomyces nodosus* has been given to this organism. The antibiotic material was shown to contain two closely related substances that were given the names amphotericins A and B. The B compound is the more active and in purified form is being used for its broad-spectrum activity against a number of deep-seated and systemic infections caused by yeastlike fungi. It does not exhibit any activity against bacteria, protozoa or viruses.

As its name indicates, this compound is an amphoteric substance that at its isoelectric point is water-soluble. It also is insoluble in methanol, chloroform and ether. Dutcher *et al.*[90] have tentatively assigned the molecular formula $C_{46}H_{73}NO_{20}$ to amphotericin B and indicate that it is a heptaene. Its structure is not known, but it is interesting to note that mycosamine, the same sugar that has been found in nystatin, is glycosidically linked in amphotericin B. It also contains carboxyl and lactone groups.

Amphotericin B is very poorly absorbed from the gastrointestinal tract, and so its preferred route of administration is intravenous infusion. Since aqueous solutions deteriorate rapidly and should not be used after 24 hours, it is available only as the dry powder that is to be dissolved in 5 per cent dextrose solution just before use. The dry powder, as well as any solution made for a day's use, should be stored in a refrigerator and protected from light.

Category—antifungal.

Usual dose—intravenous infusion, 250 mcg. per kg. of body weight in 500 ml. of 5 per cent Dextrose Injection over a period of 6 hours.

Usual dose range—100 to 250 mcg. per kg. every 2 to 4 days to 1.5 mg. per kg. every other day for 4 to 8 weeks.

OCCURRENCE
Amphotericin B for Injection U.S.P.

Candicidin N.F., Candeptin®. The macrolide polyene antibiotic candicidin was isolated in 1953 by Lechevalier *et al.*[91] from a strain of *Streptomyces griseus*. Although its potent antifungal property had been known for some time, it was not until 1964 that it became available for medicinal use in the United States. It is recommended for use in the treatment of monilia infections of the vaginal tract. Its chemistry is not yet well known but it is a heptaene macrolide closely related to amphotericin B. It is available as a 3-mg. vaginal tablet and as a vaginal ointment containing 3 mg. of candicidin per 5 g. of ointment.

Category—Local antifungal.

Application—vaginal, 3 mg. inserted twice daily for 14 days.

OCCURRENCE
Candicidin Ointment N.F.
Candicidin Suppositories N.F.

THE POLYPEPTIDES

Among the most powerful bactericidal antibiotics are those possessing a polypeptide structure. Many of them have been isolated but, unfortunately, their clinical use has been limited by their undesirable side-reactions, particularly renal toxicity. The chief source of the medicinally important members of this class has been various species of the genus *Bacillus*. A few have been isolated from other bacteria but have not gained a place in medical practice. Three medicinally useful polypeptide antibiotics have been isolated from a *Streptomyces* species.

Polypeptide antibiotics are of three main types: neutral, acidic and basic. It had been presumed that the neutral compounds such as the gramicidins possessed cyclopeptide structures and thus had no free amino or carboxyl groups. It has been shown that the neutrality is due to the formylation of a terminal amino group and that the neutral gramicidins are linear rather than cyclic. The acidic compounds have free carboxyl[92] groups, indicating that at least part of the structure is noncyclic. The basic compounds have free amino groups and, similarly, are noncyclic at least in part. Some, like the gramicidins, are active against gram-positive organisms only; others, like the polymyxins, are active against gram-negative organisms and thus have achieved a special place in antibacterial therapy. Significant comments about the biosynthesis and structure-activity relationships of peptide antibiotics have been published by Bodanszky and Perlman.[93]

Tyrothricin N.F. Tyrothricin is a mixture of polypeptides usually obtained by extraction of cultures of *Bacillus brevis*. It was isolated in 1939 by Dubos[94] in a planned search to find an organism growing in soil that would have antibiotic activity against human pathogens. Having only limited use in therapy now, it is of historical interest as the first in the series of modern antibiotics. Tyrothricin is a white to slightly gray or brownish-white powder with little or no odor or taste. It is practically insoluble in water and is soluble in alcohol and in dilute acids. Suspensions for clinical use can be prepared by adding an alcoholic solution to calculated amounts of distilled water or isotonic saline solutions.

Tyrothricin is a mixture of two groups of antibiotic compounds, the gramicidins and the tyrocidines. Gramicidins are the more active components of tyrothricin, and this fraction, occurring in 10 to 20 percent quantities in the mixture, may be separated and used in topical preparations for the antibiotic effect. Five gramicidins, A_1, A_2, B_1, B_2, and C, have been identified. Their structures have been proposed and confirmed through synthesis by Sarges and Witkop.[95] It may be noted that the gramicidins A differ from the gramicidins B by having a tryptophan moiety substituted by an L-phenylalanine moiety. In gramicidin C, a tyrosine moiety substitutes for a tryptophan moiety. In both of the gramicidin A and B pairs, the only difference is the amino acid located at the end of the chain having the neutral formyl group on it. If that amino acid is valine, the compound is either valine-gramicidin A or valine-gramicidin B. If that amino acid is isoleucine, the compound is isoleucine-gramicidin, either A or B.

$$HC = O$$
$$|$$
$$\overset{OH}{\underset{|}{(CH_2)_2}}$$
$$|$$
L-Val-Gly-L-Ala-D-Leu-L-Ala-D-Val-L-Val-D-Val-L-Try-D-Leu-L-Try-D-Leu-L-Try-D-Leu-L-Try-NH$$

Valine-gramicidin A

$$HC = O$$
$$|$$
$$\overset{OH}{\underset{|}{(CH_2)_2}}$$
$$|$$
L-Ileu-Gly-L-Ala-D-Leu-L-Ala-D-Val-L-Val-D-Val-L-Try-D-Leu-L-Try-D-Leu-L-Try-D-Leu-L-Try-NH$$

Isoleucine-gramicidin A

$$HC = O$$
$$|$$
$$\overset{OH}{\underset{|}{(CH_2)_2}}$$
$$|$$
L-Val-Gly-L-Ala-D-Leu-L-Ala-D-Val-L-Val-D-Val-L-Try-D-Leu-L-Phel-D-Leu-L-Try-D-Leu-L-Try-NH$$

Valine-gramicidin B

$$HC = O$$
$$|$$
$$\overset{OH}{\underset{|}{(CH_2)_2}}$$
$$|$$
L-Ileu-Gly-L-Ala-D-Leu-L-Ala-D-Val-L-Val-D-Val-L-Try-D-Leu-L-Phel-D-Leu-L-Try-D-Leu-L-Try-NH$$

Isoleucine-gramicidin B

Tyrocidine is a mixture of tyrocidines A, B, C and D whose structures have been determined by Craig and co-workers.[95]

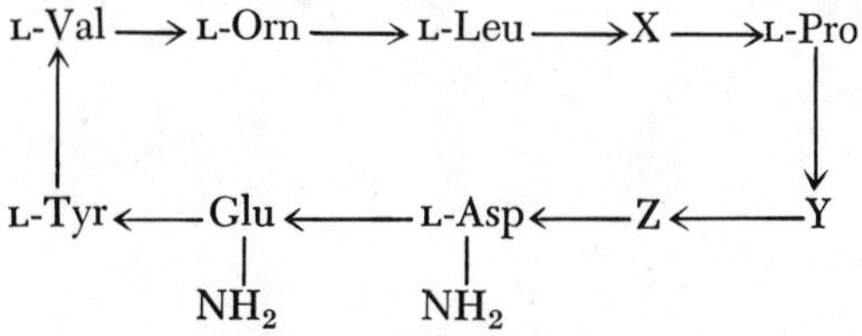

$$\text{L-Val} \rightarrow \text{L-Orn} \rightarrow \text{L-Leu} \rightarrow \text{X} \rightarrow \text{L-Pro}$$

$$\text{L-Tyr} \leftarrow \text{Glu} \leftarrow \text{L-Asp} \leftarrow \text{Z} \leftarrow \text{Y}$$

$$\text{NH}_2 \qquad \text{NH}_2$$

	X	Y	Z
Tyrocidine A	D-Phe	D-Phe	D-Phe
Tyrocidine B	D-Phe	L-Try	D-Phe
Tyrocidine C	D-Try	L-Try	D-Phe
Tyrocidine D	D-Try	L-Try	D-Try

The synthesis of tyrocidine A has been reported by Ohno *et al.*[96]

Gramicidin-S (referring to "Soviet gramicidin" has been isolated from a strain of *B. brevis* and has antibiotic and chemical properties much like the tyrocidines. Its structure has been determined and it has been synthesized. It is not used in medicine in the United States.

Tyrothricin and gramicidin are effective primarily against gram-positive organisms. Their use is restricted to local applications. The ability of tryothricin to cause lysis of erythrocytes makes it unsuitable for the treatment of systemic infections. Its applications should avoid direct contact with the bloodstream through open wounds or abrasions. It is ordinarily safe to use tyrothricin in troches for throat infections, as it is not absorbed from the gastrointestinal tract.

Category—antibacterial.

Application—topically, to the skin and mucous membrane in dosage forms which contain from 0.05 to 0.3 per cent of tyrothricin.

OCCURRENCE

Gramicidin N.F.

Bacitracin U.S.P. The organism from which Johnson, Anker and Meleney[97] produced bacitracin in 1945 is a strain of *Bacillus subtilis*. The organism had been isolated from débrided tissue from a compound fracture in 7-year-old Margaret Tracy, hence the name bacitracin. Production of bacitracin is now accomplished from the licheniformis group (Sp. *Bacillus subtilis*). Like tyrothricin, the first useful antibiotic obtained from bacterial cultures, bacitracin is a complex mixture of polypeptides. So far, at least 10 polypeptides have been isolated by countercurrent distribution technics: A, A', B, C, D, E, F_1, F_2, F_3 and G. It appears that the commercial product known as bacitracin is a mixture principally of A with smaller amounts of B, D, E and F.

The official product is a white to pale-buff powder that is odorless or nearly so. In the dry state, bacitracin is stable, but it rapidly deteriorates in aqueous solutions at room temperature. Because of its hygroscopic nature, it must be stored in tight containers, preferably under refrigeration. The stability of aqueous solutions of bacitaricin is affected by pH and temperature. Slightly acidic or neutral solutions are stable for as long as 1 year if kept at a temperature of 0 to 5°. If the pH rises above 9, inactivation occurs very rapidly. For greatest stability, the pH of a bacitracin solution is best adjusted at 4 to 5 by the simple addition of acid. The salts of heavy metals precipitate bacitracin from its solutions, with resulting inactivation. In addition to being soluble in water, bacitracin is soluble in low-molecular-weight alcohols but is insoluble in many other organic solvents, including acetone, chloroform and ether.

The principal work on the chemistry of the bacitracins has been directed toward bacitracin A, the component in which most of the antibacterial activity of crude bacitracin resides. The structure shown on page 382 is that proposed by Stoffel and Craig[98] but it has not yet been confirmed by synthesis.

The chemistry of the other bacitracins has been worked on only to a limited extent. While there is evidence of considerable similarities in structure to bacitracin A among the other members of the group, there is considerable difficulty in fixing the dissimilarities that do exist.

The activity of bacitracin is measured in units. The potency per mg. is not less than 40 U.S.P. Units except for material prepared for parenteral use which has a potency of not less than 50 Units per mg. It is a bactericidal antibiotic that is active

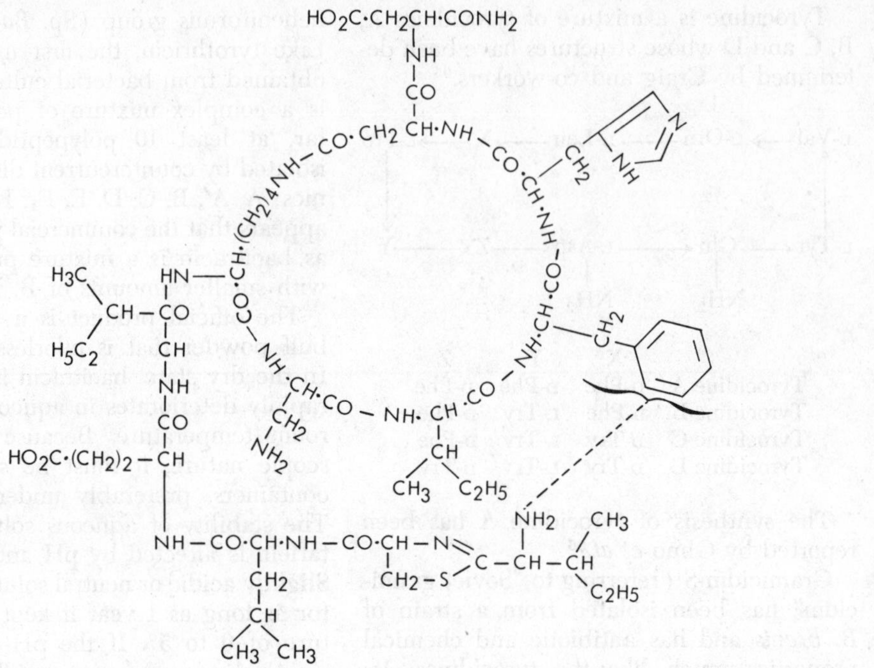

Bacitracin A

against a wide variety of gram-positive organisms, some gram-negative organisms and a few others. Although bacitracin has found its widest use in topical preparations for local infections, it is quite effective in a number of systemic and local infections when administered parenterally. It is not absorbed from the gastrointestinal tract, so oral administration is without effect except for the treatment of amebic infections within the alimentary canal.

Category—antibacterial.

Usual dose—intramuscular, 10,000 to 20,000 units three to four times a day.

Usual dose range—intramuscular, 30,000 to 100,000 units daily.

For external use—topically to the skin, as a 500-units-per g. ointment 2 or 3 times daily.

OCCURRENCE

Bacitracin Ointment U.S.P.
Sterile Bacitracin U.S.P.
Zinc Bacitracin U.S.P.

Polymyxin B Sulfate U.S.P., Aerosporin® Sulfate. Polymyxin was discovered in 1947 almost simultaneously in three laboratories in America and Great Britain.[99, 100, 101]

As often happens when similar discoveries are made in widely separated laboratories, differences in nomenclature referring both to the antibiotic-producing organism and the antibiotic itself appeared in references to the polymyxins. Since it now has been shown that the organisms first designated as *Bacillus polymyxa* and *B. aerosporus* *Greer* are identical species, the one name, *B. polymyxa,* is used to refer to all of the strains that produce the closely related polypetides called polymyxins. Other organisms (see colistin, for example) also produce polymyxins. Identified so far are polymyxins A, B_1, B_2, C, D_1, D_2, M, colistin A (polymyxin E_1), colistin B, (polymyxin E_2), circulins A and B, and polypeptin. The known structures of this group and their properties have been reviewed by Vogler and Studer.[102] Of these, polymyxin B as the sulfate is usually used in medicine because, when used systemically, it causes less kidney damage than the others.

Polymyxin B sulfate is a nearly odorless, white to buff-colored powder. It is freely soluble in water and slightly soluble in alcohol. Its aqueous solutions are slightly

$$
\begin{array}{c}
\text{NH}\!-\!\text{CO} \\
\text{C}_6\text{H}_5\text{CH}_2\!-\!\text{CH} \quad \text{CH-CH}_2\text{CH}_2\text{NH}_2 \\
\text{CO} \qquad \text{NH} \\
\text{NH} \qquad \text{CO} \qquad \text{CH}_2\text{CH}_2\text{NH}_2 \qquad \text{CH}_2\text{CH}_2\text{NH}_2 \qquad \text{CH}_3 \\
(\text{H}_3\text{C})_2\,\text{CHCH}_2\!-\!\text{CH} \quad \text{CH-NHCO}\!-\!\text{C}\!-\!\text{NH}\!-\!\text{CO}\!-\!\text{CH}\!-\!\text{NH}\!-\!\text{CO}\!-\!\text{CH}\!-\!\text{NH}\!-\!\text{CO}\!-\!(\text{CH}_2)_4\!-\!\text{CHCH}_2\text{CH}_3 \\
\text{CO} \qquad \text{CH}_2 \qquad \text{H} \qquad \text{CHOHCH}_3 \\
\text{NH} \qquad \text{CH}_2 \\
\text{H}_2\text{NCH}_2\text{CH}_2\!-\!\text{CH} \quad \text{NH} \\
\text{CO} \qquad \text{CO} \\
\text{NH} \qquad \text{CH-CHOHCH}_3 \\
\text{H}_2\text{NCH}_2\text{CH}_2\!-\!\text{CH} \quad \text{NH} \\
\text{CO}
\end{array}
$$

Polymyxin B_1

acidic or nearly neutral (pH 5 to 7.5) and, when refrigerated are stable for at least 6 months. Alkaline solutions are unstable. Polymyxin B has been shown by Hausman and Craig,[103] who used countercurrent distribution technics, to contain two fractions that differ in structure only by one fatty acid component. Polymyxin B_1 contains (+)-6-methyloctan-1-oic acid (isopelargonic acid), a fatty acid isolated from all of the other polymyxins. The B_2 component contains iso-octanoic acid, $C_8H_{16}O_2$, of undetermined structure. The structural formula for polymyxin B has been proved by the synthesis accomplished by Vogler *et al.*[104]

Polymyxin B sulfate is useful against many gram-negative organisms. Its main use in medicine has been in topical applications for local infections in wounds and burns. For such use it is frequently combined with bacitracin, which is effective against gram-positive organisms. Polymyxin B sulfate is poorly absorbed from the gastrointestinal tract, so oral administration of it is of value only in the treatment of intestinal infections such as pseudomonas enteritis or those due to Shigella. It may be given parenterally by intramuscular or intrathecal injection for systemic infections. The dosage of polymyxin is measured in U.S.P. Units. One mg. contains not less than 6,000 U.S.P. Units.

Category—antibacterial.

Usual dose—intramuscular, 2,500 to 5,000 units per kg. of body weight 4 times a day; intravenous infusion, 5,000 to 10,000 units per kg. of body weight in 200 to 500 ml. of 5 per cent Dextrose Injection over a period of 60 to 90 minutes twice a day.

Usual dose range—intramuscular, up to 1,500,000 units daily; intravenous, up to 2,000,000 units daily.

For external use—topically, as an ointment containing 20,000 units in each g. or as a solution containing 10,000 to 25,000 units in each ml. in a wet dressing.

OCCURRENCE

Polymyxin B Sulfate Ointment U.S.P.
Polymyxin B Sulfate Tablets N.F.
Sterile Polymyxin B Sulfate U.S.P.
Neomycin Sulfate, Polymyxin B Sulfate and Gramicidin Ophthalmic Solution N.F.
Neomycin Sulfate, Polymyxin B Sulfate and Zinc Bacitracin Ointment N.F.

Colistin Sulfate N.F., Coly-mycin®. In 1950, Koyama and co-workers[105] isolated an antibiotic from *Aerobacillus colistinus* (*B. polymyxa* var. *colistinus*) that has been given the name colistin. It had been used in Japan and in some European countries for a number of years before it was made available for medicinal use in the United States. It is especially recommended for the treatment of refractory urinary tract infections caused by gram-negative organisms such as *Aerobacter, Bordetella, Escherichia, Klebsiella, Pseudomonas, Salmonella* and *Shigella*.

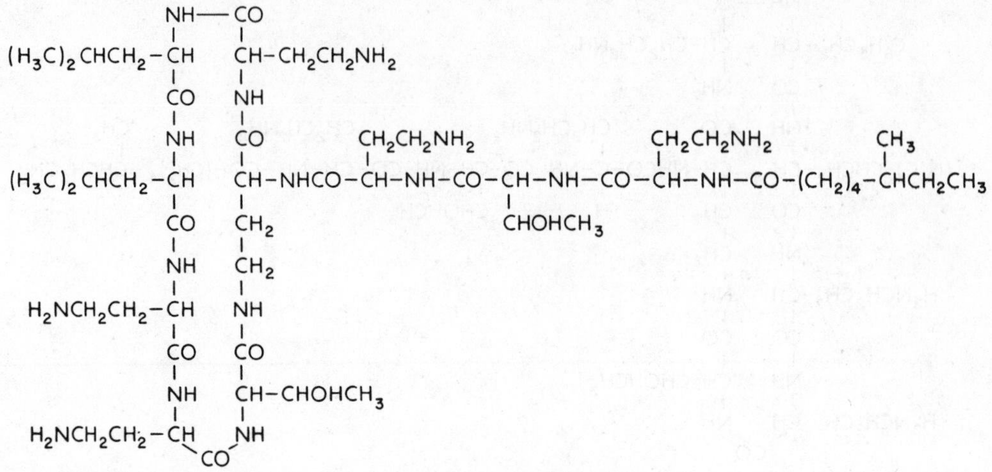

$$
\begin{array}{c}
\text{NH} \text{—} \text{CO} \\
| \quad\quad | \\
(H_3C)_2\,CHCH_2 - CH \quad\quad CH - CH_2CH_2NH_2 \\
| \quad\quad\quad | \\
CO \quad\quad NH \\
| \quad\quad\quad | \\
NH \quad\quad CO \quad\quad CH_2CH_2NH_2 \quad\quad\quad CH_2CH_2NH_2 \quad\quad CH_3 \\
| \quad\quad\quad | \quad\quad\quad\quad | \quad\quad\quad\quad\quad | \quad\quad\quad | \\
(H_3C)_2\,CHCH_2 - CH \quad CH - NHCO - CH - NH - CO - CH - NH - CO - CH - NH - CO - (CH_2)_4 - CHCH_2CH_3 \\
| \quad\quad\quad | \quad\quad\quad\quad\quad\quad\quad CHOHCH_3 \\
CO \quad\quad CH_2 \\
| \quad\quad\quad | \\
NH \quad\quad CH_2 \\
| \quad\quad\quad | \\
H_2NCH_2CH_2 - CH \quad NH \\
| \quad\quad\quad | \\
CO \quad\quad CO \\
| \quad\quad\quad | \\
NH \quad\quad CH - CHOHCH_3 \\
| \quad\quad\quad | \\
H_2NCH_2CH_2 - CH \quad NH \\
\diagdown CO
\end{array}
$$

Colistin A (Polymyxin E_1)

Chemically, colistin is a polypetide that has been reported by Suzuki *et al.*[106] to be heterogeneous with the major component being colistin A. They proposed the structure above for colistin A, which may be noted to differ from polymyxin B_1 only by the substitution of D-leucine for D-phenylalanine as one of the amino-acid fragments in the cyclic portion of the structure. Wilkinson and Lowe[107] have corroborated the structure and have shown colistin A to be identical with polymyxin E_1. Some additional confusion in nomenclature for this antibiotic exists, as Koyama *et al.* originally named the product colimycin, and that name is still used. Particularly, it has been the basis for variants used as brand names such as Coly-Mycin®, Colomycin®, Colimycine® and Colimicina®.

Two forms of colistin have been made, the sulfate and methanesulfonate, and both forms are available for use in the United States. The sulfate is used to make an oral pediatric suspension; the methanesulfonate is used to make an intramuscular injection. In the dry state, the salts are stable, and their aqueous solutions are relatively stable at acid pH from 2 to 6. Above pH 6, solutions of the salts are much less stable.

Category—antibacterial (intestinal).

Usual dose—3 to 5 mg. per kg. of body weight in 3 divided doses.

OCCURRENCE

Colistin Sulfate for Oral Suspension N.F.

Sodium Colistimethate U.S.P.; Coly-Mycin M®; Pentasodium Colistinmethanesulfonate; Sodium Colistimethanesulfonate. In colistin, four of the terminal amino groups of the α, γ-aminobutyric acid fragment may be readily alkylated. In sodium colistimethate, the methanesulfonate radical is the attached alkyl group and, through each of them, a sodium salt may be made. This provides a highly water-soluble compound that is very suitable for injection. In the injectable form, it is given intramuscularly and is surprisingly free from toxic reactions as compared with Polymyxin B. Sodium colistimethate does not readily induce the development of resistant strains of microorganisms, and no evidence of cross resistance with the common broad-spectrum antibiotics has been shown. It is used for the same conditions as those mentioned for colistin.

Category—antibacterial.

Usual dose—intramuscular, the equivalent of 1 mg. of the base per kg. of body weight 2 to 4 times a day.

Usual dose range—1.5 to 5 mg. per kg. daily.

OCCURRENCE

Sodium Colistimethate for Injection U.S.P.

THE ACTINOMYCINS

One of the largest groups of chemically related antibiotics are the actinomycins, which were first isolated from actinomycetes in 1940 by Waksman and Woodruff.[108] Since that time, many naturally occurring and biosynthetically modified actinomycins have been isolated from various genera of the actinomycetes and particularly from species of *Streptomyces*. The nomenclature of actinomycins has been complicated by the use of both letters and Roman numerals to refer to the various members of the group. In addition, numerical subscripts are used as notations to refer to compounds that belong to the same actinomycin complex. Thus, one sees references to actinomycins A, C, D, F, X, etc., to actinomycins C_1, C_2 and C_3 and to actinomycins I, II, III and IV.

These antibiotics have attracted great interest because of their potent antitumor effects and their successful use in the treatment of several different human neoplasms. Of less interest, because of their cytotoxic effect, is their bacteriostatic activity, particularly against gram-positive organisms. A review of their importance in the treatment of tumors has been published.[109] Their potential usefulness has stimulated interest in other cytotoxic antibiotics, such as puromycin, for cancer therapy.

The actinomycins are yellow or red peptide-containing derivatives of phenoxazine. The chemistry of these compounds has been determined largely through the efforts of Brockmann[110, 111] and Johnson[112] and their colleagues. The chromophore of the molecule is 2-amino-4,6-dimethyl-3-oxophenoxazine-1,9-dicarboxylic acid (called actinosin), and this nucleus appears in all of the naturally occurring actinomycins. Through both the 1 and the 9 carboxyl groups there are attached cyclic polypeptides which may have varying amino acid components to produce the various actinomycins. The formula shown for dactinomycin (actinomycin C_1 actinomycin D; also called actinomycin IV) illustrates the arrangement typical of these compounds.

It may be noted that in dactinomycin the component amino acids in the cyclic peptides attached to carbon atom 1 (group A) and to carbon atom 9 (group B) are identical in nature and arrangement. In the actinomycin C series, the only variations that occur are in the two D-valine components (subgroup 4). Actinomycin C_3 contains two D-alloisoleucine components in place of the two D-valine components. The structure of actinomycin C_2 contains one D-valine component (most probably in group A) and one D-alloisoleucine component. Changes in other subgroups lead to other series of actinomycins. For example, in the actinomycin X series, changes in the L-proline components (subgroup 3) occur. In the actinomycin E series, changes in the L-N-methylvaline component (subgroup 1) occur. In the actinomycin F series, additional sarcosine components are found in subgroup 3 substituted for proline components, with the rest of the structure as found in actinomycins C_2 or C_3. It is obvious that the potential for variety in actinomycins is large. In addition to the naturally occurring amino acids found in the C series, other amino acids may be inserted into the actinomycin molecule by providing particular substrates in biosynthetic procedures or by total synthesis. Several of the actinomycins have been prepared by totally synthetic procedures.

In addition to changes in the cyclopeptide structures, modifications of the chromophoric nucleus have been made, chiefly dealing with changes at the 2 position. Desaminoactinomycin (2-hydroxyactinomycin) and the 2-chloro- and the 2-dialkylaminoactinomycins are inactive. Monoalkylation of the 2-amino group results in compounds with diminished activity. Reduction of the aromatic ring system also leads to loss of activity. Animal studies indicate

Dactinomycin

differences in the carcinolytic activities and toxicities of the various actinomycins, but these differences have not appeared to be significant in therapy. Further variations in the structures of actinomycins are very likely to be made, and it will be interesting to see if improvements in activity will result.

The cytotoxic effect of actinomycins is due to an inhibition of protein synthesis. It has been shown that they bind reversibly to the guanidine moiety of DNA, with a resulting inhibition of DNA-dependent RNA and RNA-polymerase synthesis. The biochemistry of the actinomycins may be studied in the review by Reich.[113]

Dactinomycin U.S.P., Cosmegen®; Actinomycin C_1, Actinomycin D, Actinomycin IV. The one actinomycin released for therapy in the United States has the structure shown previously. It is obtained from *Streptomyces parvullus* which, unlike other Streptomyces species, elaborates dactinomycin in a nearly pure form. The use of this drug is restricted to hospitalized patients, for treatment of a few kinds of cancer: Wilms's tumor, rhabdomyosarcoma, and some carcinomas of the testes and uterus.

Dactinomycin is a bright red, crystalline powder that is slightly hygroscopic and affected by light and heat. It is soluble in water at 10° and slightly soluble in water at 37°. It is freely soluble in alcohol and very slightly soluble in ether. In handling, caution should be observed to avoid inhalation of particles or exposure to the skin. Because of its cytotoxic effects, side-reactions are frequent and may be severe. The toxic effects usually are reversible upon withdrawal of therapy.

Category—antineoplastic.

Usual dose—intravenous, 500 mcg. daily for 5 days.

OCCURRENCE
Dactinomycin for Injection U.S.P.

UNCLASSIFIED ANTIBIOTICS

Among the many hundreds of antibiotics that have been evaluated for activity are a number that have gained significant clinical attention but which do not fall into any of the previously considered groups. Some of these have quite specific activities against a narrow spectrum of microorganisms. Some have found a useful place in therapy as substitutes for other antibiotics to which resistance has developed. Among the most exciting recent developments has been the successful use of antibiotics for the treatment of various cancers. Since the members of this miscellaneous group of antibiotics bear no significant chemical relationships to other antibiotics, they are considered as individual compounds.

Griseofulvin U.S.P., Fulvicin®, Grisactin®, Grifulvin®. Although griseofulvin was reported in 1939 by Oxford *et al.*[114] as an antibiotic obtained from *Penicillin griseofulvum Dierckx*, it was not until 1958 that its use for the treatment of fungal infection in man was demonstrated successfully. Previously, it had been used for its antifungal action in plants and animals. Its release in the United States in 1959, 20 years after its discovery, as a potent agent for the treatment of ringworm infections re-emphasizes the need for the broad screening of drugs to find their potential uses.

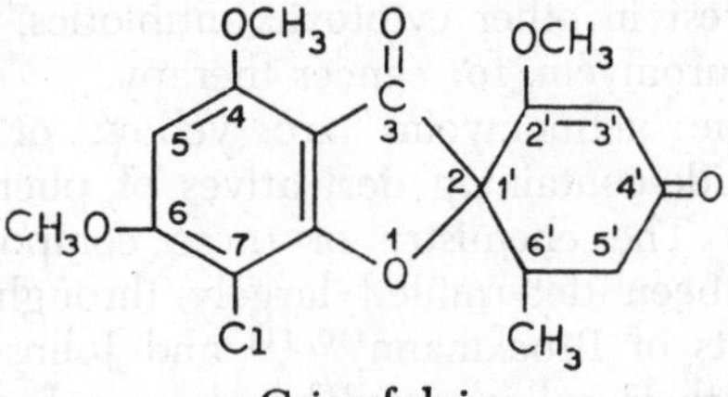

Griseofulvin

The structure of griseofulvin was determined by Grove *et al.*[115] to be 7-chloro-2', 4,6-trimethoxy-6'β-methylspiro[benzofuran-2 (3H), 1'-[2]cyclohexene]3,4'-dione. It is a white, bitter, thermostable powder that may occur also as needlelike crystals. It is relatively soluble in alcohol, chloroform and acetone. In the dry state it is stable for at least 20 months.

Since its introduction, griseofulvin has provided startling cures for infections due to trichophytons and microspora resulting in refractory ringworm infections of the body, the nails and the scalp (tinea corporis, tinea unguium and tinea capitis) and athlete's foot (tinea pedis). In the treatment of these infections it is administered orally and is absorbed from the gastrointestinal tract. Following systemic circulation, it is concentrated in the keratin of growing skin, nails and hair. As new tissue develops, the fungistatic action of the griseofulvin prevents the growth of the organism in it. The old tissue continues to support viable fungi, so the drug must be continued until exfoliation of the old tissue is complete. In the case of infected nails, therapy may need to be continued for months because of the slow rate of growth. Griseofulvin does not cause many adverse side-effects, but careful observation of patients receiving it is indicated. It is not active against bacteria and other fungi or yeasts.

A number of methods for the synthesis of griseofulvin have been developed that have permitted the synthesis of some structural analogs. None of these has shown activity superior to griseofulvin. The mode of action of griseofulvin is unknown, and little fundamental work has been published concerning possible mechanisms of its inhibitory effects. Of interest is the effect crystal size has on absorption of the orally administered powder. "Microsize" griseofulvin may be administered in significantly smaller doses than the conventional size powder to obtain the same effect. The *U.S.P.* specifies that the official product is the "Microsize" powder.

Category—antifungal.

Usual dose—250 mg. two times a day.

Range—500 mg. to 1 g. daily.

OCCURRENCE

Griseofulvin Capsules U.S.P.
Griseofulvin Tablets U.S.P.

Viomycin Sulfate U.S.P., Vinactane® Sulfate, Viocin® Sulfate. One of the major infectious diseases for which a truly effective chemotherapeutic agent still is sought is tuberculosis. Occasionally, an antibiotic is found to possess antituberculous activity, but few possess sufficient quality to warrant their introduction into medical use. Viomycin is one of these few antibiotics. It has been isolated from Streptomyces species *floridae* and *puniceus* and *Actinomyces vinaceus*.

Viomycin sulfate is an odorless powder that varies in color from white to slightly yellow. It is very soluble in water, forming solutions ranging in pH from 4.5 to 7.0. It is insoluble in alcohol and in other organic solvents. Since it is slightly hygroscopic, it should be stored in tightly closed containers. Viomycin itself is a strongly basic polypeptide of undetermined chemical structure. At least two components have been obtained from S. *vinaceus* and have been named vinactins A and B. A closely related substance, identified as vinactin C, also has been found to be present. Vinactin A appears to be the major component of viomycin.

Early work by Haskell et al.[116] and Mayer et al.[117] showed that viomycin had no free α-amino carboxy groups and, on vigorous acid hydrolysis, yielded carbon dioxide, ammonia, urea, L-serine, α,β-diaminopropionic acid, β-lysine and a guanidino compound. Based on further evidence, Dyer et al.[118] have suggested the structure below for viomycin.

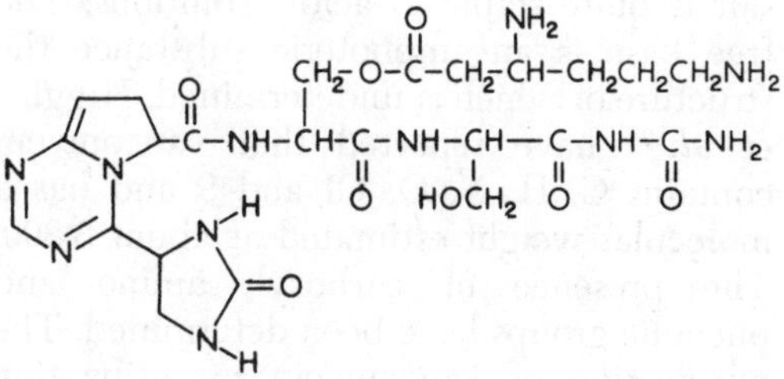

Viomycin

As an antitubercular drug, viomycin is less active than streptomycin and is more active than aminosalicylic acid. Its main use is as an adjunct in the treatment of infections that are resistant to streptomycin and isoniazid. No cross resistance with these drugs has been observed. It is administered intramuscularly in aqueous or normal saline solutions. It may produce a number of toxic reactions, and its use should be accompanied by careful observation.

Category—antibacterial (tuberculostatic).

Usual dose—intramuscular, the equivalent of 1 g. of viomycin 2 times a day every third day.

Usual dose range—4 to 14 g. weekly.

OCCURRENCE

Sterile Viomycin Sulfate U.S.P.

Vancomycin Hydrochloride U.S.P., Vancocin®. The isolation of vancomycin from *Streptomyces orientalis* was described in 1956 by McCormick *et al.*[119] The organism was originally obtained from cultures of an Indonesian soil sample and subsequently has been obtained from Indian soil. It was introduced in 1958 as an antibiotic active against gram-positive cocci, particularly streptococci, staphylococci and pneumococci. It is recommended for use when infections have not responded to treatment with the more common antibiotics or when the infection is known to be caused by a resistant organism. Vancomycin has not exhibited cross resistance with any other known antibiotic. Perkins[120] states that vancomycin interferes with mucopeptide biosynthesis, perhaps in a manner similar to that of penicillin.

Vancomycin hydrochloride is a free-flowing, tan to brown powder that is relatively stable in the dry state. It is very soluble in water and insoluble in organic solvents. The salt is quite stable in acidic solutions. The free base is an amphoteric substance the structure of which is undetermined. Higgins *et al.*[121] have reported that vancomycin contains C, H, N, O, CI and S and has a molecular weight estimated at about 3,300. The presence of carboxyl, amino and phenolic groups have been determined. The purification of vancomycin by utilization of its chelating property to form a copper complex has been reported by Marshall.[122] Present knowledge of the structure of vancomycin indicates that it contains two glucose, two aspartic acid, one N-methyl-leucine, four 3-chloro-4-hydroxybenzyl, one 2-hydroxybenzyl and one 3-methyl-4-keto-hexanoic acid components per molecule. Removal of the glucose units produces a compound, aglucovancomycin, that retains

about three fourths of the activity of vancomycin.

Vancomycin HCI is always administered intravenously, either by slow injection or by continuous infusion. In short-term therapy, the toxic side-reactions are usually slight, but continued use may lead to impairment of auditory acuity and to phlebitis and skin rashes.

Category—antibacterial.

Usual dose—intravenous infusion in 100 to 200 ml. of an isotonic solution over a period of 30 minutes 4 times daily.

Usual dose range—1 to 2 g. daily.

OCCURRENCE
Sterile Vancomycin Hydrochloride U.S.P.

Novobiocin, Albamycin®, Cardelmycin®, Cathomycin®, Streptonivicin. In the search for new antibiotics, three different research groups independently isolated novobiocin from Streptomyces species. It was first reported in 1955 as a product from S. *spheroides* and from S. *niveus*. It is currently produced from cultures of both species. Until the common identity of the products obtained by the different research groups was ascertained, confusion in the naming of this compound existed. Its chemical identity has been established as 7-[4-(carbamoyloxy)tetrahydro-3-hydroxy-5-methoxy-6,6-dimethylpyran-2-yloxy]-4-hydroxy-3-[4-hydroxy-3-(3-methyl-2-butenyl)-benzamido]-8-methylcoumarin by Shunk *et al.*[123] and Hoeksema, Caron and Hinman[124] and confirmed by Spencer *et al.*[125, 126]

Chemically, novobiocin has a unique structure among antibiotics although, like a number of others, it possesses a glycosidic sugar moiety. The sugar in novobiocin, devoid of its carbamate ester, has been named noviose and is an aldose having the con-

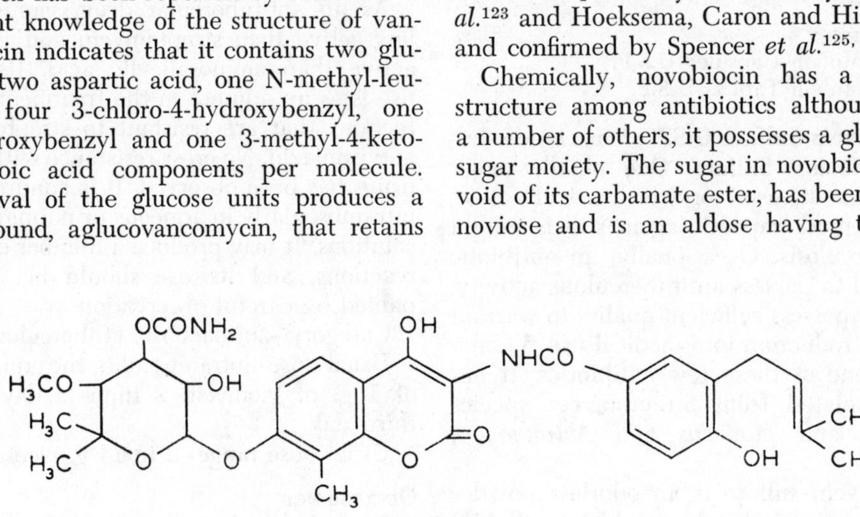

Novobiocin

figuration of L-lyxose. The aglycon moiety has been termed novobiocic acid.

Novobiocin is a pale-yellow, somewhat photosensitive compound that crystallizes in two chemically identical forms having different melting points. It is soluble in methanol, ethanol and acetone but is quite insoluble in less polar solvents. Its solubility in water is affected by pH. It is readily soluble in basic solutions, in which it deteriorates, and is precipitated from acidic solutions. It behaves as a diacid, forming two series of salts. The enolic hydroxyl group on the coumarin moiety behaves as a rather strong acid and is the group by which the commercially available sodium and calcium salts are formed. The phenolic —OH group on the benzamido moiety also behaves as an acid but is weaker than the former. Disodium salts of novobiocin have been prepared. The sodium salt is stable in dry air but decreases in activity in the presence of moisture. The calcium salt is quite water insoluble and is used to make aqueous oral suspensions. Because of its acidic characteristics, novobiocin combines to form salt complexes with basic antibiotics. Some of these salts have been investigated for their combined antibiotic effect, but none has been placed on the market, as no advantage is offered by them.

The antibiotic activity of novobiocin is exhibited chiefly against gram-positive organisms and *Proteus vulgaris*. Because of its unique structure, it appears to exert its action in a manner (still unknown) different from other anti-infectives. It may be that its ability to bind magnesium causes an intracellular deficiency of that ion which is necessary for the maintenance of the integrity of the cell membrane.[127, 128] Although resistance to novobiocin can be developed in microorganisms, cross resistance with other antibiotics is not developed. For this reason, the medical use of novobiocin is most generally reserved for the treatment of infections, particularly staphylococcal, resistant to other antibiotics and the sulfas.

Usual dose (of the sodium salt)—oral, the equivalent of 250 mg. of novobiocin every 6 hours; intravenous and intramuscular, 500 mg. every 12 hours.

Usual dose range—1 to 2 g. daily in 4 divided doses.

OCCURRENCE
Calcium Novobiocin N.F.
Calcium Novobiocin Oral Suspension N.F.
Sodium Novobiocin N.F.
Sodium Novobiocin Capsules N.F.
Sodium Novobiocin for Injection N.F.

Cycloheximide, Actidione®, Acti-dione®. This antibiotic is a glutarimide derivative that was isolated from *Streptomyces griseus* by Whiffen *et al.*[129] In 1946 and is produced, usually along with other antibiotics (e.g., streptomycin), by several species of Streptomyces. The chemical structure of cycloheximide, which has been known for some time, is β-[2-(3,5-dimethyl-2-oxycyclohexyl)-2-hydroxyethyl]glutarimide. The stereochemical structure as shown is that reported by Schaeffer and Jain[130, 131] and by Starkovsky and Johnson.[132]

Cycloheximide

Cycloheximide inhibits the growth of many fungi and yeasts, but its toxic effects have caused it to be withheld from clinical use in spite of its success in the treatment of some systemic mycotic infections. It has been used successfully as an antifungal agent for plants and has attracted attention as a potential anticancer agent. Cycloheximide and its acetoxy derivative inhibit protein synthesis in much the same manner as chloramphenicol,[133] but, unlike the last named, they are highly specific in their effects on yeast and mammalian cells and without effect on bacteria.

Cycloserine U.S.P., Oxamycin® Seromycin®, D-(+)-4-Amino-3-isoxazolidinone. One of the simplest structures to possess antibiotic action is the antitubercular substance, cycloserine. It has been isolated from three different species of Streptomyces: S. *orchidaceus,* S. *garyphalus* and S. *lavendulus.* Its structure has been determined by Kuehl *et al.*[134] and Hidy *et al.*[135] to be D-4-amino-3-isoxazolidone. No doubt the compound exists in equilibrium with its enol form.

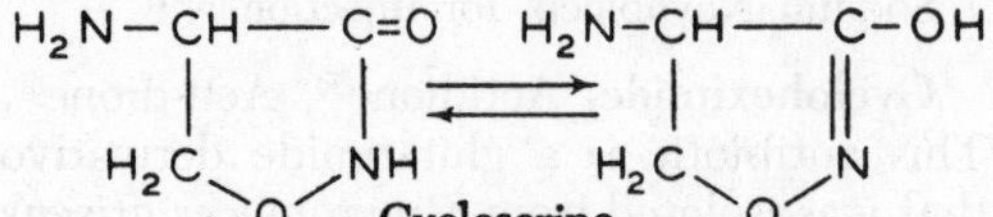

In aqueous solutions, cycloserine will form a dipolar ion that, on standing, will dimerize to 2,5-bis-(aminoxymethyl)-3,6-diketopiperazine:

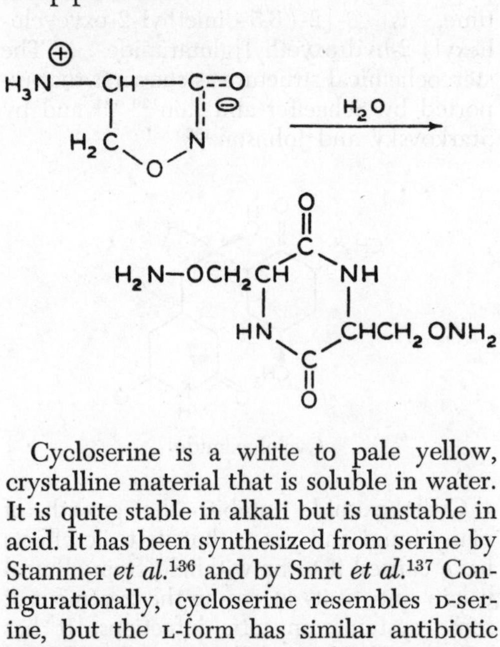

Cycloserine is a white to pale yellow, crystalline material that is soluble in water. It is quite stable in alkali but is unstable in acid. It has been synthesized from serine by Stammer *et al.*[136] and by Smrt *et al.*[137] Configurationally, cycloserine resembles D-serine, but the L-form has similar antibiotic activity. Most interesting is the observation that the racemic mixture is more active than either enantiomorph, indicating that the isomeric pair act on each other synergistically.

Although cycloserine exhibits antibiotic activity in vitro against a wide spectrum of both gram-negative and gram-positive organisms, its relatively weak potency and frequent toxic reactions limit its use to the treatment of tuberculosis. It is recommended for cases who fail to respond to other tuberculostatic drugs or are known to be infected with organisms resistant to other agents. It is usually administered orally in combination with other drugs, commonly isoniazid.

Category—antibacterial (tuberculostatic).

Usual dose—250 mg. twice a day.

Usual dose range—250 mg. to 1 g. daily.

OCCURRENCE

Cycloserine Capsules U.S.P.

Lincomycin Hydrochloride U.S.P., Lincocin®, Methyl 6,8-Dideoxy-6-(1-methyl-*trans*-4-propyl-L-2-pyrrolidine-carboxamido)-1-*thio*-D-erythro-α-D-galacto-octopyranoside Monohydrochloride. In 1962, Mason *et al.*[138] reported the isolation of a new antibiotic from *Streptomyces lincolnensis var. lincolnensis* to which the name lincomycin was given. It was found to be effective for the treatment of infections caused by gram-positive organisms, with a spectrum similar to that of erythromycin. Its general characteristics have led to its acceptance as a chemotherapeutic agent for infections due to streptococci (except S. *fecalis*), staphylococci and pneumococci, particularly when such infections are resistant to other antibiotics. Like other antibiotics, lincomycin may cause the development of resistant strains, particularly among staphylococci. It appears to be well tolerated and not to induce allergic reactions other than mild skin reactions. Its mechanism of action has not been worked out in detail, but it appears to act by inhibition of protein synthesis.

The chemistry of lincomycin has been reported by Hoeksema *et al.*[139] who have assigned the structure given below. That structure has been confirmed by Slomp and MacKellar.[140] The molecule contains one basic function, the pyrrolidine nitrogen,

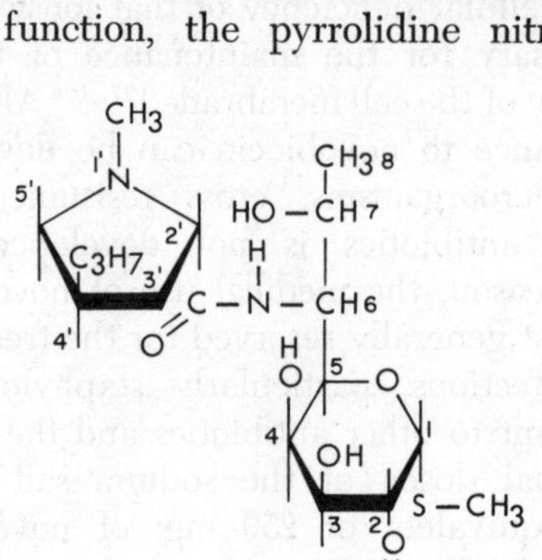

Lincomycin

that shows an apparent pKa of 7.6 and by which water soluble acids salts may be formed. When subjected to hydrazinolysis, lincomycin is cleaved at its amide bond into *trans*-L-4-*n*-propylhygric acid (the pyrrolidine moiety) and methyl α-thiolincosaminide (the sugar moiety). Lincomycin-related antibiotics have been reported by Argoudelis *et al*.[141] to be produced by *S. lincolnensis*. These new antibiotics differ in structure at one or more of three positions of the lincomycin structure: (1) the N-methyl of the hygric acid moiety is substituted by a hydrogen; (2) the *n*-propyl group of the hygric acid moiety is substituted by an ethyl group; and (3) the thiomethyl ether of the α-thiolincosamide moiety is substituted by a thioethyl ether. Magerlain *et al*.[142] have shown that increased lipophilia of lincomycin analogs resulting from larger alkyl groups on the 4 position causes increased penetration of the molecule to the site of action and greater activity.

Lincomycin hydrochloride occurs as the nonohydrate, a white crystalline solid that is stable in the dry state. It is readily soluble in water and alcohol and its aqueous solutions are stable at room temperature. It is slowly degraded in acid solutions but is well absorbed from the gastrointestinal tract. Forist *et al*.[143] have reported a half-life of 39 hours for a 0.1 N HCl solution. Lincomycin diffuses well into peritoneal and pleural fluids and into bone, but not into cerebrospinal fluid. It is excreted in the urine and the bile. It is available in capsule form for oral administration and in ampules and vials for parenteral administration.

Category—antibacterial.

Usual dose—oral, the equivalent of 500 mg. of lincomycin 3 to 4 times a day; intramuscular, the equivalent of 600 mg. of lincomycin 1 to 2 times a day; intravenous infusion, the equivalent of 600 mg. of lincomycin in 8 to 12 hours.

OCCURRENCE

Lincomycin Hydrochloride Capsules U.S.P.
Lincomycin Hydrochloride Injection U.S.P.

Clindamycin Hydrochloride Hydrate, Cleocin®. In 1967, Magerlein *et al*.[142] reported that replacement of the 7-hydroxyl

group of lincomycin by chlorine resulted in a compound with enhanced antibacterial activity. Research and clinical tests on that compound resulted in its release in the United States in 1970 for use in the treatment of infections due to susceptible streptococci, pneumococci and staphylococci.

Clindamycin inhibits protein synthesis in

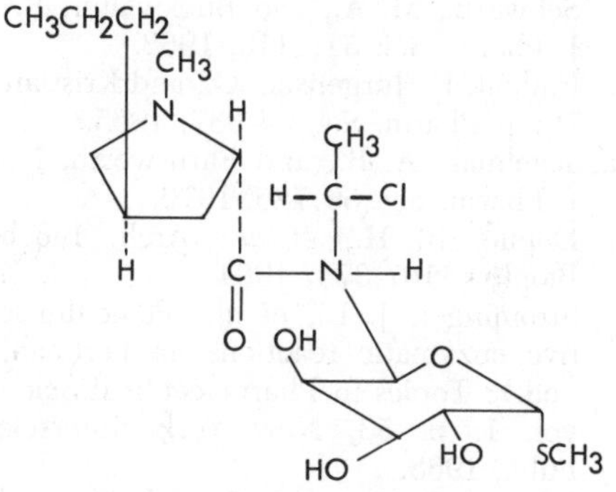

Clindamycin

the bacterial cell. In action, it may be either bacteriostatic or bactericidal, depending on the concentration of the antibiotic and the sensitivity of the pathogen. It may show cross resistance in organisms to lincomycin, but no reports of cross resistance with other antibiotics have appeared.

It is rapidly absorbed after oral administration, achieving peak serum levels (2.5 mcg./ml.) in about 45 minutes. The average biologic half-life is 2.4 hours. It is available in a capsule dosage form. The usual dose is the equivalent of 150 mg. of clindamycin base 4 times a day and the usual dose range is the equivalent of 150 mg. to 450 mg. of clindamycin base 4 times a day.

REFERENCES

1. Fleming, A.: Brit. J. Exp. Path. 10:226, 1929.
2. Vuillemin, P.: Assoc. franc. avance sc. Part 2:525-543, 1889.
3. Waksmann, S. A.: Science 110:27, 1949.

4. Benedict, R. G., and Langlykke, A. F.: Ann. Rev. Microbiol. 1:193, 1947.

5. Baron, A. L.: Handbook of Antibiotics, p. 5, New York, Reinhold, 1950.

6. Clarke, H. T., *et al.*: The Chemistry of Penicillin, p. 454, Princeton, N. J., Princeton Univ. Press, 1949.

7. Sheehan, J. C., and Henery-Logan, K. R.: J. Am. Chem. Soc. 81:3089, 1959.

8. Batchelor, F. R., *et al.*: Nature 183:257, 1959.

9. Sheehan, J. C., and Ferris, J. P.: J. Am. Chem. Soc. 81:2912, 1959.

10. Schwartz, M. A., and Buckwalter, F. H.: J. Pharm. Sci. 51:1119, 1962.

11. Finholt, P., Jürgensen, G., and Kristiansen, H.: J. Pharm. Sci. 54:387, 1965.

11a. Segelman, A. B., and Farnsworth, N. R.: J. Pharm. Sci. 59:726, 1970.

12. Depue, R. H., *et al.*: Arch. Biochem. Biophys. 107:374, 1964.

13. Strominger, J. L., *et al.*: Penicillin-sensitive enzymatic reactions, *in* Perlman, D. (ed.): Topics in Pharmaceutical Sciences, vol. 1, p. 53, New York, Interscience Publ., 1968.

14. Behrens, O. K., *et al.*: J. Biol. Chem. 175:793, 1948.

15. Stedman, R. J., *et al.*: J. Med. Chem. 7:251, 1964.

15a. Sutherland, R., and Robinson, O. P. W.: Brit. Med. J. 2:804, 1967.

16. Acred, P., *et al.*: Nature 215:25, 1967.

17. Abraham, E. P.: Penicillins and cephalosporins—their chemistry in relation to biological activity, *in* Perlman, D. (ed.): Topics in Pharmaceutical Sciences, vol. 1, p. 1, New York, Interscience Publ., 1968.

18. Van Heyningen, E.: Cephalosporins, *in* Harper, N. J., and Simmonds, A. B. (eds.): Advances in Drug Research, vol. 4, p. 1, New York, Academic Press, 1967.

19. Woodward, R. B., *et al.*: J. Am. Chem. Soc. 88:852, 1966.

20. Schatz, A., *et al.*: Proc. Soc. Exp. Biol. Med. 55:66, 1944.

21. Dyer, J. R., and Todd, A. W.: J. Am. Chem. Soc. 85:3896, 1963.

22. Dyer, J. R., *et al.*: J. Am. Chem. Soc. 87:654, 1965.

23. Gale, E. F.: Promotion and Prevention of Synthesis in Bacteria, p. 23, Syracuse, N. Y., Syracuse Univ. Press, 1968.

24. Umezawa, H., *et al.*: J. Antibiot. [A] 10:181, 1957.

25. Tatsuoka, S., *et al.*: J. Antibiot. [A] 17:88, 1964.

26. Hichens, M., and Rinehart, K. L., Jr.: J. Am. Chem. Soc. 85:1547, 1963.

27. Nakajima, M.: Tetrahedron Letters 623, 1968.

28. Umezawa, S., *et al.*: J. Antibiot. 21:162, 367, 424, 1968.

29. Aoki, T., *et al.*: J. Antibiot. [A] 12:98, 1959.

30. Morikubo, Y.: J. Antibiot. [A] 12:90, 1959.

31. Umezawa, S., *et al.*: J. Antibiot. [A] 12:114, 1959.

32. Waksman, S. A., and Lechevalier, H. A.: Science 109:305, 1949.

33. Waksman, S. A. (ed.): Neomycin, Its Nature and Practical Applications, Baltimore, Williams & Wilkins, 1958.

34. Rinehart, K. L., Jr., *et al.*: J. Am. Chem. Soc. 84:3218, 1962.

35. Huettenrauch, R.: Pharmazie 19:697, 1964.

36. Coffey, G. L., *et al.*: Antibiot. Chemother. 9:730, 1959.

37. Haskell, T. H., *et al.*: J. Am. Chem. Soc. 81:3482, 1959.

38. DeJongh, D. C., *et al.*: J. Am. Chem. Soc. 89:3364, 1967.

39. Weinstein, M. J., *et al.*: J. Med. Chem. 6:463, 1963.

40. Cooper, D. J., *et al.*: J. Infect. Dis. 119:342, 1969.

41. Maehr, H., and Schaffner, C. P.: J. Am. Chem. Soc. 89:6788, 1968.

42. Ehrlich, J., *et al.*: Science 106:417, 1947.

43. Controulis, J., *et al.*: J. Am. Chem. Soc. 71:2463, 1949.

44. Long, L. M., and Troutman, H. D.: J. Am. Chem. Soc. 71:2473, 1949.

45. Brock, T. D.: Chloramphenicol, *in* Schnitzer, R. J., and Hawking, F. (eds.): Experimental Chemotherapy, vol. 3, p. 119, New York, Acad. Press, 1964.

46. Muxfelt, H., *et al.*: J. Am. Chem. Soc. 90:6534, 1968.

47. Korst, J. J., *et al.*: J. Am. Chem. Soc. 90:439, 1968.

48. Leeson, L. J., Krueger, J. E., and Nash, R. A.: Tetrahedron Letters, No. 18:1155, 1963.

49. Stephens, C. R., *et al.*: J. Am. Chem. Soc. 78:4155, 1956.

50. Rigler, N. E., *et al.*: Anal. Chem. 37:872, 1965.

51. Benet, L. Z., and Goyan, J. E.: J. Pharm. Sci. 55:983, 1965.

52. Albert, A.: Nature 172:201, 1953.

53. Jackson, F. L.: Mode of Action of Tetracyclines, *in* Schnitzer, R. J., and Hawking, F. (eds.): Experimental Chemotherapy, vol. 3, p. 103, New York, Acad. Press, 1964.

54. Maxwell, J. H.: Biochim. biophys. acta 138:337, 1967.
55. Barrett, G. C.: J. Pharm. Sci. 52:309, 1963.
56. Boothe, J. H.: Antimicrob. Agents Chemother 1962: 213.
57. McCormick, J. R. D., et al.: J. Am. Chem. Soc. 82:3381, 1960.
58. Blackwood, R. K., et al.: J. Am. Chem. Soc. 83:2773, 1961.
59. Blackwood, R. K., and Stephens, C. R.: J. Am. Chem. Soc. 84:4157, 1962.
60. Hirokawa, S., et al.: Z. Krist. 112:439, 1959.
61. Takeuchi, Y., and Buerger, M. J.: Proc. Nat. Acad. Sci. U.S. 46:1366, 1960.
62. Cid-Dresdner, H.: Z. Krist. 121:170, 1965.
63. Gottstein, W. J., et al.: J. Am. Chem. Soc. 81:1198, 1959.
64. Conover, L. H.: U.S. Patent 2,699,054. Jan. 11, 1955.
65. Bunn, P. A., and Cronk, G. A.: Antibiot. Med. 5:379, 1958.
66. Gittinger, W. C., and Weiner, H.: Antibiot. Med. 7:22, 1960.
67. Remmers, E. G., et al.: J. Pharm. Sci. 53:1452, 1534, 1964.
68. ———: J. Pharm. Sci. 54:49, 1965.
69. Duggar, B. B.: Ann. N. Y. Acad. Sci. 51:177, 1948.
70. Finlay, A. C., et al.: Science 111:85, 1950.
71. Hochstein, F. A., et al.: J. Am. Chem. Soc. 75:5455, 1953.
72. McCormick, J. R. D., et al.: J. Am. Chem. Soc. 79:4561, 1957.
73. Stephens, C. R., et al.: J. Am. Chem. Soc. 80:5324, 1958.
74. Schach von Wittenau, M., et al.: J. Am. Chem. Soc. 84:2645, 1962.
75. Wiley, P. F.: Research Today (Eli Lilly and Co.) 16:3, 1960.
76. Miller, M. W.: The Pfizer Handbook of Microbial Metabolites, New York, Mc-Graw-Hill, 1961.
77. Morin, R., and Gorman, M.: Kirk-Othmer Encyl. Chem. Technol., ed. 2, 12;637, 1967.
78. McGuire, J. M., et al.: Antibiot. Chemother. 2:821, 1952.
79. Wiley, P. F., et al.: J. Am. Chem. Soc. 79: 6062, 1957.
80. Celmer, W. D.: J. Am. Chem. Soc. 87: 1801, 1965.
81. Wilhelm, J. M., et al.: Antimicrob. Agents Chemother. 1967:236.
82. Sobin, B. A., et al.: Antibiotics Annual 1954-1955, p. 827, New York, Medical Encyclopedia, 1955.
83. Hochstein, F. A., et al.: J. Am. Chem. Soc. 82:3227, 1960.
84. Celmer, W. D.: J. Am. Chem. Soc. 87: 1797, 1965.
85. Golding, B. T., et al.: Tetrahedron Letters 3551, 1966.
86. Lechevalier, H. A., et al.: Antibiot. Chemother. 11:640, 1961.
87. Hazen, E. L., and Brown, R.: Proc. Soc. Exp. Biol. Med. 76:93, 1951.
88. Ikeda, M., et al.: Tetrahedron Letters 3745, 1967.
89. Gold, W., et al.: Antibiotics Annual 1955-1956, p. 579, New York, Medical Encyclopedia, 1956.
90. Dutcher, J. D., et al.: Antibiotics Annual 1956-1957, p. 866, New York, Medical Encyclopedia, 1957.
91. Lechevalier, H. A., et al.: Mycologia 45: 155, 1953.
92. Sarges, R., and Witkop, B.: J. Am. Chem. Soc. 86:1861, 1964.
93. Bodanszky, M., and Perlman, D.: Science 163:352, 1969.
94. Dubos, R. J.: J. Exp. Med. 70:1, 1939.
95. Paladini, A., and Craig, L. C.: J. Am. Chem. Soc. 76:688, 1954; King, T. P., and Craig, L. C.: J. Am. Chem. Soc. 77:6627, 1955.
96. Ohno, M., et al.: Bull. Soc. Chem. Japan 39:1738, 1966.
97. Johnson, B. A., et al.: Science 102:376, 1945.
98. Stoffel, W., and Craig, L. C.: J. Am. Chem. Soc. 83:145, 1961.
99. Benedict, R. G., and Langlykke, A. F.: J. Bact. 54:24, 1947.
100. Stansly, P. J., et al.: Bull. Johns Hopkins Hosp. 81:43, 1947.
101. Ainsworth, G. C., et al.: Nature 160:263, 1947.
102. Vogler, K., and Studer, R. O.: Experientia 22:345, 1966.
103. Hausmann, W., and Craig, L. C.: J. Am. Chem. Soc. 76:4892, 1954.
104. Vogler, K., et al.: Experientia 20:365, 1964.
105. Koyama, Y., et al.: J. Antibiot. [A] 3:457, 1950.
106. Suzuki, T., et al.: J. Biochem. 54:414, 1963.
107. Wilkinson, S., and Lowe, L. A.: J. Chem. Soc. 1964:4107.
108. Waksman, S., and Woodruff, H. B.: Proc. Soc. Exp. Biol. Med. 45:609, 1940.
109. Waksman, S. (ed.): Ann. N. Y. Acad. Sci. 89:283, 1960.
110. Brockmann, H.: Fortschr. Chem. organ. Naturst. 18:1, 1960.

111. Brockmann, H.: Ann. N. Y. Acad. Sci. 89:323, 1960.
112. Johnson, A. W.: Ann. N. Y. Acad. Sci. 89:336, 1960.
113. Reich, E.: Cancer Res. 23:1428, 1963.
114. Oxford, A. E., *et al.*: Biochem. J. 33:240, 1939.
115. Grove, J. F., *et al.*: J. Chem. Soc. 1952: 3977.
116. Haskell, T. H., *et al.*: J. Am. Chem. Soc. 74:599, 1952.
117. Mayer, R. L., *et al.*: Experientia 10:335, 1954.
118. Dyer, J. R., *et al.*: Tetrahedron Letters No. 10:585, 1965.
119. McCormick, M. H., *et al.*: Antibiotics Annual 1955-1956, p. 606, New York, Medical Encyclopedia, 1956.
120. Perkins, H. R.: Biochem. J. 111:195, 1969.
121. Higgins, H. M., *et al.*: Antibiotics Annual 1957-1958, p. 906, New York, Medical Encyclopedia, 1958.
122. Marshall, F. J.: J. Med. Chem. 8:18, 1965.
123. Shunk, C. H., *et al.*: J. Am. Chem. Soc. 78:1770, 1956.
124. Hoeksema, H., *et al.*: J. Am. Chem. Soc. 78:2019, 1956.
125. Spencer, C. H., *et al.*: J. Am. Chem. Soc. 78:2655, 1956.
126. ———: J. Am. Chem. Soc. 80:140, 1958.
127. Brock, T. D.: Science 136:316, 1962.
128. ———: J. Bacteriol. 84:679, 1962.
129. Whiffen, A. J., *et al.*: J. Bacteriol. 52: 610, 1946.
130. Schaeffer, H. J., and Jain, V. K.: J. Pharm. Sci. 50:1048, 1961.
131. ———: J. Pharm. Sci. 52:639, 1963.
132. Starkovsky, N. A., and Johnson, F.: Tetrahedron Letters No. 16:919, 1964.
133. Ennis, H. L., and Lubin, M.: Science 146:1474, 1964.
134. Kuehl, F. A., Jr., *et al.*: J. Am. Chem. Soc. 77:2344, 1955.
135. Hidy, P. H., *et al.*: J. Am. Chem. Soc. 77:2345, 1955.
136. Stammer, C. H., *et al.*: J. Am. Chem. Soc. 77:2346, 1955.
137. Smrt, J.: Experientia 13:291, 1957.
138. Mason, D. J., *et al.*: Antimicrob. Agents Chemother 1962:554.
139. Hoeksema, H., *et al.*: J. Am. Chem. Soc. 86:4223, 1964.
140. Slomp, G., and MacKellar, F. A.: J. Am. Chem. Soc. 89:2454, 1967.
141. Argoudelis, A. D., *et al.*: J. Am. Chem. Soc. 86:5044, 1964.
142. Magerlein, B. J., *et al.*: J. Med. Chem. 10:355, 1967.
143. Forist, A. A., *et al.*: J. Pharm. Sci. 54: 476, 1965.

SELECTED READING

Childress, S. J.: Chemical Modification of Antibiotics *in* Rabinowitz, J. L., and Myerson, R. M. (eds.): Topics in Medicinal Chemistry, vol. 1, pp. 109-136, New York, Interscience, 1967.

Cline, D. L. J.: Chemistry of tetracyclines, Quart. Rev. 22:435, 168.

Evans, R. M.: The Chemistry of the Antibiotics Used in Medicine, New York, Pergamon Press, 1965.

Gale, E. F.: Mechanisms of antibiotic action, Pharmacol. Rev. 15:48, 1963.

Garrod, L. P.: and O'Grady, F.: Antibiotic and Chemotherapy, ed. 2, Baltimore, Williams and Wilkins Co., 1968.

Goldberg, H. S.: Antibiotics, Their Chemistry and Non-medical Uses, Princeton, N. J., Van Nostrand, 1959.

Gottlieb, D., and Shaw, P. D. (eds.): Antibiotics, vols. 1 and 2, New York, Springer-Verlag, 1967.

Herold, M., and Gabriel, Z. (eds.): Antibiotics—Advances in Research, Production and Clinical Use, Washington, D. C., Butterworth, 1966.

Hoover, J. R. E., and Stedham, R. S.: The β-lactam antibiotics, *in* Burger, A. (ed.): Medicinal Chemistry, ed. 3, p. 371, New York, Wiley-Interscience, 1970.

Korzybski, T., Kowszyk-Gindfinder, S., and Kurylowicz, W.: Antibiotics, New York, Pergamon Press, 1967.

Perlman, D.: Antibiotics, *in* Burger, A. (ed.): Medicinal Chemistry, ed. 3, p. 305, New York, Wiley-Interscience, 1970. Chemother. 1966:670.

Price, K. E., *et al.*: Structure- Activity relationships of the penicillins, Antimicrob. Agents Chemother.: 670, 1966.

15

Central Nervous System Depressants

T. C. Daniels, Ph.D.,

Emeritus Professor of Pharmaceutical Chemistry, School of Pharmacy,
University of California

and

E. C. Jorgensen, Ph.D.,

Professor of Chemistry and Pharmaceutical Chemistry, School of Pharmacy,
University of California

INTRODUCTION

The agents described in this chapter produce depressant effects on the central nervous system as their principal pharmacologic action. These include the general anesthetics, hypnotic-sedatives, central nervous system depressants with skeletal muscle relaxant properties, tranquilizing agents and anticonvulsants. The general anesthetics (e.g., ether) and hypnotic-sedatives (e.g., phenobarbital) produce a generalized or nonselective depression of central nervous system function and overlap considerably in their depressant properties. Many sedatives, if given in large enough doses, produce anesthesia. The central depressant properties of a group of skeletal muscle-relaxing agents (e.g., meprobamate) resemble closely the depressant properties of the hypnotic-sedatives. The tranquilizing agents (e.g., reserpine, chlorpromazine) exert a more selective action and, even in high doses, are incapable of producing anesthesia. The anticonvulsants also act in a more selective fashion, modifying the brain's ability to respond to seizure-evoking stimuli.

The analgesics, another group of agents which selectively depress central nervous system function, are discussed separately in Chapter 24.

The brain possesses a unique and specialized mechanism for excluding many substances presented to it by the circulation.[1] This *blood-brain barrier* appears to involve a complex interplay of anatomic, physiologic, and biochemical factors. The capillary endothelium and surrounding glial cells play an important role among the anatomic components that may selectively bar or admit substances to the functional areas of the brain. There are few metabolic reserves in the central nervous system, and a substantial flow of nutrients and oxygen must be supplied continuously. Cerebral circulation is large, but, in spite of this, only minute amounts of exogenous substances are accepted by the brain. The concept of the blood-brain barrier is used comprehensively to describe all phenomena which either hinder *or* facilitate the penetration of substances into the central nervous system. Penetration may occur by many different mechanisms: dialysis, ultrafiltration, osmosis, Donnan equilibrium, lipoid solubility, active transport, or diffusion due to concentration differences created by special tissue affinities or metabolic activity. In general, however, lipoid-soluble, nonionized molecules pass most readily into the central nervous system, whether their ultimate pharmacologic effect is depression or stimulation. Except for the relatively few active transport systems involving ionic molecules, weak acids or weak bases pass into the brain when their acid or base strengths are such that a high proportion exists as the nonionized lipoid-soluble form at the pH of the plasma (pH 7.4). For this reason, metabolically induced changes in plasma pH, such as those produced by respiratory acidosis or alkalosis, may strongly influence the effects of drugs on the central nervous system. Penetration of the brain by weak acids such as phenobarbital and acetazolamide is increased under conditions of hypercapnia (plasma pH 6.8), and decreased with hypocapnia (plasma pH 7.8). The converse would be true of weak bases, such as amphetamine.

GENERAL ANESTHETICS

General anesthesia is the controlled, reversible depression of the functional activity of the central nervous system, producing loss of sensation and consciousness. The relief of pain through general anesthesia during surgery was first carried out by Crawford Long (Georgia, 1841), who used ether during the removal of a cyst. However, it was the use of nitrous oxide anesthesia by Horace Wells (Connecticut, 1844) during extraction of a tooth that excited in William Morton, a dental associate, awareness of the possibilities of anesthesia during surgery. Morton, while a student at Harvard Medical School, learned of the anesthetizing properties of ethyl ether from his chemistry instructor, Professor Charles Jackson. Morton then persuaded the professor of surgery, J. C. Warren, to allow him to administer ether as a general anesthetic during surgery. The success attending this led to the rapid introduction of ether anesthesia for surgical operations. The word anesthesia, signifying insensibility, was coined by Oliver Wendell Holmes in a letter to Morton shortly after his successful demonstration. Chloroform was introduced in Edinburgh in 1847, and the search for new and better anesthetics has continued to this day.

The stages of anesthesia developed are related to functional levels of the central nervous system successively depressed and are present to varying degrees for all agents capable of producing general anesthesia.

STAGE I (CORTICAL STAGE): Analgesia is produced, consciousness remains, but the patient is sleepy as the higher cortical centers are depressed.

STAGE II (EXCITEMENT): Loss of consciousness results, but depression of higher motor centers involving the brain stem and the cerebellum leads to excitement and delirium.

STAGE III (SURGICAL ANESTHESIA): Spinal cord reflexes are diminished in activity, and skeletal muscle relaxation is obtained. This stage, in which most operative procedures are performed, is further divided into Planes i-iv, mainly differentiated on the basis of increasing somatic muscle relaxation and decreased respiration.

STAGE IV (MEDULLARY PARALYSIS): Respiratory failure and vasomotor collapse occur, due to depression of vital functions of the medulla and the brain stem.

General anesthesia may be produced by a variety of chemical types and routes of administration. Inhalation of gases or the vapors from volatile liquids is by far the most frequently used, although the intravenous and the rectal routes ("fixed anesthetics") are also used. The manner in which the general anesthetics of widely varying structure act to depress central nervous system function is unknown. Theories derived from the relatively high lipid solubility of most members of this class, the size of anesthetic molecules and their participation in the formation of hydrate microcrystals are discussed in Chapter 2.

HYDROCARBONS

The saturated hydrocarbons possess an anesthetic effect which increases from methane to octane and then decreases. However, toxicity is too high in these to be useful, and only hydrocarbons with unsaturated character are used.

Cyclopropane U.S.P., Trimethylene. Cyclopropane was introduced for use as a general anesthetic in 1934 and is the most potent gaseous anesthetic agent currently in use. Following premedication with depressant drugs such as morphine or barbiturates, surgical anesthesia may be obtained with concentrations of about 15 volume per cent cyclopropane and 85 volume per cent oxygen. Induction of anesthesia is rapid, requiring only 2 to 3 minutes. Cyclopropane is rapidly eliminated in the lungs. The high potency of the anesthetic, which allows use of high oxygen concentrations, is particularly advantageous in providing adequate tissue oxygenation.

Cyclopropane is a colorless gas, b.p. −33°, which liquefies at 4 to 6 atmospheres pressure. It is flammable and forms an explosive mixture with air in a concentration range from 3.0 to 8.5 per cent; in oxygen, 2.5 to 50 per cent. The solubility is about 1 volume of gas in 2.7 volumes of water; it is freely soluble in alcohol.

Category—general anesthetic (inhalation).
Application—by inhalation as required.

Ethylene N.F., Ethene, $H_2C=CH_2$. This unsaturated hydrocarbon is a colorless gas which produces rapid induction of anesthesia (6 to 8 deep breaths) but requires the high concentration of 90 volume per cent ethylene, 10 volume per cent oxygen (90:10 mixture). The danger of anoxia may be reduced if premedication permits use of an 80:20 mixture.

In 1908 it was observed that the contaminant in illuminating gas that prevented greenhouse carnations from opening was ethylene. Subsequent evaluation in experimental animals showed its ability to produce anesthesia and led to its introduction into clinical use in 1923.

Ethylene is an excellent anesthetic with a minimum of unpleasant side-effects. However its explosive nature when mixed with oxygen is an important disadvantage. Nitrous oxide-ethylene mixtures are violently explosive. Ethylene is a colorless gas, b.p. −104°, soluble in water, 1:9 by volume at 25°.

Category—general anesthetic.
Application—by inhalation as required.

HALOGENATED HYDROCARBONS

The anesthetic potency of the lower molecular weight hydrocarbons is increased as hydrogen is successively replaced by halogen. Thus, anesthetic potency increases in the order methane, methyl chloride, dichloromethane, chloroform, carbon tetrachloride. However, the favorable factors of general increase in anesthetic potency and decrease in flammability are counterbalanced by a general increase in toxicity which has limited the anesthetic applications of the halogenated hydrocarbons.

Chloroform N.F., trichloromethane, $CHCl_3$. (See Chap. 4 for physical and chemical properties.) Chloroform was introduced as an obstetric anesthetic in 1847. It possesses almost 5 times the potency of ether. It is toxic to the liver and the kidney, and sudden cardiac arrest may occur during the induction phase of anesthesia with chloroform. For these reasons it is seldom used except in the tropics where its lower volatility, b.p. 61°, is a special advantage.

Ethyl Chloride N.F., Chloroethane, CH_3-$CH_2\ Cl$. Ethyl chloride is capable of producing rapid induction of anesthesia, followed by rapid recovery after administration ceases. Surgical anesthesia may be obtained with 4 volumes per cent of the vapor. However, its potential for producing liver damage and cardiac arrhythmias has limited its application in anesthesia to occasional use as an induction anesthetic before another agent is used.

It is also used to produce local anesthesia of short duration. When sprayed on the unbroken skin, rapid evaporation freezes the tissues and allows short, minor operations to be performed.

Ethyl chloride is a gas, b.p. 12°, available under pressure in the liquid form. It is flammable, and explosive when mixed with air.

Category—local anesthetic.
For external use—topically, as spray on intact skin.

Trichlorethylene U.S.P., Chlorylen®, Trilene®, 1,1,2-trichloroethene. Trichloroethyl-

$$Cl_2C=C\overset{H}{\underset{Cl}{}}$$

ene is effective as a general anesthetic by inhalation, but side-effects (cardiac irregularities and hepatic damage) limit its use to analgesic effects. Inhalation of 0.25 to

0.75 volume per cent in air produces analgesia suitable for minor surgery and obstetrics. Trichloroethylene produces relief from the intense pain of trigeminal neuralgia (tic douloureux). For this purpose it is available in fabric-covered sealed glass tubes. One of these, containing 1 ml., is crushed in a tumbler, and the vapor is inhaled until the chloroformlike odor disappears.

Trichloroethylene is a volatile liquid, b.p. 88°. It is practically insoluble in water and is miscible with ether, alcohol and with chloroform. It is nonexplosive and nonflammable.

Category—analgesic.

Application—by inhalation, as required.

Halothane U.S.P., Fluothane®, 2-Bromo-2-chloro-1,1,1-trifluoroethane. Halothane, a

$$
\begin{array}{ccc}
\text{F} & \text{Cl} \\
| & | \\
\text{F}-\text{C}-\text{C}-\text{H} \\
| & | \\
\text{F} & \text{Br}
\end{array}
$$

Halothane

general anesthetic with potency estimated at 4 times that of ether, was introduced in 1956. Experience with the chemical inertness and low toxicity of fluorinated hydrocarbons containing the CF_3 or CF_2 groupings, and used as refrigerants, led to the development of halothane as an anesthetic.[2] The presence of a trifluoromethyl group, and bromine, chlorine and hydrogen atoms on a single carbon atom, produced an asymmetric molecule, not yet separated into its diasteriomeric forms, with physical properties and anesthetic potency close to those of chloroform ($CHCl_3$), but with much lower toxicity. The solubility and vapor pressure of halothane were found to be in a desirable range for the potential production of anesthesia, as proposed by Ferguson's Principle (see Chap. 2). In addition, the high electronegativity of the fluorine atom stabilizes the C–F bonds of CF_3, but tends to weaken the adjacent C–C and C–halogen bonds. As a result, the major metabolic products are chloride and bromide ions, and trifluoroacetic acid (CF_3COOH).

The induction period is very rapid, surgical anesthesia being produced in 2 to 10 minutes. Recovery is equally rapid following removal of anesthetic. Side-effects produced by halothane are hypotension and bradycardia. Liver necrosis, in some cases similar to that induced by chloroform and carbon tetrachloride, has been observed. An impurity, 2,3-dichloro-1,1,1,4,4,4-hexafluorobutene-2, has been found in low concentrations in freshly opened bottles of halothane.[3] Its concentration is increased in the presence of copper, oxygen and heat. This impurity has proved to be acutely toxic to dogs in anesthetic concentrations and produces degenerative changes in the lungs, the liver and the kidney of rats. It is possible that the degradation product may be responsible for liver damage in the United States where copper vaporizers are widely used. Halothane is a volatile, noninflammable liquid, b.p. 50°, which is given by inhalation. Anesthesia may be induced with 2 to 2.5 per cent, vaporized by a flow of oxygen. The compound is sensitive to light and is distributed in brown bottles, stabilized by the addition of 0.01 per cent thymol.

Category—general anesthetic (inhalation).

Application—by inhalation as required.

ETHERS

Ether U.S.P., Ethyl Ether, Diethyl Ether, $CH_3CH_2OCH_2CH_3$. Ether was the first (1842) of the general anesthetics used in surgical anesthesia, and because of the wealth of knowledge and experience concerning its effects in each plane of anesthesia, it is still one of the safest. The prolonged induction time may be avoided by the initial use of a more rapidly acting agent (e.g., vinyl ether, nitrous oxide), followed by a gradual change to ether at the proper concentration for maintenance. Ether is flammable and forms explosive mixtures with air and oxygen. It tends to form peroxides and further degradation products on standing in contact with the air. (Physical properties and standards for ether are described in Chapter 5.)

Category—general anesthetic (inhalation).

Application—by inhalation as required.

Vinyl Ether N.F., Divinyl Oxide, Vinethene®, $CH_2=CH-O-CH=CH_2$. Vinyl ether alone is useful for induction anesthesia, with production of onset in about 30 seconds, or for short operative procedures. It produces extensive liver damage on prolonged use.

Vinyl ether is a volatile liquid, b.p. 28° to 31°, with about the same explosive hazard as ether. (Physical properties and standards are described in Chapter 5.)

Category—general anesthetic.

Application—by inhalation as required.

Fluroxene N.F., Fluoromar®, 2,2,2-Trifluoroethyl Vinyl Ether; CF_3—CH_2—O—CH=CH$_2$. Fluroxene is a volatile liquid, b.p. 42.7°, with an anesthetic potency similar to that of diethyl ether. It was introduced in 1960 and has been used as the anesthetic agent in a variety of minor surgical operations, or as an induction anesthetic for other anesthetics. Fluroxene is flammable in air in a concentration range of 4 to 12 per cent and should not be used in the presence of an open flame or when diathermy or cautery apparatus is used.

Surgical anesthesia is produced by concentrations of 3 to 8 per cent.

Category—general anesthetic.

Application—by inhalation as required.

Methoxyflurane N.F., Penthrane®, 2,2-Dichloro-1,1-difluoroethyl Methyl Ether, Methyl β-dichloro-α-difluoroethyl ether; $CHCl_2$—CF_2—O—CH_3. Methoxyflurane is a volatile liquid, b.p., 101°, whose vapors produce general anesthesia with a slow onset and a fairly long duration of action. It was introduced in 1962. It is nonflammable in any concentration in air or oxygen and produces anesthesia at concentrations of 1.5 to 3 per cent when vaporized by a rapid flow of oxygen. Methoxyflurane is a stable compound, even in the presence of bases.

Category—general anesthetic.

Application—by inhalation as required.

ALCOHOLS

Alcohols, most notably ethanol, have been long known and used for their ability to depress certain higher centers of the central nervous system. As the series is ascended, hypnotic activity of the normal alcohols reaches a maximum at 6 or 8 carbons and then declines as the alkyl chain is further lengthened. None of the unsubstituted alcohols possesses sufficient potency for use as a general anesthetic. Some halogenated alcohols, particularly those bearing 3 bromine or chlorine atoms on a single carbon atom (e.g., tribromoethanol, trichloroethanol), are potent hypnotics and are capable of producing basal anesthesia.

Tribromoethanol N.F., Avertin®, Ethobrom®, 2,2,2-Tribromoethanol, CBr_3CH_2-OH. Tribromoethanol was introduced as a basal anesthetic in 1926. It is a white, crystalline material, soluble in water (3.4:100), alcohol or organic solvents. The drug usually is supplied in solution with amylene hydrate of such strength that 1 ml. of solution contains 1 g. of tribromoethanol. A distilled-water solution should produce an orange-red color (pH 5) when tested with Congo red solution. If the color is blue or violet (pH 3), decomposition of the drug has occurred. The decomposition products (dibromovinyl alcohol, hydrobromic acid,

$$\underset{\text{Tribromoethanol}}{\overset{\displaystyle Br}{\underset{\displaystyle Br}{Br-C-CH_2OH}}} \longrightarrow \underset{\substack{\text{Dibromovinyl}\\\text{alcohol}}}{\overset{\displaystyle Br}{Br-C=CHOH}} \rightleftarrows$$

$$\underset{\text{Dibromoacetaldehyde}}{\overset{\displaystyle Br}{Br-CH-CHO}} + HBr$$

dibromoacetaldehyde) cause severe irritation, which militates against the use of such a solution.

For rectal administration in the production of general anesthesia, 2.5 ml. of Tribromoethanol Solution (1 g. in 1 ml.) is diluted to 100 ml. with warm purified water just before use. The usual rectal dose is 60 to 80 mg. of tribromoethanol per kg. of body weight. This corresponds to 2.4 to 3.2 ml. of the diluted solution per kg. of body weight. The total dose of tribromoethanol should not exceed 8 g. for women, 10 g. for men. The use of tribromoethanol in the production of basal anesthesia has

TABLE 27. ULTRASHORT-ACTING BARBITURATES USED TO PRODUCE GENERAL ANESTHESIA

GENERAL STRUCTURE

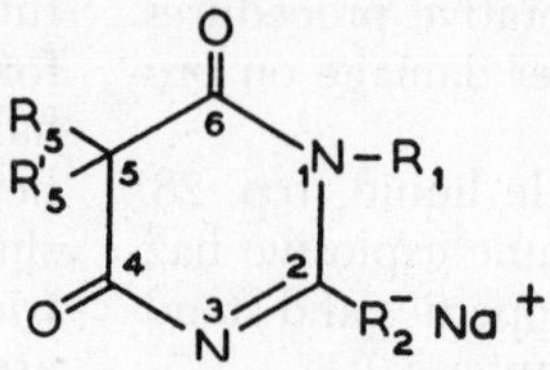

$$\begin{array}{c} R_5 \\ R'_5 \end{array}\!\!\!\overset{O}{\underset{O}{\big|}}\!\!\!\begin{array}{c} N-R_1 \\ R_2^- \; Na^+ \end{array}$$

Generic name Proprietary name	Substituents				Anesthetic dose (in mg.)
	R_5	R'_5	R_1	R_2	
Sodium Hexobarbital Evipal Sodium	CH_3-	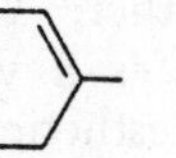 (cyclohexenyl)	CH_3	O	200-400
Sodium Methohexital Brevital Sodium	$CH_2\!=\!CH\!-\!CH_2-$	$CH_3CH_2C\!\equiv\!C\!-\!\underset{\underset{CH_3}{\big|}}{CH}-$	CH_3	O	70-100
Sodium Thiamylal Surital Sodium	$CH_2\!=\!CH\!-\!CH_2-$	$CH_3CH_2CH_2\underset{\underset{CH_3}{\big|}}{CH}-$	H	S	75-150
Sodium Thiopental U.S.P. Pentothal Sodium	CH_3CH_2-	$CH_3CH_2CH_2\underset{\underset{CH_3}{\big|}}{CH}-$	H	S	50-75

decreased because of the difficulty of reversing an overdose, once administered.

A 1 per cent aqueous solution with 5 per cent dextrose is used intravenously for general anesthesia.

Category—general anesthetic.

OCCURRENCE

Tribromoethanol Solution N.F.

ULTRASHORT-ACTING BARBITURATES

The ultrashort-acting barbiturates (Table 27), as their sodium salts, may be administered intravenously or by retention enema for the production of surgical anesthesia. Anesthesia begins rapidly (in less than 1 minute) and is usually of short duration. Intravenous barbiturate anesthesia provides smooth induction, with rapid passage through the excitement stage, absence of salivary secretions, fair muscular relaxation, nonexplosive properties, and rapid depletion of the agent from the central nervous system, leading to a short and uncomplicated period of postoperative recovery. Disadvantages include potent respiratory depression, tissue-irritating properties on extravasation, laryngospasm, and a precipitous fall in blood pressure if the intravenous barbiturates are administered too rapidly.

The rapid onset and brief duration of action of the ultrashort-acting barbiturate anesthetics have been considered to be due to their high lipid solubility, enabling free passage of the lipoid cellular membrane of the blood-brain barrier, followed by rapid loss to the peripheral lipoidal storage areas.[4] Metabolism of the barbiturates occurs too slowly to have any significant influence on the onset and duration of anesthesia. Thiopental, for example, is metabolized at the rate of 10 to 15 per cent per hour. However, the rate at which fat concentrates thiopental following its intravenous injection is too slow to account for the rate at which the central nervous system is depleted.[5] Instead, the lean body tissues (e.g., muscle), which are well perfused with blood, provide the initial pool and rapidly take up most of the thiopental lost by the brain. Redistribution to body fat and metabolic degradation appear to occur more slowly and do not account for the rapid recovery from thiopental anesthesia.

The production of surgical anesthesia by barbiturates for short operations is not without dangers, chiefly because of variations in individual susceptibility to respiratory depression and to other side-effects. In general, the ultrashort-acting barbiturates produce a greater degree of respiratory depression for a given degree of skeletal muscular relaxation than do the inhalation anesthetics. The agents commonly used are those with a very short duration of action.

Sodium Hexobarbital, Evipal® Sodium, Sodium 5-(1-Cyclohexenyl)-1,5-dimethyl-barbiturate. Sodium hexobarbital was introduced in 1934 for intravenous use in the production of anesthesia of short duration. Consciousness is lost in less than a minute, and with ordinary doses is restored in from 15 to 30 minutes. The usual dose is 2 to 4 ml. of a 10 per cent solution. Both the powder and the solution are sensitive to air; a discolored solution, even if freshly prepared, should be discarded.

Sodium Methohexital, Brevital® Sodium, Sodium α-(±)-1-Methyl-5-allyl-5-(1-methyl-2-pentynyl)barbiturate. Sodium methohexital was introduced in 1960 as an intravenously administered ultrashort-acting barbiturate. Induction of anesthesia with methohexital is as rapid as with thiopental, and recovery is more rapid, perhaps due to a faster metabolism. The drug has no muscle-relaxant properties; for surgical procedures requiring muscle relaxation, it requires supplementation with a gas anesthetic and a muscle relaxant. Sodium methohexital is supplied in crystalline form together with anhydrous sodium carbonate and is administered only by the intravenous route. Anesthesia is induced with 7 to 10 ml. of a 1 per cent solution.

OCCURRENCE
Sodium Methohexital for Injection N.F.

The isomeric β-(±)-analog also is a potent central nervous system depressant but produces a greater degree of central stimulant and skeletal muscle activity during the recovery period.

Sodium Thiamylal, Surital® Sodium, Sodium 5-Allyl-5-(1-methylbutyl)-2-thiobar-biturate. Sodium thiamylal was introduced in 1952 as an ultrashort-acting intravenous anesthetic. It is available in vials as a sterile powder admixed with anhydrous sodium carbonate as a buffer. Onset of anesthesia occurs in 20 to 60 seconds and lasts for 10 to 30 minutes after the last injection. The initial intravenous dose is 3 to 6 ml. of a 2.5 per cent solution at the rate of 1 ml. every 5 seconds. Maintenance, 0.5 to 1 ml. as required; the maximum total dose should not exceed 1 g.

A nonsterile form of sodium thiamylal, characterized by the green dye which it contains, is available for rectal instillation. The dose of a 5 to 10 per cent solution is determined by the physician.

OCCURRENCE
Sodium Thiamylal for Injection N.F.

Sodium Thiopental U.S.P., Pentothal® Sodium, Sodium 5-Ethyl-5-(1 methylbutyl)-2-thiobarbiturate. Sodium thiopental has been the most widely used of the intravenous barbiturate anesthetics. Onset is rapid (about 30 seconds) and duration brief (10 to 30 minutes). The usual intravenous dose for induction of anesthesia is 2 to 3 ml. of a 2.5 per cent solution, at the rate of 1 ml. every 5 seconds. Maintenance, 0.5 to 2 ml. as required. Ampuls containing the sterile white to yellowish-white powder also contain anhydrous sodium carbonate as a buffer.

Sodium thiopental is available as a non-sterile powder containing a green dye and is intended for rectal application; 45 mg. per kg. of body weight, in 10 per cent solution; range, 25 to 45 mg. per kg.

Category—general anesthetic (intravenous).

OCCURRENCE
Sodium Thiopental for Injection, U.S.P.

MISCELLANEOUS

Ketamine Hydrochloride, Ketalar®, 2-(*o*-Chlorophenyl)-2-methylaminocyclohexanone Hydrochloride. Ketamine hydrochloride was introduced in 1970 as an anesthetic agent, with rapid onset and short duration of action on parenteral administration. Unlike the ultrashort-acting barbiturates, which are sodium salts of acids, ketamine is solubilized for parenteral administration as the hydrochloride of a weakly basic amine. Anesthesia is produced within 30 seconds after intravenous administration, with the effect lasting for 5 to 10 minutes. Intramuscular doses bring on surgical anesthesia within 3 to 4 minutes, with a duration of action of 12 to 25 minutes. It may also be used as an induction anesthetic, prior to the use of other anesthetics, or administered together with volatile anesthetics.

Ketamine was developed as a structural analog of phencyclidine [1-(1-phenylcyclohexyl)piperidine], a parenteral anesthetic. Like the parent compound, ketamine has produced disagreeable dreams or hallucinations during the brief period of awakening and reorientation. Other untoward effects, including moderate increase in blood pressure, are minimal.

Ketamine hydrochloride is a water-soluble white crystalline powder. The drug is available at concentrations of 10 mg. per ml. or 50 mg. per ml. The usual initial intravenous dose is 2 mg. per kg. of body weight, followed by maintenance doses of 1 mg. per kg. The usual intramuscular dose is about 10 mg. per kg.

Nitrous Oxide U.S.P., nitrogen monoxide; N_2O, is useful for the rapid induction of anesthesia. For surgical anesthesia, a concentration of 80 to 85 per cent is required. The 15 to 20 per cent oxygen which may be used with nitrous oxide provides borderline oxygenation of tissues, and the gas is not recommended for prolonged administration.

Category—general anesthetic (inhalation).

Application—by inhalation as required.

SEDATIVES AND HYPNOTICS

Historically, the first sedative-hypnotic was ethanol, obtained by fermentation of a variety of carbohydrates. The opium poppy provided the limited armamentarium of the early physician with a second source of a depressant drug. The introduction of inorganic bromides as sedative-hypnotics and anticonvulsants in the 1850's was followed shortly by the development of the effective depressants chloral, paraldehyde, sulfonal and urethan. The recognition of the depressant properties of the barbiturates in 1903 was followed by a variety of related sedative-hypnotics which possess many properties in common.

A characteristic shared by all of the sedative-hypnotic drugs is the general type of depressant action on the cerebrospinal axis. In their clinical applications they differ mainly in the time required for onset of depression and in the duration of the effect produced. The degree of depression depends largely upon the potency of the agent selected, the dose used and the route of administration. All sedative-hypnotic drugs are capable of producing depression ranging from slight sedation, a condition in which the patient is awake but possesses decreased excitability, to sleep. In sufficiently high doses, depression of the central nervous system continues, and many sedative-hypnotic agents may produce surgical anesthesia which resembles that brought about by the volatile anesthetics. The same sequelae of events, including a stage of excitement due to depression of the higher cortical centers, proceeds into surgical anesthesia with both sedative-hypnotics and anesthetics. However, the dangers attending use of anesthetic doses of these drugs largely limit their use to the production of sedation and sleep. The longer-acting central nervous system depressants are usually selected for the production of sedation. The The situations in which such sedation is useful include[6]: (1) sudden, limited stressful situations involving great emotional strain, (2) chronic tension states created by disease or sociologic factors, (3) hypertension, (4) potentiation of analgesic drugs, (5) the control of convulsions, (6) adjuncts to anesthesia, (7) narcoanalysis in psychiatry.

The hypontic dose is used to overcome insomnia of many types. Sedative-hypnotics with a short to moderate duration of action are useful in relieving the insomnia of individuals whose high level of activity during the day makes it difficult for them to decrease their activities as a prelude to sleep. Once asleep, they have no more need for the drug, and the shorter-acting agents provide little after-depression. For others, who for reasons of health, external disturbances or psychic abnormalities awake frequently during the night, the longer-acting agents are more useful.

STRUCTURE-ACTIVITY RELATIONSHIPS

Although the sedative-hypnotic drugs include many chemical types, they have certain common physicochemical and structural features. The polar portion of the molecule is one of the most water-solubilizing of the nonionic functional groups. These include, with their Hansch π values[7] as measures of polar character (see Chap. 2), the unsubstituted barbituric acid nucleus

$$\left(\begin{array}{c} \text{-C-CONH} \\ | \quad | \\ \text{CONHCO ,} \end{array} \quad \pi = -1.35 \right)$$

acyclic dirureides (-CONHCONHCO-, $\pi = -1.68$), amides (-CONH$_2$, $\pi = -1.71$), alcohols (-OH, $\pi = -1.16$), carbamates (-OCONH$_2$, $\pi = -1.16$), and sulfones (-SO$_2$CH$_3$, $\pi = -1.26$). These polar groups are attached to a nonpolar moiety, usually alkyl, aryl or haloalkyl, so that the partition coefficient between a lipid and an aqueous phase (octanol-water) for the resulting molecule is close to 100 (log P = 2). In general, the potency of many classes of sedative-hypnotic drugs varies in a parabolic fashion, with a maximum close to a partition coefficient value of log P = 2.0.

It appears that these molecules have the proper solubility characteristics to be absorbed from the gastrointestinal tract, to be transported in the aqueous body fluids and to be sufficiently lipophilic so that they readily penetrate the central nervous system where their nonionic surfactant characteristics may serve to distort essential lipoprotein matrices thus depressing function.

In addition to their solubility character-istics most of the useful sedative-hypnotic drugs possess structural features which resist the rapid metabolic attack which their partition behavior would normally facilitate (see Chap. 3).

Thus, tertiary alcohols, which are resistant to metabolic oxidation, are generally more effective than primary or secondary alcohols, which are rapidly oxidized, conjugated and excreted. Amides and carbamates are generally hydrolyzed slowly as compared with esters. Sulfones require cleavage of a carbon-sulfur bond for further metabolic oxidation to occur, and they are generally excreted unchanged.

BARBITURATES

The barbiturates are the most widely used of the sedative-hypnotic drugs. Barbital, the first member of the class, was introduced in 1903; the method of synthesis of the thousands of analogs prepared since has undergone little change.

Diethylmalonate reacts with alkyl halides in the presence of sodium alkoxides to form the intermediate monoalkyl malonic ester. This may be allowed to react with a different alkyl halide to form a dialkyl malonic ester, which may be condensed with urea in the presence of a sodium alkoxide to form the sodium salt of 5,5-dialkylbarbituric acid. In an acid environment, the free 5,5-dialkyl-barbituric acid is formed (see p. 404).

If thiourea is used in place of urea in the condensation, thiobarbiturates which contain a sulfur atom attached to the 2-carbon atom are formed. The use of N-methylurea in this condensation leads to the 1-methyl-barbiturates.

All of the barbiturates are colorless, crystalline solids that melt at from 96° to 205°. They are not very soluble in water but form sodium salts that are quite soluble. Solutions of the latter are usually rather strongly alkaline and often hydrolyze enough to give precipitates of the barbiturate. Any admixture with acidic substances will be almost certain to give such a precipitate, an incompatibility that frequently must be considered in the dispensing laboratory. The alkalinity of the solutions for parenteral injections can be overcome largely by appro-

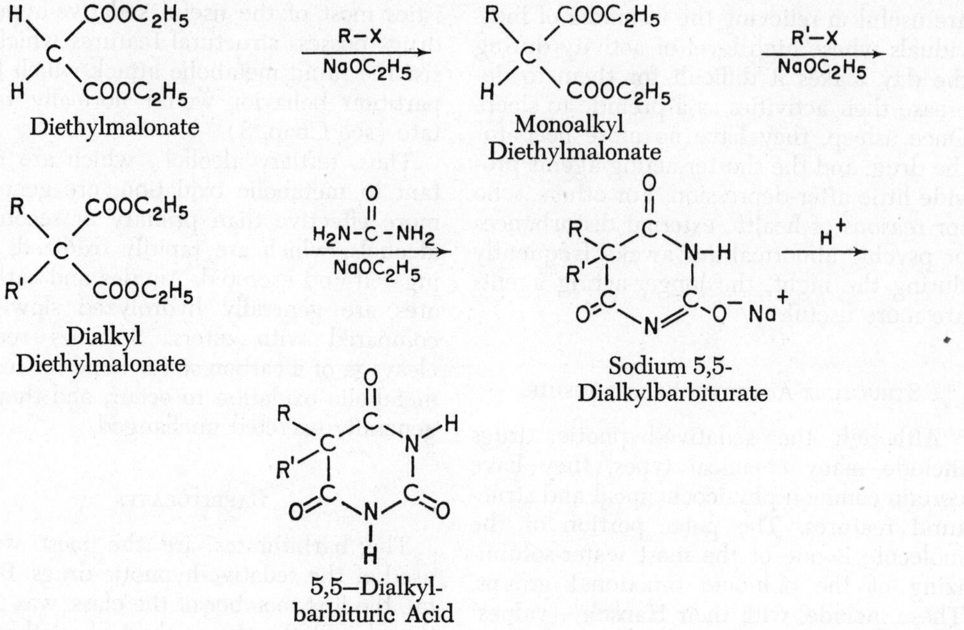

Diethylmalonate

$$\xrightarrow[\text{NaOC}_2\text{H}_5]{\text{R}-\text{X}}$$

Monoalkyl
Diethylmalonate

$$\xrightarrow[\text{NaOC}_2\text{H}_5]{\text{R}'-\text{X}}$$

Dialkyl
Diethylmalonate

$$\xrightarrow[\text{NaOC}_2\text{H}_5]{\text{H}_2\text{N}-\overset{\text{O}}{\overset{\|}{\text{C}}}-\text{NH}_2}$$

Sodium 5,5-
Dialkylbarbiturate

$$\xrightarrow{\text{H}^+}$$

5,5—Dialkyl-
barbituric Acid

priate buffering, principally by the use of sodium carbonate.

Many of the names of the barbituric acid derivatives end with the suffix "al." This has been used to denote hypnotics since the introduction of chloral hydrate, in 1896, and has been applied to a wide variety of chemical types (e.g., Sulfonal, Carbromal, Bromural, Veronal, Luminal).

The action of an ideal hypnotic would be exerted only on cells in the psychic center of the brain and on the centers of pain perception, with no effect on those of motor control, of the automatic process, such as respiration and circulation, or on any other functions or glands. Therefore, the ideal agent would bring about sleep and freedom from pain without interfering with other normal processes; the effect should be of sufficient duration for the purpose intended, and there should be no undesirable secondary reactions. While no such agent has yet been discovered, the barbiturates appear to approach closest to these criteria, although chloral, paraldehyde, codeine and a few others seem to be of advantage under differing circumstances.

The mechanism by which hypnotics bring about the desired selective depression is not

well understood, but the barbiturates are believed to act on the brain stem reticular formation to reduce the number of nerve impulses ascending to the cerebral cortex. No one knows how narcotics in general affect the cell activities unselectively or why some of them act only on particular cells. The Meyer-Overton law, that the depressant efficiency of any agent is measured by lipid solubility, is accepted generally as applying to hypnotics. However, it is readily apparent from a knowledge of the physiologic action of thousands of compounds that other factors also must be considered. For a general discussion of these factors relating to narcosis, see Chapter 2.

Variations in properties among the barbiturates involve chiefly the dose required, the length of time after administration before the effects are observed, duration of action, ratio of therapeutic to toxic or fatal dose and extent of accumulation. Since the dosage can be regulated and the margin of safety is usually satisfactory, the main considerations are promptness and duration of action, chiefly the latter. These factors, together with the structures and the usual doses of the currently distributed barbiturates, are compiled in Table 28.

Structure-Activity Relationships

Major findings in regard to structure-activity relationships are as follows:

Both hydrogen atoms in position 5 of barbituric acid must be replaced for maximal activity. This is likely due to the susceptibility to rapid metabolic attack[7] of C–H bonds in such a position.

Increasing the length of an alkyl chain in the 5-position enhances the potency up to 5 or 6 carbon atoms; beyond that, depressant action decreases and convulsant action may result. This is probably due to the excess over ideal lipophilic character.

Branched, cyclic or unsaturated chains in the 5-position generally produce a briefer duration of action than do normal saturated chains containing the same number of carbon atoms. This appears to be due to a combination of decreased lipophilic character and increased ease of metabolic conversion to a more polar, inactive metabolite.

Compounds with alkyl groups in the 1 or 3 position may have a shorter onset and duration of action. The N-methyl group results in a barbiturate which is a weaker acid (e.g., hexobarbital, $pK_a = 8.4$) compared with the usual $pK_a = 7.6$ for barbiturates without the N-methyl substituent. The weaker acid is largely in the nonionic lipid soluble form (plasma pH 7.4), which readily enters the central nervous system and rapidly equilibrates into peripheral fatty stores.

Replacement of oxygen by sulfur on the 2-carbon shortens the onset and duration of action. Thiobarbiturates, although little different from barbiturates in acid strength (e.g., thiopental, $pK_a = 7.4$), are much more lipid-soluble in the un-ionized form than are the corresponding oxygen analogs. Rapid movement into and out of the central nervous system, as well as ease of metabolic attack, accounts for the rapid onset and short duration of action.

These phenomena are discussed further in the following paragraphs.

It may be noted from Table 28, that the total number of carbon atoms contained in the groups substituted in the 5-position of barbituric acid is closely related to the duration of action. The compounds with the most rapid onset and shortest duration of action, following oral administration, are those with the most lipophilic substituents, totaling 7 to 9 carbon atoms. This may be related to rapid absorption and distribution to the central nervous system, followed by rapid loss to neutral storage sites, such as peripheral fat. Conversely, the barbiturates with the slowest onset and longest duration of action contain the most polar side chains—either the 4 carbon atoms contributed by two ethyl groups, or an ethyl and a phenyl group (e.g., phenobarbital). The phenyl group attached to a polar substituent has a water solubility greater than that expected of its 6 carbon content, apparently due to the polarizability of its *pi* electrons. In a barbiturate, for example, the lipophilic character of the benzene ring ($\pi = +1.77$) is between that of a 3 and 4 carbon aliphatic chain (n = propyl, $\pi = +1.5$; n-butyl, $\pi = +2.0$).[7] The barbiturates with an intermediate duration of action have 5 alkyl substituents of intermediate polarity (5 to 7 carbon atoms total).

Lipophilia is somewhat reduced by branched chains and unsaturation, but the total carbon content of the groups in the 5-position provides a good first approximation of duration of action. The long-acting, relatively polar phenobarbital, for example, both enters and leaves the central nervous system very slowly as compared with the more lipophilic thiopental. In addition, the lipoidal barriers to drug metabolizing enzymes lead to a slower metabolism for the more polar barbiturates, phenobarbital being metabolized to the extent of only about 10 per cent per day.

Pharmacologic Properties

The effects following administration of the barbiturates pursue about the same course, regardless of the compound used. Therapeutic doses in the smaller amounts calm nervous conditions of any origin and in larger amounts cause a dreamless sleep in from 20 to 60 minutes after oral administration and almost immediately if given intravenously. In some patients, there may be considerable excitement before sedation is

TABLE 28. BARBITURATES USED AS SEDATIVES AND HYPNOTICS

GENERAL STRUCTURE

A. Long Duration of Action (6 or more hours)

Generic name / Proprietary name	Substituents R5	R'5	R1	Sedative dose (in mg.)	Hypnotic dose (in mg.)	Usual onset of action (in min.)
Barbital / Veronal	C_2H_5	C_2H_5	H	—	300	30-60
Mephobarbital N.F. / Mebaral	C_2H_5	(phenyl)	CH_3	30-100*	100	30-60
Metharbital N.F. / Gemonil	C_2H_5	C_2H_5	CH_3	50-100*	—	30-60
Phenobarbital U.S.P. / Luminal	C_2H_5	(phenyl)	H	15-30*	100	20-40

* Daytime sedative and anticonvulsant.

B. Intermediate Duration of Action (3-6 hours)

Generic name / Proprietary name	Substituents R5	R'5	R1	Sedative dose (in mg.)	Hypnotic dose (in mg.)	Usual onset of action (in min.)
Allylbarbituric acid / Sandoptal	$CH_2{=}CHCH_2{-}$	$(CH_3)_2CHCH_2{-}$	H	—	200-600	20-30
Amobarbital U.S.P. / Amytal	$CH_3CH_2{-}$	$(CH_3)_2CHCH_2CH_2{-}$	H	20-40	100	20-30
Aprobarbital N.F. / Alurate	$CH_2{=}CHCH_2{-}$	$(CH_3)_2CH{-}$	H	20-40	40-160	—
Sodium Butabarbital N.F. / Butisol Sodium	$CH_3CH_2{-}$	$CH_3CH_2\overset{CH_3}{\underset{}{CH}}{-}$	H	15-30	100	20-30
Butallylonal / Pernocton	$CH_2{=}\overset{}{\underset{Br}{C}}{-}CH_2{-}$	$CH_3CH_2\overset{CH_3}{\underset{}{CH}}{-}$	H	—	200	—
Butethal / Neonal	$CH_3CH_2{-}$	$CH_3CH_2CH_2CH_2{-}$	H	—	100-200	30-60
Allobarbital / Dial	$CH_2{=}CHCH_2{-}$	$CH_2{=}CHCH_2{-}$	H	30	100-300	15-30
Probarbital / Ipral	$CH_3CH_2{-}$	$(CH_3)_2CH{-}$	H	50	130-390	20-30
Butalbital N.F. / Lotusate	$CH_2{=}CHCH_2{-}$	$CH_3\overset{CH_3}{\underset{}{CH}}CH_2{-}$	H	50	120	20-30
Vinbarbital N.F. / Delvinal	$CH_3CH_2{-}$	$CH_3CH_2CH{=}\overset{CH_3}{\underset{}{C}}{-}$	H	30	100-200	20-30

TABLE 28. (*Continued*)

C. Short Duration of Action (less than 3 hours)

Generic name Proprietary name	R₅	R′₅	R₁	Sedative dose (in mg.)	Hypnotic dose (in mg.)	Usual onset of action (in min.)
Cyclobarbital Phanodorn	CH_3CH_2-	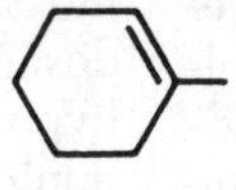 (cyclohexenyl)	H	—	100-300	15-30
Cyclopentenylallyl- barbituric acid Cyclopal	$CH_2{=}CHCH_2-$	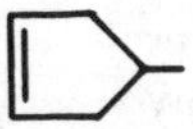 (cyclopentenyl)	H	50-100	100-400	15-30
Heptabarbital Medomin	CH_3CH_2-	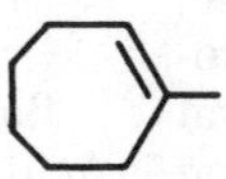 (cycloheptenyl)	H	50-100	200-400	20-40
Hexethal Ortal	CH_3CH_2-	$CH_3(CH_2)_5-$	H	50	200-400	15-30
Sodium Pentobarbital *U.S.P.* Nembutal Sodium	CH_3CH_2-	$CH_3CH_2CH_2\overset{\underset{\displaystyle CH_3}{\mid}}{C}H-$	H	30	100	20-30
Secobarbital U.S.P. Seconal	$CH_2{=}CHCH_2-$	$CH_3CH_2CH_2\overset{\underset{\displaystyle CH_3}{\mid}}{C}H-$	H	15-30	100	20-30

initiated. Still larger doses produce a form of anesthesia, and in all cases, up to this stage, there is little disturbance of other functions, such as respiration, circulation, metabolism or the action of smooth muscle. The effects last for ½ to 12 hours, depending on the compound, the dosage and the stage to which the narcosis has been carried. The patient awakens refreshed but may not be as alert as usual and may experience some lassitude for a time. In the presence of pain, these drugs have very little analgesic action but may potentiate other compounds that do have such effect. Excretion is principally by the kidneys, where they appear partly unchanged, partly oxidized in the side chain, and partly conjugated. Some of the compounds, especially those containing sulfur, are destroyed almost entirely in the body, probably in the liver. For a more detailed description of barbiturate metabolism, see Chapter 3.

Untoward reactions are uncommon except with very large doses. The chief one of inconvenience is the appearance in some individuals of delayed effects, extreme depression, excitement or even mania. Occasionally, some persons are hypersensitive and experience dermatologic lesions as manifested by wheals, angioneurotic edema or scarletinal-like rashes. The respiration and the circulation are depressed slightly by anesthetic doses, but the temperature and basal metabolism are scarcely affected. Relatively large amounts cause profound and prolonged coma, a marked fall in blood pressure and eventual paralysis of the respiratory center. The remedial measures used are persistent administration of central nervous system stimulants (see Chap. 16), various supportive procedures and artifical respiration or oxygen if necessary. Even in therapeutic doses, there is an occasional fatal collapse due to peripheral paralysis of the blood vessels. The margin between therapeutic and toxic doses is comparatively large, and poisoning would be of little importance were it not for the fact that the barbiturates are widely used, and their ready availability leads to frequent accidental or deliberate overdosage.

Habituation to the barbiturates is widespread and well recognized. It is not so well known that these widely used drugs are capable of producing a primary addiction.[8] Tolerance to increased doses develops slowly, but physical dependence may develop fairly rapidly. Oral ingestion of about 800 mg. daily of the potent, short-acting barbiturates for a period of 8 weeks will result in mild to moderate withdrawal symptoms in most individuals. The average daily dose for the barbiturate addict is about 1.5 g. Abrupt withdrawal of the drug from an addicted individual will frequently result in delirium and grand-mal-like convulsions. Severe withdrawal symptoms, including insomnia, nausea, cramps, vomiting, orthostatic hypotension, convulsive seizures and visual and auditory hallucinations, may continue for days. The extent of mental, emotional and neurologic impairment together with the severity of the withdrawal reactions have caused barbiturate addiction to be classed by some as a public health and medical problem more serious than morphine addiction.

The barbiturates are used chiefly as sedatives and hypnotics in a wide variety of conditions. Selection of a barbiturate with the appropriate onset time and duration of action is desirable.[6] The disorders for which sedation is indicated vary from a state of "overwrought nerves" to a violent mania. The barbiturates have the advantage over the bromides in that the action may be enhanced to more profound states by increasing the dose. They are indicated in any type of insomnia that is not due to pain and, even in cases where pain is present, may advantageously be combined with analgesics. They also are applied to suppress a variety of convulsions with origins in the central nervous system, including those from tetanus, meningitis, chorea, epilepsy, eclampsia, insulin overdosage and poisoning by strychnine and similar drugs. In epilepsy, phenobarbital or a similar long-acting barbiturate will diminish the number and the severity of the attacks.

Barbiturates also are employed to produce anesthesia, either as premedication before other agents, or to act as the sole anesthetic. In providing preliminary sedation, the short or the intermediate-acting barbiturates are superior in some respects to morphine, and it often may be of advantage to combine the two. In somewhat larger doses, the ultrashort-acting barbiturates contribute to the narcosis, thus diminishing the amount of volatile anesthetic required and reducing the undesirable side-

effects of the latter. In combination with morphine and scopolamine, the barbiturates are used in producing obstetric amnesia.

Some of the more frequently used barbiturates are described briefly in the following sections. For the structures, the usual dosages required to produce sedation and hypnosis, the times of onset and the duration of action see Table 28.

Barbiturates With a Long Duration of Action (six hours or more)

Barbital, Veronal®, 5,5-Diethylbarbituric Acid. Barbital is used orally as a hypnotic-sedative with a duration of about 8 to 12 hours. It is also available as the more water-soluble salt, sodium barbital (Veronal Sodium).

Mephobarbital N.F., Mebaral®, 5-Ethyl-1-methyl-5-phenylbarbituric Acid. Mephobarbital produces sedation of long duration, but is a relatively weak hypnotic. It is used orally in the prevention of grand mal and petit mal epileptic seizures.

Category—anticonvulsant; sedative.

Usual dose range—anticonvulsant, 400 to 600 mg. daily; sedative, 32 to 100 mg. 3 or 4 times daily.

Occurrence
Mephobarbital Tablets N.F.

Metharbital N.F., Gemonil®, 5-5-Diethyl-1-methylbarbituric Acid. Metharbital produces less sedation than phenobarbital and is most often used in the control of epileptic seizures of the grand mal, the petit mal, the myoclonic or the mixed type.

Category—anticonvulsant.

Usual dose—initial, 100 mg. up to 3 times daily.

Usual dose range—100 to 800 mg. daily.

Occurrence
Metharbital Tablets N.F.

Phenobarbital U.S.P., Luminal®, 5-Ethyl-5-phenylbarbituric Acid. Phenobarbital is a long-acting hypnotic-sedative, more potent than barbital but slower in onset of action, requiring about 1 hour. The duration of action is 10 to 16 hours. It is also effective in the prevention of epileptic seizures, being more effective in the grand mal than in the petit mal types. *Sodium Phenobarbital U.S.P.*, a salt that is more water soluble, is available for either oral use or parenteral use by the subcutaneous, the intramuscular and the intravenous routes.

Category—anticonvulsant; hypnotic; sedative.

Usual dose—anticonvulsant and sedative, 15 to 30 mg. 3 or 4 times daily; hypnotic, 100 mg. at bedtime.

Usual dose range—15 to 200 mg. daily.

OCCURRENCE

Phenobarbital Elixir U.S.P.
Phenobarbital Tablets U.S.P.
Sodium Phenobarbital U.S.P.
Sodium Phenobarbital Injection U.S.P.
Sterile Sodium Phenobarbital U.S.P.

Barbiturates With an Intermediate Duration of Action (3 to 6 hours)

Amobarbital U.S.P., Amytal®, 5-Ethyl-5-isopentylbarbituric Acid. The doses of amobarbital range as follows: sedation, 16 to 50 mg.; hypnosis or preanesthetic medication, 100 to 200 mg.; anticonvulsant, 200 to 400 mg. *Sodium Amobarbital U.S.P.* is the water-soluble salt used for oral, rectal, intramuscular or subcutaneous administration; the usual dose for hypnosis is 100 mg.

OCCURRENCE

Amobarbital Elixir N.F.
Amobarbital Tablets U.S.P.
Sodium Amobarbital U.S.P.
Sodium Amobarbital Capsules U.S.P.
Sterile Sodium Amobarbital U.S.P.

Aprobarbital N.F., Alurate®, 5-Allyl-5-isopropylbarbituric Acid. Aprobarbital is used orally in a dose that ranges from 20 to 160 mg., as a sedative and hypnotic.

Sodium Butabarbital N.F., Butisol® Sodium, Sodium 5-*sec*-butyl-5-ethylbarbiturate. Sodium butabarbital is used orally as a sedative at a dose of 8 to 60 mg.; as a hypnotic, 100 to 200 mg. Sedation is sustained for about 5 to 6 hours.

OCCURRENCE

Sodium Butabarbital Capsules N.F.
Sodium Butabarbital Elixir N.F.
Sodium Butabarbital Tablets N.F.

Butalbital N.F., Lotusate®, 5-Allyl-5-iso-butylbarbituric Acid. Talbutal was introduced in 1955 as a hypnotic-sedative with intermediate duration of action. The sedative dose is 30 to 50 mg., the hypnotic dose 120 mg.

Vinbarbital N.F., Delvinal®, 5-Ethyl-5-(1-methyl-1-butenyl)barbituric Acid. Vinbarbital was introduced in 1950 as a barbituric acid derivative with intermediate duration of action. Both vinbarbital and the water-soluble sodium salt, Sodium Vinbarbital, Delvinal Sodium, are administered orally at a dose of 30 mg. for sedation and 100 to 200 mg. for hypnosis. The sodium salt is used for intravenous administration also.

OCCURRENCE

Sodium Vinbarbital Injection N.F.
Vinbarbital Capsules N.F.

Barbiturates With a Short Duration of Action (less than 3 hours)

Heptabarbital, Medomin®, 5-(1-Cyclohepten-1-yl)-5-ethylbarbituric Acid. Heptabarbital was introduced in 1954 as a short-acting barbiturate, capable of producing sedation (50 to 100 mg.) and hypnosis (200 to 400 mg.).

Sodium Pentobarbital U.S.P., Nembutal®, Sodium 5-ethyl-5-(1-methylbutyl)barbiturate. Sodium pentobarbital is a short acting barbiturate, used as a hypnotic at a usual oral or intravenous dose of 100 mg. The usual dose range for oral administration is 15 to 200 mg. daily, for intravenous administration 50 to 200 mg. daily.

Pentobarbital Calcium is better suited for the preparation of compressed tablets, and for use when it is desirable to avoid sodium.

OCCURRENCE

Sodium Pentobarbital Capsules U.S.P.
Sodium Pentobarbital Elixir U.S.P.
Sodium Pentobarbital Injection U.S.P.

Secobarbital U.S.P., Seconal®, 5-Allyl-5-(1-methylbutyl) barbituric Acid. Secobarbital is the free acid form, used for its hypnotic effect in a usual adult oral dose of 100 mg. at bedtime.

Sodium Secobarbital U.S.P. is used for hypnosis in either an oral or rectal dose of 100 mg. The sodium salt is also used to produce hypnosis or as an adjunct to anesthesia in a usual parenteral dose of 100 mg.

OCCURRENCE

Secobarbital Elixir U.S.P.
Sodium Secobarbital Capsules U.S.P.
Sterile Sodium Secobarbital U.S.P.
Sodium Secobarbital and Sodium Amobarbital Capsules N.F.

NONBARBITURATE SEDATIVE-HYPNOTICS

Many drugs varying widely in their chemical structures are capable of producing sedation and hypnosis that closely resembles that of the barbiturates. The same factors are important in selecting either a nonbarbiturate or a barbiturate sedative-hypnotic, and these are principally the time required for onset of the depressant effect, the duration of the effect, and the incidence and the nature of undesirable side-effects.

Acyclic Ureides

The acyl derivatives of urea are referred to as ureides and constitute a class of compounds which includes the barbiturates and many substitution products of pyrimidine and purine. Either or both of the amino groups of urea may be acylated by mono-basic and polybasic acids, thus forming either acyclic or cyclic ureides, respectively. A barbiturate may be described as a cyclic diureide formed by diacylation of urea by the dibasic malonic acid. An acyclic monoureide (acylurea) has the structure $R-CO-NH-CO-NH_2$; an acyclic diureide (diacylurea), $R-CO-NH-CO-NH-CO-R$. The structural relationship between the barbiturates and acyclic ureides is shown

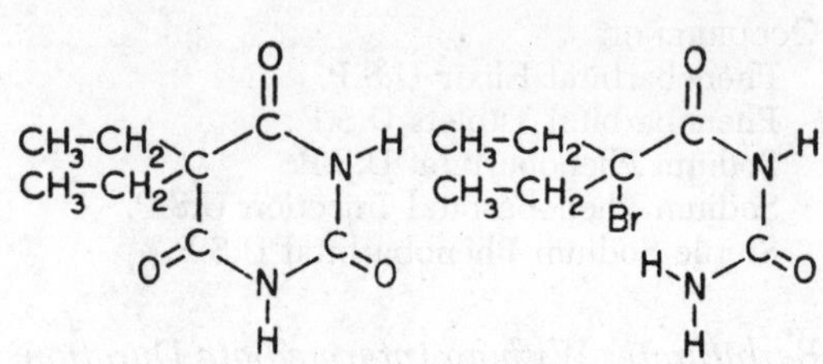

Barbital   Carbromal

by comparing the formulas of barbital and carbromal.

In all ureides at least one of the nitrogen atoms is flanked by two carbonyl groups, resulting in an acidic hydrogen atom ($-CO-NH-CO-$) which will form a water-soluble salt in the presence of alkali hydroxides.

TABLE 29. ACYCLIC UREIDE SEDATIVE-HYPNOTICS

GENERAL STRUCTURE

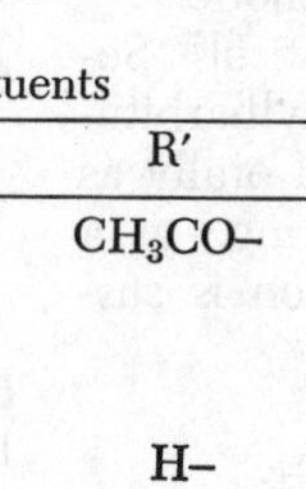

| Generic Name | Substituents | | Usual Oral Dose | |
Proprietary name	R	R'	Sedation (mg.)	Hypnosis (mg.)
Acetylcarbromal Sedamyl	$(CH_3CH_2)_2C-$ Br	CH_3CO-	250-500	—
Bromisovalum Bromural	$(CH_3)_2CHCH-$ Br	H–	300	600-900
Ectylurea Levanil	CH_3CH_2C- HCCH_3	H–	150-300	300-600

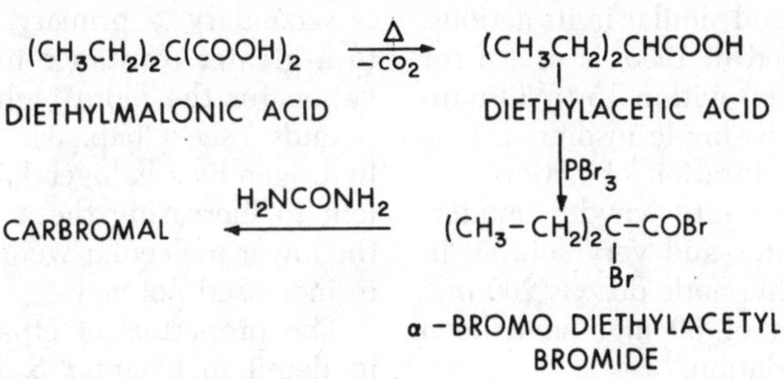

$$(CH_3CH_2)_2C(COOH)_2 \xrightarrow[-CO_2]{\Delta} (CH_3CH_2)_2CHCOOH$$

DIETHYLMALONIC ACID   DIETHYLACETIC ACID

$$\downarrow PBr_3$$

CARBROMAL $\xleftarrow{H_2NCONH_2}$ $(CH_3-CH_2)_2\underset{Br}{C}-COBr$

α–BROMO DIETHYLACETYL
BROMIDE

Glutethimide

The acyclic ureides are prepared from the corresponding alkyl or dialkylmalonic acids which are decarboxylated to form the substituted acetic acids. The acid halides of these, or their α-halo derivatives may be condensed with urea to form the desired ureide. The synthesis of carbromal is shown to illustrate these reactions.

The acyclic ureides possess weak depressant activity, and relatively high doses are required to produce hypnosis. Their main application has been as daytime sedatives in the treatment of simple anxiety and nervous tension. Although they are sometimes called "tranquilizers," they do not fit this classification as defined in this chapter and they have no appreciable effect against the major psychoses. The structures and the usual doses of the acyclic ureides are presented in Table 29. Properties of some members of this group are described separately.

Bromisovalum, Bromural®, 2-Bromoisovalerylurea. Bromisovalum is used as a sedative-hypnotic. It is slightly soluble in cold water but dissolves readily in hot water, alcohol, or in alkaline solutions. It is said to be metabolized or excreted unchanged in about 3 to 4 hours, a factor which contributes to its low toxicity. The usual sedative dose is 300 mg.; the hypnotic dose is 600 to 900 mg.

Ectylurea, Levanil®, 2-Ethyl-*cis*-crotonylurea. Ectylurea is a typical member of the acyclic ureide group, capable of producing mild sedation at a dose of 150 to 300 mg. 3 or 4 times daily.

Amides and Imides

A group of cyclic amides and imides, some bearing a close structural relationship to the barbiturates, have proved to be effective as sedative-hypnotic drugs.

Glutethimide N.F., Doriden®, 2-Ethyl-2-phenylglutarimide. Glutethimide was introduced in 1954 as a sedative hypnotic drug, closely related in action to the barbiturates. Its hypnotic effects begin about 30 minutes after administration and last for about 4 to 8 hours. The drug is useful for the induction of sleep in cases of simple insomnia and has been used as a daytime sedative to relieve anxiety-tension states. Glutethimide is a white powder, soluble in alcohol but practically insoluble in water. The usual oral dose of this structural analog of phenobarbital is 125 to 250 mg. for sedation, 500 mg. to 1 g. for hypnosis.

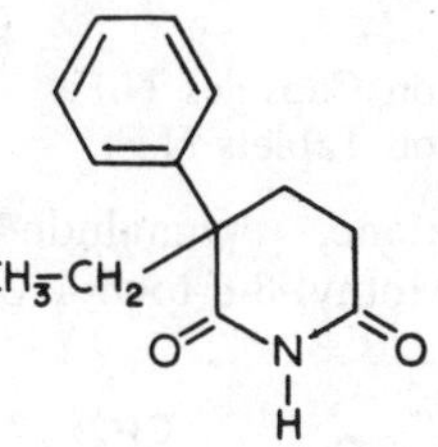

Numerous reports for addiction to glutethimide have been published,[9, 10] including epileptic seizures during withdrawal.

OCCURRENCE
Glutethimide Tablets N.F.

Methyprylon N.F., Noludar®, 3,3-Diethyl-5-methyl-2,4-piperidinedione. Methyprylon is a sedative-hypnotic, structurally related

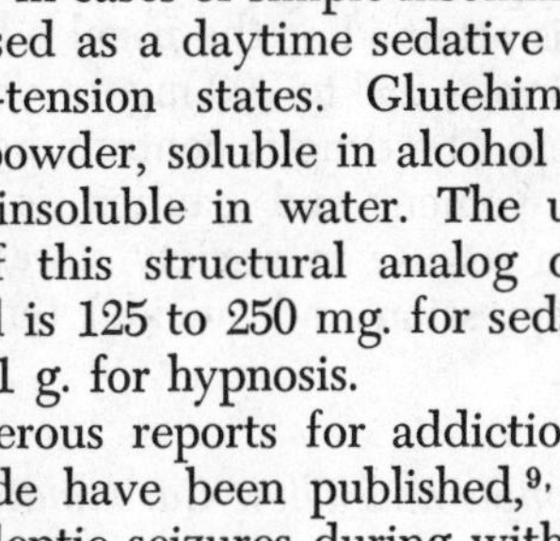

Methyprylon

to the barbiturates and similar in its actions. The drug, introduced in 1955, is useful for the induction of sleep within 15 to 30 minutes in patients with simple insomnia. It is intermediate in its duration of action.

Methyprylon is a white powder, moderately soluble in water and very soluble in alcohol. The usual hypnotic dose is 200 mg. taken orally. Doses of 50 mg. have been used for daytime sedation.

OCCURRENCE
Methyprylon Capsules N.F.
Methyprylon Tablets N.F.

Methaqualone, Quaalude®, Parest®, Sopor®; 2-Methyl-3-*o*-tolyl-4(3H)-quinazo-

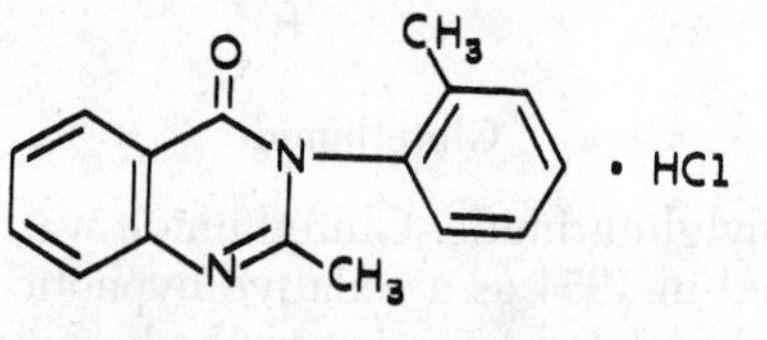

Methaqualone Hydrochloride

linone hydrochloride. Methaqualone is a sedative-hypnotic drug, introduced in 1965. Sedation is attained at the usual adult oral dose of 75 mg.; 150 to 300 mg. of drug produces sleep. The drug is contraindicated in pregnant women, and caution is recommended in its use in anxiety states where mental depression and suicidal tendencies may exist.

Alcohols and Their Carbamates

Ethanol has played a prominent role as a sedative and hypnotic for centuries. However, the feeling of stimulation which precedes that of depression has been recognized more widely. Because of the many problems associated with the use of alcohol, such as the development of chronic alcoholism on continued use, other depressant drugs have been favored for sedative-hypnotic use.

The hypnotic activity of the normal alcohols increases as the molecular weight and the lipid solubility increase, reaching a maximum depressant effect at 8 carbons (see Chap. 2). Branching of the alkyl chain increases activity, and the order of potency in an isomeric series of alcohols is tertiary

> secondary > primary. This may be due to a greater resistance to metabolic inactivation for the more highly branched compounds (see Chap. 3). Replacement of a hydrogen by a halogen has an effect equivalent to increasing the alkyl chain and, for the lower molecular weight alcohols, results in increased potency.

The properties of ethanol are described in detail in Chapter 5. It is a depressant drug whose apparent stimulation is produced as a result of the increased activity of lower centers freed from control by the depression of higher inhibitory mechanisms.

n-Butanol has been used clinically for the relief of pain, presumably taking advantage of its weak sedative-hypnotic properties. Some higher alcohols and their derivatives used for their central depressant effects are described below.

Amylene Hydrate, N.F., Tertiary Pentylalcohol, 2-Methyl-2-butanol. Amylene hy-

$$CH_3-CH_2-\underset{\underset{CH_3}{|}}{\overset{\overset{CH_3}{|}}{C}}-OH$$

drate is a tertiary alcohol, synthesized by the hydration of amylene [$CH_3-CH=C-(CH_3)_2$] in the presence of sulfuric acid. It has mild hypnotic properties in a dose of 1 to 4 g. and is primarily used as a solvent which contributes its depressant effect to the basal anesthetic, tribromoethanol (Avertin).

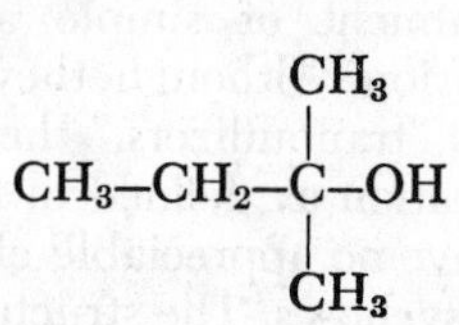

Amylene hydrate is a colorless liquid having a burning taste and a characteristic camphorlike odor. It is soluble in water (1:8) and miscible with most organic solvents.

Category—pharmaceutical necessity, ingredient of Tribromoethanol Solution N.F.

Chlorobutanol U.S.P., Chloretone®, 1,1,1-Trichloro-2-methyl-2-propanol. Chlorobu-

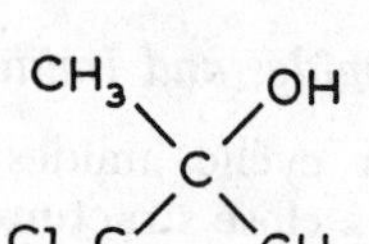

Chlorobutanol

tanol possesses the sedative-hypnotic properties characteristic of the higher alcohols. Before the development of more selective agents, its depressant properties were wildly used for the prevention of motion sickness. It is now used most often as an antibacterial preservative. It also has some local anesthetic activity. The sedative-hypnotic dose varies from 0.3 to 1.0 g.

See Chapter 9 for the properties and pharmaceutical applications of chlorobutanol.

Ethchlorvynol N.F., Placidyl®, 1-Chloro-3-ethyl-1-penten-4-yn-3-ol. Ethchlorvynol is a colorless to yellow liquid with a pungent odor. It darkens on exposure to light and to air.

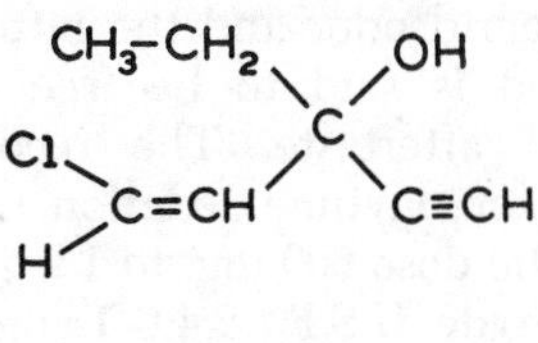

Ethchlorvynol

Ethchlorvynol was introduced as a mild hypnotic-sedative in 1955. It has a fairly rapid onset of action and a duration of about 5 hours. The drug is most useful in the induction of sleep for patients with simple insomnia and for use as a daytime sedative. Physical dependence has been reported following excessive intake. The oral hypnotic dose is 500 mg.; for sedation, 100 to 200 mg.

OCCURRENCE
Ethchlorvynol Capsules N.F.

Meparfynol, Dormison®, Methyl Ethyl Ethynyl Carbinol, 3-Methyl-4-pentyn-3-ol. Meparfynol is a mild sedative-hypnotic which shows an onset of action in from 30 to 60 minutes. The oral dose ranges from 250 to 750 mg. An unpleasant aftertaste and a low incidence of exfoliative dermatitis are

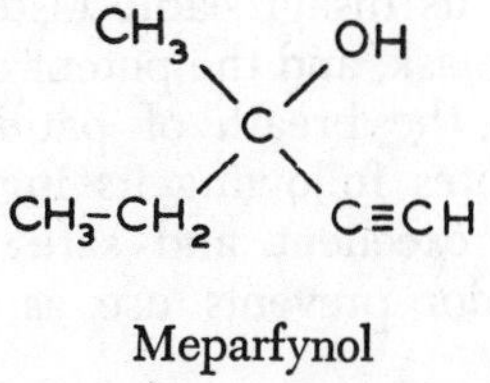

Meparfynol

the principal side-effects reported. Some hepatotoxicity has been demonstrated in experimental animals.

Urethan N.F., Ethyl carbamate, Urethane, $CH_3CH_2O-CO-NH_2$. Urethan was once used as a hypnotic agent in a dose of 2 to 4 g. Due to its weak and uncertain action as a central depressant, it is now rarely used in this way.

Ethinamate N.F., Valmid®, 1-Ethynyl-cyclohexanol Carbamate. Ethinamate was

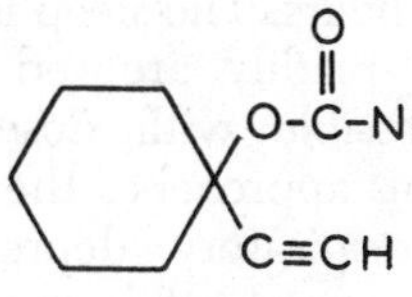

Ethinamate

introduced as a sedative-hypnotic in 1955. The onset of its depressant effects requires about 20 to 30 minutes. The drug is metabolized rapidly and the duration of action is short, lasting less than 4 hours. Tolerance and physical dependence have been observed on prolonged use of large doses.

Usual dose—500 mg.

Usual dose range—500 mg. to 1 g.

OCCURRENCE
Ethinamate Tablets N.F.

Aldehydes, Ketones and Derivatives

Members of this group were among the first of the organic hypnotics. Chloral was introduced in 1869 in the mistaken belief that it would be converted to the anesthetic chloroform in the body. However, instead of undergoing the haloform reaction of the test tube, chloral is reduced in the body to trichloroethanol, which may be responsible for most of chloral's depressant properties (see Chap. 3). Analogs and derivatives, such as chloral betaine and petrichloral, have been introduced to reduce or eliminate the disadvantages of chloral.

Chloral Hydrate U.S.P., Noctec®, Somnos®, Chloral, Trichloroacetaldehyde Monohydrate, $CCl_3CH(OH)_2$. Chloral is a reliable and safe hypnotic, useful in inducing sleep where insomnia is not due to pain, for the drug is a poor analgetic. The parent aldehyde, chloral, (CCl_3CHO, note that the synonym for the hydrate is a misnomer) is an oily liquid which in the presence of water yields the crystalline hydrate.

Chloral hydrate occurs as colorless or white crystals having an aromatic, penetrating and slightly acrid odor and a bitter caustic taste. It is very soluble in water (1:0.25) and in alcohol (1:1.3).

In the usual oral dose (500 mg. to 1 g.) chloral hydrate causes sedation in 10 to 15 minutes. Sleep occurs within an hour and lasts for 5 to 8 hours. The sleep is light, and the patient is readily aroused. Complete anthesia is possible with doses of 6 g. or over, but this approaches the dose causing marked respiratory depression, and chloral hydrate is not safely used for anesthesia. Alcohol synergistically increases the depressant effect of chloral and a mixture of the two ("knockout drops," "Mickey Finn") is a very potent depressant, although the chloral alcoholate formed ($CCl_3-CHOH-O-C_2H_5$, a hemiacetal) is no more hypnotic than is the hydrate. Chloral hydrate causes local irritation and may cause nausea, vomiting and diarrhea, particularly if taken with inadequate fluids.

Category—hypnotic.

Usual dose—500 mg. at bedtime.

Usual dose range—250 mg. to 2 g. daily.

OCCURRENCE

Chloral Hydrate Capsules U.S.P.

Choral Hydrate Syrup U.S.P.

Chloral Betaine N.F., Beta-Chlor®. Chloral betaine is a chemical complex of chloral

$$CCl_3CH(OH)_2 \cdot (CH_3)_3\overset{+}{N}CH_2OO^-$$

hydrate and betaine with a hypnotic and sedative potency equal to that of the chloral hydrate which it contains. The 870 mg. tablets contain 500 mg. of chloral hydrate. The complex is tasteless, and is said to produce gastric irritation infrequently.

Category—sedative.

Usual dose—870 mg. 1.74 g. 15 to 30 minutes before bedtime:

OCCURRENCE

Chloral Betaine Tablets N.F.

Petrichloral, Periclor®, Pentaerythritol Chloral. Petrichloral, the hemiacetal be-

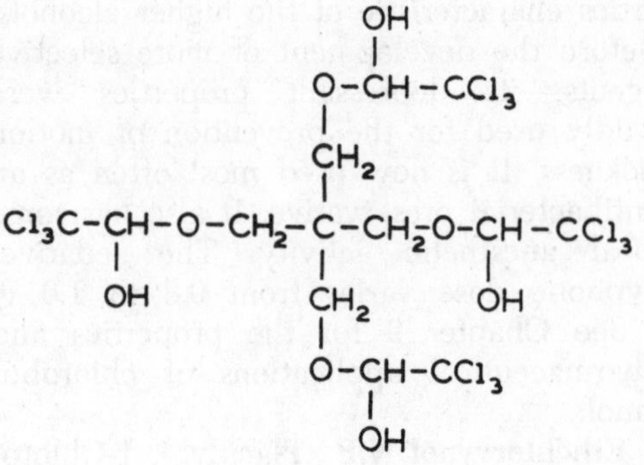

Petrichloral

tween pentaerythritol and chloral, was introduced in 1955 as a sedative-hypnotic. It lacks the acrid odor and the bitter taste of chloral and is said to be free of gastric upset and aftertaste. The recommended oral dose for daytime sedation is 300 mg., the hypnotic dose 600 mg. to 1.2 g.

Paraldehyde U.S.P., 2,4,6-Trimethyl-*s*-trioxane, Paracetaldehyde.

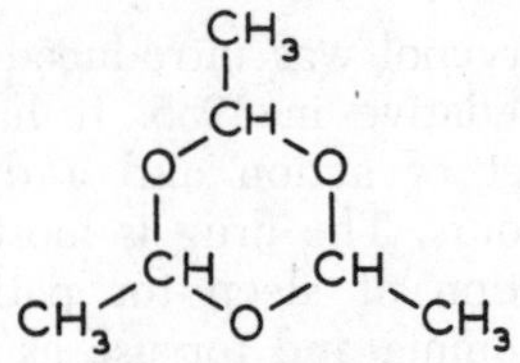

Paraldehyde

Paraldehyde, in 1882, was the second of the synthetic organic compounds to be introduced for use as a sedative-hypnotic.

Paraldehyde is a colorless liquid with an odor which is not pungent or unpleasant but with a disagreeable taste. The drug is more potent and toxic than ethanol but less so than chloral hydrate. With the usual oral or rectal dose of 10 ml., sleep is induced within 10 to 15 minutes. The sleep is a natural one and is accompanied by little change in respiration or circulation.

The chief objections to the use of paraldehyde are its disagreeable taste, which is difficult to mask, and the potent odor which appears on the breath of patients within a few minutes following its ingestion. Although an excellent and safe depressant drug, its odor prevents use as a daytime sedative, and the more acceptable barbiturates have largely replaced paraldehyde in routine use as a soporific. It is used most frequently in delirium tremens and in the treatment of psychiatric states characterized by excitement, where drugs must be given over long periods.

Paraldehyde may oxidize to form acetic acid on storage. Stored samples have been found to consist of up to 98 per cent of acetic acid. Since oxidation occurs more rapidly in opened, partially filled containers, the drug should not be dispensed from a container that has been opened for longer than 24 hours. Refrigeration is indicated to retard oxidation. The U.S. Public Health Service has recommended that its hospitals stock the drug only in single-dose ampules for injection and in the oral capsule form.

The usual oral or rectal dose is 10 ml., the usual intramuscular dose is 5 ml. The oral dose may range from 5 to 30 ml.

OCCURRENCE
Sterile Paraldehyde U.S.P.

Miscellaneous

Many of the antihistaminic drugs (see Chap. 23) possess a significant degree of central nervous system depression as their principal side-effect. In certain cases, this has been selected as the therapeutic effect. For example, promethazine (Phenergan) has been used as a preoperative sedative. Several of the antihistaminic drugs, principally methapyrilene, make up the depressant component of a large number of proprietary sleeping preparations.

CENTRAL RELAXANTS
(CENTRAL NERVOUS SYSTEM DEPRESSANTS WITH SKELETAL MUSCLE RELAXANT PROPERTIES)

The search for drugs capable of diminishing skeletal muscle tone and involuntary movement has led to the introduction during recent years of a variety of agents capable of a relatively weak, centrally mediated muscle relaxation. However, these agents have been of more signficance in the therapeutic application of their mild depressant properties on the central nervous system. This depressant effect has been variously described as "tranquilization," "ataraxia," "calming" and "neurosedation." At therapeutic levels, this has been shown to resemble more closely the well-known sedation produced by the sedative-hypnotics (e.g., amobarbital), and to be quite different from the depressant effects on the central nervous system produced by the tranquilizing agents (e.g., reserpine, chlorpromazine) (see Tables 31, 32).

The skeletal muscle relaxation is produced by drugs of this group in a manner completely different from that of curare and its analogs, which act at the neuromuscular junction. These centrally acting muscle relaxants block impulses at the interneurons of polysynaptic reflex arcs, mainly at the level of the spinal cord. This is demonstrated by the abolishment or the diminution of the flexor and crossed extensor reflexes which possess one or more interneurons between the afferent (sensory) and the efferent (motor) fibers.

The knee-jerk response, which acts through a monosynaptic reflex system and therefore possesses no interneurons, is unaffected by these drugs.

The skeletal muscle relaxation produced by this centrally mediated mechanism can be applied therapeutically by employing those members of this class which produce muscle relaxation without excessive sedation. The therapeutic applications of the skeletal muscle relaxant effect include relief in the variety of conditions in which painful muscle spasm may be present, such as bursitis, spondylitis, disk syndromes, sprains, strains and low back pain.

The major sites of the sedative effects of these drugs are the brain stem and subcortical areas. The ascending reticular formation, which receives and transmits some sensory stimuli, transmits and maintains a state of arousal. When the passage of stimuli is blocked at the level of the ascending reticular formation, response to sensory stimuli is reduced, and depression, ranging from sedation to anesthesia, may occur. The barbiturates[11] and other sedative-hypnotics, as well as meprobamate[12] and its analogs, are capable of producing inhibition of this arousal system. Suppression of polysynaptic reflexes at the spinal level is not sufficient to account for depression of the arousal system.

The depressant effect of members of this class has been applied in producing mild hypnosis in case of simple insomnia, or as an adjunct to psychotherapy in the management of anxiety and tension states associated or unassociated with physical ills, such as hypertension and cardiovascular disorders, in which excessive excitation should be avoided.

Meprobamate, a member typical of this class, has been shown[13, 14] to produce withdrawal symptoms similar to those of the barbiturates. Prolonged administration of overdosages of any drug of this class with significant sedative action may lead to withdrawal symptoms (convulsions, tremor, abdominal and muscle cramps, vomiting or sweating) if medication is stopped abruptly.

As a general class, members of this group may be described as mild sedatives with moderate to weak skeletal muscle-relaxing properties.

GLYCOLS AND DERIVATIVES

In 1945, during a study on potential preservatives for penicillin during production and processing, F. M. Berger observed the muscle-relaxant effects of aryl glycerol ethers in experimental animals. Following a study of a series of structural analogs, 3-*o*-toloxy-1,2-propanediol (mephenesin) was introduced in 1948 as a skeletal-muscle relaxant. However, the duration of action was too brief. Since metabolic attack of the terminal hydroxy group occurred in part (see Chap. 3), structural analogs which protected this functional group, including the carbamate (mephenesin carbamate) were prepared. Further structural alterations demonstrated that the aromatic nucleus was not a requisite for activity, and, in an attempt to prolong muscle relaxant activity in the aliphatic series, 2-methyl-2-propyl-1,3-propanediol dicarbamate (from which the generic name meprobamate was derived) was synthesized.

The drug not only showed the desired longer acting skeletal-muscle-relaxant properties but was shown to be an effective central nervous system depressant, producing a sedative effect in experimental animals and in man. The drug was marketed in 1955 under the trade name Miltown (its development originated in Milltown, a village in New Jersey) and was widely promoted for its "tranquilizing" properties. Since then a number of related glycols and their carbamate derivatives, as well as structurally unrelated compounds, have been shown to possess in varying degrees the properties of skeletal-muscle relaxation mediated through the blockade of polysynaptic reflexes and a mild depression of the central nervous system.

Mephenesin, Tolserol®, 3-*o*-(Tolyloxy)-1,2-propanediol, Mephenesin produces mus-

$$CH_3$$

$$O-CH_2-CH-CH_2-O-R$$
$$\qquad\qquad OH$$

Mephenesin: R = H

$$O$$
$$\|$$
Mephenesin Carbamate: R = $-C-NH_2$

cular relaxation and has mild sedative properties, both of short duration. The drug is used for relief of tremors of parkinsonism and acute alcoholism, or in any condition in which muscle spasm is present. The drug is also used in the temporary relief of states of anxiety and tension. It is an odorless white powder, with a bitter taste. The drug is sparingly soluble in water, freely soluble in alcohol. The usual oral dose is 1 to 3 g., 3 to 5 times a day. Intravenous, 30 to 150 ml. of a 2 per cent solution at a rate of 6 to 7 ml. per minute.

Category—skeletal muscle relaxant.

Mephenesin Carbamate, Tolseram®, 2-Hydroxy-3-*o*-tolyloxypropyl Carbamate. Mephenesin carbamate has been available since 1954 for the same uses as mephenesin, from which it is derived by carbamate formation. It possesses the same potency but a longer duration of action than mephenesin. The oral dose is 1 to 3 g., given 3 to 5 times a day.

Glyceryl Guaiacolate N.F., Dilyn®, Robitussin®, 3-*o*-(Methoxyphenoxy)-1,2-propanediol. Glyceryl guaiacolate is used alone for its sedative action in anxiety and tension states. It is most often used in combination with antihistamines, analgesics and vasoconstrictors in cough medicines for its expectorant action.

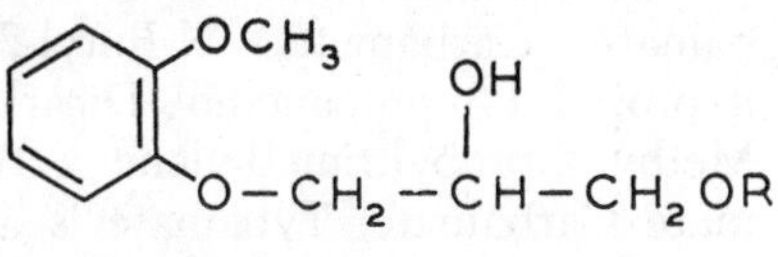

Glyceryl Guaiacolate: R = H

$$\text{Methocarbamol:} \quad R = -\overset{\displaystyle O}{\overset{\displaystyle \|}{C}}-NH_2$$

Category—expectorant.
Usual dose—100 mg. every 3 or 4 hours.

OCCURRENCE
Glyceryl Guaiacolate Syrup N.F.

Methocarbamol N.F., Robaxin®, 3-(o-Methoxyphenoxy)-1,2-propanediol 1-Carbamate. Methocarbamol was introduced in 1957 for the relief of skeletal-muscle spasm. Peak plasma concentrations of methocarbamol are reached more slowly (1 hour) than for mephenesin (30 minutes) but are more sustained. The drug is administered by the oral (1.5 to 2 g. 4 times daily), the intramuscular and the intravenous (1 g. at intervals of 8 hours) routes. Preparations of methocarbamol for parenteral use contain polyethylene glycol as a solvent. This is contraindicated in patients with impaired renal function, since it increases urea retention and acidosis in such patients.

OCCURRENCE
Methocarbamol Injection N.F.
Methocarbamol Tablets N.F.

Chlorphenesin Carbamate, Maolate®, 3-p-Chlorophenoxy-2-hydroxypropyl carbamate. This close analog of mephenesin carbamate and of methocarbamol was introduced in 1967 as a muscle relaxant and mild

sedative. The usual oral dose is 400 to 800 mg. 3 times daily.

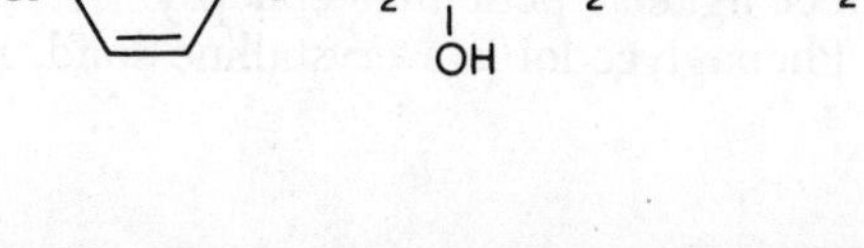

Mephenoxalone, Trepidone®, Lenetran®, 5-(o-Methoxyphenoxymethyl)-2-oxazolidinone. Mephenoxalone is prepared by the

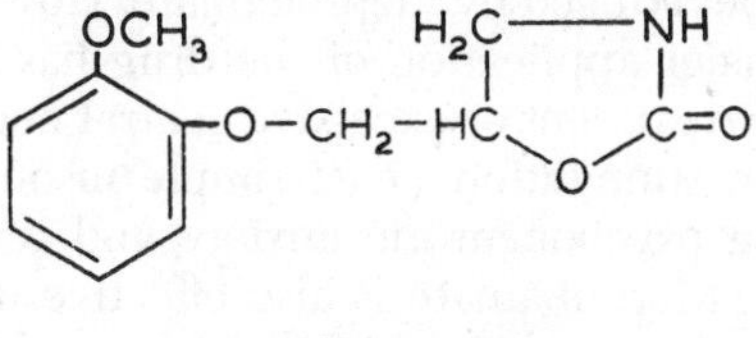

Mephenoxalone

fusion of 3-(o-methoxyphenoxy)-1,2-propanediol (glyceryl guaiacolate) with urea. The compound is the first of a new class of 5-aryloxymethyl-2-oxazolidinones, introduced in 1961, which are reported to possess pharmacologic properties similar to the propanediol-type sedatives and is recommended for the treatment of neuroses associated with anxiety and tension. It is also said to be effective as a centrally acting skeletal muscle relaxant. The drug is largely metabolized (less than 0.5% excreted unchanged) but the metabolic products have not been identified. The drug is administered orally (400 mg. four times daily).

Metaxalone, Skelaxin®, 5-(3,5-Dimethylphenoxymethyl)-2-oxazolidinone. Metaxa-

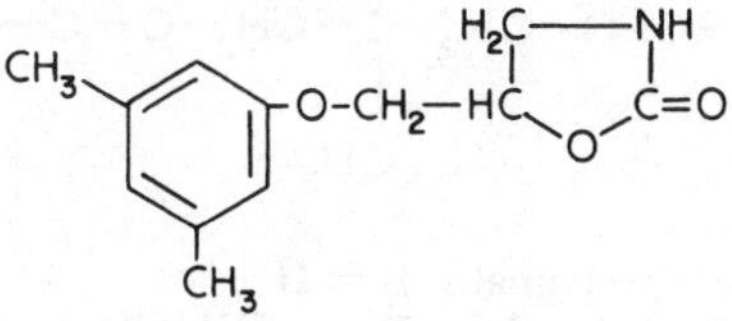

Metaxalone

lone, a structural analog of mephenoxalone, is recommended primarily for its muscle relaxant properties in the treatment of acute muscle spasm of local origin, such as sprains, strains, fractures, dislocations and trauma to tendons and ligaments.

The usual adult oral dose ranges from 2.4 to 3.2 g. daily in divided doses.

Meprobamate U.S.P., Equanil®, Miltown®, 2-Methyl-2-propyltrimethylene Dicarbamate, 2-Methyl-2-propyl-1,3-propanediol Dicarbamate. Meprobamate produces skeletal-muscle relaxation by interneuronal blockade at the spinal level. The duration of its muscle relaxant effect is 8 to 10 times longer than that produced by mephenesin, but of the same type. Certain types of abnormal motor activity and muscle spasm may be reduced by meprobamate. However, the major application of the drug has been in the treatment of excessive central nervous system stimulation (e.g., simple insomnia) and in psychoneurotic anxiety and tension states. Meprobamate is also effective in the prevention of attacks of the petit mal form of epilepsy and in the treatment of alcoholism.

The drug is a white powder, possessing a bitter taste. It is relatively insoluble in water (0.34% at 20°) and freely soluble in alcohol and most organic solvents. The drug is stable in the presence of dilute acid or alkali.

Category—minor tranquilizer.

Usual dose—oral or intramuscular, 400 mg. every 3 or 4 hours.

Usual dose range—1 to 2.4 g. daily.

OCCURRENCE

Meprobamate Injection U.S.P.
Meprobamate Oral Suspension N.F.
Meprobamate Tablets U.S.P.

$$H_2N-\underset{\underset{O}{\|}}{C}-O-CH_2-\underset{\underset{CH_3}{|}}{\overset{\overset{CH_2-CH_2-CH_3}{|}}{C}}-CH_2-O-\underset{\underset{O}{\|}}{C}-NHR$$

Meprobamate: R = H
Carisoprodol: R = $-CH(CH_3)_2$
Tybamate: R = $-(CH_2)_3CH_3$

Carisoprodol, Soma®, Rela®, N-Isopropyl-2-methyl-2-*n*-propyl-1,3-propanediol Dicarbamate. This N-isopropyl derivative of meprobamate was introduced in 1959 for therapeutic application of its centrally mediated skeletal-muscle-relaxant properties. It is recommended for use in acute skeletomuscular conditions characterized by pain, stiffness and spasticity. Carisoprodol is said to modify pain perception without affecting peripheral pain reflexes. Drowsiness is the principal side-effect of the drug.

The usual oral dose of carisoprodol is 350 mg., 4 times daily. Peak blood levels are reached 1 to 2 hours after ingestion. The compound is a bitter, odorless, white crystalline powder. It is soluble in water to the extent of 30 mg. in 100 ml. at 25°.

Tybamate N.F., Solacen®, Tybatran®, 2-(Hydroxymethyl)-2-methylpentyl Butylcarbamate Carbamate, N-Butyl-2-methyl-2-*n*-propyl-1,3-propanediol Dicarbamate, 2-Methyl-2-propyltrimethylene Butylcarbamate Carbamate. Tybamate is an N-butyl substituted analog of meprobamate introduced in 1965 for the treatment of psychoneurotic anxiety and tension states. The usual oral dose is 250 to 500 mg., 3 or 4 times daily.

OCCURRENCE

Tybamate Capsules N.F.

Mebutamate, Capla®, 2-*sec*-Butyl-2-methyl-1,3-propanediol Dicarbamate; 2,2-Dicarbamoyloxmethyl - 3 - methylpentane. Mebutamate is employed as a mild anti-

$$H_2N-\underset{\underset{O}{\|}}{C}-O-CH_2-\underset{\underset{CH_3}{|}}{\overset{\overset{CH_3-CH-CH_2-CH_3}{|}}{C}}-CH_2-O-\underset{\underset{O}{\|}}{C}-NH_2$$

hypertensive agent and, like related carbamates, is a central nervous system depressant. (See Chap. 21.)

Phenaglycodol, Ultran®, 2-*p*-Chlorophenyl-3-methyl-2,3-butanediol. Phenagly-

$$Cl-\underset{}{\bigcirc}-\underset{\underset{CH_3}{|}}{\overset{\overset{OH}{|}}{C}}-\underset{\underset{CH_3}{|}}{\overset{\overset{CH_3}{|}}{C}}-OH$$

Phenaglycodol

codol is an interneuronal blocking agent, introduced in 1957 and related in its structure and pharmacology to mephenesin and meprobamate. It is a mild sedative with weak muscle-relaxing properties, used in anxiety-tension states and as an adjunct in the treatment of simple neuroses. Like meprobamate, it displays some protective effect against petit mal epilepsy.

Phenaglycodol is a crystalline solid, relatively insoluble in water and quite soluble in alcohol. The usual oral dose is 300 mg., 3 or 4 times daily.

Styramate, Sinaxar®, 2-Hydroxy-2-phenyl-ethyl Carbamate. Styramate, the carbamate of an ethylene glycol, is related in structure and pharmacology to mephenesin carbamate but possesses a somewhat longer duration of action in the reduction of spasm in various musculoskeletal disorders. It was

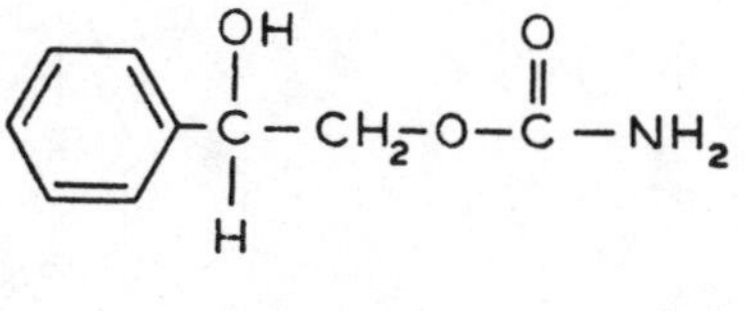

Styramate

introduced in 1958 and is used orally in an average dose of 200 to 400 mg., 4 times daily, in relieving muscle spasm. Sedation is an expected side-effect.

Hydroxyphenamate; Listica®, 2-Hydroxy-2-phenylbutyl Carbamate. Hydroxyphena-

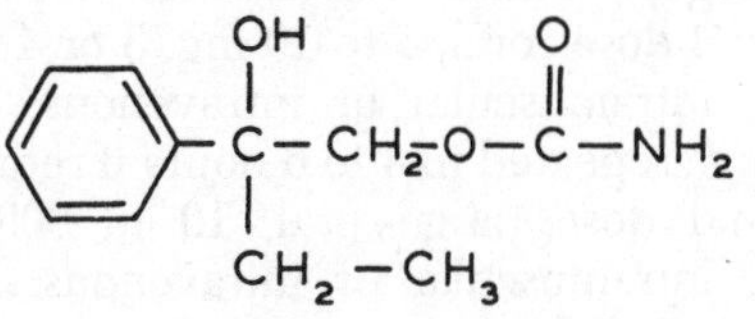

Hydroxyphenamate

mate was introduced in 1961 for use in the suppression of symptoms of anxiety and tension. The most commonly encountered side-effect is drowsiness. The usual adult oral dose is 600 to 800 mg. per day, in divided doses.

Oxanamide, Quiactin®, 2-Ethyl-2,3-epoxy-caproamide, 2-Ethyl-3-propylglycidamide. Oxanamide, the epoxide derived from a 1,2-

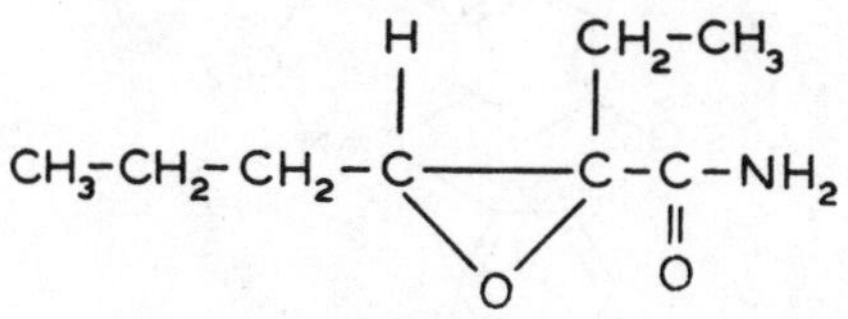

Oxanamide

glycol, was introduced in 1958 as a calming agent in the management of psychoneuroses and moderate excited states characterized by anxiety and tension. Like other central muscle relaxants, it is an internuncial neuronal blocking agent which produces skeletal-muscle relaxation by inhibition of polysynaptic reflexes. Large doses produce a hypnotic response similar to that of the short-acting barbiturates. The usual oral dose is 400 mg., 4 times daily.

Phenyramidol, Analexin®, 2-(β-Hydroxy-β-phenylethylamino)pyridine Hydrochlo-

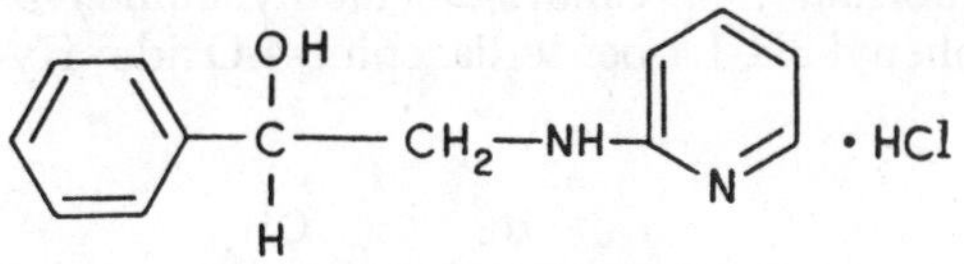

Phenyramidol

ride. Phenyramidol was introduced in 1960 as a muscle relaxant with analgetic activity. It may be considered a nitrogen isostere of the 1,2-glycol derivatives (e.g., styramate) which produce skeletal muscle relaxation, as does phenyramidol, by interneuronal blockade of spinal polysynaptic transmission. Phenyramidol is used for the relief of pain associated with muscle strain or spasm. The usual oral dose is 200 to 400 mg. every 4 hours.

BENZODIAZEPINE DERIVATIVES

The initial synthesis of the 1,4-benzo-diazepines resulted from an attempt to prepare 2-methylamino-6-chloro-4-phenyl-

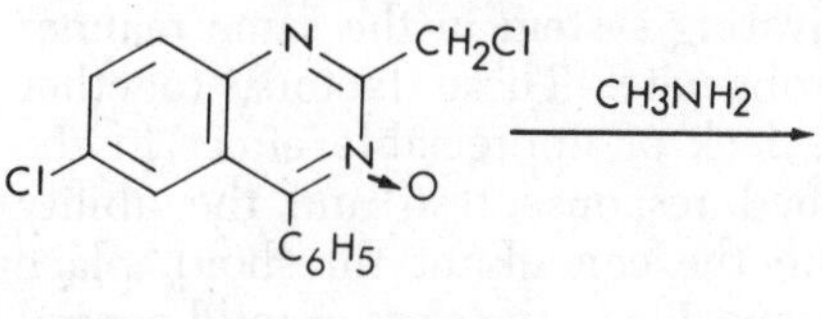

6-Chloro-2-chloromethyl-4-phenylquinazoline-3-oxide

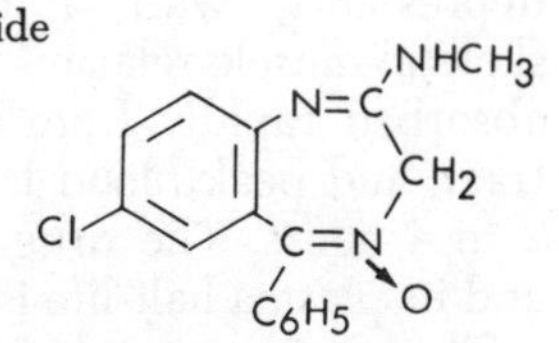

Chlordiazepoxide

quinazoline-3-oxide by the treatment of 6-chloro-2-chloromethyl-4-phenylquinazoline-3-oxide with methylamine. The unexpected ring enlargement reaction instead produced the benzodiazepine chlordiazepoxide, which was shown to possess sedative, muscle relaxant, and anticonvulsant properties much like those of the barbiturates. The drug and its congeners have been widely used for the relief of aniety, tension, apprehension and related neuroses, and of agitation during withdrawal from alcohol.

Chlordiazepoxide Hydrochloride U.S.P., Librium®, 7-Chloro-2-(methylamino)-5-phenyl-3H-1,4-benzodiazepine-4-Oxide Hydrochloride. Chlordiazepoxide was introduced in 1960 for use in the treatment of anxiety and tension. It has been shown to block spinal reflexes at one tenth the dose required of meprobamate and is therefore a moderately effective skeletal-muscle relaxant. Doses larger than those necessary to block the spinal reflexes depress the reticular activating system in the same manner as meprobamate. These factors, together with the lack of appreciable effect in the conditioned response test, and the ability to elevate the convulsant threshold, place chlordiazepoxide in the class of mild central depressants, with a centrally mediated skeletal-muscle-relaxant effect. The drug is absorbed rapidly from the gastrointestinal tract, and peak blood levels are reached in 2 to 4 hours. The drug is excreted slowly, and its plasma half-life is 20 to 24 hours.

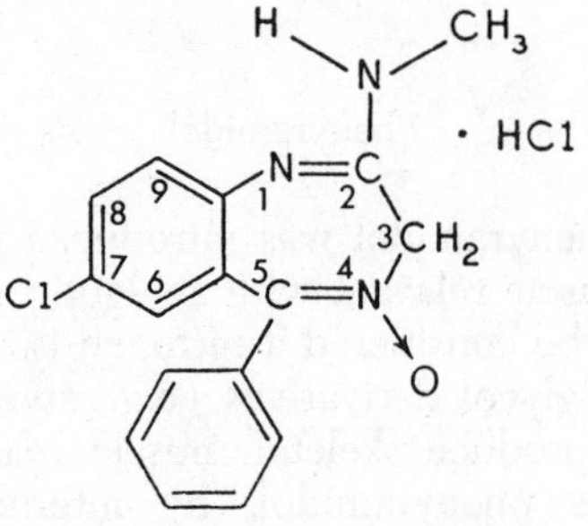

The major metabolites of chlordiazepoxide are a lactam, an amino acid ("open lactam") resulting from ring opening of the lactam, and a small amount of a conjugate of the amino acid, all primarily excreted in the urine.

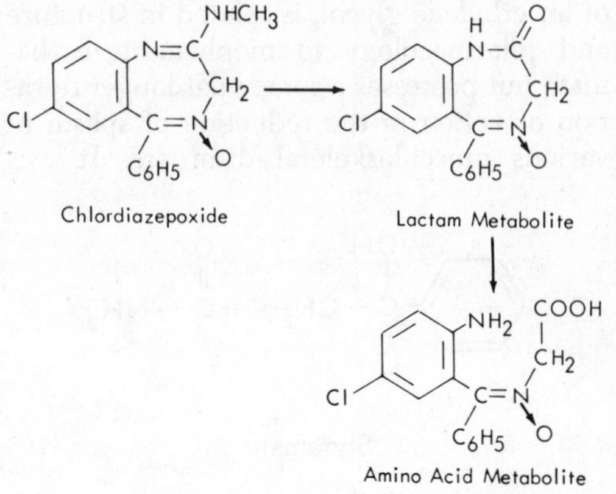

Chlordiazepoxide is a colorless crystalline substance, light sensitive and highly soluble in water but unstable in aqueous solution. Doses ranging from 50 to 300 mg. per day have been used in the treatment of acute agitation such as delirium tremens.

Category—minor tranquilizer.

Usual dose—oral, 5 to 10 mg. 3 or 4 times daily; intramuscular or intravenous, 50 to 100 mg., repeated in 4 to 6 hours if required.

Usual dose range—oral, 10 to 300 mg. daily; intramuscular or intravenous, 25 to 300 mg. in 6 hours.

OCCURRENCE

Chlordiazepoxide N.F.

Sterile Chlordiazepoxide Hydrochloride U.S.P.

Chlordiazepoxide Tablets N.F.

Chlordiazepoxide Hydrochloride Capsules U.S.P.

Diazepam N.F., Valium®, 7-Chloro-1,3-dihydro-1-methyl-5-phenyl-2H-1,4-benzo-

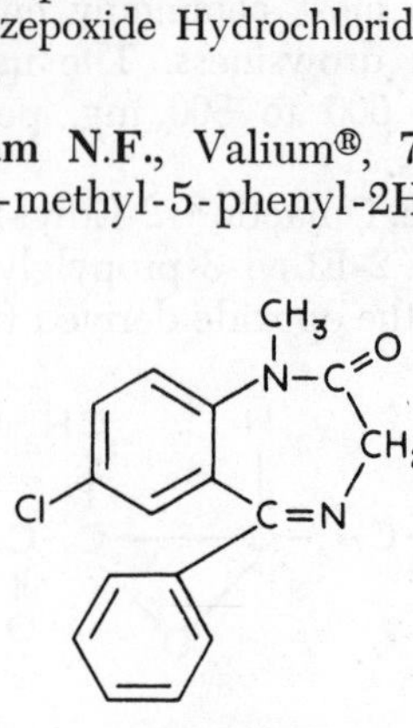

Chlordiazepoxide Hydrochloride

Diazepam

diazepin-2-one. Diazepam is a substituted benzodiazepine, introduced in 1964, which is related in structure and pharmacology to chlordiazepoxide. Diazepam is used for the control of anxiety and tension states, the relief of muscle spasm, and for the management of acute agitation during withdrawal from alcohol. Diazepam is metabolized in the liver, one of the metabolites being the 3-hydroxy derivative, oxazepam, which is also active as a sedative and muscle relaxant.

Category—minor tranquilizer.

Usual dose range—oral, 2 to 10 mg. 2 to 4 times daily; intramuscular or intravenous, 2 to 10 mg., repeated in 3 to 4 hours if needed, but no more than 300 mg. in an 8-hour period.

OCCURRENCE

Diazepam Injection N.F.
Diazepam Tablets N.F.

Oxazepam N.F., Serax®, 7-Chloro-1,3-dihydro-3-hydroxy-5-phenyl-2H-1,4-benzo-

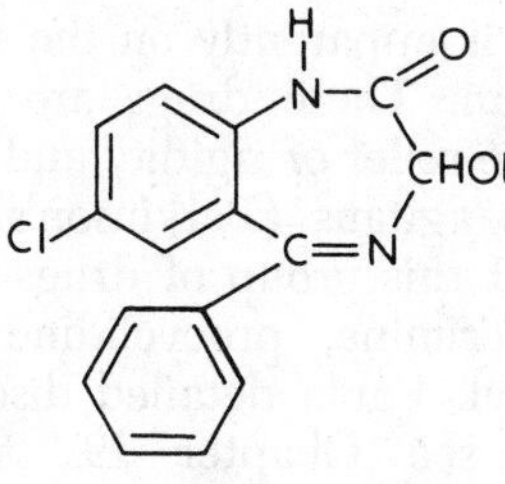

Oxazepam

diazepin-2-one. Oxazepam is a benzodiazepine derivative introduced in 1965 for use in the relief of psychoneuroses characterized by anxiety and tension. It is said to show a lower incidence of side-effects and reduced toxicity, perhaps due to the ease of conjugation of the 3-hydroxy group, and elimination as the glucuronide which is the major metabolite.

The dosage should be individualized, but the usual dose ranges from 10 to 15 mg. 3 or 4 times daily.

Category—minor tranquilizer.

OCCURRENCE

Oxazepam Capsules N.F.

Flurazepam Hydrochloride, Dalmane®, 7-Chloro-1-(2-diethylaminoethyl)-5-(2-fluoro-

phenyl)-1,3-dihydro-2H-1,4-benzodiazepine-2-one Dihydrochloride. Flurazepam is a pale-yellow, crystalline compound, freely soluble in alcohol and in water.

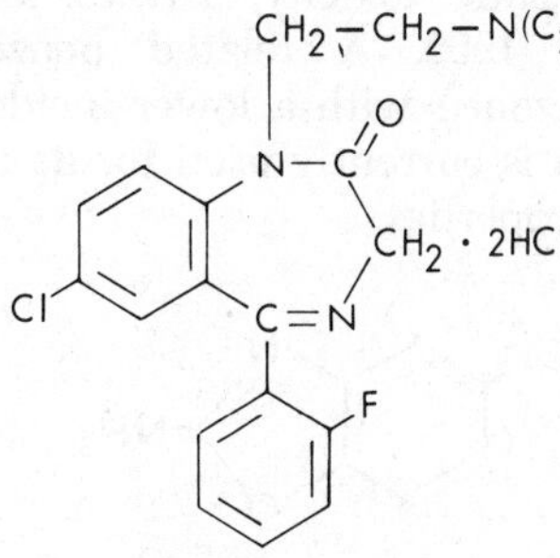

Flurazepam Hydrochloride

Flurazepam was introduced in 1970 as a hypnotic drug, useful in types of insomnia characterized by difficulty in falling asleep, and by early awakening. Although flurazepam is chemically related to the other benzodiazepines which show antianxiety and psychostimulant properties, the potent hypnotic effect appears to be unique to flurazepam. The drug is reported to provide 7 to 8 hours of restful sleep, during which normal dreaming activity, as characterized by rapid eye movements, is maintained.

The usual oral dose of 15 to 30 mg. is rapidly absorbed, and induces sleep in about 20 minutes. The drug is rapidly converted to a glucuronide and/or sulfate conjugate, which are eliminated in the urine.

MISCELLANEOUS

Several compounds of unique structure have been found to inhibit polysynaptic reflexes of the spinal cord, as well as to depress higher centers. Like previous central relaxants, their application as skeletal-muscle relaxants or sedative drugs has depended upon which effect predominates.

Certain derivatives of benzoxazole have been shown to inhibit transmission of nervous impulses through polysynaptic reflex arcs, thus acting as skeletal-muscle-relaxing agents in the same manner as mephenesin and related compounds.

The first compound of this type, zoxazolamine (Flexin), was introduced as a muscle relaxant in 1956. The uricosuric effect of the drug was recognized, and, in 1958, zoxazolamine was recommended for treatment of gout. A high frequency of side-effects which included, in a small percentage of cases, serious hepatic toxicity, led to its withdrawal in 1962. A related benzoxazole (chlorzoxazone) with a lower incidence of side-effects is currently used for its muscle-relaxing properties.

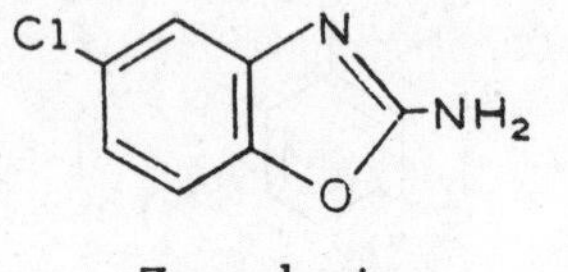

Zoxazolamine

Chlorzoxazone, Paraflex®, 5-Chloro-2-benzoxazolinone. Chlorzoxazone was introduced in 1958 as a skeletal-muscle relaxant used for reduction of painful muscle spasms

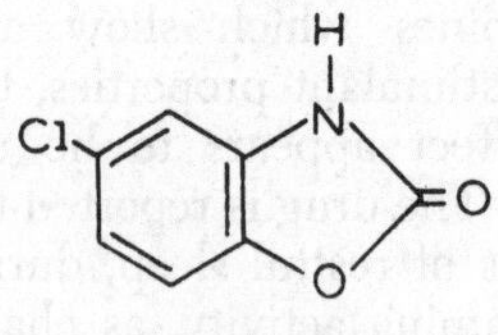

Chlorzoxazone

in medical and orthopedic disorders such as bursitis, myositis, sprains and strains, and acute or chronic back pain. A low incidence of liver damage and gastrointestinal disturbances, as side effects, has been reported.

The usual oral dose of chlorzoxazone ranges from 250 to 750 mg., 3 or 4 times a day.

Chlormezanone, Trancopal®, 2-(4-Chlorophenyl)-3-methyl-4-metathiazanone-1,1-dioxide. Chlormezanone was introduced in 1958 as a skeletal-muscle relaxant, useful

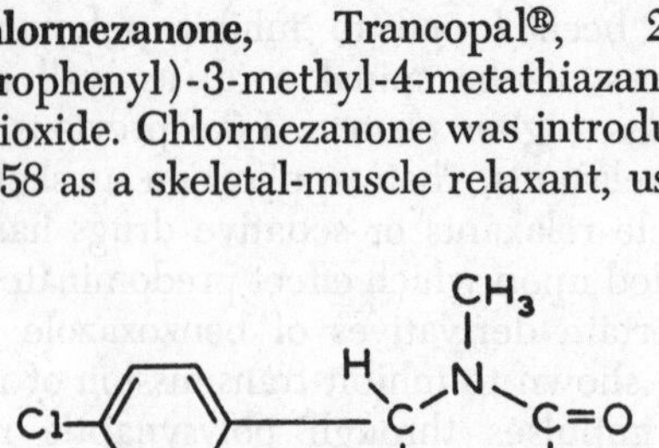

Chlormezanone

in the treatment of conditions characterized by muscle spasm. Its mild depressant effect on the central nervous system resembles that of meprobamate, and the drug is recommended for use in anxiety and tension states.

Chlormezanone is a white crystalline powder with solubility in water of less than 0.25 per cent and less than 1 per cent in alcohol. The usual oral dose is 100 mg., 3 or 4 times daily.

A large number of compounds which act as peripheral cholinergic blocking (parasympatholytic) agents possess a central depressant effect which produces a reduction of voluntary muscle spasm. They produce no inhibition of transmission through peripheral neuromuscular pathways (no blockade of polysynaptic reflexes) and therefore the muscle relaxation produced is unlike that produced by the agents discussed as "central relaxants" or central nervous system depressants with skeletal muscle relaxant properties. The inhibition by the "centrally acting cholinergic blocking agents" is apparently on the extrapyramidal system. These drugs are used primarily in the relief of rigidity and spasticity of paralysis agitans (Parkinson's disease). Examples of this group of drugs are benztropine, cycrimine, procyclidine and trihexyphenidyl. For a detailed discussion of this class, see Chapter 19, Autonomic Blocking Agents.

TRANQUILIZING AGENTS

The introduction of two drugs, one an alkaloid from a small Asian shrub, the other a synthetic compound related to the antihistamines, has led to revolutionary advances in the understanding and the treatment of certain types of mental disease. The Rauwolfia alkaloids were first made generally available to Western medicine in 1953 and were followed shortly by chlorpromazine in 1954. These drugs possess a variety of pharmacologic properties, but their unique depressant effects on the central nervous system have been widely used in the treatment of serious mental and emotional disorders which are characterized by

TABLE 30. A PHARMACOLOGIC COMPARISON OF CENTRAL RELAXANTS, SEDATIVE-HYPNOTICS AND TRANQUILIZERS[15, 16, 17]

ACTION	CENTRAL RELAXANTS	SEDATIVE HYPNOTICS	TRANQUILIZERS	
	Meprobamate	Barbiturates	Reserpine	Chlorpromazine
Adrenergic blocking (central)	no	no	yes	yes
Cholinergic blocking (peripheral)	no	no	no	yes
Antihistaminic	no	no	no	yes
Anesthesia	yes	yes	no	no
Arousal	difficult	difficult	easy	easy
Addiction liability	yes	yes	no	no
Ataxia	yes	yes	no	no
Convulsant threshold	raised	raised	lowered	lowered
Excitement	present	present	absent	absent
Lethal dose	respiratory depression	respiratory depression	convulsions (muscle spasticity)	convulsions

varying degrees of excitation. The Rauwolfia alkaloids, chlorpromazine and its analogs, and other chemically and pharmacologically dissimilar compounds whose use in mental disorders followed, have been described by a variety of names, among those "ataraxic," "neurosedative," "calming." The designation most widely used for the depressant drugs has been "tranquilizing agent."

This term may be applied more specifically to represent those agents capable of exerting a unique type of selective central nervous system depression. They act differently from the barbiturates and other sedatives which act by producing a general central nervous system depression. The action of the tranquilizers is believed to take place primarily in the paleocortex and the subcortical areas of the brain. They give strong sedation without producing sleep and produce a state of indifference and disinterest. They are effective in reducing excitation, agitation, aggressiveness and impulsiveness which are not controlled by the ordinary sedative-hypnotics (e.g., phenobarbital) and central relaxants (central depressant drugs with skeletal-muscle-relaxing properties, e.g., meprobamate).

A comparison of some pharmacologic properties of the central relaxants, sedative-hypnotics and tranquilizers displays close similarity for the depressant effects of the central relaxants and the sedative-hypnotics, and a unique set of properties for the tranquilizers (see Table 30).

In addition to the analogs of the Rauwolfia alkaloids and chlorpromazine, a group of diphenylmethane derivatives and miscellaneous compounds structurally related to chlorpromazine have been shown to be best described as tranquilizing agents. Members of this group do not produce true anesthesia, even in high dose, and arousal is easy from the sleep which may be induced. They possess no demonstrated addiction liability, produce no ataxia, muscle tone is increased at high doses, and the convulsant threshold is lowered. No excitement stage precedes hypnosis, and toxic doses produce convulsions. In contrast, most of these properties are opposite or significantly different from both central relaxants and sedative-hpynotics.

Certain biogenic amines, the catecholamines norepinephrine and dopamine, and the indoleamine serotonin, are known to occur in appreciable quantities in both the central and the peripheral nervous systems. Much is known about their roles as neurotransmitting substances in the peripheral nervous system (see Chap. 17), but very little is known of their function in the central nervous system. By analogy to their

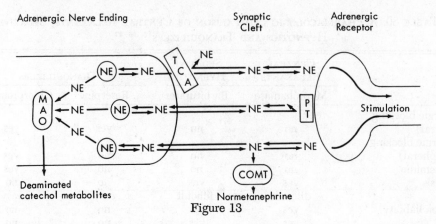

Figure 13

peripheral function, it is presumed that imbalances in the normal pattern of synthesis, distribution, and metabolism of these amines in the central nervous system may lead to changes in brain function resulting in marked alterations of mood and behavior.[18]

Norepinephrine is present in many parts of the brain, with the highest concentration in the hypothalamic area. In contrast, epinephrine occurs in the brain in very low concentration compared to norepinephrine. Dopamine is present in highest concentration in the basal ganglia, and in lower concentrations elsewhere in the brain. Serotonin occurs both in the brain and in peripheral tissues in appreciable amount. Norepinephrine, about which most is known in terms of peripheral function, serves as a general model for theories of behavioral changes induced by drug-neurohormonal interactions. In general, drugs that produce high levels of available norepinephrine in in the central nervous system produce excitation or stimulation. Drugs that enhance the depletion and inactivation of norepinephrine in the CNS produce sedation or depression.

Norepinephrine is synthesized from tyrosine, with the intermediate formation of 3,4-dihydroxyphenylalanine (dopa) and 3,4-dihydroxyphenethylamine (dopamine). Hydroxylation of dopamine at the β-carbon produces norepinephrine. A wide variety of structural analogs can serve as substrates for one or more steps in the synthetic pathway, e.g., L-α-methyl-*m*-tyrosine, L-α-methyldopa. These yield "false transmitters"

which are generally much weaker in their neurohormonal actions. Norepinephrine is stored within the nerve in intraneuronal granules (see Fig. 13), and may be released intracellularly by the action of the sedative-hypotensive reserpine alkaloids or the related synthetic benzquinolizine derivatives. The released intracellular norepinephrine may be inactivated, mainly by mitochondrial monoamine oxidase, forming deaminated catechol metabolites, such as 3,4-dihydroxymandelic acid, before leaving the cell. This depletion of a physiologically active form of norepinephrine is associated with the sedative-hypotensive properties of the reserpine alkaloids. Monoamine oxidase inhibitors (see Chap. 16), which block the intracellular inactivation of norepinephrine, act as stimulant drugs.

Norepinephrine is discharged from neuronal endings in its physiologically active form, either by nerve impulses or by the action of some sympathomimetic drugs. It is presumed to produce its stimulant effect by either direct action as a neurohormone on central adrenergic receptors, or as a regulator of synaptic transmission by mediating the release of other chemical transmitters, such as acetylcholine. Some centrally acting sympathomimetic drugs may exert a direct effect on such receptors. The phenothiazine tranquilizers are thought to act by blocking the effective interaction of norepinephrine with its receptors. Norepinephrine released to the synaptic cleft is inactivated by cellular re-uptake, or by enzymatic methylation of the 3-hydroxyl group by catechol O-methyl transferase, to form the less active normetanephrine. The tricyclic antidepressants (see Chap. 16) are thought to elevate mood by inhibition of the cellular re-uptake of norepinephrine, thus prolonging its existence and action within the synaptic cleft.

The central roles of dopamine and serotonin are less well defined. Serotonin is synthesized by decarboxylation of 5-hydroxytryptophan and, like norepinephrine, exists in the neuron in free and bound forms. Serotonin is metabolized by monoamine oxidases, forming 5-hydroxyindoleacetic acid.

THE RAUWOLFIA ALKALOIDS AND SYNTHETIC ANALOGS

Rauwolfia serpentina and other Rauwolfia species have been widely used in India for centuries in a variety of ailments including snakebite, dysentery, cholera, fevers, insomnia and insanity. A gradually increasing literature from India that emphasized the effectiveness of plant extracts and dried root powder in the reduction of elevated blood pressure culminated in a publication by Vakil in 1949.[19] This led, in 1950, to trial of the crude drug in the United States for treatment of hypertension. Hypotensive and sedative effects which developed slowly were observed.

Concurrent with the medical interest, Swiss chemists during 1947 to 1951 studied the structures of the crystalline alkaloids from *Rauwolfia serpentina* reported by Indian chemists in 1931[20] but found in these only moderate sedative and hypotensive activity. However, pharmacologic tests revealed the potent activity of the crude drug was concentrated in the noncrystalline "oleoresin fraction," and from this was isolated reserpine, the major active constituent. Animal studies demonstrated the unique sedative effects of reserpine, a quiet and subdued state being gradually developed, often leading to sleep from which the animals could be aroused readily. Unlike the sedative-hypnotics, large doses did not cause deep hypnosis and anesthesia. Together with this sedative effect, blood pressure was gradually lowered.

At the same time (1952) that enthusiastic clinical reports on the hypotensive effect and the unique sedation produced by oral use of the powdered root were being presented, the crystalline alkaloid, reserpine, was made available. Clinical reports on the antihypertensive effects of reserpine noted the sedative effect and suggested use in the treatment of psychiatric states of agitation and anxiety. Although used in India for at least 5 centuries in treatment of the mentally disturbed (called in some areas pagal-ka-dawa, or "insanity remedy"), trial in psychotherapy outside of India was delayed until 1954 when the powdered whole root was used with moderate success in a wide variety of mental disorders characterized by excitement (mania) rather than depression.[21, 22] A better understanding of the type of patient responsive to Rauwolfia therapy has led to its favorable application as a psychotherapeutic-sedative in the management of patients with anxiety or tension psychoneuroses and in those chronic psychoses involving anxiety, compulsive aggressive behavior and hyperactivity. The introduction of the first of the phenothiazine tranquilizing agents, chlorpromazine, in 1954, has somewhat limited the application of the Rauwolfia alkaloids for psychotherapeutic treatment. The alkaloids and their synthetic analogs are widely used for their hypotensive effects, either alone in cases of mild or labile hypertension or in combination with the more potent hypotensive agents (e.g., ganglionic blocking agents) for the management of essential hypertension (see Chap. 21 for detailed descriptions of these drugs).

It has been proposed that reserpine and related Rauwolfia alkaloids produce both their sedative and hypotensive effects by the release and the depletion of body amines such as serotonin, norepinephrine and hydroxytryamine.[23] In support of this concept, it has been shown that only those Rauwolfia alkaloids which produce a sedative response affect brain serotonin levels.[24] A reserpine analog, the *m*-dimethylaminobenzoic acid ester of methyl reserpate, which has a more prolonged effect on levels of brain norepinephrine than on brain serotonin, demonstrates depressant effects which closely

TABLE 31. RAUWOLFIA ALKALOIDS AND SYNTHETIC ANALOGS

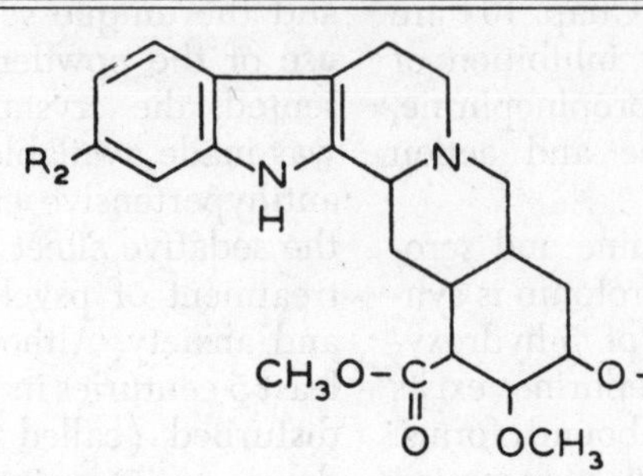

R_2 ... N—H ... N ... CH_3O-C ... $O-R_1$ ... O ... OCH_3

Generic name Proprietary name	Substituents		Usual oral dose (mg./day)
	R_1	R_2	Psychoses
Reserpine U.S.P. Sandril, Serpasil	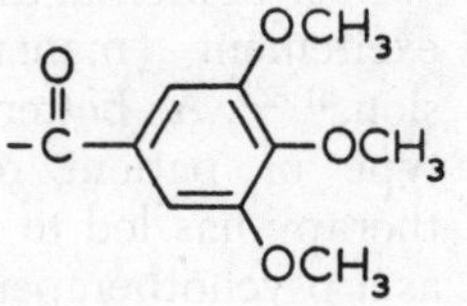 (3,4,5-trimethoxybenzoyl) $-C(=O)-$ with OCH_3, OCH_3, OCH_3	$-OCH_3$	3-5
Rescinnamine N.F. Moderil	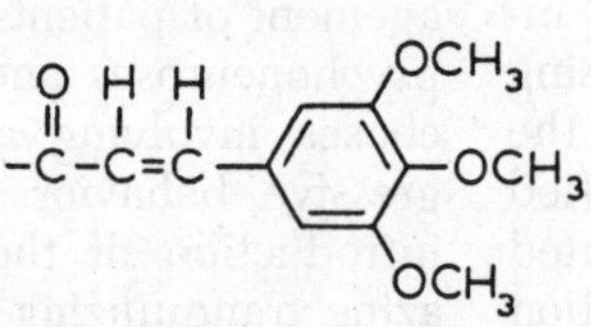 $-C(=O)-CH=CH-$ (3,4,5-trimethoxycinnamoyl) OCH_3, OCH_3, OCH_3	$-OCH_3$	3-12
Deserpidine Harmonyl	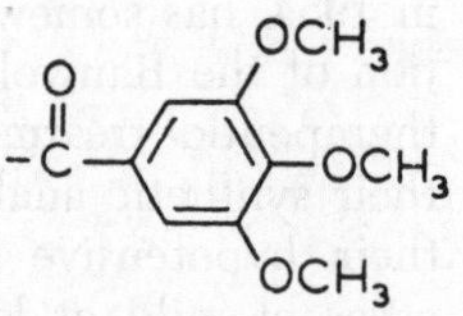 $-C(=O)-$ with OCH_3, OCH_3, OCH_3	H	2-3

parallel changes in the level of serotonin but not norepinephrine.

The structures of Rauwolfia compounds and their doses used in the treatment of psychoses are presented in Table 31.

BENZOQUINOLIZINE DERIVATIVES

The slow onset of action, the prolonged duration, and the signficant side-effects produced by the Rauwolfia alkaloids and their derivatives encouraged a search for compounds with similar tranquilizing properties. Many partial structures of reserpine have been synthesized and tested, but the most promising lead has come from an unrelated program.

Emetine is a potent but toxic amebicidal alkaloid not present in Rauwolfia species.

BENZOQUINOLIZINE PORTION

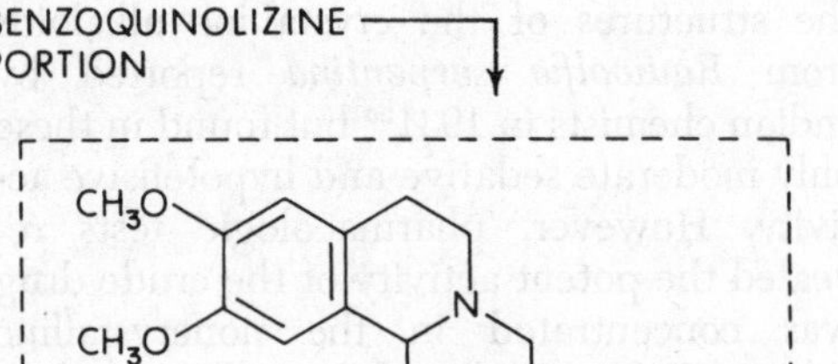

CH_3O ... CH_3O ... N ... CH_2CH_3 ... CH_2 ... HN ... OCH_3 ... OCH_3

Emetine

Approaches to its chemical synthesis yielded a substituted benzoquinolizine, *Tetrabenazine,* as an intermediate which was found to have pharmacologic properties closely resembling those of reserpine; it caused release and depletion of brain amines and was effective in depressing the conditioned avoidance response in experimental animals.

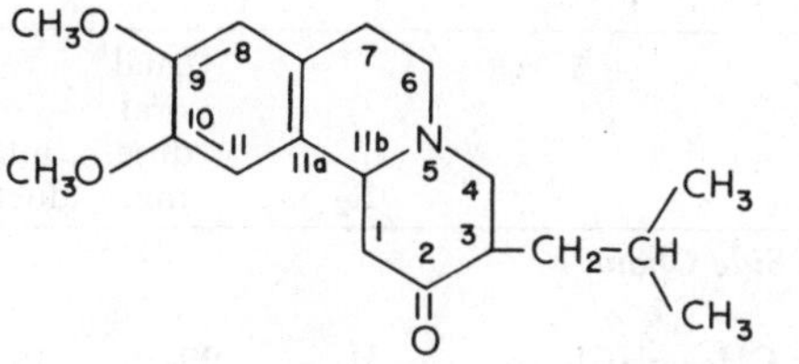

Tetrabenazine

It is less potent than reserpine. However, it has a briefer onset and duration of action than reserpine, being rapidly metabolized.

Benzquinamide, Quantril®, N,N-Diethyl-2-acetoxy-9,10-dimethoxy-1,2,3,4,6,7-hexa-

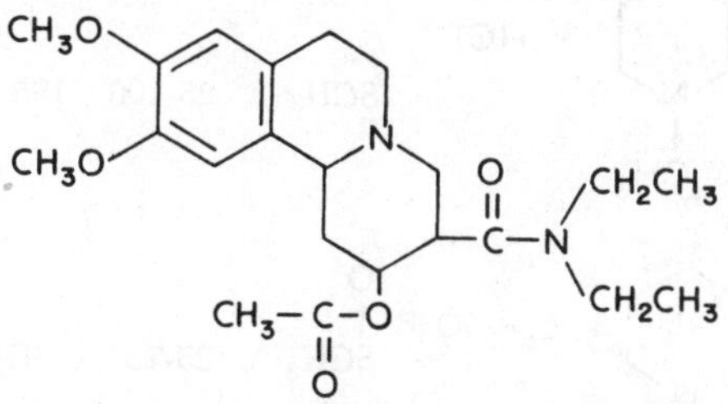

Benzquinamide

hydro-11bH-benzo[a]quinolizine-3-carboxamide. Benzquinamide, like reserpine and tetrabenazine disrupts the conditioned avoidance response in experimental animals. However, unlike the other analogs it produces no measurable changes in the level of brain amines. This indicates the possi-

bility that the tranquilizing effect may be separated from the amine-depletion effect and from resulting side-effects.

PHENOTHIAZINE DERIVATIVES

During World War II a number of phenothiazine derivatives were prepared in the Paris laboratories of the French pharmaceutical manufacturer Rhone Poulenc. Among these was a series of 10-(2-dimethylaminoalkyl)phenothiazines which on pharmacologic screening were found to possess strong antihistaminic properties (see Chap. 23). 10-(2-Dimethylamino)propylphenothiazine (promethazine) was studied extensively and among its diverse pharmacologic properties were a sedative effect alone and a potentiating effect on the sedative action of the barbiturates. Structural analogs of promethazine were prepared in an attempt to develop derivatives with a more marked central depressant action; among these chlorpromazine was synthesized by Charpentier in 1950.

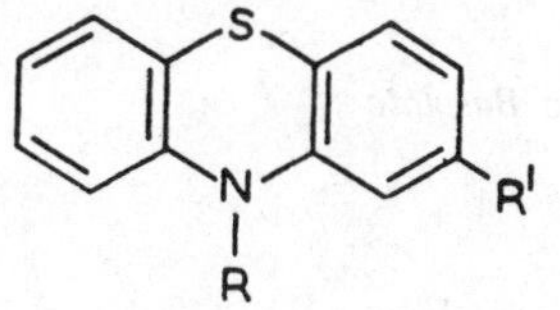

Promethazine: $R = CH_2CH(CH_3)N(CH_3)_2$
$R' = H$
Chlorpromazine: $R = CH_2CH_2CH_2N(CH_3)_2$
$R' = Cl$

A general synthesis for phenothiazine derivatives of this type is shown below.

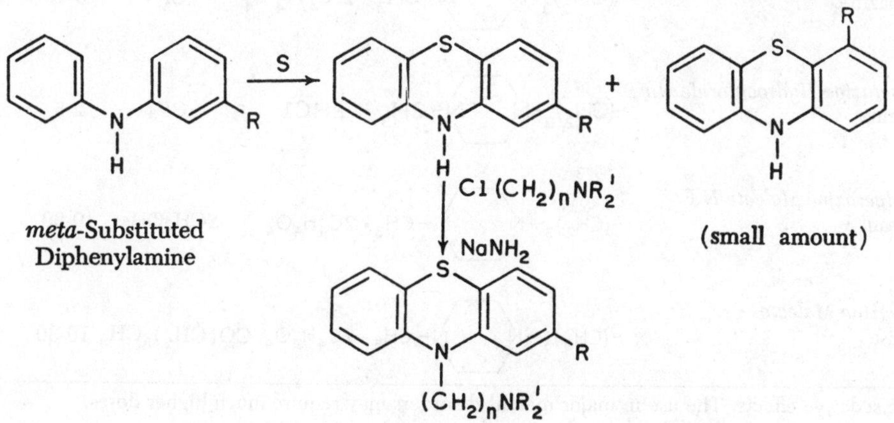

meta-Substituted Diphenylamine

(small amount)

TABLE 32. PHENOTHIAZINE DERIVATIVES (AMINO-PROPYL SIDE CHAIN)

Generic name / Proprietary name	R_{10}	R_2	Usual* oral dose mg.	Year of intro-duction
Propyl Dialkylamino Side Chain				
Promazine Hydrochloride N.F. Sparine	$-(CH_2)_3N(CH_3)_2 \cdot HCl$	H	25-50	1955
Chlorpromazine Hydrochloride U.S.P. Thorazine	$-(CH_2)_3N(CH_3)_2 \cdot HCl$	Cl	10-25	1954
Triflupromazine Hydrochloride N.F. Vesprin	$-(CH_2)_3N(CH_3)_2 \cdot HCl$	CF_3	10-25	1957
Alkyl Piperidyl and Pyrrolidinyl Side Chain				
Thioridazine Hydrochloride U.S.P. Mellaril		SCH_3	25-100	1959
Mesoridazine Besylate Serentil		SCH_3	25-75	1970
Methdilazine Hydrochloride N.F. Tacaryl		H	8	1960
Propyl Piperazine Side Chain				
Prochlorperazine Maleate U.S.P. Compazine	$-(CH_2)_3-N\!\!\!\!\diamond\!\!\!\!N-CH_3 \cdot 2C_4H_4O_4$	Cl	5-10	1957
Trifluoperazine Hydrochloride N.F. Stelazine	$-(CH_2)_3-N\!\!\!\!\diamond\!\!\!\!N-CH_3 \cdot 2HCl$	CF_3	2-5	1959
Thiethylperazine Maleate N.F. Torecan	$-(CH_2)_3-N\!\!\!\!\diamond\!\!\!\!N-CH_3 \cdot 2C_4H_4O_4$	SCH_2CH_3	10-30	1961
Butaperazine Maleate Repoise	$-(CH_2)_3-N\!\!\!\!\diamond\!\!\!\!N-CH_3 \cdot 2C_4H_4O_4$	$CO(CH_2)_3CH_3$	10-30	1968

* For sedative effects. The use in major mental disorders may require much higher doses.

TABLE 32. PHENOTHIAZINE DERIVATIVES (AMINO-PROPYL SIDE CHAIN) (*Continued*)

Generic name Proprietary name	R_{10}	R_2	Usual[*] oral dose mg.	Year of intro- duction
Perphenazine N.F. Trilafon	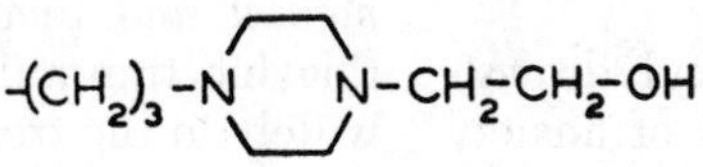 $-(CH_2)_3-N\diagup\diagdown N-CH_2-CH_2-OH$	Cl	2-8	1957
Fluphenazine Hydrochloride N.F. Permitil, Prolixin	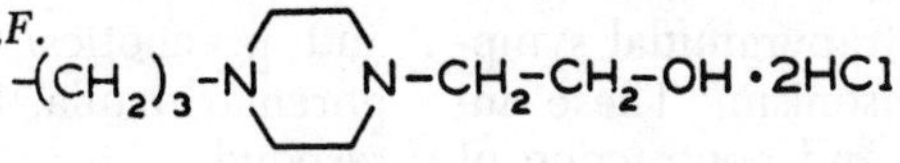 $-(CH_2)_3-N\diagup\diagdown N-CH_2-CH_2-OH\cdot 2HCl$	CF$_3$	0.25-1	1960
Thiopropazate Hydrochloride N.F. Dartal	$-(CH_2)_3-N\diagup\diagdown N-CH_2-CH_2O-\overset{O}{\overset{\|}{C}}CH_3\cdot 2HCl$	Cl	2-5	1957
Acetophenazine Maleate N.F. Tindal	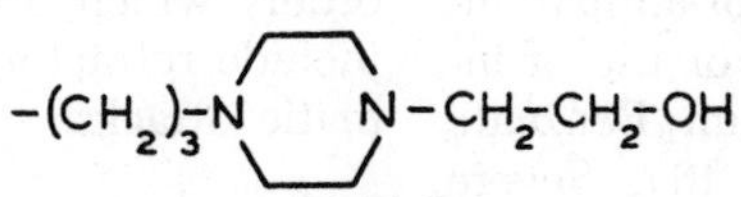 $-(CH_2)_3-N\diagup\diagdown N-CH_2-CH_2-OH$ $2C_4H_4O_4$	$-\overset{O}{\overset{\|}{C}}-CH_3$	40-80	1961
Carphenazine Maleate N.F. Proketazine	$-(CH_2)_3-N\diagup\diagdown N-CH_2CH_2OH\cdot 2C_4H_4O_4$	$-\overset{O}{\overset{\|}{C}}-CH_2CH_3$	25-100	1963

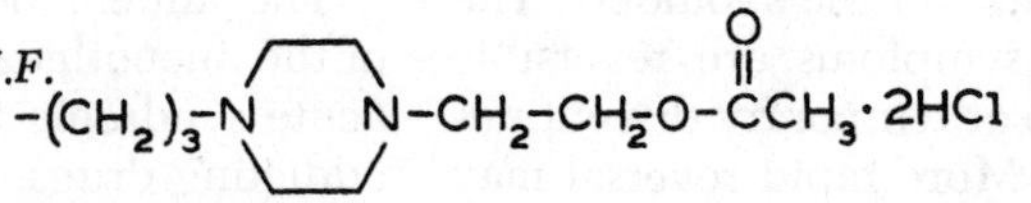

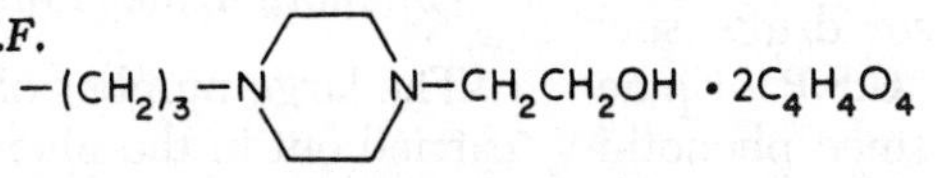

Branched Propyl Dialkylamino Side Chain

Generic name Proprietary name	R_{10}	R_2	Usual[*] oral dose mg.	Year of intro- duction
Trimeprazine Tartrate N.F. Termaril	$-CH_2CHCH_2N(CH_3)_2\cdot\tfrac12 C_4H_6O_6$ $\quad\quad\overset{\|}{CH_3}$	H	2.5-5	1958
Methotrimeprazine N.F. Levoprome	$-CH_2CHCH_2N(CH_3)_2\cdot HCl$ $\quad\quad\overset{\|}{CH_3}$	OCH$_3$	10-20(I.M.)	1968

[*] For sedative effects. The use in major mental disorders may require much higher doses.

Between 1951 and 1954 chlorpromazine was used extensively in Europe under the name of Largactil. The drug was first used in 1951 in combination with meperidine and promethazine and external cooling to lower body temperature to produce "artificial hibernation" for surgery. Its synergistic effect with analgetics was applied in 1952 and in the same year it was first reported to be useful in the quieting and the control of hyperactive psychotic patients. Chlorpromazine was introduced in the United States in 1954 gaining rapid and widespread acceptance for its use as an antiemetic for the potentiation of anesthetics, analgetics and sedatives and in the treatment of major mental and emotional disorders. Since 1954 a number of analogs varying principally in the nature of the aminopropyl side chain and the substituent in the 2-position of the phenothiazine ring have been introduced (see Table 32).

Like the Rauwolfia alkaloids, chlorpromazine and its analogs produce a central reduction of sympathetic outflow at the level of the hypothalamus. The resulting decrease in peripheral sympathetic tone which would lead to reserpinelike side-effects if unopposed is counterbalanced by the pronounced atropinelike (cholinergic-blocking) properties of the phenothiazine derivatives. Therefore, side-effects are a mixture of those produced by central adrenergic and peripheral cholinergic blockade. These include dry mouth, dizziness, blurred vision, orthostatic hypotension and tachycardia.

Central effects of sedation and drowsiness, desirable in the treatment of anxiety and agitation, are undesirable in other applications. At high doses, side-effects occur frequently, including extrapyramidal symptoms resembling parkinsonism. These include tremor, spasticity and contraction of muscles of the head and the shoulders. The parkinsonism-like symptoms are reversible on lowered dosage or temporary discontinuation of the drug. More rapid reversal may be achieved by administration of antiparkinsonism drugs (*see* Chap. 19), or use of intravenous Caffeine and Sodium Benzoate Injection U.S.P. (*see* Chap. 16). Severe hypotension may also occur, calling for immediate supportive measures, including the use of intravenous vasopressor drugs, such as Levarterenol Bitartrate U.S.P. Epinephrine should not be used, since phenothiazine derivatives reverse its action, resulting in a further lowering of blood pressure. Jaundice has been reported in about 1 per cent of the patients receiving chlorpromazine, and a much smaller percentage have developed agranulocytosis. Ocular changes including opacities of the lens and cornea (cataract formation) have been observed in patients receiving large doses of the phenothiazines for an extended period.

The therapeutic applications of the phenothiazine tranquilizing agents may be grouped into three major areas:

1. *Antiemetic effect.* These agents are the best available for the treatment or the prevention of emesis which is drug-induced (e.g., nitrogen mustards), due to infections or toxicoses, or postoperative. They are generally less effective in the prevention of motion sickness. Certain derivatives, but not all, are recommended for the relief of nausea and vomiting during pregnancy.

2. *Potentiation of the effects of anesthetics, analgetics and sedatives.* As an adjuvant in surgical procedures, the phenothiazine tranquilizing agents reduce apprehension by their sedative effects. They also potentiate the anesthetics, potent (narcotic) analgetics and sedatives, permitting their use in a smaller dose which results in decreased respiratory depression.

3. *Treatment of moderate and severe mental and emotional states.* The phenothiazine tranquilizing drugs are used most widely in the treatment of mental and emotional disorders. Anxiety, tension and agitation are reduced in both psychoneurotics and psychotics. Selected cases of schizophrenia, mania, toxic and senile psychoses respond.

The antiemetic and tranquilizing effects of the phenothiazines combine to relieve the acute syndrome following withdrawal from addicting drugs. Miscellaneous somatic disorders which are relieved by these drugs include refractory hiccups and severe asthmatic attacks.

Structure-Activity Relationships for the Phenothiazine Tranquilizing Agents

The large number of structural variations carried out in the phenothiazine series have resulted in the establishment of some fairly consistent patterns of relationships between structure and activity.[25]

The nature and the position of substituents on the phenothiazine nucleus strongly influence activity. Replacement of the hydrogen in position 2 (R_2) by chlorine (chlorpromazine), trifluoromethyl (triflupromazine) or a dimethylsulfonamido group (thioperazine) results in increased activity. The trifluoromethyl analog is generally more potent than the chloro compound, but this is accompanied usually by an increase in extrapyramidal symptoms. Tranquilizing activity is retained with a variety of 2-substituents such as thioalkyl (thioridazine, thiethylperazine) and acyl groups (butaperazine, acetophenazine, carphenazine). The 2-thioalkyl derivatives are said to produce fewer extrapyramidal side-effects. A ring substituent in positions 1,3,4 or simultaneous substitution in both aromatic rings results in loss of tranquilizing activity.

The three-carbon side chain connecting the nitrogen of the phenothiazine ring and the more basic side-chain nitrogen is optimal for tranquilizing activity. Compounds with a two-carbon side chain (amino-ethyl side chain) still possess a moderate central depressant activity, but their antihistaminic and antiparkinsonism effects generally predominate. If the side chain is altered significantly in its length and polarity, tranquilizing activity is lost, although compounds of this type (dimethoxinate, Chap. 24) show antitussive properties.

Branching at the β-position of the side chain (R_3) with a small group such as methyl greatly reduces tranquilizing activity but may enhance antihistaminic and antipruritic effects (trimeprazine). This has been attributed to steric repulsion between the methyl group at the β-position, and the 1,8-*peri* hydrogens of the phenothiazine rings, resulting in a decrease in the coplanarity of the benzene rings.[26] It would also be expected that branching at the β-position would slow the rate of metabolic attack on the side chain. Antipruritic activity is retained if the branching on the β-carbon is part of a ring (methdilazine). Side-chain substitution with a large or polar group such as phenyl, dimethylamino or hydroxyl results in loss of tranquilizing activity. The importance of the side chain is further emphasized by the fact that stereospecificity exists, the levo isomer being far more active than the dextro isomer for β-methyl derivatives.

Substitution of the piperazine group (prochlorperazine, trifluoperazine) in place of the terminal dimethylamino moiety on the side chain increases potency but usually results in an increase in extrapyramidal symptoms. Substitution of $-CH_2CH_2OH$ for the terminal methyl group on piperazine (perphenazine, fluphenazine) results in a slight increase in potency, which is also true for the readily hydrolyzed O-acetyl deriva-

tive (thiopropazate). In general, the dimethylamino compounds seem more likely to produce a parkinsonlike syndrome (tremors, rigidity, salivation), while the piperazine derivatives produce, in addition, dyskinetic reactions, generally involving the muscles of the face and the neck. Skin and liver disorders and blood dyscrasias have been associated with the dimethylamino- and alkylpiperidyl-types to a greater extent than with the piperazine derivatives.

Quaternization of the side chain nitrogen in any of the phenothiazine derivatives results in a decrease in lipid solubility, leading to decreased penetration of the central nervous system and virtual loss of central effects.

Products

The structures of the phenothiazine derivatives containing the amino-propyl side chain attached to the nitrogen atom of the phenothiazine ring are presented in Table 32.

Chlorpromazine Hydrochloride U.S.P., Thorazine® Hydrochloride, 2-Chloro-10-[3-(dimethylamino)propyl]phenothiazine Hydrochloride. Chlorpromazine hydrochloride is used orally or parenterally in the treatment of nausea and vomiting, to potentiate the effects of anesthetics, analgetics, hypnotics and sedatives, and in a variety of mental and emotional disturbances. The free base (Chlorpromazine U.S.P.) is available in suppository form. Chlorpromazine hydrochloride is a white crystalline powder, very soluble in alcohol or water.

Category—major tranquilizer.

Usual dose—oral, 25 mg. 4 times daily; intramuscular or intravenous, 25 mg. repeated in 2 to 4 hours if needed.

Usual dose range—oral, 10 mg. to 1 g. daily; intramuscular or intravenous, 25 mg. to 1 g. daily.

OCCURRENCE

  Chlorpromazine Hydrochloride Injection
    U.S.P.
  Chlorpromazine Hydrochloride Syrup
    U.S.P.
  Chlorpromazine Hydrochloride Tablets
    U.S.P.
  Chlorpromazine Suppositories
    U.S.P.

Thioridazine Hydrochloride U.S.P., Mellaril®, 10[2-(1-Methyl-2-piperidyl)ethyl]-2-(methylthio)phenothiazine Monohydrochloride. Thioridazine inhibits psychomotor function, and its use ranges from the treatment of minor conditions of anxiety and tension, to the more severe psychoneuroses and psychoses. It shows minimal antiemetic activity and minimal extrapyramidal stimulation. In addition to tablets, the drug is available in a concentrated solution containing 30 mg. per ml. The solution is light sensitive and should be dispensed in amber bottles. The free base is water insoluble, and dilutions should be made in acidified water, or suitable acidic juices.

Category—tranquilizer.

Usual dose—25 to 50 mg. 3 or 4 times daily.

Usual dose range—20 to 800 mg. daily.

OCCURRENCE

Thioridazine Hydrochloride Solution U.S.P.

Thioridazine Hydrochloride Tablets U.S.P.

Prochlorperazine Maleate U.S.P., Compazine® Dimaleate, 2-Chloro-10-[3-(4-methyl-1-piperazinyl)propyl]phenothiazine Dimaleate. Prochlorperazine is capable of producing the same effects as chlorpromazine in a much smaller dose. Except for a very low potentiating effect on other central depressant drugs, the two drugs are similar in their applications and side-effects.

Prochlorperazine maleate (a white crystalline powder, practically insoluble in water and alcohol) is administered orally in a usual dose of the equivalent of 5 mg. of prochlorperazine 3 or 4 times a day. Range—5 to 50 mg. daily. The usual effective dose for institutionalized psychotics ranges between 75 and 125 mg. daily. The water-soluble ethanedisulfonate salt **Prochlorperazine Edisylate U.S.P.** (Compazine® Edisylate) is used by the oral or intramuscular routes. Dose—5 mg., range—5 to 100 mg. daily. The free base **Prochlorperazine N.F.** (Compazine®) is available in suppository form in a rectal dose of 2.5, 5 and 25 mg.

OCCURRENCE

Prochlorperazine Edisylate Injection U.S.P.

Prochlorperazine Edisylate Syrup U.S.P.

Prochlorperazine Maleate Tablets U.S.P.

Prochlorperazine Suppositories N.F.

Promazine Hydrochloride N.F., Sparine Hydrochloride®, 10-[3-(Dimethylamino)-propyl]phenothiazine Hydrochloride. Promazine is used in the management of acute neuropsychiatric agitation but is less potent than chlorpromazine. The hydrochloride salt is a white to slightly yellow crystalline solid that oxidizes on prolonged exposure to air and acquires a blue or pink color. It is available for oral, intramuscular or intravenous administration; however, the oral route is preferred because of the production of orthostatic hypotension on parenteral administration. The total daily dose ranges from 25 to 300 mg., with a maximum recommended dose of 1 g.

OCCURRENCE

Promazine Hydrochloride Injection N.F.

Promazine Hydrochloride Syrup N.F.

Promazine Hydrochloride Tablets N.F.

Triflupromazine Hydrochloride N.F., Vesprin®, 10[3-(Dimethylamino)propyl]-2-(trifluoromethyl)phenothiazine Hydrochloride. Triflupromazine is a potent tranquilizing agent used for the treatment of anxiety and tension, and for the management of psychotic disorders. It is also employed for the control of nausea and vomiting, and as an adjunct to the potent analgesics and general anesthetics. Adverse reactions are the same as for other phenothiazine tranquilizers, and including parkinsonism, hypotension, liver damage, and blood dyscrasias. The adult dose ranges from 30 to 150 mg. daily in divided doses. Liquid forms of the drug should be protected from exposure to light.

OCCURRENCE

Triflupromazine Hydrochloride Tablets N.F.

Methdilazine Hydrochloride N.F., Tacaryl®, 10-[(1-Methyl-3-pyrrolidyl)methyl] phenothiazine Hydrochloride. Methdilazine has a very low potency as a tranquilizing agent and is used clinically for its effective antihistaminic and antipruritic activity. See Chapter 23 for a description of uses and forms available.

Trifluoperazine Hydrochloride N.F., Stelazine® Hydrochloride, 10-[3-(4-Methyl-1-

piperazinyl)propyl]-2-(trifluoromethyl)phenothiazine Dihydrochloride. Trifluoperazine is a relatively highly potent drug in the control of acute and chronic psychoses marked by hyperactivity. Extrapyramidal symptoms occur frequently at doses required for control of psychoses. The usual oral dose ranges from 1 to 4 mg. twice daily for minor anxiety reactions. The daily adult dose for patients with psychoses may range from 15 to 30 mg. Intramuscular injections of 1 or 2 mg. are given usually at intervals of 4 hours or longer when rapid control of symptoms is necessary.

Category—major tranquilizer.

OCCURRENCE

Trifluoperazine Hydrochloride Injection N.F.
Trifluoperazine Hydrochloride Tablets N.F.

Thiethylperazine Maleate N.F., Torecan®, 2-(Ethylthio)-10-[3-(4-methyl-1-piperazinyl)propyl]phenothiazine Maleate. Thiethylperazine is a potent tranquilizing agent which may also be used as an antiemetic and for the treatment of vertigo. The tranquilizing properties are similar to chlorpromazine but achieved with a smaller dose. The compound may give a higher incidence of extrapyramidal reactions but less agranulocytosis and jaundice than chlorpromazine. The usual oral dose ranges from 10 to 30 mg. daily, in divided doses. The initial dose may be by intramuscular injection or by suppository in the semiconscious or actively vomiting patient.

Category—antiemetic.

OCCURRENCE

Thiethylperazine Maleate Tablets N.F.

Butaperazine Maleate, Repoise®, 1-{10-[3-(4-methyl-1-piperazinyl)propyl]phenothiazin-2-yl}-1-butanone Dimaleate. Butaperazine is reported to be effective in the management of chronic schizophrenic patients who are under close psychiatric supervision. It is also used for the control of all forms of psychomotor agitans, mania, hallucinations, anxiety and tension. Adverse reactions are similar to those of other major tranquilizers but it is said to give a higher incidence of extrapyramidal reactions and a lower incidence of undesirable sedation.

The initial oral dose of 5 to 10 mg., three times daily, is increased by 5 to 10 mg. every 3 or 4 days until maximum response is achieved. The maximum dose recommended is 100 mg. daily. Maintenance dosage is usually ¼ to ½ the dose required during the acute stage.

Perphenazine N.F., Trilafon®, 2-Chloro-10-{3-[4-(2-hydroxyethyl)piperazinyl]propyl}phenothiazine, 4-[3-(2-Chlorophenothiazin-10-yl)propyl]-1-piperazineethanol. Perphenazine is a major tranquilizing agent of relatively high potency. Its uses include acute and chronic schizophrenia and the manic phase of manic-depressive psychoses. The drug is also used as an antiemetic. Adverse reactions are similar to those of the other phenothiazine tranquilizers. Significant autonomic side effects, such as blurred or double vision, nasal congestion, dryness of the mouth, and constipation, are infrequent at doses below 24 mg. per day.

Perphenazine is administered intramuscularly, intravenously, orally, or rectally. The usual oral dose for simple anxiety and tension is 2 to 8 mg., three times daily; for hospitalized adults with psychoses, the usual dose is 8 to 16 mg., 2 to 4 times daily.

Category—major tranquilizer.

OCCURRENCE

Perphenazine Injection N.F.
Perphenazine Tablets N.F.

Fluphenazine Hydrochloride, N.F., Permitil®, Prolixin®, 4-[3-[2-(Trifluoromethyl)-phenothiazin-10-yl]propyl]-1-piperazineethanol Dihydrochloride, 10-{3-[4-(2-Hydroxyethyl)piperazinyl]propyl}-2-trifluoromethylphenothiazine Dihydrochloride. Fluphenazine is the most potent of the currently available phenothiazine tranquilizers on a milligram basis. It is effective in the control of major psychotic states marked by hyperactivity but displays a high incidence of extrapyramidal side-effects at the dose required. The oral adult dose in the treatment of major psychoses may range from an initial dose of 0.5 mg. per day to 2 mg. per day in hospitalized patients.

A solution of the enanthic acid ester of fluphenazine (**Fluphenazine Enanthate, N.F.**) in sesame oil containing 1.5 per cent of benzyl alcohol is useful in the treatment of chronic schizophrenia. A single dose of 12.5 to 25 mg. may be given by the parenteral route (subcutaneous or intramuscular), with the therapeutic effect lasting for 1 to 3 weeks.

Category—major tranquilizer.

OCCURRENCE
Fluphenazine Hydrochloride Injection N.F.
Fluphenazine Hydrochloride Tablets N.F.
Fluphenazine Enanthate N.F.
Fluphenazine Enanthate Injection N.F.

Thiopropazate Hydrochloride, N.F., Dartal® Hydrochloride, 2-Chloro-10-{3-[4-(2-acetoxyethyl)piperazinyl]propyl}phenothiazine Dihydrochloride; 4-[3-(2-Chlorophenothiazin-10-yl)propyl]-1-piperazineethanol Acetate Dihydrochloride. Thiopropazate is a relatively potent agent in the control of major psychoses; extrapyramidal side-effects are common at the therapeutic dose level. The usual initial oral dose is 10 mg. three times daily, being adjusted at 10 mg. intervals each 3 or 4 days, depending on patient response. The maximal recommended daily dose is 100 mg.

Category—major tranquilizer.

OCCURRENCE
Thiopropazate Hydrochloride Tablets N.F.

Acetophenazine Maleate N.F., Tindal®, 10{3-[4-(2-Hydroxyethyl)-1-piperazinyl]-propyl}phenothiazin-2-yl Methyl Ketone Dimaleate. Acetophenazine, a 2-acyl phenothiazine derivative, is a homolog of carphenazine, and is also similar in structure to butaperazine. The compound is a relatively potent tranquilizing agent and, like others of its class, may be preferable for treatment of nonhospitalized patients because of its low incidence of agranulocytosis and jaundice. The usual adult oral dose is 20 mg. 3 times a day.

Category—major tranquilizer.

OCCURRENCE
Acetophenazine Maleate Tablets N.F.

Carphenazine Maleate N.F., Proketazine®, 1-{10-(3-[4-(2-Hydroxyethyl)-1-piperazinyl]propyl)phenothiazin-2-yl}-1-propanone Dimaleate. Proketazine is a comparatively short-acting antipsychotic agent used in the management of schizophrenic psychotic reactions. Like other phenothiazine

tranquilizers, it may produce adverse reactions including extrapyramidal symptoms and blood dyscrasias. The adult dose varies from 12.5 to 50 mg. 3 times a day, with a maximum dose of 400 mg. daily.

Category—major tranquilizer.

OCCURRENCE
Carphenazine Maleate Tablets N.F.

Trimeprazine Tartrate N.F., Temaril®, (±)-10-[3-(Dimethylamino)-2-methylpropyl]phenothiazine Tartrate. Trimeprazine is identical in structure to promazine except for the 2-methyl substituent in the propyl chain. The antiemetic, tranquilizing and barbiturate-potentiating effects are relatively weak as compared with other phenothiazine derivatives of this type. However, the antihistaminic effects and, particularly, the antipruritic effect are pronounced, even at low doses. (See Chap. 23.)

Methotrimeprazine N.F., Levoprome®, (−)-2-Methoxy-10-(3-dimethylamino-2-methylpropyl)phenothiazine Hydrochloride. Methotrimeprazine is a unique phenothi-

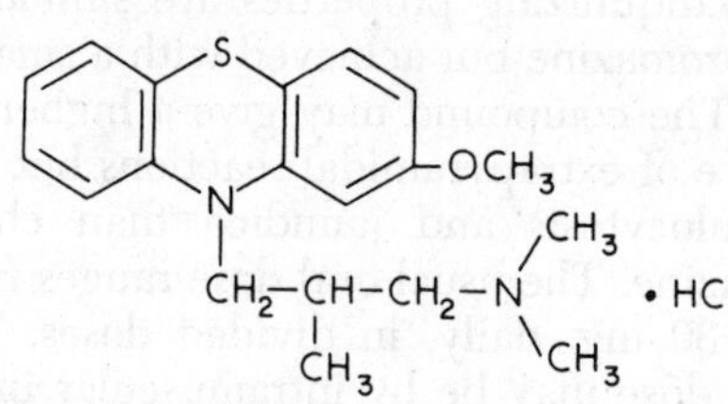

Methotrimeprazine

azine derivative, in that it acts as a potent *analgesic*. By intramuscular injection, 15 mg. of methotrimeprazine produces relief of pain equivalent to that produced by 10 mg. of morphine sulfate. (See Chap. 24.)

Mesoridazine Besylate, Serentil®, 10-[2-(1-Methyl-2-piperidyl)ethyl]-2-(methylsul-

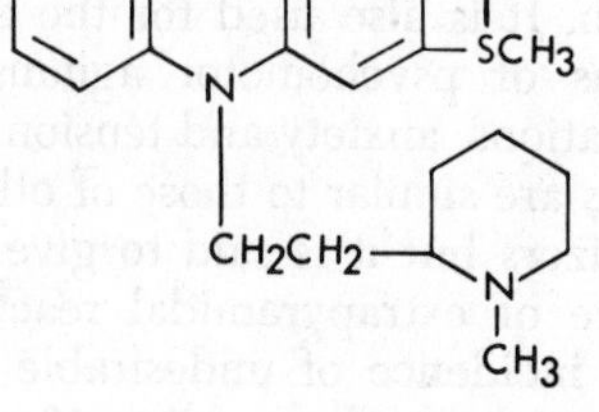

Mesoridazine

TABLE 33. PHENOTHIAZINE DERIVATIVES (AMINO-ETHYL SIDE CHAIN)

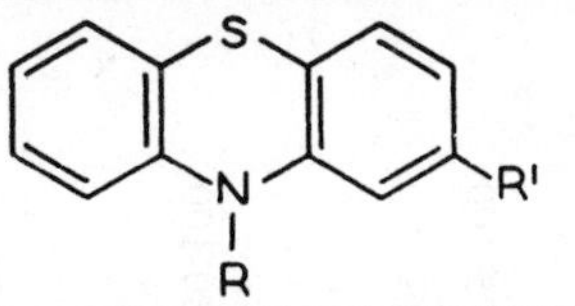

Generic name Proprietary name	R	R'	Usual sedative dose mg.	Date of introduction
Promethazine Hydrochloride U.S.P. Phenergan	$-CH_2-CH(CH_3)N(CH_3)_2 \cdot HCl$	H	25-50	1951
Pyrathiazine Hydrochloride Pyrrolazote	$-CH_2-CH_2-N$ 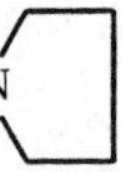 ⟩ $\cdot HCl$	H	25-50	1949
Diethazine Hydrochloride Diparcol	$-CH_2CH_2N(C_2H_5)_2 \cdot HCl$	H	25-50	1949
Ethopropazine Hydrochloride Parsidol	$-CH_2CH(CH_3)N(C_2H_5)_2 \cdot HCl$	H	25-50	1954
Propiomazine Hydrochloride N.F. Largon	$-CH_2CH(CH_3)N(CH_3)_2 \cdot HCl$	$-\underset{\underset{O}{\|\|}}{C}-CH_2CH_3$	20*	1960

* Intramuscular for nighttime or preoperative sedation.

finyl)phenothiazine Benzensulfonate. Mesoridazine differs from thioridazine in the presence of the sulfoxide group at the 2-position of the phenothiazine ring. It is reported to be as effective as thioridazine and chlorpromazine in one half the dose. The usual oral dose as a tranquilizing agent ranges from 25 mg. twice a day to 75 mg. three times a day.

Piperacetazine, Quide®, 2-Acetyl-10-{3-[4-(2-hydroxyethyl)piperidino]propyl}phenothiazine. Piperacetazine differs from acetophenazine by the presence of a piperidine ring, rather than a piperazine ring, in the side chain. Its uses and side effects are similar to those of the other phenothiazine

Piperacetazine

tranquilizers. The usual oral dose is 10 to 30 mg. per day.

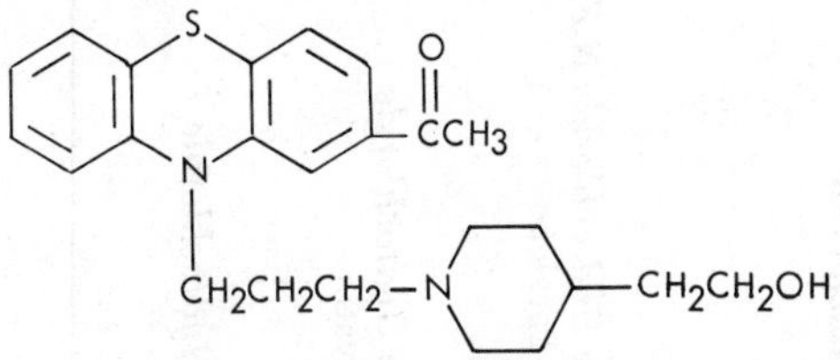

In addition to the chlorpromazinelike compounds which have 3 carbon atoms separating the heterocyclic and aliphatic or alicyclic nitrogen atoms, derivatives in which 2 carbon atoms separate the nitrogens (e.g., promethazine) show marked central depressant properties. Some of these compounds are used therapeutically in other ways (e.g., the antihistamines, promethazine and pyrathiazine, the antiparkinson agents, diethazine and ethopropazine), but newer analogs (e.g., the sedative propiomazine) reflect the recognition that the sedation produced by this group is like that of chlorpromazine and unlike that of the sedative-hypnotics (e.g., phenobarbital, meprobamate). The sedative side-effects of members of this series are frequently applied therapeutically in the relief of excited emotional states. In contrast to the phenothiazine derivatives with amino-propyl (C_3) side chains, tranquilizing potency in the amino-ethyl series is generally *reduced* by substitution in the 2-position of the phenothiazine nucleus. For example, 2-chloro-

TABLE 34. DIPHENYLMETHANE DERIVATIVES

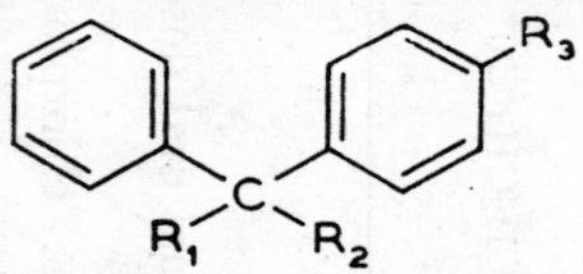

Generic name Proprietary name	R_1	R_2	R_3	Usual oral dose mg.	Year of introduction
Hydroxyzine Hydrochloride N.F. Atarax, Vistaril	H	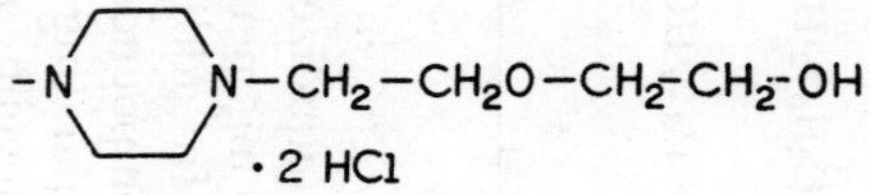 $-N$(piperazine)$-CH_2-CH_2O-CH_2-CH_2-OH$ · 2 HCl	Cl	25–100	1956
Buclizine Hydrochloride Softran	H	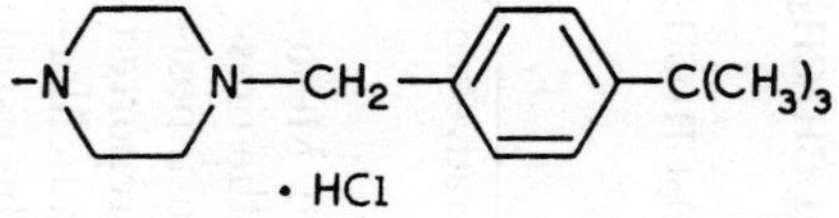 $-N$(piperazine)$-CH_2-$(phenyl)$-C(CH_3)_3$ · HCl	Cl	50–100	1956
Chlorcyclizine Hydrochloride N.F. Perazil, Di-Paralene	H	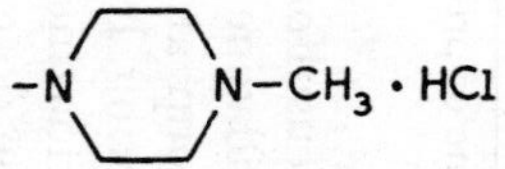 $-N$(piperazine)$-CH_3$ · HCl	Cl	25–50	1950
Captodiamine Hydrochloride Suvren, Covatin	H	$-S-(CH_2)_2N(CH_3)_2$ · HCl	$-S-(CH_2)_3CH_3$	50–100	1960–61
Benactyzine Hydrochloride Suavitil	OH	$-C(=O)-O-(CH_2)_2N(C_2H_5)_2$ · HCl	H	1–3	1957

promethazine is less effective than promethazine.

The structures of representative members of this class are presented in Table 33. These drugs are discussed in more detail in Chapter 19, Autonomic Blocking Agents, and Chapter 23, Histamine and Antihistaminic Agents.

Promethazine Hydrochloride U.S.P., Phenergan®, 10-(2-Dimethylaminopropyl) phenothiazine Hydrochloride. Promethazine is employed primarily for its antihistaminic effect (see Chap. 23), but is also used for its sedative and antiemetic properties, and for its potentiating effect on analgesics and other central nervous system depressants. The average adult oral dose for sedation is 25 mg. taken before retiring.

Propiomazine Hydrochloride N.F., Largon®, (±)-1-(10-[2-(Dimethylamino)propyl]phenothiazin-2-yl)-1-propanone Hydrochloride. The sedative effect of propiomazine is utilized to provide night-time, presurgical or obstetrical sedation of short duration and ready arousal. The drug enhances the effect of other central nervous system depressants; therefore, the dose of such agents should be reduced in the presence of propiomazine. Propiomazine is given by either the intravenous or intramuscular route in doses of 10 to 40 mg.; the solution contains 20 mg. per ml., together with preservatives and buffer salts.

Category—sedative.

OCCURRENCE

Propiomazine Hydrochloride Injection N.F.

DIPHENLYMETHANE DERIVATIVES

The diphenylmethane derivatives are a group of drugs with diverse pharmacologic actions. However, the sedative properties shown by many members of this series more closely resemble the sedation produced by the tranquilizing agents than by the sedative-hypnotics. Table 34 lists those compounds whose sedative or tranquilizing properties have led to their use as psychotherapeutic or calming agents in the treatment of a variety of emotional or mental disorders characterized by tension, anxiety and agitation. In general, this group of drugs is not effective against the major psychoses. Some possess significant antihistaminic properties (hydroxyzine, buclizine, chlorcyclizine, captodiamine), while others (benactyzine) show pharmacologic properties and side-effects which are primarily anticholinergic. This combination of anticholinergic, antihistaminic and tranquilizing properties has also been noted for the phenothiazine derivatives.

Products

Hydroxyzine Hydrochloride N.F., Atarax® Hydrochloride, Vistaril®, 1-(p-Chlorobenzhydryl)-4-[2-(2-hydroxyethoxy)ethyl] piperazine Dihydrochloride; 2-[2-[4-(p-Chloro-α-phenylbenzyl)-1-piperazinyl]ethoxy]ethanol Dihydrochloride. Hydroxyzine hydrochloride is useful for the management of neuroses with agitation and anxiety as characterizing features. It is of little use in frank psychoses or depressive states. In addition to its sedative effects, the drug possesses antihistaminic properties useful in the management of acute and chronic urticaria and other allergic states (see Chap. 23). Anticholinergic properties have been demonstrated pharmacologically.

Hydroxyzine hydrochloride is a white solid, very soluble in water and in ethanol. The usual oral dose for its calming effect ranges from 25 to 100 mg., 3 or 4 times daily. Intramuscular or intravenous administration may be used for emergencies where rapid onset of response is necessary.

Category—minor tranquilizer.

OCCURRENCE

Hydroxyzine Hydrochloride Injection N.F.
Hydroxyzine Hydrochloride Syrup N.F.
Hydroxyzine Hydrochloride Tablets N.F.

The salt of hydroxyzine with pamoic acid (1,1'-methylene bis-[2-hydroxy-3-naphthalene carboxylic acid]) is **Hydroxyzine Pamoate N.F.** (Vistaril®), used orally for the central depressant and antihistaminic properties of hydroxyzine.

OCCURRENCE

Hydroxyzine Pamoate Capsules N.F.
Hydroxyzine Pamoate Oral Suspension N.F.

Buclizine, chlorcyclizine and captodiamine resemble hydroxyzine in indications, side-reactions and potency.

Benactyzine Hydrochloride, Suavitil®, 2-Diethylaminoethyl Benzilate Hydrochloride. Benactyzine is an anticholinergic compound with about one fourth the peripheral activity of atropine. In a dose which produces little peripheral effect, benactyzine is useful in the management of psychoneurotic disorders characterized by anxiety and tension. The usual oral dose of benactyzine hydrochloride ranges from 1 to 3 mg., three times daily.

MISCELLANEOUS

There is also another group of tranquilizing agents derived by isosteric replacement of one or more groups or atoms in the structure of the phenothiazine tranquilizing agents. The compounds thus derived possess many clinically useful pharmacologic properties in common with phenothiazine tranquilizers.

Chlorprothixene N.F., Taractan®, *cis*-2-Chloro-9-(3-dimethylaminopropylidene)-thioxanthene. Chlorprothixene, an isostere of chlorpromazine in which nitrogen is replaced with a methylene group, was released in 1961 for use as a psychothera-

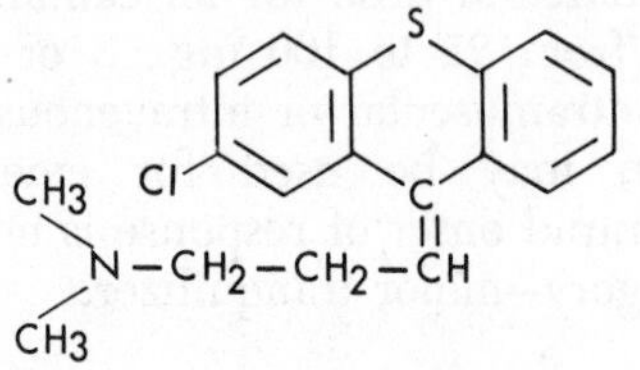

Chlorprothixene

peutic drug. It appears to be effective in the treatment of schizophrenia, and in psychotic and severe neurotic conditions characterized by anxiety and agitation, thus resembling the phenothiazines. In addition, it is claimed to exert some benefit in depressive states. Chlorprothixene potentiates the effect of sedatives, has a hypotensive effect and shows antihistaminic and antiemetic properties.

The oral dose may range from 30 to 60 mg. per day. The intramuscular dose is 12.5 to 25 mg. for antiemetic effects and ranges

from 75 to 200 mg. per day for the agitated patient.

Category—major tranquilizer.

OCCURRENCE
Chlorprothixene Tablets N.F.

Thiothixene, Navane®, *cis*-N,N-Dimethyl-9-[3-(4-methyl-1-piperazinyl)propylidene]-

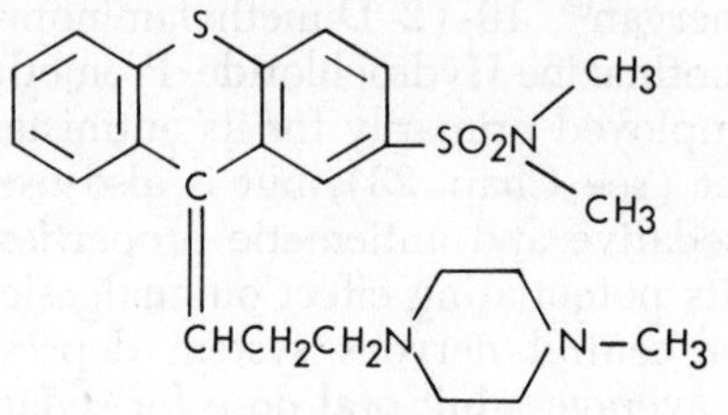

Thiothixene

thioxanthene-2-sulfonamide. Thiothixene, a thioxanthene derivative related to chlorprothixene, was introduced in 1967 as an antipsychotic agent useful in the management of schizophrenia and other psychotic states. It also shows antidepressant properties. Thiothixene is similar in its actions to the phenothiazine tranquilizers and may potentiate the actions of the central nervous system depressants including anesthetics, hypnotics, and alcohol. At higher dosage levels it may produce extrapyramidal symptoms and orthostatic hypotension.

The usual daily adult dose ranges from 8 to 25 mg. in divided doses. The maximum recommended daily dose is 60 mg.

Cyproheptadine Hydrochloride N.F., Periactin®, 1-Methyl-4(5-dibenzo[a,e]cyclo-heptenylidene)piperidine Hydrochloride; 4-(5H-Dibenzo[a,d]cyclohepten-5-ylidene)-1-methylpiperidine Hydrochloride.

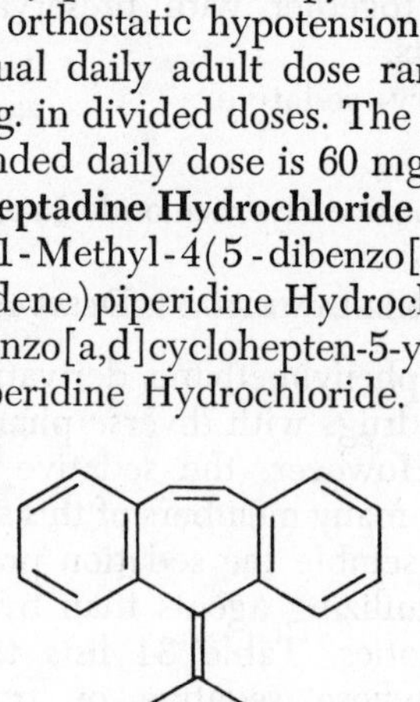

Cyproheptadine Hydrochloride

Cyproheptadine, a potent serotonin and histamine antagonist, is used primarily to relieve itching, and in the treatment of both chronic and acute allergic conditions (see Chap. 23). The usual adult dose ranges from 12 to 16 mg. a day in divided doses.

Fluorobutyrophenones

A series of related fluorobutyrophenones, derived from studies in Europe on potential analgesics,[27] were found to be effective in the management of major psychoses. The first compound of the series, haloperidol, was introduced in the United States in 1967.

Haloperidol N.F., Haldol®, 4-[4-(*p*-Chlorophenyl)-4-hydroxypiperidino]-4'-fluorobutyrophenone. Haloperidol is a major tranquilizer. It is used in the management of the agitated states, as well as mania, aggressiveness, and hallucinations associated with acute and chronic psychoses including schizophrenia and psychotic reactions in adults with organic brain damage.

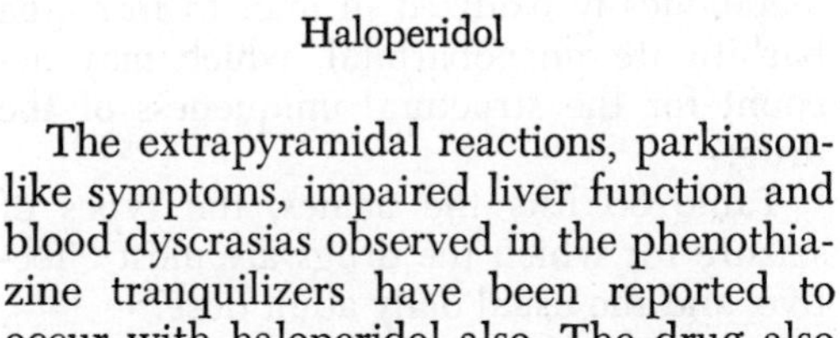

Haloperidol

The extrapyramidal reactions, parkinson-like symptoms, impaired liver function and blood dyscrasias observed in the phenothiazine tranquilizers have been reported to occur with haloperidol also. The drug also potentiates the actions of central nervous system depressant drugs such as analgesics, anesthetics, barbiturates and alcohol.

The initial oral dose is 1 to 2 mg. administered 2 or 3 times daily; the maximum recommended daily dose is 15 mg. The onset of action occurs in 5 to 15 minutes, and duration of action is from 3 to 5 hours.

Category—major tranquilizer.

OCCURRENCE
Haloperidol Solution N.F.
Haloperidol Tablets N.F.

Droperidol, Inapsine®, 1-{1-[3-(*p*-Fluorobenzoyl)propyl]-1,2,3,6-tetrahydro-4-pyridyl}-2-benzimidazolinone. Droperidol, a fluorobutyrophenone tranquilizer, is used together with the potent narcotic analgesic,

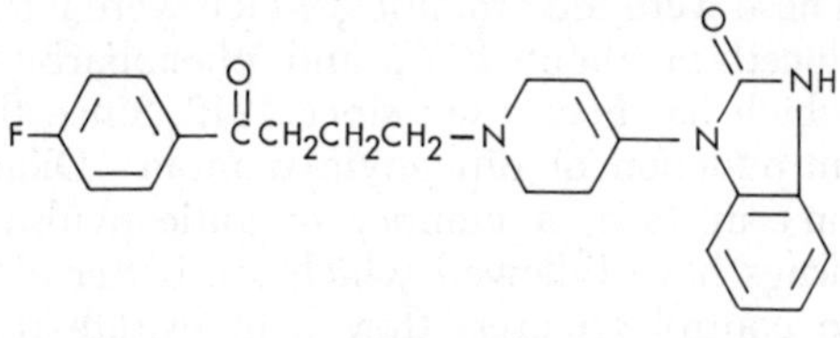

Droperidol

fentanyl (Sublimaze) (see Chap. 24). The combination (Innovar) is administered by the intramuscular or intravenous routes for preanesthetic sedation and analgesia, and as an adjunct to the induction of anesthesia. For sedation and analgesia, the usual intramuscular dose is 0.5 to 2 ml., each ml. containing 0.05 mg. of fentanyl and 2.5 mg. of droperidol. As an adjunct to the induction of anesthesia, the usual intravenous dose is 1 ml. for each 20 to 25 pounds of body weight.

A related compound, *trifluperidol*, is under clinical study. It appears to be useful in the treatment of schizophrenia and paranoia, at a usual dose range of 1 to 2.5 mg. daily.

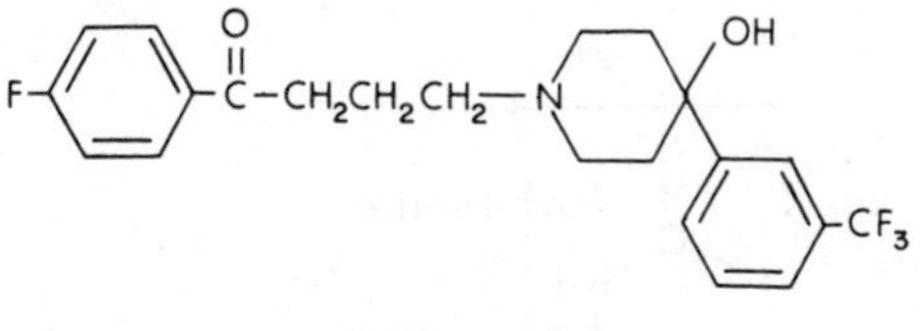

Trifluperidol

ANTICONVULSANT DRUGS

The primary use of anticonvulsant drugs is in the prevention and the control of epileptic seizures. In most cases treatment is symptomatic; only in a few cases suitable for surgery is a cure possible.

The disease affects approximately from 0.5 to 1.0 per cent of the population. Since symptomatic treatment is frequently lifelong, and a feeling of inferiority and self-consciousness often causes a withdrawal from society, this disease constitutes a major public health problem.

Until recently, only two useful drugs were available which could depress the motor cortex (prevent convulsions) as well as the sensory cortex (produce sleep). These were the bromides, which were introduced in about 1857, and phenobarbital, which has been used since 1912. Since the introduction of diphenylhydantoin (Dilantin) in 1938, a number of anticonvulsant drugs have followed which are better able to control seizures; they demonstrate that sedation may be dissociated from anticonvulsant activity in the various types of epilepsy. Each type of epilepsy may be distinguished by clinical and electroencephalographic patterns, and each responds differently to the various classes of anticonvulsant drugs. The major types of epileptic seizures are:

1. *Grand Mal.* Sudden loss of consciousness followed by general muscle spasms

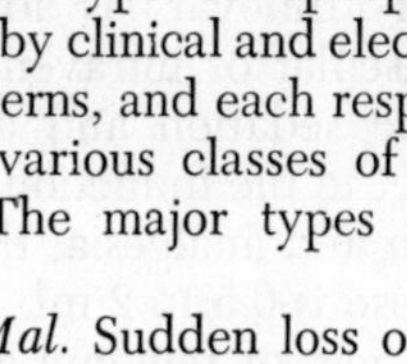

Structure common to anticonvulsant drugs.

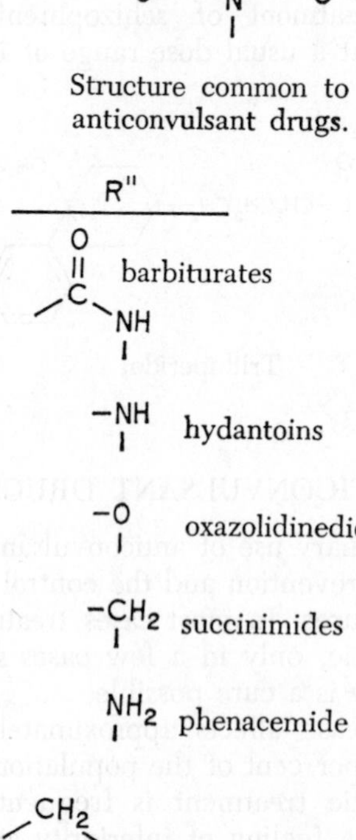

R''	
$\underset{C}{\overset{O}{\underset{\diagdown NH}{\parallel}}}$	barbiturates
$-\underset{\mid}{NH}$	hydantoins
$-\underset{\mid}{O}$	oxazolidinediones
$-\underset{\mid}{CH_2}$	succinimides
$\underset{\mid}{NH_2}$	phenacemide
$\underset{CH_2}{\overset{CH_2}{\diagup}}\diagdown CH_2$	glutarimides

lasting for an average of 2 to 5 minutes. The frequency and the severity of attacks are variable.

2. *Petit Mal.* Sudden, brief loss of consciousness with minor movements of the head, the eyes and the extremities, lasting for about 5 to 30 seconds. The patient is immediately alert, ready to continue normal activity. There may be many episodes in a day; the highest incidence is found in children.

3. *Psychomotor Seizures.* Automatic, patterned movements lasting from 2 to 3 minutes occur. Amnesia is common, with often no memory of the incident remaining. This state is sometimes confused with psychotic behavior.

Compounds being tested in the laboratory for anticonvulsant activity are assayed for protection against convulsions induced both chemically and electrically. Clinically useful drugs are usually effective in elevating the threshold to seizures produced by the central nervous system stimulant pentylenetetrazol (Metrazol; see Chap. 16) or by electroshock.

The drugs acting as selective depressants of convulsant activity have a common structural feature, as shown.

The single exception to this structural pattern among currently useful anticonvulsant drugs is primidone (Mysoline). However, it is known[28, 29] that primidone is metabolically oxidized in man to form the barbiturate phenobarbital, which may account for the structural uniqueness of the drug.

Table 35 lists the names, the types of seizure for which the drugs are most effective, and the usual daily adult dose.

BARBITURATES

Of the commonly employed barbiturates, only phenobarbital, mephobarbital (Mebaral) and metharbital (Gemonil) show the selective anticonvulsant activity which makes them useful in the symptomatic treatment of epilepsy. The mechanism by which these drugs reduce the excitability of the motor cortex is unknown. The structures of these barbiturates, all members of the long-acting class, are listed in Table 28.

TABLE 35. DRUGS USED IN THE TREATMENT OF EPILEPSY

DRUG	TYPES OF SEIZURE	AVERAGE ADULT DOSE, DAILY (GM.)
I. Barbiturates		
Phenobarbital	Grand mal	0.09-0.13
Mephobarbital	Grand mal	0.40-0.60
Metharbital	Grand mal	0.10-0.30
II. Hydantoins		
Diphenylhydantoin	Grand mal*	0.30-0.40
Mephenytoin	Grand mal*	0.20-0.60
Ethotoin	Grand mal*	2.0 -3.0
III. Oxazolidinediones		
Trimethadione	Petit mal	1.0 -2.0
Paramethadione	Petit mal	0.6 -1.2
IV. Succinimides		
Phensuximide	Petit mal	1.0 -3.0
Methsuximide	Petit mal*	0.3 -1.2
Ethosuximide	Petit mal	1.0 -1.5
V. Miscellaneous		
Primidone	Grand mal*	0.5 -1.5
Amino-glutethimide	General	0.75-1.5
Phenacemide	General	2.0 -3.0

* Some effectiveness against psychomotor seizures.

TABLE 36. THE ANTICONVULSANT HYDANTOIN DERIVATIVES

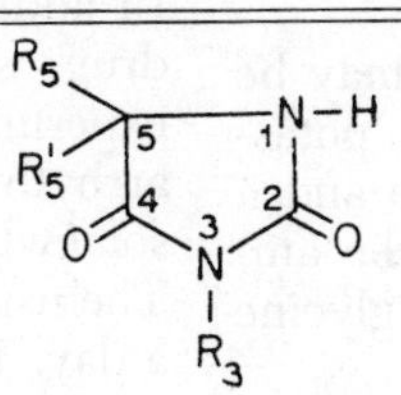

Generic name Proprietary name	Substituents			Year of Introduction
	R_5	R'_5	R_3	
Phenylethylhydantoin Nirvanol	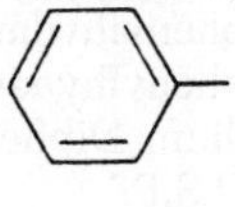 (phenyl)	CH_3-CH_2-	H	1917
Diphenylhydantoin U.S.P. Dilantin, Diphentoin	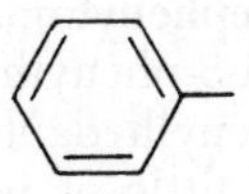 (phenyl)	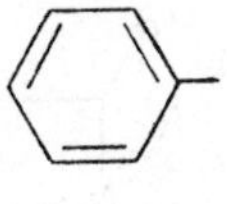 (phenyl)	H	1938
Mephenytoin Mesantoin	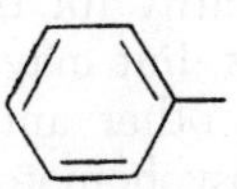 (phenyl)	CH_3-CH_2-	CH_3-	1947
Ethotoin Peganone	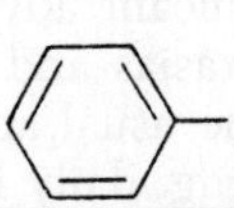 (phenyl)	H	CH_3-CH_2-	1957

The anticonvulsant activity of the barbiturates is not related to the sedation they produce, for protection from convulsions is often shown at nonsedating dose levels. The three anticonvulsant barbiturates show a high degree of effectiveness in grand mal but are of little or no benefit in petit mal and psychomotor epilepsy. Average adult daily doses are presented in Table 35.

HYDANTOINS

As cyclic ureides, related in structure to the barbiturates, many hydantoins were synthesized as potential hynotics following the introduction of the barbiturates in 1903. The first of the hydantoins, nirvanol, was introduced as a hypnotic and anticonvulsant in 1914 but has since been replaced by less toxic analogs. Following systematic pharmacologic and clinical studies, 5-5-diphenylhydantoin was reported[30] in 1938 to be the least hypnotic and most strongly anticonvulsant of the related compounds studied. The hydantoins are most effective against grand mal; psychomotor attacks are sometimes controlled. These drugs are ineffective against petit mal.

The 5,5-disubstituted hydantoins may be prepared by the reaction between potassium cyanide, ammonium carbonate and a ketone, or by the condensation of the ammonium salt of α,α-disubstituted glycine and phosgene.

The cyclic ureide structure exists in equilibrium with its enolic form, 2,4-dihydroxy-5,5-disubstituted imidazole. Sodium, or other metal salts of the acidic, 2-hydroxyl group may be formed in alkaline solution; the insoluble free-acid form is formed in the presence of acid.

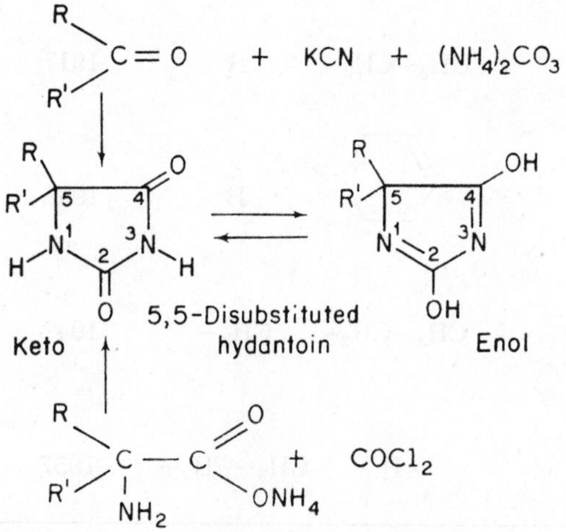

The names, the structures and the years of introduction for the anticonvulsant hydantoins are listed in Table 36. The average daily doses are listed in Table 35.

Diphenylhydantoin U.S.P., Dilantin®, Diphentoin®, 5,5-Diphenyl-2,4-imidazolidinedione, 5,5-Diphenylhydantoin. Diphenylhydantoin is an anticonvulsant with little or no sedative properties. It is most effective in controlling grand mal seizures when used alone or in combination with phenobarbital. Various untoward effects have been observed; these include dizziness, skin rashes, itching, tremors, fever, vomiting, blurred vision, difficult breathing and hyperplasia of the gums.

Diphenylhydantoin is a white powder, practically insoluble in water, and slightly soluble in alcohol. The water-soluble and somewhat hygroscopic sodium salt is available as **Sodium Diphenylhydantoin, U.S.P.** In addition to its use as an anticonvulsant drug, sodium diphenylhydantoin is used experimentally for the control of cardiac arrhythmia. Its aqueous solution is usually somewhat turbid due to partial hydrolysis. The usual oral dose is 100 mg. up to 4 times a day. The parenteral dose is 100 to 200 mg.

OCCURRENCE

Sterile Sodium Diphenylhydantoin U.S.P.
Diphenylhydantoin Capsules U.S.P.
Diphenylhydantoin Oral Suspension N.F.
Diphenylhydantoin Tablets U.S.P.
Sodium Diphenylhydantoin Capsules
U.S.P.

Mephenytoin, Mesantoin®, 3-Methyl-5-ethyl-5-phenylhydantoin. Mephenytoin, like diphenylhydantoin, is an anticonvulsant with little or no sedative effect. It is used primarily for the control of grand mal seizures, but may also be used in conjunction with other anticonvulsants for the control of psychomotor and jacksonian seizures. Significant adverse reactions include blood dyscrasias and skin rashes.

The usual adult dose ranges from 200 to 600 mg. daily in divided doses.

Ethotoin, Peganone®,3-Ethyl-5-phenylhydantoin. Ethotoin is used primarily for the control of grand mal epilepsy. Adverse reactions are like those of related hydantoins: blood dyscrasias, skin rash, ataxia, and gum hypertrophy. The drug may be used alone, but it is most frequently used in combination with other anticonvulsants.

The usual adult dose ranges from 400 to 600 mg. in 4 to 6 divided doses, taken after food.

OXAZOLIDINEDIONES

The oxazolidine-2-4-diones, compounds isosterically related to the hydantoins by substitution of an oxygen for nitrogen were first tested as hypnotics in 1938.[31] The most active anticonvulsant drugs of this type (3,5,5-trialkyloxazolidine-2,4-diones) may be synthesized by condensation of an ester

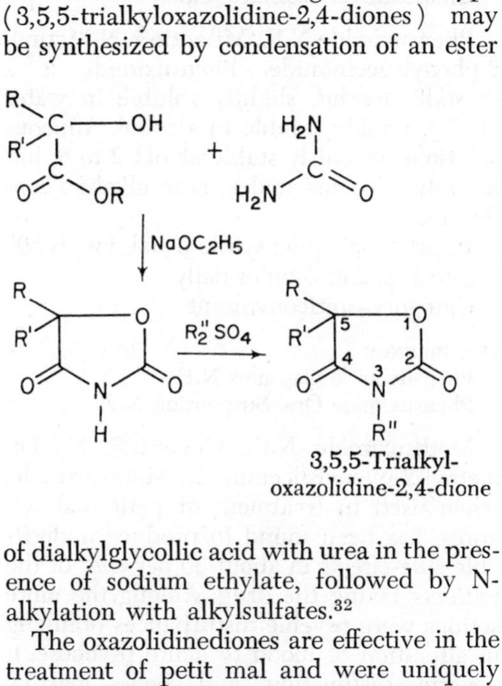

of dialkylglycollic acid with urea in the presence of sodium ethylate, followed by N-alkylation with alkylsulfates.[32]

The oxazolidinediones are effective in the treatment of petit mal and were uniquely so when first introduced in 1946. They are ineffective against grand mal but are used in conjunction with other drugs in the treatment of mixed types of seizures, such as combined petit mal and grand mal epilepsy.

Trimethadione U.S.P., Tridione®, 3,5,5-Trimethyl-2,4-oxazolidinedione. Trimethadione was introduced in 1946 for use in the treatment and the prevention of epileptic seizures of the petit mal type. Toxic effects

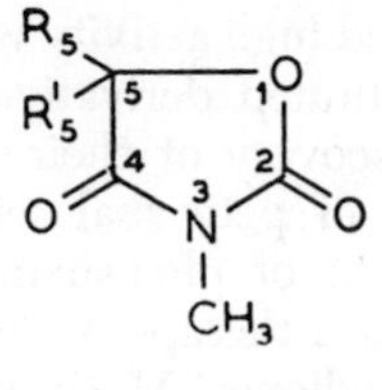

| Trimethadione | $R_5 = R'_5 = CH_3$ |
| Paramethadione | $R_5 = CH_3; R'_5 = C_2H_5$ |

experienced with the drug include gastric irritation, nausea, skin eruptions, sensitivity to light, disturbances of vision and dizziness and drowsiness. Infrequent reports of aplastic anemia and nephrosis indicate the need for routine blood and urine examinations.

Trimethadione is a white, granular substance with a weak camphorlike odor. It is soluble in water or alcohol, giving a slightly acidic solution.

Category—anticonvulsant.

Usual dose—300 to 600 mg. 3 or 4 times daily.

Usual dose range—900 mg. to 2.4 g. daily.

OCCURRENCE
Trimethadione Capsules U.S.P.
Trimethadione Solution U.S.P.
Trimethadione Tablets N.F.

Paramethadione U.S.P., Paradione®, 5-Ethyl-3-5-dimethyl-2,4-oxazolidinedione. Paramethadione, introduced in 1947, has the same use and side-effects as trimethadione, although individual variation may show one to be effective in patients in which the other is ineffective.

The compound is an oily liquid, slightly soluble in water but readily soluble in alcohol.

Category—anticonvulsant.

Usual dose—300 mg. 3 to 4 times a day.

Usual dose range—300 mg. to 2.1 g. daily.

OCCURRENCE
Paramethadione Capsules U.S.P.
Paramethadione Solution U.S.P.

SUCCINIMIDES

Extensive screening for anticonvulsant activity among aliphatic and heterocyclic amides revealed high activity within a series of α,N-disubstituted derivatives of succinimide. The discovery of their usefulness in the treatment of petit mal seizures led to the introduction of phensuximide (Milontin) in 1953 as a therapeutic companion to the oxazolidinediones. Methsuximide (Celontin) followed in 1958, and ethosuximide (Zarontin) in 1960.

The substituted succinimides are prepared by reaction of the derivative of succinic acid with ammonia or an alkyl amine. The preparation of phensuximide from α-phenylsuccinic acid is shown.[33]

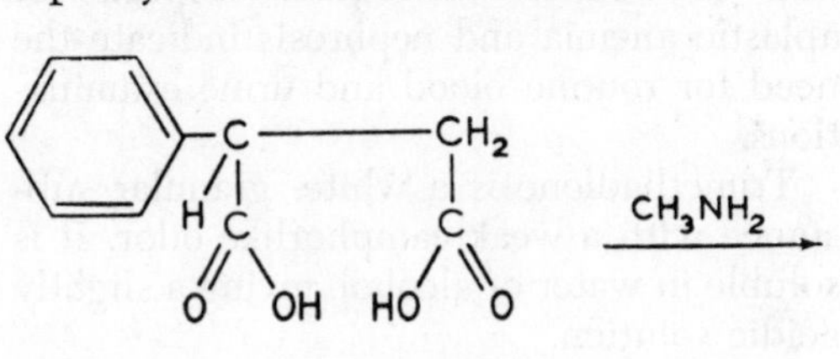

α-Phenylsuccinic Acid

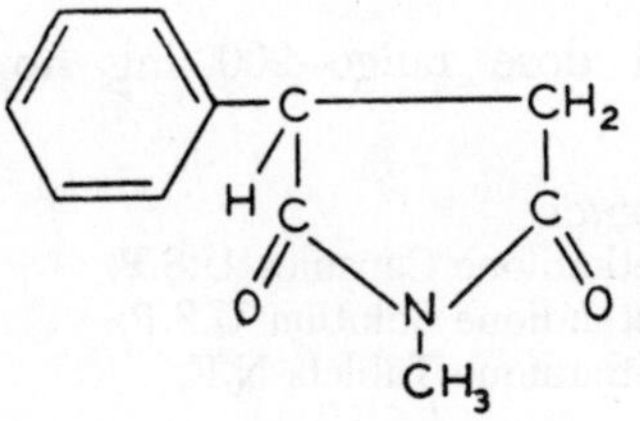

Phensuximide

The succinimides appear to be less potent than the oxazolidinediones and to possess significant side effects. Periodic blood and urine studies are advisable during treatment. These drugs are moderately effective in the control of petit mal seizures but are ineffective against grand mal. They are administered orally.

Ethosuximide U.S.P., Zarontin®, 2-Ethyl-2-methylsuccinimide. Ethosuximide has been shown to be effective in pure petit mal but less effective in mixed petit mal. Early reports show that the development of refractiveness to treatment may be slower with ethosuximide. Cases of agranulocytosis, cytopenia and bone marrow depression have been reported. The oral dose of 250-mg. capsules must be adjusted according to pa-

Phensuximide   R= [phenyl] , R' = H, R" = CH$_3$

Methsuximide   R= [phenyl] , R' = CH$_3$, R" = CH$_3$

Ethosuximide   R = C$_2$H$_5$- , R' = CH$_3$, R" = H

tient response. The usual daily dose is 250 mg. 2 or 3 times daily.

Category—anticonvulsant.

OCCURRENCE
Ethosuximide Capsules U.S.P.

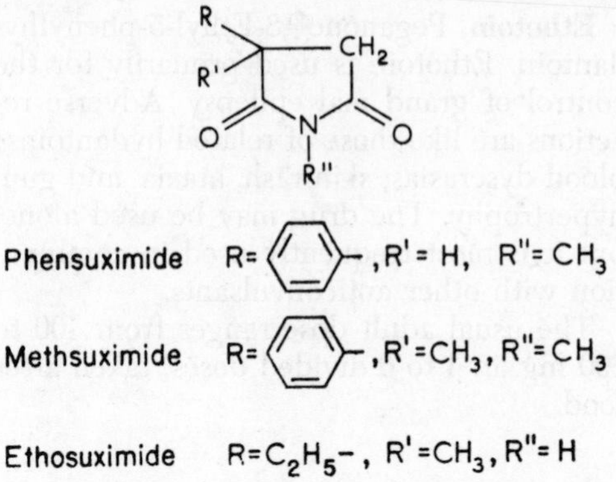

Phensuximide N.F., Milontin®, N-Methyl-2-phenylsuccinimide. Phensuximide is a crystalline solid, slightly soluble in water (0.4%), readily soluble in alcohol. Aqueous solutions are fairly stable at pH 2 to 8, but hydrolysis occurs under more alkaline conditions.

In petit mal epilepsy the usual dose is 500 mg. to 1 g., 2 or 3 times daily.

Category—anticonvulsant.

OCCURRENCE
Phensuximide Capsules N.F.
Phensuximide Oral Suspension N.F.

Methsuximide N.F., Celontin®, N,2-Dimethyl-2-phenylsuccinimide. Methsuximide, when used in treatment of petit mal seizures, has been found to produce undesirable side-effects in about 30 per cent of the patients taking the drug. Among the more serious were psychic disturbances (ranging in alteration of mood to acute psychoses), hepatic disfunction and bone marrow aplasia.

The drug has shown usefulness in a significant number of cases of psychomotor seizures. Physical properties are like those of phensuximide.

The usual oral dose is 300 mg. daily, increased gradually up to 1.2 g. daily.

Category—anticonvulsant.

OCCURRENCE
Methsuximide Capsules N.F.

MISCELLANEOUS

Several compounds of miscellaneous structure show useful anticonvulsant properties. All possess structural features common to the class, but each chemical type, as yet, is represented by a single member.

Primidone U.S., Mysoline®, 5-Ethyldihydro-5-phenyl-4,6-(1H,5H)-pyrimidinedione; 5-phenyl-5-ethylhexahydropyrimidine-

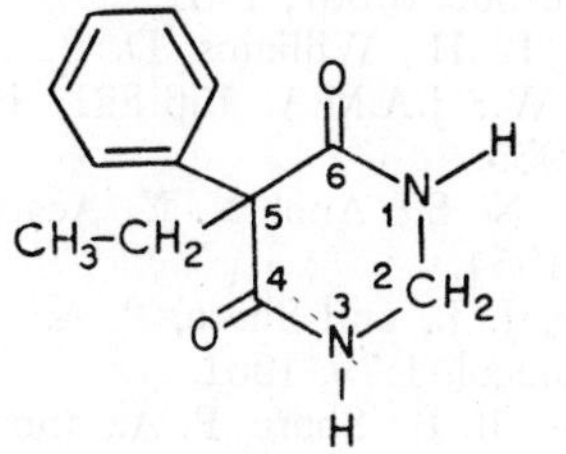

Primidone

4,6-dione. Primidone, a 2-deoxy analog of phenobarbital, was synthesized in 1949[34] and introduced in 1954 for use in the control of grand mal and psychomotor epilepsy. It is prepared by the reductive desulfurization of 5-ethyl-5-phenylthiobarbituric acid. The high incidence (20%) of drowsiness, and the anticonvulsant effects of primidone may be due to its oxidation to phenobarbital. This has been shown to occur to the extent of about 15 per cent in man.[29]

Although in general the toxicity of primidone is low, a few cases of megaloblastic anemia have been associated with its use.

Primidone is a white, odorless, crystalline powder. It has low solubility in water (1:2,000) and in alcohol (1:200).

Usual dose—250 to 500 mg. 3 times daily.

Usual dose range—125 mg. to 2 g. daily.

OCCURRENCE

Primidone Oral Suspension U.S.P.
Primidone Tablets U.S.P.

Phenacemide, Phenurone®, Phenylacetylurea. Phenacemide was introduced in 1951 for the treatment of psychomotor, grand mal and petit mal epilepsies and in mixed seizures. Serious side-effects associated with the use of phenacemide include personality changes (suicide attempts and toxic psychoses), fatalities attributed to liver damage, and bone marrow depression. Its

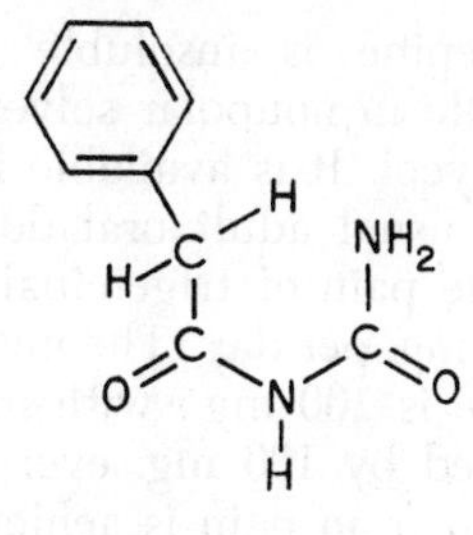

Phenacemide

unique effectiveness in the control of psychomotor seizures may indicate its use in spite of these hazards, but only after other drugs have been found to be ineffective.

Phenacemide is an odorless and tasteless, white, crystalline solid. It is very slightly soluble in water and slightly soluble in alcohol. The usual oral dose is 1.5 to 2.5 g. daily. The common dosage form is 500 mg. tablets.

Carbamazepine, Tegretol®, 5H-Dibenz-(b,f)-azepine-5-carboxamide. Carbamazepine contains the dibenzazepine ring system of the psychotherapeutic drug imipramine. It differs from imipramine in having the double bond in the 10,11-position of the ring, and in having the carboxamide side chain rather than the dimethylamino propyl group.

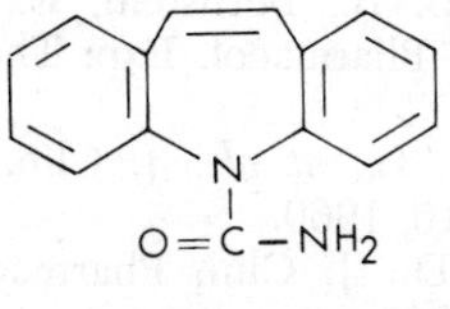

Carbamazepine

Carbamazepine was introduced in 1968 for relief of pain of trigeminal neuralgia (*tic douloureux*). The anticonvulsant drug diphenylhydantoin has proved useful in some cases in the treatment of trigeminal neuralgia, perhaps by producing an increased threshold to the sensory discharge of the trigeminal nerve; carbamazepine is thought to act in a similar way.

Carbamazepine is an effective anticonvulsant agent that is particularly useful in the control of psychomotor epilepsy, and is under investigation in the treatment of epilepsy.

Carbamazepine is insoluble in water, readily soluble in nonpolar solvents and in propylene glycol. It is available in 200-mg. tablets. The usual adult oral dose for the control of the pain of trigeminal neuralgia is 400 to 800 mg. per day. The initial recommended dose is 100 mg., with meals, with dose increased by 100 mg. every 12 hours until freedom from pain is achieved.

One of the most important factors for persons susceptible to seizures is the control of living conditions. States favorable for prevention of seizures include dehydration, systemic acidosis, adequate oxygen and freedom from stress. Therefore, adjuncts to anticonvulsant therapy include drugs that produce acidosis, such as glutamic acid, and those that produce both acidosis and dehydration, such as the carbonic anhydrase inhibiting diuretics, e.g., acetazolamide (Diamox), ethoxzolamide (Cardrase) (see Chap. 20).

REFERENCES CITED

1. Roth, L. J., and Barlow, C. F.: Science 134:22, 1961.
2. Suckling, C. W.: Brit. J. Anaesth. 29:466, 1957.
3. Cohen, E. N., *et al.*: Science 141:899, 1963.
4. Brodie, B. B., Bernstein, E., and Mark, L. C.: J. Pharmacol. Exp. Ther. 105:421, 1952.
5. Price, H. L., *et al.*: J. Clin. Pharmacol. Ther. 1:16, 1960.
6. Friend, D.: J. Clin. Pharmacol. Ther. 1 (6):5, 1960.
7. Hansch, C., *et al.*: J. Med. Chem. 11:1, 1968.
8. Goodman, L. S., and Gilman, A.: The Pharmacological Basis of Therapeutics, p. 290, New York, Macmillan, 1970.
9. Rogers, G. A.: Am. J. Psychiat. 115:551, 1958.
10. Bonnet, H., *et al.*: J. Med. Lyon 39:924, 1958.
11. Domino, E. F.: J. Pharmacol. Exp. Ther. 115:449, 1955.
12. Schallek, W., Kuehn, A., and Seppelin, D. K.: J. Pharmacol. Exp. Ther. 118:139, 1956.
13. Swinyard, E. A., and Chin, L.: Science 125:739, 1957.
14. Essig, C. F., and Ainslie, J. D.: J.A.M.A. 164:1382, 1957.
15. Berger, F. M.: Ann. N. Y. Acad. Sci. 67: 685, 1957.
16. Jacobsen, E.: J. Pharm. Pharmacol. 10: 282, 1958.
17. Burbridge, T. N.: A Pharmacologic Approach to the Study of the Mind, Springfield, Thomas, 1959.
18. Schildkraut, J. J., and Kety, S. S.: Science 156:21, 1967.
19. Vakil, R. J.: Brit. Heart J. 11:350, 1949.
20. Siddiqui, S., and Siddiqui, R. H.: J. Indian Chem. Soc. 8:667, 1931.
21. Noce, R. H., Williams, D. B., and Rapaport, W.: J.A.M.A. 156:821; 1954; 158: 11, 1955.
22. Kline, N. S.: Ann. N. Y. Acad. Sci. 59: 107, 1954.
23. Burns, J. J., and Shore, P. A.: Ann. Rev. Pharmacol. 1:79, 1961.
24. Brodie, B. B., Shore, P. A., and Pletscher, A.: Science 123:992, 1956.
25. Gordon, M., Craig, R. N., and Zirkle, C. L.: Molecular Modification in the Development of Phenothiazine Drugs, Molecular Modification in Drug Design, Advances in Chemistry Series 45, Am. Chem. Soc., Washington, D.C., 1964.
26. Bloom, B. M., and Laubach, G. D.: Ann. Rev. Pharmacol. 2:69, 1962.
27. Janssen, P. A. J., *et al.*: J. Med. Pharm. Chem. 1:281, 1959.
28. Butler, T. C., and Waddell, W. S.: Proc. Soc. Exp. Biol. Med. 93:544, 1956.
29. Plaa, G. L., Fujimoto, J. M., and Hine, C. H.: J.A.M.A. 168:1769, 1958.
30. Merritt, H. M., and Putnam, T. J.: Arch. Neurol. Psychiat. 39:1003, 1938; Epilepsia 3:51, 1945.
31. Erlenmeyer, H.: Helv. chim. acta 21: 1013, 1938.
32. Spielman, M. A.: J. Am. Chem. Soc. 66: 1244, 1944.
33. Miller, C. A., and Long, L. M.: J. Am. Chem. Soc. 73:4895, 5608, 1951; 75:373, 6256, 1953.
34. Bogue, J. Y., and Carrington, H. C.: Brit. J. Pharmacol. 8:230, 1953.

SELECTED READING

Braceland, F. J. (ed.): The Effect of Pharmacologic Agents on the Nervous System, Baltimore, William & Wilkins, 1959.
Brodie, B. B., Sulser, F., and Costa, E.: Psychotherapeutic drugs, Ann. Rev. Med. 12: 349, 1961.
Burger, A. (ed.): Drugs Affecting the Central Nervous System, Medicinal Research, Vol. 2, New York, Marcel Dekker, 1968.

Delgado, J. N., and Isaacson, E. I.: Anticonvulsants, *in* Burger, A. (ed.): Medicinal Chemistry, ed. 3, p. 1886, New York, Wiley-Interscience, 1970.

Domino, E. F.: Human pharmacology of tranquilizing drugs, Clin. Pharmacol. Ther. 3:599, 1962.

Gordon, M.: Psychopharmacological Agents, Academic Press, New York, vols. 1 and 2, 1964 and 1965.

Jucker, E.: Some new developments in the chemistry of psychotherapeutic agents, Angewandte Chemie, Internat. Ed. 2:492, 1963.

McIlwain, H.: Chemotherapy and the Central Nervous System, London, Churchill, 1955.

Mautner, H. G., and Clemson, H. C.: Hypnotics and Sedatives, *in* Burger, A. (ed.): Medicinal Chemistry, ed. 3, p. 1365, New York, Wiley-Interscience, 1970.

Patel, A. R.: General Anesthetics, *in* Burger, A. (ed.): Medicinal Chemistry, ed. 3, p. 1314, New York, Wiley-Interscience, 1970.

Toman, J. E. P., and Goodman, H. S.: Anticonvulsants, Pharmacol. Rev. 28:409, 1948.

Zirkle, C. L., and Kaiser, C.: Antipsychotic Agents, *in* Burger, A. (ed.): Medicinal Chemistry, ed. 3, p. 1410, New York, Wiley-Interscience, 1970.

16

Central Nervous System Stimulants

T. C. Daniels, Ph.D.,

Emeritus Professor of Pharmaceutical Chemistry, School of Pharmacy,
University of California

and

E. C. Jorgensen, Ph.D.,

Professor of Chemistry and Pharmaceutical Chemistry, School of Pharmacy,
University of California

The central nervous system is a complex network of subunits which act as conducting pathways between peripheral receptors and effectors, enabling man to respond to his environment. It also adjusts behavior to the quality and the intensity of stimuli and coordinates activities, providing a unified set of actions. Drugs which have in common the property of increasing the activity of various portions of the central nervous system are called central nervous system stimulants.

Until recently, the major therapeutic applications of central nervous system stimulants were their use as respiratory stimulants and analeptics. Respiratory stimulation may be brought about not only by the action of drugs directly upon the respiratory center of the medulla but by pH changes in the blood which supplies the center. Carbonic acid is most effective in this manner, and carbon dioxide, in some cases, is the respiratory stimulant of choice. Stimulation of the chemoreceptor of the carotid body (cyanide), afferent impulses from sensory stimuli (ammonia inhalation) and higher centers (visual stimuli) may affect respiration through the respiratory center.

Analeptics, agents used to lessen narcosis brought about by excess of depressant drugs, often stimulate a variety of other centers as well. The vasomotor center, which maintains the constriction of the blood vessel walls, is frequently affected. Many analeptic drugs are also pressor drugs because their stimulation of the vasomotor center produces an increase in vasoconstriction. The resulting increased peripheral resistance to blood flow causes an elevation of blood pressure.

Stimulation of the emetic center, also located in the medulla, is not infrequent with therapeutic doses of many drugs. A few drugs, such as apomorphine, apparently exert a selective effect on the emetic chemoreceptor trigger zone of the medulla and may be used as emetics in the treatment of poisoning.

In the sense that an effect on the "appetite control center" may be classed as a central-stimulant effect, drugs classed as anorexigenic agents and used to decrease appetite in the control of obesity are included in this discussion.

There are many drugs of varying pharmacologic class which, in addition to their desired effects, elicit a pronounced stimulatory effect on the central nervous system. Examples are found in the local anesthetics (cocaine), parasympatholytics (atropine), sympathomimetics (many, including ephedrine, amphetamine, etc.), and as important toxic effects in high doses, salicylates, local anesthetics and many others.

In 1955, Goodman and Gilman[1] summarized the status of central stimulants.

Although the central nervous system stimulants are sometimes dramatic in their pharmacological effects, they are relatively unimportant from a therapeutic point of view. It is not possible to stimulate the nervous system over a long period of time, for heightened nervous activity is followed by depression, proportional in degree to the intensity and duration of the stimulation. Consequently, therapeutic excitation of the central nervous system is usually of brief duration and is reserved for emergencies characterized by severe central depression.

This statement was based on the respiratory stimulant and analeptic properties of available agents such as picrotoxin, metrazol, strychnine, nikethamide, camphor, and caffeine and the related xanthines. This view-point has been altered by the recognition that a more moderate and prolonged degree of central stimulation may be achieved in the treatment of patients with mental depression. This group of drugs is now called *Psychomotor Stimulants* and may be subdivided into the *Central Stimulant Sympathomimetics*, the *Antidepressants* (*Monoamine Oxidase Inhibitors, Tricyclic Antidepressants*), and a group of compounds of *miscellaneous* chemical and pharmacologic class.

ANALEPTICS

The following drugs are used chiefly as analeptics to counteract respiratory depression and coma resulting from overdosage of central depressant agents.

Picrotoxin N.F., cocculin, is the active principle from the seed ("fishberries") of the shrub *Anamirta cocculus*. It is a molecular compound easily separated into the component dilactones, the active picrotoxinin[2] and inactive picrotin.[3]

Picrotoxin is a powerful central nervous system stimulant. In its ability to reverse the central depression resulting from the administration of high doses of barbiturates, the drug is one of the most effective of the central nervous system stimulants (analeptics). Small doses are effective in reversing mild hypnotic intoxication such as that caused by overdoses of barbiturates, chlorobutanol, paraldehyde and tribromoethanol.

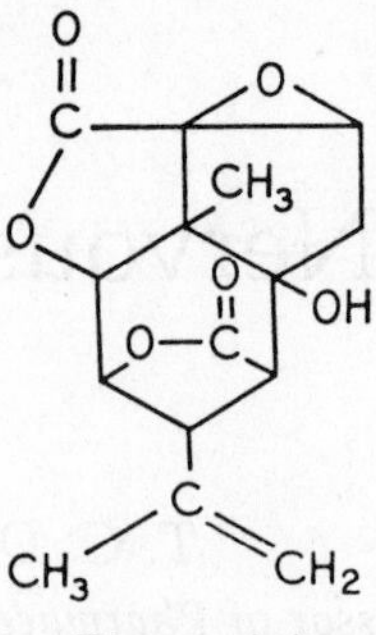

Picrotoxinin 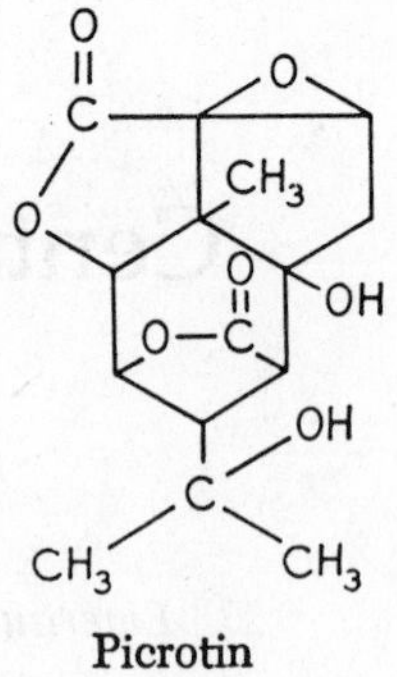   **Picrotin**

Seriously poisoned patients in coma receive 1 to 2 mg. per minute until corneal and swallowing reflexes return, followed by spaced intravenous injections. Due to the delay in twitching response, overdose leading to convulsions is possible, and a complex program of management is essential in addition to use of the analeptic. Some studies have shown that conservative, supportive therapy of barbiturate poisoning has resulted in a lower incidence of deaths than treatment involving analeptics such as picrotoxin and pentylenetetrazol.

Picrotoxin is a crystalline powder, stable in air but affected by light. One gram of picrotoxin dissolves in about 350 ml. of water to form a neutral solution. It is more readily soluble in dilute acid or alkali and is sparingly soluble in ether and chloroform. It is available as an injectable solution, usually containing 3 mg. per ml.

Category—central and respiratory stimulant.

Usual dose—intravenous: to be determined by the physician according to the needs of the patient.

OCCURRENCE
Picrotoxin Injection N.F.

Pentylenetetrazol, N.F., Metrazol®, 6,7,8,9-Tetrahydro-5H-tetrazoloazepine, 1,5-Pentamethylenetetrazole. Pentylenetetrazol acts rapidly to produce a stimulation of the cen-

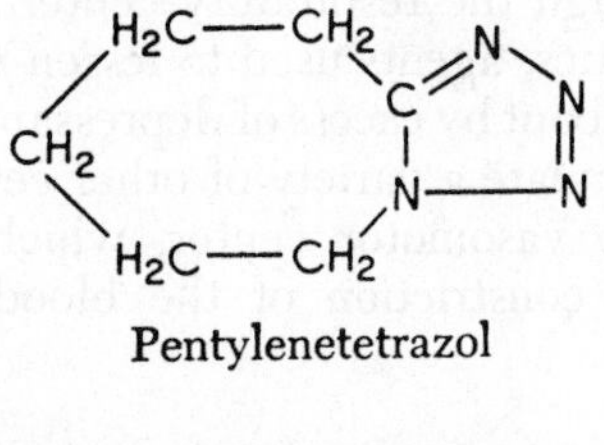

Pentylenetetrazol

tral nervous system, chiefly at the medullary and cerebral levels, but also on the spinal cord. The effect is produced within 1 minute of intravenous injection, resulting in a series of clonic convulsions of comparatively short duration. It has been used for shock therapy in the treatment of mental disorders, particularly in schizophrenia and depressive psychoses. Although more effective than camphor and, due to its rapid metabolic destruction in the liver, safer than insulin, it has been largely replaced by the simpler and safer electrical methods.

Pentylenetetrazol is used as an antidote for poisoning by depressant drugs, although it is not as potent as picrotoxin.

Pentylenetetrazol, in subeffective doses, is mixed with therapeutic doses of barbiturates to counteract excessive central nervous system depression if an overdose is taken accidently or with suicidal intent. The stimulant action of pentylenetetrazol has also been applied in geriatric patients, to improve emotional responsiveness.

Pentylenetetrazol is synthesized from cyclohexanone and hydrazoic acid. The drug is crystalline solid, freely soluble in water and most organic solvents.

Category—central stimulant.

Usual dose—intravenous and subcutaneous: initial, 500 mg., then adjust dosage and repeat at 30-minute intervals as needed.

Usual dose range—100 mg. to 2 g.

OCCURRENCE
Pentylenetetrazol Injection N.F.

Nikethamide, N.F., Coramine®, Nikorin®, N,N-Diethylnicotinamide. Nikethamide is a respiratory stimulant, acting by reflex stimulation of chemoreceptors of the carotid body, together with stimulation of medul-

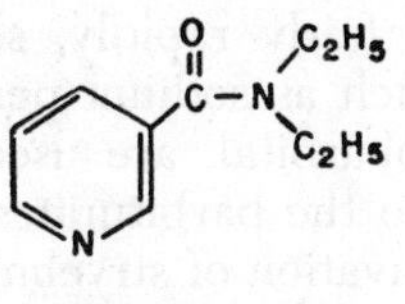

Nikethamide

lary centers. It has an intermediate central stimulant effect, resembling that of the amphetamines rather than the more potent picrotoxin or pentylenetetrazol.

It is a viscous, high-boiling oil which is miscible with water, alcohol and ether. Nikethamide is administered by injection, to counteract respiratory depression following administration of depressant drugs such as the barbiturates.

Category—central and respiratory stimulant.

Usual dose range—intramuscular and intravenous: 375 mg. to 3.75 g. repeated as needed.

OCCURRENCE
Nikethamide Injection N.F.

Ethamivan N.F., Emivan®, N,N-Diethylvanillamide, 3-Methoxy-4-hydroxybenzoic

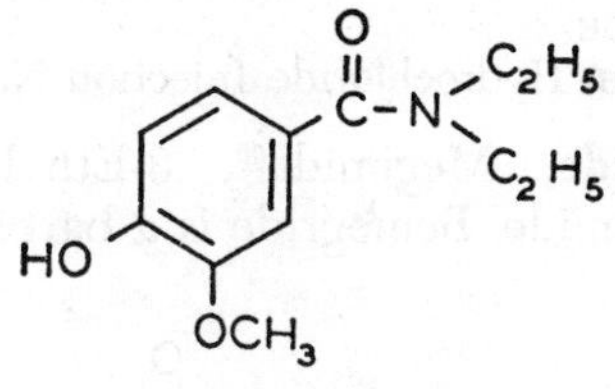

Ethamivan

Acid Diethylamide. Ethamivan is an analeptic drug useful as an adjunctive agent in the treatment of severe respiratory depression. Continuous intravenous infusion maintains an increase in both rate and depth of respiration in such patients. The drug produces general stimulation of the central nervous system, and excessively high doses may lead to convulsions.

Category—central and respiratory stimulant.

Usual dose range—intravenous, 0.5 to 5 mg. per kg. of body weight, given slowly as a single injection; may be followed with continuous intravenous infusion at the rate of 10 mg. per minute, as determined by response of the patient.

OCCURRENCE
Ethamivan Injection N.F.

Doxapram Hydrochloride N.F., Dopram®, 1-Ethyl-3,3-diphenyl-4-(2-morpholinoethyl)-2-pyrrolidinone Hydrochloride hydrate. Doxapram is used to stimulate respiration in patients with postanesthetic respiratory depression, and to hasten arousal during this period.

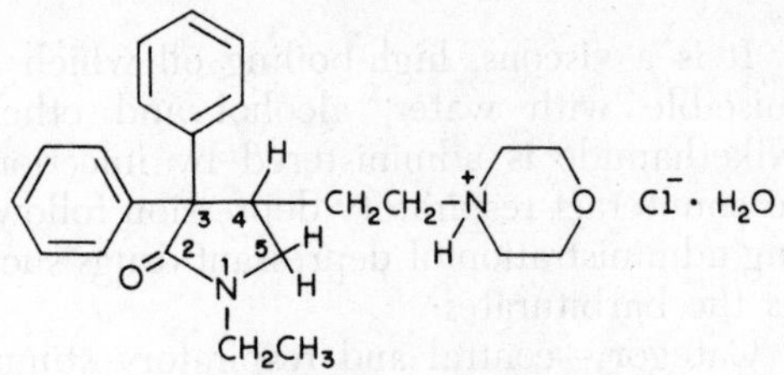

Doxapram Hydrochloride

Category—respiratory stimulant.

Usual dose range—intravenous, 500 mcg. to 1 mg. per kg. of body weight given as a single injection, or 1.5 to 2.0 mg. per kg. of body weight given as injections of 500 mcg. to 1.0 mg. per kg. of body weight at 5-minute intervals.

OCCURRENCE

Doxapram Hydrochloride Injection N.F.

Bemegride, Megimide®, 3-Ethyl-3-methylglutarimide. Bemegride is a barbiturate

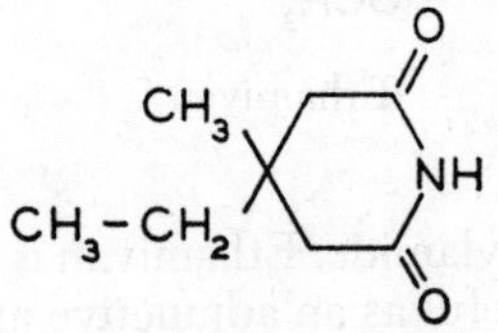

Bemegride

antagonist, useful for the treatment of barbiturate intoxication and for rapid termination of barbiturate anesthesia. Its central stimulant effects are not limited to reversal of barbiturate depression, for it has been used in the treatment of glutethimide poisoning. In overdose it causes central stimulation resembling that of other analeptics. Bemegride is administered intravenously, the usual dose being 50 mg. at intervals of 3 to 5 minutes until reversal of depression occurs or until signs of toxicity intervene. The compound is a crystalline solid, soluble in water.

Lobeline, α-Lobeline Hydrochloride, 2-(2-Phenyl-2-hydroxyethyl)-6-(benzoylmethyl)-N-methylpiperidine Hydrochloride. Lobeline is a respiratory stimulant recommended for respiratory emergencies. As the hydrochloride, it is available for parenteral administration in ampules containing 3 mg. or 10 mg. per ml. The usual subcutaneous

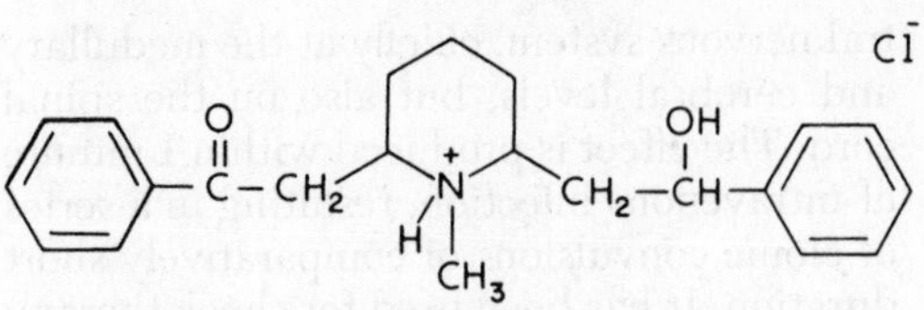

Lobeline

or intramuscular dose is 10 to 20 mg. the intravenous dose is 3 to 6 mg.

The physiologic effects of lobeline are similar to those of nicotine, and lobeline, usually as the sulfate salt, is used in preparations to discourage tobacco smoking.

Strychnine Sulfate. Strychnine, an ex-

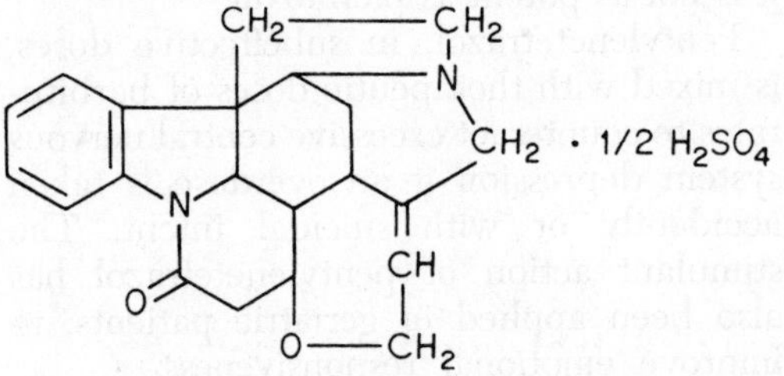

Strychnine sulfate

tremely poisonous alkaloid isolated from the ripe seed of *Strychnos nux vomica*, is a central stimulant, acting predominantly on the spinal cord; larger doses extend stimulation to the medulla. Although once widely used as a "tonic," stimulation of the gastrointestinal tract and increased voluntary muscle tone is not significant at therapeutic levels, and strychnine in such tonics has mainly contributed its extremely bitter taste.

An overdose of strychnine produces exaggerated skeletal muscle contractions, followed by convulsions. Therapy for strychnine poisoning calls for short-acting intravenous barbiturates. Since strychnine is metabolized fairly rapidly, short-acting barbiturates, such as sodium pentobarbital or sodium amobarbital, are used to avoid toxic reaction to the barbiturates following metabolic inactivation of strychnine.

Strychnine sulfate is a white crystalline solid, soluble to the extent of 1 g. in 35 ml. of water, 85 ml. of alcohol and 220 ml. of chloroform.

Flurothyl N.F., Indoklon®, Bis(2,2,2-trifluoroethyl) Ether, $CF_3CH_2-O-CH_2CF_3$.

Flurothyl is an inhalant which may be used in place of electroshock therapy in depressive disorders. Convulsions usually occur after 4 to 6 inhalations of the vapor.

Flurothyl is a colorless liquid, b.p. 63.9°, with a mild ethereal odor. It is insoluble in water. The drug is supplied in 2-ml. and 10-ml. ampuls.

Category—central stimulant (convulsant).
Application—1 ml. by special inhalation.

PURINES

Purines occur widely distributed among natural products (e.g., in uric acid, coffee, tea, cocoa, nucleic acids and enzymes). The 2,6-dihydroxylated purines, or xanthine derivatives, are caffeine, theobromine and theophylline. The world-wide use of stimulating drinks containing one or more of these principles causes them to assume added significance.

By progressive oxidation or the reverse, reduction, the relationship of uric acid, xanthine and purine can be observed.

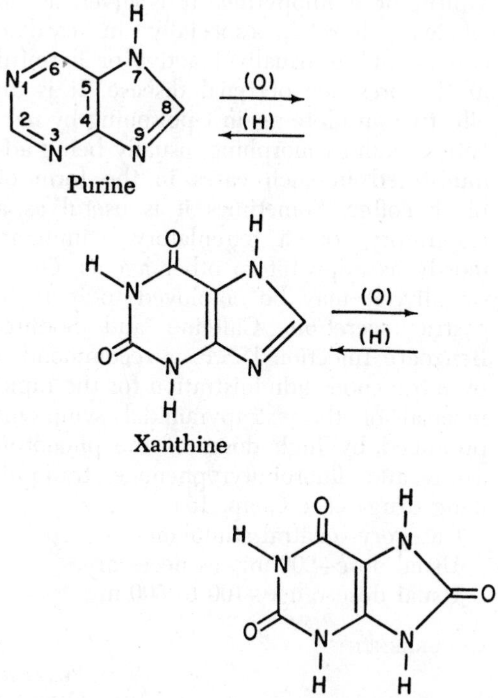

TABLE 37. XANTHINE ALKALOIDS

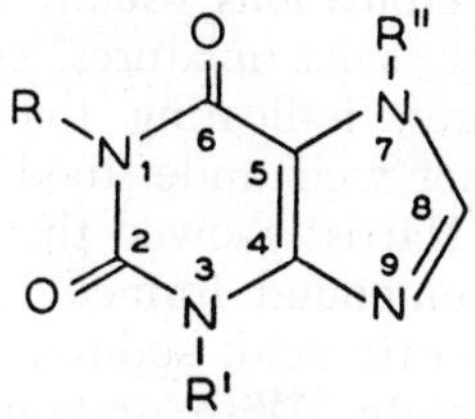

Xanthine

(R, R' & R" = H)

Compound	R	R'	R"	Common Source
Caffeine	CH$_3$	CH$_3$	CH$_3$	Coffee, Tea
Theophylline	CH$_3$	CH$_3$	H	Tea
Theobromine	H	CH$_3$	CH$_3$	Cocoa

Table 37 summarizes the structural relationships of xanthine alkaloids. The relative pharmacologic potencies of the xanthines are summarized[1] in Table 38.

In therapeutics, caffeine is the drug of choice among the three xanthines for obtaining a *stimulating effect* on the *central nervous system*. This stimulant action is almost physiologic in nature and helps to combat fatigue and sleepiness. Apparently, little tolerance is built up toward caffeine

Purine

Xanthine

Uric Acid

TABLE 38. RELATIVE PHARMACOLOGIC POTENCIES OF THE XANTHINES

Xanthine	C.N.S. Stimulation	Respiratory Stimulation	Diuresis	Coronary Dilatation	Cardiac Stimulation	Skeletal Muscle Stimulation
Caffeine	1*	1	3	3	3	2
Theophylline	2	2	1	1	1	3
Theobromine	3	3	2	2	2	1

* 1 = most potent.

stimulation therefore, habitual coffee drinkers continue to experience stimulation from day to day. Ordinarily, caffeine is not of value in other conditions, in spite of its other pharmacologic actions, because of excessive stimulation at the dose necessary to elicit other effects.

The xanthine alkaloids have poor water solubility, and this has prompted the use of numerous solubilizers. Alkali salts of organic acids (sodium acetate, sodium benzoate and sodium salicylate) are often used to solubilize caffeine.

These combinations usually are referred to as double salts, mixtures, combinations or complexes, indicating that their true nature is not well understood. Studies by Blake and Harris[4] showed that there is no chemical compound formed between caffeine and citric acid, sodium benzoate or sodium acetate. With sodium salicylate, some hydrogen bonding between the hydroxyl of the salicylate and the carbonyl of caffeine was observed.

Caffeine U.S.P., 1,3,7-Trimethylxanthine. Caffeine occurs as a white powder, or as white, glistening needles. The alkaloid is odorless and possesses a bitter taste. Caffeine is soluble in water (1:50), alcohol (1:75) or chloroform (1:6) but is less soluble in ether. Its solubility is increased in hot water (1:6 at 80°) or hot alcohol (1:25 at 60°).

Caffeine is a very weak base and does not form salts which are stable in aqueous or alcoholic solutions. The official Citrated Caffeine N.F. is not a true salt of caffeine, being merely a mixture of equal parts of citric acid and caffeine. This mixture is more soluble in water (1:4) than the caffeine alone.

A cup of coffee or tea (the prepared beverage) contains about 60 mg. of caffeine.

In physiologic action, caffeine and its relatives theobromine and theophylline are qualitatively alike. The primary effect from therapeutic doses is a stimulation of the central nervous system, beginning in the psychic center and progressing downward, with little or no reversal by continued or larger doses. There is also a direct stimulation of all muscles, partly central and partly peripheral, an increase in diuresis and vaso-dilation by direct action on particular vessels. The dominant effect on the psychic center causes increased flow of thought, lessens drowsiness and mental fatigue, relieves headache and gives a sense of comfort and well-being. In combination with the action on muscles, it brings about a condition in which more work can be done before fatigue sets in, and this is performed more rapidly and accurately, although later there may be impairment of these qualities for some time.

Caffeine often is employed in headaches of certain kinds, such as in neuralgia, rheumatism, migraine and in those due to fatigue, most often combined with other analgesics, especially acetanilid, phenacetin, aspirin or aminopyrine. It is given as an efficient diuretic, especially in cardiac edema, but is usually inactive or harmful in the presence of renal disease. It is an effective antidote against poisoning by narcotics, such as morphine, usually being administered in such cases in the form of black coffee. Sometimes it is useful as a respiratory or a circulatory stimulant, mostly as adjuvant to other agents. Occasionally, it may be employed to increase gastric secretion. Caffeine and Sodium Benzoate Injection U.S.P. is recommended by intravenous administration for the rapid reversal of the extrapyramidal symptoms produced by high doses of the phenothiazine and fluorobutyrophenone tranquilizing drugs (see Chap. 15).

Category—central stimulant.

Usual dose—200 mg. as necessary.

Usual dose range—100 to 500 mg.

OCCURRENCE

	PER CENT CAFFEINE
Caffeine and Sodium Benzoate Injection U.S.P.	45-52
Citrated Caffeine N.F.	48-52
Citrated Caffeine Tablets N.F.	45-55
Aspirin, Phenacetin, and Caffeine Tablets N.F.	
Codeine Phosphate, Aspirin, Phenacetin, and Caffeine Tablets N.F.	
Ergotamine Tartrate and Caffeine Suppositories N.F.	
Ergotamine Tartrate and Caffeine Tablets N.F.	

PSYCHOMOTOR STIMULANTS

The psychomotor stimulants are used to elevate the mood or improve the outlook of patients with mental depression. They may be divided into the general classes: *Central Stimulant Sympathomimetics,* the *Antidepressants* (*Monoamine Oxidase Inhibitors, Tricyclic Antidepressants*), and *Miscellaneous.*

CENTRAL STIMULANT SYMPATHOMIMETICS

The sympathomimetic agents are discussed in Chapter 17 in terms of peripheral or autonomic effects. Certain of the sympathomimetics, by virtue of structural features and physical properties, exert a significant stimulant effect on the central nervous system. The structures of these compounds are presented in Table 39.

These agents vary in their intensity of central stimulant activity.[5] The most potent stimulating drugs are amphetamine and its N-methyl analog, methamphetamine. Steric differences are of considerable importance, for the *dextro* insomers are 10 to 20 times more stimulating than the *levo* isomers. The branched methyl group (amphetamine, methamphetamine) or similar substitution, such as incorporation of the nitrogen in a ring system (methylphenidate, phenmetrazine), is an important feature of these central stimulants, presumably providing resistance to enzymatic inactivation by steric protection of the amino group. These compounds are attacked by monoamine oxidase at a slower rate than those which do not possess branching in the chain connecting the aromatic nucleus and amino group. The parent β-phenylethylamine is not useful as a central stimulant. An increase in the number of carbons in the branching side chain also decreases activity; 1-phenyl-2-aminobutane and 1-phenyl-2-aminopentane show a low order of central stimulation as compared with amphetamine, as does the α,α-dimethyl analog, mephentermine. N-alkyl substitution by groups larger than methyl also decreases activity. Substitutions which increase the hydrophilic character decrease central stimulant activity. A hydroxyl group in the 2-position (phenylpropanolamine, ephedrine) or in the aromatic nucleus (hydroxyamphetamine) results in sympathomimetic amines with distinctly less central stimulant activity than their nonhydroxylated analogs. Reduction of the aromatic ring, or its replacement by an alkyl group, produces compounds with little or no stimulating action on the central nervous system, although many retain peripheral activity and serve as vasoconstrictors, useful as nasal decongestants.

Those stimulants which are most potent (amphetamine, methamphetamine) may be used as analeptics in reversing the profound central depression due to anesthetic, narcotic and hypnotic drugs. Narcoleptic patients show considerable relief from attacks of sleep and cataplexy and are often improved by the potent amphetamine analogs. These have also been used in conjunction with the centrally depressant atropine analogs (Chap. 19) in the treatment of postencephalitic parkinsonism and are most effective in providing relief from oculogyric crises. A centrally mediated decrease in appetite brought about by amphetamine and its analogs has caused these to be used as anorexigenic agents, as adjuncts to dietary control in the management of obesity. In addition to the potent central stimulants amphetamine and methamphetamine, several related sympathomimetic amines are advocated for use as anorexigenic agents and are said to decrease appetite, with a lowered degree of the central stimulant effects leading to restlessness and insomnia. These include benzphetamine, diethylpropion, phenmetrazine, phendimetrazine, phentermine and chlorphentermine (Table 39).

The use of amphetamine and methamphetamine for fatigue and as "pep pills" has been long recognized and has led to frequent abuse.[6] The amphetamines have not been shown to produce true addiction, although tolerance to larger doses, without increased effect, and habituation occur.

The sympathomimetic amines with central stimulant properties have been used in the treatment of those psychogenic disorders related to depressive states. They are most effective in the treatment of mild

Table 39. Sympathomimetics With Significant Central Stimulant Activity

Generic Name	Base Structure		
Amphetamine	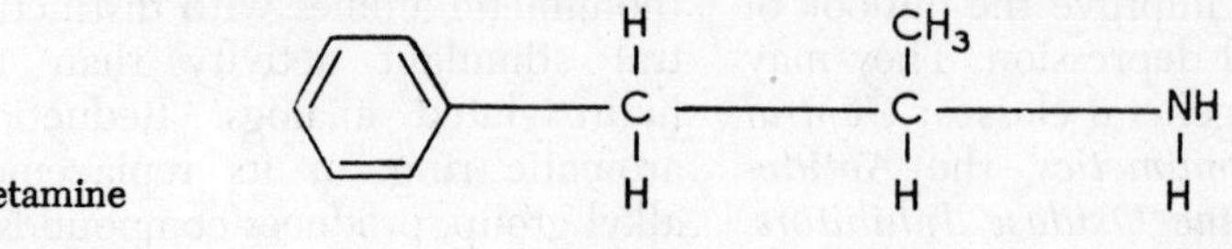		
Methamphetamine	H	H	CH_3
Phentermine	H	CH_3	H
Benzphetamine	H	H	CH_3 / $CH_2C_6H_5$
Diethylpropion	O*	H	C_2H_5 / C_2H_5
Phenylpropanolamine	OH	H	H
Ephedrine	OH	H	CH_3
Chlorphentermine	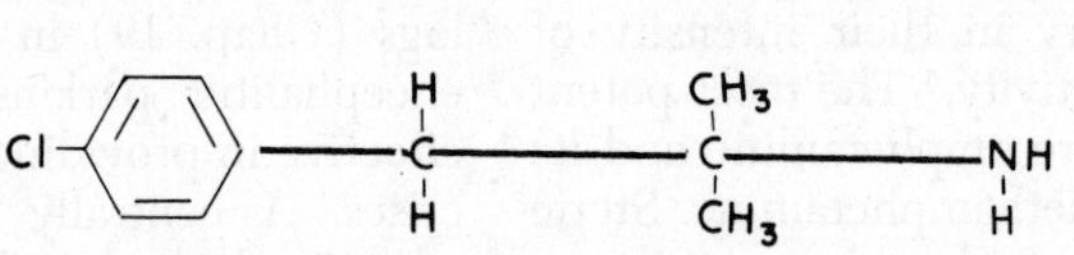		
Phenmetrazine	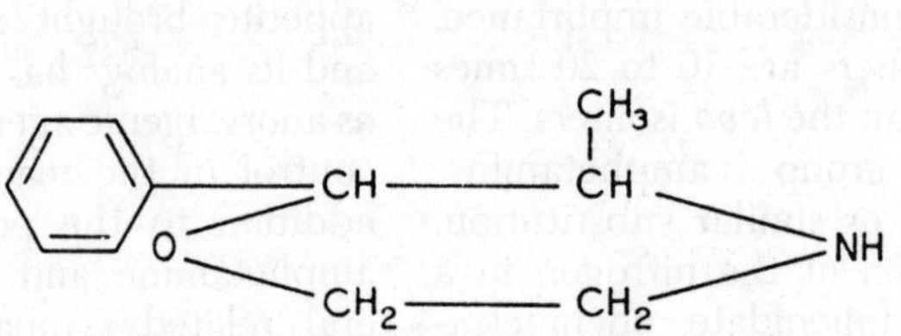		
Phendimetrazine	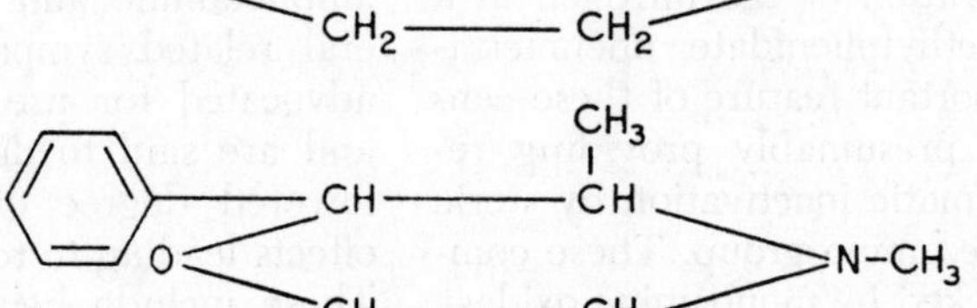		
Methylphenidate	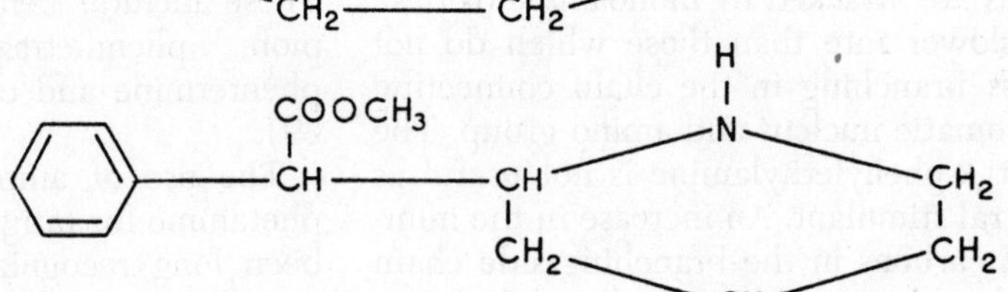		
Pipradrol	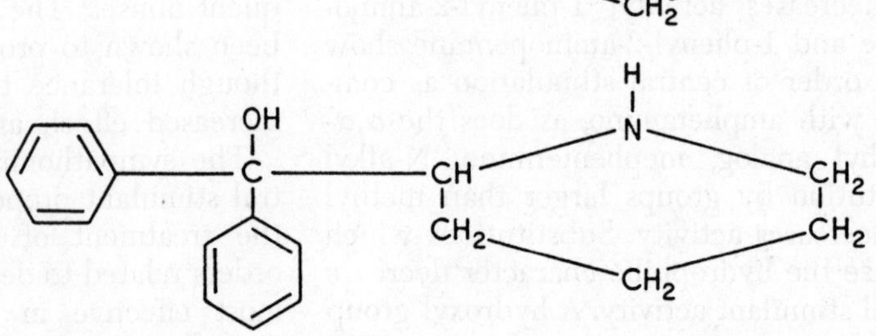		

* Carbonyl.

depression, apathy, mood disturbances and chronic exhaustion. The stimulants with activities less than amphetamine and more than caffeine, such as methylphenidate and pipradrol (Table 39), have been introduced specifically as mild cortical stimulants for improving mood and behavior and increasing motor and mental activity in patients with neuroses and psychoses characterized by depression.

Amphetamine shows little or no contracting effect on chronically denervated tissue, nor does it produce a pressor response in animals in which pre-treatment with reserpine has depleted stores of catecholamines. Therefore, amphetamine is thought to act primarily by releasing catecholamines such as norepinephrine and not to exert a direct adrenergic effect, at least in peripheral tissues.

Products

Amphetamine, Benzedrine®, (±)-1-phenyl-2-aminopropane, is a colorless liquid. The free base and the carbonate salt have been used as nasal decongestants. The generic name was derived from one chemical designation *a*lpha *m*ethyl *ph*enyl*et*hyl *amine;* the proprietary name from an alternate chemical name: *benz*yl *m*ethyl carbinam*ine*. Two salts, *Amphetamine Sulfate N.F.* (Benzedrine Sulfate) and *amphetamine phosphate* (Raphetamine Phosphate), are used as analeptic agents, in the treatment of narcolepsy, as adjuncts to treatment of alcoholism, to decrease appetite in the management of obesity, and in depressive conditions characterized by apathy and psychomotor retardation. Amphetamine is a fairly strong base, pK_a 9.77, which is slightly soluble in water, and readily soluble in alcohol, ether and aqueous acids. The sulfate salt is given orally, the usual dose being 5 mg. 3 times a day. The phosphate salt is more soluble and is used both orally and by intramuscular or intravenous injection. As an analeptic, parenteral administration in doses of 10 to 50 mg. every 30 to 60 minutes is used.

OCCURRENCE
Amphetamine Sulfate Injection N.F.
Amphetamine Sulfate Tablets N.F.

Dextroamphetamine Sulfate, U.S.P., (+)-Dexedrine® Sulfate, (+)-α-methylphenethylamine sulfate is the *dextro* isomer with the same actions and uses as the racemic amphetamine sulfate but possessing a greater stimulant activity. It was introduced in 1944. The usual oral dose for simple depression is 5 to 15 mg., 2 or 3 times a day; for narcolepsy, 10 to 50 mg. in divided doses. The appetite depressant daily dose is 15 to 30 mg. in 3 divided doses, taken 30 to 60 minutes before meals.

Category—central stimulant.
Usual dose—2.5 to 5 mg. 1 to 3 times a day.
Usual dose range—5 to 50 mg. daily.

OCCURRENCE
Dextroamphetamine Sulfate Elixir U.S.P.
Dextroamphetamine Sulfate Tablets U.S.P.

Dextroamphetamine Phosphate N.F., Monobasic (+)-α-Methylphenethylamine Phosphate. The phosphate salt is used for the same purpose as the sulfate, for its central stimulant effects. Aqueous solutions have a pH between 4 and 5.
Category—central stimulant.
Usual dose—5 mg. every 4 to 6 hours.
Usual dose range—5 to 10 mg.

OCCURRENCE
Dextroamphetamine Phosphate Tablets N.F.

Levamfetamine Succinate, Cydril®. The *levo*-isomer of amphetamine is used at an oral dose of 7 mg. 3 times daily as an appetite depressant. The drug may act directly on the brain, or may act indirectly by increasing the emptying time of the stomach by relaxing smooth muscle of the gastrointestinal tract. It produces less central stimulation than the *dextro* isomer, but is slightly more potent as a pressor agent.

Methamphetamine Hydrochloride, U.S.P., Desoxyn®, Drinalfa®, Methedrine®, Desoxyephedrine Hydrochloride, (+)-N,α-Dimethylphenethylamine Hydrochloride. This N-methyl analog of amphetamine was introduced in 1944 and shows a slightly greater central stimulant effect and slightly less circulatory action than the parent compound. It has the same range of therapeutic applications. The drug occurs as colorless or white crystals with a bitter taste, m.p. 170-175°. The hydrochloride is soluble in water (1:2), alcohol (1:3), and chloroform

(1:5), and is insoluble in ether. The free base is a fairly strong base, $pK_a = 9.86$, which is readily soluble in ether.

Category—central stimulant.

Usual dose—2.5 to 5 mg. 1 to 3 times a day.

Usual dose range—2.5 to 50 mg. daily.

OCCURRENCE

Methamphetamine Hydrochloride Tablets U.S.P.

Phentermine, Ionamine®, Wilpo®, 1-Phenyl-2-methyl-2-aminopropane; α-α-Dimethylphenethylamine. Phentermine was introduced in 1959 as an agent to lessen appetite in the management of obesity. The free base is bound to an ion exchange resin for delayed release into the gastrointestinal tract. Phentermine is also available as the hydrochloride salt.

The usual oral dose of phentermine resin is 15 to 30 mg. before breakfast. The usual dose of the hydrochloride salt is 8 mg. taken before mealtime.

Chlorphentermine Hydrochloride, Pre-Sate®, 1-(4-Chlorophenyl)-2-methyl-2-aminopropane Hydrochloride; p-Chloro-α,α-

Chlorphentermine Hydrochloride

dimethyl-β-phenylethylamine Hydrochloride. Chlorphentermine is the p-chloro analog of phentermine, introduced in 1965 for treatment of obesity. Depression of appetite is said to occur without significant CNS stimulation, or effect on blood pressure and heart rate. As with other sympathomimetic amines, chlorphentermine should not be taken by patients with glaucoma or those receiving monoamine oxidase inhibitors. The recommended adult daily dose is 1 tablet (65 mg. as the free base, 75 mg. as the hydrochloride) before the morning meal.

Benzphetamine Hydrochloride N.F., Didrex®, (+)-1-Phenyl-2-(N-methyl-N-benzylamino)propane Hydrochloride, (+)-N-Benzyl-N,α-dimethylphenethylamine Hydrochloride. Benzphetamine is an anorexigenic agent introduced in 1960. At the oral dose of about 75 mg. a day in divided doses, little restlessness, anxiety, insomnia and other symptoms of excess central stimulation are said to occur.

Category—anorexiant.

OCCURRENCE

Benzphetamine Hydrochloride Tablets N.F.

Diethylpropion Hydrochloride N.F., Tenuate®, Tepanil®, 2-(Diethylamino)propiophenone Hydrochloride, 1-Phenyl-2-diethylaminopropanone-1 Hydrochloride, was introduced in 1959 for the suppression of appetite in the management of obesity. Central and cardiovascular stimulation appears to be minimal at the recommended oral dose of 25 mg., 3 or 4 times daily.

OCCURRENCE

Diethylpropion Hydrochloride Tablets N.F.

Phenylpropanolamine Hydrochloride N.F., Propadrine®, ($\pm$)-Norephedrine Hydrochloride, ($\pm$)-1-Phenyl-2-amino-1-propanol Hydrochloride. The main uses for phenylpropanolamine are for its peripheral effects as nasal decongestant and bronchial dilator. It possesses a central stimulant effect which is less than that of ephedrine. (See Chap. 17.)

Ephedrine Sulfate, U.S.P.; ($-$)-*erythro*-α[1-(Methylamino)ethyl]benzyl Alcohol Sulfate. Ephedrine is available as a number of salts. In addition to the peripheral effects discussed in Chapter 17, ephedrine produces a moderate effect in stimulating the central nervous system, and it has been used as an analeptic and respiratory stimulant in the treatment of poisoning by depressant drugs. Its stimulant properties are noticeable and undesirable side-effects, especially in children, when used as a bronchodilator in the treatment of mild forms of asthma.

Phenmetrazine Hydrochloride, N.F., Preludin®, ($\pm$)-3-Methyl-2-phenylmorpholine Hydrochloride, was introduced in 1956 as an appetite suppressant with the side-effects of nervousness, euphoria and insomnia attributable to central nervous stimulation being much less than with amphetamine.

Category—anorexic.

Usual dose range—25 to 75 mg. per day,

in divided doses, administered 1 hour before meals.

OCCURRENCE
Phenmetrazine Hydrochloride Tablets
N.F.

Phendimetrazine Tartrate, Plegine®, (+)-3,4-Dimethyl-2-phenylmorpholine Bitartrate, was introduced in 1961 as an anorexigenic agent. It appears to possess the same degree of effectiveness and the same order of central stimulation as its close analog, phenmetrazine. The usual oral dose is 35 mg. taken 1 hour before meals.

Methylphenidate Hydrochloride, U.S.P., Ritalin® Hydrochloride, Methyl-α-Phenyl-2-piperidineacetate Hydrochloride. Methylphenidate is a mild cortical stimulant used in the treatment of depressive states since 1956. Because it does not possess pronounced pressor effects when given orally (but does when administered parenterally), it may be used to counteract the oversedation of hypertensive patients receiving the Rauwolfia alkaloids. It is used as a mild analeptic and for its stimulating effects in the treatment of psychoses characterized by depression. In emergency situations where rapid cortical stimulation is desired, parenteral dosage is determined by the degree of depression and may range up to 30 to 50 mg. every 30 minutes in counteracting overdosage of sedatives.

Category—central stimulant.

Usual dose—oral or parenteral, 10 mg. three times daily.

Usual dose range—20 to 60 mg. daily.

OCCURRENCE
Methylphenidate Hydrochloride for
Injection U.S.P.
Methylphenidate Hydrochloride Tablets
U.S.P.

MONOAMINE OXIDASE INHIBITORS

In 1952 the independent observations were made that iproniazid produced central stimulation in patients being treated for tuberculosis[7] and also inhibited the enzyme monoamine oxidase.[8] These properties were not shared by the related antitubercular agent isoniazid. In 1957 it was noted[9] that pretreatment of animals with iproniazid reversed the usual depressant effect of reserpine, producing instead central stimulation. At the same time it was observed that the reserpine-induced depletion of serotonin and norepinephrine in the brain was prevented by pretreatment with iproniazid.

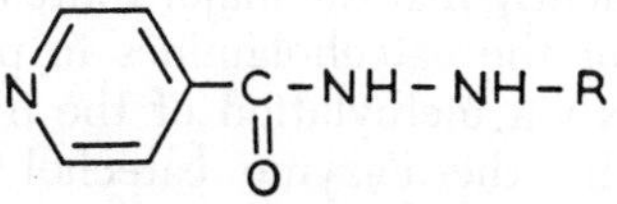

Iproniazid; R=CH(CH₃)₂
Isoniazid; R=H

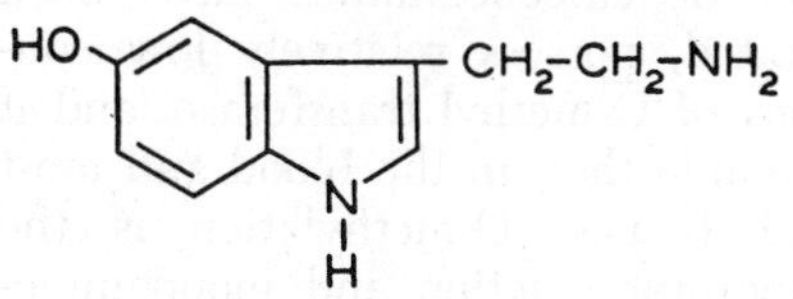

Serotonin

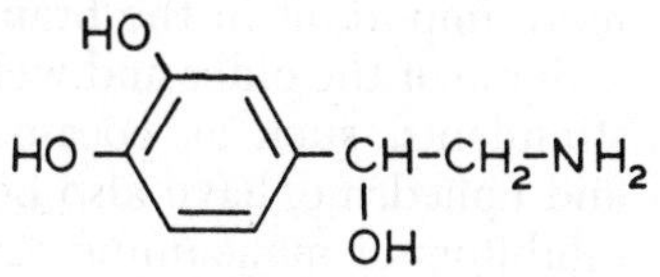

Norepinephrine

These observations prompted the successful clinical re-examination of iproniazid as a central stimulant in the treatment of mental depression.[10] It has been proposed that the clinical antidepressant actions of iproniazid and related compounds which inhibit monoamine oxidase are due to the decreased metabolic destruction of brain amines such as norepinephrine and serotonin. In experimental animals (rabbit, rat, mice, monkey) both serotonin and norepinephrine levels increase after treatment with monoamine oxidase inhibitors, and in a variety of species concentration of brain norepinephrine seems to be best correlated with excitation.[11] However, the metabolic destruction of other amines whose physiologic function in the central nervous system has not been as well explored is prevented by the monoamine oxidase inhibitors. Examples include phenylethylamine, tyramine and its ortho and meta isomers 3,4-dihydroxyphenylethylamine and tryptamine. One or more of these or others not yet discovered may contribute to the pharmacologic effects of the monoamine oxidase inhibitors.[12]

The hypothesis[13] that oxidative deamination of the catecholamines by monoamine oxidase represents the major metabolic pathway in the brain was a result of the above observations. It has been demonstrated[14] subsequently that the major route of metabolism for the catecholamines in peripheral tissue is via methylation of the 3-hydroxyl group by the enzyme catechol-O-methyl transferase (see Chap. 3). However, brain and heart tissue, where potentiation and protection of catecholamines have been demonstrated, possess relatively low concentrations of O-methyl transferase, and it seems possible that, in the blood and most peripheral tissues, O-methylation is the more important reaction and monoamine-oxidase-mediated oxidative demethylation more important in the brain and the heart.

Some of the older and well-known central stimulants, such as cocaine, amphetamine and ephedrine, have also been shown to be inhibitors of monoamine oxidase. However, they are relatively weak and short-acting enzyme inhibitors when compared with potent members of the hydrazine series and related compounds, and it is generally believed that the portion of their pharmacologic effects brought about by monoamine oxidase inhibition is small.

The monoamine oxidase inhibitors are used in the treatment of psychotic patients with mild to severe depression. Responsive patients show an increased sense of well-being, increased desire and ability to communicate, elevation of mood, increased physical activity and mental alertness as well as improvement in appetite. These drugs are used in the milder depressive states in place of electroshock, or in conjunction with electroshock in the management of more serious depressions. Because of their slow onset of action, they are of no value in psychiatric emergencies.

The monoamine oxidase inhibitors, by an unknown mechanism, reduce the frequency and the severity of anginal attacks. They do not improve the pathologic condition, and care must be taken against overexertion following the decrease in warning signals provided by anginal pains.

Because of their enzyme-inhibiting properties, they potentiate and prolong the actions of many drugs such as amphetamines, caffeine, barbiturates and local anesthetics.

Toxic side-effects to the monoamine oxidase inhibitors are large in number, and some are of a serious nature. Such side-effects, including hepatic toxicities and visual disturbances, have resulted in the removal from distribution of some of the earlier agents (e.g., iproniazid, pheniprazine, etryptamine) as less toxic drugs were developed. The term psychic energizer, which has been applied to them, implies a use in enhancing normal behavior. The newness of these agents and the seriousness of toxicities demonstrated during a relatively short period of use would dictate against their application in this manner.

In addition to the hydrazine derivatives, a number of nonhydrazines have been shown to possess potent monoamine-oxidase-inhibiting properties. The hydrazines (e.g., phenelzine, nialamide, isocarboxazid) have a slow onset of response, 2 or 3 weeks often being required before any degree of improvement in the mentally depressed state is noted. Tranylcypromine, a nonhydrazine monoamine oxidase inhibitor, frequently produces a response within several days.

Some currently available monoamine oxidase inhibitors are listed by name and structure in Table 40.

Structure-Activity Relationships

The in-vitro potency of a large series of hydrazine derivatives in inhibiting the metabolism of serotonin by rat-liver homogenate has been used to establish structure-activity relationships for the property of monoamine oxidase inhibition.[15] Analeptic properties were tested by measuring arousal of mice from a reserpine-induced stupor. Maximum monoamine oxidase and analeptic activities were shown by compounds with the amphetaminelike structure. For example, pheniprazine, a nitrogen isostere of methamphetamine, was one of the most potent agents tested, showing strong enzyme-inhibiting properties, and analeptic activity comparable with amphetamine's.

Nuclear substitution (methoxy, methyl,

TABLE 40. MONOAMINE OXIDASE INHIBITORS

Generic Name Proprietary Name	STRUCTURE
Phenelzine Sulfate U.S.P. Nardil	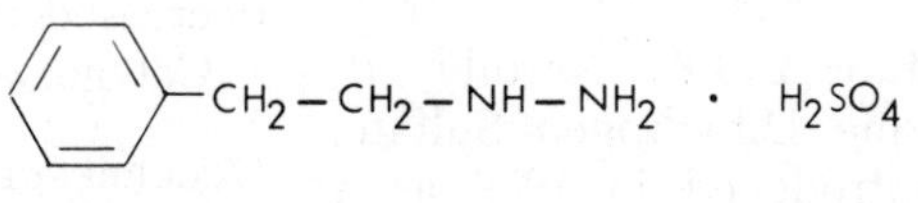 ⬡—$CH_2-CH_2-NH-NH_2$ · H_2SO_4
Nialamide N.F. Niamid	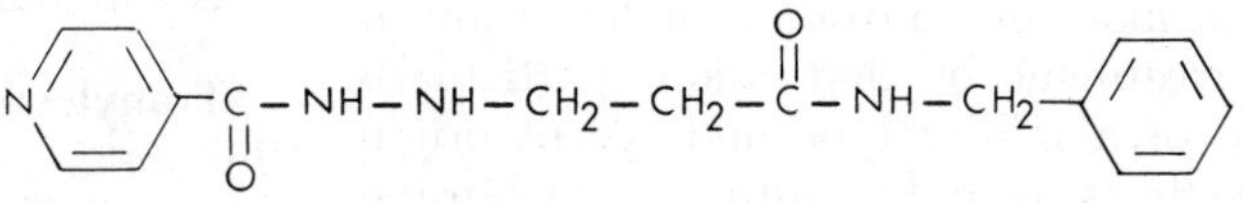 ⬡(N)—$C-NH-NH-CH_2-CH_2-\overset{O}{\overset{\|}{C}}-NH-CH_2$—⬡
Isocarboxazid N.F. Marplan	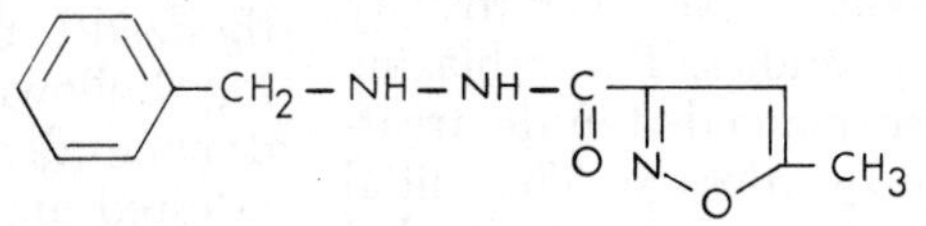 ⬡—$CH_2-NH-NH-C$⟍(isoxazole ring)—CH_3
Tranylcypromine Sulfate N.F. Parnate	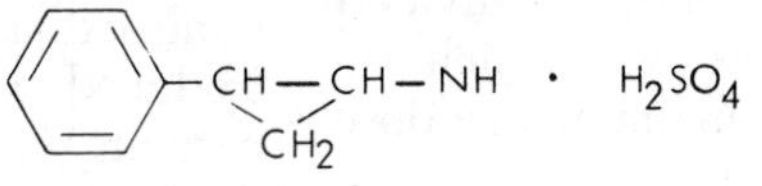 ⬡—$CH-CH-NH$ · H_2SO_4 with CH_2
Pargyline Hydrochloride N.F. Eutonyl	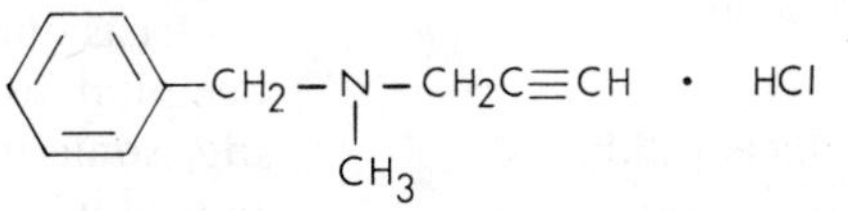 ⬡—$CH_2-\underset{CH_3}{N}-CH_2C\equiv CH$ · HCl

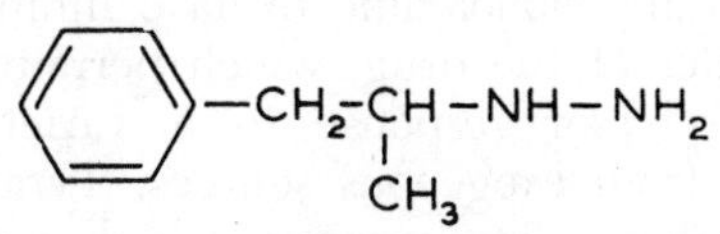

⬡—$CH_2-\underset{CH_3}{CH}-NH-NH_2$

Pheniprazine

⬡—$CH_2-\underset{CH_3}{CH}-NH-CH_3$

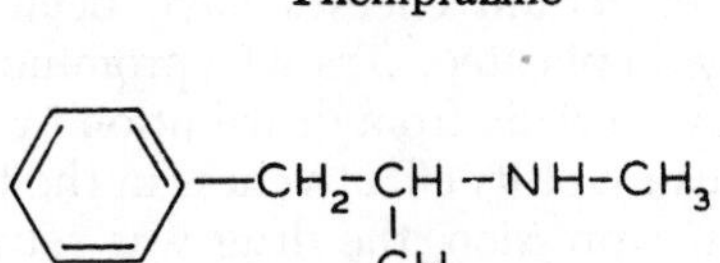

Methamphetamine

hydrogenation) reduced both enzyme-inhibitory and analeptic properties of pheniprazine, just as they do with amphetamine. Both N-acylation and N-alkylation of the hydrazines that have been tested decreased enzyme-inhibitory and analeptic potency. Replacement of the phenyl ring by several heterocyclic ring systems reduced enzyme-inhibitory properties; analeptic properties were absent. An increase or a decrease in chain length between aryl and hydrazinyl groups caused variations in analeptic and enzyme-inhibiting properties which demonstrated that these were separable. For example, iproniazid, benzylhydrazine, α-phenylethylhydrazine and γ-phenylisobutylhydrazine showed significant monoamine oxidase-inhibiting properties, both in vivo and in vitro, but were without significant analeptic effect in mice.

Compounds possessing amino or aliphatic acids as acyl groups attached to an N-isopropyl hydrazine nucleus have shown high monoamine-oxidase-inhibiting activity.[16] Among the more potent hydrazides were those in which the acyl group was provided by the amino acids leucine, serine, glutamic acid and tyrosine.

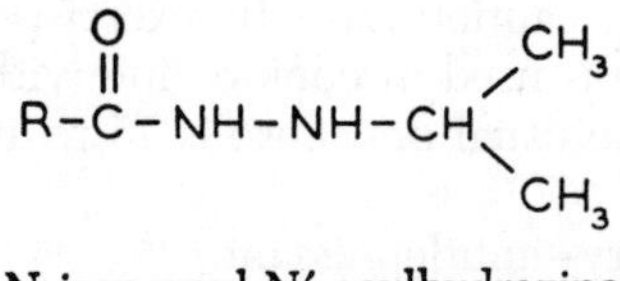

$$R-\overset{O}{\overset{\|}{C}}-NH-NH-CH\overset{CH_3}{\underset{CH_3}{\diagup}}$$

N-isopropyl-N'-acylhydrazines

It appears that the central stimulant properties of the hydrazines may be produced by structurally separable features of (1) direct stimulation of the central nervous system (amphetaminelike), and (2) monoamine-oxidase inhibition in the brain (iproniazidlike), preventing the metabolic destruction of central excitatory amines.

Products

Phenelzine Sulfate U.S.P., Nardil®, β-Phenylethylhydrazine Dihydrogen Sulfate. Phenelzine was introduced in 1959 as a potent monoamine oxidase inhibitor and in the treatment of depression, particularly those of a milder type and recent origin. Side-effects include postural hypotension, constipation and edema. No liver toxicity was found in early studies. Phenothiazine tranquilizers are recommended in the treatment of accidental overdosage. The initial oral dose is usually 45 mg. a day in divided doses. Following improvement, which requires about a week, the maintenance dose is about 15 mg. a day.
Category—antidepressant.

OCCURRENCE
Phenelzine Sulfate Tablets U.S.P.

Nialamide N.F., Niamid®, Isonicotinic Acid 2-[2-(Benzylcarbamoyl)ethyl]hydrazide, N-Benzyl-β-(isonicotinoylhydrazine)-propionamide. Nialamide was introduced in 1959 for use in the treatment of depression and in reducing the frequency and the severity of attacks of angina pectoris. The drug has a slow onset of action and may require days for an initial response to a change in dose level. Central stimulation with nervousness, restlessness and insomnia are side-effects, as well as blurred vision and dry mouth. Hepatic damage has not been noted in early use. Nialamide has been shown to prolong the central depressant effects of barbiturates and to augment the hypotensive effects of chlorothiazide. Therefore, caution must be exercised when nialamide is used in conjunction with other drugs. The usual oral dose is 75 to 100 mg. per day.
Category—antidepressant.

OCCURRENCE
Nialamide Tablets N.F.

Isocarboxazid N.F., Marplan®, 5-Methyl-3-isoxazolecarboxylic Acid 2-Benzylhydrazide, 1-Benzyl-2-(5-methyl-3-isoxazolylcarbonyl)hydrazine. Isocarboxazid was introduced in 1959 for the treatment of mental depression and in the symptomatic relief on angina pectoris. Side-effects include a low incidence of altered liver function. The usual initial dose is 30 mg. daily, this being decreased for maintenance to 10 to 20 mg.
Category—antidepressant.

OCCURRENCE
Isocarboxazid Tablets N.F.

Tranylcypromine Sulfate N.F., Parnate Sulfate®, (±)-trans-2-Phenylcyclopropylamine Sulfate. Tranylcypromine was synthesized[17] in 1948 as an amphetamine analog. Following the introduction of the hydrazine derivatives, tranylcypromine was retested and found to be a potent monoamine oxidase inhibitor. The drug was introduced in 1961 for the treatment of patients with psychoneurotic and psychotic depression.

Occasional hypertensive crises, severe occipital headache and intracranial bleeding, sometimes resulting in death, were reported with patients using tranylcypromine. These side-effects have been related to the long-lasting monoamine oxidase inhibitory properties of the drug, which permitted a strong pressor response to a variety of amines from exogenous sources. Tyramine and related amines present in high concentration in certain cheeses have been particularly implicated. Tranylcypromine was withdrawn briefly from distribution in 1964, but, because of its effectiveness in the treatment of depression, the drug was returned for restricted use in the treatment of hospitalized cases of severe depression or in cases outside the hospital in which other medication has been found ineffective. It is not to be used in patients over 60 years of age or with a history of hypertension or other cardiovascular disease.

Tranylcypromine sulfate is available as 10-mg. tablets.
Category—antidepressant.

Usual dose range—initial: 10 mg. in the morning and afternoon daily for 2 weeks; if no response appears, increase dosage to 20 mg. in the morning and 10 mg. in the afternoon daily for another week; maintenance: 10 to 20 mg. per day.

OCCURRENCE
Tranylcypromine Sulfate Tablets N.F.

Pargyline Hydrochloride N.F., Eutonyl®, N-Methyl-N-(2-propynyl)benzylamine Hydrochloride. Pargyline is a monoamine oxidase inhibitor which possesses hypotensive and stimulant properties. The postural hypotension common as a side-effect in the monoamine oxidase inhibitors is emphasized in pargyline, and it is recommended for the treatment of hypertension rather than for use in depressed states. (See Chap. 21.)

As with other monoamine oxidase inhibitors, patients receiving pargyline should not receive sympathomimetic amines, such as amphetamine, ephedrine and their analogs; foods that contain pressor amines, such as aged cheese containing tyramine; drugs that cause a sudden release of catecholamines, such as parenteral reserpine; tricyclic antidepressants, such as imipramine, or other monoamine oxidase inhibitors.

Category—antihypertensive.

Occurrence
Pargyline Hydrochloride Tablets N.F.

TRICYCLIC ANTIDEPRESSANTS

Following the discovery of the therapeutic value of the phenothiazine derivative chlorpromazine in the treatment of psychiatric disorders, many structural analogs were tested. Among these, the dibenzazepine derivative imipramine was found to be of therapeutic value in the treatment of depressive states, a condition in which chlorpromazine is not effective. A group of compounds, related in structure and pharmacological effects, is now available for the treatment of depression: imipramine (Tofranil), desipramine (Norpramin, Pertofrane), amitriptyline (Elavil), nortriptyline (Aventyl), and protriptyline (Vivactil). Because the initial compounds introduced as antidepressant drugs were related in structure by their similar three-ring systems, these drugs are frequently designated the *tricyclic antidepressants.*. Unlike the hydrazine derivatives, they do not inhibit monoamine oxidase. They rarely produce stimulation and excitement and may produce mild sedation like that of the phenothiazine tranquilizers. However, unlike the phenothiazine

tranquilizers, they are effective in the treatment of emotional and psychiatric disorders in which the major symptom is depression.

Depressed individuals, particularly those involved with endogenous depression rather than exogenous or reactive depressions, may respond with an elevation of mood, increased physical activity, mental alertness and an improved appetite. Many of the side-effects, such as dryness of mouth, tachycardia, constipation and sweating, are due to the atropinelike anticholinergic properties of these drugs which also aggravate glaucoma and cause urinary retention. Dangerous synergistic effects may occur when monoamine oxidase inhibitors are administered with imipramine and related compounds. For this reason, it is recommended that a period of at least 7 days should be allowed before changing from a monoamine oxidase inhibitor to an imipramine-type compound, or vice versa.

Structure-Activity Relationships

A number of tricyclic ring systems, if appropriately substituted, may possess antidepressant properties. These include the 5H-dibenz[*b,f*]azepine (dibenzazepine, iminostilbene); the related ring system with the 10,11-double bond reduced, 10,11-dihydro-5H-dibenz[*b,f*]azepine (dihydrodibenzazepine, iminodibenzyl); the ring system without a heteroatom, 5H-dibenzo[*a,d*]cycloheptene (dibenzocycloheptene), the corresponding reduced system, 10,11-dihydro-5H-dibenzo[*a,d*]cycloheptene (dihydrodibenzocycloheptene); the sulfur-bridged analog, thioxanthene; and the parent structure, originally related only to tranquilizing action, the phenothiazine.

Relationships between structure and antidepressant activity have been developed in a pharmacologic screening test based on the reversal of depression produced in the rat by reserpinelike benzoquinolizine compounds.[18] The following generalizations relate to derivatives of the ring structures: dibenzazepines, dibenzocycloheptenes, thioxanthenes, and phenothiazines.

Variations in R^1 (side chain). Activity is restricted to compounds having two or three carbons in the side chain. Compounds lack-

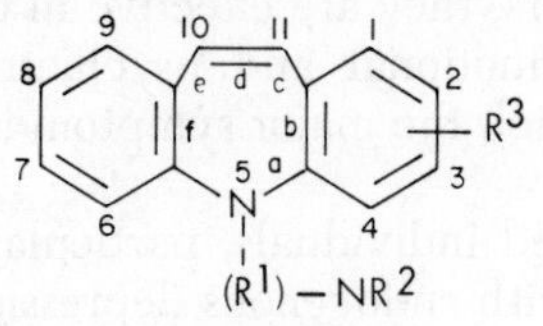

5H-Dibenz[b,f]azepine

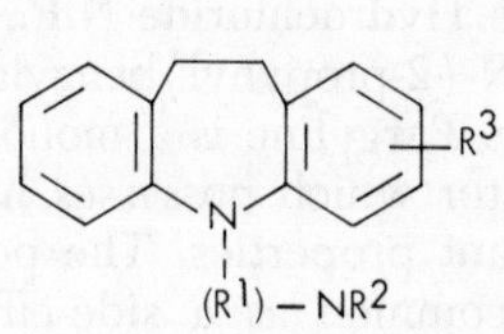

10,11-Dihydro-5H-dibenz[b,f]azepine

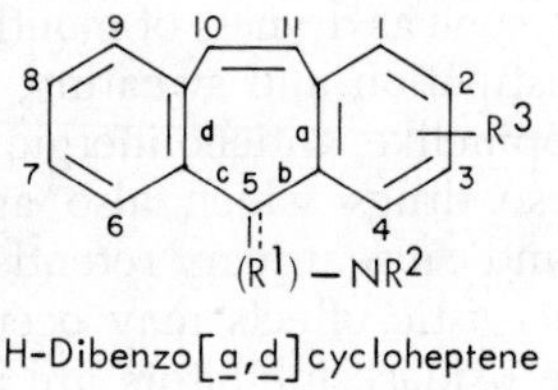

5H-Dibenzo[a,d]cycloheptene

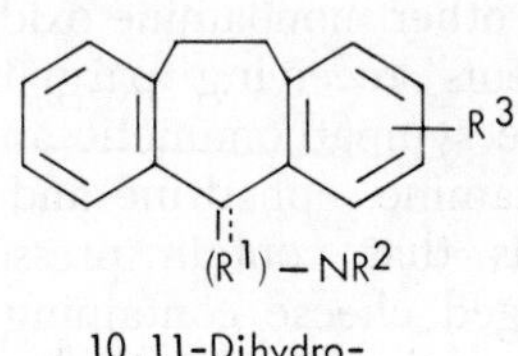

10,11-Dihydro-5H-dibenzo[a,d]cycloheptene

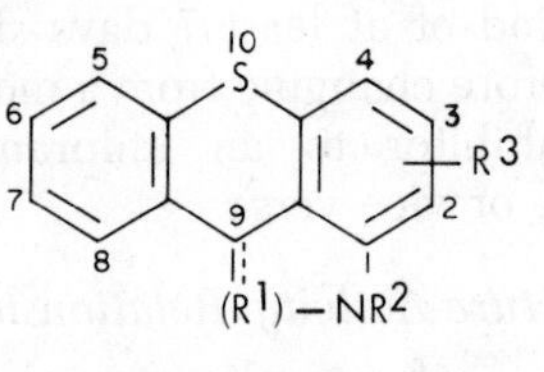

Thioxanthene

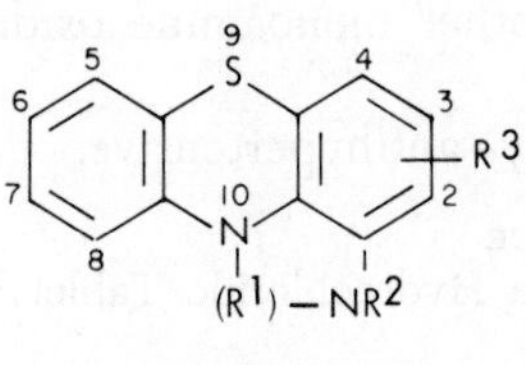

Phenothiazine

ing the side chain, or with branched chains and chains containing more than four carbons are inactive.

Variations in R² (N-substituents). Activity is confined to methyl-substituted or unsubstituted amines. Ethyl or higher alkyl groups on the side-chain nitrogen result in compounds that are inactive, and show toxicity that increases with increasing length of the side chain. Almost all antidepressant compounds are primary and secondary amines. The antidepressant action of some tertiary amines has been attributed to the rapid formation of their secondary analogs in the body. Generally, the tertiary amines show sedative properties. Thus, imipramine exerts a weak tranquilizing action. Amitriptyline is even more pronounced in this respect, and triflupromazine is a potent tranquilizer. Imipramine and amitriptyline were revealed as antidepressants in pharmacological tests, only if the secondary amine active metabolites were allowed to accumulate by repeated administration of the parent drug.

Variations in R³ (ring substituents). A number of ring-substituted compounds are active (e.g., 3-chloro, 10-methyl, 10,11 dimethyl) provided that they contain the aminoethyl or aminopropyl side chain.

Variations in the 10,11-bridge. The bridge in the 10,11-position may be formed by $-CH_2CH_2-$ (dihydrodibenzazepine) or by $-CH=CH-$ (dibenzazepine). Thus, when a dibenzazepine is active, the corresponding 10,11-dihydro compound is also active. In the case of desipramine, activity is also preserved if the C-2 bridge is replaced by an S-bridge (the related phenothiazine derivative, desmethylpromazine), or is left out altogether (a diphenylamine). This suggests that the 10,11-bridge is not vital for antidepressant activity.

Variations in Ring Systems. The ring nitrogen of desipramine can be replaced by carbon to yield the active dihydrodibenzocycloheptene, nortriptyline. Of 20 phenothiazines tested, all were inactive except desmethylpromazine and desmethyltriflupromazine. Several appropriately substituted thioxanthenes were active, as were a number of related tricyclic ring structures.

Removal of one benezene ring (bicyclic ring structures) resulted in loss of activity.

Imipramine Hydrochloride, U.S.P., Tofranil®, 5-[3-(Dimethylamino)propyl]-10,11-dihydro-5H-dibenz[*b,f*]azepine Hydrochloride. Imipramine was introduced in 1959 for the treatment of mental depression. In its clinical effect and pharmacology it shows some similarity to the phenothiazine derivatives to which it is chemically related in that it has mild tranquilizing properties. Unlike the phenothiazines, it is effective as an antidepressant agent. Its effects on behavior and mood resemble those of the amphetamines and the monoamine oxidase inhibitors, although imipramine does not inhibit monoamine oxidase. The drug is a potent parasympatholytic and displays prominent atropinelike side-effects.

Imipramine is most useful in treating endogenous depression. In treating depressions accompanied by anxiety, the drug is sometimes used together with a phenothiazine tranquilizing agent. Severe toxic reactions have occurred when imipramine was taken concurrently or immediately after the administration of monoamine oxidase inhibitors.

Imipramine undergoes metabolic N-demethylation to form the antidepressant

Imipramine: R = CH₃
Desipramine: R = H

metabolite desmethylimipramine (desipramine), which is then slowly demethylated to form the primary amine desdimethylimipramine. Both imipramine and desipramine are hydroxylated in the 2-position, followed by O-glucuronide formation.

Imipramine is available as 10-mg. and 25-mg. tablets for oral use or in ampuls containing 25 mg. for intramuscular administration. Small crystals may form in some ampuls, but this has no influence on therapeutic effectiveness. The crystals redissolve

when the ampuls are immersed in hot tap water for 1 minute.

Category—antidepressant.

Usual dose—oral or intramuscular, 25 mg. 3 to 4 times daily.

Usual dose range—50 to 200 mg. daily.

OCCURRENCE

Imipramine Hydrochloride Injection U.S.P.
Imipramine Hydrochloride Tablets U.S.P.

Desipramine Hydrochloride, N.F., Norpramin®, Pertofrane®, 5-(3-Methylaminopropyl)-10,11-dihydro-5H-dibenz[*b,f*]azepine Hydrochloride. Desipramine is a metabolite of imipramine which demonstrates similar antidepressant activity and was introduced in 1964. Although it is produced relatively slowly in the body by N-demethylation of imipramine, its subsequent metabolism and excretion is even slower, permitting accumulation. The slow onset of action of imipramine encouraged the theory that the parent compound might be exerting its antidepressant effect through a metabolite. Desipramine appears to have a somewhat shorter onset of action than imipramine, but it is somewhat less potent. The therapeutic range of effectiveness in the treatment of depressive states is the same as for imipramine, as are the side-effects and the precautions for use. Atropinelike side-effects are common, and the concomitant or prior use of monoamine-oxidase-inhibiting compounds is not recommended.

Category—antidepressant.

Usual dose—150 mg. per day in divided doses.

Usual dose range—50 to 200 mg. daily.

OCCURRENCE

Desipramine Hydrochloride Capsules N.F.
Desipramine Hydrochloride Tablets N.F.

Studies of ring analogs of the phenothiazine tranquilizing drugs included a series in which the sulfur bridge of phenothiazine was replaced by an ethylene bridge, and the ring nitrogen was replaced by carbon. Antidepressant activity was noted as well as retention of tranquilizing properties to a slight extent in a member of the series, amitriptyline, and the compound was introduced as an antidepressant drug in 1961.

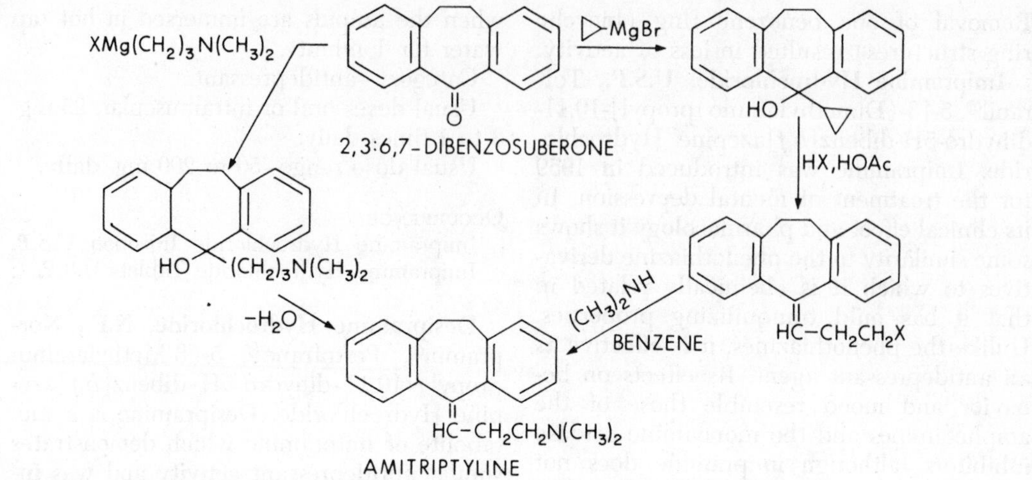

2,3:6,7-DIBENZOSUBERONE

AMITRIPTYLINE

Its metabolite, nortriptyline, was introduced in 1964.

Alternate methods have been developed for addition of the dimethylaminopropylidine side chain to 2,3:6,7-dibenzosuberone in syntheses of amitriptyline.[19, 20]

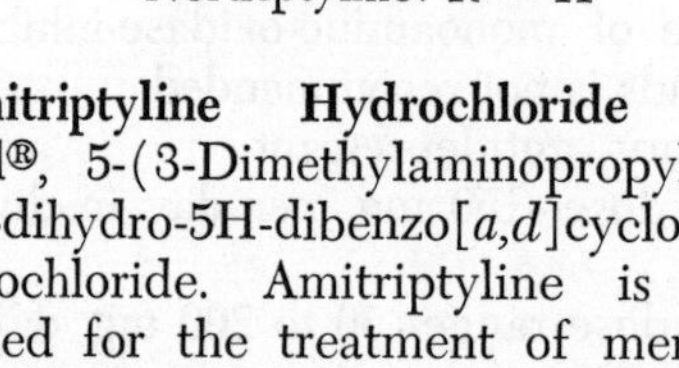

Amitriptyline: R = CH₃
Nortriptyline: R = H

Amitriptyline Hydrochloride U.S.P., Elavil®, 5-(3-Dimethylaminopropylidene)-10,11-dihydro-5H-dibenzo[a,d]cycloheptene Hydrochloride. Amitriptyline is recommended for the treatment of mental depression. It also has a tranquilizing component of action which is useful in cases in which anxiety accompanies depression. The sedation effect of amitriptyline is manifested quickly; however, the antidepressant effect may vary in onset from about 4 days to 6 weeks. Generally, improvement in mood and behavior is seen in 2 to 3 weeks after the start of medication.

Minor side-effects reflecting amitriptyline's anticholinergic activity are common. These include dryness of mouth, blurred vision, tachycardia and urinary retention. Amitriptyline is contraindicated in the pres-

ence of glaucoma and in patients with cardiovascular complications. The drug should not be administered with a monoamine oxidase inhibitor, since serious potentiation of side-effects may occur. Such combinations have caused cardiovascular collapse, impaired consciousness, hyperpyrexia, convulsions, and death.

Metabolic alteration of amitriptyline occurs by monodemethylation of the side chain nitrogen, and hydroxylation of the 10-position, forming *cis* and *trans* isomers of 10-hydroxyamitriptyline. Aromatic hydroxylation, rupture of the ethylene bridge, and oxidative deamination also occur. Nortriptyline is not excreted in the urine in appreciable amount after administration of amitriptyline.

Amitriptyline is available as 10-mg. and 25-mg. coated tablets, and as an injection for intramuscular use, with 10 mg. per ml.

The usual oral daily dose range is 30 mg. to 300 mg. The parenteral solution is used only for initial therapy at doses of 20 mg. to 30 mg., 4 times a day.

Category—antidepressant.

OCCURRENCE
Amitriptyline Hydrochloride Injection U.S.P.
Amitriptyline Hydrochloride Tablets U.S.P.

Nortriptyline Hydrochloride N.F., Aventyl®, 5-(3-Methylaminopropylidene)-10,11-dihydro-5H-dibenzo[a,d]cycloheptene Hydrochloride. Nortriptyline is the N-demethylated metabolite of amitriptyline. It possesses antidepressant and tranquilizing properties like those of the parent drug. The anticholinergic side-effects of nortriptyline are reported to be less than those of amitriptyline; however, they are still significant enough to preclude use in patients with glaucoma and urinary retention. The drug should not be used concurrently with monoamine oxidase inhibitors or before an interval of 1 to 2 weeks following termination of monoamine oxidase inhibitor therapy.

Nortriptyline undergoes N-demethylation, as well as hydroxylation in the 10-position, forming *cis* and *trans* isomers of 10-hydroxynortriptyline, which are excreted in the urine as conjugates.

Nortriptyline is administered orally in capsule or liquid form for the treatment of mental depression, anxiety-tension states and psychosomatic disorders. The usual adult dose is 20 to 100 mg. daily in divided doses.

Nortriptyline is available as 10-mg. or 25-mg. capsules or in a liquid preparation containing 10 mg. in 5 ml.

Category—antidepressant.

OCCURRENCE
Nortriptyline Hydrochloride Capsules N.F.

Protriptyline Hydrochloride, Vivactil®, 5-(3-Methylaminopropyl)-5H-dibenzo[*a,d*] cycloheptene Hydrochloride. Protriptyline, an isomer of nortriptyline containing an

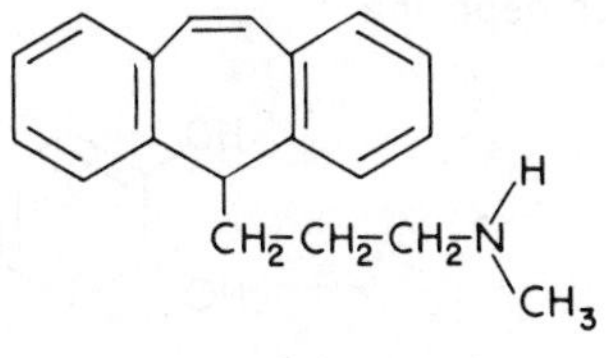

Protriptyline

endocyclic rather than exocyclic double bond, was introduced in 1967 as a selective antidepressant agent. It is used for the treatment of mental depression in patients under close medical supervision, and is said to produce little sedation. Because of potentially serious drug interactions, protriptyline should not be used in patients receiving monoamine oxidase inhibitors, guanethidine, or other hypotensive agents.

The average dose ranges from 15 to 40 mg. a day divided into 3 or 4 doses.

Doxepin Hydrochloride, Sinequan® Hydrochloride, N,N-Dimethyl-3-(dibenz[*b,e*] oxepin-11(6H)-ylidene)propylamine. Doxepin was introduced in 1969 as an anti-

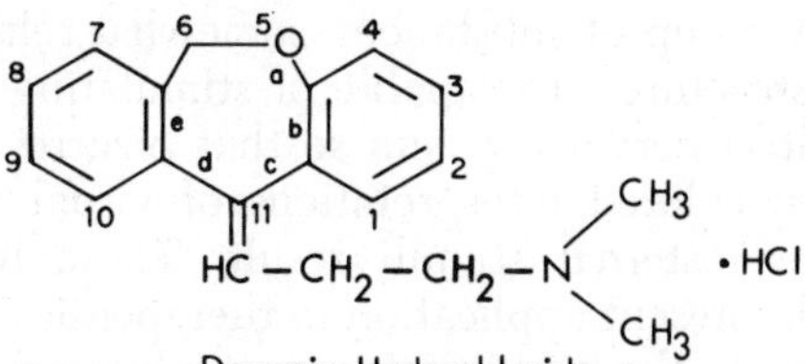

Doxepin Hydrochloride

depressant drug useful in the treatment of mild to moderate endogenous depression. It differs in structure from amitriptyline by the presence of an oxygen atom in the central ring, which leads to the formation of *cis* and *trans* isomers. The *cis* isomer is more active than the *trans*. As with related tricyclic antidepressant agents, the drug-induced elevation of mood may be accompanied by atropinelike anticholinergic side effects such as dryness of mouth, and by sedation.

Doxepin is available as 10 mg., 25 mg. and 50 mg. tablets for oral use. The usual daily dose is 50 to 200 mg. in divided doses.

MISCELLANEOUS PSYCHOMOTOR STIMULANTS

Deanol Acetamidobenzoate, Deaner®, the *p*-Acetamidobenzoic Acid Salt of 2-Dimethylaminoethanol. Deanol base, dimethyl-

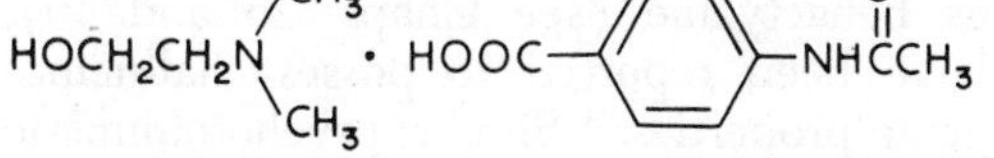

Deanol Acetamidobenzoate

aminoethanol, is the nonquaternized precursor to choline. The salt was introduced in 1958 for use in the treatment of a variety of mild depressive states and for alleviation of behavior problems and learning difficulties of school-age children. It has been proposed that deanol penetrates the central nervous system, there serving as a precursor to choline and acetylcholine. The drug is of low toxicity, and side-effects are relatively mild. These include headache, constipation, muscle tenseness and twitching, insomnia and postural hypotension. The oral dose ranges from 25 to 150 mg. per day and is expressed in terms of the free base.

HALLUCINOGENS

A group of substances, somewhat related in structure, are capable of stimulating the central nervous system so that bizarre and even colored interpretations of visual and other external stimuli result. These have little present application in therapeutics but provide the opportunity to produce model psychoses in man under controlled conditions. The effects of drugs in these artificially induced psychoses have been studied, and considerable impetus has been provided for testing the hypothesis relating abnormal metabolism of body components to spontaneous mental diseases.

Many hallucinogenic compounds possess close structural and pharmacologic relationships to the endogenous chemical mediators that are known to act in the peripheral nervous system and one believed to act in the central nervous system. These are acetylcholine and the β-arylethylamines, epinephrine, norepinephrine, and serotonin. Most hallucinogens can be referred to as adrenergic in function, that is, related to the β-arylethylamines. However, compounds affecting the parasympathetic nervous system are also known to produce psychotomimetic effects, possibly by producing an imbalance of acetylcholine activity in the central nervous system. The glycollic acid esters, such as benactyzine (see Chaps. 15 and 19), have been reported to possess hallucinogenic properties.[21] Similar psychotomimetic

$$CH_3COCH_2CH_2N(CH_3)_3^+$$

Acetylcholine

$$(C_6H_5)_2CCOCH_2CH_2N(C_2H_5)_2$$
$$|$$
$$OH$$

Benactyzine

effects may be produced by a variety of atropinelike anticholinergic compounds, such as scopolamine, that are sufficiently lipophilic to penetrate the central nervous system.

Some hallucinogenic compounds related to the β-arylethylamines can be visualized as arising from the faulty metabolism of normal body constituents. This has led to the hypothesis that some mental disorders may be due to the production of psychotoxins, such as bufotenine from serotonin, and adrenolutin, from epinephrine.

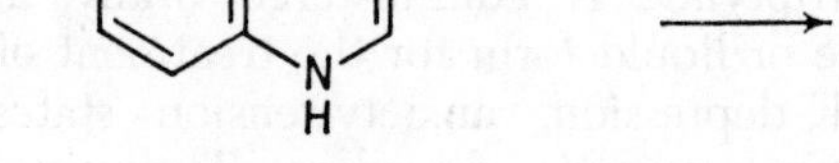

Serotonin

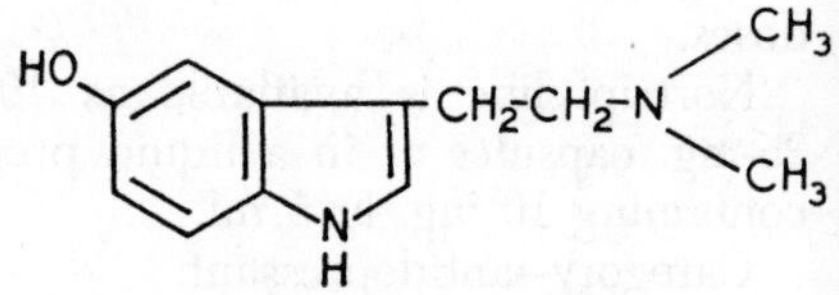

Bufotenine

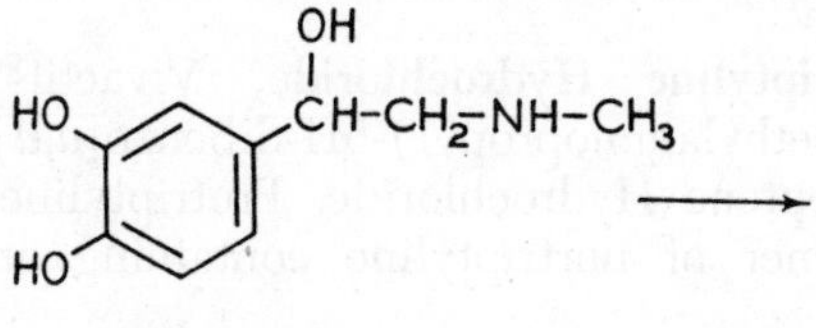

Epinephrine

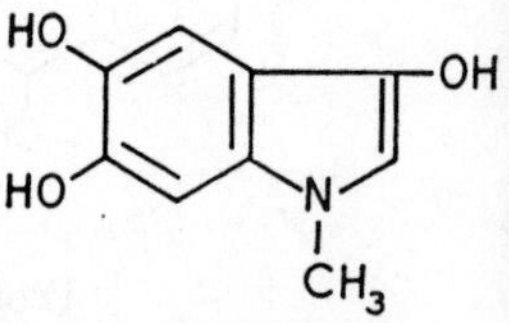

Adrenolutin

Serotonin does not cause behavioral changes after ingestion or injection, since it does not effectively penetrate the central nervous system. A precursor of serotonin, 5-hydroxytryptophan, is capable of central nervous system penetration; particularly in the presence of a monoamine oxidase inhibitor, it produces elevated brain serotonin levels and excited behavior.

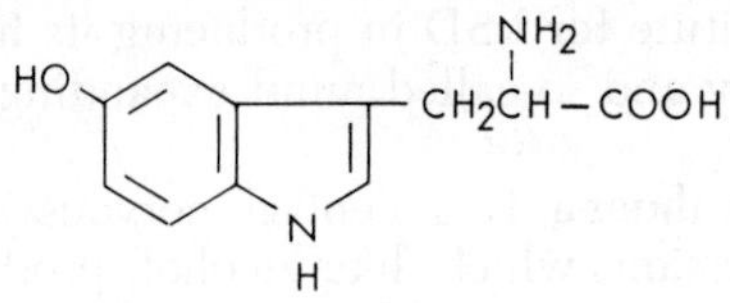

5-Hydroxytryptophan

A number of hallucinogens possess the 3-(β-aminoethyl)indole structure in common with serotonin but, in addition, are N,N-dimethyl derivatives. This tertiary amine structure facilitates penetration of the central nervous system by its increased lipophilic character and appears to delay metabolic oxidative deamination reactions, thus prolonging the existence of active amines in the body.

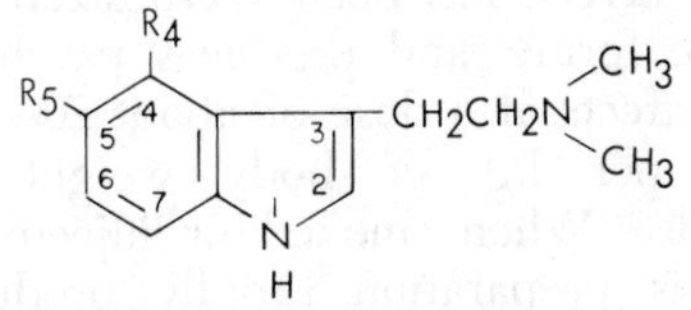

Dimethyltryptamine: $R_4 = R_5 = H$
Bufotenine: $R_4 = H$; $R_5 = OH$
Psilocybin: $R_4 = OPO_2OH$; $R_5 = H$
Psilocyn: $R_4 = OH$; $R_5 = H$

Dimethyltryptamine, DMT, N,N-dimethyltryptamine, and N,N-diethyltryptamine (DET) are hallucinogenic if given by injection. Their psychotomimetic effects have a duration of less than one hour, and are accompanied by pronounced sympathomimetic side effects.

Bufotenine, 5-hydroxy-3-(β-dimethylaminoethyl)indole, is the N,N-dimethyl derivative of serotonin. It occurs naturally in the secretions of the skin of a toad (L. *bufo*) and in the seeds of a plant, *Piptadenia peregrina,* used in the form of a snuff by some South American Indians. Bufotenine is hallucinogenic after injection.

Psilocybin, 4-phosphoryloxy-N,N-dimethyltryptamine, is the active hallucinogenic principle of the mushroom *Psilocybe mexicana.* The mushroom is used in religious ceremonies by Mexican Indians. Psilocybin resembles mescaline and lysergic acid diethylamide in its pharmacologic properties. It has a short onset and duration of action, the peak effect being reached about 2 minutes

after administration. Psilocyn, 4-hydroxy-N,N-dimethyltryptamine, which is the hydrolysis product of psilocybin, is also hallucinogenic.

Lysergic Acid Diethylamide, LSD, a potent hallucinogen, contains both the indole ethylamine (indole nucleus, C_4, C_5, N—CH_3) and the phenylethylamine (benzene ring, C_{10}, C_5 N—CH_3) units within its structure.

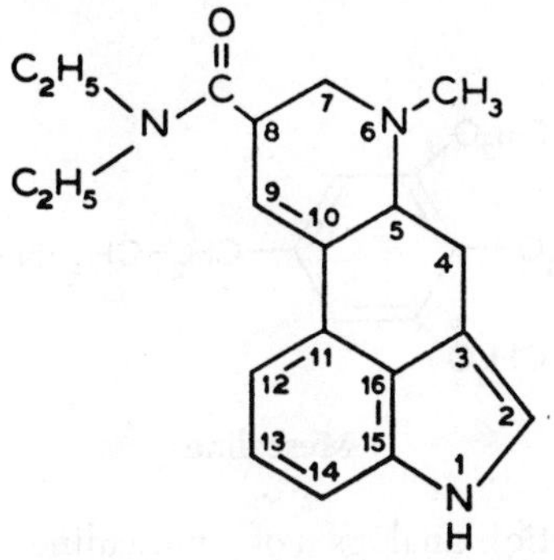

Lysergic Acid Diethylamide

The usual hallucinogenic dose of LSD ranges from 100 to 400 micrograms, taken orally, usually absorbed on an inert support such as sucrose in sugar cube form, or lactose in tablets or capsules. Clammy skin, anxiety and a slight clouding of consciousness occur about 20 to 45 minutes after ingestion, followed in about 15 minutes by the major psychic effects, which last for about 6 hours. With intensities varying widely depending on the individual and his surroundings, effects include disturbances of perception, hallucinations characterized by vivid patterns of color, excitation, euphoria, and loss of personal identity. A phenothiazine tranquilizer, such as chlorpromazine, may be used orally or intramuscularly to terminate the effects of LSD. Prolonged psychic disturbances may occur after use of the drug, frequently including flashes of color. Prolonged psychotic episodes, including schizophrenia, have been reported after discontinuance of use of LSD.

Several β-phenethylamine derivatives, structurally related to norepinephrine and to amphetamine, act rapidly and intensely on the central nervous system. Sympathomimetic side effects, such as elevated blood pressure and pupillary dilatation, are greater than with LSD.

Mescaline, 3,4,5-Trimethoxyphenethylamine, isolated from the stem parts (mescal buttons, peyote) of the cactus *Lophophora williamsii*, has long been used for its hallucinogenic effect during religious ceremonies of certain Indian tribes in Mexico and the southwestern United States.

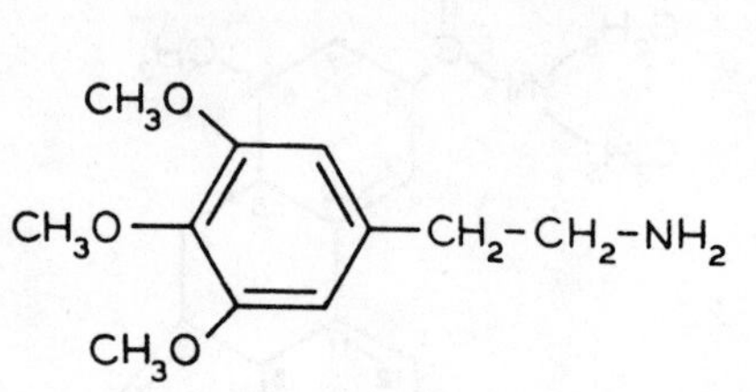

Mescaline

Synthetic analogs of mescaline that possess the branched methyl side chain of amphetamine show greatly enhanced potency as central nervous system stimulants and hallucinogens and are used illegally for these effects. Thus, 1-(2,5-dimethoxy-4-methylphenyl)-2-aminopropane (DOM), also called STP (Serenity, Tranquility, Peace), is about 100 times as potent as mescaline, although still only one thirtieth as potent as LSD. A related synthetic analog, 3,4-methylenedioxyamphetamine (MDA), has also been used illegally as a

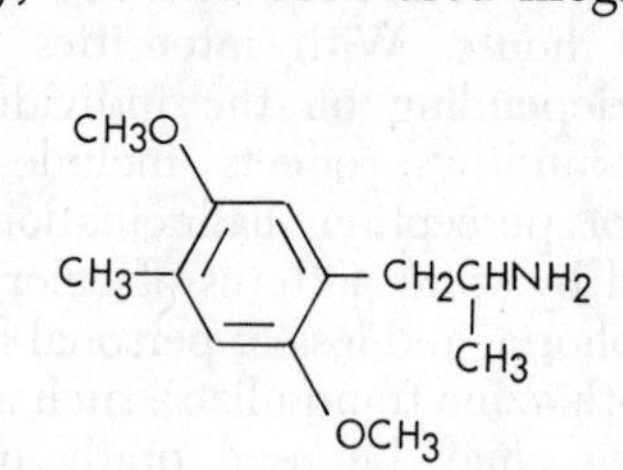

1-(2,5-Dimethoxy-4-methylphenyl)-2-aminopropane
(DOM, STP)

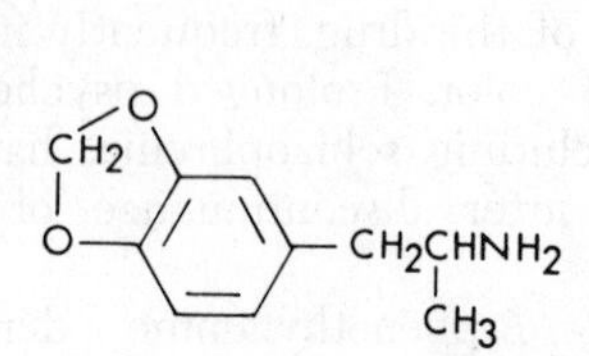

3,4-Methylenedioxyamphetamine
(MDA)

substitute for LSD in producing its hallucinatory and so-called mind expanding properties.

Marihuana is a central nervous system depressant, which, like alcohol, produces a disinhibited, excited state resulting from depression of higher centers of the brain. The drug is widely (although illegally) used. The active components are present in the leaves and especially concentrated in the resin exuded from the flowering tops of the female Indian hemp plant, *Cannabis sativa L.* The highly potent resin is known as hashish in the Middle East, or charas in India. The term marihuana generally refers to the mixed leaves and flowering tops of cannabis; it is usually smoked. The major compound, $(-)$-Δ^1-*trans*-tetrahydrocannabinol (THC), has been synthesized in the racemic form[22] and produces psychotomimetic effects at a dose of about 200 micrograms per kg. of body weight when smoked.[23] When smoked or injected, any cannabis preparation rapidly produces a maximal excitatory effect as the THC rapidly enters the central nervous system; the latency period for onset of effects may be 45 to 60 minutes after ingestion. The effect quickly passes from excitation to sedation as concentrations decrease in the central nervous system, owing to redistribution of the highly lipophilic drug to peripheral lipoidal tissues.

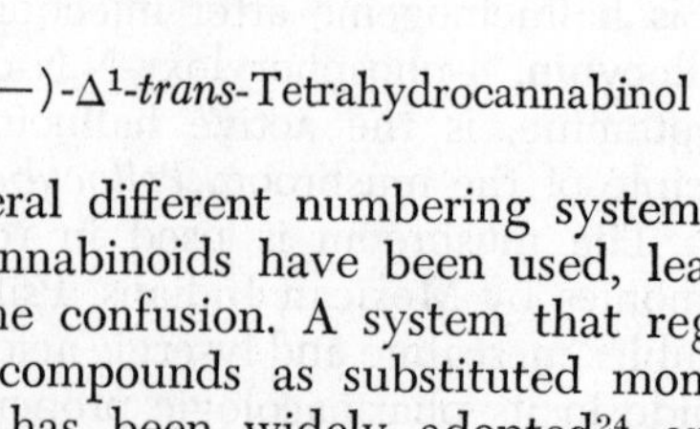

$(-)$-Δ^1-*trans*-Tetrahydrocannabinol

Several different numbering systems for the cannabinoids have been used, leading to some confusion. A system that regards these compounds as substituted monoterpenes has been widely adopted[24] and is used here.

Widespread abuse of the hallucinogenic

drugs led, in 1966, to the inclusion under Food and Drug Administration control of dimethyltryptamine, *d*-lysergic acid diethyl amide, mescaline and its salts, peyote, psilocybin, and psilocyn.

REFERENCES CITED

1. Goodman, A., and Gilman, L.: The Pharmacological Basis of Therapeutics, ed. 3, pp. 324 and 340, New York, Macmillan, 1965.
2. Conroy, H.: J. Am. Chem. Soc. 79:5551, 1957.
3. Holker, J. S. E., Robertsen, A., and Taylor, J. H.: J. Am. Chem. Soc. 80:2987, 1958.
4. Blake, M., and Harris, H. E.: J. Am. Pharm. A. (Sci. Ed.) 41:521, 1952.
5. Lands, A. M.: First Symposium on Chemical-Biological Correlation, p. 73-119, National Academy of Sciences, Washington, 1951.
6. Leake, C. D.: The Amphetamines, Their Actions and Uses, Springfield, Ill., Thomas, 1958.
7. Selikoff, I. J., Robitzek, E. H., and Ornstein, G. G.: Am. Rev. Tuberc. 67:212, 1953.
8. Zeller, E. A., *et al.*: Experientia 8:349, 1952
9. Shore, P. A., and Brodie, B. B.: Proc. Soc. Exp. Biol. Med. 94:433, 1957.
10. Loomer, H. P., Saunders, J. C., and Kline, N. S.: Psychiat. Res. Rep. 8:129, 1958.
11. Spector, S., Shore, P. A., and Brodie, B. B.: J. Pharmacol. Exp. Ther. 128:15, 1960.
12. Jepson, J. B., *et al.*: Biochem. J. (London) 74:5P, 1960.
13. Shore, P. A., *et al.*: Science 126:1063, 1957.
14. Axelrod, J., Senohi, S., and Witkop, B.: J. Biol. Chem. 233:697, 1958.
15. Biel, J. H., Nuhfer, P. A., and Conway, A. C.: Ann. New York Acad. Sc. 80:568, 1959.
16. Zeller, P., *et al.*: Ann. N. Y. Acad. Sci. 80: 555, 1959.
17. Burger, A., and Yost, W: J. Am. Chem. Soc. 70:2198, 1948.
18. Bickel, M. H., and Brodie, B. B.: Int. J. Neuropharmacol. 3:611, 1964.
19. Jucker, E.: Chimia, 15:267, 1961.
20. Hoffsommer, R. D., Taub, D., and Wendler, N. L.: Org. Chem. 27:4134, 1962.
21. Abood, L. G., *in* Burger, A. (ed.): Drugs Affecting the Central Nervous System, Vol. 2, p. 136, New York, M. Dekker, 1968.
22. Mechoulam, R., and Gaoni, Y.: J. Am. Chem. Soc. 87:3273, 1965.
23. Isbell, H.: Psychopharmacologia 11:184, 1967.
24. Mechoulam, R., and Gaoni, Y.: Progr. Chem. Org. Nat. Prod. 25:175, 1967.

SELECTED READING

Biel, J. H.: Some Rationales for the Development of Antidepressant Drugs, *in* Molecular Modification in Drug Design, Advances in Chemistry Series, no. 45, Am. Chem. Soc. Applied Pub., Washington, D.C., 1964.

Burger, A.: Hallucinogenic Agents, *in* Burger, A. (ed.): Medicinal Chemistry, ed. 3, p. 1511, New York, Wiley-Interscience, 1970.

Burns, J. J., and Shore, P. A.: Biochemical effects of drugs, Ann. Rev. Pharmacol. 1:79, 1961.

Hoffer, A., and Osmond, H.: The Hallucinogens, New York, Academic Press, 1967.

Kaiser, C., and Zirkle, C. L.: Antidepressant Drugs, *in* Burger, A. (ed.): Medicinal Chemistry, ed. 3, p. 1476, New York, Wiley-Interscience, 1970.

Rice, L. M., and Dobbs, E. C.: Analeptics, *in* Burger, A. (ed.): Medicinal Chemistry, ed. 3, p. 1402, New York, Wiley-Interscience, 1970.

Whitelock, O. V. (ed.): Amine Oxidase Inhibitors, Ann. N. Y. Acad. Sci. 80:551, 1959.

17

Adrenergic Agents

Edward E. Smissman, Ph.D.,
Professor of Medicinal Chemistry, The School of Pharmacy,
University of Kansas, Lawrence, Kansas

INTRODUCTION

The nervous system with its complex terminology is necessary background to an understanding of the action of the medicinal agents to be discussed in this and the following two chapters. Figure 14 is a simplified illustration of nerve structure. The impulse received at the receptor site is carried to the central nervous system, and the release of a chemical mediator at the synapse gives rise to an action at the effector site.

This action of the nervous system illustrates external stimulus—external action; however, in some cases the effector organ is activated by nervous impulses originating in the brain. If such action occurs at will, it is a voluntary action and is mainly action of skeletal muscles; this is controlled by the *somatic nerves.* Breathing, circulation of blood, glandular secretion, etc., are involuntary actions and are controlled by the autonomic nervous system. Formerly, this system was classified anatomically into sympathetic and parasympathetic divisions. This classification has largely been replaced by the physiologic classification based on the chemical mediator released at the synapse. Therefore, in the *adrenergic nerve,* norepinephrine is the mediator released at the synapse, and in the *cholinergic nerve* acetylcholine is released.

Adrenergic nerves consist only of the postganglionic sympathetic fibers. Cholinergic nerves comprise (a) all postganglionic parasympathetic fibers, (b) all preganglionic fibers, (c) somatic motor nerves to skeletal muscles, and (d) some postganglionic sympathetic fibers such as those to sweat glands and certain peripheral blood vessels (see Fig. 15).

Virtually every involuntary muscle or gland is innervated by both sympathetic and parasympathetic fibers. Stimulation of the sympathetic nerves brings about a liberation of norepinephrine at the neuromuscular or the neurovisceral junction. Stimulation of the parasympathetic nerve causes a liberation of acetylcholine. The autonomic drugs are usually classified as adrenergic or cholinergic, depending on the action they elicit; therefore, the terms adrenergic and sympathomimetic action are used synonymously as are cholinergic and parasympathomimetic.

Some effector cells that respond to adrenergic agents are stimulated, while others are depressed. Ahlquist[1] postulated the existence of two types of adrenergic receptors to explain the dual effects. He termed the receptors responsible for excitatory actions α-types and those with mainly inhibitory action, β-types. The ratio of α- to β-receptors in a given organ determines the type of response.

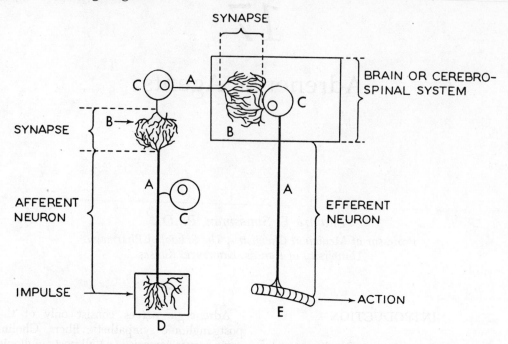

SYNAPSE

BRAIN OR CEREBRO-SPINAL SYSTEM

SYNAPSE

AFFERENT NEURON

EFFERENT NEURON

IMPULSE

ACTION

Fig. 14. A. Axon. B. End brush. C. Cell. D. Receptor dendron. E. Effector organ.

The distinction between the two types of receptor sites is made on the basis of their response to three different agents. The α-receptor is most sensitive to epinephrine, less to norepinephrine and least to isoproterenol. The β-receptor is most responsive to isoproterenol, less to epinephrine, and least to norepinephrine.

Epinephrine

Norepinephrine

Isoproterenol

Table 41 gives the observed effects on various organs by the stimulation of α- and β-receptor sites.

The recognition of chemical mediation in nervous processes was due mainly to the work of Barger and Dale.[2] They observed that the structure and the chemical properties of Adrenalin (epinephrine), a secretion of the adrenal medulla, were similar to those of certain putrefactive amines and caused the same effect as stimulation of the postganglionic sympathetic nerves. They coined the word *sympathomimetic* as a term for this adrenergic action. For many years it was thought that epinephrine was the secretion from the adrenal medulla and also the chemical mediator in postganglionic sympathetic nerves, which was sometimes called sympathin. Contrary to previously accepted theories, epinephrine is almost certainly not a mediator at sympathetic endings. Norepinephrine and epinephrine are present in most visceral organs but only the norepinephrine content is correlated with the number of sympathetic nerve fibers to the organs.

The evidence for two types of adrenergic receptors is excellent, and any postulated mode of action of adrenergics will have to explain this phenomenon. Electrophysiologically, Bülbring and coworkers[3] have shown that catecholamines have a dual action on

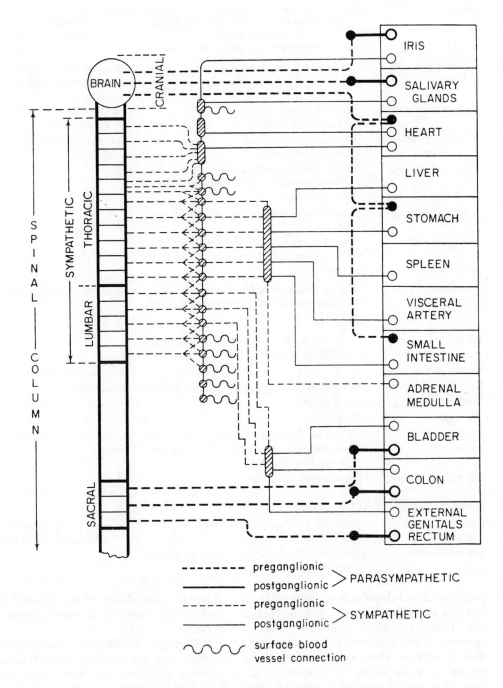

FIG. 15. Diagram of autonomic nervous system.

TABLE 41. DIFFERENTIATION IN THE EFFECTS OF NOREPINEPHRINE,
EPINEPHRINE, AND ISOPROTERENOL

Organs	NOREPINEPHRINE AND EPINEPHRINE: α-RECEPTORS α-SYMPATHOMIMETIC ACTIONS	EPINEPHRINE AND ISOPROTERENOL: β-RECEPTORS
Vascular system heart	Myocardial ectopic excitation	Myocardial contraction-augmentation; cardial acceleration
muscular vessels	Slight decrease in bloodflow Vasoconstriction	Strong increase in bloodflow. Vasodilation
brain vessels (human)	Decrease in bloodflow Vasoconstriction	Increase in bloodflow. Vasodilation
splanchnic area	Strong decrease in bloodflow Vasoconstriction (regulation of blood pressure)	
splenic capsule	Contraction (regulation of blood pressure)	
renal vessels	Strong decrease in bloodflow (regulation of glomerular bloodflow in resp. to diuresis)	
cutaneous vessels	**Strong decrease in bloodflow Vasoconstriction (temperature regulation)**	Slight increase of bloodflow
Pilomotor response	Contraction. Raising of hairs (temperature regulation)	
Bronchial tree		Bronchial relaxation
Intestine	Relaxation of intestinal smooth muscle	Relaxation
Ureter	Contraction	
Vas deferens	Contraction	
Uterus	Excitation. Uterine contractions (depending on condition of uterus; promoted by estrogens)	Inhibition of uterine contractions
Dilator muscle of iris	Contraction (mydriasis)	
Membrana nictitans	Contraction	
Carbohydrate metabolism	Increase in blood-sugar level (glycogenolysis, in liver)	Increase in blood sugar level (glycogenolysis, in muscle)
Fat metabolism	Mobilization of fat (shift from depots to liver)	

tissues—a direct depolarizing contractile effect and an indirect, hyperpolarizing relaxation. These roughly correspond to α and β agonism. It has also been observed that the smooth muscles that usually relax in response to epinephrine (β response) are those that are always in a state of semi-contraction, and those that contract (α response)are those that are normally fully relaxed. However interesting these correlations, they unfortunately do not bring us closer to a knowledge of biochemical detail and/or the receptors involved in these reactions.

Bülbring's realization that catecholamines may be involved in metabolic reactions, however, has been most useful. This early realization was based on the facts that epinephrine fails to produce hyperpolarization in the presence of iodoacetate, that an increase in metabolic rate produced by a rise in temperature causes hyperpolarization, and that glycogen must be present for hyperpolarization to occur. When Sutherland observed in 1960 that epinephrine activated the enzyme phosphorylase, this was regarded as an important part of its metabolic function. It rapidly became apparent, however, that the really important effect was the enhanced formation of adenosine-3′,5′-phosphate from ATP (stimulation of adenyl cyclase). This phenomenon is now associated with the β response to epinephrine. Sutherland[4] has offered an authoritative review summarizing the now voluminous evidence for the involvement of cyclic 3′,5′-AMP in response to catecholamines.

Because of the work of Barger and Dale and later workers, there now are available compounds related to epinephrine that are more or less selective in their action, the β-phenylethylamine structure being altered in 4 major ways:

1. Phenolic hydroxyl substitution in the aromatic nucleus
2. Substitutions on the beta carbon atom
3. Substitutions on the alpha carbon atom
4. Substitution on the amino group

STRUCTURE AND ACTIVITY RELATIONSHIPS OF PHENYLETHYLAMINE ANALOGS

In this series, studied by Barger and Dale, definite conclusions were reached concerning the relationship of structure to pressor activity (production of a rise in blood pressure). The following generalizations can be made.

1. Optimum activity is obtained when the aromatic nucleus is separated from the amino nitrogen by two carbon atoms, AR—C—C—N. Thus, it was found that aniline had no pressor activity, benzylamine had slight activity, α-phenylethylamine was more active, and β-phenylethylamine was the most active. If the carbon chain between the aromatic ring and the amino nitrogen was extended to 3 carbons, the activity decreased.

2. The primary amines are generally more active and less toxic than the corresponding methylated secondary amines. Norepinephrine is less toxic and somewhat more active than epinephrine. Alkyl groups larger than methyl may show varied effects; epinephrine, an N-methylated secondary amine, is a vasopressor agent, but replacement of the N-methyl with N-isopropyl produces a depressor agent.

Dialkylation of the amino group, forming a tertiary amine, decreases the activity. There is usually a simultaneous increase in toxicity, yet such compounds may possess an adequate therapeutic ratio and worthy medicinal properties.

3. The compounds having a methyl group on the alpha carbon have a lower pressor activity than the parent compound and are more toxic; however, this is offset in that the resulting compounds are orally active and have a stimulating action on the central nervous system.

4. In general, an alcoholic hydroxyl group attached to the carbon bearing the aromatic ring renders the drug less toxic and, in many cases, augments the pressor activity. If the secondary alcohol is converted to a ketone, the resulting compound is less active.

The alcoholic hydroxyl group, as a part of the structure, appreciably decreases but does not abolish completely the stimulating effect on the central nervous system (compare ephedrine and desoxyephedrine). The desoxycompounds, as the free bases or as carbonates, are more volatile than the parent hydroxylic compounds and can be used in inhalers.

5. Phenolic compounds have greater intensity of action than the nonphenolic compounds. Monohydroxyl substitution is most effective in the following order: meta, para and ortho. Dihydroxy-substitution is best in the 3,4-position of the aromatic nucleus.

6. Compounds with 3 carbon atoms in the side chain such as ephedrine are much more active when given orally than epinephrine and other compounds having only 2 carbon atoms in the side chain. Increas-

ing the length of the chain beyond 3 carbon atoms affects the pressor activity adversely[5]; however, certain desirable effects, such as bronchodilator effect without increase in blood pressure, may appear. If an alkyl substituent on the alpha carbon is greater than methyl, the pressor activity can be expected to decrease and the toxicity increase. The resulting compound may cause a decrease in blood pressure. Some exceptions to this generalization are known.

The length of action of adrenergic agents is also based on their structural features. The ability of the body to inactivate or detoxify these compounds depends on the length and the branching of the aliphatic chain and on the degree of oxidation of the parent compound.

There are numerous routes of inactivation for the sympathomimetic amines. A review of this subject lists the various modes of in-vivo inactivation.[6] Examples of deaminations, O-methylation, N-demethylation, O-demethylation, and hydroxylation have all been observed for inactivation of epinephrine and related compounds.

Amine oxidase, an enzyme capable of oxidative deamination, is widely distributed in the animal organism. This enzyme promotes the oxidation of primary amines to

TABLE 42. STRUCTURAL RELATIONSHIPS AND PRINCIPAL USES OF
EPINEPHRINE AND RELATED COMPOUNDS

COMPOUND	AR	|CH—CH$_2$—NH|		USE
Epinephrine	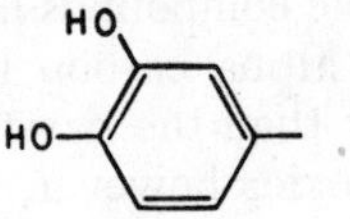	OH	CH$_3$	Vasoconstrictor Bronchodilator Cardiac stimulant
Norepinephrine	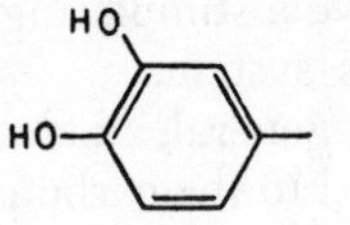	OH	H	Vasoconstrictor Blood pressure maintenance
Isoproterenol	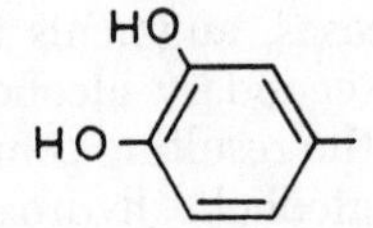	OH	i—C$_3$H$_7$	Bronchodilator
Sympatol	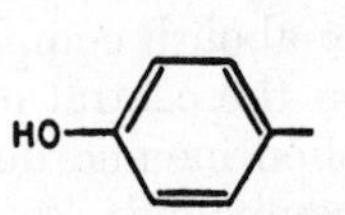	OH	CH$_3$	Nasal decongestant Blood pressure maintenance
Phenylephrine	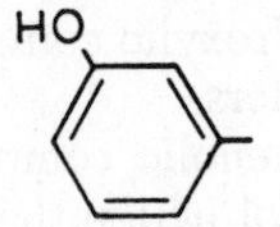	OH	CH$_3$	Nasal decongestant
Methylepinephrine	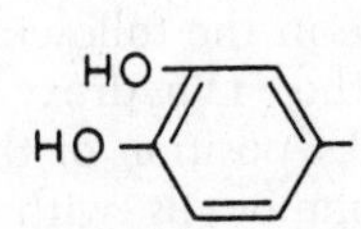	OH	$\left(\begin{array}{c}-N-CH_3\\ \vert \\ CH_3\end{array}\right)$	Hypoglycemic
Metaproterenol	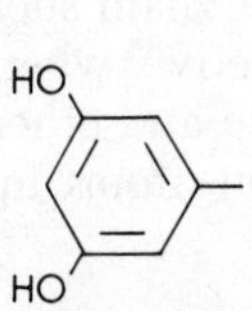	OH	i—C$_3$H$_7$	Bronchodilator

give aldehydes and ammonia; secondary amines and tertiary amines give aldehydes and primary and secondary amines, respectively. This enzyme system is responsible for the complete and rapid degradation of substances such as β-phenylethylamine, isoamylamine and partly for tyramine. Secondary amines, such as ephedrine, have greater stability by virtue of the fact that the deamination is much slower when the amino nitrogen is attached to a secondary carbon, and little deamination occurs.[7]

Phenolases, enzymes capable of oxidizing hydroxylated aromatic systems to quinoid systems, can deactivate phenolic systems rapidly. For example, tyrosinase will oxidize a compound containing a *p*-phenolic hydroxyl group to a catechol derivative. The catechol system is then further oxidized to an *o*-quinoid system. Axelrod[8] found that metanephrine (3-O-methylepinephrine) and normetanephrine (3-O-methylnorepinephrine) in conjugated form occur normally in rat urine, and that the excretion of these compounds is increased when the parent drugs are administered to rats. Metanephrine is only one five-hundredth as active as epinephrine as a pressor agent.

O-Methylation has been proposed as the principal pathway in the metabolism of catecholamines. In man, approximately 68 per cent of administered epinephrine is O-methylated to metanephrine and 23 per cent is deaminated or oxidized; the remainder is excreted as unchanged and conjugated catecholamines. After O-methylation, aminoxidase action occurs to some extent.[9]

For structural relationships and uses see Table 42.

EPINEPHRINE AND RELATED COMPOUNDS

Epinephrine U.S.P., Adrenalin®, Suprarenin®, Suprarenalin®, (−)-3,4-Dihydroxy-α-[(methylamino)methyl]benzyl Alcohol. This compound is a white, odorless, crystalline substance which is light-sensitive. In the official product norepinephrine is present. Initially, epinephrine was isolated from the medulla of the adrenal glands of animals used for food. Although synthetic epinephrine became available soon after the structure of the hormone had been elucidated, the synthetic (±)-base has not been used widely in medicine because the natural levorotatory form is about 15 times as active as the racemic mixture. The steady increase in the price of glands has largely offset the economic advantage of the natural process, especially in Europe. Synthetic epinephrine is prepared as follows:

CATECHOL + CHLOROACETYL CHLORIDE $Cl-\overset{O}{\overset{\|}{C}}-CH_2Cl$ $\xrightarrow{POCl_3}$

1. CH_3NH_2
2. H_2/cat

(±)−EPINEPHRINE

(±)-Epinephrine is resolved into the active (−)-isomer and the less active (+)-isomer by preparing the (+)-tartaric acid salt and separating the diastereoisomeric salts by fractional crystallization. The (+)-isomer can be racemized and the resolution process repeated.

Epinephrine is important historically as the first hormone to be isolated and the first to be synthesized. It arises biogenetically from phenylalanine which is hydroxylated to produce tyrosine and dopa (3,4-dihydroxyphenylalanine). Dopa is decarboxylated to yield dopamine which is then hydroxylated at the β-carbon to give norepinephrine, the latter being N-methylated to epinephrine. (See p. 480 for reactions.)

Because of its catechol nucleus, it is oxidized easily and darkens slowly on exposure to air. Dilute solutions are partially stabilized by the addition of chlorobutanol and by reducing agents, e.g., sodium bisulfite. As the free amine, it is available in oil solution for intramuscular injection

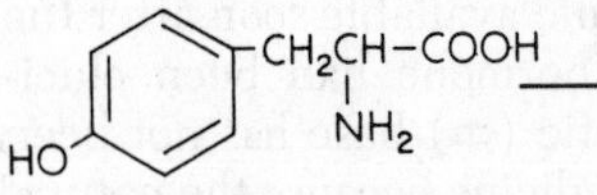

Tyrosine

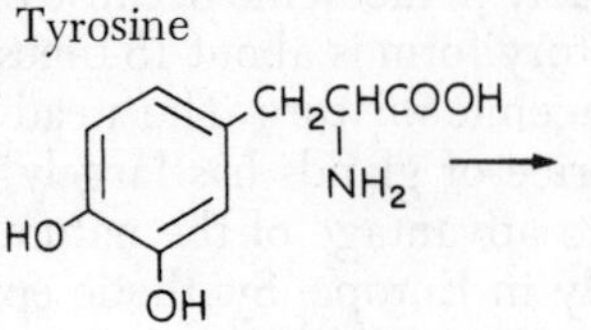

Dopa

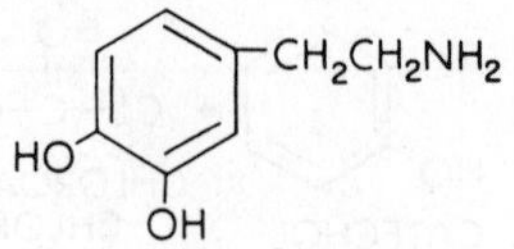

Dopamine

and in glycerin solution for inhalation. Like other amines, it forms salts with acids; for example, those now used include the hydrochloride, the borate and the bitartrate. The bitartrate has the advantage of being less acid and, therefore, is used in the eye because its solutions have a pH close to that of lacrimal fluid. Epinephrine is destroyed readily in alkaline solutions by aldehydes, weak oxidizing agents and oxygen of the air.

Epinephrine is used to cause a rise in blood pressure from stimulation of the vasoconstrictor mechanism of the systemic vessels and of the accelerator mechanism of the heart. Local application is limited but is of value as a constrictor in hemorrhage or nasal congestion. One of its major uses is to enhance the activity of local anesthetics. It is used by injection to relax the bronchial muscle in asthma and in anaphylactic reactions. It is given intravenously to treat acute circulatory collapse. Forms of administration are aqueous or oil solutions, ointment and suppositories.

Epinephrine has the following disadvantages: short duration of action; decomposition of its salts in solution; vasoconstrictive action frequently followed by vasodilation; and inactivity on oral administration.

Category—adrenergic.

Usual dose—oral inhalation, a 0.1 to 1 per cent solution applied as a fine mist as required; intramuscular, 300 to 500 mcg. in a 0.1 per cent solution, or 1 mg. (0.5 ml.) in a 0.2 per cent oil suspension, repeated

as required; subcutaneous, 300 to 500 mcg. in a 0.1 per cent solution, repeated as required.

Usual dose range—200 mcg. to 1 mg. (0.2 to 1 ml.) in a 0.1 per cent solution, or 1 to 3 mg. (0.5 to 1.5 ml.) in a 0.2 per cent oil suspension, repeated as required.

OCCURRENCE

Epinephrine Inhalation U.S.P.
Epinephrine Injection U.S.P.
Epinephrine Solution U.S.P.
Sterile Epinephrine Suspension U.S.P.
Epinephrine Bitartrate U.S.P.
Epinephrine Bitartrate Ophthalmic Solution U.S.P.

Levarterenol Bitartrate U.S.P., Levophed®, $(-)$-α-(Aminomethyl)-3,4-dihydroxybenzyl Alcohol Bitartrate, $(-)$, Norepinephrine Bitartrate. Levarterenol differs

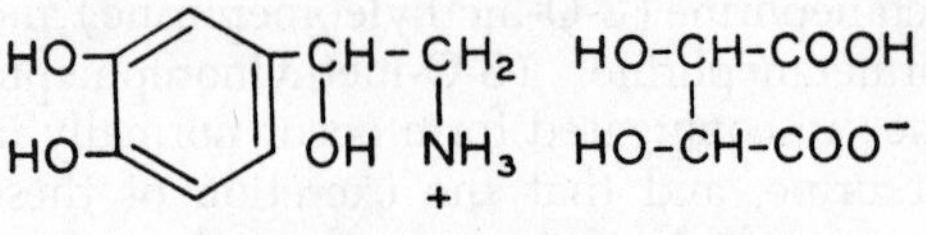

Levarterenol Bitartrate

from epinephrine in that it is a primary amine rather than a secondary amine. A synthesis similar to that of epinephrine can be utilized except that hexamethylenetetramine is used in place of methylamine. Another synthesis starts with protocatechuic aldehyde. The racemic mixture can be resolved through the formation of the $(+)$-acid tartrates. It has been shown that the

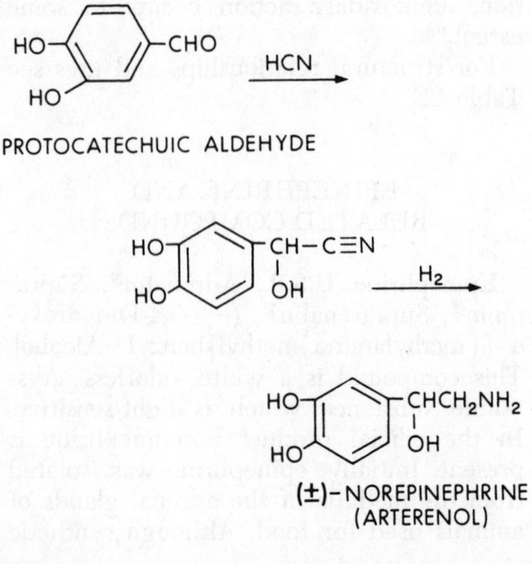

PROTOCATECHUIC ALDEHYDE

(±)-NOREPINEPHRINE
(ARTERENOL)

(−)-isomer is 27 times as potent as the (+)-form.[10]

The bitartrate is a white, crystalline powder which is soluble in water (1:2.5) and in alcohol (1:300). Solutions of the hydrochloride of norepinephrine are comparable with those of epinephrine hydrochloride with regard to stability. The bitartrate salt is available as a more stable injectable solution. It has a pH of 3 to 4 and is preserved by using sodium bisulfite. It is used to maintain blood pressure in acute hypotensive states resulting from surgical or nonsurgical trauma, central vasomotor depression and hemorrhage.

Its action differs from that of epinephrine, since the latter raises blood pressure by increasing cardiac output but has an over-all vasodilator action, while norepinephrine raises blood pressure by peripheral vasoconstriction.

Category—adrenergic (vasopressor).

Usual dose range—intravenous infusion, 8 mg., the equivalent of levarterenol, in 500 ml. of 5 per cent Dextrose Injection, at a rate adjusted to maintain blood pressure.

OCCURRENCE

Levarterenol Bitartrate Injection U.S.P.

Isoproterenol Hydrochloride U.S.P., Aludrine® Hydrochloride, Isuprel® Hydrochloride, 3,4-Dihydroxy-α-[(isopropylamino)methyl]benzyl Alchohol Hydrochloride, Isopropylarterenol Hydrochloride. Isoproterenolium Chloride.

Isoproterenol Hydrochloride

This compound is a white, odorless, slightly bitter, crystalline powder. It is soluble in water (1:3) and in alcohol (1:50). A 1 per cent solution in water is slightly acidic (pH 4.5 to 5.5). It gradually darkens on exposure to air and light. Its aqueous solutions become pink on standing.

The presence of a larger alkyl group on the nitrogen atom almost eliminates circulatory effects except palpitation of the heart. It is related to epinephrine and norepinephrine in many of its actions, the most important difference being its bronchiospasmolytic activity. When given sublingually or by inhalation, the drug is of value in the symptomatic treatment of mild and moderately severe asthma. It is effective in some asthmatics who do not respond to epinephrine or theophylline combinations.

Besides palpitation of the heart, other side-effects on the administration of isoproterenol are nausea, headache, nervousness and weakness.[11] The use of the drug must be supervised carefully.

The drug should not be given parenterally because of its intense stimulation of the heart muscle; also, it should not be given with epinephrine but may be alternated with it.

Category—adrenergic (bronchodilator).

Usual dose—sublingual, 10 to 15 mg. 3 to 4 times daily; by aerosol inhalation, 125 to 250 mcg. of a 0.25 per cent solution, repeated at 1-,5-, and 10 minute intervals as necessary; intramuscular or subcutaneous, initial, 200 mcg., followed by 200 mcg. to 1 mg. repeated as necessary; intravenous, initial, 10 to 20 mcg., followed by 10 to 200 mcg. repeated as necessary; infusion, 2 mg. in 500 ml. of 5 per cent Dextrose Injection at a rate adjusted to maintain blood pressure.

Usual dose range—sublingual, 10 to 60 mg. daily; by aerosol inhalation, 125 to 750 mcg. in 2 hours; parenteral, 10 to 200 mcg.

OCCURRENCE

Isoproterenol Hydrochloride Inhalation U.S.P.
Isoproterenol Hydrochloride Injection U.S.P.
Isoproterenol Hydrochloride Tablets U.S.P.

Isoproterenol Sulfate U.S.P., Isonorin® Sulfate, Medihaler-Iso®, Norisodrine® Sulfate, 3,4-Dihydroxy-α-[(isopropylamino)methyl]benzyl Alcohol Sulfate, 1-(3′,4′-Dihydroxyphenyl)-2-isopropylaminoethanol Sulfate. This compound is a white odorless, slightly bitter crystalline powder. It is slightly soluble in alcohol and freely soluble in water and is hygroscopic. A 1 per cent solution in water is acidic (pH 3.5 to 4.5). Its aqueous solutions become pink on standing. It is used for the same purpose as is the corresponding hydrochloride.

Category—adrenergic (bronchodilator).

Usual dose—by aerosol inhalation, 0.5 ml. of a 0.5 per cent solution.

OCCURRENCE

Isoproterenol Sulfate Inhalation U.S.P.
Isoproterenol Sulfate Aerosol N.F.

Phenylephrine Hydrochloride U.S.P., Neo-Synephrine® Hydrochloride, Isophrin® Hydrochloride, (−)-*m*-Hydroxy-α-[(methylamino)methyl]benzyl Alcohol Hydrochloride, Phenylephrinium Chloride. This compound is a white, odorless, crystalline, slightly bitter powder which is freely soluble in water and in alcohol. It is relatively stable in alkaline solution and is unharmed by boiling for sterilization.

Phenylephrine can be synthesized by several methods. The following is a method not previously discussed:

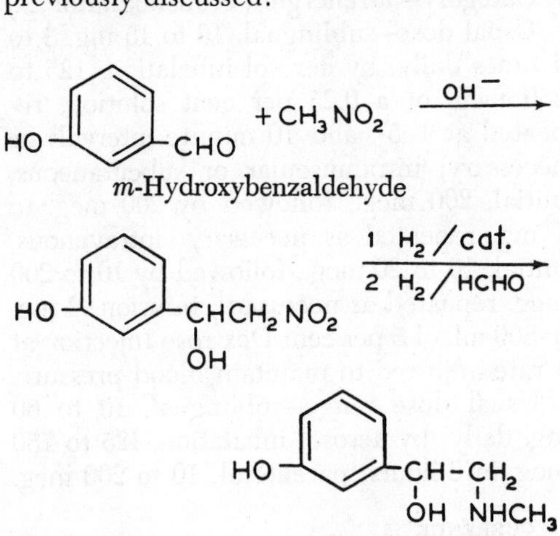

m-Hydroxybenzaldehyde

(±) Phenylephrine

The racemic mixture is resolved by preparing the (+)-camphorsulfonic acid salt.

The duration of action is about twice that of epinephrine. It is a vasoconstrictor and is active when given orally. It is relatively nontoxic and, when applied to mucous membrane, reduces congestion and swelling by constricting the blood vessels of the mucous membranes. It has little central nervous stimulation and finds its main use in the relief of nasal congestion. It is also used as a mydriatic agent, as an agent to prolong the action of local anesthetics and

to prevent a drop in blood pressure during spinal anesthesia.

Category—adrenergic.

Usual dose—subcutaneous or intramuscular, 5 mg., repeated in 10 minutes if required; intraveous infusion, 10 to 20 mg. in 500 ml. of an isotonic infusion solution at a rate adjusted to maintain blood pressure.

Usual dose range—1 to 20 mg.

For external use—topically: 0.1 ml. of a 0.125 to 10 per cent solution to the conjuctiva; 0.2 to 0.6 ml. of a 0.125 to 1 per cent solution to mucous membranes 4 to 6 times daily.

OCCURRENCE

Phenylephrine Hydrochloride Injection U.S.P.
Phenylephrine Hydrochloride Ophthalmic Solution U.S.P.
Phenylephrine Hydrochloride Solution U.S.P.

Sympatol Tartrate, Synephrine® Tartrate, (±)-1-(p-Hydroxyphenyl)-2-methylaminoethanol Tartrate. This compound is a

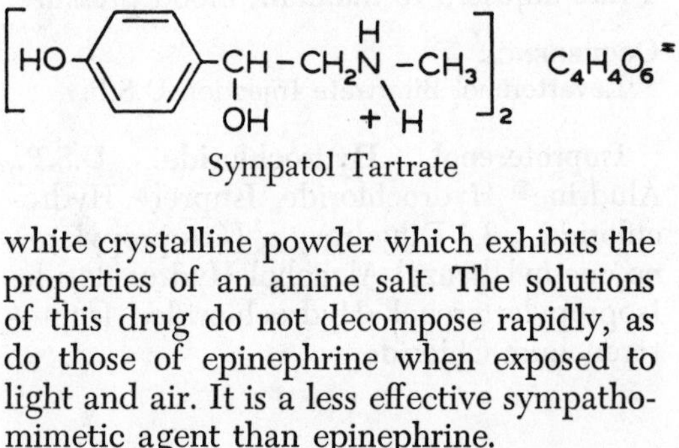

Sympatol Tartrate

white crystalline powder which exhibits the properties of an amine salt. The solutions of this drug do not decompose rapidly, as do those of epinephrine when exposed to light and air. It is a less effective sympathomimetic agent than epinephrine.

It has been used for chronic hypotension, collapse due to shock, and other disorders resulting in hypotension.

Isoetharine, Dilabron®, N-isopropylethylnorepinephrine.

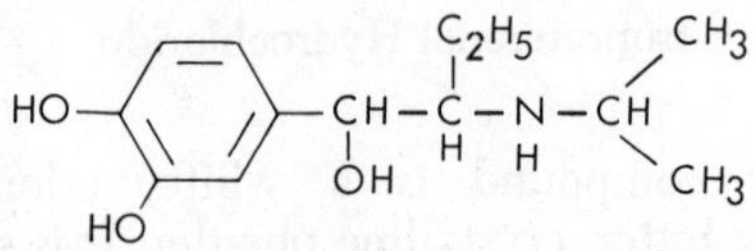

Isoetharine

This compound is a synthetic sympathomimetic amine, largely beta-active in effect. Its bronchodilating properties closely

simulate those of epinephrine but without significant pressor effects. It thus may be safer than epinephrine for hypertensive patients and severely ill patients in whom such effects are undesirable. Isoetharine is particularly adapted for use in children because of its relative lack of adverse effects, especially central nervous system excitation. It also may be of value in diabetic asthmatics due to its lack of glycogenolytic activity.

Metaproterenol, Alupent®, 3,-Dihydroxy-α-[(isopropylamino)methyl]benzyl Alcohol. This is a recently studied sym-

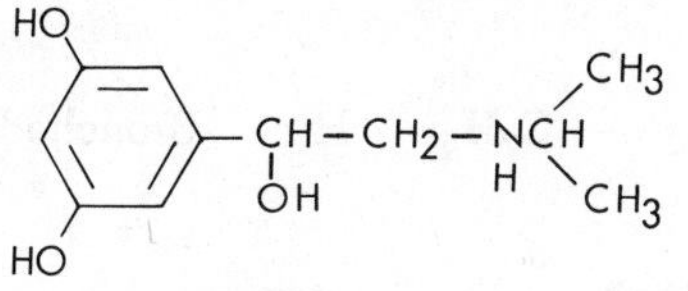

Metaproterenol

pathomimetic amine differing chemically from isoproterenol only in that both hydroxyl groups of the phenyl nucleus are in the meta position. Aerosol isoproterenol and oral Alupent in the doses used in the study produced similar bronchodilator response in asthmatics. Aerosol isoproterenol had its optimal effect earlier, but its duration of action was shorter. Subjective and objective side reactions were more prominent with isoproterenol when the two drugs were given under the same conditions.

EPHEDRINE AND RELATED COMPOUNDS

A comparison of structures of the compounds in this series is given in Table 43.

Ephedrine N.F., (−)-*erythro*-α-[(1-Meth-

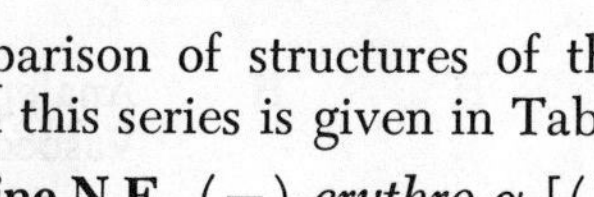

Ephedrine

ylamino)ethyl]benzyl Alcohol. Ephedrine is an alkaloid which can be obtained from the stems of various species of Ephedra. The drug Ma Huang, containing ephedrine, was known to the Chinese in 2800 B.C., but

the active principle, ephedrine, was not isolated until 1885. The pure alkaloid was obtained and named ephedrine by Nagai.[12] It was first synthesized in 1920 by Späth and Göhring.[13] Chen and Schmidt[14] investigated the pharmacology of ephedrine, and this led to wide interest in its use as a medicinal agent.

Ephedrine has 2 asymmetric carbon atoms; thus there are 4 optically active forms. The *erythro* racemate is called ephedrine, and the *threo* racemate is known as *pseudo*ephedrine (ψ-ephedrine). Natural

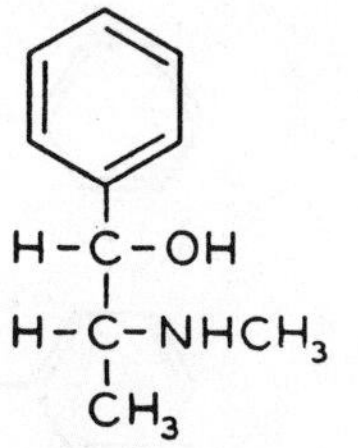

Erythro form
Ephedrine

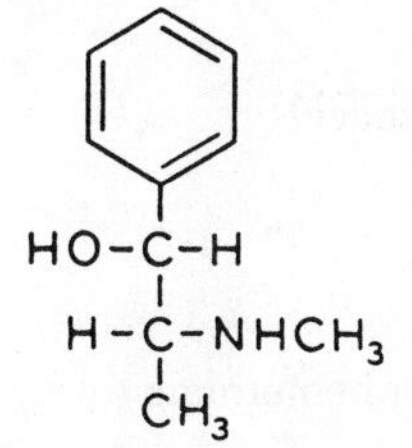

Threo form
ψ-Ephedrine

ephedrine is D(−) and is the most active of the 4 isomers as a pressor amine. Table 44 lists the relative pressor activity of isomers of ephedrine. Racemic ephedrine, racephedrine, is used for the same purpose as the optically active alkaloids. *Pseudo*ephedrine can be partially epimerized by treatment with 20 per cent hydrochloric acid to give a mixture of ephedrine and *pseudo*-ephedrine.

Lapidus[15] has shown that D(−)ψ-ephedrine can block the action of D(−)-ephedrine. He proposes that both compounds can occupy the same three sites on a receptor surface and that a competitive inhibition will be established.

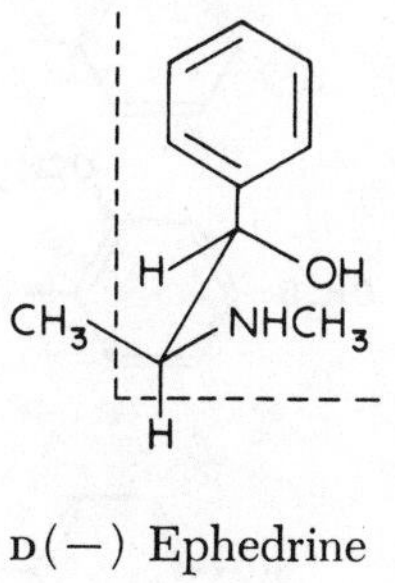

D(−) Ephedrine 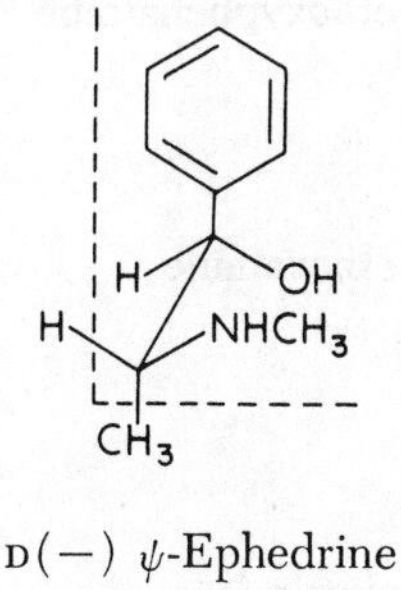   D(−) ψ-Ephedrine

TABLE 43. STRUCTURAL RELATIONSHIPS AND PRINCIPAL USES OF EPHEDRINE AND RELATED COMPOUNDS

Compound	AR—	—CH—	—CH—	—N		Principal Uses
Ephedrine	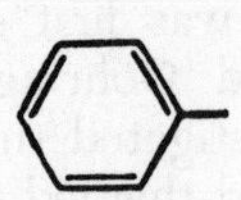 (phenyl)	OH	CH$_3$	CH$_3$	H	Vasopressor Analeptic Antiasthmatic
Phenylpropanolamine	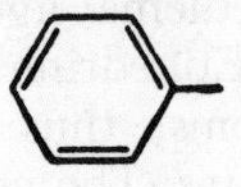 (phenyl)	OH	CH$_3$	H	H	Vasopressor Nasal decongestant
Etafedrin	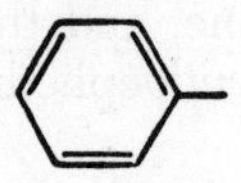 (phenyl)	OH	CH$_3$	C$_2$H$_5$	CH$_3$	Bronchodilator
Nethacol	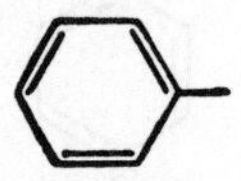 (phenyl)	OH	CH$_3$	C$_2$H$_5$	H	Bronchodilator
Mephentermine	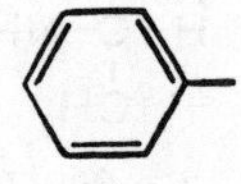 (phenyl)	H	$\left(-\overset{\text{CH}_3}{\underset{\text{CH}_3}{\text{C}}}-\right)$	CH$_3$	H	Vasoconstrictor
Metaraminol	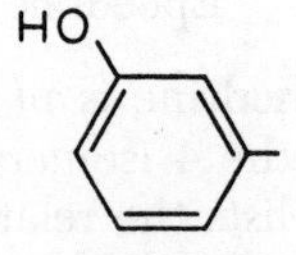 (3-HO-phenyl)	OH	CH$_3$	H	H	Nasal decongestant
Hydroxyamphetamine	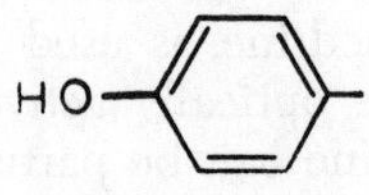 (4-HO-phenyl)	H	CH$_3$	H	H	Mydriatic Nonstimulating vasopressor
Ethylnorepinephrine	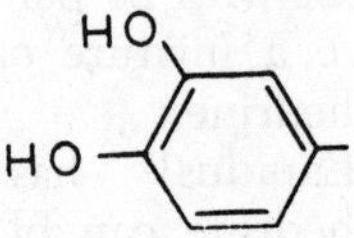 (3,4-diHO-phenyl)	OH	C$_2$H$_5$	H	H	Bronchodilator
Levonordefrin	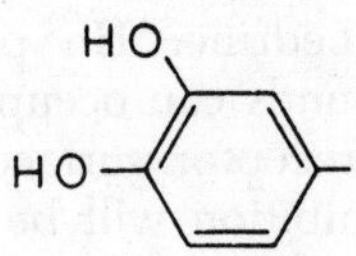 (3,4-diHO-phenyl)	OH	CH$_3$	H	H	Analeptic Vasoconstrictor
Methoxyphenamine	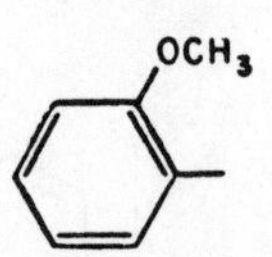 (2-OCH$_3$-phenyl)	H	CH$_3$	CH$_3$	H	Bronchodilator
Methoxamine	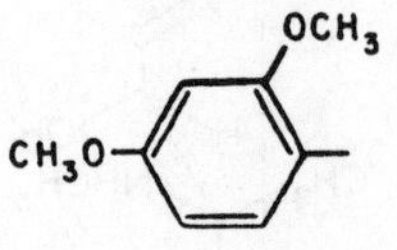 (2-OCH$_3$, 5-CH$_3$O-phenyl)	OH	CH$_3$	H	H	Vasoconstrictor
Nylidrin	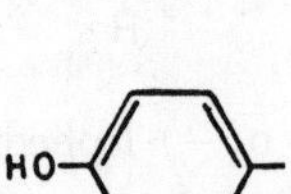 (4-HO-phenyl)	OH	CH$_3$	CH–CH$_3$ $\underset{\text{CH}_2}{\overset{}{|}}$ CH$_2$	H	Peripheral vasodilator

TABLE 44.

ISOMER			RELATIVE PRESSOR ACTIVITY
D	(−)	Ephedrine	36
DL	(±)	Ephedrine	26
L	(+)	Ephedrine	11
L	(+)	*Pseudo*ephedrine	7
DL	(±)	*Pseudo*ephedrine	4
D	(−)	*Pseudo*ephedrine	1

Partial biosynthesis of ephedrine can yield predominantly the D(−) form. The first step, acyloin formation is biologically stereochemically controlled. The second step is an example of asymmetric induction: control of the entering group by the asymmetric center present in the molecule.[17]

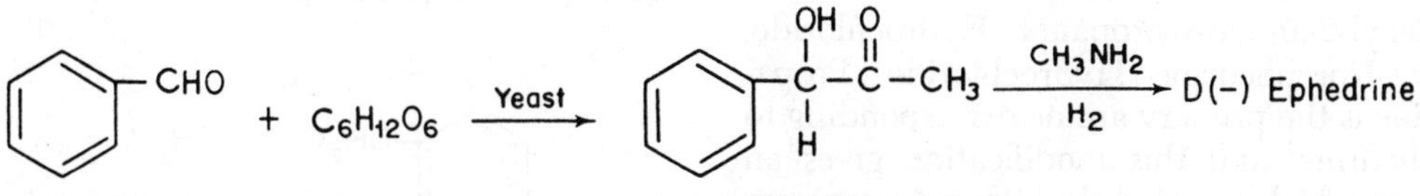

Benzaldehyde   Glucose   (−)-1-Phenyl-1-hydroxy-propanone-2

The ephedrine alkaloid occurs as a waxy solid and as crystals or granules and has a characteristic pronounced odor. Because of its instability in light, it decomposes gradually and darkens. It may contain up to one half molecule of water of hydration. It is soluble in alcohol, water (5%), some organic solvents and liquid petrolatum. The free alkaloid is a strong base with a pKa of 9.6. An aqueous solution of the free alkaloid has a pH above 10.

Ephedrine simulates epinephrine in physiologic effects but its pressor action and local vasoconstrictor action are of greater duration. It causes more pronounced stimulation of the central nervous system than does epinephrine, and it is effective when given orally or systemically. On inspection of the ephedrine molecule, it should be noted that there is an α-methyl group, in contrast with the absence of such a group in epinephrine. It has been concluded that an α-methyl group in the β-phenylethylamine structure confers oral activity by virtue of the greater in-vivo stability of such a molecule. Thus, β-phenylethylamine is in-

active when given orally, but amphetamine (1-phenyl-2-aminopropane) is effective by mouth.

Ephedrine and its salts are used orally, intravenously, intramuscularly and topically in a variety of conditions such as allergic disorders, colds, hypotensive conditions and narcolepsy. It is employed locally to constrict the nasal mucosa and cause decongestion, to dilate the pupil or the bronchi and to diminish hyperemia. Systemically, it is effective for asthma, hay fever, urticaria, low blood pressure and the alleviation of muscle weakness in myasthenia gravis.

Category—adrenergic (bronchodilator).

Usual dose (sulfate salt)—oral or parenteral, 25 mg. 4 to 6 times daily.

Usual dose range (sulfate salt)—oral or parenteral, 100 to 200 mg. daily.

For external use (sulfate salt)—topically, as a 1 to 3 per cent solution to mucous membranes 4 to 6 times daily.

OCCURRENCE

Ephedrine Hydrochloride N.F.
Ephedrine Sulfate U.S.P.
Ephedrine Sulfate Capsules U.S.P.
Ephedrine Sulfate Injection U.S.P.
Ephedrine Sulfate Solution U.S.P.
Ephedrine Sulfate Syrup U.S.P.
Ephedrine Sulfate and Phenobarbital
    Capsules N.F.
Ephedrine Sulfate Tablets N.F.
Theophylline, Ephedrine Hydrochloride, and
    Phenobarbital Tablets N.F.

Pseudoephedrine Hydrochloride N.F., Sudafed®, (+)-*threo*-α-[(1-Methylamino)-ethyl]benzyl Alcohol Hydrochloride, Isoephedrine Hydrochloride. The hydrochloride salt is a white, crystalline material, soluble in water, in alcohol, and in chloroform. Pseudoephedrine, like ephedrine, is a useful bronchodilator, but is much less active in increasing blood pressure. However, it should be used with caution in hypertensive individuals.

Category—adrenergic.

Usual dose—30 mg. 3 times daily.

Usual dose range—30 to 60 mg. 3 or 4 times daily.

OCCURRENCE

Pseudoephedrine Hydrochloride Syrup N.F.
Pseudoephedrine Hydrochloride Tablets N.F.

Phenylpropanolamine   Hydrochloride N.F., Propadrine® Hydrochloride, (±)-1-

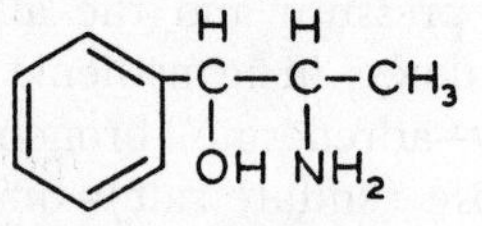

Phenylpropanolamine

Phenyl-2-amino-1-propanol Hydrochloride, (±)-Norephedrine Hydrochloride. Propadrine is the primary amine corresponding to ephedrine, and this modification gives an agent which has slightly higher vasopressor action and lower toxicity and central stimulation action than has ephedrine. It can be used in place of ephedrine for most purposes and is used widely as a nasal decongestion agent. For the latter purpose it is applied locally to shrink swollen mucous membranes; its action is more prolonged than that of ephedrine. It also is stable when given orally.

Propadrine is available as the racemic mixture synthetically, but the (−) form has been found as a minor alkaloid in Ephedra species.

Category—adrenergic (vasoconstrictor).

The usual dose is 25 to 50 mg. orally at intervals of 3 to 4 hours. (It is used topically as a 0.66 per cent jelly and as a 1 per cent solution.)

Etafedrine Hydrochloride, Nethamine® Hydrochloride,   2 - Methylethylamino - 1 - phenyl-1-propanol Hydrochloride, *Levo*-N-ethylephedrine Hydrochloride. This compound differs from ephedrine, in that it is

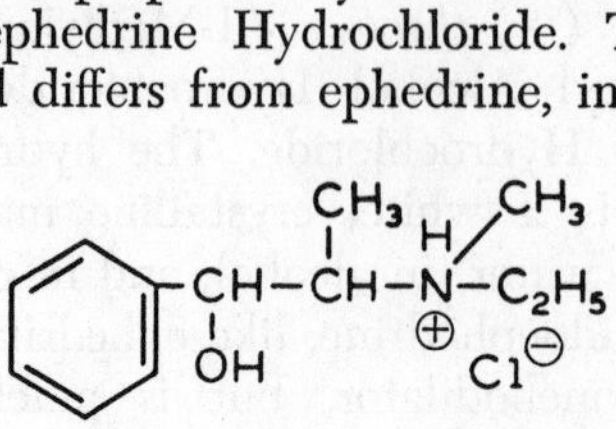

Etafedrine Hydrochloride

a tertiary amine. This difference also alters its physiologic properties. It is claimed to be an effective smooth muscle relaxant but has only slight effect as a vasopressor and a central nervous system stimulant.

Mephentermine, Wyamine®, N,α,α-Trimethylphenethylamine. This compound is a clear, colorless to pale-yellow liquid with a fishy odor. It is very soluble in alcohol and practically insoluble in water. It is a sympathomimetic amine which is sufficiently volatile to produce vasoconstriction of the congested nasal mucous membrane, being about one half as active in this respect as amphetamine. Mephentermine is less toxic than amphetamine, and in pressor action it resembles ephedrine rather than epinephrine; cerebral stimulant action is much less than that of amphetamine.

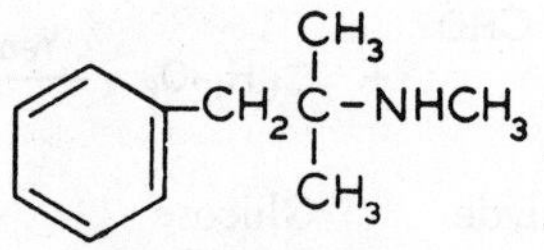

Mephentermine

Mephentermine is used only by inhalation as a nasal decongestant to provide temporary relief in the treatment of acute or chronic rhinitis, allergic or nonallergic rhinitis, vasomotor rhinitis and sinusitis. It is applied to the nasal mucosa by means of an inhaler containing a total of 250 mg. of the drug. Two inhalations through each nostril are recommended as a single dose.

Mephentermine Sulfate U.S.P., Wyamine® Sulfate, N,α,α-Trimethylphenethylamine Sulfate. The sulfate is a white crystalline powder with a faint fishy odor. It is soluble 1:20 in water and 1:50 in alcohol. A 1 per cent solution in water is acidic, pH 5.5 to 6.2.

It exhibits pressor amine properties and is used topically as a nasal decongestant. It may be injected parenterally as a vasopressor agent in acute hypotensive states.

Category—adrenergic (vasopressor).

Usual dose—oral, 12.5 to 25 mg. once or twice daily; intramuscular or intravenous, the equivalent of 15 to 30 mg. of mephentermine; by infusion, 150 mg. in 500 ml. of an isotonic solution at a rate adjusted to maintain the blood pressure.

Usual dose range—the equivalent of 12.5 to 80 mg. of mephentermine or mephentermine sulfate, repeated as required.

OCCURRENCE
Mephentermine Sulfate Injection N.F.
Mephentermine Sulfate Tablets U.S.P.

Metaraminol Bitartrate U.S.P., Aramine® Bitartrate, (−)-α-(1-Aminoethyl)-*m*-hy-

Usual dose—intravenous, the equivalent of 500 mcg. of metaraminol; by infusion, the equivalent of 50 mg. of metaraminol in 500 ml. of an isotonic solution at a rate adjusted to maintain blood pressure; intramuscular or subcutaneous, the equivalent of 2 to 5 mg. of metaraminol.

Usual dose range—intravenous, the equivalent of 500 mcg. to 5 mg. of metaraminol;

Metaraminol Bitartrate

droxybenzyl Alcohol Tartrate, (−)-*m*-Hydroxynorephedrine Bitartrate. Metaraminol bitartrate is freely soluble in water and 1:100 in alcohol. This compound is a potent vasopressor with prolonged duration of action. It is employed as a nasal decongestant in the symptomatic relief of nasal edema accompanying the common cold, rhinitis, sinusitis and nasopharyngitis.

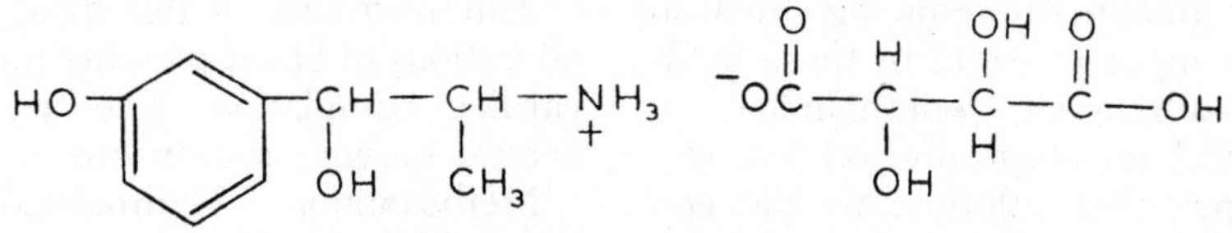

Metaraminol bitartrate is useful for parenteral administration in hypotensive episodes during surgery, for sustaining blood pressure in patients under general or spinal anesthesia and for the treatment of shock associated with trauma, septicemia, infectious diseases and adverse reactions to medication. It does not produce central nervous system stimulation.

It is synthesized from an acyloin prepared by a fermentative process by the following method:

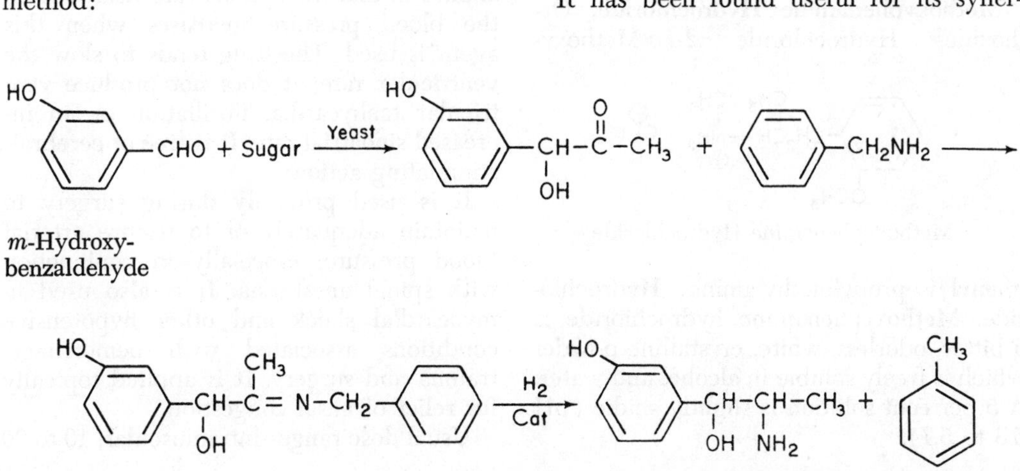

by infusion, 15 to 100 mg. per 500 ml.; intramuscular or subcutaneous, the equivalent of 2 to 12 mg. of metaraminol.

OCCURRENCE
Metaraminol Bitartrate Injection U.S.P.

Hydroxyamphetamine Hydrobromide U.S.P., Paredrine® Hydrobromide, (±)-*p*-(2-Aminopropyl)phenol Hydrobromide, 1-

Hydroxyamphetamine Hydrobromide

(p-Hydroxyphenyl)-2-aminopropane Hydrobromide. This compound is a white crystalline material which is very soluble in water (1:1) and in alcohol (1:2.5).

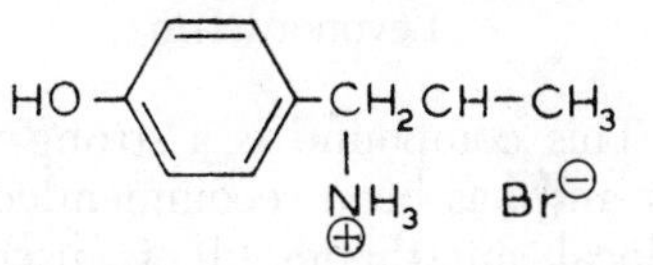

It has been found useful for its syner-

m-Hydroxybenzaldehyde

Metaraminol

gistic action with atropine in producing mydriasis. A more rapid onset, more complete dilation and more rapid recovery are observed with a mixture of atropine and hydroxyamphetamine hydrobromide than with atropine alone.

Hydroxyamphetamine has little or no ephedrinelike central-nervous-system-stimulating action but retains the ability to shrink the nasal mucosa. Its actions as a bronchodilator or as an appetite-reducing agent are too weak to make it useful in these fields.

Category—adrenergic (ophthalmic).

For external use—topically, 0.1 ml. of a 0.25 to 1 per cent solution, to the conjunctiva, repeated as required.

OCCURRENCE

Hydroxyamphetamine Hydrobromide Ophthalmic Solution U.S.P.

Levonordefrin N.F., Cobefrin®, (−)-α-(1-Aminoethyl)-3,4-dihydroxybenzyl Al-

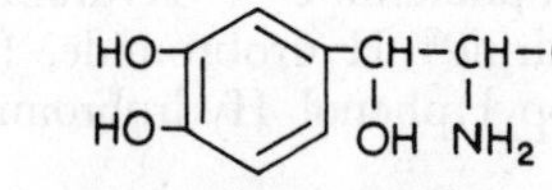

Levonordefrin

cohol. This compound is a strong vasoconstrictor and has been recommended for use with local anesthetics. It is used as an analeptic which is very active by oral administration.

The structure has the catechol nucleus of epinephrine but the side chain of norephedrine.

Category—vasoconstrictor.

Methoxyphenamine Hydrochloride, Orthoxine® Hydrochloride, 2-(o-Methoxy-

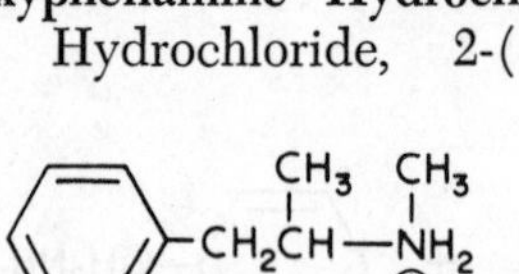

Methoxyphenamine Hydrochloride

phenyl)isopropylmethylamine Hydrochloride. Methoxyphenamine hydrochloride is a bitter, odorless, white, crystalline powder which is freely soluble in alcohol and water. A 5 per cent solution is slightly acidic (pH 5.3 to 5.7).

It is a sympathomimetic compound whose predominant actions are bronchodilation and inhibition of smooth muscle. Its effects on blood vessels are slight, its pressor effect being considerably less than that of ephedrine or epinephrine. Methoxyphenamine is useful as a bronchodilator in the treatment of asthma and also is effective in allergic rhinitis, acute urticaria and gastrointestinal allergy.

Administration of this drug produces no alterations in blood pressure and only slight cardiac stimulation. The actions on the central nervous system are minor.

Methoxamine Hydrochloride U.S.P., Vasoxyl® Hydrochloride, α-(1-Aminoethyl)-2,5-dimethoxybenzyl Alcohol Hydrochloride; 2-Amino-1-(2,5-dimethoxyphenyl)-

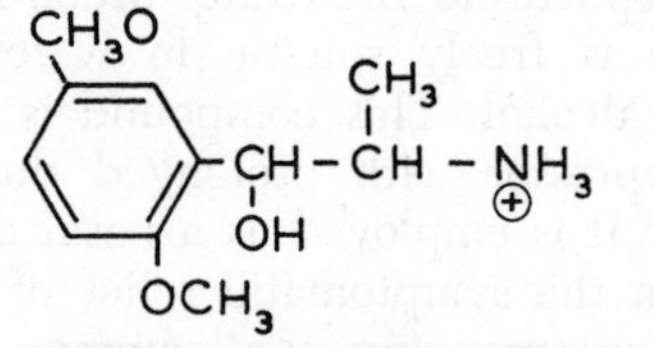

Methoxamine Hydrochloride

propanol Hydrochloride. This compound is a white, platelike crystalline substance with a bitter taste. It is odorless or has only a slight odor. It is soluble in water 1:2.5 and in alcohol 1:12. A 2 per cent solution in water is slightly acidic (pH 4.0 to 5.0), and it is affected by light.

Methoxamine hydrochloride is a sympathomimetic amine that exhibits the vasopressor action characteristic of other agents of this class, but is unlike most pressor amines in that the cardiac rate decreases as the blood pressure increases when this agent is used. The drug tends to slow the ventricular rate; it does not produce ventricular tachycardia, fibrillation or an increased sinoatrial rate. It is free of cerebral-stimulating action.

It is used primarily during surgery to maintain adequately or to restore arterial blood pressure, especially in conjunction with spinal anesthesia. It is also used in myocardial shock and other hypotensive conditions associated with hemorrhage, trauma and surgery. It is applied topically for relief of nasal congestion.

Usual dose range—intramuscular, 10 to 20

mg., intravenous, 5 to 10 mg.; infusion, 60 mg. in 500 ml. of an isotonic solution at a rate adjusted to maintain blood pressure.

OCCURRENCE
Methoxamine Hydrochloride Injection U.S.P.

Nylidrin Hydrochloride N.F., Arlidin®, *p*-Hydroxy-α-[1-[(1-methyl-3-phenyl-

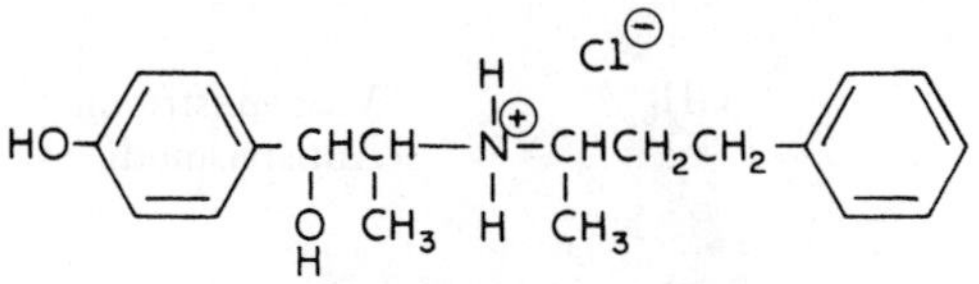

Nylidrin Hydrochloride

propyl)amino]ethyl]benzyl Alcohol Hydrochloride. This compound is a white, odorless, practically tasteless, crystalline powder. It is soluble in water 1:65 and in alcohol 1:40. A 1 per cent solution in water is acidic (pH 4.5 to 6.5).

Nylidrin is an epinephrine-ephedrine type compound that acts as a peripheral vasodilator. It is indicated in vascular disorders of the extremities that may be benefited as the result of increased blood flow. It is administered orally or parenterally by subcutaneous or intramuscular injection.

Category—peripheral vasodilator.

Usual dose—oral, 6 mg. 3 times daily; parenteral, 5 mg. one or more times daily.

Usual dose range—oral, 3 to 12 mg.; parenteral, 2.5 to 5 mg.

OCCURRENCE
Nylidrin Hydrochloride Injection N.F.
Nylidrin Hydrochloride Tablets N.F.

Isoxsuprine Hydrochloride N.F., Vasodilan®, *p*-Hydroxy-α-[1-[(1-methyl-2-phenoxyethyl)amino]ethyl]benzyl Alcohol Hydrochloride.

This compound is used as a vasodilator for symptomatic relief in peripheral vascu-

Isoxsuprine Hydrochloride

lar disease and cerebrovascular insufficiency. It acts without ganglionic blockade, neural mediation, hormonal or other undesirable effects. It may be used safely in patients with coronary artery diseases, diabetes, and asthma.

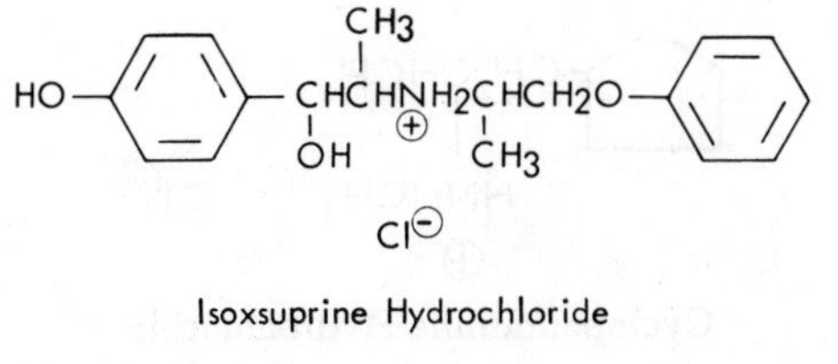

Category—vasodilator.

Usual dose range—10 mg. to 20 mg. 3 or 4 times daily.

OCCURRENCE
Isoxsuprine Hydrochloride Tablets N.F.

ALIPHATIC AMINES

In the classic work of Barger and Dale[2] in 1910, the pressor action of the aliphatic amines was described, but only since the early 1940's have these agents become of pharmaceutical importance. An investigation in 1944[18] determined the influence of the location of an amino group on an aliphatic carbon chain and the effect of branching of the carbon chain carrying an amino group on pressor action. Optimal conditions were found in compounds of 7 to 8 carbon atoms with a primary amino group in the 2-position. Branching of the chain increases pressor activity. Table 45 includes some of the more important members of this group.

A series of secondary β-cyclohexylethyl- and β-cyclopentylethylamines have been shown to have sympathomimetic activity.[19]

Tuaminoheptane, Tuamine®, 2-Amino-

$$CH_3CHCH_2CH_2CH_2CH_2CH_3$$
$$|$$
$$NH_2$$

Tuaminoheptane

heptane. This compound is a colorless to pale-yellow liquid. It is freely soluble in alcohol and sparingly soluble in water. A 1 per cent solution in water is alkaline (pH 11.5). It is a vasoconstrictor and a sympathomimetic amine. Inhalation of the vapors is an effective treatment of acute rhinologic conditions and is very useful when prolonged and repeated medication is required. It should be used with caution by those who have cardiovascular disease.

The dosage is one or two gentle inhalations through each nostril, repeated at hourly intervals if necessary.

TABLE 45. ALIPHATIC ADRENERGIC AMINES AND PRINCIPAL USE

$$\underset{\displaystyle R-CH_2-CH-N-R'}{\overset{\displaystyle CH_3 \quad H}{\big|\big|}}$$

Compound	R	R'	Use
Tuaminoheptane	$CH_3CH_2CH_2CH_2-$	H	Nasal decongestant
Methylhexaneamine	CH_3CH_2CH- $\quad\quad\;\; \underset{CH_3}{\big|}$	H	Vasopressor in spinal anesthesia
Isometheptene	$CH_3C=CHCH_2-$ $\;\;\underset{CH_3}{\big|}$	CH_3	Vasoconstrictor Antispasmodic
Cyclopentamine	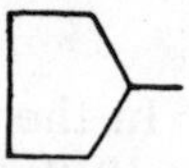	CH_3	Nasal decongestant
Propylhexedrine	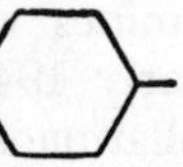	CH_3	Nasal decongestant

Tuaminoheptane Sulfate N.F., Tuamine® Sulfate, 1-Methylhexylamine Sulfate. This compound is a white, odorless powder. It is soluble in alcohol and freely soluble in water. A 1 per cent solution in water is slightly acidic (pH 5.4).

The vasoconstrictive effects of a 1 per cent solution of tuaminoheptane sulfate exceed those of a similar concentration of ephedrine, and the duration of effect is greater than that of ephedrine. For topical use a 1 per cent solution of tuaminoheptane sulfate may be applied to the mucous membranes of infants and adults and usually is adequate for routine treatment.

Category—adrenergic.

Application—to the nasal mucosa as a 0.5 to 2 per cent solution.

OCCURRENCE
Tuaminoheptane Sulfate Solution N.F.

Methylhexaneamine, Forthane®, 2-

$$\underset{\displaystyle \quad\;\; CH_3 \quad\;\; NH_2}{\overset{\displaystyle CH_3CH_2CHCH_2CHCH_3}{\big|\big|}}$$
Methylhexaneamine

Amino-4-methylhexane. This compound is a colorless to pale-yellow liquid with an ammonialike odor. It is readily soluble in alco-

hol and is very slightly soluble in water. Methylhexaneamine is a volatile sympathomimetic amine, the physiologic properties and uses of whose salts are the same as those of other vasoconstrictor drugs. Its pressor action is more prolonged than that of epinephrine. Soluble salts of the base produce mydriasis after local installation.

Methylhexaneamine is used as an inhalant for its local vasoconstrictor action on the nasal mucosa. This treatment produces temporary relief of nasal congestion and is used as an adjunct in the treatment of allergic or infectious rhinitis and sinusitis. The drug is supplied in the form of the carbonate which releases the volatile amine when the inhaler is used. Each inhaler contains methylhexaneamine carbonate equivalent to 250 mg. of the free amine. One or two inhalations through each nostril is the recommended dose, to be repeated at intervals of not less than one-half hour.

Cyclopentamine Hydrochloride N.F., Clopane® Hydrochloride, N,α-Dimethylcyclopentaneethylamine Hydrochloride, 1-

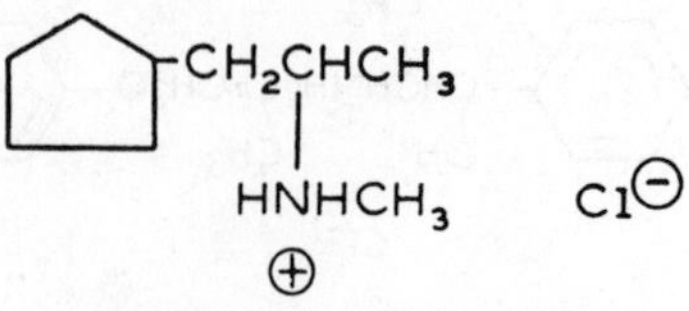

Cyclopentamine Hydrochloride

Cyclopentyl-2-methylaminopropane Hydrochloride. This white, bitter, crystalline powder is a sympathomimetic agent with uses and actions characteristic of other pressor amines. It is soluble in water 1:1 and in alcohol 1:2. Its effects are similar to ephedrine, but it produces only slight cerebral excitation. Orally, it is more effective than ephedrine. Presently, cyclopentamine is used by topical application for the temporary relief of nasal congestion.

Too frequent application topically should be avoided to prevent side-effects such as increased blood pressure, nervousness, nausea and dizziness.

Category—adrenergic (vasoconstrictor).

Usual dose—intramuscular, 25 mg.; intravenous, 5 to 10 mg.; intranasal, 1 or 2 drops of a 0.5 to 1 per cent solution every 3 or 4 hours.

OCCURRENCE

Cyclopentamine Hydrochloride Injection N.F.

Cyclopentamine Hydrochloride Solution N.F.

Propylhexedrine N.F., Benzedrex®, N,α-Dimethylcyclohexaneethylamine. This material is a clear, colorless liquid, with a characteristic fishy odor. Propylhexedrine is very soluble in alcohol and very slightly soluble in water. It volatilizes slowly at

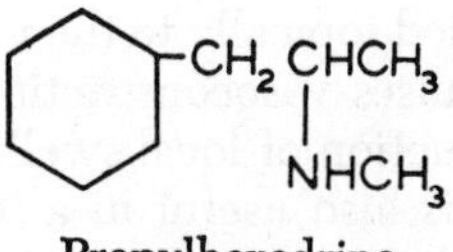

Propylhexedrine

room temperature and absorbs carbon dioxide from air. Its uses and actions are similar to those of other volatile sympathomimetic amines. It produces vasoconstriction and a decongestant effect on the nasal membranes but has only about one half the pressor effect of amphetamine and produces decidedly less effect on the nervous system. Therefore its major use is for local shrinking effect on nasal mucosa in the symptomatic relief of nasal congestion caused by the common cold, allergic rhinitis or sinusitis.

Hydroaromatic compounds can be prepared by reduction of the corresponding aromatic amines: thus, methamphetamine yields propylhexedrine on reduction.[20]

Category—adrenergic (vasoconstrictor).

Application—by inhalation as needed.

OCCURRENCE

Propylhexedrine Inhalant N.F.

Isometheptene Hydrochloride, Octin® Hydrochloride, 2-Methylamino-6-methyl-5-

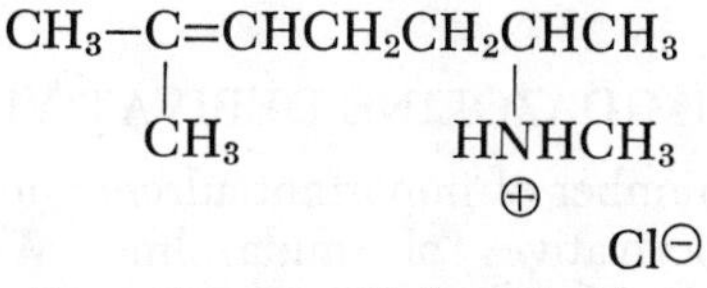

Isometheptene Hydrochloride

heptene Hydrochloride. This compound exhibits antispasmodic and vasoconstrictor properties. Its antispasmodic effect is caused by stimulation of sympathetic (inhibitory) nerve endings rather than by inhibition of parasympathetic endings, as with atropine. Isometheptene resembles epinephrine in that it produces moderate peripheral vasoconstriction, an increase in the contractile force of the myocardium and a transient increase in blood pressure. Other effects include a slight bronchodilation, mydriasis, respiratory stimulation and a shrinkage of nasal and pharyngeal mucosa.

This drug is administered orally or intramuscularly. The usual oral dose for adults is 15 to 20 drops of a 10 per cent solution (containing 100 mg. per ml.) every half hour for a total of 4 doses. By the intramuscular route 50 to 100 mg. is injected.

Isometheptene Mucate, Octin® Mucate, 2-Methylamino-6-methyl-5-heptene Mucate.

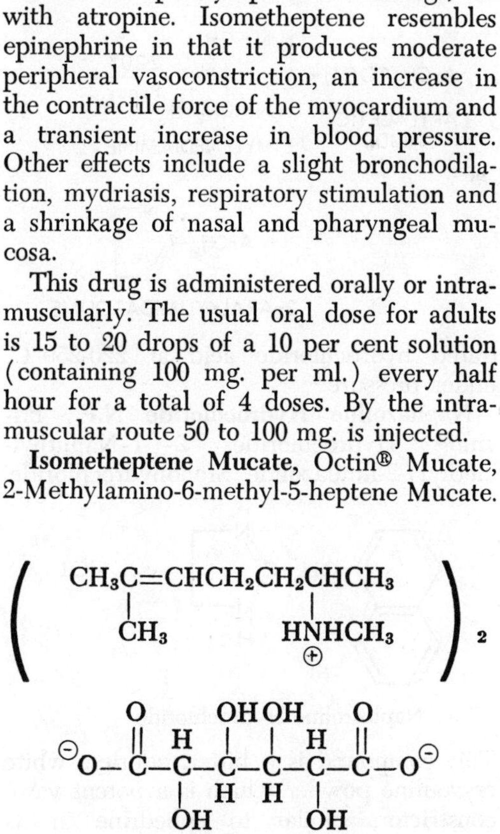

Isometheptene Mucate

This agent has the same actions and uses as the hydrochloride salt. Because it is not used by the parenteral route, it rarely causes hypertension.

Isometheptene mucate is administered orally or rectally; the usual oral dose for adults is 120 mg. every half hour for a total of 4 doses. Alternatively, one suppository containing 250 mg. may be inserted into the rectum; this procedure may be repeated in 1 hour if necessary.

IMIDAZOLINE DERIVATIVES

A number of important adrenergic agents are derivatives of imidazoline. While 2-benzylimidazoline (Priscoline) is a vasodilator and sympatholytic agent,[21] introduction of a hydroxyl group into the para position of the benzenoid ring converts the compound into a potent pressor agent.[22]

Compounds of this type may be prepared conveniently by the reaction of ethylenediamine with an aryl acetic acid in concentrated hydrochloric acid at 220-250°C. under pressure.[23]

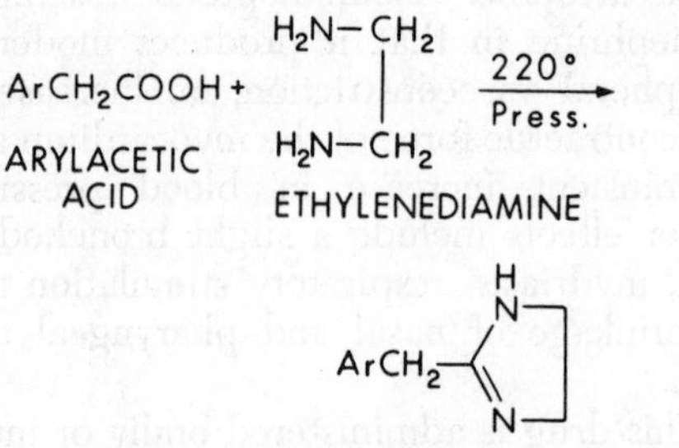

$$ArCH_2COOH + \begin{array}{c} H_2N-CH_2 \\ | \\ H_2N-CH_2 \end{array} \xrightarrow[\text{Press.}]{220°}$$

ARYLACETIC ACID   ETHYLENEDIAMINE

2-ARALKYLIMIDAZOLINE

Naphazoline Hydrochloride N.F., Privine® Hydrochloride, 2-(1-Naphthylmethyl)-2-imidazoline Monohydrochloride.

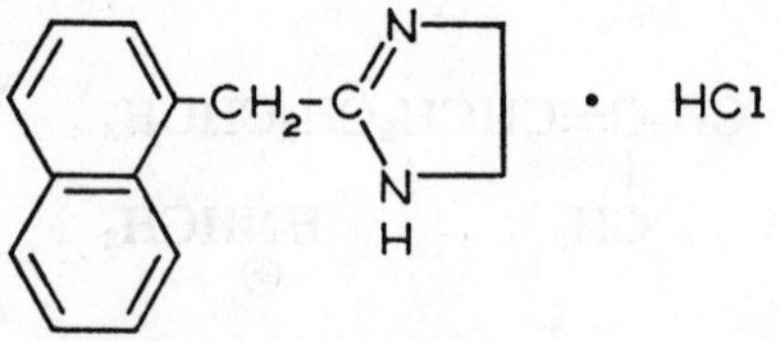

Naphazoline Hydrochloride

This compound is a bitter, odorless white crystalline powder which is a potent vasoconstrictor, similar to ephedrine in its action. It is freely soluble in water and in alcohol. When applied to nasal and ocular mucous membranes it causes a prolonged reduction of local swelling and congestion. It is of value in the symptomatic relief of disorders of the upper respiratory tract. In acute nasal congestion, excessive use of vasoconstrictors may delay recovery. A rebound congestion of the mucosa is sometimes caused by naphazoline hydrochloride but can be alleviated by discontinuing all nasal medication.

Category—adrenergic (vasoconstrictor).

Application—to the nasal mucosa, 2 drops of a 0.05 per cent solution no more frequently than every 3 hours.

OCCURRENCE
Naphazoline Hydrochloride Solution N.F.

Tetrahydrozoline Hydrochloride N.F., Tyzine® Hydrochloride, Visine® Hydrochloride, 2-(1,2,3,4-Tetrahydro-1-naphthyl)-

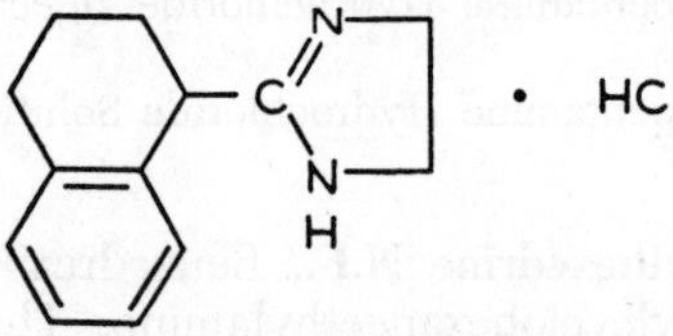

Tetrahydrozoline Hydrochloride

2-imidazoline Monohydrochloride. This compound is closely related to naphazoline hydrochloride in its pharmacologic action. When applied topically to the nasal mucosa, the drug causes vasoconstriction, which results in reduction of local swelling and congestion. It is also useful in a 0.05 per cent solution (Visine) as an ocular decongestant. When used 2 or 3 times daily, there is no influence on pupil size. It does not appear to increase intraocular pressure; however, its use in the presence of glaucoma is not recommended.

Category—adrenergic (vasoconstrictor).

Application—applied as a 0.05 or 0.1 per cent solution as nasal drops or spray at 4-hour intervals; to the eye as a 0.05 per cent solution.

OCCURRENCE
Tetrahydrozoline Hydrochloride
    Ophthalmic Solution N.F.
Tetrahydrozoline Hydrochloride Solution
    N.F.

TABLE 46. COMPARISON OF SOME VASOPRESSOR AGENTS

Product Trade names or generic Titles	Stability	Detoxi-cation	Central Stimulation	Tendency to Arrhythmias	Cardiac Effects Rate	Output	Pressure Systol.	Diastol.	Dose I.M. (mg.)	I.V. (mg.)	Action I.V. Onset (min.)	Dura-tion (min.)	I.M. Onset (min.)	Dura-tion (min.)	Effect on Respiration
Methoxamine Vasoxyl	Stable	Destroyed or eliminated	None	None	Slowed	No increase	Elevated	Elevated	10-15	5	1-2 max. 5	60-120	15	Several hours	Slight to none
Epinephrine Adrenalin Suprarenin	Unstable	Oxidized	Tremor and excitement	Ventricular fibrillation	Accelerated	Increased	Elevated markedly	Lowered	1	0.1	½-1	2-3	3-4	5-10	No increase reflex apnea
Norepinephrine Arterenol Levophed	Readily oxidized	Oxidized	Less than epinephrine	Ventricular fibrillation	Slowed occasional acceleration	Slight increase	Elevated	Elevated	Necro-tizing	2-4 mcg.	½-1	2-3			None
Ephedrine	Base decomp. HCl stable	Slowly destroyed	Less than epinephrine	Auricular fibrillation	Accelerated	Increased	Elevated	Slightly elevated	50	25	5-10	10-20	10-20	60-120	Slight increase
Desoxyephedrine Methamphetamine 'Methedrine'	Stable	Destroyed or eliminated	Marked	Auricular fibrillation	Accelerated	Increased	Elevated	Elevated	15-20	5-10	1	Several hours	5-15	Several hours	Slight increase
Phenylephrine Neosynephrine	Oxidizes easily	Oxidized	Slight	None	Slowed	Increased	Elevated	Elevated	5	0.5	1-2	15	10-15	45-90	Slight reflex apnea
Mephentermine Wyamine		Oxidized	None	None	Unchanged, occasional bradycardia	No increase	Elevated	Elevated	35	15	1-2	15-20	15	30	Slight to none
Cyclopentamine Clopane	Stable		¼ less than ephedrine	None	Variable	Increased	Elevated	Elevated	25	5-10	3-5	10-15	5-15	35-45	None
2-Methylaminoheptane Oenethyl	Stable		Slight	Paroxysmal tachycardia, premature ventricular contractions	Slightly increased	Increased	Elevated	Slightly elevated	100	25-50	1-2 max. 5	30	2-3 max. 10	30-60	None

Xylometazoline Hydrochloride N.F., Otrivin® Hydrochloride, 2-(4-*t*-Butyl-2,6-dimethylbenzyl)-2-imidazoline Hydrochloride. This compound is used as a nasal vaso-

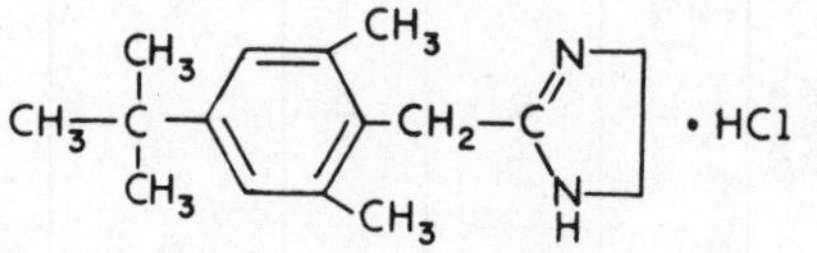

Xylometazoline Hydrochloride

constrictor. Its duration of action is approximately 4 to 6 hours.

Category—adrenergic (vasoconstrictor).

Application—nasally, 2 or 3 drops of a 0.05 per cent or 0.1 per cent solution every 4 to 6 hours.

OCCURRENCE
Xylometazoline Hydrochloride Solution N.F.

Oxymetazoline Hydrochloride N.F., Afrin® Hydrochloride, 6-*t*-Butyl-3-(2-imidazolin-2-ylmethyl)-2,4-dimethylphenol Monohydrochloride, 2-(4-*t*-Butyl-2,6-dimethyl-3-hydroxybenzyl)-2-imidazoline Hydrochloride. This compound, closely related to xylometazoline, is a long-acting vasoconstrictor. It is used as a topical aqueous nasal decongestant in a wide variety of disorders of the upper respiratory tract.

Category—adrenergic (vasoconstrictor).

Application—nasally, 2 to 4 drops of a 0.05 per cent solution instilled into each nostril twice daily.

OCCURRENCE
Oxymetazoline Hydrochloride Solution N.F.

REFERENCES

1. Ahlquist, R. P.: Am. J. Physiol. 153:586, 1948.
2. Barger, G., and Dale, H. H.: J. Physiol. 41:19, 1910.
3. Bülbring, E.: Adrenergic Mechanisms (A Ciba Foundation Symposium), Boston, Little, Brown and Co., 1960.
4. Sutherland, E. W., and Robison, G. A.: Pharmacol. Rev. 18:145, 1966.
5. Hartung, W. H.: Ind. Eng. Chem. 37: 126, 1945.
6. Iisalo, E.: Acta pharmacol. toxicol. 19 (Suppl. 1):1962.
7. Beyer, K. H., and Lee, W. V.: Pharmacol. Exp. Ther. 74:155, 1942.
8. Axelrod, J.: Science 126:593, 1958.
9. ——: Science 140:499, 1963.
10. Ludena, F. P., Ananenko, E., Siegmund, O. H., and Miller, L. C.: J. Pharmacol. Exp. Ther. 95:155, 1949.
11. Gay, L. N., and Long, J. W.: J.A.M.A. 139:452, 1949.
12. Nagai, T.: Pharm. Z. 32:700, 1887.
13. Späth, E., and Göhring, R.: Monatsh. 41: 319, 1920.
14. Chen, K. K., and Schmidt, C. F.: J. Pharmacol. Exp. Ther. 24:339, 1924.
15. Lapidus, J. B., Tye, A., Patil, P., and Modi, B. A.: J. Med. Chem. 6:76, 1963.
16. Saari, W. S., Raab, A. W., and Engelhardt, E. L.: J. Med. Chem.: 11:1115, 1968.
17. Menshikov, G. P., and Rubinstein, M. M.: J. Gen. Chem. U.S.S.R. 13:801, 1943, *in* Chem. Abstr. 39:1172, 1945.
18. Rohrmann, E., and Shonle, H.: J. Am. Chem. Soc. 66:1517, 1944.
19. Lands, A. M., Lewis, J. R., and Nash, V. L.: J. Pharmacol. Exp. Ther. 83:253, 1945.
20. Zenitz, B. L., Machs, E. B., and Moore, M. L.: J. Am. Chem. Soc. 69:1117, 1947.
21. Ahlquist, R. P., Huggins, R. A., and Woodbury, R. A.: J. Pharmacol. Exp. Ther. 89:271, 1947.
22. Scholz, C. R.: Ind. Eng. Chem. 37:120, 1945.
23. Deutsch. Rep. Pat. 687, 196; Eng. Pat. 514, 411.

SELECTED READING

Barlow, R. B.: Introduction to Chemical Pharmacology, ed. 2, p. 282, New York, Wiley, 1964.
Belleau, B.: Steric effects in catecholamine interactions with enzymes and receptors. Pharmacol. Rev. 18:131, 1966.
Bloom, B. M., and Goldman, I. M., *in* Harper and Simmonds (eds.): Advances in Drug Research, New York, Academic Press, 1966.
Burger, A.: Drugs Affecting the Peripheral Nervous System, New York, Dekker, 1967.
Triggle, D. J.: Chemical Aspects of the Autonomic Nervous System, New York, Academic Press, 1965.
——: Adrenergic Hormones and Drugs, *in* Burger, A: Medicinal Chemistry, ed. 3, p. 1235, New York, Wiley-Interscience, 1970.

18

Cholinergic Agents and Related Drugs

Ole Gisvold, Ph.D.,
Professor of Medicinal Chemistry, College of Pharmacy,
University of Minnesota

INTRODUCTION

The autonomic nervous system is composed of two parts, viz., the sympathetic and the parasympathetic. Acetylcholine mediates (regulates) in the transmission of nerve impulses in the preganglionic fibers and the ganglia of the former and in both the preganglionic and the postganglionic fibers, the ganglia and the neuromuscular junctions in the latter. It also mediates at the neuromuscular junctions in the voluntary nervous system.

Acetylcholine, like norepinephrine, is stored in vesicles where these can function both by normal release and in greater quantities to meet any such demands.[1]

The acetylcholine system composed of acetylcholine, acetylcholinesterase and choline acetylase is present and similarly functional in excitable membranes of the axon, the nerve terminal and the pre- and postsynaptic excitable membranes. These membranes, whose variations of shape, structure and organization are virtually infinite, have the special ability of changing rapidly and reversibly their permeability to ions, i.e., K^+ and Na^+. These ions are carriers of bioelectric currents for conduction along the axon and transmission across junctions. Conduction velocities vary from 0.1 to 100 meters per second. Such changes in membrane potential can increase the membrane pH by as much as 0.2. This change is sufficient to alter the permeability of the membrane and trigger the pulse of sodium ions and the nerve impulse. Acetylcholine exerts reversible conformational changes in the receptor protein only in the excitable membrane. In most axonal membranes this leads to depolarization, with a reverse of charge and a propagation of the stimulus. These effects seem to be most pronounced on the glycerophosphate groups that have been proposed as the polar gates of the membrane. Acetylcholine can trigger 10^5 impulses per hour in and along the axon. These impulses (electrical) in a desheathed axon can be reversibly blocked by curare, atropine, decamethonium, etc., applied at a node of Ranvier. The barrier effects of the sheath of the axons that render it impermeable to many quaternary compounds including acetylcholine can be reduced by snake venoms (phospholipase activity) or detergents.

Only in the presence of cholinergic agents such as physostigmine, (q.v.) does acetylcholine leak from the excitable membrane, presumably because of the increased (above normal) amounts of acetylcholine.

The standard conception of cholinergic transmission is that at certain synapses the action potential in the presynaptic nerve fiber causes the quantal release from its terminations of acetylcholine which then diffuses across the synaptic gap to excite the specific postsynaptic acetylcholine receptors.

The receptor sites involved in normal ganglionic transmission may be both nicotinic and muscarinic, as acetylcholine has both nicotinic and muscarinic activities (q.v.). The biologic role of the muscarinic receptor sites as yet has not been elucidated. While the action of acetylcholine on the nicotinic receptor sites normally initiates the potential changes leading to the discharge of impulses from the ganglion cell, its action on muscarinic receptors may serve to modify ganglionic transmission.

The major sites of activity of acetylcholine are (1) the postganglionic parasympathetic (muscarinic) receptor, (2) the autonomic ganglia and (3) the skeletal neuromuscular junction. The first is blocked by atropine, the second by hexamethonium and the third by decamethonium or curare. The above data might imply that the receptor sites are different but the agonist acetylcholine can adopt a conformation to effectively interact with the respective receptor site. On the other hand, it is possible that all the receptor sites per se are very similar but the immediate adjacent areas are different.

The postganglionic fibers of the parasympathetic nerve system are called cholinergic, and drugs that mimic the action of acetylcholine are called cholinergic. True cholinergic drugs are those whose qualitative mode of action is the same as that of acetylcholine. False or indirect cholinergic drugs such as phyostigmine, DFP, etc., are those that inhibit the hydrolysis of acetylcholine by cholinesterase. A third group represented by pilocarpine is proposed to act directly on the receptor cells; however, the molecular architecture conforms to the requirements of the true drugs and conceivably it also could act as a true drug.

The activity of acetylcholine is in part regulated by its hydrolysis by cholinesterase* to yield inactive choline and acetic acid

* Cholinesterase is present at motor end-plates and at all synapses whether central or peripheral, mammalian or fish, vertebrate or invertebrate and in the envelope and not the axoplasm of the axon of the squid.

and by the resynthesis of acetylcholine from these products via acetyl coenzyme A (q.v.).

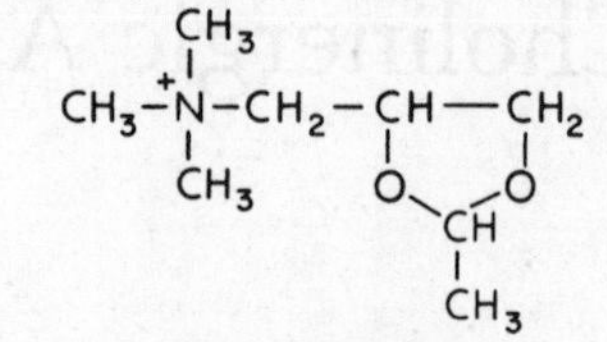

Bovets Acetal

Acetylcholine has the following dimensions and, with the exception of Bovets acetal, has the greatest cholinergic activity. A cationic quaternary nitrogen (head) con-

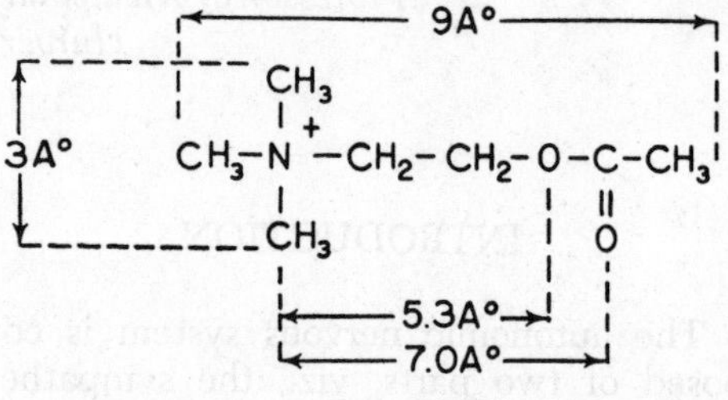

taining at least two CH_3 groups and a hydrocarbon (tail) are the minimum requirements for cholinergic activity.[2] Activity is increased if the tail contains a keto or ether oxygen atom approximately 7 Å from one of the methyl groups, and a second oxygen atom at 5.3 Å increases the activity still more. Last but not least, the over-all size and structure of the molecule are also of great importance.

Free rotation about the C—C and C—O bonds of the acetylcholine permits a great number of possible conformers. N.M.R. studies of acetylcholine in deuterium oxide suggest the following conformation that is similar to that found in the crystal lattice in that the N—C—C—O system is in a gauche arrangement but differs in the CH_2 —O—CO— CH_3 grouping has the normal conformer populations of a primary ester.

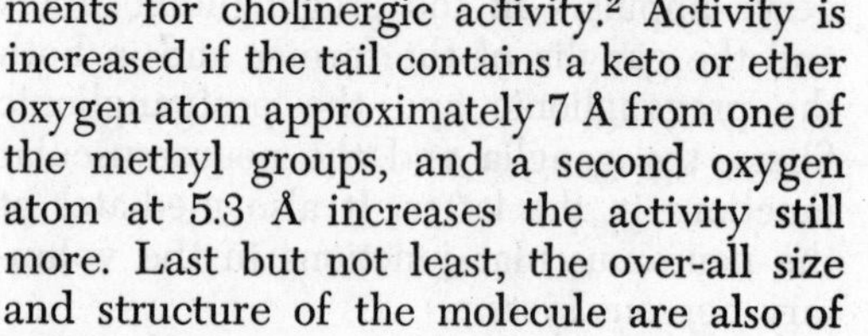

The exact conformation adopted by acetylcholine in vivo to exert its nicotinic and muscarinic activities (q.v.) is not known, although one closely related to those compounds that have significant nicotinic and muscarinic activities may be possible.

The primary portion of the acetylcholine molecule and related compounds may be designated as follows:

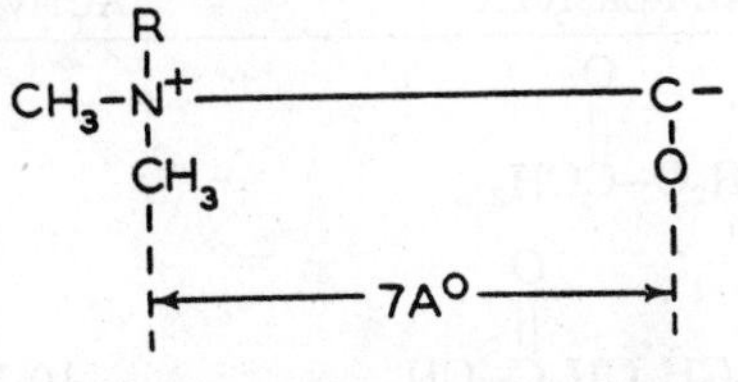

The remainder of the molecule may be designated as secondary. However, even though the term secondary is used, these structures may exert profound effects on the type of activity, the degree of activity, drug distribution, duration of action, metabolism, etc.

Table 47 has been compiled to illustrate some of the above generalizations.[2, 3]

The acetylcholine molecule is tailored to fit two sites, i.e., the receptor cell where it exerts its activity in the transmission of a nerve impulse and a specific site on the surface of the cholinesterase enzyme molecule. This requires a rather marked degree of specificity. However, it is well established in many cases that small changes can be made with the retention of varying degrees of activity. For example, a small methyl group placed on the β-carbon atom of the choline portion of acetylcholine exerts at least two effects, i.e., it slows the rate of hydrolysis by cholinesterase to give a drug with a longer half-life, probably due primarily to stereochemical effects, and it lowers but does not destroy qualitative biologic activity.

Small groups replacing a methyl-group on the N of acetylcholine have little effect on the congeners as substrates for AChE. Even N-benzylnoracetylcholine is hydrolyzed at 40 per cent of the rate of acetylcholine. Replacing of all the methyl groups with ethyl groups to give triethylacetylcholine (which also has intrinsic activity) does not appreciably reduce the rate of hydrolysis by AChE. However, when these three ethyl groups are present as in a quinuclidinum moiety, inhibition of AChE is obtained.[4]

Some of the effects of acetylcholine are similar to those of nicotine and muscarine. Muscarine has a ganglionic stimulant action that has a slower induction period and is of longer duration than the nicotinic activity produced by nicotine and some related nicotine type stimulants. The high specificity of L(+) and D(−) isomers of muscarine and its great sensitivity to inhibition by atropine but not by hexamethonium indicate that the postganglionic receptors sensitive to muscarine are similar to the receptors of postganglionic parasympathetic effector sites.[5]

In regard to the activity of acetylcholine, one should note the following phenomena: (1) its action (called muscarinic) in the vertebrate heart is mimicked by pilocarpine and muscarine and is blocked by atropine, and (2) its action (called nicotinic) at vertebrate neuromuscular junctions and at autonomic ganglia is mimicked by low concentrations of nicotine and is blocked by tetraethylammonium ions. Advantage can be taken of this knowledge, for example, in the treatment of myasthenia gravis. Neostigmine or physostigmine (q.v.) are combined with atropine, which prevents the undesirable potentiation of the action of acetylcholine on the digestive tract and at other points where its action is muscarinic, while allowing the other agents to potentiate the nicotinelike action at neuromuscular junctions where too little acetylcholine apparently is responsible for the crippling affliction.

4-Ketoamyltrimethylammonium chloride and ethoxycholine bromide (see Table 47) have, respectively, one twelfth and one hundred and sixtieth the nicotinic activity of acetylcholine. Thus the carbonyl group makes a greater contribution to nicotinic activity than does the ether oxygen. The preferred conformation of nicotine to exert maximum nicotinic activity is probably one in which the N of the pyrrolidine ring is 4.76Å from the pyridine ring N and its unshared pair of electrons in the non-protonated state. This is very close to the situation in acetylcholine in which the negatively charged carbonyl oxygen atom is 4.93Å from the quaternary nitrogen atom when the acyl group is in a 120° orientation to the ether oxygen-carbon bond.[6]

Acetylcholinesterase (AChE), an enzyme, is found in nervous tissue and red blood cells. Although it has a greater specificity for the hydrolysis of acetylcholine, it can

TABLE 47

NAME	STRUCTURAL FORMULA	ACTIVITY[a]
Acetylcholine	$(CH_3)_3\overset{+}{N}-CH_2CH_2O-\overset{O}{\overset{\|}{C}}CH_3$	1
4-Ketoamyltrimethyl ammonium chloride	$Cl^-(CH_3)_3\overset{+}{N}CH_2CH_2CH_2\overset{O}{\overset{\|}{C}}-CH_3$	10-12
3-Ketoamyltrimethyl ammonium iodide	$I^-(CH_3)_3\overset{+}{N}CH_2CH_2-\overset{O}{\overset{\|}{C}}-CH_2CH_3$	160
Ethoxycholine bromide	$Br^-(CH_3)_3\overset{+}{N}CH_2CH_2-O-CH_2CH_3$	160
n-Propionyl choline bromide	$Br^-(CH_3)_3\overset{+}{N}-CH_2CH_2-O-\overset{O}{\overset{\|}{C}}-CH_2CH_3$	105
Chloracetylcholine chloride	$Cl^-(CH_3)_3\overset{+}{N}-CH_2CH_2-O-\overset{O}{\overset{\|}{C}}-CH_2Cl$	960
β-Carbomethoxyethyltrimethyl ammonium bromide	$Br^-(CH_3)_3\overset{+}{N}-CH_2CH_2-\overset{O}{\overset{\|}{C}}-O-CH_3$	15
Amyltrimethylammonium chloride	$Cl^-(CH_3)_3\overset{+}{N}-CH_2CH_2CH_2CH_2CH_3$	70
4-Methylamyltrimethyl ammonium bromide	$Br^-(CH_3)_3\overset{+}{N}CH_2CH_2CH_2\overset{}{\underset{CH_3}{C}}HCH_3$	1,370
4-Hydroxyamyltrimethyl ammonium chloride	$Cl^-(CH_3)_3\overset{+}{N}CH_2CH_2CH_2\overset{}{\underset{OH}{C}}H-CH_3$	1,500
Carbamyl choline chloride	$Cl^-(CH_3)_3\overset{+}{N}CH_2CH_2O-\overset{O}{\overset{\|}{C}}-NH_2$	80
Acetyl β-methyl choline chloride	$(CH_3)_3\overset{+}{N}CH_2\overset{}{\underset{CH_3}{C}}H-O-\overset{O}{\overset{\|}{C}}-CH_3$	1,100

[a] Equiactive molar ratios (acetylcholine 1) when tested for depressant action on the spontaneous beat of the isolated heart of the mollusk genus mercenaria. This action is nicotinelike and of a type such as is found at autonomic ganglia, for it is blocked by tetraethylammonium ions but not by curare alkaloids.

hydrolyze a greater variety of ester bonds and even other linkages; thus, in some important ways, the groups in the "catalytic center" of AChE resemble other hydrolytic enzymes, especially chymotrypsin and trypsin. AChE is most active at pH 7.5 to 8.5 and chemical data at pH values 6.5 and 9.5 suggest that the catalytic center contains an acid group, probably tyrosine OH, and a basic group, probably histidine. A serine OH group and an anionic site also are very important parts of this "catalytic center."

Substrates and anticholinesterase agents can induce alterations in the conformation of this enzyme. Such changes may be essential and provide an explanation for some of its reactions.[7]

Butyrylcholinesterase (BChE) of serum inhibits substrate specificities different from those inhibited by AChE.

AChE from the electric eel has a minimal molecular wt. of about 240,000 and may exist as an aggregate of about 20 million. The minimal mol. wt. of BChE is about 80,000 and this enzyme also may occur as an aggregate. The "catalytic center activity" (molecules of substrate hydrolyzed per minute per catalytic center) is estimated to be 7×10^5 for AChE from the electric eel, 3×10^5 for AChE from bovine erythrocytes and 9×10^4 for BChE from horse serum. In the pH range in which it is catalytically active, AChE is an anion having an isoelectric point of about 4.

Below are schematic diagrams of the mechanism of hydrolysis by AChE as suggested by Krupa.[8]

The esteratic site is depicted by AH (probably the OH of tyrosine), OH of serine, and B:, a basic group, probably histidine. The anionic site is unoccupied in the acetyl enzyme (and probably also in the carbamoyl enzyme, q.v.).

CHOLINERGIC AGENTS

Acetylcholine exerts a powerful stimulation of the parasympathetic nerve system. Attempts have been made to utilize it as a cholinergic agent,[2, 9] though its duration of action is too short for sustained effects, due to rapid hydrolysis by cholinesterase.

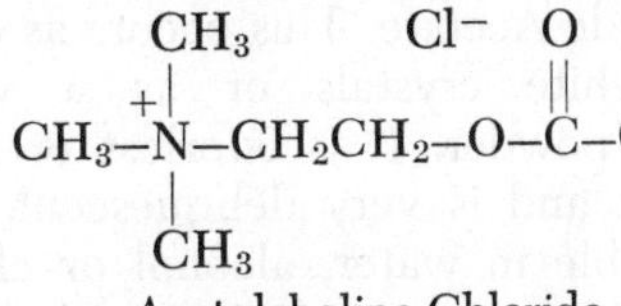

$$CH_3-\overset{\overset{\displaystyle CH_3}{|}}{\underset{\underset{\displaystyle CH_3}{|}}{N^+}}-CH_2CH_2-O-\overset{\overset{\displaystyle O}{\|}}{C}-CH_3 \quad Cl^-$$

Acetylcholine Chloride

Acetylcholine is a cardiac depressant and an effective vasodilator. The stimulation of the vagus and the parasympathetic nervous system produces a tonic action on smooth muscle and induces a flow from the salivary and the lachrymal glands. Its vasodilator action is observed to be primarily on the arteries and the arterioles, with distinct effect on the peripheral vascular system. It is ineffective orally but is used parenterally in salt form for Raynaud's disease, postoperative distention, paralytic ileus, paroxysmal tachycardia and tobacco or alcohol amblyopia.

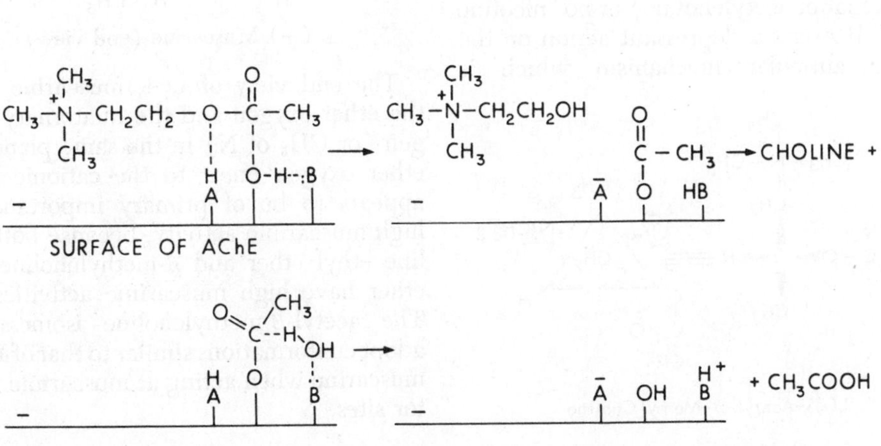

PRODUCTS

Acetylcholine Chloride, Miochol®, is a hygroscopic powder that is available in a stable solution in ampuls containing 20, 50, 100 or 200 mg. per ml. An external application also may be used for varicose ulcers, dermatosis, similar disorders.

Methacholine Chloride N.F, Mecholyl® Chloride, Acetyl-β-Methylcholine Chlo-

$$CH_3C-O-CHCH_2\overset{+}{N}-CH_3 \ Cl^-$$

ride, (2-Hydroxypropyl)trimethylammonium Chloride Acetate. This occurs as colorless or white crystals or as a white, crystalline powder. It is odorless or has a slight odor and is very deliquescent. It is freely soluble in water, alcohol or chloroform, and its aqueous solution is neutral to litmus and has a bitter taste.

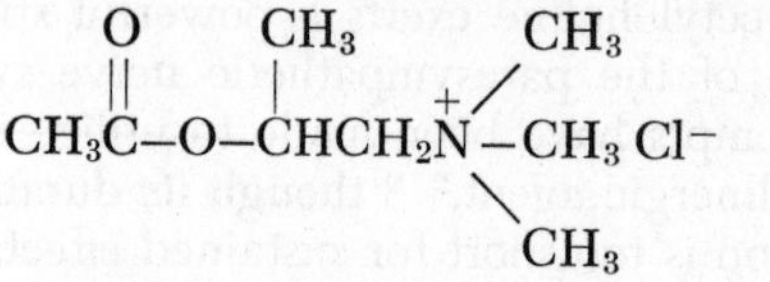

It is rapidly hydrolyzed in basic solution. Solutions are relatively stable to heat and will keep for at least 2 or 3 weeks when refrigerated to delay growth of molds.

Methacholine may be prepared from trimethylacetonylammonium chloride $(CH_3)_3$-$NCI-CH_2-CO-CH_3$, which is catalytically reduced in absolute alcoholic solution, using hydrogen with platinum oxide and a trace of ferric chloride. The product then is acetylated by heating with acetic anhydride.

Unlike acetylcholine, the methyl derivative has sufficient stability in the body to give sustained parasympathetic stimulation, and this action is accompanied by little ($\frac{1}{100}$ that of acetylcholine) or no "nicotine effect." It exerts a depressant action on the cardiac auricular mechanism which is

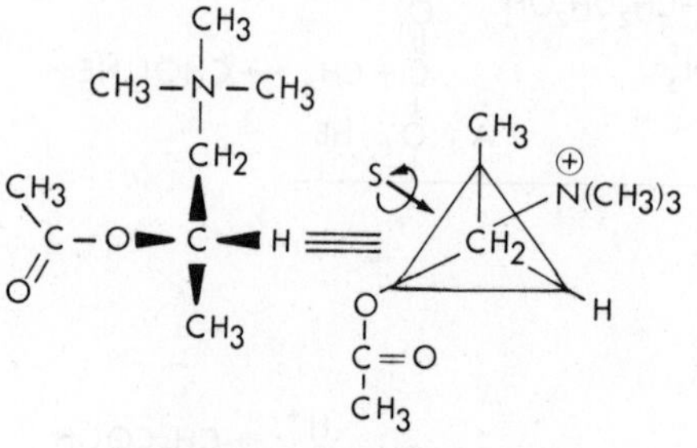

L(+)S-Acetyl-β-Methyl Choline

blocked by quinidine, a stimulation of gastrointestinal peristalsis and a general vasodilation followed by a fall in blood pressure. All of these effects are intensified and prolonged by physostigmine and neostigmine, which inhibit cholinesterase, but are rapidly and completely blocked by atropine.

It should be noted here that the (+) isomers of acetyl-α- and acetyl-β-methylcholine are respectively about 10 and 250 times more active as muscarinic than their corresponding enantiomorphs. (*See* Nature 189:671, 1961.) Thus the spatial relationships in (+)acetyl-β-methylcholine more closely resemble those in (+) muscarine than those found in the (−) isomer.

Acetyl-L(+)s-β-methylcholine is hydrolyzed by acetylcholinesterase, whereas the D-(−)-R-isomer is not. Acetyl-D(−)R-β-methylcholine is a weak inhibitor of the hydrolysis of the L-isomer and of acetylcholine by acetylcholinesterase[10] from the electric organ of Electrophorus (but not from rat serum esterase) at that concentration of substrate at which ACh is being hydrolyzed predominantly by nonspecific esterase.

In the following formula, which is a preferred conformation for L(+)s-muscarine, 3.0 Å distance separates the N from the ether oxygen and a 5.7Å distance separates the N from the alcohol group. The former distance is, therefore, very close to that found in the preferred conformation of acetylcholine (q.v.).

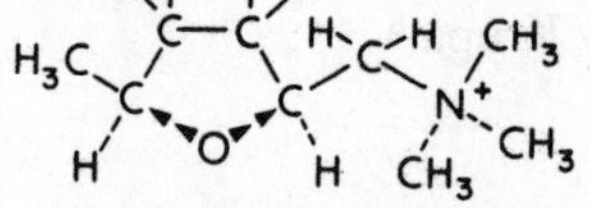

L (+) Muscarine (end view)

The end view of L(+)muscarine shows the ether oxygen and the quaternary nitrogen (or CH_3 of N) in the same plane. The ether oxygen (next to the cationic head) appears to be of primary importance for high muscarinic activity, because both choline ethyl ether and β-methylcholine ethyl ether have high muscarinic activities.[10, 11] The acetyl-β-methylcholine isomers may adopt conformations similar to that of L(+)-muscarine when acting at muscarinic receptor sites.

Muscarine and muscarinelike compounds are weak inhibitors in vitro of the hydrolysis of acetylcholine by acetylcholinesterase; thus, their muscarinic activities might be a measure of their interaction with a muscarinic receptor site.

L(+)-Acetyl-β-methylcholine's rate of hydrolysis is about 54 per cent of acetylcholine in the biophase of the muscarinic receptors, and this rate probably compensates for the decreased association of this molecule (due to the β-methyl group) with the muscarinic receptor site. This may account for the fact that acetylcholine and L(+)-acetyl-β-methylcholine have equimolecular muscarinic potencies. D(−)-Acetyl-β-methylcholine weakly inhibits acetylcholinesterase and slightly reinforces the muscarinic activity of the L(+) isomer in the (±) acetyl-β-methylcholine.

The D(+)- and the L(−)-acetyl-α-methylcholine weakly inhibits acetylcholin-acetylcholinesterase at 78 per cent and 97 per cent the rate of acetylcholine.

Methacholine chloride may be administered subcutaneously for paroxysmal auricular tachycardia, although it is inferior to quinidine, for Raynaud's disease, scleroderma, chronic ulcers and vasospastic conditions of the extremities.

Category—cholinergic.

Usual dose—subcutaneous, initial, 10 mg.; then 25 mg. may be given 10 to 30 minutes later.

Usual dose range—10 to 40 mg.

OCCURRENCE

Sterile Methacholine Chloride N.F.

Methacholine Bromide N.F., Mecholyl® Bromide, Acetyl-β-methylcholine Bromide, (2-Hydroxypropyl)trimethylammonium Bromide Acetate. This compound is very similar to the chloride in all of its properties, but it is somewhat less hygroscopic. It occurs as a white, crystalline powder with a faint, fishy odor. The pH of a freshly prepared solution (1 in 20) is about 5. Therefore, for oral use it may have an advantage in tablet form, but for injection or ion transfer, the chloride is preferable.

Category—cholinergic.

Usual dose—200 mg. 2 or 3 times a day.

Usual dose range—200 to 600 mg.

OCCURRENCE

Methacholine Bromide Tablets N.F.

Carbachol U.S.P., Doryl®, Choline Chloride Carbamate. Carbachol is a white or faintly yellow, crystalline solid or powder that is odorless or has a slight, aminelike odor and is hygroscopic. It is soluble 1:1 in water and 1:50 in alcohol. The aqueous solution is neutral to litmus paper. It may be prepared by the interaction of trimethylamine with β-chloroethyl carbamate. It differs from acetylcholine chloride in having the carbamyl group in place of the acetyl group

$$(CH_3)_3N + ClCH_2-O-CO-NH_2 \rightarrow$$
$$\overset{+}{(CH_3)_3NCH_2CH_2OCO-NH_2Cl^-}$$

and is more stable to hydrolysis. It can be noted here that the N-methyl and N,N-dimethylcarbamoyl cholines are highly stable compounds in aqueous solution and can be hydrolyzed only when boiled for several hours in alkaline solution.

It is absorbed less readily from the gastrointestinal tract than is methacholine, and effects after oral administration occur there chiefly. The nicotine effect is somewhat greater (3 to 14 times), and all actions are prolonged, so that carbachol is inclined to be more toxic than the others after injection. Atropine abolishes the muscarinelike effect, but the antidotal action is much slower, and greater doses of atropine are required.

As a remedy, it can be used for the same purposes as others of the class. It is excellent in the pains of peripheral vascular disease of the vasospastic type, for urinary retention after operation and for abdominal distention due to stasis, and its effects outlast those of other choline derivatives. However, because of its toxicity it is limited usually to those cases that do not respond to safer drugs. It is often employed to reduce the intraocular tension of glaucoma when response cannot be obtained with pilocarpine, physostigmine, neostigmine or methacholine.

Category—cholinergic (ophthalmic).

For external use—topically, 0.1 ml. of 0.75

to 3 per cent solution instilled in the conjunctival sac, 2 to 3 times a day.

OCCURRENCE

Carbachol Ophthalmic Solution U.S.P.

Bethanechol Chloride U.S.P., Urecholine® Chloride, β-Methylcholine Carbamate Chloride, (2-Hydroxypropyl)trimethylammonium Chloride Carbamate, Carbamylmethylcholine Chloride. This is a urethane

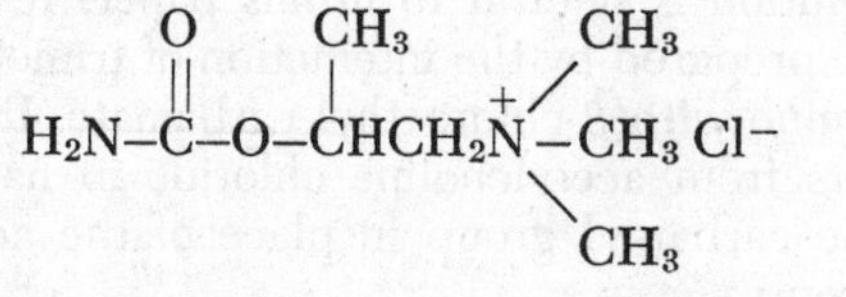

$$H_2N-\overset{\overset{\textstyle O}{\|}}{C}-O-\overset{\overset{\textstyle CH_3}{|}}{C}HCH_2\overset{+}{N}-CH_3\; Cl^-$$
$$\overset{|}{\underset{CH_3}{}}\quad\underset{CH_3}{\diagdown}$$

of β-methylcholine chloride, occurring as a white, crystalline solid with an aminelike odor. It is soluble in water 1:1 and 1:10 in alcohol but nearly insoluble in most organic solvents. An aqueous solution has a pH of 5.5 to 6.3.

It is similar in activity to methacholine chloride but has slight ganglionic stimulating action and resists hydrolysis by cholinesterase. Administration of the drug is associated with low toxicity and no serious side-effects. Its duration of action is about 1 hour.

Category—cholinergic.

Usual dose—oral, 10 to 30 mg. 3 times daily, subcutaneous, 2.5 mg. 3 times a day.

Usual dose range—oral, 30 to 120 mg. daily; subcutaneous, 2.5 to 30 mg. daily.

OCCURRENCE

Bethanechol Chloride Injection U.S.P.
Bethanechol Chloride Tablets U.S.P.

INDIRECT REVERSIBLE CHOLINERGIC AGENTS

Early studies suggested that carbamates inhibited cholinesterase reversibly, because inhibition could be reversed by washing, dialysis, dilution or adding of substrate. Carbamylation by the inhibitor was rejected; steric and not electronic factors are of predominant importance in determining potency of carbamates, because electron-withdrawing substituents weaken potency —somewhat the reverse of expectation for a simple carbamylation mechanism. Furthermore, the rates of carbamylation are

relatively slow. That carbamylation does occur was shown by the fact that acetylcholinesterase (AChE) activity that had been inhibited by dimethyl-carbamylfluoride, neostigmine and pyridostigmine respectively recovered at the same rate. The half life of the dimethylcarbamyl acetylcholinesterase (human) was about 35 minutes. The much longer persistence of the pharmacologic effect must, therefore, be caused by other factors, i.e., drug deposition, etc.[12] The half life for the methyl carbamates is 19 minutes at 38° and for the carbamates 2 minutes. Recovery is fairly rapid in water, and more rapid in the presence of hydroxylamine or choline, whereas pyridine-2-aldoxime methiodide (2-PAMI) had no effect.

Acetylcholinesterase (AChE) can be inhibited by carbamylating agents in vivo to increase the level of acetylcholine. Carbamylation of the serine OH of AChE is effected by

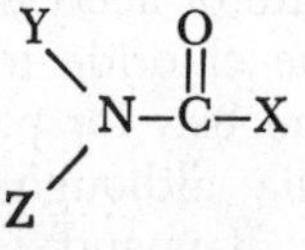

$$\underset{Z}{\overset{Y}{\diagdown}}N-\overset{\overset{\textstyle O}{\|}}{C}-X$$

where Y and Z = H, CH₃, φ, etc., and

$$X \;=\; F, \quad Cl, \quad (CH_3)_3\overset{+}{N}-CH_2CH_2O-;$$

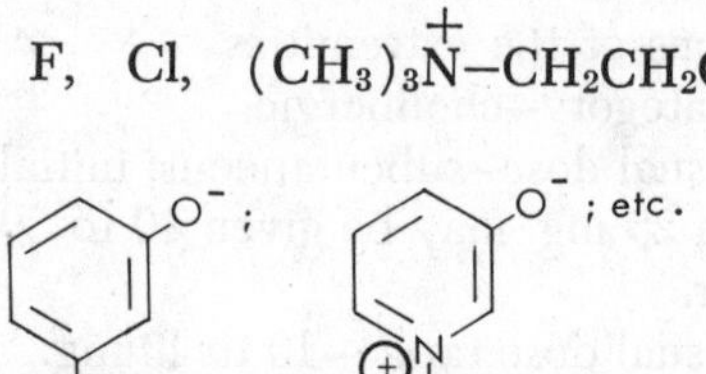

The rate of carbamylation of AChE decreases in the following order: R–OCONH₂ >ROCONHCH₃ >ROCON(CH₃)₂. The rate of cleavage of the resulting carbamylated AChE is the reverse of the carbamylation rate.[13]

It is interesting to note that carbamylation of AChE by dimethylcarbamyl fluoride is markedly accelerated by tetraethylammonium ions. All tetra-alkylammonium ions[14] slow the hydrolysis of dimethylcarbamyl AChE.

Methanesulfonyl fluoride and alkyl methanesulfonates inhibit AChE by formation of AChE methylsulfonate CH₃SO₃-AChE. This ester is not hydrolyzed by water,

H₂NOH or 2-PAM-I but is hydrolyzed by pyridine-3-aldoxime methiodide or phenyl-3-pyridyl-anti-ketoxime methiodide.[15]

New kinetic evidence[16] indicates that carbamylation is preceded by reversible complex formation, i.e., $E + CX \rightleftharpoons ECX$, where E is the AChE, C the carbamate group and X the remainder of the molecule. Subsequently, $ECX \rightarrow EC + X$ and, finally, $EC \rightarrow E + C$. An over-all expression of carbamate participation with AChE would be:

$$E + CX \underset{K_1}{\overset{K_1}{\rightleftharpoons}} ECX \overset{K_2}{\underset{X}{\rightarrow}} EC \overset{K_3}{\rightarrow} E + C.$$

Thus, carbamates are substrates for AChE, with high affinity, low carbamylation rates, and even lower decarbamylation rates. The differences in anti-AChE activity of a number of carbamates was due to the differences in complexing ability; therefore, one cannot neglect the complex formation step. It should follow that the inhibition of AChE by carbamates is a summation of the complex formation and the rate of hydrolysis of the carbamylated AChE or the decarbamylated step.

PRODUCTS

Physostigmine U.S.P. is an alkaloid usually obtained from the dried ripe seed of *Physostigma venenosum*. It occurs as a white, odorless, microcrystalline powder that is slightly soluble in water and freely soluble in alcohol, chloroform and the fixed oils. This alkaloid as the free base is quite sensitive to heat, light, moisture and bases and readily undergoes decomposition. When used topically to the conjunctiva it is better tolerated than its salts. Its liposolubility properties permit adequate absorption from the appropriate ointment bases.

Physostigmine is a competitive inhibitor of acetylcholinesterase when acetylcholine is simultaneously present. A noncompetitive inhibition is observed when the enzyme is preincubated with physostigmine.

Category—cholinergic (ophthalmic).

For external use—topically, as a 0.25 per cent ointment to the conjunctiva 1 to 4 times daily.

Physostigmine Salicylate U.S.P., Eserine Salicylate. This is the salicylate of an alkaloid usually obtained from the dried ripe seed of *Physostigma venenosum*. It may be prepared by neutralizing an ethereal solution of the alkaloid with an ethereal solution of salicylic acid. Excess salicylic acid is removed from the precipitated product by washing it with ether. The salicylate is less deliquescent than the sulfate.

Physostigmine Salicylate

It occurs as a white, shining, odorless crystal, or white powder that is soluble in water (1:75), alcohol (1:16) or chloroform (1:6) but is much less soluble in ether (1:250). Upon prolonged exposure to air and light, the crystals turn red in color. The red color may be removed by washing the crystals with alcohol, although this causes loss of the compound as well. Aqueous solutions are neutral or slightly acidic in reaction and take on a red coloration after a period of time. The coloration may be taken as an index of the loss of activity of physostigmine solutions. To guard against decomposed solutions, only enough should be made at one time for about a week's use. If it is necessary to sterilize the solution, it can be done by bacteriologic filtration. Physostigmine salicylate solutions are incompatible with the usual alkaloidal reagents, with alkalies and with iron salts.

Physostigmine in solution first hydrolyzes to methylcarbamic acid and eserinol (a phenol) which is oxidized readily to the red colored compound rubreserine. The addition of sulfite will prevent the oxidation of the phenol, eserinol (see epinephrine), and no red color will develop. However, hydrolysis does take place, and the physostigmine is inactivated. Solutions are most stable at pH 6 and never should be sterilized by heat. The addition of ascorbic acid (1:1,000) stabilizes up to 6 months under average conditions.[17]

Physostigmine effectively inhibits cholinesterase at about 10^{-6} M concentration. Its cholinesterase-inhibiting properties vary with pH (see Fig. 16).

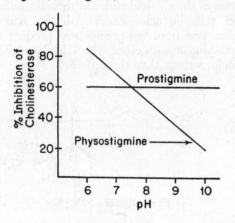

FIGURE 16

Physostigmine's cationic properties, greatest below pH 7.5, make a marked contribution to its activity. It is chiefly neutral at pH 8.5 or higher.

The ophthalmic effect (miotic) of physostigmine and related compounds is due to contraction of the ciliary body. This promotes drainage through the canal of Schlemm and thereby decreases intraocular pressure. For this reason, physostigmine is used in the treatment of glaucoma by direct instillation of a 0.1 to 1 per cent solution in the eye. It is directly antagonistic to atropine in the eye and is sometimes used to help restore the pupil to normal size following atropine dilatation. Physostigmine also causes *stimulation of the intestinal musculature* and because of this is used in conditions of depressed intestinal motility. In gaseous distention of the bowel, due to a number of causes, physostigmine often aids in the evacuation of gas as well as restoring normal bowel movement. It is administered by injection for this purpose. Much research has been done to find synthetic drugs with a physostigminelike action. This has resulted in compounds of the neostigmine type (q.v.), which, at least for intestinal stimulation, are superior to physostigmine.

Recently, acetylcholine was isolated and positively identified in brain tissue for the first time. Physostigmine depressed pole jump behavior in conditioned rats, and these effects suggest a central acetylcholinesterase inhibition. Hyoscyamine (q.v.) blocked these central effects of physostigmine.[18]

Category—cholinergic (ophthalmic).

For external use—topically to the conjunctiva, 0.1 ml. of 0.025 to 0.5 per cent solution 1 to 4 times a day.

Physostigmine Sulfate U.S.P. occurs as a white, odorless, microcrystalline powder that is deliquescent in moist air. It is soluble in water 1:4, 1:0.4 in alcohol and 1:1,200 in ether.

Category—cholinergic (ophthalmic).

For external use—topically, 0.1 ml. to the conjunctiva as 0.25 to 0.5 per cent solution 1 to 4 times a day.

Neostigmine Bromide U.S.P., Prostigmine® Bromide, (*m*-Hydroxyphenyl)trimethylammonium Bromide Dimethylcarbamate, Dimethylcarbamic Ester of 3-Hydroxyphenyltrimethylammonium Bromide.

A method of preparation is from dimethylcarbamyl chloride and the potassium salt of 3-hydroxyphenyldimethylamine. Methyl bromide readily adds to the tertiary amine, forming the stable quaternary amine (see formula for neostigmine bromide). It occurs as an odorless, white, crystalline powder having a bitter taste. It is soluble in water (1:05) and is soluble in alcohol. The crystals are much less hygroscopic than are those of neostigmine methylsulfate and thus may be used in tablets. Solutions are stable and may be sterilized by boiling. Aqueous solutions are neutral to litmus.

Use of physostigmine as prototype of an indirect parasympathomimetic drug led to the development of stigmine in which a trimethylamine group was placed para to a dimethyl carbamate group in benzene. However, activity was obtained when these groups were placed meta to each other; neostigmine, a more active and useful drug, was obtained. Although physostigmine contains a methyl carbamate grouping, greater stability toward hydrolysis was obtained with a dimethyl carbamate group as in neostigmine.[19] The meta substituents are more stable than the para.

Of neostigmine that reaches the liver, 98 per cent is metabolized in 10 minutes. Its transfer from plasma to liver cells and then to bile is probably passive in character. Since cellular membranes permit the passage of plasma proteins synthesized in liver into the blood stream through capillary walls or lymphatic vessels, they may not present a barrier to the diffusion of quaternary amines such as neostigmine. Possibly the rapid hepatic metabolism of neostigmine provides a downhill gradient for the continual diffusion of this compound.[20] A certain amount may be hydrolyzed slowly by plasma cholinesterase.

Neostigmine has a mechanism of action quite similar to that of physostigmine. Prostigmine effectively inhibits cholinesterase at about 10^{-6} M concentration. Its activity does not vary with pH and at all ranges exhibits similar cationic properties (see Fig. 16). There may be a direct action of the drug on tissues innervated by cholinergic nerves, but this has not yet been confirmed.

The uses of neostigmine are similar to those of physostigmine, but they differ in that there are greater miotic activity, fewer and less unpleasant local and systemic manifestations and greater stability. Most frequent application is to prevent atony of the intestinal, skeletal and bladder musculature. An important use is in the treatment of myasthenia gravis. The bromide is used orally.

Category—cholinergic.

Usual dose—15 to 30 mg. 3 to 6 times daily.

Usual dose range—15 to 375 mg. daily.

OCCURRENCE
Neostigmine Bromide Tablets U.S.P.

Neostigmine Methylsulfate U.S.P., Pro-

stigmine® Methylsulfate, (*m*-Hydroxyphenyl)trimethylammonium Methylsulfate Dimethylcarbamate, Dimethylcarbamic Ester of 3 - Hydroxyphenyltrimethylammonium Methylsulfate. Neostigmine is prepared, as in the method previously described, and the quaternary amine is made with methylsulfate. This compound is an odorless, white crystalline powder with a bitter taste. It is very soluble in water and is soluble in alcohol. Solutions are stable and can be sterilized by boiling. The compound is too hygroscopic for use in a solid form and thus always is used in injection. Aqueous solutions are neutral to litmus.

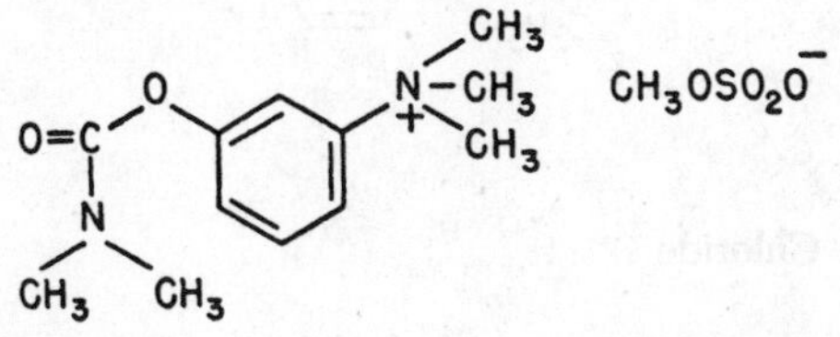

The methylsulfate is used for the same conditions as the bromide. By subcutaneous or intramuscular injection, it prevents postoperative distention.

Category—cholinergic.

Usual dose—intramuscular or subcutaneous, 500 mg. 4 to 8 times daily.

Usual dose range—1 to 5 mg. daily.

OCCURRENCE
Neostigmine Methylsulfate Injection U.S.P.

Pyridostigmine Bromide U.S.P., Mestinon® Bromide, 3-Hydroxy-1-methylpyridinium Bromide Dimethylcarbamate, Pyridostigminium Bromide. This occurs as a white,

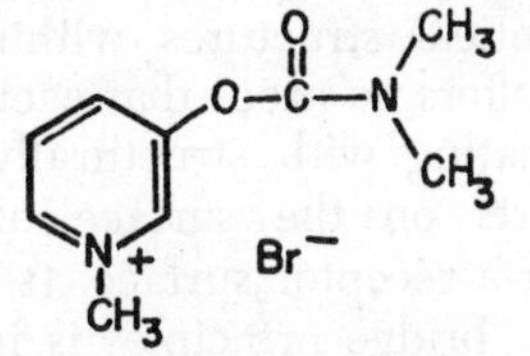

hygroscopic crystalline powder having an agreeable characteristic odor. It is freely soluble in water, alcohol and in chloroform.

Pyridostigmine bromide is about one fifth as toxic as neostigmine. It appears to function in a manner like that of neostigmine but is said really to inactivate pseudocholinesterase rather than cholinesterase. This agent is used primarily to treat myasthenia gravis. It has a longer period of duration and less muscarinic effect on the gastrointestinal tract.

Category—cholinergic.

Usual dose—60 to 120 mg. 3 to 6 times daily.

Usual dose range—60 mg. to 1.5 g. daily.

OCCURRENCE
Pyridostigmine Bromide Syrup U.S.P.
Pyridostigmine Bromide Tablets U.S.P.

Demecarium Bromide U.S.P., Humorsol®, (*m*-hydroxyphenyl)trimethylammonium bromide decamethylenebis[methylcarbamate], is the diester of (m-hydroxyphenyl)trimethylammonium bromide with decamethylene bis(methylcarbamic acid) and thus is comparable to a bis-prostigmine molecule.

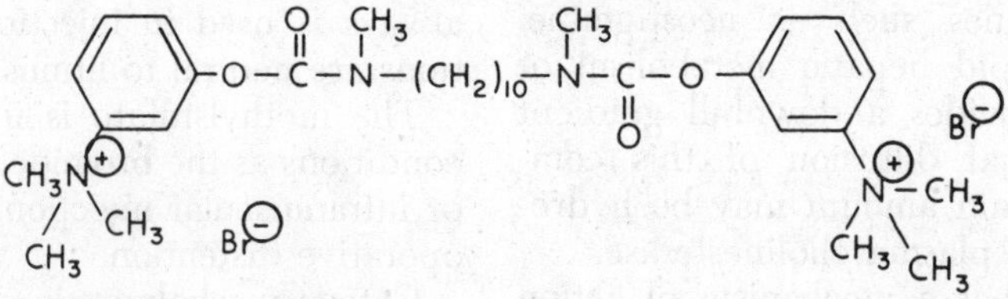

OCCURRENCE
Demecarium Bromide Ophthalmic Solution U.S.P.

Ambenonium Chloride N.F., Mytelase® Chloride, [Oxalylbis(iminoethylene)]bis [(*o*-chlorobenzyl)diethylammonium Chloride]. This compound is a white, odorless powder, soluble in water and in alcohol, slightly soluble in chloroform, and practically insoluble in ether and in acetone. Ambenonium chloride is a cholinergic drug

It occurs as a slightly hygroscopic powder that is freely soluble in water or alcohol. Aqueous solutions are neutral, stable and may be sterilized by heat. Its efficacy and toxicity are comparable to those of other potent anticholinesterase inhibitor drugs. It is useful in the management of glaucoma and accommodative convergent strabismus. Although demecarium bromide possesses a bis type molecule, symmetrical molecules such as this or even the presence of 2 or more identical structures within a given molecule offers no proof that each structure is participating with structurally identical counterparts on the surface of a given enzyme or a receptor surface. It is possible that some "bridge principle" is involved.

Category—cholinergic (ophthalmic).

For external use—topically, 0.03 to 0.06 ml. of a 0.125 to 0.25 per cent solution twice weekly, to one or two times daily, to the eye.

used in the treatment of myasthenia gravis. This condition is characterized by a pathologic exhaustion of the voluntary muscles, which is caused by either an underproduction of acetylcholine or an overproduction and activity of cholinesterase at the myoneural junction. Ambenonium chloride acts by suppressing the activity of cholinesterase. It possesses a relatively prolonged duration of action and causes fewer side effects in the gastrointestinal tract than the other anticholinesterase drugs. The dosage requirements vary considerably, and the dosage must be individualized according to the response and tolerance of the patient.

Category—cholinergic.

Usual dose—initial, 5 mg., gradually increased as required up to 5 to 25 mg. 3 or 4 times daily.

Usual dose range—5 to 50 mg.

OCCURRENCE
Ambenonium Chloride Tablets N.F.

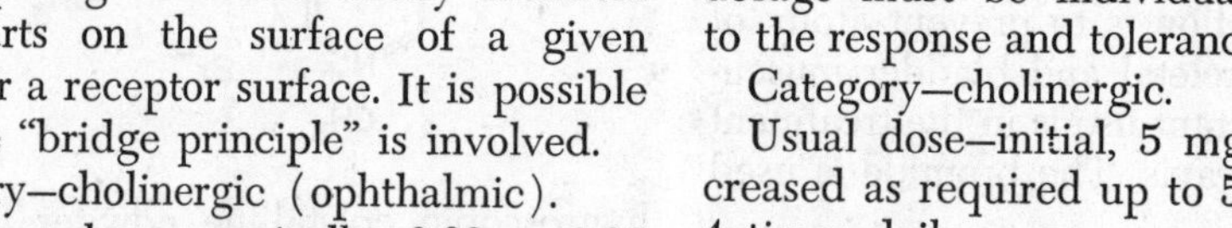

Ambenonium Chloride

IRREVERSIBLE INDIRECT CHOLINERGIC AGENTS

Cholinesterase can be inhibited irreversibly by a group of phosphate esters that are highly toxic (MLD for humans is 0.1 to 0.01 mg./kg.) and are called nerve poisons. They permit acetylcholine to accumulate in concentrations above normal. A general type formula for such compounds is as follows:

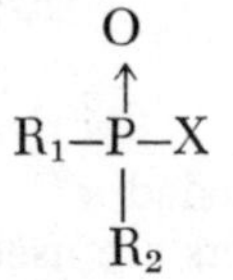

$$
\begin{array}{l}
\text{O} \\
\uparrow \\
R_1-P-X \\
| \\
R_2
\end{array}
\quad
\begin{array}{l}
R_1 = \text{alkoxyl} \\
R_2 = \text{alkoxyl, alkyl or} \\
\quad\quad\text{tertiary amine} \\
X = \text{F or C}\equiv\text{N}
\end{array}
$$

and some of the better known ones are:

Parathion $(EtO)_2P(S)O(p\text{-}NO_2C_6H_4)$ is inactive for cholinesterase in vitro and in vivo; its metabolite, paraoxon $(EtO)_2$-$P(O)O(p\text{-}NO_2C_6H_4)$, is very active and it also is inactivated by the liver. As is the case with some other biologically active substances, the route of administration may influence the quantitative effects. In the case of some cholinesterase inhibitors, the availability of the drug for metabolism by the liver is a major factor in its toxicity. The data given in Table 48 tend to support these conclusions.[21]

The stability of these esters permits of affinity labeling[22] studies in this and other

NAME	R_1	R_2	X
Isoflurophate	$O-CH(CH_3)_2$	$OCH(CH_3)_2$	F
Sarin	OCH_2CH_3	$-CH_3$	F
Soman	$OCH-C(CH_3)_3$ $\quad\mid$ $\quad CH_3$	$-CH_3$	F
Tabun	OCH_2CH_3	$N(CH_3)_2$	$C\equiv N$

TABLE 48. INFLUENCE OF THE ROUTE OF ADMINISTRATION ON TOXICITY* †

		LD50, MG/KG BODY WEIGHT (95% FIDUCIAL LIMITS)			
		HEPATIC ROUTES		PERIPHERAL ROUTES	
COMPOUND	MOLECULAR WEIGHT	INTRA-PERITONEAL	ORAL	SUB-CUTANEOUS	INTRA-VENOUS
Physostigmine salicylate	413.45	ca. 1.0	5.50	1.12	0.46
Neostigmine methylsulfate	334.39	0.62	>5.0	0.66	0.47
Azodrin	223.17	8.91	14.4	8.71	ca. 9.2
Bidrin	237.20	11.8	20.0	11.5	ca. 9.9
Chlorfenvinphos	359.59	87.0	398	339	87
Ciodrin	314.28	70.8	186.2	15.1	4.5
Phosdrin	224.14	2.51	12.30	1.18	0.68
Paraoxon	275.21	2.29	12.80	ca. 0.6	0.59
Parathion	291.27	15.1	25.7	21.4	17.4

* 24-Hour median lethal doses of cholinesterase inhibitors following administration by different routes in female mice.

CHEMICAL FORMULAE OF THE ORGANOPHOSPHORUS CHOLINESTERASE INHIBITORS EXAMINED

COMPOUND	R	R'	R"
Azodrin	Me	Me	$-CO\cdot NH\cdot Me$
Bidrin	Me	Me	$-CO\cdot NMe,$
Chlorfenvinphos	Et	2,4-dichlorophenyl	Cl
Ciodrin	Me	Me	$-CO\cdot O\cdot CH(Me)Ph$
Phosdrin	Me	Me	$-CO\cdot OMe$
Paraoxon	$(EtO)_2P(O)O(p\text{-}NO_2C_6H_4)$		
Parathion	$(EtO)_2P(S)O(p\text{-}NO_2C_6H_4)$		

† Natoff, I. L.: J. Pharm. Pharmacol. 19:612, 1967.

enzymes that have the serine OH as part of their sites. They combine irreversibly with the esteratic site (HG-, which probably is the OH of the amino acid residue serine) of cholinesterase as follows.[23]

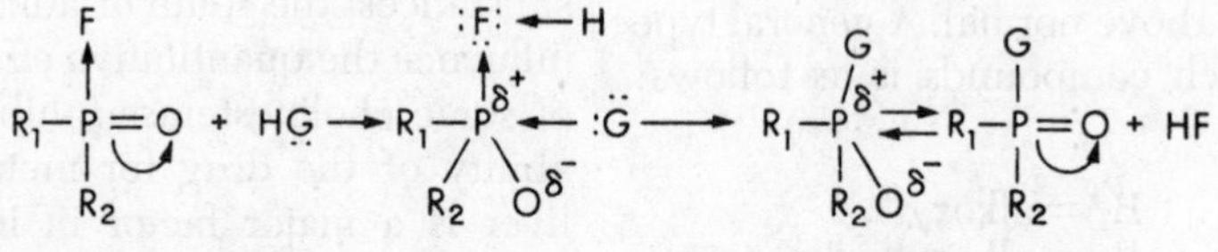

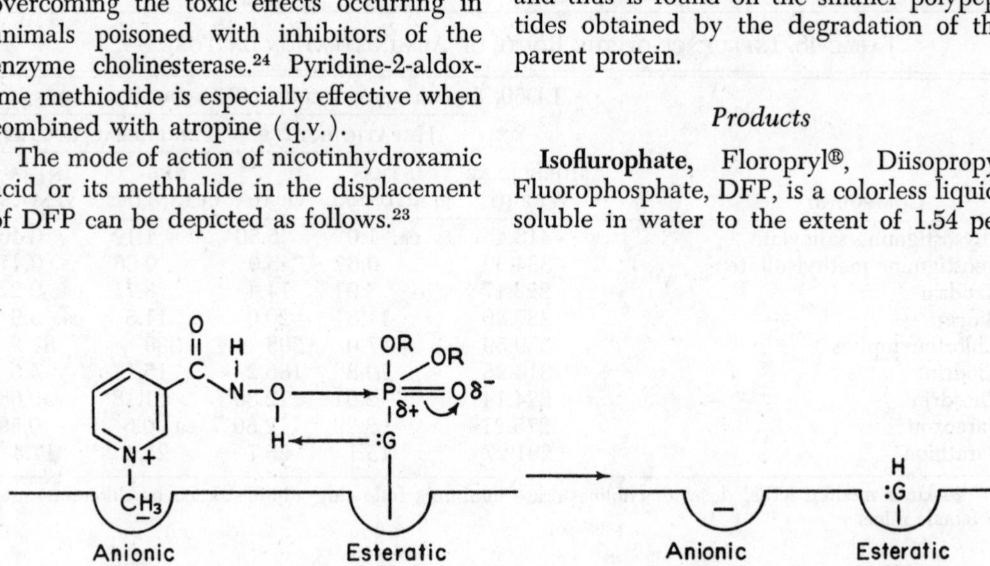

It is interesting to note that when R is ethyl, the phosphate group is displaced readily by many nucleophilic agents, i.e., amines, alcohols, mercaptans, etc. However, when R is isopropyl, a stronger nucleophilic agent

$$R_1-P=O \ + \ HG \ \longrightarrow \ R_1-P \xrightarrow{\quad} :G \ \longrightarrow \ R_1-P \rightleftarrows R_1-P=O \ + \ HF$$

The above postulations have been supported by the finding that a nucleophilic attacking agent such as hydroxylamine or hydroxamic acid can displace and reverse the activity of the phosphate ester. This led to the development of more effective displacing agents such as nicotinhydroxamic acid, its methhalide, pyridine-2-aldoxime methiodide, monoisonitrosoacetone and diacetonylmonoxime. The last three compounds have been reported to be effective in overcoming the toxic effects occurring in animals poisoned with inhibitors of the enzyme cholinesterase.[24] Pyridine-2-aldoxime methiodide is especially effective when combined with atropine (q.v.).

The mode of action of nicotinhydroxamic acid or its methhalide in the displacement of DFP can be depicted as follows.[23]

such as hydroxylamine is needed.

Some of the phosphate esters are used as insecticidal agents and must be handled with extreme caution because they also are very toxic to humans.

DFP and possibly related compounds also can phosphorylate the OH of the serine residue found as a functional group at the active site in other enzymes such as trypsin, alpha-chymotrypsin, etc. The derivative from DFP is stable to proteolytic enzymes and thus is found on the smaller polypeptides obtained by the degradation of the parent protein.

Products

Isoflurophate, Floropryl®, Diisopropyl Fluorophosphate, DFP, is a colorless liquid, soluble in water to the extent of 1.54 per cent at 25° to give a pH of 2.5. It is soluble in alcohol and to some extent in peanut oil.

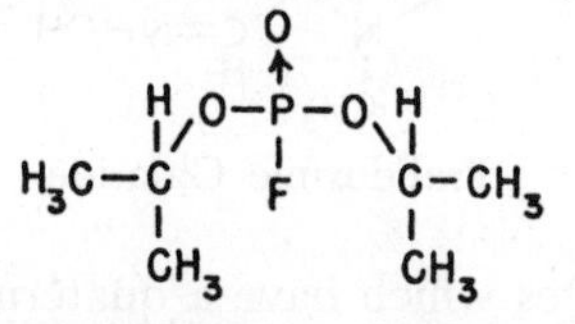

Diisopropyl Fluorophosphate

It is stable in the latter for a period of 1 year but decomposes in water in a few days. Solutions in peanut oil can be sterilized by autoclaving. The compound should be stored in hard glass, since continued contact with soft glass is said to hasten decomposition, as evidenced by a discoloration.

It must be handled with extreme caution. Avoid contact with eyes, nose, mouth and even the skin, because it can be absorbed readily through intact epidermis and more so through mucous tissues, etc.

Because DFP irreversibly[25] inhibits cholinesterase (q.v.), its activity lasts for days or even weeks. During this period new cholinesterase may be synthesized in plasma, the erythrocytes and other cells.

DFP has been used clinically in the treatment of myasthenia gravis and glaucoma. A combination of atropine sulfate and magnesium sulfate has been found to give protection in rabbits against the toxic effects of DFP. One counteracts the muscarine, and the other the nicotine effect of the drug.[26]

Isofluorophate is official in the *N.F.* as an ophthalmic ointment (0.025 per cent), applied topically to the conjunctiva once every 3 days to 3 times a day and in the *U.S.P.*, as an ophthalmic solution (0.1 per cent) to be used in the treatment of glaucoma, 1 to 3 drops 3 times a day to once every 3 days.

Echothiophate Iodide U.S.P., Phospholine® Iodide, S-Ester of (2-Mercaptoethyl)trimethylammonium Iodide with O, O-Diethyl Phosphorothioate. This occurs as a white, crystalline, hygroscopic solid that

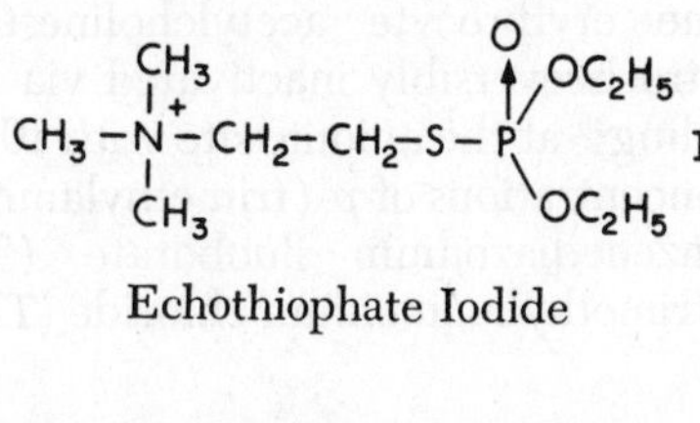

Echothiophate Iodide

has a slight mercaptanlike odor. It is soluble in water 1:1, and 1:25 in dehydrated alcohol; aqueous solutions have a pH of about 4.

Echothiophate Iodide is a long lasting cholinesterase inhibitor of the irreversible type such as isoflurophate. However, unlike the latter, it is a quaternary salt, and thus, when applied locally, its distribution in tissues is limited, which can be very desirable.

Category—cholinergic (ophthalmic).

For external use—topically, 0.1 ml. of a 0.06 to 0.125 per cent solution to the conjunctiva, once or twice daily.

OCCURRENCE
Echothiophate Iodide for Ophthalmic
  Solution U.S.P.

Hexaethyltetraphosphate, HETP; and **Tetraethylpyrophosphate.** These two substances are compounds that also show anticholinesterase activity. HETP was developed by the Germans during World War II and is used as an insecticide against aphids. It has been reported that some HETP being sold in the United States does not have this structure but is a mixture, in which the active constituent is tetraethylpyrophosphate.

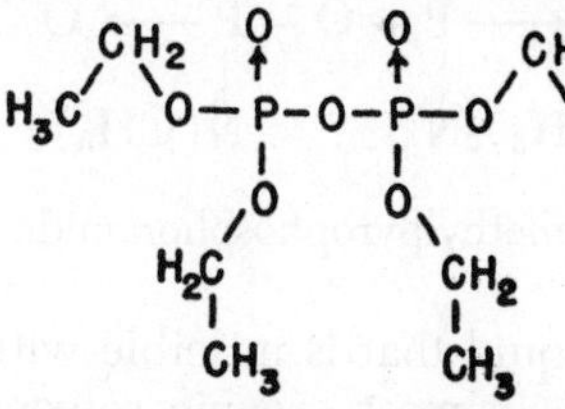

Tetraethylpyrophosphate

When used as insecticides, these compounds have the advantage of being hydrolyzed rapidly to the relatively nontoxic water-soluble compounds phosphoric acid and ethyl alcohol. Fruit trees or vegetables sprayed with this type of compound retain no harmful residue after a period of a few days or weeks, depending on the weather conditions. The disadvantage of their use comes from their very high toxicity, which results mainly from their anticholinesterase activity. Workers spraying with these agents should use extreme caution that none of the vapors are breathed and that none of the vapor or liquid comes in contact with the eyes or skin.

Parathion; Thiophos; Niran; Alkron; O,-O-Diethyl-O-*p*-nitrophenyl Thiophosphate; Diethyl-*p*-nitrophenyl Monothiophosphate.

$$S \leftarrow \overset{\overset{\displaystyle O-C_2H_5}{|}}{\underset{\underset{\displaystyle O-C_2H_5}{|}}{P}}-O-\!\!\!\bigcirc\!\!\!-NO_2$$

Parathion

This compound is a yellow liquid that is freely soluble in aromatic hydrocarbons, ethers, ketones, esters and alcohols but practically insoluble in water, petroleum ether, kerosene and the usual spray oils. It is decomposed at pH's higher than 7.5. Parathion is used as an agricultural insecticide. It is highly toxic, the effects being cumulative. Special precautions are necessary to prevent skin contamination or inhalation. A fuller discussion of pesticides may be found in the literature.[27]

Octamethylpyrophosphoramide; OMPA; Pestox III; Schradan; *Bis*(bisdimethylaminophosphonous) Anhydride. This compound is

$$\begin{array}{ccc}(CH_3)_2N & & N(CH_3)_2 \\ | & & | \\ O \leftarrow P - O - P \longrightarrow O \\ | & & | \\ (CH_3)_2N & & N(CH_3)_2\end{array}$$

Octamethylpyrophosphoramide

a viscous liquid that is miscible with water and soluble in most organic solvents. It is not hydrolyzed by alkalies or water but is hydrolyzed by acids. It is used as a systemic insecticide for plants, being absorbed by the plants without appreciable injury, but insects feeding on the plant are incapacitated. The compound has been used as an experimental cholinergic drug.

Pralidoxime Chloride U.S.P., Protopam® Chloride, 2-Formyl-1-methylpyridinium Chloride, 2-PAM Chloride, 2-Pyridine Aldoxime Methyl Chloride. It occurs as a white, nonhygroscopic crystalline powder that is soluble in water, 1 g. in less than 1 ml.

Pralidoxime chloride is used as an antidote for poisoning by parathion and related pesticides. It may be effective against some

$$\underset{\underset{\displaystyle CH_3}{|}}{\overset{\displaystyle \bigcirc}{\underset{\displaystyle \overset{+}{N}}{}}}C\!=\!N\!-\!OH \quad Cl^-$$

Pralidoxime Chloride

phosphates which have a quaternary nitrogen. It also is an effective antagonist for some carbamates such as neostigmine methyl sulfate and pyridostigmine bromide.

In addition, pralidoxime effects depolarization at the neuromuscular junction, it is anticholinergic, cholinomimetic, it inhibits cholinesterase, potentiates the depressor action of acetylcholine in nonatropinized animals and potentiates the pressor action of acetylcholine in atropinized animals.

It is not effective against a number of amide and carbamate type pesticides such as Dimetan, Dimetilan and Pyrolan; it is contraindicated in cases involving poisoning by Sevin. It is not effective against eserine in the mouse.

The mode of action of pralidoxime is analagous to that described previously for nicotinhydroxamic acid or its methhalide.

The biologic half-life of 2-PAM chloride in man is about 2 hours and its effectiveness is a function of its concentration in plasma that reaches a maximum of 2 to 3 hours after oral administration. Concentrations of 4 and 8 μg/ml. of 2-PAM chloride in the blood plasma of rats significantly decreases the toxicity of sarin by factors of 2 and 2.5 respectively.[28]

Pralidoxime chloride, a quaternary ammonium compound, is most effective by intramuscular, subcutaneous or intravenous administration. Treatment of poisoning by an anticholinesterase will be most effective if given within a few hours. Little will be accomplished if the drug is used more than 36 hours after parathion poisoning has occurred.

Aldoxime analogs of arecoline were less toxic and much less effective reactivators than the quaternary aldoximes.[29]

Bovine erythrocyte acetylcholinesterase is in vitro irreversibly inactivated via affinity labeling[30] at the anionic site with 10^{-5} to 10^{-7} concentrations of *p*-(trimethylammonium)benzenediazonium fluoborate (Tdf). Phenyltrimethylammonium chloride (TMA), a reversible inhibitor, protects against Tdf inactivation.

Category—cholinesterase reactivator.

Usual dose—intravenous, 1 g. injected over a period of not less than 2 minutes; infusion, 1 g. in 250 ml. of Sodium Chloride Injection over a period of 30 minutes, repeat in 1 hour, if required; oral, 1 g., repeated in 3 hours if required.

OCCURRENCE

Sterile Pralidoxime Chloride U.S.P.
Pralidoxime Chloride Tablets U.S.P.

DRUGS ACTING DIRECTLY ON CELLS

Pilocarpine Hydrochloride U.S.P., pilocarpine monohydrochloride, is the hydrochloride of an alkaloid obtained from the dried leaflets of *Pilocarpus jaborandi* or *P. microphyllus* where it occurs to the extent of about 0.5 per cent together with other alkaloids with a total of 1 per cent.

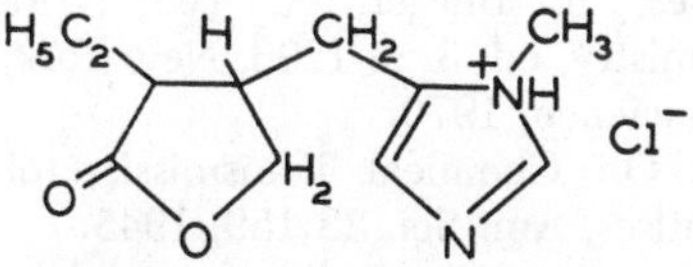

Pilocarpine Hydrochloride

It occurs as colorless, translucent, odorless, faintly bitter crystals that are soluble in water (1:0.3), alcohol (1:3) and chloroform (1:360). It is hygroscopic and affected by light; its solutions are acid to litmus and may be sterilized by autoclaving. Alkalies saponify its ester group to give the corresponding inactive hydroxy acid (pilocarpic acid).

Pilocarpine has a physostigminelike action but appears to act by direct cell stimulation rather than by disturbance of the cholinesterase-acetylcholine relationship as is the case with physostigmine.

Recent evidence supports the view that pilocarpine mimics the action of muscarine to stimulate ganglia through receptor site occupation similar to acetylcholine.[31] Its over-all molecular architecture and the interatomic distances of its functional groups in certain conformations are similar to mus-

carine, i.e., about 4 Å from the tertiary $N—CH_3$ nitrogen to the ether oxygen or carbonyl oxygen, and are compatible with this concept.

Outstanding pharmacologic effects are the production of copious sweating, salivation and gastric secretion. It also exerts a miotic effect upon the eye and, because it causes a drop in intraocular pressure, it is used as a 0.5 to 6 per cent solution (i.e., of the salts) in treating glaucoma. Secretion in the respiratory tract is noted following therapeutic doses, and therefore the drug sometimes is used as an expectorant.

Category—cholinergic (ophthalmic).

For external use—topically, 0.1 ml. of a 0.5 to 6 per cent solution to the conjunctiva 1 to 6 times daily.

OCCURRENCE

Pilocarpine Hydrochloride Ophthalmic Solution U.S.P.

Pilocarpine Nitrate U.S.P., Pilocarpine Mononitrate. This salt occurs as shining, white crystals which are not hygroscopic but are light sensitive. It is soluble in water (1:4), soluble in alcohol (1:75) and insoluble in chloroform and in ether. Aqueous solutions are slightly acid to litmus and may be sterilized in the autoclave. The alkaloid is incompatible with alkalies, iodides, silver nitrate and the usual alkaloidal precipitants.

Category—cholinergic (ophthalmic).

For external use—topically, 0.1 ml. of a 0.5 to 6 percent solution to the conjunctiva 1 to 6 times daily.

OCCURRENCE

Pilocarpine Nitrate Ophthalmic Solution U.S.P.

REFERENCES

1. Barker, L. A., *et al.*: Biochem. Pharmacol. 16:2181, 1967.
2. Welsh, H. H., and Taub, R.: Science 112:47, 1950.
3. Welsh, H. H.: Am. Scientist 38:239, 1950.
4. Thomas, T., and Roufogalis, B.: Mol. Pharmacol. 3:103, 1967.
5. Gyermeck, L., *et al.*: Am. J. Physiol. 204: 68, 1963; Unna, K., and Murayama, S.: J. Pharmacol. Exp. Ther. 140:183, 1963.
6. Kier, L.: Mol. Pharmacol. 4:70, 1968.

7. Kitz, R. J., and Kremzner, L. T.: Mol. Pharmacol. 4:104, 1968.
8. Krupa, R.: Can. J. Biochem. 42:667, 1964.
9. Schueler, F. W., and Keasling, H. H.: Am. Sci. 113:512, 1951.
10. Hoskin, F.: Proc. Soc. Exp. Biol. Med. 113:320, 1963.
11. Beckett, A., *et al.*: J. Pharm. Pharmacol. 15:362, 1963.
12. Wilson, I. B.: Ann. N. Y. Acad. Sci. 135:177, 1968.
13. Wilson, I., *et al.*: J. Biol. Chem. 236, 1498, 1961.
14. Metzger, H., and Wilson, I.: J. Biol. Chem. 238:3432, 1963.
15. Alexander, J., *et al.*: J. Biol. Chem. 238:741, 1963.
16. O'Brien, R. D., *et al.*: Mol. Pharmacol. 2:593, 1966; O'Brien, R. D.: Ibid. 4:121, 1968.
17. Swallow, W.: Pharm. J. 66:11, 1951.
18. Rosecrans, J. A., *et al.*: Int. J. Neuropharmacol. 7:127, 1968.
19. Aeschlimann, J. A., and Reinert, M.: J. Pharmacol. Exp. Ther. 43:413, 1931.
20. Calvey, T. H.: Biochem. Pharmacol. 16:1989, 1967.
21. Natoff, I. L.: J. Pharm. Pharmacol. 19:612, 1967.
22. Oosterbaan, R. A., and Cohen, J. A., Goodwin, T. W., Harris, J. I., and Hartley, B. S., (eds.): *in* Structure and Activity of Enzymes, p. 87, New York, Academic Press, 1964.
23. Wilson, B., and Meislich, E. K.: J. Am. Chem. Soc. 74:4628, 1953; 77:4286, 1955.
24. Ellin, R. I.: Am. Chem. Soc. 80:6588, 1958. Also see Wills, J. H.: J. Med. Pharm. Chem. 3:353, 1961.

25. Tenn, J. G., and Tomarelli, R. C.: Am. J. Ophth. 35:46, 1952.
26. McNamara, P., *et al.*: J. Pharmacol. Exp. Ther. 87:281, 1946.
27. DuBois, K.: Bull. Am. Soc. Hosp. Pharm. 9:168, 1952; Fleck, E. E.: Ibid. 9:174, 1952; Heyroth, F. F.: Ibid. 9: 178, 1952.
28. Zvirblis, P., and Kondritzer, A.: J. Pharmacol. Exp. Ther. 157:432, 1967.
29. Wells, J. N., *et al.*: J. Pharm. Sci. 56:1190, 1967.
30. Wofsy, L., and Michaeli, D.: Proc. Nat. Acad. Sci. 58:2296, 1967; Wofsy, L., *et al.*: Biochem. 1:1031, 1962.
31. Jones, A.: J. Pharmacol. Exp. Ther. 141:195, 1963.

SELECTED READING

Ariens, E. J., and Simonis, A. M.: Cholinergic and anticholinergic drugs, Ann. N. Y. Acad. Sci. 144:842-868, 1967.

Crossland, J.: Chemical Transmission in the Central Nervous System, J. Pharm. Pharmacol. 12:1-36, 1960.

Gearien, J. E.: Cholinergics and Anticholinesterases, *in* Burger, A. (ed.): Medicinal Chemistry, ed. 3, p. 1296, New York, Wiley-Interscience, 1970.

Loewi, O.: Chemical Transmission of Nerve Impulses, Am. Sci. 33:159, 1945.

Nachmansohn, D.: Role of acetylcholine in neuromuscular transmission, N. Y. Acad. Sci. 135:136-149, 1966.

———: Molecular Biology, New York, Acad. Press, 1960.

O'Brien, R. D.: Toxic Phosphorus Esters, New York, Acad. Press, 1960.

19

Autonomic Blocking Agents and Related Drugs

Taito O. Soine, Ph.D.,
Professor of Medicinal Chemistry, and Chairman of the
Department of Medicinal Chemistry, College of Pharmacy,
University of Minnesota

INTRODUCTION

The autonomic nervous system already has been considered in previous chapters from the standpoint of stimulants, i.e., *adrenergic agents* (Chap. 17) and *cholinergic agents* (Chap. 18). To complete the picture of drug action on this nervous system it is desirable to examine the inhibitory drugs, i.e., *adrenergic blocking agents* and *anticholinergics*. It is important to note that, whereas the normal physiology provides a neurohumoral transmitter substance to evoke a stimulant action, there is no comparable substance to provide an inhibitory action. However, there are many synthetic and plant-produced compounds that provide such an inhibitory action. These compounds act by blocking synaptic transmission in either the sympathetic or the parasympathetic innervations. Moreover, the blocking action by these inhibitors is usually quite specific as to its locus. Thus, blocking can occur at the ganglionic nerve fiber terminations (or synapses) in either the sympathetic or the parasympathetic ganglia (ganglion blocking action) or it can occur at the postganglionic nerve fiber terminations of either system (nerve-muscle junction blocking action). The specificity of action can be attributed to a number of factors among which the anatomic charac-teristics of the action site and the character of the transmitter substance undoubtedly play a major role. These blocking agents are considered to have access more readily to the nerve-muscle junctions than to the respective ganglia and, with respect to the ganglia, the sympathetic ganglia are considered less accessible than the parasympathetic.

It is important to note that all preganglionic fibers in both systems as well as the postganglionic fibers in the parasympathetic system release acetylcholine as the chemical mediator of nervous response. This is also true of the fibers in the voluntary nervous system which, having no ganglia, can be looked upon as preganglionic fibers, although it must be emphasized that this nervous system is not to be confused with the autonomic nervous system. All acetylcholine-producing nerve fibers are classed as *cholinergic*, and drugs which produce a response similar to that produced by stimulation of the nerve fibers are termed *cholinergic drugs*. Postganglionic fibers of the sympathetic system release norepinephrine* (and possibly some epinephrine) as the chemical mediator and are said to be *adrenergic* in nature. Drugs producing the effects of this transmitter substance are

---

* This neurohumoral transmitter substance formerly was thought to be epinephrine (Adrenalin) only.

TABLE 49. AUTONOMIC BLOCKING AGENTS

SITE OF BLOCKING ACTION	NEUROHUMORAL TRANSMITTER SUBSTANCE	TYPE OF BLOCKING ACTION	EXAMPLE OF DRUG
Sympathetic			
Ganglion	Acetylcholine	Anticholinergic	Hexamethonium
Postganglionic synapse	Norepinephrine (and epinephrine)	Antiadrenergic	Dibenamine
Parasympathetic			
Ganglion	Acetylcholine	Anticholinergic	Hexamethonium
Postganglionic synapse	Acetylcholine	Anticholinergic	Atropine
*Voluntary**			
Neuromuscular junction	Acetylcholine	Anticholinergic	Curare

* Included as a matter of convenience and because of certain similarities with the ganglionic blocking agents. These drugs are not autonomic blocking agents.

adrenergic drugs. It is obvious, then, that those drugs which block the activity resulting from acetylcholine are *anticholinergics,* and those which block activity resulting from norepinephrine are *antiadrenergics* (or more commonly, *adrenergic blocking agents* or *adrenolytics*). Table 49 briefly summarizes some of the essential points concerning the autonomic nervous system and its neurohumoral transmitter substances, together with examples typical of drugs acting on the various synaptic loci. Further considerations of these autonomic and related drugs will be facilitated by considering the following major groups:

1. *Cholinergic Blocking Agents* acting at the postganglionic terminations of the parasympathetic nervous system (see below).

2. *Cholinergic Blocking Agents* acting at the ganglia of *both* the parasympathetic and the sympathetic nervous systems (see p. 561).

3. *Adrenergic Blocking Agents* acting at the postganglionic terminations of the sympathetic nervous system (see p. 568).

4. *Cholinergic Blocking Agents* acting at the neuromuscular junction of the voluntary nervous system (see p. 584).

In addition to the more or less well-defined categories of inhibitory action exemplified by the preceding groups as well as in Table 49, there are others that will be treated in connection with the drugs whose action they resemble most closely. For example, papaverine and its congeners will be discussed in connection with the antispasmodic anticholinergics, although their mechanism of action is not the same.

Cholinergic Blocking Agents Acting at the Postganglionic Terminations of the Parasympathetic Nervous System

These blocking agents are also known as *anticholinergics, parasympatholytics,* or *cholinolytics.* One might be more specific in stating that members typical of this group are "antimuscarinics." This term derives from the action of acetylcholine at the postganglionic synapse, i.e., on the receptor substances of the target cell at the myoneural junction of the postganglionic parasympathetic fibers which is imitated by the alkaloid, muscarine. Thus, any drug that opposes this specific action is an antimuscarinic drug.

THERAPEUTIC ACTIONS

Because organs controlled by the autonomic nervous system are doubly innervated by both the sympathetic and the parasympathetic systems, it is believed that there is a continual state of dynamic balance between the two systems. Theoretically, one should achieve the same end-result by stimulation of one of the systems or by blockade of the other and, indeed, in some cases this is true. Unfortunately, in most cases there is a limitation on this type of generalization, and the results of antimuscarinic blocking of the parasympathetic system are no exception. However, there are three predictable and clinically useful results from blocking the muscarinic effects of acetylcholine. These are:

1. *Mydriatic effect* (dilation of pupil of the eye) and *cycloplegia* (a paralysis of the ciliary structure of the eye, resulting in a paralysis of accommodation for near vision).

2. *Antispasmodic effect* (lowered tone and motility of the gastrointestinal tract and the genitourinary tract).

3. *Antisecretory effect* (reduced salivation (*antisialogogue*), reduced perspiration (*anhydrotic*) and reduced acid and gastric secretion).

These three general effects of parasympatholytics can be expected in some degree from any of the known drugs, although in some cases it is necessary to administer rather heroic doses to demonstrate the effect. The mydriatic and cycloplegic effects, when produced by topical application, are not subject to any great undesirable side-effects due to the other two effects, because of limited systemic absorption. This is not the case with the systemic antispasmodic effects obtained by oral or parenteral administration, and it has been stated by Bachrach[1] that no drug with effective blocking action on the gastrointestinal tract is free of undesirable side-effects on the other organs. The same is probably true of the antisecretory effects. Perhaps the most commonly experienced obnoxious effects from

the oral use of these drugs under ordinary conditions is dryness of the mouth, mydriasis and urinary retention.

Mydriatic and cycloplegic drugs are generally prescribed or used in the office by ophthalmologists. The principal purpose is for refraction studies in the process of fitting glasses. This permits the physician to examine the eye retina for possible discovery of abnormalities and diseases as well as to provide controlled conditions for the proper fitting of glasses. Because of the inability of the iris to contract under the influence of these drugs, there is a definite danger to the patient's eyes unless they are protected from strong light by the use of dark glasses. These drugs also are used to treat inflammation of the cornea (keratitis), inflammation of the iris and the ciliary organs (iritis and iridocyclitis), and inflammation of the choroid (choroiditis). Interestingly, a dark-colored iris appears to be more difficult to dilate than a light-colored one and may require more concentrated solutions. A caution in the use of mydriatics is advisable because of their demonstrated effect in raising the intraocular pressure. The pressure rises because pupil dilation tends to cause the iris to restrict drainage of fluid through the canal of Schlemm by crowding the angular space, thus leading to increased intraocular pressure. This is particularly the case with glaucomatous conditions which should be under the care of a physician.

It is well to note at this juncture that atropine is used widely as an antispasmodic because of its marked depressant effect on parasympathetically innervated smooth muscle. Indeed, atropine is the standard by which other similar drugs are measured. It is to be noted also that the action of atropine is a blocking action on the transmission of the nerve impulse, rather than a depressant effect directly on the musculature. Therefore, its action is termed *neurotropic* in contrast with the action of an antispasmodic such as papaverine, which appears to act by depression of the muscle cells and is termed *musculotropic*. Papaverine is the standard for comparison of musculotropic antispasmodics and, while not strictly a parasympatholytic, will be treated together with its synthetic analogs later in this chapter. The synthetic antispasmodics appear to combine neurotropic and musculotropic effects in greater or lesser measure, together with a certain amount of ganglion blocking activity in the case of the quaternary derivatives.

Because of the widespread use of anticholinergics in the treatment of various gastrointestinal complaints, it is desirable to examine the pharmacologic basis on which this therapy rests. Smooth muscle spasm, hypermotility and hypersecretion, individually or in combination, are associated with many painful ailments of the gastrointestinal tract. Among these are peptic ulcer, ulcerative colitis, gastritis, regional enteritis, pylorospasm, cardiospasm and functional diarrhea. Although the causes have not been clearly defined, there are many who feel that emotional stress is the underlying common denominator to all of these conditions rather than a simple malfunction of the cholinergic apparatus. On the basis of Selye's original work on stress and Cannon's classic demonstration of the disruptive effects on normal digestive processes of anger, fear and excitement, stress is considered as being causative. The excitatory (parasympathetic) nerve of the stomach and the gut is intimately associated with the hypothalamus (the so-called "seat of feelings") as well as with the medullary and the sacral portions of the spinal cord. It is believed that emotions arising or passing through the hypothalamic area can transmit definite effects to the peripheral neural pathways such as the vagus and other parasympathetic and sympathetic routes. This is commonly known as a *psychosomatic reaction.* The stomach appears to be influenced by emotions more readily and more extensively than any other organ, and it does not strain the imagination to establish a connection between emotional effects and malfunction of the gastrointestinal tract. Individuals under constant stress are thought to develop a condition of "autonomic imbalance" due to repeated overstimulation of the parasympathetic pathways. The result is little rest and gross overwork on the part of the muscular and the secretory cells of the stomach and other viscera.

One of the earlier hypotheses advanced for the formation of ulcers proposed that strong emotional stimuli could lead to a spastic condition of the gut with accompanying anoxia of the mucosa due to prolonged vasoconstriction. The localized ischemic areas, combined with simultaneous hypersecretion of hydrochloric acid and pepsin, could then provide the groundwork for peptic ulcer formation by repeated irritation of the involved mucosal areas. Lesions in the protective mucosal lining would, of course, then permit the normal digestive processes to attack the tissue of the organ. Other studies[2] have tended to implicate hydrochloric acid as the causative agent because it is known that ulcer patients secrete substantially higher quantities of the acid than do normal people and also that ulcers can be induced in dogs with normal stomachs if the gastric acidity level is raised to the level of that found in ulcer patients. Nervous influence is thought to be basic to the hypersecretion of acid resulting in duodenal ulcers, whereas humoral or hormonal influences are believed to be responsible for excessive secretion in the case of gastric ulcers.

The condition of overstimulation of the parasympathetic nervous supply (vagus) to the stomach is sometimes termed *parasympathotonia.* Reduction of this overstimulated condition can be achieved by surgery (surgical vagotomy) or by the use of anticholinergic drugs (chemical vagotomy), resulting in inhibition of both secretory and motor activity of the stomach. Although anticholinergic drugs can exert an antimotility effect, there is some question as to whether they can correct disordered motility or counteract spontaneous "spasms" of the intestine. In addition, although these drugs can (in adequate dosage) diminish the basal secretion of acid, there is said to be little effect on the acid secreted in response to food or to insulin hypoglycemia. Bachrach,[1] in a critical review in 1958, suggested that none of the anticholinergics had any material advantages over the naturally occurring atropine or belladonna group, and that the anticholinergic group, as a whole, left much to be desired for the management of gastrointestinal ailments. In spite of this, new anticholinergics have appeared on the market regularly but, hopefully, research will eventually provide drugs about which there will be no question of efficacy. For the present, the most rational therapy seems to be a combination of bland diet to reduce acid secretion, antacid therapy, reduction of emotional stress, and administration of anticholinergic drugs. Most of the anticholinergic drugs on the market are offered either as the chemical alone or in combination with a central nervous system depressant such as phenobarbital or with one of the tranquilizers in order to reduce the central nervous system contribution to parasympathetic hyperactivity. Some clinical findings tend to show that phenobarbital is preferable to the tranquilizers.[3] Whereas combinations of anticholinergics with sedatives are considered rational, there is not complete agreement on combinations with antacids. This is based on the fact that anticholinergic drugs affect primarily the fasting phases of gastrointestinal secretion and motility and are most efficient if administered at bedtime and well before mealtimes. Antacids neutralize acid largely present in the between-meal, digestive phases of gastrointestinal activity and are of more value if given after meals.

In addition to the antisecretory effects of anticholinergics on hydrochloric acid and gastric secretion described above there have been some efforts to employ them as *antisialogogues* (to suppress salivation) and *anhydrotics* (to suppress perspiration). Although these studies are interesting, it would appear that the matter needs additional study and less toxic compounds.

Paralysis agitans or parkinsonism, first described by the English physician James Parkinson in 1817, is another condition that is often treated with the anticholinergic drugs. It is characterized by tremor, "pill rolling," cog-wheel rigidity, festinating gait, sialorrhea and masklike facies. Fundamentally, it represents a malfunction of the extrapyramidal system, with possible involvement of the substantia nigra and the globus pallida of the basal ganglia. It is probable that subtle degenerative changes secondary to cerebral arteriosclerosis (or of unknown cause) are responsible. The changes responsible for parkinsonism apparently are never reversed, and chemotherapy is of necessity palliative. The usefulness of the belladonna group of alkaloids was an empiric discovery of Charcot. The several synthetic preparations were developed in an effort to retain the useful antitremor and antirigidity effects of the belladonna alkaloids while at the same time reducing undesirable side-effects. Incidentally, it was also discovered rather empirically that antihistamine drugs (e.g., diphenhydramine) sometimes reduced tremor and rigidity. Although the mechanism of action of these drugs is obscure, it obviously reflects a central mechanism of action. The activity is confined to those compounds that can pass the blood-brain barrier, i.e., tertiary amines and not quaternary ammonium compounds. There are some postulations to the effect that acetylcholine is a neurohumoral agent in the central nervous system as well as peripherally and that anticholinergics can block its action in either locus. In this context it is worth noting that injections of tremorine (1,4-dipyrrolidino-2-butyne) or its active metabolite oxotremorine (1-(2-pyrrolidono)-4-pyrrolidino-

Tremorine: $R = H_2$
Oxotremorine: $R = O$

2-butyne) have been shown to increase the brain acetylcholine level in rats up to 40 per cent.[4] This increase coincides roughly with the onset of tremors similar to those observed in parkinsonism. The mechanism of acetylcholine increase in rats is uncertain but has been shown not to be due to acetylcholinesterase inhibition or to activation of choline acetylase. However, the tremors are stopped effectively by administration of the tertiary amine type anticholinergic but not by the quaternaries.

Although many compounds have been introduced for treatment of the syndrome, there is apparently a dire need for compounds that will provide more potent action with fewer side-effects and, also, will provide a wide assortment of replacements for those drugs that seem to lose their efficacy with the passing of time. Since its introduction, L-dopa seems to have great promise.

The chemical classification of anticholinergics acting at the postganglionic terminations of the parasympathetic nervous system is complicated somewhat by the fact that some of them, especially the quaternary ammonium derivatives, also act at a ganglionic level and, in high enough dosage, the latter group will also act at voluntary synapses. However, the following classification will serve to delineate the major chemical types that are encountered:

1. Aminoalcohol derivatives
   A. Esters
      a. Solanaceous alkaloids
      b. Synthetic analogs of the solanaceous alkaloids
         a. Esters of tropine (and scopine)
         b. Esters of other aminoalcohols
   B. Ethers
   C. Carbamates
2. Aminoalcohols
3. Aminoamides
4. Diamines
5. Amines, miscellaneous
6. Papaveraceous alkaloids and their synthetic analogs*

STRUCTURE-ACTIVITY CONSIDERATIONS

It will be apparent from the following discussion of the development of anticholinergics, both natural and synthetic, that a wide variety of compounds possess such activity. Also apparent will be the fact that the development of such compounds has been largely empiric and based principally on atropine as a natural prototype. The structural permutations have resulted in compounds that do not always have obvious ancestral relationships to the parent molecule. Nevertheless, modern contributions are beginning to indicate that there

* Although these are not anticholinergics acting at the postganglionic terminations of the parasympathetic system, they are included here as a matter of convenience because of their employment as antispasmodics.

is not such a chaotic situation with respect to structure-activity relationships as may appear at first glance. For example, it is quite generally agreed that the activity of atropine-related anticholinergics is a competitive one with acetylcholine (see p. 514). Thus, if one utilizes the terminology of Ariëns,[5] it is assumed that the cholinomimetic agent (i.e., acetylcholine) possesses both affinity and intrinsic activity—it can be bound to the receptor site and can elicit the characteristic mimetic response. The anticholinergic agent, on the other hand, has the necessary affinity to bind firmly to the receptor but is unable to bring about an effective response, i.e., it has no intrinsic activity. The blocking agent, in sufficient concentration, effectively competes for the receptor sites and prevents acetylcholine from binding thereon, thus preventing nerve activity.

There are several ways in which the structure-activity relationships could be considered, but in this discussion we shall follow, in general, the considerations of Long *et al.*[6] who based their postulations on the *l*-hyoscyamine molecule as being one of the most active anticholinergics and, therefore, having an optimal arrangement of groups.

The Cationic Head. Most authors consider that the anticholinergic molecules have a primary point of attachment to cholinergic sites through the so-called *cationic head,* i.e., the positively charged nitrogen. In the case of the quaternary compounds there is no question of what is implied, but in the case of tertiary amines one assumes, with good reason, that the cationic head is achieved by extensive protonation of the amine at physiologic pH. The nature of the substituents on this cationic head is critical insofar as a mimetic response is concerned but is far less critical for blocking action. It is undoubtedly true that a cationic head is far better than none at all; yet, it is possible to obtain a typical competitive block *without a cationic head.* That this is the case has been shown by Ariëns and his coworkers[7] in the case of the so-called *carbocholines* typified by benzylcarbocholine. These compounds show a typical competitive action with acetylcholine, although they are less

$$C_6H_5 \qquad\qquad CH_3$$
$$| \qquad\qquad\qquad |$$
$$HO-C-COOCH_2CH_2-C-CH_3$$
$$| \qquad\qquad\qquad |$$
$$C_6H_5 \qquad\qquad CH_3$$

Benzylcarbocholine

effective than the corresponding compounds possessing a cationic head.

The Hydroxyl Group. It has already been mentioned that a suitably placed alcoholic hydroxyl group in an anticholinergic usually enhances the activity over a similar compound without the hydroxyl group. The position of the hydroxyl group with respect to the nitrogen appears to be fairly critical with the diameter of the receptive area being estimated at about 2 to 3 Å. It is assumed that the hydroxyl group contributes to the strength of binding, probably by hydrogen bonding to an electron-rich portion of the receptor surface.

The Esteratic Group. Many of the highly potent compounds possess an ester grouping, and it may be a necessary feature for the most effective binding. This is reasonable in view of the fact that the agonist (i.e., acetylcholine) possesses a similar function for binding to the same site. That it is not necessary for activity is amply illustrated by the several types of compounds not possessing such a group (e.g., ethers, amino-alcohols, diamines, etc.). However, by far the greater number of active compounds possess this grouping. It is probable that it attaches to the receptor area at a positive site, similarly to acetylcholine, and may be necessary for maximal activity.

Cyclic Substitution. It will be apparent from an examination of the active compounds discussed in the following sections that at least one cyclic substituent (phenyl, thienyl, etc.) is a feature of the molecule. Aromatic substitution seems to be the most used in connection with the acidic moiety in esters. However, it will be noted that virtually all of the acids employed are of the aryl-substituted acetic acid variety. Use of aromatic acids per se leads to low activity as anticholinergics but with potential activity as local anesthetics. The question of the superiority of the cyclic species used

(i.e., phenyl, thienyl, cyclohexyl, etc.) appears not to have been explored in depth, although phenyl rings seem to predominate. Substituents on the aromatic rings seem to contribute little to activity.

In connection with the apparent need for a cyclic group it is instructive to consider the postulations of Ariëns in this respect. He points out that the "mimetic" molecules, richly endowed with polar groups, undoubtedly require a complementary polar receptor area for effective binding. As a consequence, it is implied that a relatively nonpolar area surrounds such sites. Thus, by increasing the binding of the molecule in this peripheral area by means of introducing flat nonpolar groups (e.g., aromatic rings) it should be possible to achieve compounds with excellent affinity but not possessing intrinsic activity. That this postulate is consistent with most anticholinergics, whether they possess an ester group or not, is quite obvious.

Stereochemical Requirements. It is instructive to consider the stereochemical implications inherent in the competitive process. Although one cannot examine such relationships with acetylcholine because of the lack of an asymmetric center, the problem has been examined by Ellenbroek[8] through various esters of beta-methylcholine. The results are summarized in Table 50 and show quite conclusively the tremendous (320-fold) effect of an S over the R configuration in the agonist molecule (i.e., acetyl beta-methylcholine) indicating a rather precise stereochemical requirement. In contrast, the benzilate esters of the isomeric beta-methylcholines show only a small difference in competitive antagonistic activity (ratio = 5/6; S/R), indicating that the stereochemical requirement is small. Now, if the stereochemical difference is removed from the choline moiety and introduced into the acidic portion, once again a significant difference (ratio = 100/1; R/S) in activity is noted in the R and the S forms of cyclohexylphenylglycolate esters of choline. These findings are further reinforced by Ellenbroek's findings[8] on the comparative blocking activities of the four possible stereoisomers of the cyclohexylphenylglycolate esters of beta-methylcholine as

TABLE 50. STEREOISOMERS AND BIOLOGICAL ACTIVITY OF CHOLINE ESTERS

est/org.	pD₂ ± P₉₅		config.	activity ratio	config.		pD₂ ± P₉₅	est/org.
	7.0						7.0	
11/7	6.8 ± 0.14		S_B	320	R_B		4.1 ± 0.23	7/4
	pA₂ ± P₉₅						pA₂ ± P₉₅	
28/9	8.0 ± 0.14		S_B	5/6	R_B		8.1 ± 0.10	31/10
26/5	8.6 ± 0.18						8.6 ± 0.18	26/5
19/18	9.6 ± 0.26		R_A	25	S_A		8.2 ± 0.14	24/23

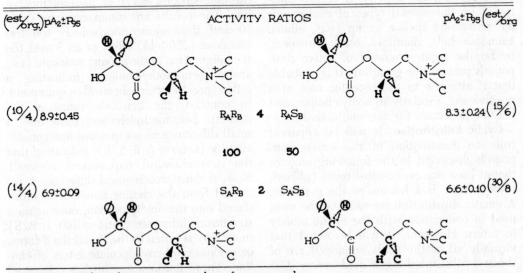

est/org = number of estimations/number of organs used.

pD₂ and pA₂ = (in Mols.) the negative logarithms of doses, of the agonists and the antagonists, respectively, that induce a certain standard response. The 95 per cent confidence limits (P₉₅) are also given.

ϕ = phenyl (—C₆H₅)　　H = cyclohexyl (—C₆H₁₁)

(Ariens, E. J., in Advances in Drug Research, vol. 3, p. 237, New York, Academic Press, 1966)

TABLE 51. STEREOISOMERS AND BIOLOGICAL ACTIVITY OF CHOLINE ESTERS OF PHENYL CYCLOHEXYL GLYCOLIC ACID

(est/org.) pA₂ ± P₉₅		ACTIVITY RATIOS			pA₂ ± P₉₅ (est/org)
(10/4) 8.9 ± 0.45		R_A R_B　4　R_A S_B			8.3 ± 0.24 (15/6)
		100	50		
(14/4) 6.9 ± 0.09		S_A R_B　2　S_A S_B			6.6 ± 0.10 (30/8)

est/org = number of estimations/number of organs used.

pD₂ and pA₂ = (in Mols.) the negative logarithms of doses, of the agonists and the antagonists, respectively, that induce a certain standard response. The 95 per cent confidence limits (P₉₅) are also given.

ϕ = phenyl (—C₆H₅)　　H = cyclohexyl (—C₆H₁₁)

(Ariens, E. J., in Advances in Drug Research, vol. 3, p. 237, New York, Academic Press, 1966) summarized in Table 51. Similar relationships have been noted for the beta-methylcholine esters of α-methyltropic acid. Thus, one is drawn to the conclusion that, for blocking activity, the structural requirements are low for the aminoalcohol portion and high for the acidic portion. As a consequence, it may be assumed that, for antagonistic action, the function of the alcoholic moiety becomes mainly one of hindering the approach of the agonist molecule to the receptor, although it is probable that the cationic head contributes significantly to the binding process.

Long *et al.*,[6] after considering the implications of the above discussion, arrived at several postulations as to the character of the receptor site. These are beyond the scope of this text but the reader is urged to consult the original paper for further information.

AMINOALCOHOL ESTERS

Solanaceous Alkaloids

Prominent among the parasympatholytics are the solanaceous alkaloids which are represented by (−)-hyoscyamine, atropine [(±)-hyoscyamine] and scopolamine (hyoscine). These alkaloids are found principally in henbane (*Hyoscyamus niger*), deadly nightshade (*Atropa belladonna*) and jimson weed (*Datura stramonium*). There are certain other alkaloids that are members of the solanaceous group (e.g., apoatropine, noratropine, belladonnine, tigloidine, meteloidine) but are not of sufficient therapeutic value to be considered in this text. However, it should be mentioned that apoatropine is a more powerful antispasmodic than atropine but never has been used clinically in this country.

The crude drugs containing these alkaloids have been used since early times for their marked medicinal properties, which depend largely on inhibition of the parasympathetic nervous system and stimulation of the higher nervous centers. Belladonna, probably as a consequence of the weak local anesthetic activity of atropine, has been used topically for its analgesic effect on

hemorrhoids, certain skin infections and various itching dermatoses. The application of sufficient amounts of belladonna or of its alkaloids results in mydriasis. Internally, the drug causes diminution of secretions, increases the heart rate (by depression of the vagus nerve), depresses the motility of the gastrointestinal tract and acts as an antispasmodic on various smooth muscles (ureter, bladder and biliary tract). In addition, it stimulates the respiratory center directly. The very multiplicity of actions exerted by the drug causes it to be looked upon with some disfavor, because the physician seeking one type of response unavoidably obtains the others. The action of scopolamine-containing drugs differs from those containing hyoscyamine and atropine in that there is no central nervous system stimulation, and a narcotic or sedative effect predominates. The use of this group of drugs is accompanied by a fairly high incidence of reactions due to individual idiosyncrasies, death from overdosage usually resulting from respiratory failure. The official compendia have recognized a variety of products of all three crude drugs, such as tinctures, fluid extracts, extracts, ointments and even a liniment and a plaster (e.g., belladonna). A complete treatment of the pharmacology and the uses of these drugs is not within the scope of this text, and the reader is referred to the several excellent pharmacology texts which are available. However, the introductory pages of this chapter have reviewed briefly some of the more pertinent points in connection with the major activities of these drug types.

Structural Considerations. All of the solanaceous alkaloids are esters of the bicyclic aminoalcohol, 3-hydroxytropane, or of related aminoalcohols.

The structural formulae on page 522 show the piperidine ring system in the commonly accepted chair conformation because this form has the lowest energy requirement. However, the alternate boat form can exist under certain conditions, because the energy barrier is not great. Inspection of the 3-hydroxytropane formula also indicates that, even though there is no optical activity because of the plane of symmetry, two stereoisomeric forms (tropine and pseu-

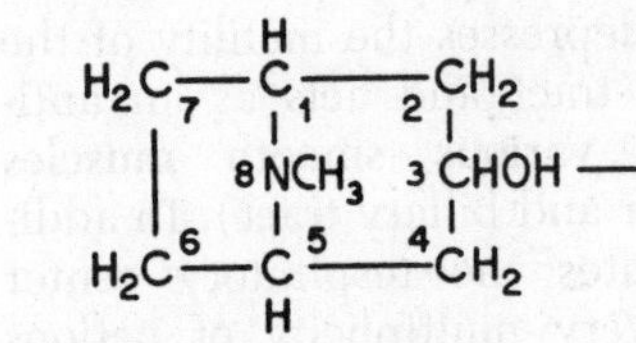

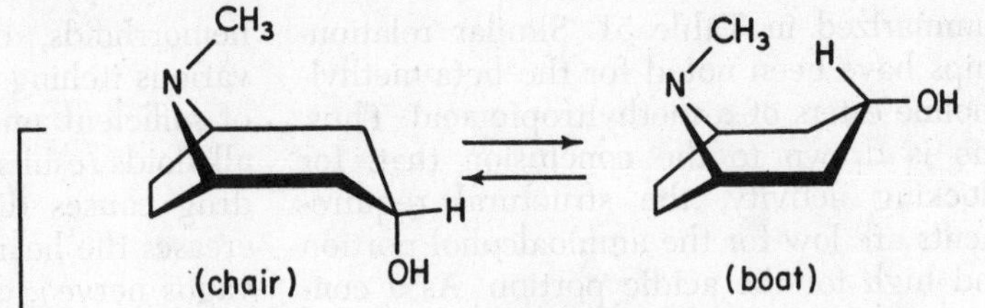

$H_2C \overset{7}{C} \overset{H}{\underset{1}{C}} \overset{2}{CH_2}$
$8 NCH_3 \quad 3 CHOH$
$H_2C \underset{6}{C} \overset{H}{\underset{5}{C}} \overset{4}{CH_2}$

3–Hydroxytropane
(Tropine or Pseudotropine)

(chair) (boat)

TROPINE
(3α–hydroxytropane or
3α–tropanol)

(chair) (boat)

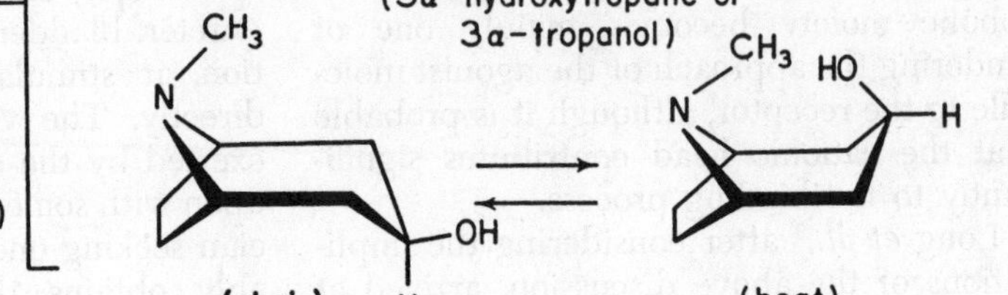

PSEUDOTROPINE
(3β–hydroxytropane or
3β–tropanol)

$\begin{array}{ccc} \overset{H}{C}-\overset{H}{C} & CH_2 \\ O & NCH_3 & CHOH \\ \underset{H}{C}-\underset{H}{C} & CH_2 \end{array}$

6:7-Epoxy-3-Hydroxytropane
(Scopine)

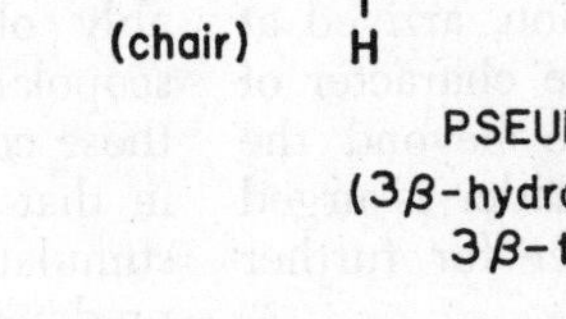

SCOPINE
(6:7β–epoxy–3α–hydroxytropane
or 6:7β–epoxy–3α–tropanol)

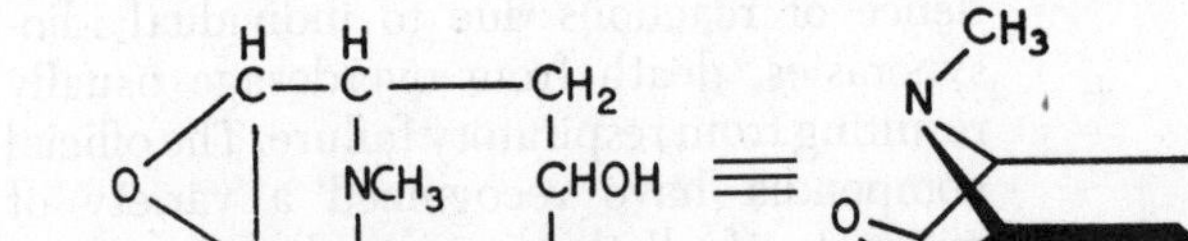

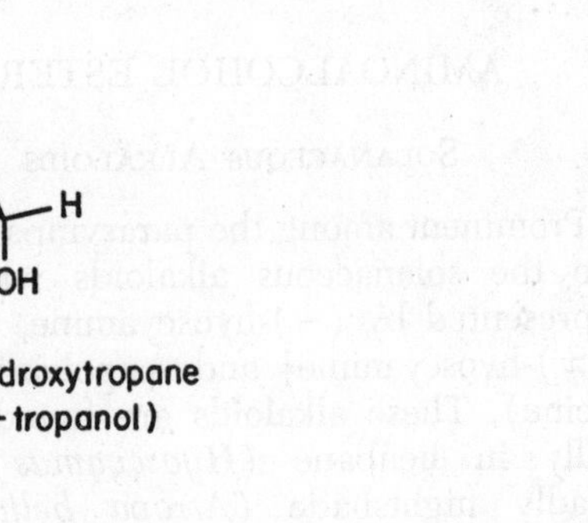

$\begin{array}{cc} \overset{H}{C} & \\ H_2C-\overset{C}{} & CH_2 \qquad CH_2OH \\ NCH_3 & CH-O-CO-CH \\ H_2C-\underset{C}{} & CH_2 \\ \underset{H}{} & \end{array}$

ATROPINE
(or Hyoscyamine)

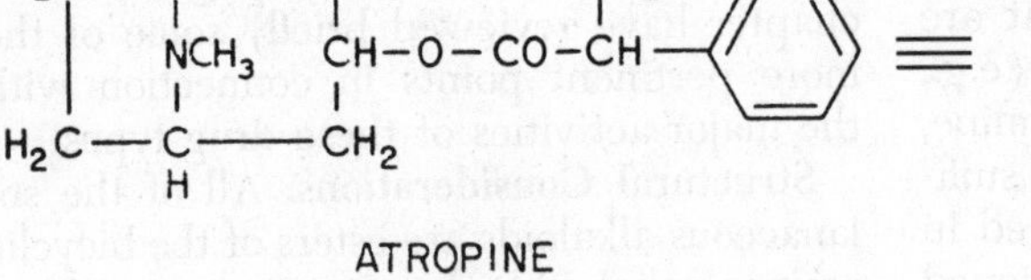

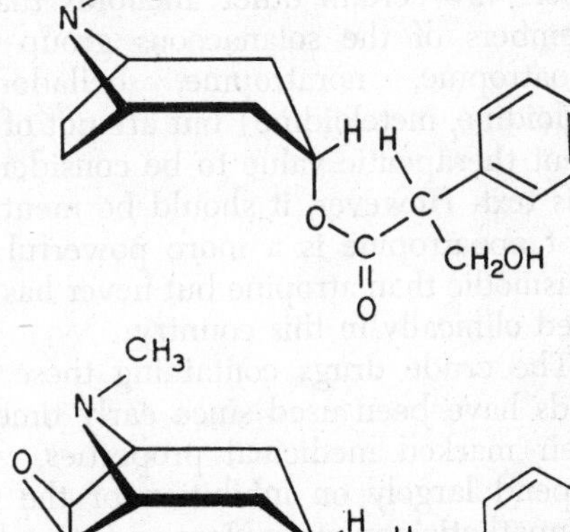

$\begin{array}{cc} \overset{H}{C}-\overset{H}{C} & CH_2 \qquad CH_2OH \\ O \quad NCH_3 & CH-O-CO-CH \\ \underset{H}{C}-\underset{H}{C} & CH_2 \end{array}$

SCOPOLAMINE

dotropine) can exist because of the rigidity imparted to the molecule through the ethane chain across the 1,5 positions. In tropine the axially oriented hydroxyl group, trans to the N-bridge, is designated as *alpha,* and the alternate cis equatorially oriented hydroxyl group is *beta.* The aminoalcohol derived from scopolamine, namely *scopine,* has the axial orientation of the 3-hydroxyl group but, in addition, has a *beta*-oriented epoxy group bridged across the 6,7-positions as shown. Of the several different solanaceous alkaloids known it has already been indicated that (−)-hyoscyamine, atropine and scopolamine are the most important. Their structures are indicated at right, but it can be pointed out that antimuscarinic activity is associated with all of the solanaceous alkaloids that possess the tropinelike axial orientation of the esterified hydroxyl group. It will be noted in studying the formulae on p. 522 that tropic acid is, in each case, the esterifying acid. It also will be apparent that tropic acid contains an easily racemized asymmetric carbon atom, the moiety accounting for optical activity in these compounds in the absence of racemization. The proper enantiomorph is necessary for high antimuscarinic activity, as illustrated by the potent (−)-hyoscyamine in comparison with the weakly active (+)-hyoscyamine. The racemate, atropine, has an intermediate activity. The marked difference in antimuscarinic potency of the optical enantiomorphs apparently does not extend to the action on the central nervous system, inasmuch as both seem to have the same degree of activity.[9]

Products

Atropine N.F., 1αH, 5αH-Tropan-3α-ol, (±)-Tropate (Ester). Atropine is the tropine ester of tropic acid (see above) and is optically inactive. It possibly occurs naturally in various *Solanaceae,* although some claim with justification that whatever atropine is isolated from natural sources results from racemization of (−)-hyoscyamine during the isolation process. Conventional methods of alkaloid isolation are used to obtain a crude mixture of atropine and hyoscyamine from the plant material.[10] This crude mixture is racemized to atropine by refluxing in chloroform or by treatment with cold dilute alkali. Because atropine is made by the racemization process, an official limit is set on the hyoscyamine content by restricting atropine to a maximum levorotation under specified conditions.

Synthetic methods for preparing atropine take advantage of Robinson's synthesis, employing modifications to improve the yield of tropinone. Tropinone may be reduced under proper conditions to tropine, which is then used to esterify tropic acid. Other acids may be used in place of tropic acid to form analogs, and numerous compounds of this type have been prepared which are known collectively as *tropëines.* The most important one, homatropine, will be considered in the section on synthetic anticholinergics.

Atropine occurs in the form of optically inactive, white, odorless crystals possessing a bitter taste. It is not very soluble in water (1:460; 1:90 at 80°) but is more soluble in alcohol (1:2; 1:1.2 at 60°). It is soluble in glycerin (1:27), in chloroform (1:1) and in ether (1:25).* Saturated aqueous solutions are alkaline in reaction (approximate pH = 9.5). The free base is useful when nonaqueous solutions are to be made, such as in oily vehicles and ointment bases.

Category—anticholinergic.

Usual dose—250 mcg. three times a day.

Atropine Sulfate U.S.P., Atropine sulfate is prepared by neutralizing atropine in acetone or ether solution with an alcoholic solution of sulfuric acid, care being exercised to prevent hydrolysis.

The salt occurs as colorless crystals or as a white crystalline powder. It is efflorescent in dry air and should be protected from light to prevent decomposition.

Atropine sulfate is freely soluble in water (1:0.5), in alcohol (1:5; 1:2.5 at boiling point) and in glycerin (1:2.5). Aqueous solutions of atropine are not very stable, although it has been stated[11] that solutions may be sterilized at 120° (15 lb. pressure) in an autoclave if the pH is kept below 6.

---

* In this chapter a solubility expressed as 1:460 indicates that 1 g. is soluble in 460 ml. of the solvent at 25°. Solubilities at other temperatures will be so indicated.

Sterilization probably is best effected by the use of aseptic technic and a bacteriologic filter. The above reference suggests that no more than a 30-day supply of an aqueous solution should be made, and, for small quantities, the best procedure is to use hypodermic tablets and sterile distilled water. Kondritzer and his coworkers[12, 13] have studied the kinetics of alkaline and proton catalyzed hydrolyses of atropine in aqueous solution. The region of maximum stability lies between pH 3 and approximately 5. They also have proposed an equation to predict the half life of atropine undergoing hydrolysis at constant pH and temperature.

The action of atropine or its salts is the same. It produces a mydriatic effect by paralyzing the iris and the ciliary muscles and for this reason is used by the oculist in iritis and corneal inflammations and lesions. Its use is rational in these conditions because one of the first rules in the treatment of inflammation is rest, which, of course, is accomplished by the paralysis of muscular motion. Its use in the eye (0.5% to 1% solutions or gelatin disks) for fitting glasses is widespread. Atropine is administered in small doses before general anesthesia to lessen oral and air-passage secretions and, where morphine is administered with it, it serves to lessen the respiratory depression induced by morphine. Its ability to dry secretions also is utilized in the commonly used "rhinitis tablets" for symptomatic relief in colds. In cathartic preparations, atropine or belladonna acts as an antispasmodic to lessen the smooth muscle spasm (griping) often associated with catharsis.

A more recent use for atropine sulfate has emerged following the development of the organic phosphates which are potent inhibitors of acetylcholinesterase. Atropine is a specific antidote to prevent the "muscarinic" effects of acetylcholine accumulation such as vomiting, abdominal cramps, diarrhea, salivation, sweating, bronchoconstriction and excessive bronchial secretions.[14] It is used intravenously but does not protect against respiratory failure due to depression of the respiratory center and the muscles of respiration.

Category—anticholinergic.

Usual dose—500 mcg. oral, intravenous or subcutaneous, 3 or 4 times a day.

Usual dose range—300 mcg. to 4 mg. daily.

For external use—topically, to the conjunctiva, as a 1 per cent ointment or 0.1 ml. of a 0.5 to 3 per cent solution 3 to 5 times daily.

OCCURRENCE
Atropine Sulfate Injection U.S.P.
Atropine Ophthalmic Solution U.S.P.
Atropine Sulfate Tablets U.S.P.
Morphine and Atropine Sulfates
    Tablets N.F.

Atropine Tannate, Atratan®. This salt of atropine was developed as a means of slowing down the absorption of atropine and to provide a more sustained release of the alkaloid. It is indicated for relief of smooth muscle spasm and pain resulting from ureteral colic. Post-instrumentation therapy in urology and renal colic are other indications. It occurs in the form of 1-mg. tablets, with a dose, in renal colic, of 1 to 2 tablets every 4 hours. The dose depends somewhat on the severity of the condition.

Atropine Oxide Hydrochloride, X-tro®, Atropine N-oxide Hydrochloride. This derivative of atropine has also been known as *genatropine hydrochloride* (cf. *genoscopolamine,* p. 526) and is designed to release atropine slowly upon administration. It has the same action as atropine and the same side-effects. It is administered orally in doses of 500 mcg. to 1 mg. and is marketed as capsules with or without phenobarbital.

Hyoscyamine is a levorotatory alkaloid obtained from various solanaceous species. One of the commercial sources is Egyptian henbane (*Hyoscyamus muticus*), in which it occurs to the extent of about 0.5 per cent. One method for extraction of the alkaloid utilizes *Duboisia* species.[15] Usually, it is prepared from the crude drug in a manner similar to that used for atropine and is purified as the oxalate. The free base is obtained easily from this salt.

It occurs as white needles which are sparingly soluble in water (1:281), more soluble in ether (1:69) or benzene (1:150) and very soluble in chloroform (1:1) or alcohol. It is official as the sulfate. The principal reason for the popularity of the hydrobromide has been its nondeliquescent nature. The salts have the advantage over the free base in being quite water soluble.

As mentioned previously, hyoscyamine is the levo-form of the racemic mixture which is known as atropine and, therefore, has the same structure. The dextro-form does not exist naturally but has been synthesized. Comparison of the activities of $(-)$-hyoscyamine, $(+)$-hyoscyamine and $(\pm)$-hyoscyamine (atropine) was carried out by Cushny in 1903. He found that the peripheral action on the iris, the cardiac vagus and glands is 12 to 18 times stronger with the levo-isomer than with the dextro-isomer and twice as strong as with the racemate. More recently, Long *et al.*[6] have shown a *levo* to *dextro* activity ratio of 1:110 to 1:250, depending on the assay procedure. All of the isomers behave the same with respect to the central nervous system. A preparation consisting principally of $(-)$-hyoscyamine malate is on the market under the trade name of Bellafoline. It has been promoted extensively on the bases of less central activity and greater peripheral activity than atropine possesses.

The uses of hyoscyamine are essentially those of atropine, although it is said to be better suited for topical application (i.e., eye) than atropine, because its effect is not as prolonged.

Hyoscyamine Sulfate N.F., $1\alpha H, 5\alpha H$-Tropan-3α-ol $(-)$-Tropate (Ester) Sulfate (2:1), Levsin® Sulfate. This salt is a white, odorless, crystalline compound of a deliquescent nature. It is affected by light. It is soluble in water (1:0.5) and alcohol (1:5) but almost insoluble in ether. Solutions of hyoscyamine sulfate are acidic to litmus.

This drug is used as an anticholinergic in the same manner and for the same uses as atropine and hyoscyamine (q.v.), but possesses the disadvantage of being deliquescent.

Category—anticholinergic.

Usual dose range—125 to 250 mcg. 3 or 4 times a day.

OCCURRENCE

Hyoscyamine Sulfate Tablets N.F.

Hyoscyamine Hydrobromide N.F. This

levorotatory salt occurs as white, odorless crystals or as a crystalline powder which is affected by light. It is not deliquescent. The salt is freely soluble in water, alcohol and chloroform but only slightly soluble in ether. The solutions, when freshly prepared, are neutral to litmus.

The uses are virtually the same as those cited for atropine and hyoscyamine, although it is believed that there is less central effect than with atropine.

Category—anticholinergic.

Usual dose—250 mcg. to 1.0 mg.

Scopolamine, Hyoscine. This alkaloid is found in various members of the *Solanaceae* (e.g., *Hyoscyamus niger, Duboisia myoporoides, Scopolia* sp. and *Datura metel*). It usually is isolated from the mother liquor remaining from the isolation of hyoscyamine.

The name *hyoscine* is the older name for this alkaloid, although *scopolamine* is more popular in this country. Scopolamine is the levo-component of the racemic mixture which is known as *atroscine*. Scopolamine is racemized readily when subjected to treatment with dilute alkali.

The alkaloid occurs in the form of a levorotatory, viscous liquid which is only slightly soluble in water but very soluble in alcohol, chloroform or ether. It forms crystalline salts with most acids, the hydrobromide being the most stable and the most popularly accepted. An aqueous solution of the hydrobromide, containing 10 per cent of mannitol (Scopolamine Stable), is said to be less prone to decomposition than unprotected solutions.

Scopolamine Hydrobromide U.S.P., Hyoscine Hydrobromide. This salt occurs as white or colorless crystals or as a white granular powder. It is odorless and tends to effloresce in dry air. It is soluble in water (1:1.5) or alcohol (1:20), only slightly soluble in chloroform and insoluble in ether.

Scopolamine gives the same type of depression of the parasympathetic nervous system as does atropine but it differs markedly from atropine in its action on the higher nerve centers. Whereas atropine stimulates the central nervous system, causing restlessness and talkativeness, scopolamine acts as a narcotic or sedative. In this capacity, it has found a use in the treatment of parkinsonism, although its value is depreciated by the fact that the effective dose is very close to the toxic dose. A sufficiently large dose of scopolamine will cause an individual to sink into a restful, dreamless sleep for a period of some 8 hours, followed by a period of approximately the same length in which the patient is in a semiconscious state. During this time, the patient does not remember events that take place. When scopolamine is administered with morphine, this temporary amnesia is termed "twilight sleep." It has been taken advantage of in obstetric and gynecologic procedures to promote loss of memory about events during labor or during preoperative or postoperative gynecologic care.

Another use for scopolamine, largely developed under stress of World War II, is for motion sickness.[16] While it is not the final answer to the treatment or prevention of travel ills, it is one of the useful remedies. For prevention, a dose of 600 to 700 mcg. is administered 30 minutes to 1 hour before the trip commences. The effect lasts for only a few hours.

Category—anticholinergic.

Usual dose—oral or parenteral, 600 mcg. 3 or 4 times daily.

Usual dose range—300 mcg. to 4 mg. daily.

For external use—topically, 0.1 ml. of a 0.2 to 0.5 per cent solution, to the conjunctiva 3 to 5 times daily.

OCCURRENCE

Scopolamine Hydrobromide Injection U.S.P.
Scopolamine Hydrobromide Tablets U.S.P.

Genoscopolamine is scopolamine-N-oxide, formed by treating the alkaloid with hydrogen peroxide. It occurs as a white, crystal-

Genoscopolamine

line powder or in the form of one of its salts (e.g., the hydrobromide).

Because the compound gradually is converted in the body to scopolamine, it exerts the effects of scopolamine but is said to be much less toxic. It is used for the same purposes as scopolamine. It is marketed in the form of "Pellets" and is given in doses of 1 to 2 mg. daily in divided doses, the dose being increased gradually to 3 or 4 mg. over a period of 6 to 8 days.

SYNTHETIC ANALOGS OF THE SOLANACEOUS ALKALOIDS*

Although the naturally occurring alkaloids are potent parasympatholytics, they retain the undesirable attribute of having a wide spectrum of activity. Thus, efforts on the part of the physician to elicit one desired reaction unavoidably result in some measure of undesirable side-effects. For this reason, the synthesis of compounds possessing one or another of the desirable actions without the others has been an active field of investigation. This ideal specificity of action may be an unattainable goal in view of the mode of action of these drugs, but it does seem reasonable to expect that the undesirable side-effects can be minimized, in reference to the desired action. The goals of research efforts with respect to synthetic amino-esters reflect the principal useful activities of the parasympatholytic anticholinergics (see p. 515), namely antispasmodic, antisecretory, mydriatic and cycloplegic actions.

Because the antispasmodic effect is possibly the most important one, it is found that a large proportion of the research in the field has been directed toward emphasizing spasmolysis and minimizing mydriasis and antisecretory effects. The hope of finding a useful synthetic antispasmodic, embodying both the neurotropic and the musculotropic types of effect (see p. 515), has occupied the attention of many investigators. In spite of the volume of research on compounds of this type, it can be said safely that the most desirable synthetic compound has yet to be found, although some of the compounds on the market today have certain desirable attributes.

Work in another field of synthetic modification has been directed toward develop-

* Roman numerals in parentheses following the names in this, discussion refer to the entries in Table 52, which lists the common antispasmodics with their formulas.

ing active mydriatics and cycloplegics.

Efforts at synthesis started with rather minor deviations from the atropine molecule, but today they have departed rather markedly from any close resemblance to the rigid atropine type aminoalcohol. Indeed, one finds that the acid portions of the aminoesters also have changed markedly over the years. One of the major developments in the field of aminoalcohol esters was the successful introduction of the quaternary ammonium derivatives as contrasted with the previously utilized tertiary amines. Although there are some outstanding tertiary amine type esters in use today, and more are being developed constantly, it would appear that the quaternaries as a group represent the most popular type that thus far has been developed. The following discussion will treat first on those compounds (esters of tropine or tropëines) which represent only minor modifications and then proceed on to those which represent more radical changes from the original atropine prototype (esters of aminoalcohols other than tropine). Following each general discussion, the principal useful compounds of each type will be given alphabetically, with a brief monograph on their individual properties.

Esters of Tropine (and Scopine)

In tribute to the remarkable specificity of action exhibited by atropine (I) it is well to point out that few aminoalcohols have been found that will impart the same degree of neurotropic activity as that exhibited by the combination of tropine with tropic acid. In a like manner, the tropic acid portion is highly specific for the anticholinergic action, and substitution by other acids results in decreased neurotropic potency, although the musculotropic action may increase. Early attempts to modify the atropine molecule retained the tropine portion of the molecule and substituted various acids for tropic acid. In this way a series of *tropëines* was built up which gave a number of active compounds. The only one which has survived to the present day is homatropine (II) or mandelyl tropëine, although the reader may be interested in the other tropëines which were reviewed ably by von Oettingen.[17]

Besides changing the acid residue, the other changes have been directed toward the quaternization of the nitrogen. Examples of this type of compound are atropine methylnitrate (IIIa), atropine methylbromide (IIIb), scopolamine methylbromide (IVa), scopolamine methylnitrate (IVb), homatropine methylbromide (V) and anisotropine methylbromide (VI). Quaternization in these compounds may or may not decrease activity. Ariëns[18] ascribes decreased activity, especially where the groups attached to nitrogen are larger than methyl, to a possible decrease in affinity for the anionic receptor site. This decreased affinity he attributes to a combination of greater electron repulsion by such groups and greater steric interference to approach of the cationic head to the anionic site. Decreases in activity are apparent in comparing atropine with methyl atropine salts and scopolamine with methylscopolamine salts. In general, however, the effect of quaternization is much greater in reduction of parasympathomimetic action than of parasympatholytic action. This may be due partially to the additional blocking at the parasympathetic ganglion induced by quaternization, which could serve to offset the decreased affinity at the postganglionic site. However, it also is to be noted that quaternization increases the curariform activity of these alkaloids and aminoesters, a usual consequence of quaternizing alkaloids. Another disadvantage in converting an alkaloidal base to the quaternary form is that the quaternized base is more poorly absorbed through the intestinal wall, with the consequence that the activity becomes erratic and, in a sense, unpredictable. The reader will find Brodie and Hogben's[19] comments on the absorption of drugs in the dissociated and the undissociated states of considerable interest, although space limitations do not permit expansion on the topic in this text. Briefly, however, they point out that bases (such as alkaloids) are absorbed through the lipoidal gut wall only in the undissociated form, which can be expected to exist in the case of a tertiary base, in the small intestine. On the other hand, quaternary nitrogen bases cannot revert to an undissociated form even in basic media and, presumably, would have difficulty passing through the gut wall. That quaternary compounds can be absorbed indicates that other

TABLE 52. STRUCTURAL RELATIONSHIPS OF SYNTHETIC ANTICHOLINERGICS

AMINOALCOHOL ESTERS

$$R_2-\underset{\underset{R_3}{|}}{\overset{\overset{R_1}{|}}{C}}-COOR_4$$

Compound	R_1	R_2	R_3	R_4	Name
I	$-C_6H_5$	$-CH_2OH$	$-H$	Tropine	Atropine
II	$-C_6H_5$	$-OH$	$-H$	Tropine	Homatropine
III	$-C_6H_5$	$-CH_2OH$	$-H$		(a) Atropine Methylnitrate ($X^- = NO_3^-$) (b) Atropine Methylbromide ($X^- = Br^-$)
IV	$-C_6H_5$	$-CH_2OH$	$-H$		(a) Methscopolamine Bromide ($X^- = Br^-$) (b) Methscopolamine Nitrate ($X^- = NO_3^-$)
V	$-C_6H_5$	$-OH$	$-H$	Same as in compound III ($X^- = Br^-$)	Homatropine Methylbromide
VI	$-C_3H_7(n)$	$-C_3H_7(n)$	$-H$	Same as V	Anisotropine Methylbromide
VII	$-C_6H_5$	$-OH$	$-H$		Eucatropine
VIII	$-C_6H_5$	$-CH_2OH$	$-H$	$-CH_2C(CH_3)_2N(C_2H_5)_2$	Amprotropine
IX	$-C_6H_5$	$-C_6H_{11}°$	$-H$	$-CH_2CH_2N(C_2H_5)_2$	Trasentine-H
X	$-C_6H_5$	$-C_6H_5$	$-H$	$-CH_2CH_2N(C_2H_5)_2$	Adiphenine
XI		$-C_6H_5$	$-H$	$-CH_2CH_2N(CH_3)_2$	Cyclopentolate
XII	$-C_6H_5$	$-C_6H_{11}$	$-OH$		Oxyphencyclimine
XIII	$-C_6H_5$	$-C_6H_{11}$	$-OH$	$-CH_2CH_2\overset{+}{N}(C_2H_5)_2$ $\underset{CH_3 \quad Br^-}{|}$	Oxyphenonium Bromide
XIV	$-C_6H_{11}$			$-CH_2CH_2-N$	Dihexyverine
XV	$-C_5H_9°$	†	$-OH$	$-CH_2CH_2\overset{+}{N}(C_2H_5)_2$ $\underset{CH_3 \quad Br^-}{|}$	Penthienate Bromide

° C_5H_9 indicates cyclopentyl and C_6H_{11} indicates cyclohexyl.

† This structure will be designated by $-C_4H_3S$ in the rest of the table.

AMINOALCOHOL ESTERS (*Continued*)

Compound	R_1	R_2	R_3	R_4	Name
XVI	$-C_6H_5$	$-CH(CH_3)CH_2CH_3$	$-H$	$-CH_2CH_2\overset{+}{N}(CH_3)(C_2H_5)_2$ Br^-	Valethamate Bromide
XVII	$-C_6H_{11}$	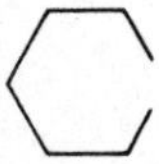		$-CH_2CH_2N(C_2H_5)_2$	Dicyclomine
XVIII	$-C_6H_5$	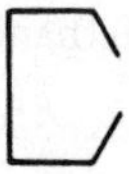		$-CH_2CH_2N(C_2H_5)_2$	Caramiphen
XIX	$-C_6H_5$	$-C_6H_5$	$-H$	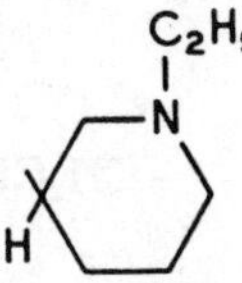	Piperidolate
XX	$-C_6H_5$	$-C_6H_5$	$-OH$	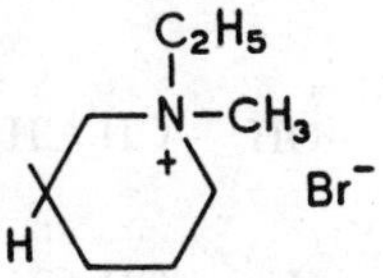	Pipenzolate Methyl-bromide
XXI	$-C_6H_5$	$-C_6H_5$	$-OH$	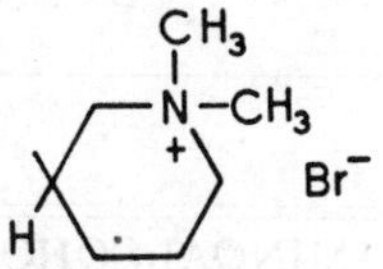	Mepenzolate Methyl-bromide
XXII	$-C_6H_5$	$-C_6H_5$	$-OH$	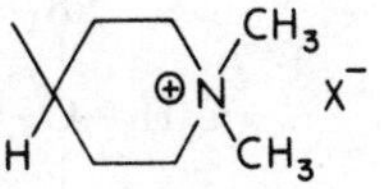	Parapenzolate Bromide ($X^- = Br^-$)
XXIII	$-C_6H_5$	$-CH(CH_2CH_3)-CH_3$	$-H$	Same as in compound XXII	Pentapiperide Methylsulfate ($X = CH_3SO_4^-$)
XXIV	$-C_6H_5$	$-C_6H_5$	$-OH$	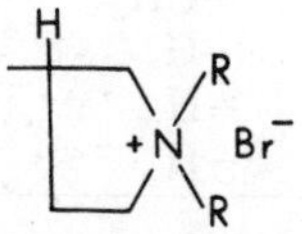	Benzilonium Bromide ($R = -C_2H_5$)
XXV	$-C_6H_5$	$-C_5H_9$	$-OH$	Same as in compound XXIV	Glycopyrrolate ($R = -CH_3$)
XXVI	$-C_6H_5$	Same as XV	$-OH$	Same as in compound XXIV	Heteronium Bromide ($R = -CH_3$)
XXVII	$-C_6H_5$	$-C_6H_5$	$-OH$	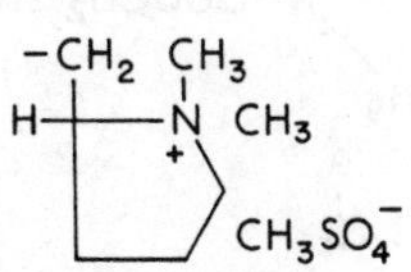	Poldine Methylsulfate
XXVIII	$-C_6H_5$	$-C_6H_5$	$-OH$	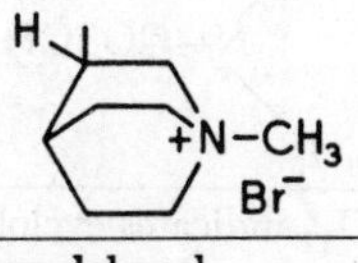	Clidinium Bromide

C_5H_9 indicates cyclopentyl and C_6H_{11} indicates cyclohexyl.

Compound	R_1	R_2	R_3	R_4	Name
XXIX	$-C_6H_5$	$-C_6H_{11}$	$-OH$	$-CH_2C \equiv CCH_2N(C_2H_5)_2$	Oxybutynin Chloride
XXX	$-C_6H_5$	$-C_6H_5$	$-H$	Same as X except the general formula should read, 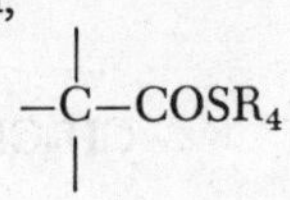 $$-\overset{\textstyle\vert}{\underset{\textstyle\vert}{C}}-COSR_4$$	Thiphenamil Hydrochloride
XXXI	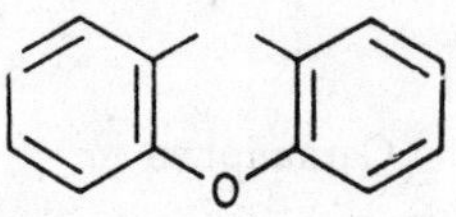		$-H$	$-CH_2CH_2\overset{+}{N}(C_2H_5)_2$ $\quad CH_3 \quad Br^-$	Methantheline Bromide
XXXII	$R_1 + R_2 =$ same as in compound XXI		$-H$	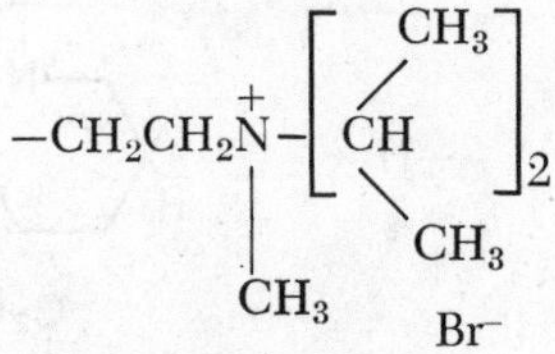 $-CH_2CH_2\overset{+}{N}-\left[\underset{\textstyle CH_3}{\overset{\textstyle CH_3}{CH}}\right]_2$ $\quad CH_3 \quad Br^-$	Propantheline Bromide
XXXIII	$-C_6H_5$	$-C_6H_5$	$-OH$	$-CH_2CH_2N(C_2H_5)_2$	Benactyzine
XXXIV	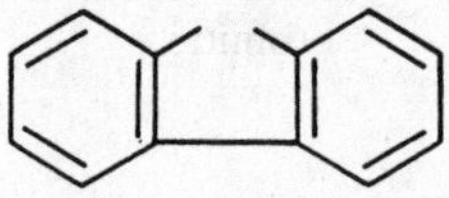		$-H$	$-CH_2CH_2N(C_2H_5)_2$	Aminocarbofluorene

AMINOALCOHOL ETHERS

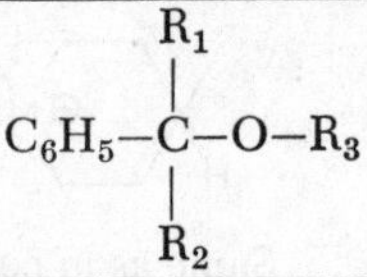

$$C_6H_5-\overset{\textstyle R_1}{\underset{\textstyle R_2}{C}}-O-R_3$$

Compound	R_1	R_2	R_3	Name
XXXV	$-C_6H_4Cl$ (*p*)	$-CH_3$	$-CH_2CH_2N(CH_3)_2$	Chlorphenoxamine
XXXVI	$-C_6H_4CH_3$ (*o*)	$-H$	$-CH_2CH_2N(CH_3)_2$	Orphenadrine
XXXVII	$-C_6H_5$	$-H$	$-CH_2CH_2N(CH_3)_2$	Diphenhydramine
XXXVIII	$-C_6H_5$	$-H$	$-$Tropine (See I)	Benztropine

AMINOALCOHOL CARBAMATES

Compound	Structural Formula	Name
XXXIX	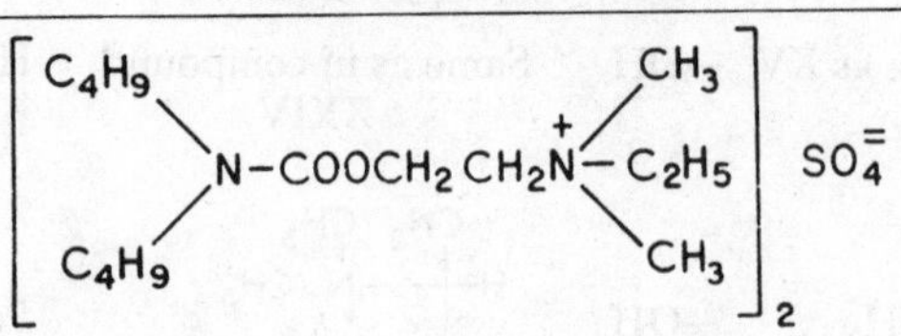 $$\left[\underset{\textstyle C_4H_9}{\overset{\textstyle C_4H_9}{N}}-COOCH_2CH_2\overset{+}{\underset{\textstyle CH_3}{\overset{\textstyle CH_3}{N}}}-C_2H_5\right]_2 SO_4^=$$	Dibutoline Sulfate
XL	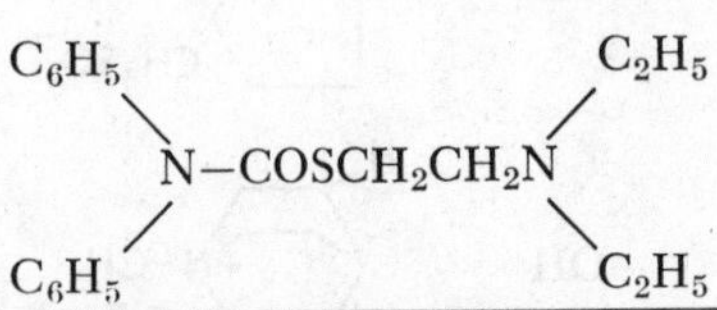 $$\underset{\textstyle C_6H_5}{\overset{\textstyle C_6H_5}{N}}-COSCH_2CH_2\underset{\textstyle C_2H_5}{\overset{\textstyle C_2H_5}{N}}$$	Phencarbamide

C_5H_9 indicates cyclopentyl and C_6H_{11} indicates cyclohexyl.

AMINOALCOHOLS

$$R_2-\underset{\underset{OH}{|}}{\overset{\overset{R_1}{|}}{C}}-R_3$$

COMPOUND	R_1	R_2	R_3	NAME
XLI	$-C_6H_5$	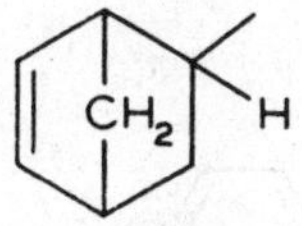	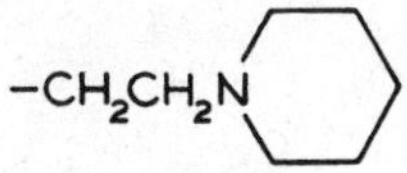 $-CH_2CH_2N$	Biperiden
XLII	$-C_6H_5$	$-C_5H_9$	Same as above	Cycrimine
XLIII	$-C_6H_5$	$-C_6H_{11}$	Same as above	Trihexyphenidyl
XLIV	$-C_6H_5$	$-C_6H_{11}$	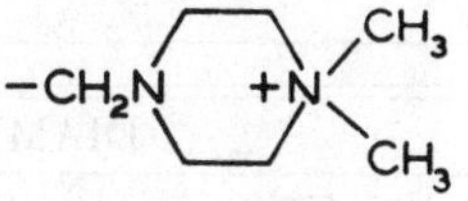	Hexocyclium Methosulfate
XLV	$-\underset{\underset{C_6H_5}{\mid}}{CH}-CH(CH_3)_2$	$-H$	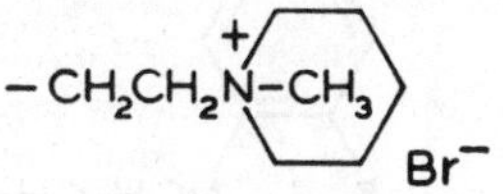	Mepiperphenidol
XLVI	$-C_6H_5$	$-C_6H_{11}$	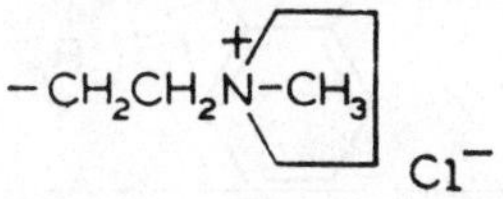	Tricyclamol Chloride
XLVII	$-C_6H_5$	$-C_6H_{11}$	$-CH_2CH_2\overset{+}{N}(C_2H_5)_3$ Cl^-	Tridihexethyl Chloride
XLVIII	$-C_6H_5$	$-C_6H_{11}$	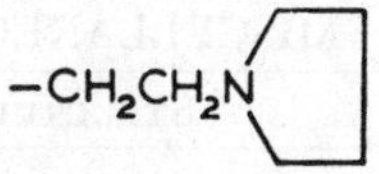	Procyclidine
XLIX	$-C_4H_3S$	$-C_4H_3S$	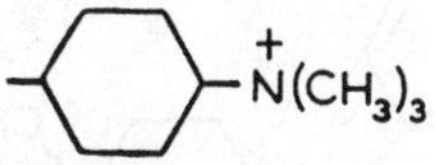	Thihexinol Methylbromide

AMINOAMIDES

$$R-\underset{\underset{C_6H_5}{\mid}}{\overset{\overset{C_6H_5}{\mid}}{C}}-CONH_2$$

COMPOUND	R	NAME
L	$-CH_2\underset{\underset{CH_3}{\mid}}{CH}-N(CH_3)_2$	Aminopentamide
LI	$-CH_2CH_2-\underset{\underset{C_2H_5}{\mid}}{\overset{+}{N}}(CH_3)_2$ Br^-	Ambutonium Bromide

C_5H_9 indicates cyclopentyl and C_6H_{11} indicates cyclohexyl.

COMPOUND	R	NAME
LII	$-CH_2CH_2-\overset{+}{N}-CH_3 \quad I^-$ $[CH(CH_3)_2]_2$	Isopropamide Iodide
LIII	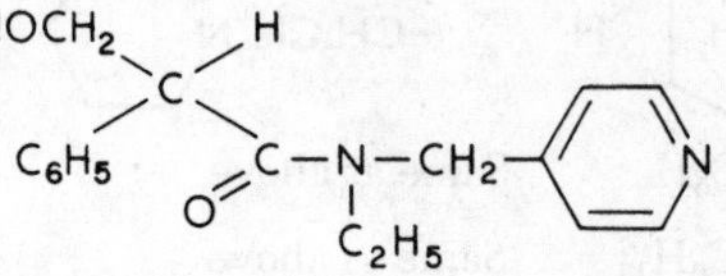	Tropicamide (entire structure as represented to the left)

DIAMINES

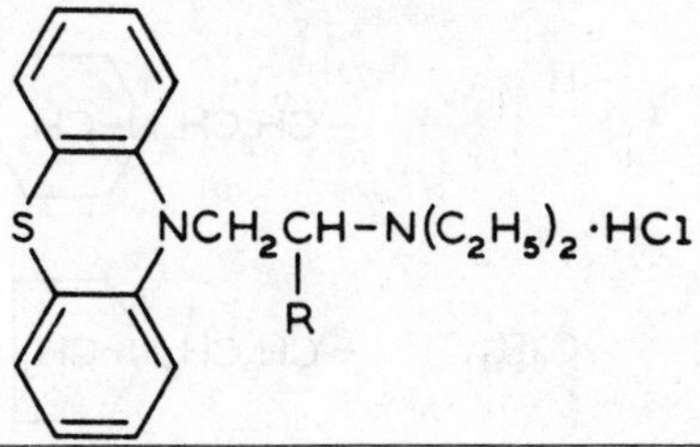

COMPOUND	R	NAME
LIV	—H	Diethazine Hydrochloride
LV	—CH₃	Ethopropazine Hydrochloride

MISCELLANEOUS AMINES

COMPOUND	STRUCTURE	NAME
LVI	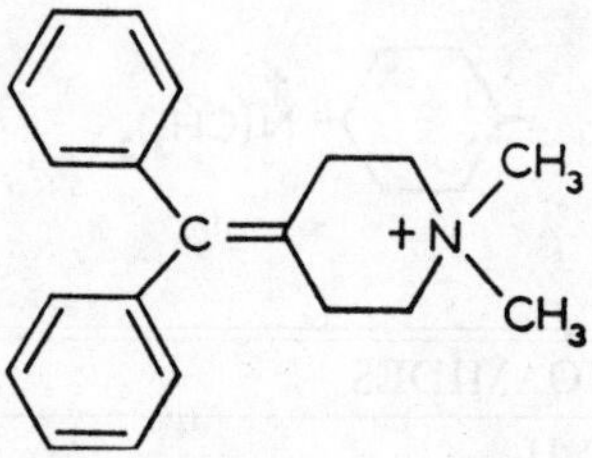	Diphemanil Methylsulfate

less efficient mechanisms for absorption probably prevail. The comments of Cavallito[20] are interesting in this respect. Asher,[3] in connection with a long-term clinical study on anticholinergic compounds, states that "Observations concerning the synthetic tertiary amine derivatives were deleted since it was found that, in general, these drugs were quite weak when compared clinically with the drugs of the quaternary ammonium series."

Products

Anisotropine Methylbromide, Valpin®, 8-Methyl-tropinium Bromide 2-Propylvalerate. This compound occurs as a crystalline white powder which is quite soluble in water and alcohol. Its actions are quite similar to those of atropine, to which it is structurally related. These actions are inhibition of gastrointestinal and urinary tract motility together with an antisecretory action affecting salivation, perspiration, etc. The promotional literature claims a high specificity for the inhibition of gastrointestinal motility, but these claims may be subject to dispute. Unfortunately, studies that have been carried out on this drug have not been well controlled.[21] A low incidence of side-effects has been claimed, although concomitantly, convincing evidence of therapeutic results was lacking. It is probable that, in common with virtually all anticholinergics, it is necessary to elicit some of the characteristic side-effects (dry mouth, urinary retention, blurring of vision, etc.) in order to obtain therapeutic levels. If given in sufficient dose, this drug produces the usual side-effects just mentioned. It is contraindicated in glaucoma and is to be used with great caution in cardiac disease and obstructive conditions of the genitourinary and the gastrointestinal tracts.

It is promoted for use as a spasmolytic in the management of gastrointestinal spasm, peptic ulcer and any other disorders that are associated with hypermotility and respond to anticholinergic therapy. It is administered orally as an elixir or tablets (with or without phenobarbital) in an adult dose of 10 mg. 3 or 4 times a day before meals and on retiring.

Homatropine Hydrobromide U.S.P. Chemically, this compound is the hydrobromide of mandelyl tropëine and is essentially atropine with the tropic acid residue replaced by mandelic acid.

It may be prepared by evaporating tropine (obtained from tropinone) with mandelic and hydrochloric acids. The hydrobromide is obtained readily from the free base by neutralizing with hydrobromic acid. The hydrochloride may be obtained in a similar manner.

The hydrobromide occurs as white crystals, or as a white, crystalline powder which is affected by light. It is soluble in water (1:6) and in alcohol (1:40), less soluble in chloroform (1:420) and insoluble in ether.

Solutions are incompatible with alkaline substances, which precipitate the free base, and also with the common alkaloidal reagents. As in the case of atropine, solutions are sterilized best by filtration through a bacteriologic filter, although it is claimed that autoclaving has no deleterious effect.[22]

It is used topically in therapy to paralyze the ciliary structure of the eye (cycloplegia) and to effect mydriasis. It behaves very much like atropine but is weaker and less toxic. In the eye, it acts more rapidly but less persistently than atropine. The dilatation of the pupil takes place in about 15 to 20 minutes, and the action subsides in about 24 hours. By utilizing a miotic, such as physostigmine (q.v.), it is possible to restore the pupil to normality in a few hours. The drug is used in concentrations of 1 to 2 per cent in aqueous solution or in the form of gelatin disks (lamellae).

Category—anticholinergic (ophthalmic).

For external use—topically, 0.1 ml. of a 2 to 5 per cent solution to the conjunctiva, repeated in 5 minutes.

OCCURRENCE
Homatropine Hydrobromide Ophthalmic Solution U.S.P.

Homatropine Methylbromide N.F., Novatropine®, Mesopin®, 8-Methylhomatropinium Bromide. This compound is the tropine methylbromide ester of mandelic acid. It may be prepared from homatropine by treating it with methyl bromide, thus forming the quaternary compound.

It occurs as a white, odorless powder having a bitter taste. It is affected by light. The compound is readily soluble in water and in alcohol but is insoluble in ether. The pH of a 1 per cent solution is 5.9 and of a 10 per cent solution of 4.5. Although a solution of the compound yields a precipitate with alkaloidal reagents, such as mercuric-potassium-iodide test solution, the addition of alkali hydroxides or carbonates does not cause a precipitate as is the case with nonquaternary nitrogen salts (e.g., atropine, homatropine).

Homatropine methylbromide is said to be less stimulating to the central nervous system than atropine, while retaining virtually all of its parasympathetic depressant action. It is used orally, in a manner similar to atropine, to reduce oversecretion and relieve gastrointestinal spasms.

Category—anticholinergic.

Usual dose—2.5 to 5 mg. 4 times daily.

Homatropine Methylbromide Tablets N.F.

Methscopolamine Bromide N.F., Pamine® Bromide, N-Methylscopolammonium Bromide, Scopolamine Methylbromide. This compound may be made by treating either scopolamine or norscopolamine with methyl bromide.[23]

It is a crystalline, colorless compound, freely soluble in water, slightly soluble in alcohol and insoluble in acetone and chloroform. The drug is a potent parasympatholytic and is distinguished especially by its ability to inhibit the secretion of acid gastric juice through a depression of the vagus innervation of the stomach. This is in some contrast to methantheline bromide wherein the principal activity seems to be toward inhibition of the motility of the gastrointestinal tract. The effect of methscopolamine bromide is claimed to be specifically on the parasympathetic nervous system, although it is to be noted that blocking of the sympathetic system will occur with very large doses. The drug is said not to possess the depressant effect usually associated with scopolamine, and the manufacturer deliberately has avoided the use of the term "scopolamine methylbromide" in the promotion of the drug to further de-emphasize the connection between the tertiary amine and the quaternary salt. According to Kirsner and Palmer,[24] methscopolamine bromide is one of the more effective antisecretory drugs, although they point out that it, too, has atropinelike side-effects when administered in large doses. The drug is also promoted for use as an antisialogogue and anhydrotic.

Perhaps the principal use of the drug is in the medical management of peptic ulcer, gastric hyperacidity and gastric hypermotility. Because of its atropinelike effect on secretions, it is of use in excessive salivation and sweating. Dryness of the mouth and blurred vision are the most common side-effects encountered. It is supplied in the form of 2.5 mg. tablets, with or without 15 mg. of phenobarbital or in a protracted action form with 7.5 mg. per capsule. The usual form for injection is a solution containing 1 mg. per ml.

Category—anticholinergic.

Usual dose—oral 2.5 mg. 4 times daily; parenteral, 500 mcg. up to 4 times daily.

Usual dose range—oral, 2.5 to 5 mg.; parenteral, 250 mcg. to 1 mg.

Methscopolamine Bromide Injection N.F.
Methscopolamine Bromide Tablets N.F.

Methscopolamine Nitrate, Methine®, Scopolamine Methylnitrate. This compound may be made by treating the methylbromide salt with silver nitrate. It is a crystalline, colorless compound that is freely soluble in water and has the same uses and indications as has the corresponding methylbromide salt.

The usual adult oral dosage of this drug (introduced in 1954) is 2 to 4 mg. given one-half hour before each meal and at bedtime. Each dose lasts for about 6 hours after being absorbed into the blood. The total daily dose, in general, should not exceed 12 mg. For prolonged action it is also supplied in the form of sustained action tablets. When used parenterally it is given in an adult dose of 250 to 500 mcg. in the form of an injection (5 mg. in 10 ml.) subcutaneously or intramuscularly, repeated 3 or 4 times daily until oral therapy is instituted. It is supplied as 2 mg. tablets, plain or sustained action, or in the form of the injectable solution.

Methylatropine Bromide. This derivative is the methyl bromide addition compound of atropine and is exactly the same type of salt as is found in homatropine methylbromide (q.v.). It occurs as white crystals that are soluble in water (1:1) or alcohol. Its activity is less than that of atropine, but it has been used in concentrations of 0.5 to 2 per cent as a mydriatic because of the short recovery period.

Methylatropine Nitrate, Metropine®. This compound, introduced as a mydriatic in 1903, is a synthetic quaternary derivative of atropine. It is prepared by first making the methyl bromide derivative of atropine, which then is treated with an equivalent amount of silver nitrate.

It occurs in the form of white crystals

which are freely soluble in water. It is precipitated by most of the common alkaloidal reagents, but, being a quaternary ammonium salt, it is not precipitated by alkalies.

It closely resembles atropine in its actions, although it is stated that it is more toxic.[25] When used, it may be given in approximately the same doses as atropine. There seems to be some confusion in the literature as to its toxicity (compared with atropine); Sollmann[26] stated a ratio of 1:50 and Graham and Lazarus[25] a ratio of 3:1.

In the eye, it is used in concentrations varying from 1 to 5 per cent for mydriasis and cycloplegia. It also has been used as an antispasmodic in the treatment of pyloric stenosis.

Esters of Other Aminoalcohols

It has already been pointed out that the stereochemical arrangement in the rigid atropine molecule lends itself to high activity, presumably because of a good fit of its prosthetic groups with the receptor site. Therefore, one might come to the conclusion that any deviation from this arrangement might reduce the activity substantially, if not remove it completely. However, early studies employing the empiric idea of structural dissection (so successful with local anesthetics) led to the conclusion that, even though atropine did seem to have a highly specific action, the tropine portion was nothing more than a highly complex aminoalcohol and was susceptible to simplification. The accompanying formula shows the portion of the atropine molecule (enclosed in the curved dotted line) believed

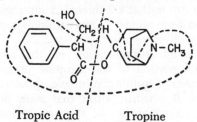

Tropic Acid   Tropine

to be responsible for its major activity. This group is sometimes called the "spasmophoric" group and compares with "anesthesiophoric" group obtained by similar dissection of the cocaine molecule (q.v.). The validity of this conclusion has been amply

borne out by the many active compounds having only a simple diethylaminoethyl residue replacing the tropine portion.

Eucatropine (VII) may be considered as a conservative approach to the simplification of the aminoalcohol portion, in that the bicyclic tropine has been replaced by a monocyclic aminoalcohol and, in addition, mandelic acid replaces tropic acid.

One of the earliest compounds to be prepared utilizing a simplified noncyclic aminoalcohol was amprotropine (VIII), which was prepared by Fromherz[27] in 1933 and for many years was widely used as a gastrointestinal antispasmodic but has been displaced by much more active compounds. In this particular case, the tropic acid residue was retained, but the bulk of modern research on antispasmodics of this nature has been directed toward compounds in which both the acid and the aminoalcohol portions have ben modified. Aromatic acids, particularly with a carbonyl or hydroxy function (e.g., mandelic and benzilic), were shown early to be among the most highly active acids to be employed. Table 53 illustrates the general situation with respect to substitution and spasmolytic potency of several different compounds. It will be noted that, starting with a simple acetyl ester (which is spasmogenic) the activity increases with increasing aromatic substitution. An enhancing effect is apparent when hydroxylation of the acetyl carbon is employed, although the dangers of broad generalizations are noted in the decreased activity of I vs. H. Likewise, two phenyl groups appear to be maximal, inasmuch as a sharp drop in activity is noted when three phenyls are employed. This is caused possibly by steric hindrance in the triphenylacetyl moiety. Comparison of compounds E and H would indicate also that enhancement of action results from bonding the phenyl groups together into the fluorene moiety (XXXIV), a compound which enjoyed some commercial success under the trade name of Pavatrine. However, it has recently been withdrawn from the market.

It is evident that the acid portion (corresponding to tropic acid) should be somewhat bulky in nature, especially when the

TABLE 53

$$R_2-\underset{\underset{R_3}{|}}{\overset{\overset{R_1}{|}}{C}}-COOCH_2CH_2N(C_2H_5)_2$$

COMPOUND	STRUCTURE			SPASMOLYTIC POTENCY[†]	
	R_1	R_2	R_3	Acetylcholine pD[‡]	Relative Potency %
A	H	H	H	Stimulates	
B	H	H	OH	4.0-4.3	1-2
C	Phenyl	H	H	5.0-5.3	10-20
D	Phenyl	H	OH	5.3-5.7	20-50
E§	Phenyl	Phenyl	H	6.0	100
F‖	Phenyl	Phenyl	OH	7.6	4,000
G	Phenyl	Phenyl	Phenyl	5.0	10
H¶	**Fluorene-9-carboxylic			6.8	600
I	††Fluorene-9-hydroxy-9-carboxylic			6.7	500

* Adapted from a table by Lands, A. M., *et al.*: J. Pharmacol. & Exper. Therap. 100:19, 1950.
† All esters were tested as the hydrochlorides on rabbit small intestine (isolated segments).
‡ Logarithm of the reciprocal of the ED$_{50}$.
§ Trasentine.　　‖ WIN 5606.　　¶ Pavatrine.

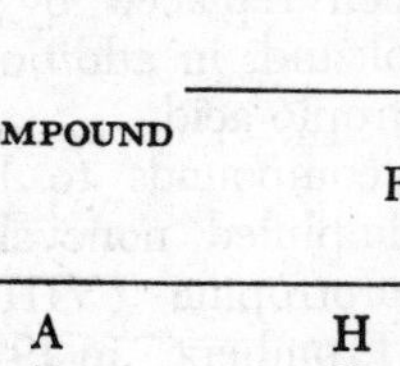

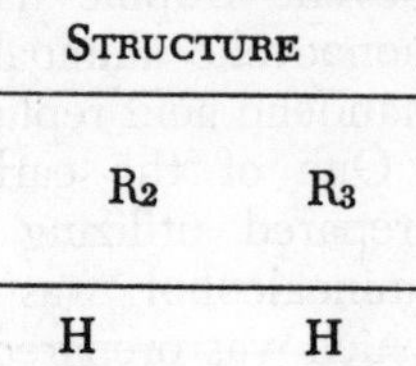

aminoalcohol portion is a simple one. This is an indication for the need of at least one portion of the molecule to have the space-occupying, umbrellalike shape which leads to firm binding at the receptor site area.

Although aromatic character in the acids utilized seems to be desirable, activity is retained and sometimes enhanced by reduction of one of the phenyl groups, e.g. in Trasentine-H (IX), which is more active than adiphenine (X). Other compounds with reduced rings are cyclopentolate (XI), oxphencyclimine (XII), oxyphenonium bromide (XIII), dihexyverine (XIV) and penthienate bromide (XV). In the case of valethamate bromide (XVI) a reduced ring is replaced by a *sec*-butyl group, which may suggest that, whereas steric bulkiness is requisite, a ring is non-essential. Particularly interesting is dicyclo-mine (XVII), which has no phenyl groups and possesses only alicyclic rings. The dicyclomine structure has a feature that has been employed also in caramiphen (XVIII) and dihexyverine (XIV), namely that of incorporating the acetyl carbon into a reduced ring structure which is either cyclohexyl or cyclopentyl.

Penthienate bromide (XV) and heteronium bromide (XXVI) are interesting compounds employing the principle of bioisosterism in that a 2-thienyl group replaces the isosteric phenyl group with no apparent loss in effectiveness.

Although simplification of the aminoalcohol portion of the atropine prototype has been a guiding principle in most research, it is worth noting that many of the more recent entries in the anticholinergic field have employed cyclic aminoalcohols. Among the earlier introductions with this feature were piperidolate (XIX), pipenzolate (XX) and mepenzolate (XXI), all of which employed the 3-piperidol type aminoalcohol in which a return to the cyclic structure is evident. Closely related to the preceding compounds are parapenzolate bromide (XXII) and pentapiperide methylsulfate (XXIII) derived from 4-piperidol, both of which are reminiscent of eucatropine (VII). Subsequently, a number of manufacturers have explored the possibilities of utilizing the pyrrolidinols as the aminoalcohol moiety. In this approach, 3-pyrrolidinol esters such as benzilonium bromide (XVIV), glycopyrrolate (XXV) and heteronium bromide (XXVI) have received considerable attention. Differing from the 3-pyrrolidinol esters but still retaining the 1,2 relationship of alcoholic hydroxyl to amino nitrogen is the 2-hydroxymethyl-pyrrolidine type seen in poldine methylsulfate (XVII). Another type of cyclic aminoalcohol is apparent in oxyphencyclimine (XII) in which a partially hydrogenated pyrimidine ring is utilized. A particular feature of this ring system is that it may be looked upon as an amidine and, indeed, it is hinted that this may account for its unusually long duration of action.[28] Clidinium bromide (XXVIII) employs a still more complex aminoalcohol, i.e., quinuclidinol. Being a bicyclic aminoalcohol, this compound represents the completion of the cycle of research from complex bicyclic aminoalcohol (tropine) to simple acyclic aminoalcohol (2-diethylaminoethanol) to complex bicyclic aminoalcohol (quinuclidinol)! Note may be made also of an anticholinergic employing an acetylenic linkage in the aminoalcohol, i.e., oxybutynin (XXIX), a drug still under experimental investigation. Finally, it should be mentioned that a thioaminoalcohol derivative, namely thiphenamil hydrochloride (XXX), also has been shown to be an active anticholinergic. This compound differs from adiphenine (X) only in the presence of a sulfur atom in place of an oxygen atom and has an activity slightly greater than that of adiphenine.

An important feature to be found in many of the synthetic anticholinergics is that they employ quaternization of the nitrogen, presumably to enhance activity. The initial synthetic compound of this type produced was methantheline (XXXI), which served as a forerunner for many others (XIII, XV, XVI, XX-XXVIII, XXXII). These combined anticholinergic activity of the antimuscarinic type with

some ganglionic blocking activity to reinforce the parasympathetic blockade. However, it must be noted that quaternization also introduces the possibility of sympathetic blockade and blocking of voluntary synapses (curariform activity) as well, and these can become evident with sufficiently high doses. Perhaps the most serious drawback to the quaternary nitrogen compounds is their erratic absorption, which has already been alluded to in conjunction with the esters of tropine (p. 527). Suitably active tertiary amine compounds seem to have an advantage over the quaternaries in the matter of gastrointestinal absorption. In particular, if the anticholinergic is to be used for central effects (e.g., benactyzine (XXX) or the parkinsonlytics (q.v.)), the quaternaries would have difficulty in passing the blood-brain barrier, whereas the tertiary amines reasonably may be expected to pass this lipoidal barrier in the undissociated form.

It is probable that, with the modern high degree of development in synthesizing potent anticholinergics, many of the earlier compounds (even those in current use) would not have survived the screening tests if discovered today.

Products

Many active compounds have been discussed above, most of which have been on the market. Those currently in use are described in the following monographs.

Adiphenine Hydrochloride, Trasentine® Hydrochloride, 2-Diethylaminoethyl Diphenyl Acetate. This is prepared by the interaction of either diphenyl acetic acid chloride or diphenylketene on diethylaminoethanol. The base is then converted to the hydrochloride.

It occurs as white crystalline needles that are stable but will hydrolyze in solution on standing. The salt is soluble in water or alcohol but insoluble in the organic solvents. A 5 per cent aqueous solution is neutral to litmus.

It possesses about one twenty-fifth the activity of atropine but is equal to papaverine as an antispasmodic. Its toxicity is approximately the same as that of atropine but, inasmuch as the toxic dose is so much higher than the therapeutic dose, it is a quite safe drug. In addition, it has few side-reactions of the atropine type. Its action is selective on smooth muscle, which it serves to relax. This spasmolytic action has been demonstrated on most smooth muscle organs, e.g., gastrointestinal tract, gall-bladder, urinary bladder.

Clinically, it is used for the treatment of gastrointestinal spasm due to gastric or duodenal ulcer, pylorospasm, cholecystitis and similar disorders and for the spasm associated with duodenal or biliary-tract disease. In common with most of the other antispasmodics, it is effective in relieving those cases of dysmenorrhea due to uterine hypertonicity. The effect on the ureters may be employed occasionally to promote spontaneous passage of calculi into the bladder. Other conditions, too numerous to mention, may be found in the literature cited by the manufacturers.

The oral dose may vary from 225 to 450 mg. daily, although larger doses have been employed without harmful results. A 50-mg. dose is suggested for parenteral use by intramuscular or intravenous route. In addition, a rectal suppository also is available for use when nausea and vomiting prevent the use of the oral route. The oral tablets are 75 mg.

Benactyzine Hydrochloride, Phobex®, Suavitil®, 2-Diethylaminoethyl Benzilate Hydrochloride. Although this compound is an anticholinergic, its activity in this respect is so low that it is not used as a peripherally acting agent. Its low incidence of side-effects in the small doses used to bring out its central effects has led to its employment as a psychotherapeutic agent. It is discussed with central depressant drugs (Chap. 15).

Clidinium Bromide N.F., Quarzan® Bromide, 3-Hydroxy-1-methylquinuclidinium Bromide Benzilate. The preparation of this compound is described by Sternbach and Kaiser.[29, 30] It occurs as a white or nearly white, almost odorless, crystalline powder which is optically inactive. It is soluble in water and in alcohol but only very slightly soluble in ether and in benzene.

This anticholinergic agent is marketed in a combination form only, with the minor tranquilizer chlordiazepoxide (Librium), the resultant product being known as Librax. The rationale of the combination for the treatment of gastrointestinal complaints is the use of an anxiety-reducing agent together with an anticholinergic based on the recognized contribution of anxiety to the development of the diseased condition. It is suggested for peptic ulcer, hyperchlorhydria, ulcerative or spastic colon, anxiety states with gastrointestinal manifestations, nervous stomach, irritable or spastic colon, etc. The combination capsule contains 5 mg. of chlordiazepoxide hydrochloride and 2.5 mg. of clidinium bromide. It is, of course, contraindicated in glaucoma and other conditions that may be aggravated by the parasympatholytic action, such as prostatic hypertrophy in the elderly male which could lead to urinary retention. The usual recommended dose for adults is 1 or 2 capsules 4 times a day before meals and at bedtime.

Category—anticholinergic.

Usual dose—2.5 mg. 3 times a day.

Usual dose range—2.5 to 5 mg. 3 or 4 times a day.

Cyclopentolate Hydrochloride U.S.P., Cyclogyl® Hydrochloride, 2-(Dimethylamino)ethyl 1-Hydroxy-α-phenylcyclopentaneacetate Hydrochloride. This compound, together with a series of closely related compounds, was synthesized by Treves and Testa[31] It is a crystalline, white, odorless solid, m.p. 137° to 141° C. It is very soluble in water, easily soluble in alcohol and only slightly soluble in ether. A 1 per cent solution has a pH of 5.0 to 5.4.

This drug is used only for its effects on the eye, where it acts as an ophthalmic parasympatholytic. It produces cycloplegia and mydriasis quickly when placed in the eye. Its primary field of usefulness is in refraction studies. However, it can be used as a mydriatic in the management of iritis, iridocyclitis, keratitis and choroiditis. Although it does not seem to affect intraocular tension significantly, it is desirable to be very cautious with patients with high intraocular pressure and also with elderly patients with possible unrecognized glaucomatous changes.

The drug has one half the antispasmodic activity of atropine and has been shown to be nonirritating when instilled repeatedly into the eye. If not neutralized after the refraction studies, the effect is usually gone in 24 hours. Neutralization with a few drops of pilocarpine nitrate solution, 1 to 2 per cent often results in complete recovery in 6 hours.

The drug is supplied as a ready-made ophthalmic solution in concentrations of either 0.5 or 1 per cent, and also in the form of a gel for better application to the eye.

Category—anticholinergic (ophthalmic).

For external use—topically, 0.1 ml. of 0.5 to 2 per cent solution to the conjunctiva, 3 to 5 times daily.

OCCURRENCE
Cyclopentolate Hydrochloride Ophthalmic Solution U.S.P.

Dicyclomine Hydrochloride N.F., Bentyl® Hydrochloride, 2-(Diethylamino)ethyl [Bicyclohexyl]-1-carboxylate Hydrochloride. The synthesis of this drug is described by Tilford and his co-workers.[32] In common with similar salts, this drug is a white crystalline compound that is soluble in water.

It is reported to have one eighth of the neurotropic activity of atropine and approximately twice the musculotropic activity of papaverine. Again, this preparation has minimized the undesirable side-effects associated with the atropine-type compounds. It is used for its spasmolytic effect on various smooth-muscle spasms, particularly those associated with the gastrointestinal tract. It is also useful in dysmenorrhea, pylorospasm and biliary dysfunction.

The drug, introduced in 1950, is marketed in the form of capsules (10 mg.), with or without 15 mg. of phenobarbital, and also in the form of a syrup, with or without phenobarbital. For parenteral use (intramuscularly) it is supplied as a solution containing 20 mg. in 2 ml.

Category—anticholinergic.

Usual dose range—10 to 20 mg. 3 to 4 times daily.

OCCURRENCE
Dicyclomine Hydrochloride Capsules N.F.
Dicyclomine Hydrochloride Syrup N.F.

Eucatropine Hydrochloride U.S.P., Euphthalmine Hydrochloride, 1,2,2,6-Tetramethyl-4-piperidyl Mandelate Hydrochloride. Eucatropine is very similar to β-eucaine. The salt is an odorless, white, granular powder that is soluble in water and stable in air.

The action of eucatropine closely parallels that of atropine, but the former is much weaker. It is used topically in a 0.1-ml. dose as a mydriatic in 2 per cent solution or in the form of small tablets. However, the use of concentrations of from 5 to 10 per cent is not uncommon. Dilation, with little impairment of accommodation, takes place in about 30 minutes, and the eye returns to normal in from 2 to 3 hours.

Category—anticholinergic (ophthalmic), repeated in 5 minutes.

For external use—topically, 0.1 ml. of a 2 to 5 per cent solution to the conjunctiva, repeated in 5 minutes.

Glycopyrrolate N.F., Robinul®, 3-Hydroxy 1,1-dimethylpyrrolidinium Bromide α-Cyclopentylmandelate. The drug occurs as a white crystalline powder that is soluble in water or alcohol but is practically insoluble in chloroform or ether.

Glycopyrrolate is a typical anticholinergic and possesses, at adequate dosage levels, the atropinelike effects characteristic of this group. It has a spasmolytic effect on the musculature of the gastrointestinal tract as well as the genitourinary tract. It diminishes gastric and pancreatic secretions and diminishes the quantity of perspiration and saliva. Its side-effects are typically atropinelike also, i.e., dryness of the mouth, urinary retention, blurred vision, constipation, etc.[33] Because of its quaternary ammonium character it rarely causes central nervous system disturbances, although, in sufficiently high dosage, it can bring about ganglionic and myoneural junction block.

The drug is used as an adjunct in the management of peptic ulcer and other gastrointestinal ailments associated with hyperacidity, hypermotility and spasm. In common with other anticholinergics its use does not preclude dietary restrictions or use of antacids and sedatives if these are indicated.

Category—anticholinergic.

Usual dose—1 mg. 3 times a day.
Usual dose range—1 to 2 mg.

OCCURRENCE
Glycopyrrolate Tablets N.F.

Mepenzolate Bromide N.F., Cantil®, 3-Hydroxy-1,1-dimethylpiperidinium Bromide Benzilate. This compound may be prepared by the method of Biel *et al.*[34] utilizing the transesterification reaction with 1-methyl-3-hydroxypiperidine and methyl benzilate. The resulting base is quaternized with methyl bromide to give a white, crystalline product, m.p. 234° to 236° C. The product is water soluble. It is structurally similar to pipenzolate methylbromide (q.v.) except for the replacement of a methyl group with an ethyl on the nitrogen in the latter.

It has an activity of about one half that of atropine in reducing acetylcholine-induced spasms of the guinea pig ileum, although some reports rate it as equal to atropine in effectiveness and duration of action. It is specifically promoted for a claimed "markedly selective action on the colon." The selective action on colonic hypermotility is said to relieve pain cramps, and bloating and to help curb diarrhea. The evidence for this specific action is conflicting. Bachrach,[1] for example, in a survey of the anticholinergic literature questions the specificity of these drugs "for any particular gastrointestinal organ, function, or segment of the gastrointestinal tract."

The drug, introduced in 1956, is marketed in 25-mg-tablets, either plain or with 15 mg. of phenobarbital.

Category—anticholinergic.
Usual dose—25 mg. 4 times a day.
Usual dose range—25 to 50 mg.

OCCURRENCE
Mepenzolate Bromide Tablets N.F.

Methantheline Bromide N.F., Banthine® Bromide, Diethyl(2-hydroxyethyl)methyl-ammonium Bromide Xanthene-9-carboxylate. Methantheline may be prepared according to the method outlined by Burtner and Cusic,[35] although this reference does not show the final formation of the quaternary salt. The compound from which the quaternary salt is prepared was in the series of esters from which aminocarbofluorene was selected as the best spasmolytic agent.

It is a white, slightly hygroscopic, crystalline salt which is soluble in water to produce solutions with a pH of about 5. Aqueous solutions are not stable and hydrolyze in a few days. The bromide form is preferable to the very hygroscopic chloride, although the latter is slightly more active in doses of equal weight.

This drug, introduced in 1950, is a potent anticholinergic agent and acts at the *ganglia* of the sympathetic and the parasympathetic systems, as well as at the myoneural junction of the postganglionic cholinergic fiber. Methantheline has no action at the effector site of the sympathetic system, because norepinephrine instead of acetylcholine is the mediator of the nervous transmission at this point. Therefore, a given dose of the drug acts more potently on the parasympathetic subdivision, because it has two points to block at instead of the single one at the ganglion in the sympathetic system. It differs from atropine, because atropine acts only at the effector site of the parasympathetic system and has no effect at the ganglia.

Among the conditions for which methantheline is indicated are gastritis, intestinal hypermotility, bladder irritability, cholinergic spasm, pancreatitis, hyperhydrosis and peptic ulcer, all of which are manifestations of parasympathotonia. The last indication (peptic ulcer) has been responsible for much of the publicity accorded the drug. The parasympathetic system is represented in its gastric innervation by the vagus nerve, and, prior to the introduction of such a drug as methantheline, the surgical procedure of vagotomy had been shown to give relief to peptic ulcer patients. The drug is, in effect, a nonsurgical vagotomy that can be withdrawn whenever desired, in common with other quaternary anticholinergic agents.

Side-reactions are atropinelike (mydriasis, cycloplegia, dryness of mouth), and the drug is contraindicated in glaucoma. High overdosage may bring about a curarelike action, a not too surprising fact when it is considered that acetylcholine is the mediating factor for neural transmission at the somatic myoneural junction. This side-effect can be counteracted with neostigmine methylsulfate.

The drug is marketed as 50-mg. tablets for oral use with or without phenobarbital (15 mg.) and for parenteral use, 50-mg. ampuls are supplied.

Category—anticholinergic.

Usual dose—50 mg. 4 times daily.

Usual dose range—50 to 100 mg.

OCCURRENCE

Methantheline Bromide Tablets N.F.

Oxyphencyclimine Hydrochloride N.F., Daricon®, Vio-Thene®, (1,4,5,6-Tetrahydro-1-methyl-2-pyrimidinyl)methyl α-Phenyl-cyclohexaneglycolate Monohydrochloride. The synthesis of this compound is described by Faust *et al.*[36] The product is a white, crystalline compound which is sparingly soluble in water (1.2 g./100 ml. at 25° C.). It has a bitter taste.

This compound, introduced in 1958, is promoted as a peripheral anticholinergic-antisecretory agent with little or no curare-like activity and with little or no ganglionic-blocking activity. That these activities are absent is probably due to the tertiary character of the compound, which is in some-what marked contrast with the quaternaries that have dominated the anticholinergic scene for the most part and potentiate anti-cholinergic activity due to a coupling of antimuscarinic action with ganglion-blocking action. Also, the tertiary character of the nitrogen should promote its better in-testinal absorption, as previously outlined (see p. 527). Another feature of the compound is its relatively long duration of action (12 hours.) which is suggested as being in some way related to the amidine-type structure to be found in the aminoalcohol portion of the molecule.[28] Perhaps the most significant activity of this compound is its marked ability to reduce both the volume and the acid content of the gastric juices,[37] a desirable action in view of the more recent hypotheses pertaining to peptic ulcer ther-apy. Another important feature of this com-pound is its low toxicity in comparison with many of the other available anticholinergics.

Oxyphencyclimine is suggested for use in peptic ulcer, pylorospasm and functional bowel syndrome. The therapeutic range given for the compound is 5 to 25 mg. twice a day, and the average adult dose is 10 mg. twice a day. For children, in the manage-ment of enuresis, pylorospasm and infant colic, the suggested dose is 0.2 mg./kg. twice a day. It is contraindicated, as are other anticholinergics, in patients with pros-tatic hypertrophy and glaucoma. The drug is supplied in the form of 10-mg. tablets plain or in combination with 25 mg. of hydroxyzine as the drug know as Enarax.

Category—anticholinergic.

Usual dose—10 mg. twice daily.

Usual dose range—10 to 50 mg. daily, in divided doses.

OCCURRENCE

Oxyphencyclimine Hydrochloride Tablets N.F.

Oxyphenonium Bromide, Antrenyl® Bro-mide, Diethyl (2-hydroxyethyl)methylam-monium Bromide α-Phenylcyclohexylglyco-late. This compound is prepared in a manner similar to methantheline.[38] It occurs as colorless crystals and is easily soluble in water to form solutions neutral to litmus.

It has an action very similar to that of methantheline and is used for the same pur-poses, namely, the management of peptic ulcer and the control of spasm due to para-sympathotonia. The side-effects of the drug are atropinelike in nature and are usually a result of the administration of high doses.

The drug, introduced in 1952, may be administered orally, intramuscularly or sub-cutaneously. For oral use it is supplied as drops containing 1 mg. per drop, a syrup containing 1.25 mg. per ml. and as 5 mg. tablets. The usual oral dosage is 10 mg. 4 times daily for adults. For quick action the injection (20 mg. per 10 ml.) is used paren-terally in a dose of 1 or 2 mg. every 6 hours.

Pentapiperide Methylsulfate, Quilene®; 4-Hydroxy-1,1-dimethylpiperidinium Meth-ylsulfate *dl*-3-Methyl-2-phenylvalerate; 3-Methyl-2-phenylvaleric acid-1-methyl-4-pi-peridyl ester. This compound is prepared according to the method of Martin *et al.*[39] It occurs as white crystals and is water soluble.

This anticholinergic is promoted for use as adjunctive therapy in the management of peptic ulcer. In vitro its antispasmodic activity is equal to that of atropine in vivo, but, because of its quaternary nature, it is less readily absorbed from the gastrointestinal tract. When administered orally to rats, antispasmodic activity is about one fifth that of atropine and its antisecretory activity is about one fourth that of atropine on a milligram-for-milligram basis. However, when given subcutaneously in rats, its activity is two and one half times that of atropine, although this difference is minimized by the fact that it is used orally. It is equivalent to propantheline bromide when given orally for the antispasmodic and antisecretory effects. It is said to offer greater protection than propantheline bromide against gastric ulcers induced in rats by the standard Shay method.

This drug, in common with other anticholinergics, is contraindicated in glaucoma, pyloric obstruction or stenosis, gastric retention, known or suspected obstructive disease of the gastrointestinal tract and urinary bladder neck obstruction, prostatic hypertrophy, stenosing peptic ulcer, megaesophagus, organic cardiospasm and, for that matter, in any patient with hypersensitivity to the drug. Caution is recommended, as with all anticholinergics, in cases of cardiac decompensation, coronary insufficiency and in tachycardias and arrythmias aggravated by vagal blockade. Similarly, although no teratogenesis has been shown in animal studies, it should be used with caution in pregnancy and should not be prescribed for children under 12 years of age because safe conditions for its use have not been established.

The usual side effects of anticholinergic therapy are evident with this drug, with dryness of the mouth (xerostomia) being the most common and other effects such as dilation of pupils, urinary retention, tachycardia, palpitation, nervousness, etc. being observed with less regularity.

The drug is administered to adults in a dose of 1 or 2 tablets (10 mg.) three or four times daily (before meals and at bedtime). On occasion, for overnight control, an extra tablet may be indicated at bedtime. It is recommended that individualization of the dose be adopted for maximal effect, since the optimal dose could be as high as 30 mg. 4 times daily.

Penthienate Bromide N.F., Monodral® Bromide, Diethyl(2-hydroxyethyl)methylammonium Bromide α-Cyclopentyl-2-thiopheneglycolate. This compound is synthesized according to the method of Blicke and Tsao.[40]

It occurs as a freely water-soluble (1:5), white, crystalline powder. The pH of a 1 per cent solution is 5.5 to 7.0. It is a potent anticholinergic featuring inhibition of gastric secretion and gastrointestinal motility. The spasmolytic effect is said to be more specific to the upper portion of the gastrointestinal tract than to the lower portion, although it has been used clinically in spastic colon. Its antisecretory action is made use of in hyperhidrosis, but this is secondary to its gastric antisecretory action. Therefore, its principal usefulness lies as an adjunct to the treatment of peptic ulcer in combination with other forms of medical management. The main side-effect in therapeutic doses is xerostomia (dryness of the mouth). Severe toxic effects have not been encountered. Contraindications are similar to those of other anticholinergics (glaucoma, prostatic hypertrophy, etc.).

The drug is available in the form of 5-mg. tablets or as an elixir containing 0.5 mg. per ml.

Category—anticholinergic.

Usual dose—5 mg. 3 or 4 times daily.

Usual dose range—2.5 to 10 mg.

OCCURRENCE
Penthienate Bromide Tablets N.F.

Pipenzolate Methylbromide, Piptal®, 1-Ethyl-3-piperidyl Benzilate Methylbromide. This compound is prepared in a manner exactly analogous to Cantil (q.v.). It exists as a white or light cream-colored powder, highly soluble in water.

The principal activity of this compound introduced in 1955, is as a peripheral atropinelike anticholinergic with an activity encompassing both an antispasmodic and a gastric antisecretory effect. The spasmolytic effect appears to extend to the bilary tract as well as the lower gastrointestinal tract. Its spasmolytic activity is said to be equal to that of atropine but is more active in depressing gastric secretion.

Clinically, pipenzolate methylbromide is used mostly for adjunctive treatment of peptic ulcer in combination with proper dietary measures. Its spasmolytic action on the lower gastrointestinal tract suggests its usefulness in ileitis, the irritable colon syndrome and other functional disorders of the gastrointestinal tract. For no apparent reason, it appears to be better than average in the treatment of certain of the postgastrectomy disturbances.

Its side-effects are atropinelike in nature, with drying of the mouth being the commonest complaint. Although no serious toxic reactions have been noted, precaution should be exercised in administering it to patients with glaucoma, prostatic hypertrophy, etc. However, Asher[3] suggests that it is particularly useful for older patients because of its low incidence of side-effects (especially urinary retention) as compared with the stronger anticholinergics.

It is available in the form of 5-mg. tablets with an average dose of 5 mg. 3 times a day before meals and 5 to 10 mg. at bedtime. However, the dosage should be individualized for each patient.

Piperidolate Hydrochloride, Dactil®, 1-Ethyl-3-piperidyl Diphenylacetate Hydrochloride. This compound is synthesized according to the method of Biel and coworkers,[34] utilizing the conventional acylation of the appropriate aminoalcohol with diphenylacetyl chloride in the presence of triethylamine. The hydrochloride exists as white crystals which are water soluble.

The principal activity of this compound, introduced in 1954, seems to be antimuscarinic in nature with little or no action on ganglia or voluntary muscle innervations. Its central action is neglible. Its antimuscarinic activity is about 1/100 that of atropine. The specificity as a spasmolytic for the smooth musculature of the gastrointestinal tract seems to be its main action, and it is termed a "visceral eutonic" (an agent producing normal tone of a viscus) in the promotional literature. There is little or no effect on gastric secretion and, in therapeutic doses, it seems to have little action on the biliary tract musculature. The rapid effect on gastrointestinal motility (within 10 to 20 minutes) is attributed to a local anesthetic effect.

Its clinical usefulness has been as an adjunctive for management of functional gastrointestinal disorders characterized by spasm and hypermotility associated with pain. The upper gastrointestinal tract seems to be affected more by the drug than the lower tract and, whereas it is useful for gastroduodenal spasm, pylorospasm and cardiospasm, it is of little value for colonic spasm. It is *not* intended for use in peptic ulcer. The drug is also promoted for relief of spasm of biliary sphincter and biliary dyskinesia.

It is supplied in capsules of 50 mg. plain or with 16 mg. of phenobarbital for oral use. The average dose is 50 mg. 4 times daily.

Pipethanate Hydrochloride, Sycotrol®, 2-(1-Piperidino)ethyl Benzilate Hydrochloride. This compound is closely related to benactyzine hydrochloride and, rather than being used as an anticholinergic, is proposed for use in the management of anxiety and tension of somatic origin. Because its principal therapeutic emphasis has been placed on its use in relieving anxiety states it is discussed in Chapter 15.

Poldine Methylsulfate, Nacton®, 2-(Hydroxymethyl)-1,1-dimethylpyrrolidinium Methylsulfate Benzilate. This compound occurs as a water-soluble, creamy-white crystalline powder.

It has, qualitatively, the same atropinelike actions as other anticholinergics both as to the therapeutically desirable effects and the undesirable side-effects.[33] It is promoted for the same purposes as, for example, glycopyrrolate and has the same side-effects and precautions concerning its use. Its principal use is as an adjunct in the management of peptic ulcer and related conditions.

It is administered orally in a 4-mg. dose 3 or 4 times daily before meals and at bedtime. The dosage is adjusted individually, with an average maximum tolerated dosage of about 30 mg. daily. It is marketed as 4 mg. tablets with or without 15 mg. of butabarbital sodium.

Propantheline Bromide U.S.P., Proban-thine® Bromide, (2-Hydroxyethyl)diisopropylmethylammonium Bromide Xanthene-9-carboxylate.

The method of preparation of this compound is exactly analogous to that for methantheline bromide (q.v.).

It is a white, water-soluble, crystalline substance with properties quite similar to those of methantheline.

Its chief difference from methantheline is in its potency, which has been estimated variously as being from 2 to 5 times as great. This greater potency is reflected in its smaller dose. For example, instead of a 50-mg. initial dose, a 15-mg. initial dose is suggested for propantheline bromide. It is available in 15-mg. sugar-coated tablets and in the form of a powder (30 mg.) for preparing parenteral (I.V. or I.M.) solutions.

Category—anticholinergic.

Usual dose—oral, 15 mg. 4 times daily; parenteral, 15 to 30 mg. 4 times daily.

Usual dose range—oral, 30 to 75 mg. daily; parenteral, 30 to 240 mg. daily.

Occurrence
Propantheline Bromide Tablets U.S.P.
Sterile Propantheline Bromide U.S.P.

Thiphenamil Hydrochloride, Trocinate®, 2-(Diethylamino)ethyl Diphenylthiolacetate Hydrochloride. This compound occurs as water-soluble colorless needles or large prisms. Aqueous solutions are practically neutral to litmus.

This drug possesses primarily a papaverinelike (see p. 515) direct spasmolytic effect on the smooth muscle of the gastrointestinal, the biliary and the genitourinary tracts. Because of its structural relationship to the atropine-type esters, it may be expected to have some atropinelike effects.[21] However, these are minimal, and the usual side-effects of such drugs occur only rarely with thiphenamil. It has only a small effect on gastric and salivary secretions and is lacking in mydriatic action.

Thiphenamil finds use in the treatment of gastric and genitourinary hypermotility and spasm. It is administered orally in a usual adult dose of 200 mg. (300 mg. maximum) 4 times daily. It is supplied as tablets with or without 16 mg. of phenobarbital.

Valethamate Bromide N.F., Murel®, Diethyl(2-hydroxyethyl)methylammonium Bromide 3-Methyl-2-phenylvalerate. The synthesis of the compound is mentioned by Krause *et al.*[41]

This compound, introduced in 1958, is claimed to have activity of three different types in the same molecule, namely, anticholinergic, musculotropic and ganglion-blocking effects. Its anticholinergic effect is equal to that of atropine, with fewer side-effects. Its direct action on muscle (musculotropic effect) is about 15 times that of papaverine, and its ganglion-blocking effect is equal to that of tetraethylammonium bromide (q.v.). It has little effect on salivary secretion and has less effect than atropine on the gastric secretions, with little or no decrease in gastric secretory volume in humans. Its antispasmodic effect extends to the gastrointestinal, biliary and the genitourinary tracts. Probably the outstanding feature of this drug is its low incidence of side-effects, including dry mouth which is the chief complaint.[42]

The drug is suggested as adjunctive therapy in peptic and duodenal ulcers, various functional gastrointestinal complaints, genitourinary tract conditions, and biliary tract conditions associated with spasm and pain. Caution should be exercised in administering the drug to patients with glaucoma and prostatic hypertrophy, as with other anticholinergics.

It is available as 10-mg. tablets plain or with 15 mg. of phenobarbital as well as in 40-mg. sustained-action tablets with or without 30 mg. of phenobarbital. For parenteral use it is available in an injection containing 10 mg. per ml.

Category—anticholinergic.

Usual dose range—intramuscular and intravenous, 10 to 20 mg. every 4 to 6 hours up to a maximum dosage of 60 mg. in a 24-hour period.

Occurrence
Valethamate Bromide Injection N.F.

AMINOALCOHOL ETHERS

The aminoalcohol ethers thus far introduced have been used as antiparkinsonism drugs rather than as conventional anticholinergics (i.e., as spasmolytics, mydriatics, etc.). In general, they may be considered as closely related to the antihistaminics and, indeed, do possess antihistaminic properties of a substantial order. Comparison of chlorphenoxamine (XXXV) and orphenadrine (XXXVI) with the antihistaminic diphenhydramine (XXXVII) illustrates the close similarity of structure. The use of diphenhydramine in parkinsonism has been cited earlier (see p. 517). Benztropine (XXXVIII) may also be considered as a structural relative of diphenhydramine, although the aminoalcohol portion is tropine and, therefore, more distantly related than XXXV and XXXVI. In the structure of XXXVIII, a 3-carbon chain intervenes between the nitrogen and oxygen functions, whereas in the others a 2-carbon chain is evident. However, the rigid ring structure possibly orients the nitrogen and oxygen functions into more nearly the 2-carbon chain interprosthetic distance than is apparent at first glance. This, combined with the flexibility of the alicyclic chain, would help to minimize the distance discrepancy.

PRODUCTS

Benztropine Mesylate U.S.P., Cogentin® Methanesulfonate, 3α-(Diphenylmethoxy)-1αH,5αH-tropane Methanesulfonate. The compound occurs as a white, colorless, slightly hygroscopic, crystalline powder. It is very soluble in water, freely soluble in alcohol and very slightly soluble in ether. The pH of aqueous solutions is about 6. It is prepared by the method of Phillips[43] by interaction of diphenyldiazomethane and tropine.

Benztropine mesylate combines anticholinergic, antihistaminic and local anesthetic properties of which the first is the applicable one in its use as an antiparkinsonism agent. It is about as potent as atropine as an anticholinergic and shares some of the side-effects of this drug such as mydriasis, dryness of mouth, etc. Of importance, however, is the fact that it does not produce central stimulation but, on the contrary, exerts the characteristic sedative effect of the antihistamines and, for this reason, pa-

tients using the drug should not engage in jobs that require close and careful attention.

Tremor and rigidity are relieved by benztropine mesylate, and it is of particular value for those patients who cannot tolerate central excitation (e.g., aged patients). It also may have a useful effect in minimizing drooling, sialorrhea, masklike facies, oculogyric crises and muscular cramps.

The usual caution that is exercised with any anticholinergic in glaucoma, prostatic hypertrophy, etc., is observed with this drug.

Category—anticholinergic.

Usual dose—oral, intramuscular, or intravenous: 1 to 2 mg., once or twice daily.

Usual dose range—500 mcg. to 8 mg.

OCCURRENCE
Benztropine Mesylate Injection U.S.P.
Benztropine Mesylate Tablets U.S.P.

Chlorphenoxamine Hydrochloride, Phenoxene®, 2-(p-Chloro-α-methyl-α-phenylbenzyloxy)-N,N-dimethylethylamine Hydrochloride. This compound is made according to the synthesis described by Arnold[44] in the patent literature. It occurs in the form of colorless needles which are soluble in water. Aqueous solutions are stable.

This drug was originally introduced in Germany as an antihistaminic.[45] However, it is stated that this close relative of diphenhydramine (Benadryl) has its antihistaminic potency lowered by the para-Cl and the α-methyl group present in the molecule.[46] At the same time, the anticholinergic action is increased. The drug has an oral LD_{50} of 410 mg. in mice, indicating a substantial margin of safety. It was introduced to U.S. medicine in 1959.

It is indicated for the symptomatic treatment of all types of Parkinson's disease and is said to be especially useful when rigidity and impairment of muscle contraction are evident. It is not as useful against tremor, and combined therapy with other agents may be necessary. The drug has proved to be very useful either on its own or as a replacement for orphenadrine (effect tends to wear off) in counteracting akinesia, adynamia, mental sluggishness and lack of mobility in patients with paralysis agitans.

Although it would be well to use the drug

with caution in cases of glaucoma, Doshay and Constable[46] suggest that chlorphenoxamine may be the drug of choice in patients with paralysis agitans who also suffer from glaucoma. This is on the basis of their lack of finding objective pupillary changes and no complaints of blurred vision. The chief complaints are drowsiness, indigestion and dryness of the mouth.

The drug is available in 50-mg. tablets, and the usual dose is 50 mg. 3 times a day, although some patients may require 2 tablets 4 times a day. In general, however, the dose should not exceed 100 mg. 3 times a day.

Orphenadrine Citrate N.F., Norflex®, N, N-Dimethyl-2-[(o-methyl-α-phenylbenzyl)-oxy]ethylamine Citrate. This compound is synthesized according to the method in the patent literature.[47] It occurs as a white, bitter-tasting crystalline powder. It is sparingly soluble in water, slightly soluble in alcohol, and insoluble in chloroform, in benzene and in ether. The hydrochloride salt is marketed as Disipal®.

Although this compound, introduced in 1957, is closely related to diphenhydramine structurally, it has a much lower antihistaminic activity and a much higher anticholinergic action. Likewise, it lacks the sedative effects characteristic of diphenhydramine. Pharmacologic testing indicates that it is not primarily a peripherally acting anticholinergic because it has only weak effects on smooth muscle, the eye and on secretory glands. However, it does reduce voluntary muscle spasm by a central inhibitory action on cerebral motor areas, a central effect similar to that of atropine.

The drug is used for the symptomatic treatment of Parkinson's disease. Although it is not effective in all patients, it appears from the literature that about one half of the patients are benefited. It relieves rigidity better than it does tremor, and in certain cases it may accentuate the latter. The drug combats mental sluggishness, akinesia, adynamia and lack of mobility, but this effect seems to be diminished rather rapidly on prolonged use. It is best used as an adjunct to the other agents such as benztropine, procyclidine, cycrimine and trihexyphenidyl in the treatment of paralysis agitans.

The drug has a low incidence of side-effects, which are the usual ones for this group, namely, dryness of mouth, nausea, mild excitation, etc.

Category—skeletal muscle relaxant.

Usual dose—oral, 100 mg. twice daily; parenteral, 60 mg. every 12 hours.

AMINOALCOHOL CARBAMATES

There is only one aminoalcohol carbamate (XXXIX) of importance in therapy as an anticholinergic, although a thiocarbamate (phencarbamide, XL) is under investigation. One should recall the fact that the nonalkylated carbamates, carbachol and bethanechol chloride, are actually parasympathomimetics. It is only through the minimization of the polar characteristics of the unsubstituted NH_2 by insertion of lipophilic alkyls that the blocking action becomes evident. Undoubtedly, the same receptor site is operative in the case of both stimulant and inhibitory molecules, except that the latter has additional hydrophobic binding capacity on sites adjacent to the receptor.

Dibutoline Sulfate, Dibuline® Sulfate, *Bis*[(dibutylcarbamate) of Ethyl(2-hydroxyethyl)dimethylammonium] Sulfate. This compound is prepared by the method described by Swan and White.[48] It was introduced for therapeutic use in 1952.

This product is a very hygroscopic water- and benzene-soluble colorless powder. Because of its hygroscopic nature, it is marketed usually in the form of an aqueous solution. Aqueous solutions (5%) are quite stable under storage conditions for at least 1 year at room temperature but are hydrolyzed by boiling. Therefore, sterilization is best effected by means of bacteriologic filtration. However, it is noteworthy that the compound will inhibit growth of staphylococci, pneumococci and α-hemolytic streptococci in ocular infections in a concentration of 0.1 to 0.3 per cent, and a 0.5 per cent solution sterilizes a 24-hour broth culture of the same organisms.[49] Another interesting feature of dibutoline is that it is a surface-active agent, which, apparently, is a result of balancing the highly polar quaternary ammonium group in choline with nonpolar groups at the carbamyl end of the molecule.[50]

Dibutoline has an atropinelike action on mammalian ocular muscles, rather than the muscarinic and nicotinelike effects of the other choline esters. This reversal of the action of carbamylcholine by the addition of nonpolar groups is stated by Swan and White to be unparalleled in autonomic pharmacology. In addition to its eye effects, it also has the ability to decrease the frequency of the contractions in the gastrointestinal tract or in larger doses to inhibit contractions entirely.[51]

It has been used in the eye as a mydriatic and cycloplegic in concentrations of 5 per cent and is claimed to be superior to homatropine for these purposes. Glaucoma is a contraindication, although it is believed that dibutoline has less tendency to raise intraocular pressure than has homatropine.

Because it is both antisecretory and antispasmodic it is used in the treatment of peptic ulcer as an adjunct to other treatment. (q.v.). In addition to this, it offers relief in the other spastic conditions of the gastrointestinal tract (pylorospasm, irritable colon syndrome, etc.) as well as mitigating pain associated with biliary and genitourinary tract spasm. Its side-effects are atropinelike, namely, dryness of the mouth and dilatation of the pupil. In common with other agents it is contraindicated in glaucoma, prostatic hypertrophy, etc. One of its principal advantages is its low toxicity and wide margin of safety.

The drug is administered only by the parenteral route, subcutaneously or intramuscularly, in a dose of 25 mg. It is supplied in the form of an injection containing 125 mg. in each 5 ml.

AMINOALCOHOLS

The development of aminoalcohols as parasympatholytics has taken place during the last several years, most of the research being directed toward finding useful parkinsonlytics. All the useful compounds have had the general characteristic of possessing rather bulky groups around the hydroxyl function, together with a cyclic amino function (e.g., XLI, XLII, XLIII). This is reminiscent of the bulky groups in the acids that were found to be desirable in the aminoester type of anticholinergic (q.v.). It serves to emphasize the fact that the ester group, per se, is not a necessary adjunct to activity, provided that other polar groupings such as the hydroxyl can substitute as a prosthetic group for the carboxyl function. Another structural feature common to all aminoalcohol anticholinergics, with the notable exception of hexocyclium, is the γ-aminopropanol arrangement with 3 carbons intervening between the hydroxyl and amino functions. All of the aminoalcohols used for paralysis agitans are tertiary amines and, because the desired locus of action is central, quaternization of the nitrogen destroys the antiparkinsonism properties. However, quaternization of these aminoalcohols has been utilized to enhance the anticholinergic activity to produce antispasmodic and antisecretory compounds such as hexocyclium (XLIV), mepiperphenidol (XLV), tricyclamol chloride (XLVI) and tridihexethyl chloride (XLVII). The marked difference in activity by simple quaternization is shown vividly by comparison of procyclidine (XLVIII) with its methochloride, tricyclamol chloride (XLVI). The former is a useful drug in parkinsonism, but the latter has very little value. However, there is not such a great disparity in their action as spasmolytics on smooth muscle, both being active, but with the greater activity being found in the quaternized form. Among the quaternized aminoalcohols, the special feature of bioisosterism is noted in thihexinol methylbromide (XLIX), wherein 2-thienyl groups have been effectively substituted for the isosteric phenyl groups without loss of activity.

PRODUCTS

Biperiden N.F., Akineton®, α-5-Norbornen-2-yl-α-phenyl-1-piperidinepropanol. The drug consists of a white, practically odorless, crystalline powder. It is practically insoluble in water and only sparingly soluble in alcohol although it is freely soluble in chloroform. Its preparation is described by Haas and Klavehn.[52]

Biperiden, introduced in 1959, has a relatively weak visceral anticholinergic but a strong nicotinolytic action in terms of its ability to block nicotine-induced convulsions. Therefore, its neurotropic action is rather low on intestinal musculature and blood vessels, but it has a relatively strong musculotropic action, about equal to papaverine, in comparison with most synthetics. Its action on the eye, although mydriatic, is much less than that of atropine. These weak anticholinergic effects serve to add to its usefulness in Parkinson's syndrome by minimizing side-effects.

The drug is used in all types of Parkinson's disease (postencephalitic, idiopathic, arteriosclerotic) and helps to eliminate akinesia, rigidity and tremor. It is also used in drug-induced extrapyramidal disorders by eliminating symptoms and permitting continued use of tranquilizers. Biperiden is also of value in spastic disorders not related to parkinsonism, such as multiple sclerosis, spinal cord injury and cerebral palsy. It is contraindicated in all forms of epilepsy.

It is usually taken orally in tablet form but the free base form is official to serve as a source for the preparation of Biperiden Lactate Injection N.F. which is a sterile solution of biperiden lactate in water for injection prepared from biperiden base with the aid of lactic acid. It usually contains 5 mg. per ml.

Category—anticholinergic.

Usual dose—intramuscular, 2 mg. (as the lactate) which may be repeated every half hour until relief is obtained, but not more than 4 consecutive doses should be given in a 24-hour period; intravenous, 5 mg. (as the lactate) given slowly, which may be repeated once in a 24-hour period.

OCCURRENCE
Biperiden Lactate Injection N.F.

Biperiden Hydrochloride N.F., Akineton® Hydrochloride,α-5-Norbornen-2-yl-α-phenyl-1-piperidine-propanol Monohydrochloride. It is a white, optically inactive, crystalline, odorless powder which is slightly soluble in water, ether, alcohol and chloform and sparingly soluble in methanol.

Biperiden hydrochloride has all of the actions described for biperiden above. The hydrochloride is used for tablets, because it is better suited to this dosage form than is the lactate salt. As with the free base and the lactate salt, xerostomia (dryness of the mouth) and blurred vision may occur.

Category—anticholinergic.

Usual dose range—2 mg. 3 to 4 times daily.

OCCURRENCE
Biperiden Hydrochloride Tablets N.F.

Cycrimine Hydrochloride N.F., Pagitane® Hydrochloride, α-Cyclopentyl-α-phenyl-1-piperidinepropanol Hydrochloride. This drug is made by the procedure of Denton *et al.*[53] It occurs as a white, odorless, bitter solid which is sparingly soluble in alcohol (2:100), and only slightly soluble in water (0.6:100). A 0.5 per cent solution in water is slightly acidic (pH 4.9-5.4).

Cycrimine is a potent antispasmodic of the neurotropic type with an activity of about one fourth to one half that of atropine sulfate. It has little or no effect against spasms induced by histamine. It is slightly more toxic than atropine sulfate.

The drug has been introduced as an aid in the treatment of paralysis agitans (Parkinson's disease). It is well known that this disease is a difficult one to treat and, naturally, the greater the selection of drugs the more likely the physician is to find one suitable to a specific case. Magee and DeJong[54] described a series of patients with paralysis agitans who were treated with the drug. Beneficial results were obtained in 46 per cent of the cases treated. This percentage was superior to that obtained by standard medications.

The drug is supplied as 1.25-mg. and 2.5-mg. tablets. Dosage is quite variable and is adjusted to the individual. The postencephalitic group of patients was reported to be able to tolerate the largest doses (up to 30 and 50 mg. per day), whereas arteriosclerotic and idiopathic types exhibited adverse effects with the larger doses. According to Magee and DeJong, the best procedure is to start with a small initial dose and increase the dose by 2.5 mg. (or smaller) increments to the point of tolerance. The side-effects are the usual atropinelike ones encountered with this group of compounds, namely, drying of the mouth, blurring of vision and epigastric distress.

Category—anticholinergic.

Usual dose—initial, 1.25 mg. 3 times daily; maintenance, to be determined by the physician.

Usual dose range—1.25 to 5 mg.

OCCURRENCE

Cycrimine Hydrochloride Tablets N.F.

Hexocyclium Methylsulfate, Tral®, N-(β-Cyclohexyl-β-hydroxy-β-phenylethyl)-N'-methylpiperazine Dimethylsulfate. The preparation of this compound, introduced in 1957, is described by Weston in the patent literature.[55] It occurs as white crystals soluble in water to about 50 per cent but only slightly soluble in chloroform and insoluble in ether.

Hexocyclium is primarily an antisecretory and antispasmodic drug. At therapeutic dose levels its effects are said to be limited to blocking postganglionic cholinergic nerves. It has, in sufficient dosage, the same effects as are expected from other anticholinergics, such as dilation of pupils, inhibition of salivation, relaxation of the bladder musculature—in other words, atropinelike effects. Large doses produce the expected ganglionic blockade and curariform activity characteristic of quaternary ammonium anticholinergics. In spite of its quaternary character it is said to be absorbed readily from the gastrointestinal tract and has a prompt onset of action, with a duration of 3 to 4 hours.

It is used as adjunctive therapy for peptic ulcer and other gastrointestinal complaints where anticholinergic action is of benefit.[56]

The side-effects are the usual ones for anticholinergics, such as drying of the mouth (xerostomia), infrequent palpitations and occasional blurring of the vision. These side-effects can be minimized by proper adjustment of the dosage. It should be used with caution in the usual contraindications for this type of drug (glaucoma, prostatic hypertrophy, etc.).

The usual dose is 25 mg. 4 times a day orally or the same effect is obtained by use of sustained release tablets (Gradumets) of 50 mg. twice a day. It is supplied in 25-mg. plain tablets or in sustained-action tablets of 50 and 75 mg.

Mepiperphenidol Bromide, Darstine® Bromide, 1-(3-Hydroxy-5-methyl-4-phenylhexyl)-1-methylpiperidinum Bromide. The preparation of this compound may be effected by the synthesis described by Schultz.[57] It occurs as a white, crystalline compound soluble in water (63.9:100). A 5 per cent solution has a pH of about 6.

This drug is a potent anticholinergic agent and is claimed to have a selective action toward the gastrointestinal tract. According to Kirsner and Palmer,[24] the gastric antisecretory effect is not as great as that of certain other anticholinergics. However, McCarthy and his co-workers[58] have shown that the antisecretory effect of mepiperphenidol is not due entirely to its anticholinergic effect, inasmuch as inhibitory effect is observed in the denervated gastric pouch as well as in the innervated one. It is a ganglionic-blocking agent and is said to be approximately equal to tetraethylammonium chloride in this respect. Side-effects are chiefly atropinelike; however, they are much less frequent.

It is promoted as an adjunct in the treatment of peptic ulcer to inhibit the motility of the gastrointestinal tract, which is believed by some to be the cause of ulcer pain. It is used also for other, less specific spasms of gastrointestinal origin. Also, it is used in hyperhidrosis and as an antisialogogue.

The dose is from 50 to 100 mg. 3 times daily, with the final dosage schedule depending on the reaction of the patient. It is supplied in the form of 50-mg. tablets.

Procyclidine Hydrochloride, Kemadrin®, 1-Cyclohexyl-1-phenyl-3-pyrrolidino-1-propanol Hydrochloride. This compound is prepared by the method of Adamson[59] or Bottorff[60] as described in the patent literature. It occurs as white crystals which are moderately soluble in water (3:100). It is more soluble in alcohol or chloroform and is almost insoluble in ether.

Although procyclidine, introduced in 1956, is an effective peripheral anticholinergic and, indeed, is used for peripheral effects similarly to its methochloride (see tricyclamol chloride), its clinical usefulness lies in its ability to relieve spasticity of voluntary muscle by its central action. Therefore, it has been employed with success in the treatment of Parkinson's syndrome.[61] It is said to be as effective as cycrimine and trihexyphenidyl and is used for reduction of muscle rigidity in the postencephalitic, the arteriosclerotic and the idiopathic types of the disease. Its effect on tremor is not predictable and probably should be supplemented by combination with other similar drugs.

The toxicity of the drug is low, but side-effects are noticeable when the dosage is high. At therapeutic dosage levels dry mouth is the most common side-effect. The same care should be exercised with this drug as with all other anticholinergics when administered to patients with glaucoma, tachycardia or prostatic hypertrophy.

The oral dosage is variable and depends on the age of the patient and the cause of disease. The usual initial dose is 7.5 mg. daily administered in 3 divided doses after meals. The dose may be increased to 15 to 30 mg. (in some cases 45 to 60 mg.) per day without undue side-effects. The drug is marketed in the form of 5-mg. tablets.

Thihexinol Methylbromide N.F., [4-(Hydroxydi-2-thienylmethyl)cyclohexyl]-trimethylammonium Bromide. The preparation of this compound is described by Villani[62] in the patent literature. It occurs as a white to creamy white, crystalline powder which is sparingly soluble in water and in chloroform. It is soluble in alcohol but is only very slightly soluble in acetone and ether.

This anticholinergic agent appears to find use mainly in combination with a hydrophilic agent known as polycarbophil which the manufacturer claims to be useful for the control of excessive fecal fluidity and too frequent evacuation. The anticholinergic, through an antispasmodic action, reduces the frequency of evacuation and the polycarbophil is claimed to have an "extraordinary capacity" for free fecal water. Uncontrolled tests of the combination suggested that it had some value in acute diarrheas and in the irritable-colon syndrome. However, diarrheas of an organic

nature and those resulting from acute bacterial or viral causes were much less responsive. Conservative opinion[63] suggests that although the combination, known as Sorboquel®, may be useful in mild diarrheas and in the irritable-colon syndrome, even here it is doubtful that the effect in the latter is greater than that achieved with placebos or sedatives. The combination product is marketed in tablet form containing 15 mg. of the anticholinergic and is obtainable either with or without neomycin (150 mg.). The combination without neomycin is given in a dose of 1 tablet 4 times daily; the neomycin-containing product is to be taken in doses of 4 to 6 tablets daily.

The usual contraindications for anticholinergics, such as glaucoma, bladder neck obstruction, prostatic hypertrophy, etc., are to be observed.

Category—anticholinergic.

Usual dose—15 mg. 4 times a day.

Tricyclamol Chloride, Elorine® Chloride, 1-Cyclohexyl-1-phenyl-3-pyrrolidino-1-propanol Methylchloride. The procedure for preparing the aminoalcohol from which this drug is derived is exactly the same as that utilized for the synthesis of procyclidine (q.v.). The methochloride is prepared readily from the aminoalcohol by treating the base with an equimolecular amount of methyl chloride.

Pharmacologically, the drug is a parasympatholytic, and by blocking vagal stimulation of the gastrointestinal tract it effectively reduces the motility of the tract. Although it has an antisecretory action, it is not as effective in reducing gastric acidity as, for example, propantheline or epoxymethamine tropate.[21] Tricyclamol chloride is used for the treatment of peptic ulcer, irritable colon, colitis and various spastic intestinal conditions. It has an advantage over some of the other drugs used in these conditions in that it apparently may be administered over relatively long periods of time without untoward reactions. The drug has approximately the same toxicity as atropine but does not have the central excitatory properties of the latter.

The dosage of this drug is variable, ranging between 50 and 500 mg. although 100 to 150 mg. is an average dose. Usually, patients are started on 50 mg. 4 times a day and the dose increased until satisfactory results are obtained. If indicated, the dose may be given 6 times daily. The side-effects of the drug are atropinelike, with dryness of the mouth and blurring of the vision predominating. It is supplied in the form of 50- and 100 mg. capsules and 50-mg tablets.

Tridihexethyl Chloride N.F., Pathilon® Chloride, (3-Cyclohexyl-3-hydroxy-3-phenylpropyl)triethylammonium Chloride. The preparation of this compound as the corresponding bromide is described by Denton and Lawson.[64] It occurs in the form of a white bitter, crystalline powder possessing a characteristic odor. The compound is freely soluble in water and alcohol, the aqueous solutions being nearly neutral in reaction.

Although this drug, introduced in 1958, has ganglion-blocking activity, it is said that its peripheral atropinelike activity predominates; therefore, its therapeutic application has been based on the latter activity. It possesses the antispasmodic and the antisecretory activities characteristic of this group but, because of its quaternary character, is valueless in the Parkinson syndrome.

The drug is useful for adjunctive therapy in a wide variety of gastrointestinal diseases such as peptic ulcer, gastric hyperacidity and hypermotility, spastic conditions such as spastic colon, functional diarrhea, pylorospasm and other related conditions. Because its action is predominantly antisecretory it is most effective in gastric hypersecretion rather than in hypermotility and spasm. It is best administered intravenously for the latter conditions.

The side-effects usually found with effective anticholinergic therapy occur with the use of this drug. These are dryness of mouth, mydriasis, etc. As with other anticholinergics, care should be exercised when administering the drug in glaucomatous conditions, cardiac decompensation and coronary insufficiency. It is contraindicated in patients with obstruction at the bladder neck, prostatic hypertrophy, stenosing gastric and duodenal ulcers or pyloric or duodenal obstruction.

The drug may be administered orally or parenterally. Oral therapy is preferable. The drug is supplied in 25-mg. tablets and as powder for injection (10 mg. in 1 ml.).

Category—anticholinergic.

Usual dose—25 mg. 3 times a day and 50 mg. at bedtime.

Usual dose range—25 to 75 mg. up to 4 times a day.

OCCURRENCE

Tridihexethyl Chloride Tablets N.F.

Trihexyphenidyl Hydrochloride U.S.P., Artane® Hydrochloride, Tremin® Hydrochloride, Pipanol®, α-Cyclohexyl-α-phenyl-1-piperidinepropanol Hydrochloride. This compound was synthesized by Denton and his co-workers.[65] It occurs as a white, odorless, crystalline compound that is not very soluble in water (1:100). It is more soluble in alcohol (6:100) and chloroform (5:100) but only slightly soluble in ether and benzene. The pH of a 1 per cent aqueous solution is about 5.5 to 6.0.

Introduced in 1949, it is approximately one half as active as atropine as an antispasmodic, but it has milder side-effects, such as mydriasis, drying of secretions and cardioacceleration. It has a good margin of safety, although it is about as toxic as atropine. It has found a place in the treatment of parkinsonism and is claimed also to provide some measure of relief from the mental depression often associated with this condition. However, it does exhibit some of the side-effects typical of the belladonna-type preparation, although it is said that these often may be eliminated by adjusting the dose carefully. According to Doshay and Constable,[46] trihexyphenidyl is a superior agent in all types of parkinsonism, and its most striking effect is the reduction of rigidity.

The drug is administered orally either 3 or 4 times a day near mealtimes. Usually, the beginning of treatment is carried out with a small (1 mg.) dose, followed by 2 mg. the next day. Thereafter, the dose is raised by 2-mg. increments until a total daily dosage of 6 to 10 mg. is being taken. However, doses up to 15 mg. per day may be required in the postencephalitic group. It is available in tablets or as an elixir.

Category—anticholinergic.

Usual dose—2 mg. 3 to 4 times daily.
Usual dose range—1 to 15 mg. daily.

OCCURRENCE
Trihexyphenidyl Hydrochloride
Elixir U.S.P.
Trihexyphenidyl Hydrochloride
Tablets U.S.P.

AMINOAMIDES

The aminoamide type of anticholinergic, from a structural standpoint, represents the same type of molecule as the aminoalcohol group with the important exception that the polar amide group replaces the corresponding polar hydroxyl group. Aminoamides retain the same bulky structural features as are found at one end of the molecule or the other in all of the active anticholinergics. One of the first to be introduced was aminopentamide sulfate (L) which was soon followed by ambutonium bromide (LI) and isopropamide iodide (LII). However, the last two possess the additional quaternary ammonium feature.

Another amide-type structure is that of tropicamide (LIII), formerly known as bistropamide, a compound having some of the atropine features.

PRODUCTS

Aminopentamide Sulfate, Centrine®, 4-Dimethylamino-2,2-diphenylvaleramide Sulfate. This compound may be prepared according to the method of Cheney.[66] It occurs as a white, odorless, crystalline powder. It is freely soluble in water as well as in alcohol, and the aqueous solutions are quite acidic (2.5% solution has pH 1.3 to 2.2).

This drug was introduced in 1953. The pharmacologic study was carried out by Hoekstra.[67] It was ascertained that the drug was about one half as active as atropine and one fifth as active as papaverine, with less effect on the eye and the salivary glands. However, its action is predominantly atropinelike, and because of this it decreases tone and motility of the gastrointestinal tract, decreases gastric secretion and also causes mydriasis and dry mouth.

It is used in the treatment of duodenal or gastric ulcer, hypertrophic gastritis and pylorospasm. In common with most anticholinergics it is used primarily as an adjunct to other treatments.

The drug is administered orally to adults in a usual initial dose of 0.5 mg. 3 or 4 times daily but with the proviso that the dosage be adjusted to the individual, based on his response and the severity of side-effects. Dosage forms available are a solution containing 0.7 mg. per ml. for oral use and 0.5-mg. tablets.

Ambutonium Bromide, (3-Carbamoyl-3,-3-diphenylpropyl)ethyldimethylammonium Bromide, 4-Dimethylamino-2,2-diphenylbutyramide Ethyl Bromide. This compound may be prepared by the method of Janssen *et al.*[68] and occurs as an odorless, white or nearly white, crystalline powder with a bitter taste. It is soluble in water (1 g./3 ml. at 25°) to give a slightly alkaline solution. A 1 per cent solution has a pH of 8.0 to 8.5.

According to the report by Judge and his co-workers[69] concerning the pharmacology of this drug, it was noted that 10 to 25 mg. orally significantly reduced gastric acidity and secretory volume without significant side-effects. They felt that it was superior to atropine with respect to its effects on both gastric motility and gastric secretion. In spite of the favorable reports, the manufacturer cautions that the drug may have side-effects similar to any other anticholinergic, and it is necessary to adjust the dose carefully to avoid them. The drug is indicated for relief of hypersecretion and hypermotility accompanying gastric and duodenal ulcer as well as for the other common complaints of the gastrointestinal tract treated with anticholinergics. It is used in combination with aluminum hydroxide gel.

Isopropamide Iodide N.F., Darbid®, 3-Carbamoyl-3,3-diphenylpropyl)diisopropylmethylammonium Iodide. This compound may be made according to the method of Janssen and his co-workers.[68] It occurs as a white to pale yellow crystalline powder with a bitter taste and is only sparingly soluble in water but is freely soluble in chloroform and alcohol.

This drug, introduced in 1957, is a potent anticholinergic producing atropinelike effects peripherally. Even with its quaternary nature it does not cause sympathetic blockade at the ganglionic level except in high-level dosage. Its principal distinguishing feature is its long duration of action. It is said that a single dose can provide antispasmodic and antisecretory effects for as long as 12 hours.

It is used as adjunctive therapy in the treatment of peptic ulcer and other conditions of the gastrointestinal tract associated with hypermotility and hyperacidity. It has the usual side-effects of anticholinergics (dryness of mouth, mydriasis, difficult urination) and is contraindicated in glaucoma, prostatic hypertrophy, etc.

Category—anticholinergic.

Usual dose—5 mg. twice a day.

Usual dose range—10 mg. to 20 mg. daily.

OCCURRENCE

Isopropamide Iodide Tablets N.F.

Tropicamide U.S.P., Mydriacyl®, *N*-Ethyl-2-phenyl-*N*-(4-pyridylmethyl)hydracryl-amide. The preparation of this compound is described in the patent literature.[70] It occurs as a white or practically white, crystalline powder which is practically odorless. It is only slightly soluble in water but is freely soluble in chloroform and in solutions of strong acids. The pH of ophthalmic solutions ranges between 4.0 and 5.0, the acidity being achieved with nitric acid.

This drug is an effective anticholinergic for ophthalmic use where mydriasis is produced by relaxation of the sphincter muscle of the iris, allowing the adrenergic innervation of the radial muscle to dilate the pupil. Its maximum effect is achieved in about 20 to 25 minutes and lasts for about 20 minutes, with complete recovery being noted in about 6 hours. Its action is more rapid in onset and wears off more rapidly than that of most other mydriatics. To achieve mydriasis either the 0.5 or 1.0 per cent concentration may be used, although cycloplegia is achieved only with the stronger solution. Its uses are much the same as those described in general for mydriatics (see p. 515), but opinions differ as to whether the drug is as effective as homatropine, for example, in achieving cycloplegia. For mydriatic use, however, in ex-amination of the fundus and treatment of acute iritis, iridocyclitis and keratitis it is quite adequate, and, because of its shorter duration of action, it is less prone to initiate a rise in intraocular pressure than are the more potent longer lasting drugs. However, as with other mydriatics, pupil dilation can lead to increased intraocular pressure. In common with other mydriatics it is contraindicated in cases of glaucoma, either known or suspected, and should not be used in the presence of a shallow anterior chamber. Thus far, allergic reactions and/or ocular damage have not been observed with this drug.

Category—anticholinergic (ophthalmic).

For external use—topically, 0.1 ml. of a 0.5 to 1 per cent solution to the conjunctiva, repeated in 5 minutes.

OCCURRENCE

Tropicamide Ophthalmic Solution U.S.P.

DIAMINES

The diamines in this classification number only two, both being derivatives of phenothiazine and both being utilized for Parkinson's syndrome. An inspection of their formulae (LIV, LV) shows that they are closely related to certain of the anti-histaminics (q.v.) and they do, indeed, have an antihistaminic activity. Although they have peripheral effects of an anticholinergic nature, their usefulness lies in the fact that they can bring about depression of the central nervous system. The ability to reach this area of the body is a consequence of their tertiary amine nature which permits them to exist in an undissociated form (see p. 527).

PRODUCTS

Diethazine Hydrochloride, Diparcol®, 10-(2-Diethylaminoethyl)phenothiazine. This drug is prepared according to the method of Charpentier and others.[71] It occurs as a white, crystalline compound with a burning taste and imparts a numbness to the tongue. It is water soluble (1:5) as well as soluble in alcohol (1:6) and chloroform

(1:5) but is insoluble in ether. The pH of a 10 per cent aqueous solution is 5.0 to 5.3.

This drug is said to possess both parasympatholytic and sympatholytic properties and is actually an antihistaminic. It is suggested for use in parkinsonism but has not met with too enthusiastic a reception, if the reports of Doshay and Constable,[72] Schwab[73] and Raffle[74] are any indication. It appears to be unsuited to all forms of the disease except the mild postencephalitic and the early idiopathic forms. Its toxicity militates against its widespread use. Dry mouth occurs with ordinary therapeutic doses. It is marketed abroad, but it has not yet been accepted for use in the U.S.

Ethopropazine Hydrochloride, Parsidol®, 10 - (2 - Diethylaminopropyl) phenothiazine Hydrochloride. The compound is prepared in a number of ways, among which is the patented method of Berg and Ashley.[75] It occurs as a white crystalline compound with a poor solubility in water at 20° C. (1:400) but greatly increased solubility at 40° C. (1:20). It is soluble in ethanol and chloroform but almost insoluble in ether, benzene and acetone. The pH of an aqueous solution is about 5.8.

This close relative of diethazine (q.v.) was introduced to therapy in 1954. It has similar pharmacologic activities and has been found to be especially useful in the symptomatic treatment of parkinsonism. In this capacity it has value in controlling rigidity, and it also has a favorable effect on tremor, sialorrhea and oculogyric crises. It is often used in conjunction with other parkinsonlytics for complementary activity.

Side-effects are common with this drug but not usually severe. Drowsiness and dizziness are the most common side-effects at ordinary dosage levels, and as the dose increases xerostomia, mydriasis, etc., become evident. It is contraindicated in conditions such as glaucoma because of its mydriatic effect.

The daily oral dosage for optimal effect may vary from 100 mg. to 1 g., although it is seldom necessary to give more than 700 mg. per day. The usual initial dose is 10 mg. 4 times a day with an increase of 10 mg. per dose every 2 or 3 days until an optimum dosage level is found. It is supplied in 10-, 50- and 100-mg. tablets.

AMINES, MISCELLANEOUS

The miscellaneous group contains only one useful compound (LVI). This compound has the bulky group characteristic of the usual anticholinergic molecule.

Diphemanil Methylsulfate N.F., Prantal® Methylsulfate, 4-Diphenylmethylene-1,1-dimethylpiperidinium Methylsulfate. This compound may be prepared by two alternative syntheses as outlined by Sperber and co-workers.[76] It was introduced in 1951.

The drug is a white, crystalline, odorless compound that is sparingly soluble in water (50 mg./ml.), alcohol and chloroform. The pH of a 1 per cent aqueous solution is between 4.0 and 6.0.

The methylsulfate radical was chosen as the best anion because the chloride is hygroscopic and because the bromide and iodide ions have exhibited toxic manifestations under clinical usage.

As mentioned previously, it is a potent cholinergic blocking agent. In the usual dosage range it acts as an effective parasympatholytic by blocking nerve impulses at the parasympathetic ganglia but does not invoke a sympathetic ganglionic blockade. It is claimed to be highly specific in its action upon those innervations that have to do with gastric secretion and motility. Although this drug is capable of producing atropinelike side-effects, these are not a problem because in the doses used they are reported to occur very rarely. The highly specific nature of its action on the gastric functions makes it useful in the treatment of peptic ulcer, and its lack of atropinelike effects makes this use much less distressing than is the case with some of the other similarly used drugs. In addition to its action in gastric hypermotility, it is valuable in hyperhydrosis in low doses (50 mg. twice daily) or topically.

The drug is not well absorbed from the gastrointestinal tract, particularly in the presence of food, so it is desirable to administer the oral doses between meals. In addition to the regular tablet form the drug also is supplied in a so-called "repeat action" tablet that has an enteric coated tablet embedded in an ordinary tablet and gives about 8 hours of activity. It also is supplied as a parenteral injection.

Category—anticholinergic.

Usual dose—oral, 100 mg. every 4 to 6 hours; intramuscular and subcutaneous, 25 mg. 4 times daily.

Usual dose range—oral, 50 to 200 mg.

OCCURRENCE

Diphemanil Methylsulfate Injection N.F.
Diphemanil Methylsulfate Tablets N.F.

PAPAVERINE AND RELATED COMPOUNDS

Previously, it was pointed out that papaverine is in actuality not a parasympatholytic. However, it exerts an antispasmodic effect and for that reason is customarily considered together with the solanaceous alkaloids.

Modern pharmacologic technics have shown that papaverine is an antagonist of the noncompetitive type (see p. 284) in contrast with the competitive type of antagonism shown by atropine and its congeners. A noncompetitive antagonism indicates that papaverine attaches to receptors other than those recognized as being involved with the natural agonist, acetylcholine. Thus, it interferes with the mechanism of muscle contraction somewhere other than the acetylcholine receptor. However it does not inactivate the contractile elements of the muscle, because it is still possible to obtain a response after papaverine under certain conditions. Perhaps a more precise way of expressing the spasmolysis induced by a drug such as papaverine is that it does not interfere with the induction of the stimulus but rather with the response in the effector system. Because of its nonspecific action (i.e., with respect to the acetylcholine receptor) it is often called a nonspecific antagonist. This is sometimes referred to as a musculotropic type of spasmolysis, in contrast with the so-called neurotropic action of atropine and its congeners. Such nonspecific antagonists act against a great variety of smooth muscle spasmogens (e.g., para-

sympathomimetics, sympathomimetics, histamine, barium chloride, etc.). Thus, regardless of the type of smooth muscle, papaverine acts as a spasmolytic although its effectiveness is greater in some muscles than in others. This is evident in its principal application as a coronary blood vessel relaxant rather than as a general spasmolytic. As a consequence of its noncompetitive mode of action one expects to observe no atropinelike side effects from this type of spasmolytic, an expectation borne out by experience. The absence of such effects is a desirable characteristic of papaverine-type compounds, but, unfortunately, these compounds do not compare in potency to the atropine congeners.

Papaverine (see formula on page 559) is the principal naturally occurring member of this group that is of any therapeutic consequence as an antispasmodic.

Papaverine Hydrochloride N.F. This alkaloid was isolated first by Merck (1848) from opium, in which it occurs to the extent of about 1 per cent. Its structure was elucidated by the classic researches of Goldschmiedt, and its synthesis was effected first by Pictet and Gam in 1909.

Previous to World War II, papaverine had been obtained in sufficient quantities from natural sources. However, as a result of the war, the United States found itself early in 1942 without a source of opium and, therefore, of papaverine. Consequently, the commercial synthesis of papaverine took on a new significance, and methods soon were developed to synthesize the alkaloid on a large scale.[77]

Papaverine itself occurs as an optically inactive, white, crystalline powder. It possesses one basic nitrogen and forms salts quite readily. The most important salt is the hydrochloride, which is official. The hydrochloride occurs as white crystals or as a crystalline, white powder. It is not optically active, is odorless and has a slightly bitter taste. The compound is soluble in water (1:30), alcohol (1:120) or chloroform. It is not soluble in ether. Aqueous solutions are acid to litmus and may be sterilized by autoclaving. Unless properly handled and stored, extemporaneous solutions of papaverine salts deteriorate rapidly.

Because of the antispasmodic action of papaverine on blood vessels, it has become extremely valuable for relieving the arterial spasm associated with acute vascular occlusion. It is useful in the treatment of peripheral, coronary and pulmonary arterial occlusions. Administration of an antispasmodic is predicated on the concept that the lodgement of an embolus causes an intense reflex vasospasm. This vasospasm affects not only the artery involved but also the surrounding blood vessels. Relief of this neighboring vasospasm is imperative in order to prevent damage to these vessels and to limit the area of ischemia. Thus, it appears to increase collateral circulation in the affected area rather than act on the occluded vessel.

Other than its antispasmodic action on the vascular system, it is used for bronchial spasm and visceral spasm. In the latter type of spasm, it is not advisable to administer morphine simultaneously because it opposes the relaxing action of papaverine.

Category—smooth muscle relaxant.

Usual dose—oral, 100 mg.; intramuscular, 30 mg.

Usual dose range—oral, 60 to 200 mg.; intramuscular, 30 to 60 mg.

OCCURRENCE

Papaverine Hydrochloride Injection N.F.
Papaverine Hydrochloride Tablets N.F.

Because papaverine is a musculotropic drug, it has provided the starting point for synthetic analogs in which it has been hoped that a neurotropic activity could be combined with its musculotropic action. This combination of activities would be desirable, if possible, without the introduction of any atropinelike side-effects. Comparing the results of this research with those of atropine analogs, it appears that the use of the latter has proved to be more successful. However, there are a number of important developments in this field, and several active drugs have been developed as a result of these research activities.

It will be of advantage to consider the developments in this field of activity as falling into two general categories: (1) changes in the peripheral groups of papaverine without extensive nuclear changes, e.g., replacing methoxyl groups, altering the methylene group, nuclear alkylation; (2) changes involving reduction and opening of the nitrogen ring.

1. It is apparent that one of the especially easily altered peripheral groups of papaverine is the methoxyl group, of which there are 4. These have been changed to various alkoxyl groups, among which are the ethoxyl and the methylenedioxy groups. This research has shown that, as far as papaverine itself is concerned, the most active compound is one in which there are 4 ethoxyls replacing the 4 methoxyls. This the commercially available compound, ethaverine (Perparin) (I)* which has 3 times the activity of papaverine with only one third to one half of its toxicity. However, no definite statement can be made as to the relative desirabilities of the methoxyl, the ethoxyl and the methylenedioxy groups, inasmuch as they vary in the active compounds. Thus, ethaverine has ethoxyl groups, eupaverine (II) has methylenedioxy groups, and dioxyline (III) has a combination of 1 ethoxyl group with 3 methoxyl groups. It is not entirely certain that the alkoxyl groups are necessary for activity, although they seem to be present in most of the accepted compounds. Activity is known to reside in both 1-phenyl-3-methyl isoquinoline (IV) and 1-benzyl-3-methyl isoquinoline (V) but, on the other hand, spasmo*genic* properties are found in 1-phenyl-3-methyl-6,7-methylenedioxy-3,4-dihydro-isoquinoline (VI).

Alterations of the methylene group in the benzyl residue have not been of any outstanding importance. If anything, alteration of this group, as by introducing a hydroxyl or a carbonyl group, makes for more toxic compounds as exemplified by papaverinol (VII) and papaveraldine (VIII). Indeed, the presence of the methylene group is not even essential to high activity. This was shown by Kreitmair[78] who found that, qualitatively, the muscle-relaxing action was the same with or without a $-CH_2-$ group, although if a second $-CH_2-$ group was inserted the action was reversed, (i.e., stimulant). Quantitatively, the two com-

* The Roman numerals following the names of various compounds in this discussion refer to the corresponding numerals in Table 54.

TABLE 54

COMPOUND	STRUCTURE			NAME
	R_1	R_2	R_3	
I	$-OC_2H_5$	$-H$	$-CH_2-$ (3,4-diethoxyphenyl)	Ethaverine
II	methylenedioxy ($O-CH_2-O$)	$-CH_3$	$-CH_2-$ (methylenedioxyphenyl)	Eupaverine
III	$-OCH_3$	$-CH_3$	$-CH_2-$ (ethoxy-methoxyphenyl)	Dioxyline
IV	$-H$	$-CH_3$	phenyl	1-Phenyl-3-methyl-isoquinoline
V	$-H$	$-CH_3$	$-CH_2-$ phenyl	1-Benzyl-3-methyl-isoquinoline
VI*	methylenedioxy ($O-CH_2-O$)	$-CH_3$	phenyl	1-Phenyl-3-methyl-6, 7-methylenedioxy-3, 4-dihydroisoquinoline
VII	$-OCH_3$	$-H$	$-\overset{OH}{\underset{H}{C}}-$ (dimethoxyphenyl)	Papaverinol
VIII	$-OCH_3$	$-H$	$-\overset{O}{C}-$ (dimethoxyphenyl)	Papaveraldine
IX	methylenedioxy ($O-CH_2-O$)	$-CH_3$	(methylenedioxyphenyl)	Neupaverine
X	methylenedioxy ($O-CH_2-O$)	$-CH_3$	phenyl	1-Phenyl-3-methyl-6, 7-methylenedioxyisoquinoline

* This differs from all others in having the double bond at the 3 to 4 position in the isoquinoline nucleus saturated.

TABLE 54 (*Continued*)

COMPOUND	STRUCTURE			NAME
	R₁	R₂	R₃	
XI	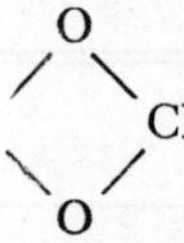 O–CH₂–O	—CH₃	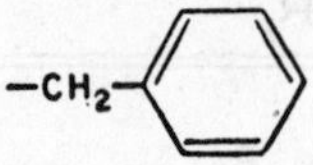 —CH₂–⬡	1-Benzyl-3-methyl-6, 7-methyl-enedioxyisoquinoline
XII	 O–CH₂–O	—CH₃	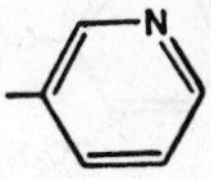 (pyridyl ring)	1-(β-Pyridyl)-3-methyl-6, 7-methylenedioxyisoquinoline
XIII	 O–CH₂–O	—CH₃	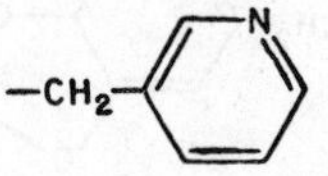 —CH₂–(pyridyl ring)	1-(β-Picolyl)-3-methyl-6, 7-methylenedioxyisoquinoline
XIV	—OCH₃	—H	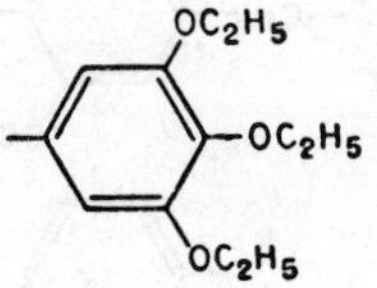 –⬡(OC₂H₅, OC₂H₅, OC₂H₅)	Octaverine
XV	—OCH₃	—CH₃	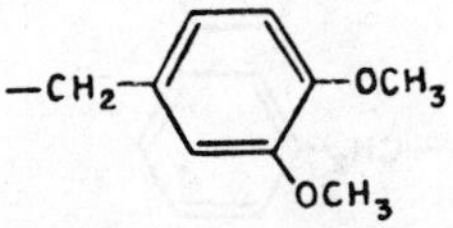 —CH₂–⬡(OCH₃, OCH₃)	3-Methylpapaverine

pounds (IX and X) without the —CH₂— group were about 10 times as active as the corresponding compounds (II and XI) with such a group. Repetition of his experiments with 1-(β-pyridyl)-3-methyl-6,7-methylene-dioxyisoquinoline (XII) and 1-phenyl-3-methylisoquinoline (IV) showed that they also had a grater activity than the similarly constructed (−CH₂−)-containing compounds (XIII and V) and a lesser toxicity. Fodor[79] substantiated the observation concerning the nonessential nature of the methylene group, as illustrated by the adoption of octaverine (XIV) as an active antispasmodic.

The reader will note that in the examples of active compounds the majority bear a methyl group in position 3 of the isoquinoline nucleus. This feature stems from the work of Kreitmair and also of Fodor, certain aspects of which have already been discussed. One of Fodor's compounds, the 3-methyl homolog of papaverine (XV), was

the most potent of his series. An example of an active compound in this series is dioxyline (III) which is the 3-methyl homolog of papaverine and possesses an ethoxyl instead of a methoxyl group in position 4 of the benzyl group.

2. Blicke[80] has pointed out that compounds based on a close similarity to papaverine very likely would have some of the defects peculiar to papaverine. Among these defects are the low water solubility of the salts, the tendency of the salts to produce acidic solutions by hydrolysis because of a feebly basic nitrogen and poor absorption of the compounds because of precipitation of the free bases due to hydrolysis. These factors are, of course, of greater importance in parenteral than in oral medication.

Recognizing these limitations and having observed that tetrahydropapaverine showed qualitatively the same type of action as papaverine, workers began to investigate the open chain models of tetrahydropapaverine. Inspection of the following formulas shows the logical progression from papaverine to the *bis-β*-phenylethylamine type of compound.

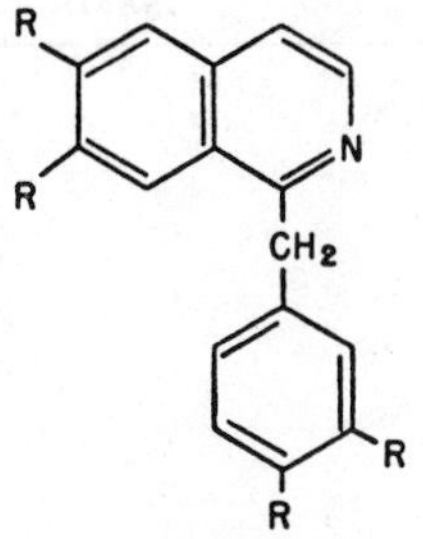

Papaverine
(R=OCH₃)

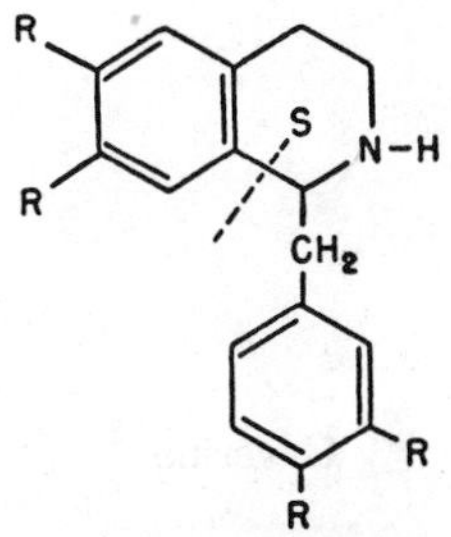

Tetrahydropapaverine
(R=OCH₃)

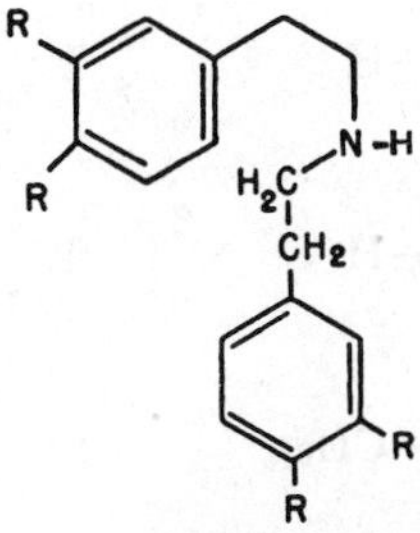

Bis-β-phenyl-
ethylamine
(R=H)

Rosenmund and his co-workers[81, 82, 83] have been among the most active in carrying out this type of permutation and early demonstrated that *bis-β*-phenylethylamine (A)* itself had a slight but unmistakable activity. From this point, it has been natural for many other workers to extend the studies and develop this activity into a more potent one.

It was found that activity was retained by compounds in which the phenyl rings were substituted with alkoxy groups (B), although it was apparent that these groups were not necessary for maximum activity. Methylation of the phenyl groups in the para position (C) was found to be advantageous in potentiating the activity. Re-

* The letters following certain compounds in this discussion identify these compounds as those in Table 55 (p. 560).

placement of the phenyl groups by one (D) or two α-thienyl groups (E) or by cyclohexyl groups (F) did not result in loss of activity. The latter type of compounds is exemplified in *bis-*(β-cyclohexylethyl)methylamine hydrochloride (G) which formerly was marketed under the trade name of Cyverine. Alkylation of the nitrogen increased the water solubility as well as the physiologic activity. In this particular series, the *n*-hexyl compound (H) was found to exhibit a greater activity than papaverine. Substitution of alkyl (I), aralkyl (J), aryl (K) or alkoxyl groups (L) on the carbon atoms adjacent to the nitrogen in *bis-β*-phenylethylamine produced active compounds also. Finally, one of the important findings was that the optimum chain length was not 2 carbons (ethyl) but 3 carbons (propyl). From a study of a great number of these compounds, *bis-*(γ-phenylpropyl)-ethylamine (M) was selected as the best all around compound. It has been recognized by the *N.F.* as Alverine and it is said to be 2.3 times as active as papaverine.

As early as 1933, it was known that both saturated and unsaturated acyclic amines had spasmolytic properties. In addition, they had sympathomimetic properties. The best compound in this group was 2-methyl-amino-6-methyl-5-heptene which is commercially obtainable under the generic name of isometheptene (N) (see p. 491). According to Issekutz,[84] this compound has a direct paralyzing effect on smooth muscle of the intestine in a manner similar to papaverine and also stimulates sympathetic nerve endings to thus inhibit intestinal functions.

In conclusion, it is well to point out that some of the sympathomimetic amines possess specialized antispasmodic properties toward the bronchi and are used as bronchodilators. Among this group, we find ephedrine, isoproterenol and epinephrine. However, the mechanism of action here is not a parasympatholytic or muscle-depressant action but may be characterized as an overstimulation of the sympathetic system which simulates in many ways the paralysis of the parasympathetic system.

TABLE 55

$$R_1(CH_2)_nCH-N-CH(CH_2)_nR_2$$

with R_3 on the left CH, R_3 on the right CH, and R_4 on the central N.

COMPOUND	STRUCTURE R₁	R₂	R₃	R₄	n	COMMERCIAL NAME
A	C_6H_5-	C_6H_5-	$-H$	$-H$	1	
B	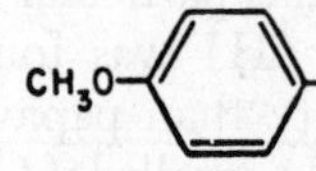 CH_3O- (phenyl)	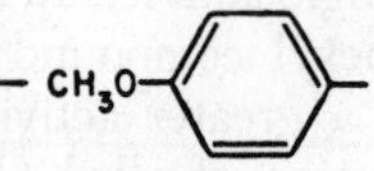 CH_3O- (phenyl)	$-H$	$-H$	1	
C	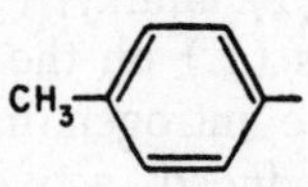 CH_3- (phenyl)	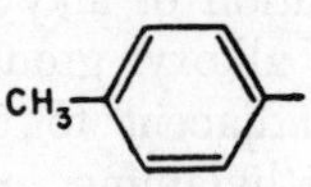 CH_3- (phenyl)	$-H$	$-H$	1	
D	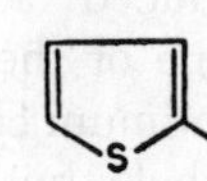 thienyl	C_6H_5-	$-H$	$-H$	1	
E	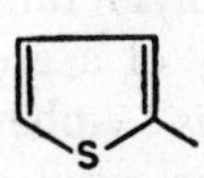 thienyl	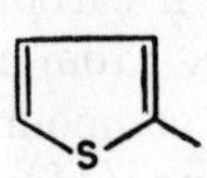 thienyl	$-H$	$-H$	1	
F	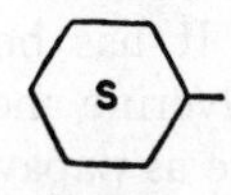 thienyl	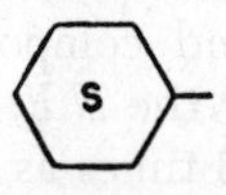 thienyl	$-H$	$-H$	1	
G	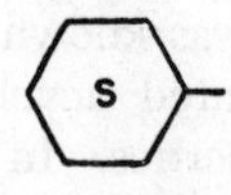 thienyl	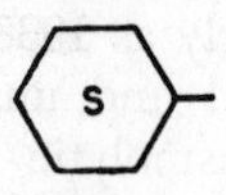 thienyl	$-H$	$-CH_3$	2	Cyverine
H	C_6H_5-	C_6H_5-	$-H$	$-C_6H_{13}-n$	1	
I	C_6H_5-	C_6H_5-	CH_3-, C_2H_5-, etc.	$-H$	1	
J	C_6H_5-	C_6H_5-	$-CH_2-$ 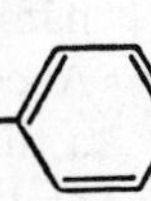 (phenyl)	$-H$	1	
K	C_6H_5-	C_6H_5-	$-C_6H_5$	$-H$	1	
L	C_6H_5-	C_6H_5-	$-OCH_3$	$-H$	1	
M	C_6H_5-	C_6H_5-	$-H$	$-C_2H_5$	2	Alverine
N			$(CH_3)_2C=CHCH_2CH_2CHCH_3$ with $-N(H)(CH_3)$			Octin

Ethaverine Hydrochloride, Isovex®, 6,7-Diethoxy-1-(3,4-diethoxybenzyl)isoquinoline. This well known derivative of papaverine is synthesized in exactly the same way as papaverine, but intermediates that bear ethoxyl groups instead of methoxyl groups are utilized.[85]

The hydrochloride, which has a melting point between 186° and 188°, is soluble to the extent of 1 g. in 40 ml. of water at room temperature. The aqueous solutions are acidic in reaction, with a 1 per cent solution having a pH of 3.6 and a 0.1 per cent solution having a pH of 4.6.

The pharmacologic action of ethaverine is quite similar to that of papaverine, although its effect is said to be longer in duration. It is used as an antispasmodic in doses of 30 to 60 mg. Its effect in angina pectoris appears to be somewhat questionable on the basis of results of Voyles and his co-workers[86] using a daily dose of 400 mg.

Dioxyline Phosphate, Paveril® Phosphate, 6,7-Dimethoxyl-1-(4′-ethoxy-3′-methoxybenzyl)-3-methylisoquinoline phosphate. This may be prepared according to the usual Bischler-Napieralski isoquinoline synthesis followed by dehydrogenation.[87]

This compound is related quite closely to papaverine and gives the same type of antispasmodic action as papaverine, with less toxicity. By virtue of the lesser toxicity, it can be given in larger doses than papaverine if desired, although usually the same dosage regimen can be followed as with the natural alkaloid.

The drug is useful for mitigating the reflex vasospasm that already has been described for papaverine (p. 556), during peripheral, pulmonary or coronary occlusion. The indications are the same as for papaverine.

The dosage is usually 200 mg. 3 or 4 times daily. It is supplied in the form of tablets (100 and 200 mg.) for oral use.

Alverine Citrate N.F., Spacolin®, Gamatran®, Profenil®, Proverine®, N-ethyl-3,3′-diphenyldipropylamine Citrate. This compound is prepared by the interaction of γ-phenylpropyl bromide (or chloride) with ethylamine in the presence of a base.[88]

As pointed out previously, alverine is 2.3 times more active as an antispasmodic than papaverine. It occurs as a white to off-white powder with a sweet odor and a slightly bitter taste. It is slightly soluble in water and in chloroform, sparingly soluble in alcohol and very slightly soluble in ether. The acute and the chronic toxicities of this compound are very low, and it appears that prolonged use has no deleterious effects. The drug has both an anticholinergic and an antibarium (musculotropic) action.

It is indicated in those conditions where it is desired to relieve smooth muscle spasms. Specifically, it is directed toward various kinds of spasms of the gastrointestinal tract, hyperemesis gravidarum, spasms of the ureter due to inflammation or gravel and also to circulatory spasms.

The tablets have a slight local anesthetic action and are to be swallowed and not chewed.

Category—anticholinergic.

Usual dosage—120 mg. 1 to 3 times a day.

OCCURRENCE

Alverine Citrate Tablets N.F.

Cholinergic Blocking Agents Acting at the Ganglionic Synapses of Both the Parasympathetic and the Sympathetic Nervous Systems

These compounds commonly are called "ganglionic blocking agents" in allusion to their ability to block transmission of impulses through the autonomic ganglia (sympathetic and parasympathetic). Although the gross effect of all blocking agents, i.e., failure of nervous transmission, is common to all types, this does not imply that the effect is necessarily achieved by the same mechanism in all cases. On the contrary, certain classifications arise when one considers the effects of these blocking agents on the electrical events in the ganglia

which are associated with nerve impulse transmission. Among others, Paton and Perry[89] have shown interesting electrical changes at the ganglia during impulse transmission. The receptor cell membrane, in common with membranes of other cells, is polarized (outside positive with respect to inside). They have demonstrated that the action of acetylcholine is to produce a temporary depolarization of this membrane (an effect known as end-plate potential) which causes a response by the cell. Such a depolarizing effect, very similar to that at ganglia, has been noted at the neuromuscular junction as well. Therefore, it is not unreasonable to expect the ganglionic blocking agents to be involved in one way or another with these electrical events at the ganglionic synapse. Van Rossum[90, 91] has reviewed the mechanisms of ganglionic synaptic transmission, the mode of action of ganglionic stimulants, and the mode of action of ganglionic blocking agents. He has conveniently classified the blocking agents in the following manner:

Depolarizing Ganglionic Blocking Agents. These blocking agents are actually ganglionic stimulants. Thus, in the case of nicotine, it is well known that small doses give an action similar to the natural neuroeffector, acetylcholine, an action known as the "nicotinic effect of acetylcholine." However, larger amounts of nicotine bring about a ganglionic block, characterized initially by depolarization followed by a typical competitive antagonism. In order to conduct nervous impulses the cell must be able to carry out a polarization and depolarization process, and if the depolarized condition is maintained without repolarization, it is obvious that no nerve conduction occurs. Acetylcholine, itself, in high concentrations will bring about an autoinhibition. There are a number of compounds which cause this type of ganglionic block but they are not of therapeutic significance. However, the remaining classes of ganglionic blocking agents have therapeutic utility.

Nondepolarizing Competitive Ganglionic Blocking Agents. Compounds in this class possess the necessary affinity to attach to the receptor sites that are specific for acetylcholine but lack the intrinsic activity necessary for impulse transmission, i.e., they are unable to effect depolarization of the cell. Under experimental conditions, in the presence of a fixed concentration of blocking agent of this type, a large enough concentration of acetylcholine can offset the blocking action by competing successfully for the specific receptors. When such a concentration of acetylcholine is administered to a ganglion preparation, it appears that the intrinsic activity of the acetylcholine is as great as it was when no antagonist was present, the only difference being in the larger concentration of acetylcholine required. It is evident, then, that such blocking agents are "competitive" with acetylcholine for the specific receptors involved and either the agonist or the antagonist can displace the other if present in sufficient concentration. Drugs falling into this class are tetraethylammonium salts, azamethonium, hexamethonium, and trimethaphan. Mecamylamine possesses a competitive component in its action but is also noncompetitive—a so-called "dual antagonist."

Nondepolarizing Noncompetitive Ganglionic Blocking Agents. These blocking agents produce their effect, not at the specific acetylcholine receptor site, but at some point further along the chain of events that is necessary for transmission of the nervous impulse. When the block has been imposed, increase of the concentration of acetylcholine has no effect, and, thus, apparently acetylcholine is not acting competitively with the blocking agent at the same receptors. Theoretically, a pure noncompetitive blocker should have a high specific affinity to the noncompetitive receptors in the ganglia, and it should have very low affinity for other cholinergic synapses, together with no intrinsic activity. Among the drugs that possess activity of this type are chlorisondamine chloride and trimethidinium sulfate. Mecamylamine, as mentioned before, has a noncompetitive component but is also competitive.

Finally, it may have occurred to the student that a rather obvious classification, i.e., specific ganglionic blocking, has been overlooked. Thus, one might expect to find such drugs as "parasympathetic ganglionic blockers" or "sympathetic ganglionic blockers." Such specificity of action toward the ganglia has not been widely studied, and there appear to be some discrepancies in the results thus far reported. On a limited number of drugs, Garrett[92] has shown that tetraethylammonium salts and azamethonium are nondiscriminating, whereas hexamethonium shows some selective sympathetic blocking action. None of the commonly used ganglionic blockers that was tested showed a selectivity toward the parasympathetic ganglia, although an experimental compound known as MG 624 [triethyl-(4-stilbenehydroxyethyl)ammonium iodide] showed such action.

The first ganglionic blocking agents employed in therapy were tetraethylammonium chloride and bromide (I).* Although one might assume that curariform activity would be a deterrent to their use, it has been shown that the curariform activity of the tetraethyl compound is less than 1 per cent that of the corresponding tetramethyl ammonium compound. A few years after the introduction of the tetraethyl ammonium compounds, Paton and Zaimis[93] investigated the usefulness of the *bis*-trimethyl-ammonium polymethylene salts:

+ $N(CH_3)_3$ | $(CH_2)_n$ $2Br^-$ | $N(CH_3)_3$ +	$n = 5$ or 6, active as ganglionic blockers (feeble curariform activity) $n = 9$ to 12, weak ganglionic blockers (strong curariform activity)

As shown above, their findings indicate that there is a critical distance of about 5 to 6 carbon atoms between the onium centers for good ganglion blocking action. Interestingly enough, the pentamethylene and the hexamethylene compounds are effective antidotes for counteracting the curare effect of the decamethylene compound. Hexamethonium (II), as the bromide and the chloride, emerged from this research as a clinically useful product. Pentolinium tartrate (III), another symmetric *bis*-quaternary with onium groups spaced 5 carbons apart, but with the important difference of incorporating the onium heads in a heterocyclic moiety, represents one of the most useful in the group of symmetric compounds largely replacing hexamethonium. Deviation from the symmetric arrangements resulted in other useful *bis*-quaternaries such as chlorisondamine chloride (IV) and trimethidinium methosulfate (V). Although all of these compounds were well absorbed and predictable in action following parenteral injection, this was not the case following oral administration, with unpredictable and erratic absorption being the rule. This poor absorption picture is largely due to the

* Compounds referred to by Roman numerals in this section are in Table 56 (p. 564).

completely ionic character of the products (see, however, p. 532). Consequently, parenteral administration of these compounds is usually desirable if predictable effects are to follow. Nevertheless, some of the newer ones are used orally in spite of somewhat erratic results. Trimethaphan camphorsulfonate (VI), a monosulfonium compound, bears some degree of similarity to the quaternary ammonium types because it, too, is a completely ionic compound. Although it produces a prompt ganglion blocking action on parenteral injection, its action is evanescent, and it is used only for a specialized purpose (q.v.). Almost simultaneously with the introduction of chlorisondamine (now removed from the market), announcement was made of the powerful ganglionic blocking action of mecamylamine (VII), a secondary amine *without* quaternary ammonium character. As expected, the latter compound showed uniform and predictable absorption from the gastrointestinal tract as well as a longer duration of action. The action was similar to that of hexamethonium.

Other drugs of a nonquaternary nature that show a marked ganglionic-blocking action but have not been marketed in this country are 1,2,2,6,6-pentamethylpiperidine (Pempidine, Perolysen, Tenormal) and 2,2,6,6-tetramethylpiperidine hydrochloride (M and B 4500). These are extremely potent drugs with the latter being about twice as potent as the former. Even with the discovery of these potent nonquaternaries there has been a persistent search among the quaternaries, particularly for effective hypotensive agents. Among the more recent of these is bretylium tosylate (see also p. 568), which is not properly a ganglionic blocker but rather a selective blocker of the peripheral sympathetic nervous system in which it selectively accumulates. It has been introduced abroad commercially as Darenthin but seems to lack the qualities necessary for long term hypotensive therapy in spite of its lack of activity on the parasympathetic system. Another onium salt that has seen clinical acceptance abroad is phenacyl homatropinium chloride (Trophenium), a powerful ganglionic blocker. A related compound is the 4-diphenylmethyl

TABLE 56. STRUCTURES OF GANGLIONIC BLOCKING AGENTS

COMPOUND	STRUCTURE	NAME
I	$(C_2H_5)_4 \overset{+}{N}$ X^-	Tetraethylammonium Chloride (X=Cl) " " Bromide (X=Br)
II	$(CH_3)_3\overset{+}{N}-(CH_2)_6-\overset{+}{N}(CH_3)_3$ $2X^-$	Hexamethonium Chloride (X=Cl) " " Bromide (X=Br)
III	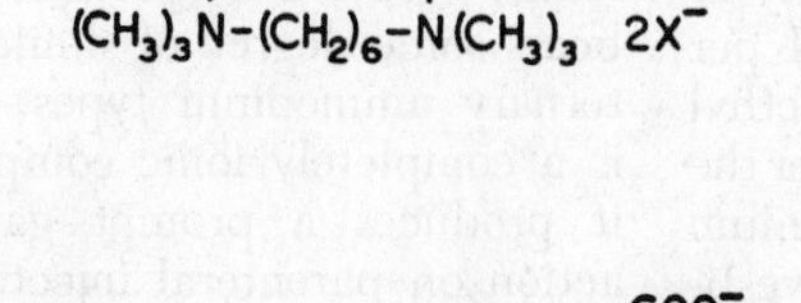 $2\begin{array}{l}COO^-\\(CHOH)_2\\COOH\end{array}$	Pentolinium Tartrate
IV	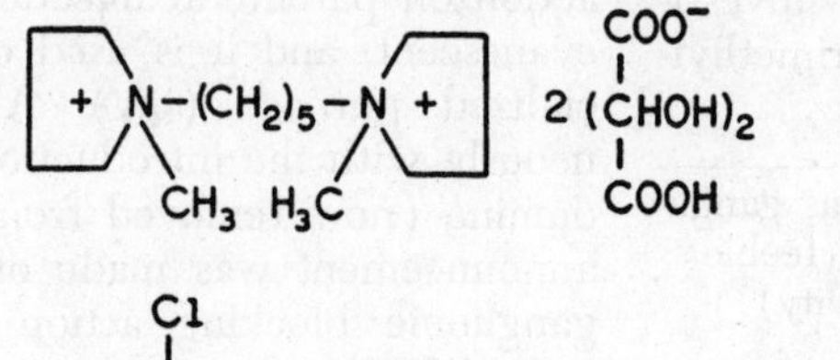	Chlorisondamine Chloride
V	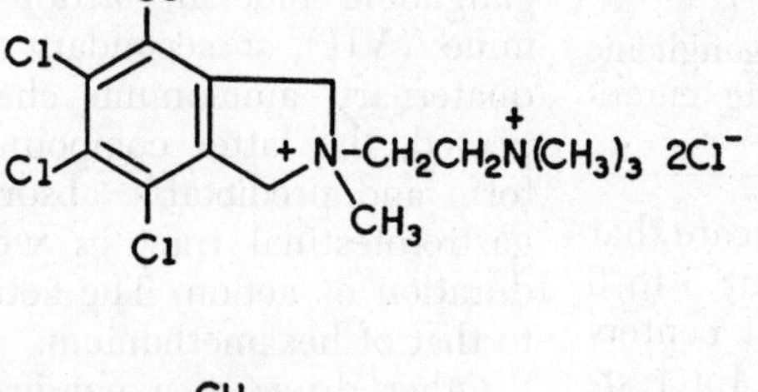	Trimethidinium Methosulfate
VI	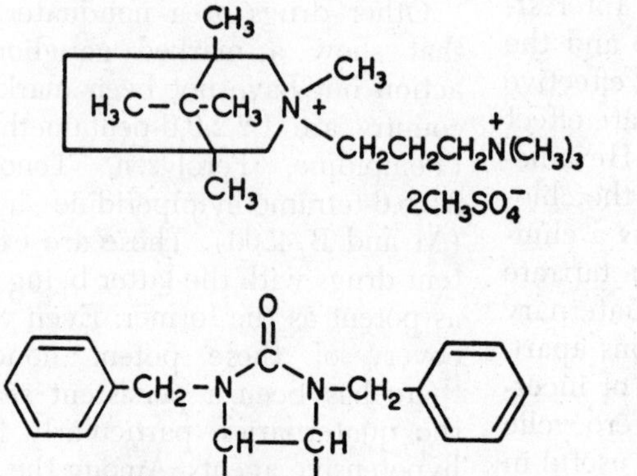	Trimethaphan Camphorsulfonate
VII	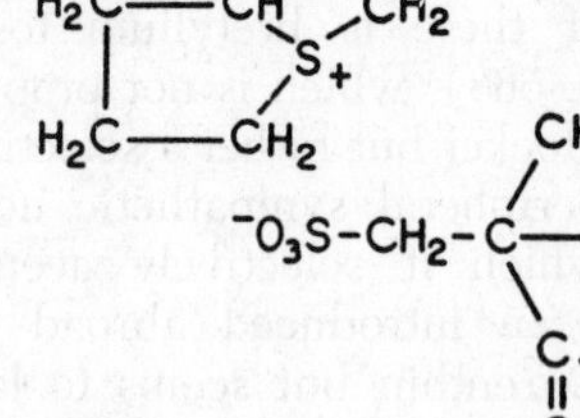	Mecamylamine Hydrochloride

quaternary derivative of atropine known as gastropin, which has a marked ganglionic blocking action coupled with only slight parasympathetic paralyzing action. In summary, although the number of ganglionic blocking agents of the onium type is large and continually increasing it appears that little has been accomplished in correlating structure with activity, and it seems likely that a single common mechanism of action is unlikely to emerge.

Drugs of this class have a limited usefulness as diagnostic and therapeutic agents in the management of peripheral vascular diseases (e.g., thromboangiitis obliterans, Raynaud's disease, diabetic gangrene, etc.). However, the principal therapeutic application has been in the treatment of hypertension through blockade of the sympathetic pathways. Unfortunately, the action is not specific, and the parasympathetic ganglia, unavoidably, are blocked simultaneously to a greater or lesser extent, causing visual disturbances, dryness of the mouth, impotence, urinary retention, constipation and the like. Constipation, in particular, probably due to unabsorbed drug in the intestine (poor absorption), has been a drawback because the condition can proceed to a paralytic ileus if extreme care is not exercised. For this reason, cathartics or a parasympathomimetic (e.g., pilocarpine nitrate) are frequently administered simultaneously. Another serious side-effect is the production of orthostatic (postural) hypotension, i.e., dizziness when the patient stands up in an erect position. Prolonged administration of the ganglionic blocking agents results in their diminished effectiveness due to a build-up of tolerance, although some are more prone to this than others. Because of the many serious side-effects this group of drugs has been largely abandoned by researchers seeking effective hypotensive agents.

In addition to the side-effects mentioned above, there are a number of contraindications to the use of these drugs. For instance, they are all contraindicated in disorders characterized by severe reduction of blood flow to a vital organ (e.g., severe coronary insufficiency, recent myocardial infarction, retinal and cerebral thrombosis, etc.) as well as in situations where there have been large reductions in blood volume. In the latter case, the contraindication is based on the fact that the drugs block the normal vasoconstrictor compensatory mechanisms necessary for homeostasis. A potentially serious complication, especially in older patients with prostatic hypertrophy, is urinary retention. These drugs should be used with care or not at all in the presence of renal insufficiency, glaucoma, uremia and organic pyloric stenosis.

PRODUCTS

Tetraethylammonium Chloride, Etamon® Chloride. This compound exists in the form of white deliquescent crystals which are freely soluble in water, alcohol, chloroform and acetone but only slightly soluble in benzene. However, the drug is prepared in the form of a 50 per cent aqueous solution (pH 5.8 to 6.5) from which the dosage forms are prepared. The pH of a 10 per cent aqueous solution is 6.48, and the solution is stable to heating for prolonged periods.

This drug was introduced to produce blockade of the autonomic ganglia, both sympathetic and parasympathetic, obliterating vasoconstrictive tone. The action was originally observed by Burn and Dale in 1915, but therapeutic application was delayed until 1947. Although the blockade of sympathetic vasospasm results in an increased blood supply to the affected regions and decreased arterial pressure, the parasympathetic blockade causes a number of characteristic side-effects. These are the usual side-effects associated with parasympathetic blockade, such as dryness of the mouth, disturbance of vision, urinary retention, etc.

The clinical usefulness of the drug as a diagnostic and therapeutic aid in the treatment of peripheral vascular diseases and related vascular difficulties is well established. However, it is of limited usefulness because of its oral ineffectiveness, its short duration of activity and its side-effects. It has been used successfully in Buerger's disease, trench foot, immersion foot, various causalgias and related peripheral vascular diseases. As a diagnostic agent it finds use as a provocative test for pheochromocytoma when the blood pressure is normal and for differential diagnosis of headache due to high blood pressure. In spite of its effectiveness in reducing blood pressure it is not useful in treating hypertension because of its short duration of action. It should be used with caution in elderly patients because of the exaggerated hypotensive response.

The drug is used intravenously and intramuscularly. The intravenous dose is 2 to 5 ml. of a 10 per cent solution but not to exceed 7 mg. per kg. body weight. The intramuscular dose is approximately 3 times the intravenous dose. The blocking action of the drug is effectively counteracted by intravenous administration of 0.5 to 1 mg. of neostigmine.

Hexamethonium Chloride, Hexamethylene-*bis*(trimethylammonium Chloride). As the bromide, it is known as hexamethonium bromide.

It occurs in the form of a colorless, odorless, crystalline, slightly hygroscopic powder. It is freely soluble in water, ethyl alcohol and methyl alcohol, but insoluble in acetone, benzene, chloroform and ether. The pH of a 1 per cent aqueous solution is 6.56, and that of a 10 per cent solution is 6.0. The solutions are stable to heating and keeping. A 20 per cent solution of sodium hydroxide causes a white precipitate which is soluble on heating.

This drug is a powerful autonomic ganglion blocking agent, inhibiting transmission of impulses through both sympathetic and parasympathetic ganglia. By virtue of this action, it brings about a marked and prolonged drop in the blood pressure.

The drug has been used in the treatment of severe malignant hypertension, where its effect is quite dramatic. It is not necessarily promoted for the treatment of benign essential hypertension, although positive results have been obtained. It is of limited use in the treatment and the management of the peripheral vascular diseases mentioned under tetraethylammonium chloride.

Oral administration is far less suitable as a mode of administration when compared with the parenteral route, being slower and less dependable. For this reason the oral route is not recommended in the therapy or the diagnosis of peripheral vascular disease. The parenteral dose is 50 to 100 mg. of the salt, repeated every 6 hours as necessary, for peripheral vascular disease or for hypotensive effect. The oral dose for hypotensive effect ordinarily should not exceed 3 g., although 4 to 5 g. may be tolerated. The usual method of administration is to start with 125 mg. 4 times a day and then to raise the dosage slowly to the point of tolerance. It is supplied in the form of 250-mg. tablets or in 10-ml. vials containing 30 mg. of hexamethonium ion per milliliter.

Pentolinium Tartrate, Ansolysen® Tartrate, Pentamethylene-1,1'-*bis*(1-methylpyrrolidinium) Bitartrate. This compound occurs as a white, crystalline powder which is odorless and nonhygroscopic. It is readily soluble in water but sparingly soluble in alcohol and is insoluble in ether and chloroform. Aqueous solutions have an acidic reaction, with a 1 per cent solution having a pH range of 3.0 to 4.0 and a 10 per cent solution having a pH of about 3.5. Aqueous solutions are stable to autoclaving.

Pentolinium tartrate is useful as an orally active blocking agent for treatment of moderate to severe hypertension. In common with the other ganglionic blocking agents it finds use in the treatment and the diagnosis of peripheral vascular diseases. It also exhibits side-effects of the same types found with other parasympathetic blocking agents.

The usual starting dose of 20 mg. is given after breakfast and then every 8 hours. To secure optimal results, the dose may be raised by 20-mg. increments to the desired maintenance dose, which is just short of that causing postural hypotension. The drug is also administered intramuscularly or subcutaneously but not by the intravenous route. The last route is not used because of the slow dissociation of the bitartrate ion which makes the response to the drug unpredictable. It is supplied in the form of 20-, 40-, 100-, or 200-mg. tablets or as a 1 per cent solution for parenteral injection.

Trimethidinium Methosulfate N.F., Ostensin®, 3-[3-(Dimethylamino)propyl]-1,3,8,8-tetramethyl-3-azoniabicyclo[3.2.1]octane *Bis*(methyl sulfate). This compound occurs as a white, hygroscopic powder and may have a slight camphoraceous odor. It is soluble in water and in alcohol, very slightly soluble in acetone, and insoluble in ether. It is used for the management of moderate and severe hypertension. It is preferred when the side-effects, such as dry mouth and constipation, are objectionable with the other ganglionic blockers, although mydriasis is a characteristic side-effect. However, this usually can be controlled by the use of 2.5 to 5.0 mg. of pilocarpine nitrate with each dose. Oral use of pilocarpine nitrate in a similar dosage is also recommended for prophylaxis against constipation. The drug is contraindicated in the presence of organic pyloric stenosis, bleeding peptic ulcer and marked cerebral arteriosclerosis. It is to be used with caution in cases of severe renal disease.

Category—antihypertensive.

Usual dose—initial, 20 mg. 2 times a day, increasing by 20-mg. increments every third day until the desired response is obtained.

Usual dose range—40 mg. to 300 mg. daily in divided doses.

OCCURRENCE

Trimethidinium Methosulfate Tablets N.F.

Trimethaphan Camsylate U.S.P., Arfonad® Camsylate, (+)-1,3-Dibenzyldecahydro-2-oxoimidazo[4,5-*c*]thieno[1,2-*α*]thiolium 2-Oxo-10-bornanesulfonate. The drug consists of white crystals or is a crystalline powder with a bitter taste and a slight odor. It is soluble in water and alcohol but only slightly soluble in acetone and ether. The pH of a 1 per cent aqueous solution is 5.0 to 6.0.

This ganglionic blocking agent is used only for certain neurosurgical procedures where excessive bleeding obscures the operative field. Certain craniotomies are included among these operations. The action of the drug is a direct vasodilation, and because of its evanescent action, it is subject to minute-by-minute control. On the other hand, this type of fleeting action makes it useless for hypertensive control. In addition, it is ineffective when given orally, and the usual route of administration is intravenous.

The hypotensive action during operative procedures is produced by continuous intravenous infusion of a 0.1 per cent solution at such a rate as to give the desired hypotension.

Category—antihypertensive.

Usual dose—intravenous infusion, 500 mg. in 500 ml. of isotonic solution at a rate adjusted to maintain blood pressure.

Usual dose range—0.2 to 5 mg. per minute.

OCCURRENCE

Trimethaphan Camsylate Injection U.S.P.

Mecamylamine Hydrochloride U.S.P., Inversine® Hydrochloride, N,2,3,3-Tetramethyl-2-norbornanamine Hydrochloride. The drug occurs as a white, odorless, crystalline powder. It has a bittersweet taste. It is freely soluble in water and chloroform, soluble in isopropyl alcohol, slightly soluble in benzene and practically insoluble in ether. The pH of a 1 per cent aqueous solution ranges from 6.0 to 7.5, and the solutions are stable to autoclaving.

This secondary amine has a powerful ganglionic blocking effect which is almost identical with that of hexamethonium. It has an advantage over most of the ganglionic blocking agents in that it is readily and smoothly absorbed from the gastrointestinal tract. This makes it quite suitable for oral administration. It has a longer duration of action than hexamethonium, and the same effect can be obtained with lower doses. Although tolerance is built up to the drug on prolonged administration, this effect is less pronounced than that with hexamethonium and pentolinium. As with other ganglionic blocking agents, this drug is capable of producing the undesirable side-effects associated with parasympathetic blockade, although they are of less intensity than with most of the others. It is probably the drug of choice among the ganglion blockers.

It is used for the treatment of moderate to severe hypertension and is occasionally effective in malignant hypertension. The dosage is highly individualized and depends on the severity of the condition and the patient response. It is supplied in 2.5- and 10-mg. tablets.

Category—antihypertensive.

Usual dose—initial, 2.5 mg. twice daily, increased by 2.5 mg. increments at intervals of not less than 2 days as required; maintenance, 7.5 mg. 3 times daily.

Usual dose range—2.5 to 60 mg. daily.

OCCURRENCE
Mecamylamine Hydrochloride Tablets U.S.P.

Adrenergic Blocking Agents Acting at the Postganglionic Terminations of the Sympathetic Nervous System

The drugs falling into this group have been termed *antisympathetics, sympatholytics, adrenolytics* and *adrenergic blocking agents*. The earlier classification of these agents separated them into adrenolytics or sympatholytics, on the basis that the former were those drugs that block response to endogenous or exogenous circulating epinephrine, whereas the latter blocked response to adrenergic nerve stimulation. This distinction is now thought to be somewhat artificial, and it appears that the differences are simply quantitative rather than qualitative in nature. Furthermore, terms implying "lysis" (e.g., of nerve ending, effector cell or mediator) are not accurate or meaningful, and current usage favors the term "adrenergic blocking agents." However, this term has been modified in view of the current classification of adrenergic receptors as α or β, based on Ahlquist's suggestions[94] (see p. 473). Most α-receptors are stimulatory in nature and β-receptors inhibitory, but this is not invariably so. Thus, we may distinguish further between the so-called "α-blockers" and the "β-blockers." Historically, only the α-blockers were the investi-

gatively and clinically useful ones before propranolol (III), the first officially accepted β-blocker, appeared recently on the American market. Although the α-blockers have dominated the investigative and therapeutic scene since the recognition by Dale[95] in 1906 of the α-adrenergic blocking action of the ergot alkaloids, it may well be that a shift in the ratio of importance will occur in view of the many active programs today concerning β-blocking agents.

Another class of sympathetic blocking agents differs from those discussed above in that they *prevent* the release of the adrenergic transmitter substance at sympathetic nerve endings, rather than blocking the effects of the released transmitter at the effector cell. Indeed, they have no blocking effect on circulating or injected epinephrine or norepinephrine, and the end-organs of the sympathetic fibers remain sensitive to these catecholamines. Among these drugs are xylocholine, bretylium, debrisoquin (Declinax)[96] and guanethidine, all of which effectively prevent release of epinephrine and/or norepinephrine from the nerve terminus. In addition, guanethidine (and

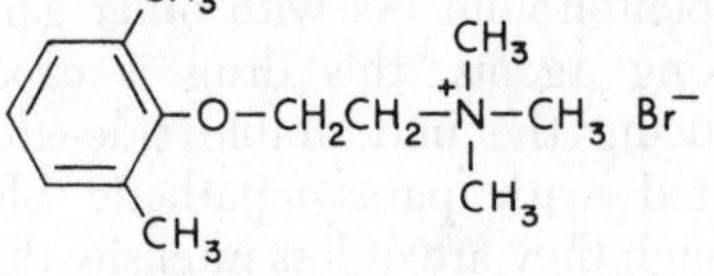

Xylocholine Bromide

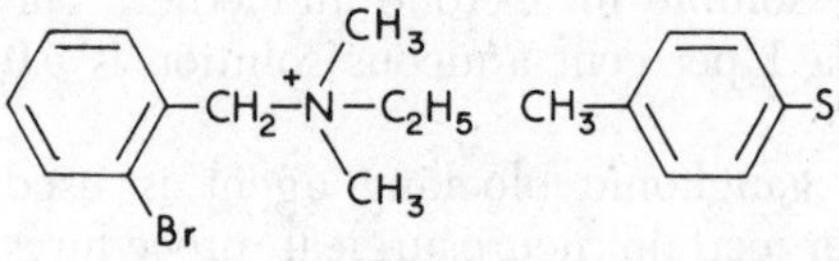

Bretylium Tosylate

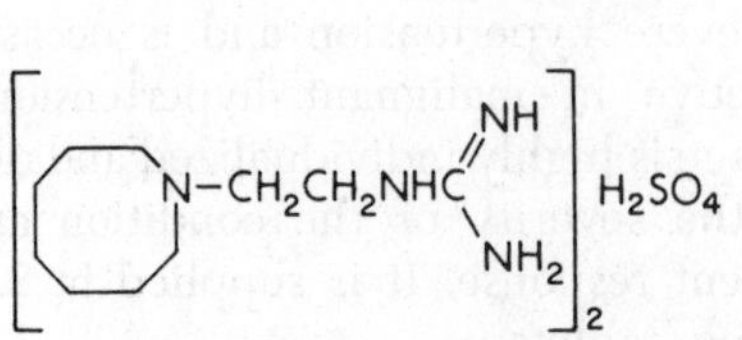

Guanethidine Sulfate

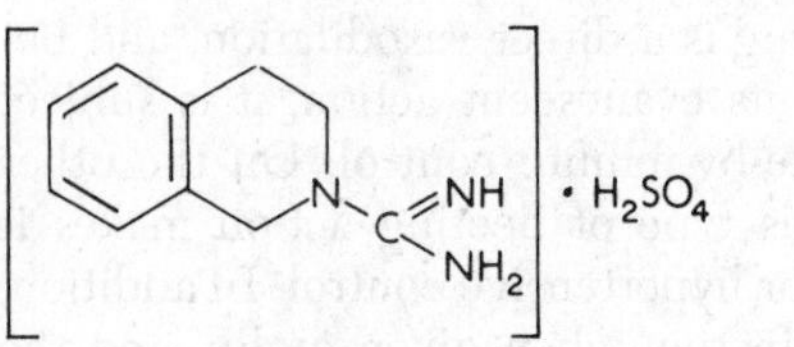

Debrisoquin Sulfate

debrisoquin to a lesser extent) promote the loss of tissue stores of norepinephrine. Radioactive studies on tagged bretylium have suggested that its high specificity of action may be due to a preferential accumulation in sympathetic nerve tissue.[97] In common with other drugs of this type it is virtually without effect on the central nervous system or the parasympathetic ganglia. With the exception of guanethidine,[98] these drugs are not used clinically in the U.S., but research is being pursued actively. Further discussion of guanethidine may be found in Chapter 21, in conjunction with the antihypertensive agents.

α-ADRENERGIC BLOCKING AGENTS

Adrenergic blocking agents of the α-type can be classified as either competitive or non-competitive antagonists of epinephrine and/or norepinephrine. In the competitive group are found the ergot alkaloids, yohimbine, some imidazolines and, probably, the benzodioxanes. In common with the characteristics of competitive cholinergic blocking agents (see p. 518), the competitive adrenergic blockers apparently possess suitable affinity for adrenergic α-receptors but lack the intrinsic activity characteristic of the natural neuroeffector. Therefore, the blocking action of a given concentration of the antagonist can be offset, under suitable experimental conditions, by increased concentrations of the agonist to provide its full intrinsic activity. The principal representatives of the noncompetitive type have traditionally been the β-haloethylamines. The usual block induced by these agents and termed an "irreversible competitive antagonism" has been characterized by two phases, an initial phase that could be competitively reversed by sufficient agonist (e.g., norepinephrine) and a more slowly developing second phase that was insurmountable by added agonist regardless of the dose and thus could be characterized as noncompetitive. The reversible phase has commonly been ascribed to the presence of "spare receptors," i.e., although enough receptors were noncompetitively blocked to bring about a lack of response with the usual doses of agonist, there were assumed

to be extra non-blocked receptors (the so-called spare receptors) sufficient to give a full response if enough agonist were added. The noncompetitive aspect of the block is thought to be due to alkylation of the receptor to form a covalent bond (see p. 36). The spare receptor hypothesis with respect to adrenergic agents recently has been challenged by Moran *et al.*[99] who concluded from their experiments that the above described agonist-antagonist behavior could be explained adequately without invoking spare receptors.

Some of the adrenergic blocking agents have not been investigated sufficiently to determine which classification applies. Among these are the dibenzazepines and the hydrazinophthalazines.

It might be concluded that the more specific action of an adrenergic blocking agent would confer certain desirable attributes to such drugs over the less discriminatory action of the ganglionic blocking agents. However, this theoretical advantage has not been realized in clinical practice (especially for the treatment of hypertension) for a number of reasons. They have been too short-acting or too ineffectual in

TABLE 57. ANTAGONISM OF EPINEPHRINE TOXICITY IN RATS

COMPOUND	APPROXIMATE RELATIVE ACTIVITY (Chlorpromazine = 100).
Chlorpromazine HCl	100
Dihydroergocornine (Methane sulfonate)	180
Ergonovine maleate	25
Ergotoxine ethane sulfonate	41
D-Lysergic acid diethylamide (LSD)	6.6
2-Brom-D-Lysergic acid diethylamide (Br.LSD)	8.6
Piperoxan HCl	8.7
Phentolamine HCl	140
Phenindamine tartrate	26
Phenoxybenzamine	67
Promazine HCl	51
Tolazoline HCl	3.5
Yohimbine HCl	3.5
Reserpine	0
Win 13,645	125
Win 14,020	100

some cases and, on the other hand, those with a high activity have been too potent and have produced unpleasant side-effects. Furthermore, not all of the drugs in this category have the same spectrum of activity. Their effect is selective with respect to the tissues upon which the blockade reaction is exercised. For further details of the pharmacology of these drugs the excellent review by Nickerson should be consulted.[100]

An insight into the relative protective effects of numerous adrenergic blocking agents against epinephrine toxicity in the rat (the most sensitive test animal) is given by the study of Luduena and co-workers.[101] Table 57 shows the relative ED_{50} when the antagonist is injected simultaneously with 2.7 times the LD_{50} of epinephrine.

A suitable classification for the α-adrenergic blocking agents is on the basis of their chemical structures. These are conveniently grouped as follows:
1. Ergot and the ergot alkaloids
2. The yohimbine group
3. Benzodioxanes
4. Beta-haloalkylamines
5. Dibenzazepines
6. Imidazolines
7. Miscellaneous agents

ERGOT AND THE ERGOT ALKALOIDS

Ergot consists of the dried sclerotium of *Claviceps purpurea*, Fam. *Hypocreaceae*, a fungus that develops on the rye plant. However, it has been shown that hosts other than rye can produce a comparable ergot.

Recorded accounts of the poisonous nature of ergot extend back to early times; and, in the late 17th century, it was identified as the cause of the medieval gangrenous scourge known as St. Anthony's fire. The gangrenous conditions were shown clearly to result from the ingestion of ergot-infected rye-grain products. The oxytocic action of ergot was recognized as early as the 16th century, and it was used by midwives for years prior to its acceptance by the medical profession. Modern acceptance is based largely on the extensive researches conducted during the past half century. Ergotoxine, isolated in 1906, and ergotamine, isolated in 1920, for many years were thought to be the principal alkaloids present. Since then, the former has been shown to be nonhomogeneous and composed of equal parts of three bases: ergocornine, ergocristine and ergocryptine. In 1933, sensibamine was reported as a new base, only to be shown later to be a mixture of equal parts of ergotamine and ergotaminine,* similarly, ergoclavine (1934) has been shown to be a mixture of ergosine and ergosinine. In 1935, an active water-soluble alkaloid was reported simultaneously by four research groups and is the alkaloid now known as ergonovine (ergometrine in Great Britain). In addition to the alkaloids just described, which are obtained from rye grain ergot, some new alkaloids have been isolated from the bases produced by artificial cultivation of the ergot fungus and related microorganisms. Among these are agroclavine, elymoclavine, penniclavine, etc., some of which possess interesting oxytocic actions and all of which show certain structural resemblances to the lysergic acid ring structure of the older alkaloids. Although they may be of future pharmaceutical interest, they are not used medicinally at present and will not be considered further in this text.

Structural studies, largely due to Jacobs and Craig[102, 103, 104] as well as to Stoll and his co-workers,[105] have shown that the active alkaloids are all amides of lysergic acid, whereas the inactive diastereoisomeric counterparts are similarly derived from *iso*lysergic acid. The only difference between the two acids is the configuration of the substituents at position 8 of the molecule. The structure of ergonovine is the simplest of these alkaloids, being the amide of lysergic acid derived from (+)-2-aminopropanol. The other alkaloids are of a more complex polypeptidelike structure in which the common structural elements are (1) lysergic acid, (2) ammonia and (3) proline. These are coupled with various combina-

---

* See Table 58. The addition of "*in*" to the suffix indicates the pharmacologically inactive diastereoisomer, which differs from the active one by the configuration of the groups at position 8.

tions of (4) pyruvic or dimethylpyruvic acid and (5) phenylalanine, leucine, or valine (see Table 58). The total synthesis of the key fragment, lysergic acid, was reported by Kornfeld and his co-workers in 1954,[106] confirming all structural assignments that had been made previously.

The isomers of ergonovine (A) have been prepared for pharmacologic study. Only the propanolamides of (+)-lysergic acid were found to be active. The optical configuration of the amino alcohol did not seem to be important to pharmacologic activity. Other partially synthetic derivatives of (+)-lysergic acid have been prepared, with two of them showing notable activity: methylergonovine (B), the amide formed from (+)-lysergyl chloride and 2-aminobutanol,

and the N-diethyl amide of (+)-lysergic acid (C). The latter compound, also known as LSD, has an oxytocic action comparable with that of ergonovine and, in addition, has been known to cause, in very small doses (30 to 50 mcg.), marked psychic changes combined with hallucinations and colored visions. The most recent active synthetic derivative of lysergic acid is the 1-methyl butanolamide (D), known generically as methysergide. The hydrogenation of the C_9 to C_{10} double bond in the lysergic acid portion of the ergot alkaloids, other than ergonovine, enhances the adrenergic blocking activity as assayed against the constrictive action of circulating epinephrine upon the seminal vesicle of an adult guinea pig. Comparative activities are demon-

TABLE 58. ERGOT ALKALOIDS*

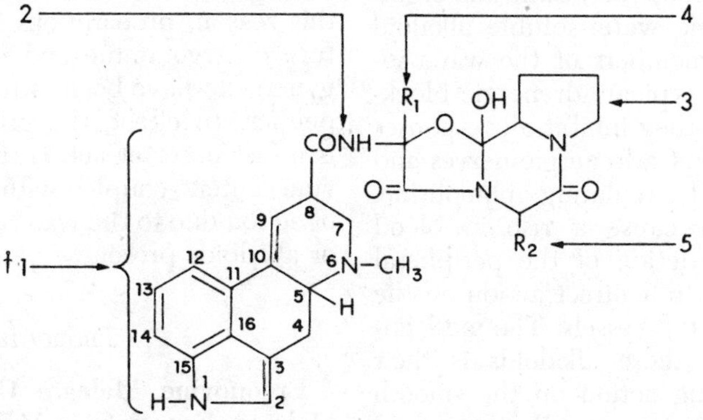

	R_1	R_2
Ergotamine Group		
Ergotamine	$-CH_3$	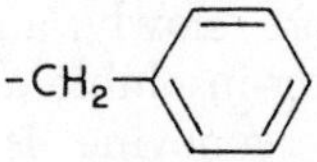 $-CH_2-\langle phenyl \rangle$
Ergosine	$-CH_3$	$-CH_2CH(CH_3)_2$
Ergotoxine Group		
Ergocristine	$-CH(CH_3)_2$	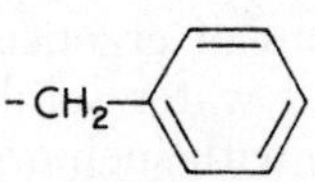 $-CH_2-\langle phenyl \rangle$
Ergocryptine	$-CH(CH_3)_2$	$-CH_2CH(CH_3)_2$
Ergocornine	$-CH(CH_3)_2$	$-CH(CH_3)_2$

* Each of the listed alkaloids has an inactive diastereoisomer derived from *iso*-lysergic acid which, in the above formulae, differs only in that the configuration of the hydrogen and the carboxyl groups at position 8 is interchanged. The nomenclature also differs, in that the suffix "*in*" is added to the name, e.g., ergotaminine instead of ergotamine. However, in the case of ergonovine, the diastereoisomer is named "ergometrinine" because this derives from the name of ergonovine commonly used in England, i.e., ergometrine.

† The numbers refer to the discussion in the text above, indicating the constituent fragments of the alkaloidal molecule.

		R_1	R_2	R_3
A.	Ergonovine	—H	—CH(CH₃)CH₂OH	—H
B.	Methyl-			
	ergonovine	—H	—CH(C₂H₅)CH₂OH	—H
C.	LSD	—C₂H₅	—C₂H₅	—H
D.	Methysergide	—H	—CH(C₂H₅)CH₂OH	—CH₃

strated in the following results recorded by Brugger.[107]

Ergotamine	1	Dihydroergotamine	7
Ergocornine	2	Dihydroergocornine	25
Ergocristine	4	Dihydroergocristine	35
Ergocryptine	4	Dihydroergocryptine	35

Pharmacologically, the ergot alkaloids may be placed in two classes: (1) the water-insoluble, polypeptidelike group comprising ergocryptine, ergocornine, ergocristine (ergotoxine group), ergosine and ergotamine and (2) the water-soluble alkaloid ergonovine. The members of the water-insoluble group are typical adrenergic blocking agents in that they inhibit all responses to the stimulation of adrenergic nerves and block the effects of circulating epinephrine. In addition, they cause a rise in blood pressure by constriction of the peripheral blood vessels due to a direct action on the smooth muscle of the vessels. The most important action of these alkaloids is their strongly stimulating action on the smooth muscle of the uterus, especially the gravid or puerperal uterus. This activity develops more slowly and lasts longer when the water-insoluble alkaloids are used than when ergonovine is administered. Toxic doses or the too frequent use of these alkaloids in small doses are responsible for the symptoms of ergotism. These alkaloids are rendered water soluble by preparing salts of them with such organic acids as tartaric, maleic, ethylsulfonic or methylsulfonic.

Ergonovine has little or no activity as an adrenergic blocking agent and, indeed, has many of the pharmacologic properties (produces mydriasis in the rabbit's eye, relaxes isolated strips of gut, constricts blood vessels) of a sympathomimetic drug. It does not raise the blood pressure when injected intravenously into an anesthetized animal. It possesses a strong, prompt, oxytocic action. Although ergonovine exerts a constrictive effect upon peripheral blood vessels, no cases have been reported yet of ergotism due to its use. It is highly active orally and causes little nausea or vomiting. It usually is dispensed as a salt of an organic acid, such as maleic, tartaric or hydracrylic acid.

Some physicians feel that none of the individual ergot alkaloids offers the advantages of mixtures of the alkaloids. For this reason, preparations containing a mixture of ergotamine and ergonovine (Neogynergen) have been introduced into therapeutics. In effect, the principal advantage is a fast onset of action due to the ergonovine content, coupled with a longer duration of action due to the water-insoluble alkaloid or alkaloids present.

PRODUCTS

Ergonovine Maleate U.S.P., Ergotrate® Maleate, Ergometrine Maleate. This water-soluble alkaloid was isolated, as indicated, from ergot, in which it occurs to the extent of 0.2 mg. per gram of ergot. The several research groups which isolated the alkaloid almost simultaneously named the alkaloid according to the dictates of each. Thus, the names, ergometrine, ergotocin, ergostetrine and ergobasine were assigned to this alkaloid. To clarify the confusion, the Council on Pharmacy and Chemistry of the American Medical Association adopted a new name, ergonovine, which is in general use today. Of course, commercial names differ from the Council-accepted name, the principal one being Ergotrate (ergonovine maleate).

Isolation of the alkaloid is based on the difference in its water solubility from that of the accompanying free alkaloids. An extract is made with an immiscible solvent of the crude alkalinized ergot. The solvent is removed from the extract, and the residue is dissolved in acetone. Upon dilution of the acetone solution with water, only the ergonovine remains in solution and is recovered easily.

The free base occurs as white crystals which are quite soluble in water or alcohol and levorotatory in solution. It readily forms crystalline, water-soluble salts, behaving in this respect as a mono-acidic base. The nitrogen involved in salt formation obviously is not the one in the indole nucleus, since it is far less basic than the other nitrogen. The official salt is the maleate. It is said to be a convenient form in which to crystallize the alkaloid and is also quite stable.

Ergonovine maleate occurs in the form of a light-sensitive, white or nearly white, odorless, crystalline powder. It is soluble in water (1:36) and in alcohol (1:20) but is insoluble in ether and in chloroform.

Ergonovine has a powerful stimulating action on the uterus and is used for this effect. Since it seems to exercise a much greater effect on the gravid uterus than on the nongravid one, it is used safely in small doses with ample effect. Some physicians utilize oxytocics of this kind during the first and the second stages of labor in the mistaken notion that delivery is hastened thereby. This practice is a possible source of danger to both mother and fetus. During the third stage of labor, these drugs should not be used until at least after presentation of the head and preferably after passage of the placenta. Ordinarily, 200 mcg. of ergonovine is injected at this stage to bring about prompt and sustained contraction of the uterus. The effect lasts about 5 hours and prevents excessive blood loss. It also lowers the incidence of uterine infection. A continued effect may be obtained by further administration of the alkaloid, either orally or parenterally.

Category—oxytocic.

Usual dose—oral, 200 mcg. 3 or 4 times a day; intramuscular or subcutaneous, 200 mcg., repeated after 2 to 4 hours if required.

Usual dose range—400 mcg. to 1.6 mg. daily.

Occurrence

Ergonovine Maleate Injection U.S.P.
Ergonovine Maleate Tablets U.S.P.

Ergotamine Tartrate U.S.P.; Gynergen®. Ergotamine, one of the insoluble ergot alkaloids, is obtained from the crude drug by the usual isolation methods.

It occurs as colorless crystals or as a white to yellowish white crystalline powder. It is not especially soluble in water (1:500) or in alcohol (1:500) although the aqueous solubility is increased with a slight excess of tartaric acid.

Previous to the discovery of ergonovine, ergotamine was the ergot drug of choice as a uterine stimulant, either orally or parenterally. Because it offered no advantage over ergonovine except for a more sustained action and, in addition, was more toxic, it fell into disuse. However, it has been employed for a new use, i.e., as a specific analgesic in the treatment of migraine headache, in which capacity it is reasonably effective. Cafergot, a combination of ergotamine tartrate and caffeine, is an available product. It is of no value in other types of headaches and sometimes fails to abort migraine headaches. It has no prophylactic value. It is customary to administer 250 mcg. subcutaneously to determine whether idiosyncrasy to the drug exists. In the event that no sensitivity is shown, the full dose is injected. Oral or sublingual administration may be resorted to, but they are much less effective than the parenteral route. Care should be exercised in its continued use to prevent signs of ergotism.

Category—analgesic (specific in migraine).

Usual dose—2 mg. orally followed by 1 mg. every half hour, if required, to a maximum of 6 mg. per attack; 250 to 500 mcg. intramuscularly or subcutaneously, repeated in 1 hour if necessary; rectally, 2 mg., repeated in 1 hour if required.

Usual dose range—oral or rectal, 2 to 10 mg. weekly; intramuscular or subcutaneous, 250 mcg. to 1 mg. weekly.

Occurrence

Ergotamine Tartrate Injection U.S.P.
Ergotamine Tartrate Tablets U.S.P.

Ergotamine Tartrate and Caffeine Suppositories N.F.

Ergotamine Tartrate and Caffeine Tablets N.F.

Dihydroergotamine; D.H.E.45. This compound is produced by the hydrogenation of the easily reducible C_9 to C_{10} double bond in the lysergic acid portion of the ergotamine molecule. The compound is marketed only in ampul form as a solution of dihydroergotamine methanesulfonate.

Dihydroergotamine, although very closely related to ergotamine, differs significantly from the latter in its action. For all practical purposes, the uterine action is lacking. However, the adrenergic blocking action is stronger. Nausea and vomiting are at a minimum, as is its cardiovascular action. One of its principal uses has been in the relief of migraine headache in a manner similar to ergotamine, over which it excels not only in decreased toxicity but also in a higher percentage of favorable results. According to authorities, good results are obtainable in about 75 per cent of the cases treated. Because the drug is not very effective orally, it usually is administered subcutaneously, intravenously or intramuscularly in a dose of 1 or 2 ml. (1 or 2 mg.) which may be repeated in 1 hour if necessary.

Methylergonovine Maleate U.S.P., Methergine®, N-[α-(Hydroxymethyl)propyl]-D-lysergamide. This compound occurs as a white to pinkish tan microcrystalline powder which is odorless and has a bitter taste. It is only slightly soluble in water and alcohol and very slightly soluble in chloroform and ether.

It is very similar to ergonovine in its pharmacologic actions. It is said to be about one to three times as powerful as ergonovine in its action. The action of methylergonovine is quicker and more prolonged than that of ergonovine. It has been shown to be relatively nontoxic in the doses used. It is marketed as 200 mcg. tablets and as ampuls.

Category—oxytocic.

Usual dose—oral, 200 mcg. 3 or 4 times daily; intramuscular or intravenous, 200 mcg., repeated after 2 to 4 hours if required.

Usual dose range—200 to 800 mcg. daily.

OCCURRENCE

Methylergonovine Maleate Injection U.S.P.

Methylergonovine Maleate Tablets U.S.P.

Hydergine® is composed of the hydrogenated alkaloids in the ergotoxine group. It is an equiproportional mixture of dihydroergocornine, dihydroergocristine and dihydroergocryptine obtained by hydrogenation of the double bond present in the lysergic acid nucleus of the natural ergot alkaloids. The hydrogenated alkaloids are known as the D-H alkaloids of ergot. In hydergine, these are solubilized by forming the methansulfonate salt. Each 1-ml. (0.3 mg.) ampul contains 100 mcg. of each D-H alkaloid.

As mentioned before, natural alkaloids of ergot have the ability to stimulate the contraction of smooth muscle (blood vessel and uterus), whereas hydrogenation of these alkaloids brings about a direct contrast in physiologic action. The D-H alkaloids do not stimulate smooth muscle but have a vasodilation effect, have a stronger adrenergic blocking action, have a central sedative action, show a central dampening of pressoreceptor reflexes and cause a mild slowing of the heart rate. Due to these effects Hydergine has been used in the treatment of hypertension and peripheral vascular disease, such as Raynaud's and Buerger's disease, frostbite, chilblains, thrombophlebitis, gangrene and varicose ulcers.

Methysergide Maleate N.F., Sansert®, N-[1-Hydroxymethyl)propyl]-1-methyl-D-lysergamide Maleate (1:1). This drug was introduced in 1962 and possesses the structure (D) given on page 572. It occurs as a white to yellowish white, crystalline powder that is practically odorless. It is only slightly soluble in water and alcohol and very slightly soluble in chloroform and ether.

Although it is closely related in structure to methylergonovine it does not possess the potent oxytocic action of the latter. It has been shown to be a potent serotonin antagonist and has found its principal utility in the prevention of migraine headache, but the exact mechanism of prevention has not been elucidated. It has only a weak adrenolytic activity.

The drug produces a variety of untoward side effects, but many of them are mild and disappear with continued use. Some of the most common of these effects are nausea, epigastric pain, dizziness, restlessness, drowsiness, leg cramps and psychic effects. Numerous other effects, including some blood dyscrasias, have been reported. Patients receiving this drug should be under the close supervision of the physician and should report any untoward effects to him immediately. Contraindications are pregnancy (residual oxytocic action), peripheral vascular disturbance, severe hypertension, thrombophlebitis, peptic ulcer, renal disease and coronary artery disease.

Methysergide is used prophylactically in the management of all types of migraine headache in patients in whom frequency and severity of headache warrants continuous therapy. It is of no value in treating tension headache nor is it useful in treating the acute phase of a migraine attack.

Category—antiadrenergic (migraine prophylactic).

Usual dose—4 to 8 mg. daily in divided doses, preferably with meals.

OCCURRENCE
Methysergide Maleate Tablets N.F.

THE YOHIMBINE GROUP

The alkaloids in this group are obtained from the bark of *Corynanthe johimbe* K. Schum and from related trees. Yohimbine itself has been isolated from *Rauwolfia serpentina*. The chemical structure of yohimbine, as well as the related alkaloids corynanthine and alpha-yohimbine (rauwolscine), has the essential difference from reserpine that the configuration of the hydrogen at C-3 is opposite (there are, of course, other differences). These differences are shown in the accompanying figure.

Although the adrenergic blocking action of yohimbine has been known since the original work of Raymond-Hamet in 1925, the drug has had only a limited use as a laboratory tool and has found little employment as a blocking agent in therapy. A principal deterrent was a strychninelike

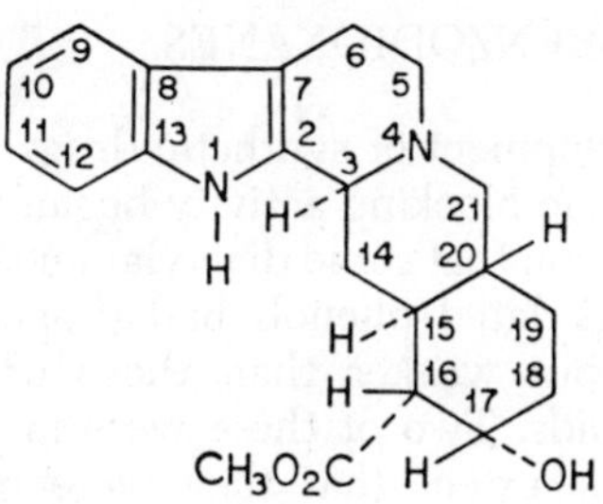

Yohimbine = As shown.
Corynanthine = Reverse configurations of substituents at C-16.
α-Yohimbine = Same as that for Corynanthine except reverse configuration of hydrogen at C-20.

central stimulation. Some of the derivatives of yohimbine have been prepared, such as the ethyl, allyl, butyl, phenyl, etc., and they appear to have similar properties. Of these, only ethyl yohimbine has been studied because of its relatively low toxicity, but most studies have been superficial. The isomer, corynanthine, also appears to be less toxic and, indeed, is more potent than yohimbine as a blocking agent, but it has not been studied thoroughly. Huebner and co-worker[108] also have reported that various esters of yohimbine, such as the benzoate, the anisate the veratrate and the 3,4,5-trimethoxybenzoate, and some of its isomers possess hypotensive and adrenergic blocking action and that they have lost the central effects that are undesirable in yohimbine.

Yohimbine, Quebrachine, Aphrodine, Corynine. The free base occurs as water-insoluble but alcohol-soluble colorless needles. It is a monoacidic base and readily forms salts, of which the hydrochloride is the most important. It is water soluble (1:120) and gives practically neutral solutions which may be heat sterilized.

The drug was used formerly for its aphrodisiac properties, which depended on a dilating effect on the blood vessels of the genital organs. At times it also has been employed for local anesthesia, mydriasis, angina pectoris and arteriosclerosis. However, its use in human therapy has been largely abandoned.

BENZODIOXANES

The development of synthetic drugs having adrenergic blocking activity began with the observation that some dialkylaminoalkyl ethers of alkylated phenols had properties similar to but weaker than those of the ergot alkaloids. Two of these were investigated to the extent that they were given names in addition to their laboratory numbers. They were: 1-Methoxy-2-β-diethylaminoethoxy-3-allylbenzene (β-diethylaminoethylether of 6-allylguaiacol), named gravitol, and β-dimethylaminoethylether of 3-methyl-6-isopropylphenol (β-dimethylaminoethylether of thymol), named tastromine.[109]

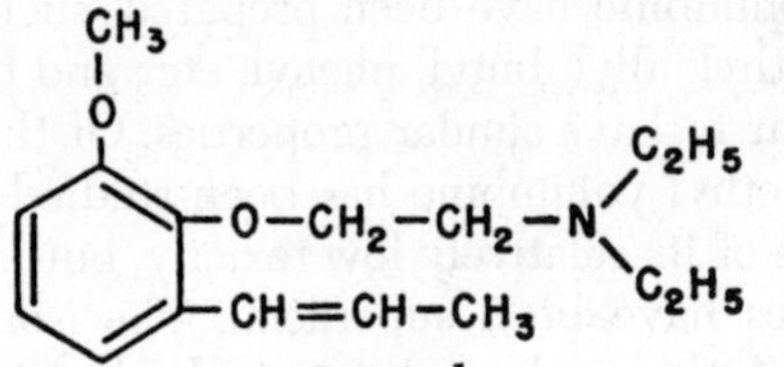

Gravitol

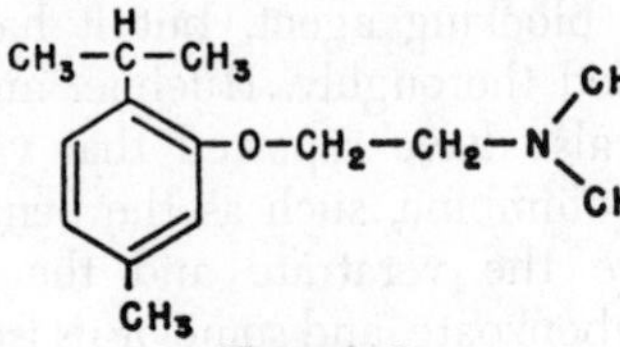

Tastromine

The solid lead furnished by these compounds prompted the syntheses and the testing of hundreds of analogous compounds and the variants of these suggested by the fertile minds of a number of groups of pharmaceutical chemists and pharmacologists. The earliest work was done in the laboratories of the Pasteur Institute in France under the direction of Ernest Fourneau and his colleagues.[110]

As a direct result of this activity a number of potent antagonists to circulating epinephrine were synthesized, based upon the nucleus of gravitol, which is a catechol derivative. The more successful compounds resulted from the inclusion of the oxygen atoms of catechol within a ring structure, thus forming a fused ring heterocyclic named 1,4-benzodioxane.

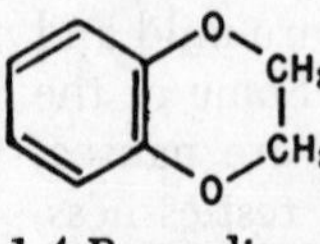

1,4-Benzodioxane 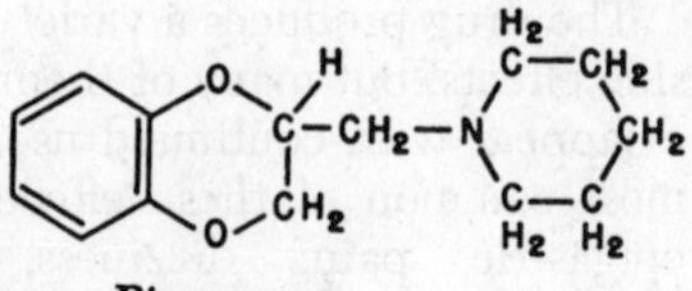   Piperoxan

The only clinically useful compound resulting from these studies was piperoxan, formerly on the market as Benodaine but no longer available. Another, named prosympal (2-diethylaminomethyl-1,4-benzodioxane), has been studied experimentally but never marketed in the U.S. For the most part the benzodioxanes are effective against responses to circulating sympathomimetic amines, although some are effective against responses to both circulating mediators and sympathetic nerve activity.[100] Piperoxan falls into the former category, whereas prosympal represents the latter. The use of these drugs in essential hypertension is disappointing for the reason that they do not inhibit the adrenergic cardioaccelerator nerves to the heart. The resultant increased heart activity tends to mask any hypotensive effects.

β-HALOALKYLAMINES

Although dibenamine (N,N-dibenzyl-β-chloroethylamine), the prototype of these

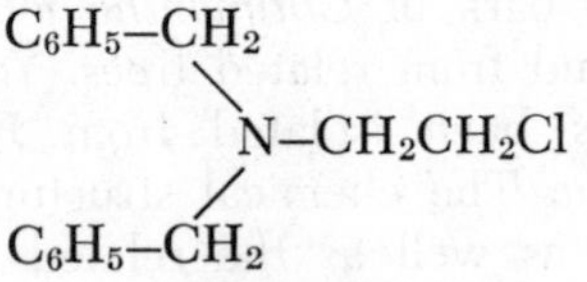

Dibenamine

compounds, was characterized in 1934 by Eisleb[111] incidental to a description of some other synthetic intermediates, it was the report of Nickerson and Goodman[112] in 1947 on the pharmacology of the compound that revealed the powerful adrenergic blocking properties. The blockade produced by this group of compounds seems to be the most complete and specific of the entire group of blocking agents. The differences in activity of the members of this group differ only quantitatively, being qualitatively the same. When given in adequate doses, they produce a slowly developing prolonged adrenergic blockade which is not overcome by massive doses of epinephrine (10 mg./kg. I.V.). Although a large number of compounds related to dibenamine have been synthesized, only a few have reached the stage of general distribution and clinical trial. Much of the early work done with this group was confined to dibenamine but this has been largely supplanted by the more orally useful and potent phenoxybenzamine. The mass of pharmacologic data accumulated with respect to this class of compounds has led to the establishment of certain structural requirements that are necessary for activity. Ullyot and Kerwin[113] state that most of the presently known effective compounds may be defined broadly by the following formula:

$$\begin{array}{c} R' \\ \diagdown \\ N-CH_2CH_2X \\ \diagup \\ R'' \end{array}$$

R′ = Aralkyl (Benzyl, Phenethyl, etc.)
　 = Phenoxyalkyl (β-phenoxyethyl, etc.)
R″ = Alkyl, Alkenyl, dialkylamino-alkyl, aralkyl, β-phenoxylethyl, etc.
X = Halogen, sulfonic acid ester.

However, not all compounds answering this description or broad generalization will be active. The degree of effectiveness depends on the character of R′, R″ and X. Furthermore, ring substitution either increases or decreases activity. Likewise, substitution on the β-haloethyl side chain has an effect on activity. In all cases of good blocking, X is readily ionizable and capable of displacement by an intramolecular cyclization mechanism to form an immonium ion:

$$\begin{array}{c} R' \\ \diagdown \\ R'' \end{array} N-CH_2CH_2X \;\rightleftharpoons\; \begin{array}{c} R' \quad CH_2 \\ \diagdown \overset{+}{N} \diagdown \quad \Big| \quad X^- \\ \diagup \quad \diagup \\ R'' \quad CH_2 \end{array}$$

Early reports that quaternary salts derived from effective blockers are also effective have not been confirmed, and this is considered to be a good argument for the existence of the immonium ion as the active intermediate. A possible sequence of events following oral ingestion of these blockers is

that they exert their effect by alkylating a "receptor substance" (i.e., the α-receptors) and that the process requires the formation of an intermediate immonium ion to act as the active alkylating species. However, others have suggested that the noncyclized drug can possibly concentrate in the fat depots of the body and be slowly released to the plasma to account for the long duration of activity. Belleau[114] subscribes in part to the above hypotheses but holds to the belief that the slow recovery from blockade can be ascribed to alkylation of phosphate or carboxylate anions by the ethyleneimmonium ions leading to labile esters (carboxylate anions are found in proteins and phosphate anions in nucleotides). He suggests that two phases are involved in the establishment of adrenergic blockade:

1. Attraction of the ethyleneimmonium ion by the receptor and retention by weak forces.

2. Reaction chemically of the ethyleneimmonium ion with the receptor and slow hydrolysis (or fast, depending on adjacent basic groups) to regenerate the receptor anions.

Belleau also postulates a reasonable sequence of events in the establishment of blockade to fit the blocking agent into the so-called "phenethylamine mold," which is necessary if a common basic structural pattern is to relate stimulator and blocker.

The recent work of Moran et al.[99] reinforces the suggestions of Belleau. It, further, provides experimental evidence in support of the belief that during the developing phase of the block the ethyleneimmonium ion is acting in a typically competitive fashion in that it can be displaced by sufficient agonist and that receptor alkylation is a more slowly developing process.* The original concept of "spare receptors" to account for the seemingly competitive phase of block has been seriously challenged by these studies. Although a consideration of all of the interesting postulations of Belleau[114, 115] and of Moran et al.[99] are not within the scope of this text, the inquiring reader will find them interesting and provocative reading.

---

* However, certain of the haloethylamines have so rapid a rate of alkylation that no reversible competitive phase can be detected.

Phenoxybenzamine Hydrochloride, Dibenzyline® Hydrochloride, N-(2-Chloroethyl)-N-(1-methyl-2-phenoxyethyl)benzylamine Hydrochloride. The compound exists in the form of colorless crystals which are

Phenoxybenzamine Hydrochloride

insoluble in water but soluble in warm propylene glycol. It slowly hydrolyzes in neutral and basic solutions but is stable in acid solutions and suspensions.

The action of phenoxybenzamine has been described as representing a "chemical sympathectomy" because of its selective blockade of the excitatory responses of smooth muscle and of the heart muscle. Its antipressor action is not confined to antagonizing epinephrine alone, because it is also effective against other sympathomimetic amines. However, it is characteristic that the activity of this blocking agent is slow in developing and, before full blockade is developed, its action can be reversed by large doses of epinephrine. However, once the blockade is developed there is no known drug that will reverse it. The principal effects following administration are an increase in peripheral blood flow, increase in skin temperature and a lowering of blood pressure. It has no effect on the parasympathetic system and has little effect on the gastrointestinal tract. The most common side-effects are miosis, tachycardia, nasal stuffiness and postural hypotension, which are all related to the production of adrenergic blockade.

The drug is employed (and is superior to other drugs of this class, e.g., dibenamine) for all peripheral vascular disease characterized by excessive vasospasm. These conditions include Raynaud's syndrome, acrocyanosis, causalgia, chronic vasospastic ulceration and the effects following frostbite. Its value in Buerger's disease and intermittent claudication is only fair. Likewise, it is of limited value in the management of hypertension because, although it reduces

blood pressure, it brings about excessive postural hypotension as well as other undesirable side-effects. The drug is contraindicated in those conditions where a drop in blood pressure is dangerous.

The drug is administered orally in doses which must be individualized to the patient and are based on patient response and development of side-effects. Usually, 10 to 20 mg. once daily may be given to begin therapy. After 4 days of this therapy the dose is increased by 10-mg. increments and even smaller increments until satisfactory results are obtained. The usual maintenance dose is 20 to 60 mg. once or twice daily. Two weeks are needed for development of the full blockade. It is supplied in 10-mg. capsules.

DIBENZAZEPINES

A group of compounds somewhat related to the β-haloalkylamines was reported on by Wenner[116] and also by Randall and Smith.[117] These were the so-called "dibenzazepines," a name derived from the term *azepine*, which denotes a ring containing 6 carbon atoms and 1 nitrogen atom. The particular azepines that were active were the dibenzazepines with the fused benzene rings in positions reminiscent of those in the potent dibenamine, as illustrated by the formula given for azapetine, the most active hypotensive agent of the series. There are many similarities between

Azapetine

the action of the dibenzazepines and the benzodioxanes (as well as imidazolines). Principally, they establish a blockade that is reversible by administration of sufficient epinephrine or other sympathomimetic amines (the so-called "labile" type of blockade). Moore and co-workers[118] have shown that this type of drug exerts a direct vasodilatory action. They noted that the epinephrine reversal effect of azapetine was at least that of tolazoline and that it lasted longer. The combination of direct vasodilatation and blockage of the vasconstrictive response of smooth muscle to circulating epinephrine makes the group effective in the treatment of peripheral disorders in which vasospasm is the predominant cause of restricted blood flow. Administration of the drugs, e.g., azapetine, results in increases in skin temperature and peripheral blood flow as well as a small decrease in blood pressure. Parasympathetic effects are not serious when the drug is used.

Azapetine Phosphate, Ilidar® Phosphate, 6-Allyl-6,7-dihydro-5H-dibenz[c,e]azepine Phosphate. This drug has the general pharmacologic activities that have been described briefly above. Indeed, presently it is the only member of its class that is being used clinically, and practically all reports dealing with dibenzazepines are concerned with azapetine. It has found employment successfully in the treatment of Raynaud's disease, acrocyanosis, causalgia, ulcers of the extremities from chronic peripheral vasospasm or frostbite after-effects. It has not been useful in organic occlusive forms of peripheral vascular disease such as Buerger's disease, acute arterial occlusion, etc. It has been employed in moderate to severe hypertension, but as yet its value is not established.

The drug is administered orally in an initial test dose of 25 mg. 3 times a day for 7 days. If the drug is tolerated, the usual dose of 50 to 75 mg. 3 times daily is worked up to on the basis of individual response. It is supplied in the form of 25-mg. tablets.

IMIDAZOLINES

In 1939, Hartmann and Isler [119] reported on the pharmacology of the first known member of this class, namely, tolazoline. Their report noted the fact that it was an active depressor agent but failed to recognize the adrenergic blocking action. This was first noted by Schnetz and Fluch in 1940.[120] It has a relatively short duration of action compared with the β-haloalkyl-amines and, by possessing an "equilibrium" (labile) type of blockade, is more closely related to the benzodioxanes than to the haloalkylamines. It blocks both circulating epinephrine and sympathetic nerve stimulation. A more recent member of this group, phentolamine, was reported on in 1952 by Roberts and his co-workers[121]; it was said to antagonize the vasoconstrictor effects of epinephrine about 6 times as effectively as tolazoline. Nevertheless, on the whole, these are relatively weak adrenergic blocking agents. Indeed, there is some evidence of sympathomimetic activity, with perhaps the most important effect being tachycardia due to cardiac stimulation. This renders this group ineffective as hypotensive agents (see also benzodioxanes, p. 576). The gastrointestinal tract is affected by a parasympathomimetic stimulation due to the drug, and unpleasant gastric symptoms (nausea, diarrhea, pain, etc.) may result. A histaminelike side-effect also is noticed which produces a direct peripheral vasodilation of peripheral blood vessels. This effect, which may be termed a musculotropic effect, reinforces the neurotropic (adrenergic nerve block) effect and results in a useful degree of peripheral vasodilation. Another result of the histaminelike action is increased gastric secretion.

PRODUCTS

Tolazoline Hydrochloride, Priscoline® Hydrochloride, Benzazoline Hydrochloride, Tolazolinium Chloride, 2-Benzyl-2-imidazoline Hydrochloride.

Tolazoline Hydrochloride

The synthesis of tolazoline is described by Scholz.[122] The drug occurs as a white or creamy white, bitter, crystalline powder possessing a slight aromatic odor. It is freely soluble in water and alcohol. A 2.5 per cent aqueous solution is slightly acidic (pH 4.9 to 5.3). It is only slightly soluble in ether and ethyl acetate but is soluble in chloroform.

As described in the introduction to these drugs, this drug has the ability not only to block circulating epinephrine but also to block sympathetic nerve activity. In addition, its direct histaminelike activity gives it a vasodilating property unlike that of other adrenergic blocking agents. For this reason, it finds its chief use in the treatment of peripheral vascular disorders in which vasospasm is a prominent factor. Likewise, it is of value where angiospasm is a factor and finds use in the treatment of acrocyanosis, arteriosclerosis obliterans, Buerger's disease, Raynaud's disease, frostbite sequelae, thrombophlebitis, etc.

It is used orally and parenterally in a usual dose of 50 mg. 4 times a day with a usual dosage range of 25 to 75 mg.

Phentolamine Mesylate U.S.P., Regitine® Methanesulfonate, *m*-[*N*-(2-Imidazolin-2-ylmethyl)-*p*-toluidino]phenol Monomethanesulfonate. This compound may be made

Phentolamine Mesylate

by the procedure of Urech and co-workers.[123] It occurs as a white, odorless, bitter powder which is freely soluble in alcohol and very soluble in water. Aqueous solutions are slightly acidic (pH 4.5-5.5) and deteriorate slowly. However, the chemical itself is stable when protected from moisture and light. The stability and the solubility of this salt of phentolamine are superior to those of the hydrochloride and account for the use of the methanesulfonate (mesylate) rather than the hydrochloride for parenteral injection. The imidazoline ring structure is susceptible to degradation by means of a base-catalyzed hydrolytic mechanism with concurrent ring opening. The kinetics of this type of ring opening are discussed by Stern and co-workers[124] in conjunction with another of the imidazolines (i.e., naphazoline).

This adrenergic blocking agent is used parenterally in the diagnosis and the surgical management of pheochromocytoma.

Category—antiadrenergic.

Usual dose—intramuscular or intravenous, 5 mg.

OCCURRENCE

Phentolamine Mesylate for Injection U.S.P.

Phentolamine Hydrochloride N.F., Regitine® Hydrochloride, *m*-[*N*-(2-Imidazolin-2-ylmethyl)-*p*-toluidino]phenol Monohydrochloride. It occurs as a white or slightly grayish, odorless, bitter powder. It is slightly soluble in alcohol and sparingly soluble in water. Its solutions in water are slightly acidic (pH 4.5 to 5.5) and foam when shaken. It is affected by light, and its aqueous solutions are unstable.

This salt of phentolamine is suitable for oral administration and is used as a potent adrenergic blocking agent to block circulating epinephrine as well as sympathetic nerve stimulation. The fact that it suppresses the pressor response to levarterenol as well as to injected epinephrine makes it of value in the control of the hypertension produced by pheochromocytoma. Similarly, it is of value wherever it is necessary to increase blood flow to the extremities and where adrenergic blocking will be effective. Although the physical characteristics (solubility and stability) of the hydrochloride prevent it from being used parenterally (necessary for diagnosis of pheochromocytoma as well as surgical management) it is used to prevent hypertension in such patients until surgical removal is possible.

Category—antiadrenergic.

Usual dose—50 mg. 4 to 6 times daily.

Usual dose range—50 to 100 mg.

OCCURRENCE

Phentolamine Hydrochloride Tablets N.F.

β-ADRENERGIC BLOCKING AGENTS

In contrast to the long-known α-blocking agents, the literature on β-blocking agents is of relatively recent origin and dates only from 1958 when Powell and Slater[125] reported on the specific adrenergic β-receptor blocking action of dichloroisoproterenol (DCI) (I*). Up to the time of discovery of DCI no agent was known that would block adrenergic stimuli that produced stimulation of the heart and inhibition of several types of smooth muscle. Powell and Slater[125] demonstrated that DCI blocked the inhibitory effects of sympathomimetic amines on blood vessels, the uterus, the intestine and the tracheobronchial system and also showed a depressant effect on the frog heart and decreased the inotropic† and chronotropic‡ effects of epinephrine. Within the same year, Moran and Perkins[126] confirmed these actions and clearly demonstrated a highly specific blockade in the heart of the dog and the rabbit. Furthermore, they demonstrated a direct sympathomimetic action of DCI in the dog leading to an increase in both frequency and force of heart contraction. However, administration of successively larger doses led to profound cardiac depression. It was Moran and Perkins who suggested that Ahlquist's long neglected classification[94] of adrenergic receptors into α- and β-types be applied to this new blocking agent. Thus, DCI was classed as a β-adrenergic blocking agent, whereas all of the previously known adrenergic blockers were α-blocking agents. Numerous other workers[127, 128, 129] soon confirmed the fact that DCI possesses both agonistic and antagonistic properties. Unfortunately, DCI, because of its partial agonist character, was of little prospective value as a drug. Black et al.,[130] in 1962, reported on the adrenergic blocking properties of pronethalol (II) (nethalide, Alderlin), a compound possessing the same type of β-blocking action as DCI but with a considerably lower sympathomimetic potency. Although pronethalol showed promise in clinical trial, it was shown to produce tumors in animals and was quickly withdrawn. To replace it, Black et al.,[131] in 1964, introduced propranolol (III), which was

* Roman numerals refer to entries in Table 59, p. 582.

† Affecting the *force* or *energy* of muscular contractions.

‡ Affecting the time or rate, applied especially to nerves whose stimulation or agents whose administration affects the *rate* of contraction of the heart.

10 times more potent than pronethalol and did not have the tumor-producing propensity of the latter. Propranolol has since appeared on the American market under the trade name of Inderal. MJ 1999 (IV),[132] an experimental β-adrenergic blocking product of another American pharmaceutical company, has long been expected on the market but at this writing is still under experimental observation. Other active β-blocking agents being considered abroad are H 13/57 (V),[133] H 13/62 (VI),[133] Kö 592 (VII)[134] and butoxamine (VIII),[135] but none of these has achieved acceptance in the U.S. Present use of propranolol in the U.S. is restricted to its cardiac effects and limited to the control of arrythmias. There are some indications that it might prove useful in the anginal syndrome but this is not a currently approved usage.[136]

Besides the specific β-receptor blockade produced by drugs such as propranolol there also appears to be a nonspecific "quinidinelike" component to their action. This was first noted by Lucchesi et al.[137] in connection with the reversal of digitalis-induced arrhythmias by DCI; later the same group[138] made similar observations with regard to propranolol. Numerous observations by others support the idea of a nonspecific antiarrhythmic effect and are well summarized by Lucchesi et al.[138] Thus, it appears that β-blocking agents may be of value in the control of cardiac rhythm disorders that are not due to adrenergic mechanisms. However, some authorities hold that antiarrhythmic agents should preferably not be β-adrenergic blocking agents, in order to avoid inhibition of the sympathetic control of the heart and, possibly bronchoconstriction. Vargaftig et al.[139] have critically evaluated three methods for the study of adrenergic β-blocking and antiarrhythmic agents with a view toward providing suitable test procedures for the eventual production of the ideal antiarrhythmic agent.

The study of β-adrenergic blocking agents is still in its infancy. Thus far all of the active blocking agents have been derived from modifications of isoproterenol (IX), a potent β-receptor agonist. Some of the modifications that have been successful

TABLE 59. BETA-ADRENERGIC BLOCKING AGENTS

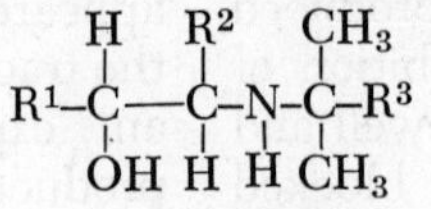

$$\begin{array}{c} \text{H} \quad \text{R}^2 \quad \text{CH}_3 \\ | \quad\;\; | \quad\;\; | \\ \text{R}^1\text{-C--C--N--C--R}^3 \\ | \quad\; | \;\; | \quad | \\ \text{OH} \;\; \text{H} \;\; \text{H} \;\; \text{CH}_3 \end{array}$$

No.	NAME	R¹	R²	R³
I	Dichloroisoproterenol	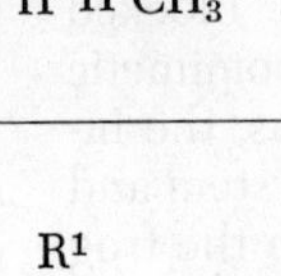	H–	H–
II	Pronethalol	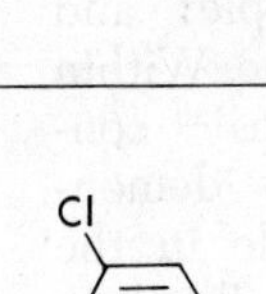	H–	H–
III	Propranolol	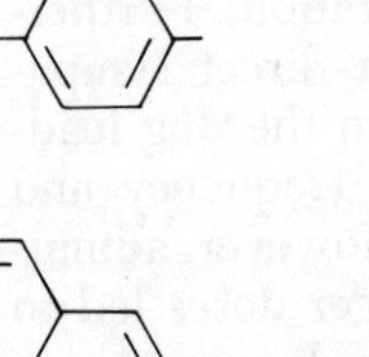	H–	H–
IV	MJ 1999		H–	H–
V	H 13/57	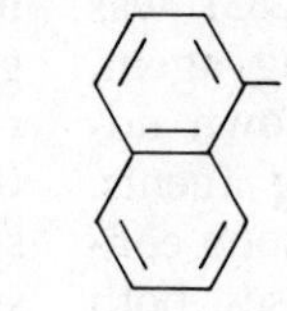	H–	H–
VI	H 13/62	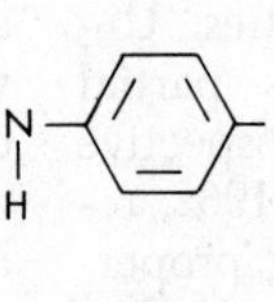	CH₃–	H–
VII	Kö 592	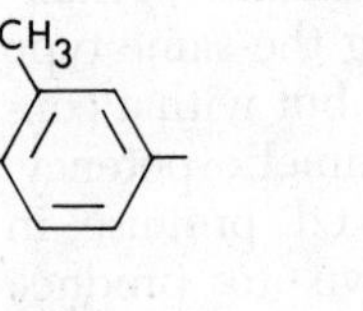	H–	H–

TABLE 59. BETA-ADRENERGIC BLOCKING AGENTS (*Continued*)

No.	NAME	R¹	R²	R³
VIII	Butoxamine	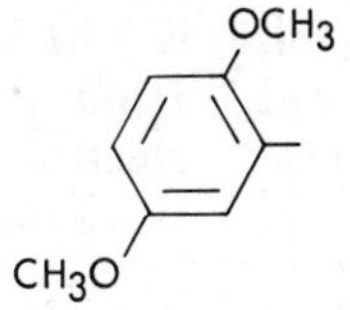 OCH₃ / CH₃O substituted benzene ring	CH₃–	CH₃–
IX	Isoproterenol*	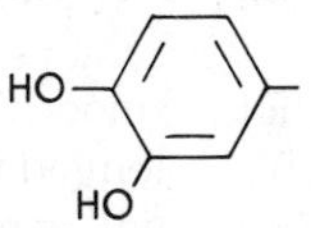 HO / HO substituted benzene ring	H–	H–

* Included to show structural relation.

are shown in Table 59 although this is only a fraction of the compounds that have been so far made and tested. It is of interest also to note that the resolution of the active blocking agents possessing asymmetric centers has shown that the (−)-isomer is the most active, this also being true of the agonists from which they have been derived.[140, 141, 142] It is not within the scope of this text to review the entire spectrum of structure-activity studies thus far undertaken but the interested reader will find a wealth of information in the excellent presentations of Ariëns,[143] Biel *et al.*[144] and Ghouri *et al.*[145]

PRODUCTS

Propranolol Hydrochloride, Inderal®, 1-(Isopropylamino)-3-(1-naphthyloxy)propanol Hydrochloride. This compound is a colorless, crystalline solid, soluble in water or ethanol and insoluble in nonpolar solvents such as ether, benzene and ethyl acetate. Its preparation is described in the patent literature[146] and the separation of its optical isomers is described by Howe *et al.*[147]

This drug is an effective and potent β-adrenergic blocking agent and is claimed to represent an entirely new approach to the control of cardiac arrhythmias, i.e., the suppression of cardiac adrenergic stimuli by selective blocking of the β-adrenergic receptors in the myocardium. It appears that the myocardium is endowed primarily with receptors of the β-type. However, a direct

quinidinelike antiarrhythmic effect of the drug cannot be ignored (see p. 581) and may, in fact, be as important as the adrenergic blocking action in certain arrhythmias.

Side-effects due to reduction of resting sympathetic nervous activity may be hypotension and/or marked bradycardia resulting in vertigo, syncopal attacks or orthostatic hypotension. Occasional nausea, vomiting, light-headedness, mild diarrhea, constipation and mental depression may also be noted. The incidence of side effects is low and these effects are usually transient. However, the physician using this drug, as well as the pharmacist dispensing it, should be familiar with the basic modern concepts of adrenergic receptors and with the pharmacology of specific blocking agents.

Contraindications to the use of the drug are given by the manufacturer as: (1) bronchial asthma, (2) allergic rhinitis during the pollen season, (3) sinus bradycardia and greater than second degree or total heart block, (4) cardiogenic shock, (5) right ventricular failure secondary to pulmonary hypertension, (6) congestive heart failure *unless* the failure is secondary to a tachyarrhythmia treatable with the drug, (7) the use of anesthetics that produce myocardial depression (e.g., ether and chloroform), and (8) the use of adrenergic-augmenting psychotropic drugs (including monoamine oxidase inhibitors), and during the two week withdrawal period from such drugs. Furthermore, because the safety of usage during pregnancy has not been ascertained its use should be attended with good judgment concerning relative risks to the mother and the fetus.

The drug is available in 10 mg. tablets or as an injection containing 1 mg. per ml. The oral route of administration is preferred and it is given before meals. The usual dose for control of arrhythmias is 10 to 30 mg. 3 or 4 times daily; for hypertrophic subaortic stenosis 20 to 40 mg. 3 or 4 times daily; preoperatively (together with an α-blocker) for pheochromocytoma 60 mg. daily in divided doses. The usual intravenous dose is 1 to 3 mg. administered slowly (not over 1 mg. per min.) under monitoring with an electrocardiograph.

Cholinergic Blocking Agents Acting at the Neuromuscular Junction of the Voluntary Nervous System

Once again it must be pointed out that these blocking agents are treated in this chapter simply as a matter of convenience and not because they are considered as autonomic blocking agents. The principal point of similarity is that neuromuscular junctions of the voluntary system are mediated by acetylcholine and that these blockers have some points in common with some of the ganglion blocking agents which are certainly classed as autonomic blocking agents. The therapeutically useful compounds in this group are sometimes referred to as possessing "curariform" activity in reference to the original representatives of the class which were obtained from curare. Since then, synthetic compounds have been prepared with a similar activity. Although all of the compounds falling into this category, natural and synthetic alike, bring about substantially the same end-result, i.e., voluntary muscle relaxation, there are some significant differences in the mechanisms whereby this is brought about. Basically, the mechanisms involved are quite similar to those already encountered in the discussion on ganglionic blocking agents. Thus, the following types of neuromuscular junction blockers have been noted.

Depolarizing Blocking Agents. Drugs in this category are known to bring about a depolarization of the membrane of the muscle end-plate. This depolarization is quite similar to that produced by acetylcholine itself at ganglia and neuromuscular junctions (i.e., its so-called "nicotinic" effect), with the result that the drug, if it is in sufficient concentration, eventually will produce a block. It has been known for years that either smooth or voluntary muscle, when challenged repeatedly with a depolarizing agent, will eventually become insensitive. This phenomenon is known as *tachyphylaxis* or *desensitization* and is convincingly demonstrated under suitable experimental conditions with repeated applications of acetylcholine itself, the results indicating that within a few minutes the end-plate becomes insensitive to acetylcholine. The previous statements may imply that a blocking action of this type is quite clear-cut, but under experimental conditions it is not quite so clear and unambiguous because a block that initially begins with depolarization may regain the polarized state even before the block. Furthermore, a depolarization induced by increasing the potassium ion concentration does not prevent impulse transmission. For these and other reasons it is probably best to consider the blocking action as a desensitization until a clearer picture emerges. The drugs falling into this classification are decamethonium and succinylcholine.

Competitive Blocking Agents. There is no depolarization accompanying the block by these agents. It is thought that these agents successfully compete with acetylcholine for the receptor sites but, importantly, are unable to effect the necessary depolarization characteristic of the natural neuroeffector. Thus, by decreasing the effective acetylcholine-receptor combinations the end-plate potential becomes too small to initiate the propagated action potential. The action of these drugs is quite analogous to that of atropine at the muscarinic receptor sites of acetylcholine. All experiments suggest that the agonist (acetylcholine) and the antagonist compete on a one-to-one basis for the end-plate receptors. Drugs falling into this classification are tubocurarine, dimethyltubocurarine, gallamine, laudexium and benzoquinonium.

Mixed Blocking Agents. It has already been intimated that pure classifications of the blocking agents may be difficult. Because of this, some authorities believe that there are mixed types of blockers which possess both depolarizing and competitive components in the blocking action. Indeed, decamethonium and succinylcholine, while commonly classed as producing a depolarizing block, show evidence of some typical competitive action as well. Other examples could be cited, but, for the purposes of this discussion, it will be sufficient to recognize that such mixed types of action can and probably do occur.

CURARE AND CURARE ALKALOIDS

Originally *curare* was a term used to describe collectively the very potent arrow poisons used since early times by the South American Indians. The arrow poisons were prepared from numerous botanic sources and often were mixtures of several different plant extracts. Some were poisonous by virtue of a convulsant action and others by a paralyzant action. It is only the latter type that is of value in therapeutics and is ordinarily spoken of as "curare."

Chemical investigations of the curares were not especially successful because of the difficulties attendant on the obtaining of authentic samples of curare with definite botanic origin. It was only in 1935 that King was able to isolate a pure crystalline alkaloid, which he named *d*-tubocurarine chloride, from a curare of unknown botanic origin.[148] It was shown to possess, in great measure, the paralyzing action of the original curare. Wintersteiner and Dutcher,[149] in 1943, also isolated the same alkaloid. However, they showed that the botanical source was *Chondodendron tomentosum* (Fam. *Menispermaceae*) and thus provided a known source of the drug.

Following the development of quantitative bioassay methods for determining the potency of curare extracts, a purified and standardized curare was developed and marketed under the trade name of Intocostrin® (Purified Chondodendron Tomentosum Extract) the solid content of which consisted of almost one-half (+)-tubocurarine solids. More recently, (+)-tubocurarine chloride and its dimethyl ether have appeared on the market.

PRODUCTS

Tubocurarine Chloride U.S.P., (+)-Tubocurarine Dichloride, Tubarine®. This alkaloid is prepared from crude curare by a process of purification and crystallization and it occurs as the pentahydrate.

The structural considerations* have been investigated, principally by King and Wintersteiner and Dutcher. The structure given is that proposed by King.[148]

Tubocurarine Chloride
(R = H, R' = CH₃, X = Cl)
Dimethyltubocurarine Iodide
(R = R' = CH₃, X = I)

* In a recently published paper (Everett, A. J., *et al.*: Revision of the structures of (+)-tubocurarine chloride and (+)-chondrocurine, Chemical Communications, 1970, p. 1020) the authors indicate that the traditional structure for (+)-tubocurarine as given above is incorrect and that it is more adequately represented by the given formula where R = R' = H. In other words, this implies that the structure is not a bis-quaternary but is rather a mono-quaternary combined with a tertiary amine moiety. If the structure proposed is, indeed, correct (as seems likely), it should provide some food for thought to the theoreticians who have been proposing mechanisms for the blocking action of (+)-tubocurarine on the basis of its supposed diquaternary character.

Tubocurarine chloride occurs as a white or yellowish-white to gray or light tan, odorless crystalline powder, which is soluble in water. Chemically, it is a quaternary salt and solutions of it are stable to sterilization by heat.

This drug is of value for its paralyzing action on voluntary muscles, the site of action being the neuromuscular junction. Its action is inhibited or reversed by the administration of acetylcholinesterase inhibitors such as neostigmine or by edrophonium chloride (Tensilon Chloride) which acts by a competitive mechanism. Such inhibition of its action is necessitated in respiratory embarrassment due to overdosage. It is often necessary to use artificial respiration as an adjunct until the maximum curare action has passed. The drug is inactive orally because of inadequate absorption through lipoidal membranes in the gastrointestinal tract and, when used therapeutically, is usually injected intravenously.

Tubocurarine, in the form of a purified extract, was first used in 1943 as a muscle relaxant in shock therapy of mental disorders. By its use the incidence of bone and spine fractures and dislocations resulting from the convulsions due to shock were reduced markedly. Following this, it was employed as an adjunct in general anesthesia to obtain complete muscle relaxation, a usage that persists to this day. Prior to its use, satisfactory muscle relaxation in various surgical procedures (e.g., abdominal operations) was obtainable only with "deep" anesthesia using the ordinary general anesthetics. Tubocurarine permits a lighter plane of anesthesia with no sacrifice in the muscle relaxation so important to the surgeon. A reduced dose of tubocurarine is administered with ether, because ether itself has a curarelike action.

Another recognized use of tubocurarine is in the diagnosis of myasthenia gravis, because, in minute doses, it causes an exaggeration of symptoms by accentuating the already deficient acetylcholine supply. It has been used to a limited extent in the treatment of spastic, hypertonic and athetoid conditions, but one of its principal drawbacks has been its relatively short duration of activity. When used intramuscularly its action lasts longer than when given by the intravenous route although this characteristic has not made it useful in the above conditions. Tubocurarine has found increasing use with intravenous sodium thiopental anesthesia; however, care should be exercised, since the usual concentration of tubocurarine (3 mg. per ml.) may cause undesirable dilution and precipitation of the thiobarbiturate because of high acidity. To aid in overcoming this, more concentrated solutions (15 mg. per ml.) with a lower acidity may be used. Because of the high potency, adequate care should be taken to prevent confusion with the less potent solution.

Category—skeletal muscle relaxant.

Usual dose—intravenous, 6 to 9 mg., followed in 5 minutes by 3 to 6 mg. more if necessary.

OCCURRENCE

Tubocurarine Chloride Injection U.S.P.

Dimethyl Tubocurarine Iodide N.F.; Metubine® Iodide; Dimethyl Ether of (+)-Tubocurarine Iodide. This drug is prepared from natural crude curare by extracting the curare with methanolic potassium hydroxide. When the extract is treated with an excess of methyl iodide the (+)-tubocurarine is converted to the diquaternary dimethyl ether and crystallizes out as the iodide (*see* tubocurarine chloride). Other ethers besides the dimethyl ether also have been made and tested. For example, the dibenzyl ether was one third as active as tubocurarine chloride and the diisopropyl compound had only one half the activity. This is compared with the dimethyl ether which has two or three times the activity of tubocurarine chloride. It is only moderately soluble in cold water but more so in hot water. It is easily soluble in methanol but insoluble in the water-immiscible solvents. Aqueous solutions have a pH of from 4 to 5, and the solutions are stable unless exposed to heat or sunlight for long periods of time. As is the case with tubocurarine chloride, the acidity of the solution causes a transient precipitation of barbiturates (e.g., sodium thiopental) when used in intravenous anesthesia.

The pharmacologic action of this compound is the same as that of tubocurarine chloride, namely, a competitive blocking effect on the motor end-plate of skeletal muscles. However, it is considerably more potent than the latter and has the added advantage of exerting much less effect on the respiration. The effect on respiration is not a significant factor in therapeutic doses. Accidental overdosage is counteracted best by forced respiration.

The drug is used for much the same purposes as tubocurarine chloride but in a smaller dose. The dose ranges from 2 to 8 mg. The exact dosage is governed by the physician and depends largely on the depth of surgical relaxation.

It is marketed in the form of a parenteral solution in ampuls.

Category—skeletal muscle relaxant.

Usual dose range—intravenous, initial, 1.5 to 8 mg. given over a 60-second period; maintenance, 500 mcg. to 1 mg. every 25 to 90 minutes.

OCCURRENCE

Dimethyl Tubocurarine Iodide Injection N.F.

SYNTHETIC COMPOUNDS WITH CURARIFORM ACTIVITY

Almost 100 years have passed since Brown and Fraser[150] laid the basis for the study of quaternary ammonium salts with respect to their structure and pharmacologic activity. Since that time many quaternary salts have been investigated in an effort to find potent, easily synthesized curare substances. It has been found that the curarelike effect is a common property of all "onium" compounds. In the order of decreasing activity they are:

$$(CH_3)_4N^+ > (CH_3)_3S^+$$
$$(CH_3)_4P^+ > (CH_3)_4As^+ > (CH_3)_4Sb^+$$

Even ammonium, potassium and sodium ions and other ions of alkali metals have been shown to exhibit a certain amount of curare action.[151] Thus far, however, it has been impossible to establish any quantitative relationships between the magnitude or the mobility of the cation and the intensity of action. Possibly the only exception to the rule that "onium" compounds are necessary for curarelike activity has been the demonstrated activity of the *Erythrina* alkaloids which are known to contain a tertiary nitro-

gen. Indeed, they seem to lose their potency when the nitrogen is quaternized. Whether or not two quaternary groups are necessary still appears to be open to question in view of the reportedly high activity of certain monoquaternary compounds, naturally occurring and synthetic. However, there is a strong feeling that the presence of two or more such quaternary groups permits higher activity by virtue of a more firm attachment at the site of action. [152, 153]

Curare, until relatively recent times, remained the only useful curarizing agent, and it, too, suffered from a lack of standardization. The history of the development of curare into a reliable, standardized product (Intocostrin), together with the isolation and the structural characterization of (+)-tubocurarine chloride, has been mentioned already (q.v.). The establishment of the structure of (+)-tubocurarine chloride led other workers to hope for activity in synthetic substances of less complexity. The quaternary ammonium character of the curare alkaloids, coupled with the known activity of the various simple "onium" compounds, hardly seemed to be coincidental, and it was natural for research to follow along these lines.

One of the first approaches to the synthesis of this type of compound was based on the assumption that the highly potent effect of tubocurarine chloride was a function of some optimum spacing of the two quaternary nitrogens. Indeed, the bulk of the experimental work tends to suggest an optimum distance of 15 Angstroms between quaternary nitrogen atoms in most of the bis-quaternaries for maximum curariform activity. However, other factors may modify this situation. [154, 155] Bovet and his co-workers[156, 157, 158] were the first to develop synthetic compounds of significant potency. In the order of development, some of the potent compounds turned out by Bovet *et al.* in 1946-1947 are shown on page 588.

The third compound (C), although less potent than the first two, still had a potent action and had the advantage of producing less side-effects. It was marketed in 1951 as Flaxedil (gallamine triethiodide).

In 1948, another series of even simpler compounds was described independently

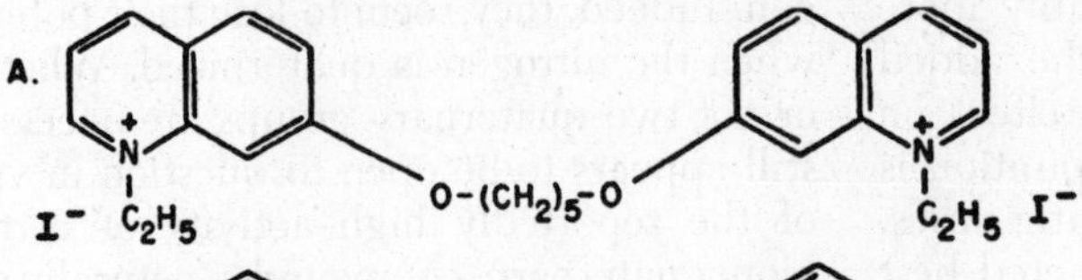

A.

I^- C_2H_5 $O-(CH_2)_5-O$ C_2H_5 I^-

3381 R.P.
Head-drop dose = 0.25 mg./kg.

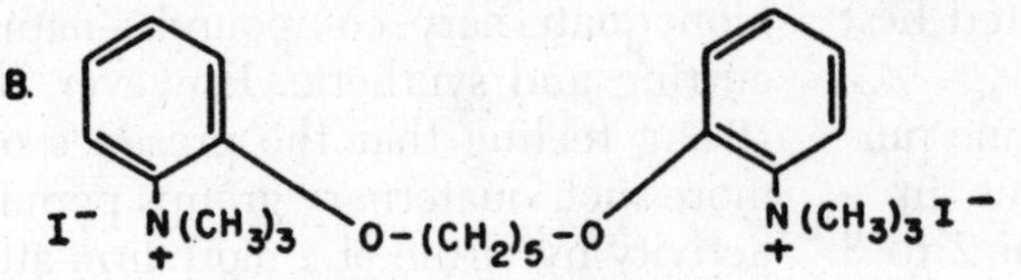

B.

I^- $N(CH_3)_3$ $O-(CH_2)_5-O$ $N(CH_3)_3$ I^-

3565 R.P.
Head-drop dose = 0.2 mg./kg.

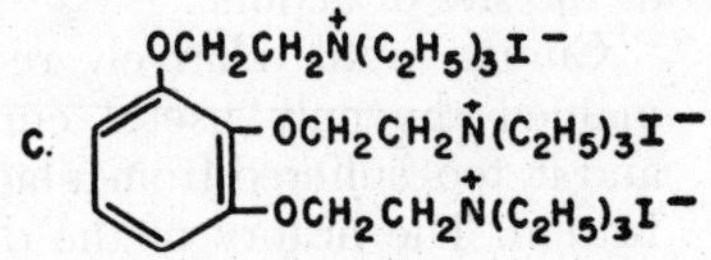

C.

$OCH_2CH_2\overset{+}{N}(C_2H_5)_3 I^-$
$OCH_2CH_2\overset{+}{N}(C_2H_5)_3 I^-$
$OCH_2CH_2\overset{+}{N}(C_2H_5)_3 I^-$

3697 R.P., 2559 F; Flaxedil
Head-drop dose = 0.5 mg./kg.

by Barlow and Ing[159] and by Paton and Zaimis.[160] These were the *bis*-trimethyl-ammonium polymethylene salts (formula below), and certain of them were found to possess a potency greater than that of (+)-tubocurarine chloride itself. Both groups concluded that the decamethylene compound was the best in the series and that the shorter chain lengths exhibited only feeble activity. The accompanying formula

$$Br^-(CH_3)_3\overset{+}{N}-(CH_2)_n-\overset{+}{N}(CH_3)_3 \, Br^-$$

n = 2 to 5, feebly active
    7 to 9, rise in activity
    9 to 12, constant activity
        at high level
    13, slight drop in activity

shows the conclusions of Barlow and Ing with respect to these compounds. The commercially obtainable preparation known as

Cavallito and his co-workers[153] introduced another type of quaternary ammonium compound with a high curarelike activity. This type is represented by the following formula and may be designated as ammonium-alkylaminobenzoquinones. The distance between the "onium" centers is the same as in other less-active "onium" compounds without the quinone structure, and for this reason, they speculate that the quinone itself may be involved in the activity. This appears to be reasonable in view of the fact that even the monoquaternary compounds and the corresponding nonquaternized amines show a significant curarelike activity. One compound (benzoquinonium chloride) was selected from this study and was marketed as Mytolon Chloride but has been withdrawn (n = 3, R_1 = C_2H_5, R_2 = $CH_2C_6H_5$).

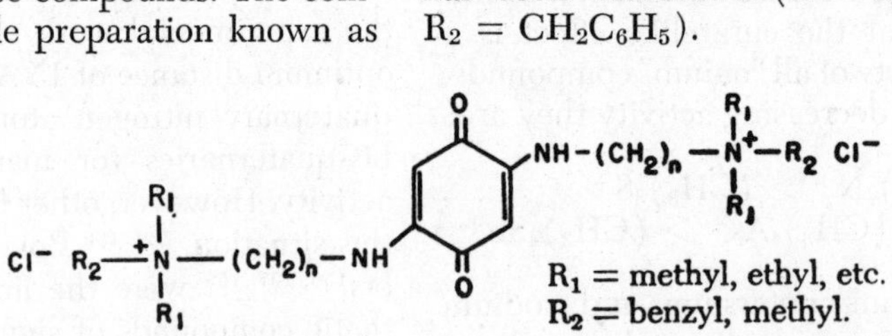

R_1
Cl^- R_2—$\overset{+}{N}$—$(CH_2)_n$—NH ... $NH-(CH_2)_n-\overset{+}{N}-R_2$ Cl^-
R_1 R_1

R_1 = methyl, ethyl, etc.
R_2 = benzyl, methyl.

decamethonium (represented by the formula above where n = 10, salt may be I or Br) represents the decamethylene compound. An interesting finding is that the shorter chain compounds, such as the pentamethylene and the hexamethylene compounds, are effective antidotes for counteracting the blocking effect of the decamethylene compound.

Other laboratories engaged in research toward new and better neuromuscular blocking agents produced the stilbazoline quaternary ammonium salts,[161] the *Erythrina* alkaloids,[162] and laudexium methylsulfate[163, 164] (old name, laudolissin). The last named compound was largely tubocurarinelike in its action, in keeping with the fact that *bis*-onium compounds in which

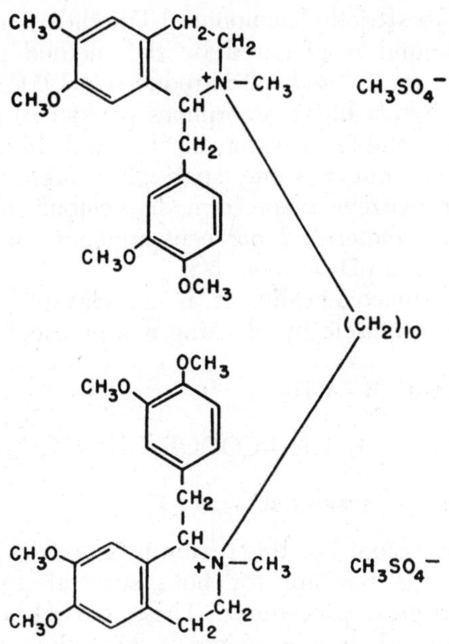

Laudexium Methylsulfate

the onium head forms part of a heterocyclic system are tubocurarinelike rather than decamethoniumlike in action.[165] It has not been marketed in the U.S. The English group of Stenlake *et al.* have carried out interesting studies on what may be termed

"poly-onium" compounds.[166, 167] These are linear compounds with varying methylene chain lengths between 3,4, and 5 onium centers. It was found that the number of onium heads influenced potency, reversibility and duration of action, whereas the length of the methylene chain between the onium centers determined whether the action would be tubocurarinelike or decamethoniumlike. For maximum tubocurarine type action a chain length of 5 to 6 methylene units seemed optimal with a mixed type of action becoming predominant as the chain was lengthened beyond that. An interesting area of research has been found in the dicarboxylic acid *bis-β*-tertiaryaminoalkyl amides (I) and their quaternary salts[168] (II). Whereas they possess little or no curarelike activity at the dose levels usually used, nevertheless, they exert a powerful activity in prolonging the duration of neuromuscular block produced by succinylcholine chloride (III) which is a related curarelike agent.[169] In this respect, the *bis-β*-piperidinoethyl succinamide (IV) was about 5 times as active as the *bis-β*-dimethylaminoethylsuccinamide *bis*-methiodide (V).[170] However, the *bis-β*-piperidinoethylsuccinamide when converted to the *bis*-methiodide was virtually without activity as a potentiator of suc-

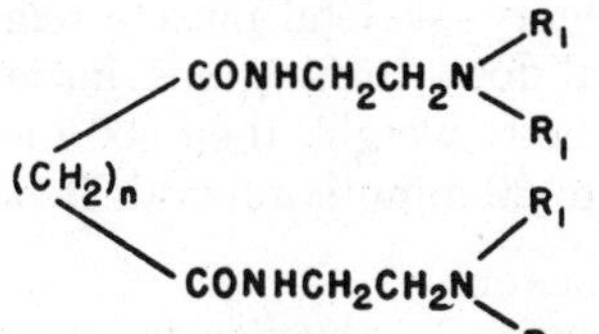

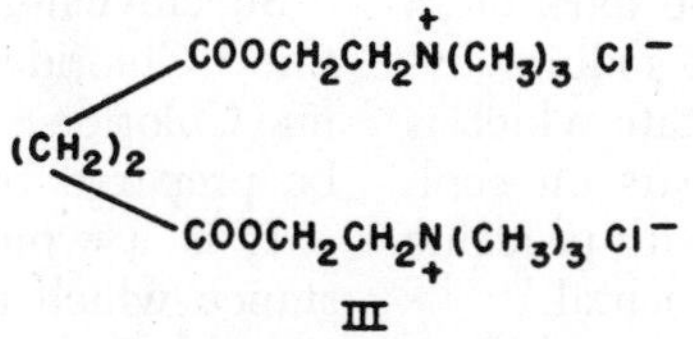

III
(Succinylcholine Chloride)

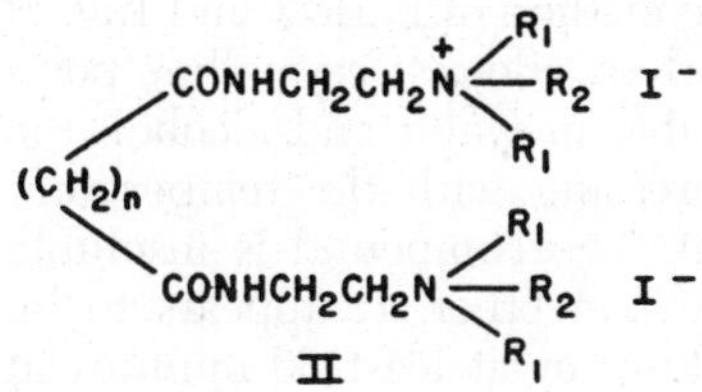

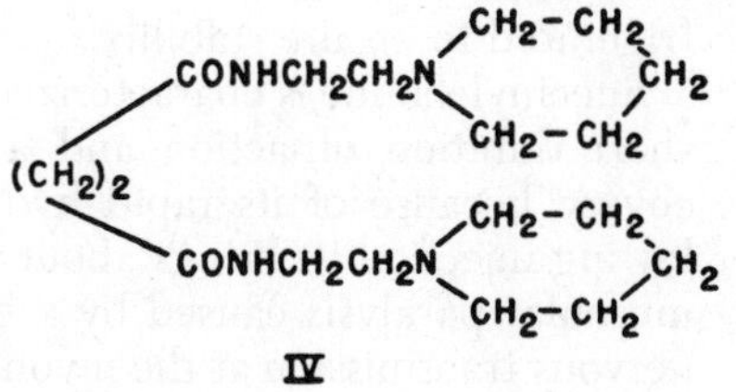

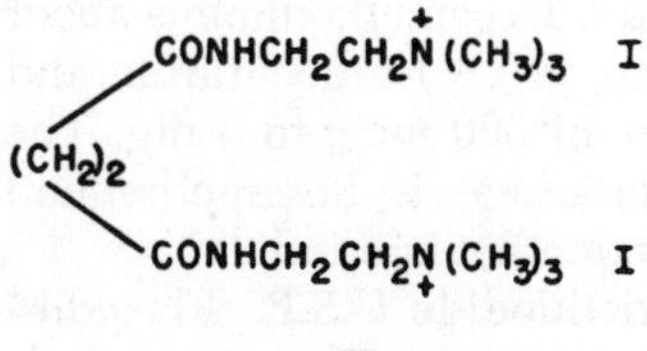

cinylcholine. (See structure on p. 589.)

In connection with the useful drug that has emerged from this research, i.e., succinylcholine chloride, it can be said that virtually every possible modification has been made but the originally discovered parent compound still remains the best. However, studies on the probable conformation in which succinylcholine chloride acts at the receptor site have led to some other interesting (and active) compounds.

$$\begin{array}{l} \overset{+}{CH_2}-COOCH_2CH_2\overset{+}{N}(CH_3)_3 \\ | \qquad\qquad\qquad\qquad\qquad 2Cl^- \\ CH_2-COOCH_2CH_2N(CH_3)_3 \\ \qquad\qquad\qquad\qquad\quad + \end{array}$$

"Eclipsed"

Among the first were the dicholine esters of maleic and fumaric acid, prepared with the objective of determining whether succinylcholine acts in the "eclipsed" or the "staggered" conformation.[171] With the same objective in mind, Burger and Bedford[172] and McCarthy *et al.*[173] have prepared dicholine esters of the *cis*- and *trans*-cyclopropane dicarboxylic acids. In all cases it was evident that the "staggered" conformation was the most effective.

Decamethonium Bromide, Syncurine®, Decamethylene - *bis* - (trimethylammonium Bromide). This compound is prepared according to the method of Barlow and Ing.[159]

It is a colorless, odorless crystalline powder. It is soluble in water and alcohol, the solubility increasing with the temperature of the solvent. The compound is insoluble in chloroform and ether. It appears to be stable to boiling for at least 30 minutes in physiologic saline, either in the dark or in the sunlight. Twenty per cent sodium hydroxide causes a white precipitate which is soluble on heating but reappears on cooling. Solutions are compatible with procaine hydrochloride and sodium thiopental.

The drug is used as a muscle relaxant, especially in combination with the anesthetic barbiturates. Decamethonium is about 5 times as potent as (+)-tubocurarine and is used in a dose of 500 mcg. to 3 mg. The antidote to overdosage is hexamethonium bromide or pentamethonium iodide.

Gallamine Triethiodide U.S.P., Flaxedil® Triethiodide,   [*v*-Phenenyltris (oxyethyl-

ene)]tris[triethylammonium] Triiodide. This compound is prepared by the method of Bovet *et al.*[158] and was introduced in 1951. It is a slightly bitter, amorphous powder. It is freely soluble in water, alcohol and dilute acetone, but it is only sparingly soluble in ether, benzene, chloroform, dry acetone and ethyl acetate. A 2 per cent aqueous solution has a pH of about 5.8.

Pharmacologically, it is a relaxant of skeletal muscle by blocking neuromuscular

$$\begin{array}{l} (CH_3)_3\overset{+}{N}CH_2CH_2OOC-CH_2 \\ \qquad\qquad\qquad\qquad\qquad | \qquad\qquad\quad + \quad 2Cl^- \\ \qquad\qquad\qquad\qquad CH_2-COOCH_2CH_2\overset{+}{N}(CH_3)_3 \end{array}$$

"Staggered"

transmission. For this reason it is used as a muscular relaxant for both surgical and nonsurgical procedures. These have been mentioned in the general discussion of curare. It has an advantage over (+)-tubocurarine in that it exerts little or no effect on the autonomic ganglia, and it is readily miscible with the thiobarbiturate solutions used in anesthesia.

The drug is contraindicated in patients with myasthenia gravis, and it should also be borne in mind that the drug action is cumulative, as with curare. The antidote for gallamine triethiodide is neostigmine.

Category—skeletal muscle relaxant.

Usual dose—intravenous, initial, 1 mg. per kg. of body weight, then 500 mcg. to 1 mg. at 30- to 60-minute intervals if needed.

OCCURRENCE
Gallamine Triethiodide Injection U.S.P.

Succinylcholine Chloride U.S.P., Anectine® Chloride, Sucostrin® Chloride, Choline Chloride Succinate. This compound may be prepared by the method of Phillips.[174]

It is a white, odorless, crystalline substance which is freely soluble in water. It is stable in acidic solutions but unstable in alkali. The aqueous solutions should be refrigerated to ensure stability.

Succinylcholine is characterized by a very short duration of action and a quick recovery because of its rapid hydrolysis following injection. It brings about the typical muscular paralysis caused by a blocking of nervous transmission at the myoneural junction. Large doses may cause a temporary respiratory depression in common with other similar agents. Its action, in contrast with that of (+)-tubocurarine, is not antagonized by neostigmine, physostigmine or edrophonium chloride. These anticholinesterase drugs actually prolong the action of succinylcholine, and on this basis it is believed that the drug probably is hydrolyzed by cholinesterases. The brief duration of action of this curarelike agent is said to render an antidote unnecessary if the other proper supportive measures are available. However, succinylcholine has a disadvantage in that its action cannot be terminated promptly by the usual antidotes. This difficulty has led to further research, in an effort to overcome it.

It is used as a muscle relaxant for the same indications as other curare agents. It may be used for either short or long periods of relaxation, depending on whether one or several injections are given. In addition, it is suitable for the continuous intravenous drip method.

Succinylcholine chloride should not be used with sodium thiopental because of the high alkalinity of the latter or, if used together, should be administered immediately following mixing. However, separate injection is preferable.

Category—skeletal muscle relaxant.

Usual dose—intravenous, 30 mg.; infusion, 2.5 mg. of a 0.1 to 0.2 per cent solution per minute.

Usual dose range—10 to 40 mg.

OCCURRENCE

Sterile Succinylcholine Chloride U.S.P.
Succinylcholine Chloride Injection U.S.P.

Hexafluorenium Bromide, Mylaxen®, Hexamethylene-1,6-bis(9-fluorenyldimethylammonium Dibromide). This compound occurs as a white, water-soluble crystalline

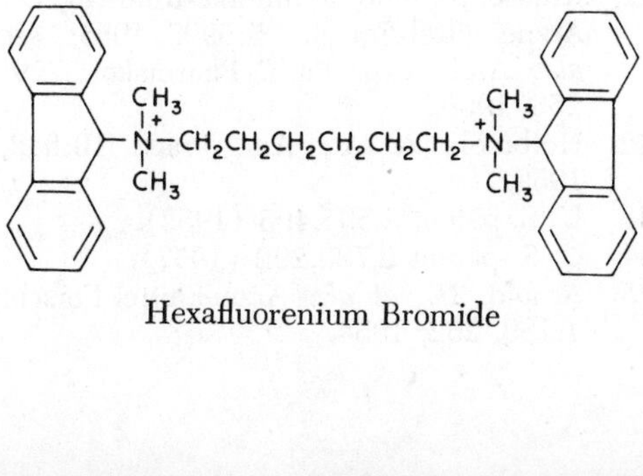

Hexafluorenium Bromide

material. Aqueous solutions are stable at room temperature but are incompatible with alkaline solutions. Hexafluorenium bromide is used clinically to modify the dose and extend the duration of action of succinylcholine chloride.

In most cases the dose of succinylcholine chloride can be lowered to about one fifth the usual total dose. The combination produces profound relaxation and facilitates difficult surgical procedures. Its principal mode of action appears to be suppression of the enzymatic hydrolysis of succinylcholine chloride, thereby prolonging its duration of action.

Edrophonium Chloride U.S.P., Tensilon® Chloride, Ethyl(m-hydroxyphenyl)dimethylammonium Chloride.

This compound has the following structure:

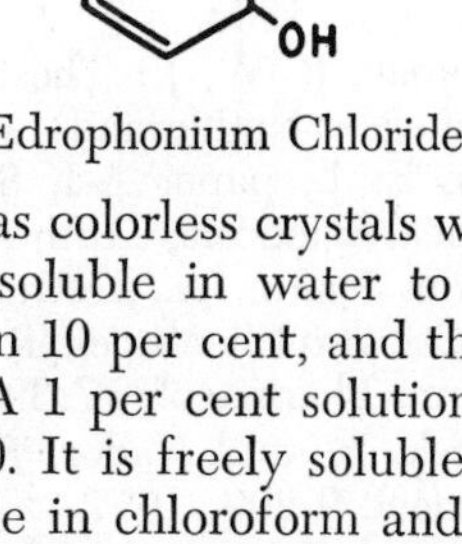

Edrophonium Chloride

It occurs as colorless crystals with a bitter taste. It is soluble in water to the extent of more than 10 per cent, and the solutions are stable. A 1 per cent solution has a pH of 4.0 to 5.0. It is freely soluble in alcohol but insoluble in chloroform and ether.

It is a specific anticurare agent and acts within 1 minute to alleviate overdosage of tubocurarine, dimethyl tubocurarine, or gallamine triethiodide. The drug also is used to terminate the action of any one of the above drugs when the physician so desires. However, it is of no value in terminating the action of the depolarizing blocking agents such as decamethonium, succinylcholine, etc. because it acts in a competitive manner.

Edrophonium chloride is related structurally to neostigmine methylsulfate and because of this has been tested as a potential diagnostic agent for myasthenia gravis.[175] It has been found to bring about a rapid increase in muscle strength without significant side-effects. The drug usually is administered in a 1-ml. intravenous dose, and the test is completed within 2 minutes.

Category—antidote to curare principles.

Usual dose—intravenous, 10 mg., repeated in 5 to 10 minutes if necessary.

Usual dose range—5 to 40 mg. in one episode.

OCCURRENCE

Edrophonium Chloride Injection U.S.P.

REFERENCES

1. Bachrach, W. H.: Am. J. Digest. Dis. 3:743, 1958.
2. Dragstedt, L. R.: J.A.M.A. 169:203, 1959.
3. Asher, L. M.: Am. J. Digest. Dis. 4:250, 1959.
4. Holmstedt, B., *et al.*: Biochem. Pharmacol. 14:189, 1965.
5. Ariens, E. J., *et al.*, *in* Ariens, E. J. (ed.): Molecular Pharmacology, vol. 1, pp. 137, 200, New York, Academic Press, 1964.
6. Long, J. P., *et al.*: J. Pharmacol. Exp. Ther. 117:29, 1956.
7. Ariens, E. J., *et al.*, *in* Molecular Pharmacology, vol. 1, p. 258, New York, Academic Press, 1964.
8. Ellenbroek, B. W. J.: Thesis, University of Nijmegen, Netherlands, 1964, through Ariens, E. J., paper A-I, Scientific and Technical Symposia, 112th Annual Meeting, Am. Pharm. A., Detroit, Michigan, 1965. *See also* Ellenbroek, B. W. J., *et al.*: J. Pharm. Pharmacol. 17:393, 1965.
9. Gyermek, L., and Nador, K.: J. Pharm. Pharmacol. 9:209, 1957.
10. Chemnitius, F.: J. prakt. Chem. 116:276, 1927; *see also* Hamerslag, F.: The Chemistry and Technology of Alkaloids, p. 264, New York, Van Nostrand, 1950.
11. J. Am. Pharm. A. (Pract. Ed.) 8:377, 1947.
12. Zvirblis, P., *et al.*: J. Am. Pharm. A. (Sci. Ed.) 45:450, 1956.
13. Kondritzer, A. A., and Zvirblis, P.: J. Am. Pharm. A. (Sci. Ed.) 46:531, 1957.
14. Rodman, M. J.: Am. Prof. Pharm. 21:1049, 1955.
15. Ralph, C. S., and Willis, J. L.: Proc. Roy. Soc. [N. S. Wales] 77:99, 1944.
16. J. Am. Pharm. A. (Pract. Ed.) 7:109, 1946.
17. Von Oettingen, W. F.: The Therapeutic Agents of the Pyrrole and Pyridine Group, p. 130, Ann Arbor, Edwards, 1936.
18. Ariens, E. J., *et al.*: *in* E. J. Ariens (ed.): Molecular Pharmacology, Vol. 1, p. 205, New York, Academic Press, 1964.
19. Brodie, B., and Hogben, C. A. M.: J. Pharm. Pharmacol. 9:345, 1957.
20. Cavallito, C. J., and O'Dell, T. B.: J. Am. Pharm. A. 47:169, 1958.
21. The Medical Letter 5:86, 1963.
22. Pittenger, P. S., and Krantz, J. C.: J. Am. Pharm. A. 17:1081, 1928.
23. U. S. patent 2,753,288 (1956).
24. Kirsner, J., and Palmer, W.: J.A.M.A. 151:798, 1953.
25. Graham, J. D. P., and Lazarus, S.: J. Pharmacol. Exp. Ther. 70:165, 1940.
26. Sollmann, T.: A Manual of Pharmacology, ed. 8, p. 384, Philadelphia, Saunders, 1957.
27. Fromherz, K.: Arch. exp. Path. Pharmakol. 173:86, 1933.
28. Nash, J. B., *et al.*: J. Pharmacol. Exp. Ther. 122:56A, 1958.
29. Sternbach, L. H., and Kaiser, S.: J. Am. Chem. Soc. 74:2219, 1952.
30. U.S. patent 2,648,667 (1953).
31. Treves, G. R., and Testa, F. C.: J. Am. Chem. Soc. 74:46, 1952.
32. Tilford, C. H., *et al.*: J. Am. Chem. Soc. 69:2902, 1947.
33. The Medical Letter 4:30, 1962.
34. Biel, J. H., *et al.*: J. Am. Chem. Soc. 77: 2250, 1955; *see also* Long, J. P., and Keasling, H. K.: J. Am. Pharm. A. (Sci. Ed.) 43:616, 1954.
35. Burtner, R. R., and Cusic, J. W.: J. Am. Chem. Soc. 65:1582, 1943.
36. Faust, J. A., *et al.*: J. Am. Chem. Soc. 81:2214, 1959.
37. Steigmann, F., *et al.*: Am. J. Gastroent. 33:109, 1960.
38. Swiss patent 259,958 (1949); *see also* Chem. Abstr. 44:5910, 1950.
39. Brit. patent 781,382 (1957); U.S. patent 2,987,517 (1961).
40. Blicke, F. F., and Tsao, M. U.: J. Am. Chem. Soc. 66:1645, 1944; *see also* U. S. patent 2,541,634 (1951).
41. Krause, D., and Schmidtke-Ruhman, D.: Arzneimittel-Forsch. 5:599, 1955; *see also* Arch. exp. Path. Pharmakol. 229: 258, 1956.
42. Holbrook, A. A.: Am. Pract. 10:842, 1959.
43. U. S. patent 2,595,405 (1952).
44. U. S. patent 2,785,202 (1957).
45. Arnold, H., *et al.*: Arzneimittel-Forsch. 4:189, 262, 1954.

46. Doshay, L. J., and Constable, K.: J.A.M.A. 170:37, 1959.

47. U. S. patent 2,567,351 (1951).

48. Swan, K. C., and White, N. G.: U. S. patents 2,408,898 and 2,432,049.

49. ———: Arch. Ophth. 31:289, 1944.

50. ———: J. Pharmacol. Exp. Ther. 80: 285, 1944.

51. Peterson, C. G., and Peterson, D. R.: Gastroenterology 5:169, 1945.

52. Haas, H., and Klavehn, W.: Arch. exp. Path. Pharmakol. 226:18, 1955.

53. Denton, J. J., et al.: J. Am. Chem. Soc. 72:3795, 1950.

54. Magee, K., and DeJong, R.: J.A.M.A. 153:715, 1953.

55. U. S. patent 2,907,765 (1959).

56. Kasich, A. M., and Fein, H. D.: Am. J. Dig. Dis. 3:12, 1958.

57. U. S. patent 2,665,278 (1954).

58. McCarthy, J., et al.: J. Pharmacol. Exp. Ther. 108:246, 1953.

59. U. S. patent 2,891,890 (1959); see also Adamson et al.: J. Chem. Soc. 52, 1951.

60. U. S. patent 2,826,590 (1958).

61. Schwab, R. S., and Chafetz, M. E.: Neurology, 5:273, 1955.

62. U. S. patent 2,764,519 (1956).

63. The Medical Letter 3:66, 1961.

64. Denton, J. J., and Lawson, V. A.: J. Am. Chem. Soc. 72:3279, 1950; see also U. S. patent 2,698,325.

65. Denton, J. J., et al.: J. Am.: Chem. Soc. 71:2053, 1949.

66. Cheney, L. C., et al.: J. Org. Chem. 17: 770, 1952.

67. Hoekstra, J. B., et al.: J. Pharmacol. Exp. Ther. 110:55, 1954.

68. Janssen, P., et al.: Arch. intern. pharmacodyn. 103:82, 1955.

69. Judge, R. D., et al.: J. Lab. Clin. Med. 47:950, 1956.

70. U. S. patent 2,726,245 (1955).

71. U. S. Patents 2,530,451 and 2,607,773; see also Charpentier, P.: Compt. rend. Acad. Sci. 225:306, 1947; Huttrer, C. P.: Enzymologia 12:293, 1948.

72. Doshay, L. J., and Constable, K.: Neurology 1:68, 1951.

73. Schwab, R. S.: Postgrad. Med. 9:52, 1951.

74. Raffle, R. B.: Practitioner 168:62, 1952.

75. U. S. patent 2,607,773 (1952).

76. Sperber, N., et al.: J. Am. Chem. Soc. 73:5010, 1951.

77. Wilson, C. O., and Gisvold, O. (eds.): Textbook of Organic Medicinal and Pharmaceutical Chemistry, ed. 3, p. 364, Philadelphia, Lippincott, 1956.

78. Kreitmair, H.: Arch. exp. Path. Pharmakol. 164:509, 1932.

79. Von Fodor, G.: Chem. Abstr. 32:2124, 1938. See also Ber. deutsch. chem. Ges. 71:541, 1938; 76:1216, 1943.

80. Blicke, F. F.: Ann. Rev. Biochem. 13: 549, 1944.

81. Buth, W., et al.: Ber. deutsch. chem. Ges. 72:19, 1939.

82. Külz, F., et al.: Bericht. deutsch. chem. Ges. 72:2161, 1939.

83. Külz, F., and Rosenmund, K. W.: Klin. Wschr. 17:345, 1938.

84. Issekutz, B. V., Jr.: Arch. exp. Path. Pharmakol. 177:388, 1935.

85. Weijlard, J., et al.: J. Am. Chem. Soc. 71:1889, 1949.

86. Voyles, C., et al.: J.A.M.A. 153:12, 1953.

87. U. S. patent 2,728,769.

88. Külz, F., et al.: Ber. deutsch. chem. Ges. 72:2165, 1939.

89. Paton, W. D. M., and Perry, W. L. M.: J. Physiol. 112:49P, 1951. See also J. Physiol. 114:47P.

90. Van Rossum, J. M.: Int. J. Neuropharmacol. 1:97, 1962.

91. ———: Int. J. Neuropharmacol. 1:403, 1962.

92. Garrett, J.: Arch. int. pharmacodyn. 144: 381, 1963.

93. Paton, W. D. M., and Zaimis, E. J.: Brit. J. Pharmacol. 4:381, 1949.

94. Ahlquist, R. P.: Am. J. Physiol. 154:585, 1948.

95. Dale, H. H.: J. Physiol. 34:163, 1906.

96. Abrams, W. B., et al.: J. New Drugs 4:268, 1964.

97. Boura, A. L. A., et al.: Lancet 2:17, 1959. See also Boura, A. L. A., and Green, A. F.: Brit. J. Pharmacol. 14:536, 1959.

98. Maxwell, R. A., et al.: Experientia 15: 267, 1959. See also Nature 180:1200, 1957.

99. Moran, J. F., Triggle, C. R., and Triggle, D. J.: J. Pharm. Pharmacol. 21:38, 1969.

100. Nickerson, M.: Pharmacol. Rev. 1:27, 1949.

101. Luduena, F. P., et al.: Arch. int. pharmacodyn. 122:111, 1959.

102. Jacobs, W. A., and Craig, L. C.: J. Am. Chem. Soc. 60:1701, 1938.

103. Craig, L. C., et al.: J. Biol. Chem. 125: 289, 1938.

104. Uhle, F. C., and Jacobs, W. A.: J. Organic Chem. 10:76, 1945.

105. Stoll, A.: Chem. Rev. 47:197, 1950.

106. Kornfeld, E. E., *et al.*: J. Am. Chem. Soc. 76:5256, 1954; *see also* J. Am. Chem. Soc. 78:3087, 1956.

107. Brugger, J.: Helv. physiol. pharmacol. acta 3:117, 1945.

108. Huebner, C. F., *et al.*: J. Am. Chem. Soc. 77:469, 1955.

109. Annan, S.: Ber. ges. Physiol. 53:430, 1930 (abstr.).

110. Fourneau, E., *et al.*: J. pharm. chim. 18: 185, 1933. See also Benoit, G., and Bovet, M. D.: J. pharm. chim. 22:544, 1935.

111. Eisleb, O.: U. S. patent 1,949,247. *See also* Chem. Abstr. 28:2850, 1934.

112. Nickerson, M., and Goodman, L. S.: J. Pharmacol. Exp. Ther. 89:167, 1947.

113. Ullyot, G. E., and Kerwin, J. F.: Medicinal Chemistry, vol. 2, p. 234, New York, Wiley, 1956.

114. Belleau, B.: Canad. J. Phys. 36:731, 1958. See also J. Med. Pharm. Chem. 1:327, 1959.

115. Belleau, B.: Ann. N.Y. Acad. Sci. 139: 580, 1967.

116. Wenner, W.: J. Organic Chem. 16:1475, 1951.

117. Randall, L. O., and Smith, T. H.: J. Pharmacol. Exp. Ther. 103:10, 1951.

118. Moore, P. E., *et al.*: J. Pharmacol. Exp. Ther. 106:14, 1952.

119. Hartmann, M., and Isler, H.: Arch. exp. Path. Pharmakol. 192:141, 1939.

120. Schnetz, H., and Fluch, M.: Z. klin. Med. 137:667, 1940.

121. Roberts, G., *et al.*: J. Pharmacol. Exp. Ther. 105:466, 1952.

122. Scholz, C. R.: Ind. Eng. Chem. 37:120, 1945.

123. Urech, E. A., *et al.*: Helv. chim. acta 33: 1386, 1950.

124. Stern, M. J., *et al.*: J. Am. Pharm. A. 48:641, 1959.

125. Powell, C. E., and Slater, I. H.: J. Pharmacol. Exp. Ther. 122:480, 1958.

126. Moran, N. C., and Perkins, M. E.: *Ibid.* 124:223, 1958.

127. Furchgott, R. F.: Pharmacol. Rev. 11: 429, 1959.

128. Dresel, P. E.: Canad. J. Biochem. Physiol. 38:375, 1960.

129. Fleming, W. W., and Hawkins, D. F.: J. Pharmacol. Exp. Ther. 129:1, 1960.

130. Black, J. W., and Stephenson, J. S.: Lancet, 2:311, 1962.

131. Black, J. W., *et al.*: *Ibid.* 1:1080, 1964.

132. Larsen, A. A., and Lish, P. M.: Nature, London, 203:1283, 1964.

133. Corrodi, H. *et al.*: J. Med. Chem. 6:751, 1963.

134. Engelhardt, A.: Arch. Exp. Path. Pharmak. 250:245, 1965.

135. Burns, J. J., and Lemburger, L.: Fed. Proc. 24:298, 1965.

136. Dornhorst, A. C.: Ann. N.Y. Acad. Sci. 139:968, 1967.

137. Lucchesi, B. R., and Hardman, H. F.: J. Pharmacol. Exp. Ther. 132:372, 1961.

138. Lucchesi, B. R., *et al.*: Ann. N.Y. Acad. Sci. 139:940, 1967.

139. Vargaftig, B., and Coignet, J. L.: European J. Pharmacol. 6:49, 1969.

140. Pratesi, P., *et al.*: J. Chem. Soc. p. 2069, 1958.

141. LaManna, A., and Ghislandi, V.: Farmaco (Pavia), Ed. Sci. 19:377, 1964.

142. Howe, R.: Biochem. Pharmacol. 12: suppl. 85, 1963.

143. Ariëns, E. J.: Ann. N.Y. Acad. Sci. 139: 606, 1967.

144. Biel, J. H., and Lum, B. K. B.: Progr. Drug Res. 10:46, 1966.

145. Ghouri, M. S. K., and Haley, T. J.: J. Pharm. Sci. 58:511, 1969.

146. Belgian patent 640,312 (1964).

147. Howe, R., and Shanks, R. G.: Nature, London 210:1336, 1966.

148. King, H.: J. Chem. Soc. 1381, 1935. *See also* 265, 1948.

149. Wintersteiner, O., and Dutcher, J. D.: Science 97:467, 1943.

150. Brown, A. C., and Fraser, T.: Tr. Roy. Soc. Edinburgh 25:151, 693, 1868-1869.

151. Craig, L. E.: Chem. Rev. 42:285, 1948.

152. Phillips, A. P., and Castillo, J. C.: J. Am. Chem. Soc. 73:3949, 1951.

153. Cavallito, C. J., *et al.*: J. Am. Chem. Soc. 72:2661, 1950.

154. ——: J. Am. Chem. Soc. 76:1862, 1954.

155. Macri, F. J.: Proc. Soc. Exp. Biol. Med. 85:603, 1954.

156. Bovet, D., Courvoisier, S., Ducrot, R., and Horclois, R.: Compt. rend. Acad. sci. 223:597, 1946.

157. Bovet, D., Courvoisier, S., and Ducrot, R.: Compt. rend. Acad. sci. 224:1733, 1947.

158. Bovet, D., Depierre, F., and de Lestrange, Y.: Compt. rend. Acad. sci. 225:74, 1947.

159. Barlow, R. B., and Ing, H. R.: Nature, London 161:718, 1948.

160. Paton, W. D. M., and Zaimis, E. J.: Nature, London 161:718, 1948.

161. Phillips, A. P., and Castillo, J. C.: J. Am. Chem. Soc. 73:3949, 1951.

162. Taylor, E. P., and Collier, H. O. J.: Nature, London 167:692, 1951.
163. ———: J. Chem. Soc. 142-145, 1952.
164. Wylie, W. D.: Lancet 263:517, 1952.
165. Stenlake, J. B. *in* Ellis, G., and West, G. B. (eds.): Progress in Medicinal Chemistry, Vol. 3, pp. 1-44, London, Butterworths, 1963.
166. Edwards, D., *et al.*: J. Med. Pharm. Chem. 3:369, 1961.
167. Carey, F. M., *et al.*: J. Pharm. Pharmacol. (Suppl.) 16:89T, 1964.
168. Phillips, A. P.: J. Am. Chem. Soc. 73: 5822, 1951.
169. Bourne, J. G., Collier, H. O. J., and Somers, G. F.: Lancet 262:1225, 1952.
170. Phillips, A. P.: J. Am. Chem. Soc. 74: 4320, 1952.
171. McCarthy, J. F., *et al.*: J. Pharm. Sci. 52:1168, 1963.
172. Burger, A., and Bedford, G. R.: J. Med. Chem. 6:402, 1963.
173. McCarthy, J. F.: *et al.*: J. Med. Chem. 7:72, 1964.
174. Phillips, A. P.: J. Am. Chem. Soc. 71: 3264, 1949.
175. Osserman, K. E., and Kaplan, L. I.: J.A.M.A. 150:265, 1952.

SELECTED READING

A.

Ariëns, E. J., Simonis, A. M., and van Rossum, J. M.: Drug-Receptor Interactions: Interaction of One or More Drugs with One Receptor System *in* Ariëns, E. J. (ed.): Molecular Pharmacology, vol. 1, p. 119-286, New York, Academic Press, 1964.

Bachrach, W. H.: Anticholinergic Drugs, Am. J. Dig. Dis. 3:743, 1958.

Barlow, R. B.: Atropine and Related Compounds *in* Introduction to Chemical Pharmacology, ed. 2, pp. 214-240, New York, Wiley, 1964.

Bebbington, A., and Brimblecombe, R. W.: Muscarinic Receptors in the Peripheral and Central Nervous Systems, *in* Harper, N. J., and Simmonds, A. B. (eds.): Advances in Drug Research, vol. 2, p. 143-172, New York, Academic Press, 1965.

Burtner, R. R.: Antispasmodics: derivatives of carboxylic acids *in* Suter, C. M. (ed.): Medicinal Chemistry, vol. 1, p. 151, New York, Wiley, 1951.

Cannon, J. G., and Long, J. P.: Postganglionic Parasympathetic Depressants (Cholinolytic or Atropinelike Agents) *in* Burger, A. (ed.): Drugs Affecting the Peripheral Nervous System, p. 133-148, New York, Dekker, 1967.

Friedman, A. H., and Everett, G. M.: Pharmacological Aspects of Parkinsonism, *in* Garattini, S., *et al.* (eds.): Advances in Pharmacology, vol. 3, p. 83, New York, Academic Press, 1964.

Lichtin, J. L.: The autonomic nervous system, Am. Prof. Pharm. 23:56, 144, 1957.

———: Parasympathetic blocking agents. Am. Prof. Pharm. 25:163, 1959.

Rama Sastry, B. V.: Anticholinergics: Antispasmodic and Antiulcer Drugs, *in* Burger, A. (ed.): Medicinal Chemistry, ed. 3, p. 1544, New York, Wiley-Interscience, 1970.

Triggle, D. J.: Chemical Aspects of the Autonomic Nervous System, p. 106-122, New York, Academic Press, 1965.

B.

Barlow, R. B.: Antagonists at Autonomic Ganglia *in* Introduction to Chemical Pharmacology, ed. 2, pp. 160-184, New York, Wiley, 1964.

Cavallito, C. J.: Molecular Modifications among Antihypertensive Agents *in* Gould, R. F. (ed.): Molecular Modification in Drug Design, p. 77, Washington, D.C., Am. Chem. Soc., Advances in Chemistry Ser. No. 45, 1964.

Cavallito, C. J., and Gray, A. P.: Chemical Nature and Pharmacological Action of Quaternary Ammonium Salts, *in* Jucker E. (ed.): Progress in Drug Research, p. 135, Basle, Birkhäuser, 1960.

Corcoran, A. C.: The Choice of Drugs in the Treatment of Hypertension *in* Modell, W. (ed.): Drugs of Choice 1964-1965, p. 437, St. Louis, Mosby, 1964.

Gyermek, L.: Ganglionic Stimulant and Depressant Agents *in* Burger, A. (ed.): Drugs Affecting the Peripheral Nervous System, p. 149-326, New York, Dekker, 1967.

Kharkevich, D. A.: Ganglion-blocking and Ganglion-stimulating Agents, p. 1-367, New York, Pergamon Press, 1967.

Lichtin, J. L.: Ganglion blocking agents, Am. Prof. Pharm. 24:223, 1958.

Nador, K.: Ganglienblocker, *in* Jucker, E. (ed.): Progress in Drug Research, vol. 2, p. 297, Basle, Birkhäuser, 1960.

Rice, L. M., and Dobbs, E. C.: Ganglionic Stimulants and Blocking Agents, *in* Burger, A. (ed.): Medicinal Chemistry, ed. 3, p. 1600, New York, Wiley-Interscience, 1970.

Van Rossum, J. M.: Classification and Molecular Pharmacology of Ganglionic Blocking Agents, I. Mechanism of Ganglionic Synaptic Transmission and Mode of Action of Ganglionic Stimulants. II. Mode of Action of Competitive and Non-competitive Ganglionic Blocking Agents, Int. J. Neuropharmacol. 1:97, 1962.

Volle, R. L.: Interactions of Cholinomimetic and Cholinergic Blocking Drugs at Sympathetic Ganglia, *in* Koelle, G. B., *et al.* (eds.): Pharmacology of Cholinergic and Adrenergic Transmission, p. 85, New York, Macmillan, 1965.

C.

Barlow, R. B.: Antagonists at Adrenergic Receptors *in* Introduction to Chemical Pharmacology, ed. 2, pp. 319-343, New York, Wiley, 1964.

Belleau, B.: Mechanism of drug action at receptor surfaces. I. Introduction. General interpretation of the adrenergic blocking activity of β-haloalkylamines, Canad. J. Phys. 36:731, 1958.

———: An analysis of Drug-Receptor Interactions, *in* Brunnings, K. J. (ed.): Modern Concepts in the Relationship Between Structure and Pharmacological Activity, vol. 7, p. 75, New York, Macmillan, 1963.

Bloom, B. M., and Goldman, I. M.: The Nature of Catecholamine-Adenine Mononucleotide Interactions in Adrenergic Mechanisms *in* Harper, N. J., and Simmonds, A. B. (eds.): Advances in Drug Research, vol. 3, p. 121-170, New York, Academic Press, 1966.

Boura, A. L. A., and Green, A. F.: Adrenergic Neurone Blocking Agents *in* Cutting, W. C. (ed.): Annual Review of Pharmacology, vol. 5, p. 183, Palo Alto, Annual Reviews, Inc., 1965.

Chapman, N. B. and Graham, J. D. P.: Synthetic Postganglionic Sympathetic Depressants *in* Burger, A. (ed.): Drugs Affecting the Peripheral Nervous System, p. 473-519, New York, Dekker, 1967.

Comer, W. T., and Gomoll, A. W.: Antihypertensive Agents, *in* Burger, A. (ed.): Medicinal Chemistry, ed. 3, p. 1019, New York, Wiley-Interscience, 1970.

Copp, F. C.: Adrenergic Neurone Blocking Agents *in* Harper, N. J., and Simmonds, A. B. (eds.): Advances in Drug Research, vol. 1, p. 161-189, New York, Academic Press, 1964.

Gifford, R. W., and Moyer, J. H.: Vasodilator Drugs for the Treatment of Peripheral Vascular Disturbances *in* Modell, W. (ed.): Drugs of Choice 1964-1965, p. 481, St. Louis, Mosby, 1964.

Graham, J. D. P.: 2-Halogenoalkylamines, *in* Ellis, G. P., and West, G. B. (eds.): Progress in Medicinal Chemistry, p. 132, London, Butterworths, 1962.

Iversen, L. L.: The Inhibition of Noradrenaline Uptake by Drugs *in* Harper, N. J., and Simmonds, A. B. (eds.): Advances in Drug Research, vol. 2, p. 1-46, New York, Academic Press, 1965.

Lichtin, J. L.: Adrenergic blocking agents, Am. Prof. Pharm. 24:115, 1958.

Schlittler, E., *et al.*: Antihypertensive Agents *in* Jucker, E. (ed.): Progress in Drug Research, vol. 4, p. 295, New York, Interscience, 1962.

Triggle, D. J.: 2-Halogenoethylamines and Receptor Analysis *in* Harper, N. J., and Simmonds, A. B. (eds.): Advances in Drug Research, vol. 2, p. 173-189, New York, Academic Press, 1965.

Ullyot, G. E., and Kerwin, J. F.: β-Haloethylamine adrenergic blocking agents, *in* Blicke, F., and Suter, C. M. (eds.): Medicinal Chemistry, vol. 2, p. 234, New York, Wiley, 1956.

D.

Barlow, R. B.: Antagonists at the Neuromuscular Junction *in* Introduction to Chemical Pharmacology, ed. 2, pp. 121-139, New York, Wiley, 1964.

Bovet, D., Bovet-Nitti, F., and Marini-Bettolo, G. B.: Curare and Curare-like Agents, Amsterdam, Elsevier, 1959.

Bowman, W. C.: Mechanisms of Neuromuscular Blockade, *in* Ellis, G. P., and West, G. B. (eds.): Progress in Medicinal Chemistry, vol. 2, p. 88, London, Butterworths, 1962.

Carrier, J. O.: Curare and curareform drugs, *in* Burger, A. (ed.): Medicinal Chemistry, ed. 3, p. 1581, New York, Wiley-Interscience, 1970.

De Reuck, A. V. S. (ed.): Curare and Curare-like Agents (Ciba Foundation Study Group No. 12), Boston, Little, 1962.

Lewis, J. J., and Muir, T. C.: Drugs Acting at Nerve-Skeletal-Muscle Junctions *in* Burger, A. (ed.): Drugs Affecting the Peripheral Nervous System, p. 327-364, New York, Dekker, 1967.

Paton, W. D. M., and Waud, D. R.: Neuromuscular Blocking Agents, Brit. J. Anaesth. 34:251, 1962.

Thesleff, S., and Quastel, D. M. J.: Neuromuscular Pharmacology *in* Cutting, W. C. (ed.): Annual Review of Pharmacology, vol. 5, p.

263, Palo Alto, Annual Reviews, Inc., 1965.

Waser, P. G.: Nature of the Cholinergic Receptor *in* Brunings, K. J. (ed.): Modern Concepts in the Relationship Between Structure and Pharmacological Activity, vol. 7, p. 101, New York, Macmillan, 1963.

————: The Molecular Distribution of ¹⁴C-Decamethonium In and Around the Motor End-Plate and Its Metabolism in Cats and Mice *in* Koelle, G. B., *et al.* (eds.): Pharmacology of Cholinergic and Adrenergic Transmission, vol. 3, p. 129, New York, Macmillan, 1965.

————: Autoradiographic Investigations of Cholinergic and other Receptors in the Motor End Plate, *in* Harper, N. J., and Simmonds, A. B. (eds.): Advances in Drug Research, vol. 3, p. 81-120, New York, Academic Press, 1966.

20
Diuretics

T. C. Daniels, Ph.D.,
Emeritus Professor of Pharmaceutical Chemistry, School of Pharmacy,
University of California

and

E. C. Jorgensen, Ph.D.,
Professor of Chemistry and Pharmaceutical Chemistry, School of Pharmacy,
University of California

INTRODUCTION

The kidney is the organ mainly responsible for maintaining an internal environment compatible with life processes. Its primary function is the regulation of the volume and composition of the body fluids, which it accomplishes by the elimination of variable amounts of water and selective ions such as Na^+, K^+, H^+, Cl^-, HPO_4^{--} and SO_4^{--}. The extracellular fluids, which comprise about 15 per cent of normal body weight, are influenced directly by changes in kidney function. The fluid within the cell (intracellular) is under osmotic equilibrium with extracellular fluid, and changes in extracellular fluid composition lead to changes in internal cellular fluid composition and function. A diuretic substance increases the excretion of urine by the kidney, thereby decreasing body fluids, especially the extracellular fluids.[1]

The pH of the body fluids is maintained by the excretion of anions such as HPO_4^{--} and, through the mediation of carbonic anhydrase, the synthesis from carbon dioxide and water of carbonic acid, which dissociates to H^+ and HCO_3^- ions. The renal tubular cells are able to synthesize ammonia by the deamination of amino acids, and in this way also the acid-base balance is maintained.

The kidney also serves as an excretory organ for the elimination of water-soluble substances present in excess of body needs. Thus, urinary excretion controls plasma concentrations of many nonelectrolytes that are end products of normal body metabolism, such as urea and uric acid, as well as metabolically solubilized derivatives of foreign molecules, such as glucuronides and sulfate esters of phenols.

The main functional unit is the nephron, and there are approximately one million nephrons in each kidney. It will be noted (Fig. 17) that each nephron has three functional parts:

1. The renal corpuscle, consisting of a cluster or tuft of capillaries, known as the glomerulus, which is enclosed in *Bowman's capsule*. The blood enters the nephron through the afferent arteriole under high capillary pressure, and portions of the dissolved substances are filtered through the walls of the capillaries and the epithelium of Bowman's capsule into the lumen of the capsule.

2. The renal tubule, which in turn may be divided into three segments: (a) the *proximal convoluted tubule,* (b) the *Loop of Henle* and (c) the *distal convoluted tubule.*

3. The *collecting tubule*, which leads to the renal pelvis, the ureter and, finally, to a larger collecting duct emptying into the bladder, where the urine is stored.

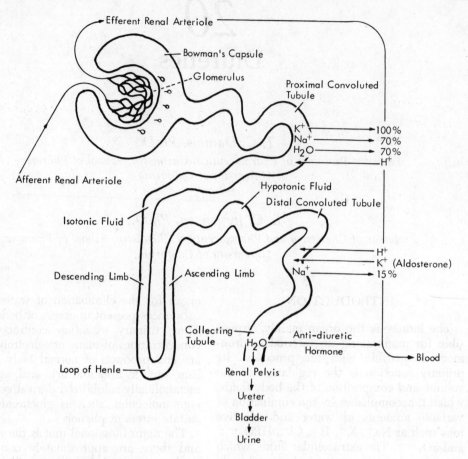

Fig. 17. The nephron—the functional unit of the kidney.

The glomerular filtrate which enters Bowman's capsule has the same general composition as the blood plasma, except that substances with a molecular weight of 67,000 or more do not pass through the filtering membrane. In this way, serum albumin and globulin are retained and are not present in the ultrafiltrate of the normal kidney.

The glomerular filtrate in the normal adult is formed at the rate of approximately 120 ml. per minute, or more than 7 liters per hour. Since the total extracellular fluid in the average adult amounts to approximately 12.5 liters, it will undergo complete filtration in less than 2 hours. As the filtrate passes down the renal tubule (nephron), the epithelial cells of the tubule reabsorb most of the water and solutes (over 99%), returning them to the bloodstream. Thus, it requires more than 100 ml. of glomerular

filtrate to produce 1 ml. of urine. Approximately 70 per cent of the water and sodium ions and essentially all of the potassium ions are reabsorbed in the proximal tubules (Fig. 17). The fluid entering the loop of Henle is isotonic.[2] It becomes increasingly more concentrated as it passes through the descending limb, either by the loss of water,[3] or by the addition of solutes from the capillaries.[3a] The fluid passing through the ascending limb becomes increasingly more dilute, as about half of the remaining sodium and chloride ions are reabsorbed. Most of the remaining sodium ions may be reabsorbed in the distal tubule by a cation exchange mechanism under the control of the adrenal cortex hormone, aldosterone. Sodium from the renal tubular fluid is exchanged for potassium ions and, to a lesser extent, hydrogen ions from the blood. Reabsorption of water from the hypotonic fluid of the distal convoluted tubule takes place primarily under the influence of the antidiuretic hormone, vasopressin (ADH), of the posterior pituitary.

Most of the clinically useful diuretics increase the excretion of sodium ion (as chloride or bicarbonate) by decreasing reabsorption of the ion in the renal tubules. Any compound capable of interfering with the reabsorption of water and solutes from the glomerular filtrate will give a diuretic response. It is also evident that an increase in the rate of glomerular filtration due to increased blood pressure or blood flow will increase urine formation. Likewise, inhibition of the antidiuretic hormone (vasopressin, ADH) of the posterior pituitary will produce diuresis. Since most of the reabsorption takes place in the proximal tubules, it is not surprising to find that the most potent diuretics, i.e., thiazides, mercurials, act primarily on the proximal tubules. Unfortunately, they also tend to produce excessive loss of potassium, which may lead to hypokalemia, a major complication in diuretic therapy. For this reason a potassium-sparing diuretic such as triamterene may be used in combination with another diuretic which produces potassium loss, such as a thiazide.[4]

At the present time, only a small number of diuretics (e.g., the aldosterone inhibitors) that act primarily on the distal tubules are known. Understandably, these compounds have a weaker diuretic action and are commonly used in combination with a thiazide or other agent acting on the proximal tubules.

The exact mechanism for the reabsorption of the normal electrolytes in the renal tubules remains to be determined, but it is generally believed that the electrolyte transport process may involve a "carrier molecule" to which the electrolyte has been attached through ion exchange. A number of specific transport mechanisms may be involved.

The diuretics are generally employed for the treatment of all types of edema and, therefore, may be properly regarded as lifesaving drugs. The main diseases associated with edema are congestive heart failure, premenstrual tension, edema of pregnancy, renal edema and cirrhosis with ascites. Diuretics are also used for the treatment of edema induced by the administration of ACTH and the other corticosteroids. In addition, diuretics are frequently employed as adjuncts in the treatment of hypertension, where the extracellular fluid may be at a relatively normal level. Some diuretics, notably the benzothiadiazines, potentiate the action of the antihypertensive drugs, including the ganglionic blocking agents. Moreover, diuresis alone may serve a useful purpose in the treatment of hypertension by decreasing the extracellular fluid and thereby lowering blood volume and pressure.

The presently used diuretics may be conveniently divided into the following classes:

1. Water and osmotic agents
2. Acidifying salts
3. Mercurials
4. α,β-Unsaturated ketones
5. Purines, pyrimidines and related heterocyclic compounds
6. Sulfonamides
   A. Inhibitors of carbonic anhydrase
   B. Benzothiadiazines ("thiazides") and related heterocyclic compounds
7. Endocrine antagonists

WATER AND OSMOTIC AGENTS

Any compound that is poorly reabsorbed by the renal tubules and is present in a concentration in excess of the concentration of electrolytes and dissolved substances in the body fluids will cause water and electrolytes to pass into the more concentrated solution and be excreted. Thus, diuresis and a mobilization of edema fluids take place.

The ingestion of large amounts of water and the resultant rapid dilution of blood increases urinary excretion by inhibiting the antidiuretic hormone (ADH), but the electrolyte concentration is not affected. There is no net loss of water from the tissues, since it is retained in proportion to the concentration of the electrolytes present.

In edematous states sodium salts are retained, and this produces an increase in the body fluids; therefore, low or salt-free diets are used to restrict sodium intake. The substitution of potassium salts for sodium has little value and may be harmful by developing hyperkalemia.

A number of nonelectrolytes, such as urea and certain sugars, may be used as osmotic diuretics.

Urea U.S.P., Ureaphil®, Carbamide. Urea is poorly reabsorbed by the renal tubules

$$H_2N-\overset{\overset{\displaystyle O}{\|}}{C}-NH_2$$

Urea

and therefore serves as an osmotic diuretic. When administered in large amounts, electrolytes are excreted, and a mobilization of edema fluid takes place. Oral administration has been successful in the treatment of cardiac edema and nephrosis, but requires doses of up to 20 g. 2 to 5 times daily.

Sterile lyophilized urea (Urevert®) is available for the preparation of solutions containing 4 or 30 per cent of urea in 10 per cent invert sugar solutions (Travert®). The 30 per cent solution (Urevert for Injection®) is used to control cerebral edema, or for the symptomatic relief of headache and vomiting due to increased intracranial pressure. It is also used in conjunction with surgical treatment of narrow angle closure glaucoma. The 4 per cent urea and 10 per cent invert sugar solution may be employed as a diuretic. A dose of 1 g. of urea per kg. of body weight reduces intracranial and intraocular pressure and produces diuresis. The lyophilized urea and invert sugar solutions should be freshly prepared for intravenous use.

Category—diuretic.

Usual dose—intravenous infusion, 100 mg. to 1 g. per kg. of body weight daily, as a 30 per cent solution in Dextrose Injection at a rate not exceeding 6 ml. per minute.

Occurrence

Sterile Urea U.S.P.

Glucose, sucrose and mannitol (e.g., Osmitrol, an injectable solution of mannitol) have been used as osmotic diuretics. For this purpose they are administered intravenously and must be used in large doses

in the order of 50 ml. of a 50 per cent solution.

The osmotic diuretics have obvious disadvantages and have been largely replaced by more effective agents, except for special indications, such as the early treatment of acute reduction in renal blood flow, the reduction of intraocular pressure prior to eye surgery for glaucoma, or the enchancement of urinary excretion of intoxicants such as barbiturates.

Mannitol U.S.P., Osmitrol®, D-Mannitol, is a hexahydroxy alcohol which is essentially

$$HOCH_2CH-\overset{\overset{\displaystyle OH}{|}}{CH}-\overset{\overset{\displaystyle OH}{|}}{CH}-\overset{\overset{\displaystyle}{|}}{\underset{\underset{\displaystyle OH}{|}}{CH}}-\overset{\overset{\displaystyle}{|}}{\underset{\underset{\displaystyle OH}{|}}{CH}}CH_2OH$$

D-Mannitol

not metabolized. It is filtered by the glomerulus, but only negligible amounts are reabsorbed by the tubules. Therefore, it is used as a diagnostic agent to measure glomerular filtration rates (see Chap. 29) and to produce osmotic diuresis in various edematous states, including those which fail to respond to thiazide preparations. It is administered intravenously, the normal adult dose being 50 to 100 g. within a 24-hour period; the maximum recommended dose is 200 g. Approximately 80 per cent is excreted in 12 hours.

The usual solution for injection contains 12.5 g. in 50 ml.

Category—diuretic.

Usual dose range—intravenous infusion, 50 to 200 g. daily.

Occurrence

Mannitol Injection U.S.P.

Mannitol and Sodium Chloride Injection U.S.P.

ACIDIFYING SALTS

Acidifying salts, such as ammonium chloride and ammonium nitrate, elicit a weak diuretic response by producing an excess of the anion (Cl^-, NO_3^-) in the glomerular filtrate. This is made possible by the ammonium cation conversion to urea, which is a neutral compound. Accompanying the excess anion, there is an increase in Na^+ output, but a greater increase in H^+ concentration leading to an acidification of the urine. There remains an excess of hydrogen ions, leading to systemic acidosis. The same general mechanism is involved when calcium chloride or calcium nitrate is administered, except that the calcium cation (Ca^{++}) is depleted by deposition in the bone or is excreted as the phosphate.

The use of acidifying salts alone for diuresis is quite unsatisfactory, for the kidney is able to develop rapidly a compensating mechanism to neutralize the acids formed in the glomerular filtrate by increasing the formation of ammonia. Thus, in a comparatively short time (1 to 2 days) the acidifying salts lose their ability to produce acidosis and no longer serve effectively as diuretics. Other disadvantages of the acidifying salts are: gastric irritation, which may lead to anorexia, nausea and vomiting; the ammonium salts may produce hyperammonemia; and acidosis may lead to renal insufficiency.

At the present time, the acidifying agents (principally ammonium chloride) are used primarily in conjunction with the mercurial diuretics which they potentiate.

MERCURIALS

Prior to 1950, the mercurial compounds were the only effective diuretics available for use. Although they have since been largely replaced by newer orally effective agents, they are still regarded as useful for the treatment of severe edematous states. Under optimal conditions, they are approximately four times as effective as the thiazides in increasing the excretion of sodium (as chloride).

The mercurial diuretics have a number of undesirable properties. They are not well absorbed from the gastrointestinal tract, but when administered parenterally usually give a rapid and reliable onset of diuresis. Toxic reactions are comparatively uncommon, but local irritation, necrosis, hypersensitivity reactions and electrolyte disturbances may occur. Mercurialism also may develop following prolonged use, especially with the mercurials used orally.

The continued use of the mercurial diuretics may be attributed to the following desirable properties: With the exception of the new potassium-sparing diuretics, including the aldosterone antagonists, the organic mercurials produce less potassium loss than most other classes of diuretics and supplementary potassium administration is normally unncessary. They do not significantly alter the excretion of potassium, ammonium, bicarbonate or phosphate ions and therefore may be used without producing a marked disturbance of the electrolyte balance of the body fluids. Carbohydrate metabolism is unaltered and thus they are free from the danger of producing hyperglycemia and the onset of diabetes. Finally, uric acid elimination is unchanged, which avoids the possibility of hyperuricemia.

Mercurous chloride was first used as a diuretic by Paracelsus in the 16th century, and its use alone or in combination with other agents continued until comparatively recent times. The organic mercurial diuretics originated in 1919 following the observation that a new compound (Merbaphen, Novasural) introduced for the treat-

Merbaphen

ment of syphilis elicited a pronounced diuretic response. However, Merbaphen, with the mercury attached directly to the aromatic ring, was found to be too toxic for general use as a diuretic, and was later found that related compounds with the mercury attached to an aryl group were likewise too toxic.[1]

Structural variations led in 1924 to the less toxic, and still useful, mersalyl.

Mersalyl

All of the mercurials in clinical use at the present time are close structural analogs, in which an alkoxy-mercuripropyl group is attached to a mono- or dicarboxylic acid or the amide derivative of an acid.[5] The following compound illustrates the general structural features of the mercurial diuretics:

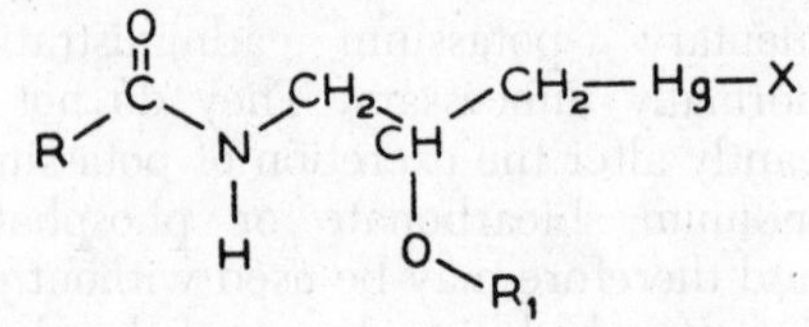

R = aliphatic, alicyclic, aromatic, heterocyclic groups, usually carrying a carboxyl function, and attached to the 3 carbon chains through amide, urea, ether or carbon-carbon linkages.

$R_1 = -H, -CH_3, -C_2H_5$ or $-CH_2CH_2-OCH_3$

$X = -OH, -Cl, -Br, -O_2CCH_3, -SCH_2-CO_2H, -SCH_2(CHOH)_4CH_2OH$

The general method for the synthesis of the mercurial diuretics involves the mercuration of an alkene:

$$RCH_2CH = CH_2 \xrightarrow[\text{R}_1\text{OH}]{\text{Hg(OCOCH}_3)_2}$$

The R_1 substituent is determined by the solvent in which the mercuration reaction is carried out, being hydroxyl if the solvent is water, and methoxy or ethoxy in the corresponding alcohol. The acetoxy group on mercury is replaceable by a variety of groups (X), theophylline being most often used.

STRUCTURE-ACTIVITY RELATIONSHIPS

Diuretic activity for the organic mercurials requires a hydrophilic group (e.g., $RCONH-$) attached not less than three carbon atoms distant from the mercury.[6] Compounds with a shorter chain show little or no activity.

Each of the 3 groups (R, R_1 and X) influences the diuretic activity and the toxicity of the molecule. Of the 3 variants (R) has the greatest and (R_1) the least influence.[5] The (X) group does not increase the activity of the molecule per se, but when theophylline represents (X) there is improved absorption from the site of injection and an enhanced diuretic response due to a potentiating effect. Theophylline also lowers the tissue irritation and therefore is commonly used in combination with the mercurial diuretics.[6, 7, 8] Ammonium chloride administered concurrently with the mercurials increases the diuretic response.

MODE OF ACTION

The mechanism of action of the mercurial diuretics remains to be established, but the primary site of action has recently been shown to be the proximal renal tubules.[9] Their action is believed to be due primarily to a combination of mercury ions with sulfhydryl groups attached to the renal enzymes responsible for the production of energy necessary for tubular reabsorption. When these specific enzymes are blocked, there is a marked increase in the amount of sodium chloride and water excreted, due to interference with the reabsorption process. Administration of dimercaprol (BAL) or other related *vicinal* dithiols with mercurial diuretics prevents blocking of the enzymes, and diuresis does not occur, because the dithiols have a greater affinity for ionic mercury than does the enzyme.

There is uncertainty as to whether the organic mercurials act in the mono or the divalent form, that is, as intact molecules ($R-Hg^+$) or, by a splitting of the molecule to give mercuric ions (Hg^{++}) at the site of action.[10] In support of the mercuric ion postulate, it has been observed that all active mercurial diuretics are acid labile, yielding mercuric ion. Acid stable compounds were inactive as diuretics.[1] If mercuric ions are responsible for the diuretic response, it must be assumed that the un-

TABLE 60. MERCURIAL DIURETICS

$$R-CH_2-CH-CH_2-Hg-X$$
$$\overset{|}{O-R_1}$$

Generic Name Proprietary Name	R	R_1	X	ADMINISTERED
Meralluride U.S.P. Mercuhydrin	$HO_2C-CH_2CH_2CONHCONH-$	$-CH_3$	Theophylline	IM
Sodium Mercaptomerin U.S.P. Thiomerin	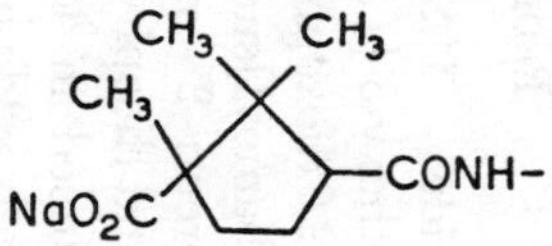 (cyclopentane structure with CH_3, CH_3, NaO_2C, $CONH-$)	$-CH_3$	$-S-CH_2-CO_2Na$	S. C.
Chlormerodrin N.F. Neohydrin	$H_2N-CO-NH-$	$-CH_3$	$-Cl$	Orally
Mercurophylline N.F. Mercuzanthin	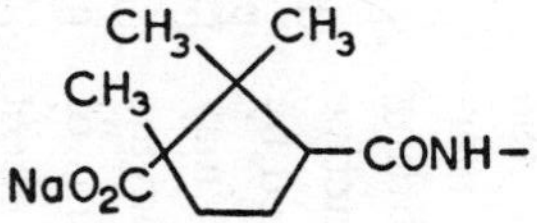 (cyclopentane structure with CH_3, CH_3, NaO_2C, $CONH-$)	$-CH_3$	Theophylline	IM and Orally
Mercurophylline	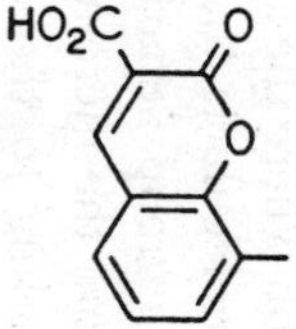 (coumarin structure with HO_2C)	$-CH_3$	Theophylline	Orally
Merethoxylline procaine Dicurin	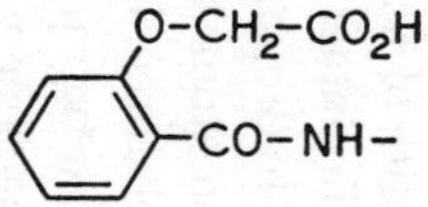 (benzene structure with $O-CH_2-CO_2H$, $CO-NH-$)	$-CH_2-CH_2-O-CH_3$	Theophylline + Procaine	IM and S. C.
Mersalyl Salyrgan	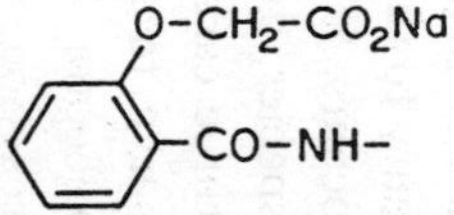 (benzene structure with $O-CH_2-CO_2Na$, $CO-NH-$)	$-CH_3$	Theophylline	IM and IV

A. Action due to intact molecule

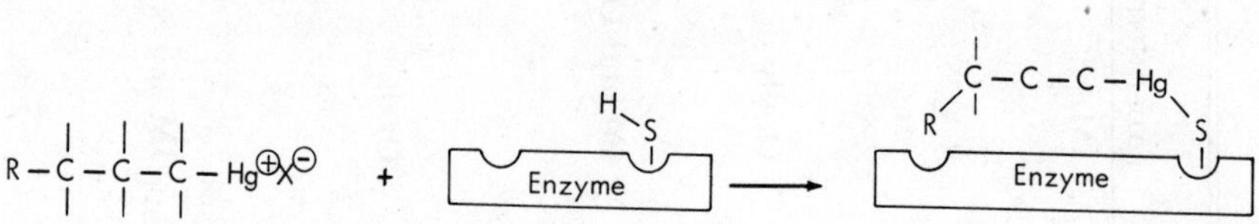

B. Action due to mercuric ion

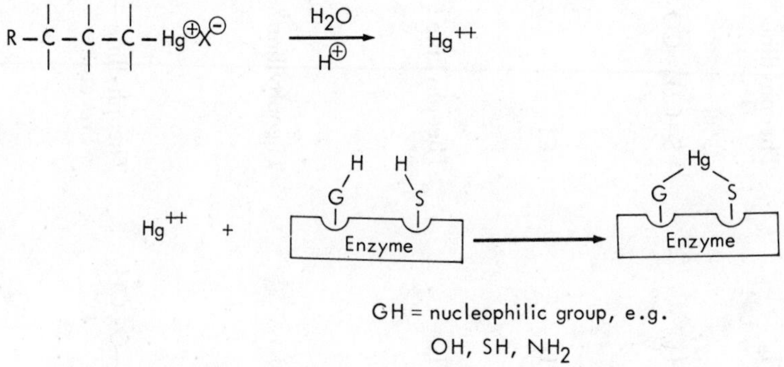

GH = nucleophilic group, e.g.

OH, SH, NH$_2$

C. Combination of BAL with mercuric ion

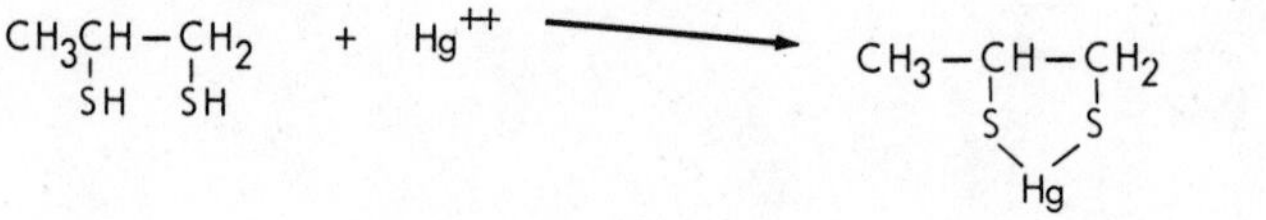

Stable Cyclic Mercurial

dissociated molecule contributes to the lower toxicity and serves to carry the mercury to the site of action where the splitting occurs. If the intact molecule is the active species, the hydrophilic group, separated by three carbon atoms from the mercury, may serve to reinforce binding to the enzyme receptor. The two proposed mechanisms of reaction, together with the blocking effect of BAL, are shown above.

Table 60 gives the structural features and the modes of administration of the mercurial diuretics.

PRODUCTS

Meralluride U.S.P., Mercuhydrin®, N - [[2-Methoxy-3-[(1,2,3,6-tetrahydro - 1,3-dimethyl - 2,6 dioxopurin - 7 - yl) - mercuri]-propyl]carbamoyl]succinamic Acid, 1-(3′-hydroxymercuri-2′- methoxypropyl)-3-succinylurea and theophylline. Meralluride was first introduced in 1943 and has received comparatively wide use since that time. The free acid is only sparingly soluble in water, and the injection is prepared by making a slightly alkaline solution with sodium hydroxide. The solution contains 130 mg. of

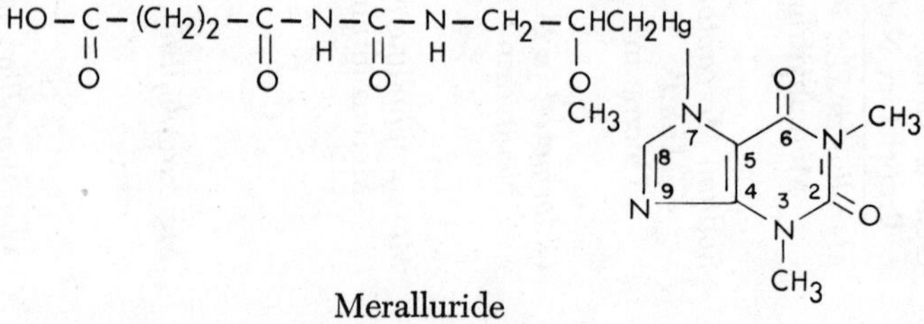

Meralluride

meralluride (equivalent to 39 mg. of mercury) and 48 mg. of theophylline per ml. The solution is administered IM. The usual adult dose is 1 to 2 ml. of the solution twice weekly. Suppositories containing 600 mg. of the drug in the free acid form are available for adjunctive maintenance therapy.

Category—diuretic.

Usual dose—parenteral, 1 ml. of the Injection [equivalent to 39 mg. of mercury and 43.6 mg. of anhydrous theophylline (48 mg. of hydrous theophylline)] once or twice a week.

Usual dose range—1 to 2 ml.

OCCURRENCE
Meralluride Injection U.S.P.

Sodium Mercaptomerin U.S.P., Thiomerin® Sodium, [[[3-(3-Carboxy-1,2,2-trimethylcyclopentanecarboxamido) - 2 - methoxy - propyl]thio]mercuri]acetate, Disodium N-[3-(Carboxymethylthiomercuri)-2-methoxy-

Usual dose range—25 to 250 mg. daily to weekly.

OCCURRENCE
Sterile Sodium Mercaptomerin U.S.P.
Sodium Mercaptomerin Injection U.S.P.

Chlormerodrin N.F., Neohydrin®, 3-Chloro(2-methoxy-3-ureidopropyl)mercury.

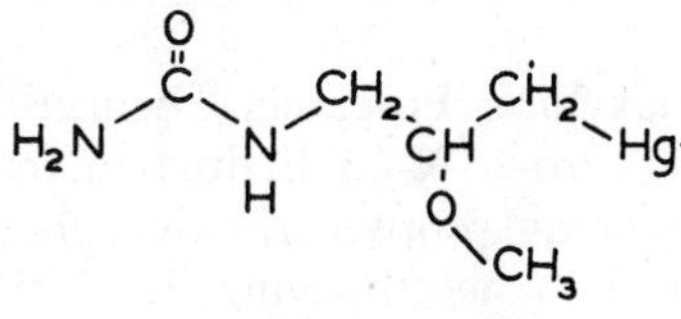

Chlormerodrin

Chlormerodrin is a white powder, appreciably soluble in water (1.1%), which gives a bitter metallic taste. It was first introduced in 1952 and has the advantage of being an orally effective compound. It may be used alone or for maintenance therapy in conjunction with a parenterally administered mercurial to reduce the number of

Mercaptomerin

propyl]-α-camphoramate. This compound occurs as a white hygroscopic powder or amorphous solid that is freely soluble in water and soluble in alcohol. A 2 per cent solution is neutral to litmus.

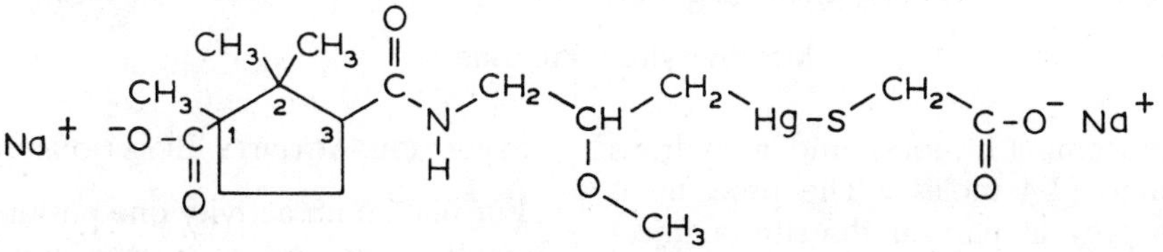

Mercaptomerin was first introduced in 1946 and is one of the most widely used mercurial diuretics. It gives less local irritation at the site of injection and is less toxic to the heart than other mercurials. Because of its low local irritation effect, it may be administered subcutaneously. Mercaptomerin is heat- and light-sensitive; therefore, it should be stored in lightproof containers under refrigeration.

Mercaptomerin may also be administered rectally in the form of suppositories for maintenance therapy. The suppositories contain 500 mg. of the drug.

Category—diuretic.

Usual dose—parenteral, 125 mg. once daily.

injections necessary for control. It is less effective than the parenterally administered compounds and may produce gastrointestinal irritation. The drug is supplied in 18.3-mg. tablets.

Category—diuretic.

Usual dose range—55 to 110 mg. daily.

OCCURRENCE
Chlormerodrin Tablets N.F.

Mercurophylline, Sodium 3-{[3-(hydroxymercuri)-2-methoxypropyl]carbamoyl}-1,2,2-trimethylcyclopentanecarboxylate and theophylline. Mercurophylline consists of the sodium salt of the mercurial compound and theophylline in equimolar proportion. Diuresis develops slowly following oral administration and does not reach its maximum for 48 hours, although the total diuretic response may approach that produced by injection.

Usual dose—oral, 200 mg. daily; intra-

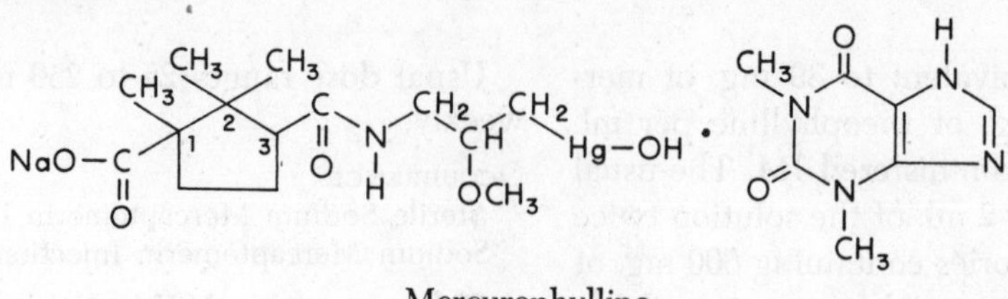

Mercurophylline

muscular, 135 mg. (in 1 ml.) once or twice a week.

Merethoxylline Procaine, Dicurin® Procaine, Dehydro-2-[N-(3'-hydroxymercuri-2'-methoxyethoxy)propylcarbamyl]phenoxyacetic acid (merethoxylline), 2-diethylaminoethyl p-aminobenzoate (procaine) and theophylline. The compound is a mixture consisting of the organic mercurial (1

sociated with the transport mechanism responsible for the reabsorption of electrolytes in the kidney tubules. Since the α,β-unsaturated ketones react readily and reversibly with sulfhydryl groups, derivatives of arylphenoxyacetic acid containing the α,β-unsaturated carbonyl group were studied and found to show a high order of diuretic activity.[11]

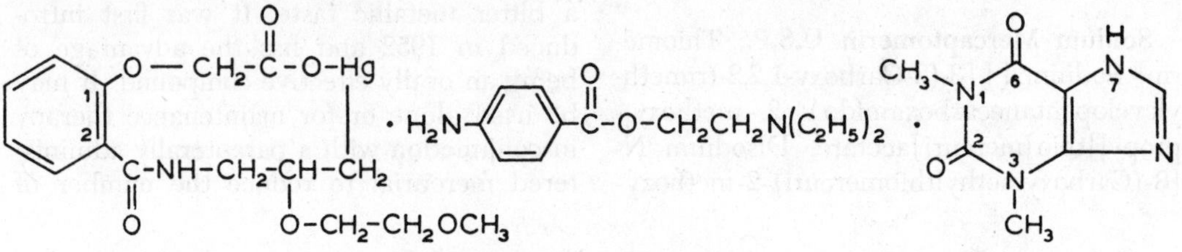

Merethoxylline Procaine

mole), procaine (1 mole) and anhydrous theophylline (1.4 moles). The procaine is added to prevent pain at the site of injection; theophylline, to improve absorption and reduce irritation. The compound can be used both subcutaneously and IM. It was first introduced in 1953.

Merethoxylline procaine solution is made to contain 100 mg. of the organic mercurial per ml. The average adult dose is 2 ml. daily.

α,β-UNSATURATED KETONES

The diuretic properties of the mercurials are postulated to be dependent on the blocking of an essential sulfhydryl group as-

STRUCTURE-ACTIVITY RELATIONSHIPS [1, 11]

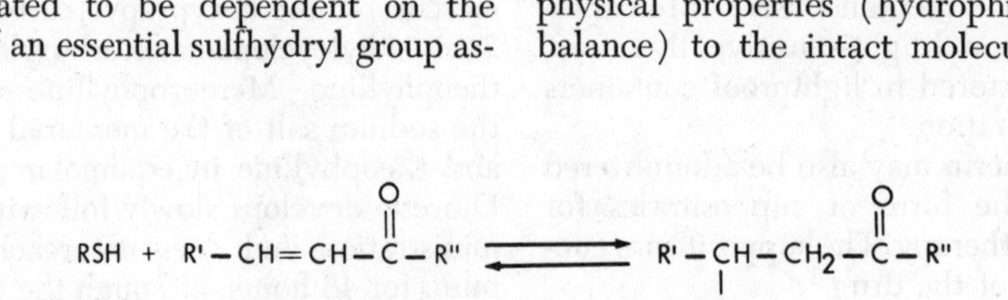

For maximum activity one position in the aromatic ring *ortho* to the unsaturated ketone must be substituted with a halogen or methyl group. Disubstitution in the 2,3-positions increases the activity; additional substitution in the ring may lower the activity. It is of interest that the first organic mercurial diuretic (Merbaphen) was a derivative of a chlorophenoxyacetic acid, which suggests that such acids may serve to increase the activity of the functional groups of these two classes of diuretics (i.e., labile mercury bonding, reactive α,β-unsaturated carbonyl groups) and contribute physical properties (hydrophilic-lipophilic balance) to the intact molecules favoring their transport to receptor sites for inter-action and blocking of the essential sulf-hydryl transport mechanism.

The structure of the α,β-unsaturated car-bonyl group influences the activity. Maxi-mum activity is shown if the β-position of the unsaturated ketone is unsubstituted. Higher alkyl groups substituted for ethyl in the α-position lower the activity.

The unsaturated ketone group must be *para* to the oxyacetic acid group for maxi-mum activity. The *ortho* and *meta* isomers are much less active.

All the compounds in the series that gave a significant diuretic response were also highly reactive in vitro with sulfhydryl-con-taining compounds. Ethacrynic acid was the most active member of the series.

Ethacrynic Acid U.S.P., Edecrin®, [2,3-Dichloro-4-(2-methylenebutyryl)phen-oxy]acetic Acid.

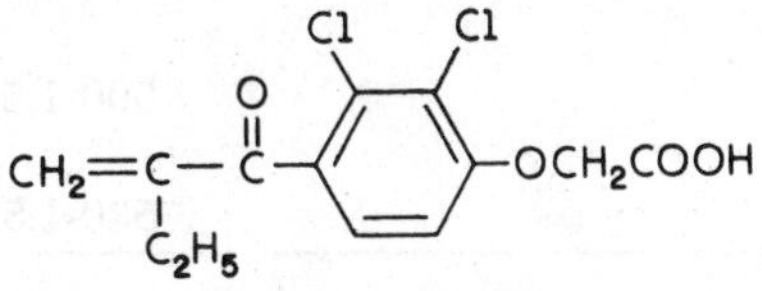

Ethacrynic Acid

Ethacrynic acid is a potent saluretic agent which may produce a marked diuretic re-sponse in cases of refractory edema. Under optimal conditions it has a natriuretic effect at least five times greater than the thia-zides[12] and is equivalent in activity to the mercurials.[13] Unlike the mercurials, the di-uretic activity is not influenced by metabolic alkalosis or acidosis. Like the thiazides, it may lower uric acid excretion and cause hy-peruricemia but (unlike the thiazides) it has little influence on carbohydrate metab-olism and blood glucose levels.[1] It has been suggested[13] that ethacrynic acid inhibits sodium reabsorption primarily at two sites in the nephron: (1) in the ascending limb of the loop of Henle, and (2) in the distal tubule where urinary dilution occurs. Chlo-ride excretion is increased to a greater ex-tent than sodium excretion, which may lead to systemic alkalosis.

Ethacrynic acid may produce hypoalbu-minemia, hypochloremia, hyponatremia, hypokalemia and metabolic alkalosis.[14]

Transient and permanent deafness has been reported following treatment with ethacry-nic acid in patients with uremia. The cause of deafness is unknown, but it should be used with caution in uremic patients.[15]

Ethacrynic acid is especially useful in the treatment of refractory edema and may be used in combination with a potassium-spar-ing diuretic.

Category—diuretic.

Usual dose—50 mg. twice daily or 2 times every other day.

Usual dose range—50 to 400 mg. daily.

OCCURRENCE
Ethacrynic Acid Tablets U.S.P.
Sodium Ethacrynate for Injection U.S.P.

PURINES, PYRIMIDINES AND RELATED HETEROCYCLIC COMPOUNDS

Purine and pyrimidine bases are consti-tuents of the nucleic acids. The purine bases, which consist of fused pyrimidine and imidazole rings, occur in nature primarily as oxidized derivatives. The 2,6-dihydroxy-purine is xanthine, and the N-methylated xanthines, by virtue of their widespread oc-currence in plant materials used by man, especially the traditional beverages (tea, coffee, cocoa), have long been recognized for their diuretic properties. The 3 most common naturally occurring xanthines are caffeine, theobromine and theophylline, and

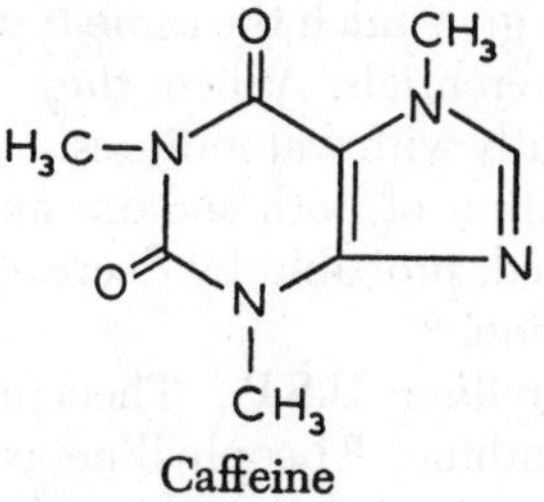

Caffeine

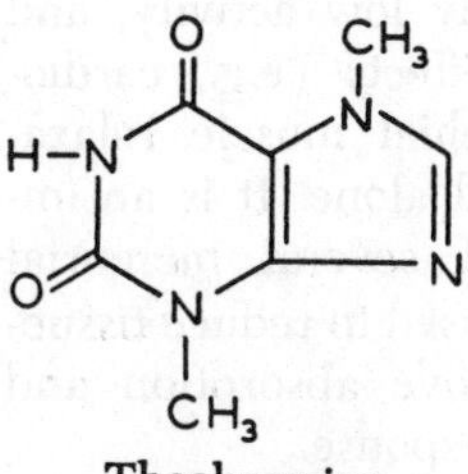

Theobromine

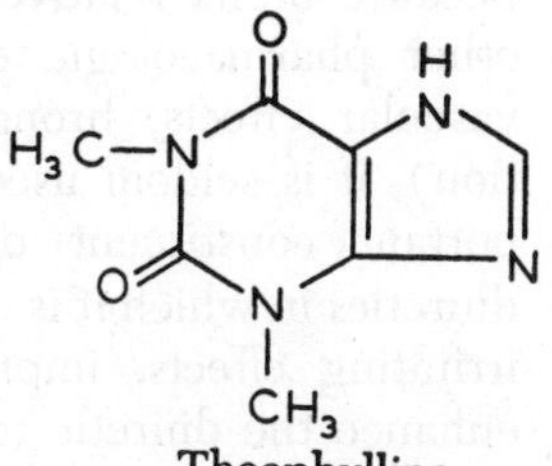

Theophylline

TABLE 61. PARTIAL LIST OF PURINE (XANTHINE) DERIVATIVES AND COMBINATIONS

| GENERIC NAME | PROPRIETARY NAME | DOSAGE FORMS | | | AVERAGE DOSE RANGE mg. |
		INJEC-TION	SUPPOSI-TORIES	TAB-LETS	
Theophylline U.S.P.	Theocin			X	100-200
Aminophylline U.S.P.	Lixaminol	X	X	X	200-500
Theophylline Sodium Acetate N.F.	Theocin Soluble			X	200-300
Theophylline Sodium Glycinate N.F.	Glynazan Glytheonate	Aerosol Elixir	X	X	300-1,000
Theophylline Calcium Salicylate	Phyllicin			X	250-300
Theophylline-Meglumine	Glucophylline		X	X	150-500
Oxytriphylline	Choledyl			X	200-400
Dyphylline	Neothylline			X	100-200
Theobromine N.F.	--			X	300-500
Theobromine Sodium Acetate	Thesodate			X	500-750
Theobromine Sodium Salicylate	Diuretin			X	500-1,000
Theobromine Calcium Salicylate	Theocalcin			X	500-1,500
Theobromine Calcium Gluconate	Calpurate			X	500-1,500

the diuretic potency increases in the order named.

These compounds are neither potent nor reliable diuretics, and their principal use has been in conjunction with the mercurial diuretics. The mode of action of the purine (xanthine) diuretics is not known, but they appear to give much the same type of action as the mercurials. When they are given, concurrently with water diuresis, the urinary concentration of both sodium and chloride is increased, probably by decreased tubular reabsorption.[10]

Theophylline U.S.P., Theocin®, 1,3-Dimethylxanthine. Theophylline is the most active diuretic of the xanthine alkaloids, but because of its relatively low activity, and other pharmacologic effects (e.g., cardiovascular effects, bronchial muscle relaxation), it is seldom used alone. It is an important constituent of several mercurial diuretics in which it is used to reduce tissue-irritating effects, improve absorption and enhance the diuretic response.

OCCURRENCE
Theophylline, Ephedrine Hydrochloride, and Phenobarbital Tablets N.F.
Theophylline Tablets N.F.
Theophylline Sodium Acetate N.F.
Theophylline Sodium Acetate Tablets N.F.
Theophylline Sodium Glycinate N.F.
Theophylline Sodium Glycinate Tablets N.F.

Aminophylline U.S.P., Theophylline Eth-

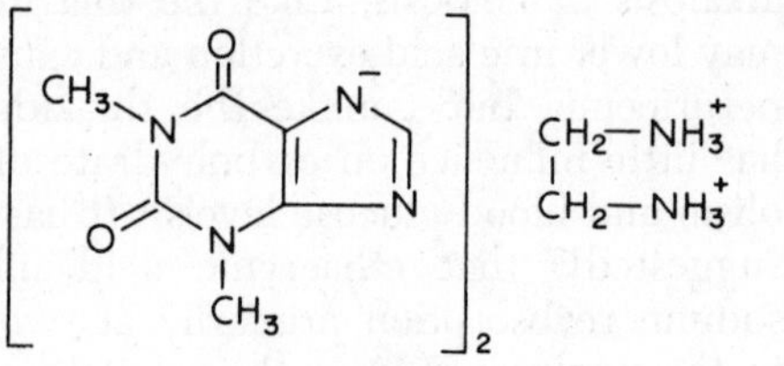

Aminophylline

ylenediamine. Aminophylline is the principal purine salt employed as a diuretic. Its prinicipal advantage is increased water solubility (1 g. dissolves in about 5 ml. of water), which enables it to be administered orally, parenterally, or by rectal suppositories. In addition to its diuretic effect, aminophylline is useful as a peripheral vasodilator, a myocardial stimulant for the relief of pulmonary edema and as an antiasthmatic agent. Table 61 lists some xanthine derivatives, most of these consisting of salts possessing a higher solubility than the parent xanthine.

OCCURRENCE

Aminophylline Injection U.S.P.
Aminophylline Suppositories U.S.P.
Aminophylline Tablets U.S.P.

PYRIMIDINES

Since pyrimidines generally are used as intermediates in the synthesis of purines, these heterocycles were examined for diuretic activity. In 1907 5-methyluracil was reported to produce marked diuresis when administered orally to dogs.[16] In searching for more effective diuretics in this series, Papesch and Schroeder[17] also noted that the pyrimidine portion of the purine molecule produced diuresis in experimental animals, and from this observation, aminometradine, 1-allyl-3-ethyl-6-aminouracil, was developed and introduced as an orally active diuretic. The compound produced gastric irritation

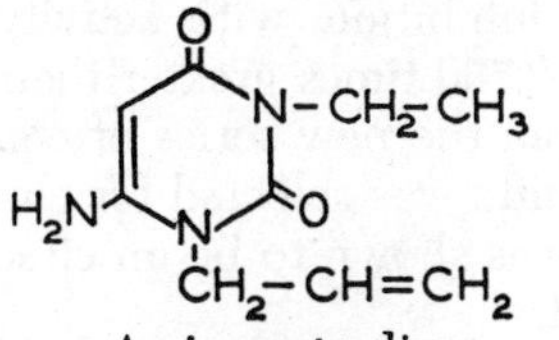

Aminometradine

and nausea in a high percentage (20 to 30%) of patients and has been replaced in therapy by the isomeric amisometradine (Rolicton) which shows a reduced incidence of gastrointestinal disturbance with the same degree of diuretic effectiveness.

Amisometradine, Rolicton®, 6-Amino-3-methyl-1-(2-methylallyl)uracil.

This occurs as a white crystalline powder that is slightly soluble in water and freely soluble in alcohol.

Amisometradine is an orally effective diuretic. It may be used for continuous therapy without losing its effectiveness through acquired tolerance. The diuretic effect of

amisometradine appears to be the same as that of the purines, in which there is increased excretion of sodium chloride, due

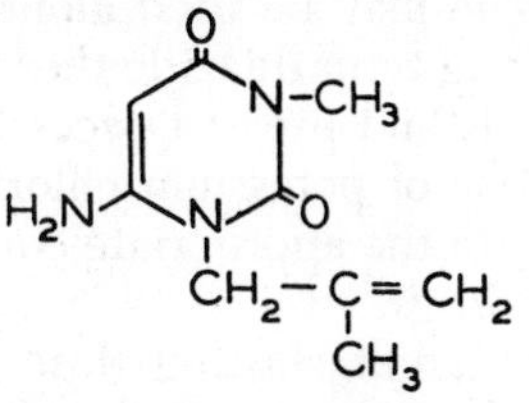

Amisometradine

probably to decreased tubular resorption. Like the purines, amisometradine may be used as an adjunct to mercurial therapy.

The average adult dose is approximately 800 mg. per day, divided into 2 doses of 400 mg. each.

PTERIDINES

The pteridine ring system consists of fused pyrimidine and pyrazine rings. Derivatives of this heterocyclic system show unique diuretic properties.

Triamterene U.S.P., Dyrenium®, 2,4,7-

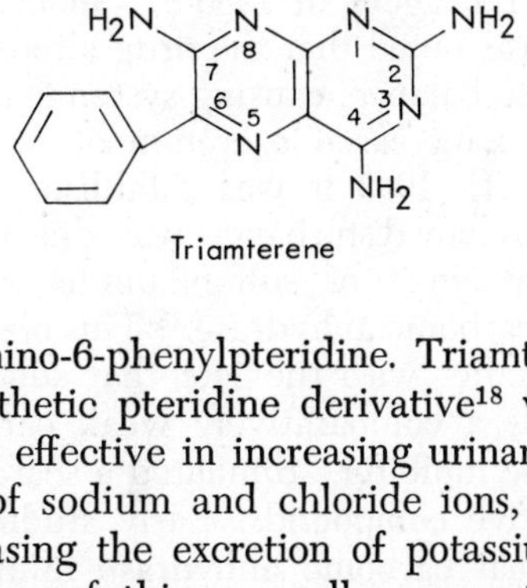

Triamterene

Triamino-6-phenylpteridine. Triamterene is a synthetic pteridine derivative[18] which is orally effective in increasing urinary excretion of sodium and chloride ions, without increasing the excretion of potassium ions. Because of this, as well as reports of increased effectiveness when aldosterone secretion is elevated, it was proposed originally that triamterene acts as an aldosterone antagonist. However, direct interference with the action of the mineralocorticoids has not been demonstrated, and available data indicate that triamterene acts by interfering with the processes of cation exchange in the distal renal tubule.[20]

An increase in blood urea nitrogen levels has been observed with triamterene. This is believed to be due to a reduced glomerular filtration rate, and, therefore, use of the drug is contraindicated in the presence of renal disease and hepatitis.

Triamterene may be used alone or as an adjunct to long-term thiazide therapy to improve diuresis and prevent excessive potassium loss. Use of potassium chloride is not necessary with the appropriate combination of the two drugs.

The usual adult starting dose of triamterene is 100 mg. twice daily after meals. Maintenance therapy is usually 100 mg. daily or every other day, with the peak diuretic effect occurring 2 to 8 hours after administration.

Category—diuretic.

Usual dose—100 mg. once daily.

Usual dose range—100 to 300 mg. daily.

OCCURRENCE

Triamterene Capsules U.S.P.

SULFONAMIDES

INHIBITORS OF CARBONIC ANHYDRASE

Sulfanilamide was first introduced as an antibacterial agent in 1936-37. Soon thereafter it was noted that the drug altered the electrolyte balance, causing systemic acidosis due to increased excretion of bicarbonate.[21, 22] In 1940 it was established that the electrolyte disturbance was due to the inhibitory effect of sulfanilamide on the enzyme carbonic anhydrase.[23] This observation, together with the fact that sulfanilamide was a comparatively weak carbonic anhydrase inhibitor, stimulated a search for more active compounds. Early studies revealed that carbonic anhydrase inhibition was limited to those compounds in which the amide nitrogen is free. The mono- and disubstituted derivatives of the sulfamyl ($-SO_2NH_2$) group are inactive. It has been proposed that the structural similarity of the unsubstituted sulfamyl group and carbonic acid permits competitive binding by

the sulfonamide at the active site of the carbonic anhydrase enzyme.

CARBONIC ACID

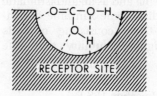

SULFONAMIDE INHIBITION

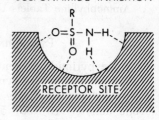

In 1950 Roblin and Clapp[24] synthesized a series of heterocyclic sulfonamides to test as carbonic anhydrase inhibitors. These workers set out to prepare more acidic sulfonamides in the hope that the more highly ionized compounds would bind more strongly to the carbonic anhydrase enzyme and thus be more active inhibitors. It was found that the degree of dissociation of the compounds roughly paralleled the carbonic anhydrase inhibition, with activity ranging as high as 2,500 times greater than sulfanilamide. From the new series of compounds, acetazolamide was selected for clinical trial and later was shown to be an effective diuretic agent.

The mode of action of the carbonic anhydrase inhibitors appears to be reasonably well established. Carbonic anhydrase is known to catalyze the hydration of carbon dioxide (produced metabolically in the renal tubules) to carbonic acid and likewise its reverse dissociation to carbon dioxide and water. The carbonic acid formed ionizes to give bicarbonate and hydrogen ions, as illustrated in the following equations:

$$HOH + CO_2 \underset{\text{Inhibition blocks this step}}{\overset{\text{Carbonic anhydrase}}{\rightleftharpoons}} H_2CO_3 \rightleftharpoons H^+ + HCO_3^-$$

The hydrogen ions formed exchange for sodium ions and to a lesser extent, for potassium ions in the renal tubules, or they may combine with bicarbonate ions to produce carbonic acid and carbon dioxide, thereby propagating the cycle. Inhibition of carbonic anhydrase reduces the concentration of hydrogen ions in the renal tubules and leads to increased excretion of sodium and bicarbonate ions (decreased reabsorption), thereby producing diuresis. There may also be a significant loss of potassium, leading to hypokalemia; chloride excretion is not greatly altered. The normally acidic urine becomes alkaline, and hydrogen ion is retained, which may lead to systemic acidosis. When acidosis occurs, the carbonic anhydrase inhibitors are no longer effective as diuretics, and administration of the drug must be interrupted until the acid-base balance has returned to normal.

The ability of the carbonic anhydrase inhibitors to develop systemic acidosis makes this class of drugs useful as adjuncts to anticonvulsant therapy in epilepsy, since acidosis reduces, or may prevent, epileptic seizures.

Another important clinical use for the carbonic anhydrase inhibitors is in the treatment of glaucoma. Acetazolamide and other carbonic anhydrase inhibitors produce a partial depression of aqueous humor formation, thus reducing the high intraocular pressure associated with this disease. With the exception of the miotics, the carbonic anhydrase inhibitors are considered to be the most useful agents for the treatment of glaucoma, a disease which is estimated to occur in at least 2 per cent of the population over 40 years of age and is responsible for blindness in more than 325,000 persons in this country.

A large number of sulfonamide compounds have been studied for diuretic activity, and many are reported to have a high order of effectiveness as carbonic anhydrase inhibitors. The following compounds belonging to this class are in present use.

Acetazolamide U.S.P., Diamox®, 5-Acetamido-1,3,4-thiadiazole-2-sulfonamide. It occurs as a white to faintly yellowish white, crystalline powder that is slightly soluble in water or alcohol.

Acetazolamide was introduced in 1953 as the first member of the series of carbonic anhydrase inhibitors. It is absorbed following oral administration to give peak levels in the blood plasma in about 2 hours, and the duration of its diuretic action is from 8 to 12 hours. The compound is well tolerated and may be used alone in mild or moderate cases of edema or in conjunction with a mercurial diuretic. When used alone, it may rapidly lose its effectiveness due to systemic acidosis, in which case interruption of therapy is necessary until the acid-base balance is restored. In addition to its diuretic effect, acetazolamide is a useful agent for the treatment of glaucoma and epilepsy. The usual range in the treatment of epilepsy is 375 mg. to 1 g. once a day. When used together with other anticonvulsants, the initial dose should not exceed 250 mg. The drug is available as 250-mg. tablets and as a syrup containing 50 mg. per ml. Sterile Sodium Acetazolamide, U.S.P., is available for intravenous administration when the oral route is impractical.

Category—carbonic anhydrase inhibitor.

Usual dose—250 mg. 2 to 4 times daily.

Usual dose range—250 mg. to 1 g. daily.

OCCURRENCE

Acetazolamide Tablets U.S.P.

Sterile Sodium Acetazolamide U.S.P.

Methazolamide U.S.P., Neptazane®, N-(4-Methyl-2-sulfamoyl-Δ^2-1,3,4-thiadiazolin-5-ylidene)acetamide,5-Acetylimino-4-methyl-Δ^2-1,3,4-thiadiazoline-2-sulfonamide. Methazolamide is a more active carbonic anhydrase inhibitor than the parent compound acetazolamide, and it is claimed to show better penetration into the brain and the eye than acetazolamide; hence, it is recommended primarily for use in the treatment of glaucoma and epilepsy. The drug is available as 50-mg. tablets for oral administration. Like acetazolamide, it can produce an electrolyte imbalance leading to acidosis.

Category—carbonic anhydrase inhibitor.

Usual dose—50 to 100 mg. 2 to 3 times daily.

Usual dose range—100 to 600 mg. daily.

OCCURRENCE

Methazolamide Tablets U.S.P.

Ethoxzolamide U.S.P., Cardrase®, 6-Ethoxybenzothiazole-2-sulfonamide. Ethoxzolamide is approximately 2 times more active

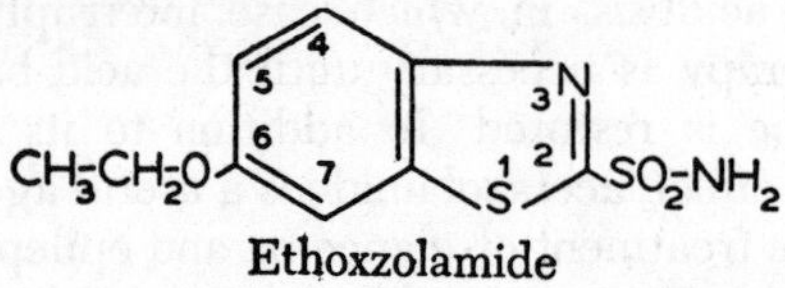

Ethoxzolamide

as a carbonic anhydrase inhibitor than acetazolamide. Following oral administration, maximum plasma levels are attained in approximately 2 hours, and the duration of action is from 8 to 12 hours. Like acetazolamide, the drug may be used as an adjunct in the treatment of glaucoma and epilepsy, but its principal use is as a diuretic, either alone or in conjunction with a mercurial diuretic. The properties of ethoxzolamide closely resemble those of acetazolamide.

Category—carbonic anhydrase inhibitor.

Usual dose—125 mg. 2 to 4 times daily.

Usual dose range—62.5 mg. to 1 g. daily.

OCCURRENCE

Ethoxzolamide Tablets U.S.P.

Dichlorphenamide U.S.P.; Daranide®, 1,2-Dichloro-3,5-disulfamylbenzene. Dichlorphenamide is a carbonic anhydrase inhibitor recommended primarily as an adjunct for the treatment of glaucoma. Like

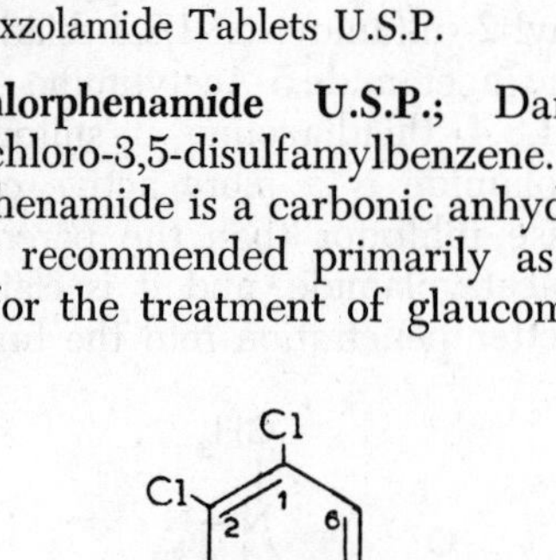

Dichlorphenamide

other drugs of this class, it reduces intraocular pressure by inhibiting aqueous humor formation. Normally, it is used in conjunction with miotic agents such as pilocarpine, physostigmine, etc., and is claimed to be effective when other therapy, including miotics, has failed or is poorly tolerated. Since it is a carbonic anhydrase inhibitor, it is able to produce a disturbance of the acid-base balance, leading to systemic acidosis, but this is not usually experienced in the dosage recommended.

Category—carbonic anhydrase inhibitor.

Usual dose—25 to 50 mg. 1 to 3 times daily.

Usual dose range—50 to 300 mg. daily.

OCCURRENCE

Dichlorphenamide Tablets U.S.P.

BENZOTHIADIAZINE DERIVATIVES

In a study of aromatic sulfonamides as diuretics, Novello and Sprague[25] observed an unexpected high order of activity in the benzene 1,3-disulfonamides, and that activity was enhanced by certain substituents on the ring; chlorine, amino and acylamino groups gave a marked increase in the activity, as did the methyl substituent. Higher alkyl groups decreased activity. 1,4-Disulfonamides were less active. In a further study of the chemistry of the benzene disulfonamides it was observed that when the acylamino group occupied a position ortho to an unsubstituted sulfamyl ($-SO_2NH_2$) group, the compound could be cyclized to give a new type of diuretic of still greater interest. Chlorothiazide is formed by ring closure (elimination of water) of 3-chloro-4,6-disulfamyl formanilide.

By employing other acylated amines, analogs substituted in the 3-position are obtained.

Chlorothiazide was first introduced in 1958, and since that time several closely related analogs have been released for use. The heterocyclic benzothiadiazines are potent orally effective diuretics. Their potency approaches that of parenteral meralluride. Unlike most diuretic drugs, tolerance is not a problem. The benzothiadiazines are closely related chemically, but it has been observed that minor changes in structure

3-Chloro-4,6-disulfamyl formanilide Chlorothiazide

may have a large influence on activity. Thus, saturation of the thiadiazine ring of chlorothiazide gives dihydrochlorothiazide, which is approximately 10 times more active and less toxic than the parent compound.[26]

Mode of Action

The diuretic effect of the benzothiadiazines is due largely to their ability to inhibit the renal tubular reabsorption of sodium and chloride ions and, to a lesser extent, potassium and bicarbonate ions. The same general electrolyte excretion pattern is elicited by the organic mercurial diuretics. In comparing the differences in action of chlorothiazide and other potent diuretics, Beyer[27] has noted the following:

(1) Chlorothiazide gives an increase in urinary pH from 6.0 (control phase) to 7.96 (drug phase) which is not typical of the mercurials, but is typical of acetazolamide and other carbonic anhydrase inhibitors. The latter compounds do not give the marked chloruresis typical of the benzothiadiazines; (2) Chlorothiazide like acetazolamide is a carbonic anhydrase inhibitor, but they differ so markedly in their effect on the kidney (renotropic), their mode of action appears to be quite different. Under conditions of ammonium chloride acidosis, chlorothiazide produces a larger excretion of chloride than sodium and under conditions of sodium bicarbonate alkalosis, it produces a greater excretion of sodium and bicarbonate than chloride ions. (3) Chlorothiazide retains its activity under conditions of experimental alkalosis whereas the organic mercurial diuretics are ineffective under the same conditions; (4) Unlike the mercurials, chlorothiazide and other benzothiadiazines are not inactivated by 2,3-dimercaptopropanol (BAL); and (5) the primary action of chlorothiazide is a saluretic effect with or without a substantial increase in bicarbonate elimination.

Evaluation of the Benzothiadiazines

The several benzothiadiazines in clinical use differ from the parent compound, chlorothiazide, primarily in their activity, toxicity and duration of action. The following observations, based on accumulated clinical experience have been made[28]: (1) Patients resistant to one benzothiadiazine derivative are likely to be resistant to others. (2) Each drug has a "ceiling dose" above which sodium loss does not increase. However, doses much lower than the "ceiling dose" are usually adequate. (3) The diuretic potency varies markedly, as reflected in the average dose for each of the compounds, which varies from 2 mg. to 500 mg. daily. (4) The doses necessary to produce equivalent loss of sodium and water may also produce comparable loss of potassium and bicarbonate. (5) To avoid hypokalemia the supplemental administration of potassium or concurrent administration of a potassium-sparing diuretic is recommended for all members of the series. Due to the observed increase in the incidence of small-bowel ulceration and stenosis following thiazide and potassium therapy,[29, 30] the Food and Drug Administration has ordered certain preparations that contain potassium salts to go on prescription-order status. The regulation applies to any capsule and coated or uncoated tablet that supplies 100 mg. or more of potassium per unit dose, and to liquid preparations that contain potassium salts and supply 20 mg. or more potassium per ml. (6) The side-effects, which may be potentially serious, are common to all of the benzothiadiazines.

The side-effects of the benzothiadiazines may include nausea, anorexia and headaches, hyperuricemia and, in rare instances, leukopenia and rash. Gastrointestinal distress, jaundice, photosensitization, acute

TABLE 62. CHLOROTHIAZIDES AND ANALOGS

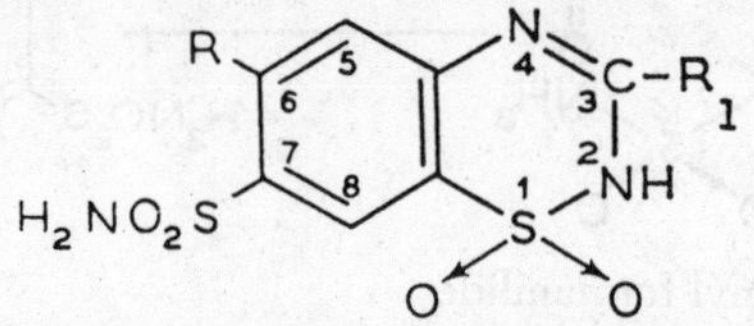

GENERIC NAME	PROPRIETARY NAME	R	R_1	APPROXIMATE DAILY DOSE mg.
Chlorothiazide N.F.	Diuril	—Cl	H	500
Benzthiazide N.F.	Exna	—Cl	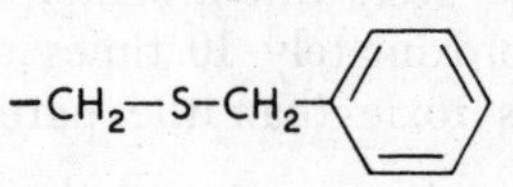 $-CH_2-S-CH_2-$⟨benzene⟩	50-150

glomerulonephritis and pancreatitis have been observed. These drugs may also precipitate onset of diabetes and gout. They decrease the responsiveness to the catechol amines and increase the skeletal muscle paralysis of (+)-tubocurarine.[28]

The benzothiadiazines, like other diuretics, are employed primarily for the treatment of edemas of both pathologic or drug-induced origin. They have the important advantage of potentiating the effect of antihypertensive agents such as reserpine, veratrum alkaloids, hydralazine and the ganglionic blocking agents, thus serving as

TABLE 63. HYDROCHLOROTHIAZIDE AND ANALOGS

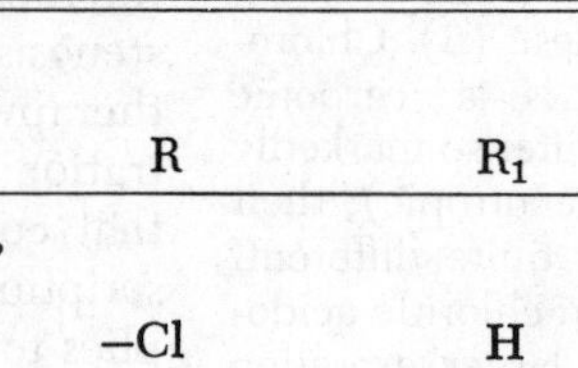

GENERIC NAME	PROPRIETARY NAME	R	R_1	R_2	APPROXIMATE DAILY DOSE mg.
Hydrochlorothiazide U.S.P.	Hydrodiuril, Esidrex, Oretic	—Cl	H	H	50
Hydroflumethiazide N.F.	Saluron,	$-CF_3$	H	H	50
Bendroflumethiazide N.F.	Naturetin, Benuran	$-CF_3$	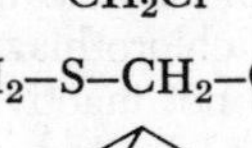 $-CH_2-$⟨benzene⟩	H	5
Trichlormethiazide N.F.	Naqua, Metahydrin	—Cl	$-CHCl_2$	H	4
Methyclothiazide N.F.	Enduron	—Cl	$-CH_2Cl$	CH_3	5
Polythiazide N.F.	Renese	—Cl	$-CH_2-S-CH_2-CF_3$	$-CH_3$	4
Cyclothiazide N.F.	Anhydron	—Cl	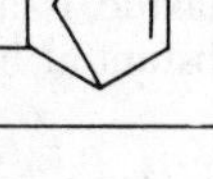 ⟨norbornene⟩	H	2

useful adjuncts for the treatment of hypertension. Employed alone, the benzothiadiazines are useful antihypertensive agents, but they are used more commonly in combination with other antihypertensives.

The mechanism of the antihypertensive effect is not known but is generally assumed to be related to the diuretic and naturetic properties. Interestingly, removal of the sulfamyl group gives compounds devoid of diuretic properties, yet retaining the antihypertensive activity. Thus, 7-chloro-3-methyl-2H-1,2,4,-benzothiadiazine-1,1-dioxide (diazoxide, for structure see numbering of ring, Table 62) has been reported[31] to be an effective antihypertensive agent without diuretic properties; useful drugs may result from this observation.

The benzothiadiazine diuretics may be divided conveniently into two general types: (1) the analogs of chlorothiazide (Table 62) and (2) the analogs of hydrochlorothiazide (Table 63). They differ only with respect to the 3,4-positions of the thiadiazine ring, which is unsaturated in the chlorothiazide group.

Structure-Activity Relationships of Benzothiadiazines

Optimum diuretic activity has thus far been associated with the following structural features: (1) The benzene ring must

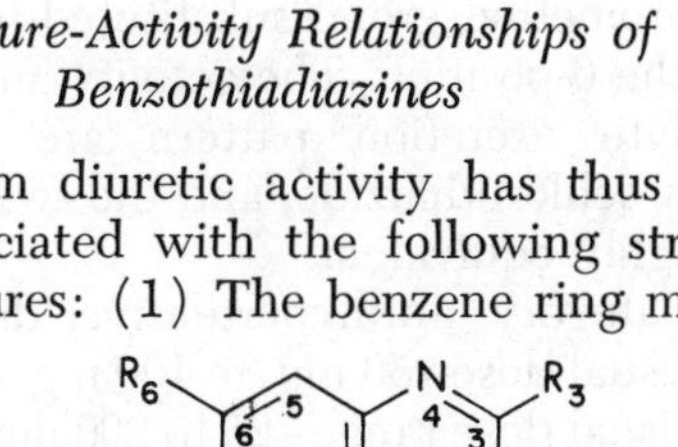

have a sulfamyl group, preferably unsubstituted, at position 7 and a halogen or halogenlike group (e.g., CF_3) at position 6. (2) Saturation of the 3,4 double bond generally produces increased activity. (3) Lipophilic substituents at position 3 enhance activity, as do lower alkyl groups, such as methyl at position 2. (4) Position 1 of the heterocyclic ring may be

$$SO_2 \text{ or } C=O,$$

but higher activity is associated with the sulfur heterocycle.

Products

Chlorothiazide N.F., Diuril®, 6-Chloro-2H-1,2,4-benzothiadiazine-7-sulfonamide 1,1-dioxide. This compound occurs as a white

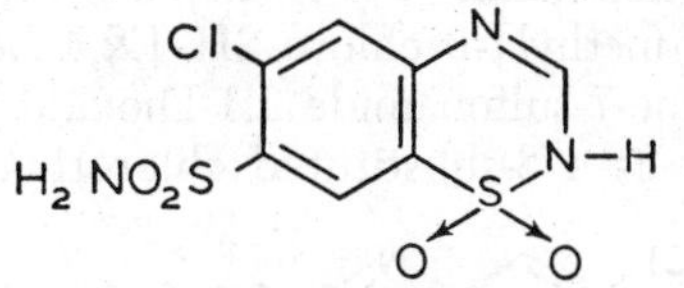

Chlorothiazide

crystalline powder that is very slightly soluble in water.

Chlorothiazide was first introduced in 1958 and is the original benzothiadiazine diuretic. It gives primarily a saluretic effect but may also produce significant loss of potassium and bicarbonate. To prevent the development of hypokalemia, adequate amounts of potassium chloride or a potassium-sparing diuretic normally are given with chlorothiazide or with other thiazide diuretics. Hypochloremic alkalosis also may develop, and this may be treated by a temporary discontinuance of the drug or by giving appropriate amounts of ammonium chloride. Chlorothiazide is a potent diuretic and is used for the treatment of all types of edemas. It is also used alone and as an adjunct in the management of hypertension. Unlike other antihypertensive agents, chlorothiazide and other benzothiadiazine analogs lower blood pressure only in hypertensive and not in normotensive individuals. The onset of action is comparatively rapid (approximately 2 hours) and lasts from 6 to 12 hours. The drug normally maintains its effectiveness with prolonged administration.

Chlorothiazide shows no effect on intraocular pressure or on the rate of aqueous humor formation. Unlike the related carbonic anhydrase inhibitors, it is not used in the treatment of glaucoma.

Sodium chlorothiazide, Lyovac Diuril®, is available for parenteral administration.

Category—diuretic, antihypertensive.

Usual dose—antihypertensive, 250 mg. 3 times daily; diuretic, 500 mg. once or twice daily.

Usual dose range—antihypertensive, 250 to 500 mg.; diuretic, 500 mg. to 1 g.

Chlorothiazide Oral Suspension N.F.
Chlorothiazide Tablets N.F.
Sodium Chlorothiazide for Injection N.F.

Benzthiazide N.F., Exna®, 3-[(Benzyl-thio)methyl]-6-chloro-2H-1,2,4-benzothia-diazine-7-sulfonamide 1,1-Dioxide. Benzthi-azide is a 3-substituted chlorothiazide and

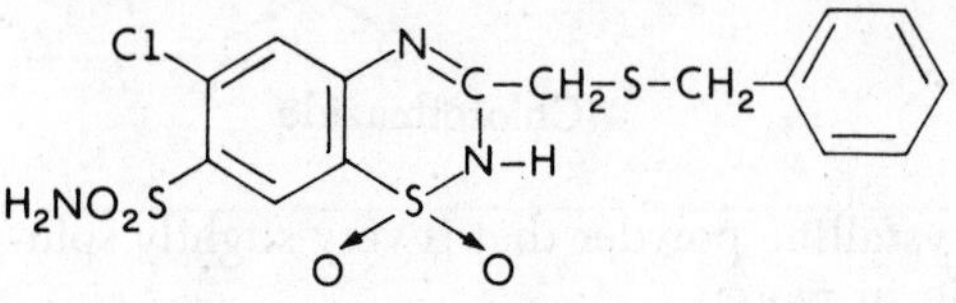

Benzthiazide

is significantly more active than the parent compound. It is reported to be approximately 85 per cent as active as hydrochloro-thiazide and to give a similar electrolyte excretion pattern. Hypokalemia and other electrolyte imbalances may occur on prolonged use.

Category—diuretic; antihypertensive.

Usual dose range—diuretic: initial, 50 to 200 mg. daily; maintenance, 50 to 150 mg. daily; antihypertensive: initial, 25 to 50 mg. twice daily; maintenance, adjust to the response of the patient with a maximal dose of 50 mg. 3 times daily.

Occurrence
Benzthiazide Tablets N.F.

Hydrochlorothiazide U.S.P., Hydrodiu-ril®, Esidrex®, Oretic®, 6-Chloro-3,4-dihy-dro-2H-1,2,4-benzothiadiazine-7-sulfona-mide 1-1-Dioxide.

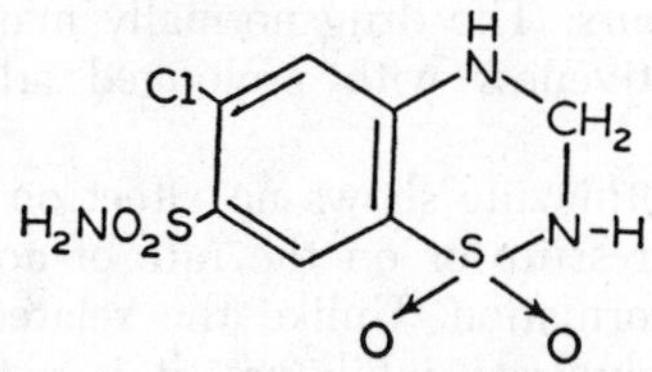

Hydrochlorothiazide

Reduction of the 3-4-position of the thia-diazine ring increases activity of the benzo-thiadiazines by approximately 10 times and also lowers the toxicity. Hydrochlorothia-zide is the parent compound belonging to the class of reduced benzothiadiazines. Qualitatively, the metabolic properties of

hydrochlorothiazide are similar to chloro-thiazide, but the reduced compound has the important advantage of causing less loss of potassium and bicarbonate and, therefore, less tendency to produce an electrolyte imbalance. Another advantage is that it is better tolerated and produces fewer side-effects (nausea, anorexia, headache, restlessness and constipation) than chlorothiazide.

Category—diuretic.

Usual dose—50 mg. once or twice daily.

Usual dose range—25 to 200 mg. daily.

Occurrence
Hydrochlorothiazide Tablets U.S.P.

Hydroflumethiazide N.F., Saluron®, 3,4-Dihydro-6-(trifluoromethyl)-2H-1,2,4-ben-zothiadiazine-7-sulfonamide 1,1-Dioxide. Hydroflumethiazide differs in structure from hydrochlorothiazide by having a tri-

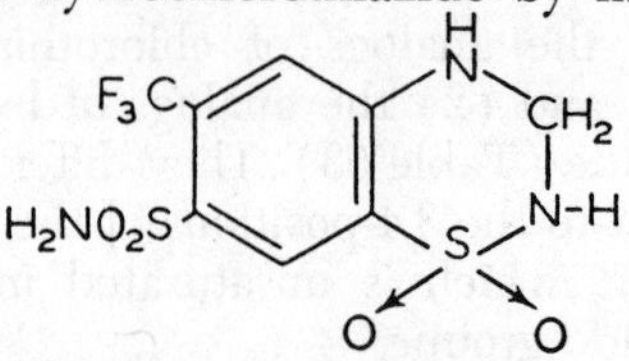

Hydroflumethiazide

fluoromethyl group substituted for chlorine in the 6-position. The activity and the electrolyte excretion pattern are similar to hydrochlorothiazide, and the two drugs are roughly equivalent.

Category—antihypertensive; diuretic.

Usual dose—50 mg. to 100 mg. daily.

Usual dose range—25 to 200 mg.

Occurrence
Hydroflumethiazide Tablets N.F.

Bendroflumethiazide N.F., Naturetin®, Benuran®, 3-Benzyl-3,4-dihydro-6-(trifluoro-methyl)-2H-1,2,4-benzothiadiazine-7-sulfon-

Bendroflumethiazide

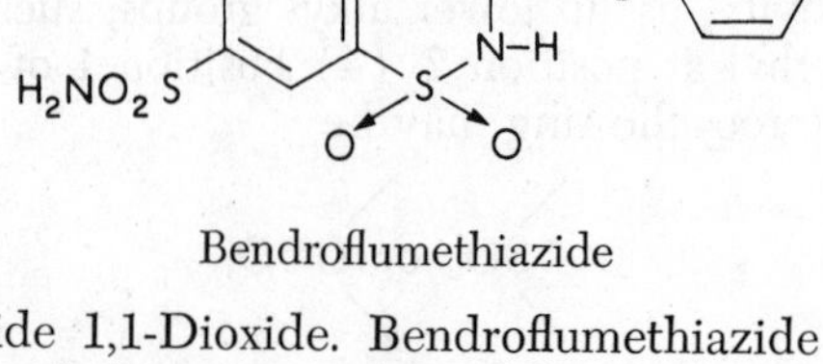

amide 1,1-Dioxide. Bendroflumethiazide is one of the most potent diuretic and anti-hypertensive agents available for use. It incorporates in its structure a reduced thia-diazine ring and a benzyl substitution on the 3-position, both of which enchance the activity. The trifluoromethyl group is substi-tuted for chlorine in the 6-position, which may lower the toxicity. The benzyl substi-tution on the 3-position enhances activity and gives a longer duration of action (ap-proximately 18 hours), but this may not be important clinically.

Qualitatively, bendroflumethiazide is sim-ilar to hydrochlorothiazide, but it is re-ported to have the advantage of producing less loss of potassium and bicarbonate and fewer side-effects. For long-term therapy, potassium chloride is recommended as a supplement to avoid hypokalemia.

Category—diuretic; antihypertensive.

Usual dose range—diuretic: initial, 5 to 20 mg. daily; maintenance, 2.5 to 5 mg. daily; antihypertensive: initial, 5 to 20 mg. daily; maintenance, 2.5 to 15 mg. daily.

OCCURRENCE

Bendroflumethiazide Tablets N.F.

Trichlormethiazide N.F., Naqua®, Meta-hydrin®, 6-Chloro-3-(dichloromethyl)-3,4-dihydro-2H-1,2,4-benzothiadiazine-7-sulfon-amide 1,1-Dioxide. Trichlormethiazide dif-

Trichlormethiazide

fers in structure from hydrochlorothiazide by the substitution of a dichloromethyl group for hydrogen in the 3-position, which increases the diuretic potency by approxi-mately 10 times. The compound is excreted more slowly than hydrochlorothiazide, but the difference in duration of action has not been shown to be clinically important. Similar diuretic responses are reported fol-lowing the administration of 8 mg. of tri-chlormethiazide and 75 mg. of hydro-chlorothiazide.

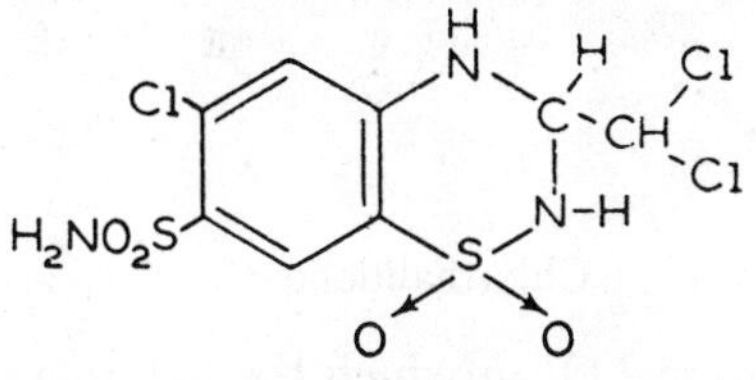

Category—diuretic; antihypertensive.

Usual dose range—initial, 2 to 4 mg. twice daily, then 2 to 4 mg. once daily.

OCCURRENCE

Trichlormethiazide Tablets N.F.

Methyclothiazide N.F., Enduron®, 6-Chloro-3-(chloromethyl)-3,4-dihydro-2-methyl-2H-1,2,4-benzothiadiazine-7-sulfon-amide 1,1-Dioxide. Methyclothiazide is the first of the benzothiadiazines to be substi-tuted in the 2-position. It is well absorbed and develops a diuretic response within 2 hours following administration and a maxi-mum response at about 6 hours. The diu-

Methyclothiazide

retic response continues for 24 hours or more; therefore, a continuous therapeutic effect may be obtained from a single daily dose. Serum potassium levels are reported to be comparatively unchanged, and the predominant effects are diuresis, chloruresis and natriuresis. Urinary pH is not signifi-cantly altered.

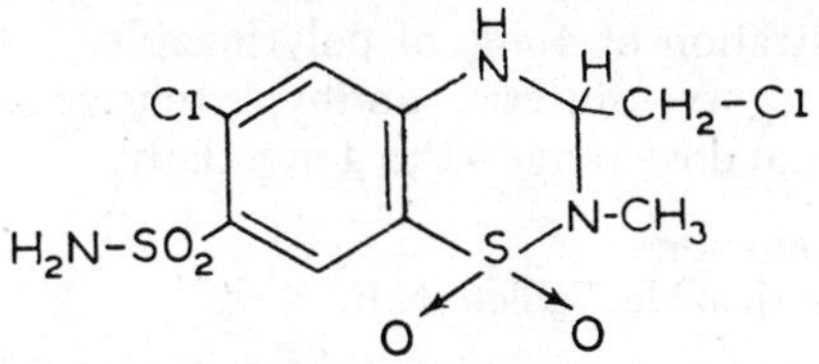

Methyclothiazide is a potent oral diu-retic, and, like other benzothiadiazines, it potentiates the effects of ganglionic block-ing and other antihypertensive agents.

Category—diuretic; antihypertensive.

Usual dose range—2.5 to 10 mg. once daily, 10 mg. being the maximum single dose.

OCCURRENCE

Methyclothiazide Tablets N.F.

Polythiazide N.F., Renese®, 6-Chloro-3,4-dihydro-2-methyl-3-[[(2,2,2-trifluoroethyl)-thio]methyl]-2H-1,2,4-benzothiadiazine-7-sulfonamide 1,1-Dioxide. The principal ef-fect of polythiazide is on the renal excretion of sodium and chloride with a lesser effect

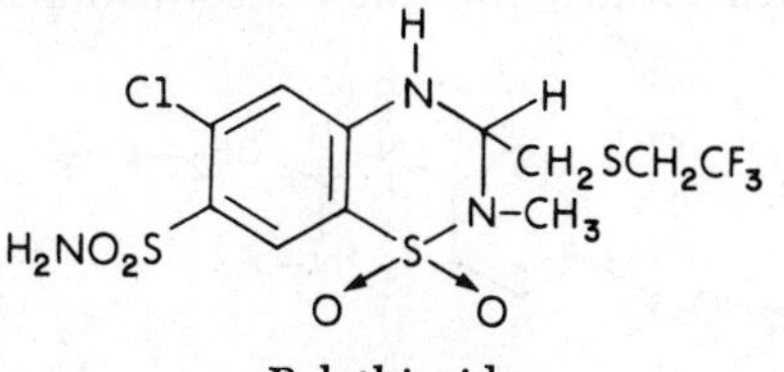

Polythiazide

on the excretion of potassium and bicarbonate. Substitution of both a methyl group in the 2-position and a trifluoroethylthiomethyl group in the 3-position results in an equivalent increase in urinary sodium excretion at about one tenth the dose of hydrochlorothiazide. The drug is excreted slowly, due to binding by plasma proteins and reabsorption by the distal tubules. This may be responsible for its long duration of action. Sodium excretion levels have been reported to be elevated 72 hours after administration of 4 mg. of polythiazide.

Category—diuretic; antihypertensive.

Usual dose range—1 to 4 mg. daily.

OCCURRENCE

Polythiazide Tablets N.F.

Cyclothiazide N.F., Anhydron®, 6-Chloro-3,4-dihydro-3-(5-norbornen-2-yl)-2H-1,2,4-benzothiadiazine-7-sulfonamide 1,1-Dioxide.

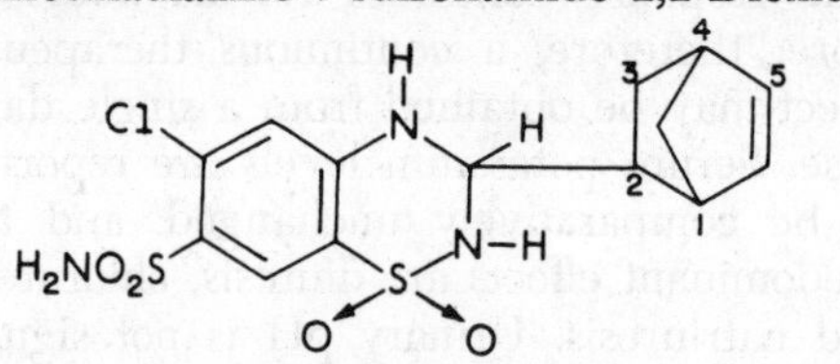

Cyclothiazide

Cyclothiazide is a potent orally effective diuretic with a lipophilic terpene substituent in the 3-position.

Category—diuretic; antihypertensive.

Usual dose range—diuretic: initial, 1 to 2 mg. daily; maintenance, 1 to 2 mg. every other day or 2 or 3 times a week; antihypertensive: 2 mg. 1 to 3 times daily.

OCCURRENCE

Cyclothiazide Tablets N.F.

QUINAZOLINONE DERIVATIVES

Quinethazone U.S.P., Hydromox®, 7-Chloro-2-ethyl-1,2,3,4-tetrahydro-4-oxo-6-quinazoline Sulfonamide. Quinethazone differs structurally from the benzothiadiazines

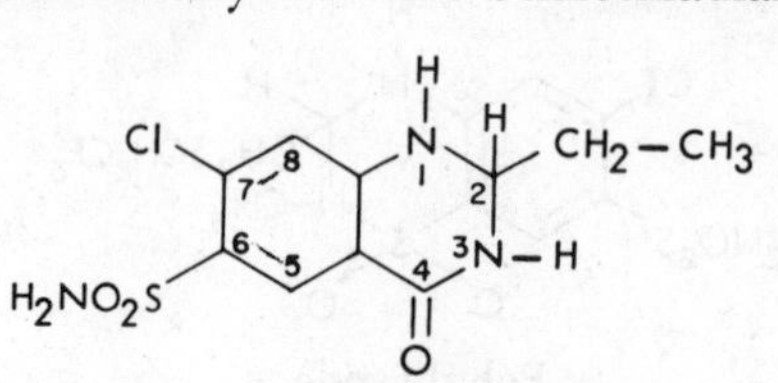

Quinethazone

in having the ring sulfone (S-dioxide) replaced with a carbonyl group. The drug is a potent, long-acting, orally effective diuretic. The compound has the same order of potency as hydrochlorothiazide, with a duration of action between 24 and 72 hours.

Category—diuretic.

Usual dose—50 to 100 mg. once daily.

Usual dose range—50 to 200 mg. daily.

OCCURRENCE

Quinethazone Tablets U.S.P.

PHTHALIMIDINE DERIVATIVES

Chlorthalidone U.S.P., Hygroton®, 2-Chloro-5-(1-hydroxy-3-oxo-1-isoindolinyl)-benzenesulfonamide. Chlorthalidone contains a sulfamyl and carboxamide group in each ring of the benzophenone system (II), but exists primarily in the tautomeric

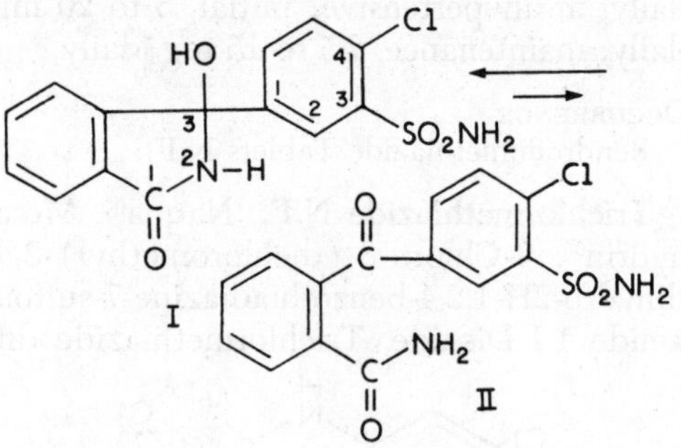

Chlorthalidone

lactam form (I), in which the heterocyclic nucleus may be named as an isoindoline or as a phthalimidine. Chlorthalidone is a potent, long-acting, orally effective diuretic, and antihypertensive agent. The compound is unique in that it is the only diuretic making use of the phthalimidine ring system but is clearly related structurally to other diuretic sulfonamides. Average therapeutic doses are reported to give primarily a saluretic effect with minimal loss of potassium and bicarbonate. The mode of action of chlorothalidone is not established, but the electrolyte excretion pattern is similar to that given by the benzothiadiazines. Chlorthalidone has no effect on either renal circulation or glomerular filtration, and the diuretic response is believed to be due to interference with the renal tubular reabsorption of sodium and chloride, thereby promoting loss of salt and water. Compared on a weight basis (orally), it is 1.8 times as potent as meralluride intramuscularly and gives a duration of action up to 60 hours. The compound is concentrated in the kidney, and a large portion of the drug is eliminated unchanged.

Category—diuretic.

Usual dose—100 mg. once daily.

Usual dose range—50 to 200 mg. daily or every other day.

OCCURRENCE

Chlorthalidone Tablets U.S.P.

ANTHRANILIC ACID DERIVATIVES

Furosemide U.S.P., Lasix®, 4-Chloro-*N*-furfuryl-5-sulfamoylanthranilic acid.

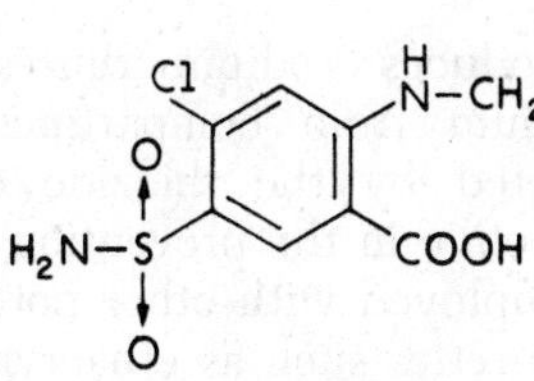

Furosemide

Furosemide is a highly active saluretic agent that produces a rapid diuretic response of comparatively short duration (6 to 8 hours). It inhibits the reabsorption of sodium throughout the renal tubules, which may account for its high potency and effectiveness in cases of reduced glomerular filtration in which the thiazides and other diuretics fail. In ceiling dosages furosemide shows 8 to 10 times the saluretic effect of the thiazides.[2] Like the thiazides, furosemide promotes potassium excretion and is commonly used with potassium supplementation or a potassium-sparing diuretic. Other side-effects may include hypochloremic alkalosis, hyperuricemia, and hyperglycemia.[32] Furosemide has a blood-pressure-lowering effect similar to that of the thiazides. In addition to the 40-mg. tablets for oral use, the drug is available as a sterile solution in 2-ml. ampules, each containing 20 mg.

Category—diuretic.

Usual dose—40 to 80 mg. once daily.

Usual dose range—40 to 200 mg. daily.

OCCURRENCE

Furosemide Tablets U.S.P.

ENDOCRINE ANTAGONISTS

Aldosterone is a potent antidiuretic hormone secreted by the adrenal cortex. In congestive heart failure, nephrosis and other pathologies associated with edema, there is increased secretion of aldosterone which is believed to be responsible for the retention of salt and water. The biologically active corticosteroids, such as cortisone, hydrocortisone, desoxycorticosterone, aldosterone, and closely related analogs tend to increase retention of salt and water by increasing the renal tubular reabsorption of sodium and chloride and to promote the excretion of potassium. The corticosteroids differ widely in their activity to produce retention of salt and water, but the most potent of the compounds is aldosterone, which is at least 1,000 times more active than hydrocortisone and is believed to play an important role in maintaining the normal electrolyte balance.

Aldosterone occurs as a hemiacetal in equilibrium with a hydroxy aldehyde form, as shown in the following structures:

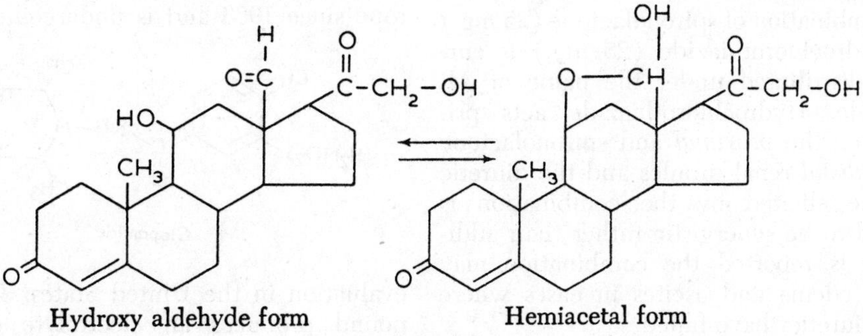

Hydroxy aldehyde form   Hemiacetal form

Aldosterone

Antagonists of aldosterone decrease the amount of sodium and chloride reabsorbed by the renal tubules, thereby promoting diuresis—a process of competitive inhibition. Aldosterone antagonists in combination with other diuretics are able to restore the electrolyte balance and are potentially most useful agents. Spironolactone is the only aldosterone inhibitor available for use at the present time, but equally or more effective compounds of this class may be anticipated in the future.

Spironolactone U.S.P., Aldactone®, 17-Hydroxy-7α-mercapto-3-oxo-17α-pregn-4-ene-21-carboxylic Acid γ-Lactone 7-Acetate.

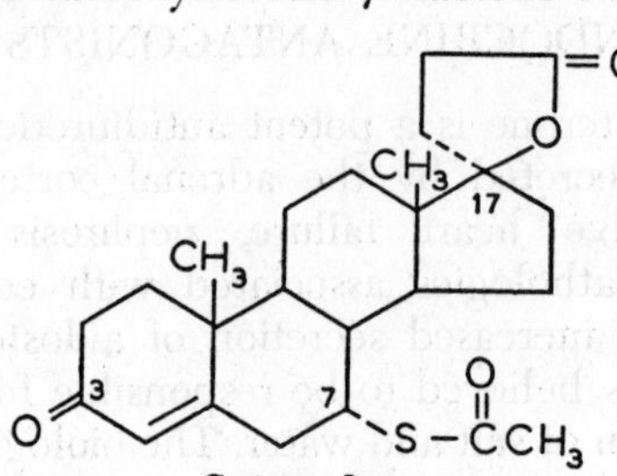

Spironolactone

Spironolactone is a synthetic steroid in which the side chain on the C-17 carbon of 4-androsten-3-one is replaced by a 5-membered lactone ring, and the 7α-position is substituted with an acetylthio group. Spironolactone has some structural similarity to aldosterone and blocks the latter's effect of promoting reabsorption of sodium and loss of potassium in the distal renal tubules. Unlike other diuretics, spironolactone does not produce loss of potassium. The compound is sometimes effective when used alone, but a slow onset of action (3 to 7 days) and variable diuresis cause it to be used normally in combination with other diuretics, such as the mercurials and the benzothiadiazines.

A combination of spironolactone (25 mg.) and hydrochlorothiazide (25 mg.) is currently distributed under the name of Aldactazide. Hydrochlorothiazide acts primarily on the *proximal* and spironolactone on the *distal* renal tubules and the diuretic response elicited by the combination is claimed to be synergistic rather than additive. It is reported the combination may control edema and ascites in cases where other diuretics have failed.

Category—diuretic.
Usual dose—25 mg. 2 to 4 times daily.

OCCURRENCE
Spironolactone Tablets U.S.P.

COMPOUNDS UNDER INVESTIGATION

Amiloride, Colectril®, N-Amidino-3,5-diamino-6-chloropyrazinecarboxamide Hydrochloride. Amiloride, an aminopyrazine derivative, is a long-acting potassium-sparing

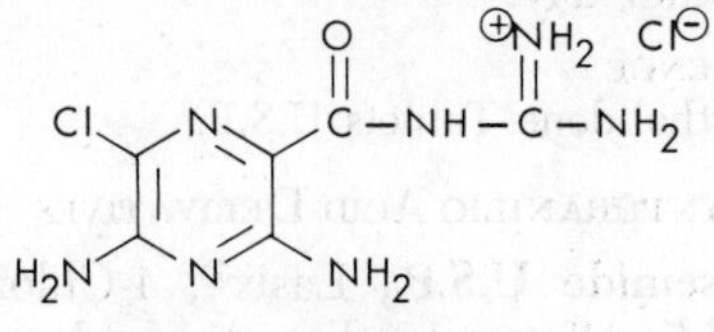

Amiloride

diuretic that produces sodium diuresis, withdrawing sodium from compartments not usually affected by the thiazide diuretics.[33] It is effective in the prevention of kaluresis when employed with other potassium-depleting diuretics such as ethacrynic acid. It may also produce an increase in blood urea nitrogen levels in patients with impaired renal function or with diabetes. The compound appears to be a useful diuretic for treatment of refractory edema, and especially in patients with hypokalemia. Amiloride potentiates the diuretic effect of the thiazides and produces no change in uric acid clearance. The maximum diuretic response to amiloride is produced in 4 to 6 hours, with a duration of action of from 10 to 12 hours.[34] The usual adult oral dose ranges from 5 to 40 mg. daily.

Clopamide, Aquex®, 1-(4-Chloro-3-sulfamylbenzamido)-2,6-*cis*-dimethylpiperidine. Clopamide has been marketed in Europe since 1963 and is undergoing clinical

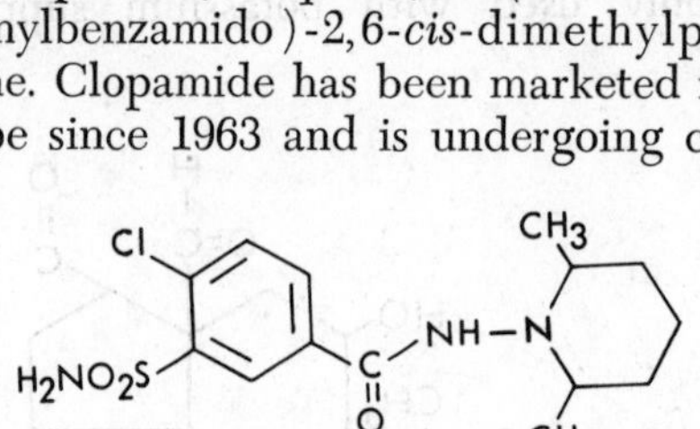

Clopamide

evaluation in the United States. The compound produces an electrolyte excretion pattern similar to that of the thiazides.[35] It shows weak carbonic anhydrase inhibiting activity but does not alter the acid-base balance. Dosages of from 25 to 50 mg. produce an increase in excretion of sodium, potassium and chloride ions in both normal subjects and those with various states of fluid retention. Like the thiazides, clopamide is a hypotensive agent, and also induces biochemical changes, such as hyperuicemia, hypergylcemia and hypokalemia.[13]

Clorexolone, Nefrolan®,5-Chloro-2-cyclohexyl-1-oxo-6-sulfamylisoindoline. Clorexolone is a potassium-sparing long-acting diuretic, chemically related to chlorthalidone.

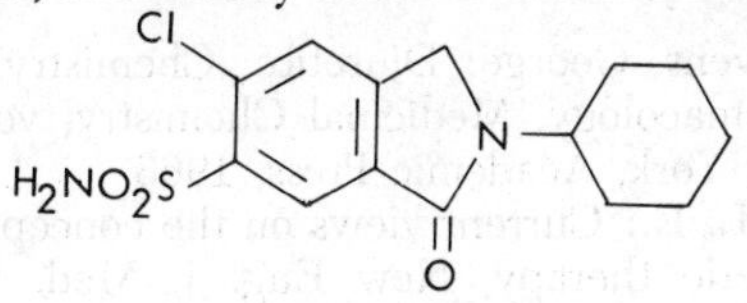

Clorexolone

Its electrolyte excretion pattern is similar to that of the thiazides, except that at therapeutic dosage levels of 25 to 100 mg. per day it produces a pronounced diuresis without causing hypokalemia. The natriuretic effect persists for more than 48 hours. Clorexolone increases the blood serum uric acid levels but appears to have no significant effect on carbohydrate metabolism and is, therefore, less likely to cause hypergylcemia.[13, 36] The compound is approximately two and one-half times more active than hydrochlorothiazide.[37]

Mefruside, N-(4-chloro-3-sulfamylbenzenesulfonyl)-N-methyl-2-aminomethyl-2-methyltetrahydrofuran. Mefruside, a 1,3-disulfonamide derivative, is a long acting potent diuretic that differs in action from the related dichlorphenamide in that it acts primarily as a saluretic agent and does not increase bicarbonate excretion. Mefruside lactone, the main metabolic product, is several times more active than is the parent compound.[38] The (−)-enantiomers of me-

fruside and its lactone are more active than are the (+)-forms.

During recent years a large number of new compounds have been evaluated for their diuretic action, and a small percentage of these have been introduced as therapeutic agents. The ideal diuretic agent, one that effectively promotes sodium and water excretion without producing an electrolyte imbalance or disturbing carbohydrate metabolism and uric acid elimination, remains to be discovered. However, advances in renal physiology and biochemistry have led to a better understanding of the basic problem, and advances in pharmacology and medicinal chemistry have increased understanding of the modes of diuretic action and the rational design of more selective agents. Although many highly effective diuretics are now available, new agents which more nearly approach the ideal may be anticipated.

REFERENCES

1. Sprague, J. M.: Diuretics; *in:* Rabinowitz, J. L., and Myerson, R. M. (eds.): Topics in Medicinal Chemistry, vol. 2, New York, Interscience Pub., John Wiley and Sons, 1968.
2. Early, L. E.: New Eng. J. Med. 276:966, 1967.
3. Kleeman, C. R., and Fichman, M. P.: New Eng. J. Med. 277,1300, 1967.
3a. de Rouffinac, C., and Morel, F.: J. Clin. Invest. 41:474, 1969.
4. Early, L. E., and Orloff, J.: Ann. Rev. Med. 15:149, 1964.
5. Sprague, J. M.: Ann. N. Y. Acad. Sci. 71: 328, 1958.
6. Kessler, R. H., Lozano, R., and Pitts, R. F.: J. Clin. Invest. 36:656, 1957.
7. Friedman, H. L.: Ann. N. Y. Acad. Sci. 65:461, 1957.
8. Batteman, R. C., Unterman, D., and DeGraff, A. C.: J.A.M.A. 140:1268, 1949.

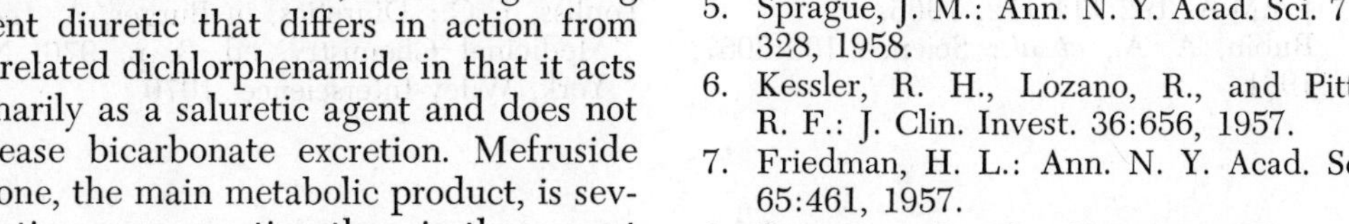

Mefruside

Mefruside Lactone
(metabolic product)

9. Vander, A. J., Malvin, R. L., Wilde, W. S., and Sullivan, L. P.: Am. J. Physiol. 195: 558, 1958.

10. Mudge, G. H., and Wiener, I. N.: Ann. N. Y. Acad. Sci. 71:344, 1958.

11. Schultz, E. M., *et al.*: J. Med. Pharm. Chem. 5:660, 1962.

12. Early, L. E., *et al.*: J. Clin. Invest. 43: 1160, 1964.

13. Hutcheon, D. E.: Am. J. Med. Sci. 253: 620, 1967.

14. Kirkendall, W. M., and Stern, J. H.: Am. J. Cardiol. 22:162, 1968.

15. Schwartz, F. D.: Lancet 1:77, 1969.

16. Levene, P. A.: Biochem. J. 4:316, 1907.

17. Papesch, V., and Schroeder, E. F.: J. Org. Chem. 16:1879, 1951.

18. Pachter, I. J., and Nemeth, P. E.: J. Org. Chem. 28:1187, 1963.

19. Ginsberg, D. J., Load, A., and Gabuzda, G. J.: New Eng. J. Med. 271:1229, 1964.

20. Liddle, G. W.: Ann. N. Y. Acad. Sci. 139:466, 1966.

21. Southworth, H.: Proc. Soc. Exp. Biol. Med. 26:586, 1937.

22. Strauss, M. B., and Southworth, H.: Bull. Johns Hopkins Hosp. 63:41, 1938.

23. Mann, T., and Keilin, D.: Nature 146: 164, 1940.

24. Roblin, R. O., Jr., and Clapp, J. W.: J. Am. Chem. Soc. 72:4289, 1950.

25. Novello, F. C., and Sprague, J. M.: J. Am. Chem. Soc. 79:2028, 1957.

26. Friend, D. G.: Clin. Pharmacol. Ther. 1:5, 1960.

27. Beyer, K. H.: Ann. N. Y. Acad. Sci. 71: 363, 1958.

28. The Medical Letter 2 (No. 15):57, 1960; 3 (No. 9):36, 1961.

29. Boley, S. J., *et al.*: J.A.M.A. 192:93-98, 1965.

30. Abbruzzese, A. A., and Gooding, C. A.: J.A.M.A. 192:111-112, 1965.

31. Rubin, A. A., *et al.*: Science 133:2067, 1961.

32. Laragh, J. H., *et al.*: Ann. N. Y. Acad. Sci. 139:304, 1966.

33. Baer, J. E., *et al.*: J. Pharmacol. Exp. Ther. 157:472, 1967.

34. Hitzenberger, G., *et al.*: Clin. Pharmacol. Exp. Ther. 9:71, 1967.

35. Juker, E., *et al.*: Arzn. Forsch. 13:269, 1963.

36. Lant, A. F., *et al.*: Clin. Pharmacol. Ther. 7:196, 1966.

37. Russell, R. R., *et al.*: Clin. Pharmacol. Ther. 10:265, 1969.

38. Schlossman, K.: Arzn. Forsch. 17:688, 1967.

SELECTED READING

de Stevens, George: Diuretics: Chemistry and Pharmacology, Medicinal Chemistry, vol. 1, New York, Academic Press, 1963.

Early, L. E.: Current views on the concepts of diuretic therapy, New Eng. J. Med. 276: 966, 1967.

Grollman, A. (ed.): New diuretics and antihypertensive agents, Ann. N. Y. Acad. Sci. 88:771-1020, 1960.

Hess, H.-J.: Diuretic Agents, *in* Cain, C. K. (ed.): Ann. Reports Med. Chem., 1967 Academic Press, New York, 1968.

Hutcheon, D. E.: The pharmacology of the established diuretic drugs, Am. J. Med. Sci. 253:620, 1967.

Kleeman, C. R., and Fickman, M. P.: The regulation of renal water metablolism, New Eng. J. Med. 277:1330, 1967.

Pitts, R. F., *et al.*: Chlorothiazide and other diuretics, Ann. N. Y. Acad. Sci. 71:371-478, 1958.

Sprague, J. M.: Diuretics, *in* Rabinowitz, J. L., and Myerson, R. M. (eds.): Topics in Med. Chem., vol. 2, New York, Wiley-Interscience, 1968.

Topliss, J. G.: Diuretics, *in* Burger, A. (ed.): Medicinal Chemistry, ed. 3, p. 976, New York, Wiley-Interscience, 1970.

21
Cardiovascular Agents

Robert F. Doerge, Ph.D.,
Professor of Pharmaceutical Chemistry,
Chairman of the Department of Pharmaceutical Chemistry, School of Pharmacy,
Oregon State University

INTRODUCTION

Cardiovascular agents are used for their action on the heart or on other parts of the vascular system so that they modify the total output of the heart or the distribution of blood to certain parts of the circulatory system. These include cardiotonic drugs (see Chap. 25), vasodilators, hypotensive drugs, drugs that modify cardiac rhythm, antihypercholesterolemic drugs and sclerosing agents. Of course, there are other classes of drugs which do not necessarily have a direct action on the cardiovascular system but are of considerable value in the treatment of cardiac disease. These include the diuretics and the anticoagulants.

This chapter is concerned with drugs which have a direct action on the cardiovascular system, and, in addition, it covers classes of drugs that affect certain constituents of the blood. The latter include the anticoagulant drugs, the hypoglycemic agents, the thyroid hormones and the antithyroid drugs.

VASODILATORS

This group of drugs acts primarily on the vascular system and includes the esters of nitrous and nitric acid, and certain alkaloids. There are other drugs useful in the treatment of hypertension, such as chlorothiazide and its congeners and certain autonomic blocking agents. These are discussed in Chapters 19 and 20.

ESTERS OF NITROUS AND NITRIC ACIDS

Inorganic acids, like organic acids, will form esters with an alcohol. Pharmaceutically, the important ones are the bromide, the chloride, the nitrite and the nitrate. The chlorides and the bromides are thought of conventionally as chloro- or bromo- compounds, and they are discussed with those types of compounds. Hydrogen cyanide forms the organic cyanides or nitriles (R—CN). Sulfuric acid forms organic sulfates, of which methyl sulfate and ethyl sulfate are the most common.

Nitrous acid, HNO_2, esters may be formed readily from an alcohol and nitrous acid. The usual procedure is to mix sodium nitrite, sulfuric acid and the alcohol. Organic nitrites are generally very volatile liquids that are only slightly soluble in water but soluble in alcohol. Preparations containing water are very unstable, due to hydrolysis.

The organic nitrates and nitrites and the inorganic nitrites have their primary utility in the prophylaxis and treatment of angina pectoris. They have a more limited application in treating asthma, gastrointestinal spasm and certain cases of migraine headache. Nitroglycerin (glyceryl trinitrate) was one of the first members of this group to be introduced into medicine and still remains an important member of the group. By varying the chemical structure of the organic nitrates, differences in speed of onset, duration of action and potency can be obtained. It is interesting to note that although the number of nitrate ester groups may vary from 2 to 6 or more, depending on the compound, there is no direct relationship between the number of nitrate groups, and the level of activity. It appears that the higher the oil:water partition coefficient, the greater the potency. The orientation of the groups within the molecule may also affect potency.

A long held theory explains the beneficial action of these compounds as the result of coronary vasodilation. In fact, laboratory studies have shown that the compounds increase blood flow in experimental animals, but in patients with coronary artery disease, vasodilation of the coronary arteries may not always be accompanied by an increase in coronary flow. There is evidence that these compounds decrease myocardial work and oxygen consumption, so that probably they possess a dual action.

PRODUCTS

Amyl Nitrite N.F., Isopentyl Nitrite. Amyl nitrite [$(CH_3)_2CHCH_2CH_2ONO$], is a mixture of isomeric amyl nitrites but is principally isoamyl nitrite. It may be prepared from amyl alcohol and nitrous acid by several procedures. It usually is dispensed in ampul form and used by inhalation, or orally in alcohol solution. Currently, it is recommended in treating cyanide poisoning; although not the best, it does not require intravenous injections.

Amyl nitrite is a yellowish liquid having an ethereal odor and a pungent taste. At room temperature it is volatile and inflammable. Amyl nitrite vapor forms an explosive mixture in air or oxygen. Inhalation of the vapor may involve definite explosion hazards if a source of ignition is present, as both room and body temperatures are within the flammability range of amyl nitrite mixtures with either air or oxygen. It is nearly insoluble in water but is miscible with organic solvents. The nitrite also will decompose into valeric acid and nitric acid.

Category—vasodilator.

Usual dose—0.3 ml. by inhalation, as needed.

OCCURRENCE
Amyl Nitrite Inhalant N.F.

Glyceryl Trinitrate, Nitroglycerin, Glonoin, is the trinitrate ester of glycerol and is official in tablet form in the *U.S.P.* It is prepared by carefully adding glycerin to a mixture of nitric and fuming sulfuric acids. This reaction is exothermic and the reaction mixture must be cooled to 10° to 20°.

The ester is a colorless oil with a sweet, burning taste. It is only slightly soluble in water, but it is soluble in organic solvents.

$$\begin{array}{c}
CH_2OH \\
| \\
CHOH \\
| \\
CH_2OH
\end{array}
\xrightarrow[H_2SO_4]{3HNO_3}
\begin{array}{c}
CH_2ONO_2 \\
| \\
CHONO_2 \\
| \\
CH_2ONO_2
\end{array}
+ 3H_2O$$

Nitroglycerin is used extensively as an explosive in dynamite. A solution of the ester, if spilled or allowed to evaporate, will leave a residue of nitroglycerin. To prevent an explosion, the ester must be decomposed by the addition of alkali. It has a strong vasodilating action and, since it is absorbed through the skin, is prone to cause headaches among workers associated with its manufacture. In medicine, it has the action typical of nitrites but its action is developed more slowly and is of longer duration. Of all the known coronary vasodilator drugs, nitroglycerin is the only one capable of stimulating the production of coronary collateral circulation and the only one able to prevent experimental myocardial infarction by coronary occlusion.

Previously, the nitrates were thought to be hydrolyzed and reduced in the body to nitrites, which then lowered the blood pressure. However, this is not the case.[1] The action depends on the intact molecule.

Category—vasodilator (coronary).

Usual dose—sublingual, 400 mcg., repeated as needed.

Usual dose range—300 mcg. to 10 mg. daily.

OCCURRENCE

Nitroglycerin Tablets U.S.P.

Trolnitrate Phosphate; Metamine®, Nitretamin®; Triethanolamine Trinitrate Biphosphate is a white, stable, amine salt.

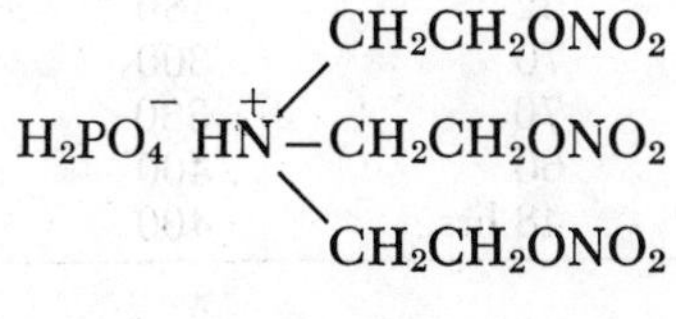

$$H_2PO_4^- \ \overset{+}{H}N \overset{CH_2CH_2ONO_2}{\underset{CH_2CH_2ONO_2}{-CH_2CH_2ONO_2}}$$

It is available in 2-mg. tablets and possesses the usual properties and therapeutic uses of the other nitrates. Trolnitrate phosphate is suggested for the prevention and the management of angina pectoris. It usually is given 4 times daily, and the full therapeutic effect is not obtained until after the first few days of treatment. There have been very few reported cases of side-effects or a tolerance from use of the drug.

Diluted Erythrityl Tetranitrate N.F., Cardilate®, Erythrol Tetranitrate, Tetranitrol, is the tetranitrate ester of erythritol and nitric acid, and it is prepared in a manner analogous to that used for nitroglycerin. The result is a solid, crystalline material. This ester also is very explosive and is diluted with lactose or other suitable inert diluents to permit safe handling; it is slightly soluble in water and is soluble in organic solvents.

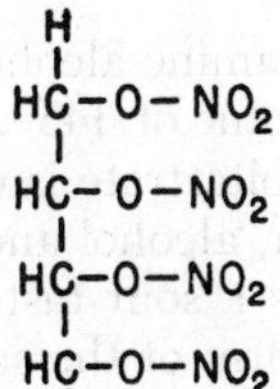

$$\begin{array}{c} H \\ | \\ HC-O-NO_2 \\ | \\ HC-O-NO_2 \\ | \\ HC-O-NO_2 \\ | \\ HC-O-NO_2 \\ | \\ H \end{array}$$

Erythrityl tetranitrate requires slightly more time than nitroglycerin to develop its action and this is of longer duration. It is useful where a mild, gradual and prolonged vascular dilation is wanted and is used in the treatment of, and as a prophylaxis against, attacks of angina pectoris and to reduce blood pressure in arterial hypertonia.

Category—vasodilator.

Usual dose—oral, initial, 10 mg. of erythrityl tetranitrate 3 times daily; sublingual, initial, 5 mg. of erythrityl tetranitrate 3 times daily. These doses may be increased in 2 or 3 days if required.

OCCURRENCE

Erythrityl Tetranitrate Tablets N.F.

Diluted Pentaerythritol Tetranitrate N.F., Peritrate®, Pentritol®, 2,2-Bis-(hydroxymethyl)-1,3-propanediol Tetranitrate, PETN. This compound is a white, crystalline material with a melting point of 140°. It is insoluble in water, slightly soluble in alcohol and readily soluble in acetone. The drug is a nitric acid ester of the tetrahydric alcohol, pentaerythritol, and is a powerful explosive.

For this reason it is diluted with lactose or mannitol or other suitable inert diluents to permit safe handling.

$$\begin{array}{c} O_2NOCH_2 \quad\quad CH_2ONO_2 \\ \diagdown C \diagup \\ O_2NOCH_2 \quad\quad CH_2ONO_2 \end{array}$$

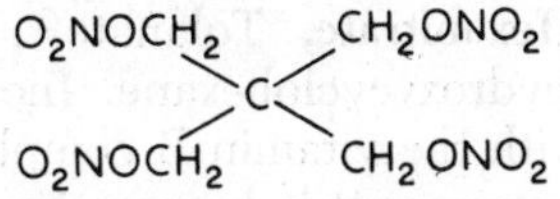

It relaxes smooth muscle of smaller vessels in the coronary vascular tree. It is used prophylactically to reduce the severity and frequency of anginal attacks.

Category—vasodilator.

Usual dose—10 mg. of pentaerythritol tetranitrate 3 or 4 times daily.

Usual dose range—10 to 20 mg. of pentaerythritol tetranitrate.

OCCURRENCE

Pentaerythritol Tetranitrate Tablets N.F.

Mannitol Hexanitrate, Nitranitol®, Mannitol Nitrate, is prepared by the nitration of mannitol. It, too, is an explosive com-

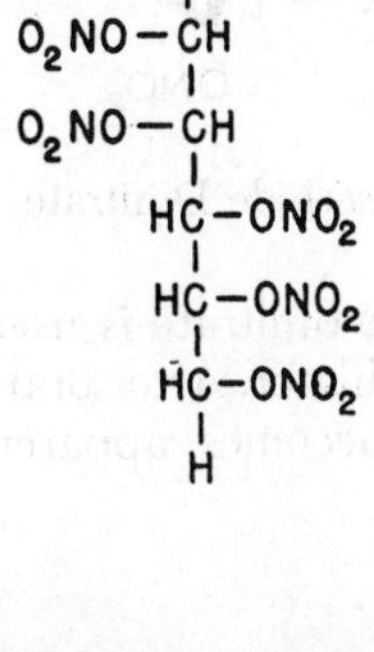

$$\begin{array}{c} H \\ | \\ HC-ONO_2 \\ | \\ O_2NO-CH \\ | \\ O_2NO-CH \\ | \\ HC-ONO_2 \\ | \\ HC-ONO_2 \\ | \\ HC-ONO_2 \\ | \\ H \end{array}$$

TABLE 64. RELATION BETWEEN SPEED AND DURATION OF ACTION OF SODIUM
NITRITE AND CERTAIN INORGANIC ESTERS

COMPOUND	ACTION BEGINS (MINUTES)	MAXIMUM EFFECT (IN MINUTES)	DURATION OF ACTION (IN MINUTES)
Amyl Nitrite	¼	½	1
Nitroglycerin	2	8	30
Sodium Nitrite	10	25	60
Erythrityl Tetranitrate	15	32	180
Mannitol Hexanitrate	15	70	300
Pentaerythritol Tetranitrate	20	70	330
Inositol Hexanitrate	30	60	400
Trolnitrate Phosphate	..	48 hrs.	400

pound and is used in medicine diluted with nine parts of carbohydrate.

It has physiologic properties and uses similar to those of nitroglycerin and erythrityl tetranitrate. Table 64 gives the relation between these inorganic esters and sodium nitrite as to speed of action and duration.

The usual dose is 15 to 30 mg.

Inositol Hexanitrate, Tolanate®, is nitrated hexahydroxycyclohexane. Inositol is associated with the vitamin B complex and is a normal component in human tissue. The solid nitrated compound is used in tablet form but is usually available commercially admixed with mannitol, 1 to 13.

It is very stable in tablet form and has the advantage over other nitrates of long duration of activity, slight effects on capillary fragility and decreased tendency to cause headache, and less tolerance is observed to develop.

Isosorbide Dinitrate, Isordil®, Sorbitrate®, 1,4:3,6-dianhydrosorbitol 2,5-dinitrate, occurs as a white, crystalline powder. Its water solubility is about 1 mg. per ml.

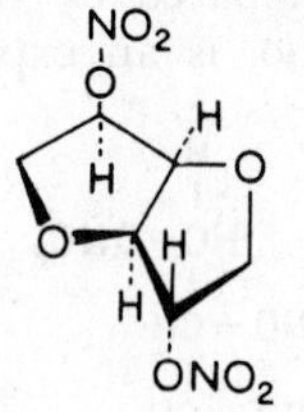

Isosorbide Dinitrate

Isosorbide dinitrate is used as a coronary artery vasodilator. After oral administration, the effect becomes apparent in about 15 minutes and lasts about 4 to 5 hours. If given sublingually, the effect begins in about 2 minutes, with a shorter duration of action than when given orally. The average dose is 10 mg. taken about 30 minutes before meals and at bedtime. The dose range is from 5 mg. to 20 mg.

MISCELLANEOUS VASODILATORS

Nicotinyl Alcohol Tartrate; Roniacol®; β-Pyridylcarbinol Bitartrate or 3-Pyridine-methanol Tartrate (the alcohol corresponding to nicotinic acid).

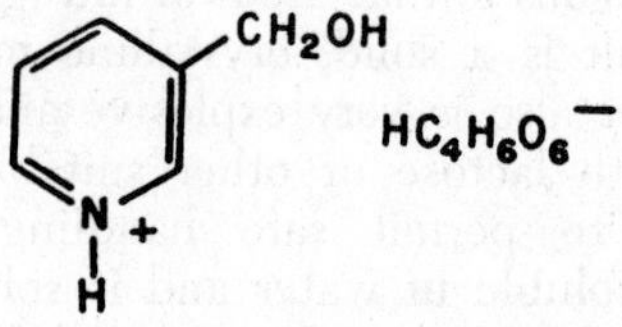

Nicotinyl Alcohol Tartrate

The free amine-alcohol is a liquid having a boiling point of 145°. It forms salts with acids. The bitartrate is crystalline and soluble in water, alcohol and ether. An aqueous solution has a sour taste, partly due to the bitartrate form of the salt.

In 1950, it was introduced as a vasodilator, following the lead that nicotinic acid is a vasodilator. The action of Roniacol is peripheral vasodilation similar to that of nicotinic acid. There is a direct relaxing effect on peripheral blood vessels, producing a longer action with less flushing than does nicotinic acid. It is given orally in tablets or as an elixir. Medicinal use includes the treatment of vascular spasm,

Raynaud's disease, Buerger's disease, ulcerated varicose veins, chilblains, migraine, Ménière's syndrome and most conditions requiring a vasodilator.

Another use is in the treatment of dermatitis herpetiformis. This came about because both sulfapyridine and niacin were found to be effective.

The usual dose is 50 to 200 mg.

Dipyridamole, Persantin®, 2,6-bis(di-2-hydroxyethylamino) - 4,8 - dipiperidinopyrimido[5,4,-d]pyrimidine, is used for coronary and myocardial insufficiency. It is a yellow, crystalline powder, with a bitter taste. It is soluble in dilute acids, methanol or chloroform.

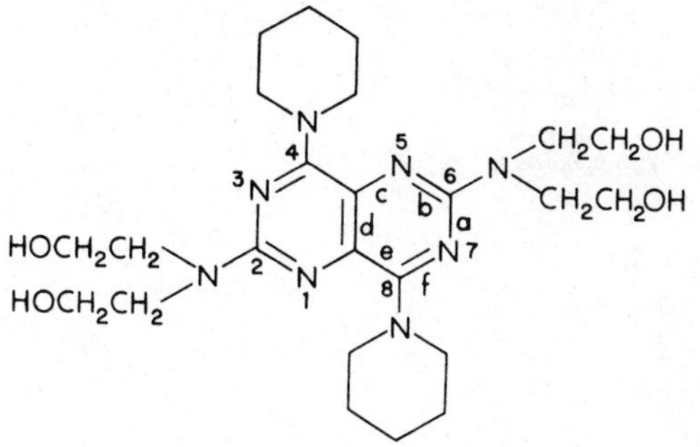

Dipyridamole

The recommended oral dose is 25 to 50 mg. 2 or 3 times daily before meals. Optimum response may not be apparent until the third or fourth week of therapy. Dipyridamole is available in 25-mg. sugar-coated tablets and in 2-ml. ampuls (5 mg. per ml.) for intravenous administration.

Cyclandelate, Cyclospasmol®, 3,5,5-Trimethylcyclohexyl Mandelate. This compound was introduced in 1956 for use especially in peripheral vascular disease in which there is vasospasm. It is a white to off-white crystalline powder, practically insoluble in water and readily soluble in alcohol and in other organic solvents. Its actions are similar to those of papaverine.

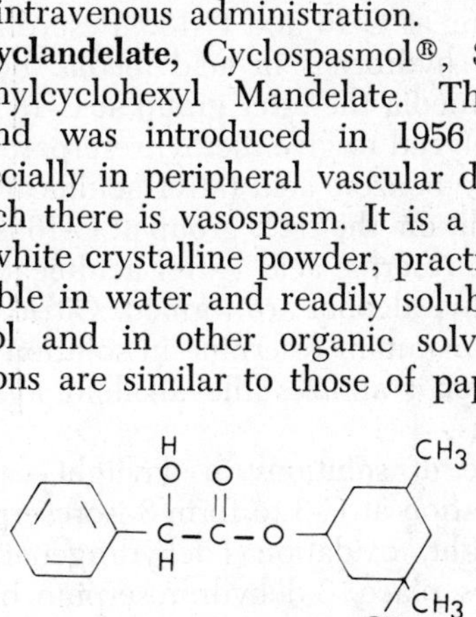

Cyclandelate

When cyclandelate is effective, the improvement in peripheral circulation usually occurs gradually and treatment often must be continued over long periods. At the maintenance dose of 100 mg. four times daily, there is little incidence of serious toxicity. At higher doses, as high as 400 mg. four times daily, which may be needed initially, there is a greater incidence of unpleasant side effects such as headache, dizziness, and flushing. It must be used with caution in patients with glaucoma. The oral dosage forms are 200-mg. capsules and 100-mg. tablets.

ANTIHYPERTENSIVE AGENTS

Progress in the treatment of hypertensive disease has come only in recent years. As a result of this progress in the past 20 years, the mortality rate from hypertension has been markedly reduced. In most cases, physicians now can reduce the high blood pressure and its accompanying symptoms through the use of drugs.

Many compounds have been made and tested which cause a precipitous but extremely brief fall in blood pressure, but, of course, these are not useful for therapy. The desired action is a slow reduction of blood pressure with prolonged effect. Further, increased doses should cause a more prolonged effect rather than a more pronounced fall in blood pressure. Finally, the drugs should be active after oral administration because undoubtedly they would be used for extended periods.

The first drugs of value in the treatment of hypertension were adrenergic blocking agents; by blocking the action of epinephrine and norepinephrine it was hoped that contraction of the smooth muscle of the vascular walls could be blocked. The hopes were not fulfilled because the duration of action was far too short and the side-effects generally precluded long-term therapy. The clinical importance of these drugs now lies in the value for the treatment of peripheral vascular disease and for diagnosis of pheochromocytoma. (See Chap. 19.)

The cholinergic agents, which act as antagonists to the adrenergic agents, cause peripheral vasodilatation and could possibly serve as hypotensive drugs. However, they cause a sharp fall in blood pressure which is of short duration, and none is presently in use as an antihypertensive drug. Some of the ganglionic blocking agents serve in a useful but somewhat limited capacity as antihypertensive drugs (see Chap. 19). In addition, several of the benzothiadiazine diuretics, either alone or in combination with other drugs, have proved to be useful in the treatment of hypertension (see Chap. 20).

The antihypertensive agents to be considered in this chapter include the Rauwolfia alkaloids, the veratrum alkaloids and an unclassified group.

Rauwolfia Alkaloids

Powdered Rauwolfia Serpentina N.F., Raudixin®, Rauserpa®, Rauvel®, is the powdered whole root of *Rauwolfia serpentina* (Benth). It is a light-tan to light-brown powder, sparingly soluble in alcohol and only slightly soluble in water. It contains the total alkaloids, of which reserpine accounts for about 50 per cent of the total activity. Orally, 200 to 300 mg. is roughly equivalent to 500 mcg. of reserpine. It is used in the treatment of mild, labile hypertension or in combination with other hypotensive agents in severe hypertension.

Category—antihypertensive.

Usual dose—initial, 200 mg. daily for 1 to 3 weeks; maintenance, 50 to 300 mg. daily.

Occurrence

Rauwolfia Serpentina Tablets N.F.

Alseroxylon, Rauwiloid®, is a fat-soluble alkaloidal fraction obtained from the whole root of *Rauwolfia serpentina*. Reserpine is the most potent alkaloid in the fraction. It is given orally in a dose of 2 to 4 mg. daily in the treatment of hypertension.

Reserpine U.S.P., Serpasil®, Reserpoid®, Rau-Sed®, Sandril®; this is a white to light-yellow crystalline alkaloid practically insoluble in water, obtained from various species of Rauwolfia. In common with other compounds with an indole nucleus, it is susceptible to decomposition by light and oxidation, especially when in solution. In the

dry state discoloration occurs rapidly when exposed to light, but the loss in potency is usually small.[2] In solution there may be

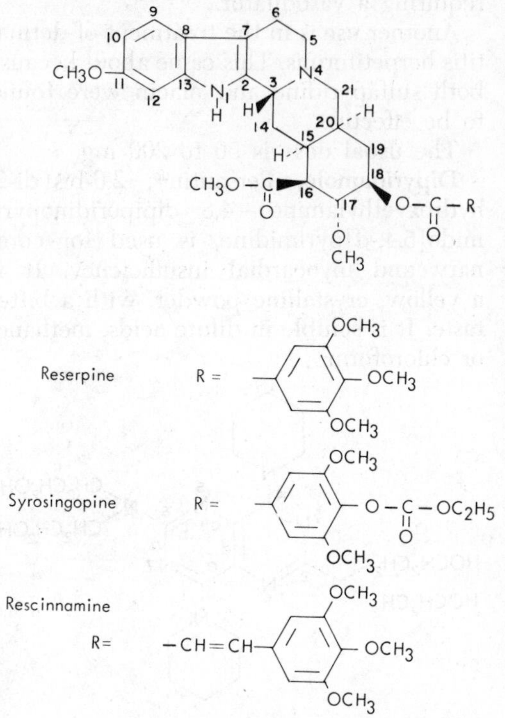

Reserpine

$$R = \text{(3,4,5-trimethoxybenzoyl group)}$$

Syrosingopine

$$R = \text{(ethoxycarbonyloxy trimethoxybenzoyl group)}$$

Rescinnamine

$$R = -CH=CH- \text{(3,4,5-trimethoxyphenyl group)}$$

breakdown when exposed to light, especially in clear glass containers, with no appreciable color change; thus, color change cannot be used as an index of the amount of decomposition.

There are several possible points of breakdown in the reserpine molecule. Hydrolysis may occur at C-16 and C-18.[3] Reserpine is stable to hydrolysis in acid media, but in alkaline media the ester group at C-18 may be hydrolyzed to give methyl reserpate and trimethoxybenzoic acid (after acidification). If, in addition, the ester group at C-16 is hydrolyzed, reserpic acid (after acidification) and methyl alcohol are formed. Citric acid helps to maintain reserpine in solution and in addition stabilizes the alkaloid against hydrolysis.

Storage of solutions in daylight causes epimerization at C-3 to form 3-isoreserpine. In daylight, oxidation (dehydrogenation) also takes place, 3-dehydroreserpine being formed. It is green in solution, but, as the oxidative process progresses, the color disappears and, finally, a strongly orange color appears. Oxidation of solutions takes place in the dark at an increasing rate with increased amounts of oxygen and at an even faster rate when exposed to light. Sodium metabisulfite will stabilize the solutions if kept protected from light, but when exposed to light it actually oxidizes the reserpine so that the solutions are less stable than if the metabisulfite were absent. Nordihydroguaiaretic acid (NDGA) aids in stabilizing solutions when protected from light, but in daylight the degradation is retarded only slightly. Urethan in the solution stabilizes it in normally filled ampuls but affords no protection in daylight.

Reserpine is effective orally and parenterally in the treatment of hypertension. After a single intravenous dose the onset of antihypertensive action usually begins in about 1 hour. After intramuscular injection the maximum effect occurs within approximately 4 hours and lasts about 10 hours. When given orally, the maximum effect occurs within about 2 weeks and may persist up to 4 weeks after the final dose. When used in conjunction with other hypotensive drugs in the treatment of severe hypertension the daily dose varies from 0.5 to 2 mg.

Category—antihypertensive.

Usual dose—oral, 250 mcg. once or twice daily; intramuscular, 2.5 mg. 1 to 3 times daily.

Usual dose range—oral, 100 mcg. to 1 mg. daily; intramuscular, 250 mcg. to 10 mg. daily.

OCCURRENCE

Reserpine Injection U.S.P.
Reserpine Tablets U.S.P.

Syrosingopine N.F., Singoserp®, methyl reserpate ester of syringic acid ethyl carbonate, is closely related to reserpine, the only difference being the acid used to esterify the hydroxyl group at C-18. It is less toxic and less potent than reserpine and possesses about the same therapeutic index. It is effective in the control of some cases of mild hypertension but must be used with other hypotensive agents in the treatment of severe hypertension.

Category—antihypertensive.

Usual dose—initial, 1 to 2 mg. daily; maintenance, 500 mcg. to 3 mg. daily.

OCCURRENCE

Syrosingopine Tablets N.F.

Deserpidine, Harmonyl®, is 11-desmethoxyreserpine.[4] It differs from reserpine only in the absence of a methoxyl group at C-11. Deserpidine is claimed to have a more rapid onset of action than reserpine, is less potent and causes less depression. The initial dose in the treatment of mild hypertension is 250 mcg. 3 or 4 times daily until the blood pressure has been controlled; then it is reduced to a maintenance dose.

Rescinnamine N.F., Moderil®, is the 3,4,5 trimethoxycinnamic acid ester of methyl reserpate. It differs from reserpine only in the acid used to esterify the hydroxyl group at C-18.

The dose must be adjusted carefully, as is true with the other Rauwolfia alkaloids.

Category—antihypertensive.

Usual dose—initial, 500 mcg. once or twice daily for up to 2 weeks; maintenance, 250 mcg. daily.

Usual dose range—250 mcg. to 2 mg. daily.

OCCURRENCE

Rescinnamine Tablets N.F.

VERATRUM ALKALOIDS

The veratrum alkaloids used in medicine are a family of chemically related substances obtained chiefly from the roots and the rhizomes of *Veratrum viride* and *Veratrum album*. The alkaloids have a complex polycyclic nucleus and include a group of esters of tertiary amines and a group of esters of secondary amines. Preparations of veratrum alkaloids may vary greatly in the relative amounts of the various alkaloids and thus may vary in the pharmacologic response that they elicit. In general, the free alkaloids are soluble in organic solvents but practically insoluble in water. Salts of the alkaloids are water soluble and, when in such solutions, are stable at room temperature to light and air but are precipitated by alkaline or alcoholic solutions.

After oral administration, the alkaloids usually act within 2 hours, with a duration of action for 4 to 6 hours. Much of the activity is lost upon oral administration; from 5 to 20 times as much as is administered by intramuscular injection must be given for equivalent effects. After intramuscular administration, the onset of action occurs in 1 to 1½ hours and lasts from 3 to 6 hours. After intravenous administration, the onset occurs within a few minutes and lasts from 1 to 3 hours. The distribution and the metabolic fate of the alkaloids is not well understood. Only a small part of the dose is excreted in the urine. The alkaloids have been used orally in conjunction with other hypotensive agents—Rauwolfia alkaloids, ganglionic blocking agents, etc.

The available preparations of the veratrum alkaloids fall into 3 general groups: (1) powdered whole root and rhizome, (2) mixtures of partially purified alkaloids and (3) the pure alkaloids.

Veratrum viride; Vertavis®; this is a biologically standardized powdered preparation of the crude drug. It is available in oral tablets of 5 or 10 Craw units. (It is assayed by cardiac arrest in the crustacean, *Daphnia magna.*) The total daily dose varies from 20 to 80 Craw units. As with the other veratrum preparations, it is usually given in divided doses after meals and at bedtime to minimize the tendency toward nausea.

Alkavervir; Veriloid®; this is a mixture of alkaloids from *V. viride.* It is a light-yellow powder, practically insoluble in water but freely soluble in alcohol. It is given orally, by intravenous injection or by intramuscular injection. The injectable solutions contain 0.25 per cent of acetic acid to solubilize the alkaloids, with the intramuscular injection containing 1 per cent of procaine hydrochloride to lessen pain after injection.

Veratrone® is a similar preparation.

Cryptenamine; Unitensin®; this is a mixture of alkaloids from *V. viride* and is used as the tannate and the acetate salts. The acetate salts are more water-soluble and are used in parenteral solutions. Cryptenamine tannates are a tan powder, slightly soluble in water but freely soluble in alcohol. The dose of both salts is expressed in terms of the equivalent amount of free alkaloids. One mg. is equivalent to 130 C.S.R. units (Carotid Sinus Reflex unit based on biologic

assay on the dog). The starting oral dose is 2 mg. twice daily.

Protoveratrines A and B; Veralba®; this is a mixture of 2 alkaloids obtained from *V. album.* The mixture is a white crystalline powder. It is stable in light and air in solution of pH 4 to 6 but is rapidly destroyed in basic and alcoholic solutions. The usual starting oral dose is 500 mcg. after each meal and at bedtime. Parenteral administration is used only to control hypertensive crises.

Protoveratrine A and B Maleates; Provell® Maleate; this mixture of the maleate salts of 2 alkaloids is obtained from *V. album.* It is a white to light-tan powder, freely soluble in alcohol or water. The maleate salts have the same actions and uses as the parent alkaloids. On a weight basis, about 13 per cent more of the maleate salts than of the free alkaloids must be given to provide equivalent dosage; the dosage must be carefully individualized for each patient. The average oral total daily dose is from 1 to 2.5 mg., given at intervals of 4 to 6 hours, preferably after meals and at bedtime.

Protoveratrine A, Protalba®, is a purified alkaloid from the rhizomes of *V. album.* It differs structurally from protoveratrine B by a single hydroxyl group. It is only slightly more potent than protoveratrine B when given parenterally but is considerably more potent when given orally. The usual oral starting dose is 200 mcg. 4 times daily, after meals and at bedtime.

UNCLASSIFIED ANTIHYPERTENSIVE AGENTS

Guanethidine Sulfate U.S.P., Ismelin® Sulfate, [2-(Hexahydro-1(2H)-azocinyl)-ethyl]guanidine Sulfate, is a white crystalline material which is very soluble in water. It is chemically unrelated to previously introduced antihypertensive agents and produces a gradual, prolonged fall in blood pressure. Usually 2 to 7 days of therapy are

Guanethidine Sulfate

required before the peak effect is reached, and usually this peak effect is maintained for 3 or 4 days; then, over a period of 1 to 3 weeks the blood pressure returns to pretreatment levels. Because of this slow onset and prolonged duration of action only a single daily dose need be given.

Category—antihypertensive.

Usual dose—initial, 25 to 50 mg. once daily.

Usual dose range—10 to 100 mg. daily.

OCCURRENCE

Guanethidine Sulfate Tablets U.S.P.

Hydralazine Hydrochloride N.F., Apresoline® Hydrochloride, is 1-hydrazinophthalazine monohydrochloride. It occurs as yellow crystals and is soluble in water to the extent of about 3 per cent. A 2 per cent aqueous solution has a pH of 3.5 to 4.5.

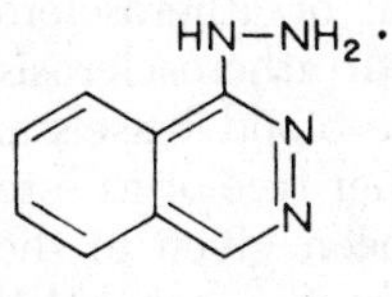

Hydralazine Hydrochloride

Hydralazine is useful in the treatment of moderate to severe hypertension. It is frequently used in conjunction with less potent antihypertensive agents, because when used alone in adequate doses there is a frequent occurrence of side-effects. In combinations it can be used in lower and safer doses. Its action appears to be centered on the smooth muscle of the vascular walls, with a decrease in peripheral resistance to blood flow. This results in an increased blood flow through the peripheral blood vessels. Also of importance is its unique property of increasing renal blood flow, an important consideration in patients with renal insufficiency.

Hydralazine (as the hydrochloride) is readily absorbed after oral administration. As mentioned above, the therapeutic benefits, when the drug is administered alone, are often limited by the development of unpleasant side-effects such as nausea, palpitation and headache.

Category—antihypertensive.

Usual dose range—oral, initial, 10 mg. 4 times daily; maintenance, up to 100 mg. 4 times daily. Intramuscular or intravenous, 20 to 40 mg. repeated as required.

OCCURRENCE

Hydralazine Hydrochloride Injection N.F.

Hydralazine Hydrochloride Tablets N.F.

Methyldopa U.S.P., Aldomet®, (−)-3-(3,4-Dihydroxyphenyl)-2-methylalanine or α-

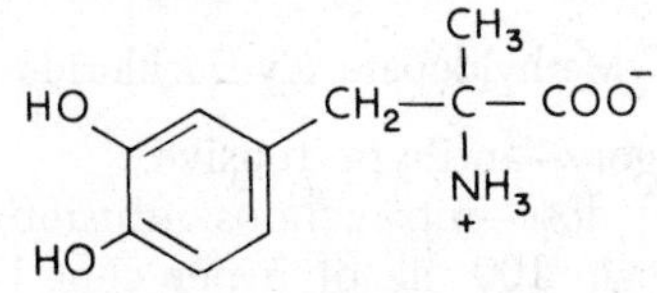

Methyldopa (zwitter ion)

Methyldopa. This compound was investigated as part of a program to develop antagonists to the biochemical synthesis of pressor amines that may be implicated in the development of hypertension, e.g., serotonin, dopamine and norepinephrine. While such inhibition by methyldopa has been demonstrated in man, it has not been established definitely as being responsible for the antihypertensive effect. There is other evidence that indicates that the decarboxylated compound, α-methyldopamine, acts as a catecholamine-releasing and depleting agent.[5]

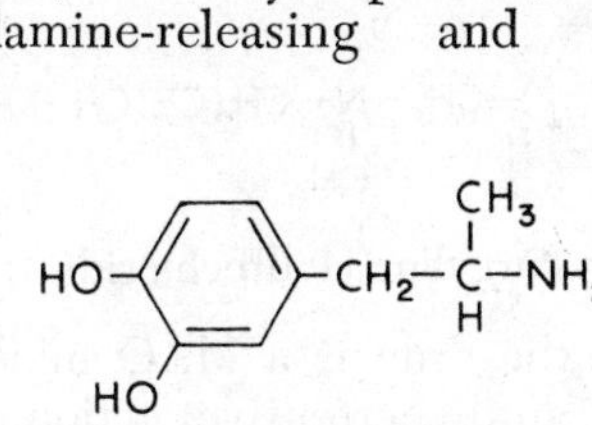

α-Methyldopamine

Methyldopa is recommended for patients with moderate to severe hypertension.

Category—antihypertensive.

Usual dose—250 mg. 3 times daily.

Usual dose range—500 mg. to 2 g. daily.

OCCURRENCE

Methyldopa Tablets U.S.P.

Methyldopate Hydrochloride U.S.P., Aldomet® Ester Hydrochloride, (−)-3-(3,4-Dihydroxyphenyl)-2-methylalanine Ethyl Ester Hydrochloride. Methyldopa, suitable for oral use, is a zwitter ion and is not soluble enough for parenteral use. This problem was solved by making the ester, leaving the amine free to form the water-soluble hydrochloride salt. It is supplied as a stable, buffered solution, protected with antioxidants and chelating agents.

$$HO-\underset{HO}{\diagdown}\hspace{-1em}\bigcirc\hspace{-1em}-CH_2-\underset{\underset{NH_3}{\overset{\overset{CH_3}{|}}{|}}}{C}-COOC_2H_5 \quad Cl^-$$

Methyldopate Hydrochloride

Category—antihypertensive.

Usual dose—intravenous infusion, 250 to 500 mg. in 100 ml. of 5 per cent Dextrose Injection over a period of 30 to 60 minutes 4 times daily.

Usual dose range—100 mg. to 4 g. daily.

OCCURRENCE
Methyldopate Hydrochloride Injection U.S.P.

Pargyline Hydrochloride N.F., Eutonyl®, N-Methyl-N-(2-propynyl)benzylamine Hydrochloride. This drug, introduced in 1963, is a nonhydrazine monoamine oxidase (MAO) inhibitor which is effective in lowering systolic and diastolic blood pressure without depressing the patient.

$$\bigcirc\hspace{-1em}-CH_2-\underset{\underset{CH_3}{|}}{N}-CH_2C\equiv CH \cdot HCl$$

Pargyline Hydrochloride

Although the drug is a MAO inhibitor, its mode of antihypertensive action has not been clearly established. Its action as a MAO inhibitor may cause a potentiation of the action of other drugs and substances; these include barbiturates, adrenergic amines, antihistamines, caffeine, alcohol and tyramine (from aged cheese).

Pargyline hydrochloride is well absorbed from the gastrointestinal tract and is administered orally. Little metabolism of the drug occurs and it is excreted in the urine, largely unchanged.

Category—antihypertensive.

Usual dose—initial, 10 to 25 mg. once daily; then increase daily dose in increments of 10 mg. once a week until the desired response is obtained.

OCCURRENCE
Pargyline Hydrochloride Tablets N.F.

Mebutamate, Capla®, 2-*sec*-Butyl-2-methyl-1,3-propanediol Dicarbamate. This compound is a meprobamate congener (see Chap. 15). It is a bitter, white crystalline powder, slightly soluble in water and freely soluble in alcohol. It has been used with some degree of success in the treatment of mild hypertension. When used as the sole medication, it is not effective in the treatment of severe hypertension. It is used orally in a dose of 300 mg. 3 or 4 times daily. The side effects are similar to those of meprobamate.

ANTIHYPERCHOLESTEROLEMIC DRUGS

Within recent years cholesterol has been believed to play an important role in the development of atherosclerosis in man. In patients with atherosclerosis, the fatty deposits in the blood vessels are high in cholesterol, either free or as esters. Further impetus has been given to the indictment of cholesterol by the fact that the incidence of atherosclerosis in Americans is significantly higher than in persons in other countries where the national diet is lower in cholesterol and saturated fats.[6]

The range of serum cholesterol in normal individuals is 190 to 250 mg. per cent, with approximately 30 per cent in the free state and 70 per cent as cholesterol esters.[7] Three basic methods have been used in attempts to bring serum cholesterol levels within this range. These are: (1) diminish cholesterol absorption from the gastrointestinal tract; (2) increase the metabolism and the biliary excretion of cholesterol; and (3) inhibit endogenous liver synthesis of cholesterol.

Dihydrocholesterol, sitosterol and stigmasterol have been used, in the expectation that, because these sterols were not absorbed, the enzymatic reabsorption of cholesterol would be blocked.[8] It was found that dihydrocholesterol was itself absorbed. In tests using sitosterol and stigmasterol in the diet of experimental animals, there was found a reduction in serum cholesterol for several weeks, but then the levels returned to the original values. Thus it appears that there was an increased endogenous production of cholesterol to counterbalance the early inhibition (see Chap. 25).

Since fat is the major source of the precursors of cholesterol formed in the body, it seemed probable that modification of the fat content of the diet may modify cholesterol synthesis. This is indeed the case. The use of unsaturated fats from vegetable sources appears to lead to a decreased serum cholesterol, and evidence is accumulating to indicate there may also be an effect on the arterial concentration of cholesterol.

Many drugs to control cholesterol levels are being investigated.[9] These include clofibrate, the aluminum salt of nicotinic acid, modifications of the thyroxine molecule, lipotropic agents, estrogens and others.

PRODUCTS

Clofibrate N.F., Atromid-S®, Ethyl 2-(p-Chlorophenoxy)-2-methylpropionate. Clofibrate is a stable, colorless to pale-yellow liquid with a faint odor and a characteristic taste. It is soluble in organic solvents, but insoluble in water.

Clofibrate is absorbed from the gastrointestinal tract and is rapidly hydrolyzed by serum enzymes to the free acid and is extensively bound to serum proteins. The biologic half-life is of the order of 12 hours. It appears to inhibit cholesterol biosynthesis in the liver at the point prior to mevalonic

$$Cl-\text{C}_6\text{H}_4-O-\overset{\overset{\displaystyle CH_3}{|}}{\underset{\underset{\displaystyle CH_3}{|}}{C}}-\overset{\overset{\displaystyle O}{\|}}{C}-O-C_2H_5$$

Clofibrate

acid. There is also considerable evidence that it is effective in lowering serum triglycerides.

Clofibrate is well tolerated by most patients, the most common side effects being nausea and, to a lesser extent, other gastrointestinal distress. The dosage of anticoagulants, if used in conjunction with this drug, should be reduced by one third to one half, depending on the individual response, so that the prothrombin time may be kept within the desired limits.

Category—anticholesterolemic.

Usual dose—500 mg. 4 times daily.

OCCURRENCE

Clofibrate Capsules N.F.

Sodium Dextrothyroxine N.F., Choloxin®, Sodium D-3-[4-(4-Hydroxy-3,5-diiodophenoxy)-3,5-diiodophenylalanine], Sodium D-3,3′,5,5,′-Tetraiodothyronine. This compound occurs as light yellow to buff-colored powder. It is stable in dry air, but discolors on exposure to light; for this reason it should be stored in light-resistant containers. It is very slightly soluble in water, slightly soluble in alcohol and insoluble in acetone, in chloroform and in ether.

The hormones secreted by the thyroid gland have marked hypocholesterolemic activity along with their other well-known actions. With the finding that not all active thyroid principles possessed the same degree of physiologic actions, a search was made for congeners that would cause a decrease in serum cholesterol without other effects such as angina pectoris, palpitation and congestive failure. D-Thyroxine has resulted from this search. However, at the dosage required, the L-thyroxine contamination must be minimal, otherwise it will exert its characteristic actions. One route to optically pure (at least 99 per cent pure) D-thyroxine is the use of an L-aminoacid oxidase from snake venom which acts only on the L-isomer and makes separation possible.

The mechanism of action of D-thyroxine appears to be stimulation of oxidative catabolism of cholesterol in the liver. The catabolic products are bile acids which are conjugated with glycine or taurine and excreted via the biliary route into the feces. Cholesterol biosynthesis is not inhibited by the drug and abnormal metabolites of cholesterol do not accumulate in the blood. There is also a decrease in serum levels of triglycerides, but this is less consistent than the decrease of cholesterol.

D-Thyroxine potentiates the action of anticoagulants such as warfarin or dicoumarol; thus, dosage of the anticoagulants should be reduced by one third if used concurrently and then further modified, if necessary to maintain the prothrombin time within the desired limits. Also, it may increase the dosage requirements of insulin or of oral hypoglycemic agents if used concurrently with them.

Category—antihypercholesterolemic.

Usual dose—initial, 1.0 to 2.0 mg. daily; maintenance, 4.0 to 8.0 mg. daily.

Usual dose range—1.0 to 8.0 mg. daily.

OCCURRENCE

Sodium Dextrothyroxine Tablets N.F.

Aluminum Nicotinate, Nicalex®, is aluminum hydroxy nicotinate and some free nicotinic acid.[10] Studies have shown that nicotinic acid in high doses is an effective agent for reduction of elevated serum cholesterol levels. However, at these high doses, the side-effects of nicotinic acid limit its usefulness. Aluminum nicotinate is reported to overcome largely these side-effects and to be as effective as nicotinic acid in producing and maintaining lower serum cholesterol levels in hypercholesterolemic patients.[11]

SCLEROSING AGENTS

Several different kinds of irritating agents have been used for the obliteration of varicose veins. These are generally called sclerosing agents and include invert sugar solutions, dextrose, ethyl alcohol, iron salts, quinine and urea hydrochloride, fatty acid salts (soaps) and certain sulfate esters. Many of these preparations contain benzyl alcohol which acts as a bacteriostatic agent and relieves pain after injection.

PRODUCTS

Sodium Morrhuate Injection U.S.P. is a sterile solution of the sodium salts of the fatty acids of cod liver oil. The salt (a soap) was introduced first in 1918 as a treatment for tuberculosis and, in 1930, it was reported to be useful as a sclerosing agent. Sodium morrhuate is not a single entity, although morrhuic acid, $C_9H_{13}NO_3$ has been known for years. Sodium morrhuate is a mixture of the sodium salts of the saturated and unsaturated fatty acids from cod liver oil.

The preparation of the free fatty acids of cod liver oil is carried out by saponification

with alkali and then acidulation of the resulting soap. The free acids are dried over anhydrous sodium sulfate before being dissolved in an equivalent amount of sodium hydroxide solution. Sodium morrhuate is obtained by careful evaporation of this solution. The result is a pale-yellowish, granular powder having a slight fishy odor.

Commercial preparations are usually 5 per cent solutions, which vary in properties and in color from light yellow to medium yellow to light brown. They are all liquids at room temperature and have congealing points that range from −11° to 7°. A bacteriostatic agent, not to exceed 0.5 per cent, and ethyl or benzyl alcohol to the extent of 3 per cent, may be added.

Category—sclerosing agent.

Usual dose—intravenous, by special injection, 1 ml. of 5 per cent solution to localized area.

Usual dose range—0.5 to 5 ml.

Sodium Tetradecyl Sulfate, Sodium Sotradecol®, is a colorless waxy solid. A 5 per cent aqueous solution varies from pH 6.5 to 9.0 and is clear and colorless.

$$CH_3CH-CH_2-CH(CH_2)_2\underset{\substack{|\\C_2H_5}}{CH}(CH_2)_3-CH_3$$
$$\underset{CH_3}{|}\qquad \underset{\substack{O\\SO_3Na}}{|}$$

Chemically, this is sodium 7-ethyl-2-methyl-4-hendecanol sulfate. Sodium Sotradecol is the medicinal grade of Tergitol-4 and is not less than 85 per cent pure. It is an anionic surfactant which also possesses useful sclerosing properties. It is used in an aqueous solution of 1 to 5 per cent.

Quinine and Urea Hydrochloride is a double salt of quinine and urea hydrochlorides, which has been used as a sclerosing agent in a 1 ml. dose of a 5 per cent solution.

ANTIARRHYTHMIC DRUGS

Quinidine and procainamide are the most effective drugs for use in the treatment of cardiac rate and rhythm. There are, in addition, a number of other drugs which are unrelated either chemically or pharmacologically which are also of some value. These include procaine, quinacrine, quinine, papaverine, digitalis preparations and others.

Quinidine is an isomer of quinine and is obtained from various cinchona species. It has much the same physical and chemical properties as quinine; however, it is dextrorotatory, whereas quinine is levorotatory. The molecule contains two nitrogens, that in the side chain being the more basic. It thus forms two series of salts.

PRODUCTS

Quinidine Sulfate U.S.P., Quinidinium Sulfate. This salt crystallizes from water as

Quinidine

the dihydrate, in the form of fine, needlelike, white crystals. It has a bitter taste and is light-sensitive. Aqueous solutions are nearly neutral or slightly alkaline. It is soluble to the extent of 1 per cent in water and is more highly soluble in alcohol or chloroform.

Commercial quinidine sulfate may contain up to 20 per cent dihydroquinidine sulfate. (The vinyl group at C-3 is converted to an ethyl group.) Studies indicate that dihydroquinidine has a greater antifibrillating action than quinidine, but is more toxic.[12]

When quinidine is administered intramuscularly the peak effect, as measured by the prolongation of the QT interval on the electrocardiogram, is 1 to 1½ hours; given orally in the same dose, the peak effect occurred in 2 to 2½ hours.[13] The duration of effect is about the same in both cases, but the intramuscular injection gives a greater peak effect. From 2 hours after administration the activity curves fall off at almost the same rate. When compared with the same oral dose of quinine, the cardiac response is qualitatively the same but of lesser magnitude and shorter duration. A study of the

effect of gastric acidity on the absorption of quinidine sulfate from the gastrointestinal tract showed a consistently higher plasma level in the group with achlorhydria than in the normal subjects, but the difference between the mean values for each group was not statistically significant.[14] However, it has been pointed out that there appears to be no correlation between plasma levels of quinidine and cardiac response.[15]

Category—cardiac depressant (antiarrhythmic).

Usual dose—100 to 200 mg. 1 to 4 times daily.

Usual dose range—100 mg. to 4 g. daily.

OCCURRENCE
Quinidine Sulfate Capsules U.S.P.
Quinidine Sulfate Tablets U.S.P.

Quinidine Gluconate U.S.P., Quinidinium Gluconate; occurs as an odorless, white powder with a very bitter taste. In contrast with the sulfate salt, it is freely soluble in water. This is important, because there are emergencies when the condition of the patient and the need for a rapid response may make the oral route of administration inappropriate. The high water-solubility of the gluconate salt along with a low irritant potential makes it of value when an injectable form is needed in these emergencies. Quinidine salts have been given intravenously for a prompt response, but this route is rather risky, so that the intramuscular route is usually used when the oral route is inadvisable. Quinidine dihydrochloride has been used parenterally, but this causes painful inflammatory induration at the injection site.[16] Quinidine hydrochloride dissolved in water with urea and antipyrine as solubilizers has been used successfully, but within a few months the solution turns brown, and crystallization occurs.[13] Quinidine sulfate in propylene glycol has been used satisfactorily.[17]

Quinidine gluconate forms a stable aqueous solution. When used for injection, it usually contains 80 mg. per ml., equivalent to 50 mg. of quinidine or 60 mg. of quinidine sulfate.

Category—cardiac depressant (antiarrhythmic).

Usual dose—intramuscular or intravenous,

400 mg., repeated every 2 hours, if needed. Usual dose range—300 mg. to 2.4 g. daily.

OCCURRENCE
Quinidine Gluconate Injection U.S.P.

Quinidine Polygalacturonate, Cardioquin®. This is formed by reacting quinidine and polygalacturonic acid in a hydroalcoholic medium. It contains the equivalent of approximately 60 per cent quinidine. This salt is only slightly ionized and slightly soluble in water, but studies have shown that although equivalent doses of quinidine sulfate give higher peak blood levels earlier, a more uniform and sustained blood level is achieved with the polygalacturonate salt.[18]

In many patients, the local irritant action of quinidine sulfate in the gastrointestinal tract causes pain, nausea, vomiting and especially diarrhea and often precludes oral use in adequate doses. It has been reported that in studies with the polygalacturonate salt no evidence of gastrointestinal distress was encountered. It is available as 275-mg. tablets. Each tablet is the equivalent of 200 mg. of quinidine sulfate or 166 mg. of free alkaloid.

Procainamide Hydrochloride U.S.P., Procainamidium Chloride; Pronestyl® Hydrochloride; *p*-Amino-N-[2-(diethylamino)-ethyl]benzamide Monohydrochloride. It is

the amide form of procaine hydrochloride (see Chap. 22) in that the amide group (·CO·NH) replaces the ester group (CO·O). The amide occurs as a white to tan crystalline powder, soluble in water but insoluble in alkaline solutions. Its aqueous solutions have a pH of about 5.5. Hydrolysis in water to the corresponding acid and amine is less likely than with procaine hydrochloride. This stability permits its use orally.

A kinetic study of the acid-catalyzed hydrolysis of procainamide has shown it to be unusually stable to hydrolysis in the pH range 2 to 7, even at elevated temperatures.[19]

Procainamide hydrochloride appears to have a direct depressant action on the ventricular muscle. It is used for the treatment of ventricular arrhythmias and extrasystoles and to correct cardiac arrhythmias during anesthesia. Advantages over quinidine and procaine include oral administration, less toxicity and reliable activity. It was learned first that the procaine hydrochloride was useful in treating arrhythmias. However, it must be given intravenously and is hydrolyzed by plasma enzymes to the acid and diethylaminoethanol. This aminoalcohol is still effective, but it has hypotensive effects and is quickly removed from the bloodstream. A study of the physiologic disposition and cardiac effects of procainamide showed the drug to be relatively stable in the body, since it is not affected by the enzyme which catalyzes the hydrolysis of procaine. It is absorbed rapidly and completely from the gastrointestinal tract. After the steady state is reached the plasma levels in man decrease at 10 to 20 per cent per hour.[20] Some of the drug is destroyed in the tissues, but the greater part is excreted in the urine. A study of the intramuscular use of the drug showed that the efficacy by this route was similar to that for oral or intravenous doses.[21] Appreciable serum levels were achieved in 5 minutes with the peak at 15 to 60 minutes; significant amounts were present after 6 hours. Higher serum levels and slower rate of decline were noted in patients with renal insufficiency.

Category—cardiac depressant (antiarrhythmic).

Usual dose—oral or intramuscular, 500 mg. to 1 g. 4 to 6 times daily; intravenous, 50 to 100 mg. per minute up to a total of 1 g.

Usual dose range—oral or intramuscular, 500 mg. to 6 g. daily; intravenous, 50 mg. to 1 g. daily.

OCCURRENCE
Procainamide Hydrochloride Capsules U.S.P.
Procainamide Hydrochloride Injection U.S.P.

ANTICOAGULANTS

Retardation of clotting is important in blood transfusions, to avoid thrombosis after operation or from other causes, to prevent recurrent thrombosis in phlebitis and pulmonary embolism and to lessen the propagation of clots in the coronary artery. This retardation may be accomplished by agents that inactivate the thrombin (heparin) or those substances that prevent the formation of prothrombin in the liver—the coumarin derivatives and the phenylindanedione derivatives.

Although heparin (q.v.) is a useful anticoagulant, it has limited applications. Many of the anticoagulants in use today were developed following the discovery of bishydroxycoumarin, an anticoagulant that is present in spoiled sweet clover. These compounds are orally effective, but there is a lag period of 18 to 36 hours before they significantly increase the clotting time. Heparin, in contrast, produces an immediate anticoagulant effect following intravenous injection. A major disadvantage of heparin is that the only effective therapeutic route is parenteral.

Bishydroxycoumarin functions by reversibly competing with vitamin K (q.v.) in the role that it plays in the clotting of blood. This lengthens the clotting time by decreasing the prothrombin concentration in the blood (in vivo), presumably by interfering with its production in liver, because prothrombin activity in vitro is not retarded. The molecule is symmetrical, and only one half of the molecule serves as a primary functional portion in the competition for vitamin K. The other half is secondary, and its contribution to activity is determined by its physical, chemical and structural properties which play a role in absorption, distribution, deposition, excretion, metabolism, and secondary binding forces at the enzymatic site. This is well illustrated by the drug warfarin (q.v.), which is many times more active than bishydroxycoumarin. The differences in toxicity, and manifestations other than fundamental function, may vary in the case of other modifications of structure.

The discovery of bishydroxycoumarin and related compounds as potent reversible* competitors of vitamin K led to the development of antivitamin K compounds such as

phenindione, which was designed in part according to metabolite-antimetabolite concepts. The active compounds of the phenylindanedione series are characterized by a phenyl, a substituted phenyl or a diphenylacetyl group in the 2-position. Another requirement for activity is a keto group in the 1 and 3 positions, one of which may form the enol tautomer. A second substituent, other than hydrogen, at the 2-position prevents this keto-enol tautomerism and the resulting compounds are ineffective as anticoagulants.

The activity of bishydroxycoumarin and related compounds and the phenindione types can be reversed by the proper amounts of vitamin K_1†, menadione, etc.

Out of hundreds of active anticoagulants the following are accepted for clinical use.

Products

Protamine Sulfate U.S.P. has an anticoagulant effect, but it counteracts the action of heparin if used in the proper amount and is used as an antidote for the latter in cases of overdosage. It is administered intravenously in a dose depending on the circumstances.

Category—antidote to heparin.

Usual dose—intravenous, 50 mg., repeated as required.

Occurrence
Protamine Sulfate Injection U.S.P.
Protamine Sulfate for Injection U.S..P.

Bishydroxycoumarin U.S.P., Dicumarol®, 3,3'-Methylenebis(4-hydroxycoumarin), is a white or creamy-white, crystalline powder

Bishydroxycoumarin

with a faint, pleasant odor and a slightly bitter taste. It is practically insoluble in water or alcohol, slightly soluble in chloroform and is dissolved readily by solutions

---

* At high levels Dicumarol is not reversed by vitamin K.

† Vitamin K_1 is considerably more effective than menadione.

of fixed alkalies. The effects after administration require 12 to 72 hours to develop and persist for 24 to 96 hours after discontinuance.

Bishydroxycoumarin is used alone or as an adjunct to heparin in the prophylaxis and treatment of intravascular clotting. It is employed in postoperative thrombophlebitis, pulmonary embolus, acute embolic and thrombotic occlusion of peripheral arteries and recurrent idopathic thrombophlebitis. It has no effect on an already formed embolus but may prevent further intravascular clotting. Since the outcome of acute coronary thrombosis is largely dependent on extension of the clot and formation of mural thrombi in the heart chambers with subsequent embolization, bishydroxycoumarin has been used in this condition. It also has been administered to arrest impending gangrene and frostbite. The dose, after determination of the prothrombin clotting time, is 0.2 to 0.3 g., depending on the size and the condition of the patient, the drug being given orally in the form of capsules or tablets. On the second day and thereafter, it may be given in amounts sufficient to maintain the prothrombin clotting time at about 30 seconds. If hemorrhages should occur, 50 to 100 mg. of menadione sodium bisulfite is injected, supplemented by a blood transfusion.

Category—anticoagulant.

Usual dose—initial, 200 to 300 mg. daily, then 50 to 100 mg. daily, according to prothrombin-time determinations.

OCCURRENCE

Bishydroxycoumarin Tablets U.S.P.
Bishydroxycoumarin Capsules N.F.

Ethyl Biscoumacetate N.F., Tromexan®, ethyl bis(4-hydroxy-2-oxo-2H-1-benzopyran-3-yl)acetate, 3,3′-(Carboxymethylene) bis(4-hydroxycoumarin) Ethyl Ester, is slightly soluble in alcohol and insoluble in water.

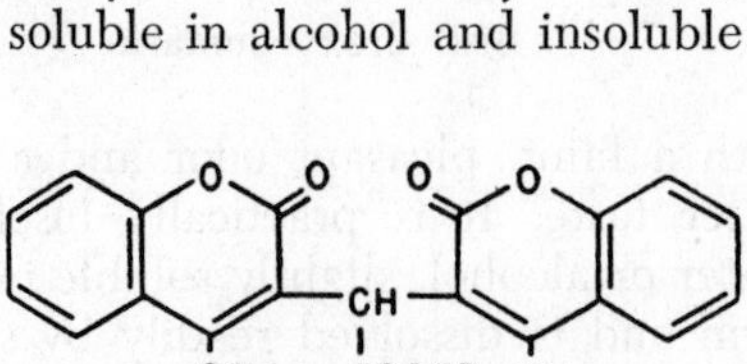

It is an anticoagulant similar to bishydroxycoumarin but is claimed to give more rapid

action, is absorbed and excreted more rapidly and gives less cumulative effect. The same precautions must be exercised as with similar agents.

Category—anticoagulant.

Usual dose—initial, 1.2 to 1.8 g., then 300 to 900 mg. daily, in accordance with prothrombin time determinations.

OCCURRENCE

Ethyl Biscoumacetate Tablets N.F.

Sodium Warfarin U.S.P., Coumadin® Sodium, Warcoumin®, 3-(α-acetonylbenzyl)-4-hydroxycoumarin sodium salt, is a white, odorless, crystalline powder, having a slightly bitter taste; it is slightly soluble in chloroform, soluble in alcohol or water. A 1 per cent solution has a pH of 7.2 to 8.5.

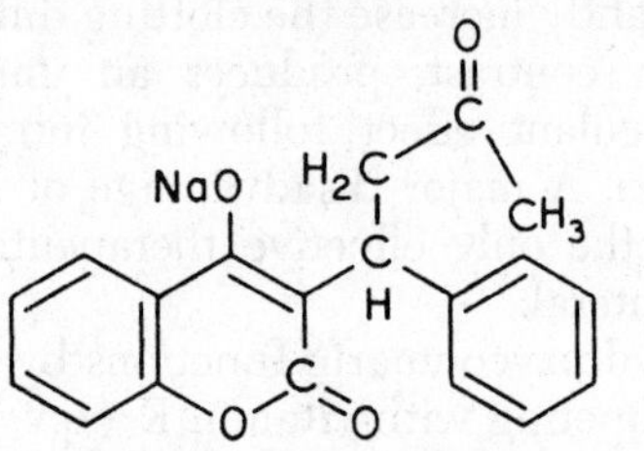

Sodium Warfarin

By virtue of its great potency warfarin at first was considered unsafe for use in humans and was utilized very effectively as a rodenticide, especially against rats. However, when used in the proper dosage level, it can be used in humans, especially by the intravenous route.

Category—anticoagulant.

Usual dose—oral, intramuscular or intravenous: initial, 50 mg., then 5 to 10 mg. daily, in accordance with prothrombin time determinations.

Usual dose range—initial, 25 to 75 mg.; maintenance, 2 to 10 mg. daily.

OCCURRENCE

Sodium Warfarin for Injection U.S.P.
Sodium Warfarin Tablets U.S.P.

Potassium Warfarin N.F.; Athrombin-K®, 3-(α-Acetonylbenzyl)-4-hydroxycoumarin Potassium Salt. Potassium warfarin is readily absorbed after oral administration, with a therapeutic hypoprothrombinemia being produced in 12 to 24 hours after administration of 40 to 60 mg. This salt is therapeutically interchangeable with sodium warfarin.

Category—anticoagulant.

Usual dose range—initial, 25 to 50 mg., then 2.5 to 10 mg. daily, in accordance with prothrombin time determinations.

OCCURRENCE

Potassium Warfarin Tablets N.F.

Acenocoumarol, Sintrom®, is 3-(α-acetonyl-4-nitrobenzyl)-4-hydroxycoumarin and differs from warfarin only in that the benzyl group contains a 4-nitro group. It is claimed to be the most active anticoagulant used clinically. Usual dose is 16 to 28 mg. the first day; 8 to 16 mg. the second day; and 2 to 10 mg. as a maintenance dose.

Phenprocoumon N.F., Liquamar®, 3-(α-Ethylbenzyl)-4-hydroxycoumarin. This drug has been shown to possess marked and prolonged anticoagulant activity.

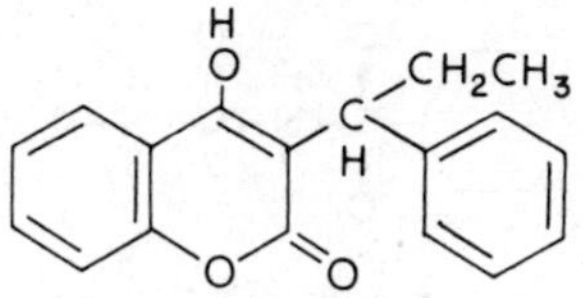

Phenprocoumon

Category—anticoagulant.

Usual dose range—initial, 21 mg. the first day, 9 mg. the second day, and 3 mg. the third day; maintenance, 1 to 4 mg., daily, according to prothrombin level.

OCCURRENCE

Phenprocoumon Tablets N.F.

Phenindione, Hedulin®, Danilone®, 2-phenyl-1,3-indandione, is an oral anticoagulant specifically designed to function as an antimetabolite for vitamin K. It is a pale-

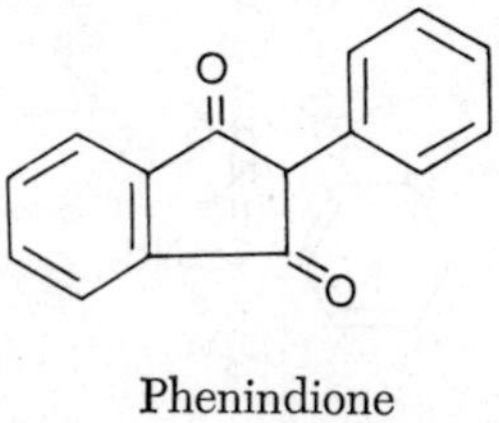

Phenindione

yellow crystalline material that is slightly soluble in water but very soluble in alcohol. It is more prompt-acting than bishydroxycoumarin. The initial dose of 200 to 300 mg., divided between morning and bedtime,

gives therapeutic prothrombin levels in 18 to 24 hours, with a return to normal on discontinuance in 24 to 48 hours. Maintenance dose is 50 to 100 mg. daily in two divided doses. The rapid elimination is presumed to make the drug safer than others of the class.

Bromindione, Halinone®, 2-(p-bromophenyl)-1,3-indandione is a potent long-acting oral anticoagulant. This is the p-bromo congener of phenindione; bromination increases the potency and also increases the duration of action. The increase in potency and in duration of action makes possible a single dose once daily, with a continued and stable effect through each 24-hour period. On a weight basis it is reported to be about 30 times as potent as bishydroxycoumarin and 3 times as potent as warfarin.[22] Twelve to 18 mg. of the drug induced therapeutic hypoprothrominemia within 28 to 34 hours. Treatment was resumed on the third day with doses in the range of 2 to 5 mg. daily.

Anisindione N.F., Miradon®, 2-(p-methoxyphenyl)-1,3-indandione, 2-(p-anisyl)-1,3-indandione is a p-methoxy congener of phenindione. It is a white crystalline powder, slightly soluble in water, tasteless, and well absorbed after oral administration.

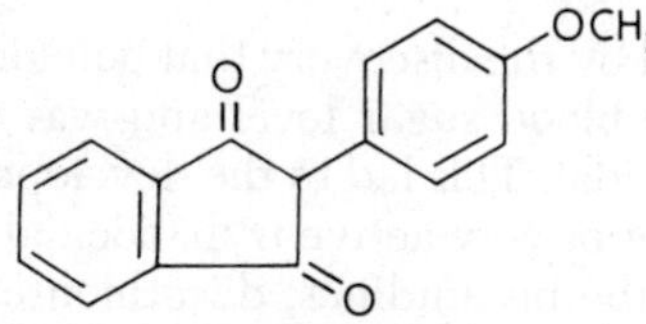

Anisindione

In instances where the urine may be alkaline, an orange color may be detected. This is due to metabolic products of anisindione and is not hematuria.

Category—anticoagulant.

Usual dose—initial, 300 mg. the first day, 200 mg. the second day, 100 mg. the third day; maintenance, 75 to 100 mg. daily.

Usual dose range—maintenance, 25 to 250 mg. daily, as determined by prothrombin time determinations.

OCCURRENCE

Anisindione Tablets N.F.

Diphenadione N.F., Dipaxin®, is 2-(diphenylacetyl)-1,3-indandione and therefore differs from phenindione only in the nature of the group at the 2-position. It occurs as yellow crystals or crystalline powder that are odorless, slightly soluble in alcohol and practically insoluble in water; however, it is lipo-soluble.

Category—anticoagulant.

Usual dose—initial, 20 to 30 mg. the first day, 10 to 15 mg. the second day; maintenance, 2.5 to 5 mg. daily.

Usual dose range—2.5 to 30 mg.

OCCURRENCE

Diphenadione Tablets N.F.

SYNTHETIC HYPOGLYCEMIC AGENTS

The discovery that certain organic compounds will lower the blood sugar level is not a recent one. In 1918 guanidine was shown to lower the blood sugar level. The discovery that certain trypanosomes need much glucose and will die in its absence was

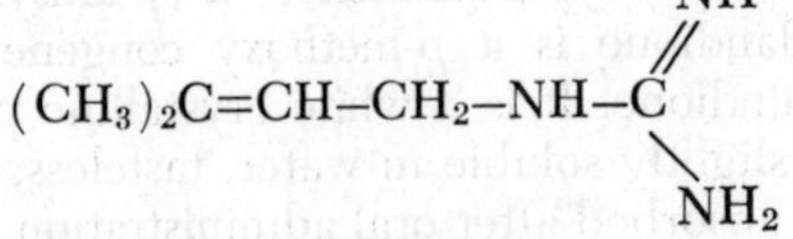

$$(CH_3)_2C=CH-CH_2-NH-C\overset{\displaystyle NH}{\underset{\displaystyle NH_2}{\big<}}$$

Galegine

followed by the discovery that galegine lowered the blood sugar level and was weakly trypanocidal. This led to the development of a number of very active trypanocidal agents such as the bisamidines, diisothioureas, bisguanidines, etc. Synthalin (trypanocidal at 1:250,000,000) and pentamidine are outstanding examples of very active trypanocidal agents. Synthalin lowers the blood sugar level in normal, depancreatized and completely alloxanized animals. This may

be due to a reduction in the oxidative activity of mitochondria resulting from inhibition of the mechanisms which simultaneously promote phosphorylation of adenosine diphosphate and stimulate oxidation by diphosphopyridine nucleotide in the citric acid cycle. Hydroxystilbamidine Isethionate U.S.P. is used as an antiprotozoan agent.

In 1942, *p*-aminobenzenesulfonamidoisopropylthiadiazole (an antibacterial sulfonamide) was found to produce hypoglycemia. These results stimulated the research for the development of synthetic hypoglycemic agents, several of which are in use today. These may be divided into two groups—the sulfonylureas and the biguanides.

SULFONYLUREAS

The sulfonylureas may be represented by the following general structure:

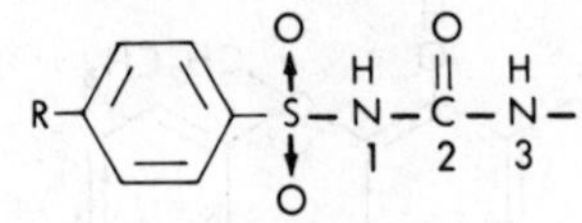

$$R-\hspace{-1mm}\bigcirc\hspace{-1mm}-\overset{\displaystyle O}{\underset{\displaystyle O}{\overset{\displaystyle \uparrow}{\underset{\displaystyle \downarrow}{S}}}}-\overset{H}{\underset{1}{N}}-\overset{O}{\overset{\|}{\underset{2}{C}}}-\overset{H}{\underset{3}{N}}-R'$$

These are urea derivatives with an aryl-sulfonyl group in the 1-position and an aliphatic group at the 3-position. R' must be of a certain size so that it confers lipophilic properties on the molecule. The methyl group gives an inactive compound, ethyl some activity and maximal activity results with alkyl groups containing 3 to 6 carbon atoms, as in chlorporpamide, tolbutamide and acetohexamide. Aryl groups at R' generally give toxic compounds. The R group on the aromatic ring primarily influences the duration of action of the compound. Tolbutamide disappears quite rapidly from the blood stream through being metabolized

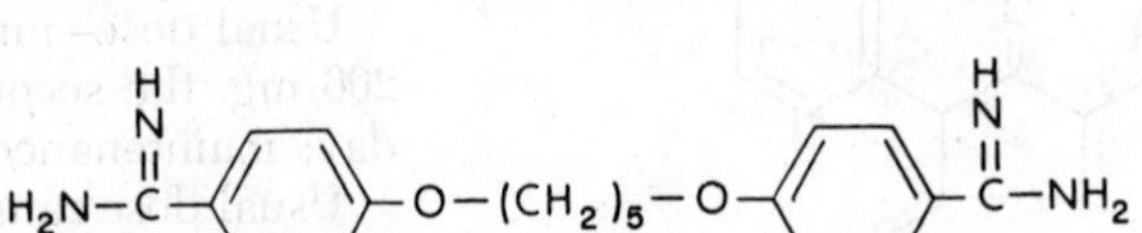

$$H_2N-\overset{H}{\overset{N}{\overset{\|}{C}}}-\hspace{-1mm}\bigcirc\hspace{-1mm}-O-(CH_2)_5-O-\hspace{-1mm}\bigcirc\hspace{-1mm}-\overset{H}{\overset{N}{\overset{\|}{C}}}-NH_2$$

Pentamidine

$$H_2N-\overset{H}{\overset{N}{\overset{\|}{C}}}-N-(CH_2)_{10}-\overset{H}{N}-\overset{H}{\overset{N}{\overset{\|}{C}}}-NH_2$$

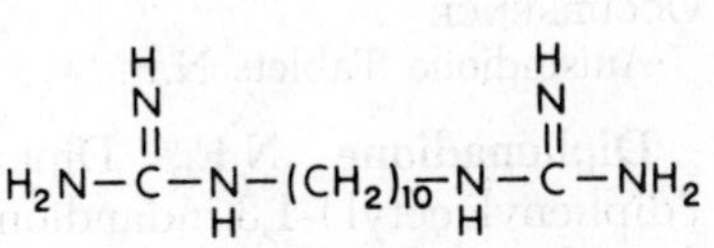

Synthalin

to the inactive carboxy compound which is rapidly excreted. On the other hand, chlorpropamide does not undergo extensive metabolic attack before excretion and it persists in the blood for a much longer time.

The mechanism of action of the sulfonylureas is to increase the release of insulin from the functioning beta cells of the intact pancreas. In the absence of the pancreas, they have no significant effect on blood glucose. They may have other actions, such as inhibition of glycogenolysis in the liver, but these are still uncertain. This group of drugs is of most value in the diabetic patient whose disease had its onset in adulthood. Accordingly, the group of sulfonylureas is not indicated in the juvenile onset diabetic.

Tolbutamide U.S.P., Orinase®, 1-Butyl-3-(*p*-tolylsulfonyl)urea, occurs as a white crystalline powder that is insoluble in water and soluble in alcohol or aqueous alkali. It is stable in air.

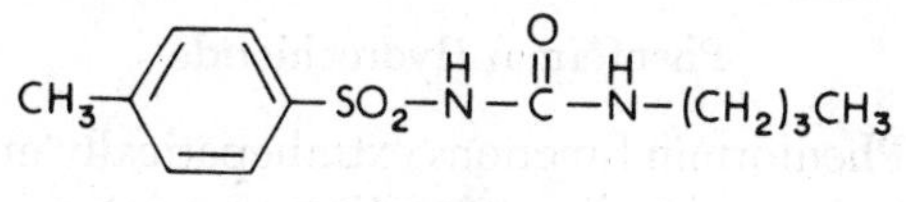

Tolbutamide

Tolbutamide is absorbed rapidly in responsive diabetic patients. The blood sugar level reaches a minimum after 5 to 8 hours. It is oxidized rapidly in vivo to 1-butyl-3-p-carboxyphenylsulfonylurea, which is inactive. The metabolite is freely soluble at urinary pH; however, if the urine is strongly acidified, as in the use of sulfosalicylic acid as a protein precipitant, a white precipitate of the free acid may be formed.

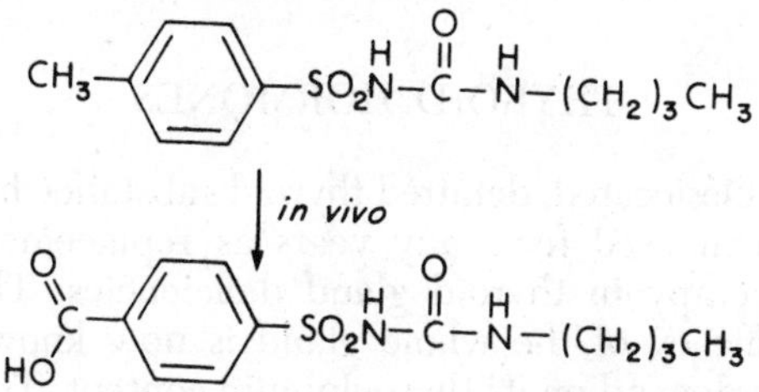

Tolbutamide should be used only where the diabetes patient is an adult or shows maturity onset in character, and the patient should adhere to dietary restrictions.

Category—hypoglycemic.

Usual dose—500 mg. twice daily.

Usual dose range—500 mg. to 2 g. daily.

OCCURRENCE

Tolbutamide Tablets U.S.P.

Sodium Tolbutamide U.S.P., Orinase® Diagnostic, 1-Butyl-3-(*p*-tolylsulfonyl)urea Monosodium Salt. Sodium tolbutamide is a white crystalline powder, freely soluble in water, soluble in alcohol and in chloroform and very slightly soluble in ether.

This water-soluble salt of tolbutamide is used intravenously for the diagnosis of mild diabetes mellitus and of functioning pancreatic islet cell adenomas. The sterile dry powder is dissolved in sterile water for injection to make a clear solution which then should be administered within one hour. The main route of breakdown is to butylamine and sodium *p*-toluenesulfonamide.

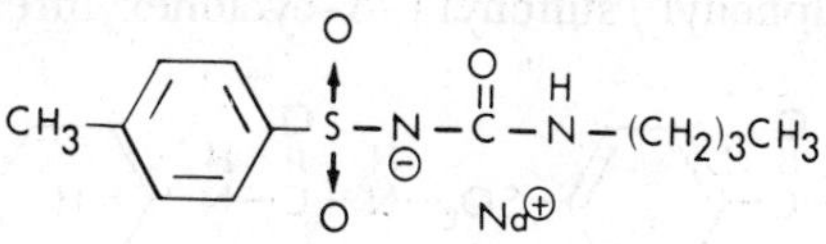

Sodium Tolbutamide

Category—diagnostic aid (diabetes).

Usual dose—intravenous, the equivalent of 1 g. tolbutamide.

OCCURRENCE

Sterile Sodium Tolbutamide U.S.P.

Chlorpropamide U.S.P.; Diabinese®; 1-[(*p*-Chlorophenyl)sulfonyl]-3-propylurea. Chlorpropamide is a white crystalline powder, practically insoluble in water, soluble in alcohol and sparingly soluble in chloroform. It will form water-soluble salts in basic solutions. This drug is more resistant to conversion to inactive metabolites than is tolbutamide and, as a result, has a much longer duration of action. After control of the blood sugar levels the maintenance dose is usually on a once-a-day schedule.

Category—hypoglycemic.

Usual dose—100 to 250 mg. once to twice daily.

Usual dose range—100 to 750 mg. daily.

OCCURRENCE

Chlorpropamide Tablets U.S.P.

Tolazamide U.S.P., Tolinase®, 1-(Hexahydro-1H-azepin-1-yl-3-(*p*-tolylsulfonyl)-urea. This agent is an analog of tolbutamide and is reported to be effective, in general, under the same circumstances where tolbutamide is useful. However, tolazamide appears to be more potent than tolbutamide, and is nearly equal in potency to chlorpropamide. In studies with radio-active tolazamide, investigators found that 85 per cent of an oral dose appears in the urine as metabolites which are more soluble than tolazamide itself.

Category—hypoglycemic.
Usual dose—250 mg. once or twice daily.
Usual dose range—100 mg. to 1 g. daily.

OCCURRENCE
Tolazamide Tablets U.S.P.

Acetohexamide N.F., Dymelor®; 1-[(*p*-Acetylphenyl) sulfonyl]-3-cyclohexylurea.

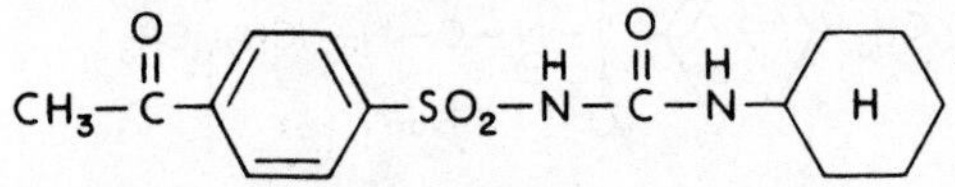

Acetohexamide

Acetohexamide is chemically and pharmacologically related to tolbutamide and chlorpropamide. Like the other sulfonylureas, acetohexamide lowers the blood sugar, primarily by stimulating the release of endogenous insulin.[23]

Acetohexamide is metabolized in the liver to a reduced form—the α-hydroxyethyl derivative. This metabolite, the main one in humans, possesses hypoglycemic activity. Acetohexamide is intermediate between tolbutamide and chlorpropamide in potency and duration of effect on blood sugar levels.

Category—hypoglycemic.
Usual dose—250 mg. to 1.0 g. daily.
Usual dose range—250 mg. to 1.5 g. daily.

OCCURRENCE
Acetohexamide Tablets U.S.P.

BIGUANIDES

Biguanides of the following type structure are effective hypoglycemic agents.

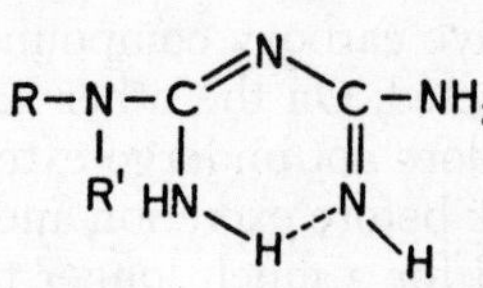

R may be alkyl or aralkyl. When alkyl, the activity is greatest when R is *n*-amyl, and good activity is obtained when R is benzyl or β-phenethyl. Activity is retained when the phenyl ring is replaced by a pyridyl, thienyl or furanyl group. R′ preferably should be H, although activity is present in some cases where it is a methyl group.

Phenformin Hydrochloride U.S.P., DBI®, 1-Phenethylbiguanide Monohydrochloride. This oral hypoglycemic agent is completely unrelated to the hypoglycemic sulfonylureas in chemical structure and mode of action.

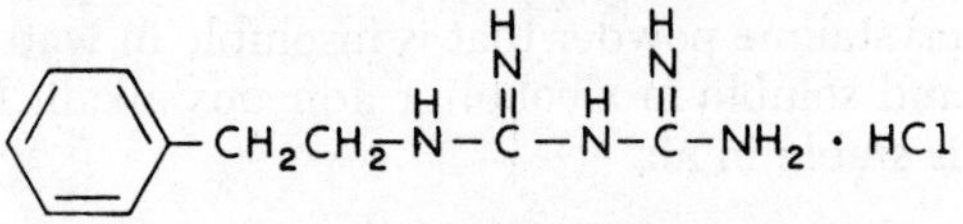

Phenformin Hydrochloride

Phenformin functions extrahepatically and acts as an insulin-supporting or reinforcing agent. The main use proposed for this drug is in maturity-onset diabetes mellitus, especially in patients who have no other complications and cannot be controlled by diet alone.

Category—hypoglycemic.
Usual dose—25 to 50 mg. once to 3 times daily.
Usual dose range—25 to 300 mg. daily.

OCCURRENCE
Phenformin Hydrochloride Tablets U.S.P.

THYROID HORMONES

Desiccated, defatted thyroid substance has been used for many years as replacement therapy in thyroid gland deficiencies. The efficacy of the whole gland is now known to depend on its thyroglobulin content. This is an iodine-containing globulin. Thyroxine was obtained as a crystalline derivative by Kendall of the Mayo Clinic in 1916. It showed much the same action as the whole thyroid substance. Later thyroxine was synthesized by Harington and Barger in England. Later studies showed that an even more potent iodine-containing hormone existed, which is now known as triiodothyronine. There is now evidence that thyroxine may be the storage form of the hormone, while triiodothyronine is the circulating form. Another point of view is that, in the blood, thyroxine is more firmly bound to the globulin fraction than is triiodothyronine, which can then enter the tissue cells.

Products

Sodium Levothyroxine U.S.P., Synthroid® Sodium, Letter®, Levoroxine®, Sodium L-3-[4-(4-Hydroxy-3,5-diodophenoxy)-3,5-diio-

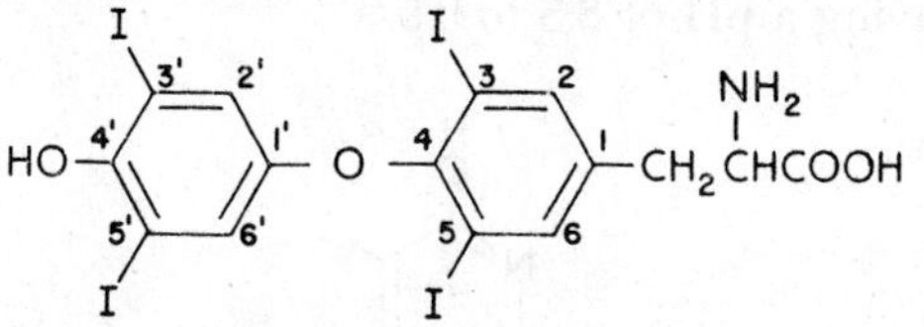

Sodium Levothyroxine

dophenyl]alanine, Sodium L-3,3′,5,5′-Tetra-iodothyronine. This compound is the sodium salt of the levo isomer of thyroxine, which is an active physiologic principle obtained from the thyroid gland of domesticated animals used for food by man. It is also prepared synthetically. The salt is a light-yellow, tasteless, odorless powder. It is hygroscopic but stable in dry air at room temperature. It is soluble in alkali hydroxides, 1:275 in alcohol and 1:500 in water to give a pH of about 8.9.

Sodium levothyroxine is used in replacement therapy of decreased thyroid function (hypothyroidism). In general, 100 mcg. of sodium levothyroxine is clinically equivalent to 30 to 60 mg. of Thyroid U.S.P.

Category—thyroid hormone.

Usual dose—100 to 400 mcg. once daily.

Usual dose range—25 mcg. to 1 mg. daily.

Occurrence

Sodium Levothyroxine Tablets U.S.P.

Sodium Liothyronine U.S.P., Cytomel®, sodium L-3-[4-(4-hydroxy-3-iodophenoxy)-3,5-diiodophenyl]alanine, is the sodium salt of L-3,3′,5-triiodothyronine. It occurs as a light tan, odorless, crystalline powder slightly soluble in water or alcohol and has a specific rotation of +18° to +21° in acid (HCl) alcohol.

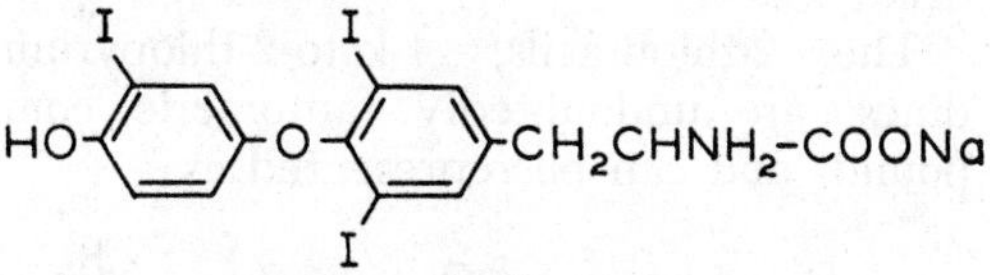

Sodium Liothyronine

Liothyronine occurs in vivo together with levothyroxine; it has the same qualitative activities as thyroxin but is more active. It is absorbed readily from the gastrointestinal tract, is cleared rapidly from the bloodstream and is bound more loosely to plasma proteins than is thyroxin, probably due to the less acidic phenolic hydroxyl group.

Its uses are the same as those of levothyroxine, including treatment of metabolic insufficiency, male infertility and certain gynecologic disorders.

Category—thyroid hormone.

Usual dose—the equivalent of 25 to 75 mcg. of liothyronine once daily.

Usual dose range—5 to 100 mcg. daily.

Occurrence

Sodium Liothyronine Tablets U.S.P.

ANTITHYROID DRUGS

When hyperthyroidism exists (excessive production of thyroid hormones), the condition usually requires surgery, but prior to surgery the patient must be prepared by preliminary abolition of the hyperthyroidism through the use of antithyroid drugs. Thiourea and related compounds show an antithyroid activity, but they are too toxic for clinical use. The more useful drugs are 2-thiouracil derivatives and a closely related 2-thioimidazole derivative. All of these appear to have a similar mechanism of action, i.e., prevention of the iodination of the precursors of thyroxine and triiodothyronine. The main difference in the compounds lies in their relative toxicities.[24]

Thiourea 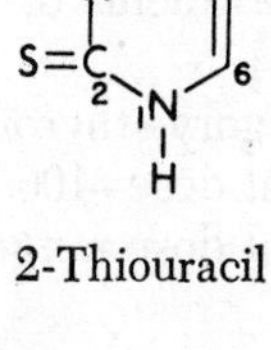   2-Thiouracil

These compounds are well absorbed after oral administration and are excreted in the urine.

The 2-thiouracils, 4-keto-2-thiopyrimidines, are undoubtedly tautomeric compounds and can be represented as

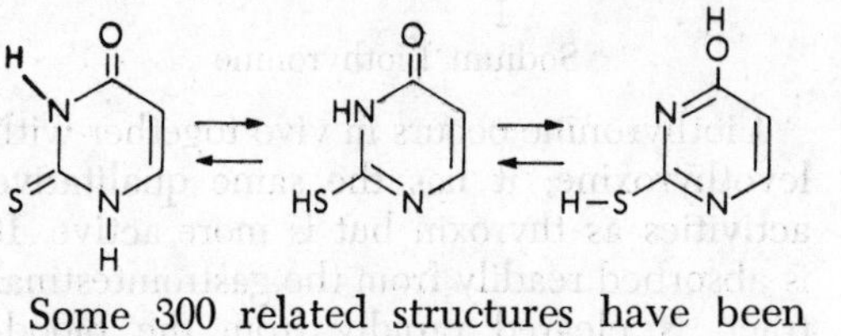

Some 300 related structures have been evaluated for antithyroid activity, but, of these, only the 6-alkyl-2-thiouracils and closely related structures possess useful clinical activity. The most serious side-effect of thiouracil therapy is agranulocytosis.

Products

Propylthiouracil U.S.P.; Propacil®; 6-Propyl-2-thiouracil. Propylthiouracil is a stable, white crystalline powder with a bitter taste. It is slightly soluble in water but

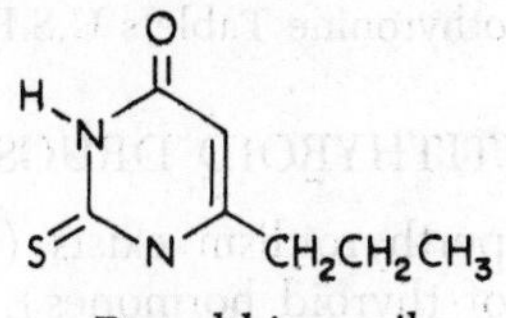

Propylthiouracil

is readily soluble in alkaline solutions (salt formation).

This drug is useful in the treatment of hyperthyroidism. There is a delay in appearance of its effects, because propylthiouracil does not interfere with the activity of thyroid hormones already formed and stored in the thyroid gland. This lag period may vary from several days to weeks, depending on the condition of the patient. The need for three equally spaced doses during a 24-hour period is often stressed, but there is now evidence that a single daily dose is as effective as multiple daily doses in the treatment of most hyperthyroid patients.[25]

Category—thyroid inhibitor.

Usual dose—100 mg. 3 times daily.

Usual dose range—50 to 500 mg. daily.

OCCURRENCE

Propylthiouracil Tablets U.S.P.

Methylthiouracil N.F., Methiacil®, Thimecil®, 6-methyl-2-thiouracil is a white, odorless crystalline powder with solubilities similar to those of propylthiouracil. It should be stored in well-closed, light-resistant containers. The action and the uses are similar to those of propylthiouracil.

Category—thyroid suppressant.

Usual dose—50 mg. 4 times daily.

Sodium Iothiouracil, Itrumil® Sodium, sodium 5-iodo-2-thiouracil is a white to off-white, odorless, crystalline powder. It is water soluble, with a two per cent solution having a pH of 8.5 to 9.5.

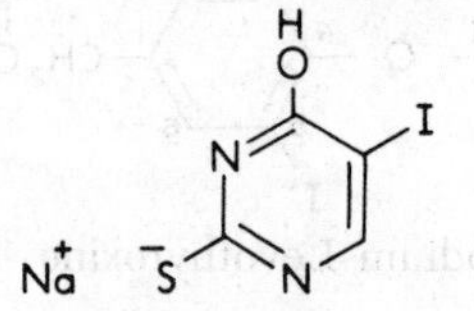

Sodium Iothiouracil

The iodinated thiouracil compounds were investigated in an effort to obtain a drug with the combined effects of iodine and thiouracil. The special hope was that the action of the iodine would counteract some of the undesirable effects of thiouracil, namely, the glandular hyperplasia and hypervascularity induced by thiouracil. Sodium iothiouracil is catabolized to iodine (or iodide) and thiouracil, which are excreted separately. Studies have shown that the clinical response is different from that produced with equivalent amounts of thiouracil and Lugol's solution or potassium iodide.[26]

In preparation for surgery, the oral dose is initially 300 mg. daily in divided doses, then is adjusted as needed to control the symptoms of hyperthyroidism. Itrumil Sodium is available as 50-mg. pink, scored tablets.

Methimazole U.S.P., Tapazole®, 1-methylimidazole-2-thiol occurs as a white to off-white crystalline powder with a characteristic odor and is freely soluble in water. A 2 per cent aqueous solution has a pH of 6.7

to 6.9. It should be packaged in well-closed, light-resistant containers.

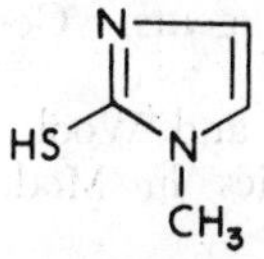

Methimazole

Methimazole is indicated in the treatment of hyperthyroidism. It is more potent than propylthiouracil. The side-effects are similar to those of propylthiouracil. As with other antithyroid drugs, patients using this drug should be under medical supervision. Also, similar to the other antithyroid drugs, methimazole is most effective if the total daily dose is subdivided and given at 8-hour intervals.

Category—thyroid inhibitor.
Usual dose—10 mg. 3 times daily.
Usual dose range—5 to 60 mg. daily.

OCCURRENCE
Methimazole Tablets U.S.P.

REFERENCES CITED

1. Krantz, J. C., *et al.*: J. Pharmacol Exp. Ther. 70:323, 1940.
2. Leyden, A. F., *et al.*: J. Am. Pharm. A. (Sci. Ed.) 45:771, 1956.
3. Weis-Fogh, O.: Pharm. acta helv. 35:442, 1960.
4. MacPhillamy, H. B., *et al.*: J. Am. Chem. Soc. 77:4335, 1955.
5. Udenfriend, S., *et al.*: Biochem. Pharmacol. 8:419, 1962.
6. Dock, W. D., *et al.*: Bull. N. Y. Acad. Med. 31:198, 1955.
7. Oaks, W., *et al.*: Arch. Int. Med. 104:527, 1959.
8. Curran, G. L.: Am. Pract. 7:1412, 1956.
9. Brit. Med. J. 1964:1181 (Nov. 7.)
10. Miale, J. P.: Curr. Ther. Res. 7:392, 1965.
11. Parsons, W. B., Jr., and Flinn, J. H.: J.A.M.A. 165:234, 1957.
12. Scott, C. C. *et al.*: J. Pharmacol. Exp. Ther. 84:184, 1945.
13. Riseman, J. E. F., *et al.*: Arch. Int. Med. 71:460, 1943.
14. Mankin, J. W.: J. Lab. Clin. Med. 41:929, 1953.
15. Blinder, H., *et al.*: Arch. Int. Med. 86-917, 1950.
16. Riseman, J. E. F., *et al.*: Am. Heart J. 22:219, 1941.
17. Gold, H., *et al.*: J.A.M.A. 145:637, 1951; Brass, H.: J. Am. Pharm. A. (Pract. Ed.) 4:310, 1943.
18. Halpern, A., *et al.*: Antibiot. Chemother. 9:97, 1959.
19. Marcus, A. D., and Taraszka, A. J.: J. Am. Pharm. A. (Sci. Ed.) 46:28, 1957.
20. Brodie, B. B. *et al.*: J. Pharmacol. Exp. Ther. 102:5, 1951.
21. Bellet, S., *et al.*: Am. J. Med. 13:145, 1952.
22. Singer, M. M., *et al.*: J.A.M.A. 179:150, 1962.
23. Council on Drugs: J.A.M.A. 191:127, 1965.
24. McClintock, J. C., *et al.*: Surg. Gynec. Obstet. 112:653, 1961.
25. Greer, M. A., *et al.*: New Eng. J. Med. 272:888, 1965.
26. Catz, B., and Starr, P.: Trans. Am. Goiter A. 1952:53.

SELECTED READING

Anderson, G. W.: Antithyroid compounds, Med. Chem. 1:1, 1951.
Arora, R. B., and Mathur, C. N.: Brit. J. Pharmacol. 20:29, 1963.
Astwood, E. B.: Chemotherapy of Hyperthyroidism, Harvey Lectures, Ser. 40, 195, 1944-1945, Lancaster, Pa., Science Press, 1945.
Bach, F. L.: Antilipemic Agents, *in* Burger, A. (ed.): Medicinal Chemistry, ed. 3, p. 1123, New York, Wiley-Interscience, 1970.
Bender A. D.: Antihypertensive agents, Topics in Med. Chem. 1:177, 1967.
Comer, W. T., and Gomall, A. W.: Antihypertensive Agents, *in* Burger, A. (ed.): Medicinal Chemistry, ed. 3, p. 1019, New York, Wiley-Interscience, 1970.
Davis, C. S., and Halliday, R. P.: Cardiac Drugs, *in* Burger, A. (ed.): Medicinal Chemistry, ed. 3, p. 1078, New York, Wiley-Interscience, 1970.
Divald, S., and Joullié, M. M.: Coagulants and Anticoagulants, *in* Burger, A. (ed.): Medicinal Chemistry, ed. 3, p. 1092, New York, Wiley-Interscience, 1970.
Friend, D. G.: Drugs for peripheral vascular disease, Clin. Pharmacol. Ther. 5:666, 1964.
Grunwald, F. A.: Hypoglycemic Agents, *in* Burger, A. (ed.): Medicinal Chemistry, ed. 3, p. 1172, New York, Wiley-Interscience, 1970.

Ingram, G. I. C.: Anticoagulant therapy, Pharmacol. Rev. 13:279, 1961.

Jorgensen, E. C.: Thyroid Hormones and Anti-thyroid Drugs, *in* Burger, A. (ed.): Medicinal Chemistry, ed. 3, p. 838, New York, Wiley-Interscience, 1970.

Owen, W. R.: Efficacy of drugs in lowering blood cholesterol, Med. Clin. N. Am. 48:347, 1964.

Pinter, K. G., and Van Itallie, T. B.: Drugs and atherosclerosis, Ann. Rev. Pharmacol. 6:251, 1966.

Schlittler, E., *et al.*: Progr. Drugs Res. 4:295, 1962.

Seidensticker, J. F., and Hamwl, G. J.: Oral hypoglycemic agents, Geriatrics 22:112, 1967.

Selenkow, H. A., and Wool, M. S.: Thyroid hormones, Topics in Mod. Chem. 1:242, 1967.

Slater, J. D. H.: Oral hypoglycaemic drugs, Progr. Med. Chem. 2:187, 1962.

Wien, R.: Hypotensive agents, Progr. Med. Chem. 1:34, 1961.

22

Local Anesthetic Agents

Robert F. Doerge, Ph.D.,
Professor of Pharmaceutical Chemistry,
Chairman of the Department of Pharmaceutical Chemistry, School of Pharmacy,
Oregon State University

INTRODUCTION

Local anesthetics are found in several different chemical classes of organic compounds. There is a wide variety of structures which possess local anesthetic activity and are of value in medicine. In general the useful compounds can be divided into three groups:

1. Hydroxy-compounds
2. Esters
3. A miscellaneous group of compounds, most of which contain one or more nitrogen atoms

Hydroxy-compounds that are useful are predominantly in the aromatic series, but alicyclic alcohols such as menthol and cyclohexanol possess a little activity. Phenol, as well as resorcinol and the cresols, has been mentioned previously as having local anesthetic properties. Note, however, that toxicity increases and activity decreases with methoxy- and hydroxy-substitution on the aromatic nucleus. Aliphatic alcohols with no aryl (aromatic) groups in the molecule are of no practical value as local anesthetics.

Benzyl alcohol produces useful anesthesia without the disadvantages of phenol (q.v.). Here an alkyl group separates the aromatic ring from the hydroxy group. This is also the case with saligenin, β-phenethyl alcohol and α-phenylcinnamyl alcohol.

The miscellaneous group of local anesthetics includes those compounds that are unrelated chemically yet exhibit anesthetic properties. Many compounds in this group have been studied and are recorded in the literature, but relatively few are used in medicine for this purpose. Examples are dibucaine—an amide; phenacaine—an amidine; diperodon—a urethane; and quinine—a cupreine. Methyl and ethyl chloride may also be mentioned in this group; these compounds produce their effects by freezing the tissue.

Amides and imides may be considered as having the group —NH— substituted for the —O— of an ester:

AR—CO—O—R	an ester
AR—CO—NH—R	a substituted amide
AR—CO—NR'—CO—AR	a substituted imide

Active amides in which R represents a dialkylaminoalkyl group are benzamide, *p*-aminobenzamide, cinnamide and naphthoamide (see dibucaine). Phthalimides possess some degree of local anesthetic properties.

Urethanes or carbamates, when properly substituted, are effective compounds. In the structure

$$R^1$$
$$\diagdown$$
$$N-CO-O-R^3$$
$$\diagup$$
$$R^2$$

it can be seen that a wide variety of compounds is possible (see diperodon).

The ester group of compounds is by far the most important class. The possible number of compounds is large because of the many alcohols and acids which are available, but the most productive field of investigation has been the aminoalcohol esters of benzoic and *p*-aminobenzoic acids. From the tremendous amount of work which has been done in the synthesis and the testing of esters for local anesthetic activity it has been found that three criteria must be met for the compound to show a high degree of activity:

1. The ester must contain nitrogen, in the alcohol, the acid or in both.
2. The acid must be aromatic.
3. The alcohol is usually aliphatic, either open chain or alicyclic.

In 1884, Koller[1] observed that cocaine hydrochloride produced anesthesia in the eye. The work of Willstätter on the elucidation of the cocaine molecule then led to the synthesis of hundreds of compounds possessing local anesthetic properties. On hydrolysis, cocaine gives benzoic acid, methyl alcohol and ecgonine. Ecgonine contains an alcoholic hydroxyl group and a carboxyl group and has the properties of a tertiary amine. On examination, ecgonine can be seen to be a combined piperidine and pyrrolidine ring. That portion of the cocaine formula enclosed by a dotted line has been

found to represent the anesthesiophoric group. By 1890, ethyl *p*-aminobenzoate had been prepared, showing that esters with nitrogen on the acid group possess local anesthetic properties. Soon thereafter, Einhorn prepared orthoform (1909) and procaine (1906). During this same period, studies were made on esters of basic alcohols with benzoic acid, for example, amylocaine (1904) (see Table 66).

COCA AND THE COCA ALKALOIDS

Coca leaves (*Erythroxylon coca*) and a few other related species of coca contain a number of important alkaloids of which (−)-cocaine is the most important from a therapeutic standpoint. For commercial purposes, the leaves from South America (Bolivia and Peru) and Java are used most. Among the alkaloids present in the leaves are (−)-cocaine, cinnamoylcocaine, the truxillines and tropacocaine. These alkaloids, with the exception of tropacocaine, are closely related, because acid or alkaline hydrolysis yields (−)-ecgonine and methyl alcohol as common products. The other product of hydrolysis is an acid, which differs in the case of each alkaloid. Tropacocaine, as noted, does not yield (−)-ecgonine on hydrolysis but instead gives pseudotropine together with benzoic acid. Therefore it is closely related to atropine. Table 65 shows the different constituent portions of the various alkaloids.

Coca leaves have long been chewed by the South American Indians to prevent hunger and to increase endurance. They are highly habit-forming. Because of this and its toxic nature, the leaf is little used in medicine.

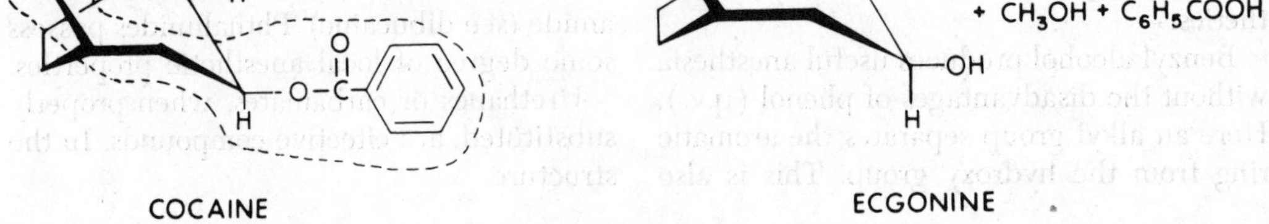

COCAINE → ECGONINE + CH$_3$OH + C$_6$H$_5$COOH

TABLE 65. CONSTITUENT PORTIONS OF VARIOUS ALKALOIDS

ALKALOID	BASIC RESIDUE	ACIDIC RESIDUE
(−)-Cocaine	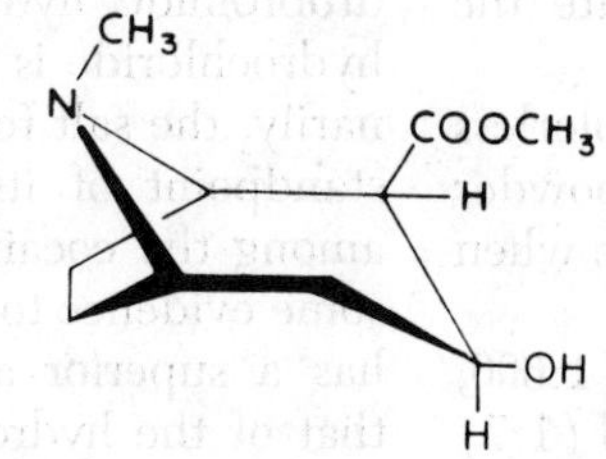	C_6H_5COOH Benzoic acid

(−)-Ecgonine methyl ester

ALKALOID	BASIC RESIDUE	ACIDIC RESIDUE
Cinnamoylcocaine	(−)-Ecgonine methyl ester	C_6H_5−CH=CH−COOH Cinnamic acid
α-Truxilline	(−)-Ecgonine methyl ester (2 molecules)	C_6H_5−CH—CH−COOH HOOC−CH—CH−C_6H_5 α-Truxillic acid
β-Truxilline	(−)-Ecgonine methyl ester (2 molecules)	C_6H_5−CH—CH−COOH C_6H_5−CH—CH−COOH β-Truxillic acid
Tropacocaine	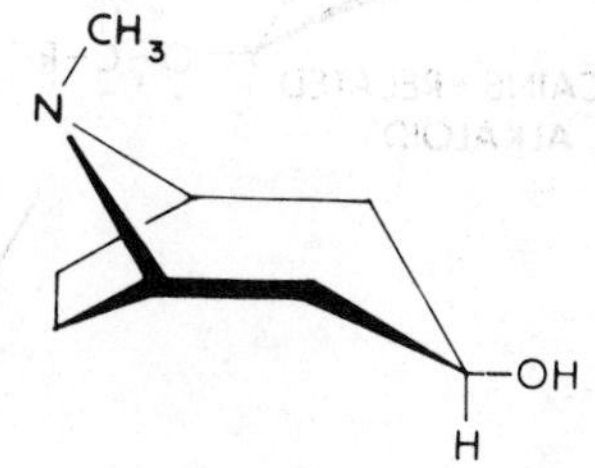	C_6H_5COOH Benzoic acid

Pseudotropine

PRODUCTS

Cocaine N.F. Cocaine is an alkaloid that is obtained from the leaves of *Erythroxylon coca* Lamarck and other species of Erythroxylon (Fam. Erythroxylaceae), or by synthesis from ecgonine or its derivatives. Chemically, it is methylbenzoylecgonine. It occurs to the extent of approximately 1 per cent in South American leaves and to the equivalent (as other derivatives) of about 2 per cent in Java leaves. It was first discovered by Gaedecke (1855), and it was rediscovered by Niemann (1859) and named cocaine. Its constitution was elucidated by the combined work of many

researchers, among whom Willstätter and his co-workers figured quite prominently.

Since cocaine does not exist naturally to any great extent in any leaves other than those from South America, the usual method of manufacture is not a simple isolation of the active constituent. Commercial methods aim first to isolate the cocaine and related alkaloids from the crude leaves, following which the alkaloids are hydrolyzed either to (−)-ecgonine or, by the use of methyl alcohol hydrogen chloride, to (−)-ecgonine methyl ester. Following purification of the ecgonine or its derivative by crystallization, (−)-cocaine is synthesized by methylation and benzoylation in the case of (−)-ecgonine or by simple benzoylation in the case of (−)-ecogonine methyl ester. The reactions shown below illustrate the manufacture of cocaine.

Cocaine occurs as levorotatory, colorless crystals or as a white, crystalline powder which numbs the lips and the tongue when applied topically.

It is slightly soluble in water (1:600; 1:270 at 80°), more soluble in alcohol (1:7) and quite soluble in chloroform (1:1) and in ether (1:3.5). The crystals are fairly soluble in olive oil (1:12) but less soluble in mineral oil (1:80 to 100). Because of its solubility characteristics, it is used principally where oily solutions or ointments are indicated.

Cocaine is basic, with a pKa of 8.4 at 15°, and readily forms crystalline salts, such as benzoate, borate, citrate, hydrochloride, hydrobromide, hydriodide and salicylate. The hydrochloride is official in the *U.S.P.* Ordinarily, the salt form is useful only from the standpoint of its solubility in water, but among the cocaine salts there seems to be some evidence to indicate that the salicylate has a superior anesthetic action (4 times that of the hydrochloride).[2]

Category—local anesthetic (narcotic).

Application—topical to mucous membrane as a 1 per cent solution.

Cocaine Hydrochloride U.S.P.; Cocaini Hydrochloridum P.I. This cocaine salt occurs as colorless crystals, or as a white, crystalline powder. It is soluble in water

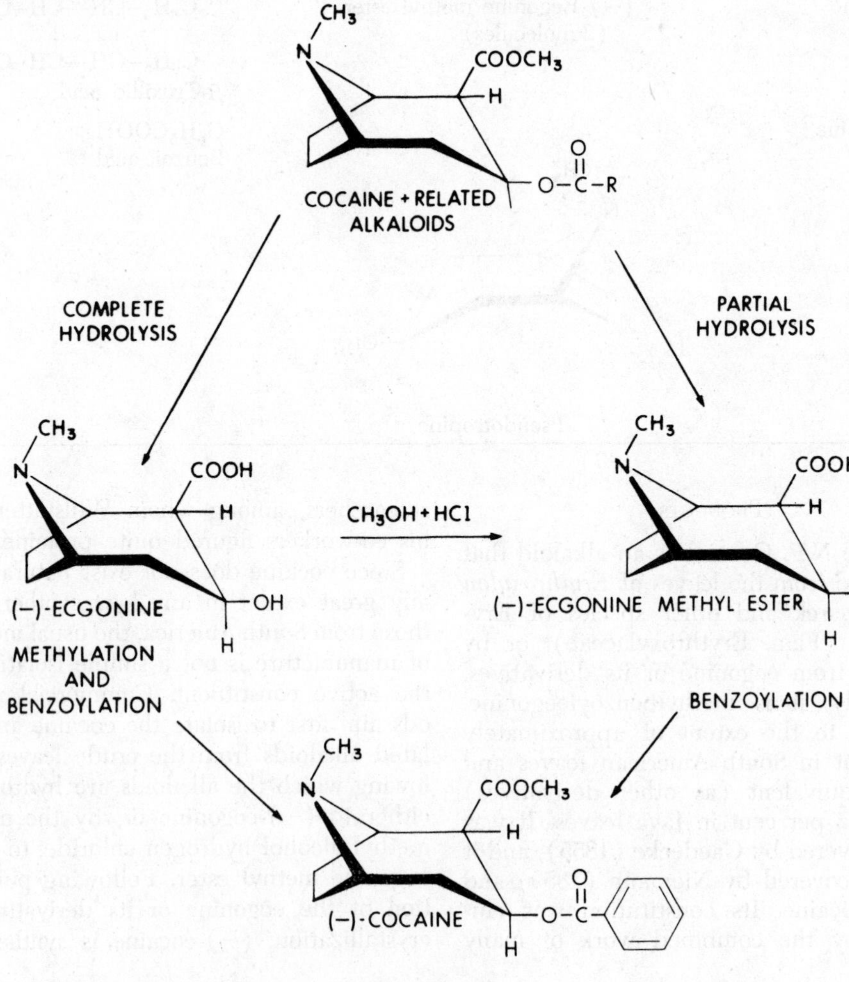

(1:0.5), alcohol (1:3.5), chloroform (1:15) or glycerin. It is insoluble in ether.

Aqueous solutions of cocaine hydrochloride are stable if not subjected to elevated temperatures or stored for prolonged periods. Up to 10 per cent of potency may be lost by autoclaving at 120° for 15 minutes.[3] Bacteriologic filtration is a better method of sterilization. It is sometimes necessary that the solution be as near neutrality as possible. At pH's near 7 the solution becomes very unstable and should be used soon after preparation and should not be autoclaved. Solutions are incompatible with alkalies, the usual alkaloidal precipitants and triethanolamine, silver nitrate, sodium borate, calomel and mercuric oxide.

The local anesthetic properties of cocaine were demonstrated first by Wohler, in 1860. However, Koller first used the drug, in 1884, as a topical anesthetic in the eye. Its toxicity has prevented cocaine from being used for anything other than topical anesthesia and, even in this capacity, it is desirable to limit its use for fear of causing systemic reactions and addiction. It is absorbed easily from mucous membranes, as is evidenced by the quick response obtained by addicts when cocaine ("snow") is snuffed into the nostrils. It is worthy of note that cocaine does not penetrate through the intact skin. As a local anesthetic, it is used topically as solutions of 2 to 5 per cent applied to mucous membranes such as those of the eye, the nose, and the throat. The anesthesia produced by such concentrations lasts approximately a half hour. Its use in ophthalmology is marred by the fact that not infrequently it causes corneal damage with resultant opacity. For *nose* and *throat* work, concentrations higher than 10 per cent are rarely used, 4 per cent being a common concentration. In any case, the total amount of cocaine used should not exceed 100 mg. More detailed summaries of its uses may be consulted in the literature.[4]

The difficulties in regard to sterilization coupled with the addictive character of cocaine have resulted in much research aimed at the production of more desirable local anesthetics. Since the discovery of procaine by Einhorn in 1905 (q.v.), a host of compounds possessing local anesthetic properties has been synthesized.

Category—topical anesthetic.

For external use—topically as a 2 to 10 per cent solution to mucous membranes.

OCCURRENCE

Cocaine Hydrochloride Tablets N.F.

Tropacocaine. This alkaloid occurs in Javanese coca, or it may be prepared synthetically by the esterification of benzoic acid with pseudotropine. The latter is the commercial procedure.

It[*] is marketed as the hydrochloride, which occurs as colorless needles, slightly soluble in alcohol but readily soluble in water to form a neutral solution.

The compound, although more irritant than cocaine, is reported to have approximately the same anesthetic action with one half the toxicity. Also, it is more stable. It is used principally in spinal anesthesia.

SYNTHETIC COMPOUNDS

The anesthesiophoric group may be represented more simply by the following structure.

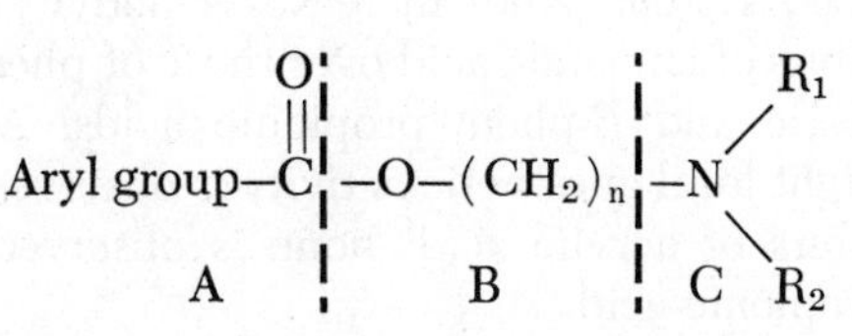

$$\underset{A}{\text{Aryl group}-\overset{\overset{\displaystyle O}{\|}}{C}-\underset{B}{O-(CH_2)_n}-\underset{C}{N}\overset{R_1}{\underset{R_2}{}}$$

The A portion or acid group includes practically every reasonably available aromatic acid. Benzoic and *p*-aminobenzoic acids have been proved to be best, and in Table 66 are listed acids of which esters have been prepared and tested. Examination of numbers 2 to 23 indicates how the kind and the position of substituents have been varied on the benzene ring. Except for esters of *m*-aminobenzoate, β-napthoic and cinnamic acids, none of the others has produced a useful local anesthetic. Because of the voluminous literature concerning local anesthetics, the varied types of organic esters and the individual evaluation methods reported by investigators, comparison is almost impossible. A few rules, not too well-established, may be cited from the researches thus far:

1. Introduction of a methylene group (CH₂), as in phenylacetic acid, decidedly decreases the activity. Note that the following structure is practically inactive:

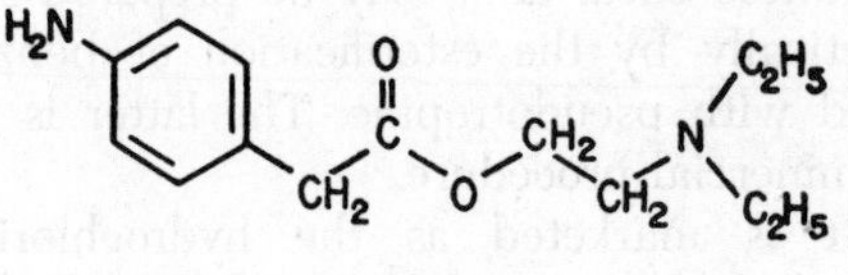

2. The carboxyl group must be conjugated with an aryl group. Structures such as the following have no local anesthetic action:

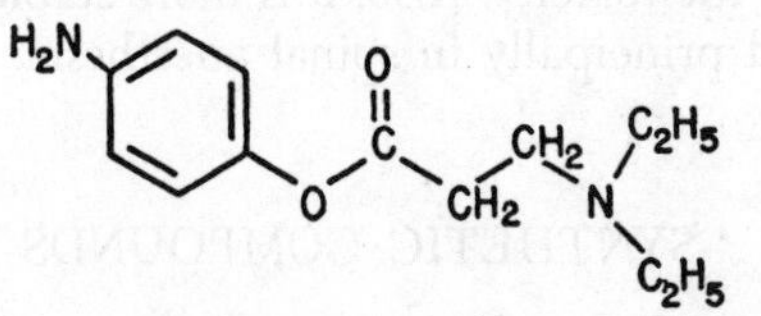

3. The carbonyl group must be conjugated with an aromatic nucleus or other related system. Note increase of activity for esters of cinnamic acid over those of phenylacetic and β-phenylpropionic acids. Also, slight local anesthetic activity is observed in esters of acrylic acid; none is observed in propionic acid.

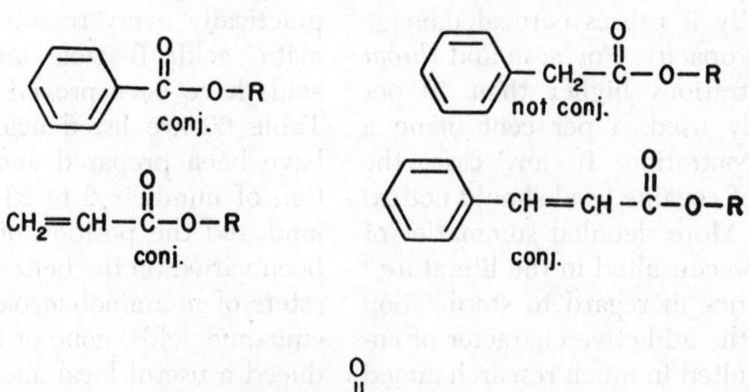

4. The esters of *p*-aminobenzoic acid are more effective than are the esters of benzoic, *p*-hydroxybenzoic and *p*-alkoxybenzoic acid.

5. *p*-Alkoxybenzoates increase in activity as the alkyl group increases in molecular weight.

6. The addition of an amino group in the meta position of *p*-alkoxybenzoates increases activity but also increases toxicity.

7. An alkyl group on the aromatic amino nitrogen increases anesthetic potency and toxicity (tetracaine). In cases of the simplest esters of *p*-aminobenzoic acid (benzocaine), when the amino group has a hydrogen replaced by an alkyl group, the activity is lost.

8. The most desirable position of the amino group on the aromatic ring is para to a carboxy group.

9. Most esters of heterocyclic acids possess some local anesthetic properties, except for α-picolinic, nicotinic and quinolinic acids. They decrease in the following order: 30, 31, 28, 33 (see Table 66).

10. Esters of thiolbenzoates have about the same order of action and toxicity as benzoates but cause some dermatitis.

The B portion or alkyl chain of the alcohol involved in the ester has been the subject of considerable study. Local anesthetics in current use attest to the varied structures that have been found desirable for Part B. The chain not only may be branched and contain two tertiary amino groups, but may include one of the following general types:

STRUCTURES USED IN B PORTION

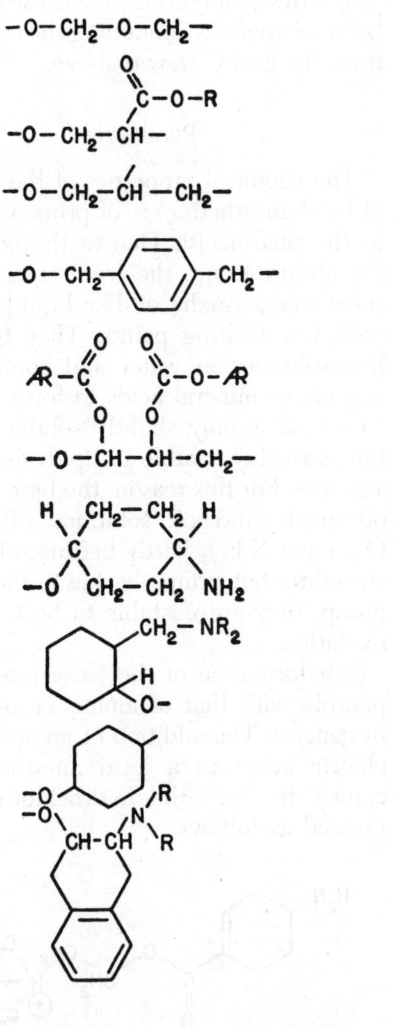

$$-O-CH_2-O-CH_2-$$

In general, the group $-CH_2CH_2CH_2-$ (propylene) provides the most active compounds, with $-CH_2CH_2-$ (ethylene) next. The group $-CH_2-$ makes the ester too irritant. Only a few compounds have been prepared where B is $(CH_2)_4$ or larger. However, indications are that there is slight increase in activity in chains containing more than 4 carbon atoms.

The C-portion represents the amine part of the amino alcohols used to esterify the aromatic acids. Here again, variation of structures is observed among the local anesthetics in use today. The advantage of the aliphatic amino nitrogen is its ability to form salts with inorganic acids, thereby providing water-soluble compounds. On this point, some workers have claimed variations in anesthetic activity, depending upon the acid salt. Procaine borate (q.v.) is an example, but there is some question as to whether or not it has any advantage over the hydrochloride. Another variation of the acid is employed, for example, in procaine nitrate to avoid incompatibility with silver salts. Butacaine is used as the sulfate because it is more soluble than the hydrochloride.

TABLE 66. SOME AROMATIC ACIDS THAT
HAVE BEEN ESTERIFIED FOR LOCAL
ANESTHETIC STUDIES

1. Benzoic Acid
2. Phenylacetic Acid
3. Phenylhydroxyacetic Acid
4. Phenylmethylacetic Acid
5. α-Phenylpropionic Acid
6. p-Toluic Acid
7. Piperonylic Acid
8. p-Hydroxybenzoic Acid
9. p-(β-Hydroxyethyl) benzoic Acid
10. p-(β-Ethoxyethyl) benzoic Acid
11. p-Hydroxy-m-aminobenzoic Acid
12. Alkoxybenzoic Acid
13. Aryloxybenzoic Acid
14. p-Ethylmercaptobenzoic Acid
15. p-Halobenzoic Acid
16. p-Aminobenzoic Acid
17. m-Aminobenzoic Acid
18. p-Alkylaminobenzoic Acid
19. p-Dialkylaminobenzoic Acid
20. p-Aminomethylbenzoic Acid
21. p-(Diethylaminoethyl) benzoic Acid
22. p-(Diethylaminoethoxy) benzoic Acid
23. p-Phenylbenzoic Acid
24. α-Naphthoic Acid
25. β-Naphthoic Acid
26. Phthalic Acid
27. 3-Aminophthalic Acid
28. Picolinic Acid
29. Nicotinic Acid
30. α-Carboxypyrrole
31. α-Carboxythiophene
32. α-Carboxyfuran
33. α-β-Dicarboxylquinoline
34. p-Aminothiolbenzoic Acid
35. Cinnamic Acid

The amino group always has one or both of its hydrogens substituted. Very little information is available on nonsubstituted amino-alcohol esters except that some are reported to be unstable. The β-aminoethyl

STRUCTURES USED FOR C PORTION

Pyrrolidino

Pyrrolino

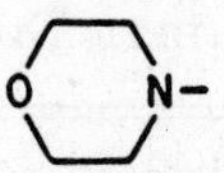

Morpholino

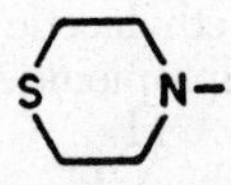

Thiomorpholino

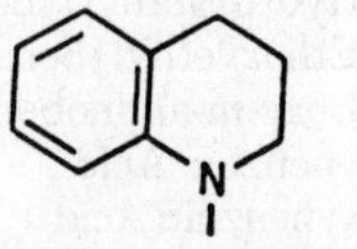

Tetrahydroquinolino

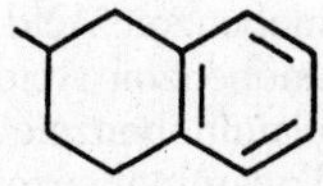

Tetrahydronaphthyl

p-aminobenzoate has been reported as having no anesthetic effect in a 5 per cent solution. The types that are most common are those in which R_1 and R_2 are the same, usually lower alkyl groups (C_2 to C_5). In general, the anesthetic property increases with the size of the alkyl groups, the maximum being at C_3 to C_4. The groups R_1 and R_2 do not have to be the same, but no advantage has been observed when this is true. R_1 and R_2 also may be identical and unsaturated groups or may be hydrogen and an alkyl group (naepaine). Very satis-

factory compounds are possible when R_1 and R_2, together with the nitrogen, form an aliphatic heterocyclic ring. The piperidine ring has proved to elicit strong anesthetic properties (piperocaine). Also studied have been compounds containing the cyclic structures for Part C shown above.

PROPERTIES

The chemical properties of the ester type of local anesthetics are of prime importance to the pharmacist. Due to the presence of the amino group, the free esters are basic substances, usually oil-like liquids or solids with low melting points. They form alkaline solutions in water and combine with organic or mineral acids to form salts. The free base is only slightly soluble in water but is readily soluble in lipids and organic solvents. For this reason, the base is used in ointments and oil solutions (Benzocaine Ointment N.F.). Partly because of the ester structure, but primarily due to the *p*-amino group, they are unstable to heat, light and oxidation.

Salt formation of the basic ester is comparable with that of ammonia and amines in general. The addition of an acid (hydrochloric acid) to a local anesthetic (procaine) to form the hydrochloride is expressed as follows:

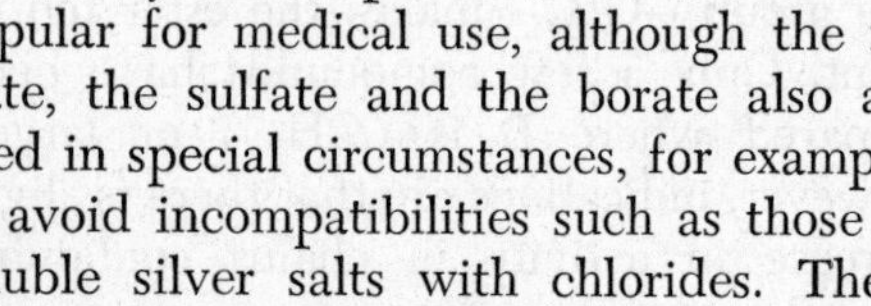

The hydrochloride, which occurs as a white, crystalline powder, is the salt most popular for medical use, although the nitrate, the sulfate and the borate also are used in special circumstances, for example, to avoid incompatibilities such as those of soluble silver salts with chlorides. These acid salts are very soluble in water or alcohol but are only slightly soluble in lipids and organic solvents.

In aqueous solution, they ionize as do the ammonium salts, producing a solution having a pH of 4 to 6. This varies, of course, with the local anesthetic base and the acid used. The free base is liberated when alkaline substances, such as hydroxides, carbonates and bicarbonates or any other normal salt that is alkaline, are added to a solution of the salt. Since local anesthetics are alkaloidlike, they often are precipitated by the alkaloid reagents.

MODE OF ACTION

The precise mode of action of local anesthetics is unknown. More information must first be obtained on the mechanism of the transmission of nerve impulses. It is felt that the function of a local anesthetic is not one of competition with a natural nerve substance. However, it is a blocking of the nerve impulse that occurs within the nerve cell. Although sensory fibers are blocked earlier than motor nerve fibers, it may be due to their relative size. A local anesthetic appears to anesthetize small fibers first. If the actions depend on the passage of the local anesthetic compound into the cell, this observation would support it.

The compounds also may function according to the Meyer-Overton law, since

ministered, gain entrance into the body by absorption and are destroyed by the liver.

Most of the compounds that have a local anesthetic action do not possess the property of vasoconstriction. However, cocaine does possess this property in sufficient degree so that actual contact with nervous tissue is prolonged. This lack of constriction of blood vessels speeds absorption and explains the short duration of anesthesia obtained from the synthetic compounds. In 1903, Braun suggested the addition of epinephrine hydrochloride or other vasoconstrictors to solutions of local anesthetics. This is now standard practice and concentrations of 1:25,000, 1:50,000 or 1:100,000 are the rule. Not only is anesthesia prolonged, but the rate of absorption is reduced, thus allowing time for detoxication before a toxic concentration can be built up in the circulation.

From time to time, some investigators endeavor to prepare a compound having local anesthetic and vasoconstrictive properties. Some are recorded in the literature, but, actually, the combination of these qualities is of questionable value. An example[5] is "Epicaine," a combination of a pressor residue (A) and an anesthetic residue (B).

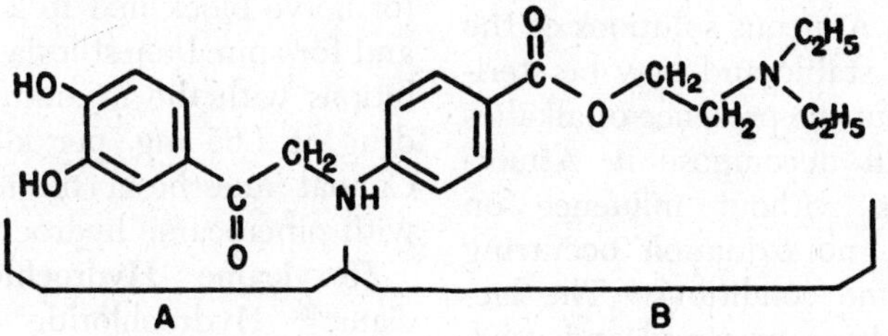

A   B

the free base is lipid soluble. Perhaps the base is liberated by the slightly alkaline fluids in the tissues.

It has been observed that a neutral or basic solution of procaine, obtained by using sodium bicarbonate, has a greater effectiveness than does a solution of the hydrochloride. This indicates that neutralization of the acid salt is, perhaps, a preliminary step to action.

Most local anesthetic drugs, when ad-

BENZOIC ACID DERIVATIVES

Eucaine Hydrochloride; Betaeucaine Hydrochloride; Betacaine, Benzamine. Eucaine was one of the first synthetic local anesthetics developed by Harries[6] soon after the elucidation of the cocaine molecule. It is one of the compounds resulting from the early attempts at simplification while retaining the cyclic structure of the alcohol moiety. On examination of the cocaine molecule,

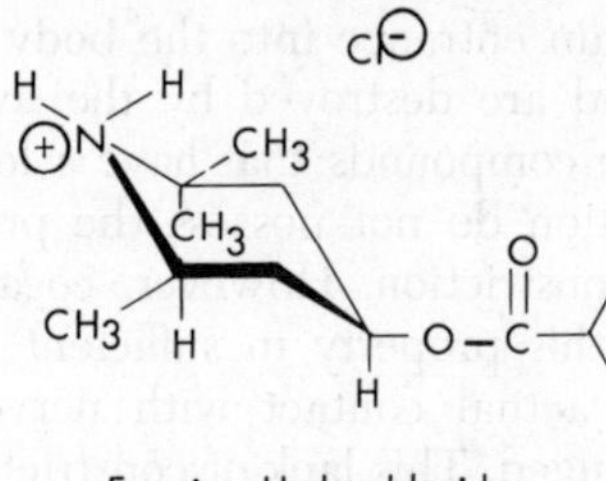

Eucaine Hydrochloride

that portion related to eucaine is observed readily.

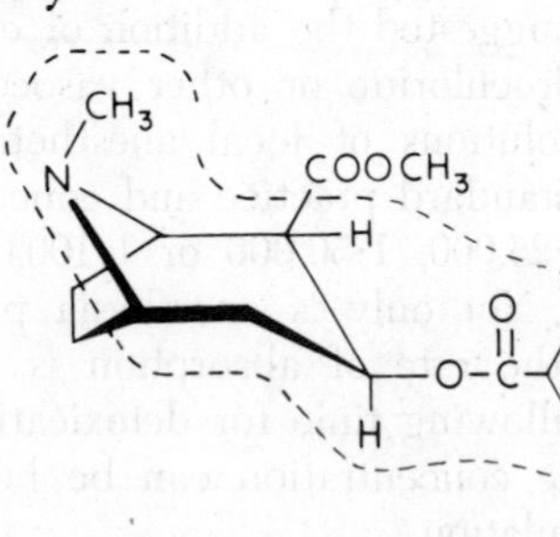

The part enclosed by a dashed line is essentially the basis of eucaine. The development of this local anesthetic confirmed in part the essential portion of the cocaine molecule responsible for local anesthetic action.

The hydrochloride of eucaine is an odorless, white, crystalline powder, stable in air. It is affected by light and soluble in water (1:30), alcohol (1:35) and chloroform but insoluble in ether. Aqueous solutions of the hydrochloride are stable and may be sterilized by boiling, but the presence of alkalies or carbonates will decompose it. Atmospheric oxygen is without influence on eucaine solutions, no oxidation occurring even under extreme conditions.[7] The lactate salt also has been prepared and used.

It has the advantage over cocaine of not being habit forming, not being a mydriatic and not being a vasoconstrictor, although it does come under stringent legal control. The action is little slower than that of cocaine, but it is less irritating. Since it is more toxic than procaine, procaine has become the drug of choice. Eucaine hydrochloride is applied in 1 to 5 per cent solutions by injection, instillation or perfusion.

Piperocaine Hydrochloride, Metycaine® Hydrochloride, the salt of 3-(2-methylpi-peridino)propyl benzoate, is a fine, white, odorless, crystalline powder that is soluble in water (1:15) or alcohol (1:4.5), freely soluble in chloroform and on the tongue gives a slightly bitter taste followed by a sense of numbness.

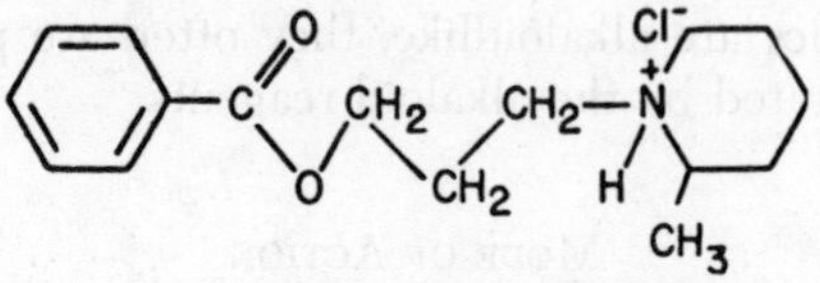

Piperocaine Hydrochloride

This is an ester of benzoic acid with the basic alcohol 3-(2-methylpiperidino) propanol. The compound contains an asymmetric carbon atom and is used as a racemic mixture. The solubility is similar to that of the hydrochlorides of other basic esters. Aqueous solutions are slightly acid; they are stable and can be sterilized by autoclaving at 115°.

Piperocaine hydrochloride is used as a local anesthetic topically, for infiltration anesthesia and as a spinal anesthetic. It is one third as toxic as cocaine. It is recommended for application to the eye in 2 to 4 per cent solutions; to the nose and the throat in 2 to 10 per cent solutions; for infiltration in 0.5 to 1.0 per cent solutions; for nerve block in 1 to 2 per cent solutions and for spinal anesthesia in 1.5 per cent solutions with the maximum quantity of the drug at 1.65 mg. per kilogram of weight. Caudal anesthesia often is accomplished with piperocaine hydrochloride.

Hexylcaine Hydrochloride N.F., Cyclaine® Hydrochloride is 1-(cyclohexyl-amino)-2-propanol benzoate (ester) hydrochloride. It is soluble in water (1:7) and

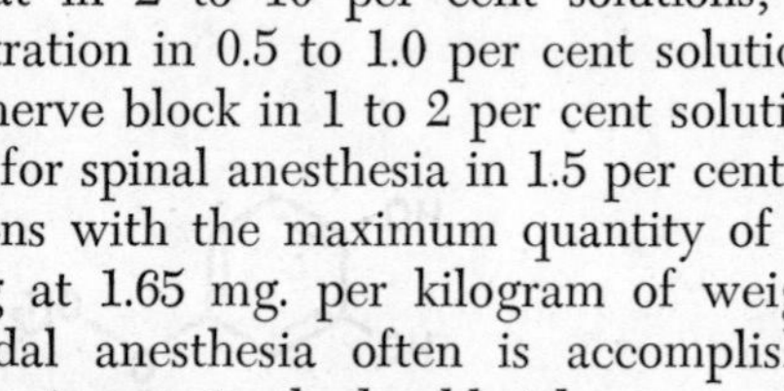

Hexylcaine Hydrochloride

freely soluble in alcohol and in chloroform. Solutions are stable to boiling and autoclaving. A 1 per cent solution has a pH of 4.4.

Hexylcaine has about the same toxicity as procaine and piperocaine; topical anesthesia is like cocaine and butacaine; for nerve block anesthesia, it is between butacaine and tetracaine. It has been used successfully in spinal anesthesia in a 2.5 per cent solution in 10 per cent glucose.

Category—local anesthetic.

Usual dose—topically or by injection, according to site and condition.

OCCURRENCE

Hexylcaine Hydrochloride Injection N.F.
Hexylcaine Hydrochloride Solution N.F.

Meprylcaine Hydrochloride N.F., Oracaine® Hydrochloride, is 2-methyl-2-(propylamino)-1-propanol benzoate (ester) hydrochloride. It is freely soluble in alcohol or water or in chloroform. A 2 per cent aqueous solution has a pH of 5.7, and is reported that such a solution can be sterilized by autoclaving without decomposition. Studies have shown that it is hydrolyzed in human serum 8 to 10 times as rapidly as procaine hydrochloride.[8] It is used primarily in dentistry in a 2 per cent solution as an infiltration and nerve block anesthetic.

Category—local anesthetic.

OCCURRENCE

Meprylcaine Hydrochloride and
Epinephrine Injection N.F.

Isobucaine Hydrochloride N.F., Kincaine® Hydrochloride, 2-(isobutylamino)-2-methyl-1-propanol benzoate (ester) hydrochloride, is a white crystalline solid, freely soluble in water but only sparingly soluble in isopropyl alcohol.

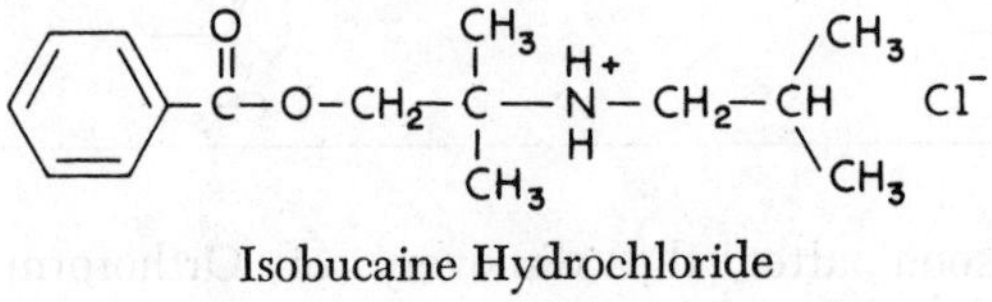

Isobucaine Hydrochloride

The pH of a 2 per cent solution is about 6.

Category—local anesthetic (dental).

OCCURRENCE

Isobucaine Hydrochloride and
Epinephrine Injection N.F.

Parethoxycaine Hydrochloride, Intracaine® Hydrochloride, is 2-diethylamino-ethyl *p*-ethoxybenzoate hydrochloride. It is an example of *p*-alkoxy substitution in the local anesthetic compounds of the ester type. It occurs as a white, crystalline water-soluble powder. Solutions become cloudy on exposure to air; for this reason solutions made extemporaneously from the sterile crystals should be used shortly after preparation.

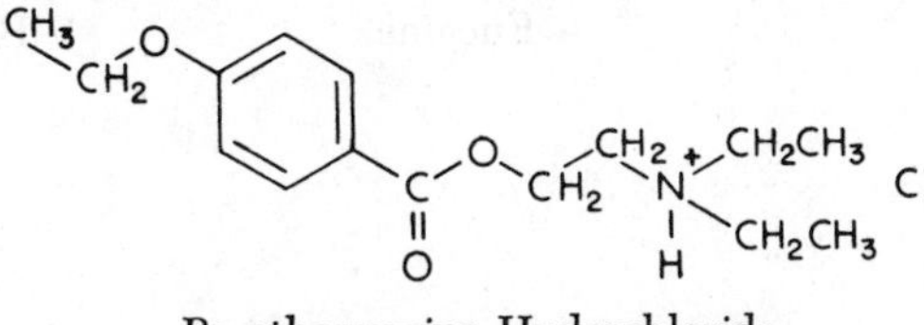

Parethoxycaine Hydrochloride

It is reported to possess about the same degree of potency as procaine; the uses are comparable with those of procaine. It is available in crystals for making solutions for injection, as 2 and 5 per cent ointment, and 2 and 5 per cent solution for injection.

Cyclomethycaine Sulfate, Surfacaine® Sulfate, is 3-(2-methylpiperidino)propyl *p*-cyclohexyloxybenzoate sulfate. It differs from piperocaine by having a cyclohexyloxy-group in the *p*-position of the benzoic acid moiety.

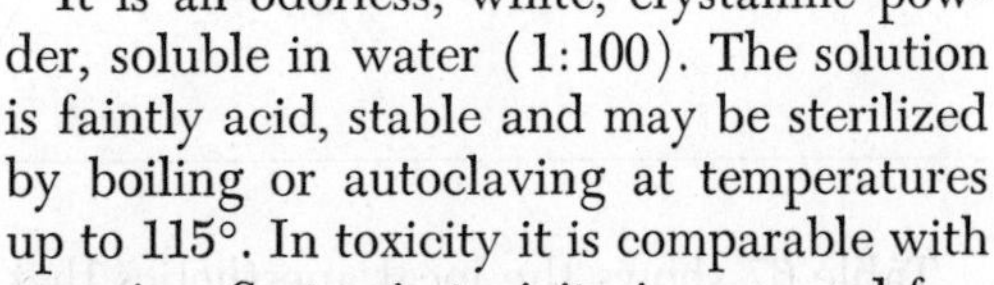

Cyclomethycaine
Sulfate

It is an odorless, white, crystalline powder, soluble in water (1:100). The solution is faintly acid, stable and may be sterilized by boiling or autoclaving at temperatures up to 115°. In toxicity it is comparable with procaine. Systemic toxicity is rare, and few allergic reactions have been observed.

Cyclomethycaine sulfate is an effective local anesthetic on damaged or diseased skin and on rectal mucous membrane. It is useful topically on burns (sunburn), abrasions and mucous membranes, and is applied most commonly in 0.25 to 1 per cent ointments or suppositories.

TABLE 67. BENZOIC ACID DERIVATIVES

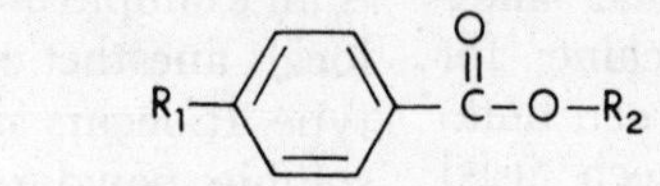

PROPRIETARY NAME	GENERIC NAME	R₁	R₂
	Eucaine	H	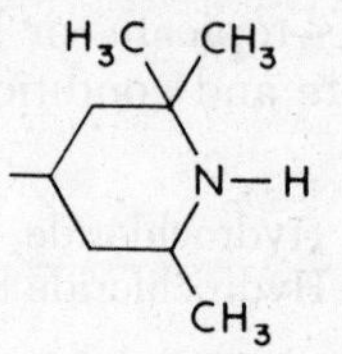
Metycaine	Piperocaine	H	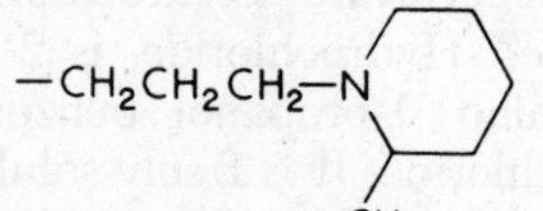 $-CH_2CH_2CH_2-N$ (piperidine ring with CH_3)
Cyclaine	Hexylcaine	H	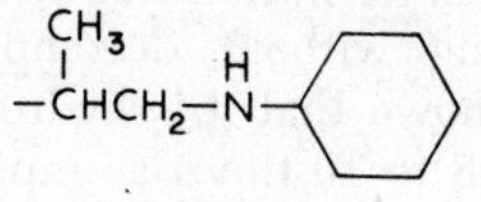 $-CHCH_2-N$ with CH_3 and H (cyclohexyl)
Oracaine	Meprylcaine	H	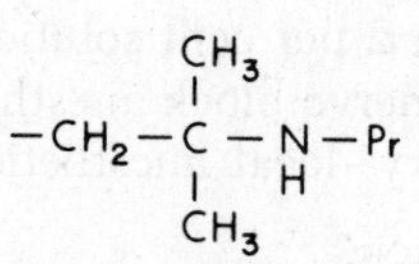 $-CH_2-C(CH_3)_2-N(H)-Pr$
Kincaine	Isobucaine	H	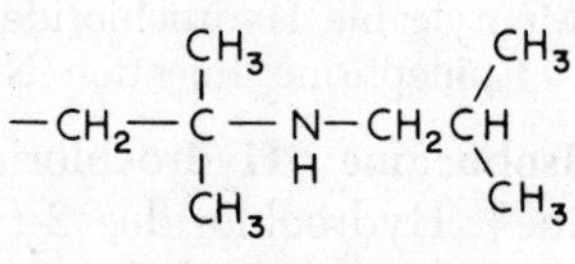 $-CH_2-C(CH_3)_2-N(H)-CH_2CH(CH_3)_2$
Intracaine	Parethoxycaine	C_2H_5O	$-CH_2CH_2N(C_2H_5)_2$
Surfacaine	Cyclomethycaine	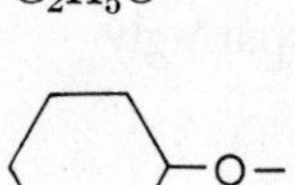 (cyclohexyl)—O—	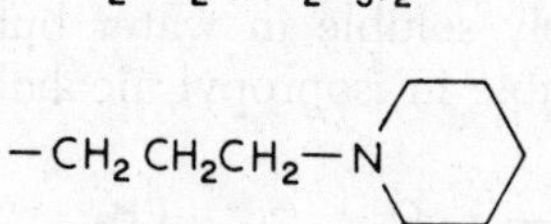 $-CH_2CH_2CH_2-N$ (piperidine ring with CH_3)

Table 67 shows the local anesthetics that are esters of benzoic acid.

p-AMINOBENZOIC ACID DERIVATIVES

Benzocaine N.F.; Anesthesin®; Orthesin; Parathesin®; Ethyl p-Aminobenzoate. The ester, ethyl aminobenzoate, was prepared soon after the discovery of Orthoform (q.v.).

$$C_6H_5CH_3 \xrightarrow[H_2SO_4]{HNO_3} p-NO_2C_6H_4CH_3 \xrightarrow{(O)}$$

$$p-NO_2C_6H_4COOH \xrightarrow[H_2SO_4]{C_2H_5OH} p-NO_2C_6H_4COOC_2H_5$$

$$\xrightarrow{H} p-NH_2C_6H_4COOC_2H_5$$

It occurs as a white, odorless, crystalline powder, stable in air. It is soluble in alcohol (1:5), ether (1:4), chloroform (1:2), fixed oils (1:40), glycerin, propylene glycol or mineral acids and insoluble in water (1:2,500) and alkaline solutions.

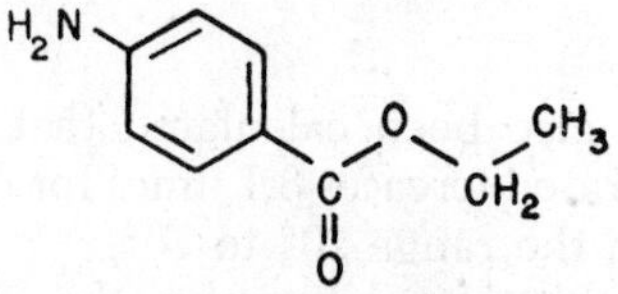

Ethyl Aminobenzoate

Benzocaine is destroyed when boiled with water, but its oil solutions may be boiled without change. Benzocaine forms a water-soluble complex with caffeine, which is much less subject to hydrolysis than benzocaine.[9] It forms a sticky mass with resorcinol and forms colored mixtures with bismuth subnitrate. Benzocaine has also been found to be incompatible with certain agents in a throat lozenge formulation. This was due to the reactivity of the aromatic amine group with aldehydes and with citric acid.[10] A hydrochloride is formed readily. However, it cannot be used because it is too irritant, because aqueous solutions are extensively hydrolyzed to an acid reaction.

Benzocaine, propyl aminobenzoate, butyl aminobenzoate and Orthoform represent a class of slightly soluble local anesthetics that are unsuited for injection. They are all absorbed slowly. They are nonirritant and nontoxic. Anesthesia on abraded skin and on mucous membrane is effective and long lasting.

Although ethyl aminobenzoate is nearly insoluble in aqueous media, there is sufficient absorption through abraded surfaces and mucous membranes so that it acts almost entirely on the nerve terminals. Cocaine, in contrast, is thought to affect the nerve trunk. Benzocaine is used for its local anesthetic action in powders up to 20 per cent, in ointments up to 20 per cent, in suppositories and for various types of surface pain and itching. Use is made of it in throat lozenges and in ulcerative conditions of the digestive tract. Often it is used orally to prevent nausea and vomiting with compounds such as aminophylline.

Category—local anesthetic.

For external use—topical, as a 1 to 20 per cent ointment to the skin.

	PER CENT
OCCURRENCE	BENZOCAINE
Benzocaine Ointment N.F.	5.0
Antipyrine and Benzocaine Solution N.F.	
Trimethobenzamide and Benzocaine Suppositories N.F.	

Propaesin, Propyl Aminobenzoate, is the ester from propyl alcohol and p-aminobenzoic acid. It has properties and uses very similar to those of benzocaine. This compound is present in some proprietary suppositories.

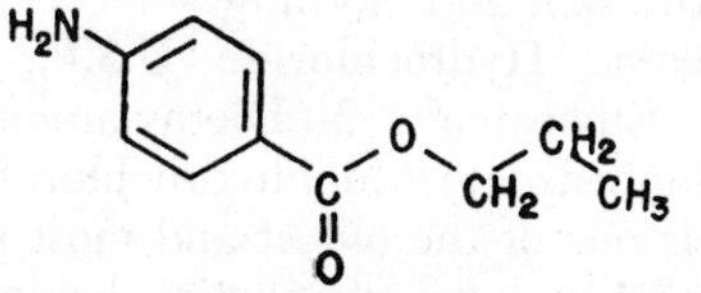

Propaesin

Butyl Aminobenzoate N.F., Butesin®, butyl p-aminobenzoate, is the butyl analog of benzocaine, using butyl alcohol in place of ethyl alcohol. It occurs as a white, crys-

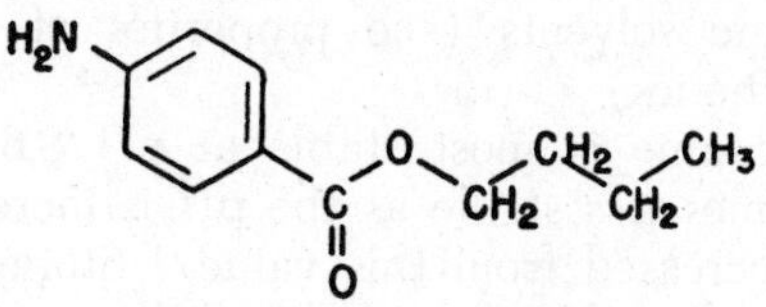

Butyl Aminobenzoate

talline powder that is tasteless and odorless. It is insoluble in water and miscible with absolute alcohol and the incompatibilities are the same as those for ethyl aminobenzoate. Butyl aminobenzoate is used in the same way as benzocaine and is claimed to be more effective; however, it is more toxic.

Isobutyl p-aminobenzoate is used in Diothoid® Suppositories. The amyl p-aminobenzoate compound is an ingredient in Ultracain® Ointment used for burns.

Category—local anesthetic.

Butamben Picrate, Butesin® Picrate, (n-butyl p-aminobenzoate)$_2$ trinitrophenol, is thought to contain 2 moles of butyl aminobenzoate and 1 mole of picric acid. It occurs as a yellow, amorphous powder that is slightly soluble in water (1:2,000) and

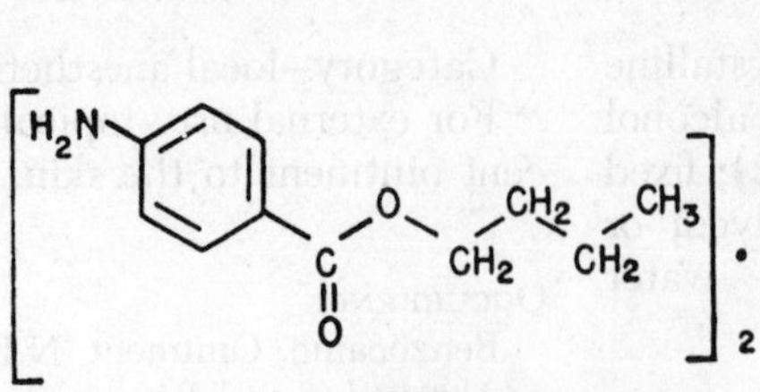

Butamben Picrate

soluble in fixed oils or organic solvents. Note that this is the reverse of all the other local anesthetics.

In a saturated aqueous solution, it is used in the eye, and, as a 1-per cent ointment, for burns and denuded areas of the skin. A disadvantage is that the yellow color will stain the skin and clothing.

Procaine Hydrochloride U.S.P.; Novocain®; Ethocaine®; 2-(Diethylamino)ethyl *p*-Aminobenzoate Monohydrochloride. Procaine is one of the oldest and most used of the synthetic local anesthetics, having been developed by Einhorn in 1906.

The free ester is an oil, but it is isolated and used as the hydrochloride salt. It occurs as an odorless, white crystalline powder that is stable in air, soluble in water (1:1), alcohol (1:30) but much less soluble in organic solvents (see properties of local anesthetics).

Procaine is most stable at pH 3.6 and becomes less stable as the pH is increased or decreased from this value.[11] Storage of buffered solutions at room temperature resulted in the following amounts of hydrolysis.

pH	Amount of Hydrolysis
3.7-3.8	0.5 to 1% in 1 year
4.5-5.5	1.0 to 1.5% in 1 year
7.5	1% in 1 day

Dosage forms are generally regarded as being satisfactory for use as long as not more than 10 per cent of the active ingredient has been lost and there is no increase in toxicity. The following times for a 10 per cent loss in potency at 20° are based on kinetic studies on procaine solutions.

pH	Time in Days
3.6	2,300
5.0	1,200
7.0	7

It has also been calculated that the hydrolysis rate increases 3.1 times for each rise of 10° in the range 20° to 70°.

The following data show the effect on buffered 2 per cent procaine hydrochloride solutions of autoclaving at 15 pounds pressure for 2 hours.[12]

pH Before and After Autoclaving	Per Cent of Original Assay
2.4	97.5
2.6	97.9
2.8	98.1
3.0	98.4
3.2	98.5
3.4	98.5
3.6	98.3
3.8	98.2
4.0	97.8

It was early found that neutral or slightly alkaline solutions of procaine had certain physiologic advantages over acid solutions in that there was less pain on injection, there was less pain after injection and less tissue damage, and, most important, the rate of onset of anesthesia was quicker, and a smaller quantity of procaine could be used.[13] Thus there is much in favor for the use of alkaline solutions. However, these solutions are very unstable and cannot be sterilized by autoclaving. The problem is even greater if epinephrine is to be used because it is still more unstable than procaine in alkaline solutions.

Although it has been recommended that procaine solutions be sterilized by using moist heat at 100° for 30 minutes on 3 successive days or at 75° to 80° for longer periods, it has been shown that autoclaving for the usual period at 120° is preferable to prolonged sterilization at 100°.[14] Between 80° and 120° the hydrolysis rate of procaine increases 2.5 times for each increase of 10°, while for killing bacterial spores the 10° increases are 4.6 times from 80° to 90°, 5.6 times from 90° to 100°, and 15 times over 100°.[15] Thus it can be seen readily that a short heat treatment at a higher temperature is more effective bacteriologically and less destructive chemically than prolonged heating at lower temperatures.

The procaine molecule is subject to oxidative decomposition also, but this is not a function of the ester linkage but of the aromatic amine portion. This type of breakdown can be controlled by nitrogen flushing of the solutions and by the addition of an antioxidant. The aromatic amine group will also undergo the diazo reaction with nitrous acid and α-naphthol, forming a red dye. This is characteristic of the local anesthetics having a primary aromatic amine group. In solutions with glucose, this functional group is responsible for the formation of procaine N-glucoside. There is no significant change in the clinical results, but there is the possibility of interference with the assay. It has been found that procaine forms a soluble complex with sodium carboxymethyl cellulose.[16] However, sodium chloride displaces it to a great extent, so, here again, there probably would be no problem physiologically, but there may be an analytical problem.

Procaine hydrochloride solutions are not effective on intact skin or mucous membrane, but they act promptly when used by infiltration. This action can be prolonged by the concurrent use of epinephrine to slow the release into the bloodstream where the procaine is rapidly hydrolyzed and inactivated.

In 1940, procaine hydrochloride was first used intravenously for pruritus associated with jaundice and has since been administered for the pain of burns, arthritis and many other conditions. It is eliminated completely from the blood in 20 minutes after infusion. A 0.1 per cent solution is available and is used in a dosage of 4 mg. per kilogram per 20 minutes.

It has been observed that procaine hydrochloride solutions with penicillin form the insoluble procaine penicillin (q.v.). It is this low solubility of about seven parts per thousand that accounts for the prolonged action of this penicillin salt. The penicillin is slowly released from the intramuscular depot.

Procaine is also used to form a sparingly soluble salt of heparin.

Category—local anesthetic.

Usual dose—infiltration, 50 ml. of a 0.5 per cent solution; peripheral nerve block, 25 ml. of a 1 or 2 per cent solution; epidural, 25 ml. of a 1.5 per cent solution.

Usual dose range—up to 160 ml. of a 0.5 per cent solution, 70 ml. of a 1 per cent solution, or 30 ml. of a 2 per cent solution.

OCCURRENCE

Procaine Hydrochloride Injection U.S.P.
Sterile Procaine Hydrochloride U.S.P.
Procaine Hydrochloride and Levonordefrin Injection N.F.
Procaine and Phenylephrine Hydrochlorides Injection N.F.
Procaine and Propoxycaine Hydrochlorides and Levarterenol Bitartrate Injection N.F.
Procaine and Propoxycaine Hydrochlorides and Levonordefrin Injection N.F.
Procaine and Tetracaine Hydrochlorides and Levarterenol Bitartrate Injection N.F.
Procaine and Tetracaine Hydrochlorides and Levonordefrin Injection N.F.
Procaine, Tetracaine and Phenylephrine Hydrochlorides Injection N.F.

Procaine Borate, Borocaine®, is a borate salt prepared by refluxing procaine base with boric acid in acetone. The final product (a complex) contains 1 mole of procaine base to 5 moles of metaboric acid (HBO_2), resulting in crystals containing about 51 per cent of procaine base. The boric acid evidently loses 1 molecule of water during the refluxing to form the metaboric acid complex.

Procaine borate solution (pH 8.4) has a greater local anesthetic action than a solution of procaine hydrochloride (pH 5.6), yet at pH 8 they are of equal potency. This would indicate that the efficiency of the borate salt is dependent on its alkalinity rather than being due to any specific effect of the borate radical.[17]

Procaine Nitrate possesses the single advantage over other procaine salts of not forming a precipitate with soluble silver salts. Other salts that are on the market but possess no advantages are procaine butyrate

(Probutylin), and procaine ascorbate (Scorbacaine).

Isocaine is marketed as the hydrochloride of di-isopropylaminoethyl *p*-aminobenzoate, which, for all pharmaceutic purposes, has the same properties as procaine. A product utilizing it is Proctocaine.

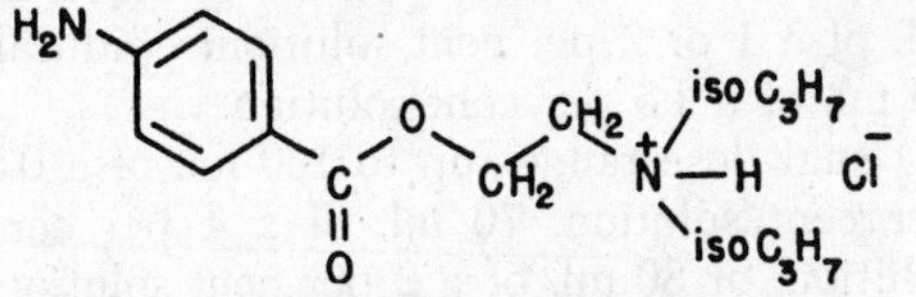

Isocaine Hydrochloride

Chloroprocaine Hydrochloride N.F., Nesacaine® Hydrochloride, 2-(Diethylamino)ethyl 4-Amino-2-chlorobenzoate Monohydrochloride. This compound is similar in chemical structure and pharmacologic activity to procaine hydrochloride. Its potency is about twice that of procaine, while the toxicity of the two compounds is similar. In the pH range of 4 to 8 it is hydrolyzed faster than procaine. Chloroprocaine also is hydrolyzed by plasma about 4 times faster than procaine; thus the possibility of the accumulation of toxic amounts in the patient is limited.[18] With therapeutic doses the concentration of liberated cleavage product is innocuous. Chloroprocaine hydrochloride is used by injection and is supplied in 1, 2 and 3 per cent solutions.

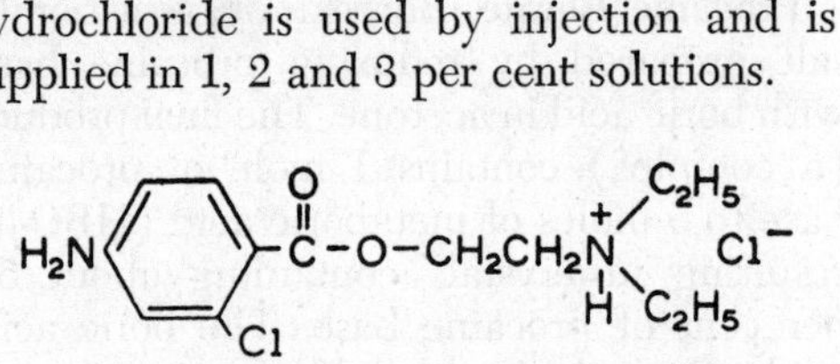

Chloroprocaine Hydrochloride

Category—local anesthetic.

Usual dose—infiltration, 100 ml. of 0.5 per cent solution; peripheral nerve block, 50 ml. of 1 per cent solution.

Usual dose range—up to 1 g. as a 0.5 to 3 per cent solution.

OCCURRENCE

Chloroprocaine Hydrochloride
   Injection N.F.

Benoxinate Hydrochloride; Dorsacaine® Hydrochloride; 2-Diethylaminoethyl 4-Amino-3-*n*-butoxybenzoate Hydrochloride. It is a white, odorless, crystalline powder that is soluble in water, alcohol and chloroform but insoluble in ether. The melting point is 157° to 160°. The crystals are stable in air

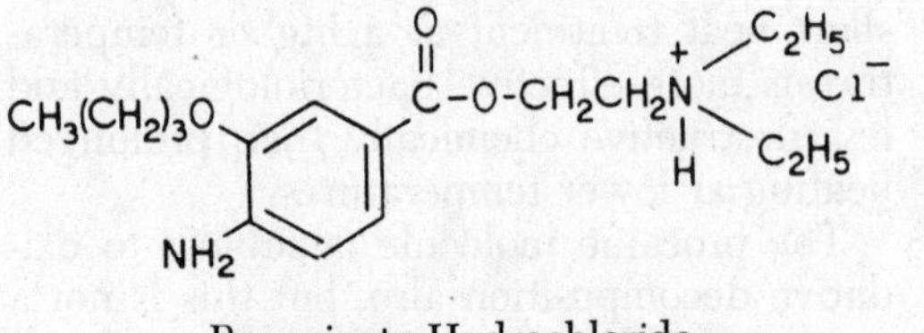

Benoxinate Hydrochloride

and not affected by light or heat. An aqueous solution possesses a pH of 4.5 to 5.2.

The chemical properties are similar to those of procaine except that the 3-butoxy group appears to stabilize the molecule to hydrolysis. This is in marked contrast with the 2-chloroprocaine which is hydrolyzed more readily than procaine. From kinetic studies the following times have been calculated for a 10 per cent loss in potency when stored at 20°.[19]

pH	Time in Days
3.5	4800
4.0	8800
4.5	5300
5.0	1860
5.5	590
6.0	186
6.5	59
7.0	19
7.5	5.9

Besides being a local anesthetic, this agent possesses bacteriostatic properties. It is employed primarily in ophthalmology as a 0.4 per cent solution. The duration of anesthesia is between 20 and 30 minutes. Applications do not cause any significant irritation, constriction or dilation of the pupil, nor any noticeable light sensitivity or symptoms indicating that absorption into the system has taken place.

Propoxycaine Hydrochloride N.F.; Blockain® Hydrochloride; Ravocaine® Hydrochloride; 2-(Diethylamino)ethyl 4-Amino-2-propoxybenzoate   Monohydrochloride.

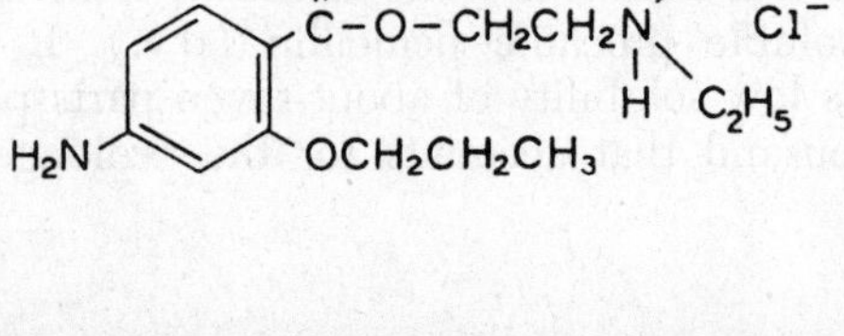

This compound is a white, odorless crystalline solid. It is soluble in water to the extent of at least 20 per cent. The pH of a 1 per cent solution is 5.5, which when adjusted to pH 7 does not precipitate. It appears that the 2-propoxy group labilizes the ester group in much the same way as the 2-chloro group in chloroprocaine, because the solutions are not to be autoclaved.

Propoxycaine hydrochloride is used by injection as a 0.5 per cent solution without vasoconstrictors, for nerve block and filtration anesthesia. The duration of anesthesia is claimed to be twice as long as that with procaine.

Category—local anesthetic.

OCCURRENCE
Propoxycaine Hydrochloride Injection N.F.

Tetracaine N.F., Pontocaine®, 2-(dimethylamino)ethyl *p*-(butylamino)benzoate, is a white to light yellow, waxy solid. It is only very slightly soluble in water, but is used because of its solubility in lipid substances. It must be protected from light.

Tetracaine is so named because it contains two groups of 4 (tetra) carbon atoms each. It differs from the other local anesthetics in having a substituent (butyl) replace a hydrogen on the *p*-amino nitrogen of *p*-aminobenzoic acid.

Category—local anesthetic.

For external use—topically as a 0.5 per cent ointment to the conjunctiva.

OCCURRENCE
Tetracaine Ophthalmic Ointment N.F.

Tetracaine Hydrochloride U.S.P., Pontocaine® Hydrochloride, 2-(dimethylamino)-ethyl *p*-(butylamino)benzoate monohydro-

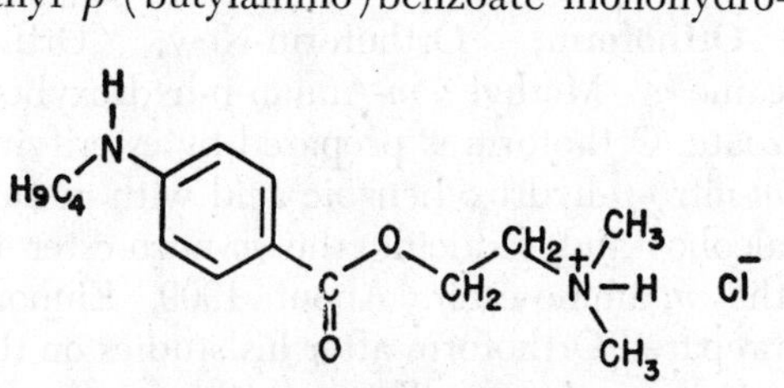

Tetracaine Hydrochloride

chloride, is very soluble in water and soluble in alcohol. Its solutions are more stable to hydrolysis than are procaine solutions. The enzymatic hydrolysis rate in human plasma is about one third that of procaine.[18] An aqueous solution may be sterilized by boiling. If it is made neutral or slightly basic, decomposition results.

Tetracaine resembles procaine for infiltration anesthesia and approaches the effectiveness of cocaine for topical use.

Category—local anesthetic; topical anesthetic.

Usual dose—subarachnoid, 2 ml. of a 0.5 per cent solution.

Usual dose range—1 to 3 ml.

For external use—topically, 0.1 ml. of a 0.5 per cent solution to the conjunctiva repeated at 5 to 10 minute intervals if needed; 1 ml. of a 2 per cent solution in nose and throat.

OCCURRENCE
Tetracaine Hydrochloride Injection U.S.P.
Sterile Tetracaine Hydrochloride U.S.P.

Butacaine Sulfate, Butyn® Sulfate, 3-Di-*n*-butylaminopropyl *p*-Aminobenzoate Sulfate. This compound is similar to procaine except that the amino alcohol used in synthesis is di-*n*-butylaminopropanol. The sul-

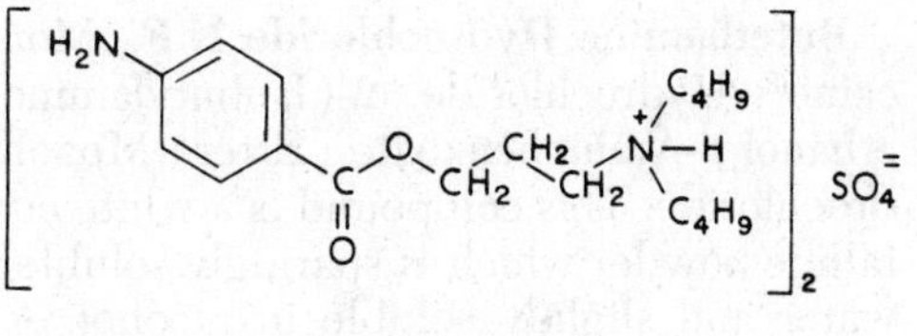

Butacaine Sulfate

fate is an odorless, white, crystalline powder that is tasteless and has the same solubilities and incompatibilities as procaine. In this case, the sulfate salt is prepared and used because of the low solubility of the hydrochloride salt. It resembles cocaine in action and, intravenously, has the same toxicity; however, it is more toxic on subcutaneous use. Today, it largely has replaced cocaine because it possesses all the advantages and none of the disadvantages of the natural alkaloid. Solutions may be sterilized by boiling and are effective on mucous membrane and in the eye. It is the local anesthetic of choice for use in the eye, since it has few side-effects and does not cause mydriasis. Usually, a 2 per cent solution is applied topically for use in the eye, the nose and the throat.

TABLE 68. *p*-AMINOBENZOIC ACID DERIVATIVES

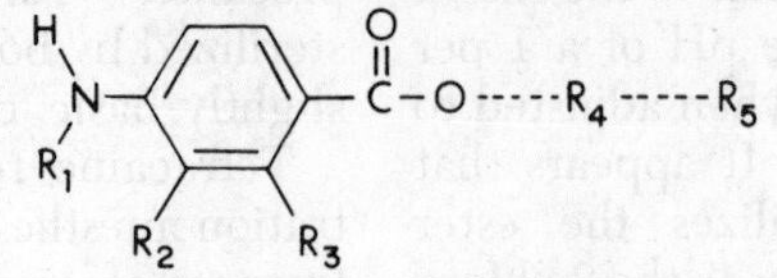

PROPRIETARY NAME	GENERIC NAME	R_1	R_2	R_3	R_4	R_5
Anesthesin	Benzocaine	H	H	H	$-C_2H_5$	————
	Propaesin	H	H	H	$-CH_2CH_2CH_3$	————
Butesin	Butyl Amino-benzoate	H	H	H	$-CH_2CH_2CH_2CH_3$	————
Novocaine	Procaine	H	H	H	$-CH_2CH_2-$	$-N(C_2H_5)_2$
	Isocaine	H	H	H	$-CH_2CH_2-$	$-N(iC_3H_7)_2$
Nesacaine	Chloroprocaine	H	H	Cl	$-CH_2CH_2-$	$-N(C_2H_5)_2$
Dorsacaine	Benoxinate	H	BuO—	H	$-CH_2CH_2-$	$-N(C_2H_5)_2$
Blockain	Propoxycaine	H	H	PrO—	$-CH_2CH_2-$	$-N(C_2H_5)_2$
Pontocaine	Tetracaine	Bu	H	H	$-CH_2CH_2-$	$-N(C_2H_5)_2$
Butyn	Butacaine	H	H	H	$-CH_2CH_2CH_2-$	$-N[C_4H_9 (n)]_2$
Monocaine	Butethamine	H	H	H	$-CH_2CH_2-$	$-N$ $\overset{H}{\underset{C_4H_9 (i)}{\diagup}}$

Butethamine Hydrochloride N.F., Mono-caine® Hydrochloride. 2-(Isobutylamino)-ethanol *p*-Aminobenzoate (Ester) Monohy-drochloride. This compound is a white crys-talline powder which is sparingly soluble in water and slightly soluble in alcohol. A 1 per cent solution has a pH of 5 and is stable in air. Its anesthetic and toxic activities are greater than those of procaine, to the extent that a 1.5 per cent solution is equivalent to a 2 per cent solution of procaine. Butetha-mine hydrochloride solutions are used by injection and are not recommended for topical use. Usually, 1 to 2 per cent solutions are used. It is used a great deal for nerve block anesthesia in dentistry.

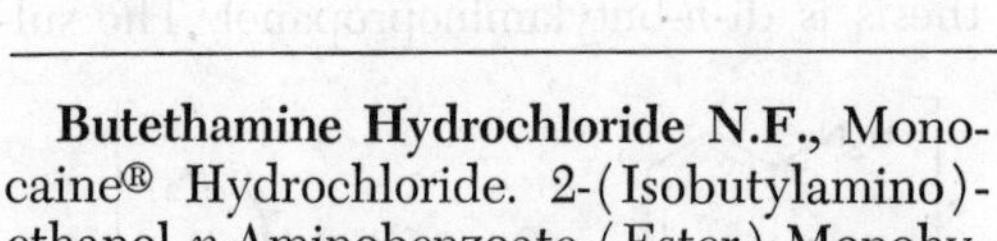

Butethamine Hydrochloride

Category—local anesthetic (dental).

Usual dose—local injection 1.8 to 2.2 ml. of a 1.5 per cent solution with epinephrine

1:100,000 or a 2 per cent solution with epinephrine 1:50,000.

OCCURRENCE

Butethamine Hydrochloride and Epinephrine Injection N.F.

Table 68 shows the local anesthetics that are esters of *p*-aminobenzoic acid.

m-AMINOBENZOIC ACID DERIVATIVES

Orthoform; Orthoform-New; Ortho-caine®; Methyl *m*-Amino-*p*-hydroxyben-zoate. Orthoform is prepared by esterifying *m*-nitro-*p*-hydroxybenzoic acid with methyl alcohol and reducing the *m*-nitro-ester to the *m*-amino-ester. About 1909, Einhorn prepared Orthoform after his studies on the cocaine molecule. This was the forerunner of esters from *p*-aminobenzoic acid, which have become very useful as local anesthet-ics. The original "Orthoform" as synthesized by Einhorn was a *p*-amino-*m*-hydroxy com-pound.

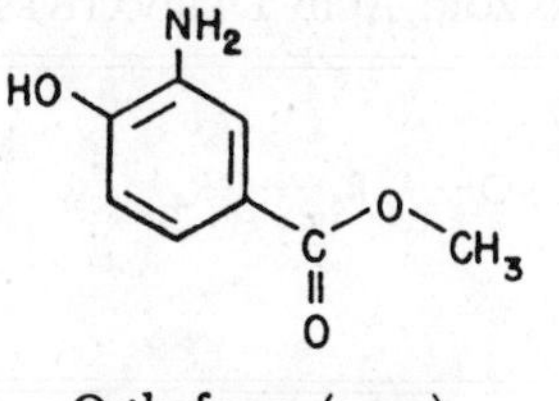

Orthoform (new)

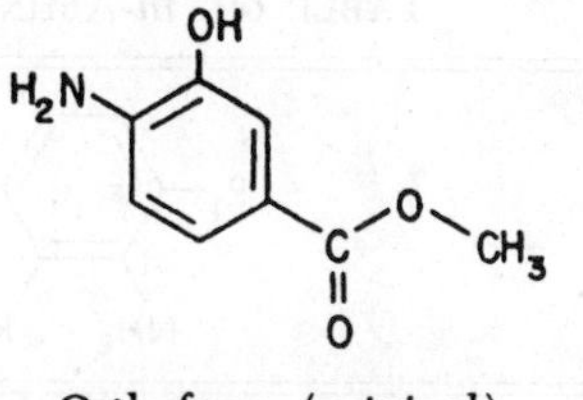

Orthoform (original)

The ester occurs as a fine, white, crystalline powder that is tasteless and odorless; it is insoluble in water but soluble in organic solvents.

The use of Orthoform as an anesthetic is restricted principally to powders and ointments because of its low order of solubility. It is ineffective on unbroken skin.

Metabutethamine Hydrochloride N.F., Unacaine® Hydrochloride, is 2-(isobutylamino)ethanol *m*-aminobenzoate (ester) monohydrochloride. It differs from other local anesthetics in being an ester of *m*-aminobenzoic acid. It appears to have a better ratio of potency to toxicity than butethamine or procaine.

It occurs as a white, crystalline salt soluble in water (1:10). Aqueous solutions are a slightly yellowish color when first prepared and deepen on standing. It is stable to heat sterilization and autoclaving. There is no precipitate formed with penicillin, as in the case of procaine.

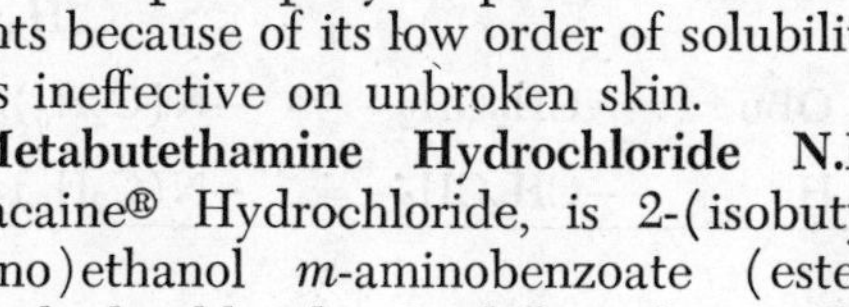

It is used primarily for infiltration and block anesthesia in dentistry.

Category—local anesthetic (dental).

OCCURRENCE

Metabutethamine Hydrochloride and
  Epinephrine Injection N.F.

Metabutoxycaine Hydrochloride, Primacaine® Hydrochloride is 2-(diethylamino)-ethyl 3-amino-2-butoxybenzoate hydrochloride. It occurs as a white crystalline solid and is very soluble in water and in alcohol. Solutions can be sterilized by autoclaving.

This compound differs from the procaine series by having the amino group in the *m*-position. It has been found to have

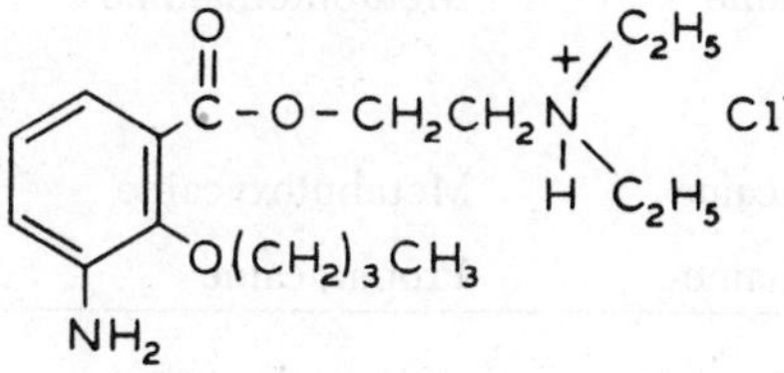

Mebutoxycaine Hydrochloride

greater anesthetic potency than procaine and is used in a 1.5 per cent solution. It is a short-acting local anesthetic for dental use.

Proparacaine Hydrochloride U.S.P., Ophthaine®, is 2(diethylamino)ethyl 3-amino-4-propoxybenzoate monohydrochloride. The

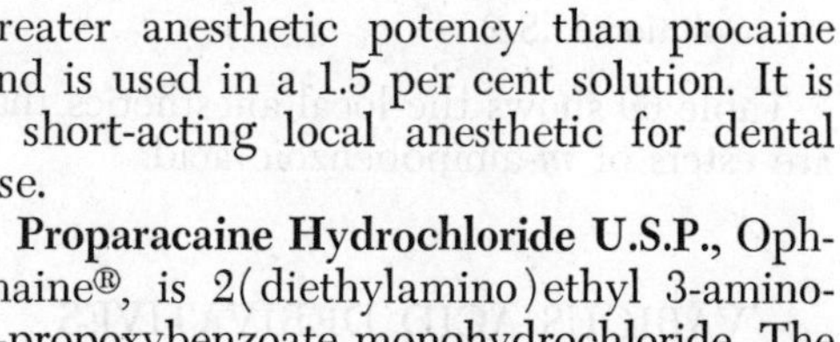

molecule is similar in structure to metabutoxycaine except that the alkoxy group is in the *p*-rather than the *o*-position. It is used topically in ophthalmology as a 0.5 per cent aqueous solution, with glycerin as a stabilizer and chlorobutanol and benzalkonium chloride as preservatives. It is slightly more potent than an equal amount of tetracaine. Solutions will discolor in the presence of air, and when discolored should not be used.

Category—topical anesthetic (ophthalmic).

For external use—topically, 0.05 ml. of a 0.5 per cent solution to the conjunctiva, repeated at 5 to 10 minute intervals if needed.

TABLE 69. *m*-AMINOBENZOIC ACID DERIVATIVES

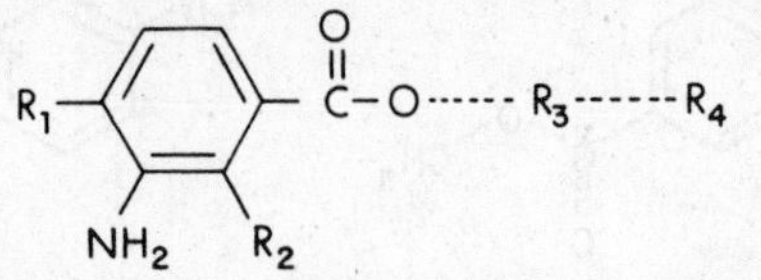

PROPRIETARY NAME	GENERIC NAME	R_1	R_2	R_3	R_4
	Orthoform-New	HO	H	$-CH_3$	————
Unacaine	Metabutethamine	H	H	$-CH_2CH_2-$	$-N\begin{smallmatrix}H\\ \\C_4H_9\ (i)\end{smallmatrix}$
Primacaine	Metabutoxycaine	H	OBu	$-CH_2CH_2-$	$-N(C_2H_5)_2$
Ophthaine	Proparacaine	OPr	H	$-CH_2CH_2-$	$-N(C_2H_5)_2$

OCCURRENCE

Proparacaine Hydrochloride Ophthalmic Solution U.S.P.

Table 69 shows the local anesthetics that are esters of *m*-aminobenzoic acid.

VARIOUS ACID DERIVATIVES

γ-Diethylaminopropyl Cinnamate Hydrochloride; Apothesine® Hydrochloride.

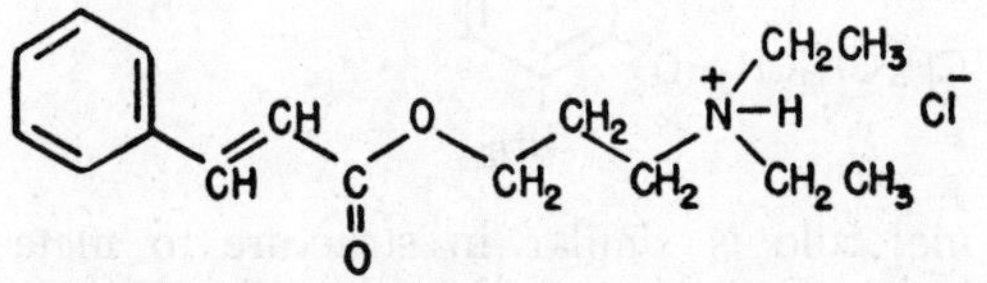

This is an ester of cinnamic acid and an amino alcohol similar to that used in procaine, γ-diethylaminopropanol. Introduced in 1916, it was the first local anesthetic developed in the United States. It occurs as odorless, white crystals that are soluble in water and alcohol. This structure is the only usable ester type of local anesthetic in which the aryl group is not directly attached to the carboxyl group. If β-phenylpropionic acid is used, the ester has no local anesthetic effect; therefore, the conjugated double bond appears to be necessary.

The ester is slightly more toxic than pro-caine. Its solutions may be sterilized by boiling, and it is used in 0.5 to 2 per cent concentration, similarly to the way procaine is used.

Biphenamine Hydrochloride is β-diethylaminoethyl 3-phenyl-2-hydroxybenzoate hydrochloride. It is a white, crystalline ma-

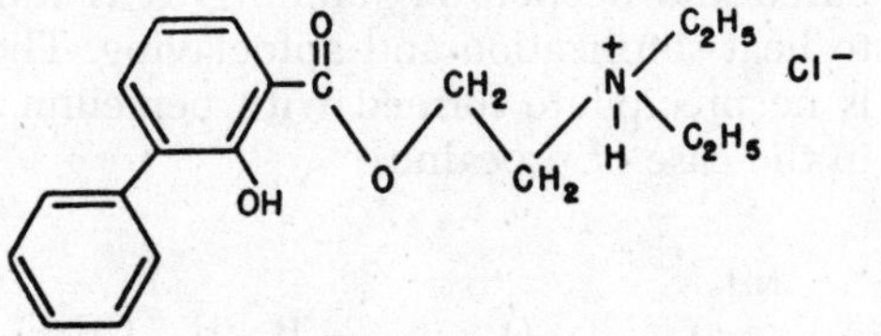

terial which is water soluble and has the usual properties of salts of aminoesters. It possesses marked fungicidal and bactericidal properties and is used for this purpose as a 1 per cent concentration in a shampoo (Alvinine).

Piridocaine Hydrochloride, Lucaine® Hydrochloride, is β-(2-piperidyl)ethyl *o*-aminobenzoate hydrochloride. It occurs as a white, crystalline, odorless, nonhygroscopic powder. It is soluble in water (2.5%), and at 37° a 4 per cent solution may be obtained. In aqueous solutions, the pH range is 6.2 to 6.8. Alkalies will precipitate the free base from aqueous solutions. A solution of the local anesthetic may be autoclaved at 15 pounds pressure (120°) for 20 minutes. Usually, the compound and its solutions are stable to light, but prolonged exposure to direct sunlight should be avoided.

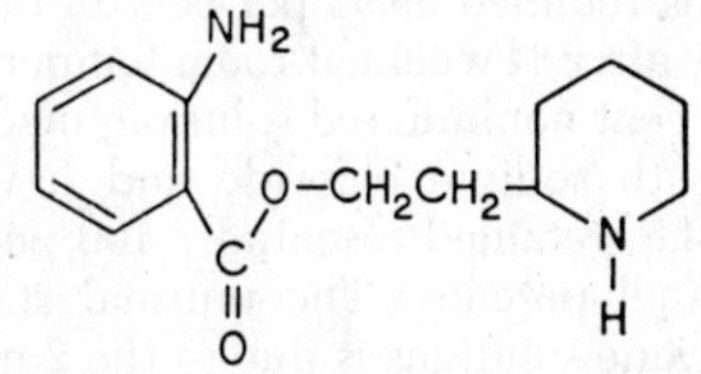

Piridocaine hydrochloride solutions are twice as effective on the rabbit's cornea as cocaine hydrochloride and eight times as

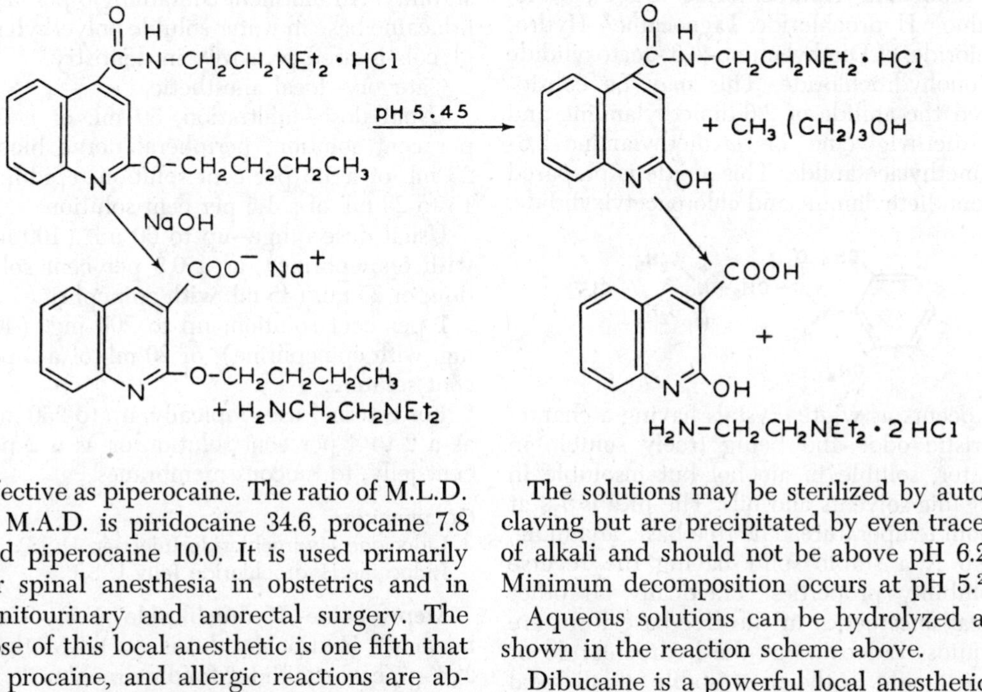

effective as piperocaine. The ratio of M.L.D. to M.A.D. is piridocaine 34.6, procaine 7.8 and piperocaine 10.0. It is used primarily for spinal anesthesia in obstetrics and in genitourinary and anorectal surgery. The dose of this local anesthetic is one fifth that of procaine, and allergic reactions are absent. It is supplied as crystals (lyophilized) in ampuls of 20 and 30 mg.

AMIDES

Dibucaine N.F., Nupercaine®, 2-butoxy-N-[(2-diethylamino)ethyl]cinchoninamide, is a colorless, somewhat hygroscopic powder which darkens on exposure to light. It is used in solution or as an ointment.

Category—local anesthetic.

Application—topically as a 0.5 per cent cream or a 1 per cent ointment several times daily.

OCCURRENCE

Dibucaine Cream N.F.
Dibucaine Ointment N.F.

Dibucaine Hydrochloride U.S.P., Nupercaine® Hydrochloride, Percaine®, 2-butoxy-N-[(2-diethylamino)ethyl]cinchoninamide, monohydrochloride, is a white, hygroscopic, crystalline powder that is freely soluble in water or alcohol. It is prepared from 2-hydroxycinchoninic acid by action of phosphorus pentachloride, condensation of the product with diethylethylenediamine and finally heating with sodium butylate.

The solutions may be sterilized by autoclaving but are precipitated by even traces of alkali and should not be above pH 6.2. Minimum decomposition occurs at pH 5.[20]

Aqueous solutions can be hydrolyzed as shown in the reaction scheme above.

Dibucaine is a powerful local anesthetic, being about five times as active as cocaine by injection and at least twenty times as strong when applied to the cornea, but it is also much more poisonous, both subcutaneously and intravenously.

Category—local anesthetic.

Usual dose—subarachnoid, 10 ml. of a 0.067 per cent (1 in 1500) solution or 1.5 ml. of a 0.5 per cent (1 in 200) solution.

Usual dose range—5 to 18 ml. of a 0.067 per cent (1 in 1500) solution or 0.5 to 2 ml. of a 0.5 per cent (1 in 200) solution.

OCCURRENCE

Dibucaine Hydrochloride Injection U.S.P.

Lidocaine U.S.P., Xylocaine®, is 2-(diethylamino)-2',6'-acetoxylidide. It is a white to off-white crystalline powder with a characteristic odor. It is stable in air. It is practically insoluble in water and is used because of its good solubility in lipid materials.

Category—topical anesthetic.

For external use—topically, as 2.5 per cent ointment or a solution to mucous membranes.

OCCURRENCE
Lidocaine Ointment U.S.P.

Lidocaine Hydrochloride U.S.P., Xylocaine® Hydrochloride, Lignocaine® Hydrochloride, 2-Diethylamino-2',6'-acetoxylidide Monohydrochloride. This may be considered the anilide of 2,6-dimethylaniline and N-diethylglycine or ω-diethylamino-2,6-dimethylacetanilde. This amide is prepared from diethylamine and chloroacetylxylidide.

It occurs as white crystals having a characteristic odor and being freely soluble in water, soluble in alcohol but insoluble in organic solvents and oils. The pKa is 7.9 at room temperature. The free base, an amide, also is a stable solid having the reverse solubility properties. Practically all other local anesthetics in the free-base form are liquids. Lidocaine as the hydrochloride or free amide is the most stable of all local anesthetics known to date. The ester-type local anesthetics are often prone to hydrolysis in aqueous solution, and solutions of dibucaine (an amide) must be stored in alkali-free glass containers.

Thus the clinical advantages over procaine and other ester-type local anesthetics probably would not have led to extensive use of lidocaine had it not been for its extreme resistance to hydrolysis. Two-per cent solutions buffered at pH 7.3 retained 99.95 per cent of original potency after autoclaving at 115° for 3 hours; however, the solutions became turbid on heating, which was attributed to a change of the

dissociation constant of water at the elevated temperature, because, on cooling, the solution became clear again.[21] The same solutions retained 99.98 per cent of original potency after 84 weeks at room temperature; a 2 per cent nonbuffered solution, made isotonic with sodium chloride and having a pH of 4.8, retained essentially 100 per cent of original potency. The unusual stability of lidocaine solutions is due to the 2 methyl groups ortho to the amide linkage.

In application, lidocaine is similar to procaine but has about twice the potency, a more rapid onset of action, and greater stability. An ointment containing 5 per cent lidocaine base in water-soluble polyethylene glycols is used primarily in dentistry.

Category—local anesthetic.

Usual dose—infiltration, 50 ml. of a 0.5 per cent solution; peripheral nerve block, 25 ml. of a 1.5 per cent solution; epidural, 15 to 25 ml. of a 1.5 per cent solution.

Usual dose range—up to 60 ml. (100 ml. with epinephrine), as a 0.5 per cent solution; or 27 ml. (45 ml. with epinephrine) as a 1 per cent solution; up to 200 mg. (400 mg. with epinephrine), or 20 ml. of a 2 per cent solution.

For external use—topically, up to 250 mg. as a 2 to 4 per cent solution or as a 2 per cent jelly, to mucous membranes.

OCCURRENCE
Lidocaine Hydrochloride Injection U.S.P.
Lidocaine Hydrochloride Jelly U.S.P.

Mepivacaine Hydrochloride N.F., Carbocaine® Hydrochloride, is (±)-1-Methyl-2',6'-pipecoloxylidide Monohydrochloride. It occurs as a white, crystalline solid. It is an analog of lidocaine and shows the same stability pattern; solutions are highly resistant to hydrolysis and can be autoclaved several times with no appreciable breakdown.

Mepivacaine Hydrochloride

The drug is used as the racemic mixture,

since the two optical isomers have been tested and found to have the same toxicity and potency.[22] Like lidocaine, it may be used with epinephrine but it also has the property of satisfactory action without it.

Category—local anesthetic.

OCCURRENCE

Mepivacaine Hydrochloride Injection N.F.
Mepivacaine Hydrochloride and
Levonordefrin Injection N.F.

Pyrrocaine Hydrochloride N.F., Endo-caine® Hydrochloride, Dynacaine® Hydrochloride, is 1-pyrrolidinoaceto-2′,6′-xylidide (pyrrolidino-2,6-dimethylacetanilide) mon-ohydrochloride.[23] In general, it resembles lidocaine in both its chemical properties and its pharmacologic actions. It is reported that in equianesthetic doses it is about one half as toxic as lidocaine. In addition, the epinephrine requirement for localization of activity appears to be less than that of lidocaine.

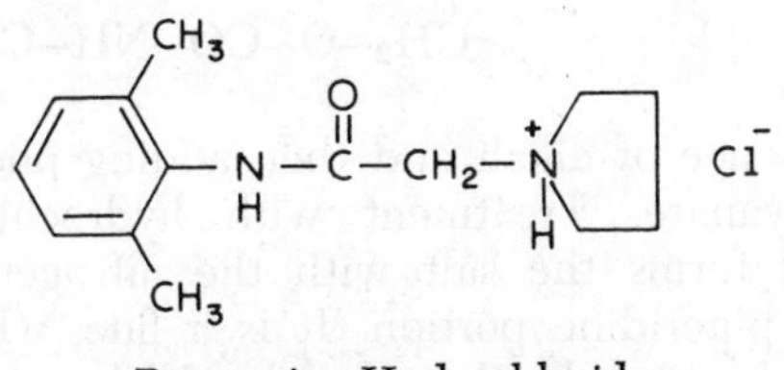

Pyrrocaine Hydrochloride

Category—local anesthetic (dental).

Usual dose—infiltration, 1 ml. of a 2 per cent solution; nerve block, 1.5 to 2 ml. of a 2 per cent solution.

OCCURRENCE

Pyrrocaine Hydrochloride and
Epinephrine Injection N.F.

Prilocaine Hydrochloride N.F., Citanest®, 2-(Propylamino)-o-propionotoluidide Monohydrochloride, Propitocaine Hydrochloride. Prilocaine Hydrochloride, originally named propitocaine hydrochloride, is a white, crystalline powder, freely soluble in water and in alcohol, slightly soluble in chloroform and very slightly soluble in acetone. This is another local anesthetic related in structure to lidocaine, mepivacaine and pyrrocaine. Solutions in ampuls and vials are stable to autoclaving. Prilocaine is effective without the addition of epinephrine or other vasoconstrictors; therefore, it is useful when such agents are contraindicated.

Category—local anesthetic.

Usual dose range—therapeutic nerve block, 3 to 5 ml. of a 1 or 2 per cent solution; infiltration, 20 to 30 ml. of a 1 or 2 per cent solution; regional anesthesia, peridural, and caudal, 15 to 20 ml. of a 3 per cent solution or 20 to 30 ml. of a 1 or 2 per cent solution; nerve block in dentistry, 0.5 to 5.0 ml. of a 4 per cent solution.

OCCURRENCE

Prilocaine Hydrochloride Injection N.F.

Digammacaine is 1-benzamido-1-phenyl-3-piperidinopropane and is used as a local anesthetic in erythromycin injections because of its compatibility, stability and prolonged anesthetic action.

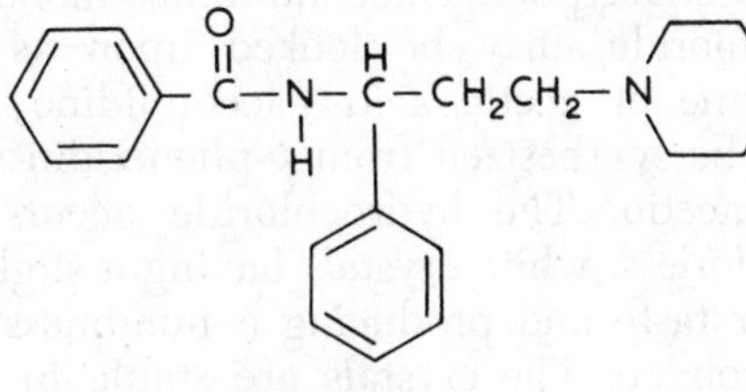

Digammacaine

Oxethazaine,2,2′-(2-hydroxyethylimino)-bis[N-(α,α-dimethylphenethyl)-N-methylacetamide], has local anesthetic properties. It is used orally along with aluminum hydroxide gel in the symptomatic treatment of chronic gastritis. The onset of action is too slow and the duration of action is too long for the drug to be used for topical application or infiltration, but it is reported to have an effective local action when taken

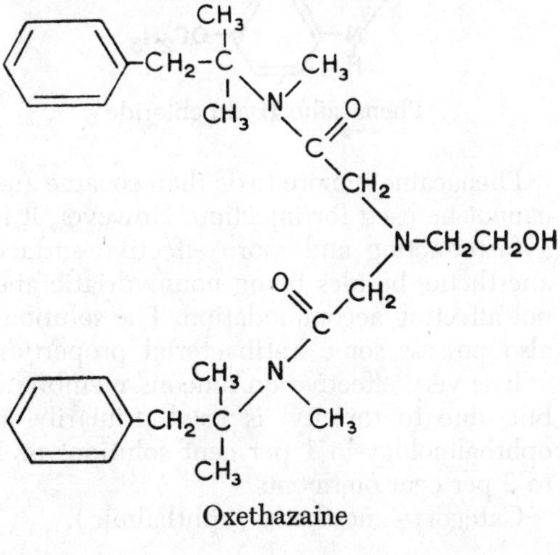

Oxethazaine

orally. Oxaine contains 2 per cent of the free base of oxethazaine in suspension with aluminum hydroxide gel. Oxethazaine is a very weak base, so that when taken orally it is only partially converted to the salt form, leaving much of the dose as free base to penetrate nerve endings and to produce mucosal anesthesia. The recommended dosage is 10 to 20 mg. suspended in 5 to 10 ml. of aluminum hydroxide gel, given 4 times a day, 15 minutes before meals and at bedtime.

AMIDINES

Phenacaine Hydrochloride N.F., N,N'-bis(*p*-ethoxyphenyl)acetamidine monohydrochloride, may be looked upon as an amidine of acetic acid (acetamidine). It may be synthesized from *p*-phenetidin and phenacetin. The hydrochloride occurs as as odorless, white crystals having a slightly bitter taste and producing a numbness on the tongue. The crystals are stable in air, sparingly soluble in water (1:50), freely soluble in chloroform or alcohol and insoluble in ether. Aqueous solutions may be sterilized by boiling. The salt is very susceptible to decomposition by alkalies and has the same incompatibilities as the other local anesthetics.

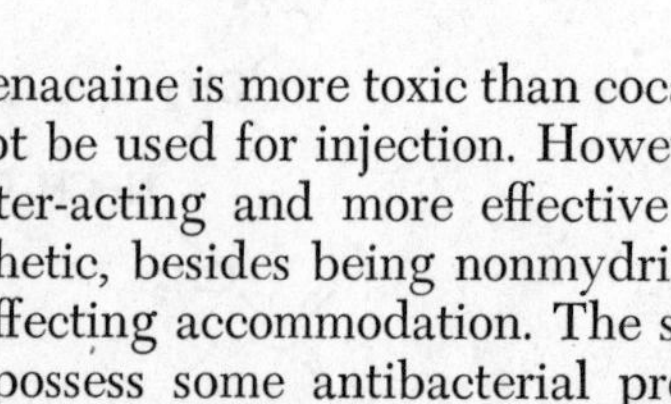

Phenacaine Hydrochloride

Phenacaine is more toxic than cocaine and cannot be used for injection. However, it is a faster-acting and more effective surface anesthetic, besides being nonmydriatic and not affecting accommodation. The solutions also possess some antibacterial properties.

It is very effective on mucous membrane but, due to toxicity, is used primarily in ophthalmology in 1 per cent solutions or 1 to 2 per cent ointments.

Category—anesthetic (ophthalmic).

Application—to the conjunctiva as a 1 per cent solution or as a 1 to 2 per cent ointment.

URETHANES

Diperodon Hydrochloride, Diothane® Hydrochloride, is 3-(1-piperidyl)-1,2-propanediol diphenylurethan hydrochloride. The base is made by combining glycerol monochlorohydrin with piperidine in the

$$
\begin{array}{ccc}
CH_2Cl & & CH_2{-}NC_5H_{10} \\
| & & | \\
CH_2OH & \xrightarrow{C_5H_{10}NH} & CHOH \\
| & & | \\
CHOH & & CH_2OH \\
& & \downarrow C_6H_5NCO \\
& & CH_2{-}NC_5H_{10} \\
& & | \\
& & CH{-}O{-}CO{-}NH{-}C_6H_5 \\
& & | \\
& & CH_2{-}O{-}CO{-}NH{-}C_6H_5
\end{array}
$$

presence of alkali and then adding phenyl isocyanate. Treatment with hydrochloric acid forms the salt with the nitrogen of the piperidine portion. It is a fine, white, odorless powder that is more soluble in alcohol than in water; when applied to the tongue, it is tasteless but produces a sense of numbness.

The free base is relatively unstable to heat, but aqueous solutions of the hydrochloride salt at pH 4.5 to 4.7 will retain at least 99 per cent of their potency after storage for 18 months at room temperature. Samples autoclaved at 100° for 18 hours had a loss of only 2 per cent of the original potency.[24] Diperodon hydrochloride solutions decompose by the sequence of reactions shown on page 673. Lowering the pH to about 4.5 stabilizes the solution by shifting equilibrium (1) to the left. When compared with other ester-type local anesthetics, its solutions are much more stable under ordinary storage conditions and at normal autoclaving temperatures and times.[25] Furthermore, any significant deterioration under adverse conditions is evidenced by discoloration, precipitation or both.

It is used as a local anesthetic in about

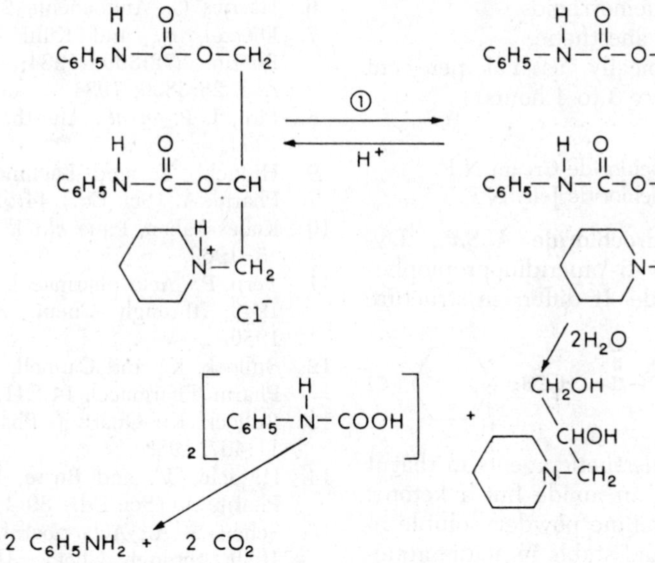

$$C_6H_5 \overset{H}{\underset{|}{N}} - \overset{O}{\underset{||}{C}} - O - CH_2$$

$$\xrightarrow[H^+]{①}$$

$$C_6H_5 - \overset{H}{\underset{|}{N}} - \overset{O}{\underset{||}{C}} - O - CH_2$$

$$C_6H_5 \overset{H}{\underset{|}{N}} - \overset{O}{\underset{||}{C}} - O - CH$$

$$C_6H_5 - \overset{H}{\underset{|}{N}} - \overset{O}{\underset{||}{C}} - O - CH$$

$$N - CH_2 \quad Cl^-$$

$$^+HCl$$

$$\bigg[C_6H_5 - \overset{H}{\underset{|}{N}} - COOH \bigg]_2 \quad + \quad$$

$$2H_2O$$

$$CH_2OH$$
$$CHOH$$
$$N - CH_2$$

$$2\ C_6H_5NH_2 \ + \ 2\ CO_2$$

the same way as cocaine and procaine, but it is claimed that the effects last much longer. After intravenous injection, it is just about as toxic as cocaine; hence, it should not be injected except in small amounts. It is recommended for the relief of pain and irritation in abrasions of the skin and the mucous membranes, following hemorrhoidectomy and in cases of nonoperable hemorrhoids and is applied in 0.5 to 1.0 per cent solution.

It is available in a 1 per cent cream and in solutions of 0.5 and 1 per cent.

MISCELLANEOUS

Dimethisoquin Hydrochloride N.F.; Quotane® Hydrochloride; 3-Butyl-1-[(2-dimethylamino)ethoxy]isoquinoline Monohydrochloride. In chemical structure, this differs from most other local anesthetics. The hydrochloride occurs as white crystals that are

$$CH_2CH_2CH_2CH_3$$

$$O - CH_2CH_2 \overset{+}{\underset{|}{N}} \overset{CH_3}{\underset{CH_3}{}} \quad Cl^-$$

Dimethisoquin Hydrochloride

soluble in water (1:8), alcohol (1:3) and chloroform (1:2).

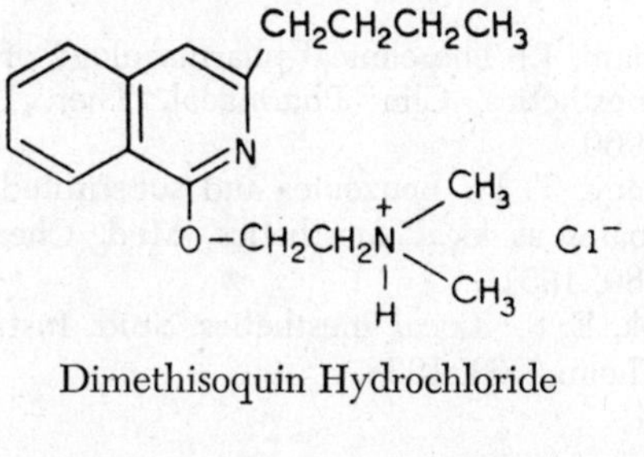

Dimethisoquin is a safe, effective compound for general application as a topical anesthetic. It is available as a water-soluble ointment (0.5%) for dry dermatologic conditions and in lotion form for more moist skin surfaces.

Category—local anesthetic.

Application—topically, to the skin as a 0.5 per cent ointment or lotion twice to 4 times daily.

OCCURRENCE
Dimethisoquin Hydrochloride Lotion N.F.
Dimethisoquin Hydrochloride Ointment N.F.

Pramoxine Hydrochloride N.F., Tronothane® Hydrochloride, 4-[3-(*p*-butoxyphenoxy)propyl]morpholine hydrochloride, is a white, odorless crystalline powder freely soluble in alcohol or water. The molecule contains the butoxy group of dibucaine and dyclonine but differs by having another ether linkage.

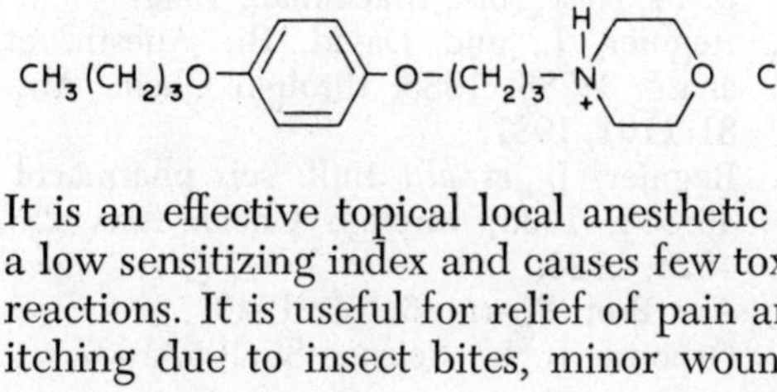

$$CH_3(CH_2)_3O - \bigcirc - O - (CH_2)_3 - \overset{H}{\underset{+}{N}} \bigcirc O \quad Cl^-$$

It is an effective topical local anesthetic of a low sensitizing index and causes few toxic reactions. It is useful for relief of pain and itching due to insect bites, minor wounds and lesions, and hemorrhoids.

Category—local anesthetic.

Application—topically, as a 1 per cent cream or jelly every 3 to 4 hours.

OCCURRENCE

Pramoxine Hydrochloride Cream N.F.
Pramoxine Hydrochloride Jelly N.F.

Dyclonine Hydrochloride U.S.P., Dyclone®, 4'-Butoxy-3-piperidinopropiophenone Hydrochloride. It differs in structure

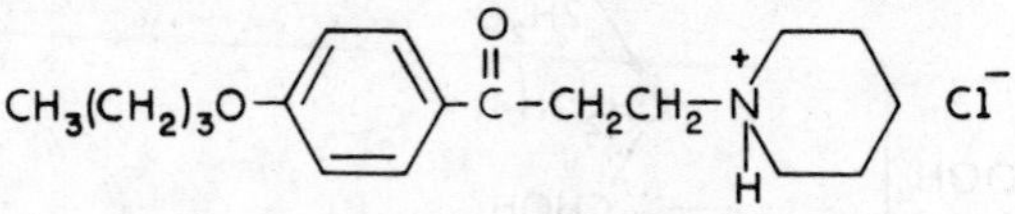

$$CH_3(CH_2)_3O-\text{⟨benzene⟩}-\overset{\overset{O}{\|}}{C}-CH_2CH_2-\overset{+}{\underset{H}{N}}\text{⟨piperidine⟩}\quad Cl^-$$

from most local anesthetic agents in that it is not an ester or an amide but a ketone. It is a white crystalline powder, soluble in alcohol or water and stable in acidic aqueous solutions if not autoclaved.

Dyclonine is used as a topical anesthetic on the skin or mucous membranes. It is effective for anesthetizing the mucous membrane of the mouth, pharynx, trachea, esophagus, and urethra prior to various endoscopic procedures. Anesthesia usually occurs in 5 to 10 minutes after application and persists for 20 minutes to 1 hour. When a 0.5 per cent solution is instilled into the conjunctiva, it produces local anesthesia without causing miosis or mydriasis. The anesthesia is rapid in onset, and normal sensitivity of the cornea usually returns in 30 to 50 minutes.

Category—topical anesthetic.

For external use—topically, as a 0.5 to 1 per cent solution, to the mucous membranes.

OCCURRENCE

Dyclonine Hydrochloride Solution U.S.P.

REFERENCES CITED

1. Silverman, M. M.: Magic in a Bottle, p. 74, New York, Macmillan, 1941.
2. Regnier, J., and David, R.: Anesth. et analg. 1:285, 1935; through Chem. Abs. 31:1101, 1937.
3. Regnier, J., et al.: Bull. sci. pharmacol. 40:353, 1933; through Chem. Abs. 27: 4628, 1933.
4. Am. Prof. Pharm. 6:163, 1940.
5. Osborne, R. L.: Science 85:105, 1937.
6. Harries, C.: Ann. chemie 296:328, 1897.
7. Dietzel, R., and Kühl, G. W.: Arch. Pharm. 272:369, 1934; through Chem. Abs. 28:3836, 1934.
8. Piro, J. P., et al.: Anesth. Analg. 33:391, 1954.
9. Higuchi, T., and Lachman, L.: J. Am. Pharm. A. (Sci. Ed.) 44:521, 1955.
10. Kabasakalian, P., et al.: J. Pharm. Sci. 58: 45, 1969.
11. Terp, P.: Acta pharmacol. toxicol. 5:353, 1949; through Chem. Abs. 44:6576c, 1950.
12. Bullock, K., and Cannell, J. S.: Quart. J. Pharm. Pharmacol. 14:241, 1941.
13. Bullock, K.: Quart. J. Pharm. Pharmacol. 11:407, 1938.
14. Higuchi, T., and Busse, L. W.: J. Am. Pharm. A. (Sci. Ed.) 39:411, 1950.
15. Schou, S. A.: Acta pharm. intern. 1:117, 1950; through Chem. Abs. 45:10486g, 1951.
16. Kennon, L., and Higuchi, T.: J. Am. Pharm. A. (Sci. Ed.) 45:157, 1956.
17. Fosdick, L. S., et al.: Proc. Soc. Exp. Biol. Med. 27:529, 1930; through Chem. Abs. 24:580, 1930.
18. Foldes, F. F., et al.: J. Am. Chem. Soc. 77:5149, 1955.
19. Willi, A. V.: Pharm. acta helv. 33:635, 1958.
20. Mørch, J.: Dansk. tids. farm. 27:173, 1953; through Chem. Abs. 47:12760d, 1953.
21. Bullock, K., and Grundy, J.: J. Pharm. Pharmacol. 7:755, 1955.
22. Sadove, M., and Wessinger, G. D.: J. Int. Coll. Surg. 34:573, 1960.
23. Schlesinger, A., and Gordon, S. M.: U. S. patent 2,949,470, August 16, 1960.
24. Cook, E. S., et al.: J. Am. Pharm. A. (Sci. Ed.) 24:269, 1935.
25. Cook, E. S., and Ryder, T. H.: J. Am. Pharm. A. (Sci. Ed.) 26:222, 1937.

SELECTED READING

Adriani, J.: The clinical pharmacology of local anesthetics, Clin. Pharmacol. Ther. 1:645, 1960.

Carney, T. P.: Benzoates and substituted benzoates as local anesthetics, Med. Chem. 1: 280, 1951.

Cook, E. S.: Local anesthetics, Stud. Inst. Div. Thom. 2:63, 1938.

Daniels, T.: Synthetic drugs—local anesthetics, Ann. Rev. Biochem. 12:462, 1943.

Geddes, I. C.: Chemical structure of local anesthetics, Brit. J. Anaesth. 34:229, 1962.

Lofgren, Nils: Studies on Local Anesthetics, Stockholm, University of Stockholm, 1948.

Moore, M. B.: Synthetic organic local anesthetics, J. Am. Pharm. A. (Sci. Ed.) 33:193, 1944.

Wiedling, S., and Tegner, C.: Local anesthetics, Progr. Med. Chem. 3:332, 1963.

23

Histamine and Antihistaminic Agents

Robert F. Doerge, Ph.D.,
Professor of Pharmaceutical Chemistry, Chairman of the Department of
Pharmaceutical Chemistry, School of Pharmacy,
Oregon State University

HISTAMINE

Histamine, or β-imidazolylethylamine, was identified in 1907.[1] It has been shown that the compound is widespread in nature, being found in ergot and other plants and in all the organs and the tissues of the human body in very small amounts.[2] Histidine, a naturally occurring amino acid, can be decarboxylated to form histamine, and this may be the source of histamine in the body.

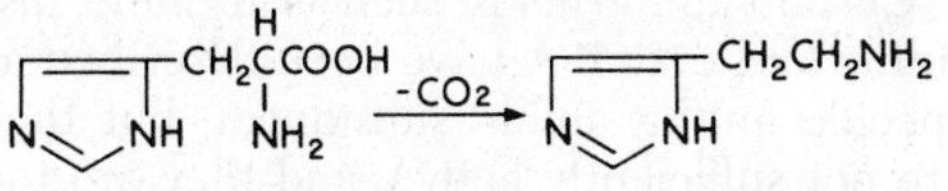

This drug causes a wide variety of physiologic responses, almost every tissue of the body being affected by it to some extent. It produces a strong vasodilatation of the capillaries and in large doses may cause an increase in their permeability, so that fluid and plasma proteins may escape into the extracellular fluid and lead to edema. In the lungs, histamine acts on smooth muscle, producing bronchiolar constriction. It has a stimulating action on certain excretory glands, causing an increase in the secretion of acid in the stomach. The lacrimal and nasal secretory glands also are stimulated.

There is much circumstantial evidence that histamine plays an important part in human allergy.[3] Researchers in allergy and anaphylaxis have pointed out that the symptoms are similar regardless of the sensitizing agent used.[4] This has led to the assumption that some common substance could be responsible for the symptoms. It is believed now that this substance is histamine and/or some closely related substances. Very little histamine is absorbed from the gastrointestinal tract, even after large doses, but small parenteral doses will cause an intense response. This response is similar to many allergic manifestations. However, it should be pointed out that the positive identification of histamine in the blood of human beings has not been made. Thus, the term "histaminelike substance" is used by many writers.

Histamine Phosphate U.S.P. is 4-(2-aminoethyl)imidazole bis(dihydrogen phosphate). The crystals are stable in air but affected by light; they melt at about 140° and readily dissolve in water (1 g. in 4 ml.) to form an acid solution. Histamine is less important as a remedial agent than in its theoretic role in allergic reactions. Attempts have been made to use injections for desensitizing against allergic reactions, histamine cephalgia, migraine and similar disorders. Results have been inconclusive. In order to increase the antigenicity of histamine, it was coupled with despeciated horse-serum globulin by using a

677

histamine derivative which could be diazotized.[5] This preparation had great theoretic possibilities, but the clinical results were not very encouraging.

The chief use of histamine is to diagnose impairment of the acid-producing cells of the stomach. Normally, it is the most powerful stimulant that is available, and the absence of acid after injection is considered proof that the acid-secreting glands are nonfunctional, a condition that is particularly symptomatic of pernicious anemia. The wheal caused by intradermal injection of a 1:1,000 solution had been suggested as a diagnostic test of local circulation; in normal individuals the wheal appears in about 2.5 minutes, and any delay is considered a sign of vascular disease.

Category—diagnostic aid (gastric secretion indicator).

Usual dose—subcutaneous, 800 mcg. (the equivalent of 300 mcg. of histamine base).

Usual dose range—800 mcg. to 2 mg. (the equivalent of 300 to 700 mcg. of histamine base).

OCCURRENCE

Histamine Phosphate Injection U.S.P.

Betazole Hydrochloride U.S.P., Histalog®, 3-(2-Aminoethyl)pyrazole Dihydrochloride.

$$\begin{array}{c} \text{CH}_2\text{CH}_2\overset{+}{\text{NH}}_3 \\ \text{N}-\text{H} \qquad \text{2 Cl}^- \\ \overset{+}{\text{N}} \\ \text{H} \end{array}$$

Betazole Hydrochloride

This drug is an analog of histamine which retains the ability to stimulate gastric secretion but with much less tendency to the other effects usually observed after the use of histamine. It is a water-soluble, white, crystalline, nearly odorless powder. The pH of a 5 per cent solution is about 1.5. A dose of 5 mg. is comparable in gastric secretory response to 10 mcg. of histamine base. Most clinicians usually give a subcutaneous dose of 50 mg. to all patients of normal weight.

Category—diagnostic acid (gastric secretion indicator).

Usual dose—subcutaneous, 50 mg.

Usual dose range—40 to 60 mg.

OCCURRENCE

Betazole Hydrochloride Injection U.S.P.

ANTIHISTAMINIC AGENTS

Earlier Drugs Used

If histamine or a "histaminelike substance" were the cause of allergic symptoms, an obvious approach to the problem would be to find agents that would destroy or counteract the effects of histamine. It was found that an enzyme, histaminase or diaminoxidase, would destroy histamine, and attempts were made to use it therapeutically. An enzymatic histaminase extract derived from the intestinal mucosa of the hog was introduced as Torantil. The clinical results were disappointing, probably because histaminase is inhibited or actually destroyed before the site of action can be reached.[6]

Many other drugs have been used in past years. In fact, for a long time symptomatic treatment was in vogue, using drugs such as epinephrine, ephedrine, phenylpropanolamine and aminophylline. These drugs do not block the action of histamine, nor do they inactivate it, but they exert an action that is diametrically opposed to some of the actions of histamine.

Certain amino acids, such as arginine, histidine and cysteine, have been described as specific antihistamine substances, but they are not sufficiently active, and they are too toxic for use in animals and man in doses adequate to block the effects of histamine.[7]

The origin of antihistamine research as we know it today was in the Pasteur Institute in France in 1937.[8] Sympathomimetic and sympatholytic substances were being investigated for their ability to antagonize the action of histamine on the isolated intestine. In this study, the phenolic ethers were found to be better than any previously tested compound. The compound designated as F 929 was the most active of the group. It is of great historical importance but of no clinical value because of its toxicity.

Because of the proved activity of various amines of the alkyl aromatic series (sympathomimetic amines), it also was decided to investigate a series of phenylethylenediamines.[9] These two original studies led to the development of the many effective antihistaminics that are available today.

GENERAL FORMULA

All of the more active antihistaminic agents which have been developed may be represented by the following general formula, in which X represents nitrogen, oxygen or carbon connecting the side chain to the nucleus. The chemical structures of

$$R-X-\overset{|}{\underset{|}{C}}-\overset{|}{\underset{|}{C}}-N\diagdown$$

these drugs vary greatly otherwise, yet the prominent compounds exert similar pharmacologic and therapeutic action. However, there is considerable difference in potency, in both animal experiments and clinical tests.

MODE OF ACTION

Antihistaminic agents may be defined as "those drugs which are capable of diminishing or preventing several of the pharmacologic effects of histamine and which do so by a mechanism other than the production of pharmacologic response diametrically opposed to those produced by histamine."[10] The mode of action may be considered as a competition, in tissue, between the antihistaminic agent and histamine for a receptive substance. The combining of the antihistaminic agent with the receptive substance at the site of action prevents histamine from exerting its characteristic effect on the tissue. It is known that the antihistamines do not combine with histamine to neutralize its action, nor do they prevent its liberation from the cells. Also, no indication has been found that these drugs activate enzymes which catalyze the breakdown of histamine (i.e., histaminase). Also, it is held generally that antihistaminics do not interfere with the antigen-antibody reaction.[11] None of the presently known antihista-

minics has any significant influence on histamine-induced gastric secretion.[10]

It would be well to point out that the use of antihistaminic drugs is no more than the treatment of symptoms as they arise and in no way changes the fundamental cause of the allergic state.[12] They must be used as long as exposure to the allergen persists. They relieve symptoms for from 4 to 24 hours following each dose, and seldom do they relieve all symptoms completely. Certain of the drugs are of more benefit in some cases than in others, and the pattern may be different for individual patients. In one case, a drug may be effective in small doses, and in another it may have no desirable action no matter what size dose is used. Differences also are found in the capacity of individual patients to tolerate the drugs. These facts justify to some extent the large number of the drugs available.

The beneficial action of the antihistaminic compounds is most apparent in seasonal hay fever. Other conditions which have been relieved include serum sickness, urticaria, motion sickness and the nausea of pregnancy. Antihistamines, particularly in conjunction with analgesic agents, are widely used in the treatment of the common cold. However, there is considerable difference of opinion as to their value. [13, 14]

OVERLAPPING ACTIVITIES AND SIDE REACTIONS

Seldom, if ever, is a drug found which has only one action on the intact organism. This is true of the antihistamines, which exhibit, in varying degrees, local anesthetic, adrenergic blocking, antispasmodic, sympathomimetic, analgesic, cholinergic blocking and quinidinelike actions.[9] It has been reported that tinea pedis (athlete's foot) responds to topical application of various antihistamine creams.[15] A study of the fungistatic activity of a number of antihistamines has shown that there is little correlation between their antihistaminic potency and their effectiveness as fungistatic agents.[16]

Undesirable side-reactions vary with the individual drug and the individual patient. Most of the antihistamines are somewhat depressant to the central nervous system and cause sedation.[3] However, some of the drugs have a stimulating action. In many cases the reaction wears off on continued use of the drug. Other undesirable but not necessarily limiting side-effects which have been reported include muscular weakness, dizziness, gastric irritation, loss of appetite, diarrhea, dryness of the mouth and throat, palpitation and nervousness. In some cases, the side-effects can be overcome by the use of a different antihistaminic agent.

TESTING

The antihistamine drugs are assayed by observing whether or not they will diminish or block one or more of the physiologic actions of histamine. The methods available for the determination of the potency of potential antihistaminic substances include: (1) the evaluation of the protective effect of the antihistaminic against lethal doses of intravenously administered histamine in the guinea pig; (2) the abolition of histamine-induced spasm of an isolated strip of the guinea pig ileum; (3) the protection of the guinea pig against histamine aerosol (inhalation of the histamine).

Next, the drug must be evaluated clinically. This procedure may sometimes be disappointing, for some of the highly active drugs are not always superior therapeutically. Another weak spot in the clinical evaluation is the lack of quantitative, objective methods of evaluation, and in many cases the criteria for evaluation are not comparable.

SALT FORMATION

Salt formation occurs at the nitrogen in the aliphatic chain of the antihistaminic. The nitrogen occurs as a tertiary amine and can be represented as

$$R \overset{..}{\underset{\cdot \times}{N}} \times CH_3$$
$$CH_3$$

This leaves a pair of electrons of the nitrogen free to react with a proton. When an acid is added to form a salt, the hydrogen ion is combined to form a substituted ammonium ion, such as

$$\left[\begin{array}{c} CH_3 \\ \cdot \times \\ R \underset{\times}{N} \colon H \\ \cdot \times \\ CH_3 \end{array} \right]^+$$

These salts can be converted to the free amine by the addition of a base, such as ammonium hydroxide or sodium hydroxide. The free amine is only slightly soluble in water but is readily soluble in lipids and organic solvents.

Hydrochloric acid is used commonly to form a water-soluble salt of various amines, but in the field of antihistaminics, we have, in addition, the use of various dicarboxylic organic acids, such as succinic, fumaric, maleic, malic and tartaric acids, and the tricarboxylic citric acid. The salt with the antihistamine base is usually an acid salt of the dicarboxylic acid, that is, only one of the acidic hydrogens enters into salt formation with the amino group.

Sometimes there is a noticeable difference in the potency of the various salts of a particular antihistaminic agent. This may be due to increased solubility or absorption in the animal. For example, chlorpheniramine maleate appears to be more active than the hydrochloride. In a study of the salts of diphenhydramine, it was found that the acid succinate and acid oxalate had an order of antihistaminic action similar to that of the hydrochloride, but the acid succinate appeared to be less toxic.[17]

Another important reason for the use of various salts is the difference in their stability in air. Thus, in a study of some furan antihistaminic agents, it was found that the hydrochloride was very hygroscopic, the salts with maleic, tartaric, oxalic and malic acids were less hygroscopic, while the salt with fumaric acid was stable in air. In a study of the salt formation of chlorothen, it was found that the dihydrogen citrate was nonhygroscopic, gaining less than 0.5 per cent in weight when exposed in a closed vessel to an atmosphere saturated with water vapor for 24 hours.[18]

The various dosage forms that are available for many of the antihistaminics include capsules, sustained-release capsules, plain compressed tablets, sugar-coated tablets, enteric-coated tablets, sustained-action tablets, elixirs, syrups, suppositories, nose drops, ointments and creams. Solutions for intramuscular or intravenous injections are available if a rapid effect is needed in an emergency.

STRUCTURE AND ACTIVITY RELATIONSHIPS

From a study of the activity of the various antihistaminics that have been synthesized and examined pharmacologically, it is possible to make some general statements summarizing the structural requirements for optimum activity. Most of the antihistaminics are either ethanolamine derivatives, ethylenediamine derivatives or propylamine derivatives. To have maximum activity, it is necessary that in both types of derivatives the terminal N atom should be a tertiary amine, and it appears that the dimethylamine derivatives have a better therapeutic index than the corresponding diethylamine derivatives. It is of interest to note that this is in contrast with many antispasmodics and local anesthetics in which the diethylamine derivatives are the better drugs (e.g., aminocarbofluorene and procaine). It should be noted that the terminal N atom may be part of a heterocyclic structure, as in antazoline and in chlorcyclizine, and still result in a compound of high antihistaminic value.

The chain between the O and N atoms or the N and N atoms should be the ethylene group $-CH_2CH_2-$. A longer chain or a branched chain gives a less active compound. However, in the promethazine molecule an isopropyl group separates the two N atoms, but it may be that the phenothiazine portion of the molecule alters the picture.

Substitution of a chlorine or a bromine, especially in the *p*-position, of a phenyl group enhances the potency (compare bromopheniramine, chlorpheniramine and pheniramine). In compounds that exhibit optical isomerism the dextrorotatory isomer possesses most of the antihistaminic activity. Geometric isomers (*cis* and *trans*) may also show this phenomenon.

In the ethanolamine series, the most effective group attached to the O atom has been found to be benzhydryl[19]; whereas, in the ethylenediamine series, several different groups on the second N of the chain have led to active compounds. In the active compound, the groups are either isocyclic or heterocyclic aromatic ring systems, one of which may or may not be separated from the N atoms by a methylene group $-CH_2-$. If one or both of the ring systems is hydrogenated, the activity is lost. In many of the effective antihistaminics, one of the aromatic groups is α-pyridyl,

with the second substituent on the N atom being the benzyl group,

a substituted benzyl group or one of the isosteres of the benzyl group. Included here are pyrilamine, tripelennamine, methapyrilene and thenyldiamine.

It appears that, in order to be an effective antihistaminic, the nucleus should have a minimum of two aryl or aralkyl groups or their equivalent in a polycyclic ring system.

Several physical properties of antihistamines have been studied in an effort to relate them to activity of the various drugs.[20] Ionization constants, solubilities at pH 7.4 and 37.5°, and relative surface activities at pH 7 of 16 commercially available antihistamines were determined, but no direct correlation could be made between these physical properties and the antihistaminic effect. It may be that steric factors are more important in determining antihistaminic action in view of the fact that certain stereoisomers have been found to be more active than others (e.g., dexchlorpheniramine, dexbrompheniramine and triprolidine).

ETHANOLAMINE DERIVATIVES

F 929; Thymoxyethyldiethylamine. This

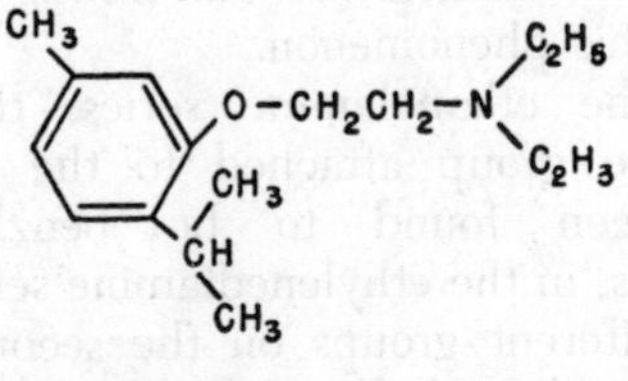

compound is of historical importance only. It was synthesized and investigated by Fourneau and his co-workers for its action on anaphylactic shock in guinea pigs and was the first compound discovered to have a specific antagonism to histamine. However, this and other related phenolic ethers were too toxic for clinical use.

Diphenhydramine Hydrochloride U.S.P., Benadryl®, 2-(Diphenylmethoxy)-*N,N*-dimethylethylamine Hydrochloride. This is

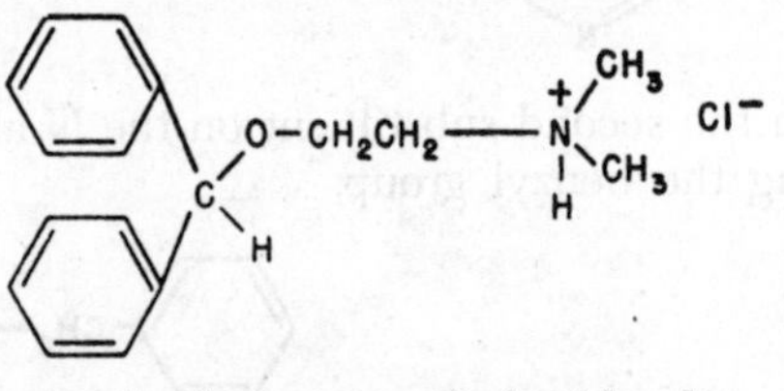

Diphenhydramine Hydrochloride

available as the hydrochloric salt, which is a stable, white, crystalline powder, soluble in water (1:1) or alcohol (1:2) and in chloroform (1:2) and possessing a bitter taste. The drug has a pKa of 9, and a 1 per cent aqueous solution has a pH of about 5.

A synthesis which has been reported in the patent literature can be represented as shown below, starting with diphenylmethane.[21] It has been reported that changes in the benzhydryl portion of the molecule increase the activity.[22]

Rearrangement of the phenyl rings to give fluorene or dihydroanthracene derivatives yields compounds with little activity. If the dimethylamino group is replaced by a piperidino group, the potency is unchanged, but if the morpholino group is used, the potency is decreased by one half, but the toxicity is unchanged.

As an antihistaminic agent, diphenhydramine is recommended in various allergic conditions and, to a lesser extent, as an anti-

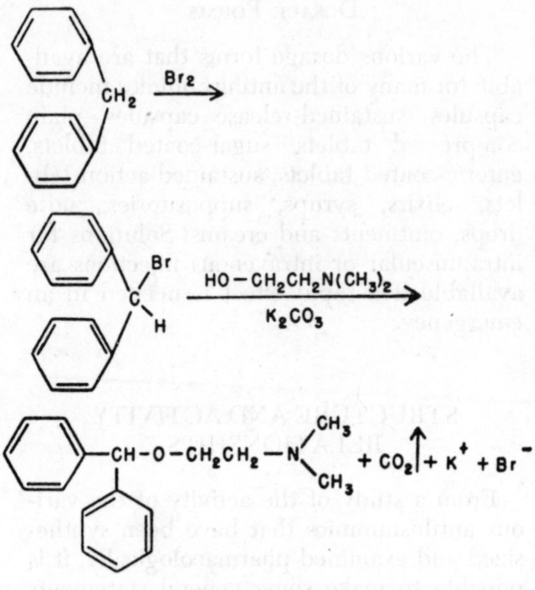

Diphenhydramine

spasmodic. Conversion to the quaternary ammonium salt does not alter the antihistamine action greatly but does increase the antispasmodic action.

It is useful either orally or intravenously in the treatment of urticaria, hay fever, bronchial asthma, vasomotor rhinitis and some dermatoses. The most common side-effects is drowsiness. In the treatment of "colds," it is finding use as a constituent of cough mixtures.

Category—antihistaminic.

Usual dose—oral, 25 to 50 mg. 3 to 4 times daily; intramuscular and intravenous, 10 to 50 mg.

Usual dose range—oral and parenteral, 10 to 400 mg. daily.

OCCURRENCE

Diphenhydramine Hydrochloride Capsules U.S.P.

Diphenhydramine Hydrochloride Elixir U.S.P.

Diphenhydramine Hydrochloride Injection U.S.P.

Dimenhydrinate U.S.P., Dramamine®, 8-Chlorotheophylline-2-(Diphenylmethoxy)-*N,N*-dimethylethylamine Compound.

The 8-chlorotheophyllinate salt of diphenhydramine[23] is recommended for the nausea of motion sickness and for hyperemesis gravidarum (nausea of pregnancy).

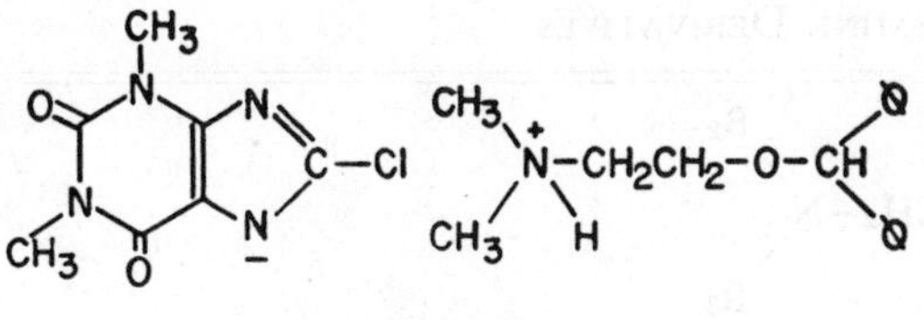

Dimenhydrinate

It is white, crystalline, odorless powder which is slightly soluble in water and freely soluble in alcohol.

The usual dosage schedule is 50 mg. taken one half hour before meals, and, if used for the prevention of motion sickness, 50 mg. also is taken one half hour before beginning the trip; where necessary, the same dosage schedule may be followed using 100-mg. doses.

Category—antinauseant.

Usual dose—50 mg. 4 times daily.

Usual dose range—25 to 600 mg. daily.

OCCURRENCE

Dimenhydrinate Syrup U.S.P.

Dimenhydrinate Tablets U.S.P.

Bromodiphenhydramine Hydrochloride N.F., Ambodryl® Hydrochloride, 2-[(*p*-Bromo-α-phenylbenzyl)oxy]-N,N-dimethylethylamine Hydrochloride. This drug is a white to pale buff, crystalline powder and is freely soluble in water and in alcohol and soluble in isopropyl alcohol.

Bromodiphenhydramine differs from diphenhydramine by having a bromine atom substituted in the para position on one of the benzene rings. Based on the protection of guinea pigs against the lethal effects of histamine aerosols, bromodiphenhydramine was found to be twice as effective as diphenhydramine.

Category—antihistaminic.

Usual dose—25 mg. 3 times daily.

Usual dose range—25 to 50 mg.

OCCURRENCE

Bromodiphenhydramine Hydrochloride Capsules N.F.

Bromodiphenhydramine Hydrochloride Elixir N.F.

Doxylamine Succinate, Decapryn® Succinate, 2-[α[(2-Dimethylamino)ethoxy]-α-methylbenzyl]pyridine Bisuccinate. This antihistaminic is related closely in structure and activity to diphenhydramine. It is available as the acid succinate salt, which is soluble in water (1:1), alcohol (1:2) and

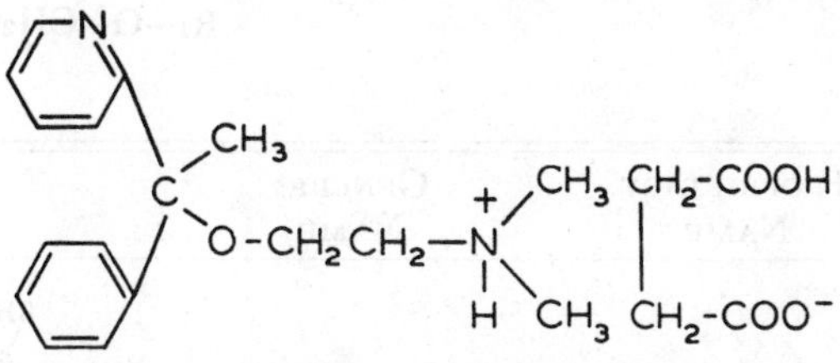

Doxylamine Succinate

chloroform (1:2). A 1 per cent aqueous solution has a pH of about 5.

It is a highly effective drug but has pronounced sedative properties, especially when large doses are required.[24, 25]

The usual dose is 25 mg. up to 4 times daily.

Carbinoxamine Maleate N.F., Clistin® Maleate, 2-[p-chloro-α-[2-(dimethylamino)-ethoxy]benzyl]pyridine bimaleate is a white crystalline powder that is very soluble in water and freely soluble in alcohol and in chloroform. The pH of a 1 per cent solution is between 4.6 and 5.1.

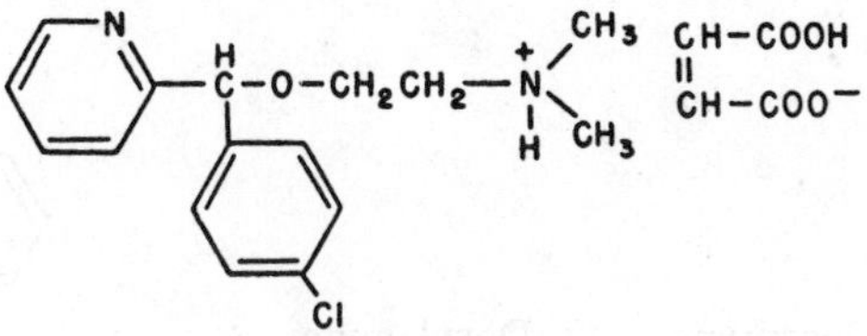

Carbinoxamine Maleate

This can be considered as being related to chlorpheniramine maleate (q.v.). It differs only in that the carbon atom bearing the aromatic nuclei is separated from the remainder of the side chain by an oxygen atom. It is of the same order of activity as chlorpheniramine maleate. There is reported to be a minimum of side-effects when used in ordinary therapeutic dosage levels.

Category—antihistaminic.

Usual dose—4 mg. 3 or 4 times daily.

Usual dose range—4 to 8 mg.

OCCURRENCE

Carbinoxamine Maleate Elixir N.F.

Carbinoxamine Maleate Tablets N.F.

TABLE 70. ETHANOLAMINE DERIVATIVES

$$R_1-O-CH_2CH_2-N\begin{cases}R_2\\R_3\end{cases}$$

PROPRIETARY NAME	GENERIC NAME	R_1	R_2	R_3
	F 929	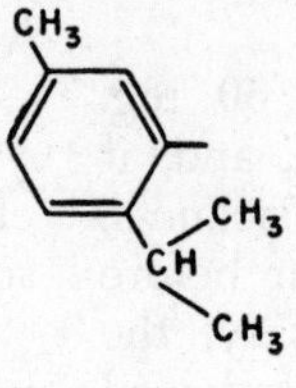 (2,4-dimethyl-α-methylbenzyl)	$-CH_2CH_3$	$-CH_2CH_3$
Benadryl	Diphenhydramine	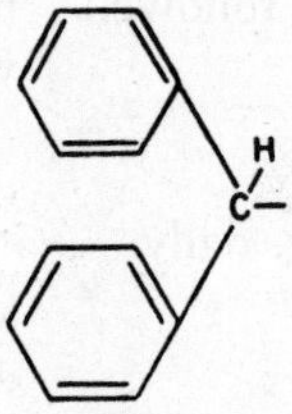 (diphenylmethyl)	$-CH_3$	$-CH_3$
Ambodryl	Bromodiphen-hydramine	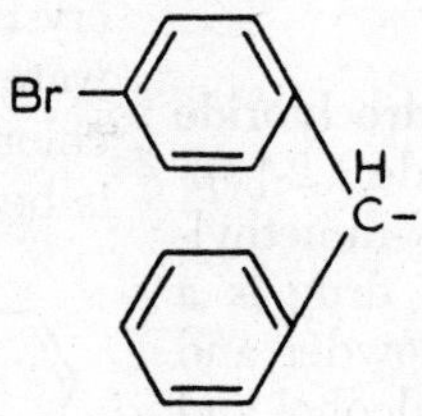 (p-bromo-diphenylmethyl)	$-CH_3$	$-CH_3$
Decapryn	Doxylamine	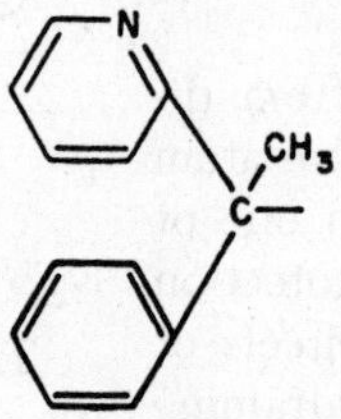 (1-phenyl-1-(2-pyridyl)ethyl)	$-CH_3$	$-CH_3$
Bristramin	Phenyltoloxamine	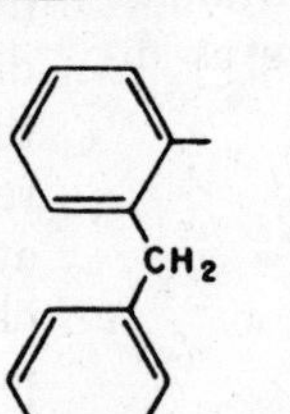 (o-methylbenzyl-phenyl)	$-CH_3$	$-CH_3$
Clistin Twiston	Carbinoxamine Rotoxamine	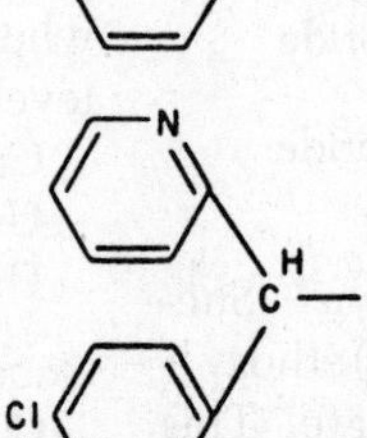 (p-chlorophenyl-2-pyridylmethyl)	$-CH_3$	$-CH_3$

Rotoxamine Maleate, Twiston®. This levorotatory isomer of carbinoxamine malleate is available as plain and sustained-action tablets.

The structural relationships of the ethanolamine derivatives are shown in Table 70.

Ethylenediamine Derivatives

2325 RP, N′-Phenyl-N′-ethyl-N,N-dimethylethylenediamine. This compound was synthesized and investigated by Halpern in the laboratory of the Rhone-Poulenc Society of Chemical Manufacturing in France and was found to be of higher antihistaminic order than any previously reported drug. Although this compound was too toxic to be used clinically, many effective present-day antihistaminic agents are related to it. Replacement of the N′-ethyl group by a benzyl group yields Antergan, which was the first useful drug in this series.[26]

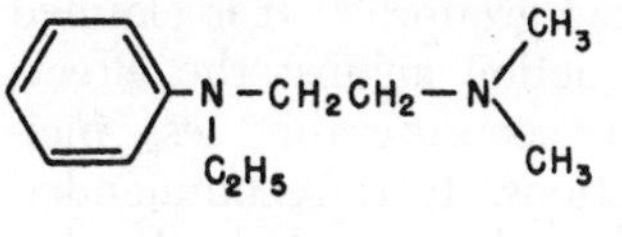

2325RP

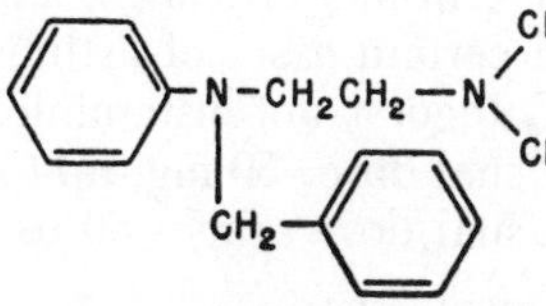

Antergan

Methaphenilene Hydrochloride, Diatrin® Hydrochloride, N′-Phenyl-N′(2-thenyl)-N-dimethylethylenediamine Hydrochloride. This compound contains the isosteric 2-thenyl group in place of the benzyl group in Antergan. It is reported to be an effective antihistaminic.[27]

The adult dosage is 50 mg. 4 times daily.

Pyrilamine Maleate N.F., Neo-Antergan® Maleate, 2-{[(2-Dimethylamino)ethyl](*p*-methoxybenzyl)amino}pyridine Bimaleate, Pyranisamine Maleate. This drug differs from Antergan by having a 2-pyridyl group in place of the phenyl group and by having a methoxy group in the para position of the benzyl radical.[28] It is soluble in water (1:0.5), alcohol (1:3) and chloroform

(1:2). A 5 per cent solution in water has a pH of 5. The free base is liberated as an oil by adding sodium hydroxide to an aqueous solution. It is available as the acid maleate

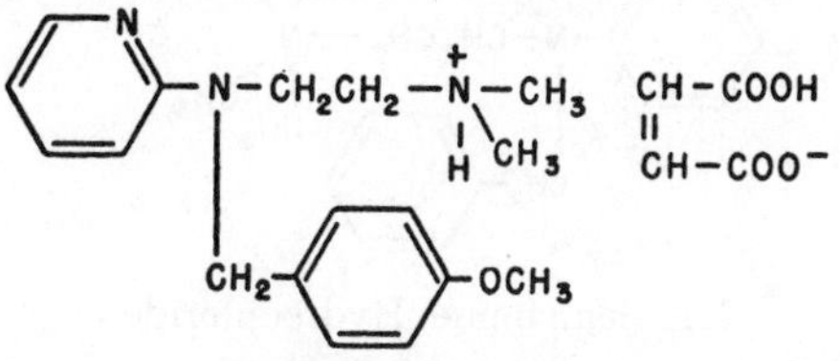

Pyrilamine Maleate

salt, a white crystalline powder with a bitter taste. No special precautions are necessary in handling and storage.

Successful results have been obtained in the treatment of hay fever, urticaria, allergic rhinitis and allergic drug reactions. It is less useful in the treatment of asthma, atopic and contact dermatitis and eczema. It is recommended that the drug be taken with food and that the tablets not be chewed because of the pronounced local anesthetic action.

Category—antihistaminic.

Usual dose—25 mg. up to 4 times daily.

Usual dose range—25 to 50 mg.

OCCURRENCE

Pyrilamine Maleate Tablets N.F.

Tripelennamine Citrate U.S.P., Pyribenzamine® Citrate, 2-{benzyl[2-(dimethylamino)ethyl]amino}pyridine dihydrogen citrate occurs as a white crystalline powder. A 1 per cent solution has a pH of about 4.25. It is freely soluble in water and in alcohol. The advantage of this salt over the hydrochloride is that it is more palatable for oral administration in liquid dosage forms. Thirty mg. of the citrate salt are equivalent to 20 mg. of the hydrochloride salt because of the difference in molecular weights.

Category—antihistaminic.

Usual dose—75 mg. 1 to 3 times daily.

Usual dose range—37.5 to 900 mg. daily.

OCCURRENCE

Tripelennamine Citrate Elixir U.S.P.

Tripelennamine Hydrochloride U.S.P., Pyribenzamine® Hydrochloride, Stanzamine®, 2-{Benzyl[2-(dimethylamino)ethyl]amino}pyridine Hydrochloride. This antihistaminic agent differs from pyrilamine only in not having a methoxy group on the benzyl radical.[29]

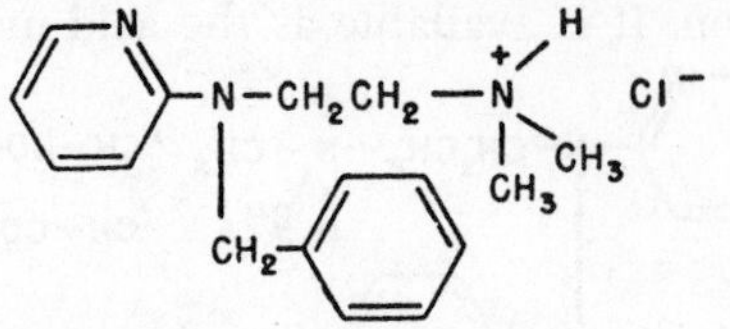

Tripelennamine Hydrochloride

A published synthesis,[30] starting with 2-aminopyridine and 2-dimethylaminoethyl bromide, is as follows:

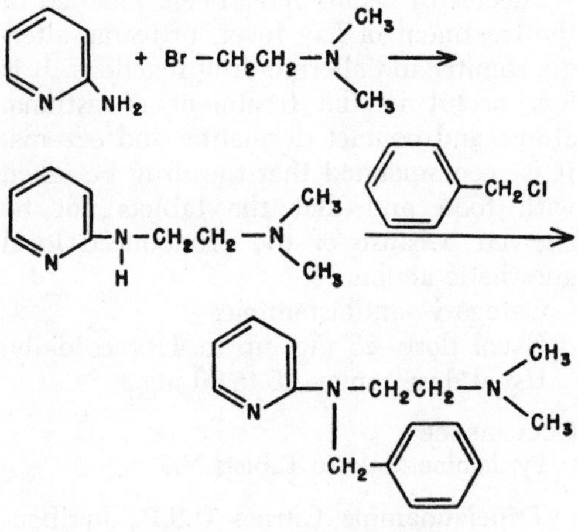

Tripelennamine

It is available as the hydrochloride salt, which is a white, crystalline powder that slowly darkens on exposure to light. The drug is soluble to the extent of 1 g. in 1 ml. of water or in 6 ml. of alcohol and is insoluble in ether or benzene. The drug has a pKa of about 9 and an aqueous solution has a pH of about 6.5.

On the basis of clinical experience, tripelennamine appears to be as effective as diphenhydramine and may have the advantage of fewer and less severe side-reactions. It is well absorbed when given orally in 50 mg. doses.

Category—antihistaminic.

Usual dose—oral, 50 mg. 1 to 3 times daily; parenteral, 25 mg. twice daily.

Usual dose range—oral, 50 to 600 mg. daily; parenteral, 25 to 100 mg. daily.

OCCURRENCE

Tripelennamine Hydrochloride Injection U.S.P.

Tripelennamine Hydrochloride Tablets U.S.P.

Methapyrilene Hydrochloride N.F., Histadyl®, Thenylpyramine Hydrochloride, Semikon®, 2-{[2-(Dimethylamino)ethyl]-2-thenylamino}pyridine Monohydrochloride. This occurs as a white crystalline powder

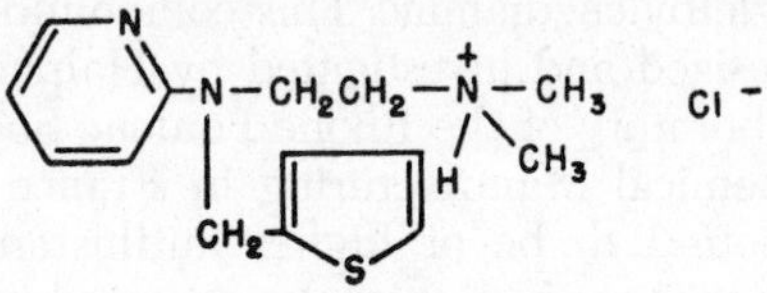

Methapyrilene Hydrochloride

that is soluble in water (1:0.5) alcohol (1:5) and in chloroform (1:3) and its aqueous solutions have a pH of about 5 to 6. This drug differs from tripelennamine in having a 2-thenyl (thiophene-2-methylene) group in place of the benzyl group.[31-34] It is claimed to have marked action against the effects of histamine with considerably less incidence of side-actions. It is recommended in the treatment of all suspected allergies, especially hay fever, allergic rhinitis, acute and chronic urticarias, allergic dermatitis and certain cases of asthma.

Category—antihistaminic.

Usual dose—50 mg. up to 4 times daily.

Usual dose range—50 to 100 mg.

OCCURRENCE

Methapyrilene Hydrochloride Capsules N.F.

Methapyrilene Hydrochloride Tablets N.F.

Chlorothen Citrate N.F., Tagathen®, Pyrithen®, 2-[(5-Chloro-2-thenyl)[2-(dimethylamino)ethyl]amino]pyridine Dihy-

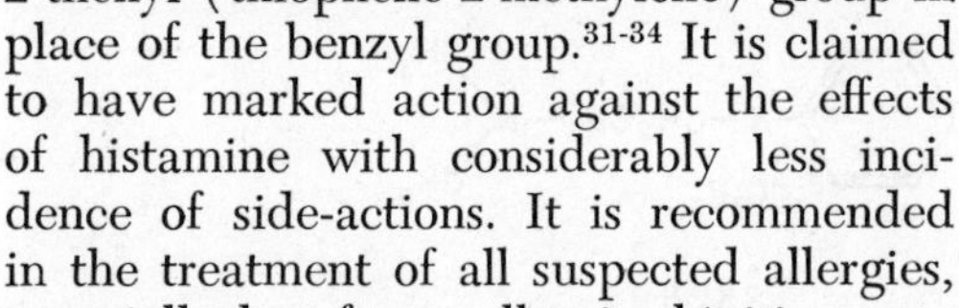

Chlorothen Citrate

drogen Citrate, Chloromethapyrilene Citrate. This is a white crystalline powder that is soluble in water (1:35) and slightly soluble in alcohol. Its solutions are acid to litmus. Chlorothen is similar in structure to tripelennamine, the difference being that the

TABLE 71. ETHYLENEDIAMINE DERIVATIVES

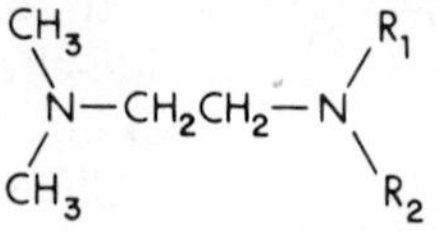

$$CH_3-N(CH_3)-CH_2CH_2-N(R_1)(R_2)$$

PROPRIETARY NAME	GENERIC NAME	R_1	R_2
2325 RP		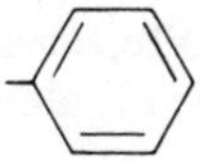 phenyl	—CH$_2$CH$_3$
Antergan		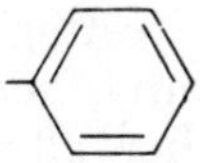 phenyl	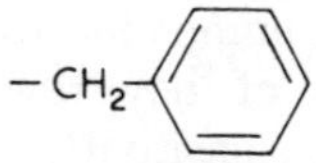 —CH$_2$—phenyl
Diatrin	Methaphenilene	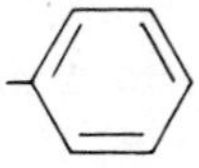 phenyl	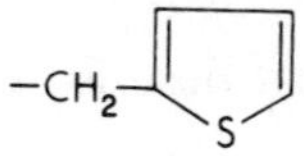 —CH$_2$—thienyl
Neo-Antergan	Pyrilamine	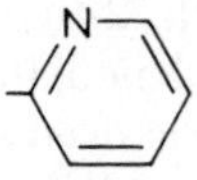 pyridyl	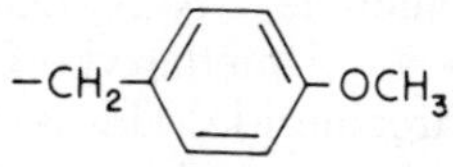 —CH$_2$—C$_6$H$_4$—OCH$_3$
Pyribenzamine Stanzamine	Tripelennamine	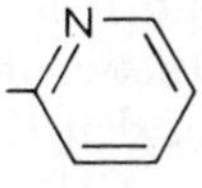 pyridyl	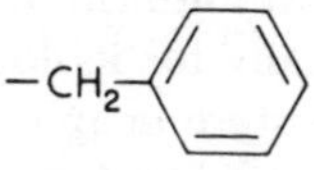 —CH$_2$—phenyl
Histadyl	Methapyrilene	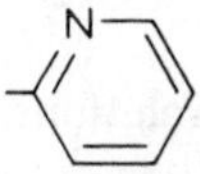 pyridyl	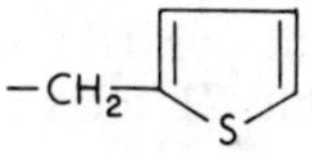 —CH$_2$—thienyl
Tagathen	Chlorothen	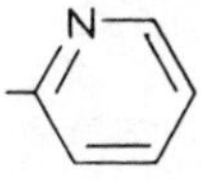 pyridyl	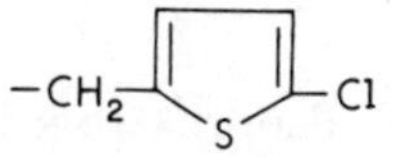 —CH$_2$—thienyl—Cl
Neohetramine	Thonzylamine	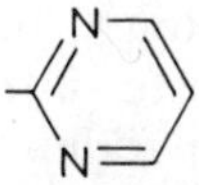 pyrimidyl	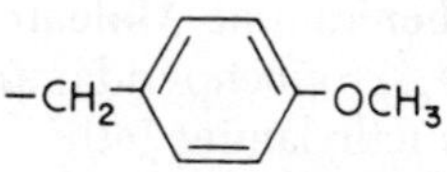 —CH$_2$—C$_6$H$_4$—OCH$_3$
	Zolamine	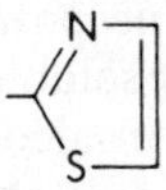 thiazolyl	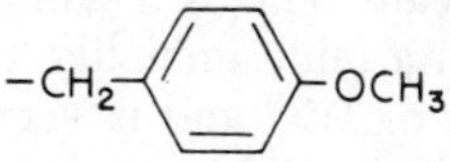 —CH$_2$—C$_6$H$_4$—OCH$_3$

benzyl group of triplennamine is replaced by the 5-halothenyl group. It is reported that the introduction of the halogen yields a compound which is more active and less toxic than the nonhalogenated compound.[18] However, bromination of the pyridyl moiety or introduction of a second halogen atom into the thenyl nucleus decreases the activity. Using equal doses, chlorothen protects against histamine shock twice as long as tripelennamine.

Category—antihistaminic.

Usual dose—25 mg. every 3 or 4 hours, should not exceed 150 mg. in a 24-hour period.

OCCURRENCE

Chlorothen Citrate Tablets N.F.

Thonzylamine Hydrochloride, Neohetramine® HCl, Anahist®, Resistab®, 2-{[(2-Dimethylamino)ethyl](p-methoxybenzyl)-amino}pyrimidine Hydrochloride. This compound is weaker than other members of the antihistaminic group of drugs[35]; however, its outstanding advantage is that it causes less drowsiness than the other drugs.

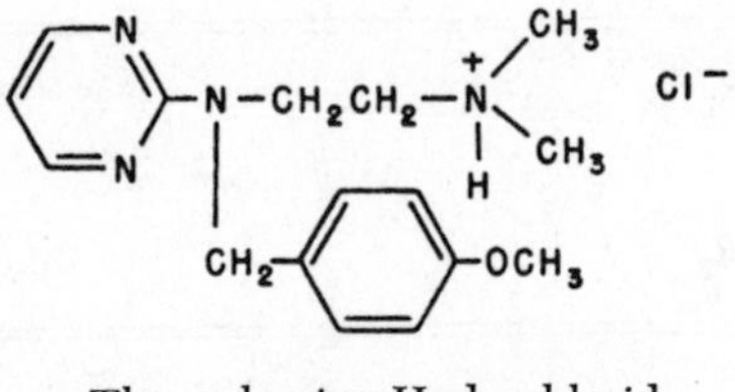

Thonzylamine Hydrochloride

A 2 per cent aqueous solution has a pH of about 5.5. The drug is a white, crystalline powder, soluble in water (1:1), alcohol (1:6) and chloroform (1:4).

It is recommended for use in treating the symptoms of hay fever, urticaria, serum and drug sensitivities and other allergic conditions

The usual dose is 50 mg. up to 4 times daily.

Zolamine, N′,N′-Dimethyl-N-(2-thiazolyl)-N-(p-methoxybenzyl)ethylenediamine Hydrochloride. This compound is used currently as an antihistaminic agent. It is also used topically for its local anesthetic effect. It is an isostere of pyrilamine in that the S of the thiazolyl moiety replaces a CH=CH group of the pyridyl residue.

See Table 71 for structure relationships of the ethylenediamine derivatives.

Propylamine Derivatives

Pheniramine Maleate, Trimeton®, Inhiston®, Prophenpyridamine Maleate, 2-{α[2-(Dimethylamino)ethyl]benzyl}pyridine Bimaleate. This is a white crystalline powder with a faint aminelike odor. It is soluble in water (1:5) and is very soluble in alcohol.

This antihistamine resulted from a study of the effect of replacement of the nitrogen or oxygen function between the nucleus and the side chain by a carbon linkage.[36]

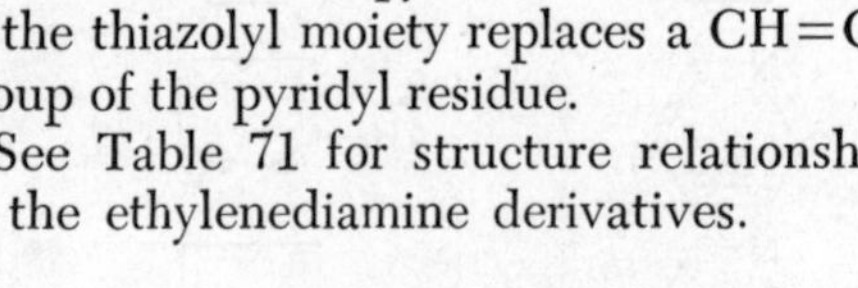

Pheniramine Maleate

In general, compounds of this series are not as active as those derived from ethylenediamine, but this is compensated by lowered toxicity and decreased incidence of side-effects. This usual adult dose is 20 to 40 mg. 3 times daily.

Chlorpheniramine Maleate U.S.P., Chlor-Trimeton®, Chlorprophenpyridamine Maleate, (±)2-{p-Chloro-α-[2-(dimethylamino)ethyl]benzyl}pyridine Bimaleate. This is

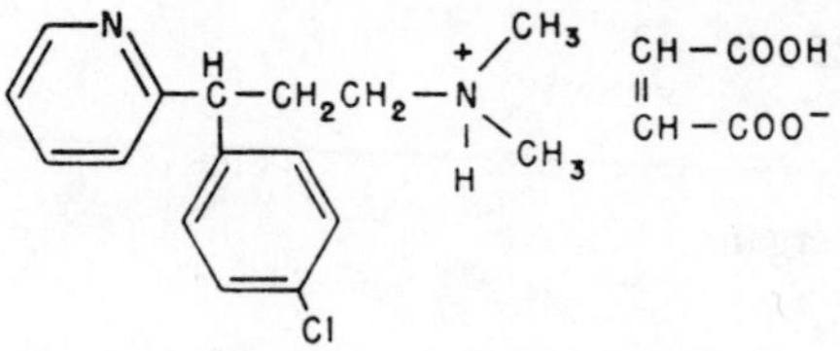

Chlorpheniramine Maleate

one of the most potent oral antihistamines that is available today. Chlorination of pheniramine in the p-position of the phenyl ring gave a 20-fold increase in potency with no appreciable change in toxicity. Chlorpheniramine is reported to have a therapeutic index of 50 as compared with 1 for tripelennamine and 4 for pheniramine.[37]

Chlorpheniramine maleate is a white, crystalline solid which is soluble in water (1:4), alcohol (1:10) and chloroform (1:10). It has a pKa of about 9.2 and an aqueous solution has a pH between 4 and 5.

Category—antihistaminic.

Usual dose—2 to 4 mg. 3 to 4 times daily.

Usual dose range—2 to 24 mg. daily.

Occurrence

Chlorpheniramine Maleate Injection N.F.
Chlorpheniramine Maleate Elixir U.S.P.
Chlorpheniramine Maleate Tablets U.S.P.

Dexchlorpheniramine Maleate N.F., Polaramine® Maleate, (+)-2-{p-Chloro-α-[2-(dimethylamino)ethyl]benzyl}pyridine Bimaleate. Chlorpheniramine exists as a racemic mixture. This mixture has been resolved, and studies on animals showed that the antihistaminic activity exists predominantly in the dextrorotatory enantiomorph.[38] The acute toxicity of the (+)-isomer was no greater than that of the racemic mixture.

Category—antihistaminic.

Usual dose—2 mg. 3 or 4 times daily.

Usual dose range—1 to 2 mg.

TABLE 72. PROPYLAMINE DERIVATIVES

PROPRIETARY NAME	GENERIC NAME	
Trimeton	Pheniramine	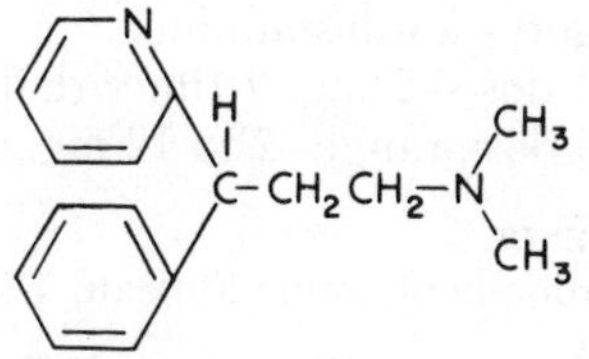
Chlor-Trimeton Polaramine	Chlorpheniramine Dexchlorpheniramine	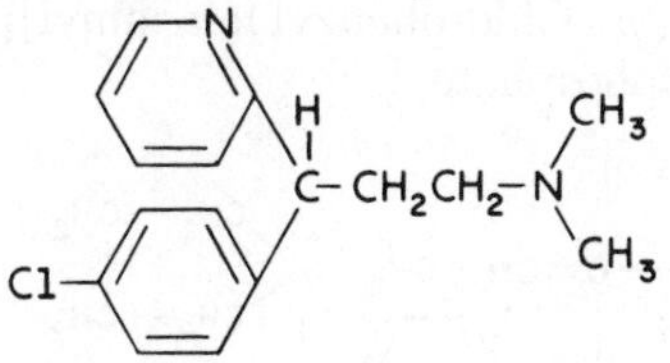
Dimetane Disomer	Brompheniramine Dexbrompheniramine	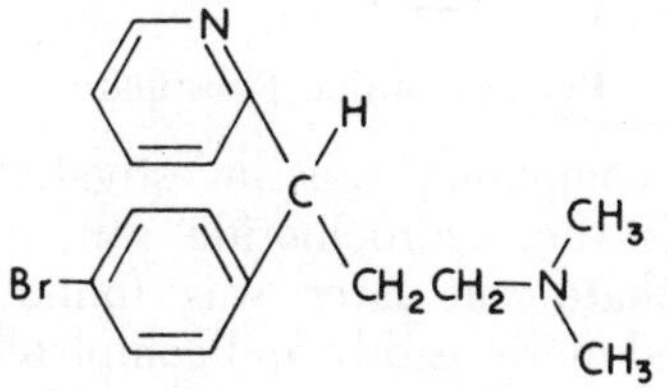
Pyronil	Pyrrobutamine	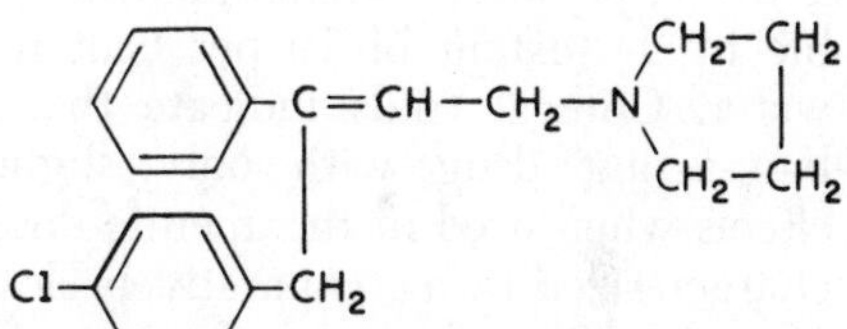
Actidil	Triprolidine	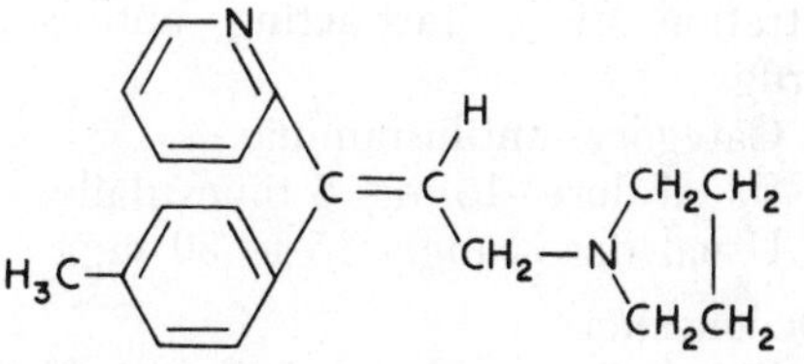

OCCURRENCE

Dexchlorpheniramine Maleate Syrup N.F.

Dexchlorpheniramine Maleate Tablets N.F.

Brompheniramine Maleate N.F., Dimetane®, Parabromdylamine Maleate, (±)-{p-Bromo-α-[2-(diethylamino)ethyl]benzyl}pyridine Bimaleate. This drug differs from chlorpheniramine by the substitution of a bromine atom for the chlorine atom.[39] Its actions and uses are similar to those of chlorpheniramine.

Category—antihistaminic.

Usual dose—oral, 4 mg. 1 to 4 times daily.

Usual dose range—4 to 8 mg.

OCCURRENCE

Brompheniramine Maleate Elixir N.F.

Brompheniramine Maleate Tablets N.F.

Dexbrompheniramine Maleate N.F., Disomer®, (+)-2-[p-Bromo-α-[2-(dimethylamino)ethyl]benzyl]pyridine Bimaleate. It is a white crystalline powder and the pH of a 1 per cent solution is about 5. Like the chlorine congener, the antihistaminic activity exists predominantly in the dextrorotatory isomer.

Category—antihistaminic.

Usual dose—2 mg. 4 times daily.

Usual dose range—2 to 12 mg. daily.

OCCURRENCE

Dexbrompheniramine Maleate Tablets N.F.

Pyrrobutamine Phosphate N.F., Pyronil®, 1 - [γ - (p - Chlorobenzyl) cinnamyl] pyrrolidine Diphosphate.

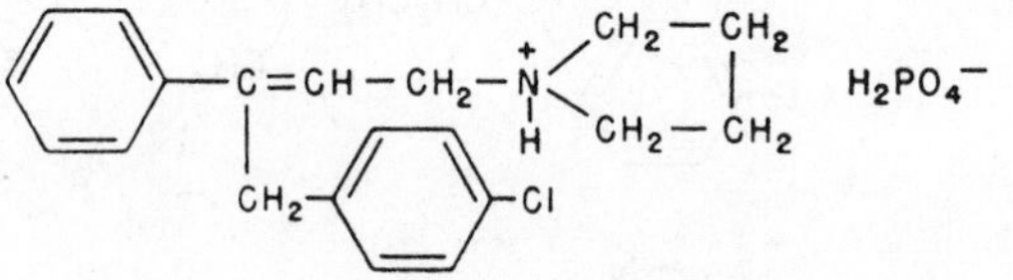

Pyrrobutamine Phosphate

This compound was investigated originally as the hydrochloride salt, but the diphosphate salt later was found to be absorbed more readily and completely. It is a white, crystalline material which is soluble to the extent of 10 per cent in warm water. Clinical trials indicate that it is a long-acting drug with only slight side-effects when used in therapeutic doses. It is characterized by a comparatively slow onset of action. The slowness of onset of action can be overcome by the concurrent administration of a fast-acting antihistaminic drug.

Category—antihistaminic.

Usual dose—15 mg. 3 times daily.

Usual dose range—15 to 30 mg.

OCCURRENCE

Pyrrobutamine Phosphate Tablets N.F.

Triprolidine Hydrochloride, N.F., Actidil®, *trans*-2-[3-(1-Pyrrolidinyl)-1-*p*-tolylpropenyl]pyridine Monohydrochloride. Tri-

Triprolidine Hydrochloride

prolidine was synthesized and tested as an antihistamine in the course of studies on a series of phenylpyridylallylamines.[40] The activity is confined mainly to the geometric isomer in which the pyrrolidinomethyl group is *trans* to the 2-pyridyl group. The potency is of the same order as that of chlorpheniramine. In contrast with most antihistamines, which are bitter tasting, triprolidine is claimed to be virtually tasteless. The peak effect occurs in about 3½ hours after oral administration, and the duration of effect is about 12 hours.

Category—antihistaminic.

Usual dose—2.5 mg. 2 or 3 times daily.

OCCURRENCE

Triprolidine Hydrochloride Tablets N.F.

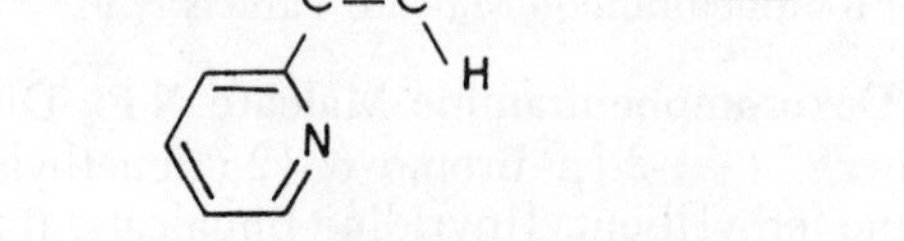

The structural relationships of the propylamine derivatives are shown in Table 72.

PHENOTHIAZINE AND RELATED STRUCTURE DERIVATIVES

Promethazine Hydrochloride U.S.P., Phenergan® Hydrochloride, Promethazine Hydrochloride, 10-[(2-Dimethylamino)propyl]phenothiazine Monohydrochloride. A search for effective antimalarials among phenothiazine derivatives and their subsequent testing as antihistaminic agents led to the discovery that the bridged structure, in which the aromatic rings on one of the nitrogen atoms were joined through a sulfur atom, had good activity. The most active histamine antagonist in the series was promethazine.[41] It occurs as a white to faint yellow, crystalline powder that is very soluble in water, in hot absolute alcohol and in chloroform.

In addition to its antihistaminic action it possesses an antiemetic effect, a tranquilizing action and a potentiating action on analgesic and sedative drugs.

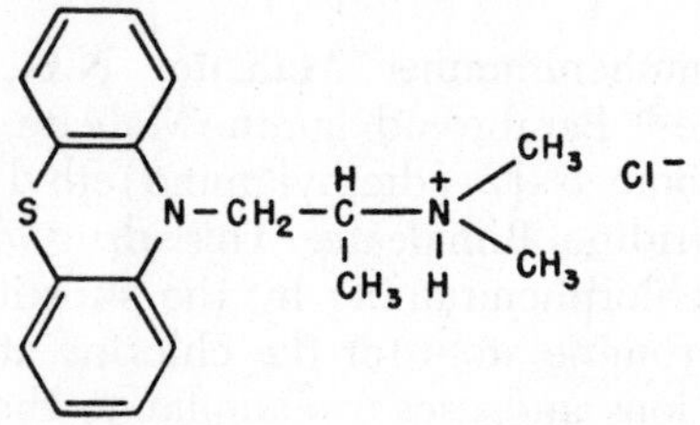

Promethazine Hydrochloride

It has been reported to be as much as 7 times more potent than certain other antihistaminics and to have approximately 3 times the duration of action.[42] It was found that a single dose given at night is enough to control symptoms in the majority of cases. A common side-effect is drowsiness, but this actually becomes an advantage if the drug is given at bedtime. The next day the soporific effect has worn off, but the antihistaminic effect still persists.

Category—antihistaminic.

Usual dose—oral or parenteral, 25 mg., repeated after 4 hours if required.

Usual dose range—5 to 75 mg. daily.

OCCURRENCE

Promethazine Hydrochloride Injection U.S.P.

Promethazine Hydrochloride Syrup U.S.P.

Promethazine Hydrochloride Tablets U.S.P.

Pyrathiazine Hydrochloride, Pyrrolazote®, 10-[2-(1-Pyrrolidyl)ethyl]phenothiazine Hydrochloride. This compound was

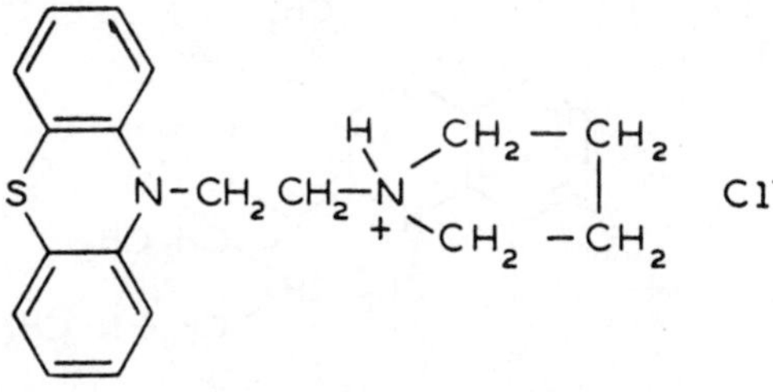

Pyrathiazine Hydrochloride

found to be the most active of a series of N-pyrrolidinoethylphenothiazines. It is about as potent as tripelennamine but less toxic and, like promethazine, it is a long-acting drug. It is interesting to note that the 2-pyrrolidino analog of promethazine is relatively inactive. Oxidation of the sulfur to sulfoxide or sulfone results in a great loss of activity.

In hay fever due to pollen sensitization, pyrathiazine has been shown to be effective in alleviating symptoms in about 80 per cent of the cases. It is also effective in relieving perennial vasomotor rhinitis, urticaria and bronchial asthma. The usual dose is 25 to 50 mg. 3 or 4 times a day.

Trimeprazine Tartrate N.F., Temaril®, (±)-10-[3-(Dimethylamino)-2-methylpropyl]phenothiazine Tartrate. This compound

was synthesized during the investigation of phenothiazine derivatives for physiologic activity.[43] Its antihistaminic action has been reported to be from 1½ to 5 times that of promethazine. Clinical studies have shown it to have a pronounced antipruritic action. This action may not be related to its histamine-antagonizing action.

Category—antipruritic.

Usual dose—2.5 mg. 4 times daily.

Usual dose range—10 to 40 mg. daily.

OCCURRENCE

Trimeprazine Tartrate Syrup N.F.

Trimeprazine Tartrate Tablets N.F.

Methdilazine N.F., Tacaryl®, 10-[(1-Methyl-3-pyrrolidinyl)methyl]phenothiazine. This compound is a light tan, crystalline powder, practically insoluble in water. Methdilazine, as the free base, is used in chewable tablets. These may cause some local anesthesia of the buccal mucosa if not chewed and swallowed promptly. The activity is that of methdilazine hydrochloride.

Category—antipruritic.

Usual dose—7.2 mg. (equivalent to 8 mg. of methdilazine hydrochloride) 2 to 4 times daily.

OCCURRENCE

Methdilazine Tablets N.F.

Methdilazine Hydrochloride N.F., Tacaryl® Hydrochloride, 10-[(1-Methyl-3-pyrrolidinyl)methyl]phenothiazine Mono-

Methdilazine Hydrochloride

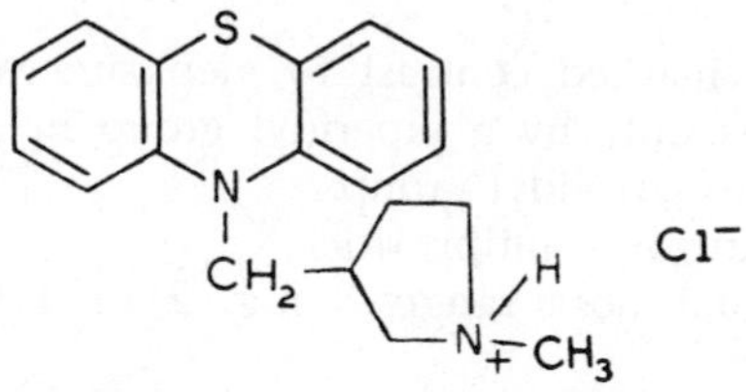

hydrochloride. On the basis of animal studies, methdilazine is rapidly and completely absorbed from the gastrointestinal tract.[44] Peak tissue levels are obtained within 30 minutes, whether given orally or by subcutaneous injection. The drug is excreted chiefly as a metabolite which is thought to be the sulfoxide. It is excreted fairly rapidly; no residual methdilazine was found in the tissues 24 hours after a single dose. This

TABLE 73. PHENOTHIAZINE AND RELATED STRUCTURE DERIVATIVES

PROPRIETARY NAME	GENERIC NAME	
Phenergan	Promethazine	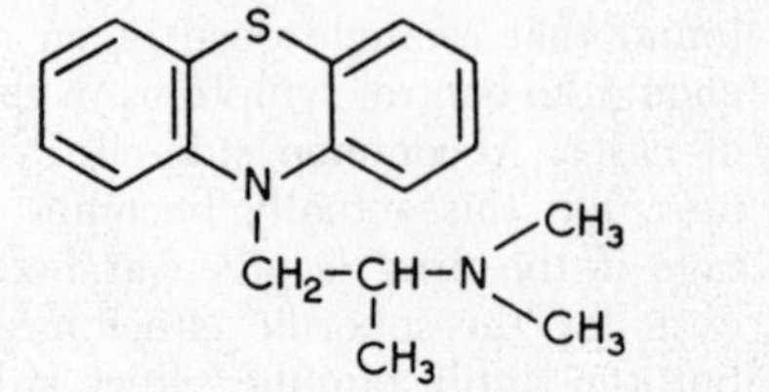
Pyrrolazote	Pyrathiazine	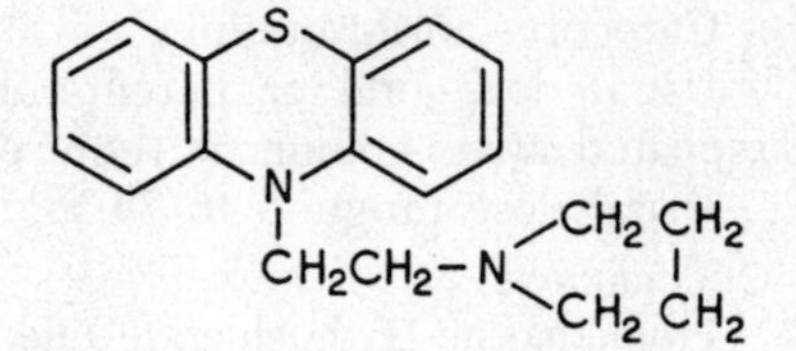
Temaril	Trimeprazine	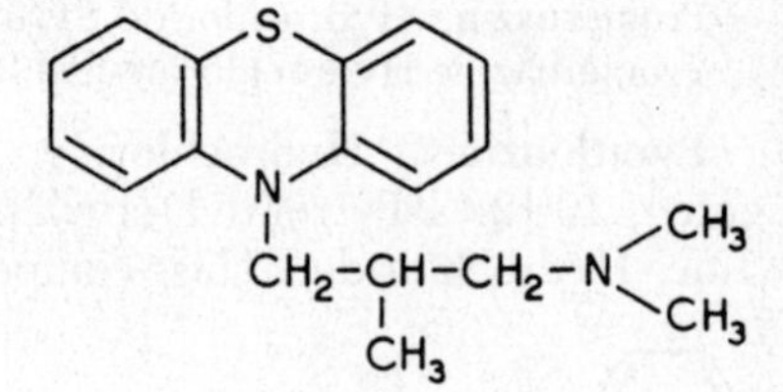
Tacaryl	Methdilazine	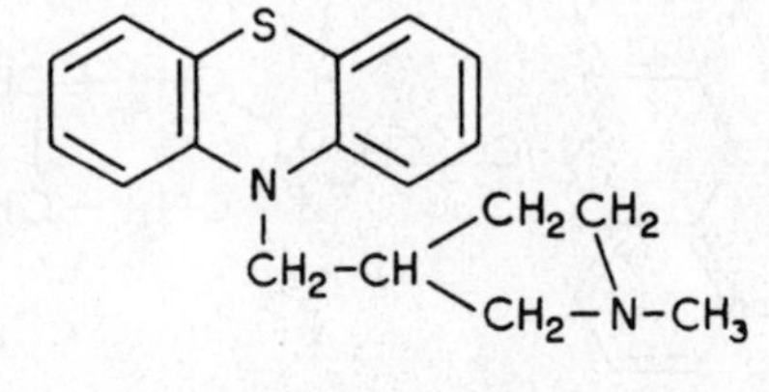

is in marked contrast to mepazine which differs only by a piperidyl group in place of the pyrrolidyl group.

Category—antipruritic.

Usual dose range—8 mg. 2 to 4 times daily.

OCCURRENCE

Methdilazine Hydrochloride Syrup N.F.
Methdilazine Hydrochloride Tablets N.F.

MISCELLANEOUS STRUCTURES

Cyclizine Hydrochloride U.S.P., Marezine® Hydrochloride, 1-(Diphenylmethyl)-4-methylpiperazine Monohydrochloride. This drug occurs as a light-sensitive, white, crystalline powder having a bitter taste. It is soluble in water (1:115), alcohol (1:115)

and chloroform (1:75). It is not used as an antihistaminic but has been found to be of value in the prophylaxis and treatment of motion sickness.

The lactate salt is used for intramuscular injection because of the limited water-solubility of the hydrochloride. The injection should be stored in a cold place because it may develop a slight yellow tint if stored at room temperature for several months. This does not indicate any loss in biologic potency.

Category—antinauseant.

Usual dose—oral, 50 mg. three times daily.

Usual dose range—50 to 200 mg. daily.

OCCURRENCE

Cyclizine Hydrochloride Tablets U.S.P.

Chlorcyclizine Hydrochloride N.F., Diparalene® Hydrochloride, Perazil®, 1-(p-Chloro-α-phenylbenzyl)-4-methylpiperazine Monohydrochloride. This occurs as a light-

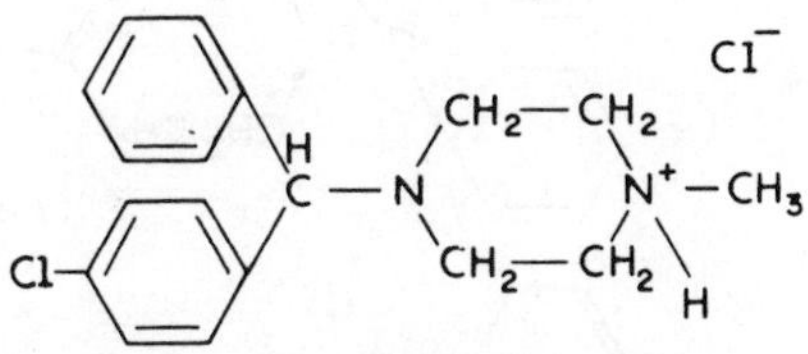

Chlorcyclizine Hydrochloride

sensitive, white, crystalline powder that is soluble in water (1:12) alcohol (1:11) and in chloroform (1:4). A 1 per cent aqueous solution has a pH between 4.8 and 5.5.

In a study of a series of related compounds for antihistaminic activity, chlorcyclizine was found to have the best therapeutic index. It is reported to be 4 times as active and one half as toxic as diphenhydramine and to have a long duration of action.[45]

Chlorcyclizine is distinguished by a piperazine ring rather than the ethylenediamine grouping common to most antihistaminics now commercially available. Disubstitution, or halogen in the 2- or 3-positions of one of the benzhydryl rings, results in a compound less potent than chlorcyclizine.[46]

In a study of the human pharmacology of chlorcycline, it was found to elicit a smaller percentage of drowsiness and other side-reactions, such as dryness of mouth, nausea and headache, than tripelennamine.[47] It is recommended as an adjunct to the basic therapeutic methods in the treatment of allergic conditions.[48] The action is prolonged, 1 oral daily dose of 50 mg. usually being sufficient for relief of symptoms. However, it must be noted that the interval between administration of the drug and the onset of its action may be as long as 2 hours. Chlorcyclizine is indicated in the symptomatic relief of urticaria, hay fever, certain types of vasomotor rhinitis and sinusitis and certain types of asthma.

Category—antihistaminic.
Usual dose—50 mg. up to 4 times daily.
Usual dose range—25 to 100 mg.
OCCURRENCE
Chlorcyclizine Hydrochloride Tablets N.F.

Meclizine Hydrochloride U.S.P., Bonine®, 1-(p-Chloro-α-phenylbenzyl)-4-(m-methylbenzyl)piperazine Dihydrochloride. Meclizine differs from chlorcyclizine by having an N-m-methylbenzyl group in place of the N-methyl group. Meclizine hydrochloride occurs as a bland tasting, white or slightly yellowish crystalline powder. It is soluble

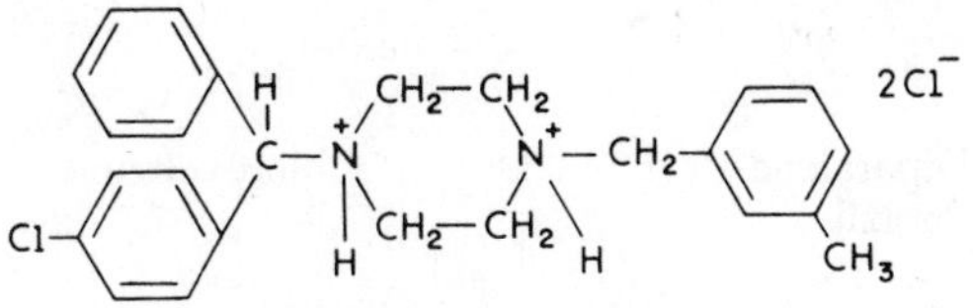

Meclizine Hydrochloride

in water to the extent of 0.1 per cent and in alcohol to the extent of about 4 per cent.

This compound is a potent antihistaminic which is characterized by a slow onset of action and a long duration of action.[49] The action of a single oral dose extends over a period of 9 to 24 hours. Although it is a potent antihistaminic, its primary use is as an antinauseant in the prevention or treatment of motion sickness and, also, in the treatment of nausea and vomiting associated with vertigo and radiation sickness.

Category—antinauseant.
Usual dose—25 to 50 mg. once daily.
Usual dose range—25 to 100 mg. daily.

OCCURRENCE
Meclizine Hydrochloride Tablets U.S.P.

Diphenylpyraline Hydrochloride, Hispril®, Diafen®; 4-Diphenylmethoxy-1-methylpiperidine Hydrochloride. This occurs as

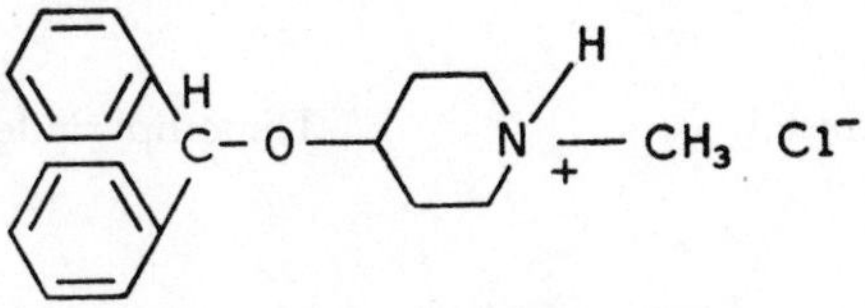

Diphenylpyraline Hydrochloride

a white or slightly off-white crystalline powder which is soluble in water or alcohol. It can be synthesized by refluxing 1-methyl-4-piperindol and benzhydryl bromide in xylene.[50]

It is a potent antihistaminic agent which

TABLE 74. MISCELLANEOUS STRUCTURES

PROPRIETARY NAME	GENERIC NAME	
Marezine	Cyclizine	
Diparalene Perazil	Chlorcyclizine	
Bonine	Meclizine	
Hispril Diafen	Diphenylpyraline	
Thephorin	Phenindamine	
Forhistal	Dimethpyrindene	
Allercur	Clemizole	

TABLE 74. MISCELLANEOUS STRUCTURES (*Continued*)

PROPRIETARY NAME	GENERIC NAME	
Antistine	Antazoline	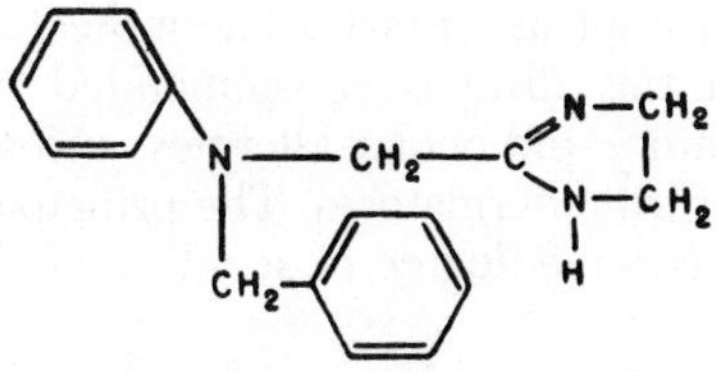
Periactin	Cyproheptadine	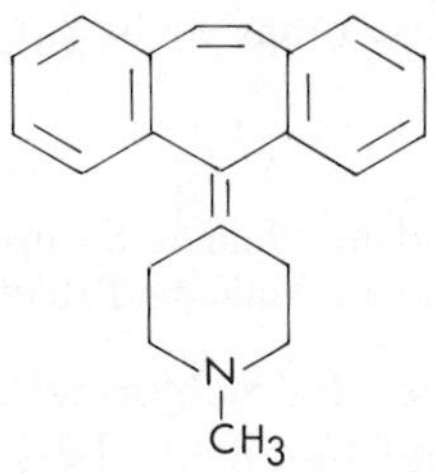

is structurally related to diphenhydramine. At ordinary doses it causes only a low incidence of side-reactions. It has been reported that many patients who previously had been treated unsuccessfully with other antihistamines responded to treatment with diphenylpyraline.[51] The drug is available as 2-mg. tablets and as 5-mg. sustained release capsules.

Phenindamine Tartrate N.F., Thephorin®, 2,3,4,9-Tetrahydro-2-methyl-9-phenyl-1H-indeno-[2,1-c]pyridine Bitartrate. This compound seemingly is unrelated to the conventional antihistamines, but the usual dimethylaminoethyl side chain is apparent if one considers the open-ring structure.

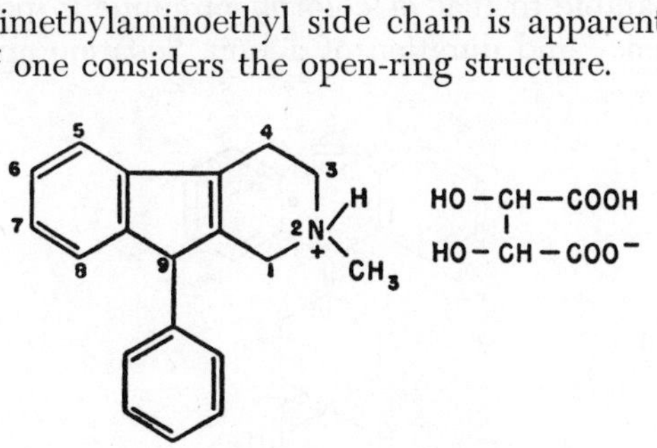

Phenindamine Tartrate

This is somewhat analogous to the imidazoline side chain of antazoline.[52] In a comparison of relative toxicities, it was found that phenindamine was less toxic than diphenhydramine.[53] See Table 74 for structural comparison with other antihistaminic agents.

Phenindamine is a stable, white, crystalline powder, soluble up to 3 per cent in water; a 2 per cent aqueous solution has a pH of about 3.5. It is most stable in the pH range 3.5 to 5.0 and is unstable in solutions of pH 7 or higher. Oxidizing substances should not be combined with it, nor should it be heated because this may cause isomerization to an inactive form.

Unlike the other antihistamines in common use, it does not produce drowsiness and sleepiness; on the contrary, it has a mildly stimulating action in some patients and as a result may cause insomnia when taken just before bedtime.[54]

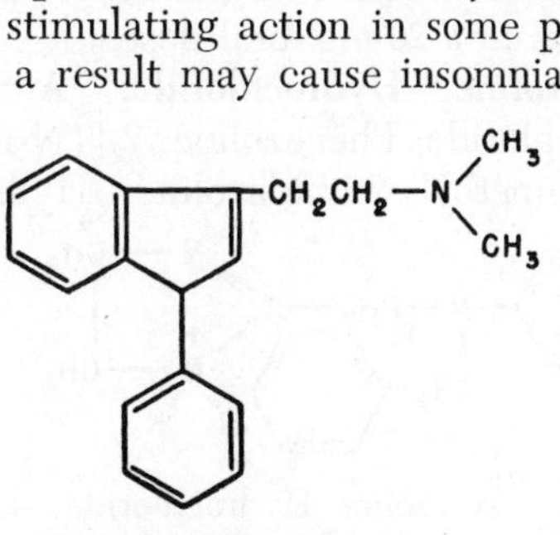

Category—antihistaminic.

Usual dose—25 mg. up to 4 times daily as needed.

Usual dose range—25 to 50 mg.

OCCURRENCE

Phenindamine Tartrate Tablets N.F.

Dimethindene Maleate N.F., Forhistal® Maleate, 2-{1-[2-[2-(Dimethylamino)ethyl]-inden-3-yl]ethyl}pyridine Maleate, in contrast with many of the antihistamines, is free of unpleasant taste. The indications for which this drug is recommended include respiratory and ocular allergies, and allergic and pruritic dermatoses. The principal side-effect is some degree of sedation or drowsiness.

Category—antihistaminic.

Usual dose range—1 to 2 mg. 1 to 3 times daily.

OCCURRENCE
Dimethindene Maleate Syrup N.F.
Dimethindene Maleate Tablets N.F.

Clemizole Hydrochloride, Allercur®, Reactrol®, 1-*p*-Chlorobenzyl-2-(1-pyrrolidinyl-methyl)benzimidazole Hydrochloride. This

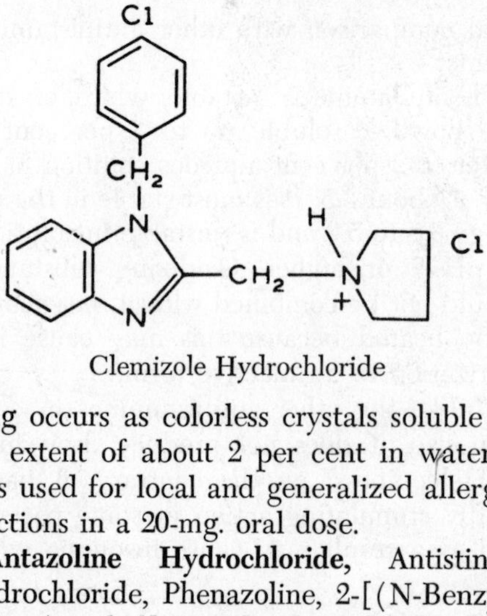

Clemizole Hydrochloride

drug occurs as colorless crystals soluble to the extent of about 2 per cent in water.[55] It is used for local and generalized allergic reactions in a 20-mg. oral dose.

Antazoline Hydrochloride, Antistine® Hydrochloride, Phenazoline, 2-[(N-Benzyl-anilino)methyl]-2-imidazoline Hydrochlo-

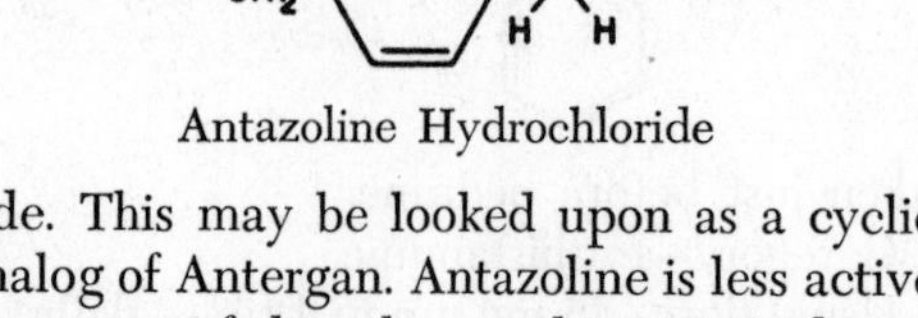

Antazoline Hydrochloride

ride. This may be looked upon as a cyclic analog of Antergan. Antazoline is less active than most of the other antihistaminic drugs, but it is characterized by the lack of local irritation. The effects of imidazoline deriva-

tives are highly varied. A slight change of the antazoline structure yields the sympathomimetic drug naphazoline.[56] Given orally, its action compares favorably with that of tripelennamine and diphenhydramine.[57]

Antazoline hydrochloride is a white, crystalline material which is sparingly soluble in alcohol or water and practically insoluble in ether. It has a pKa of about 10 and a 1 per cent solution has a pH of about 6.

The usual dose is 50 to 100 mg.

Antazoline Phosphate N.F., Antistine® Phosphate, 2-[(N-Benzylanilino)methyl]-2-imidazoline Dihydrogen Phosphate. This salt is soluble in water and less irritating when applied locally for allergies of the eye.

A 2 per cent aqueous solution has a pH of about 4.5.

Category—antihistaminic.

Application—1 or 2 drops of a 0.5 per cent solution in each eye every 3 or 4 hours.

OCCURRENCE
Antazoline Phosphate Ophthalmic Solution N.F.

Cyproheptadine Hydrochloride, N.F., Periactin® Hydrochloride, 4-(5H-Dibenzo-[*a,d*]cyclohepten-5-ylidene)-1-methylpiperidine Hydrochloride. This compound possesses both an antihistamine and an antiserotonin activity. Animal experiments have shown it to have antihistaminic activity comparable to that of chlorpheniramine in potency and duration of action. Sedation ap-

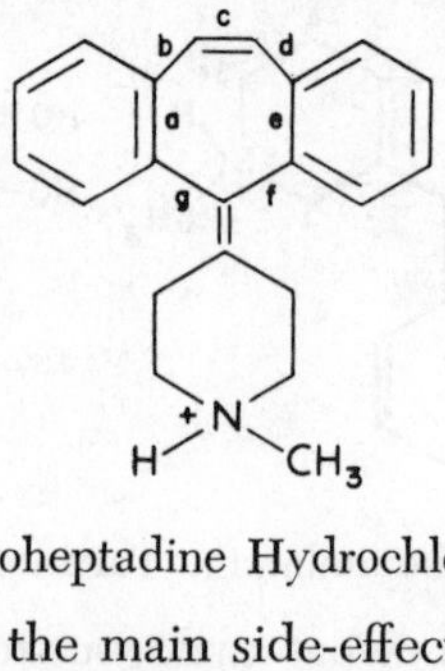

Cyproheptadine Hydrochloride

pears to be the main side-effect, and this is usually brief, disappearing after 3 or 4 days treatment.

Cyproheptadine Hydrochloride is reported to be effective in the treatment of various itching skin conditions.

Category—antihistaminic; antipruritic.

Usual dose—4 mg. 3 or 4 times daily.

Usual dose range—4 to 20 mg. daily.

OCCURRENCE

Cyproheptadine Hydrochloride Syrup N.F.

Cyproheptadine Hydrochloride Tablets N.F.

REFERENCES CITED

1. Windaus, A., and Vogt, W.: Ber. deutsch. chem. Ges. 40:3691, 1907.
2. Best, C. H., et al.: J. Physiol. 62:397, 1927.
3. Feinburg, S. M.: J.A.M.A. 132:702, 1946.
4. Dale, H. H.: Lancet 1:1233, 1929.
5. Fell, N. H.: U. S. patent 2,376,424, May 22, 1945; through Chem. Abstr. 39:4722, 1945.
6. Council on Pharmacy and Chemistry: J.A.M.A. 115:1019, 1940.
7. Rocha, E., and Silva, M.: J. Pharmacol. Exp. Ther; 80:399, 1944.
8. Bovet, D., and Staub, A. M.: Compt. rend. Soc. biol. 124:547, 1937.
9. Bovet, D., and Bovet-Nitti, F.: Medicaments du systeme nerveux vegetaif, p. 741, Basel, Karger, 1948.
10. Loew, E. R.: Physiol. Rev. 27:542, 1947.
11. Fischel, E. .E: Proc. Soc. Exp. Biol. Med. 66:537, 1947.
12. Strauss, W. T.: J. Am. Pharm. A. (Pract. Ed.) 9:728, 1948.
13. Brewster, J. M.: U. S. Navy Med. Bull. 49:1, 1949.
14. Buchan, R. F., et al.: Indust. Hygiene Occup. Med. 4:32, 1951.
15. Carson, L. E., and Campbell, C. C.: Science 111:689, 1950.
16. Mitchell, R. B., et al.: J. Am. Pharm. A. (Sci. Ed.) 41:472, 1952.
17. Winder, C. V., et al.: J. Pharmacol. Exp. Ther. 87:121, 1946.
18. Clapp, R. C., et al.: J. Am. Chem. Soc. 69:1549, 1947.
19. Loew, E. R., et al.: J. Pharmacol. Exp. Ther. 86:1, 1946.
20. Lordi, N. G., and Christian, J. E.: J. Am. Pharm. A. (Sci. Ed.) 45:300, 1956.
21. Rieveschl, G., Jr.: U. S. Patent 2,421,714, June 3, 1947; through Chem. Abstr. 41: 5550h, 1947.
22. McGavack, T. H., et al.: J. Allergy 22:31, 1951.
23. Cusic, J. W.: U. S. patent 2,499,058, February 28, 1950; through Chem. Abstr. 44: 4926g, 1950.
24. Brown, E. A., et al.: Ann. Allergy 6:1, 1948.
25. MacQuiddy, E. L.: Nebraska M. J. 34: 123, 1948.
26. Halpern, B. N.: Arch. int. pharmacodyn. 68:339, 1942.
27. Ercoli, N., et al.: Arch. Biochem. 13:487, 1947.
28. Bovet, D., et al.: Compt. rend. Soc. biol. 138:99, 1944.
29. Mayer, R. O., et al.: Science 102:93, 1945.
30. Huttrer, C. P., et al.: J. Am. Chem. Soc. 68:1999, 1946.
31. Clark, J. H., et al.: J. Org. Chem. 14:216, 1949.
32. Kyrides, L. P., et al.: J. Am. Chem. Soc. 69:2239, 1947.
33. Leonard, F., and Solmssen, U. V.: J. Am. Chem. Soc. 70:2064, 1948.
34. Weston, A. W.: J. Am. Chem. Soc. 69: 980, 1947.
35. Biel, J. H.: J. Am. Chem. Soc. 71:1306, 1949.
36. Sperber, N., et al.: J. Am. Chem. Soc. 73: 5752, 1951.
37. Tislow, R., et al.: Fed. Proc. 8:338, 1949.
38. Roth, F. E., and Gavier, W. W.: J. Pharmacol. Exp. Ther. 124:347, 1958.
39. Sperber, N., et al.: U. S. patent 2,676,964, April 27, 1954; through Chem. Abstr. 49: 6316f. 1955.
40. Green, A. F.: Brit. J. Pharmacol. 8:171, 1953. Wellcome Foundation Ltd.: Brit. patent 719,276, December 1, 1954; through Chem. Abstr. 50:1090b, 1956.
41. Halpern, B. N.: Bull. Soc. Chem. Biol. 39:309, 1947.
42. Bain, W. A., et al.: Lancet 2:47, 1949.
43. Jacob, R. M., and Robert, J. G.: U. S. patent 2,837,518, June 3, 1958; through Chem. Abstr. 52:16382d, 1958.
44. Weikel, J. H., et al.: Toxicol. & Applied Pharmacol. 2:68, 1960.
45. Roth, L. W., et al.: Arch. intern. pharmaçodyn. 80:378, 1949.
46. Baltzly, R., et al.: J. Org. Chem. 14:775, 1949.
47. Jaros, S. H., et al.: Ann. Allergy 7:458, 1949.

48. Roth, L. W., *et al.*: Arch. intern. pharma-codyn. 80:466, 1949.
49. P'An, S. Y., *et al.*: J. Am. Pharm. A. (Sci. Ed.) 43:653, 1954.
50. Knox, L. H., and Kapp, R.: U. S. patent 2,479,843, August 23, 1949; through Chem. Abstr. 44:1144a, 1950.
51. Maxwell, M. J.: Lancet 2:828, 1958.
52. Huttrer, C. P.: Experientia 5:53, 1949.
53. Lehman, G.: J. Pharmacol. Exp. Ther. 92:249, 1948.
54. Criep, L. H.: Journal-Lancet 68:55, 1948.
55. Finkelstein, M., *et al.*: J. Am. Pharm. A. (Sci. Ed.) 49:18, 1960.
56. Nickerson, M.: J. Pharmacol. Exp. Ther. 95:27, 1949.
57. Friedlander, A. S., and Friedlander, S.: Ann. Allergy 6:23, 1948.

SELECTED READING

Friend, D. G.: The antihistamines. Clin. Pharmacol. Ther. 1:5, 1960.
Haley, T. J.: The antihistaminic drugs, J. Am. Pharm. A. (Sci. Ed.) 37:383, 1948.
Idson, B.: Antihistamine drugs, Chem. Rev. 47:307, 1950.
——: Chemical Industries Week, p. 7, March 31, 1951.
Leonard, F., and Huttrer, C. P.: Histamine Antagonists, Washington, D. C., Nat. Res. Council, 1950.
Wilhelm, R. E.: The new anti-allergic agents, Med. Clin. N. Amer. 45:887, 1961.
Witiak, D. T.: Antiallergic Agents, *in* Burger, A. (ed.): Medicinal Chemistry, ed. 3, p. 1643, New York, Wiley-Interscience, 1970.

24
Analgesic Agents

Taito O. Soine, Ph.D.,
Professor of Medicinal Chemistry, and Chairman of the
Department of Medicinal Chemistry, College of Pharmacy,
University of Minnesota

and

Robert E. Willette, Ph.D.,
Associate Professor of Medicinal Chemistry,
School of Pharmacy,
University of Connecticut

INTRODUCTION

Man's struggle to relieve pain began with the origin of mankind. Ancient writings, both serious and fanciful, dealt with secret remedies, religious rituals, and other methods of pain relief. Slowly, there evolved the present, modern era of synthetic analgesics.*

Tainter[1] has divided the history of analgesic drugs into 4 major eras, namely:

1. The period of discovery and use of naturally occurring plant drugs;

2. The isolation of pure plant principles, e.g., alkaloids, from the natural sources and their identification with analgesic action;

3. The development of organic chemistry and the first synthetic analgesics;

4. The development of modern pharmacologic technics, making it possible to undertake a systematic testing of new analgesics.

The discovery of morphine's analgesic activity by Sertürner, in 1806, ushered in the second era. It continues today only on a small scale. Wöhler introduced the third era indirectly with his synthesis of urea in 1828. He showed that chemical synthesis could be used to make and produce drugs. In the third era, the first synthetic analge-

sics used in medicine were the salicylates. These originally were found in nature (methyl salicylate, salicin) and then were synthesized by chemists. Other early, man-made drugs were acetanilid (1886), phenacetin (1887), and aspirin (1899).

These early discoveries were the principal contributions in this field until the modern methods of pharmacologic testing initiated the fourth era. The effects of small structural modifications on synthetic molecules now could be assessed accurately by pharmacologic means. This has permitted systematic study of the relationship of structure to activity during this era. The development of these pharmacologic testing procedures, coupled with the fortuitous discovery of meperidine by Eisleb and Schaumann,[2] has made possible the rapid strides in this field today.

The consideration of synthetic analgesics, as well as the naturally occurring ones, will be facilitated considerably by dividing them into 2 groups: morphine and related compounds and the antipyretic analgesics.

It should be called to the reader's attention that there are numerous drugs which, in addition to possessing distinctive pharmacologic activities in other areas, may also possess analgesic properties. The analgesic property exerted may be a direct effect or may be indirect but is subsidiary to some other more pronounced effect. Some ex-

---

* An analgesic may be defined as a drug bringing about insensibility to pain without loss of consciousness. The etymologically correct term "analgetic" may be used in place of the incorrect but popular "analgesic."

TABLE 75. SYNTHETIC DERIVATIVES OF MORPHINE

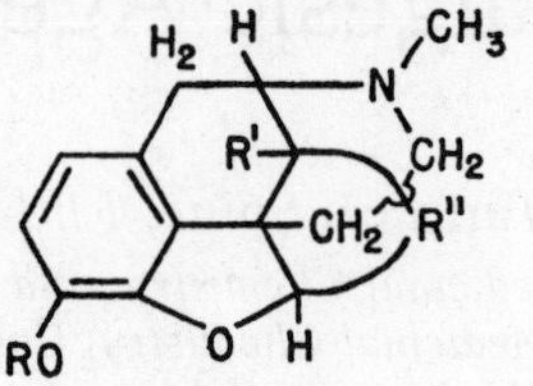

Compound *Proprietary Name*	R	R′	R″	Principal Use
Morphine	H	H	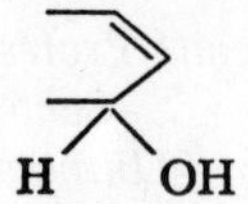 (allyl group with H and OH)	Analgesic
Codeine	CH_3	H	Same as above	Analgesic and to depress cough reflex
Ethylmorphine *Dionin*	C_2H_5	H	Same as above	Ophthalmology
Diacetylmorphine (Heroin)	CH_3CO	H	$-O-\overset{O}{\overset{\|}{C}}-CH_3$	Analgesic (prohibited in U.S.)
Hydromorphone (Dihydromorphinone) *Dilaudid* *Hymorphan*	H	H	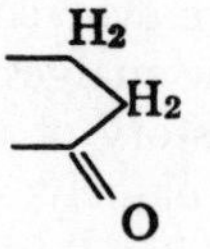 H_2 H_2 O	Analgesic
Hydrocodone (Dihydrocodeinone) *Dicodid*	CH_3	H	Same as above	Analgesic and to depress cough reflex
Oxymorphone (Dihydrohydroxy- morphinone) *Numorphan*	H	OH	Same as above	Analgesic
Oxycodone (Dihydrohydroxy- codeinone)	CH_3	OH	Same as above	Analgesic and to depress cough reflex
Dihydrocodeine *Paracodin*	CH_3	H	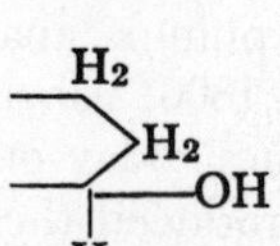 H_2 H_2 OH H	Depress cough reflex
Dihydromorphine	H	H	Same as above	Analgesic
Dihydrodesoxy- morphine-D (Desomorphine)	H	H	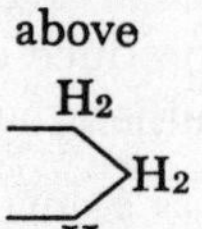 H_2 H_2 H_2	Analgesic
Dihydrodesoxy- codeine-D (Desocodeine)	CH_3	H	Same as above	Analgesic

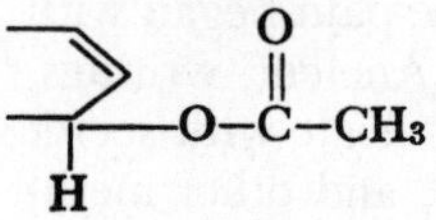

TABLE 75. SYNTHETIC DERIVATIVES OF MORPHINE (*Continued*)

COMPOUND *Proprietary Name*	R	R'	R''	PRINCIPAL USE
Methyl dihydro- morphinone *Metopon*	H	H	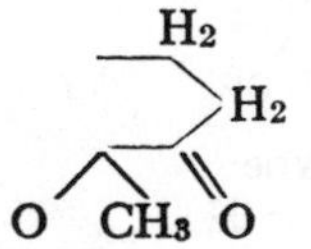	Analgesic

(Structure for R'': showing H_2 and H_2 groups with O, CH₃, O and CH₃)

 amples of these, which are discussed elsewhere in this text, are: sedatives (e.g., barbiturates); muscle relaxants (e.g., mephenesin, methocarbamol, carisoprodol, phenyramidol chlorzoxazone); tranquilizers (e.g., meprobamate, phenaglycodol, chlormezanone) etc. These types will not be considered in this chapter.

MORPHINE AND RELATED COMPOUNDS

The discovery of morphine early in the 19th century and the demonstration of its potent analgesic properties led directly to the search for more of these potent principles from plant sources. In tribute to the remarkable potency and action of morphine, it has remained alone as an outstanding and indispensable analgesic from a plant source.

It is only since 1938 that synthetic compounds rivaling it in action have been found, although many earlier changes made on morphine itself gave more effective agents.

Modifications of the morphine molecule will be considered under the following headings:

1. Early changes on morphine prior to the work of Small, Eddy and their co-workers.

2. Changes on morphine initiated in 1929 by Small, Eddy and co-workers[3] under the auspices of the Committee on Drug Addiction of the National Research Council and extending to the present time.

3. The researches initiated by Eisleb and Schaumann,[2] in 1938, with their discovery of the potent analgesic action of meperidine, a compound departing radically from the typical morphine molecule.

4. The researches initiated by Grewe, in 1946, leading to the successful synthesis of the morphinan group of analgesics.

Early Morphine Modications. Morphine is obtained from **opium,** which is the partly dried latex from incised unripe capsules of *Papaver somniferum.* Opium contains numerous alkaloids (as meconates and sulfates), of which morphine, codeine, noscapine (narcotine) and papaverine are therapeutically the most important. Other alkaloids, such as narceine, also have been tested medicinally but are not of great importance. The action of opium is due principally to its morphine content. As an analgesic, opium is not as effective as morphine because of its slower absorption, but it has a greater constipating action and is thus better suited for antidiarrheal preparations (e.g., paregoric). Opium, as a constituent of Dover's powders and Brown Mixture, also exerts a valuable expectorant action which is superior to that of morphine.

Two types of basic structures usually are recognized among the opium alkaloids, i.e., the *phenanthrene* (morphine) type and the *benzyl-isoquinoline* (papaverine) type (see p. 702).

The pharmacologic actions of the two types of alkaloids are dissimilar. The morphine group acts principally on the central nervous system as a depressant and stimulant, whereas the papaverine group has little effect on the nervous system but has a marked antispasmodic action on smooth muscle.

Clinically, the depressant action of the morphine group is the most useful property, resulting in an increased tolerance to pain, a sleepy feeling, a lessened perception to external stimuli, and a feeling of well-being (euphoria). Respiratory depression, central

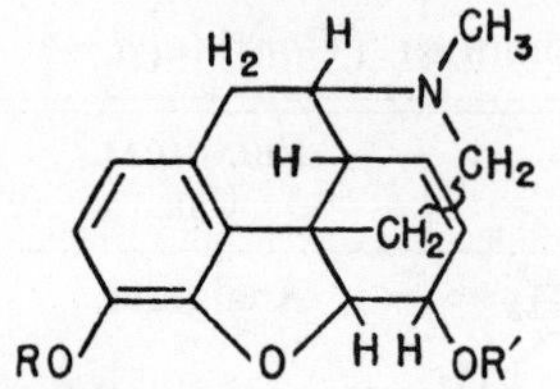

Phenanthrene Type
(Morphine, R & R′ = H)

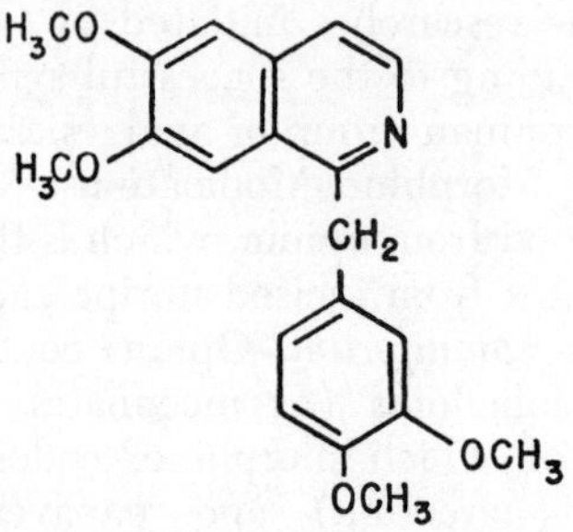

Benzyl-Isoquinoline Type
(Papaverine)

in origin, is perhaps the most serious objection to this type of alkaloid, aside from its tendency to cause addiction. The stimulant action is well illustrated by the convulsions produced by certain members of this group (e.g., thebaine).

Prior to 1929, the derivatives of morphine that had been made were primarily the result of simple changes on the molecule, such as esterification of the phenolic or alcoholic hydroxyl group, etherification of the phenolic hydroxyl group and similar minor changes. The net result had been the discovery of some compounds with greater activity than morphine but also greater toxicities and addiction tendencies. No compounds had been found that did not possess in some measure the addiction liabilities of morphine.*

Some of the compounds that were in common usage prior to 1929 are listed in Table 75, together with some other more recently introduced ones. All have the morphine skeleton in common.

* The term "addiction liability," or the preferred term "dependence liability," as used in this text, indicates the ability of a substance to develop true addictive tolerance and physical dependence and/or to suppress the morphine abstinence syndrome following withdrawal of morphine from addicts.

Among the earlier compounds is codeine, the phenolic methyl ether of morphine which also had been obtained from natural sources. It has survived as a good analgesic and cough depressant, together with the corresponding ethyl ether which has found its principal application in ophthalmology. The diacetyl derivative of morphine, heroin, has been known for a long time; it has been banished for years from the United States and is being used in decreasing amounts in other countries. Among the reduced compounds were dihydromorphine and dihydrocodeine and their oxidized congeners, dihydromorphinone (hydromorphone) and dihydrocodeinone (hydrocodone). Derivatives of the last two compounds possessing a hydroxyl group in position 14 are dihydrohydroxymorphinone, or oxymorphone, and dihydrohydroxycodeinone, or oxycodone. These represent the principal compounds that either had been on the market or had been prepared prior to the studies of Small, Eddy and co-workers.† It is well to note that no really systematic effort had been made to investigate the structure-activity relationships in the molecule, and only the easily changed peripheral groups had been modified.

Morphine Modifications Initiated by the Researches of Small and Eddy. The avowed purpose of Small, Eddy and co-workers,[3] in 1929, was to approach the morphine problem from the standpoint that:

1. It might be possible to separate chemically the addiction property of morphine from its other more salutary attributes. That this could be done with some addiction-producing compounds was shown by the development of the nonaddictive procaine from the addictive cocaine.

2. If it were not possible to separate the addictive tendencies from the morphine molecule, it might be possible to find other synthetic molecules without this undesirable property.

Proceeding on these assumptions, they first examined the morphine molecule in an exhaustive manner. As a starting point, it

† The only exception is oxymorphone; this was introduced in the U. S. in 1959 but is mentioned here because it obviously is closely related to oxycodone.

TABLE 76. SOME STRUCTURAL RELATIONSHIPS IN THE MORPHINE MOLECULE

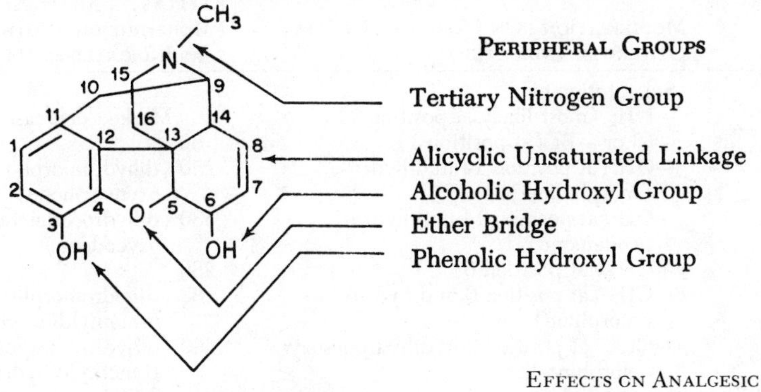

PERIPHERAL GROUPS OF MORPHINE	MODIFICATION (ON MORPHINE UNLESS OTHERWISE INDICATED)	EFFECTS ON ANALGESIC ACTIVITY* (MORPHINE OR ANOTHER COMPOUND AS INDICATED $= 100$)
Phenolic Hydroxyl	$-OH \rightarrow -OCH_3$ (codeine)	15
	$-OH \rightarrow -OC_2H_5$ (ethylmorphine)	10
	$-OH \rightarrow -OCH_2CH_2-N\underset{}{\bigcirc}O$ (pholcodine)	1
Alcoholic Hydroxyl	$-OH \rightarrow -OCH_3$ (heterocodeine)	500
	$-OH \rightarrow -OC_2H_5$	240
	$-OH \rightarrow -OCOCH_3$	420
	$-OH \rightarrow =O$ (morphinone)	37
	†$-OH \rightarrow =O$ (dihydromorphine to dihydromorphinone)	600 (dihydromorphine vs. dihydromorphinone)
	†$-OH \rightarrow =O$ (dihydrocodeine to dihydrocodeinone)	390 (dihydrocodeine vs. dihydrocodeinone)
	†$-OH \rightarrow -H$ (dihydromorphine to dihydrodesoxymorphine-D)	1000 (dihydromorphine vs. dihydrodesoxymorphine-D)
Ether Bridge	‡$=C-O-CH- \rightarrow =C-OH\ HCH-$ (dihydrodesoxymorphine-D to tetrahydrodesoxymorphine)	13 (dihydrodesoxymorphine-D vs. tetrahydrodesoxymorphine)
Alicyclic Unsaturated Linkage	$-CH=CH- \rightarrow -CH_2CH_2-$ (dihydromorphine)	120
	†$-CH=CH- \rightarrow -CH_2CH_2-$ (codeine to dihydrocodeine)	115 (codeine vs. dihydrocodeine)
Tertiary Nitrogen	$\diagdown N-CH_3 \diagdown \rightarrow N-H$ (normorphine)	5
	$\diagdown N-CH_3 \diagdown \rightarrow N-CH_2CH_2-C_6H_5$	1400
	§ $\diagdown N-CH_3 \diagdown \rightarrow N-R$	Reversal of activity (morphine antagonism). R = propyl, isobutyl, allyl, methallyl.
	$\diagdown N-CH_3 \rightarrow \overset{CH_3}{\underset{CH_3}{\diagdown \overset{+}{N} \diagup}}\ Cl^-$	1 (strong curare action)
	Opening of nitrogen ring (morphimethine)	Marked decrease in action.

(Table 76 continues on p. 704)

TABLE 76. SOME STRUCTURAL RELATIONSHIPS IN THE MORPHINE MOLECULE (*Continued*)

PERIPHERAL GROUPS OF MORPHINE	MODIFICATION (ON MORPHINE UNLESS OTHERWISE INDICATED)	EFFECTS ON ANALGESIC ACTIVITY* (MORPHINE OR ANOTHER COMPOUND AS INDICATED = 100)
Nuclear Substitution	Substitution of:	
	$-NH_2$ (most likely at position 2)	Marked decrease in action.
	$-Cl$ or $-Br$ (at position 1)	50
	$-OH$ (at position 14 in dihydromorphinone)	250 (dihydromorphinone vs. oxymorphone)
	$-OH$ (at position 14 in dihydrocodeinone)	530 (dihydrocodeinone vs. oxycodone)
	#$-CH_3$ (at position 6)	280
	#$-CH_3$ (at position 6 in dihydromorphine)	33 (dihydromorphine vs. 6-methyldihydromorphine)
	#$-CH_3$ (at position 6 in dihydrodesoxymorphine-D)	490 (dihydrodesoxymorphine-D vs. 6-methyldihydrodesoxymorphine)
	#$-CH_2$ (at position 6 in dihydrodesoxymorphine-D)	600 (dihydrodesoxymorphine-D vs. 6-methylenedihydrodesoxymorphine)

* Per cent ratio of the ED_{50} of morphine (or other compound indicated) to the ED_{50} of the compound as determined in mice. These conclusions have been adapted from data in references 3 and 4. For a wealth of additional tabular material the reader is urged to consult the original references.

† These represent cases in which, for various reasons, a direct comparison with morphine itself cannot be made. The alternative has been to compare the effect of modifying the group in a pair of compounds where the changes can be made. It is felt that the direction of change in analgesic activity at least can be determined in this way.

‡ See, however, discussion of N-methylmorphinan, p. 712.

§ Although many of these derivatives possess morphine antagonism, it has been shown that many of them also possess analgesic activity in their own right. Indeed, the ability to antagonize morphine in the rat is used as a screening method to assure low addiction potential in man.[44]

Not included in the studies of Small *et al*: See Ref. 8.

offered the advantages of ready availability, proven potency, and ease of alteration. In addition to its addictive tendency, it was hoped that other liabilities, such as respiratory depression, emetic properties, and gastrointestinal tract and circulatory disturbances, could be minimized or abolished as well. Since early modifications of morphine (e.g., acetylation or alkylation of hydroxyls, quaternization of the nitrogen, and so on) caused variations in the addictive potency, it was felt that the physiologic effects of morphine could be related, at least in part, to the peripheral groups.

It was not known if the actions of morphine were primarily a function of the peripheral groups or of the structural skeleton. This did not matter, however, because modification of the groups would alter activity in either case. These groups and the effects on activity by modifying them are listed in Table 76. The results of these and earlier studies[4] have not, in all cases, shown quantitatively the effects of simple modifications on the analgesic action of morphine. However, they do indicate in which direction the activity is apt to go. The studies are far more comprehensive than Table 76 indicates, and the conclusions depend on more than one pair of compounds in most cases.

Unfortunately, these studies on morphine did not provide the answer to the elimination of addiction potentialities from these compounds. In fact, the studies suggested that any modification bringing about an increase in the analgesic activity caused a concomitant increase in addiction liability.

The second phase of the studies, engaged in principally by Mosettig and Eddy,[3] had to do with the attempted synthesis of substances with central narcotic and, especially, analgesic action. It is obvious that the morphine molecule contains in its makeup certain well-defined types of chemical structures. Among these are the phenanthrene nucleus, the dibenzofuran nucleus and, as a variant of the latter, carbazole. These synthetic studies, although extensive and interesting, failed to provide significant findings and will not be discussed further in this text.

One of the more useful results of the investigations was the synthesis of 5-methyldihydromorphinone* (see Table 75). Although it possessed addiction liabilities, it was found to be a very potent analgesic with a minimum of the undesirable side-effects of morphine, such as emetic action and mental dullness.

More recently, the high degree of analgesic activity demonstrated by morphine congeners in which the alicyclic ring is either reduced or methylated (or both) and the alcoholic hydroxyl at position 6 is absent has prompted the synthesis of related compounds possessing these features. These include 6-methyldihydromorphine and its dehydrated analog 6-methyl-Δ^6-desoxymorphine or methyldesorphine,[6] both of which have shown promise. Also of interest are compounds reported by Rapoport and his coworkers[7]: morphinone; 6-methylmorphine; 6-methyl-7-hydroxy-, 6-methyl- and 6-methylenedihydrodesoxymorphine. In analgesic activity in mice, the last-named compound proved to be 82 times more potent, milligram for milligram, than morphine. Its therapeutic index (T.I.$_{50}$) was 22 times as great as that of morphine.[8]

The structure-activity relationships of 14-hydroxy morphine derivatives have been reviewed recently, and several new compounds have been synthesized.[9] Of these, the dihydrodesoxy compounds possessed the highest degree of analgesic activity. Also, esters of 14-hydroxycodeine derivatives have shown very high activity.[10] For example, in rats, 14-cinnamyloxycodeinone was 177 times more active than morphine.

In 1963, Bentley and Hardy[11] reported the synthesis of a novel series of potent analgesics derived from the opium alkaloid thebaine. In rats the most active members of the series (I, R_1 = H, R_2 = CH_3, R_3 = isoamyl; and I, R_1 = $COCH_3$, R_2 = CH_3, R_3 = n-C_3H_7) were found to be several thousand times stronger than morphine.[12]

* The location of the methyl substituent was originally assigned to position 7.[5]

These compounds exhibited marked differences in activity of optical isomers, as well as other interesting structural effects. It was postulated that the more rigid molecular structure might allow them to fit the receptor surface better. Extensive structural and pharmacologic studies have been reported.[13] Some of the N-cyclopropylmethyl compounds are the most potent antagonists yet discovered and are currently being studied very intensively.

As indicated in Table 76, replacement of the N-methyl group in morphine by larger alkyl groups not only lowers analgesic activity but confers morphine-antagonistic properties on the molecule (see p. 714). In direct contrast to this effect, the N-phenethyl derivative has 14 times the analgesic activity of morphine. This enhancement of activity by N-aralkyl groups has wide application, as will be shown later.

It has been observed that the morphine antagonists, such as nalorphine, are also strong analgesics.[14] The similarity of the ethylenic double bond and the cyclopropyl group has prompted the synthesis of N-cyclopropylmethyl derivatives of morphine and its derivatives.[15] This substituent confers strong narcotic antagonistic activity in most cases, with variable effects on analgesic potency. The dihydronormorphinone derivative had only moderate analgesic activity.

Morphine Modifications Initiated by the Eisleb and Schaumann Research. In 1938 Eisleb and Schaumann[2] reported the fortuitous discovery that a simple piperidine derivative, now known as meperidine, possessed analgesic activity. It was prepared as an antispasmodic, a property it shows as well. As the story is told, during the pharmacologic testing of meperidine in mice, it was observed to cause the peculiar erection of the tail known as the Straub reaction. Because the reaction is characteristic

TABLE 77. COMPOUNDS RELATED TO MEPERIDINE

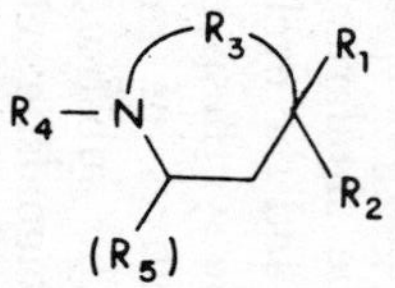

(R$_5$ = H except in trimeperidine, where it is CH$_3$ (see p. 708.).)

COMPOUND	R$_1$	R$_2$	R$_3$	R$_4$	NAME (If Any)	ANALGESIC ACTIVITY* (Meperidine = 1)
A-1	—C$_6$H$_5$	—COOC$_2$H$_5$	—CH$_2$CH$_2$—	—CH$_3$	Meperidine	1.0
A-2	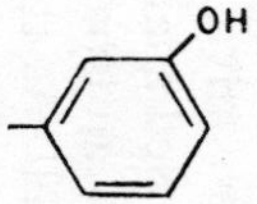 C$_6$H$_4$(OH)	—COOC$_2$H$_5$	—CH$_2$CH$_2$—	—CH$_3$	Bemidone	1.5
A-3	—C$_6$H$_5$	—COOCH(CH$_3$)$_2$	—CH$_2$CH$_2$—	—CH$_3$	Properidine	15
A-4	—C$_6$H$_5$	—C(=O)—C$_2$H$_5$	—CH$_2$CH$_2$—	—CH$_3$		0.5
A-5	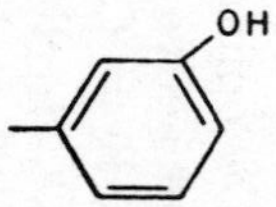 C$_6$H$_4$(OH)	—C(=O)—C$_2$H$_5$	—CH$_2$CH$_2$—	—CH$_3$	Ketobemidone	6.2
A-6	—C$_6$H$_5$	—O—C(=O)—C$_2$H$_5$	—CH$_2$CH$_2$—	—CH$_3$		5
A-7	—C$_6$H$_5$	—O—C(=O)—C$_2$H$_5$	—CH$_2$CH(CH$_3$)—	—CH$_3$	Alphaprodine Betaprodine	5 14

TABLE 77. COMPOUNDS RELATED TO MEPERIDINE (*Continued*)

COMPOUND	STRUCTURE				NAME (If Any)	ANALGESIC ACTIVITY[*] (Meperidine = 1)
	R_1	R_2	R_3	R_4		
A-8	$-C_6H_5$	$-O-\underset{\parallel O}{C}-C_2H_5$	$-CH_2\overset{CH_3}{\underset{\vert}{CH}}-$	$-CH_3 (R_5 = CH_3)$	Trimeperidine	7.5
A-9	$-C_6H_5$	$-COOC_2H_5$	$-CH_2CH_2-$	$-CH_2CH_2C_6H_5$	Pheneridine	2.6
A-10	$-C_6H_5$	$-COOC_2H_5$	$-CH_2CH_2-$	$-CH_2CH_2-\!\!\left\langle\!\!\bigcirc\!\!\right\rangle\!\!-NH_2$	Anileridine	3.5
A-11	$-C_6H_5$	$-COOC_2H_5$	$-CH_2CH_2-$	$-(CH_2)_3-NH-C_6H_5$	Piminodine	55[†]
A-12	$-C_6H_5$	$-O-\underset{\parallel O}{C}-C_2H_5$	$-CH_2CH_2-$	$-CH_2CH_2CHC_6H_5$; $O-\underset{\parallel O}{C}-C_2H_5$		1880[†]
A-13	$-C_6H_5$	$-COOC_2H_5$	$-CH_2CH_2-$	$-CH_2CH_2C(C_6H_5)_2$; CN	Diphenoxylate	None
A-14	$-C_6H_5$	$-COOC_2H_5$	$-CH_2CH_2CH_2-$	$-CH_3$	Ethoheptazine	1
A-15	$-C_6H_5$	$-O-\underset{\parallel O}{C}C_2H_5$	$-\overset{}{\underset{\vert CH_3}{CH}}-$	$-CH_3$	Prodilidine	0.3
A-16	$-H$	$-N-\underset{\parallel O}{C}C_2H_5$; $\vert$ C_6H_5	$-CH_2CH_2-$	$-CH_2CH_2C_6H_5$	Fentanyl	940

[*] Ratio of the ED_{50} of meperidine to the ED_{50} of the compound in mg./kg. administered subcutaneously in mice, based on data in References 4, 27, 32, 33.
[†] In rats. See Reference 21.

of morphine and its derivatives, the compound then was tested for analgesic properties and was found to be about one fifth as active as morphine. This discovery led not only to the finding of an active analgesic but, far more important, it served as a stimulus to research workers. The status of research in analgesic compounds with an activity comparable with that of morphine was at a low ebb in 1938. Many felt that potent compounds could not be prepared, unless they were very closely related structurally to morphine. However, the demonstration of high potency in a synthetic compound that was related only distantly to morphine spurred the efforts of various research groups.[16, 17]

The first efforts, naturally, were made upon the meperidine type of molecule in an attempt to enhance its activity further. It was found that replacement of the 4-phenyl group by hydrogen, alkyl, other aryl, aralkyl, and heterocyclyl groups reduced analgesic activity. Placement of the phenyl and ester groups at the 4-position of 1-methylpiperidine also gave optimum activity. Several modifications of this basic structure are listed in Table 77.

Among the simplest changes to increase activity is the insertion of a *m*-hydroxyl group on the phenyl ring. It is in the same relative position as in morphine. The effect is more pronounced on the keto compound (A-4) than on meperidine (A-1). Ketobemidone is equivalent to morphine in activity and was widely used.

More significantly, Jensen and co-workers[18] discovered that replacement of the carbethoxyl group in meperidine by acyloxyl groups gave better analgesic, as well as spasmolytic, activity. The "reversed" ester of meperidine, the propionoxy compound (A-6), was the most active, being 5 times as active as meperidine. These findings were validated and expanded upon by Lee *et al.*[19] In an extensive study of structural modifications of meperidine, Janssen and Eddy [20] concluded that the propionoxy compounds were always more active, usually about 2-fold, regardless of what group was attached to the nitrogen.

Lee[21] had postulated that the configuration of the propionoxy derivative (A-6)

more closely resembled that of morphine, with the ester chain taking a position similar to that occupied by carbons 6 and 7 in morphine. His speculations were based on space models and certainly did not reflect the actual conformation of the nonrigid meperidine. However, he did arrive at the correct assumption that introduction of a methyl group into position 3 of the piperidine ring in the propionoxy compound would yield two isomers, one with activity approximating that of desomorphine and the other with lesser activity. One of the two diastereoisomers (A-7), betaprodine, has an activity in mice of about 9 times that of morphine and 3 times that of A-6. Beckett *et al.*[22] have established it to be the *cis* (methyl/phenyl) form. The *trans* form, alphaprodine, is twice as active as morphine. Resolution of the racemates shows one enantiomer to have the predominant activity. In man, however, the sharp differences in analgesic potency are not so marked. The *trans* form is marketed as the racemate. The significance of the 3-methyl group is not clear, as the 3-ethyl isomers show little differences in activity, whereas the *trans* 3-allyl isomer is more active than the *cis* isomer.[23] This may reflect a change in preferred conformations.

Until only the last few years it appeared that a small substituent, such as methyl, attached to the nitrogen was optimal for analgesic activity. This was believed to be true not only for the meperidine series of compounds but for all the other types as well. It is now well established that replacement of the methyl group by various aralkyl groups can increase activity markedly.[20] A few examples of this type of compound in the meperidine series are shown in Table 77. The phenethyl derivative (A-9) is seen to be about 3 times as active as meperidine (A-1). The *p*-amino congener, anileridine (A-10) is about 4 times more active. Piminodine, the phenylaminopropyl derivative (A-11), has 55 times the activity of meperidine in rats and in clinical trials is about 5 times as effective in man as an analgesic.[24] The most active meperidine-type compounds to date are the propionoxy derivative (A-12), which is nearly 2,000 times as active as meperidine, and the N-phenethyl analog of betaprodine, which is over 2,000 times as active as morphine.[22] Diphenoxylate (A-13), a structural hybrid of meperidine and methadone types, lacks analgesic activity although it reportedly suppresses the morphine abstinence syndrome in morphine addicts.[25, 26] It is quite effective as an intestinal spasmolytic and is used for the treatment of diarrhea. Several other derivatives of it have been studied.[27]

Another manner of modifying the structure of meperidine with favorable results has been the enlargement of the piperidine ring to the 7-membered hexahydroazepine (or hexamethylenimine) ring. As was the case in the piperidine series, the most active compound was the one containing a methyl group on position 3 of the ring adjacent to the quaternary carbon atom in the propionoxy derivative, that is, 1,3-dimethyl-4-phenyl-4-propionoxyhexahydroazepine, to which the name proheptazine has been given. In the study by Eddy and co-workers, previously cited, proheptazine was one of the more active analgesics included and had one of the highest addiction liabilities. The higher ring homolog of meperidine, ethoheptazine, has been marketed. Though originally thought to be inactive,[28] it is almost equivalent to codeine as an analgesic in man and has the advantages of being free of addiction liability and having a low incidence of side-effects.[29]

Contraction of the piperidine ring to the 5-membered pyrrolidine ring has also been successful. The lower ring homolog of alphaprodine, prodilidene (A-15), is an effective analgesic, 100 mg. being equivalent to 30 mg. of codeine, with the same advantages as ethoheptazine.[30]

A more unusual modification of the meperidine structure may be found in fentanyl (A-16), in which the phenyl and the acyl groups are separated from the ring by a nitrogen. It is a powerful analgesic, 50 times stronger than morphine in man, with minimal side-effects.[31] Its short duration of action makes it well suited for use in anesthesia.[32] It is marketed for this purpose in combination with a neuroleptic, droperidol.

It should be recalled by the reader that when the nitrogen ring of morphine is opened, as in the formation of morphimethines, the analgesic activity virtually is abolished. On this basis, the prediction of whether a compound would or would not have activity without the nitrogen in a cycle would be in favor of lack of activity or, at best, a low activity. The first report indicating that this might be a false assumption was based on the initial work of Bockmuehl and Ehrhart[33] wherein they claimed that the type of compound represented by B-1 in Table 78 possessed analgesic as well as spasmolytic properties. The Hoechst laboratories in Germany followed up this lead during World War II by preparing the ketones corresponding to these esters. Some of the compounds they prepared with high activity are represented by formulas B-2 through B-7. Compound B-2 is the well-known methadone. In the meperidine and bemidone types, the introduction of a *m*-hydroxyl group in the phenyl ring brought about slight to marked increase in activity, whereas the same operation with the methadone type compound brought about a marked decrease in action. Phenadoxone (B-8), the morpholine analog of methadone, has been marketed in England. The piperidine analog, dipanone (q.v.), has been under study in this country after successful results in England.

Methadone was first brought to the attention of American pharmacists, chemists and allied workers by the Kleiderer report[34] and by the early reports of Scott and Chen.[35] Since then, much work has been done on this compound, its isomer known as isomethadone, and allied compounds. The report by Eddy, Touchberry and Lieberman[36] covers most of the points concerning the structure-activity relationships of methadone. It was demonstrated that the levo isomer (B-3) of methadone (B-2) and the levo isomer of isomethadone (B-4) were about twice as effective as racemic mixtures. It is also of interest that all structural derivatives of methadone had a greater activity than the corresponding structural derivatives of isomethadone. In other words, the superiority of methadone over isomethadone seems to hold even through the derivatives. Conversely, the methadone series of compounds was always more toxic than the isomethadone group.

TABLE 78. Compounds Related to Methadone*

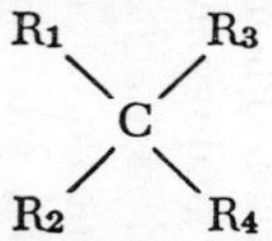

Central structure: R_1, R_3 and R_2, R_4 attached to C.

COMPOUND	STRUCTURE				NAME	ISOMER, SALT	ANALGESIC ACTIVITY† (Methadone = 1)
	R_1	R_2	R_3	R_4			
B-1	$-C_6H_5$	$-C_6H_5$	$-COO-Alkyl$	$-CH_2CH_2N(CH_3)_2$		—	0.17
B-2	$-C_6H_5$	$-C_6H_5$	$-\overset{\displaystyle \parallel O}{C}-C_2H_5$	$-CH_2\overset{\displaystyle \mid CH_3}{CH}N(CH_3)_2$	Methadone	(±)-HCl	1.0
B-3		Same as in B-2			Levanone	(−)-bitartr.	1.9
B-4	$-C_6H_5$	$-C_6H_5$	$-\overset{\displaystyle \parallel O}{C}-C_2H_5$	$-\overset{\displaystyle \mid CH_3}{CH}CH_2N(CH_3)_2$	Isomethadone	(±)-HCl	0.65
B-5	$-C_6H_5$	$-C_6H_5$	$-\overset{\displaystyle \parallel O}{C}-C_2H_5$	$-CH_2CH_2N(CH_3)_2$	Normethadone	HCl	0.44
B-6	$-C_6H_5$	$-C_6H_5$	$-\overset{\displaystyle \parallel O}{C}-C_2H_5$	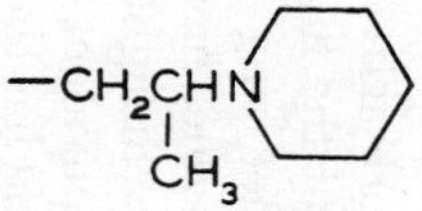 $-CH_2\overset{\displaystyle \mid CH_3}{CH}N\langle\text{piperidine}\rangle$	Dipanone	(±)-HCl	0.80
B-7	$-C_6H_5$	$-C_6H_5$	$-\overset{\displaystyle \parallel O}{C}-C_2H_5$	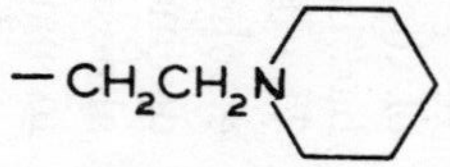 $-CH_2CH_2N\langle\text{piperidine}\rangle$	Hexalgon	HBr	0.50

TABLE 78. COMPOUNDS RELATED TO METHADONE* (Continued)

COMPOUND	R_1	R_2	R_3	R_4	NAME	ISOMER, SALT	ANALGESIC ACTIVITY[†] (Methadone = 1)
B-8	$-C_6H_5$	$-C_6H_5$	$-C(=O)-C_2H_5$	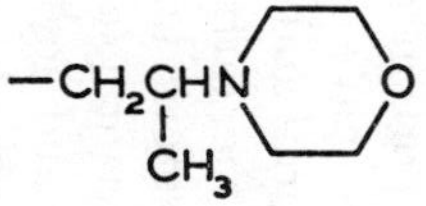 morpholine $-CH_2CHN-$, CH_3	Phenadoxone	(±)-HCl	1.4
B-9	$-C_6H_5$	$-C_6H_5$	$-CHC_2H_5-O-C(=O)CH_3$	$-CH_2CHN(CH_3)_2$, CH_3	Alphacetylmethadol	α, (±)-HCl	1.3
B-10	Same as in B-9			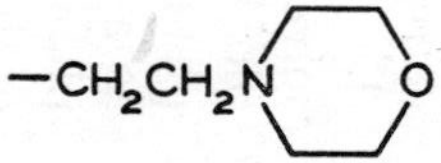 morpholine $-CH_2CH_2N-$	Betacetylmethadol	β, (±)-HCl	2.3
B-11	$-C_6H_5$	$-C_6H_5$	$-COOC_2H_5$	morpholine	Dioxaphetyl butyrate	HCl	0.25
B-12	$-C_6H_5$	$-C_6H_5$	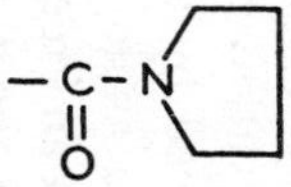 pyrrolidine $-C(=O)N$	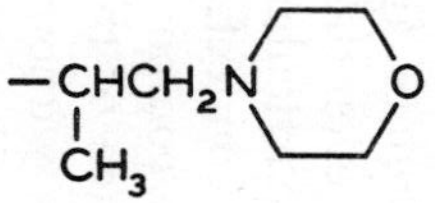 $-CHCH_2N-$, CH_3	Racemoramide	(±)-base	3.6
B-13	Same as in B-12				Dextromoramide	(+)-base	13
B-14	$-C_6H_5$	$-CH_2C_6H_5$	$O-C(=O)-C_2H_5$	$-CHCH_2N(CH_3)_2$, CH_3	Propoxyphene	(+)-HCl	0.21

* Table adapted from Janssen, P. A. J.: Synthetic Analgesics, Part 1, New York, Pergamon Press, **1960**.

† Ratio of the ED$_{50}$ of methadone to the ED$_{50}$ of the compound in mg./Kg. administered subcutaneously to mice as determined by the hot-plate method.

More extensive permutations, such as replacement of the propionyl group (R_3 in B-2) by hydrogen, hydroxyl or acetoxyl, led to decreased activity. In a series of amide analogs of methadone, Janssen and Jageneau[37] synthesized racemoramide (B-12), which is more active than methadone. The (+)-isomer, dextromoramide (B-13), is the active isomer and has been marketed. A few of the other modifications that have been carried out, together with the effect on analgesic activity relative to methadone, are described in Table 78, which comprises most of the methadone congeners that are or were on the market. It can be assumed that much deviation in structure from these examples will result in varying degrees of activity loss. Particular attention should be called to the two phenyl groups in methadone and the sharply decreased action resulting by removal of one of them. It is believed that the second phenyl residue helps to lock the $-COC_2H_5$ group of methadone in a position to simulate again the alicyclic ring of morphine, even though the propionyl group is not a particularly rigid group. However, in this connection it is interesting to note that the compound with a propionoxy group in place of the propionyl group (R_3 in B-2) is without significant analgesic action.[17] In direct contrast with this is (+)-propoxyphene (B-14), which is a propionoxy derivative with one of the phenyl groups replaced by a benzyl group. In addition, it is an analog of isomethadone (B-4), making it an exception to the rule. This compound is about equal to codeine in analgesic activity, possesses few side-effects and has little or no addiction liability.[38] Replacement of the dimethylamino group in (+)-propoxyphene with a pyrrolidyl group gives a compound that is nearly three fourths as active as methadone and possesses morphinelike properties.

Morphine Modifications Initiated by Grewe. Grewe, in 1946, approached the problem of synthetic analgesics from another direction when he synthesized the tetracyclic compound which he first named morphan and then revised to N-methylmorphinan. The relationship of this compound to morphine is obvious.

N-Methylmorphinan

N-Methylmorphinan differs most significantly from the morphine nucleus in the lack of the ether bridge between carbon atoms 4 and 5. Because this compound has been found to possess a high degree of analgesic activity, it suggests the nonessential nature of the ether bridge. The 3-hydroxyl derivative of N-methylmorphinan (racemorphan) was on the market and had an intensity and duration of action that exceeded that of morphine. The original racemorphan was introduced as the hydrobromide and was the (±)-or racemic form as obtained by synthesis. Since then, realizing that the levorotatory form of racemorphan was the actively analgesic portion of the racemate, the manufacturers have successfully resolved the (±)-form and have marketed the levo-form as the tartrate salt (levorphanol). The dextroform has also found use as a cough depressant (see dextromethorphan). The ethers and acylated derivatives of the 3-hydroxyl form also exhibit considerable activity. The 2- and 4-hydroxyl isomers are, not unexpectedly, without value as analgesics. Likewise, the N-ethyl derivative is lacking in activity and the N-allyl compound, levallorphan, is a potent morphine antagonist.

Eddy and co-workers[39] have reported on an extensive series of N-aralkylmorphinan derivatives. The effect of the N-aralkyl substitution was more dramatic in this series than it was in the case of morphine or meperidine. The N-phenethyl and N-p-aminophenethyl analogs of levorphanol were about 3 and 18 times, respectively, more active than the parent compound in analgesic activity in mice. The most potent member of the series was its N-β-furylethyl analog, which was nearly 30 times as active as levorphanol or 160 times as active as morphine. The N-acetophenone analog, levophenacylmorphan, was once under clinical investigation. In mice, it is about 30 times more active than morphine, and in man a 2-mg. dose is equivalent to 10 mg. of morphine in its analgesic response.[40] It has a much lower physical dependence liability than morphine.

The N-cyclopropylmethyl derivative of 3-hydroxymorphinan (cyclorphan), reported recently,[15] is a potent morphine antagonist capable of precipitating morphine withdrawal symptoms in addicted monkeys, indicating that it is nonaddicting. Clinical studies have indicated that it is about 20 times stronger than morphine as an analgesic but has some undesirable side-effects, primarily hallucinatory in nature.

Inasmuch as removal of the ether bridge and all the peripheral groups in the alicyclic ring in morphine did not destroy its analgesic action, May and co-workers[41] synthesized a series of compounds in which the alicyclic ring was replaced by one or two methyl groups. These are benzomorphan derivatives. They may be represented by the formula

The trimethyl compound (II, $R_1 = R_2 = CH_3$) is about 3 times more potent than the dimethyl (II, $R_1 = H$, $R_2 = CH_3$). The N-phenethyl derivatives have almost 20 times the analgesic activity of the corresponding N-methyl compounds. Again, the more potent was the one containing the 2-ring methyls (II, $R_1 = CH_3$, $R_2 = CH_2$-$CH_2C_6H_5$). Deracemization proved the levoisomer of this compound to be more active, being about 20 times as potent as morphine in mice. The ($\pm$)-form, phenazocine, was on the market but has been removed.

Chignell and his co-workers[42] have demonstrated an extremely significant difference between the two isomeric N-methyl benzomorphans in which the alkyl in the 5 position is *n*-propyl (R_1) and the alkyl in the 9 position is methyl (R_2). These have been termed the α-isomer and the β-isomer and have the groups oriented as indicated. The

α-isomer *(cis)* β-isomer *(trans)*

cis-isomer has been shown to possess analgesic activity (in mice) equal to that of morphine but has little or no capacity to suppress withdrawal symptoms in addicted monkeys. On the other hand, the *trans*-isomer has one of the highest analgesic potencies among the benzomorphans but is quite able to suppress morphine withdrawal symptoms. Further separation of properties is found between the enantiomers of the cis-isomer. The ($+$)-isomer has weak analgesic activity but a high physical dependence capacity. The ($-$)-isomer is a stronger analgesic without the dependence capacity, and possesses antagonistic activity.[43] The same was found true with the 5,9-diethyl and 9-ethyl-5-phenyl derivatives. The ($-$)-*trans*-5,9-diethyl-isomer was similar except it had no antagonistic properties. This demonstrates that it is possible to divorce analgesic activity comparable to morphine from addiction potential. The fact that N-methyl compounds have shown some antagonistic properties is of great interest as well.

An extensive series of the antagonist-type analgesics in the benzomorphans has been reported.[44] Of these, pentazocine (II, $R_1 = CH_3$, $R_2 = CH_2CH=C(CH_3)_2$) and cyclazocine (II, $R_1 = CH_3$, $R_2 = CH_2$-cyclopropyl) have proved to be the most interesting. Pentazocine has about half the analgesic activity of morphine, with a lower incidence of side-effects.[45] Its addiction liability is much lower, approximating that of propoxyphene.[46] It is currently available in parenteral and tablet form. Cyclazocine is a strong morphine antagonist, showing about 10 times the analgesic activity of morphine.[47]

It was mentioned previously that replacement of the N-methyl group in morphine by larger alkyl groups lowered analgesic activity. In addition, these compounds were found to counteract the effect of morphine and other morphinelike analgesics and are thus known as *narcotic antagonists*. The reversal of activity increased as the size of the group increased, with the allyl group being maximal. This property was found to be true not only in the case of morphine but with other analgesics as well. N-allylnormorphine (nalorphine) and levallorphan, the corresponding allyl analog of levorphanol, are the two narcotic antagonists presently on the market. Of those currently under investigation, naloxone, N-allylnoroxymorphone, has shown the most promise. It appears to be a true antagonist with no morphine- or nalorphine-like effects. It also blocks the effects of other antagonists. These drugs are used to prevent, diminish or abolish many of the actions or the side-effects encountered with the narcotic analgesics. Some of these are respiratory and circulatory depression, euphoria, nausea, drowsiness, analgesia and hyperglycemia. They are thought to act by competing with the analgesic molecule for attachment at the receptor site. As indicated previously, the observation that these narcotic antagonists, which are devoid of addiction liability, are also strong analgesics has spurred considerable interest in them.[14] The N-cyclopropylmethyl compounds mentioned are the most potent antagonists, but appear to produce psychotomimetic effects and may not be useful as analgesics.

Much research, other than that described in the foregoing discussion, has been carried out by the systematic dissection of morphine to give a number of interesting fragments. These approaches have not produced important analgesics yet; therefore, they are not discussed in this chapter. However, the interested reader may find a key to this literature from the excellent reviews of Eddy,[4] Bergel and Morrison,[17] and Lee.[21]

Structure-Activity Relationships

Several reviews on the relationship between chemical structure and analgesic action have been published.[4, 25, 48-53] Only the major conclusions will be considered here, and the reader is urged to consult these reviews for a more complete discussion of the subject.

From the time Small and co-workers started their studies on the morphine nucleus to the present, there has been much light shed on the structural features connected with morphinelike analgesic action. In a very thorough study made for the United Nations Commission on Narcotics in 1955, Braenden and co-workers[49] found that the features possessed by all known morphinelike analgesics were:

1. A tertiary nitrogen, the group on the nitrogen being relatively small;

2. A central carbon atom, of which none of the valences is connected with hydrogen;

3. A phenyl group or a group isosteric with phenyl, which is connected to the central carbon atom;

4. A 2-carbon chain separating the central carbon atom from the nitrogen for maximal activity.

From the foregoing discussion it is evident that a number of exceptions to these generalizations may be found in the structures of compounds that have been synthesized in the last several years. Eddy[25] has discussed the more significant exceptions.

In regard to the first feature mentioned above, extensive studies of the action of normorphine have shown it to possess analgesic activity in the order of morphine. In man, it is about one fourth as active as morphine when administered intramuscularly but was slightly superior to morphine when administered intracisternally. On the basis of the last-mentioned effect, Beckett and his co-workers[54] postulated that N-dealkylation was a step in the mechanism of analgesic action. This has been questioned.[55] It is clear, from the previously discussed N-aralkyl derivatives, that a small group is not necessary.

Several exceptions to the second feature have been synthesized. In these series, the central carbon atom has been replaced by a tertiary nitrogen. They are related to methadone and have the following structures:

III

IV

Diampromide (III) and its related anilides have potencies that are comparable to those of morphine;[56] however, they have shown addiction liability and have not appeared on the market. The closely related cyclic derivative fentanyl (A-16, Table 77), is used in surgery. The benzimidazoles, such as etonitazene (IV), are very potent analgesics, but show the highest addiction liabilities yet encountered.[57]

Possibly an exception to feature 3, and the only one that has been encountered, may be the cyclohexyl analog of A-6 (Table 77), which has significant activity.

Eddy[25] mentions two possible exceptions to feature 4.

As a consequence of the many studies on molecules of varying types that possess analgesic activity, it became increasingly apparent that activity was associated not only with certain structural features but also with the size and the shape of the molecule. The hypothesis of Beckett and Casy[58] has dominated thinking for a number of years in the area of stereochemical specificity of these molecules. They noted initially that the more active enantiomers of the methadone and thiambutene type analgesics were related configurationally to R-alanine. This suggested to them that a stereoselective fit at a receptor could be involved in analgesic activity. In order to depict the dimensions of an analgesic receptor, they selected morphine (because of its semi-rigidity and high activity) to provide them with information on a complementary re-

ceptor. The features that were thought to be essential for proper receptor fit were:

1. A basic center able to associate with an anionic site on the receptor surface;
2. A flat aromatic structure, coplanar with the basic center, allowing for van der Waal's bonding to a flat surface on the receptor site to reinforce the ionic bond; and
3. A suitably positioned projecting hydrocarbon moiety forming a 3-dimensional geometric pattern with the basic center and the flat aromatic structure.

These features were selected, among other reasons, because they are present in N-methylmorphinan which may be looked upon as a "stripped down" morphine, i.e., morphine without the characteristic peripheral groups (except for the basic center). Inasmuch as N-methylmorphinan possessed significant activity of the morphine type, it was felt that these three features were the fundamental ones determining activity and that the peripheral groups of morphine acted essentially to modulate the activity.

In accord with the above postulations, Beckett and Casy,[58] proposed a complementary receptor site (see Fig. 18) and suggested ways[59, 60] in which the known active molecules could be adapted to it. Subsequent to their initial postulation it was demonstrated that natural (−)-morphine was related configurationally to methadone and thiambutene, a finding that lent weight to the hypothesis. Fundamental to their proposal, of course, was that such a receptor was essentially inflexible and that a lock-and-key type situation existed.

Although the above hypothesis appeared to fit the facts quite well and was a useful hypothesis for a number of years, it now appears that certain anomalies exist which cannot be accommodated by it. For example, the more active enantiomer of α-methadol is not related configurationally to R-alanine, in contrast to the methadone and thiambutene series. This is also true for the carbethoxy analog of methadone (V) and for diampromide (III) and its analogs. Another factor that was implicit in considering a proper receptor fit for the morphine molecule and its congeners was that the phenyl ring at the 4-position of the piperidine moiety should be in the axial

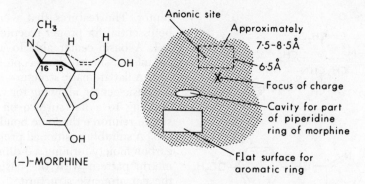

FIG. 18. Diagram of the surface of the analgesic receptor site with the corresponding lower surface of the drug molecule. The 3-dimensional features of the molecule are shown by the bonds: —, - - -, and — represent in front of, behind, and in the plane of the paper respectively. (Gourley, D. R. H., *in* Jucker, E. (ed.): Progress in Drug Research, Vol. 7, p. 36, Basel, Birkhauser, 1964)

orientation for maximum activity. The fact that structure VI has only an equatorial phenyl group, yet possesses activity equal to that of morphine would seem to cast doubt on the necessity for axial orientation as a receptor-fit requirement.

In view of the difficulty of accepting Beckett and Casy's hypothesis as a complete picture of analgesic-receptor interaction, Portoghese[61, 62] has offered an alternative hypothesis. This hypothesis is based in part on the established ability of enzymes and other types of macromolecules to undergo conformational changes[63, 64] on interaction with small molecules (substrates or drugs). The fact that configurationally unrelated analgesics can bind and exert activity is interpreted as meaning that more than one mode of binding may be possible at the same receptor. Such different modes of binding may be due to differences in positional or conformational interactions with the receptor. The manner in which the hypothesis can be adapted to the methadol anomaly is illustrated in Figure 19. Portoghese, after considering activity changes in various structural types (i.e., methadones, meperidines, prodines, etc.) as related to the identity of the N-substituent, noted that in certain series there was a parallelism in direction of activity when identical changes in N-substituents were made. In others there appeared to be a nonparallelism. He has interpreted parallelism and nonparallelism, respectively, as being due to similar and to dissimilar modes of binding. As viewed by this hypothesis, while it is still a requirement that analgesic molecules be bound in a fairly precise manner, it nevertheless liberalizes the concept of binding in that a response may be obtained by two different molecules binding stereoselectively in two different precise modes at the same receptor. A schematic representation of such different possible binding modes is shown in Figure 20. This representation will

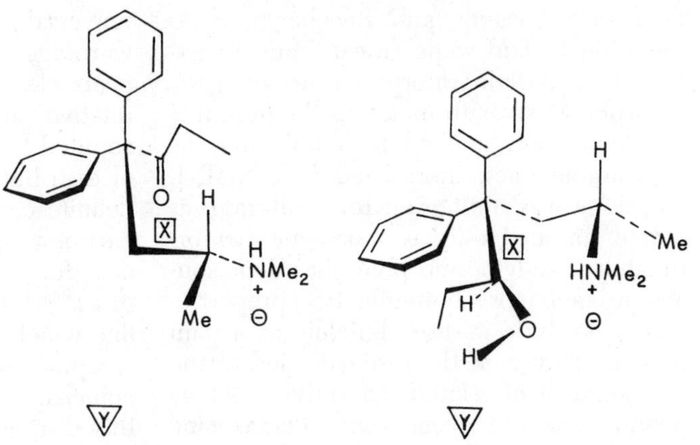

FIG. 19. An illustration of how different polar groups in analgesic molecules may cause inversion in the configurational selectivity of an analgesic receptor. A hydrogen bonding moiety is denoted by x. y represents a site which is capable of being hydrogen bonded.

(6R)

(6S)

aid in visualizing the meaning of *similar* and *dissimilar* binding modes. If two different analgesiophores* bearing identical N-substituents are positioned on the receptor surface so that the N-substituent occupies essentially the same position, a similar pharmacologic response may be anticipated. Thus, as one proceeds from one N-substituent to another the response should likewise change, resulting in a parallelism of

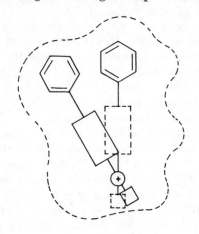

FIG. 20. A schematic illustration of two different molecular modes of binding to a receptor. The protonated nitrogen is represented by ⊕. The square denotes an N-substituent. The anionic sites lies directly beneath ⊕.

* The analgesic molecule less the N-substituent, i.e., the portion of the molecule giving the characteristic analgesic response.

effect. On the other hand, if two different analgesiophores are bound to the receptor so that the N-substituents are not arranged identically, one may anticipate nonidentical responses on changing the N-substituent, i.e., a nonparallel response. From the preceding statements, as well as the diagram, it is not to be implied that the analgesiophore necessarily will be bound in the identical position within a series. They do, however, suggest that, in series with parallel activities, the pairs being compared will be bound identically to produce the parallel effect. Interestingly, when binding modes are similar he has been able to demonstrate the existence of a linear free energy relationship. There also is the possibility that more than one receptor is involved.

Although this hypothesis is new, it appears to embrace virtually all types of analgesic molecules presently known,† and it will be interesting to see whether it is of further general applicability as other molecules with activity are devised.

Another of the highly important developments in structure-activity correlations has been the development of highly active analgesics from the N-allyl type derivatives that once were thought to be only morphine antagonists and devoid of analgesic properties. Serendipity played a major role in this

† Two possible exceptions are 4-propionoxy-4-cyclohexyl-1-methylpiperidine and 1-tosyl-4-phenyl-4-ethylsulfone piperidine. (Helv. chim. acta 36: 819, 1953)

discovery: Lasagna and Beecher,[65] in attempting to find some "ideal" ratio of antagonist (N-allylnormorphine) to analgesic (morphine) so as to maintain the desirable effects of morphine while minimizing the undesirable ones, discovered that N-allylmorphine was, milligram for milligram, as potent an analgesic as morphine. Unfortunately, N-allylnormorphine has depersonalizing and psychotomimetic properties which preclude its use clinically as a pain reliever. However, the discovery led to the development of related derivatives such as pentazocine and cyclazocine. Pentazocine appears to have achieved some notable success in providing an addiction-free analgesic, although it is not totally free of some of the other side effects of morphine. The pattern of activity in these and other N-allyl and N-cyclopropyl derivatives indicates that the potent antagonists possess psychotomimetic activity, whereas the weak antagonists do not. It is from this latter group that useful analgesics, such as pentazocine, have been found.

What structural features are associated with antagonist-like activity has become uncertain. The N-allyl and dimethylallyl substituent does not always confer antagonist properties. This is true in the meperidine and thevinol series. Demonstration of antagonistlike properties by specific isomers of N-methyl benzomorphans has raised still further speculation. The exact mechanisms by which morphine and the narcotic antagonists act are not clearly defined, and a great amount of research is presently being carried on. Recent reviews and symposia may be consulted for further discussions of these topics.[66, 67, 67a]

A further problem also is demonstrated in the testing for analgesic activity. As noted above, the analgesic activity of the antagonists was not apparent from animal testing but was observed only in man. Screening in animals can be used to assess the antagonistic action, which indirectly indicates possible analgesic properties in man.[68]

It has been customary in the area of analgesic agents to attribute differences in their activities to structurally related differences in their receptor interactions. This rather universal practice continues in spite of early warnings and recent findings. It now appears clear that much of the differences in relative analgesic potencies can be accounted for on the basis of pharmacokinetic or distribution properties.[67a] For example, a definite correlation was found between the partition coefficients and the I.V. analgesic data for 17 agents of widely varying structures.[68a] Usual test methods do not help define which structural features are related to receptor and which to distribution phenomena. Studies directed toward making this distinction are using the measurement of actual brain and plasma levels[68b, 68c] or direct injection into the ventricular area,[68a] and are providing valuable insight in regard to the designing of new and more successful agents.

PRODUCTS*

Morphine. This alkaloid was isolated first in 1803 by Derosne, but the credit for isolation generally goes to Serturner (1803) who first called attention to the basic properties of the substance. Morphine, incidentally, was the first plant base isolated and recognized as such. Although intensive research was carried out with respect to the structure of morphine, it was only in 1925 that Gulland and Robinson[69] postulated the currently accepted formula. The total synthesis of morphine finally was effected by Gates and Tschudi[70] in 1952, thus confirming the Gulland and Robinson formula.

Morphine is obtained only from opium, in which it occurs in amounts varying from 5 to 20 per cent (*U.S.P.* requires not less than 9.5%). It is isolated by various methods, of which the final step is usually the precipitation of morphine from an acid solution by using excess ammonia. The precipitated morphine then is recrystallized from boiling alcohol.

The free alkaloid occurs as levorotatory,

* In General Circular No. 253, March 10, 1960, the Treasury Department, Bureau of Narcotics, Washington 25, D.C. has published an extensive listing of narcotics of current interest in the drug trade. This listing will be much more extensive than the following monographic coverage of compounds primarily of interest to American pharmacists.

odorless, white, needlelike crystals possessing a bitter taste. It is almost insoluble in water (1:5,000,* 1:1,100 at boiling point), ether (1:6,250) or chloroform (1:1,220). It is somewhat more soluble in ethyl alcohol (1:210, 1:98 at boiling point). Because of the phenolic hydroxyl group, it is readily soluble in solutions of alkali or alkaline earth metal hydroxides.

Morphine is a mono-acidic base and readily forms water-soluble salts with most acids. Thus, because morphine itself is so poorly soluble in water, the salts are the preferred form for most uses. Numerous salts have been marketed, but the ones in use are principally the sulfate and, to a lesser extent, the hydrochloride. Morphine acetate, which is freely soluble in water (1:2.5), has been used to a limited extent in liquid antitussive combinations.

Many writers have pointed out the "indispensable" nature of morphine, based on its potent analgesic properties toward all types of pain. It is properly termed a narcotic analgesic. However, because it causes addiction so readily, it should be used only in those cases where other pain-relieving drugs prove to be inadequate. It controls pain caused by serious injury, neoplasms, migraine, pleurisy, biliary and renal colic and numerous other causes. It often is administered as a preoperative sedative, together with atropine to control secretions. With scopolamine, it is given to obtain the so-called "twilight sleep." This effect is used in obstetrics, but care is exercised to prevent respiratory depression in the fetus. It is worthy of note that the toxic properties of morphine are much more evident in the very young and in the very old than in middle-aged people.

Morphine Hydrochloride. This salt may be prepared by neutralizing a hot aqueous suspension of morphine with diluted hydrochloric acid and then concentrating the resultant solution to crystallization.

It occurs as silky, white, glistening needles or cubical masses or as a crystalline, white powder. The hydrochloride is soluble

* In this chapter a solubility expressed as (1:5,000) indicates that 1 g. is soluble in 5,000 ml. of the solvent at 25°. Solubilities at other temperatures will be so indicated.

in water (1:17.5, 1:0.5 at boiling point), alcohol (1:52, 1:46 at 60°) or glycerin, but it is practically insoluble in ether or chloroform. Solutions have a pH of approximately 4.7 and may be sterilized by boiling.

Its uses are the same as those of morphine.

The usual oral and subcutaneous dose is 15 mg. every 4 hours as needed, with a suggested range of 8 to 20 mg.

Morphine Sulfate U.S.P. This morphine salt is prepared in the same manner as the hydrochloride, i.e., by neutralizing morphine with diluted sulfuric acid.

It occurs as feathery, silky, white crystals, as cubical masses of crystals or as a crystalline, white powder. Although it is a fairly stable salt, it loses water of hydration and darkens on exposure to air and light. It is soluble in water (1:16, 1:1 at 80°), poorly soluble in alcohol (1:570, 1:240 at 60°) and insoluble in chloroform or ether. Aqueous solutions have a pH of approximately 4.8 and may be sterilized by heating in an autoclave.

Category—narcotic analgesic.

Usual dose—parenteral, 10 mg. 4 to 6 times daily, as needed.

Usual dose range—5 to 75 mg. daily.

OCCURRENCE

Morphine and Atropine Sulfates Tablets N.F.
Morphine Sulfate Tablets U.S.P.
Morphine Injection U.S.P. (other salts may be used also)

Codeine N.F. Codeine is an alkaloid which occurs naturally in opium, but the amount present is usually too small to be of commercial importance. Consequently, most commercial codeine is prepared from morphine by methylating the phenolic hydroxyl group. The methylation methods usually are patented procedures and make use of reagents such as diazomethane, dimethyl sulfate and methyl iodide.

It occurs as levorotatory, colorless, efflorescent crystals or as a white, crystalline powder. It is light sensitive. Codeine is slightly soluble in water (1:120) and sparingly soluble in ether (1:50). It is freely soluble in alcohol (1:2) and very soluble in chloroform (1:0.5).

Codeine is a mono-acidic base and readily forms salts with acids, the most important salts being the sulfate and the phosphate. The acetate and the methylbromide derivatives have been used to a limited extent in cough preparations. The free base is used little as compared with the salts, its greatest use being in Terpin Hydrate and Codeine Elixir N.F.

The general pharmacologic action of codeine is similar to that of morphine but, as previously indicated, it does not possess the same degree of analgesic potency. Lasagna[71] comments on the status of the drug as follows:

"Despite codeine's long use as an analgesic drug, it is amazing how little reliable information there is about its efficacy, particularly by the parenteral route."

There are studies that indicate that 30 to 120 mg. of codeine are considerably less efficient parenterally than 10 mg. of morphine and the usual side-effects of morphine—respiratory depression, constipation, nausea, etc.—are apparent. Codeine is less effective orally than parenterally, and it has been stated by Houde and Wallenstein[72] that 32 mg. of codeine is about as effective as 650 mg. of aspirin in relieving terminal cancer pain. However, it also has been recognized that combinations of aspirin and codeine act additively as analgesics, thus giving some support to the common practice of combining the two drugs.

Codeine has a reputation as an antitussive, depressing the cough reflex, and is used in many cough preparations. It is one of the most widely used morphinelike analgesics. It is less addicting than morphine and in the usual doses respiratory depression is negligible, although an oral dose of 60 mg. will cause such depression in a normal person. It is probably true that much of codeine's reputation as an antitussive rests on subjective impressions rather than on objective studies. The average 5 ml. dose of Terpin Hydrate and Codeine Elixir contains 10 mg. of codeine. This preparation and many like it have been sold over the counter as exempt narcotic preparations. However, abuse of these preparations has led to their being placed on a prescription-only status in many states.

A combination of codeine and papaverine (Copavin) was advocated by Diehl[73] for the prophylaxis and treatment of common colds. When administered at the first signs of a cold, it was claimed to have aborted the cold in a significant percentage of the cases.

Category—analgesic (narcotic); antitussive.

Usual dose—analgesic, 30 mg. every 4 hours; antitussive, 5 to 10 mg. every 4 hours.

Usual dose range—analgesic, 15 to 60 mg.

OCCURRENCE

Terpin Hydrate and Codeine Elixir N.F.

Codeine Phosphate U.S.P. This salt may be prepared by neutralizing codeine with phosphoric acid and precipitating the salt from aqueous solution with alcohol.

Codeine phosphate occurs as fine, needle-shaped, white crystals or as a white, crystalline powder. It is efflorescent and is sensitive to light. It is freely soluble in water (1:2.5, 1:0.5 at 80°) but less soluble in alcohol (1:325, 1:125 at boiling point). Solutions may be sterilized by boiling.

Because of its high solubility in water as compared with the sulfate, this salt is used widely. It is often the only salt of codeine stocked by pharmacies and is dispensed, rightly or wrongly, on all prescriptions calling for either the sulfate or the phosphate.

Category—antitussive; narcotic analgesic.

Usual dose—analgesic: oral or parenteral, 30 mg. 4 to 6 times daily as required; antitussive: oral, 10 mg. 6 to 8 times daily as necessary.

Usual dose range—5 to 300 mg. daily.

OCCURRENCE

Codeine Phosphate Injection U.S.P.
Codeine Phosphate Tablets U.S.P.
Codeine Phosphate, Aspirin, Phenacetin and Caffeine Tablets N.F.

Codeine Sulfate N.F. Codeine sulfate is prepared by neutralizing an aqueous solution of codeine with diluted sulfuric acid and then effecting crystallization.

It occurs as white crystals, usually needlelike, or as a white, crystalline powder. The salt is efflorescent and light sensitive. It is soluble in water (1:30, 1:6.5 at 80°), much less soluble in alcohol (1:1,280) and insoluble in ether or chloroform.

This salt of codeine is prescribed frequently but is not as suitable as the phosphate for liquid preparations. Solutions of the sulfate and the phosphate are incompatible with alkaloidal reagents and alkaline substances.

Category—analgesic (narcotic); antitussive.

Usual dose—analgesic, 30 mg. every 4 hours; antitussive, 5 to 10 mg. every 4 hours.

Usual dose range—analgesic, 15 to 60 mg.

OCCURRENCE
Codeine Sulfate Tablets N.F.

Ethylmorphine Hydrochloride N.F., Dionin®. This synthetic compound is analogous to codeine, but instead of being the methyl ether it is the ethyl ether. Ethylmorphine may be prepared by treating an alkaline alcoholic solution of morphine with diethyl sulfate. The hydrochloride is obtained from the free base by neutralizing it with diluted hydrochloric acid.

The salt occurs as a microcrystalline, white or faintly yellow, odorless powder. It has a slightly bitter taste. It is soluble in water (1:10) and in alcohol (1:25) but only slightly soluble in ether and in chloroform.

The systemic action of this morphine derivative is intermediate between those of codeine and morphine. It has analgesic qualities and sometimes is used for the relief of pain. As a depressant of the cough reflex, it is as effective as codeine and, for this reason, is found in some commercial cough syrups. However, the chief use of this compound is in ophthalmology. By an irritant dilating action on vessels, it stimulates the vascular and lymphatic circulation of the eye. This action is of value in chemosis (excessive edema of the ocular conjunctiva), and the drug is termed a *chemotic*.

Category—chemotic.

Application—topically, as a 1 to 5 per cent solution in the eye.

Diacetylmorphine Hydrochloride, Heroin Hydrochloride, Diamorphine Hydrochloride. Although heroin is much more potent than morphine as an analgesic, its sale and use is prohibited in the United States because of its intense addiction liability. It is available in some European countries where it has a limited use as an antitussive and as an analgesic in terminal cancer patients.

Hydromorphone Hydrochloride N.F., Dilaudid®, Hymorphan®, Dihydromorphinone Hydrochloride. Hydromorphone is a synthetic derivative of morphine. It is prepared by the catalytic hydrogenation and dehydrogenation of morphine under acidic conditions, using a large excess of platinum or palladium.

The free base has properties similar to those of morphine. Its hydrochloride occurs as a light-sensitive, white, crystalline powder which is soluble in water (1:3), sparingly soluble in alcohol and practically insoluble in ether.

This compound is of German origin and was introduced in 1926. It is a substitute for morphine (5 times as potent) but has approximately equal addicting properties and a shorter duration of action. Because its narcotic effect is not as great in proportion to its analgesic effect, it possesses the advantage over morphine of causing less daytime drowsiness. It has a powerful effect upon the cough reflex and occasionally is incorporated into cough syrups for coughs that are hard to control. It is used in about one fifth the dose of morphine for any of the indications of morphine.

Category—analgesic (narcotic).

Usual dose—oral and subcutaneous, 2 mg. every 4 hours as necessary.

Usual dose range—1 to 4 mg.

OCCURRENCE
Hydromorphone Hydrochloride Injection N.F.
Hydromorphone Hydrochloride Tablets N.F.

Hydrocodone Bitartrate U.S.P., Dicodid®, Mercodinone®, Dihydrocodeinone Bitartrate. This drug is prepared by the catalytic rearrangement of codeine or by hydrolyzing dihydrothebaine. It occurs as fine, white crystals or as a white, crystalline powder. It is soluble in water (1:16), slightly soluble in alcohol and insoluble in ether. It forms acidic solutions and is affected by light. The hydrochloride is also available.

Hydrocodone has a pharmacologic action midway between those of codeine and morphine, with 15 mg. being equivalent to 10 mg. of morphine in analgesic power. Although it has been shown to possess more addiction liability than codeine, it has been said to give no evidence of dependence or addiction when used over long periods of time. Its principal advantage is in the lower incidence of side-effects encountered with its use. It is more effective than codeine as an antitussive and is used primarily for this purpose. It is on the market in many cough preparations as well as in tablet and parenteral forms. It has also been marketed in an ion exchange resin complex form under the trade name of Tussionex. The complex has been shown to release the drug at a sustained rate and is said to produce effective cough suppression over a 10- to 12-hour period.

Although this drug found extensive use in antitussive formulations for many years, recently it has been placed under more stringent narcotic regulations, and it is being replaced gradually by codeine or dextromethorphan in most over-the-counter cough preparations.

Category—antitussive.

Usual dose—5 to 10 mg. 3 to 4 times daily as needed.

Usual dose range—5 to 50 mg. daily.

OCCURRENCE

Hydrocodone Bitartrate Syrup U.S.P.
Hydrocodone Bitartrate Tablets U.S.P.

Methyldihydromorphinone, Metopon®. As indicated previously (p. 705), methyldihydromorphinone was prepared during the studies of Small *et al.*[3]

Methyldihydromorphinone hydrochloride is freely soluble in water but only sparingly soluble in alcohol. It is slightly soluble in the immiscible solvents. Dilute aqueous solutions have a pH of about 5. It has been estimated that the analgesic action of this derivative is substantially greater than that of morphine with no greater toxicity or addiction liability. Indeed, investigators have found that placement of confirmed morphine addicts on a methyldihydromorphinone regimen failed to control adequately the morphine withdrawal symptoms. Taking the lack of control of withdrawal symptoms as an indication of its addictive potentialities, it would appear that methyldihydro-

morphinone is, therefore, less liable to produce addiction than is morphine. The drug is very effective orally and, because it elicits practically no emetic action, it is suitable for prolonged administration. Its use is probably limited by its difficult and expensive synthesis. It is marketed only for oral administration in the form of 3-mg. capsules, with each 3-mg. dose being approximately equal in effect to 10 mg. of morphine. The dose of the hydrochloride is from 3 to 6 mg., and the dose is to be repeated only on the recurrence of pain. Around-the-clock medication is to be avoided, because it is conducive to habituation. The drug is suggested primarily for the relief of pain in the treatment of conditions such as inoperable cancer. A narcotic form is required to obtain the drug.

Oxymorphone Hydrochloride N.F., Numorphan®, (−)-14-Hydroxydihydromorphinone Hydrochloride. Oxymorphone, introduced in 1959, is prepared by cleavage of the corresponding codeine derivative. It is used as the hydrochloride salt, which occurs as a white, crystalline powder freely soluble in water and sparingly soluble in alcohol. In man, oxymorphone is as effective as morphine in one eighth to one tenth the dosage, with good duration and a slightly lower incidence of side-effects.[74] It has high addiction liability. It is used for the same purposes as morphine, such as control of postoperative pain, pain of advanced neoplastic diseases as well as other types of pain that respond to morphine. Because of the risk of addiction it should not be employed for relief of minor pains that can be controlled with codeine. It is also well to note that it has poor antitussive activity and is not used as a cough suppressant.

It may be administered orally, parenterally (i.v., i.m., s.c.) or rectally and for these purposes is supplied as a solution for injection (1.0 and 1.5 mg. per ml.), suppositories (2 and 5 mg.) and in tablets (10 mg.).

Category—analgesic (narcotic).

Usual dose—oral, 10 mg. every 4 to 6 hours, with a maximum dose of 40 mg. per day; subcutaneously and intramuscularly, 1.0 to 1.5 mg. every 4 to 6 hours as needed;

intravenous, 0.5 mg. initially, repeated in 4 to 6 hours if necessary.

OCCURRENCE
Oxymorphone Hydrochloride Injection N.F.
Oxymorphone Hydrochloride Tablets N.F.

Oxycodone Hydrochloride, Dihydrohydroxycodeinone Hydrochloride. This compound is prepared by the catalytic reduction of hydroxycodeinone, the latter compound being prepared by hydrogen peroxide (in acetic acid) oxidation of thebaine. This derivative of morphine occurs as a white, crystalline powder which is soluble in water (1:10) or alcohol. Aqueous solutions may be sterilized by boiling. Although this drug is almost as likely to cause addiction as morphine, it has been introduced in the United States in Percodan® as a mixture of its hydrochloride and terephthalate salts in combination with aspirin, phenacetin and caffeine.

It is used as a sedative, an analgesic and a narcotic. Because it is believed to exert a physostigminelike action, it is used externally in the eye in the treatment of glaucoma and related ocular conditions. To depress the cough reflex, it is used in 3- to 5-mg. doses and as an analgesic in 5- to 10-mg. doses. For severe pain, a dose of 20 mg. is given subcutaneously.

Dihydrocodeine Bitartrate, Paracodin®. Dihydrocodeine is obtained by the reduction of codeine. The bitartrate salt occurs as white crystals which are soluble in water (1:4.5) and only slightly soluble in alcohol. Subcutaneously, 30 mg. of this drug is almost equivalent to 10 mg. of morphine as an analgesic, giving more prompt onset and negligible side-effects. It has addiction liability. It is available in parenteral and 10-mg. tablet forms. As an analgesic and antitussive, the usual dose is 10 to 30 mg.

Normorphine. This drug may be prepared by N-demethylation of morphine.[75] It is still undergoing investigation and evaluation of its pharmacologic properties at the present time. In man, it is about one fourth as active as morphine in producing analgesia but has a much lower physical dependence capacity. It does not show the sedative effects of morphine in single doses but does so cumulatively. Normorphine

suppresses the morphine abstinence syndrome in addicts, but after its withdrawal it gives a slow onset and a mild form of the abstinence syndrome.[76] It may prove useful in the treatment of narcotic addiction.

Concentrated Opium Alkaloids, Pantopon®, consists of a mixture of the total alkaloids of opium. It is free of nonalkaloidal material, and the alkaloids are said to be present in the same proportions as they occur naturally. The alkaloids are in the form of the hydrochlorides, and morphine constitutes 50 per cent of the weight of the material.

This preparation is promoted as a substitute for morphine, the claim being that it is superior to the latter, due to the synergistic action of the opium alkaloids. This synergism is said to result in less respiratory depression, less nausea and vomiting and an antispasmodic action on smooth muscle. According to several authorities, however, the superiority to morphine is overrated, and the effects produced are comparable with the use of an equivalent amount of morphine. The commercial literature suggests a dose of 20 mg. of Pantopon to obtain the same effect as is given by 15 mg. of morphine.

Solutions prepared for parenteral use may be slightly colored, a situation which does not necessarily indicate decomposition.

Apomorphine Hydrochloride N.F. When morphine or morphine hydrochloride is heated at 140° under pressure with strong (35%) hydrochloric acid, it loses a molecule of water and yields a compound known as apomorphine.

The hydrochloride is odorless and occurs

Apomorphine

as minute, glistening, white or grayish-white crystals or as a white powder. It is light sensitive and turns green on exposure to air and light. It is sparingly soluble in water (1:50, 1:20 at 80°) and in alcohol (1:50) and is very slightly soluble in ether or chloroform. Solutions are neutral to litmus.

The change in structure from morphine to apomorphine causes a profound change in its physiologic action. The central depressant effects of morphine are much less pronounced, and the stimulant effects are enhanced greatly. In particular, the vomiting center is stimulated; therefore, the emesis produced by apomorphine is purely central in origin. It is administered subcutaneously to obtain emesis. It is ineffective orally. Apomorphine is one of the most effective, prompt (10 to 15 minutes) and safe emetics in use today. However, care should be exercised in its use because it may be depressant in already depressed conditions.

Category—emetic.

Usual dose—subcutaneous, 5 mg.

OCCURRENCE
Apomorphine Hydrochloride Tablets N.F.

Meperidine Hydrochloride U.S.P., Demerol® Hydrochloride, Ethyl 1-Methyl-4-phenyl-isonipecotate Hydrochloride, Ethyl 1-Methyl-4-phenyl-4-piperidinecarboxylate Hydrochloride. This is a fine, white, odorless, crystalline powder that is very soluble in water, soluble in alcohol and sparingly soluble in ether. It is stable in the air at ordinary temperature, and its aqueous solution is not decomposed by a short period of boiling. The free base may be made by heating benzyl cyanide with *bis*(β-chloroethyl)methylamine, hydrolyzing to the corresponding acid and esterifying the latter with ethyl alcohol.[2]

Meperidine first was synthesized in order to study its spasmolytic character, but it was found to have analgesic properties in far greater degree. The spasmolysis is due primarily to a direct papaverinelike depression of smooth muscle and, also, to some action on parasympathetic nerve endings. In therapeutic doses, it exerts an analgesic effect which lies between those of morphine and codeine, but it gives little tendency toward hypnosis. It is indicated for the relief of pain in the majority of cases for which morphine and other alkaloids of opium generally are employed, but it is especially of value where the pain is due to spastic conditions of intestine, uterus, bladder, bronchi, and so on. Its most important use seems to be in lessening the severity of labor pains in obstetrics and, with barbiturates or tranquilizers, to produce amnesia in labor. In labor, 100 mg. is injected intramuscularly as soon as contractions occur regularly, and a second dose may be given after 30 minutes if labor is rapid or if the cervix is thin and dilated (2 to 3 cm. or more). A third dose may be necessary an hour or two later, and at this stage a barbiturate may be administered in small dose to ensure adequate amnesia for several hours. Meperidine possesses addiction liability. There is a development of psychic dependence in those individuals who experience a euphoria lasting for an hour or more. The development of tolerance has been observed, and it is significant that meperidine can be successfully substituted for morphine in addicts who are being treated by gradual withdrawal. Furthermore, mild withdrawal symptoms have been noted in certain persons who have become purposely addicted to meperidine. The possibility of dependence is great enough to put it under the Federal narcotic laws. Nevertheless, it remains as one of the more widely used analgesics.

Category—narcotic analgesic.

Usual dose—oral or parenteral, 50 to 100 mg. 4 to 6 times daily as necessary.

Usual dose range—25 to 500 mg. daily.

OCCURRENCE
Meperidine Hydrochloride Injection U.S.P.
Meperidine Hydrochloride Syrup N.F.
Meperidine Hydrochloride Tablets U.S.P.

Alphaprodine Hydrochloride N.F., Nisentil® Hydrochloride, (±)-1,3-Dimethyl-4-phenyl-4-piperidinol Propionate Hydrochloride. This compound is prepared according to the method of Ziering and Lee.[77]

It occurs as a white, crystalline powder, which is freely soluble in water, alcohol and chloroform but insoluble in ether.

The compound is an effective analgesic, similar to meperidine and has been found to be of special value in obstetric analgesia. It appears to be quite safe for use in this capacity, causing little or no depression of respiration in either mother or fetus.

Category—analgesic (narcotic).

Usual dose—subcutaneous, 20 to 40 mg.; intravenous, 20 mg.

Usual dose range—subcutaneous, 20 to 60 mg.; intravenous, 20 to 30 mg.

OCCURRENCE

Alphaprodine Hydrochloride Injection N.F.

Anileridine N.F., Leritine®, ethyl 1-(*p*-aminophenethyl)-4-phenylisonipecotate is prepared by the method of Weijlard *et al.*[78] It occurs as a white to yellowish white crystalline powder which is freely soluble in alcohol but only very slightly soluble in water. It is oxidized on exposure to air and light. The injection is prepared by dissolving the free base in phosphoric acid solution.

Anileridine is more active than meperidine and has the same usefulness and limitations. Its dependence capacity is less and is considered a suitable substitute for meperidine.

Category—analgesic (narcotic).

Usual dose—subcutaneous or intramuscular, 25 to 50 mg. of anileridine, present as the phosphate, repeated every 6 hours, if necessary.

Usual dose range—25 to 75 mg.

OCCURRENCE

Anileridine Injection N.F.

Anileridine Hydrochloride N.F., Leritine® Hydrochloride, ethyl 1-(*p*-aminophenethyl)-4-phenylisonipecotate hydrochloride, is prepared as cited above for anileridine except that it is converted to the dihydrochloride by conventional procedures. It occurs as a white or nearly white crystalline, odorless powder which is stable in air. It is freely soluble in water, sparingly soluble in alcohol and practically insoluble in ether and chloroform.

This salt has the same activity as that cited for anileridine (see above).

Category—analgesic (narcotic).

Usual dose—25 mg. of anileridine, present as the dihydrochloride, repeated every 6 hours if necessary.

Usual dose range—25 to 50 mg.

OCCURRENCE

Anileridine Hydrochloride Tablets N.F.

Piminodine Esylate N.F., Alvodine®, Ethyl 1-(3-Anilinopropyl)-4-phenylisonipecotate Monoethanesulfonate, Ethyl 4-Phenyl-1-[3-(phenylamino)propyl]-piperidine-4-carboxylate Ethanesulfonate. This drug is somewhat more effective as an analgesic than morphine, being about 5 times more so than meperidine in man. Although it has addiction liability, it has a lower incidence of side-effects and is suggested for use in any condition in which meperidine or morphine is indicated. It is available in tablet and parenteral forms.

Category—analgesic (narcotic).

Usual dose range—oral: 25 to 50 mg. every 4 to 6 hours; intramuscularly or subcutaneously, 10 to 20 mg. every 4 hours as needed, depending on degree of pain and the patient's response.

OCCURRENCE

Piminodine Esylate Injection N.F.

Piminodine Esylate Tablets N.F.

Diphenoxylate Hydrochloride N.F., Lomotil®, Ethyl 1-(3-Cyano-3,3-diphenylpropyl)-4-phenylisonipecotate Monohydrochloride. It occurs as a white, odorless, slightly water-soluble powder with no distinguishing taste.

Although this drug has a strong structural relationship to the meperidine-type analgesics it has very little, if any, such activity itself. Its most pronounced activity is its ability to inhibit excessive gastrointestinal motility, an activity reminiscent of the constipating side-effect of morphine itself. Investigators have demonstrated the possibility of addiction,[25, 26] particularly with large doses, but virtually all studies using ordinary dosage levels show nonaddiction. Its safety is reflected in its classification as an exempt narcotic, with, however, the warning that it may be habit forming. To discourage possible abuse of the drug, the commercial product (Lomotil) contains a subtherapeutic dose (25 mcg.) of atropine sulfate in each 2.5-mg. tablet and in each 5 ml. of the liquid which contains a like amount of the drug.

It is indicated in the oral treatment of diarrheas resulting from a variety of causes. The usual initial adult dose is 5 mg. 3 or 4 times a day, with the maintenance dose usually being substantially lower and being individually determined. Appropriate dosage schedules for children are available in the manufacturer's literature.

The incidence of side-effects is low, but the drug should be used with caution, if at all, in patients with impaired hepatic function. Similarly, patients taking barbiturates concurrently with the drug should be observed carefully, in view of reports of barbiturate toxicity under these circumstances.

Category—antidiarrheal (narcotic).

Usual dose—5 mg. 4 times a day.

Usual dose range—5 to 40 mg. per day.

OCCURRENCE

Diphenoxylate Hydrochloride and Atropine Sulfate Solution N.F.

Diphenoxylate Hydrochloride and Atropine Sulfate Tablets N.F.

Ethoheptazine Citrate N.F.;[79] Zactane Citrate®, Ethyl Hexahydro-1-methyl-4-phenyl-1H-azepine-4-carboxylate Dihydrogen Citrate, 1-Methyl-4-carbethoxy-4-phenylhexamethylenimine Citrate. It is effective orally against moderate pain in doses of 50 to 100 mg., with minimal side-effects. Parenteral administration is limited, due to central stimulating effects. It appears to have no addiction liability, but toxic reactions have occurred with large doses. A double blind study in man rated 100 mg. of the hydrochloride salt equivalent to 30 mg. of codeine, and found that the addition of 600 mg. of aspirin increased analgesic effectiveness.[29] It is available as a 75-mg. tablet and in combination with 600 mg. of aspirin (Zactirin).

Category—analgesic.

Usual dose—75 mg. 3 to 4 times a day.

Usual dose range—75 to 150 mg.

OCCURRENCE

Ethoheptazine Citrate Tablets N.F.

Fentanyl Citrate, Sublimaze®, N-(1-phenethyl-4-piperidyl)propionanilide citrate, occurs as a crystalline powder, soluble in water (1:40) and methanol, and sparingly soluble in chloroform.

This novel anilide derivative has demonstrated analgesic activity 50 times that of morphine in man.[31] It has a very rapid onset (4 minutes) and short duration of action. Side effects similar to those of other potent analgesics are common—in particular, respiratory depression and bradycardia. It is used primarily as an adjunct to anesthesia. For use as a neuroleptanalgesic in surgery, it is available in combination with the neuroleptic droperidol as Innovar®. It has dependence liability.

Methadone Hydrochloride U.S.P., Dolophine®, 6-(Dimethylamino)-4,4-diphenyl-3-heptanone Hydrochloride. It occurs as a white, crystalline powder with a bitter taste. It is soluble in water, freely soluble in alcohol and chloroform, and insoluble in ether.

The synthesis of methadone is effected in several ways. The method of Easton and co-workers[80] is noteworthy in that it avoids the formation of the troublesome isomeric intermediate aminonitriles. The analgesic effect and other morphinelike properties are exhibited chiefly by the (−)-form. Aqueous solutions are stable and may be sterilized by heat for intramuscular and intravenous use. Like all amine salts, it is incompatible with alkali and salts of heavy metals. It is somewhat irritating when injected subcutaneously.

The toxicity of methadone is 3 to 10 times greater than that of morphine, but its analgesic effect is twice that of morphine and 10 times that of meperidine. It has been placed under Federal narcotic control because of its high addiction liability.

Methadone is a most effective analgesic, used to alleviate many types of pain. It can replace morphine for the relief of withdrawal symptoms. It produces less sedation and narcosis than does morphine and appears to have fewer side-reactions in bedridden patients. In spasm of the urinary bladder and in the suppression of the cough reflex, methadone is especially valuable.

The levo-isomer, levanone, does not produce euphoria or other morphinelike sensations and has been advocated for the treatment of addicts.[81] Methadone itself is being used quite extensively in addict treatment, although not without some controversy.[81a] It will suppress withdrawal effects and can be withdrawn more easily. Large doses are often used to "block" the effects of heroin during treatment.

Category—narcotic analgesic.

Usual dose—oral or parenteral, 5 to 10 mg. 3 to 6 times daily.

Usual dose range—2.5 to 50 mg. daily; up to 200 mg. daily for treatment of heroin addiction.

OCCURRENCE

Methadone Hydrochloride Injection U.S.P.

Methadone Hydrochloride Tablets U.S.P.

Propoxyphene Hydrochloride U.S.P., Darvon®, (+)-α-4-Dimethylamino-1,2-diphenyl-3-methyl-2-butanol Propionate Hydrochloride. This drug was introduced into therapy in 1957. It may be prepared by the method of Pohland and Sullivan.[82] It occurs as a bitter, white, crystalline powder which is freely soluble in water, soluble in alcohol, chloroform and acetone but practically insoluble in benzene and ether. It is the α-(+)-isomer, the α-(−)-isomer and β-diastereoisomers being far less potent in analgesic activity. The α-(−)-isomer, levopropoxyphene, is an effective antitussive (see p. 733).

In analgesic potency, propoxyphene is approximately equal to codeine phosphate and has a lower incidence of side-effects. It has no antidiarrheal, antitussive or antipyretic effect, thus differing from most analgesic agents. Although it is able to suppress morphine abstinence syndrome in addicts, there is no other evidence to indicate that it possesses addiction liabilities. It is not very effective in deep pain and appears to be no more effective in minor pain than aspirin. Its widespread use in dental pain seems justified, since aspirin is reported to be relatively ineffective. It is classified as non-addictive and is not controlled by Federal narcotic law. It does give some euphoria in high doses and has been abused. Indiscriminate refilling of the drug should be avoided if misuse is suspected.

Category—analgesic.

Usual dose—60 mg. 3 or 4 times daily as needed.

Usual dose range—30 to 500 mg. daily.

OCCURRENCE

Propoxyphene Hydrochloride Capsules U.S.P.

Propoxyphene Hydrochloride and Aspirin Capsules N.F.

Levorphanol Tartrate N.F., Levo-Dromoran® Tartrate, (−)-3-Hydroxy-N-methylmorphinan Bitartrate. The basic studies in the synthesis of this type of compound were made by Grewe, as already pointed out (p. 712). Schnider and Grüssner synthesized the hydroxymorphinans, including the 3-hydroxyl derivative, by similar methods. The racemic 3-hydroxy-N-methylmorphinan hydrobromide (racemorphan, (±)-Dromoran) was the original form in which this potent analgesic was introduced. This drug is prepared by resolution of racemorphan. It should be noted that the levo compound is available in Europe under the original name Dromoran. As the tartrate, it occurs in the form of colorless crystals. The salt is sparingly soluble in water (1:60) and is insoluble in ether.

The drug is used for the relief of severe pain and is in many respects similar in its actions to morphine except that it is from 6 to 8 times as potent. The addiction liability of levorphanol is as great as that of morphine, and, for that reason, caution should be observed in its use. It is claimed that the gastrointestinal effects of this compound are significantly less than those experienced with morphine. Nalorphine (q.v.) is an effective antidote for overdosage. Levorphanol is useful for relieving severe pain originating from a multiplicity of causes, e.g., inoperable tumors, severe trauma, renal colic, biliary colic. In other words, it has the same range of usefulness as morphine. It is supplied in ampuls, in multiple-dose vials and in the form of oral tablets. The drug requires a narcotic form.

Category—analgesic (narcotic).

Usual dose—oral and subcutaneous, 2 mg.

Usual dose range—1 to 3 mg.

OCCURRENCE

Levorphanol Tartrate Injection N.F.

Levorphanol Tartrate Tablets N.F.

Pentazocine N.F.; Talwin®, 1,2,3,4,5,6-hexahydro-*cis*- 6,11-dimethyl-3-(3-methyl-2-butenyl)-2,6-methano-3-benzazocin-8-ol, *cis*-2-dimethylallyl-5,9-dimethyl-2′-hydroxy-6,7-benzomorphan, occurs as a white, crystalline powder which is readily soluble in water and sparingly soluble in alcohol.

Pentazocine in a parenteral dose of 30

mg. or an oral dose of 50 mg. is about as effective as 10 mg. of morphine in most patients. There is now some evidence that the analgesic action resides principally in the (−)-isomer, with 25 mg. being approximately equivalent to 10 mg. of morphine sulfate.[83] Occasionally, doses of 40 to 60 mg. may be required. At the lower dosage levels, it appears to be well tolerated, although some degree of sedation occurs in about one third of those persons receiving it. The incidence of other morphinelike side-effects is as high as with morphine and other narcotic analgesics. In patients who have been receiving other narcotic analgesics, large doses of pentazocine may precipitate withdrawal symptoms. It shows an equivalent or greater respiratory depressant activity. Pentazocine has given rise to a few cases of possible dependence liability. It is not under narcotic control but its abuse potential should be recognized and close supervision of its use maintained. Nalorphine or levallorphan cannot reverse its effects, although naloxone can, and methylphenidate is recommended as an antidote for overdosage or excessive respiratory depression.

Pentazocine as the lactate is available in vials containing the equivalent of 30 mg. of base per ml., buffered to pH 4 to 5. It should not be mixed with barbiturates. Tablets of 50 mg. (as the hydrochloride) are also available for oral administration.

Category—analgesic.

Usual dose—parenteral, 30 mg. (as the lactate) every 3 to 4 hours.

Usual dose range—20 to 60 mg.

OCCURRENCE
Pentazocine Lactate Injection N.F.

Cyclazocine, *cis*-2-Cyclopropylmethyl-5, 9-dimethyl-2′-hydroxy-6,7-benzomorphan, is a potent narcotic antagonist that has shown a high degree of analgesic activity in man. It does possess considerable side-effects which precludes its usefulness as an analgesic. It has found use in the treatment of narcotic addiction. By voluntary treatment with cyclazocine, addicts are deprived of the euphorogenic effects of heroin. Its dependence liability is lower, and the effects of withdrawal develop more slowly and are milder. Tolerance develops to the side-effects of cyclazocine but not to its antagonist effects.[84] A usual maintenance dose of 4 mg. is obtained by gradually increasing doses. The effects are long lasting and are not reversed by other antagonists such as nalorphine.

Methotrimeprazine N.F., Levoprome®, (−)-10-[3-(dimethylamino)-2-methylpropyl]-2-methoxyphenothiazine, is a phenothiazine derivative, closely related to chlorpromazine, which possesses strong analgesic activity. An intramuscular dose of 15 to 20 mg. is equal to 10 mg. of morphine in man. It has not shown any dependence liability and appears not to produce respiratory depression. The most frequent side-effects are similar to those of phenothiazine tranquilizers, namely, sedation and, orthostatic hypotension. These often result in dizziness and fainting, limiting the use of methotrimeprazine to nonambulatory patients. It is to be used with caution along with antihypertensives, atropine, and other sedatives. It shows some advantage in cases in which addiction and respiratory depression are problems.[85]

Category—analgesic.

Usual dose—intramuscular, 10 to 30 mg. every 4 to 6 hours.

Usual dose range—5 to 40 mg.

OCCURRENCE
Methotrimeprazine Injection N.F.

NARCOTIC ANTAGONISTS

Nalorphine Hydrochloride U.S.P., Nalline® Hydrochloride, N-allylnormorphine hydrochloride, may be prepared according to the method of Weijlard and Erickson.[75] It occurs in the form of white or practically white crystals that slowly darken on exposure to air and light. It is freely soluble in water (1:8) but is sparingly soluble in alcohol (1:35) and is almost insoluble in chloroform and ether. The phenolic hydroxyl group confers water solubility in the presence of fixed alkali. Aqueous solutions of the salt are acid, having a pH of about 5.

Nalorphine has a directly antagonistic effect against morphine, meperidine, methadone and levorphanol. However, it has little antagonistic effect toward barbiturate or general anesthetic depression.

Perhaps one of the most striking effects is on the respiratory depression accompanying morphine overdosage. The respiratory minute volume is quickly returned to normal by intravenous administration of the drug. However, it does have respiratory depressant activity itself, which may potentiate the existing depression. It affects circulatory disturbances in a similar way, reversing the effects of morphine. Other effects of morphine are affected similarly. It is interesting to note that morphine addicts, when treated with the drug, exhibit certain of the withdrawal symptoms associated with abstinence from morphine. Thus, it may be used as a diagnostic test agent to determine narcotic addiction. Chronic administration of nalorphine along with morphine prevents or minimizes the development of dependence on morphine. As pointed out earlier, it has been found to have strong analgesic properties but is not acceptable for such use, due to the high incidence of undesirable psychotic effects.

Nalorphine hydrochloride is administered intravenously, intramuscularly or subcutaneously. The intravenous route gives the quickest results, and the dose may be repeated in 10 to 15 minutes if necessary. Doses up to 40 mg. have been used without untoward results in severe cases of poisoning.

Category—antidote to narcotic overdosage.

Usual dose—parenteral, 5 mg., repeated 2 times at 3-minute intervals if needed.

Usual dose range—2 to 10 mg. per dose.

OCCURRENCE
Nalorphine Hydrochloride Injection U.S.P.

Levallorphan Tartrate U.S.P., Lorfan®, (−)-N-Allyl-3-hydroxymorphinan Bitartrate. This compound occurs as a white or practically white, odorless, crystalline powder. It is soluble in water (1:20), sparingly soluble in alcohol (1:60) and practically insoluble in chloroform and ether. Levallorphan resembles nalorphine in its pharmacologic action, being about 5 times more effective as a narcotic antagonist. It

has been found also to be useful in combination with analgesics such as meperidine, alphaprodine and levorphanol to prevent the respiratory depression usually associated with these drugs.

Category—antidote to narcotic overdosage.

Usual dose—parenteral, 1 mg., repeated twice at 3-minute intervals if needed.

Usual dose range—500 mcg. to 2 mg., repeated if needed.

OCCURRENCE
Levallorphan Tartrate Injection U.S.P.

Naloxone, Narcan®, N-Allyl-14-hydroxynordihydromorphinone, N-Allylnoroxymorphone, is currently being investigated as a clinically useful narcotic antagonist. It lacks not only the analgesic activity shown by other antagonists but most of the other side-effects. It is almost 7 times more active than nalorphine in antagonizing the effects of morphine. It shows no withdrawal effects after chronic administration. The duration of action is about 4 hours. It is presently undergoing trials for the treatment of heroin addiction. With adequate doses of naloxone, the addict will not receive any effect from heroin. However, it can be given to an addict only after a detoxification period. Its long-term usefulness is currently limited, because of its short duration of action. Long-acting forms or alternate antagonists are being investigated.

ANTITUSSIVE AGENTS

Cough is a protective, physiologic reflex that occurs in health as well as in disease. It is very widespread and commonly ignored as a mild symptom. However, in many conditions it is desirable to take measures to reduce excessive coughing. It should be stressed that many etiologic factors cause this reflex; and in a case where a cough has been present for an extended period of time, or accompanies any unusual symptoms, the person should be referred to a physician. Cough preparations are widely advertised and often sold indiscriminately; so it is the obligation of

TABLE 79. Non-narcotic Antitussive Agents

Compound *Proprietary Name*	Structural Formula	Action
Noscapine (Narcotine) *Nectadon*	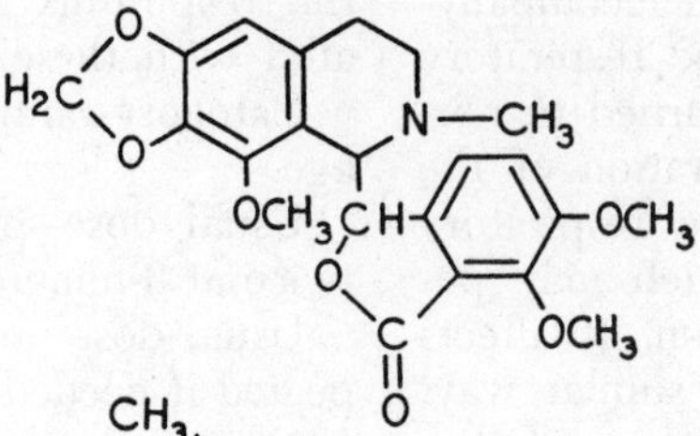	Central action; Bronchodilation
Dextromethorphan *Romilar*	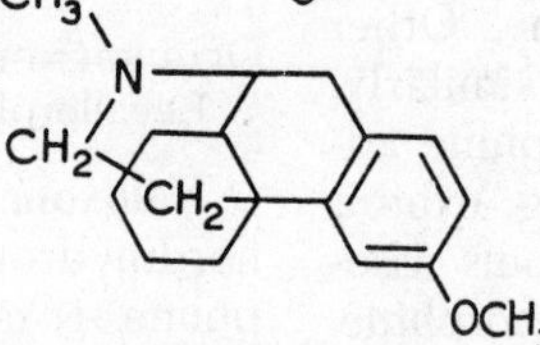	Central action
Dextro-Methadone-S (Sulfamethadone)	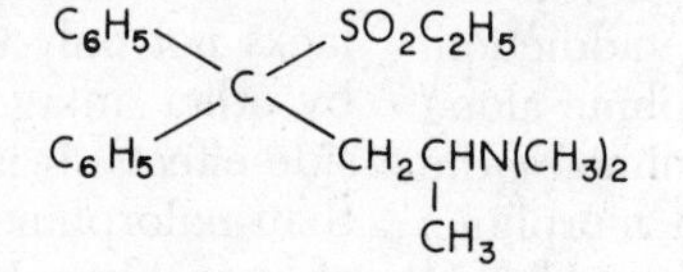 C_6H_5 C $SO_2C_2H_5$ / C_6H_5 $CH_2CHN(CH_3)_2$ CH_3	Central action
Chlophedianol *ULO*	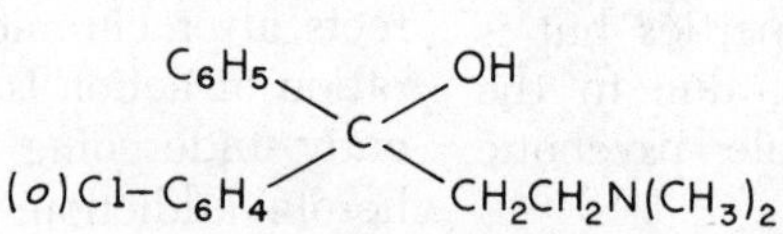 C_6H_5 C OH / $(o)Cl-C_6H_4$ $CH_2CH_2N(CH_3)_2$	Spasmolysis; Bronchodilation (?)
Hoe 10682	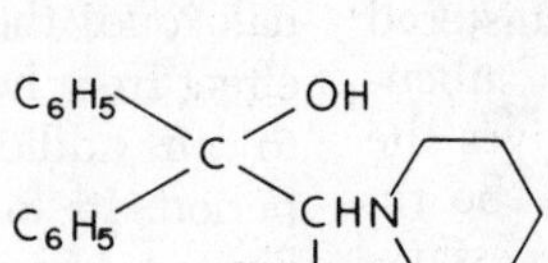 C_6H_5 C OH / C_6H_5 CHN CH_3	
Levopropoxyphene *Novrad*	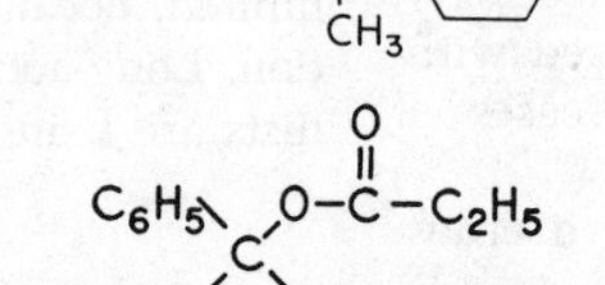 C_6H_5 $O-C-C_2H_5$ / $C_6H_5CH_2$ $CHCH_2N(CH_3)_2$ CH_3	Central action
Isoaminile	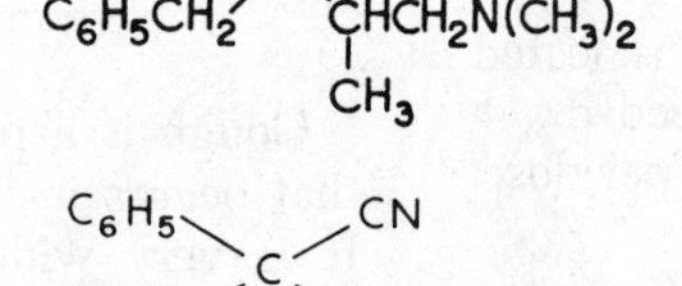 C_6H_5 C CN / $(CH_3)_2CH$ $CH_2CHN(CH_3)_2$ CH_3	Bronchodilation
KAT-256	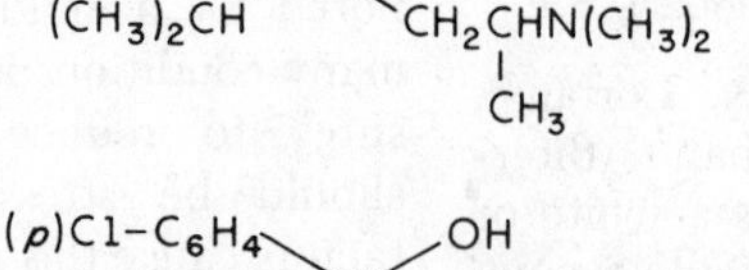 $(p)Cl-C_6H_4$ C OH / CH_3 $CHCH_2N(CH_3)_2$ CH_3	Central action
Oxolamine	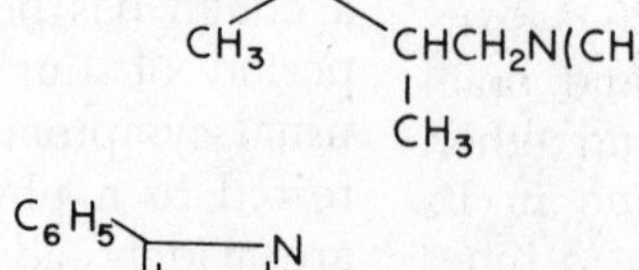 C_6H_5 ... $CH_2CH_2N(CH_3)_2$	Spasmolysis; Bronchodilation

TABLE 79. NON-NARCOTIC ANTITUSSIVE AGENTS (*Continued*)

COMPOUND *Proprietary Name*	STRUCTURAL FORMULA	ACTION
Benzonatate (Benzononatine) *Tessalon, Ventussin*	$C_4H_9NH-\langle\ \rangle-\overset{O}{\underset{\parallel}{C}}-O-(CH_2CH_2O)_nCH_3$ $n = 9$ (average)	Local anesthetic
Caramiphen Ethanedisulfonate	$\left[C_6H_5-\overset{O}{\underset{\parallel}{C}}-O-CH_2CH_2N(C_2H_5)_2\right]_2 \cdot \begin{smallmatrix}CH_2SO_3H\\ \mid\\ CH_2SO_3H\end{smallmatrix}$	Central action Bronchodilation
Carbetapentane Citrate	$C_6H_5-\overset{O}{\underset{\parallel}{C}}-(OCH_2CH_2)_2N(C_2H_5)_2 \cdot C_6H_8O_7$	Bronchodilation
Oxeladin *Pectamol*	$C_6H_5-\overset{O}{\underset{\parallel}{C}}-(OCH_2CH_2)_2N(C_2H_5)_2 \cdot C_6H_8O_7$ ... CH_2 CH_2 / CH_3 CH_3	Bronchodilation (?)
Dimethoxanate *Cothera*	$S\ N-\overset{O}{\underset{\parallel}{C}}-(OCH_2CH_2)_2N(CH_3)_2 \cdot HCl$	
Pipazethate *Theratuss*	$S\ N-\overset{O}{\underset{\parallel}{C}}-(OCH_2CH_2)_2N$	Spasmolysis; Bronchodilation
Sodium Dibunate, R = Na Ethyl Dibunate R = Et *Neodyne*	$(CH_3)_3C$ ... $C(CH_3)_3$ ... SO_3R	
Meprotixol	S ... $CH_2CH_2CH_2N(CH_3)_2$ ... OCH_3	

the pharmacist to warn the public of the inherent dangers.

Among the agents used in the symptomatic control of cough are those which act by depressing the cough center located in the medulla. These have been termed anodynes, cough suppressants and centrally acting antitussives. Until recently, the only effective drugs in this area were members of the narcotic analgesic agents. The more important and widely used ones are morphine, hydromorphone, codeine, hydrocodone, morpholinoethylmorphine (pholcodine), methadone and levorphanol, which were discussed in the foregoing section.

In recent years, several compounds have been synthesized that possess antitussive activity without the addiction liabilities of the narcotic agents. Some of these act in a similar manner through a central effect. In a new hypothesis for the initiation of the cough reflex, Salem and Aviado[86] proposed that bronchodilation is an important mechanism for the relief of cough. Their hypothesis suggests that irritation of the mucosa initially causes bronchoconstriction, and this in turn excites the cough receptors. Many of these compounds are summarized in Table 79, together with the mechanism of action(s) attributed to them.

Chappel and von Seemann[87] have pointed out that most antitussives of this type fall into two structural groups. The larger group represented in Table 79 has those that bear a structural resemblance to methadone, i.e., dextromethorphan through Melipan. The other group has large, bulky substituents on the acid portion of an ester, usually connected by means of a long, ether-containing chain to a tertiary amino group. The notable exceptions shown in Table 79 are benzonatate and sodium dibunate. Noscapine could be considered as belonging to the first group.

It should be pointed out that many of the cough preparations sold contain various other ingredients in addition to the primary antitussive agent. The more important ones include: antihistamines, useful when the cause of the cough is allergic in nature, although some antihistaminic drugs, e.g., diphenhydramine, have a central antitussive action as well; sympathomimetics, which are quite effective, due to their bronchodilatory activity, the most useful being ephedrine, methamphetamine, phenylpropanolamine, homarylamine, isoproterenol and isoöctylamine; parasympatholytics, which help to dry secretions in the upper respiratory passages; and expectorants. It is not known if these drugs potentiate the antitussive action, but they usually are considered as adjuvant therapy.

The more important drugs in this class will be discussed in the following section. For a more exhaustive coverage of the field the reader is urged to consult the excellent review of Chappel and von Seemann.[87]

Products

Some of the narcotic antitussive products have been discussed previously with the narcotic analgesics (q.v.).

Noscapine N.F., (−)-Narcotine, Nectadon®. This opium alkaloid was isolated, in 1817, by Robiquet. It is isolated rather easily from the drug by ether extraction. It makes up 0.75 to 9 per cent of opium. Present knowledge of its structure is due largely to the researches of Roser.

Noscapine occurs as a fine, white or practically white, crystalline powder which is odorless and stable in the presence of light and air. It is practically insoluble in water, freely soluble in chloroform, and soluble in acetone and benzene. It is only slightly soluble in alcohol and ether.

With the discovery of its unique antitussive properties, the name of this alkaloid was changed from narcotine to noscapine. It was realized that it would not meet with widespread acceptance as long as its name was associated with the narcotic opium alkaloids. The selection of the name "noscapine" was probably due to the fact that a precedent existed in the name of (±)-narcotine, namely "gnoscopine."

Although noscapine had been used therapeutically as an antispasmodic (similar to papaverine), antineuralgic and antiperiodic, it had fallen into disuse. It had also been used in malaria, migraine and other conditions in the past in doses of 100 to 600 mg. Newer methods of testing for antitussive compounds were responsible for revealing the effectiveness of noscapine in this respect. In addition to its central action, it has been shown to exert bronchodilation effects.

Noscapine is an orally effective antitussive, approximately equal to codeine in effectiveness. It is free of the side-effects usually encountered with the narcotic antitussives and, because of its relatively low toxicity, may be given in larger doses in order to obtain a greater antitussive effect. Although it is an opium alkaloid, it is devoid of analgesic action and addiction liability. However, it is still listed as a narcotic under Federal control. It is available in various cough preparations.

Category—antitussive.

Usual dose—15 mg. up to 4 times daily.

Usual dose range—15 to 30 mg.

Dextromethorphan Hydrobromide N.F., Romilar® Hydrobromide, (+)-3-Methoxy-17-methyl-9α,13α,14α-morphinan Hydrobromide. This drug is the O-methylated (+)-form of racemorphan left after the resolution necessary in the preparation of levorphanol. It occurs as practically white crystals, or as a crystalline powder, possessing a faint odor. It is sparingly soluble in water (1:65), freely soluble in alcohol and chloroform and insoluble in ether.

It possesses the antitussive properties of codeine without the analgesic, addictive, central depressant and constipating features. Ten milligrams is suggested as being equivalent to a 15-mg. dose of codeine in antitussive effect.

It affords an opportunity to note the specificity exhibited by very closely related molecules. In this case, the (+)- and (−)-forms both must attach to receptors responsible for the suppression of cough reflex, but the (+)- form is apparently in a steric relationship such that it is incapable of attaching to the receptors involved in analgesic, constipative, addictive and other actions exhibited by the (−)- form. It is rapidly replacing many older antitussives, including codeine, in prescription and nonprescription cough preparations.

Category—antitussive.

Usual dose range—15 to 30 mg. 1 to 4 times daily.

OCCURRENCE

Dextromethorphan Hydrobromide Syrup N.F.

Terpin Hydrate and Dextromethorphan Hydrobromide Elixir N.F.

Levopropoxyphene Napsylate N.F., Novrad®, (−)-α-4-(Dimethylamino)-3-methyl-1,2-diphenyl-2-butanol Propionate (ester) 2-Naphthalenesulfonate (salt). This compound, the levoisomer of propoxyphene, does not possess the analgesic properties of the (+)-form but is equally effective as an antitussive, 50 mg. being equivalent to 15 mg. of codeine.[88] Side-effects are infrequent. Levopropoxyphene is used in suspension in the form of its 2-naphthalenesulfonate salt ("napsylate") which is effective and has the advantage of being virtually tasteless.

Category—antitussive.

Usual dose—50 to 100 mg. every 4 hours.

OCCURRENCE

Levopropoxyphene Napsylate Capsules N.F.

Benzonatate N.F., Tessalon®, Ventussin®, 2,5,8,11,14,17,20,23,26-Nona-oxaoctacosan-28-yl p-(Butylamino)benzoate. This compound was introduced in 1956. It is a pale-yellow, viscous liquid insoluble in water and soluble in most organic solvents. It is chemically related to p-aminobenzoate local anesthetics except that the aminoalcohol group has been replaced by a methylated polyethylene glycol group (see Table 79).

Benzonatate is said to possess both peripheral and central activity in producing its antitussive effect. It somehow blocks the stretch receptors thought to be responsible for cough. Clinically, it is not as effective as codeine but produces far fewer side-effects and has a very low toxicity. It is available in 50- and 100-mg. capsules ("perles") and ampuls (5 mg./ml.).

Category—antitussive.

Usual dose—100 mg. 3 times a day.

Usual dose range—100 to 200 mg.

OCCURRENCE

Benzonatate Capsules N.F.

Chlophedianol, ULO®, 1-o-Chlorophenyl-1-phenyl-3-dimethylaminopropan-1-ol. This compound, which first was described as an antispasmodic, was found to be an effective antitussive agent.[89] It is useful in doses of

20 to 30 mg. given 3 to 5 times daily, with a duration of effect for a single dose lasting up to 5 hours. It has a low incidence of side-effects. It is available in several combinations (15 mg./ml.) (Acutuss, Ulogesic).

Caramiphen Edisylate, 2-Diethylaminoethyl 1-Phenylcyclopentane-1-carboxylate Ethanedisulfonate. Caramiphen occurs in the form of water- and alcohol-soluble crystals. The antitussive activity of this compound is less than that of codeine. It has been shown to have both central and bronchodilator activity. The incidence of side-effects is lower than with the narcotic antitussives. It is currently marketed as a combination under the tradename of Tuss-Ornade®, both in a liquid form and in a sustained release form.

Carbetapentane Citrate N.F., 2-[2-(Diethylamino)ethoxy]ethyl 1-Phenylcyclopentanecarboxylate Citrate (1:1). This salt is a white, odorless, crystalline powder, which is freely soluble in water, slightly soluble in alcohol and insoluble in ether. It is similar to caramiphen chemically and is said to be equivalent to codeine as an antitussive. Introduced in 1956, it is well tolerated and has a low incidence of side-effects. It is available as a syrup (7.25 mg./5 ml.) in combination (Tussar-2).

Category—antitussive.

Usual dose—15 to 30 mg. 3 to 4 times daily.

The tannate is also available (Rynatan, Rynatuss) and is said to give a more sustained action.

Dimethoxanate Hydrochloride, Cothera®, 2-(Dimethylaminoethoxy)ethyl Phenothiazine-10-carboxylate Hydrochloride. It is claimed that this is an effective antitussive with 25 mg. being equivalent to 15 mg. of codeine and with less side-effects. The recommended dose is 25 to 50 mg.

Pipazethate; Theratuss®; 2'-(2-Piperidinoethoxy)ethyl 10H-pyrido [3,2-b] [1,4] benzothiazine-10-carboxylate or 1-Azaphenothiazine-10-carboxylate. Pipazethate has been reported to give effective relief, being somewhat less potent than codeine and with a reportedly low incidence of side-effects.[90] The recommended dose is 20 to 40 mg.

THE ANTIPYRETIC ANALGESICS

The growth of this group of analgesics was related closely to the early belief that the lowering or "curing" of fever was an end in itself. Drugs bringing about a drop in temperature in feverish conditions were considered to be quite valuable and were sought after eagerly. The decline of interest in these drugs coincided more or less with the realization that fever was merely an outward symptom of some other, more fundamental, ailment. However, during the use of the several antipyretics, it was noted that some were excellent analgesics for the relief of minor aches and pains. These drugs have survived to the present time on the basis of the analgesic rather than the antipyretic effect. Although these drugs are utilized today principally for the alleviation of minor aches and pains, they are also employed extensively in the symptomatic treatment of rheumatic fever, rheumatoid arthritis and osteoarthritis. The dramatic effect of salicylates in reducing the inflammatory effects of rheumatic fever is time-honored, and, even with the development of the corticosteroids, these drugs are still of great value in this respect. It has been reported that the steroids are no more effective than the salicylates in preventing the cardiac complications of rheumatic fever.[91]

The analgesic drugs that fall in this category have been disclaimed by some as not deserving the term "analgesic" because of the low order of activity in comparison with the morphine-type compounds. Indeed, Fourneau has suggested the name "antalgics" to designate this general category and, in this way, to make more emphatic the distinction from the true analgesics. Two of the principal features distinguishing these minor analgesics from the true analgesics are: the low activity for a given dose and the fact that higher dosage does not give any significant increase in effect.

Considerable research has continued in an effort to find new nonsteroidal anti-inflammatory agents. Long-term therapy with the corticosteroids is often accompanied by various side-effects. Although several new agents have been introduced for use in rheumatoid arthritis, aspirin remains the agent of choice.

Discussion of these drugs will be facilitated by considering them in their various chemical categories.

SALICYLIC ACID DERIVATIVES

Historically, the salicylates were among the first of this group to achieve recognition as analgesics. Leroux, in 1827, isolated salicin, and Piria, in 1838, prepared salicylic acid. Following these discoveries, Cahours (1844) obtained salicylic acid from oil of wintergreen (methyl salicylate); and Kolbe and Lautermann (1860) prepared it synthetically from phenol. Sodium salicylate was introduced in 1875 by Buss, followed by the introduction of phenyl salicylate by Nencki, in 1886. Aspirin, or acetylsalicylic acid, was first prepared in 1853 by Gerhardt but remained obscure until Felix Hoffmann discovered its pharmacologic activities in 1899. It was tested and introduced into medicine by Dreser, who named it *aspirin* by taking the "a" from acetyl and adding it to "spirin," an old name for salicylic or spiric acid, derived from its natural source of spirea plants.

The pharmacology of the salicylates and related compounds has been reviewed extensively by Smith.[92, 93] Salicylates, in general, exert their antipyretic action in febrile patients by increasing heat elimination of the body through the mobilization of water and consequent dilution of the blood. This brings about perspiration, causing cutaneous dilation. This does not occur with normal temperatures. The antipyretic and analgesic actions are believed to occur in the hypothalamic area of the brain. It is also thought by some that the salicylates exert their analgesia by their effect on water balance, reducing edema usually associated with arthralgias. Aspirin has been shown to be particularly effective in this respect.

For an interesting account of the history of aspirin and a discussion of its mechanisms of action, the reader should consult an article on the subject by Collier,[94] as well as the reviews by Smith.[92, 93]

The possibility of hypoprothrombinemia and concomitant capillary bleeding in conjunction with salicylate administration accounts for the inclusion of menadione in some salicylate formulations. However, there is some doubt as to the necessity for this measure. A more serious aspect of salicylate medication has been the possibility of inducing hemorrhage due to direct irritative contact with the mucosa. Alvarez and Summerskill have pointed out a definite relationship between salicylate consumption and massive gastrointestinal hemorrhage from peptic ulcer.[95] Barager and Duthie,[96] on the other hand, in an extensive study find no danger of increase in anemia or in development of peptic ulcer. Levy[97] has demonstrated with the use of labeled iron that bleeding does occur following administration of aspirin. The effects varied with the formulation. It is suggested by Davenport[98] that back-diffusion of acid from the stomach is responsible for capillary damage.

The salicylates are readily absorbed from the stomach and the small intestine, being quite dependent on the pH of the media. Absorption is considerably slower as the pH rises (more alkaline), due to the acidic nature of these compounds and the necessity for the presence of undissociated molecules for absorption through the lipoidal membrane of the stomach and the intestines. Therefore, buffering agents administered at the same time in *excessive* amounts will decrease the rate of absorption. In small quantities, their principal effect may be to aid in the dispersion of the salicylate into fine particles. This would help to increase absorption and decrease the possibility of gastric irritation due to the accumulation of large particles of the undissolved acid and their adhesion to the gastric mucosa. Levy and Hayes[99] have shown that the absorption rate of aspirin and the incidence of gastric distress were a function of the dissolution rate of its particular dosage form. A more rapid dissolution rate of calcium and buffered aspirin was believed to account for faster absorption. They also established that significant variations exist in dissolution rates of different nationally distributed brands of plain aspirin tablets. This may account for some of the conflicting reports and opinions concerning the relative advantages of plain and buffered aspirin tablets. Lieberman and co-workers[100] have also shown that buffering is effective in raising the blood levels of aspirin. In a measure of the antianxiety effect of aspirin by means of electroencephalograms (EEG), differences between buffered, brand name, and generic aspirin preparations were found.[101]

Potentiation of salicylate activity by virtue of simultaneous administration of *p*-aminobenzoic acid or its salts has been the basis for the introduction of numerous products of this kind. Salassa and his co-workers have shown this effect to be due to the inhibition both of salicylate metabolism and of excretion in the urine.[102] This effect has been proved amply, provided that the ratio of 24 g. of *p*-aminobenzoic acid to 3 g. of salicylate per day is observed. However, there is no strong evidence to substantiate any significant elevation of plasma salicylate levels when a lesser quantity of *p*-aminobenzoic acid is employed.

The derivatives of salicylic acid are of two types (I and II (a, b)):

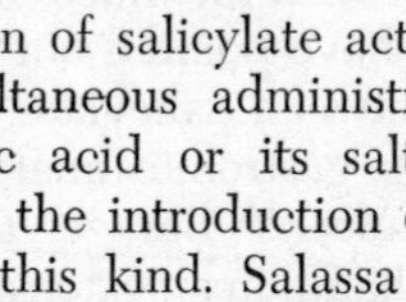

I

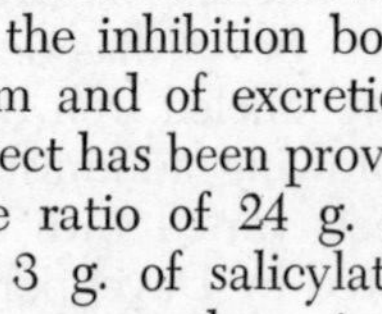

IIa 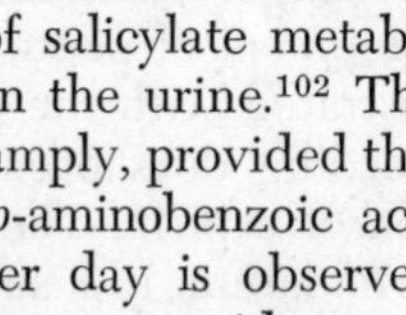   **IIb**

Type I represents those which are formed by modifying the carboxyl group (e.g., salts, esters or amides). Type II (a and b) represents those which are derived by substitution on the hydroxyl group of salicylic acid. The derivatives of salicylic acid were introduced to prevent the gastric symptoms and the undesirable taste inherent in the common salts of salicylic acid. Hydrolysis of type I takes place to a greater extent in the intestine, and most of the type II compounds are absorbed into the bloodstream (see aspirin).

Compounds of Type I

The alkyl and aryl esters of salicylic acid (type I) are used externally, primarily as counterirritants, where most of them are well absorbed through the skin. This type of compound is of little value as an analgesic.

A few inorganic salicylates are used internally when the effect of the salicylate ion is intended. These compounds vary in their irritation of the stomach. To prevent the development of pink or red coloration in the product, contact with iron should be avoided in the manufacture.

Sodium Salicylate U.S.P. may be prepared by the reaction, in aqueous solution, between 1 mole each of salicylic acid and sodium bicarbonate; upon evaporating to dryness, the white salt is obtained.

Generally, the salt has a pinkish tinge or is a white, microcrystalline powder. It is odorless or has a faint, characteristic odor, and it has a sweet, saline taste. It is affected by light. The compound is soluble in water (1:1), alcohol (1:10) and glycerin (1:4).

In solution, particularly in the presence of sodium bicarbonate, the salt will darken on standing (see salicylic acid). This darkening may be lessened by the addition of sodium sulfite or sodium bisulfite. Also, a color change is lessened by using recently boiled distilled water and dispensing in amber-colored bottles. Sodium salicylate forms a eutectic mixture with antipyrine and produces a violet coloration with iron or its salts. Solutions of the compound must be neutral or slightly basic, to prevent precipitation of free salicylic acid. However, the U.S.P. salt forms neutral or acid solutions.

This salt is the one of choice for salicylate medication and usually is administered with sodium bicarbonate to lessen gastric distress, or it is administered in enteric-coated tablets. The use of sodium bicarbonate[103] is ill-advised, since it has been shown to decrease the plasma levels of salicylate and to increase the excretion of free salicylate in the urine.

Category—analgesic.

Usual dose—600 mg. 4 to 6 times daily.

Usual dose range—300 mg. to 4 g. daily.

Sodium Salicylate Tablets U.S.P.

Choline Salicylate, Arthropan®. This salt of salicylic acid is extremely soluble in water. It is claimed to be absorbed more rapidly than aspirin, giving faster peak blood levels. It is used in conditions where salicylates are indicated in a recommended dose of 0.87 to 1.74 g. 4 times daily.

Other salts of salicylic acid that have found use are those of ammonium, lithium and strontium. They offer no distinct advantage over sodium salicylate.

Methyl Salicylate U.S.P., Wintergreen Oil, Sweet Birch Oil. Methyl Salicylate is produced synthetically or is obtained by maceration and subsequent distillation with steam from the leaves of *Gaultheria procumbens* Linné (Fam. Ericaceae) or from the bark of *Betula lenta* Linné (Fam. Betulaceae).

It is prepared synthetically through the esterification of salicylic acid with methyl alcohol in the presence of sulfuric acid.

Methyl salicylate is a colorless, yellowish or reddish, fragrant, oily liquid, slightly soluble in water and soluble in alcohol. It usually is labeled to indicate whether it was prepared synthetically or distilled from natural sources.

Only the carboxyl group of salicylic acid has reacted with methyl alcohol. Therefore, the hydroxyl group reacts with ferric chloride T.S. to produce a violet color. It is readily saponified by alkalies and reacts like the other salicylates.

It is used most often as a flavoring agent and in external medication as a rubefacient in liniments. In water and hydroalcoholic solutions it is absorbed rapidly, thus penetrating deeply into the tissue and exerting also a systemic action. It is generally applied as a lotion in 10 to 25 per cent concentration.

Internal use is limited to small quantities due to its toxic effects in large doses. The average lethal dose is 10 ml. for children and 30 ml. for adults. In veterinary practice, it finds some use as a carminative.

Category—rubefacient; pharmaceutic aid (flavor for Aromatic Cascara Fluidextract).

For external use—topically, in lotions and solutions in 10 to 25 per cent concentration.

Aromatic Cascara Fluidextract U.S.P.

Carbethyl Salicylate, Sal-Ethyl Carbonate®, Ethyl Salicylate Carbonate, is an ester of ethyl salicylate and carbonic acid and thus is a combination of a type I and type II compound.

It occurs as white crystals, insoluble in water and in diluted hydrochloric acid, slightly soluble in alcohol or ether and readily soluble in chloroform or acetone. The insolubility tends to prevent gastric irritation and makes it tasteless.

In action and uses it resembles aspirin and gives the antipyretic and analgesic effects of the salicylates. The pharmaceutic forms are powder, tablet and a tablet containing aminopyrine.

The usual dose is 1.0 g.

The Salol Principle

Nencki introduced salol in 1886 and by so doing presented to the science of therapy the "Salol Principle." In the case of salol, two toxic substances (phenol and salicylic acid) were combined into an ester which, when taken internally, will slowly hydrolyze in the intestine to give the antiseptic action of its components (q.v.). This type of ester is referred to as a "Full Salol" or "True Salol" when both components of the ester are active compounds. Examples are guaiacol benzoate, β-naphthol benzoate, and salol.

This "Salol Principle" can be applied to esters of which only the alcohol or the acid is the toxic, active or corrosive portion, and this type is called a "Partial Salol."

Examples of a "Partial Salol" containing an active acid are ethyl salicylate, and methyl salicylate. Examples of a "Partial Salol" containing an active phenol are creosote carbonate, thymol carbonate and guaiacol carbonate (see phenols, Chap. 10).

Although a host of the "salol" type of compounds have been prepared and used to some extent, none is presently very valuable in therapeutics, and all are surpassed by other agents.

Phenyl Salicylate, Salol. Phenyl salicylate occurs as fine, white crystals or as a white, crystalline powder with a characteristic taste and a faint, aromatic odor. It is insoluble in water (1:6700), slightly soluble in glycerin, soluble in alcohol (1:6), ether, chloroform, acetone or fixed and volatile oils.

Damp or eutectic mixtures form readily with many organic materials, such as thymol, menthol, camphor, chloral hydrate and phenol.

Salol is insoluble in the gastric juice but is slowly hydrolyzed in the intestine into phenol and salicylic acid. Because of this fact, coupled with its low melting point (41-43°), it has been used in the past as an enteric coating for tablets and capsules. However, it is not efficient as an enteric coating material, and its use has been superseded by more effective materials.

It also is used externally as a sun filter (10 per cent ointment) for sunburn prevention.

Salicylamide N.F., Salicin®, *o*-Hydroxybenzamide. This is a derivative of salicylic acid that has been known for almost a century and is currently enjoying a revival.[104]

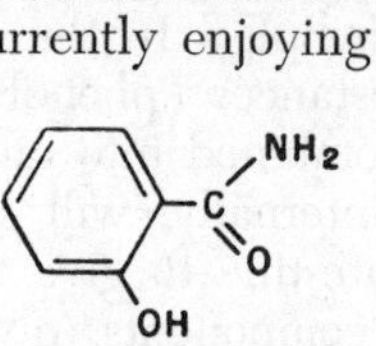

Salicylamide

It is readily prepared from salicyl chloride and ammonia. The compound occurs as a nearly odorless, white, crystalline powder. It is fairly stable to heat, light and moisture. It is slightly soluble in water (1:500), soluble in hot water, alcohol (1:15) and propylene glycol, and sparingly soluble in chloroform and ether. It is freely soluble in solutions of alkalies. In alkaline solution with sodium carbonate or triethanolamine, decomposition takes place, resulting in a precipitate and yellow to red color.

Salicylamide is said to exert a moderately quicker and deeper analgesic effect than does aspirin. Long-term studies on rats revealed no untoward symptomatic or physiologic reactions. Its metabolism is different from that of other salicylic compounds, and it is not hydrolyzed to salicylic acid.[92] Its analgesic and antipyretic activity is probably no greater than that of aspirin and possibly less. However, it can be used in place of salicylates and is particularly useful for those cases where there is a demonstrated sensitivity to salicylates. It is excreted much more rapidly than other salicylates, which probably accounts for its lower toxicity, and thus does not permit high blood levels.

The dose for simple analgesic effect may vary from 0.3 to 1 g. administered 3 times daily; but for rheumatic conditions the dose may be increased to 2 to 4 g. 3 times a day. However, gastric intolerance may limit the dosage. The usual period of this higher dosage should not extend beyond 3 to 6 days. Drops (60 mg./ml.) are also available (Liquiprin®).

Category—analgesic.

Usual dose—300 to 600 mg. every 3 or 4 hours; child's dose, 75 mg. for each year of age, repeated every 3 or 4 hours with a maximum single dose of 300 mg.

OCCURRENCE

Salicylamide Oral Suspension N.F.
Salicylamide Tablets N.F.

Aspirin U.S.P., Aspro®, Empirin®, Acetylsalicylic Acid. Aspirin was introduced into medicine by Dreser in 1899. It is prepared by treating salicylic acid, which was first prepared by Kolbe in 1874, with acetic anhydride.

The hydrogen atom of the hydroxyl group in salicylic acid has been replaced by the acetyl group; this also may be accomplished by using acetyl chloride with salicylic acid or ketene with salicylic acid.

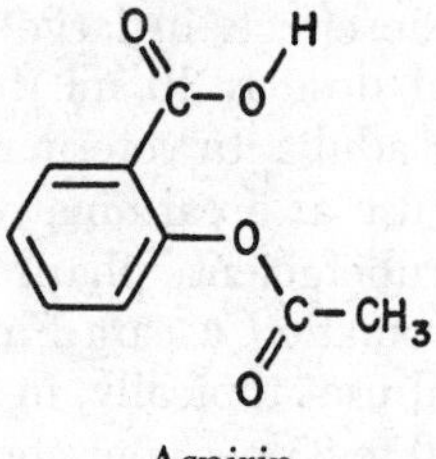

Aspirin

Aspirin occurs as white crystals or as a white, crystalline powder. It is slightly soluble in water (1:300) and soluble in alcohol (1:5), chloroform (1:17) and ether (1:15). Also, it dissolves easily in glycerin. Aqueous solubility may be increased by using acetates or citrates of alkali metals, although these are said to decompose it slowly.

It is stable in dry air, but in the presence of moisture, it slowly hydrolyzes into acetic and salicylic acids. Salicylic acid will crystallize out when an aqueous solution of aspirin and sodium hydroxide is boiled and then acidified.

Aspirin itself is sufficiently acid to produce effervescence with carbonates and, in the presence of iodides, to cause the slow liberation of iodine. In the presence of alkaline hydroxides and carbonates, it decomposes, although it does form salts with alkaline metals and alkaline earth metals. The presence of salicylic acid, formed upon hydrolysis, may be confirmed by the formation of a violet color upon the addition of ferric chloride solution.

Aspirin is not hydrolyzed appreciably on contact with weakly acid digestive fluids of the stomach, but on passage into the intestine, is subjected to some hydrolysis. However, most of it is absorbed unchanged. The gastric mucosal irritation of aspirin has been ascribed by Garrett[105] to salicylic acid formation, the natural acidity of aspirin, or the adhesion of undissolved aspirin to the mucosa. He has also proposed the nonacidic anhydride of aspirin as a superior form for oral administration. Davenport[98] concludes that aspirin causes an alteration in mucosal cell permeability, allowing back-diffusion of stomach acid which damages the capillaries. A number of proprietaries (e.g., Bufferin) employ compounds, such as sodium bicarbonate, aluminum glycinate, sodium citrate, aluminum hydroxide or magnesium trisilicate, to counteract this acid property. One of the better antacids is Dihydroxy-aluminum Aminoacetate N.F. Aspirin has been shown to be unusually effective when prescribed with calcium glutamate. The more stable, nonirritant calcium acetylsalicylate is formed, and the glutamate portion (glutamic acid) maintains a pH of 3.5 to 5.

Preferably, dry dosage forms (i.e., tablets, capsules or powders) should be used, since aspirin is somewhat unstable in aqueous media. In tablet preparations, the use of acid-washed talc has been shown to improve the stability of aspirin.[106] Also, it has been found to break down in the presence of phenylephrine hydrochloride.[107] Aspirin in aqueous media will hydrolyze almost completely in less than 1 week. However, solutions made with alcohol or glycerin do not decompose as quickly. Citrates retard hydrolysis only slightly. Some studies have indicated that sucrose tends to inhibit hydrolysis. A study of aqueous aspirin suspensions has indicated sorbitol to exert a pronounced stabilizing effect.[108] Stable liquid preparations are available that use triacetin, propylene glycol or a polyethylene glycol. Aspirin lends itself readily to combination with many other substances but tends to soften and become damp with methenamine, aminopyrine, salol, antipyrine, phenol or acetanilid.

Aspirin is one of the most widely used compounds in therapy and, until recently, was not associated with untoward effects. Allergic reactions to aspirin now are observed commonly. Asthma and urticaria are the most common manifestations and, when they occur, are extremely acute in nature and difficult to relieve. Like sodium salicylate, it has been shown to cause congenital malformations when administered to mice.[109] Pretreatment with sodium pentobarbital or chlorpromazine resulted in a significant lowering of these effects.[110] The reader is urged to consult the excellent review by Smith for an account of the pharmacologic aspects of aspirin.[92, 93]

Practically all salts of aspirin, except those of aluminum and calcium, are unstable for pharmaceutic use. This salt appears to have fewer undesirable side-effects and to induce analgesia faster than aspirin.

A timed-release preparation (Measurin®) of aspirin is available. It does not appear to offer any advantages over aspirin except for bedtime dosage.

Aspirin is used as an antipyretic, analgesic and antirheumatic, usually in powder, capsule, suppository or tablet form. Its use in rheumatism has been reviewed and it is said to be the drug of choice over all other salicylate derivatives.[111, 112] There is some anesthetic action when applied locally, especially in powder form in tonsilitis or pharyngitis, and in ointment form for skin itching and certain skin diseases. In the usual dose, 52 to 75 per cent is excreted in the urine, in various forms, in a period of 15 to 30 hours. It is believed that analgesia is due to the unhydrolyzed acetylsalicylic acid molecule.[92-94] A widely used combination is aspirin, phenacetin and caffeine (Tablets N.F.), known as APC.

Category—analgesic; antipyretic; antirheumatic.

Usual dose—oral or rectal, 600 mg. 4 to 6 times daily as needed.

Usual dose range—oral, 300 mg. to 8 g. daily; rectal, 300 mg. to 2 g. daily.

OCCURRENCE
Aspirin Capsules N.F.
Aspirin Suppositories U.S.P.
Aspirin Tablets U.S.P.
Aspirin, Phenacetin and Caffeine Tablets N.F.
Codeine Phosphate, Aspirin, Phenacetin, and Caffeine Tablets N.F.

Aluminum Aspirin N.F., Aspirin Dulcet®, Hydroxybis(salicylato)aluminum Diacetate. This salt of aspirin may be prepared by thoroughly mixing aluminum hydroxide

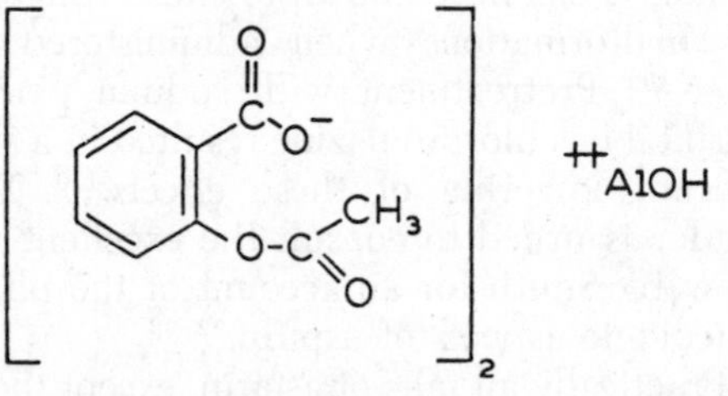

gel, water and acetylsalicylic acid, maintaining the temperature below 65°. Aluminum aspirin occurs as a white to off-white powder or granules and is odorless or has only a slight odor. It is insoluble in water and organic solvents, is decomposed in aqueous solutions of alkali hydroxides and carbonates and is not stable above 65°. It offers the advantages of being free of odor and taste and possesses added shelf-life stability. It is available in a flavored form for children (Dulcet).

Category—analgesic.

Usual dose—670 mg., equivalent to about 600 mg. of aspirin, every 4 hours as necessary. Children's dose, 75 mg. every 4 hours.

OCCURRENCE
Aluminum Aspirin Tablets N.F.

Calcium Acetylsalicylate, Soluble Aspirin, Calcium Aspirin. This compound is pre-

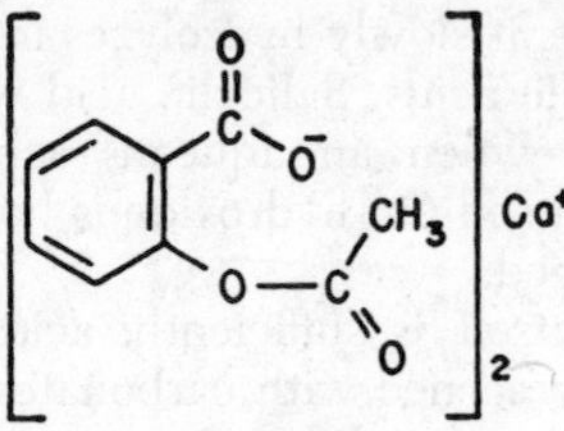

pared by treating acetylsalicylic acid with calcium ethoxide or methoxide in alcohol or acetone solution. It is readily soluble in water (1:6) but only sparingly soluble in alcohol (1:80). It is more stable in solution than aspirin and is used for the same conditions.

Calcium aspirin is marketed also as a complex salt with urea, calcium carbaspirin (Calurin), which is claimed to give more rapid salicylate blood levels and to be less irritating than aspirin, although no clear advantage has been shown.

The usual dose is 0.5 to 1.0 g.

Sodium Gentisate, Gentasal®. Gentisic acid is used as the sodium salt in tablet form. Its mode of action may be a hyaluronidase-inhibiting effect. It does not increase prothrombin time, cause tinnitus or give rise to aural symptoms. Also, salts of resorcyclic acid (2,6-dihydroxybenzoic acid) have been shown to be effective in rheumatism. The usual dose is 500 mg.

THE N-ARYLANTHRANILIC ACIDS

The most recent advances in the search for non-narcotic analgesics appear to be centered in the N-arylanthranilic acids. Their outstanding characteristic is that they are primarily nonsteroidal anti-inflammatory agents and, secondarily, that some possess analgesic properties.

Mefenamic Acid, Ponstel®, N-(2,3-Xylyl)-anthranilic acid, (a), occurs as an off-white, crystalline powder that is insoluble in water and slightly soluble in alcohol. It appears to be the first genuine antiphlogistic analgesic discovered since aminopyrine. Because it is believed that aspirin and aminopyrine owe their general-purpose analgesic efficacy to a combination of peripheral and central effects,[113] a wide variety of arylanthranilic acids were screened for antinociceptive (analgesic) activity if they showed significant anti-inflammatory action. It has become evident that the combination of both effects is a rarity among these compounds. The actual mechanism of analgesic action is unknown at present and no relationship to lipid-plasma distribution, partition coefficient, or pK_a has been noted. The interested reader, however, will find additional information with respect to anti-bradykinin and anti-UV-erythema activities of these compounds together with speculations on a receptor site in the literature.[114]

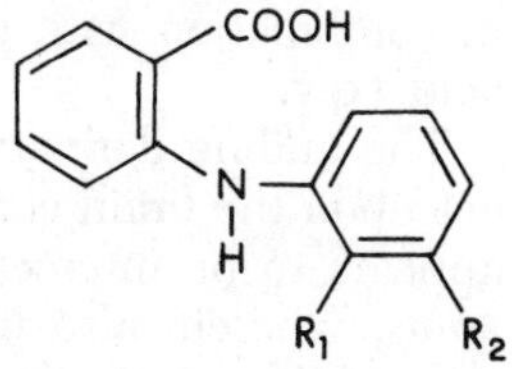

It has been shown[115] that mefenamic acid in a dose of 250 mg. is superior to 600 mg. of aspirin as an analgesic and that doubling the dose gives a sharp increase in efficacy. A study[116] examining this drug with respect to gastrointestinal bleeding indicated that it has a lower incidence of this side-effect than has aspirin. Diarrhea, drowsiness and headache have accompanied its use. The possibility of blood disorders has prompted limitation of its administration to 7 days. It is not recommended for children or during pregnancy.

Flufenamic Acid, Arlef®, N-(*m*-Trifluoromethylphenyl)anthranilic Acid, (b), is similar in activity to mefenamic acid. It appears to have superior anti-inflammatory activity and less analgesic activity. It is currently under investigation.

(a) $R_1 = R_2 = CH_3$
(b) $R_1 = H$
$R_2 = CF_3$

Arylacetic Acid Derivatives

Indomethacin N.F., Indocin®, 1-(*p*-chlorobenzoyl)-5-methoxy-2-methylindole-3-acetic acid, occurs as a pale yellow to yellow-tan crystalline powder which is soluble in ethanol and acetone and practically insoluble in water. It is unstable in alkaline solution and sunlight. It shows polymorphism, one form melting at about 155° and the other at about 162°. It may occur as a mixture of both forms with a melting range between the above melting points.

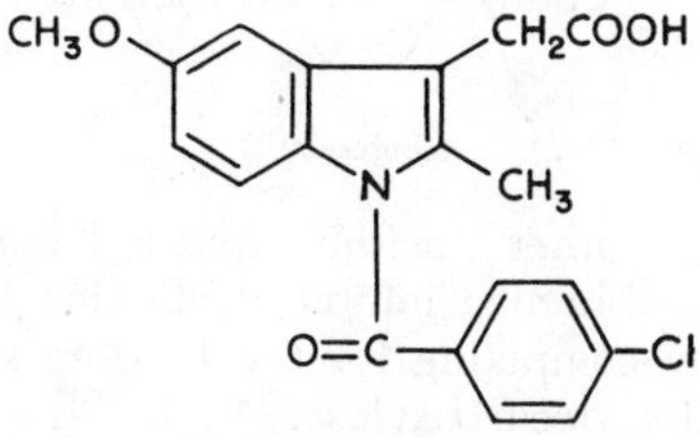

Indomethacin

Since its introduction in 1965, it has been widely used as an anti-inflammatory analgesic in rheumatoid arthritis, spondylitis, and osteoarthritis, and to a lesser extent in gout. Although both its analgesic and anti-inflammatory activities have been well established, it appears to be no more effective than aspirin.[117]

The most frequent side-effects are gastric distress and headache. It has also been associated with peptic ulceration, blood disorders, and possible deaths. The side-effects appear to be dose-related and sometimes can be minimized by reducing the dose. It is not recommended for use in children because of possible interference with resistance to infection. Like many other acidic compounds, it circulates bound to blood protein, requiring caution in the concurrent use of other protein-binding drugs.

Indomethacin is recommended only for those patients by whom aspirin cannot be tolerated, and in place of phenylbutazone in long-term therapy, for which it appears to be less hazardous than corticosteroids or phenylbutazone.

Category—anti-inflammatory (nonsteroid).

Usual dose—25 or 50 mg. 2 to 3 times daily.

OCCURRENCE
Indomethacin Capsules N.F.

Ibufenac, *p*-isobutylphenylacetic acid, is one of a new group of arylacetic acid derivatives which were developed after the discovery of indomethacin's activity. Ibufenac has shown about 2 to 4 times the anti-inflammatory, analgesic and antipyretic properties of aspirin in animal testing. It is being tested in man.

$$CH_3{-}CHCH_2{-}\underset{\text{Ibufenac}}{\bigcirc}{-}CH_2CO_2H$$

Like other acidic anti-inflammatory agents, ibufenac inhibits histamine formation by competing for the binding site on histadine decarboxylase.[118]

Namoxyrate, Namol®, 2-(4-biphenyl)butyric acid dimethylaminoethanol salt, is another new phenylacetic acid derivative under investigation. Namoxyrate shows high analgesic activity, being about 7 times that of aspirin and nearly as effective as codeine. It has high antipyretic activity but appears

$$\bigcirc{-}\bigcirc{-}\underset{\underset{CH_2CH_3}{|}}{CHCO_2H}$$

Namoxyrate

to be devoid of anti-inflammatory activity. These effects are peripheral. The dimethylaminoethanol increases its activity by increasing intestinal absorption. The ester of these two components is less active.[119]

ANILINE AND *p*-AMINOPHENOL DERIVATIVES

The introduction of aniline derivatives as analgesics is based on the discovery by Cahn and Hepp, in 1886, that aniline (C-1)* and acetanilid (C-2) both have powerful antipyretic properties. The origin of this group from aniline has led to their being called "coal tar analgesics." Acetanilid was introduced by these workers because of the known toxicity of aniline itself. Aniline brings about the formation of methemoglobin, a form of hemoglobin that is incapable of functioning as an oxygen carrier. The acyl derivatives of aniline were thought to exert their analgesic and antipyretic effects by first being hydrolyzed to

* See Table 80.

aniline and the corresponding acid, following which the aniline was oxidized to *p*-aminophenol (C-3). This is then excreted in combination with glucuronic or sulfuric acid (q.v.).

The aniline derivatives do not appear to act upon the brain cortex; the pain impulse appears to be intercepted at the hypothalamus, wherein also lies the thermoregulatory center of the body. It is not clear if this is the site of their activity, because most evidence suggests that they act at peripheral thermoceptors. They are effective in the return to normal temperature of feverish individuals. Normal body temperatures are not affected by the administration of these drugs.

It is significant to note that, of the antipyretic-analgesic group, the aniline derivatives show little if any anti-inflammatory activity.

Table 80 shows some of the types of aniline derivatives that have been made and tested in the past. In general, any type of substitution on the amino group that reduces its basicity results also in a lowering of its physiologic activity. Acylation is one type of substitution that accomplishes this effect. Acetanilid (C-2) itself, although the best of the acylated derivatives, is toxic in large doses but when administered in analgesic doses is probably without significant harm. Formanilide (C-4) is readily hydrolyzed and too irritant. The higher homologs of acetanilid are less soluble and, therefore, less active and less toxic. Those derived from aromatic acids (e.g., C-5) are virtually without analgesic and antipyretic effects. One of these, Salicylanilide N.F. (C-6), is used as a fungicide and antimildew agent. Exalgin (C-7) is too toxic.

The hydroxylated anilines (*o, m, p*), better known as the aminophenols, are quite interesting from the standpoint of being considerably less toxic than aniline. The *para* compound (C-3) is of particular interest from two standpoints, namely, it is the metabolic product of aniline, and it is the least toxic of the three possible aminophenols. It also possesses a strong antipyretic and analgesic action. However, it is too toxic to serve as a drug and, for this reason, there were numerous modifications

TABLE 80. SOME ANALGESICS RELATED TO ANILINE

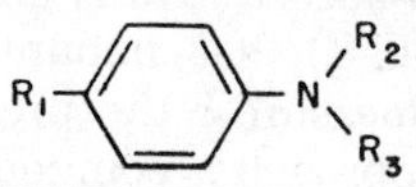

$$R_1 - \text{C}_6\text{H}_4 - N\begin{smallmatrix}R_2\\R_3\end{smallmatrix}$$

COMPOUND	STRUCTURE			NAME (If Any)
	R_1	R_2	R_3	
C-1	–H	–H	–H	Aniline
C-2	–H	–H	$-\overset{\text{O}}{\underset{\|\|}{C}}-CH_3$	Acetanilid
C-3	–OH	–H	–H	p-Aminophenol
C-4	–H	–H	$-\overset{}{C}-H$, $\|\|$ O	Formanilid
C-5	–H	–H	$-C-C_6H_5$, $\|\|$ O	Benzanilid
C-6	–H	–H	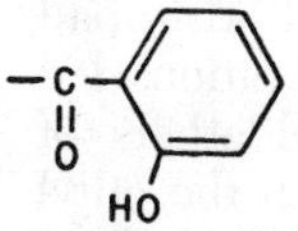 $-C(=O)-C_6H_4-OH$	Salicylanilide (not an analgesic, but is an antifungal agent)
C-7	–H	–CH₃	$-C-CH_3$, $\|\|$ O	Exalgin
C-8	–OH	–H	$-C-CH_3$, $\|\|$ O	Acetaminophen
C-9	$-OCH_3$	–H	–H	Anisidine
C-10	$-OC_2H_5$	–H	–H	Phenetidine
C-11	$-OC_2H_5$	–H	$-C-CH_3$, $\|\|$ O	Phenacetin
C-12	$-OC_2H_5$	–H	$-C-CHCH_3$, $\|\|$ O OH	Lactylphenetidin
C-13	$-OC_2H_5$	–H	$-C-CH_2NH_2$, $\|\|$ O	Phenocoll
C-14	$-OC_2H_5$	–H	$-C-CH_2OCH_3$, $\|\|$ O	Kryofine
C-15	$-OC-CH_3$, $\|\|$ O	–H	$-C-CH_3$, $\|\|$ O	p-Acetoxyacetanilid
C-16	$-OC(=O)-C_6H_4-OH$	–H	$-C-CH_3$, $\|\|$ O	Phenetsal
C-17	$-OCH_2CH_2OH$	–H	$-C-CH_3$, $\|\|$ O	Pertonal

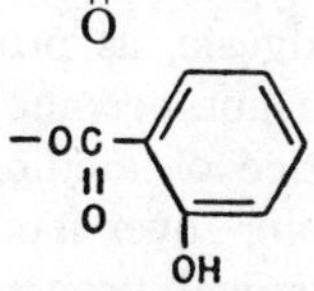

attempted. One of the first was the acetylation of the amine group to provide N-acetyl-*p*-aminophenol (acetaminophen) (C-8), a product which retained a good measure of the desired activities. Another approach to the detoxification of *p*-aminophenol was the etherification of the phenolic group. The best known of these are anisidine (C-9) and phenetidine (C-10), which are the methyl and ethyl ethers, respectively. However, it became apparent that a free amino group in these compounds, while promoting a strong antipyretic action, was also conducive to methemoglobin formation. The only exception to the preceding was in compounds where a carboxyl group or sulfonic acid group had been substituted on the benzene nucleus. In these compounds, however, the antipyretic effect also had disappeared. The above considerations led to the preparation of the alkyl ethers of N-acetyl-*p*-aminophenol of which the ethyl ether was the best and is now the official phenacetin (C-11). The methyl and propyl homologs were undesirable from the standpoint of causing emesis, salivation, diuresis and other reactions. Alkylation of the nitrogen with a methyl group has a potentiating effect on the analgesic action but, unfortunately, has a highly irritant action on mucous membranes.

The phenacetin molecule has been modified by changing the acyl group on the nitrogen with sometimes beneficial results. Among these are lactylphenetidin (C-12), phenocoll (C-13) and kryofine (C-14). None of these, however, is in current use.

Changing the ether group of phenacetin to an acyl type of derivative has not always been successful. *p*-Acetoxyacetanilid (C-15) has about the same activity and disadvantages as the free phenol. However, the salicyl ester (C-16) exhibits a diminished toxicity and an increased antipyretic activity. Pertonal (C-17) is a somewhat different type in which glycol has been used to etherify the phenolic hydroxyl group. It is very similar to phenacetin. None of these is presently on the market.

With respect to the fate in man of the types of compounds discussed above, Brodie and Axelrod[120] point out that acetanilid and phenacetin are metabolized by two different routes. Acetanilid is metabolized primarily to N-acetyl-*p*-aminophenol and only a small amount to aniline, which they showed to be the precursor of phenylhydroxylamine, the compound responsible for methemoglobin formation. Phenacetin is mostly de-ethylated to N-acetyl-*p*-aminophenol, whereas a small amount is converted by deacetylation to *p*-phenetidine, also responsible for methemoglobin formation. With both acetanilid and phenacetin, the N-acetyl-*p*-aminophenol formed is believed to be responsible for the analgesic activity of the compounds.

Acetanilid, Antifebrin, Phenylacetamid. Acetanilid is the monoacetyl derivative of aniline, prepared by heating aniline and acetic acid for several hours.

It can be recrystallized from hot water and occurs as a stable, white, crystalline compound. It is slightly soluble in water (1:190) and easily soluble in hot water, acetone, chloroform, glycerin (1:5), alcohol (1:4) or ether (1:17).

Acetanilid is a neutral compound and will not dissolve in either acids or alkalies.

It is prone to form eutectic mixtures with aspirin, antipyrine, chloral hydrate, menthol, phenol, pyrocatechin, resorcinol, salol, thymol or urethane.

It is definitely toxic in that it causes formation of methemoglobin, affects the heart and may cause skin reactions and a jaundiced condition. Nevertheless, in the doses used for analgesia, it is a relatively safe drug. However, it is recommended that it be administered in intermittent periods, no period exceeding a few days.[121]

The analgesic effect is selective for most simple headaches and for the pain associated with many muscles and joints.

The usual dose is 200 mg.

A number of compounds related to acetanilid have been synthesized in attempts to find a better analgesic, as previously indicated. They have not become very important in the practice of medicine, for they have little to offer over acetanilid. The physical and chemical properties are also much the same. Eutectic mixtures are formed with many of the same compounds.

Phenacetin U.S.P., Acetophenetidin, *p*-Acetophenetidide.

Phenacetin may be synthesized in several steps from *p*-nitrophenol.

It occurs as stable, white, glistening crystals, usually in scales, or a fine, white, crystalline powder. It is odorless and has a slightly bitter taste. It is very slightly soluble in water (1:1,300), soluble in alcohol (1:15) and chloroform (1:15), but only slightly soluble in ether (1:130). It is sparingly soluble in boiling water (1:85).

In general properties and incompatibilities, such as being decomposed by acids and alkalies, it is similar to acetanilid. Phenacetin forms eutectic mixtures with chloral hydrate, phenol, aminopyrine, pyrocatechin or pyrogallol.

It is used widely as an analgesic and antipyretic, having essentially the same actions as acetanilid. It should be used with the same cautions because the toxic effects are the same as those of acetaminophen, the active form to which it is converted in the body. Some feel there is little justification for its continued use.[120] In particular, a suspected nephrotoxic action[122] has been the basis for the present warning label requirements by the Food and Drug Administration, i.e., "This medication may damage the kidneys when used in large amounts or for a long period of time. Do not take more than the recommended dosage, nor take regularly for longer than 10 days without consulting your physician." Some recent evidence suggests that phenacetin may not cause nephritis to any greater degree than aspirin, with which it is most often combined.[123]

Category—analgesic.
Usual dose—300 mg. 4 to 6 times a day.
Usual dose range—300 mg. to 2 g. daily.

Occurrence
Phenacetin Tablets U.S.P.
Aspirin, Phenacetin and Caffeine
   Tablets N.F.
Codeine Phosphate, Aspirin, Phenacetin, and
   Caffeine Tablets N.F.

Acetaminophen N.F., Tempra®, Tylenol®, 4'-Hydroxyacetanilide. This may be prepared by reduction of *p*-nitrophenol in glacial acetic acid, acetylation of *p*-aminophenol with acetic anhydride or ketene, or from *p*-hydroxyacetophenone hydrazone. It occurs as a white, odorless, slightly bitter crystalline powder. It is slightly soluble in water and ether, soluble in boiling water (1:20), alcohol (1:10), and sodium hydroxide T.S.

Acetaminophen has analgesic and antipyretic activities comparable with acetanilid and is used in the same conditions. Although it possesses the same toxic effects as acetanilid, they occur less frequently and with less severity; therefore, it is considered somewhat safer to use. However, the same cautions should be applied. The new Food and Drug Administration warning label reads: "Warning: Do not give to children under three years of age or use for more than 10 days unless directed by a physician."[122]

It is available in several nonprescription forms and, also, is marketed in combination with aspirin and caffeine (Trigesic).

Category—analgesic.
Usual dose—650 mg. every 4 hours.
Usual dose range—325 mg. to 650 mg.; not to exceed 2.6 g. per 24-hour period.

Occurrence
Acetaminophen Elixir N.F.
Acetaminophen Tablets N.F.

The Pyrazolone and Pyrazolidinedione Derivatives

The simple doubly unsaturated compound containing 2 nitrogen and 3 carbon atoms in the ring, and with the nitrogen atoms neighboring, is known as pyrazole. The reduction products, named as are other rings of 5 atoms, are pyrazoline and pyrazolidine. Several pyrazoline substitution products are used in medicine. Many of these are derivatives of 5-pyrazolone.

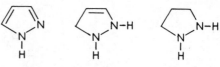

Pyrazole   Pyrazoline   Pyrazolidine

5-Pyrazolone   3,5-Pyrazolidinedione

Some can be related to 3,5-pyrazolidinedione.

Ludwig Knorr, a pupil of Emil Fischer, while searching for antipyretics of the quinoline type, in 1884, discovered the 5-pyrazolone now known as antipyrine. This discovery initiated the beginnings of the great German drug industry that dominated the field for approximately 40 years. Knorr, although at first mistakenly believing that he had a quinoline-type compound, soon recognized his error, and the compound was interpreted correctly as being a pyrazolone. Within 2 years, the analgesic properties of this compound became apparent when favorable reports began to appear in the literature, particularly with reference to its use in headaches and neuralgias. Since then, it has retained some of its popularity as an analgesic, although its use as an antipyretic has declined steadily. Since its introduction into medicine, there have been over 1,000 compounds made in an effort to find others with a more potent analgesic action combined with a lesser toxicity. That antipyrine remains as one of the useful analgesics today is a tribute to its value. Many modifications of the basic compound have been made. The few derivatives and modifications on the market are listed in Tables 81 and 82. Phenylbutazone, although analgesic itself, was originally developed as a solubilizer for insoluble aminopyrine. It is being used at the present time for the relief of many forms of arthritis, in which capacity it has more than an analgesic action in that it also reduces swelling and spasm by an anti-inflammatory action.

Products

Antipyrine N.F., Felsol®, Phenazone, 2,3-Dimethyl-1-phenyl-3-pyrazolin-5-one. This was one of the first important drugs to be made (1887) synthetically.

Antipyrine and many related compounds are prepared by the condensation of hydrazine derivatives with various esters. Antipyrine itself is prepared by the action of ethyl acetoacetate on phenylhydrazine and subsequent methylation.

It consists of colorless, odorless crystals or a white powder with a slightly bitter taste. It is very soluble in water, alcohol or chloroform, less so in ether, and its aqueous solution is neutral to litmus paper. However, it is basic in nature, which is due primarily to the nitrogen at position 2.

Locally, antipyrine exerts a paralytic action on the sensory and the motor nerves, resulting in some anesthesia and vasoconstriction, and it also exerts a feeble antiseptic effect. Systemically, it causes results that are very similar to those of acetanilid, although they are usually more rapid. It is readily absorbed after oral administration, circulates freely and is excreted chiefly by the kidneys

TABLE 81. DERIVATIVES OF 5-PYRAZOLONE

Compound *Proprietary Name*	R_1	R_2	R_3	R_4
Antipyrine (Phenazone)	$-C_6H_5$	$-CH_3$	$-CH_3$	$-H$
Aminopyrine (Amidopyrine) 　*Pyramidon*	$-C_6H_5$	$-CH_3$	$-CH_3$	$-N(CH_3)_2$
Dipyrone (Methampyrone) 　*Dimethone, Pydirone*	$-C_6H_5$	$-CH_3$	$-CH_3$	$-NCH_2SO_3Na$ 　　　$\mid$ 　　CH_3

TABLE 82. DERIVATIVES OF 3,5-PYRAZOLIDINEDIONE

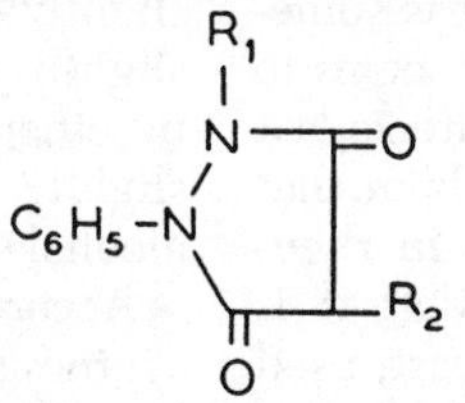

Compound Proprietary Name	R_1	R_2
Phenylbutazone Butazolidin	$-C_6H_5$	$-C_4H_9$ (n)
Oxyphenbutazone Tandearil	$-C_6H_4(OH)$ (p)	$-C_4H_9$ (n)
Sulfinpyrazone Anturane	$-C_6H_5$	$-CH_2CH_2S-C_6H_5$ $\downarrow$ O

without having been changed chemically. Any abnormal temperature is reduced rapidly by an unknown mechanism, usually attributed to an effect on the hypothetic heat-regulating center of the nervous system, although this appears doubtful. It has a higher degree of anti-inflammatory activity than aspirin, phenylbutazone, and indomethacin. It also lessens perception to pain of certain types, without any alteration in the mentality or motor functions, which differs from the effects of morphine. Very often it produces unpleasant and possibly alarming symptoms, even in small or moderate doses. These are giddiness, drowsiness, cyanosis, great reduction in temperature, coldness in the extremities, tremor, sweating and morbilliform or erythematous eruptions; with very large doses there are asphyxia, epileptic convulsions, and collapse. Treatment for such untoward reactions must be symptomatic. It is probably less likely to produce collapse than acetanilid and is not known to cause the granulocytopenia that sometimes follows aminopyrine.

Antipyrine has been employed in medicine less often in recent years than formerly. It is administered orally to reduce pain and fever in neuralgia, the myalgias, migraine, other headaches, chronic rheumatism and neuritis but is less effective than salicylates and more toxic. When used orally it is given in 300-mg. dose. It sometimes is employed in motor disturbances, such as the spasms of whooping cough or epilepsy. Occasionally, it is applied locally in 5 to 15 per cent solution for its vasconstrictive and anesthetic effects in rhinitis and laryngitis and sometimes as a styptic in nosebleed.

Category—local anesthetic.

Application (as Antipyrine and Benzocaine Solution N.F.)—instilled into ear canal from 3 times daily to every 1 to 2 hours as needed.

OCCURRENCE
Antipyrine and Benzocaine Solution N.F.

The great success of antipyrine in its early years led to the introduction of a great many derivatives, especially salts with a variety of acids, but none of these has any advantage over the parent compound. Currently in use is the compound with chloral hydrate (Hypnal).

Aminopyrine, Amidopyrine, Aminophenazone, 2,3-Dimethyl-4-dimethylamino-1-phenyl-3-pyrazolin-5-one. It is prepared from nitrosoantipyrine by reduction to the 4-amino compound followed by methylation.

It consists of colorless, odorless crystals that dissolve in water and the usual organic solvents. It has about the same incompatibilities as antipyrine.

It has been employed as an antipyretic and analgesic, as is antipyrine, but is somewhat slower in action. However, it seems to be much more powerful, and its effects last longer; indeed, some authorities claim that it is as effective as the salicylates in rheumatic fever and even maintain that it is practically specific. It was, in the past, used as routine treatment for measles and at one time was considered by some as the best agent for affording relief in diabetes insipidus, where it lowered the urinary output if not administered repeatedly. The usual dose is 300 mg. for headaches, dysmenorrhea, neuralgia, migraine and other like disorders, and it may be given several times daily in rheumatism and other conditions that involve continuous pain.

One of the chief disadvantages of therapy with aminopyrine is the possibility of producing agranulocytosis (granulocytopenia). It has been shown that this is caused by drug therapy with a variety of substances, including mainly aromatic compounds, but particularly with aminopyrine; indeed, a number of fatal cases have been traced definitely to this drug. The symptoms are a marked fall in leukocytes, absence of granulocytes in the blood, fever, sore throat, ulcerations on mucous surfaces, and prostration, with death in the majority of cases from secondary complications. The treatment is merely symptomatic with penicillin to prevent any possible superimposed infection. The condition seems to be more or less an allergic reaction, because only a certain small percentage of those who use the drug are affected, but great caution must be observed to avoid susceptibility. Many countries have forbidden or greatly restricted its administration, and it has fallen more or less into disfavor.

Dipyrone, Dimethone®, Pydirone®, methampyrone, occurs as a white, odorless, crystalline powder possessing a slightly bitter taste. It is freely soluble in water (1:1.5) and sparingly soluble in alcohol.

It is used as an analgesic, an antipyretic and an antirheumatic. The recommended dose is 300 mg. to 1 g. orally and 500 mg. to 1 g. intramuscularly or subcutaneously.

Phenylbutazone U.S.P., Butazolidin®, 4-Butyl-1,2-diphenyl-3,5-pyrazolidinedione.

This drug is a white to off-white, odorless, slightly bitter-tasting powder. It has a slightly aromatic odor and is freely soluble in ether, acetone and ethyl acetate, very slightly soluble in water, and is soluble in alcohol (1:20).

According to the patents describing the synthesis of this type of compound, it can be prepared by condensing n-butyl malonic acid or its derivatives with hydrazobenzene to get 1,2-diphenyl-4-n-butyl-3,5-pyrazolidinedione. Alternatively, it can be prepared by treating 1,2-diphenyl-3,5-pyrazolidinedione, obtained by a procedure analogous to the above condensation, with butyl bromide in 2 N sodium hydroxide at 70° or with n-butyraldehyde followed by reduction utilizing Raney nickel catalyst.

The principal usefulness of phenylbutazone lies in the treatment of the painful symptoms associated with gout, rheumatoid arthritis, psoriatic arthritis, rheumatoid spondylitis and painful shoulder (peritendinitis, capsulitis, bursitis and acute arthritis of the joint). Because of its many unwelcome side-effects, this drug is not generally considered to be the drug of choice but should be reserved for trial in those cases that do not respond to less toxic drugs. It should be emphasized that, although the drug is an analgesic, it is not to be considered as one of the simple analgesics and is not to be used casually. The initial daily dosage in adults ranges from 300 to 600 mg., divided into 3 or 4 doses. The manufacturer suggests that an average initial daily dosage of 600 mg. per day administered for 1 week should determine whether the drug will give a favorable response. If no results are forthcoming in this time, it is recommended that the drug be discontinued to avoid side-effects. In the event of favorable response, the dosage is reduced to a minimal effective daily dose, which usually ranges from 100 to 400 mg.

The drug is contraindicated in the presence of edema, cardiac decompensation, a history of peptic ulcer or drug allergy, blood dyscrasias, hypertension and whenever renal, cardiac or hepatic damage is present. All patients, regardless of the history given, should be careful to note the occurrence of black or tarry stools which might be indicative of reactivation of latent peptic ulcer and is a signal for discontinuance of the drug. The physician is well advised to read the manufacturer's literature and warnings thoroughly before attempting to administer the drug. Among the precautions the physician should take with regard to the patient are to examine the patient periodically for toxic reactions, to check for increase in weight (due to water retention) and to make periodic blood counts to guard against agranulocytosis.

Category—antirheumatic.

Usual dose—100 mg. 1 to 4 times daily.

Usual dose range—100 to 600 mg. daily.

OCCURRENCE

Phenylbutazone Tablets U.S.P.

Oxyphenbutazone N.F., Tandearil®, 4-Butyl-1-(*p*-hydroxyphenyl)-2-phenyl-3,5-pyrazolidinedione. This drug is a metabolite of phenylbutazone and has the same effectiveness, indications, side-effects and contraindications. Its only apparent advantage is that it causes acute gastric irritation less frequently.

Category—antiarthritic; anti-inflammatory (nonsteroid).

Usual dose—400 mg. daily in divided doses.

Usual dose range—100 to 600 mg.

OCCURRENCE

Oxyphenbutazone Tablets N.F.

Sulfinpyrazone U.S.P., Anturane®, 1,2-Diphenyl-4-[2-(phenylsufinyl)ethyl]-3,5-pyrazolidinedione. Unlike phenylbutazone, this is a potent uricosuric agent and is used primarily for the prevention of attacks of acute gouty arthritis. It has only a weak analgesic and anti-inflammatory action so that pain relief must be obtained by administration of phenylbutazone or other analgesics. Salicylates, however, are contraindicated as they antagonize its uricosuric action. Gastric distress is the most common side-effect and, like the other pyrazolones, it should be taken with milk or food. The initial daily dose is 50 mg. 4 times daily, gradually increased over a week until a daily dose of 400 to 800 mg. is reached. A lower dosage may be satisfactory after the blood urate level is reduced. The same precautions employed with phenylbutazone should be followed.

Category—uricosuric.

Usual dose—100 mg. 4 times daily.

Usual dose range—200 mg. to 800 mg. daily.

OCCURRENCE

Sulfinpyrazone Tablets U.S.P.

The pharmacology of these and other analogs has been reviewed extensively.[124]

QUINOLINE DERIVATIVES

The quinoline group of compounds has been the source of the two drugs cinchophen and neocinchophen, both of which are antipyretics as well as analgesics. In addition, they possess the ability of bringing about the excretion of uric acid from the system. Cinchophen was introduced to the medical profession in 1906 and soon became a widely used ingredient of arthritis remedies and simple analgesic preparations. However, it was recognized early that the drug was quite dangerous, and this finding stimulated considerable research in this field. The principal modification to emerge from this research was neocinchophen which, although less toxic than cinchophen, still had a potential for causing fatal hepatic disease. Since the emergence of much less toxic uricosuric agents the use of the cinchophens has been abandoned.

REFERENCES CITED

1. Tainter, M. L.: Ann. N. Y. Acad. Sci. 51:3, 1948.
2. Eisleb, O., and Schaumann, O.: Deutsche med. Wschr. 65:967, 1938.
3. Small, L. F., Eddy, N. B., Mosettig, E. and Himmelsbach, C. K.: Studies on Drug Addiction, Supplement No. 138 to the Public Health Reports, Washington, D.C., Supt. Doc., 1938.
4. Eddy, N. B., Halbach, H., and Braenden, O. J.: Bull. W.H.O. 14:353-402, 1956.
5. Stork, G., and Bauer, L.: J. Am. Chem. Soc. 75:4373, 1953.
6. U. S. patent 2,831,531; through Chem. Abstr. 52:13808, 1958.
7. Rapoport, H., Baker, D. R., and Reist, H. N.: J. Org. Chem. 22:1489, 1957;

Chadha, M. S., and Rapoport, H.: J. Am. Chem. Soc. 79:5730, 1957.

8. Okun, R., and Elliott, H. W.: J. Pharmacol. Exp. Ther. 124:255, 1958.

9. Seki, I., Takagi, H., and Kobayashi, S.: J. Pharm. Soc. Jap. 84:280, 1964.

10. Buckett, W. R., Farquharson, M. E., and Haining, C. G.: J. Pharm. Pharmacol. 16: 174, 68T, 1964.

11. Bentley, K. W., and Hardy, D. G.: Proc. Chem. Soc. 220, 1963.

12. Lister, R. E.: J. Pharm. Pharmacol. 16: 364, 1964.

13. Bentley, K. W., and Hardy, D. G.: J. Am. Chem. Soc. 89:3267, 1967.

14. Telford, J., Papadopoulos, C. N., and Keats, A. S.: J. Pharmacol. Exp. Ther. 133:106, 1961.

15. Gates, M., and Montzka, T. A.: J. Med. Chem. 7:127, 1964.

16. Schaumann, O.: Arch. exp. Path. Pharmakol. 196:109, 1940.

17. Bergel, F., and Morrison, A. L.: Quart. Revs. (London) 2:349, 1948.

18. Jensen, K. A., Lindquist, F., Rekling, E., and Wolffbrandt, C. G.: Dansk. tids. farm. 17:173, 1943; through Chem. Abstr. 39:2506, 1945.

19. Lee, J., Ziering, A., Berger, L., and Heineman, S. D.: Jubilee Volume—Emil Barell, p. 267, Basel, Reinhardt, 1946; J. Org. Chem. 12:885, 894, 1947; Berger, L., Ziering, A., and Lee, J.: J. Org. Chem. 12:904, 1947; Ziering, A., and Lee, J.: J. Org. Chem. 12:911, 1947.

20. Janssen, P. A. J., and Eddy, N. B.: J. Med. Pharm. Chem. 2:31, 1960.

21. Lee, J.: Analgesics: B. Partial structures related to morphine *in* American Chemical Society: Medicinal Chemistry, vol. 1, pp. 438-466, New York, Wiley, 1951.

22. Beckett, A. H., Casy, A. F., and Kirk, G.: J. Med. Pharm. Chem. 1:37, 1959.

23. Ziering, A., Motchane, A., and Lee, J.: J. Org. Chem. 22:1521, 1957.

24. Groeber, W. R., et al.: Obstet. Gynec. 14:743, 1959.

25. Eddy, N. B.: Chem. & Ind. (London), p. 1462, Nov. 21, 1959.

26. Fraser, H. F., and Isbell, H.: Bull. Narcotics 13:29, 1961.

27. Janssen, P. A. J., et al.: J. Med. Pharm. Chem. 2:271, 1960.

28. Blicke, F. F., and Tsao, E.: J. Am. Chem. Soc. 75:3999, 1953.

29. Cass, L. J., et al.: J.A.M.A. 166:1829, 1958.

30. Batterman, R. C., Mouratoff, G. J., and

Kaufman, J. E.: Am. J. Med. Sci. 247:62, 1964.

31. Finch, J. S., and DeKornfeld, T. J.: J. Clin. Pharmacol. 7:46, 1967.

32. Yelnosky, J., and Gardocki, J. F.: Toxic. Appl. Pharmacol. 6:593, 1964.

33. Bockmuehl, M., and Ehrhart, G.: German patent 711,069.

34. Kleiderer, E. C., Rice, J. B., and Conquest, V.: Pharmaceutical Activities at the I. G. Farbenindustrie Plant, Höchstam-Main, Germany. Report 981, Office of the Publication Board, Dept. of Commerce, Washington, D.C., 1945.

35. Scott, C. C., and Chen, K. K.: Fed. Proc. 5:201, 1946; J. Pharmacol. Exp. Ther. 87:63, 1946.

36. Eddy, N. B., Touchberry, C., and Lieberman, J.: J. Pharmacol. Exp. Ther. 98: 121, 1950.

37. Janssen, P. A. J., and Jageneau, A. H.: J. Pharm. Pharmacol. 9:381, 1957; 10: 14, 1958. See also Janssen, P. A. J.: J. Am. Chem. Soc. 78:3862, 1956.

38. Cass, L. J., and Frederik, W. S.: Antibiotic Med. 6:362, 1959, and references cited therein.

39. Eddy, N. B., Besendorf, H., and Pellmont, B.: Bull. Narcotics, U.N. Dept. Social Affairs 10:23, 1958.

40. DeKornfeld, T. J.: Curr. Res. Anesth. 39:430, 1960.

41. Murphy, J. G., Ager, J. H., and May, E. L.: J. Org. Chem. 25:1386, 1960, and references cited therein.

42. Chignell, C. F., Ager, J. H., and May, E. L.: J. Med. Chem. 8:235, 1965.

43. May, E. L., and Eddy, N. B.: J. Med. Chem. 9:851, 1966.

44. Archer, S., et al.: J. Med. Chem. 7:123, 1964.

45. Cass, L. J., Frederik, W. S., and Teodoro, J. V.: J.A.M.A. 188:112, 1964.

46. Fraser, H. F., and Rosenberg, D. E.: J. Pharmacol. Exp. Ther. 143:149, 1964.

47. Lasagna, L., DeKornfeld, T. J., and Pearson, J. W.: J. Pharmacol. Exp. Ther. 144: 12, 1964.

48. deStevens, G. (ed.): Analgetics, New York, Academic Press, 1965.

49. Braenden, O. J., Eddy, N. B., and Halbach, H.: Bull. W.H.O. 13:937, 1955.

50. Leutner, V.: Arzneimittelforschung 10: 505, 1960.

51. Janssen, P. A. J.: Brit. J. Anaesth. 34: 260, 1962.

52. Beckett, A. H., and Casy, A. F., *in* Ellis, G. P., and West, G. B. (eds.): Progress

in Medicinal Chemistry, vol. 2, pp. 43-87, London, Butterworth, 1962.

53. Mellet, L. B., and Woods, L. A., *in* Progress in Drug Research, vol. 5, pp. 156-267, Basel, Birkhäuser, 1963.

54. Beckett, A. H., Casy, A. F., and Harper, N. J.: Pharm. Pharmacol. 8:874, 1956.

55. Lasagna, L., and DeKornfeld, T. J.: J. Pharmacol. Exp. Ther. 124:260, 1958.

56. Wright, W. B., Jr., Brabander, H. J., and Hardy, R. A., Jr.: J. Am. Chem. Soc. 81:1518, 1959.

57. Gross, F., and Turrian, H.: Experientia 13:401, 1957; Fed. Proc. 19:22, 1960.

58. Beckett, A. H., and Casy, A. F.: J. Pharm. Pharmacol. 6:986, 1954.

59. Beckett, A. H.: J. Pharm. Pharmacol. 8:848, 860, 1958.

60. ———: Pharm. J., p. 256, Oct. 24, 1959.

61. Portoghese, P. S.: J. Med. Chem. 8:609, 1965.

62. ———: J. Pharm. Sci. 55:865, 1966.

63. Koshland, D. E., Jr.: Proc. First Intern. Pharmacol. Meeting, 7:161, 1963, and references cited therein.

64. Belleau, B.: J. Med. Chem. 7:776, 1964.

65. Lasagna, L., and Beecher, H. K.: J. Pharmacol. Exp. Ther. 112:356, 1954.

66. Martin, W. R.: Pharmacol. Rev. 19:463, 1967.

67. Soulairac, A., Cahn, J., and Charpentier, J. (eds.): Pain, New York, Academic Press, 1968.

67a. Willette, R. E.: Am. J. Pharm. Educ. 34:662, 1970.

68. Archer, S., and Harris, L. S., *in* Jucker, E. (ed.): Progress in Drug Research, vol. 8, p. 262, Basel, Birkhäuser, 1965.

68a. Kutter, E., *et al.*: J. Med. Chem. 13:801, 1970.

68b. Portoghese, P. S., *et al.*: J. Med. Chem. 14:144, 1971.

68c. ———: J. Med. Chem. 11:219, 1968.

69. Proc. Manchester Lit. Phil. Soc. 69:79, 1925.

70. Gates, M., and Tschudi, G.: J. Am. Chem. Soc. 74:1109, 1952; 78:1380, 1956.

71. Lasagna, L.: Pharmacol. Rev. 16:47, 1964.

72. Houde, R. W., and Wallenstein, S. L.: Minutes of the 11th Meeting, Committee on Drug Addiction and Narcotics, National Research Council, 1953, p. 417.

73. Diehl, H. S.: J.A.M.A. 101:2042, 1933.

74. Eddy, N. B., and Lee, L. E.: J. Pharmacol. Exp. Ther. 125:116, 1959.

75. Weijlard, J., and Erickson, A. E.: J. Am. Chem. Soc. 64:869, 1942.

76. Fraser, H. F., *et al.*: J. Pharmacol. Exp. Ther. 122:359, 1958; Cochin, J., and Axelrod, J.: J. Pharmacol. Exp. Ther. 125:105, 1959.

77. Ziering, A., and Lee, J.: J. Org. Chem. 12:911, 1947.

78. Weijlard, J., *et al.*: J. Am. Chem. Soc. 78:2342, 1956.

79. Council on Drugs, Special Report, J.A.M.A. 172:1518, 1960.

80. Easton, N. R., Gardner, J. H., and Stevens, J. R.: J. Am. Chem. Soc. 69:2941, 1947. See also reference 34.

81. Freedman, A. M.: J.A.M.A. 197:878, 1966.

81a. The Medical Letter 11:97, 1969.

82. Pohland, A., and Sullivan, H. R.: J. Am. Chem. Soc. 75:4458, 1953.

83. Forrest, W. H., *et al.*: Clin. Pharmacol. Ther. 10(4):468, 1969.

84. Jasinski, D. R., Martin, W. R., and Sapira, J. D.: Clin. Pharmacol. Ther. 9:215, 1968.

85. The Medical Letter 9:49, 1967.

86. Salem, H., and Aviado, D. M.: Am. J. Med. Sci. 247:585, 1964.

87. Chappel, C. I., and von Seemann, C., *in* Ellis, G. P., and West, G. B. (eds.): Progress in Medicinal Chemistry, vol. 3, pp. 133-136, London, Butterworth, 1963.

88. Chernish, S. M.: Annals Allergy 21:677, 1963.

89. Chen, J. Y. P., Biller, H. F., and Montgomery, E. G.: J. Pharmacol. Exp. Ther. 128:384, 1960.

90. Amler, A. B., and Rothman, C. B.: J. New Drugs 3:362, 1963.

91. Five Year Report, Brit. Med. J. 2:1033, 1960.

92. Smith, P. K.: Ann. N. Y. Acad. Sci. 86:38, 1960.

93. Smith, M. J. H., and Smith, P. K. (eds.): The Salicylates. A Critical Bibliographic Review, New York, Wiley, 1966.

94. Collier, H. O. J.: Sci. Amer. 209:97, 1963.

95. Alvarez, A. S., and Summerskill, W. H. J.: Lancet 2:920, 1958.

96. Barager, F. D., and Duthie, J. J. R.: Brit. Med. J. 1:1106, 1960.

97. Leonards, J. R., and Levy, G.: Abstracts of the 116th Meeting of the American Pharmaceutical Association, p. 67, Montreal, May 17-22, 1969.

98. Davenport, H. W.: New Eng. J. Med. 276:1307, 1967.

99. Levy, G., and Hayes, B. A.: New Eng. J. Med. 262:1053, 1960.

100. Lieberman, S. V., *et al.*: J. Pharm. Sci. 53:1486, 1492, 1964.

101. Pfeiffer, C. C.: Arch. Biol. Med. Exp. 4:10, 1967.

102. Salassa, R. M., Bollman, J. M., and Dry, T. J.: J. Lab. Clin. Med. 33:1393, 1948.

103. Smith, P. K., *et al.*: J. Pharmacol. Exp. Ther. 87:237, 1946.

104. The Medical Letter 6:14, 1964.

105. Garrett, E. R.: J. Am. Pharm. A. (Sci. Ed.) 48:676, 1959.

106. Gold, G., and Campbell, J. A.: J. Pharm. Sci. 53:52, 1964.

107. Troup, A. E., and Mitchner, H.: J. Pharm. Sci. 53:375, 1964.

108. Blaug, S. M., and Wesolowski, J. W.: J. Am. Pharm. A. (Sci. Ed.) 48:691, 1959.

109. Obbink, H. J. K.: Lancet 1:565, 1964.

110. Goldman, A. S., and Yakovac, W. C.: Proc. Soc. Exp. Biol. Med. 115:693, 1964.

111. Anon.: Brit. Med. J. 2:131T, 1963.

112. The Medical Letter 8:7, 1966.

113. Winder, C. V.: Nature 184:494, 1959.

114. Scherrer, R. A., Winder, C. V., and Short, F. W.: Ninth National Medicinal Chemistry Symposium of the American Chemical Society, June 21-24, 1964, University of Minnesota, Minneapolis, p. 11a.

115. Cass, L. J., and Frederik, W. S.: J. Pharmacol. Exp. Ther. 139:172, 1963.

116. Lane, A. Z., Holmes, E. L., and Moyer, C. E., J. New Drugs 4:333, 1964.

117. The Medical Letter 10:37, 1968.

118. Adams, S. S., Hebborn, P., and Nicholson, J. S.: J. Pharm. Pharmacol. 20:305, 1968.

119. Emele, J. F., and Shanaman, J. E.: Arch. Int. Pharmacodyn. Ther. 170:99, 1967.

120. Brodie, B. B., and Axelrod, J.: J. Pharmacol. Exp. Ther. 94:29, 1948; 97:58, 1949. See also Axelrod, J.: Postgrad. Med. 34:328, 1963.

121. Bonica, J. J., and Allen, G. D., *in* Modell, W. (ed.): Drugs of Choice 1970-1971. p. 210, St. Louis, Mosby, 1970.

122. The Medical Letter 6:78, 1964.

123. Brown, D. M., and Hardy, T. L.: Brit. J. Pharmacol. Chemother. 32:17, 1968.

124. Burns, J. J., *et al.*: Ann. N. Y. Acad. Sci. 86:253, 1960; Domenjoz, R.: Ann. N. Y. Acad. Sci. 86:263, 1960.

SELECTED READING

American Chemical Society, First National Medicinal Chemistry Symposium, pp. 15-49, 1948.

Archer, S., and Harris, L. S.: Narcotic antagonists, *in* Jucker, E. (ed.): Progress in Drug Research, vol. 8, pp. 262-320, Basel, Birkhäuser, 1965.

Barlow, R. B.: Morphine-like analgesics *in* Introduction to Chemical Pharmacology, pp. 39-56, Wiley, New York, 1955.

Beckett, A. H., and Casy, A. F.: The testing and development of analgesic drugs *in* Ellis, G. P., and West, G. B. (eds.): Progress in Medicinal Chemistry, vol. 2, pp. 43-87, London, Butterworth, 1963.

Bergel, F., and Morrison, A. L.: Synthetic analgesics, Quart. Rev. (London) 2:349, 1948.

Berger, F. M., *et al.*: Non-narcotic drugs for the relief of pain and their mechanism of action, Ann. N. Y. Acad. Sci. 86:1-310, 1960.

Braenden, O. J., Eddy, N. B., and Halbach, H.: Relationship between chemical structure and analgesic action, Bull. W.H.O. 13:937, 1955.

Brümmer, T.: Die historische Entwicklung des Antipyrin und seiner Derivative, Fortschr. Therap. 12:24, 1936.

Chappel, C. I., and von Seemann, C.: Antitussive drugs *in* Ellis, G. P., and West, G. B. (eds.): Progress in Medicinal Chemistry, vol. 3, pp. 89-145, London, Butterworth, 1963.

Chen, K. K.: Physiological and pharmacological background, including methods of evaluation of analgesic agents, J. Am. Pharm. A. (Sci. Ed.) 38:51, 1949.

Clouet, D. H.: Narcotic Drugs: Biochemical Pharmacology, New York, Plenum Press, 1971.

Collins, P. W.: Antitussives, *in* Burger, A. (ed.): Medicinal Chemistry, ed. 3, pp. 1351-1364, New York, Wiley-Interscience, 1970.

Coyne, W. E.: Nonsteroidal Antiflammatory Agents and Antipyretics, *in* Burger, A. (ed.): Medicinal Chemistry, ed. 3, pp. 953-975, New York, Wiley-Interscience, 1970.

deStevens, G. (ed.): Analgetics, New York, Academic Press, 1965.

Eddy, N. B.: Chemical structure and action of morphine-like analgesics and related substances, Chem. & Ind. (London), p. 1462, Nov. 21, 1959.

Eddy, N. B., Halbach, H., and Braenden, O. J.: Bull. W.H.O. 14:353-402, 1956; 17:569-863, 1957.

Fellows, E. J., and Ullyot, G. E.: Analgesics: A. Aralkylamines, *in* American Chemical Society, Medicinal Chemistry, vol. 1, pp. 390-437, New York, Wiley, 1951.

Gold, H., and Cattell, M.: Control of Pain, Am. J. Med. Sci. 246 (5):590, 1963.

Greenberg, L.: Antipyrine: A Critical Bibliographic Review, New Haven, Hillhouse, 1950.

Gross, M.: Acetanilid: A Critical Bibliographic Review, New Haven, Hillhouse, 1946.

Hellerbach, J., Schnider, O., Besendorf, H., Dellmont, B., Eddy, N. B., and May, E. L.: Synthetic Analgesics: Part II. Morphinans and 6,7-Benzomorphans, New York, Pergamon, 1966.

Jacobson, A. E., May, E. L., and Sargent, L. J.: Analgetics, *in* Burger, A. (ed.): Medicinal Chemistry, ed. 3, pp. 1327-1350, New York, Wiley-Interscience, 1970.

Janssen, P. A. J.: Synthetic Analgesics: Part I. Diphenylpropylamines, New York, Pergamon, 1960.

Janssen, P. A. J., and van der Eycken, C. A. M.: *in* Burger, A. (ed.): Drugs Affecting the Central Nervous System, pp. 25-85, New York, Marcel Dekker, 1968.

Lasagna, L.: The clinical evaluation of morphine and its substitutes as analgesics, Pharmacol. Rev. 16:47-83, 1964.

Lee, J.: Analgesics: B. Partial structures related to morphine, *in* American Chemical Society, Medicinal Chemistry, vol. 1, pp. 438-466, New York, Wiley, 1951.

Martin, W. R.: Opioid Antagonists, Pharmacol. Rev. 19:463-521, 1967.

Mellet, L. B., and Woods, L. A.: Analgesia and addiction, *in* Progress in Drug Research, vol. 5, pp. 156-267, Basel, Birkhäuser, 1963.

Portoghese, P. S.: Stereochemical factors and receptor interactions associated with narcotic analgesics, J. Pharm. Sci. 55:865, 1966.

Reynolds, A. K., and Randall, L. O.: Morphine and Allied Drugs, Toronto, Univ. Toronto Press, 1957.

Salem, H., and Aviado, D. M.: Antitussive drugs, Am. J. Med. Sci. 247 (5):585, 1964.

Winder, C. A.: Nonsteroid anti-inflammatory agents, *in* Jucker, E. (ed.): Progress in Drug Research, vol. 10, pp. 139-203, Basel, Birkhäuser, 1966.

25

Steroids, Nonsteroidal Estrogens and Cardiac Glycosides

Ole Gisvold, Ph.D.,

Professor of Medicinal Chemistry, College of Pharmacy,
University of Minnesota

INTRODUCTION

A number of very important compounds that have been found in both plants and animals contain a perhydro-1,2-cyclopentanophenanthrene carbon skeleton as part of their molecules. Some of these compounds are sterols, bile acids, sex hormones, corticoids, cardiac glycosides, certain saponins and certain toad poisons.

STEROLS

A group of tetracyclic, unsaturated, high molecular weight, secondary alcohols that are closely related in structure occur in all orders of plants and animals, with the possible exception of some bacteria. They are isolated from the unsaponifiable portion of fats. They are insoluble in water, but are soluble in the fat solvents. They crystallize in the form of colorless leaflets, platelets or needles, depending on the solvent from which they are crystallized. They are somewhat waxy in nature. Sterols form loose associations with each other, their own esters and derivatives, the bile acids, the saponins and with unrelated substances such as urea, substituted ureas, glycerol, maleic and citraconic anhydrides and solvents of crystallization. These many associations increase the difficulties in obtaining absolutely pure sterols. Furthermore, upon this property of sterols to enter into these associations may depend, in part, their role in physiologic processes. With many saponins they form crystalline, nonhemolytic, insoluble products.* Therefore, by using digitonin, a digitalis saponin, it is possible to separate out most sterols from unsaponifiable mixtures. The insoluble products can be resolved readily into their component parts in one of several ways.[1] Sterols, even in minute amounts give characteristic color reactions[2] with certain reagents. These tests can be used both qualitatively and quantitatively.

An examination of the structural formula for cholesterol as a representative sterol reveals that it has 8 asymmetric centers; C_3, C_8, C_9, C_{10}, C_{13}, C_{14}, C_{17} and C_{20} with 2^8 or 256 possible isomers. In cholestanol (double bond between C_5 and C_6 in cholesterol reduced), an additional asymmetric center at C_5 is introduced which increases the number of isomers possible to 512. Stepwise reduction of Δ^4-cholestenone and Δ^5-cholestenone, obtained from cholesterol, yields cholestanol, epicholestanol, coprostanol and epicoprostanol. These 4 alcohols differ only in the orientation of the groups attached to C_3 and C_5.

The naturally occurring substances that have a steroidal (cyclopentanoperhydrophenanthrene) nucleus are related closely

---

* This is a property of the secondary alcohol group when it is *cis* to the methyl group at C_{10}. See the formula for cholesterol.

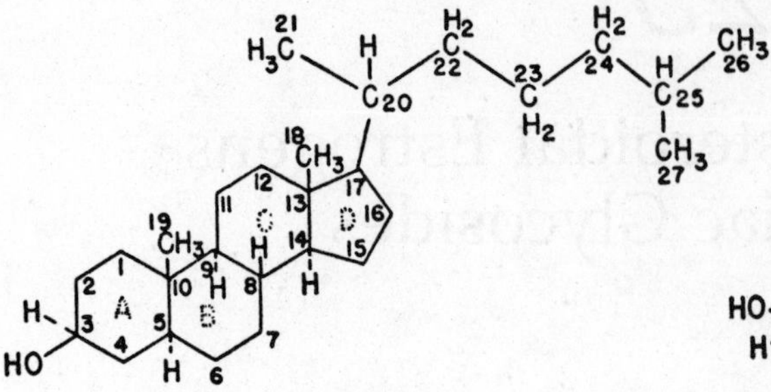

Cholestanol

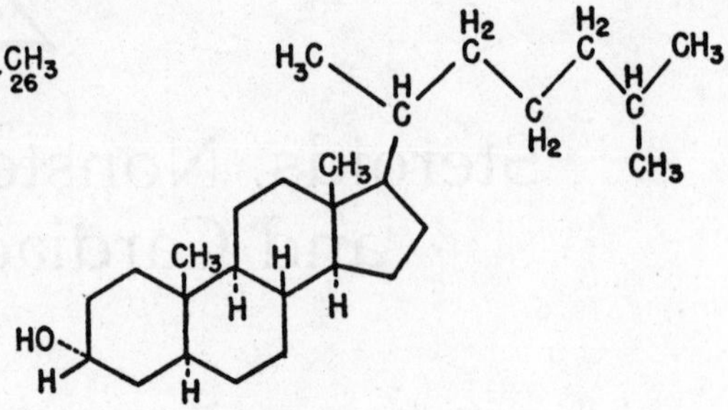

Epicholestanol

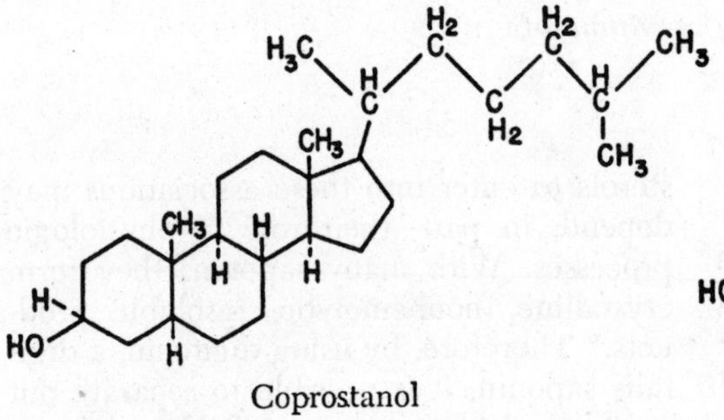

Coprostanol

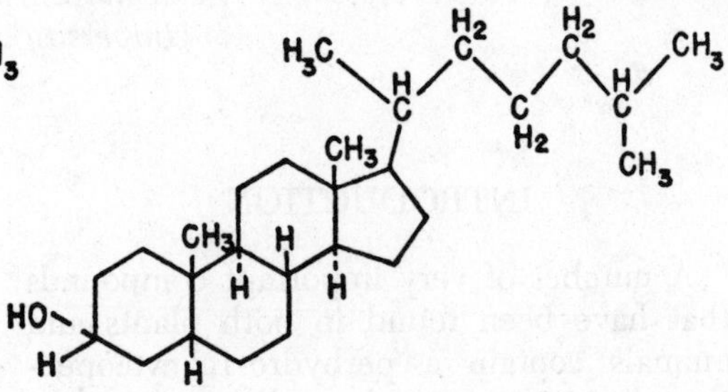

Epicoprostanol

in structure insofar as this polycyclic ring system is concerned, i.e., ring D is *trans* to ring C and ring C in turn is *trans* to ring B. In the saturated series, both the *cis* and the *trans* fusions of rings A and B are found. Naturally occurring compounds are known in which the C_3—OH is normal (*cis* to the methyl group at C_{10}) and in which it is *epi* (*trans* to the methyl group at C_{10}).

Specific attention should be called to the fact that the addenda to all steroid nuclei that are depicted with a dotted line have been assigned the α-configuration; and those with a solid line, the β-configuration. Although these are conventional and useful, the terms equatorial (e) and axial (a) are more descriptive in terms of the actual positions occupied by the addenda. An equatorial addendum is approximately horizontal in the plane of the ring and may carry either an α or a β designation. An axial addendum is vertical and parallel with the central axis of the ring. It may be above or below the plane of the ring and also may carry an α or a β designation.

The rings found in all saturated steroids most probably have the all-chair conforma-

tion; thus, cholestanol and coprostanol can be depicted as shown in the structures which are given on page 757. The addenda in these formulae are depicted to show the axial and the equatorial relationships; thus, a comparison can be made with the conventional alpha and beta designations.

ZOOSTEROLS

Products

Cholesterol U.S.P., Cholest-5-en-3β-ol; Cholesterin. Cholesterol, the principal sterol of animals, occurs in the tissues of the most diverse phyla. It appears to be present in all metazoa. Cholesterol has been identified in very small amounts as a natural constituent of several plants, and radioactive cholesterol has been converted to the C-27 steroidal sapogenins diosgenin and kryptogenin by *Dioscorea spiculiflora* seedlings.[3] The cholesterol of the bloodstream may be of endogenous or exogenous origin. It may be free or combined as esters or as a component of giant molecular cholesterol-pro-

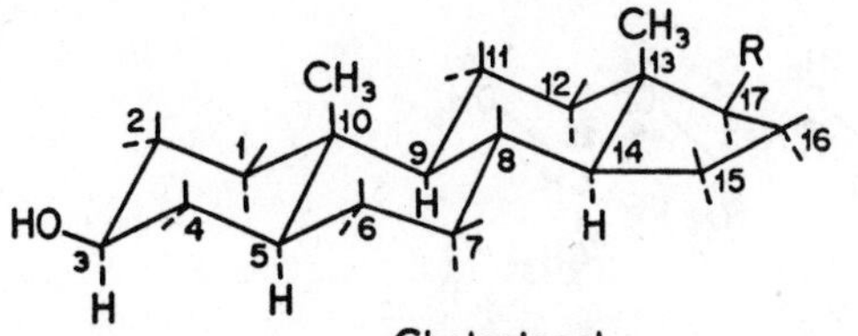

Cholestanol

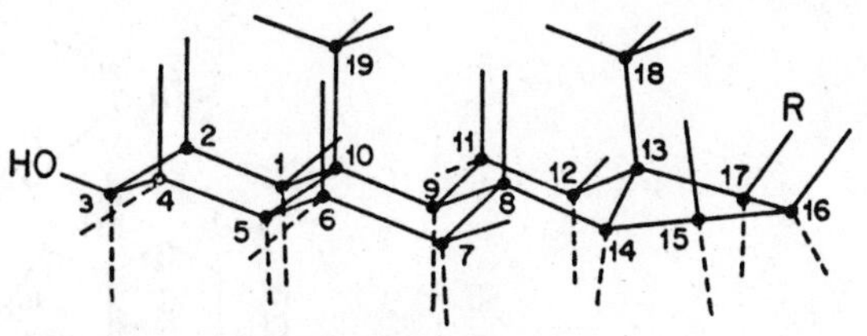

Cholestanol

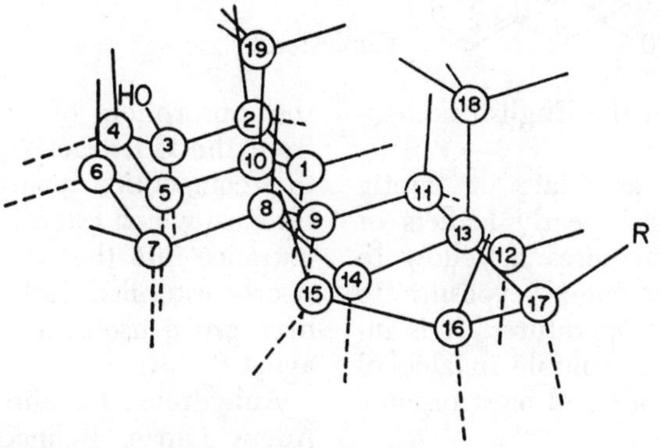

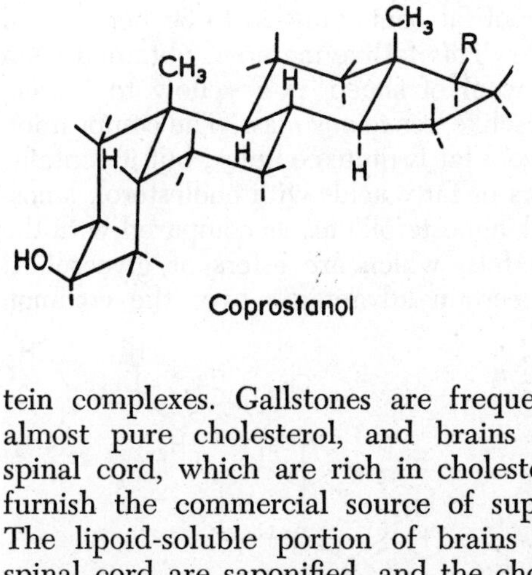

Cholestanol

Coprostanol

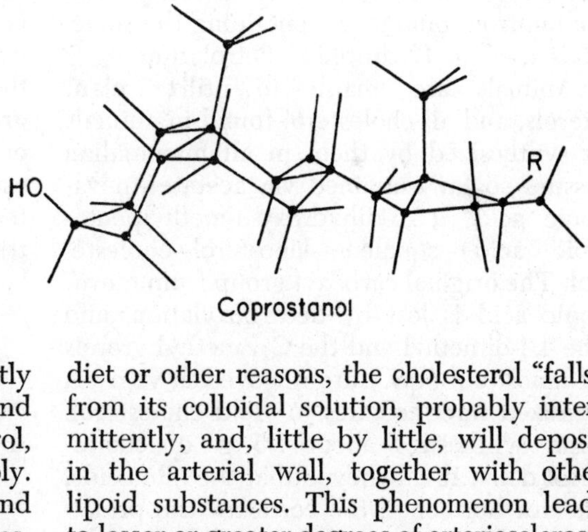

Coprostanol

tein complexes. Gallstones are frequently almost pure cholesterol, and brains and spinal cord, which are rich in cholesterol, furnish the commercial source of supply. The lipoid-soluble portion of brains and spinal cord are saponified, and the cholesterol, which is concentrated in the nonsaponifiable portion, is crystallized from organic solvents, such as alcohol, ethers and the like. Cholesterol may contain very small amounts of 7-dehydrocholesterol, which is unstable upon prolonged exposure to light and oxygen and imparts a yellow color to it. This slight color change does not reduce the usefulness of cholesterol.

In some individuals, if the cholesterol content of the blood exceeds a possible critical level (above 220 mg.%) due to the type of diet or other reasons, the cholesterol "falls" from its colloidal solution, probably intermittently, and, little by little, will deposit in the arterial wall, together with other lipoid substances. This phenomenon leads to lesser or greater degrees of arteriosclerosis (atherosclerosis). There is no proof that atherosclerosis can be prevented by reducing the oral ingestion or absorption of exogenous cholesterol.

Cholesterol, which is perhaps the most important of all sterols, was isolated first from gallstones by Paulleitier de la Salle, in about 1770. In 1815, Chevreul[4] showed that cholesterol was unsaponifiable, and he called it cholesterin (chole, bile; stereos, solid). In 1859, Berthelot[5] established its alcoholic nature, and since then it has been

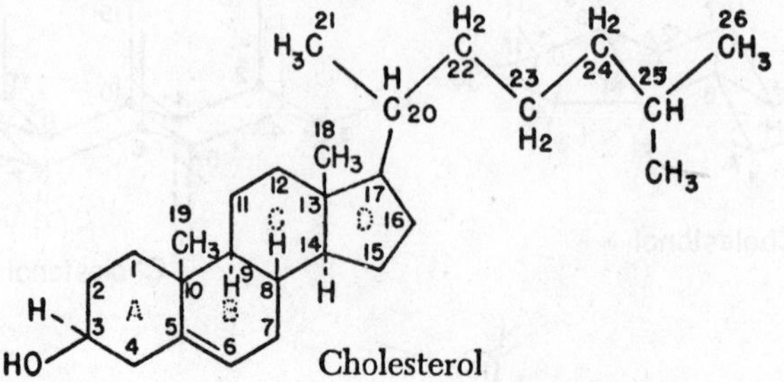

Cholesterol

properly designated in the English nomenclature as cholesterol.

Cholesterol occurs as white or faintly yellow, almost odorless, pearly leaflets or granules. It usually acquires a yellow to pale-tan color on prolonged exposure to light or to elevated temperatures. It is insoluble in water and is soluble in alcohol (1:100) in vegetable oils and most organic solvents.

Cholesterol is stable under normal storage conditions. Because it is an unsaturated secondary alcohol, it may possibly add iodine from ointments containing the same. It is used in Hydrophilic Petrolatum.

Animals are unable to utilize plant sterols, and all cholesterol found in animals is synthesized by them in all mammalian tissues so far examined via acetate—mevalonic acid* (3,5-dihydroxy-3-methylpentanoic acid)—squalene—lanosterol—cholesterol. The original carboxyl group from mevalonic acid is lost by decarboxylation, and the 4,4-dimethyl and the C_{14} methyl groups of lanosterol (q.v.) are removed in vivo via oxidized intermediates to yield cholesterol. Man synthesizes about 1.5 g. of cholesterol daily. It is believed that the bile acids, the sex hormones, the corticometric principles and other steroids are synthesized by the body from cholesterol. Cholesterol labeled with ^{14}C can be transformed by the isolated adrenal gland into the corticoids. Part of the cholesterol is excreted in the feces as coprostanol and much of the remainder as bile acids. Coprostanol differs from cholestanol in that in the former rings A and B are *cis* and the latter they are *trans*.

Cholestane-3β,5α,6β-triol (CT) inhibits

* Mevalonic acid is incorporated into carotenoids, rubber, etc. (See Bonner, J., et al.: J. Biol. Chem. 234:1081, 1959.)

the absorption of exogenous cholesterol from the G.I. tract. This inhibition is due to a competition for the enzyme cholesterol esterase which esterifies cholesterol prior to entrance into the lymphatic system. CT is poorly esterified and poorly absorbed and may prove useful as an antiatherosclerotic agent.[5a]

Anhydrous Lanolin U.S.P., Wool Fat, Adeps Lanae, Refined Wool Fat. Lanum is the purified, anhydrous, fatlike substance from the wool of sheep. **Lanolin U.S.P., Hydrous Wool Fat, Adeps Lanae Hydrosus,** is wool fat containing 25 to 30 per cent of water. This fatlike material, obtained from the wool of sheep, is a yellow to brown, greaselike, tenacious mass. The composition of wool fat is quite complex, but it contains esters of fatty acids with cholesterol, lanosterol, agnosterol, etc., as compared with the true fats, which are esters of glycerol. It has certain advantages over the common

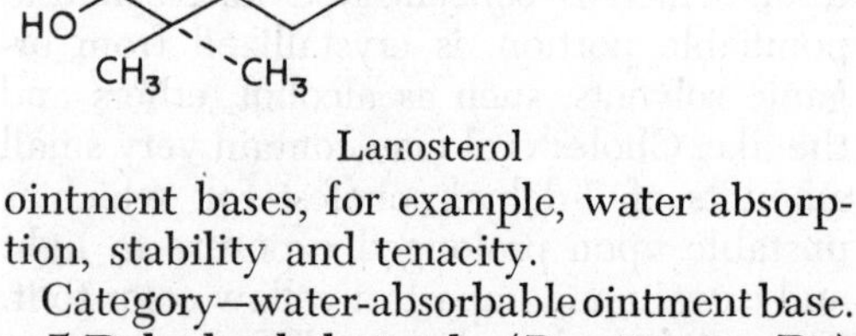

Lanosterol

ointment bases, for example, water absorption, stability and tenacity.

Category—water-absorbable ointment base.

7-Dehydrocholesterol (Provitamin D_3) occurs in small amounts in the skins of animals. It no doubt also occurs along with cholesterol, because the latter, upon irradiation, has antirachitic properties. It occurs in

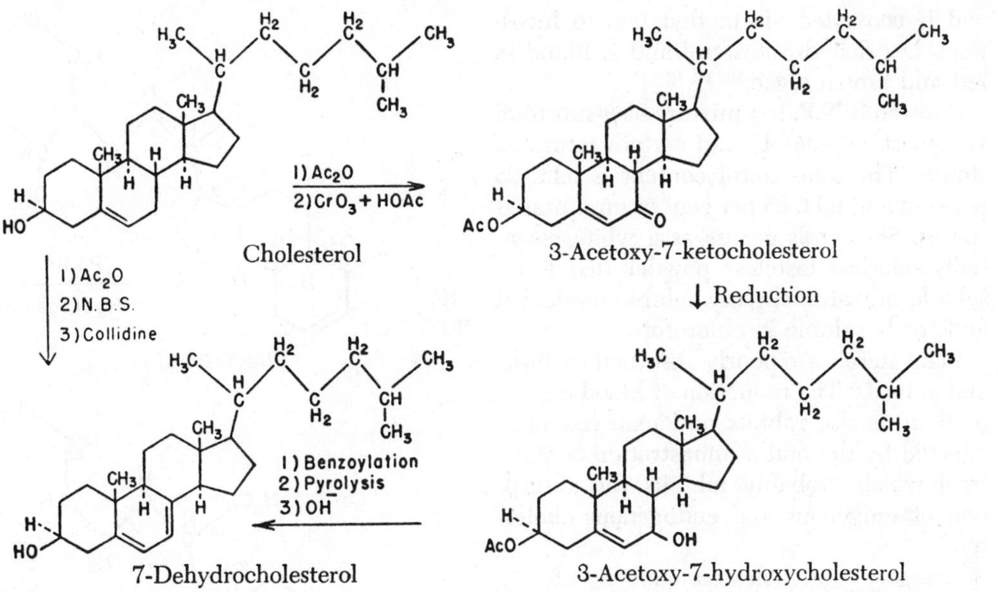

Cholesterol

1)Ac₂O
2)CrO₃ + HOAc →

3-Acetoxy-7-ketocholesterol

↓ Reduction

3-Acetoxy-7-hydroxycholesterol

7-Dehydrocholesterol

1)Ac₂O
2)N.B.S.
3)Collidine

1)Benzoylation
2)Pyrolysis
3)OH⁻

such small amounts in natural sources that they do not furnish a source of supply. However, it can be synthesized from cholesterol as shown above.

7-Dehydrocholesterol resembles the sterols in most of its properties, except that it undergoes a number of photochemical changes under the influence of ultraviolet light, with the generation of a number of products, one of which is vitamin D_3 (see Chap. 28). 7-Dehydrocholesterol is unstable in clear glass containers under normal storage conditions, and it gives rise to degradation products that are very complex in character.

PHYTOSTEROLS

In the years following the isolation of cholesterol from gallstones, cholesterol-like substances, even though found in plants, were designated as cholesterins or cholesterols. However, Hesse,[6] in 1878, pointed out that the sterol isolated from peas by Beneke,[7] in 1862, was not identical with cholesterol; therefore, he proposed the term "phytosterol" for those sterols occurring only in plants. Because this term embraced all sterols obtained from the entire plant kingdom, and inasmuch as all plant sterols isolated did not exhibit identical physical prop-

erties, some investigators chose to name sterols after the names of the plants from which they were isolated, for example, ergosterol from ergot, stigmasterol from physostigma, sitosterol from wheat, etc. Because the physical properties of many of the phytosterols are very similar to those of the sterol isolated from wheat germ by Burian[8] in 1897, many phytosterols frequently are designated as sitosterols.

The sitosterols are the most predominantly occurring phytosterols of the Spermatophytes. The name "sitosterol" has been used to designate not only the mixture of sterols obtained by Burian, in 1897, from wheat germ, but also those of other phytosterols which, either pure or in a mixture, have physical properties closely resembling those of Burian's phytosterol mixture. The sterol from wheat germ now has been resolved into α-, β- and γ-sitosterols, dihydrositosterol, α-tritisterol, β-tritisterol, stigmasterol, ergosterol and possibly others. β-Sitosterol occurs as the greater part of the sterols of cottonseed oil and can be prepared somewhat readily in a pure state.[9]

Cholesterol occurs in very small amounts in most plants and in large amounts in Chlorophyta. C-21 steroids appear to be widely distributed. 24-Methylenecholesterol is an intermediate derived from cholesterol and is converted via methylation to fucosterol, i.e., dehydrositosterol and is found in red and brown algae.[10, 11, 12]

Sitosterols N.F. is a mixture of β-sitosterol (stigmast-5-en-3β-ol) and certain saturated sterols. The total sterol content is n.l.t. 95 per cent and n.l.t. 85 per cent of unsaturated sterols. Sitosterols occurs as a white, essentially odorless tasteless powder that is insoluble in water, slightly soluble in alcohol and freely soluble in chloroform.

Plant sterols are poorly absorbed by birds and animals. The reduction of blood cholesterol in chicks, rabbits and man has been effected by the oral administration of sitosterol which probably inhibits the absorption of exogenous and endogenous cholesterol.

Category—anticholesterolemic agent.

Usual dose—3 g. 3 times a day before meals.

Usual dose range—9 to 30 g. daily.

OCCURRENCE
Sitosterols Suspension N.F.

Cytellin®, sitosterols, is a lightly flavored aqueous suspension of beta sitosterol with a small portion of dihydro beta sitosterol. Each 15 ml. contains 3 g. of Cytellin. Its use and actions are the same as those of Sitosterols N.F.

Dose—15 ml. immediately before each meal.

Stigmasterol was isolated first by Windaus and Hauth, in 1906, from the sterols found in *Physostigma venenosum*.[13] They were able to effect a separation because of the insolubility of stigmasterol tetrabromide. It also occurs in quite small amounts in the sterol fractions of many plants. Stigmasterol is important because it contains a double bond in the side chain and thus furnishes a somewhat more readily manipulatable starting material for the synthesis of some of the sex hormones and corticoids. Stigmasterol can be obtained in commercial quantities from soybean oil.

The relationship in structure between stigmasterol and one of the sitosterols can be shown as follows:

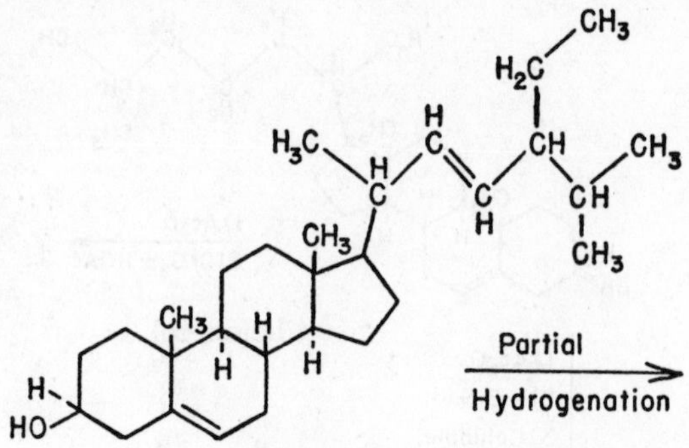

Stigmasterol

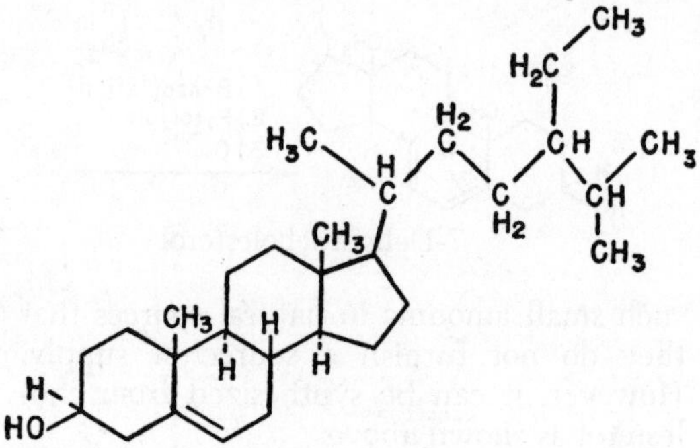

Partial Hydrogenation →

β-Sitosterol

In 1889, Tanret[14] succeeded in preparing from ergot ergosterol that was almost free from the difficultly separable, contaminating fungus sterols. Gerard (1892–1895) demonstrated that ergosterol occurs in many cryptogams, such as brown algae, slime fungi, bacteria, mucors, penicillia and lichens. He proposed the rule that ergosterol was characteristic of the cryptogams. Molds and yeast furnish the commercial source of supply of ergosterol. The type of culture media used and the species of fungi grown determine the percentage of ergosterol produced. Thus, for example, *Saccharomyces Carlsbergensis* yields about 2 per cent of ergosterol, based on the dry weight of the organisms. Even though ergosterol is characteristic of the cryptogams, it has been isolated in very small amounts from some phaenerogams.

Ergosterol undergoes a number of photochemical changes under the influence of ultraviolet light, with the generation of a number of products, one of which is vitamin D_2 (See Chap. 28). Under normal storage conditions, ergosterol in clear glass

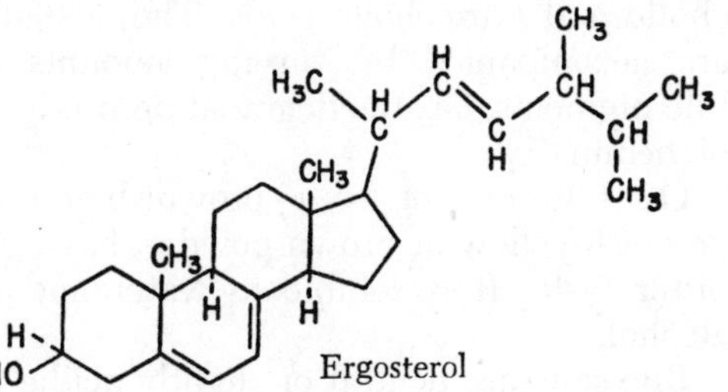

Ergosterol

containers is unstable and gives rise to degradation products that are very complex in character.

BILE SALTS

The bile of man and a number of animals contains a group of closely related acids, the bile acids, conjugated by amide linkages to the common amino acid glycine and the rare amino acid taurine, $H_2N-CH_2-CH_2-SO_3H$. These conjugates are present as their sodium salts and are called bile salts. These bile salts can be hydrolyzed by alkali to give, in the case of ox bile, 5 to 6 per cent of cholic acid and 0.6 to 0.8 per cent of desoxycholic acid, together with smaller amounts of chenodesoxycholic acid and

lithocholic acid. These acids contain the ring systems found in the sterol coprostanol and are not precipitable by digitonin, even though they have an alcohol group at C_3. This is due to the fact that C_3OH is *trans* to the methyl group at C_{10}.

There are no free bile acids in human bile. The bile acids are conjugated in the liver via cholyl-CoA with glycine (80 to 90%) and taurine (10 to 15%). It is interesting to note that of ^{14}C cholesterol administered to humans 90 per cent is excreted as bile acids. In the in-vivo synthesis of cholic acid from cholesterol the first hydroxyl appears in the 7 α-position followed by a 12α-hydroxyl ("12α-hydroxylase"). Then the side chain is oxidized.[15]

These acids are stable under normal storage conditions. They are insoluble in water, whereas their sodium salts are both water- and alcohol-soluble. The secondary alcohol groups can be converted by mild chromic acid oxidation to the corresponding ketones. These keto acids usually are called the dehydro bile acids, i.e., dehydrocholic acid.

Desoxycholic acid possesses to a marked degree the capacity to form stable molecular complexes, the choleic acids. The molar ratios of the bile acid and the solvent vary

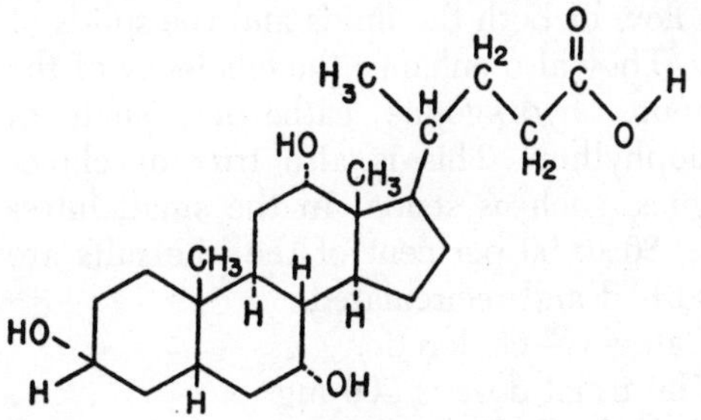

Cholic Acid

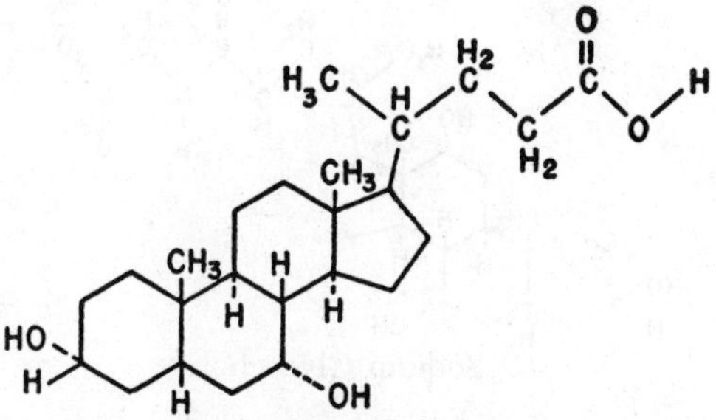

Chenodesoxycholic Acid

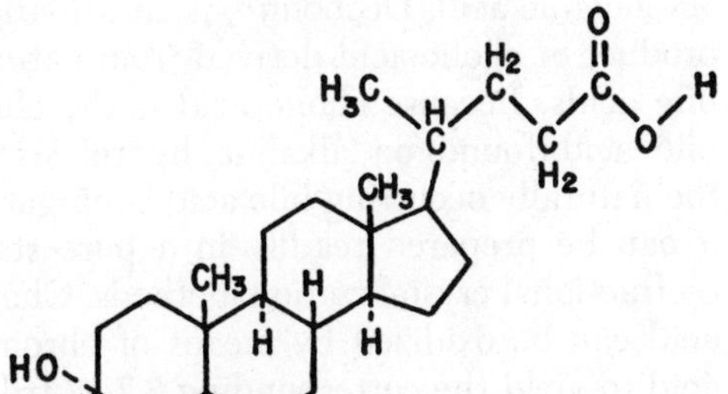

Lithocholic Acid

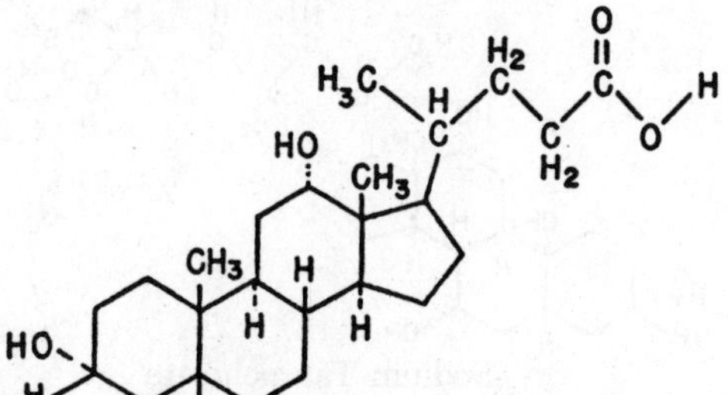

Desoxycholic Acid

with the type of solvent, such as ether, xylene, acetic acid, palmitic acid, stearic acid and even paraffins. Although choleic acids dissociate in dilute solutions in organic solvents until a point of equilibrium is reached, they will dissolve in aqueous alkali without dissociation. If a biologically active material is administered as a solution of the sodium salt of the choleic acid, any effect in the organism appears to be attributable to the liberation and the dissociation of the free choleic acid.

The bile acids are suitable raw materials for the synthesis of some of the corticoids, male and female sex hormones.

Products

Ox Bile Extract, Powdered Oxgall Extract; Extractum Fellis Bovis. Ox Bile Extract contains an amount of the sodium salts of ox bile acids equivalent to not less than 45 per cent of cholic acid.

It usually is prepared by extracting dried ox bile with absolute alcohol. Ether is added to the alcoholic extract in order to precipitate the desired bile salts from some of the nondesirable ether and alcohol soluble constituents.* This solid extract is composed chiefly of the sodium salts of glyco-

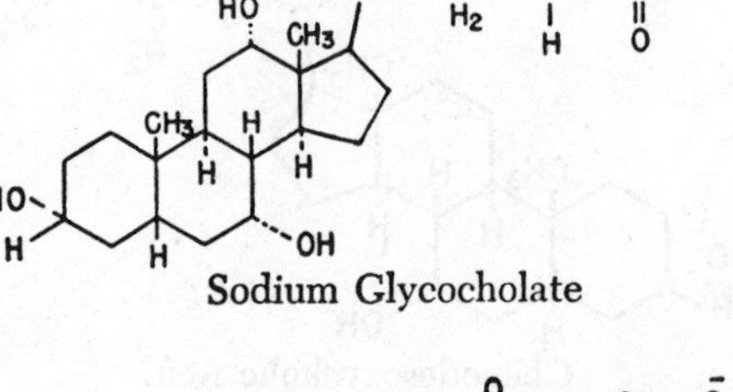

Sodium Glycocholate

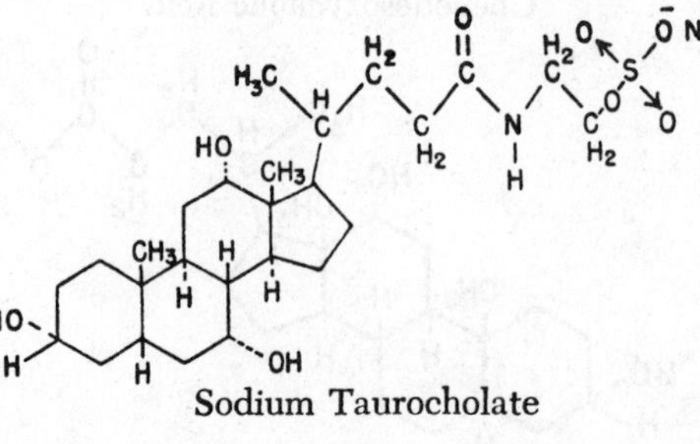

Sodium Taurocholate

* These include cholesterol, lecithin, bile pigments.

cholic and taurocholic acids. They usually are accompanied by varying amounts of bile pigments, i.e., the degradation products of hemin.

Ox bile extract is a brownish-yellow, greenish-yellow or brown powder, having a bitter taste. It is soluble in water and in alcohol.

Bile salts are neutral or slightly acidic to litmus and behave as anionic detergents (see Surface-active Agents). Because sodium glycocholate is the sodium salt of a weak acid, it may be converted by mineral acids to the water-insoluble glycocholic acid. Bile salts are stable under normal storage conditions, and their aqueous solutions can be sterilized by boiling.

Because the bile salts are surface-active agents, they emulsify water-insoluble fats, fat-soluble vitamins A, D, E and K and other lipoids and thus promote their absorption through the intestinal mucosa. The emulsified fats are hydrolyzed more readily by the lipolytic enzyme lipase and, therefore, are more readily assimilable.

It is known that when the flow of bile from the gallbladder into the intestines is obstructed, the absorption of vitamin K is imperfect, and a vitamin-K deficiency may result (see Chap. 28). Bile salts stimulate the secretory activity of the liver, increasing the flow of both the fluids and the solids of bile. They also enhance the efficiency of the resinous hydragogue cathartics, such as podophyllum. This is also true of cholagogues, such as senna. In the small intestine, 80 to 90 per cent of the bile salts are absorbed and recirculated.

Category—choleretic.

The usual dose is 300 mg.

Dehydrocholic Acid N.F., 3,7,12-trioxo-5β-cholanic acid, Decholin®, is an oxidation product of cholic acid derived from natural bile acids. Because cholic acid is the chief bile acid found on alkaline hydrolysis of the naturally occurring bile acid conjugates, it can be prepared readily in a pure state by fractional crystallization methods. Cholic acid can be oxidized by means of chromic acid to yield the corresponding 3,7,12-trike-tocholanic acid, which commonly is called dehydrocholic acid.

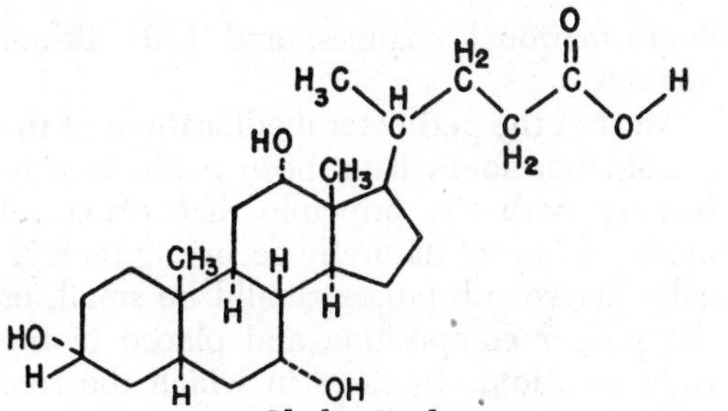

Cholic Acid
3,7,12-Trihydroxycholanic Acid

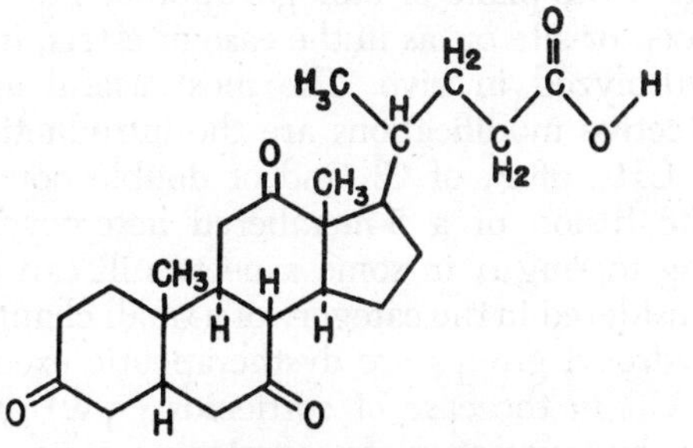

Dehydrocholic Acid
3,7,12-Triketocholanic Acid

Some of the bile acids occur naturally in small quantities as the corresponding dehydroacids. This fact stimulated research in this field, with the production and the use of dehydrocholic acid.

Dehydrocholic acid occurs as a white, odorless, fluffy powder with a bitter taste. It is insoluble in water and soluble in alcohol (1:100). It is stable under normal storage conditions and can be sterilized by boiling.

Dehydrocholic acid increases the secretion of a thin, free-flowing bile by the liver, thus facilitating drainage throughout the entire biliary tract. It is not cholagogic in action (stimulates evacuation of the gallbladder). Therefore, it is used to alleviate biliary stasis and functional insufficiency of the liver, to outline the bile ducts at operation and in cholecystography to accelerate the appearance of the gallbladder shadow and to hasten removal of residual tetraiodophenolphthalein from the biliary tract. Dehydrocholic acid produces diuresis in edema of cardiac origin but is less effective than the mercurials. It potentiates the diuretic effect of mercurials. It may be used with antispasmodics, if so desired.

Category—choleretic.
Usual dose—500 mg. 3 times a day.
Usual dose range—250 to 750 mg.

OCCURRENCE
Dehydrocholic Acid Tablets N.F.

Sodium Dehydrocholate, Decholin® Sodium, the sodium salt of dehydrocholic acid, is a fine, colorless, crystalline powder with a very bitter taste. It is soluble in water or alcohol, and its aqueous solution is alkaline to litmus. Its actions and uses are the same as those of dehydrocholic acid; however, because it is soluble in water, it can be administered intravenously when necessary.

Sodium Dehydrocholate Injection N.F. is available in various concentrations for intravenous injection.

Category—choleretic; diagnostic agent (circulation time).

Usual dose—intravenous, 3 to 5 ml. of a 20 per cent solution.

Ketocholanic Acids, Ketochol®, contains chiefly dehydrocholic acid together with small amounts of dehydrodesoxycholic, dehydrochenodesoxycholic and dehydrolithocholic acids. Its uses and actions are essentially the same as those of dehydrocholic acid.

Tocamphyl, Syncuma®, Gallogen®. To-

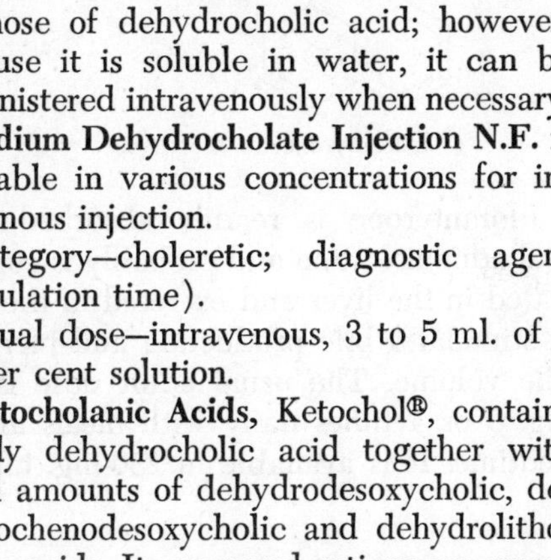

Tocamphyl

camphyl is 1-(*p*, α-dimethylbenzyl)camphorate and diethanolamine. It occurs as a pale yellow to amber, unctuous mass and is soluble in water and in alcohol.

The acid component of this amine salt stimulates secretion by the liver of an increased volume of bile that contains less solids per unit volume than does normal bile.

Usual dose—75 mg. 3 times a day.

Florantyrone, Zanchol®, γ-Oxo-8-fluoranthenebutyric Acid. Florantyrone, a synthetic compound only remotely related in structure to the bile salts, possesses pharmacologic actions and clinical uses similar to those of dehydrocholic acid. It occurs as fine platelets which are soluble in bases, in methanol and in ethanol.

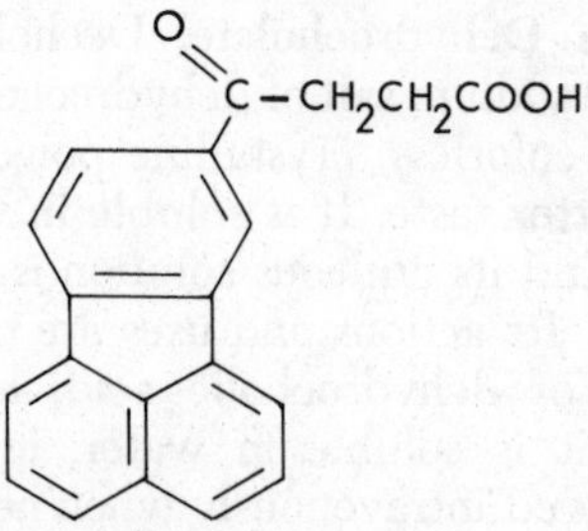

Florantyrone

Florantyrone is readily absorbed after oral administration and probably is concentrated in the liver and excreted in the bile. It stimulates bile production and increases bile volume. The usual adult dose is 250 mg. 3 or 4 times daily with meals and at bedtime. It is available in 250-mg. tablets.

SEX HORMONES

PERIMETER MODIFICATIONS

In the past decade a vast number of structural modifications of the steroid hormones have been reported in the literature. These have been prepared in order to (1) increase biologic activities, (2) increase oral activity, (3) increase duration of action, (4) effect a separation of biologic activities, (5) decrease the requirement for some essential perimeter functional group of the parent hormone, and (6) obtain better solubility properties.

These ends have been accomplished by (1) protection of some essential group against metabolic attack or attack by intestinal bacteria; (2) prevention of some position against metabolic attack in vivo; (3) prevention of the conversion in vivo of one steroid hormone into another steroid hormone; (4) alteration of physical properties, and by employing (5) ester derivatives, (6) enol ethers, (7) acetals and ketals, (8) possible electron-attracting effects, (9)

conformational changes, and (10) 19-nor analogs.

Most of the perimeter modifications of the steroid hormones have been made in conformity with the principle that structural modifications of naturally occurring biologically active substances should be small, of the proper composition and placed in the right positions. In cases in which the rule is violated in regard to size, one may postulate that the modification, in its position, is out of the plane of biologic approach to the receptor site or, as in the case of esters, it is hydrolyzed in vivo. The most useful and effective modifications are the introduction of CH_3, of F, of Cl, and of double bonds. The fusion of a 5-membered heterocyclic ring to ring A in some aspects still can be considered in the category of a small change. Hydroxyl groups are dystherapeutic except at C_{16} in the case of corticoids (q.v.) and 17α-hydroxyesters of progesterone. Removal of the C_{19}-CH_3 group has proved very useful in some cases. The rule concerning change in structure appears to be violated in the case of the 17α-substituted testosterones that have progestational activity; however, this apparent violation may be compensated, since testosterone, by the intrauterine assay, has pronounced progestational activity. A second case is that of the doisynolic acids (q.v.) that are effective active estrogens.

Although the bioassays of the steroid hormones usually are measured by some single specific biologic response, the hormonal action observed may not be the result of one hormone, i.e., the estrogens may interact with their metabolic products, progesterone, testosterone and the adrenocorticoids to bring about metabolic alterations of the uterus, etc. (i.e., growth).

STEROID ESTERS

The effects of esterification of steroid hormones may be multiple in character, depending on the nature of the hormone and the acyl radical. Factors affected include:
1. Stability
2. Solubility
3. Rate of absorption
4. Rate of hydrolysis

A. As paralleling activity; e.g., polyestradiol phosphate is a water-soluble high molecular weight polyester with marked antienzymatic activity; it is absorbed rapidly from the site of injection and has prolonged activity[16] (up to 200 days also by IV administration[17]) in rodents and humans, due to its rate of hydrolysis[18] in vivo

B. As not paralleling activity; e.g., hydrolysis of certain testosterone esters in vitro does not parallel the duration of action in vivo[19]

5. Activity

A. By creation of an active compound (ester may be active per se)

B. By conversion of an inactive to an active compound; e.g., 17-hydroxyprogesterone vs. its esters, which have greatly prolonged and fairly strong activity orally, with very little antiestrogenic activity.[20]

6. Oral activity, which may be enhanced by

A. Sulfates* (estrone sulfate) and phosphates (q.v.), as soluble salts exert enough desorptive properties so that those enzymes (some of bacterial origin in the gut) responsible for their metabolic alteration cannot act upon them.

B. C_3-enol esters; e.g., cyclopentyl propionate of dihydroprogesterone (see also esters of testosterone).

Esterification can in some cases alter the biologic activity of the steroid, e.g., DOCA diethyl acetate biologically more resembles the activity of a glucocorticoid.

Steroid Ethers

The degree of oral activity conferred upon the C_3-keto steroid enol ethers is greatly dependent upon the structure of the parent steroid and the R group. Enhanced oral activity is obtained with methyl testosterone, 17α-acetoxyprogesterone, cortisone acetate, desoxycorticosterone, testosterone and progesterone. In the case of the last two steroids, the degree of oral activity is still very low compared to the subcutaneous activity of the parent steroids and therefore they are not clinically useful orally. No sig-

nificant increase in oral activity is obtained in the case of very active steroids such as 6-methyl-17-acetoxyprogesterone or 17-ethinyl-19-nortestosterone.

Whereas small daily subcutaneous doses† of methyltestosterone - 3 - cyclopentylenol ether were only slightly effective, the subcutaneous administration of a single, relatively large dose (1 or 5 mg.) resulted in a more prolonged intense stimulation of the sex accessory organs in castrated male rats than that obtained by comparable doses of methyltestosterone[22] (MT). The MT enol ether is stored in body fats whereas MT is not. The enol ethers may be active per se.[23]

The biologic activities of the enol ethers may differ from those of the parent steroid. Thus, the benzyl enol ether of testosterone phenylpropionate parenterally is weakly androgenic but has very effective antitumoral activity in the human breast. This also is true of the Δ^2-enol ethers of 5α-dihydro testosterone. The C_3-cyclopentyl enol ether of 17α-acetoxy progesterone (quinoxygesterone acetate) orally produces a secretory endometrium in the Clauberg assay and maintains pregnancy in ovariectomized rabbits at doses comparable with those of subcutaneously administered progesterone. Quinoxygesterone acetate has a relative lack of ability to inhibit estrus, ovulation, mating, fertilization and delivery at term.

17β-Enol ethers such as testosterone 17-cyclopentenyl ether are active orally and parenterally. Orally 19-nortestosterone 17-cyclopentenyl ether had a LA/SV = 10.8 and LA/VP = 31.7, with MT as the standard.‡ Subcutaneously its LA activity was equal to 19-nortestosterone propionate.

Steroid enol ethers are very sensitive to hydrolysis, especially below pH 7, and special precautions must be taken in their storage before use.

Steroid 17β-yl acetal derivatives such as that of 5α-androstan-17β-ol-3-one are active orally and parenterally. This also is true of cyclopentanone ethyl testosteronyl acetal which orally is twice as androgenic as MT

---

* Steroid sulfates are biosynthesized by liver, human adrenocortical tissue and human ovarian tissue.[21]

† Also true of other C_3-enol ethers.

‡ LA = Levator ani (muscle). SV = seminal vesicles. VP = ventral prostate. See text, p. 794, for explanation.

and by injection has a prolonged activity comparable to that of the most potent esters.

Acetals and ketals (stable in vivo) markedly enhance the activity and the duration[24] of activity of the $17\alpha,16\alpha$-dihydroxy corticoids.

2'-Tetrahydropyranyloxy ethers of estradiol,[25] estrone and some androgens[26] have yielded derivatives with greatly enhanced

TABLE 83. COMPARATIVE ACTIVITIES* OF THE TETRAHYDROPYRANYL ETHERS (THP)†

	SUBCUTANEOUS	ORAL
Estrone 3-THP	18	400
17β-Estradiol	280	130
17β-Estradiol 3-THP	4.0	1,600
17β-Estradiol 17-THP	280	2,000
17β-Estradiol 3,17-bis THP		28
Ethinyl estradiol		2,340
Ethinylestradiol 3,17-bis THP		450
Ethinylestradiol 3-methyl ether	35	1,350
Ethinylestradiol-17-THP-3-methyl ether	412	200

* Based on estrone as standard ($= 100$).
† These THP ethers are acid labile.
Data from Reference 69.

oral activities versus subcutaneous activities in some cases (see Table 83).

Certain labile ethers such as the 17-2-chloro and 2-fluoro ethyl ethers of estradiol 3-methyl ether had 4 times the estrogenic activity of estrone orally; subcutaneously, however, they had 0.5 and 0.05 times the activity of estrone.[27]

The duration of action of 17α-ethinylestradiol-3-cyclopentyl ether's uterotropic activity by injection was more than $3\frac{1}{2}$ times that of ethinylestradiol. Orally it was 2 to 3 times as active as ethinylestradiol and in rodents was effective as a pituitary suppressant and contraceptive in lower doses than mestranol.

The anterior lobe of the pituitary gland secretes proteinoid gonadotropic substances that stimulate the development of the gonads of both the male and the female of the species. The gonads, in turn, under this stimulation, continuously synthesize the sex hormones, which have the ring systems found in the sterols and the bile acids. In the case of the human female, two gonadotropic hormones are involved, i.e., follicle-stimulating and luteinizing hormones. At puberty, the follicle-stimulating hormone stimulates maturation of the graafian follicles, in which the egg cells, or ova, develop. Under this developmental stimulation, the cells around the follicle produce the follicular hormone (an estrogenic compound), which is absorbed and circulated by the bloodstream throughout the body. After ovulation, the luteinizing hormone stimulates development of the corpus luteum from the graafian follicle. Under this developmental stimulation, the corpus luteum produces progesterone, which is absorbed and circulated through the body by the bloodstream. If the ovum has been fertilized, pregnancy ensues, and the corpus luteum grows for several months and attains a large size. However, if pregnancy does not occur, the corpus luteum degenerates, menstrual hemorrhage occurs, and the cycle is repeated.

ESTROGENS

Besides β-estradiol,* which is the original estrogenic substance produced by the ovaries, estrogenic substances have been found in other sources: e.g., in the placenta, the adrenals, the testes of the stallion, pregnancy urine of women and mares, stallion urine, urine of men, palm-kernel oil, female willow flowers, suitable extracts of peat, brown coal, lignite, coal tar and petroleum, bile, feces and yeast. The pregnancy urine of women contains 10,000 mouse units per liter, whereas mare's pregnancy urine contains 10 times this amount. Stallion urine contains 170,000 m. u. per liter. This is not as paradoxical as it may appear at first, for the steroids excreted in the urine are products of metabolism and not functioning hormones, and it is only in the case of equines

* The Greek symbols used here conform to the most recent structural configurations. Thus, the historical α-estradiol becomes the configurational β-estradiol.

that males show a higher excretion of female sex hormones than females. The estrogenic substances found in the female pregnancy urine are oxidation products of β-estradiol.

Pueraria mirifica (Fam. Butea superba), a legume, contains miroestrol that has 1/6 the activity of estradiol. Coumestrol (6',7-dihydroxybenzofuro [3',2',3,4] coumarin) a coumarin from several clover forage crops had 1/2,500 the activity of stilbestrol that was effective at the 0.0001-mg. level. The synthetic isoflaven, 7,4'-diacetoxy-2-methyl-4-ethyl-Δ³-isoflaven is almost as active as stilbestrol.[28] Gentistein, biochanin, daidzein and formononetin are isoflavones found in forage crops, and, even though they have a low order of estrogenic activity, the quantities consumed by sheep, etc., have led to a high incidence of infertility (q.v.) associated with cystic glandular hyperplasia of the endometrium.[29]

Antifertility Effects

In rats and rabbits, estrogens (steroid and nonsteroid) play an important role in nidation[30] (the implantation of the fertilized ovum in the endometrium of the pregnant uterus). Nidation is delayed if the amount of estrogen is below the threshold level for implantation. Optimal levels (0.3 to 1 μg. of estrone daily for rats) induced nidation in rats ovariectomized 3 days after coitus only if sufficient quantities of progesterone were available to ensure the presence of viable blastocysts. Higher levels of estrogens (0.5 mg./kg. of estrone subcutaneously) inhibited nidation in rats that were treated during the preimplantation period (postcoital days 1, 2, 3 and 4). 3,4-Bis-*p*-hydroxy-diphenyl-2-hexanone was approximately 5 times as active as estrone, which was 100 per cent effective in rats in daily doses of 50 mcg./kg. subcutaneously. Although high estrogenic activity gave high antifertility activity in some cases, in others antifertility effects were 10 to 15 times as great as was expected on the basis of the estrogenicity.[31] The level of estrogen at the time of nidation may affect the viability of the embryo of placentation at later stages of pregnancy. A single injection of 100 mcg.

estradiol cyclopentylpropionate immediately after coitus caused a total death of rabbit embryos 8 days after mating.[32]

In other studies[33] ethinylestradiol (EE) 50 mcg./kg. given orally before mating prevented pregnancy in 100 per cent of the mice treated. EE inhibited pituitary gonadotropin release and subsequent blockage of ovulation. The following findings also are significant with reference to the antifertility effects of estrogens.

Estrogens, i.e., EE and mestranol (80 to 100 mcg.), markedly inhibit the levels of the follicle stimulating hormone (FSH). This may well be a control of the synthesis of FSH by the pituitary through a typical feedback repressor mechanism, since small doses (10 to 20 mcg.) of mestranol stimulate FSH levels. Long-term high doses of mestranol depress at peak levels of luteinizing hormone (LH) mid term of the estrus cycle, whereas 50 to 80 mcg. of mestranol do not consistently bring about this effect, an activity that is difficult to understand. The repression of the FSH levels leads to inhibition of follicle developments—a most important contribution to the inhibition of ovulation by steroid hormones.

Estrogens can maintain the life of the corpus luteum. Time studies showed that EE given after ovulation had taken place did not inhibit ova transport, ova nidation or formation of deciduomata.

Ethynyl-estradiol 3-methyl ether administered from day 1 in hamsters suppressed follicular development early in the cycle.[34]

Estrogens when given alone in appropriate doses will inhibit ovulation in the human female; however, they do not induce the development of a normal secretory endometrium.

Synthetic estrogens such as dimethylstilbestrol have antiestrogenic activity[35] when applied locally to the vagina in doses much smaller than the estrogenic dose. Evidence to date does not indicate clearly whether the antifertility effects of estrogens are related to their estrogenic or their antiestrogenic activities.

Products

Estradiol N.F.; Dihydrotheelin; Oestra-diol; Estra-1,3,5(10)-triene-3,17β-diol; β-Estradiol; the Follicular Hormone; the Female Sex Hormone. β-Estradiol is believed to be the follicular hormone first synthesized by the graafian follicle of the ovary. Because it is produced continuously in such small quantities by the ovary and is promptly utilized and excreted in a changed form, the ovaries of cattle and hogs cannot be used as commercial sources of supply. It can be prepared by the reduction of estrone, which is obtained commercially from pregnant mare's urine or from stallion urine. It has been synthesized from Δ^4-androstene-3,17-diol.[36] Ring A in the steroids can be aromatized.

In 1922, Frank[37] discovered in the fluid of the graafian follicle an active estrogenic substance, that is, a substance which, when injected into rodents whose ovaries had been removed previously, produced estrus, or sexual heat. Allen[38] confirmed this work in 1922. Aschheim and Zondek,[39] in 1927, showed that the urine of pregnant women was active in the above test and presented a potential source of supply. Doisy *et al.*,[40] in 1929, isolated from the urine of pregnant women, a very active crystalline substance that was later known as estrone.

β-Estradiol actually was extracted from sows' ovaries,[41] in which it occurs to the extent of 6 mg. per ton or about 1 part in 150,000,000. It also has been isolated from urine of pregnant mares[42] and women[43] and from the human placenta.[44]

Estradiol occurs as odorless, white or creamy-white, small crystals or crystalline powder. It is almost insoluble in water, soluble in alcohol and sparingly soluble in vegetable oils. It is stable in air.

Estradiol and its related products found in urine differ in part from all other naturally occurring steroidal compounds in that ring A is aromatic; therefore, the OH group at C_3 is phenolic and exhibits all the properties of phenols; i.e., it will form sodium salts with alkali hydroxides and so on. In the dry state, it is quite stable under normal storage conditions and can be sterilized. Because it is phenolic, its stability will be comparable with phenol when in contact with moisture, peroxides of fats and other oxidizing agents.

Estrone

β-Estradiol

Estra-1,3,5(10)-triene-3,17-β-diol

Allen and Doisy,[45] in 1923, developed a convenient method of bioassay in which the determination is made of the minimum amount of hormone required to give full estrous cycle in castrated female mice or rats. This response is detected by microscopic examination of a stained smear of the vaginal epithelium. This cycle takes place in 4 stages, i.e., diestrus (epithelial nucleated cells), proestrus (large nucleated cells accompanied by epithelial cells and cornified cells), estrus (only cornified cells) and postestrus (vaginal epithelium invaded by leukocytes which make up practically the entire smear). This assay gave rise to mouse and rat units as a measure of potency of a preparation. However, because a number of variables (e.g., solvent, numbers of doses, the route of administration and others) did not afford completely uniform results, a crystalline estrone standard was adopted as the International Unit. This I.U. is equal to 0.1 mcg. of crystalline estrone and is about equal to 1 m.u. A rat unit is 4 or 5 or even 10 times greater than the mouse unit. β-Estradiol has 120,000 I.U. per milligram, whereas α-estradiol is quite inert.

Estradiol is synthesized by ovarian follicles from childhood to the menopause. The peak of secretion of estradiol is at the time of ovulation, and this is due no doubt to the liberation of a large amount of follicular fluid. In the immature female, estradiol accelerates the growth of secondary sex organs, i.e., the vagina, the fallopian tubes and the uterus; the development of the breasts (mammary glands); the development of the genital tract and the distribution of hair. At sexual maturity, it plays a role in the development of the uterine wall for reception of the fertilized ovum and arouses sex instinct, i.e., estrus (heat, rut). A deficiency of estradiol in the immature female delays the above developments. In the mature female, estradiol maintains the normal size and functional capacity of the uterus, the fallopian tubes and the vagina. Estradiol preserves the normal epithelium of the vulva and the vagina and the muscular activity of the myometrium and the fallopian tubes. It promotes the growth of the duct tissues of the breast, especially during pregnancy. It acts on the pituitary so as to control the secretion of the gonadotropic and other hormones. Estradiol exerts a powerful influence on the psychic attitudes and general health of the adult woman. It plays a role in the maintenance of a normal condition of nasal and oral mucous membranes. It also sustains the basal metabolic rate and has certain vasodilator actions. Estradiol plays a role in the nutrition of unborn and prematurely born children. It exhibits some corticoid activity, particularly that concerned with fluids and salts.

It follows that estradiol therapy is indicated in cases where a deficiency of this hormone is known. Such therapy will bring about a normal state of being. For example, estradiol can be used in the treatment of sexual infantilism and senile vaginitis, and for smoother transition during the menopause, for inhibition of the pituitary in certain overactive states and in treatment of juvenile vaginitis due to gonococcus. In the last-named condition, estradiol effects a brief conversion of the juvenile epithelium of the adult type, which has secretions of a lower pH, is a more rugged kind of tissue and is markedly more resistant to the infection. Estradiol increases muscular strength, bodily vigor and mental acumen.

The symptoms of the menopause and the percentages of their incidence have been reported.[46]

β-Estradiol is oxidized in the body and excreted in the urine as estrone, estriol, equilin[47] and equilenin sulfates and glu-

Equilin

Equilenin
(3-Hydroxy-17-keto-estra-1,3,5(10),6,8 pentaene)

curonates and as the free compounds. All these excretory products have estrogenic activity that is qualitatively equal to that of β-estradiol but differs quantitatively. Estrone, estriol and equilenin have one third, one sixty-sixth[48] and one three-hundred-sixtieth the activities of β-estradiol, respectively. Orally, estradiol has one-fifteenth to one-twentieth its subcutaneous activity.

Category—estrogen.

Usual dose range—implantation, 25 mg. repeated when necessary; intramuscular, 220 mcg. to 1.5 mg. 2 or 3 times a week; oral, 200 to 500 mcg. 1 to 3 times a day.

OCCURRENCE
Estradiol Pellets N.F.
Sterile Estradiol Suspension N.F.
Estradiol Tablets N.F.

Estradiol Benzoate N.F.; Beta-estradiol Benzoate; 1,3,5(10)-Estratriene-3,17β-diol 3-benzoate. This is prepared by the benzoylation of β-estradiol at the C_3-OH.

Estradiol benzoate occurs as a white or creamy white, crystalline powder that is odorless and stable in air. It is almost insoluble in water, soluble in alcohol, slightly soluble in ether and sparingly soluble in vegetable oils.

Of the more common and simple esters, such as the acetate, the propionate and the benzoate, which prolong the activity of estradiol when administered intramuscularly, the benzoate is the most desirable

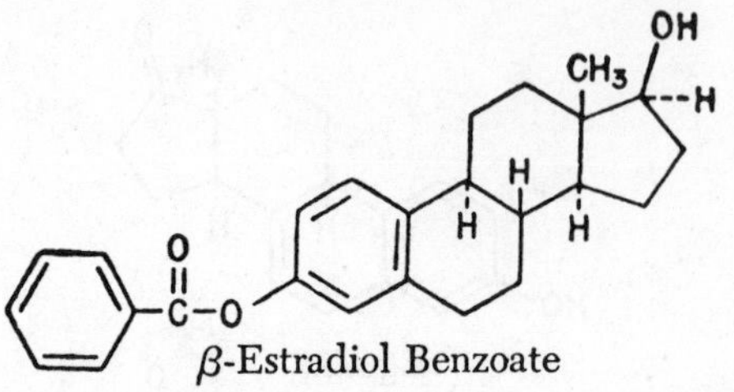

β-Estradiol Benzoate

because it is absorbed more slowly. Orally, these compounds have a marked decrease in activity due to destruction by the intestinal bacteria and the liver.

Category—estrogen.

Usual dose range—intramuscular, initial, 1.0 to 1.66 mg. 2 to 3 times a week for 2 or 3 weeks; maintenance, 330 mcg. to 1.0 mg. twice weekly.

OCCURRENCE
Estradiol Benzoate Injection N.F.

Estradiol Dipropionate N.F.; 1,3,5(10)-Estratriene-3,17β-diol Dipropionate. This occurs as a small, white crystals or crystalline powder that is insoluble in water, sparingly soluble in vegetable oils and soluble in alcohol. It has the same action and uses as other estrogen products. It is available in oil solution for intramuscular injection.

Category—estrogen.

Usual dose range—intramuscular, initial, 1 to 5 mg. every 1 to 2 weeks; maintenance, 1 to 2.5 mg. every 10 days to 2 weeks.

OCCURRENCE
Estradiol Dipropionate Injection N.F.

Estradiol Cypionate N.F., Depo®-Estradiol, estradiol 17-β-(3-cyclopentyl) propionate, occurs as a white crystalline solid that is insoluble in water, soluble in alcohol 1:50 and freely soluble in chloroform. When dissolved in cottonseed oil, this derivative of beta estradiol furnishes an estrogen with prolonged action (3 to 4 weeks) for intramuscular injection.

Category—estrogen.

Usual dose range—intramuscular, initial, 1 to 5 mg. weekly for 2 to 3 weeks; maintenance, 2 to 5 mg. every 3 to 4 weeks.

OCCURRENCE
Estradiol Cypionate Injection N.F.

Estradiol Valerate U.S.P., Delestrogen®, estra-1,3,5-(10)-triene-3,17β-diol 17-valerate, estradiol-17-valerate, is a white, crystalline powder, odorless or having a faint, fatty-type odor. It is practically insoluble in water, soluble in methanol and castor oil and sparingly soluble in sesame and peanut oils.

Category—estrogen.

Usual dose—intramuscular, 5 to 30 mg. every 2 weeks.

Usual dose range—5 to 40 mg. every 1 to 3 weeks.

OCCURRENCE
Estradiol Valerate Injection U.S.P.

Estrone N.F.; 3-Hydroxyestra-1,3,5(10)-trien-17-one; Oestrone; Theelin; Estrogenic Hormone; Estrin; Folliculin; Thelykinin. Estrone is obtained from pregnant mare's urine and from stallion urine.[49] Because much of the estrone is conjugated as the sulfate,[50] the urine first must be hydrolyzed before extraction with organic solvents. The estrone is purified[51] by the use of Girard's reagent,[52] adsorption technics and fractional crystallization processes. It also can be synthesized from suitable steroid starting materials such as $\Delta^{1, 4, 6}$-androstatriene-3,17-dione via vapor phase aromatization of ring

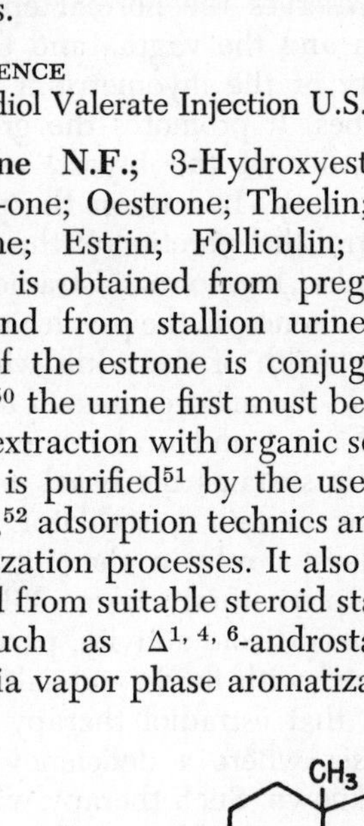

Estrone
(3-Hydroxy-17-keto-estra-1,3,5(10)-triene)

A with the loss of the C_{10}—CH_3 group and subsequent reduction of the Δ^6 double bond or by total synthesis.

Estrone occurs as small, white crystals or as a white, crystalline powder that is odorless and is stable in air. It is slightly soluble in water but is soluble in alcohol, in vegetable oils and in solutions of fixed alkali hydroxides.

The maximum activity of estrone can be obtained when it is administered intramuscularly. Orally, its activity is reduced greatly because intestinal bacteria and the liver oxidize it to produce substances with lowered activities.

Category—estrogen.

Usual dose—intramuscular, 1 mg. one or more times a week as required.

Usual dose range—200 mcg. to 5 mg. per week.

OCCURRENCE
Estrone Injection N.F.

3,16α-Dihydroxyestratriene-1,3,5:10 has one fourth the activity of estradiol, whereas the corresponding 16β-isomer is inactive.[53]

Estriol, Theelol, is 3,16α,17β-Trihydroxy estra-1,3,5(10)-triene. It is obtained from the same natural sources as estrone. It was isolated first by Doisy[54] and others, in 1930, and was secured as the glucuronate in 1935.[55] Estriol can be converted to estrone and the estradiols by the following reactions.

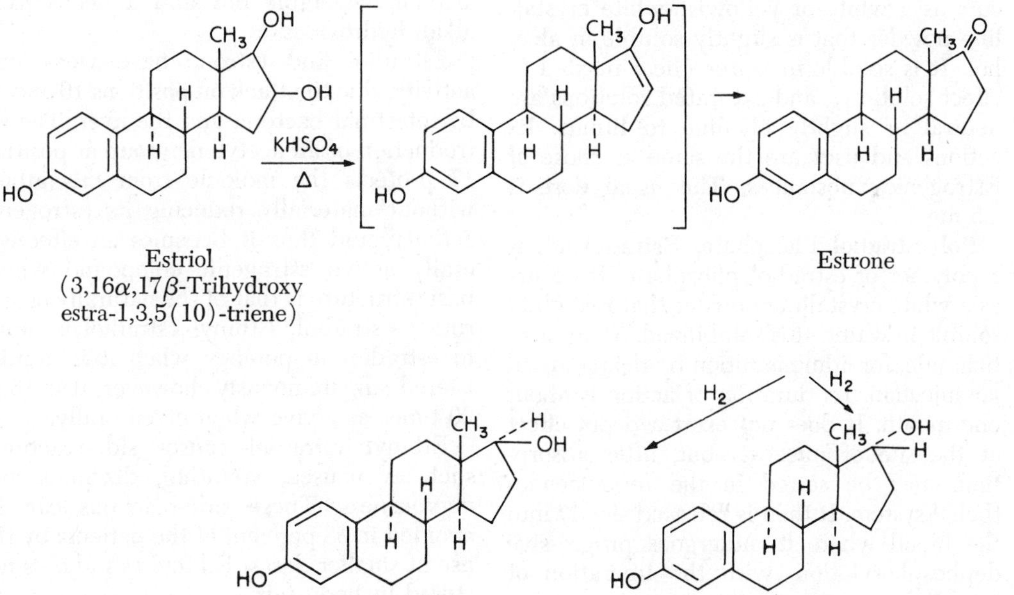

Pentovis®, Colpovis®, 3-cyclopentyloxy-oestra - 1,3,5(10) - triene - 16α,17β - diol, is claimed to be especially suitable when interference with the menstrual cycle is not desired and when the production of uterine hemorrhage in postmenopausal women is to be avoided.[56]

Conjugated Estrogens U.S.P., Premarin®, Amnestrogen®, Conestron®, Estrifol®, Konogen®, are amorphous preparations containing the naturally occurring, water-soluble, conjugated forms of the mixed estrogens obtained from the urine of pregnant mares. This preparation contains chiefly estrone sulfate, together with small amounts of other equine estrogens and relatively large quantities of nonestrogenic material.

Premarin is prepared by extracting concentrated pregnant mare's urine with butyl alcohol and subsequently removing the nondesirable alkali-soluble constituents from

Estriol
(3,16α,17β-Trihydroxy
estra-1,3,5(10)-triene)

Estrone

β-Estradiol

α-Estradiol

Estriol occurs as a white, odorless, microcrystalline powder. It is practically insoluble in water, soluble in alcohol and vegetable oils.

Estriol is used orally (capsules) for its estrogenic activity, which is 1/66 that of β-estradiol.

The usual dose is 60 to 120 mcg.

the butyl alcohol with tenth normal sodium hydroxide. After the butyl alcohol is removed, the solid residue is treated with acetone. The acetone solution is concentrated and then mixed with an excess of ether to precipitate the desirable constituents as an amorphous reddish brown to almost white powder. This powder is soluble in water and alcohol but is insoluble in oils and ether.

The mixture is used orally for its estrogenic activity. Apparently, the sulfates of the estrogenic substances are not oxidized as readily by intestinal bacteria and the liver as are the free estrogens, since they may be absorbed and circulated in the bloodstream more readily.

Category—estrogen.

Usual dose—oral, 1.25 to 2.5 mg. 1 to 3 times daily; intramuscular or intravenous, 20 mg.

Usual dose range—oral, 1.25 to 30 mg. daily; intramuscular or intravenous, 20 to 40 mg. daily.

OCCURRENCE

Conjugated Estrogens for Injection U.S.P.
Conjugated Estrogens Tablets U.S.P.

Piperazine Estrone Sulfate, Ogen®, Sulestrex®, Piperazinium Estronesulfate, occurs as a white or yellowish-white crystalline powder that is slightly soluble in alcohol. It is soluble in water (heat needed to effect solution), and saturated solutions are neutral or slightly alkaline to litmus. Its actions and uses are the same as those of Estrogenic Substances. The usual dose is 1.5 mg.

Polyestradiol Phosphate, Estradurin®, is a polymer of estradiol phosphate. It occurs as a white crystalline powder that is slightly soluble in water. It is solubilized, using nicotinamide, for administration by deep muscular injection. Its duration of action is about one month. It does not exert a depot effect at the site of injection but, after absorption, may be stored in the reticuloendothelial system. It then is released slowly into the blood where it undergoes progressive dephosphorylation, with the liberation of estradiol.

It is used only in the palliation of prostatic carcinoma. The dose is 40 mg. every 2 to 4 weeks.

Ethinyl Estradiol U.S.P., 19-nor-17α-pregna-1,3,5(10)-trien-20-yne-3,17-diol, 17-ethynylestradiol, Estinyl®, Eticylol®, Lynorol®, Oradiol®, Orestralyn®, Feminone®, is made by treating estrone with potassium tertiary amylate in acetylene or potassium acetylide in liquid ammonia as follows.[57]

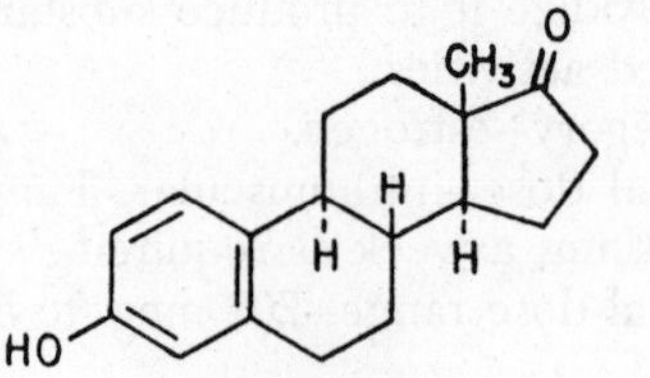

Estrone

KC≡CH
↓ NH₃

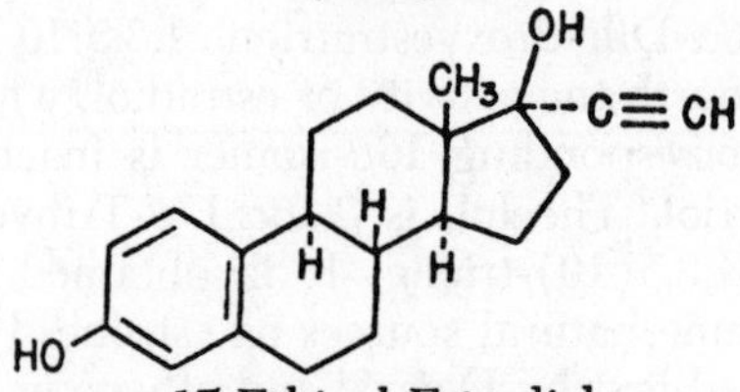

17-Ethinyl Estradiol

Ethinyl Estradiol occurs as a white to creamy-white odorless, crystalline powder that is insoluble in water but is soluble in alcohol, vegetable oils and aqueous fixed alkali hydroxides.

Estradiol and estrone have weak oral activity, due to attack on positions 16 and 17 by intestinal bacteria and the liver. The introduction of an acetylene group at position 17 protects the molcule from this attack without materially reducing its estrogenic activity, and thus it becomes an effective, orally active estrogenic compound whose basic structure is that of the naturally occurring β-estradiol. Ethinyl estradiol is equal to estradiol in potency when it is administered subcutaneously; however, it is 15 to 20 times as active when given orally.

Ethinyl estradiol causes side-reactions, such as nausea, vomiting, dizziness and nervousness. These side-reactions can be avoided in 93 per cent of the patients by the use of smaller doses. Ethinyl estradiol is not stored in body fats.

Ethinyl estradiol is most effective in sequential antiovulatory preparations at 100 mcg. dose levels.[58]

Category—estrogen.

Usual dose—50 mcg. 1 to 3 times a day.
Usual dose range—20 mcg. to 3 mg. daily.

OCCURRENCE

Ethinyl Estradiol Tablets U.S.P.
Ethinyl Estradiol and Dimethisterone Tablets N.F.

Structure-activity relationships in the estrogens (estrane derivatives) are much narrower than with other steroid hormones. The following groups or changes lead to a considerably less estrogenic activity; 11α, 11β,[59] 6α, 6β, or 7α-hydroxy groups; 1, 2 dimethyl,[60] or 1,2 or 6β-methyl[61] groups; a $\Delta^{9,\,11}$-double bond; 18 nor[62]; 3 or 17 desoxy estrogens.

The C_{18}—CH_3 group appears to be necessary for estrogenic activity, probably because it stabilizes the *trans* C, D configuration and an ethyl or *n*-propyl group can replace the CH_3 without great reduction in activity.[63] This perimeter change increases some activities of the progestins and androgens (q.v.). 8-Isoestrone (rings B and C *cis*) is half as active as estrone. D-homoestrone[64] (ring D 6-membered) had 1/30 the estrogenic activity of estrone (contrast with D-homotestosterone). Of the stereoisomers of equilenin, the *d*-isomer (natural isomer) is most active in rats (30 γ) whereas the *l*-isomer is only 1/13 as active.

16-Ketoestradiol and its 3-methyl and 3-ethyl ethers act as spindle or metaphase poisons and hinder mitosis or cell division in chick embryo fibroblast-primitive cells that give rise to connective tissue. The 3-methyl ether of 16-ketoestradiol has only 1/400 the estrogenic activity of diethylstilbestrol; however, it has significantly greater mitosis or cell division inhibiting properties.[65]

The 3-glycolic acid ether of estradiol is almost inactive.[66]

d-2,3,4-Trimethoxyestra-1,3,5(10)-triene-17β-ol had 40 times the analgesic activity of morphine with no estrogenic activity. The I.V. acute LD_{50} in mice and rats was about 180 mg./kg. Abdominal surgery in cats and dogs was possible after the administration of 3 to 5 mg./kg.[67]

ANTIESTROGENIC AGENTS

Clomiphene Citrate, Clomid®, is 1[*p*-diethylaminoethoxy)phenyl]-1,2-diphenyl-2-chloroethylene (*trans* form) and is available as a citrate salt. In men it stimulates the pituitary–Leydig cell axis to produce an increase in plasma LH and testosterone levels. In females it stimulates an increased output

of ketosteroids, estrogen, FSH and LH. Ovulation was induced with clomiphene in patients receiving adequate estrogen to suppress ovulation. It appears that clomiphene has antiestrogenic activities and can displace estradiol from its receptor sites. It, therefore, would have affinity for the receptor site, with little or no intrinsic activity.

NONSTEROIDAL ESTROGENS

Doisynolic acid (1-ethyl-2-methyl-7-hydroxy-1,2,3,4,9,10,11,12-octahydrophenanthryl-2-carboxylic acid) and **bisdehydrodoisynolic acid** (1-ethyl-2-methyl-7-hydroxy-1,2,3,4-tetrahydrophenanthryl-2-carboxylic acid) have been obtained by the fusion with potassium hydroxide of estrone and equilenin respectively.[68]

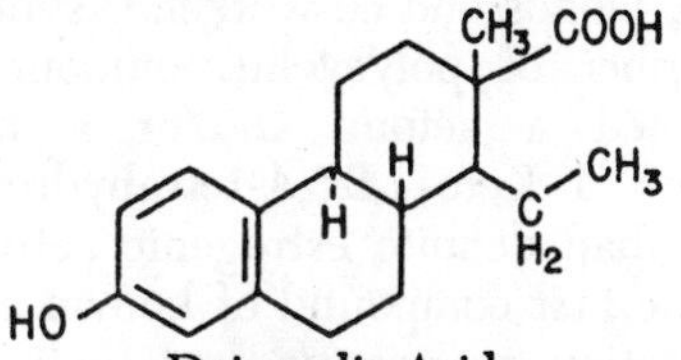

Doisynolic Acid

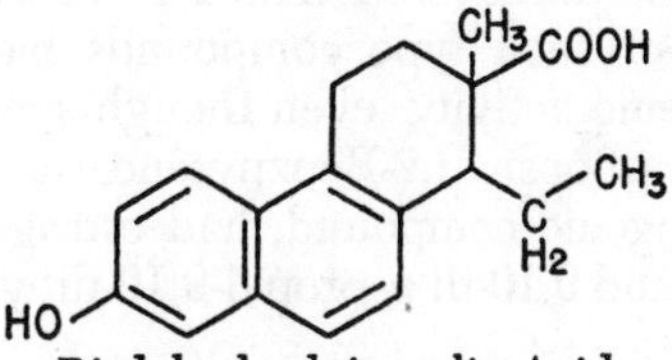

Bisdehydrodoisynolic Acid

Both of the above-mentioned compounds have been synthesized.[68] These compounds exhibit high estrogenic activity. Doisynolic acid, when bioassayed by the Kahnt-Doisy method, using pure estrone as a standard, gave the typical estrus response in doses of 0.8 to 0.9 mcg. It is very effective when administered orally.[69]

The following open-chain analog of 7-methylbisdehydrodoisynolic acid ethyl ester showed appreciable estrogenic activity.

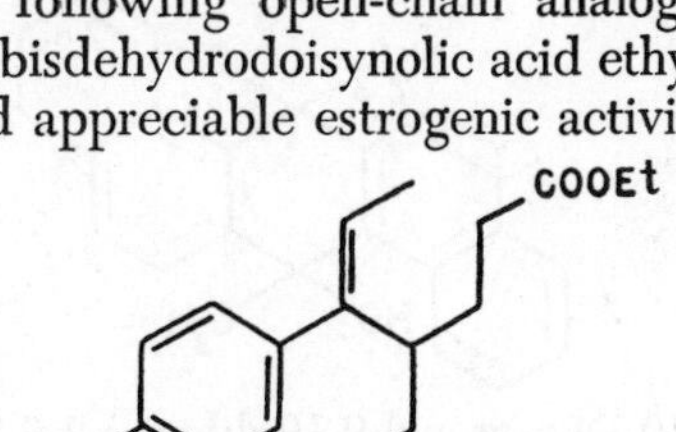

It is conceivable that the estrogenic activity of **Methallenestril**; Vallestril®; can be attributed to a structural relationship to bis-dehydrodoisynolic acid when it is depicted as follows:

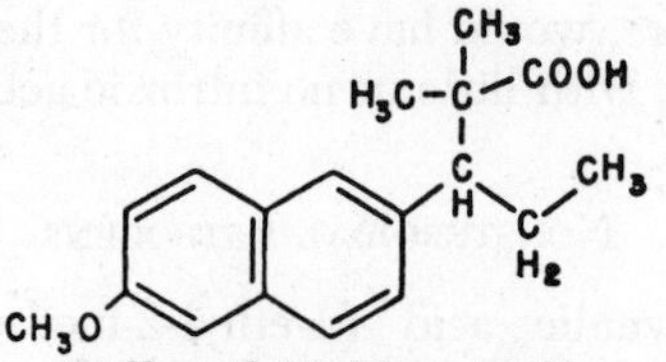

Vallestril (Horeau Acid)
[3-(6-Methoxy-2-naphthyl)-
2,2-dimethylpentanoic Acid]

Methallenestril has a high estrogenic activity and a low incidence of untoward side-reactions. It is available in 3-mg. tablets.

Because estrone contained a polycyclic structure, a keto and a phenolic hydroxyl group, Dodds and co-workers[70] synthesized a number of polycyclic compounds that contained a ketone and/or a hydroxyl group. 1-Keto-1,2,3,4-tetrahydrophenanthrene had definite estrogenic activity and was the first compound of known chemical constitution to possess such activity.[70] It soon was discovered that a relatively wide range of such type compounds had some estrogenic activity, even though some were hydrocarbons. 1,2-Benzpyrene, a potent carcinogenic compound, had estrogenic activity, and 9,10-di-n-propyl-9,10-dihydro-1,2, 5,6-dibenzanthracene-9,10-diol was more active than estrone.[71]

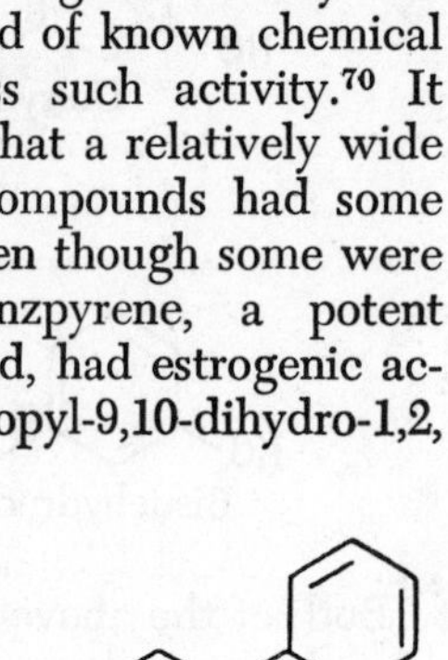

These results were followed by attempts to determine which portions of the molecule were responsible for the estrogenic activity. It was shown that the phenanthrene nucleus was not necessary and that estrogenic activity of a low order was demonstrated in the following type compounds: p,p'-dihydroxy-diphenyl, p,p'-dihydroxydiphenylmethane, di-(p-hydroxyphenyl)alkyl or dialkyl methane, 4,4'-dihydroxybenzophenone and the naturally occurring substance phloretin.[72] Furthermore, activity was present to a lesser degree in compounds that had only one phenolic hydroxyl group, e.g., 4-hydroxy-phenylcyclohexane.[73]

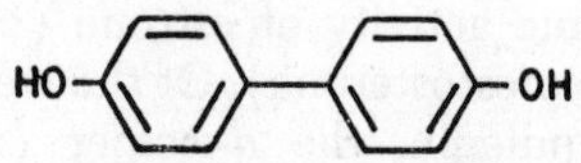

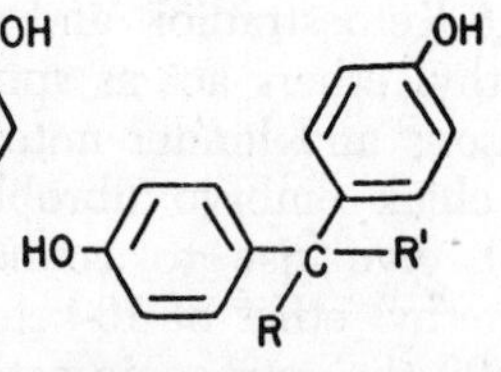

p-p'-Dihydroxydiphenyl

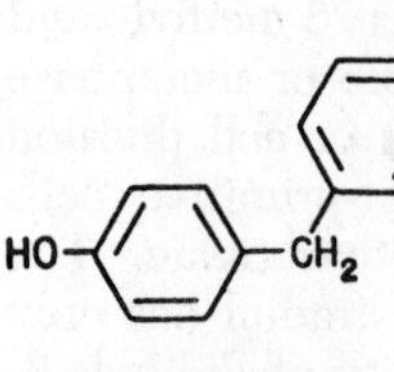

p-p'-Dihydroxy-   Di-(p-hydroxyphenyl)-
diphenyl Methane   alkyl or dialkyl methane
where R = CH₃, Et or Pr
and R' = H or CH₃.

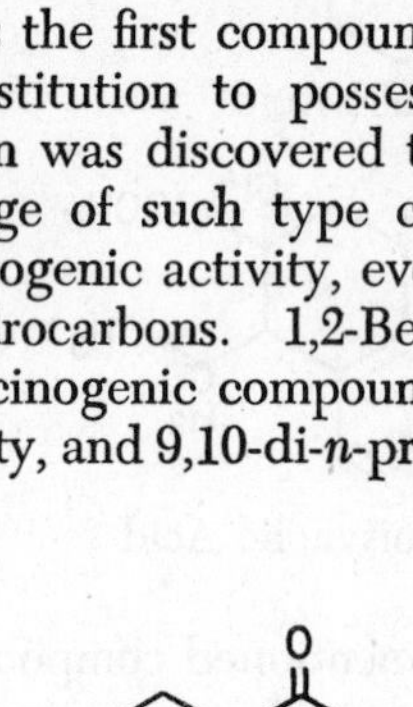

1-Keto-1,2,3,4-tetra-   1,2-Benzpyrene
hydrophenanthrene

4,4'-Dihydroxybenzophenone

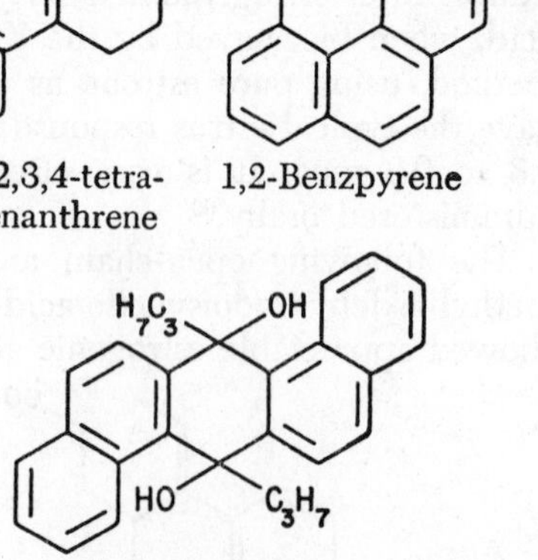

9,10-Di-n-propyl-9,10-dihydro-1,2,5,6-di-
benzanthracene-9,10-diol

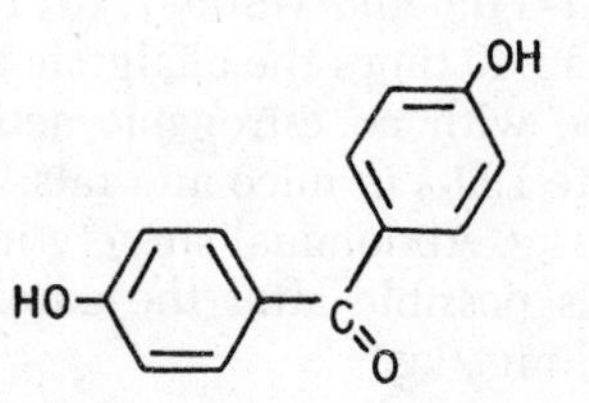

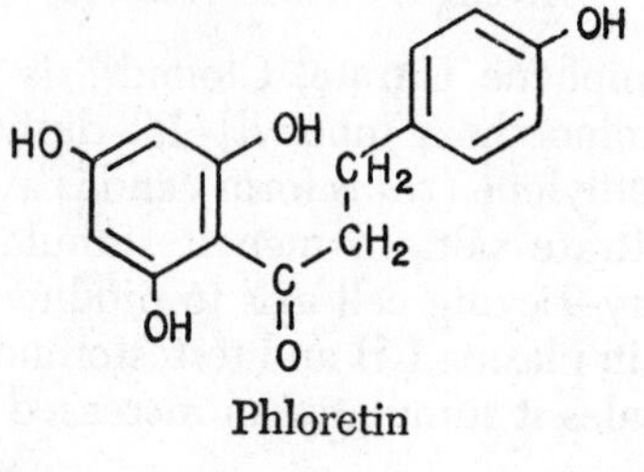

Phloretin

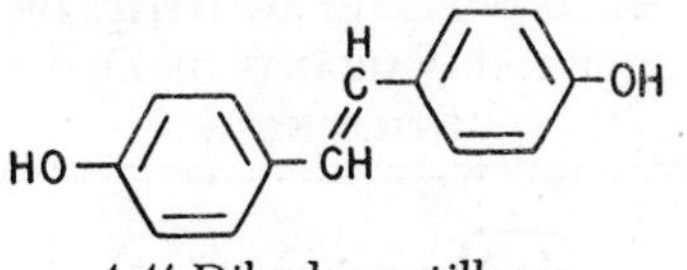

4,4′-Dihydroxystilbene

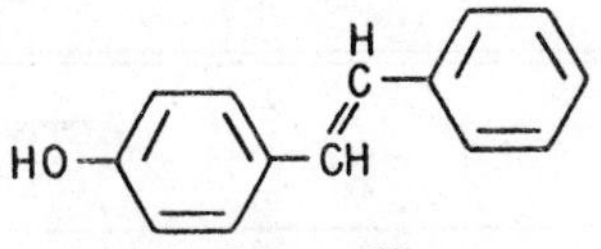

4-Hydroxystilbene

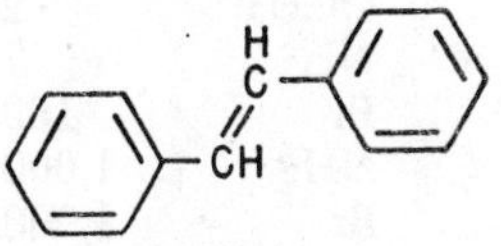

Stilbene

Extension of these studies led to the preparation of 4,4′-dihydroxystilbene, 4-hydroxystilbene and stilbene, which produced full estrus response in castrated mice in 10, 10 and 25 mg. doses, respectively. The two former compounds were, therefore, 10 times

as active as the bis-phenolic compounds described above. Because activity was present in still simpler molecules, such as *p*-hydroxyphenylethyl alcohol, i.e., 100 mg. giving a 60 per cent response, the demethylation of anethole by means of potassium hydroxide was tried in order to prepare a simple phenol with a short unsaturated side chain. This compound, whose structure was supposedly *p*-hydroxypropenylbenzene, or anol, exhibited a very high degree of activity. Later it was shown that some batches of anol were highly active while others were not. Anol that was completely inactive at the 10 mg. level was converted to a highly active substance when heated for 15 hours at 180° to 200° with alcoholic potassium hydroxide.[74] It was suspected that a polymerized substance was the active product, and the demethylation of di-anethole with potassium hydroxide yielded di-anol, which gave a 100 per cent response in castrated female mice at the 100-mcg. level.[75]

These and other studies stimulated efforts to prepare estrogenic compounds bearing a close relationship to estrone (whose structure was now known) but which would be

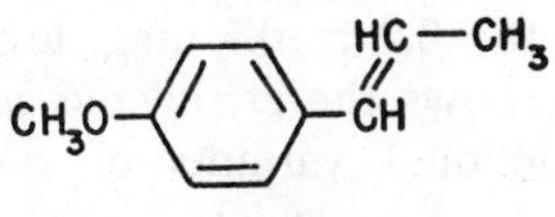

Anethol

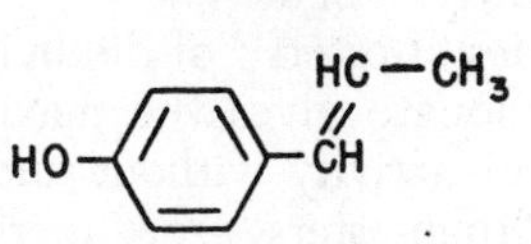

Anol

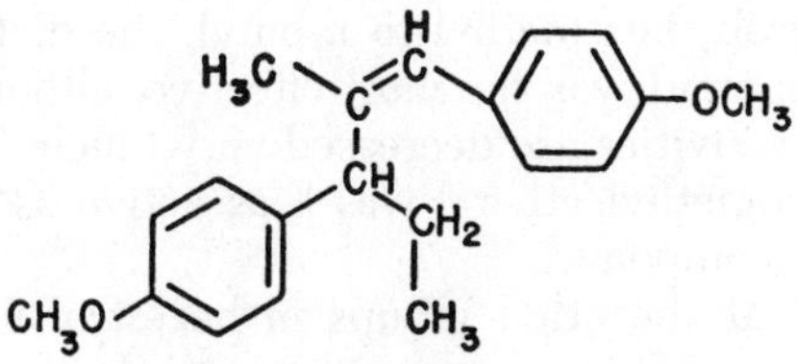

Di-anethole

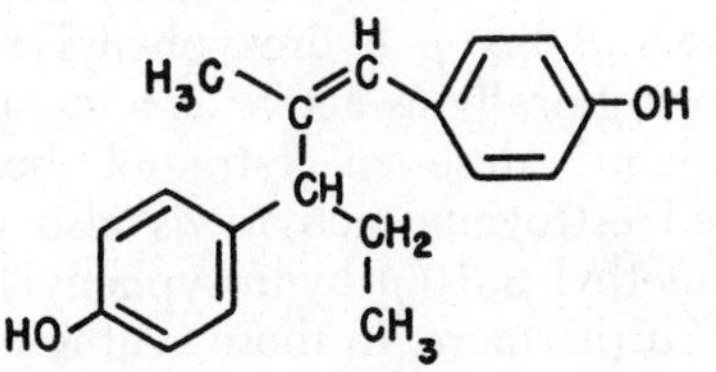

Di-anol

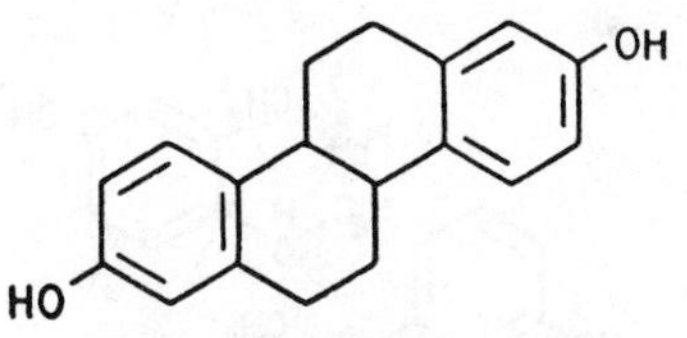

Dihydroxyhexahydrochrysene
(Active at the 1 mg. level)

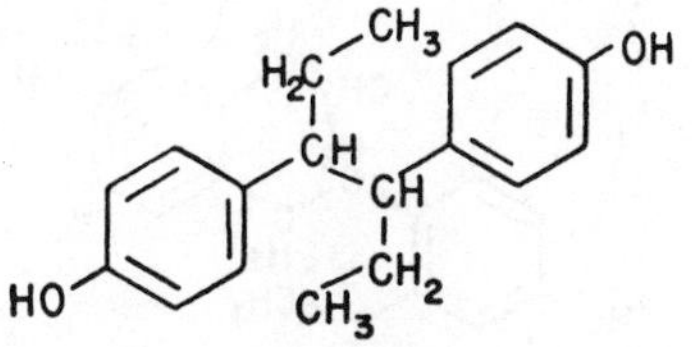

Hexestrol

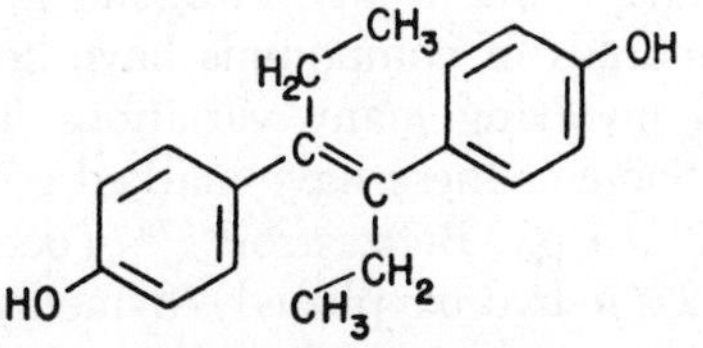

Diethylstilbestrol
(Active at 0.3 and 0.4 mcg. levels)

capable of ready synthesis. These studies culminated in the synthesis of diethylstilbestrol, hexestrol and dihydroxyhexahydrochrysene[76] (see p. 775).

Later[77] it was shown that hexestrol could be found in the products obtained by the heating of anol with potassium hydroxide; this accounts for the marked activity obtained.

Studies have shown that the following compounds, whose structures can be depicted as being related structurally to es-

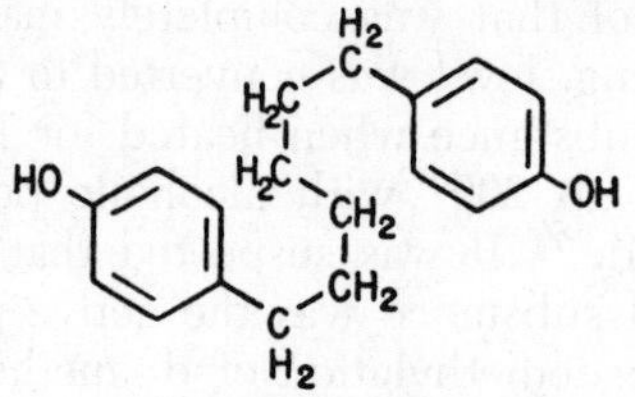

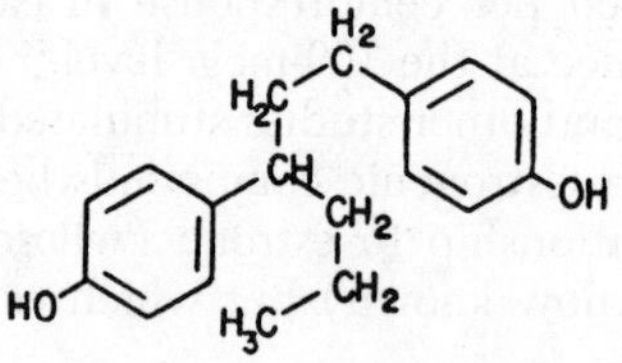

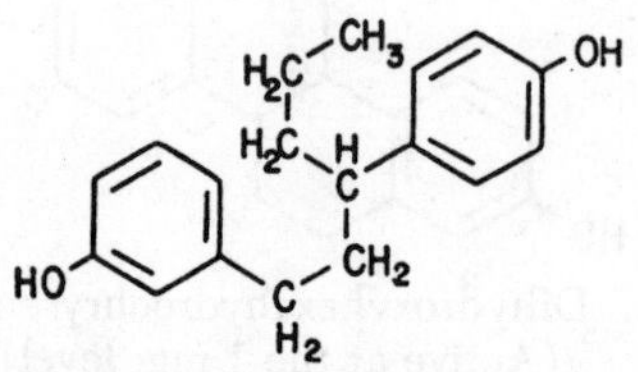

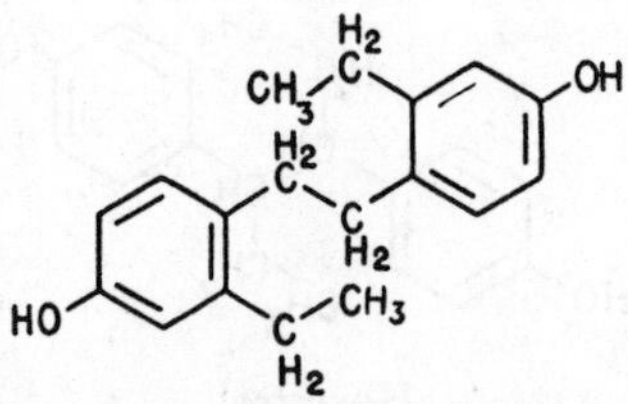

trone, have little if any estrogenic activity.[78]

Hundreds of compounds have been prepared involving many variations in structure. Some of these have marked estrogenic activity, e.g., Benzestrol[78, 79] (octofollin) and 2-(*p*-hydroxyphenyl)-3-methyl-4-(or 6)-hydroxyindene.[80] The latter compound is

TABLE 84. ESTROGENIC ACTIVITIES OF SOME FUNCTIONAL VARIANTS OF DIETHYL STILBENE[*]

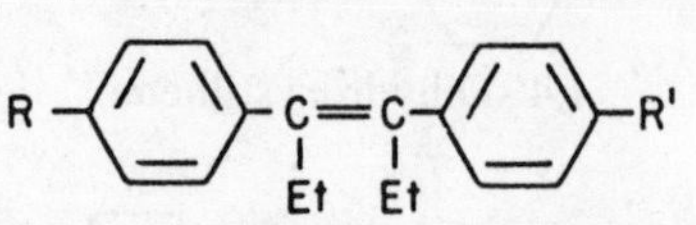

R	R'	ACTIVITY[†] IN MCG.
HO	HO	0.3
HO	OCH$_3$	2.5
HO	NH$_2$	7.5
HO	Br	100.0
OCH$_3$	NH$_2$	1,000.0
OCH$_3$	Br	1,000.0
OCH$_3$	COOH	Inactive at 1,000
H	H	Inactive at 1,000

[*] After Rubin, M., and Wishinsky, H.: J. Am. Chem. Soc. 66:1948, 1944.

[†] These weights are equivalent to 1 rat unit as assayed by the Allen-Doisy method.

active at the 0.2 to 0.5 mcg. level in rats. Table 84 shows the estrogenic activities of some functional variants of diethyl stilbene.[81]

Table 85 shows the relationship between activity and the varying alkyl groups in *p,p'*-dihydroxystilbene.[82]

Of a series of esters[83] of diethylstilbestrol, the dipropionate gives the maximum prolongation of activity without undue loss in activity. Mono-esters are superior to the more easily prepared diesters.[84]

Of a series[85] of dialkylated diethylstilbestrols, i.e., methyl to *n*-butyl, the diethyl compound was the most effective, although the activities are decreased very much. The monomethyl ether[86] was ⅛ as active as the free compound.

That the ethyl groups in hexestrol need not be of a nonpolar character for estrogenic activity is demonstrated by the water-soluble β-bis (*p*-hydroxyphenyl) succinic acid being orally as active as a comparable intravenous dose of estradiol benzoate. Marked estrogenic activity is also present in β-methyl-β,β-(*p*-hydroxyphenyl)valeric acid. Furthermore, in those synthetic estro-

TABLE 85. RELATIONSHIP BETWEEN ACTIVITY AND ALKYL GROUPS IN P,P′-DIHYDROXYSTILBENE*

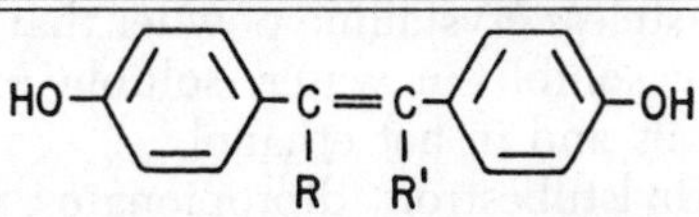

R	R′	Dose in mcg.	Per cent response in ovariectomized rats
H	H	5,000.0	80
		10,000.0	100
H	Ethyl	100.0	50
Methyl	Methyl	20.0	80
		30.0	100
Methyl	Ethyl	0.5	30
		1.0	100
Ethyl	Ethyl	0.3	80
		0.4	100
Ethyl	*n*-Propyl	1.0	trace
		10.0	100
n-Propyl	*n*-Propyl	10.0	75
		100.0	100
n-Butyl	*n*-Butyl	10.0	0
		100.0	40
Monohydroxy diethylstilbene		100.0	trace
		1,000.0	100

* After Dodds, E. C.: Nature 142:34, 1938.

gens containing a carboxyl group, both aromatic rings are not necessary, since the following open chain analog of 7-methyl bisdehydrodoisynolic acid ethyl ester showed appreciable estrogenic activity.

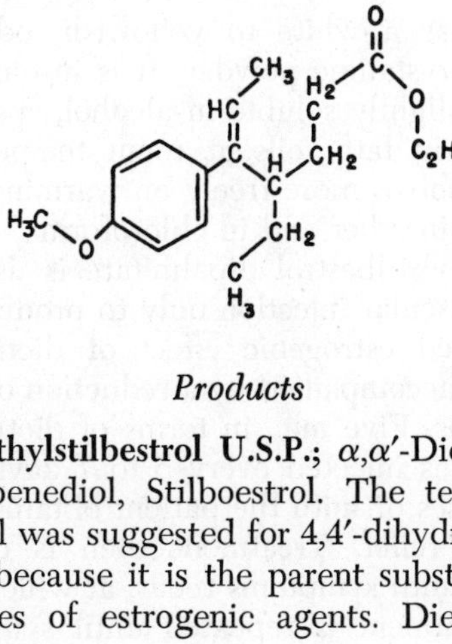

Products

Diethylstilbestrol U.S.P.; α,α'-Diethyl-4,-4′-stilbenediol, Stilboestrol. The term stilbestrol was suggested for 4,4′-dihydroxystilbene because it is the parent substance of a series of estrogenic agents. Diethylstilbestrol was first prepared by Dodds[76] as follows:

Anisaldehyde $\xrightarrow{\text{KCN}}$ Anisoin $\xrightarrow[\text{100°, 24 hrs.}]{\text{Zn dust, AcOH}}$

Desoxyanisoin $\xrightarrow[\text{EtI}]{\text{EtONa}}$ Ethyldesoxy anisoin $\xrightarrow{\text{EtMgBr}}$

3,4-Bis-(*p*-anisyl)-3-hexanol $\xrightarrow[\text{or KHSO}_4]{\text{PBr}_3}$

Diethylstilbestrol dimethyl ether $\xrightarrow[\text{205°}]{\text{KOH}}$ Diethylstilbestrol

Because of the marked interest in this highly successful estrogenically active compound, it has been prepared in a number of ways, one of which involved a 3-step synthesis as follows.[87]

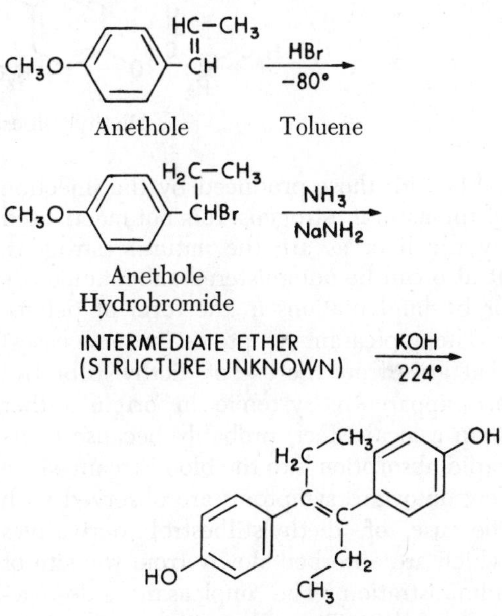

Diethylstilbestrol (*Trans* Form)

Diethylstilbestrol occurs as a white, odorless, crystalline powder. It is almost insoluble in water and is soluble in alcohol, chloroform, ether, fatty oils or dilute alkali hydroxides.

Because diethylstilbestrol is a phenol, it exhibits the stability properties of the phenols in general. Aqueous neutral or alkaline solutions of diethylstilbestrol kept for 2 weeks at room temperature or in the cold become yellow in color and lose their activity.[88] Air or oxidizing agents accelerate this decomposition.

Diethylstilbestrol can exist in a *cis* or a *trans* form. The *trans* form or *trans*-4,4′-dihydroxy-α,α'-diethylstilbene is more active than the lower melting *cis* isomer, which has a melting point of 142°. The official drug is the *trans* form, which has a melting point of 169° to 172°.

A 3-dimensional molecular model of *trans* diethylstilbestrol, made with the Fischer-Hirschfelder-Taylor type models, resembles estradiol quite closely in its spatial configuration.

Diethylstilbestrol when administered orally produces hormonelike effects compa-

This ester of *trans*-diethylstilbestrol can be prepared in the usual way for the preparation of esters of phenols. It is a white, odorless, tasteless crystalline powder that is very slightly soluble in water, soluble in vegetable oils and in hot ethanol.

Diethylstilbestrol dipropionate is absorbed more slowly than diethylstilbestrol. Therefore, a more prolonged estrogenic effect can be obtained, accompanied by fewer side-reactions.

Category—estrogen.

Usual dose range—100 mcg. to 1 mg. daily.

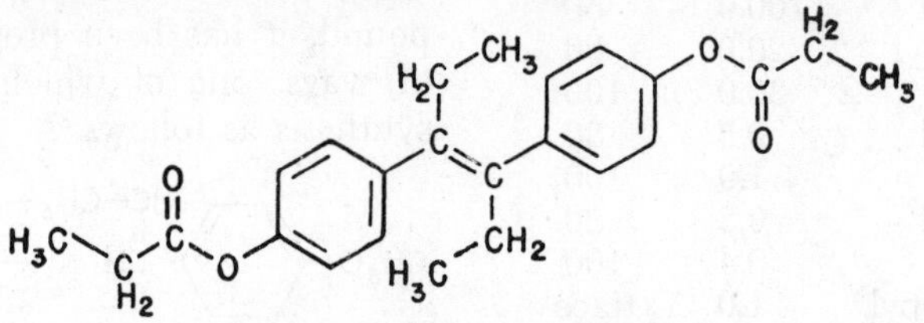

Diethylstilbestrol Dipropionate

rable with those produced by the injection of the natural estrogens. It is not inactivated by the liver as are the natural estrogens. It also can be administered subcutaneously or by implantations in the form of pellets.

The unpleasant symptoms (in some cases) that arise from the use of diethylstilbestrol are apparently systemic in origin rather than a local effect, probably because of its rapid absorption into the bloodstream, since few untoward symptoms are observed with the use of diethylstilbestrol derivatives which are absorbed slowly from the site of administration. The unpleasant side-reactions of the stilbene derivatives are nausea, vomiting and headache.

Category—estrogen.

Usual dose—oral, intramuscular or vaginal, 100 mcg. to 1 mg. once daily.

Usual dose range—oral, 100 mcg. to 15 mg. daily; intramuscular, 100 mcg. twice a week to 5 mg. 3 times daily.

OCCURRENCE

Diethylstilbestrol Injection U.S.P.
Diethylstilbestrol Suppositories U.S.P.
Diethylstilbestrol Tablets U.S.P.

Diethylstilbestrol Dipropionate N.F.; α, α'-Diethyl-4,4′-stilbenediol Dipropionate.

OCCURRENCE

Diethylstilbestrol Dipropionate Tablets N.F.

Diethylstilbestrol Dipalmitate, Stilpalmitate®, can be prepared in the usual way for the preparation of esters of phenols. It

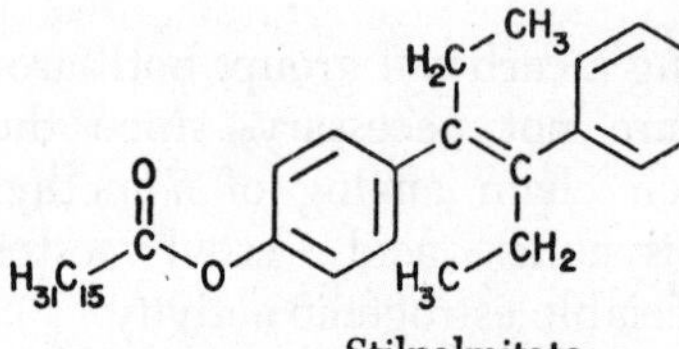

Stilpalmitate

occurs as a white to yellowish, odorless, waxy, crystalline powder. It is insoluble in water, slightly soluble in alcohol, sparingly soluble in fatty oils at room temperature but dissolves more freely on warming. It is soluble in ether and in chloroform.

Diethylstilbestrol dipalmitate is used by intramuscular injection only to produce the prolonged estrogenic effect of diethylstilbestrol accompanied by a reduction of side-reactions. Five mg., in terms of diethylstilbestrol, is injected every 5 to 7 days for 3 to 5 doses or until the patient obtains satisfactory relief. Treatment then is discontinued until symptoms recur, at which time the treatment is repeated until symptoms are relieved again. The duration of symptomatic relief between dosage periods will vary from 4 to 12 weeks.

Monomestrol; Mestilbol; Diethylstilbestrol Monomethyl Ether; α,α′-Diethyl-4′-methoxy-4-stilbenol. Monomestrol is prepared by the methylation of diethylstilbestrol or by the partial demethylation of the dimethyl ether of diethylstilbestrol.

Monomestrol occurs as a white, odorless, crystalline powder that is insoluble in water and soluble in alcohol, acetone or ether, dilute aqueous solutions of sodium or potassium hydroxide or in vegetable oils. It is more stable than diethylstilbestrol.

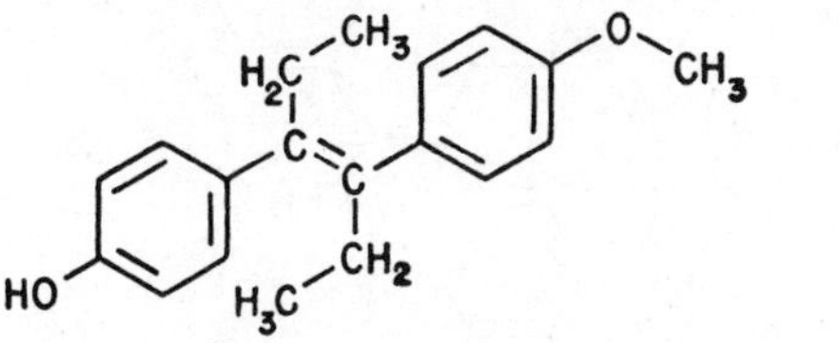

Monomestrol

Monomethylation of diethylstilbestrol has increased somewhat the duration of action and reduced the side-effects of diethylstilbestrol. These changes are accompanied by a reduction to one eighth of the activity of the parent compound.[81]

Monomestrol is effective orally and parenterally. The oral dose ranges from 100 mcg. to 5 mg. daily, and the parenteral dose in oil ranges from 10 to 25 mg. biweekly.

Hexestrol, 4,4′-(1,2-Diethylethylene)diphenol, *Meso*-3,4-di-*p*-hydroxyphenyl-*n*-hexane; *p,p′*-(1,2-Diethylethylene)diphenol. Hexestrol originally was prepared by Dodds and co-workers[77] by the demethylation of anethole with potassium hydroxide and by the reduction of diethylstilbestrol (see synthesis of diethylstilbestrol). Kharasch and Kleiman,[87, 89] by a three-step reaction, were able to synthesize hexestrol in good yield.

Hexestrol can exist in 3 forms, i.e., D-, L- or the *meso* forms. The *meso* form is more effective than the lower melting DL- form. It occurs as a white, odorless crystalline powder that is insoluble in water but soluble in alcohol and vegetable oils. Hexestrol has about the same order of activity as diethylstilbestrol; however, the claim is made that the incidence of side-reactions is lower.

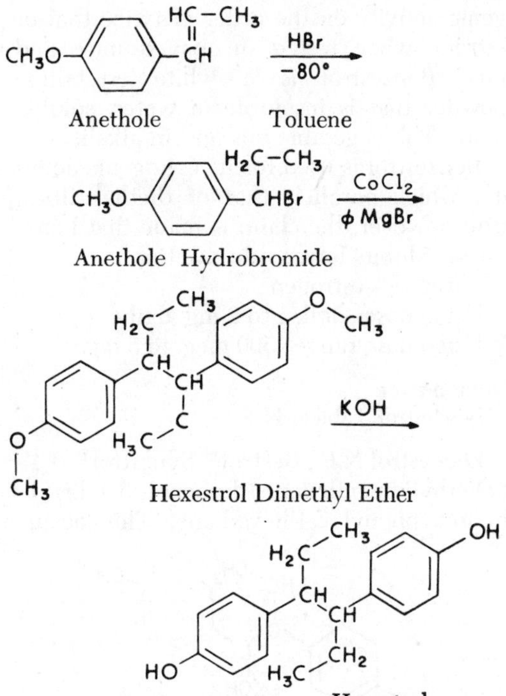

Anethole   Toluene

Anethole Hydrobromide

Hexestrol Dimethyl Ether

Hexestrol

The *meso* form, in which the two ethyl groups are *cis* in the Fischer projectional type formula, would adopt a more nearly *trans* type of orientation in vitro and in vivo. Thus, the whole molecule could adopt a conformation more closely related to that of the *trans* isomer of diethylstilbestrol which is the most active isomer.

The dosage range is 0.2 to 3 mg. daily.

Benzestrol N.F. is 4,4′-(1,2-diethyl-3-methyltrimethylene)diphenol. It was the most active member of a series[53, 79] of trialkyl derivatives of 1,3-di-(*p*-hydroxyphenyl)propanes.

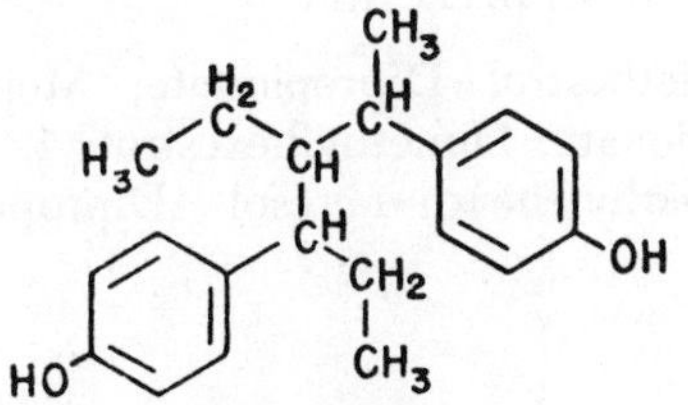

Benzestrol

Because benzestrol has 3 asymmetric carbon atoms, it can exist in 8 possible isomeric forms, one of which is more active[78] than the others. This isomer has an estrogenic activity on the order of twice that of estrone when tested in spayed mice and rats. Benzestrol is a white crystalline powder that is insoluble in water, soluble in alcohol, vegetable oils and in alkali.

Benzestrol is used for its estrogenic activity, which parallels that of diethylstilbestrol; however, the claim is made that benzestrol exhibits fewer side-reactions.

Category—estrogen.

Usual dose—oral, 1 to 2 mg. daily.

Usual dose range—500 mcg. to 5 mg.

OCCURRENCE

Benzestrol Tablets N.F.

Dienestrol N.F., Restrol®, Synestrol®, 4,4'-(Diethylidineethylene)diphenol, 3,4-bis(*p*-hydroxyphenyl)2,4-hexadiene. This occurs

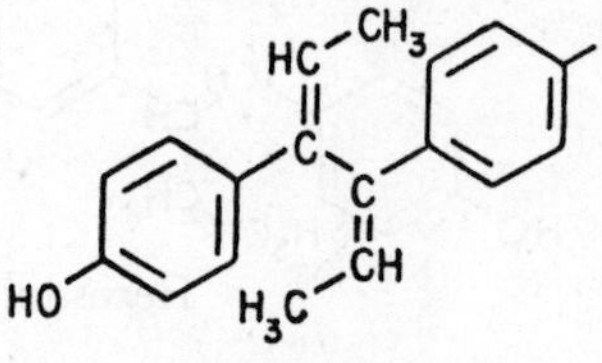

Dienestrol

as a crystalline powder that is insoluble in water but soluble in alcohol and slightly soluble in fatty oils. It is an orally effective estrogen that is claimed to be well tolerated.

Category—estrogen.

Usual dose—500 mcg. daily.

Usual dose range—100 mcg. to 1.5 mg.

Application—vaginal, 5 g. of a 0.01 per cent cream once or twice daily for 7 to 14 days, then once every 48 hours for 7 to 14 days.

OCCURRENCE

Dienestrol Cream N.F.

Dienestrol Tablets N.F.

Promethestrol Dipropionate; Meprane® Dipropionate; Dimethylhexestrol; 4,4'-(1,2-Diethylethylene)di-*o*-cresol Dipropionate.

The estrogenic activity of this compound shows that the phenyl groups of hexestrol can be substituted without a significant loss of activity. The propionic acid ester is used for more prolonged duration of action. The average maintenance dose is 1 mg.

Chlorotrianisene N.F.; Tace®; Chlorotris-(*p*-methoxyphenyl)ethylene. It occurs as a white, odorless crystalline powder that is very slightly soluble in water, 280 mg. in 100 ml. of alcohol and is freely soluble in chloroform.

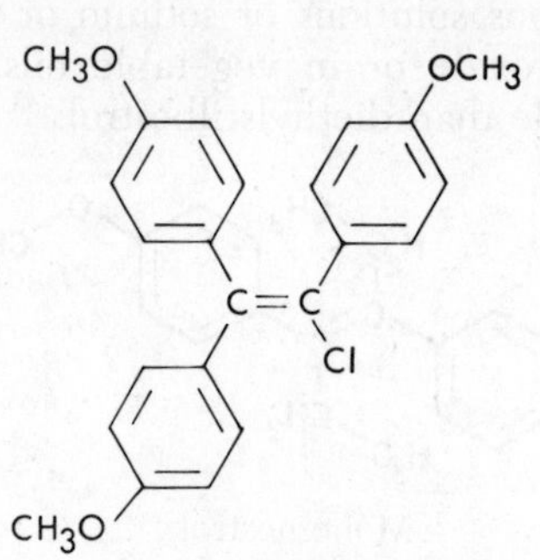

Chlorotrianisene

The use of a true solution of a water-insoluble medicinal agent in a digestible oil for oral administration enables the development of smaller particles, if not a molecular state, in the gut as the oil is digested.

Category—estrogen.

Usual dose—24 mg. daily.

Usual dose range—12 to 50 mg.

OCCURRENCE

Chlorotrianisene Capsules N.F.

PROGESTINS

Frankel,[90] as early as 1903, showed that the corpus luteum was essential for the maintenance of pregnancy. Corner and Allen,[91] in 1930, prepared active extracts of corpora lutea capable of functionally re-

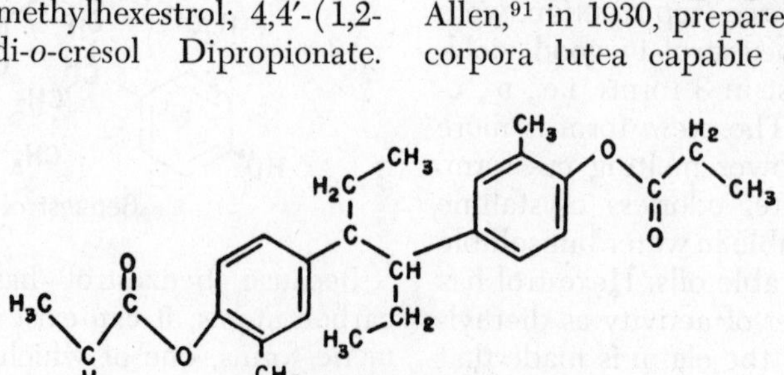

Promethestrol Dipropionate

placing the organ. Butenandt,[92] in 1934, isolated progesterone from the corpus luteum, and Beall[93] isolated it from the adrenal glands in 1938. It is the precursor to the adrenal cortex hormones.

Progesterone is thought to be the original progestational hormone produced by the corpus luteum. It also is produced by the adrenal glands and the placenta. In the event of pregnancy, progesterone is secreted throughout gestation, the first 3 months by the corpus luteum and the remainder of the time by the placenta.

The level of production of progesterone is not uniform, since it is a time cycle hormone. It has the following biologic functions and effects.

1. Proliferation of the endometrium (i.e., it stimulates the maturation of the uterine mucosa in order to prepare it for reception of a fertilized ovum)
2. Maintenance and protection of successful pregnancy by
    A. Preventing uterine motility
    B. Inhibiting ovulation (antiovulatory)
3. Suppression of secretion of pituitary gonadotropin
4. Antiestrogenic
5. Increase of carbonic anhydrase in the uterus
6. Maintenance of deciduomata in the estrone-primed histamine-treated uteri of spayed rats
7. Stimulation of gland formation in the breasts
8. Thermogenic (when administered by injection)
9. Induction of the copulatory reflex in estrogen-primed ovariectomized guinea pig
10. Suppression of ovulation
11. Antagonist of the salt-retaining properties of desoxycorticosterone
12. CNS depressant (sleep or anesthesia) (see also pregnanedione)

Because progesterone may exert both antiestrogenic and antiandrogenic activity it may protect the sensitive hypothalamic pituitary–gonad system of the developing fetus against possible damaging effects of increases in androgen and/or estrogen concentrations. The latter steroids, when present in excess quantities during the first few days of life, seriously inhibit sexual development and maturation.[94]

It was early recognized that the activities of progesterone should be useful for the treatment of a number of conditions. However, progesterone differed from other steroids in that (1) relatively large doses were required to produce a physiologic effect, (2) being a time cycle hormone, the where, the when, and the how of its delivery become a difficult problem in logistics, and (3) it has been a frustrating problem of how to administer a continuous supply which will maintain continuous optimal blood and tissue levels. Drug distribution also may complicate matters in some cases, e.g., where antiovulatory activity may be due to a local effect only and, therefore, the presence of the progestin in other body tissues may not be desired. Other factors that were encountered included the following:

1. Progesterone probably is a precursor to other steroids in vivo.
2. Progesterone (and other progestins) are synergized by estrogens.
3. If the tissue is not estrogen-primed, no amount of progesterone can be effective, and
4. Faulty cycle of progesterone production or metabolism can lead to numerous disorders such as amenorrhea, habitual abortion, premature delivery, etc.

The first breakthrough in the field of progestins came with the synthesis of 19-norprogesterone, which was followed by norethisterone and norethynodrel. These compounds enabled physiologists for the first time to reproduce those effects attributed to endogenous progesterone which they had never been able to duplicate with injections of progesterone. For example, a true progestin should be able to bring about indefinite postponement of ovulation and, therefore, of menstruation. Progesterone itself—at least, in the ways we know how to use it—fails to meet this requirement. The 19-nor progestins provided gynecologists with precise control of the menstrual cycle, permitting them to adjust irregular menses, stop excessive uterine bleeding and produce anovulatory cycles which often put a stop to dysmenorrhea, amenorrhea, endometriosis and infertility.

TABLE 86. ANTIOVULATORY ACTIVITY OF SOME STEROIDS, IN WOMEN

	MIN. EFF. OVULATION-INHIBITING DOSE		$\dfrac{SC}{O}$	ORAL/SC RATIO AS PROGESTINS
	S.C.	O		
17-Acetoxyprogesterone	0.05	5	0.01	0.1
6α-Methyl-17-acetoxy progesterone	0.05	0.4	0.13	0.2
19-Nor-17α-ethyl testosterone	0.2	2	0.10	0.6
19-Nor-17α-ethinyl testosterone	0.3	0.25	1.2	1.4
17α-Ethinyl-5(10)- estrenolone	5	0.2	25.0	5.4

S.C. = Subcutaneous. O = Oral.

Developments following the above-mentioned 19-nor steroids led to the synthesis of a large number of congeners, some of which offered good potential as useful progestins. The results of their bioassays showed that not all the progestins had an increase in all the biologic activities of progesterone. This enabled the development of progestins that could be used for specific purposes, e.g., antiovulatory, maintenance of pregnancy, etc. Some of the new progestins exhibited activities and effects which may or may not be desirable, e.g., estrogenic, anabolic, androgenic; adrenal atrophy, etc. It also is significant that it is difficult to compare potency in animals to activity in humans. For example, there may be a positive Clauberg assay for proliferation of the endometrium of the uteri of rabbits but nonmaintenance of pregnancy in humans or animals (q.v.); also, 17α-acetoxyprogesterone, subcutaneously, has powerful antiovulatory activity in humans.

Although there are no sure criteria for predicting the antiovulatory activity of progestins in women, Pincus points out that, in a limited series of steroids, oral subcutaneous ratios in the ovulation inhibition test in the rabbit appear thus far to be best correlated with ovulation-inhibiting activity in women. See data in Table 86, which show the androstane derivatives to be the most antiovulatory in women.[95, 96]

The antiovulatory activity of some of the progestins is tabulated in Table 87.

TABLE 87.

	ORAL[*]	S.C.[†]	S.C.[‡]
Progesterone	0.0017	0.2	1
Norethisterone	1	1	3.3-5
Norethynodrel	0.25		0.2
17α-Acetoxyprogesterone	0.01		7-20
6α-Methyl-17-acetoxy-progesterone	0.03		20
Δ⁶-Dehydro-17α-acetoxyprogesterone	3	2	2
6-Chloro-Δ⁶-dehydro-17α-acetoxyprogesterone	35	8	40
6-Methyl-Δ¹-dehydro-17α-acetoxyprogesterone	10		
19-Norprogesterone	0.02	30	150
19-Nor-17α-methyl-testosterone	1		
19-Nor-17α-ethyl-testosterone			5
3-Desoxy-19-nor-17α-ethyltestosterone	1		
3-Desoxy-19-nor-17α-ethinyltestosterone	3		
Δ3,6-Pregnadien-3β-,17α-21-trioltriacetate			63
3-Methoxy-17α-ethinylestradiol	0.3	1	

[*] Oral Standard, Norethisterone = 1.
[†] Subcutaneous Standard, Norethisterone = 1.
[‡] Standard, Progesterone = 1.

Data from Kincl, F., and Dorfman, R.: Acta endocrinol. 42 (Suppl. 73):3, 1963; Steroids, vol. 2, 521, 1963.

Although some of the progestins are more active subcutaneously than orally, in a few of them (e.g., 19-norprogesterone and 6-chloro-6-dehydro-17α-methylprogesterone) this is reversed. No adequate explanation has been offered to account for 17-α-ethinyl-5(10)-estrenolone's good oral activity versus its low subcutaneous activity and inactivity in the intrauterine McGinty assay. In general, there is poor correlation between the subcutaneous activity and the intrauterine (McGinty) activity.

For some progestational activity* only one oxygen function need be present as a 17β-methyl ketone or a 17β-OH with a 17 side chain (q.v.). The fact that progestational activity is present in some of the 17β-OH steroids is not yet explained, and this phenomenon represents one of the more significant apparently anomalous cases in the steroid hormones. In some cases even a 17α-side chain is not necessary, for testosterone, administered locally in the intrauterine Clauberg assay, is almost as potent as progesterone itself.

The minimum requirements for oral activity in those congeners of progesterone are those that are substituted in the 17α-position by

$$-O-\overset{\overset{\displaystyle O}{\|}}{C}-R \ (R = CH_3 \text{ to } C_6H_{13});$$

R (CH_3 to $C_{10}H_{21}$); Br; OR or those that contain a 6α-CH_3, Cl or F, and the contribution of those substituents is enhanced by a Δ^6 double bond. The function of the above substituents is to prevent metabolic reduction of the C_3 and C_{20} carbonyls and the Δ^4-double bond and metabolic hydroxylation at C_6. 19-Norprogesterone has marked oral activity. 3-Desoxyprogesterone has 1/12 the activity of progesterone by subcutaneous administration. However, 3-desoxy-17α-acetoxyprogesterone and 17α-acetoxyprogesterone each have 1/7 the activity of norethisterone by the buccal route. 17α-acetoxy-$\Delta^{3,\,5}$-pregnadien-20-one has the same activity as norethisterone. 3β-Fluoro-Δ^5-pregnene-17α-ol-20-one acetate orally is 3 times as active as 17α-acetoxyprogesterone. 3-Des-

* de Winter, M. S., *et al.*: Chem. and Ind. 905, 1959.

oxynorethisterone (Lynestrenol®) orally is highly progestational, being 1½ times as active as norethisterone.

In the derivatives of testosterone oral progestational activity is present in those steroids that have a 17α-alkyl or modified alkyl substituents such as CH_3, C_2H_5,

$$CH_2-C=CH_2, \ C\equiv CH$$
$$\overset{|}{CH_3}$$

and $-C\equiv C-Cl$. This oral activity may be enhanced by a 6α-Cl, F or CH_3, and their contribution may be increased by a Δ^6 double bond. This also is true of the 19-nor congeners. Note 19-norethisterone is 10 times more active orally than ethisterone. The C_3-enol ethers and esters contribute to oral activity.

Inhibition of ovulation[97, 98] in experimental animals may be effected by an active corpus luteum, progesterone or other progestins.

The complete mechanism or sequence of events by which the antifertility progestins inhibit ovulation is not unambiguously established. Orally or subcutaneously these steroids suppress the plasma levels of the gonadotrophin luteinizing hormone (LH) and 20α-hydroxypregn-4-ene-3-one (20α-OH). Some evidence[99] suggests that the progestins effect this suppression at the pituitary level, in that direct injection of chlormadinone into the pituitary reduced the levels of gonadotrophins (i.e., LH) in plasma. Other evidence[100] suggests that progestins act at the hypothalamic level or at a higher brain center to suppress the synthesis or release of luteinizing hormone releasing factor LRF and FSHRF (q.v.), which act in the pituitary. Time factors and the possibility of back tracking (steroid diffusion) from the site of injections might account for the above experimental results.

The activities of the antifertility steroids can be blocked or overcome by the administration of LRF or LH or electrical stimulation of the median eminence of the tuber cinereum of the hypothalamus. This may be the result of increased sympathetic activity. Brain-norepinephrine-depleting agents such as reserpine exert antiovulatory effects, probably by causing insufficiency in sympathetic nervous system activity leading to the hypothalamus-hypophyseal system. Some women on certain tranquilizers stop ovulating.

The sensitivity of the pituitary to the stimulating effect of the cyclically discharged hypothalamic gonadotrophin-releasing factor is increased by estrogen, and in this way an ovulation-inducing augmentation of gonadotrophin (LH) secretion is obtained.

The antiovulatory activity of a steroid is not necessarily related to its progestational activity or its antigonadotrophic activity as determined in a parabiosis assay. Nevertheless, many potent progestational steroids also have potent antiovulatory activity. 6-Chloro-Δ^6-dehydro-17α-acetoxyprogesterone has as much antiovulatory as progestational activity. It is equally active orally and subcutaneously, i.e., 10 mcg. produces 50 per cent inhibition of ovulation in rabbits. However, 40 mg. are needed for significant gonadotrophin inhibition in the parabiotic rat. Also, 6α-chloro-Δ^1-dehydro-17α-acetoxyprogesterone is antiovulatory at the 0.3 mg. level but did not inhibit the pituitary at 20 mg. It is of interest that 6 mcg. of 3-methoxyethinyl estradiol inhibited the pituitary gonadotrophin but 1 mg. was ineffective as an antiovulatory agent in rabbits. It has some antiovulatory activity in women. It may be required in small quantities (0.075 to 0.1 mg.) in order to obtain antiovulatory[95, 96] activity with progestins in women and in order to effect adequate menstrual cycle regulation without escape or breakthrough bleeding. The use of progestins with estrogens leads to a suppression of the levels of the gonadotrophins, i.e., LH and FSH, particularly their mid-cycle peaks. Combinations appear to act synergistically in their antiovulatory effects. Because some of the progestogens have antiestrogenic activities, proportionate doses of estrogens are needed to overcome the effects. Thus by the use of proper progestin/estrogen* combination for 20 days from day

5 of a cycle one obtains an ovulation-free cycle of approximately normal length, since the cessation of medication by the 24th day results in withdrawing bleeding after a latent period of one or more days. Superficially this artificial cycle is indistinguishable from a normal ovulatory cycle except for a tendency to a reduction in the quantity of menstrual effluent as well as in the number of days of duration of flow. With the prolonged use of 19-norethisterone, 17α-ethyl-19-norethisterone, 17α-ethyl-19-nortestosterone or 17α-ethinyl-estr-5,10-enolone in women there was no significant decline in 17-ketosteroid output but a marked reduction in pregnanediol excretion, indicating the reduction of progesterone secretion to preovulatory levels. After 7 to 9 days of treatment with these four progestins, a rapid stimulation of the endometrium† takes place, after which the response of the endometrial glands fails whereas the stroma cells continue to be stimulated, so that, by the last day of treatment, a predecidual state ordinarily is attained. There is no inhibition of the basic ovogenetic process by these progestins; only the formation of large follicles and corpora lutea is inhibited. Although the ova may develop to the brink of maturation independent of pituitary hormone secretion, growth and rupture of the follicles in which they reside are gonadotrophin dependent.

In Table 88 are listed the compositions of some of the antifertility (antiovulatory) preparations in use today.

Progesterone and its congeners are biologically assayed for a number of activities. These are: (1) proliferation of a secretory endometrium (progestational activity), by the Clauberg assay or its equivalent; (2) progestational activity, intrauterine, McGinty assay, (3) antiovulatory activity, (4)

---

* Although both steroid and synthetic estrogens such as diethylstilbestrol are effective, ethinylestradiol and its 3-methyl ether are preferred because of the low dosage needed and for best control

of menstrual bleeding (See Molecular Modification in Drug Design, Advances in Chemistry Series 45, Am. Chem. Soc. p. 201, 1964). The use of progestin or estrogen alone for antiovulatory activity leads to an irregular succession of menstrual cycles. Estrogens alone do not induce a desirable normal secretory endometrium.

† The condition of the endometrium is analogous to that found in pregnancy and, therefore, is frequently said to be in a pseudopregnant state.

TABLE 88. ANTIFERTILITY PREPARATIONS

PROPRIETARY NAME	COMBINATION			
	PROGESTIN	MG.	ESTROGEN	MCG.
Anovlar	Norethisterone acetate	4	Ethinyl estradiol	50
Enovid E	Norethynodrel	2.5	Mestranol	100
Enovid	Norethynodrel	5	Mestranol	75
Enovid	Norethynodrel	9.85	Mestranol	150
Demulen	Ethynodiol diacetate	1	Ethinyl estradiol	50
Gynovlar	Norethisterone acetate	3	Ethinyl estradiol	50
Lyndiol	Lynestrenol	5	Mestranol	150
Norinyl-1+80	Norethindrone	1	Mestranol	80
Norinyl-1	Norethindrone	1	Mestranol	50
Norinyl-2	Norethindrone	2	Mestranol	100
Norinyl-10 mg.	Norethindrone	10	Mestranol	60
Norlestrin	Norethisterone acetate	1	Ethinyl estradiol	50
Norlestrin	Norethisterone acetate	2.5	Ethinyl estradiol	50
Ortho-Novum	Norethindrone	2	Mestranol	100
Ortho-Novum-10	Norethindrone	10	Mestranol	60
Ortho-Novum 1/50	Norethindrone	1	Mestranol	50
Ortho-Novum 1/80	Norethindrone	1	Mestranol	80
Ovral	Norgestrol	0.5	Ethinyl estradiol	50
Ovulen	Ethynodiol diacetate	1	Mestranol	100
Volidan	Megestrol acetate	4	Ethinyl estradiol	50

SEQUENTIAL

Norquen	Mestranol 80 mcg. 14 tablets Mestranol 80 mcg. + Norethindrone 2 mg. 6 tablets
Oracon	Ethinyl estradiol 0.1 mg. 16 tablets Ethinyl estradiol 0.1 mg. + Dimethisterone 25 mg. 5 tablets
Ortho-Novum-SQ	Mestranol 80 mcg. 14 tablets Mestranol 80 mcg. + Norethindrone 2 mg. 6 tablets

gonadotrophin inhibition, (5) antiestrogenic activity, (6) uterine carbonic anhydrase activity, and (7) deciduomata test. The usual reference standards for the assays are progesterone, for subcutaneous only, and ethisterone and norethisterone for both oral and injection methods. The oral and the subcutaneous progestational activities of ethisterone are the same, i.e., about one tenth of progesterone's S.C. activity. This is true of norethisterone also; however, its activity is about equal to progesterone's S.C. activity.

Steroids that show progestational activity as measured by the Clauberg assay may not always maintain pregnancy in spayed ani-

TABLE 89. PROGESTATIONAL ACTIVITY OF PROGESTINS

PROGESTIN	REFERENCE STANDARDS					
	PROGESTERONE		ETHISTERONE		NORETHISTERONE	
	O	S.C.	O	S.C.	O	S.C.
Progesterone		1*				
Ethisterone		0.1	1*	1*		
Norethisterone		0.5-1	5-10	5-10	1*	1*
Norethynodrel					0.1	0.1
Dimethisterone			12			
Ethynodiol Diacetate						
3-Desoxynorethisterone (Lynestrenol)					1.5	
21-Fluoroprogesterone †		2				
17α-Hydroxyprogesterone Acetate‡		4-10	2-10			
17α-Hydroxyprogesterone Caproate		4-10				
Provera (orally)		12				
Medroxyprogesterone Acetate U.S.P. (Provera) (6α-Methyl-17α-acetoxyprogesterone)		50	12-25			
6α-Methyl-17α-acetoxy-21-fluoro-progesterone§		35-50				
Chlormadinone (Lormin) (6-Dehydro-6-chloro-17α-acetoxyprogesterone)		35-40				
Chlormadinone (orally)		35-40				
Megestrol (orally)		16				
Megestrol (6-Dehydro-6-methyl-17α-acetoxyprogesterone)		25				
6,16α-Dimethyl-6-dehydro-17α-acetoxy-progesterone			130			
19-Norprogesterone		5-10				
17α-Acetoxy-Δ³,⁵-pregnadien-20-one						1
3β-Fluoro-Δ⁵-pregnene-17α-ol-20-one						
(±)-13β-Ethyl-17α-ethinyl-17β-hydroxy-gon-4-en-3-one (Norgestrel)		9				30-80
Norgestrel (orally)		3				
6-Methyl-6-dehydro-17-methylprogesterone (Medrogesterone)		4				

* Reference Standards
　Ethisterone O=S.C.
　Norethisterone O=S.C.
† Cl, Br much less active

‡ Not clinically useful in women
§ Twice as active orally as 6α-methyl-17α-acetoxyprogesterone

mals (e.g., ethinylestrenol, dimethisterone, ethinylestrenolone, etc.).

Table 89 lists the progestational (Clauberg assay) activity of some progestins.

Products

Progesterone   N.F.,   Pregn-4-ene-3,20-dione, Progestin, Corpus Luteum Hormone,

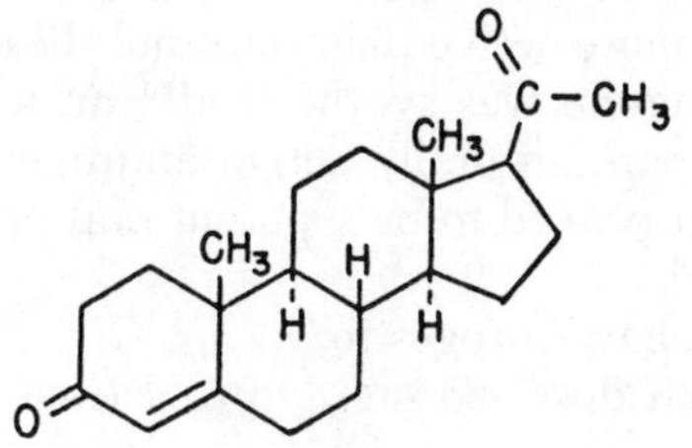

Progesterone

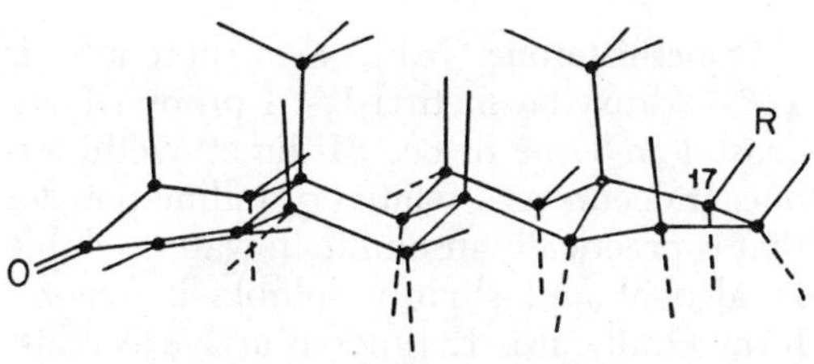

Progesterone R = CH₃ CO⁻

Oestrogenic Hormone, is Δ⁴-pregnen-3,20-dione. Progesterone occurs as a white, crystalline powder that is colorless and stable in air. It is practically insoluble in water but is soluble in alcohol and acetone and is sparingly soluble in vegetable oils.

Progesterone is sensitive to alkalies and to light and should be stored in tight, light-resistant containers. It can be sterilized by means of heat.

Progesterone is excreted chiefly as the reduction product pregnandiol [pregnan-diol-3α,20α]. It has been found in the pregnancy urine of man, cows, mares and monkeys.

Progesterone is prepared on a commercial scale from cholesterol, stigmasterol[101] and the sapogenins. Δ⁵-Pregnenol-3-one-20 is obtained as a by-product of the technical oxidation of cholesteryl acetate dibromide or in excellent yields from diosgenin. It is converted to progesterone by an Oppenauer oxidation as follows.

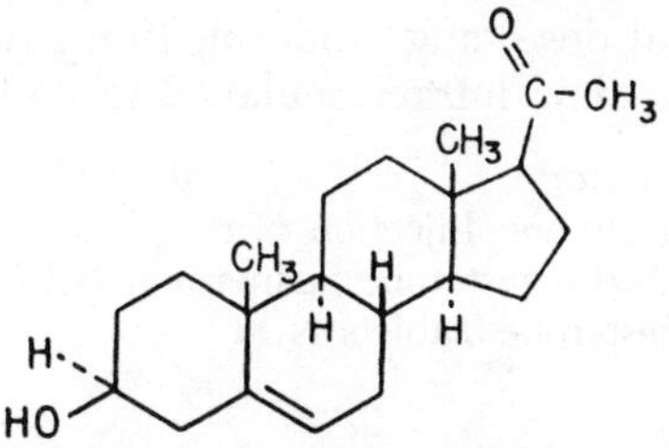

Δ⁵-Pregnenol-3-one-20

Al. tert. butoxid
& acetone

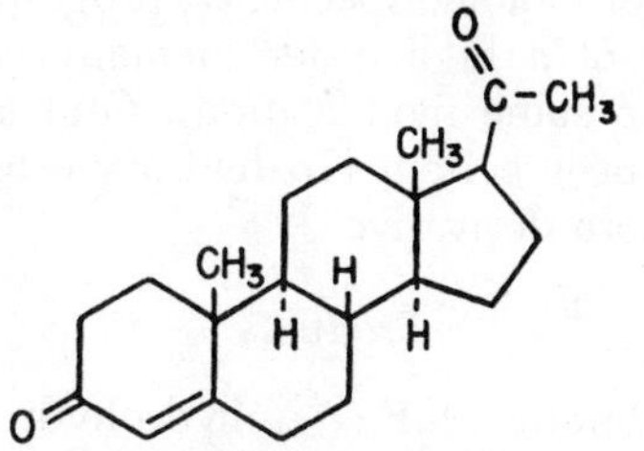

Progesterone

In the synthesis of testosterone from one of the sapogenins,[102] one of the intermediates, 4β-bromopregnanedione-3-20, is dehydrohalogenated to give progesterone (see the synthesis of testosterone).

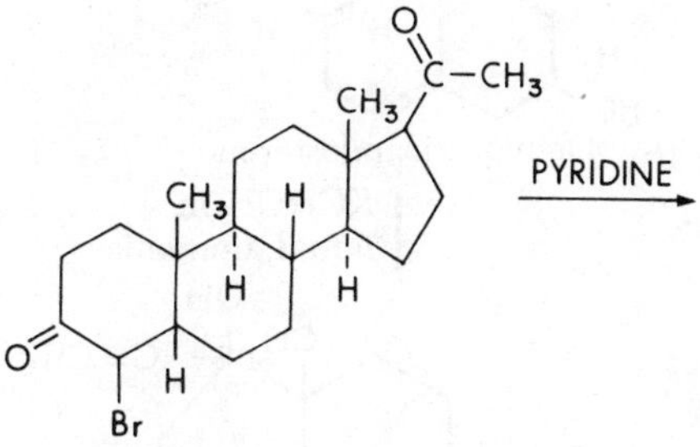

PYRIDINE

4β-Bromopregnanedione-3-20

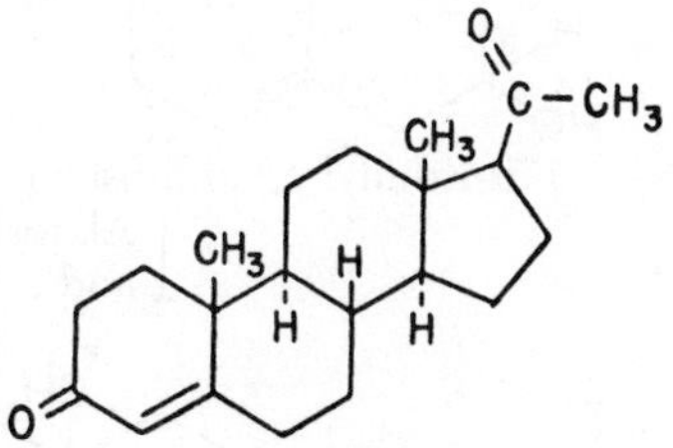

Progesterone

Progesterone is administered intramuscularly in order to obtain the maximum effect. Orally its activity is of a low order.

Category—progestin.

Usual dose range—buccal, 10 mg. up to 4 times a day; intramuscular, 2 to 25 mg.

OCCURRENCE

Progesterone Injection N.F.
Sterile Progesterone Suspension N.F.
Progesterone Tablets N.F.

MODIFIED PROGESTINS

Although progesterone exhibits a high degree of biologic specificity, progestational activity of a high order in many cases is found in some modifications. Oral activity is present in certain 17α-hydroxy esters and a 21-fluoro derivative.

Products

Ethisterone N.F.; Anhydrohydroxyprogesterone; 17-Hydroxy-17α-pregn-4-en-20-yn-3-one; 17-Ethinyltestosterone.

Ethisterone occurs as white or slightly yellow crystals or as a crystalline powder

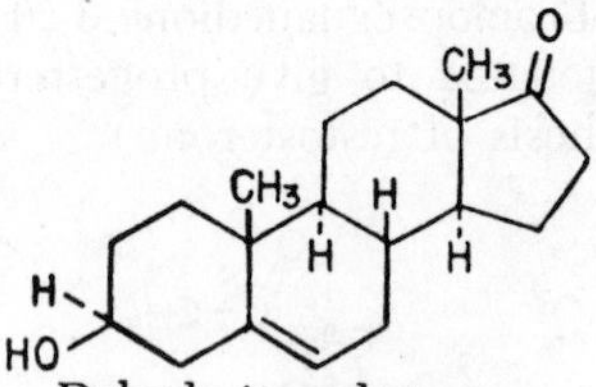

Dehydroisoandosterone

KC≡CH in
liquid ammonia

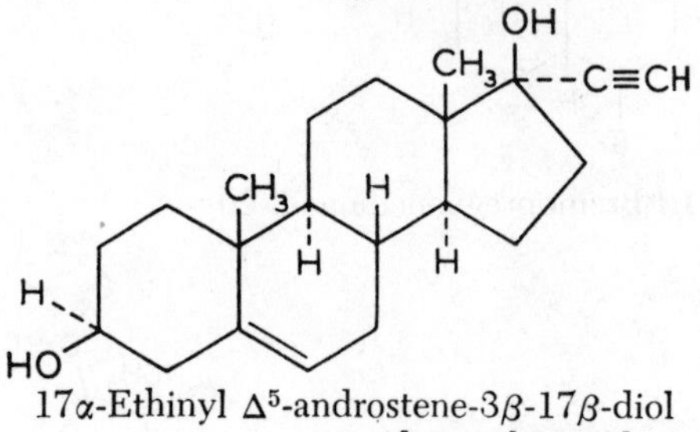

17α-Ethinyl Δ⁵-androstene-3β-17β-diol

Al. tert. butoxide
and acetone

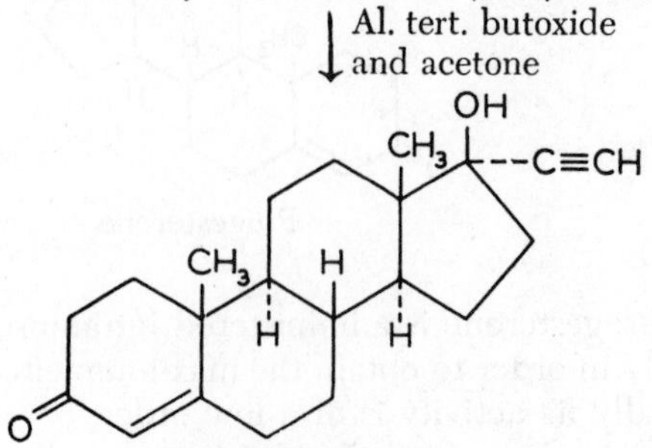

Anhydrohydroxyprogesterone

that is odorless, stable in air but is affected by light, particularly in the presence of alkali. It is practically insoluble in water and slightly soluble in alcohol and ether and in vegetable oils.

Ethisterone's progestational activity is the same both orally and subcutaneously and has about 1/10 the activity of progesterone administered subcutaneously.

After 17-ethinylestradiol was shown to be orally more active than estradiol, 17-ethinyltestosterone was synthesized[103] in an effort to prepare an orally active androgen; however, it proved to be a potent oral progestin instead.

Category—progestin.
Usual dose—25 mg., up to 4 times a day.

OCCURRENCE
Ethisterone Tablets N.F.

Dimethisterone N.F., Secrosterone®, is 17β-hydroxy-6α-methyl-17-(1-propynyl)androst-4-en-3-one or 6α, 21-dimethylethisterone. It occurs as a white crystalline powder that is practically insoluble in water, soluble in alcohol and slightly soluble in chloroform. Orally, it is 12 times as active as ethisterone in the Clauberg assay. It has no anabolic, androgenic, estrogenic or mineral corticoid effects. It is progestational in women and is antiovulatory only in relatively high dosages.

Category—progestin.
Usual dose—25 mg. daily for 5 days of each month.

OCCURRENCE
Ethinyl Estradiol and Dimethisterone Tablets N.F.

17α-Ethinylandrost-5-ene-3β,17β-diol, its 3-monoesters (but not the diesters), and related compounds possess outstanding activity as pituitary inhibitors. However, the free diols and their 3-monoacetates also had a high estrogenic activity because ring A is aromatized in vivo. The 3-cyclohexylpropionate of 17α-ethinylandrost-5-ene-3β,17β-diol (Ethandrostate) had a more favorable activity ratio (defined as the ratio of pituitary inhibition to estrogenicity).[104]

19-Norprogesterone is 6 to 10 times as active as progesterone and ½ as active orally as norethisterone in the Clauberg assay. Its antiovulatory activity[105] is 30 times that of progesterone S.C. and orally it is 150 times progesterone's S.C. activity. It is inactive in the parabiosis gonadotrophin assay. Subcutaneously it equals progesterone's antiestrogenic activity,[106] whereas orally it has twice this activity. A few mcg. per day absorbed from a subcutaneously implanted pellet in mice produced an effective antiluteinizing effect, as evidenced by absolute sterility that lasted 176 days, and, after removal of the pellet, litters were produced in 24 to 42 days. However, 18,19-dinorprogesterone is feebly active.[107]

The outstanding activity found in 19-norprogesterone also is present in its congener ($\pm$)-18-methyl-19-norprogesterone (19-norprogesterone that has an ethyl group at 13 in lieu of a methyl group), which is 10 times as active as progesterone in the Clauberg assay.[108]

Ethynodiol Diacetate U.S.P., 19-nor-17α-pregn-4-en-20-yne-3β,17-diol diacetate, 17α-ethinyl-4-estrene-3,17-diol diacetate, occurs as a white crystalline powder that is insoluble in water, soluble in alcohol and very soluble in chloroform. It is more active orally than by injection as an antiovulatory agent in rabbits and has high oral progestational activity. It has minimal side-effects, and 1 to 2 mg. plus 0.1 mg. of 17α-ethinyl estradiol methyl ether (EEME) is effective as an antiovulatory agent in women.[109] It is more effective than norethynodrel + EEME.

Category—progestin.

Usual dose—1 mg. once daily.

Norethindrone U.S.P., Norethisterone, Norlutin®,19-nor-17α-ethinyltestosterone,[110] has both oral and subcutaneous progestational activity 5 to 10 times that of ethisterone, and suppression of ovulation activity orally is 4 times that of 17α-ethinyl-5(10)-estrenolone. It has powerful antiestrogenic activity orally or subcutaneously in rabbits (q.v.) and does not maintain pregnancy. It has some estrogenic and androgenic activities.[111] However, highly purified norethindrone has no estrogenic activity.[112] Clinically it is a potent progestin. It is recommended to control menstrual disorders.

Category—progestin.

Usual dose—1 to 10 mg. once daily.

Usual dose range—1 to 40 mg. daily.

OCCURRENCE
Norethindrone Tablets N.F.

Norethindrone with Mestranol, Ortho-Novum®, is available as tablets containing 2 mg. of norethindrone with 100 mcg. of mestranol or 10 mg. of norethindrone with 60 mcg. of mestranol. Mestranol, ethinylestradiol-3-methyl ether, is the estrogen component and is utilized with the antiovulatory progestins to effect adequate menstrual cycle regulations without escape or breakthrough bleeding. The 10-mg. tablet may exert recognizable anabolic and androgenic activities.

Norethindrone Acetate N.F., Norlutate®, 17-Hydroxy-19-nor-17α-pregn-4-en-20-yn-3-one Acetate. Norethindrone acetate occurs as a white, crystalline powder, soluble in ether and in alcohol, very soluble in chloroform and practically insoluble in water. It is about twice as potent as the nonesterified norethindrone; otherwise, its actions are the same as those of norethindrone. It is also used in combination with the estrogen, ethinyl estradiol, as a pregnancy test (Gestest), and for ovulation control (Norinyl, Ortho-Novum).

Category—progestin.

Usual dose—10 mg. daily.

Usual dose range—5 to 15 mg.

OCCURRENCE
Norethindrone Acetate Tablets N.F.

Norethynodrel, U.S.P., 17-hydroxy-19-nor-17α-pregn-5(10)-en-20-yn-3-one, is an isomer of norethindrone. The double bond in the 5(10) position is obtained directly when for example estradiol is subjected to a Birch reduction. Its progestational activity is one tenth that of norethisterone orally or subcutaneously, and its antiovulatory activity orally and subcutaneously is about one fourth that of norethisterone (q.v.). It has some estrogenic activity,[113] 5 to 7 per cent that of estrone. At high dosage levels it caused masculinization of the offspring in rats.[114] Norethynodrel does not interfere with follicular development, but, apparently, it prevents the release of sufficient ovulating hormone (LH) to induce ovulation in hamsters and rats.[115]

Category—progestin.

Usual dose—2.5 to 10 mg. once daily.

Usual dose range—2.5 to 30 mg. daily.

Norethynodrel with Mestranol, Enovid®, is available as tablets containing 5 mg. of norethynodrel and 75 mcg. of mestranol or 2.5 mg. of norethynodrel and 100 mcg. of mestranol.

It is used primarily to control fertility (q.v.).

Norvinodrel, 17α-vinyl-5-(10)-estrene-17 β-ol-3-one, when compared to pure norethynodrel, was found to have qualitatively the same biologic activities. It had higher progestational activity (Clauberg assay) and high gonadotrophin inhibitory activity.[116]

Norgestrol, (±)-13β-ethyl-17α-ethinyl-17β-hydroxygon-4-en-3-one, is a homolog of norethisterone. Although it previously had been established that the C-18 methyl was essential, a change in the size of the methyl group at this position gave a remarkable increase in activity, with an ethyl group being the most outstanding.[117] Norgestrol is about 9 and 80 times the activity of progesterone and norethisterone respectively in the Clauberg assay by the subcutaneous route. Orally it is ½ to ⅛ as active as by the subcutaneous route. It is about 3 times more potent than progesterone by the intravenous route. It has 10 times the antiestrogenic activity of norethisterone. It has endometrial proliferating properties and supports pregnancy in spayed female rats. Norgestrol, like progesterone, lacks antifertility effects in rats unless estrogens are added.

Norgestrol, 500mcg., is used together with 50 mcg. of ethinyl estradiol for antifertility effects that act primarily through the mechanism of gonadotrophin suppression to inhibit ovulation. This activity may be supplemented by a possible second mechanism such as the development of a cervical mucus hostile to sperm penetration and migration.

Activity resides primarily in the enantiomorph corresponding in absolute configuration to naturally occurring steroids.

The major metabolite has been identified as optically pure 13β-ethyl-17α-ethinyl-5β-gonan-3α,17β-diol.

Esters of 17α-hydroxyprogesterone are somewhat more active than progesterone.[118]

Hydroxyprogesterone Caproate U.S.P., Delalutin®, 17-hydroxypregn-4-ene-3,20-dione hexanoate, 17α-hydroxyprogesterone hexanoate, is available as an oil solution to be administered by intramuscular injection.

Subcutaneously it has 4 to 10 times the progestational activity of progesterone or norethisterone and orally 0.2 to 1 times the activity of norethisterone. It could be active per se because the unesterified precursor is nonprogestational in humans, rats and rabbits. Estrogens are needed to maintain pregnancy (1:3000).

Category—progestin.

Usual dose—cyclic therapy: intramuscular, 250 mg. every 4 weeks; continuous therapy, 250 mg. once a week.

Usual dose range—125 mg. to 1 g. monthly.

OCCURRENCE

Hydroxyprogesterone Caproate Injection U.S.P.

A 6α-(equatorial) methyl group[119] probably increases activity and duration of action by protecting the Δ⁴-double bond and the 3-keto group from metabolic attack (reduction) and by protecting from metabolic attack (hydroxylation) at position 6. By combining this favorable change with a favorable 17α-hydroxy ester change a remarkable increase in activity[120] was obtained. This synergism is not always the case, as in the 11-dehydro-17α-methyl progesterone.[121, 122]

Medroxyprogesterone Acetate U.S.P., Provera®, 17-hydroxy-6α-methylpregn-4-ene-3,20-dione 17-acetate, 6α-methyl-17α-acetoxyprogesterone, is an off-white, odorless, crystalline powder that is stable in air. It is insoluble in water, freely soluble in chloroform and sparingly soluble in alcohol and in methanol.

Medroxyprogesterone acetate orally is 12 to 25 times as active as ethisterone and subcutaneously 50 times as active as progesterone in the Clauberg assay, with no androgenic and low estrogenic activities and some corticoid activity.

Category—progestin.

Usual dose—oral, 10 mg. daily; intramuscular, 50 mg. weekly.

Usual dose range—cyclic therapy: oral, 2.5

to 10 mg. daily for 5 to 10 days during second half of cycle; continuous therapy: oral, 10 to 40 mg. daily; intramuscular, 50 to 100 mg. weekly or every 2 weeks.

OCCURRENCE

Sterile Medroxyprogesterone Acetate Suspension U.S.P.

Medroxyprogesterone Acetate Tablets U.S.P.

A 6α-(equatorial) chloro or fluoro atom probably functions similar to a 6α-methyl group. It also can increase lipid solubility, and it should be noted that it is alpha to the 3 keto Δ⁴-double bond system where it can exert its electron-attracting effects. Therefore, advantage, especially in the case of chlorine, has been taken of the size and the lipophilic, electronegative, conformational and enzyme-blocking effects of this atom to increase biologic (progestational) activities.

Chlormadinone Acetate N.F., 6-chloro-17-hydroxypregna-4,6-diene-3,20-dione acetate, 6-dehydro-6-chloro-17α-acetoxyprogesterone. It occurs as an off-white to pale yellow, fluffy, crystalline powder that is practically insoluble in water and is soluble in alcohol, in chloroform and in ether.

This progestin represents a case in which several favorable changes are cumulative. It was used in combination with mestranol for ovulation control (C-Quens). Chlormadinone acetate is orally and subcutaneously about 35 to 40 times as active as progesterone S.C. in the Clauberg assay. Antiovulatory activity[124] is orally 35 times and subcutaneously 8 times that of norethisterone. Ten micrograms exert significant antiovulatory effects, whereas 40 mg. are required for significant gonadotropic inhibition. It has ⅓ the oral antiestrogenic activity[125] of norethisterone and has no estrogenic, androgenic or thermogenic effects. In humans it has high pure progestational activity (1.5 to 2 mg. daily), and must be supplemented with estrogens for antifertility use and to produce painless bleeding.

Although chlormadinone acetate can inhibit ovulation in combined contraceptive regimens, extremely low doses have been used for human contraception which do not achieve their effect by suppression of ovulation in the majority of women. The antifertility effects possibly may be due to the inhibition of implantation which also may be an activity in part of other 17-acetoxy progestins.

Category—progestin.

OCCURRENCE

Chlormadinone Acetate and Mestranol Tablets N.F.

The C₆-methyl analog, 17α-acetyl-6-dehydro-6-methylprogesterone (B.D.H. 1298), Megestrol® Acetate, though less active, has comparable properties and 2.5 mg. + 0.1 mg. of mestranol daily from day 5 of the cycle exerts antiovulatory activity in women.[126] It is a tasteless, odorless white crystalline compound that is insoluble in water, soluble 1:55 in alcohol and soluble in chloroform. It has 100 to 200 times the progestational activity of dimethisterone, i.e., 0.005 mg./g., one half the antiestrogenic activity of progesterone, 10 times the antiovulatory activity of 19-norethynodrel and no virilizing effects upon the developing fetus.

Megestrol acetate implanted in a dimethylpolysiloxane(DPS) membrane exerts an activity 6 to 25 times that when administered by the usual routes. DPS membranes permit a uniform predictable diffusion rate of steroid hormones.

6-Dehydro-16-methylene-17α-acetoxyprogesterone (DMAP) orally was 13 times as active[127] as progesterone (Clauberg assay) subcutaneously. Clinical oral studies with humans show a good parallelism of activity with the results of animal studies.

Orally in the Clauberg assay 6,16α-dimethyl-6-dehydro-17α-acetoxyprogesterone[128] was 130 times as active as ethisterone, with no virilizing effects on the fetus. C₂₁-Fluorination reduced this activity to 100 times that of ethisterone. However, diminishing returns were obtained in the 1,6-bis-dehydro-6-chloro-17α-acetoxyprogesterone. In this series chlorine is more effective than fluorine.[129]

As a second example one can cite the case of 6α-methyl-17α-acetoxyprogesterone and 6α-methyl-17α-acetoxy-21-fluoroprogesterone, both of which subcutaneously are 50 times as active as progesterone; however, the latter has twice the oral activity of the former.[130]

The 21-fluoro atom in 21-fluoroprogesterone among other contributions no doubt prevents hydroxylation of the C_{21}—CH_3 group. It has 2 times (subcutaneously) and 5 times (orally) the activity of progesterone. This contribution decreases progressively from chlorine to iodine (inactive), and oral activity is present only when the halogen is fluorine.[130] 21-Fluoroprogesterone produced an inhibition of the uterotropic or vaginal keratinizing action of estrone, equal to or greater than that of 9α-fluoro-11β-hydroxyprogesterone.

The acetophenone (Deladroxone®, Dihydroxyprogesterone Acetophenide) and the 2-acetofuran derivatives of 16α,17α-dihydroxyprogesterone were about 5 times as active as progesterone in preventing pregnancy in approximately 100 per cent of the mice tested.[131] They have a prolonged duration of activity when parenterally administered and they are devoid of estrogenic, androgenic or glucocorticoid activities. Deladroxone is equal to norethisterone in promoting secretory endometrium and in maintaining pregnancy in the ovariectomized rat and does not cause masculinization of the female rat fetus. Parenterally it has a long duration of activity. Deladroxate® a combination of Deladroxone (100 mg.) and estradiol enanthate (10 or 20 mg.), a long-acting injectable estrogen, have been used by injection to control ovulation.[132]

The C_3-cyclopentyl enol ether of progesterone orally appears to be 10 times as active as progesterone in the Clauberg test without eliciting the narcotic effect of progesterone.[133]

3β,17α - Diacetoxy - 6 - chloropregna - 4, 6-diene-20-one by the subcutaneous route of administration was 20 times as active as progesterone and about equal to chlormadinone in the Clauberg assay. It maintains pregnancy in ovariectomized rabbits, it has no anabolic or androgenic activity, it does not suppress gonadotrophic activity in the parabiotic rat, it has 1/50 the antiovulatory activity of chlormadinone acetate but does not maintain pregnancy in ovariectomized rats or induce uterine deciduoma formation in rats. The differences in activity in rabbits and rats may be due to different metabolic patterns.[134]

STEROIDS OF UNNATURAL CONFIGURATION

Although the 17 α- and 14 β, 17 α-progesterones[135] were inactive, their corresponding 19-nor congeners[136] were 8 times as active as progesterone in the subcutaneous Clauberg assay. The 8α-progesterone was about one half as active as progesterone and the 9β,10α-isomer was 5 times as active as progesterone.[137] The 6-dehydro-9β,10α-progesterone (q.v.) is used as a progestational agent. It is obvious that the isomeric changes that alter the conformational characteristics of progesterone, with retention and increase of activity, differ from those in the case of 19-norprogesterone. The total over-all molecular architecture in both cases leads to good activity as compared to progesterone. Whether such increases in activity are due to greater interaction at the receptor site, lowered rate of metabolism, or other factors has not yet been established.

One might postulate that the C-19 angular methyl group in α-position permits a greater interaction of the C_3-keto group with its receptor site. The same effect could be obtained with the 19-norprogestins.

Dydrogesterone N.F., Gynorest®, Duphaston®, 9β,10α-pregna-4,6-diene-3,20-dione, 6-dehydro-9β,10α-progesterone, markedly differs in its structure and conformational aspects as compared to progesterone; nevertheless, it has progestational activity: 0.2 mg. S.C. and 5 mg. orally[138] produced a maximum secretory proliferation of the rabbit endometrium. Orally it is 0.4 times as active as is progesterone subcutaneously. Subcutaneously it maintains pregnancy in spayed rats, whereas orally, added estrogens also are necessary. It has no androgenic, antiovulatory, estrogenic or thermogenic effects, no masculinizing effects on the female fetus and is not a precursor to other steroids.

Category—progestin.

Usual dose—10 to 20 mg. daily in divided doses.

Usual dose range—10 to 30 mg.

OCCURRENCE

Dydrogesterone Tablets N.F.

Certain progestins also exhibit species differences in activity; for example, ethinyl testosterone is ineffective in monkeys (200 mg. daily for 3 to 5 kg.), whereas women responded quite regularly to 60 to 80 mg. for 50 to 60 kg. On the other hand, ethinyl-19-nortestosterone gave quantitative reproducible effects in rabbits, monkeys and women. As a second example, 17α-hydroxyprogesterone caproate that has very little antiestrogenic activity will not maintain pregnancy in spayed rats but will in castrated rabbits.[139]

Probably in-vivo biologic transformations account for the observations that 17α-ethyl-19-nortestosterone, 17α-methyl-19-nortestosterone and 17α-ethinyl-$\Delta^{5(10)}$-estren-17-ol-3-one have marked progestational activity in rabbits and humans, and that the first two steroids also have androgenic activity and the latter has estrogenic activity in man.

Progesterone, some of its congeners and some other steroids have antiestrogenic activity (see Table 90).

A-Nor Progestins

Although the A-norsteroid hormones are inactive, 6-chloro-6-dehydro-17α-acetoxy-A-norprogesterone orally and subcutaneously is 4 times as active as progesterone in the Clauberg assay. This may also be a case where certain favorable perimeter groups overcome a dystherapeutic modification, i.e., the A-nor alteration.

17α-Acetoxy-6-dimethylaminoethyl-21-fluoro-3-ethoxypregna-3,5-dien-20-one hydrochloride is more analgesic than codeine or meperidine but less than morphine.[140] Its activity is not antagonized by nalorphine. It gives a positive Straub-tail effect in mice, and, therefore, it could have addicting properties.[140]

Certain 5β-pregnanes exhibit significant anesthetic activity. Of a series of congeners prepared, epipregnanolone (3α-hydroxy-5β-pregnan-20-one) was the most potent—ED_{50} in mice was 2.5 mg./kg. Sodium hydroxydione succinate, Viadril®, sodium 21-hydroxypregnane-3,20-dione succinate, has been used for its CNS effects, which induce hypnosis and general anesthesia with little analgesia.[141] This drug has now been withdrawn.

The sedative or mild tranquilizing activity of 11β-hydroxy-11α-methyl-5β-preg-

TABLE 90. Antiestrogenic Steroids*

	S.C.	†	Oral	†
Progesterone	250	36		
2α,17α-Dimethyl-17β-hydroxy-5α-androstan-3-one	2	30		
6α,16α-Dimethyl progesterone	5	60		
17α-Ethyl-19-nortestosterone	8	54	40	43
6α-Fluoroprogesterone	10	34	500	25
17α-Ethinyl-19-nortestosterone	16	56	32	40
17α-Acetoxyprogesterone	25	39	0	
17α-Methyl-19-nortestosterone	32	46	40	44
17α-Methyltestosterone	32	35	250	
17α-Acetoxyprogesterone			50	21
6-Chloro-Δ^6-dehydro-17α-acetoxyprogesterone	50	40	100	40
6α-Methyl-17α-acetoxyprogesterone	150	30	500	24
3-Deoxy-17α-Acetoxyprogesterone	0			
Estradiol-17α	125	22		

* Data from Dorfman, R., and Kincl, F.: Steroids 1:207, 1963.
† Per cent inhibition of estrogen administered.
Figures are the minimum dose in mcg.

nane-3,20-dione in humans was about equivalent to that of an equal dose of meprobamate.[142]

Androgens

In the case of the male, there is one gonadotrophic hormone which functions in stimulating the development of the testes. Under this stimulation, the interstitial cells, lying outside of the seminiferous tubules and not concerned with spermatogenesis, continuously produce very small quantities of the male sex hormone, testosterone, which is absorbed and circulated by the bloodstream throughout the body. The term androgenic hormones is applied to those substances whose activities are comparable with those of testosterone.

Androgens and related steroid production are stimulated by the administration of ACTH, whereas their production by the adrenal and perhaps the gonads may be inhibited by amphenone (3,3-*bis*(*p*-aminophenyl)-2-butanone).

Androgenic substances also occur in the urine of human males and females and are found in the urine of horses, both male and female.

Testosterone and its congeners are assayed biologically for their androgenic (AG), myotropic (MT) and anabolic (AB) activities. The androgenic activity is usually determined by the stimulation of growth of the ventral prostate (VP) or seminal vesicles (SV) of castrated male rodents and, less often, by the promotion of comb growth in capons. The ventral prostate assay may give a more sensitive indication of androgenic potency at dose levels roughly ⅔ those required for the seminal vesicle response.[143] Sometimes the SV and the VP values are averaged. It should be noted that the degree of androgenic activity may not be the same in castrated rats and by the chick comb growth in capons, when compared to a suitable reference standard.

The myotropic activity, which is independent of the androgenic activity, is a measure of the growth-promoting properties of the levator ani (LA) muscle in castrated young rats.[144] Even though the levator ani muscle is found only in male rats and

certain steroids stimulate its growth more than they do that of skeletal muscle, it is not considered by some to be a male sex characteristic because (1) the uterus responds in the same way as the levator ani, (2) certain androgens, chiefly anabolic, stimulate the growth of the levator ani to a greater extent than they do that of the seminal vesicles or the ventral prostate while, with others, the degree of stimulation is opposite, chiefly androgenic, and (3) clinical results with certain steroids gave preferential anabolic activity, in conformity with the results of bioassays in rats.[145]

Anabolic activity is based on nitrogen-retention[146] (or excretion) properties or the change (increase) in body weight. Anabolic androgenic ratios based on nitrogen retention properties may yield estimates clearly different from those derived from an increase in weight in the levator ani muscle assay. The latter assay is based on a particular muscle response (weight increase) of the perineal complex rather than to the responses of the over-all muscle mass. Thus AB/AG is greater than MT/AG for some pyrazole and isoxazole conjugated steroids (q.v.) and for norethandrolone. In some cases AB/AG is similar to MT/AG, as in the case of methandrostenolone and oxymetholone. Still others show M.T. activity but have no AB activity in rats and exhibit estrogenic properties.[147] Lastly, two androgens may have about equal LA values but their VP and SV values may differ. Thus, using testosterone as a standard, 17α-methyldihydrotestosterone had LA=107, VP=254 and SV=78, compared to 97, 10 and 29 respectively for 2-formyl-17α-methyl-5α-androst-2-en-17β-ol.[148]

Because androgenic values are obtained in several ways some variables are also introduced, so that the AG value used in AB/AG and MT/AG should be clearly defined.

Differences in assay procedures used by various investigators account for the marked variability of the estimations of the myotropic activity of a given steroid.

Some androgens that appear to have greater anabolic than androgenic activity in rats are potent androgens for man (e.g., 19-nortestosterone, 17α-ethyl-19-nortestosterone and methylandrostenediol). Some anabolic steroids that are highly active androgens for rodents are practically inactive in monkeys and man. These cases of species difference may be due to major differences in species metabolism of steroids.

Although it appears that anabolic and androgenic activities are separable by certain molecular manipulations, all of the clinically available anabolic steroids have varying degrees of androgenic activity.[149]

The usual reference standards are testosterone, testosterone propionate or testosterone acetate for subcutaneous assays only and methyltestosterone for both oral and subcutaneous assays. Testosterone and methyltestosterone administered by injection had about equal potency for VP, SV and LA activities.[150] Testosterone propionate was about 1.7 times as active as testosterone for its VP or SV activities[151] subcutaneously. Methyltestosterone orally is 0.4 times as active as testosterone subcutaneously. There can be significant differences in the LA/VP of AB(N)/VP values, whether the steroid is administered orally or parenterally.[152] On a w/w dose basis testosterone propionate subcutaneously is about 4 times more androgenic than myotropic[153] at the dose needed to produce a minimum response in each type test. Testosterone was about 0.6 as anabolic as it was androgenic[154] and slightly more myotropic (LA) than androgenic (VP).

The basic functions[155] of the male sex hormone, i.e., testosterone, are to induce normal development of the male reproductive tract and to maintain the secondary male sex characteristics and behavior patterns. It plays a role in the development of penis, seminal vesicles and prostate gland and in the descent of the testes. The accessory characteristics affected are depth of voice, distribution of facial and body hair and male type of skeletal muscular development.

The anabolic or nitrogen retaining properties of testosterone are associated with the synthesis of protein in muscle and, probably, other tissues in the body. Thus, body weight, muscular strength and endurance are measurably increased. Testosterone confers a sense of well-being, favors mental equilibrium and measurably increases the resistance of the central nervous system to fatigue.

Testosterone and some of its congeners exert a favorable effect[156] on the calcium balance (i.e., Ca retention), with some retention of potassium in greater amounts than sodium.

By suitable structural modifications analogs of testosterone have been made in which varying degrees of separation of androgenic from anabolic or myotropic activities have been effected. These activities in some cases quantitatively exceed those of testosterone. Thus, steroids are available for use primarily to promote rapid protein anabolism, such as in patients recovering from chronic undernutrition, the aging male, etc. Others can be used for the treatment of sexual infantilism in the male, male climacteric, etc. The androgens also can be used as antagonists for estrogens in females for the treatment of certain kinds of cancer and for the suppression of lactation, breast engorgement and ovarian dysfunctions such as dysmenorrhea, menorrhagia, etc.

Because the androgenic activity has not been completely separated from the myotropic (anabolic) activity caution should be exercised in the use of these steroids in pregnant women and growing children.* In the latter the less desirable androgenic activity will accelerate bone maturation to a rate more rapid than that of linear growth, thus inducing premature closure of epiphyses, which results in a reduction in ultimate adult height. Some anabolic androgens have found considerable clinical use in the treatment of osteogenesis imperfecta and of osteoporotic conditions that respond to the calcium and nitrogen retention activities of these steroids.[157]

Table 91 is a recent evaluation of four parameters (biological) of a number of androgens. Complete separation of anabolic from androgenic activity has not been achieved; however, some have high anabolic and low androgenic activities. Also, a separation between anabolic and antigonadotropic activities has not been obtained.

Androsterone (3α-hydroxy-5α-androstan-17-one) administered subcutaneously has

* Stimulation of growth with anabolic steroids is an important clinical use of these agents.

TABLE 91. MINIMUM TESTED DOSE OF VARIOUS ANDROGENS WITH SIGNIFICANT EFFECT
ON FOUR TISSUES

	TESTES*	SEMINAL VESICLES†	VENTRAL PROSTRATE†	LEVATOR ANI†
	(miocrograms per rat per day, subcutaneously)			
Oxymetholone	100	400	1000	100
Oxandrolone	100	400*	1000*	200
Methandrostenolone	100	1000	1000	200
Nandrolone phenpropionate	≤ 40‡	100*	200*	≤ 40
Norbolethone	≤ 40	100*	200*	≤ 40
Ethylestrenol	100	400*	1000*	100
Bolasterone	≤ 40	100	100	≤ 40
Norethandrolone	≤ 40‡	200*	400*	≤ 40
Oxymesterone	≤ 40	200	400	≤ 40
Methenolone acetate	≤ 40‡	100	200*	≤ 40
Chlorotestosterone acetate	100	400	2000*	100
Testosterone	≤ 40‡	≤ 40‡	≤ 40	≤ 40
Stanozolol	≤ 40	200*	400	100

* Weight decrease
† Weight increase
‡ Indicates biphasic effects on organ within dosage range studied

hypocholesterolemic properties and reduces the level of serum triglycerides. It is noteworthy that serum levels of androsterone in the premenopausal female are as high as levels in the male.

3α-Methoxy-17α-methyl-5α-androstan-17-ol has an oral hypocholesterolemic activity[158] similar to that of parenterally administered androsterone, with androgenic activity (weak) of the same order as androsterone.

Androgens such as testosterone and nandrolone administered to starved female rats caused a marked increase[159] in plasma unesterified fatty acid concentrations, similar to the effects of the growth hormone.

Products

Testosterone N.F., 17β-Hydroxy-4-androsten-3-one, Androlin®, Testrone®, occurs in such small amounts in testes that Laqueur,[160] in 1935, first isolated it by working with large quantities of bull testes. About 90 to 270 mg. are present in a ton of testes.[161] Therefore, this is not a practical source of supply. In fact, there is no natural source rich enough in testosterone to be of preparative value. However, testosterone is synthesized in commercial quantities from the androgenic excretory products found in

stallion urine, cholesterol and some of the sapogenins, i.e., diosgenin and sarsasapogenin as shown on page 798.[162]

Testosterone is a white, odorless and tasteless, crystalline powder. It is insoluble in water but soluble in alcohol, vegetable oils and organic solvents. It is stable in air under normal storage conditions. It is sensitive to alkali.

Category—androgen.

Usual dose—buccal, 10 mg. daily; implantation, 300 mg.; intramuscular, 25 mg. daily or 2 times a week depending on condition being treated.

OCCURRENCE
Testosterone Pellets N.F.
Sterile Testosterone Suspension N.F.

Testosterone is inactive orally and must be administered intramuscularly. In the latter case, it is absorbed quite rapidly. It is thought to be inactivated in the intestinal tract by the bacteria that are present; they possibly convert it to feebly active compounds.

Testosterone is converted in the body and excreted in the urine as androsterone, dehydroisoandrosterone and isoandrosterone. Androsterone is 1/7 to 1/10 as active as testosterone, and dehydroisoandrosterone

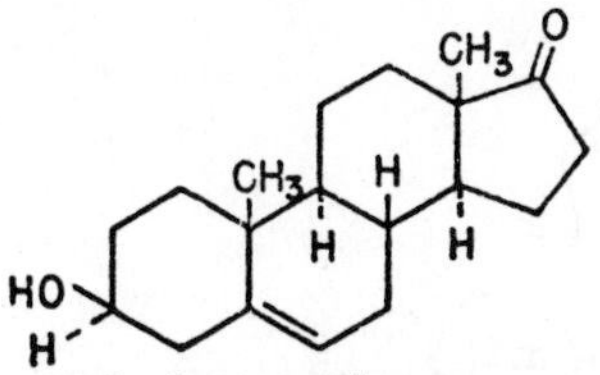

Dehydroisoandrosterone

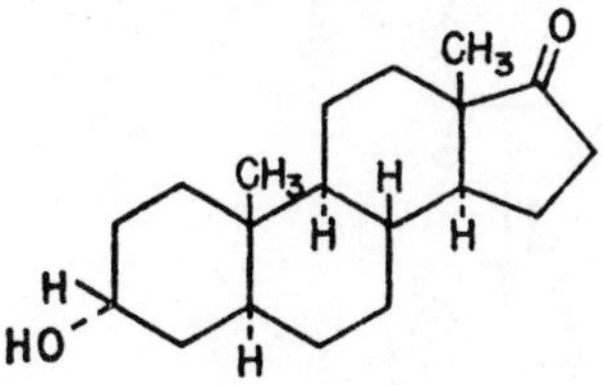

Androsterone

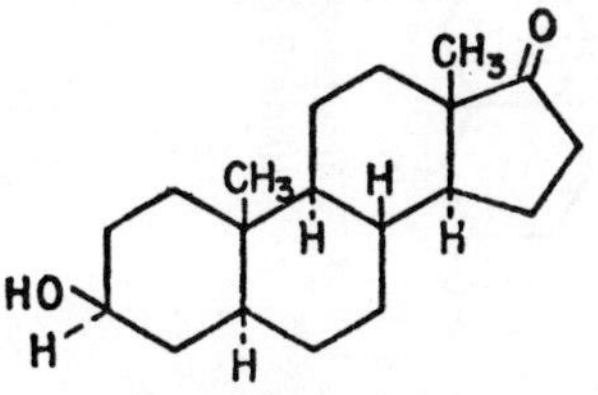

Isoandrosterone

and isoandrosterone are 1/20 as active as testosterone. It is interesting to note the marked difference in the physiologic activity of these two compounds that differ only in the spatial configuration of the hydrogen and hydroxyl group at C_3.

For the relationships in structure to activity in this series which have been established[163] see Table 92.

From a study of Table 92 one can conclude that for maximum androgenic activity, in the saturated compounds, rings A and B must be *trans*, an OH at C_{17} must be β, and the oxygen function at C_3 may be ketonic or an α OH group. A Δ^4 or Δ^5 double bond is not significant. However, it should be noted that 17α-methyl-Δ^4-androstene-$3\beta,17\beta$-diol is as androgenic as methyltestosterone, and the former is probably converted into the latter in vivo.

The double bond originally present in testosterone apparently contributes very little to the activity.

As in the case of other steroids, it is possible to determine the effect that a single change in the structure or derivative will have on biologic activity, and in general the following can be observed.

The minimum structural requirements for maximum androgenic and anabolic activities are found in 5α-androstane-17β-ol or 17α-methyl-5α-androstane-17β-ol.

For high anabolic activity a high electron density at C_2 and/or C_3 is necessary. This may be effected by a double bond at C_1, C_2 or C_3; a C_3 keto, a $C_3\alpha$-OH, a C_3 keto as an enol or enolate anions where a $\Delta^2\pi$ bond is present. The introduction of more than one Sp^2 hybridized carbon atom into ring A results in a pronounced flattening of ring A from a chair to a more planar conformation. This may be significant for anabolic activity. Anabolic activity is increased by a 2-hydroxymethylene group; a C_2-methylene; a 2-formyl or a 2-aminomethylene group; the 19-nor congeners and the following heterocyclic rings fused to ring A: (3,2-c)-pyrazole; (2,3-d)-isoxazole and (3,2-c)-isoxazole. The following modifications when utilized to good advantage also contribute to anabolic activity: these are, methyl groups at 2α and 7α; an ethyl group at 17α; fluorine at 2α, 9α, 3 or 4; 2 oxa analog and acyl, enol ether or -yl acetals of the 17β-OH.

The perimeter modifications that increase or contribute to retention of androgenic ac-

TABLE 92. ANDROGENIC ACTIVITIES OF SOME ANDROGENS

COMPOUND	MICROGRAMS EQUIVALENT TO AN INTERNATIONAL UNIT
Testosterone (17β-ol)	15
Epitestosterone (17α-ol)..	400
17α-Methyltestosterone ..	25–30
17α-Ethyltestosterone	70–100
17α-Methylandrostane-3α, 17β-diol	35
17α-Methylandrostane-3-one-17β-ol	15
Androsterone	100
Epiandrosterone	700
Androstane-3α, 17β-diol ..	20–25
Androstane-3α, 17α-diol ..	350
Androstane-3β, 17β-diol ..	500
Androstane-17β-ol-3-one ..	20
Androstane-17α-ol-3-one ..	300
Δ^5-Androstene-3α, 17β-diol	35
Δ^5-Androstene-3β, 17β-diol	500
Androstanedione-3, 17 ...	120–130
Δ^4-Androstenedione	120

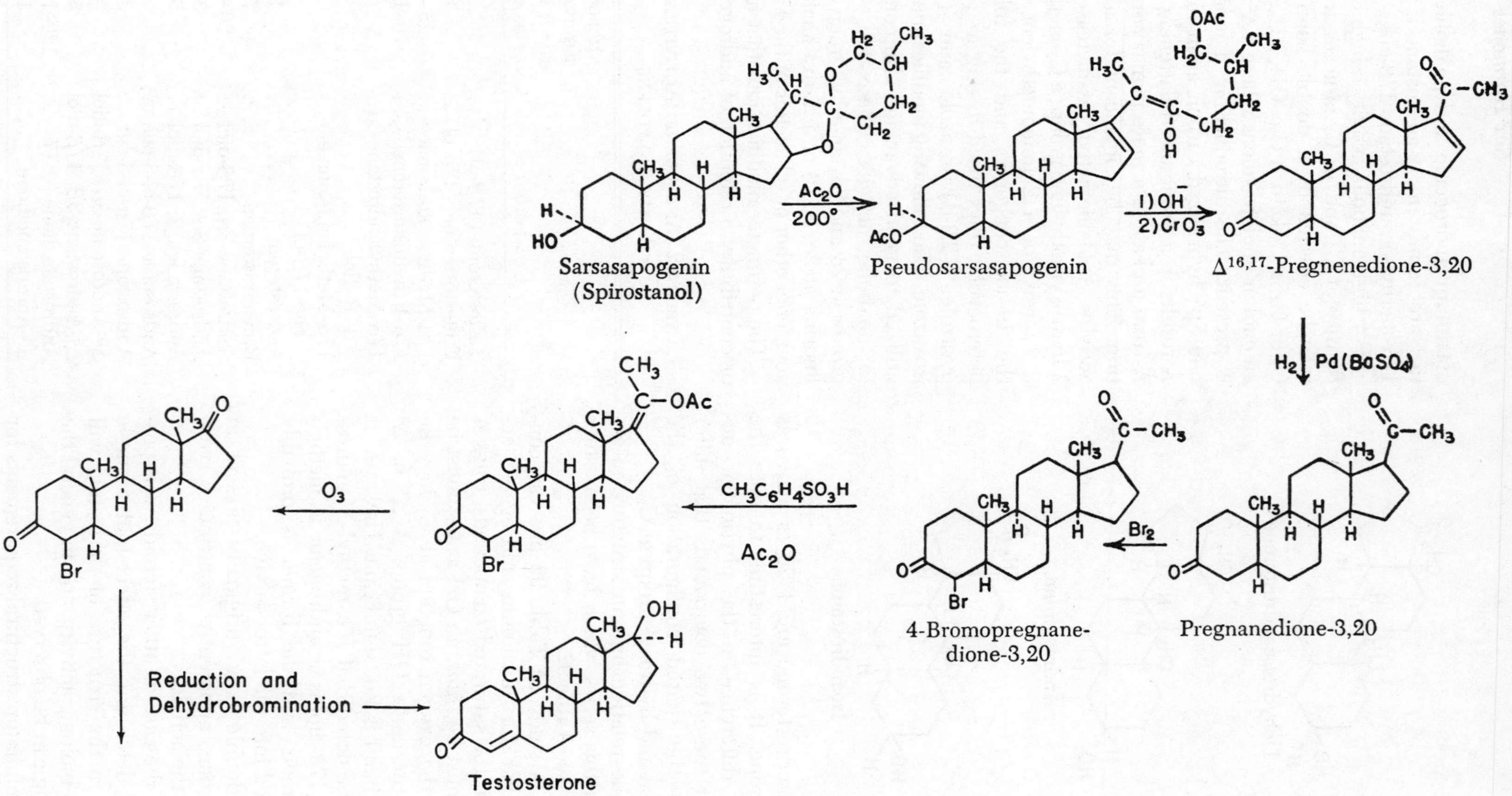

CH₃

$\Delta^{16,17}$-Pregnenedione-3,20

$\xrightarrow[\text{H}_2]{\text{Pd(BaSO}_4)}$

Pregnanedione-3,20

CH₃

1)OH⁻ 2)CrO₃

Pseudosarsasapogenin

OAc

$\xrightarrow[\text{200°}]{\text{Ac}_2\text{O}}$

Sarsasapogenin (Spirostanol)

HO

Br₂

4-Bromopregnane-dione-3,20

Br

CH₃C₆H₄SO₃H

Ac₂O

CH₃—C—OAc

Br

O₃

CH₃

Br

Reduction and Dehydrobromination

OH

Testosterone

tivity of 5α-androstane-17β-ol are as follows: (1) C$_3$α-OH or C$_3$-keto group; (2) double bond at C$_2$; (3) methyl groups at 2α-, 7α-, 6β-, 17α; (4) ethyl group for methyl group at C$_{13}$; (5) halogens C$_4$-Cl, C$_4$-F, 9α-F; (6) C$_3$-enol ethers; (7) 17β-enol ethers; (8) 17β-yl acetals; (9) 17β-OH esters such as propionate, dichloroacetate, etc.; (10) 2-hydroxymethyl + a double bond[164] at C$_2$; and (11) 2-formyl + a double bond at C$_2$.

Dihydrotestosterone, 17β-hydroxy-5α-androstan-3-one when compared to testosterone as a standard of 100 has VP = 268, SV = 158 and LA = 152, and these values[165] for 17α-methyldihydrotestosterone are 254, 78 and 107.

3α, 17β-Dihydroxy-5α-androstane has potent androgenic activity (two-thirds that of testosterone), whereas the 3β-isomer is weakly androgenic.

In the case of multiple changes the above and other changes may be additive, whereas this may not always be the case.

The amount and the duration of activity of testosterone esters may possibly be correlated with their rates of hydrolysis.[166] In general, esters from *n*-acids (most rapidly hydrolyzed) are less active than those from 2-alkyl esters (more slowly hydrolyzed), and the 2,2-dialkyl esters in some animal tests are inactive, possibly because the rate of hydrolysis is too low to yield enough of the biologically active testosterone. Solubility factors also are significant. The high efficacy of these derivatives implies that they may protect testosterone from metabolism until it reaches the site of action. In rats aqueous suspensions of 4-*t*-butylphenoxyacetate and 4-chlorophenoxyacetate gave maxima of activity 3 times that of the propionate, and the former showed a useful level of activity for over 120 days. 4-Alkyl cyclohexanecarboxylate's peak activity was maintained better than the best fatty acids.

Testosterone dichloroacetate by subcutaneous ventral prostate assay was over 10 times as androgenic as testosterone.[167] Two 9α-fluoro derivatives also had a greater androgenic activity than testosterone.[168]

The C$_3$-cyclohexylenol ether of testosterone acetate orally is more than 5 times as active as the parent compound as both an

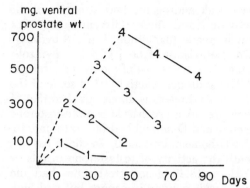

FIG. 21. Duration of action and activity of some esters of testosterone.

1. Propionate; 2. Phenyl propionate; 3. *p*-Hydroxyphenylpropionate; 4. *p*-Hexoxyphenyl propionate.

Data in the above figure show the marked effect exerted by certain esters on the duration of action and intensity of activity which is much greater than can be obtained even by huge doses of testosterone. (Data from Diczfalusy, E.: Acta endocr. 35:59, 1960.)

androgen and a myotropic agent and thus is equal to testosterone given parenterally.[169]

Orally the 2'-tetrahydropyranyl ether of 2α-methyl-5α-dihydrotestosterone was 88, 82 and 395 per cent as active as methyltestosterone on the end points ventral prostate (VP), seminal vesicles (SV) and levator ani (LA) respectively.[170] By subcutaneous injection it had low activity and the parent steroid showed only weak activity orally.

2α-Fluorotestosterone has no androgenic activity in the male rat. In the female rat it produced a marked increase in weight, yet effected nearly 100 per cent inhibition of the mammary fibroadenoma that had grown resistant to the action of testosterone propionate.

Testosterone and certain other derivatives of androstane are converted to estrogens by pregnant mares, ovariectomized-adrenalectomized women and by placental, ovarian, adrenal and stallion testicular tissues. An intermediate in this conversion is believed to be a C$_{19}$ hydroxylated compound. 19-Hydroxy-Δ^4-androstene-3,17-dione in vitro is converted relatively efficiently to estrone, whereas 19-nor-Δ^4-androstene-3,17-dione or Δ^{1,4}-androstadiene-3,7-dione is not.

2-Hydroxymethylene-17α-methyldihydrotestosterone (Adroyd®, Anasterone®) orally has weak androgenic and strong anabolic activities, and these differences are of a much greater magnitude[171] in 17β-hydroxy-17α-methylandrostano-[3,2-c] pyrazole, which is 35 times that of methyltestosterone.

The androgenic activity[172] of some of the D-homoandrogens is about equal to that of their precursors, such as D-homotestosterone acetate and D-homodihydrotestosterone. 18-Nor-D-homoandrostane-3,17α-dione has one tenth the activity of testosterone in rats.[173] D-Homo-18:19-bisnortestosterone had no androgenic activity in capons but had high myotropic (anabolic) activity.[133]

Testosterone Propionate U.S.P., 17β-hydroxyandrost-4-en-3-one propionate, is prepared by esterifying the OH at C_{17} with propionic acid. It occurs as white or creamy-white crystals or as a crystalline powder that is odorless and stable in air. It is insoluble in water, freely soluble in alcohol, ether and other organic solvents and soluble in vegetable oils. It is sensitive to alkali, a property of the C_3 ketone conjugated with the double bond.

Testosterone is used as the propionate to increase and prolong its activity when injected intramuscularly. It is weakly active orally.

Category—androgen.

Usual dose—intramuscular, 25 mg. 3 times a week.

Usual dose range—25 to 125 mg. weekly.

OCCURRENCE
Testosterone Propionate Injection U.S.P.

Testosterone Cypionate U.S.P., Depo®-Testosterone, 17β-Hydroxyandrost-4-ene-3-one Cyclopentanepropionate, Testosterone Cyclopentanepropionate, is a white or creamy-white crystalline powder that is freely soluble in alcohol, chloroform, slightly soluble in water and soluble in vegetable oils. When cottonseed oil solutions of this derivative of testosterone are used intramuscularly, its duration of action in animals or man is 7 to 30 days.

Category—androgen.

Usual dose range—intramuscular, 100 to 400 mg. once or twice monthly.

OCCURRENCE
Testosterone Cypionate Injection U.S.P.

Testosterone Enanthate U.S.P., Delatestryl®, testosterone heptanoate, occurs as a white or creamy-white crystalline powder that is insoluble in water, soluble in vegetable oils, and 1:0.3 of ether.

Testosterone enanthate dissolved in sesame oil is used intramuscularly for prolonged (2 to 4 weeks) androgenic effects.

Category—androgen.

Usual dose range—intramuscular, 100 to 400 mg. once or twice monthly.

OCCURRENCE
Testosterone Enanthate Injection U.S.P.

Dromostanolone Propionate N.F., Drolban®, 17β-hydroxy-2α-methyl-5α-androstan-3-one propionate, is reported to be active anabolically and weakly androgenic. Subcutaneously LA/VP + SV = 4(2/0.5), with T.P. as the standard.[174] Prolonged use can exert androgenic effects.

Dromstanolone is recommended for use in the treatment of advanced or metastatic carcinoma of the breast.

Category—antineoplastic.

Usual dose—intramuscular, 100 mg. 3 times a week.

OCCURRENCE
Dromostanolone Propionate Injection N.F.

A 2α-(equatorial) methyl group such as in 2α-methyl dihydrotestosterone propionate lowers androgenic activity but increases (twice) the anabolic activity. The 2α-methyl androgens are more effective than testosterone or dihydrotestosterone as potent inhibitors of the development of a transplantable rat mammary tumor, probably because the methyl group protects the C_3 keto group and prevents conversion of the androgen into estradiol or estrone in vivo.[175] Even if they were converted to estrogens in vivo, 2-methylestrone has only 1/200 the tumor-stimulating effect of estrone.

Stanolone, Neodrol® is androstane-17β-ol-3-one or dihydrotestosterone and occurs as a white crystalline powder that is practically insoluble in water and soluble 6:100 in alcohol. It is considerably more stable than testosterone. Its LA/VP = 0.57(1:5/2.7); LA/SV = 1(1.5/1.5) subcutaneously, with testosterone as a standard.[176]

When used as an anabolic and tumor suppressant, some of the disadvantages of testosterone are minimized. It is used in a saline suspension that when injected has a duration of action slightly more prolonged than that of testosterone propionate in oil.

Methyltestosterone N.F., Oreton®-Methyl, Metandren®, 17β-Hydroxy-17-methylandrost-4-en-3-one.

Methyltestosterone occurs as white or creamy-white crystals or as a crystalline powder that is odorless and stable in air but is affected by light. It is insoluble in water, soluble in alcohol, ether and other organic solvents and sparingly soluble in vegetable oils. It is sensitive to alkali. It is prepared from dehydroisoandrosterone as follows:

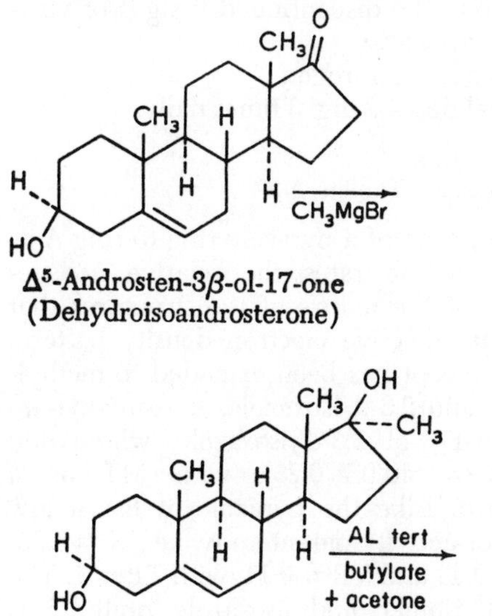

Δ⁵-Androsten-3β-ol-17-one
(Dehydroisoandrosterone)

Methyltestosterone

Methyltestosterone possesses the same action as testosterone; however, orally it is two fifths as active as testosterone administered by injection. By the sublingual route, ½ to ⅓ the oral dose is used. Apparently, one of the points of attack by intestinal bacteria to inactivate testosterone is the OH group at C_{17}. The small methyl group replacing the hydrogen prevents this attack, but the stereochemical specificity has not been altered sufficiently to prevent it from possessing a relatively high androgenic activity. Methyltestosterone has a more pronounced corticometric activity.

Hepatic toxicity occurs with the introduction of a 17α-alkyl group and this activity is increased in the 19-nor congeners.[177]

It may be used orally, sublingually or topically.

Category—androgen.

Usual dose—oral, 10 mg. 3 times a day; buccal, 5 mg. up to 4 times a day.

Usual dose range—oral 10 to 50 mg. per day; buccal, 5 to 25 mg. per day.

OCCURRENCE

Methyltestosterone Tablets N.F.

Fluoxymesterone U.S.P., Halotestin®, 9α-fluoro-11β,17β-dihydroxy-17-methylandrost-4-en-3-one, 9α-fluoro-11β-hydroxy-17α-methyltestosterone, has both anabolic and androgenic activity about 10 times that of methyltestosterone. It is orally effective and is used for the same purposes as methyltestosterone. The 11β-hydroxy group is present, probably because it is convenient to introduce a 9α-fluoro group via the 9:11 epoxide and HF. A 9α-fluoro atom usually increases the biologic activity of the parent steroid hormone (q.v.).

Category—androgen.

Usual dose—2 to 4 mg. 1 to 3 times daily.

Usual dose range—2 to 30 mg. daily.

OCCURRENCE

Fluoxymesterone Tablets U.S.P.

Bolasterone, Myagen®, is 7α,17α-Dimethyltestosterone whose oral LA/SV = 4.3 (13/3) and N/VP = 3.2 (4.2/1.3), with MT as the standard.[178] Expressing human metabolic balance effects as differences between pretreatment and post-treatment nitrogen excretion values, bolasterone was 25, 80 and 2 times the activity of fluoxymesterone, norethandrolone and methandrostenolone, respectively.[179]

A 7α-methyl group exerts a modest increase in the LA and SV values of testosterone, more so with methyltestosterone and markedly with 19-normethyltestosterone, i.e., LA/SV = 2 (41/18). The latter compound is active in rats in daily doses of 1 mcg. orally or 0.1 mcg. subcutaneously, which begins to approximate the physiologic dose of the potent natural estrogens.

Methandrostenolone N.F., Dianabol®, Primobolan®, is 17β-hydroxy-17-methylandrosta-1,4-dien-3-one, whose oral[180] N/VP = 3.4 (1.2/0.35); LA/VP = 5 (0.3/0.06) and LA/SV = 2.5 (0.3/0.12), with MT as the standard.[176]

Methandrostenolone is used orally as an anabolic agent, with mild androgenic effects over a prolonged period of time.

Category—androgen.

Usual dose—2.5 to 5 mg. daily.

OCCURRENCE
Methandrostenolone Tablets N.F.

Oxymetholone N.F., Adroyd®, Anadrol®, is 17β-hydroxy-2-(hydroxymethylene)-17α-methyl-5α-androstan-3-one or 2-hydroxymethylene-17α-methyldihydrotestosterone. It occurs as a white to creamy white crystalline powder that is practically insoluble in water, sparingly soluble in alcohol and freely soluble in chloroform. Its oral N/VP = 8.75 (1.75/0.2), with MT as the standard.[181]

The 2-hydroxymethylene is one of the perimeter modifications[182] that increases electron density in the area, which causes marked depression of the androgenic activity together with an increase in anabolic activity. It is used as an anabolic agent.

Category—androgen.

Usual dose—oral, 5 to 10 mg. daily.

Usual dose range—5 to 30 mg.

OCCURRENCE
Oxymetholone Tablets N.F.

Stanozolol N.F., Winstrol®, is 17-methyl-2'H-5α-androst-2-eno[3,2-c]pyrazol-17β-ol—oral N/VP = 30 (10/0.33), with MT as a standard;[181] parenterally, AB/AG = 2 (0.05/0.025), with T.P. as a standard; LA/VP + SV = 0.8 (1.2/1.5) by injection with testosterone as a standard.[176]

By employing nitrogen balance studies[183] in suitable patients the steroid protein activity index (SPAI) of a number of androgens was compared. These were: testosterone pro-

Stanozolol

pionate +6; norethandrolone +8; 19-nortestosterone +9; 4-hydroxymethyltestosterone +11; methandrostenolone +16; oxandrolone +17; and stanozolol +24. Antianabolic or catabolic adrenal cortical steroids all have negative values by this assay.

Stanozolol is used as an anabolic agent but should be discontinued if signs of virilization appear.

Category—androgen.

Usual dose—2 mg. 3 times daily.

OCCURRENCE
Stanozolol Tablets N.F.

The fusion of a pyrazole ring to ring A as above demonstrates the relative nonspecificity of the nature of the structures that have an effective electron density pattern. This concept has been extended to methylandrostano[2,3-d]isoxazole, 17α-methyl-androstan-17β-ol[2,3-d]isoxazole, whose oral N/VP = 40(9.7/0.25), with MT as a standard. Like the pyrazole, it has a low order of activity parenterally, i.e., N = 0.25; LA = 0.43 and VP = 0.11, with TP = 1. The [3,2-c] isomer androisoxazole orally[176] is less active, i.e., N/VP = 7(1.5/0.22) or 1.7.

Androgenic activity is decreased, with the retention or increase of the anabolic activity, by the introduction of a formyl or hydroxymethyl in the 2 position. Thus 2-formyl-5α-androst-2-en-17β-ol had LA/SV + VP = 5-10(1/0.1-0.2) subcutaneously, with testosterone as the standard. 2-Hydroxymethyl-5α-androst-2-en-17β-ol and its 17α-methyl congener had about the same activity orally as subcutaneously and a LA/SV + VP = 11 (2.2/0.2) or (4.5/0.40), with testosterone and methyltestosterone as the standards.[182]

Ethylestrenol, Orgabolin®, is 17α-ethyl-17β-hydroxyestr-4-ene—oral N/VP = 8(1.7/0.21); LA/SV = 20, with MT as a standard. It is used as an anabolic steroid.

Methyl Androstenediol, Stenediol®, Methostan®, Methanabol®, Mestendiol®, is 17-methylandrost-5-en-3β,17β-diol. It is a close relative of methyltestosterone, differing primarily in the position of the unsaturation and the 3 position having an alcohol group. These changes have decreased the male hormone effect and increased the corticoid and anabolic properties.

In the dosage used, this steroid promotes a tissue-building action, with the retention of nitrogen and an increase in body proteins. It is available for intramuscular injection or for oral and sublingual use.

Testolactone, Testololactone, Teslac®, is 17α-oxa-*D*-homo-1,4-androstadiene-3,17-dione. It is a white, odorless crystalline solid which is slightly soluble in water and

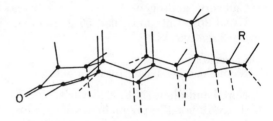

Testolactone

soluble in ethanol. The aqueous suspension for intramuscular administration is intended for use in the palliative management of advanced or disseminated mammary cancer in postmenopausal women when hormonal therapy is indicated. It possesses no androgenic activity. The usual dosage is 100 mg. 3 times per week.

19-NOR ANDROGENS

With the loss of the C_{19}-CH_3 via Birch's synthesis[184] to give 19-nortestosterone, full anabolic activity was retained; however, most (85%) of the androgenic activity was lost. According to Edgren,[153] 19-nortestosterone's LA/VP = 15 (60/4), which was about the same as for its β-phenylpropionate 16 (220/14) S.C., with T.P. as the standard; moreover, some compounds related to 19-nortestosterone consistently produced significant myotropic (LA) effects at dose levels that were below the threshold for an androgenic response (VP), whereas some testosterone derivatives showed no such separation, with the exception of stanozolol

and the enanthate of 1-methyl-Δ^1-4,5α-dihydrotestosterone.

19-Nordihydrotestosterone[185] exhibits 8 times the antiestrogenic activity of 19-nortestosterone when assayed by the inhibition of the uterotropic activity of estrone in the immature mouse. Oral potent anti-estrogenic activity[185] also is present in others, such as 17α-methyl-19-nordihydrotestosterone, 17α-methyl-19-norandrostan-3β, 17β-diol and 19-nor-17α-methyl-$\Delta^{4, 9}$-androstadien-11β-ol-3-one.

The following formula is presented to permit a better mental image of the conformational formula of the 19-nor steroids.

R = OH = 19-Nortestosterone

Even though 19-nortestosterone and some of its congeners (norethandrolone, etc.) showed a favorable anabolic/androgenic ratio and, in some cases, androgenic activity less than the standards, the 19-nor modification as a limiting factor when in conjunction with certain other modifications markedly increases androgenic activity, with a less favorable anabolic/androgenic ratio. For example, 7α,17α-dimethyl-19-nortestosterone; LA/SV = 2 + (41/18).

19-Nortestosterone is more effective than testosterone in man and it is more androgenic in man than in animals.

17α-Methyl-19-nortestosterone has LA/VP of 2(0.6/0.3) subcutaneously in rats, whereas in the chick its activity was equal to that of methyltestosterone.[89]

Nandrolone Phenpropionate N.F., Durabolin®, is 17β-hydroxyestr-4-en-3-one hydrocinnamate, 19-nortestosterone phenylpropionate, whose subcutaneous LA/SV = 16(0.6 to 1/0.04 to 0.06), with 19-nortestosterone as the standard.[186] Nandrolone phenpropionate is administered by injection as a long-acting anabolic steroid, but virilization may occur with prolonged use of high doses.

Category—androgen.

Usual dose—intramuscular, 25 to 50 mg. each week.

OCCURRENCE

Nandrolone Phenpropionate Injection N.F.

Nandrolone Decanoate N.F., 17β-hydroxyestr-4-en-3-one decanoate, or 19-nortestosterone decanoate, occurs as a fine, white to creamy white crystalline powder that is practically insoluble in water but is soluble in chloroform, alcohol, and in vegetable oils. It is used for its very prolonged anabolic effects, which after 4 to 6 weeks, are more pronounced than the androgenic effects.[187]

Category—androgen.

Usual dose—intramuscular, 50 to 100 mg. every 3 to 4 weeks.

OCCURRENCE

Nandrolone Decanoate Injection N.F.

Norethandrolone N.F., Nilevar®, is 17-hydroxy-19-nor-17α-pregn-4-en-3-one, 17α-Ethyl-19-nortestosterone, whose oral N/VP = 20(3.9/0.19), with MT as a standard,[181] and LA/VP or SV = 7(7/1), with T as the standard.[176] This is a better anabolic androgenic ratio than 17α-methyl-19-nortestosterone, whose LA/VP = 2(0.6/0.3), and interesting because methyltestosterone has a more favorable ratio than ethyltestosterone.

Norethandrolone is used for its anabolic effects both by oral and intramuscular injection. At high dosage levels definite androgenic effects are obtained.

Category—androgen.

Usual dose—10 mg. 2 or 3 times daily.

OCCURRENCE

Norethandrolone Tablets N.F.

(I)-13β-Ethyl-17α-ethynyl-17β-hydroxygon-4-en-3-one, is a homolog of 19-norethisterone where the C_{18} methyl is replaced by an ethyl. The value of 20 for its LA/VP (340/17) ratio was obtained with T.P. as a standard.[153] It produced acceleration of body growth at a dose of 300 mcg., and was more effective than 19-nortestosterone or testosterone propionate.

Norbolethone, Genabol®, ($\pm$) $13\beta,17\beta$-Diethyl-17β-hydroxygon-4-en-3-one, has 3.4 times the myotropic and 0.15 times the androgenic activity of T.P. Its oral nitrogen retention activity in castrate rats was about $3 \times$ M.T. and 4.2 times that of stanozolol which, in turn, was more active than methandrostenolone, oxymetholone, ethylestrenol or norethandrolone. Norbolethone was 6 times more active subcutaneously than orally, and parenterally was about equal to T.P.[188]

Oxandrolone N.F., Anavar®, is 17β-hydroxy-17α-methyl-2-oxa-5α-androstan-3-one, whose intramuscular LA/AG = 5(1/0.2), with MT as a standard. It is a white crystalline solid that is soluble in alcohols and glycols and is stable to air, heat and light. Orally it has ⅛ to ¼ the androgenic activity of methyltestosterone. The anabolic potency as measured by its nitrogen-sparing activity in humans is 6 times that of methyltestosterone and 2 to 3 times that of norethandrolone. It has no estrogenic, antiestrogenic, progestational or antiprogestational activities.

Oxandrolone is used for its anabolic effects.

The hetero oxygen at C_2 furnishes a means to obtain a high electron density at C_2 without significantly altering the conformation of ring A. This lowers androgenic but increases anabolic activity.

Category—androgen.

Usual dose range—initial, 5 to 10 mg. daily; maintenance, 2.5 to 5 mg. daily.

OCCURRENCE

Oxandrolone Tablets N.F.

In 1959 several 17-alkyl-3-deoxy-19-nortestosterones were reported to have progestational activities of an order equal to the parent steroid.[189] 5α-Androstan-17β-ol had 60 per cent of the androgenic activity of testosterone in the chick inunction assay.[190] The Δ^1, Δ^2, Δ^3 and Δ^4 congeners subcutaneously in the rat had LA/VP + SV of 3 (1/0.5), 3 (1.5/0.5), 2 (0.8/0.4) and 1(0.1/0.1) respectively.[191] In general, 17α-alkylation increases anabolic and decreases androgenic activity so that for 17α-methyl-5α-androst-2-en-17β-ol LA/SV + VP = 4-10 (2-5/0.5) orally.[191] This also is true[192] of 2-methyl-5α-androst-2-en-17β-ol, LA/VP = 0.9 (1.17/1.27), and its 17α-methyl derivative, LA/VP = 5 (3.1/0.64) subcutaneously, with testosterone as the standard and, orally, the LA/VP = 10 (10.4/0.97), with methyltestosterone as the standard.[176] In general, a 17α-ethyl group induces larger anabolic/androgenic ratios.

Androst-2-en-17β-ol in olive oil when injected had an activity equal to that of testosterone phenylpropionate in amount and duration. Its myotropic and androgenic activities are respectively 1.5 and 0.5 those of testosterone.[193]

ADRENAL CORTEX HORMONES

It long has been known that adrenalectomy in the experimental animal will cause death in a few days. Adrenalectomy also reduces glycosuria, increases amino acid metabolism, deranges fat metabolism and forms ketone bodies in diabetic animals. It also has been known for some time that suitable extracts of the adrenal cortex will delay or prevent death in adrenalectomized animals. Thomas Addison, in 1855, described a syndrome, now usually known as Addison's disease, which is caused by diminished adrenal cortex activity that, for example, may be due to a disease, such as adrenal tumors or tuberculosis of the adrenal gland. Addison's disease is characterized by extreme weakness, anorexia, anemia, nausea and vomiting, pigmentation of the skin, low blood pressure and mental depression.

The most important deficiency symptoms which follow adrenalectomy and are susceptible to quantitative estimations are: (1) disturbance of the Na^+, K^+, Cl^- and water balance (increases excretion of Na^+ Cl^- and water; retention of K^+), (2) increase of the urea content of the blood, (3) asthenia (inefficiency of muscle), (4) disturbance of carbohydrate metabolism (decrease in liver glycogen, decreased resistance to insulin) and (5) reductions of resistance to various traumata (cold, mechanical or chemical shock).

Most methods of assay depend on the alleviation of such deficiency symptoms in adrenalectomized animals. However, some make use of normal animals. A properly prepared adrenal cortex extract, when administered to an adrenalectomized animal, will abolish the above symptoms.

The adrenals of a normal adult secrete about 20 to 30 mg. of hydrocortisone and 20 to 200 mcg. of aldosterone per day. Adrenocorticoid production reaches a maximum about the age of 30 and then gradually declines.

The well-known mediation of corticoid synthesis by ACTH in the adrenals is repressed by agents, such as puromycin, chloramphenicol and cycloheximide that inhibit protein synthesis. Elevated levels of corticoids also repress their own synthesis in the adrenals probably by a primary effect on the rate of synthesis of a regulator protein or some cofactor. Thus a feedback mechanism in the adrenals may be involved to control the levels of the corticoids.[194] This is independent of the control of the ACTH by the pituitary. There also appears to be regulatory mechanisms in which other naturally occurring steroids participate at the cellular sites of biosynthesis.[195]

Cushing's disease is the result of hyperadrenalism in man.

Conn's syndrome is a disease characterized by hypernatremia, polyuria, alkalosis and hypertension accompanied by a high excretory level of aldosterone (q.v.). This condition is said to be due to the inability of the adrenals to perform the 17α-hydroxylation.

In the adrenogenital syndrome, the adrenals are unable to carry out 21-hydroxylation, with the resultant accumulation of 17α-hydroxyprogesterone which may be degraded to androgens and cause masculinization or excreted as pregnane-3α, 17α, 20α-triol. The deficiency of hydrocortisone under these conditions leads to greater ACTH stimulation of the adrenals.

The so-called sodium excretion factor (S.E.F.) has been identified as allopregnane-3β, 16α-diol-20-one and is a causative agent in the loss of sodium chloride. This steroid is an abnormal metabolite produced as a variant in the adrenogenital syndrome.

Since the discovery by Hench *et al.* that cortisone gave dramatic results in the treatment of rheumatoid arthritis, a tremendous amount of research has been conducted in an effort to prepare corticoids having high anti-inflammatory activity to be used for the above-mentioned and many other inflammatory conditions. This entailed new methods of assay, the most important of which measures the prevention of the deposition of granulation tissue when cotton pellets (weight 6 to 8 mg.) are placed in the ventrolateral subcutaneous connective tissue.

Recent evidence suggests that adrenal corticosteroids have a role in maintaining physiologic phosphorylase levels and in the conversion of inactive to active phosphorylase, possibly at the level of the phosphorylase kinase reaction—this, because administered hydrocortisone following adrenalectomy results in restoration of diminished inactive liver phosphorylase levels and a restoration of the diminished hyperglycemic effects of administered epinephrine and cyclic AMP.[195a]

In some cases the topical anti-inflammatory activity of the corticoids can be correlated with a vasoconstrictor activity. The anti-inflammatory activity of the glucocorticoids is believed to function as an antagonist of a stress-activated microcirculatory dilator, histamine, and thus inhibit relaxation of microvascular sphincters. The dilator tends to open capillaries and the glucocorticoids tend to close them. In the absence of glucocorticoids, mild stress activation of dilator synthesis leads to excessive opening of capillaries, while greater activation leads to pooling and circulatory failure. Glucocorticoids in low doses restore the normal distribution of open and closed vessels. With larger doses of glucocorticoids the number of capillaries carrying blood is markedly reduced.[196]

Hydrocortisone and its many anti-inflammatory congeners have the common property of exerting some stabilization of the release of acid phosphatase and beta-glucuronidase from lysosomes. This may be a general stabilizing effect on lysosomes and play a role in their anti-inflammatory activities.[196a]

The term "glucocorticoid" has been coined to describe the glycogen deposition activity of the adrenal corticoids, and the term "mineralocorticoid" for their effect on the electrolyte balance. The primary effect of the glucocorticoids is exerted on existing enzyme systems so as to direct metabolites toward carbohydrate formation, and induction of enzyme synthesis is in the nature of a secondary effect.[197]

The specific glycoprotein (α-2-GP) is not detectable in the serum of mature normal rats or adrenalectomized rats following trauma.[198] Increased levels of α-2-GP can be detected in the serum of mature normal rats 8 to 12 hours following trauma to connective tissue and in the serum of adrenalectomized rats under the same conditions only after the administration of hydrocortisone. Without concomitant trauma, hydrocortisone administered by various routes did not induce an α-2-GP response. Serum levels of α-2-GP quantitatively reflect the severity of an inflammatory response. The synthesis and concentration of α-2-GP in the serum of traumatized animals is a parameter of glucocorticoid activity apart from that of liver glycogen deposition or thymolysis. An immunoassay can be utilized to detect the α-2-GP levels in serum. Only those corticoids having an oxygen function at C_{11} play a role in the synthesis of α-2-GP.

Corticosteroids also increase the blood lipids, suppress lymphatic tissue growth (thymus; spleen, lymph nodes), cause a drop in circulating lymphocytes and eosinophils and inhibit the effect of hyaluronidase on membrane permeability.

Mineralocorticoid activity (sodium retention) appears to regulate sodium transport across anuran skin and urinary bladder by stimulating DNA-dependent synthesis of RNA and de novo synthesis of proteins. The effectiveness of inhibition of the response to aldosterone is correlated with inhibition of RNA and protein synthesis. The effect of a single dose of aldosterone on RNA synthesis is rapid and short lived. Levels of membrane Na^+-K^+-dependent ATPase activity associated with sodium transport do not appear to be regulated by aldosterone.[199]

Up to 1959, 46 crystalline steroids[200] were isolated from the adrenal cortex. Some of these were first known as cortin, at which time an amorphous fraction was described. Of the crystalline steroids, 7 give varying degrees of relief in adrenal cortical insufficiency. Of those more recently isolated, aldosterone (q.v.) is of outstanding significance.

One crystalline compound, i.e., corticosterone, is effective in all the standard tests, even though it may not be as active as some others, such as desoxycorticosterone and its acetate and the amorphous fraction in certain specific tests.

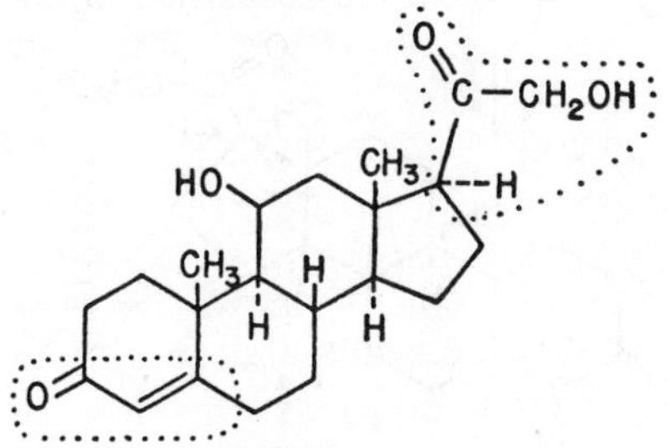

Corticosterone

Desoxycorticosterone

17-Hydroxycorticosterone
(Compound F, Hydrocortisone)

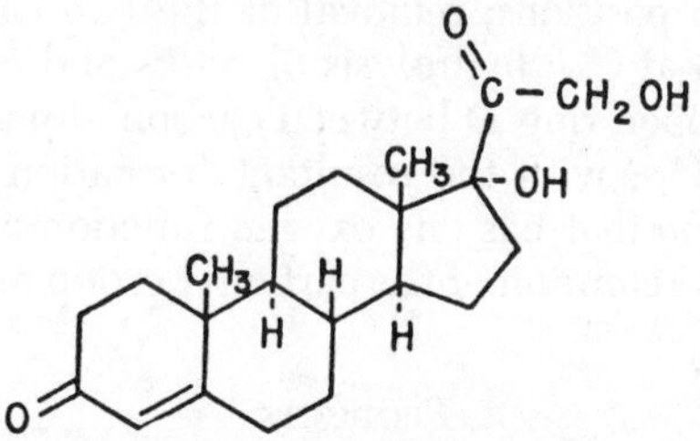

17-Hydroxy-11-desoxycorticosterone

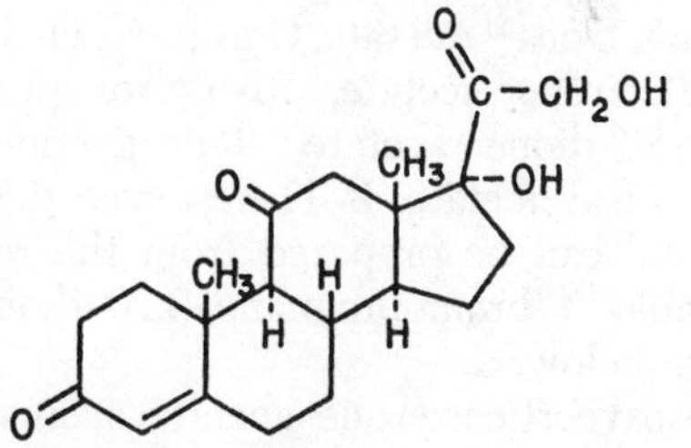

17-Hydroxy-11-dehydrocorticosterone
(Cortisone)

The two structures encircled with a dotted line are essential for all known activity and are apparently the only ones necessary for the control of the salt balance. The $C_{11}-$ OH is essential for carbohydrate activity, which is enhanced if an OH also is present at C_{17}.[201]

11-Dehydro-17-hydroxycorticosterone is 3 times as active as 11-dehydrocorticosterone when tested for glycogen deposition activity in adrenalectomized mice. 11-Dehydro-17-hydroxycorticosterone is 12 times as active as corticosterone and 13 times as active as 11-desoxycorticosterone in the cold test employing adrenalectomized rats.[202]

All adrenocorticoids are synthesized because isolation from natural sources is not practical and would not supply the demand. Starting materials for their syntheses are sterols such as stigmasterol and ergosterol, cholic acid and the sapogenins such as sarsasapogenin, diosgenin, etc. The latter are

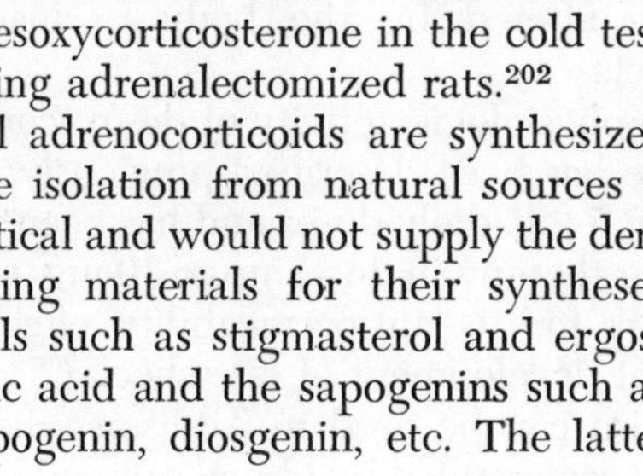

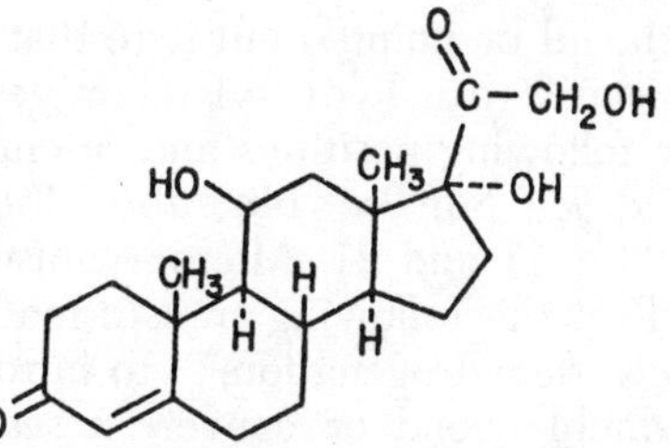

11-Dehydrocorticosterone

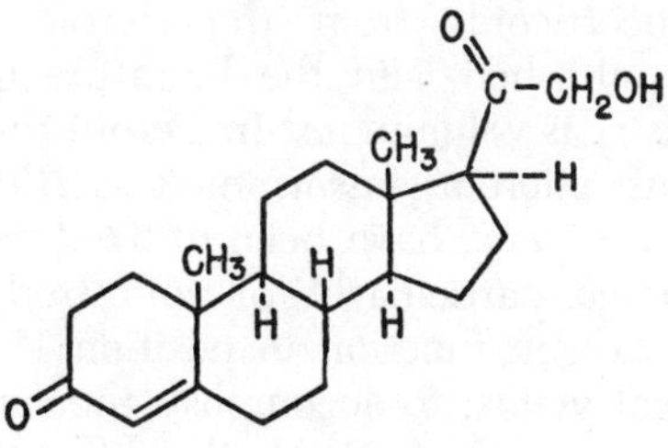

particularly useful because the side chain at position 17, when subjected to a 2-step reaction, can be degraded readily to give very good yields of the pregnane carbon skeleton with a $CH_3C(O)-$ (methylketone) at the 17 position and a double bond at the 16,17 positions.[162] For example, one of these intermediates is $\Delta^{16, 17}$-pregnene-3,20-dione which can be utilized readily for the preparation of progesterone and other steroid hormones.[162]

Many routes for the syntheses of the

adrenocorticoids from the steroids have been published, and the literature of this chemistry is voluminous. In steroid transformations, microorganisms such as *Rhizopus arrhizus*,[203] etc., have been utilized to great advantage, particularly for the introduction of an oxygen function in position 11 in 90 per cent yields; to accomplish the same by chemical methods is both difficult and expensive.

It should be pointed out here that microorganisms[204] can hydroxylate progesterone in the following positions and orientations: 6β; 7α; 7β; 8β; 9α; 14α; 15α; 15β; 16α; 11α; 11β; 17 and 21. Microorganisms also can effect the following transformations of steroids: dehydrogenation,[205] to introduce a new double bond or convert a secondary alcohol to a ketone; reduction of a double bond or keto group; epoxidation at 9:11 or 14:15 positions; removal of the two carbon chain at C_{17}; hydrolysis of esters and ethers and open ring D between carbon atoms C_{13} and C_{17}, with the resultant formation of a lactone that has one oxygen function at C_{13} and a retention of the carbonyl group at C_{17}.

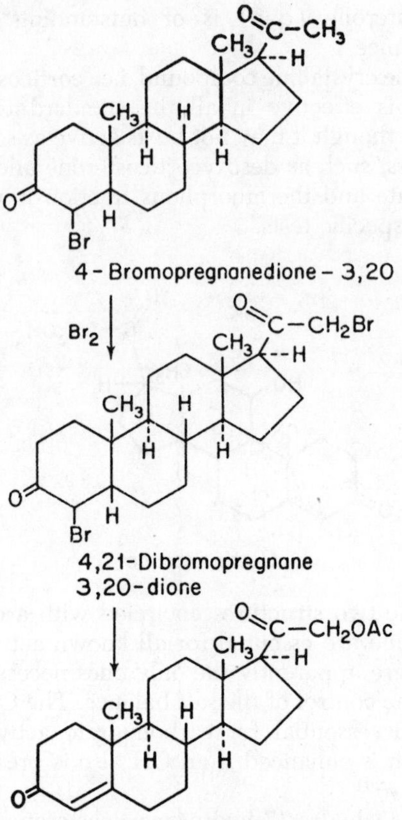

4 - Bromopregnanedione - 3,20

4,21-Dibromopregnane 3,20-dione

Desoxycorticosterone Acetate

PRODUCTS

Desoxycorticosterone Acetate U.S.P., Percorten®, Doca® Acetate, Cortate®, 11-deoxycorticosterone acetate, 21-hydroxypregn-4-ene-3,20-dione acetate, 4-pregnene-3,20-dione-21-ol acetate, is 11-desoxycorticosterone and can be prepared from the readily available 4-bromopregnane-3,20-dione as shown below.

Desoxycorticosterone acetate occurs as a white or creamy-white crystalline powder, odorless and stable in air. It is practically insoluble in water, sparingly soluble in alcohol or acetone and slightly soluble in vegetable oils.

The acetate group at C_{21} is so unstable that it can be saponified readily by means of potassium bicarbonate. Desoxycorticosterone is very unstable in alkali and is oxidized very readily, as is evidenced by the fact that it will reduce Tollen's reagent in the cold. It should be stored in well-closed, light-resistant containers.

Desoxycorticosterone acetate is about 3 times as active as the free compound be-

cause, when administered by intramuscular injection or crystal implantation, it is absorbed more slowly and apparently less readily altered by the body to inactive products.

The physiologic activity of desoxycorticosterone has been described previously. It is used in 5-mg. daily doses and has been used in burn therapy in doses up to 10 mg. every 4 hours for its antipermeability effect, to prevent development of late shock.[206]

It also can be used in Addison's disease when accompanied by liberal salt and water intake. It is useful in surgical shock both preoperatively and postoperatively.

Desoxycorticosterone activity differs from that of extracts of the adrenal cortex in being concerned chiefly with salt and water metabolism.

Category—adrenocortical steroid (salt-regulating).

Usual dose—intramuscular, 1 to 5 mg. daily.

Usual dose range—intramuscular, 1 to 10 mg. daily.

OCCURRENCE

Desoxycorticosterone Acetate Injection U.S.P.

Desoxycorticosterone Acetate Pellets N.F.

Desoxycorticosterone Pivalate N.F., Percorten® Pivalate, 21-hydroxypregn-4-ene-3, 20-dione pivalate, deoxycorticosterone trimethylacetate, occurs as a white or creamy-white crystalline powder, odorless, stable in air, insoluble in water, slightly soluble in alcohol and in vegetable oils. This ester is used for its prolonged duration[166] of mineralocorticoid activity (q.v.).

Category—mineralcorticoid.

Usual dose range—intramuscular, 50 to 100 mg., repeated in not less than 30 days.

OCCURRENCE

Sterile Desoxycorticosterone Pivalate Suspension N.F.

Pregnanedione is a potent C.N.S. depressant[207] and it has been called a "steroid anesthetic." In mice it induces a deep sleep whose duration continues until the steroid is metabolized. Progesterone given in large doses to humans induces depression, drowsiness and even sleep.[208]

Aldosterone, Electrocortin®, 11β,21-Dihydroxy-3,20-diketo-4-pregnene-18-ol, has been isolated from the amorphous fraction of both beef and hog adrenal cortex extracts.[209] Its structure has been proved by synthesis.

Aldosterone

It melts at 164° to 169° and has an $[\alpha]_D$ of +161 in chloroform. It is less stable than most corticoids, particularly in the presence of bases.

Using the toad bladder for in-vitro tests, aldosterone was active at physiologic concentrations (3.3×10^{-10}M to 10^{-7}M) to increase the permeability to sodium of the mucosal surface of the single layer of epithelial cells[210] (through which sodium has been shown to enter passively). Other steroids having sodium-retaining activity in vivo also were active in this test and their activities could be inhibited by progesterone, aldactone (q.v.) etc.

Aldosterone is soluble in alcohol, chloroform, ethyl acetate and aqueous acetone. It has 30 to 100 times the mineralocorticoid activity of desoxycorticosterone and is administered orally or intramuscularly for Addison's disease.

Dose—100 to 300 mcg.

The mineralocorticoid activity of desoxycorticosterone and some other corticoids can be altered by structural modifications. Salt retention properties are increased by the following: 9α-halo; 6α-chloro; 2-methyl; 6β-methyl; a double bond in the 6,7-position and in the 19-nor compounds. Some of these effects are additive, and corticoids have been prepared that are equal to or more active than aldosterone—for example, 2-methyl-9α-fluorohydrocortisone has 100 times the salt-retaining activity of DOCA. Salt retention properties can be lowered by the following: 6α-methyl; 16α-methyl; 16α- or β-hydroxyl or a double bond in the 1,2-position.

Adrenal Cortex Extract is an extract of adrenal glands, from domesticated animals used as food by man, containing the cortical steroids essential for the maintenance of life in adrenalectomized animals.

Adrenal Cortex Injection is a clear, colorless or nearly colorless aqueous alcoholic solution of a mixture of the adrenocorticoids obtained from healthy domestic animals used for food by man, and it is used for its adrenocorticoid activity.

Cortisone, Cortone®, 17-hydroxy-11-dehydrocorticosterone, 11-dehydro-17-hydroxycorticosterone, Compound E, is used primarily as the 21-acetate because of its increased stability and possibly to obtain more prolonged activity. This corticometric principle was used first by Hench *et al.*, in 1948, with dramatic results in the treatment of rheumatoid arthritis.[211]

Cortisone Acetate U.S.P., Cortogen® Acetate, Cortone® Acetate, 17,21-dihydroxy-pregn-4-ene-3,11,20-trione 21-acetate, is a white or practically white, odorless, crystalline powder that is insoluble in water, soluble in alcohol 1:350 and chloroform 1:4. Its stability is less than that of desoxycorticosterone or corticosterone due to the presence of an additional hydroxyl group at position 17. It is more stable than cortisone.

Cortisone acetate and cortisone have about equal activity. When administered by injection, a more prolonged action is obtained and smaller doses can be used.

Although cortisone originally gave dramatic results in the treatment of rheumatoid arthritis, it has been used in a number of other conditions with good to excellent remission of clinical symptoms. Cessation of usage of cortisone causes a recurrence of symptoms in varying degrees and periods of time. Beneficial effect to often dramatic results are obtained in rheumatoid arthritis, rheumatoid spondylitis, Still's disease, psoriatic arthritis, rheumatic fever, lupus erythematosus, allergic disorders (i.e., bronchial asthma, hay fever, angioneurotic edema, drug sensitization, serum sickness), inflammatory eye diseases, pemphigus, exfoliative dermatitis and panhypopituitarism. Highly encouraging results have been obtained in acute gouty arthritis, ulcerative colitis, regional enteritis, nephrotic syndrome, scleroderma, dermatomyositis, psoriasis, periarteritis nodosa, pulmonary granulomatosis and alcoholism.[212]

Adverse hormonal reactions occur to a greater or lesser degree in some cases, and these reactions are the chief obstacle to the continued use of cortisone in large or relatively large doses.

Category—adrenocortical steroid (anti-inflammatory).

Usual dose—oral, 25 mg. 4 times a day; intramuscular, 100 mg. daily.

Usual dose range—oral or intramuscular, 10 to 400 mg. daily.

For external use—topically, 0.1 ml. of a 0.5 to 2.5 per cent suspension to the conjunctiva, 6 to 12 times daily.

OCCURRENCE

Cortisone Acetate Ophthalmic Suspension U.S.P.

Cortisone Acetate Tablets U.S.P.
Sterile Cortisone Acetate Suspension U.S.P.

Hydrocortisone U.S.P., Cortef®, Cortril®, Hydrocortone®, 11β,17,21-trihydroxypregn-4-ene-3,20-dione, was isolated by Kendall, in 1936, from the adrenal cortex. Today it is produced synthetically by procedures used to synthesize cortisone.

Hydrocortisone is a white, crystalline powder that is considerably more soluble than its acetate in such media as water (28 times), plasma (35 times) and synovial fluid (6 times). It is soluble 1:40 in alcohol, insoluble in water and slightly soluble in chloroform. Its stability properties parallel those of cortisone.

It is significant to note that hydrocortisone and not cortisone is the chief product found when: (1) ACTH is perfused through isolated beef adrenal glands; (2) ACTH is administered or in patients with hyperfunctioning adrenal cortical tumors (found in urine); (3) corticosterone is perfused through adrenal glands (hydrocortisone only is yielded).

Although hydrocortisone structurally is related very closely to cortisone, there are quite significant differences in their potencies. The onset of and subsequent improvement in cases of rheumatoid arthritis is more striking and more uniformly rapid with hydrocortisone than with cortisone acetate when these compounds are administered orally in similar doses.

Hydrocortisone or its acetate is distinctly superior to cortisone when injected intraarticularly. It is effective when injected directly into rheumatoid arthritic and osteoarthritic joints. Hydrocortisone is absorbed more rapidly than most if not all of its esters. Unfortunately, large initial suppressive doses of hydrocortisone given systemically will produce endocrine disorders of the same order as cortisone.[213]

Category—adrenocortical steroid (anti-inflammatory).

Usual dose—oral, 10 mg., 3 to 4 times daily.

Usual dose range—10 to 300 mg. daily.

For external use—topically, 0.5 to 2.5 per cent ointment 2 or 3 times daily.

OCCURRENCE

Hydrocortisone Cream U.S.P.
Hydrocortisone Lotion N.F.
Hydrocortisone Ointment U.S.P.
Hydrocortisone Tablets U.S.P.

Hydrocortisone Acetate U.S.P., Hydrocortone® Acetate, Cortef® Acetate, Cortril® Acetate, hydrocortisone 21-acetate, is a more stable form of hydrocortisone. It occurs as a white, odorless, crystalline powder that is insoluble in water, soluble 1:230 of alcohol and 1:200 of chloroform.

Category—adrenocortical steroid (anti-inflammatory).

Usual dose—intra-articular, 25 mg. at each site every 2 weeks.

Usual dose range—10 to 50 mg. at each site 1 to 4 times a month.

For external use—topically, 0.5 to 2.5 per cent ointment or suspension.

OCCURRENCE

Hydrocortisone Acetate Ointment U.S.P.
Hydrocortisone Acetate Ophthalmic Ointment U.S.P.
Hydrocortisone Acetate Ophthalmic Suspension U.S.P.
Sterile Hydrocortisone Acetate Suspension U.S.P.

Hydrocortisone Sodium Succinate U.S.P., Solu-Cortef®, hydrocortisone 21-sodium succinate, occurs as a white or nearly white, odorless, hygroscopic, amorphous solid that is very soluble in water and in alcohol. It is suitable for intravenous or intramuscular injection in the management of emergencies amenable to intense corticoid therapy.

Category—adrenocortical steroid (anti-inflammatory).

Usual dose—intravenous or intramuscular, the equivalent of 100 to 250 mg. of hydrocortisone.

Usual dose range—100 to 1 g. daily.

OCCURRENCE

Hydrocortisone Sodium Succinate for Injection U.S.P.

Hydrocortisone Sodium Phosphate U.S.P., Hydrocortone® Phosphate, Hydrocortisone 21-(Disodium Phosphate). This derivative occurs in a white to light yellow hygroscopic powder which is freely soluble in water. The solution, buffered at pH 7.5 to 8.5, is used for intravenous or intramuscular injection.

Category—adrenocortical steroid (anti-inflammatory).

Usual dose—intravenous or intramuscular, the equivalent of 100 to 250 mg. of hydrocortisone.

Usual dose range—100 to 1 g. daily.

OCCURRENCE

Hydrocortisone Sodium Phosphate Injection U.S.P.

Hydrocortisone Cypionate, Cortef® Fluid, is hydrocortisone 21-cyclopentylpropionate and occurs as a water-insoluble, white, tasteless, odorless solid. It is used orally in doses expressed in terms of hydrocortisone for slower absorption from the gastrointestinal tract.

Hydrocortamate Hydrochloride, Magnacort®, hydrocortisone 21-dimethylaminoacetate; is used topically for its anti-inflammatory properties in a 0.5 per cent ointment. The dimethylamino group together with the ester group impart favorable physical properties for good absorption and distribution. It is claimed to be more than twice as potent as hydrocortisone or its acetate.

MODIFIED ADRENAL CORTEX HORMONES

During the synthesis of 11β-OH corticoids, intermediates containing a 9α-halogen were obtained by the action of halogen acids upon the 9,11 epoxide or hypohalous acids upon the $\Delta^{9,\ 11}$-unsaturated compound. Biologic testing of the 9α-halo compounds led to the important discovery that these new derivatives exhibited pronounced glucocorticoid activity.[214] This discovery provided the stimulus for further modifications such as the introduction of new double bonds, methyl groups and hydroxyl groups. It is important to note here that these modifications are small and do not represent an unusual departure from established concepts of structure-activity relationships that previously have been utilized. This is an excellent demonstration of the way in which small modifications can be utilized, singly or in combinations, in several positions in a large molecule that contains several functional groups, to obtain selective enhancement or suppression of one or more of the biologic activities of the parent steroid. Each substituent exerts its own characteristic effect on the biologic activity of the total molecule, an effect that is largely independent of the presence of other activity-enhancing or -modifying groups.

The introduction of a 9α-halo substituent increases the glucocorticoid, anti-inflammatory, anti-arthritic and salt-retention activities. In general, these activities increase with the decreasing atomic weight of the halogen.

The 9α-halo substituent has an axial position and literally is on the opposite side of the steroid molecule that is involved with its biologic activity. Therefore, it cannot interfere with its function. On the other hand, its ability to withdraw electrons from the $C_{11}\beta$-(axial) OH group increases its acidity so it can exert a stronger binding effect at the site of activity.[*] The halogen also may increase liposolubility to facilitate better absorption and distribution and it may prevent the reduction of the biologically important Δ^4-3 keto group in ring A. An analogous situation exists in the 12α-(axial) halocorticoids.[215] However, a 12α-halo group may prevent the reduction in vivo of a C_{11} keto group and thus account for the lower activity of such compounds.

Products

Fludrocortisone Acetate, Alflorone® Acetate, F-Cortef® Acetate, Florinef® Acetate, 9α-fluorohydrocortisone, is a white, odorless, crystalline powder that is very slightly soluble in water and sparingly soluble in alcohol. Although its anti-inflammatory activity is greatly increased, it also has increased salt-retaining properties and, therefore, is recommended for topical use as an anti-inflammatory agent in the treatment of anogenital pruritus, dermatitis, eczema, sunburn, etc.

The salt-retaining activity due to the 9α-fluoro atom is enhanced by a 2-methyl group (q.v.) so that 2 methyl-9α-fluorohydrocortisone is 100 times as active as desoxycorticosterone.[216]

[*] Other electro-negative ($-I$ effect) groups, such as $-OH$, OCH_3, OC_2H_5, exhibit similar effects. Hertz, J. E., *et al.*: J. Am. Chem. Soc. 78:2017, 1956.

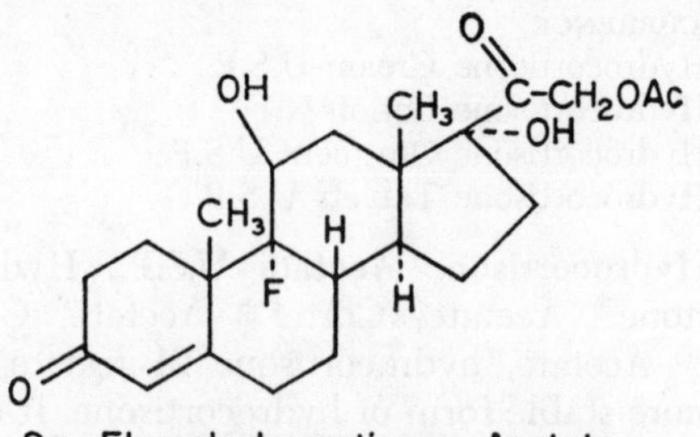

9α – Fluorohydrocortisone Acetate

A double bond in the 1,2-position[217] can be introduced by dehydrobromination of the 2β-(axial) bromo intermediate or very effectively by the use of microorganisms such as *Fusarium solani, Corynebacterium simplex,* etc. The introduction of a Δ^1-double bond increases the glycogenic and anti-inflammatory activity of the parent corticoid 3 to 4 times and does not increase the salt retaining activity.

Prednisone U.S.P., Meticorten®, Deltasone®, Deltra®, 17,21-dihydroxypregna-1,4-diene-3,11,20-trione, Δ^1-dehydrocortisone, a white, odorless, crystalline powder, is very

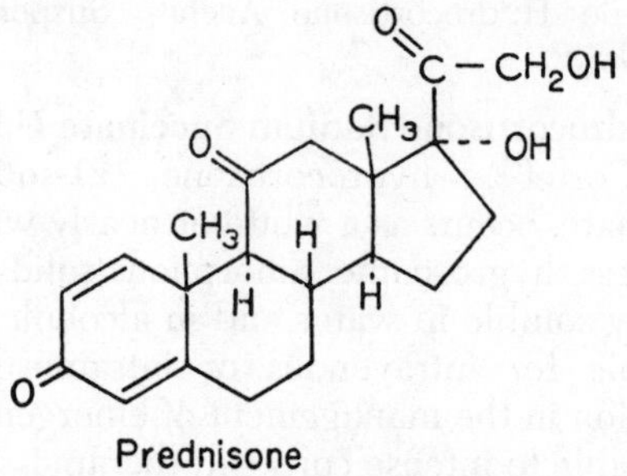

Prednisone

slightly soluble in water, soluble 1:150 in alcohol and 1:200 in chloroform.

Category—adrenocortical steroid (anti-inflammatory).

Usual dose—5 mg. 2 to 4 times a day.

Usual dose range—5 to 80 mg. daily.

OCCURRENCE
Prednisone Tablets U.S.P.

Prednisolone U.S.P., Meticortilone®, Delta Cortef®, Hydeltra®, 11β,17,21-trihydroxypregna-1,4-diene-3,20-dione, Δ^1-dehydrohydrocortisone, occurs as a white, odorless, crystalline powder that is very slightly soluble in water, soluble 1:30 in alcohol, 1:180 in chloroform and soluble in methanol.

Category—adrenocortical steroid (anti-inflammatory).

Usual dose—oral, 5 mg. 2 to 4 times daily.

Usual dose range—5 to 80 mg. daily.

OCCURRENCE

Prednisolone Tablets U.S.P.

Prednisolone Acetate U.S.P., Sterane®, prednisolone 21-acetate, occurs as a white, odorless, crystalline powder that is practically insoluble in water, soluble 1:120 in alcohol and slightly soluble in chloroform.

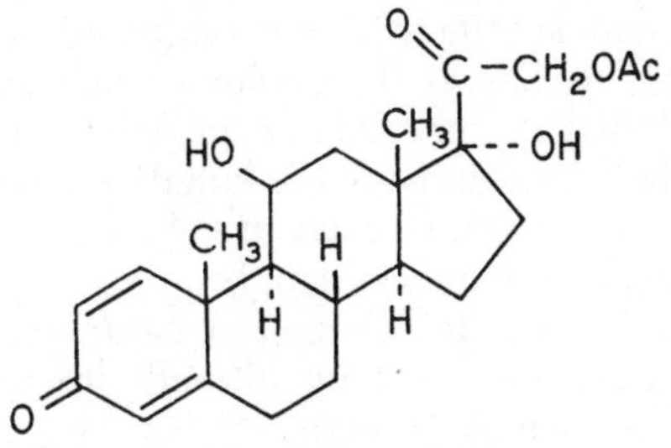

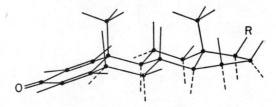

Prednisolone Acetate

Category—adrenocortical steroid (anti-inflammatory).

Usual dose—intra-articular, 10 to 25 mg. at each site every 2 weeks; intramuscular, 5 mg. 4 times a day.

Usual dose range—intra-articular, 5 to 50 mg. 1 to 4 times monthly; intramuscular, 5 to 80 mg. daily.

OCCURRENCE

Sterile Prednisolone Acetate Suspension U.S.P.

Prednisolone Butyl Acetate, Hydeltra-T.B.A.®, prednisolone *tert*-butyl acetate, $\Delta^{1,4}$-pregnadiene-3,20-dione-11β,17α-21-triol *tert*-butyl acetate, is long-acting, due to its low solubility and very slow rate of hydrolysis (q.v.). It therefore is suitable as a suspension for injection only in intra-synovial and soft tissues. It is used in the treatment of inflammation of the joints for bursitis, etc.

Dose—4 to 20 mg.

Prednisolone Sodium Phosphate U.S.P., Hydeltrasol®, prednisolone 21-(disodium phosphate), is a white, water-soluble derivative of prednisolone for a rapid-acting, short-acting corticoid for parental use. It

also is applied topically as an ointment or solution. Aqueous solutions of this disodium salt have a pH of 8.5. The monosodium salt has a pH of 4.5.

Category—adrenocortical steroid (anti-inflammatory).

Usual dose—intramuscular or intravenous, 20 mg.

Usual dose range—intramuscular or intravenous, 10 to 200 mg.

For external use—topically, 0.1 ml. of a 0.5 per cent solution 6 to 12 times a day, to the eye.

OCCURRENCE

Prednisolone Sodium Phosphate Injection U.S.P.

Prednisolone Sodium Phosphate Ophthalmic Solution U.S.P.

Prednisolone Succinate N.F., 11β,17,21-trihydroxypregna-1,4-diene-3,20-dione 21-(hydrogen succinate), the 21 acid succinate of prednisolone, occurs as a fine, creamy white powder with friable lumps. It is very slightly soluble in water and freely soluble in alcohol. Its sodium salt is very soluble in water.

Category—glucocorticoid.

Usual dose—intravenous (as the sodium salt), the equivalent of 25 or 50 mg. of prednisolone given over a 1-minute period, repeated every 3 or 4 hours for 4 doses.

The following formula is presented to permit a better mental image of the nature of those steroids that have a 3 keto 1,4-diene system.

A 3Keto–1,4–diene Steroid

Methyl groups have been introduced in the 2α-(equatorial)[218] and 6α-(equatorial)[219] positions, with the resultant considerable increase in glycogenic, anti-inflammatory and muscle work activity in the rat. The 2α-methyl derivatives markedly increased sodium retention whereas the

6α-methyl derivatives lowered sodium retention. In man and the dog there is no significant increase in glucocorticoid activity with 2α-methylhydrocortisone; however, glycogenic activity (eosinopenia) in man persisted more than twice as long as with hydrocortisone.[219, 220] This duration of action is probably due to the slower rate of reduction of the Δ^4-3 keto group, due to the presence of the 2α-methyl group.*

In the rat the 6α-(equatorial) methyl group[219, 220] doubled the glucocorticoid activity and the anti-inflammatory activity with a marked drop in sodium retention. It should be noted that the 6α-methyl group is in the same general plane as the B ring and could interfere with any biological reduction of the Δ^4-double bond.

Methylprednisolone N.F., Medrol®, 11β, 17,21-trihydroxy-6α-methylpregna-1,4-diene-3,20-dione, 6α-methyl prednisolone, occurs as a white to practically white, odorless crystalline powder that is sparingly soluble in alcohol, slightly soluble in chloroform and insoluble in water. Results to date indicate that it rarely induces peptic ulcer. It has pronounced selective enhancement of corticoid activities.

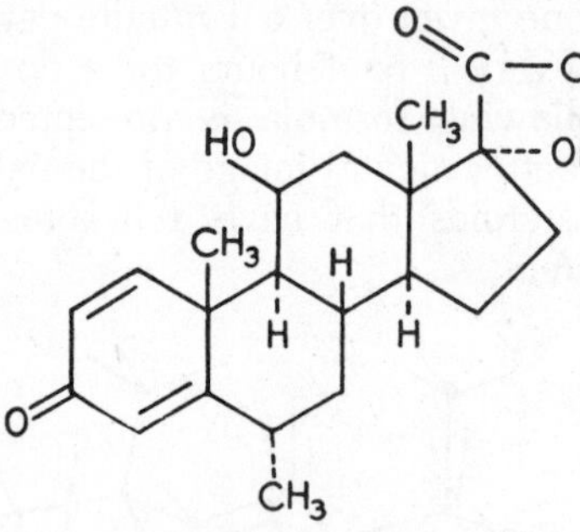

Methylprednisolone

Category—glucocorticoid.
Usual dose—4 mg. 4 times a day.
Usual dose range—2 to 60 mg. daily.

Occurrence
Methylprednisolone Tablets N.F.

* β-Methylacetylcholine furnishes an excellent example of how a methyl group can be in close proximity to a functional group (ester) without complete loss of activity or loss of enzymatic hydrolysis.

Methylprednisolone Acetate N.F., Depo-Medrol®; methylprednisolone 21-acetate, has properties very similar to those of methylprednisolone.
Category—glucocorticoid.
Usual dose—intra-articular or intramuscular, 40 mg.
Usual dose range—10 to 80 mg.

Occurrence
Sterile Methylprednisolone Acetate Suspension N.F.

Methylprednisolone Sodium Succinate N.F., Solu-Medrol®, methylprednisolone 21-(sodium succinate), is a white, or nearly white, odorless, hygroscopic amorphous solid that is very soluble in water and in alcohol. This derivative is suitable for intravenous or intramuscular use.
Category—glucocorticoid.
Usual dose—intravenous or intramuscular, the equivalent of 40 mg. of methylprednisolone every 6 to 24 hours.
Usual dose range—the equivalent of 10 to 125 mg. of methylprednisolone.

Occurrence
Methylprednisolone Sodium Succinate for Injection N.F.

The additive favorable (glucocorticoid) effect of a $\Delta^{1,2}$-double bond and a 6α-methyl group is enhanced to a marked degree by a 9α-fluoro atom: 6α-methyl-9α-fluoro-11β,17α-dihydroxy-21-acetoxy-1,4-pregnadiene-3,20-dione has 120 times the glucocorticoid activity of hydrocortisone when administered parenterally and 190 times by oral administration. However, it has appreciable sodium-retaining properties.[221]

Fluprednisolone N.F., Alphadrol®, is 6α-fluoro-11β,17α,21-trihydroxypregna-1,4-diene-3,20-dione or 6α-fluoroprednisolone. It occurs as a white to off-white, odorless, crystalline powder that is practically insoluble in water, sparingly soluble in alcohol and slightly soluble in chloroform. In man it is 10 times more potent than hydrocortisone on a milligram-for-milligram basis by hyperglycemic-eosinopenic assay.
Category—glucocorticoid.
Usual dose range—750 mcg. to 5.25 mg. 1 to 4 times a day.

OCCURRENCE
Fluprednisolone Tablets N.F.

Fluorometholone N.F., Oxylone®, 9α-fluoro-11β,17α-dihydroxy-6α-methylpregna-1,4-diene-3,20-dione,[222] occurs as a white to yellowish white crystalline powder that is practically insoluble in water, slightly soluble in alcohol and is very slightly soluble in chloroform. It is orally equal to hydrocortisone in man; topically it is 40 times as active, and clinically it is used topically for its anti-inflammatory properties for certain dermatoses, pruriti, etc. Although it has no C_{21}—OH group, the loss in activity due to its absence is offset by the positive contributions of the 9α-fluoro and the 1:2 double bond.* A similar case is that of 21-fluoro-9α-fluoroprednisolone which is 10 to 15 times more potent than hydrocortisone.

Category—glucocorticoid.

For external use—topically, as a 0.025 per cent cream 1 to 3 times a day.

OCCURRENCE
Fluometholone Cream N.F.

A 16α-methyl (away from the top face of the molecule in ring D) markedly increases the anti-inflammatory activity in the rat and the anti-arthritic activity in man with complete suppression of salt retention in these compounds prepared. This also is true of a 16β-methyl group.[223]

Dexamethasone U.S.P., Deronil®, Decadron®, Hexadrol®, 9α-Fluoro-11β,17,21-trihydroxy-16α-methylpregna-1,4-diene-3,20-dione, 9α-Fluoro-16α-methylprednisolone. It occurs as a white, odorless crystalline powder that is sparingly soluble in alcohol, slightly soluble in chloroform, practically insoluble in water and is stable in air. Dexamethasone has 28 to 40 times the anti-inflammatory properties of hydrocortisone in man, with no salt retention.[224] In rats subcutaneously antigranuloma, thymus involution and liver glycogen activities are 104, 47 and 90 times respectively those of hydrocortisone acetate.[225] It is another example of the utilization of the combination of a number of favorable small changes to obtain additive effects.

* It should be noted also that 21-deoxytriamcinolone is 4 times as active as hydrocortisone with no sodium retention.

Category—adrenocortical steroid (anti-inflammatory).

Usual dose—750 mcg. 2 to 4 times daily.

Usual dose range—500 mcg. to 5 mg. daily.

OCCURRENCE
Dexamethasone Aerosol N.F.
Dexamethasone Elixir N.F.
Dexamethasone Tablets U.S.P.

Dexamethasone Sodium Phosphate U.S.P., dexamethasone 21-(disodium phosphate) is a water-soluble (pH 8.5) ester of dexamethasone (1:2). It is slightly soluble in alcohol, freely soluble in water and aqueous solutions 1 in 100 have a pH between 7.5 and 10. It occurs as a white or slightly yellow crystalline powder that is very hygroscopic.

Category—adrenocortical steroid (anti-inflammatory).

Usual dose—intravenous or intramuscular, the equivalent of 2 to 4 mg. of dexamethasone phosphate 6 to 8 times a day.

Usual dose range—2 to 50 mg. daily.

OCCURRENCE
Dexamethasone Sodium Phosphate Cream N.F.
Dexamethasone Sodium Phosphate Ophthalmic Ointment N.F.
Dexamethasone Sodium Phosphate Ophthalmic Solution N.F.
Dexamethasone Sodium Phosphate and Neomycin Sulfate Cream N.F.

Paramethasone Acetate N.F., Haldrone®, 6α-fluoro-11β,17,21-trihydroxy-16α-methylpregna-1,4-diene-3,20-dione 21-acetate occurs as a fluffy, practically white, crystalline powder that is insoluble in water, soluble in ether and chloroform. It has 45 to 50 times the anti-inflammatory and 8 to 9 times the glucogenic activity of prednisolone in animal studies. It has some sodium excretion properties and topically in man is 10 times as anti-inflammatory as hydrocortisone in skin induced irritation by croton oil.

For oral use 2 mg. of paramethasone is equivalent to 5 mg. of prednisolone or 20 mg. of hydrocortisone.

A 16α-methyl group[226] may hinder reduction in vivo of the C_{20} keto group and thus prolong the activity of the parent steroid.

Surprisingly, this also appears to be true of a 16β-methyl group.[223]

Category—glucocorticoid.

Usual dose range—initial, 4 to 12 mg. daily in 3 or 4 divided doses; maintenance, 1 to 8 mg. daily in divided doses.

OCCURRENCE
Paramethasone Acetate Tablets N.F.

Betamethasone N.F., Celestone®, is 9α-fluoro - 11β,17,21 - trihydroxy - 16β-methyl - pregna-1,4-diene-3,20-dione and, thus, is isomeric with dexamethasone at C-16. It represents one of the few instances where a significant contribution to activity is obtained by a perimeter substituent in both the alpha and the beta modifications.

It occurs as a white to practically white, odorless, crystalline powder that is insoluble in water, sparingly in alcohol and slightly soluble in chloroform.

Category—glucocorticoid.

Usual dose—initial, 2.4 to 4.8 mg. daily in divided doses; maintenance, 600 mcg. to 1.2 mg. daily.

Usual dose range—600 mcg. to 8.4 mg. daily.

Application—topically, as a 0.2 per cent cream applied to the skin 2 or 3 times daily.

OCCURRENCE
Betamethasone Cream N.F.
Betamethasone Tablets N.F.

Betamethasone Acetate N.F., Celestone® Acetate, is the 21-acetate of Betamethasone N.F. It occurs as a white to creamy white, odorless powder that is practically insoluble in water, soluble in alcohol and chloroform. This ester provides a more satisfactory form of betamethasone for injection purposes.

Category and Dose—*See* Betamethasone Acetate and Betamethasone Sodium Phosphate Injection N.F.

Betamethasone Sodium Phosphate N.F. is the 21-phosphate ester of Betamethasone N.F. as the disodium salt. It occurs as a white to practically white, hygroscopic powder that is freely soluble in water and practically insoluble in acetone and chloroform. Aqueous solutions have a pH of about 8.5. Like other 21-sodium phosphate corticoids it is rapid acting upon injection.

Category—glucocorticoid.

Sterile Betamethasone Sodium Phosphate and Betamethasone Acetate Suspension N.F. is a mixture containing in each 1 ml. the equivalent of 3 mg. of betamethasone acetate and 3 mg. of betamethasone present as the disodium phosphate.

This combination permits an initial rapid acting effect followed by a more prolonged effect. This preparation is used intramuscularly for systemic corticosteroid therapy or by direct injection into joints, bursae, tendon sheaths, skin and other soft tissues for local anti-inflammatory effects.

Category—glucocorticoid.

Usual dose—intramuscular 1 ml. repeated at intervals of 3 days to 1 week; intra-articular, 0.25 to 2 ml. depending on the size of the joint.

Betamethasone Valerate N.F., Valisone®, is 9-fluoro-11β,17,21-trihydroxy-16β-methyl-pregna-1,4-diene-3,20-dione 17-valerate or the 17-valerate of betamethasone. It occurs as a white to practically white powder that is practically insoluble in water, freely soluble in chloroform, soluble in alcohol and slightly soluble in ether.

Category—glucocorticoid.

Application—topically as a cream, containing the equivalent of 0.1 per cent of betamethasone, to the affected area 1 to 3 times a day.

OCCURRENCE
Betamethasone Valerate Cream N.F.

A 16α-hydroxy (below the plane of the ring D) eliminates or remarkably reduces the sodium-retaining activity[227] of the parent steroid without a commensurate reduction of the glucocorticoid and anti-inflammatory activities which are less than those of the parent compound. But, in the case of 9α-fluoro-16α-hydroxy-1-dehydrohydrocortisone and 9α-fluoro-16α-hydroxyhydrocortisone, conversion to certain 16α,17α-ketals and acetals yielded compounds with remarkable liver glycogen and anti-inflammatory activities, accompanied by excretion or retention of sodium. To account for their activities versus those of the parent compound,[228] the ketal or acetal, it is believed, remain intact in vivo.

Triamcinolone U.S.P., Aristocort® Kenacort®, 9α-fluoro-11β,16α,17,21-tetrahydroxy-pregna-1,4-diene-3,20-dione, 9α-fluoro-16α-hydroxy-Δ¹-hydrocortisone, when administered orally or topically to humans has anti-inflammatory properties that are 10 to 50 per cent greater than those of prednisolone.

In rats by the subcutaneous route, liver glycogen deposition activity is about 3.5 times that of prednisolone acetate which is 5 times that of hydrocortisone.[225] Undesirable side-reactions are less than with prednisolone. It causes some loss of sodium, usually in the first 7 to 14 days.

Category—adrenocortical steroid (anti-inflammatory).

Usual dose—initial, 4 mg. 1 to 4 times daily.

Usual dose range—1 to 32 mg. daily.

OCCURRENCE
Triamcinolone Tablets U.S.P.

Triamcinolone Acetonide U.S.P., Kenalog®, Aristocort® Acetonide, is 9α-fluoro-16α,17α-isopropylidene dioxy-Δ¹-hydrocortisone and in rats subcutaneously its antigranuloma, thymus involution and liver glycogen deposition activities are 48.5, 37.7 and 108 times respectively those of hydrocortisone acetate. Its duration of action also exceeded that of triamcinolone.

Category—adrenocortical steroid (anti-inflammatory).

Usual dose range—intra-articular, 2.5 to 15 mg.; intramuscular, 40 to 80 mg. once a week.

For external use—topically, as a 0.1 per cent ointment, cream, or lotion 2 or 3 times daily.

OCCURRENCE
Triamcinolone Acetonide Aerosol N.F.
Triamcinolone Acetonide Cream U.S.P.
Triamcinolone Acetonide Ointment U.S.P.
Sterile Triamcinolone Acetonide Suspension U.S.P.

Triamcinolone Diacetate N.F., Aristocort®, 9α-Fluoro-11β,16α,17,21-tetrahydroxypregna-1,4-diene-3,20-dione 16,21-Diacetate. This derivative possesses anti-inflammatory properties similar to those of the acetomide.

Category—glucocorticoid.

Usual dose—4 mg. daily.
Usual dose range—4 to 30 mg. daily.

OCCURRENCE
Sterile Triamcinolone Diacetate Suspension N.F.
Triamcinolone Diacetate Syrup N.F.

Fluocinolone is 6α,9α-Difluoro-16α-hydroxyprednisolone and subcutaneously in rats its antigranuloma, thymus involution and liver glycogen activities are, respectively, 19.7, 8.5, and 44.2 times those of hydrocortisone.[225]

Fluocinolone Acetonide N.F., is 6α,9α-difluoro-16α-hydroxyprednisolone 16α,17α-acetonide and subcutaneously in rats its antigranuloma, thymus involution and liver glycogen activities are, respectively, 446, 263 and 138 times those of hydrocortisone acetate.[225] Topically it is 40 to 100 times as effective as hydrocortisone. It has proved effective in the treatment of a variety of dermatoses such as psoriasis that had been resistant to other systemic and topically active steroids.

Category—glucocorticoid.

For external use—topically, as a 0.01 to 0.2 per cent cream, ointment, or solution.

Flurandrenolide N.F., Cordan®, is 6α-fluoro-11β,16α,17,21-tetrahydroxypregn-4-ene-3,20-dione 16,17-acetonide or 6α-fluoro-16α-hydroxyhydrocortisone 16,17-acetonide. A 0.05 per cent solution has greater anti-inflammatory potency than a 1 per cent solution of hydrocortisone. It is available in 0.05 per cent preparations for topical application.

Category—glucocorticoid.

For external use—topically, as a 0.025 per cent or 0.05 per cent cream or ointment to the affected area 2 or 3 times daily.

OCCURRENCE
Flurandrenolide Cream N.F.
Flurandrenolide Ointment N.F.

Dichlorisone Acetate, Diloderm®, Disoderm®, is the 9α,11β-dichloro analog of prednisolone and occurs as a white crystalline powder that is very slightly soluble in water and slightly soluble in alcohol.

This steroid is used as an aerosol spray or applied topically for its anti-inflammatory effects, where it is 4 times as effective as is hydrocortisone.

This corticoid is unique in that the oxygen function at position 11 thought to be essential for significant glucocorticoid and anti-inflammatory activity is replaced by a chlorine atom. However, if enough favorable groups are ultilized some perimeter essential groups can be deleted. q.v.

A 6β-(axial, up)[228] chloro substituent decreases the thymolytic and anti-inflammatory activities and increases sodium retention. A 6α (equatorial, in the plane of the ring) chloro substituent enhances the thymolytic and anti-inflammatory activities, with moderate sodium retention to sodium loss.[229]

Multiple α-halosubstituted prednisolones have yielded highly active corticoids[230]; thus, 16α-chloro-9α-fluoro; 6α,9α,16α-trifluoro; and 16α-chloro-6α,9α-difluoroprednisolones have respectively 313, 480 and 1100 times the anti-inflammatory and 360, 425 and 1030 times the glucocorticoid activity of hydrocortisone.

A 12α-fluoro atom also increases the glucocorticoid activity of hydrocortisone, for probably much the same reasons as a 9α-fluoro atom.[231] However, 2α-fluorohydrocortisone is less active (about one third) than hydrocortisone in liver glycogen and anti-inflammatory assays.[232]

The 3'-phenyl[3,2-d]-3'H-1',2',3'-triazole function is an effective activity-enhancing group; thus, 11β,17α,21-trihydroxy-6,16α-dimethyl-20-oxo-3'-phenyl-4,6-pregnadiene[3,2-d]-3'H-1',2',3'-triazole-21-acetate had 190 times the anti-inflammatory activity of hydrocortisone.[233] This also is true of the [3,2-c] pyrazoles; thus, 9α-fluoro-6,16α-dimethyl-4,6-pregnadiene-11β,17α,21-triol-3,20-dione-[3,2-c]-2'-phenylpyrazole has 2,000 times the systemic granuloma activity of hydrocortisone or about 12.5 times that of dexamethasone.[234] 9α-Fluoro-16α-methyl-4-pregnene-11β,17α,21-triol-3,20-dione-[3,2-c]-2'-p-fluorophenylpyrazole has 500 times the systemic granuloma activity of hydrocortisone.[235] In the above examples the 2'-phenyl or 2'-p-fluorophenyl group markedly enhances the activity of the parent [3,2-c] pyrazole which if unsubstituted is only 10 times as active as hydrocortisone; a 2'-

methyl group gives only 1.5 times the activity of hydrocortisone.

19-Norhydrocortisone is less active (glycogenic) than hydrocortisone; however, 19-nordesoxycorticosterone has twice the sodium-retaining properties of desoxycorticosterone. (See other 19 norsteroid hormones).[236]

The substitution of Cl or F for the 17α-hydroxyl of cortisone almost obliterates adrenocorticoid activity as measured by the eosinophil test in the mouse.[237]

21-Aldehyde hydrates[238] are as active as the parent compounds in the case of cortisone and hydrocortisone, probably because of the reduction in vivo to the 21-hydroxyl. However, the corresponding 9α-chloro and 9α-fluoro compounds have 1/30 the glycogenic and the salt-retaining activities of the parent 21-hydroxy corticoid. The 9α-halo substituent probably inhibits the in-vivo reduction of the 21-aldehyde group.[239]

Not all the essential perimeter functional groups found in hydrocortisone are necessary to equal or exceed its anti-inflammatory and glucocorticoid activities when certain perimeter modifications have been made. Examples are: (1) Fluorometholone (q.v.); (2) 21-fluoro-9α-fluoroprednisolone, which is 10 to 15 times as active as hydrocortisone, with no salt retention; (3) Delmeson (Hoechst) 21-desoxy-6α-methyl-9-fluoroprednisolone; (4) 9α-fluoro-11β,17α-dihydroxy-6α-methyl-3,20-dioxopregna-1,4-dien-16α-acetic acid-γ-lactone, with 72 times hydrocortisone's anti-inflammatory activity and 63 times its glucocorticoid activity, with no salt retention;[240] (5) 16α,17α-isopropylidene dioxy-6α-methylpregna-1,4-diene-3,20-dione, which is almost as active[241] as prednisolone and topically is equal to hydrocortisone and therefore is presumed to be active per se, and (6) 6,16α-dimethyl-4,6-pregnadiene-11β,17α-diol-3,20-dione-[3,2-c]-2'-p-fluorophenylpyrazole, which has 468 times the systemic granuloma activity of hydrocortisone, or about 3 times that of dexamethasone.[234]

14α-Hydroxyhydrocortisone is more active than hydrocortisone and has been found to be an anti-inflammatory agent in rheumatoid arthritis.[242]

The modification of the steroid nucleus has produced a compound having new activities: Δ^5-androsten-3β,16α-diol (Cetadiol®) is claimed to have a tranquilizing effect in acute alcohol withdrawal symptoms and appears to lower blood pressure in hypertension caused by mental factors.[243] 3-(3-Oxo-17β-hydroxy-4-androsten-17α-yl) propionic-γ-lactone and its 19-nor homolog reversibly block the action of aldosterone and desoxycorticosterone.[244] This blocking action is the first example of a modified steroid that has antimetabolite properties, although some of the steroids (q.v.) do lead to some salt loss and therefore possibly also could have antimetabolite properties.

Although the modified corticoids showed selective enhancement or suppression of one or more biologic activities, many of them, such as 1-dehydrocortisone and 1-dehydrohydrocortisone and others, cause decrease in body thymus, spleen and adrenal weights per unit of body weight as well as increases in the relative weights of liver, kidney, testes, thyroid and pituitary in young intact male rats and in these activities may exceed the parent steroid. Other undesirable side-effects which may develop on prolonged use of the corticoids and modified corticoids include fat deposition, hirsutism, hyperadrenalism and rounding of the face as well as adverse effects on nitrogen-calcium balance, gastric secretion (leading to ulcer formation) and the central nervous system.

In some cases of the modified corticoids there is a high degree of correlation between anti-inflammatory activity and antirheumatic activity in man. Low glucocorticoid activity with high anti-inflammatory activity has yet to be established.

At present, complete separation of anti-inflammatory and glycogenic activities has not been accomplished in the modifications of the corticoids. However, effective ·inhibition of liver-glycogen deposition without antagonism of the antigranuloma activity of hydrocortisone has been demonstrated in rats, by the simultaneous administration of Δ^1-testololactone and hydrocortisone. Testololactone has no androgenic, estrogenic, progestational or corticoid activities.[225]

Very often a study of the products resulting from the in-vivo metabolic alteration of a drug, an essential metabolite, etc., can be utilized to good advantage in the preparation of new medicinal agents. This appears to be so in the case of the steroids. Some androgens, estrogens, progesterone, adrenal corticoids and related steroids are oxygenated by the following mammalian tissues: dog liver, bovine adrenals, rat liver, human adrenal mitochondria and human placenta. Hydroxyl groups are introduced at 2α, 2β, 6α, 6β* and 16α-positions, as the case may be. A keto group has been introduced at C_{16} in estradiol and estrone and at C_6 in progesterone. The metabolic attack at these positions is prevented by the introduction of a methyl group or a halogen atom, with resultant increase in one or more of the activities in many cases.

The C_3-propyl enol ether of cortisone acetate has the same anti-inflammatory activity as prednisone, whereas its parenteral activity is decreased.[133]

Effective mineralocorticoid antagonistic activity has been demonstrated with progesterone[245] 15-oxygenated progesterones,[246] and 18-hydroxyprogesterone.[247]

Some steroids on local application have no activity, whereas after systemic administration they have high activity. In some cases the biologic properties may vary with the mode of administration.

THE BIOGENESIS OF STEROID HORMONES

It might be well to point out here the biogenetic relationship of the steroid hormones. No attempt will be made to postulate the significance of these relationships at this early date in this field. Where cholesterol has been proved to be the precursor to the steroid hormones in the adrenal cortex, the side chain has been cleaved at C_{20} to yield Δ^5-pregnen-3-ol-20-one (pregnenolone). It is rapidly converted to progesterone and the latter is hydroxylated by the appro-

---

* 6β-Hydroxycorticosterone is found in human urine.

priate hydroxylases at positions C_{17}, C_{21}, C_{11} and C_{18} to give the adrenal cortical hormones or their precursors that have been described previously in this chapter. In the case of the C_{11} and the C_{17} hydroxylations, the $C_{11}-OH$ is beta and the $C_{17}-OH$ is alpha. Desoxycorticosterone, for example, has been shown to be converted to aldosterone by the adrenal gland.

Testosterone is synthesized[248] in the gonads according to the following sequence: Cholesterol → Pregnenolone → Progesterone → 17α-Hydroxyprogesterone → Δ^4-Androstene-3,17-dione → Testosterone. Testosterone also is elaborated in small amounts by the adrenals and the human ovarian tissue. Some Δ^4-androstene-3,17-dione is hydroxylated in vivo at position 11.

Carbon 19 must be eliminated as one step in the synthesis (in vivo) of the estrogens, and the initial step is via the C_{19}-hydroxylated steroid, because 19-hydroxy-Δ^4-androstene-3,17-dione and 19-oxo(aldehyde)-Δ^4-androstene-3,17-dione are converted relatively efficiently to estrone. The aldehyde is oxidized to the acid, and C_{19} carbon is lost as CO_2. Subsequent dehydrogenation is followed by aromatization of ring A.[249] Radioactive estrone has been found in mare's urine when testosterone-4 ^{14}C had been administered.[250]

Biologic hydroxylations[251] of the steroids are effected by liver, adrenal, gonadal and placental tissues as well as with microbial systems (q.v.). In adrenal preparations and microbiologic systems molecular oxygen is utilized to effect direct substitution with inversion to produce the hydroxylated steroid. Liver tissue has been shown to contain 3 distinct hydroxylating systems for C_{19} steroids, including 2β, 6β and 16α. Adrenals contain 3 hydroxylating systems for C_{19} steroids that are 6α, 6β and 11β.

CARDIAC GLYCOSIDES

Digitalis purpurea and other Digitalis species and the cardiac glycosides have a long and interesting history. Whole leaf preparations were first investigated and written records indicate that as early as the 16th century physicians were praising the medicinal properties of *Digitalis purpurea*. However, inconsistencies in the therapeutic results obtained, together with some fatalities, caused the removal of digitalis from the London Pharmacopoeia. However, after William Withering in 1785 published a treatise entitled "An Account of the Foxglove and Some of Its Medical Uses," *Digitalis purpurea* was readmitted to the London Pharmacopoeia.

Since that time there has been a gradual increase in the use of digitalislike substances and today the cardiac glycosides are referred to as venerable and irreplaceable drugs. A rather large number of naturally occurring substances are related closely in structure and activity to the active compounds of foxglove, i.e., digitalis. These compounds, because of their glycoside structures and cardiac action, often are called the cardiac glycosides or heart poisons. They are characterized by the highly specific and powerful action which they exert upon the cardiac muscle. In the proper dosage, many of them are very valuable in the treatment of congestive heart failure (q.v.) due to heart disease or other causes. They stimulate the heart to greater contractile activity and restore the original tonicity. Thus ventricles are more completely emptied, blood flow is increased in the heart and through all tissues, heart muscle nourishment is increased and becomes more efficient and the heart size is decreased.

In patients with atrial fibrillation and rapid ventricular rate digitalis dramatically slows the ventricular rate, probably by two mechanisms—one vagal (increased force of the heart's systolic action) and one extravagal (increased dose of digitalis).

Cardiac glycosides also exert a positive diuretic action and a positive inotropic effect when applied directly to cardiac tissue. It is significant that certain steroids can produce positive inotropic action directly upon cardiac tissue.

These active glycosides are found in a number of plant families, such as *Apocynaceae, Scrophulariaceae, Liliaceae* and *Ra-*

nunculaceae. Some of the so-called African arrow poisons are prepared from the *Strophanthus* species, *Apocynum cannabinum, Periploca graeca, Gomphocarpus* and other plants. The sea onion or squill, *Scillia maritima,* also has proved a useful drug. Squill is one of the oldest drugs used in medicine. Squill was used in ancient times by the Egyptians, Greeks and Romans for the treatment of many illnesses, and they worshiped the plant as a general protector against evil. The earliest mention of squill is found in a medical prescription contained in the Papyrus Ebers (about 1500 B.C.). G. L. B. van Swieten, in 1764, emphasized the importance of squill as a remedy against dropsy and used, with clear insight, fresh squill because he found it to be especially active. However, Withering's introduction of the foxglove as a cardiac remedy resulted in the disuse of squill, even though F. Home, in 1780, discovered that squill was cardioactive. Nevertheless, squill and its preparations were always mentioned in pharmaceutical textbooks.

Assay Methods

The physiologic potency of the cardiac glycosides and tinctures of the drugs that contain them may be determined experimentally by injecting an aqueous alcoholic solution of the glycoside into the bloodstream of pigeons or cats or in the lymph sac of frogs. The smallest dose or amount of the substance necessary to produce systolic standstill of the heart or cardiac arrest is determined, and the so-called minimum lethal dose (M.L.D.) is calculated per kilogram of cat or per gram of frog. The pigeon assay is official in the *U.S.P.* and is more economical than the cat assay and more reliable than the frog assay.

The relative digitoxin sensitivities of man, dog, cat, rabbit, rat and toad differ considerably, with a total range of 1,000. The toxicity values found in the cat assay can be utilized directly in man. Approximately 5 cat LD's of any glycoside or aglycone when injected intravenously into a patient with cardiac failure always produces evidence of digitalization of varying duration without appearance of toxic symptoms.

Also, approximately 3 cat LD's of any of the glycosides can be suggested as the initial dose given I.V. or by mouth (if well absorbed) as a means of lowering the ventricular rate of atrial fibrillation.

The utilization of LD values for therapeutic application must be integrated with the mode of administration and the duration of action (q.v.).

The value of the results of the methods of assay described above is questionable. Absorption from the lymph sac of the frog cannot be correlated with the absorption from the human gastrointestinal tract. Further, in the case of the cat and pigeon assays, the active glycosides usually can exert their full effect, whereas, when administered orally to humans or cats, marked differences in the rate and degree of absorption are found.

It has been postulated[252] that the chloroform/aqueous methanol partition coefficient of some purified cardiac glycosides (see Table 93) can be correlated with the

TABLE 93

DRUG	DISTRIBUTION COEFFICIENT $CHCl_3$/16% aq. Me OH
Acetyldigoxin	98
Digitoxin	96.5
Lanatoside C	16.2
Gitalin	90.5
Digoxin	81.5
Gitoxin	Very low

rate and the degree to which the glycosides are absorbed orally in cats.* Furthermore, it was shown in these absorption studies that particle size also plays a marked role. These results can be briefly summarized by stating that high partition coefficient favors good absorption, whereas a low partition coefficient favors poor absorption. Gitalin, which has a high partition coefficient, is well absorbed; however, unlike the other glycosides mentioned, it induces nausea in cats. Gitoxin was so insoluble, in both aqueous methanol and chloroform, that its partition coefficient was not determined. The oral absorption in cats of gitoxin was very poor. One can add to the above data the well-known fact that ouabain and strophanthin are absorbed so

* See Figs. 22, 23, 24 and 25.

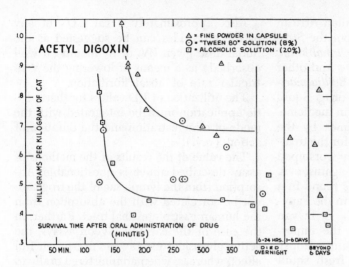

FIG. 22. Unanesthetized cats were given single, oral doses of acetyl digoxin in 10 ml. of 20 per cent alcohol or 8 per cent "Tween 80" in water by somach tube, followed by 30 ml. of water. Capsules containing finely ground powder were placed in the mouth and washed down with 20 to 30 ml. of water in 5-ml. portions.

poorly and erratically that they are not recommended to be used orally. A check on their solubilities indicates that they are poorly soluble in chloroform but quite soluble in water. Therefore, they would have very low partition coefficients and fit the aforementioned general postulates.

It thus appears that cardiac glycosides are passively [252a] absorbed from the G.I. tract and that nonpolar cardiac glycosides are able to diffuse readily across the intestinal mucosa whereas polar glycosides are not.

It is well known that the various species of *Digitalis,* particularly *Digitalis purpurea* and *D. lanata,* yield mixtures of glycosides. Furthermore, the composition of these mixtures is not constant, particularly if the plant is grown in different habitats (Tables 94 and 95 illustrate this point).[253]

Leaves carefully and rapidly dried under specified conditions (usually at about 50°) chiefly contain the native glycosides and an enzyme that can partially degrade these glycosides to give the desglucoglycosides. The enzyme is active at pH's from 7 to 1 and in concentrations of ethanol or methanol at least up to 35 per cent. It is inactivated by 70 per cent or greater concentrations of alcohol. Thus, tinctures prepared from carefully dried leaves contain the glycosides chiefly as native glycosides. Whole leaf preparations should yield the generation of the desglucoglycosides in the G. I. tract.

It becomes apparent immediately from the tables and the previous discussion that wide differences could exist between intravenous assay potencies and oral absorption when whole leaf preparations or tinctures are used. The amount of activity obtained upon the oral administration of such preparations would be correlated with the kinds and percentages of the different glycosides that are present.

Similar studies have been conducted using humans, with essentially the same results.[254]

Another factor of importance is the time required for a material to develop its full effect. All of the effects likely to occur from

TABLE 94. *Digitalis purpurea**

No.	SOURCE	EXTRACT OF 1 KG. OF DRIED LEAF	
		DIGITOXIN (MG.)	GITOXIN (MG.)
1.	Thuringia	5	420
2.	Black Forest	500	200
3.	Vauges	630	000
4.	Swiss Jura	150	700
5.	Vauges	490	50
6.	Harz	130	260
7.	Unknown Commercial	270	130
8.	Unknown Commercial	180	540
9.	U.S.A. Commercial	330	290
10.	Unknown Commercial	210	700

Data from Movitt, E.[253]

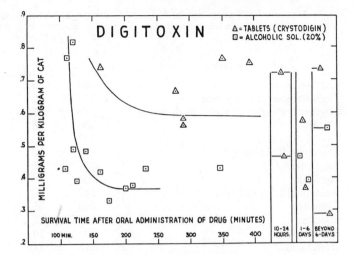

Fig. 23. Unanesthetized cats were given single, oral doses of digitoxin in 10 ml. of 20 per cent alcohol, followed by 30 ml. of water by stomach tube. The tablets were given by mouth followed by 20 to 30 ml. of water in 5-ml. portions to induce swallowing.

a single injection of ouabain or crystalline strophanthin take place within 2 hours. A single injection of digitoxin develops its full effect in about 6 hours. Digifoline, lanatoside C and digoxin stand between 2 and 6 hours. Ouabain is the drug usually used in an emergency for an immediate effect. Also, it follows that digitoxin dosage by intravenous injection need not be repeated in less than 4- or 5-hour intervals. The duration of action of digitoxin is greater than that of ouabain; i.e., two thirds of the effect of ouabain wears off in 4 days, whereas less than one third of the effect of digitoxin wears off in the same length of time.[255] Digoxin and lanatoside C are intermediate in their duration of action.

The cardiac glycoside genins are short acting, presumably because in part they are not bound to heart tissue—a property at present attributed only to the glycosides

(glycoside linkage) with one or more sugar residues. The sugar residue is not essential for activity, as both the free genin and its monoside are promptly effective in the isolated heart.

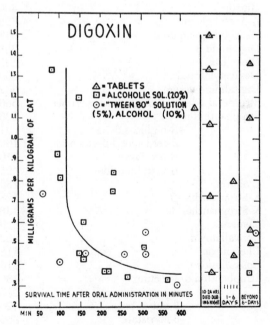

TABLE 95. *Digitalis lanata**

No.	Source	% OF GLYCOSIDES LANATOSIDE A	% OF GLYCOSIDES LANATOSIDE C
1.	Neighborhood of Vienna	48	37
2.	Near Basel	55	39
3.	Near Vienna	34	52
4.	From France	25	62

* Data from Movitt, E.[253]

Fig. 24. Unanesthetized cats were given single oral doses of digoxin in 10 ml. of 20 per cent alcohol or in aqueous "Tween 80" solution, followed by 30 ml. of water by stomach tube. Tablets were placed in the mouth and washed down with 20 to 30 ml. of water in 5-ml. portions.

TABLE 96. M.L.D. IN CATS OF A NUMBER OF COMPOUNDS*

COMPOUND	ASSAY IN CATS M.L.D. (MCG./KG.)		SUGARS OR OTHER RESIDUES
Digitoxigenin	459	± 36	None
Digitoxigenin-β-D-glucoside	124.7	± 12	Glucose
Digitoxigenin-β-tetraacetyl-D-glucoside	1,184	± 127	Tetraacetyl glucose
Digitoxin	330	± 8	3-Digitoxose
Acetyl digitoxin	447		3-Digitoxose + AcOH
Desacetyldigilanide-A	469	± 13	3-Digitoxose + glucose
Digilanide-A	361	± 17	3-Digitoxose + glucose + AcOH
Neriifolin	196		Thevetose
Acetylneriifolin cerberin	147		Thevetose + AcOH
Thevetin	920	± 35	Thevetose + 2-glucose
Somalin	288	± 17	Cymarose
Odoroside-H	200	± 10	Digitalose
Honghelin	213	± 81	Thevetose
Digoxigenin	441	± 42	None
Digoxigenin-β-D-glucoside	100	± 5	Glucose
Digoxigenin-β-tetraacetyl-D-glucoside	1,057	± 147	Tetraacetyl glucose
Digoxin	230	± 10	3-Digitoxose
Acetyldigoxin	380		3-Digitoxose + AcOH
Desacetyl digilanide C	228		3-Digitoxose + glucose
Digilanide C	232	± 18	3-Digitoxose + glucose + AcOH
Gitoxin	400	± 12	3-Digitoxose
α-Acetyl gitoxin	525		3-Digitoxose + AcOH
Deacetyldigilanide B	548	± 21	3-Digitoxose + glucose
Digilanide B	388	± 28	3-Digitoxose + glucose + AcOH
Oleandrin	197		Oleandrose + C_{16}-OAc
Desacetyloleandrin	300		Oleandrose
Digitalinum verum	1,332	± 193	Digitalose + glucose
Strospesid (desglucodigitalinum verum)	586		Digitalose
Strophanthidin	306	± 39	None
" -β-D-glucoside	91.3	± 2.46	Glucose
" -β-tetraacetyl-D-glucoside	1,166	± 125	Tetraacetyl glucose
" -β-D-xyloside	110	± 4.39	Xylose
" -β-L-arabinoside	95	± 3	Arabinose
Cymarin	130	± 3	Cymarose
K-Strophanthin-β	128		Cymarose + glucose
K-Strophanthoside	186		Cymarose + 2-glucose
Cymarol	99		Cymarose
Convallatoxin	80	± 2	Rhamnose
Cheirotoxin	118		Lyxose + glucose
Convalloside	215		Rhamnose + glucose
Strophanthidin-3-acetate	186.6	± 24.6	Acetic acid
" -3-propionate	257		Propionic acid
" -3-butyrate	426		Butyric acid
" -3-myristate	983		Myristic acid
" -3-benzoate	2,720		Benzoic acid
Perilogenin	719		None
Periplogenin-β-D-glucoside	125		Glucose
Periplocymarin	154		Cymarose
Emicymarin	138		Digitalose
Periplocin	121		Cymarose + glucose
Uzarin	5,080	± 437	2-Glucose
Hellebrigenin	76.9	± 5.5	None
Hellebrigenin monoacetate	66.2	± 2.9	Acetic acid

TABLE 96. (*Concluded*)

COMPOUND	ASSAY IN CATS M.L.D. (MCG./KG.)			SUGARS OR OTHER RESIDUES
Hellebrin	100			Glucose + rhamnose
Desglucohellebrin	86			Rhamnose
Hellebrigenol monoacetate	77			None
Scillarenin	125			None
Scillaridin (anhydroscillarenin)	1,660			None
Proscillaridin A (desglucoscillaren A)	157			Rhamnose
Scillaren A	146			Rhamnose Glucose
Scillirosidin	70	±	3	None
Acetyl sillirosidin	201	±	8	AcOH
Scilliroside	130	±	5	Glucose
Sarmentogenin	458	±	32	None
Sarmentocymarin	210	±	8	Sarmentose
Sarmentoside A	112			Fucose
Ouabain	120	±	2	Rhamnose
Calotropin	120	±	2	Methylreductinic acid
Calotoxin	112			Hydroxymethylreductinic acid

* Data from Chen, K. K., *et al.*[257]; Uhle, F. C., and Elderfield, R. C.[257]; Elderfield, R. C., *et al.*[257]

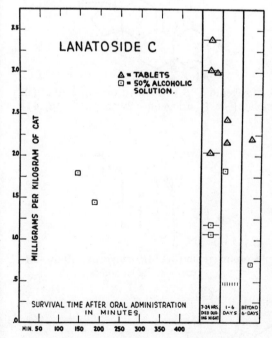

FIG. 25. Unanesthetized cats were given single, oral doses of lanatoside C in 10 ml. of 50 per cent alcohol, followed by 30 ml. of water. Tablets were placed in the mouth and washed down with 20 to 30 ml. of water in 5-ml. portions.

Cardiac activity is not entirely limited to the glycosides, for some of the *Erythrophloeum* alkaloids and the bufagins and bufotoxins of the toads exhibit the same type of activity.[256] Table 96 gives the M.L.D. in cats of a number of naturally occurring cardiac glycosides, some synthetic glycosides and their acetates, strophanthidin and its acetate, a few other aglycons together with some squill glycosides.[257]

STRUCTURE-ACTIVITY RELATIONSHIPS

The aglycon portion of the cardiac glycosides contains the essential structures for biologic activity (q.v.). The sugar residue or residues or other derivatives that have no biologic activity per se, such as a C_{16}-OAc, in many cases markedly increase activity (q.v.) (as much as 35 times[258] in the case of gitoxigenin versus its 16-formate (gitaloxigenin)) via contributions of a secondary nature.

With the exception of the squill glycosides and hellebrin from *Radix Helleborus niger,* the cardiac glycosides are substituted etiocholanyl-butenolides or 21-hydroxy-norcholenic acid lactones and etioallocholanylbutenolides or 21-hydroxy norallocholenic acid lactones. The type and extent of substitution, together with certain stereochemical configurations in the aglycons,[259] determines very markedly the degree of activity in the cardiac glycosides. Most cardiac aglycons contain the basic nucleus found in car-

danolid (cardogenan, see below). In addition to the cardanolid nucleus, the following structural features are the minimum requirements for marked cardiac activity; an axial 3β-hydroxyl group; $C_{14}OH$ (β orientated and a double bond in the 21–23 position) (see digitoxigenin).

NOTE: See Helv. chim. acta 34:1680, 1951 and Chemistry and Industry, 1951, on nomenclature in the field of steroids. Stoll, A.: Helv. chim. acta 34:214, 1950 proposes the term diganolide because the six-membered ring in the squill glycosides also produces cardioactive compounds. See also I.U.P.C.A. Definitive rules for nomenclature of steroids, J. Am. Chem. Soc. 82:5577, 1960, which are used here and elsewhere in this text.

5β-Cardanolide or Cardogenan

14-Hydroxy-5β-Cardanolide
(All chair conformation)

Digitoxigenin [3β,14β-Dihydroxy-5β-card-20(22)-enolide] or [Cardogenin-(20:22)-diol-(3β,14)]

Uzarigenin [3β,14-Dihydroxy-5α-card-20(22)-enolide]

Corotoxigenin [3β,14β-Dihydroxy-19-oxo-5β-card-20(22)-enolide]

Sarmentogenin [3β,11α,14β-Trihydroxy-5β-card-20(22)-enolide]

An appreciable loss in activity is found in the $C_{3\alpha}$-OH epimers. Activity[258] is still retained when the $C_{3\beta}$-OH is oxidized to a keto group, e.g. Δ^4-digitoxigenone (4-anhydroperiplogenone) (0.7). Although no examples are found in the cardenolides, in the case of the bufadienolides activity, though of a lower order, is retained by the replacement of the $C_{14\beta}$-OH by a 14,15-β-epoxide.

Recently[260] significant activity (3×10^{-6} g./ml.) was obtained with 14-deoxy-14β H-uzarigenin in isolated frog's heart (Straub's preparation). Thus a 14β-OH is not indispensable for the specific cardiotonic activity. The decreased activity of uzarigenin is due to the *trans* fusion of rings A and B.

Even though AB *trans* fusion lowers activity greatly, a Δ^4-double bond does not alter activity significantly (scillarenin (8) versus bufalin (7.3)); however, 5-anhydro-digitoxigenin (0.8) is less than half as active as digitoxigenin (2.2). The β-orientation of the side chain is necessary for maximum activity: the allo compounds formed by enzymatic activity, in which this side chain is α-oriented, are inactive. The double bond[261] in the lactone side chain is necessary for marked activity, as reduction of the double bond in this ring in cymarin reduces activity to one twenty-third of its original value. Dihydroouabain is one sixteenth as active as ouabain. Under certain conditions, the double bond in the side chain can shift to the β-γ position and react with the C_{14}-OH to form a ring containing an oxygen bridge to give the iso compounds which are inactive physiologically.

Additional single oxygen functions may increase activity,* (e.g., a C_{19}-CHO in corotoxigenin), decrease activity (e.g., a $C_{5\beta}$-OH in periplogenin (1.4), a $C_{1\beta}$-OH

* Expressed in LD's per mg.: digitoxigenin = 2.2 (Chen, K. K.[257]; Tamm, C.: New Aspects of Cardiac Glycosides 3:11, 1963)

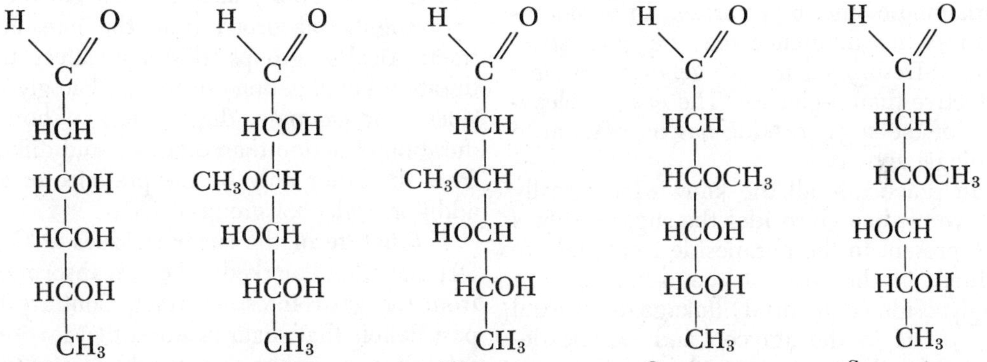

Digitoxose   Digitalose   Diginose   Cymarose   Sarmentose

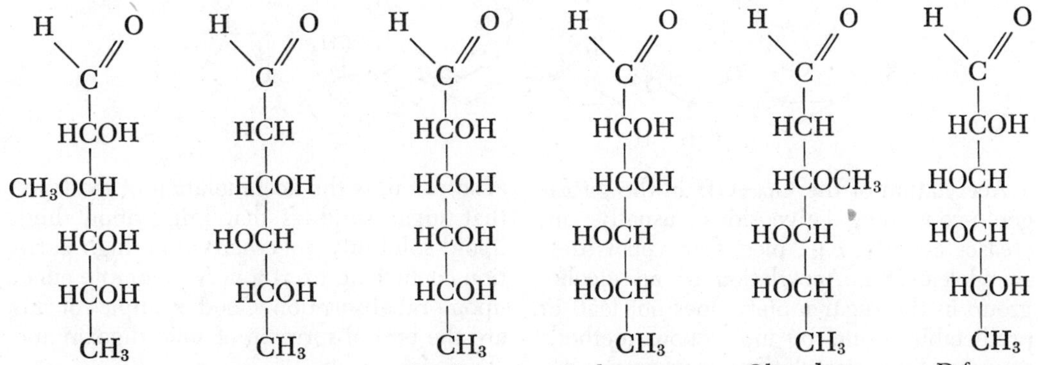

Thevetose   Boivinose   D-allo-methylose   L-rhamnose   Oleandrose   D-fucose

in acovenosigenin (1.4), a $C_{16}-\beta$-OH in gitoxigenin (0.3)), or effect essentially no change in activity (e.g., a $C_{12}-\beta$-OH in digoxigenin (2.3) and $C_{11}-\alpha$-OH in sarmentogenin (2.2)). In the case of 2 or more additional oxygen functions, activity may be increased (e.g., C_{19}—CHO + $C_5-\beta$-OH in strophanthidin (3.9); C_{11} keto + $C_{12}-\beta$-OH in caudoside (7.1); $C_{11}-\alpha$-OH + C_{12} keto in sinoside (8.5)), whereas activity is decreased in caudosine) (1.0), C_{11}—keto + $C_{12}-\alpha$-OH.

The sugars on page 827 have been found in the cardiac glycosides.

The nature of the sugar residue or residues at C_3—OH may or may not affect markedly the cardiac activity when measured by intravenous assay methods (see Table 96). Generally speaking, the monosides are the most active on a milligram-for-milligram basis when compared to the di-, tri- and tetrasides. A variation of the sugar residue in the monoside of a given aglycon also may affect the potency in the following decreasing order: D-glucose, D-diginose, L-rhamnose and D-cymarose. In some instances this difference may be quite striking. 2-Desoxy-sugars in general are less effective than D-glucose. The ease of cleavage of the sugar residue has no relation to biologic activity.

In practically all the known biologically active cardiac glycosides the sugar residues are present in the pyranoside form and are joined, in the case of the D-sugars, in a β-glycoside (equatorial) linkage to the axial $C_3-\beta$-OH in the aglycon and can be depicted by the type glycoside as follows.

glycoside, e.g., digoxin versus acetyl digoxin; and penta-acetylgitoxin[262] versus gitoxin. These differences in activity in oral absorption may well be a function of the partition coefficient, which is increased upon acetylation.

Intraduodenally, penta-acetylgitoxin is absorbed[262] more rapidly than digitoxin; however, it is deacetylated quickly in vivo, and blood levels are reduced in a relatively short time.

Formylation of the $C_{16\beta}$-OH of gitoxigenin increases the activity considerably more than does acetylation. Compare gitoxigenin (0.3) to its 16-acetate oleandrigenin (4.6) and 16-formate gitaloxigenin (10.3).

It should be noted here that the reduction in activity* of a 14,15-β-epoxide in marinobufagen ($3\beta,5\beta$-dihydroxy-14,15β-epoxy-bufadienolide) (0.7) is markedly reversed by a 16β-acetate as in cinobufagin ($3\beta,16\beta$-dihydroxy-14,15β-epoxy-bufadienolide-16-acetate) (5.0).

The C_{16}—OH group in gitoxin so adversely affects the solubility in water that gitoxin is very slightly absorbed from the intestinal tract. Alcohol groups also may affect the duration (cumulation) of the cardiac glycosides. For example, digoxin has a shorter duration of action than digitoxin and differs from the latter only by the presence of an additional alcohol group at C_{12}.

The nature of the sugar residues at C_3—OH may affect markedly the oral absorption from the gastrointestinal tract. Contrary to past beliefs that sugar residues that favored solubility in water also would favor oral

Acetylation of the C_{16}—OH in the gitoxigenin-containing glycosides usually increases activity, e.g., oleandrin versus desacetyloleandrin. Acetylation of an alcohol group in the sugar moiety does not lead to predictable results by intravenous methods of assay. Such acetylations may exert a favorable effect on oral absorption of the solid

absorption, is the accumulation of evidence that sugar residues that bring about high lipoid solubility, together with a high partition coefficient, exert a very favorable effect upon oral absorption. Good examples of this are the oral absorption of solid digoxin and

* Expressed in LD's per mg.: digitoxigenin = 2.2.

lanatoside C versus acetyldigoxin, and digoxin in solution versus lanatoside C.

The sugar residues also prevent the rapid inactivation in vivo that takes place with the free genins—i.e., via final epimerization and conjugation at C_3–OH as the glucuronide.

In humans, intravenously administered digitoxigenin and digoxigenin act almost immediately and have the same kind of action as do digitoxin and digoxin. Also, they reach a maximum effect more quickly; the degree of effect is less and duration of effect is shorter. Activities suggesting CNS effects were observed in the case of digitoxigenin. Orally 9 mg. of digoxigenin produced a negligible effect.[262a]

That the specificity of the 5-membered unsaturated lactone is not so rigid is revealed by the more active squill glycosides that have a six-membered, double-unsaturated lactone.

It is interesting to note that a C_3–OAc in strophanthidin acetate gives a more active compound than strophanthidin itself.

Under proper storage conditions the cardiac glycosides are stable over long periods whether they be a purified form or in the original powdered leaf. A tincture freshly prepared from properly dried leaves and U.S.P. Digitalis Reference Standard[263] (1942) contained, in 1958, large amounts of "genuine glycosides." By this term is meant the unchanged, initial cardiac glycosides as they pre-exist in the fresh plant, carefully dried. Comparative standardization of tinctures of known age by both the cat and the frog methods shows a rapid decrease in the apparent toxicity (activity) as determined by the frog method, without a corresponding decrease as determined by the cat method. This indicates some change that affects the absorption of the original active glycosides. Because digitoxin is absorbed slowly and irregularly from the frog lymph sac, it is thought that some of the genuine glycosides may be desugared partially in the tincture upon standing to give substances less absorbable from frog lymph sac.

Some of these cardio-active glycosides, such as strophanthin and those obtained from digitalis, have a 2-desoxy sugar directly linked to the C_3–OH. These 2-desoxy sugars are cleaved very easily even by weak hydrogen ion concentrations.[264] However, the glucose residue in the genuine glycosides of digitalis is cleaved more readily. The sugar residues in such glycosides as ouabain and convallatoxin are fairly resistant to hydrolysis.

The side chain at C_{17} can be isomerized by enzymatic means and by bases as weak as sodium bicarbonate to give the allo compounds, which are inactive.[265] The side chain, because of its lactone character, can be saponified by alkali. Because this side chain is an α,β-butenolide, it gives an aldehydic acid upon alkaline saponification and, thus, gives a positive Legal's test. The side chain plays a role in the Baljet and Raymond reactions, which have been used as bases for quantitative colorimetric procedures.[266]

The *U.S.P.* color test for digitoxin is essentially the Keller-Kiliani test. The concentrated sulfuric acid layer becomes red-brown to violet-brown (aglycons), and the acetic acid layer becomes blue (digitoxose).

The trend in cardiac therapy is toward the increased use of single purified cardiac glycosides or purified mixtures of these glycosides. This is done in order to standardize more accurately the product or drug and the treatment of the patient.

ACTIVITY AND MODE OF ACTION

Classical congestive heart failure basically appears to be the inability of the myofibril to utilize phosphate bond energy.[267] Thus, there is low cardiac output even though oxygen consumption, ATP and phosphocreatine concentrations and utilization of substrates (glucose, etc.) are normal or increased. Cardiac decompensation results in the inability of the heart to maintain adequate circulation. It is characterized by dyspnea and venous engorgement.

In humans, chronic hypertensive and valvular disease may lead to diminution of work capacity per unit weight of myocardial tissue. If the work capacity falls below a critical value, the result is congestive heart failure. Hypertrophy of the myocardium may be the most important compensatory mechanism in heart failure. Heart mass may be increased by six times and effective work by as much as twenty times, and such increased demand may be accompanied by anoxia of the myocardium. Failing hearts have diminished myosin and actomyosin concentrations, and the physiochemical changes in the properties of the contractile proteins may be a causative factor in congestive heart failure. Myosin and actin were found to have increased molecular weights and molecule lengths as compared to the normal (q.v.).

The cardiac glycosides exert a positive inotropic action, and the increase in the force of myocardial contraction that they induce is considered to be their outstanding pharmacologic contribution to the relief of cardiac failure. Indeed, the cardiac glycosides are the pharmacologic specific for heart failure. They are the only drugs that increase the efficiency of the failing heart, i.e., increase the force of contraction of the myocardium without at the same time increasing oxygen consumption. Cardiac slowing is secondary to this positive inotropic action and is an indirect or reflex action, in which vagal tone increases. In the tachycardia of cardiac failure, the cardiac glycosides may cause some direct sinus (SA node) slowing that is independent of the positive inotropic effect, but may not be clinically significant. Cardiac slowing by the cardiac glycosides may result from a direct action on the AV node also.

Because cardiac glycosides increase ventricular myocardium contractility and, consequently, cardiac output, their use enables the failing heart to handle an increased venous return. This simultaneously reduces venous engorgement, the elevated venous pressure and the need for increased sympathetic vasoconstriction in the venous system. Increased cardiac output reverses the sequence that leads to edema accumulation. The increased rate of glomerular filtration, the reduced rate of tubular resorption of sodium, and the negative sodium balance result in obligatory loss of intercellular water (i.e., diuresis) and reduction of edema. This also relieves pulmonary edema and thus assists in the relief of dyspnea and cyanosis.

Usually the cardiac glycosides, in thera-peutic as well as toxic doses, primarily depress or block AV conduction.* Simultaneously other types of heart tissue with excitability, conductivity,† automaticity‡ and refractoriness also are affected. For example, the His-Purkinje (HP) system as an origin of ventricular arrhythmias may demonstrate increased or decreased automaticity at low levels of digitalis. Furthermore, if the dose is increased, spontaneous pacemaker activity located in the HP system might appear as a first sign of digitalis administration and prior to any significant effect of the drug on AV transmission. As early as 1938 it was demonstrated that cardiac glycosides directly enhanced the contractile force of cat papillary muscle independent of other biologic effects. Thus, the effect of the cardiac glycosides depends on the tissue and/or electrophysiologic property that is predominantly affected by the glycoside.

The response to the cardiac glycosides may be modified by vagal stimulation§ to release acetylcholine, causing alteration of catecholamine‖ levels, change in local ionic concentrations, acidosis, hypercapnia and hypoxia. These factors may exert considerable effects and, until corrected, present new hazards. Thus the effectiveness and predictability—hence the safety—of cardiac glycoside therapy are further complicated by the increase and overlapping of biologic alterations.

---

* AV block provides the main evidence of overdose of cardiac glycosides when the myocardium is unaffected by disease (e.g., in suicidal attempts or accidental poisoning in children).

† Conductivity or conduction (from cell to cell) is possible probably because of the structure peculiar to heart tissue and described as interdigitation.

‡ The ability of each cell to initiate an impulse that leads to conduction.

§ Patients receiving cardiac glycosides become sensitive to carotid sinus massage or vagal stimulation of other types, and a positive response may occasionally provide the earliest evidence of cardiac glycoside toxicity by depression of SA or AV conduction or by ventricular ectopic beats or rhythms. Ectopic beats are located away from the normal position on the EKG.

‖ Catecholamines counteract the antiadrenergic as well as the vagal effects of digitalis and there appears to be no evidence that the maximal inotropic effect of digitalis can be further augmented by catecholamine or sympathetic stimulation.

In atrial fibrillation the counting of the ventricular rate is dose related in most cases, because the slowing of the ventricular response in atrial fibrillation is mediated via the effect of digitalis on conduction through the junctional tissue. However, in the management of congestive heart failure in patients with normal SA node rhythm (no atrial fibrillation), the basic difficulty lies in assessing contractility clinically. Measurement of the cardiac output alone does not necessarily provide evidence for the enhancement of the force of contraction. The clinician therefore has the choice of (a) digitalizing until toxic symptoms appear and then reducing the dose, on the assumption that a full therapeutic effect is directly proportional to the dosage and thus is not maximally obtained until toxic levels are reached, or (b) using the dose, calculated statistically from the study of a large number of patients, that may be expected to produce a good therapeutic effect. However, although such dose levels no doubt exert significant contractility they may not be "fully digitalizing," and patients with refractory heart failure may not receive the maximum potential therapeutic benefits. The great potency of the cardiac glycosides necessitates close patient supervision in order to optimize the dose levels.

Toxicity

Arrhythmia is usually the principal effect of digitalis overdosage. The arrhythmias and their combinations are many and almost all known arrhythmias can be produced by digitalis. Increased ventricular irritability is very common and is manifested by ventricular premature contractions, ventricular tachycardia and ventricular fibrillation. Depression of AV node and P–R prolongation are the most common findings and, although not dangerous in themselves, are always an indication that digitalis dosage should be reduced. Second degree block is quite common; complete block is very serious.

Gastrointestinal side effects are anorexia, nausea, vomiting, and, less commonly, diarrhea. They are far more common during initial oral digitalization but can occur with parenteral use. Gitoxin in the whole leaf is not absorbed and can contribute to local G. I. effects.

The nervous system effects of digitalis are both central and peripheral. Many different manifestations have been described. In addition to the classical visual changes, an organic mental syndrome and even stupor can result from central effects, and a transient but striking peripheral neuropathy can result. That the most liposoluble of the cardiac glycosides exert the greatest C.N.S. effects is no doubt due in part to drug distribution properties.

Overdosage with the cardiac glycosides may cause death, and the lethal effects are almost always specifically related to the cardiac toxicity of these compounds.

Structure of Muscle[268]

The unit structure in both striated and cardiac muscle is the long muscle fiber cell which is 10 to 100 μ in diameter and a few millimeters to several centimeters in length. It is enclosed in a membrane or sarcolemma. The muscle fibers are composed of bundles of myofibrils (the contractile units of the cell), which, in turn, are composed of filaments that run parallel to the length of the muscle fiber cell, and these compose series of structurally similar units called sarcomeres (see Fig. 26). The myofilaments of the myofibrils overlap each other and are surrounded by fluid called sarcoplasm.

The thick zones in the diagram are chiefly myosin (extractable preferentially), and the thin lines are chiefly actin (also extractable). Myosin, the main structural protein of muscle, has a molecular weight of 400,000 to 450,000 and the molecule is 1600 Å long and 50 Å wide.

Myosin has a large number of anionic sites that can combine with cations. It has a separate binding site for actin, which suggests that actin may be an allosteric effector for myosin. ATP is the only anion that combines with myosin only in the presence of K^+ or Mg^{++} or both. Myosin ATP has myosin ATPase activity only in the presence of cations, i.e., K^+, Na^+, or Ca^{++}, and is said to be Ca^{++}-dependent.

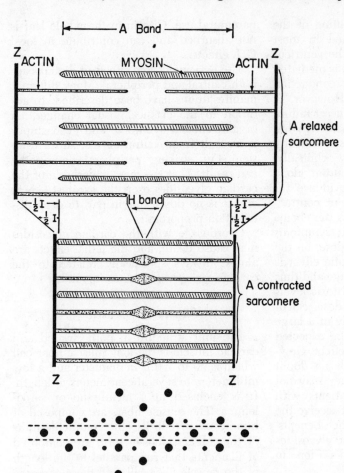

FIG. 26. Diagram of part of one sarcomere of striated muscle.

The small diagram (below) illustrates a cross section through the A band and shows two fine myofilaments between a pair of coarse ones. (From Ham, A.: Histology, ed. 6, p. 556, Philadelphia, Lippincott, 1969)

Actin, another important muscle protein, has a molecular weight of 70,000 and can exist in two forms—F-actin, which is fibrous, and G-actin, which is globular. It contains some Ca^{++}. Additional cations reduce electrostatic repulsion between actin molecules and permit aggregation. Myosin greatly accelerates polymerization of actin, an effect that can be inhibited or reversed by high salt concentrations (e.g., KCl) or the removal of salts. When myosin, F-action, ATP and Mg^{++} are mixed together in the proper ratios and at the proper pH and ionic strength, syneresis and catalytic hydrolysis of ATP take place.

Proteinlike substances called tropomyosin* and troponin* inhibit the above described interaction of actin and myosin and, thus, ATPase activity located within the cross bridge, and suppress the energy needed for contractile energy. Ca^{++} supposedly inhibits their suppressant activity by combining with them. Ca^{++} in less than $10^{-7}M$ terminates or markedly weakens contraction.†

Recently it has been reported that digoxin causes an increase in extractable cardiac actomyosin ATPase, probably by participation in its formation from F-actin and myosin.[269] Cardiac actomyosin ATPase activity is tightly coupled to the contraction process (q.v.) which has never been observed in the absence of ATP hydrolysis.

Autoradiographic studies using tritiated digoxin in therapeutic doses indicate that digoxin is localized in the A band of the sarcomere where actin and myosin interact and the most intense development of grains in the A bands occurred at the time of maximum cardiac contractile force.

* Are these synonymous with the relaxing factor (q.v.)?

† Ca^{++} reputedly has other functions in the contraction of muscle (q.v.).

Cardioactive glycosides do not inhibit cardiac actomyosin ATPase but do inhibit a Na-K ATPase (q.v.). They have no effect upon ATP synthesis.

The myofibrils are 0.5 to 2.0 mμ in diameter and are enclosed in and separated from each other longitudinally and transversely by the sarcoplasmic reticulum* (S.R.)† which, in part, resembles a network of tubules whose ends form a saclike cavity. The S.R. also is surrounded (bathed) by the fluid sarcoplasm. The S.R. is formed by a membrane that has active enzymes such as an ATPase on its surface, and electrical signals can be transmitted on its surface. The S.R. can take in substances to be metabolized (e.g., glycogen), pass out metabolites (e.g., lactic acid) and distribute enzymes. The enlarged portions (sacs) of the S.R. that adjoin on opposite sides of the T system (q.v.) to form the triad structure of the muscle fiber are the reservoirs, or sinks, for Ca ions. This portion of the S.R. rapidly can remove (i.e., extract) Ca ions from the environment of the myofibrils, thus inducing relaxation. Its intimate proximity to the T-system (tubules) which is implicated in the conduction of intracellular stimuli suggests that it may be a source from which Ca^{++} is released to the myofibrils to trigger contraction.

T channels are a second communication system; they are transverse to the myofilaments and are intimately in contact with the S.R. The membranes of these T-channels are an extension of the fiber membrane (sheath or sarcolemma), and they extend deep into the interior of the fiber at each Z line‡ or, in some muscles, at each junction between the A and the I bands.§ Because the T-channels and the S.R. are intimately in contact they can function together in ion movements, electrical conduction, transportation of metabolites and metabolic products, etc.

* Highly developed in rapidly acting skeletal muscles and less so in other striated muscles.
† S.R. is a system similar to the endoplasmic reticulum (E.R.) which performs similar functions in many other kinds of cells.
‡ Fish and frog muscle.
§ Mammals, birds and reptiles.

Contraction-Relaxation Cycle of Muscle Cells

It may be assumed that the triadic junction (q.v.) at the Z line and the I band constitutes a zonula occludens, so that membrane depolarization can progress from the sarcolemma inward along the T-channel into the cisternal‖ elements (sacs of the S.R., q.v.). Then, with a minute delay, some mechanism releases Ca ions which need to diffuse over a distance of only one micron or so to reach the myosin in the A bands of the fibrils. This Ca^{++} catalyzes the interaction of actin and myosin (filaments) to form actomyosin,¶ giving rise to contraction (q.v.). The source of energy for this contraction comes from ATP $\rightarrow$ ADP + P$_i$. Calcium then is removed via the Ca pump to end myofilament interaction and produce relaxation. The enzyme involved is a Ca-transport ATPase. The energy for Ca-ion uptake comes from ATP $\rightarrow$ ADP + P$_i$. Regeneration of ATP comes from ADP and creatine phosphate.

Two theories[270] have been proposed to explain muscle contraction. (1) The flexible polyelectrolyte theory postulates one filament in which attraction and repulsion of charged groups on an α-helix structure (as in myosin) are responsible for contraction and relaxation. In the resting state, a balance exists between entropic force tending to shorten the filament and electrostatic forces tending to keep the molecular chain extended. Contraction or collapse of such a chain (polyelectrolyte) is thought to be the result of a discharge (adsorption) of the fiber by the quadrivalent anions of ATP. The contracted (*random coil*) actomyosin ATP complex,[271] which now is considered an active actomyosin ATPase, leads to the hydrolysis of ATP $\rightarrow$ ADP + energy. This energy is utilized to cause reversal of the contraction and desorption of ADP, and the resting state is restored. Ca ions are needed as a cofactor in the above transformations, and their possible transitory removal by relaxing factor (q.v.), Ca^{++} and,

‖ Relatively large cistern-like structures.
¶ Actomyosin binds[248] cardiac glycosides probably via the myosin portion. One mg. of actomyosin bound 1 μg of lanatoside A.

possibly, K^+ may cause myosin to change its structure charge to accommodate them and thus produce structural changes of the actomyosin complex to enhance its contractility.[272] (2) The two filament, or sliding mode, theory proposes that, on excitation, the thin filaments in the I band slide past the relatively fixed thick (chiefly myosin) filaments in the A band. When the muscle is stretched passively, the I band filaments slide out again. The width of the A band remains essentially constant at different degrees of stretch and contraction, and the changes in sarcomere length are due primarily to changes in the width of the I band. Cross links between the filaments, i.e., interaction between ATPase sites on the dense myosin filament in the A band and ATP bound to actin in the thin I band filament, mediate the sliding. Ionized Ca may act either by establishing or strengthening cross links between the filaments or by increasing ATPase activity and, thereby, enhancing the probability of interaction.

The cross bridges effect a temporary but firm union between the sets of fibrils (actomyosin complex). The energy needed for the above process is derived by the cleavage of ATP, physically located within the myosin cross bridge, to ADP, P_i and free energy by an ATPase—possibly myosin ATPase or actomyosin ATPase, also located within the cross bridge. Shortening, tension, contraction and work in any muscle are then the consequence of the continuous making and breaking of cross links between the two contractile proteins (myosin and actin) in a ratchetlike fashion dependent on a continuous breakdown and synthesis of ATP.

Electrolyte Factors[273]

Cardiac tissue contains about the same amounts of Ca and Mg, less K and more NaCl and water than skeletal muscle. K, Na and Ca in proper proportions and concentrations are required for normal cardiac excitation and contraction. Na is involved in the process of excitability and depolarization, K determines the magnitude and the resting potential, and there is much evidence that Ca combines with the contractile proteins actin and myosin as an essential aspect of muscle excitation-contraction coupling. Ca ions are localized (bound and free) at the Z line (sacs of the S.R., q.v.) in the relaxed state of skeletal muscle but, some with depolarization and muscle shortening, are found in the A band where myosin and actin overlap. It long has been known that Ca and the cardiac glycosides act very synergistically to produce an increase in myocardial contractility, myocardial contracture or systolic arrest. The cardiac glycosides may act at the triad to permit during depolarization a greater reduction in binding affinity for Ca, and, thus, more Ca would be available for the contractile proteins. Cardiac glycosides do not improve muscle contractility at extremely low concentrations of intracellular Ca. Although some reports conclude that therapeutic concentrations of cardiac glycosides do not induce uptake of extracellular Ca by heart atria, others have shown that beating rabbit atria take up Ca in the presence of ouabain ($4 \times 10^{-7}M$). This uptake can be inhibited by an excess of K. Digitalis in therapeutic concentrations has been re-

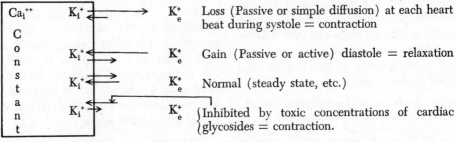

Ca_i^{++} = Intracellular Ca^{++}, K_i^+ = Intracellular K^+
K_e^+ = Extracellular K^+.

ported (Holland, W. C.: Cir. Res. 11:85, 1964) to increase Ca exchange.

Potassium is localized in the A bands of cardiac and striated muscle. Upon excitation of muscle a redistribution of K occurs and it moves to the I bands and accumulates at the cell surface. The diagram on p. 834 indicates that when the Ca_i/K_i ratio is increased contraction can be expected and the reverse when it is lowered. The duration of contraction in either case obviously is dependent in part on the half life of these ratios.

Evidence to date indicates that the cardiac glycosides can inhibit K influx,[*][†][‡] exert a positive inotropic effect and cause contracture of heart muscle. Thus, it is possible that there is competition[§] for K or its function or action in the Ca_i/K_i balance which, in turn, tips the balance in favor of Ca_i and contraction. Research on this mode of action of the cardiac glycosides (however, specifically brought about) was stimulated by the early discoveries that hearts of digitalized patients[275] contained less K at autopsy than did those of nondigitalized patients and that skeletal muscle[276] lost K when it was soaked in solutions containing 1 or 2 μg. of ouabain per ml.

Recently,[277] using therapeutic concentrations, i.e., 2.5 to 5×10^{-8} M of ouabain, positive inotropic effects of cardiac glyco-

sides have been demonstrated on isolated guinea pig ventricles with no change in the K_i/K_e equilibrium states compared to the controls.[278]

An ATPase that splits ATP is apparently a component of the "transport ATPase" that plays a role in the "active" transport of cations through the cell membrane. It has now been identified in particulate fractions prepared from a very large number of tissues and in a wide range of animals. In every case the ATPase activity appears to be associated with that fraction which is thought to contain fragments of the original cell membrane. Its presence in microsomes from heart tissue has been reported. Good correlation between ATPase activity and the amount of pumping activity has been detected. The richest sources of "transport ATPase" are mammalian kidney and brain, and the electric organs of electric eels and torpedoes.

The "transport ATPase" is Na and K-dependent and is highly sensitive to cardiac glycosides[279] in the range of 10^{-8} to more than 10^{-4}M. Tissues vary widely in their sensitivity to cardiac glycosides. Low concentrations of cardiac glycosides act slowly and appear to be cumulative, whereas higher concentrations reach a maximum of activity in a few minutes.

An ATPase (an ATP hydrolyzing enzyme system) "that has been isolated" from microsomes of heart[280] homogenates and from the S.R.[||] contains two active sites, one that is Mg-dependent and is inhibited by ethanol, guanidine and, to some extent, by amobarbital and a second that is Mg-, Na- and K-dependent.[¶] The latter site is specifically inhibited by ouabain, p-hydroxymercuribenzoate (POMB) and other SH inhibiting agents such as N-ethylmalemide, iodacetamide and mersalyl. Cysteine and mercaptoethanol will inhibit the activity of POMB but only partially reverse the inhibition previously established by POMB. The presence of guanidine (which inhibits the Mg-dependent site) increases the sen-

---

[*] In general, when an "uphill" transfer of cations such as is present in red blood cells has been found, cardiac glycosides can depress it in a wide range of species and tissues.[274] The more the cation composition of the cell differs from that of its environment, the greater the effect of the cardiac glycosides.

[†] After digitalis administration, especially in larger doses, there is a loss of K from the myocardium (Regan, T., *et al.*: J. Clin. Invest. 35:1220, 1956; Daggett, W. H., *et al.*: Fed. Proc. 23:357, 1964).

[‡] This inhibition of K influx in the case of red blood cells can be reversed or decreased by raising the external concentration of K (q.v.). Passive K loss (efflux) is not effected by cardiac glycosides (q.v.).

[§] It should be noted that a relatively small change of Ca can produce marked changes of balance of the more abundant K, thus, the affinity of the hypothetical site must be higher for Ca than for K. Frog ventricles took up twice as much Ca when perfused with a K-free solution than they did from a solution containing both K and Ca.

[||] Also found in many other body tissues.

[¶] The Mg, Na and K dependent ATPase may be the same as the above described transport ATPase that is Na and K dependent.

sitivity of heart ATPase transport system to ouabain and sulfhydryl inhibitors.[281]

It is significant to note that the general structural features (q.v.) necessary for high cardiac activity of the cardiac glycosides also are necessary for the inhibition of the activity of the above described ATPase. Such activity could establish a high intracellular Na^+ concentration, making increased ions available for exchange. This would cause release of Ca^{++}, slow the return of Ca^{++} to its storage area (pool) (largely because of high Na^+ concentrations there) and block the re-entry of K^+, with consequent cellular K^+ depletion and, finally (in the same or other regions), accumulation of K^+ at the outer membrane boundary, with excess K^+ conductance even in the phase of metabolic slowdown.[281a]

The Ca-ATPase transport that plays an active role in the movement of Ca from the sarcoplasm surrounding the filaments into the S.R. (sarcotubular system) is not inhibited by cardiac glycosides.

Perfusion of hearts with cardiac glycosides in 10^{-7} M concentration, resulted in subcellular distribution with highest concentrations in the microsomes, very significant concentrations in the sarcoplasmic reticulum fragments (SRF) and least in mitochondria. In microsomes the following concentrations were found: proscillaridin 13.50, digitoxin 2.97, ouabain 2.20, digoxin 1.79 and convallatoxol 1.72 picomoles/mg. of protein. About half of these quantities were found in mitochondria.[281b]

The exact role of the cardiac glycosides at the molecular level is yet to be specifically elucidated. The preceding discussion indicates a possible multiple role for these highly biologically active substances.

Relaxing Factor

Prolonged extraction of muscle fibers with glycerin removes a functioning cell membrane and a relaxing factor. The latter can prevent the precipitation of a suspension of myofibrils, and it induces a marked lengthening or relaxation of glycerinated fibers that previously have been shortened under the influence of ATP. Its activities

are inhibited by the proper concentration of Ca ions. This might imply that its role is related to its capacity to compete for or "pull" selectively some of the Ca ions from the contractile proteins. It should be noted that the relaxing factor or ETDA do not remove from the myofibrils Ca^{++} that is essential for actomyosin ATPase activity.

Ouabain causes an increased tension of actomyosin (from glycerol-extracted muscle fibers) in the presence of relaxing factor and ATP. Ca ions are more effective than ouabain which is ineffective in the absence of relaxing factor.[282] These data are difficult to explain.

PRODUCTS

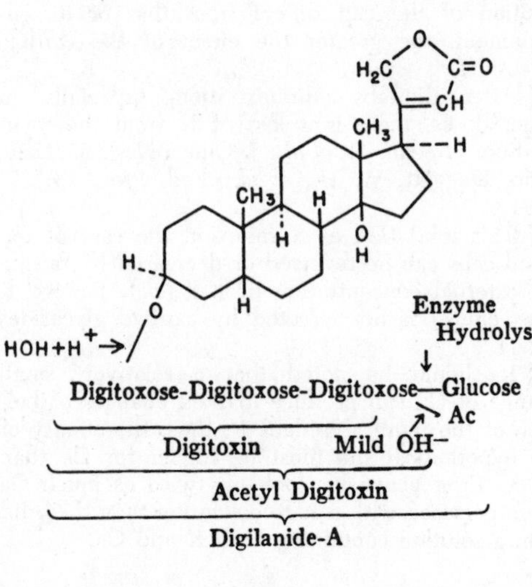

Digitoxin U.S.P. is a cardiotonic glycoside obtained from *Digitalis purpurea* Linné and *Digitalis lanata* Ehrh. and other suitable species of Digitalis, and when properly dried contains n.l.t. 90 per cent of digitoxin. Digitoxin was described first as a pure or nearly pure crystalline substance "digitaline crystallisée," in 1869, by Nativelle.[283] However, the careful chemical and pharmacologic investigation of digitoxin was not published until 1920 by Cloetta.[284] Digitoxin per se pre-exists in *Digitalis purpurea* as purpurea glycoside A[285] and differs from the latter in that the sugar moiety has lost a molecule of glycose by hydrolysis. It also pre-exists in *Digitalis lanata* as digilanide A or lanatoside-A, that is, monoacetyl purpurea glycoside A. These relationships can be expressed diagrammatically as shown on page 836.

Digitoxin is a white or pale buff, odorless, microcrystalline powder. It is insoluble in water and very slightly soluble in ether. One g. dissolves in about 40 ml. of chloroform and in about 150 ml. of alcohol.

Digitoxin is absorbed rapidly and almost completely, whereas the other digitalis glycosides may vary from very rapid and complete to very slow and quite incomplete to essentially no absorption (q.v.). However, digitoxin is especially cumulative, more so than the other digitalis glycosides. Given orally, digitoxin is 1,000 times as active as digitalis leaf.[286] Digitoxin has no selective affinity for the cardiac tissue of man or experimental animals. Only 1 per cent of the dose of digitoxin reaches the heart.[287] The liver, the kidney and the gastrointestinal tract had higher concentrations than the heart.[288]

After I.V. injection and equilibration of digitoxin, the amount of the glycoside retained by the myocardium continuously declined and was always less than the initial levels. Therefore minute amounts are required regularly for a positive inotropic effect. For example, unchanged digitoxin found in the atrial appendage of a patient undergoing cardiac surgery 3 hours after I.V. administration of 0.5 mg. of ^{14}C-digitoxin was 1.1 mμ per 100 g. of tissue or 1.4×10^{-8} M. About 90 per cent of the digitoxin was found in the soluble supernatant fractions (cytoplasm), 5.4 per cent in the microsomes, 1.9 per cent in mitochondria and 1.6 per cent in the nuclei. The metabolites of digitoxin (digoxin, etc.) were present in cardiac muscle in larger quantities than was digitoxin. It appears that digoxin has a somewhat greater affinity for cardiac muscle than does digitoxin. Digitoxin is loosely and reversibly bound to serum proteins.

The half life of digitoxin in man is 9 days. Approximately 80 per cent of an I.V. dose is metabolized and excreted over a period of 3 weeks, and 30 days are required for the complete elimination of a single dose. About 17 per cent is excreted in the feces and the remainder by way of the kidney.

Of the total activity in a 24-hour period from a subject receiving ^{14}C-digitoxin, 74 per cent was in the form of metabolites.

Spironolactone, in rats, antagonizes the activities of digitoxin in the heart and at various receptor sites, an effect owing not merely to the restriction of potassium elimination. Spironolactone and norbolethone also inhibit the characteristic neuromuscular disturbances and the lethal effects produced by digoxin, digitalin, gitalin and proscillaridin but not ouabain, strophanthin K and digitoxigenin. Clinically these antagonizing effects have not been appraised; however, it should be noted that digitalis and spironolactone are frequently administered conjointly to cardiac patients.[288a] Spironolactone also has some antiandrogenic activity.[288b]

Category—cardiotonic.

Usual dose—digitalization, 1.5 mg. divided over 24 to 48 hours; maintenance, 100 mcg. once daily.

Usual dose range—digitalization 1 to 1.5 mg. in 1 or 2 days; maintenance, 100 to 200 mcg. daily.

OCCURRENCE
Digitoxin Injection N.F.
Digitoxin Tablets U.S.P.

Acetyldigitoxin N.F., Acylanid®, is obtained by the enzymatic hydrolysis of Lanatoside A. Acetyldigitoxin[289] can be isolated or demonstrated to be present in properly prepared (short-time enzyme favoring) extracts of *Digitalis lanata, D. furruginea, D. lutea, D. mertonensis* and *D. siberica* where its precursor is Lanatoside A. Two forms—alpha and beta—have been reported. The former is Acylanid. These forms are interconvertible by boiling in methanol.

Acetyldigitoxin occurs as a white microcrystalline hygroscopic powder that is less soluble in water than digitoxin but is more liposoluble. Alcoholic or hydroalcoholic solutions slowly lose the acetyl group by hydrolysis.

Acetyldigitoxin is absorbed orally equally as well as digitoxin. It is claimed to be less toxic and have a greater margin of safety than digitoxin. It is dissipated twice as fast as digitoxin (14% daily versus 7 % daily).

Category—cardiotonic.

Usual dose—initial; rapid digitalization, 1.8 mg. in one dose or several divided doses given within 24 hours; slow digitalization, 2.4 mg. in divided doses given over a period of 2 to 6 days; maintenance; 150 mcg. per day.

Usual dose range—rapid digitalization, 1.6 to 2.2 mg.; slow digitalization, 1.8 to 2.4 mg.; maintenance, 100 to 200 mcg.

OCCURRENCE

Acetyldigitoxin Tablets N.F.

Gitoxin was the second glycoside isolated from *Digitalis purpurea* by Cloetta[290] in 1926 and by Kraft[291] in 1912. It is very poorly absorbed (q.v.).

Digoxin U.S.P., Lanoxin®, is a cardiotonic glycoside from the leaves of *Digitalis lanata* Ehrh. (*Fam.* Scrophulariaceae) and when dried contains n.l.t. 96 per cent of digoxin. Digoxin was isolated first from *Digitalis lanata,* in 1930, by Smith,[292] who at the same time also isolated gitoxin from this same plant. Digoxin per se pre-exists in *Digitalis lanata* as lanatoside-C or digilanide-C[285] and differs from the latter in that the sugar moiety has lost a molecule of glucose and acetic acid by hydrolysis.

Digoxin occurs as colorless to white crystals or as a white, crystalline powder. Digoxin is insoluble in water, in chloroform, and in ether. It is freely soluble in pyridine and soluble in dilute alcohol. $[\alpha]^D = 13.6°$ to $+ 14.2°$.

Digoxin is less cumulative and its duration of action is shorter than that of digitoxin. In an emergency it is claimed that it can be used intravenously in dilute (8%) alcohol. Digoxin administered orally is reported to be absorbed up to 65 to 85 per cent of dose. However, in solution it is rapidly and almost completely absorbed (q.v.). This difference in absorption should be taken into consideration in the size of the dose when this drug is administered in these two forms.

Digoxin has a half life of 44 hours and approximately 10 per cent of the dose is detected as metabolites. Eighty per cent is excreted in the urine and 12 per cent in the feces over a period of 7 days.

Even though it is less toxic and shorter acting than digitoxin, digoxin given in one daily morning dose tends to result in overdigitalization in the early part of the day and underdigitalization at night. Considerably better results are obtained when the daily dose is divided into 3 equal amounts in order to reduce the maximum and raise the minimum levels of the effects of the drug.

Category—cardiotonic.

Usual dose—oral: initial, 2 to 3 mg. divided over 24 hours; maintenance, 250 mcg. 2 or 3 times daily; intravenous; initial, 500 mcg. to 1.5 mg. divided over 12 hours; maintenance, 250 mcg. once or twice daily.

Usual dose range—oral, 750 mcg. to 3 mg. in 1 day; intramuscular or intravenous, 500 mcg. to 2 mg. in one-half day.

OCCURRENCE

Digoxin Elixir U.S.P.
Digoxin Injection U.S.P.
Digoxin Tablets U.S.P.

Acetyl digoxin has been isolated as the chief glycoside from the first year's growth of fresh *Digitalis lanata* leaves grown in Minnesota when enzyme-favoring technics were employed.[293]

Acetyldigoxin, Novadigal®, obtained from *Digitalis lanata* is digoxin containing an acetyl group at C_4 of the terminal digitoxose residue. It is absorbed[294] enterally better than digoxin, and absorption is claimed by some to be close to 100 per cent. It also is claimed to be better tolerated than other orally useful cardiac glycosides.

Two to 4 hours after oral administration in rats greatest quantities are found in the blood and heart. Absorption can be detected 30 minutes after oral administration. The heart contained 25 to 33 per cent more digoxin than β-acetyldigoxin, and in about 50 per cent of the test cases the hearts had ½ to ⅓ as much digoxigenin as β-acetyldigoxin. Some digoxin is found in blood after oral administration. Therefore, partial deacetylation starts during or soon after the absorption of the glycoside.[295]

α-Acetyldigoxin (the acetyl group at C_3 of the terminal digitoxose residue) appears to have no advantage over digoxin. It is marketed as Lanatilin. References can be found in Kraupp, O., *et al.*: Pharmacol. 1:345, 1968.

Lanatoside-C N.F. is a glycoside[285] obtained from the leaves of *Digitalis lanata* Ehrh. (*Fam.* Scrophulariaceae).

Lanatoside-C occurs as colorless or white crystals or as a white crystalline powder. It is soluble 1:20,000 in water, 1:45 in alcohol, 1:20 in methanol and 1:2,000 in chloroform.

Alcoholic and hydroalcoholic solutions are slowly deacetylated upon standing.

Lanatoside-C is less cumulative, and its duration of action is shorter than that of digitoxin. In an emergency, it is claimed that it can be used intravenously. Lanatoside-C is absorbed poorly (about 10 per cent) from the gastrointestinal tract, both in the solid state and in solution (q.v.).

The relationship of digoxin to lanatoside-C can be expressed diagrammatically as shown below.

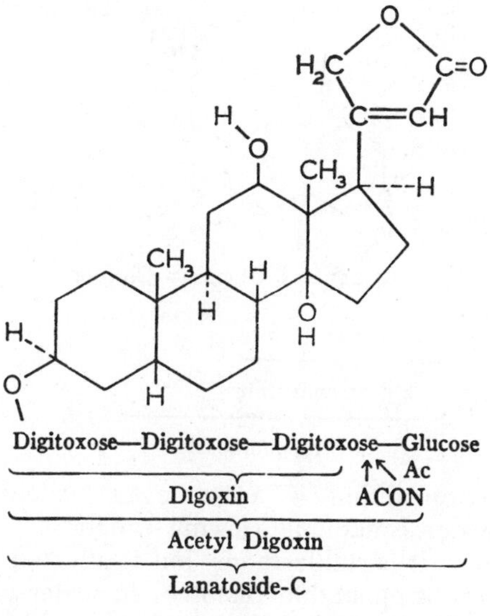

Digitoxose—Digitoxose—Digitoxose—Glucose

Digoxin ACON

Acetyl Digoxin

Lanatoside-C

Category—cardiotonic.

Usual dose—oral: initial, 7.5 to 10 mg.; maintenance, 0.5 to 1.5 mg.

OCCURRENCE
Lanatoside C Tablets N.F.

Deslanoside U.S.P., Cedilanid-D®, Desacetyl-lanatoside C, is obtained by alkaline deacetylation of lanatoside C. It occurs as odorless white crystals or as a white, crystalline powder that is hygroscopic, absorbing about 7 per cent water. It is soluble 1:300 in alcohol, very slightly soluble in chloroform, and, although it is called water insoluble, it is more soluble than lanatoside-C.

The physical properties of deslanoside render it useful for intramuscular and intravenous use. It acts rapidly, i.e., in 10 to 30 minutes, has a quick dissipation rate with a wide margin of safety and low toxicity and is well tolerated. Orally, it is probably absorbed poorly.

Category—cardiotonic.

Usual dose—intramuscular or intravenous, for digitalization, 1.6 mg. in 1 or 2 doses.

Usual dose range—intramuscular or intravenous, for digitalization, 1 to 1.6 mg. in one-half day.

OCCURRENCE
Deslanoside Injection U.S.P.

The metabolic products of digoxin are produced in the following order: digoxigenin bisdigitoxoside, digoxigenin monodigitoxoside, digoxigenin and its glucuronide and epidigoxigenin and its glucuronide. Digitoxin also is converted into the above metabolic products via digoxin, digitoxigenin bisdigitoxoside, digitoxigenin monodigitoxoside and digitoxigenin. In addition digitoxigenin and epidigitoxigenin and their glucuronides are formed. All the metabolic products are active with the exception of the epigenins and their conjugates. Other unknown metabolites have be detected.[296, 297, 298]

The more polar glycosides and those containing more polar sugars such as lanatoside-C and -A are excreted more rapidly and in greater amounts into bile than are the less polar glycosides such as digitoxin. They are less readily reabsorbed than digitoxin. Although digitoxin is excreted into bile in lesser amounts, it is retained in greater amounts by the enterohepatic circulation as a result of passive reabsorption of the drug across the lipid membrane. The enterohepatic system has the highest concentration of the more liposoluble cardiac glycosides, and other body organs in general reflect the biologic half-life of the glycosides of this system. The high concentration gradient favors the continuous supply of glycoside from the enterohepatic system to organs of lower concentration. In time only 6 to 10 per cent of digitoxin appears to be excreted in the urine unchanged.

Desoxy sugar residues such as D-digitoxosyl, D-cymarosyl and L-thevetosyl can be cleaved in vivo by animal enzymes but at a rate slower than the metabolism of the genins that are generated,[299] and it appears that the aglycon is protected from metabolic attack by a sugar residue at C_3. In the case of digitoxin, digoxin and gitoxin, the digitoxose residues are progressively cleaved to finally yield the genin in which the C_3-OH is epimerized (inactive genin) and conjugated as the glucuronide. The glucose residue of digilanide-A is not cleaved in vivo (heart or liver slices) but is cleaved by cellulolytic bacteria in the intestinal tract.[287]

It is interesting to note that the extremely digitoxin-sensitive guinea pig shows the highest detoxification rates among all species so far studied.[287]

Gitalin N.F. (Amorphous Gitaligin) is a glucosidal constituent of *Digitalis purpurea* prepared according to the method of Kraft. It contains considerable amounts of digitoxin and gitoxin.[288] Oral absorption studies in cats (q.v.) indicate that this glycoside has nauseating properties much greater than those of digoxin, acetyl digoxin, lanatoside-C or digitoxin. It is a white or slightly buff-colored, amorphous powder which is readily soluble in chloroform, ether and alcohol and is slowly soluble in 600 parts of cold water. It is marketed as 0.75-mg. tablets that equal 1 cat unit.

Category—cardiotonic.

Usual dose—initial: rapid digitalization, 2.5 mg. followed by 750 mcg. every 6 hours until therapeutic effect or toxicity develops, with a total dose of about 6 mg. being given in 24 hours; slow digitalization, 1.5 mg. daily for 4 to 6 days; maintenance: 500 mcg. daily.

Usual dose range—maintenance, 250 mcg. to 1.25 mg. daily.

OCCURRENCE
Gitalin Tablets N.F.

Strophanthin (k-β-Strophanthin). A glycoside or a mixture of glycosides obtained from *Strophanthus kombe* Oliver (*Fam.* Apocynaceae). Of the plants that yield cardio-active substances, the various *Strophanthus* species, particularly, are characterized by a relatively very high glycoside content. Although strophanthin is isolated readily from *Strophanthus kombe*, 90 per cent of the cardio-active constituents are present as k-strophanthoside,[285] which can be isolated by careful handling during isolation. k-Strophanthoside is very soluble in water and in methyl or ethyl alcohols. Prior to the isolation of this glycoside, k-β-strophanthin was thought to be the chief glycoside from *Strophanthus kombe*. Cymarin is a glycoside found in Canadian hemp. The following formula shows some relationships among these three compounds.

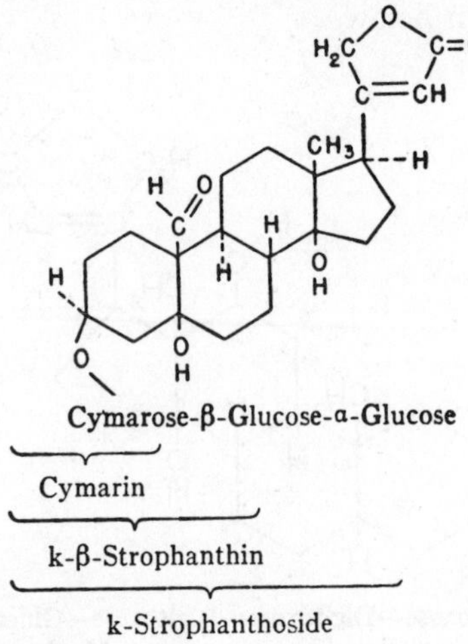

Cymarose-β-Glucose-α-Glucose

Cymarin

k-β-Strophanthin

k-Strophanthoside

Strophanthin is a white or yellowish powder containing varying proportions of water. It is stable in air, but is affected by light. Strophanthin dissolves in water and in diluted alcohol, but is less soluble in dehydrated alcohol. It is nearly insoluble in chloroform, in ether, and in benzene.

Strophanthin is used intravenously in emergencies for its cardiac effects. It is absorbed too poorly to be used orally.

The usual dose (intravenous) is 600 mcg.

Ouabain U.S.P., G-Strophanthin, is a glycoside obtained from the seeds of *Strophanthus gratus* (Wall. et Hook.) Baillon (*Fam.* Apocynaceae).

Ouabain occurs as white, odorless, crystals, or as a crystalline powder. It is stable

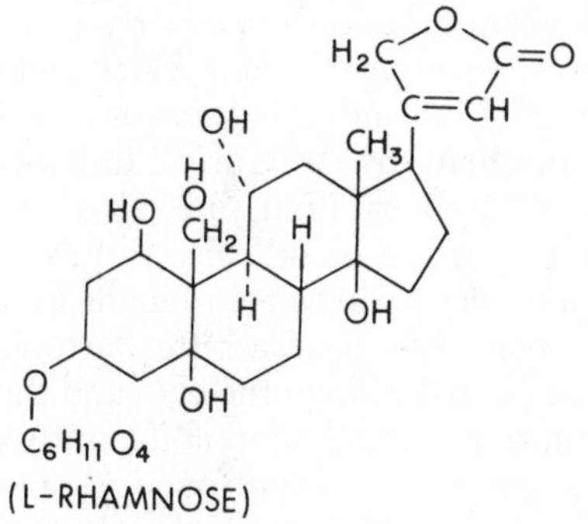

(L-RHAMNOSE)

in air, but is affected by light. One gram of ouabain dissolves slowly in about 75 ml. of water, and in about 100 ml. of alcohol. $[\alpha]^D = -31°$ to $-32.5°$.

Ouabain inhibits energy dependent cation and amino-acid transport by Ehrlich ascites cells without marked effects on cellular respiration. At concentrations of 6×10^{-4} or higher, ouabain also inhibits cell division of Ehrlich ascites cells even in mitosis.[300] This activity is reversible.

Ouabain is absorbed so slowly and so irregularly from the alimentary canal that the oral administration of the drug is not to be recommended and is even considered unsafe. It is used intravenously for its cardiac effects in emergencies.

Category—cardiotonic.

Usual dose—intravenous, for digitalization, 500 mcg. to 1 mg., divided over 24 hours.

OCCURRENCE

Ouabain Injection U.S.P.

Scillaren is a mixture of the natural glycosides, scillaren-A and scillaren-B, occurring in fresh squill, *Urginea maritima*, in the proportions in which they exist in the fresh crude drug, i.e., about 2 parts of scillaren-A to 1 part of scillaren-B. It occurs as a white or yellowish-white, odorless, granular powder, possessing a very bitter taste; it is soluble in absolute alcohol (1:5) and sparingly soluble in water (1:3,000).

Extensive attempts by early workers to isolate the native glycosides of squill resulted only in purified extracts. Thus, the scillipicrin, scillitoxin and scillin of Merk,[301] 1879; the scillain of von Jarmersted[302]; the apparently crystalline substances, scillipicrin, scillenin and scillimarin, of Waliszewski,[303] 1894, and the substances, scillitin and scilli diuretin, by Kopazewski,[304] 1914, were not characterized as pure compounds. Furthermore, the three earlier workers failed to make definite statements concerning the physiologic activity. The successful isolation of the genuine, active, pure cardio-active principles of fresh squill was accomplished by A. Stoll[285] and his co-workers, in 1933 and 1935, after extensive investigations. His work was assisted by the discovery that the toxicity of the preparations ran parallel to the intensity of the so-called Liebermann's cholesterol reaction. Active and partially purified extracts of squill first show a strong red color, which changes rapidly to blue and then to a bluish green when the above color reaction is applied.

The structural formula below is proposed for scillaren-A and its related degradation products.

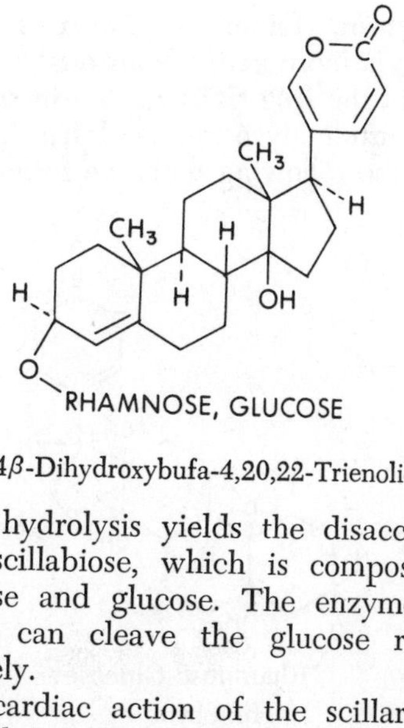

RHAMNOSE, GLUCOSE

* 3β,14β-Dihydroxybufa-4,20,22-Trienolide.

Acid hydrolysis yields the disaccharide called scillabiose, which is composed of rhamnose and glucose. The enzyme scillarinase can cleave the glucose residue selectively.

The cardiac action of the scillarens is essentially similar to that of digitalis, but this action is apparently less persistent than that of digitalis. The squill glycosides produce copious diuresis and are often active where digitalis and strophanthin fail to act, or act insufficiently, or where intolerance to digitalis exists. Because of its high therapeutic index and rapid elimination, it maintains compensation in those cases where prolonged treatment is necessary.

Scillaren-B is the amorphous component of the natural mixture of the glycosides occurring in squill, *Urginea maritima.* It is a fine, white or slightly yellowish white, odorless, granular powder, possessing a very bitter taste. It is freely soluble in water, soluble in alcohol, and in methyl alcohol, 1:5. It is marketed as 500 mcg. in 1-ml. ampuls and is used the same as scillaren. Scillaren B is more water soluble and more active in the frog assay than scillaren A. The ratio of activity is about 3 to 5. Scillaren B is more stable than scillaren A.

Urginin is a mixture of two water-insoluble glycosides, urginin-A and urginin-B, derived from squill, in the proportions in which they exist in the drug, i.e., about equal parts of each. It occurs as a pale-yellow, granular powder possessing a slight characteristic odor and an extremely bitter taste. It is soluble in alcohol, and 1 cat unit is equal to 200 mcg. per kg. It is marketed as 1.0-mg. tablets.

Hellebrin, found in *Radix Hellebori nigri,*[305] is more active than ouabain when tested in the frog. It is related in structure to the squill glycosides and has been assigned the following tentative formula.

Rhamnose-Glucose
Hellebrin

Some of the cardiac glycosides[306] such as cymarin, apocannoside and calotropin have cytotoxic properties, i.e., they were found to show significant inhibitory activity in cell cultures of human carcinoma of the nasopharynx.

SAPONINS

Saponins are glycosides that are found in plants and have the property of forming colloidal aqueous solutions which foam upon shaking. Like other glycosides, saponins vary in chemical structure. Most saponins, even in high dilution, are able to effect hemolysis of red blood cells. In this respect, they are very toxic to cold-blooded animals. In general, they have a bitter taste and are extremely irritating to the eyes and the nose. The more commonly and abundantly occurring saponins are those found in soapbark, soaproot, snakeroot, *Smilax,* cacti and many plants that grow in the desert regions of the southwestern U.S. and in Mexico. Many saponins occur as admixtures and, because some of these are so closely related in physical properties, they are very difficult to separate from one another. Therefore, in a number of instances, the crude mixtures are hydrolyzed to facilitate separation and study of the aglycons.

The digitalis saponins form solid molecular complexes with the higher alcohols, phenols and thiophenols. Digitonin has proved particularly useful in the field of steroidal chemistry, where it forms insoluble complexes that are readily isolated with those steroids that have a C_3—OH that is normal. Those in which the C_3—OH is *epi* form complexes that are soluble. Thus, a twofold usefulness is gained, i.e., determination of the configuration about C_3—OH and a means of purification, because these complexes can be resolved readily into their respective components.

Sarsasapogenin, whose structure and chemistry have been intensively investigated, is one of a large number of sapogenins whose structures are related very closely. These sapogenins have proved exceedingly useful as starting materials for the synthesis of all corticoids and steroid hormones.

Examples of these types of sapogenins can be illustrated by sarsasapogenin and diosgenin.

Table 97 gives a number of steroidal saponins together with their aglycons and sugar components. Some of the aglycons

TABLE 97. BOTANICAL SOURCES, AGLYCONS, FORMULAS AND SUGAR COMPONENTS OF A NUMBER OF SAPONINS

SAPONIN	BOTANICAL SOURCE	AGLYCON	FORMULA	SUGAR
Digitonin	*Digitalis purpurea*	Digitogenin	$C_{27}H_{44}O_5$	4-Galactose xylose
Gitonin	*Digitalis purpurea*	Gitogenin	$C_{27}H_{44}O_4$	3-Galactose pentose
Tigonin	*Digitalis lanata*	Tigogenin	$C_{27}H_{44}O_3$	2-Glucose 2-Galactose rhamnose
Sarsasaponin	*Radix sarsaparillae*	Sarsasapogenin	$C_{27}H_{44}O_3$	3-Glucose rhamnose
Amolonin	*Chlorogalum pomeridianum*	Tigogenin	$C_{27}H_{44}O_3$	3-Glucose 2-Rhamnose galactose
Trillin	*Trillium erectum*	Diosgenin	$C_{27}H_{42}O_3$	Glucose

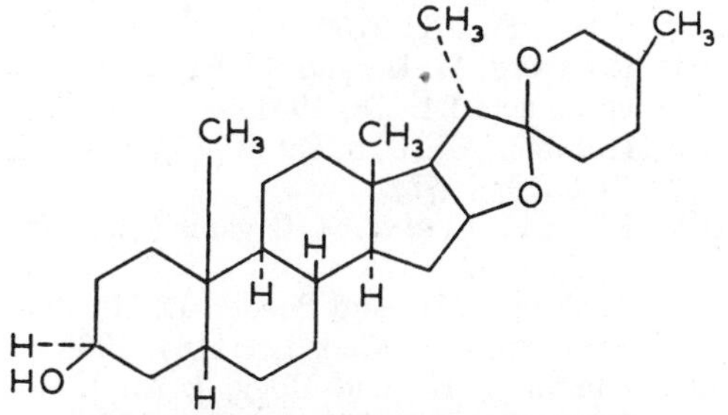

Sarsasapogenin [Spirostanol—(3β)]
22β-Spirostan-3β-ol

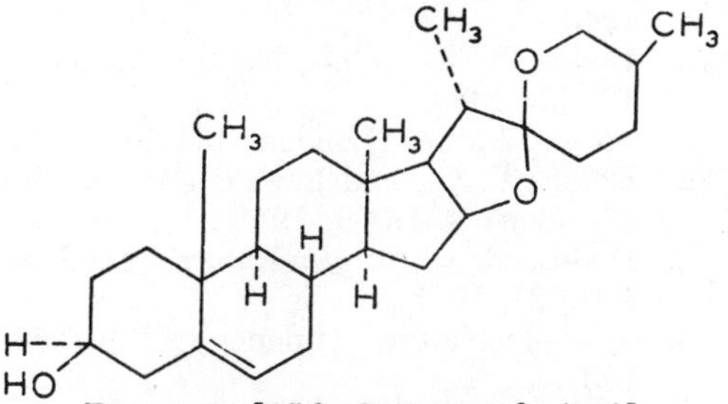

Diosgenin [Δ^{5,6}—Spirostenol—(3β)]
(22β-Spirost-5-en-3β-ol)

shown exist in glycosidal combination and are found in other plants.

REFERENCES CITED

1. Windaus, A.: Z. physiol. Chem. 65:110, 1910; 215:59, 1933.
   Lifschutz, I.: Biochem. Z. 282:441, 1935.
   ———: J. Biol. Chem. 132:471, 1940.
2. Fieser, L. F., and Fieser, M.: Steroids, Baltimore, Reinhold, 1959.
3. Gawienowski, A., and Gibbs, C.: Steroids 12(4):545, 1968.
4. Chevreul, M.: Ann. Chim. phys. (Ser. 1) 95:5, 1815.
5. Berthelot, M.: Ann. Chim. phys. (Ser. 3) 56:51, 1859.
5a. Ito, M., *et al.*: J. Lipid Res. 10:694, 1969.
6. Hesse, O.: Ann. Chem. 192:175, 1878.
7. Beneke, G. M.: Ann. Chem. 122:249, 1862.
8. Burian, R.: Mschr. Chem. 18:551, 1897.
9. Wallis, E. S., and Charkravorty, P. N.: J. Organic Chem. 2:335, 1938.
10. Morisake, N., and Kyosuke, T.: Steroids 12:41, 1968.
11. Tschesche, K.: Bull. Soc. Chim. France, 1219, 1965.
12. Patterson, G. W.: Comp. Biochem. Physiol. 24:501, 1968.
13. Windaus, A., and Hauth, A.: Ber. deutsch. chem. Ges. 39:4378, 1906.
14. Tanret, C.: Compt. rend. Acad. sci. 108:98, 1889.
15. Bergstrom, S.: Proc. 4th Int. Cong. Biochem. 4:161, 1959.
16. Tillinger, K. G., and Westman: Acta endocrinol. 25:113, 1957.
17. Diczfalusy, E., *et al.*: Acta endocrinol. (Suppl.) 25:24, 1956.
18. ———: Acta endocrinol. 21:321, 1956.
19. Schenck, M., and Junkmann, K.: Arch. exp. Path. Pharmak. 227:210, 1955.
20. Junkmann, K.: Rec. Progr. Horm. Res. 13:389, 1957.
21. Wallace, E., and Silberman, N.: J. Biol. Chem. 239:2809, 1964.
22. Meli, A., *et al.*: Endocrinol. 74:79, 1964.

23. ———: Steroids 1:287, 1963.
24. Lerner, L. J., *et al.*: Ann. N. Y. Acad. Sci. 116:1071, 1964.
25. Cross, A. D., *et al.*: Steroids 4:423, 1964.
26. ———: Steroids 4:229, 1964.
27. ———: Steroids 5:557, 1965.
28. Mecheli, R. A., *et al.*: J. Med. Chem. 5: 321, 1962.
29. Bruch, M. G.: Biochem. J. 88:55, 1963; Bickoff, E. M., *et. al.*: Agr. Food Chem. 6:536, 1958; 12:537, 1964.
30. Nutting, E. F., and Meyer, R. K.: J. Endocrinol. 29:235, 243, 1964.
31. Saunders, F. J., and Rorig, K.: Fertil. Steril. 15:202, 1964.
32. Chang, M. C., and Yanagimachi, R.: Fertil. Steril. 16:281, 1965.
33. Whatnick, A. S., *et al.*: Proc. Soc. Exp. Biol. Med. 116:343, 1964.
34. Greenwald, G. S.: Acta endocrinol. 47: 10, 1964; see also France, E. S., and Pincus, G.: Endocrinol. 75:359, 1964.
35. Emmens, C. W., *et al.*: Acta endocrinol. (Suppl.) 90:61, 1964.
36. Wilds, A., and Djerassi, C.: J. Am. Chem. Soc. 68:2155, 1946.
37. Frank, R. T.: J.A.M.A. 78:181, 1922.
38. Allen, E.: Am. J. Anat. 30:297, 1922.
39. Aschheim, S., and Zondek, B.: Klin. Wschr. 6:1322, 1927.
40. Doisy, E. A., Veler, C. D., and Thayer, S.: Am. J. Physiol. 90:329, 1929; Butenandt, A.: Naturwiss. 17:878, 1929.
41. MacCorquodale, D. W., *et al.*: J. Biol. Chem. 115:435, 1936.
42. Wintersteiner, O., *et al.*: Proc. Soc. Exp. Biol. Med. 32:1087, 1935.
43. Huffman, M. N., *et al.*: J. Biol. Chem. 134:591, 1940.
44. ———: J. Biol. Chem. 133:567, 1940.
45. Allen, E., and Doisy, E. A.: J.A.M.A. 81: 819, 1923.
46. Hawkinson, L. F.: J.A.M.A. 111:39, 1938.
47. Girard, A. A., *et al.*: Compt. rend. Acad. sci. 194:98, 909, 981, 1020, 1932.
48. Mazer, C., and Shecter, F. R.: J.A.M.A. 112:1925, 1939.
49. Cross, A. D., *et al.*: Steroids 4:425, 1964.
50. Schacther, B., and Marrian, G. F.: J. Biol. Chem. 126:633, 1938.
51. Westerfield, W. W., *et al.*: J. Biol. Chem. 126:181, 1938.
52. Girard, A., and Sandulesco, G.: Helvet. chim. acta 19:1095, 1936.
53. Stuart, A. H., *et al.*: J. Am. Chem. Soc. 65:1579, 1943.
54. Doisy, E. A., *et al.*: Proc. Soc. Exp. Biol. Med. 28:88, 1930; Marrian, G. F.: J. Soc. Chem. Ind. (London) 49:515, 1930.
55. Cohen, S. L., and Marrian, G. F.: Biochem. J. 30:57, 1936.
56. Sharman, A.: J. Endocrinol. 29:209, 1964.
57. Inhoffen, H. H., and Hohlweg, W.: Naturwiss. 26:96, 1938.
58. Jackson, J. L., *et al.*: Fertil. Steril. 19: 649, 1968.
59. Hogg, J. A., *et al.*: J. Am. Chem. Soc. 80: 2220, 1958.
60. Iriarte, J., and Ringold, H. J.: Tetrahedron 3:28, 1958.
61. Djerassi, C., *et al.*: J. Organic Chem. 24: 311, 1959.
62. Johns, U. F.: J. Am. Chem. Soc. 80:6456, 1958.
63. Bachmann, W., and Holmes, D.: J. Am. Chem. Soc. 62:2750, 1940.
64. Goldberg, M. U., and Studer, S.: Helvet. chim. acta 24:478, 1941.
65. Huffman, M.: U. S. Pat. 2,779,773 (C. A. 61:8819, 1957).
66. Kincl, F. A., *et al.*: J. Organic Chem. 22: 1127, 1957.
67. Axelrod, L. R., and Baeder, D. H.: Proc. Soc. Exp. Biol. Med. 121:1184, 1966.
68. Hunter, J. H., and Hogg, J. A.: J. Am. Chem. Soc. 68:1676, 1946; Meischer, K., *et al.*: Helvet. chim. acta. 30:1422, 1947.
69. Noel, C. J.: Nature (London) 157:132, 1946.
70. Dodds, E. C., *et al.*: Nature (London) 131:56, 1933.
71. ———: Nature (London) 131:205, 1933.
72. Dodds, E. C., and Lawson, W.: Nature (London) 137:996, 1936.
73. Dodds, E. C., *et al.*: Nature (London) 139:627, 1937.
74. ———: Nature (London) 139:1068, 1937.
75. ———: Nature (London) 141:78, 1938.
76. ———: Nature (London) 141:34, 247, 1938.
77. ———: Nature (London) 142:1121, 1938.
78. Blanchard, E. W., and Stebbins, R. B.: Endocrinology 32:307, 1943; Baker, B. R.: J. Am. Chem. Soc. 65:1572, 1943.
79. Blanchard, E. W., and Stebbins, R. B.: Endocrinology 36:297, 1945; Stuart, A. H.: J. Am. Chem. Soc. 68:729, 1946.
80. Solmssen, U. V.: J. Am. Chem. Soc. 65: 2370, 1943.
81. Rubin, M., and Wishinsky, H.: J. Am. Chem. Soc. 66:1948, 1944.

82. Dodds, E. C., *et al.*: Nature (London) 142:34, 1938.

83. ———: Nature (London) 142:211, 1938.

84. Ried, E. E., and Wilson, E.: J. Am. Chem. Soc. 64:1625, 1942.

85. Sondern, C., *et al.*: Endocrinology 28:849, 1941.

86. Wilds, A. L., and Biggerstaff, W. R.: J. Am. Chem. Soc. 67: 789, 1945.

87. Kharasch, M. S., and Kleinman, M.: J. Am. Chem. Soc. 65:11, 1943.

88. Vanderlinde, R. E., *et al.*: J. Am. Chem. Soc. 77:4176, 1955.

89. Kharasch, M. S., *et al.*: J. Organ. Chem. 10:401, 1945.

90. Frankel, L.: Arch. Gynak. 68:438, 1903.

91. Allen, W. M.: Am. J. Physiol. 92:612, 1930: Allen, W. M., and Corner, G. W.: Proc. Soc. Exp. Biol. Med. 27:403, 1930; Hisaw, F. L., *et al.*: Proc. Soc. Exp. Biol. Med. 27:400, 1930.

92. Butenandt, R., *et al.*: Ber. deutsch. chem. Ges. 67:1440, 1934.

93. Beall, D., and Reichstein, T.: Nature (London) 142:479, 1938.

94. Dorfman, R. I.: Anat. Rec. 157:547, 1967.

95. Pincus, G., *et al.*: Ann. N. Y. Acad. Sci. 71:677, 1961.

96. Pincus, G.: Modern Trends in Endocrinology, p. 235, New York, Harper, 1961.

97. Kincl, F., and Dorfman, R.: Acta endocrinol. 42:3, 1963.

98. Sawyer, C., and Kawakami, M.: Control of Ovulation *in* Villee, C. A. (ed.): Harvard U. Proc. Conf. Endicott House, p. 79, 1961.

99. Docke, F., *et al.*: J. Endocrinol. 41:353, 1968.

100. Schally, A. V., *et al.*: J. Clin. Endocrinol. 28:1747, 1968; Exley, D., *et al.*: J. Physiol. 195:697, 1968.

101. Butenandt, A., *et al.*: Ber. deutsch. chem. Ges. 67:1611, 1934; Fernholz, E.: Ber. deutsch. chem. Ges. 67:2027, 1934.

102. Marker, R., and Krueger, J.: J. Am. Chem. Soc. 62:2543, 3349, 1940.

103. Kathol, J., *et al.*: Naturwiss. 25:682, 1937; Ruzicka, L., *et al.*: Helv. chim. acta 20:1280, 1937.

104. Beyler, A. L., and Clinton, R. O.: Proc. Soc. Exp. Biol. Med. 92:404, 1956; Clinton, R. O., *et al.*: J. Organ. Chem. 22: 473, 1957.

105. Kincl, F., and Dorfman, R.: Acta endocrinol. 42:3, 1963.

106. Dorfman, R., and Kincl, F.: Steroids 1: 185, 1963.

107. Nelson, N. A., and Garland, R. B.: J. Am. Chem. Soc. 79:6313, 1957.

108. Hughes, G. A., *et al.*: Steroids 8:947, 1966.

109. Pincus, G.: Science 138:439, 1962.

110. McGinty, D. A., and Djerassi, C.: Ann. New York Acad. Sci. 71:500, 1958; Djerassi, C., *et al.*: J. Am. Chem. Soc. 76: 4092, 1954.

111. Edgren, R., *et al.*: Steroids 2:319, 1963.

112. Kincl, F., and Dorfman, R.: Proc. Soc. Exp. Biol. Med. 119:340, 1965.

113. Drill, V. A.: J. Endocrinol. XVII Proc. 1962.

114. Kincl, F., and Dorfman, R.: Acta endocrinol. 41:275, 1962.

115. Greenwald, G. S.: Acta endocrinol. 47: 10, 1964; France, E. S., and Pincus, G.: Endocrinol. 75:359, 1964.

116. de Ruggieri, P., *et al.*: Steroids 5:73, 1965.

117. Edgren, R. A., *et al.*: Steroids 2:321, 1963; Internat. J. Fert. 11:389, 1966; DeJongh, D. C., and Hribar, J. D.: Steroids 11:649, 1968.

118. Davis, E., and Wied, G.: J. Clin. Endocrinol. 15:923, 1955.

119. Babcock, J. C., *et al.*: J. Am. Chem. Soc. 80:2904, 1958.

120. *Ibid.*

121. Heusser, H., *et al.*: Helvet. chim. acta. 32:270, 1949; 33:2229, 1950; 35:2437, 1952.

122. Engel, R., *et al.*: J. Am. Chem. Soc. 78: 6153, 1956.

123. Provest Symposium, Int. J. Fertil. 8:589, 1963.

124. Kincl, F., and Dorfman, R.: Acta endocrinol. 41:274, 1962; Rubio, B.: Fertil. Steril. 14:254, 1963.

125. Dorfman, R., and Kincl, F.: Steroids 1: 185, 1963.

126. Hartly, F.: J. Endocrinol. Vol. 24, p. xvi of Proc. 1962.

127. Cekan, Z., *et al.*: Steroids 4:415, 1964.

128. Graber, R. P., *et al.*: J. Med. Chem. 7: 540, 1964.

129. Bowers, A., *et al.*: J. Am. Chem. Soc. 81: 5991, 1959.

130. Dodson, R. M., *et al.*: J. Am. Chem. Soc. 82:2322, 1960; Engel, C. R., and Noble, R. L.: Endocrinology 61:318, 1957.

131. Lerner, L., *et al.*: Int. J. Fertil. 9:547, 1964.

132. Taymor, M. L., *et al.*: Fertil. Steril. 15: 653, 1964.

133. Birch, A. J., and Smith, H.: J. Chem. Soc. 4909, 1956. Ercoli, A., and Gardi, R.: J. Am. Chem. Soc. 82:746, 1960.
134. Revesz, C., et al.: Steroids 10:291, 1967.
135: Plattner, P. A., et al.: Helv. chim. acta 31:249, 1948.
136. Ehrenstein, M., et al.: Endocrinol. 60: 681, 1957; Miyake, T., and Rooks, W. H. in Dorfman, R. I. (ed.) Methods in Hormone Research, Vol. 5, p. 59, 1966.
137. Scholer, H.: Acta endocrinol. 35:188, 1960.
138. Bishop, P., et al.: Acta endocrinol. 40: 203, 1962.
139. Suchowsky, G., and Junkman, K.: Acta endocrinol. 28:129, 1958.
140. Craig, C. R.: J. Pharmacol. Exp. Ther. 164:371, 1968.
141. Gyermek, L., et al.: J. Med. Chem. 11: 117, 1968; Int. J. Neuropharmacol. 6: 191, 1967.
142. Fonken, G. S., and Wechter, W. J.: J. Med. Chem. 11:633, 1968.
143. ———: Acta endocrinol. 39:68, 1962.
144. Hershberger, L. G., et al.: Proc. Soc. Exp. Biol. Med. 83:175, 1953.
145. Overbeek, G.: Symposium on Anabolic Therapy, March 21, 1963, Detroit, Michigan.
146. Arnold, A., et al.: Proc. Soc. Exp. Biol. 102:184, 1959; Endocrinol. 72:409, 1963.
147. Potts, G., et al.: Endocrinol. 67:849, 1960.
148. Dorfman, R., and Kincl, F.: Endocrinol. 72:260, 1963.
149. Boris, A., and Ng, C.: Steroids 9:299, 1966.
150. Dorfman, R., and Kincl, F.: Endocrinol. 72:259, 1963.
151. ———: Steroids 3:109, 1964.
152. Arnold, A., et al.: Endocrinol. 72:408, 1963.
153. Edgren, R.: Acta endocrinol. (Suppl.) 87:3, 1963.
154. Junkmann, K., and Suchowsky, G.: Arzneimittelforsch. 12:214, 1962; Arnold, A., et al.: Acta endocrinol. 44:490, 1963.
155. Male Sex Hormone Therapy, A Clinical Guide, p. 10, Bloomfield, N. J., Schering Corp., 1941.
156. Heaney, R. P.: Am. J. Med. 23:188, 1962.
157. Anderson, J. A.: Acta endocrinol. (Suppl.) 63:54, 1961.
158. Ramney, R., and Saunders, F.: Proc. Soc. Exp. Biol. Med. 116:596, 1964.
159. Laron, Z., et al.: Acta endocrinol. 45: 427, 1964.
160. Laquer, E., et al.: Z. physiol. Chem. 233: 281, 1935.
161. Koch, F. C.: Bull. N. Y. Acad. Med. 14: 655, 1938.
162. Marker, R.: J. Am. Chem. Soc. 62:2543, 1940.
163. Koch, F. C.: Ann. Rev. Biochem. 13:284, 1944.
164. Dorfman, R., and Kincl, F.: Steroids 3: 173, 1964.
165. ———: Endocrinol. 72:259, 1963.
166. Gould, D., et al.: J. Am. Chem. Soc. 79: 4472, 1957.
167. Kincl, F., and Dorfman, R.: Steroids 3: 109, 1964.
168. Lyster, S. C., et al.: Endocrinology 58: 781, 1956.
169. Sydnor, K.: Endocrinology 62:322, 1958.
170. Cross, A. D., et al.: Steroids 4:229, 1964.
171. Clinton, R. O., et al.: Proc. Soc. Exp. Biol. Med. 81:1513, 1959.
172. Goldberg M. W., and Wydler, E.: Helvet. chim. acta 26:1142, 1943.
173. Johnson, W. S., et al.: J. Am. Chem. Soc. 75:4866, 1953.
174. Ringold, H. J., et al.: J. Am. Chem. Soc. 81:429, 1959.
175. Ringold, H. J.: J. Am. Chem. Soc. 81: 427, 1959.
176. Dorfman, R., and Kincl, F.: Endocrinol. 72:259, 1963.
177. Carter, A. C.: Cancer Chemother. Rep. 11:121, 1961.
178. Stucki, J. C., et al.: Proc. Int. Congress on Hormonal Steroids, Milan, 2:119, 1962.
179. Korst, D., et al.: Clin. Pharmacol. Ther. 4:734, 1963.
180. Arnold, A., et al.: Endocrinol. 72:410, 1963.
181. ———: Endocrinol. 72:408, 1963.
182. Orr, J. C., et al.: J. Med. Chem. 6:166, 1963; Kincl, F., and Dorfman, R.: Steroids 3:121, 1964; Endocrinol. 72:261, 1963.
183. Albanese, A. A.: Ann. Prog. Rep. 1962 Burke Foundation Rehabilitation Center, White Plains, N. Y. 1963.
184. Birch, A. J.: J. Chem. Soc. 367, 1950; Stafford, R. O., et al.: Proc. Soc. Exp. Biol. Med. 86:322, 1954.
185. Bowers, A., et al.: J. Am. Chem. Soc. 79: 4556, 1957; Iriarte, J., et al.: J. Am. Chem. Soc. 81:436, 1959.
186. Kincl, F., and Dorfman, R.: Steroids 3: 119, 1963.

187. de Visser, J., and Overbeek, G.: Acta endocrinol. 35:405, 1960.
188. Tomarelli, R. M., and Berhart, F. W.: Steroids 4:451, 1964.
189. de Winter, M. S., et al.: Chem. Ind. (London) 905, 1959.
190. Dorfman, R., and Kincl, F.: Steroids 3:173, 1964.
191. Bowers, A., et al.: J. Med. Chem. 6:156, 1963; Edwards, J. A., and Bowers, A.: Chem. Ind. Dec. 2, 1961.
192. Cross, A. D., et al.: J. Med. Chem. 6:162, 1963.
193. Cekan, Z., et al.: Steroids 5:113, 1965.
194. Morrow, L. B., et al.: Endocrinol. 80:833, 1967.
195. Dorfman, R. I., et al.: Proc. 3rd Asia and Oceania Congress of Endocrinol. p. 171, 1967.
195a. Schaeffer, L. D., et al.: Biochim. biophys. acta 192:292, 1969.
196. Schayer, R. W.: Perspect. Biol. Med. 10:409, 1967.
196a. Symons, A. M., et al.: Biochem. Pharmacol. 18:2581, 1969.
197. Ray, P. D., et al.: J. Biol. Chem. 239:3396, 1964.
198. Bogden, A. E., et al.: Endocrinol. 82:1093, 1968.
199. Forte, L., and Landon, E. J.: Biochim. biophys. acta 157:303, 1968; Edelman, I. S., et al.: Am. J. Physiol. 213:954, 1967.
200. Reichstein, T., and Shoppee, C. W.: Vitamins & Hormones 1:352, 1943; Reichstein, T., et al.: Erg. Vit. Hormonforsch. 1:334, 1938; Wettstein, A.: Proc. 4th Int. Cong. Biochem. 4:233, 1959.
201. Lews, R. H., et al.: Science 94:348, 1941; Thorn, G. W., and Clinton, M.: Science 96:343, 1943.
202. Dorfman, R. I., et al.: Endocrinology 38:189, 1946.
203. Peterson, D. H., and Murray, H. C.: J. Am. Chem. Soc. 74:1871, 1952; U. S. Patent 2,602,769; Peterson, D. H.: Proc. 4th Int. Cong. Biochem. 4:83, 1959.
204. Florey, K.: Chimia 8:81, 1954; Hanc, O., and Riedl-Tumova, E.: Pharmazie 9:877, 1954; Waksman, S. A. (ed.): Perspectives and Horizons in Microbiology, p. 121, New Brunswick, N. J., Rutgers Univ. Press, 1955.
205. Herzog, H. L., et al.: J. Am. Chem. Soc. 77:4184, 1955.
206. Gordon, E. S.: J.A.M.A. 114:2549, 1940.
207. Selye, H.: Anesthesiol. 4:36, 1943.
208. Merryman, W., et al.: J. Clin. Endocrinol. 14:1567, 1954.
209. Harman, R. E., et al.: J. Am. Chem. Soc. 76:5035, 1955.
210. Sharp, G., and Leaf, A.: Nature 202:1185, 1964.
211. Hench, P. S., et al.: Proc. Staff Meet. Mayo Clin. 24:181, 1949.
212. Gibson, A., et al.: Postgrad. Med. 10:1, 1951.
213. Boland, E. W.: J. Am. Pharm. A. (Pract. Ed.) 13:540, 1952.
214. Fried, J., and Sabo, E. F.: J. Am. Chem. Soc. 76:1455, 1954; Liddle, G. W., et al.: Science 120:496, 1954.
215. Taub, B., et al.: J. Am. Chem. Soc. 78:2912, 1956.
216. Byrnes, W. W., et al.: Proc. Soc. Exp. Biol. Med. 91:67, 1956.
217. Herzog, H. L., et al.: Science 121:176, 1955.
218. Swingle, W. W., et al.: Endocrinology 60:658, 1957.
219. Hogg, J. A., et al.: J. Am. Chem. Soc. 78:6213, 1956; Bowers, A., and Ringold, H. J.: J. Am. Chem. Soc. 80:3091, 1958; Sarett, L. H., et al.: J. Am. Chem. Soc. 81:1235, 1959.
220. Liddle, G. W., and Richard, J. E.: Science 123:324, 1956.
221. Spero, G., et al.: J. Am. Chem. Soc. 79:1515, 6213, 1957.
222. West, K.: Metabolism 7:441-456, 1958.
223. Pearlman, P., et al.: J. Am. Chem. Soc. 80:6687, 1958; Taub, D., et al.: J. Am. Chem. Soc. 82:4012, 1960.
224. Arth, E. G., et al.: J. Am. Chem. Soc. 80:3161, 1958.
225. Lerner, L. J., et al.: Ann. N. Y. Acad. Sci. 116:1071, 1964.
226. Boland, E. W.: California Med. 88:417, 1958; Oliveto, E. P., et al.: J. Am. Chem. Soc. 80:4428, 4431, 6687, 1958; Arth, E. G., et al.: J. Am. Chem. Soc. 81:1235, 1959.
227. Bernstein, S., et al.: J. Am. Chem. Soc. 81:1256, 1959; 78:5693, 1956.
228. Fried, J., Borman, A.: Vitamins & Hormones 16:358, 1958; J. Am. Chem. Soc. 80:2338, 1958.
229. Ringold, H. J., et al.: J. Am. Chem. Soc. 80:6464, 1958.
230. Kagan, F., et al.: J. Med. Chem. 7:748, 751, 1964.
231. Hogg, J. A.: Sixth Nat. Med. Chem. Symposium A. C. S., Madison, Wis., June 23-25, 1958; Taub, et al.: J. Am. Chem.

Soc. 78:2912, 1956; Bowers, H., and Ringold, H. J.: Tetrahedron 3:14, 1958.

232. Weiss, M. J., *et al.*: J. Am. Chem. Soc. 81:1262, 1959.

233. Mrozik, H., *et al.*: J. Med. Chem. 7:584, 1964.

234. Fried, J., *et al.*: J. Am. Chem. Soc. 85: 236, 1963.

235. Tishler, M., *et al.*: J. Am. Chem. Soc. 85: 120, 1963.

236. Fried, J., and Borman, A.: Vitamins & Hormones, 16:363, 1958; Sandoval, A., *et al.*: J. Am. Chem. Soc. 77:148, 1955; Hogg, J. A., *et al.*: J. Am. Chem. Soc. 80: 2226, 1958; Zaffroni, A., *et al.*: J. Am. Chem. Soc. 80:6110, 1958.

237. Herzog, H., *et al.*: J. Am. Chem. Soc. 82: 369, 1960.

238. Leanza, W. J., *et al.*: J. Am. Chem. Soc. 76:1691, 1954.

239. Fried, J., and Borman, A.: Vitamins & Hormones 16:316, 1958.

240. Pike, J., *et al.*: J. Org. Chem. 28:2502, 1963.

241. Bianchi, C., *et al.*: J. Pharm. Pharmacol. 13:355, 1961.

242. Laubach, G. D., *et al.*: J. Am. Chem. Soc. 77:4685, 1955.

243. Huffman, M.: C. & E. N., V. 34, June 18, p. 2982, 1956.

244. Kakawa, C. M., *et al.*: Science 126:1015, 1957.

245. Landau, R. L., *et al.*: J. Clin. Endocrinol. Metab. 15:1194, 1955.

246. Tweit, R., and Kagawa, C.: J. Med. Chem. 7:524, 1964.

247. Kagawa, C., and Pappo, R.: Endocrinol. 74:999, 1964.

248. Dorfman, R.: Proc. 4th Int. Cong. Biochem. 4:178, 1959.

249. *Ibid.*, p. 189.

250. Heard, R., *et al.*: J. Endocrinol. 57:201, 1955.

251. Dorfman, R.: Proc. 4th Int. Cong. Biochem. 4:186, 1959.

252. White, W. F., and Gisvold, O.: J. Am. Pharm. A. (Sci. Ed.) 41:42, 1952.

252a. Greenberger, N. J., *et al.*: J. Pharmacol. Exp. Ther. 167:265, 1969.

253. Movitt, E. R.: Digitalis and Other Cardiotonic Drugs, p. 72, New York, Oxford, 1946.

254. Gold, H., *et al.*: J. Pharmacol. Exp. Ther. 69:177, 1940; 98:337, 1950; Batherman, R. C., and De Graff, A. C.: Fed. Proc. 4:112, 1945.

255. Gold, H.: Connecticut M. J. 9:13, 1945.

———: J. Am. Pharm. A. (Pract. Ed.) 8:494, 1947.

———: Therapeutic Notes, p. 227, 1946.

———: Tile and Till 33:79, 1947.

256. Research Today 2 (no. 3), Fall 1945.

257. Chen, K. K., *et al.*: J. Am. Pharm. A. (Sci. Ed.) 25:579, 1936.

———: J. Am. Pharm. A. (Sci. Ed.) 26: 214, 1937.

———: J. Am. Pharm. A. (Sci. Ed.) 27: 113, 1938.

———: J. Am. Pharm. A. (Sci. Ed.) 31: 236, 1942.

———: J. Pharmacol. Exp. Ther. 77:401, 1943.

Chen, K. K., and Elderfield, R. C.: J. Pharmacol. Exp. Ther. 76:81, 1942.

Chen, K. K.: Ann. Rev. Physiol. 7:681, 1945.

Uhle, F. C., and Elderfield, R. C.: J. Organic Chem. 8:162, 1943.

Elderfield, R. C., *et al.*: J. Am. Chem. Soc. 69:2235, 1947.

258. Chen, K. K.: *in* New Aspects of Cardiac Glycosides, vol. 3, p. 27, New York, Pergamon, 1963.

259. Tschesche, R., and Bohle, K.: Ber. deutsch. chem. Ges. 69:2443, 1936.

260. Shigei, T., and Mineshita, S.: Experientia 24:466, 1968.

261. Jacobs, W. A., and Hoffman, A.: J. Biol. Chem. 74:787, 1927.

———: Physiol. Rev. 13:241, 1933.

Stoll, A., and Hoffman, A.: Helvet. chim. acta 18:401, 1935.

262. Megges, R., and Repke, K. *in* New Aspects of Cardiac Glycosides, vol. 3, p. 271, New York, Pergamon, 1963.

262a. Gold, H., *et al.*: J. Clin. Pharmacol. 9: 148, 1969.

263. Gisvold, O.: J. Am. Pharm. A. 47:600, 1958.

264. Jacobs, W. A., and Bigelow, N. M.: J. Biol. Chem. 13:241, 1933.

265. Jacobs, W. A., and Hoffman, A.: Physiol. Rev. 13:241, 1933.

266. Anderson, R. C., and Chen, K. K.: J. Am. Pharm. A. (Sci. Ed.) 35:353, 1946.

Bell, F. K., and Krantz, J. C.: J. Am. Pharm. A. (Sci. Ed.) 35:260, 1946.

Bell, F. K., and Krantz, J. C.: J. Pharmacol. Exp. Ther. 87:198, 1946.

267. Hdjdu, S., and Leonard, E.: Phys. Rev. 11:173, 1959.

268. Huxley, H., and Hanson, J.: *In* Structure and Function of Muscle, New York, Acad. Press, 1960; Porter, K. R., and Franzini-

Armstrong, C.: Sci. Am. 212(3):72, 1965.

269. Fritz, P. J., *et al.*: Pharmacol. 2:32, 1968.
270. Morales, M.: Rev. Mod. Physics 31:426, 1959; See also Podalsky, R.: Mech. Basis of Muscular Contraction, Fed. Proc. 21: 964, 1962; Remington, Introduction of Muscle Mechanics, Fed. Proc. 21:954, 1962.
271. Weber: Biochem. biophys. acta 12:150, 1963.
272. Waser, P.: New Aspects of Cardiac Glycosides, vol. 3, p. 173, 1963.
273. Holland, W., and Klein, R.: Chemistry of Heart Failure, p. 63, Springfield, Ill., Thomas, 1960.
274. For Reference see Tuttle, R., *et al.*: J. Pharmacol. Exp. Ther. 133:281, 1961.
275. Calhoun, R., and Harrison, T.: J. Clin. Invest. 10:139, 1931.
276. Cattell, Mc.: J. Pharmacol. 62:459, 1938.
277. Holland, W., and Sekul, A.: J. Pharmacol. Exp. Ther. 133:228, 1961; Lee, K.: *In* New Aspects of Cardiac Glycosides, vol. 3, p. 185, 1963.
278. Walker, J., and Weatherall, M.: Brit. J. Pharmacol. 23:66, 1964.
279. Bonting, S., *et al.*: Arch. Biochem. Biophys. 98:413, 1962; Auditore, J., and Murray, L.: Arch. Biochem. Biophys. 99: 372, 1962; ———: Arch. int. Pharmacodyn. 145:137, 1963.
280. Schwartz, A.: Cardiovascular Res. Center. Bull. p. 73, Spring, 1964.
281. Schwartz, A., *et al.*: Biochem. Pharmacol. 13:337, 1964.
281a. Hecht, H. H.: Med. Clin. N. Am. 54:221, 1970.
281b. Dutta, S., *et al.*: J. Pharmacol. Exp. Ther. 194:10, 1968.
282. Lee, K.: J. Pharmacol. Exp. Ther. 137: 186, 1962; New Aspects of Cardiac Glycosides, vol. 3, p. 185, 1963.
283. Nativelle, C. A.: J. pharm. chim. 9:255, 1869.
284. Cloetta, M.: Arch. exp. Path. Pharmakol. 88:113, 1920. Windaus, A., and Stein, G.: Ber. deutsch. chem. Ges. 61:2436, 1928.
285. Stoll, A.: J. Am. Pharm. A. (Sci. Ed.) 27:761, 1938.
286. Gold, H., *et al.*: J. Pharmacol. Exp. Ther. 32:187, 1944.
287. Greenwald, G. S.: Acta endocrinol. 47: 59, 1964.
288. Dimond, E. G.: Digitalis, pp. 65 and 86, Springfield, Ill., Thomas, 1957. For References see Repke, K., *in* New Aspects of

Cardiac Glycosides, vol. 3, p. 61, 1963.
288a. Selye, H., *et al.*: Science 164:842, 1969; Brit. J. Pharmacol. 37:485, 1969.
288b. Steelman, S. L., *et al.*: Steroids 14:449, 1969.
289. Gisvold, O.: J. Am. Pharm. A. 47:594, 1958.
290. Cloetta, M.: Arch. exp. Path. Pharmakol. 112:261, 1926.
291. Kraft, F.: Arch. Pharm. 250:126, 1912. Windaus, A., *et al.*: Ber. deutsch. chem. Ges. 61:1847, 1928.
292. Smith, S.: J. Chem. Soc. p. 508, 1930; 23, 1931.
293. Hopponen, R. E., and Gisvold, O.: J. Am. Pharm. A. (Sci. Ed.) 51:146, 1952. Krishnamurty, G. G.: J. Am. Pharm. A. (Sci. Ed.) 51:152, 1952.
294. Benthe, H. F., Arzneimittelforsch. 15: 486, 1965.
295. Forster, W., and Schulzeck, S.: Biochem. Pharmacol. 17:489, 1968.
296. Wright, S. E., *et al.*: J. Biol. Chem. 220: 431, 1956; 232:315, 1958.
297. For references see Repke, K., *in* New Aspects of Cardiac Glycosides, p. 57, 1963.
298. Wright, S. E., and Cox, E.: J. Pharmacol. Exp. Ther. 126:117, 1959.
299. Greenwald, G. S.: Acta endocrinol. 47: 55, 1964.
300. Mayhew, E., and Levinson, C. J.: Cell. Physiol. 72:73, 1968.
301. Merk, E.: Pharm. Ztg. 24:286, 1879.
302. von Jarmersted, E.: Arch. exp. Path. Pharmakol. 11:22, 1879.
303. Waliszewski, S.: l'Union Pharm. 34:251, 1894.
304. Kopaczewski, W.: Compt. rend. Acad. sci. 158:1520, 1914. ———: Biochem. Z. 66:501, 1914.
305. Karrer, W.: Helvet. chim. acta 26:1353, 1943.
306. Kupchan, S. M., *et al.*: J. Med. Chem. 7: 803, 1964.

SELECTED READING

STEROID HORMONES

Applezweig, H.: Steroid Drugs, New York, McGraw-Hill, 1962-1964.
Babcock, J. C.: Synthetic Progestational Agents, *in* Molecular Modification of Drug Design, Adv. in Chem. Series 45, Washington, D.C., Am. Chem. Soc., 1964.
Burrows, H.: Biological Action of Sex Hormones, London, Cambridge, 1945.

Complete Manual of Therapy With the Meti-steroids, Bloomfield, N. J., Schering Corp., 1956.

Counsel, R. E., and Klimstra, P. D.: Androgens and Anabolic Agents *in* Burger, A. (ed.): Medicinal Chemistry, ed. 3, pp. 923-952, New York, Wiley-Interscience, 1970.

Deghenghi, R., and Manson, A. S.: Estrogens and Progestational and Contraceptive Agents *in* Burger, A. (ed.): Medicinal Chemistry, ed. 3, pp. 900-922, New York, Wiley-Interscience, 1970.

Diczfalusy, E. The Mode of Action of Contraceptive Drugs, Am. J. Obstet. Gynec. 100: 136-163, 1968.

Dorfman, R. I., and Shipley, R. A.: Androgens, New York, Wiley, 1956.

Ercoli, A., *et al.*: Steroidal Ethers *in* Research Progress in Organic-Biological and Medicinal Chemistry, vol. 1, p. 156, 1964.

Fieser, L. F., and Fieser, M.: Steroids, Baltimore, Reinhold, 1959.

Fisch, C., and Surawicz, B.: Digitalis, N. Y., Grune and Stratton, 1969.

Fried, J., and Borman, A.: Synthetic derivatives of cortical hormones, Vitamins & Hormones 16:304, 1958.

Gordon, E. S. (ed.): A Symposium on Steroid Hormones, Madison, Wis., Univ. Wis. Press, 1950.

Hecht, H. H.: The cellular action of digitalis compounds, Med. Clin. N. Am. 54:221, 1970.

Heftmann, E., and Mosettig, E.: Biochemistry of Steroids, New York, Reinhold, 1960.

Maas, J. M.: The oral contraceptives, Clin. Med. 77:14, 1970.

Martin, L., and Pecile, A.: Hormonal Steroids, New York, Academic Press, 1964.

Miyake, T., and Roaks, W. H.: The relation between the structure and physiological activity of progestational steroids, Methods Hormone Res. 5:59, 1966.

Petrow, V.: Current aspects of fertility control, Chemistry in Britain 6:167, 1970.

Pincus, G.: Clinical Control of Fertility *in* Molecular Modification in Drug Design, p. 177, Washington, D.C., Am. Chem. Soc., 1964.

Pincus, G.: Progestational agents and control of fertility, Vitamins & Hormones 17:307, 1959.

Pincus, G., and Pearlman, W.: The Intermedi-ate Metabolism of the Sex Hormones, *in* Harris, R. S., and Thimann, K. V.: Vitamins & Hormones, pp. 293-343, New York, Acad. Press, 1943.

Reichstein, T., and Shoppee, C. W.: The Hormones of the Adrenal Cortex, *in* Harris, R. S., and Thimann, K. V.: Vitamins & Hormones, pp. 345-413, New York, Acad. Press, 1943.

Rodig, O. R.: The Adrenal Cortex Hormones *in* Burger, A. (ed.): Medicinal Chemistry, ed. 3, pp. 878-899, New York, Wiley-Interscience, 1970.

Rudel, H. W., and Kincl, F. A.: The biology of anti-fertility steroids, Acta endocrinol. Suppl. 105, Vol. 51, pp. 7-45, 1966.

Segaloff, A.: Androgenic-Anabolic Agents *in* Molecular Modification in Drug Design, p. 204, Washington, D.C., Am. Chem. Soc., 1964.

West, K. M.: Relative eosinopenic and hyperglycemic potencies of glucocorticoids, Metabolism 7:441-456, 1958.

Wolff, H. P.: Aldosterone in clinical medicine, Acta endocrinol. 124:65-86, 1967.

CARDIAC GLYCOSIDES

Chen, K. K., and Henderson, F. G.: Pharmacology of 64 glycosides and aglycones, J. Pharmacol. Exp. Ther. 111:365, 1954.

Dimond, E. G.: Digitalis, Springfield, Ill., Thomas, 1957.

Elderfield, R. C.: The chemistry of the cardiac glycosides, Chem. Rev. 17:187, 1935.

Jacobs, W. A.: Chemistry of the cardiac glycosides, Physiol. Rev. 13:222, 1933.

Movitt, E. R.: Digitalis and Other Cardiotonic Drugs, New York, Oxford, 1946.

Stoll, A.: The cardiac glycosides of digitalis, Chem. Ind. 1558, 1959.

Stoll, A.: The Cardiac Glycosides, London, Pharm. Press, 1937.

Tamm, Ch.: The stereochemistry of the cardiac glycosides in relation to biological activity, Proc. 1st Int. Pharmacological Meeting, Stockholm (1961), Vol. 3, New York, Pergamon, 1963.

Weise, M. H.: Digitalis, Leipzig, Thieme, 1936.

Wilbrandt, W. (ed.): New Aspects of Cardiac Glycosides, Proc. 1st Int. Pharmacological Meeting, Stockholm (1961), Vol. 3, New York, Pergamon, 1963.

26

Carbohydrates

Ole Gisvold, Ph.D.,
Professor of Medicinal Chemistry, College of Pharmacy,
University of Minnesota

INTRODUCTION

Carbohydrates are synthesized by chlorophyll-containing plants from carbon dioxide and water in the presence of sunlight via a very complex route (demonstrated in the case of the alga *Chlorella pyrenoidosa*[*]). One of the carbohydrates that is synthesized is glucose; in many cases, this is converted immediately into the osmotically less effective polysaccharide, starch. Because the empirical formula for glucose was shown to be CH_2O, it was designated as a carbohydrate, which literally means hydrated carbon (from the French, *hydrate de carbone*). Therefore, other naturally occurring substances that, before or after hydrolysis, exhibited similar chemical and physical properties also were called carbohydrates.

As a result of a vast amount of research in this field, it can be concluded that carbohydrates may be defined as polyhydroxy aldehydes or ketones or condensation products thereof. It may be inferred that such a definition does not hold to the empirical formula $C_nH_{2n}O_n$. This is due to the fact that in some rare and important desoxy sugars that have been found in nature, such as 2-desoxyribose, digitoxose, cymarose, rhamnose, this relationship of C, H and O does not exist.

Carbohydrates furnish a source of energy for both plants and animals. In the form of cellulose, together with lignin, they function as the supporting tissues of plants. Their degradation products, by enzymatic means, are used in the synthesis of fats, proteins and many other naturally occurring products. Sugars, modified or in combination, give rise to or are present in important naturally occurring substances such as vitamin C, riboflavin, coenzymes I and II, yeast and muscle nucleic acids, glycosides, saponins, etc.

CLASSIFICATION

Carbohydrates may be classified according to the number of sugar residues[†] present in the molecule. These divisions in turn may be subdivided further into the number of carbon atoms in each residue and into those containing a potential aldehyde or ketone group. Thus, for example, the following classification is possible:

Monosaccharides or Monoses.[‡]

TRIOSES ($C_3H_6O_3$). For example, the aldotriose D-glyceric aldehyde, or glycerose, and the ketotriose dihydroxyacetone, both

[*] See Bassham, J. A.: The path of carbon photosynthesis, Sci. Am., June, 1962.

[†] The smallest units or molecules into which the carbohydrate can be resolved by acid hydrolytic methods.

[‡] It is recommended that Emil Fischer's designations *d-* and *l-* be denoted by small capitals D and L in order to prevent confusion with optical rotation.

851

of which occur as an intermediate in muscle glycolysis and yeast fermentation of glucose.

TETROSES ($C_4H_8O_4$). For example, the aldotetroses D- and L-erythrose and the aldoketose D-erythrulose, all of which have been prepared synthetically.

PENTOSES ($C_5H_{10}O_5$).

Aldopentoses. L-Arabinose occurs naturally in arabans, hemicelluloses and other vegetable gums. D-Xylose occurs in wood gum, mucilages, hemicelluloses, xylans and other wood products. D-Ribose is found in some nucleic acids.

HEXOSES ($C_6H_{12}O_6$).

Aldohexoses. D-Glucose, the most abundant sugar of the monosaccharides, occurs chiefly in starch, cellulose, glycogen, sucrose and lactose. D-Galactose occurs in galactans, gums, mucilages, hemicelluloses and lactose. D-Mannose can be prepared from the reserve polysaccharide found in date seeds and vegetable ivory. It also is found in cherry gum and in several other polysaccharides.

Ketohexoses. D-Fructose can be prepared from the polysaccharide inulin, which is found in dahlias and artichokes. It also is found in sucrose. L-Sorbose is found as a fermentation product of the sugar alcohol sorbitol.

Oligosaccharides.

DISACCHARIDES.

Reducing Disaccharides. Maltose (4-D-glucopyranosyl-α-D-glucopyranoside) is an end-product resulting from the action of diastase upon starch. Cellobiose (4-D-glucopyranosyl-β-D-glucopyranoside) is one of the hydrolytic products of cellulose. Gentiobiose (6-D-glucopyranosyl-β-D-glucopyranoside) has been prepared by the hydrolysis of the glycoside amygdalin. Melibiose (6-D-glucopyranosyl-α-D-glucopyranoside) is obtained upon partial hydrolysis of the trisaccharide raffinose. Lactose (4-D-glucopyranosyl-β-D-galactopyranoside) is found in milk.

Nonreducing Disaccharides. Sucrose (1-α-D-glucopyranosyl-β-D-fructofuranoside) is obtained on a commercial scale from sugar cane and sugar beets. It occurs very abundantly in nature.

TRISACCHARIDES.

Nonreducing. Raffinose, which yields ga-lactose, glucose and fructose upon hydrolysis, is found in such plants as cotton, barley, lotus and eucalyptus. Melecitose, upon hydrolysis, yields two molecules of glucose and one of fructose. It is found in Persian manna, *Larix decidua* and the manna from Pseudatsuga. Gentianose, as the name implies, has been found in the roots of *Gentiana lutea.* Upon hydrolysis, it yields two molecules of glucose and one of fructose.

TETRASACCHARIDES.

Nonreducing. Stachyose yields two molecules of galactose and one molecule each of glucose and fructose upon hydrolysis. It is found in *Stachys sieboldi* and in several leguminous seeds.

Polysaccharides or Nonsugars.

PENTOSANS [($C_5H_8O_4$)n]. Arabans that yield only arabinose upon hydrolysis are rare. On the other hand, certain gums, such as cherry gum and gum arabic, yield chiefly arabinose upon hydrolysis. Xylans that yield chiefly xylose upon hydrolysis have been obtained from esparto cellulose, deciduous trees and straw.

HEXOSANS [($C_6H_{10}O_5$)n]. The glucosans starch, cellulose and glycogen are outstanding examples of hexosans that are composed of glucose units. The fructosan inulin is found as a reserve carbohydrate in many composites. It can be prepared from the roots of the dahlia and the artichoke, and it yields fructose upon hydrolysis. Mannans that yield mannose upon hydrolysis are found in vegetable ivory, date seeds and fenugreek seeds. Galactans are rare.

MIXED POLYSACCHARIDES.

Hemicellulose, a not too satisfactory term, is used to designate the members of a group of high molecular weight carbohydrates that occur as constant companions of cellulose but differ from it by being considerably more soluble and more easily broken down. This group includes those cell wall polysaccharides that can be extracted by either cold or hot alkali and hydrolyzed to the sugars or uronic acids. They are composed of short chain hexosans and pentosans. Upon hydrolysis, hemicelluloses yield glucose, galactose, mannose, xylose, arabinose, rhamnose and the uronic acids, especially glucuronic and galacturonic. Hemicelluloses are found in almost all plants. They are most abundant in young tissues, succulents, nonwoody plant parts and some seeds.

Gums and mucilages are very closely related to the hemicelluloses both in composition and function. They are perhaps more soluble in water than the hemicelluloses, although some of them merely swell up but do not dissolve. Upon hydrolysis, they yield galactose, arabinose, xylose and glucuronic and galacturonic acids.

Pectins, upon hydrolysis, yield chiefly galacturonic acid. They are found as an integral part of the cell framework and structural parts of green plant tissues, such as green leaves, stems and more particularly the fleshy fruits and roots. They are found chiefly in the middle lamella; however, in plants of a succulent nature, they also may diffuse into the cell proper.

CONFIGURATION OF SUGARS

Through the brilliant researches of Emil Fischer, the spatial configurations of the alcohol groups in glucose and a number of other simple sugars were elucidated. The carbon atoms in the simple aldoses are designated by arabic numerals, the carbon atom of the aldehyde group having been assigned position 1. In the case of the ketoses, the carbonyl group which is one carbon removed from the end of the chain is given position 2. For example, writing the structural formulas for glucose and fructose in the form of a straight chain, they can be numbered as shown below.

The formula for D-glucose shows that car-

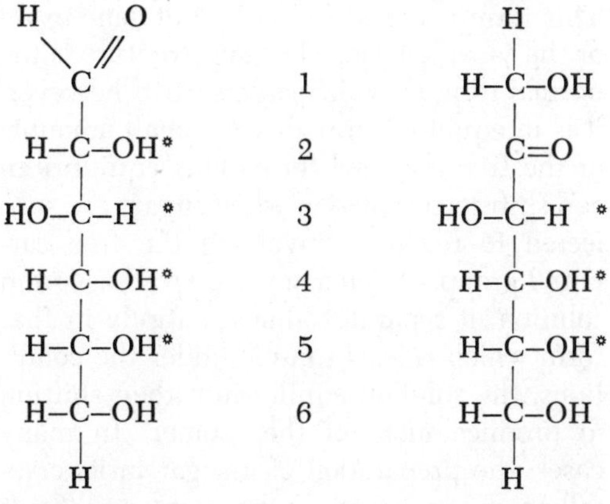

bon atoms 2, 3, 4 and 5 are asymmetric and, therefore, the number of isomers of glucose would be 2^4 or 16. In the case of keto-hexoses, such as fructose, carbon atoms 3, 4 and 5 are asymmetric; therefore, the number of isomers would be 2^3 or 8. Of the 16 possible isomers of glucose, 8 have mirror images and, being identical in physical and chemical properties except for their effect upon polarized light, are enantiomorphic. Therefore, 8 of the isomers would be the D-form, and 8 of them the L-form. Unfortunately, the designations *d-* and *l-* in the sugar series were introduced by Fischer not to denote the sign of rotation of polarized light but to indicate arbitrarily the two possible enantiomorphs of glucose. Therefore, other sugars that were structurally related to *d-* or *l-*glucose, in so far as the asymmetry of their carbon atoms were concerned, likewise were called *d-* or *l-*sugars, irrespective of their sign of rotation. Thus, *d-*fructose is levorotatory. A convenient method for assigning sugars into their D- and L-forms has been suggested by Wohl and Freudenberg. The configurations of D- and L-glyceric aldehyde can be illustrated as follows:

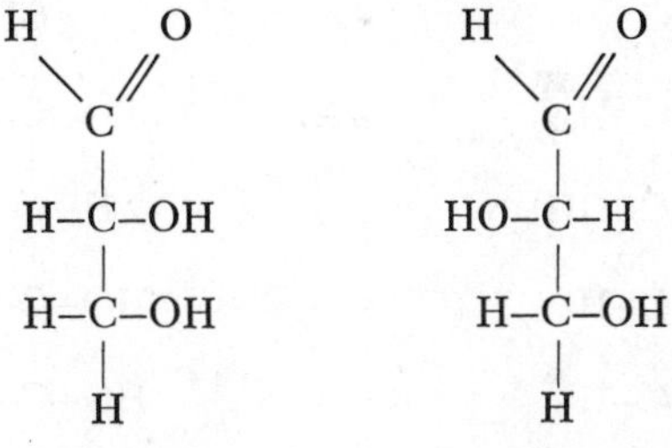

```
   H   O              H   O
    \  //              \  //
     C                  C
     |                  |
  H—C—OH            HO—C—H
     |                  |
  H—C—OH             H—C—OH
     |                  |
     H                  H
  D-Glyceric         L-Glyceric
  Aldehyde           Aldehyde
```

These compounds are the simplest aldose sugars that contain an asymmetric carbon atom. The one in which the asymmetric hydroxyl group is to the right is called D-glyceric aldehyde, and its mirror image, where this hydroxyl group is to the left, is called L-glyceric aldehyde. Following this type of nomenclature, the D-forms of all the sugars must have the hydroxyl group to the right on the carbon atom adjacent to the terminal primary alcohol group or the asymmetric carbon atom farthest removed

```
   H   O
    \  //
     C          1
     |
  H—C—OH*       2
     |
 HO—C—H  *      3
     |
  H—C—OH*       4
     |
  H—C—OH*       5
     |
  H—C—OH        6
     |
     H
  D-Glucose
```

```
     H
     |
  H—C—OH        1
     |
     C=O        2
     |
 HO—C—H  *      3
     |
  H—C—OH*       4
     |
  H—C—OH*       5
     |
  H—C—OH        6
     |
     H
  D-Fructose
```

from the carbonyl group. The rest of the molecule must be arranged according to its actual configuration. The converse is true of the L-series. All hydroxyls when assigned to the right are called "D" hydroxyls, and the configurations of the *carbon atoms* containing such hydroxyls are called "D". When the hydroxyls are to the left, the "L" terminology is employed. For example, writing the structural formulas for D-glucose, D-fructose, L-glucose and L-fructose in the form of a straight chain, these positions can be indicated by the letters D and L or + and −.

$$
\begin{array}{ll}
\underset{\text{H}}{}\quad\underset{\text{O}}{} & \\
\text{C} & \text{CH}_2\text{OH} \\
\text{H–C–OH} \quad \text{D}+ & \text{C=O} \\
\text{HO–C–H} \quad \text{L}- & \text{HO–C–H} \\
\text{H–C–OH} \quad \text{D}+ & \text{H–C–OH} \\
\text{H–C–OH} \quad \text{D}+ & \text{H–C–OH} \\
\text{CH}_2\text{OH} & \text{CH}_2\text{OH} \\
\text{D-Glucose} & \text{D-Fructose}
\end{array}
$$

$$
\begin{array}{ll}
\underset{\text{H}}{}\quad\underset{\text{O}}{} & \\
\text{C} & \text{CH}_2\text{OH} \\
\text{HO–C–H} \quad \text{L}- & \text{C=O} \\
\text{H–C–OH} \quad \text{D}+ & \text{H–C–OH} \\
\text{HO–C–H} \quad \text{L}- & \text{HO–C–H} \\
\text{HO–C–H} \quad \text{L}- & \text{HO–C–H} \\
\text{CH}_2\text{OH} & \text{CH}_2\text{OH} \\
\text{L-Glucose} & \text{L-Fructose}
\end{array}
$$

Although the usual type of an optically active compound exhibits a fixed degree of specific rotation, such is not the case with D-glucose and a number of other sugars. D-Glucose, crystallized under certain conditions, when dissolved in water, gave an initial rotation that slowly changed until an equilibrium of +52.7 was obtained. The initial rotation may be +19.7 or +113.4, depending upon the method whereby the crystalline glucose is obtained. This phenomenon is called mutarotation and is shown by all the simple sugars and oligosaccharides that have a potential carbonyl group that is not in a glycoside linkage. Therefore, glucose and other simple sugars and some oligosaccharides can exist in two forms. These forms involve an asymmetric center because of their effect on polarized light. This new asymmetric center can be generated only by a hemiacetal formation from the carbonyl group and one of the hydroxyl groups in the remainder of the molecule. In the case of D-glucose this can be expressed as follows:

$$
\begin{array}{ccc}
\text{HO}\quad\text{H} & \text{H}\quad\text{O} & \text{H}\quad\text{OH} \\
\text{H–C} & \text{C} & \text{C} \\
\text{H–C–OH} & \text{H–C–OH} & \text{H–C–OH} \\
\text{HO–C–H} \;\rightleftarrows\; & \text{HO–C–H} \;\rightleftarrows\; & \text{HO–C–H} \\
\text{H–C–OH} & \text{H–C–OH} & \text{H–C–OH} \\
\text{C} & \text{H–C–OH} & \text{H–C} \\
\text{CH}_2\text{OH} & \text{CH}_2\text{OH} & \text{CH}_2\text{OH} \\
\beta\text{-D-glucose} & \text{D-glucose} & \alpha\text{-D-glucose}
\end{array}
$$

This 6-membered ring structure in the sugar series is called the pyranose, amylene oxide or the δ-oxide form. Hudson suggested that the form of D-glucose which has the strongest dextrorotatory power be designated as α-D-glucose and the other β-D-glucose. In the L-series the same suggestion was made, thus α-L-glucose is the most optically active form. Under certain conditions, some simple sugars have a 5-membered oxide ring structure. Some oligosaccharides, polysaccharides, coenzymes and other forms have the 5-membered oxide ring structure. This form is called furanose, butylene oxide or the γ-oxide form. The ring structure is the normal form in which sugars exist; however, it is in equilibrium with very small amounts of the free carbonyl form. This equilibrium can shift to completion when sugars are subjected to reactions involving the free carbonyl group. When a sugar crystallizes from solution, it separates almost entirely in that form which is least soluble under the conditions, the solution equilibrium then shifting to produce more of this isomer. In many cases, the preparation of a sugar in its crystalline α- and β-forms becomes a difficult matter because it is necessary to find conditions under which each form will crystallize. The form in which a sugar exists also influences its degree of sweetness (see lactose).

The cyclic structures of the simple sugars can be visualized better from the following formulas, now described as lactal rings.

A   B
α-D-glucose

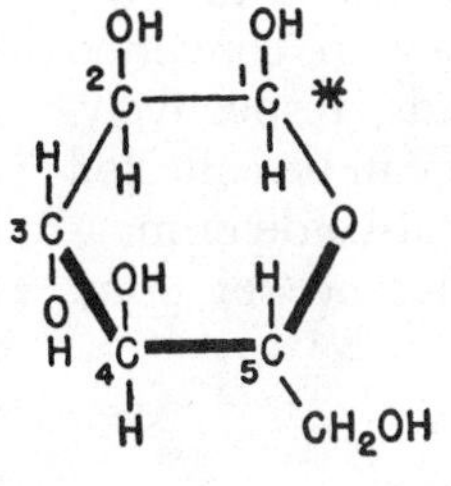

C
α-D-glucose   or α-D-glucose

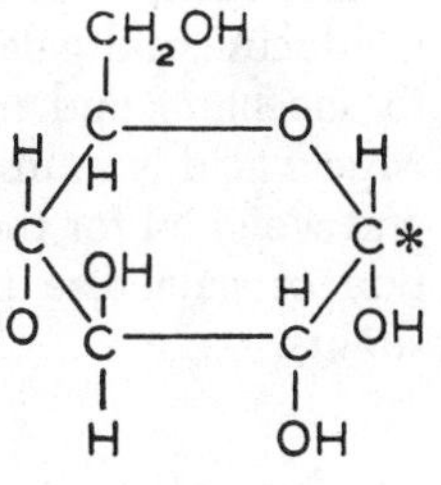

The reducing group of glucose is represented by the position marked with an asterisk, and those addenda which are coplanar are shown by the thickened lines. By comparing formula B with formula A, it will be seen that the hydroxyl groups and the hydrogen atoms attached to the central chain correspond in each formula and may be read off directly from A to B in the same order. In formulas B and C, the hydroxyls at C_1 and C_2 are in the same plane, the hydroxyl at C_3 and the hydrogen at C_4 are in the same plane, the hydrogens at C_4 and C_5 are in the same plane and the hydroxyl at C_4 and the oxygen bridge are in the same plane. The primary alcohol group can be visualized as a side chain, such as is present in benzyl alcohol. For example, its oxidation to a carboxyl group in plants and animals gives rise to the naturally occurring and important uronic acids.

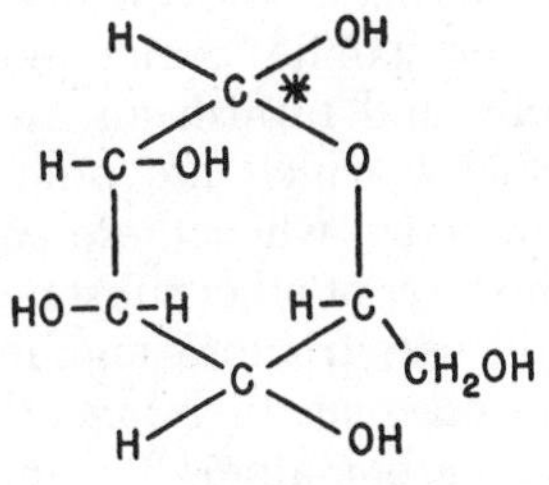

The chemical properties of the carbohydrates are a function of the primary alcohol group when it exists per se, the uncombined secondary alcohol groups, the alcohol group generated by ring formation and the carbonyl group existing in aqueous solution in equilibrium with the ring structure.

The alcohol groups can be esterified with organic acids, such as acetic, propionic, butyric and with the inorganic acids, nitric, phosphoric and sulfuric. They can be condensed with acetone, acetaldehyde, benzaldehyde to produce the isopropylidene (acetone), ethylidene and benzylidine derivatives. They can be converted to the methyl and ethyl ethers.

The preferred conformational formula for glucose is one in which the C_6—CH_2OH group and the secondary alcohol groups occupy equatorial positions. Such type conformational formulae can be utilized to better advantage to explain the state and the properties of the carbohydrates.

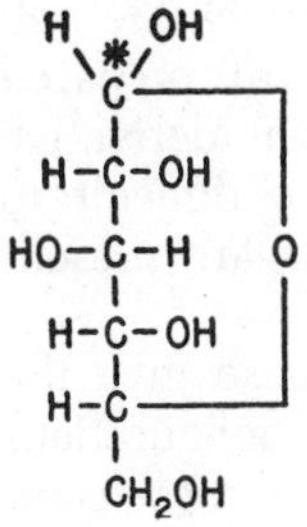

Under the proper conditions, the terminal alcohol group can be oxidized to a carboxyl group to give the uronic acids. The alcohol group generated by the ring structure can be etherified selectively to give the glycosides (q.v.).

The carbonyl group can be reduced catalytically or with sodium amalgam to give one sugar alcohol in the case of the aldoses and a mixture of two sugar alcohols in the case of the ketoses.

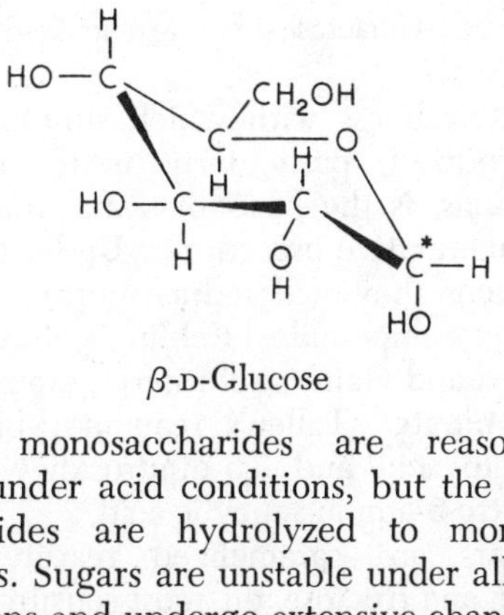

β-D-Glucose

The monosaccharides are reasonably stable under acid conditions, but the oligosaccharides are hydrolyzed to monosaccharides. Sugars are unstable under alkaline conditions and undergo extensive changes in structure. The monosaccharides and the reducing oligosaccharides can be oxidized carefully by chemical methods and by fermentation methods to give the corresponding "onic acids." Thus, glucose is converted to gluconic acid on a commercial scale (see calcium gluconate). More vigorous oxidation under acid conditions converts monosaccharides and the residues of the oligosaccharides into dicarboxylic acids. In the case of glucose and galactose, they are converted to saccharic and mucic acids, respectively. Mucic acid is so insoluble it can be collected and weighed for the quantitative determination of galactose in the presence of other sugars.

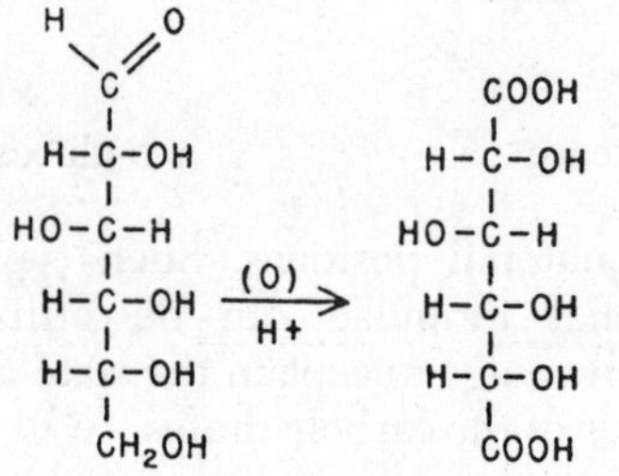

D-Glucose   Saccharic Acid

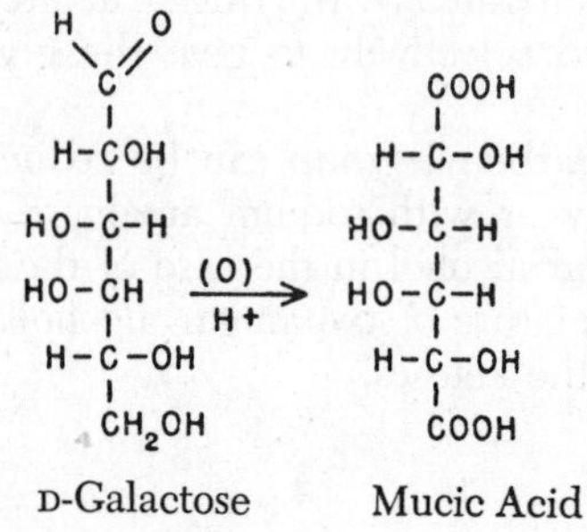

D-Galactose   Mucic Acid

The readiness with which simple sugars are oxidized, particularly under alkaline conditions, is the basis of some qualitative and quantitative procedures. Under alkaline conditions they will reduce cupric salts to cuprous compounds (Fehling's, Benedict's, Barfoed and Haine's solutions), ammoniacal silver nitrate (Tollen's reagent), picric to picramic acid and 3,5-dinitrosalicylic acid to 3-nitro-5-amino-salicylic acid.

Sugars are caramelized readily when heated, and fructose, the most sensitive sugar in this respect, will caramelize above 70°.

Dehydrating agents, such as 12 per cent hydrochloric acid, will convert the pentoses quantitatively to furfural, which is of great commercial significance. The aldohexoses yield some hydroxymethylfurfural, and fructose yields levulinic acid, which is produced now on a commercial scale.

The monosaccharides and the reducing oligosaccharides yield oximes, hydrazones and osazones. The osazones, which are yellow, crystalline compounds, are prepared readily and furnish an excellent derivative of sugars which are otherwise difficult to characterize when these sugars are present in mixtures of other substances.

Glucose, fructose and mannose give the same osazone; therefore, the configurations about carbon atoms 3, 4 and 5 in these compounds must be the same.

PRODUCTS

Dextrose U.S.P., D (+)-Glucopyranose, Grape Sugar, D-Glucose, Glucose. Dextrose is a sugar usually obtained by the hydrolysis of starch. It can be either α-D-glucopyranose or β-D-glucopyranose or a mixture of the two. A large amount of the dextrose of commerce, whether crystalline or syrupy, usually is obtained by the acid hydrolysis of corn starch, although other starches can be used.

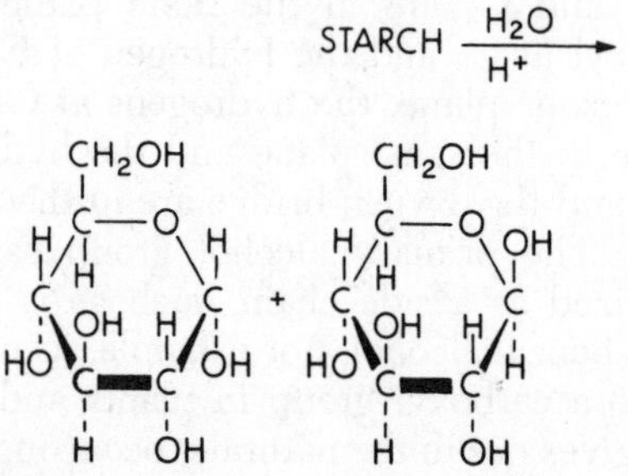

$$\text{STARCH} \xrightarrow[\text{H}^+]{\text{H}_2\text{O}}$$

α-D-Glucopyranose β-D-Glucopyranose

Although some free glucose occurs in plants and animals, most of it occurs in starches, cellulose, glycogen and sucrose. It also is found in other polysaccharides, oligosaccharides and glycosides.

Dextrose occurs as colorless crystals or as a white, crystalline or granular powder. It is odorless, and has a sweet taste. One g. of dextrose dissolves in about 1 ml. of water and in about 100 ml. of alcohol. It is more soluble in boiling water and in boiling alcohol.

Aqueous solutions of glucose can be sterilized by autoclaving.

Glucose can be used as a ready source of energy in various forms of starvation. It is the sugar found in the blood of animals and in the reserve polysaccharide glycogen which is present in the liver and muscle. It can be used in solution intravenously to supply fluid and to sustain the blood volume temporarily. It has been used in the management of the shock which may follow the administration of insulin used in the treatment of schizophrenia. This, because a "hypoglycemia" results from the use of insulin in this type of therapy, and the "hypoglycemic" state can be reversed by the use of dextrose intravenously. When dextrose is used intravenously, its solutions (5 to 50%) usually are made with physiologic salt solution or Ringer's solution. The dextrose used for intravenous injection must conform to the *U.S.P.* requirements for dextrose.

OCCURRENCE

Dextrose Injection U.S.P.
Dextrose & Sodium Chloride Injection U.S.P.
Sodium Chloride and Dextrose Tablets N.F.

Liquid Glucose U.S.P. Glucose is a product obtained by the incomplete hydrolysis of starch. It consists chiefly of dextrose (D-glucose, $C_6H_{12}O_6$), with dextrins, maltose and water. This glucose usually is prepared by the partial acid hydrolysis of cornstarch and, hence, the common name corn syrup and other trade names refer to a product similar to liquid glucose. The official product contains not more than 21 per cent of water.

Liquid glucose is a colorless or yellowish, thick, syrupy liquid. It is odorless, or nearly so, and has a sweet taste. Liquid glucose is very soluble in water, but is sparingly soluble in alcohol.

Liquid glucose is used extensively as a food (sweetening agent) for both infants and adults. It is used in the massing of pills, in the preparation of pilular extracts and for other similar uses. It is not to be used intravenously.

Calcium Gluconate U.S.P. The gluconic acid used in the preparation of calcium gluconate can be prepared by the electrolytic oxidation of glucose as follows.

D-Glucose   Calcium Gluconate

Gluconic acid is produced on a commercial scale by the action of a number of fungi, bacteria and molds upon 25 to 40 per cent solutions of glucose. The fermentation is best carried out in the presence of calcium carbonate and oxygen to give almost quantitative yields of gluconic acid. A number of organisms can be used, for example, *Acetobacter oxydans, A. aceti, A. rancens, B. gluconicum, A. xylinum, A. roseus* and *Penicillium chrysogenum*. The fermentation is complete in 8 to 18 days.

Calcium gluconate occurs as a white, crystalline or granular powder without odor or taste. It is stable in air. Its solutions are neutral to litmus paper. One g. of calcium gluconate dissolves slowly in about 30 ml. of water, and in 5 ml. of boiling water. It is insoluble in alcohol and in many other organic solvents.

Calcium gluconate will be decomposed by the mineral acids and other acids that are stronger than gluconic acid. It is incompatible with soluble sulfates, carbonates, bicarbonates, citrates, tartrates, salicylates and benzoates.

Calcium gluconate fills the need for a soluble, nontoxic well-tolerated form of calcium that can be employed orally, intramuscularly or intravenously. Calcium therapy is indicated in conditions such as parathyroid deficiency (tetany), general calcium deficiency, and when calcium is the limiting factor in decreased clotting time of the blood. It can be used both orally and intravenously.

Category—calcium replenisher.

Usual dose—1 g. 3 or more times a day; intravenous, 1 g. one or more times daily.

Usual dose range—oral or intravenous, 3 to 15 g. daily.

OCCURRENCE

Calcium Gluconate Tablets U.S.P.
Calcium Gluconate Injection U.S.P.

Calcium Glucoheptonate, Calcium Gluceptate, is a sterile aqueous approximately neutral solution of the calcium salt of glucoheptonic acid, a homolog of gluconic acid. Each ml. represents 90 mg. of Ca. Its uses and actions are the same as those of calcium gluconate.

Ferrous Gluconate N.F., Iron (2+) Gluconate, Fergon®, occurs as a fine yellowish gray or pale greenish-yellow powder with a slight odor like that of burnt sugar. One gram of this salt is soluble in 10 ml. of water; however, it is nearly insoluble in alcohol. A 5 per cent aqueous solution is acid to litmus.

Ferrous gluconate can be administered orally or by injection for the utilization of its iron content.

Category—iron supplement.
Usual dose—300 mg. 3 times a day.
Usual dose range—200 to 600 mg.

OCCURRENCE

Ferrous Gluconate Tablets N.F.

Glucuronic acid occurs naturally as a component of many gums, mucilages, hemicelluloses and in the mucopolysaccharide portion of a number of glycoproteins. It is used by animals and humans to detoxify such substances as camphor, menthol, phenol, salicylates and choral hydrate. None of the above can be used to prepare glucuronic acid for commercial purposes. It is prepared by oxidizing the terminal primary alcohol group of glucose or a suitable de-

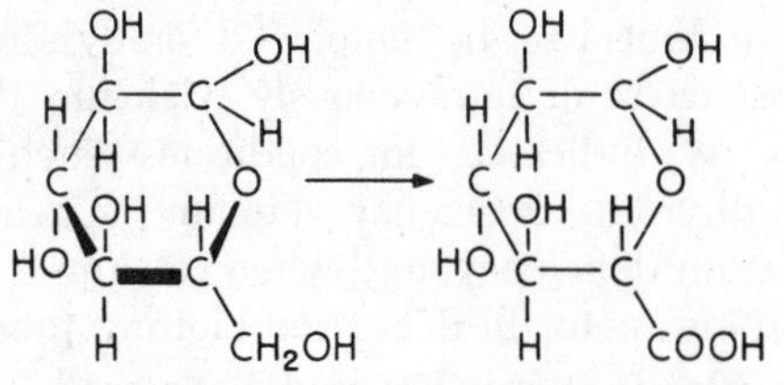

D-Glucose　　D-Glucuronic Acid

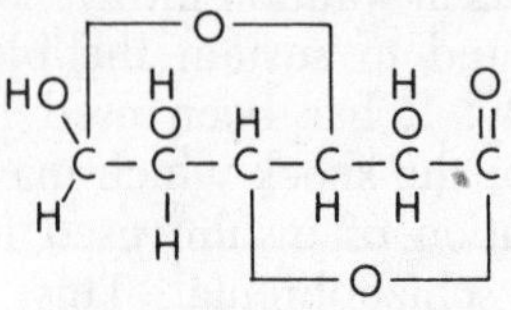

rivative thereof, such as 1,2-isopropylidine-D-glucose. It is a white, crystalline, solid that is water-soluble and stable. It exhibits both aldehydic and acidic properties. It also may exist in a lactone form and as such is marketed under the name "Glucurone," an abbreviation of glucuronolactone.

Glucuronolactone

An average of 60 per cent effectiveness was obtained in the relief of certain arthritic conditions by the use of glucuronic acid. A possible rationale for the effectiveness of glucuronic acid in the treatment of arthritic conditions is based upon the fact that it is an important component of cartilage, nerve sheath, joint capsule tendon and joint fluid and intercellular cement substances. The dose is 0.5 to 1.0 g. orally 4 times a day or 3 to 5 ml. of a 10 per cent buffered solution given intramuscularly.

Fructose N.F., D(−)fructose, levulose β-D(−)fructopyranose, is a sugar usually obtained by hydrolysis of aqueous solutions of sucrose and subsequent separation of fructose from glucose. It occurs as colorless crystals or as a white or granular powder that is odorless and has a sweet taste. It is soluble 1:15 in alcohol and is freely soluble in water. Fructose is considerably more sensitive to heat and decomposition than is glucose and this is especially true in the presence of bases.

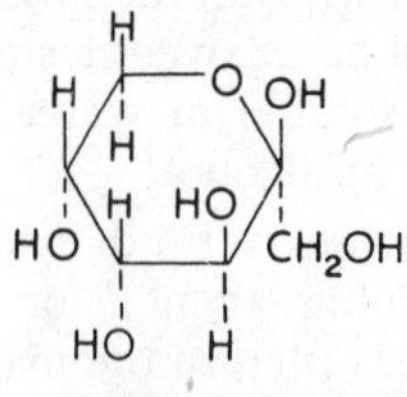

D-Fructose

Fructose (a 2-ketohexose) can be utilized to a greater extent than glucose by diabetics and by patients who must be fed by the intravenous route.

Category—fluid and nutrient replenisher (as intravenous solution).

Fructose Injection N.F.
Fructose and Sodium Chloride Injection N.F.

Lactose U.S.P., saccharum lactis, milk sugar, is a sugar obtained from milk. Lactose is a by-product of whey, which is the portion of milk that is left after the fat and the casein have been removed for the production of butter and cheese. Cows' milk contains 2.5 to 3 per cent of lactose, whereas, in other mammals, it is present in 3 to 5 per cent concentrations. Although common lactose is a mixture of the alpha and beta forms, the pure beta form is sweeter than the slightly sweet-tasting mixture.

Lactose occurs as white, hard, crystalline masses or as a white powder. It is odorless, and has a faintly sweet taste. It is stable in air, but readily absorbs odors. Its solutions are neutral to litmus paper. One g. of lactose dissolves in 5 ml. of water, and in

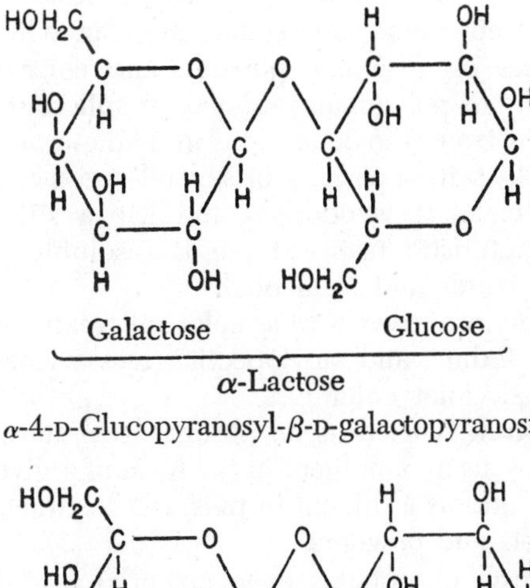

Galactose · Glucose

α-Lactose

α-4-D-Glucopyranosyl-β-D-galactopyranoside

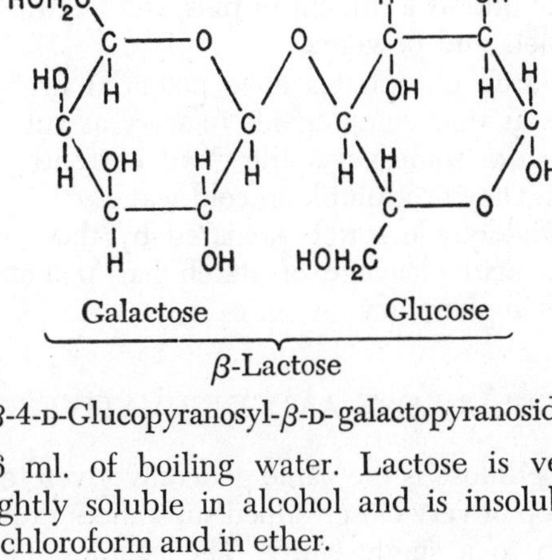

Galactose · Glucose

β-Lactose

β-4-D-Glucopyranosyl-β-D-galactopyranoside

2.6 ml. of boiling water. Lactose is very slightly soluble in alcohol and is insoluble in chloroform and in ether.

Lactose is hydrolyzed readily in acid solutions to yield one molecule each of D-glucose and D-galactose. It reduces Fehling's solution.

Lactose is used as a diluent in tablets and powders and as a nutrient for infants.

β-Lactose when applied locally to the vagina brings about a desirable lower pH. The lactose probably is fermented, with the production of lactic acid.

Maltose or malt sugar, 4-D-glucopyranosyl-α-D-glucopyranoside, is an end-product of the enzymatic hydrolysis of starch by the enzyme diastase. It is a reducing disaccharide that is fermentable and is hydrolyzed by acids or the enzyme maltose to yield 2 molecules of glucose.

Maltose is a constituent of malt extract and is used for its nutritional value for infants and adult invalids.

Malt Extract is a product obtained by extracting malt, the partially and artificially germinated grain of one or more varieties of *Hordeum vulgare* Linné (Fam. Gramineae). Malt extract contains maltose, dextrins, a small amount of glucose and amylolytic enzymes.

Malt extract is used in the brewing industry because of its enzyme content which converts starches to fermentable sugars. It also is used in infant feeding for its nutritive value and laxative effect.

The usual dose is 15 g.

Dextrins are obtained by the enzymatic (diastase) degradation of starch. These degradation products vary in molecular weight in the following decreasing order: amylodextrin, erythodextrin and achroodextrin. Lack of homogeneity precludes the assignment of definite molecular weights. With the decrease in molecular weight, the color produced with iodine changes from blue to red to colorless.

Dextrin occurs as a white, amorphous powder that is incompletely soluble in cold water but freely soluble in hot water.

Dextrins are used extensively as a source of readily digestible carbohydrate for infants and adult invalids. They often are combined with maltose or other sugars.

Sucrose U.S.P., Saccharum, Sugar, Cane Sugar, Beet Sugar. Sucrose is a sugar obtained from *Saccharum officinarum* Linné (*Fam.* Gramineae), *Beta vulgaris* Linné (*Fam.* Chenopodiaceae), and other sources. Sugar cane (15 to 20% sucrose) is expressed, and the juice is treated with lime to neutralize the plant acids. The water-soluble proteins are coagulated by heat and are removed by skimming. The resultant liquid is decolorized by means of charcoal and con-

Sucrose
1-α-D-Glucopyranosyl-β-D-fructofuranoside

centrated. Upon cooling, the sucrose crystallizes out. The mother liquor, upon concentration, yields more sucrose and brown sugar and molasses.

Sucrose occurs as colorless or white crystalline masses or blocks, or as a white, crystalline powder. It is odorless, has a sweet taste, and is stable in air. Its solutions are neutral to litmus. One g. of sucrose dissolves in 0.5 ml. of water and in 170 ml. of alcohol.

Sucrose does not respond to the tests for reducing sugars, i.e., reduction of Fehling's solution and others. It is hydrolyzed readily, even in the cold, by acid solutions to give one molecule each of D-glucose and D-fructose. This hydrolysis also can be effected by the enzyme invertase. Sucrose caramelizes at about 210°.

Sucrose is used in the preparation of syrups and as a diluent and sweetening agent in a number of pharmaceutic products, e.g., troches, lozenges and powdered extracts. In a concentration of 800 mg. per ml., sucrose is used as a sclerosing agent.

OCCURRENCE
Syrup U.S.P.

Invert sugar, Travert®, is a hydrolyzed product of sucrose (invert sugar) prepared for intravenous use.

STARCH AND DERIVATIVES

Starch U.S.P., amylum, cornstarch, consists of the granules separated from the grain of *Zea mays* Linné (*Fam.* Gramineae). Corn, which contains about 75 per cent dry weight of starch, is first steeped with sulfurous acid and then milled to remove the germ and the seed coats. It then is milled with cold water, and the starch is collected and washed by screens and flotation. Starch is a high molecular weight carbohydrate composed of 10 to 20 per cent of a hot-water-soluble "amylose" and 80 to 90 per cent of a hot-water-insoluble "amylopectin." Amylose is hydrolyzed completely to maltose by the enzyme β-amylase, whereas amylopectin is hydrolyzed only incompletely (60%) to maltose. The glucose residues are in the form of branched chains in the amylopectin molecule. The chief linkages of the glucose units in starch are α-1,4, since β-amylase hydrolyzes only alpha linkages and maltose is 4-D-glucopyranosyl-α-D-glucopyranoside.

Starch occurs as irregular, angular, white masses or as a fine powder, and consists chiefly of polygonal, rounded, or spheroidal grains from 3 to 35 microns in diameter and usually with a circular or several-rayed central cleft. It is odorless and has a slight characteristic taste. Starch is insoluble in cold water and in alcohol.

Amylose gives a blue color on treatment with iodine, and amylopectin gives a violet to red-violet color.

Starch is used as an absorbent in starch pastes, as an emollient in the form of a glycerite and as a diluent in pills, solid extracts, tablets and powders.

Starch phosphates have potential usefulness as thickening agents to serve as substitutes for some vegetable gums currently in use. They are soluble in cold water.

Dialdehyde starch prepared by the periodic acid cleavage of starch has potential uses in pharmacy.

CELLULOSE AND DERIVATIVES

Cellulose is the name generally given to a group of very closely allied substances rather than to a single entity. The celluloses are anhydrides of β-glucose, possibly existing as long chains that are not branched, consisting of 100 to 200 β-glucose residues. These chains may be cross linked by residual valences (hydrogen bonds) to produce the supporting structures of the cell walls of plants. The cell walls found in cotton, pappi on certain fruits and other sources are the purest forms of cellulose; however, because they are cell walls, they enclose varying amounts of substances that are proteinaceous, waxy, fatty. These, of course, must be removed by proper treatment in order to obtain pure cellulose. Cellulose from almost all other sources is combined by ester linkages, glycoside linkages and other combining forms with encrusting substances, such as lignin, hemicelluloses, pectins. These can be removed by steam under pressure, weak acid or alkali solutions, and sodium bisulfite and sulfurous acid. Plant celluloses, especially those found in wood, can be resolved into β-cellulose, which is soluble in 17.5 per cent sodium hydroxide, and alkali-insoluble α-cellulose. The cellulose molecule can be depicted in part as shown below.

It occurs as a fine, white odorless crystalline powder that is insoluble in water, in dilute alkalies and in most organic solvents.

Category—tablet diluent.

Methylcellulose U.S.P., Syncelose®, Cellothyl®, Methocel®, is a methyl ether of cellulose whose methoxyl content varies between 26 and 33 per cent. A 2 per cent solution has a centipoise range of n.l.t. 80 and n.m.t. 120 per cent of the labeled amount when such is 100 or less, and n.l.t. 75 and n.m.t. 140 per cent of the labeled amount for viscosity types higher than 100 centipoises.

Methyl and ethyl cellulose ethers (Ethocel®) can be prepared by the action of methyl and ethyl chlorides or methyl and ethyl sulfates, respectively, on cellulose that has been previously treated with alkali. Purification is accomplished by washing the reaction product with hot water. The degree of methylation or ethylation can be controlled to yield products that vary in their viscosities when they are in solution. Eight viscosity types of methylcellulose are produced commercially and have the following centi-

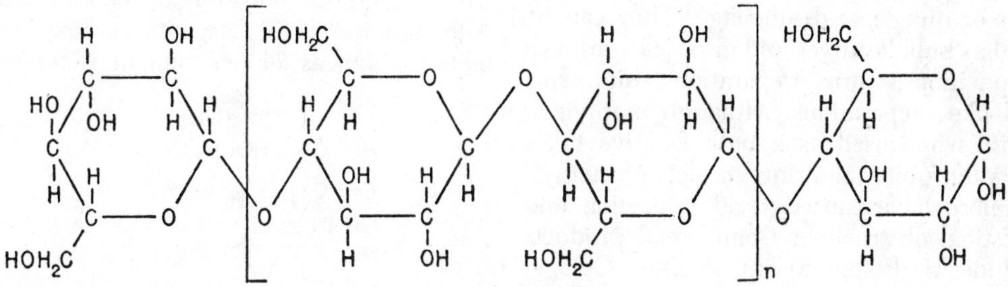

Cellulose

Purified Cotton U.S.P., is the hair of the seed of cultivated varieties of *Gossypium hirsutum* Linné, or of other species of *Gossypium* (*Fam.* Malvaceae), freed from adhering impurities, deprived of fatty matter, bleached and sterilized.

Category—surgical aid.

Microcrystalline Cellulose N.F. is purified, partially depolymerized cellulose prepared by treating alpha cellulose, obtained as a pulp from fibrous plant material, with mineral acids.

poise ranges: 10, 15, 25, 100, 400, 1,500 and 4,000, respectively. Other intermediate viscosities can be obtained by the use of a blending chart.[1] The ethylcelluloses have similar properties.

Methylated celluloses of a lower methoxy content are soluble in cold water, but, in contrast to the naturally occurring gums, they are insoluble in hot water and are precipitated out of solution at or near the boiling point. Solutions of powdered methylcellulose can be prepared most readily by first mixing the powder thoroughly with one fifth to one third of the required water as hot water (80° to 90°) and allowing it to macerate for 20 to 30 minutes. The remaining water then is added as cold water. With the increase in methoxy content, the solubility in water decreases until complete water insolubility is reached.

Methylcellulose resembles cotton in appearance and is neutral, odorless, tasteless and inert. It swells in water and produces a clear to opalescent, viscous, colloidal solution. Methylcellulose is insoluble in most of the common organic solvents. On the other hand, aqueous solutions of methylcellulose can be diluted with ethanol.

Methylcellulose solutions are stable over a wide range of pH (2 to 12) with no apparent change in viscosity. The solutions do not ferment and will carry large quantities of univalent ions, such as iodides, bromides, chlorides and thiocyanates. However, smaller amounts of polyvalent ions, such as sulfates, phosphates, carbonates and tannic acid or sodium formaldehyde sulfoxylate, will cause precipitation or coagulation.

The methylcelluloses are used as substitutes for the natural gums and mucilages, such as gum tragacanth, gum karaya, chondrus or quince seed mucilage. They can be used as bulk laxatives and in nose drops, eye preparations, burn preparations, ointments and like preparations. Although methylcellulose when used as a bulk laxative takes up water quite uniformly, tablets of methylcellulose have caused fecal impaction and intestinal obstruction. Commercial products include Hydrolose Syrup, Anatex, Cologel Liquid, Premocel Tablets and Valocall. In general, methylcellulose of the 1,500 or 4,000 cps. viscosity type is the most useful as a thickening agent when used in 2 to 4 per cent concentrations. For example, a 2.5 per cent concentration of a 4,000 cps. type methylcellulose will produce a solution with a viscosity obtained by 1.25 to 1.75 per cent of tragacanth.

Category—suspending agent.

Ethylcellulose N.F. is an ethyl ether of cellulose containing n.l.t. 45 per cent and n.m.t. 50 per cent of ethoxy groups and is prepared (q.v.) from ethyl chloride and cellulose. It occurs as a free-flowing stable white powder that is insoluble in water, glycerin and propylene glycol but is freely soluble in alcohol, ethyl acetate or chloroform. Aqueous suspensions are neutral to litmus. Films prepared from organic solvents are stable, clear, continuous, flammable and tough.

Category—tablet binder.

Hydroxypropyl Methylcellulose N.F., propylene glycol ether of methyl cellulose, contains a degree of substitution of n.l.t. 19 and n.m.t. 30 per cent as methoxyl groups (OCH_3), and n.l.t. 3 and n.m.t. 12 per cent as hydroxypropyl groups (OC_3H_6-OH). It occurs as a white, fibrous or granular powder that swells in water to produce a clear to opalescent, viscous, colloidal solution.

Category—suspending agent.

Oxidized Cellulose U.S.P., Oxycel®, Hemo-Pak®, Novocell®, when thoroughly dry contains not less than 16 nor more than 24 per cent of carboxyl groups. Oxidized cellulose is cellulose in which a part of the terminal primary alcohol groups of the glucose residues have been converted to carboxyl groups. Therefore, the product is possibly a synthetic polyanhydrocellobiuronide. Although the *U.S.P.* accepts carboxyl contents as high as 24 per cent, it is reported

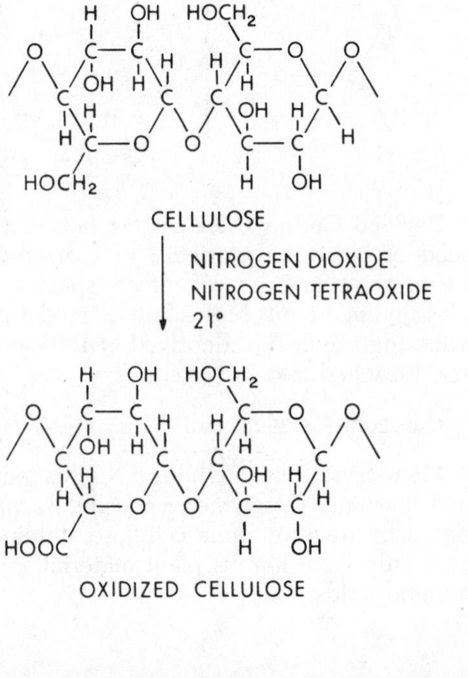

CELLULOSE

NITROGEN DIOXIDE
NITROGEN TETRAOXIDE
21°

OXIDIZED CELLULOSE

that products which contain 25 per cent carboxyl groups are too brittle (friable) and too readily soluble to be of use. Those products which have lower carboxyl contents are the most desirable. Oxidized cellulose is slightly off-white in color, is acid to the taste and possesses a slight, charred odor. It is prepared by the action of nitrogen dioxide, or a mixture of nitrogen dioxide and nitrogen tetroxide, upon cellulose fabrics at ordinary temperatures.[2] Because cellulose is a high molecular weight carbohydrate composed of glucose residues joined 1,4- to each other in their beta forms, the reaction must be as shown (see p. 862) on the cellulose molecule in part.

The oxidized cellulose fabric, such as gauze or cotton, resembles the parent substance. It is insoluble in water and in acids but is soluble in dilute alkalies. In weakly alkaline solutions, it swells and becomes translucent and gelatinous. When wet with blood, it becomes slightly sticky and swells, forming a dark brown, gelatinous mass. Oxidized cellulose cannot be sterilized by autoclaving. Special methods are needed to render it sterile.

Oxidized cellulose has noteworthy hemostatic properties. However, when it is used in conjunction with thrombin, it should be neutralized previously with a solution of sodium bicarbonate. It is used in various surgical procedures in much the same way as gauze or cotton, by direct application to the oozing surface. Except when used for hemostasis, it is not recommended as a surface dressing for open wounds. Oxidized cellulose implants in connective tissue, muscle, bone, serous and synovial cavities, brain, thyroid, liver, kidney and spleen were absorbed completely in varying lengths of time, depending on the amount of material introduced, the extent of operative trauma and the amount of blood present.[3]

Category—local hemostatic.

Sodium Carboxymethylcellulose U.S.P., Natulose®, CMC®, Sodium Tylose®, Thylose®, Sodium Cellulose Glycolate, is the sodium salt of a polycarboxymethyl ether of cellulose, containing, when dried, 6.5 to 9.5 per cent of sodium. It is prepared by treating alkali cellulose with sodium chloro-

acetate. This procedure permits a control of the number of $-OCH_2COONa$ groups that are to be introduced. The number of $-OCH_2COONa$ groups introduced is related to the viscosity of aqueous solutions of these products. C.M.C. is available in various viscosities, i.e., 5 to 2,000 centipoises in 1 per cent solutions. Therefore, high molecular weight polysaccharides containing carboxyl groups have been prepared whose properties in part resemble those of the naturally occurring polysaccharides, whose carboxyl groups contribute to their pharmaceutic and medicinal usefulness.

Sodium carboxymethylcellulose occurs as a hygroscopic white powder or granules. Aqueous solutions may have a pH between 6.5 and 8. It is easily dispersed in cold or hot water to form colloidal solutions that are stable to metal salts and pH conditions from 2 to 10. It is insoluble in alcohol and organic solvents.

It can be used as an antacid but is more adaptable for use as a nontoxic, nondigestible, unabsorbable, hydrophilic gel as an emollient type bulk laxative. Its bulk-forming properties are not as great as those of methylcellulose; on the other hand, its lubricating properties are superior, with little tendency to produce intestinal blockage.[4]

It is used in a usual dose of 1.5 g. 3 times daily with meals.

Category—suspending agent.

<small>Occurrence</small>
Sodium Carboxymethylcellulose Tablets
    N.F.

Pyroxylin U.S.P., soluble guncotton, is a product obtained by the action of nitric and sulfuric acids on cotton, and consists chiefly of cellulose tetranitrate $[C_{12}H_{16}O_6$ $(NO_3)_4]$. The glucose residues in the cellulose molecule contain 3 free hydroxyl groups which can be esterified. Two of these 3 hydroxyl groups are esterified to give the official pyroxylin, and, therefore, it is really a dinitrocellulose or cellulose dinitrate which conforms to the official nitrate content.

Pyroxylin occurs as a light-yellow, matted mass of filaments, resembling raw cotton in appearance, but harsh to the touch. It is exceedingly flammable and decomposes when exposed to light, with the evolution of nitrous vapors and a carbonaceous residue. Pyroxylin dissolves slowly but completely in 25 parts of a mixture of 3 volumes of ether and 1 volume of alcohol.

In the form of collodion and flexible collodion, it is used for coating purposes per se or in conjunction with certain medicinal agents.

	PER CENT
OCCURRENCE	PYROXYLIN
Collodion U.S.P.	4.0
Flexible Collodion U.S.P.	3.8

Cellulose Acetate Phthalate U.S.P. is a partial acetate ester of cellulose which has been reacted with phthalic anhydride. One carboxyl of the phthalic acid is esterified with the cellulose acetate. The finished product contains about 20 per cent acetyl groups and about 35 per cent phthalyl groups. In the acid form it is soluble in organic solvents and insoluble in water. The salt form is readily soluble in water. This combination of properties makes it useful in enteric coating of tablets because it is resistant to the acid condition of the stomach but is soluble in the more alkaline environment of the intestinal tract.

Category—tablet coating agent.

MISCELLANEOUS CARBOHYDRATES

Psyllium Hydrophilic Mucilloid, Metamucil®, contains about 50 per cent of a highly purified hemicellulose that is obtained from the outer seed coat of blonde psyllium seed *Plantago ovata* Forsk. The remaining 50 per cent is anhydrous dextrose, together with small amounts of sodium bicarbonate, monobasic potassium phosphate, citric acid and benzyl benzoate.

Mucilose is a highly purified hemicellulose obtained from the outer seed coat of *Plantago loeflingii.*

Plancello tablets and other commercially available products utilize the same hemicelluloses.

The mucilages, which are hemicelluloses, are found in the outer seed coats of the various species of Plantago, for example, *Plantago ovata, P. pyllsium, P. lanceolata, P. loeflingii* and *P. indica* have similar com-

positions and properties. They yield[5] D-galacturonic acid, D-galactose, L-arabinose and D-xylose upon hydrolysis. Their composition depends upon the way in which they are prepared, if, for example, by water extraction methods or, in some cases, mechanical methods. The mucilage content may be as high as 20 per cent of the dry weight of the seed.

The mucilages from Plantago species are used to provide a soft, bland bulk that exerts a gentle, stimulating effect upon the bowel. These mucilages are not digestible and do not interfere with the assimilation of food or vitamins. They are used in spastic constipation, mucous colitis, atonic colitis and ulcerative colitis conditions.

Plantago Ovata Coating, Konsyl®, consists primarily of the mucilaginous layers of *P. ovata* seeds (blonde psyllium).

Acacia, U.S.P., gum arabic, is the dried gummy exudation from the stems and the branches of *Acacia Senegal* (Linné) Willdenow or of some other African species of Acacia (*Fam.* Leguminosae). Gum arabic is a mixture of the calcium, magnesium and potassium salts of a mixed polysaccharide called arabic acid that has a molecular weight of 1,000 to 1,200. Upon mild acid hydrolysis (0.01N sulfuric acid), arabic acid yields L-arabinose, L-rhamnose, and 3-D-galactoside-L-arabinose plus a branched chain nucleus that appears to be homogenous and is composed of galactose and glucuronic acid residues in the proportions of 3 to 1.[6] The carboxyl groups of the glucuronic acid residues are responsible for the acidic properties of gum arabic.

Acacia occurs as speroidal tears, or as a white to yellowish white powder. It is insoluble in alcohol but is almost completely soluble in twice its weight of water at room temperature, the resulting solution flowing readily. It is acid to litmus.

Acacia is used as a demulcent and as an emulsifying agent and vehicle in the preparation of emulsions, pills and troches. It can be used to restore the normal colloidal osmotic pressure in the treatment of nephrotic edema, in which case sodium arabinate is administered intravenously as a 6 per cent solution in physiologic salt solution. It also can be used intravenously in the treatment of posthemorrhagic shock.

Category—suspending agent.

OCCURRENCE	PER CENT ACACIA
Acacia Syrup N.F.	10.0

Tragacanth U.S.P., gum tragacanth, is the dried gummy exudation from *Astragalus gummifer* Labillardiere, or other Asiatic species of Astragalus (*Fam.* Leguminosae). This plant gum is a complex mixture of a polyuronide (tragacanthic acid), a galacto-araban and a glycosidic substance, probably steroidal in nature.[7] Tragacanthic acid has a branched chain structure similar to that of arabic acid. The repeating residue is galacturonic acid (pyranose and linked C_1 to C_4 as in pectic acid) that is terminated by D-xylopyranose and the rare L-fucopyranose sugars. The galactoaraban contains arabinose and galactose.

Tragacanth occurs as white or weak yellow, flattened, lamellated, frequently curved fragments or in straight or spirally twisted linear pieces or as a white to yellowish-white powder. One gram of tragacanth in 50 ml. of water swells and forms a smooth, nearly uniform, stiff, opalescent mucilage free from cellular fragments.

Tragacanth is used in hand lotions, as a demulcent, as an emulsifying agent and as a suspending agent for heavy powders in liquid preparations. It also is used as an adhesive in pills, troches and similar products.

Category—suspending agent.

Sterculia Gum, gum karaya, is the dried gummy exudation from certain species of Sterculia or Cochlospermum. It occurs in tears of variable size or in irregular broken pieces. It has a pale-yellow to pinkish-brown color and has a mucilaginous and slightly acetous taste. Sterculia gum swells in water and is insoluble in alcohol.

Gum karaya is chiefly a polygalactouronide together with galactose and L-rhamnose and a ketohexose.

Category—cathartic (bulk).

Agar U.S.P., agar-agar, is the dried hydrophilic, colloidal substance extracted from *Gelidium cartilagineum* (Linné) Gaillon (*Fam.* Gelidiaceae), *Gracilaria confervoides* (Linné) *Greville* (*Fam.* Sphaerococcaceae), and from related red algae (*Class* Rhodophyceae). A number of agars, either identical in structure or closely related in structure, can be extracted with hot water from *Gelidium cartilagineum, Gelidium latifolium, Gelidium crinale* and *Gracilaria confervoides.* Agar[8] is composed chiefly of the calcium salt of the sulfuric acid ester of a linear polygalactose, in which the repeating unit is composed of D-galactopyranose residues terminated at the reducing end by one residue (molecule) of L-galactose. The D-galactose residues (molecules) are joined to each other by 1,3-glycosidic linkages, whereas the reducing end member, L-galactose, is attached to the chain through position 4. The L-galactose is esterified at carbon atom six with sulfuric acid. A tentative formula, in part to satisfy the above relationships, can be depicted as follows, even though the sulfur content is higher than that found in related agars.

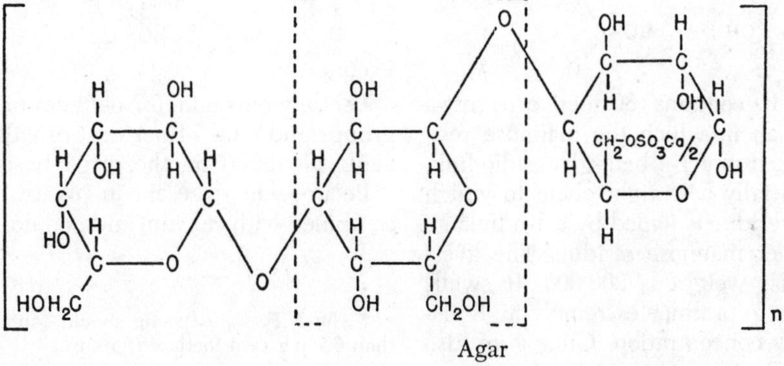

Agar

Agar occurs unground in bundles consisting of thin, membranous agglutinated pieces, as cut, flaked or granulated pieces or as a white to yellowish-white or pale-yellow powder. Agar is soluble in boiling water but insoluble in cold water, alcohol and organic solvents.

Agar forms gels with water and can absorb at least 5 times its weight of water. This property of gel formation, together with its indigestibility, is the basis of the use of agar to furnish bulk in the treatment of chronic constipation. It also is used in conjunction with mineral oil and phenolphthalein. Agar has been used extensively in culture media in bacteriology.

Agar can be used orally in doses of 4 to 16 g.

Category—suspending agent.

Chondrus, Irish-moss, carrageen, is the dried bleached plant of *Chondrus crispus* (Linné) Stackhouse, or of *Gigartina mamillosa* (Goodenough et Woodward) J. Agardh (*Fam.* Gigartinaceae). These sea algae, like *Gigartina stellata,* yield a polysaccharide composed chiefly of galactose residues that are esterified with potassium acid sulfate.[9] This polysaccharide has muciliaginous properties. Kondremul is a representative product.

Guar Gum, Jaguar, is the refined endosperm of the guar seed (*Cyamopsis tetra-*

contains 5 per cent of protein. It is edible and can be used in pharmaceuticals and foods where such an agent is indicated. A suitable preservative usually is necessary.

Pectin N.F. is a purified carbohydrate product obtained from the dilute extract of the inner portion of the rind of citrus fruits or from apple pomace. It consists chiefly of partially methoxylated polygalacturonic acids. Pulped apple pomace or cull lemons are washed first and then cooked with a dilute mineral or organic acid, usually in the presence of sulfurous acid. These operations liberate any combined pectin and bring it into solution quite free of soluble sugars. The mixture is filtered, and the filtrate is concentrated to a syrup which may be marketed as such or used in the preparation of a solid product. When lemon albedo extracts are used, the pectin is precipitated with aluminum sulfate and ammonia. The aluminum is removed from the dried precipitate by washing with acidified alcohol. Solid pectin also may be prepared by the addition of alcohol to the syrup described above to effect a precipitation of the pectin, or the syrup may be dried as a thin film on a drum. Pectin is composed of galacturonic acid residues, with each fourth residue being esterified with methyl alcohol.[*][10] The pectin molecule has a rodlike structure[11] and can be depicted in part as follows:

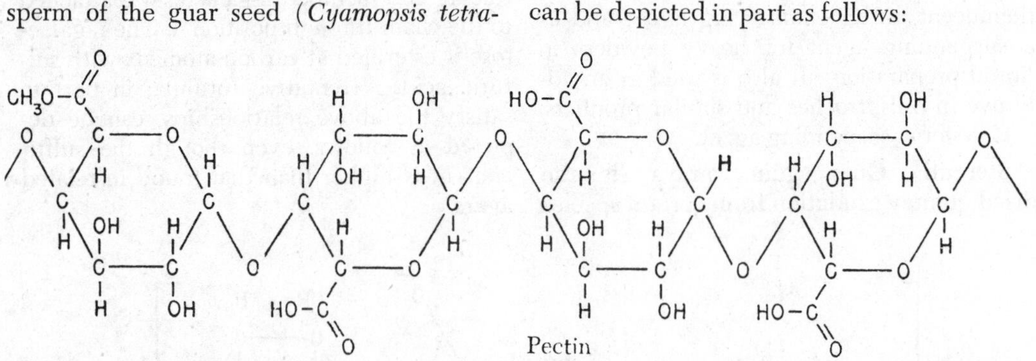

Pectin

gonaloba). It contains 80 per cent of a galactomannan in which the mannose residues are linked in a 1-4 beta-glycosidic linkage as essentially a straight chain to which a galactose residue is joined by a 1-6 linkage to every other mannose residue. The average molecular weight is 200,000. It swells in cold water to produce extremely high viscosity at low concentration. Guar gum also

Pectin yields n.l.t. 6.7 per cent of methoxy groups and n.l.t. 74 per cent of galacturonic acid calculated on the dried basis.

Pectin, when present in plants, usually is combined with calcium and magnesium.

* The *N.F.* states pectin should contain not less. than 6.7 per cent methoxyl groups.

Pectins occur in leaves, stems, roots and fleshy fruits. The pectin content of the fleshy portion of many fruits and green leaves may be as high as 20 or 30 per cent of the total dry weight. Dried sugar-beet pulp and the albedo (white portion) of orange peelings often contain 50 per cent pectins. Lemon peels and apple pulp are also rich in pectin.

Pectin occurs as a coarse or fine powder, yellowish white in color, almost odorless, and with a mucilaginous taste. Pectin is almost completely soluble in 20 parts of water at 25°, forming a viscous, opalescent, colloidal solution which flows readily and is acid to litmus paper. It is insoluble in alcohol or in diluted alcohol, and in other organic solvents. Pectin dissolves in water more readily if first moistened with alcohol, glycerin, or simple syrup, or if first mixed with 3 or more parts of sucrose.

Jellies are made possible by the fact that dilute aqueous solutions of pectin can be induced to set into a clear firm jell in the presence of a medium that contains a high concentration of sugar and a definite, optimal concentration of hydrogen ions. The methyl ester linkages in pectin are saponified readily by means of alkali, even at room temperature, to yield a gel or semigel. Acidification of such saponified pectin gives pectic acid which separates to form a voluminous, colorless, gelatinous precipitate. Such treatment or conditions destroy the usefuless of pectin. Heating acidified pectin solutions for prolonged periods of time also produces pectic acid.

Pectin usually is sold according to its "jelly grade," which refers to the number of pounds of sucrose that 1 lb. of pectin can "carry" in a jelly of standard acidity and water content. Therefore, a "100 jelly grade" means that 1 lb. of such a pectin will produce a satisfactory jelly with 100 lbs. of sugar, when the finished mixture contains 65 per cent of sugar and the proper acidity.* Pure pectins usually have a jelly grade of 180 to 300. The commercial grade of pectin is standardized to a value of not more than 150. This standardization is accomplished by the addition of dextrose or other sugars, and some preparations contain sodium citrate or other buffer salts.

* One per cent tartaric acid.

Pectin is not digested by humans but passes into the intestinal tract, where it appears to be of value in the treatment of intestinal disorders, such as diarrhea and dysentery. Pectin in the form of pastes appears to be bacteriostatic and, therefore, it is used in the treatment of indolent ulcers and deep wounds.

Category—protective.

OCCURRENCE	PER CENT PECTIN
Pectin with Kaolin Mixture N.F.	1

Sodium Alginate N.F. is the purified carbohydrate product extracted from giant brown seaweeds by the use of dilute alkali. The horsetail kelp, *Laminaria digitata*, and the broad-leaf or sugar kelp, *Laminaria saccharina*, are harvested as raw materials from the European and the American coasts of the Atlantic ocean. The giant kelp, *Macrocystis pyrifera*, is collected from the Pacific coast of the United States for the production of sodium alginate. The chopped and shredded seaweed first is leached with acid water to remove soluble salts and free the alginic acid from its calcium and magnesium salts. It then is digested with alkali or alkali carbonates to dissolve the alginic acid that is present in the seaweed. The resulting alkaline aqueous solution is bleached or decolorized. Acidulation of this solution yields insoluble alginic acid, which is subjected to further purification by washing with acid water and alcohol. The alginic acid is reconverted into the sodium salt as a final step in its preparation.

Alginic acid is a polymannuronide in which the D-mannuronic acid residues have a beta linkage. Therefore, it has a free carboxyl group for each sugar residue.

Sodium alginate occurs as a nearly odorless and tasteless, coarse or fine powder, yellowish white in color. It is soluble in water to give a viscous, colloidal solution. It is insoluble in the common organic solvents.

Solutions of sodium alginate yield precipitates with polyvalent cations. Acidulation of solutions of sodium alginate yield voluminous gelatinous precipitates.

Sodium alginate is used because of its properties as an emulsifying, suspending, jellying, thickening and body agent. It is used as a stabilizer for ice cream and is used in other food products. It can be used in greaseless lubricating jellies, in hand lotions and in other pharmaceuticals which require thickening agents.

Category—suspending agent.

Sodium Heparin U.S.P., Heparin is a mixture of active principles, having the property of prolonging the clotting time of blood in man or other animals. The existence of an anticoagulant material in liver was first demonstrated by McLean, in 1916, while working in Howell's laboratory at Johns Hopkins. In 1918, Howell and Holt named the substance heparin (Gr., *hepar*—liver) to indicate its origin and source. Since then it has been shown to be present in animal tissue of practically all types but predominantly in lung and liver tissue.

Heparin may be prepared commercially from lung and liver, employing the procedure of Kuizenga and Spaulding[12] combined with suitable methods for purifying the isolated heparin. The sodium salt is a white, amorphous, hygroscopic powder that is soluble 1:20 in water; but poorly soluble in alcohol. A 1 per cent aqueous solution has a pH of 5 to 7.5. It is stable to heat and solutions may be sterilized by autoclaving.

Sodium Heparin has not less than 120 U.S.P. Heparin Units in each mg. and 10 per cent plus or minus the potency stated on the label.

Heparin is a member of the mucopolysaccharides such as chondroitin sulfates A, B, and C, keratosulfate, hyaluronic acid and heparitin sulfate, and the last named is the most closely related to heparin. Heparin has a molecular weight of 10,000 to 12,000 (some 40 to 50 monosaccharide residues) plus 1.25 sulfate groups per residue.

Heparin is a polymer composed of alternating D-glucuronic acid and 2-amino-2-deoxy-D-glucose units linked glycosidically in an 2-D-(1→4) manner.[12a] Essentially all the amino groups of the hexoseamine are sulfated.

The "Toronto unit" represents the anticoagulant activity of 10 mcg. of the pure crystalline barium salt of heparin. It can be converted to the sodium salt, with a potency of 130 units per mg.

Heparin, as previously indicated, is a blood anticoagulant and has little other biologic action. The exact mechanism whereby it prevents coagulation is not known and has been the object of much pharmacologic investigation. It is not active orally and usually is administered intravenously because intramuscular or subcutaneous injection might cause bleeding at the site of injection. Recently, however, heparin has been administered subcutaneously in Pitkin's menstruum,[13] which, to be most effective, consists of gelatin (18%), dextrose (8%), glacial acetic acid (0.5%) and water (q.s.). The purpose of this vehicle is to slow down and equalize the release of the heparin into the tissues. Ordinarily, heparin acts quickly following injection, its action being apparent in approximately 10 minutes. Its effect is over in approximately 1 to 2 hours, i.e., the blood coagulation time returns to normal (2 to 6 minutes for blood from skin puncture or 4½ minutes for blood withdrawn from a vein. The usual clotting time which it is desirable to attain with heparin is from 15 to 20 minutes. The effect of heparin may be terminated rapidly by infusion of fresh whole blood. The fast onset of action and short duration of action are in contrast with the delayed but longer lasting effect of bishydroxycoumarin (q.v.), a synthetic blood anticoagulant. Note should be made of the fact that digitalis inhibits the action of heparin and should not be administered during heparinization.

The anticoagulant effect, of course, is the basis for its use medicinally. Its greatest use has been in the prevention and arrest of thrombosis. It has been used successfully in mesenteric thrombosis, cavernous sinus thrombosis, thrombosis of the central vein of the retina and routinely as a postsurgical preventive of thrombotic conditions. It does not destroy a thrombus, but it prevents the thrombus from increasing in size and in so doing reduces the possibility of an embolism. It has been used experimentally with encouraging results in the treatment of frostbite to prevent gangrene.[14] It may be used in two ways for blood transfusions, i.e., by putting it in the flask into which the blood is to be drawn or by heparinizing the donor before withdrawal of the blood.

Heparin is administered intravenously in two ways: (1) the intermittent dose method and (2) the continuous drip method. In the intermittent dose method, a dose of 50 mg. is repeated every 4 hours until a total of 250 mg. per day have been given. The continuous drip method is to be preferred; it consists of a slow infusion of a heparin containing solution into the vein adjusting the flow according to the observed clotting time. A solution containing 100 to 200 mg. of heparin in each 1,000 ml. of 5 per cent dextrose or physiologic saline solution is used for the latter method.

Category—anticoagulant.

Usual dose—intravenous or subcutaneous, as indicated by clotting time determinations.

Usual dose range (of the Injection)— intravenous, 5,000 to 10,000 U.S.P. Units every 4 to 6 hours; infusion, 5,000 to 40,000 units per liter at a rate of 1 ml. per minute; subcutaneous, 10,000 to 20,000 units twice daily.

OCCURRENCE

Anticoagulant Heparin Solution U.S.P.
Sodium Heparin Injection U.S.P.

Highly branched polyanhydroglucose molecules whose molecular weights range from 1,000 to 200,000 are obtained when glucose is condensed with an acid catalyst under vacuum. These synthetic polysaccharides can be sulfated, and the number of sulfate groups per glucose residue ranges from 0.6 to 3. These derivatives can inhibit the depolymerization action of ribonucleic acid by ribonuclease. At maximum sulfation this inhibition is 1,000 times as active as heparin, which is the best-known inhibitor to date. The T2 bacteriophage that attacks *E. coli* also can be inhibited by these sulfated synthetic polysaccharides.

GLYCOSIDES

Because a number of plant constituents yielded glucose and an organic hydroxide upon hydrolysis, the term "glucoside" was introduced as a generic term for these substances. The fact that a number of plant constituents yielded sugars other than glucose led to the suggestion of the less specific general term "glycoside." When the nature of the sugar residue is known, more specific terms can be used where desired, such as glucoside, fructoside, rhamnoside and others, respectively. The nonsugar portion of the glycoside generally is referred to as the aglycon or genin.

Two general types of glycosides are known, viz., the nitrogen glycosides and the conventional type glycoside. The conventional type glycoside has an acetal structure and can be illustrated by the simplest type in which methyl alcohol is the aglycon or organic hydroxide. Two forms of this as well as all other glycosides are possible, viz., alpha and beta, because of the asymmetry centering about carbon atom 1 of the sugar residue that contains the acetal structure. It is thought that all naturally occurring glycosides are of the beta variety because the

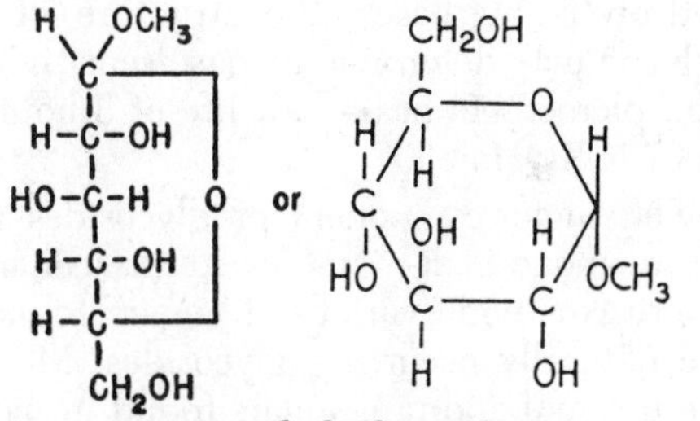

α-Methyl Glucoside

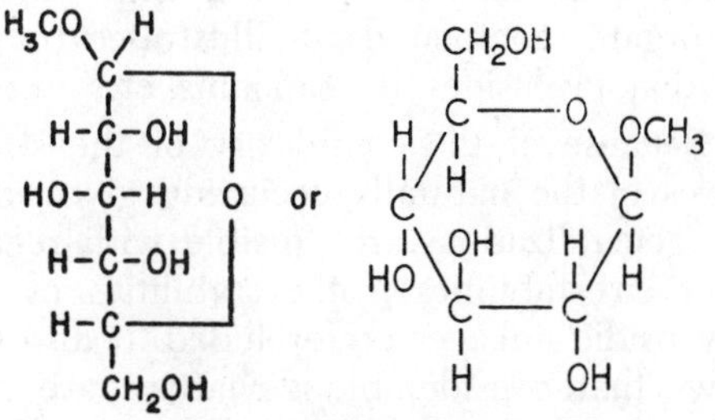

β-Methyl Glucoside

enzyme emulsin, which cannot hydrolyze synthetic alpha glycosides, hydrolyzes naturally occurring glycosides. Some of the beta-glycosides also are hydrolyzed by amygdalase, cellobiase, genitobiase and the phenol-glycosidases. The alpha glycosides are hydrolyzed by maltase, mannosidase and trehalase.

Glycosides usually are hydrolyzed by acids and are relatively stable toward alkalies. Some glycosides are much more resistant to hydrolysis than others. For example, those glycosides that contain a 2-desoxy sugar (see cardiac glycosides) are easily cleaved by weak acids, even at room temperature. On the other hand, most of the glycosides containing the normal type sugars are quite resistant to hydrolysis, and of these some may require rather drastic hydrolytic measures. The drastic treatment required for the hydrolysis of some glycosides causes chemical changes to take place in the aglycon portion of the molecule; these changes present problems in the elucidation of their structures. On the other hand, those glycosides that are very easily hydrolyzed present problems in regard to isolation and storage. Examples of the latter are the cardiac glycosides.

Although most glycosides are stable to hydrolysis by bases, the structure of the aglycon may determine its base sensitivity;[15] e.g., picrocrocin has a half life of 3 hours in 0.007 N KOH at 30°.

The sugar component of glycosides may be a mono-, di-, tri- or tetrasaccharide. There is a wide variety of sugars found in the naturally occurring glycosides. Most of the unusual and rare sugars found in nature are components of glycosides.

The aglycons or nonsugar portions of glycosides are represented by a wide variety of organic compounds, as illustrated by the cardiac glycosides, the saponins, etc.

Because of the complexity of the structures of the naturally occurring glycosides, no generalizations are possible with regard to their stabilities if the stabilities of the glycosidic linkages are excluded. It also follows that considerable deviations are met with in their solubility properties. Many glycosides are soluble in water or hydroalcoholic solutions because the solubility properties of the sugar residues exert a considerable effect. Some glycosides, such as the phytosterolins and cardiac glycosides,

are slightly soluble or insoluble in water. In these cases, the steroid aglycon is markedly insoluble in water and offsets the solubility properties of the sugar residues. Most glycosides are insoluble in ether. Some glycosides are soluble in ethyl acetate, chloroform or acetone.

Glycosides occur widely distributed in nature. They are found in varying amounts in seeds, fruits, roots, bark and leaves. In some cases, two or more glycosides are found in the same plant, e.g., cardiac glycosides and saponins. Glycosides often are accompanied by enzymes that are capable of synthesizing or hydrolyzing them. This phenomenon introduces problems in the isolation of glycosides because the disintegration of plant tissues, with no precautions to inhibit enzymatic activity, leads, in some cases, to partial or complete hydrolysis of the glycosides.

Most glycosides are bitter to the taste, although there are many that are not. Glycosides per se or their hydrolytic products furnish a number of drugs, some of which are very valuable. Some plants that contain the cyanogenetic type of glycoside present an agricultural problem. Cattle have been poisoned by eating plants which are rich in the cyanogenetic type of glycoside.

Phlorhizin (Phlorizin) is a glucoside obtained from the root of the apple, pear,

Phlorhizin

Phloretin

cherry and the like. Upon hydrolysis it yields glucose and the dihydrochalcone, phloretin.

Phlorhizin occurs as a white or yellowish crystalline powder that is sparingly soluble in water and soluble (1:4) in alcohol.

When administered to man or animals, phlorhizin produces glycosuria of renal origin. When administered by injection, it is used as a means of testing the functional activity of the kidney.

Rutin is the 3-rhamnoglycoside of 5,7,3′,4′-tetrahydroxyflavonol obtained from buckwheat, *Fagopyrum moench*, or other sources. It has been assigned the following formula in which the sugar residue, rutinose, is 6-D-glucopyranosyl-β-L-rhamnopyranoside.

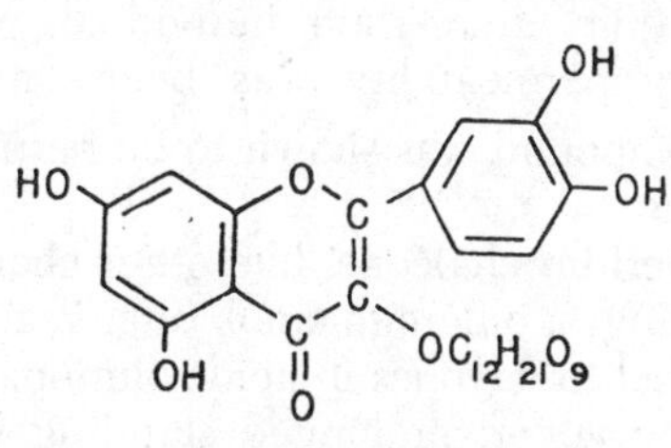

Rutin

Rutin was isolated first, in 1842, from *Ruta graveolens*. It has been isolated subsequently from 40 species of plants under many names. Stein first isolated it, in 1853, from *Sophora japonica*, a Chinese ornamental pagoda tree, whose flower buds, "Chinese yellow berries," when dry, may yield as much as 22 per cent of rutin.[16] This source also yields toxic alkaloids and other pigments that are used to dye garments.

Rutin occurs in varying concentrations in yellow pansies, fresh forsythia flowers[17] (1 to 4%), hydrangea, garden rue, violets, elder flowers, bright tobacco leaves (0.4%) and green buckwheat[18] (3.2 to 8%). Rutin is found only in flue-cured tobacco leaves, and the best yields are obtained from leaves of high quality U. S. Grade C3L or better. Because of its high content, green buckwheat furnishes the best commercial source of supply of rutin.

Rutin occurs as greenish yellow, needle-shaped crystals that are difficultly soluble in water (0.15 g. per liter at 20°), soluble 1:650 in alcohol, acetone or ethyl acetate and insoluble in chloroform, ether or the hydrocarbons. Because rutin contains a catechol group, it is unstable to light, oxidizing agents and alkalies or alkali bicarbonates.

Category—bioflavonoid.

Usual dose—20 mg. (3 to 4 times a day).

Szent-Gyorgyi *et al.*[19] reported, in 1936, that lemon juice or extracts of Hungarian red pepper were beneficial in a number of cases of purpura hemorrhagica which had failed to respond to large doses of ascorbic acid. Because a factor other than ascorbic acid was thought to be responsible for this activity, it was given the name vitamin P, to indicate its effect on the permeability of the capillaries. A crystalline flavanone glycoside fraction called "citrin" was isolated from lemon juice and shown to be physiologically active. "Citrin" subsequently was shown to contain largely hesperidin together with smaller amounts of eriodictyol glycoside (eriodictin).[20]

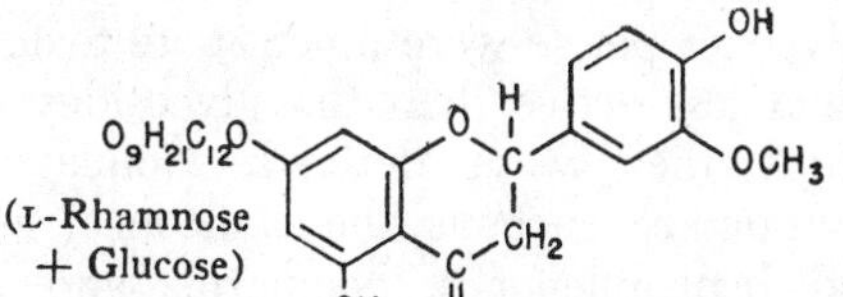

Hesperidin

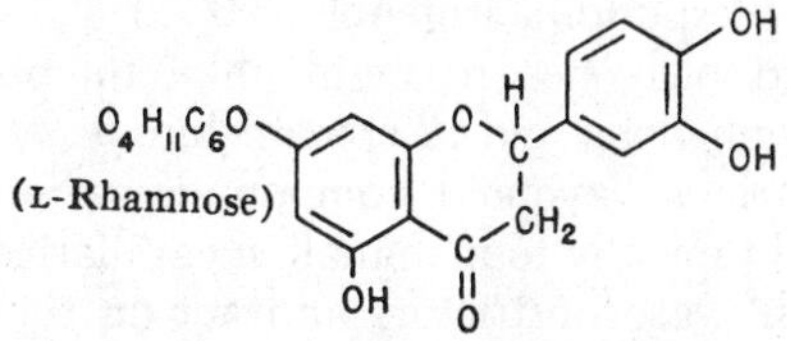

Eriodictin

As a result of chromatographic studies of citrin, Robezieks[21] detected the presence of a third pigment that had quercitrinlike properties. This discovery may explain the observation that citrin brings about a definite lowering of blood pressure.[22] Therefore, sources of rutin, a well-known flavanol glycoside, were sought in order to test it for its activity.

Preliminary clinical data indicate that rutin may decrease capillary fragility favorably in a high percentage of cases when it is administered over a period of many weeks or months. Increased capillary fragility may be encountered in patients with hypertension, diabetes mellitus or hereditary telangiectasia and as a complication of salicylate, thiocyanate and arsenical therapy. Rutin appears to be of value in certain cases of glaucoma by decreasing intraocular pressure when it is used with miotics.

Quercetin, the aglycon of rutin, is more effective than rutin in correcting capillary fault in hypertensive subjects. Also, it is effective in those cases that are refractory to rutin.

Reports that rutin decreased mortality or hastened the recovery of Roentgen-ray irradiated animals stimulated others[23] to test the "Vitamin P-Like" activity of a number of flavonoids for their capillary permeability modifying effects. No clear-cut conclusions could be drawn between structures with relation to activity, and the original publication should be consulted for specific details. Suffice it to say that a number of compounds were considerably more active in the tests performed than was rutin. The aglycones per se were much more toxic and much less active than the glycosides from which they were derived. Nonflavonoid compounds, such as the chalcones, aceto- and benzophenones, coumarins and even phloroglucinol, showed some activity. The most active compounds were phosphorylated hesperidin, irigenol (5,6,7,3',4',5-hexahydroisoflavone), naringin, phloretin, phloroacetophenone and *dl*-epicatechin.

Various flavonoid compounds, when applied topically to mammalian capillaries, are potent vasoconstrictors and act on the precapillary sphincters, thus shunting the flow of blood into the larger vessels. Therefore, an indirect action decreases capillary permeability. The basic mode of action is proposed as an inhibition of the conversion of ADP (adenosine diphosphate) to ATP (adenosine triphosphate), a change necessary for muscle relaxation.

Phosphorylated hesperidin prevents fertilization of the ova in about 80 per cent of female rats when this compound is given orally or intraperitoneally. Its mode of action is attributed to its powerful, nontoxic hyaluronidase-inhibiting activity. Its activity in humans parallels that in rats.[24]

The activity of phosphorylated hesperidin in regard to alteration of blood-vessel permeability apparently prevents the accumulation of cholesterol. Therefore, it may be of some value in the treatment of arteriosclerosis.

Hesperidin and other citrus bioflavonoids, together with ascorbic acid, when taken by athletes, decreased the number of bruises and hastened healing.[25]

Hesperidin Methyl Chalcone. Investigations upon pure substances[26] indicated that the hesperidin and eriodictyol[27] of "citrin" did not have the pronounced effect on capillary permeability as originally reported for "citrin." On the other hand, the water-soluble yellow pigment from crude orange hesperidin was found to increase capillary permeability in certain individuals whose capillary permeability was below normal. This compound was shown to be hesperidin chalcone.

Hesperidin chalcone, like other chalcones containing a phloroglucinol ring, readily is converted to flavones in acid solution, when heated or upon prolonged standing in the dry state. Conversely, flavones are converted to chalcones by the action of alkali. Methylation of one of the phenolic groups in the phloroglucinol portion of the chalcones stabilizes or prevents ring closure to form the flavones.

The stabilized hesperidin methyl chalcone has lost none of the capillary permeability properties shown to be present in hesperidin chalcone.

Hesperidin methyl chalcone is a deep yellow powder which is readily soluble in water, quite soluble in alcohol and acetone and slightly soluble in ether.

Parenteral administration of hesperidin methyl chalcone causes a striking depression of blood pressure. Orally it is claimed to increase capillary permeability (vitamin P(?) activity). For uses see rutin.

Hesperidin methyl chalcone is used in 50-mg. capsules.

Good sources of substances that appear to restore abnormally increased permeability of capillaries are grapes, lemons, oranges, rose hips, blue plums, prunes and black currants. Moderately good sources are grapefruit, apricots, cherries, blackberries and blueberries. Little or no activity was found in tomatoes, cabbage, cauliflower, lettuce,

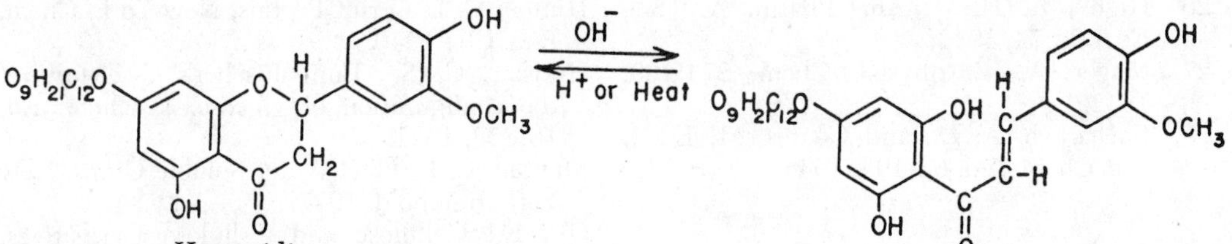

Hesperidin

$\rightleftharpoons$ (OH^- / H^+ or Heat)

Hesperidin Chalcone

turnips, carrots, potatoes and parsnips. Although some of the above sources contain much vitamin C, there is no constant relationship between the vitamin C content and vitamin P activity.[28] This is particularly true of grapes.

↓ Methylation

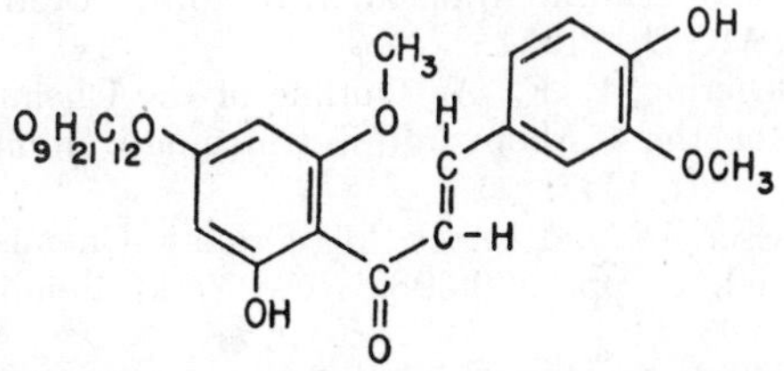

Hesperidin Methyl Chalcone

REFERENCES CITED

1. Bergy, G. A.: Am. Prof. Pharm. 18:340, 1952.
2. Kenyon, W. O., et al.: J. Am. Chem. Soc. 64:121, 127, 1943.
   ————: Textile Res. J. 16:1, 1946.
3. Frantz, V. K.: Ann. Surg. 118:116, 1943. Grahzit, O. M.: Fed. Proc. 5:222, 1946.
4. Blythe, R. H., et al.: J. Am. Pharm. A. (Sci. Ed.) 38:59, 1949.
5. Nelson, W., and Percival, E.: J. Chem. Soc. p. 58, 1942.
   Mullan, J., and Percival, E.: J. Chem. Soc. p. 1501, 1940.
   Anderson, E., and Krznarich, P. W.: J. Biol. Chem. 111:549, 1935.
6. Smith, F.: J. Chem. Soc. p. 1724, 1939; p. 1935, 1940.
   Jackson, J., and Smith, F.: J. Chem. Soc. pp. 74 & 79, 1940.
7. James, S. P., and Smith, F.: J. Chem. Soc. p. 739, 1945.
8. Percival, E., et al.: J. Chem. Soc. p. 1615, 1937.
   Jones, W., and Peat, S.: J. Chem. Soc. p. 225, 1942.
9. Butler, R. M.: Biochem. J. 28:759, 1934; 29:1025, 1935.
10. Schneider, G., et al.: Ber. deutsch. chem. Ges. 70B:1617, 1937; 69B:309, 2530, 1936.
11. Owens, H. S., et al.: J. Am. Chem. Soc. 68:1628, 1946.
12. Kuizenga, M. H., and Spaulding, L. B.: J. Biol. Chem. 148:641, 1943.
12a. Wolfrom, M. L., et al.: J. Org. Chem. 29: 3284, 1964.
13. Loewe, L., et al.: J.A.M.A. 130:386, 1226, 1946.
14. Lange, K., and Loewe, L.: Surg. Gynec. Obstet. 82:256, 1946.
15. Ballou, C. C.: Adv. Carb. Chem. 9:59, 1954.
16. Harrisson, J., et al.: J. Am. Pharm. A. (Sci. Ed.) 39:556, 1950.
17. Couch, J. F., et al.: J. Am. Chem. Soc. 69:572, 1947.
18. Griffith, J. Q., et al.: Proc. Soc. Exp. Biol. Med. 55:228, 1944.
   Sando, E., and Lloyd, J. U.: Biol. Chem. 58:737, 1924.
   Couch, J. F., et al.: Science 103:197, 1946.
19. Szent Gyorgyi, A., et al.: Deutsche med. Wschr. 62:1326, 1936.
20. Szent Gyorgyi, A., and Brucher, V.: Nature, London 138:1057, 1936.
   Rusnyak, I., and Szent Gyorgyi, A.: Nature, London 138:27, 1936.
21. Robezieks, I.: T. Vitaminforsch. 8:27, 1938.
22. Armentane, L.: Z. ges. exp. Med. 102: 219, 1937.
23. Haley, J. T., and Rhodes, B. M.: J. Am. Pharm. A. (Sci. Ed.) 40:179, 1951.
24. Martin, G. J., and Beiler, J. M.: Science 115:402, 1952.
   Sieve, B. F.: Science 116:373, 1952.
25. Miller, M. J.: Injury to athletes, Med. Times 88:313, 1960.

26. Higby, R. H.: J. Am. Pharm. A. (Sci. Ed.) 32:74, 1943.
27. Mager, A.: Z. physiol. Chem. 274:109, 1942.
28. Bacharach, A. L., and Coates, M. E.: J. Soc. Chem. Ind. 63:198, 1944.

SELECTED READING

Couch, J. F., *et al.*: Chemistry, Pharmacology and Clinical Application of Rutin, U.S.D.A. AIC-291, 1951.

Degering, E. F.: An Outline of the Chemistry of the Carbohydrates, Cincinnati, John S. Swift, 1943.

Fieser, L., and Fieser, M.: Organic Chemistry, ed. 3, pp. 350-398, New York, Reinhold, 1956.

Griffith, G. C., and Silverglade, A.: Symposium on Heparin, Am. J. Card. 14:1-54, 1964.

Haworth, H. W.: The Constitution of Sugars, New York, Longmans, 1929.

Hinton, C. L.: Fruit Pectins, New York, Chem. Pub. Co., 1940.

Hudson, C. S.: Emil Fischer's discovery of the configuration of glucose, J. Chem. Ed. 18:353, 1941.

Mantell, C. L.: The Water Soluble Gums, New York, Reinhold, 1947.

Ott, E.: Cellulose and Cellulose Derivatives, New York, Interscience, 1947.

Percival, E. G. V.: Structural Carbohydrate Chemistry, London, Muller, 1950.

Pigman, W. W.: Advances in Carbohydrate Chemistry, vols. 1-6, New York, Acad. Press, 1945-1951.

Pigman, W. W., and Goep, R. M.: Chemistry of the Carbohydrates, New York, Acad. Press, 1948.

Smith, F., and Montgomery, R.: Chemistry of Plant Gums and Mucilages, New York, Reinhold, 1959.

Stanek, J.: The Monosaccharides, New York, Acad. Press, 1963.

Wolfrom, M. L., Raymond, A. L., and Heuser, E.: The Carbohydrates, *in* Gilman, H.: Organic Chemistry, vol. II, ed. 2, New York, Wiley, 1943.

27

Amino Acids, Proteins, Enzymes and Hormones With Proteinlike Structure

Ole Gisvold, Ph.D.,
Professor of Medicinal Chemistry, College of Pharmacy,
University of Minnesota

AMINO ACIDS

Table 98 gives the formula, the symbol (abbreviation) and some properties of those amino acids that have been obtained by the hydrolysis of proteins.

The amino acids are colorless, crystalline solids that dissolve in varying degrees in water; the notably less soluble compounds are cystine, tyrosine, thyroxine and tryptophan. Except proline, they are mostly insoluble in alcohol, and all of them are insoluble in ether. From hydrolytic mixtures, they sometimes can be separated by differences in solubility in various solvents and by fractional distillation of their methyl esters. Most of them have a high melting point, which is consistent with their structure as salts, and usually decompose at the same time. All of them except glycine are optically active, but the direction and the amount of rotation vary considerably with the solvent, the pH, the temperature and other factors. The rotation of some of the salts is opposite in direction to that of the free base.

It is now more proper to use the prefixes D and L when discussing the α-amino acids to show their relationship to the D and L lactic acids. All the naturally occurring amino acids derived from proteins have the L-configuration. Thus, the formerly designated d-alanine is now more properly called L(+)-alanine and l-leucine more properly L(−)-leucine, etc.

The α-amino carboxyl $(-CH(NH_2)-COOH)$ portion of the amino acids can exist in water solutions largely as neutral dipolar ions, also called "zwitterions" or "amphions," and, if completely equal, would make no contribution to migration in an electric field. When they are not equal they can be made so by adjusting the pH of the solution until no migration takes place. The pH at this stage is called the isoelectric point. At the isoelectric point the solubility of both amino acids and proteins is lowest.

In an electric field, an amino acid such as glycine, alanine, etc., in acid solutions will migrate to the cathode and in basic solutions to the anode.

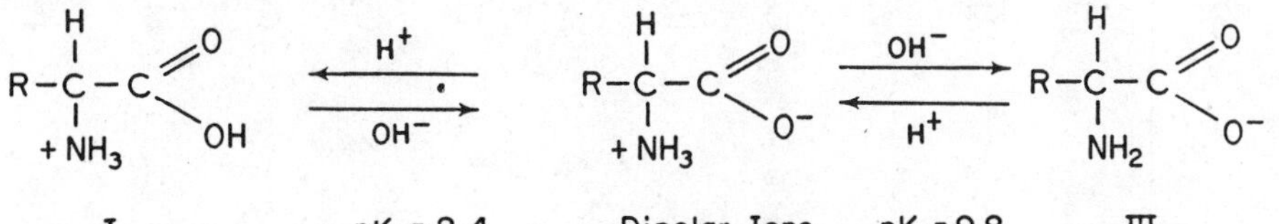

875

TABLE 98. NATURALLY OCCURRING AMINO ACIDS

NAME	SYMBOL	FORMULA	Iso Electric Point	M.P.	$[\alpha]_D$	SOLUBILITY IN WATER
Glycine	Gly.	H_2NCH_2COOH	5.97	230		1:4
Alanine	Ala.	$CH_3CH(NH_2)COOH$	6.00	297	+ 3.5	1:5
Valine	Val.	$(CH_3)_2CHCH(NH_2)COOH$	5.96	315	+ 6.5	1:50
Leucine	Leu.	$(CH_3)_2CHCH_2CH(NH_2)COOH$	6.02	295	− 10.5	1:45
Isoleucine	Ileu.	$CH_3CH_2CH(CH_3)CH(NH_2)COOH$	5.98	280	+ 9.5	1:26
Serine	Ser.	$HOCH_2CH(NH_2)COOH$	5.68	228	− 6.8	easily
Threonine	Thr.	$CH_3CH(OH)CH(NH_2)COOH$	5.60	285	− 6.8	soluble
Cysteine	CySH	$HSCH_2CH(NH_2)COOH$	5.05			easily
Cystine	CySSCy.	$(-SCH_2CH(NH_2)COOH)_2$	4.8	260	− 225	1:8,840
Methionine	Met.	$CH_3SCH_2CH_2CH(NH_2)COOH$	5.74	283	− 7	soluble
Proline	Pro.	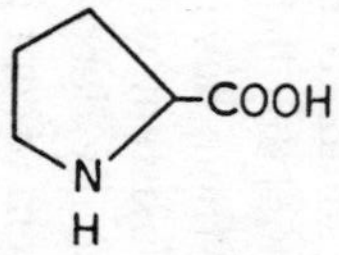	6.3	206–209	− 80	easily

TABLE 98. NATURALLY OCCURRING AMINO ACIDS (*Continued*)

NAME	SYMBOL	FORMULA	ISO ELECTRIC POINT	M.P.	$[\alpha]_D$	SOLUBILITY IN WATER
Hydroxyproline	Hypro.	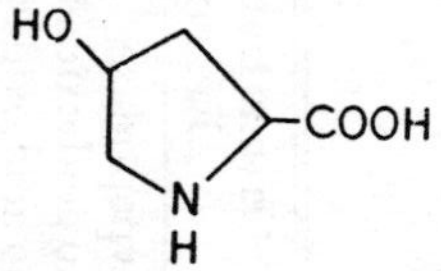 (structure)	5.83	260–270	− 80	easily
Phenylalanine	Phe.	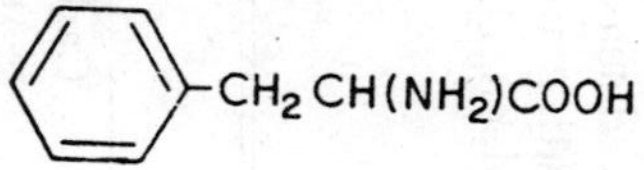 $CH_2CH(NH_2)COOH$	5.48	283	− 35	1:35
Tyrosine	Tyr.	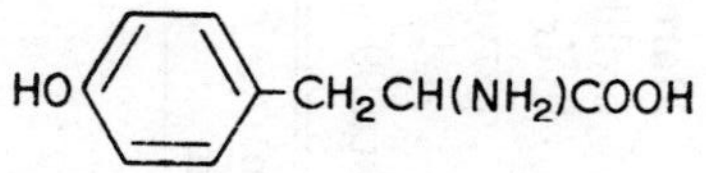 HO—$CH_2CH(NH_2)COOH$	5.66	290–299	− 8.5	slightly
Tryptophane	Try.	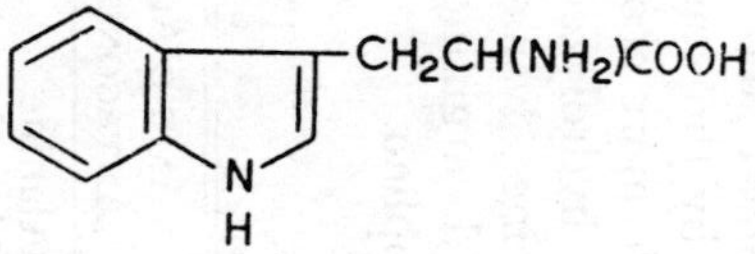 $CH_2CH(NH_2)COOH$ (indole)	5.89	289	− 32	slightly
Aspartic acid	Asp.	$HOOCCH_2CH(NH_2)COOH$	2.77	270	+ 4.36	1:250
Glutamic acid	Glu.	$HOOCCH_2CH_2CH(NH_2)COOH$	3.22	206	+ .2	1:100
Lysine	Lys.	$H_2NCH_2CH_2CH_2CH_2CH(NH_2)COOH$	9.74	224	+ 15.3	easily
Arginine	Arg.	$H_2NC(=NH)NH_2CH_2CH_2CH_2CH(NH_2)COOH$	10.76	207	+ 11.4	easily
Histidine	His.	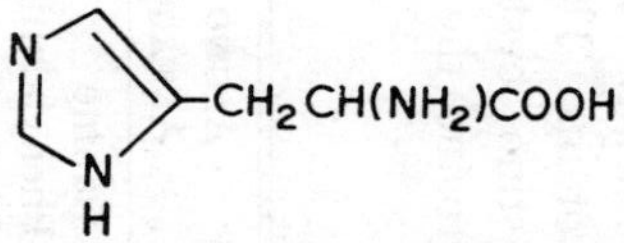 $CH_2CH(NH_2)COOH$ (imidazole)	7.59	277	− 40	freely

Glycine has two acidity constants, $pk_1 = 2.4$ for the carboxyl group and $pk_2 = 9.8$ for the amino group. This also is true for alanine, leucine and valine. The positive charge of I tends to repel a proton from the carboxyl group so that I is more strongly acidic than acetic acid ($pk = 4.76$). The pk_2 value for III is less than methylamine because of the electron-withdrawing effect of the carboxyl group (see structure on page 875).

A study of Table 98 shows that in addition to the α-amino carboxyl groups some amino acids have other polar groups such as $-OH$; $-NH_2$; $-COOH$; SH; phenolic $-OH$; guanidine, etc., which make their contribution to the chemical and the physical properties of the respective amino acids or to the proteins in which they are present. For example, the amino and the α-amino groups can be blocked by formaldehyde to permit the conventional titration of the carboxyl groups (Sorenson's formal titration).

Although all of the naturally occurring amino acids have been synthesized, and a number of them are available by the synthetic route, others are available more economically by isolation from hydrolyzed proteins. The latter are leucine, lysine, cystine, cysteine, glutamic acid, arginine, tyrosine, the prolines and tryptophan.

An essential amino acid may be defined as one which cannot be synthesized by the animal body at a rate commensurate to meet its demands for normal growth. If this definition is correct, it is interesting to note that phenylalanine, tryptophan, histidine and methionine also are effective in the D-forms.

Application of some of the concepts utilized in the preparation of antimetabolites, i.e., one small change at a time, has resulted in the preparation of a number of antimetabolites of the naturally occurring amino acids. Table 99 contains examples of amino acid antimetabolites.

In a number of instances the amino acid antimetabolite was incorporated into protein molecules, thus indirectly altering certain living processes, in such a way as usually to arrest eventually the growth of the organism. This phenomenon is sometimes called lethal synthesis. This approach has given similar results with dipeptides.

L-O-Ethylthreonine is incorporated into tRNA in *E. coli* and inhibits growth. It also decreases the incidence of mortality in chickens infected with virus responsible for Marek's disease, a neoplasticlike lymphoproliferative malady that affects primarily liver, spleen, gonads, kidneys and peripheral nerves.

TABLE 99

AMINO ACID ANTAGONIST	AMINO ACID ANTAGONIZED	OTHER INHIBITORY EFFECTS
D-Alanine	L-Alanine	Carboxypeptidase
D-Phenylalanine	L-Phenylalanine	D-Amino acid oxidase
α-Methy-L-methionine	L-Methionine	D-Amino acid oxidase
α-Methyl-L-glutaric acid	L-Glutaric acid	Glutamic decarboxylase
Ethionine	Methionine	
α-Methyldopa	Dopa	Dopa decarboxylase
Allyl glycine	Methionine	Growth of *E. coli*
Propargylglycine	Methionine	Growth of *E. coli*
2-Amino-5-heptenoic acid	Methionine	Growth of *E. coli*
2-Thienylalanine	Phenylalanine	Growth of yeast
p-Fluorophenylalanine	Phenylalanine	Incorporation of phenylalanine into protein molecules
L-O-Methyl threonine	Isoleucine	Competitive incorporation of leucine into proteins
4-Oxalysine	Lysine	Growth of *E. coli, L. casei*, etc.
6-Methyltryptophan	Tryptophan	
5,5,5-Trifluoronorvaline	Leucine, Methionine	Growth of *E. coli*, etc.
3-Cyclohexene-l-glycine	Isoleucine	Inhibits *E. coli*
O-Carbamyl-L-serine	L-Glutamine	Inhibits *E. coli, S. lactis*

N-Dichloracetyl-DL-serine and DL-threonine recently have effected regression of some tumors in mice and rats. The serine derivative is particularly effective against sarcoma 37 and in some cases causes complete regression. Its toxicity in mice is low.

PRODUCTS

Aminoacetic Acid N.F., Glycocoll®, Glycine, contains not less than 98.5 per cent and not more than 101.5 per cent of $C_2H_5NO_2$. It occurs as a white, odorless, crystalline powder, having a sweetish taste. It is insoluble in alcohol but soluble in water (1:4) to make a solution that is acid to litmus paper.

Category—nutrient.

Usual dose—30 g. daily, in divided doses.

Application—irrigating solution as a 1.5 per cent solution.

OCCURRENCE

Aminoacetic Acid Sterile Solution N.F.

Diglycine Hydrochloride, Glyco-HCl®, is supplied in capsules containing 194 mg. equivalent to 10 minims of diluted hydrochloric acid for achlorhydria.

Diglycine Hydriodide Glyco-HI®, consists of capsules used as expectorant and in iodide therapy. The dose is 1 to 2 capsules during or after meals.

Dihydroxyaluminum Aminoacetate N.F., Basic Aluminum Glycinate, Alglyn®, Aspogen®, Alzinox®, Doraxamin®, Robalate®, Alminate®, Dimothyn®, may be represented by the formula $H_2NCH_2COOAl(OH)_2$. It is a white, odorless water-insoluble powder with a faintly sweet taste and is employed as a gastric antacid in the same way as aluminum hydroxide gel. Over the latter, it is claimed to have the advantages of more prompt, greater and more lasting buffering action. Also, it is said to have less astringent and constipative effects because of its smaller content of aluminum. However, medical authorities are not yet satisfied that any of these claims are justified. The compound is furnished in powder, magma, or in tablets containing 500 mg.

Category—antacid.

Usual dose—500 mg. to 1 g. 4 times daily.

Usual dose range—500 mg. to 2 g.

OCCURRENCE

Dihydroxyaluminum Aminoacetate Magma N.F.

Dihydroxyaluminum Aminoacetate Tablets N.F.

Histidine Monohydrochloride, a salt of histidine, contains not less than 21.5 per cent and not more than 22.2 per cent of N, calculated on the moisture-free basis, corresponding to not less than 98 per cent of $C_6H_9N_3O_2.HCl$. It consists of small, glistening, colorless crystals that are odorless, possess a salty taste and dissolve in water (1:8) to produce a solution that is acid to litmus. The specific rotation in about 2.5 per cent solution is +9.3 to +11.2°, although histidine itself is levorotatory. Histidine occurs in the urine of pregnant women. It is prepared most conveniently from blood corpuscles by hydrolysis and precipitation with mercuric sulfate.

It has been used in the treatment of peptic ulcer and intestinal ulcerations.

Glutamic Acid Hydrochloride N.F., Acidulin®, Glutamicol®, is essentially a pure compound that occurs as a white crystalline powder soluble 1:3 in water and insoluble in alcohol. It has been used in place of glycine in the treatment of muscular dystrophies with rather unpromising results. It also is combined (8 to 20 g. daily) with anticonvulsants for the petit mal attacks of epilepsy, a use which appears to depend on change in pH of the urine.

The hydrochloride, which releases the acid readily, has been recommended under a variety of names for furnishing acid to the stomach in the achlorhydria of pernicious anemia and other conditions.

Category—acidifier (gastric).

Usual dose range—600 mg. to 1.8 g. during meals.

OCCURRENCE

Glutamic Acid Hydrochloride Capsules N.F.

Methionine N.F., Amurex®, Diameth®, Meonine®, Metione®, DL-2-amino-4-(methylthio)butyric acid, occurs as white, crystalline platelets or powder with a slight, characteristic odor; it is soluble in water (1:30) and a 1 per cent solution has a pH of 5.6 to 6.1. It is insoluble in alcohol. In recent years, the racemic compound has been produced in ever increasing quantities and at considerably reduced cost. The human body needs proteins that furnish methionine in order to prevent pathologic accumulation of fat in the liver, a condition which can be counteracted by administration of the acid or proteins that provide it. Methionine also has a function in the synthesis of choline, cystine, lecithin and, probably, creatine. Deficiency not only limits growth in rats but also inhibits progression of tumors.

In therapy, methionine has been employed in the treatment of liver injuries caused by poisons such as carbon tetrachloride, chloroform, arsenic and trinitrotoluene. While many physicians are enthusiastic about its value under such circumstances, this action has not been established satisfactorily. It also is administered to treat eclampsia and shock, probably on the basis of protecting the liver, and as a supplement to a high protein diet. It usually is administered orally in doses of 3 to 6 g. daily, but much more has been employed. It also has been given by intravenous drip as a 3 per cent solution in amounts up to 10 g. daily.

Category—lipotropic.

Usual dose—3 g. one to 3 times a day.

Usual dose range—3 to 20 g. per day.

OCCURRENCE

Methionine Capsules N.F.

Methionine Tablets N.F.

Methionine is available in combination with other ingredients in various commercial preparations. It has been combined as a lipotropic with choline, inositol and other agents in Cholimeth Fortified, Chol-nine, C.M.I., Lipothyn, Lipotropic Capsules, Lipovite and Methischol. It is used as a choleretic in Timagol, where it is combined with dehydrocholic acid, inositol and methenamine; for vitamin therapy in Amvitol, Meovite, and Rydiamin; and as antinauseant with pentobarbital and vitamins in Nidoxitol.

Cysteine has been used in solution for application to ulcers and slow-healing wounds and once was recommended in Addison's disease. Iodogorgoic acid has been administered in Graves' disease, but there is no indication that better effects are ob-

tained than with the same amount of inorganic iodides. Tryptophan and tyrosine are added sometimes to other agents in supplementary feeding, although the indications for their use are still obscure.

Arginine Hydrochloride, Argivene®, has pharmacologic uses similar to those of sodium glutamate. It is used intravenously.

Arginine Glutamate, Modumate®, is the L(+)-arginine salt of L(+)glutamic acid. It is administered intravenously because of its ability to prevent or relieve symptoms of ammoniemia which most commonly is caused by hepatic insufficiency. Glutamic acid is an ammonia acceptor. Hepatic arginase cleaves arginine to urea and ornithine which also is an ammonia acceptor. These amino acids, unlike their sodium or potassium salts, do not introduce additional cations into the blood stream. Although use of these amino acids may be urgent in ammonia intoxication to relieve coma and to prevent cerebral damage, they should be considered supplementary to other forms of treatment.

These amino acids are available for injection (intravenous) as 250 mg./ml. in 100 ml. containers.

Aminocaproic Acid N.F., Amicar®, 6-aminohexanoic acid, occurs as a fine, white crystalline powder that is freely soluble in water, slightly soluble in alcohol and practically insoluble in chloroform.

Aminocaproic acid is a competitive inhibitor of plasminogen activators such as streptokinase and urokinase. It is effective because it is an analog of lysine whose position in proteins is attacked by plasmin. To a lesser degree it also inhibits plasmin (fibrinolysin). Lowered plasmin levels lead to more favorable amounts of fibrinogen, fibrin and other important clotting components.

Aminocaproic acid has been used in the control of hemorrhage in certain surgical procedures. It is of no value in controlling hemorrhage due to thrombocytopenia or other coagulation defects or vascular disruption, e.g., bleeding ulcers, functional uterine bleeding, post-tonsillectomy bleeding, etc. Since it inhibits the dissolution of clots, it may interfere with normal mechanisms for maintaining the patency of blood vessels.

Aminocaproic acid is well absorbed orally. Plasma peaks occur in about two hours. It is excreted rapidly, largely unchanged.

Category—hemostatic.

Usual dose—oral and intravenous, initial, 5 g. followed by 1 to 1.25 g. every hour to maintain a plasma level of 13 mg. per 100 ml; no more than 30 g. per 24 hour period is recommended.

OCCURRENCE

Aminocaproic Acid Injection N.F.
Aminocaproic Acid Tablets N.F.

Acetylcysteine, Mucomyst®, is the N-acetyl derivative of L-cysteine. It is used primarily to reduce the viscosity of the abnormally viscid pulmonary secretions in patients with cystic fibrosis of the pancreas (mucoviscidosis) or various tracheobronchial and bronchopulmonary diseases.

Acetylcysteine is more active than cysteine and its mode of action in reducing the viscosity of mucoprotein solutions, including sputum, may be by opening the disulfide bonds in the native protein.

Acetylcysteine is most effective in 10 to 20 per cent solutions with a pH of 7 to 9. It is used by direct instillation or by aerosol nebulization. It is available as a 20 per cent solution of the sodium salt in 10 and 30 ml. containers. An opened vial of acetylcysteine must be covered, stored in a refrigerator and used within 48 hours.

Levodopa, Larodopa®, Dopar®, Levopa® is (−)-3-(3,4-dihydroxyphenyl)-L-alanine. It occurs as a colorless, crystalline material. It is slightly soluble in water and insoluble in alcohol. Levodopa is a precursor of dopamine and has been found to be of value in the treatment of Parkinson's disease. Dopamine does not cross the blood-brain bar-

Levodopa

rier and, therefore, is ineffective. Levodopa does cross the blood-brain barrier and pre-

sumably is metabolically converted to dopamine in the basal ganglia. The dose must be carefully determined for each patient.

PROTEIN HYDROLYSATES

Exhaustive hydrolysis of simple proteins (proteolysis) ultimately results in liberation of the constituent amino acids. Incomplete proteolysis yields a mixture of these amino acids with some dipeptides, polypeptides and derived proteins of higher molecular weight. The hydrolysis can be accomplished by alkalies, but these agents bring about destruction of methionine, cystine and perhaps others and racemization of most of the amino acids. The hydrolysis can be performed by acids, a method that has the advantage of leaving no peptides but which results in loss of tryptophan. It also can be brought about by proteolytic enzymes, a process that is slow and seldom complete. In the manufacture of amino acid mixtures, the raw materials often are pure proteins, such as casein and lactalbumin, but the most convenient ones are good sources of protein, such as yeast, liver, beef, vegetables.

The resulting protein hydrolysates are employed to supplement the diet in cases of protein deficiencies and have a limited but, as yet, undetermined value in the treatment of disease. The lack of adequate protein may result from a number of conditions, but the case is not always easy to diagnose. It may arise from insufficient intake, normally about 1 g. or less per kg. of weight; from temporarily increased demands, as in pregnancy; from impaired digestion or absorption in the intestinal tract due to disease conditions; from poor assimilation because of diseases of the liver or other organs; from increased decomposition or loss of amino acids or proteins during diseases, such as fevers, leukemia or hemorrhage, or after operations, burns, fractures or shock. Under the last set of conditions, the loss of protein may be very large, as much as 150 g. per day. There seems little need for any therapy other than an adequately selected diet in the majority of cases, but, when there is an allergy towards excessive protein, the hydrolysates may be indicated. The dose under such circumstances will depend upon the desired purpose but, because the amino acids are normal food constituents, may be practically unlimited.

PRODUCTS

Protein Hydrolysate Injection U.S.P., Protein Hydrolysates (Intravenous), Amigen®, Aminosal®, Hyprotigen®, Parenamine®, Travamin®. Protein Hydrolysate Injection is a sterile solution of amino acids and short-chain peptides which represent the approximate nutritive equivalent of the casein, lactalbumin, plasma, fibrin, or other suitable protein from which it is derived by acid, enzymatic, or other method of hydrolysis. It may be modified by partial removal and restoration or addition of one or more amino acids. It may contain dextrose or other carbohydrate suitable for intravenous infusion. Not less than 50 per cent of the total nitrogen present is in the form of α-amino nitrogen. It is a yellowish to red amber transparent liquid that has a pH of 4 to 7.

Parenteral preparations are employed for the maintenance of a positive nitrogen balance in cases where there is interference with ingestion, digestion or absorption of food. In such cases, the material to be injected must be nonantigenic and must not contain pyrogens or peptides of high molecular weight. Injection may result in untoward effects, such as nausea, vomiting, fever, vasodilatation, abdominal pain, twitching and convulsions, edema at the site of injection, phlebitis and thrombosis. Sometimes these reactions are due to inadequate care in cleanliness or too rapid administration.

Category—fluid and nutrient replenisher.

Usual dose—intravenous, 500 ml. of a 5-per-cent solution.

Usual dose range—250 to 1500 ml.

Protein Hydrolysate (Oral), Aminoat®, Caminoids®, are oral protein hydrolysates obtained in a similar manner to those for intravenous use. The same kinds of proteins are used and may be hydrolyzed by any one of the three methods. Usually, they are available in powdered form, flavored or unflavored. Most often, the oral form is recommended for diets of infants allergic to milk and in the treatment of peptic ulcer.

PROTEINS AND PROTEINLIKE SUBSTANCES

The subject of proteins and proteinlike substances is a very complex one, so much so that it is very difficult to discuss this subject adequately in a textbook of this kind. Therefore, this subject will receive only a very limited and condensed treatment. However, one interested in medicinal chemistry would do well to become oriented in this area, because to a large extent living processes are primarily dependent upon the activities of proteins and proteinlike substances which may function per se or in conjunction with certain atoms or organic molecules. In many instances the mode of action of medicinal agents is associated with proteins and proteinlike substances or coenzymes.

Mulder in 1839 first used the term "protein," which is derived from the Greek word *proteios,* meaning primary or holding in first place.

Proteins may be found in intracellular and extracellular substance or as integral components of cell as walls in both animals and plants. True enzymes, viruses, antigens and antibodies are proteins, and a number of hormones are low molecular weight proteins. Proteins of all kinds are composed of one or more polypeptide chains. The polypeptide is composed of naturally occurring amino acids (see Amino Acids) joined in amide linkages. For example, the case in which the amino acid residues are in the form of a fully extended chain can be represented as shown in Figure 27, together with the bond lengths and bond angles. The number of different polypeptides that are theoretically possible, containing all or part of these amino acids, is almost infinite. If one utilized 16 amino acids (the number found in the B chain of insulin, which is a relatively low molecular weight protein), 6×10^{56} different arrangements are possible. If one had only one molecule each of all these possible kinds, the total weight would

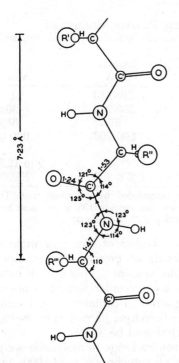

FIG. 27. A diagrammatic representation of a fully extended polypeptide chain with the bond lengths and the bond angles derived from crystal structures and other experimental evidence.[1] (Corey and Pauling, 1953)

be about 1,000 billion times that of the earth. Polyeptide chains can assume different shapes, and this, added to the above, would give a figure so vast that a mathematical brain might conclude that, from a statistical point of view, life is completely improbable.

Not all the 20 naturally occurring amino acids occur in all proteins. In addition, the relative proportions of the amino acids in proteins vary.

SHAPES

In addition to the very abridged definition that proteins and proteinlike substances are polypeptides, it can be said that each member of this group can be characterized by the kind, the number and the sequential arrangement of the amino acids. In addition, in the native state, shape and structure are very important to the function of the polypeptide chain in living matter. Proteins and proteinlike substances vary in their shapes. Figure 28 shows silhouettes of the molecules of the blood proteins at a magnification of 1-million-fold. At this magnification each red blood cell would be about 30 feet in diameter, and a man would be about 1,200 miles tall.

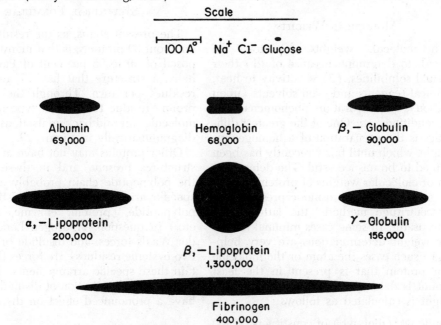

FIG. 28. Relative dimensions of protein molecules in the blood.

TABLE 100. SIZE AND SHAPE OF SOME PROTEIN MOLECULES

PROTEIN	SHAPE	CROSS SECTION IN Å	LENGTH IN Å	MOL. WT.
γ-Globulin	Cylindrical	19 and 57	230-240	1.6×10^5
Yeast ribose	Rodlike	15-17	200-260	30,000
Hemoglobin	Cylindrical	33.5	57	68,000
Fibrinogen	Rodlike	30-35	600-700	400,000 to 500,000
Influenza virus	Spherical	97-114		
Bushy stunt virus	Spherical	25-27		9 to 12.8 million
Tobacco mosaic virus*	Rodlike	15.2-17.8	270-280	17,000,000

* Now shown to be composed of a nucleic acid core surrounded by a protein coat. This coat is composed of several homogeneous protein subunit molecules that contain 158 amino acid residues, the sequence of which has been elucidated recently by W. Stanley and co-workers.

Table 100 illustrates the size and the shape of some protein molecules.

The data given for some proteins indicate that at the isoelectric point each long chain is closely coiled on itself; however, this coiling is not so uniform that the individual macromolecules can be regarded as disks, but the van der Waals forces between the side chains are sufficient to preserve a certain 2-dimensional compactness in the structure. On moving away from the isoelectric point, the repulsive effect of the intramolecular electric charges leads to a partial or complete unfolding. This can be detected by a rise in the surface viscosity.

MOLECULAR WEIGHTS

The molecular weights of proteins are difficult to determine because of (1) their unusual solubilities, (2) sensitivity to heat, chemical reagents and even solvents (upon dilution), (3) hydration phenomena, (4) pH conditions, etc. One of the greatest difficulties is the preparation of a homogenous sample, which until fairly recently has been claimed to be unsuccessful. The determination of molecular weights of proteins is usually accomplished by osmotic pressure and ultracentrifuge methods, the latter being quite useful. In some cases minimal molecular weight determinations are very helpful. In such cases the atom or the molecule of a protein that is present in the least amount is determined. Then the molecular weight is calculated as follows:

$$\frac{\text{Atomic wt. (mol. wt.) of constituent} \times 100}{\text{Per cent of constituent in protein}}$$

N = number of atoms or molecules in a molecule of protein. For example, hemoglobin contains 4.2 per cent of heme, therefore its molecular weight is about 69,000. Egg albumin contains 4.2 per cent of tyrosine; therefore, its molecular weight (minimal) should be 34,496.

Proteins range in molecular weight, from that of cytochrome C, 13,000, to tobacco mosaic virus 17,000,000. Examples of a few others are albumin, 34,500; hemoglobin, 68,000; serum globulin, 150,000; visual purple, 270,000; and the group of hemocyanins, from 380,000 to 6,800,000.

STRUCTURAL FORMULAE

The present status, as the result of work on about 30 proteins, is that many are composed of 90 to 95 per cent of Pauling's α-helix, a structure that has 3.7 amino acid residues per turn. Through the aid of a proline residue the α-helix type of protein molecule can fold back on itself, as depicted diagrammatically in Figure 29.

Other proteins may not have any α-helix structures present, and in these proteins the polypeptide chain probably assumes a specific arrangement characteristic of the polypeptide (protein, enzyme, hormone, etc.) in question. Hydrogen bonding, van der Waals forces, and disulfide bridges (of two cysteine residues) are forces that maintain these specific arrangements. The number and the distribution of disulfide bridges have a pronounced effect on the ability of

$\times$ N = Minimal mol. wt. of protein

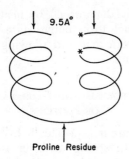

FIG. 29. * to * = 5.1 Å and 3.6-3.7 amino acid residues per turn.

hydrogen-bonded structures in their vicinity to maintain a reasonably constant structural form. Where the α-helix is present, this structure is maintained by hydrogen bonding. (See dotted line in Fig. 30.) The environment, i.e., solvent, pH, salts, temperature, etc., also will determine to a certain extent the arrangement of the polypeptide in question. Furthermore, one probably can assume that the kind, the number and the sequential arrangement of the amino acid residues in the polypeptide chain also will exert an effect on the specific arrangement. It should be made clear that not necessarily all the functional groups of the tails of the amino acid residues are exposed or readily detected.

To explain how water plays such an important role in the structure and activity of proteins and polypeptides, it has been proposed[2] recently that the polypeptide chains of naturally occurring substances might occur in a hexagonal arrangement.

Ribonuclease can be depicted diagrammatically to show how the folds in this polypeptide molecule are quite firmly fixed by

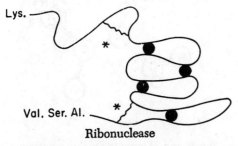

Ribonuclease

* These bonds ruptured by urea
● Disulfide bridges

intramolecular disulfide bridges and how the ends of the polypeptide chain are free to assume a more random arrangement.

SOLUBILITY

The solubility of proteins and proteinlike substances is governed by their molecular weights and those factors described under the discussion of structural formulae.

In Table 101 is given the solubilities of some proteins and proteinlike substances.

DENATURATION

Denaturation can be considered as an intramolecular structural change of the protein molecule which does not involve the rupture of covalent bonds and is due to reversible and irreversible rearrangements of nonionic bonds, such as hydrogen bonds, and of cohesive forces such as van der Waals. The rupture of native disulfide bonds or the formation of new disulfide bonds is usually considered as a more extensive state of denaturation. Denaturation, which also can be likened to an unfolding or loosening process, can be confirmed in some cases by the detection of previously masked groups such as -SH, imidazole and -NH$_2$ that are the functional groups of the tails of some of the amino acid residues. Other criteria for the detection of denaturation are, (2) decreased solubility, (3) increased digestibility by proteolytic enzymes, (4) decrease of the diffusion constant and increase of the viscosity of the protein solution, (5) loss of enzymatic properties if the protein is an enzyme, (6) modification of antigenic properties.

The degree or the extent of denaturation that takes place under any specified set of conditions is in turn dependent upon the protein per se. These physical or chemical conditions may be exceedingly mild on the one hand to relatively drastic on the other. These conditions include (1) heat, (2) strong acids, (3) strong bases, (4) alcohols, (5) surface-active agents, (6) urea, (7) high pressure, (8) surface forces, (9) ultraviolet light, (10) inorganic salts, (11) distilled water, (12) acetone (in some cases), (13) freezing and thawing (in some cases),

(14) membranes used in dialysis, (15) adsorbents (charcoal, etc.).

The rate, the ease and the degree of reversal of a denatured protein will depend on (1) the extent of denaturation that has been effected, (2) the nature of the protein and (3) the conditions employed. It is believed that most, if not all, denatured proteins can be obtained again in the native state. If the albumin in the egg white is coagulated (denatured) by heat, it can be brought back to its native state by dissolving it in acetic acid and allowing it to stand for some time.

HYDROLYSIS AND DEGRADATION OF PROTEINS

Proteins are nonspecifically hydrolyzed by boiling with strong acids or strong alka-

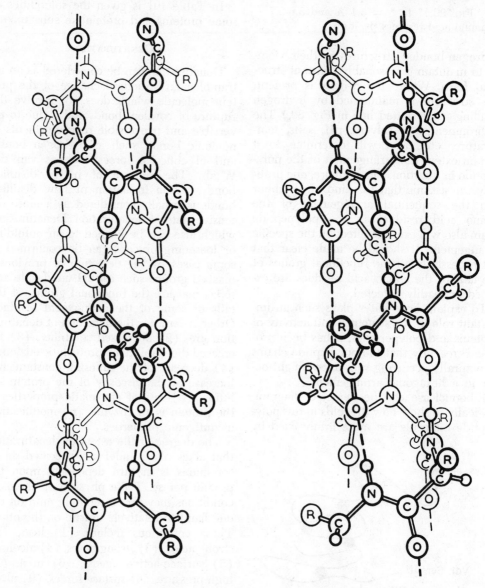

FIG. 30. Left-handed and right-handed α-helices. The R and H groups on the α-carbon atom are in the correct position corresponding to the known configuration of the L-amino acids in proteins. (L. Pauling and R. B. Corey, unpublished drawings.)

lies. The proteolytic enzymes hydrolyze only certain peptide bonds and give lower molecular weight polypeptides that can be used for structural studies. Sangers and Edman's[3] end member methods of analysis have been used to great advantage to study the amino acid sequence in proteins and polypeptides. They also have been used for the quantitative determination of C-terminal amino acids residue.[4]

CLASSIFICATION AND CRITERIA OF PURITY OF PROTEINS

The brief preceding discussion on the molecular weights, the structure, the shape, composition, the solubility, etc., of proteins precludes anything but an arbitrary classification of proteins and related substances. Before a substance can be classified properly, first it should be prepared in a pure state. Probably proteins are characterized more often by their biologic activity rather than by physical chemical criteria. The following criteria are usually needed to establish homomolecularity: (1) crystallinity, (2) constant solubility at a given temperature independent of the amount of excess protein added, (3) osmotic pressure in different solvents, (4) diffusion rate, (5) electrophoretic mobility, (6) dielectric constant, (7) chemical analysis.

Conjugated proteins are proteins that are intimately associated with certain metals, i.e., Fe, Cu, etc., or organic molecules, some of which are usually called coenzymes, or with other substances such as carbohydrates, nucleic acids, lipids, phosphoric acid, lecithin and heme, of which the conjugated proteins are often called glycoproteins, nucleoproteins, lipoproteins, phosphoproteins, lecithoproteins and chromoproteins or hemoglobins respectively.

Table 101 illustrates the way some of the proteins are classified according to their solubilities.

CHEMICAL PROPERTIES OF PROTEINS

The chemical properties of proteins are governed by their structures and by the kind, the number and the sequential arrangement of the amino acid residues. The polar groups of the amino acid residues make the greatest contributions to the chemical properties of proteins. In the native state some of these polar groups are "buried" and will not respond to their usual chemical properties when the native state is maintained. However, certain denaturation conditions can expose these buried

TABLE 101. SIMPLE (TRUE) PROTEINS

CLASS	CHARACTERISTICS	OCCURRENCE
ALBUMINS	Soluble in water, coagulable by heat and reagents	Egg albumin, lactalbumin, serum albumin, leucosin of wheat, legumelin of legumes
GLOBULINS	Insoluble in water, soluble in dilute salt solution, coagulable	Edestin of plants, vitelline of egg, serum globulin, lactoglobulin, amandin of almonds, myosin of muscles
PROLAMINES	Insoluble in water or alcohol, soluble in 60-80% alcohol, not coagulable	Found only in plants, e.g., gliadin of wheat, hordein of barley, zein of corn and secalin of rye
GLUTELINS	Soluble only in dilute acids or bases, coagulable	Found only in plants, e.g., glutenin of wheat and oryzenin of rice
PROTAMINES	Soluble in water or ammonia, strongly alkaline, not coagulable	Found only in the sperm of fish, e.g., salmine from salmon
HISTONES	Soluble in water, but not in ammonia, predominantly basic, not coagulable	Globin of hemoglobin, nucleohistone from nucleoprotein
ALBUMINOIDS	Insoluble in all solvents	In keratin of hair, nails and feathers; collagen of connective tissue; chondrin of cartilage; fibroin of silk and spongin of sponges

polar groups so that they can be detected by chemical methods.

By selecting the proper chemical reagents some of the polar groups can be manipulated without disturbing the native state of the protein per se. For example, an exposed $-NH_2$ group can be selectively acetylated by ketene. Further acetylation with this reagent will esterify the phenolic $-OH$ group. The free $-SH$ groups will combine readily with some of the heavy metals such as Hg (supposedly a specific reagent in some cases for the $-SH$ group), As, Bi, Ag, etc. The $-COOH$ group can be esterified with diazomethane. Sorensen's formal titration (see p. 878) can be used to estimate the $-COOH$ content of a protein. The imidazoline ring exists as a resonance hybrid and can combine with zinc. The free amino groups can be diazotized to liberate nitrogen, which is the basis for the Van Slyke method for their determination. Under acid conditions the $-NH_2$ group will take on a + charge, $-NH_3^+$, and serve as a cationic site. Under basic conditions the $-COOH$ will take on a negative charge and serve as an anionic site. The acidity constant of a protein (polypeptide) is dependent upon a number of factors: i.e., structures, kind, number and sequential arrangements of amino acid residues. Thus many factors can affect these constants. In the case of a simple dipeptide, such as glycylglycine, they are as follows:

$$\overset{+}{H_3N}-CH_2-\underset{\underset{O}{\|}}{C}-N-CH_2-COOH$$
$$H$$

pK$_2$ 8.2 \qquad\qquad pK$_1$ 3.1

Color Tests of Proteins

Proteins respond to the following color tests: (1) biuret, a pink to purple color with an excess of alkali and a small amount of copper sulfate; (2) ninhydrin, a blue color when boiled with ninhydrin (triketohydrindene hydrate) that is intensified by the presence of pyridine; (3) Millon's test for tyrosine, a brick-red color or precipitate when boiled with mercuric nitrate in an excess of nitric acid; (4) Hopkins-Cole test

for tryptophan, a violet zone with a salt of glyoxylic acid and stratified over sulfuric acid; and (5) xanthoproteic test, a brilliant orange zone when a solution in concentrated nitric acid is stratified under ammonia.

Almost all so-called alkaloidal reagents will precipitate proteins in slightly acid solution.

The qualitative identification of the amino acids found in proteins, etc., has been simplified very greatly by the application of paper chromatographic technics to the proper hydrolysate of proteins and related substances. End-member degradation technics for the detection of the sequential arrangements of the amino acid residues in polypeptides (proteins, hormones, enzymes, etc.) have been developed to such a high degree with the aid of paper chromatography that very small samples of the polypeptides can be utilized. These technics, together with statistical methods, have led to the elucidation of the sequential arrangement of the amino acid residues in oxytocin, vasopressin, insulin, hypertensin, glucagon, corticotropins, etc.

Some of these hormones have been synthesized by utilizing improved synthetic methods.

Products

Gelatin U.S.P. is a protein obtained by the partial hydrolysis of collagen, an albuminoid found in bones, skins, tendons, cartilage, hoofs and other animal tissues. The products seem to be of great variety, and, from a technical standpoint, the raw material must be selected according to the purpose intended. The reason for this is that collagen usually is accompanied in nature by elastin and especially by mucoids, such as chondromucoid, which enter into the product in small amount. The raw materials for official gelatin, and also that used generally for a food, are skins of calf or swine and bones. First, the bones are treated with hydrochloric acid to remove the calcium compounds and then are digested with lime for a prolonged period, which converts most other impurities to a soluble form. The fairly pure collagen is extracted with hot water at a pH of about 5.5, and the aqueous solution of gelatin is concentrated, filtered and cooled to a stiff gel. Calf skins are treated in about the same way, but those from hogs are not given any lime treatment. The product derived from an acid-treated precursor is known as Type A and exhibits an isoelectric point between pH 7 and 9, while that where alkali is used is known as Type B and exhibits an isoelectric point between pH 4.7 and 5. The minimum of gel strength officially is that a 1 per cent solution, kept a 0° for 6 hours, must show no perceptible flow when the container is inverted.

Gelatin occurs in sheets, shreds, flakes or coarse powder. It is white or yellowish, has a slight but characteristic odor and taste and is stable in dry air but subject to microbial decomposition if moist or in solution. It is insoluble in cold water but swells and softens when immersed and gradually absorbs 5 to 10 times its own weight of water. It dissolves in hot water to form a colloidal solution; it also dissolves in acetic acid and in hot dilute glycerin. Gelatin commonly is bleached with sulfur dioxide, but that used medicinally must have not over 40 parts per million of sulfur dioxide. However, a proviso is made that for the manufacture of capsules or pills it may have certified colors added, may contain as much as 0.15 per cent of sulfur dioxide and may have a lower gel strength.

Gelatin is used in the preparation of capsules and the coating of tablets and, with glycerin, as a vehicle for suppositories. It also has been employed as a vehicle for other drugs when slow absorption is required. When dissolved in water, the solution becomes somewhat viscous, and such solutions are used to replace the loss in blood volume in cases of shock. This is accomplished more efficiently now with blood plasma, which is safer to use. In hemorrhagic conditions, it sometimes is administered intravenously to increase the clotting of blood or is applied locally for the treatment of wounds.

The most important value in therapy is as an easily digested and adjuvant food. It fails to provide any tryptophan at all and is lacking notably in adequate amounts of other essential acids; approximately 60 per cent of the total acids consist of glycine and the prolines. Nevertheless, when supplemented, it is very useful in various forms of malnutrition, gastric hyperacidity or ulcer, convalescence and general diets of the sick. It is specially recommended in the preparation of modified milk formulas for feeding infants.

Category—suspending agent.

Special Intravenous Gelatine Solution, is a 6 per cent sterile, pyrogen-free, nonantigenic solution in isotonic chloride, the gelatin being specially prepared from beef-bone collagen. It is odorless, clear, amber colored, slightly viscous above 29° but a gel at room temperature and has a pH of 6.9 to 7.4. It is employed as an infusion colloid to support blood volume in various types of shock and thus is a substitute for plasma and whole blood. It is contraindicated in kidney ailments and must be used with care in cardiac disease. Any typing of blood must be done before injection because gelatin interferes with proper grouping. The semisolid preparation is warmed to 50° before use, and about 500 ml. is injected at the rate of not over 30 ml. per minute. It gives adequate protection for 24 to 48 hours.

Absorbable Gelatin Film, Gelfilm®, is a sterile, nonantigenic, absorbable, water-insoluble gelatin film. The gelatin films are prepared from a solution of specially prepared gelatin-formaldehyde combination by spreading on plates and drying under controlled humidity and temperature. The film is available as light yellow, transparent, brittle sheets 0.076 to 0.228 mm. thick. Although insoluble in water, they become rubbery after being in water for a few minutes.

Absorbable gelatin film is used primarily in surgical closures and for repair of defects in such tissues as the dura mater and the pleura.

Absorbable Gelatin Sponge U.S.P., Gelfoam®, is a sterile, absorbable, water-insoluble, gelatin-base sponge that is a light, nearly white, nonelastic, tough, porous matrix. It is stable to dry heat at 150° for 4 hours. It absorbs 50 times its own weight of water or 45 of oxalated whole blood.

It is absorbed in 4 to 6 weeks when it is

used as a surgical sponge. When applied topically to control capillary bleeding, it should be moistened with sterile isotonic sodium chloride solution or thrombin solution.

Category—local hemostatic.

Beef Extract is the residue from beef broth obtained by extracting fresh, sound, lean beef by cooking with water and evaporating the broth at low temperature, usually in a vacuum, until a thick, pasty residue is obtained. It consists of a yellowish- to dark-brown, slightly acid, pasty mass, having an agreeable, meatlike odor and taste. It is soluble in 10 times its weight of water to yield a nearly clear solution, free from sediment. At least 90 per cent of the solids are soluble in dilute alcohol, and these alcohol-soluble solids must contain at least 6 per cent of nitrogen. The total solids, comprising about 75 per cent of the extract, must contain not more than 0.35 per cent of amide ammonia.

Beef extract, as well as similar products, such as Valentine's Meat Extract, contains natural salts, meat bases, including xanthine, and noncoagulable protein, chiefly gelatin and proteoses. In the small doses used, the amount of nutriment is insignificant, but the meaty flavor and odor are excellent stomachics.

Gastric Mucin is prepared by digesting the lining of hogs' stomachs with pepsin and precipitating from the supernatant liquid with 60 per cent alcohol. It is a white to yellow powder or brownish-yellow granules and forms a viscous, gray, opalescent solution with water. It probably contains proteoses and peptones derived from the glycoprotein mucin. It is employed in a dose of 2.5 g. every 2 hours in the treatment of gastric ulcer and hyperchlorhydria, where it acts as an emollient but does not effectively neutralize gastric acidity in man. In Mucotin® Almagucin® and Trimucolan®, the antacid effect is increased by adding aluminum hydroxide and magnesium trisilicate.

Milk products are marketed as dietary supplements, especially with added vitamins or as condensed or dried milk for infant feeding.

Casein, the chief protein of milk, is obtained readily in fairly pure form by precipitating the fat-free serum with acid. It is a white or yellowish, granular powder that is insoluble in water but soluble in dilute mineral acids or alkalies. In the dispensary, it has some value in emulsions and creams, but its chief importance in medicine is for making protein hydrolysates, because hydrolysis of the compound gives all of the essential amino acids in satisfactory amounts; such proteins are described as "adequate." Solutions of casein that are highly purified have been injected for nonspecific protein therapy. A calcium salt (Casec) is recommended as an accessory food in high protein diets and to be used for infants in summer diarrhea. The sodium salt also has been recommended as an adjunct food.

Nonspecific Proteins. The intravenous injection of foreign protein is followed by fever, muscle and joint pain, sweating and decrease and then increase in leukocytes; it even can result in serious collapse. The results have been used in the treatment of various infections, originally the chronic form. The method is presumed to be of value in acute and chronic arthritis, peptic ulcer, certain infections of the skin and eye, some vascular diseases, cerebrospinal syphilis, especially dementia paralytica, and in other diseases. Since a fever is necessary in this system, the original program has developed into the use of natural fevers, such as malaria, of external heat and similar devices. However, the highly purified proteins of milk still are recommended for some diseases; they are available commercially as Activin, Caside, Clarilac, Bu-Ma-Lac, Lactoprotein, Mangalac, Nat-i-lac, Neo-lacmanese and Proteolac. Mucosol is a purified beef peptone, and Omniadin is a similar purified bacterial protein. Synodal contains nonspecific protein with lipoids, animal fats and emetine hydrochloride and is designed for the treatment of peptic ulcer. One of the favorite agents of this class has been typhoid vaccine.

Venoms. Cobra (Naja) Venom Solution, from which the hemotoxic and proteolytic principles have been removed, has been credited with virtues due to toxins and has been injected intramuscularly as a non-narcotic analgesic in doses of 1 ml. daily. Snake Venom Solution of the water moccasin is employed subcutaneously in doses of 0.4 to 1.0 ml. as a hemostatic in recurrent epistaxis, thrombocytopenic purpura and as a prophylactic before tooth extraction and minor surgical procedures. Stypven® from the Russell viper is used topically as a hemostatic and as thromboplastic agent in Quick's modified clotting time test. Ven-Apis®, the purified and standardized venom from bees, is furnished in graduated strengths of 32, 50 and 100 bee-sting units. It is administered topically in acute and chronic arthritis, myositis and neuritis.

Nucleoproteins. The nucleoproteins previously mentioned are found in the nuclei of all cells and also in the cytoplasm. They can be deproteinized by a number of methods. Those compounds that occur in yeast usually are treated by grinding with a very dilute solution of potassium hydroxide, adding picric acid in excess and precipitating the nucleic acids with hydrochloric acid, leaving the protein in solution. The nucleic acids are purified by dissolving in dilute potassium hydroxide, filtering, acidifying with acetic acid and finally precipitating with a large excess of ethanol.

The nucleic acids prepared in some such manner differ in a few respects according to source, but they seem to be remarkably alike in chemical composition. They are slightly soluble in cold water, more readily soluble in hot water and easily soluble in dilute alkalies with the production of salts, from which they can be reprecipitated by acids. Neutral solutions of the sodium salts from thymonucleic acids set to a jelly on cooling, but those from yeast do not. All of them can be hydrolyzed to nucleotides, and these in turn to nucleosides and phosphoric acid. The nucleosides further hydrolyze to D-ribose or 2-deoxyribose and derivatives of pyrimidine: adenine and guanine, which are purines, and cytosine (6-amino-2-hydroxypyrimidine), uracil (2,6-dihydroxypyrimidine), thymine (5-methyluracil), and 5-methylcytosine. The nucleotides are known as adenylic acid, guanylic acid, cytidylic acid, uridylic acid, thymylic acid, 5-methylcytidylic acid and their corresponding deoxy-congeners.

Adenylic acid is found in muscle in the free state and in combination with addi-

Adenylic Acid

tional phosphoric acid as adenyl diphosphate (ADP) and as adenyl triphosphate (ATP). During muscular exertion, the last compound is hydrolyzed enzymatically to adenylic acid or the diphosphate to furnish phosphoric acid and energy during metabolism, and regeneration of the triphosphate takes place in the muscle by further enzymatic action.

Both the nucleotides and the deoxynucleotides are the residues of the polynucleotide RNA and DNA is a polydeoxynucleotide. DNA transmits genetic information in organisms that contain it. RNA carries instructions from the genes to the sites where it directs the assembly of proteins.

Phosphentaside, My-B-Den®, is a preparation of the sodium salt of adenosine-5-monophosphate, more commonly known as adenylic acid, a specific nucleotide. It is one of the constituents of yeast cozymase and apparently of immense importance in carbohydrate metabolism of the animal. It is presumed to be effective in treating the complications of varicose veins (pruritis, dermatitis and ulceration), pruritis of other kinds and possibly bursitis. Phosphentaside is furnished in tablets for oral administration, a solution for intramuscular injection and a solution for sustained action. The dose is 20 mg. 5 to 7 times a day for 4 to 7 days, then decreased.

BLOOD PRODUCTS

The blood, as the circulating medium of the body, has numerous functions. It distributes nutriments, transfers oxygen and carbon dioxide between the lungs and the tissues, collects and delivers waste materials, acts as a buffer to acids and alkalies, regulates isotonicity by adding or subtracting water and distributes many agents that are needed at different points, for example, immune bodies, other protective agents, hormones and enzymes of various types. Therefore, its composition will change markedly during altering conditions, although it is fairly constant under normal circumstances. About 45 per cent consists of corpuscles that can be separated by centrifuging, and of these only about 0.2 per cent are other than erythrocytes. The 55 per cent of removed plasma contains about 7 to 8.7 per cent of solids, of which a small portion (less than 1%) can be removed by clotting to produce the defibrinated plasma, which is called the serum.

The serum contains some inorganic salts and simple organic compounds, but the total solids are chiefly protein, mostly albumin and the rest nearly all globulin. The inorganic anions are mainly chloride, sulfate and phosphate, and the cations are magnesium, calcium, sodium, potassium, copper and iron. Some of the simple organic compounds are urea, uric acid, creatine, creatinine, dextrose, cholesterol, fatty acids, lecithin, bile pigments, carbonates, ascorbic acid and other vitamins, hormones, phenols and lactic acid. In the serum also are many of the substances that are associated with development of immunity, such as opsonins and antitoxins. In addition, the plasma contains the protein fibrinogen, which is converted by coagulation to insoluble fibrin, and the separated serum has an excess of the clotting agent thrombin.

The serum globulins can be obtained by salting out with magnesium sulfate or half saturation with ammonium sulfate. They can be subdivided by various solubility methods into euglobulin and pseudoglobulin, but more particularly by electrophoresis into alpha, beta and gamma globulins that contain most of the antibodies. The serum albumins, which are in about double the amount of globulins, can be isolated by completely saturating the filtrate with ammonium sulfate and probably can be separated into several individual albumins. It has been found that regeneration of all of these proteins in the serum is best accomplished by feeding whole blood or liver.

The clotting process is a very important one to prevent excessive bleeding in cases of operation or wounding. The time re-quired for this is normally about 5 minutes, and any prolongation beyond 10 minutes is considered to be abnormal and to require treatment. Thrombin, the agent or enzyme that is partly responsible for bringing about the clotting process, is contained in the blood in the inactive form, prothrombin, and calcium salts seem to be necessary to set it free in the active state. However, this action cannot occur in the blood because the prothrombin is present combined with an inhibiting agent that is called antiprothrombin. Antiprothrombin can be removed only by the action of a substance that usually is termed thromboplastin or thrombokinase. This compound apparently is contained in the thrombocytes, and also probably is present in many tissues that are not directly accessible to the blood. When blood is shed, the thrombocytes disintegrate to liberate thromboplastin, which releases prothrombin, and calcium salts finally activate the prothrombin to thrombin. The last then converts fibrinogin to fibrin to complete the process. Fibrinogen, which can be prepared in a relatively pure state, resembles globulins in that it is insoluble in water but dissolves in dilute salt solution and can be precipitated again by increasing the amount of salt. The change during the formation of fibrin is not understood and could be a denaturation. The antiprothrombin is assumed to be identical with heparin, a compound obtained from connective tissue and particularly from the walls of the liver. It has no effect on the activity of thrombin, but antithrombins are known to exist, such as hirudin of leeches. Little is known about the composition of thromboplastin, which can be obtained from disintegrating tissue cells of various types, but it is probable that several substances can assume the role of releasing prothrombin.

The human erythrocytes are specialized cells in the form of biconcave disks, although occasionally they may assume a cup-shape, and are 7.5 to 9.5 μ in diameter. The number per cu. mm. averages about 5,000,000, but this is quite variable with sex (men have more than women), with age, with individuals, with conditions of living and particularly in the presence of disease.

The erythrocytes contain 32 to 35 per cent

hemoglobin, about 60 per cent water and the rest as stroma. The last can be obtained, after hemolysis of the corpuscles by dilution, through the process of centrifuging and is found to consist of lecithin, cholesterol, inorganic salts and a protein, stromatin. Hemolysis of the corpuscles, or laking as it sometimes is called, may be brought about by hypotonic solution, by fat solvents, by bile salts which dissolve the lecithin, by soaps or alkalies, by saponins, by immune hemolysins and by hemolytic serums, such as those from snake venom and numerous bacterial products. Determination of the concentration of hemoglobin in the blood is an important diagnostic procedure. The normal amount is 14 to 16 per cent and usually is taken as 15.6 for purposes of calculating the percentage of normal in the color index, the volume index or the saturation index.

Hemoglobin, Hb, is a conjugated protein, the prosthetic group being heme (hematin) and the protein (globin) which is composed of four polypeptide chains, usually in identical pairs. The total molecular weight is about 66,000, including four heme molecules. The molecule has an axis of symmetry and therefore is composed of identical halves with an over-all ellipsoid shape of the dimensions $55 \times 55 \times 70$Å.

Normal human beings synthesize and incorporate into Hb four distinct but related polypeptide chains designated as α, β, γ and δ respectively. Hb molecules contain two α chains together with two β, γ or δ chains. In hemoglobin HbA[*] the chains[5] are called α and β. The gross structure may be represented[6] as

$$HbA = He\ \alpha_2{}^A\beta_2{}^A$$

where He is heme; the superscript denotes the hemoglobin that is the source of the chain and the subscript has the usual chemical designation. Correspondingly, normal human fetal hemoglobin is

$$HbF = He\ \alpha_2{}^F\gamma_2{}^F.$$

The β^A chain[7] contains 146 amino acid residues, with the C terminus being histidine and the N terminus being valine, the amino acid sequence determined with a fair degree of certainty. The chain contains 8

[*] A = Adult human hemoglobin. Hemoglobin F = fetal hemoglobin; Hemoglobin S = sickle-cell hemoglobin; etc. (*See* Blood 8:386, 1953)

sections of the α- helix structure that differ in their number of amino acid residues. The conformation and the structures of the non-helical areas are less well known. The $\alpha_2{}^A$ chain contains 141 amino acid residues whose sequence is quite well established, with the C terminus being valine and the N terminus being arginine.

In the complete structure, the two pairs of polypeptide chains assume a tetrahedral configuration in which the irregular parts mesh together very well, with little free space in the interior. There is relatively little contact between like chains but more between unlike ones, with subsequently greater bond strength. There are no disulfide bridges in globin which has an isoelectric point of about 7.5.

There are other hemoglobins, and the subject is too comprehensive[7] to be discussed here.

Heme is linked to the α^A and β^A chains through two histidyl residues, one close to and the other more distant from the iron atom of heme. The proximal histidyl residue is at position 87 in the α-chain and 92 in the β-chain. The distal histidyl residue is 58 in the α-chain and 63 in the β-chain. The latter residues appear to be involved with the heme group in carrying out oxygen transport.[7] The environment of the heme groups and the linkage of heme to globin is becoming better understood. X-ray data suggest that the heme groups, although in crevices, are near the surface of the molecule and spatially distant from each other. Heme-heme interaction thus must be transmitted through the globin.

Iron in the heme of hemoglobin (ferrohemoglobin) is in the ferrous state and can combine reversibly with oxygen to function as a transporter of oxygen.

$$\text{Hemoglobin} + \text{oxygen}\ (O_2) \rightleftharpoons \text{oxyhemoglobin}$$

In this process, the formation of a stable oxygen complex, the iron remains in the ferrous form because the heme moiety lies within a cover of hydrophobic groups of the globin. Both Hb and O_2 are magnetic, whereas HbO_2 is dimagnetic because the unpaired electrons in both molecules have become paired. When oxidized to the ferric state (hemiglobin, methemoglobin or ferrihemoglobin) this function is lost. Carbon monoxide will combine with hemoglobin to form carboxyhemoglobin (carbonmonoxyhemoglobin) to inactivate it.

The other corpuscles of the blood consist of leukocytes and thrombocytes (platelets). The latter are important for carrying the thromboplastin and to the pathologist for diagnostic purposes. The leukocytes or white corpuscles are classified by the cytologist into numerous subdivisions, but the most important types are granulocytes or true leukocytes (65%) and the lymphocytes (30%). They are nucleated and have ameboid motion, and the general composition is quite varied, including albumins, globulins, nucleoproteins, glycogen, phospholipids and cholesterol. They are mobilized during infection to counteract the bacteria, which they engulf and destroy (phagocytosis).

The defense armamentarium of the body also includes the formation of antibodies that act against antigens. All of these antibodies are apparently proteins that usually are concentrated in the globulin fraction of the serum, and it has been suggested that they are globulins modified in structure in some way as to fit only the special needs. They include agglutinins, which cause clumping of the antigen cells or molecules, hemolysins and other cytolysins, which bring about destruction of cells, preciptins, which form precipitates with antigens, and antitoxins, which specifically counteract toxic products from organisms. These terms are used rather loosely and probably are more or less overlapping. The fixation of antigen-complement-antibody is an irreversible process. It is used in the Wassermann test to detect syphilitic organisms.

The practical aspects of the facts about antibody formation are utilized in the production of serums and vaccines for counteracting infections.

According to agglutination tests, human blood may be classified into four groups which are mutually incompatible, although the first of these sometimes has been called the universal donor type. Unfortunately, they have been given different names and one must know which system is being employed. The international system, probably most often used, is here noted, with those of Jansky and of Moss, respectively, in parentheses. The types and approximate percentage of incidence are: O(I,IV) 43, A(II,II) 42, B(III,III) 10, AB(IV,I) 5.

PRODUCTS

Blood-Group Specific Substances A and B N.F., is a sterile, isotonic solution of the polysaccharide-amino acid complexes capable of reducing the titer of the anti-A and anti-B isoagglutinins of group-O donor blood. The anti-A substance is from a precipitate of a tryptic digest of hog gastrin mucin, and anti-B is a similar precipitate of the glandular portion of equine gastric mucosa. They are added to group-O blood just prior to administration or at the time of collection and storage, in order that the treated blood may be used in recipients of other groups. One transfusion unit (10 ml.) can reduce the titer of 500 ml. of blood to one fourth of its original titer.

Discovery of the Rh factor in the blood was extremely significant. This is an agglutinogen which was given its name because the original work was done on rhesus monkeys. About 86 per cent of human beings have this factor in the erythrocytes and are said to be Rh-positive, although more recent work has shown that there are at least 5 or 6 subtypes. The Rh-negative person will develop reactions if he receives a transfusion of Rh-positive blood; also, those persons who are positive may show a sensitivity if one of the subtype factors is missing. Besides the necessity of determining if the patient is Rh-negative before transfusion, this factor is important in marriage, because union of a positive father and a negative mother is apt to result in miscarriage or stillbirth or in fatal erythroblastosis fetalis in the child.

Category—blood neutralizer (isoagglutinins, group O blood).

Usual dose-intravenous, one transfusion unit (10 ml.) in approximately 500 ml. of group O blood.

Whole Blood U.S.P. Whole blood (Human) is blood that has been drawn from a selected donor under rigid aseptic conditions. It contains citrate ion or heparin as an anticoagulant. It should be stored at 1° to 6° held constant within a 2° range, except during shipment, when the temperature may be between 1° and 10°. If it is over 21 days old it should not be used.

Only those persons may serve as a source of Citrated Whole Human Blood who are in physical condition to give blood and are free of those diseases transmissible by transfusion of blood, as far as can be determined from the donor's personal history and from such physical examination and clinical tests as appear necessary for each donor on the day upon which the blood is drawn. It is used extensively for a variety of purposes but mostly to give replacements in case of disease or loss.

Allergic reactions that sometimes develop include edema, bronchial asthma and skin rashes. Incompatibilities in unmatched blood will vary with the types, but the symptoms usually include chills, nausea, vomiting, fever and pain in the region of the kidneys.

Category—blood replenisher.

Usual dose—intravenous, 1 unit (500 ml.), repeated as necessary.

Usual dose range—as needed to replenish blood volume.

Hemoglobin is available commercially as brownish red scales that are soluble in water. It has been used as a treatment for various anemias in doses of 200 to 500 mg., but the amount of iron is very small and probably no more assimilable than inorganic forms, and intravenous injection is likely to produce reactions. The dried and powdered erythrocytes have been recommended for local application to wounds to stimulate healing. A preparation of globin from the erythrocytes also has been employed to promote diuresis in kidney diseases.

Normal Human Plasma is the sterile plasma obtained by pooling approximately equal amounts of the liquid portion of citrated whole blood from 8 or more adult humans who are free from any disease which is transmissible by blood transfusion at the time of drawing the blood. Each bleeding is drawn under aseptic conditions into individual sterile centrifuge bottles already containing a suitable anticoagulant,

usually 50 ml. of a sterile 4 per cent solution of sodium citrate in isotonic sodium chloride solution for each 500 ml. of whole centrifugation, and transferred to a pool by means of a closed system. It must comply with the requirements of the National Institutes of Health, and it may be stored as such, frozen or dried. The liquid plasma, which contains 5 per cent of dextrose to stabilize it, is a faintly yellow or amber liquid with a slight opalescence and no odor, in the absence of an odorous preservative. Frozen plasma, made by promptly freezing normal human plasma must be kept in this condition at not above −18° until liquefied for use by placing in a water bath at 37°. Dried plasma, made by freezing the whole plasma without dextrose and drying in a vacuum until it contains not more than 1 per cent of moisture (lyophilization), has a light-yellow to deep-cream color, is microscopically of a honeycomb-like structure and shows no evidence of fusion. It can be restored to the original volume by dissolving in 0.1 per cent of citric acid in water for injection. The outside label for any of these products must bear the name; the volume of the original plasma; the manufacturer's lot number, name, address and license number and the expiration date. The liquid and the dried products are preserved at 15° to 30°.

Plasma is used under the same conditions as whole blood but is preferred in most cases, except where the corpuscles are needed, such as in anemia. It has the advantages that it can be preserved more easily, it can be administered in more concentrated form, it is less likely to give reactions and, in the desiccated form, it can be transported more readily.

Category—blood-volume supporter.

Usual dose—intravenous, 500 ml., repeated as necessary.

Usual dose range—500 to 1,500 ml.

Normal Human Serum Albumin U.S.P., Normal Serum Albumin (Human), is a sterile solution of the serum albumin component of blood from healthy human donors and is prepared by a fractionation process.

The nearly odorless, moderately viscous, clear, brown solution contains 5 or 25 g. of serum albumin in each 100 ml., osmotically equivalent to 100 or 500 ml., respectively, of normal human serum plasma and should be stored between 2° and 10°. N.L.T. 96 per cent of its total protein is albumin. It contains no added preservative but may contain stabilizing agents. Expiration date is 3 to 10 years. It is used to replace blood in the treatment of shock, to reduce edema and to raise the serum protein level in hypoproteinemia.

Category—blood-volume supporter.

Usual dose—intravenous, a volume equivalent to 25 g. of albumin.

Usual dose range—volumes equivalent to 25 to 125 g. of albumin.

Plasma Protein Fraction U.S.P., Plasma Protein Fraction (Human) is a sterile solution of selected proteins derived from the blood plasma of adult human donors. It contains 4.5 to 5.5 g. of protein per 100 ml., of which about 83 to 90 per cent is albumin, and the remainder is alpha and beta globulins. It may contain suitable stabilizing agents. Temperatures between 2° and 8° are preferable for storage with an expiration date of 5 years; 3 years when stored between 8° and 30°.

It occurs as a transparent, nearly colorless or slightly brownish liquid and may develop a slight, granular or flaky deposit during storage.

Category—blood-volume supporter.

Usual dose—intravenous infusion, 500 ml. at a rate not exceeding 8 ml. per minute.

Usual dose range—250 to 1,500 ml.

Packed Human Blood Cells U.S.P., Packed Red Blood Cells (Human), is whole blood from which plasma has been removed. It may be prepared from whole blood at any time prior to its expiration date; however, when a centrifuge is used to effect its preparation, whole blood over 6 days old should not be used. It should be stored at a temperature between 1° and 6°, held constant within a 2° range. Its expiration date is not later than that of the whole blood from which it is derived and in any case, it should be used not later than 24 hours from the time the hermetic seal is broken. It is identified with respect to the donor's blood group and Rh type.

Category—blood replenisher.

Usual dose—intravenous, the equivalent of 1 unit (500 ml.) of whole blood repeated as necessary.

Iodinated I 131 Serum Albumin U.S.P., Radio-iodinated (^{131}I) Serum Albumin (Human), is a sterile, clear, colorless, buffered, isotonic solution containing in each ml. at least 10 mg. of radio-iodinated normal human serum albumin adjusted to provide n.m.t. 1 millicurie of radioactivity per ml. and should contain n.l.t. 95 per cent and n.m.t. 105 per cent of the labeled amount of ^{131}I. Other chemical forms of radioactivity should not exceed 3 per cent of the total radioactivity. Its pH. is 7.0 to 8.5.

The half-life of ^{131}I is 8.08 days, and this rate of decay should be considered in calculating the intravenous dose. In blood volume determinations this is 3 to 60 microcuries; in cardiac output determinations it is 5 to 60 microcuries.

Normal human serum albumin is iodinated with ^{131}I so as to introduce not more than 1 gram-atom of iodine for each gram-molecule (60,000 g.) of albumin.

This preparation should be properly shielded, stored at 2° to 10° and not used 4 weeks after date of standardization.

Category—diagnostic aid (blood volume and cardiac output determination).

Usual dose range—intravenous, 5 to 60 microcuries.

Iodinated I 125 Serum Albumin U.S.P. conforms in all its specifications, use and dose to that of Iodinated I 131 Serum Albumin U.S.P., with the exception of the half life of I 125, which is 60 days.

Category—diagnostic aid (blood volume and cardiac output determination).

Usual dose—intravenous, 5 to 60 microcuries.

Immune Serum Globulin U.S.P., Immune Serum Globulin (Human), is a sterile solution of globulins which contains those antibodies normally present in adult human blood. Each lot is prepared from an original plasma or serum pool represented by the venous or placental blood of at least 1,000 humans. The protein content, 15 to 18 g. per 100 ml., contains n.l.t. 90 per cent of globulin.

It is a transparent or slightly opalescent liquid, colorless or with a brownish color, due to denatured hemoglobin; it should be stored at 2° to 8° and is stable for 3 years.

Category—passive immunizing agent.

Usual dose—prophylactic intramuscular, 0.02 ml. per kg.

Measles Immune Globulin U.S.P. is prepared from immune serum globulin that complies, after dilution if necessary, with the measles antibody requirements of the U.S. Public Health Service.

Category—passive immunizing agent.

Usual dose—intramuscular—prophylactic, 0.25 ml. per kg. of body weight; modification, 0.02 to 0.05 ml. per kg.

Pertussis Immune Globulin U.S.P. is a sterile solution of globulins derived from the blood plasma of adult human donors who have been immunized with pertussis vaccine.

Category—passive immunizing agent.

Usual dose—intramuscular, prophylactic, 1.25 to 2.5 ml. at 1-week intervals for 1 or 2 weeks; therapeutic, 1.25 to 2.5 ml. at 1-day intervals for 1 or 2 days.

Tetanus Immune Globulin U.S.P. is a sterile solution of globulins derived from the blood plasma of adult human donors who have been immunized with tetanus toxoid.

Category—passive immunizing agent.

Usual dose—intramuscular, prophylactic, 250 units as a single dose; therapeutic, 3,000 to 10,000 units as a single dose.

The above products all have properties, storage requirements, and expiration dates the same or similar to those for Immune Serum Globulin U.S.P.

Antihemophilic Human Plasma U.S.P., Antihemophilic Plasma (Human), is normal human plasma (both the dried and the frozen forms) that have been processed not more than 6 hours after withdrawal of blood from the donor to prevent the denaturation of the antihemophilic globulin component. The frozen form is not pooled, is stored at −18° or lower and is stable for 1 year. The dried form is pooled, is stored at 2° to 10° and is stable for 5 years.

Category—antihemophilic.

Usual dose—intravenous infusion, 250 ml. daily.

Usual dose range—250 to 500 ml. daily.

PLASMA SUBSTITUTES

For supporting the circulation volume after hemorrhage and shock, whole blood is the best agent. However, its use is accompanied by serious disadvantages, chief of which are deficiency in available amounts, deterioration of supplies, the necessity for typing and matching and the danger of homologous serum jaundice. The use of plasma obviates the first disadvantage but still may carry the virus of hepatitis. Human serum albumin has some advantages but is very expensive, and the supply is limited. Animal albumin, which can be produced inexpensively, is not used because it cannot be rendered nonantigenic easily and always presents the possibility of transmitting tuberculosis and undulant fever. Human globulin from red cells in the processing of plasma is stable, and its successful modification for safe use is being studied now. In times of emergency, whole blood should be reserved for the most severe injuries, and the less serious cases should be treated with blood derivatives or with the so-called plasma substitutes.

A large number of substitutes for blood derivatives have been proposed and used. Before World War I, reliance was placed on crystalloids, such as sodium chloride and dextrose; they are very effective in emergencies, but the effects are lost too rapidly to be of value generally. During the war and after, acacia was used in 6 per cent solution, but apparently it is stored in various organs to interfere with normal functions and has the potentiality of producing allergic reactions. Other gums, such as algin and laminarin, also have been proposed. Pectin from citrus fruits, human ascitic fluid, methylcellulose, glutamic acid polypeptide, sodium glycerol polysuccinate, casein, isinglass, polyvinyl alcohol and others have been given consideration, and some of them still are being studied.

The efforts in recent times have been to develop colloidal materials with about the same molecular weight as serum albumin (70,000). Any such substance must have the desirable characteristics of human plasma, be completely metabolized without producing toxic or allergic reactions in the recipient, be sufficiently stable to permit storage and distribution and be available and economic. None has received sufficient clinical trial yet to warrant endorsement, but a few are undergoing tests.

PRODUCTS

Polyvinylpyrrolidone, P.V.P., Plasdone®, Kollidon®, Periston®, is a synthetic polymer derived from acetylene which was introduced in Germany during World War II. The product is nonantigenic and can be made to have a molecular weight of 40,000 to 50,000, and a 3 per cent solution has about the same colloidal osmotic pressure as plasma. It has been found that a 3.5 per

$$\left[\begin{array}{c} \\ N \\ | \\ -CH-CH_2- \end{array} \right]_n$$

Polyvinylpyrrolidone

n = 350 to 450

cent solution in saline is most effective. The preparation Plasmosan® has added salts to simulate plasma.

Dextran, Expandex®, Gentran®, Plavolex®, is a biosynthetic polysaccharide that has been known for years as an objectionable slime in the sugar house. It can be obtained by the fermentation of sucrose by *Leuconostoc mesenteroides* and similar organisms, and its high molecular weight can be reduced by partial acid hydrolysis to an average of 75,000. In the powdered form it is stable at room temperature. It is a white to light-yellow, tasteless, odorless, amorphous, freely water-soluble solid. It is presumed to be a mixture of polymers of α-D-glucopyranose, a small percentage of which consists of branched chains that are undesirable. It is administered as a 6 per cent solution in normal saline. Toxic reactions have been observed, but these seem to be trivial or are absent when it is given under anesthesia.

COAGULANTS

The use of coagulants or hemostatics to aid in the normal process of coagulation is of greatest importance. In this process, as previously described, there are four necessary factors: calcium salts, thrombin, thromboplastin and fibrinogen. If one of the first three is deficient or lacking in the body, administration or topical application of this substance might be expected to speed up clotting. In addition, other compounds have been introduced that seem to promote coagulation or that favor it mechanically.

PRODUCTS

Thrombin N.F. is a sterile protein substance prepared from prothrombin of bovine origin through interaction with added thromboplastin in the presence of calcium. It is a white or grayish, amorphous substance dried from the frozen state, best preserved at 2° and 8°, preferably at the lower limit and is stable for 3 years. The coagulating power is expressed in units, each of which is the amount necessary to coagulate 1 ml. of standard fibrinogen in 15 seconds.

Category—local hemostatic.

Application—topically as a powder or as a solution containing 1,000 units per ml.

Fibrinogen U.S.P., Fibrinogen (Human), is a sterile fraction of normal human plasma, dried from the frozen state, which in solution has the property of being converted into insoluble fibrin when thrombin is added. It is a white or grayish amorphous, water-soluble protein that is best stored at 2° to 8° and is stable for 5 years.

Fibrinogen is administered for certain extensive surgical procedures to increase plasma fibrinogen levels.

Category—coagulant (clotting factor).

Usual dose—intravenous, 2 g.

Usual dose range—2 to 8 g.

Carbazochrome Salicylate, Adrenosem® Salicylate, is adrenochrome monosemicarbazone sodium salicylate complex. It occurs as a fine, orange-red odorless powder with a sweetish saline taste and is soluble in alcohol and in water. Solutions have a pH of 6.7 to 7.3

Carbazochrome salicylate is used to prevent bleeding or oozing from a vascular bed of small blood vessels. It is proposed to control bleeding in a number of surgical procedures and is used orally or intramuscularly, but with unproved effectiveness.

Cotarnine Chloride, Stypticin®, is a pale-yellow, hygroscopic powder that is freely soluble in water. It was formerly used in various forms of hemorrhage.

ENZYMES AND PROTEINLIKE HORMONES

The term "enzyme" might very well be reserved for those naturally occurring substances that are protein or proteinlike and function both in vivo. and in vitro as catalytic agents. They are responsible for most of the chemical changes that occur in cells. These chemical changes may or may not produce physical changes in the cell itself. Some enzymes are functional as true simple proteins; others require in addition a co-factor.

Many enzymes have been obtained in a crystalline state; however, this may or may not be a measure of homogeneity, activity or purity. Some of the noncrystalline enzyme preparations no doubt have been obtained in a high state of purity. Because enzymes and conjugated enzymes are proteins, their structures, physical and chemical properties, shapes, etc., are analogous to those of the proteins, and the student is referred to these discussions under proteins. From these discussions it follows that the activity of enzymes is dependent upon temperature, pH, solvent, inorganic salts, concentration of enzyme, concentration of the substrate, etc. Conditions that effect the denaturation of proteins usually have an adverse effect upon the activity of the enzyme. Because proteins can be reversibly denatured, the same might be expected of enzymes; however, the kind, the number and the sequential arrangement of the amino acid residues in the enzyme (polypeptide) molecule may alter the rate or the extent.

Enzymes vary in molecular weights from 12,700 for ribonuclease to about 1,000,000 for L-glutamate dehydrogenase. Examples of the molecular weights of other enzymes are lysozyme (eggwhite) 17,000; papain 20,700; trypsin 23,800; carboxypeptidase 34,400; pepsin 34,500; rennin 40,000, α-amylase (pancreas) 45,000; alcohol dehydrogenase (liver) 73,000; lactate dehydrogenase (yeast) 100,000; luciferase 100,000; aldolase (liver) 160,000; catalase (ox liver) 248,000; urease (soybean) 480,000.

According to Dixon and Webb,[8] 659 enzymes have been reported, and others may yet be discovered. No cell or tissue contains all known enzymes. These authors have tabulated 659 enzymes.

RELATION OF STRUCTURE TO FUNCTION

Evidence is accumulating to support the "active center hypothesis" of biologically active proteins and the hypothesis that many portions of these proteins are expendable.

When only short-range forces are considered, in some cases at least, only a small portion of the enzyme molecule is involved in catalysis, e.g., urease molecular weight 600,000 versus the substrate urea molecular weight 60.

One possibility is that the configurational specificity of the active center may consist of a series of contiguous residues along the peptide chain; the other, that a number of widely separated amino acid residues are held in juxtaposition by the secondary structure of the molecule. In both cases these centers are stabilized by the molecule as a whole. Evidence of the former is supported by the fact that ribonuclease in the presence of 8M urea (which completely denatures, i.e., destroys, intact hydrogen-bonded secondary structures) is as active as the enzyme in the native state.

The activation of many enzymes by denaturation points toward the importance of secondary and tertiary structures in relation to function. This appears to be true even in the case of urease. Mild peptic hydrolysis of ribonuclease only cleaved a tetrapeptide from the C-terminal portion of the enzyme. The progressive cleavage of this tetrapeptide was correlated with a progressive loss of enzyme activity. However, ultraviolet studies indicated a rupture of a hydrogen bond involving tyrosine. Therefore, no clear-cut evidence was available as to whether one or both processes were correlated with activity.

In some cases surprisingly extensive modification fails to alter enzymatic activity. For example, when papain is treated with swine kidney leucine aminopeptidase—one third of the papain molecule (60 amino acid residues)—still fully active can be recovered. Pepsin can cleave 11 amino acid residues from the C-terminal end of ACTH, with retention of activity. Even 24 amino acid residues can be cleaved from ACTH, with retention of activity. In hog β-corticotropin (ACTH) apparently the last 15 or the 39 amino acid residues are not necessary for activity. However, activity is destroyed if the N-terminal residue is altered.

Evidence is accumulating that some structural features in many enzymes and hormones appear to be absolutely expendable.

The above discoveries were made in experiments carried out in vitro; thus, even if only a small part of the molecule is directly responsible for its catalytic role (enzymatic activity), the rest of the molecule may have in vivo other useful or even essential attributes and functions: i.e., (1) orientation of the active protein to structural features in the cell and to other enzymes in the cell, (2) determination of the thermodynamic activity of the enzyme in the cell under different physiologic conditions, (3) its net charge, as influenced by interaction with small molecules at sites other than the active center, and (4) simply the ability to stay inside a semipermeable membrane.

SPECIFICITY OF ENZYMES

One of the outstanding properties of enzymes is their specificity for certain substrates. Of course, this could be expected if there is to be any order to the biologic processes that take place in living cells; however, even though this is true, it is remarkable when one considers the large number of different enzymes that are present in a single cell. The physical, chemical, structural, conformational, antipodal and optical properties of the substrate are factors that effect the specificity of enzymes and conjugated enzymes. In the former only the "active center" of the polypeptide

(enzyme) is involved, whereas in the latter the coenzyme also is involved. In some cases the specificity is very narrow, whereas in others this is not the case.

In some cases the "active center" on the surface of the enzyme (protein) is complementary to the substrate molecule in a strained configuration corresponding to the "activated" complex for the reaction catalyzed by the enzyme. The substrate molecule is attracted to the enzyme and is caused by the forces of attraction to assume the strained state, with conformational changes,[*] which favors the chemical reaction; that is, the activation energy requirement of the reaction is decreased by the enzyme to such an extent as to cause the reaction to proceed at an appreciably greater rate than it would in the absence of the enzyme. If in all cases the enzyme were completely complementary in structure to the substrate, then no other molecule would be expected to compete successfully with the substrate in combination with the enzyme, which in this respect would be similar in behavior to antibodies. However, in some cases an enzyme complementary to a strained substrate molecule might attract more strongly to itself a molecule resembling the strained substrate molecule than it would the strained substrate molecule itself; e.g., the hydrolysis of benzoyl-L-tyrosylglycine amide was practically inhibited by an equal amount of benzoyl-D-tyrosylglycine amide. This illustration also might serve to illustrate a type of antimetabolite activity.

While there is evidence that the active site may be mobile,[†] it is obviously not infinitely so. If it were, structural and stereochemical specificity would not be observed.

Recent attempts to correlate[‡] the function of a protein molecule with its structure have been based on introducing a molecule "reporter group," sensitive to

[*] This resembles the "Rack" theory,[9] or the "induced fit" theory.[10]

[†] The conformation of the active state of α-chymotrypsin may vary with hydrogen ion concentration (Mukatis, W., and Niemann, C.: Proc. Nat. Acad. Sci. 51:397, 1964).

[‡] See Koshland, D. E., et al.: Proc. Nat. Acad. Sci. U.S. 52:1017, 1964. C & EN Nov. 2 p. 42, 1964.

changes in its environment, into a specific position on a protein molecule. The reporter group can then transmit a signal to an appropriate detector.

Many studies dealing with specificity of enzymes have been carried out with the proteolytic enzymes because substrates (di-, tri-, etc., peptides) were prepared readily. Antipodal specificity in this area can be illustrated as follows: L-leucylglycylglycine is cleaved by amino peptidase, whereas D-leucylglycylglycine is not. 4-Alanylglycylglycine is slowly cleaved by this enzyme to show that at the "active center" of aminopeptidase it is a matter of closeness of approach that is one of the factors concerned with hydrolysis.

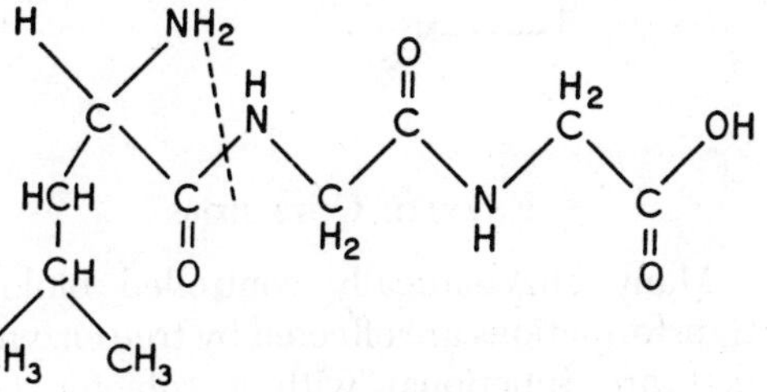

Normal cleavage
L-Leucylglycylglycine

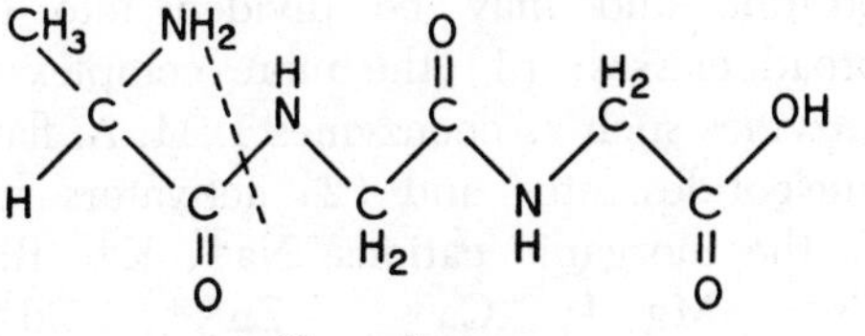

Slow cleavage
D-Alanylglycylglycine

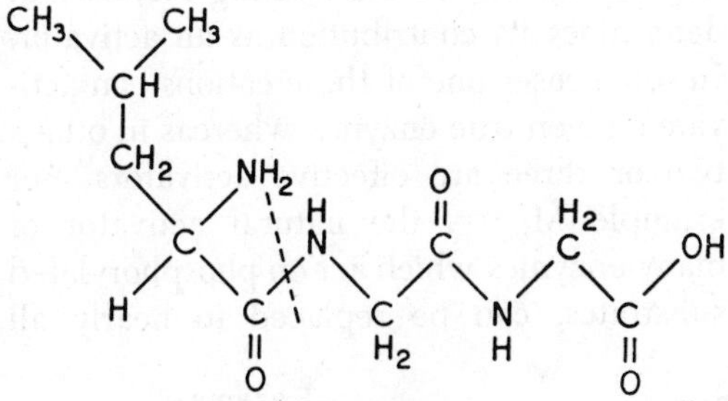

No cleavage
D-Leucylglycylglycine

As an example where apparently larger "active centers" are involved in enzymatic activity one can site the slow hydrolysis of

L-leucyl-L-tyrosine versus a 50- to 200-fold increase in the hydrolysis of acyl-L-leucyl-L-tyrosine by carboxypeptidase.

The polar groups of the amino acid residues of the polypeptide chain of the enzyme are in a number of cases an important part of the "active centers" because, when the polar groups are carefully blocked without further alteration of the native state, enzymatic activity ceases.

Although the above type substrates yield much useful information, they suffer one defect, i.e., they are conformationally indeterminate. Thus the discovery[11] of a pair of conformationally constrained model substrates of α-chymotrypsin in D- and L-3-carbomethoxydihydroisocarbostyril was a significant contribution toward the solution of the nature of the "active site" and/or its immediate environs.

The −SH group probably is found in more enzymes as a functional group than are the other polar groups. It should be noted that in some cases, e.g., urease, the less readily available SH groups are necessary for biologic activity and cannot be detected by the nitroprusside test that is used to detect the freely reactive SH groups.

A free −OH group of the tyrosyl residue is necessary for the activity of pepsin. Both the −OH of serine and the imidazole portion of histidine appear to be necessary parts of the active center of certain hydrolytic enzymes such as trypsin and chymotrypsin and furnish the electrostatic forces involved in the proposed mechanism[12] (shown on page 902), in which E denotes enzyme, the other symbols being self-evident.

These two groups, i.e., −OH and =NH, could be located on separate peptide chains in the enzyme so long as the specific 3-dimensional structure formed during activation of the zymogen brought them near enough to form a hydrogen bond. The polarization of the resulting structure would cause the serine oxygen to be the nucleophilic agent which attacks the carbonyl function of the substrate. The complex is stabilized by the simultaneous "exchange" of the hydrogen bond from the serine oxygen to the carbonyl oxygen of the substrate.

The intermediate acylated enzyme is written with the proton on the imidazole

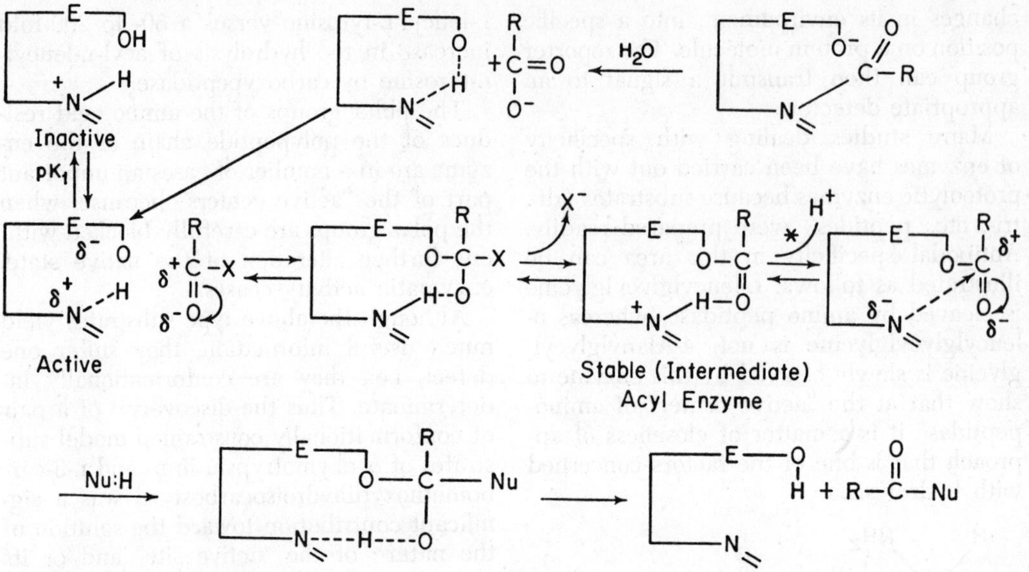

Inactive
pK_1

Active

Stable (Intermediate)
Acyl Enzyme

Nu:H

nitrogen. The deacylation reaction involves the loss of this positive charge simultaneously with the attack of the nucleophilic reagent (abbreviated Nu:H).

A possible alternative route to deacylation would involve the nucleophilic attack of the imidazole nitrogen on the newly formed ester linkage of the postulated acyl intermediate, leading to the formation of the acyl imidazole. The latter is unstable in water, hydrolyzing rapidly to give the product and regenerated active enzyme.

The reaction of alkyl phosphate in such a scheme may be written in an entirely analogous fashion, except that the resulting phosphorylated enzyme would be less susceptible to deacylation through nucleophilic attack.

The following diagrammatic scheme[13] has been proposed to explain the function of the active thiol ester site of papain. This ester site is formed and maintained by the folding energy the enzyme (protein) molecule.

ENZYME COFACTORS

Many enzymatically controlled biologic transformations are effected by true enzymes that are functional with a cofactor (coenzyme, prosthetic group, etc.). These cofactors are exceedingly diversified in composition and may be divided into two broad classes: (1) the more complex coenzymes such as coenzymes 1, 11, A, flavin nucleotides, etc., and (2) activators such as the inorganic cations, Na^+, K^+, Rb^+, Cs^+, Mg^{++}, Ca^{++}, Zn^{++}, Cd^{++}, Cr^{+++}, Cu^{++}, Mn^{++}, Fe^{++}, Co^{++}, Ni^{++}, Al^{+++}, and NH_4^+. The size of the ion may be one of the limiting factors that determines its contribution as an activator. In some cases one of these cations can activate a given true enzyme, whereas in others two or three are effective activators. For example Mg^{++}, the natural activator of many enzymes which act on phosphorylated substrates, can be replaced in nearly all

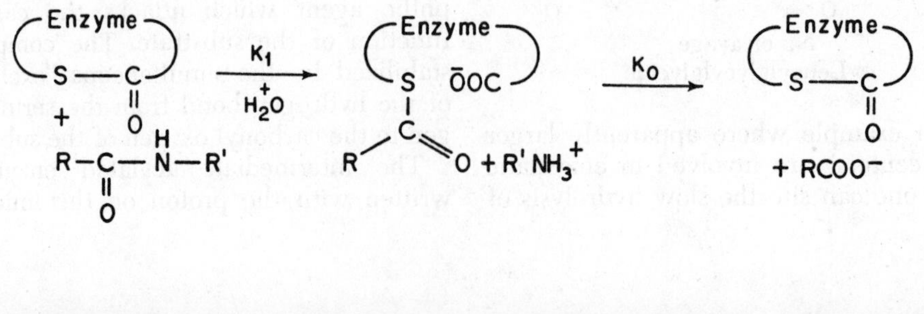

Enzyme
S—C
+
O
R—C—N—R'
O

$\xrightarrow{K_1}$ H_2O

Enzyme
S $\bar{O}OC$
C
R O + $R'NH_3^+$

$\xrightarrow{K_0}$

Enzyme
S—C
O
+ $RCOO^-$

cases by Mn^{++} but usually not by any other metal.

Some of these cations can antagonize each other, e.g., Na^+ versus K^+, Mg^{++} versus Ca^{++}, and Mn^{++} versus Zn^{++}.

In some cases a cation may enhance the activity of a coenzyme by a secondary catalytic action, as in the case of co-decarboxylase.

One or more of the same cofactors may be associated for each molecule of true enzyme as the case may be.

ZYMOGENS AND KINASES

Nature has provided an ingenious way to synthesize a reserve supply of inactive precursors of enzymes, hormones, etc. (e.g., trypsinogen; pepsinogen). These substances are released and activated as warranted by the need or demand. Activation may be effected by simple conditions such as certain pH states, inorganic ions, bile salts, glutathione, etc. These activators, which do not cause peptide cleavage, probably effect a subtle reorganization of the tertiary structure of the polypeptide molecule so that those structures that are more intimately involved in the active sites become functional. In other cases, activation that leads to similar final results may be effected by the hydrolysis of one or more peptide linkages by a proteolytic enzyme or autocatalytically (q.v.). The number of peptide linkages cleaved and the number of amino acid residues removed are specific for the zymogen in question.

The activators often are called kinases[14] and if some control is to be exercised in vivo one would expect some degree of specificity by a given activator to be of paramount importance.

SECRETION OF ENZYMES

Exportable proteins (enzymes) such as amylase, ribonuclease, chymotrypsin(ogen), trypsin (ogen), insulin, etc. are synthesized on the ribosomes.[15] They pass across the membrane of the endoplasmic reticulum into the cisternae and directly into smooth vesicular structure* which effect further

* Condensing vacuoles.

transportation. They are finally stored in highly concentrated form within membrane bound granules. In the case of the digestive glands these granules are called zymogen granules whose exportable protein content may reach a value of 40 per cent of the total protein of the gland cell. In the above sequences the newly synthesized exportable protein (enzymes) is not free in the cell sap. The stored exportable proteins are released into the extracellular milieu in the case of the digestive enzymes and into adjacent blood capillaries in the case of hormones. The release of these proteins is initiated (triggered) by specific inducers: for example, cholinergic agents (but not epinephrine) and Ca^{++} effect a discharge of amylase, lipase, etc. into the medium; increase in glucose level stimulates the secretion of insulin, etc. This release of the reserve enzymes and hormones is completely independent of the synthetic process as long as the stores in the granules are not completely depleted. Energy—oxidative phosphorylation—does not play an important role in these releases. Electron microscopic studies indicate a fusion of the zymogen granule membrane with the cell membrane so that a direct opening of the granule into the extracellular lumen of the gland is formed.

ANTIENZYMES

Familiar examples of antienzymes are true immune bodies which today are believed to be protein molecules.[16]

PRODUCTS

Pancreatin N.F., Panteric®, is a substance obtained from the fresh pancreas of the hog or of the ox and contains a mixture of enzymes, principally pancreatic amylase (amylopsin), protease and pancreatic lipase (steapsin). It converts not less than 25 times its weight of N.F. Potato Starch Reference Standard into soluble carbohydrates, and not less than 25 times its weight of casein into proteoses. Pancreatin of a higher digestive power may be brought to this standard by admixture with lactose, or with sucrose containing not more than 3.25 per cent of starch, or with pancreatin of lower digestive power. The time allowed in the assay, which is fundamentally important and should have been included in the rubic, is 5 minutes for starch and 1 hour for casein. Pancreatin is a cream-colored, amorphous powder having a faint, characteristic, but not offensive, odor. It is slowly but incompletely soluble in water and insoluble in alcohol. It acts best in neutral or faintly alkaline media, and excessive acid or alkali renders it inert. Pancreatin can be prepared by extracting the fresh gland with 25 per cent alcohol or with water and subsequently precipitating with alcohol. Besides the enzymes mentioned, it contains some trypsinogen, which can be activated by enterokinase of the intestines, chymotrypsinogen, which is converted by trypsin to chymotrypsin, and carboxypeptidase.

Pancreatin is used largely for the predigestion of food and for the preparation of hydrolysates. It originally was introduced as an emulsifying agent but, with the availability of more effective substances for this purpose, is not used much now in this way. The value of its enzymes orally must be very small because they are digested by pepsin and acid in the stomach, although some of them may escape into the intestines without change. Even if they are protected by enteric coatings, it is doubtful if they could be of great assistance in digestion. Pancreatin has been recommended in the official dose of 500 mg. in gastric and intestinal achylia, but it is probably of little value. It also has been applied topically for removing the false membrane of diphtheria and dead tissue in cancers and for the treatment of tuberculous abscesses.

Category—digestive aid.

Usual dose—500 mg. to 1 g.

Trypsin Crystallized N.F. is a proteolytic enzyme crystallized from an extract of the pancreas gland of the ox, *Bos taurus*. It contains n.l.t. 2,500 N.F. Trypsin Units per mg. It occurs as a white to yellowish-white, odorless, crystalline or amorphous powder, and 500,000 N.F. Trypsin Units are soluble in 10 ml. of water or saline T.S.

Category—proteolytic.

Usual dose range—intramuscular, 12,500 to 25,000 N.F. units daily; buccal, 12,500 N.F. units 4 times daily; topically, as a wet dressing of a solution containing 10,000 N.F. units per ml.

Trypsinogen is composed of a single polypeptide chain of 229 amino acid residues that has 6 disulfide bridges at 13-143; 31-47; 115-216; 122-189; 154-168 and 179-203. Histidine at position 46 and serine at position 183 are involved at the active site.[17] Trypsinogen is activated to give trypsin by cleavage of the peptide bond Lys^6-Ile^7 via trypin, enterokinase or peptidase A from Penicillium janthinellum. The loss of the terminal hexapeptide is accompanied by a calcium dependent conformational change probably essential for trypsin's proteolytic properties. The $-NH_2$ of the terminal isoleucine residue is essential for activity.

Trypsinogen cleaved by trypsin or enterokinase yields a hexapeptide and active trypsin that has an active site histidine near the amino terminus and an active site serine close to the carboxyl terminus. One seventh of the pancreatic juice protein is trypsin. Trypsin hydrolyzes only ester or peptide bonds in which the amino acid contributing carboxyl moiety is lysine or arginine.

Trypsin is inhibited by a variety of polypeptide substances occurring in soy bean, egg white, potato and beef parotid gland. There are two serum globulins, alpha-1-globulin (90) and alpha-2-globulin, and two pancreatic inhibitors and these appear to be competitive. Trypsin of serum may be the factor most directly responsible for the cause and course of pancreatitis.

Trypsin has been used for a number of conditions in which its proteolytic activities relieve certain inflammatory states, liquefy tenacious sputum, etc.; however, the many side reactions encountered, particularly when it is used parenterally, militate against its use.[18]

Pepsin is a substance containing a proteolytic enzyme obtained from the glandular layer of the fresh stomach of the hog. It occurs as lustrous, transparent or translucent scales; as a granular or spongy mass, ranging in color from weak yellow to light brown; or as a fine, white to weakly yellow, amorphous powder. It is free from offensive odor, has a slightly acid or salty taste and is not more than slightly hygroscopic, although it dissolves freely in water to give an opalescent solution. Dry pepsin is not injured by heating to 100°; in solution, it is destroyed rapidly by alkalies, by temperatures in excess of 70° and by other proteolytic enzymes.

The enzyme itself can be obtained from materials such as the official article, by dialysis against dilute acid in concentrated solution. By the aid of low temperatures, it has been crystallized in six-sided pyramids that are soluble in strong acids or alkalies but are insoluble at a pH of 3. Its activity in the official assay is about 1:20,000.

Pepsinogen is a single polypeptide chain of about 36 amino acid residues and is stabilized by 3 disulfide bridges. Pepsinogen can be activated spontaneously at acid pH values and is accompanied by the cleavage from the N-terminal end of a peptide chain of 40 to 42 amino acid residues. Pepsinogen cleaved by pepsin yields five small peptides, a polypeptide inhibitor, and pepsin, which has 4 basic residues (from an initial 19) and 36 free carboxyl groups. The latter supposedly account for the stability and proteolytic activity of pepsins in 0.1 N HCl. The isoelectric point of pepsin is about 1 and that of pepsinogen is about 3.7.

Pepsin has the widest range of proteolytic specificity, preferentially in report to residues from phenylalanine, tyrosine, and tryptophan, but also with those from leucine, valine, cysteic acid and glutamic acid. By changing the pH conditions, pepsin can catalyze transpeptidation reactions.[19]

Pepsin has been used locally to remove dead tissue in cancer, sloughing ulcers, the false membrane of diphtheria or in wounds. Internally it can be employed where there is a deficiency of the natural enzyme, but it is not entirely necessary to proteolytic digestion because the intestinal and pancreatic enzymes can function entirely unaided by pepsin, and administration of the latter can accomplish very little. Furthermore, many of the preparations are devoid of activity which is destroyed so easily by admixture with reagents and by various other conditions. A more logical use of pepsin is in preparation of predigested foods, when it must be added alone in an active form with a trace of acid.

Plasmin. Plasmogen in plasma and, pos-sibly, eosinophils is converted to plasmin by activators found in most tissues and by urokinase from urine. In spontaneous fibrinolysis, a proactivator present in human blood, milk, tears, saliva, amniotic fluid and most tissues is converted to an activator. Streptokinase from bacteria can activate proactivators. Plasmin is a proteolytic enzyme that under physiological and pathological conditions can dissolve a fibrin clot. It is inhibited by D.F.P. and, therefore, may contain a serine residue at its active site. In vitro it splits a variety of proteins whereas in vivo it exhibits considerable specificity for fibrin. Circulating plasmin is neutralized by antiplasmin. Since fibrinolysis often occurs without demonstrable levels of plasmin in the blood, plasminogen may be intimately associated with fibrinogen and is activated in the interstices of the clot and thus effect fibrinolysis. There also exists the possibility of a competition between the above factors in favor of a plasmin-fibrin interaction complex.

Spontaneous and abnormal fibrinolysis is due to a number of conditions. Nonspecific stresses include fasting, exercise, childbirth, hemorrhage, electroshock therapy, pyrogens, cirrhosis of the liver, prostatic surgery, etc. Obstetrical conditions account for a large percentage of the cases of abnormal fibrinolysis. The levels of fibrinogen and inhibitors of plasmogen activation appear to parallel estrogen levels.

The use of a urokinase-activated plasmin led to a significant increase in the survival rate in hyaline membrane disease resulting from lack of plasminogen activator.

Actase and Thrombolysin which are mixtures of streptokinase, activator, and free plasmin, have been used in thromboembolic disease. Side reactions are common and consideration should be given to the antiplasmin and antistreptokinase levels which vary markedly with the population.

Urokinase is a more promising agent in that it appears to be nontoxic and nonallergenic, it elicits no antibodies and it is a competitive inhibitor.

Rennin, Rennet®, Chymosin®, Seriparium®, is the partially purified milk-curdling enzyme obtained from the glandular layer of the fourth stomach of the calf. It occurs as a yellowish white, hygroscopic powder or as yellow grains or scales, having a characteristic and slightly salty taste and a peculiar, not unpleasant, odor. It is partly soluble in water and in diluted alcohol. Rennin is sold as junket tablets for the preparation of thickened foods from milk, and large quantities are employed in the cheese industry. Occasionally, it is administered to assist in the digestion of casein, especially in infants.

The clotting of milk represents a special case of limited proteolysis where an individual protein component of a large complex of different proteins is especially susceptible to proteolysis. This activity transforms the remaining part of the protein complex into an insoluble form.

Papain, Papayotin®, Papoid®, Carotid®, Papase®, the dried and purified latex of the fruit of *Carica papaya* L. (*Fam.* Caricaceae), has the power of digesting protein in either acid or alkaline media; it is best at a pH of from 4 to 7, and at 65 to 90°. The activity is increased greatly by the presence of glutathione, hydrogen sulfide or cyanides. These reagents cleave an EnS-SCyst disulfide linkage to generate an active SH on papain. It occurs as light brownish gray to weakly reddish brown granules or as a yellowish-gray to weakly yellow powder. It has a characteristic odor and taste and is incompletely soluble in water to form an opalescent solution. The commercial material is prepared by evaporating the juice, but the pure enzyme also has been prepared and crystallized.

Papain[20] is a single polypeptide chain of 212 residues, mol. wt. 23,000. The chain is folded into two distinct parts that are divided by a cleft at whose surface the active site, containing a cysteine residue at position 25 and a histidine residue at position 106, resides. Tryptophan, at position 128 also may play a role in the catalytic process of papain. Other groups in the active site region are Gln 19, Asp 105, and Asp 160. The latter are at about 10 Å from the SH group and may participate in the binding of substrates. The total α-helix content is about

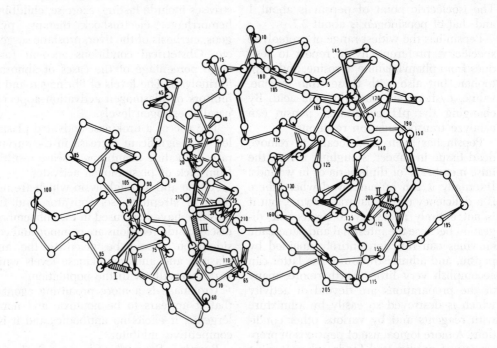

FIG. 31. Perspective drawing of the main chain conformation of papain. The circles represent the α-C atoms of the 212 residues. (Drenth, J., et al.: Structure of papain, Nature (Lond.) 218:929, 1968)

20 per cent and involves residues 26-41, 50-56, 69-78, and 116-126. A β-structure involves residues 163-172. The conformation of the remainder of the peptide chain is irregular. Three disulfide bridges are present at positions 22-159, 43-152 and 100-186 and one free SH at position 25.

The chief use of papain is to tenderize beef commercially, in which case the crude substance is smeared on or mixed intimately before cooking; about 250,000 pounds are used annually in this way in the United States. It also is employed in the brewing industry to make "chill-proof" beer, where it removes the proteins that would precipitate on cooling, and in the tanning industry to prepare hides. In medicine, it has been used locally in various conditions similar to those for which pepsin is employed. It has the advantage of activity over a wider range of conditions, but it is often much less reliable. Intraperitoneal instillation of a weak solution has been recommended to counteract a tendency to adhesions after abdominal operations, and several enthusiastic reports have been made about its value under these conditions. Papain has been reported to cause allergies in persons who handle it, especially those who are exposed to inhalation of the powder.

Bromelin is a somewhat similar proteolytic enzyme from the pineapple, *Ananas comosus* (L.) Merr., and can be prepared from the juice by precipitating with ammonium sulfate or by alcohol. Its activity is greatest at a pH of from 3 to 4. It has been suggested as an anthelmintic because it has the power of digesting living worms in a test tube.

Chymotrypsin N.F., Chymar®, is a proteolytic enzyme crystallized from an extract of the pancreas gland of the ox, *Bos taurus*. Each mg. contains n.l.t. 1,000 N.F. Chymotrypsin units.

It occurs as a white to yellowish white odorless, crystalline or amorphous powder; 100 mg. is soluble in 10 ml. of water or in 10 ml. of saline T.S.

Both intramuscular and oral administration of chymotrypsin give measurable blood levels lasting about one hour. It is relatively stable for one half hour in human intestinal juice. Systemic benefits from chymotrypsin

are dubious and the advisability of using parenteral chymotrypsin is open to serious question.[21]

Category—proteolytic.

Usual dose range—intramuscular, 2,500 to 5,000 N.F. units 1 to 3 times a day.

Application—(Sterile Chymotrypsin N.F.) for zonule lysis, 150 to 500 N.F. units as a 1:5,000 solution.

OCCURRENCE
Sterile Chymotrypsin N.F.
Chymotrypsin Injection N.F.

Bovine Chymotrypsinogen A (proteolytically inactive, mol. wt. 25,000) is a linear polypeptide cross linked by 5 disulfide bridges.

Activation of this zymogen proceeds stepwise. Initial cleavage of the Arg^{15}-Ile^{16} peptide bond forms π-chymotrypsin. Chymotrypsin then effects three successive hydrolytic cleavages. The first occurs at the Leu^{13}-Ser^{14} pepside bond to form a Ser-Arg dipeptide and δ-chymotrypsin. The latter suffers two chymotryptic splits at Tyr^{145}-Thr^{146} and Asp^{147}-Ala^{148} to yield a second dipeptide and α-chymotrypsin. The latter, composed of 242 amino acid residues, consists of three peptide chains, A, B, and C, which are linked by 5 disulfide bridges and is the enzyme used clinically and in most biochemical studies involving chymotrypsin. The cleavage of the Arg^{15}-Ile^{16} peptide bond can be preceded by the other three cleavages to give "neochymotrypsinogens" that are not active. The trypsin cleavage of the Arg^{15}-Ile^{16} peptide bond leads to the active enzyme. No change in the estimated α-helix content occurs in the conversion of the zymogen to active enzyme. Chymotrypsinogen is more stable and less completely folded than α-chymotrypsin. The NH_2 of the N-terminal amino acid residue plays a role in controlling the conformation of the enzyme.

Chain A (see below) is linked to chain B, which has 130 residues, amino terminal isoleucine, carboxyl terminal tyrosine;

$$\overset{+}{H_3N} \ Cy\overset{|}{S}\cdot Gly\cdot Val\cdot Pro\cdot Ala\cdot Ileu\cdot Val\cdot Pro\text{-}\cdot$$

$$GluNH_2\cdot Leu\cdot Ser\cdot Gly\cdot LeuCO_2.$$

Chain A of α-Chymotrypsin

chain C has 100 residues, amino terminal alanine, carboxyl terminal asparagine. α-Chymotrypsin contains a single active OH (detected by D.F.P. + hydrolysis) of a serine residue found in chain C. A histidine residue found in chain B and essential for activity implies that the "active center" of α-chymotrypsin is an interchain feature[22] dependent on the tertiary or 3-dimensional structure of α-chymotrypsin.

Iodination of the tyrosine residue least reactive toward iodine inactivates chymotrypsin but not its ability to react with diisopropyl fluorophosphate.[23]

Until recently it was believed that α-chymotrypsin cleaved only those substrates whose stereochemical configurations were the same as those found in the L-α-amino acid residues found in proteins; however, in several instances the D-antipodes were found to be the more reactive,[12, 25] whereas in others the D-antipode was a fully competitive inhibitor of the hydrolysis of the L-antipode.[25] It is evident that the two complexes differ primarily only in the way in which the two antipodes are oriented at the active site (for a more extensive discussion see Niemann, C.: Science, 143:1287, 1964).

In some cases it is established that α-chymotrypsin-catalyzed hydrolysis proceeds via an acyl enzyme intermediate; however, rigid proof is lacking that such an intermediate always is involved.[26]

α-Chymotrypsin is an endopeptidase with a considerably wider spectrum than trypsin. It rapidly hydrolyzes peptide bonds where the amino acid residues R′ are derived from tyrosine, phenylalanine or tryptophan and R″ can be a peptide nitrogen, $-NH_2$; $-NH.NH_2$; $-NHOH$ or an OR group. Hydrolysis is slower when R′ is leucyl, methionyl, asparaginyl or glutaminyl. It can

$$\begin{array}{ccc} O & H & H & R \\ \| & | & | & \| \\ R-C-N-C-C-R'' \\ & & | \\ & & R' \end{array}$$

function under alkaline conditions and has powerful milk clotting properties, a process involving proteolysis. Many proteins are extensively hydrolyzed by α-chymotrypsin.

The activity of α-chymotrypsin is not limited[27] to the above type peptide and a more inclusive type formula is as follows: $R_1CHR_2R_3$ where R_1=H, X, OH, OAc, Ac, NH_2, NH_3^+, NHAc, peptide, or α-N-acylated peptide residue; R_2=H, R, ArR−; or Ar, and R_3=COOH or its derivatives, with retention of the −C=O− group. Not all combinations are suitable substrates.

α-Chymotrypsin in solution under the proper conditions can yield β- and γ-chymotrypsins which are not interconvertible or reversed to the α-form; however, all three have the same enzymatic activity.

α-Chymotrypsin can be inhibited by the benzoyl derivatives of phenylalanine, glycine and methionine, and benzoyl-D-phenylalanine is about twice as potent an inhibitor as the L-isomer. β-Phenylpropionic acid is a strong inhibitor of α-chymotrypsin.

α-Chymotrypsin, Alpha Chymar®, is used in 1:5,000 concentration as an adjunct in ophthalmic surgery where it exerts a selective lytic action on the zonule fibers of the lens. It is especially useful in cataract surgery.

Chymotrypsin in a variety of dosage forms is used to counteract the local sequelae of trauma—inflammation, edema, hematoma and pain. The anti-inflammatory activity of prednisolone is synergised by chymotrypsin.[28] It may be injected or used buccally.

Chymoral® consists of one part chymotrypsin and six parts trypsin. Administered orally, it causes marked reduction of sputum viscidity in patients with respiratory diseases.[29]

Diastase, Taka®-Diastase, is derived from the action of a fungus, *Aspergillus oryzae* Cohn (*Eurotium O.* Ahlburg), on rice hulls or wheat bran. It is a yellow, hygroscopic, almost tasteless powder that is freely soluble in water and can solubilize 300 times its weight of starch in 10 minutes. It is employed in doses of 0.3 to 1.0 g. in the same conditions as malt diastase. Taka-Diastase is combined with alkalies as an antacid in Takazyme®, with vitamins in Taka-Combex® and in other preparations.

Alpha Amylase, Buclamase®, Fortizyme®, is a carbohydrase with a molecular weight of about 45,000. It is slightly acidic, water soluble and contains one atom of Ca per mol. which is essential for its activity and protects it from chemical and proteolytic degradation. It catalyzes the hydrolysis of α-1-4 glucosidic linkages, e.g., in starch, glycogen or their degradation products. The exact mode of action and the character of the active site is yet to be established. No organic prosthetic molecule is required. The role of calcium probably deals with the forming of a tight intramolecular metal-chelate structure that maintains the secondary and tertiary structure of the protein and the proper configuration for its hydrolytic activity.

α-Amylase is proposed for use in the treatment of soft tissue inflammation and edema associated with traumatic injury, localized inflammations, postoperative tissue reactions and connective tissue disorders. Limited experimental evidence indicates that α-amylase opposes the increased capillary permeability associated with induced inflammation, and control of the permeability factor may be significant in anti-inflammatory action. Other evidence is inconclusive and no standard anti-inflammatory response has been demonstrated experimentally.

α-Amylase is available as 10 mg. tablets for buccal administration.

Hyaluronidase for Injection N.F., Alidase®, Wydase®, Hyazyme®, Premdase®, Diffusin®, is a sterile, dry, soluble enzyme product prepared from mammalian testes and capable of hydrolyzing the mucopolysaccharide hyaluronic acid. It contains n.m.t. 0.25 mcg. of tyrosine for each N.F. Hyaluronidase Unit. Hyaluronidase is supplied only in the dry form because it is unstable in solution. Hyaluronic acid, an essential component of tissues, limits the spread of fluids and other extracellular material, and, because the enzyme destroys this acid, injected fluids and other substances tend to spread farther and faster than normal when administered with this enzyme. Hyaluronidase may be used to increase the spread and consequent absorption of hypodermoclysis solutions, to diffuse local anesthetics, especially in nerve blocking, and to increase diffusion and absorption of other injected materials, such as penicillin. It also enhances local anesthesia in surgery of the eye and is useful in glaucoma because it causes a temporary drop in intraocular pressure.

Hyaluronidase is practically nontoxic, but caution must be exercised in the presence of infection, because the enzyme may cause a local infection to spread, through the same mechanism; it never should be injected in an infected area. Sensitivity to the drug is rare.

The activity of hyaluronidase is determined by measuring the reduction in turbidity that it produces on a substrate of native hyaluronidate and certain proteins, or by measuring the reduction in vicosity that it produces on a buffered solution of sodium or potassium hyaluronidate. Each manufacturer defines his product in turbidity or viscosity units, but they are not the same because they measure different properties of the enzyme.

Category—spreading agent.

Usual dose—hypodermoclysis, 150 N.F. Units.

Streptokinase-Streptodornase, Varidase®, is a mixture containing streptokinase and streptodornase. The former acts to activate an enzyme in the blood that reacts on fibrin and brings about dissolution of blood clots and fibrinous exudates. The latter acts in a similar way to dissolve constituents of pus and has no effect on living cells. The mixture is used locally to remove dead tissue in surgery and before making skin grafts. It is recommended also in the treatment of hemothorax, hematoma, empyema, osteomyelitis, draining sinuses, tuberculous abscesses, infected wounds or ulcers, severe burns and other chronic suppurations. It is supplied in vials containing 100,000 units of streptokinase and 25,000 units of streptodornase.

Proteolytic enzymes in a stable, sterile solution for intramuscular injection, Protamide®, is used in the therapy of tabes dorsalis and herpes zoster. The dose is 1.3 ml. twice a week in tabes and daily for 2 to 4 days in herpes.

Plant Protease Concentrate, Ananase®, is a mixture of proteolytic enzymes obtained from the pineapple plant. It is proposed for use in the treatment of soft tissue inflammation and edema associated with traumatic injury, localized inflammations and postoperative tissue reactions. The swelling that accompanies inflammation may possibly be caused by occlusion of the tissue spaces with fibrin. If this be true, sufficient amounts of Ananase would have to be absorbed and reach the target area after oral administration to act selectively on the fibrin. This is yet to be firmly established and its efficacy as an anti-inflammatory agent is inconclusive. On the other hand, an apparent inhibition of inflammation has been demonstrated with irritants such as turpentine and croton oil (granuloma pouch technic).

Ananase is available in 50,000 unit tablets for oral use.

Ficin, the proteolytic enzyme from *Ficus laurifolia* has the remarkable power of digesting living helminths and has been used in the treatment of whipworm (trichuriasis) infestation. The dose of 60 ml. of the sap (latex) does not need a subsequent cathartic.

Fibrinolysin and desoxyribonuclease are available in Elase®, which rapidly lyses fibrinous material in serum, clotted blood and purulent exduates but does not appreciably attack living tissue. It is used topically in surgical wounds, burns, chronic skin ulcerations, sinus tracts, abscesses, etc.

Pancreatic Dornase, Dornavac®, is a deoxyribonuclease obtained from beef pancrease. It partially degrades deoxyribonucleoprotein (extracellular only) in a few minutes. This reduces the viscosity of secretions containing these substances. Thus, expectoration of pulmonary secretions in certain bronchopulmonary infections is facilitated. Pancreatic dornase may be useful as an adjunct to other supportive measures in treating patients with tracheobronchitis, bronchiectasis, lung abscess, atelectasis, unresolved pneumonia, and the respiratory complications of cystic fibrosis of the pancreas.

Pancreatic dornase is a freeze-dried powder of the purified enzyme. It is dissolved in isotonic sodium chloride solution immediately prior to use by inhalation or irrigation. Its potency is expressed in units that are a measure of the rate at which it reduces the viscosity of thymus deoxy-ribonucleic acid. One unit causes a drop of one viscosity unit in 10 minutes at 30°C., when the flow-time of water is one viscosity unit.

Dornavac® is available as a powder 100,000 units with 2 ml. of sterile diluent.

L-Asparaginase, Kidd reported in 1953 that when normal guinea-pig serum was injected into mice and rats bearing certain transplanted lymphomas, a marked regression, even of established tumors, occurred. This activity has been shown to be present in the sera of all the members of the superfamily Cavioida only and not in the sera of horses, rabbits, humans, etc. Recently it was shown that the effective agent in these sera is L-asparaginase, an enzyme that hydrolyzes asparagine.[30] Certain tumors such as lymphosarcoma in dogs, acute lymphoblastic leukemia in man, Lorenz's Lymphoma 2 in mice and the Murphy Sturm lymphosarcoma of rats have a requirement for the nutritionally nonessential amino acid, L-asparagine. It appears that these asparagine-dependent cells lack an asparagine synthetase that in normal cells (and in cancer cells unresponsive to asparaginase) converts aspartic acid to asparagine. Similarly, in some microorganisms mutation may cause the loss of an enzyme or the production of a defective enzyme, leading to an absolute requirement for a substance that normally would be synthesized. This is in contrast to the behavior of some malignant cells which revert to a primitive state, in the relative increase in self-sufficiency.

L-Asparaginase is widely distributed in nature, however, some asparaginases, owing to their short half life in blood, are less effective than others in suppressing leukemias. Fortunately, *E. coli,* which can be cultivated in massive quantities, produces an asparaginase, EC-2, whose half life in plasma is 2.5 hours (half life of L-asparaginase guinea pig serum is 19 hours) and is suitable for use. L-Asparaginase (EC-2) has 100 per cent specificity for L-asparagine and specificity of a much lower order (5 to 8% or less) for other L-amino acid amides. Its optimum activity is at pH 7, range 6 to 8.4; optimal pH for guinea pig serum L-asparaginase is 9.6. Both asparaginases have a molecular weight of about 150,000.

L-Asparaginase depresses both plasma and tissue levels of asparagine. In addition to its antitumor effects it can interfere with the development of other rapidly growing tissues such as the regenerating liver of the rat and the rabbit embryo and, in the latter case, can prevent the appearance of malformed newborn.

Some L-asparaginase-sensitive tumors are made resistant by treatment with subeffective doses of the enzyme. The resistant variant possesses L-asparagine synthetase activity and no longer requires exogenous L-asparagine for growth in vitro.

In dogs lymphosarcoma that is asparagine dependent and resembles some human cancers readily and rapidly responds to asparaginase treatment. In humans the gross manifestations of the acute lymphoblastic type of leukemia disappear for various periods on treatment with asparaginase, and promising clinical results have been obtained in some cases. This activity appears to be more nearly specific for the malignant cells, which is not the case with certain corticoids and drugs that induce remissions of this leukemia.

Penicillinase, Neutrapen®, is an enzyme preparation for use in treating reactions to penicillin. Penicillinase is produced and obtained by fermentation from cultures of a strain of *Bacillus cereus*. Its production by *E. coli* and many strains of staphylococci contributes significantly to the resistance of some, but not all, penicillin-insensitive microorganisms.

Penicillinase rapidly hydrolyzes penicillin G to inactivate both its bacteriostatic and its allergenic effects. It acts less rapidly against methicillin, nafcillin, and oxacillin. A single intramuscular injection of 100,000 to 800,000 units of penicillinase reduces blood levels of penicillin G to zero. Its duration of action is 4 to 7 days and is most useful in the treatment of reactions caused by long-acting injectable forms of penicillin G. Anaphylactic reactions to penicillin should first be controlled by giving epinephrine and antihistamines followed by the use of corticosteroids.

Penicillinase is a protein (mol. wt. about 50,000) and a foreign one to animals and humans, and therefore must be regarded as potentially allergenic and capable of producing anaphylactic reactions.

Neutrapen® is available as a powder 800,000 units per ml. in 2 ml. containers for intramuscular injection.

INSULIN

Insulin is a polypeptide hormone that is synthesized and stored in the beta cells of the Langerhan islets which comprise 1 per cent of the pancreas and are dispersed throughout this organ. Insulin is synthesized by ribosomes in the endoplasmic reticulum. Synthesis of the B chain (q.v.) begins at the N-terminal end, as with other proteins such as hemoglobin, lysozyme or ribonuclease. Recent evidence suggests the synthesis of a single polypeptide chain containing both the A and B chains in the form of a zymogen, proinsulin (q.v.). The A and B chain sections (q.v.) are probably paired via a special enzyme. Adequate reserves of insulin are stored in beta secretory protein granules, which contain 10 per cent of insulin. These granules, containing the insulin, fuse with the cell membrane, with the simultaneous liberation of insulin which enters the portal vein and passes through the liver, where large amounts are trapped before the remainder is delivered to the systemic circulation. The half life of insulin in plasma is 40 minutes. In most cases exogenous insulin is weakly antigenic. No insulin antibodies were found in thousands of persons who had never received insulin. The release of insulin is probably triggered by certain levels of glucose in the blood (in the case of rat pancreas, 50 mg. per 100 ml.) or a metabolic product of glucose and the insulin levels in the blood. Secretin and ACTH can directly stimulate the secretion of insulin. Other factors such as glucagon cause an increase in plasma insulin probably via indirect mechanisms i.e., release of glucose.

"Clinical" insulin that has been crystallized 5 times and then subjected to countercurrent distribution (2-butanol: 1% dichloroacetic acid) yielded about 90 per cent insulin-A, with varying amounts of insulin-B, together with other minor components. A and B differ by an amide group and have the same activity. End member analysis, sedimentation and diffusion studies indicate a molecular weight of about 6,000. The value of 12,000 for the molecular weight of insulin containing trace amounts of zinc (obtained by physical methods) is probably a bimolecular association product through the aid of zinc.

The extensive studies of Sanger[32] and others employing modern methods have elucidated the amino acid sequence in the two polypeptide chains, e.g., the phenylalanyl (B) and the glycyl (A) chains of insulin, together with the positions of the amide groups. Two of the disulfide bridges are

the A and B chains are synthesized in vivo separately and are subsequently combined to form insulin.

Careful cleavage with carboxypeptidase of the C-terminal alanyl residue of the B chain gives the fully active dealanine-insulin. Further cleavage to the dealanine-deamido-insulin reduces the activity by one third, but dealanine-deasparagine-insulin is inactive. The first six amino acid residues of the B chain are supposed to be unnecessary for the hypoglycemic action of insulin.[35]

During the acidic extraction of the pancreas there is some partial degradation of

$$
\begin{array}{l}
\overset{\text{amide}}{|} \quad \overset{\quad\quad\text{S-S}\quad\quad}{\boxed{\quad\quad\quad}} \qquad\qquad\qquad \overset{\text{amide}}{|} \quad\quad \overset{\text{amide}}{|} \quad\quad \overset{\text{amide}}{|} \\
\text{Gly·Ileu·Val·Glu·Glu·Cy·Cy·Ala·Ser·Val·Cy·Ser·Leu·Tyr·Glu·Leu·Glu·Asp·Tyr·Cy·Asp} \\
\qquad\qquad\qquad\qquad\qquad\quad | \qquad\qquad\qquad\qquad\qquad\qquad\qquad\qquad\qquad\qquad\qquad | \\
\qquad\qquad\qquad\qquad\qquad\quad \text{S} \qquad\qquad\qquad\qquad\qquad\qquad\qquad\qquad\qquad\qquad\qquad \text{S} \\
\overset{\text{amide}}{|} \qquad\qquad\qquad\quad | \qquad\qquad\qquad\qquad\qquad\qquad\qquad\qquad\qquad\qquad\qquad \text{S} \\
\text{Phe·Val·Asp·Glu·His·Leu·Cy·Gly·Ser·His·Leu·Val·Glu·Ala·Leu·Tyr·Leu·Val·Cy·Gly·Glu·-} \\
\qquad | \\
\text{amide} \\
\text{Arg·Gly·Phe·Phe·Tyr·Thr·Pro·Lys·Ala}
\end{array}
$$

intermolecular (between the two chains), and the third is intramolecular in the glycyl chain. The diagrammatic structural formula that has been proposed for beef insulin is given above.

Pig, sheep, horse and whale insulin differ from beef insulin in the 8 to 10 positions of the glycyl chains as follows: pig = Thr., Ser., Ileu.; sheep = Ala., Gly., Val.; horse = Thr., Gly., Ileu.; and whale = Thr., Ser., Ileu.

Human insulin differs in having threonine instead of alanine in positions A_5 and B_{30}, and isoleucine instead of valine in position A_{10}.

The insulins of other species also differ to a minor degree.[33]

The A and B chains of human, bovine and sheep insulin have been synthesized in a few weeks by the peptide synthesis on solid supports. The A and B chains have been combined to form insulin in 60 to 80 per cent yields, with a specific activity comparable to that of the natural hormone.[34] This lends support to the suggestion that

the quite acid labile primary β-amide group of the C-terminal asparagine residue (A_{21}) to give the almost fully active deamido-insulin.[36]

Insulin is inactivated when the phenolic hydroxyl, carboxyl, imidazole and amide groups are esterified or altered chemically. On the other hand, this is not the case with the amino and the aliphatic hydroxyl groups and the guanidine residue. Insulin also is inactivated when the disulfide bridges are broken by oxidation, treatment with cysteine, sodium bisulfite, sulfurous acid or by in vivo enzymes. In vivo, further degradation is effected by peptide hydrolysis, particularly in liver, kidney, pancreas, testes and placenta.

There is no complete agreement on the 3-dimensional structure of insulin; however, some believe that, on solution, the part of the polypeptide chains between the disulfide bridges exists as an α-helix, whereas the head and tail portions may exist between a folded and an unfolded state.

The particle size of insulin in solution

depends on pH, temperature, ionic strength, concentration, and foreign ions. In low concentrations and at pH below 2 the molecular weight is 6,000 which dimerizes as the concentration and pH is raised to 2. Above pH 4, complexes with metals such as Zn^{++} are formed and between pH 4 and pH 7 these complexes precipitate as a highly amorphous aggregate. At pH 7 zinc insulin can take up 2.2 per cent of zinc. The carboxyl groups and one of the two imidazole groups are involved with zinc.[37]

Insulin occurs as an amorphous powder, although it can be crystallized. It is insoluble in neutral water, aqueous alcohols, acetone and ether. It is insoluble at a pH of 4.5 to 7. It keeps best at a pH 5.7 and is soluble in acidified alcohol and dilute acid. The isoelectric point is at pH 5.3 and to 5.5. Insulin can be heated for ½ hour at 80° and at pH 4 to 7.2 or at 75° in 0.1 to 0.2N HCl or boiled for 20 minutes in 0.2N HCl with retention of activity.[38]

All the official insulin preparations are stable up to 36 months when stored at 5°C., with the exception of Isophane Insulin suspension which suffers about a 25 per cent loss after 30 months. At 25°C. Protamine Zinc Insulin and Isophane Insulin Suspension were stable for 36 months. The latter suffered some aggregation that was difficult to disperse. Insulin Zinc Suspension, Prompt Insulin Zinc Suspension and Extended Insulin Zinc Suspension retained potency for 24 months and an excess of 30 months was required before losses in potency were significant. However, Insulin Injection lost 10 per cent of its potency after 18 months and color change occurred after 24 months at 25°C. Higher temperatures were more deleterious. Ultraviolet light breaks down the cystine groups, with slower decomposition of the tyrosine residues. Photo-oxidation leads to preferential breakdown of the imidazole groups of histidine. Insulin is rapidly decomposed in alkaline solution.[39]

Insulin is inactivated in vivo by (1) an immunochemical system in the blood of insulin treated patients, (2) reduction of the disulfide bonds (probably by glutathione) and (3) by insulinase (a proteolytic enzyme) that occurs in liver. Pepsin and chymotrypsin will hydrolyze some peptide bonds that lead to inactivation. It is inactivated by reducing agents such as sodium bisulfite, sulfurous acid, hydrogen, etc.

The isolation of insulin suitable for clinical use can be outlined briefly as follows[40]: To 1 kg. of finely ground fresh beef pancreas 20 to 30 ml. of 10M sulfuric acid is added followed by the addition of 1,500 ml. of ethanol. After stirring and standing 4 to 12 hours the filtrate is concentrated to 1/10 its volume (at 25° to 30°). After filtration the insulin is precipitated by ammonium sulfate, the precipitate is dissolved and adjusted to pH 5 with acetic acid. The resulting precipitate can be dissolved in 0.1N HCl in slight excess and used clinically.

Very small quantities of insulin can be isolated by precipitation with antibodies.[41] Repeatedly recrystallized insulins are not homogeneous and may contain some of the insulin antagonist glucagon, deamidoinsulin, inactive insulinlike proteins and antigenic substances. The most purified insulins are isolated by counter-current distribution chromatographic methods or gel filtration.[42]

Insulin is inactivated when the phenolic, hydroxyl, carboxyl and imidazole groups are blocked by acetylation, esterification, etc., or the disulfide bridges are broken by reduction, oxidation or treatment with cysteine. On the other hand, this is not true of the amino and aliphatic hydroxyl groups. Insulin will form complexes with certain metals such as zinc, nickel, cobalt and cadmium. Insulin will crystallize when the pH of a solution of highly purified amorphous preparation in a zinc-ion-containing buffer is adjusted to a pH close to its isoelectric point.

Insulin appears to play a role in the following: (1) transport of glucose through the cell membrane; (2) formation of glucose-6-phosphate from glucose and ATP + hexokinase; (3) oxidations in the tricarboxylic acid cycle; (4) oxidative phosphorylation (formation of "energy rich" phosphate bonds); (5) glucose uptake, enhanced in muscle, adipose tissue and liver; (6) glycogen synthesis in muscle and adipose tissue; (7) protein synthesis in muscle, adipose tissue and liver, and (8) fatty acid

TABLE 102. INSULIN PREPARATIONS

NAME	PARTICLE SIZE (Microns)	ACTION	COMPOSITION	DOSE (U.S.P. Units*)	ZN (mg. per 100 U.S.P. U.)	N (mg. per 100 U.S.P. U.)	pH	DURATION (Hours)
Insulin Injection† U.S.P.	....	Prompt	Insulin + Zn	5-100				5-7
Prompt Insulin Zinc Suspension† U.S.P.	2§	Rapid	Insulin + ZnCl$_2$ + buffer	10-80	0.20-0.25	-0.7	7.1-7.5	12
Insulin Zinc Suspension† U.S.P.	10-40 (70%) 2 (30%)§	Intermediate	Insulin + ZnCl$_2$ + buffer	10-80	0.20-0.25	0.7	7.1-7.5	18-24
Extended Insulin Zinc Suspension† U.S.P.	10-40	Long-acting	Insulin + ZnCl$_2$ + buffer	10-80	0.20-0.25	0.7	7.1-7.5	~24
Globin Zinc Insulin Injection† N.F.	....	Intermediate	‖Globin + ZnCl$_2$ + insulin	10-80	0.25-0.35	1.50	3.4-3.8	18-24
Protamine Zinc Insulin Suspension† U.S.P.	....	Long-acting	#Protamine + insulin + Zn	10-80	0.20-0.25	1.25	7.1-7.4	36
Isophane Insulin Suspension† U.S.P.	30	Intermediate	Protamine** ZnCl$_2$ insulin buffer	10-80	0.016-0.004	0.085	7.1-7.4	24-48

* One U.S.P. Insulin Unit = 1 International Unit = 0.0167 mg.; 1 mg. of insulin = 24 units.
† Clear or almost clear.
‡ Turbid.
§ Amorphous.
‖ Globin (3.6 to 4.0 mg. per 100 U.S.P. Units of insulin) prepared from beef blood.
Protamine (1.0 to 1.5 mg. per 100 U.S.P. Units of insulin) from the sperm or the mature testes of fish belonging to the Genus Oncorhynchus or Salmo.
** Protamine (0.3 to 0.6 mg. per 100 U.S.P. Units of insulin) (q.v.).

synthesis, increased in liver and adipose tissue.[43]

Of the above activities, the role of insulin in the reversible transport of glucose through cell membranes appears to be of paramount importance. This transport is stereoselective in character and L-glucose is not transported. Competitive inhibition between various pairs of hexoses and pentoses lends additional support for the existence of a transport carrier in the membrane. Glucose uptake by cells even in the presence of adequate insulin is reduced by respiratory fuels such as acetate and pyruvate. Insulin does not effect the transport of phosphorylated glucose through the cell membrane.

A deficiency of insulin leads to hyperglycemia, followed by glucosuria, disturbance of electrolyte and water balance, increased fat metabolism with defective fat metabolism (β-hydroxybutyric acid, acetoacetic acid and acetone and acidosis), and increased amino acid metabolism with increased nitrogen excretion. Thus, decrease in blood volume, collapse of the peripheral blood vessels, failure of kidney function and, finally, shock occur.

Unmodified insulin has a rapid onset of activity and a short duration of action. Insulin containing zinc (2 per cent) and, when prepared in larger crystalline particles (q.v.), has decreased solubility and slower and more prolonged activity. An insoluble amorphous small particle (2μ) that is prompt acting also can be prepared. Each of these can be used alone or in combination to obtain the desired effect. The discovery that insulin would combine with certain basic proteins such as protamine, histone, globin and kyrin to form complexes which had a more prolonged duration of action led to the development of Protamine Zinc Insulin. In some cases one injection daily of such a preparation would suffice to control the diabetic condition. More controlled medication can be obtained by combining insulin and protamine zinc insulin in the ratio of 2 to 1. The development of NPH insulin is an improvement over the latter combination. It is composed of a suspension of insulin, protamine and zinc.

Insulin that is slightly esterified is active and not very strongly antigenic, and is slightly bound by antibodies that are present. Sulfated insulins containing only one sulfonate and six sulfate (of threonine and serine) groups per molecule and very few O-tyrosyl sulfate groups were clinically useful. In insulin-resistant diabetics sulfated insulin was more effective, by 5 to 10 times, per hypoglycemic unit than unaltered insulin over a period of 5 to 32 months.[44] Sulfation of tyrosyl (and histidine) residues leads to inactivity.

Intraperitoneal administration of tranylcypromine to mice caused a rapid and marked stimulation of insulin secretion, which was followed at 60 to 90 minutes by profound hypoglycemia. The conversion of lactate to blood glucose also was markedly depressed. Other monoamine oxidase inhibitors tested had no effect on insulin secretion or blood glucose.[45]

Diazoxide can cause an inhibition of insulin secretion and hyperglycemia in man.[46]

Proinsulin (Porcine). A single polypeptide chain composed of the A and B chains of insulin joined together by a polypeptide containing 33 amino acid residues and containing the 3 disulfide bridges of insulin has been isolated and characterized.[31] The potency of this proinsulin is about 3 I.U. per mg. by the mouse-convulsion assay and about 6 I.U. per mg. by the radioimmunoassay. Digestion with trypsin yields desalanine insulin (insulin minus the terminal alanine residue of the B chain), with potencies of 23 I.U. per mg. by the mouse-convulsion assay and 26 I.U. by radioimmunoassay. Insulin has a potency of 25 I.U. per mg. Thus the possibility exists that insulin is synthesized in vivo as a single polypeptide chain precursor which is subsequently converted in a zymogen-like (q.v.) manner.

Six proinsulin molecules combine with 2 zinc ions to form a complex with a molecular weight of 55,000, which corresponds to a hexamer. If this should be found to take place in vivo also, it may have some implications for the mechanism of conversion of proinsulin to insulin in the beta cell. Six mols of insulin also combine with 2 zinc ions to form a complex that probably is analogous to that established for the hex-amer unit of crystalline zinc-insulin.

CORTICOTROPIN *

"Clinical" adrenocorticotropin (ACTH) or corticotropin is a mixture of very closely related polypeptides (8, more or less) that are obtained from the (fresh) anterior lobe of the pituitary gland of mammals (usually hogs or sheep).

In the case of sheep corticotropin, the major component was α-corticotropin, whereas in the hog it was designated β-corticotropin.† The sequential arrange-ment[47] of the amino acid in β-corticotropin was determined, and, through recent syn-thetic efforts, together with bioassays[48] the following formula that contains 39 amino acid residues has been proposed for β-corticotropin. It contains one amide group, free amino groups of the lysine and phenolic

Ser·Tyr·Ser·Met·Glu·His·Phe·Arg·Tyr·Gly·-
Lys·Pro·Val·Gly·Lys·Lys·Arg·Arg·Pro·Val·-
Lys·Val·Tyr·Pro·Asp·Gly·Ala·Glu·Asp·Glu·-
Leu·Ala·Glu·Ala·Phe·Pro·Leu·Glu·Phe

Porcine β-corticotropin

groups of the tyrosine residues, and serine occupies the N-terminus position. Its molec-ular weight is 4,566.

α-Corticotropin appears to differ from β-corticotropin in three minor respects, i.e., sequences at 31 to 32; at 25 to 28; and the presence of two amide groups instead of one. Controlled pepsin[49] digestion of β-corticotropin cleaved the peptide bonds be-tween 28-29, 30-31 and the 31-32 residues to yield polypeptides containing 28, 30 and 31 amino acid residues, respectively. These fragments are as active as β-corticotropin in the Sayers assay, and the one with 28 amino acid residues was fully effective in rheumatoid arthritis.

Mild acid hydrolysis of β-corticotropin readily cleaved bonds 24 and 25. The re-sulting substances with 24 and 25 amino acid residues would appear at present to be the smallest active degradation products.

* Now generally accepted name, see J.A.M.A. 147:326, 1951.

† Sometimes called corticotropin A.

A polypeptide containing the first 20 amino acid residues (serine→valine) of β-corticotropin has been synthesized and shown to have full ACTH activity.[50] One with 17 amino acids[51] (serine→arginine) had 5 per cent, and one with 19 amino acids (serine→proline) had 50 per cent of the adrenal stimulating activity of ACTH in ani-mals, as measured by the increase in ad-renal weight. In man, the latter had 80 per cent of the activity[52] of ACTH.

A synthetic 25 amino acid polypeptide analog of corticotropin, in which D-serine replaced the N-terminal L-serine, norleu-cine replaced methionine at position 4 and valinamide replaced asparagine at position 125, exhibited high activity in humans with no apparent side effects.[53]

It should be noted that a polypeptide containing the first 13 amino acid residues (serine→valine) has MSH properties (q.v.), thus accounting for this activity of ACTH. It also is the smallest fragment that exhibits some ACTH activity.

The terminal $-NH_2$[54] group and the methionine residue[55] are not essential for activity.

All reactions involving peptide cleavages at the amino end (Ser.) of the β-cortico-tropin molecule lead to inactive fragments. Both β-corticotropin and the above deg-radation fragments have a small amount of intermedin activity that is believed to be inherent in these substances.

Evidence for the existence of a precor-ticotropin substance (inactive) that has ACTH activity after acid treatment has been reported.[56]

Acetone-dried anterior lobes of the pitu-itary glands are first extracted with glacial acetic acid at 70° or with boiling 40 per cent acetic acid in methanol. A crude pre-cipitated preparation obtained by the addi-tion of ether is purified via oxycellulose ad-sorption from 0.1N acetic acid and eluted by 0.1N hydrochloric acid and countercur-rent distribution (e.g., n-butanol vs. 0.5% trichloroacetic acid). Three main fractions were obtained: i.e., 12 per cent α, 49 per cent β, 19 per cent γ and 20 per cent δ. All these fractions except the δ showed equal biologic activity at about 100 U.S.P. units per mg. by the Sayers test.

β-Corticotropin is soluble in water, physiologic saline solutions, glacial acetic acid, and aqueous alcohol, especially if acidic. It is insoluble in acetone, ether and petroleum ether. It is basic (free amino groups) and forms salts with acids. Its isoelectric point is in the range of pH 5.5 to 6.0. It is stable in the dry state, and activity persists after boiling for 16 hours in 0.1N hydrochloric acid. Therefore, it can be sterilized by the usual technics if the solutions are less than pH 7. It is more sensitive to alkali, and boiling for 20 minutes in 0.1N sodium hydroxide is quite deleterious; however, deamidation without loss of activity occurs at a pH 9 (NaHCO₃ 0.1M) at 25° for 18 to 22 hours. Trypsin and chymotrypsin inactivate β-corticotropin with 30 minutes; on the other hand, it is stable to pepsin (granular) for 1 hour and to carboxypeptidase for 40 to 60 hours.

The various corticotropins lose some of their potencies when subjected to air oxidation, especially in the presence of iron; however, activity can be restored with reducing agents such as hydrogen sulfide.

Corticotropins can bind Mn, Co, Ni, Zn and Cu, and only in the case of Zn and Cu does a relationship exist to biologic activity.

Corticotropin Injection U.S.P., ACTH Injection, Adrenocorticotropin Injection, is a sterile preparation of the principle or principles derived from the anterior lobe of the pituitary of mammals used for food by man. It occurs as a colorless or light straw-colored liquid, or soluble amorphous solid by drying such liquid from the frozen state. It exerts a tropic influence on the adrenal cortex. The solution has a pH range of 3.0 to 7.0 and is used for its adrenocorticotropic* activity.

Category—adrenocorticotropic hormone.

Usual dose—intramuscular, 20 U.S.P. Units 4 times a day.

Usual dose range—intramuscular, 40 to 160 U.S.P. Units daily.

Porcine β-corticotropin is about 3 times more active subcutaneously than by the I.V. route.[57]

ACTH increases the weight of the adrenals and thus promotes the production of the adrenal cortical hormones with their subsequent biologic effects (q.v.) It causes an increased release of ascorbic acid† from the adrenals into the blood. Red blood cell production is enhanced, metabolic rate is elevated and the level of both free fatty acids and degradation products of the fatty acids in the blood serum of fasted rats is elevated by ACTH. It plays a role in the production of milk in the mammary glands. The administration of ACTH can cause the thymus to shrink and decrease the eosinophils‡ of the blood. Porcine ACTH has activities of about equal potency to human ACTH, with the former being more steroidogenic.[58]

Repository Corticotropin Injection U.S.P., Corticotropin Gel, Depo® ACTH Purified Corticotropin, ACTH Purified, is corticotropin in a solution of partially hydrolyzed gelatin to be used intramuscularly for a more uniform and prolonged maintenance of activity.

Category—adrenocorticotropic hormone.

Usual dose—intramuscular, initial, 40 U.S.P. Units once or twice daily.

Usual dose range—10 to 160 U.S.P. Units daily.

Sterile Corticotropin Zinc Hydroxide Suspension U.S.P. is a sterile suspension of corticotropin, adsorbed on zinc hydroxide and contains n.l.t. 45 and n.m.t. 55 mcg. of zinc for each 20 U.S.P. corticotropin units. Because of its prolonged activity due to slow release of corticotropin, an initial dose of 40 U.S.P. units can be administered intramuscularly, followed by a maintenance dose of 20 units, 2 or 3 times a week.

Category—adrenocorticotropic hormone.

Usual dose—intramuscular, 40 U.S.P. Units once or twice daily.

Usual dose range—10 to 160 Units daily.

NEUROHYPOPHYSEAL HORMONES

Vasopressin and oxytocin are synthesized in the pericarya of cells in the supraoptic nucleus of the hypothalamus. They are combined with a protein (sometimes called neurophysin) to form neurosecretory gran-

---

* In addition to the stimulation of the production of adrenal corticoids, androgen and related steroid production is increased.

† Basis for ACTH assays.

‡ Basis for ACTH assays.

ules which pass down the hypophyseal stalk, probably by way of the axons, to the posterior lobe of the pituitary. There they are stored and from there the controlled release takes place. They are released in the general circulation in association with a transport protein (neurophysin) to prevent metabolic attack until they reach their targets. They are deactivated in the liver and kidneys in non-pregnant women and by the enzyme oxytocinase (q.v.) of the serum in pregnant women. Oxytocin release in pregnant women can be effected by stimulation of the nipples and preparation for breast feeding.

The oxytocins and vasopressins have multiple activities—uterine-contracting, milk-ejecting, vasopressor and antidiuretic properties—that are qualitatively similar. However, the nature of the amino acid residues and their sequential order determine the quantitative potencies, which may vary greatly in the above activities. In recently adopted nomenclature[*] the names of all analogs have been derived from the term vasopressin. Thus, oxytocin is Ile[3] Leu[8]-vasopressin, and beef vasopressin is Arg[8]-vasopressin, etc. Vasopressin therefore, should contain the following amino acids in the order given, with no name given to the eighth amino acid residue: Cys.Tyr.Phe.-Glu(NH_2) . Asp . (NH_2) . Cys . Pro . AA . Gly.NH_2 (AA, an amino acid). This nomenclature was adopted, no doubt, to systematize and simplify relationships in these polypeptides.

These naturally occurring neurohypophyseal hormones that have been assigned a trivial name with the suffix -tocin have significant oxytocic blood depressor (fowl) and milk-ejecting activities, with very low antidiuretic activity (rat) with the exception of arginine vasotocin, which also has significant rat antidiuretic activity.

A large number of peptides[†] closely related in structure to oxytocin and vasopressin have been synthesized and evaluated biologically. Table 103 shows the activities

[*] I.U.P.A.C.–I.U.B. tentative rules, J. Biol. Chem. 241:527, 1966.

[†] For nomeclature, *see* Vitamins and Hormones 22:266, 1964.

For extensive discussion on structure-activity relationships, ibid., p. 270.

For assays, ibid., p. 262.

of some of these analogs and demonstrates the species differences in these assays.[59] Activities are expressed in I.U. per mg.

Table 103 gives a number of SAR features of some of the naturally occurring neurohypophyseal hormones and a few selected congeners (out of hundreds). Nine L-amino acids joined in peptide linkages together with a 20-member cyclic disulfide structure appear to be the minimum requirements for significant biological activities.

Replacement of one of the S atoms of the –S–S– bridge in desamino[1]-oxytocin with a CH_2 group gave an analog that had 60 I.U. per mg. of oxytocic activity, 25 I.U. per mg. of avian depressor activity and 1 I.U. per mg. of antidiuretic activity. Thus, the –S–S–bridge is not an absolute requirement for activity in the oxytocins.[60]

The two sulfur atoms can be replaced by selenium atoms: e.g.: 1-deamino-1,6-L-selenocystine-oxytocin has avian depressor activity of 985, oxytocic activity of 1,100, rat pressor activity of 4.8 and rat antidiuretic activity of 29 U.S.P. units per mg.

The length of the side chain (i.e., a tripeptide) is critical. D-Amino acids are dystherapeutic. It is apparent from Table 103 that certain limited changes in the composition of certain amino acid residues are not significantly dystherapeutic; e.g., Ileu[8]-oxytocin (also desamino-oxytocin, q.v.) has enhanced activities with a 10:1 ratio of antidiuretic: pressor activity, as compared with oxytocin which has a 1:1 ratio approximately. Although the loss of the amino group in the 1-hemicystine residue led to some advantages, acylation, carbamylation or alkylation of this amino group markedly reduces activities or even produces antagonists[61] of oxytocin without intrinsic activity. On the other hand, some analogs that have amino acid residues or short peptide chains attached to the terminal amino group of oxytocin have distinctly prolonged effects and have been designated hormonogens. Such a derivative may be defined as one whose biologic effects, particularly in vivo, are wholly or largely due to its conversion to the active hormone via enzyme peptide cleavage. Leucyl and phenylalanyl residues are relatively readily cleaved—thus, potency is high but duration of action is less marked.

TABLE 103.

Neurohypophyseal Peptides	Source	Oxytocic, Uterus			Depressor (Chicken blood)	Milk Ejecting (Rabbit)	Blood Pressure (Rat)	Anti-diuresis (Rat)
		Rat (In vitro)	Cat (In situ)	Human (In situ)				
Oxytocin ($Ille^3$-Leu^8-vasopressin)	cattle	415-500	450	450	360-510	360-450	3-7	1-5
Arginine vasotocin (Ile^3-Arg^8-vasopressin)	synthetic	37-160			75-285	80-270	64-245	74-250
Isotocin (Ile^3-Ser^4-Ile^8-vasopressin)		130-150	250	335	320	300	0.06	0.18
Mesotocin (Ile^3-Ile^8-vasopressin)	frog	289			498	328	6	1.1
Glumitocin (Ile^3-Ser^4-Glu^8-vasopressin)	Raia species	8	563					
Valyloxytocin (Val^3-Leu^8-vasopressin)		59	226	277	58	207		
Isoleucine^8-oxytocin (Ile^3-Ile^8-vasopressin)		289	563	365	498	328		
Aspargine^4-oxytocin (Ile^3-Asp-Leu^8-vasopressin)		108 / 830	335	150	202 / 975	300		
Desamino^1-oxytocin (β-Mpr^1-Ile^3-Leu^8-vasopressin)	synthetic	360	900	1,030	400	541	2	15-19
1-Deamino-^{4}-valine-oxytocin	synthetic	350			800	350	0	5
Arginine-vasopressin (Arg^8-vasopressin)	cattle	9-20			42-60	51-70	300-500	300-400
Lysine-vasopressin (Lys^8-vasopressin)	hog	4-7.5			28-48	31-60	243-290	150-250
β-Mpr^1-Arg^8-vasopressin β-Mpr-β-mercaptopropionyl								1300-200
1-Deamino-8-arginine-vasopressin	synthetic	27±4			150±4		370±20	1300±200
4-Deamino-8-arginine-vasopressin	synthetic	5			25		38	760
1-Deamino-4-decarboxamido-8-arginine vasopressin	synthetic	3.3			9.9		10.7	1020±67
1-Deamino-4-decarboxamido-8-lysine-vasopressin	synthetic	1.5			12.6		3.5	729±26

All values —I.U. per. mg.

Glycyl residues are more resistant to cleavage—thus, potency is rather low but duration of action is particularly long.

O-Methylation of the tyrosyl residue or its replacement by H,CH$_3$ or Et leads to general dystherapeutic effects, a loss of pressor activity in the rat (but not in rabbits or cats) and, in some cases produce antagonists. 2-0-Methyltyrosine oxytocin has high affinity and low or no intrinsic activity and is inhibitory in the rat uterus and pressor assays.

2-Phenylalanine oxytocin has avian depressor activity of 60 to 70 I.U. per mg., oxytocic activity of 30 I.U. per mg. and no pressor activity. Studies with the rat uterus indicate that this analog has a high affinity for the receptor site but a low intrinsic activity. 2-Leucine oxytocin has a lower affinity but the same intrinsic activity as the 2-phenylalanine analog.

The 3 position is one of the sites in the native neurohypophyseal peptides in which changes in the amino acid residue have developed in the course of evolution. Therefore, a considerably greater tolerance to changes of the amino acid residue at this position might be expected than appears to be the case.

Position 4 appears to be a favored position for evolution changes (see isotocin and gumitocin versus the other 4 naturally occurring hormones). Also, 4-decarboxamido-[60, 62] and deamino-4-decarboxamido-oxytocins have significant activities: depressor (fowl) 108 and 193, oxytocic (rat) 72 and 93, and milk-ejecting (rabbit) 225 and 266 I.U. per mg. respectively. For a 4 valyl analog, see 1-deamino-4-valine oxytocin (Table 103) which has high activity. 4-Serine and 4-alanine oxytocin also have relatively high orders of activity. 4-Leucine-oxytocin, the CH$_2$ homolog of valine, has very low activities. However, it is noteworthy that it has a potent natriuretic and diuretic effect: It inhibits antidiuretic hormone activity, reversing the free-water reabsorption induced by vasopressin.

Changes in the 4 position of lysine vasopressin seem to reduce the vasopressin like properties quite severely; e.g., rat pressor activity of 4-alanyl-lysine vasopressin is 1.6 I.U. per mg., antidiuretic activity is 30 I.U.

per mg., rat pressor activity of 4-seryl lysine vasopressin is 3.3 I.U. per mg., antidiuretic activity is 69 I.U. per mg.

The asparagine residue at position 5 carries great structural specificity, as does, also, though to a somewhat lessor extent, the glycine amide at position 9. 5-D-Carboxamido-oxytocin has low activity.

Five natural neurohypophyseal hormones contain different amino acid residues at positions 8, i.e., Leu, Ile, Lys, Arg, and Glu. Therefore, this position, with its specific amino acid, is of great importance in respect to the biologic activities of these hormones (See Table 103, and also Table 111 in ref. 59). In some synthetic analogs an 8-alanine residue together with a 1-deamino also is effective; e.g., 1-deamino-8-alanine oxytocin has 314 I.U. of oxytocic activity, 453 I.U. of avian depressor activity and 8 I.U. per mg. of rat pressor activity.

Substitution of suitable residues at this position and perhaps also at position 3 leads to more or less intermediate types of biologic activities.

The phenolic OH and the amide groups at positions 5 and 9 contribute to affinity and intrinsic activity. The amide group at position 4 has an affinity role. The amino group at position 1 is not essential. The side chain plays an outstanding role in the activity of these hormones; its absence leads to weak oxytocic and milk-ejecting activity (3 and 1 I.U. per mg. respectively).

Substitution of other amino acid residues for the leucine residue in position of 8 of oxytocin results in analogs that possess a significant degree of biologic activities.[63]

Very high antidiuretic activity together with very low activities of other types are found in 1-deamino-4-decarboxamido-arginine vasopressin and 1-deamino-4-decarboxamido-lysine vasopressin (see Table 103). These congeners offer attractive possibilities for clinical application as antidiuretic agents.[64]

Oxytocin

Oxytocin, Syntocinon®, a uterine-contracting, milk-ejecting principle (hormone) is a cyclic octapeptide amide that can be extracted readily from the fresh posterior lobe of the pituitary gland by means of aqueous acetic acid (0.1 to 1.0%). It has been synthesized,[65] and the synthetic product is identical in all respects with natural oxytocin. Reduction of the disulfide bridge yields an open chain biologically active octapeptide amide rendered inactive when the sulfhydryl groups are blocked with benzyl groups.

$$
\begin{array}{c}
Cys.-Tyr. \\
S \quad 1 \quad 2 \quad 3Ileu. \\
| \quad\quad | \\
S \quad 6 \quad 5 \quad 4Glu.-Amide \\
Cys.-Asp.-amide \\
/ \\
7Pro. \\
\backslash \quad 8 \quad\quad 9 \\
Leu.-Gly.-amide
\end{array}
$$

Oxytocin

Oxytocin is soluble in water, aqueous acetic acid, alcohols or acetone and insoluble in ether or petroleum ether. It is stable in the dry state. It is comparatively stable at a pH 3.8 to 4.4. When heated at pH 5, considerable loss is encountered. It forms a biologically active crystalline flavianate that melts at 190° to 200° C. Specific rotation of oxytocin is $[\alpha]_D 21.5 = 26.1 = (c. 0.53, H_2O)$. The distribution coefficient between 0.05 per cent acetic acid and s-butyl alcohol is $K = 0.35$. Its isoelectric point is at pH 7.7.

Pure synthetic[66] and naturally occurring oxytocin have the following activities per mg.: avian depressor 507; oxytocic (rat uterus) 486; milk ejecting (rabbit) 410; pressor (rat) 3.1 and antidiuretic (rat) 2.7.

Oxytocin Injection U.S.P. is a sterile solution in water for injection of an oxytocic principle prepared by synthesis or obtained from the posterior lobe of the pituitary of healthy, domestic animals used for food by man. pH 2.5 to 4.5; expiration date, 3 years.

Oxytocin preparations are widely used with or without amniotomy to induce and stimulate labor. Although injection is the usual route of administration the sublingual route is extremely effective. Sublingual and intranasal spray routes of administration also will stimulate milk let-down.

Category—oxytocic.

Usual dose—intramuscular, 1 ml. (10 U.S.P. Units/ml.) repeated in 30 minutes if necessary; intravenous infusion, 1 ml. in 1,000 ml. of isotonic solution at a rate of 0.5 to 2 ml. per minute.

Usual dose range—intramuscular, 0.3 to 1 ml.

Valyl Oxytocin

Valyl[3] oxytocin differs from oxytocin only in that the isoleucine residue is replaced by a valine residue. It is more active in vivo, e.g., cat uterus in situ or on the milk-ejecting pressure in the lactating rabbit than it is in vitro, e.g., isolated rat uterus compared with oxytocin whose activities are the same in vivo as in vitro. Valyl oxytocin also has small undesirable pressor effect (spinal cat) and very slight antidiuretic effect (nonanesthetized rat) and thus is similar to oxytocin.

Desamino[1]-Oxytocin

Loss of the free amino group of the 1-cystine residue of oxytocin results in an enhancement of its avian depressor (733 units/mg.), rat oxytocic (684 units/mg.) and antidiuretic activities (15 units/mg.) and a decrease in its rat pressor activity (1.1 units/mg.). The oxytocic activity in the human uterus (in situ) is 1,030 ± 300 units/mg., which is about twice that of oxytocin. Desamino[1]-oxytocin is not inactivated by the serum of pregnant women that contains an enzyme, serum oxytocinase, which can cleave the peptide linkage between the cysteine[1] and tryosine[2] residues. However, plasma from women in labor can reduce its activity—probably via reduction of the disulfide linkage, because activity can be restored by suitable oxidation technics.[67] This mechanism of inactivation also appears to take place in the liver and kidneys followed by aminopeptidase degradation (and called tissue oxytocinase) and also may be present in placental tissue. In a species such as the cat that has no serum oxytocinase, desamino[1]-oxytocin also is about twice as active as oxytocin in the cat uterus (in situ) assay. This may imply that desamino[1]-oxytocin per se is more active than oxytocin. Furthermore, in equipotent doses desamino[1]-oxytocin and oxytocin have the same duration of action. Milk-ejecting activity in lactating rabbits is not appreciably affected.[68] The above activity ratios suggest this compound be tested for diabetes insipidus and obstetrics.

Arginine Vasotocin

A cyclic octapeptide amide containing the cyclic pentapeptide amide moiety of oxytocin with the side chain of arginine-vasopressin was found to possess oxytocic (75 units per mg.), avian depressor (150 units per mg.) and pressor activities (125 units per mg.). The high pressor activity probably is in part a function of arginine in the side chain. The avian depressor and oxytocic activities probably are a function of the polypeptide ring structure.

It is interesting to note that arginine vasotocin seems to be the natural hormone in all cold-blooded vertebrates except the elasmobranchs.

Citrulline Oxytocin

The substitution of citrulline for the 8-leucine oxytocin reduces oxytocic and milk-ejecting activities, with some increase in vasopressor and antidiuretic activities, which, however, are far lower than those of argenine vasotocin.[69]

Other synthetic analogs of oxytocin that possess a more selective response than the native hormone are likely to find clinical application. For example, 4-asparagine oxytocin has one-fifth to three-fifths the oxytocin activity but only one-hundredth to one-fiftieth the vasopressin activity of oxytocin. 3-Alloisoleucine oxytocin is very selective for milk let-down because its ratio of milk let-down to uterus activity (125 I.U. per mg. and 24 I.U. per mg.) is 5:1 compared to that of oxytocin, which is 1:1.

D-Leucine oxytocin,[70] glycyl oxytocin,[66, 71] carbamoyloxytocin,[72] l-hemi-D-cystine oxytocin,[73] leucyloxytocin,[74] N-methylhemicystine oxytocin[74] and 5-alanine oxytocin[62, 75] have activities of a low order or are inactive.

Carbamoylated derivatives of oxytocin, assayed on isolated rat uterus, act as specific inhibitors[76] of oxytocin without intrinsic activity.

Isotocin is Ser^4-Ile^8-oxytocin isolated from the pituitaries of the teleost fishes and has 120 I.U. per mg. of oxytocic activity in the isolated rat uterus assay.[77]

Serum oxytocinase can be inhibited by desthio-oxytocin and S,S'-dibenzyl oxytocin.

Two particularly potent inhibitors of oxytocin have been developed[78] i.e., l-L-penicillamine oxytocin, where β,β-dimethyl cysteine replaces cysteine, and l-deamino-penicillamine oxytocin. These inhibitors are effective in vivo and in vitro, have no intrinsic activities and exhibit competitive and reversible properties. They have no inhibitory effect on the responses to hypotensin or bradykinin or on spontaneous contractions.

VASOPRESSINS

The antidiuretic vasodepressor principles (hormones) of vasopressins, arginine from beef and lysine from hogs) are cyclic octapeptide amides that can be extracted readily[79] from the fresh posterior lobe of the

$$S\text{————————}S$$
$$|\qquad\qquad\qquad|$$
Cys-Tyr-Phe-Glu-Asp-Cys-Pro-Arg·Gly·amide
$$\qquad\qquad|\quad|$$
amide amide

Vasopressin (Beef) (Argenine vasopressin)

$$S\text{————————}S$$
$$|\qquad\qquad\qquad|$$
Cys·Tyr·Phe-Glu-Asp-Cys·Pro·Lys·Glyamide
$$\qquad\qquad|\quad|$$
amide amide

Vasopressin (Hog) (Lysine vasopressin)

pituitary gland by means of aqueous acetic acid (0.1 to 1.0%). They have been synthesized, and the synthetic products are identical in all respects with the natural hormones.[80]

It will be noted that the eighth amino acid residues, i.e., arginine and lysine, respectively, differ from that of oxytocin, which has leucine. Both of these vasopressins differ from oxytocin in that they contain phenylalanine instead of isoleucine.

The vasopressins resemble quite closely oxytocin in their solubilities and stability properties. Pharmacologically effective doses can be dissolved in physiologic saline solutions which can be sterilized by heating in steam and by filtration with some loss of activity. Arginine vasopressin's isoelectric point is at pH 10.85, which is very high and due to the guanidine group (very basic) of the arginine residue, and at this pH the phenolic group of tyrosine will exist as the phenoxide ion. Arginine vasopressin can be fractionally precipitated from an acetic acid (98%) solution of oxytocin and arginine vasopressin by the addition of ether. It can be purified further by electrophoretic methods or by countercurrent distribution between 0.06M *p*-toluene sulfonic acid and 2-butanol (K = 0.85) or 2-butanol and 0.1M acetic acid (K = 0.11), and these values also hold for the synthetic product.

Lysine vasopressin, when purified by countercurrent distribution methods using *n*-butyl alcohol and 0.09 M *p*-toluenesulfonic acid, has a distribution coefficient K = 0.66. It can also be purified by 0.08M *p*-toluene sulfonic acid and *s*-butyl alcohol. Its specific rotation is $[\alpha]_D^{20} = -47.5°$ (c— 0.99, H₂O).

Arginine and lysine vasopressins are identical in biologic activity, i.e., antidepressor and antidiuretic activities of about 350 to 400 U.S.P. units per mg., together with uterine-contracting, milk-ejecting, and avian vasodepressing effects equal to 5, 20 and 15 per cent, respectively, of the activity found in pure oxytocin. However, lysine vasopressin has only one seventh the antidiuretic potency of the arginine vasopressin when assayed I.V. in dogs.

Arginine and lysine vasopressin have one unambiguously physiologic role in the economy of the organism, i.e., the facilitation of water conservation by the kidney. Their physiologic effect is to inhibit the excretion of "osmotically free" water in the distal sections of the nephrons. It is possible that they also play a subtle role in the regulation of smooth muscle tone, especially in the arterioles, and, thereby, in the maintenance of blood pressure after hemorrhage and shock, although the pressor effects for which they were named require dosages that are in the pharmacologic range.

Vasopressin exerts its pressor effects in part by potentiating normally released catecholamines.[81] It should be noted that, in rat assays, a minimum pressor effect requires approximately 100 times as much arginine vasopressin as is needed for a minimum antidiuretic effect. No physiochemical methods are available for the detection of the minute amounts of vasopressin (5×10^{-11} to 2.5×10^{-11} g.) which elicit a significant antidiuretic response in the rat. For this reason it is difficult to be sure that an antidiuretic effect is due specifically to vasopressin.[82]

Vasopressin has marked corticotropin-releasing activity (CRF). In a limited number of congeners tested, CRF activity is not realted to pressor or antidiuretic activity. The CRF pressor ratio was largest in 1-deamino-4-decarboxamide-8-lysine-vasopressin—approximately 98 times that of lysine vasopressin.[83]

The vasopressins are not recommended as pressor agents. They are used by injection to treat diabetes insipidus. The significantly prolonged antidiuretic response of 1-glycyl-8-lysine vasopressin might render it useful in this condition (rat pressor 15 I.U. per mg. and antidiuretic 10 to 40 I.U. per mg. dose dependent).

Lysine vasopressin is inactivated by the enzymatic cleavage of the terminal glycine-amide residue to yield a —COOH group.

Vasopressin Injection U.S.P. is a sterile solution in water for injection of the water-soluble pressor principle of the posterior lobe of the pituitary of healthy domestic animals used for food by man, or prepared by synthesis. Each ml. possesses a pressor activity equal to 20 U.S.P. Posterior Pituitary Units. Expiration date, 3 years.

Category—posterior pituitary hormone (antidiuretic).

Usual dose—intramuscular, 0.5 ml.

Usual dose range—0.25 to 2 ml.

Vasopressin Tannate, Pitressin® Tannate, is a water-insoluble tannate of vasopressin administered intramuscularly (1.5 to 5 pressor units daily) for its prolonged duration of action due to the slow release of vasopressin. It is particularly useful for patients that have diabetes insipidus and never should be used intravenously.

Felypressin, 2-(phenylalanine)-8-lysine

vasopressin, has relatively small antidiuretic activity and little oxytocic activity. It has considerable pressor (i.e., vasoconstrictor) activity which, however, differs from that of epinephrine, i.e., following capillary constriction in the intestine it lowers the pressure in the vena portae, whereas epinephrine raises the portal pressure. Felypressin also causes an increased renal blood flow in the cat, whereas epinephrine brings about a fall in renal blood flow. Felypressin is 5 times more effective a vasopressor than lysine vasopressin and is recommended in surgery to minimize blood flow, especially in obstetrics and gynecology.

Selective pressor activity[84] also has been obtained with other vasopressin analogs. Thus Orn[8]-vasopressin has a pressor activity almost equal to Arg[8]-vasopressin but has about one fifth its antidiuretic activity in the rat assay. It has only one fiftieth the oxytocic activity of oxytocin.

On the other hand, the pressor and antidiuretic activities of 8-L-citrulline vasopressin was about one tenth that of vasopressin.[85]

Thus, the spacing and basic character of the side chain of residue 8 is significant. A reduction in basicity leads to a considerable reduction in pressor and antidiuretic activities.

Highly purified arginine vasopressin has been used to great advantage to produce uterine blanching by direct injection into the uterus during gynecologic surgery.

Desamino-8-lysine vasopressin is more active than 8-lysine vasopressin in the antidiuretic (300 I.U. per mg. versus 203 I.U. per mg.) and less active (126 I.U. per mg. versus 243 I.U. per mg.) in the pressor tests. It has approximately the same activity (31 I.U. per mg.) as the parent hormone in the milk let-down and is more active in the uterus (12 I.U. per mg. versus 5 I.U. per mg.) and avian (61 I.U. per mg. versus 48 I.U. per mg.) depressor assays. It has a more favorable antidiuretic (126 I.U. per mg.) to pressor (301 I.U. per mg.) ratio than has arginine vasopressin (380 I.U. per mg.: 429 I.U. per mg.).

In general, the pressor effects of vasopressin do not complicate control of diuresis, but, in specific instances, this greater selectivity might be advantageous.

L-Dab[8]-vasopressin* has about one-half of the pressor (rat) and antidiuretic (rat) and about two-thirds the uterotropic (rat) activities of Lys[8]-vasopressin. D-Dab[8]-vasopressin's activities are: antidiuretic 90-150 I.U. per mg., pressor 6.3 I.U. per mg. and uterotonic 0.1 I.U. per mg. Thus, it has been possible to effect a separation of the antidiuretic and pressor activities by the use of D-amino acid. The reverse of the above separation has been effected in Orn[8]-oxytocin (103 to 2.15 I.U.); Phen[2]-Orn[8]-oxytocin (120 to 0.55 I.U.), to a lesser extent in Orn[8]-vasopressin (360 to 88) and, to a still lesser extent, in Phe[2]-Lys[8]-vasopressin (felypressin, PLV®†) (55 to 20 I.U.). The latter analog has been used clinically to enhance the duration of action of general and local anesthetics.[86] It also has been used to restore blood pressure during hemorrhagic shock, in which its primary effect is exerted on the capacitance of the vessels.

PARATHYROID HORMONE

Parathyroid hormone is a single chain polypeptide that has 83 amino acid residues and a molecular weight of about 9,500. It contains one tyrosine, one tryptophan, two methionine and no cysteine residues. Varying degrees of inactivation are effected by oxidation (methionine residues), tyrosinase (tyrosine residue), 2-hydroxy-5-nitrobenzyl bromide (tryptophan residue), and esterification via methanol or acetylation. The intact molecule possesses some areas of ordered 3-dimensional structure and other large areas that are not so ordered.

Parathyroid hormone is synthesized and stored in the parathyroid gland, which can synthesize this hormone very rapidly. The calcium level of the serum provides a highly effective negative feedback control of this synthesis. The hormone has a half life of about 20 minutes, although the effect of the hormone may last for 12 to 24 hours. This is a common pattern, for the effects of many hormones persist long after the hormone itself has disappeared from the blood.

Parathyroid hormone plays a role of prime

* Dab = α, γ-diaminobutyric acid.
† Felypressin-Sandoz.

importance in calcium homeostasis. It causes a rise in serum calcium by mobilization of calcium from the deep bone stores rather than from the recently deposited surface calcium. Activity is similar to that of vitamin D, and the two appear to act synergistically. This activity is accompanied by an increase in urinary phosphate and a decrease in serum phosphate, beginning in 30 to 60 minutes after administration. These effects may persist for 12 to 24 hours and are accompanied by an increase in RNA synthesis. The levels of succinic dehydrogenase and glucose utilization also are elevated.

Although the mode of action at the molecular level has not been established, parathyroid hormone causes rapid activation of adenyl cyclase in renal and skeletal cells. This may accelerate the conversion of ATP to cyclic AMP which acts as a secondary messenger to promote the release and synthesis of the specific lysosomal enzymes involved in the mobilization of calcium from bone.

Parathyroid Injection U.S.P. is a sterile solution in water for injection of the water-soluble principle or principles of the parathyroid glands that have the property of increasing the calcium content of the blood. Each ml. contains n.l.t. 100 U.S.P. Parathyroid Units. One unit equals one one-hundredth of the amount of Parathyroid Injection required to raise the Ca content of 100 ml. of the blood serum of normal dogs 1 mg. in 16 to 18 hours.

Category—blood-calcium regulator.

Usual dose—intramuscular, 40 U.S.P. Units twice daily.

Usual dose range—20 to 100 Units daily.

MELANOCYTE-STIMULATING HORMONE

Melanocyte-stimulating Hormone (MSH)[87] or intermedin occurs in the anterior lobe of the pituitary gland and, in the case of amphibia, the intermediate lobe of the hypophysis. It is probably a mixture of at least 2 polypeptides designated as α-MSH (about 75%) and β-MSH (about 25%). β-MSH (molecular weight 2,177) from hogs has been assigned the following structure.[88] α-MSH has more amino acid residues. Bovine

Asp·Glu·Gly·Pro·Tyr·Lys·Met·Glu·His· Phe·Arg·Tyr·Gly·Ser·Pro·Pro·-Lys·Asp·

MSH (molecular weight 2,135) differs from that proposed for hog β-MSH only at position 2, where a seryl residue replaces a glutamyl. It should be noted that the amino acid sequence 7 to 13 of β-MSH is identical with that between 4 and 10 in ACTH, and this no doubt accounts for the melanocyte-stimulating activity of ACTH.

The solubility and the stability of these principles closely approximates that of ACTH, and they accompany ACTH during the isolation. However, they can be separated from ACTH and each other by the very useful countercurrent distribution technics. The β-MSH component has an isoelectric point of 5.2 and $[\alpha]^D = 97.5$. Bovine MSH is reported to have a distribution coefficient lower and an isoelectric point higher (pH 7) than hog β-MSH.

β-MSH is inactivated by trypsin and chymotrypsin.

In amphibia and some fish MSH plays a role in the pigmentation of the skin, and the darkening of the melanophores in the skins of frogs is employed as a biologic index for the activity of the hormone preparation. In humans it appears to play a role in shortening the time of adaptation to darkness and in increasing sensitivity to light.

HUMAN CHORIONIC GONADOTROPHIN (HCG)

HCG is a glycoprotein that is synthesized in the cytotrophoblast by the placenta.[89] During the first trimester of pregnancy women produce 100,000 units per day, after which the quantities produced drop rapidly. Estrogens stimulate the anterior pituitary to produce placentotrophin, which stimulates HCG production. HCG causes ovulation in the female and the expulsion of spermatozoa in the male. It is more active than LH, probably because of the difference in their biologic half-lives. HCG is stable in 0.2 per cent aqueous solution for 8 days at pH 3.0 to 11.0. It is inactivated by nuraminidase (a receptor-destroying enzyme) and by influenza virus. In certain cases of infertility, treatment with LH followed by the administration of HCG effected ovulation and, in many cases, multiple births.

HCG that was isolated from normal pregnancy urine and urine from patients with trophoblastic tumors had a minimum mol. wt. of 27,000 to 30,000. Selected hydrolytic technics yielded 8 moles of sialic acid and two glycopeptides. Each of the latter contained a multiple branched mixed polysaccharide—one composed of Fuc, Gal_4, Man_5, and $GluNAc_5$ and the other of Gal_2, Man_5, and $GlNAc_5$. The total amino acid composition (about 66%) was Lys_7, $Hist_3$, Arg_{11}, Asp_{12}, $Thre_{12}$, Ser_{16}, Glu_{13}, Pro_{20}, Gly_9, Ala_9, Cys_{14}, Val_{13}, $Meth_3$, $Ileu_4$, Leu_{10}, Tyr_4, Phe_4 Try_1. The carbohydrate moiety was linked to the polypeptide chain via a N-acetylglucosaminyl-asparagine linkage. Biologic activity appears to be related to the sialic acid content.

Releasing Factors

The hypothalamus synthesizes, primarily in the SME region, substances called releasing factors that are transported via the hypophysial portal vessels to the anterior lobe of the pituitary where they effect the release of hormones such as luteinizing hormone, LH, follicle stimulating hormone, FSH, corticotrophin, thyrotropin, growth hormone, melanocyte stimulating hormone and prolactin inhibiting factor. Because they all exert analogous type activities their abbreviated terms are: LRF, FSH-RF, CRF, TRF, PRF, MIF and PIF respectively. Because they exhibit a high degree of specificity of activity to release hormones into the circulatory system they might more suitably be referred to as luteinizing hormone-releasing hormone (LRH or LH-RH), follicle-stimulating hormone-releasing hormone (FRH or FSH-RH), corticotropin-releasing hormone (CRH), etc. To date these substances have not been completely characterized. Some are small polypeptides slightly larger than vasopressin and others are slightly larger polypeptides. Porcine TRF is not a polypeptide.

These releasing factors in minute amounts are capable of releasing from the anterior lobe of the pituitary gland up to 200 or more times their own weight of the respective hormones. This is called a multiplier effect and supports the concept of their being designated as hormones or "neurohumors."

The amounts of the releasing factors may be determined by the activity of the sympathetic nervous system. Adrenergic blocking agents or the norepi-depleting tranquilizer drugs can reduce the amounts of circulating LH to the extent that ovulation can be inhibited. It therefore appears that norepi and the sympathetic nervous system —in part, at least—initiate the release of the releasing factors at the hypothalamic level. This conclusion is supported by the fact that LRF is active in normal female and in ovariectomized rats in which LH release has been inhibited by lesions in the median eminence. These lesions block the participation of the sympathetic nervous system. Therefore, LRF acts directly on the anterior pituitary to release LH. LRF injected directly into the anterior lobe of the pituitary could effect ovulation, whereas systemic administration of the same dose was without effect.

LRF in appropriate dose levels can overcome gonadal steroid inhibition of LH release.

Lactogenic Hormone

Lactogenic hormone (prolactin, luteotropin) is a polypeptide produced by the anterior pituitary that stimulates lactation in mammals. It also has been found in blood, placental tissue and urine. Prolactin obtained from sheep pituitary glands is a single polypeptide chain of 211 amino acid residues (molecular weight 24,100) containing the threonine as the N-terminal amino acid. It is soluble at a pH of 8.0, and a 1 per cent solution at pH 7.6 is stable (biologically) at 100° C. for 20 minutes. Its isoelectric point is pH 5.73, and its specific rotation is −40.5°.

Growth Hormone

The growth hormone has been isolated from the anterior pituitary gland of the sheep, pig, horse, fish, monkey and human and is a polypeptide of considerably greater molecular weight than the other polypeptide peptide hormones. There is a species difference in molecular weight (25,000 for humans and monkeys and 46,000 for the bovine hormone). Biologic activity is not entirely specific, i.e., the bovine hormone, is active in fish and rats, and the human hormone is active in rats. Bovine growth hormone has been investigated most extensively to date.

The growth hormone of the anterior pituitary produces a much greater (about 5-fold) protein anabolic response than testosterone propionate in the hyophysectomized rat. However, in the castrated animal, the responses produced by the two hormones are identical for the first 14 days, after which the protein anabolic effect of the growth hormones continues at a slower rate, while that of the androgen ceases. A "high" dose of the growth hormone gives a response similar to that of the androgen, but on cessation of the injections there is a tremendous loss in body weight and nitrogen (protein).

HYPERTENSIN

Hypertensin I (Angiotensin I or angiotonin) is a decapeptide that is slowly released by the action of renin (a proteolytic enzyme extracted from kidneys) upon angiotensinogen synthesized in the liver and contained in plasma. Hypertensin I is cleaved, probably only on passing through the lungs, by an angiotensinase (an endopeptidase) to give hypertensin II (angiotensin), whose half life in blood is 3 minutes. In a single passage through peripheral vascular beds 50 to 60 per cent disappears and the remainder passes through the pulmonary circulation without loss.[90]

All amino acids have the L-configuration, and the hypertensins differ very slightly, depending on the blood source.

Asp.Arg.Val.Tyr.X.His.Pro.Phe.His.Leu.
 1   2   3   4 5 6   7   8   9 10
                               Hypertensin I
Asp.Arg.Val.Tyr.X.His.Pro.Phe.
                          Hypertensin II
X = Isoleucyl residue (from horse or hog blood)[91]
X = Valyl residue (from beef blood)
Human hypertensin I is the same as hyper-

tensin I obtained from the horse[92] and has been synthesized.

Hypertensin II has been found in the blood of many human beings with essential hypertension as well as in dogs with experimental renal hypertension.

Hypertensin II is the most powerful pressor substance discovered to date. It is reported to release norepinephrine from tissue and the adrenal.[93]

Hypertensins I and II are soluble in water, particularly at pH values below 7, and are soluble in aqueous alcohol (up to 80 per cent). They can be sterilized at pH 6.5 in a boiling water bath for 15 minutes.

Hypertensin I (isoleucyl) per se has no in-vitro pressor activity; however, in vivo its pressor activity is equal to hypertensin II, which is about 3 to 5 times that of noradrenalin. When hypertensin I is injected I.V. in intact anesthetized animals, a maximum rise in blood pressure occurs in 1 minute and returns to normal in 3 minutes. Therefore, hypertensin I must be converted rapidly to hypertensin II in vivo. Hypertensin II (isoleucyl) also has high smooth muscle contracting (myotropic, oxytocic) activity. The site of action of hypertensin II is different from that of the pressor amines (noradrenaline, etc.).

Hypertensin II, with the end-member aspartyl residue as an amide, has about twice the pressor activity of hypertensin II and is 2 to 4 times as active as valine hypertensin I. Minimum structural requirements for maximum biologic activity of hypertensin II are phenylalanine's free carboxyl group, phenyl group of phenylalanine, tyrosine's phenolic group, proline in position 7, and histidine in position 6. Histidine's contribution is of a sterochemical nature and its charge is not of crucial significance.[94]

Conversion[95] of the C-terminal carboxyl group to a carboxamide group decreased the biologic activity to one thirtieth. Even esterification is dystherapeutic, with activity being decreased to one tenth.

Asparagine can replace the aspartyl residue, and the high activity[96] of β-aspartyl- and deamino-angiotensin[97] in which the $-NH_2$ terminus has been modified suggests the comparative unimportance of these modifications of this section of the molecule.

However, D-aspartyl[1]-valyl[5] angiotensin II was 50 per cent more active than bovine angiotensin II on rat blood pressure with a 2- and 3-fold increase in duration.[98] These derivatives are more stable than angiotensin to the action of enzymes (one possibly an aminopeptidase) in serum.

Modification[99] of both the $-NH_2$ terminus and the C terminus, such as in D-aspartyl[1]-phenylalanyl[8] amide angiotensin II, decreased the biologic activity to one three hundredth, with a 3-fold increase in duration of action.

The aspartyl $-NH_2$ terminal residue can be replaced by a glycyl residue or removed altogether without reducing the activity[100] by more than 50 per cent. Lysine can replace the arginyl residue, leucine the valyl residue, or valine the isoleucyl residue with essentially no loss in pressor or myotropic (oxytocic) activity. L-Al[3]-L-Ileu[5]-hypertensin II has 68 per cent of the pressor activity of Ileu[5]-hypertensin II, thus a branched or straight chain in position 3 does not carry the specificities[101] as at a position 5. On the other hand, if leucine replaces the isoleucyl residue or alanine the prolyl residue, activities are lost. When the aspartyl residue is replaced with arginine, 15 per cent of the pressor and 70 per cent of the myotropic activities are retained; thus, one activity can be reduced to a much greater extent than the other.

Assuming a helical structure for hypertensin II, a model would show the necessary groups such as tyrosine's phenolic group and the phenyl group of phenylalanine all on one side and all the unnecessary groups on the opposite side. Treatment with 10 per cent urea drops the activity to 50 per cent, and this indicates that hydrogen bonding is a force responsible for the orientation of the amino acid residues that leads to maximum activity.

It has been shown[120] that hypertensin in vitro in aqueous solution exists in a random conformation as a monomer between pH 2.5 and 8.6 and that a helical conformation[103] is improbable. However, its conformation when combined with its cellular receptor is not known.

A number of clinical observations support the view that angiotensin is by far superior to catecholamines in the mediation of shock of various origins. Angiotensin effects incorporation of iron into red blood cells and has an effect qualitatively similar to that of vasopressin on water and salt excretion in diabetes insipidus.

Aspartyl[1]-amide-valyl[5]-angiotensin II, Hypertensin® is available for clinical use. It has a higher ratio of oxytocic:pressor activity.

Angiotensin Amide N.F., N-[1-[N-[N-[N-(N.²L-Asparaginyl-L-arginyl)-L-valyl]-L.tyrosyl]-L-valyl-L-histidyl]-L-prolyl]-3-phenyl-L-alanine, Aspamide.Arg.Val.Tyr.Val.His.Pro.Phe., aspartyl[1]-amide-valyl[5]-angiotensin II, is an octapeptide that occurs as a white to slightly off-white amorphous powder that is soluble in water and insoluble in chloroform.

Angiotensin amide has about twice the pressor activity of hypertensin II.

Category—vasopressor.

Usual dose range—intravenous infusion, 3 to 10 mcg. per minute as determined by blood pressure response.

OCCURRENCE

Angiotensin Amide for Injection N.F.

GLUCAGON

Glucagon U.S.P., a hyperglycemic-glycogenolytic factor (HGF) that is a proteinlike hormone that contains 29 amino acid residues, occurs in the α-cells of the pancreas. It has been isolated from the amorphous fraction of a commercial insulin (4% glucagon) and prepared in a crystalline state and reported to have the following structural formula in which all the amino acids have the L-configuration.

amide
|
His·Ser·Glu·-Gly·Thr·Phe·Thr·Ser·Asp·Tyr·
Ser·Lys·Tyr·Leu·Asp·Ser·Arg·Arg·Ala·Glu.
|
amide
Asp·Phe·Val·Glu·Tyr·Leu·Met·Asp·Thr.
| |
amide   amide

Glucagon's solubility is 50 mcg. per ml. in most buffers between pH 3.5 and 8.5. It is soluble 1 to 10 mg. per ml. in the pH ranges

2.5 to 3.0 and 9.0 to 9.5. Solutions of 0.2 mg./ml. at pH 2.5 to 3.0 are stable for at least several months at 4° if sterile. Loss of activity via fibril formation occurs readily at high concentrations of glucagon at room temperature or above at pH 2.5. The isoelectric point appears to be at pH 7.5 to 8.5. Because it has been isolated from commercial insulin its stability properties should be comparable with those of insulin.

Glucagon causes a transient rise in blood sugar concentration[104] in most of the mammals, in birds and in reptiles probably via the activation of liver phosphorylase, causing increased glycogenolysis. This is accompanied by a transient loss of liver glycogen that is restored in 24 hours due to the release of insulin and the corticoid hormones. The need for glucagon is greatest in time of glucose need and apparently is the hyperglycemic factor in the pancreatic effluent of man and animals made hypoglycemic with insulin, phloridzin or induced starvation. Induced hyperglycemia is accompanied by a rapid fall in glucagon to baseline levels, which returns to normal upon a reduction of glucose levels. Thus glucagon is one of the hormones of glucose whose function may be concerned with maintaining the maximal hepatic glucose output. Thus, the vital glucose-dependent tissues of the brain are one of the beneficiaries of glucagon's activities.[105]

Glucagon can affect the clinical course of diabetes in man, as well as hypoglycemia due to α-cell carcinoid tumor, schizophrenia, peptic ulcer, etc.

Glucagon has been used for the emergency treatment of insulin overdosage, for the smooth termination of insulin shock in psychiatric patients, in the treatment of "spontaneous" hypoglycemia and of hypoglycemia due to islet-cell tumors. It also can be used diagnostically for liver function in glycogenosis, hyperinsulinism, functional hypoglycemia and in Addison's disease where the blood sugar rise in response to glucagon is smaller than in normal subjects.[106]

Category—hyperglycemic.

Usual dose—parenteral, 0.5 U.S.P. Glucagon Unit, repeated in 20 minutes if necessary.

Usual dose range—0.5 to 1 unit repeated as necessary.

OCCURRENCE

Glucagon for Injection U.S.P.

BRADYKININ

The term bradykinin was coined by Rocha e Silva from the Greek *bradys* and *kinein,* slow moving, because this factor's contracting effects on the intestine were developed more slowly than are those of histamine. He and his co-workers showed that the venom of the Brazilian snake *Bothrops jararaca,* when added to blood or serum, released, within a minute, a factor that induced the contraction of guinea-pig ileum and produced a profound fall in the arterial blood pressure of rabbits. Subsequently, bradykinin also was shown to increase capillary permeability to blood plasma (to cause a local swelling) and to induce intense pain at local sites.

Bradykinin, a nonapeptide, is Arg.Pro.Pro. Gly.Phe.Ser.Pro.Phe.Arg. It occurs in blood as the precursor bradykininogen (a globulin) from which it is released via peptide cleavage by proteolytic enzymes, i.e., kallikrein (of blood), snake venoms and trypsin. Bradykinin also has been isolated from the skin of *Rana temporaria.* Its duration of action—about 10 minutes—is terminated by a kininase, which is a carboxypeptidase that cleaves the arginine residue at the 9 position. In addition to bradykinin, kallidin, lysylbradykinin* (a decapeptide) and methionyl-lysyl-bradykinin[107]† have been identified in plasma.

Kinins from other sources are derivatives of bradykinin, e.g., phyllokinin (from the skin of *Phyllomedusa rohdei,* a Brazilian amphibian) which is bradykinyl-isoleucyl-tyrosine-O-sulfate.[108] Upon trypsin digestion bradykinin is obtained. Glycyl bradykinin is derived from the main peptide of wasp venom by the action of trypsin. Additional examples are hornet kinin, urine kinin, ortho kinin and clostrokinin. These kinins have biologic activities similar to bradykinin.

* One half uterus, same guinea pig capillary permeability.
† One third uterus, same guinea pig capillary permeability.

The generation of bradykinin is not a simple process and safeguards against excessive release and spread appear to be present. This suggests a local use in local injuries and the term *local hormones* appears appropriate. This term implies that such a hormone may be produced in many places to act locally and is needed only occasionally at a concentration that would be toxic if widely distributed. It is postulated that the functions of bradykinin are pain (as a warning), increase of local blood supply arising from vasodilation, the bringing of appropriate nutritive and immunizing support to cells combating injury, and production of vascular permeability which allows antibodies and white cells to pass out of the blood into the tissues. These activities are accompanied by an inflammatory response.

Bradykinin is one of the most powerful vasodilator substances known, i.e., 0.05 to 0.5 μg. per kg. I.V. produces a fall in blood pressure in all mammals so far investigated.

By suitable substitutions in the polypeptide chain one might conclude that both the phenylalanine and arginine residues and the terminal carboxyl group play essential interaction roles in the biologic activities of bradykinin. The glycine and serine residues are spacers with no bulk at the glycyl residue and a limited bulk and a nonessential OH at the seryl residue permissible.[109] Three prolyl residues because of their rigid ring structures impose severe restrictions on the over-all conformation of the polypeptide molecule. Only the prolyl residue at the 3 position can be replaced by the flexible alanyl residue with the retention of full activity.[110] Significant activity is retained when one of the arginine residues is replaced by a large side chain containing some basic character such as in the case of lysine, ornithine, histidine and citrulline. Also, the aromatic ring of phenylalanine can be substituted in the p-position by OH, OCH_3 and F with the retention of significant activity.[111] In the case of 8-*p*-fluoro-L-phenylalanine bradykinin the activity is increased by 50 per cent, and Gly[6], Tyr (Me)[8] bradykinin has 1.2 times the uterotropic and ing support to cells combating injury, and

2 times the capillary permeability activity of bradykinin.[112]

Some bradykinin derivatives and analogs have both intrinsic and inhibitory activities. Examples are acetyl bradykinin and acetyl Gly[6], Tyr (Me)[8]-bradykinin. D-Pro[2,3,7]-bradykinin has low intrinsic activity, with 28 per cent inhibition at 0.0001 μg with 0.01 μg bradykinin. No analogs prepared to date have the much desired practically useful inhibitor properties.

Although the kinins per se are not used directly, kallikrein enzyme preparations which release bradykinin from an inactive precursor in vivo are used in the treatment of Raynaud's disease, claudication and circulatory diseases of the eyegrounds.

Butylated hydroxanisole in concentrations as low as 8×10^{-9} mole per liter can inhibit detectably the contraction of smooth muscle elicited by bradykinin. Other phenolic antioxidants and some tranquilizers also act as inhibitors.[113]

THYROCALCITONIN

Thyrocalcitonin (TCT), a polypeptide hormone produced by the thyroid gland, contains 32 amino acids in a single polypeptide chain. It has an intrachain disulfide bridge between the N-terminal cysteine residue and a cysteine residue at position seven.[114] This hormone can be extracted using 0.1 NHCl from the thyroid glands of all species tested including rat, goat, pig, guinea pig, dog, cow, monkey and man. Its activity can be destroyed by pepsin, trypsin or boiling with excess acid or alkali. It is stable at pH 5 or below.

Thyrocalcitonin inhibits calcium resorption from bone; changes in plasma phosphate usually parallel the changes in plasma calcium. It may function to prevent the dangerous effects of hypercalcemia. TCT can inhibit the PTH-induced resorption of bone in tissue culture and in vivo. It is active in the absence of PTH and its hypocalcemic effects are not the result of an inhibition of PTH secretion. Responsiveness to TCT is best demonstrated in young animals; the younger the animal the greater the effect. This might explain some of the poor effects observed thus far in man, since most trials have involved the elderly.

Bioassays of TCT-containing substances are based upon their hypocalcemic activity in Holtzerman rats that have been maintained on a low calcium diet for 4 days. A positive response, i.e., a decrease in plasma calcium, is measured one to one and one half hours after administration of the test substance. This demonstrates the rapid activity of TCT.

At the molecular level very little is known about the mode of action of TCT.

Clinically, it is hoped that TCT will be a useful therapeutic agent in the treatment of osteoporosis and other bone disorders. Some positive results have been obtained in hypercalcemia of malignancy and in an infant with idiopathic hypercalcemia. TCT is most effective in those states where bone resorption is most active and, therefore, should have some therapeutic potential in such conditions as Paget's disease.

Induced hypercalcemia triggers a rapid increase in the secretion of TCT, whereas a hypocalcemic state of serum induces an increase in the levels of PTH.

Thyrotropin, Thyptropar®, thyroid stimulating hormone TSH appears to be a glycoprotein (molecular weight 26,000 to 30,000) containing glucosamine, galactosamine, mannose and fucose, whose homogeneity is yet to be established. It is produced by the basophil cells of the anterior lobe of the pituitary gland. TSH enters the circulation from the pituitary, presumably traversing cell membranes in the process. After exogenous administration it is widely distributed and disappears very rapidly from circulation. Some evidence suggests that the thyroid may directly inactivate some of the TSH via an oxidation mechanism that may involve iodine. TSH thus inactivated can be reactivated by certain reducing agents. TSH regulates the production by the thyroid gland of thyroxine which stimulates the metabolic rate. Thyroxine feedback mechanisms regulate the production of TSH by the pituitary gland.

The decreased secretion of TSH from the pituitary is a part of a generalized hypopituitarism that leads to hypothyroidism. This type of hypothyroidism can be distinguished from primary hypothyroidism by the administration of TSH in doses sufficient to increase the uptake of radioiodine or to elevate the blood or plasma protein-bound iodine (PBI) as a consequence of enhanced secretion of hormonal iodine (thyroxine).

It is of interest that massive doses of vitamin A inhibit the secretion of TSH.

TSH is used as a diagnostic agent to differentiate between primary and secondary hypothyroidism. Its use in hypothyroidism due to pituitary deficiency has limited application; other forms of treatment are preferable.

Dose, intramuscular or subcutaneous, 10 U.S.P. Units.

THE PROSTAGLANDINS

In 1930 Kurzrok and Lieb showed that human semen effected a strong contraction of the uteri of women with a history of childless marriages whereas a relaxation of the uteri from fertile women was obtained. A few years later Goldblatt in England and von Euler in Sweden showed that human seminal plasma exerted a strong smooth-muscle stimulating effect. The factor (now known to be composed of several structurally related compounds) responsible for the above effects was named prostaglandin because it was supposed that it was secreted by the prostrate.

Although the prostaglandins are found in the highest concentrations in seminal fluids being formed in the seminal vesicles and in the vesicular gland, they also are found in low concentrations in lung, iris, stomach, intestinal mucosa, brain and human endometrial curettings and, to a lesser extent, in the uterus, thymus, heart, liver, kidney and human decidua. Of the 13 prostaglandins that have been found in human and sheep seminal plasma, the PGEs occur in the greatest amounts and $PGF_{2\alpha}$ is the one that is most commonly distributed both in various organs and in various animals.

The prostaglandins are liposoluble acids derived in vivo from the unsaturated eicosanoic acids. Thus PGE_1 (perhaps the most important member) is derived from the all *cis*-8,11,14-eicosatrienoic acid. The prostaglandins are derivatives of prostanoic acid and at the present time all the naturally occurring prostaglandins are congeners of 15(S)-hydroxy-13-trans-prostenoic acid, which may be depicted as follows. An oxy-

gen function—either an α-OH or a keto group—also always is present at position 9. Some members may contain an 11α-OH group. Thus the structure of PGE$_1$ first isolated from sheep glands is as follows and is $(-)$ 11α, 15(S)-dihydroxy-9-oxo-13-

PGE$_1$

trans-prostenic acid. The subscript utilized in prostaglandin nomenclature indicates the number of double bonds present in the molecule in the side chains attached to the 5 membered ring. Thus PGE$_2$ is $\triangle$5,6 *cis* PGE$_1$ and PGE$_3$ is $\triangle$5,6,17,18 di*cis* PGE$_1$. Because the PGEs are betahydroxy ketones they readily dehydrate to yield the alpha-beta unsaturated compounds (double bond in the 10,11 position) which are called the PGAs; the one derived from PGE$_1$ is called PGA$_1$. The PGAs therefore in some instances might be artifacts generated during isolation of the PGEs. The letters E, F, A and B are used to designate the differences found in the composition of the 5-membered ring in the natural prostaglandins.

The $\triangle$10,11 double bond in the PGAs can rearrange to the $\triangle$8,12 position to give the PGBs; the one from PGA$_1$ is called PGB$_1$; PGE$_2$ is $\triangle$5,6-PGE$_1$, and PGE$_3$ is $\triangle$17,18-PGE$_2$. The PGFs differ from the PGEs only in that the former contain an α-OH at the 9 position in lieu of a keto group. The PGFs do not dehydrate readily as do the PGEs. Thus PGF$_{1a}$ would be $(-)$ 9α,11α,15 (S)-trihydroxy 13-transprostenoic acid. The α in PGF$_{1a}$ signifies the position of the OH group.

The prostaglandins demonstrate a great variety of activities, many of which have been demonstrated in man. Not all the prostaglandins exhibit the same spectrum of activities from both a qualitative and a quantitative standpoint. The activities of PGE$_1$ probably have been the most extensively investigated.

The prostaglandins (chiefly PGE$_1$ and PGF$_{2a}$) have the ability to lower blood pressure (an important activity). Units of activity have been proposed; thus, 1 Euler unit ($= 4.5\ \mu g$ of PGE$_1$) will lower the blood pressure of rabbits about 30 per cent. PGE$_1$ administered I.V. uniformly lowers arterial pressure in dog, cat, rabbit, guinea pig, rat, mouse and chick. Stimulation (contraction) of various smooth muscles (the basis of some assays) in the intestine of rabbit and guinea pig and rat stomach fundus by PGE$_1$ is not inhibited by antimuscarinics (atropine), antihistamines, and α- and β-adrenergic blockers. PGE$_1$ in small doses can enhance the activities of acetylcholine, vasopressin, angiotensin, serotonin, bradykinin and procaine on rat stomach fundus, and of bretylium, dichloroisoproterenol, dibenamine, hexamethonium, epi, norepi and tyramine on selected tissues. On the other hand, PGE$_1$ can cause relaxation of the rat duodenum and can antagonize the action of epi and norepi on the isolated rat stomach fundus.

The contractile action of PGE$_1$ seems to be coupled with oxidative metabolism and membrane depolarization, and release of bound Ca^{++} may be involved.

The prostaglandins E$_1$, E$_2$, and A$_1$ are powerful inhibitors of gastric secretion, and exert significant antiulcer effects.

In humans all three PGEs (most active), PGA$_1$, PGB, and the 19 OH analogs decrease tonus, frequency and amplitude of spontaneous contraction of uterine strips. The PGEs cause a relaxation of the human myometrium, an activity that is enhanced at ovulation. On the other hand, PGF$_{1a}$ and PGF$_{2a}$ cause contraction.

At low doses PGE$_1$ I.V. stimulates free fatty acid (FFA) mobilization via the sympathetic nervous system, probably as a compensatory reaction to lowered blood pressure. At high doses FFA mobilization is reduced by inhibition of lipolysis enough to overcome the stimulation of FFA mobilization by epi and norepi.* PGE_1 probably acts at some point in the lipolytic process before the activation of the lipase system by cyclic AMP. It should be noted that the PGEs in particular can inhibit epi-stimulated cyclic AMP formation and the enzyme adenyl cyclase which forms cyclic AMP has been found in all tissues except possibly the erythrocytes. PGE_1 competitively inhibits the stimulation of lipolysis in rat adipose tissue by theophylline or vasopressin. The lipolytic activity of glucagon, histamine and the growth hormone also is altered by the prostaglandins.

PGE_1 has hyperglycemic effects that may be mediated by a reflex release of adrenal epi in response to lowered blood pressure. It is known that several hypotensive drugs cause hyperglycemia in dogs.

The prostaglandins are very potent substances; 0.05 to 0.1 μg per kg. per minute I.V. caused a flushing of the face and a headache of a pulsating character. These effects may be of C.N.S. origin. PGE administered intraventricularly to cats causes catatonia and transitory stupor. The prostaglandins are distributed throughout all regions of the central nervous system and, when one considers their occurrence, release via nerve stimulation and profound biologic effects in the C.N.S., the probability of their participation in some unknown way at central synapses is suggested.

$PGF_2\alpha$, unlike most other prostaglandins, is pressor in rats and dogs. In Rhesus monkeys $PGF_2\alpha$ effects a regression of the corpus luteum and a shedding of the uterine endometrium, to start the menstrual cycle even after an ovum has been fertilized.

The application of prostaglandins as

* Epi and norepi raise plasma FFA and glycerol by mobilizing them from adipose tissue. A triglyceride lipase called the hormone sensitive lipase appears to be rate-limiting in the lipolytic process. Cyclic $3',5'$=AMP plays an important role in mediating the lipolytic actions of catecholamines and agents by activating this lipase. Glucose metabolism has a key position in the control of re-esterification by providing the obligatory α-glycerophosphate, which cannot be formed from glycerol in adipose tissue.

medicinal agents seems only a matter of time.

REFERENCES CITED

1. Corey, R. B., and Pauling, L.: Proc. Roy. Soc. London (ser. B) 141:10, 1953, see also Ad. Protein Chem., p. 147, 1957.
2. C. E. N., p. 53, 1964.
3. Edman, P.: Acta chem. scandinav. 4:283, 1950.
4. Fox, S., et al.: J. Am. Chem. Soc. 77:3119, 1953.
5. Rhinesmith, H., et al.: J. Am. Chem. Soc. 80:3358, 1958.
6. Jones, R., et al.: J. Am. Chem. Soc. 81:3161, 1959.
7. Schroeder, W.: Ann. Rev. Biochem. 32:301, 1963; Progr. Chem. Natural Products 17:322, 1959.
8. Dixon, M., and Webb, E. C.: Enzymes, pp. 185-227, New York, Acad. Press, 1958. Also see pp. 672-785 in 2nd ed. 1964.
9. Lumry, R., and Eyring, H.: J. Phys. Chem. 58:110, 1954.
10. Koshland, D.: in The Enzymes, vol. 1, p. 305, New York, Acad. Press, 1959.
11. Hein, G., and Nieman, C.: J. Am. Chem. Soc. 84:4487, 4495, 1962.
12. Cunningham, L. W.: Science 125:1145, 1957.
13. Smith, E.: J. Biol. Chem. 233:1392, 1958.
14. Neurath, H.: The activation of zymogens, Adv. Protein Chem. 12:320, 1957.
15. Schramm, M.: Ann. Rev. Biochem. 36:307, 1967.
16. Isliker, H.: The chemical nature of antibodies, Adv. Protein Chem. 12:387, 1957.
17. Ann. Rev. Biochem. 36:55-76, 1967.
18. Colman, R. W.: Clin. Pharmacol. Ther. 6:602, 1965.
19. Fruton, J. S., et al.: Proc. Nat. Acad. Sci. 47:759, 1961.
20. Drenth, J., et al.: Nature 218:929, 1968.
21. Colman, R. W.: Clin. Pharmacol. Ther. 6:602, 1965.
22. Shibata, K., et al.: Biochim. biophys. acta 81:323, 1964.
23. Dube, S. K., et al.: J. Biol. Chem. 239:1809, 1963.
24. Cohen, S. C., et al.: J. Am. Chem. Soc. 84:4163, 1962; Rapp, J., and Niemann, C.: J. Am. Chem. Soc. 85:1896, 1963.
25. Huang, H., and Niemann, C.: J. Am. Chem. Soc. 73:1514, 1951.
26. See References found in ref. 28.

27. Niemann, C.: Science 143:1289, 1964.
28. Wohlman, A., *et al.*: Canad. J. Physiol. Pharmacol. 47:301, 1969.
29. Bruce, R. A., and Quinton, K. C.: Brit. Med. J. 1:282, p. 62.
30. Lloyd, J. O., *et al.*: Sci. Am. 219:34, 1968; Adamson, R. H., and Fabro, S.: Cancer, Chemother. Rep. 52:617, 1968; Broome, J. D.: Trans. N. Y. Acad. Sci. 30:690, 1968.
31. Chance, R. E., *et al.*: Science 161:165, 1968.
32. For references to Sanger's studies, see Ann. Rev. Biochem. 27:58, 1958.
33. Randle, P. J.: *in* Pincus, G., *et al.* (eds.): The Hormones, vol. 4, p. 483, New York, Acad. Press, 1964.
34. Katsoyannis, P. G.: Science 154:1509, 1966.
35. Smith, E. L., *et al.*: Biochim. biophys. acta 29:207, 1958.
36. Slobin, L. I., and Carpenter, F. N.: Biochem. 2:22, 1963; Mirsky, I. A., and Kawamura, K.: Endocrinol. 78:1115, 1966.
37. Klostermeyer, H., and Humbel, R. E.: Angew. Chem. 5:810, 1966.
38. Piper, H. A., *et al.*: J. Biol. Chem. 58:321, 1923.
39. Storvick, W. O., and Henry, H. J.: Diabetes 17:499, 1968.
40. Somogyi, M., *et al.*: J. Biol. Chem. 60:31, 1924.
41. Taylor, K. W., *et al.*: Biochem. biophys. acta 100:521, 1965.
42. Klostermeyer, H., and Humbel, R. E.: Angew. Chem. 5:807, 1966.
43. Randle, P. J.: *in* Pincus, G., *et al.* (eds.): The Hormones, vol. 4, p. 520, New York, Acad. Press, 1964.
44. Moloney, P. J., *et al.*: J. New Drugs 4:258, 1964.
45. Bressler, R., *et al.*: Diabetes 17:617, 1968.
46. Seltzer, H. S., *et al.*: Diabetes 18:19, 1969.
47. Bell, P. N.: J. Am. Chem. Soc. 76:5565, 1954.
48. Schwyzer, R., and Sieber, P.: Nature 199:172, 1963.
49. Bell, P. H., *et al.*: J. Am. Chem. Soc. 78:5059, 1956.
50. Hofmann, K., *et al.*: J. Am. Chem. Soc. 84:1054, 4475, 1962; Lebovitz, H. E., and Engel, F. L.: Endocrinol. 75:831, 1964.
51. Li, C. H.: Sci. Am. 209:46, 1963.
52. Li, C. H., *et al.*: J. Am. Chem. Soc. 84:2460, 1962.
53. Rose, L. T., *et al.*: Clin. Pharmacol. Ther. 9:740, 1968.
54. Dixon, H., and Weitkamp, L.: Biochem. J. 84:462, 1962.
55. Hofmann, K., *et al.*: J. Am. Chem. Soc. 85:1546, 1963.
56. Dorfman, R. I., *et al.*: Acta Endocrinol. 55:43, 1967.
57. White, W. F., and Landmann, W. A.: J. Am. Chem. Soc. 77:1711, 1955.
58. Blair, A. J., *et al.*: Can. J. Physiol. Pharmacol. 42:391, 1964.
59. Hope, D., and Vigneaud, V. du: J. Biol. Chem. 237:3146, 1962.
60. Rudinger, J., and Jost, K.: Experientia 20:570, 1964.
61. Bisset, G., *et al.*: Nature 199:69, 1963; J. Physiol. 169:12-40P, 1963.
62. Vigneaud, V. du, *et al.*: J. Biol. Chem. 239:472, 1964.
63. Baxter, J. W. M., *et al.*: Biochem. 8:3592, 1969.
64. Gillessen, D., and duVigneaud, V.: J. Med. Chem. 13:346, 1970.
65. Vigneaud, V. du, *et al.*: J. Am. Chem. Soc. 75:4879, 1953; 76:3115, 1954; Science 123:967, 1956.
66. Drabarek, S.: J. Am. Chem. Soc. 86:4477, 1964.
67. Rucanski, B., *et al.*: Science 160:81, 1968.
68. Chan, W. Y., and Vigneaud, V. du: Endocrinol. 71:977, 1962; Vigneaud, V. du, *et al.*: J. Biol. Chem. 239:472, 1964.
69. van Dyke, H. B., *et al.*: Endocrinol. 73:637, 1963.
70. Schneider, C., and Vigneaud, V. du: J. Am. Chem. Soc. 84:3005, 1972.
71. Vigneaud, V. du, *et al.*: Proc. Soc. Exp. Biol. Med. 104:653, 1960.
72. Risset, G. W., *et al.*: Nature 199:29, 1963.
73. Hope, D., *et al.*: J. Am. Chem. Soc. 85:3686, 1963.
74. Jost, K., *et al.*: Coll. Czeck. Chem. Comm. 28:2021, 1960.
75. Vigneaud, V. du, *et al.*: J. Biol. Chem. 238:P.C. 1560, 1963.
76. Smyth, D., *et al.*: Nature 199:69, 1963.
77. Johl, A., *et al.*: Biochim. biophys. acta 69:193, 1963.
78. Chan, W. Y., *et al.*: Endocrinol. 81:1267, 1968.
79. Schally, A. V., *et al.*: Arch. Biochem. Biophys. 107:332, 1964.
80. Vigneaud, V. du: J. Am. Chem. Soc. 80:3355, 1958.
81. Traber, D. L., *et al.*: Arch. int. Pharmacodyn. 168:288, 1967.

82. Schwartz, I. L., and Livingston, L. M.: Vitamins and Hormones 22:261, 1964.
83. Arimura, A., *et al.*: Endocrinol. 84:579, 1969.
84. Berde, B., *et al.*: Experientia 20:42, 1964, Bodansky, M., *et al.*: J. Am. Chem. Soc. 86:4452, 1964.
85. Bodanszky, M., and Birkhimer, C.: Am. Chem. Soc. 84:4943, 1962.
86. Bosomworth, P. P.: J. New Drugs 5:308, 1965.
87. For other names *see* Li, C. H.: Adv. Protein Chem. 12:269, 1957 (including the proposed one, melanotropin).
88. Geschwind, I. I., *et al.*: J. Am. Chem. Soc. 78:4494, 1956.
89. van Hell, H., *et al.*: Acta Endocrinol. 59:89-138, 1968; Bahl, O.: J. Biol. Chem. 244:567, 575, 1969.
90. Ng, K., and Vane, J. R.: Nature 218:144, 1968.
91. Schwartz, H.: J. Am. Chem. Soc. 79:5697, 1957; Science 125:886, 1957.
92. Arakawa, K., et al.: Biochem. J. 104:900, 1967; Bull. Chem. Soc. Japan 41:433, 1968.
93. Peach, M., and Ford, G.: J. Pharmacol. Exp. Ther. 162:92, 1968.
94. Hofmann, K., *et al.*: J. Am. Chem. Soc. 90:1654, 1968.
95. Schwyzer, R.: Pure Appl. Chem. 6:265, 1963.
96. Riniker, B., *et al.*: Angew. Chem. 74:469, 1962.
97. Arakawa, K., *et al.*: J. Am. Chem. Soc. 84:1424, 1962.
98. Regoli, B., *et al.*: Biochem. Pharmacol. 12:637, 1963.
99. Hess, H. J., and Constantine, J.: J. Med. Chem. 7:602, 1964.
100. Schwyzer, R.: Pure and Appl. Chem. 6:284, 1963.
101. Khosla, M. C., *et al.*: Biochem. 6:754, 1967.
102. Scheraga, H., *et al.*: Biochem. 2:1327, 1963.
103. Smeby, R., *et al.*: Biochim. Biophys. Acta 58:550, 1962.
104. Staub, A., *et al.*: Science 117:628, 1953.
105. Unger, R. H., and Eisentraut, A. M.: Diabetes 13:563, 1964; Ohneda, A., *et al.*: Diabetes 18:1, 1969.
106. Foa, P. P.: *in* Pincus, G., *et al.* (eds.): The Hormones, vol. 4, p. 550, New York, Acad. Press, 1964; Lawrence, A. M.: Med. Clin. N. Am. 54:183, 1970.
107. Elliott, D. F., and Lewis, G. P.: Biochem. J. 95:437, 1965.
108. Anastasi, A., *et al.*: Brit. J. Pharmacol. Chemother. 27:479, 1966.
109. Suzuki, K., *et al.*: Chem. Pharm. Bull. 14:211, 1966; Bodanszky, M., *et al.*: J. Am. Chem. Soc. 85:991, 1963.
110. Schroeder, E.: Annalen der Chem. 679:207, 1964.
111. Nicolaides, E. D., *et al.*: J. Med. Chem. 6:524, 1963.
112. Stewart, J. M.: Fed. Proc. 27:63, 1968.
113. Posati, L. P., and Pallansch, M. J.: Science 168:121, 1968.
114. Potts, J. T., *et al.*: Proc. Nat. Acad. Sci. 59:1321, 1968; Rittel, W., *et al.*: Helv. chim. acta 51:924, 1968; Arnaud, C. D., and Tsao, H.: Biochem. 8:449, 1969.

SELECTED READING

Albanese, A.: Protein and Amino Acid Metabolism, New York, Acad. Press, 1959.
Bergstrom, S.: Prostaglandins, Science 157:382, 1967.
———: The prostaglandins, Pharmacol. Rev. 20:1, 1968.
Boyer, P., *et al.*: The Enzymes, vols. 1-8, New York, Acad. Press, 1959-1964.
Collier, H.: Bradykinin and its allies, Endeavour 100:14, 1968.
Colman, R. W.: Proteolytic enzymes in clinical medicine, Clin. Pharmacol. Ther. 6:598, 1965.
Dixon, M.: Enzymes, New York, Acad. Press, 1958-1964.
Eisenstein, H. B.: The Biochemical Aspects of Hormone Action, Boston, Little, 1964.
Fox, S.: Introduction to Protein Chemistry, New York, Wiley, 1957.
Greenstein, J. P., and Winitz, M.: Chemistry of the Amino Acids, New York, Wiley, 1961.
Haurowitz, F.: The Chemistry and Function of Proteins, New York, Academic Press, 1963.
Hofmann, K.: Structure-Function Relations in the Corticotropin Series, Pharmaceutical Chemistry, p. 245, London, Butterworths, 1963.
Horton, E. W.: Hypotheses on physiological roles of prostaglandins, Physiol. Rev. 49:122, 1969.
Klostermeyer, H., and Humbel, R. E.: The chemistry and biochemistry of insulin, Angew. Chem. Int. Ed. 5:807, 1966.
Knobil, E., and Hotchkiss, J.: Growth hormone, Ann. Rev. Physiol. 26:47, 1964.

Knowles, J. R.: On the mechanism of action of pepsin, Phil. Trans. Roy. Soc. Lond. B 257: 135, 1970.

Law, H. D.: Polypeptides of medicinal interest, Progr. Med. Chem. 4:86, 1965.

Litwack, G., and Kritchevsky, D.: Actions of Hormones on Molecular Processes, New York, Wiley, 1964.

Matsuzaki, F., and Raben, M. S.: Growth hormone, Ann. Rev. Pharmacol. 5:137, 1965.

Neuberger, A.: Symposium on Protein Structure, New York, Wiley, 1958.

Neurath, H.: The Proteins, vol. 1, 1963, vol. 2, 1964, vol. 3, 1965, New York, Acad. Press.

Perutz, M. F.: The Hemoglobin Molecule, Sci. Am. Nov. 211:64, 1964.

Pickles, V. R.: The prostaglandins, Biol. Rev. 42:614, 1967.

Rieser, P.: Insulin, Membranes and Metabolism, Baltimore, The Williams and Wilkins Co., 1967.

Sahyun, M.: Outline of Amino Acids and Proteins, New York, Reinhold, 1948.

Schram, M.: Secretion of enzymes and other macromolecules, Ann. Rev. Biochem. 36: 307, 1967.

Schroeder, W. A.: The Chemical Structure of the Normal Hemoglobins, Progress in the Chemistry of Organic Natural Products 17: 322, 1959, Springer Verlag.

Schwartz, I. L., and Livingston, L. M.: Cellular and molecular aspects of the antidiuretic action of vasopressins and related peptides, Vitamins and Hormones 22:261, 1964.

Schwyzer, R.: Chemical structure and biological activity in the field of polypeptide hormones, Pure and Applied Chem. 6:265-295, 1963.

Skeggs, L. T., et al.: Biochemistry and kinetics of the renin-angiotensin system, Fed. Proc. 26:42, 1967.

Sneath, P. H. A.: Relations between chemical structure and biological activity in peptides, J. Theoret. Biol. 12:157, 1966.

Steiner, R. F.: The Chemical Foundations of Molecular Biology, New York, van Nostrand, 1965.

Stewart, J. M.: Structure activity relationships in bradykinin analogues, Fed. Proc. 27:63, 1968.

Summer, J. B., and Myrbäck, K.: The Enzymes, New York, Acad. Press, 1950-1952.

Traketellis, A. C., and Schwartz, G. P.: Insulin, *in* Progress in the Chemistry of Organic Natural Products, Zechmeister, L. (ed.), Wien, Springer Verlag., p. 121, 1968.

Vickery, H. B., and Schmidt, C. L. A.: The history of the discovery of the amino acids, Chem. Rev. 9:169, 1931.

Von Euler, V. S.: Prostaglandins, Clin. Pharmacol. Therap. 9:228, 1968.

Von Euler, V. S., and Eliasson, R.: Prostaglandins, New York, Acad. Press, 1967.

Walter, R., et al.: Chemistry and structure-activity relations of the antidiuretic hormone, Am. J. Med. 42:653, 1967.

28

Vitamins

Ole Gisvold, Ph.D.,
Professor of Medicinal Chemistry, College of Pharmacy,
University of Minnesota

INTRODUCTION

In 1905, Pekelharing and, in 1906, Hopkins pointed out that in addition to proteins, fats, carbohydrates and minerals a small amount of milk was necessary to maintain animal life. Hopkins concluded that milk contained "accessory food factors." The word "vitamine" first was coined by Funk, in 1912, to describe the substance that was present in rice polishings (Eijkman's antiberiberi factor) and in foods that cured polyneuritis in birds and beriberi in man. Because the antiberiberi substance contained nitrogen, it was thought to be an amine. This, is addition to its being necessary for life, gave rise to the term "vitamine," in which the prefix vita means life. Later investigations revealed the presence of other "accessory food factors" or "vitamines" that did not contain nitrogen and hence Drummond's suggestion to drop the terminal "e" was accepted.

Symptoms or diseases in humans, that we know today are due to nutritional deficiencies of vitamins, have been described or known for centuries. Some of these have been described under the names of beriberi, scurvy, rickets, pellagra and night blindness.

All known naturally occurring vitamins are synthesized by plants, with the exception of the vitamins D and vitamin A; however, precursors (provitamins) in these two cases also are synthesized by plants. In some cases, some animals and birds can synthesize some of the vitamins and provitamins, such as vitamin C and 7-dehydrocholesterol.

Vitamins are classified arbitrarily according to their solubility in water and fats, e.g., fat-soluble A, D, E and K and the water-soluble B_1, B_2, B_6, B_{12}, C, nicotinic acid, folic acid, pantothenic acid, biotin, inositol and p-aminobenzoic acid.

Vitamins can be administered in doses far exceeding the daily requirement with no apparent untoward effects. The intake of the water-soluble vitamins in excess of that needed by the body is excreted in the urine. The amounts of vitamins found in the urine furnish a means of measuring the vitamin reserves in the body. The fat-soluble vitamins usually are stored in the liver and, thus, the body is able to conserve these factors.

LIPID-SOLUBLE VITAMINS

THE VITAMINS A

About 1913, McCollum and Davis[1] and Osborne and Mendel[2] showed that rations of purified casein, carbohydrates, various salt mixtures and lard induced an apparent normal growth in experimental animals for periods that varied from 70 to 120 days, after which time little or no increase in body weight could be induced. The resumption of growth occurred quite promptly upon addition to the diet of the ether extract of egg or butter. The factor responsible for this growth was called "fat-soluble A" to distinguish it from the "water-soluble B." Therefore, it was called vitamin A. Further work showed this factor to be present in cod-liver oil[3] but not in lard, olive, corn, cottonseed, linseed, soybean or almond oils.[4] McCollum and Davis,[4] in 1914, showed that this factor could be concentrated in the non-saponifiable portion of butter oil. It was shown to be absent from cereal grains and seeds, whereas alfalfa and cabbage leaves were found to be excellent sources of the vitamin. Furthermore, ether extracts of spinach leaf or clover were shown to be rich in vitamin A. In about 1919, Steenbock[5] pointed out that the vitamin A potency of certain plant sources seemed to run parallel with the amount of yellow, fat-soluble pigments present in them. He prepared[6] nonsaponifiable concentrates from carrots, alfalfa and yellow corn. He suggested that vitamin A activity might be associated with the "carotenoid pigments." Because cod-liver oil concentrates (nonsaponifiable portion) are essentially colorless but very potent in vitamin A activity, Steenbock stated that the vitamin A of animals might be a colorless or leuco form of carotene. In 1928, von Euler[7] noted that substances that were rich in vitamin A gave certain chemical color tests similar to those given by carotene. He demonstrated that carotene was active when fed to vitamin A deficient rats. Moore, in 1930, showed that ingested carotene is converted to vitamin A by the rat.[8] This established the relationship to the active yellow carotenes of plants and the nearly colorless, highly active vitamin concentrates from liver oils. Karrer,[9] who had previously determined the constitution of beta-carotene, suspected a structural relationship between certain carotenes and vitamin A. He degraded, by ozonization, an impure preparation of vitamin A obtained from halibut-liver oil and obtained geronic acid in an amount that indicated the presence of one beta-ionone ring. Two years later, the carbon skeleton of vitamin A was established by the synthesis, starting from

beta-ionone, of perhydrovitamin A.[10] Knowing that beta-carotene has a conjugated system of 11 double bonds and that vitamin A also contained a system of 5 conjugated double bonds and a primary alcohol group, Karrer proposed the structural formula for vitamin A below.

Two numbering systems to indicate the positions of the double bonds are reported in the literature (see formulas for vitamin A [all *trans*] and neovitamin A.)

Vitamin A (all *trans*)
(Retinol)

Neovitamin A (Δ^4-*cis* or 11-mono-*cis*
Vitamin A

For steric reasons the number of isomers of vitamin A most likely to occur would be limited. These are: All-*trans*, 9-*cis* (Δ^3-*cis*), 13-*cis* (Δ^5-*cis*) and the 9,13-di-*cis*. A *cis* linkage at double bond 7 or 11 encounters steric hindrance. The 11-*cis* isomer is twisted as well as bent at this linkage; nevertheless this is the only isomer that is active in vision (q.v.). Some of these have been prepared in a crystalline state.[11]

Most liver oils contain vitamin A and neovitamin A in the ratio of 2 to 1.

The biologic activity[12] of the isomers of vitamin A acetate in terms of U.S.P. units* per gram are as follows: Vitamin A, all *trans* 2,907,000; neovitamin A 2,190,000; Δ^3-*cis* 634,000; $\Delta^{3,5}$-di-*cis* 688,000 and $\Delta^{4,6}$-di-*cis* 679,000. In the case of the isomers of vitamin A aldehyde,[12] the following values have

* U.S.P. and I.U. units are the same, i.e., 0.3 mcg. of vitamin A alcohol (Retinol).

been reported; all *trans* 3,050,000; neo (Δ^5-*cis*) 3,120,000; Δ^3-*cis* 637,000; $\Delta^{3,\,5}$-di-*cis* 581,000 and $\Delta^{4,6}$-di-*cis* 1,610,000.

Disregarding stereochemical variations, a number of compounds with structures corresponding to vitamin A, its ethers and its esters have been prepared.[13] These compounds, as well as a synthetic vitamin A acid, possess biologic activity. Some of these compounds have been prepared.[14, 15, 16]

Although fish-liver oils are used as such for their vitamin A content, purified or concentrated forms of vitamin A are of great commercial significance. These are prepared in three ways: (1) saponification of the oil and concentration of the vitamin A in the nonsaponifiable matter by solvent extraction, the product is marketed as such; (2) molecular distillation of the nonsaponifiable matter, from which the sterols have previously been removed by freezing, giving a distillate of vitamin A containing 1,000,000 to 2,000,000 I.U. per gram; (3) subjecting the fish oil to direct molecular distillation to recover both the free vitamin A and vitamin A palmitate and myristate.

Pure crystalline vitamin A occurs as pale yellow plates or crystals. It melts at 63 to 64° and is insoluble in water but soluble in alcohol, the usual organic solvents and the fixed oils. It is unstable in the presence of light and oxygen and in oxidized or readily oxidized fats and oils. It can be protected by the exclusion of air and light and by the presence of antioxidants.

Like all substances that have a polyene structure, vitamin A gives color reactions with many reagents, most of which are either strong acids or chlorides of polyvalent metals. An intense blue color (Carr-Price) is obtained with vitamin A in dry chloroform solution upon the addition of a chloroform solution of antimony trichloride. This color reaction has been studied intensively and is the basis of a colorimetric assay for vitamin A.[17]

The 11-mono-*cis* isomer of vitamin A (Retinol) can be oxidized with manganese dioxide[18] to give biologically active neoretinene b.

The chief source of natural vitamin A is fish-liver oils, which vary greatly in their content of this vitamin (see Table 104). It

Neoretinene b.
(Retinal)

occurs free and combined as the biologically active esters, chiefly of palmitic and some myristic and dodecanoic acids. It also is found in the livers of animals, especially those which are herbivorous. Milk and eggs are fair sources of this vitamin. The provitamins A, e.g., beta, alpha, and gamma carotenes and cryptoxanthin, are found in green parts of plants, carrots, red palm oil, butter, apricots, peaches, yellow corn, egg yolks and other similar sources. The provitamins aphanin and myxoxanthin are found in sea algae. The carotenoid pigments are utilized poorly by humans, whereas animals differ in their ability to utilize these compounds. These carotenoid pigments are provitamins A because they are converted, in part in the liver, to the active vitamin A. This, because cleavage of the molecule at the double bond in the middle and conversion of this terminal carbon atom to a primary alcohol group can give rise in the case of β-carotene to 2 molecules of vitamin A, whereas in the other 3 carotenoids only 1 molecule of vitamin A is possible by this transformation.

TABLE 104. VITAMIN A CONTENT OF SOME FISH-LIVER OILS

SOURCE OF OIL	ANIMAL	POTENCY (I.U./ Gm.)
Halibut, liver	*Hippoglossus hippoglossus*	60,000
Percomorph, liver	Percomorph fishes (mixed oils)	60,000
Shark, liver	*Galeus zygopterus*	25,500
Shark, liver	*Hypoprion brevirostris* and other varieties	16,500
Burbot, liver	*Lota maculosa*	4,880
Cod, liver	*Gadus morrhua*	850

These carotenoids have only 1 ring (see formula for β-carotene) at the end of the polyene chain that is identical with that found in β-carotene and is necessary and found in vitamin A.

Oils or lipids enhance the absorption of carotene, which is poorly absorbed (10 per cent) when ingested in dry vegetables.

The conjugated double bond systems found in vitamin A and β-carotene are necessary for activity, for when these compounds are partially or completely reduced, activity is lost. The ester and methyl ethers of vitamin A have a biologic activity on a molar basis equal to vitamin A. Vitamin A acid is biologically active but is not stored in the liver. The rat can utilize it as efficiently as vitamin A in general tissue functions other than the role it plays in vision. Vitamin A acid cannot be reduced by the rat to vitamin A or Vitamin A aldehyde. Kitol, a substance obtained from the nonsaponifiable portion of whale oil after the removal of vitamin A, has no activity as such but upon pyrolysis yields some vitamin A. Highly purified kitol analyzes for the formula $C_{40}H_{58}(OH)_2$.

Vitamin A often is called the "growth vitamin" because a deficiency of it in the diet causes a cessation of growth in young rats.* A deficiency of vitamin A is manifested chiefly by a degeneration of the mucous membranes throughout the body. This degeneration is evidenced to a greater extent in the eye than in any other part of the body and gives rise to a condition known as xerophthalmia. The eyes become hemorrhagic, incrusted, infected and this is accompanied by a dryness of the conjunctivae. It is associated with atrophy of the para-ocular glands, with metaplasia of the epithelium of the glands and conjunctivae and with loss of sensation in the conjunctiva, especially affecting the cornea. At first, the dryness may be associated with photophobia. Triangular white spots (Bitot's spots) appear in the palpebral fissure; irregular wrinkled patches of xerosis are found over the bulbar conjunctivae and later over the cornea. A characteristic light-brown pigmentation develops in the con-

* This property is used as the basis for the quantitative assay of vitamin A-containing substances.

junctiva. The superficial layers of the cornea degenerate. In the final stage, corneal softening or keratomalacia develops, with resulting severe impairment of vision. Permanent blindness is likely to follow. In the earlier stages of vitamin A deficiency, there may develop a night blindness (nyctalopia) which can be cured by vitamin A. Night blindness can be defined as the inability to see in dim light. Vitamin A deficiency that leads to night blindness and total blindness is especially common in India, China, Yucatan, Java, Sumatra and the Malay States. It has been observed in prisoners, slaves and workmen and the people of Denmark during World War I.

"Dark adaptation" or "visual threshold" is a more suitable description than "night blindness" when applied to many subclinical cases of vitamin A deficiency. The visual threshold at any moment is just that light intensity required to elicit a visual sensation. Dark adaption is the change which the visual threshold undergoes during a stay in the dark after an exposure to light. This change may be very great. After exposure of the eye to daylight, a stay of 30 minutes in the dark results in a decrease in the threshold by a factor of a million. This phenomenon is used as the basis to detect subclinical cases of vitamin A deficiencies. These tests vary in their technic, but, essentially, they measure visual dark adaptation after exposure to bright light and compare it with the normal.[19]

Advanced deficiency of vitamin A gives rise to a dryness and scaliness of the skin, accompanied by a tendency to infection. Characteristic lesions of the human skin due to vitamin A deficiency usually occur in sexually mature persons between the ages of 16 and 30 and not in infants. These lesions appear first on the anterolateral surface of the thigh and on the posterolateral portion of the upper forearms and later spread to adjacent areas of the skin. The lesions consist of pigmented papules, up to 5 mm. in diameter, at the site of the hair follicles.

In the rat, marked vitamin A deficiency leads to a degeneration of the testes of the male, accompanied by sterility in 3 months. In the female rat, the vaginal mucous membrane becomes permanently cornified, and the ovaries also are affected.

Vitamin A regulates the activities of osteoblasts and osteoclasts, influencing the shape of the bones in the growing animal. The teeth also are affected. In vitamin A deficiency states, a long overgrowth occurs. Overdoses of vitamin A in infants for prolonged periods of time led to irreversible changes in the bones, including retardation of growth, premature closure of the epiphyses and differences in the lengths of the lower extremities. Thus, a close relationship exists between the functions of vitamins A and D with regard to cartilage, bones and teeth.[20]

The tocopherols exert a sparing and what appears to be a synergistic action[21] with vitamin A.

Blood levels of vitamin A decrease very slowly, and a decrease in dark adaptation was observed in only 2 of 27 volunteers (maintained on a vitamin A free diet) after 14 months, at which time blood levels had decreased from 88 I.U. per 100 ml. of blood to 60 I.U.

The mode of action of vitamin A in the deficiency symptoms described above is not definitely known. It has been demonstrated that vitamin A stimulates the production of mucus by the basal cells of the epithelium whereas in its absence keratin can be formed. Vitamin A plays a role in the biosynthesis of glycogen and some steroids, and increased quantities of the coenzymes Q (q.v.) are found in the livers of vitamin-deficient rats. However, vitamin A also plays a role in vision[22] that can be explained as follows. The modern duplicity theory considers the vertebrate retina as a double sense organ in which the rods are concerned with colorless vision at low light intensities and the cones with color vision at high light intensities. A dark-adapted, excised retina is rose-red in color and, when it is exposed to light, its color changes to chamois, to orange, to pale yellow, and finally upon prolonged irradiation it becomes colorless. The rods contain photosensitive visual purple (rhodopsin) which, when acted upon by light of a definite wavelength, is converted to visual yellow and initiates a series of chemical steps necessary to vision. Visual purple is a conjugated, carotenoid protein having a molecular weight of about 40,000 and 1 prosthetic group per molecule. It has an absorption maximum of about 510 mμ. The prosthetic group is retinene (neoretinene b or retinal) which is joined to the protein through a protonated Shiff's base linkage. The function of retinene in visual purple is to provide an increased absorption coefficient in visible light and thus sensitize the protein which is denatured. This process initiates a series of physical and chemical steps necessary to vision. The protein itself differs from other proteins by having a lower energy of activation, which permits it to be denatured by a quantum of visible light. Other proteins require a quantum of ultraviolet light to be denatured. The bond between the pigment and the protein is much weaker when the protein is denatured than when it is native. The denaturation process of the protein is reversible and takes place more readily in the dark to give rise, when combined with retinene, to visual purple. The effectiveness of the spectrum in bleaching visual purple runs fairly parallel with its absorption spectrum (510 mμ) and with the sensibility distribution of the eye in the spectrum at low illuminations. It has been calculated that for man to see a barely perceptible flash of light, in a dark-adapted eye there need be transformed photochemically only 1 molecule of visual purple in each 5 to 14 rod cells. In vivo, visual purple is constantly re-formed as it is bleached by light, and, under continuous illumination, an equilibrium between visual purple,* visual yellow,† and visual white‡ is maintained. If an animal is placed in the dark, the regeneration of visual purple continues until a maximum concentration is obtained. Visual purple in the eyes of an intact animal may be bleached by light and regenerated in the dark an enormous number of times.

Visual purple has been found in the whole vertebrate series from Petromyzon to man. In cattle the dry retina contains 14 per cent of rhodopsin and in the frog 35 per cent of rhodopsin. It is not distributed

---

* Protein combined with retinene.
† Protein denatured plus free retinene.
‡ Protein plus vitamin A.

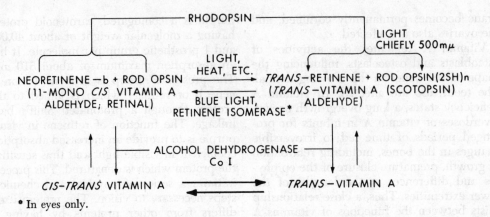

* In eyes only.

evenly over the retina. It is missing in the fovea, and in the regions outside of the fovea its concentration undoubtedly increases to a maximum in the region about 20° off center, corresponding to the high density of rods in this region. Therefore, to see an object best in the dark, one should not look directly at it.

The diagram shown above represents some of the changes that take place in the visual cycle involving the rhodopsin system in which the 11-mono-*cis* isomer of vitamin A is functional in the aldehyde form.

Temperature controlled studies led to results that can be depicted as follows.

Reaction	Product	Color
Photochemical stereisomerization of chromophore (reversible)	Rhodopsin (11-Mono-*cis*-vitamin-A aldehyde)	Red
Thermal rearrangement of the receptor site of opsin	Lumirhodopsin (all-*trans* vitamin A-aldehyde)	Red to Violet
Hydrolysis of chromophore from opsin	Metarhodopsin (all-*trans* vitamin A-aldehyde)	Orange to Red
	Opsin + (all-*trans* vitamin A-aldehyde)	Yellow

In these studies the meta pigments were found to be stable below −20° C. but, on warming, will hydrolyze in the dark.

Three additional visual pigments are known: (1) iodopsin, composed of cone opsin and retinene₁; (2) porphyropsin, composed of retinene₂ and rod opsin; (3) cyanopsin, composed of retinene₂ and cone opsin.

Only neoretinene b (retinal) (Δ^4- or 11-mono-*cis*) can combine with opsin (scotopsin) to form rhodopsin. The isomerization of *trans*-retinene may take place in the presence of blue light. However, vision continues very well in yellow, orange and red light in which no isomerization takes place. The neoretinene b (retinal) under these circumstances is replaced by an active form of vitamin A from the bloodstream which, in turn, obtains it from stores in the liver. The isomerization of *trans*-vitamin A in the body to *cis-trans*-vitamin A seems to keep pace with long-term processes such as growth, since vitamin A, neovitamin A and neoretinene b (retinal) are equally active in growth tests in rats.

The sulfhydryl groups (2 for each retinene molecule isomerized) exposed on the opsin molecule play a role of a cathode nature in the transmission of impulses in the phenomenon of vision.

Pure vitamin A has the activity of 3,500,-000 I.U. per gram. The average daily requirements for vitamin A are 1,500 I.U. for infants and 5,000 I.U. for adults. The therapeutic adult dose is 25,000 Units. Moderate to massive doses of vitamin A have been used in pregnancy, lactation, acne, abortion of colds, removal of persistent follicular hyperkeratosis of the arms, persistent and abnormal warts, corns and calluses and similar conditions. Phosphatides or the tocopherols enhance the absorption of vitamin A.

Vitamin A U.S.P. is a product that contains retinol (vitamin A alcohol) or its esters from edible fatty acids, chiefly acetic and palmitic acids and whose activity is n.l.t. 95 per cent of the labeled amount. If it is retinol, its potency is n.l.t. 1660 U.S.P. units per mg. (0.3 mcg. of vitamin A alcohol equals 1 U.S.P. Unit). Retinol acetate has 1,450 units and the palmitate has 910 units per mg.

Vitamin A occurs as a yellow to red, oily liquid; it is nearly odorless or has a fishy odor and is unstable to air and light. It is insoluble in water or glycerin and is soluble in absolute alcohol, vegetable oils, ether and chloroform.

Category—antixerophthalmic vitamin.

Usual dose—prophylactic, 1.5 mg. (5,000 units) daily; therapeutic, 7.5 mg. (25,000 units).

Usual dose range—prophylactic, 1.2 to 2.4 mg. (4,000 to 8,000 units); therapeutic 7.5 to 60 mg. (25,000 to 200,000 units) daily.

OCCURRENCE

Vitamin A Capsules U.S.P.

VITAMIN A₂

Vitamin A₂ is found in vertebrates which live or at least begin their lives in fresh water. It also is found in the marine fishes of the family Labridae (wrasses) and in turtles. Vitamin A₂ exhibits chemical, physical and biologic properties very similar to vitamin A. The crystalline vitamin A₂, phenylazobenzoate, was orange colored and melted at 76 to 77°. The regenerated alcohol was obtained as an orange-yellow oil which showed two absorption maxima in the ultraviolet, one at 351 mμ and a subsidiary peak at 287 mμ. A chloroform solution of antimony trichloride gave a blue color with a single absorption maximum at 693 mμ. Vitamin A₂ has the structural formula shown below.

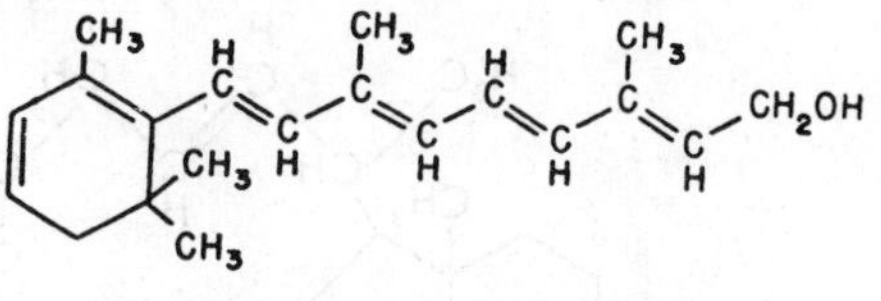

Vitamin A₂ (all trans)
3-Dehydroretinol or dehydroretinol

Vitamin A₂ has a biologic potency of 1,300,000 U.S.P. units per gram which is approximately 40 per cent of the activity of crystalline vitamin A acetate. It is stored in the liver of the rat but does not appear to be converted to vitamin A in vivo.

In the rods (of the retinas) of fresh water vertebrates—lampreys, fresh water fishes and certain amphibia larvae—rhodopsin is replaced by the purple, light-sensitive pigment porphyropsin (q.v.). The retina of the tortoise contains only cones and therefore would have cyanopsin (q.v.).

Beta-carotene, a precursor to vitamin A, is synthesized by microorganisms in sufficient quantities to provide a commercial source of supply. Beta-carotene is used as a coloring agent in butter, in animal food supplements, etc.

THE VITAMINS D

The first full description of rickets was published in a treatise by Francis Glisson, in 1650, an English professor at Cambridge. This deficiency disease is most common in northern countries, especially in large cities, whereas it is fairly rare in southern countries. The idea that rickets was connected with nutrition began to develop in the last quarter of the 19th century. In about 1890, Palm,[23] an English medical man, first pointed out that rickets is prevalent where there is little sunlight and quite rare wherever sunshine is abundant. He recommended the use of sunlight in the treatment of rickets. J. Raczynski,[24] in 1913, exposed puppies to sunlight to see if he was able to increase the amount of mineral substances in their bones and, from his results, concluded that sunlight plays a principal role in the etiology of rickets. In 1919-1920, Huldschinsky[25] reported that ultraviolet light cured rickets in children. Hess and Unger,[26] in 1921, confirmed this observation

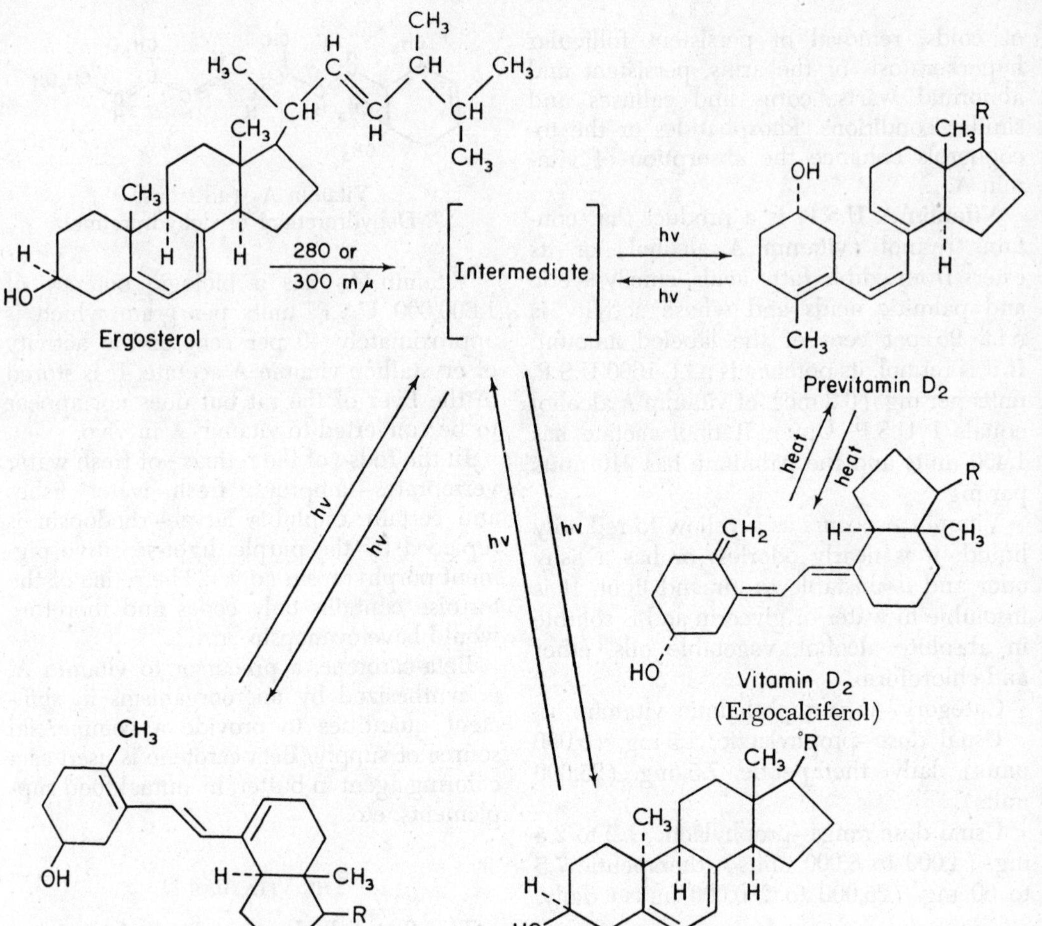

Ergosterol

Intermediate

$\xrightarrow{\text{280 or } 300 \text{ m}\mu}$

Previtamin D$_2$

Vitamin D$_2$
(Ergocalciferol)

Tachysterol$_2$

Lumisterol$_2$

and demonstrated that sunshine would exert the same effect. Although cod-liver oil had long been used and, as early as 1848, recommended for adult rickets and later for rickets in children, it was not accepted by many in the field of medicine until laboratory experiments with animals demonstrated its value in this deficiency disease.

In 1919, Mellanby[27] prevented rickets in puppies by the inclusion of cod-liver oil or butterfat in the diet. McCollum et al.[28] (1921-1922) reported that this factor was distinct from vitamin A. It also was recognized by these and other workers[29] that calcium and phosphorus were also necessary in the prevention and cure of rickets. McCollum et al.[30] (1922) furnished clear evidence of the existence of vitamin D as an entity distinct from vitamin A.

During the period from 1921 to 1924, reports from laboratories showed that irradiation of the body with light rays of short wave length in the region of 300 mμ from sunlight, mercury vapor quartz lamps or from carbon arc lamps promoted growth and improved ossification in children and experimental animals when the ratio of calcium and phosphorus in the diet was not conducive to normal development of bones.

Hess[31] and Steenbock,[32] working independently, in 1924, reported that the irradiation of foods, including oils, confers antirachitic properties upon them. It was shown also that the irradiation of highly purified phytosterol or cholesterol gave an antirachitic product.[33] Phytosterol and cholesterol when purified through their dibromides[34] could not be activated. This indicated the presence of small quantities of activatable substances in what previously was thought to be pure cholesterol and phytosterol. Because ergosterol* was thought to be one of the impurities that possibly might be activatible, it was irradiated, and the resultant irradiation product was shown to have exceedingly potent antirachitic properties.[35]

The course of the irradiation of ergosterol is not a simple one. The solvent employed, the time of exposure and the wavelength of light employed determine the nature and the amounts of the end-products obtained. Under the best conditions, nearly 50 per cent of a vitamin D (now designated as vitamin D_2) is obtained, together with tachysterol$_2$, and lumisterol$_2$ accounting for most of the remainder. Some suprasterols I_2 and II_2 and toxisterol$_2$ are formed. When ultraviolet rays of about 300 mμ are employed, a predominance of lumisterol$_2$ and vitamin D_2 are obtained, whereas shorter rays of about 280 mμ give rise to a predominance of tachysterol$_2$ and vitamin D_2.

Ergosterol and other 7-dehydrosterols also can be converted into vitamins D by treating them with low velocity electrons, electrons of high frequency, alternating current of high frequency, x-rays, radium emanations and cathode rays.

Products

Ergocalciferol† **U.S.P.** Irradiated Ergosta-5,7,22-trien-3β-ol. Vitamin D_2. Calciferol. The history and preparation of this vitamin have been described (q.v.).

Vitamin D_2 is a white, odorless, crystalline compound that is soluble in fats and in the usual organic solvents, including alcohol. It is insoluble in water.

Vitamin D_2 is oxidized slowly in oils by oxygen of the air, probably through the fat peroxides that are formed. Vitamin A is much less stable under the same conditions.

The structures of vitamin D_2 and tachysterol have been elucidated by Windaus and Thiele[36] and by Grundmann.[37]

* Detected by absorption spectrum data before and after irradiation and by its sensitivity toward oxidation as compared to the purified (through dibromide) and nonpurified cholesterol and phytosterol.

† Accepted nomenclature by I.U.P.A.C.

Pure vitamin D_2 will protect rats from rickets in daily doses of 0.015 mg. However, it was soon shown that, rat unit for rat unit, vitamin D_2 or irradiated ergosterol was not as effective as cod-liver oil for the chick. Therefore, the vitamin D of cod-liver oil must differ from vitamin D_2.

One microgram equals 40 U.S.P. units.

Category—antirachitic vitamin.

Usual dose—daily, in rickets, prophylactic, 10 mcg. (400 U.S.P. vitamin D units); therapeutic, 30 mcg.; in hypocalcemic tetany, 1.25 to 5 mg.

OCCURRENCE

Ergocalciferol Capsules U.S.P.
Ergocalciferol Solution U.S.P.

Cholecalciferol† **U.S.P.**, activated 5,7-cholestadien-3β-ol, vitamin D_3, activated 7-dehydrocholesterol, occurs as white, odorless crystals that are soluble in fatty oils, alcohol and many organic solvents. It is insoluble in water. The irradiation of 7-dehydrocholesterol,[38] when carried out under conditions similar to those used for the irradiation of ergosterol, gave analogous end-products which have been identified as vitamin D_3, lumisterol$_3$ and tachysterol$_3$.

Monochromatic light of 296.7 mμ activates 7-dehydrocholesterol to a greater degree than light of any other wavelength.[39]

Vitamin D_3 also occurs in tuna- and halibut-liver oils, from which it has been prepared[40] as a crystalline 3,5-dinitrobenzoate indistinguishable[41] from that obtained by the irradiation of 7-dehydrocholesterol. One milligram of crystalline vitamin D_3 has the activity of 40,000 I.U.[42] and, therefore, has the same activity as vitamin D_2 in rats. Vitamin D_3 is more effective for the chick; however, both vitamins have equal activity for humans.

Vitamin D_3 exhibits stability comparable to that of vitamin D_2.

Epimerization of the C_3–OH in vitamin D_2 or D_3 or conversion of C_3–OH to a ketone group greatly diminishes the activity but does not destroy it completely. Ethers and esters that cannot be cleaved in the body have no vitamin D activity. Inversion of the hydrogen at C_9 in ergosterol and other 7-dehydrosterols prevents the normal course of irradiation.

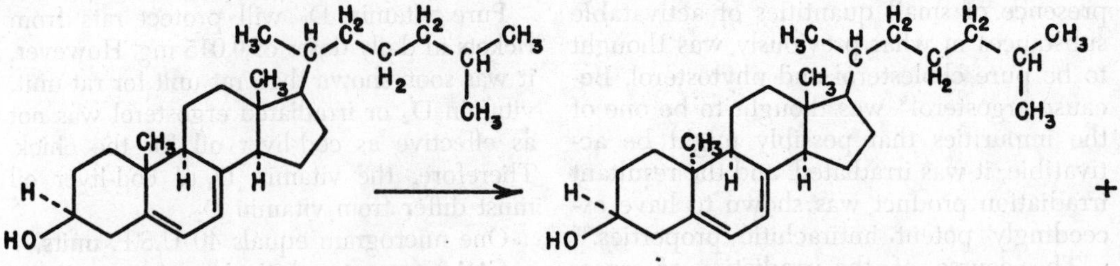

7-Dehydrocholesterol

Lumisterol$_3$

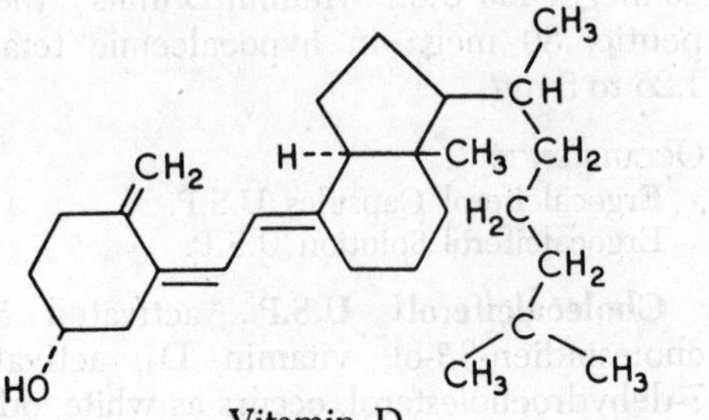

+

Vitamin D$_3$

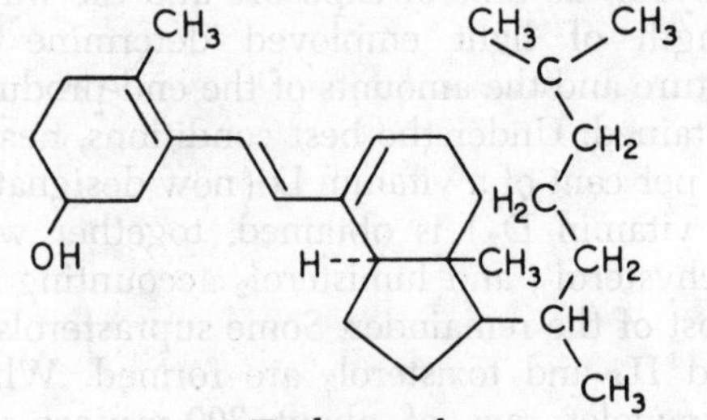

Tachysterol$_3$
(Cholecalciferol)

The structure of primary importance in the vitamins D is the unsaturated conjugated portion of the molecule because the 2,1'-*cis* isomer of 1-cholestanylidene-2'-(5'methoxy-2'-methylene-1'-cyclohexylidene)ethane is almost as active as crystalline vitamin D$_2$ in rats.[43]

Crystalline vitamin D$_4$ has been obtained from the irradiation products of 22-dihydro-ergosterol.[44] It is one half to three fourths as active as vitamin D$_2$ in rats but more effective in chicks. The crude product obtained by the irradiation of 7-dehydrositosterol[45] is one fortieth to one twentieth as active as crude irradiated ergosterol. Irradiated 7-dehydrostigmasterol is $\frac{1}{25}$ to $\frac{1}{400}$ as active as irradiated ergosterol,[46] and irradiated 7-dehydrocampesterol[47] has $\frac{1}{10}$ the activity of irradiated 7-dehydrocholesterol. The tachy-

sterols are feebly active. The loss or partial degradation of the side chain at C_{17} or a side chain at C_{17} containing a carboxyl group leads to inactive or freely active compounds. A C_3 keto, halo or SH group leads to inactive or feebly active compounds that are analogous to vitamin D$_2$. When the C_3–OH is epimeric, 90 per cent of the activity is lost.

Ergosterol occurs as the characteristic sterol of the cryptograms and in very small amounts in the phanerograms. 7-Dehydrocholesterol has been found in the skins of animals, in birds, molluscs, snails and others.[47] 7-Dehydrocholesterol is prepared synthetically as are the other 7-dehydrosterols.

Although fish-liver oils serve as excellent sources for vitamin D, they vary greatly in

TABLE 105. VITAMIN D CONTENT OF SOME FISH-LIVER OILS

SOURCE OF OIL	ANIMAL	POTENCY (I.U./g.)
Bluefin tuna, liver	*Thunnus thynnus*	40,000
Yellowfin tuna, liver	*Neothunnus macropterus*	10,000
Halibut, liver	*Hippoglossus hippoglossus*	1,200
Burbot, liver	*Lota maculosa*	640
Cod, liver	*Gadus morrhua*	100
Shark, liver		50

their vitamin D content and in their activity for rats and chicks. Bills[48] assayed the liver oils from 25 species of fish and found that, although many of these assayed essentially like cod-liver oil, the liver oil of the bluefin tuna is only about one sixth as potent, rat unit for rat unit, as cod-liver oil. The non-saponifiable fractions of the oils showed the vitamin D of tuna-liver oil to be one seventh as potent for chicks, rat unit for rat unit, as the cod-liver oil vitamin D. Bills states that certain liver oils in this series contain an antirachitic factor which is more effective, rat unit for rat unit, for the chick than cod-liver oil. Apparently, the New England and European oils assay like cod, whereas the California and Oriental species of tuna oils are definitely inferior in relative effectiveness in the chick.

Rickets, classically defined, is a lack of calcification of the hypertrophic cartilage zone and osteoid, with a consequent elongation and widening of the epiphyseal cartilage plate. The changes in the bones in rickets result in gross manifestations recognizable clinically[53] in enlargement of the wrists, knees and ankles, bowed legs, beading of the ribs, the rachitic rosary, Harrison's groove and craniotabes. In children and some experimental animals (e.g., dogs) a deficiency of vitamin D, with or without a great distortion of the diet, including variations in the calcium and phosphorus ratios, is sufficient to produce rickets that is reversed by rather wide ranges of vitamin D dosages. Rats do not develop rickets in the absence of vitamin D when calcium and phosphorus are in a favorable ratio in the diet. In humans favorable ratios of calcium and phosphorus (2:1 or 1:2) require minimal amounts of vitamin D. The chick always requires the presence of vitamin D for adequate intestinal calcium absorption.

Other factors also play a role in the metabolism of calcium and phosphorus. An excess of acid in the diet causes a depletion through excretion of calcium and phosphorus and leads to rickets. Correct amounts of acidity in the duodenum favor absorption, while alkalinity leads to precipitation and excretion.

Because vitamin D plays a role in the absorption of calcium and phosphorus from the intestinal tract, it follows that the level of these substances in the bloodstream would be affected also. When vitamin D is given, the blood serum values tend to become normal, regardless of the type of diet. With rather wide ranges of vitamin D dosage, these values remain normal; however, with massive doses of vitamin D, the calcium content of the blood becomes excessive and the phosphate content may be depressed if it is high.

In some unknown manner, vitamin D decreases to a normal level the abnormally high phosphatase content of the blood serum found in rachitic animals and humans. This enzyme is believed to be concerned with the deposition of calcium phosphate in the bones.

The absorption of calcium from the intestine is in part a vitamin D-dependent process in which the parathyroid hormone plays no role. Vitamin D also plays a dominant role in the resorption of calcium from bones. At physiologic doses vitamin D and dihydrotachysterol exert equal effects on both intestinal calcium absorption and bone resorption. However, with increasing doses vitamin D exerts a greater effect on calcium resorption from bone. This may explain the partial effectiveness of the clinical administration of large amounts of dihydrotachysterol rather than vitamin D to patients with vitamin D-resistant rickets. No direct effects of vitamin D on intestinal absorption of phosphate have been found. There is the possibility of vitamin D interaction in the renal reabsorption of phosphate. The end result of these activities is to maintain the serum levels of calcium and perhaps phosphate at some relatively constant level. When the physiologic activities of vitamin D are integrated with those of parathyroid hormone in regard to bone resorption and renal phosphate excretion, the organism is provided with the sophisticated mechanism for an exact and delicate homeostatic control of the levels of calcium and phosphorus in its internal environment.

The administration of vitamin D to a rachitic subject starts calcification at the line of provisional calcification. The first histologic evidence of repair is the presence along the diaphyseal border of degenerated cartilage cells. This effect is visible at the end of 24 hours and is accompanied by extensive vascular penetration within 48 hours. The penetration of blood vessels permits the deposition of the bone-forming salts. There is, thus, produced the so-called line test* for healing. The mass of irregular cartilage cells becomes arranged in short, orderly, parallel columns of a few cells, osteoid material is formed and repair takes place to a remarkable degree. There is no fundamental pathologic condition in the rachitic bone which prevents its calcification.

It recently has been established that all the known in-vivo effects of vitamin D, even when given in doses 4,000 times a physiologic dose, can be suppressed by actinomycin D (AD). AD in 1×10^{-6} concentration inhibits DNA-directed mRNA synthesis. Small physiologically effective doses of vitamin D localize predominantly in the nucleus of the intestinal mucosa in amounts that appear to be too low for cofactor type activity. Vitamin D stimulates mRNA synthesis in the intestinal mucosa. A lag in the Vitamin D-mediated activities occurs after administration of vitamin D and is not due to a lack of the vitamin in the target organs. These and other data suggest that vitamin D may mediate in the synthesis of appropriate enzyme system or systems that play a role in (promote or facilitate) the intestinal absorption of calcium. This concept is supported in part by the nature of vitamin D-resistant rickets, a disease that is almost always inherited and is usually congenital. Therefore, vitamin D may act at the gene level to affect the utilization of DNA-coded information—a link in its biologic response.[49] The fact that the chick has a high structural specificity for vitamin D_3 over vitamin D_2 might also suggest that the vitamins D do not have classical cofactor activity.

25-Hydroxycholecalciferol[50] has been isolated as a major metabolite of cholecalciferol. It is formed in chicks, rats, hogs and humans and was found in human plasma and porcine plasma. The biologic activity of this metabolite is 1.4 times that of vitamin D_3 in the curing of rickets in rats. It can elevate the calcium level in serum and can stimulate calcium transport by everted intestinal sacs. In addition, when administered orally to vitamin D deficient rats, it stimulated calcium transport within 8 to 10 hours, whereas vitamin D had a 20-hour lag. It was the predominant form of vitamin D in the target tissues (intestine and bone) after truly physiologic doses (10 I.U.) of the vitamin. A metabolite of 25-hydroxycholecalciferol that is over twice as active as 25-hydroxycholecalciferol in stimulating the transport of calcium has been reported.[50a] Its onset of action is rapid.

Massive doses† of vitamin D result in a blood level of calcium and phosphorus above normal. This leads to an increased rate of calcification; the structures most affected are the tubules of the kidney, the blood vessels, the heart, the stomach and the bronchi. There is evidence of irritation and degeneration in these tissues and in the liver. The animals under these conditions lose weight rapidly, have intense diarrhea and die in from 5 to 14 days. With smaller doses, death is delayed or the animal survives.

The international unit of vitamin D is equivalent to 0.025 mcg. of crystalline vitamin D_2.

Dosage range same as Ergocalciferol.

Dihydrotachysterol U.S.P.; A.T. 10, Hytakerol®, Dihydrotachysterol₂, 9,10-Seco-5,7, 22-ergostatrien-3β-ol. Reduction by sodium and alcohol of the 3,5-dinitro-4-methylbenzoic acid ester of tachysterol₂, followed by saponification, leads to the production of dihydrotachysterol,[51] which can be obtained in a crystalline form.

It occurs as colorless or white crystals or white, crystalline odorless powder. It is soluble in alcohol, freely soluble in chloroform, sparingly soluble in vegetable oils and practically insoluble in water.

Dihydrotachysterol has slight antirachitic activity.[52] It causes an increase of the calcium concentration in the blood, an effect for which tachysterol is only one tenth as active.

Dihydrotachysterol (A.T. 10, meaning antitetanus) is used in infantile tetany[53] and

---

* In the assay of vitamin D-containing materials using vitamin D-depleted rats.

† 1,000 times the therapeutic dose.

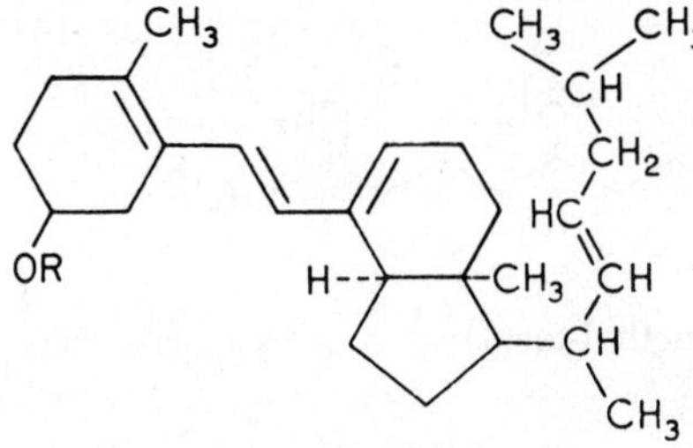

R=3,5-Dinitro-4-methyl Benzoyl Group
R=H=Tachysterol₂

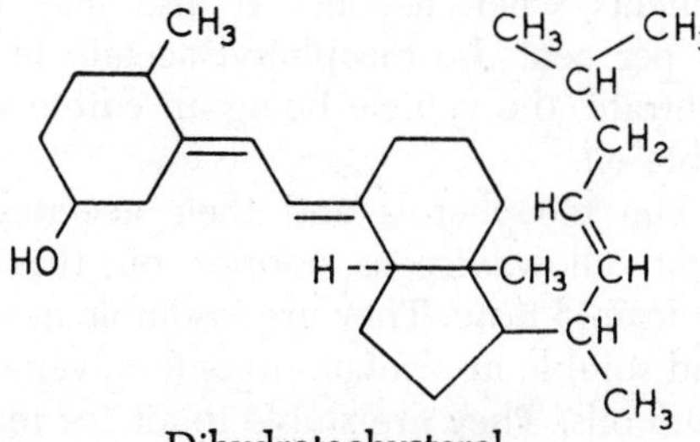

Dihydrotachysterol₂

in postoperative (hypoparathyroid) tetany, in which conditions it increases the calcium content of the blood serum through absorption from the gut. Both vitamin D_2 and dihydrotachysterol are of equal value in hypoparathyroidism. Dihydrotachysterol has greater activity than vitamin D for parathyroid tetany.

Category—blood-calcium regulator.

Usual dose range—initial, 750 mcg. to 2.5 mg. once a day; maintenance, 250 mcg. to 1.75 mg. once a week.

25-Hydroxydihydrotachysterol₃, prepared recently, has weak antirachitic activity, but it is a more important bone-mobilizing agent and is more effective than dihydrotachysterol₃. Also, it is more effective in increasing intestinal calcium transport and bone mobilization in thyroparathyroidectomized rats. Its activity suggests that it may be the drug of choice in the treatment of hypoparathyroidism and similar bone diseases.[58a]

The Vitamins E

In 1922, Evans and Bishop[54] showed that rats maintained on certain diets did not produce offspring, although normal growth occurred and the rats seemed normal in other respects. Therefore, they postulated that some factor, unrelated to the known dietary essentials, controlled fertility and they temporarily called this factor X. Later Sure[55] noted the same phenomenon and proposed the name "vitamin E" for the factor essential for reproduction. This condition was corrected by the addition of lettuce, whole wheat, dry cereals, corn and other foods to the diet. It soon was shown that the unsaponifiable portions of certain natural oils would exert the same effect.

Much difficulty was encountered in preparing a pure fraction from the unsaponifiable portions of oils. Evans et al.,[56] however, succeeded in isolating from wheat-germ oil an active substance in the form of a crystalline monoallophanate, which melted at 158° to 160°.

$$ROH + HCNO \rightarrow ROOCNH_2$$
$$ROOCNH_2 + HCNO \rightarrow RCOOCNHCONH_2$$

The regenerated alcohol was an oil and agreed with the formula $C_{29}H_{50}O_2$. Single doses of 1 to 3 mg. produced litters in 50 per cent of pregnant rats that were maintained on a diet deficient in this or similar factors. This factor was given the name α-tocopherol, from *tokos* meaning "child" and *pherein* "to bear" and the ending *-ol* indicating an alcohol. A second allophanate which melted at 144.6° was obtained. The regenerated substance was also an oil and agreed with the formula $C_{28}H_{48}O_2$. It was called β-tocopherol and originally was reported to be biologically active in single doses of 8 mg. In addition to α- and β-tocopherols, γ-tocopherol was isolated as the allophanate, with a melting point between 138° and 140°, from cottonseed oil.[57] Although the activity of this compound originally was reported to be equal to that of β-tocopherol, reinvestigations have shown it to be only one one-hundreth as active as α-tocopherol.[58] A fourth tocopherol, i.e., δ-tocopherol,[59] has been isolated from soybean oil, in which it comprises 30 per cent of the tocopherols of this oil. Its biologic activity is equal to that of γ-tocopherol.

The absolute configuration of α-tocopherol has been established and is shown on page 950.[60]

Karrer[61] postulated that the tocopherols could be conceived as end-products of the condensation of di- and trimethylhydro-

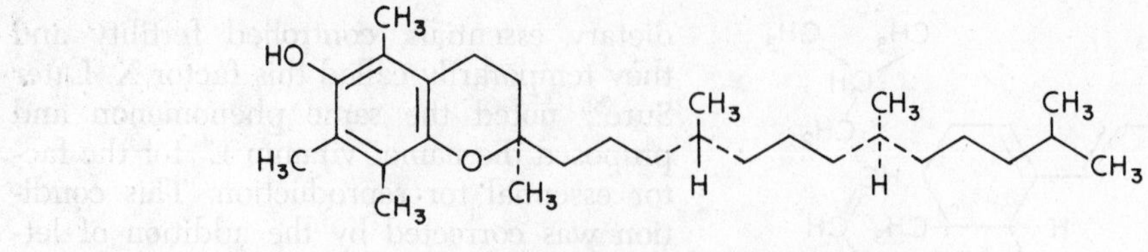

α-Tocopherol (5,7,8-trimethyl-tocol)

quinones and phytol. Fernholz[62] was the first to suggest that the correct formula for α-tocopherol should contain a chroman ring. α-Tocopherol was synthesized[63] in a number of different laboratories by condensing durohydroquinone and phytol, phytyl bromide or phytadiene. When the halide is used, no solvent or catalyst is needed; however when the alcohol or diene is used, it is more advantageous to use a catalyst and a solvent. One of these syntheses can be depicted as shown below.

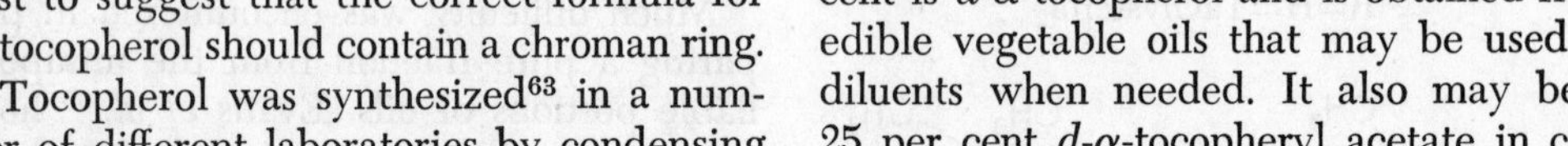

Trimethyl Hydroquinone   Phytyl Halide   Intermediate α-Tocopherol

Ring closure of the above intermediate involves the addition of the hydroxyl group to the double bond in the side chain in accordance with Markownikoff's rule. The nature of the ring structure, i.e., coumaran or chroman in the final product, is determined by the type of substitution at the carbon atom in the allylic compound. Dialkyl substitution at the carbon atom gives rise to the chroman ring, whereas, if hydrogens are present, the product will be a coumaran.

Degradative and synthetic studies comparable with those used in connection with α-tocopherol have been applied to β-, γ- and δ-tocopherols, and they have been shown to have the structures pictured on the following page.

Vitamin E N.F. may consist of *d-* or *dl-*α-tocopherol or their acetates or their succinates, 97.0 to 100 per cent pure. It also may be mixed tocopherols concentrate containing not less than 34 per cent of total tocopherols of which not less than 50 per cent is *d-*α-tocopherol and is obtained from edible vegetable oils that may be used as diluents when needed. It also may be a 25 per cent *d-*α-tocopheryl acetate in concentrate, the vehicle being an edible vegetable oil.

The tocopherols and their acetates are light yellow, viscous, odorless oils that have an insipid taste. They are insoluble in water and soluble in alcohol, organic solvents and fixed oils. They are stable in air for reason-

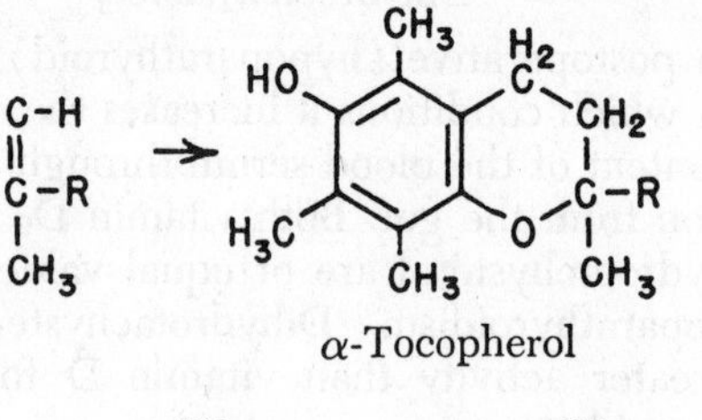

R equals the above radical

able periods of time, but oxidized slowly by air. They are oxidized readily by ferric salts, mild oxidizing agents and by air in the presence of alkali. They are inactivated rapidly by exposure to ultraviolet light; however, not all samples behave alike in this respect because traces of impurities apparently affect the rate of oxidation very much. The tocopherols have antioxidant properties for fixed oils in the following decreasing order of effectiveness: δ-, γ-, β- and α-.[64] In the process of acting as antioxidants, the tocopherols are destroyed by the accumulating fat peroxides that are decomposed by them. They are added to Light

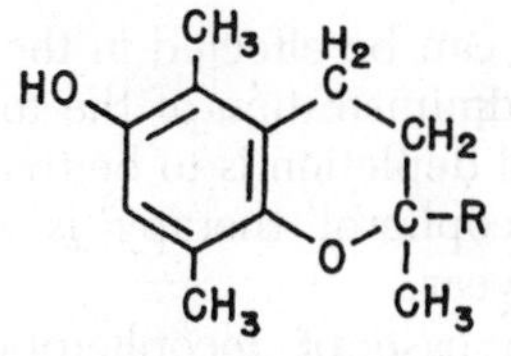

β-Tocopherol
(*p*-Xylotocopherol or 5,8-Dimethyltocol)

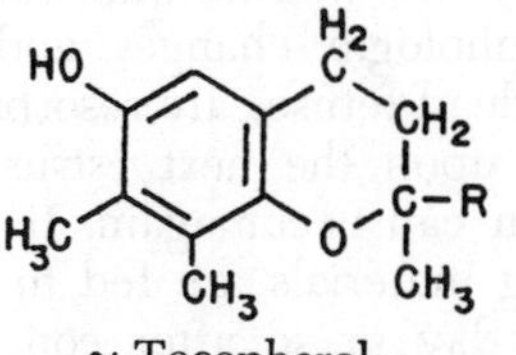

γ-Tocopherol
(*o*-Xylotocopherol or 7,8-Dimethyltocol)

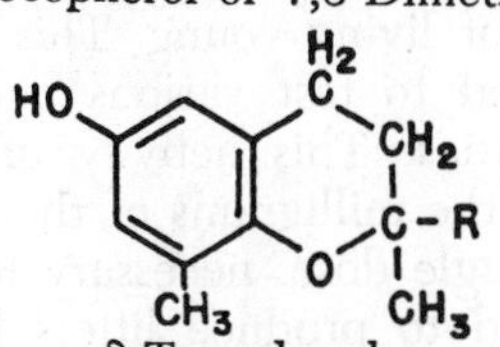

δ-Tocopherol
(8-Methyltocol)

Mineral Oil N.F. and Mineral Oil U.S.P. because of their antioxidant property. The tocopherols can be converted to the acetates and benzoates, respectively, which are oils and are as active as the parent compounds and have the advantage of being more stable toward oxidation.

l-α-Tocopherol is absorbed from the gut more rapidly than the *d*-form; however, the absorption of the mixture of *d*- and *l*-α-tocopherol was considerably higher (about 55% av.) than was to be expected from the data obtained after administration of the single compounds.[65] No marked differences were noted in the distribution in various tissues and the metabolic degradation of *d*- and *l*-α-tocopherols.[65] The liver is an important storage site where the tocopherols are enriched in the mitochondria and microsomes. High concentrations were found particularly in the adrenals, heart, nerves and uterus.[66]

The tocopherols exert a sparing and what appears to be a synergistic action with vitamin A.[67]

d-Alpha Tocopheryl Acid Succinate contains not less than 96 per cent of $C_{33}H_{54}O$.

It occurs as a white crystalline powder that has little or no taste or odor and is stable in air. It is insoluble in water, slightly soluble in aqueous alkali and is soluble in alcohol, acetone, chloroform and vegetable oils. This solid derivative is more convenient to handle than the parent oily compound or its oily esters.

The tocopherols are especially abundant in wheat germ, rice germ, corn germ, other seed germs, lettuce, soya and cottonseed oils. All green plants contain some tocopherols, and there is some evidence that some green leafy vegetables and rose hips contain more than wheat germ. It probably is synthesized by leaves and translocated to the seeds. All 4 tocopherols have been found in wheat-germ oil. α-, β-, and γ-Tocopherols have been found in cottonseed oil. Corn oil contains predominantly γ-tocopherol and thus furnishes a convenient source for the isolation of this, a difficult member of the tocopherols to prepare. δ-Tocopherol is 30 per cent of the mixed tocopherols of soya bean oil.

d-α-Tocopherol is about 1.36 times as effective as *dl*-α-tocopherol in rat antisterility bio-assays. β-Tocopherol is about one half as active as α-tocopherol, and the γ- and δ-tocopherols are only 1/100 as active as α-tocopherol. Meta-xylo-tocopherol, prepared synthetically, is active at the 3-mg. level, which is about equal to α-tocopherol. The esters of the tocopherols, such as the acetate, propionate and butyrate, are more active than the parent compound.[68] This is also true of the phosphoric acid ester of (±)-δ-tocopherol when it is administered parenterally.[69] The ethers of the tocopherols are inactive. The oxidation of the tocopherols to their corresponding quinones also leads to inactive compounds. Replacement of the methyl groups by ethyl groups leads to decreased activity. The introduction of a double bond in the 3,4-position of α-tocopherol reduces its activity by about two thirds. Reduction of the size of the long alkyl side chain or the introduction of double bonds in this side chain markedly reduced activity. Many synthetic compounds[70] have been tested for their activity; forty of these, distributed among a half dozen chemical classes, show some activity, though of a low order.

The exact role that the tocopherols play in the animal body is not known. They apparently play an essential part in the metabolism of skeletal muscle in all species of mammals that have been investigated and in ducklings. Tocopherols are concerned with contractile rather than with the resting metabolism of muscle.[71] Chicks deficient in vitamin E develop encephalomalacia. A reduction of egg hatchability and increased embryonic abnormalities occur in eggs obtained from hens deficient in vitamin E.

Evidence indicates that the tocopherols may play a role as an antioxidant[72] in vivo, especially for the polyunsaturated fatty acids, and as components of the cytochrome C reductase portion of the terminal respiratory chain.

α-Tocopherol appears to be concerned with the biogenesis of coenzyme Q and to be necessary for the maintenance of α-ketoglucuronidate and succinate oxidation.[73]

It is worthy of note that several antioxidants appear to substitute for the tocopherols in several physiologic situations. Thus, diphenyl para-phenylenediamine (DPPD) can almost replace α-tocopherol in rats, and methylene blue was able to prevent resorption in gestating rats.[74]

More evidence is accumulating to support the concept that some of the activity of the tocopherols is associated with their antioxidant properties. However, their biologic activities are in reverse order to their antioxidant activities in vitro.

It is interesting and perhaps significant that the adminstration of 100 mg. of α-tocopherol quinone, which was without effect in nonpregnant animals, brought about resorption of the fetus and hemorrhage in pregnant mice. This effect was not prevented by giving α-tocopherol acetate; vitamin K was the only compound tested that was effective in reversing the effect of α-tocopherol quinone.

Male rats maintained on a diet deficient in the tocopherols suffer a degeneration of the germinal epithelium of the testicular seminal tubules with the resultant complete liquefaction of the sperm nuclei. Although recovery can be effected in the early stages by the administration of the tocopherols, if advanced depletion is to be treated successfully, tocopherol therapy is required for about 1 year.

In the case of tocopherol-depleted females, the normally fertilized eggs start their journey to the uterus, but within a few days the embryos, which have become anchored to the uterine wall, exhibit visible and morphologic changes and gradually die. The dead fetuses are resorbed; this has no effect upon the next estrus cycle and conception can occur again. If tocopherol-containing materials are fed to a depleted female a day or so after conception, the pregnancy will be terminated by the birth of a litter of living young. This procedure can be used to test various materials for their activities. This activity often is expressed as the milligrams of the substance, fed in a single dose, necessary to cure the sterility and to produce litters in 50 percent of the animals used. The litter of a depleted female cannot be saved by the administration of tocopherols after the fifth day of gestation.

Nurslings reared on a tocopherol-deficient female rat develop paralytic symptoms that can be prevented by the administration of tocopherols directly to the nurslings. This observation led to the use of this vitamin in muscular dystrophies and related conditions in humans, for example, fibrositis, which is defined as an inflammatory reaction of fibrous connective tissue and may occur either as a primary or secondary disease. Many common diseases show evidence of primary fibrositis but masquerade under other terms, i.e., muscular rheumatism, lumbago, tenonitis, periarticular fibrositis, torticollis, myositis and panniculitis. Some muscular disorders are being treated effectively by α-tocopherol.

Evidence to date would indicate that vitamin E therapy is useful for (1) intermittent claudication of moderate severity, (2) fat malabsorption syndromes (e.g., fibrocystic disease of the pancreas, sprue), (3) supplementation of diet, for prematures on artificial foods, and (4) diets containing large amounts of unsaturated fats.[73] It also may be useful in habitual abortion in some cases.[75]

In doses of 10 to 30 mg. daily, it appears to relieve menopausal symptoms in many cases; however, its value in the relief of a number of heart conditions is controversial and inconclusive.

The International Unit is 1 mg. of synthetic racemic α-tocopherol acetate. This is the average amount which, when administered orally, prevents resorption in gestating rats deprived of vitamin E. One rat unit is the smallest amount of vitamin E which, when given per os daily to resorption-sterile female rats for the entire period of gestation (21 days), results in the birth of at least one living young in 50 per cent of the animals.

Category—vitamin E supplement.

Usual dose range—prophylactic: from 5 to 30 International Units of vitamin E; therapeutic: to be determined by the physician according to the needs of the patient.

The Vitamins K

In 1929, Dam,[76] using a special fat-free diet, reported experimentally induced bleeding tendencies in chicks, accompanied by a tendency toward delayed blood clotting. Later, he[77] and others[78] furnished strong evidence for the existence of a vitaminlike organic factor present in fresh cabbage or an ether extract of alfalfa or putrefied fish meal, cereals, hog livers and other sources that would cure an experimentally induced condition in chicks.* This condition is char-

acterized by subcutaneous, intramuscular and abdominal hemorrhages. Dam[79] proposed the name vitamin K (koagulations vitamin) for this new factor found in the unsaponifiable portion of certain fats.

Vitamin K_1 was obtained in a crystalline form, in 1937, from the unsaponifiable portion of alfalfa fat.[80] Its yellow color, together with oxidation-reduction potential measurements, reductive acetylation and a broad absorption band with strong absorption in the region 240 to 275 mμ with a rather fine structure revealing sharp maxima at 243, 249, 260 and 270 mμ, indicated that vitamin K_1 contained a quinone structure of the 1,4 type. It was known that phthiacol and several natural pigments, such as lapachol, whose structures were known, exhibited some antihemorrhagic activity. These data, together with degradative studies, led to the proposal of a structure for vitamin K_1 that was verified by its synthesis in several laboratories.[81]

Alfalfa, chestnut leaves and spinach are excellent sources of vitamin K. It also occurs in hog-liver fat, hempseed, tomatoes, kale, soybean oil. In most plants, it appears to be confined to the green leafy parts. The highest concentrations of antihemorrhagic agents are present in certain microorganisms which may be 11 to 38 times as active as alfalfa. These microorganisms are *Bacillus cereus*, *B. cereus var. mycoides*, *B. subtilis*, *Proteus vulgaris*, *Myobacterium tuberculosis*, *Sarcina lutea* and *Staphylococcus aureus*. Other microorganisms produce little if any antihemorrhagic agents. Putre-

n = 4 = Vitamin $K_2(30)$

n = 5 = Vitamin $K_2(35)$

* Vitamin K deficiency can be induced in chickens, ducklings and goslings when maintained on a vitamin-K-free diet because, even though the intestinal bacteria can synthesize antihemorrhagic agents, absorption from the lower portion of the intestine is minimal in birds. This is not true of mammals, except in cases of faulty absorption.

fied fish meal was used to prepare antihemorrhagic agents for clinical trial.

The factor that previously was reported to be vitamin K_2 or 2-methyl-3-difarnesyl-1,4-naphthoquinone[82] now has been shown to be 2-methyl-3-(all-*trans*-farnesylgeranyl

geranyl)-1,4-naphthoquinone[83] m.p. 54°, for which the designation vitamin $K_{2\ (35)}$ has been proposed. Vitamin K_2 or 2-methyl-3-(all-*trans*-difarnesyl)-1,4-naphthoquinone also has been isolated in smaller amounts from microorganisms and is designated as Vitamin $K_{2\ (30)}$ or farnoquinone.* These factors have chemical, physical and biologic properties similar to vitamin K_1. Other isoprenologs of the vitamin K_2 type have been found in nature.

* A trivial name acceptable to the I.U.P.A.C. in biochemical papers.

Products

Phytonadione U.S.P., Mephyton®, Konakion®, 2-Methyl-3-phytyl-1,4-naphthoquinone, Vitamin K_1†, is described as a clear, yellow, very viscous, odorless or nearly odorless liquid.

Pure vitamin K_1[83a] is a yellow, crystalline solid that melts at 69°. It is insoluble in water, slightly soluble in alcohol, soluble in vegetable oils and in the usual fat solvents.

† The trivial name phylloquinone is acceptable in biochemical papers by I.U.P.A.C.

Table 106. Antihemorrhagic Agents and Their Effective Doses

Antihemorrhagic Agent	Effective Dose in Micrograms*
2-Methyl-3-Alkyl and 3-β-Alkenyl-1,4-Naphthoquinones	
2-Methyl-3-phytyl-1,4-naphthoquinone	1.0
2-Methyl-3-farnesyl-1,4-naphthoquinone	5.0
2-Methyl-3-β-γ-dihydrophytyl-1,4-naphthoquinone	8.0
2-Methyl-3-geranyl-1,4-naphthoquinone	25.0
2-Alkyl	
2-Methyl-1,4-naphthoquinone	0.3
Naphthoquinone Oxides	
Vitamin K_1 oxide	1.2
2-Methyl-1,4-naphthoquinone oxide	5.0
Water-Soluble Inorganic Esters	
Sodium 2-methyl-1,4-naphthohydroquinone diphosphate	0.5
Sodium 2-methyl-1,4-naphthohydroquinone disulfate	2.0
Esters and Ethers	
2-Methyl-1,4-naphthohydroquinone dibenzoate and diacetate	1.0
Dimethyl ether	5.0
Monomethyl ether	1.0
Dibenzyl ether	7.0
Reduction Products of Vitamin K_1 and 2-Methylnaphthoquinone	
5,8-Dihydrovitamin K_1	4.0
2-Methyl-5,8-dihydro-1,4-naphthoquinone	6.0
2-Methyl-5,8,9,10-tetrahydro-1,4-naphthoquinone	8.0
Methylnaphthols, Methyltetralones and Related Compounds	
2-Methyl-1-naphthol	1.0
3-Methyl-1-naphthol	0.6
3-Methyl-1-tetralone	1.0
2-Methyl-1-tetralone	0.6
2-Methyl-1-naphthylamine	5.0
Naphthohydroquinones	
2-Methyl-1,4-naphthohydroquinone	0.5

* The minimum amount of each compound given orally in one dose (dissolved in 0.1 ml. of peanut oil) that will reduce the clotting times of 60 to 80 per cent of vitamin-K-deficient chicks to less than 10 minutes in a period of 18 hours. (Tarbel, D. S., *et al.*: J: Biol. Chem. 137:659, 1941)

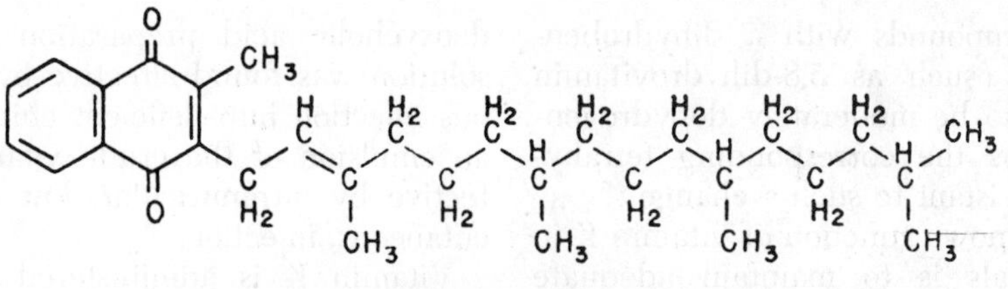

Vitamin K$_1$
(2-Methyl-3-phytyl-1,4-naphthoquinone)

It is unstable toward light, oxidation, strong acids and halogens. It easily can be reduced to the corresponding hydroquinone, which, in turn, can be esterfied. Vitamin K$_1$ gives the following color reaction.[83b] To a few milligrams of material in 1 to 2 ml. of methanol, 1 ml. of sodium methylate (2 to 3 g. of sodium in 50 ml. of methanol) is added. If interfering pigments are practically absent, the mixture slowly develops a distinct purple color, provided that sufficient vitamin K$_1$ is present. The color soon changes to a reddish purple and finally to a reddish brown. Certain synthetic α-naphthoquinones that have at least one allyl group in the quinoid ring give intense and transient blue or purple colors.

A large number of compounds have been tested for their antihemorrhagic activity, and the compounds in Table 106 have been chosen because of their pronounced activity.

Significant biologic activity is manifested in compounds with the following structure when:

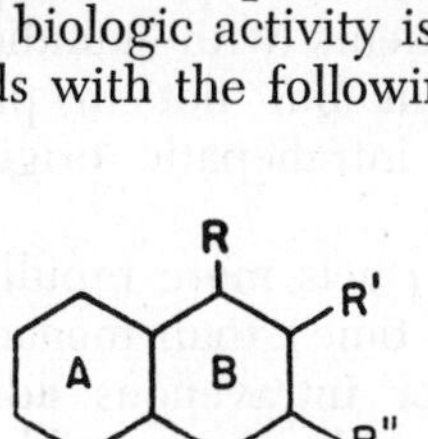

1. Ring A is aromatic or hydro-aromatic.
2. Ring A is not substituted.
3. Ring B is aromatic or hydro-aromatic.
4. R equals OH, CO, OR, OAc (the R in OR equals methyl or ethyl).
5. R' equals methyl.
6. R" equals H, sulfonic acid, dimethylamino or an alkyl group containing 10 or more than 10 carbon atoms. A double bond in the β, γ-position of this alkyl group enhances potency, whereas, if the double bond is further removed, it exerts no effect.

Isoprenoid groups are more effective than straight chains. In the case of the vitamin K$_{2\,(30)}$ type compounds (q.v.) the 6',7'-mono-*cis* isomer is significantly less active than the all-*trans* or the 18',19'-mono-*cis* isomer. This also was true of the vitamin K$_{2\,(20)}$ isoprenolog. A vitamin K$_{2\,(25)}$ isoprenolog was 20 per cent more active than vitamin K$_1$.[83]

7. R''' equals H, OH, NH$_2$, CO, OR, Ac (the R in OR equals methyl or ethyl).

Decreased antihemorrhagic activity is obtained when:

1. Ring A is substituted.
2. R' is an alkyl group larger than a methyl group.
3. R" is a hydroxyl group.
4. R" contains a hydroxyl group in a side chain.

It is interesting to note that, if ring A is benzenoid in character, the introduction of sulfur in place of a —CH=CH— in this ring in 2-methylnaphthoquinone permits the retention of some antihemorrhagic activity.[84] This might indicate that, in the process of exerting vitamin K activity, the benzenoid end of the molecule must fit into a pocket carefully tailored to it. That the other end is not so closely surrounded is shown by the retention of activity on changing the alkyl group in the 2-position.

Although marked antihemorrhagic activity is found in a large number of compounds, the possibility exists that they may be converted in the body to a vitamin K$_1$ type compound. The esters of the hydroquinones may be hydrolyzed, and the resulting hydroquinone may be oxidized to the quinone. The methyl tetralones, which are very active, possibly could be dehydrogenated to the methylnaphthols, which are hydroxylated, and the latter product converted to the biologically equivalent quinone. Compounds with a dihydrobenzenoid ring (such as 5,8-dihydrovitamin K_1) appear to be moderately dehydrogenated, whereas the corresponding tetrahydrides are resistant to such a change.

The only known function of vitamin K in higher animals is to maintain adequate plasma levels of the protein prothrombin (factor II), and three other essential clotting factors: VII (proconvertin), IX (autoprothrombin II) and X (Stuart-Prower Factor). It has been reported that[85] vitamin K's role is the stimulation of mRNA via a repressor protein. Others[86] believe that vitamin K is required for the synthesis of Factor VII from a polypeptide precursor, and, therefore, its activity as an inducer of Factor VII synthesis is not at the level of mRNA but by some other mechanism. Still others have[87] proposed that vitamin K is involved in the removal of the properly folded protein (Factor VII) from the precursor peptide on the ribosome with simultaneous disulfide bond formation, but a later report does not support these results.[88]

It follows that any condition which does not permit the full utilization of the antihemorrhagic agents or the production of prothrombin would lead to an increase in the amount of time in which the blood will clot or to hemorrhagic conditions. Some of these conditions are: (1) faulty absorption caused by a number of conditions, e.g., obstructive jaundice, biliary fistulas, intestinal polyposis, chronic ulcerative colitis, intestinal fistula, intestinal obstruction and sprue; (2) damaged livers or primary hepatic diseases, such as atrophy, cirrhosis or chronic hepatitis; (3) insufficient amounts of bile or abnormal bile in the intestinal tract and (4) insufficient amounts of vitamin K.

Bile of a normal composition is necessary to facilitate the absorption of vitamin K from the intestinal tract. The bile component principally concerned in the absorption and transport of fat-soluble vitamin K from the digestive tract is thought to be deoxycholic acid. The molecular compound of vitamin K with deoxycholic acid was effective upon oral administration to rats with biliary fistula. A crude vitamin K-deoxycholic acid preparation in aqueous solution was found effective by subcutaneous injection into deficient chicks, whereas an emulsion of the crude vitamin was effective by intramuscular, but not by subcutaneous, injection.

Vitamin K is administered in conjunction with bile salts or their derivatives in pre- and postoperative jaundiced patients to bring and maintain a normal prothrombin level in the blood.

In the average infant, the birth values of prothrombin content are adequate, but during the first few days of life they appear to fall rapidly, even dangerously low, and then slowly recover spontaneously. This transition period was and is a critical one because of the numerous sites of hemorrhagic manifestations, traumatic or spontaneous, that may prove serious if not fatal. This condition now is recognized as a type of alimentary vitamin K deficiency. The spontaneous recovery is due perhaps to the establishment of an intestinal flora capable of synthesizing vitamin K after ingestion of food. However, administration of vitamin K orally effects a prompt recovery.

Vitamin K can be used to diagnose liver function accurately. The intramuscular injection of 2 mg. of 2-methyl-1,4-naphthoquinone has led to response in prothrombin index in patients with jaundice of extrahepatic origin but not in patients with jaundice of intrahepatic origin, e.g., cirrhosis.

Vitamin K_1 acts more rapidly (effect on prothrombin time) than menadione within 2 hours after intravenous administration. However, no difference could be detected after 2 hours.[83]

The menadiones are much less active than vitamin K_1 in normalizing the prolonged blood-clotting times caused by Dicumarol and related drugs.[83]

Vitamin K_1 is the drug of choice for humans because of its low toxicity. Its duration of action is longer than that of menadione and its derivatives.

Category—prothrombogenic vitamin.

Usual dose—oral, 10 mg. once daily; intramuscular, 5 mg. repeated as needed.

Usual dose range—oral or intramuscular, 1 to 50 mg. daily.

Phytonadione Injection U.S.P.
Phytonadione Tablets U.S.P.

Menadione N.F., 2-Methyl-1,4-naphtho-quinone, Menaphthone, Thyloquinone. Menadione can be prepared very readily by the oxidation of 2-methyl-naphthalene with chromic acid. It is a bright yellow, crystalline powder and is nearly odorless. It is affected by sunlight. Menadione is practically insoluble in water; it is soluble in vegetable oils, and 1 g. of it is soluble in about 60 ml. of alcohol. The *N.F.* has a caution that menadione powder is irritating to the respiratory tract and to the skin, and an alcoholic solution has vesicant properties.

On a mole for mole basis, menadione is equal to vitamin K_1 in activity and can be used as a complete substitute for this vitamin. It is effective orally, intravenously and intramuscularly. If given orally to patients with biliary obstruction, bile salts or their equivalent should be administered simultaneously in order to facilitate absorption. It can be administered intramuscularly in oil when the patient cannot tolerate an oral product, has a biliary obstruction or where a prolonged effect is desired.

[14]C-labeled menadiol diacetate in small physiologic doses is converted in vivo to a vitamin $K_{2(20)}$, and the origin of the side chain probably is via mevalonic acid. This suggests that menadione may be an intermediate or a provitamin K.[83]

It has been shown that 2-methyl-1,4-naphthoquinone inhibits the glucose metabolism of the parasite causing schistosomiasis and, in conjunction with subcurative doses of fuadin, eliminates parasites from the intestinal blood vessels of mice.

Menadione in oil is three times more effective than a menadione suspension in water. More of menadione than of vitamin K_1 is absorbed orally, but 38 per cent of the former is excreted by the kidney in 24 hours whereas only very small amounts of the latter are excreted by this route in 24 hours. In rats menadione in part is reduced to the hydroquinone and excreted as the glucuronide 19 per cent and the sulfate 9.3 per cent.

Category—vitamin K supplement.

Usual dose—oral and parenteral, 2 mg. daily.
Usual dose range—2 to 5 mg.

Menadione Injection N.F.
Menadione Tablets N.F.

Menadione Sodium Bisulfite N.F., 2-Methyl-1,4-napthoquinone Sodium Bisulfite, Menadione Bisulfite, Hykinone®, is prepared by adding a solution of sodium bisulfite to menadione.

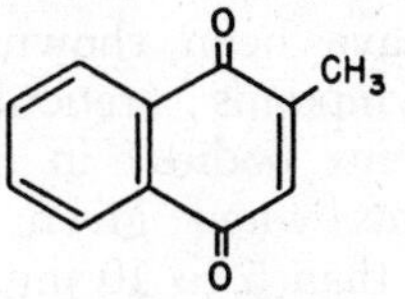

Menadione

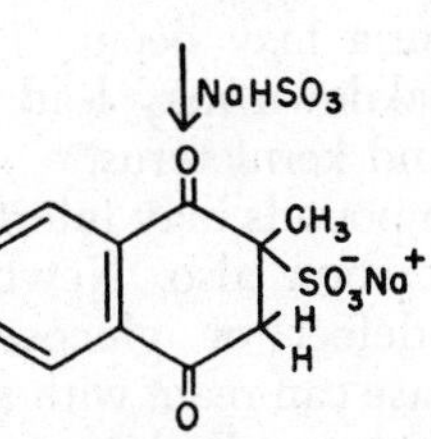

Menadione Sodium Bisulfite

Menadione sodium bisulfite occurs as a white, crystalline, odorless powder. One gram of it dissolves in about 2 ml. of water, and it is slightly soluble in alcohol. It decomposes in the presence of alkali to liberate the free quinone.

Category—form of vitamin K.
Usual dose—intravenous and subcutaneous 2 mg. daily.

Menadione Sodium Bisulfite Injection N.F.

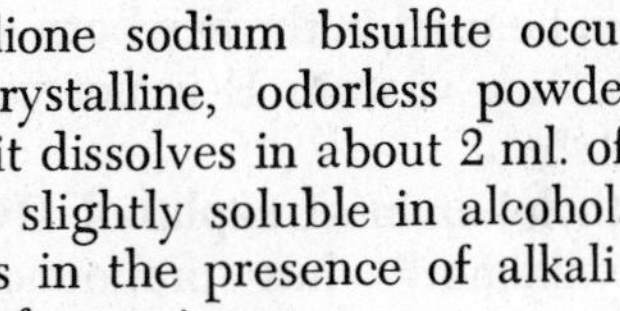

Menadiol Sodium Diphosphate U.S.P., Synkayvite®, Tetrasodium 2-methyl-1,4-naphthalenediol bis(dihydrogen phosphate), tetrasodium 2-methyl-naphthohydroquinone diphosphate, kappadione, is a white hygroscopic powder very soluble in water, giving solutions that have a pH of 7 to 9. It is available in ampuls for use subcutaneously, intramuscularly or intravenously and in tablets for oral administration.

Category—prothrombogenic vitamin.

Usual dose—parenteral, 5 mg. daily.

Range—5 to 75 mg. daily.

OCCURRENCE

Menadiol Sodium Diphosphate Injection U.S.P.

Menadiol Sodium Diphosphate Tablets N.F.

Menadione bisulfite and menadiol diphosphate have been shown to produce hemolytic symptoms (reticulocytosis, increase in Heinz bodies) in newborn premature infants when given in excessive doses (more than 5 to 10 mg. per kg.) In severe cases overt hemolytic anemia with hemoglobinuria may occur. The increased red cell breakdown may lead to hyperbilirubinemia and kernicterus.

These compounds may interfere with bile pigment secretion also. Newborns with a congenital defect of glucose-6-phosphate dehydrogenase can react with severe hemolysis even with small doses of menadione derivatives. However, small nonhemolyzing doses can be used in the newborn, and combination with vitamin E is not considered essential.[89]

2-Methyl-4-amino-1-naphthol Hydrochloride, Synkamin®, has pronounced antihemorrhagic activity (equal to menadione) and was introduced as a water-soluble drug. It is a strong antimicrobial agent for a number of pathogenic bacteria and fungi, as well as saprophytic bacteria, yeast and fungal despoilers of foods, beverages and other products. It is active in concentrations ranging from 10 to 300 ppm., depending on the organism. Administration may be parenterally or orally in 1 to 5 mg. doses.

Oxides. The oxides[90] of menadione, vitamin K and their compounds can be prepared by the action of hydrogen peroxides,

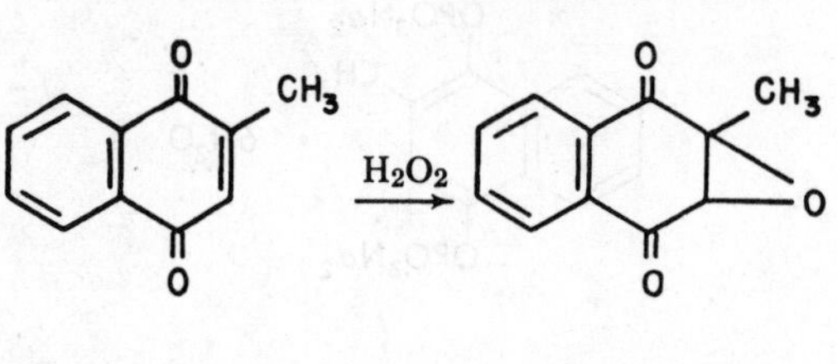

in alkaline solution, upon the parent compound. For example, 2-methyl-1,4-naphthoquinone oxide can be prepared as shown above. This type of compound appears to counteract the hemorrhagic effect of bishydroxycoumarin more effectively than menadione or vitamin K_1.

A new series of naphthoquinones structurally related to phytonadione (phylloquinone) has been isolated from a *Streptomyces* species. They differ from one another in the number of saturated isoprene units in the side chain at position 3. They are called menaquinones (abbreviated as MK_n) and have 9 isoprene units in the side chain. The most unsaturated member would be MK_9 and have the following formula. Progressive saturation would give $MK_9(2\text{-H})$,

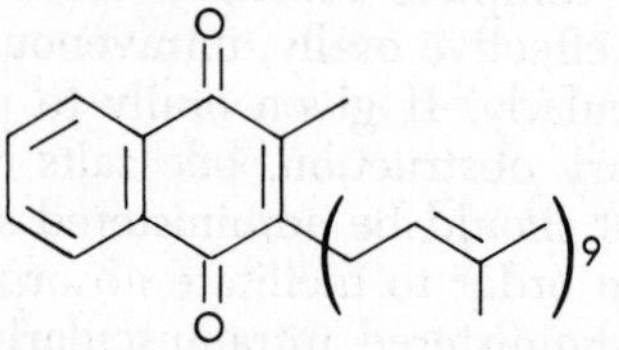

$MK_9(4\text{-H})$, $MK_9(6\text{-H})$, and $MK_9(8\text{-H})$ which were found in the *Streptomyces* species. Phylloquinone by such nomenclature would be $MK_4(6\text{-H})$. These compounds no doubt have vitamin K-like activity and may participate in important oxidation-reduction reactions in vivo.[90a]

Ubiquinones. The term ubiquinone is synonomous with the term coenzyme Q.

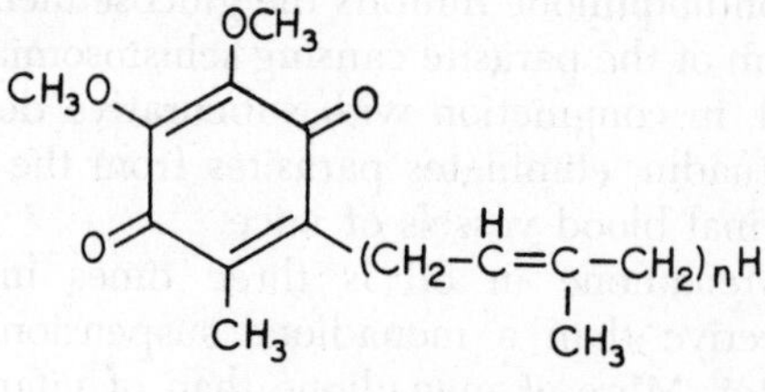

Ubiquinone Q_{10}; n=10
Q_6-Q_9; n=6-9

They are 2,3-dimethoxy-5-methyl-benzoquinones with a prenyl side chain at position 4 where n = 6 to 10.

$$(CH_2CH=C-CH_2)_nH$$
$$\overset{|}{CH_3}$$

$$H_2O_2 \longrightarrow$$

Ubiquinones occur in representatives of all vertebrate classes,[91] invertebrates, higher plants, the algae, a wide range of bacteria and in all species of fungi and yeasts so far examined.[92] Ubiquinone-10 has probably the widest distribution in nature. The total amount of ubiquinone-10 in a 70-kg. human may be about 500 to 1,500 mg. and only 0.01 per cent of the body content is normally eliminated via the urine in 24 hours.[93] In general, one homolog occurs much in excess over the other homologs. For example, beef heart, spinach, *Pseudomonas dinitrificans,* etc., contain ubiquinone-10, *E. Coli* ubiquinone-8, *Saccharomyces cerevisiae* ubiquinone-6, etc.[94] Some molds[92] contain chiefly dihydroubiquinone-10 (terminal isoprene unit reduced). It may be significant that, in the case of the rat and the chick, dietary ubiquinone-7 or -10 accumulates in liver but not in cardiac muscle where ubiquinone-9 normally is present.[91] This suggests specificity at their functional sites (receptor sites).

Ubiquinols (dihydroubiquinones) also are found in nature and may function as antioxidants in mitochondria and are as efficient as α-tocopherol in inhibiting lipid peroxidation.[94]

Shikimic acid-1,2-^{14}C is incorporated into ubiquinone and mevalonic acid-2-^{14}C into the prenyl side chain.[95]

The ubiquinones (and plastoquinones) are tightly bound and selectively localized within the lipoprotein membrane system of cells as highly water-insoluble complexes. Thus, ubiquinones are concentrated in mitochondria and plastoquinones in chloroplasts in the same plant cells.

Ubiquinone-10 is an orange crystalline solid[96] that melts at 48° to 49°C., absorption max. in alcohol at 275 mμ E% 163, and 410 mμ, E% 8. It is stable to boiling alcohol KOH in the presence of pyrogallol and is isolated with the hexane-soluble nonsaponifiables. It is sparingly soluble in water and soluble in alcohol, acetone, and many other organic solvents. It can be reduced by leucomethylene blue on paper.

Ubiquinone-9,2,3-dimethoxy-5-methyl-6-solanesylbenzoquinone[97] synthesized from solanesol and 2,3-dimethoxy-5-methylhydroquinone melts at 42° to 43.5°C. and has properties similar to those of ubiquinone-10.

Probably there is at least one quinone associated with every major type of electron transport system, be it of animal, plant or microbial origin—in animal mitochondria, chloroplasts, chromophores of photosynthetic bacteria and particulate electron transport systems isolated from various bacterial cells. Although ubiquinone has been reported in microsomes of liver and adrenal glands, nuclear fraction from liver cells and rods of the retina, there is no evidence to date of its quinone function in subcellular fractions other than mitochondria, chloroplasts, and bacterial particles. The requirement for ubiquinone and its analogs is best demonstrated by the succinic dehydrogenase complex, a particulate fraction from mitochondria, that contains only the succinic dehydrogenase flavoprotein, nonheme iron proteins, ubiquinone, and cytochromes b and c_1. Activity of this complex that is lost upon extraction with acetone can be restored only by the members of the ubiquinone group that have a side chain containing more than 10 carbon atoms. Soluble NADH ubiquinone reductase (a flavoprotein enzyme) from beef heart submitochondrial particles can utilize the higher isoprenologs of ubiquinone as electron acceptors.[98] Kinetic studies strongly suggest that ubiquinone-10 can function rapidly enough to be on the main path of electron transport from substrate to oxygen in *Acetobacter xylinum.*[99]

Hexahydroubiquinone is orally both prophylactic and curative in the dystrophic rabbit and is somewhat superior to vitamin E.

Hexahydroubiquinone-4 prevented in chicks exudative diathesis, which is characterized by edema, hemorrhage and anemia. It also maintained levels of hemoglobin and packed cell volume comparable with those of chicks on a practical ration and reduced the levels of cathepsin to nearly normal. Similar activities are reported for turkey poults maintained on similar diets. Hexahydroubiquinone-4 is believed to exhibit qualitatively activity similar to ubiquinone-10. Rabbits, calves and rats maintained on certain vitamin E-deficient diets also suffer a deficiency of ubiquinone-10. Added vitamin E to such diets does not alleviate all the expected deficiency symptoms that are relieved by added ubiquinone-10. In addition to the vitamin activities described above, the ubiquinones also exhibit antioxidant properties in vivo. Some activities previously ascribed to the vitamins E may be those of the ubiquinones.

Plastoquinones. These quinones have been found in all plants[100] investigated—34 plant families—from algae to angiosperms, with the exception of some red algae. Plastoquinone-A_9, 2,3-dimethyl-5-solanesylbenzoquinone synthesized[101] from 2,3-dimethylhydroquinone and solanesol melts at 43.5-45°, λ max (iso-octane) 253 mμ ($E_{1\,cm}^{1\%}$ 239) and 261 mμ ($E_{1\,cm}^{1\%}$ 222). It is a yellow crystalline solid that is sparingly soluble in water and soluble in hexane. It occurs by far to the greatest extent and is the most common homolog of this type quinone found in most plant material in which oxygen-evolving photosynthesis occurs. Plastoquinone-A_4 occurs in horse chestnut leaves. Plastoquinone A is a requirement for the Hill reaction* and photosynthetic phosphorylation in chloroplasts. Plastoquinone-C, i.e., plastoquinone A_9 with an OH in the second isoprenoid unit in the prenyl side chain, appears in small amounts to support [102] the activity of plastoquinone-A_9.

The plastoquinones appear to be concentrated in the chloroplasts where they are tightly bound within the lipoprotein membrane as a water-insoluble complex.

WATER-SOLUBLE VITAMINS

Thiamine† Hydrochloride U.S.P., Thiamine Monohydrochloride, Thiamin Chloride, Vitamin B_1 Hydrochloride, Vitamin B_1, Aneurine Hydrochloride. Thiamine hydrochloride, the first water-soluble vitamin to be obtained in a crystalline form, was

* Hill reaction activity (photoreduction of ferricyanide or indophenol, combined with evolution of cyanide).

† The name thiamine was suggested because this vitamin contained sulfur. See J.A.M.A. 109:952, 1937.

Thiamine Hydrochloride

isolated from rice bran by Jansen and Donath,[103] in 1926. They reported its formula to be $C_6H_{10}ON_2$. Later, in 1932, a more nearly correct formula, $C_{12}H_{18}ON_4$-SCl_2, was given by Windaus et al.[104] Several laboratories have contributed to the elucidation of its structural formula. Williams, et al.,[105] working with very large quantities, were able to isolate sufficient quantities of thiamine hydrochloride for degradative studies.

The complete elucidation of the structure of vitamin B_1, along with its synthesis, was accomplished, by 1936, in at least three different laboratories. Cline et al.[106] have developed a successful commercial method for its synthesis.

Many natural foods have been shown to contain moderate quantities of vitamin B_1. The germ of cereals, brans, egg yolks, yeast extracts, peas, beans and nuts are good sources of B_1. It is too costly to isolate the crystalline vitamin on a commercial scale, and all vitamin B_1 so marketed is prepared synthetically. However, concentrates from rice, bran, yeast and other sources that contain other water-soluble vitamins in a concentrated form in addition to the vitamin B_1 are marketed.

Thiamine hydrochloride occurs as small, white crystals or as a crystalline powder; it has a slight, characteristic yeastlike odor. The anhydrous product, when exposed to air, will absorb rapidly about 4 per cent of water. One gram is soluble in 1 ml. of water and in about 100 ml. of alcohol. It is soluble in glycerin. An aqueous solution, 1 in 20, has a pH of 3. Aqueous solutions 1:100 have a pH of 2.7 to 3.4.

Thiamine hydrochloride, because it contains a free amino group, a quaternary amino group and a pyrimidine ring, gives a positive test with some of the alkaloidal reagent test solutions, such as mercuric potassium iodide, trinitrophenol, iodine and bichloride of mercury.

Thiamine hydrochloride is sensitive toward alkali.[107] The addition of 3 moles of sodium hydroxide per mole of thiamine hydrochloride reacts as shown below.

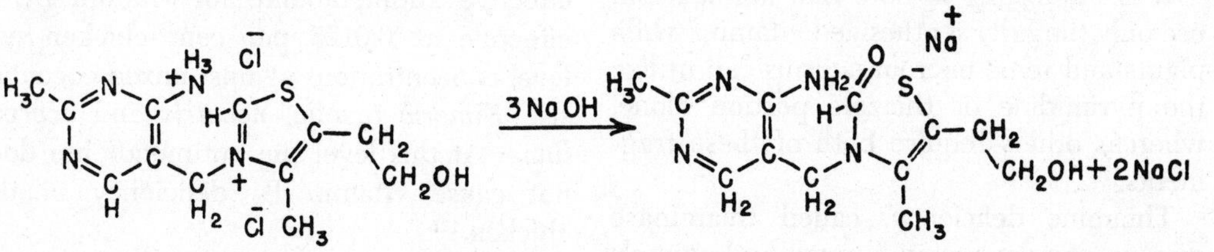

The amino group in thiamine will react readily with aldehydes to form Schiff bases, and these compounds are inactive. The amino group apparently labilizes the pyrimidine ring so that it undergoes a coupling reaction; thus, highly colored com-

$$\text{Thiamine HCl} \xrightarrow{\text{3 NaOH}} \text{product} + 2\,\text{NaCl}$$

It generally is agreed that the stability of thiamine hydrochloride in aqueous solutions decreases as the pH is increased above 5. The following substances, which are capable of increasing the pH of thiamine preparations above the danger point, include sodium bicarbonate, sodium salicylate, sodium barbiturates, sodium citrate, aminophylline and sodium sulfa drugs.

At a pH of 5 to 6, thiamine is cleaved readily by sulfites as follows:

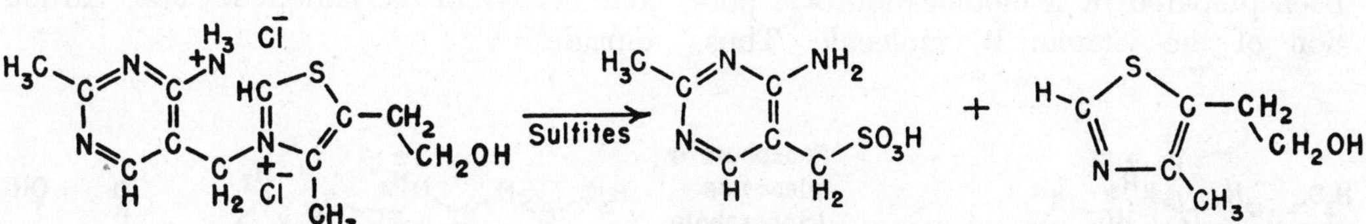

pounds can be made and have been used as a basis of colorimetric assays. Thiamine is quite stable in acid media, and cleavage does not occur even at a pH of 4 or less when heated at 120° for 20 minutes. In the dry state, it can be heated at 100° for 24 hours without diminishing its potency. However, thiamine nitrate is used in many products today because of its greater stability.

Many methods of assay[111] have been de-

$$\text{Thiamine} \xrightarrow{\text{Sulfites}} \text{2-Methyl-5-sulfonmethyl-6-aminopyrimidine} + \text{4-Methyl-5-hydroxyethyl-thiazole}$$

2-Methyl-5-sulfonmethyl-6-aminopyrimidine

4-Methyl-5-hydroxyethyl-thiazole

Thiamine[108] hydrochloride is oxidized readily by air,[109] hydrogen peroxide, permanganate or alkaline potassium ferricyanide[110] to give thiochrome, which exhibits a vivid blue fluorescence and furnishes the basis for the quantitative colorimetric assay of this vitamin. This assay is official in the *U.S.P.*

veloped for thiamine.

A large number of compounds[112] related in structure to vitamin B_1 have been prepared and tested for their activity. The following groups have been found necessary for activity.[113]

1. The amino group on the pyrimidine ring.

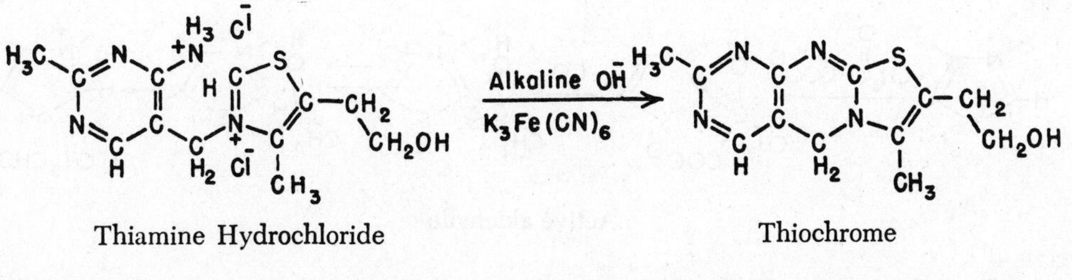

$$\text{Thiamine Hydrochloride} \xrightarrow[\text{K}_3\text{Fe(CN)}_6]{\text{Alkaline OH}^-} \text{Thiochrome}$$

Thiamine Hydrochloride

Thiochrome

2. A 5-β-hydroxyethyl group on the thiazole ring.

3. A hydrogen in position 2 on the thiazole ring.

4. A methylene bridge between the thiazole and the pyrimidine rings.

It is interesting to note that animals can use only the fully synthesized vitamin, while plants and some microorganisms can utilize the pyrimidine or thiazole portion alone, whereas others require both of these fragments.

Thiamine deficiency, called thiaminase disease, can develop in humans and animals with the growth of certain bacteria such as *Bacillus thiaminolyticus* in their intestinal tracts. These bacteria can decompose thiamine and effect an exchange between the thiazole portion of thiamine and bases such as pyridine, aniline, etc. Flavonoids (such as isoquercetin in sweet potato leaves), phenols (including catechol derivatives), quinones and other compounds have thiamine decomposing activity.

Potent antithiamine compounds have been prepared by a modification of a portion of the vitamin B$_1$ molecule. Thus,

This condition can be prevented or alleviated by administering vitamin B$_1$ in the ratio of 1 mole to 40 moles of pyrithiamine.

1-(4-Amino-2-n-propyl-1-pyrimidinylmethyl)-2-picolinium chloride hydrochloride (Amprolium, Mepyrium), also an effective antimetabolite for vitamin B$_1$, is effective at 0.0125 per cent chicken-feed level concentrations against mixed coccidiosis (*Eimeria tenella, necatrix* and *acervulina*). At this level the antimetabolite does not cause vitamin B$_1$ deficiency in the chicken.[115]

Vitamin B$_1$ occurs in the free state in plants. In yeast, some of the vitamin B$_1$ is present as the pyrophosphate known as cocarboxylase. In mammalian tissue, all of the vitamin B$_1$ occurs as the pyrophosphate. Cocarboxylase has been isolated in a crystalline form[116] and also has been synthesized from vitamin B$_1$ by both the use of enzymes[117] and by chemical (in vitro) methods.[118] In yeast, cocarboxylase, together with carboxylase, decarboxylates pyruvic acid to yield acetaldehyde and carbon dioxide.

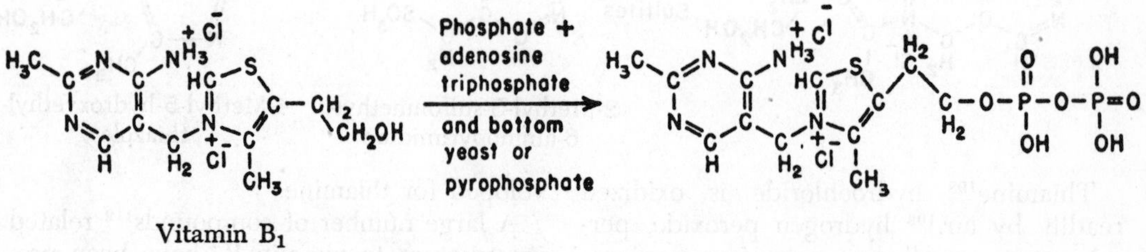

Vitamin B$_1$

pyrithiamine[114] [2-methyl-4-amino-5-pyrimidinylmethyl-(2-methyl-3-hydroxyethyl)-pyridinium bromide], for example, inhibits the growth of fungi and, when given to mice, will induce signs of vitamin B$_1$ deficiency.

The C$_2$ position of the thiazolium ring plays a major role in the functions of thiamine as a cofactor. Thus the decarboxylation of pyruvic acid via "active aldehyde" proceeds as follows.[119]

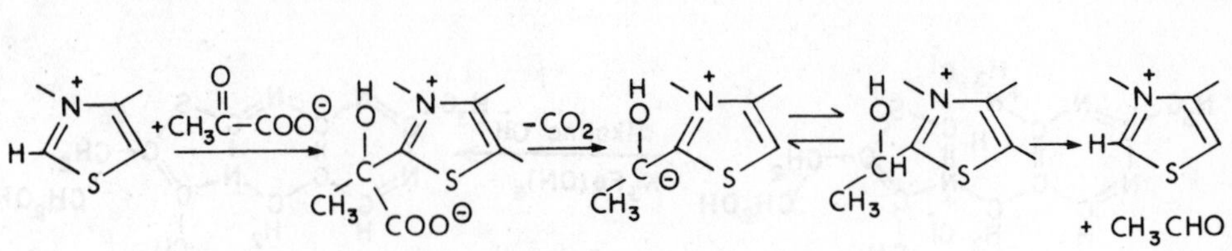

Active aldehyde

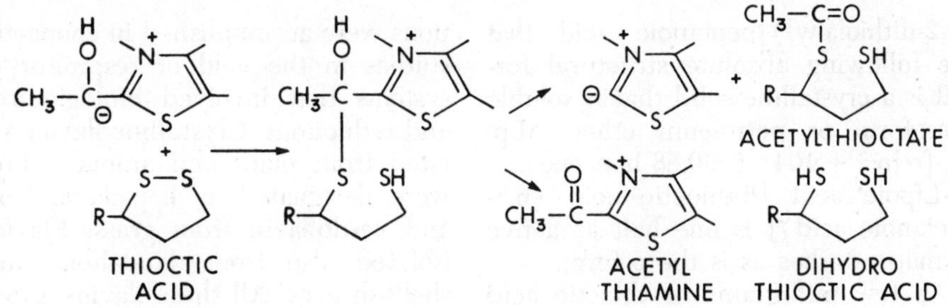

THIOCTIC ACID

R=−(CH₂)₄COOH

ACETYLTHIOCTATE

ACETYL THIAMINE   DIHYDRO THIOCTIC ACID

The active aldehyde above in the presence of thioctic acid is most probably involved in the oxidative decarboxylation of pyruvate to form acetyl thioctate or 2-acetyl thiamine, both of which are capable of acetylating Co ASH (q.v.).

The parenteral administration of thiamine hydrochloride in sufficiently large doses (100 mg. fatal in one case) has produced peripheral circulatory collapse and a shock syndrome similar to that produced by vasodepressor and allergenic agents. In animals (rats and dogs) thiamine monophosphate is about one half as active in this respect. Sodium pyruvate furnishes some protection against these effects.[120]

Because vitamin B_1 plays a role in the metabolism of carbohydrates, and also possibly of amino acids and fats, the source of energy in each cell is affected. Those tissues and organs which utilize the greatest amount of carbohydrates will be affected the most. In the adult, the condition known as beriberi is characterized by polyneuritis, muscular atrophy, cardiovascular changes, serous effusions and generalized edema. These manifestations vary so greatly in number and severity and order of appearance from one patient to another that beriberi has come to be classified into several types, i.e., (1) dry beriberi; (2) wet beriberi; (3) cardiac, pernicious or acute beriberi and (4) mixed beriberi.[121]

The signs and symptoms of clinical beriberi are loss of strength, fatigue, headache, insomnia, nervousness, dizziness, dyspnea, loss of appetite, dyspepsia, tachycardia and tenderness of the calf muscles.

The following systems are affected by a marked deficiency of vitamin B_1.

1. Skin and other epithelial tissues—atrophy, scaling, dermatitis, pigmentation, ulceration and cornification.

2. Nervous system—a neuritis accompanied by pain, paresthesia, weakness and paralysis; degeneration of the spinal cord and mental disturbances.

3. Alimentary tract—anorexia, stomatitis, glossitis, atrophy of the tongue, achlorhydria, diarrhea, loss of tone of gastrointestinal tract and ulceration of the intestine.

4. Hematopoietic system—macrocytic, microcytic and hypochromic anemias.

5. Cardiovascular system—hemorrhage, easy bruising, edema, nutritional heart disease and enlargement of the heart.

Category—enzyme cofactor vitamin.

Usual dose—oral or parenteral, prophylactic, 2 mg. daily; therapeutic, 10 to 15 mg. 2 or 3 times daily.

Usual dose range—2 to 100 mg. daily.

OCCURRENCE

Thiamine Hydrochloride Injection U.S.P.
Thiamine Hydrochloride Tablets U.S.P.

Thiamine Mononitrate U.S.P., Thiamine Nitrate, Vitamin B_1 Mononitrate, is a colorless compound that is soluble in water 1:35 and slightly soluble in alcohol. Two per cent aqueous solutions have a pH of 6.0 to 7.1. This salt is more stable than the chloride hydrochloride in the dry state, is less hygroscopic and is recommended for multivitamin preparations and the enrichment of flour mixes.

+-α-Lipoic Acid is the cyclic disulfide,

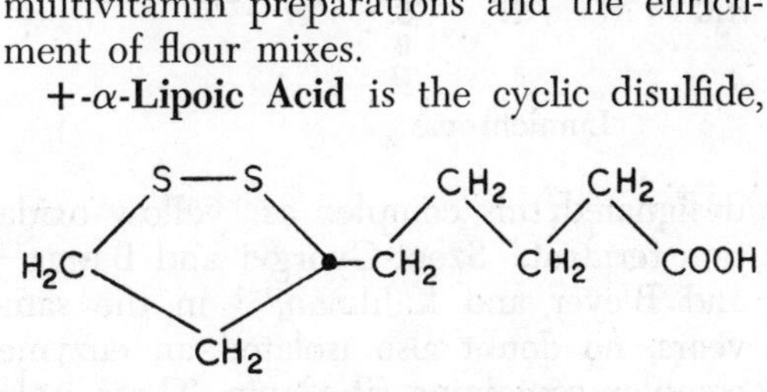

+ − α − Lipoic Acid

5-[3-(1,2-dithiolanyl)]pentanoic acid that has the following absolute structural formula. It is a crystalline solid that is soluble in chloroform or petroleum ether. M.p. 46-48°; $[\alpha]_D^{23} + 104°$ (c. 0.88 benzene).

DL-α-Lipoic acid [6-thioctic acid (6,8-dithiooctanoic acid)] is one half as active in enzymatic studies as is the D form.

Only a few milligrams of thioctic acid were obtained from over 100 tons of liver. This factor is so potent that one part in 10 billion parts of culture media can be detected by a test with microorganisms. Evidence has been presented to show that α-lipoic acid can be conjugated by chemical or enzymatic methods through an amide linkage to the pyrimidine portion of thiamine pyrophosphate to form lipothiamide pyrophosphate (LTPP).

The enzymes and enzyme complexes participating in lipoic acid-dependent reactions are pyruvate dehydrogenase complex–oxoglutarate dehydrogenase complex, lipoic acid reductase-transacetylase or -succinylase, lipoamide oxidoreductase, lipoamidase and lipoic acid activating system.

Lipoic acid is frequently superior to BAL in the treatment of poisoning by heavy metals such as As, Pb, Hg and Se.[122]

Riboflavin* U.S.P., Riboflavine, Lactoflavin, Vitamin B_2, Vitamin G. In the years 1932 and 1933, riboflavin was isolated from milk and egg white and was called lactoflavin and ovoflavin by Ellinger and Koschara,[123] and by Kuhn *et al.*[124] It was isolated as a coenzyme enzyme-complex, from yeast, by Warburg and Christian,[125] who

tions were accomplished in connection with studies in the field of respiratory-enzyme systems that involved biologic oxidations and reductions. Crystalline flavins were isolated from plant and animal sources and were designated as hepaflavin from liver and verdoflavin from grass. Flavins were isolated also from dandelions, malt and shellfish eyes. All these flavins appeared to be identical with riboflavin.

Although the successful isolation of crystalline riboflavin was not accomplished until 1932 and 1933, interest in this pigment dates back to about 1881 in connection with the color[128] in the whey of milk. Osborne and Mendel[129] in 1913, showed that a water-soluble factor was present in milk that was necessary for the growth of young rats. The presence of a substance with the same activity was demonstrated by researchers to be present in germinating wheat, egg yolk, yeast, liver, in crude casein and widely distributed in the plant and animal kingdom. Attempts to obtain this growth factor for rats in a pure form resulted only in the preparation of very active fractions.

Kuhn *et al.*,[130] in 1933, demonstrated that lactoflavin could be substituted in diets deficient in the rat-growth factor to obtain normal growth, and, thus, they established its vitamin nature and synonymity with vitamin B_2. Kuhn[131] showed that the irradiation of lactoflavin in neutral and alkaline media yielded chloroform-soluble, colored fragments, i.e., lumichrome and lumiflavin, from which a sugarlike side chain was lost.

Lactoflavin

neutral or acid
ultra violet light

alkali irradiation

Lumichrome   Lumiflavin

designated this complex as "yellow oxidation ferment." Szent-Gyorgyi and Banga[126] and Bleyer and Kahlman,[127] in the same years, no doubt also isolated an enzyme-complex containing riboflavin. These isola-

* Term Riboflavine recommended by I.U.P.A.C.

These studies, together with other data,[132] furnished clues that led to the synthesis of riboflavin. The chloroform-soluble fragments furnish a basis for the colorimetric estimation of riboflavin.

Riboflavin has been synthesized by a

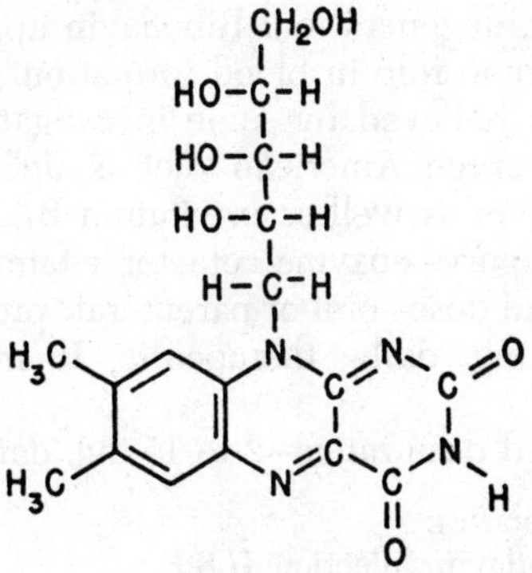

CH2OH structure:

$$CH_2OH$$
$$HO-C-H$$
$$HO-C-H$$
$$HO-C-H$$
$$H-C-H$$

Riboflavin
6,7-Dimethyl-9-(D-1'-Ribityl)-isoalloxazine

naphthoate are also effective solubilizing agents for riboflavin.

When dry, riboflavin is not appreciably affected by diffused light; however, as previously mentioned, it deteriorates in solution in the presence of light, and this deterioration is very rapid in the presence of alkalies. This deterioration can be retarded by buffering on the acid side.

A large number of flavins have been synthesized by varying the substitution in the benzene ring and the nature of the polyhydroxy side chain at position 9. The following flavins have about one half of the biologic activity of riboflavin itself: (1) 6-methyl-9-(D-1'-ribityl)isoalloxazine, (2) 7-methyl-9-(D-1'-ribityl)isoalloxazine, (3) 6-ethyl-7-methyl-9-(D-1'-ribityl)isoalloxazine.

The antipode of riboflavin, 6,7-dimethyl-9-(L-1'-ribityl)isoalloxazine, has one third the biologic activity of riboflavin.

Some flavins, do not produce a permanent growth in rats; in fact, after a temporary increase in weight, growth is arrested, and the animal finally dies. Examples of these flavins are: (1) 6,7-dimethyl-9-(L-1'-arabityl)isoalloxazine, (2) 6,7-dimethyl-9-(D-1'-arabityl)isoalloxazine, (3) 7-ethyl-9-(D-1'-ribityl)isoalloxazine.

Riboflavin has been shown to exist as the phosphate in a coenzyme in combination with a number of enzymes, as enzyme-coenzyme complexes that function in a number of biologic oxidation-reduction systems. It also is present as the active component of the coenzyme of the D-amino acid oxidase which is flavin adeninedinucleotide (F.A.D.) composed of riboflavin, pyrophosphoric acid, D-ribose and adenine. Such

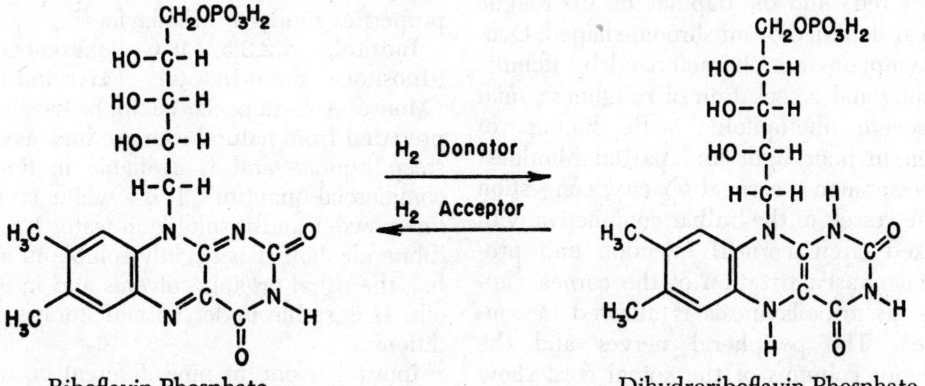

number of methods, some of which are used commercially.[132]

Riboflavin is found distributed widely in nature. The best sources are yeast, rice polishings, wheat germ, turnip greens, liver, milk and kidneys. Excellent sources are lean beef, eggs, spinach, beet greens, oysters, veal, and other foods.

Riboflavin is a yellow to orange-yellow, crystalline powder with a slight odor. It is soluble in water 1:3,000 to 1:20,000 ml., the variation in solubility being due to difference in internal crystalline structure, but it is more soluble in an isotonic solution of sodium chloride. A saturated aqueous solution has a pH of 6. It is less soluble in alcohol and insoluble in ether or chloroform. Benzyl alcohol (3%), gentisic acid (3%), urea in varying amounts and niacinamide are used to solubilize riboflavin when relatively high concentrations of this factor are needed for parenteral solutions. Gentisic ethanol amide and sodium-3-hydroxy-2-

$$CH_2OPO_3H_2$$
H2 Donator → / H2 Acceptor ←
$$CH_2OPO_3H_2$$

Riboflavin Phosphate

Dihydroriboflavin Phosphate

enzyme-coenzyme complexes also are called flavoproteins. The riboflavin portion of the enzyme-coenzyme complex is thought to act as a hydrogen transporting agent, oxygen, cytochromes and others being the acceptors.

The hydrogen donators may be coenzymes I and II or other suitable substrates. They are reduced in the presence of certain dehydrogenases and suitable substrates, such as alcohol, acetaldehyde, glucose. Riboflavin phosphate may accept hydrogen from α-ketoglutarate, fumarate and pyruvate in the presence of certain enzymes (dehydrogenases). D-Amino acids and xanthine also may act as hydrogen donors in the presence of the appropriate enzymes (oxidases). The hydrogen acceptor in many cases is the cytochrome system, although oxygen and certain other substrates also may be effective.

In some cases evidence indicates that certain cations such as Cu, Mo and Fe facilitate the interaction of reduced flavoproteins with one-electron acceptors such as cytochrome and ferricyanide. Flavoprotein metal complexes are called metalloproteins.[133]

A deficiency of riboflavin causes a cessation of growth in rats, together with the following symptoms: ophthalmia, dermatitis, falling hair, incrustations of the skin, ulceration of the corners of the mouth and inflammation of the gums.

A deficiency of riboflavin in humans is characterized by cheilosis, seborrheic accumulations in the nasolabial folds, a glossitis (tongue a purplish-red or magenta color, whereas, in nicotinic acid deficiency, it is a fiery red) and the papillae on the tongue appear flattened or mushroom-shaped. Ocular symptoms are characterized by itching, burning and a sensation of roughness, mild or severe photophobia with dimness of vision in poor light and partial blindness (less common), corneal opacity, congestion of the vessels of the bulbar conjunctiva with marked circumcorneal injection and progressive vascularization of the cornea (appears as ariboflavinosis is allowed to continue). The peripheral nerves and the posterior columns of the spinal cord show myelin degeneration. Riboflavin appears to play some role in blood formation.

It is believed by some investigators that the average American diet is deficient in riboflavin as well as in vitamin B_1.

Category—enzyme cofactor vitamin.

Usual dose—oral or parenteral: prophylactic, 2 mg. daily; therapeutic, 10 mg. once daily.

Usual dose range—2 to 15 mg. daily.

OCCURRENCE
Riboflavin Injection U.S.P.
Riboflavin Tablets U.S.P.

Riboflavin Phosphate (Sodium), riboflavin 5'-phosphate sodium, flavin mononucleotide as the commercial product is soluble 68 mg./ml. and has a pH about 5 to 6. This derivative of riboflavin is fully active biologically. It is more sensitive to U.V. light than is riboflavin.

Methylol Riboflavin, Hyflavin® is a mixture of methylol derivatives of riboflavin by the action of formaldehyde on riboflavin in weakly alkaline solution. The number of methylol groups formed in the ribityl moiety varies from 1 to 3. It possesses the activity of riboflavin and is recommended for parenteral therapy.

Lyxoflavin, 6,7-dimethyl-9-(D-1'-lyxityl)-isoalloxazine, has been isolated from human heart myocardium. It is devoid of riboflavin activity in rats by the standard rat assay; however, a basal diet supplemented with lyxoflavin induced weight gains in rats, chicks and pigs.[134]

Lyxoflavin occurs as orange needles that melt at 283° to 284° (dec.); $[\alpha]23_D = -49°$ (c. o. 26 in 0.05 N sodium hydroxide). Lyxoflavin exhibits other chemical and physical properties similar to riboflavin.[135]

Inositol, 1,2,3,5/4,6-Cyclohexanehexol, i-Inositol, meso-Inositol (Myo-Inositol) (Mouse Anti-alopecia Factor). Inositol is prepared from natural sources, such as corn steep liquors, and is available in limited commercial quantities. It is a white, crystalline powder and is soluble in water 1:6 and dilute alcohol. It is slightly soluble in alcohol, the usual organic solvents and in fixed oils. It is stable under normal storage conditions.

Inositol is one of nine different *cis-trans*

isomers of hexahydroxycyclohexane and usually is assigned the following configuration.

$$\begin{array}{c}
\text{OH} \quad \text{OH} \\
| \quad | \\
\text{C} \quad \text{C} \\
\text{H} / | \quad | \backslash \text{OH} \\
| / \text{H} \quad \text{H} | \\
\text{C} \quad \text{C} \\
| \backslash \text{OH} \quad \text{H} / | \\
\text{HO} \backslash | \quad | / \text{H} \\
\text{C} \text{——} \text{C} \\
| \quad | \\
\text{H} \quad \text{OH}
\end{array}$$

Inositol has been found in most plants and animal tissues. It has been isolated from cereal grains, other plant parts, eggs, blood, milk, liver, brain, kidney, heart muscle and other sources. The concentration of inositol in leaves reaches a maximum shortly before the time that the fruit ripens. Good sources of this factor are fruits, especially citrus fruits,[136] and cereal grains. Large amounts of inositol are found in certain yeasts and molds. [137]

Inositol occurs free and combined in nature. In plants, it is present chiefly as the well-known phytic acid which is inositol hexaphosphate. It is also present in the phosphatide fraction of soybean as a glycoside.[138] In animals, much of it occurs free.

Inositol in the form of phosphoinositides is almost as widely distributed as inositol, and these forms are, in some cases, more active metabolically. Phosphatidylinositol (monophosphoinositide) is the most widely distributed of the inositides and the chief

$$\text{HO}\underset{\text{HO}\quad\text{OH}}{\overset{\text{OH}}{\bigcirc}}\text{O-}\overset{\overset{\text{O}}{\|}}{\underset{\underset{\text{O}^-}{|}}{\text{P}}}\text{-O-CH}_2\text{-CH-CH}_2\text{OOCR}'$$
$$\qquad\qquad\qquad\qquad\qquad |$$
$$\qquad\qquad\qquad\qquad\text{OOCR}$$

fatty acid residue is stearic acid. Di- and triphosphoinositides, of which the former contains an additional phosphate residue at position 4 and the latter two additional phosphate residues at positions 4 and 5 of phosphatidylinositol, have been found in the brain. Possibly other more complex inositides exist.

Inositol lipids occur in all mammalian tissues which have been investigated, and phosphatidylinositol is present in many tissues as about 2 to 8 per cent of the total lipid phosphorus.

The phosphoinositides may be involved in the transport of certain cations and have as yet undetermined functions.

Eastcott,[139] in 1928, showed that bios I was the well-known, naturally occurring optically inactive inositol. In 1941, Woolley[140] showed that mice maintained on an inositol-deficient diet ceased growing, lost their hair and finally developed a severe dermatitis. In rats, a denudation about the eyes, called "spectacle eye," takes place in the absence of inositol in the diet. These symptoms also are accompanied by the development of a special type of fatty liver containing large amounts of cholesterol.

Inositol has been shown to be an essential growth factor for a wide variety of human cell lines in tissue culture. It is considered a characteristic component of seminal fluid, and the content is an index of the secretory activity of the seminal vesicles.

Phytin (calcium-magnesium salt of inositol hexaphosphate), inositol hexa-acetate, cephalin[141] and methylinositol (Mytilitol) are active in mice.

Evidence is accumulating to indicate that inositol will reduce elevated blood cholesterol levels. This, in turn, may prevent or mitigate cholesterol depositions in the intima of blood vessels in man and animals and, therefore, be of value in atherosclerosis.

Biotin, Coenzyme R,[142] Vitamin H,[143] Anti-Egg White Injury Factor[144] or S or Skin Factor. In 1901, Wilders[145] observed that wort or similar extracts, in addition to fermentable sugars and inorganic salts, were necessary for growth and fermentation of yeast but not wild yeasts. The name bios was provisionally given (to a mixture of subtances) to designate this growth stimulant activity, and biotin, for one specific substance. Kogl et al.[146] spent 5 years in order to isolate 70 mg. of the methyl ester of biotin. Two parts per 100 billion of crystalline biotin stimulated the growth of yeast. Because of its marked activity, it can be detected in a few milligrams of natural substances.

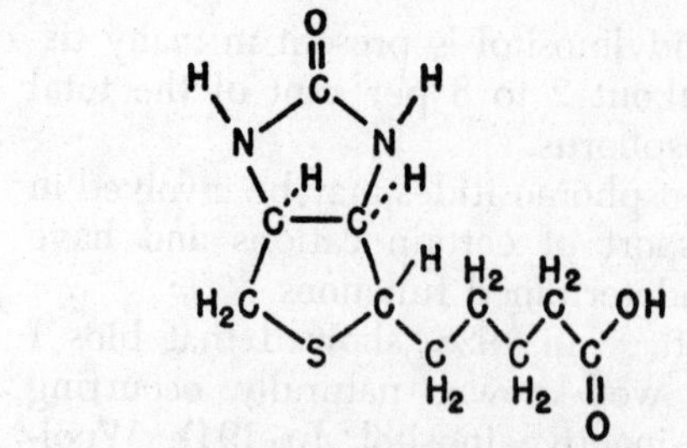

(±)-Biotin (Racemates)
2'-Keto-3,4-imidazolido-2-tetra-
hydrothiophene-*n*-valeric Acid

Biotin occurs both free and combined, even if in minute quantities in some cases. Small amounts are present in all higher animals. The highest concentrations are found in liver, kidney, eggs and yeast as a water-insoluble, firmly-bound complex. Considerable quantities are found both free and combined in vegetables, grains, nuts. Alfalfa, string beans, spinach and grass are fair sources, while beets, cabbage, peas and potatoes are low in biotin. Peaches and raspberries are high in free biotin. The bound form seems to be as well utilized as the free form.

Biotin occurs as a white, crystalline powder that melts at 230 to 232°. It is optically active, $[\alpha]_D = +92$ for a 0.3 per cent solution in 0.1 N sodium hydroxide or +57 for chloroform solution of the methyl ester. It is stable in the dry state and in acid solutions but is slowly inactivated in alkali and is very rapidly destroyed by oxidizing agents. The methyl ester is as active biologically as the free acid.

Curative tests on rats fed avidin have shown that synthetic (+)-biotin and natural biotin are equal in activity,[147, 148] (±) Oxybiotin has been synthesized,[149] and its microbiologic activity for certain organisms has been shown to be equal to that of (±)-biotin.[149, 150] It is also less active in animals. It seems likely that oxybiotin and biotin have identical spatial configurations and that the two compounds differ from one another only in the nature of the one hetero atom.

Numerous observations indicate that biotin functions as a carboxylation cofactor[151] via 1'-N-carboxy biotin (CO_2-biotin) that is formed as follows:

$$\text{Biotin-enzyme} + HCO_3^- \xrightarrow{Mg^{++}} CO_2\text{-biotin enzyme} + ADP + Pi.$$

The oxygen for ATP cleavage is derived from bicarbonate and appears in the Pi.

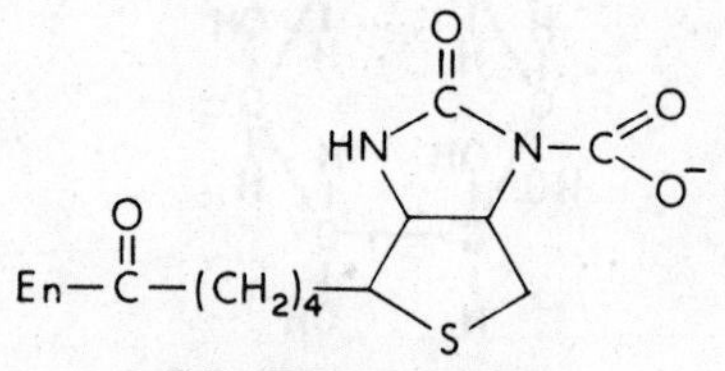

CO_2-Biotin enzyme

Purified preparations of acetyl CoA carboxylase contained biotin (1 mole of biotin per 350,000 g. of protein (enzyme)). It catalyzed the first step in palmitate synthesis as follows:

$$CH_3COSCoA + HCO_3^- + ATP \xrightarrow{Mg^{++}} {}^-OOCCH_2COSCoA + ATP + Pi.$$

Other enzymes with which biotin appears to be intimately associated in carboxylation are beta-methylcrotonyl CoA carboxylase, propionyl CoA carboxylase, pyruvate carboxylase and methylmalonyl-oxalacetic transcarboxylase.

Biotin also is joined in an amide linkage to the epsilon amino group of a lysine residue of carbamyl phosphate synthetase (CPS) to form biotin-CPS which participates with 2 ATP, HCO_3^- and glutamine in the synthesis of carbamyl phosphate. This takes place stepwise as follows:

(1) Biotin CPS + ATP + HCO^-_3 ⇌ carbonic phosphoric anhydride biotin CPS (CPA biotin CPS) + ADP;

(2) CPA biotin CPS ⇌ ^-OOC biotin CPS + Pi;

(3) ^-OOC biotin CPS + glutamine ⇌ H_2NOC biotin CPS + ATP ⇌ biotin CPS + carbamyl-phosphate + ADP.

Carbamyl phosphate can participate in amino acid metabolism and some nucleic acid syntheses.

Biotin deficiency in mammals develops only* when raw egg white is added to the biotin-deficient diet. The first symptoms of biotin deficiency in rats is a characteristic

* Because intestinal bacteria can synthesize biotin as well as some other factors of the B complex.

dermatitis around the eyes called "spectacle eye," which progresses to general alopecia and a scaly dermatitis. Raw egg white contains a substance called avidin which binds biotin in a nonabsorbable form. The avidin-biotin complex is excreted in the feces. Biotin causes the development in rats of fatty livers that are characterized by a high cholesterol content. The fatty liver effect of biotin can be prevented by the simultaneous feeding of lipocaic (an internal secretion of the pancreas) or inositol.

When large amounts of raw egg white (induced biotin deficiency) were fed to humans, the following symptoms were observed: a maculosquamous dermatitis of the neck, hands, arms and legs; striking ashy pallor of the skin and mucous membranes; diminution in hemoglobin and erythrocytes; a rise in serum cholesterol; atrophy of the papillae on the tongue; muscle pains; hyperesthesia; lassitude and depression.

Certain cases of baldness in men are caused by seborrheic conditions and can be improved by biotin. Encouraging results were obtained in severe cases of seborrhea that seemed to be related to the skin disease called psoriasis.

Biocytin. Certain extracts from natural products, particularly those from controlled autolysis of actively metabolizing yeasts, showed a higher biotin content when assayed with *L. casei* than when *L. arabinosus* was used. This led to the belief that biotin existed in some complex that was not utilized by *L. arabinosus*. This was substantiated by the isolation and characterization of a compound called biocytin (from the Greek *kutos,* cell) from yeast, where it occurs in much larger quantities (two and

PANTOTHENIC ACID

Pantothenic Acid (Chick Antidermatitis Factor). In 1933, R. J. Williams *et al.*[153] reported that extracts of very diverse tissues representing many different biologic groups, i.e., chordates, arthropods, echinoderms, molluscs, annelids, platyhelminthes, myxomycetes, bacteria, fungi, molds, algae and spermatophytes, all contain a substance that was capable of stimulating to a very marked degree the growth of Gebrude Mayer yeast. This property of growth stimulation was a convenient way with which to trace the concentration and isolation of the active substance. By this method, the active substance can be detected quantitatively when only 5 parts per 10 billion are present. This was fortunate because liver, one of the richest sources of this factor, contains only 40 parts per million. The active concentrates from diversified sources, even though they contained some impurities, behaved so nearly uniformly that the same factor was thought to be present in all of them. The active substance appeared to be an acid and, since its occurrence was so widespread, it was called "pantothenic acid," from the Greek meaning "from everywhere."

In 1938, Williams *et al.*[154, 155] reported the isolation of one tenth of an ounce of crude pantothenic acid from 500 lbs. of raw liver, employing a very complicated process accompanied by 700 fractionations of an alkaloidal salt of the acid. He recognized that β-alanine was one cleavage product, and the other appeared to be dihydroxyvaleric acid. In the same year, Elvehjem *et al.*[156] pointed out that the chick antidermatitis factor and pantothenic acid were similar in chemical properties. At the same time,

Biocytin

one half times) than biotin. It has been assigned the structural formula given above, as ε-N-biotinyl-L-lysine.[152]

Jukes[157] showed that the chick antidermatitis factor and *calcium pantothenate* supplied by Williams had the same biologic activity. In 1940, Williams and Major reported the synthesis of pantothenic acid as shown in the following equations.[158]

$$CH_3 \quad O$$
$$H_3C-\underset{\underset{H}{|}}{C}-\overset{//}{C} \quad + \quad CH_2O \quad \longrightarrow \quad H_3C-\underset{\underset{CH_2OH}{|}}{\overset{CH_3}{|}}-\overset{//}{\underset{H}{C}} \quad \xrightarrow[\text{CaCl}_2]{\text{KCN}} \quad H_3C-\underset{\underset{CH_2OH}{|}}{\overset{CH_3}{|}}-\underset{\underset{H}{|}}{\overset{OH}{|}}-CN$$

$$\xrightarrow[\text{(H}^+)]{\text{HOH}} \quad H_3C-\underset{\underset{CH_2OH}{|}}{\overset{CH_3}{|}}-\underset{\underset{H}{|}}{\overset{OH}{|}}-COOH \quad \longrightarrow \quad H_3C-\underset{\underset{OH}{|}}{\overset{CH_3}{|}}-\underset{\overset{|}{O}}{\overset{H}{\underset{|}{C}}}-C=O$$

$$(\pm)\text{-}\alpha\text{-}\gamma\text{-Dihydroxy-}\beta,\beta\text{-dimethyl Butyric Acid}$$
(center/right)
$$(\pm)\text{-}\alpha\text{-Hydroxy-}\beta,\beta\text{-dimethyl-}$$
$$\gamma\text{-butyrolactone}$$
$$(\text{Natural Form } [\alpha]\ 27 = -49.8)$$

$$H_3C-\underset{\underset{HC}{\overset{|}{\underset{H}{}}}}{\overset{CH_3}{|}}-\underset{\underset{O}{}}{\overset{H}{\underset{|}{C}}}-C=O \quad + \quad H_2NCH_2CH_2COOH \longrightarrow H_3C-\underset{\underset{CH_2OH}{|}}{\overset{CH_3}{|}}-\underset{\underset{OH}{|}}{\overset{H}{C}}-\overset{O}{\overset{||}{C}}-\underset{\overset{|}{H}}{N}-\overset{H_2}{\underset{}{C}}-\overset{H_2}{\underset{}{C}}-\overset{//}{\underset{OH}{C}}$$

$$(\pm)\text{-Lactone} \qquad \beta\text{-Alanine}$$

$(\pm)$-Pantothenic Acid (the (+)-form is the naturally occurring one) N-(α, γ-Dihydroxy-β,β-dimethyl) butyryl-β-aminopropionic Acid

The dextrorotatory biologically active form of pantothenic acid has the (R)-configuration.[158a]

Pantothenic acid occurs as an odorless, white, microcrystalline powder. It is very soluble in water and is unstable in acid and alkaline solutions.

Products

Calcium Pantothenate U.S.P., Dextro Calcium Pantothenate, is a slightly hygroscopic, white, odorless, bitter-tasting powder that is stable in air. It is insoluble in alcohol, soluble 1:3 in water, and aqueous solutions have a pH of about 9 and $[\alpha]^D = +25°$ to $+27.5°$. Autoclaving calcium pantothenate at 120° for 20 minutes may cause a 10 to 30 per cent decomposition. Some of the phosphates of pantothenic acid that occur naturally in coenzyme are quite stable to both acid and alkali, even upon heating.[159]

Usual dose—10 mg. once daily.

Usual dose range—10 to 50 mg. daily.

OCCURRENCE

Calcium Pantothenate Tablets U.S.P.

Racemic Calcium Pantothenate U.S.P.

is recognized to provide a more economical source of this vitamin. Other than containing n.l.t. 45 per cent of the dextrorotatory biologically active form, its properties are very similar to Calcium Pantothenate U.S.P.

The richest sources[160] of pantothenic acid are liver, yeast, cereal brans, leafy vegetables, dairy products and eggs. It is produced by various molds and microorganisms in the soil and elsewhere by green plants after they develop their capacity for photosynthesis. In some natural sources, such as liver, pantothenic acid occurs in a bound form from which it can be released by enzymatic activity.

Levorotatory pantothenic acid is biologically inactive. Ethyl pantothenate and ethyl monoacetylpantothenate are active in rats and chicks but are not utilized by microorganisms. In certain tests, fragments of the pantothenic acid molecule are active. For example, β-alanine stimulates the growth of yeast and certain strains of bacteria and is partially effective for the rat but ineffective in chicks.

A large number of compounds[161] have been prepared that are related in structure to pantothenic acid. With few exceptions, all compounds made had little or no activity. Some of the most effective pantothenic acid antagonists are (+)- and (−)-pantoyltaurine and $(+)$- and $(-)$-pantoyltaurine amide. $(+)$-Pantoyltaurine is about 32 times more active in this respect than the $(-)$-form for certain bacteria. This antagonistic effect is reversed by pantothenic acid. This type of antagonism is believed to be a competitive inhibition of the conversion of pantothenic acid to coenzyme A. Pantoyltaurine inhibits the growth of a wide variety of microorganisms which require pantothenic acid for growth, and phenyl pantothenone markedly inhibits the growth of malaria organisms.

Pantoyltaurine constitutes the first case of an effective chemotherapeutic agent being designed in accordance with the concept of competitive analog-metabolite growth inhibition. Rats were protected from 10,000 lethal doses of a virulent strain of streptococcus and less completely from 1,000,000 lethal doses by frequent subcutaneous doses of pantoyltaurine. Sulfonamide-resistant streptococci were just as sensitive to pantoyltaurine as the nonresistant strains.

Utilizing a combination of enzyme and microbial assays, it has been shown that pantothenic acid exists in the free state in plasma and almost exclusively in the form of coenzyme A in tissue cells. Some bacteria require preformed coenzyme A, whereas others can utilize pantothenic acid and still others pantothenic acid intermediates.

Coenzyme A tentatively has been assigned the following structural formula.

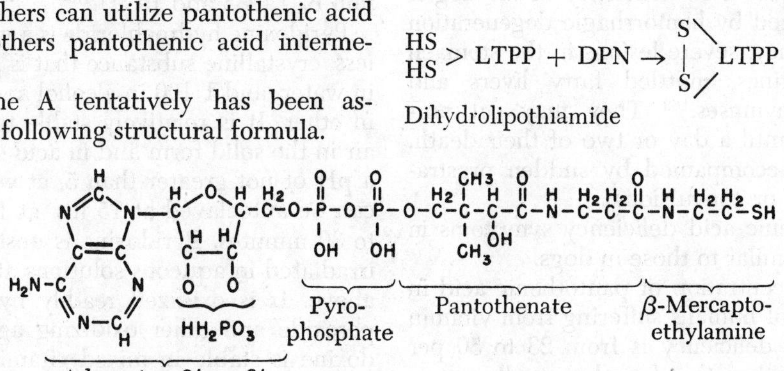

between acetyl donors and acetyl acceptors in the presence of suitable true enzymes. Examples of direct acetyl donors are acetyl phosphate (bacteria) and the adenosine triphosphate-acetate system (mammalian). Acetyl-CoA can be generated from pyruvate, acetaldehyde, fatty acids, β-keto fatty acids and acetate through the above direct acetyl donors. Acetyl acceptors are choline, p-aminobenzoic acid, sulfonamides, glycine in the presence of acetylase. Oxalacetic acid accepts an acetyl group in the presence of the condensing enzyme to form citric acid which initiates the cycles by which both carbohydrates and fatty acids are metabolized aerobically.

The conversion of CoASH to CoAS~COCH$_3$ probably takes place via acetyl lipothiamide as follows:

$$CH_3COCOOH + \begin{array}{c} S\backslash \\ | \quad LTPP \rightleftharpoons \\ S/ \end{array}$$

Lipothiamide

$$\begin{array}{c} CH_3COS \\ HS \end{array} > LTPP + CO_2$$

Acetyl Lipothiamide

$$\begin{array}{c} CH_3COS \\ HS \end{array} > LTPP + CoASH \rightleftharpoons$$

$$CoAS \sim COCH_3 + \begin{array}{c} HS \\ HS \end{array} > LTPP$$

$$\begin{array}{c} HS \\ HS \end{array} > LTPP + DPN \rightarrow \begin{array}{c} S\backslash \\ | \quad LTPP + DPNH_2 \\ S/ \end{array}$$

Dihydrolipothiamide

Adenosine 2′ or 3′ Phosphate   Pyrophosphate   Pantothenate   β-Mercaptoethylamine

Coenzyme-A

Coenzyme A

Coenzyme A represented by CoASH can function as

$$CoA-S-\overset{\overset{\displaystyle O}{\|}}{C}-CH_3,$$

acetyl-CoA, an energy-rich compound in biologic transformations. Acetyl-CoA (active acetate) appears to act as an acetyl transfer

It is interesting to note that in a 24-hour period an adult rat produces acetic acid equal to 1 per cent of its body weight. It is believed that coenzyme A participates in the utilization of most if not all of this acetic acid.

Chicks maintained on a diet deficient in pantothenic acid develop a dermatitis that has been characterized as follows:

External manifestations appear chiefly at the eyes, the corners of the mouth and upon the legs and the feet. The feathering is retarded and very ruffled because birds peck at themselves continuously. The crusty scabs at the corners of the mouth gradually enlarge and often involve the margins of the skin around the nostrils and underneath the lower mandible.[162]

Feather depigmentation occurs in black fowl.

Rats, dogs and hogs suffer severe pantothenic acid deficiency symptoms. Rats cease to grow after 3 to 4 weeks, show adrenal hemorrhages, scant, coarse fur with rusty spots, inflammation of the nose and blood-caked whiskers. Piebald and black rats also show a characteristic depigmentation of the fur, a condition which in most cases can be brought to normal by sufficiently large doses of pantothenic acid. In this induced achromotrichia, a highly individualistic response with regard to restoration of color was obtained upon pantothenic acid therapy, and it is possible that other factors may be involved.

Adult female rats failed to reproduce when they were deficient in pantothenic acid.

Pantothenic acid deficiency in the dog is characterized by hemorrhagic degeneration of the kidney, severe lesions in the stomach and intestines, mottled fatty livers and mottled thymuses.[163] They may eat normally up until a day or two of their death, which is accompanied by sudden prostration, coma or convulsions.

Pantothenic acid deficiency symptoms in hogs are similar to those in dogs.

The concentration of pantothenic acid in the blood of humans suffering from vitamin B-complex deficiency is from 23 to 50 per cent lower than that found normally.

Panthenol, the alcohol analog of pantothenic acid, exhibits both qualitatively and quantitatively the vitamin activity of pantothenic acid. It is considerably more stable than pantothenic acid in solutions with pH values of 3 to 5, but of about equal stability at pH 6 to 8. It appears to be more readily absorbed from the gut, particularly in the presence of food.

PYRIDOXOL

Pyridoxine Hydrochloride U.S.P., 5-Hydroxy-6-methyl-3,4-pyridinedimethanol Hydrochloride, (Vitamin B_6 Hydrochloride, Rat Antidermatitis Factor). In 1935, P. Gyorgy[164] showed that "rat pellagra" was not the same as human pellagra but that it resembled a particular disease of infancy known as "pink disease" or acrodynia. This "rat acrodynia" is characterized by a symmetric dermatosis affecting first the paws and the tips of the ears and the nose. These areas become swollen, red and edematous, with ulcers developing frequently around the snout and on the tongue. Thickening and scaling of the ears is noted, and there is a loss of weight, with fatalities occurring in from 1 to 3 weeks after the appearance of the symptoms. Gyorgy was able to cure the above conditions with a supplement obtained from yeast which he called "vitamin B_6." In 1938, this factor was isolated from rice paste and yeast in a crystalline form in a number of laboratories.[165] A single dose of about 0.1 mg. produced healing in 14 days in a rat having severe vitamin B_6 deficiency symptoms.

Chemical tests, electrometric titration determinations and absorption spectrum studiest gave clues as to its composition. These were substantiated by the synthesis of vitamin B_6 (1938 and 1939).[166]

Pyridoxine hydrochloride is a white, odorless, crystalline substance that is soluble 1:5 in water, and 1:100 in alcohol and insoluble in ether. It is relatively stable to light and air in the solid form and in acid solutions at a pH of not greater than 5, at which pH it can be autoclaved at 15 lbs. at 120° for 20 to 30 minutes. Pyridoxine is unstable when irradiated in aqueous solutions at pH 6.8 or above. It is oxidized readily by hydrogen peroxide and other oxidizing agents. Pyridoxine is stable in mixed vitamin preparations to the same degree as riboflavin and nicotinic acid. A 1 per cent aqueous solution has a pH of 3.

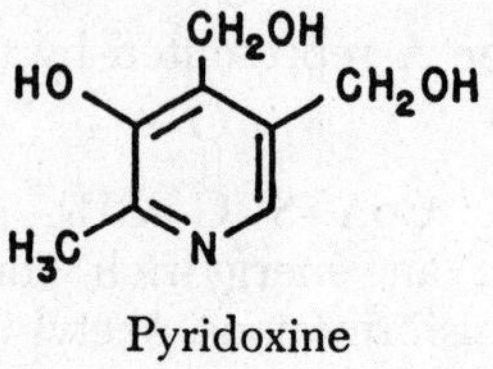

Pyridoxine

The pK$_1$ values for pyridoxine, pyridoxal and pyridoxamine are 5.00, 4.22 and 3.40, respectively, and their pK$_2$ values are 8.96, 8.68 and 8.05, respectively.[167]

Rats[168] that have been deprived of vitamin B$_6$ over long periods of time develop fits that are epileptiform in character and show several successive stages, i.e., violent stage, helpless condition and, finally, a comatose condition. Others have reported that severe microcytic hypochromic anemia developed in puppies[169] and dogs[170] when vitamin B$_6$ was apparently the only missing factor in the diet. This anemia was cured by supplementing the diet with this vitamin. The symptoms of vitamin B$_6$ deficiency in chicks[171] are slow growth, depressed appetite and inefficient utilization of food, followed in some cases by spasmodic convulsions and death.

Spies,[172] in 1939, showed that patients maintained on a pellagra-producing diet, supplemented with nicotinic acid, vitamin B$_1$ and riboflavin, still showed nervousness, insomnia, irritability, abdominal pains, weakness and difficulty in walking. These symptoms were alleviated by the administration of vitamin B$_6$. Vitamin B$_6$ apparently plays a role in the cure of cheilosis, thought to be due to riboflavin deficiency.

The intravenous injection of synthetic pyridoxine alleviated muscular dystrophy and Parkinson's syndrome in a number of cases. Large doses (50 to 100 mg.) had a sedative effect[173] in normal persons, epileptics and patients with deficiency diseases. It reduced or completely abolished the seizures in persons with idiopathic epilepsy.

Vitamin B$_6$ has been used with success for nausea and vomiting of pregnancy (30 to 100 mg. daily) and in the treatment of acne and possibly the prevention of atherosclerosis. In the latter cases it may play a role in the conversion of linoleic to arachidonic acid. Rats on a vitamin B$_6$-deficient diet develop enlarged fatty livers. The cheilosis associated with pellagra, sprue, celiac disease and digestive upset has been treated successfully with vitamin B$_6$.

Di- and triacetyl pyridoxine[174] are as effective as free pyridoxine in experimental animals, i.e., rats, whereas the benzoate is inactive.

Evidence[175] has accumulated that pyridoxine alone is not responsible for all the activity found in naturally occurring materials. Pyridoxamine and pyridoxal have been isolated and their structures proved by synthesis.[176] All three compounds have been isolated and their structures proved by synthesis.[176] All three compounds have about equal activity for the rat, whereas

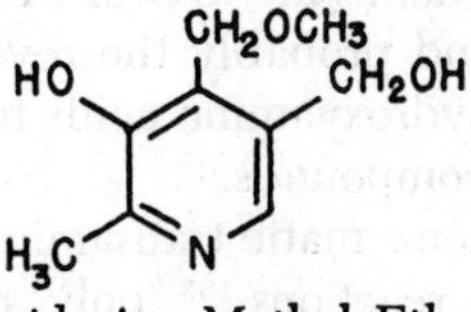

Pyridoxine Methyl Ether

$$\downarrow NH_3 \quad 120\text{-}140°$$

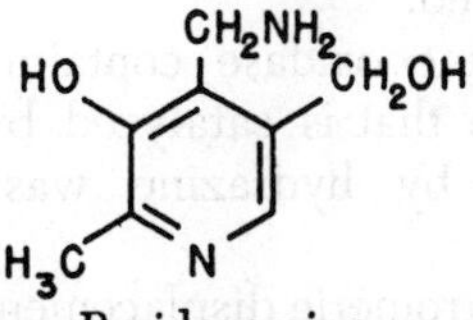

Pyridoxamine

there is a great variation in their activities for different microorganisms. These compounds are inactivated rapidly[177] by exposure to light, especially ultraviolet light, particularly at a pH of 7 or higher. In 0.1 N acid they are comparatively stable to light; however, pyridoxamine is destroyed fairly rapidly by direct exposure to sunlight.

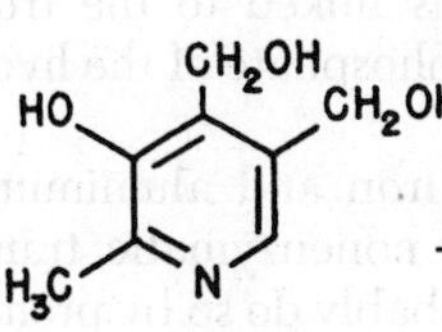

Pyridoxine

$$\xrightarrow{\text{KMnO}_4}$$

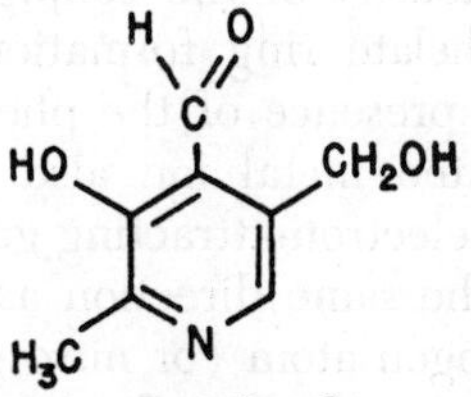

Pyridoxal

Oxygen, apparently, does not play a role in this light sensitivity.

Pyridoxamine, pyridoxine and pyridoxal,

as the corresponding 5-hydroxymethyl phosphates,[178] function as cotransaminase in biologic transaminations[179] and in the decarboxylation[180, 181] of certain amino acids, such as tyrosine, lysine, arginine, ornithine, aspartic and glutamic acids. Other biological transformations[182] of amino acids in which pyridoxal can function are racemization, elimination of the α-hydrogen together with a β-substituent (i.e., OH or SH) or a γ-substituent, and probably the reversible cleavage of β-hydroxyamino acids to glycine and carbonyl compounds.

For nonenzymatic transamination and dehydration reactions,[183] only the aldehyde and the phenolic hydroxyl groups in the proper orientation on the pyridine nucleus are required.

An amine oxidase containing pyridoxal phosphate that is catalyzed by Cu^{++} and inhibited by hydrazine was isolated in plants.[184]

An electromeric displacement of electrons from bonds a, b or c would result in the release of a cation (H, R′ or COOH) and subsequently lead to the variety of reactions observed with pyridoxal. The extent to which one of these displacements predominates over others depends on the structure of the amino acid and the environment (pH, solvent, catalysts, enzymes, etc.). When the above mechanism applies in vivo, the pyridoxal component is linked to the true enzyme through the phosphate of the hydroxymethyl group.

Metals such as iron and aluminum that markedly catalyze nonenzymatic transaminations in vitro probably do so by promoting the formation of the Schiff base and maintaining planarity of the conjugated system through chelate ring formation which requires the presence of the phenolic group. This chelated metal ion also provides an additional electron-attracting group that operates in the same direction as the heterocyclic nitrogen atom (or nitro group), thus increasing the electron displacements from the alpha carbon atom as shown on this page.

Kinetic studies have shown that imidazole catalyzes 3-hydroxypyridine-4-aldehyde in transamination reactions via the following

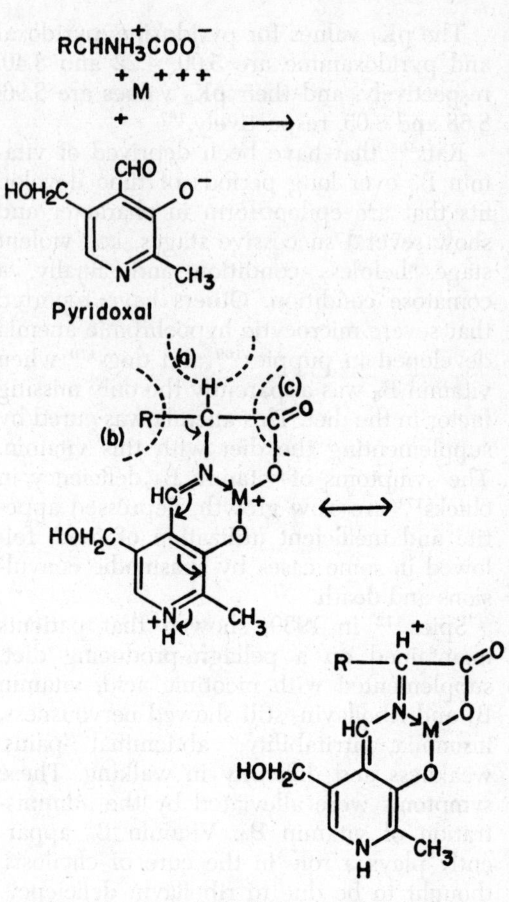

Pyridoxal

intermediate.[185] Imidazole catalysis also has

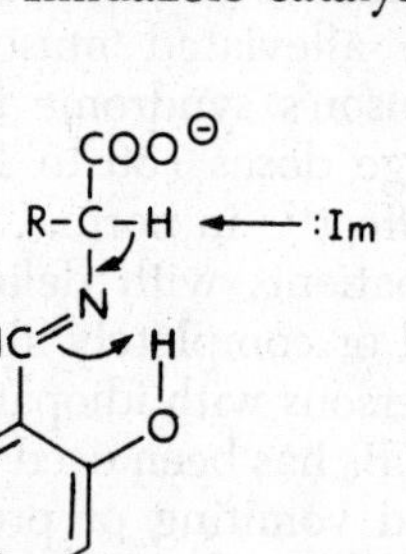

been demonstrated with pyridoxal.[186]

The growth inhibitory effects of compounds related to pyridoxal, pyridoxamine and pyridoxine vary greatly with the test organism. For example, 2-ethyl-3-hydroxy-4-formyl-5-hydroxymethylpyridine antagonizes pyridoxal for yeast but has some (1 to 35%) growth-promoting properties for lactic acid bacteria.[187] 5-Desoxypyridoxal, 5-desoxypyridoxamine and 4-desoxypyridoxine are examples of potent vitamin B_6 inhibitors.[188]

Coenzymatic studies of simple homologs of pyridoxal at the 2 position suggests that the methyl group at the 2 position plays an important spatial role in interaction with the coenzyme binding site of the apoenzyme such that variations in this group lead to variations in the conformation of the substrate binding site and catalytic site of the holoenzyme. On the other hand, no catalytic role in the reactions catalyzed by pyridoxal phosphate proteins is due to the homologs at the 2 position.[189]

Monoamine oxidase inhibitors of the hydrazine type, e.g., isoniazid, inhibit pyridoxal phosphokinase (which converts pyridoxal to its phosphate) in the brain via pyridoxal. Such inhibition if great enough can induce seizures in man and animal.

Category—enzyme cofactor vitamin.

Usual dose—oral or parenteral: prophylactic, 2 mg. once daily; therapeutic, 5 to 150 mg. daily.

Usual dose range—2 to 150 mg. daily.

OCCURRENCE

Pyridoxine Hydrochloride Injection U.S.P.
Pyridoxine Hydrochloride Tablets U.S.P.

NICOTINIC ACID

Niacin N.F., Nicotinic Acid, 3-Pyridinecarboxylic Acid. Nicotinic acid first was prepared in 1867, by the oxidation of nicotine.[190] It was isolated by Funk,[191] in 1913, from yeast concentrates and about the same time by Suzuki *et al.*[192] from rice polishings in connection with antineuritic studies in which it failed to have antineuritic activity. Interest in nicotinic acid and its derivatives in biologic processes lagged until Warburg and Christian[193] showed that nicotinic acid amide was obtained upon the hydrolysis of a coenzyme which they isolated from red blood cells of horse blood. This coenzyme is now known as coenzyme II. Kuhn and Vetter[194] isolated nicotinic acid amide from heart muscle, and von Euler *et al.*[195] isolated it from cozymase. These findings again focused attention upon the possible value of nicotinic acid and its amide in the nutrition of experimental animals. Some growth responses were obtained in rats and pigeons on certain diets when given these factors.

However, the magnitudes were not sufficiently great to warrant classifying the factors as vitamins.

Pellagra* has been known nearly 2 centuries, and the term has its origin from the Italian *"pelle agara"* meaning rough skin. By 1930, due to the excellent work of Goldberger *et al.*, pellagra in humans had been established definitely as a deficiency disease, and the protective factor was associated with the more heat stable fraction of the vitamin B complex. Although pellagralike conditions have been produced in some experimental animals, it was shown finally that only black tongue, a dietary deficiency disease in dogs, was comparable with pellagra in humans. From liver extracts, which were efficacious in treating pellagra in humans and black tongue in dogs, nicotinic acid amide was isolated.[196] This compound, as well as nicotinic acid, cured black tongue in dogs. Nicotinic acid soon was tested and shown to be successful in the treatment of human pellagra.[197]

Nicotinic acid can be prepared by the oxidation of nicotine with nitric acid.[198] It

Nicotine   Nicotinic Acid

also can be prepared by the oxidation of other beta substituted pyridines, such as β-picoline, 3-ethylpyridine, 3,3'-dipyridyl, 3-phenylpyridine. The pyridine-polycarboxylic acids, with the exception of those having carboxyl groups in the 4-position, that have a carboxyl group in the 3-position undergo decarboxylation, thermal or acidic, to give nicotinic acid. The pyridinepolycarboxylic acids are obtained by the oxidation of quinoline, quinaldine, etc.

Nicotinic acid occurs as white crystals or as a crystalline powder. It is odorless or it

* Usually prevalent in people who eat certain simplified diets, such as the salt pork, maize and molasses diet of some Negroes and poor white people of southern United States.

may have a slight odor. One gram of nicotinic acid dissolves in 60 ml. of water. It is freely soluble in boiling water, in boiling alcohol and in solutions of alkali hydroxides and carbonates but is almost insoluble in ether. A 1 per cent aqueous solution has a pH of 6.

Nicotinic acid is stable under normal storage conditions. It sublimes without decomposition.

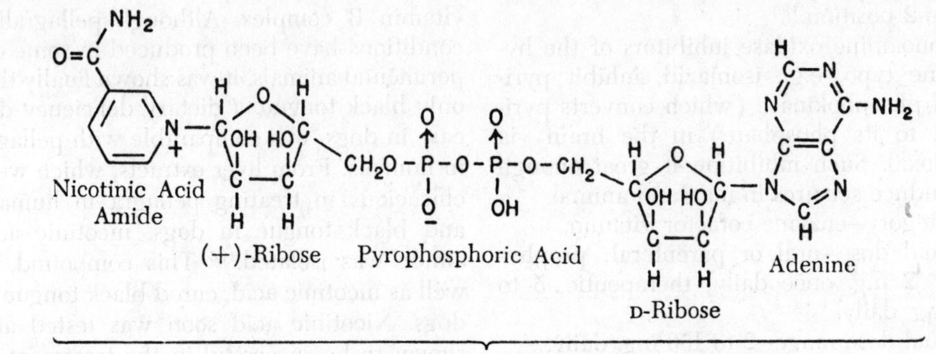

pounds that are capable of oxidation or hydrolytic conversion to these substances in the body possess anti-black tongue activity.[200] Examples of these are ethyl nicotinate, nicotinic acid N-methylamide, nicotinic acid N-diethylamide, beta picoline and nicotinuric acid.

Nicotinic acid has been shown to exist as the amide as a component of coenzymes I and II. The latter differs from the former

Nicotinic Acid Amide   (+)-Ribose   Pyrophosphoric Acid   D-Ribose   Adenine

Coenzyme I; Cozymase, Codehydrase

The most satisfactory chemical assay for nicotinic acid seems to be that which depends on the reaction with cyanogen bromide and aniline to produce a yellow color.[199] The biologic methods of importance use the cure of induced pellagra in dogs or pigs and the growth of various bacteria.

Excellent sources of nicotinic acid, its amide or its coenzymes, are pork, lamb, and beef livers, hog kidneys, brewers and bakers yeasts, pork, beef tongue, hearts, lean meats, wheat germ, peanut meal, green peas. These factors are distributed quite widely

only in that it contains an additional molecule of phosphoric acid. Nevertheless, they possess remarkable specificity in relation to the dehydrogenases* with which they will function. In most cases, a substance (together with its dehydrogenase) will react with one of the coenzymes but not with the other. The nicotinic acid portion of these coenzymes acts as hydrogen acceptors and donors in a number of biologic oxidation and reduction systems in the metabolism of carbohydrates, amino acids and fats.

Coenzymes I and II accept hydrogen or are reduced in the presence of certain spe-

Substrate (Alcohol)   Coenzyme I   (Acetaldehyde)   Reduced Coenzyme I

in varying concentrations in animal tissues and in plants.

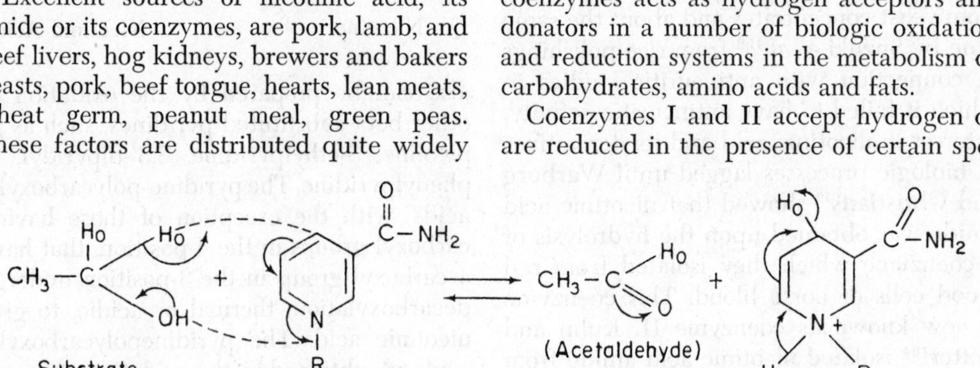

It appears probable that, in addition to nicotinic acid and its amide, only those com-

cific dehydrogenases and suitable substrates, such as alcohol, acetaldehyde, hexose phos-

* Lactic acid, malic acid, glycerophosphoric acid, glutamic acid, glucose, Robison ester are some.

phates, phosphoglyceric aldehyde, isocitrate, oxalacetate. These coenzymes act as hydrogen transporting agents in aerobic systems and will reduce riboflavin phosphate in the presence of a suitable enzyme. They also will act as reducing agents in anaerobic systems, and many of the above dehydrogenations are reversible. The function of coenzyme I, for example, in an anaerobic system can be depicted as follows:

$$\text{Triosephosphoric Acid} + D^T\text{–}C_0 \rightleftarrows D^T\text{–}C_0H_2 + \text{Phosphoglyceric Acid}$$

$$\text{Pyruvic Acid} + D^L\text{–}C_0H_2 \rightleftarrows D^L\text{–}Co + \text{Lactic Acid}$$

Key: D^T = Triosedehydrogenase
$\quad\ \ D^L$ = Lactic Acid Dehydrogenase

Co = Coenzyme I
CoH_2 = Reduced or Dihydrocoenzyme I

It should be noted that certain substituted pyridines will exchange with the nicotinamide moiety of coenzyme I [diphosphopyridine nucleotide (D.P.N.)] to form coenzyme I (D.P.N.) analogs by the use of D.P.N.-ases such as pig brain D.P.N.-ase. The substituted pyridines that have been used are isoniazid, iproniazid and pyridine substituted in the 3 position by the following groups: –COCH$_3$; –CONHOH; –CONHNH$_2$; –CHNOH; –COC$_6$H$_5$; –NH$_2$; –NHCOCH$_3$; –COC$_3$H$_7$ and –CSNH$_2$. The 4-substituted pyridine D.P.N. analogs are inactive coenzymes. In the case of the 3-substituted analogs, some are inactive, e.g., –NH$_2$; –NHCOCH$_3$, whereas some others are active, e.g., –COCH$_3$; –COC$_3$H$_7$; –CSNH$_2$; –CONHNH$_2$ and –CHNOH; however, the activity (rate of reduction) varies greatly, depending on the source (animal, part of the animal or microorganism) of the dehydrogenase. The rate of reduction also is dependent upon the inductive effect conferred upon position 4 of the pyridine ring by the 3 substituent.[201] The biologic implications of the above phenomena should be of distinct interest to the medicinal chemist.

It has been shown[202] that corn contains 3-acetylpyridine, which inhibits the full utilization of the nicotinic acid normally present in corn. It also has been shown that some of the nicotinic acid is bound via its carboxyl group so that it is not available to

man. This, in part, explains why people who consume corn as a major portion of their diet develop pellagra.

Studies on nutrition have indicated that coenzyme I (cozymase) could replace nicotinic acid amide completely and be more effective on a mole for mole basis. The administration of nicotinic acid has been shown almost invariably to increase the level of cozymase in the blood of both normal persons and pellagrins. The cozymase content of liver and muscle rises and falls with the nicotinic acid intake. Evidence is accumulating that indicates a direct relationship between carbohydrate metabolism, insulin and nicotinic acid (in the form of coenzymes I and II). This is because pellagra signs and a deficiency of cozymase of the blood have been shown to be associated with diabetics.

Pellagra in humans is manifested by very complex symptoms.[197] In brief, these are characterized as follows: (1) A typical dermatitis may develop anywhere, usually following a bilaterally symmetric pattern, and may be accompanied by pigmentation. (2) A characteristic glossitis appears, and in the early stages the tip and the lateral margins of the tongue are reddened and swollen. As the involvement of the mucous membranes increases, swelling and reddening become more intense. Deeply penetrating ulcers are common, and their surfaces often are covered with a thick, gray membrane filled with Vincent's organisms and debris. Other mucous membranes, such as those of the alimentary tract, urethra, vagina, etc., may be affected similarly. (3) Diarrhea is present. (4) There is anorexia. (5) The patient exhibits weakness and lassitude. (6) Nausea and vomiting are present. (7) There are various types of psychoses, such as mental confusion, loss of memory, disorientation and confabulation. Frequently, these are accompanied by excitement, mania, depression and delirium.

In many cases, pellagrins also suffer from a deficiency of other vitamins, particularly vitamin B_1, riboflavin and vitamin B_6. This condition serves to complicate the clinical picture. The administration of all the known vitamins except one, such as nicotinic acid, for example, permits a more accurate study of the deficiency symptoms due to a lack of this factor in the diet. Nicotinic acid, when administered for the usual case of pellagra, 500 mg. daily in 50-mg. doses, brings about a prompt and dramatic relief of the symptoms described above. Large amounts of nicotinic acid often are followed by sensations of heat and tingling of the skin accompanied by flushing and a rise in skin temperature which has no clinical significance. Nicotinic acid amide does not produce this effect.

Category—vitamin B-complex component.

Usual dose—oral requirement: 20 mg. daily; therapeutic oral or parenteral, 50 mg. 3 to 10 times a day.

OCCURRENCE
Niacin Injection N.F.
Niacin Tablets N.F.

Niacinamide U.S.P., Nicotinamide, Nicotinic Acid Amide. Nicotinamide is prepared by the amidation of esters of nicotinic acid or by passing ammonia gas into nicotinic acid at 320° C.

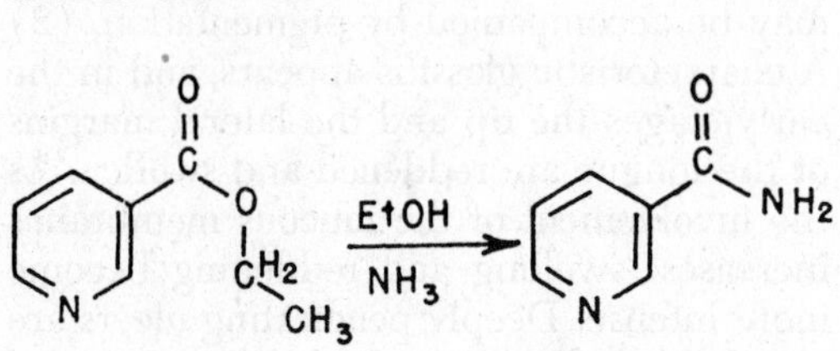

Ethyl Nicotinate   Nicotinamide

Nicotinamide is a white, crystalline powder that is odorless or nearly so and has a bitter taste. One gram is soluble in about 1 ml. of water, 1.5 ml. of alcohol and in about 10 ml. of glycerin. Aqueous solutions are neutral to litmus. For occurrence, action and uses see nicotine acid.

Niacinamide hydrochloride recently has been made available. It is more stable in solution and more compatible with thiamine chloride in solution.

Usual dose—oral or parenteral: prophylactic, 20 mg. once daily; therapeutic, 50 mg. 3 to 10 times a day.

Usual dose range—20 to 500 mg. daily.

OCCURRENCE
Niacinamide Injection U.S.P.
Niacinamide Tablets U.S.P.

FOLIC ACID

Folic Acid U.S.P., Folacin®, Folvite®, N-{p-{[(2-Amino-4-hydroxy-6-pteridinyl)methyl]amino}benzoyl}glutamic Acid, Pteroylglutamic Acid.* Folic acid was defined originally as the active principle required for the growth of a *Streptococcus lactis* R.† under specified conditions. It was called folic acid by Williams *et al.,*[103] who showed that it was present in leaves and foliage and in a particular spinach, from which they produced very active concentrates when they worked with tons of spinach. This name is not wholly justified, since it also is found in whey, mushrooms, liver, yeast, bone marrow, soybeans, urine and fish meal.

The structural formula for folic acid has

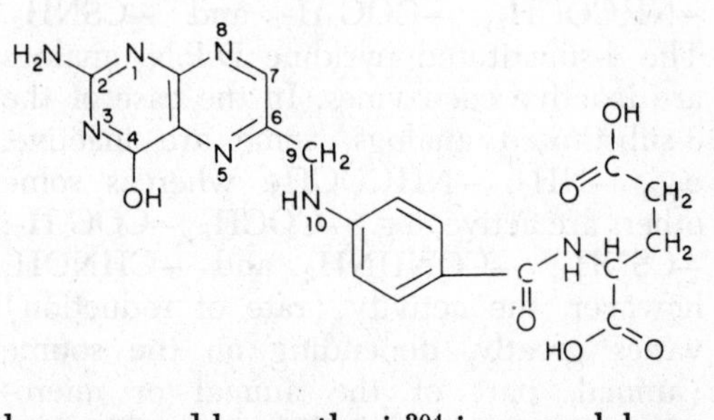

been proved by synthesis[204] in many laboratories.

Folic acid occurs as a yellow or yellowish-orange powder that is only slightly soluble in water (1 mg. per 100 ml.). It is insoluble in the common organic solvents. The sodium salt is soluble (1:66) in water.

Aqueous solutions of folic acid or its sodium salt are stable to oxygen of the air, even upon prolonged standing. These solutions

* Pteroylmonoglutamic acid accepted by I.U.P.-A.C.

† An organism which sours milk.

can be sterilized by autoclaving at a pressure of 15 pounds per square inch in the usual manner. Folic acid in the dry state and in very dilute solutions is decomposed readily by sunlight or ultraviolet light. Although folic acid is unstable in acid solutions, particularly below a pH of 6, the presence of liver extracts has a stabilizing effect at lower pH levels than is otherwise possible. Iron salts do not materially affect the stability of folic acid solutions. The water-soluble vitamins that have a deleterious effect on folic acid are listed in their descending order of effectiveness as follows: riboflavin, thiamine hydrochloride, ascorbic acid, niacinamide, pantothenic acid and pyridoxine. This deleterious effect can be overcome to a considerable degree by the inclusion of approximately 70 per cent of sugars in the mixture.

Folic acid in foods is destroyed more readily by cooking than are the other water-soluble vitamins. These losses range from 46 per cent in halibut to 95 per cent in pork chops and from 69 per cent in cauliflower to 97 per cent in carrots.

The pteridine portion of the molecule is unique, whereas the *p*-aminobenzoic and glutamic acids are simple and well known. Hopkins, as early as 1889,[205] recognized that the pigments of butterfly wings were purine derivatives. Wieland,[206] in 1925, proposed the generic term pterins for the colored and colorless pigments that occur in butterfly wings. In 1935, Bavarian children collected 250,000 cabbage butterflies, from which sufficient pigment was obtained to elucidate the constitution of xanthopterin,* which has been shown to be as follows:

Xanthopterin

The fermentation† *L. casei* factor contains three glutamic acid residues and is pteroly - γ - glytamyl - γ - glutamyl - glutamic

* Found in human urine.

† Fermentation product of a diphthoid bacterium that was isolated more easily than other related

acid. Both factors have the same pteridine nucleus attached to *p*-aminobenzoic acid. Since a chemical name was too long for general usage, a name for the basic nucleus indicating its pterine nature was proposed. Thus, the name "pteroylglutamic acid" was proposed for the liver *L. casei* factor. The basic structure was proposed as "pteroic acid" and chemically is 4[< (2-amino-4-hydroxy-6-pteridyl) methyl > amino] benzoic acid; it was synthesized in the same way as pteroylglutamic acid.

Vitamin B_c conjugate is pteroylhexaglutamylglutamic acid, and the enzyme vitamin B_c conjugate is a peptidase, and, since it does not hydrolyze vitamin B_c conjugate methyl ester, it can be identified as a pteroylglutamylcarboxypeptidase. Most of the pteroylglutamic acid in food exists in a conjugated form.

It appears that a folinic acid and not folic acid is the natural factor in digests of liver and that folic acid arises from a folinic acid‡ during the isolation of folic acid. Folinic acid, Folinic acid-S.F.[207] and leucovorin,[208] a growth factor for *Leuconostoc citrovorum*, appear to be identical.[209] Folinic acid-S.F.§ was prepared from folic acid by formylation, reduction and autoclaving‖ or the action of dilute alkali and has been assigned the structural formula shown as follows.

Folinic Acid-S.F. (Leucovorin)
(5-Formyl-5,6,7,8-tetrahydrofolic Acid)

factors and, therefore, employed to a large extent for proof of structure work.

‡ Very mild acid hydrolysis alters folinic acid (as measured by biologic activities) but forms a compound with folic acid activity.

§ The letters S.F. are used to differentiate for the time being between this synthetic factor and unidentified, biologically similarly active naturally occurring factor or factors.

‖ Rearranges the formyl group from position 10 to position 5.

Folinic acid-S.F. is very sensitive to acid; however, it is stable at $90°$ in 0.1 N sodium hydroxide for 6 hours; whereas, after 22 hours, there is a loss of about 65 per cent. It is a colorless compound that decomposes at 240 to $250°$.

It also can be prepared by the formylation of tetrahydrofolic acid. Folinic acid-S.F. is an effective antianemic substance for humans, a growth factor for chicks, prevents the toxicity of aminopterin for the mouse and is more effective than folic acid in preventing the toxicity of methyl folic acid for *L. casei*. Tetrahydrofolic acid is also biologically active and reverses the effect of 4-aminofolic acid in mice.

It appears that one or more of the above mentioned formyl derivatives (probably a folinic acid similar or identical to folinic acid-S.F.) is the coenzyme that is involved in the introduction of a single carbon unit into purines, pyrimidines and probably histidine.

Tetrahydrofolic acid (THFA) can accept formaldehyde to form N^5,N^{10}-methylene THFA reductase to N^5-methyl THFA. The latter can donate the N^5 methyl group to homocysteine to form methionine in the presence of B_{12}-enzyme (DPNN, FAD, ATP, Mg++) and to regenerate THFA.[210]

Tetrahydrofolic acid (THFA) also can accept a formimino ($-CH=NH$) group at its N_5 position from such substrates as formiminoglycine (FIG) and formiminoglutamic acid (FIGLU). This derivative may then participate in a number of ways, one of which is as follows[211]

FIGLU + THFA → 10 formyl − THFA + Glutamic Acid

A formimino derivative of THFA + NH_3 also may be formed that may function in certain formimino transfer reactions catalyzed by enzymes.

In the presence of an adequate supply of preformed pteroylglutamic acid, enterococci and certain lactobacilli were relatively insensitive to the sulfonamides. *p*-Aminobenzoic acid showed a competitive type of antagonism. Therefore, it can be concluded that in these experiments the primary point of inhibition was the synthesis of pteroylglutamic acid and related compounds by

means of *p*-aminobenzoic acid. *p*-Aminobenzoyl-L-glutamic acid was 8 to 10 times as active on a molar basis as *p*-aminobenzoic acid in antagonizing the inhibition of *Lactobacillus arabinosus* by sulfanilamide. Sulfonamides inhibit the growth of those bacteria that synthesize their own supply of folic acid but not those that cannot synthesize this factor but require it as a preformed substance. Some purines and thymine are also products of enzyme systems in which *p*-aminobenzoic acid functions. These compounds, when added to the media, in some cases render the bacteria insensitive to sulfonamides.[212]

Some pernicious anemia patients[213] in relapse fail to utilize the conjugate, whereas the normal individual appears to be able to utilize this form. The free vitamin is excreted in the urine of the latter after administration of the conjugate, whereas, in the former, this does not occur. Spies[214] believes that folic acid (vitamin B_c) is liberated by enzymes more readily in some of the anemias than in others. Relatively large amounts of folic acid are required in eliciting a favorable hemopoietic response, as compared with the small quantity contained in equally effective liver extract.

In macrocytic anemias, folic acid duplicates the effects of liver therapy, i.e., an increase in reticulocytes, red and white cells, platelets, normoblasts and hemoglobin. The decrease of bone marrow megaloblasts and the return of vigor and appetite are the physical manifestations. Folic acid performs a specific function in the maturation of the various cells of the bone marrow and has other obvious profound effects on the human body.

In contrast with the excellent hematopoietic activity in pernicious anemia, several reports indicate the failure of therapy with pteroylglutamic acid to arrest the progress or to prevent the subsequent development of neurologic symptoms and signs. Neurologic manifestations may develop rapidly many weeks after the blood values have been restored to and maintained at normal levels. Recently, dihydrobiopterin has been shown to be the phenylalanine hydroxylation cofactor and to play a role in the enzymatic conversion of phenylalanine to tyrosine. It is the first un-conjugated pteridine shown to have a meta-bolic role.[215] It should be noted that 5-methyl-5,6-dihydrofolate[216] is effective in methionine biosynthesis.[217]

Folic acid can be administered orally or parenterally in the treatment of a number of macrocytic anemias, including sprue,[218] macrocytic anemias of pregnancy, those of gastrointestinal origin and those associated with pellagra and similar states.[219]

Five to six times as much folic acid is ex-creted in sweat as in the urine in humans; this amounts to 3.8 to 23.8 mcg. per day.

Category—hematopoietic vitamin.

Usual dose—100 mcg. daily; therapeutic: oral or intramuscular, 250 mcg. to 1 mg. daily.

OCCURRENCE

Folic Acid Injection U.S.P.
Folic Acid Tablets U.S.P.

Calcium Leucovorin, Calcium Folinate-S.F., Calcium N-[p-{(-Amino-5-formyl-5,6, 7,8 - Tetrahydro - 4 - hydroxy - 6 - pteridinyl) - methyl}amino]benzoylglutamate, Calcium 5-Formyl-5,6,7,8-tetrahydrofolate, occurs as a yellowish-white or yellow, odorless, micro-crystalline powder that is insoluble in alco-hol and very soluble in water.

The chemistry and the function of leuco-vorin have been described previously. It can be used as an antidote for folic acid antagonists.

Dose—intramuscular 6 mg.
Range—200 mcg. to 15 mg.

Antimetabolites of Folic Acid

A large number of structural modifica-tions[220] of pteroylglutamic acid have been synthesized and, in general, the variations fall into the following classes: (1) changes in the substituents in the 2- and 4-positions; (2) replacement of the pteridine moiety by other cyclic systems; (3) alteration of the 9, 10 configuration; (4) replacement of glutamic acid by other amino acids; (5) re-placement of *p*-aminobenzoic acid by posi-tion isomers and by sulfanilic acid; (6) substitution in the benzene ring of the *p*-aminobenzoic acid. Many of these analogs competitively inhibit the activity of pteroyl-glutamic acid, formyl folic acid, folinic acid

and others, both in microorganisms and in animals. Death may result from this inhibi-tory effect. This type of inhibition has proved useful in the detection of folinic acid. Some of these analogs have been used in studies involving leukemia and neo-plasms, both in experimental animals and in man.[221]

Methotrexate U.S.P., 4-Amino-10 methyl-folic Acid, occurs as an orange-brown odor-less, crystalline powder that is practically insoluble in water, alcohol or chloroform but is soluble in alkali hydroxides.

Category—antineoplastic.

Usual dose—oral, intramuscular or intra-venous, 2.5 to 5 mg. once daily or every other day.

Usual dose range—2.5 to 10 mg. daily or every other day.

OCCURRENCE

Methotrexate for Injection U.S.P.
Methotrexate Tablets U.S.P.

THE COBALAMINS

About 100 years ago, Addison described pernicious anemia—a condition that was uniformly fatal. In 1926 Minot and Mur-phy[222] demonstrated that the disease could be kept in check by feeding about one half pound of whole liver daily to per-nicious anemia patients. Although the years that followed saw the advent of extracts of liver that could be used orally or by injec-tion, the isolation of the pure principle was very elusive. The isolation studies were complicated by the fact that man was the only satisfactory experimental subject. How-ever, in 1947, M. Shorb's discovery[223] that a growth stimulating factor for *L. lactis D*, which was present in liver extracts used for the treatment of pernicious anemia, fur-nished a simple, rapid, biologic assay method for tracing the concentration and isolation of an antipernicious anemia fac-tor. In April 1948, Rickes *et al.*[224] isolated minute amounts of a red, crystalline com-pound from clinically active liver fractions, which was also highly effective in promot-ing the growth of *L. lactis*. This compound was called vitamin B_{12} and in single doses, as small as 3 to 6 mcg., produced positive

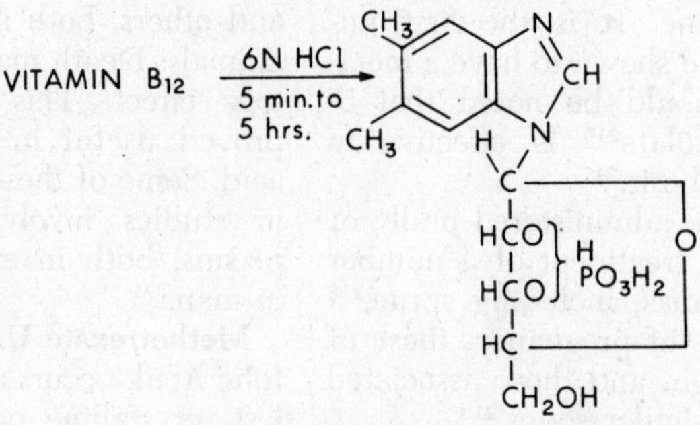

VITAMIN B₁₂ → (6N HCl, 5 min. to 5 hrs.)

2′- or 3′-Phosphoryl-5:6-dimethyl benziminazole-1-α-D ribofuranoside
α-Ribazole Phosphate

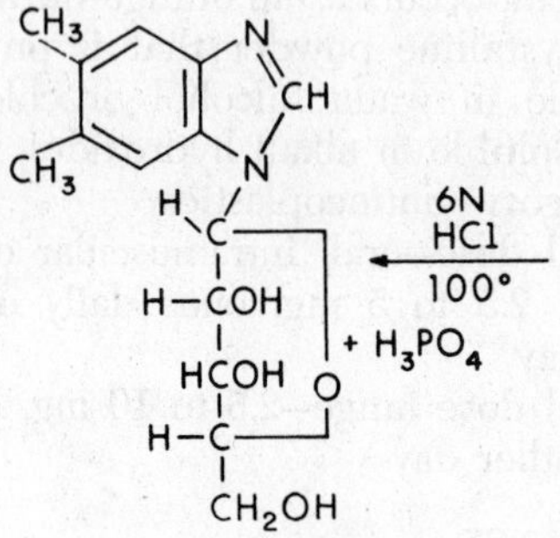

← (6N HCl, 100°, + H₃PO₄)

α-Ribazole
1-α-D Ribofuranoside-5,-6-dimethylbenzimidazole

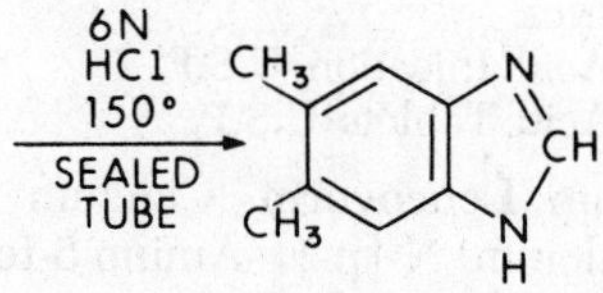

→ (6N HCl, 150°, SEALED TUBE)

5,6-Dimethylbenzimidazole

hematologic activity in patients having addisonian pernicious anemia. Evidence indicates that its activity is comparable with Castle's extrinsic factor and that it can be stored in liver.

The term animal protein factor (A.P.F.) was coined to describe an essential metabolite necessary for the growth of chicks maintained on vegetable diets. This factor was shown to be present in cow manure and chick feces as a result of bacterial activity, and it was shown that it had antianemic properties. Vitamin B₁₂ also is found in commercial fermentation processes of antibiotics, such as *Streptomyces griseus*, *S. olivaceus*, *S. aureofaciens*, sewage, milorganite and others. Some of these fermentations furnish a commercial source of vitamin B₁₂.

Vitamin B₁₂ (cyanocobalamin, q.v.) occurs in nature as a cofactor (q.v.) which originally was isolated as cyanocobalamin and vitamin B₁₂b (hydroxocobalamin, q.v.).

Stepwise degradation[225, 226, 227] of vitamin B₁₂ yielded the fragments shown above, together with cobinamide (q.v.).

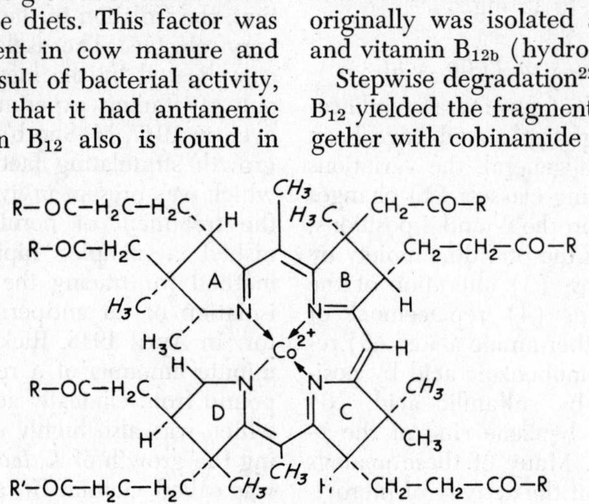

Cobyrinic Acid

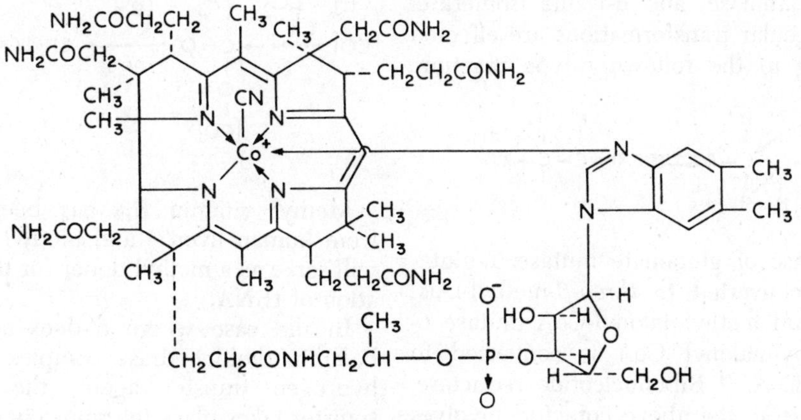

The chemical structure diagram showing cyanocobalamin with labels:
NH₂COCH₂CH₂, CH₃, CH₃, CH₂CONH₂, NH₂COCH₂, CH₃, CN, CH₂CH₂CONH₂, CH₃, CH₃, N, N, Co³⁺, N, N, NH₂COCH₂, CH₃, CH₃, CH₃, CH₃, CH₂CH₂CONH₂, CH₃, CH₃, N, N, CH₃, CH₃, CH₂CH₂CONHCH₂CH—O—P—O, O⁻, HO, H, H, CH₂OH

Cyanocobalamin

The macro-ring system devoid of periph-eral substituents and found in vitamin B_{12} is named corrin, and its derivatives are cor-rinoids that contain central cobalt atom when found in nature. When the corrin ring contains the peripheral groups shown in the structure on page 982 and $R = R' = OH$, it is called cobyrinic acid which contains the parent hydrocarbon.

In *cobyric acid* $R = NH_2$ and $R' = OH$; in *cobinic acid* $R = OH$ and $R' = $ D-$(-)$-1-amino-propan-2-ol; and in *cobinamide* $R = NH_2$ and $R' = $ 1-aminopropan-2-ol.

In *cobamide* the OH of *cobinamide* is es-terified with 3'-phospho-D-ribofuranose. Thus cobalamin is 5,6-dimethylbenzimid-azolylcobamide, and its CN ligand is cyano-cobalamin, and the Co is Co^{3+} and has a coordination[228] number of 6.

In contrast to the porphyrins, the metal atom in vitamin B_{12} is held so tightly[229] that it has not yet been possible to remove it without destroying the molecule. The Co atom participates in the resonance of the corrin ring system.

Bases other than 5,6-dimethylbenzimida-zole such as imidazole, benzimidazole, naph-thimidazole or purine also can be incorpo-rated into cobamide.

Pseudovitamin B_{12} contains an adenine residue instead of 5,6-dimethylbenzimida-zole.[230, 231] It is active for certain micro-organisms but inactive for chicks. Other vitamin B_{12}-like substances, active for cer-tain microorganisms, have been found.

Under the usual conditions, in the ab-sence of cyanide ions only the hydroxo form of cobalamin is isolated from natural sources, and this was called vitamin B_{12b}. It has good depot properties[232] and can be used that way for its B_{12} properties. Hy-droxocobalamin will react with chloride, bromide, nitrite, thiocyanate, cyanate, cya-nide and other ions to form the correspond-ing cobalamins. Advantage is taken of this property to pretreat the crude sources with cyanide to facilitate its isolation and to en-sure a more uniform assay due to the in-creased stability of the cyanocobalamin.

Cobalamin exists in a coenzyme[229] form in the livers of man and animals. A 5'-deoxyadenosyl residue is linked covalently to carbon 5 of the 5'-deoxyadenosyl residue and the Co atom of cobalamin, and, there-fore, the compound is Co-$(5'-\beta$-deoxyadeno-syl)cobalamin and the Co is Co^{3+}. It also is called dimethylbenzimidazolyl cobama-mide coenzyme (DBC coenzyme). This coenzyme is extremely sensitive to light and readily cleaved by cyanide to yield cyano-cobalamin, adenine and *erythro*-3,4-dihy-droxy-1-penten-5-al. Acid cleavage yields hydroxocobalamin and the last two frag-ments. Other related coenzymes also have been isolated.[233, 234]

Co-5'-deoxyadenosylcorrinoid can func-tion as a prosthetic group with enzymes to transfer hydrogen and form a new carbon-hydrogen bond. Examples are the glutamate-and methylmalonyl-CoA mutase reactions, dioldehydrase, glycerol dehydrase, ethanolamine deaminase, and β-lysine isomerase. Intramolecular transformations are effected according to the following type reaction.

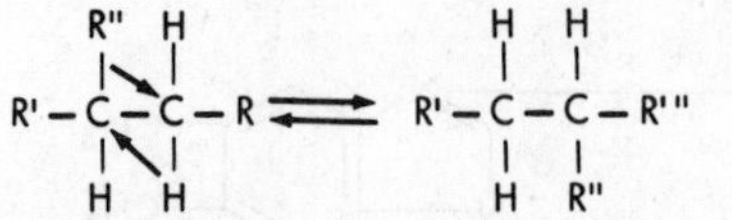

$$R'-\underset{\underset{H}{|}}{\overset{\overset{R''}{|}}{C}}-\underset{\underset{H}{|}}{\overset{\overset{H}{|}}{C}}-R \rightleftharpoons R'-\underset{\underset{H}{|}}{\overset{\overset{H}{|}}{C}}-\underset{\underset{R''}{|}}{\overset{\overset{H}{|}}{C}}-R'''$$

In the case of glutamate mutase, L-glutamate is converted to *threo*-β-methyl-L-aspartate and methylmalonyl-CoA mutase L-(R)-methylmalonyl CoA is converted to succinyl CoA.[228] Ribonucleotide reductase, also requiring the above cofactor, involves inter- and not intramolecular reactions. The hydrogen donor and the hydrogen acceptor are different compounds. This enzyme catalyzes the reduction of ribonucleoside triphosphates to the corresponding 2'-deoxyribonucleoside triphosphates. The hydrogen donors are dihydrolipoate, etc. Schematically this can be depicted as follows:

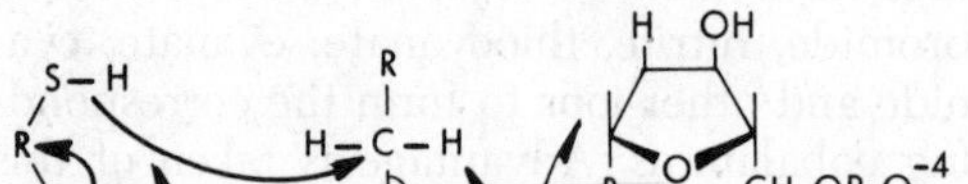

Exchange   Reduction

A Co-corrinoid participates as a Co-methyl-corrinoid (Cofactor) in the methylation of homocysteine,[235] the formation of methane and the synthesis of acetate from CO_2. The methyl group of N^5-methyl-H_4-folate is transferred to a reduced cobamide enzyme complex to form a CH_3 cobamide enzyme complex holoenzyme which in turn can donate its methyl group as a methyl carbonium ion to homocysteine to form methionine under the influence of methionine synthetase. The following scheme depicts the role of cobamide in the synthesis of acetate from CO_2:

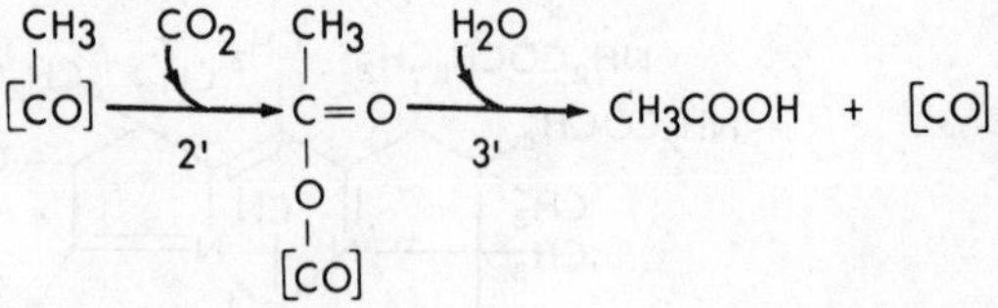

Methyl vitamin B_{12} has been isolated from human liver[236] and methyl cobamide will serve as a methyl donor for the methylation of tRNA.

In the case where 5'-deoxyadenosylcobalamin dioldehydrase complex acts as a hydrogen transfer agent, the hydrogen transfer takes place intermolecularly. In the conversion of 1,2-propanediol to propionaldehyde, hydrogen is transferred from C_1 of the substrate to C-5' of enzyme-bound 5'-deoxyadenosylcobalamin, with the transfer of hydrogen from C-5' of the cobalamin to C-2 of the substrate.

Several mechanisms have been proposed to explain these transformations, i.e., carbanion,[237] organometallic chemistry[238] and hydride ion types.[239]

One of the most unusual features of vitamin B_{12} is that it can be reduced to B_{12s} Co(I) complex, one of the most powerful nucleophiles known.[*][240] This enormously nucleophilic character might account for the minuteness of the amounts of this vitamin needed in vivo.

Products

Cyanocobalamin U.S.P., vitamin B_{12}, is a cobalt-containing substance usually produced by the growth of suitable organisms or obtained from liver. It occurs as dark red crystals or as an amorphous or crystalline powder. The anhydrous form is very hygroscopic and may absorb about 12 per cent of water. One gram is soluble in about 80 ml. of water. It is soluble in alcohol but insoluble in chloroform and in ether.

Vitamin B_{12} loses about 1.5 per cent of its activity per day when stored at room temperature in the presence of ascorbic acid; whereas, vitamin B_{12b} is very unstable[241] (completely inactivated in one day). This loss in activity is accompanied by a release

[*] B_{12s} reacts with methyl iodide 50 million **times** faster than with cyanide.

of cobalt and a disappearance of color. The greater stability of vitamin B_{12} is attributed to the increased strength of the bond between cobalt and the benzimidazole nitrogens by cyanide. Unusual resonance energy is imputed to the cobalt-cyanide complex, giving a positive charge to the cobalt atom and thereby strengthening the Co-N bond. The protective action of certain liver extracts of vitamin B_{12b} toward ascorbic acid and its sodium salt is, no doubt, due to the presence of copper and iron. Iron salts will protect vitamin B_{12b} in 0.001 per cent concentration. Catalysis of the oxidative destruction of ascorbate by iron is well-known.[242] On exposure to air, liver extracts containing B_{12} lose most of the B_{12} activity in 3 months. The most favorable[242] pH for a mixture of cyanocobalamin and ascorbic acid appears to be 6 to 7. Niacinamide can stabilize[243] aqueous parenteral solutions of cyanocobalamin and folic acid at a pH of 6 to 6.5. However, it is unstable in B complex solution. Cyanocobalamin is stable in solutions of sorbitol and glycerin but not in dextrose or sucrose.

Aqueous solutions of vitamin B_{12} are stable to autoclaving for 15 minutes at 121°. It is almost completely inactivated in 95 hours by 0.015 N sodium hydroxide or 0.01 N hydrochloric acid. The optimum pH for the stability of cyanocobalamin is 4.5 to 5.0. Cyanocobalamin is stable in a wide variety of solvents.

The extreme lack of toxicity of vitamin B_{12} indicates that the cyano group is tightly bound within the coordination complex. No deaths or toxic symptoms were produced by the intraperitoneal or intravenous administrations of 1,600 mg./kg. This dose level corresponds to 112,000,000 times the daily human dose of 1 mcg. The thiocyanocobalamin[244] compound had biologic (in bacteria, rats and man) activity and toxicity of the same order of vitamin B_{12}. Other cobalamins are also biologically active.

The degradation products of vitamin B_{12}, i.e., α-ribazole and 5,6-dimethyl benzimidazole, have growth promoting properties when tested in rats. 1,2-Diamino-4,5-dimethylbenzene also has similar activity. The growth promoting activity of 0.25 mcg.

of vitamin B_{12}, approximates 400 mcg. of α-ribazole and 2 to 5 mg. of 5,6-dimethylbenzimidazole and 1,2-diamino-4,5-dimethylbenzene, respectively.

Cyanocobalamin plays a role in the maturation of erythrocytes via its function in the synthesis of thymidine, which, in turn, is an essential component of nucleic acids. The above activity is potentiated by folic acid, which also is a limiting factor in the synthesis of the nucleic acids. Cyanocobalamin is essential for the treatment of pernicious anemia. It also plays a role in certain neurologic disorders and is a growth factor for children.

In cases of pernicious anemia, cyanocobalamin is very effective when administered intramuscularly in small doses. Its oral absorption is enhanced markedly by Castle's intrinsic factor, a purified preparation of which was reported to be active orally at 1 mg. per day.[245]

Certain other cyanocobalamin polypeptide complexes[246] are also active. Very large doses[247] of cyanocobalamin are orally effective in pernicious anemia.

Probably both active and passive mechanisms play a role in the absorption of the cobalamins from the G.I. tract.[248] The active mechanism is operative in the ileum; passive absorption occurs probably along the entire length of the small intestine, especially with large quantities of the vitamin. Sixty to 80 per cent of the first 2 mg. of B_{12} is absorbed via the intrinsic factor mechanism, about 1 per cent of the remainder is absorbed by diffusion. In the active absorption process, a carrier glycoprotein (Castle's gastric intrinsic factor, mol. wt. about 50,000) binds to vitamin B_{12} and delivers it to receptor sites on the brush border of the ileal mucosal cell. In the presence of calcium and alkaline pH, it attaches to these receptor sites, from which it is absorbed.

Eighty to 85 per cent of cyanocobalamin is bound by serum proteins and stored in the liver.

Category—hematopoietic vitamin.

Usual dose—maintenance: intramuscular, 100 mcg. once a month; therapeutic: intramuscular, 100 mcg. once a week.

Usual dose range—30 mcg. to 1 mg. per dose.

OCCURRENCE

Cyanocobalamin Injection U.S.P.

Cyanocobalamin Co 57 Capsules U.S.P. contain cyanocobalamin in which some of the molecules contain radioactive cobalt (Co 57). Each mcg. of this cyanocobalamin preparation has a specific activity of n.l.t. 0.5 microcurie.

The *U.S.P.* cautions that in making dosage calculations one should correct for radioactive decay. The radioactive half-life of Co 57 is 270 days.

Category—diagnostic aid (pernicious anemia).

Usual dose—the equivalent of 0.5 to 1 microcurie.

Cyanocobalamin Co 57 Solution U.S.P. has the same potency, dosage and use as described under Cyanocobalamin Co 57 capsules U.S.P. It is a clear, colorless to pink solution that has a pH range of 4.0 to 5.5.

Cyanocobalamin Co 60 Capsules N.F. is the counterpart of Cyanocobalamin Co 57 capsules in potency dosage and use. It differs only in its radioactive half-life, which is 5.27 years.

Category—diagnostic aid (pernicious anemia).

Usual dose—0.5 to 2 mcg. containing not more than 1 microcurie.

Usual dose range—the equivalent of 0.5 to 1 microcurie.

Cyanocobalamin Co 60 Solution N.F. has the same potency, dosage and use as Cyanocobalamin Co 60 capsules. It is a clear, colorless to pink solution that has a pH range of 4.0 to 5.5.

The above four preparations must be labeled "Caution—Radioactive Material" and "Do not use after 6 months from date of standardization."

Cobalamin Concentrate N.F., derived from Streptomyces cultures or other cobalamin-producing microorganisms contains 500 mcg. of cobalamin per g. of concentrate.

A cyanocobalamin zinc tannate complex can be used as a repository form for the slow release of cyanocobalamin when it is administered by injection.

Category—vitamin B_{12} supplement.

AMINOBENZOIC ACID

Aminobenzoic Acid. In 1940,[249] it was shown that p-aminobenzoic acid was an essential factor for the growth of bacteria. It also was observed that it possessed an anti-sulfanilamide activity[250] in in-vitro experiments. These facts directed attention to the possibility that p-aminobenzoic acid might have vitamin properties. Rats,[251] which had developed a definite graying of fur when maintained on the basal ration G H–I, could be restored to normal by the administration of p-aminobenzoic acid. p-Aminobenzoic acid also was shown to be an essential factor in the maintenance of life and growth of the chick.[228] Since these original developments in this field, various claims[252] have been made for the chromotrichial value of p-aminobenzoic acid in rats, mice, chicks, minks and humans. However, in the case of experimental achromotrichia in rats, it has been shown that, in sufficiently large doses, pantothenic acid will effect a cure and that biotin eliminates the scattered gray hairs or "stippling" remaining after the pantothenic acid treatment. The problem of nutritional achromotrichia is a complex one that may involve several vitamin or vitamin-like factors and is complicated by the synthesis and absorption from the intestinal tract of a number of factors produced by bacteria.

p-Aminobenzoic acid is a white, crystalline substance that occurs widely distributed over the plant and animal kingdom. It occurs both free and combined[253] and has been isolated[254] from yeast, of which it is a natural constituent. It is soluble 1:170 in water, 1:8 in alcohol and freely soluble in alkali.

p-Aminobenzoic acid is thought to play a role in melanin formation and to influence or catalyze tyrosine activity.[255] It inhibits oxidative destruction of epinephrine and stilbestrol, counteracts the graying of fur attributable to hydroquinone in cats and mice, exhibits antisulfanilamide activity and detoxifies the toxic effects of carbarsone and other pentavalent phenylarsonates.[256]

p-Aminobenzoic acid, when given either parenterally or in the diet to experimental animals, will protect them against otherwise fatal infections of epidemic or murine typhus, Rocky Mountain spotted fever and tsutsugamushi disease.[257] These diseases have been treated clinically with most encouraging results by maintaining blood levels of 10 to 20 mg. per cent for Rocky Mountain spotted fever and tsutsugamushi diseases. The mode of action of *p*-aminobenzoic acid in the treatment of the above diseases appears to be rickettsiostatic rather than rickettsiocidal, and the immunity mechanisms of the host finally overcome the infection.

p-Aminobenzoic acid appears to function as a coenzyme in the conversion of certain precursors to purines.[258] It is also a component of the folic acid molecule (see Folic Acid). It has been used as an antirickettsial agent.

ASCORBIC ACID

Ascorbic Acid U.S.P., Vitamin C, Cevitamic Acid®, Cebione®. The disease scurvy, which now is known as a condition due to a deficiency of ascorbic acid in the diet, has considerable historical significance.[259] For example, in the war between Sweden and Russia (most likely the march of Charles XII into the Ukraine in the winter of 1708–1709) almost all of the soldiers of the Swedish army became incapacitated by scurvy. But further progress of the disease was stopped by a tea prepared from pine needles. The Iroquois Indians cured Jacques Cartier's men in the winter of 1535–1536 in Quebec by giving them a tea brewed from an evergreen tree. Many of Champlain's men died of scurvy when they wintered near the same place in 1608–1609. During the long siege of Leningrad, lack of vitamin C made itself particularly felt, and a decoction made from pine needles played an important role in the prevention of scurvy. It is somewhat common knowledge that sailors on long voyages at sea were subject to the ravages of scurvy. The British used supplies of limes to prevent this, and the sailors often were referred to as limeys.

Holst and Frolich,[260] in 1907, first demonstrated that scurvy could be produced in guinea pigs. A comparable condition cannot be produced in rats.

Although Waugh and King[261] (1932) isolated crystalline vitamin C from lemon juice and showed it to be the antiscorbutic factor of lemon juice, Szent-Gyorgyi[262] had isolated the same substance from peppers, in 1928, in connection with his biologic oxidation-reduction studies. At the time, he failed to recognize its vitamin properties and re-

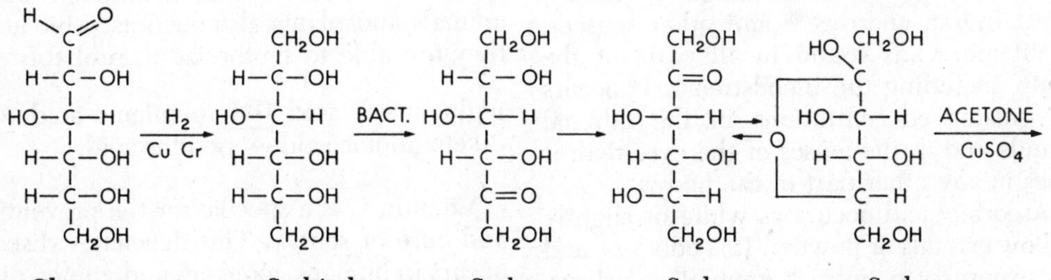

D-Glucose   D-Sorbitol   L-Sorbose (Inverted Formula)   L-Sorbose   L-Sorbose (Furanose Form)

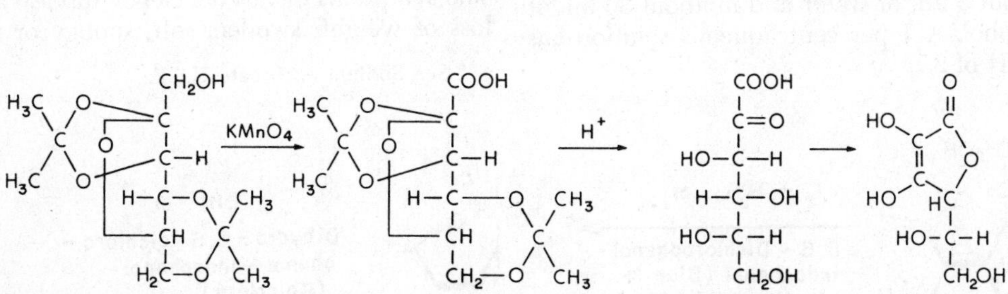

Diacetone Sorbose   Diacetone Sorburonic Acid   2-Keto-L Gulonic Acid   L-Ascorbic Acid

ported it as a hexuronic acid because some of its properties resembled those of sugar acids. Hirst, *et al.*,[263] suggested that the correct formula should be one of a series of possible tautomeric isomers and offered basic proof that the formula now generally accepted is correct. The first synthesis of L-ascorbic acid (vitamin C) was announced almost simultaneously by Haworth and Reichstein,[264] in 1933. Since that time, ascorbic acid has been synthesized in a number of different ways; the one shown on page 987 has proved commercially feasible.[265]

Although ascorbic acid occurs in relatively large quantities in some plants, fruits and other natural sources, its isolation is tedious, difficult and uncertain. It is, therefore, not prepared in a crystalline state from natural sources; however, concentrates have been and are prepared for human use.

Vitamin C is distributed very widely in the active tissues of higher plants. It is formed rapidly in germinating seeds and apparently reaches a high concentration in rapidly growing stem or root tips, green leaves and seeds. Almost all fresh fruits and tubers contain significant amounts of this vitamin. This is especially true of the citrus fruits, peppers, paprikas, tomatoes, rose hips, blackberries, green English walnuts, West Indian cherries[266] and other sources.

Vitamin C is found in all parts of the body, including the bloodstream. It occurs in greater concentrations in the adrenal glands and in the lenses of the eyes than it does in any other part of the body.

Ascorbic acid occurs as white or slightly yellow crystals or powder. It is odorless, and on exposure to light it gradually darkens. One gram of ascorbic acid dissolves in about 3 ml. of water and in about 30 ml. of alcohol. A 1 per cent aqueous solution has a pH of 2.7.

Vitamin C was given the name ascorbic acid[267] because it exhibited acid properties and would cure scurvy. The enolic groups impart acidity[268] to the molecule so that the preparation of salts* is possible; these can be regenerated by treatment with a stronger acid, i.e., hydrochloric acid. Vitamin C shows marked reducing properties but no color with Schiff's reagent. It is quantitatively reversibly oxidized in aqueous solution by iodine or by 2,6-dichlorophenol-indophenol. The iodine reaction is employed in the official assay, whereas the dye is used in the estimation of ascorbic acid in natural extracts and ascorbic acid tablets.

Ascorbic acid is reasonably stable in the dry state. In solution, it is oxidized slowly under acid conditions but is oxidized rapidly in alkaline conditions. The above oxidations are catalyzed by metals, such as iron, copper and manganese.

Ascorbic acid exhibits no mutarotation, it gives a color with ferric chloride and with basic lead acetate yields a precipitate that can be decomposed with hydrogen sulfide. Ascorbic acid can be sterilized in the presence of phenol when heated in an autoclave for short periods of time.

Man, the other primates, the guinea pig and a few microorganisms are not able to synthesize ascorbic acid. Although other animals and plants also need ascorbic acid, they are able to synthesize it, probably via

D-glucuronic acid $\xrightarrow{\text{TPN}}$ L-gulonic acid $\xrightarrow{\text{DPN}}$ 3-ketogulonic acid → ascorbic acid.

Vitamin C is a specific for the prevention and cure of scurvy. This deficiency disease occurs in humans, monkeys and guinea pigs. Rats do not need ascorbic acid in their diets. The symptoms of this deficiency disease are loss of weight; swollen, soft, spongy or ul-

* See Sodium Ascorbate U.S.P.

L-Ascorbic Acid

$I_2 + H_2O$ or
2,6 – Dichlorophenol-indophenol (Blue in alkali, red in acid)

2 HI or
Dihydro – 2,6 – dichloro – phenol – indophenol (Colorless)

Dehydroascorbic Acid

cerated gums; loose carious teeth; hemorrhages; necrosis of the bones; swollen joints; edema; hardening of the skin and often perifollicular or petechial hemorrhages; sometimes bloody conjunctiva and, occasionally, anemia. The tendency to bleed, capillary fragility, accompanied by ready injury to the vascular system is a general one. The joints become painful.

In the absence of vitamin C, there is a loss in the development and maintenance of intercellular substances. This involves the collagen of all fibrous tissues and of all nonepithelial cement substances, such as intracellular material of the capillary wall, cartilage, dentin and bone matrices. Because vitamin C is oxidized and reduced readily, it is possible that its function in the development and maintenance of intercellular substances in the above tissues may be a respiratory one. In glandular tissues, there seems to be some correlation between the high concentration of ascorbic acid and the general rate of metabolism. The fact that the vitamin C content of the adrenal cortex is depleted rapidly when the cortex is stimulated by ACTH indicates that vitamin C plays a role in metabolic processes, possibly including the synthesis of the cortical hormones. The development of arthritic lesions in the scorbutic guinea pig is corrected by cortisone or ascorbic acid but not by ACTH or desoxycorticosterone.[269]

The amino acids phenylalanine and tyrosine are not metabolized completely in vitamin C-deficient individuals. Under these conditions they are metabolized only partly and are excreted in the urine as homogentisic, *p*-hydroxyphenylpyruvic and hydroxyphenyllactic acids. It appears that vitamin C plays the role of a coenzyme in the metabolism of tyrosine through its deaminated product,[270] because scorbutic liver slices cannot metabolize this amino acid in the absence of this vitamin. Vitamin C in adequate amounts delays the oxidation of epinephrine by the body.

Although ascorbic acid is the only compound found in nature that has vitamin C properties, a large number of closely related compounds have been prepared and tested for their activities. From these studies, it can be concluded that in addition to the two enolic groups the D-configuration[271] of the fourth carbon atom (essential configuration for dextrorotatory lactone) and the L-configurations of the fifth carbon atom are necessary for maximum antiscorbutic activity. The terminal primary alcohol group also contributes to the activity, because 6-desoxy-L-ascorbic acid is one third as active as L-ascorbic acid.

D-Ascorbic, D-glucoascorbic and D-galactoascorbic acids are inactive in guinea pigs.[272] 2-Keto-L-gulonic acid when fed to guinea pigs possesses practically no antiscorbutic properties, and, therefore, no regeneration of 2,3-diketo-L-gulonic acid appears likely in the animal body. The simultaneous loss of reducing and antiscorbutic properties when the lactone ring is opened further suggests that the reducing action of ascorbic acid is in some way associated with its biologic function.

Vitamin C is excreted continuously in the urine, and ingestion of amounts above those required to saturate the tissues will also be excreted. This provides a measure of the vitamin C reserve. The titration of the ascorbic acid in the urine with indophenol, after a standard dose of vitamin C has been given, provides a direct, rapid and simple method of evaluating the condition of the body tissues in terms of vitamin C reserve. The vitamin C level also can be estimated by measuring capillary strength or by direct titration of ascorbic acid in the blood or spinal fluid. Others suggest that low plasma ascorbic acid levels indicate only the degree of saturation and are a poor index of deficiency. Patients placed on a diet deficient in vitamin C exhibited the following: (1) 10 days, plasma fell to a low level; (2) 30 days, plasma level was zero; (3) 13 weeks, first clinical evidence of scurvy; (4) 132 days, hyperkeratotic papules developed; (5) 141 days, wounds failed to heal and (6) 162 days, perifollicular hemorrhages of scurvy developed; ascorbic acid value of white cell platelets fell to zero. Loss of weight occurred, accompanied by lowered blood pressure.

Large doses of vitamin C have been of distinct value in the treatment of hay fever, for the relief of heat cramps and heat prostration in workers in an extremely hot environment and as a detoxifying agent for arsenicals. Ascorbic acid is nontoxic even in massive doses.

The daily human requirement is usually the amount of ascorbic acid needed to maintain at least 1 mg. per cent in the plasma.

The U.S.P. unit and the I.U. are equivalent to 50 mcg. of pure ascorbic acid.

Category—antiscorbutic vitamin.

Usual dose—daily, oral or parenteral: requirement, 60 mg. once daily; therapeutic, 250 mg. twice daily.

Usual dose range—40 mg. to 1 g. daily.

OCCURRENCE
Ascorbic Acid Tablets U.S.P.

Sodium Ascorbate U.S.P. is a white crystalline powder that is soluble 1:1.3 in water and insoluble in alcohol.

Ascorbic Acid Injection U.S.P. is a sterile solution of sodium ascorbate that has a pH of 5.5 to 7.0. It is prepared from ascorbic acid with the aid of sodium hydroxide, sodium carbonate, or sodium bicarbonate. It can be used for intravenous injection, whereas ascorbic acid is too acidic for this purpose. It is available as 100, 200, and 500 mg. in 2 ml.; 500 mg. and 1 g. in 5 ml.; and 1 g. in 10 ml.

Ascorbyl Palmitate N.F., Ascorbic Acid Palmitate (ester), is the C_6-palmitic acid ester of ascorbic acid. It occurs as a white to yellowish white powder that is very slightly soluble in water and in vegetable oils. It is freely soluble in alcohol. Ascorbic acid has some antioxidant properties and is a very effective synergist for the phenolic antioxidants such as propylgallate, hydroquinone, catechol, and nordihydroguaiaretic acid (N.D.G.A.) when they are used to inhibit oxidative rancidity in fats, oils and lipids. Long chain fatty acid esters of ascorbic acid are more soluble and suitable for use with lipids than is ascorbic acid.

Category—preservative (antioxidant).

REFERENCES CITED

1. McCollum, E. V., and Davis, M.: J. Biol. Chem. 15:167, 1913.
2. Osborne, T. B., and Mendel, L. B.: J. Biol. Chem. 15:311, 1913; 16:423, 1913.
3. ———: J. Biol. Chem. 17:401, 1914.
4. McCollum, E. V., and Davis, M.: J. Biol. Chem. 19:245, 1914; 21:179, 1915.
5. Steenbock, H.: Science 50:352, 1919.
6. Steenbock, H., and Boutwell, P. W.: J. Biol. Chem. 42:131, 1920; 41:149, 1920; 51:63, 1922; 47:303, 1921.
7. Euler, B., Euler, H., and Helstrom, H.: Biochem. Z. 203:370, 1928.
   Euler, B., *et al.*: Helv. chim. acta 12:278, 1929.
8. Moore, T.: Biochem. J. 24:696, 1930; 25:275, 1931.
9. Karrer, P., *et al.*: Helv. chim. acta 13: 1084, 1930.
10. ———: Helv. chim. acta 16:557, 1933.
11. Robeson, C. D., *et al.*: J. Am. Chem. Soc. 77:4111, 1955.
12. Snell, E. E., *et al.*: J. Am. Chem. Soc. 77:4134, 4136, 1955.
13. Hanze, A. R., *et al.*: J. Am. Chem. Soc. 68:1389, 1946.
    Milas, N. A.: U. S. Patents, 2,369,156; 2,369,168; 2,382,085; 2,382,086.
    Isler, O., *et al.*: Experientia 2:31, 1946.
    Karrer, P., *et al.*: Helv. chim. acta 29: 704, 1946.
    Milas, N. A., and Harrington, T. M.: J. Am. Chem. Soc. 69:2248, 1947.
    Oroshnik, W.: J. Am. Chem. Soc. 67: 1627, 1945.
    Isler, O., *et al.*: Helv. chim. acta 30: 1911, 1947.
14. Milas, N. A.: Science 103:581, 1946.
15. Arens, J. F., and van Dorp, D. A.: Nature, London 157:190, 1946.
    ———: Rec. trav. chim. 65:338,1946.
16. Isler, O., *et al.*: Helv. chim. acta 30: 1911, 1947.
17. Carr, F. H., and Price, E. A.: Biochem. J. 20:497, 1926.
    Benham, G. H.: Canad. J. Res. 22B:21, 1944.
18. Oroshnik, W.: J. Am. Chem. Soc. 78:265, 1956.
19. Pett, L. B.: J. Lab. Clin. Med. 25:149, 1939.
    Hecht, S., and Mandelbaum, J.: J.A.M.A. 112:1910, 1939.
20. McLean, F., and Budy, A.: Vitamins and Hormones 21:51, 1963.
21. Green, J.: Vitamins and Hormones 20: 485, 1962.
22. Hecht, S.: Am. Scientist 32:159, 1944.
23. Palm, T. A.: Practitioner 45:271, 321, 1890.
24. Raczynski, J.: Compt. rend. Congres. assoc. internat. pediatrie, 1st. Paris, p. 389, 1912.

25. Huldschinsky, K.: Deutsche med. Wschr. 45:712, 1919.
—: Z. orthop. Chir. 39:426, 1919-1920.
26. Hess, A. F., and Unger, L. J.: Proc. Exp. Biol. Med. 18:298, 1921.
27. Mellanby, E.: Lancet 1:407, 1919.
28. McCollum, E. V., *et al.*: Proc. Soc. Exp. Biol. Med. 18:275, 1921.
—: J. Biol. Chem. 47:507, 1921.
—: J. Biol. Chem. 50:5, 1922.
29. Hart, E. B., *et al.*: Science 52:318, 1920.
—: J. Biol. Chem. 48:33, 1921.
—: J. Biol. Chem. 53:21, 1922.
30. McCollum, E. V., *et al.*: Bull. Johns Hopkins Hosp. 33:229, 1922.
—: J. Biol. Chem. 53:293, 1922.
31. Hess, A. F., and Weinstock, M.: J.A.M.A. 83:1945, 1846, 1924.
—: J. Biol. Chem. 62:301, 1924.
—: Proc. Soc. Exp. Biol. Med. 22:5, 6, 1924.
32. Steenbock, H., and Black, A.: J. Biol. Chem. 61:405, 1924.
—: Science 60:224, 1924.
33. —: Science 64:263, 1926.
Hess, A. F., and Weinstock, M.: J. Biol. Chem. 64:181, 193, 1925.
—:J. Biol. Chem. 63:305, 1925.
34. Rosenhein, O., and Webster, T. A.: Biochem. J. 21:389, 1927.
35. Hess, A. F.: J.A.M.A. 89:337, 1927.
—: Proc. Soc. Exp. Biol. Med. 24: 461, 462, 1927.
Windaus, A.: Chem. Z. 51:113, 114, 1927.
Windaus, A., and Hess, A.: Nachr. Ges. Wiss. Gottingen, Mathphysik. Klasse 2: 175, 1927.
Rosenhein, O., and Webster, T. A.: Lancet 19:622, 1927.
Kon, S. K., *et al.*: J. Am. Chem. Soc. 50: 2573, 1928.
36. Windaus, A., and Thiele, W.: Ann. Chemie 521:160, 1935.
37. Grundmann, W.: Z. physiol. Chem. 252: 151, 1938.
38. Windaus, A., *et al.*: Ann. Chemie 533: 118, 1937.
39. Bunker, J., *et al.*: J. Am. Chem. Soc. 62: 508, 1940.
40. Brockmann, H.: Z. physiol. Chem. 241: 104, 1936; 245:96, 1937.
41. Windaus, A., *et al.*: Z. physiol. Chem. 241:100, 1936.
42. Schenk, F.: Naturwiss. 25:159, 1937.
Remp, D. G., and Marshall, I. H.: J. Nutrition 15:525, 1938.

43. Milas, N. A., and Priesing, C. P.: J. Am. Chem. Soc. 81:397, 1959.
44. Windaus, A., *et al.*: Ann. Chemie 520:98, 1935.
—: Z. physiol. Chem. 247:185, 1937.
45. Wunderlich, W.: Z. physiol. Chem. 241: 116, 1936.
Windaus, A.: Chem. Zent. 108:2787, 1937.
46. Linsert, O.: Z. physiol. Chem. 241:116, 1936.
Haslewood, G. A. D.: Biochem. J. 33: 454, 1939.
47. Bock, F., and Wetter, F.: Z. physiol. Chem. 256:33, 1938.
48. See Ann. Rev. Biochem. 6:390, 1936.
49. Norman, A. W.: Biol. Rev. 43:97, 1968.
50. Blunt, J. W.: *et al.*: Chem. Comm. 801, 1968; Biochem. 7:3317, 1968.
50a. Myrtle, J. F., and Norman, A. W.: Science 171:79, 1971; Haussler, M. R., *et al.*: Proc. Nat. Acad. Sci. 68:177, 1971.
51. Werder, F.: Angew. Chem. 51:172, 1938. I. G. Farbenindustrie: U. S. Patent 2,070,117.
52. Werder, F.: Z. physiol. Chem. 260:119, 1939.
53. MacBryde, C. M.: Surgery 16:804, 1944.
54. Evans, H. M., and Bishop, K. S.: Science 56:650, 1922.
55. Sure, B.: J. Biol. Chem. 58:693, 1924.
56. Evans, H. M., *et al.*: J. Biol. Chem. 113: 319, 1936.
57. —: J. Biol. Chem. 122:99, 1927.
58. Weisler, L., *et al.*: J. Am. Chem. Soc. 67:1230, 1945.
58a. Suda, T., *et al.*: Biochem. 9:1651, 1970.
59. Stern, M. A., *et al.*: J. Am. Chem. Soc. 69:869, 1947.
60. Mayer, H., *et al.*: Helv. chim. acta 46: 963, 1963.
61. Karrer, P., and Fritzsche, H.: Helv. chim. acta 21:520, 1938.
62. Fernholz, E.: J. Am. Chem. Soc. 60:700, 1938.
63. Smith, L. I.: Chem. Rev. 27:287, 1940.
64. Stern, M. A., *et al.*: J. Am. Chem. Soc. 69:869, 1947.
65. Weber, F., *et al.*: Biochem. Biophys. Res. Commi. 14:186, 1964.
66. Wiss, O., *et al.*: Vitamins and Hormones 20:451, 1962.
67. Green, J.: Vitamins and Hormones 20: 485, 1962.
68. Demole, V., *et al.*: Helv. chim. acta 22: 65, 1939.

69. Karrer, P., and Bussmann, G.: Helv. chim. acta 23:1137, 1940.

70. Evans, H. M., *et al.*: J. Organic Chem. 4:376, 1939.
Werder, F., and Moll, T.: Z. physiol. Chem. 254:39, 1938.
Smith, L. I.: Chem. Rev. 27:287, 1940.

71. Pappenheimer, V.: Physiol. Rev. 23:47, 1943.

72. Horwitt, M.: Vitamins and Hormones 20:556, 1962; Dam, H.: ibid., p. 538, Toppel, A.: ibid., p. 493.

73. Mark, J.: Vitamins and Hormones 20:593, 1962.

74. Moore, T., and Sharman, I. M.: Int. J. Vit. Res. 317:1964.

75. Roche Review, pp. 77-97. Dec. 1944.

76. Dam, H.: Biochem. Z. 215:475, 1929.
McFarlane: Biochem. J. 25:358, 1931.

77. ———: Nature, London 135:652, 1935.
———: Biochem. J. 29:1273, 1935.

78. Almquist, H. J., and Stokstad, E.: J. Biol. Chem. 111:105, 1935.

79. Axlerod, A. E., and Pilgrim, J. J.: Science 102:35, 1945.

80. Almquist, H. J.: J. Biol. Chem. 120:634, 1937; 125:681, 1938.

81. Fieser, L. F.: J. Am. Chem. Soc. 61:3467, 1939.
Binkley, S. B., *et al.*: J. Biol. Chem. 130:219, 433, 1939.
Klose, A., and Almquist, H. J.: J. Am. Chem. Soc. 61:1295, 1939.
———: J. Biol. Chem. 132:469, 1940.

82. Binkley, S. B., *et al.*: J. Biol. Chem. 133:707, 1940.
McKee, R. W., *et al.*: J. Am. Chem. Soc. 61:1295, 1939.

83. See Chapter, "Chemistry and Biochemistry of the K Vitamins" *in* Vitamins and Hormones 17:531, 1959.

83a. Almquist, H. J.: J. Biol. Chem. 120:634, 1937; 125:681, 1938.

83b. Almquist, H. J., and Klose, A. A.: J. Am. Chem. Soc. 61:1610, 1939; Fieser, L. F.: J. Am. Chem. Soc. 61:2213, 1939.

84. Tarbel, D. S., *et al.*: J. Am. Chem. Soc. 67:1643, 1945.

85. Olson, R. E.: Can. J. Biochem. 43:1565, 1965.

86. Babior, B. M.: Biochim. biophys. acta 123:606, 1966.
Suttie, J. W.: Arch. Biochem. biophys. 118:166, 1967.

87. Johnson, B. C., *et al.*: Life Sci. 5:385, 1966.

88. Lowenthal, J. and Simmons, E. L.: Experientia 23:421, 1967.

89. Gyorgy, P.: Vitamins and Hormones 20:600, 1962.

90. Fieser, L. F.: J. Biol. Chem. 133:391; 1940.

90a. Phillips, P. G., *et al.*: Biochem. 8:2856, 1969.

91. Diplock, A. T., and Haslewood, G.: Biochem. J. 104:1004, 1967.

92. Lavate, W. V., and Bently, R.: Arch. Biochem. Biophys. 108:287, 1964.

93. Koniuszy, F. R., *et al.*: Arch. Biochem. Biophys. 87:298, 1960.

94. Erickson, R. E., *et al.*: J. Nutrition 71:101, 1960.
Page, C. A., *et al.*: Arch. Biochem. Biophys. 85:475, 1959; 89:318, 1960; 104:169, 1964.

95. Spiller, G. H., *et al.*: Arch. Biochem. Biophys. 125:786, 1968.

96. Page, A. C., *et al.*: Arch. Biochem. Biophys. 89:318, 1960.

97. Folkers, K., *et al.*: J. Am. Chem. Soc. 81:5000, 1959.

98. Pharo, R., *et al.*: Arch. Biochem. Biophys. 125:416, 1968.
Szarkowska, L.: Arch. Biochem. Biophys. 113:519, 1966.

99. Benziman, M., and Goldhamer, H.: Biochem. J. 108:311, 1968.

100. Barr, R., and Crane, F. L.: Plant Physiol. 42:1255, 1967.

101. Folkers, K., *et al.*: J. Am. Chem. Soc. 81:5000, 1959.

102. Henninger, M. D., and Crane, F. L.: J. Biol. Chem. 241:5190, 1968.

103. Jansen, B. C. P., and Donath, W. F.: Chem. Weekblad. 23:923, 1926.

104. Windaus, A., *et al.*: J. Physiol. Chem. 204:123, 1932.
———: Nachr. Ges. Wiss. Gottingen, Mathphysik. Klasse 207:342, 1932.

105. Williams, R. R., *et al.*: J. Am. Chem. Soc. 57:1052, 1937. See also other references in Rosenberg, H. R.: Chemistry and Physiology of the Vitamins, New York, Interscience, 1945.

106. Cline, J. K., *et al.*: J. Am. Chem. Soc. 59:530, 1050, 1947.

107. Williams, R. R.: J.A.M.A. 110:730, 1938.
Williams, R. R., and Spies, T.: Vitamin B_1, p. 163, New York, Macmillan, 1938.

108. Todd, R. R., *et al.*: J. Chem. Soc., p. 1601, 1936.

109. Kinnersley, H. W., *et al.*: Biochem. J. 29:701, 1935.

110. Berger, G., *et al.*: Nature, London 136:259, 1935.

————: Ber. deutsch. chem. Ges. 68: 2257, 1935.

111. Munsell, H. E.: J.A.M.A. 110:927, 1938.

112. Williams, R. R.: Vitamin B_1 and Its Uses in Medicine, New York, Macmillan, 1938.

113. Bergel, F., and Todd, A. R.: J. Chem. Soc. 140:1504, 1937.
Price, D., and Pickel, F. D.: J. Am. Chem. Soc. 63:1067, 1941.

114. Keresztesy, J. C.: Ann. Rev. Biochem. 13:370, 1944.

115. Rogers, E. F., et al.: J. Am. Chem. Soc. 82:2974, 1960.

116. Lohman, K., and Schuster, P.: Biochem. Z. 294:188, 1937.

117. Tauber, H.: Science 86:180, 1937.
Euler, H. V., and Vestin, R.: Naturwissenschaften 25:216, 1937.

118. Tauber, H.: J. Biol. Chem. 125:191, 1938.

119. Breslow, R.: Ann. N. Y. Acad. Sci. 98: 445, 1962; Carlson, G. L., and Brown, G. M.: J. Biol. Chem. 236:2099, 1961; Miller, S. C., and Sprague, J.: Ann. N. Y. Acad. Sci. 98:401, 1962.

120. Buckley, J., et al.: J. Am. Pharm. A. 48: 404, 1959.

121. Williams, R. R., and Spies, T.: Vitamin B_1, pp. 5, 332, New York, Macmillan, 1938.

122. Goedde, H. W.: Angew. Chem. 4:846, 1965.

123. Ellinger, P., and Koschara, W.: Ber. deutsch. chem. Ges. 66:315, 1933.

124. Kuhn, R., et al.: Ber. deutsch. chem. Ges. 66:317, 1933.

125. Warburg, O., and Christian, W.: Naturwissenschaften 20:688, 980, 1932.
————: Biochem. Z. 254:438, 1932.

126. Szent-Gyorgyi, A., and Banga, I.: Ber. deutsch. chem. Ges. 246:203, 1932.

127. Bleyer, B., and Kahlman, O.: Biochem. Z. 247:492, 1933.

128. Blyth, A. W.: J. Am. Chem. Soc. 35:530, 1879.

129. Osborne, T. B., and Mendel, L. B.: J. Biol. Chem. 15:311, 1913.

130. Kuhn, R., et al.: Ber deutsch. chem. Ges. 66:1034, 1933.

131. ————: Ber. deutsch. chem. Ges. 66: 1905, 1933.

132. See Karrer, P.: The Chemistry of the Flavins, in Ergebnisse der Vitamin und Hormon Forschung, vol. 2, p. 381, 1939. See also Rosenberg, H. R.: Chemistry and Physiology of the Vitamins, p. 153, New York, Interscience, 1945.

133. Nickolas, D.: Nature 179:800, 1957;

134. Emerson, G. A., and Folkers, K.: J. Am. Chem. Soc. 73:2398, 1951; Snell, E., et al.: ibid. 79:2258, 1957.

135. Heyl, D., et al.: J. Am. Chem. Soc. 73: 3826, 1951.

136. Nelson, E. K., and Keenan, G. L.: Science 77:561, 1933.

137. Kogl, F., and Van Hasselt, W.: Z. physiol. Chem. 242:43, 1936.

138. Anderson, R. J., et al.: J. Biol. Chem. 125:299, 1938.

139. Eastcott, E. V.: J. Physiol. Chem. 28: 1180, 1928.

140. Woolley, D. W.: Science 92:384, 1940.
————: J. Biol. Chem. 136:113, 1940.
Martin, G. J., and Ansbacher, S.: Proc. Soc. Exp. Biol. Med. 48:118, 1941.

141. Woolley, D. W.: J. Nutrition 21:Supp. 17, 1941.

142. Allison, F. E., et al.: Science 78:217, 1933.

143. Gyorgy, P.: J. Biol. Chem. 131:733, 1931.

144. Parsons, H. T.: J. Biol. Chem. 90:351, 1931.
Lease, J., and Parsons, H. T.: J. Biol. Chem. 105:1, 1934.

145. Wilders, E.: Cellule 18:313, 1901.

146. Kogl, F., and Tonnis, B.: Z. physiol. Chem. 242:43, 1936.

147. Harris, S. A., et al.: Science 97:447, 1943.
————: J. Am. Chem. Soc. 66:1756, 1800, 1944.

148. Gunness, M.: J. Biol. Chem. 157:121, 1945; Ott, W., et al.: J. Biol. Chem. 157: 131, 1945.

149. Hofman, K.: J. Am. Chem. 67:1459, 1945.

150. Axelrod, A. E., and Pilgrim, J. J.: Science 102:35, 1945.

151. Mistry, S. P., and Dakshinamurti, K.: Vitamins and Hormones 22:1, 13, 1964.

152. Wright, L. D., et al.: Science 114:635, 1951.
Peck, R. L., et al.:J. Am. Chem. Soc. 74:1999, 2002, 1952.

153. Williams, R. J., et al.: J. Am. Chem. Soc. 55:2912, 1933.

154. ————: J. Am. Chem. Soc. 60:2719, 1938.

155. ————, and Major, R. T.: Science 91: 246, 1940.

156. Elvehjem, C. A., et al.: J. Biol. Chem. 124:313, 1938; 125:715, 1938.

157. Jukes, T. H.: J. Am. Chem. Soc. 61:975, 1939.
 Woolley, D. W., *et al.*: J. Am. Chem. Soc. 61:977, 1939.

158. Williams, R. J., and Major, R. T.: Science 91:246, 1940.
 ———: J. Am. Chem. Soc. 62:1784, 1940.
 Stiller, E. T., *et al.*: J. Am. Chem. Soc. 62:1785, 1940.
 Carter, H. E., and Ney, L. F.: J. Am. Chem. Soc. 63:312, 1941.

158a. Hill, R. K., and Chan, T. H.: Biochem. Biophys. Res. Comm. 38:181, 1970.

159. King, T. E., and Strong, F. M.: Science 112:562, 1950.

160. Jukes, T. H.: J. Nutrition 21:193, 1941.

161. Barnett, J. W., and Robinson, F. A.: Biochem. J. 36:357, 364, 1942.
 Shive, W., and Snell, E.: J. Biol. Chem. 158:551, 1945: 160:287, 1945.
 Woolley, D. W., and Collyer, M. L.: J. Biol. Chem. 159:271, 1945.
 Mead, J. F., *et al.*: J. Biol Chem. 163:465, 1946.

162. Kline, O., *et al.*: J. Biol. Chem. 99:295, 1932-33.

163. Shaefer, A. E., *et al.*: J. Biol. Chem. 143:321, 1942.

164. Gyorgy, P.: Nature, London 133:498, 1934.
 ———: Biochem. J. 29:741, 760, 767, 1935.

165. Lepkovsky, C.: Science 87:169, 1938.
 Gyorgy, P.: J. Am. Chem. Soc. 60:983, 1938.
 Kuhn, R., and Wendt, G.: Ber deutsch. chem. Ges. 71B:118, 1938,

166. Harris, S. A., and Folkers, K.: J. Am. Chem. Soc. 61:1245, 3307, 1939.
 Harris, S. A., and Folkers, K.: J. Am. Chem. Soc. 61:1242, 1939.
 Kuhn, R., *et al.*: Naturwissenschaften 27:469, 1939.

167. See Vitamins and Hormones 16:84, 1958.

168. Birch, T. W., *et al.*: Biochem. J. 29:2830, 1935.

169. Fouts, P. J., *et al.*: J. Nutrition 16:197, 1938.

170. Fouts, P. J., *et al.*: Am. J. M. Sci. 199:163, 1940.

171. Jukes, T. H.: Proc. Soc. Exp. Biol. Med. 42:180, 1939.

172. Spies, T. D.: J.A.M.A. 112:2414, 1939.

173. ———: Ohio M. J. 36:148, 1940.

174. Kuhn, T., and Wendt, G.: Ber. deutsch. chem. Ges. 71:118, 1938.

 Una, K.: Proc. Soc. Exp. Biol. Med. 43:122, 1940.

175. Melnick, D., *et al.*: J. Biol. Cem. 160:1, 1945.
 Snell, E. E.: J. Biol. Chem. 154:313, 1944.

176. Harris, S. A., *et al.*: J. Biol. Chem. 154:315, 1944.
 ———: J. Am. Chem. Soc. 66:2088, 1944.

177. Snell, E. E., and Cunningham, E.: J. Biol. Chem. 158:495, 1945.
 Hochberg, M., *et al.*: J. Biol. Chem. 155:129, 1944.

178. Harris, S. A., *et al.*: J. Am. Chem. Soc. 73:3436, 4693, 1951.

179. Snell, E. E., and Schlenk, F.: J. Biol. Chem. 157:425, 1945.
 ———: J. Am. Chem. Soc. 67:194, 1945.
 Wood, W. W., *et al.*: J. Biol. Chem. 170:313, 1947.

180. Gunsalus, I. C., *et al.*: J. Biol. Chem. 161:743, 1945; 170:415, 1947.

181. ———: J. Biol. Chem. 160:461, 1945.

182. Braunstein, A. E.: Enzymes 2:115, 1960.

183. Snell, E. E., *et al.*: J. Am. Chem. Soc. 76:648, 1954.

184. See Chapter, "Pyridoxal Phosphate" *in* The Enzymes 2:130, 1960.

185. Bruice, T.: Med. Chem. Symposium, June, 1964.

186. Bruice, T., *et al.*: J. Am. Chem. Soc. 85:1480, 1488, 1493, 1963.

187. Ikawa, K., and Snell, E.: J. Am. Chem. Soc. 76:637, 1954.

188. Heyl, D., *et al.*: J. Am. Chem. Soc. 75:653, 1953.

189. Morino, Y., and Snell, E. E.: Proc. Nat. Acad. Sci. 57:1692, 1967.
 Bocharov, A. L., *et al.*: Biochem. Biophys. Comm. 30:459, 1968.

190. Huber, C.: Leibig's Ann. Chem. Pharm. 141:271, 1867.

191. Funk, C.: J. Physiol. 46:173, 1913.

192. Suzuki, U., *et al.*: Biochem. Z. 43:89, 1912.

193. Warburg, O., and Christian, W.: Biochem. Z. 275:464, 1934-35.

194. Kuhn, R., and Vetter, H.: Ber. deutsch. chem. Ges. 68:2374, 1935.

195. von Euler, H., *et al.*: Z. physiol. Chem. 237:1, 1935.

196. Elvehjem, C. A., *et al.*: J. Biol. Chem. 123:137, 1938.

197. Spies, T. D., *et al.*: Ann. Int. Med. 12:1830, 1939.

198. Organic Synthesis: Coll. Vol. 1, 378, 1932.

199. Shaw, G. E., and MacDonald, C. A.: Quart. J. Pharm. Pharmacol. 9:380, 1938.

200. Elvehjem, C. A., *et al.*: J. Biol. Chem. 124:715, 1938.

201. Anderson, B. M., and Kaplan, N. O.: J. Biol. Chem. 234:1219, 1226, 1959.

202. Woolley, D. W.: J. Biol. Chem. 157:455, 1945; 162:179, 1946; 163:773, 1946.

203. Williams, R. J., *et al.*: J. Am. Chem. Soc. 63:2284, 1941; 66:267, 1944.

204. Waller, C. W., *et al.*: J. Am. Chem. Soc. 70:19, 1948.
See also Angier, R. B., *et al.*: Science 103:667, 1946.

205. Hopkins: Nature, London 40:335, 1889.
————: Chem. News 60:57, 1889.

206. Wieland, H., and Shopf: Ber. deutsch. chem. Ges. 58:2178, 1925.

207. Shieve, W., *et al.*: J. Am. Chem. Soc. 73:3067, 3247, 1951.

208. Jukes, T. H., *et al.*: J. Am. Chem. Soc. 73:3535, 1951.

209. Roth, B., *et al.*: J. Am. Chem. Soc. 74: 3247, 3252, 1952.

210. Donaldson, J. O., and Keresztesy, J. C.: J. Biol. Chem. 237:3185, 1962.

211. See Ann. Rev. Biochem. 27:287, 1958.

212. Lampsen, J., and Jones, M. J.: J. Biol. Chem. 170:133, 1947.

213. Welch, A. D., *et al.*: J. Biol. Chem. 164: 786, 1946.

214. Spies, T. D.: Scope, July, 1947, p. 15.
See also Heinle, R. W., and Welch, A. D.: Ann. New York Acad. Sci. 48: 345, 1946.

215. Kaufman, S.: Proc. Nat. Acad. Sci. 50: 1085, 1963.

216. Donaldson, K., and Keresztesy, J.: J. Biol. Chem. 237:3185, 1962.

217. Rohrbough, P.: Fed. Proc. 21:4, 1962.

218. Spies, T. D.: Ann. New York Acad. Sci. 48:313, 1946.

219. Darby, W. J., *et al.*: Science 103:108, 1946.

220. Cosulich, D. B., *et al.*: J. Am. Chem. Soc. 73:2554, 1951.

221. Thiersch, J. B., and Philips, F. S.: Am. J. M. Sci. 217:575, 1949.

222. Minot, G. R., and Murphy, W. P.: J.A.M.A., 87:470, 1926.

223. Shorb, M.: J. Biol. Chem. 169:455, 1947.

224. Rickes, E. L., *et al.*: Science 107:397, 1948.
Smith: Nature, London 162:144, 1948.
Ellis, *et al.*: J. Pharm. Pharmacol. 1:60, 1949.

225. Folkers, K., *et al.*: J. Am. Chem. Soc. 71:1854, 1949.

226. Donaldson, K., and Keresztesy, J.: J. Biol. Chem. 237:3185, 1962.

227. Kaczka, E. A., *et al.*: Science 112:354, 1950.

228. Folkers, K., *et al.*: J. Am. Chem. Soc. 73:3569, 1951.
Brink, N. G., *et al.*: Science 112:354, 1950.

229. Brot, N., *et al.*: Biochem. Biophys. Res. Comm. 18:18, 1965.

229a. Toohey, J., and Barker, H.: J. Biol. Chem. 236:560, 1961.

230. Beaven, G. H., *et al.*: J. Pharm. & Pharmacol. 1:957, 1949.
See also Buchanan, *et al.*: Chem. Ind. 22:426, 1950.
Brink, N. G., and Folkers, K.: J. Am. Chem. Soc. 72:4442, 1950.
————, *et al.*: J. Am. Chem. Soc. 72: 1886, 1950.

231. Dion, H. W., *et al.*: J. Am. Chem. Soc. 74:1108, 1952; 76:948, 1954.

232. Glass, G., *et al.*: Fed. Proc. 21:471, 1962.

233. Barker, H. A., *et al.*: Proc. Nat. Acad. Sci. 44:1093, 1958.

234. Bernhauer, K., *et al.*: Biochem. Z. 333: 106, 1960.

235. Brot, N., and Weissbach, H.: J. Biol. Chem. 241:2024, 1966; Kerwar, S. S., *et al.*: Arch. Biochem. Biophys. 116:305, 1966.

236. Walerych, W. S., *et al.*: Biochem. Biophys. Res. Comm. 23:368, 1966.

237. Hogenkamp, H.: Fed. Proc. 25:1623, 1966.

238. Retey, J., *et al.*: Biochem. Biophys. Res. Comm. 22:274, 1966.

239. Huennekens, F.: Prog. Hematology 5: 83, 1966.

240. Chem. Engr. News, Oct. 21, p. 42, 1968.

241. Trenner, N. R., *et al.*: J. Am. Pharm. A. (Sci. Ed.) 39:361, 1950; Campbell, J. A., *et al.*: J. Am. Pharm. A. 41:479, 1952.

242. Frost, D. V., *et al.*: Science 116:119, 1952.

243. Bartilucci, A., and Foss, N. E.: J. Am. Pharm. A. 43:159, 1953.

244. Buhs, R. P., *et al.*: Science 113:625, 1951.

245. Heathcote, J. G., and Mooney, F. S.: J. Pharm. Pharmacol. 10:593, 1958.

246. See Ann. Rev. Biochem. 27:303, 1958.

247. Glass, G. B., *et al.*: Science 120:74, 1954.

248. Herbert, V.: Gastroenterology, 54:110, 1968.

249. Nielsen, E., *et al.*: J. Biol. Chem. 133: 637, 1940.
Fildes, P.: Lancet 238:955, 1940.
250. Woods, D. D., and Fildes, P.: J. Soc. Chem. Ind. 59:133, 1940.
251. Ansbacher, S.: Science 93:164, 1941.
252. Emerson, G. A.: Proc. Soc. Exp. Biol. Med. 47:448, 1941.
253. Diamond, N. S.: Science 94:420, 1941.
254. Rubbo, S. D., and Gillespie, J. M.: Nature, London 146:838, 1940.
255. Wisansky, W. A., *et al.*: J. Am. Chem. Soc. 63:1771, 1941.
256. Sandground, J. H., and Hamilton, C. R.: J. Pharmacol. Exp. Therap. 78:109, 1943.
257. Am. Prof. Pharmacist 13:451, 1947.
258. Shive, W. *et al.*: J. Am. Chem. Soc. 69: 725, 1947.
259. Schick, B.: Science 98:325, 1943.
260. Holst, A., and Frolich, T.: J. Hyg. 7: 634, 1907.
261. Waugh, W. A., and King, C. C.: Science 75:357, 630, 1932.
———: J. Biol. Chem. 97:325, 1932.
Svirbely, J. L., and Szent-Gyorgyi, A.: Nature, London 129:576, 609, 1932.
———: Biochem. J. 26:865, 1932; 27: 279, 1933.
Tillmans, J., *et al.*: Biochem. Z. 250:312, 1932.
262. Szent-Gyorgyi, A.: Biochem. J. 22:1387, 1928.
263. Hirst, E. L., *et al.*: J. Soc. Chem. Ind. 2:221, 482, 1933.
Cox, E. G., and Goodwin, T. H.: J. Chem. Soc., 769, 1936.
———: Nature 130:88, 1932.
264. Haworth, W. N., *et al.*: J. Chem. Soc. p. 1419, 1933.
Reichstein, T.: Helvet. chim. acta 16: 1019, 1933.
265. Micheel, F., and Kraft, K.: Naturwissenschaften 22:205, 1934.
266. Szent-Gyorgyi, A.: Biochem. J. 28:1625, 1934.
Tuba, J., *et al.*: Science 105:70, 1947.
267. Haworth, W. N., and Szent-Gyorgyi, A.: Nature 131:23, 1933.
268. Haworth, W. N., *et al.*: J. Chem. Soc. p. 1556, 1934.
269. Hughes, C. D., *et al.*: Science 116:252, 1951.
270. Sealock, R. R., and Goodland, R. L.: Science 114:645, 1951.
271. Reichstein, T., *et al.*: Helvet. chim. acta 16:1019, 1933; 18:353, 1935.
272. Zilva, S. S.: Biochem. J. 29:1612, 2366, 1935.

SELECTED READING

Barker, H. A.: Biochemical functions of corrinoid compounds, Biochem. J. 105:1, 1967.
de Reuck, A., and O'Connor, M.: The Mechanism of Action of Water Soluble Enzymes, Boston, Little, Brown, 1961.
Hawthorne, J. N.: The Biochemistry of the Inositol Lipids, *in* Vitamins and Hormones, vol. 22, N. Y., Acad. Press, 1964.
Hogenkamp, H. P. C.: Enzymatic reactions involving corrinoids, Ann. Rev. Biochem. 37: 668, 1968.
Inhoffen, H. H., and Irmscher, K.: Progress in the Chemistry of Vitamin D *in* Progress in the Chemistry of Natural Products 17:71, 1959, Springer Verlag.
Jolly, M.: Vitamin A deficiency, a review, J. Oral. Ther. Pharmacol. 3:364, 439, 1967.
Knapp, J.: Mechanism of biotin action, Ann. Rev. Biochem. 39:757, 1970.
Morton, R., and Pitt, G.: Visual Pigments *in* Progress in the Chemistry of Natural Products, vol. 14, p. 244, 1957.
Olson, J. A.: Metabolism and function of vitamin A, Fed. Proc. 28:1670, 1969.
Reed, J. J.: Biochemistry of Lipoic Acid *in* Vitamins and Hormones, vol. 20, N. Y., Acad. Press, 1962.
Rosenberg, H. R.: Chemistry and Physiology of the Vitamins, New York, Interscience, 1945.
Sebrell, W. H.: The Vitamins, New York, Acad. Press, 1954.
Sherman, H. C., and Smith, S. L.: The Vitamins, New York, Chemical Catalog Co.,
Shils, M. E.: The Flavonoids *in* Biology and Medicine, New York, National Vitamin Foundation, 1956.
Stokstad, E. L. R., and Koch, J.: Folic acid metabolism, Physiol. Rev. 47:83, 1967.
Sure, B.: The Vitamins in Health and Disease, New York, Waverly Press, 1933.
Symposium on Vitamin B$_6$ (very comprehensive) *in* Vitamins and Hormones, vol. 22, N. Y., Acad. Press, 1964.
Symposium on Vitamin E and Metabolism *in* Vitamins and Hormones, vol. 20, N. Y., Acad. Press, 1962.
Vitamins and Hormones, New York, Acad. Press, 1960.
Wagner, A., and Falkers, K.: Vitamins and Coenzymes, N. Y., Interscience, 1964.

Weissbach, H., and Dickerman, H.: Biochemical Role of Vitamin B_{12}, Phys. Rev. 45:80, 1965.

Williams, R. R., and Spies, T. D.: Vitamin B_1 and Its Use in Medicine, New York, Macmillan, 1939.

29

Diagnostic Agents and Miscellaneous Organic Pharmaceuticals

Robert F. Doerge, Ph.D.,

*Professor of Pharmaceutical Chemistry, Chairman of the Department of
Pharmaceutical Chemistry, School of Pharmacy,
Oregon State University*

DIAGNOSTIC AGENTS

Diagnostic agents are used to detect impaired function of the body organs or to recognize abnormalities in tissue structure. Usually, these agents find no other use in medicine; however, a few are also valuable therapeutic agents. Factors that often determine the usefulness of a diagnostic agent are its solubility, mode and rate of excretion, metabolism or chemical configuration (e.g., color) and chemical composition (e.g., iodine).

Compounds used in diagnosis generally are divided into two classes. First, there are the many clinical diagnostic chemicals used to determine normal and pathologic products in urine, blood, feces and other body fluids or excrement. Also in the first group are the serologic solutions and tissue-staining dyes necessary in microscopic examination. Second, there is the group that is being discussed here, which finds application directly to or in the body and is most often intended by the use of the term "diagnostic agent." These agents are conveniently arranged into three groups: (1) radiopaque substances; (2) compounds for testing functional capacity; (3) compounds modifying a physiologic action.

Radiopaque diagnostic agents include both inorganic and organic compounds. These compounds have the property of casting a shadow on x-ray film and are also useful in fluoroscopic examination. Inorganic compounds include Barium Sulfate U.S.P., thorium oxide and bismuth oxides. Usually in suspensions, these are used in x-ray examination of the gastrointestinal tract (orally or enema) and the lungs. Organic iodinated compounds are usually considered more useful; they are more opaque and are used most in x-ray studies.

Iodine was observed to contribute opacity to x-ray in 1924 and was studied more fully by Binz, in 1935.[1] Useful iodinated compounds contain iodine in a strong covalent linkage and do not release iodide ions readily. However, their use in conditions of thyroid disease or tuberculosis should be with caution. The iodinated compounds are used primarily by two technics: systemic and retrograde.

In the systemic procedure, the agent is administered orally or intravenously and is used to examine the kidney (urography) or liver (cholecystography). The contrast medium is used in the roentgenographic visualization of accessible parts of the body, such as renal cavities, ureters, biliary tract, blood vessels, the heart and the large vessels. The patient is given a preliminary test to determine individual sensitivity by instillation of a small amount into the conjunctival sac, then a cathartic is given the night before the injection and food and liquid are withheld for at least 18 hours previous to prevent blurring of the pictures. The solution is warmed to 98°F. and injected slowly into the vein; the patient is kept under careful observation. When renal functioning is normal, good exposures are obtained in from 5 to 15 minutes. Some iodinated compounds will concentrate in the kidney or the bladder and others in the liver or the gallbladder.

The retrograde method is the introduction of the diagnostic agent by mechanical means. An iodinated compound may be introduced into the urethra, the bladder, the vagina, the lower bowel, the ulcer area or varicose veins, for example. For retrograde pyelography, the solution is diluted with normal saline to about 15 per cent and allowed to flow by gravity through a catheter that has been inserted into the ureteral orifice by means of a cystoscope, about 20 ml. being required. For the visualization of the blood vessels or the heart, a solution of special concentration (70%) is used, and the technic is more complicated. In all of these methods, mildly toxic reactions are quite frequent, but serious ones are encountered rarely, provided that the patient has been tested for susceptibility, that the injections are not repeated too often and that contraindicating diseases are not present. The most serious reactions are cyanosis and a fall in blood pressure, which lasts for less than 1 hour and can be overcome by epinephrine.

Requirements of a satisfactory radiopaque are as follows:

1. Adequate radiopacity. This usually requires an iodine content of 50 per cent or more.

2. The solution should be capable of selective concentration in certain structures, such as galibladder and kidney.

3. The solution should be retained in the area long enough for x-ray visualization; then, it should be excreted rapidly with no toxic effects.

4. High solubility is desirable, often in the range of 40 per cent.

5. It should be stable under the conditions of use (resist oxidation in vivo) as well as during storage prior to use.

6. The compound should have a low toxicity, with a minimum of pharmacodynamic activity.

Contrast media may be divided arbitrarily into those which are water-soluble and those which are not. This serves to divide them also according to their general use. The water-soluble group is used mainly for urography and also for angiography. Angiography is the term generally used to define visualization of the blood vessels using a contrast medium. It is also used to designate visualization of the heart, lymph and bile ducts. The water-insoluble group is used mainly for cholecystography, with some use in bronchography and myelography.

WATER-SOLUBLE CONTRAST MEDIA

Sodium Iodohippurate, Hippuran®, is sodium *o*-iodohippurate dihydrate containing 35 to 39 per cent of iodine based upon the anhydrous salt. It is prepared from *o*-iodobenzoic acid and glycine followed by conversion to the sodium salt. It is a white, crystalline powder having an objectionable alkaline taste and a slight odor. The crystals are soluble in water, in alcohol and in dilute alkali. Aqueous solutions are neutral or slightly alkaline to litmus.

Sodium Iodohippurate

One preparation of sodium iodohippurate available for hysterosalpingography is Medopaque H®, which contains 45 per cent of sodium *o*-iodohippurate and 1.83 per cent of carboxymethylcellulose in water.

Sodium iodohippurate has been largely replaced by newer contrast media and is mainly of historical interest. The other iodinated acetamidobenzoates owe their origin to structural modifications of this compound.

Sodium Iodohippurate I 131 Injection U.S.P., Sodium *o*-Iodohippurate. This is a sterile solution containing sodium *o*-iodohippurate in which a portion of the molecules contain radioactive iodine ^{131}I in the structure.

Category—diagnostic aid (renal function determination).

Usual dose range—intravenous, the equivalent of 2 to 35 microcuries.

Sodium Acetrizoate, Urokon® Sodium, Cystokon®, is sodium 3-acetylamino-2,4,6-triiodobenzoate. It is prepared from acetrizoic acid as follows. The salt is not isolated but is formed in solution by dissolving the acid in an equivalent amount of dilute sodium hydroxide. It contains one of the highest percentages of iodine (65.8%) of any

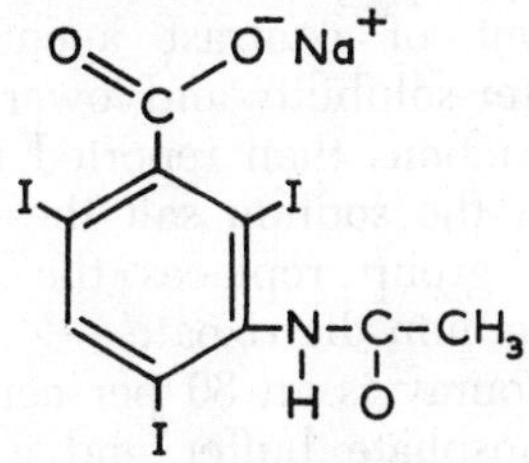

Sodium Acetrizoate

compound used in urography. Aqueous solutions are sensitive to light and, therefore, must be protected properly. Artificial light appears to have no effect.

A 30 per cent solution currently is used intravenously or in retrograde urography. This solution contains calcium ethylenediaminetetracetate as a stabilizer to maintain solution (see Chapter 12). Excretion is very rapid in the kidney, and good pictures usually are obtained. It is less toxic than many other similarly used compounds. When locally applied, there is no irritation, and the delicate mucosa tolerate it well.

The usual dosage of the 30 per cent solution in intravenous urography is 25 ml. intravenously for adults and proportionately less for children. In retrograde pyelography, 25 ml. or 15 ml. is introduced for bilateral or unilateral ureteral examination, respectively.

Sodium acetrizoate is also used in urethrography, nephrography, angiocardiography, cholangiography and cerebral angiography.

Sodium Diatrizoate Injection U.S.P., Hypaque® Sodium Injection, is sodium 3,5-diacetamido-2,4,6-triiodobenzoate in sterile solution. A 50 per cent aqueous solution is essentially neutral in reaction. The solutions

may be sterilized by autoclaving but, in common with most iodinated compounds, should be protected from light.

The 50 per cent solution which is commonly used in urographic studies may be-

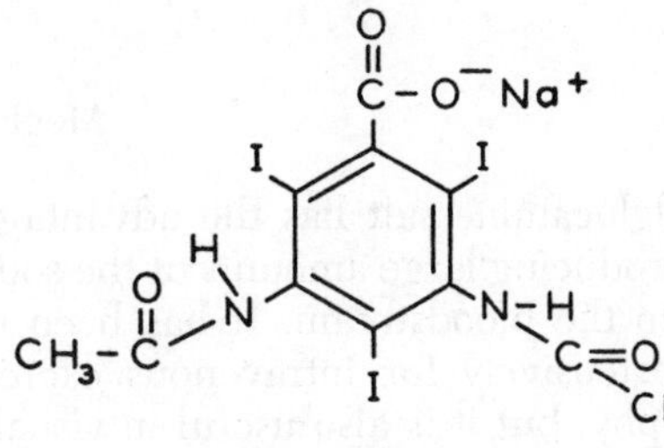

Sodium Diatrizoate

come cloudy or form a precipitate when stored at low temperatures. When warmed to 25° C., the solution should be free of haze or crystals.

This diagnostic agent is also available with a coloring agent and a surfactant for making solutions for oral administration or to be given as an enema. This powder is not intended for use in preparing solutions for parenteral administration.

A mixture of sodium and methylglucamine diatrizoate salts is used in 90 and 75 per cent aqueous solutions for use in angiocardiography only. At body temperature both concentrations are clear solutions, but, at room temperature or when chilled, crystals may form. These solutions are for use in cases which present difficult diagnostic problems, and by persons specially trained in their use.

Category—radiopaque medium.

Usual dose—excretory urography: intravenous, 30 ml. of a 50 per cent solution.

Meglumine Diatrizoate Injection U.S.P., Cardiografin®, Gastrografin®, Renografin®, is the N-methylglucamine salt of 3,5-diacetamido-2,4,6-triiodobenzoic acid in sterile solution. It is a clear, colorless to pale yellow, slightly viscous solution.

The commercially available solutions usually contain a citrate buffer, a chelating agent and bacteriostatic agents. All solutions should be protected from light. At body temperature the solutions should be clear and free of any crystals.

The action and uses of this salt are similar to those of sodium diatrizoate; however, the

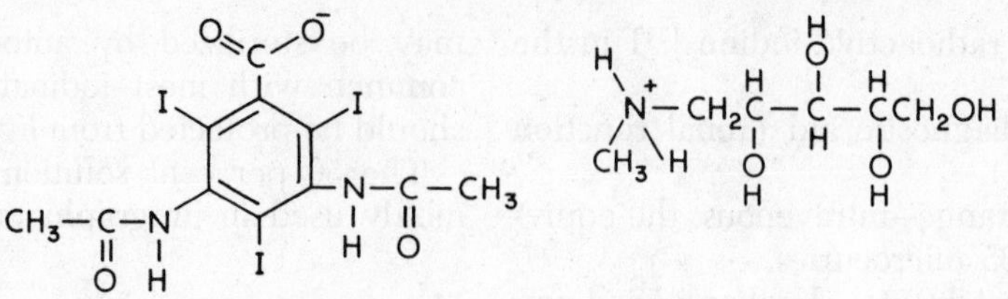

Meglumine Diatrizoate

methylglucamine salt has the advantage of not introducing large amounts of the sodium ion into the bloodstream. It has been used most extensively for intravenous excretory urography, but it is also useful in visualization of the cardiovascular system. When given orally it is only slightly absorbed and may be used instead of suspensions of barium sulfate for visualization of the gastrointestinal tract.

Category—radiopaque medium.

Usual dose—angiocardiography: intravenous or intra-arterial, 25 to 50 ml. of a 76 to 85 per cent solution; excretory urography: intravenous, 20 to 25 ml. of a 60 to 76 per cent solution.

Sodium Metrizoate, Isopaque®, is sodium 3-acetamido-2,4,6-triiodo-5-(N-methylacetamido)benzoate. This compound, closely related to sodium diatrizoate, is being investigated for use as a contrast medium.

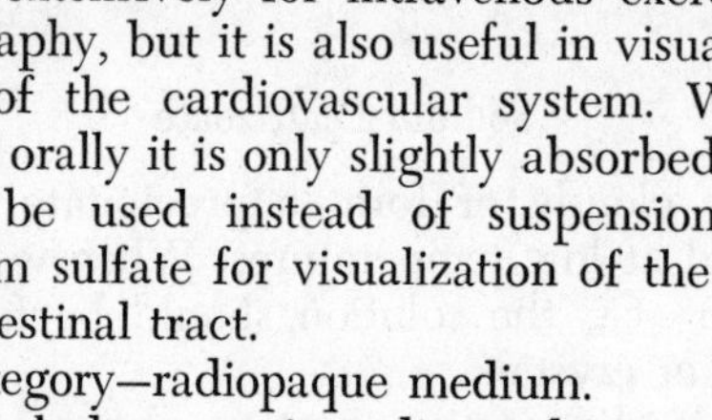

Sodium Metrizoate

Sodium Iothalamate Injection U.S.P., Angio-Conray®, Conray® 400, is sodium 5-acetamido-2,4,6-triiodo-N-methylisophthalamate in sterile aqueous solution.

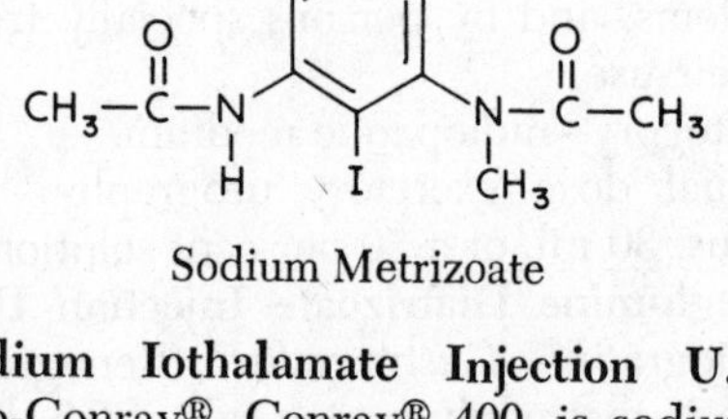

Sodium Iothalamate

Sodium iothalamate injection is a clear, pale yellow, slightly viscous liquid, pH 6.8 to 7.5.

Isothalamic acid was synthesized as part of a research project directed toward the development of contrast agents with a higher water solubility and lower incidence of toxic reactions than reported for known agents.[2] In the sodium salt the N-methylcarbamoyl group replaces the acetamido group of sodium diatrizoate.

Angio-Conray is an 80 per cent solution with a phosphate buffer and a chelating agent. It is used for intravascular angiocardiography and aortography. Conray-400 is a similar 66.8 per cent solution, and, in addition to the above indications, is also used for intravenous urography.

Category—radiopaque medium.

Usual dose—angiocardiography: intravenous or intra-arterial, 40 to 50 ml. of a 66.8 per cent or an 80 per cent solution; excretory urography: intravenous, 25 ml. of a 66.8 per cent solution.

Meglumine Iothalamate Injection U.S.P., Conray®, is the N-methylglucamine salt of 5-acetamido-2,4,6-triiodo-N-methylisophthalamic acid in sterile aqueous solution. The commercially available solution contains a phosphate buffer and a chelating agent. The solutions are sensitive to light and must be protected.

Category—radiopaque medium.

Usual dose—cerebral angiography: intra-arterial, 6 to 10 ml. of a 60 per cent solution; excretory urography: intravenous, 30 ml. of a 60 per cent solution; peripheral arteriography: intra-arterial, 20 to 40 ml. of a 60 per cent solution.

Sodium Iodipamide, Injection U.S.P., Cholografin® Sodium, sodium N,N'-adipoylbis(3-amino-2,4,6-triiodobenzoate), sodium 3,3'-(adipoyldiimino)bis[2,4,6-triiodobenzoate], is prepared by dissolving the free acid in dilute sodium hydroxide solution and adjusting to pH 6.5 to 7.7. Solutions should be protected from light and should be injected only if any cloudiness disappears when warmed and shaken.

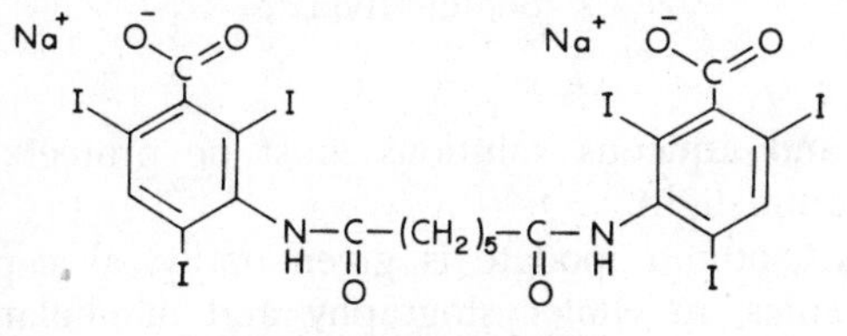

Sodium Iodipamide

Category—radiopaque medium.

Usual dose—cholecystography: intravenous, 40 ml. of 20 per cent solution administered over a period of 10 minutes.

Usual dose range—20 to 40 ml.

Meglumine Iodipamide Injection U.S.P., Cholografin® Meglumine, is the N-methylglucamine salt of 3,3'-(adipoyldiimino)bis-2,4,6-[triiodobenzoic acid], (N,N'adipoyl-bis-[3-amino-2,4,6-triiodobenzoic acid]) in sterile aqueous solution. The injection may contain a chelating agent and a phosphate buffer. The uses and precautions in handling are the same as those for the sodium salt. Meglumine iodipamide has twice the radiopaque content of sodium iodipamide, so that it is equivalent to twice the volume of the latter.

Category—radiopaque medium.

Usual dose—cholecystography: intravenous, 20 ml. of a 52 per cent solution over a period of 10 minutes.

Sodium Methiodal N.F., Skiodan® Sodium, is sodium monoiodomethanesulfonate, ICH_2SO_3Na, having an iodine content of about 52 per cent. It is a white, crystalline, odorless, powder which has a mild saline taste and a sweetish aftertaste. It is soluble in water (7:10), forming a solution neutral to litmus (pH 6 to 8), and is only slightly soluble in alcohol. Solubility in organic solvents is negligible. The salt is prone to decompose in light, turning to a yellow color (iodine). Both the solid compound and its water solutions should be kept protected from light.

Sodium methiodal is useful both by intravenous injection and by retrography. After injection the urine concentration is 4 to 6 per cent, and 75 per cent is excreted in 3 hours.

Category—diagnostic aid (radiopaque urographic).

Usual dose—intravenous, 20 g. in 50 ml.

Usual dose range—10 to 30 g.

OCCURRENCE

Sodium Methiodal Injection N.F.

Iodopyracet Injection N.F., Diodrast®, is a solution of the salt of 3,5-diiodo-4-oxo-1 (4H)-pyridineacetic acid and diethanolamine. The free acid contains not less than 61.5 per cent and not more than 63.5 per cent of iodine (I). Iodopyracet can be made in a number of ways, and the method given below is an example.

The use of ethanolamine (mono, di and tri) salts has increased greatly the water solubility over that of the sodium salts. In this case, diethanolamine is employed. This

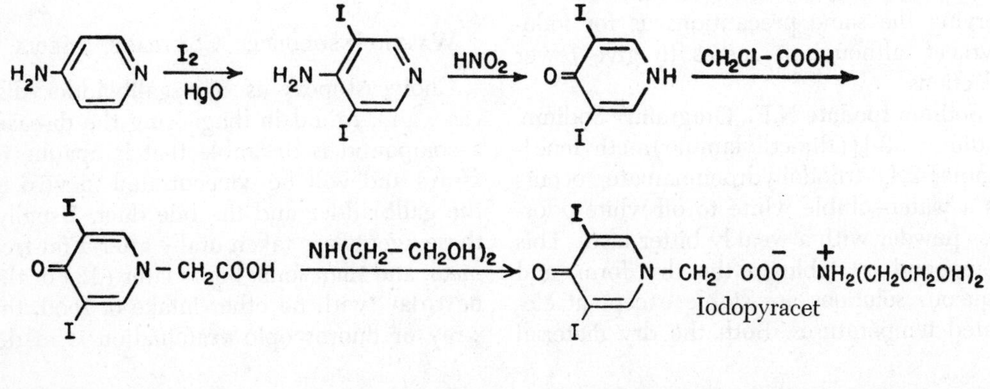

Iodopyracet

salt is very soluble in water where it forms a nearly neutral, clear and colorless solution. It is stable and may be sterilized by heat but, like most organic iodine-containing compounds, will decompose slowly on exposure to sunlight. Iodopyracet solutions have been used for many years, and their mild side reactions are well-known. It is considered one of the safest contrast mediums for intravenous use in urography. It is used also in arteriography and cholangiography and, to some extent, in retrograde pyelography.

Iodopyracet Injection N.F. (35%) is used mostly as a contrast agent for intravenous urography. The dose is warmed to body temperature and injected slowly intravenously. It sometimes is used intramuscularly or subcutaneously.

Category—diagnostic aid (radiopaque urographic).

Usual dose—intramuscular and intravenous, 20 ml.

Sodium Iodomethamate, Iodoxyl®, is disodium 1,4-dihydro-3,5-diiodo-1-methyl-4-oxo-2,6-pyridinedicarboxylate. It is a white, odorless, crystalline powder that is very soluble in water (1:1) and contains about 52 per cent of iodine.

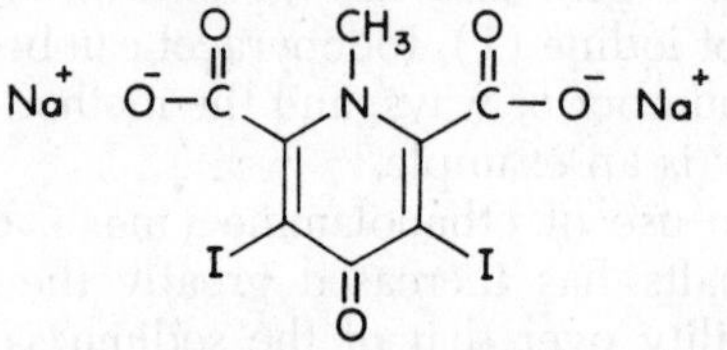

Sodium Iodomethamate

It is employed as a contrast medium in intravenous urography and retrograde pyelography, using the same technic and observing the same precautions as for iodopyracet, although it seems to give fewer reactions.

Sodium Ipodate N.F., Oragrafin® Sodium, sodium 3-[[(dimethylamino)methylene]-amino]-2,4,6-triiodohydrocinnamate, occurs as a water-soluble white to off-white odorless powder with a weakly bitter taste. This compound is stable in the dry form, and aqueous solutions are stable except at elevated temperatures. Both the dry material

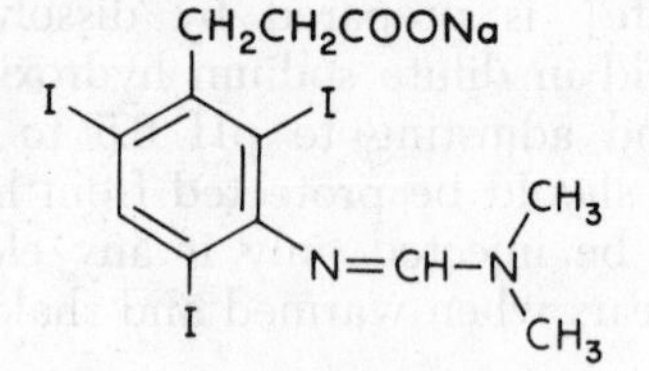

Sodium Ipodate

and aqueous solutions must be protected from light.

Sodium ipodate is given orally, as capsules, in cholecystography and in cholangiography. Maximal concentration in the hepatic and biliary ducts occurs in 1 to 3 hours in most patients and persists for about 45 minutes.

Category—diagnostic aid (radiopaque cholecystographic).

Usual dose—3 g. given 12 hours before the examination.

Usual dose range—3 to 6 g.

OCCURRENCE
Sodium Ipodate Capsules N.F.

Calcium Ipodate, N.F., Oragrafin® Calcium, calcium 3-[[(dimethylamino)methylene] amino] -2,4,6-triiodohydrocinnamate, differs from the sodium salt only in that it is almost insoluble in water and is supplied as granules with flavored sucrose. It may be administered by using an aqueous suspension.

Category—diagnostic aid (radiopaque—cholangiography and cholecystography).

Usual dose—3 to 6 g., as a single dose given 10 to 12 hours before the examination.

OCCURRENCE
Calcium Ipodate for Oral Suspension N.F.

WATER-INSOLUBLE CONTRAST MEDIA

Cholecystopexy is any gallbladder disease, and, to aid in diagnosing the disease, a compound is desirable that is opaque to x-rays and will be concentrated in vivo in the gallbladder and the bile duct. Usually, these agents are taken orally after a fat-free meal, and then some hours later (12) or the next day, with no other intake of food, the x-ray or fluoroscopic examination is made.

These diagnostic agents generally are insoluble or slightly soluble in water. They most often are used as the free organic acid. The formula below summarizes the structural modifications in this group.

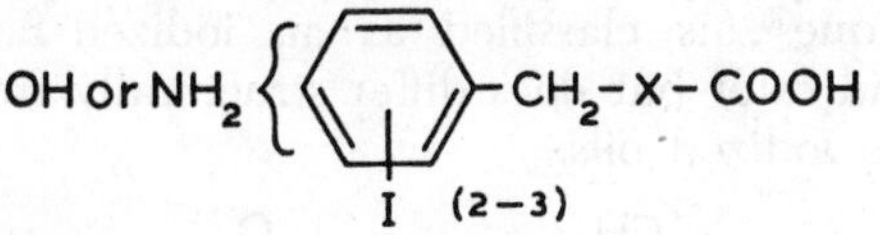

OH or NH₂ {⟨ring⟩—CH₂—X—COOH
I (2—3)

Products

Iopanoic Acid U.S.P., Telepaque®, Veripaque®, is 3-amino-α-ethyl-2,4,6-triiodohydrocinnamic acid, a cream colored solid which contains 66.68 per cent iodine. It is

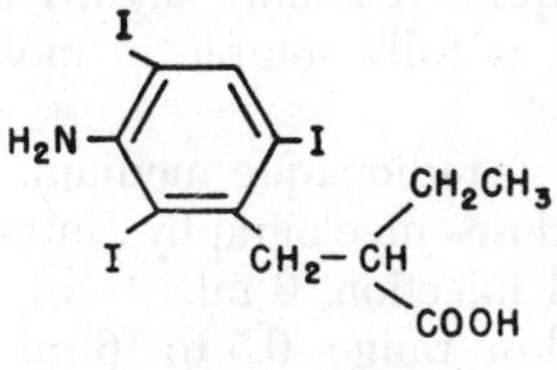

Iopanoic Acid

insoluble in water but soluble in dilute alkali and 95 per cent alcohol, as well as in other organic solvents.

In a study of derivatives, it was observed that the optimum visualization of the gallbladder was obtained when the number of carbon atoms in the alkanoic acid side chain approached five. Comparative studies with iodoalphionic acid showed iopanoic acid to be 1¼ times as effective. Also, it is about ¾ as toxic as iodoalphionic acid.

Iopanoic acid taken orally is well tolerated by the gastrointestinal tract and gives no impairment of hepatic or renal function. It is excreted in the feces and to a slight extent in the urine.

Category—radiopaque medium.
Usual dose—cholecystography, 3 g.
Usual dose range—3 to 6 g.

OCCURRENCE
Iopanoic Tablets U.S.P.

Propyliodone U.S.P., Dionosol®, is propyl 3,5-diiodo-4-oxo-1(4H)pyridineacetate and is used as a sterile aqueous or oil suspension for instillation into the trachea prior to bronchography. It occurs as a white, crystalline powder which is practically insoluble in water.

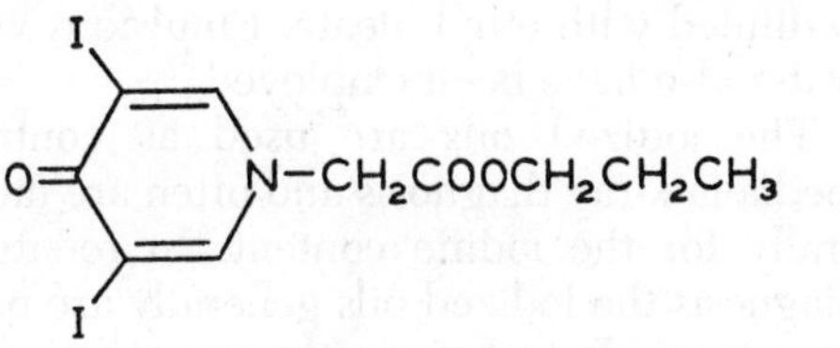

Propyliodone

Category—radiopaque medium.
Usual dose—bronchography, intrathecal, 0.2 ml. per kg. of body weight of a 50 per cent aqueous suspension or a 60 per cent oil suspension up to a maximum of 20 ml.
Usual dose range (of the sterile suspension)—10 to 20 ml.

OCCURRENCE
Sterile Propyliodone Suspension U.S.P.
Sterile Propyliodone Oil Suspension U.S.P.

2-(3,5-Diiodo-4-hydroxybenzyl)cyclohexanecarboxylic Acid, Monophen®, is a light yellowish-white solid that is insoluble in water. This acid, having 52 per cent of io-

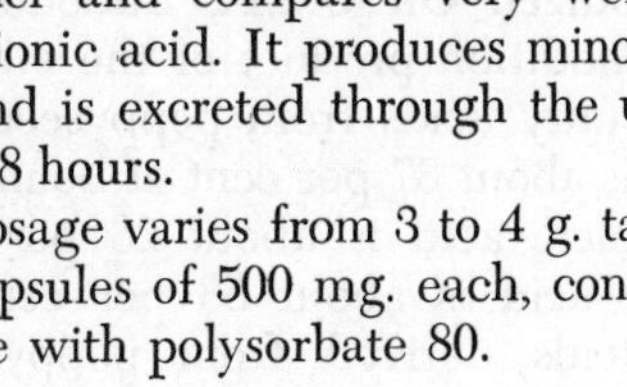

dine, has the property of opacifying the gallbladder and compares very well with iodoalphionic acid. It produces minor side-effects and is excreted through the urinary tract in 48 hours.

The dosage varies from 3 to 4 g. taken in 6 to 8 capsules of 500 mg. each, containing a mixture with polysorbate 80.

IODIZED OILS

Iodized oils are vegetable oils that have been treated with iodine, and the double bond, which is always present in an unsaturated glyceride (olein, linolein, for example), adds iodine (see Chap. 7). These iodized oils usually contain about 40 per cent of iodine. This concentration is necessary for good opacity but results in making the oil very viscous. To overcome this, simple esters (ethyl) of the unsaturated fatty acids are iodinated or the iodized oil is diluted with ethyl oleate. Emulsions with water also have been employed.

The iodized oils are used as contrast media in x-ray diagnosis and often are taken orally for the iodine content. In roentgen diagnosis the iodized oils generally are used by retrograde technic in the examination of the nasal sinuses, the bronchial tract, fistulas and the bladder, for example. They rarely are used intravenously.

Iodized Oil N.F., Lipiodol®, is an iodine addition product of vegetable oils, containing not less than 38 per cent and not more than 42 per cent of organically combined iodine (I). Any vegetable oil may be used. Iodine addition causes the oil to become more viscous, or thick, so that, in order to produce a more fluid iodized oil, the vegetable oil selected is usually a "semi-drying" oil that is moderately high in unsaturated glycerides. Long standing and exposure to air or sunlight cause iodized oil to decompose and darken, rendering it unfit for use.

Category—diagnostic aid (radiopaque—hysterosalpingographic).

Usual dose range—1 to 30 ml. by special injection, and depending on the procedure.

Iodized Poppyseed Oil, Lipiodol® Ascendent. This product contains about 10 per cent of bound iodine. It is used to visualize intradural tumors. It is available in 5-ml. vials.

Ethiodized Oil U.S.P., Ethiodol®, is an iodine addition product of the ethyl esters of the fatty acids from poppyseed oil and contains about 37 per cent of bound iodine. Since oleic acid is about 28 per cent and linoleic acid is about 58 per cent of the fatty acids derived from poppyseed oil, this would indicate that the main components of ethiodized oil are ethyl 9,10-diiodostearate and ethyl 9,10,12,13-tetraiodostearate. The prime advantage of this product over Iodized Oil N.F. and iodized poppyseed oil (Lipiodol Ascendent) is that the viscosity is much lower, being about one-fifth that of iodized poppyseed oil. This makes the injections easier to administer and makes the procedure more comfortable for the patient.

Category—radiopaque medium.

Usual dose—bronchography, 10 ml. by special injection.

Usual dose range—0.5 to 20 ml.

Iophendylate Injection U.S.P., ethyl 10-(iodophenyl)undecanoate injection, Pantopaque®, is classified as an iodized fatty acid ester but does differ structurally from the iodized oils.

Iophendylate

It is a uniform mixture of the κ and ω (10 and 11) isomers of ethyl iodophenylundecylate, occurring as a pale yellowish, odorless, viscous liquid. It is only slightly soluble in water but is fully soluble in most organic solvents.

Category—radiopaque medium.

Usual dose—myelography, intrathecal or by special injection, 6 ml.

Usual dose range—0.5 to 16 ml.

AGENTS FOR KIDNEY FUNCTION TEST

Aminohippuric Acid U.S.P., *p*-Aminohippuric Acid. This is a white crystalline powder which discolors on exposure to light.

Aminohippuric Acid

It is soluble to the extent of 1 in 100 in water or alcohol, and is readily soluble in acids or bases with salt formation occurring.

Category—pharmaceutic necessity for Sodium Aminohippurate Injection U.S.P.

Sodium Aminohippurate Injection U.S.P. is prepared by treating the free acid with an equivalent amount of sodium hydroxide and adjusting the pH to 7.0 to 7.2 with citric acid. This solution is used without isolating the sodium salt. The acid is prepared from *p*-nitrobenzoyl chloride and glycine. The *p*-nitro acid is isolated and reduced.

Solutions of sodium *p*-aminohippurate are sensitive to light.[3] The addition of 0.1 per cent of sodium bisulfite markedly retards the darkening of solutions in ampuls and prevents discoloration for at least 2 weeks in direct sunlight and 3 years in the dark or in diffused sunlight if the solution and ampuls are nitrogen-purged before filling. Dextrose should not be included in the solutions.

The sodium salt is excreted by the tubular epithelium of the kidney and by the glomerulus, thus serving as a means for measuring the effective renal plasma flow and for determining the functional capacity of the tubular excretory mechanism.

Category—diagnostic aid (renal function determination).

Usual dose—intravenous, 2 g.

Sodium Indigotindisulfonate U.S.P., sodium 5,5'-indigotindisulfonate, Indigo Carmine, occurs as a blue powder or crystal with a copper luster and is prepared from indigotin by sulfonation. This is an example of solubilizing a compound with sodium sulfonate groups. It is soluble in water (1:100), is slightly soluble in alcohol and almost insoluble in other organic solvents.

It is affected by light, but its solutions may be sterilized by autoclaving.

The dye is used in the laboratory as a coloring agent, stain and reagent. It is used to determine renal function and to locate the ureteral orifices. Normally, it appears in the urine in 10 minutes, and about 10 per cent of it is eliminated during the first hour.

Category—diagnostic aid (renal function determination).

Usual dose—intramuscular, 50 to 100 mg.; intravenous, 40 mg.

OCCURRENCE
Sodium Indigotindisulfonate Injection U.S.P.

Phenolsulfonphthalein U.S.P., α-Hydroxy-α,α-bis(*p*-hydroxyphenyl)-*o*-toluenesulfonic

Acid γ-Solutone, PSP, Phenol Red, is a red, crystalline powder that is stable in air. It is soluble in water (1:1,300), in alcohol (1:350) and almost insoluble in ether. It dissolves readily in bases. This compound may be considered as a derivative of phenolphthalein in which the CO group is replaced by an SO_2 group.

This compound, pKa 7.9, is used in the laboratory as an acid-base indicator using a

Phenolsulfonphthalein

0.02 to 0.05 per cent alcohol solution. At pH 6.8 it is yellow, and at pH 8.4 it is red. The dye is employed medicinally as a diagnostic agent for determining renal function. For this purpose, the monosodium salt is injected intravenously or intramuscularly, and the amount of phenolsulfonphthalein excreted in the urine is measured quantitatively. When kidney function is normal, the dye is excreted in a shorter time interval than when kidney function is impaired.

Category—diagnostic aid (renal function determination).

Usual dose—intramuscular or intravenous, 6 mg.

OCCURRENCE
Phenolsulphonphthalein Injection U.S.P.

AGENTS FOR LIVER FUNCTION TEST

Sodium Sulfobromophthalein U.S.P., Bromsulphalein® Sodium, disodium 4,5,6,7-tetrabromo-3',3''-disulfophenolphthalein, disodium phenoltetrabromophthalein disulfonate, is a white, crystalline, hygroscopic powder that has a bitter taste and is odorless. It is soluble in water but is insoluble in alcohol and acetone.

The bromine atoms in the compound cause it to be removed from the blood almost entirely by way of the liver. The introduction of sulfonic acid groups into compounds of this type decreases the toxicity and greatly increases the water solubility. The compound is injected intravenously, as a 5 per cent solution, and the amount remaining in the blood after a certain time interval is determined colorimetrically. The rate at which the dye is removed from the

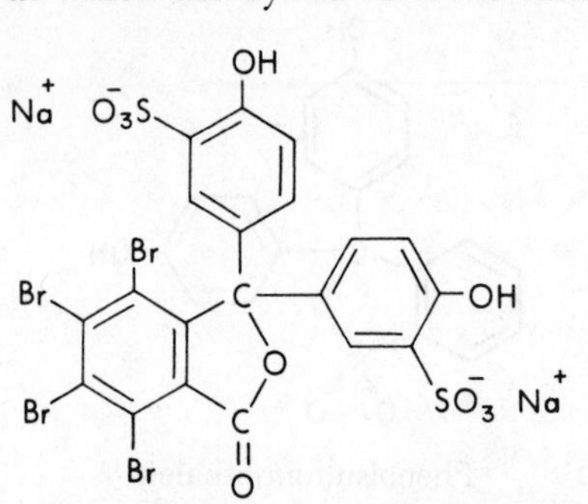

Sodium Sulfobromophthalein

blood is a measure of the hepatic function. The concentration of the dye in the bloodstream is measured at the end of 1 hour and at regular time intervals thereafter in order to determine the rate of clearance.

Category—diagnostic aid (hepatic function determination).

Usual dose—intravenous, 5 mg. per kg. of body weight.

OCCURRENCE

Sodium Sulfobromophthalein Injection U.S.P.

Rose Bengal, Tetraiodotetrachlorofluorescein, is made by reacting tetrachlorophthalic anhydride with resorcinol and iodinating the resulting product. It is used as a test for liver function. The liver almost exclusively removes the dye from the bloodstream. From 100 to 150 mg. of the dye is injected intravenously in sterile saline. A normally functioning liver will remove 50 per cent of the dye within 2 minutes. The dye is photosensitive, so the dye, its solutions and the patients receiving it should be protected from light.

This compound is also available as ^{131}I-labeled tetraiodotetrachlorofluorescein in sterile, neutral solution. A small amount of the radioactive dye is injected intrave-

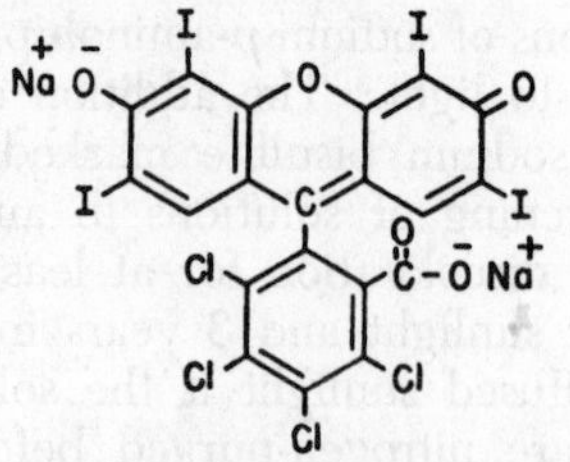

Rose Bengal

nously, then the rates of clearance from the blood by the liver and excretion into the small intestine are determined. The clearance and excretion rates are determined using standard radioisotope counting equipment. The usual intravenous dose is the equivalent of 150 microcuries, with the usual dose range being the equivalent of 10 to 200 microcuries.

OCCURRENCE

Sodium Rose Bengal I 131 Injection U.S.P.

MISCELLANEOUS DIAGNOSTIC AGENTS

Sodium Fluorescein U.S.P., resorcinolphthalein sodium, soluble fluorescein, is an orange, odorless, hygroscopic powder. It is soluble in water and sparingly soluble in alcohol.

The disodium salt forms highly fluorescent solutions when dissolved in water. The acidified solution has practically no fluorescence.

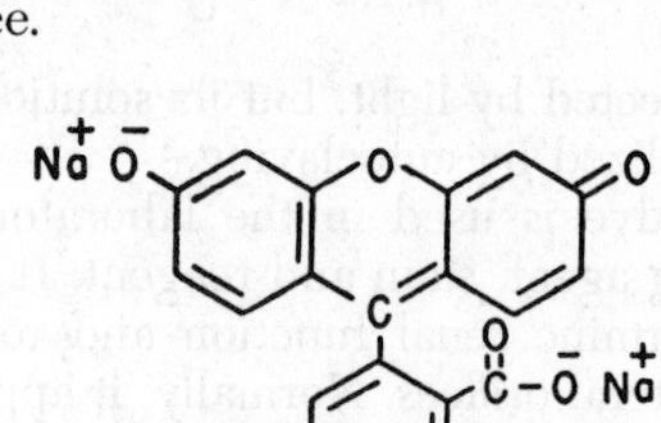

Sodium Fluorescein

Sodium fluorescein is used as an ophthalmologic diagnostic agent. For this purpose, a solution consisting of 2 per cent of the dye and 3 per cent of sodium bicarbonate is employed. Diseased or abraded areas of the cornea, such as corneal ulcers, are stained green by the solution. Foreign bodies appear with a green ring around them, while the normal cornea is not stained.

Category—diagnostic aid (corneal trauma indicator).

For external use—topically, 0.1 to 0.3 ml. of a 2 per cent solution, to the conjunctiva.

	PER CENT
OCCURRENCE	SODIUM FLUORESCEIN
Sodium Fluorescein Ophthalmic Solution U.S.P.	2

Evans Blue U.S.P., is a complex azo dye known chemically as 4,4'-bis[7-(1-amino-8-hydroxy-2,4-disulfo)naphthylazo]-3,3'-bitolyl tetrasodium salt.

It exists as blue crystals having a bronze to green luster and is soluble in water, al-

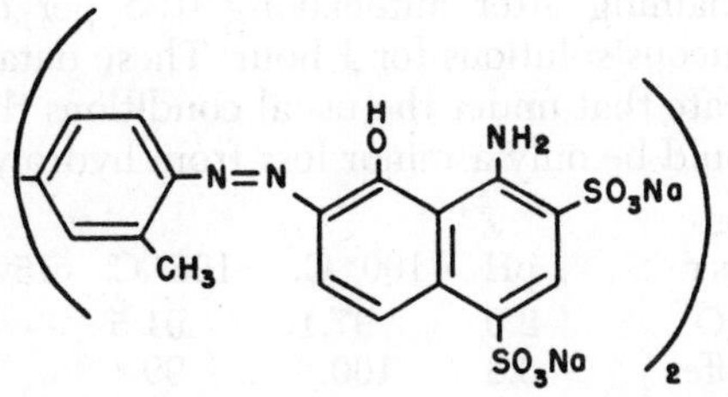

cohol, acids and alkalies. The aqueous solutions are quite stable and may be autoclaved. Saline solutions are less stable and should not be autoclaved.

Evans blue dye when injected into the bloodstream combines firmly with the plasma albumin. The color developed is directly proportional to its concentration. Spectral absorption is greatest at about 610 mμ where the photometric determination is made, and, by means of color intensity, the total blood volume may be found. This is used as a guide in replacement therapy, in shock and in hemorrhage.

Category—diagnostic aid (blood volume determination).

Usual dose—intravenous, the equivalent of 22.6 mg. of dried Evans Blue.

	PER CENT
OCCURRENCE	EVANS BLUE
Evans Blue Injection U.S.P.	0.45

Indocyanine Green U.S.P., CardioGreen®. This is a dark green to black powder which forms deep emerald-green solutions. The solutions are not stable over long periods, thus they must be made just prior to administration.

Category—diagnostic aid (blood volume and cardiac output determination; hepatic function determination).

Usual dose—intravenous, blood volume and cardiac output determination, 5 mg. in 1 ml.; hepatic function determination, 500 mcg. per kg. of body weight.

OCCURRENCE
Sterile Indocyanine Green U.S.P.

Chlormerodrin Hg 197 U.S.P., Neohydrin-197®, and **Chlormerodrin Hg 203 U.S.P.**, Neohydrin-203®, are available in sterile solution as radioactive diagnostic aids, employed for locating lesions of the brain and anatomic or functional defects of the kidneys. Mercury-197 has a shorter half-life (65 hours) than mercury-203 (46.6 days) and also a lower gamma radiation energy. Chlormerodrin Hg 203 delivers less than one half the total body radiation of radioiodinated (^{131}I) serum albumin, and is said to be diagnostically superior in locating brain lesions. Chlormerodrin Hg 197 solution contains about 1,000 microcuries per ml.; chlormerodrin Hg 203 solution contains about 250 microcuries per ml.

Category—diagnostic aid (tumor localization).

Usual dose—intravenous, 10 microcuries per kg. of body weight.

Usual dose range—up to a total of 700 microcuries.

OCCURRENCE
Chlormerodrin Hg 197 Injection U.S.P.
Chlormerodrin Hg 203 Injection U.S.P.

Azuresin N.F., Diagnex® Blue, Azure A Carbacrylic Resin, is a carbacrylic resin-dye combination that is used to diagnose achlorhydria. In the presence of acid in the gastric juice, the dye is released and absorbed from the upper intestine and then promptly excreted in the urine where it can be determined colorimetrically.

Each test unit contains two 250-mg. tablets of caffeine and sodium benzoate to be taken to stimulate gastric secretion. Histamine phosphate or betazole hydrochloride may be used in place of the caffeine and sodium benzoate.

Category—diagnostic aid (gastric secretion).

Usual dose—2 g. preceded by 500 mg. of caffeine and sodium benzoate.

Metyrapone U.S.P., Metopirone®, 2-methyl-1,2-di-3-pyridyl-1-propanone, occurs as a white to off-white crystalline powder. It has a characteristic odor. It should be protected from heat and light because of its low melting point and its light sensitivity.

Metyrapone possesses the property of selective inhibition in vivo of hydroxylation of the three principal adrenocorticoid hormones, hydrocortisone, corticosterone and

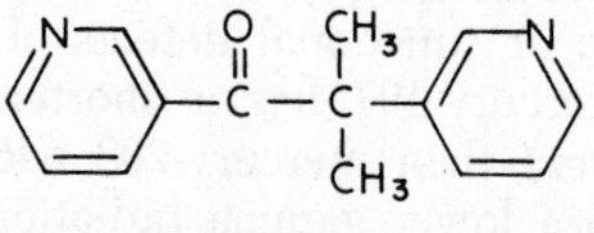

Metyrapone

aldosterone.[4] Thus, it finds use as a diagnostic tool to determine residual pituitary function in patients with hypopituitarism and, also, to evaluate a patient's ability to withstand surgery and other stresses.

Metyrapone is available as 250-mg. tablets of the base, and as ampuls with each ml. containing 100 mg. of the bitartrate salt which is equivalent of 43.8 mg. of the base.

Category—diagnostic aid (anterior pituitary function determination).

Usual dose—750 mg. every 4 hours for six doses.

OCCURRENCE
Metyrapone Tablets U.S.P.

SYNTHETIC SWEETENING AGENTS

Saccharin U.S.P., 1,2-Benzisothiazolin-3-one 1,1-dioxide. Saccharin occurs as white crystals, or a white crystalline powder. It is odorless or nearly so, but has a very pronounced sweet taste. In dilute solution, it is about 500 times as sweet as sucrose.

The following equations illustrate the preparation of saccharin:

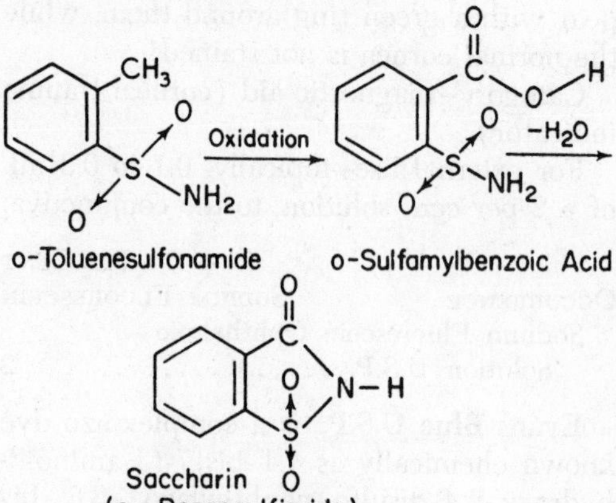

o-Toluenesulfonamide   o-Sulfamylbenzoic Acid

Saccharin
(o–Benzosulfimide)

Saccharin is relatively stable in solution from pH 3.3 to 8.0.[5] The following chart shows the per cent of unchanged saccharin remaining after autoclaving 0.35 per cent aqueous solutions for 1 hour. These data indicate that under the usual conditions there would be only a minor loss from hydrolysis.

SOL-VENT	pH	100° C.	125° C.	150° C.
H_2O	2.0	97.1	91.5	81.4
Buffer	3.3	100	99	98.1
Buffer	7.0	99.7	99.7	98.4
Buffer	8.0	100	100	100

Since it does not enter into the body's metabolism, it is employed as a sweetening agent in the diets of diabetics and others who need to restrict their intake of carbohydrates.

Category—pharmaceutic aid (flavor).

OCCURRENCE	PER CENT SACCHARIN
Aromatic Castor Oil N.F.	0.05
Aromatic Cascara Sagrada Fluidextract U.S.P.	0.2

Sodium Saccharin N.F.; Sodium 1,2-Benzisothiazolin-3-one 1,1-dioxide. The compound occurs as a white, crystalline powder.

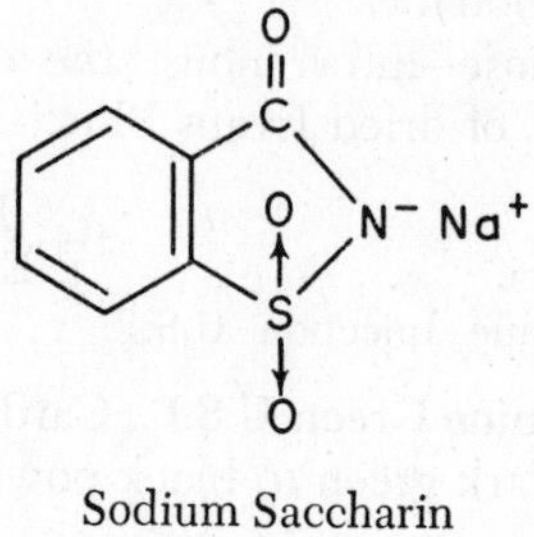

Sodium Saccharin

It is soluble in water (1:1,5) and in alcohol (1:50). Like saccharin, the sodium salt is about 500 times as sweet as sugar; and, since the salt is much more soluble in water, this is the form in which it usually is employed as a sweetening agent.

Category—non-nutritive sweetener.

	PER CENT SACCHARIN
OCCURRENCE	SODIUM
Sodium Saccharin Tablets N.F.	..
Amobarbital Elixir N.F.	0.1
Kaolin Mixture with Pectin N.F. ...	0.1

Calcium Saccharin N.F., Calcium 1,2-Benzisothiazolin-3-one 1,1-dioxide. This compound occurs as white crystals or as a white crystalline powder. One gram is soluble in 1.5 ml. of water. It is used as a sweetening agent with Calcium Cyclamate N.F. in a 1:10 ratio.

Calcium Cyclamate N.F.; Sucaryl® Calcium; Calcium Cyclohexanesulfamate. This

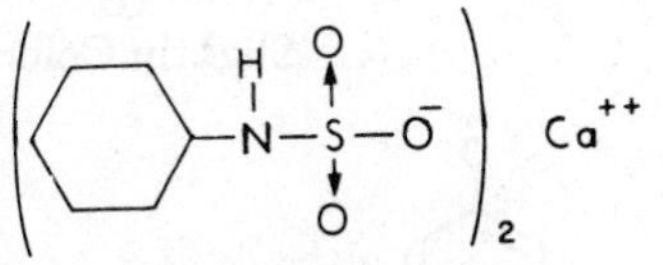

Calcium Cyclamate

compound, a derivative of sulfamic acid, is used as a sweetening agent in diets for diabetics and those who must restrict their intake of carbohydrates and sodium ions. It is essentially nontoxic, but it is recommended that intake be restricted to about 7 ml. of the 15 per cent solution per day. An excessive intake may produce a mild laxative effect. One and one-quarter ml. of this solution is equivalent to about 2 teaspoons of sucrose. In the usual concentrations the compound does not cause a bitter aftertaste.

Cyclamate salts are incompatible with any potassium salts because of the low order of solubility of potassium cyclamate. Calcium cyclamate is stable at elevated temperatures and so can be used in cooking.

Category—non-nutritive sweetener.

OCCURRENCE
Calcium Cyclamate and Calcium Saccharin Solution N.F.
Calcium Cyclamate and Calcium Saccharin Tablets N.F.

Sodium Cyclamate N.F., Sucaryl® Sodium; Sodium Cyclohexanesulfamate. This compound is used as a sweetening agent,

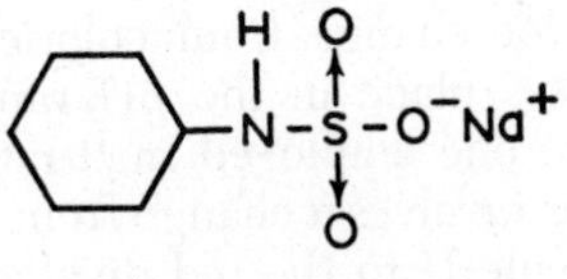

Sodium Cyclamate

being about 30 times as sweet as sucrose. It is essentially nontoxic, but excessive intake may produce a mild laxative effect. About 40 per cent is eliminated unchanged in the urine and 60 per cent in the feces. As a sweetening agent 0.125 g. is the equivalent of 1 teaspooon of sugar.

Category—non-nutritive sweetener.

OCCURRENCE
Sodium Cyclamate and Sodium Saccharin Solution N.F.
Sodium Cyclamate and Sodium Saccharin Tablets N.F.

SYNTHETIC LAXATIVES

Phenolphthalein N.F., 3,3-bis(*p*-hydroxyphenyl)phthalide, is a white or faintly yellowish-white, crystalline powder. It is soluble in alcohol (1:15), in ether (1:100) and in dilute bases but is almost insoluble in

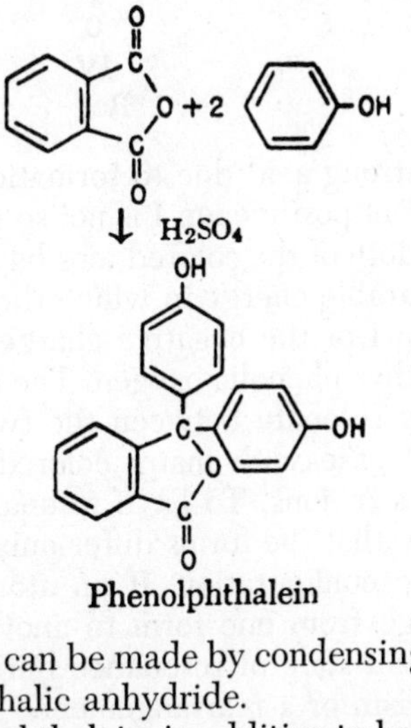

Phenolphthalein

water. It can be made by condensing phenol with phthalic anhydride.

Phenolphthalein, in addition to being used as a laxative, is one of the most commonly used indicators for the titration of weak acids with alkali. It can exist as a neutral molecule and as four different ions, depending on the pH.

The color change from colorless to red which takes place in the pH range 8.3 to 10.0 is the one employed in titrating weak acids. This involves a change from the colorless molecule II to the red divalent ion IV. That phenolphthalein becomes colorless in strong base due to the conversion to the trivalent ion V or that it turns orange-pink

tion. The colorless or almost colorless N.F. product has only about one third the laxative action of yellow phenolphthalein, a more impure product. It was thought for some time that the greater laxative action of the yellow product was due to hydroxyanthraquinones which were presumed to have been formed during the synthesis. However, more recent work has shown[6] that hydroxyanthraquinones are not present in yellow phenolphthalein nor is the laxative action of Phenolphthalein N.F. in-

| I | II | III |
| Orange-Pink | Colorless | Slightly Colored |

| IV | V |
| Red | Colorless |

in very strong acid due to formation of the monovalent positive ion I is not so generally known. Both of the colored ions have a form of comparable energy in which the positive charge in I or the negative charge in IV is on the other phenolic oxygen. The structure is said to resonate between the two forms, as is the case with many colored organic molecules or ions. To have resonance, it is necessary that the forms differ only in their electronic configuration. If an atom has to shift to go from one form to another, it is no longer a state of resonance but involves tautomerism or a rearrangement.

Phenolphthalein is used as a mild, tasteless laxative in the treatment of constipa-

creased by adding hydroxyanthraquinones to it. A number of colorless compounds have been isolated from the yellow product; these include fluoran, isophenolphthalein [3(o-hydroxyphenyl)3-(p-hydroxyphenyl) phthalide], phenolphthalein, 2-(4-hydroxybenzoyl)benzoic acid and other compounds not yet identified. None of these compounds showed a laxative action greater than that of Phenolphthalein N.F. The cause of the greater laxative action of yellow phenolphthalein is still undetermined.

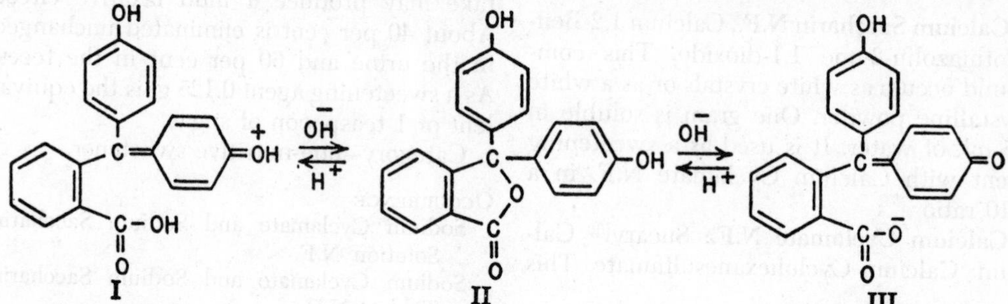

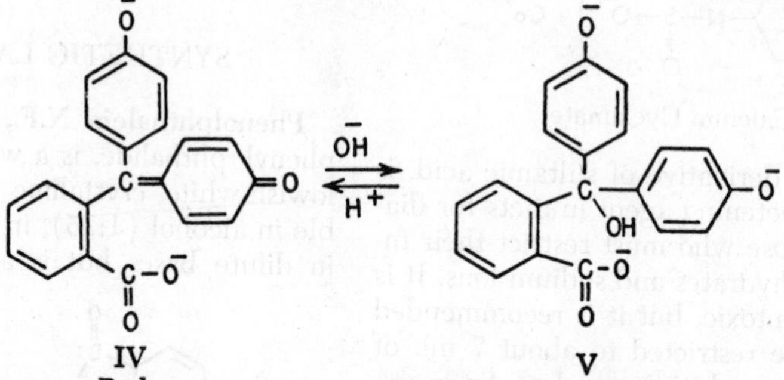

It may be combined with other drugs, such as agar-agar or mineral oil. Phenolphthalein is not well absorbed from the intestinal tract and has a low toxicity. It is found in many of the commercial laxative preparations.

Category—cathartic.

Usual dose—60 mg.

OCCURRENCE

Phenolphthalein Tablets N.F.

Oxyphenisatin Acetate; Isacen®; 3,3-Bis-(4-acetoxyphenyl)oxindole. This occurs as tasteless crystals which are insoluble in

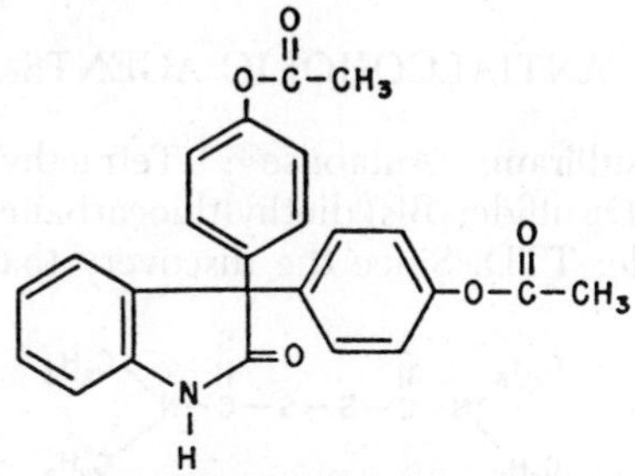

Oxyphenisatin Acetate

water or dilute hydrochloric acid, slightly soluble in alcohol and insoluble in ether. This compound is related in structure to phenolphthalein and has a similar mild purgative action. The acetylated compound gives rise to less irritation than the unacetylated compound and is completely excreted in the feces.

The usual dose is 5 mg.

Bisacodyl N.F., Dulcolax®, is 4,4'-(2-pyridylmethylene)diphenol diacetate (ester). It occurs as tasteless crystals which are practically insoluble in water and alkaline solutions. It is soluble in acids and organic solvents.

Bisacodyl appears to act directly on the colonic and rectal mucosa with little effect on the small intestine. It is recommended for use in constipation and in the preparation of patients for surgery or radiography.

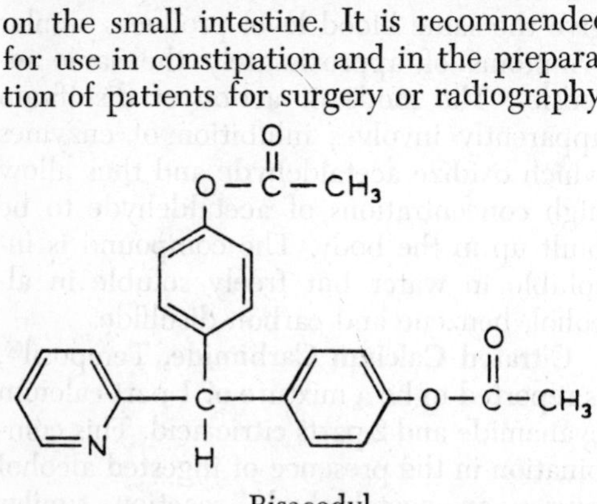

Bisacodyl

It is supplied as enteric-coated 5-mg. tablets and as 10-mg. suppositories which may be stored at normal room temperature.

The tablets must be swallowed whole, not chewed, or crushed, and should not be taken within one hour of antacids. These precautions are necessary so that the enteric coating is not disturbed until after the drug leaves the stomach. If released in the stomach, the drug may cause vomiting.

Category—cathartic.

Usual dose—oral and rectal, 10 mg.

Usual dose range—oral, 10 to 30 mg.

OCCURRENCE

Bisacodyl Suppositories N.F.

Bisacodyl Tablets N.F.

Danthron N.F., Dorbane®, is 1,8-dihydroxyanthraquinone. It is structurally related to the anthraquinone derivatives found

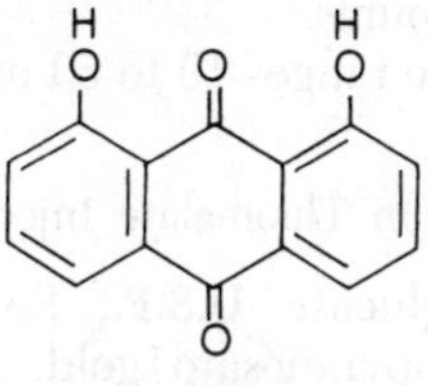

Danthron

in Cascara sagrada and other vegetable cathartics.

Danthron is administered orally at bedtime. It is frequently used in combination with a fecal softening agent, dioctyl sodium sulfosuccinate (see Chap. 12).

Category—cathartic.

Usual dose—75 to 150 mg.

OCCURRENCE

Danthron Tablets N.F.

GOLD COMPOUNDS

Gold and its compounds have been used since early times in the treatment of various diseases, including syphilis, tuberculosis and cancer. However, they have not proved to be effective therapeutic agents for these conditions. At present they are used for the treatment of lupus erythematosus and rheumatoid arthritis. Gold compounds are among the most toxic of all the metal compounds. Toxic manifestations involve skin, renal and hematologic reactions.

Gold Sodium Thiomalate U.S.P.; Myochrysine®, (Disodium mercaptosuccinato)-gold. This compound occurs as a white or yellowish-white powder that is almost insoluble in alcohol and ether but very solu-

Gold Sodium Thiomalate

ble in water. A 5 per cent aqueous solution has a pH of about 6. It is used for the treatment of rheumatoid arthritis.

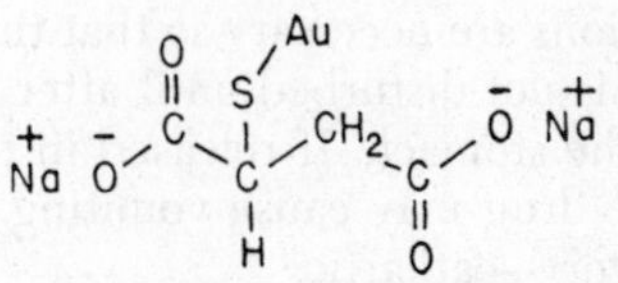

Category—antirheumatic.

Usual dose—intramuscular, 10 mg., increasing to 25 mg. and then to 50 mg. per week to a total dose of 750 mg., then in decreasing amounts.

Usual dose range—10 to 50 mg. per week.

OCCURRENCE

Gold Sodium Thiomalate Injection U.S.P.

Aurothioglucose U.S.P., Solganal®, (1-Thio-D-glucopyranosato)gold. Aurothioglucose is a water-soluble, oil-insoluble compound containing about 50 per cent of gold.

It occurs as yellow crystals with a slight mercaptanlike odor. It decomposes in water solution, so is used as a suspension in an anhydrous vegetable oil.

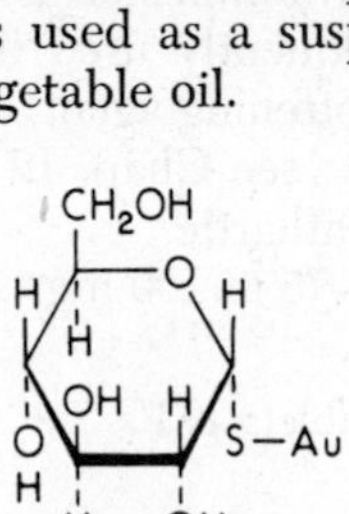

Aurothioglucose

Category—antirheumatic.

Usual dose—intramuscular, 10 mg., increased to 25 mg. and then to 50 mg. per week to a total dose of 750 mg., then in decreasing amounts.

Usual dose range—10 to 50 mg. per week.

OCCURRENCE

Aurothioglucose Injection U.S.P.

Aurothioglycanide, Lauron®, α-Auromercaptoacetanilid. This compound occurs as a

Aurothioglycanide

grayish-yellow powder which is insoluble in bases, acids, water and most organic solvents. It is used as an oil suspension in the treatment of rheumatoid arthritis.

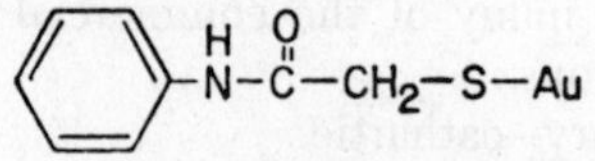

ANTIALCOHOLIC AGENTS

Disulfiram; Antabuse®; Tetraethylthiuram Disulfide; Bis(diethylthiocarbamyl)disulfide; TTD. Since the discovery that this

Tetraethylthiuram Disulfide

compound causes nausea, pallor, copious vomiting and other unpleasant symptoms when alcohol is ingested after its use, it has been proposed as a treatment for alcoholism. Up to 6 g. of the drug has been tolerated without symptoms if alcohol is not taken. However, if alcohol is taken in appreciable quantities after disulfiram, dizziness, palpitation, unconsciousness and even death may result.

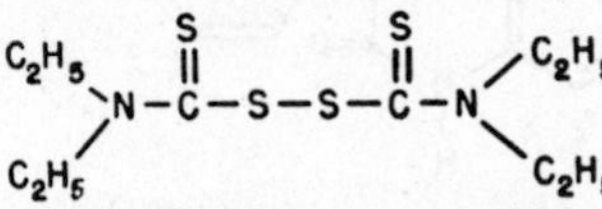

It has been observed that individuals ingesting alcohol after disulfiram have a blood acetaldehyde level 5 to 10 times greater than that obtained when the same amount of alcohol is ingested by untreated persons. The breath has a noticeable aldehyde odor. The intravenous infusion of acetaldehyde to give the same blood level produces similar symptoms of approximately the same intensity. The mode of action of disulfiram apparently involves inhibition of enzymes which oxidize acetaldehyde and thus allow high concentrations of acetaldehyde to be built up in the body. The compound is insoluble in water but freely soluble in alcohol, benzene and carbon disulfide.

Citrated Calcium Carbimide, Temposil®, is reported to be a mixture of 1 part calcium cyanamide and 2 parts citric acid. This combination in the presence of ingested alcohol causes an acetaldehyde reaction similar to that of disulfiram and appears to be a useful adjunct to other measures in the treatment of alcoholism.[7]

PSORALENS

The psoralens are furocoumarins which are widely distributed in nature. Plants containing these psoralens have been used since antiquity to produce pigmentation. The probable mechanism of action is the concentration of the psoralen in the melanocytes, which when activated by ultraviolet irradiation initiates melanin production. After an oral dose the skin becomes photosensitive in about 1 hour, reaches a peak in sensitivity in 2 hours and the effect wears off in 8 hours.[8] The results of an extensive investigation of the relationship between structure and erythematous activity of the psoralens following ultraviolet irradiation have been published.[9]

Methoxsalen, Meloxine®, 8-Methoxypsoralen, is obtained from the fruit of *Ammi*

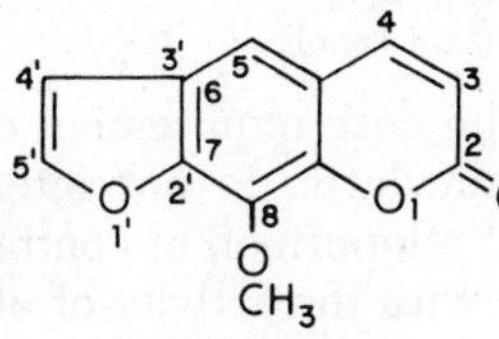

Methoxsalen

majus. Methoxsalen increases the normal response of the skin to ultraviolet radiation. Overdoses or excessive exposure early in the treatment may cause severe burning.

Trioxsalen N.F., Trisoralen®, 2,5,9-Trimethyl-7H-furo [3,2-g][1]benzopyran-7-one, 4,5′,8-Trimethylpsoralen. This is a synthetic

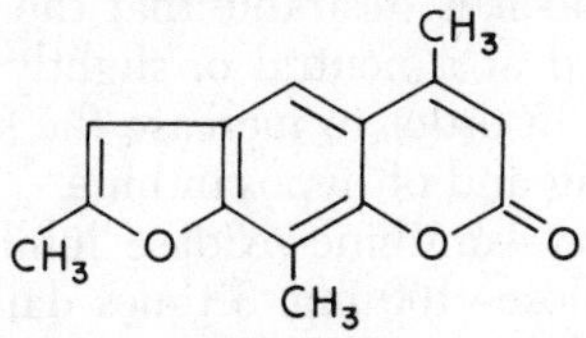

Trioxsalen

psoralen with much the same uses as methoxsalen, but it is reported to be more potent and less toxic.

Category—pigmenting agent (photosensitizer).

Usual dose—10 mg., 2 hours before exposure to sunlight.
Usual dose range—5 mg. to 10 mg.

OCCURRENCE
Trioxsalen Tablets N.F.

URICOSURIC AGENTS

Most purine derivatives in the diet are converted to uric acid and in man are excreted as such. Gout is characterized by an

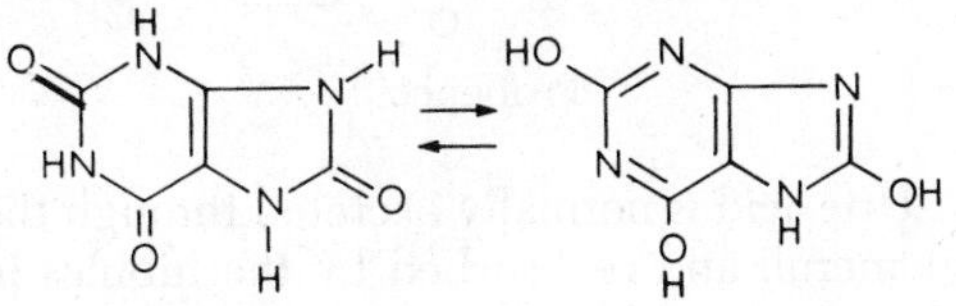

Keto   Enol
Uric Acid

error in the metabolism of uric acid, there is an elevation of serum urate and crystals form in the cartilages. Colchicine is used for acute attacks and uricosuric agents are used to aid in the excretion of the elevated levels of urates.

Colchicine U.S.P. occurs as a pale yellow powder, which is soluble in water, freely soluble in alcohol and in chloroform and is slightly soluble in ether. It darkens on exposure to light. This drug is used for the relief of acute attacks of gout. It is also useful for the prevention of acute gout when there is frequent recurrence of the attacks. The exact mechanism is not yet established, but it is known not to have any effect in uric acid metabolism. It is considered here with the uricosuric agents as a matter of convenience.

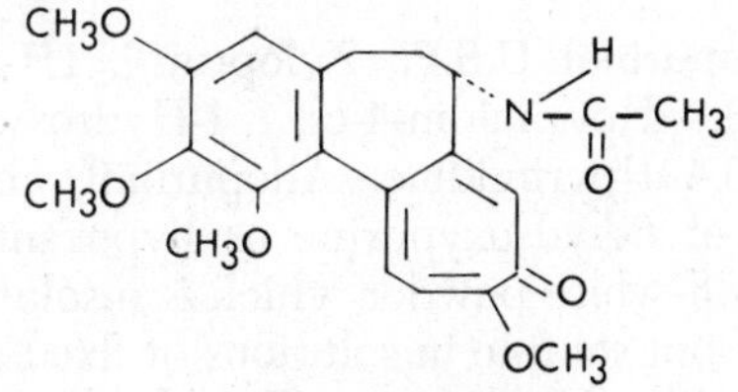

Colchicine

Category—gout suppressant.
Usual dose—therapeutic, 1 mg. every 2 hours for 6 to 8 doses, as tolerated; maintenance, 500 mcg. twice daily.

Usual dose range—1 to 8 mg. daily.

OCCURRENCE
Colchicine Tablets U.S.P.

Probenecid U.S.P., Benemid®, is *p*-(di-propylsulfamoyl)benzoic acid. It is a white, nearly odorless crystalline powder. It is soluble in dilute alkali (salt formation) but is practically insoluble in water and dilute acids.

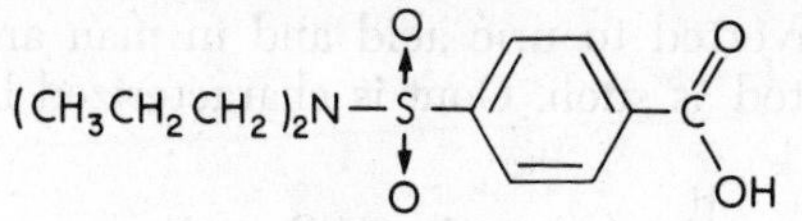

Probenecid

Uric acid is normally excreted through the glomeruli and reabsorbed by the tubules in

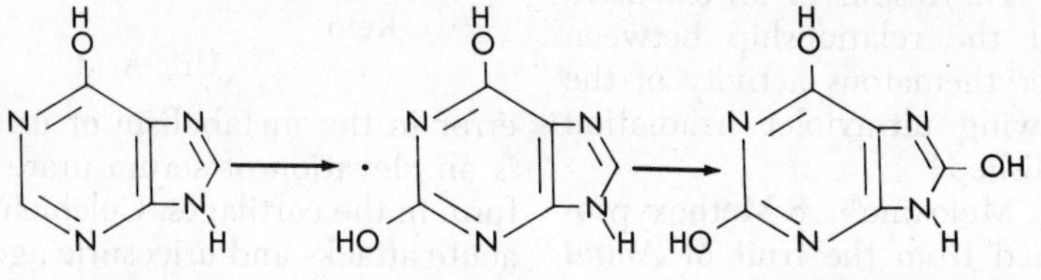

Hypoxanthine   Xanthine   Uric Acid
(Enol Form)

the kidney. Probenecid acts by interfering with this tubular reabsorption. Probenecid also inhibits excretion of compounds such as aminosalicylic acid, penicillin and sulfobromophthalein. It is rapidly absorbed after oral administration, then is metabolized and excreted slowly, mainly as the glucuronate conjugate.

Category—uricosuric.
Usual dose—500 mg. 2 to 4 times daily.
Usual dose range—500 mg. to 2 g. daily.

OCCURRENCE
Probenecid Tablets U.S.P.

Allopurinol U.S.P., Zyloprim®, 1H-Pyrazolo[3,4-d]pyrimidin-4-ol, 4-Hydroxypyrazolo[3,4-d]pyrimidine. Allopurinol, an isostere of 6-hydroxypurine or hypoxanthine, is an off-white powder which is insoluble in water but soluble in solutions of fixed alkali hydroxides. It has a pKa of about 9.4. It blocks the formation of uric acid by inhibiting xanthine oxidase, the enzyme responsible for the biotransformation of hypoxanthine to xanthine and of xanthine to uric acid. Thus, it is useful in the control of

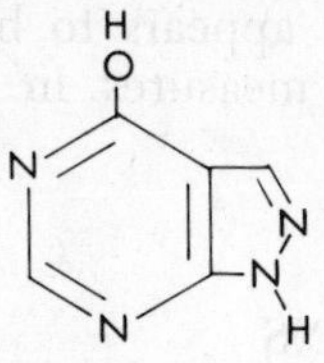

Allopurinol 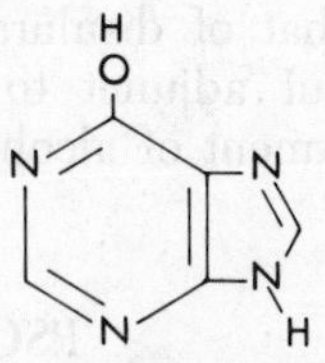   Hypoxanthine

uric acid levels associated with gout and other conditions. Allopurinol is metabolized to the corresponding xanthine isostere, alloxanthine, which in turn contributes to the inhibition of xanthine oxidase.

Allopurinol also inhibits the enzymatic oxidation of mercaptopurine, which is used as an antineoplastic antimetabolite. When the two compounds are co-administered, there may be as much as a 75 per cent reduction in the dose requirement of mercaptopurine. Salicylates do not interfere with the action of allopurinol, in contrast to their interference with the activity of other uricosuric agents.

The dosage of allopurinol required to lower serum uric acid to normal or near-normal levels varies with the severity of the disease. Divided daily doses are advisable because of the short biological half-life of the drug. While the drug is being administered, fluid intake should be adequate to produce a daily urinary output of at least 2 liters and it is desirable that the urine be maintained at a neutral or slightly alkaline pH value in order to increase the solubility of the drug and of hypoxanthine.

Category—xanthine oxidase inhibitor.
Usual dose—100 mg. 3 times daily.
Usual dose range—100 to 600 mg. daily.

OCCURRENCE
Allopurinol Tablets U.S.P.

Other uricosuric agents include sulfinpyrazone and phenylbutazone (see Chapter 24).

ANTIEMETIC AGENTS

Trimethobenzamide Hydrochloride N.F., Tigan®, N-{*p*-[2-(Dimethylamino)ethoxy]-benzyl}-3,4,5-trimethoxybenzamide Monohydrochloride. This drug is reported to block the emetic mechanism without undesirable side-effects. It is useful in nausea and vomiting associated with pregnancy, radiation therapy, drug administration and travel sickness. Effects appear within 20 to 40 minutes after administration and last for 3 to 4 hours.

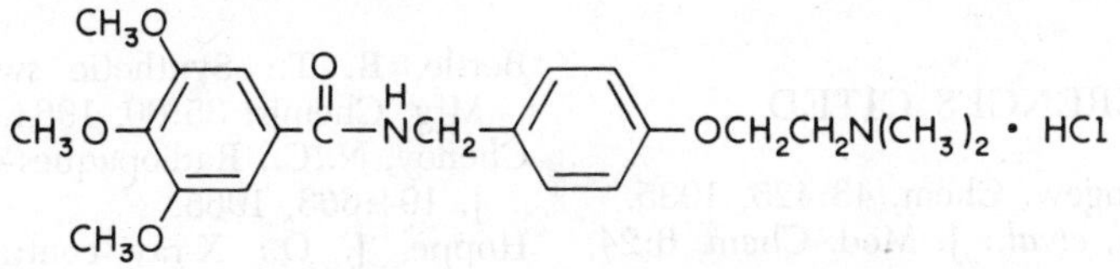

the control of vertigo and of nausea and vomiting. The soluble hydrochloride salt is used in the injectable forms and the tablets; the pamoate is used in the suspension and the free base in the suppositories.

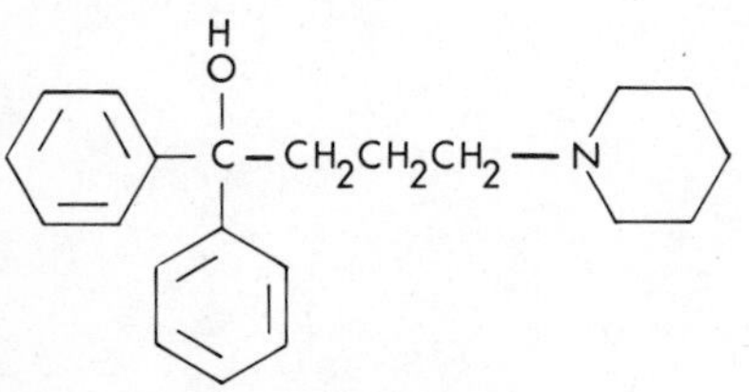

The adult dosage is 25 to 50 mg. orally or rectally 4 times daily; for acute symptoms, 20 to 40 mg. may be given 4 times daily by deep intramuscular injection.

Other antiemetics include chlorpromazine hydrochloride, prochlorperazine dimaleate

Trimethobenzamide Hydrochloride

Category—antiemetic.

Usual dose range—oral and intramuscular, 100 to 250 mg. 4 times daily; rectal, 200 mg. 3 or 4 times daily.

OCCURRENCE

Trimethobenzamide Hydrochloride Capsules N.F.

Trimethobenzamide Hydrochloride Injection N.F.

Trimethobenzamide Hydrochloride and Benzocaine Suppositories N.F.

and some of the antihistaminic agents such as cyclizine hydrochloride, dimenhydrinate and meclizine hydrochloride.

ALCOHOL DENATURANTS

Denatonium Benzoate N.F., Benzyldiethyl[(2,6-xylylcarbamoyl)methyl]ammonium Benzoate. This compound occurs as a white, crystalline powder with an intensely bitter taste. It is soluble in water and freely soluble in alcohol and in chloroform. Be-

Denatonium Benzoate

Diphenidol, Vontrol®, α,α-Diphenyl-1-piperidinebutanol. Diphenidol is useful in

Diphenidol

cause of its intensely bitter taste and good solubility in alcohol and in water it can replace brucine as a denaturant for ethyl alcohol.

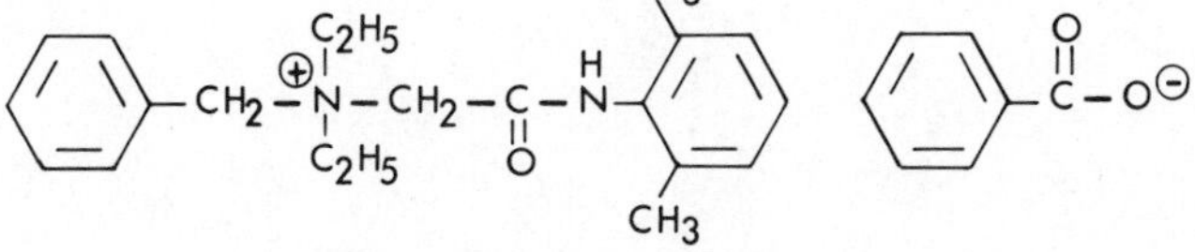

Category—alcohol denaturant.

Sucrose Octaacetate N.F., is a white, practically odorless powder with an intensely bitter taste. It is soluble in alcohol and in ether but very slightly soluble in water (1 in 1100). It is made by complete acetylation of sucrose.

Category—alcohol denaturant.

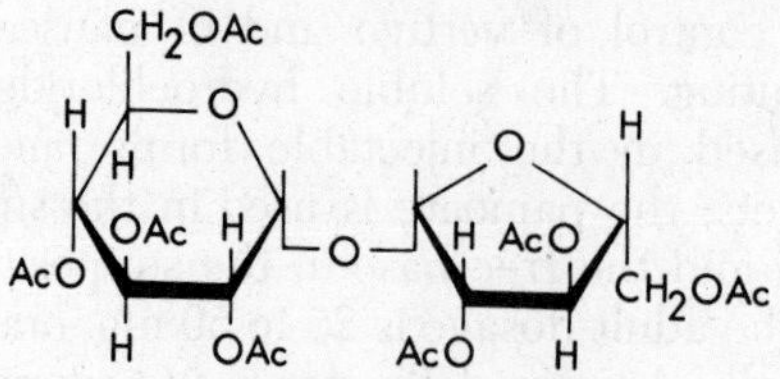

CH₂OAc structure

Sucrose Octaacetate

Other alcohol denaturants include Methyl Isobutyl Ketone N.F. (See Chapter 6.)

REFERENCES CITED

1. Binz, A.: Angew. Chem. 48:425, 1935.
2. Hoey, G. B., et al.: J. Med. Chem. 6:24, 1963.
3. Whittet, T. D., and Robinson, A. E.: Pharm. J. 1964:39 (July 11).
4. Coppage, W. S., Jr.: J. Clin. Invest. 38: 2101, 1959.
5. DeGarmo, O., et al.: J. Am. Pharm. A. (Sci. Ed.) 41:17, 1952.
6. Hubacher, M. H., and Doernberg, S.: J. Am. Pharm. A. (Sci. Ed.) 37:261, 1948.
7. Mitchell, E. H.: J.A.M.A. 168:2008, 1958.
8. Becker, S. W.: J.A.M.A. 173:1483, 1960.
9. Pattiak, M. A., et al.: J. Invest. Derm. 35: 165, 1960.

SELECTED READING

Bottle, R. T.: Synthetic sweetening agents, Mfg. Chemist 35:60, 1964.
Chenoy, N. C.: Radiopaques–a review, Pharm. J. 194:663, 1965.
Hoppe, J. O.: X-ray contrast media, Med. Chem. 6:290, 1963.
Shockman, A. T.: Radiologic diagnostic agents, Topics in Med. Chem. 1:381, 1967.

Index

Page numbers in **boldface** indicate tables; in *italics* indicate figures.

1019